CHEMISTRY

The effect of temperature on the equilibrium between pink $Co(H_2O)_6^{2+}$ and blue $CoCl_4^{2-}$ ions in aqueous solution. (See Chapter 15.)

TABLE OF ATOMIC WEIGHTS AND NUMBERS

Name	Symbol	Atomic number	Atomic weight
Actinium	Ac	89	227.028
Aluminum	Al	13	26.981539
Americium	Am	95	(243)
Antimony	Sb	51	121.75
Argon	Ar	18	39.948
Arsenic	As	33	74.92159
Astatine	At	85	(210)
Barium	Ba	56	137.327
Berkelium	Bk	97	(247)
Beryllium	Be	4	9.012182
Bismuth	Bi	83	208.98037
Boron	B	5	10.811
Bromine	Br	35	79.904
Cadmium	Cd	48	112.411
Calcium	Ca	20	40.078
Californium	Cf	98	(251)
Carbon	C	6	12.011
Cerium	Ce	58	140.115
Cesium	Cs	55	132.90543
Chlorine	Cl	17	35.4527
Chromium	Cr	24	51.9961
Cobalt	Co	27	58.93320
Copper	Cu	29	63.546
Curium	Cm	96	(247)
Dysprosium	Dy	66	162.50
Einsteinium	Es	99	(252)
Erbium	Er	68	167.26
Europium	Eu	63	151.965
Fermium	Fm	100	(257)
Fluorine	F	9	18.9984032
Francium	Fr	87	(223)
Gadolinium	Gd	64	157.25
Gallium	Ga	31	69.723
Germanium	Ge	32	72.61
Gold	Au	79	196.96654
Hafnium	Hf	72	178.49
Hassium	Hs	108	(265)
Helium	He	2	4.002602
Holmium	Ho	67	164.93032
Hydrogen	H	1	1.00794
Indium	In	49	114.82
Iodine	I	53	126.90447
Iridium	Ir	77	192.22
Iron	Fe	26	55.847
Krypton	Kr	36	83.80
Lanthanum	La	57	138.9055
Lawrencium	Lr	103	(260)
Lead	Pb	82	207.2
Lithium	Li	3	6.941
Lutetium	Lu	71	174.967
Magnesium	Mg	12	24.3050
Manganese	Mn	25	54.93805
Meitnerium	Mt	109	(267)
Mendelevium	Md	101	(258)
Mercury	Hg	80	200.59

Name	Symbol	Atomic number	Atomic weight
Molybdenum	Mo	42	95.94
Neodymium	Nd	60	144.24
Neon	Ne	10	20.1797
Neptunium	Np	93	237.048
Nickel	Ni	28	58.69
Nielsbohrium	Ns	107	(262)
Niobium	Nb	41	92.90638
Nitrogen	N	7	14.00674
Nobelium	No	102	(259)
Osmium	Os	76	190.2
Oxygen	O	8	15.9994
Palladium	Pd	46	106.42
Phosphorus	P	15	30.973762
Platinum	Pt	78	195.08
Plutonium	Pu	94	(244)
Polonium	Po	84	(209)
Potassium	K	19	39.0983
Praseodymium	Pr	59	140.90765
Promethium	Pm	61	(145)
Protactinium	Pa	91	231.0359
Radium	Ra	88	226.025
Radon	Rn	86	(222)
Rhenium	Re	75	186.207
Rhodium	Rh	45	102.90550
Rubidium	Rb	37	85.4678
Ruthenium	Ru	44	101.07
Samarium	Sm	62	150.36
Scandium	Sc	21	44.955910
Selenium	Se	34	78.96
Silicon	Si	14	28.0855
Silver	Ag	47	107.8682
Sodium	Na	11	22.989768
Strontium	Sr	38	87.62
Sulfur	S	16	32.06
Tantalum	Ta	73	180.9479
Technetium	Tc	43	(98)
Tellurium	Te	52	127.60
Terbium	Tb	65	158.92534
Thallium	Tl	81	204.3833
Thorium	Th	90	232.0381
Thulium	Tm	69	168.93421
Tin	Sn	50	118.710
Titanium	Ti	22	47.88
Tungsten	W	74	183.85
Unnilhexium	Unh	106	(263)
Unnilpentium	Unp	105	(262)
Unnilquadium	Unq	104	(261)
Uranium	U	92	238.0289
Vanadium	V	23	50.9415
Xenon	Xe	54	131.29
Ytterbium	Yb	70	173.04
Yttrium	Y	39	88.90585
Zinc	Zn	30	65.39
Zirconium	Zr	40	91.224

A value in parentheses is the mass number of the isotope of longest half-life.

Values in this table are from the 1988 Report of the International Union of Pure and Applied Chemistry (IUPAC).

S E C O N D   E D I T I O N

CHEMISTRY

STANLEY R. RADEL
The City College of the City University of New York

MARJORIE H. NAVIDI
Queens College of the City University of New York

WEST PUBLISHING COMPANY
Minneapolis/St. Paul   New York   Los Angeles   San Francisco

DEDICATION

To my wife Eva, my children Carol and Laura, and my many students at City College

STANLEY RADEL

To William and Catherine, Joseph, and Sarah, and to the chemistry students at Queens College

MARJORIE NAVIDI

INTERIOR AND COVER DESIGN Diane Beasley
COPYEDITING Luana Richards
ARTWORK Precision Graphics
CUSTOM PHOTOGRAPHY Joel Gordon Photography
COMPOSITION Black Dot Graphics
PROOFREADING Jerrold Moore
PAGE LAYOUT Diane Beasley
COVER IMAGE Joel Gordon Photography
PRODUCTION, PREPRESS, PRINTING, AND BINDING BY WEST PUBLISHING COMPANY

610 Opperman Drive
P.O. Box 64526
St. Paul, MN 55164-0526

Printed in the United States of America

01 00 99 98 97 96 95 94 8 7 6 5 4 3 2 1 0

Library of Congress Cataloging-in-Publication Data

Radel, Stanley R.
Chemistry / Stanley R. Radel, Marjorie Navidi.—2nd ed.
p. cm.
Includes index.
ISBN 0-314-02654-1
1. Chemistry. I. Navidi, Marjorie H. II. Title.
QD31.2.R33 1994
540—dc20 ♾ 93-11454
CIP

INDIVIDUAL PHOTO CREDITS FOLLOW THE INDEX.

WEST'S COMMITMENT TO THE ENVIRONMENT

In 1906, West Publishing Company began recycling materials left over from the production of books. This began a tradition of efficient and responsible use of resources. Today, up to 95 percent of our legal books and 70 percent of our college texts are printed on recycled, acid-free stock. West also recycles nearly 22 million pounds of scrap paper annually—the equivalent of 181,717 trees. Since the 1960s, West has devised ways to capture and recycle waste inks, solvents, oil, and vapors created in the printing process. We also recycle plastics of all kinds, wood, glass, corrugated cardboard, and batteries, and have eliminated the use of styrofoam book packaging. We at West are proud of the longevity and the scope of our commitment to our environment.

CONTENTS IN BRIEF

CONTENTS

CHAPTER 1

CHEMISTRY AND MEASUREMENT 1

CHAPTER 2

ATOMS, MOLECULES, AND IONS 34

SURVEY OF THE ELEMENTS 1

GROUPS 1A, 2A, AND 8A 336

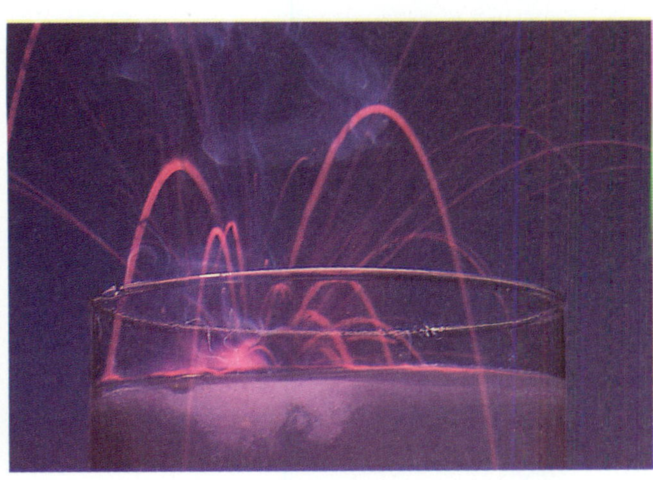

CHAPTER 9

THE CHEMICAL BOND 364

CHAPTER 12

LIQUIDS, SOLIDS, AND INTERMOLECULAR FORCES 532

CHAPTER 13

SOLUTIONS 586

CHAPTER 14

CHEMICAL KINETICS 630

CHAPTER 15

CHEMICAL EQUILIBRIUM 690

A-669
1Pt.
(473 ml.)
REAGENT A.C.S.
Ammonium
Hydroxide
A-144
500 ml.
(1.1 pt.)
REAGENT A.C.S.
Hydrochloric
Acid

CHAPTER 17

ACID–BASE EQUILIBRIA IN AQUEOUS SOLUTIONS 775

CHAPTER 18

SOLUBILITY AND COMPLEX ION EQUILIBRIA 821

SURVEY OF THE ELEMENTS 3

THE REMAINING NONMETALS 855

CHAPTER 19

FREE ENERGY, ENTROPY, AND THE SECOND LAW OF THERMODYNAMICS 905

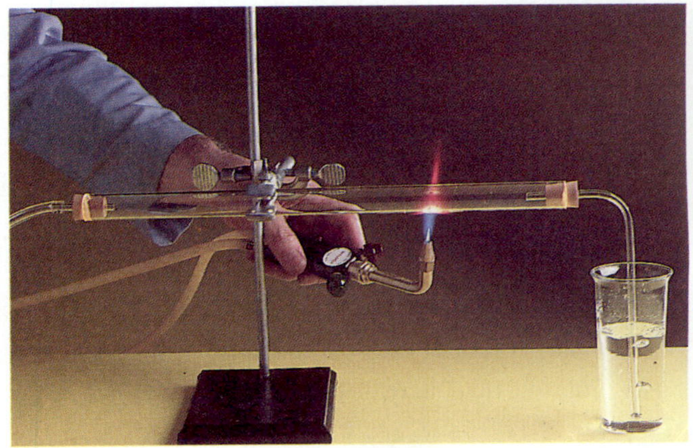

CHAPTER 20

ELECTROCHEMISTRY 937

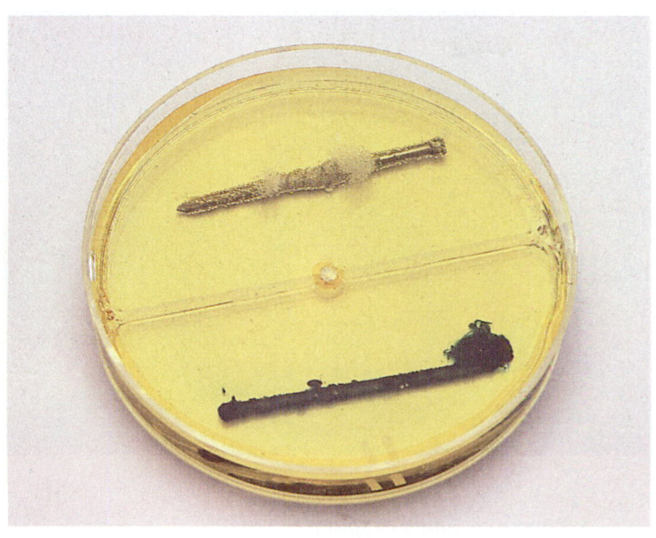

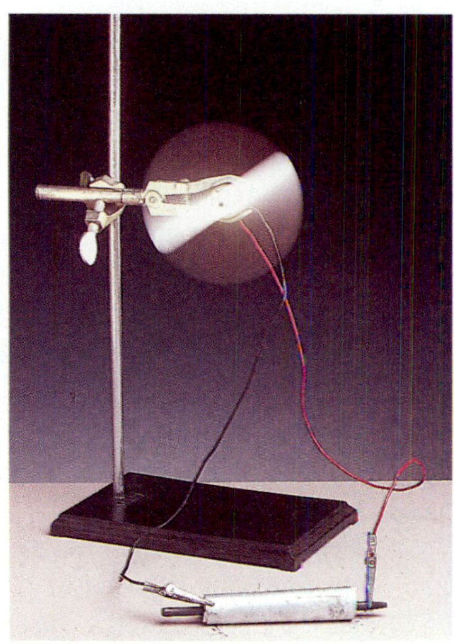

SURVEY OF THE ELEMENTS 4

METALS AND METALLURGY 1033

CHAPTER 22

NUCLEAR CHEMISTRY 1071

CHAPTER 23

ORGANIC CHEMISTRY AND THE CHEMICALS OF LIFE 1112

TEXTBOOK DEMONSTRATIONS

CHEMICAL INSIGHTS

- *Final exercises* come in a wide range of difficulty, from those that test your memory and exercise your skills to those that challenge ingenuity and provoke thought. Answer those assigned by your instructor and as many others as you have time for. Answers to exercises with blue numbers are found in Appendix D.
- *Frequently used data* are summarized on the inside back cover of this book. Here you will also find a list of pages containing important tables, figures, and other useful information.

Suggestions for Studying Chemistry

We recommend that you go over each chapter at least three times: first for orientation, second for study, and finally for review.

Orientation. Before you come to a lecture, read the chapter preview and quickly scan the contents of the material to be discussed. A few minutes of preparation time will put you in a better position to understand the lectures and to take a better set of lecture notes.

Study. Read the chapter in depth, highlight the important points, and make sure you understand each concept and example. Pay attention to the learning hints in the margin and work out each practice exercise as you come to it. If you have difficulty, go over the preceding material until the difficulty is resolved. After studying a section, turn to the end of the chapter and make sure that you have mastered the learning objectives, key terms, and important equations listed for the section. (*Note*: Your instructor may add additional learning objectives and key terms to those given in the text.)

Review. Review each chapter at least once a week. Go over the highlighted material, review the learning objectives, and work several problems.

Additional Help

Two supplements that may provide even further assistance are the *Study Guide* and the *Student Solutions Manual*. The *Study Guide* provides, for each chapter, a preliminary assessment test, a review of key concepts, a drill on key operations, further examples with solutions, and a final test. The *Student Solutions Manual*, which may not be available in every bookstore, provides detailed solutions for all practice exercises and end-of-chapter exercises. Keep in mind, however, that you should not consult the *Student Solutions Manual* until after you have made a serious attempt a solving the problem.

You are about to join thousands of students who have ventured into the fa nating world of chemistry. May you have an interesting and successful journey.

To the Instructor

The second edition of this text, like the first edition, is intended to provic solid preparation for more advanced work. It can also stand by itself as the core a a student's only chemistry course. All chapters contain examples and illustrations that relate chemistry to other disciplines and to the problems of today's world so every student should find something of value. Enthusiastic response from students and instructors using the first edition has convinced us that we accomplished our primary goal—the presentation of chemistry to beginning students in an understandable, interesting, and user-friendly manner.

Features

Chemistry is more than a collection of facts and learned laboratory skills—it is also a way of thinking. This book should help students to find that way. It is designed to be readable and understandable, without sacrificing any of the depth and thoroughness expected of a good introductory course. These goals were accomplished by paying major attention to the learning process. Each new topic is presented in a straightforward, step-by-step fashion, with liberal cross-references to preceding related material; the student is guided around pitfalls and encouraged to use common sense at every stage. Examples with solutions illustrate every procedure, and practice exercises allow for frequent self-checks. Margin remarks are used liberally to emphasize ideas that may be overlooked, add interesting comments, and relay important learning hints. Chemical principles and reactions are illustrated by an abundance of textbook demonstrations and photographs. Our own students have always been particularly appreciative of the end-of-chapter learning objectives, which provide clear and specific guidelines for study. Finally, the exercises at the end of each chapter present a copious collection of questions and problems of varying degrees of difficulty, from those that test the student's memory and emphasize routine skills, to those that challenge ingenuity and provoke thought. Many of the final exercises deal with timely issues.

Changes in the Second Edition

All suggestions from students, instructors who used the previous edition, and independent reviewers, have been carefully considered, and many of these suggestions have been incorporated into this edition. The material has been fine-tuned in some sections, completely rewritten in others, and updated to reflect recent advances in the field. Some of the major changes are as follows:

- The final exercises in each chapter have been reorganized and many new exercises have been added.
- A section dealing with the shapes of hydrocarbon molecules has been added to Chapter 10. This section provides insight into an important class of organic compounds that are used throughout the text, but were more or less taken for granted in the first edition.
- The discussion of dipole moments is now integrated with the discussion of molecular geometry.
- The oxidation number rules have been rewritten, a brief discussion of fractional oxidation numbers has been added, and there is an expanded discussion of the selection and use of oxidizing and reducing agents.
- A number of instructors and students have pointed out that the traditional method of balancing oxidation–reduction equations occasionally breaks down for reactions that occur in basic solution. A method that does away with this problem is now presented in Chapter 11.

- The chapter on kinetics (Chapter 14) has been upgraded by including a more detailed discussion of heterogeneous catalysis, including zero-order reactions on a catalytic surface. In addition, a section on the mechanism by which CFCs deplete the ozone layer has been added.
- A *Digging Deeper* section on the method of successive approximations has been added to Chapter 15.
- The section on pH control of sulfide precipitations (Chapter 18) has been rewritten to reflect recent developments in our knowledge of sulfide equilibria.
- After checking with various specialists, we decided to replace the crystal field model of complex formation (Chapter 21) with a simplified molecular orbital model. We feel that this approach is a natural extension of the molecular orbital theory developed in Chapter 10.
- In addition to updating many sections, the text and *Chemical Insights* now include a discussion of fullerenes, diamond films, supercritical fluids, atomic force microscopes, positron emission tomography, isotope enrichment by gas centrifugation, and so forth.

The second edition also includes a number of "profiles" that illustrate the diversity of careers based on chemistry. These profiles, prepared by Professor Thaddeus Ichniowski of Illinois State University, are designed to address the varied interests of the typical heterogeneous group of students that enroll in general chemistry courses.

Flexibility

It is not possible to find an order of topics that is perfect for everybody; many instructors have individual preferences for certain sequences that work best for them. The order we finally chose is a conventional one, but presented in a way that provides maximum flexibility for rearrangement. Some of the principal options are mentioned below; more detailed suggestions can be found in the *Instructor's Guide*.

Placement of Descriptive Chemistry. Surely much of the delight of chemistry lies in becoming familiar with substances in all their marvelous variety; therefore, each of the conceptual chapters, 1 through 23, is richly illustrated with descriptive material. The chemistry of the more important elements and their compounds, however, is also described systematically in four survey chapters that have been positioned to utilize and reinforce the concepts presented in the immediately preceding chapters. Each survey chapter can be taught as it appears, or taught later, or assigned for self-study. Note that a student using this book will learn a great deal of practical "bench top" chemistry even if the survey chapters are omitted.

Placement of Advanced Bonding Theory. Lewis structures (Chapter 9) and molecular shapes (Sections 10.1–10.3) will be needed right away. The more advanced bonding topics in Chapter 10 (valence bond theory and molecular orbital theory) can be deferred to later lectures where they are needed.

Placement of Redox Chemistry. An introduction to redox chemistry is given in Chapter 3. The more detailed discussion given in Chapter 11 can be deferred until just before Chapter 20, the electrochemistry chapter.

Placement of Kinetics. Instructors who prefer to teach equilibrium before kinetics can defer Chapter 14 until after Chapter 21, the chapter dealing with metals and coordination chemistry.

Placement of Nuclear Chemistry. Instructors who feel that nuclear chemistry is too important to be left to an end-of-semester rush, can present Chapter 22 as soon as the students have learned about rates and half-life in Chapter 14.

Placement of Organic Chemistry. Organic molecules are encountered in most chapters of the text, and this may be enough for students who will advance to a full organic course. For other students, however, a greater emphasis on organic chemistry may be desirable. The first three sections of Chapter 23 deal with hydrocarbons and could, in principle, be presented at any time following Chapter 10. Sections 23.1 through 23.6 would be an appropriate supplement to Survey 3, which includes inorganic carbon compounds. The remaining sections of Chapter 23 deal largely with biochemistry and require the concept of chirality, which is presented in Chapter 21.

Two other features provide further flexibility. The sections marked *Digging Deeper* contain material that can be deferred or omitted without loss of continuity. Although all of the *Digging Deeper* material is suitable for an introductory text, most of it is more difficult than the material found in the other sections. The *Chemical Insight* sections relate chemical principles to everyday matters of health, the environment, consumer products, and so forth. They make straightforward reading; their content is generally optional.

Supplements

- The *Study Guide*, by Marion Brisk, provides an assessment test, a review of key concepts, a drill on key operations, solved examples, and a final test for each chapter in the text.
- The *Solutions Manual*, by Alan Tchernoff, presents detailed solutions for all practice exercises and end-of-chapter exercises. Departmental approval is required for the sale of the *Solutions Manual* to students.
- The *Student Solutions Manual*, by Alan Tchernoff, presents detailed solutions for all practice exercises and selected end-of-chapter exercises.
- The *Laboratory Manual to Accompany Chemistry*, by A. David Baker, Lawrence F. Gries, Marjorie H. Navidi, and Thomas Strekas contains laboratory experiments keyed to chapters in the text, plus study assignments on laboratory techniques, report writing, chemical nomenclature, and chemical arithmetic. An *Instructor's Guide* and appendix list the equipment and supplies needed for each experiment.
- The *Instructor's Guide*, by Marjorie H. Navidi and Stanley R. Radel, gives suggestions for planning the course and helpful material to be used with each chapter.
- The *Test Bank*, by Paul Hunter, contains over 3000 questions in multiple-choice format. A computerized version of the *Test Bank* is available for IBM PC and Macintosh computers.

- The *Problem Supplement*, prepared by Kevin Grundy of Dalhousie University, provides additional problems with an emphasis on testing cumulative knowledge, reasoning and critical thinking skills, as well as routine concepts from a different perspective.
- *Qualitative Inorganic Analysis*, a combined text and laboratory manual by William T. Scroggins of Chabot College, introduces qualitative analysis techniques and their underlying equilibrium principles.
- *Software. Concentrated Chemical Concepts* by Trinity Software for IBM compatible computers. This computer-aided student tutorial covers the entire general chemistry course, and is highly interactive, with instant feedback and reinforcement. Another software package available is *Labsystant* from Trinity Software, which enables instructors to create electronic worksheets for quantitative laboratory data; the *Checker* program helps students find and correct calculation errors, and prints out the final results.
- *Videodisk*. West's General Chemistry videodisk contains demonstrations and experiments that are often too expensive and dangerous for the classroom. Many of the photos from the text are included.
- *Full-color Transparency Acetates* for 160 figures and diagrams from the text, and an additional 62 acetates covering the 20 textbook demonstrations.
- *Transparency Masters* for about 100 figures, diagrams, and tables from the text.

ACKNOWLEDGMENTS

The second edition of this text owes much to many individuals. Foremost among these is our science editor, Ron Pullins, who provided inspiration and invaluable guidance in all phases of the project, and whose questions, comments, and suggestions helped fine-tune the text in many ways. (Ron, it was a pleasure to work with you—your energy and enthusiasm helped us over many rough crossings.) Denise Bayko, our Developmental Editor, also deserves special mention. Her detailed analyses of the reviews, her coordination of the ancillary program, and most of all, her unfailing good humor were a constant source of amazement to us. We must also thank our Production Editors, Tom Hilt and Cliff Kallemeyn, for their calm guidance of the manuscript through the various crises that inevitably arise during a production of this magnitude. Tom and Cliff have unsurpassed eyes for detail, and we will always remember their talent for organizing and coordinating the production of this edition.

We owe a great debt of gratitude to Raymond Keywork for his expertise in setting up and orchestrating the various interesting and beautiful demonstrations found throughout the book. (Thank you, Ray, for coming out of retirement to help us with the second edition.) We also thank Mohamed Razak, of City College, and Thomas Hayden, of Queens College, for lending us equipment and providing many helpful suggestions for the demonstrations.

Once again we thank our photographer, Joel Gordon, and his assistant, Paul Hollenvack, for their many wonderful pictures and our many interesting conversations. Initially, our insistence on scientific accuracy clashed with Joel's instinct for taking a dramatic photograph. In the end, however, we learned much about each other's profession, and the result was a truly spectacular set of photographs.

Dr. Alan Tchernoff again provided superb assistance by critically reviewing every word of the manuscript. His uncanny ability to ferret out errors nearly drove us crazy. Alan contributed much to the writing and rewriting of the practice exercises and end-of-chapter exercises.

We would also like to thank our colleagues at City College and Queens College, including A. David Baker, Marion Brisk, David Brown, Harry Gafney, Denise Garland, David Gosser, Derek Lindsay, John Lombardi, Neil McKelvie, Jack Morrow, Leonard Schwartz, Tom Strekas, Sam Wilen, and Arthur Woodward, most of whom have used our book and all of whom have offered constructive criticisms. Particular thanks are due to John Arents, Norman Goldman, and Michael Weiner, with whom we discussed many of the finer points of chemistry.

Others who deserve special mention include Luana Richards, who did a wonderful job of copy editing, and Diane Beasley, who developed a beautiful design for the second edition. Still others include Thaddeus C. Ichniowski from Illinois State University, who provided profiles of *Chemists at Work*, Mary Steiner, our Promotion Manager, Paula Blair, a wonderful secretary for West Publishing, and many other members of the West staff who unfortunately are anonymous to us.

Finally, we want to thank the following people who reviewed all or part of our manuscript in its various stages:

First Edition Reviewers

David L. Adams—*Babson College*

Paul A. Barks—*No. Hennepin Community College*

Jacqueline Barton—*Columbia University*

O. T. Beachley, Jr.—*SUNY, Buffalo*

Luther K. Brice—*Virginia Polytechnic Institute*

David Brooks—*University of Nebraska, Lincoln*

Weldon Burnham—*Richland College*

Thomas J. Bydalek—*University of Minnesota, Duluth*

Donald Campbell—*University of Wisconsin, Eau Claire*

James D. Carr—*University of Virginia*

Harvey F. Carroll—*Kingsboro Community College*

Ronald J. Clark—*Florida State University*

Michael I. Davis—*University of Texas, El Paso*

Thomas C. Devore—*James Madison University*

Jimmie G. Edwards—*University of Toledo*

L. M. Epstein—*University of Pittsburgh*

Gordan J. Ewing—*New Mexico State University*

Steven L. Fedder—*Santa Clara University*

L. Peter Gold—*Pennsylvania State University*

Charles G. Haas—*Pennsylvania State University*

Arnulf P. Hagen—*University of Oklahoma*

Albert Haim—*SUNY, Stony Brook*

James C. Hill—*California State University, Sacramento*

Paul Hunter—*Michigan State University*

Stanley Israel—*University of Lowell*
Philip S. Lamprey—*University of Lowell*
Joel T. Mague—*Tulane University*
Gardiner Meyers—*University of Florida*
Ted Musgrave—*Northern Arizona University*
Robert J. Ouellette—*Ohio State University*
R. S. Perkins—*University of Southwestern Louisiana*
W. D. Perry—*Auburn University*
George Pfeffer—*University of Nebraska*
Erwin W. Richter—*University of Northern Iowa*
Don Roach—*Miami–Dade Community College*
Stephen B. W. Roeder—*San Diego State University*
R. J. Ruch—*Kent State University*
Barbara A. Sawrey—*San Diego State University*
Martha Sellers—*Northern Virginia Community College*
T. W. Sottery—*University of Maine, Portland*
Michael Strauss—*University of Vermont*
William Sweeney—*Hunter College*
Donald D. Titus—*Temple University*
Carl Trindle—*University of Virginia*
Jeanette Wasserstein—*New School for Social Research*
W. W. Wendlandt—*University of Houston, University Park*

Second Edition Reviewers

Keith O. Berry—*University of Puget Sound*
Lewis H. Brubacher—*University of Waterloo*
C. Eugene Burchill—*University of Manitoba*
B. Edward Cain—*Rochester Institute of Technology*
Harvey F. Carroll—*Kingsborough Community College of the City University of New York*
Stanley M. Cherim—*Delaware County Community College*
Ronald J. Clark—*Florida State University*
Kim Cohn—*California State University, Bakersfield*
Leslie DiVerdi—*Colorado State University*
Mark Doughty—*Concordia University*
Patricia C. Flath—*Paul Smith's College of Arts and Sciences*
Catherine I. Gall—*Fanshawe College of Applied Arts and Technology*
Nancy Gettys—*Oklahoma State University*
Peter R. Girardot—*The University of Texas at Arlington*
L. Peter Gold—*The Pennsylvania State University*

Kevin Grundy—*Dalhousie University*
Thomas E. Hagan, Jr.—*Lebanon Valley College*
Anne Harmon—*Lamar University*
Steven A. Hendrix—*The University of Tampa*
G. Hewson Hickie—*Bishop's University*
Colin D. Hubbard—*University of New Hampshire*
Donald E. Irish—*University of Waterloo*
Ronald C. Johnson—*Emory University*
Janice Kelland—*Memorial University of Newfoundland*
Robert C. Kerber—*State University of New York-Stony Brook*
Lynn Vogel Koplitz—*Loyola University*
Dorothy B. Kurland—*West Virginia Institute of Technology*
Claude R. Lassigne—*Kwantlen College*
William M. Litchman—*University of New Mexico*
Rudy Luck—*The American University*
John Martin—*Davis and Elkins College*
Lillian Martin—*Fraser Valley College*
Ronald D. Ragsdale—*The University of Utah*
Thomas W. Richardson—*North Georgia College*
Martha E. Russell—*Iowa State University of Science and Technology*
Dennis J. Sardella—*Boston College*
William Scroggins—*Chabot College*
Robert E. Smith—*Longview Community College*
Jean Stanley—*Wellesley College*
Keith Vitense—*Cameron University*
E. J. Wells—*Simon Fraser University*
Walter E. Weibrecht—*University of Massachusetts, Boston*
Daniel J. Williams—*Kennesaw State College*
Peter K. Wong—*Queensborough Community College of the City University of New York*

CHAPTER 1

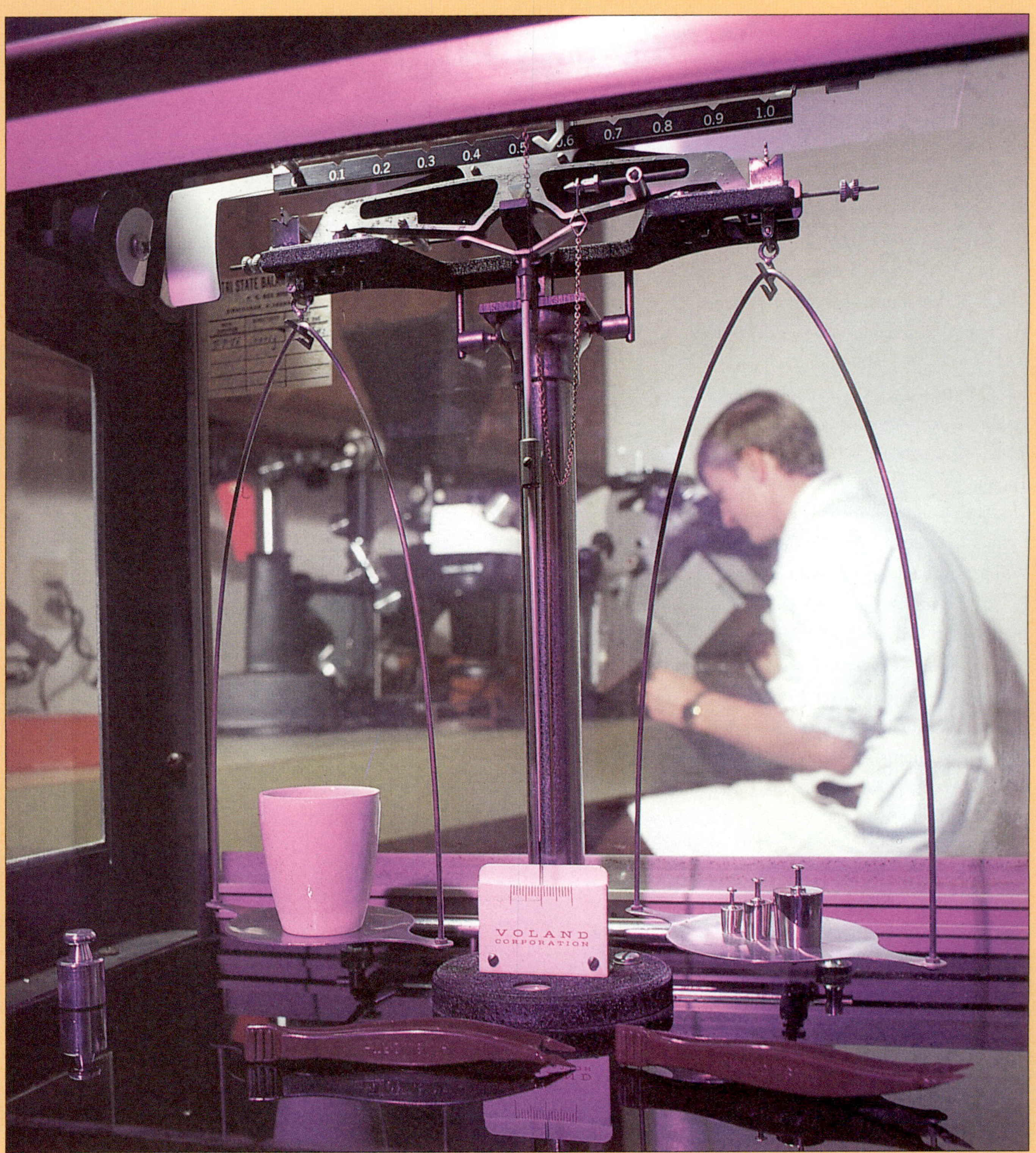

Two-pan chain balances of this type were once widely used in chemical research laboratories. They have now been supplanted by single-pan electronic balances.

CHEMISTRY AND MEASUREMENT

OUTLINE

PREVIEW

How many times a day do you measure something? Consider tape measures, rulers, postage scales, bathroom scales, thermometers, speedometers, odometers, clocks, stopwatches, and gas gauges. Measurement is an integral part of our culture, and science often progresses in leaps and bounds when new or improved methods of measurement are developed. Lord Kelvin (1824–1907), a British mathematician and physicist, wrote:

> I often say that when you can measure what you are speaking about, and express it in numbers, you know something about it; but when you cannot measure it, when you cannot express it in numbers, your knowledge is of a meager and unsatisfactory kind; it may be the beginning of knowledge, but you have scarcely, in your thoughts, advanced to the stage of science, whatever the matter may be.

Chemistry is the study of matter and its transformations. Chemists are interested in the composition, structure, and behavior of the various forms of matter, from the smallest atom to the most complex living species. They observe matter to determine its properties, and they study how various substances react with each other. Their goals are to understand why reactions take place, to develop methods for controlling them, and to create new substances for specific purposes (Figure 1.1).

Linus Pauling, winner of two Nobel prizes, has pointed out that almost all of science can be included within the definition of chemistry. A knowledge of chemical principles has long been a prerequisite for biology and medicine, and it is essential for an understanding of earth, planetary, and atmospheric sciences. Engineers and metallurgists are interested in the structural properties of matter and in the chemical principles underlying such diverse pursuits as making beer and wine, growing diamond crystals, producing electricity from nuclear reactions, and generating electrical power from the energy of the sun. Advances in astronomy have turned many astrophysicists into astrochemists, and nuclear physicists share with chemists an interest in the structure and transformations of the inner cores of atoms.

The fruits of chemical research have affected both the quality and the length of

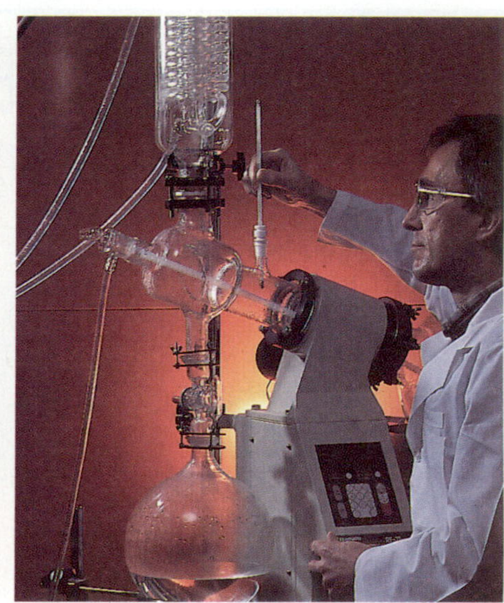

Figure 1.1
Chemists at work.

life. In ancient Rome the average human lifespan was between 20 and 30 years; today, life expectancy in much of the industrialized world is more than 70 years. A large part of this increase is a result of the development of antibiotics and other drugs for the control of disease and the alleviation of pain. Thousands of other chemical products, including plastics, steel, fertilizers, cement, and paper, contribute to making life easier, more comfortable, and more interesting.

Unfortunately, there is another side to the coin. Unknowing, careless, and sometimes unscrupulous individuals have allowed harmful pollutants and dangerous waste materials to enter the air we breathe, the water we drink, and even the food we eat. The result has been a series of environmental and health problems such as those caused by acid rain, tobacco smoke, asbestos, and mercury-contaminated fish. Many of these problems have become the focus of controversy between the victims of pollution and the groups responsible for it. These controversies often receive national attention, and a knowledge of chemical principles is almost always required to evaluate the various claims and counterclaims. Chemistry is interesting in its own right, and it may help you in your chosen field, but concern for the environment and one's health are two more very practical reasons for studying this basic science.

1.1 MATTER AND ENERGY

Our universe is one of unceasing motion and continuous change. **Matter**, by which we mean anything that has mass and occupies space, is constantly changing from one form into another. Examples of spontaneous change can be found everywhere—living organisms are born, they mature, and they eventually die; ocean water evaporates and then recondenses as rain or snow or hail; radioactive material disintegrates into simpler particles; and giant stars explode and become supernovas (Figure 1.2).

What makes it possible for matter to undergo these transformations? The answer is **energy**, the ability to do work. Energy enables objects to move and undergo change. Unlike matter, energy is intangible; we cannot touch it or pick it up like a book, but we can see and feel its effects. The universe is full of energy. Stars are vast reservoirs of radiant energy, and light energy from our star, the sun, helped to develop life on earth and continues to sustain it today. People have used wind energy for centuries to propel ships and turn windmills, and the energy of moving water has been harnessed to produce electricity. Even atoms, when split, can release energy to produce motion, change, and in the case of atomic bombs, the wholesale destruction of life and property.

Where do the matter and energy of our universe come from? Or have they always been here in one form or another? No one knows the answers to these questions. Modern theories of *cosmology,* the science that deals with the origin and evolution of the universe, assume that the contents of the universe were once in a highly condensed state and that an explosion or "big bang" occurred about 15 billion years ago. The hot products of this explosion began to rush apart and cool, and over the years they gradually evolved into the universe in which we now live. Our universe is still expanding—the most distant stars are moving away from us at speeds more than one-third the speed of light. No one knows whether this expansion will go on forever or whether gravitational attraction will someday slow it down, and perhaps even pull all the widely dispersed matter together again.

At a very early stage in the expansion—one-hundredth of a second after the big bang—the temperature is estimated to have been about 100 billion degrees Celsius. It was much too hot for atoms to exist and the universe at that time was heavily populated with less complex particles, including negatively charged *electrons,* positively charged *protons,* and uncharged *neutrons.* Electrons are the particles that flow through wires during the passage of electric current. Protons and neutrons are of special interest to us. They are the particles that ultimately fused together to form the nuclei of the various atoms so familiar to us today.

(a)

(b)

Figure 1.2

A supernova. (a) A photograph of the sky showing the blue supergiant star known as Sanduleak. (b) The same region of the sky after a stellar explosion converted Sanduleak into a supernova. In May 1987 the supernova was 200 million times brighter than the sun.

The simplest atom is the hydrogen atom, consisting of an electron moving about a proton. More complex atoms consist of two or more electrons moving about nuclei that contain protons and neutrons. Except for hydrogen, the nuclei of all naturally occurring atoms, including carbon, oxygen, nitrogen, and other substances essential for life, are produced by reactions within the stars. As the stars grow old and die, these nuclei are ejected into space where they combine with electrons and gradually coalesce to form dust grains, comets, planets, new stars, and living matter. To paraphrase Carl Sagan, a modern-day astronomer: We are all made from the dust of the stars.

PRACTICE EXERCISE 1.1

Give (a) four examples of matter and (b) four forms of energy.

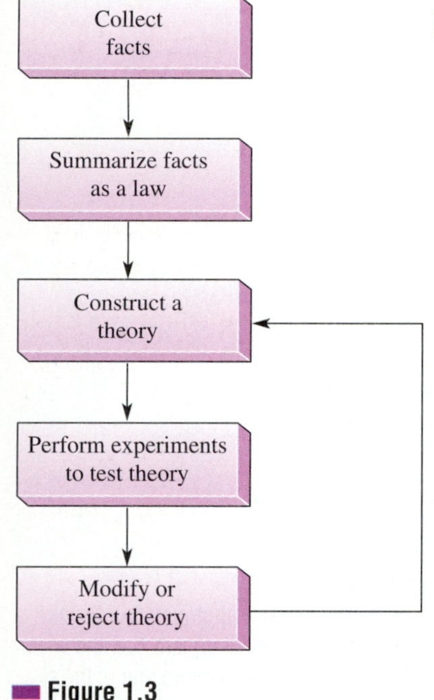

Figure 1.3
The scientific method.

1.2 MEASUREMENT AND THE SCIENTIFIC METHOD

Most scientists use some combination of intuition and reasoning to arrive at conclusions. Intuition is often responsible for new ideas, for a feeling that something is wrong and should be checked, or for just "knowing" that an idea is right. Scientific reasoning, on the other hand, is based on the rules of logic and almost always proceeds according to the **scientific method** (Figure 1.3). The steps in the scientific method are as follows: (1) the collection of facts or **data** (a single fact is a **datum**)—for example, the gases oxygen and nitrogen expand when heated; (2) a search for generalizations or **laws** to summarize these facts—all gases expand when heated; and (3) the free use of imagination to construct **theories** or models of nature that will account for the laws—gases are composed of particles that move more rapidly and tend to separate from each other when heated. A theoretical model not only offers an explanation for observed phenomena, it also allows predictions to be made about the system under study. Correct predictions build confidence in the model. Incorrect predictions mean that the model will have to be modified or even abandoned. It is interesting to note that even though a theory can be disproved, it can never be completely proved. Each step in the scientific method is tested by *experiment* and *observation*—try something and see what happens—and experiment alone determines what is right or wrong in chemistry.

Chemistry is an experimental science, and much of the data collected by chemists consists of numbers that are used in computations or subjected to mathematical analysis. Some numerical data are obtained by counting discrete items, but most data are obtained by **measurement**, that is, by finding the magnitude of some quantity in terms of a previously defined unit that serves as a standard. Measurements are usually made with instruments marked off in multiples or fractions of the basic unit (Figure 1.4). Seconds, minutes, and hours, for example, are commonly used units for time, and clocks are typical measuring instruments.

The metric system was proposed in 1670 by Gabriel Mouton, Vicar of Lyons. It was adopted by the French National Assembly in 1795.

The word *metric* is derived from the Greek *metron,* "measure."

Local customs have given rise to a variety of units, many of which measure the same quantities. Most of us have used or referred to the following volume units: gallon, quart, pint, cubic yard, cubic foot, cord, fluid ounce, cup, tablespoon, teaspoon. To discourage the proliferation of so many different units and to simplify cal-

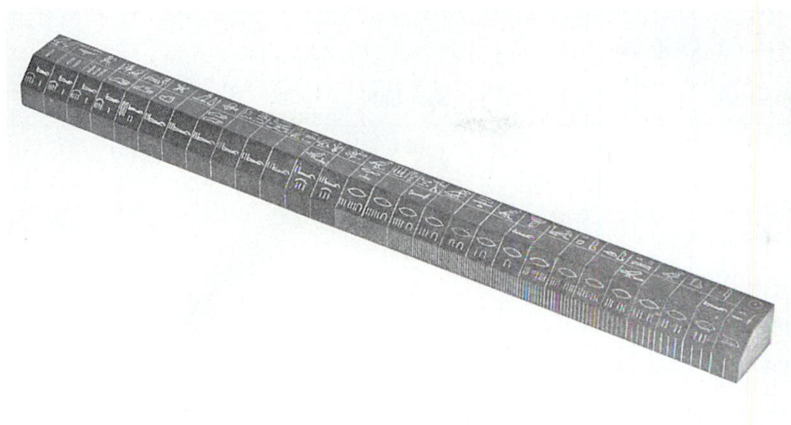

(a)

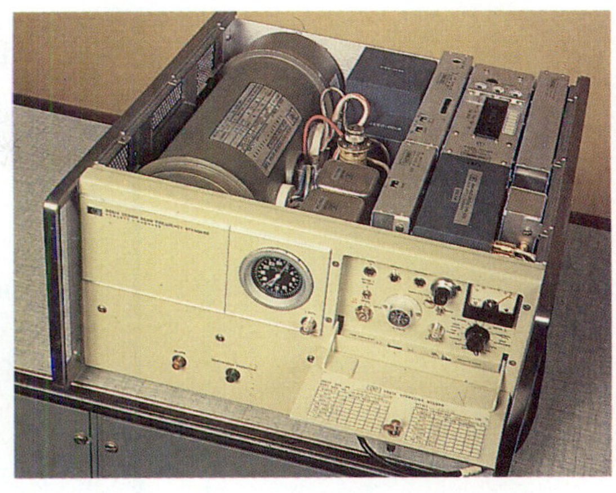

(b)

Figure 1.4

Ancient and modern measuring instruments. (a) A replica of the Egyptian royal cubit, the standard unit of length for the builders of the Great Pyramid of Cheops (ca. 3000 B.C.). A cubit is about 52.5 cm. (b) Interior view of an atomic clock, the most accurate type of clock known to date. Time scales for atomic clocks are based on periodic processes that occur in atoms and molecules.

culations, scientists have chosen to use the **metric system**, a decimal system of weights and measures originally based on the meter as the unit of length and the kilogram as the unit of mass.

The metric system is used in most countries, and it is anticipated that the United States will gradually convert to this system over the coming years. At the present time, however, many nonmetric units are still used by scientists as well as by laypersons. Thus students of the sciences must develop proficiency in metric units as well as the commonly used nonmetric units. Some metric units and their English equivalents are given in Table 1.1 and inside the back cover of this book.

TABLE 1.1 Some Metric Units and Their English Equivalents

Units of Length:	1 meter = 39.37 inches
	1 inch = 2.54 centimeters (exact)
	1 mile = 1.609 kilometers
Units of Mass:	1 kilogram = 2.205 pounds
	1 pound = 453.6 grams
	1 metric ton = 2205 pounds
Units of Volume:	1 liter = 1.057 quarts (U.S., liquid)
	1 gallon (U.S., liquid) = 3.785 liters

SI Units

A complete system of measurement can be based on a limited number of fundamental units. Four fundamental units of *length, mass, time,* and *temperature* suffice for our everyday needs. Together with three additional units of *electric current, intensity of light,* and *number of particles,* they complete a system of seven units adequate for the entire realm of physical science. Other units, called **derived units**, can be obtained by combining fundamental units. For example, the volume of an object is obtained by multiplying three lengths together; therefore, the units of

TABLE 1.2 SI Base Units

Physical Quantity	Unit	Symbol[a]
Length	meter	m
Mass	kilogram	kg
Time	second	s
Temperature	kelvin	K
Electric current	ampere	A
Luminous intensity	candela	cd
Number of particles	mole	mol

[a]Metric symbols are not followed by periods nor are they changed in the plural; 5 kilograms, for example, is written as 5 kg, not 5 kg. or 5 kgs.

volume are length × length × length or, more simply, $(\text{length})^3$. Speed units such as meters per second are obtained by combining units of length and time.

In 1960, the Eleventh General Conference on Weights and Measures recommended that the seven metric units summarized in Table 1.2 be used for all scientific work. These units, called **SI units** (from the French *Système International d'Unités,* International System of Units), are widely used, but they have not yet been fully accepted by the international scientific community. In this book we will use many SI units, but we will also use some non-SI units that chemists have found convenient to retain.

Units are defined by an international body called the General Conference on Weights and Measures.

PRACTICE EXERCISE 1.2

State the SI units of (a) length, (b) mass, and (c) time.

1.3 LENGTH AND VOLUME

The current definition of the meter in terms of the speed of light was made in 1983. The numbers in the definition might seem peculiar, but they were carefully chosen so that the actual length corresponding to the newly defined meter does not substantially differ from the length that most scientists had used all along.

The SI unit of length is the **meter** (m), defined as the distance light travels in a vacuum during 1/299,792,458 of a second. This corresponds to 39.37 inches, a distance slightly larger than 1 yard.

Fractions and multiples of SI units are named by adding appropriate prefixes. Table 1.3 lists the approved SI prefixes with their abbreviations and shows how they are used to give multiples and fractions of the meter. One-hundredth of a meter, for example, is a **centi**meter (1 cm = 10^{-2} m), a thousandth of a meter is a **milli**meter (1 mm = 10^{-3} m), and a thousand meters is a **kilo**meter (1 km = 1000 m). The **micro**meter (1 μm = 10^{-6} m), which is one-millionth of a meter, the **nano**meter (1 nm = 10^{-9} m), one-billionth of a meter, and the **pico**meter (1 pm = 10^{-12} m), one-trillionth of a meter, are small units frequently used by chemists. *Note that the fractions and multiples are all powers of 10.* This feature makes it easy to convert from one unit to another and partially accounts for the popularity of the metric system in many countries as well as in the scientific community. Table 1.4 gives the approximate dimensions of various objects in terms of the meter.

TABLE 1.3 Multiples and Fractions of the Meter

Unit[a]	Symbol	Relation to Meter
Attometer	am	10^{-18} m
Femtometer	fm	10^{-15} m
Picometer	pm	10^{-12} m
Nanometer	nm	10^{-9} m
Micrometer[b]	µm	10^{-6} m
Millimeter	mm	10^{-3} m
Centimeter	cm	10^{-2} m
Decimeter	dm	10^{-1} m
Meter	m	1 m
Dekameter	dam	10 m
Hectometer	hm	10^{2} m
Kilometer	km	10^{3} m
Megameter	Mm	10^{6} m
Gigameter	Gm	10^{9} m
Terameter	Tm	10^{12} m
Petameter	Pm	10^{15} m
Exameter	Em	10^{18} m

[a]Standard prefixes are in color. They may be used with any SI base unit listed in Table 1.2.

[b]The Greek letter mu (μ) is the only approved symbol that is not English. An older name for the micrometer is *micron*.

LEARNING HINT

Exponential notation is reviewed in Appendix A, Section A.1.

TABLE 1.4 Some Approximate Dimensions

Radius of a proton	10^{-15} m = 1 fm
Limit of scanning tunneling microscopy	10^{-11} m = 10 pm
Radius of typical atom[a]	10^{-10} m = 100 pm
Limit of optical microscopy	2×10^{-7} m = 0.2 μm
Radius of human red blood cell	10^{-5} m = 10 μm
Diameter of U.S. penny	0.019 m = 1.9 cm
Height of a 6-ft-tall human	1.83 m
Height of Mt. Everest	8840 m = 8.840 km
Earth's equatorial radius	6.378×10^{6} m = 6378 km
Solar radius	6.96×10^{8} m
Mean earth–sun distance	1.496×10^{11} m
Radius of observable universe	2×10^{26} m

[a]Atomic and molecular dimensions are often expressed in **angstroms** (Å), where 1 Å = 10^{-10} m = 100 pm. The angstrom unit is part of the metric system because it is a decimal fraction of the meter, but it is not an SI unit.

TABLE 1.5
Volume Formulas

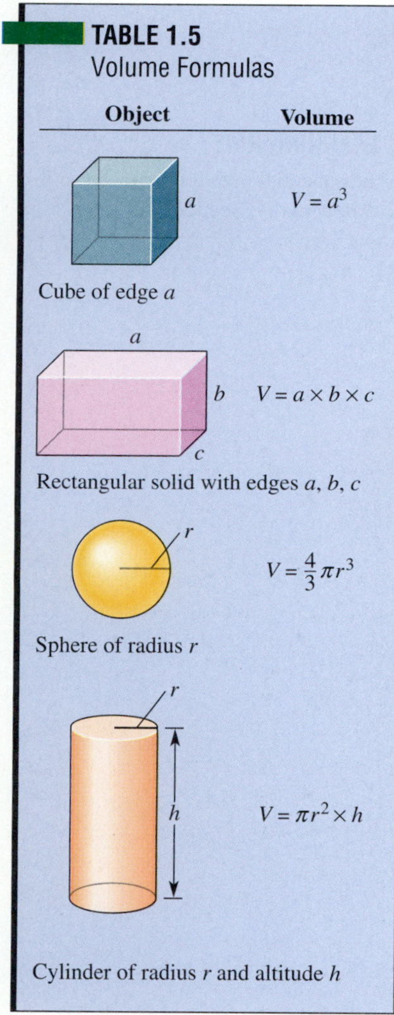

Object
Volume
a
$V = a^3$
Cube of edge a
a
b
c
$V = a \times b \times c$
Rectangular solid with edges a, b, c
r
$V = \frac{4}{3}\pi r^3$
Sphere of radius r
r
h
$V = \pi r^2 \times h$
Cylinder of radius r and altitude h

EXAMPLE 1.1

How many meters are equivalent to (a) 2 km and (b) 0.035 cm?

SOLUTION

Table 1.3 shows that

(a) 1 km = 1000 m; therefore, 2 km = 2 × 1000 m = 2000 m.

(b) 1 cm = 10^{-2} m; therefore, 0.035 cm = 0.035 × 10^{-2} m = 3.5 × 10^{-4} m.

PRACTICE EXERCISE 1.3

How many meters are equivalent to (a) 21 dm, (b) 410 nm, and (c) 0.55 μm?

The volume of an object such as a cube, sphere, or cylinder can be calculated from its linear dimensions. The formulas in Table 1.5, some of which may be familiar to you, show that volumes are obtained by multiplying three lengths together. Thus volume units are the cubes of length units. For example, the volume of a box with edges of 5 cm, 10 cm, and 6 cm is 5 cm × 10 cm × 6 cm = 300 cm^3 (300 cubic centimeters).

The volume unit most often used by the chemist is the **liter** (L), a non-SI unit defined as the volume of a cube 10 cm on edge (Figure 1.5). A liter is slightly larger than a quart (1 L = 1.057 qt), and its magnitude can be visualized by thinking of a quart container of milk or a one-liter bottle of soda. The volume of a cube of edge a is a^3 (Table 1.5); therefore,

$$1\ \text{L} = (10\ \text{cm})^3 = 1000\ \text{cm}^3$$

The SI prefixes given in Table 1.3 are also used with the liter. A milliliter (mL) is one-thousandth of a liter, so

$$1\ \text{L} = 1000\ \text{mL}$$

Comparing the two expressions for the liter shows that

$$1000\ \text{mL} = 1000\ \text{cm}^3$$

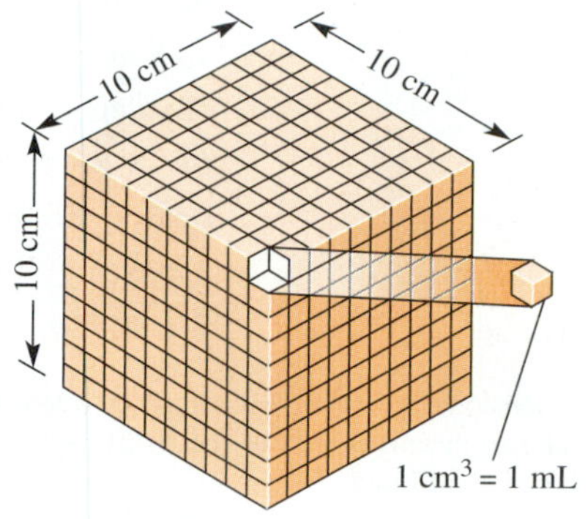

Figure 1.5
The volume of a cube 10 cm on edge is 1000 cm^3 or 1 L. Each of the 1000 smaller cubes has a volume of 1 cm^3 or 1 mL.

and dividing both sides by 1000 gives

$$1\ \text{mL} = 1\ \text{cm}^3$$

Thus *the volume units milliliter and cubic centimeter can be used interchangeably.* A milliliter is about one-fifth of a teaspoon.

Some laboratory devices for measuring liquid volumes are shown in Figure 1.6. The *graduations* (markings) etched on the glass indicate volume levels in milliliters or some other fraction of the liter.

LEARNING HINT

Since 10 centimeters is 1 decimeter (10 cm = 1 dm), it follows that 1 L = $(10\ \text{cm})^3 = (1\ \text{dm})^3 = 1\ \text{dm}^3$; that is, 1 liter is equal to 1 **cubic decimeter**. Many chemists prefer to report volumes in cubic decimeters rather than in liters.

PRACTICE EXERCISE 1.4

How many liters are equivalent to (a) 1 kL, (b) 18 mL, and (c) 500 cm^3?

1.4 RELIABILITY OF MEASUREMENTS

Scientists depend on being able to make trustworthy measurements and often devote much time to developing and refining their measuring instruments. Completely perfect measurements, however, are not possible. Even the most reliable measurement

Figure 1.6

Glassware for measuring liquids by volume. Left to right: 250-mL Erlenmeyer flask, 250-mL volumetric flask, 100-mL graduated cylinder, 1000-mL volumetric flask, 25-mL graduated cylinder, 250-mL beaker. In front: 20-mL pipet, 10-mL pipet.

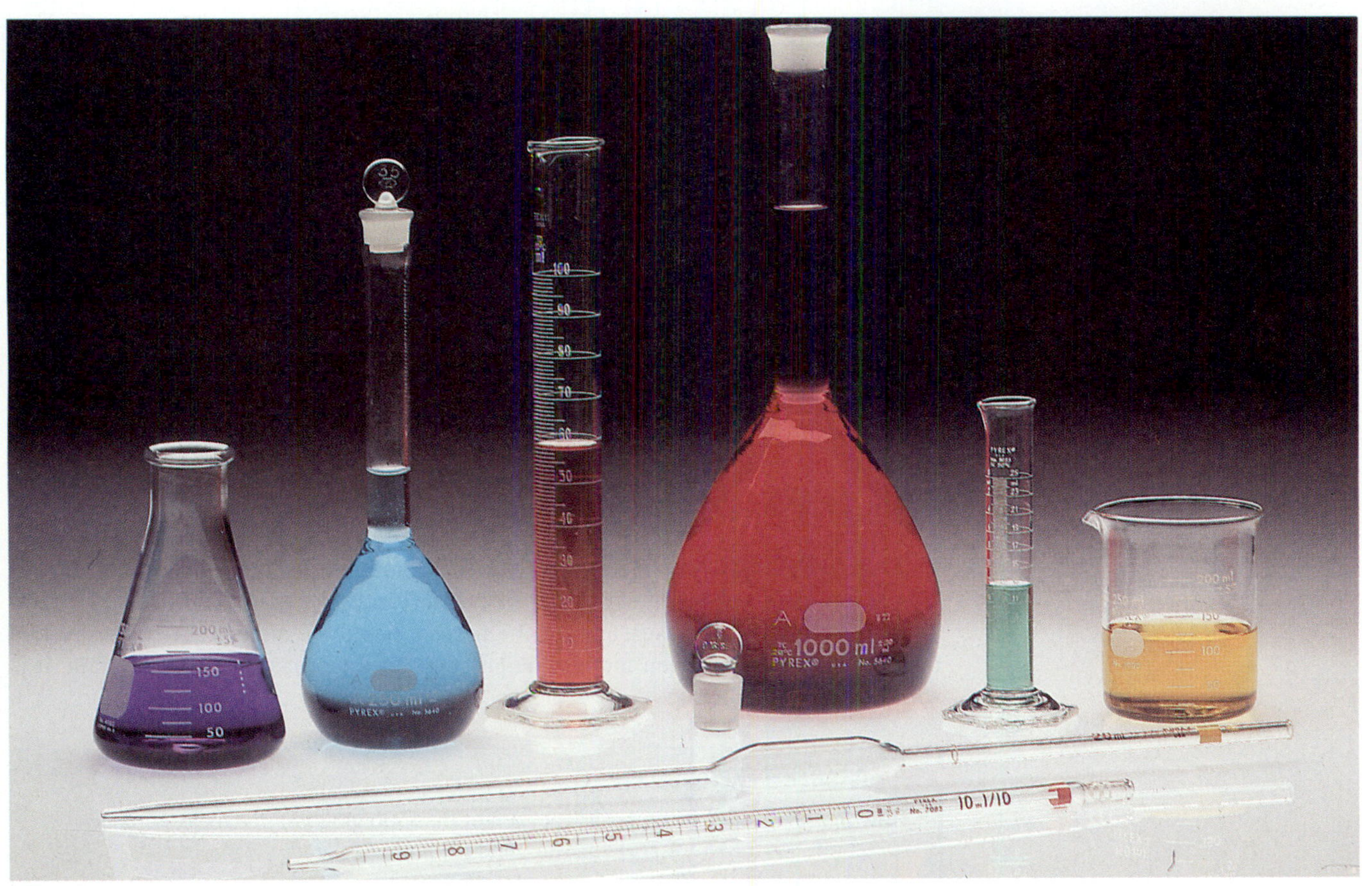

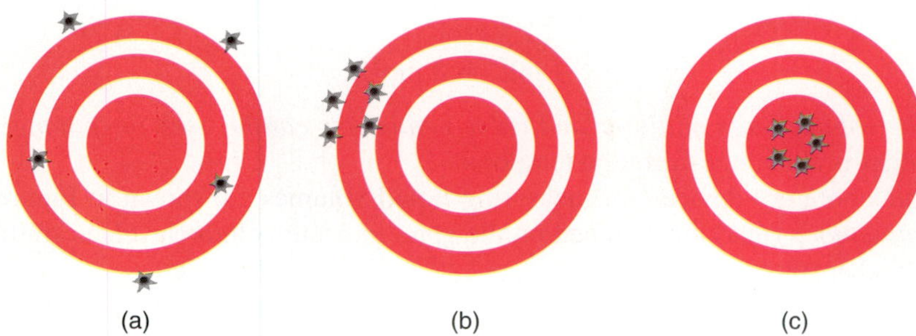

Figure 1.7

Three possible outcomes of shooting at a target. (a) The five shots were neither accurate nor precise. (b) The five shots were precise, but not accurate. (c) The five shots were accurate and precise.

contains some uncertainty in its last digit and is therefore only an approximation of the true value of the property being measured.

A reliable measurement is both accurate and precise. **Accuracy** is the extent to which a measured value coincides with the true or accepted value of the quantity measured. A volume measurement of 23.8 mL, for example, is reasonably accurate if the true volume is 23.7 mL, but it is much less accurate if the true volume is 21.4 mL. **Precision** refers to the "fineness" of a measurement as well as its reproducibility. By fineness we mean the number of digits in the measured value; a length measurement of 39.64 inches is a more precise measurement than one of 39 inches. Reproducibility is the extent to which repeated measurements agree with each other. One would probably be satisfied with the precision of a fever thermometer if three consecutive readings were 99.2, 99.3, and 99.2°F. One would not be satisfied with the precision, however, if the readings were 99.2, 97.6, and 101.4°F. It is important to remember that *precise measurements may not be accurate.* If the fever thermometer is poorly calibrated so that it consistently reads 0.5°F too high, the readings may be precise but they will not be accurate. The difference between accuracy and precision is also illustrated in Figure 1.7.

LEARNING HINT

The last reported digit in any measurement is assumed to be uncertain.

For most measurements the true value is not known, so accuracy cannot be determined. For this reason, scientists strive to make their measurements as precise as possible. The precision of a measurement is limited by the method and the instruments used. Figure 1.8 shows a block of wood measured by two rulers. The top ruler shows the block's length to be between 6 and 7 cm. There are no markings between the centimeter lines on this ruler, and the best we can do is to estimate the length as 6.7 or 6.8 cm, with the second digit uncertain. The bottom ruler provides a more precise value because it has millimeter markings. Estimating between these markings gives a length of 6.76 or 6.77 cm with the final digit again uncertain. No amount of peering at the lines could add another decimal place to this observation.

Significant Figures

The number of digits in a measured value is a rough but useful indication of its precision. Each digit obtained as a result of measurement is called a **significant figure**. The 6.77-cm length measurement in Figure 1.8 has three significant figures, while 6.8 cm has only two. Even though the last digit in a measurement is uncertain, it is still a significant figure.

PRACTICE EXERCISE 1.5

How many significant figures are in each of the following volume measurements: (a) 65 mL, (b) 173.4 mL, (c) 13.2 mL, and (d) 5 mL?

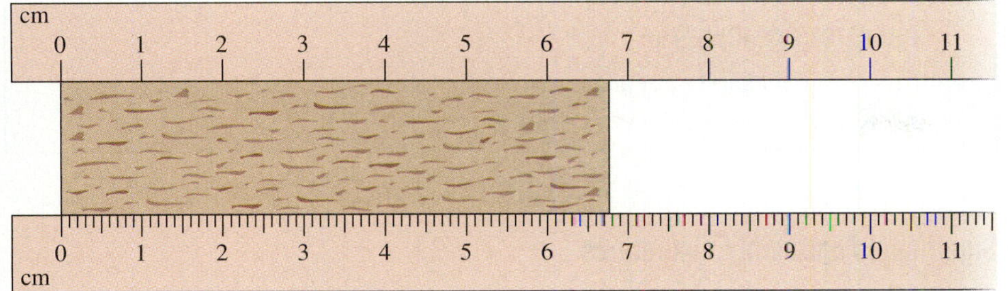

Figure 1.8

A block of wood measured by two different rulers. The top ruler shows the length to be between 6 cm and 7 cm. The bottom ruler gives a more precise measurement; it shows that the length is between 6.7 cm and 6.8 cm.

The zeros in a number warrant special attention. A zero that is the result of a measurement is significant, but zeros that serve only to mark the decimal point are not significant. The following four rules can be used to determine when a zero is significant.

1. *A zero between other significant figures is significant.* A volume of 1.05 mL, for example, has three significant figures; 2001 mL has four significant figures.

2. *Final zeros to the right of the decimal point are significant.* A mass of 6.30 g measured on a balance sensitive to 0.01 g has three significant figures because the final zero is the result of a measurement. A mass of 10.00 g measured on the same balance has four significant figures.

3. *Initial zeros are not significant.* Initial zeros, such as the two zeros in 0.028 L, serve only to show the position of a decimal point. For example, 28 mL has two significant figures—expressed as 0.028 L, it still has two significant figures. Similarly, 0.000601 m and 0.000610 m each have three significant figures.

4. *Final zeros in a number with no decimal point may or may not be significant.* If you give your age as 20 years, you mean not 19 and not 21. The zero in 20 years is significant. On the other hand, the distance from the earth to the sun is often given as 92,900,000 miles, an approximate distance accurate to only three significant figures. To avoid ambiguity, such numbers are better expressed in scientific notation (see Appendix A, Section A.1). When the mean distance from the earth to the sun is given as 9.29×10^7 miles, it is clear that only three significant figures are involved.

EXAMPLE 1.2

How many significant figures are in each of the following measurements? (a) 4.770 cm, (b) 0.0254 g, and (c) 3.02 s.

SOLUTION

(a) Rule 2 applies; 4.770 cm has four significant figures.

(b) Rule 3 applies; 0.0254 g has three significant figures.

(c) Rule 1 applies; 3.02 s has three significant figures.

PRACTICE EXERCISE 1.6

How many significant figures are in each of the following measurements? (a) 0.0060 g, (b) 7.03000 pm, (c) 0.000001 s, and (d) 1.800×10^{-6} m.

Significant Figures in Calculations

The number of significant figures in a calculated result depends on the number of significant figures in the data used for the calculation. If an employee with an actual income of $33,425 (five significant figures) were to file a tax return based on a rounded income of $30,000 (one significant figure), he or she would quickly receive a bill for the tax (and penalties) on $3425.

The following rule tells how to determine the number of significant figures in an addition or subtraction: *A sum or difference of two or more measurements has the same number of digits after the decimal point as the measurement with the least number of digits after the decimal point.*

EXAMPLE 1.3

Add 94.02 g + 61.1 g + 3.1416 g, and determine the number of significant figures in the answer.

SOLUTION

The measurement 61.1 g has the least number of digits after the decimal point (one); hence, the sum will have one digit after the decimal point.

$$\begin{array}{r|l}
94.0 & 2\ \ \ \ \text{g} \\
61.1 & \ \ \ \ \ \ \ \text{g} \\
3.1 & 416\ \text{g} \\
\hline
158.2 & 616\ \text{g}
\end{array} = 158.3 \text{ g after rounding off to one digit after the decimal point}$$

The answer has four significant figures.

Rounding off means finding a number that is closest to a given number but with fewer digits. The rules for rounding off a number are:

1. *The last digit to be retained is either kept or increased by one, whichever gives a value nearer to the original number.* The answer to Example 1.3 was rounded to 158.3 because 158.2616 is closer to 158.3 than to 158.2. Similarly, the number 3.674 would become 3.67 if one digit is dropped and 3.7 if two digits are dropped.

2. *When the digits to be dropped are a 5 or a 5 followed only by zeros, the last remaining digit is rounded to the nearest even number.* The numbers 4.775 and 6.485 round to 4.78 and 6.48. Rule 2 implies that the number 5 would be rounded up in half of such cases and rounded down in the other half.

EXAMPLE 1.4

Subtract 2.30642 s from 4.01 s and determine the number of significant figures in the answer.

SOLUTION

The measurement 4.01 s has two digits after the decimal point; therefore, the difference will have two digits after the decimal point.

$$\begin{array}{r|l} 4.01 & \quad s \\ -2.30 & 642\ s \\ \hline 1.70 & 358\ s \end{array} = 1.70 \text{ s after rounding to two digits after the decimal point}$$

The answer contains three significant figures.

PRACTICE EXERCISE 1.7

Perform the following calculations and round off each answer to the correct number of significant figures: (a) 6.8 mL + 71.35 mL, (b) 9.2241 cm − 3.43 cm, and (c) 0.0041 g − 21.33 g − 7.0844 g.

The following rule is used to determine the number of significant figures in a multiplication or division: *A product or quotient of two or more measurements has the same number of significant figures as the measurement with the least number of significant figures.*

EXAMPLE 1.5

Multiply 5.6432 by 0.020 and determine the number of significant figures in the answer.

SOLUTION

The number 5.6432 has five significant figures, while 0.020 has only two; thus, the product will have two significant figures. Hence, 5.6432 × 0.020 = 0.112864 = 0.11 after rounding to two significant figures.

EXAMPLE 1.6

During an automobile trip, a traveler pays \$39.65 for 31.8 gallons of gasoline. Calculate the average price per gallon.

SOLUTION

The average price per gallon is obtained by dividing \$39.65 by 31.8 gallons: \$39.65/31.8 gallons = \$1.25 per gallon. The answer has three significant figures because the number of gallons has three significant figures.

LEARNING HINTS

USE YOUR ELECTRONIC CALCULATOR. The number of digits shown by a calculator will vary with the model and with the calculator setting; hence, ELECTRONIC CALCULATORS DO NOT GIVE THE CORRECT NUMBER OF SIGNIFICANT FIGURES. You must round off at the end of the calculation.

When a calculation involves more than two steps, as in:

$$\frac{(A \times B \times C)}{(D + E)}$$

carry all the digits through to the end, and then round off.

Always keep in mind that the significant-figure rules are approximate and will produce misleading or contradictory results on occasion. More sophisticated (i.e., more complicated) techniques are available for obtaining unambiguous results.

$$\frac{1 \text{ inch}}{2.54 \text{ cm}} \quad \text{and} \quad \frac{2.54 \text{ cm}}{1 \text{ inch}}$$

If we multiply 100.0 cm by the factor on the left, centimeters will cancel, and the answer will be in inches:

$$100.0 \cancel{\text{cm}} \times \frac{1 \text{ inch}}{2.54 \cancel{\text{cm}}} = 39.37 \text{ inches}$$

The answer has four significant figures because 100.0 has four significant figures and 2.54 is exact.

These conversions illustrate the general rule that *a conversion factor is used as a multiplier to cancel unwanted units and replace them with new units.* This rule is a simple application of **dimensional analysis**, a powerful aid to problem solving in which the **dimensions** (units) of each quantity are included in each step of the calculation and are handled (multiplied, canceled, and raised to powers) just like numbers. Carrying units through a calculation helps in reasoning out the problem and helps ensure that the final answer is actually the quantity being sought.

Most problems involve more than one conversion. The next example shows how a series of conversion factors is used in a calculation.

EXAMPLE 1.8

The height of a 7-year-old girl is 4 ft, 1.5 in. Use the definitions 1 in = 2.54 cm and 1 ft = 12 in to convert the girl's height to meters.

SOLUTION

We will use the given definitions to find the girl's height in inches, then change inches to centimeters, and, finally, change centimeters to meters. First we convert 4 ft to inches:

$$4 \cancel{\text{ft}} \times \frac{12 \text{ in}}{1 \cancel{\text{ft}}} = 48 \text{ in}$$

Forty-eight is an exact number because 4 and 12 are exact numbers. The girl's height is 48 in + 1.5 in = 49.5 in. Now we convert inches to centimeters:

$$49.5 \cancel{\text{in}} \times \frac{2.54 \text{ cm}}{1 \cancel{\text{in}}} = 126 \text{ cm}$$

This answer is rounded off from 125.73 cm. Finally, we use the definition 1 m = 100 cm to find the height in meters:

$$126 \cancel{\text{cm}} \times \frac{1 \text{ m}}{100 \cancel{\text{cm}}} = 1.26 \text{ m}$$

PRACTICE EXERCISE 1.10

The gas tank of a car holds 17.0 gallons of gas. How many liters does the tank hold?

Conversion factors can be raised to powers in the same way that numbers are raised to powers. In the next example, a conversion factor for length is cubed to give a conversion factor for volume.

EXAMPLE 1.9

What is the volume in liters of a 10.0-cubic-foot refrigerator?

SOLUTION

Although Table 1.1 does not give the number of liters per cubic foot, we can make the conversion in steps. Our strategy will be to convert cubic feet to cubic inches, cubic inches to cubic centimeters, and cubic centimeters to liters. As in Example 1.8, each step will require a conversion factor; this time, however, we will assemble all the factors before doing the arithmetic.

Step 1: Convert cubic feet to cubic inches. Since 1 ft = 12 in, then $(1\text{ ft})^3 = (12\text{ in})^3$, and the volume in cubic inches is

$$10.0\ \cancel{\text{ft}^3} \times \frac{(12\text{ in})^3}{1\ \cancel{\text{ft}^3}}$$

Note that 1 ft^3 is the same as $(1\text{ ft})^3$.

Step 2: Convert cubic inches to cubic centimeters. Since 1 in = 2.54 cm, then $(1\text{ in})^3 = (2.54\text{ cm})^3$, and the volume in cubic centimeters is

$$10.0\ \cancel{\text{ft}^3} \times \frac{(12\ \cancel{\text{in}})^3}{1\ \cancel{\text{ft}^3}} \times \frac{(2.54\text{ cm})^3}{1\ \cancel{\text{in}^3}}$$

Step 3: Convert cubic centimeters to liters. Since 1 L = 1000 cm^3, the volume in liters is

$$10.0\ \cancel{\text{ft}^3} \times \frac{(12\ \cancel{\text{in}})^3}{1\ \cancel{\text{ft}^3}} \times \frac{(2.54\ \cancel{\text{cm}})^3}{1\ \cancel{\text{in}^3}} \times \frac{1\text{ L}}{1000\ \cancel{\text{cm}^3}}$$

$$= \left(10.0 \times 12^3 \times 2.54^3 \times \frac{1}{1000}\right)\text{L} = 283\text{ L}$$

Observe once again that each conversion factor was chosen to cancel out the preceding unit and replace it with another.

LEARNING HINTS

Use only the official symbol for an SI unit. (See Tables 1.2 and 1.3.)

Remember to square or cube the numbers when squaring or cubing the units.

PRACTICE EXERCISE 1.11

A standard football field is 100 yards long and $53\frac{1}{3}$ yards wide. Find its area in square meters.

EXAMPLE 1.10

A ball is rolling with a speed of 5.00 centimeters per second. Express this speed in meters per hour.

SOLUTION

We need to change centimeters per second (cm/s) to meters per hour (m/h). We will use one factor to convert centimeters to meters and another factor to convert seconds to hours. Note that hours must be in the denominator.

Step 1: 1 m = 100 cm; the speed in meters per second is

$$\frac{5.00\ \cancel{\text{cm}}}{1\ \text{s}} \times \frac{1\ \text{m}}{100\ \cancel{\text{cm}}}$$

Step 2: 3600 s = 1 h; the speed in meters per hour is

$$\frac{5.00\ \cancel{\text{cm}}}{1\ \cancel{\text{s}}} \times \frac{1\ \text{m}}{100\ \cancel{\text{cm}}} \times \frac{3600\ \cancel{\text{s}}}{1\ \text{h}} = 180\ \text{m/h}$$

PRACTICE EXERCISE 1.12

A fast-pitched baseball moves at a speed of 99 miles per hour. Convert this speed to meters per second.

The preceding examples illustrate the power of dimensional analysis in both simple and complex calculations. If the units do not cancel to give the desired unit, then at least one of the operations was incorrect. Note, however, that correct units do not guarantee a correct numerical answer. For example, a misplaced decimal point in Example 1.8 might have given the absurd answer of 12.6 m rather than 1.26 m for the girl's height. *Always check carefully both the arithmetic and the units for each calculation and, when possible, make sure that the result agrees with common sense.*

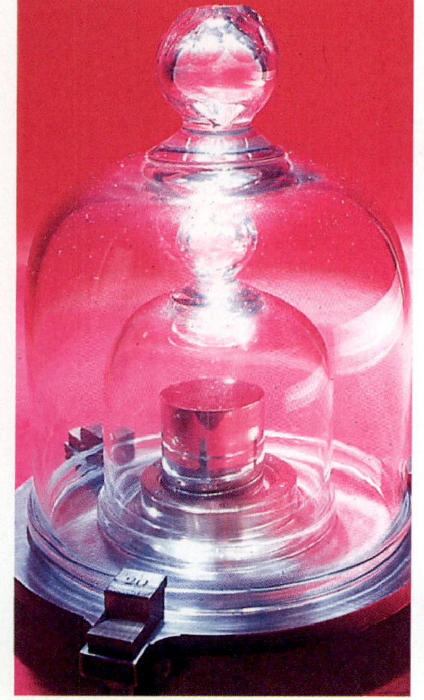

This duplicate of the international standard kilogram is called "kilogram 20." It is carefully preserved at the United States National Institute of Standards and Technology (formerly known as the National Bureau of Standards) located near Washington, D.C.

1.6 MASS, FORCE, AND WEIGHT

The **mass** of an object is a measure of the quantity of matter it contains. The SI unit of mass is the **kilogram** (kg), defined as the mass of a cylinder of platinum–iridium alloy carefully stored at the International Bureau of Weights and Measures in Sèvres, France. One kilogram is 2.205 pounds. The **gram** (g), one-thousandth of a kilogram, is a smaller and more practical mass unit for ordinary chemical work. Some approximate masses are given in Table 1.6.

EXAMPLE 1.11

How many kilograms of sugar are in a 10-lb sack?

SOLUTION

1 kg = 2.205 lb; therefore,

$$10\ \text{lb} \times \frac{1\ \text{kg}}{2.205\ \text{lb}} = 4.5\ \text{kg}$$

TABLE 1.6 Some Approximate Masses

Electron rest mass[a]	9.11×10^{-31} kg
Proton rest mass	1.67×10^{-27} kg
Carbon atom	1.99×10^{-26} kg
Water molecule	2.99×10^{-26} kg
U.S. Lincoln penny	3.11×10^{-3} kg = 3.11 g
One pound of sugar	0.4536 kg = 453.6 g
150-pound person	68.0 kg = 6.80×10^{4} g
Earth	5.977×10^{24} kg
Sun	1.993×10^{30} kg
Observable universe	1×10^{53} kg (minimum)

[a]The theory of relativity tells us that the mass of an object increases with its speed, becoming infinitely large as the speed approaches that of light. Hence, it is customary to report rest masses.

PRACTICE EXERCISE 1.13

One pound (1.00 lb) of butter is cut into 32 equal pats. What is the mass of each pat in (a) grams and (b) milligrams?

EXAMPLE 1.12

A bottle of aspirin contains 250 five-grain tablets. If one grain is equal to 64.80 milligrams, how many grams of aspirin are in the bottle? Assume that each tablet contains exactly five grains.

SOLUTION

We will use a three-step strategy: (1) use the factor 5 grains/tablet to find the mass of 250 tablets in grains; (2) use the factor 64.80 mg/grain to convert grains to milligrams; (3) convert milligrams to grams. The steps are combined as follows:

$$250 \text{ tablets} \times \underbrace{\frac{5 \text{ grains}}{1 \text{ tablet}}}_{\text{Step 1}} \times \underbrace{\frac{64.80 \text{ mg}}{1 \text{ grain}}}_{\text{Step 2}} \times \underbrace{\frac{1 \text{ g}}{1000 \text{ mg}}}_{\text{Step 3}} = 81.00 \text{ g}$$

Conversion factors such as 2.54 cm/in or 1 g/1000 mg are valid for any problem. The factor 5 grains/tablet used in the previous example is different, however, in that it is valid only for that particular problem (other aspirin preparations might contain 10 grains/tablet or 1.25 grains/tablet). Many problems can be solved by identifying quantities that are equivalent only in the context of the problem and making conversion factors from these quantities. Another illustration is given in the next example.

EXAMPLE 1.13

A certain salt solution contains 1.75 g of dissolved salt in each 35.0 mL of solution. What volume of the solution would contain 10.0 g of salt?

SOLUTION

In this problem we can equate 1.75 g of salt with 35.0 mL of solution. The resulting factors

$$\frac{1.75 \text{ g salt}}{35.0 \text{ mL solution}} \quad \text{and} \quad \frac{35.0 \text{ mL solution}}{1.75 \text{ g salt}}$$

apply only to this mixture. We choose the factor that cancels grams and leaves milliliters:

$$10.0 \text{ g salt} \times \frac{35.0 \text{ mL solution}}{1.75 \text{ g salt}} = 200 \text{ mL solution}$$

PRACTICE EXERCISE 1.14

A jewelry alloy contains 3.00 g of gold for every 2.00 g of platinum. (a) How many grams of the alloy will contain 0.500 kg of gold? (b) How many grams of platinum will be in the alloy of Part (a)?

Force and Weight

A **force** is most simply defined as any push or pull. The earth exerts a gravitational force that pulls objects downward, and if the object is attached to a scale like the one in Figure 1.9, the displacement of the indicator is a measure of the gravitational force. Experiments show that at any given location, the displacement varies directly with the mass of the object; doubling the mass, for example, produces twice the displacement. Thus the downward force f is proportional to the mass m, or

$$f = m \times g \tag{1.1}$$

where g, a quantity called the **gravitational acceleration**, can be determined experimentally. Equation 1.1 shows that the force of gravity acting on an object is not the same as its mass but is its mass multiplied by some number.

The force of gravity decreases slightly with increasing altitude; it is smaller on top of a mountain than at sea level. Hence, the value of g varies from point to point on the earth's surface. We will use a standard g value that has been defined as exactly 9.80665 m/s^2.

The gravitational force exerted on an object is called its **weight**. It is important to realize that mass and weight are different concepts. The mass of an object, that is, the amount of matter in the object, is the same on the earth and moon, but the weight or gravitational force exerted on the object is not. The following example should help clarify the distinction.

Figure 1.9

A spring scale measures weight. The object to be weighed is suspended from the bottom of a spring behind the dial; as the spring stretches, it moves the pointer. The elongation of the spring is proportional to the gravitational force on the object; the pointer moves farther for the 17.7-newton weight (1.8-kg mass) than for the 4.1-newton weight (0.42-kg mass).

EXAMPLE 1.14

Calculate the weight of a 95.0-kg astronaut (a) on the earth's surface and (b) on the moon. Assume that the gravitational acceleration on the moon is one-sixth of that on earth.

SOLUTION

(a) The mass of the astronaut is $m = 95.0$ kg. The value of g on the earth's surface is 9.81 m/s^2. Substituting these values into Equation 1.1 gives

$$\text{weight} = \text{gravitational force} = m \times g$$
$$= 95.0 \text{ kg} \times 9.81 \text{ m/s}^2 = 932 \text{ kg}\cdot\text{m/s}^2$$

(b) The mass of the astronaut is the same on the moon as on the earth. However, since the gravitational acceleration on the moon is one-sixth of that on earth, the astronaut's weight will also be one-sixth of that on earth, or 155 kg·m/s^2.

Example 1.14 shows that the SI unit of force is kg·m/s^2 (kilogram-meters per second squared). This combination of units is called a **newton** (N) in honor of the English physicist and mathematician Sir Isaac Newton (1642–1727) who discovered the law of gravitation. Because 1 N = 1 kg·m/s^2, the weight of the 95.0-kg astronaut is 932 N on the earth and 155 N on the moon. The newton is another example of a derived unit, a unit made by combining two or more other units, in this case, mass, length, and time. The unit for gravitational acceleration, meters per second squared, is also a derived unit.

TABLE 1.7 Densities of Various Substances and Mixtures[a]

Substance	Density (g/mL)
Air (dry)	0.001205
Aluminum	2.70
Blood (whole)	1.05 (37°C)
Bone	1.6
Copper	8.96
Diamond	3.513 (25°C)
Ethanol (ethyl alcohol)	0.7893
Gold	19.32
Granite	2.7
Iron	7.87
Lead	11.35
Mercury	13.55
Osmium	22.57
Quartz	2.65
Seawater	1.025 (15°C)
Sodium chloride	2.165
Water (ice)	0.917 (0°C)
Water (steam)	0.000596 (100°C)
Water (liquid)	0.99987 (0°C)
	1.00000 (3.98°C)
	0.99823 (20°C)
	0.95838 (100°C)

[a]At 20°C unless otherwise noted.

LEARNING HINT

Solutions of commercial acids and other chemicals often have specific gravity data on their labels. The **specific gravity** (sp gr) of a liquid is the ratio of its density to the density of water:

$$\text{sp gr} = \frac{\text{density of liquid}}{\text{density of water}}$$

At room temperature, the density of water is 1.00 g/mL; hence, for most practical purposes, the specific gravity of a liquid is numerically equal to its density in grams per milliliter.

PRACTICE EXERCISE 1.17

Compare the volume of 100.0 g of aluminum at 20°C with that of 100.0 g of lead at the same temperature. Refer to Table 1.7 for density data.

Most substances expand when heated, so their densities decrease with increasing temperature. Liquid water below 3.98°C is an exception. Water decreases in volume when heated between 0°C and 3.98°C, and expands when heated above 3.98°C. Water has its maximum density of 1.00000 g/mL at 3.98°C. The density of ice is less than that of liquid water, which explains why icebergs and ice cubes float in water. This very important phenomenon is discussed further in Chapter 12.

1.7 WORK, HEAT, AND OTHER FORMS OF ENERGY

Energy is the ability to do **work**, and many examples of energy were given in Section 1.1. Work, which includes tasks such as carrying a bag of groceries, hammering a nail into wood, or swimming in a 400-m relay, is done whenever force is used to move an object. In fact, physicists have formally defined work (w) as the product of a force f and the distance d through which the force operates:

$$w = f \times d \tag{1.3}$$

Because force is expressed in newtons (Section 1.6) and distance in meters, the SI unit of work and other forms of energy is the *newton-meter,* also called the **joule** (J); 1 J = 1 N·m. The joule is named in honor of James Prescott Joule (1818–1889), an English physicist whose experiments led to the discovery of the law of conservation of energy (Chapter 6).

A joule is not very much energy by human standards. Only 1 J of work is done, for example, when a medium-sized 85-g lemon is lifted 4 ft (1.2 m) against gravity. A great deal of work, however, goes on in the environment and in living things. Our surroundings are constantly shaped and reshaped by wind, rain, and flowing water. Life depends on the continuous circulation of blood and other substances, the contraction and relaxation of the heart and other muscles, and the transport of nutrients and wastes to and from the cells. Many of these processes have a chemical origin; each of them involves work, and each requires the action of a force through a distance.

Kinetic Energy and Potential Energy

The energy an object possesses by virtue of its motion is called **kinetic energy**. An object of mass m moving with a speed v has a kinetic energy (E_k) given by

$$E_k = \frac{1}{2} m\text{v}^2 \tag{1.4}$$

EXAMPLE 1.17

Calculate the kinetic energy of (a) a 5.00-kg bowling ball rolling down the lane at 10.0 m/s and (b) a hydrogen atom of mass 1.67×10^{-27} kg moving through outer space at 450 m/s.

SOLUTION

(a) Substituting m = 5.00 kg and v = 10.0 m/s into Equation 1.4 gives

$$\begin{aligned} E_k &= \frac{1}{2} m\text{v}^2 \\ &= \frac{5.00\ \text{kg} \times (10.0\ \text{m/s})^2}{2} \\ &= 250\ \text{kg·m}^2/\text{s}^2 = 250\ \text{N·m} = 250\ \text{J} \end{aligned}$$

since 1 N = 1 kg·m/s^2 and 1 J = 1 N·m.

(b) Substituting the mass and speed of the hydrogen atom into Equation 1.4 gives

$$E_k = \frac{1.67 \times 10^{-27}\ \text{kg} \times (450\ \text{m/s})^2}{2} = 1.69 \times 10^{-22}\ \text{J}$$

Even though the hydrogen atom has a greater speed than the bowling ball, its kinetic energy is less because its mass is smaller.

between the freezing and boiling points of water. Room temperature on the Celsius scale is normally taken to be 25°C, although most people would now consider this to be warm. The Celsius scale is not part of the SI system, but it is too widely used to abandon at this time. The SI system uses the Kelvin scale, which is discussed in Chapter 5.

CHAPTER REVIEW

LEARNING OBJECTIVES BY SECTION

1.1 1. Give examples of matter and energy.

1.2 1. Describe the steps of the scientific method.
2. State the two ways in which numerical data are obtained.
3. State the SI units of length, mass, and time.

1.3 1. Visualize the magnitude of a meter, a centimeter, and a millimeter.
2. Visualize the magnitude of a liter and a cubic centimeter.
3. Convert multiples and fractions of the meter (kilometers, centimeters, etc.) to meters.
4. Convert multiples and fractions of the liter to liters.

1.4 1. Distinguish between accuracy and precision.
2. State the number of significant figures in a measurement.
3. Perform calculations to the correct number of significant figures.
4. Round off a number to any specified number of digits.
5. Distinguish between systematic and random errors.

1.5 1. Use units and conversion factors in calculations.

1.6 1. Perform calculations involving units of mass.
2. Distinguish between mass and weight.
3. Calculate the weight of an object from its mass and vice versa.
4. Calculate the mass, volume, or density of an object, given two of the three quantities.

1.7 1. Give examples of work.
2. State the SI unit of energy.
3. Calculate the kinetic energy, mass, or speed of a moving object, given two of the three quantities.
4. Distinguish between potential energy and kinetic energy.
5. Describe the potential energy changes that occur when charged particles come together or move apart.
6. Distinguish between heat and temperature.
7. Convert joules to calories and vice versa.

KEY TERMS BY SECTION

1.1 Chemistry
Energy
Matter

1.2 Data
Derived unit
Law
Measurement
Metric system
Scientific method
SI units
Theory

1.3 Cubic decimeter (dm^3)
Liter (L)
Meter (m)

1.4 Accuracy
Precision
Random error
Rounding off
Significant figure
Systematic error

1.5 Conversion factor
Dimensional analysis

1.6 Density
Force
Gram (g)
Gravitational acceleration (g)
Kilogram (kg)
Mass (m)
Newton (N)
Specific gravity (sp gr)
Weight

1.7 Calorie (cal)
Celsius scale
Degree Celsius (°C)
Heat
Joule (J)
Kinetic energy
Potential energy
Temperature scale
Thermal equilibrium
Thermometer
Work

IMPORTANT EQUATIONS

1.1 $f = m \times g$

1.2 $d = \frac{m}{V}$

1.3 $w = f \times d$

1.4 $E_k = \frac{1}{2} mv^2$

FINAL EXERCISES

Answers to exercises with blue numbers are given in Appendix D.

All numerical problems should be worked out to the correct number of significant figures.

PART A. QUESTIONS AND PROBLEMS BY SECTION

Scientific Method: Measurement (Introduction; Sections 1.1 and 1.2)

1.1 Explain why a knowledge of chemistry is important to the following professions:
(a) medicine **(d)** agriculture
(b) geology **(e)** astronomy
(c) electrical engineering **(f)** computer science

1.2 Define each of the following terms:
(a) data **(d)** theory
(b) measurement **(e)** observation
(c) scientific law **(f)** experiment

1.3 Describe each step of the scientific method. Illustrate each step with examples other than those given in the text.

1.4 Discuss the statement: Theories can never be proved, only disproved.

1.5 List the seven physical quantities required for a complete system of measurement, and give the SI base unit for each one.

1.6 Suggest some reasons why the metric system of units is more popular than the English system.

1.7 What is a "derived unit"? Express each of the following physical quantities in units derived from SI base units:
(a) speed **(d)** density
(b) area **(e)** force
(c) volume **(f)** energy

1.8 Is counting a form of measurement? Explain.

Length and Volume (Section 1.3)

(Most of the problems will require a knowledge of significant figures and dimensional analysis.)

1.9 (a) State the SI units of length and volume. (b) What relation does the *liter* bear to the SI unit of volume?

1.10 Define and give symbols for the following prefixes:
(a) kilo **(d)** milli
(b) centi **(e)** mega
(c) micro **(f)** nano

1.11 A basketball player is 2.2 m tall. Express the player's height in centimeters, kilometers, and nanometers. Which of these four units do you prefer for describing a person's height?

1.12 The radius of a magnesium atom is 1.60×10^{-10} m. What is the radius in (a) picometers, (b) nanometers, and (c) angstroms?

1.13 Give the number of
(a) millimeters in a centimeter
(b) square millimeters in a square centimeter
(c) cubic millimeters in a cubic centimeter

1.14 Give the number of
(a) centimeters in a decimeter
(b) square centimeters in a square decimeter
(c) cubic centimeters in a cubic decimeter

1.15 A tennis court for doubles play is 78 ft long and 36 ft wide. Convert these measurements to meters and find the area of the court in square meters.

1.16 A room is 12.2 ft long, 10.6 ft wide, and 8.0 ft high. What is the volume of the room in cubic meters?

1.17 The diameter of a regulation tennis ball is 2.50 in. Find its volume in (a) cubic centimeters, (b) cubic decimeters, and (c) cubic meters.

1.18 Find the volume of a 750-mL wine bottle in (a) cubic centimeters, (b) cubic decimeters, and (c) cubic meters.

1.19 The radius of a silicon atom is 118 pm. Refer to Table 1.5 and estimate its volume in cubic picometers.

1.20 A bacterium has the shape of a cylindrical rod about 1.2 μm in diameter and 2.1 μm long. Refer to Table 1.5 and estimate its volume in cubic micrometers.

Reliability of Measurements (Section 1.4)

1.21 Distinguish between accuracy and precision. If you weigh an object, what might you do to evaluate the precision of your result? How could you evaluate its accuracy? Which is easier to evaluate?

1.22 An inexperienced laboratory student is trying to weigh a known 10.000 g mass. His first set of measurements is 10.184, 10.182, and 10.186 g. His second set of measurements is 10.350, 9.750, and 10.644 g. His third set of measurements is 10.001, 9.999, and 10.002 g. Which set of measurements shows (a) good accuracy and good precision, (b) poor accuracy and poor precision, and (c) poor accuracy and good precision?

1.23 State whether the errors involved in the following procedures are systematic or random:
(a) Liquid volumes are measured in a graduated cylinder whose etched markings are fuzzy.
(b) In calculating income tax, you round off each item to the nearest dollar.
(c) A laboratory worker weighs samples of a chemical on a damp day and ignores the fact that they all absorb moisture from the air.

1.24 State whether the errors involved in the following procedures are systematic or random:
(a) An experiment is supposed to be carried out at 25.0°C, but a thermostat allows the temperature to fluctuate between 24.5°C and 25.5°C.
(b) A spring scale calibrated at sea level is used for weighing an object at the top of Mount Everest.
(c) You were supposed to measure the densities of several liquids at 25°C, but you measured them at 27°C.

1.25 Which of the following numbers are obtained by measurement?
(a) 2 dozen eggs
(b) 227 g of sulfur
(c) 16.0 L of gasoline
(d) a woman is 18 years old
(e) 1000 mL/L
(f) an oil tank holds 550 gallons

1.26 Which of the following numbers are exact?
(a) a 125-lb person
(b) 6 books
(c) a crowd of 10,000 people
(d) 28.5 miles per gallon
(e) 2.54 cm/in
(f) a class of 48 students

1.27 A technician reports the following values for average numbers of red cells per cubic millimeter of blood: (a) normal males, 5,267,250; (b) normal females, 4,968,667; and (c) newborns, 7,000,000. Round these values off and express each one unambiguously to three significant figures, a reasonable precision for a conventional blood count.

1.28 How many significant figures are in each of the following measurements?
(a) .003 m
(b) 0.003 m
(c) 0.00300 m
(d) 6.022137×10^{23} atoms
(e) 2×10^{26} m
(f) 98.6°F

1.29 How many significant figures are in each of the following measurements?
(a) 22.414 L
(b) 0.040 g/L
(c) 0.0040 g/L
(d) 2.9979×10^{8} m/s
(e) 6×10^{-13} g
(f) 105 miles/h

1.30 How many significant figures are in each of the following measurements? Identify any measurement in which the number of significant figures is uncertain.
(a) 10.200 s
(b) 10,200 s
(c) 0.01020 s
(d) 10.2 million s
(e) 102,000,000 s
(f) 1.02×10^{6} s

1.31 Perform the following calculations and express each answer to the correct number of significant figures:
(a) $(3.0 \times 22.4)/1.120$
(b) $56.85 - 9.9 - 43.214$
(c) $(4.38 \times 10^{4})/(5.1 \times 10^{-3})$
(d) $22.22 + 2.71828 + 2001 + 0.03$

1.32 Perform the following calculations and express each answer to the correct number of significant figures:
(a) $15.9563 + 0.00636 + 0.03670$
(b) 4.184×1.98719
(c) $(34.0 + 2.089) \times 12$
(d) $(2.564 - 2.508)/(18.22 - 9.007)$

1.33 Perform the following calculations and express each answer to the correct number of significant figures:
(a) $(3.61 \times 10^{5}) + (2.75 \times 10^{4})$
(b) $(9.1095 \times 10^{-28}) \times (6.022 \times 10^{23})$
(c) $1.0073 + (5.486 \times 10^{-4})$
(d) $(7.01 \times 10^{7})(3.0 \times 10^{24})/(6.219 \times 10^{-16})$

1.34 Perform the following calculations and express each answer to the correct number of significant figures:
(a) $\dfrac{(3.01 \times 10^{5}) + (4.22 \times 10^{4})}{(2.33 \times 10^{8}) - (5.11 \times 10^{7})}$
(b) $\dfrac{(2.481 \times 10^{-5})(3.42 \times 10^{8})(2.2832 \times 10^{-10})}{(3.0 \times 10^{6})(5.24 \times 10^{-7})}$
(c) 4.368 L + 0.2241 L + 275.6 mL + 14.34 mL (Give your answer in liters.)
(d) (1.544 m + 2.34 cm − 6.735 dm) × 4.3 nm (Give your answer in square meters.)

Dimensional Analysis (Section 1.5)

1.35 Write two conversion factors for each of the following statements:
(a) 1 kilometer = 0.6214 mile
(b) 1 gallon = 3.785 liters

(c) 453.6 grams = 1 pound
(d) 4.184 joules = 1 calorie
(e) 1000 kilograms = 1 metric ton
(f) 1 hour = 3600 seconds

1.36 Perform the following conversions:
(a) 45.0 kcal to kJ
(b) 1.00 g/cm^3 to kg/dm^3
(c) 9.81 m/s^2 to $miles/h^2$
(d) 1 ton (2000 lb) to metric tons
(e) 2.9979×10^8 m/s to miles/h
(f) 179 lb to kg

1.37 The speed limit on many roads in the United States is 55 miles/h. What is the speed limit in (a) km/h and (b) m/s?

1.38 An atom of oxygen is moving with a speed of 482 m/s. Express this speed in (a) km/h and (b) miles/h.

1.39 The average gasoline mileage of a certain automobile is 32 miles/gal. Express this mileage in km/L.

1.40 A metric ton is 1000 kg. A short ton is 2000 lb. How many short tons are in 1.00 metric ton?

1.41 Water drips from a leaky faucet at a rate of 0.050 mL/s. Express this rate in L/h.

1.42 A homeowner buys 475 gal of fuel oil. It takes 12.3 minutes to deliver the oil from the truck to the storage tank. Calculate the delivery rate in L/s.

Mass, Force, and Weight (Section 1.6)

1.43 Distinguish between weight and mass. How are the two properties related?

1.44 Could you lose weight without losing mass? How?

1.45 Does a double-pan balance measure mass or weight? A single-pan balance? A spring scale? Explain.

1.46 What is the SI base unit of mass? Which mass unit is usually used in the chemistry laboratory? Why?

1.47 A woman is 5 ft 6 in tall and weighs 145 lb. Express (a) her height in meters, (b) her mass in kilograms, and (c) her weight in newtons.

1.48 At 25°C, one gallon of water weighs 8.230 lb. What is (a) its mass in kilograms and (b) its weight in newtons?

1.49 Add the following masses and express the total in grams: 2.125 kg, 0.4043 kg, 600.4 g, 15.25 g.

1.50 Add the following masses and express the total in grams: 3.8635 g, 50.19 cg, 2 dg, 18 mg, 5.3 mg.

1.51 If 143.5 mL of beer contains 6.0 g of alcohol, how many grams of alcohol are in 1.00 L of beer?

1.52 A 100.0-mL sample of vinegar contains 4.64 g of acetic acid. How many liters of vinegar will contain 125 g of acetic acid?

1.53 At room temperature, 100 mL of water will dissolve 71 g of Epsom salts. How many grams will dissolve in 55 mL of water?

1.54 A solution contains 0.15 g of sodium chloride (table salt) per milliliter. How many liters of this solution contain 225 g of sodium chloride?

1.55 It has been estimated that a cubic meter of seawater contains 1.30 kg of magnesium in the form of its soluble salts. How many grams of magnesium are in 200 mL (about one cupful) of seawater?

1.56 The label on a well-known cereal reports 160 mg of potassium per 1-oz serving. How many milligrams of potassium are in 2.5 servings of the cereal?

Density (Section 1.6)

1.57 A certain brass knob weighs 88.9 g. When the knob is submerged in water, it displaces a volume of 10.5 mL. Find the density of the brass in g/mL.

1.58 A cylindrical tantalum rod 5.0 mm in diameter and 3.0 cm long has a mass of 9.80 g. Find its density in g/cm^3.

1.59 Calculate and compare the mass in grams of 1.00 L of (a) water at 3.98°C, (b) ice at 0°C, and (c) steam at 100°C.

1.60 Calculate the volume in cubic centimeters of (a) 5.0 g of lead, (b) 5.0 g of water, and (c) 5.0 g of air. Assume a temperature of 20°C.

1.61 The density of mercury is 13.6 g/cm^3. Convert this density to (a) kg/dm^3 and (b) lb/ft^3.

1.62 The density of water at 3.98°C is 1.000 g/mL. Convert this density to (a) kg/L and (b) lb/ft^3.

1.63 One cubic centimeter of solid boric acid weighs 1.435 g. What size bottle (in liters) is needed to hold 1.00 kg of boric acid?

1.64 The density of tungsten metal is 19.35 g/cm^3. What is the mass in kilograms of 1.00 dm^3 of tungsten metal?

1.65 (a) Find the mass in grams of a cylindrical column of mercury 76.0 cm high and 2.00 cm^2 in cross-sectional area. (b) What volume in liters would be occupied by an identical mass of water?

1.66 The density of benzene is 0.879 g/mL at 20°C. How many metric tons of benzene are required to fill a 12,000-gal tank car?

1.67 The density of ethyl alcohol (ethanol) is 0.7893 g/mL at 20°C. Calculate its specific gravity at this temperature.

1.68 A 24-hour urine specimen has a volume of 1250 mL and weighs 1275 g. What is its specific gravity?

Heat, Work, and Energy (Section 1.7)

1.69 Define *work.* What is the SI unit of work?

1.70 Define *heat.* What is the SI unit of heat?

1.71 What is the difference between kinetic energy and potential energy? What is the SI unit of energy?

1.72 Describe what happens to the kinetic and potential energies of a pendulum as it swings back and forth under the influence of gravity. What happens to the speed of the pendulum as it swings back and forth? Assume that the total energy of the pendulum is constant.

1.73 Does the potential energy increase or decrease when (a) two positive charges are pushed together, (b) two negative charges fly apart, and (c) a positive and a negative charge are separated?

1.74 What is the difference between heat and temperature? If two objects undergo the same temperature change, will they necessarily absorb the same amount of heat? Explain.

1.75 A constant force of 15.0 N operates through a distance of 500 cm. Calculate the work done in (a) joules and (b) calories.

1.76 How many joules of work are required to lift a 5.25-lb textbook from the floor to the top of a 31-inch-high desk?

1.77 Calculate the kinetic energy in joules of (a) a standard baseball (0.64 lb) pitched at a speed of 88 miles/h and (b) an oxygen molecule of mass 5.31×10^{-26} kg moving at a speed of 500 m/s.

1.78 Calculate the kinetic energy in joules of (a) a 150-lb person running a mile in 4.00 min and (b) a neon atom with a mass of 3.35×10^{-23} g moving at a speed of 100 m/s.

1.79 The kinetic energy of a moving electron is 1.1×10^{-19} J. Its mass is 9.1×10^{-31} kg. Calculate its speed in m/s.

1.80 A hydrogen atom of mass 1.67×10^{-27} kg travels through outer space with a kinetic energy of 3.5×10^{-24} J. What is its speed in m/s?

1.81 The amount of heat required to vaporize 18.0 g of water at 100°C is 40.67 kJ. How many calories are required to vaporize 1.00 g of water at this temperature?

1.82 It takes 15.4 cal of heat to melt 1.00 g of gold at its normal melting point. How many kilojoules are required to melt 200 g of gold?

1.83 The melting point of platinum is 1772°C. Convert this to degrees Fahrenheit.

1.84 The boiling point of chloroform is 143°F. Convert this to degrees Celsius.

PART B. MISCELLANEOUS QUESTIONS AND PROBLEMS

1.85 Debate the proposition: The scientific method is useful only in science.

1.86 According to Newton's theory of gravitation, objects fall to the earth because they are acted on by the force of gravity. If you saw an object rise when you expected it to fall, would you assume that Newton's theory had been disproved? Why or why not? Discuss what it would take to overthrow an established theory.

1.87 Do you think that someday it will be possible to measure distance exactly? Explain.

1.88 Can you think of any branch of science that does not involve measuring or counting in at least some of its investigations?

1.89 A new field called *nanotechnology* deals with the construction of circuits and machines with extremely small components. What does this term imply about the size of the components? Would you expect their dimensions to be more comparable to human hairs, to blood cells, to atoms, or to protons? (Consult Tables 1.3 and 1.4.)

1.90 For thousands of years people have estimated distances using parts of their bodies as standards. The following measurements should help you in making similar estimates.

- **(a)** How many millimeters wide is each of your fingernails?
- **(b)** How many centimeters wide is the palm of your right hand?
- **(c)** How many centimeters are between the tip of your right thumb and the first joint before the thumbnail?
- **(d)** Spread your fingers as far apart as possible and measure the distance between the tip of your thumb and the tip of your little finger.
- **(e)** Use your hands and the above measurements to estimate the dimensions of this book and the dimensions of your desk. Compare your estimates with the actual dimensions as determined with a meter stick.

1.91 An old edition of the *Handbook of Chemistry and Physics* contains 1968 pages. The total thickness of the pages without the cover is 5.74 cm. Estimate the leaf thickness in (a) millimeters and (b) micrometers. Keep in mind that there are two pages on each leaf of the book.

1.92 An adult is told to take 1 teaspoonful (about 5 mL) of a liquid cough medicine every 6 hours. If the medicine contains 2 g of glyceryl guaiacolate for every 100 mL, how many grams of glyceryl guaiacolate will be taken in 1 day?

1.93 A bottle of Tedral Pediatric Suspension contains 44 mL of medication. Each teaspoonful (about 5 mL) contains 4 mg of phenobarbital, 65 mg of theophylline, and 12 mg of ephedrine hydrochloride. For frequent attacks of bronchial asthma, the label recommends 1 teaspoonful four times daily per 60 lb of body weight.

(a) How many milligrams of each of the active ingredients will a 90-lb child receive in 1 day?

(b) If one grain equals 64.80 milligrams, how many grains of phenobarbital will the child receive in 1 day?

(c) For about how many days will one bottle last?

1.94 A 3-year-old child weighing 35 lb is suffering from a middle ear infection for which ampicillin is prescribed. The *Physicians' Desk Reference* recommends that this antibiotic be administered at a daily rate of 100 mg ampicillin per kg of body weight. It also recommends that the medication be given in equally divided doses at 8-hour intervals.

(a) How many milligrams of ampicillin should be given to this child every 8 hours?

(b) The ampicillin is supplied as an oral suspension containing 250 mg ampicillin per 5 mL suspension. If 1 teaspoonful contains about 5 mL, how should the child's parent administer the medicine? Give a practical answer.

1.95 A regulation basketball weighs about 21 oz and has a circumference of about 30 in. Estimate its average density in g/cm^3. (*Hint:* The circumference of a circle is $2\pi r$, where r is the radius of the circle.)

1.96 A major source of iron is the mineral magnetite, which has a density of 5.2 g/cm^3 and contains 72.4% iron by mass. How many cubic meters of magnetite are needed to supply one metric ton (1000 kg) of iron?

1.97 How many grams of gold will have the same volume as 100.0 g of copper at 20°C?

1.98 The naked eye cannot distinguish between pure gold and gold adulterated with copper. We can tell the difference, however, by submerging the objects in water.

(a) Would an object made of pure gold displace a greater or smaller volume of water than an adulterated object of the same mass? Explain.

(b) A statuette consisting of a gold and copper alloy weighs 210 g and displaces 14.0 mL of water. Estimate the percentage by mass of gold in the statuette. Assume that volumes are additive in the formation of copper–gold alloys.

1.99 A solution is prepared by mixing 50.0 mL of ethanol with 100.0 mL of water at 20°C. Calculate the percentage by mass of ethanol in this solution.

1.100 Each gram of sugar provides about 3.9 kcal of energy to the person who eats it. How many kilojoules of energy are provided by 1.00 lb of sugar?

1.101 The text states that it takes about 1 J of energy to lift an 85-g lemon 4 ft against gravity. Perform a calculation verifying this statement. (*Hint:* consider Equations 1.1 and 1.3.)

1.102 What characteristics must a liquid have to be used in a thermometer? Could liquids other than mercury be used? Could a gas be used?

1.103 Derive the formula (see p. 27) relating Fahrenheit temperature to Celsius temperature.

1.104 At what temperature will both the Fahrenheit and Celsius scales give the same reading?

CHAPTER 2

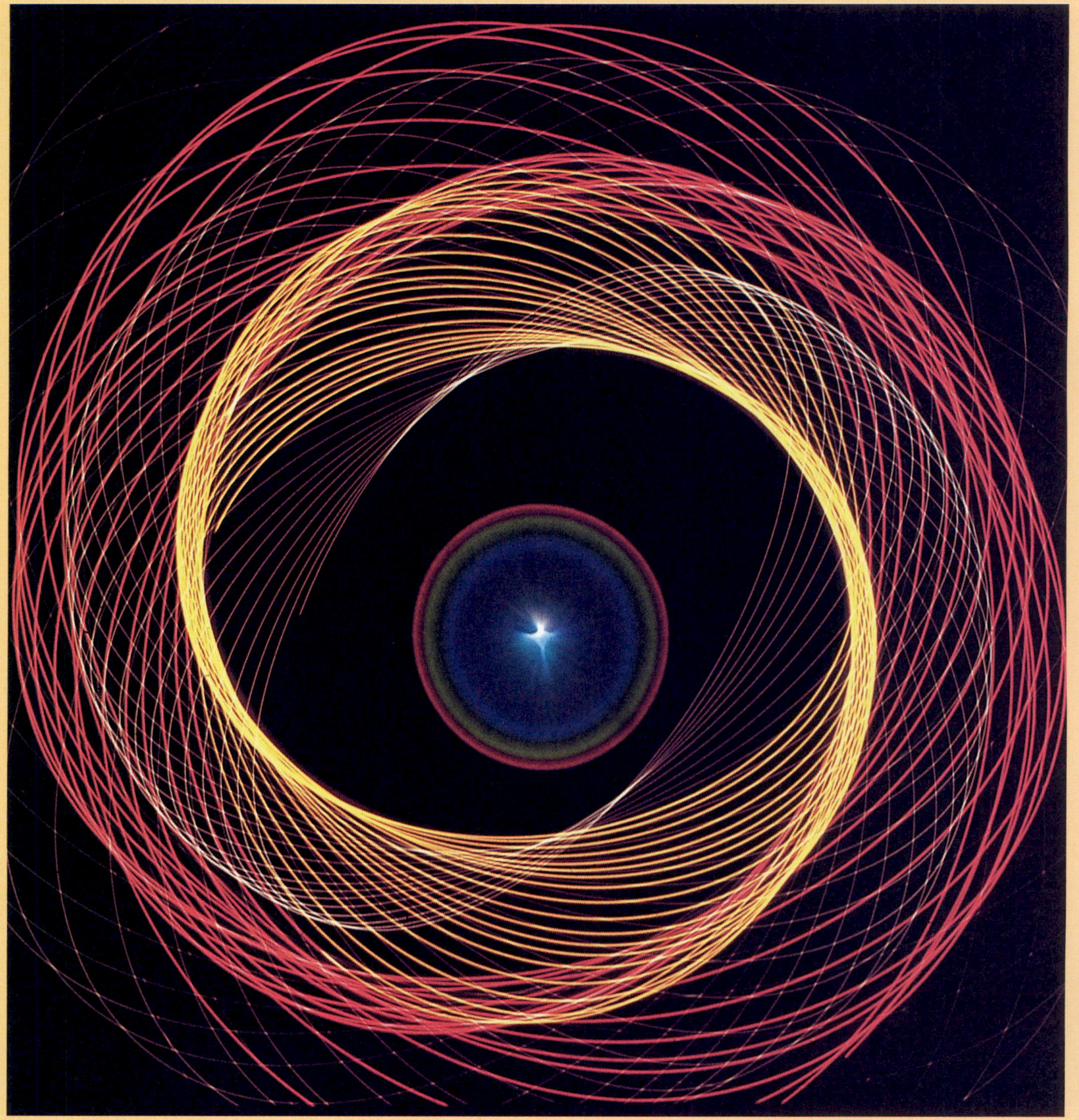

"Anatomy of an Atom," by Dr. E.R. Degginger. This photograph of orbiting points of light in a darkened room is one artist's conception of the electronic structure of an atom. The actual structure has never been observed by the human eye.

ATOMS, MOLECULES, AND IONS

OUTLINE

PREVIEW

What is matter made of? People have asked this question for thousands of years. The ancient philosopher Democritus believed that the universe consisted of an infinite number of indestructible atoms moving in empty space. He thought that atoms differed in size, weight, and shape, but that all atoms were made from the same material. Democritus had no experimental evidence to back up his ideas, and most ancient thinkers disagreed with him. Instead they sided with Aristotle, who believed that matter was continuous, infinitely divisible, and made up of four elements—earth, air, fire, and water. This view changed in the eighteenth and nineteenth centuries when the advent of modern measuring methods allowed chemists to accumulate data that could only be explained in terms of an atomic theory. In 1803 John Dalton, an English schoolmaster, published the first carefully reasoned arguments to show that atoms must exist. Today, we accept atoms as the building blocks of matter, and we also know that they are more complex than Democritus or Dalton ever imagined.

Our present concept of the **atom**—a neutral system of negatively charged electrons moving around a dense, positively charged nucleus—is a product of the twentieth century. The word *atom* comes from the Greek *atomos,* meaning "incapable of division"; the atoms originally proposed by the ancient Greek philosophers were believed to be without internal structure. Over the years, however, evidence has accumulated to suggest that atoms consist of simpler particles. Some observations, such as the ability of salt water to conduct an electric current, were explained by assuming that atoms can gain or lose an electrical charge, but the idea that atoms consist of charged particles was only conjecture until near the end of the nineteenth century. Around that time a series of discoveries and brilliant experiments presented the scientific world with the nuclear atom so familiar to us today.

2.1 THE DISCOVERY OF ELECTRONS

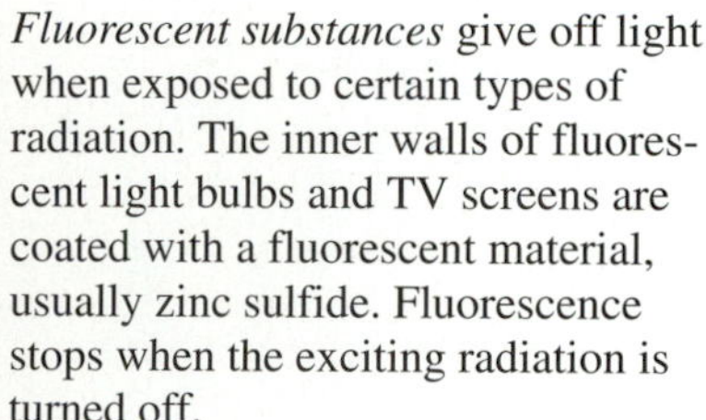

Fluorescent substances give off light when exposed to certain types of radiation. The inner walls of fluorescent light bulbs and TV screens are coated with a fluorescent material, usually zinc sulfide. Fluorescence stops when the exciting radiation is turned off.

Various investigators in the nineteenth century used **discharge tubes** for studying the conduction of electricity in gases. A discharge tube is a sealed tube that contains two metal plates called *electrodes* (Figure 2.1). One electrode, the *anode,* is given a positive charge; the other electrode, the *cathode,* a negative charge. When the difference in charge between the electrodes is sufficiently great, a beam of radiation, called a **cathode ray**, is emitted by the cathode (Figure 2.1a). The beam moves with little interference when most of the air is pumped out of the tube, and it can be detected by its impact on a fluorescent screen. A cathode ray beam always bends toward a positively charged object (Figure 2.1b), showing that it carries a negative charge.

The nature of cathode ray beams was not understood until Joseph J. Thomson, working at Cambridge University in England, demonstrated that their properties are consistent with those of a stream of negatively charged particles. Thomson originally called these particles "corpuscles," but later they came to be called **electrons**. Thomson also found that cathodes made of different materials produced identical electron beams, and he thus concluded that electrons are a fundamental and universal constituent of matter. The significance of this discovery is summarized in Thomson's own words (J. J. Thomson, "Cathode Rays," *Philosophical Magazine,* 1897):

Joseph J. Thomson working with discharge tubes.

> Thus we have in the cathode rays matter in a new state, a state in which the subdivision of matter is carried very much further than in the ordinary gaseous state: a state in which all matter—that is, matter derived from different sources such as hydrogen, oxygen, etc.—is of one and the same kind; this matter being the substance from which all the chemical elements are built up.

Thomson announced the discovery of the electron in 1897, and in 1906 he was awarded the Nobel prize for his research. His experiments were the forerunners of modern electron beam technology, the commercial development of which has had a tremendous impact on our way of life. Controlled electron beams are now used in television picture tubes, x-ray tubes, oscilloscopes, computer monitors, and electron microscopes.

Thomson was unable to determine the mass or charge of an electron separately. He did show, however, that the ratio of the electron's charge to its mass is in Equation 2.1.

Figure 2.1

The cathode ray beam in a discharge tube travels in a straight line from cathode to anode. (a) Part of the beam passes through a hole in the anode and lights up the fluorescent screen at the end of the tube. (b) The beam bends toward a positively charged plate and away from a negatively charged plate.

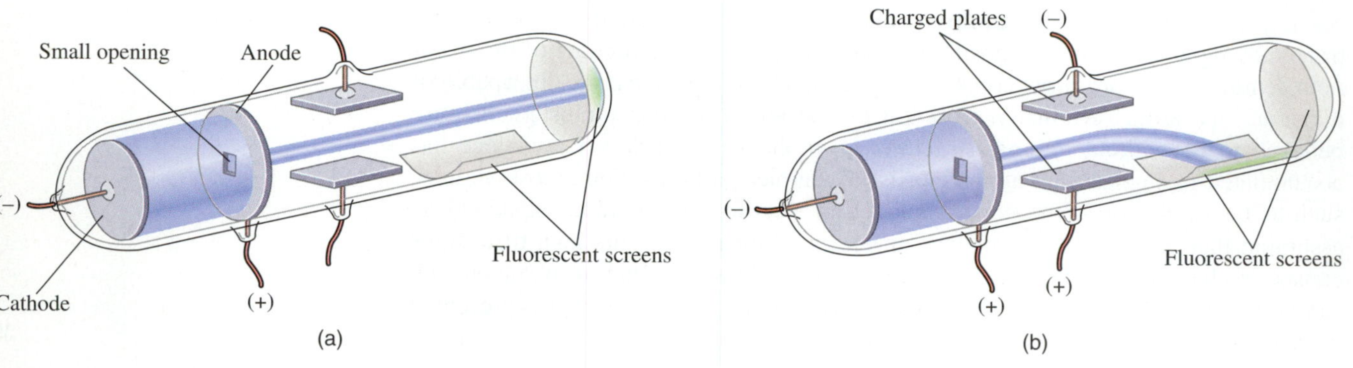

$$\frac{e}{m} = 1.759 \times 10^8 \text{ C/g} \tag{2.1}$$

where e is the magnitude (i.e., the value without a sign) of the electron charge in coulombs and m is the electron mass in grams. (The SI unit of charge is the **coulomb** (C). The usual symbol for charge is q; the symbol e is reserved for the magnitude of the electron charge.) Robert A. Millikan, working at the University of Chicago between 1909 and 1912, determined the electron charge and found that $e = 1.602 \times 10^{-19}$ C. The mass of the electron was calculated by rearranging Equation 2.1 and substituting e into the rearranged equation:

The *coulomb* is named in honor of Charles Augustin de Coulomb (1736–1806), a scientist known for his work on electricity and magnetism.

$$m = \frac{e}{1.759 \times 10^8 \text{ C/g}} = \frac{1.602 \times 10^{-19}\text{C}}{1.759 \times 10^8 \text{ C/g}} = 9.107 \times 10^{-28} \text{ g}$$

More precise values for the electron's charge and mass are given in Table 2.1.

TABLE 2.1 Masses and Charges of Some Atomic Particles

Particle	Mass (g)	Mass (u)[a]	Charge (C)[b]
Electron	9.10939×10^{-28}	5.48580×10^{-4}	-1.6022×10^{-19}
Proton	1.67262×10^{-24}	1.00728	$+1.6022 \times 10^{-19}$
Neutron	1.67493×10^{-24}	1.00866	0
Alpha particle	6.64476×10^{-24}	4.00150	$+3.2044 \times 10^{-19}$

[a]One gram equals 6.0221×10^{23} u (see Section 2.4). Note that protons and neutrons are almost 1840 times more massive than electrons.

[b]The charge of the proton is equal in magnitude but opposite in sign to the charge of the electron. The charge of the alpha particle is exactly twice the charge of the proton.

DIGGING DEEPER

How the Electron Charge and Mass Were Determined

Thomson's experiment for determining the electron's charge-to-mass ratio was the direct ancestor of mass spectroscopy, a powerful tool now universally used to determine precise masses of atoms, molecules, and molecular fragments. Thomson used two known facts: (1) an electric charge is repelled by like charges and attracted by unlike charges (Figure 2.2a), and (2) a magnetic field causes a moving electric charge to travel in a circular path (Figure 2.2b). He used a modified discharge tube similar to the one in Figure 2.3 in which the electron beam is narrowed by a series of slits so that its impact produces a small bright spot on the fluorescent screen opposite the cathode. Two oppositely charged plates are placed above and below the path of the beam. When turned on, they create an electric field that deflects the beam downward toward the positive plate. In addition, an electromagnet is positioned so that its field causes the beam to curve upward. The strengths of the electric and magnetic fields are adjustable. The deflection caused by either field or by both together can be determined by measuring the displacement of the bright spot on the glass.

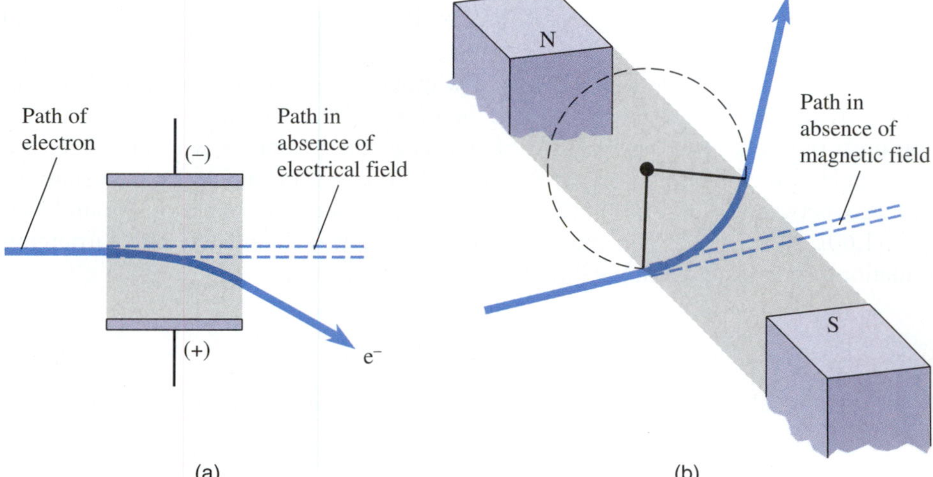

Figure 2.2
The effect of electric and magnetic fields on moving electrons. (a) In an electric field, the electron beam bends toward the positive plate and away from the negative plate. (b) In a magnetic field, the electron beam follows a circular path in the plane perpendicular to the field.

The electric field strength E at any point in space is defined as the force experienced by 1 coulomb of charge at that point. The force exerted on a charge q by a field E is given by

$$\text{electric force} = E \times q \tag{2.2}$$

(E: newtons/coulomb; q: coulombs)

The magnetic field causes the charged particle to move in a circular path. Two equations describe the magnitude of the force experienced by a charge q moving in a magnetic field of strength H:

$$\text{magnetic force} = H \times q \times \text{v} \tag{2.3a}$$

$$\text{magnetic force} = \frac{m \times \text{v}^2}{r} \tag{2.3b}$$

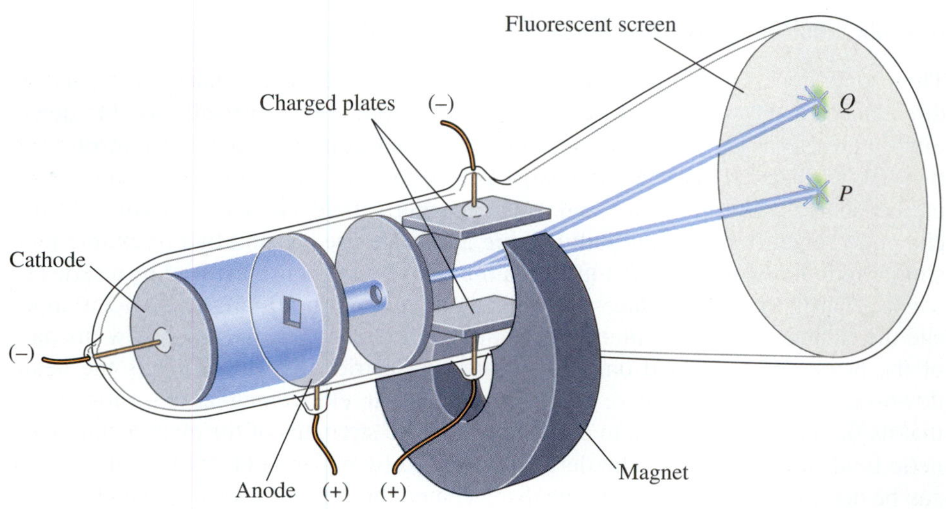

Figure 2.3
Finding the charge per unit mass (*e/m* ratio) of an electron. When there is no charge on the plates, the magnetic field causes the electron beam to curve upward so that it strikes the fluorescent screen at point *Q*. A charge on the plates creates an opposing electric field. When the electric and magnetic fields are balanced, the beam will strike the screen at point *P*.

where v is the velocity (speed) of the charged particle, m is its mass, and r is the radius of the circular path.

When the electric field is off, the magnetic field will deflect the electron beam upward so that it strikes the fluorescent detector screen at point Q in Figure 2.3. If the electric field is then turned on and adjusted to balance the magnetic field, the electrons will return to their original undeflected path, hitting the screen at point P. When both fields are exactly balanced, Equations 2.2 and 2.3a can be combined to give

$$\text{magnetic force} = \text{electric force}$$

$$H \times q \times \mathrm{v} = E \times q$$

The speed of the electrons moving through the balanced field is obtained by dividing both sides by $H \times q$:

$$\mathrm{v} = \frac{E}{H} \tag{2.4}$$

Thomson determined charge-to-mass ratio, e/m, of the electrons by equating Equation 2.3a to Equation 2.3b and substituting $\mathrm{v} = E/H$ as follows:

$$H \times q \times \mathrm{v} = \frac{m \times \mathrm{v}^2}{r}$$

$$H \times q \times \frac{E}{H} = \frac{m}{r} \times \left(\frac{E}{H}\right)^2$$

Rearrangement gives

$$\frac{q}{m} = \frac{e}{m} = \frac{E}{H^2 r} \tag{2.5}$$

(Remember that $q = e$ for electrons.) The electric field strength E, the magnetic field strength H, and the radius of curvature r in the magnetic field alone can all be measured. Substituting their values into Equation 2.5 gives the charge-to-mass ratio of the electron. Note that Thomson's method gives e/m directly; it does not provide enough information to yield separate values for e or m.

Millikan determined e by assuming that the electron charge is the smallest unit of charge and that all other charges are some multiple of the electron charge. One version of Millikan's apparatus is shown schematically in Figure 2.4. Very fine oil droplets are sprayed into a chamber and are allowed to fall between two charged plates where they can be observed with an eyepiece. The air surrounding the droplets is exposed to x-rays, a form of high-energy radiation that breaks up air particles into negative electrons and positively charged fragments. Some of the oil droplets pick up electrons, thus acquiring a negative charge; other droplets pick up the positively charged fragments. Each charged droplet is under the influence of two forces: the downward pull of gravity and the force exerted by the charged plates. The observer chooses a particular droplet and records its rate of fall with the electric field turned on. The field is then turned off and the droplet's rate of fall under gravity alone is measured. A comparison of the two rates allows the droplet charge to be calculated.

Millikan's data showed that the oil droplet charges were 1.602×10^{-19} C or some whole-number multiple of this value. No charge less than 1.602×10^{-19} C was ever found. Millikan concluded that the magnitude of the electron charge was

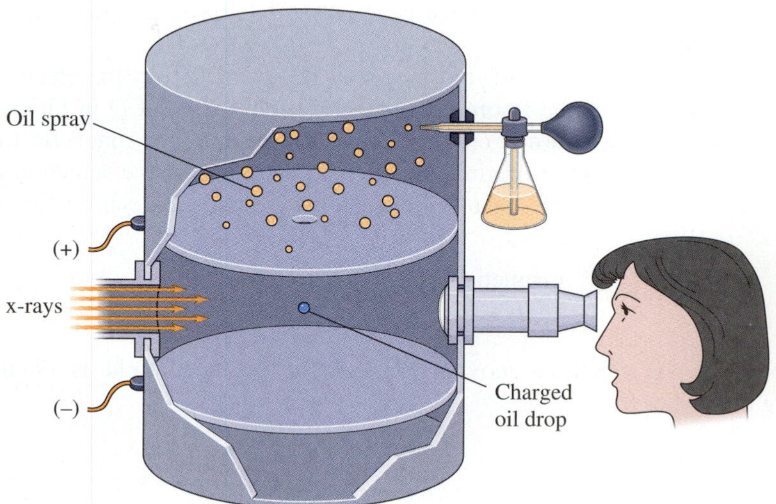

Figure 2.4
Millikan's oil drop experiment. Oil drops acquire charges that are multiples of the electron charge. The charge on any one drop is determined by comparing its rate of free fall with its rate of fall in an electric field. The lowest common denominator of the charges on a number of drops is assumed to be the charge on one electron.

equal to the smallest charge observed, that is, 1.602×10^{-19} C. Multiple charges result whenever a droplet picks up two or more electrons or when it picks up a positive fragment formed by the loss of two or more electrons.

2.2 THE NUCLEAR ATOM

Ions can be positive or negative. The ions produced by ionizing radiation such as x-rays and electron beams are usually positive.

An **ion** is an atom or group of atoms bearing an electrical charge. One way to produce ions is to bombard atoms with electrons or with high-energy radiation such as x-rays. Normally an atom is neutral because it contains as many negative charges (electrons) as positive charges. When an atom loses an electron, it is left with a net positive charge and thus becomes a positive ion. Figure 2.5 shows how a rapidly moving free electron may collide with an argon atom, knocking out an electron, and leaving a positive argon ion.

The production of positive ions is an undesirable complication in discharge tubes used for the study of electron beams. For this reason, such tubes are evacuated; the less air in the tube, the less the chance of ion formation. When a discharge tube is only partially evacuated, or when another gas such as neon or argon is introduced, collisions with the electron beam will cause some of the gas atoms to break up into

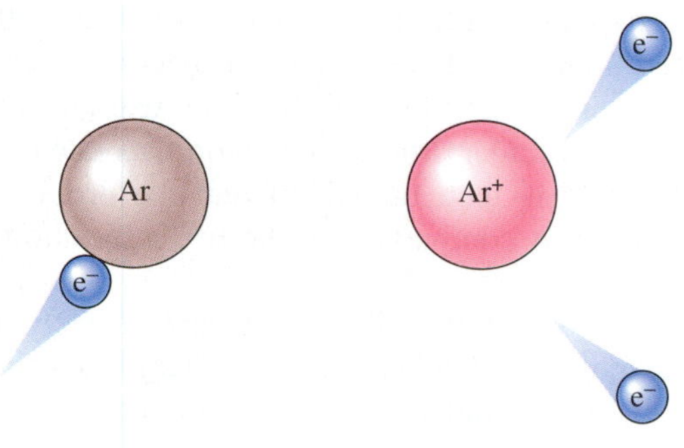

Figure 2.5
An electric discharge ionizes atoms in its path. Here one of the energetic electrons in the discharge knocks an electron out of an argon atom, leaving a positive argon ion.

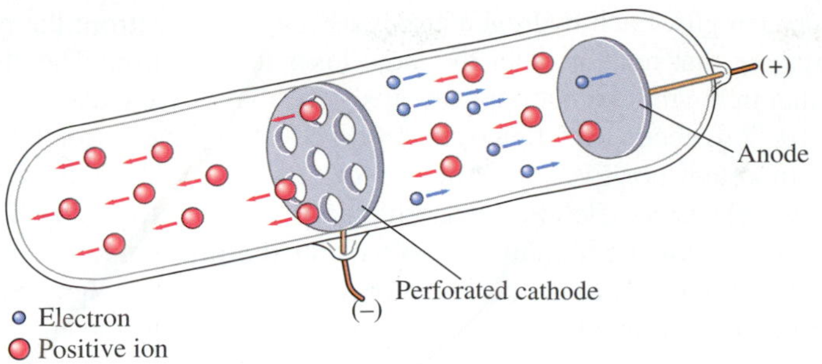

Figure 2.6
The positive ions produced in a discharge tube travel toward the cathode. If the cathode is perforated, some of the positive ions pass through and emerge as a positive beam on the other side.

positive ions and electrons. The newly released electrons will move along with the cathode-ray stream, but the positive ions will move in the opposite direction, as shown in Figure 2.6.

Beams of positive ions, called **positive rays**, were first detected and studied in 1886 by Eugen Goldstein, who observed them emerging from openings in the back of a perforated cathode similar to the one depicted in Figure 2.6. The newly formed ions are very energetic and, on recombining with electrons, release their excess energy in the form of visible light. Thus the positive rays emit a fascinating colored glow. Figure 2.7 shows the different colors exhibited by different gases. This effect is often seen in advertising displays; the familiar orange-red neon sign is essentially a discharge tube containing neon gas.

The charge and mass of the ions in a positive ion beam can be determined by methods similar to those used for electrons. The simplest (least massive) positive ion is the hydrogen ion or **proton**, which is formed when an electron beam passes

Figure 2.7
Certain gases emit distinctive colors when an electric discharge passes through them: (a) hydrogen, (b) nitrogen, and (c) mercury.

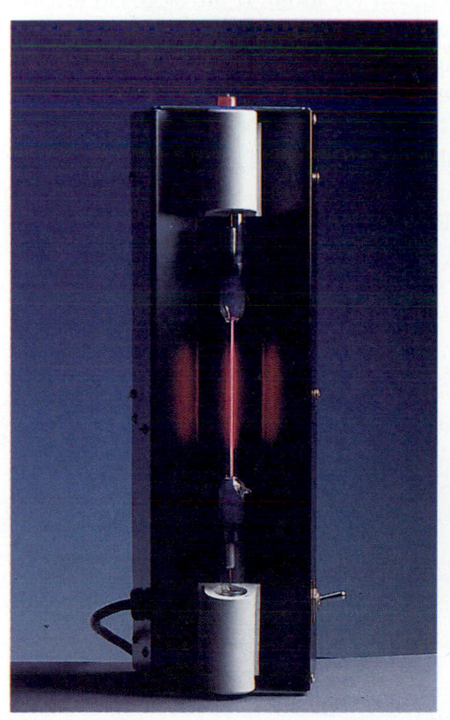
(a)

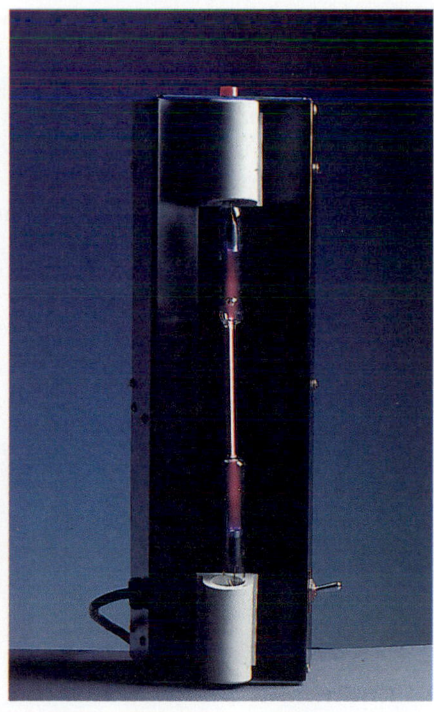
(b)

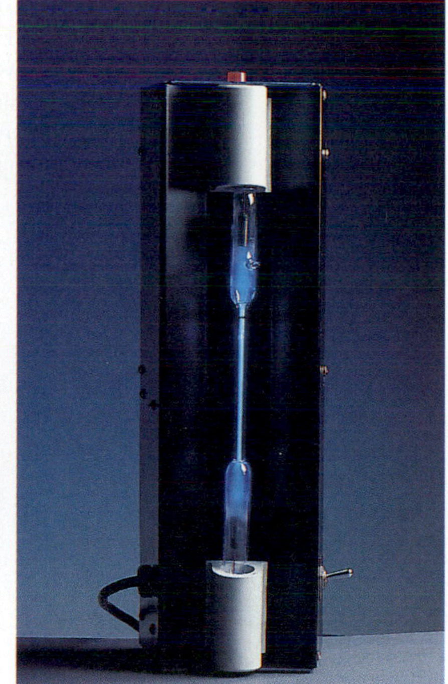
(c)

through hydrogen gas. Each hydrogen atom has only one electron; the proton is the positive fragment that remains after the atom loses this electron. The charge on the proton is equal in magnitude but opposite in sign to that of the electron, and its mass is 1.673×10^{-24} g, about 1836 times greater than that of the electron (Table 2.1).

A *radioactive* substance is one whose atoms break down spontaneously. Radioactivity is discussed in Chapter 22.

Another important positive ion is the helium ion that is formed when a helium atom loses two electrons. Helium ions, which are also emitted by radioactive substances such as radium and uranium, are often called **alpha particles**. Their properties are listed in Table 2.1. In the next section, we will see how alpha particles were used in a crucial experiment that provided us with our modern concept of the atom.

The Gold Foil Experiment

Thomson's work with cathode rays and the discovery of positive ion beams led scientists to believe that every atom contains electrons and that the removal of an electron leaves a positive ion that contains most of the atom's mass. The inner structure of the atom was still a mystery, but Thomson and other investigators initially believed that the atom's mass was uniformly distributed throughout its volume. In 1904 Thomson suggested that the atom contained electrons embedded in a jellylike sphere of positive electricity, a model of the atom that is sometimes referred to as the "raisin pudding" model.

Several years after Thomson developed his ideas about the atom, Ernest Rutherford and an associate, Hans Geiger, were working at the University of Manchester in England. They performed experiments in which an extremely thin gold metal foil was bombarded with alpha particles, and they watched for changes in direction as the alpha particles passed through the foil (Figure 2.8). Geiger initially observed that the narrow beam of alpha particles widened only slightly as it emerged from the foil; this result was consistent with Thomson's model of the atom. If the atoms in the foil consisted of electrons embedded in a homogeneous sphere of positive electricity (Figure 2.9a), then the electrical forces acting on a positively charged alpha particle would be almost evenly balanced, and the particles should pass through the foil with almost no deflection. Electrons in the atom would be pushed aside by the more massive alpha particles in much the same way that falling snowflakes are pushed aside by a fast-moving snowball.

What happened next is best described in Rutherford's own words (E. Rutherford,

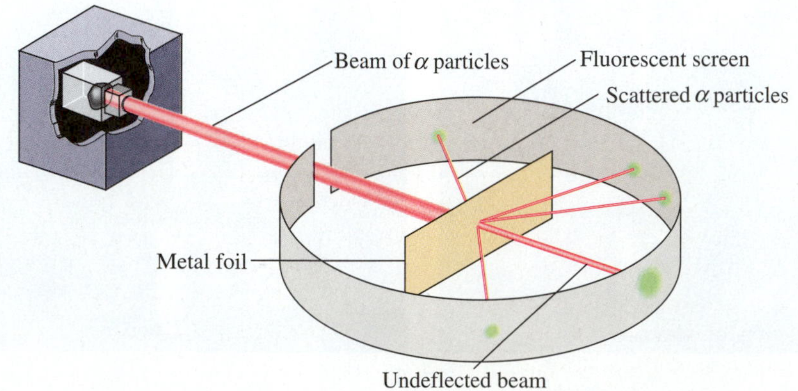

Figure 2.8
Rutherford's gold foil experiment. A beam of alpha particles is directed at a sheet of gold foil. The flashes produced by the particles reaching the fluorescent screen show that most of them travel through the foil with little or no deflection. A few, however, are scattered through large angles.

Background to Modern Science, J. Needham and W. Pagel, eds., London, Cambridge University Press, 1938):

> One day Geiger came to me and said, "Don't you think that young Marsden (Ernest Marsden, an undergraduate student at Manchester), whom I am training in radioactive methods, ought to begin a small research?" Now I had thought that too, so I said, "Why not let him see if any alpha particles can be scattered through a large angle?" I may tell you in confidence that I did not believe they would be, since we knew that the alpha particle was a very fast massive particle, with a great deal of energy, and you could show that if the scattering was due to the accumulated effect of a number of small scatterings, the chance of an alpha particle's being scattered backwards was very small. Then I remember two or three days later Geiger coming to me in great excitement and saying, "we have been able to get some of the alpha particles coming backwards." . . . It was quite the most incredible event that has ever happened to me in my life. It was almost as incredible as if you fired a fifteen-inch shell at a piece of tissue paper and it came back and hit you.

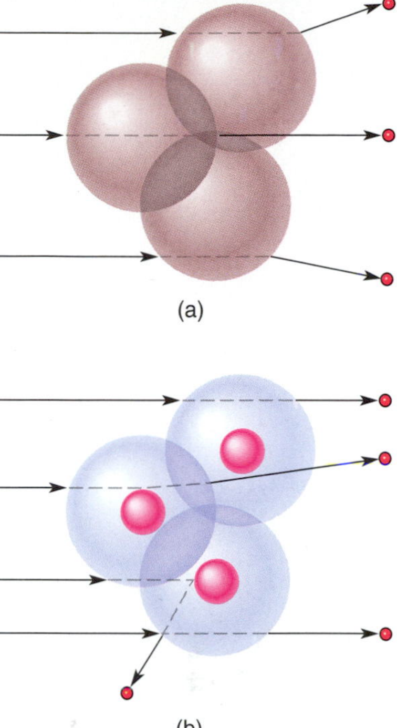

Figure 2.9
Alpha particles traveling through atoms. (a) If the atom were a sphere of diffuse positive charge with embedded electrons, it would have no mass or charge concentrated enough to strongly deflect an alpha particle. (b) The nuclear atom has its mass and charge concentrated in a small nucleus. Most alpha particles do not encounter a nucleus, but those that do are strongly deflected.

Rutherford realized that the large observed deflections could be explained only by assuming that the positive charge and most of the atom's mass are concentrated in a small core, or **nucleus**. As Figure 2.9b shows, alpha particles that collide with such a nucleus would be strongly deflected; alpha particles not encountering a nucleus would continue in nearly straight paths. Because most alpha particles pass through the atom without severe deflection, the nucleus must be very small compared to the atom as a whole. Thus Thomson's model of the atom was wrong and had to be abandoned in favor of a **nuclear atom** consisting of a small, but massive, positively charged nucleus surrounded by negative electrons.

From his deflection data, Rutherford estimated nuclear radii to be about 0.001 pm (1 picometer = 10^{-12} m). Previous estimates of atomic radii ranged from about 50 pm to about 250 pm, some one hundred thousand times larger than nuclear radii. These results showed that the atom is mostly empty space. If, for example, the nuclear diameter were expanded to the diameter of a U.S. penny, the atomic diameter would be about 2 km or $1\frac{1}{4}$ miles!

Rutherford published his work in 1911, only 3 years after receiving the Nobel prize for his previous research on radioactivity. Although he may not have realized it at the time, his discovery that the atom is not solid and substantial, but mostly empty space like the solar system, ranks among the most influential scientific achievements. It profoundly changed our way of looking at both matter and the universe.

The Rutherford Atom

Rutherford's discovery showed that atoms contain electrons moving about a small but massive nucleus. But what is in the nucleus? Rutherford and his associates believed that the nuclei of all atoms contain protons. (Recall that protons are left behind when hydrogen atoms lose their electrons.) Furthermore, because the positive charge on the proton is equal in magnitude to the negative charge on the electron, they concluded that *the number of electrons in a neutral atom must equal the number of protons in its nucleus.* This number is called the **atomic number** and is symbolized by *Z.* The atomic number of hydrogen, which has one proton in the nucleus, is 1; the atomic number of helium, with two protons, is 2. In 1913 Henry Moseley, a former student of Rutherford, used x-rays to provide experimental proof for the concept of atomic number. He also showed that the atomic number, and not the atomic mass, determines the structure and behavior of the atom.

Ernest Rutherford (right) and Hans Geiger with their apparatus for counting alpha particles.

Every atom other than hydrogen has a nuclear mass greater than the total mass of its protons. For example, alpha particles, the nuclei of helium atoms, have two protons, but the mass of each alpha particle is roughly four times that of a single proton (Table 2.1). This observation led Rutherford to suggest in 1920 that each nucleus should contain electrically neutral particles in sufficient number to account for the missing mass. These particles were given the name **neutrons**; however, they were not actually discovered until 1932. In that year James Chadwick, another of Rutherford's former students, was able to show that the bombardment of beryllium atoms with alpha particles produces a beam of neutrons. The properties of the neutron are summarized in Table 2.1. Note that the neutron's mass is similar to but slightly greater than that of a proton.

PRACTICE EXERCISE 2.1

Compare electrons, protons, and neutrons with respect to charge and mass.

2.3 THE CHEMICAL ELEMENTS AND THE PERIODIC TABLE

Earth offers an amazing variety of substances, each with its own unique characteristics. Early chemists looked for some underlying order in this abundance and proposed the existence of basic entities, called **elements**, which could be combined, like colored blocks, into an infinite variety of patterns, and thus provide all the substances we know. An element is now defined as *a substance whose atoms all have the same atomic number;* in other words, all of its atoms have the same number of protons in the nucleus. Oxygen, for example, is an element consisting only of atoms with eight protons in the nucleus ($Z = 8$). Hydrogen and carbon are elements with atomic numbers of $Z = 1$ and $Z = 6$, respectively.

LEARNING HINT

Look up the symbol of each element you read about—you will soon learn the important ones.

An atom of an element is represented in chemical notation by a one-, two-, or three-letter symbol derived from its name. The symbol for oxygen is O and the symbol for helium is He. The symbol for sodium is Na, derived from its Latin name *natrium.* The symbol for element number 104 is Unq from the name unnilquadium, which is derived from the Latin roots for 1 (un), 0 (nil), and 4 (quad). An alphabetical list of all the elements and their symbols is given on the inside front cover.

The properties of an element depend in large part on the number of electrons in each of its atoms; that is, they depend on the element's atomic number. The **periodic table** (inside front cover) shows the elements arranged in order of increasing atomic number. The 18 vertical columns in this table are called **groups**; the seven horizontal rows are called **periods**. Metallic elements are to the left of the heavy stepwise line in the periodic table; nonmetallic elements are to the right. Note that only 22 elements are nonmetals. A vast amount of information concerning the elements is summarized in this table, and we will refer to it frequently.

At present, 109 elements are known; 92 of these have been found in the earth's crust. The elements from hydrogen (H, $Z = 1$) through plutonium (Pu, $Z = 94$) occur naturally, with the exception of technetium (Tc, $Z = 43$) and promethium (Pm, $Z = 61$). These two elements and the elements from americium (Am, $Z = 95$) to meitnerium (Mt, $Z = 109$) have been synthesized in the laboratory. Naturally occurring elements vary greatly in abundance, and only eight elements account for more than 98% of the mass of the earth's crust (Table 2.2). Some naturally occurring elements, such as plutonium and neptunium (Np, $Z = 93$), are extremely rare. They can, however, be synthesized. Table 2.3 lists some elements known to be essential for life. Even though many of these are needed only in small amounts, life as we know it could not continue without them.

Figure 2.10 shows a portion of the periodic table along with some of the symbols you will need in the near future. Note that 11 elements are gases at room temperature (25°C). Six of these—helium, neon, argon, krypton, xenon, and radon—are found in Group 8A. (Group 8A is also called Group 18.) These six elements are called the **noble gases**. The other five gaseous elements are hydrogen, nitrogen, oxygen, fluorine, and chlorine. Bromine and mercury are the only liquid elements at room temperature. The remaining elements, most of which are metals, are solid.

The metals cesium (melting point = 28.4°C) and gallium (melting point = 29.8°C) will melt on a hot day.

Isotopes

The **mass number** (A) of an atom is the total number of protons and neutrons in its nucleus. A potassium atom containing 19 protons and 20 neutrons, for example, has a mass number of 19 + 20 = 39. A carbon atom with six protons and six neutrons has a mass number of 12. Although all atoms of a given element have the same atomic number, they may have different mass numbers because of differences in the number of neutrons. Atoms with the same atomic number but different mass numbers are called **isotopes**. Figure 2.11 shows three isotopes of oxygen. The atomic number of oxygen is 8, so each oxygen isotope has eight protons in the nucleus and eight electrons moving around the nucleus.

The prefix *iso* means "the same."

A given isotope is specified by adding its mass number to the name of the element, as in carbon-12, uranium-235, and strontium-90. Atomic symbols for specific isotopes are written with the mass number as a superscript on the left and the atomic number as a subscript, also on the left; for example, the symbol for carbon-12 is

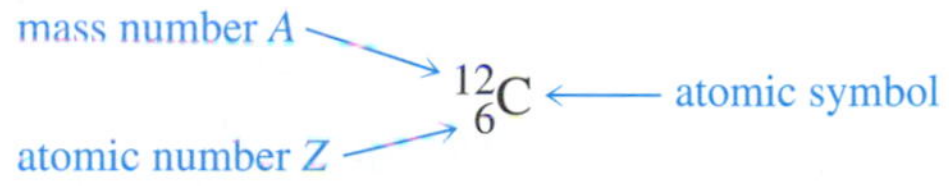

TABLE 2.2 Elemental Composition of the Earth's Crust[a]

Element	Percent by Mass	Percent by Atom Count
Oxygen	46.60	62.55
Silicon	27.72	21.22
Aluminum	8.13	6.47
Iron	5.00	1.92
Calcium	3.63	1.94
Sodium	2.83	2.64
Potassium	2.59	1.42
Magnesium	2.09	1.84

[a]All other elements account for less than 2% by mass of the total.

TABLE 2.3 Elemental Composition of the Human Body[a]

Element	Percent by Atom Count
Hydrogen	63
Oxygen	25.5
Carbon	9.5
Nitrogen	1.4
Calcium	0.31
Phosphorus	0.22
Potassium	0.06
Sulfur	0.05
Sodium	0.03
Chlorine	0.03
Magnesium	0.01

[a]All other elements account for less than 0.01% of the total number of atoms.

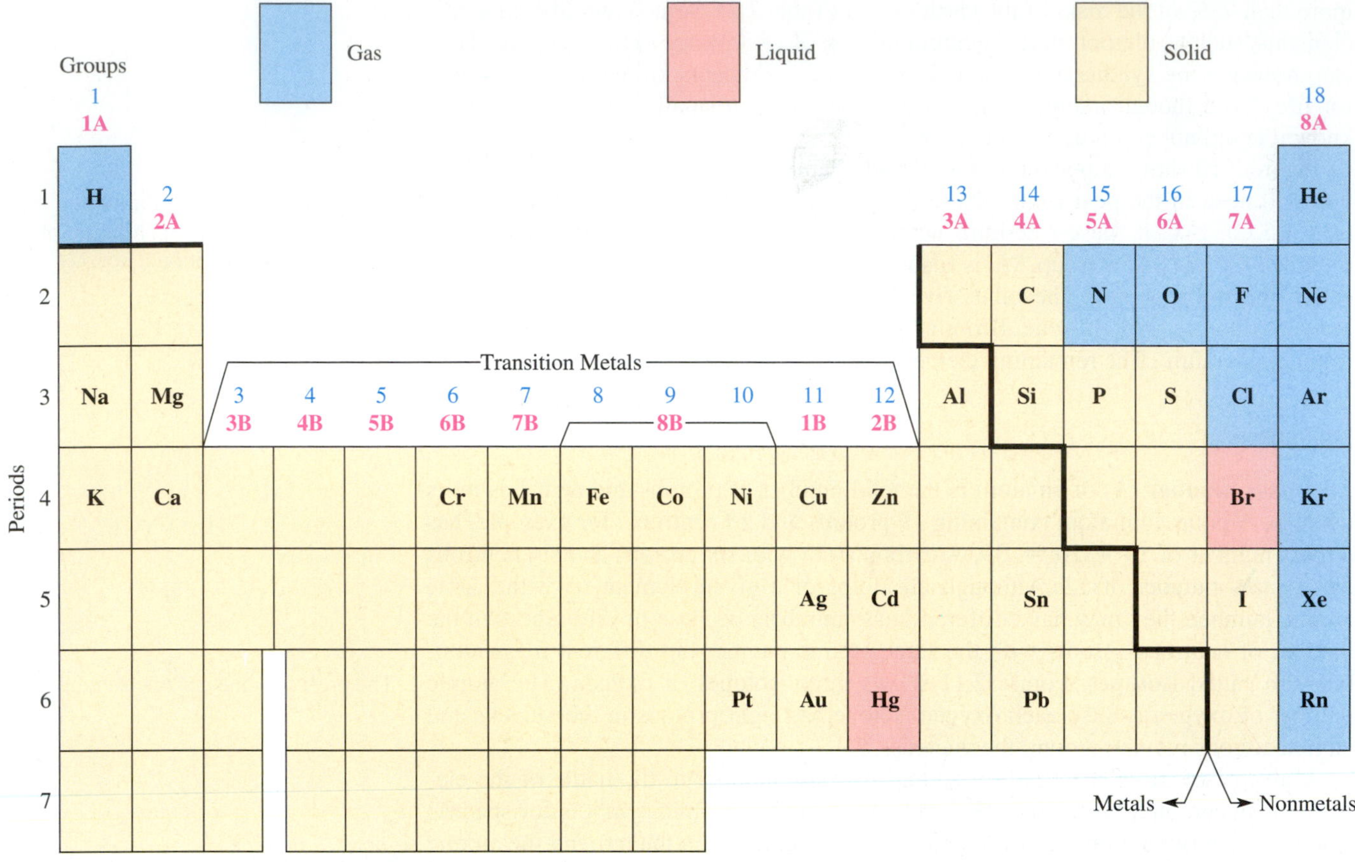

Figure 2.10

Some important elements and their symbols. Eleven elements are gases at 25°C, two are liquids, and the rest are solids.

EXAMPLE 2.1

The symbol for an isotope of uranium is $^{238}_{92}U$. How many (a) protons, (b) neutrons, and (c) electrons are in the atom?

SOLUTION

The symbol shows that the atomic number Z is 92 and the mass number A is 238.

(a) The number of protons equals the atomic number; hence, the number of protons $= Z = 92$.

(b) The number of neutrons is the mass number minus the number of protons; hence, the number of neutrons $= A - Z = 238 - 92 = 146$.

(c) The number of electrons in a neutral atom must equal the number of protons; hence, the number of electrons = number of protons = 92.

PRACTICE EXERCISE 2.2

The principal isotope of iron has 26 protons and 30 neutrons. (a) What is the mass number of this isotope? (b) Write the symbol for this isotope.

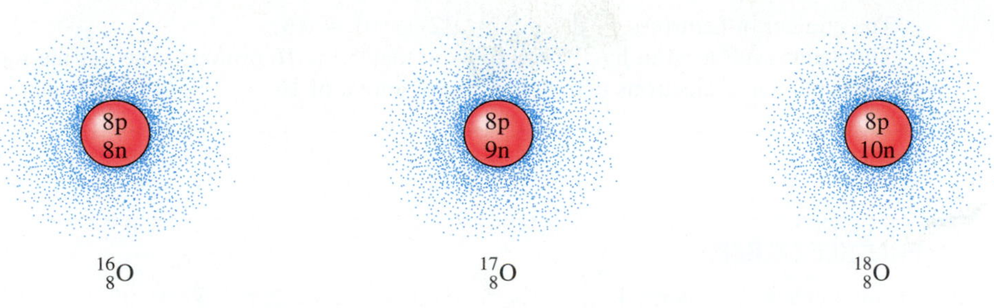

Figure 2.11

The three isotopes of oxygen. All oxygen atoms have eight protons in the nucleus and eight electrons outside the nucleus. The nucleus also contains neutrons—8 in oxygen-16, 9 in oxygen-17, and 10 in oxygen-18. (The eight electrons outside the nucleus are represented by clouds of dots because electrons move fast and change positions rapidly.)

Even though only 109 elements have been identified, the existence of isotopes gives rise to more than 1450 different atomic species. Most elements have one isotope that is much more plentiful than the others. Naturally occurring carbon, for example, consists of 98.89% carbon-12; the remaining 1.11% includes carbon-13 and trace amounts of carbon-14. Magnesium consists of 78.70% magnesium-24, 10.13% magnesium-25, and 11.17% magnesium-26. A few elements such as fluorine, sodium, and manganese have only one natural isotope. At the other extreme are xenon with 9 natural isotopes and tin with 10. Hydrogen is the only element whose isotopes have been given special names and special symbols. Most abundant (99.985%) is hydrogen, $^{1}_{1}H$, whose nucleus is a single proton. The other isotopes are *deuterium,* symbolized by D or by $^{2}_{1}H$, and *tritium,* symbolized by T or by $^{3}_{1}H$.

Symbols can also be written for ions. We are already acquainted with some of the positive ions that form when atoms lose electrons. Negative ions form when neutral atoms gain additional electrons. The symbol for an ion is written with a superscript on the right that indicates its charge and the number of electrons lost or gained. A positive superscript gives the number of electrons lost by the atom; a negative superscript gives the number of electrons gained. For example, Li^+ and Al^{3+} are symbols for positive lithium and aluminum ions formed from atoms that have lost one and three electrons, respectively. (The superscripts + and − are understood to mean 1+ and 1−.) The chloride ion, Cl^-, and sulfide ion, S^{2-}, are negative ions formed by the respective addition of one and two electrons to neutral chlorine and sulfur atoms.

LEARNING HINT

A positive ion is formed when an atom loses electrons; a negative ion is formed when an atom gains electrons.

EXAMPLE 2.2

How many protons, neutrons, and electrons are in (a) the rhodium ion, $^{103}_{45}Rh^{3+}$ and (b) the sulfide ion, $^{32}_{16}S^{2-}$?

SOLUTION

(a) The symbol for the rhodium ion shows that the atomic number Z is 45 and the mass number A is 103.

The number of protons $= Z = 45$.

The number of neutrons $= A - Z = 103 - 45 = 58$.

A neutral rhodium atom would have 45 electrons to balance the charge of the 45 protons. The rhodium ion has 3 fewer electrons than the atom, or $45 - 3 =$ 42 electrons.

(b) For the sulfide ion, $Z = 16$ and $A = 32$.

The number of protons $= Z = 16$.

The number of neutrons $= A - Z = 32 - 16 = 16$.
The neutral sulfur atom has 16 electrons to match the 16 protons. The sulfide ion has two more electrons than the atom, for a total of $16 + 2 = 18$ electrons.

PRACTICE EXERCISE 2.3

A chloride ion has 17 protons, 18 neutrons, and one negative charge. Write its symbol.

Atomic Masses

Individual atoms are too small to be picked out and weighed on a balance, and in the early days of chemistry atomic masses could be estimated only by indirect methods that involved weighing sizable samples of materials. Methods changed, however, with the development of instruments called **mass spectrometers**, which have been used for atomic mass determinations since about 1920. Improvements in mass spectrometer technology have produced precision and accuracy far exceeding that of the best older methods, most of which are now only of historical interest.

The definition of an atomic mass unit in terms of carbon-12 was a compromise accepted by the international scientific community in 1962. Prior to that time, chemists and physicists used different scales for their atomic mass determinations. The symbol u emphasizes that the new unit is a *unified* atomic mass unit.

Atomic masses are most conveniently expressed in **atomic mass units** (u or amu), where 1 atomic mass unit is defined as one-twelfth the mass of a carbon-12 atom. (In other words, the mass of a single carbon-12 atom is exactly 12 atomic mass units.) Table 2.1 gives the masses of the fundamental particles in both grams and atomic mass units. Note that the proton and neutron each have masses approximately equal to 1 u.

Chemical Atomic Weights

It is customary to use the phrase *chemical atomic weight* rather than *chemical atomic mass* for the average isotopic mass of an element. This usage has been approved by IUPAC, the International Union of Pure and Applied Chemistry.

Naturally occurring elements contain mixtures of isotopes, and mass spectrometer data can be used to find the percentage abundance as well as the mass of each isotope in a sample. Ordinary oxygen, for example, consists of 99.7587% oxygen-16, 0.0374% oxygen-17, and 0.2039% oxygen-18. Since these percentages are fairly constant from one sample of oxygen to the next, chemists find it convenient to use an average isotopic mass in their calculations. This average mass is called the **chemical atomic weight** and is found by multiplying each isotopic mass by its **fractional abundance** (the percentage abundance divided by 100%) and adding the products. The procedure is illustrated by the following calculation for oxygen:

Isotope	Isotopic Mass (u)		Fractional Abundance		Product (u)
^{16}O	15.994915	×	0.997587	=	15.9563
^{17}O	16.999133	×	0.000374	=	0.00636
^{18}O	17.99916	×	0.002039	=	0.03670
			Sum	=	15.9994

The average, 15.9994 u, is the chemical atomic weight of oxygen and is listed as such in atomic weight tables (see the inside front cover). Note carefully that the chemical atomic weight is not obtained by simply adding the isotopic masses and dividing by three. Instead, it is recognized that each isotope contributes to the average mass in proportion to its abundance. Averages obtained in this manner are called *weighted averages.*

EXAMPLE 2.3

Gallium is a metal used in transistors. It has two naturally occurring isotopes, gallium-69 and gallium-71, with masses of 68.9257 u and 70.9249 u, respectively. The percentage abundance of gallium-69 is 60.27%; the abundance of gallium-71 is 39.73%. Calculate the chemical atomic weight of gallium.

SOLUTION

The fractional abundance of each isotope is equal to its percentage abundance divided by 100%. The fractional abundance of gallium-69 is 0.6027; the abundance of gallium-71 is 0.3973. Multiplying each mass by its fractional abundance and adding the products gives

Isotope	Isotopic Mass (u)		Fractional Abundance		Product (u)
^{69}Ga	68.9257	×	0.6027	=	41.54
^{71}Ga	70.9249	×	0.3973	=	28.18
			Sum	=	69.72

Thus the chemical atomic weight of gallium is 69.72 u.

LEARNING HINT

Some students prefer to use the following formula for calculating the chemical atomic weight:

chemical atomic weight =

$$\Sigma\left(\text{isotopic mass} \times \text{fractional abundance}\right)$$

where the Greek letter sigma, Σ, means "the sum of." The sum referred to is the sum of the product for each of the isotopes.

PRACTICE EXERCISE 2.4

Lithium consists of two isotopes, lithium-6 and lithium-7, with masses of 6.01512 u and 7.01600 u, respectively. If the chemical atomic weight of lithium is 6.941 u, calculate the percentage abundance of each isotope. *Hint:* Remember that percentage abundances add up to 100%; fractional abundances add up to 1. Hence, if x equals the fractional abundance of lithium-6, then $1 - x$ equals the fractional abundance of lithium-7.

Atomic weight values show a wide range in precision (see table on the inside front cover). The atomic weight of fluorine, for example, has eight significant figures while the atomic weight of zinc has only four. The precision of a chemical atomic weight is limited by the precision of the abundance data, which may be low because of natural variations in isotopic compositions in the earth's crust. Most isotopic masses, on the other hand, have been very precisely measured.

DIGGING DEEPER

The Mass Spectrometer

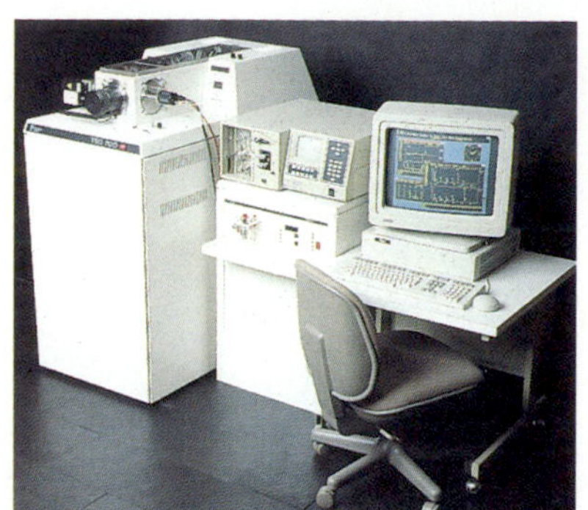

A mass spectrometer.

Mass spectrometers have gone through many modifications since the pioneering work at the beginning of the century, and various versions of these instruments are now available. In this section we describe the principles of mass spectroscopy as they apply to ions passing through a **direction-focusing mass spectrometer** (Figure 2.12).

A small sample of the element under study is vaporized. The gaseous atoms enter a partially evacuated *ionization chamber,* where a heated filament produces a stream of high-energy electrons that bombards the atoms and causes them to break down into positive ions. Naturally occurring magnesium, for example, consists of three isotopes with mass numbers 24, 25, and 26. Electron bombardment of magnesium vapor produces a positive ion beam consisting of $^{24}Mg^+$, $^{25}Mg^+$, and $^{26}Mg^+$. The ion beam, still in the ionization chamber, passes between a pair of charged plates. The ions speed up as they move toward and through the negative plate and, for this reason, the plates are called *accelerating plates.* A system of slits ultimately produces a narrow beam of rapidly moving ions.

After leaving the ionization chamber, the ions enter a *mass analyzer tube* where they are subjected to a constant magnetic field of strength *H*. This field causes the ions to move in circular paths of radius *r* such that

$$r^2 = \frac{2V}{H^2} \times \frac{m}{q} \tag{2.6}$$

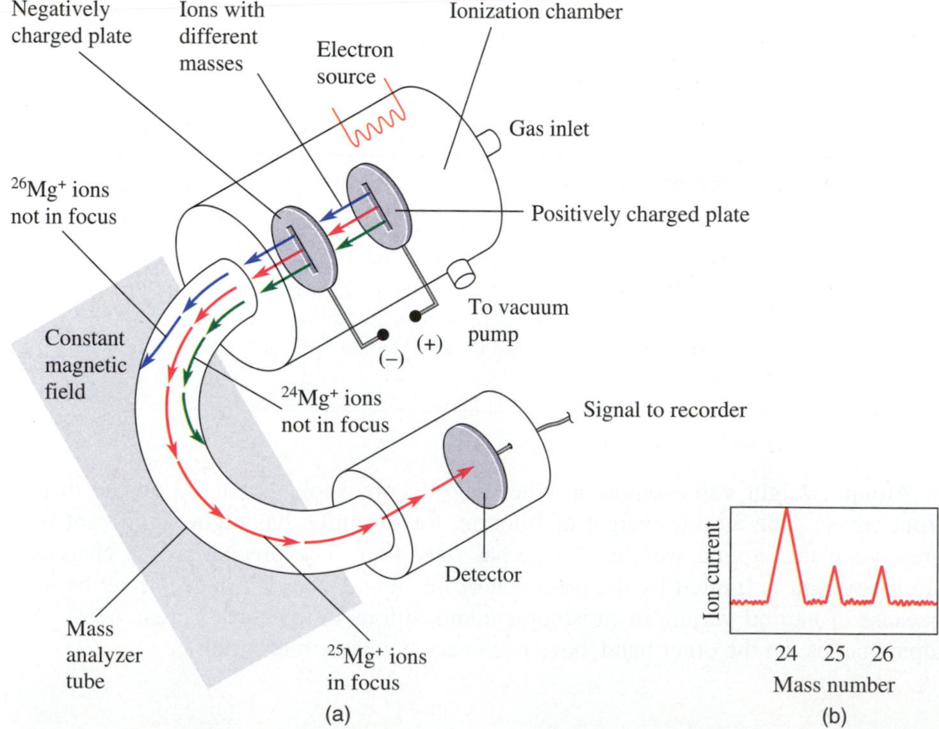

Figure 2.12
A direction-focusing mass spectrometer. (a) Ions are accelerated by a pair of charged plates in the ionization chamber, and their paths are then curved by the magnetic field. At any given voltage, only ions of a certain mass will follow the curve of the analyzer tube and reach the detector. (b) A mass spectrum is obtained when the intensity of the detector signal (ion current) is plotted against the mass number of the ion giving the signal.

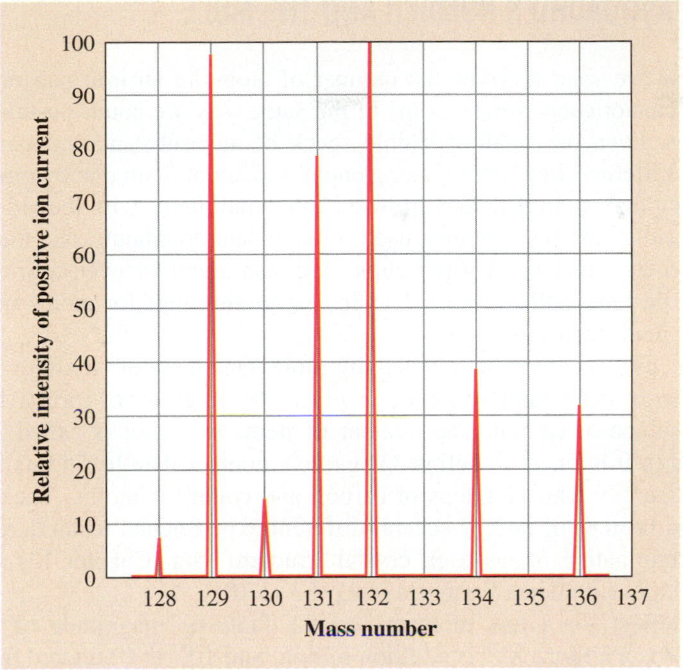

Figure 2.13
A mass spectrum showing seven isotopes of xenon. The height of the peak for each isotope is proportional to the number of its ions that reached the detector of the mass spectrometer, and is thus proportional to the relative abundance of the isotope.

where V is the voltage of the accelerating plates and m/q is the mass-to-charge ratio of the ion. Equation 2.6 and Figure 2.12 show that, for ions with the same charge, the radius of the ion beam's path increases with increasing mass. In effect, more massive ions exhibit smaller deflections, so the original ion beam breaks up into separate beams, one for each mass.

Ions that pass through the mass analyzer tube strike a *detector* that electronically measures and records an *ion current*. The ion current is proportional to the number of ions striking the detector. Note that only ions with the same value of m/q (i.e., ions following the same path) reach the detector. Ions with larger or smaller values of m/q will be deflected to the sides of the mass analyzer tube. Equation 2.6 shows that the path radius depends on both the voltage V and magnetic field strength H. The operator of a direction-focusing mass spectrometer continuously changes the voltage to successively focus ions of different m/q ratios to the same radius. In this way a positive ion detector in a fixed position can be used to scan the different ions one after the other.

If the intensity of the detector signal is graphed against the mass number of the ion giving the signal, the resulting plot is called a **mass spectrum**. Figure 2.13 shows a mass spectrum demonstrating the existence of seven xenon isotopes with mass numbers ranging from 128 to 136. This scan was obtained by continuously changing the voltage and successively focusing the ions as described above. The abundance of each ion can be calculated from the intensity of its signal, that is, from the height of its peak.

Direction-focusing mass spectrometers are extremely sensitive and have the capability of detecting individual ions. Excellent and complete spectra have been obtained in scan times of less than 1 s with submicrogram and even nanogram (1 ng $=$ 10^{-9} g) samples.

2.4 AVOGADRO'S NUMBER AND THE MOLE

Suppose that we want to know the number of atoms in an iron nail or a pinch of sulfur. We cannot count single atoms in the same way we count postage stamps or automobiles. Even the smallest visible speck of any substance contains so many atoms that a lifetime would not be long enough to count them one by one. How then can we count atoms? In everyday life we count small items with the aid of counting units such as the dozen (12 items), used for eggs and doughnuts, and the gross (144 items), used for pencils and paper clips. A dozen atoms or even a gross of atoms would still be too small to see or handle; a counting unit for atoms must contain many more items than 12 or 144.

The Latin word *mole* means "a mass"; a molecule is "a little mass." Mole is often abbreviated to mol, as in g/mol.

Chemists use a counting unit called the **mole**. One mole of anything contains the same number of items as there are atoms in exactly 12 grams of carbon-12, the most abundant isotope of carbon. The number of items in 1 mol is called **Avogadro's number** (N_A). It is easier to define Avogadro's number than to find its value, since one cannot simply take 12 grams of carbon and count the atoms. The number has nevertheless been estimated by several different experimental techniques. The most precise determination is based on crystal structure data (Chapter 12) and gives a value with eight significant figures, 6.0221367×10^{23}.

Amadeo Avogadro (1776–1856) was an Italian lawyer and scientist whose hypothesis concerning the number of molecules in a gaseous volume is discussed in Chapter 5.

Let us reflect for a few moments on the fantastic magnitude of Avogadro's number: 10^9 is a billion, 10^{18} is a billion billion, and 10^{23} is 1 hundred thousand billion billion. There are 6.022×10^{23} atoms in 12 grams of carbon-12. This means that there are more than 6 hundred thousand billion billion atoms in little more than a teaspoonful of carbon-12! Compare this with other large numbers such as the number of seconds in a normal lifetime (about 2 billion), or the number of dollars in the U.S. national debt (3.2 trillion in 1990). These are insignificant compared to Avogadro's number. You might consider Avogadro's number in terms of the earth's age, which is currently estimated at somewhat over 4 billion years, or about 10^{17} seconds. If you had started to count carbon atoms at the rate of one per second on the day the earth began, by now you would have counted only 10^{17} atoms. This corresponds to about 2 micrograms of carbon-12, a speck hardly big enough to notice.

National Mole Day is listed in *Chase's Annual Calendar of Events.* Because Avogadro's number is 6.02×10^{23}, National Mole Day is observed from 6:02 a.m. to 6:02 p.m. on October 23.

The mole is the official SI counting unit (see Table 1.2 on page 6). One mole of any substance contains Avogadro's number of particles:

1 mol of copper contains 6.022×10^{23} Cu atoms

1 mol of sodium ions contains 6.022×10^{23} Na^+ ions

1 mol of electrons contains 6.022×10^{23} electrons

EXAMPLE 2.4

How many atoms are in 2.50 mol of iron?

SOLUTION

One mole of iron contains 6.022×10^{23} atoms. The number of atoms in 2.50 mol is

$$2.50 \text{ mol Fe} \times \frac{6.022 \times 10^{23} \text{ atoms}}{1 \text{ mol Fe}} = 1.51 \times 10^{24} \text{ atoms}$$

PRACTICE EXERCISE 2.5

Use data from Table 2.1 to calculate the charge in coulombs on 1 mol of hydrogen ions (protons).

We now have enough information to determine the relationship between grams and atomic mass units. In the previous section we defined the mass of one carbon-12 atom to be exactly 12 atomic mass units; that is, there are 12 u per carbon-12 atom. Our definition of the mole tells us that there are 6.022×10^{23} atoms in 12 grams of carbon-12. When we combine these definitions,

$$\frac{12\text{ u}}{1\text{ C atom}} \times \frac{6.022 \times 10^{23}\text{ C atoms}}{12\text{ g}} = 6.022 \times 10^{23}\text{ u/g}$$

we find that there are 6.022×10^{23} atomic mass units per gram; that is, *1 gram is equal to the Avogadro number of atomic mass units.* The relation, $1\text{ g} = 6.022 \times 10^{23}\text{ u}$, allows us to convert atomic mass units into grams, and vice versa, as illustrated in the next example.

EXAMPLE 2.5

Find the mass in grams of one carbon-12 atom.

SOLUTION

One carbon-12 atom has a mass of exactly 12 u. Its mass in grams is

$$12\text{ u} \times \frac{1\text{ g}}{6.022 \times 10^{23}\text{ u}} = 1.993 \times 10^{-23}\text{ g}$$

(If you are wondering about significant figures, remember that 12 u, the mass of one carbon-12 atom, is an exact number.)

PRACTICE EXERCISE 2.6

The mass of a neutron is 1.675×10^{-24} g. Convert this into atomic mass units.

Molar Masses of Atoms

The mass of a dozen eggs is different from that of a dozen doughnuts. In the same way, the mass of a mole of sulfur atoms is different from that of a mole of iron atoms. The mass of 1 mol of a substance is called the **molar mass** ($\mathcal{M}$). Molar masses are usually, but not always, expressed in grams per mole.

Let us find the mass in grams of 1 mol of oxygen atoms. The table of atomic weights shows that the mass of one oxygen atom is 16.00 u. Two conversions are required to obtain the molar mass: First, we convert 16.00 u/atom into u/mol; then we convert u/mol into g/mol.

$$\frac{16.00\ \text{u}}{1\ \text{O atom}} \times \frac{6.022 \times 10^{23}\ \text{O atoms}}{1\ \text{mol O}} \times \frac{1\ \text{g}}{6.022 \times 10^{23}\ \text{u}} = 16.00\ \text{g/mol O}$$

Note that the mass in grams of 1 mol of oxygen atoms (16.00 g) is numerically equal to the mass in atomic mass units of one oxygen atom (16.00 u). This relation exists because we first multiplied and then divided by Avogadro's number. *For atoms, the molar mass in grams per mole is numerically equal to the atomic weight in atomic mass units.*

LEARNING HINT

The mass below each symbol in the periodic table is *both* an atomic weight in atomic mass units and a molar mass in grams per mole. For example, the mass of one hydrogen atom is 1.008 u; the mass of 1 mol of hydrogen atoms is 1.008 g.

EXAMPLE 2.6

What is the molar mass of (a) manganese and (b) potassium?

SOLUTION

The atomic weights of manganese and potassium are 54.94 u and 39.10 u, respectively (see the table of atomic weights on the inside front cover). Hence, their molar masses are (a) 54.94 g/mol and (b) 39.10 g/mol.

PRACTICE EXERCISE 2.7

Use a table of atomic weights to find the molar mass in grams of (a) magnesium and (b) gold.

A given amount of substance can be expressed as a mass (m) or as a number of moles (n). Although mass is what we measure on a laboratory balance, the number of moles is usually needed for chemical calculations. Thus it is often necessary to change mass into number of moles, and vice versa. The molar mass provides the conversion factor, as shown in the following examples.

LEARNING HINT

Some chemists prefer to use the following equation for changing moles into mass:

$$m = n\mathcal{M}$$

where m is the mass in grams, n is the number of moles, and $\mathcal{M}$ is the molar mass in grams per mole. PRACTICE—BE SURE YOU KNOW HOW TO CONVERT GRAMS INTO MOLES, AND VICE VERSA.

EXAMPLE 2.7

How many moles of iron atoms are in a 100-g sample?

SOLUTION

The atomic weight of iron to four significant figures is 55.85 u; therefore, the mass of 1 mol of iron atoms is 55.85 g. The number of moles in 100 g of iron is

$$n = 100\ \text{g Fe} \times \frac{1\ \text{mol Fe}}{55.85\ \text{g Fe}} = 1.79\ \text{mol Fe}$$

PRACTICE EXERCISE 2.8

Calculate the mass in grams of 1.55 mol of calcium atoms.

EXAMPLE 2.8

How many grams of sulfur contain the same number of atoms as 100 g of iron?

SOLUTION

A 100-g sample of iron contains 1.79 mol of atoms (Example 2.7). The mole is a counting unit, so 1.79 mol of sulfur will contain the same number of atoms as 1.79 mol of iron. The atomic weight of sulfur is 32.06 u, and its molar mass is 32.06 g/mol. The mass of 1.79 mol of sulfur atoms is

$$1.79 \text{ mol S} \times \frac{32.06 \text{ g S}}{1 \text{ mol S}} = 57.4 \text{ g S}$$

PRACTICE EXERCISE 2.9

Calculate the mass in grams of 2.00×10^{21} helium atoms.

2.5 ELEMENTS, COMPOUNDS, AND MIXTURES

Most natural materials such as air, seawater, stone, and wood consist of many different substances. A pure substance is either an **element**, whose atoms all have the same atomic number, or it is a **compound**, a substance made up of two or more elements combined in a fixed ratio. Water, one of our most abundant compounds, always consists of 88.8% oxygen and 11.2% hydrogen by mass. Carbon dioxide, a compound produced during combustion and metabolism, contains 27.3% carbon and 72.7% oxygen by mass. Aspirin, sucrose (table sugar), and sodium chloride (salt) are also compounds.

We learn to recognize familiar substances by characteristic traits or **properties** such as color, hardness, density, and combustibility. *The properties of any element or compound are the same for all samples observed under the same conditions.* Some of the properties of hydrogen, oxygen, and water are summarized in Table 2.4. Note that the properties of water are very different from those of hydrogen and oxygen. Compounds usually have properties that differ markedly from those of their constituent elements—a striking example is ordinary white table salt, a compound formed when sodium, a soft silvery white metal, combines with chlorine, a greenish yellow, highly poisonous gas.

PRACTICE EXERCISE 2.10

List three properties of (a) sodium chloride (salt) and (b) sucrose (table sugar).

Compounds should not be confused with mixtures. A **mixture** contains two or more pure substances (elements or compounds) in which each component retains its identity. Salt water is a mixture of sodium chloride and water, and air is a mixture of

TABLE 2.4 Some Properties of Oxygen, Hydrogen, and Water

Properties of Oxygen (an element)
Colorless, odorless, and tasteless gas at 25°C.
Condenses to a pale blue liquid at −183°C.
Freezes to a pale blue crystalline solid at −218.4°C.
Liquid oxygen is attracted by a magnet.
Oxygen supports combustion; that is, other substances burn in oxygen.

Properties of Hydrogen (an element)
Colorless, odorless, and tasteless gas at 25°C.
Condenses to a colorless liquid at −253°C.
Freezes to a colorless solid at −259°C.
Hydrogen is not attracted by a magnet.
Hydrogen does not support combustion, but it burns in air to form water.

Properties of Water (a compound)
Colorless, odorless, and tasteless liquid at 25°C.
Freezes to a colorless crystalline solid (ice) at 0°C.
Boils at 100°C to form a colorless gas (steam).
Water is not attracted by a magnet.
Water does not support combustion, nor will it burn in air.
Water contains 88.8% oxygen and 11.2% hydrogen by mass.

nitrogen, oxygen, and other gases. The properties of a mixture, unlike those of an element or compound, may vary from sample to sample because the component substances may be present in different proportions. Seawater samples may contain slightly more or less salt, air may contain more or less water vapor, and so forth. The components of a mixture can usually be separated from each other by simple physical operations such as sorting, sifting solids, and distilling liquids. The ingenuity of chemists is constantly being challenged by the need to find more efficient methods for identifying and separating the components of various mixtures.

Demonstration 2.1 illustrates the difference between a mixture of iron filings and sulfur, and a compound (iron sulfide) made from these elements. The elements in the mixture retain their identity and can be separated by means of a magnet. The properties of the compound are different from those of iron and sulfur, and the elements cannot be separated magnetically.

PRACTICE EXERCISE 2.11

Which of the following are pure substances and which are mixtures of substances: (a) milk, (b) carbon dioxide, (c) steel, (d) sucrose (table sugar), (e) wood, and (f) blood?

DEMONSTRATION 2.1 A MIXTURE AND A COMPOUND

Iron (gray filings), sulfur (yellow powder), and a mixture of iron filings and sulfur.

Since both components of the mixture retain their own properties, the iron is attracted to a magnet.

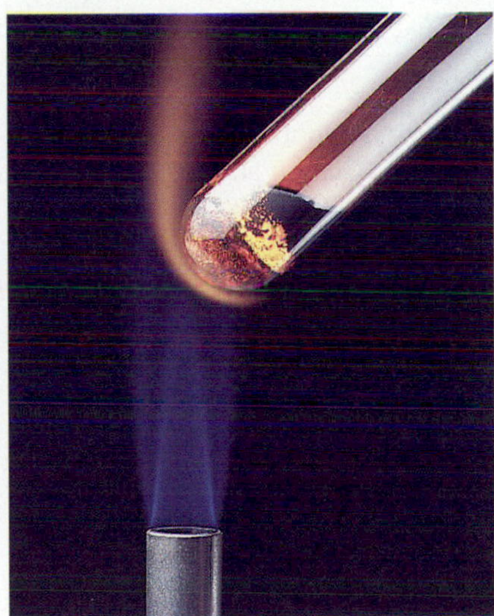

When heated, the iron–sulfur mixture begins to glow and give off heat, indicating that a chemical reaction is taking place.

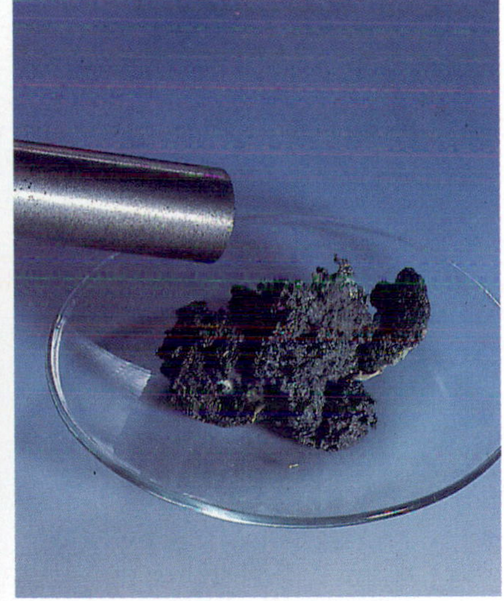

The final product is the compound iron sulfide. Iron sulfide is gray and not attracted to a magnet; its properties are different from those of iron or sulfur.

2.6 ATOMS IN COMBINATION

Thus far we have looked at individual atoms, but atoms rarely exist as separate entities—they are usually combined with other atoms. We have seen that these combinations might contain atoms of one element only, or they might be compounds containing atoms of several different elements.

Molecules

Only 109 different elements have been discovered or synthesized thus far, but there are millions of known compounds. The existence of compounds with fixed properties and constant composition led John Dalton, an English scientist and schoolteacher, to revive the ancient Greek concept of atoms and to suggest in 1803 that atoms can bind together in fixed ratios to form atomic groupings (Figure 2.14). Water consists of triatomic (three-atom) units, each of which contains two atoms of hydrogen strongly bonded to one atom of oxygen. Oxygen gas consists of diatomic (two-atom) units in which two oxygen atoms are bonded to each other. A **molecule** is a distinct, electrically neutral group consisting of a well-defined number of atoms bonded together (Figure 2.15). Water molecules contain three atoms bonded together; oxygen molecules contain two atoms. The internal forces that bond the atoms together in a molecule are generally much stronger than any forces between different molecules, and therefore each molecule is a relatively independent entity.

Monatomic: one atom
Diatomic: two atoms
Triatomic: three atoms
Polyatomic: three or more atoms

A molecule is represented by a **molecular formula** in which subscripts next to the atomic symbols show how many atoms of each kind are present. The water molecule contains two hydrogen atoms and one oxygen atom, so its formula is as follows:

$$H_2O$$

one oxygen atom
two hydrogen atoms

(An atomic symbol without a subscript, such as O in H_2O, always represents a single atom.) Carbon and oxygen form the following molecular compounds (Figure 2.15):

carbon monoxide: CO

one oxygen atom
one carbon atom

(a)

ELEMENTS

	W:		W:
Hydrogen	1	Strontian	46
Azote	5	Barytes	68
Carbon	5,4	Iron	50
Oxygen	7	Zinc	56
Phosphorus	9	Copper	56
Sulphur	13	Lead	90
Magnesia	20	Silver	190
Lime	24	Gold	190
Soda	28	Platina	190
Potash	42	Mercury	167

(b)

Figure 2.14

(a) John Dalton (1766–1844), English chemist. (b) Some of the atomic symbols Dalton used. The postulates of Dalton's atomic theory are (1) Matter is composed of small indivisible atoms. (2) The atoms of each element are identical to each other and unlike those of other elements. (3) A compound contains atoms of two or more elements bound together in a fixed proportion. (4) A chemical reaction is a rearrangement of atoms; atoms are not created or destroyed during such reactions.

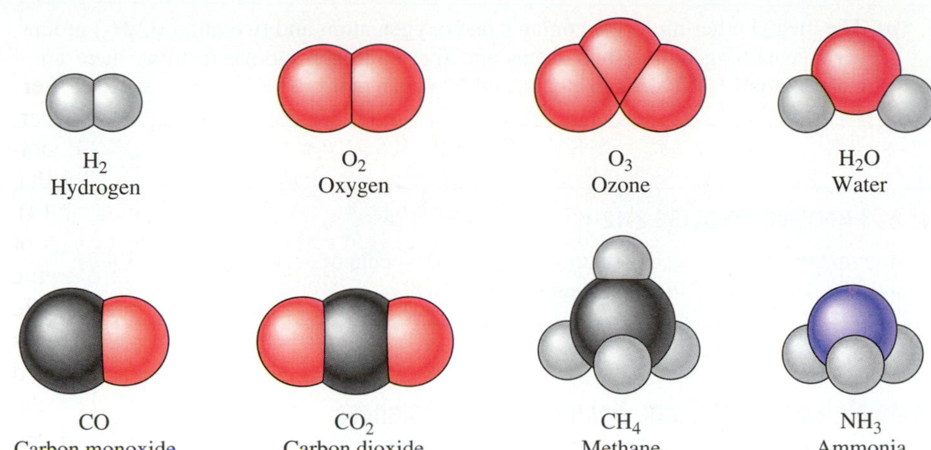

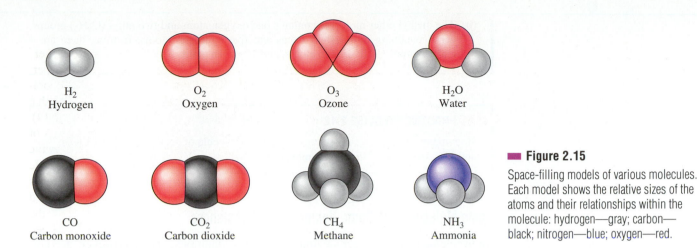

Figure 2.15
Space-filling models of various molecules. Each model shows the relative sizes of the atoms and their relationships within the molecule: hydrogen—gray; carbon—black; nitrogen—blue; oxygen—red.

carbon dioxide: CO_2

Carbon dioxide has one more oxygen atom per molecule than carbon monoxide. This difference in composition causes carbon monoxide and carbon dioxide to be different compounds with different properties.

Some molecules contain many atoms and are more complex. The cholesterol molecule, for example, has 74 atoms:

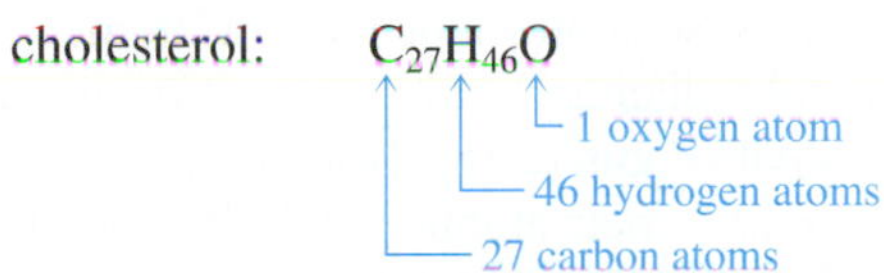

Certain conventions determine the preferred order for writing the elements in a formula, but for the time being you will just have to remember that water is written H_2O rather than OH_2. (Some of the rules for writing formulas and naming compounds are discussed in Section 2.9.) Sometimes a formula is written with parentheses to show the presence of certain groups. The urea molecule, for example, contains one carbon atom, one oxygen atom, and two NH_2 groups. Its formula is usually written as $(NH_2)_2CO$, though it can be written as CH_4N_2O. Another example is $Al(OH)_3$, which contains one aluminum atom and three OH groups, that is, three oxygen atoms and three hydrogen atoms. It is important to remember that *a subscript following a group in parentheses applies to all the atoms in the group.*

EXAMPLE 2.9

How many atoms of each element are in (a) the hydrogen peroxide molecule, H_2O_2, and (b) the diethyl ether molecule, $(C_2H_5)_2O$?

SOLUTION

(a) In H_2O_2 there are two hydrogen atoms and two oxygen atoms.

(b) The diethyl ether molecule contains one oxygen atom and two ethyl (C_2H_5) groups, each containing two carbon atoms and five hydrogen atoms. In total, there are 4 carbon atoms, 10 hydrogen atoms, and 1 oxygen atom.

PRACTICE EXERCISE 2.12

How many atoms of each element are in (a) a molecule of acetic acid, CH_3COOH, and (b) a molecule of dimethylamine, $(CH_3)_2NH$?

Some Molecular and Some Nonmolecular Substances

Some elements are molecular under ordinary conditions; most, however, are not. Look, once again, at the periodic table on the inside front cover. There are 87 metallic elements and 22 nonmetallic elements. None of the metals exist in the form of molecules. (The structure of metals will be considered more fully in later chapters.) Eight of the 22 nonmetals normally consist of diatomic molecules. These elements include hydrogen (H_2), nitrogen (N_2), oxygen (O_2), and the five members of Group 7A (Group 17)—fluorine (F_2), chlorine (Cl_2), bromine (Br_2), iodine (I_2), and astatine (At_2). The Group 7A elements are called **halogens.** Sulfur (Group 6A) consists of ring-shaped S_8 molecules (Figure 2.16) and the white form of phosphorus (Group 5A) contains P_4 molecules (Figure 2.17). The formulas and physical states for these molecular elements at 25°C are summarized in Table 2.5.

The nonmetallic elements that do not form molecules include the six **noble gases** of periodic Group 8A (Group 18). These elements—helium, neon, argon, krypton, xenon, and radon—exist as separate, independent atoms. The two most familiar forms of carbon (Group 4A), diamond and graphite, consist of atoms linked together in arrays of indefinite extent (Figure 2.18). Nonmolecular elements, both metals and nonmetals, are represented by their atomic symbols. For example, helium, carbon, and iron are written as He, C, and Fe, respectively.

As we have seen, some elements exist in more than one form. For example, some of the oxygen in the upper stratosphere occurs as ozone, a gas consisting of O_3 mol-

Figure 2.16

(a) The sulfur molecule, S_8, is a puckered ring of eight sulfur atoms. (b) Crystals of rhombic sulfur.

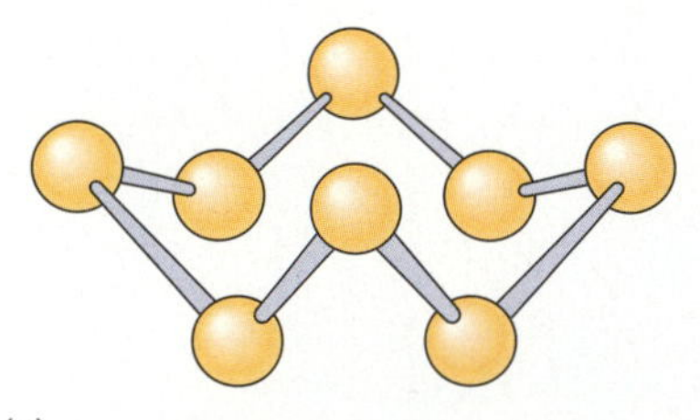

(a)

S S S S S S S S

(b)

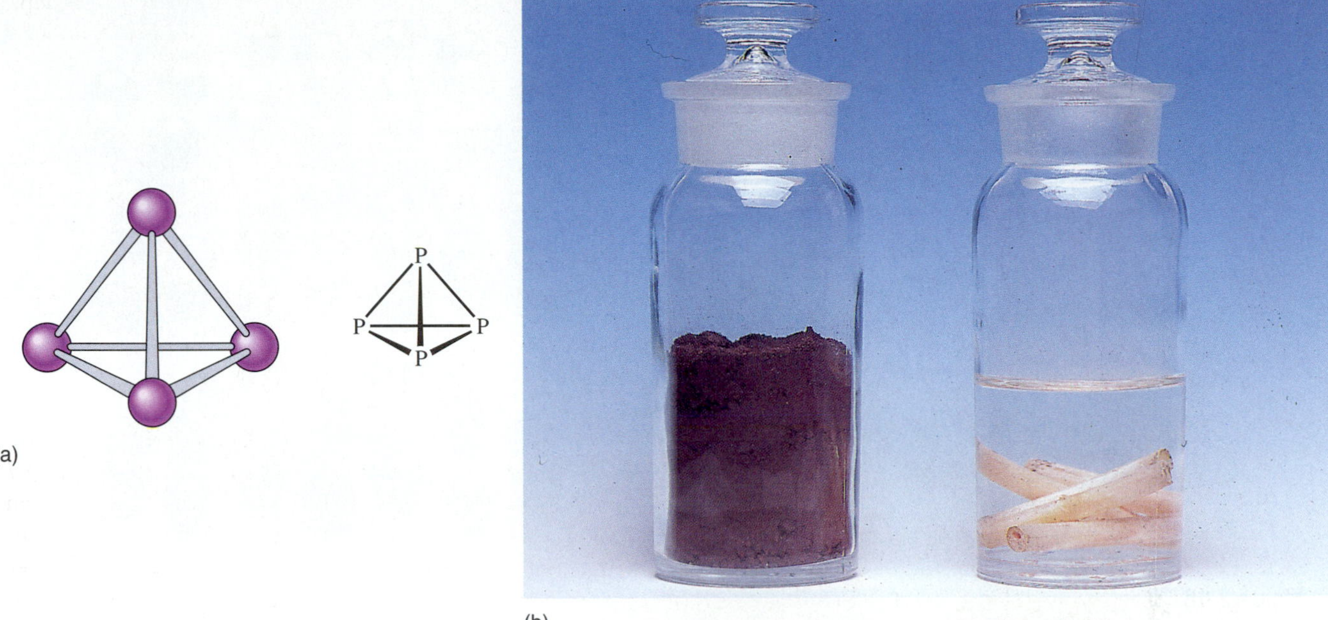

Figure 2.17

(a) White phosphorus consists of P_4 molecules. (b) Allotropic forms of phosphorus. *Left:* Red phosphorus. The atoms in this allotrope are linked together in an extended three-dimensional array. *Right:* white phosphorus is stored under water to protect it from air.

TABLE 2.5 Formulas and Physical States of Some Molecular Elements at 25°C

Gases		Liquids		Solids	
Hydrogen	H_2	Bromine	Br_2	Phosphorus (white)	P_4
Nitrogen	N_2			Sulfur	S_8
Oxygen	O_2			Iodine	I_2
Ozone	O_3				
Fluorine	F_2				
Chlorine	Cl_2				

ecules (Figure 2.15 and Table 2.5). Two or more forms of an element that exist in the same physical state (solid, liquid, or gas) are called **allotropes** or **allotropic forms**. Ordinary oxygen (O_2) and ozone (O_3) are allotropes. Diamond and graphite are nonmolecular allotropic forms of carbon. Some recently discovered carbon allotropes, called *fullerenes,* consist of large symmetrical molecules that contain 60 or more carbon atoms. (See *Chemical Insight: Buckminsterfullerenes* on page 882.) A number of other elements, including phosphorus, sulfur, and tin, exist in allotropic forms.

PRACTICE EXERCISE 2.13

List the elements that are (a) gases and (b) liquids at room temperature. Write a formula for each element on your list.

Figure 2.18

Two allotropic forms of solid carbon. (a) Each carbon atom in diamond is bonded to four others. (b) Graphite consists of flat sheets in which each carbon is bonded to three others. The bonding forces between atoms in adjacent sheets are much weaker than the bonding force within a sheet. Although there is close contact between bonded atoms (see Figure 2.15), these models show them as separate to bring out the three-dimensional pattern.

Many compounds, including water (H_2O), ethanol (C_2H_5OH), and ammonia (NH_3), are molecular; many others are nonmolecular. Solid compounds often contain atoms or ions arranged in a fixed geometrical pattern that extends throughout the solid. There are no separate molecules in such an array. For example, sodium chloride (table salt) is composed of ions. A crystal of sodium chloride consists of a cubic array of alternating sodium ions (Na^+) and chloride ions (Cl^-) arranged so that each ion has six oppositely charged ions for its nearest neighbors. Figure 2.19 shows that there are no sodium chloride molecules in the crystal. Equal numbers of sodium ions and chloride ions are present, and the formula is written as NaCl to reflect the one-to-one ratio.

Silicon dioxide, the principal ingredient of sand and the only ingredient of quartz, is a nonmolecular compound consisting of silicon and oxygen atoms. The atoms in a quartz crystal arrange themselves so that each silicon atom is bonded to four neighboring oxygen atoms and each oxygen atom in turn has two silicon neighbors (Figure 2.20). To achieve this arrangement, the total number of oxygen atoms must be twice the number of silicon atoms, so the formula for silicon dioxide is SiO_2.

2.7 MOLAR MASSES OF COMPOUNDS

Consider a single propane molecule, C_3H_8, which contains three carbon atoms and eight hydrogen atoms. Just as a dozen propane molecules would contain three dozen carbon atoms and eight dozen hydrogen atoms, a mole of propane molecules would contain 3 mol of carbon atoms and 8 mol of hydrogen atoms. Similarly, 1 mol of water, H_2O, would contain 2 mol of hydrogen atoms and 1 mol of oxygen atoms. *The mole ratio of the elements in a compound is the same as the atom ratio in the formula of the compound.*

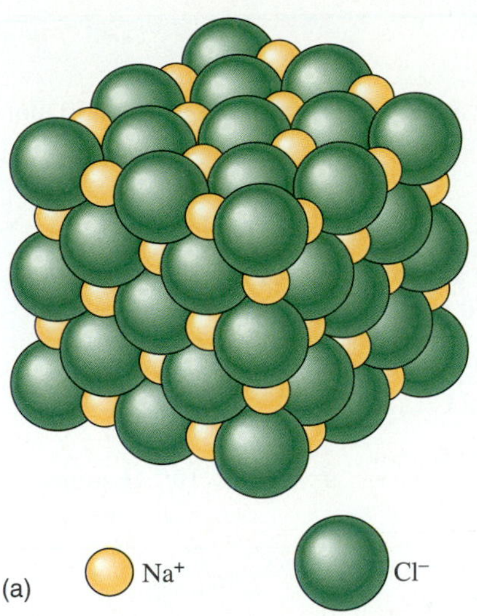

Figure 2.19

(a) Sodium chloride consists of sodium and chloride ions in an alternating cubic array. (b) Crystals of halite (sodium chloride) in a rock matrix.

EXAMPLE 2.10

How many moles of (a) carbon atoms, (b) hydrogen atoms, and (c) oxygen atoms are in exactly 5 mol of sucrose, $C_{12}H_{22}O_{11}$?

SOLUTION

One mole of sucrose contains 12 mol of carbon atoms, 22 mol of hydrogen atoms, and 11 mol of oxygen atoms. Five moles of sucrose will contain (a) $5 \times 12 = 60$ mol of carbon atoms, (b) $5 \times 22 = 110$ mol of hydrogen atoms, and (c) $5 \times 11 = 55$ mol of oxygen atoms. (Note that all numbers in this calculation are exact.)

Figure 2.20

(a) Quartz is a form of silicon dioxide (SiO_2) in which each silicon atom (black) is bonded to four oxygen atoms (red), and each oxygen atom is bonded to two silicon atoms. (b) Quartz crystals.

EXAMPLE 2.11

Ethylene consists of C_2H_4 molecules. How many atoms of each element are in 2.50 mol of ethylene?

SOLUTION

One mole of ethylene contains 2 mol of carbon atoms and 4 mol of hydrogen atoms. The number of carbon and hydrogen atoms in 2.50 mol of ethylene is:

$$2.50 \text{ mol } C_2H_4 \times \frac{2 \text{ mol C}}{1 \text{ mol } C_2H_4} \times \frac{6.022 \times 10^{23} \text{ C atoms}}{1 \text{ mol C}} = 3.01 \times 10^{24} \text{ C atoms}$$

$$2.50 \text{ mol } C_2H_4 \times \frac{4 \text{ mol H}}{1 \text{ mol } C_2H_4} \times \frac{6.022 \times 10^{23} \text{ H atoms}}{1 \text{ mol H}} = 6.02 \times 10^{24} \text{ H atoms}$$

PRACTICE EXERCISE 2.14

Calcium chloride, $CaCl_2$, consists of positive calcium ions (Ca^{2+}) and negative chloride ions (Cl^-). (a) How many moles of each ion are in 3.5 mol of calcium chloride? (b) How many calcium and how many chloride ions are in 3.5 mol of calcium chloride?

EXAMPLE 2.12

Hydrazine is a component of some rocket fuels. One mole of hydrazine contains 2 mol of nitrogen atoms and 4 mol of hydrogen atoms. What is the formula of the hydrazine molecule?

SOLUTION

The atom ratio is the same as the mole ratio; hence, one molecule of hydrazine will contain two nitrogen atoms and four hydrogen atoms. The formula is N_2H_4.

PRACTICE EXERCISE 2.15

One mole of magnesium nitride contains 3 mol of magnesium ions (Mg^{2+}) and 2 mol of nitride ions (N^{3-}). Write the formula for magnesium nitride.

The molar mass of a compound is the sum of the molar masses of the atoms in its formula. For example, the formula of sodium chloride is NaCl, and the molar mass of sodium chloride is the sum of the molar masses of the sodium and chlorine atoms, 22.990 g/mol + 35.453 g/mol = 58.443 g/mol. (Remember that the molar mass of an atom in grams per mole is numerically equal to its atomic weight in atomic mass units.) The molar mass of a compound is sometimes referred to as its **formula weight**. The molar mass of a molecular compound is often referred to as the **molec-**

Figure 2.21

Molar quantities of various substances. Clockwise from center bottom: copper (Cu, 63.5 g), iron (Fe, 55.8 g), water (H_2O, 18.0 g), sucrose ($C_{12}H_{22}O_{11}$, 342 g), copper sulfate pentahydrate ($CuSO_4 \cdot 5H_2O$, 250 g), mercury (Hg, 201 g). Center: sulfur (S_8, 257 g).

ular weight although, strictly speaking, molecular weight refers to the mass of one molecule. Some molar masses are shown in Figure 2.21.

EXAMPLE 2.13

The molecular compound urea, $(NH_2)_2CO$, is used in fertilizers to supply nitrogen. Calculate the mass of 1 mol of urea.

SOLUTION

One mole of $(NH_2)_2CO$ contains 1 mol of carbon atoms, 1 mol of oxygen atoms, 2 mol of nitrogen atoms, and 4 mol of hydrogen atoms. The molar masses of carbon, oxygen, nitrogen, and hydrogen are 12.01, 16.00, 14.01, and 1.008 g/mol, respectively. The total mass is

$$
\begin{aligned}
\text{carbon:} \quad & 1 \text{ mol} \times 12.01 \text{ g/mol} = 12.01 \text{ g} \\
\text{oxygen:} \quad & 1 \text{ mol} \times 16.00 \text{ g/mol} = 16.00 \text{ g} \\
\text{nitrogen:} \quad & 2 \text{ mol} \times 14.01 \text{ g/mol} = 28.02 \text{ g} \\
\text{hydrogen:} \quad & 4 \text{ mol} \times 1.008 \text{ g/mol} = 4.032 \text{ g} \\
& \text{Total mass of 1 mol} = 60.06 \text{ g}
\end{aligned}
$$

LEARNING HINT

In this book we will report all atomic weights and molar masses to four significant figures unless the problem or example calls for more than four significant figures.

EXAMPLE 2.14

Calculate the mass in grams of one urea molecule.

SOLUTION

One mole of urea contains 6.022×10^{23} molecules and has a mass of 60.06 g (Example 2.13). The mass of one molecule is

$$\frac{60.06 \text{ g}}{1 \text{ mol urea}} \times \frac{1 \text{ mol urea}}{6.022 \times 10^{23} \text{ molecules}} = 9.973 \times 10^{-23} \text{ g/molecule}$$

PRACTICE EXERCISE 2.16

Butane, C_4H_{10}, is a fuel used in lighters and camp stoves. Calculate (a) the molar mass of butane and (b) the mass in grams of one butane molecule.

Molar masses are often used to convert moles of a compound to grams, and vice versa, as illustrated in the following example.

EXAMPLE 2.15

Calculate the mass in grams of 2.50 mol of ammonia, NH_3.

SOLUTION

The molar masses of nitrogen and hydrogen are 14.01 and 1.008 g/mol, respectively. The mass of 1 mol of ammonia is

$$\begin{aligned} \text{nitrogen:} \quad & 1 \text{ mol} \times 14.01 \text{ g/mol} = 14.01 \text{ g} \\ \text{hydrogen:} \quad & 3 \text{ mol} \times 1.008 \text{ g/mol} = 3.024 \text{ g} \\ & \text{Total} = 17.03 \text{ g} \end{aligned}$$

The mass of 2.50 mol of ammonia is

$$2.50 \text{ mol } NH_3 \times \frac{17.03 \text{ g}}{1 \text{ mol } NH_3} = 42.6 \text{ g}$$

PRACTICE EXERCISE 2.17

Calculate (a) the number of moles and (b) the number of molecules in 1000 g of sulfur dioxide, SO_2.

2.8 IONS AND IONIC COMPOUNDS

Compounds can be classified into those that contain ions and those that do not. The former are termed **ionic**, the latter **covalent**. Sodium chloride is ionic; water and sucrose are covalent. Ionic compounds can be distinguished from covalent compounds by the fact that they can conduct an electric current when melted. Covalent substances in the liquid state are at best feeble conductors, and most do not conduct at all.

Negative ions are called **anions** (pronounced an′-ions) because they tend to migrate toward the positive electrode (anode) when an electric current is passed through a solution or a molten liquid containing the ions. Positive ions migrate toward the negative electrode (cathode), so they are called **cations** (cat′-ions). Formulas for some of the more common anions and cations are given in Table 2.6. The ionic charge is an important part of the formula and should be carefully noted.

Figure 2.22 shows a portion of the periodic table and the principal ions formed by some of the elements. Inspection of the figure shows that *positive ions are usually formed by metals while negative ions are formed by nonmetals.* We will discover in later chapters that the tendency to lose electrons is a property common to most metal atoms. Nonmetal atoms, on the other hand, tend to gain or share electrons.

LEARNING HINT

Look up the formula and charge each time you read about an ion—you will soon remember the important ones.

Certain metals such as sodium and potassium give up electrons so readily that they occur naturally only as positive ions, never as neutral atoms. Note that sodium, potassium, and all the other elements in Group 1A form 1+ ions. The elements magnesium and calcium occur naturally as the ions Mg^{2+} and Ca^{2+}; they are found in Group 2A along with other metals that form 2+ ions. Aluminum and other Group 3A metals form 3+ ions; Al^{3+} is the most important of these.

Nonmetals usually form negative ions. The halogens in Group 7A form 1− ions, oxygen and sulfur in Group 6A form 2− ions, and nitrogen in Group 5A forms a 3− ion. The noble gases in Group 8A do not readily form ions. Nonmetals rarely form positive ions, but there are two important exceptions: the hydrogen ion, H^+, and the

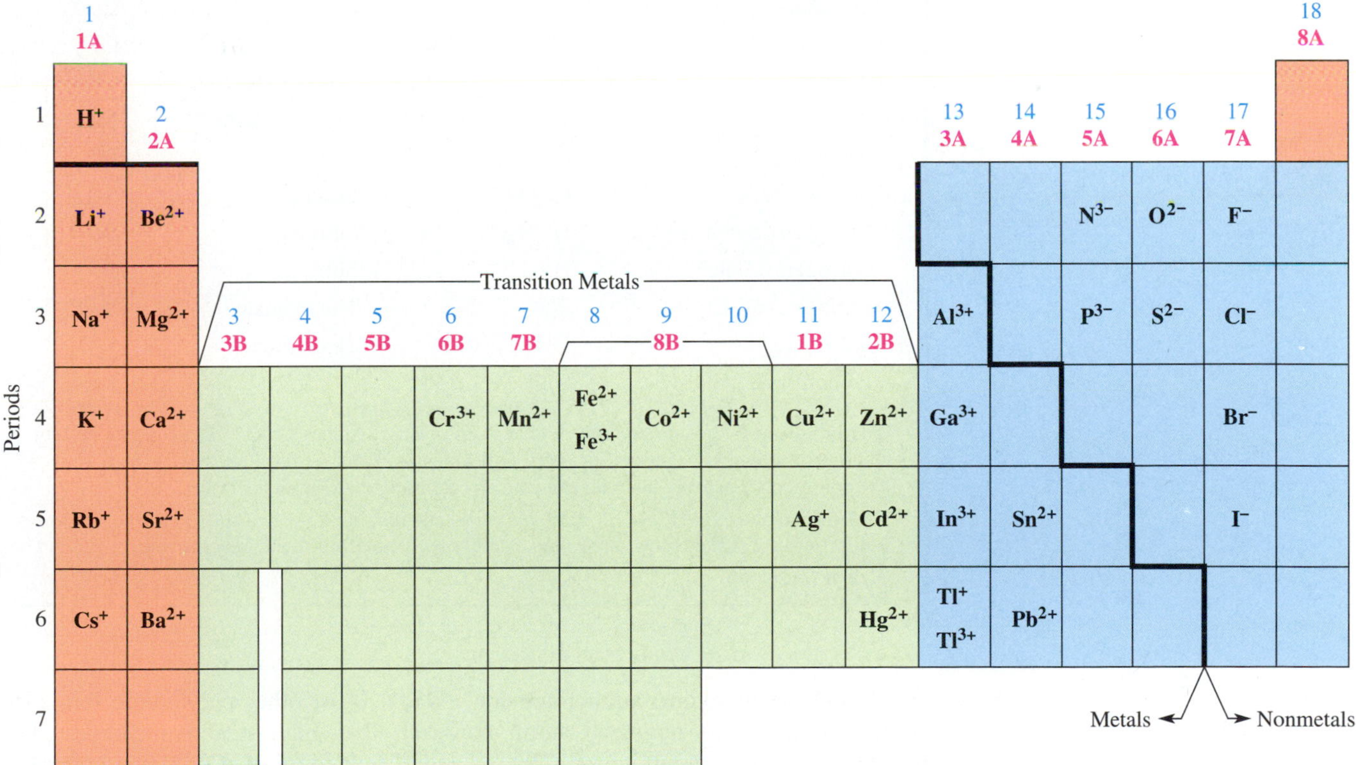

Figure 2.22

Some important monatomic (one-atom) ions. With a few exceptions, each of the A-group metal ions bears a positive charge equal to its group number. Each nonmetal ion bears a negative charge equal in magnitude to eight minus the group number.

TABLE 2.6 Formulas for Some Ions

Positive Ions (Cations)

Name	*Formula*	*Name*	*Formula*
*Aluminum	Al^{3+}	Lead	Pb^{2+}
*Ammonium	NH_4^+	Lithium	Li^+
Barium	Ba^{2+}	*Magnesium	Mg^{2+}
Cadmium	Cd^{2+}	Manganese	Mn^{2+}
*Calcium	Ca^{2+}	Mercury(II)	Hg^{2+}
Chromium(III)	Cr^{3+}	Mercury(I)	Hg_2^{2+}
Cobalt(II)	Co^{2+}	Nickel	Ni^{2+}
*Copper(II)	Cu^{2+}	*Potassium	K^+
Copper(I)	Cu^+	Silver	Ag^+
*Iron(III)	Fe^{3+}	*Sodium	Na^+
*Iron(II)	Fe^{2+}	Tin(II)	Sn^{2+}
*Hydrogen[a]	H^+	Zinc	Zn^{2+}

Negative Ions (Anions)

Name	*Formula*	*Name*	*Formula*
Acetate	CH_3COO^-	*Iodide	I^-
*Bromide	Br^-	*Nitrate	NO_3^-
*Carbonate	CO_3^{2-}	Nitrite	NO_2^-
Chlorate	ClO_3^-	Oxalate	$C_2O_4^{2-}$
*Chloride	Cl^-	*Oxide	O^{2-}
Chlorite	ClO_2^-	Perchlorate	ClO_4^-
Chromate	CrO_4^{2-}	Permanganate	MnO_4^-
Cyanide	CN^-	*Phosphate	PO_4^{3-}
Dichromate	$Cr_2O_7^{2-}$	*Sulfate	SO_4^{2-}
*Hydrogen carbonate[b]	HCO_3^-	Sulfite	SO_3^{2-}
Hydrogen sulfate[b]	HSO_4^-	*Sulfide	S^{2-}
Hydrogen sulfite[b]	HSO_3^-	Thiocyanate	SCN^-
*Hydroxide	OH^-	Thiosulfate	$S_2O_3^{2-}$
Hypochlorite	ClO^-		

[a]The proton or hydrogen ion, H^+, is formed in discharge tubes containing hydrogen gas. In solution, it combines with one or more molecules of water to form species such as H_3O^+, hydronium ion. There are no ionic compounds containing bare protons.

[b]The official name for HCO_3^- is hydrogen carbonate ion, but it is often called bicarbonate ion. Similarly, HSO_4^- and HSO_3^- are called bisulfate ion and bisulfite ion, respectively.

***Learn these now.**

polyatomic (many-atom) ammonium ion, NH_4^+. Most other polyatomic ions also contain at least one nonmetal atom; however, they have negative charges. The hydroxide ion (OH^-), nitrate ion (NO_3^-), and chromate ion (CrO_4^{2-}) are examples;

others are listed in Table 2.6. The mercury(I) ion, Hg_2^{2+}, is unusual in that it consists only of metal atoms.

Formulas of Ionic Compounds

An ionic compound contains positive and negative ions in a ratio that makes it electrically neutral. Sodium chloride, for example, contains Na^+ and Cl^- ions, each with a single charge. Neutrality is achieved when the sodium and chloride ions are present in a one-to-one ratio; hence, the formula is Na^+Cl^-, or simply NaCl. (Ionic charges are generally dropped from neutral formulas.) In calcium chloride each doubly charged Ca^{2+} ion requires the presence of two Cl^- ions, making the formula $CaCl_2$. Calcium sulfate, $CaSO_4$, contains equal numbers of Ca^{2+} and SO_4^{2-} ions.

The following examples show how the formula of an ionic compound can be deduced from the charges on the component ions.

EXAMPLE 2.16

What is the formula of an ionic solid made up of potassium ions (K^+) and sulfide ions (S^{2-})?

SOLUTION

The total charge must add up to zero. Hence there are two K^+ ions for every S^{2-} ion. The formula is $(K^+)_2S^{2-}$, or K_2S.

EXAMPLE 2.17

Use Table 2.6 to find the formula of aluminum sulfate.

SOLUTION

From the table we see that the aluminum ion is Al^{3+} and the sulfate ion is SO_4^{2-}. A total charge of zero is obtained from the two aluminum ions ($+3 \times 2 = +6$) and three sulfate ions ($-2 \times 3 = -6$). The formula is $(Al^{3+})_2(SO_4^{2-})_3$, or $Al_2(SO_4)_3$.

PRACTICE EXERCISE 2.18

Write formulas for (a) sodium carbonate, (b) magnesium chloride, and (c) iron(III) bromide.

2.9 NAMING COMPOUNDS

There are millions of chemical compounds, and it is important for each one to have a unique name. Some compounds such as water (H_2O) and ammonia (NH_3) are known by their common names, but it would be impossible to memorize a common name for every substance. For this reason, the International Union of Pure and Applied Chemistry (IUPAC) has established a special commission on nomenclature devoted to the systematic naming of chemical compounds. The official IUPAC rules can be found in the Chemical Rubber Company's *CRC Handbook of Chemistry and Physics,* which is available in most libraries. Some of the more important rules are summarized in the following sections. Additional rules are given in later chapters.

Naming Ions

1. *Positive ions (cations):*

(a) Positive ions formed from metal atoms have the same name as the metal:

Na^{+}	sodium ion
Mg^{2+}	magnesium ion
Al^{3+}	aluminum ion

(b) If a metal can form more than one positive ion, the number of electrons lost *per atom* is added as a Roman numeral, as illustrated in the middle column below:

Ion	Systematic Name	Informal Name
Fe^{2+}	iron(II) ion	ferrous ion
Fe^{3+}	iron(III) ion	ferric ion
Hg_2^{2+}	mercury(I) ion	mercurous ion
Hg^{2+}	mercury(II) ion	mercuric ion

Such ions frequently have older informal names in which the lesser and greater charges per atom are designated by the endings *-ous* and *-ic,* respectively. Informal names for the above ions are given in the right-hand column.

(c) Positive ions formed from nonmetal atoms have special names ending in *-ium:*

NH_4^{+}	ammonium ion
H_3O^{+}	hydronium ion

PRACTICE EXERCISE 2.19

Refer to a table of the elements and name the following ions: (a) K^{+}, (b) Ca^{2+}, and (c) Ga^{3+}.

PRACTICE EXERCISE 2.20

Give systematic names for the ions in each of the following pairs: (a) Cu^{+}, Cu^{2+} and (b) Cr^{2+}, Cr^{3+}.

2. *Negative ions (anions):*

(a) Monatomic (one-atom) negative ions have names ending in *-ide:*

I^-	iodide ion
S^{2-}	sulfide ion

A few polyatomic anions have special names ending in *-ide:*

OH^-	hydroxide ion
CN^-	cyanide ion
O_2^{2-}	peroxide ion

(b) Polyatomic anions containing oxygen usually have names ending in *-ite* or *-ate*. The suffix *-ite* designates an ion that contains one less oxygen than a similar ion with the suffix *-ate:*

NO_2^-	nitrite ion	SO_3^{2-}	sulfite ion
NO_3^-	nitrate ion	SO_4^{2-}	sulfate ion

The prefix *hypo-* indicates one less oxygen than *-ite;* the prefix *per-* indicates one more oxygen than *-ate:*

ClO^-	hypochlorite ion
ClO_2^-	chlorite ion
ClO_3^-	chlorate ion
ClO_4^-	perchlorate ion

(c) Negative ions derived by replacing an oxygen atom with a sulfur atom are indicated by the prefix *thio-:*

OCN^-	cyanate ion	SO_4^{2-}	sulfate ion
SCN^-	thiocyanate ion	$S_2O_3^{2-}$	thiosulfate ion

PRACTICE EXERCISE 2.21

Consult a table of the elements and name the following ions: (a) Br^-, (b) N^{3-}, and (c) O^{2-}.

PRACTICE EXERCISE 2.22

The bromite ion is BrO_2^-. Write formulas for (a) the bromate ion, (b) the hypobromite ion, and (c) the perbromate ion.

Naming Ionic Compounds

The name of an ionic compound is the cation name followed by the anion name:

K_2SO_4	potassium sulfate
NH_4Cl	ammonium chloride
$Mg(NO_3)_2$	magnesium nitrate
$Na_2S_2O_3$	sodium thiosulfate
$FeCl_3$	iron(III) chloride

The last compound is informally called ferric chloride.

PRACTICE EXERCISE 2.23

Consult Table 2.6 and use the above rules to give systematic names for the following ionic compounds: (a) $BaCl_2$, (b) CaS, (c) Mg_3N_2, (d) $FeSO_4$, (e) KCN, and (f) $Al(NO_3)_3$.

PRACTICE EXERCISE 2.24

Consult Table 2.6 and use the above rules to give formulas for the following ionic compounds: (a) ferrous chloride, (b) sodium sulfide, (c) potassium dichromate, (d) magnesium phosphate, and (e) iron(III) sulfate.

A number of unofficial names are worth remembering, some because of their common use in other fields and some because of historical interest. Some examples are as follows:

Compound	Common Name	Systematic Name
NaCl	table salt	sodium chloride
$NaHCO_3$	baking soda (sodium bicarbonate)	sodium hydrogen carbonate
NaOH	caustic soda	sodium hydroxide
CaO	lime (quicklime)	calcium oxide
$CaCO_3$	limestone	calcium carbonate

Other examples are pointed out in later chapters.

Naming Binary Molecular Compounds

A **binary compound** contains only two elements and may be ionic or molecular. Examples of binary ionic compounds were given in the previous section. Binary molecular compounds usually contain nonmetal atoms; hydrogen chloride (HCl), water (H_2O), and ammonia (NH_3) are examples. Some rules of nomenclature for binary molecular compounds are as follows:

1. (a) The element further to the left in the periodic table is usually given first in the name and in the formula:

CS_2	carbon disulfide
HCl	hydrogen chloride
H_2S	hydrogen sulfide
PCl_3	phosphorus trichloride

An exception is chlorine dioxide, ClO_2. In these examples note that the name of the first element is unchanged while the ending of the second element is changed to *-ide*.

(b) If the elements are in the same periodic group, the lower one is given first:

SO_2	sulfur dioxide
SiC	silicon carbide

2. When two elements form more than one binary molecular compound, Greek numerical prefixes (Table 2.7) are used to indicate the number of atoms of each element:

CO	carbon monoxide	N_2O	dinitrogen monoxide
CO_2	carbon dioxide	NO	nitrogen monoxide
		NO_2	nitrogen dioxide
		N_2O_5	dinitrogen pentoxide

The prefix *mono-* may be omitted. For example, N_2O is often called dinitrogen oxide rather than dinitrogen monoxide.

3. A few common names are officially recognized as chemical names:

H_2O	water	N_2H_4	hydrazine
NH_3	ammonia	PH_3	phosphine
CH_4	methane	C_6H_6	benzene

Note that, for many of these compounds, the formula is written in violation of Rule 1a; hydrogen appears as the second element rather than the first.

4. The following unofficial names are also worth remembering:

Compound	Common Name	Systematic Name
SiO_2	silica	silicon dioxide
CO_2 (solid)	Dry Ice	solid carbon dioxide
N_2O	laughing gas (nitrous oxide)	dinitrogen oxide

LEARNING HINT

Greek prefixes are *not* used in naming ionic compounds.

TABLE 2.7 Greek Prefixes Used in Naming Binary Molecular Compounds

Prefix	Number of Atoms
mono-	1
di-	2
tri-	3
tetra-	4
penta-	5
hexa-	6
hepta-	7
octa-	8
nona-	9
deca-	10

PRACTICE EXERCISE 2.25

Name the following molecular compounds: (a) SO_2, (b) SO_3, (c) PCl_3, (d) PCl_5, (e) UF_6, and (f) CCl_4.

PRACTICE EXERCISE 2.26

Give formulas for the following molecular compounds: (a) ammonia, (b) methane, (c) hydrogen bromide, (d) bromine pentafluoride, (e) carbon disulfide, (f) dinitrogen tetroxide, and (g) iodine monobromide.

PRACTICE EXERCISE 2.27

Give formulas for the following substances: (a) limestone, (b) baking soda, (c) Dry Ice, (d) silica, (e) lime, (f) laughing gas, and (g) caustic soda. *Hint:* Some of these substances are ionic, others are covalent.

CHAPTER REVIEW

Starred entries are based on the *Digging Deeper* sections.

LEARNING OBJECTIVES BY SECTION

2.1 1. Sketch a typical discharge tube.
*2. Describe how Thomson determined the electron's charge-to-mass ratio.
*3. Describe how Millikan determined the electron charge.

2.2 1. Describe how positive ions form in a discharge tube.
2. State the evidence for Rutherford's conclusion that the positive charge and most of the atom's mass are concentrated in a nucleus that is very small compared to the atom as a whole.
3. Relate the atomic number Z to the numbers of protons and electrons in a neutral atom.
4. Compare electrons, protons, and neutrons in terms of charge and mass.

2.3 1. State which elements are gases, which are liquids, and which are solids, at room temperature.
2. Given the symbol for an atom or ion, state the number of protons, neutrons, and electrons it contains.
3. Write the symbol for an atom or ion, given its mass number, atomic number, and charge.
4. State the names and write the symbols of the three isotopes of hydrogen.
5. State the exact mass of a carbon-12 atom in atomic mass units.
6. State the approximate masses of a proton and a neutron in atomic mass units.
7. Calculate chemical atomic weights from isotope mass and abundance data.
*8. Describe the operation of a direction-focusing mass spectrometer.

2.4 1. Convert moles of particles to numbers of particles, and vice versa.
2. Convert grams to atomic mass units, and vice versa.
3. Use a table of atomic weights to find the molar mass of an atom.
4. Convert grams of substance to moles of substance, and vice versa.

2.5 1. Distinguish between elements, compounds, and mixtures, and give examples of each.

2.6 1. Find the number of atoms of each element in a given formula.
2. Give examples of molecular and nonmolecular elements and compounds.
3. Name the allotropes of oxygen and carbon.

2.7 1. Calculate the number of moles of each element in a given mass of compound or in a given number of moles of compound.
2. Given the mole ratio of the elements in a compound, write the formula of the compound.
3. Calculate the molar mass of a compound from its formula.
4. Use molar masses to convert moles of compound to grams of compound, and vice versa.

2.8 1. Learn the formulas, charges, and names of the starred anions and cations in Table 2.6.
2. Learn the ionic charges associated with periodic groups 1A, 2A, 6A, and 7A.
3. Write the formula of an ionic compound when its component ions are given.

2.9 1. Write names for monatomic positive and negative ions.
2. Write names for ionic compounds and binary molecular compounds whose formulas are given.
3. Write formulas for ionic compounds and binary molecular compounds whose names are given.

KEY TERMS BY SECTION

2.1 Atom
Cathode ray
Coulomb (C)
Discharge tube
Electron

2.2 Alpha particle
Atomic number (Z)
Ion
Neutron
Nuclear atom
Nucleus
Positive ray
Proton

2.3 Atomic mass unit (u or amu)
Chemical atomic weight
*Direction-focusing mass spectrometer
Element
Fractional abundance
Group (in the periodic table)
Isotope

Mass number (A)
Mass spectrometer
*Mass spectrum
Noble gas
Period (in the periodic table)
Periodic table

2.4 Avogadro's number (N_A)
Molar mass ($\mathcal{M}$)
Mole

2.5 Compound
Mixture
Property

2.6 Allotropes
Halogen
Molecular formula
Molecule

2.7 Formula weight
Molecular weight

2.8 Anion
Cation
Covalent compound
Ionic compound

2.9 Binary compound

IMPORTANT EQUATIONS

***2.2** electric force $= E \times q$

***2.3a** magnetic force $= H \times q \times v$

***2.3b** magnetic force $= \dfrac{m \times v^2}{r}$

***2.4** $v = \dfrac{E}{H}$

***2.5** $\dfrac{q}{m} = \dfrac{E}{H^2 r}$

***2.6** $r^2 = \dfrac{2V}{H^2} \times \dfrac{m}{q}$

FINAL EXERCISES

Answers to exercises with blue numbers are given in Appendix D. Starred exercises are based on the *Digging Deeper* sections.

PART A. QUESTIONS AND PROBLEMS BY SECTION

The Discovery of Electrons; The Nuclear Atom (Sections 2.1 and 2.2)

2.1 Which discharge tube observations led Thomson to conclude that electrons are (a) negatively charged and (b) a universal constituent of matter?

2.2 When a negatively charged particle moves between oppositely charged plates, it bends toward the positive plate. Will the curvature of the particle's path increase or decrease with (a) increasing particle charge and (b) increasing particle mass?

2.3 Describe how positive rays form in a discharge tube. In which direction do they travel?

2.4 Suggest a simple experiment that would distinguish between a beam of electrons and a beam of positive ions.

***2.5** Will the force experienced by an electron moving in (a) a magnetic field and (b) an electric field increase, decrease, or remain the same with increasing electron speed?

***2.6** An electron beam moves through balanced electric and magnetic fields. If the electric field is doubled in strength, by what factor must the magnetic field be increased to restore the beam to its original path?

2.7 Describe Rutherford's gold foil experiment. Why were the observers surprised? How were the following observations finally interpreted? (a) A few alpha particles were deflected through large angles. (b) Most alpha particles were not appreciably deflected.

2.8 What is an *ion?* Protons and alpha particles are two of the many types of ions. Which neutral atoms are they related to?

2.9 The charge-to-mass ratio of an alpha particle is 4.8224×10^4 C/g. The charge on an alpha particle is 3.2044×10^{-19} C. Calculate the mass of the alpha particle.

2.10 The charge-to-mass ratio of an electron is 1836 times greater than the charge-to-mass ratio of a proton. The mass of an electron is 9.109×10^{-28} g. Calculate the mass of a proton.

The Chemical Elements; Isotopes (Section 2.3)

2.11 **(a)** What is the difference between the *atomic number* and the *mass number* of an atom?
(b) Which of these numbers is equal to the number of protons in the nucleus? The number of electrons in the neutral atom?

2.12 Does a positive ion have more or fewer electrons than the neutral atom? How about a negative ion?

2.13 What are *isotopes*? In what way do the isotopes of hydrogen differ from each other?

2.14 (a) What is an *element*?
(b) The ancient Greek philosophers considered water to be an element. Did their definition of an element agree with our modern definition? Explain.

2.15 How many *groups* are in the periodic table? How many *periods*?

2.16 Which periodic group contains only gaseous elements at room temperature (25°C)?

2.17 (a) Write a symbol for each of the following elements: iron, mercury, sodium, nitrogen, phosphorus, calcium, sulfur, copper.
(b) Name the following elements: Pb, Ag, K, Au, Ne, Zn, Ni.

2.18 Give the names and symbols for the three isotopes of hydrogen. How many protons and how many neutrons are in the nucleus of each isotope?

2.19 List the elements that are (a) gases and (b) liquids at 25°C.

2.20 What is the physical state (solid, liquid, or gas) of each of the following elements at 25°C: cesium, niobium, gallium, bromine, silicon, krypton, iodine?

2.21 Complete the following table:

Symbol	*Z*	*A*	Protons	Neutrons	Electrons
Zn	30	—	—	34	—
—	—	—	53	74	—
Eu	—	153	63	—	—

2.22 Complete the following table:

Symbol	*Z*	*A*	Protons	Neutrons	Electrons
U	—	—	—	143	92
—	—	60	23	—	—
W	—	184	—	110	—

2.23 How many protons, neutrons, and electrons are in each of the following ions?
(a) $^{208}_{82}Pb^{2+}$ (b) $^{14}_{7}N^{3-}$

2.24 How many protons, neutrons, and electrons are in each of the following ions?
(a) $^{40}_{20}Ca^{2+}$ (b) $^{3}_{1}H^{+}$

2.25 A particle contains 13 protons, 14 neutrons, and 10 electrons. Use the inside front cover to identify the element and write its complete symbol including *A*, *Z*, and charge.

2.26 Give the name and complete symbol for a particle that contains 55 protons, 68 neutrons, and 54 electrons.

2.27 Atoms with the same mass number but different atomic numbers are called *isobars*. How many protons, neutrons, and electrons are in each of the following isobars: chlorine-37, argon-37, sulfur-37?

2.28 In the following list of particles, the atomic symbol is represented by X:

$^{32}_{16}X$ $^{39}_{19}X$ $^{35}_{17}X$ $^{33}_{16}X$ $^{40}_{18}X$ $^{39}_{18}X$ $^{32}_{16}X^{2-}$

(a) Which of the particles are isotopes of the same element?
(b) Which of the particles are isoelectronic (have the same number of electrons)?
(c) Use the inside front cover to identify X for each particle.

Atomic Weights (Section 2.3)

2.29 Give the mass of each of the following particles to the nearest whole number of atomic mass units. Which of your mass values is exact?
(a) a carbon-12 atom (c) a neutron
(b) a proton (d) an alpha particle

2.30 Use data from Table 2.1 to explain why the mass of an isotope in atomic mass units is approximately the same as its mass number.

2.31 Why does the chemist use a weighted average for the chemical atomic weight?

2.32 Explain why chemical atomic weights cannot be reported with the same degree of precision for all elements.

***2.33** Briefly describe the operating principle of a direction-focusing mass spectrometer.

***2.34** Lithium consists of two isotopes with mass numbers 6 and 7. If singly positive ions of these species are in the magnetic field of a direction-focusing mass spectrometer, which will exhibit the larger radius of curvature? The larger deflection? Explain.

2.35 Beryllium, a lightweight metal used in alloys, has two isotopes with masses of 10.0135 and 9.01218 u. How many neutrons are in the nucleus of each isotope?

2.36 The two most abundant isotopes of sulfur have masses of 31.97207 and 33.96786 u. How many neutrons are in the nucleus of each isotope?

2.37 A U.S. dime weighs 2.27 g. A U.S. quarter weighs 5.67 g. Calculate the average mass of the coins in a collection containing 250 dimes and 45 quarters. Is your answer a "weighted" average or a simple average? Which kind of average is a chemical atomic weight?

2.38 Naturally occurring vanadium, the element that imparts hardness to vanadium steel, consists of 99.750% vanadium-51 and 0.250% vanadium-50. The masses of these isotopes are 50.943962 and 49.947161 u, respectively. Calculate the chemical atomic weight of vanadium.

2.39 Three naturally occurring isotopes of magnesium have masses of 23.98504, 24.98584, and 25.98259 u. Their abundances are 78.70, 10.13, and 11.17%, respectively. Calculate the chemical atomic weight of magnesium.

2.40 The percentage abundances and masses of the three isotopes of hydrogen are as follows: hydrogen-1, 99.985%, 1.007825 u; hydrogen-2 (deuterium), 0.015%, 2.014102 u; and hydrogen-3 (tritium), 0.000%, 3.016049 u. (The abundance of tritium is zero to three significant figures.) Calculate the chemical atomic weight of hydrogen.

2.41 Naturally occurring chlorine consists of chlorine-35 with an isotopic mass of 34.968852 u and chlorine-37 with an isotopic mass of 36.965903 u. The chemical atomic weight of chlorine is 35.4527 u. Determine the percentage abundance of each isotope.

2.42 Boron consists of two isotopes with masses of 10.0129 and 11.00931 u. The chemical atomic weight of boron is 10.811 u. Calculate the fractional abundance of each isotope.

2.43 The chemical atomic weight of copper is 63.546 u. Naturally occurring copper consists of 69.17% copper-63 and 30.83% copper-65. The isotopic mass of copper-63 is 62.939598 u. Calculate the isotopic mass of copper-65.

2.44 The chemical atomic weight of carbon is 12.011 u. Given that naturally occurring carbon contains 98.89% carbon-12 and 1.11% carbon-13, compute the mass of the carbon-13 atom in atomic mass units.

Avogadro's Number; Molar Masses of Atoms (Section 2.4)

2.45 Explain why (a) the dozen is not used for counting atoms and (b) the mole is not used for counting eggs.

2.46 Write out Avogadro's number showing all the zeros.

2.47 For each of the following elements, use the periodic table to find the average mass of one atom in atomic mass units and the molar mass in grams per mole: (a) silver; (b) mercury; and (c) radium.

2.48 For the elements (a) fluorine, (b) hydrogen, and (c) gold, find the average mass of one atom in atomic mass units and the mass of 1 mol of atoms in grams.

2.49 Calculate the mass in grams of (a) 2.55 mol of aluminum atoms and (b) one aluminum atom.

2.50 Calculate the mass in grams of (a) 3.44 mmol of krypton atoms, and (b) five krypton atoms.

2.51 Calculate the number of atoms in (a) 0.100 mol of iron, (b) 1.00 kg of gold, and (c) 0.250 L of mercury (density = 13.55 g/mL).

2.52 Calculate the number of atoms in 1.00 ng of (a) hydrogen, (b) carbon, and (c) uranium.

2.53 Calculate the number of moles of (a) protons, (b) neutrons, and (c) electrons in 2.35 mmol of carbon-12 atoms.

2.54 Calculate the number of (a) protons, (b) neutrons, and (c) electrons in 1.00 mol of alpha particles.

Compounds and Mixtures (Section 2.5)

2.55 Table sugar is sucrose, $C_{12}H_{22}O_{11}$. Contrast the properties of sucrose with the properties of its elements—carbon, hydrogen, and oxygen. If you need to look up some properties, try the *Handbook of Chemistry and Physics.*

2.56 Rust is predominantly iron(III) oxide, Fe_2O_3. Contrast the properties of rust with the properties of iron and oxygen. If you need to look up some properties, try the *Handbook of Chemistry and Physics.*

2.57 What is the difference between a compound and an element?

2.58 What characteristics distinguish a compound from a mixture? How would you show that (a) salt water and (b) garden soil are mixtures rather than compounds?

Molecules; Molecular and Nonmolecular Substances (Section 2.6)

2.59 What is a molecule? Do all substances consist of molecules?

2.60 Some old chemistry books refer to substances such as diamond, graphite, and silicon dioxide (see Figures 2.18 and 2.20) as giant molecules—molecular structures big enough to be visible. Is the term "giant molecule" consistent with the definition of molecule given in this textbook?

2.61 What are *allotropes?* Name the most common allotropes of (a) carbon, (b) oxygen, and (c) phosphorus.

2.62 Are the solid and liquid forms of mercury allotropes? Explain.

2.63 List the noble gases in order of increasing atomic

number. In what periodic group are they found? Do these gases consist of molecules?

2.64 List the halogens in order of increasing atomic number. In what periodic group are they found? What is the general formula of a halogen molecule?

2.65 How many atoms of each element are in the following formulas?
(a) $CaCl_2$
(b) $Mg(OH)_2$
(c) $Al(NH_4)(SO_4)_2$
(d) $CH_3(CH_2)_2CH_3$

2.66 How many atoms of each element are in the following molecules?
(a) P_4O_{10}
(b) CH_3COOH
(c) $C_7H_5(NO_2)_3$
(d) $CH_3(CH_2)_{20}CH_3$

2.67 Write a formula for each of the following elements. Does the element consist of molecules?
(a) hydrogen
(b) oxygen
(c) phosphorus (white)
(d) argon
(e) copper
(f) bromine

2.68 Write a formula for each of the following elements. Does the element consist of molecules?
(a) iron
(b) nitrogen
(c) ozone
(d) iodine
(e) helium
(f) mercury

2.69 Write a formula for each element that is a gas at 25°C.

2.70 Write a formula for each element that is a liquid at 25°C.

2.71 Write a formula for each of the following molecular compounds:
(a) ethanol
(b) acetic acid
(c) water
(d) hydrogen peroxide
(e) carbon monoxide
(f) ammonia

2.72 Write a formula for each of the following compounds. Does the compound consist of molecules, ions, or a nonmolecular extended array of atoms?
(a) carbon dioxide
(b) sodium chloride
(c) silicon dioxide
(d) methane

Molar Masses (Section 2.7)

2.73 Are the following statements true or false? Explain.
(a) One mole of H_2O contains 2 mol of hydrogen atoms.
(b) One mole of H_2O contains 1 mol of oxygen molecules.
(c) One mole of NH_3 contains 3 mol of hydrogen molecules.
(d) One mole of NH_3 contains 1 mol of nitrogen molecules.

2.74 Why might it be necessary to clarify what you mean when you say "1 mol of hydrogen," "1 mol of oxygen," or "1 mol of nitrogen"? Does a similar problem exist when referring to 1 mol of lead or 1 mol of water?

2.75 How many (a) phosphorus atoms and (b) moles of phosphorus atoms are in 1.00 mol of white phosphorus (P_4)?

2.76 **(a)** How many atoms of each element are in 1.00 mol of sulfur trioxide (SO_3)?
(b) How many moles of sulfur atoms and how many moles of oxygen atoms are in 1.00 mol of sulfur trioxide?

2.77 **(a)** How many moles of hydrogen atoms are in 2.50 mol of ammonia (NH_3)?
(b) How many hydrogen atoms are in 2.50 mol of ammonia?

2.78 **(a)** How many moles of oxygen atoms are in 1.40 mol of calcium nitrate, $Ca(NO_3)_2$?
(b) How many oxygen atoms are in 1.40 mol of calcium nitrate?

2.79 One mole of magnetite, an important iron-bearing mineral, contains 3 mol of iron atoms and 4 mol of oxygen atoms. Write the formula for magnetite.

2.80 One mole of a nitrogen oxide contains 2 mol of nitrogen atoms and 5 mol of oxygen atoms. Write the formula for the oxide.

2.81 Consult the periodic table and calculate the molar mass in grams of (a) glucose ($C_6H_{12}O_6$) and (b) sulfur trioxide (SO_3).

2.82 Find the mass of one molecule of each of the substances in the preceding exercise. Express your answer in atomic mass units and in grams.

2.83 Calculate the molar mass in grams of each of the following substances:
(a) SiC (silicon carbide; carborundum)
(b) $NaPO_3$ (sodium metaphosphate; a water softener)
(c) Al_2O_3 (aluminum oxide; corundum)
(d) $CH_3CHOHCH_3$ (isopropyl alcohol; rubbing alcohol)

2.84 Calculate the molar mass in grams of each of the following substances:
(a) H_2SO_4 (sulfuric acid)
(b) O_3 (ozone)
(c) C_2H_5OH (ethanol; ethyl alcohol)
(d) $Na_2B_4O_7 \cdot 10H_2O$ (borax; borax is a "hydrate," a compound whose structure contains water molecules in the proportion shown in the formula. Treat the 10 water molecules as part of the formula unit.)

2.85 Calculate the mass in grams of (a) 1.50 mol of silicon carbide (SiC) and (b) 6.45 mol of aluminum oxide (Al_2O_3).

2.86 Calculate the mass in grams of (a) 3.60 mol of sulfuric acid (H_2SO_4) and (b) 2.13 mol of ozone (O_3).

2.87 How many (a) moles of water and (b) molecules of water are in 20.0 g of water?

2.88 How many (a) moles of ethanol and (b) molecules of ethanol are in 750 mL of ethanol? The density of ethanol (C_2H_5OH) is 0.789 g/mL.

2.89 **(a)** How many moles of each ion are in 2.50 mol of NaCl?
(b) How many sodium and how many chloride ions are in 2.50 mol of NaCl?

2.90 Magnesium nitride (Mg_3N_2) contains Mg^{2+} and N^{3-} ions.
(a) How many moles of each ion are in 4.75 mol of Mg_3N_2?
(b) How many magnesium and how many nitride ions are in 4.75 mol of Mg_3N_2?

2.91 What mass of sodium hydroxide (NaOH) contains the same number of moles as 126 g of nitric acid (HNO_3)?

2.92 What mass of oxygen (O_2) contains the same number of moles as 52.0 g of nitrogen (N_2)?

Ionic Compounds (Section 2.8)

2.93 What property can be used to distinguish ionic compounds from covalent compounds? Suggest an experiment for determining whether the following substances are ionic or covalent: (a) methyl chloride (a liquid) and (b) potassium fluoride (a solid that melts at 860°C).

2.94 Refer to Figure 2.22.
(a) What will be the charges on monatomic ions of K, S, Br, Ba, N, Al, F, O, and Sr?
(b) Which periodic group does not normally form ions?

2.95 Find cesium, strontium, and bromine in the periodic table. Write a formula for the ionic compound that forms from (a) cesium and bromine and (b) strontium and bromine.

2.96 The oxide ion is O^{2-}. Consult Table 2.6 and write a formula for each of the following compounds:
(a) lithium oxide **(d)** iron(II) oxide
(b) magnesium oxide **(e)** iron(III) oxide
(c) aluminum oxide **(f)** zinc oxide

2.97 Consult Table 2.6 and write a formula for each of the following compounds:
(a) potassium permanganate **(d)** mercury(II) nitrate
(b) ammonium carbonate **(e)** sodium thiocyanate
(c) mercury(I) chloride **(f)** magnesium perchlorate

2.98 Consult Table 2.6 and write a formula for each of the following compounds:
(a) nickel sulfate **(d)** aluminum phosphate
(b) sodium thiosulfate **(e)** ammonium sulfide
(c) chromium(III) hydroxide **(f)** calcium hydrogen carbonate

Naming Compounds (Section 2.9)

2.99 Name the following cations:
(a) Pb^{2+} **(d)** NH_4^+
(b) Sn^{2+} **(e)** Cu^{2+}
(c) Fe^{3+} **(f)** Hg_2^{2+}

2.100 Name the following anions:
(a) I^- **(d)** OH^-
(b) SO_4^{2-} **(e)** ClO_4^-
(c) S^{2-} **(f)** PO_4^{3-}

2.101 Name the following ionic compounds:
(a) $K_2Cr_2O_7$ **(d)** Hg_2Cl_2
(b) $FeSO_4$ **(e)** $K_2C_2O_4$
(c) $KMnO_4$ **(f)** $NaClO_3$

2.102 Given that the periodate ion is IO_4^-, write formulas for (a) sodium iodate, (b) potassium iodite, and (c) calcium hypoiodite.

2.103 Name the following covalent compounds:
(a) HCl **(d)** N_2O_5
(b) H_2S **(e)** IBr
(c) H_2O **(f)** S_2F_2

2.104 Write formulas for the following compounds:
(a) carbon tetrachloride **(e)** diboron hexahydride
(b) sulfur hexafluoride **(f)** dinitrogen tetroxide
(c) diiodine heptoxide
(d) phosphorus pentachloride

PART B. MISCELLANEOUS QUESTIONS AND PROBLEMS

2.105 Sketch a typical discharge tube. Try not to refer to the diagrams in the textbook.

2.106 Some of the oil drops in Millikan's experiment absorbed positive ions from the ionized air. The charges on several of these drops were found to be 1.60×10^{-18} C, $+1.76 \times 10^{-18}$ C, and $+1.12 \times 10^{-18}$ C. Explain how these charges support the value given for the electron charge.

2.107 Compare the Rutherford model of the atom with the Thomson model of the atom. Describe the features they have in common as well as their differences.

2.108 What evidence did Rutherford have for concluding that the positive charge and most of the atom's mass are concentrated in a nucleus that is very small compared to the atom as a whole?

2.109 In what respect are the atoms of a given element alike? In what ways might they differ?

2.110 The following radioactive isotopes have been used for medical therapy or diagnosis: (a) iodine-131, (b) thallium-201, (c) technetium-99, (d) cobalt-60, and (e) sodium-24. Write complete atomic symbols including *A* and *Z* for each of these isotopes.

2.111 The textbook states that if the nuclear diameter is expanded to the diameter of a U.S. penny, the atomic diameter will be about 2 km or $1\frac{1}{4}$ miles. Verify this statement for an atom of nuclear radius 1×10^{-15} m and atomic radius 1×10^{-10} m. The diameter of a U.S. penny is 1.9 cm.

2.112 If an atom is enlarged until its diameter is equal to the length of a football field (100 yards), what would be the diameter of its nucleus in inches? Use data from the preceding exercise.

2.113 Assume a typical atom to be a sphere of radius 100 pm. If the nucleus is a sphere of radius 10^{-15} m, what percentage of the total atomic volume is occupied by the nucleus?

2.114 **(a)** Estimate the density in grams per cubic centimeter of a nucleus with two protons and two neutrons and a radius of 5×10^{-15} m.

(b) Compare your answer to Part (a) with the density of pulsar PSR1257+12, a spherical star about 12 miles in diameter and containing about 1.4 times the mass of the sun. The sun's mass is 1.991×10^{30} kg. (This pulsar is a *neutron star* in which electrons have collapsed into the nuclei to produce a dense state of matter with properties similar to those of closely packed neutrons.)

(c) The average radius of the sun is 6.96×10^{5} km. What would be its radius if it were squeezed down to the density of the neutron star in Part (b)?

2.115 Zinc has five natural isotopes. The percentage abundances and masses of four of these isotopes are as follows:

zinc-64, 48.89%, 63.9291 u
zinc-66, 27.81%, 65.9260 u
zinc-67, 4.11%, 66.9721 u
zinc-70, 0.62%, 69.9253 u

The chemical atomic weight of zinc is 65.39 u. Determine the percentage abundance and isotopic mass of the fifth isotope. What name would you give to this isotope?

2.116 What practical reasons can you think of for choosing 12 g of carbon-12 rather than some other mass of carbon-12 as the standard for the mole?

2.117 Use data from Table 2.1 to calculate the charge in coulombs on (a) 1.00 mol of electrons and (b) 1.00 mol of alpha particles.

2.118 Explain why the molar mass of an atom in grams is numerically equal to the mass of a single atom in atomic mass units.

2.119 Suggest a method for separating the following mixtures:

(a) red blood cells from blood plasma
(b) water vapor from air
(c) oxygen and nitrogen, the main components of air

2.120 Ringer's lactate is an aqueous physiological solution used for intravenous fluid therapy. One liter of the solution contains 5.96 g NaCl, 3.1 g sodium lactate ($NaC_3H_5O_3$), 0.3 g KCl, and 0.2 g $CaCl_2$.

(a) How many moles of each compound are in 1.00 L of solution?

(b) The ingredients dissolve to form Na^+, K^+, Ca^{2+}, Cl^-, and $C_3H_5O_3^-$ ions. How many moles of each ion are in 1.00 L of solution?

2.121 A 100-mL portion of a solution used for intravenous feeding contains 5.51 g of glucose. How many moles of glucose ($C_6H_{12}O_6$) are in 1.00 L of the solution?

2.122 A 1-lb (454-g) container of iodized salt contains 0.010% KI by mass. (a) How many moles of potassium iodide are in the container? (b) How many iodide ions are in the container?

2.123 The density of liquid benzene, C_6H_6, is 0.879 g/mL at 15°C. What is the volume in milliliters of 1.00 mol of benzene at this temperature?

CHEMISTS AT WORK

I KNOW YOU'RE A CHEMIST, BUT WHERE DO YOU WORK?

Chemists work everywhere. They work at all kinds of endeavors and in all kinds of positions. The majority of chemists (61.5%) work in industrial firms; 23.8% are teachers in colleges and universities; 1.4% are teachers in high schools; 8.0% work for various agencies of the federal, state, and local governments; and approximately 8.0% pursue other careers based on their scientific education.[1]

Industrial chemical firms come in all sizes. They include giant multinational corporations, intermediate-sized companies, and privately owned and operated concerns. In such companies, regardless of size, the typical chemist might function in one or more of the following areas: applied research, production support, general management, R & D (Research and Development) management, marketing, and basic research. Most chemists begin their careers in the laboratory. In giant firms, they eventually become specialists in one laboratory-type function and, in time, they often take on management responsibilities. In a smaller company, the chemist is probably a jack-of-all trades—responsible for several or all laboratory functions, involved directly with marketing and customer services, and probably responsible for some management as well.

Chemists are important contributors to the operation of government agencies at all levels. They provide a wide range of services directly related to our well-being and safety. Some chemists engage in strictly regulatory activities, making sure that environmental, health, and product safety regulations are obeyed. Other chemists perform basic and/or applied research in areas such as the development of new materials (e.g., inks and paper for currency and postage), the study and solution of environmental problems, and the utilization of agricultural products. The majority of forensic laboratories are functions of government agencies—some of the finest in the world are those of the FBI.

[1]Reported in "Today's Chemist at Work," April 1992.

Chemical scientists with all types of degrees and experiences are teachers at all levels. Frequently, it is the scientist–teacher who provides the leadership, excitement, and example that encourages people to pursue a particular career. Teachers are the foundation on which our future mutual well-being depends since they stimulate, cultivate, and challenge the world's most important raw material—students. If you elect to pursue a career based on chemistry, you will probably do so because of the example and encouragement of a teacher.

We often think of chemical careers as being pursued in industry, government, and teaching because that is where the majority of chemists practice their specialty. However, the education of a chemist is such that it can be applied to a variety of what might seem very unusual or nontraditional areas. Indeed, there are chemists who are presidents of corporations, colleges, and universities; chemists who sell and market a wide range of products, from typical materials to instruments and computer hardware; and chemists who hold elected office in various levels of government—the list of these unusual careers goes on and on.

So where do chemists work? Chemists work where their knowledge, (scientific and general), their experience, their problem-solving skills, and their capability for growth and development can be applied usefully and profitably. This can be in manufacturing, in laboratories, on college campuses, and in the business world. In these various locales, chemical scientists are engaged in the process of discovery, in teaching, in publishing, in managing operations and people, in representing and selling a product, and, really, in just about all profit-making or service-oriented activities of our society. Chemistry offers exciting careers because of the world of opportunity available.

CHAPTER 3

A mixture of powdered zinc, ammonium nitrate, and iodine is ignited by the addition of several drops of water.

CHEMICAL REACTIONS, EQUATIONS, AND STOICHIOMETRY

OUTLINE

PREVIEW

In this chapter we return to the world of ordinary dimensions, to samples of material large enough to see and touch, and to various practical questions that chemists and chemical engineers are called on to answer. A metallurgist, for example, may need to know the amount of iron in a ton of ore, agricultural chemists often test the nitrogen and phosphorus content of soils and fertilizers, and environmental chemists may require information on the sulfur contaminants in fuel, the pesticides in animal food, or the relationship between automobile exhaust and photochemical smog. The calculations necessary to obtain such information require a knowledge of chemical reactions and are carried out with the help of chemical formulas and equations.

The changes that substances undergo obey simple laws that can be verified in the laboratory. These laws provide strong support for the atomic theory on which modern chemistry is based, and they also allow us to identify elements and compounds, to analyze mixtures, and to predict the amount of each substance consumed or produced during a reaction. The study of the quantitative relationships that govern the composition of substances and their reactions is called **stoichiometry**.

Quantitative: pertaining or subject to measurement. Chemistry is a quantitative science.

Stoichiometry (stoy-key-ahm′-eh-tree): from *stoikheion* and *metron,* the Greek words for "element" and "measure," respectively.

3.1 PHYSICAL AND CHEMICAL CHANGE

A **physical change** is one in which each substance retains its identity and no new elements or compounds are formed. Examples include the melting of ice and snow, the evaporation of alcohol, and the crushing of stone. Physical changes can often be reversed, sometimes by simply changing the conditions: liquid water can be frozen by cooling, and alcohol vapor can be cooled and condensed into liquid alcohol. During these changes, water remains water, alcohol remains alcohol, and stone remains stone.

A **chemical change** or **chemical reaction** is a change in which one or more new elements or compounds are formed. Chemical reactions include such diverse transformations as a log burning in air to form hot vapor, smoke, and ashes; a seed sprouting and growing into a plant; and water decomposing into hydrogen and oxygen (Figure 3.1). The original substances in a chemical reaction are called **reactants**. The new substances are called **products**. When sulfur burns in oxygen to form sulfur dioxide, sulfur and oxygen are the reactants and sulfur dioxide is the product. The many chemical reactions that influence our lives include corrosion, decay, growth, metabolism, the burning of fuels, and the cooking of food.

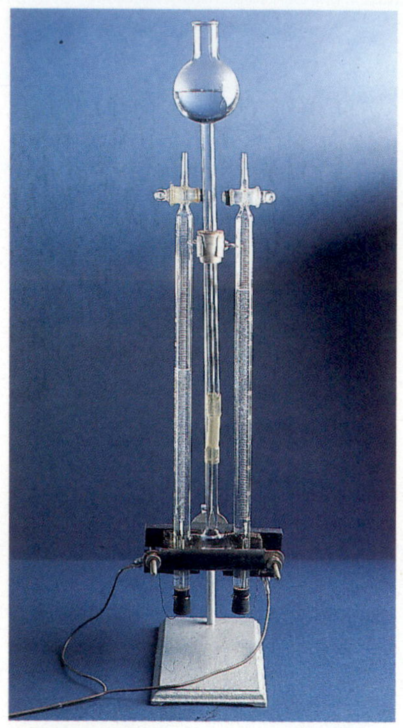

Figure 3.1
Water is decomposed by passing an electric current through it. Here the water is contained in three interconnected vertical tubes; the center tube is the reservoir. The right and left tubes contain the positive anode and negative cathode, respectively. A small amount of sulfuric acid is added to the water to help carry the current. Bubbles of oxygen gas rise from the anode, and bubbles of hydrogen gas rise from the cathode.

EXAMPLE 3.1

State whether each of the following is a chemical reaction or a physical change: (a) gold melting, (b) iron rusting, and (c) copper wire being formed from a chunk of copper metal.

SOLUTION

New substances do not form when gold melts or when copper metal is drawn into wire—these changes are physical changes. Iron forms a new substance with different properties when it rusts, so rusting is a chemical reaction.

PRACTICE EXERCISE 3.1

State whether each of the following is a chemical reaction or a physical change: (a) milk turning sour, (b) dissolved carbon dioxide bubbling out of a newly opened bottle of soda, and (c) a bronze statue gradually turning green.

PRACTICE EXERCISE 3.2

Natural gas (methane) burns in oxygen to form water vapor and carbon dioxide. Identify the reactants and products.

The properties of a substance (see Section 2.5) can be classified as chemical or physical. A **chemical property** is any chemical reaction that a substance can undergo. The ability of silver to tarnish (form a silver sulfide coating) is a chemical property of silver. The ability of hydrogen and oxygen to react and form water is a chemical property of both hydrogen and oxygen. Chemical properties can be observed only during chemical reactions, when old substances are changing into new substances.

A **physical property** is any property other than a chemical property. Physical properties include color, odor, melting and boiling points, solubility, and density. Many physical properties can be observed simply by examining the substance.

Chemical reactions are usually recognized by changes in physical properties—the properties of the products are different from those of the reactants. For example, when hydrogen burns in oxygen, the physical properties of hydrogen and oxygen are lost, and the product has the distinctive physical properties of water.

EXAMPLE 3.2

Which type of property, chemical or physical, is described by each of the following statements? (a) Air is a gas. (b) Sodium metal reacts with chlorine gas to form sodium chloride. (c) Sulfur crystals are yellow.

SOLUTION

The reaction of sodium with chlorine forms a new substance; hence, statement (b) describes a chemical property of both sodium and chlorine. Statements (a) and (c) do not describe the formation of new substances—state (solid, liquid, or gas) and color are physical properties.

PRACTICE EXERCISE 3.3

List five physical properties of water. How many chemical properties of water can you think of? List them.

3.2 LAWS OF CHEMICAL COMBINATION

Certain regularities are observed when reactant and product masses are compared in a chemical reaction. When water is decomposed by an electric current (Figure 3.1), 11.2 g of hydrogen and 88.8 g of oxygen are formed for every 100.0 g of water that breaks down. The same proportions of oxygen and hydrogen are produced by any sample of water that decomposes anywhere at any time. Two important laws are illustrated by this observation: the law of conservation of mass and the law of constant composition.

The Law of Conservation of Mass

The data for the decomposition of water—11.2 g of hydrogen and 88.8 g of oxygen from every 100.0 g of water—show that the total mass of hydrogen and oxygen produced is the same as the mass of water consumed. Other chemical reactions also occur without any measurable change in mass. The burning of a candle in a closed jar is a chemical reaction in which paraffin (candle wax) and oxygen are converted to carbon dioxide and water vapor, which remain trapped in the jar. If the jar rests on a balance pan as in Figure 3.2, and if the products are cooled to the starting temperature, the balance pointer will be at the same position before and after the reaction, showing that the total mass does not change. Another example is illustrated in Demonstration 3.1. A tube containing lead nitrate [$Pb(NO_3)_2$] dissolved in water is placed upright in a large stoppered flask containing potassium iodide (KI) dissolved

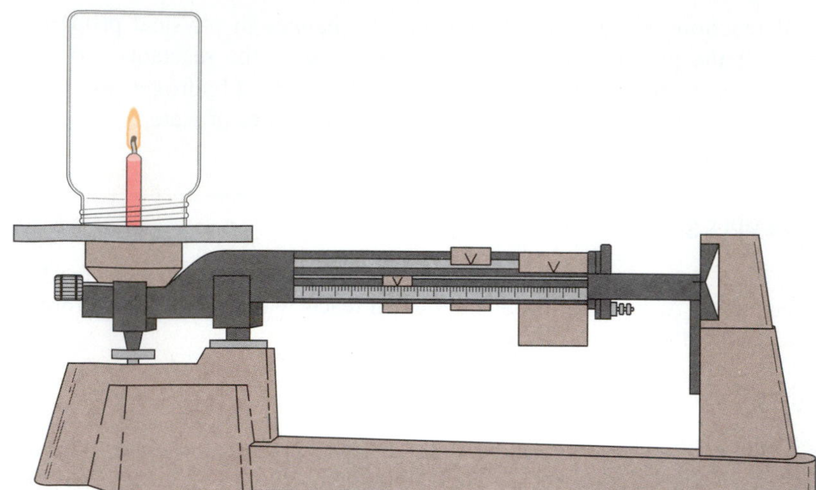

Figure 3.2
Conservation of mass. The mass of the reactants (candle wax plus oxygen) is the same as the mass of the products (carbon dioxide plus water vapor).

in water. Both solutions are clear and colorless. When the solutions are mixed by tipping the flask, a chemical reaction occurs and an insoluble yellow substance known as lead iodide (PbI_2) forms. The mass of the flask and its contents does not change during the reaction.

The results of these and many other experiments are summarized by the **law of conservation of mass**: *There is no measurable change in total mass during a chemical reaction.* Measurements of mass input and mass output on humans and animals confined to special experimental chambers show that living things also obey the law of conservation of mass.

The law of conservation of mass is explained by assuming that *atoms are not created or destroyed during a chemical reaction; they simply rearrange to form new substances.* Figure 3.3 shows how atoms of hydrogen and oxygen rearrange to form molecules of water. There is no change of mass because the same atoms are present before and after the reaction.

The theory of relativity developed by Albert Einstein predicts a small decrease in mass for reactions that evolve energy (e.g., the burning of a match) and an increase in mass when energy is absorbed. For ordinary chemical reactions, the predicted mass changes are so small that they cannot be measured by our most sensitive balances, and the conservation law is valid as worded.

DEMONSTRATION 3.1 MASS IS CONSERVED DURING A CHEMICAL REACTION

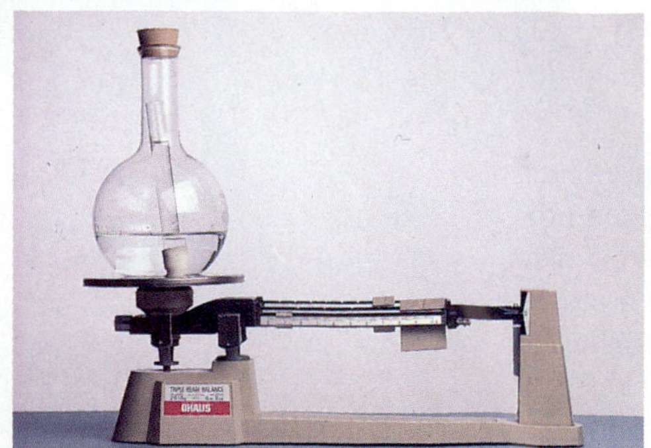

A test tube containing a solution of lead nitrate rests upright in a solution of sodium iodide. The total mass is 571 g.

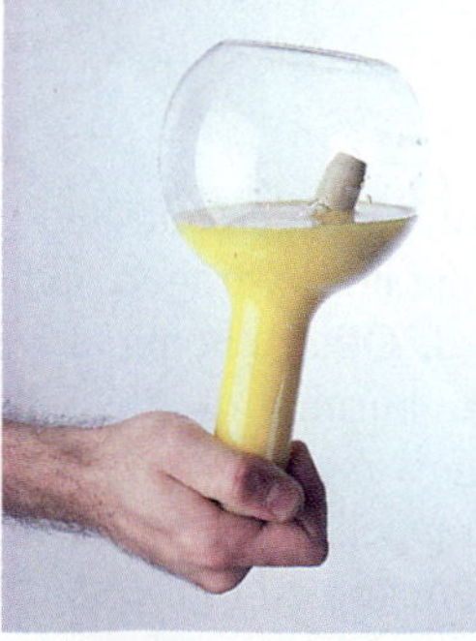

The flask is inverted to mix the solutions, which react to produce a yellow solid, lead iodide.

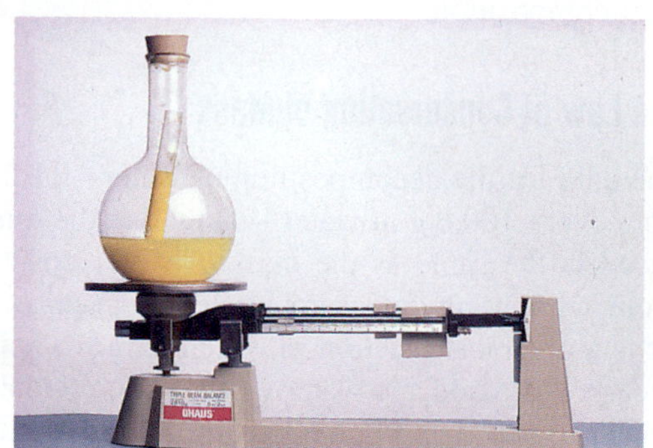

The flask is reweighed; the mass is still 571 g.

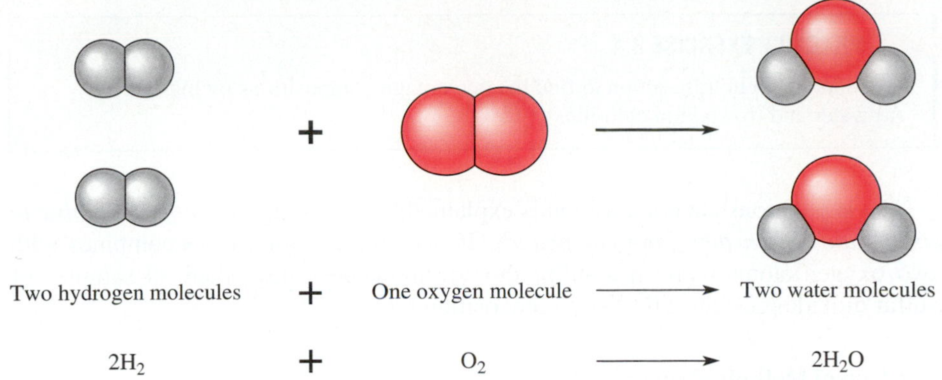

Figure 3.3

A chemical reaction is a rearrangement of atoms. When hydrogen and oxygen react to form water, the atoms in two molecules of hydrogen and one molecule of oxygen rearrange to form two molecules of water. The total mass is unchanged.

The Law of Constant Composition

The decomposition data for different samples of water indicate that water has a definite elemental composition, 88.8 g of oxygen for every 11.2 g of hydrogen. Other compounds are also found to have definite compositions. A 100.0-g sample of sodium chloride, for example, always contains 39.3 g of sodium and 60.7 g of chlorine. The composition of a compound can be found either by decomposing the compound and measuring the masses of the elements produced or by forming the compound and measuring the masses of the elements consumed. Regardless of which method is used, the result always conforms to the **law of constant composition**: *The elemental composition by mass of a given compound is the same for all samples of the compound.* An alternative version of this law is the **law of definite proportions**: *When elements combine to form a compound, they do so in a definite proportion by mass.*

The elemental makeup of a compound is customarily reported in terms of its **percentage composition**, the percent of each element by mass.

Antoine L. Lavoisier (1743–1794) discovers the law of conservation of mass. Here he is shown using an electric spark to ignite a mixture of hydrogen and oxygen.

EXAMPLE 3.3

A 30.5-mg sample of carbon combines with oxygen to form 111.8 mg of carbon dioxide. Find the percentage of (a) carbon and (b) oxygen in carbon dioxide.

SOLUTION

(a) The fraction of carbon in carbon dioxide is 30.5 mg/111.8 mg = 0.273. The percentage of carbon in carbon dioxide is the fraction of carbon multiplied by 100%:

$$\text{percent carbon} = 0.273 \times 100\% = 27.3\%$$

(b) The percentages of carbon and oxygen must add up to 100%; hence

$$\begin{aligned}\text{percent oxygen} &= 100\% - \text{percent carbon}\\ &= 100\% - 27.3\%\\ &= 72.7\%\end{aligned}$$

All samples of carbon dioxide have this composition.

PRACTICE EXERCISE 3.4

Use data given in this section to find the percentage compositions by mass of (a) water and (b) sodium chloride.

The law of constant composition is explained by assuming that *atoms combine in fixed ratios when they form compounds.* If one sulfur atom always combines with two oxygen atoms to form a sulfur dioxide molecule (SO_2), then all samples of sulfur dioxide must have the same composition by mass.

The Law of Multiple Proportions

Water and hydrogen peroxide are different compounds with different properties. Both substances, however, contain only hydrogen and oxygen atoms. When hydrogen peroxide is decomposed into its elements, 16.0 g of oxygen form for every gram of hydrogen. When water is decomposed, 8.0 g of oxygen form per gram of hydrogen. Thus hydrogen peroxide contains twice as much oxygen for a given mass of hydrogen as does water. The formulas for these substances are consistent with the decomposition data. The formula for hydrogen peroxide (H_2O_2) shows two oxygen atoms bonded to two hydrogen atoms while the formula for water (H_2O) shows only one oxygen atom bonded to two hydrogen atoms.

Now consider sulfur trioxide (SO_3) and sulfur dioxide (SO_2), two compounds that contain only sulfur and oxygen atoms. Sulfur trioxide has three oxygen atoms per sulfur atom; sulfur dioxide has two oxygen atoms per sulfur atom. The formulas suggest that for a given mass of sulfur, the masses of oxygen in sulfur trioxide and in sulfur dioxide will be in a three-to-two (3:2) ratio. This ratio has been confirmed by chemical analysis. Many other examples, including CO and CO_2, $FeCl_2$ and $FeCl_3$, and UF_3, UF_4, and UF_6, are also known.

These examples illustrate the **law of multiple proportions**: *In different compounds containing the same elements, the masses of one element combined with a fixed mass of the other element are in the ratio of small whole numbers.* John Dalton, who in 1803 explained constant composition by assuming that atoms bond together in fixed proportions, also predicted the law of multiple proportions. The prediction and subsequent verification of this law constituted the strongest experimental evidence available at that time for the atomic theory.

3.3 PERCENTAGE COMPOSITION FROM A FORMULA

The formula of a compound gives its elemental composition in terms of moles of atoms; for example, 1 mol of H_2S contains 2 mol of hydrogen atoms and 1 mol of sulfur atoms. The molar masses of the atoms can be used to calculate percentage composition by mass, as shown in the next example.

EXAMPLE 3.4

The formula for lactic acid, an ingredient of sour milk, is $C_3H_6O_3$. Calculate the percentage composition of lactic acid.

SOLUTION

The formula shows that 1 mol of lactic acid contains 3 mol of carbon atoms, 6 mol of hydrogen atoms, and 3 mol of oxygen atoms. The molar masses of carbon, hydrogen, and oxygen are 12.01, 1.008, and 16.00 g/mol, respectively; hence, the mass of 1 mol of lactic acid is

$$\begin{aligned} 3 \text{ mol C} \times 12.01 \text{ g/mol C} &= 36.03 \text{ g} \\ 6 \text{ mol H} \times 1.008 \text{ g/mol H} &= 6.048 \text{ g} \\ 3 \text{ mol O} \times 16.00 \text{ g/mol O} &= \underline{48.00 \text{ g}} \\ \text{Total mass} &= 90.08 \text{ g} \end{aligned}$$

The fraction of the molar mass supplied by carbon is 36.03 g/90.08 g = 0.4000. Note that the fraction is dimensionless. The percentage of carbon is the fraction of carbon × 100%:

$$\text{percent carbon} = 0.4000 \times 100\% = 40.00\%$$

Similarly, the percentages of hydrogen and oxygen are

$$\text{percent hydrogen} = \frac{6.048 \text{ g}}{90.08 \text{ g}} \times 100\% = 6.714\%$$

$$\text{percent oxygen} = \frac{48.00 \text{ g}}{90.08 \text{ g}} \times 100\% = 53.29\%$$

Note that the sum of the percentages is 100%.

PRACTICE EXERCISE 3.5

Chromium is obtained from the ore chromite ($FeCr_2O_4$). Calculate the percentage composition of chromite.

3.4 DETERMINATION OF FORMULAS

Thus far we have used formulas without showing where they come from. Atomic and molecular weights are obtained from mass spectrometer data, and the formulas of simple molecules can be calculated from these weights. The procedure is illustrated in the following examples.

EXAMPLE 3.5

The molecular weight of gaseous phosphorus molecules is 124 u. Determine the formula of molecular phosphorus.

SOLUTION

Phosphorus is an element, so the phosphorus molecule contains only phosphorus atoms. The atomic weight of phosphorus is 31.0 u (inside front cover). Since the molecular weight, 124 u, is four times greater than the atomic weight, there must be four atoms of phosphorus in one molecule of phosphorus. The formula is P_4.

PRACTICE EXERCISE 3.6

The molecular weight of rhombic sulfur (a form of the element sulfur) is 256.5 u. Determine the formula of rhombic sulfur.

EXAMPLE 3.6

Methane, the major component of natural gas, is a compound of carbon and hydrogen. The molecular weight of methane is 16.0 u. Find the formula of methane.

SOLUTION

The periodic table gives the atomic weight of carbon as 12.0 u and hydrogen as 1.0 u. Two atoms of carbon would have a mass of 24.0 u, more than that of the methane molecule. Hence, there can be only one atom of carbon in the molecule. The remaining mass, 16.0 u − 12.0 u = 4.0 u, is due to hydrogen. Since each hydrogen is 1.0 u, there must be four atoms of hydrogen in the molecule. Thus the formula is CH_4.

PRACTICE EXERCISE 3.7

Hydrogen sulfide, a gas with the odor of rotten eggs, contains hydrogen and sulfur. Its molecular weight is 34.08 u. Find the formula of hydrogen sulfide.

The previous examples illustrate a "common sense" method of determining formulas that works only for molecular elements and a few simple molecular compounds. Determining the formulas of more complex molecules and nonmolecular compounds requires a knowledge of both composition data and molar masses. Procedures for finding such formulas are described in the remainder of this section.

Empirical Formulas

Chemists use two kinds of formulas: molecular formulas, which we have already discussed, and empirical formulas. A **molecular formula** describes the molecule as a whole; its subscripts show the actual number of each atom in the molecule. The formulas H_2O for water and C_6H_6 for benzene are examples of molecular formulas. An **empirical formula** gives the simplest whole-number ratio of the atoms in a compound. The empirical formula of water is also H_2O, but the empirical formula of benzene, whose molecule contains six carbon and six hydrogen atoms, is simply CH.

EXAMPLE 3.7

Write the empirical formulas for (a) ethane and (b) sulfur dioxide. Their molecular formulas are C_2H_6 and SO_2, respectively.

SOLUTION

(a) The ethane molecule has two carbon atoms and six hydrogen atoms. The simplest whole-number ratio is one carbon atom for three hydrogen atoms, so the empirical formula is CH_3.

(b) The ratio of one sulfur atom for two oxygen atoms in the sulfur dioxide molecule is also the simplest ratio. Hence, the empirical formula is identical to the molecular formula, SO_2.

PRACTICE EXERCISE 3.8

Write the empirical formulas for (a) $C_6H_{12}O_6$ and (b) $Hg_2(NO_3)_2$.

Ionic compounds such as sodium chloride (NaCl) or potassium sulfate (K_2SO_4) do not contain molecules, and neither do compounds such as silicon dioxide (SiO_2), that consist of extended arrays of atoms. *The formulas of nonmolecular compounds are always empirical.*

Finding an Empirical Formula

In Section 3.3 the percentage composition of a compound was calculated from its formula. We now show how empirical formulas can be calculated from elemental composition data. The method is illustrated in the following examples.

EXAMPLE 3.8

A 13.8-g sample of a compound containing only nitrogen and oxygen produced 4.2 g of nitrogen on decomposition. Calculate the empirical formula of the compound.

SOLUTION

The sample contained 4.2 g of nitrogen and 13.8 g − 4.2 g = 9.6 g of oxygen. Nitrogen and oxygen atoms have molar masses of 14.01 and 16.00 g/mol, respectively. The number of moles of each element in the sample is

$$9.6 \text{ g O} \times \frac{1 \text{ mol O atoms}}{16.00 \text{ g O}} = 0.60 \text{ mol O atoms}$$

$$4.2 \text{ g N} \times \frac{1 \text{ mol N atoms}}{14.01 \text{ g N}} = 0.30 \text{ mol N atoms}$$

The nitrogen-to-oxygen ratio is 0.30 mol to 0.60 mol, or 1 to 2. This compound contains 1 mol of nitrogen atoms for every 2 mol of oxygen atoms. The atom ratio is the same as the mole ratio (Section 2.7); hence, the compound contains one nitrogen atom for every two oxygen atoms. The empirical formula is NO_2.

EXAMPLE 3.9

Acrolein, the eye-irritating ingredient in fumes from burning fat, contains 64.3% carbon, 7.2% hydrogen, and 28.5% oxygen by mass. Calculate the empirical formula of acrolein.

SOLUTION

When the percentage composition is given, as in this problem, it is easiest to base the calculation on a 100-g sample of the compound. The number of grams of each element in 100 g is numerically equal to the percentage of the element. For acrolein, 100 g contains 64.3 g carbon, 7.2 g hydrogen, and 28.5 g oxygen. The number of moles of each element is

$$64.3\text{ g C} \times \frac{1\text{ mol C atoms}}{12.01\text{ g C}} = 5.35\text{ mol C atoms} \qquad \frac{5.35}{1.78} = 3.01 = 3$$

$$7.2\text{ g H} \times \frac{1\text{ mol H atoms}}{1.008\text{ g H}} = 7.1\text{ mol H atoms} \qquad \frac{7.1}{1.78} = 4.0 = 4$$

$$28.5\text{ g O} \times \frac{1\text{ mol O atoms}}{16.00\text{ g O}} = 1.78\text{ mol O atoms} \qquad \frac{1.78}{1.78} = 1.00 = 1$$

The column on the right shows how to find the simplest whole-number ratio when it is not immediately apparent. Each of the molar quantities is divided by the smallest one, in this case, 1.78. The resulting numbers give the ratio 3 to 4 to 1, so the empirical formula is C_3H_4O.

The analysis of a new compound routinely includes a determination of its empirical formula. The procedure is as follows:

- ***Step 1:*** Find the mass of each element in the given sample. If percentage composition data are given, assume the sample weighs 100 g. If one mass or percentage is missing, find it by taking the difference.
- ***Step 2:*** Divide the mass of each element by its molar mass to obtain the number of moles.
- ***Step 3:*** Find the simplest whole-number mole ratio. This ratio can often be found by inspection; if not, divide all the molar quantities by the smallest one. If these quotients are not close to whole numbers, multiply them all by 2, 3, or whatever number gives a whole-number ratio.
- ***Step 4:*** Write the formula. (Remember that the atom ratio in the formula will be the same as the mole ratio determined in Step 3.)

EXAMPLE 3.10

An oxide of iron contains 69.9% iron by mass. Find its empirical formula.

SOLUTION

We will use the steps summarized above.

Step 1: Find the mass of each element. The oxide contains 69.9% iron and 100% − 69.9% = 30.1% oxygen. In a 100-g sample, there will be 69.9 g of iron and 30.1 g of oxygen.

Step 2: Find the number of moles of each element. The masses of 1 mol of iron and oxygen are 55.85 and 16.00 g, respectively.

$$69.9 \text{ g Fe} \times \frac{1 \text{ mol Fe atoms}}{55.85 \text{ g Fe}} = 1.25 \text{ mol Fe atoms}$$

$$30.1 \text{ g O} \times \frac{1 \text{ mol O atoms}}{16.00 \text{ g O}} = 1.88 \text{ mol O atoms}$$

Step 3: Find the simplest whole-number mole ratio. The ratio is not obvious from inspection, so we will divide through by the number of moles of iron.

$$\text{iron:} \quad \frac{1.25}{1.25} = 1.00 \qquad 1.00 \times 2 = 2.00 = 2$$

$$\text{oxygen:} \quad \frac{1.88}{1.25} = 1.50 \qquad 1.50 \times 2 = 3.00 = 3$$

In this case the quotients on the left-hand side, 1.00 and 1.50, have to be multiplied by 2 to obtain whole-number ratios. There are 2 mol of iron for every 3 mol of oxygen.

Step 4: Write the formula. The atom ratio in the formula is the same as the mole ratio. The formula is Fe_2O_3.

PRACTICE EXERCISE 3.9

Chloroform, an industrial solvent, contains 89.10% chlorine and 0.84% hydrogen by mass. The remainder is carbon. Calculate the empirical formula of chloroform.

PRACTICE EXERCISE 3.10

A 25.0-g sample of aluminum sulfide contains 8.98 g of aluminum. Calculate the empirical formula of aluminum sulfide.

Molecular Formulas

The molar mass of benzene (C_6H_6) is six times the molar mass of CH, its empirical formula. Similarly, the molar mass of hydrogen peroxide (H_2O_2) is twice the molar mass of HO. Because the formula of a molecular compound is always a whole-number multiple of its empirical formula, it follows that *the molar mass of a molecular compound is always a whole-number multiple of the molar mass of its empirical formula.* The next example shows how empirical formulas and molar masses are used to calculate molecular formulas.

EXAMPLE 3.11

The nitrogen compound in Example 3.8 consists of molecules with a molar mass of 92.0 g/mol. What is the molecular formula?

SOLUTION

The empirical formula of the compound in Example 3.8 was found to be NO_2. One mole of NO_2 units has a mass of 14.0 g + (2 × 16.0 g) = 46.0 g. One mole of molecules has a mass of 92.0 g, twice the mass of 1 mol of NO_2 units. Thus 1 mol of molecules contains 2 mol of NO_2 units. The molecular formula is $(NO_2)_2$, or N_2O_4.

PRACTICE EXERCISE 3.11

The molar mass of 1-butene is 56.10 g/mol. Its empirical formula is CH_2. What is the molecular formula of 1-butene?

PRACTICE EXERCISE 3.12

Nitrogen selenide, an explosive orange-red powder, contains 15.07% nitrogen and 84.93% selenium by mass. Its molar mass has been reported as 371.9 g/mol. Find the molecular formula of nitrogen selenide.

Combustion Analysis

The elemental composition of carbon-containing compounds is often established by *combustion analysis,* a procedure that involves burning a small sample of the compound in oxygen and weighing the products. If a compound contains only carbon, hydrogen, and oxygen, the products will consist of only carbon dioxide and water vapor. As shown in Figure 3.4, the gaseous combustion products pass through preweighed tubes that absorb the water and carbon dioxide separately. The tubes are then reweighed to ascertain how much of each product was formed. Laboratories that specialize in combustion analyses generally report the results in terms of percentage composition. The investigator can then convert the mass data into an empirical formula.

A modern combustion analysis is automated. The masses of CO_2 and H_2O are determined by passing the gases through an instrument called a gas chromatograph.

EXAMPLE 3.12

A 28.64-mg sample of vitamin A, a compound consisting of carbon, hydrogen, and oxygen, is burned to form 88.02 mg of carbon dioxide and 27.03 mg of water. Calculate the percentage of each element in vitamin A.

SOLUTION

To find the percentage composition, we first need to find the mass of each element in the original sample of vitamin A. After combustion, all the carbon is in the carbon dioxide, and all the hydrogen is in the water. Hence,

$$\text{mass of carbon in 28.64 mg of vitamin A} = \text{mass of carbon in 88.02 mg of } CO_2$$

$$\text{mass of hydrogen in 28.64 mg of vitamin A} = \text{mass of hydrogen in 27.03 mg of } H_2O$$

$$\text{mass of oxygen in 28.64 mg of vitamin A} = 28.64 \text{ mg} - \text{mass of carbon} - \text{mass of hydrogen}$$

Our strategy will be to (1) calculate the masses of carbon and hydrogen, (2) find the mass of oxygen by taking the difference, and (3) convert the masses to percentages.

Step 1: The molar mass of CO_2 is 12.01 g/mol + (2 × 16.00 g/mol) = 44.01 g/mol, of which 12.01 g are carbon. A 44.01-mg sample of CO_2 will contain 12.01 mg of carbon, so the mass of carbon in 88.02 mg of CO_2 is

$$\text{mass of carbon} = 88.02 \text{ mg } CO_2 \times \frac{12.01 \text{ mg C}}{44.01 \text{ mg } CO_2} = 24.02 \text{ mg C}$$

The molar mass of H_2O is (2 × 1.008 g/mol) + 16.00 g/mol = 18.02 g/mol, of which 2.016 g are hydrogen. An 18.02-mg sample of H_2O will contain 2.016 mg of hydrogen, so the mass of hydrogen in 27.03 mg of H_2O is

$$\text{mass of hydrogen} = 27.03 \text{ mg } H_2O \times \frac{2.016 \text{ mg H}}{18.02 \text{ mg } H_2O} = 3.024 \text{ mg H}$$

Step 2: The mass of oxygen in the vitamin A sample is its total mass minus the masses of carbon and hydrogen, so

$$\text{mass of oxygen} = 28.64 \text{ mg} - 24.02 \text{ mg} - 3.024 \text{ mg} = 1.60 \text{ mg}$$

Step 3: The percentage composition is

$$\text{carbon:} \quad \frac{24.02 \text{ mg}}{28.64 \text{ mg}} \times 100\% = 83.87\%$$

$$\text{hydrogen:} \quad \frac{3.024 \text{ mg}}{28.64 \text{ mg}} \times 100\% = 10.56\%$$

$$\text{oxygen:} \quad \frac{1.60 \text{ mg}}{28.64 \text{ mg}} \times 100\% = 5.59\%$$

PRACTICE EXERCISE 3.13

Vitamin A is a molecular compound with a molar mass of 286.5 g/mol. Use the results of Example 3.12 to calculate the empirical and molecular formulas of vitamin A.

PRACTICE EXERCISE 3.14

A molecular compound containing only carbon and hydrogen has a molar mass of 54.09 g/mol. When 25.00 mg of this compound is burned, 81.36 mg of CO_2 and 24.98 mg of H_2O are formed. Calculate (a) the percentage composition of the compound and (b) its molecular formula.

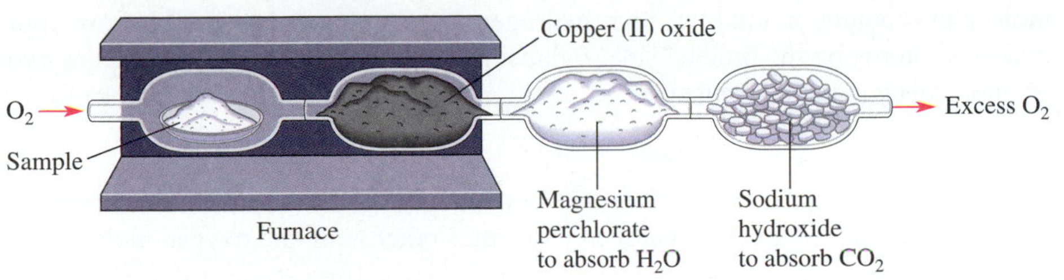

Figure 3.4
Combustion analysis. The weighed sample in the furnace is an unknown compound containing carbon, hydrogen, and oxygen. Gases from the combustion pass through copper(II) oxide (which converts traces of CO to CO_2), then through weighed tubes that absorb H_2O and CO_2. The masses of H_2O and CO_2 are found by reweighing the tubes.

3.5 CHEMICAL EQUATIONS

We now know that the apparent magic of chemical change, the vanishing of old substances and the appearance of new ones, is simply a matter of atomic rearrangement. No atoms appear or disappear in a reaction; the products contain only atoms provided by the reactants. We can use this knowledge to describe each reaction in an orderly symbolic form known as a **chemical equation**.

Formulas and equations are the language of chemistry.

A chemical equation uses formulas to represent the reactants and products, and it also accounts for all of the atoms involved in the reaction. Consider, for example, burning carbon in oxygen to form carbon dioxide gas. The "word equation" for this reaction

$$\text{carbon} + \text{oxygen} \rightarrow \text{carbon dioxide}$$

shows the reactants on the left, the products on the right, and an arrow in between that can be read as "gives," "yields," or "is transformed into." To obtain the chemical equation, the formulas C, O_2, and CO_2 are substituted into the word equation and the physical state of each substance is added in parentheses. Carbon is a solid (s), while oxygen and carbon dioxide are gases (g). The complete chemical equation is

$$C(s) + O_2(g) \rightarrow CO_2(g) \tag{3.1}$$

Equation 3.1 states that one atom of carbon combines with one molecule of oxygen (two oxygen atoms) to form one molecule of carbon dioxide. It also states that 1 mol of carbon atoms reacts with 1 mol of oxygen molecules to form 1 mol of carbon dioxide molecules. Note that the equation is **balanced** in the sense that every atom that had been in the reactants (left side) has reappeared in the products (right side). Since atoms are neither created nor destroyed in a chemical reaction, a chemical equation is not correct unless it is balanced.

As a second example, consider the reaction of hydrogen and oxygen gases to form liquid water:

$$\text{hydrogen} + \text{oxygen} \rightarrow \text{water}$$

Substituting the formulas gives

$$H_2(g) + O_2(g) \rightarrow H_2O(l) \qquad \text{(not balanced)}$$

This equation is not yet balanced. There are two oxygen atoms on the reactant side but only one on the product side. The extra atom must be accounted for, and this is done by putting the coefficient 2 before the H_2O:

$$H_2(g) + O_2(g) \rightarrow 2H_2O(l) \qquad \text{(still not balanced)}$$

LEARNING HINT

NEVER CHANGE THE SUBSCRIPTS IN A FORMULA WHEN YOU BALANCE AN EQUATION. A different subscript would indicate a different substance; the properties of water (H_2O), for example, are quite different from those of hydrogen peroxide (H_2O_2).

A **coefficient** is a number placed before a formula unit to show how many of that unit are involved; it applies to the entire unit. For example, $2H_2O$ means two water molecules, each containing two hydrogen atoms and one oxygen atom. Two water molecules contain a total of four hydrogen atoms; hence, there are now four hydrogen atoms on the product side. Balancing the hydrogen atoms will require two H_2 molecules on the reactant side:

$$2H_2(g) + O_2(g) \rightarrow 2H_2O(l) \tag{3.2}$$

Now the equation is balanced. Note that Equation 3.2 gives the same information as Figure 3.3; it states that two hydrogen molecules react with one oxygen molecule to produce two water molecules. It also states that 2 mol of hydrogen gas react with 1 mol of oxygen gas to form 2 mol of water.

The steps in writing a balanced equation are as follows:

- ***Step 1:*** Identify the reactants and products.
- ***Step 2:*** Write the reactant formulas to the left of the arrow and the product formulas on the right, with plus signs between the formulas. Indicate the physical state of each substance (s = solid, l = liquid, g = gas, and aq = aqueous, i.e., dissolved in water).
- ***Step 3:*** Choose coefficients to make the atoms balance.

LEARNING HINT

Equation writing takes practice. Frequently used formulas such as O_2, H_2, and CO_2 should be memorized.

EXAMPLE 3.13

Methane (CH_4) is the major component of natural gas. When methane burns in air, it reacts with oxygen to form carbon dioxide and water vapor. Write a balanced equation for this reaction.

SOLUTION

We will follow the steps outlined above.

Step 1: Identify the reactants and products.

$$\underbrace{\text{methane} + \text{oxygen}}_{\text{(reactants)}} \longrightarrow \underbrace{\text{carbon dioxide} + \text{water}}_{\text{(products)}}$$

Step 2: Insert the formula and physical state of each substance. Methane is CH_4; all of the substances are gaseous.

$$CH_4(g) + O_2(g) \longrightarrow CO_2(g) + H_2O(g) \qquad \text{(not balanced)}$$

Step 3: Choose the coefficients. There is one carbon atom on the left and one on the right, so the carbon atoms are balanced. There are four hydrogen atoms on the left and only two on the right. To balance them we place a 2 in front of the H_2O:

$$CH_4(g) + O_2(g) \longrightarrow CO_2(g) + 2H_2O(g) \qquad \text{(still not balanced)}$$

Now balance the oxygen. The four oxygen atoms on the right require two O_2 molecules on the left:

$$CH_4(g) + 2O_2(g) \longrightarrow CO_2(g) + 2H_2O(g) \qquad \text{(balanced)}$$

LEARNING HINT

When an element is a reactant or product, balance it last.

PRACTICE EXERCISE 3.15

Ethylene (C_2H_4) is a gas used to speed up the ripening of oranges and tomatoes. When ethylene burns in oxygen, carbon dioxide and water vapor are the products. Write a balanced equation for the combustion of ethylene.

EXAMPLE 3.14

Methanol (CH_3OH) is a fuel that has been used in spirit lamps and some picnic stoves.

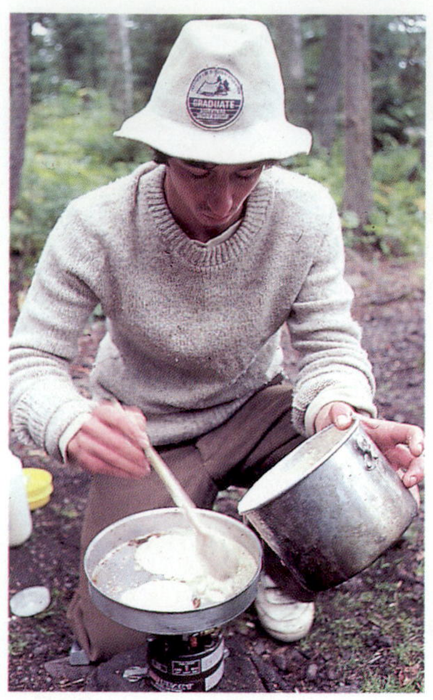

Cooking with methanol fuel

Write a balanced equation for burning liquid methanol in oxygen. The products are carbon dioxide and water vapor.

SOLUTION

The unbalanced equation is

$$CH_3OH(l) + O_2(g) \rightarrow CO_2(g) + H_2O(g)$$

It is usually best to start balancing with the molecule containing the greatest number of elements, in this case CH_3OH. There is one carbon atom on the left in CH_3OH and one carbon atom on the right in CO_2, so the carbon atoms are balanced. There are four hydrogen atoms in CH_3OH but only two on the right, so we place a 2 in front of H_2O:

$$CH_3OH(l) + O_2(g) \rightarrow CO_2(g) + 2H_2O(g) \qquad \text{(not yet balanced)}$$

Now we balance the oxygen. There are four oxygen atoms on the right: CH_3OH supplies one oxygen atom on the left and three more are needed. Rather than change the coefficient of CH_3OH, which would alter the already balanced carbon and hydrogen atoms, we change the coefficient in front of O_2. The three needed oxygen atoms will come from one and one-half ($1\frac{1}{2} = \frac{3}{2}$) O_2 molecules:

$$CH_3OH(l) + \tfrac{3}{2}O_2(g) \rightarrow CO_2(g) + 2H_2O(g)$$

The equation, although balanced, contains a fractional coefficient. Whole-number coefficients in the same ratio are obtained by multiplying each coefficient by two:

$$2CH_3OH(l) + 3O_2(g) \rightarrow 2CO_2(g) + 4H_2O(g)$$

PRACTICE EXERCISE 3.16

Balance the equation: $SO_2(g) + O_2(g) \rightarrow SO_3(g)$

The next example shows that balancing an equation can sometimes be simplified by treating a group of atoms as a single unit.

EXAMPLE 3.15

Write a balanced equation for the reaction between aqueous (water) solutions of sodium sulfate and calcium nitrate. The products are solid calcium sulfate and a solution of sodium nitrate. All of the compounds are ionic.

SOLUTION

The reactants and products are

$$\text{sodium sulfate} + \text{calcium nitrate} \rightarrow \text{sodium nitrate} + \text{calcium sulfate}$$

Formulas for the compounds can be obtained by combining ions from Table 2.6 (p. 68):

$$\text{sodium sulfate:} \quad 2Na^+ + SO_4^{2-} = Na_2SO_4$$

$$\text{calcium nitrate:} \quad Ca^{2+} + 2NO_3^- = Ca(NO_3)_2$$

$$\text{sodium nitrate:} \quad Na^+ + NO_3^- = NaNO_3$$

$$\text{calcium sulfate:} \quad Ca^{2+} + SO_4^{2-} = CaSO_4$$

The unbalanced equation is

$$Na_2SO_4(aq) + Ca(NO_3)_2(aq) \rightarrow NaNO_3(aq) + CaSO_4(s)$$

where "aq" indicates that the compounds are in aqueous solution and "s" indicates that solid calcium sulfate is formed.

The SO_4^{2-} and NO_3^- ions remain intact in this reaction and can be treated as units for the purposes of balancing the equation. Two NO_3^- ions on the left require two $NaNO_3$ formula units on the right. The balanced equation is

$$Na_2SO_4(aq) + Ca(NO_3)_2(aq) \rightarrow 2NaNO_3(aq) + CaSO_4(s)$$

PRACTICE EXERCISE 3.17

Write a balanced equation for the reaction between aqueous solutions of potassium chromate and silver nitrate. The products are solid silver chromate and a solution of potassium nitrate.

3.6 SOME CHEMICAL REACTIONS

We now look at several types of reactions in order to become better acquainted with chemical reactions and equation writing. It will become apparent as we proceed that many reactions fit into more than one category.

Combination and Decomposition Reactions

Combination reactions are those in which a compound forms from simpler substances. We have already seen equations for the formation of water from hydrogen and oxygen (Equation 3.2) and the formation of carbon dioxide from carbon and oxygen (Equation 3.1). Other examples are the formation of sodium chloride from sodium metal and chlorine gas (Demonstration 3.2):

$$2Na(s) + Cl_2(g) \rightarrow 2NaCl(s) \quad (3.3)$$

and the formation of nitrogen monoxide (NO) from nitrogen and oxygen:

$$N_2(g) + O_2(g) \rightarrow 2NO(g) \quad (3.4)$$

Nitrogen monoxide, which forms only at high temperatures, is an air pollutant that is produced in automobile engines and released along with other hot exhaust gases.

Some combination reactions involve compounds as reactants. An example is the reaction between nitrogen monoxide (NO) and oxygen to form nitrogen dioxide (NO_2), another air pollutant:

$$2NO(g) + O_2(g) \rightarrow 2NO_2(g)$$

The compound P_4O_{10} is an effective drying agent because it readily absorbs and reacts with water to form phosphoric acid:

$$P_4O_{10}(s) + 6H_2O(l \text{ or } g) \rightarrow 4H_3PO_4(aq)$$

DEMONSTRATION 3.2 A COMBINATION REACTION: $2Na(s) + Cl_2(g) \rightarrow 2NaCl(s)$

Sodium metal is freshly cut, exposing a clean metallic surface.

Sodium is added to a flask containing chlorine, a yellowish gas.

After warming, the sodium glows and sparks; a chemical reaction is under way.

The product, sodium chloride, forms fine white crystals that coat the flask walls.

EXAMPLE 3.16

Write a balanced equation for the formation of solid iron(III) chloride from its elements.

SOLUTION

Iron(III) chloride contains Fe^{3+} and Cl^- ions (Table 2.6, page 68), and its formula is $FeCl_3$. The elements are iron (Fe), a solid, and chlorine (Cl_2), a gas. The balanced equation is

$$2Fe(s) + 3Cl_2(g) \rightarrow 2FeCl_3(s)$$

A **decomposition reaction** is one in which a compound breaks down into simpler substances that may or may not be elements. Decomposition often requires an input of energy; for example, the electrolytic decomposition of water requires the energy of an electric current (Figure 3.1):

$$2H_2O(l) \xrightarrow{\text{electric current}} 2H_2(g) + O_2(g)$$

Less energy is required to decompose mercury(II) oxide (HgO). Gentle heating converts this red solid into oxygen gas and droplets of liquid mercury (Demonstration 3.3):

$$2HgO(s) \xrightarrow{\text{heat}} 2Hg(l) + O_2(g)$$

DEMONSTRATION 3.3

A DECOMPOSITION REACTION: $2HgO(s) \rightarrow 2Hg(l) + O_2(g)$

A brick-red powder, mercury(II) oxide, is heated. A lighted wood splint, barely glowing, is ready to be thrust into the test tube.

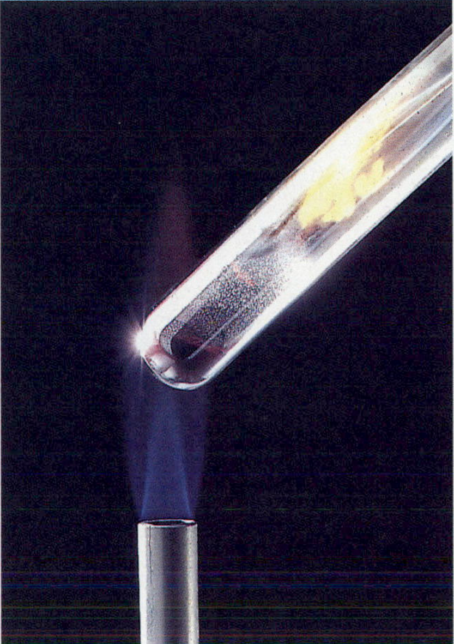

The oxide turns dark as it forms mercury vapor and oxygen. The oxygen causes the glowing splint to burst into flame.

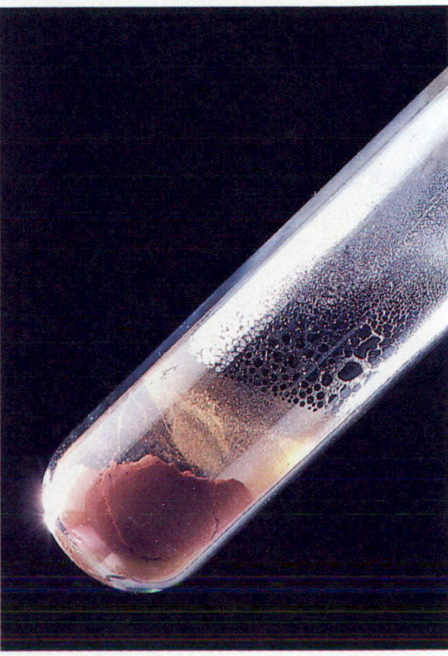

The burner is removed. The remaining oxide regains its red color. Mercury vapor condenses in droplets on the test tube walls.

PRACTICE EXERCISE 3.18

Write a balanced equation for (a) the formation of gaseous ammonia (NH_3) from its elements and (b) the decomposition of solid barium peroxide (BaO_2) into solid barium oxide (BaO) and oxygen gas.

Displacement Reactions

In a **displacement reaction** one element displaces another element from a compound; that is, one element leaves the compound and another element takes its place. The displacement of hydrogen from hydrochloric acid (aqueous HCl) by zinc metal is a reaction often used to prepare small quantities of hydrogen gas in the laboratory (Figure 3.5):

$$Zn(s) + 2HCl(aq) \longrightarrow H_2(g) + ZnCl_2(aq)$$

The other product, zinc chloride, remains in solution.

Thermite, a military incendiary used in past wars, is a finely divided mixture of aluminum and iron(III) oxide (Fe_2O_3). When ignited with a magnesium fuse, it produces a spectacular pyrotechnical display (Demonstration 3.4). Iron is displaced by aluminum; the reaction is

$$2Al(s) + Fe_2O_3(s) \longrightarrow 2Fe(l) + Al_2O_3(l)$$

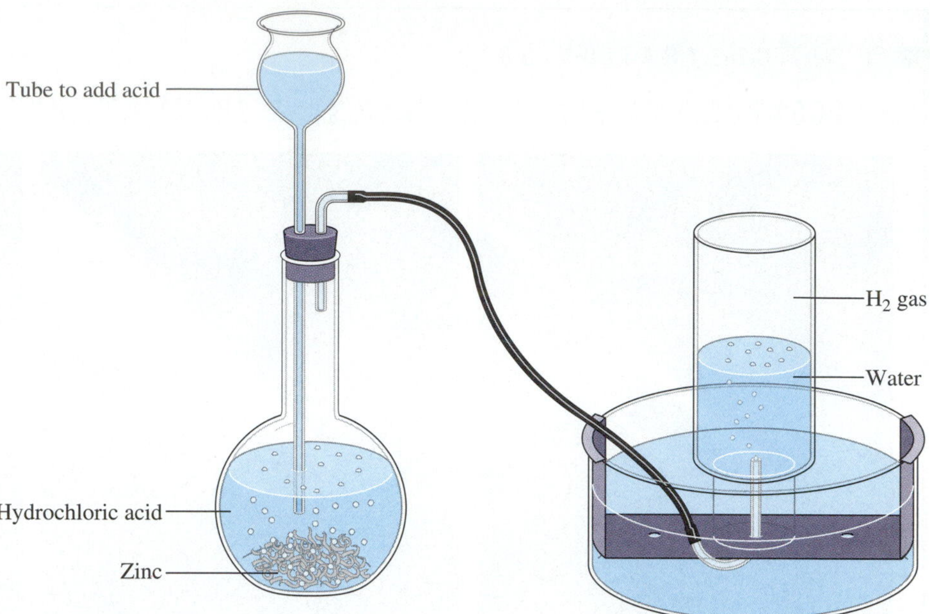

Figure 3.5
Hydrogen is prepared in the laboratory by a displacement reaction:

$$Zn(s) + 2HCl(aq) \longrightarrow ZnCl_2(aq) + H_2(g)$$

Hydrogen gas bubbling from the zinc–hydrochloric acid mixture is trapped in an inverted bottle that was initially filled with water. (Hydrogen can be collected in this way because it is not appreciably soluble in water.)

The reaction gives off a tremendous amount of heat and produces temperatures in excess of 2000°C, well above the melting point of iron. Molten iron produced by this process has been used to weld broken rails and other iron objects.

EXAMPLE 3.17

Zinc metal displaces silver from aqueous solutions of silver nitrate. The products are silver metal and a solution of zinc nitrate. Write a balanced equation for this reaction.

SOLUTION

Zinc metal and aqueous silver nitrate are the reactants; silver metal and aqueous zinc nitrate are the products. The compounds are ionic; silver nitrate is $AgNO_3$ and zinc nitrate is $Zn(NO_3)_2$ (Table 2.6, page 68). The balanced equation is

$$Zn(s) + 2AgNO_3(aq) \longrightarrow 2Ag(s) + Zn(NO_3)_2(aq)$$

PRACTICE EXERCISE 3.19

Fluorine gas (F_2) reacts with water to form oxygen gas and an aqueous solution of hydrogen fluoride (HF). Write a balanced equation for this displacement reaction.

Combustion Reactions

"Combustion" is the term commonly used for a reaction in which something burns. **Combustion reactions** are vigorous self-sustaining reactions in which substances

DEMONSTRATION 3.4

THE THERMITE REACTION: $2Al(s) + Fe_2O_3(s) \rightarrow 2Fe(l) + Al_2O_3(l)$

A clay cup is filled with thermite, a mixture of powdered aluminum and iron oxide. A magnesium strip is inserted as a fuse.

Ignition triggers a reaction that produces heat, light, and a shower of sparks, as aluminum displaces iron from its oxide.

The molten products burst through the cup and pour into the sand below.

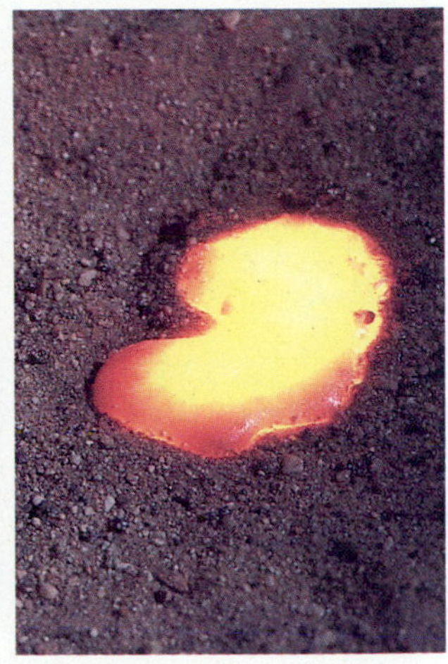

The red-hot iron lies in the sand.

combine and give off heat and light. Many combustion reactions are also combination reactions; examples include the burning of hydrogen in oxygen to form water vapor

$$2H_2(g) + O_2(g) \rightarrow 2H_2O(g)$$

and the burning of carbon in oxygen to form carbon dioxide (Equation 3.1). Many familiar combustion reactions involve **organic compounds** (compounds that contain carbon) burning in oxygen. The combustion of wood, paper, and fossil fuels such as natural gas (Example 3.13), oil, coal, and gasoline are examples. Because organic compounds always contain carbon and nearly always contain hydrogen, they form carbon dioxide and water vapor when burned completely in an unlimited supply of air.

EXAMPLE 3.18

Gasohol is a mixture of gasoline and ethanol (C_2H_5OH, also called ethyl alcohol). Write the equation for the complete combustion of ethanol in oxygen.

SOLUTION

The products will be carbon dioxide and water vapor. The unbalanced equation is

$$C_2H_5OH(l) + O_2(g) \rightarrow CO_2(g) + H_2O(g)$$

Balancing gives

$$C_2H_5OH(l) + 3O_2(g) \rightarrow 2CO_2(g) + 3H_2O(g)$$

The combustion of carbon and carbon compounds is often incomplete, and some carbon monoxide (CO, an atmospheric pollutant and toxic gas) is formed along with carbon dioxide. Carbon monoxide is especially likely to be formed at very high temperatures or when the supply of oxygen is limited. Coal, for example, is mostly carbon, and its combustion product depends on how much oxygen is present:

$$2C(s) + O_2(g) \rightarrow 2CO(g) \qquad \text{(limited oxygen)}$$

or

$$C(s) + O_2(g) \rightarrow CO_2(g) \qquad \text{(excess oxygen)}$$

Many fossil fuels contain sulfur, which produces gaseous sulfur dioxide (SO_2) on combustion. Sulfur dioxide is an atmospheric pollutant that is, to a large degree, responsible for acid rain.

PRACTICE EXERCISE 3.20

Thiophene (C_4H_4S) is a liquid used in making dyes and resins. Write a balanced equation for the combustion of thiophene in excess oxygen. One of the combustion products is SO_2.

Not all combustions require oxygen. Hydrogen reacts violently with fluorine, evolving heat and light:

$$H_2(g) + F_2(g) \rightarrow 2HF(g) \qquad (3.5)$$

Many elements and compounds react vigorously in chlorine and sulfur vapors to form chlorides and sulfides, and these reactions are also combustions. The reaction of sodium metal with chlorine gas (Demonstration 3.2) is an example.

Oxidation–Reduction Reactions Involving Oxygen

Oxygen is the most abundant element in the earth's crust, oceans, and atmosphere. Thus it is not surprising that many important compounds are compounds of oxygen, and many important reactions involve the transfer of oxygen atoms from one substance to another. As we have seen, the combustion of most fuel involves the transfer of oxygen atoms from elemental oxygen in the air to the carbon and hydrogen atoms of the fuel (see Examples 3.13, 3.14, and 3.18). When a substance has combined with oxygen, we say it has been *oxidized,* and the process is called **oxidation**. Fuels are oxidized when they are burned in air to form CO_2 and H_2O. Our bodies obtain energy from the oxidation of food.

The terms oxidation and reduction are used here in a limited sense. In later chapters we will see that these terms can also be applied to many reactions that do not involve oxygen.

An **ore** is a natural mineral product from which an element can be profitably extracted. Many metal ores are oxides (Figure 3.6), and the metals are obtained by the removal of oxygen. When a substance has lost oxygen, we say it has been *reduced,* and the process is called **reduction**. Manganese metal, for example, is obtained by the reduction of MnO_2.

(a)

(b)

Figure 3.6

Some oxide ores. (a) From left to right: magnetite (Fe_3O_4), bauxite (Al_2O_3), and hematite (Fe_2O_3). (b) From left to right: cassiterite (SnO_2) on a quartz base, four samples of rutile (TiO_2), and pyrolusite (MnO_2) crystals in a rock.

EXAMPLE 3.19

Hydrogen gas reacts with hot solid copper(II) oxide to form copper metal and water vapor. Write a balanced equation for this reaction and identify the substance oxidized and the substance reduced.

SOLUTION

The equation is

$$H_2(g) + CuO(s) \rightarrow Cu(s) + H_2O(g)$$

Hydrogen combines with oxygen; hence, it is oxidized. Copper(II) oxide loses oxygen; hence, it is reduced. Note that this reaction is also a displacement reaction because hydrogen displaces copper from the copper(II) oxide.

PRACTICE EXERCISE 3.21

Which substance is oxidized and which is reduced in the thermite reaction?

$$2Al(s) + Fe_2O_3(s) \rightarrow 2Fe(l) + Al_2O_3(l)$$

In most cases, reduction of an oxide to a metal can be accomplished by heating it with a substance such as carbon, whose affinity for oxygen is greater than that of the desired metal. ***Coke*** is one of the best and cheapest materials for this purpose. It is a form of carbon obtained by heating coal to drive out water and other volatile (easily vaporized) substances.

Tin, for example, is obtained by heating cassiterite (an ore containing SnO_2) with coke. The tin(IV) oxide is reduced to tin:

$$SnO_2(s) + 2C(s) \rightarrow Sn(l) + 2CO(g)$$

At the high temperature required for this reaction, tin is liquid and CO forms rather than CO_2.

CHEMICAL INSIGHT

OXYGEN, THE CENTRAL ELEMENT

Our discussion of chemical reactions has been heavily weighted with examples involving oxygen and its compounds. Half of the atoms in our environment are atoms of oxygen, and the chemistry of these atoms permeates all natural processes here on earth. Oxygen gas in the form of diatomic O_2 molecules comprises 20% of the atmosphere; vast amounts of the gas are dissolved in the waters of the world; and water itself is a compound containing 88.8% oxygen by mass. The rocks of the earth's crust are largely silicates, rigid networks of silicon and oxygen atoms, interspersed with various ions. The major constituent of sand is silicon dioxide (SiO_2), a nonionic compound whose continuous structure was shown in Figure 2.20, page 63. Moon rocks are also silicates, and evidence suggests that all rocky planets and moons are composed of this sort of material.

Oxygen is the atmospheric component that supports combustion. It also participates in slower oxidation reactions including those involved in metabolism, decay, and forms of corrosion such as the rusting of iron. The latter reaction is most simply represented as

$$4Fe(s) + 3O_2(g) \rightarrow \underset{\text{rust}}{2Fe_2O_3(s)}$$

Because oxygen is so prevalent, the majority of elements on or near the earth's surface exist in oxidized form, that is, in some compound containing oxygen. Many metals occur naturally as oxide ores (Figure 3.6). Metal oxides are solids, and many of them are ionic.

Nonmetal oxides are not ionic. Water, carbon dioxide, and sulfur dioxide are important examples. Carbon dioxide is the end product of food metabolism in animals. It is used commercially for carbonating beverages. Sulfur dioxide is an atmospheric pollutant formed by volcanoes, by the combustion of sulfur containing fuels, and by the roasting of sulfide ores. Nitrogen forms a number of gaseous oxides including N_2O ("laughing gas," dinitrogen monoxide), an anesthetic gas used in dentistry, and NO (nitrogen monoxide), which is produced by the action of lightning and high temperatures on atmospheric nitrogen and oxygen (Equation 3.4). Solid nonmetal oxides include SiO_2 (silicon dioxide) and P_4O_{10} (commonly referred to as phosphorus pentoxide because it was once thought to have the formula P_2O_5).

Iron, our most important structural metal, is made by reducing hematite (Fe_2O_3) or magnetite (Fe_3O_4) with coke in a blast furnace. The coke forms carbon monoxide, and the reduction can be summarized by the equations

$$2C(s) + O_2(g) \rightarrow 2CO(g)$$

$$Fe_2O_3(s) + 3CO(g) \rightarrow 2Fe(l) + 3CO_2(g)$$

$$Fe_3O_4(s) + 4CO(g) \rightarrow 3Fe(l) + 4CO_2(g)$$

Silicon has recently achieved importance as an element essential to the manufacture of computer chips. Silicon, which is not a metal, is firmly bonded to oxygen in all of its natural compounds. The element is obtained by reducing sand (SiO_2) with carbon in an electric furnace at 3000°C. At this high temperature, silicon is a liquid and the carbon forms carbon monoxide gas:

$$SiO_2(s) + 2C(s) \xrightarrow{3000^\circ C} Si(l) + 2CO(g)$$

Zinc is one of several metals found in the form of a sulfide. One way of reducing sulfide ores consists of a two-step process in which oxidation in air is followed by reduction with coke. The ore sphalerite (ZnS) is first "roasted" by heating it in air. Sulfur is oxidized by atmospheric oxygen and zinc oxide forms:

$$2ZnS(s) + 3O_2(g) \xrightarrow{\text{heat}} 2ZnO(s) + 2SO_2(g)$$

The zinc oxide is then reduced with coke in a furnace:

$$ZnO(s) + C(s) \xrightarrow{\text{heat}} Zn(g) + CO(g)$$

The zinc vapor that forms during the high-temperature reduction can be condensed directly into zinc powder, or it can be condensed first to the liquid and then cast into metal ingots.

3.7 REACTION STOICHIOMETRY

A chemist planning to carry out a reaction often needs to know something about the quantities involved. Questions such as "How much of each reactant is needed?" or "How much product will form?" can be answered with the aid of a balanced equation. Consider the combustion of propane, C_3H_8, a gaseous fuel. The equation

$$C_3H_8(g) + 5O_2(g) \rightarrow 3CO_2(g) + 4H_2O(g)$$

tells us that 1 mol of propane combines with 5 mol of oxygen to yield 3 mol of carbon dioxide and 4 mol of water. It does not matter how much propane is burned; the reactants always combine and the products always form in the same mole ratios. Mole ratios provide conversion factors that relate one quantity in the equation to another. For example, 1 mol of propane always combines with 5 mol of oxygen, so for any problem involving this reaction, the factors

$$\frac{5 \text{ mol } O_2}{1 \text{ mol } C_3H_8} \quad \text{and} \quad \frac{1 \text{ mol } C_3H_8}{5 \text{ mol } O_2}$$

relate the quantities of oxygen and propane. If we wanted to know how much oxygen is needed to burn 3.0 mol of propane, we would write

$$3.0 \text{ mol } C_3H_8 \times \frac{5 \text{ mol } O_2}{1 \text{ mol } C_3H_8} = 15 \text{ mol } O_2$$

The number of moles of propane cancels, leaving the number of moles of oxygen.

EXAMPLE 3.20

Calculate the number of moles of carbon dioxide formed when 40.0 mol of oxygen is consumed in the burning of propane.

SOLUTION

The balanced equation states that 3 mol of CO_2 is produced for every 5 mol of O_2 consumed. These quantities are stoichiometrically equivalent and can be used to convert 40.0 mol of O_2 to moles of CO_2:

$$40.0 \text{ mol } O_2 \times \frac{3 \text{ mol } CO_2}{5 \text{ mol } O_2} = 24.0 \text{ mol } CO_2$$

The answer has three significant figures because the number of moles of oxygen was given to three significant figures. The integral coefficients 3 and 5 are exact.

PRACTICE EXERCISE 3.22

Calculate the number of moles of water formed when 2.00 mol of propane is burned.

Stoichiometric problems may involve masses, volumes, and various other quantities, but in each problem a mole ratio will be central to the solution. The following four-step method can be used for solving such problems:

- ***Step 1:*** Write the balanced equation and identify the known and unknown quantities.
- ***Step 2:*** Convert the known quantity to moles.
- ***Step 3:*** Use the mole ratio in the equation to find the unknown quantity in moles.
- ***Step 4:*** Convert the number of moles in Step 3 to the final units called for in the problem.

These steps are followed in the next example.

LEARNING HINT

In a stoichiometric calculation, molar masses should have at least one more significant figure than the data given for the calculation.

EXAMPLE 3.21

Cassiterite, SnO_2, is the major ore of tin. How many kilograms of carbon are needed to convert 50.0 kg of cassiterite to metallic tin? The equation is

$$SnO_2(s) + 2C(s) \longrightarrow Sn(l) + 2CO(g)$$

SOLUTION

Step 1: Write the balanced equation and identify the known and unknown quantities. The known quantity is 50.0 kg of SnO_2. The unknown quantity is the number of kilograms of carbon:

$$\begin{array}{llll} SnO_2(s) & + \ 2C(s) & \longrightarrow Sn(l) & + \ 2CO(g) \\ 50.0\text{ kg} & \ \ ?\text{ kg} & & \end{array}$$

Step 2: Convert the known quantity, 50.0 kg = 50.0×10^3 g of SnO_2, to moles. The mass of 1 mol of SnO_2 is 118.7 g + (2 × 16.00 g) = 150.7 g.

$$50.0 \times 10^3 \text{ g } SnO_2 \times \frac{1 \text{ mol } SnO_2}{150.7 \text{ g } SnO_2} = 332 \text{ mol } SnO_2$$

Step 3: Find the number of moles of the unknown substance, in this case, carbon. The equation shows that 2 mol of carbon reduce 1 mol of SnO_2; hence,

$$332 \text{ mol } SnO_2 \times \frac{2 \text{ mol C}}{1 \text{ mol } SnO_2} = 664 \text{ mol C}$$

Step 4: Find the mass of carbon in kilograms. The molar mass of carbon is 12.01 g/mol.

$$664 \text{ mol C} \times \frac{12.01 \text{ g C}}{1 \text{ mol C}} = 7.97 \times 10^3 \text{ g C} = 7.97 \text{ kg C}$$

Steps 2, 3, and 4 of the preceding example can be summarized as follows:

$$50.0 \text{ kg } SnO_2 \rightarrow 332 \text{ mol } SnO_2 \rightarrow 664 \text{ mol C} \rightarrow 7.97 \text{ kg C}$$

This "roadmap" shows that the molar quantities serve as the bridge between the known mass of SnO_2 and the unknown mass of carbon. After you become more confident in problems of this type, you can combine Steps 2, 3, and 4 before making the final calculation. Your solution would then read

$$50.0 \times 10^3 \text{ g } SnO_2 \times \underbrace{\frac{1 \text{ mol } SnO_2}{150.7 \text{ g } SnO_2}}_{\text{Step 2}} \times \underbrace{\frac{2 \text{ mol C}}{1 \text{ mol } SnO_2}}_{\text{Step 3}} \times \underbrace{\frac{12.01 \text{ g C}}{1 \text{ mol C}}}_{\text{Step 4}}$$

$$= 7.97 \times 10^3 \text{ g C} = 7.97 \text{ kg C}$$

PRACTICE EXERCISE 3.23

How many kilograms of tin will be produced when 50.0 kg of cassiterite reacts with excess carbon?

EXAMPLE 3.22

Iron reacts with superheated steam to form hydrogen gas and the oxide Fe_3O_4:

$$3Fe(s) + 4H_2O(g) \rightarrow Fe_3O_4(s) + 4H_2(g)$$

Calculate the number of moles of hydrogen produced by 10.0 g of iron and excess steam.

SOLUTION

Step 1: The known quantity is 10.0 g Fe. The unknown quantity is the number of moles of hydrogen:

$$\underset{10.0 \text{ g}}{3Fe(s)} + 4H_2O(g) \rightarrow Fe_3O_4(s) + \underset{?\text{ mol}}{4H_2(g)}$$

Step 1 suggests the following "roadmap" for the remaining steps

$$10.0 \text{ g Fe} \rightarrow \text{mol Fe} \rightarrow \text{mol } H_2$$

Step 2: The molar mass of iron is 55.85 g/mol. The number of moles of iron in 10.0 g is

$$10.0 \text{ g Fe} \times \frac{1 \text{ mol Fe}}{55.85 \text{ g Fe}} = 0.179 \text{ mol Fe}$$

Step 3: The balanced equation shows that 4 mol of hydrogen are formed from 3 mol of iron. The number of moles of hydrogen formed from 0.179 mol Fe is

$$0.179 \text{ mol Fe} \times \frac{4 \text{ mol } H_2}{3 \text{ mol Fe}} = 0.239 \text{ mol } H_2$$

Step 4: The problem asks for the number of moles of hydrogen, so no further calculations are needed.

Like Example 3.21, the preceding problem can also be solved by combining the steps before making the final calculation:

$$10.0 \text{ g Fe} \times \underbrace{\frac{1 \text{ mol Fe}}{55.85 \text{ g Fe}}}_{\text{Step 2}} \times \underbrace{\frac{4 \text{ mol } H_2}{3 \text{ mol Fe}}}_{\text{Step 3}} = 0.239 \text{ mol } H_2$$

PRACTICE EXERCISE 3.24

Magnesium dissolves in hydrochloric acid according to the equation

$$Mg(s) + 2HCl(aq) \longrightarrow MgCl_2(aq) + H_2(g)$$

How many grams of magnesium will be dissolved by 2.25 mol of HCl?

Stoichiometric problems come in an endless variety of disguises, but they are all alike at the core. You will rarely have difficulty solving the problem if you carefully read the question and correctly determine what is given and what is called for.

EXAMPLE 3.23

The air in a closed container consists of 1.40 mol of oxygen and 7.00 mol of nitrogen plus some carbon dioxide and water vapor. A small lamp fueled by liquid methanol (CH_3OH) is lit inside the container. How many milliliters of methanol will be consumed by the time the lamp goes out? The density of methanol is 0.791 g/mL.

SOLUTION

In real life, pertinent facts are nearly always accompanied by superfluous information. Analyzing a problem includes identifying the relevant data.

The lamp goes out when the oxygen is used up. The other components of the air—nitrogen, carbon dioxide, and water vapor—do not enter into the problem.

Step 1: Methanol burns in oxygen to produce carbon dioxide and water. The known quantity is 1.40 mol of oxygen consumed; the unknown quantity is milliliters of methanol consumed:

$$\underset{?\text{ mL}}{2CH_3OH(l)} + \underset{1.40\text{ mol}}{3O_2(g)} \longrightarrow 2CO_2(g) + 4H_2O(g)$$

Our strategy for the remaining steps will be:

$$1.40 \text{ mol } O_2 \longrightarrow \text{mol } CH_3OH \longrightarrow \text{g } CH_3OH \longrightarrow \text{mL } CH_3OH$$

Step 2: The known quantity, O_2, is already given as 1.40 mol.

Step 3: Two moles of methanol react with 3 mol of oxygen. The number of moles of methanol that will burn in 1.40 mol of oxygen is

$$1.40 \text{ mol } O_2 \times \frac{2 \text{ mol } CH_3OH}{3 \text{ mol } O_2} = 0.933 \text{ mol } CH_3OH$$

Step 4: Two steps are required to find the volume of methanol. First we use the molar mass of methanol, 32.04 g/mol, to find the mass of 0.933 mol:

$$0.933 \text{ mol } CH_3OH \times \frac{32.04 \text{ g } CH_3OH}{1 \text{ mol } CH_3OH} = 29.9 \text{ g } CH_3OH$$

LEARNING HINT

ALWAYS CHECK YOUR FINAL ANSWER. Make sure that you have calculated the quantity asked for, that your answer has the correct number of significant figures, and that your calculation is correct.

Then we use the density of methanol, 0.791 g/mL, to convert mass to volume:

$$29.9 \text{ g } CH_3OH \times \frac{1 \text{ mL } CH_3OH}{0.791 \text{ g } CH_3OH} = 37.8 \text{ mL } CH_3OH$$

PRACTICE EXERCISE 3.25

Small amounts of oxygen are sometimes prepared by heating potassium chlorate:

$$2KClO_3(s) \rightarrow 2KCl(s) + 3O_2(g)$$

How many grams of potassium chlorate are required to produce 5.00 g of oxygen?

3.8 LIMITING AND EXCESS REACTANTS

Reacting substances are rarely mixed in the exact proportions given by the equation. Usually one reactant is in short supply and will be completely consumed, while portions of the other reactants will be left over (Figure 3.7). The reactant that is used up is called the **limiting reactant** or **limiting reagent**. The reactants left over are called **excess reactants** or **excess reagents**. The idea of limiting quantities is not unfamiliar. Consider this homely example: suppose you take eight packages of frankfurters (eight franks each) and twelve packages of buns (six buns each) to a cookout. You have a total of 64 frankfurters and 72 buns. The frankfurters are the limiting ingredient—only 64 hot dogs can be made. The next example uses the same principle to determine which of two reactants limits a chemical reaction.

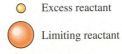

Excess reactant

Limiting reactant

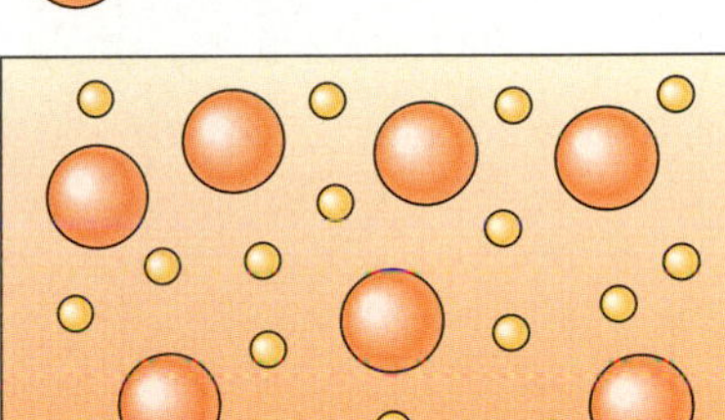

Before reaction

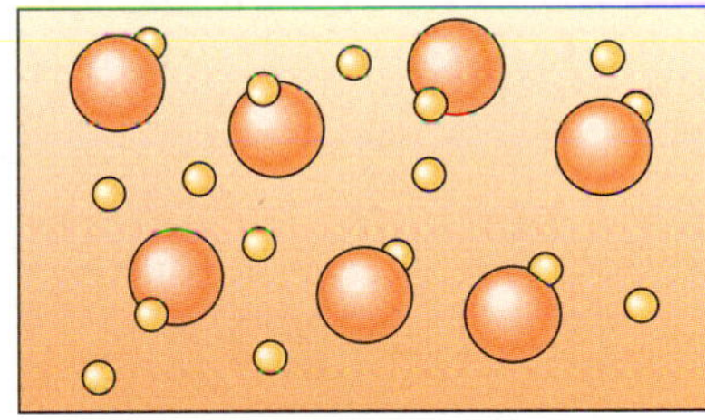

After reaction

Figure 3.7

Excess and limiting reactants. The initial mixture contains more molecules of one reactant than of the other. Since this reaction requires one molecule of each kind to form a product molecule, some of the excess reactant will still be left after the limiting reactant has been used up. The limiting reactant determines the amount of product.

EXAMPLE 3.24

Zinc metal reacts with chlorine gas according to the equation

$$Zn(s) + Cl_2(g) \rightarrow ZnCl_2(s)$$

(a) Identify the limiting and excess reactants in a mixture containing 5.00 mol Zn and 4.00 mol Cl_2. (b) How many moles of the excess reactant will remain after the reaction is completed?

SOLUTION

(a) The balanced equation states that 1 mol of chlorine reacts with 1 mol of zinc, so 4.00 mol of chlorine will react with 4.00 mol of zinc. There are 5.00 mol of zinc in the reaction mixture; hence, zinc is excess and chlorine is limiting.

(b) There will be 5.00 mol − 4.00 mol = 1.00 mol of zinc left over.

EXAMPLE 3.25

Magnesium burns in oxygen to form magnesium oxide:

$$2Mg(s) + O_2(g) \rightarrow 2MgO(s)$$

(a) Identify the limiting and excess reactants in a reaction mixture containing 2.50 mol of magnesium and 3.00 mol of oxygen. (b) How many moles of the excess reactant will be left when the reaction is over?

SOLUTION

(a) The equation states that 1 mol of oxygen reacts with 2 mol of magnesium. The amount of magnesium needed to react with 3.00 mol of O_2 is

$$3.00 \text{ mol } O_2 \times \frac{2 \text{ mol Mg}}{1 \text{ mol } O_2} = 6.00 \text{ mol Mg}$$

There are only 2.50 mol of magnesium in the reaction mixture, not enough to use up 3.00 mol of oxygen. Some oxygen will be left over: oxygen is excess and magnesium is limiting.

(b) Magnesium is the limiting reactant, so all of it will react with oxygen. The number of moles of oxygen consumed by 2.50 mol of Mg is

$$2.50 \text{ mol Mg} \times \frac{1 \text{ mol } O_2}{2 \text{ mol Mg}} = 1.25 \text{ mol } O_2$$

There were originally 3.00 mol of oxygen in the reaction mixture. The number of moles of oxygen remaining is 3.00 mol − 1.25 mol = 1.75 mol.

PRACTICE EXERCISE 3.26

White phosphorus burns in oxygen according to the equation

$$P_4(s) + 5O_2(g) \longrightarrow P_4O_{10}(s)$$

(a) Identify the limiting reactant in a reaction mixture containing 1.00 mol each of P_4 and O_2. (b) How many moles of the excess reactant will remain after the reaction is over?

EXAMPLE 3.26

Hydrogen sulfide and sulfur dioxide react on contact to form sulfur and water:

$$2H_2S(g) + SO_2(g) \longrightarrow 3S(s) + 2H_2O(l)$$

(a) If 5.00 g of H_2S gas is mixed with 5.00 g of SO_2 gas, which gas is the limiting reactant? **(b)** How many grams of the excess gas will remain at the end of the reaction?

SOLUTION

(a) We can find the limiting reactant in two steps.

Step 1: Convert the known masses to moles. The molar masses are 34.08 g/mol for H_2S and 64.06 g/mol for SO_2. The number of moles of each gas is

$$5.00 \text{ g } H_2S \times \frac{1 \text{ mol } H_2S}{34.08 \text{ g } H_2S} = 0.147 \text{ mol } H_2S$$

$$5.00 \text{ g } SO_2 \times \frac{1 \text{ mol } SO_2}{64.06 \text{ g } SO_2} = 0.0781 \text{ mol } SO_2$$

Step 2: Find the limiting reactant. This is done by assuming that one of the gases is limiting and determining if there is enough of the other gas to react with it. If there is, then our assumption is correct; if there is not, then the other gas is limiting. Let us assume that H_2S is limiting. If this is true, then all of the H_2S (0.147 mol) will be used up. The number of moles of SO_2 required to consume 0.147 mol of H_2S is obtained from the mole ratio in the balanced equation

$$0.147 \text{ mol } H_2S \times \frac{1 \text{ mol } SO_2}{2 \text{ mol } H_2S} = 0.0735 \text{ mol } SO_2$$

The amount of SO_2 in the original mixture, 0.0781 mol, is greater than 0.0735 mol. There is more than enough SO_2 to react with all of the H_2S, so our trial assumption was correct: H_2S is the limiting reagent. All of the H_2S will be consumed and some SO_2 will be left over. (If we had assumed that SO_2 was limiting, we would have found that there was not enough H_2S to react with it—the assumption would have been wrong. You should verify this by performing the calculation.)

In Example 3.26 sulfur is represented by its empirical formula S, rather than its molecular formula S_8. Using S simplifies the equation, but does not change the outcome of the calculation. (You can verify this by rewriting the equation using S_8 for elemental sulfur and reworking the example.)

(b) In Step 2 we determined that 0.0735 mol of SO_2 will react with all of the H_2S. The mixture originally contained 0.0781 mol of SO_2. If 0.0735 mol is consumed, then 0.0781 mol − 0.0735 mol = 0.0046 mol of SO_2 will remain at the end of the reaction. The number of grams of SO_2 remaining is

$$0.0046 \text{ mol } SO_2 \times \frac{64.06 \text{ g } SO_2}{1 \text{ mol } SO_2} = 0.29 \text{ g } SO_2$$

PRACTICE EXERCISE 3.27

Aqueous solutions of calcium chloride and sodium phosphate react according to the equation

$$3CaCl_2(aq) + 2Na_3PO_4(aq) \longrightarrow Ca_3(PO_4)_2(s) + 6NaCl(aq)$$

(a) Identify the limiting and excess reactants in a reaction mixture containing 50.0 g of $CaCl_2$ and 100.0 g of Na_3PO_4. (b) How many grams of the excess reactant will remain when the reaction is over?

The amount of product formed in a reaction is determined by the limiting reactant, as illustrated in the next example.

EXAMPLE 3.27

Calculate the number of grams of solid aluminum chloride ($AlCl_3$) that will form when a mixture containing 0.150 g of aluminum powder and 1.00 g of chlorine gas is allowed to react.

SOLUTION

Step 1: The balanced equation and the known and unknown quantities are

$$\begin{array}{ccc} 2Al(s) & + \ 3Cl_2(g) & \longrightarrow 2AlCl_3(s) \\ 0.150 \text{ g} & 1.00 \text{ g} & ? \text{ g} \end{array}$$

CHEMICAL INSIGHT

PHOSPHORUS, THE LIMITING ELEMENT

Living organisms grow, die, and decompose in a series of chemical reactions in which their elements are continuously recycled and reused. Six elements—carbon, oxygen, hydrogen, nitrogen, sulfur, and phosphorus—are primarily responsible for the form and structure of living tissue. All except phosphorus are abundant in the atmosphere, oceans, and lakes as N_2, O_2, H_2O, CO_2, and SO_2. Phosphorus does not occur in the atmosphere, and its salts are largely insoluble in water. Most phosphorus is recycled through living organisms by way of food chains. In a lake, for example, microorganisms feed on dead material containing phosphorus while larger organisms feed on each other. In this way, the limited supply of phosphorus is used over and over again.

Because the availability of phosphorus is limited, it often determines the total amount of life a body of water will support. As a limiting element, it also serves as a brake against runaway growth. This limiting function becomes clear when the limitation is removed. If large amounts of phosphate detergents and organic waste containing phosphorus are allowed to enter a lake, the amount of living material at first increases. Algae grow in profusion and cover the surface of the water. Eventually, however, the algae cut off sunlight from plants beneath the surface, and they begin to deplete the supply of dissolved oxygen. The death rate increases, and because decay involves oxidation, the increasing decay of dead material causes a further depletion of dissolved oxygen. The process continues until eventually everything dies, a catastrophe called *eutrophication* (Figure 3.8).

Figure 3.8

A lake in the process of eutrophication. An excess of nutrients such as phosphorus encourages an overgrowth of green algae, which depletes the oxygen supply.

The number of grams of $AlCl_3$ is calculated from the reactant that is completely consumed. The calculation is similar to those in previous examples, but there will be an extra step to determine which reactant is limiting.

Step 2: Convert the known masses to moles. The mass of 1 mol of aluminum is 26.98 g. The number of moles of aluminum is

$$0.150 \text{ g Al} \times \frac{1 \text{ mol Al}}{26.98 \text{ g Al}} = 0.00556 \text{ mol Al}$$

The mass of 1 mol of Cl_2 is $2 \times 35.45 \text{ g} = 70.90 \text{ g}$. The number of moles of chlorine is

$$1.00 \text{ g Cl}_2 \times \frac{1 \text{ mol Cl}_2}{70.90 \text{ g Cl}_2} = 0.0141 \text{ mol Cl}_2$$

Step 3: Use the mole ratios from the balanced equation to find the limiting reactant. Let us make the trial assumption that chlorine limits the reaction. The amount of aluminum that will use up 0.0141 mol of chlorine is

$$0.0141 \text{ mol Cl}_2 \times \frac{2 \text{ mol Al}}{3 \text{ mol Cl}_2} = 0.00940 \text{ mol Al}$$

There is only 0.00556 mol of aluminum in the original reaction mixture, less than the

0.00940 mol required, so some chlorine will remain. Our trial assumption was wrong: aluminum, not chlorine, is the limiting reactant. Therefore, we must use 0.00556 mol of aluminum to calculate the final mass of $AlCl_3$.

Step 4: The remainder of the calculation is based on the reaction equation, and our strategy will be:

$$0.00556 \text{ mol Al} \rightarrow \text{mol } AlCl_3 \rightarrow \text{g } AlCl_3$$

The mass of 1 mol of $AlCl_3$ is 133.3 g; the number of grams of $AlCl_3$ is

$$0.00556 \text{ mol Al} \times \frac{2 \text{ mol } AlCl_3}{2 \text{ mol Al}} \times \frac{133.3 \text{ g } AlCl_3}{1 \text{ mol } AlCl_3} = 0.741 \text{ g } AlCl_3$$

LEARNING HINT

Some chemists prefer to solve limiting reactant problems by calculating the amount of product that will form from *each* of the reactants. The *smaller* number will be correct, and the reactant that gives the smaller answer will be the limiting reactant.

PRACTICE EXERCISE 3.28

Sulfur burns in oxygen to form sulfur dioxide:

$$S(s) + O_2(g) \rightarrow SO_2(g)$$

How many grams of sulfur dioxide will form in a reaction mixture containing 100.0 g of sulfur and 200.0 g of oxygen?

3.9 THEORETICAL, ACTUAL, AND PERCENT YIELD

In Example 3.27 we calculated that 0.741 g of $AlCl_3$ would be produced when 0.150 g of aluminum powder reacts with excess chlorine gas. The amount of product calculated from a chemical equation, in this case, 0.741 g, is called the **theoretical yield**. If we were to actually do the experiment—that is, if we were to allow the mixture of aluminum and chlorine to react—we would probably collect less than 0.741 g of $AlCl_3$. For most reactions, the amount of product collected at the end of the reaction, the **actual yield**, is less than the theoretical yield. The discrepancy can be caused by various factors. The reaction itself might not be complete, or some of the reactants might be consumed by other reactions (side reactions) that were not considered in the calculation. In some cases, the product cannot be entirely separated from the final mixture, or some of it is lost in the process of purification. It is customary to report the **percent yield** of any product prepared in the laboratory:

$$\text{percent yield} = \frac{\text{actual yield}}{\text{theoretical yield}} \times 100\% \qquad (3.6)$$

EXAMPLE 3.28

The equation for the production of bromobenzene (C_6H_5Br) from bromine and benzene (C_6H_6) is

$$C_6H_6(l) + Br_2(l) \rightarrow C_6H_5Br(l) + HBr(g)$$

A 12.85-g yield of bromobenzene was obtained from 8.00 g of benzene and excess bromine. Calculate (a) the theoretical yield of bromobenzene in grams and (b) the percent yield.

SOLUTION

(a) The theoretical yield of bromobenzene is calculated in the usual way from the balanced equation. Our strategy will be:

$$8.00 \text{ g } C_6H_6 \longrightarrow \text{mol } C_6H_6 \longrightarrow \text{mol } C_6H_5Br \longrightarrow \text{g } C_6H_5Br$$

The molar masses of C_6H_6 and C_6H_5Br are 78.11 g/mol and 157.0 g/mol, respectively. The theoretical yield is

$$8.00 \text{ g } C_6H_6 \times \frac{1 \text{ mol } C_6H_6}{78.11 \text{ g } C_6H_6} \times \frac{1 \text{ mol } C_6H_5Br}{1 \text{ mol } C_6H_6} \times \frac{157.0 \text{ g } C_6H_5Br}{1 \text{ mol } C_6H_5Br}$$

$$= 16.1 \text{ g } C_6H_5Br$$

(b) The actual yield of bromobenzene was 12.85 g. The percent yield is obtained by substituting the actual and theoretical yields into Equation 3.6:

$$\text{percent yield} = \frac{\text{actual yield}}{\text{theoretical yield}} \times 100\%$$

$$= \frac{12.85 \text{ g}}{16.1 \text{ g}} \times 100\% = 79.8\%$$

PRACTICE EXERCISE 3.29

Copper(I) sulfide (Cu_2S) forms when copper is heated with excess sulfur:

$$2Cu(s) + S(s) \longrightarrow Cu_2S(s)$$

A first-year chemistry student heated 99.8 g of copper with excess sulfur. The student obtained 50.0 g of Cu_2S. Calculate the percent yield.

CHAPTER REVIEW

LEARNING OBJECTIVES BY SECTION

3.1 1. Identify the reactants and products in a chemical reaction.
2. State whether a given change is a physical change or a chemical reaction.
3. State whether a given property is a physical property or a chemical property.

3.2 1. State the laws of conservation of mass, constant composition, and multiple proportions, and explain these laws in terms of atomic theory.
2. Calculate the percentage composition of a compound given the appropriate data.

3.3 1. Calculate percentage composition from a formula.

3.4 1. Use atomic and molecular weights to calculate the formulas of simple molecules.
2. Write the empirical formula of a compound given its molecular formula.
3. Use elemental composition data and a table of atomic weights to calculate empirical formulas.
4. Use empirical formulas and molar masses to calculate molecular formulas.
5. Calculate percentage compositions and formulas from combustion data.

3.5 1. Write balanced equations for chemical reactions.

3.6 1. Write balanced equations for combination and decomposition reactions.
2. Write balanced equations for displacement reactions.
3. Predict the products and write balanced equations for the combustion of organic compounds.
4. Identify the substances oxidized and reduced in an oxidation–reduction reaction involving oxygen.
5. Write balanced equations for the reduction of oxide ores with carbon.

3.7 1. Use mole ratios to solve problems relating quantities of reactant and product.

3.8 1. Identify the limiting and excess reactants in a reaction mixture.
2. Perform stoichiometric calculations involving limiting and excess reactants.

3.9 1. Calculate the percent yield from the theoretical and actual yields.

KEY TERMS BY SECTION

3.1 Chemical property
Chemical reaction
Physical change
Physical property
Product
Reactant
Stoichiometry

3.2 Law of conservation of mass
Law of constant composition
Law of definite proportions
Law of multiple proportions
Percentage composition

3.4 Empirical formula

3.5 Balanced (as applied to a chemical equation)
Chemical equation
Coefficient (in a chemical equation)

3.6 Combination reaction
Combustion reaction
Decomposition reaction
Displacement reaction
Ore
Organic compound
Oxidation (in reactions involving oxygen)
Reduction (in reactions involving oxygen)

3.8 Excess reactant (or reagent)
Limiting reactant (or reagent)

3.9 Actual yield
Percent yield
Theoretical yield

IMPORTANT EQUATIONS

3.6 $$\text{percent yield} = \frac{\text{actual yield}}{\text{theoretical yield}} \times 100\%$$

FINAL EXERCISES

Answers to exercises with blue numbers are given in Appendix D.

PART A. QUESTIONS AND PROBLEMS BY SECTION

Physical and Chemical Change (Section 3.1)

3.1 Which of the following changes are physical changes and which are chemical reactions?
(a) A toaster filament becomes red hot.
(b) The exposed flesh of an apple turns brown.
(c) Grease catches fire and burns.
(d) A grease spot disappears when treated with cleaning fluid.

3.2 Which of the following changes are physical changes and which are chemical reactions?
(a) Heated oil smokes and turns brown.
(b) A silver bracelet becomes tarnished.
(c) A piece of paper is torn into little bits.
(d) A snowball melts.

3.3 Which type of property, physical or chemical, is described by each of the following statements?
(a) Oil floats on water.
(b) Water can be decomposed by an electric current into hydrogen and oxygen.
(c) Iron rusts.
(d) Iron melts at 1535°C.

3.4 Which type of property, physical or chemical, is described by each of the following statements?
(a) Oxygen liquefies at −183°C.
(b) Baking soda neutralizes vinegar.
(c) Sugar cubes dissolve in coffee.
(d) Gunpowder explodes when detonated.

Laws of Chemical Combination (Section 3.2)

3.5 Explain the following laws in terms of atomic theory:
(a) law of constant composition
(b) law of conservation of mass

3.6 Explain the following laws in terms of atomic theory:
(a) law of definite proportions
(b) law of multiple proportions

3.7 A 7.00-g sample of an oxide of mercury decomposes on heating to give oxygen gas and 6.48 g of liquid mercury.
(a) How many grams of oxygen are produced?
(b) Calculate the percentage composition of the oxide.

3.8 A 25.0-mg sample of sodium peroxide contains 14.74 mg of sodium. Calculate the percentage composition of sodium peroxide.

3.9 Oxygen combines with 3.05 g of iron to form 4.36 g of an iron oxide. Calculate the percentage composition of the oxide.

3.10 A 0.742-g sample of calcium metal reacts with excess chlorine to form 2.056 g of calcium chloride. Calculate the percentage composition of calcium chloride.

3.11 Show how the sulfides of copper, Cu_2S and CuS, illustrate the law of multiple proportions.

3.12 Three different compounds contain 46.7, 30.45, and 25.9% nitrogen by mass. The only other element in each of the compounds is oxygen. Show how these compounds illustrate the law of multiple proportions.

Percentage Composition from a Formula (Section 3.3)

3.13 Calculate the percent of each element in the following compounds:
(a) oxalic acid, $H_2C_2O_4$, a compound used to bleach wood
(b) stannous fluoride, SnF_2, a toothpaste additive
(c) acetone, CH_3COCH_3, an industrial solvent
(d) phosgene, $COCl_2$, a toxic war gas

3.14 Calculate the percentage composition of the following compounds:
(a) paradichlorobenzene, $C_6H_4Cl_2$, used in moth crystals
(b) thiourea, $CS(NH_2)_2$, used as a photographic fixer
(c) picric acid, $C_6H_2(NO_2)_3OH$, used in explosives
(d) cortisone, $C_{21}H_{28}O_5$, an anti-inflammatory agent

Empirical and Molecular Formulas (Section 3.4)

3.15 What is the difference between an empirical formula and a molecular formula? Under what circumstances will these formulas be the same?

3.16 In your own words, describe the steps for calculating an empirical formula from elemental composition data. What additional information is needed for calculating a molecular formula?

3.17 Write an empirical formula for each of the following substances:
(a) hydrazine, N_2H_4
(b) acetylene, C_2H_2
(c) sulfur, S_8

3.18 Write an empirical formula for each of the following substances:
(a) tetraborane, B_4H_{10}
(b) acetic acid, CH_3COOH
(c) lactic acid, $CH_3CHOHCOOH$

3.19 The molecules in pure crystalline boron have a molecular weight of 129.7 u. What is the formula of a boron molecule?

3.20 Some of the molecules in sulfur vapor have a molecular weight of 64.1 u, others have a molecular weight of 128.2 u, and still others have a molecular weight of 256.5 u. Calculate the formula of each of these molecules.

3.21 A 20.00-g sample of "king's gold," a sulfide of arsenic, contains 12.18 g of arsenic.
(a) Calculate the empirical formula of the sulfide.
(b) Is enough information given to determine its molecular formula? Explain.

3.22 The combustion of 9.29 g of phosphorus produced 21.29 g of a phosphorus oxide. Calculate the empirical formula of the oxide.

3.23 Calculate the empirical formula of the compound formed when 6.58 g of nitrogen combines with 1.42 g of hydrogen.

3.24 The percentage composition of pyrophosphoric acid is 2.27% H, 34.81% P, and 62.93% O by mass. Calculate the empirical formula of pyrophosphoric acid.

3.25 Calculate empirical formulas of the three different compounds whose percentage compositions by mass are as follows:
(a) K = 43.18%, Cl = 39.15%, O = 17.67%
(b) K = 31.90%, Cl = 28.93%, O = 39.17%
(c) K = 28.22%, Cl = 25.59%, O = 46.19%

3.26 Calculate empirical formulas for the compounds described in
(a) Exercise 3.7
(b) Exercise 3.8
(c) Exercise 3.9
(d) Exercise 3.10

3.27 The molar mass of a certain hydrocarbon (a substance containing only hydrogen and carbon) is 56.1 g/mol. Its percentage composition is 85.6% C and 14.4% H by mass. Calculate the molecular formula of the hydrocarbon.

3.28 The percentage composition of glucose is 40.00% C, 6.72% H, and 53.28% O by mass. The molar mass of glucose is 180.2 g/mol. Calculate the molecular formula of glucose.

3.29 The molar mass of D-glucuronic acid is 194 g/mol. Its percentage composition is 37.12% C, 57.69% O, and 5.188% H by mass. Calculate the molecular formula of D-glucuronic acid.

3.30 A compound contains 76.03% iodine and 23.97% oxygen by mass. Its molar mass is 333.8 g/mol. Calculate the molecular formula of the compound.

3.31 A 2.500-g sample of "red lead" (an oxide of lead) contains 2.266 g of lead. The molar mass of the oxide is 685.6 g/mol.
(a) Calculate the formula of "red lead."
(b) Can one tell from the given information whether "red lead" is a molecular compound? Explain.

3.32 Naphthalene, a widely used industrial compound, contains 93.71% C and 6.29% H by mass. Its molar mass is 128.16 g/mol.
(a) Calculate the formula of naphthalene.
(b) Can one tell from the given information whether naphthalene is a molecular compound? Explain.

3.33 The combustion of 10.00 mg of styrene, a compound consisting only of carbon and hydrogen, produced 33.80 mg of CO_2 and 6.92 mg of H_2O. Determine (a) the percentage composition and (b) the empirical formula of styrene.

3.34 Benzoic acid, a compound found in berries, contains carbon, hydrogen, and oxygen. Combustion of a 25.0-mg sample of benzoic acid produced 63.0 mg of carbon dioxide and 11.1 mg of water. Calculate (a) the percentage composition and (b) the empirical formula of benzoic acid.

3.35 A compound used as a rocket fuel contains carbon, hydrogen, and nitrogen. A 0.7148-g sample of the compound produced 1.048 g of CO_2 and 0.8573 g of H_2O in a combustion analysis. The molar mass of the compound is 60.10 g/mol. Determine the molecular formula of the rocket fuel.

3.36 The molar mass of vitamin C, a compound composed of carbon, hydrogen, and oxygen, is 176.12 g/mol. Combustion of a 35.5-mg sample produced 53.3 mg of CO_2 and 14.4 mg of H_2O. Determine the molecular formula of vitamin C.

Chemical Equations (Section 3.5)

3.37 What is meant by the phrase *balanced equation*? Explain, in terms of atomic theory, why equations should be balanced.

3.38 What is the difference between a *subscript* and a *coefficient* in a chemical equation? Can a subscript be changed when balancing an equation? Why or why not?

3.39 Write a word equation for each of the following reactions:
(a) Sugar decomposes into carbon and water vapor when it is heated.
(b) Hydrochloric acid and sodium hydroxide react to form sodium chloride and water.
(c) Green plants manufacture glucose and oxygen from carbon dioxide and water.

3.40 Balance the following equations for reactions in which compounds form or decompose:
(a) $H_2(g) + I_2(g) \longrightarrow HI(g)$
(b) $N_2O(g) \longrightarrow N_2(g) + O_2(g)$
(c) $As(s) + H_2(g) \longrightarrow AsH_3(g)$
(d) $Al(s) + O_2(g) \longrightarrow Al_2O_3(s)$
(e) $KClO_3(s) \longrightarrow KCl(s) + O_2(g)$
(f) $KClO_3(s) \longrightarrow KClO_4(s) + KCl(s)$

3.41 Balance the following equations for reactions with oxygen:
(a) $C_6H_6(l) + O_2(g) \longrightarrow CO_2(g) + H_2O(g)$
(b) $ZnS(s) + O_2(g) \longrightarrow ZnO(s) + SO_2(g)$
(c) $FeO(s) + O_2(g) \longrightarrow Fe_2O_3(s)$
(d) $CS_2(l) + O_2(g) \longrightarrow CO_2(g) + SO_2(g)$

3.42 Balance the following equations for reactions involving ionic compounds:
(a) $CaCl_2(aq) + Na_3PO_4(aq) \longrightarrow Ca_3(PO_4)_2(s) + NaCl(aq)$
(b) $AgNO_3(aq) + K_2CO_3(aq) \longrightarrow Ag_2CO_3(s) + KNO_3(aq)$
(c) $Pb(NO_3)_2(aq) + KCl(aq) \longrightarrow PbCl_2(s) + KNO_3(aq)$
(d) $NaBr(aq) + Cl_2(aq) \longrightarrow NaCl(aq) + Br_2(aq)$

3.43 Balance the following equations in which water is a reactant:
(a) $P_4O_{10}(s) + H_2O(l) \longrightarrow H_3PO_4(aq)$
(b) $XeF_6(s) + H_2O(l) \longrightarrow XeO_3(s) + HF(g)$
(c) $K(s) + H_2O(l) \longrightarrow KOH(aq) + H_2(g)$
(d) $PCl_5(s) + H_2O(l) \longrightarrow H_3PO_4(aq) + HCl(aq)$
(e) $PBr_3(l) + H_2O(l) \longrightarrow H_3PO_3(aq) + HBr(aq)$

3.44 Write balanced equations for the following reactions:
(a) Nitrogen gas plus hydrogen gas yields ammonia gas.
(b) Sodium metal reacts with water to give hydrogen gas and aqueous sodium hydroxide.
(c) Acetylene (C_2H_2) burns in oxygen to give carbon dioxide and water vapor.
(d) Silver nitrate and calcium chloride solutions give solid silver chloride and dissolved calcium nitrate.

Balance the following equations:
(a) fermentation of glucose,

$C_6H_{12}O_6(aq) \rightarrow C_2H_5OH(aq) + CO_2(g)$

(b) photosynthetic production of glucose,
$CO_2(g) + H_2O(l) \rightarrow C_6H_{12}O_6(aq) + O_2(g)$

3.46 The ultrapure silicon used in computer chips is obtained from silicon dioxide (sand) in a three-step process that uses carbon, chlorine, and hydrogen as reagents. Balance the equation for each step.

Step 1: Silicon dioxide is heated with carbon.

$SiO_2(s) + C(s) \rightarrow Si(s) + CO(g)$

Step 2: Crude silicon is converted to silicon tetrachloride.

$Si(s) + Cl_2(g) \rightarrow SiCl_4(l)$

Step 3: Silicon tetrachloride is treated with hydrogen.

$SiCl_4(l) + H_2(g) \rightarrow Si(s) + HCl(g)$

Some Chemical Reactions (Section 3.6)

3.47 Define and give at least one example of
(a) a combination reaction
(b) a decomposition reaction

3.48 Define and give at least one example of
(a) a displacement reaction
(b) a combustion reaction

3.49 Which combustion product of carbon, CO or CO_2, is more likely to form at very high combustion temperatures?

3.50 **(a)** What products usually form when an organic fuel is burned?
(b) Name and give the formulas of two common air pollutants that may be produced when an organic fuel is burned.

3.51 State whether each of the following equations represents a combination, a decomposition, or a displacement reaction.
(a) $Ca(s) + 2H_2O(l) \rightarrow H_2(g) + Ca(OH)_2(aq)$
(b) $2AgBr(s) \rightarrow 2Ag(s) + Br_2(l)$
(c) $PbO(s) + C(s) \rightarrow Pb(l) + CO(g)$
(d) $2K(s) + Cl_2(g) \rightarrow 2KCl(s)$

3.52 State whether each of the following equations represents a combination, a decomposition, or a displacement reaction.
(a) $CaO(s) + CO_2(g) \rightarrow CaCO_3(s)$
(b) $Cu(s) + 2AgNO_3(aq) \rightarrow Cu(NO_3)_2(aq) + 2Ag(s)$
(c) $NH_4Cl(s) \rightarrow NH_3(g) + HCl(g)$
(d) $C_2H_4(g) + H_2(g) \rightarrow C_2H_6(g)$

3.53 Write a balanced equation for the combustion in excess oxygen of each of the following compounds:
(a) liquid isooctane, C_8H_{18}
(b) liquid isopropyl alcohol, $CH_3CHOHCH_3$
(c) solid glucose, $C_6H_{12}O_6$
(d) liquid acetone, CH_3COCH_3

3.54 Write a balanced equation for the combustion in excess oxygen of each of the following compounds:
(a) liquid carbon disulfide, CS_2
(b) gaseous ammonia, NH_3 (*Hint:* NO(g) is a product)
(c) gaseous silane, SiH_4 (*Hint:* $SiO_2(s)$ is a product)
(d) liquid methylethyl sulfide, $CH_3SC_2H_5$

3.55 Write a balanced equation for the combustion of carbon in (a) excess oxygen and (b) limited oxygen.

3.56 Write a balanced equation for the combustion in excess oxygen of each of the following fuels:
(a) carbon
(b) liquid ethanol
(c) propane gas, C_3H_8
(d) liquid butane, $CH_3CH_2CH_2CH_3$

3.57 Germanium, an element used in making semiconductors, is obtained by heating germanium dioxide (GeO_2) with carbon.
(a) Assume carbon monoxide is also formed, and write a balanced equation for the reaction.
(b) Which substance is oxidized? Which is reduced?

3.58 Manganese metal can be obtained by heating manganese dioxide (MnO_2) with carbon monoxide.
(a) Assume that carbon dioxide is also formed, and write a balanced equation for the reaction.
(b) Which substance is oxidized? Which is reduced?

3.59 Write equations for the high-temperature reduction of each of the following ores with coke:
(a) hematite, Fe_2O_3
(b) cassiterite, SnO_2
(c) magnetite, Fe_3O_4
(d) sphalerite, ZnS (two equations)

3.60 Hydrogen, aluminum, and carbon are often used to reduce metal oxides to metals. At high temperatures they form $H_2O(g)$, $Al_2O_3(s)$, and CO(g), respectively. Write a balanced equation for each of the following reductions.
(a) Cr_2O_3 to chromium metal using aluminum
(b) Fe_2O_3 to iron metal using hydrogen
(c) Bi_2O_3 to bismuth metal using carbon

Reaction Stoichiometry (Section 3.7)

3.61 A 0.50-mol sample of solid $KClO_3$ decomposes to give solid KCl and gaseous oxygen.
(a) Write the equation for the reaction.
(b) How many moles of KCl and how many moles of oxygen will be produced?
(c) How many grams of KCl and how many grams of oxygen will be produced?

3.62 When heated with oxygen, bismuth metal reacts to form

bismuth(III) oxide, Bi_2O_3.
(a) Write the equation for the reaction.
(b) How many grams of oxygen will combine with 30.0 g of bismuth?

3.63 How many grams of oxygen are required to burn 75.0 g of ammonia? The equation is

$$4NH_3(g) + 5O_2(g) \rightarrow 4NO(g) + 6H_2O(g)$$

3.64 The *Deacon process* has been used to prepare chlorine gas from HCl. The equation is

$$4HCl(g) + O_2(g) \rightarrow 2H_2O(g) + 2Cl_2(g)$$

How many kilograms of HCl are required to produce 1.75 metric tons of Cl_2?

3.65 The equation for the combustion of glucose is

$$C_6H_{12}O_6(s) + 6O_2(g) \rightarrow 6CO_2(g) + 6H_2O(g)$$

How many (a) moles of CO_2 and (b) grams of water vapor will form when 15.0 g of glucose is burned in excess oxygen?

3.66 When chlorine gas is bubbled into hot potassium hydroxide solution, it reacts according to the equation

$$3Cl_2(g) + 6KOH(aq) \rightarrow 5KCl(aq) + KClO_3(aq) + 3H_2O(l)$$

(a) How many grams of KOH are needed to form 50.0 g of $KClO_3$?
(b) How many grams of Cl_2 will be consumed during the reaction of Part (a)?

3.67 How many grams of oxygen are required to convert 500 g of SO_2 to SO_3?

3.68 How many kilograms of NH_3 will be produced when 25.0 kg of H_2 reacts with excess N_2?

3.69 The formula of formaldehyde (a tissue preservative) is CH_2O. How many grams of formaldehyde contain 85.5 g of carbon?

3.70 The formula of nicotine is $C_{10}H_{14}N_2$. How many milligrams of nitrogen are in 50.0 mg of nicotine (the lethal dose for humans)?

3.71 A 26.73-g silver dollar is dissolved in nitric acid. When hydrochloric acid is added to the solution, all of the silver forms 31.96 g of solid silver chloride (AgCl). Calculate the percentage of silver in the coin.

3.72 How many moles of oxygen are required for the complete combustion of 0.150 mL of liquid methanol (CH_3OH)? The density of methanol is 0.791 g/mL.

Limiting Reactants (Section 3.8)

3.73 Identify the limiting and excess reactants in each of the following situations: (a) a candle burns in the open air until nothing is left of it and (b) a candle burns in a closed jar until it goes out.

3.74 The equation for the reaction of ammonia and hydrogen chloride gases is

$$NH_3(g) + HCl(g) \rightarrow NH_4Cl(s)$$

A reacting mixture contains 3.00 mol of NH_3 and 5.00 mol of HCl.
(a) Find the limiting reactant.
(b) How many moles of NH_4Cl will form?
(c) How many moles of excess reactant will remain?

3.75 Refer to the reaction in Exercise 3.66. A reacting mixture contains 6.00 mol of chlorine and 8.00 mol of potassium hydroxide.
(a) Find the limiting reactant.
(b) How many moles of $KClO_3$ will form?
(c) How many moles of excess reactant will remain?

3.76 The following reaction occurs when aluminum is heated with oxygen:

$$4Al(s) + 3O_2(g) \rightarrow 2Al_2O_3(s)$$

A reacting mixture contains 50.0 kg of aluminum and 50.0 kg of oxygen.
(a) Find the limiting reactant.
(b) How many grams of aluminum oxide will form?
(c) How many grams of excess reactant will remain?

3.77 How many grams of phosphoric acid will form when 30.0 g of P_4O_{10} is mixed with 75.0 g of water? The reaction is

$$P_4O_{10}(s) + 6H_2O(l) \rightarrow 4H_3PO_4(aq)$$

3.78 Calcium carbonate dissolves in hydrochloric acid according to the equation

$$CaCO_3(s) + 2HCl(aq) \rightarrow CaCl_2(aq) + H_2O(l) + CO_2(g)$$

(a) How many grams of calcium chloride will form if 40.0 g of $CaCO_3$ is mixed with 0.500 mol of HCl?
(b) How many grams of $CaCO_3$, if any, will remain undissolved in the reaction of Part (a)?

3.79 Hydrogen sulfide gas burns in oxygen to form sulfur dioxide and water:

$$2H_2S(g) + 3O_2(g) \rightarrow 2SO_2(g) + 2H_2O(g)$$

If a mixture containing 17.0 g of H_2S and 200 g of O_2 is ignited, how many moles of each of the four gases will be in the final mixture?

3.80 The equation for the combustion of carbon disulfide is

$$CS_2(l) + 3O_2(g) \rightarrow CO_2(g) + 2SO_2(g)$$

If a mixture containing 40.0 g of CS_2 and 30.0 g of O_2 is allowed to burn, how many grams of each of the four substances will be in the final mixture?

Percent Yield (Section 3.9)

3.81 Caustic soda (NaOH) is prepared commercially by passing an electric current through a concentrated solution of sodium chloride in water:

$$2NaCl(aq) + 2H_2O(l) \rightarrow 2NaOH(aq) + H_2(g) + Cl_2(g)$$

(a) Calculate the theoretical yield of caustic soda if 125 kg of NaCl is electrolyzed.
(b) Calculate the percent yield if the electrolysis in Part (a) produces 55.4 kg of caustic soda.

3.82 Chlorobenzene (C_6H_5Cl), a starting material in the production of aspirin and other compounds, is prepared from benzene (C_6H_6) by the following reaction:

$$C_6H_6(l) + Cl_2(g) \rightarrow C_6H_5Cl(l) + HCl(g)$$

A 10.0-kg sample of benzene treated with excess chlorine gas yields 10.4 kg of chlorobenzene. Calculate the percent yield of chlorobenzene.

3.83 Freon-12 (CCl_2F_2) is a gas that has been used as a refrigerant. It is prepared by the reaction between carbon tetrachloride and antimony trifluoride:

$$3CCl_4(l) + 2SbF_3(s) \rightarrow 3CCl_2F_2(g) + 2SbCl_3(s)$$

If the percent yield is 72.0%, how many grams of antimony trifluoride must be treated with excess carbon tetrachloride to obtain an actual yield of 25.0 grams of Freon-12?

3.84 Hydrogen chloride is prepared commercially by the reaction of sodium chloride with concentrated sulfuric acid:

$$NaCl(s) + H_2SO_4(aq) \rightarrow NaHSO_4(s) + HCl(g)$$

If the percent yield is 81.5%, how many grams of HCl will be obtained by treating 25.0 kg of NaCl with excess sulfuric acid?

3.85 Acetic acid (CH_3COOH) can be prepared by heating methanol with carbon monoxide in the presence of a catalyst:

$$CH_3OH(l) + CO(g) \rightarrow CH_3COOH(l)$$

If a mixture containing 105 g of methanol and 75.0 g of carbon monoxide is allowed to react, calculate (a) the theoretical yield and (b) the actual yield of acetic acid. Assume that the percent yield for this reaction is 88.0%.

3.86 Titanium, an important structural metal used in the aircraft industry, can be prepared by reacting titanium tetrachloride with molten magnesium:

$$TiCl_4(g) + 2Mg(l) \rightarrow Ti(s) + 2MgCl_2(l)$$

When 180 kg of $TiCl_4$ is mixed with 55.0 kg of molten magnesium, 40.4 kg of titanium is produced. Calculate (a) the theoretical yield and (b) the percent yield of titanium.

PART B. MISCELLANEOUS QUESTIONS AND PROBLEMS

3.87 How can one tell whether an observed change is a physical change or a chemical reaction?

3.88 Helium does not react with oxygen. Water does not support combustion. Do these observations describe chemical properties of helium and water? Discuss whether a listing of chemical properties should include reactions that a substance does *not* undergo.

3.89 Show how the following pairs of compounds illustrate the law of multiple proportions:
(a) $TiBr_2$ and $TiBr_4$
(b) $AgIO_3$ and $AgIO_4$
(c) NO and N_2O_3
(d) N_2O_5 and N_2O_4

3.90 The formulas of glucose and arabinose are $C_6H_{12}O_6$ and $C_5H_{10}O_5$, respectively. Could you distinguish between these two substances by finding their percentage compositions? Explain.

3.91 Why is the process for obtaining a metal from its ore often referred to as *reduction*?

3.92 Most substances evolve energy when they combine with oxygen.
(a) In what form is energy released when glucose burns in air?
(b) What forms of energy are obtained when glucose combines with oxygen in the body?

3.93 Many decomposition reactions require an input of energy.
(a) Which substance mentioned in this chapter is decomposed simply by heating?
(b) Which substance requires electrical energy for its decomposition?

3.94 Find the number of milligrams of
(a) sulfur in a 500 mg tablet of sulfadiazine ($C_{10}H_{10}N_4O_2S$), an antimicrobial drug
(b) calcium in 1.00 g of hydroxyapatite ($Ca_{10}(PO_4)_6(OH)_2$), the principal ingredient of tooth enamel

3.95 Hemoglobin, an iron-containing compound that transports oxygen from the lungs to the tissues, has an approximate molecular weight of 64,500 u and contains 0.35% iron by mass. How many iron atoms are in a hemoglobin molecule?

3.96 The isotope deuterium, ${}^2_1H$, a key atom in the development of atomic power, comprises 0.015% by mass of naturally occurring hydrogen. Estimate the mass of deuterium per metric ton (1000 kg) of water.

3.97 A compound containing only carbon and hydrogen is completely burned in air to yield 6.60 g CO_2 and 4.10 g H_2O. Calculate its empirical formula.

3.98 Hydrogen gas is passed over 10.00 g of a hot copper oxide. The oxide is completely converted to copper metal, and 2.26 g of water are formed.

(a) Find the percentage composition of the copper oxide.
(b) Find its empirical formula.
(c) Write a balanced equation for the reaction.

3.99 Compounds A, B, and C are composed of copper and oxygen. Compound A contains 79.9% copper. A 1.00-g sample of compound B contains 201 mg of oxygen. A 16.0-g sample of compound C leaves a residue of 12.8 g of copper after reduction with hydrogen. Are A, B, and C the same or different compounds?

3.100 How many grams of air (23% oxygen by mass) are required for the complete combustion of a 120-g charcoal briquet containing 95% carbon and 5% noncombustible materials?

3.101 As the earth cooled, many of its elements combined with oxygen to form the rocks and minerals of its crust. How many kilograms of oxygen are in 5.00 kg of (a) quartz (SiO_2) and (b) limestone ($CaCO_3$)?

3.102 Sodium carbonate (Na_2CO_3), also known as *washing soda,* can be prepared by heating sodium hydrogen carbonate:

$$2NaHCO_3(s) \rightarrow Na_2CO_3(s) + CO_2(g) + H_2O(g)$$

If the $NaHCO_3$ contains 8.0% unreactive impurities, how many grams of it would be needed to prepare 100.0 g of sodium carbonate? Assume that the percent yield for the reaction is 93%.

3.103 A certain mixture contains 98.0% sulfuric acid and 2.0% water by mass. Its density is 1.84 g/mL. How many grams of ammonia are required to convert all of the sulfuric acid in 35.5 mL of this mixture to ammonium sulfate? The equation for the reaction is

$$H_2SO_4(aq) + 2NH_3(g) \rightarrow (NH_4)_2SO_4(aq)$$

3.104 Carbon dioxide gas forms when acetic acid is added to sodium hydrogen carbonate:

$$CH_3COOH(aq) + NaHCO_3(s) \rightarrow CH_3COONa(aq) + CO_2(g) + H_2O(l)$$

How many grams of carbon dioxide would be produced by adding 100 g of vinegar containing 5.00% acetic acid by mass to a 5.00-g sample of baking soda (100% $NaHCO_3$)?

3.105 A minor source of elemental iodine is Chile saltpeter, which contains about 0.20% sodium iodate ($NaIO_3$) by mass. The iodine is obtained by treating the iodate with sodium hydrogen sulfite ($NaHSO_3$):

$$2NaIO_3(aq) + 5NaHSO_3(aq) + 3NaCl(aq) \rightarrow I_2(s) + 5Na_2SO_4(aq) + 3HCl(aq) + H_2O(l)$$

If the percent yield is 82%, how many kilograms of Chile saltpeter would be needed for the production of 1.00 kg of I_2?

3.106 Sodium hydrogen carbonate decomposes when it is heated:

$$2NaHCO_3(s) \rightarrow Na_2CO_3(s) + CO_2(g) + H_2O(g)$$

A 27.0-mg sample of impure sodium hydrogen carbonate loses 8.22 mg on heating. Calculate the percentage of sodium hydrogen carbonate in the original sample. Assume that the impurities do not vaporize when heated.

3.107 If the percent yield of iron is 78% when hematite (Fe_2O_3) is reduced to iron in a blast furnace and 72% when magnetite (Fe_3O_4) is reduced, which ore would yield the greater amount of iron per metric ton?

3.108 A 10.00-g sample of a certain magnesium–aluminum alloy reacted completely with hydrochloric acid (HCl) to produce 0.4500 mol of hydrogen gas. Find the mass of each metal in the alloy. The reactions are

$$Mg(s) + 2HCl(aq) \rightarrow MgCl_2(aq) + H_2(g)$$

$$2Al(s) + 6HCl(aq) \rightarrow 2AlCl_3(aq) + 3H_2(g)$$

3.109 A 24.00-g sample of a magnesium carbonate–calcium carbonate mixture was converted by strong heating into 12.00 g of mixed magnesium and calcium oxides. Find the mass of magnesium carbonate in the original mixture. The reactions are

$$MgCO_3(s) \rightarrow MgO(s) + CO_2(g)$$

$$CaCO_3(s) \rightarrow CaO(s) + CO_2(g)$$

CHAPTER 4

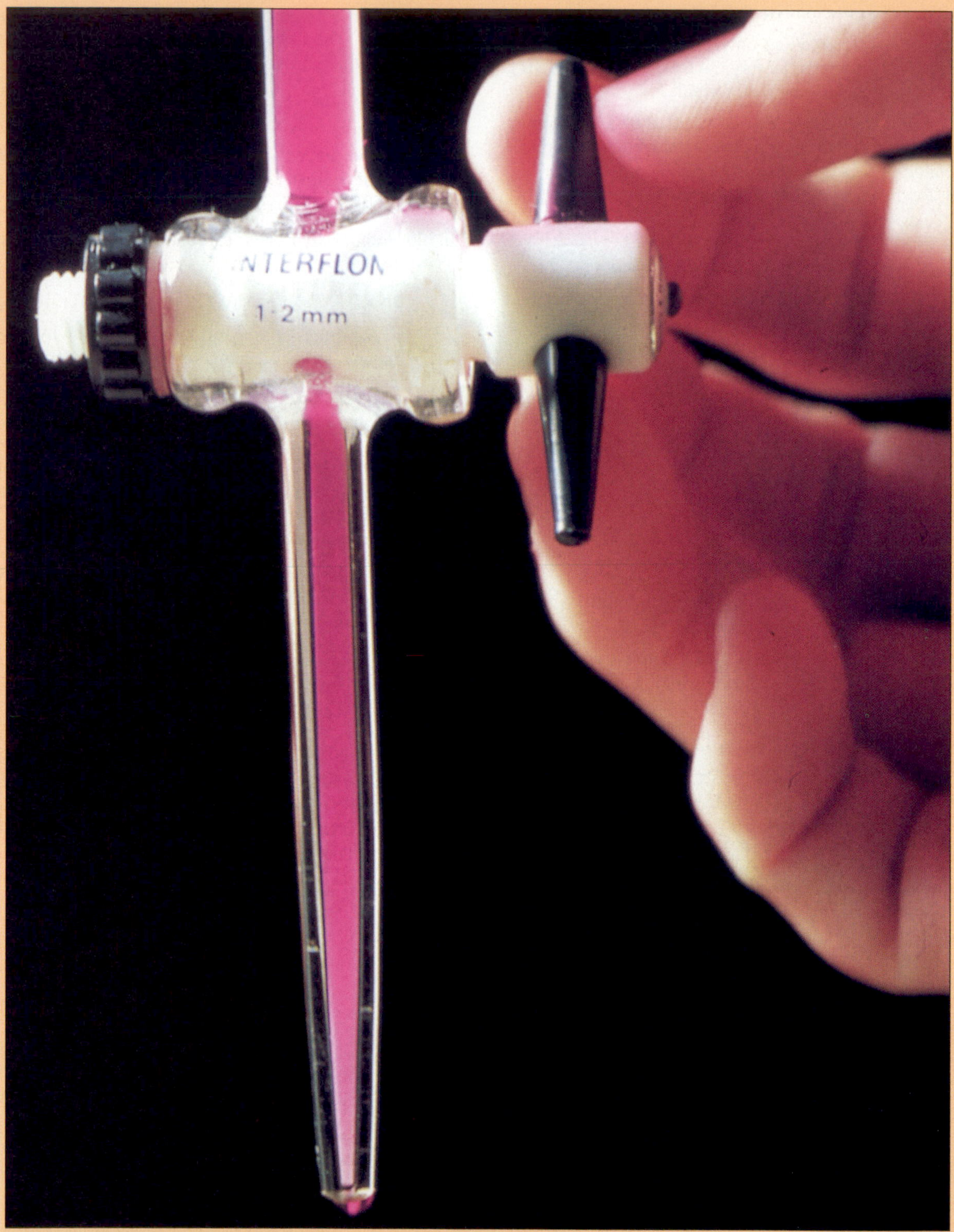

Close-up of a laboratory buret filled with purple potassium permanganate ($KMnO_4$) solution. Burets are widely used for dispensing precisely measured volumes of liquids during volumetric analyses (Section 4.6).

SOLUTIONS AND SOLUTION STOICHIOMETRY

OUTLINE

PREVIEW

Solutions, in which dissolved molecules and ions move about and collide with each other, provide the locale for most of the chemical reactions in our environment. The atmosphere is a solution of gases, and the ocean is a solution containing water, salt, and many other substances. The atmosphere and waters of primitive earth probably contained more compounds of carbon and nitrogen than the atmosphere and oceans of the earth today, and many scientists believe that the chemicals of life, and even life itself, originated from the interaction of solar energy and lightning with rich concentrations of dissolved matter.

Imagine shaking some salt and sand together. The resulting material feels like sand and tastes like salt. With a magnifying glass we can see the separate crystals of salt and sand, and when we add water, the salt dissolves, leaving the sand behind. The two components retain their individual properties; merely shaking salt and sand together does not produce a new substance. Sand and salt form a *mixture* (p. 55) whose components can be separated by physical means.

The composition of a mixture is variable; we can take 2 parts salt to 1 part sand, 3.7 parts salt to 2.8 parts sand, and so forth. Recall (see Section 2.5 and Figure 4.1) that a mixture is not a compound. Compounds have fixed compositions, and their properties are different from those of the parent elements; furthermore, compounds can be decomposed only through chemical change. Virtually all natural materials, including air, seawater, soil, and rock, are mixtures of many different elements and compounds.

4.1 SOLUTIONS

A **heterogeneous mixture** is one in which the components are not uniformly mixed. Bits of separate components can be seen either with the naked eye or under a micro-

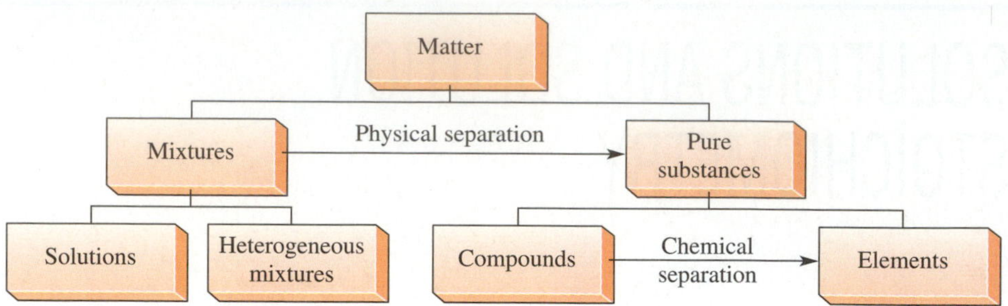

Figure 4.1
Samples of matter can be classified into mixtures and pure substances. A pure substance may be either an element or a compound. A mixture may be homogeneous (a solution) or heterogeneous.

scope (Figure 4.2). A mixture of salt and sand is heterogeneous, and wood, concrete, living tissue, and natural fluids such as blood and milk are also heterogeneous mixtures. Some heterogeneous mixtures appear uniform at first glance, but close inspection shows that some components are segregated in minute grains or droplets.

A **solution** is a **homogeneous mixture** consisting of one or more substances uniformly dispersed as separate atoms, molecules, or ions throughout another substance. Solutions can be gaseous, liquid, or solid. Air is a gaseous solution consisting of nitrogen, oxygen, and lesser amounts of other gases. Seawater is a liquid solution containing water, salt and many other dissolved substances. Alloys such as brass (copper and zinc) or pewter (tin and lead) are solid solutions. Other examples are given in Table 4.1. The physical properties of a solution are uniform throughout. Liquid and gaseous solutions may be colorless or colored, but they always appear homogeneous and clear. Turbidity or cloudiness would indicate the presence of suspended particles that are not actually dissolved (Figure 4.3).

The component that determines whether a solution is solid, liquid, or gaseous is called the **solvent**; any other component is a **solute**. In a cup of tea, for example, water is the solvent. The subtle flavor of the tea is provided by numerous solutes

TABLE 4.1 Types of Solutions

Solute	Solvent[a]	Examples
Gas	Gas	Air (oxygen in nitrogen)
Gas	Liquid	Carbonated water (CO_2 in water) Aqueous ammonia (NH_3 in water)
Gas	Solid	Hydrogen in platinum or palladium
Liquid	Liquid	Vinegar (acetic acid in water) Alcohol in water
Liquid	Solid	Mercury amalgams (mercury in silver or gold)
Solid	Liquid	Sugar or salt in water Tincture of iodine (I_2 in ethanol)
Solid	Solid	Sterling silver (copper in silver) Stainless steel (nickel and chromium in iron) Glass (sodium and calcium silicates)

[a]The solvent determines whether the solution is solid, liquid, or gaseous.

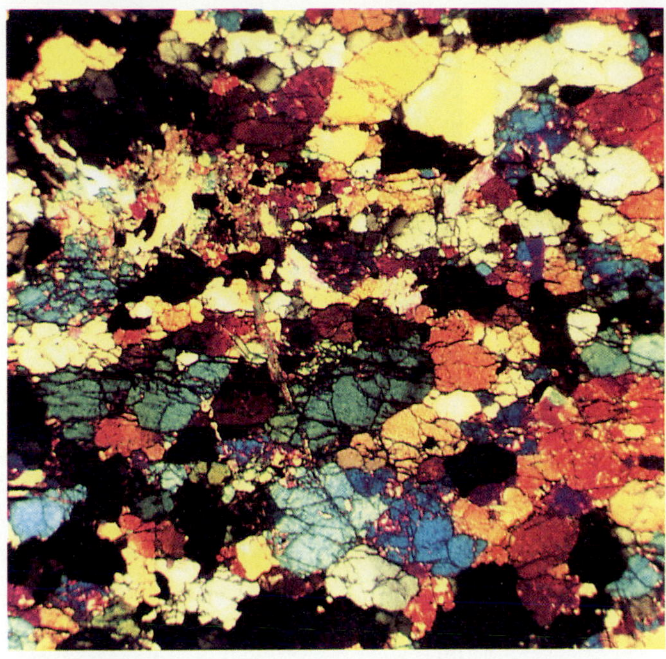

Figure 4.2

A photomicrograph of a thin section of rock containing the mineral olivine.

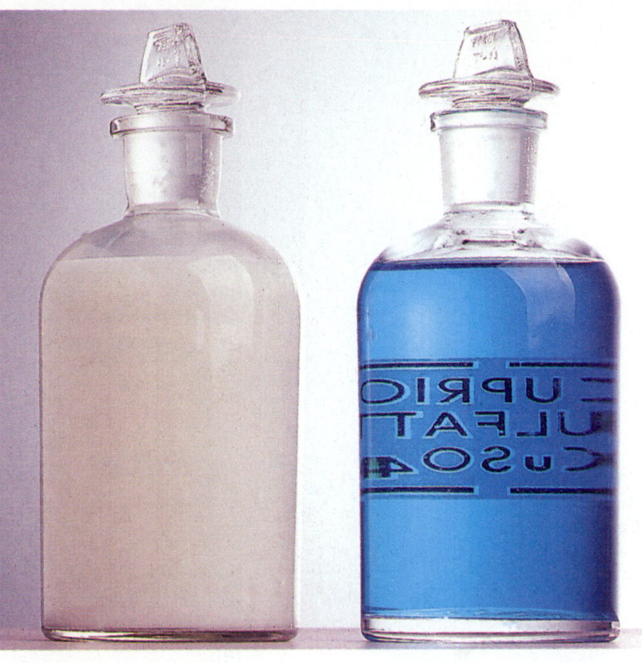

Figure 4.3

Solutions are clear; the label is clearly visible through the blue solution of copper sulfate. Suspensions are turbid; nothing can be seen through the white suspension of magnesium hydroxide.

extracted from the tea leaves and dissolved in water. The solvent is often but not always present in excess. For example, a solution containing 35 g of table sugar and 20 g of water is liquid. Water, even though it is the lesser component, is the solvent because it determines the physical state of the solution. Occasionally, the designation of solute and solvent is arbitrary. All gases dissolve in each other, so any one of the gases in a gaseous solution could be taken as the solvent. The same is true for miscible liquids (liquids that dissolve in each other in all proportions) such as ethanol and water. For such solutions the solvent is usually considered to be the component present in excess. Air contains about 78% nitrogen, so nitrogen is the solvent. In a 25% alcohol–75% water mixture, water is the solvent. In a half-and-half mixture of alcohol and water, however, either substance could be called the solvent. In this chapter we will study **aqueous solutions**, those in which the solvent is water.

Solubility

Figure 4.4a shows what happens when sucrose (table sugar) is added a few crystals at a time to a beaker of water. At first the sucrose crystals dissolve, forming an aqueous solution of sucrose in water. Eventually, however, the sucrose no longer dissolves and the added crystals then fall to the bottom of the beaker (Figure 4.4b). Although we cannot see it, there is continuous activity at the surface of the solid. Some of the sucrose molecules enter the solution, but they are immediately replaced by molecules coming out of the solution. The two processes, dissolving and crystallization (coming out of solution), occur at the same rate, so nothing appears to

(a)

(b)

Figure 4.4

(a) Solid sucrose dissolves in an unsaturated sucrose solution. (b) Solid sucrose falls to the bottom of a saturated sucrose solution.

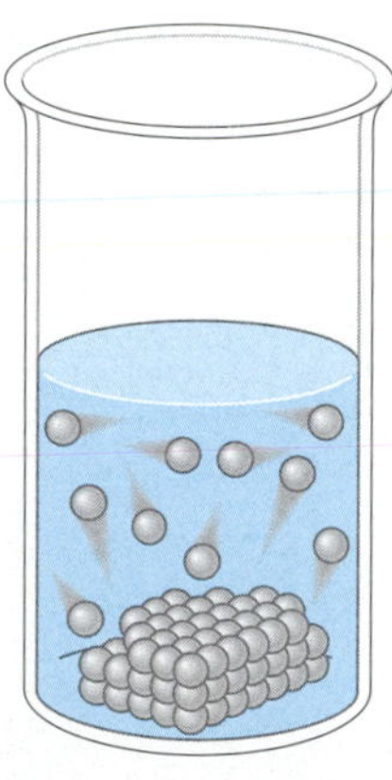

Figure 4.5

In a saturated solution, the number of molecules that leave the surface of an undissolved solute each second is equal to the number that return. The undissolved solute is in equilibrium with the dissolved solute.

The concept of dynamic equilibrium is not limited to solution chemistry. We will soon study other examples of this phenomenon.

change (Figure 4.5). The dissolved and undissolved sucrose are said to be in a state of **dynamic equilibrium**, a state in which no net change occurs because two opposing processes are going on at the same rate. Any solution in which dissolved solute is in equilibrium with undissolved solute (or would be in equilibrium if undissolved solute were present) is called a **saturated solution**. A solution containing less than the equilibrium amount of solute is said to be **unsaturated**.

The **solubility** of a substance is the amount that dissolves in a given quantity of solvent to form a saturated solution. Solubility is often expressed as grams of solute per 100 mL of solvent, but other units can also be used. Solubility varies greatly with the nature of solute and solvent, and it also varies with conditions such as temperature and pressure. The data in Table 4.2 show that sucrose is very soluble in water, other substances such as sodium chloride (NaCl) and sodium hydroxide (NaOH) are less soluble, while still other substances such as calcium carbonate ($CaCO_3$, the major component of marble) and silver sulfide (Ag_2S, silver tarnish) have very low solubilities. Compounds such as calcium carbonate and silver sulfide are often described as "insoluble," but a better term is *sparingly soluble*.

Most (but not all) solids become less soluble as the temperature decreases. Thus when a saturated solution is cooled, the excess solute often reappears in the form of solid crystals. Some solutes, however, form **supersaturated solutions** from which the excess solute does not immediately crystallize. A supersaturated solution holds more solute than the equilibrium amount. It is in a *metastable state*, an unstable state that persists as long as it is undisturbed, like a coin standing on edge. The supersaturated state is possible because solute crystals need to achieve a minimum size before they are able to grow spontaneously; smaller crystals are broken up by intermolecular collisions and fail to grow. "Seeding" the supersaturated solution with a larger solute crystal breaks this pattern and brings about immediate crystallization (see Demonstration 4.1).

DEMONSTRATION 4.1 CRYSTALLIZATION FROM A SUPERSATURATED SOLUTION

A small crystal of sodium acetate ($NaC_2H_3O_2$) is added to a supersaturated solution of sodium acetate.

Excess solute immediately begins to crystallize onto the added crystal.

The needle-like crystals grow rapidly.

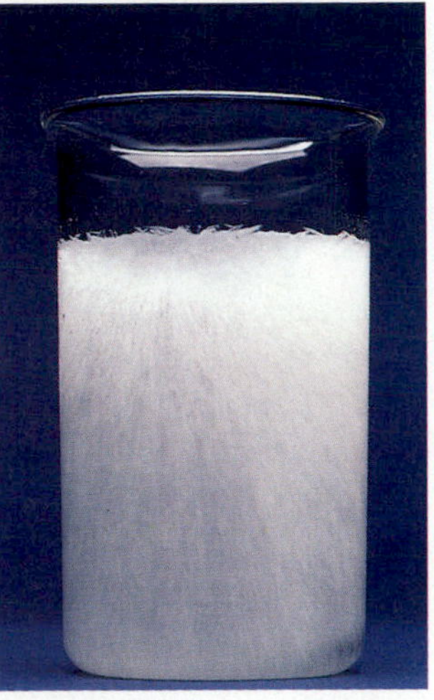

All of the excess solute has crystallized. The solution that remains between crystals is saturated with respect to sodium acetate.

TABLE 4.2 Solubilities of Various Substances in Water

Substance	Solubility (g/100 mL H_2O)	
Gases		
Ammonia (NH_3)	51.8	at 20°C
Carbon dioxide (CO_2)	0.145	at 25°C
Hydrogen chloride (HCl)	82.3	at 0°C
Nitrogen (N_2)	0.0019	at 20°C
Oxygen (O_2)	0.0043	at 20°C
Familiar Solids		
Aspirin (acetylsalicylic acid, $HC_9H_7O_4$)	0.33	at 25°C
Baking soda ($NaHCO_3$)	10	at 25°C
Caustic soda (sodium hydroxide, NaOH)	42	at 0°C
Marble ($CaCO_3$)	0.0014	at 25°C
Silver tarnish (Ag_2S)	8×10^{-16}	at 25°C
Table salt (NaCl)	35.7	at 25°C
Table sugar (sucrose, $C_{12}H_{22}O_{11}$)	211	at 25°C
Washing soda ($Na_2CO_3 \cdot 10H_2O$)	28	at 25°C

Ionic Solutions

An **ionic solution** contains ions dispersed in some solvent. Salt water, which contains Na^+ and Cl^- ions, is a familiar example. Dissolved ions move and react more or less independently of each other. Thus the properties of an ionic solution reflect the properties of its individual ions, and the source of an ion is not important once it is in solution. Figure 4.6 shows, for example, that a solution prepared by dissolving sodium nitrate and potassium chloride in water is identical to a solution prepared by dissolving sodium chloride and potassium nitrate. Both solutions contain sodium, potassium, chloride, and nitrate ions.

An electric current is a flow of electric charge.

An aqueous solution can be tested for the presence of ions using an electric circuit like the one in Figure 4.7. This circuit consists of a current source, a light bulb, and two metal rods called *electrodes* that dip into the solution. If ions are present, they will carry the current through the solution, completing the electrical circuit and causing the bulb to light. The more ions in the solution, the more current they carry, and the brighter the bulb. If the solution contains very few ions, or no ions, the bulb will not glow at all.

A substance that provides ions when dissolved in water is called an **electrolyte**. When an electrolyte such as sodium chloride (NaCl) or potassium sulfate (K_2SO_4) dissolves, the ions move apart and spread throughout the water, a process called **dissociation**. The dissolving and dissociation of ionic solids are represented by equations such as

$$NaCl(s) \rightarrow Na^+(aq) + Cl^-(aq)$$

$$K_2SO_4(s) \rightarrow 2K^+(aq) + SO_4^{2-}(aq)$$

A **strong electrolyte** is an electrolyte that is completely or almost completely ionized in aqueous solution. Ionic compounds are strong electrolytes. Solutions of strong electrolytes, if not too dilute, are good conductors of electricity and will cause the light bulb in Figure 4.7 to glow brightly.

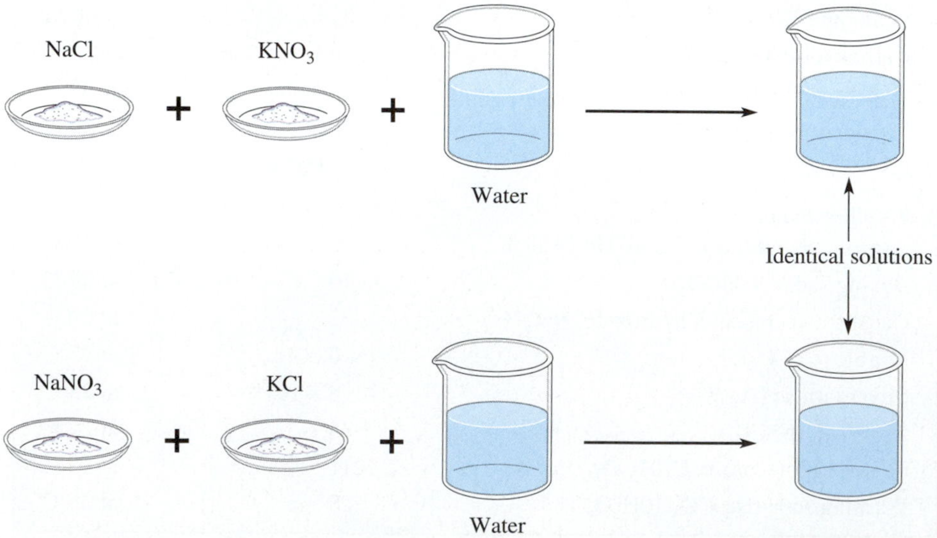

Figure 4.6

A solution prepared from NaCl and KNO_3 (top) is the same as a solution prepared from $NaNO_3$ and KCl (bottom).

Figure 4.7

A conductivity apparatus used to demonstrate the presence of ions in solution. The circuit consists of a power source, a light bulb, and electrodes that dip into the solution to be tested. If the solution conducts current, the light bulb will glow. The intensity of the glow depends on the concentration of ions in the solution.

Some molecular compounds such as hydrogen chloride (HCl) and nitric acid (HNO_3) are not ionic in the pure state but form ions when they dissolve in water. This process, called **ionization**, is also represented by equations such as

$$HCl(aq) \rightarrow H^+(aq) + Cl^-(aq)$$

$$HNO_3(aq) \rightarrow H^+(aq) + NO_3^-(aq)$$

HCl and HNO_3 are strong electrolytes because they are almost completely ionized in aqueous solution.

Weak electrolytes are molecular substances that ionize only to a limited extent in water. Acetic acid (CH_3COOH) and ammonia (NH_3) are examples. Weak electrolytes provide very few ions, and their solutions conduct only a feeble current. They will cause the light bulb in Figure 4.7 to glow dimly. The equation for the ionization of aqueous acetic acid is

$$CH_3COOH(aq) \rightleftharpoons H^+(aq) + CH_3COO^-(aq)$$

The double arrow indicates that the reaction goes in both directions; some molecules break down into ions (the forward reaction) and some ions recombine into molecules (the reverse reaction). The rate of the forward reaction very quickly becomes equal to the rate of the reverse reaction, and a dynamic equilibrium exists between the ions and molecules of acetic acid. Aqueous ammonia also forms an equilibrium mixture:

$$NH_3(aq) + H_2O(l) \rightleftharpoons NH_4^+(aq) + OH^-(aq)$$

Pure water contains so few ions that it is essentially a nonconductor of electricity. Water and molecular solutes such as ethanol (C_2H_5OH) and sucrose ($C_{12}H_{22}O_{11}$) that dissolve in water without forming ions are **nonelectrolytes**. The light bulb in Figure 4.7 remains dark when the electrodes are dipped into solutions of nonelectrolytes.

LEARNING HINT

You cannot predict which compounds are ionic and which are molecular by simply looking at their formulas. Such predictions require a more detailed knowledge of chemical bonding.

Most water conducts electricity because it is not pure. Tap water, rainwater, and groundwater all contain some ionic impurities.

CHEMICAL INSIGHT

THE WORLD'S BIGGEST SOLUTION

The oceans, which cover 70% of the earth's surface, have an average depth of 4 km and are interconnected so that water from one eventually circulates into all of them. Together they form the world's largest repository of dissolved material, a single solution with a volume of 1.37×10^9 km^3. Dissolved ions constitute about 3.5% of the mass of ocean water. Table 4.3 shows that the principal ions are sodium and chloride, but other ions are also present in significant quantities. Probably every one of earth's naturally occurring elements is present to some extent in seawater. In addition to dissolved matter, the oceans contain vast amounts of suspended debris, both organic and inorganic.

The oceans' contents are replenished by rain, runoff from the land, emissions from undersea volcanoes and fissures, and absorption of atmospheric gases. They are depleted by evaporation, precipitation (settling out) of solids on the ocean floor, and loss of gases to the atmosphere. This continuous exchange of materials causes the average concentration of dissolved material to vary slightly with location; the North Atlantic Ocean, for example, has the highest salt concentration, 3.79%. The amount of dissolved oxygen also fluctuates. Not only does oxygen enter and leave the oceans at their surface, but dissolved oxygen is continuously consumed by some organisms and produced photosynthetically by others.

The oceans are a *sink* for atmospheric carbon dioxide; that is, they absorb excess CO_2 from the atmosphere. The ratio of dissolved CO_2 to O_2 in the ocean is more than 25 times higher than in the atmosphere. Some of the dissolved CO_2 is converted to carbohydrates by plants that are then consumed and metabolized by higher life-forms. Some CO_2 is absorbed by organisms that also absorb calcium ions to form skeletons and shells of calcium carbonate. The skeletons and shells become debris on the ocean floor when the organisms die. A certain amount of this carbonate debris is redissolved by acid-secreting bacteria and converted back into CO_2.

Other substances also have oceanic cycles. Photosynthetic organisms in the ocean use silicates, phosphates, and nitrates, and many marine bacteria obtain energy by oxidizing ammonium ions, dissolved hydrogen, and hydrogen sulfide. The ocean is a solution, but unlike most laboratory solutions, it is dynamic and variable—in constant interaction with the atmosphere, the solid earth, and its own living population.

EXAMPLE 4.1

Hydrogen bromide (HBr) and hydrogen cyanide (HCN) are molecular in the pure state. In aqueous solution, hydrogen bromide is a strong electrolyte and hydrogen cyanide is a weak electrolyte. Write equations for the ionization of these substances in water.

SOLUTION The strong electrolyte HBr ionizes completely:

$$HBr(aq) \longrightarrow H^+(aq) + Br^-(aq)$$

The weak electrolyte HCN ionizes partially; ions exist in equilibrium with molecules:

$$HCN(aq) \rightleftharpoons H^+(aq) + CN^-(aq)$$

PRACTICE EXERCISE 4.1

Write equations for (a) the dissociation of $Al_2(SO_4)_3$, an ionic solid, in water; (b) the ionization of molecular HI, a strong electrolyte; and (c) the ionization of aqueous HNO_2, a weak electrolyte.

TABLE 4.3 Elemental Abundances in Seawater

Element	Milligrams per liter	Element	Milligrams per liter	Element	Milligrams per liter
Chlorine	19,000	Molybdenum	0.01	Tungsten	1×10^{-4}
Sodium	10,600	Selenium	0.004	Germanium	1×10^{-4}
Magnesium	1,300	Copper	0.003	Xenon	1×10^{-4}
Sulfur	900	Arsenic	0.003	Chromium	5×10^{-5}
Calcium	400	Tin	0.003	Beryllium	5×10^{-5}
Potassium	380	Lead	0.003	Scandium	4×10^{-5}
Bromine	65	Uranium	0.003	Mercury	3×10^{-5}
Carbon	28	Vanadium	0.002	Niobium	1×10^{-5}
Oxygen	8	Manganese	0.002	Thallium	1×10^{-5}
Strontium	8	Titanium	0.001	Helium	5×10^{-6}
Boron	4.8	Thorium	0.0007	Gold	4×10^{-6}
Silicon	3.0	Cobalt	0.0005	Praseodymium	2×10^{-7}
Fluorine	1.3	Nickel	0.0005	Gadolinium	2×10^{-7}
Nitrogen	0.8	Gallium	0.0005	Dysprosium	2×10^{-7}
Argon	0.6	Cesium	0.0005	Erbium	2×10^{-7}
Lithium	0.2	Antimony	0.0005	Ytterbium	2×10^{-7}
Rubidium	0.12	Cerium	0.0004	Samarium	2×10^{-7}
Phosphorus	0.07	Yttrium	0.0003	Holmium	8×10^{-8}
Iodine	0.05	Neon	0.0003	Europium	4×10^{-8}
Barium	0.03	Krypton	0.0003	Thulium	4×10^{-8}
Indium	0.02	Lanthanum	0.0003	Lutetium	4×10^{-8}
Aluminum	0.01	Silver	0.0003	Radium	3×10^{-11}
Iron	0.01	Bismuth	0.0002	Protactinium	2×10^{-12}
Zinc	0.01	Cadmium	0.0001	Radon	9×10^{-15}

SOURCE: *Chemistry and the Environment*, American Chemical Society, Washington, D.C., 1967.

4.2 METATHESIS REACTIONS

A **metathesis reaction** is one in which atoms or ions exchange partners. An example of a metathesis reaction involving ions is

$$Pb(NO_3)_2(aq) + 2KI(aq) \rightarrow PbI_2(s) + 2KNO_3(aq)$$

Metathesis occurs whenever ions are unable to remain dissociated, either because they form a **precipitate** (a solid that settles out of solution) or because they form a molecular substance. The molecular product may remain in solution or it may evolve as a gas. We will consider each of these cases in turn.

The word *metathesis* (meh-tath′-eh-sis) is derived from the Greek word for "to transpose." Metathesis reactions are also called *double displacement reactions* or *double replacement reactions*.

Precipitation Reactions

When sodium chloride and silver nitrate solutions are mixed, the ions momentarily present in solution are Na^+, Cl^-, Ag^+, and NO_3^-. Silver chloride (AgCl) is a sparingly soluble solid, and the silver and chloride ions combine to form a silver chloride precipitate that settles out of the solution in the form of minute white crystals (Figure 4.8):

Figure 4.8

Formation of a precipitate. The pipet contains clear silver nitrate solution; the beaker contains clear sodium chloride solution. A white precipitate of silver chloride appears where the two solutions meet.

$$Ag^+(aq) + Cl^-(aq) \rightarrow AgCl(s)$$

Sodium nitrate ($NaNO_3$) is very soluble, so Na^+ and NO_3^- ions remain in solution. These ions are examples of **spectator ions**, ions that are present in the solution but do not participate in the reaction. An equation like the one above, which does not include the spectator ions, is called a **net ionic equation**. The complete equation, which includes all of the ions, is

$$Ag^+(aq) + NO_3^-(aq) + Na^+(aq) + Cl^-(aq) \rightarrow AgCl(s) + Na^+(aq) + NO_3^-(aq)$$

or more simply

$$AgNO_3(aq) + NaCl(aq) \rightarrow AgCl(s) + NaNO_3(aq)$$

Complete equations with neutral formulas are often called *molecular equations* even when the substances are ionic. Note that the reaction between silver nitrate and sodium chloride is a metathesis reaction because the positive and negative ions in the original compounds have exchanged partners.

A molecular equation can be converted to a net ionic equation by writing out the ions on both sides and eliminating the spectator ions, as illustrated in the next example.

LEARNING HINT

The total charge is the same on both sides of an ionic equation; charges as well as atoms must be balanced.

EXAMPLE 4.2

Lead chromate precipitates when aqueous solutions of lead nitrate and potassium chromate are mixed. Write the molecular equation and the net ionic equation.

SOLUTION

The formulas of the reactants are $Pb(NO_3)_2$ and K_2CrO_4. (Refer to Examples 2.16 and 2.17, p. 69, if you need to review writing formulas of ionic compounds.) The products are solid lead chromate ($PbCrO_4$) and aqueous potassium nitrate (KNO_3). The molecular equation is

$$Pb(NO_3)_2(aq) + K_2CrO_4(aq) \rightarrow PbCrO_4(s) + 2KNO_3(aq)$$

Rewriting this equation to show the dissociated ions gives

$$Pb^{2+}(aq) + 2NO_3^-(aq) + \cancel{2K^+(aq)} + \cancel{CrO_4^{2-}(aq)} \rightarrow PbCrO_4(s) + \cancel{2K^+(aq)} + \cancel{2NO_3^-(aq)}$$

The net ionic equation

$$Pb^{2+}(aq) + CrO_4^{2-}(aq) \rightarrow PbCrO_4(s)$$

is obtained by crossing out the spectator ions common to both sides of the equation.

PRACTICE EXERCISE 4.2

Aluminum hydroxide precipitates when aqueous solutions of aluminum sulfate and sodium hydroxide are mixed. Write the molecular and net ionic equations.

A knowledge of solubilities allows us to predict whether or not a precipitation reaction will occur. Some useful solubility rules for ionic compounds are as follows:

1. Nitrates are soluble.
2. Compounds containing Li^+, Na^+, K^+, and NH_4^+ are soluble.
3. Most chlorides, bromides, and iodides are soluble. The principal exceptions are sparingly or slightly soluble halides containing silver, lead, and mercury(I) ions, such as AgBr, PbI_2, and Hg_2Cl_2.
4. Most sulfates (SO_4^{2-}) are soluble. Important exceptions are the sparingly or slightly soluble sulfates of calcium, strontium, barium, silver, and lead.
5. Most hydroxides are sparingly or slightly soluble. Important soluble hydroxides include barium hydroxide ($Ba(OH)_2$) and the hydroxides of the positive ions listed in Rule 2.
6. Most carbonates (CO_3^{2-}), phosphates (PO_4^{3-}), sulfites (SO_3^{2-}), and sulfides (S^{2-}) are sparingly soluble. The principal exceptions are those whose positive ions are listed in Rule 2.

The use of these rules is illustrated in the following examples.

The solubility rules apply at room temperature and assume the following solubility ranges:

sparingly soluble—less than 0.1 g per 100 mL of water
slightly soluble—between 0.1 and 1 g per 100 mL of water
soluble—greater than 1 g per 100 mL of water

EXAMPLE 4.3

State whether the following solids are soluble or sparingly soluble: (a) Na_2S, (b) PbS, (c) $CaCO_3$, and (d) AgCl.

SOLUTION

(a) Sodium sulfide is soluble (Rule 2).

(b) Lead sulfide is sparingly soluble (Rule 6).

(c) Calcium carbonate is sparingly soluble (Rule 6).

(d) Silver chloride is sparingly soluble (Rule 3).

PRACTICE EXERCISE 4.3

State whether the following solids are soluble or sparingly soluble: (a) Na_2SO_3, (b) Ag_3PO_4, (c) $Ba(NO_3)_2$, and (d) Na_3PO_4.

EXAMPLE 4.4

Will a precipitate form when the following ionic compounds are mixed in aqueous solution: (a) calcium chloride and potassium carbonate; (b) ammonium chloride and sodium sulfate? If so, write the molecular and net ionic equations for the precipitation reaction.

SOLUTION

(a) When calcium chloride ($CaCl_2$) and potassium carbonate (K_2CO_3) are mixed in aqueous solution, the ions momentarily present are Ca^{2+}, Cl^-, K^+, and CO_3^{2-}. The possible products of a precipitation reaction are calcium carbonate ($CaCO_3$) and potassium chloride (KCl). The solubility rules state that calcium carbonate is sparingly soluble (Rule 6) and potassium chloride is soluble (Rule 2), so a precipitate of calcium carbonate will form. The molecular equation is

$$CaCl_2(aq) + K_2CO_3(aq) \rightarrow CaCO_3(s) + 2KCl(aq)$$

Rewriting the equation to show the dissociated ions gives

$$Ca^{2+}(aq) + \cancel{2Cl^-(aq)} + \cancel{2K^+(aq)} + CO_3^{2-}(aq) \rightarrow CaCO_3(s) + \cancel{2K^+(aq)} + \cancel{2Cl^-(aq)}$$

The net ionic equation

$$Ca^{2+}(aq) + CO_3^{2-}(aq) \rightarrow CaCO_3(s)$$

is obtained by crossing out the spectator ions $K^+(aq)$ and $Cl^-(aq)$.

(b) When ammonium chloride (NH_4Cl) and sodium sulfate (Na_2SO_4) are mixed, the ions initially present are NH_4^+, Cl^-, Na^+, and SO_4^{2-}. The possible products, ammonium sulfate ($(NH_4)_2SO_4$) and sodium chloride (NaCl), are both soluble according to Rule 2, so no precipitate will form.

PRACTICE EXERCISE 4.4

Will a precipitation reaction occur when the following compounds are mixed in aqueous solution: (a) lead nitrate and potassium bromide; (b) sodium sulfide and zinc chloride? If so, write the molecular and net ionic equations.

The next example shows how the solubility rules can be used in preparing ionic compounds.

Figure 4.9

Filtering a precipitate (see Example 4.5). The mixture of solid magnesium carbonate and potassium sulfate solution is poured from the beaker into a funnel lined with filter paper. The glass rod guides the solution so that none of it drips down the outside of the beaker. The magnesium carbonate remains in the filter paper; the solution passes through.

EXAMPLE 4.5

How can magnesium carbonate be prepared from two soluble ionic compounds?

SOLUTION

Magnesium carbonate is sparingly soluble (Rule 6) and will precipitate when solutions containing magnesium ions and carbonate ions are mixed. Magnesium sulfate and potassium carbonate are both soluble (Rules 4 and 2) and can be used as reactants. The equation is

$$MgSO_4(aq) + K_2CO_3(aq) \rightarrow MgCO_3(s) + K_2SO_4(aq)$$

Potassium sulfate is soluble (Rule 2), so the ions K^+ and SO_4^{2-} remain in solution. The solid $MgCO_3$ can be separated from them and the solvent by filtration (see Figure 4.9).

PRACTICE EXERCISE 4.5

Write an equation for the preparation of silver phosphate from two soluble ionic compounds.

Gravimetric Analysis

A **gravimetric analysis** is an analytical procedure in which the amount of some substance is determined by weighing one of its reaction products. Suppose that a laboratory technician is asked to find the percentage of magnesium sulfate ($MgSO_4$) in a commercial Epsom salts preparation. The technician would begin the analysis by weighing a small portion of the mixture and dissolving it in water to give a solution containing Mg^{2+} and SO_4^{2-} along with the various soluble impurities that the preparation might contain. The amount of magnesium sulfate in this solution can then be determined by precipitating one of its ions and weighing the precipitate. Both Mg^{2+} and SO_4^{2-} form a number of sparingly soluble salts; barium sulfate is the one favored by the analytical chemist because it does not decompose when it is dried by heating and it has little tendency to absorb moisture from the air. Adding excess barium chloride ($BaCl_2$) solution to the dissolved sample will produce a precipitate of barium sulfate

$$Ba^{2+}(aq) + SO_4^{2-}(aq) \rightarrow BaSO_4(s)$$

that can be filtered, dried, and weighed (see Figure 4.10). The following example illustrates how the data would be handled.

Epsom salts is the common name of magnesium sulfate heptahydrate, a solid compound with a fixed proportion of water molecules incorporated into its structure. In the formula, the water molecules are attached with a dot: $MgSO_4{\cdot}7H_2O$.

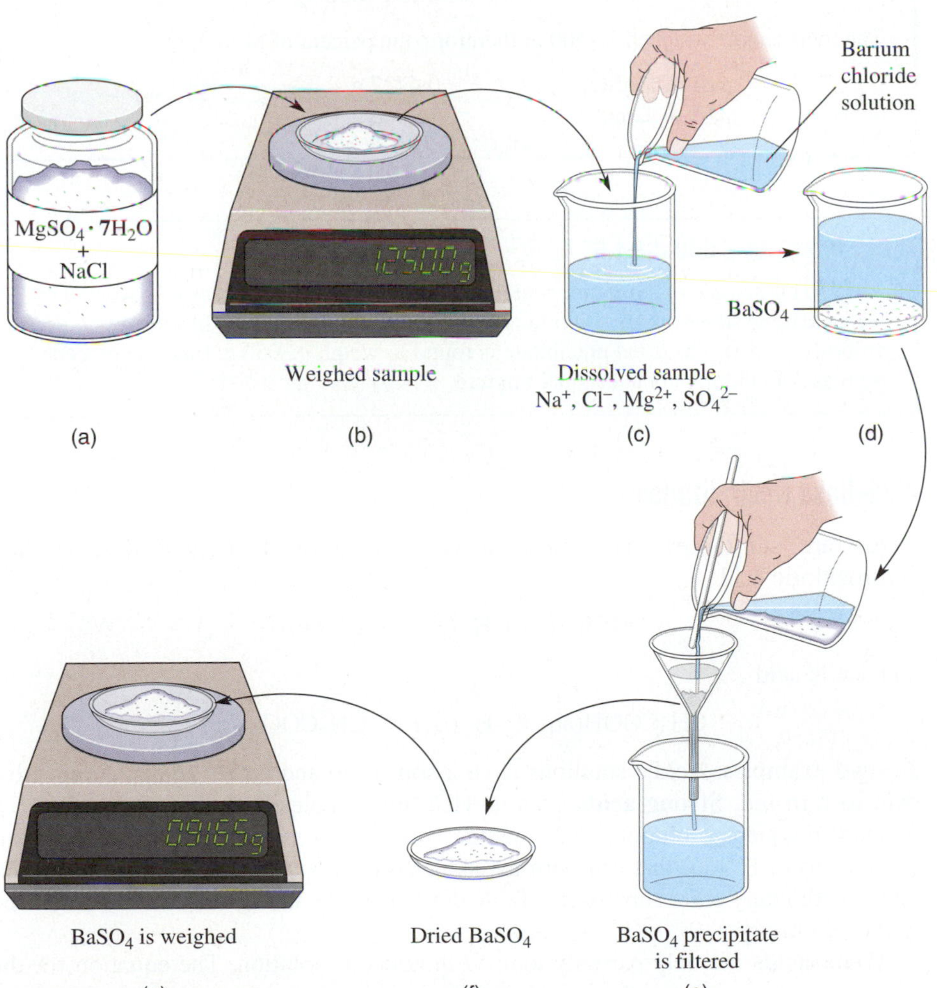

Figure 4.10
The sequence of operations in a gravimetric analysis of impure $MgSO_4{\cdot}7H_2O$. (a)–(d) A sample of the material is weighed, dissolved in water, and treated with $BaCl_2$ solution to give a precipitate of $BaSO_4$. (e)–(g) The precipitate is filtered, dried, and weighed. The reaction equation (see text) shows that the number of moles of $BaSO_4$ equals the number of moles of $MgSO_4$ in the original sample.

EXAMPLE 4.6

A 1.2500-g sample of impure Epsom salts is dissolved and treated with excess barium chloride solution. The dried barium sulfate precipitate weighs 0.9165 g. Calculate the percent by mass of $MgSO_4$ in the original mixture.

SOLUTION

The equation for the reaction

$$MgSO_4(aq) + BaCl_2(aq) \longrightarrow BaSO_4(s) + MgCl_2(aq)$$

shows that the number of moles of $BaSO_4$ is equal to the number of moles of $MgSO_4$ in the dissolved sample. The steps in the calculation are

$$0.9165 \text{ g } BaSO_4 \longrightarrow \text{mol } BaSO_4 \longrightarrow \text{mol } MgSO_4 \longrightarrow \text{g } MgSO_4 \longrightarrow \% \ MgSO_4$$

The molar masses of $BaSO_4$ and $MgSO_4$ are 233.39 and 120.37 g/mol, respectively; the first three steps give the number of grams of $MgSO_4$:

$$0.9165 \text{ g } BaSO_4 \times \frac{1 \text{ mol } BaSO_4}{233.39 \text{ g } BaSO_4} \times \frac{1 \text{ mol } MgSO_4}{1 \text{ mol } BaSO_4} \times \frac{120.37 \text{ g } MgSO_4}{1 \text{ mol } MgSO_4}$$
$$= 0.4727 \text{ g } MgSO_4$$

The total sample weighed 1.2500 g; therefore, the percent of $MgSO_4$ is

$$\frac{\text{mass of } MgSO_4}{\text{mass of sample}} \times 100\% = \frac{0.4727 \text{ g}}{1.2500 \text{ g}} \times 100\% = 37.82\%$$

PRACTICE EXERCISE 4.6

A 0.2501-g sample of a mixture containing various soluble chlorides is dissolved in water. Excess silver nitrate solution is added to precipitate all the chloride as silver chloride (AgCl). The dried precipitate is found to weigh 0.3050 g. Find the percent by mass of chloride in the original mixture.

Acid–Base Neutralization

The term *acid* is derived from the Latin word *acidus*, meaning "sour."

CAUTION! Laboratory acids and bases, both weak and strong, can cause severe injury. Never taste-test or touch them.

Acids are substances that ionize in water to form hydrogen ions, $H^+(aq)$. Hydrochloric acid

$$HCl(aq) \longrightarrow H^+(aq) + Cl^-(aq)$$

and acetic acid

$$CH_3COOH(aq) \rightleftharpoons H^+(aq) + CH_3COO^-(aq)$$

are two examples. Acidic solutions have a sour taste and cause litmus, a vegetable dye, to turn red. **Strong acids** such as HCl are completely, or almost completely, ionized in aqueous solution. Concentrated solutions of strong acids have high concentrations of $H^+(aq)$ that can burn the skin, dissolve various metals, and react violently with many other substances. Table 4.4 lists some strong acids commonly used in the laboratory.

Weak acids are only partially ionized in aqueous solution. The equation for the

TABLE 4.4 Some Strong and Weak Acids

Strong Acids		Weak Acids	
Hydrogen bromide[a]	HBr	Acetic acid	$HC_2H_3O_2$ or CH_3COOH
Hydrogen chloride[a]	HCl	Carbonic acid	H_2CO_3 (unstable)
Hydrogen iodide[a]	HI	Hydrogen fluoride[a]	HF
Nitric acid	HNO_3	Hydrogen sulfide	H_2S
Perchloric acid	$HClO_4$	Phosphoric acid	H_3PO_4
Sulfuric acid	H_2SO_4	Sulfurous acid	H_2SO_3 (unstable)

[a]Aqueous solutions of HBr, HCl, HI, and HF are called hydrobromic acid, hydrochloric acid, hydroiodic acid, and hydrofluoric acid, respectively.

ionization of acetic acid, a typical weak acid, has a double arrow to indicate that the molecules and ions are in dynamic equilibrium. Other examples of weak acids are listed in Table 4.4. The majority of acids are weak and do not produce high concentrations of $H^+(aq)$. Their action is more gentle than that of strong acids. Many are found in everyday foodstuffs. Vinegar, for example, is a dilute solution of acetic acid.

Bases are substances that produce hydroxide ion, $OH^-(aq)$, in aqueous solution. Some common bases are listed in Table 4.5. Basic solutions feel slippery, have a bitter taste, and turn litmus blue. The terms **alkali** and **alkaline** are synonyms for base and basic, respectively. **Strong bases** such as sodium hydroxide (NaOH) and potassium hydroxide (KOH) are ionic compounds that dissociate completely in solution:

The word *alkali* is derived from the Arabic *al-qaliy* meaning "ashes of saltwort," a seaside plant containing potassium hydroxide.

$$NaOH(s) \rightarrow Na^+(aq) + OH^-(aq)$$

$$KOH(s) \rightarrow K^+(aq) + OH^-(aq)$$

Concentrated solutions of strong bases are corrosive and may cause severe damage to living tissue. The active ingredient in many oven cleaners is a concentrated solution of NaOH. Less soluble hydroxides such as calcium hydroxide ($Ca(OH)_2$) and magnesium hydroxide ($Mg(OH)_2$) are also strong in the sense of being completely dissociated; however, their saturated solutions are too dilute to be harmful.

TABLE 4.5 Some Strong Bases

Chemical Name	Common Name	Formula
Barium hydroxide	—	$Ba(OH)_2$
Calcium hydroxide	Slaked lime; an aqueous solution is limewater	$Ca(OH)_2$
Magnesium hydroxide	A 30% suspension in water is milk of magnesia	$Mg(OH)_2$
Potassium hydroxide	Caustic potash	KOH
Sodium hydroxide	Caustic soda	NaOH

Ammonia is an example of a **weak base**, one that is only partially ionized in aqueous solution. Ammonia produces small amounts of $OH^-(aq)$ ion by reacting with water:

$$NH_3(aq) + H_2O(l) \rightleftharpoons NH_4^+(aq) + OH^-(aq)$$

At any instant most of the dissolved ammonia is in the molecular form.

The reaction between an acid and a base is called **neutralization**. Some examples of neutralization in aqueous solution are

$$NaOH(aq) + HCl(aq) \longrightarrow NaCl(aq) + H_2O(l)$$

$$Ca(OH)_2(aq) + 2HNO_3(aq) \longrightarrow Ca(NO_3)_2(aq) + 2H_2O(l)$$

$$Ba(OH)_2(aq) + H_2SO_4(aq) \longrightarrow BaSO_4(s) + 2H_2O(l)$$

Observe that these neutralizations are metathesis reactions in which the hydrogen and hydroxide ions combine to form the molecular product water:

$$H^+(aq) + OH^-(aq) \longrightarrow H_2O(l)$$

The disappearance of these ions means that the acidic and basic properties of the reactant solutions are lost (neutralized) during the reaction.

A **salt** is any ionic compound whose ions are neither H^+ or OH^-. Sodium chloride, NaCl, is the best-known example of a salt. Salts are produced during acid–base neutralizations; for example, sodium chloride, calcium nitrate, and barium sulfate are produced by the neutralizations whose equations are shown above. Barium sulfate is sparingly soluble, so it appears as a precipitate. The general pattern of acid–base neutralization in aqueous solution is

$$\text{acid} + \text{base} \longrightarrow \text{salt} + \text{water}$$

EXAMPLE 4.7

The compound Na_3PO_4, also known as trisodium phosphate or "TSP," is the active ingredient in various mixtures used for heavy-duty cleaning. Write an equation for the acid–base neutralization that produces Na_3PO_4.

SOLUTION

Na_3PO_4 is a salt containing Na^+ ions and PO_4^{3-} ions. The Na^+ ions will be supplied by the base NaOH. The PO_4^{3-} ions come from phosphoric acid, H_3PO_4. The equation is

$$3NaOH(aq) + H_3PO_4(aq) \longrightarrow Na_3PO_4(aq) + 3H_2O(l)$$

PRACTICE EXERCISE 4.7

Write equations for the neutralization of (a) $Fe(OH)_3$ by HNO_3 and (b) H_2SO_4 by $Mg(OH)_2$. (*Hint*: Both hydroxides are sparingly soluble.)

Acids such as H_2SO_4 and H_3PO_4 have more than one acidic hydrogen atom. If a base is added gradually to one of these acids, the acid will be neutralized in distinct steps. For example, 1 mol of sodium hydroxide added to 1 mol of sulfuric acid produces the salt sodium hydrogen sulfate, which contains the hydrogen sulfate ion, HSO_4^-:

$$NaOH(aq) + H_2SO_4(aq) \rightarrow NaHSO_4(aq) + H_2O(l)$$

This reaction is a *half-neutralization* because only half of the acidic hydrogen atoms have been converted to H_2O. More sodium hydroxide—at least 2 mol of NaOH for each mole of H_2SO_4—is required for complete neutralization:

$$2NaOH(aq) + H_2SO_4(aq) \rightarrow Na_2SO_4(aq) + 2H_2O(l)$$

EXAMPLE 4.8

How can sodium dihydrogen phosphate (NaH_2PO_4) be produced from an acid and a base?

SOLUTION

Sodium hydroxide (NaOH) will supply the Na^+ ions and phosphoric acid (H_3PO_4) will supply the $H_2PO_4^-$ ions. We want only one of the acidic hydrogens in phosphoric acid to react, so we must add no more than 1 mol of NaOH for each mole of H_3PO_4. The reaction is

$$NaOH(aq) + H_3PO_4(aq) \rightarrow NaH_2PO_4(aq) + H_2O(l)$$

Too much sodium hydroxide would produce Na_2HPO_4 or even Na_3PO_4 (see Example 4.7).

PRACTICE EXERCISE 4.8

Write equations for the preparation of (a) Na_2HPO_4 and (b) $KHSO_4$.

Reactions in Which a Gas Is Formed

Some metathesis reactions produce gases. Salts containing ammonium ion give off ammonia gas (NH_3) when treated with a strong concentrated base:

$$NH_4Cl(aq) + KOH(aq) \rightarrow KCl(aq) + NH_3(g) + H_2O(l)$$

The ammonium ion reacts with the hydroxide ion, and the net ionic equation is

$$NH_4^+(aq) + OH^-(aq) \rightarrow NH_3(g) + H_2O(l)$$

Potassium and chloride ions are spectator ions that remain in the solution.

Hydrogen sulfide (H_2S) and hydrogen cyanide (HCN) are gases that are produced when acids are added to sulfides and cyanides. The reactions of solid iron(II) sulfide and aqueous potassium cyanide with hydrochloric acid are examples:

$$FeS(s) + 2HCl(aq) \rightarrow FeCl_2(aq) + H_2S(g)$$

$$KCN(aq) + HCl(aq) \rightarrow KCl(aq) + HCN(g)$$

Hydrogen sulfide is the gas that gives rotten eggs their awful smell. Hydrogen cyanide has the odor of bitter almonds. Both of these gases are dangerously toxic,

Figure 4.11

The reaction of a strong acid with a carbonate. The seashell is approximately 98% $CaCO_3$. The cylinder contains aqueous HCl. Bubbles of CO_2 gas form as the shell dissolves in the acid.

Calcium carbonate ($CaCO_3$) is the principal ingredient of limestone, marble, coral, pearls, and shells. It is the most abundant of all carbonates.

Sodium hydrogen carbonate ($NaHCO_3$) is often referred to by its older names: sodium bicarbonate, bicarbonate of soda, and baking soda.

and strict precautions must be taken to prevent their salts from being acidified accidentally.

Hydrogen sulfide and hydrogen cyanide are weak acids, and the above equations illustrate the general rule that *a weak acid will always form by metathesis if one of its salts is treated with a strong acid.* Weak acids with small molecules, such as H_2S or HCN, tend to escape as gases. Weak acids consisting of larger molecules (e.g., acetic acid) may remain dissolved. Some of these, however, are unstable and decompose to give gases. The most important example is carbonic acid (H_2CO_3), which is formed by the action of a strong acid on a carbonate. The reaction of solid calcium carbonate with hydrochloric acid

$$CaCO_3(s) + 2HCl(aq) \rightarrow CaCl_2(aq) + H_2CO_3(aq)$$

is accompanied by the decomposition of carbonic acid into carbon dioxide and water:

$$H_2CO_3(aq) \rightarrow H_2O(l) + CO_2(g)$$

The overall reaction, the sum of the above steps, is

$$CaCO_3(s) + 2HCl(aq) \rightarrow CaCl_2(aq) + H_2O(l) + CO_2(g)$$

Calcium carbonate dissolves while carbon dioxide bubbles out of the solution (Figure 4.11). Sodium hydrogen carbonate ($NaHCO_3$) reacts in a similar way:

$$NaHCO_3(\text{s or aq}) + HCl(aq) \rightarrow NaCl(aq) + H_2O(l) + CO_2(g)$$

The action of acid on carbonates and hydrogen carbonates explains such diverse phenomena as the effectiveness of muriatic acid (an HCl solution) in cleaning stone, the neutralization of excess stomach acid with sodium bicarbonate ($NaHCO_3$), and the disintegration of marble statuary in areas of acidic rainfall. It also suggests that it is not a good idea to clean pearls ($CaCO_3$) with vinegar (an aqueous solution of acetic acid)—an experiment known to have caused much sorrow.

Sulfurous acid (H_2SO_3), formed by the action of strong acids on salts containing sulfite ion, is so unstable that its existence has never been detected:

$$Na_2SO_3(s) + 2HCl(aq) \rightarrow 2NaCl(aq) + H_2SO_3(aq)$$

H_2SO_3 decomposes instantly to give sulfur dioxide, a gas with an acrid (harsh and choking) odor:

$$H_2SO_3(aq) \rightarrow H_2O(l) + SO_2(g)$$

The overall reaction is

$$Na_2SO_3(s) + 2HCl(aq) \rightarrow 2NaCl(aq) + H_2O(l) + SO_2(g)$$

EXAMPLE 4.9

Most calcium salts are prepared from natural limestone ($CaCO_3$). What reaction would convert limestone to calcium nitrate, $Ca(NO_3)_2$?

SOLUTION

A reactant is needed to supply nitrate ions and remove carbonate ions. Nitric acid (HNO_3) is a strong acid that will convert carbonate ions into gaseous carbon dioxide while the calcium and nitrate ions remain in solution. The equation is

$$CaCO_3(s) + 2HNO_3(aq) \rightarrow Ca(NO_3)_2(aq) + H_2O(l) + CO_2(g)$$

Calcium nitrate crystals can be obtained by evaporating the water.

PRACTICE EXERCISE 4.9

Write equations for the reaction of the following solid salts with HCl: (a) NaCN, (b) NH_4HCO_3, (c) $MgCO_3$, and (d) $NaHSO_3$.

4.3 IONIC DISPLACEMENT REACTIONS

A *displacement reaction* was defined in Section 3.6 as a chemical reaction in which one element displaces another element from a compound. The reaction between iron and sulfuric acid

$$Fe(s) + H_2SO_4(aq) \rightarrow H_2(g) + FeSO_4(aq)$$

is an example of a metal displacing hydrogen from a strong acid (Figure 4.12a). Another example is the reaction of aluminum with hydrochloric acid:

$$2Al(s) + 6HCl(aq) \rightarrow 3H_2(g) + 2AlCl_3(aq)$$

In each case the metal dissolves and hydrogen gas bubbles out of the solution. These metals undergo similar reactions with other strong acids, which suggests that

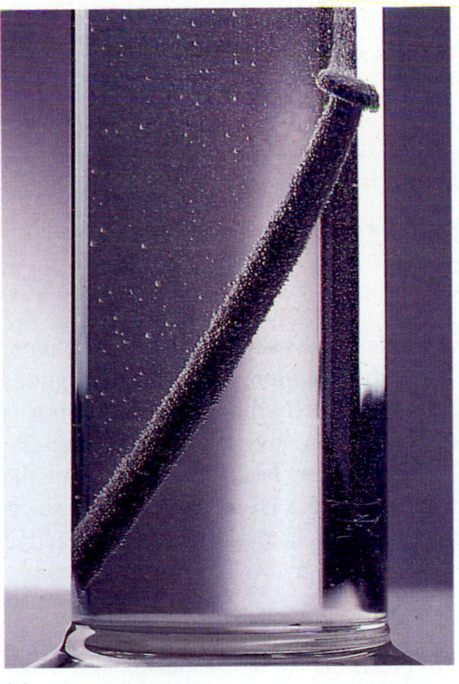

(a)

(b)

Figure 4.12

Some ionic displacement reactions of iron. (a) The iron in an iron nail displaces hydrogen from aqueous H_2SO_4. The H_2 gas bubbles out of the solution. (b) The iron displaces copper from a $CuSO_4$ solution. The reddish-brown copper metal appears as a shiny coating on the undissolved portion of the nail.

hydrogen ions are the reacting species. The net ionic equations for these displacement reactions are

$$Fe(s) + 2H^+(aq) \rightarrow H_2(g) + Fe^{2+}(aq)$$

$$2Al(s) + 6H^+(aq) \rightarrow 3H_2(g) + 2Al^{3+}(aq)$$

The negative ions are spectator ions and do not participate in the reactions.

Certain metals can also displace other metals. For example, when iron is dipped into copper sulfate solution (Figure 4.12b), the iron dissolves and solid copper precipitates:

$$Fe(s) + CuSO_4(aq) \rightarrow FeSO_4(aq) + Cu(s)$$

The sulfate ions are spectator ions; subtracting one $SO_4{}^{2-}$ ion from each side of the molecular equation leaves the net ionic equation:

$$Fe(s) + Cu^{2+}(aq) \rightarrow Fe^{2+}(aq) + Cu(s)$$

EXAMPLE 4.10

Copper displaces silver from an aqueous solution of silver nitrate. Copper nitrate, $Cu(NO_3)_2$, is one of the reaction products. Write the molecular and net ionic equations.

SOLUTION

The reactants are Cu(s) and $AgNO_3$(aq). The products are Ag(s) and $Cu(NO_3)_2$. The molecular equation is

$$Cu(s) + 2AgNO_3(aq) \rightarrow Cu(NO_3)_2(aq) + 2Ag(s)$$

The nitrate ions are spectator ions, and the net ionic equation is obtained by subtracting two $NO_3{}^-$ ions from each side of the molecular equation:

$$Cu(s) + 2Ag^+(aq) \rightarrow Cu^{2+}(aq) + 2Ag(s)$$

PRACTICE EXERCISE 4.10

Magnesium displaces hydrogen from aqueous sulfuric acid. Write the molecular and net ionic equations.

TABLE 4.6 Activity Series for Selected Metals and Hydrogen

Element		
K	Displace hydrogen from cold water	Displace hydrogen from acids
Ca	Displace hydrogen from cold water	Displace hydrogen from acids
Na	Displace hydrogen from cold water	Displace hydrogen from acids
Mg	Displace hydrogen from steam	Displace hydrogen from acids
Al	Displace hydrogen from steam	Displace hydrogen from acids
Zn	Displace hydrogen from steam	Displace hydrogen from acids
Cr	Displace hydrogen from steam	Displace hydrogen from acids
Fe	Displace hydrogen from steam	Displace hydrogen from acids
Ni		Displace hydrogen from acids
Sn		Displace hydrogen from acids
Pb		Displace hydrogen from acids
H		
Cu		
Hg		
Ag		
Pt		
Au		

NOTE: Each element is able to displace the elements below it from a compound. The metals above hydrogen are able to displace hydrogen from acid solutions containing 1 mol of H^+(aq) per liter.

In a displacement reaction the element that enters the solution is said to be *more active* than the element that is displaced from the solution. The above reactions show that iron, aluminum, and magnesium are more active than hydrogen, iron is more active than copper, and copper is more active than silver. A listing of the elements in order of their activity is called an **activity series**. Table 4.6 shows the activity series for some selected metals and hydrogen. Each element in the series is able to displace the elements below it, but not the elements above it. The most active metals such as sodium and potassium will even displace hydrogen from cold water:

$$2Na(s) + 2H_2O(l) \rightarrow 2NaOH(aq) + H_2(g)$$

Such very active metals are stored under oil to protect them from atmospheric moisture and oxygen.

EXAMPLE 4.11

Use the activity series to predict whether the following metals will dissolve in hydrochloric acid: (a) tin and (b) copper. If so, write the molecular equation.

SOLUTION

(a) Tin (Sn) is above hydrogen in the activity series and will displace hydrogen from HCl. It will dissolve according to the equation

$$Sn(s) + 2HCl(aq) \longrightarrow H_2(g) + SnCl_2(aq)$$

(b) Copper (Cu) is below hydrogen in the activity series. It cannot displace hydrogen and will not dissolve in hydrochloric acid.

PRACTICE EXERCISE 4.11

Use the activity series to predict whether (a) magnesium will react with aqueous $AgNO_3$ and (b) lead will react with aqueous $CuSO_4$. Write balanced equations for any reactions that would occur. (*Hint*: Lead sulfate is sparingly soluble.)

Displacement reactions are not confined to metals. Dissolved chlorine, for example, displaces iodide ions from a solution of sodium iodide:

$$Cl_2(aq) + 2NaI(aq) \longrightarrow I_2(s) + 2NaCl(aq)$$

The sodium ions are spectator ions, and the net ionic equation is

$$Cl_2(aq) + 2I^-(aq) \longrightarrow I_2(s) + 2Cl^-(aq)$$

4.4 SOLUTION CONCENTRATION

The **concentration** of a solution is the amount of solute in a given quantity of solvent or solution; it is a measure of the solution's "strength." Calculations involving concentration are frequently required in chemistry, and performing such calculations is not difficult provided the concepts and definitions are well understood.

Percent by Mass

The labels on commercial solutions often give concentrations as **percent by mass**, in which

$$\text{percent by mass} = \frac{\text{mass of solute}}{\text{mass of solution}} \times 100\% \qquad (4.1)$$

Consider, for example, a solution containing 10 g of sucrose in 40 g of water. The solution has a total mass of 10 g + 40 g = 50 g, and the percent by mass of sucrose is

$$\%\ \text{sucrose} = \frac{\text{mass of sucrose}}{\text{mass of solution}} \times 100\%$$

$$= \frac{10\ \text{g}}{50\ \text{g}} \times 100\% = 20\%\ \text{sucrose}$$

The percent by mass of water is 100% − 20% = 80%. This example is straightforward, but there may be more steps in the calculation if density must be used to convert volume into mass, as in the next example.

EXAMPLE 4.12

A solution is made by mixing 50.0 mL of ethanol with 50.0 mL of water. Determine the percent by mass of ethanol in this solution. The densities of ethanol and water are 0.789 g/mL and 1.00 g/mL, respectively.

SOLUTION

The mass of each component is obtained by multiplying its volume by its density (Equation 1.2, p. 22):

$$\text{mass of ethanol} = 50.0\ \text{mL} \times \frac{0.789\ \text{g}}{1\ \text{mL}} = 39.4\ \text{g}$$

$$\text{mass of water} = 50.0\ \text{mL} \times \frac{1.00\ \text{g}}{1\ \text{mL}} = 50.0\ \text{g}$$

The total mass of the solution is the mass of ethanol plus the mass of water, 39.4 g + 50.0 g = 89.4 g. The percent by mass of ethanol is obtained by substituting the mass of ethanol and the total mass into Equation 4.1:

$$\%\ \text{ethanol} = \frac{\text{mass of ethanol}}{\text{mass of solution}} \times 100\% = \frac{39.4\ \text{g}}{89.4\ \text{g}} \times 100\% = 44.1\%$$

PRACTICE EXERCISE 4.12

An aqueous solution has a density of 1.19 g/mL and contains 26.00% sulfuric acid by mass. How many grams of sulfuric acid are in 750 mL of the solution?

Percent means parts per hundred. Similar ratios used for concentration measurements include **parts per thousand** (grams per 10^3 grams), **parts per million** (**ppm**; grams per 10^6 grams), **parts per billion** (**ppb**; grams per 10^9 grams), and **parts per trillion** (**ppt**; grams per 10^{12} grams). (Note that ppt is often used to mean parts per thousand, so watch out!) These units, which usually refer to very small mass concentrations, are becoming more common as better techniques are developed for detecting trace amounts of impurities, pollutants, nutrients, and other scarce substances. Dioxin levels, for example, can now be measured down to three parts per trillion (3 ppt), and the U.S. Food and Drug Administration has advised people not to eat fish with dioxin levels above 50 ppt. (Dioxins are toxic contaminants that form during the incineration of organic material and during the synthesis of certain pesticides and defoliants.) One part per trillion is a very low concentration. It is equivalent to 1 penny in 10 billion dollars, 1 inch in 15.8 million miles, and 1 second in 31,700 years.

EXAMPLE 4.13

An Environmental Protection Agency (EPA) regulation sets the concentration limit for copper in drinking water at 1.3 ppm by mass. Calculate the number of milligrams of copper in 1.0 L of water at this concentration. Assume the density of water to be 1.0 g/mL.

SOLUTION

A copper concentration of 1.3 ppm by mass means 1.3 g of copper in one million (1.00×10^6) g of water. Each liter of water would contain

$$1.0\ \text{L} \times \frac{1000\ \text{mL}}{1\ \text{L}} \times \frac{1.0\ \text{g H}_2\text{O}}{1\ \text{mL}} \times \frac{1.3\ \text{g Cu}}{1.00 \times 10^6\ \text{g H}_2\text{O}} \times \frac{1000\ \text{mg Cu}}{1\ \text{g Cu}} = 1.3\ \text{mg Cu}$$

PRACTICE EXERCISE 4.13

Calculate the number of grams of dioxin in 100 g of fish if the dioxin level is 50 ppt.

Molarity

In chemical reactions the number of moles is usually more important than the mass, so a concentration unit based on moles is more useful than one based on mass. **Molarity** (M) is defined as the number of moles of solute per liter of solution:

$$\text{molarity} = \frac{\text{number of moles of solute}}{\text{volume of solution in liters}} \qquad (4.2a)$$

Molarity is the most convenient concentration unit for solution stoichiometry.

EXAMPLE 4.14

Calculate the molarity of a solution containing 0.750 mol of HCl in 335 mL of solution.

SOLUTION

The volume of the solution is 335 mL or 0.335 L. The molarity of HCl is found by substituting the number of moles of HCl and the volume of the solution into Equation 4.2a:

$$\text{molarity} = \frac{\text{number of moles of HCl}}{\text{volume of solution in liters}}$$

$$= \frac{0.750\ \text{mol HCl}}{0.335\ \text{L}} = 2.24\ \text{mol HCl/L}$$

The solution is 2.24 molar (2.24 M) with respect to HCl.

The molarity of a solution varies somewhat with the temperature. A rise in temperature usually causes a solution to expand, thus increasing the volume and decreasing the molarity.

PRACTICE EXERCISE 4.14

A 750-mL solution contains 1.22 mol of potassium nitrate (KNO_3). Calculate the molarity of the potassium nitrate solution.

EXAMPLE 4.15

Calculate the molarity of a solution that contains 13.5 g of sodium sulfate in 850 mL of solution. The molar mass of Na_2SO_4 is 142.0 g/mol.

SOLUTION

The number of moles of sodium sulfate in the solution is

$$13.5 \text{ g } Na_2SO_4 \times \frac{1 \text{ mol } Na_2SO_4}{142.0 \text{ g } Na_2SO_4} = 0.0951 \text{ mol } Na_2SO_4$$

The volume of the solution is 850 mL or 0.850 L. The molarity is obtained by substituting the number of moles of sodium sulfate and the volume of the solution into Equation 4.2a:

$$\text{molarity} = \frac{\text{number of moles of } Na_2SO_4}{\text{volume of solution in liters}}$$

$$= \frac{0.0951 \text{ mol } Na_2SO_4}{0.850 \text{ L}} = 0.112 \text{ mol } Na_2SO_4\text{/L}$$

The Na_2SO_4 solution is 0.112 *M*. This problem can also be done by combining the steps as follows:

$$\frac{13.5 \text{ g } Na_2SO_4}{850 \text{ mL}} \times \frac{1000 \text{ mL}}{1 \text{ L}} \times \frac{1 \text{ mol } Na_2SO_4}{142.0 \text{ g } Na_2SO_4} = 0.112 \text{ mol } Na_2SO_4\text{/L}$$

PRACTICE EXERCISE 4.15

Calculate the molarity of a solution that contains 5.25 g of $AgNO_3$ in 125 mL of solution.

The molarities of commercial solutions of acids and bases are not usually provided by the manufacturer, but they can be calculated from information on the product label.

EXAMPLE 4.16

A solution of aqueous sulfuric acid (Figure 4.13) has a density of 1.840 g/mL and contains 98.0% sulfuric acid by mass. Calculate the molarity of the solution. The molar mass of H_2SO_4 is 98.08 g/mol.

SOLUTION

The molarity is the number of moles of sulfuric acid in 1 liter of solution. Our strategy will be to (1) find the total mass (water plus acid) of 1 liter of solution, (2) find the mass of sulfuric acid in 1 liter of solution, and (3) convert the mass of sulfuric acid to moles.

Step 1: The total mass of 1 liter (1000 mL) of solution is found from its density:

$$\frac{1000 \text{ mL}}{1 \text{ L}} \times \frac{1.840 \text{ g solution}}{1 \text{ mL}} = 1840 \text{ g solution/L}$$

Step 2: The solution contains 98.0% sulfuric acid by mass. This means that 100 g of the solution contains 98.0 g of H_2SO_4. The mass of sulfuric acid in 1 liter of solution is

$$\frac{1840 \text{ g solution}}{1 \text{ L}} \times \frac{98.0 \text{ g } H_2SO_4}{100 \text{ g solution}} = 1.80 \times 10^3 \text{ g } H_2SO_4/\text{L}$$

Step 3: The number of moles of sulfuric acid in 1 liter of solution is obtained from the molar mass:

$$\frac{1.80 \times 10^3 \text{ g } H_2SO_4}{1 \text{ L}} \times \frac{1 \text{ mol } H_2SO_4}{98.08 \text{ g } H_2SO_4} = 18.4 \text{ mol } H_2SO_4/\text{L}$$

The H_2SO_4 solution is 18.4 *M*. (Try to do this problem by combining the steps as in the previous example.)

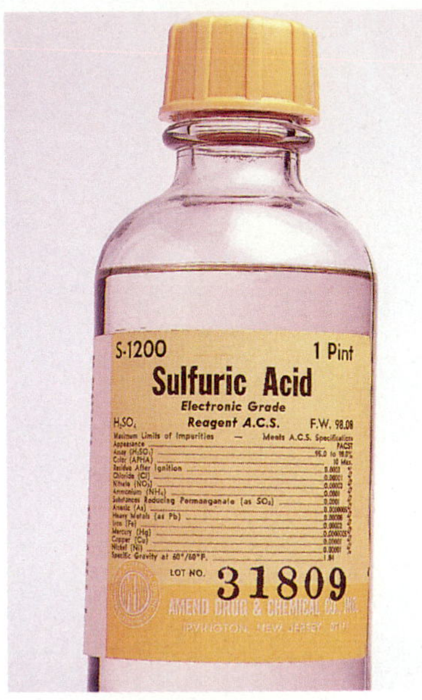

Figure 4.13

The label on a bottle of sulfuric acid describes its contents and gives the maximum concentration of each impurity. The third line "Assay (H_2SO_4)" tells us that the solution contains between 95 and 98% H_2SO_4 by mass. The last line "Specific Gravity" tells us that the density of the solution at 60°F is 1.84 times the density of water at 60°F (see Learning Hint on p. 24).

PRACTICE EXERCISE 4.16

A solution of nitric acid (HNO_3) has a density of 1.424 g/mL and contains 70.9% nitric acid by mass. Calculate the molarity of the solution.

The following example shows how molarity can be used as a conversion factor to find the number of moles of solute in a given solution.

EXAMPLE 4.17

A laboratory experiment calls for 0.300 *M* KOH solution. Calculate the number of moles of potassium hydroxide needed to make 150 mL of the solution.

SOLUTION

One liter of the solution requires 0.300 mol of KOH; 150 mL (0.150 L) of solution requires

$$0.150 \text{ L} \times \frac{0.300 \text{ mol KOH}}{1 \text{ L}} = 0.0450 \text{ mol KOH}$$

PRACTICE EXERCISE 4.17

How many milliliters of 1.50 *M* nitric acid solution will contain 0.250 mol of HNO_3?

Observe that when we used molarity as a conversion factor in Example 4.17, we actually multiplied volume by molarity to find the number of moles; in other words

$$\text{number of moles} = V\,(\text{L}) \times M\,(\text{mol/L}) \qquad (4.2b)$$

This equation, which could also have been obtained by rearranging Equation 4.2a, shows that the *number of moles of a dissolved solute is equal to the volume of the solution multiplied by the molarity of the solute*. The use of Equation 4.2b is illustrated in the following examples.

EXAMPLE 4.18

Calculate the number of grams of solute in (a) 150 mL of 0.30 *M* NaOH and (b) 150 mL of 0.30 *M* $BaCl_2$.

SOLUTION

The molar masses of NaOH and $BaCl_2$ are 40.00 and 208.2 g/mol, respectively. Each solution has a volume of 150 mL (0.150 L) and contains 0.30 mol of solute per liter. The number of moles of solute in each solution is $V \times M = 0.150 \text{ L} \times 0.30 \text{ mol/L} =$ 0.045 mol. The mass of solute in each solution is

(a) $0.045 \text{ mol NaOH} \times \frac{40.00 \text{ g NaOH}}{1 \text{ mol NaOH}} = 1.8 \text{ g NaOH}$

(b) $0.045 \text{ mol } BaCl_2 \times \frac{208.2 \text{ g } BaCl_2}{1 \text{ mol } BaCl_2} = 9.4 \text{ g } BaCl_2$

PRACTICE EXERCISE 4.18

How many grams of $NaNO_3$ are in 250 mL of a 1.45 *M* sodium nitrate solution?

Analytical procedures often require solution concentrations to be accurate to four significant figures. To make such a solution, a very pure solute must be weighed on an analytical balance. The solution is diluted to a precise volume in a *volumetric flask* like the one shown in Demonstration 4.2.

EXAMPLE 4.19

Sodium oxalate ($Na_2C_2O_4$) is a white crystalline salt used as an analytical reagent. Describe how to make 250.0 mL of 0.1000 *M* sodium oxalate solution.

SOLUTION

One liter (1000 mL) of the 0.1000 *M* solution will contain 0.1000 mol of sodium oxalate. The number of moles of sodium oxalate required for 250.0 mL is

$$250.0 \text{ mL} \times \frac{0.1000 \text{ mol } Na_2C_2O_4}{1000 \text{ mL}} = 0.02500 \text{ mol } Na_2C_2O_4$$

Sodium oxalate, a solid, is measured by weight. Its molar mass is 134.0 g/mol, and the mass of 0.02500 mol is

$$0.02500 \text{ mol } Na_2C_2O_4 \times \frac{134.0 \text{ g } Na_2C_2O_4}{1 \text{ mol } Na_2C_2O_4} = 3.350 \text{ g } Na_2C_2O_4$$

This amount of sodium oxalate is weighed out on an analytical balance and placed in a 250-mL volumetric flask as shown in Demonstration 4.2. Enough water is added to dissolve the salt; then enough additional water is added to bring the volume up to the 250-mL mark etched on the neck of the flask.

LEARNING HINT

Remember that molarity is based on the volume of *solution*, not on the volume of solvent.

DEMONSTRATION 4.2 PREPARING 250.0 mL OF 0.1000 *M* SODIUM OXALATE SOLUTION

A 3.350-g portion of sodium oxalate (0.2500 mol $Na_2C_2O_4$) is weighed and put into a 250.0-mL volumetric flask.

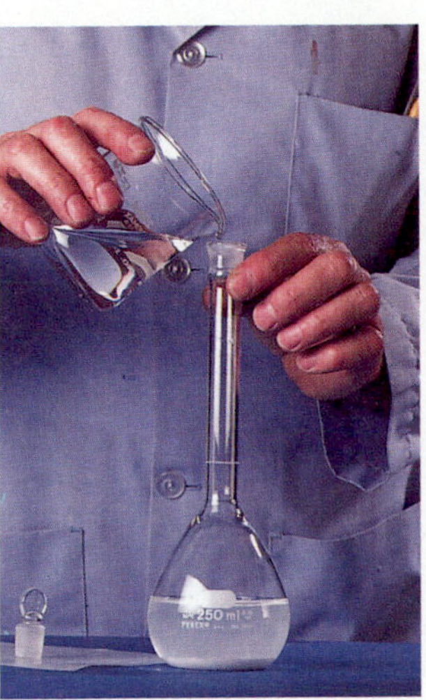

The flask is partially filled with distilled water.

The stoppered flask is inverted and shaken to dissolve the sodium oxalate.

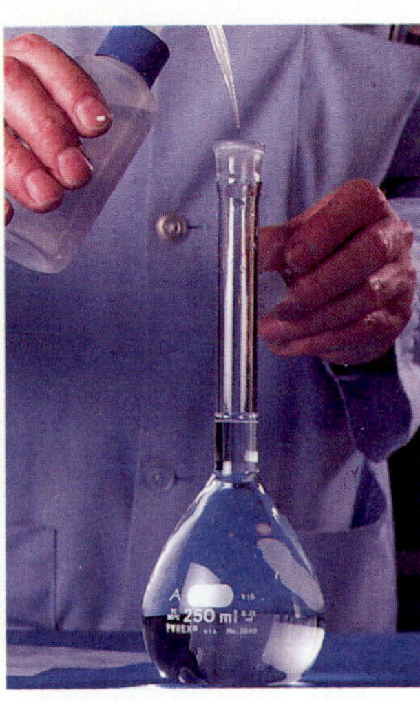

Distilled water is added until the bottom of the meniscus (curved liquid surface) reaches the etched line on the flask. (The flask will then be inverted repeatedly to mix the contents.)

PRACTICE EXERCISE 4.19

Describe how to prepare 100.0 mL of 0.250 *M* Na_2CO_3 solution.

Molarities of Ions

The molarity of an ion is often different from the molarity of its compound. For example, when 1 mol of solid aluminum sulfate dissolves in water, it dissociates into 2 mol of aluminum ions and 3 mol of sulfate ions:

$$Al_2(SO_4)_3(s) \rightarrow 2Al^{3+}(aq) + 3SO_4^{2-}(aq)$$

Thus the molarity of aluminum ions in the solution will be twice the molarity of the aluminum sulfate, and the molarity of sulfate ions will be three times the molarity of the aluminum sulfate. For example, a 0.60 *M* $Al_2(SO_4)_3$ solution will be $2 \times 0.60\ M = 1.2\ M$ in Al^{3+} and $3 \times 0.60\ M = 1.8\ M$ in SO_4^{2-}. As this illustration shows, *when an ionic compound dissolves in water, the molarity of each dissolved ion will be a small whole-number multiple of the molarity of the compound.*

PRACTICE EXERCISE 4.20

How many moles of (a) sulfate ion and (b) iron(III) ion are in 1.00 L of 0.750 *M* $Fe_2(SO_4)_3$?

EXAMPLE 4.20

Find the molarity of sodium ion in a solution obtained by mixing 300 mL of 0.450 *M* NaCl with 200 mL of 0.250 *M* Na_2CO_3.

SOLUTION

First we find the number of moles of sodium ion in each of the original solutions.

NaCl: One mole of NaCl provides 1 mol of sodium ion. The number of moles of sodium ion in 300 mL (0.300 L) of the sodium chloride solution is

$$0.300\text{ L} \times \frac{0.450\text{ mol NaCl}}{1\text{ L}} \times \frac{1\text{ mol Na}^+}{1\text{ mol NaCl}} = 0.135\text{ mol Na}^+$$

Na_2CO_3: One mol of Na_2CO_3 provides 2 mol of sodium ion. The number of moles of sodium ion in 200 mL (0.200 L) of the sodium carbonate solution is

$$0.200\text{ L} \times \frac{0.250\text{ mol Na}_2\text{CO}_3}{1\text{ L}} \times \frac{2\text{ mol Na}^+}{1\text{ mol Na}_2\text{CO}_3} = 0.100\text{ mol Na}^+$$

When the solutions are mixed, there will be 0.135 mol + 0.100 mol = 0.235 mol of sodium ion in a volume of 0.300 L + 0.200 L = 0.500 L. The final molarity of the sodium ion is obtained by substituting these values into Equation 4.2a:

$$\text{molarity} = \frac{\text{number of moles of Na}^+}{\text{volume of solution in liters}}$$

$$= \frac{0.235\text{ mol Na}^+}{0.500\text{ L}} = 0.470\text{ mol Na}^+/\text{L}$$

The solution is 0.470 *M* with respect to sodium ion.

PRACTICE EXERCISE 4.21

Find the molarity of chloride ion in a solution prepared by mixing 250 mL of 0.125 *M* $MgCl_2$ with 800 mL of 0.350 *M* $FeCl_3$.

Some Common Laboratory Solutions

Certain solutions are available in every laboratory and are almost as familiar to the chemist as water from the tap. These reagents are supplied in large bottles of concentrated solution that can be diluted to other strengths. Some frequently used solutions are listed in Table 4.7.

Hydrochloric acid (HCl) is the strong acid most commonly used in the introductory laboratory. Pure hydrogen chloride is a gas that readily dissolves in water. Concentrated hydrochloric acid is usually about 12 *M* and contains 36% HCl by mass. It has a harsh, penetrating odor and fumes in moist air. Commercial-grade hydrochloric acid, called *muriatic acid,* is used to dissolve rust and other oxides from metal and to remove scale ($CaCO_3$) from boilers and pipes. Hydrochloric acid is also found in human gastric juice, where a hydrogen ion concentration between 0.01 and 0.10 *M* is needed for the action of the digestive enzyme trypsin.

Nitric acid (HNO_3) is a strong acid. Unlike HCl, it is also an oxidizing agent

TABLE 4.7 Common Laboratory Reagents and Their Concentrations[a]

Name	Formula	Molar Mass (g/mol)	Molarity	Percent by Mass	Density (g/mL)
Acetic acid, glacial	CH_3COOH	60.05	17.4	99.5	1.05
Aqueous ammonia	NH_3	17.03	14.8	28	0.90
Hydrochloric acid	HCl	36.46	11.6	36	1.18
Nitric acid	HNO_3	63.01	16.0	71	1.42
Sulfuric acid	H_2SO_4	98.08	18.0	98	1.84

[a]Of concentrated stock solutions.

capable of dissolving substances such as copper, silver, and sulfur. Concentrated nitric acid is about 16 *M* and contains 71% HNO_3 by mass. Although pure nitric acid is colorless, concentrated solutions that stand in the light will decompose slightly and become yellowish. A solution of concentrated nitric acid must be handled with extreme care because its combined acidity and oxidizing power make it a potentially hazardous reagent.

Sulfuric acid (H_2SO_4) is the nation's leading synthetic industrial chemical as well as an important laboratory reagent. It is a strong acid with two acidic hydrogen atoms per molecule. Concentrated sulfuric acid is about 18 *M* and contains less than 4% water by mass. It is a syrupy liquid with a strong affinity for water and is a powerful dehydrating agent. It removes water from organic material such as sugar, paper, and flesh, leaving a residue of black carbon. Sulfuric acid is also an oxidizing agent, but it is not as strong in this respect as nitric acid.

Acetic acid (CH_3COOH) is a weak acid that partially ionizes in water to form hydrogen ions and acetate ions (Section 4.1). Concentrated acetic acid, also called *glacial acetic acid* because it freezes to icelike crystals, is about 17.4 *M* and contains less than 0.5% water by mass. Glacial acetic acid can burn the skin, and its vapors are dangerous to breathe. *Vinegar* is a dilute aqueous solution containing from 3 to 5% acetic acid by mass.

Ammonia (NH_3) is a weak base (Section 4.1). Pure ammonia is a colorless gas with a distinctive choking odor. It is very soluble in water, and a concentrated solution is about 15 *M*. The preferred name for a solution of ammonia in water is *aqueous ammonia*, but the name *ammonium hydroxide* is often used because the aqueous solution contains small concentrations of ammonium and hydroxide ions.

PRACTICE EXERCISE 4.22

Write formulas for (a) acetic acid, (b) hydrochloric acid, (c) sulfuric acid, (d) nitric acid, and (e) aqueous ammonia.

4.5 PREPARATION OF SOLUTIONS BY DILUTION

Concentrated laboratory reagents are often diluted to meet specific needs. A dilute solution is prepared by removing a small volume of concentrated reagent from the stock bottle and adding it to water to make a larger volume of less concentrated reagent. Adding water does not change the number of moles of solute; hence *the number of moles of solute in the diluted solution will be equal to the original number of moles in the concentrated solution.* Let V_C and M_C be the volume and molarity of the concentrated sample, and V_D and M_D the volume and molarity after dilution. The number of moles in the concentrated solution is the product of its volume (in liters) and molarity (in moles per liter): $V_C \times M_C$. The number of moles in the dilute solution is $V_D \times M_D$. Equating the two products gives

$$V_C \times M_C = V_D \times M_D \tag{4.3}$$

a relationship often referred to as the **dilution formula**.

EXAMPLE 4.21

How many milliliters of 18.0 M H_2SO_4 are required to prepare 600 mL of 6.00 M H_2SO_4 by dilution?

SOLUTION

CAUTION! Concentrated acids, especially sulfuric acid, are diluted by pouring them slowly with stirring into about half the desired volume of water. Then enough water is added to the partially diluted solution to bring it up to the final volume. Mixing water with acid generates heat, and water added directly to a concentrated acid may vaporize, spatter, and cause burns.

The diluted solution will have a volume V_D of 600 mL (0.600 L) and a concentration M_D of 6.00 mol/L. The molarity M_C of the concentrated acid is 18.0 mol/L, and its volume V_C is unknown. Substituting the known quantities into the dilution formula gives

$$V_C \times M_C = V_D \times M_D$$

$$V_C \times 18.0 \text{ mol/L} = 0.600 \text{ L} \times 6.00 \text{ mol/L}$$

and

$$V_C = \frac{0.600 \text{ L} \times 6.00 \text{ mol/L}}{18.0 \text{ mol/L}} = 0.200 \text{ L} = 200 \text{ mL}$$

The dilute solution is therefore made by removing 200 mL of concentrated acid from the stock bottle and adding it to enough water to make a final volume of 600 mL. The procedure is shown in Demonstration 4.3.

PRACTICE EXERCISE 4.23

How many milliliters of 12 M HCl are required to prepare 1.0 L of 0.10 M HCl?

4.6 SOLUTION STOICHIOMETRY

Stoichiometry is based on molar quantities, and the relation between moles, volume, and molarity (Equation 4.2b) can be used to calculate the molar quantities of dissolved substances from concentration and volume data. The use of Equation 4.2b is illustrated in the following examples.

DEMONSTRATION 4.3 PREPARING 600 mL OF A DILUTE (6 *M*) H_2SO_4 SOLUTION FROM A CONCENTRATED (18 *M*) SOLUTION

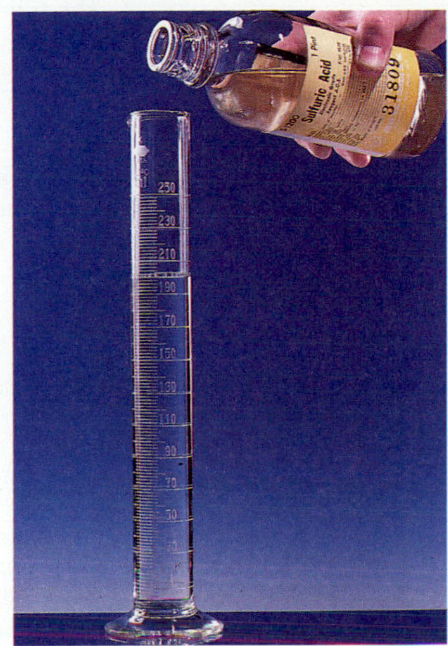

A graduated cylinder is used to measure out 200 mL of the concentrated acid.

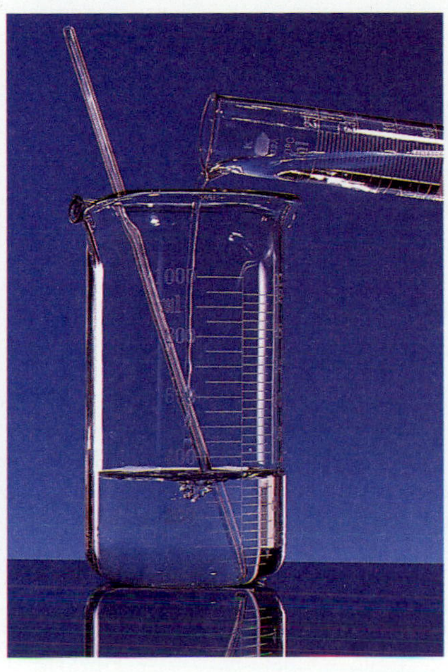
A 1-L beaker is partially filled with distilled water; the concentrated acid is slowly added with constant stirring.

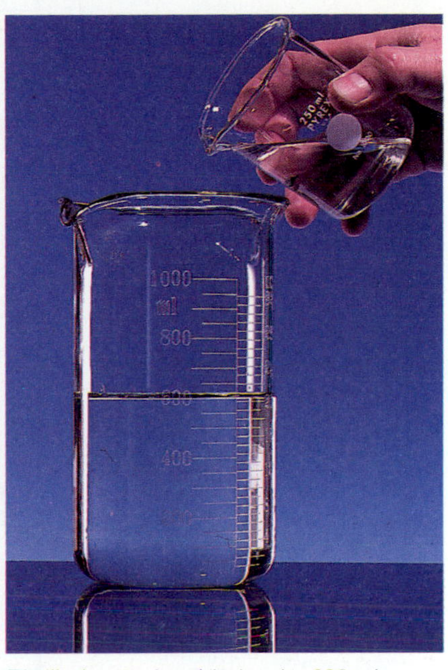

Distilled water is added to the 600-mL mark. The solution will be stirred again to make sure it is uniform.

EXAMPLE 4.22

How many milliliters of 1.25 *M* HCl are needed to neutralize 50.0 mL of 1.66 *M* NaOH?

SOLUTION

This problem can be solved by the four-step procedure outlined in Section 3.7. The balanced equation for the neutralization is

$$NaOH(aq) + HCl(aq) \longrightarrow NaCl(aq) + H_2O(l)$$

The volume of base is 50.0 mL (0.0500 L); its molarity is 1.66 mol NaOH per liter. The volume of 1.25 *M* HCl is calculated as follows:

$$V \times M \text{ (of NaOH)} \longrightarrow \text{mol NaOH} \longrightarrow \text{mol HCl} \longrightarrow \text{mL HCl}$$

$$0.0500 \text{ L NaOH} \times \frac{1.66 \text{ mol NaOH}}{1 \text{ L NaOH}} \times \frac{1 \text{ mol HCl}}{1 \text{ mol NaOH}} \times \frac{1 \text{ L HCl}}{1.25 \text{ mol HCl}}$$

$$= 0.664 \text{ L HCl} = 66.4 \text{ mL HCl}$$

PRACTICE EXERCISE 4.24

Calculate the molarity of a sulfuric acid solution if 10.0 mL of the solution is completely neutralized by 40.0 mL of 0.750 *M* sodium hydroxide solution.

EXAMPLE 4.23

Calculate the number of grams of calcium carbonate ($CaCO_3$) that will be dissolved by 1000 L of 6.00×10^{-4} *M* HCl. (This acid concentration is similar to acid concentrations found in some rain and fog.)

SOLUTION

The equation for the reaction is

$$CaCO_3(s) + 2HCl(aq) \rightarrow CaCl_2(aq) + CO_2(g) + H_2O(l)$$

The volume of acid is 1000 L, and its molarity is 6.00×10^{-4} mol/L. The molar mass of calcium carbonate is 100.1 g/mol; the mass of calcium carbonate that reacts with the acid is calculated as follows:

$$V \times M \text{ (of HCl)} \rightarrow \text{mol HCl} \rightarrow \text{mol } CaCO_3 \rightarrow \text{g } CaCO_3$$

$$1000 \text{ L HCl} \times \frac{6.00 \times 10^{-4} \text{ mol HCl}}{1 \text{ L HCl}} \times \frac{1 \text{ mol } CaCO_3}{2 \text{ mol HCl}} \times \frac{100.1 \text{ g } CaCO_3}{1 \text{ mol } CaCO_3}$$
$$= 30.0 \text{ g } CaCO_3$$

PRACTICE EXERCISE 4.25

How many milliliters of 1.25 *M* HNO_3 will react with 5.23 g of sodium hydrogen carbonate ($NaHCO_3$)?

EXAMPLE 4.24

How many milliliters of 0.250 *M* barium chloride are needed to precipitate the silver ion in 50.0 mL of 0.120 *M* silver nitrate solution? The equation is

$$2AgNO_3(aq) + BaCl_2(aq) \rightarrow 2AgCl(s) + Ba(NO_3)_2(aq)$$

SOLUTION

The volume of the silver nitrate solution is 50.0 mL (0.0500 L); its molarity is 0.120 mol/L. The volume of 0.250 *M* $BaCl_2$ solution is calculated as follows:

$$V \times M \text{ (of } AgNO_3) \rightarrow \text{mol } AgNO_3 \rightarrow \text{mol } BaCl_2 \rightarrow \text{mL } BaCl_2$$

$$0.0500 \text{ L } AgNO_3 \times \frac{0.120 \text{ mol } AgNO_3}{1 \text{ L } AgNO_3} \times \frac{1 \text{ mol } BaCl_2}{2 \text{ mol } AgNO_3} \times \frac{1 \text{ L } BaCl_2}{0.250 \text{ mol } BaCl_2}$$
$$= 12.0 \times 10^{-3} \text{ L } BaCl_2 = 12.0 \text{ mL } BaCl_2$$

PRACTICE EXERCISE 4.26

A solution is made by dissolving 5.55 g of $BaCl_2$ in water. How many milliliters of 2.00 *M* sulfuric acid are required to precipitate the barium ions according to the following equation?

$$BaCl_2(aq) + H_2SO_4(aq) \rightarrow BaSO_4(s) + 2HCl(aq)$$

Volumetric Analysis

A **volumetric analysis** is any analytical procedure in which quantitative information is obtained by measuring the *volume* of a reacting solution. It differs from a gravimetric analysis (Section 4.2) in which the mass of a product is measured. Volumetric methods of analysis are often preferred over gravimetric methods because they are quicker and easier to perform.

A volumetric analysis involves the controlled addition of one solution to another, a process called **titration**. During an acid–base titration, for example, a solution of acid is added to a solution of base (or vice versa) until the neutralization reaction is complete. A typical acid–base titration is illustrated in Figure 4.14. A precise volume of base, in this case 50.00 mL of $Ba(OH)_2$ solution, is dispensed from a *pipet* into a flask, and a few drops of an indicator solution are added to it. An **indicator** is a dye that has been chosen because it changes color at the end of the reac-

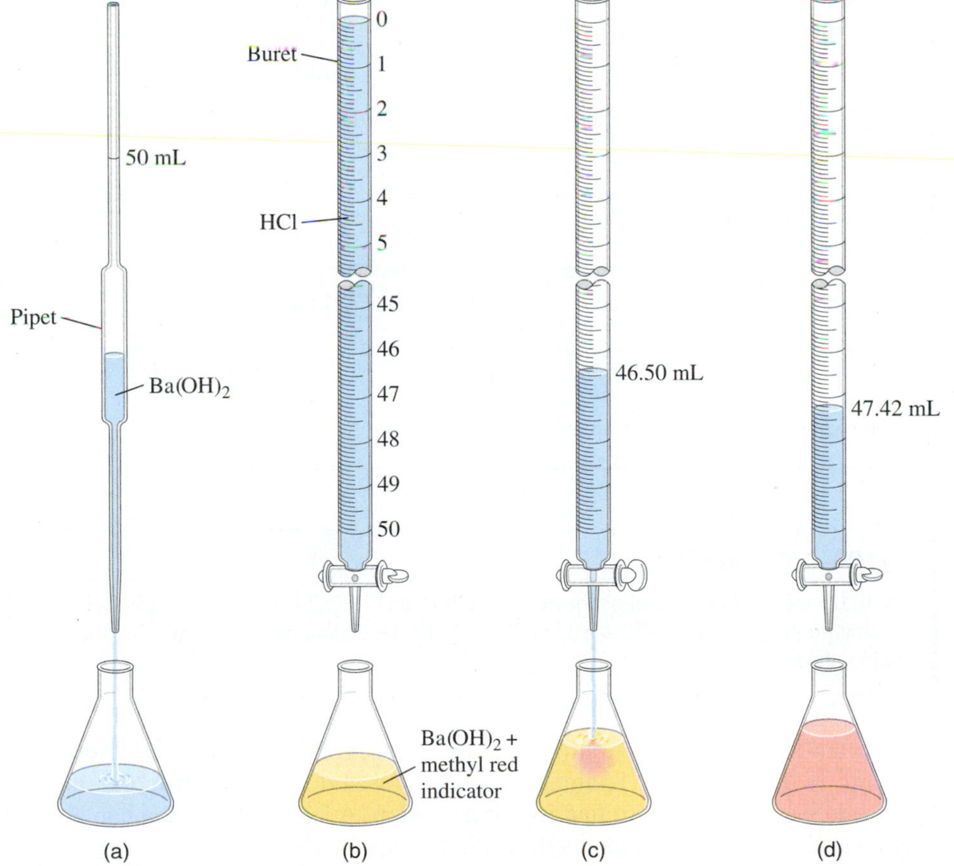

Figure 4.14

An acid–base titration. (a) A 50.00-mL portion of a $Ba(OH)_2$ solution of unknown concentration is pipetted into a flask. (b) Methyl red is added to the flask; it turns yellow in the basic solution. The buret contains standard 0.1234 *M* HCl. (c) The HCl solution is gradually added to the flask. As the endpoint nears, each new drop of acid produces a fleeting red color that disappears with mixing. (d) The endpoint is reached when one drop of acid produces a permanent red color. The buret readings show that 47.42 mL of acid was added.

tion. (Indicators are discussed more fully in Chapter 17.) Two indicators often used for acid–base titrations are methyl red (yellow in basic solutions, red in acidic solutions) and phenolphthalein (reddish pink in basic solutions, colorless in acidic solutions). The acid, in this case an HCl solution of known concentration, is added gradually to the flask from a *buret*, a long thin tube on which volume levels can be read with a high degree of precision. The contents of the flask are swirled continuously to ensure mixing. The addition of acid is stopped at the **endpoint** of the titration, that is, when the indicator changes color. If the indicator has been correctly chosen, this will also be the point at which the base has been exactly neutralized by the acid.

The next example illustrates how acid–base titration data are handled.

EXAMPLE 4.25

In the titration shown in Figure 4.14, 47.42 mL of 0.1234 *M* HCl neutralizes a 50.00-mL sample of $Ba(OH)_2$ solution. Calculate the molarity of the barium hydroxide solution.

SOLUTION

Titration problems are solved by the same method used for other stoichiometry problems. The equation for the neutralization is

$$Ba(OH)_2(aq) + 2HCl(aq) \rightarrow BaCl_2(aq) + 2H_2O(l)$$

The known quantity is 47.42 mL of 0.1234 *M* HCl solution. The volume of the $Ba(OH)_2$ solution is 50.00 mL; its molarity is unknown. First we find the number of moles of $Ba(OH)_2$:

$$V \times \text{M (of HCl)} \rightarrow \text{mol HCl} \rightarrow \text{mol Ba(OH)}_2$$

$$0.04742 \text{ L HCl} \times \frac{0.1234 \text{ mol HCl}}{1 \text{ L HCl}} \times \frac{1 \text{ mol Ba(OH)}_2}{2 \text{ mol HCl}} = 0.002926 \text{ mol Ba(OH)}_2$$

Volumetric glassware (like a buret or a pipet) is customarily calibrated at 20°C. The changes in volume that occur with changing temperature are small and can be ignored for most work.

This number of moles was present in the original 50.00 mL (0.05000 L) of solution. The molarity of the original solution is obtained using Equation 4.2a:

$$\text{molarity} = \frac{\text{number of moles of Ba(OH)}_2}{\text{volume of Ba(OH)}_2 \text{ solution in liters}}$$

$$= \frac{0.002926 \text{ mol Ba(OH)}_2}{0.05000 \text{ L}} = 0.05852 \text{ mol Ba(OH)}_2\text{/L}$$

The barium hydroxide solution is 0.05852 *M*.

PRACTICE EXERCISE 4.27

A 40.00-mL portion of a phosphoric acid solution (H_3PO_4) is titrated to a phenolphthalein endpoint with 25.54 mL of 2.111 *M* KOH. Calculate the molarity of the acid solution.

A solution whose concentration is known is called a **standard solution**. The standard HCl solution in Example 4.25 was used to **standardize** the barium hydroxide solution, that is, to find its concentration. A standard solution can also be used to analyze a solid mixture, as shown in Example 4.26.

EXAMPLE 4.26

A 0.7755-g portion of a solid mixture containing sodium hydroxide and unreactive impurities is dissolved in water and titrated with standard 0.1000 M H_2SO_4. If 34.44 mL of the acid is required to neutralize the sample, what is the percent by mass of NaOH in the mixture?

SOLUTION

The equation for the neutralization is

$$2NaOH(aq) + H_2SO_4(aq) \longrightarrow Na_2SO_4(aq) + 2H_2O(l)$$

The known quantity is 34.44 mL of 0.1000 M H_2SO_4 solution. The unknown quantity is the percentage of NaOH in the solid sample. The molar mass of NaOH is 40.00 g; the number of grams of NaOH in the sample is found as follows:

$$V \times M \text{ (of } H_2SO_4) \longrightarrow \text{mol } H_2SO_4 \longrightarrow \text{mol NaOH} \longrightarrow \text{g NaOH}$$

$$0.03444 \text{ L } H_2SO_4 \times \frac{0.1000 \text{ mol } H_2SO_4}{1 \text{ L } H_2SO_4} \times \frac{2 \text{ mol NaOH}}{1 \text{ mol } H_2SO_4} \times \frac{40.00 \text{ g NaOH}}{1 \text{ mol NaOH}}$$

$$= 0.2755 \text{ g NaOH}$$

The percent by mass of NaOH in the sample is

$$\% \text{ NaOH} = \frac{\text{mass of NaOH}}{\text{total mass}} \times 100\%$$

$$= \frac{0.2755 \text{ g}}{0.7755 \text{ g}} \times 100\% = 35.53\%$$

PRACTICE EXERCISE 4.28

A 2.9929-g sample of impure sodium carbonate is dissolved in water and titrated to a methyl red endpoint with 0.4150 M HCl. The reaction is

$$Na_2CO_3(s) + 2HCl(aq) \longrightarrow 2NaCl(aq) + H_2O(l) + CO_2(g)$$

If 33.75 mL of the acid is used for the reaction, what is the percent by mass of sodium carbonate in the sample?

CHAPTER REVIEW

LEARNING OBJECTIVES BY SECTION

4.1
1. Distinguish between a solution and a heterogeneous mixture.
2. Give examples of solid, liquid, and gaseous solutions.
3. Describe the dynamic equilibrium that exists in a saturated solution.
4. Distinguish between unsaturated, saturated, and supersaturated solutions.
5. Write equations for the dissociation of ionic compounds.
6. Write equations for the ionization of strong and weak molecular electrolytes.

4.2
1. Write molecular and net ionic equations for precipitation reactions.
2. Use solubility rules to predict whether a precipitation reaction will occur.
3. Perform the calculations required for a gravimetric analysis.
4. Write equations for complete and stepwise acid–base neutralizations.

5. Find a neutralization reaction to produce a given salt.
6. Write equations for metathesis reactions that produce NH_3, HCN, H_2S, CO_2, and SO_2 gases.

4.3 1. Write molecular and net ionic equations for displacement reactions.
2. Use the activity series to predict whether or not a displacement reaction will occur.

4.4 1. Calculate the percent by mass of each component in a solution, given the appropriate data.
2. Use parts per thousand, parts per million, parts per billion, and parts per trillion in concentration calculations.
3. Describe how to make solutions of a given molarity.
4. Calculate molarity, solution volume, or number of moles of solute, given two of the three quantities.
5. Calculate molarity, solution volume, or mass of solute, given two of the three quantities.
6. Calculate the molarity of a solution given its density and percent composition by mass.
7. Calculate the molarity of an ion in solution from the molarity of the compound supplying the ion.

4.5 1. Calculate the volume of concentrated solution needed to make a given volume of dilute solution.

4.6 1. Use molarities and solution volumes in stoichiometric calculations.
2. Describe a titration procedure.
3. Perform typical titration calculations.

KEY TERMS BY SECTION

4.1 Aqueous solution
Dissociation
Dynamic equilibrium
Electrolyte
Heterogeneous mixture
Homogeneous mixture
Ionic solution
Ionization
Nonelectrolyte
Saturated solution
Solubility
Solute
Solution
Solvent
Strong electrolyte
Supersaturated solution
Unsaturated solution
Weak electrolyte

4.2 Acid
Alkali
Base
Gravimetric analysis
Metathesis reaction
Net ionic equation
Neutralization
Precipitate
Salt
Spectator ion
Strong acid
Strong base
Weak acid
Weak base

4.3 Activity series

4.4 Concentration
Molarity (*M*)
Parts per billion (ppb)
Parts per million (ppm)
Parts per thousand
Parts per trillion (ppt)
Percent
Percent by mass

4.5 Dilution formula

4.6 Endpoint
Indicator
Standardization (of a solution)
Standard solution
Titration
Volumetric analysis

IMPORTANT EQUATIONS

4.1 $$\text{percent by mass} = \frac{\text{mass of solute}}{\text{mass of solution}} \times 100\%$$

4.2a $$\text{molarity} = \frac{\text{number of moles of solute}}{\text{volume of solution in liters}}$$

4.2b $$\text{number of moles} = V\,(\text{L}) \times M\,(\text{mol/L})$$

4.3 $$V_C \times M_C = V_D \times M_D$$

FINAL EXERCISES

Answers to exercises with blue numbers are given in Appendix D.

PART A. QUESTIONS AND PROBLEMS BY SECTION

Solutions and Mixtures (Section 4.1)

4.1 Which of the following mixtures are homogeneous and which are heterogeneous?
(a) vinegar
(b) clean air
(c) chicken soup
(d) 14-karat gold
(e) smog
(f) gasoline

4.2 Which of the following mixtures are homogeneous and which are heterogeneous?
(a) wine
(b) seawater
(c) cocoa
(d) wood
(e) sterling silver
(f) ammonia water

4.3 Give an example of
(a) a gaseous solution
(b) a liquid solution in which the solute is a solid
(c) a solid solution in which the solute is a liquid
(d) a solid solution in which the solute is a gas

4.4 Give an example of
(a) a solid solution in which the solute is a solid
(b) a liquid solution in which the solute is a liquid
(c) a liquid solution in which the solute is a gas

4.5 What will happen when a crystal of solute is added to (a) an unsaturated solution, (b) a saturated solution, and (c) a supersaturated solution?

4.6 Describe the dynamic equilibrium that exists in a saturated solution.

Ionic Solutions (Section 4.1)

4.7 Explain how electrical conductivity is used to distinguish between strong electrolytes, weak electrolytes, and nonelectrolytes. Diagram and explain the device that is used.

4.8 Use information from this chapter to identify each of the following compounds as either a nonelectrolyte, a weak electrolyte, a strong electrolyte that is an ionic compound, or a strong electrolyte that is a molecular compound:
(a) CH_3COOH
(b) $C_6H_{12}O_6$ (glucose)
(c) KOH
(d) H_2SO_4
(e) NH_3
(f) C_2H_5OH (ethanol)

4.9 Write equations to represent the dissociation in aqueous solution of the following ionic compounds:
(a) potassium carbonate
(b) sodium hydrogen carbonate
(c) iron(II) sulfate
(d) barium hydroxide
(e) calcium chloride
(f) sodium phosphate

4.10 Write equations to represent the ionization in aqueous solution of the following molecular compounds:
(a) ammonia, NH_3
(b) hydrogen chloride, HCl
(c) acetic acid, CH_3COOH
(d) nitric acid, HNO_3, a strong electrolyte
(e) hydrogen fluoride, HF, a weak electrolyte
(f) perchloric acid, $HClO_4$, a strong electrolyte

Metathesis; Precipitation (Section 4.2)

4.11 What is a *metathesis reaction*? Explain why a metathesis reaction between ions in aqueous solution must produce either a precipitate or a molecular compound.

4.12 Do the precipitation reactions described in this chapter involve changes of identity for the ions involved? If not, why are they classified as chemical reactions?

4.13 Use the solubility rules to predict whether the following solids are soluble or sparingly soluble:
(a) $Pb(NO_3)_2$
(b) Li_2CO_3
(c) Hg_2Cl_2
(d) $Mg(OH)_2$
(e) $FeSO_4$
(f) $AlPO_4$

4.14 Use the solubility rules to predict whether or not a precipitation reaction will occur when the following aqueous solutions are mixed. Write the molecular and net ionic equations for each reaction that does occur.
(a) $NaCl(aq) + Pb(NO_3)_2(aq)$
(b) $Fe_2(SO_4)_3(aq) + NaOH(aq)$
(c) $Na_2CO_3(aq) + BaCl_2(aq)$
(d) mercury(I) nitrate and sodium bromide
(e) copper(II) chloride and calcium nitrate
(f) copper sulfate and calcium chloride

4.15 The solubility of strontium hydroxide ($Sr(OH)_2$) is 0.41 g in 100 mL of water at 0°C. Will a precipitate form if 1.00 L of water containing 0.250 mol of $Sr(NO_3)_2$ is mixed with 1.00 L of water containing 0.500 mol of NaOH?

4.16 The solubility of calcium hydroxide at room temperature is 0.18 g in 100 mL of water. (a) Show that a precipitate will form if 1.00 L of water containing 0.125 mol of $Ca(NO_3)_2$ is mixed with 1.00 L of water containing 0.250 mol of NaOH. (b) How much water would you have to add to dissolve this precipitate?

4.17 A 0.3250-g sample of water contains $MgCl_2$. Addition of excess $AgNO_3$ to the sample produces 0.03500 g of AgCl. Calculate the percentage of $MgCl_2$ in the water sample.

4.18 Saccharine molecules contain sulfur atoms. All of the sulfur in a 0.590-g sample of saccharine is oxidized to sulfate ions. The addition of excess $BaCl_2$ then produces 0.750 g of $BaSO_4$. Calculate the percentage of sulfur in the saccharine sample.

4.19 The amount of chromium in an impure sample can be determined by (1) dissolving the sample in hot hydrochloric acid, (2) filtering out insoluble impurities, (3) precipitating the chromium as $Cr(OH)_3$ with aqueous ammonia, (4) filtering the precipitate, and (5) igniting (heating) the precipitate until it is converted to Cr_2O_3. A 1.4312-g sample produces 0.2748 g of Cr_2O_3 by this procedure. Calculate the percent by mass of chromium in the sample.

4.20 A sample of stainless steel can be analyzed for its nickel content by dissolving it in a hot mixture of hydrochloric and nitric acids, neutralizing the solution with aqueous ammonia, and adding an alcoholic solution of dimethylglyoxime. The nickel forms a red precipitate of nickel dimethylglyoximate, $Ni(C_4H_7N_2O_2)_2$, which can be dried and weighed. A 0.6366-g sample of stainless steel yields 0.2793 g of dry nickel dimethylglyoxime. Calculate the percent by mass of nickel in the steel.

Acid–Base Reactions (Section 4.2)

4.21 What ion do all acids produce in aqueous solution? What ion do all bases produce? What product is common to all aqueous neutralization reactions?

4.22 List some easily observable properties that are common to acids. List some properties common to bases.

4.23 Write equations for the ionization of the following weak acids in water:
(a) H_2SO_3
(b) H_2CO_3
(c) HCN

4.24 The weak base methylamine (CH_3NH_2) reacts with water to form methylammonium ion ($CH_3NH_3^+$) and hydroxide ion. Write an equation for the reaction.

4.25 Complete and balance the following neutralization equations:
(a) $Ba(OH)_2(aq) + HNO_3(aq) \rightarrow$
(b) $LiOH(aq) + H_3PO_4(aq) \rightarrow$
(c) $Cd(OH)_2(s) + H_2S(aq) \rightarrow$
(d) $Fe(OH)_3(s) + H_2SO_4(aq) \rightarrow$

4.26 Write balanced equations for the neutralization reactions that produce the following salts:
(a) NaBr (c) $Ca(NO_3)_2$
(b) sodium sulfate (d) calcium phosphate

Reactions in Which a Gas Is Formed (Section 4.2)

4.27 Write molecular and net ionic equations for at least two reactions that produce each of the following gases:
(a) CO_2 (c) H_2S
(b) SO_2 (d) HCN

4.28 Which of the following solids will release a gas when treated with hydrochloric acid? Write molecular and net ionic equations for each reaction.
(a) $Ca(HCO_3)_2$ (d) ZnS
(b) Na_2SO_4 (e) Na_2SO_3
(c) $Ca(CN)_2$ (f) $AgNO_3$

4.29 A test for aqueous ammonium ions consists of adding concentrated NaOH to the solution, heating gently, and cautiously smelling the vapor. Identify the product in the vapor. Write the equation for the reaction.

4.30 When disposing of sodium cyanide and other cyanide salts, special precautions must be taken to prevent the salts from coming into contact with acids. Explain, with the aid of equations, the reason for these precautions.

Ionic Displacement; The Activity Series (Section 4.3)

4.31 Identify the more active and the less active element in each of the following reactions:
(a) $Fe(s) + 2HCl(aq) \rightarrow FeCl_2(aq) + H_2(g)$
(b) $2Al(s) + 3CuSO_4(aq) \rightarrow Al_2(SO_4)_3(aq) + 3Cu(s)$

4.32 What is the *activity series*? List the three most active and the three least active metals in Table 4.6.

4.33 Write a molecular equation for each of the following displacement reactions:
(a) $Cr(s) + HCl(aq)$ (d) $Mg(s)$ + steam
(b) $K(s)$ + cold water (e) $Al(s) + H_2SO_4(aq)$
(c) $Zn(s) + CuSO_4(aq)$ (f) $Fe(s) + AgNO_3(aq)$

4.34 Refer to the activity series and predict whether
(a) a silver coin will dissolve in aqueous copper sulfate
(b) a copper coin will dissolve in aqueous silver nitrate
(c) aluminum foil will dissolve in an iron(II) sulfate solution
(d) tin foil will dissolve in an iron(II) sulfate solution

4.35 (a) Write the molecular and net ionic equations for the reaction of aluminum metal with lead nitrate solution.
(b) Would it be possible for lead metal to displace aluminum from aluminum nitrate solution? Explain.

4.36 Write a net ionic equation for each of the following displacement reactions:
(a) $Zn(s) + 2AgNO_3(aq)$ (c) $Cl_2(g) + 2KBr(aq)$
(b) $Ca(s) + 2HCl(aq)$ (d) $Br_2(l) + 2NaI(aq)$

Percent by Mass (Section 4.4)

4.37 The solubility of DDT has been estimated as 5.0×10^{-5} g/100 g water. Express this solubility in parts per thousand and in parts per million.

4.38 The EPA limit for benzene (C_6H_6) in drinking water is 5 ppb by mass. At this concentration, how many grams of benzene would be present in 1.5 L of drinking water (a possible day's consumption for one person)? Assume that the density of the drinking water is 1.0 g/mL.

4.39 Hydrogen peroxide disinfectant solution contains 3.0% H_2O_2 by mass and has a density of 1.0 g/mL. How many milligrams of H_2O_2 are in 10.0 mL of the solution?

4.40 A vinegar solution has a density of 1.01 g/mL and contains 5.00% acetic acid by mass. How many milliliters of this solution will contain 35.0 g of acetic acid?

4.41 A solution is prepared by dissolving 20.0 g of ethanol in 80.0 g of water. The density of the solution is 0.9687 g/mL. How many grams of ethanol are in 1.50 L of this solution?

4.42 Dry tropospheric air contains 480.0 ppm carbon dioxide by mass. (The troposphere extends from sea level to 10 km altitude.)
(a) Express this concentration in percent by mass.
(b) How many milligrams of carbon dioxide are present in 1.00 L of dry tropospheric air? Assume the average density of air to be 1.29 g/L.

4.43 (a) How many grams of a solution containing 12.0% glucose by mass can be prepared from 100 g of glucose?
(b) How many grams of water are required to prepare the solution in Part (a)?
(c) Find the volume of the solution given that its density is 1.05 g/mL.
(d) Explain how the solution could be made without weighing the water.

4.44 The concentration of magnesium in seawater is estimated to be 0.14% by mass.
(a) Express this concentration in parts per thousand.
(b) How many kilograms of seawater would have to be treated to obtain 1.00 kg of magnesium?

MOLARITY (Section 4.4)

4.45 What quantity is obtained when the molarity of the solute is multiplied by the solution volume in liters?

4.46 What quantity is obtained when the number of millimoles of solute is divided by the solution volume in milliliters?

4.47 Calculate the molarity of each of the following solutions:
(a) 45.0 g of NaCl in 250 mL of solution
(b) 40.0 g of H_2SO_4 in 2.00 L of solution
(c) 2.50 g of $Ba(OH)_2$ in 325 mL of solution
(d) 5.00 g of KNO_3 in 0.525 L of solution

4.48 Calculate the molarity of each of the following solutions:
(a) a saline solution that contains 2.00% NaCl by mass and has a density of 1.01 g/mL
(b) serum with a glucose ($C_6H_{12}O_6$) level of 119 mg/dL
(c) an ethanol solution, density 0.986 g/mL, prepared by dissolving 15.0 g of C_2H_5OH in 188 g of water

4.49 Glacial acetic acid (CH_3COOH) has a density of 1.05 g/mL and contains 99.5% acetic acid by mass. Calculate the molarity of glacial acetic acid.

4.50 A concentrated solution of aqueous ammonia (NH_3) has a density of 0.90 g/mL and contains 28.0% ammonia by mass. Calculate the molarity of the solution.

4.51 How many moles of solute are in each of the following solutions?
(a) 1.50 L of 3.25 *M* NaCl
(b) 180 mL of 1.50×10^{-3} *M* $(NH_4)_2SO_4$
(c) 25.0 mL of 6.50 *M* $Ba(NO_3)_2$

4.52 How many grams of solute are needed to prepare each of the solutions in the preceding exercise?

4.53 How many moles of solute are in each of the following solutions?
(a) 350 mL of 0.100 *M* NaOH
(b) 1.75 L of 2.00×10^{-5} *M* KBr
(c) 35.2 mL of 0.555 *M* $Al_2(SO_4)_3$

4.54 How many grams of solute are needed to prepare each of the solutions in the preceding exercise?

4.55 Give specific practical instructions for preparing 2.00 L of a 1.00 *M* aqueous glucose ($C_6H_{12}O_6$) solution.

4.56 Give specific practical instructions for preparing 500 mL of a 0.550 *M* K_2SO_4 solution.

4.57 How many milliliters of 0.15 *M* HCl will contain (a) 0.30 mol of HCl and (b) 0.30 g of HCl?

4.58 How many milliliters of 14.8 *M* H_3PO_4 will contain (a) 12.0 mol of H_3PO_4 and (b) 24.0 g of H_3PO_4?

4.59 Fluoridated water contains 2.0×10^{-5} *M* sodium fluoride (NaF). How many milligrams of sodium fluoride are in one glass (about 250 mL) of fluoridated water?

4.60 Recent guidelines from the Health and Human Services' Centers for Disease Control state that blood lead levels in excess of 70 μg/dL indicate a medical emergency for children. How many millimoles of lead would be in a 50.0-mL sample of blood at this concentration?

4.61 Battery acid has a density of 1.29 g/mL and contains 38.0% sulfuric acid by mass.
(a) What is the molarity of battery acid?
(b) How many millimoles of sulfuric acid would be in a 50-mL pipet of battery acid?

4.62 A certain vinegar solution has a density of 1.00 g/mL and contains 4.93% acetic acid by mass.
(a) How many moles of acetic acid are in 500 mL of this solution?
(b) How many milliliters of the vinegar solution will contain 15.0 g of acetic acid?

Ionic Molarities (Section 4.4)

4.63 Find the molarity of each ion in the following solutions:
(a) 3.0 *M* KNO_3
(b) 0.55 *M* $BaCl_2$
(c) 0.75 *M* $NaHCO_3$
(d) 0.75 *M* $Al_2(SO_4)_3$
(e) 0.25 *M* Na_3PO_4
(f) 1.0 *M* $Hg_2(NO_3)_2$

4.64 The 1990 EPA standard for U.S. drinking water sets 10.0 mg/L as the maximum allowable concentration of nitrate ion (NO_3^-). Find the molarity of nitrate ion at this concentration.

4.65 (a) How many grams of NaOH are needed to make 750 mL of a solution that is 1.50 *M* in OH^-?
(b) What will be the molarity of NaOH in the solution of Part (a)?

4.66 (a) How many grams of $CaCl_2$ are needed to make 400 mL of a solution that is 0.50 *M* in Cl^-?
(b) What will be the molarity of Ca^{2+} in the solution of Part (a)?

4.67 Find the molarity of (a) Cr^{3+}, (b) Na^+, and (c) SO_4^{2-} in a solution made by mixing 200 mL of 0.120 M $Cr_2(SO_4)_3$ with 150 mL of 0.100 M Na_2SO_4.

4.68 Find the molarity of each ion in a solution made by mixing 100 mL of 0.20 M $Mg(NO_3)_2$ with 300 mL of 0.010 M KCl.

Molarity and (Sections 4.4 and 4.5)

4.69 Derive the dilution formula without referring to the textbook.

4.70 Does the dilution formula remain valid if concentrations are expressed in grams of solute per liter rather than moles of solute per liter? Explain.

4.71 How many milliliters of stock solution are needed to make each of the following solutions?
- **(a)** 320 mL of 1.5 M HNO_3 from 16 M HNO_3 stock solution
- **(b)** 250 mL of 3.0 M NH_3 from 15 M stock solution
- **(c)** 1.4 L of 2.8 M CH_3COOH from 17 M stock solution

4.72 How many milliliters of the concentrated solution are needed to make each of the following solutions?
- **(a)** 500 mL of 2.00×10^{-3} M $K_2Cr_2O_7$ from 0.0150 M $K_2Cr_2O_7$
- **(b)** 1.00 L of 0.125 M H_2SO_4 from 6.00 M H_2SO_4
- **(c)** 0.750 L of 0.100 M KOH from 0.750 M KOH

4.73 A bottle of reagent-grade hydrochloric acid has the following information on its label: Assay, 37.0% HCl by mass; Density, 1.19 g/mL.
- **(a)** Determine the molarity of the acid solution.
- **(b)** How many milliliters of this solution are needed to prepare 250 mL of 0.500 M HCl by dilution?

4.74 A concentrated ammonia solution is 28.0% NH_3 by mass and has a density of 0.900 g/mL. How many milliliters of concentrated ammonia are needed to make 0.750 L of 3.00 M NH_3 by dilution?

4.75 Give practical instructions for preparing the solution described in the preceding exercise.

4.76 A concentrated sulfuric acid solution has a density of 1.84 g/mL and contains 98.0% H_2SO_4 by mass. Give practical instructions for making 250 mL of 0.100 M sulfuric acid from this solution. (REMEMBER: add acid to water; not water to acid.)

Solution Stoichiometry; Titration (Section 4.6)

4.77 Describe with words and sketches how you would titrate an unknown sulfuric acid solution against a standard solution of sodium hydroxide.

4.78 When titrating with an indicator, why must the addition of reagent be stopped as soon as the endpoint is reached?

4.79 Calculate the molarities of sulfuric acid solutions A, B, C, and D, given the following titration data:
- **(a)** 40.00 mL of 0.1000 M NaOH neutralizes 40.00 mL of solution A
- **(b)** 40.00 mL of 0.1000 M NaOH neutralizes 20.00 mL of solution B
- **(c)** 15.32 mL of 0.1000 M NaOH neutralizes 30.64 mL of solution C
- **(d)** 30.64 mL of 0.1000 M NaOH neutralizes 15.32 mL of solution D

4.80 Consider the reaction:

$$2AgNO_3(aq) + K_2CrO_4(aq) \longrightarrow Ag_2CrO_4(s) + 2KNO_3(aq)$$

- **(a)** How many milliliters of 0.10 M K_2CrO_4 must be added to excess $AgNO_3$ solution to produce 3.0×10^{-3} mol of Ag_2CrO_4?
- **(b)** How many milliliters of 0.50 M $AgNO_3$ must be added to excess K_2CrO_4 solution to produce 2.0 g of Ag_2CrO_4?
- **(c)** How many milliliters of 0.50 M $AgNO_3$ are needed to react completely with 45 mL of 0.10 M K_2CrO_4?
- **(d)** How many moles of Ag_2CrO_4 are produced in Part (c)?

4.81 When an HCl solution was titrated against 0.09850 M NaOH, it was found that 45.00 mL of the acid reacted with 40.10 mL of the base. What is the molarity of the HCl solution?

4.82 A sulfuric acid solution contains 4.90 g of H_2SO_4 per liter. A 50.0-mL portion of this solution was neutralized by 25.0 mL of a NaOH solution. Calculate the molarity of (a) the H_2SO_4 solution and (b) the NaOH solution.

4.83 A 2.122-g sample of an unknown cobalt compound is dissolved in 50.0 mL of water. The concentration of Co^{2+}(aq) is found to be 0.071 M. Calculate the percentage by mass of cobalt in the unknown compound.

4.84 All the silver in a 45.0-mL portion of silver nitrate solution is precipitated as silver iodide by 26.0 mL of a 0.250 M calcium iodide solution:

$$2AgNO_3(aq) + CaI_2(aq) \longrightarrow 2AgI(s) + Ca(NO_3)_2(aq)$$

Calculate the molarity of the silver nitrate solution.

4.85 It was found that 35.45 mL of a HCl solution was needed to dissolve 1.110 g of pure $CaCO_3$:

$$CaCO_3(s) + 2HCl(aq) \longrightarrow CaCl_2(aq) + H_2O(l) + CO_2(g)$$

What is the molarity of the HCl solution?

4.86 Potassium hydrogen phthalate, $KHC_8H_4O_4$, is useful for standardizing basic solutions because it can be obtained in a state of high purity. Its reaction with NaOH is:

$$KHC_8H_4O_4(aq) + NaOH(aq) \longrightarrow KNaC_8H_4O_4(aq) + H_2O(l)$$

A 0.4214-g sample of $KHC_8H_4O_4$ is exactly neutralized by 20.31 mL of a NaOH solution. Find the molarity of the NaOH solution.

4.87 Water containing calcium bicarbonate is often called "hard water." Hard water can be "softened" by adding calcium hydroxide:

$$Ca(HCO_3)_2(aq) + Ca(OH)_2(aq) \longrightarrow 2CaCO_3(s) + 2H_2O(l)$$

The solid calcium carbonate is removed by filtration.

(a) How many grams of $Ca(OH)_2$ are needed to soften 800 L of hard water in which the $Ca(HCO_3)_2$ concentration is 1.0×10^{-4} *M*?

(b) Would you recommend adding extra calcium hydroxide to make sure that the water really has been softened? Why or why not?

4.88 A 50.0-mL portion of an unknown $Ca(OH)_2$ solution was mixed with 100 mL of 0.200 *M* HCl solution. The resulting solution, which was still acidic, was titrated against standard base and found to be 0.0450 *M* in HCl. Find the molarity of the calcium hydroxide solution.

4.89 Dolomite is a mineral containing equal molar quantities of $MgCO_3$ and $CaCO_3$. Its formula is often written as $MgCO_3 \cdot CaCO_3$. If it requires 50.0 mL of 0.260 *M* HCl to dissolve the dolomite in a 1.00-g sample of ore, what is the percent by mass of dolomite in the ore?

4.90 A 1.450-g piece of limestone rock was ground up and treated with 25.00 mL of 1.052 *M* HCl. The resulting solution contained excess acid and required 15.85 mL of 0.09850 *M* NaOH for neutralization. Calculate the percent by mass of calcium carbonate in the rock.

PART B. MISCELLANEOUS QUESTIONS AND PROBLEMS

4.91 What is the difference between a solution and a heterogeneous mixture? Can a solution have color? Can a solution be cloudy?

4.92 Are all saturated solutions concentrated? Explain.

4.93 The solubility of sodium thiosulfate is 79.4 g in 100 mL of water at 0°C. A certain solution at this temperature contains 50 mL of water and 500 g of sodium thiosulfate. What (if anything) would happen when a crystal of sodium thiosulfate is dropped into this solution?

4.94 A saturated sugar solution and a solution of acetic acid are both in states of dynamic equilibrium. What features do these two equilibrium states have in common? How do they differ?

4.95 Is it possible that a strong electrolyte could produce a low concentration of dissolved ions? Explain.

4.96 Refer to the activity series and predict whether a lead pipe will react with aqueous solutions of (a) zinc chloride and (b) silver nitrate.

4.97 Consult the activity series and state which metal, iron or zinc, loses electrons more easily.

4.98 Does the molarity of an aqueous solution increase, decrease, or remain the same with increasing temperature? Explain.

4.99 Is the following statement true or false? Explain. The molarity of an ion provided by a dissolved ionic compound is either the same as, or greater than, the molarity of the compound that provides it.

4.100 Consider the titration in Figure 4.14. What would happen and how would the results be affected by the following errors?

(a) A few drops of acid are added after the indicator changes color.

(b) The acid is less concentrated than you thought.

(c) You forget to add the indicator.

(d) You don't swirl the flask to mix its contents thoroughly.

4.101 Stomach acid contains HCl. Write molecular and net ionic equations for the reactions that occur when excess stomach acid is neutralized by

(a) $NaHCO_3$ (bicarbonate of soda)

(b) $CaCO_3$ (Tums)

(c) $Mg(OH)_2$ (milk of magnesia)

4.102 Write balanced equations for neutralization reactions that produce the following salts:

(a) sodium hydrogen carbonate, $NaHCO_3$

(b) calcium dihydrogen phosphate, $Ca(H_2PO_4)_2$

(c) barium oxalate, BaC_2O_4 (*Hint*: oxalic acid is $H_2C_2O_4$)

(d) barium hydrogen oxalate, $Ba(HC_2O_4)_2$

4.103 A blood analysis reported a serum cholesterol level of 205 mg per 100 mL of serum. The formula for cholesterol is $C_{27}H_{46}O$. Express the serum cholesterol concentration in (a) millimoles per liter and in (b) micromoles per liter.

4.104 Normal blood plasma contains about 0.1 g of glucose per 100 mL of plasma. Estimate the total number of grams of glucose in the plasma of a 150-lb person. Assume that normal plasma volume is about 43 mL per kilogram of body mass.

4.105 A serum specimen is diluted to 12 times its original volume. If the final concentration of serum glucose is 9.09 mg/100 mL, what was the original concentration?

4.106 An average man's blood contains about 15.5 g of hemoglobin per 100 mL of blood; an average woman's about 12% less. Calculate the number of grams of hemoglobin in a 70-kg man. Assume that the total blood mass in humans is 7.7% of body mass and that the density of whole blood is 1.05 g/mL.

4.107 The volume of the oceans is estimated to be 1.37×10^9 km^3. Table 4.3 shows that the concentration of gold in seawater is about 4×10^{-6} mg per liter of seawater. (a) Estimate the total mass of gold dissolved in the oceans. (b) Look up the current price of gold and estimate the value of the dissolved gold. (*Hint*: gold is weighed in the troy unit system; 1 oz troy = 1.09714 oz avoirdupois.)

4.108 High-sulfur coal may contain as much as 3% sulfur. Assume that all the sulfur in one metric ton (1000 kg) of high-sulfur coal is oxidized to H_2SO_4, which then appears in acid rain. Calculate the number of kilograms of marble ($CaCO_3$) that this rain could destroy.

4.109 A 9.240-g sample of iron ore was treated with concentrated nitric acid to convert all of the iron to soluble $Fe(NO_3)_3$. The resulting solution was treated with NaOH solution to convert the iron(III) nitrate to sparingly soluble $Fe(OH)_3$. The solid iron(III) hydroxide was filtered out and heated to a high temperature to decompose it to iron(III) oxide (Fe_2O_3) and water vapor. The solid iron(III) oxide was found to weigh 3.766 g. What is the percent by mass of iron in the ore?

4.110 A 500-mg tablet containing an antacid mixture of aluminum hydroxide, magnesium hydroxide, and inert material was dissolved in 50.0 mL of 0.500 *M* HCl. The resulting solution, which was still acidic, required 26.5 mL of 0.377 *M* NaOH for neutralization.

(a) How many moles of OH^- were in the tablet?

(b) If the tablet contained equal masses of $Al(OH)_3$ and $Mg(OH)_2$, how many moles of each hydroxide were in it?

4.111 Jars on a laboratory shelf contain lengths of shiny wire and are labeled magnesium, tin, silver, and platinum. A student took two pieces of wire for an experiment but forgot to look at the labels. She tried to identify the wires by dropping small pieces of each one into solutions of HCl and $FeCl_2$. One wire dissolved in the HCl solution but not in the $FeCl_2$ solution. The other wire did not dissolve in either solution. Did the student have enough information to identify the wires? Explain.

4.112 Evan's Blue, a nontoxic organic dye, has been used for the determination of plasma volumes. A known quantity of the dye is injected, and samples of blood are withdrawn at appropriate intervals for determination of dye dilution.

(a) 5.0 mL of an Evan's Blue solution containing 500 mg of dye per 100 mL was injected into a vein of a patient's arm. After ten minutes a 10.0-mL blood sample was drawn from the opposite arm and found to contain 0.80 mg/100 mL of the dye. Estimate the plasma volume.

(b) The volume calculated by the Evan's Blue technique is sufficiently accurate for most clinical purposes, but it is only approximate because some of the dye leaves the plasma by diffusion into surrounding tissues. Is the true plasma volume greater than or less than that calculated in Part (a)? Explain.

4.113 The alcoholic content of beverages is often stated in terms of *proof*, which is defined as twice the percentage by volume of ethanol at 60°F.

(a) What is the proof of a Chianti Classico wine labeled 12.5% alcohol by volume?

(b) In a 70-kg man, 1 oz of 90-proof whiskey will produce a maximum ethanol concentration of 20 mg per 100 mL of blood. Medical and legal evidence of intoxication is 150 mg per 100 mL of blood. (There is often severe impairment of functional ability at concentrations well below 150 mg per 100 mL.) Estimate the number of scotch highballs, each containing 1.5 oz of 90-proof whiskey, needed to achieve this concentration.

4.114 In 1974 a team of scientists detected ethanol in Sagittarius B2, a vast cloud of dust and gas near the center of the Milky Way. They calculated that if all the ethanol in Sagittarius B2 were condensed and bottled, it would amount to 10^{28} fifths of 80-proof whiskey. (See the previous exercise for the definition of *proof*. A "fifth" is one-fifth of a gallon.) The density of ethanol is 0.794 g/mL at 60°F. Estimate the total number of kilograms of ethanol in Sagittarius B2.

4.115 If the solid NaCl in Figure 4.6 has a mass of 1.00 g, what masses of the other salts would be needed to make the two solutions identical? Assume that the volume of water is the same for each solution.

4.116 A 100.0-mL portion of 0.200 *M* NaCl solution was mixed with 200.0 mL of 0.150 *M* $AgNO_3$ solution. Determine the mass of AgCl precipitated and the final molar concentrations of the ions that remained in solution.

4.117 A 120.0-mL portion of 0.150 *M* H_2SO_4 was added to 80.0 mL of a solution containing 4.10 g $BaCl_2$. Find the mass of the $BaSO_4$ precipitate and the concentrations of the ions remaining in solution.

CHEMISTS AT WORK

THE CONGRESSMAN

A common perception is that elected office is exclusively for those whose educational background is law or political science. But this is not always the case. John W. Olver, who was elected in 1991 to the U.S. House of Representatives, is also a Ph.D. chemist.

"Chemistry was not going to be offered in high school the year I was to graduate," he said. "The only way to take the course was to do it independently, and that's what I did." He received a bachelor's degree in chemistry from Rensselaer Polytechnic Institute in 1955 and a master's degree from Tufts University in 1956. "After I completed the master's degree, I decided that I needed some practical experience and taught for 2 years at a trade school in Boston. This was an in-teresting experience because I was 19 years old and teaching a variety of chemistry courses and a course in chemical engineering to much older veterans of the Korean conflict."

John decided in 1958 that he wanted to teach in higher education and recognized that he would need a doctoral degree. In 1961, he earned a Ph.D. in chemistry from MIT and stayed there for a year to teach analytical chemistry. After that, Dr. Olver joined the faculty at the University of Massachusetts in Amherst. He recalls, "I was a typical faculty member in that I taught courses in chemistry; formed and supported a research group in electroanalytical chemistry by applying for and receiving grants from various agencies; had as many as six Ph.D. candidates in that group; and earned tenure."

There were choices to be made when he began to plan for his first sabbatical leave in 1968. One possibility was to try to find some kind of activity to enrich and expand his background in chemistry, such as applying for a Fullbright Fellowship to study in Europe. The other, and the choice that he made, was to attempt to get elected to the Massachusetts Legislature. "Prior to my election, my involvement in politics was minimal. I had done some campaigning in the Cambridge area while I was a graduate student, and I had become interested in issues related to public and social policy." Since his first election in 1968, he served for 22 years in the Massachusetts legislature, 4 years in the state House of Representatives, and 18 in the state Senate. During his tenure there, he contributed on such issues as mental retardation, health care, chronic unemployment, child care, higher education, the environment, and economic development.

JOHN W. OLVER

Dr. Olver was elected to the U.S. Congress in 1991 to fill the seat of the late Silvio O. Conte. In 1992, he was elected to his first full term. Since he has been in the U.S. House of Representatives, he has served on the committees of Education and Labor; Science, Space, and Technology; and Appropriations. "I'm not certain that my scientific education has a direct bearing on my job in Congress, although the problem-solving and numerical skills are valuable," he says. "When you're doing research, there is a high probability that problems are easily defined and you will get quantifiable results. In the legislative arena, the problems are much more complex, and there is frequently a division of thought on potential solutions, which means that there are not easy solutions."

He states that a legislator is often an educator trying to explain positions on issues to colleagues and constituents. "One of the most important skills is being able to communicate effectively."

In congress, he continues to champion "people issues." He states that "so much that we do is technologically based, and therefore we need to find ways to popularize science so that large segments of our population, especially minorities and women, are not excluded." Congressman Olver believes that science is very badly taught in the lower grades and needs to be improved. He is also concerned about the quality of higher education and the availability of appropriate financial aid for education.

"One of the differences in being in politics as compared to the classroom is that your constituents will always tell you when you're not doing a good job. In the classroom, I gave out grades. As a congressman, I receive grades from the voters every 2 years."

CHAPTER 5

Heating a hot air balloon with a propane burner. Gases expand and become less dense on warming.

GASES AND THEIR PROPERTIES

OUTLINE

PREVIEW

The molecules of a gas are relatively far apart; they move randomly and are more or less independent of each other. The rarefied nature of the gaseous state is strikingly illustrated by comparing the volume of 1 mol of liquid water at ordinary pressure and 100°C—about 19 mL—with the molar volume of steam under the same conditions—about 30,600 mL! Gases are unruly compared to solids and liquids; they cannot be scooped, poured, or shoveled, and their molecules find their way through the tiniest pores and cracks. Such properties are troublesome to anyone who wants to work safely with gases, yet these same properties are essential to their life-giving role in our biosphere. The survival of most living organisms depends on their ability to exchange gases with their environment, and an understanding of life processes requires a knowledge of the properties and behavior of gases.

The principal **states of matter** are solid, liquid, and gas. Most substances can be found in more than one state; water, for example, occurs as ice, liquid water, and steam (water vapor). Eleven elements are gaseous under ordinary conditions—the atmospheric gases nitrogen and oxygen; the noble gases helium, neon, argon, krypton, xenon, and radon; and hydrogen, fluorine, and chlorine. Many familiar compounds, including ammonia, methane, and carbon dioxide, are gases under ordinary conditions, and the vapors emitted by volatile (easily vaporized) materials such as mothballs and gasoline are also gases.

Still other states of matter can exist under special conditions. The liquid crystal, superfluid, and plasma states will be described in later chapters.

5.1 THE KINETIC THEORY OF MATTER

One of our deep-rooted scientific beliefs is that the particles making up matter—atoms, molecules, and ions—are in continuous, rapid, and chaotic motion. This belief is the basis of the **kinetic theory of matter**, a theory richly supported by a variety of experimental evidence. Some of this evidence is provided by **Brownian**

Kinetic: from the Greek *kinein*, "to move."

motion, the spontaneous random jiggling of finely divided particles suspended in a liquid. Brownian motion was discovered in 1827 by the Scottish botanist Robert Brown during his study of pollen grains in water. Each grain, when observed under a microscope, was seen to follow an erratic path as if buffeted by unseen blows (Figure 5.1). Brownian motion is explained by assuming that the grains are tossed about by constantly moving liquid molecules smashing into them from all directions. At any instant, a single grain may be hit by more molecules on one side than on the other and, as a result, the grain moves off in some direction. In effect, the incessant and random motion of the pollen grains is simply a reflection of the incessant and random motion of the much smaller liquid molecules continually bombarding the grains.

Diffusion, the spontaneous spreading of one substance through another, provides additional evidence of molecular motion. If a drop of ink is placed in a glass of water, or a cube of sugar in hot coffee, the ink and sugar spread through the liquids as shown in Figure 5.2. Gases also diffuse into each other, as evidenced by the speed with which distant odors reach our noses. To understand diffusion in terms of molecular motion, consider some water to which a drop of ink has been added. Random motion and collisions will cause more molecules of ink to move away from the concentrated drop than will return from neighboring regions where there is little ink. Similarly, more water molecules will move into the inky region than out of it. *Each substance in a mixture tends to diffuse away from regions where its concentration is high toward regions where its concentration is low*. Diffusion will continue as long as there are concentration differences within the mixture. When the concentrations become uniform, diffusion stops, but molecular motion continues unabated.

The rates of many molecular processes increase with increasing temperature. Liquids flow more easily when they are warm, and diffusion occurs more rapidly in hot fluids than in cold ones. To understand the effect of temperature on molecular motion, recall that an increase in temperature is the result of the addition of energy from a flame, an electric current, or some other source. The added energy increases the kinetic energy of the molecules, thus causing them to move more rapidly.

The Kinetic Theory of Gases

In liquids and solids, the range of molecular motion is limited by attractive forces that keep molecules together and impose a certain degree of order on their otherwise random movements. In gases, however, these attractions are very weak, and gas

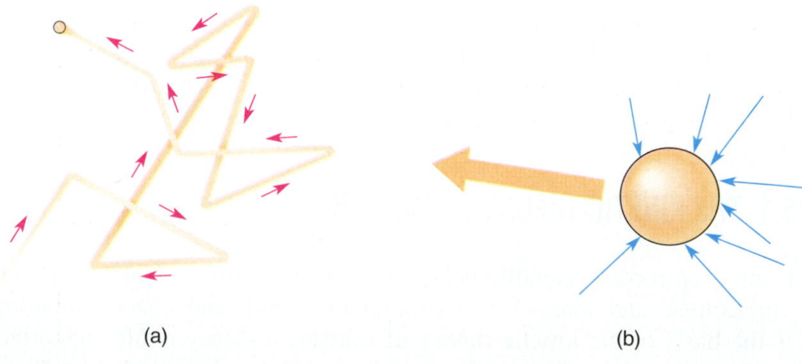

Figure 5.1

Brownian motion. (a) Sketch representing the erratic path of a minute pollen grain in water, as observed under a microscope. (b) The grain moves in a particular direction (large arrow) because more water molecules are hitting it on one side than on the other. The direction keeps changing because molecular motions are random.

(a)

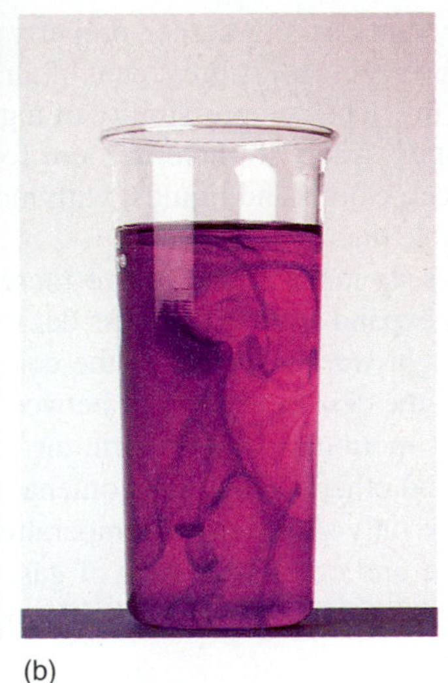
(b)

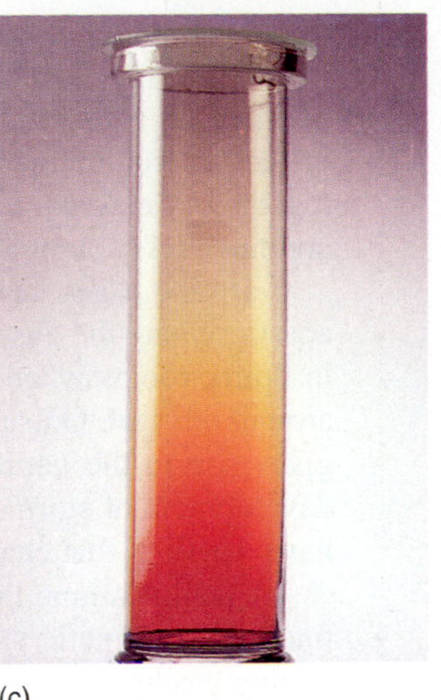
(c)

(d)

Figure 5.2

Diffusion. (a) A drop of dye is released from a pipet at the bottom of a beaker of water. (b) Over a period of time, the dye spreads throughout the water, and eventually its concentration will become uniform. (c) A few drops of liquid bromine are placed at the bottom of a cylinder, which is then sealed. Bromine vapor rises from the liquid. (d) Bromine vapor has diffused until its concentration is more or less uniform.

molecules move more or less independently of each other. Each molecule goes its own way, randomly bouncing off other molecules or off the container wall, and there is complete chaos (Figure 5.3). (The word *gas* is derived from the Greek word *khaos*, "chaos.") This independent motion accounts for the fact that gases, unlike solids and liquids, have variable volumes and expand to fill their containers: *the volume of any gas is simply the volume of its container*.

The observed properties of gases can be explained in terms of the **kinetic theory of gases**. This theory, which is an extension of the kinetic theory of matter, assumes the following for all gases:

1. *Gases consist of many molecules (or atoms) moving randomly and continuously through space*. The molecules frequently collide with each other and with the walls of their container.

2. *Gas molecules are very small compared to the distances between them*. Most of the volume occupied by a gas is empty space.

3. *Attractive forces between gas molecules are weak and usually negligible*. We know, however, that gases do liquefy when cooled or compressed. This observation shows that the attractive forces become more effective when the temperature is low and when the molecules are packed closely together.

4. *The average kinetic energy of the molecules in a gas depends only on the temperature*. Individual molecules may gain or lose energy when they collide, but the sum of their energies (i.e., the total kinetic energy of the gas) will remain unchanged as long as the temperature is constant. When the temperature increases, however, the molecules move more rapidly and their average kinetic energy also increases; the molecules collide with each other and with the container walls more frequently and with greater force. Because of its dependence on temperature, the random motion of atoms and molecules is called **thermal motion**.

Virtually all gas behavior can be explained in terms of the kinetic theory. We have already mentioned how the absence of intermolecular attractions allows gases

Figure 5.3

The kinetic molecular model of a gas. Small molecules move randomly and continuously in a relatively large volume of empty space.

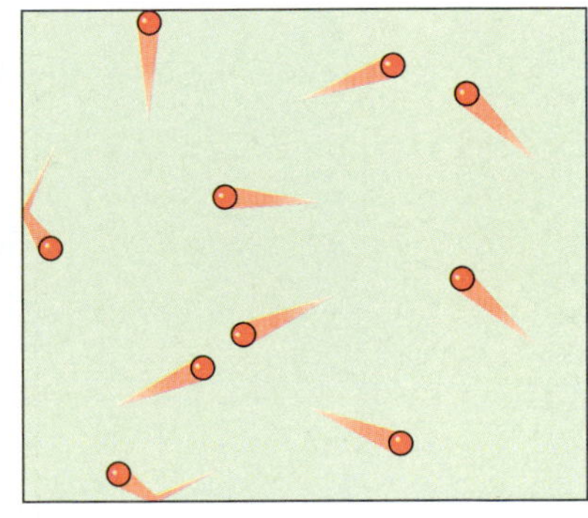

to expand and fill their containers. Another property of gases is the relative ease with which they can be compressed. A great deal of air, for example, can be pumped into a small automobile tire. The compressibility of a gas is due to the fact that most of the gas volume is empty space, so there is room for the molecules to be pushed together by external forces. Solids and liquids, with more closely packed molecules, are much less compressible than gases.

Warming a gas causes its molecules to move more rapidly; they tend to spread apart, causing the gas to expand and become less dense. We notice this phenomenon in drafty rooms where light warm air rises to the ceiling and denser cold air flows around our feet. Outside, the density difference between cool air and sun-warmed air gives rise to the vertical motion of atmospheric air masses and contributes to the development of storms and other weather phenomena. In contrast, liquids and solids have densities that change only slightly with temperature.

We will examine these and other properties of gases in more detail, and we will find that the kinetic theory of gases is a remarkable tool for understanding and predicting gas behavior.

5.2 PRESSURE

Pressure is a common word in our vocabulary; most of us are familiar with phrases such as water pressure, blood pressure, and tire pressure. **Pressure** is defined as force per unit area:

$$\text{pressure} = \frac{\text{force}}{\text{area over which the force is exerted}} \tag{5.1}$$

The SI units of force and area are newtons and square meters, respectively; hence, the SI unit of pressure is the newton per square meter (N/m^2). This unit is also called a **pascal** (Pa) in honor of Blaise Pascal (1623–1662), who investigated the effect of pressure on fluids; 1 Pa = 1 N/m^2. A pressure of 1 Pa is small by ordinary standards. A U.S. Roosevelt-type dime, for example, exerts a pressure of 88.5 Pa when it is lying flat on the ground or in the palm of your hand. Another unit, the **bar**, is defined as 10^5 Pa; 1 bar = 10^5 Pa = 10^2 kPa.

A solid object exerts pressure on the surface that supports it. This pressure results from gravitational force; for an object standing on a horizontal surface, the pressure is the object's weight divided by the area on which it rests.

EXAMPLE 5.1

An 80.0-kg man stands on snowshoes whose combined area is 2.5×10^3 cm^2. What pressure, in newtons per square meter, is exerted on the snow beneath him?

SOLUTION

The pressure is obtained by substituting the gravitational force (i.e., the man's weight) in newtons and the snowshoe area in square meters into Equation 5.1. The man's weight is given by Equation 1.1:

$$\text{weight} = m \times g = 80.0 \text{ kg} \times 9.81 \text{ m/s}^2 = 785 \text{ kg}\cdot\text{m/s}^2 = 785 \text{ N}$$

since 1 N = 1 kg·m/s^2. The snowshoe area in square meters is

$$2.5 \times 10^3 \text{ cm}^2 \times \left(\frac{1 \text{ m}}{100 \text{ cm}}\right)^2 = 0.25 \text{ m}^2$$

The pressure is calculated from Equation 5.1:

$$\text{pressure} = \frac{\text{force}}{\text{area}} = \frac{785 \text{ N}}{0.25 \text{ m}^2} = 3.1 \times 10^3 \text{ N/m}^2$$

PRACTICE EXERCISE 5.1

Calculate the pressure in newtons per square meter exerted by a cylindrical column of liquid mercury 76.0 cm high and 1.00 cm^2 in cross-sectional area. The density of mercury is 13.6 g/cm^3. (*Hint*: The volume of a cylinder is equal to its height multiplied by its cross-sectional area.)

LEARNING HINT

Assume the same column height but different cross-sectional areas (2.00 cm^2, 10.0 cm^2, etc.), and repeat the calculation in Practice Exercise 5.1. Explain why the calculated pressure does not change.

Atmospheric Pressure

Liquids and gases consist of rapidly moving molecules that constantly collide with each other and with their container walls. The force of these collisions causes liquids and gases to exert pressure in all directions, against every surface they touch. Water in a pail presses against the bottom of the pail and also against the sides. Air presses on the walls and ceilings of a room as well as on the floor. A dramatic experiment demonstrating the existence of air pressure starts with an "empty" tin can containing nothing but air. The forces on the inner and outer walls of the open can are balanced because there is air both inside and outside the can. If the air is pumped out as in Demonstration 5.1, a pressure imbalance is created, and the external (atmospheric) pressure crushes the walls of the can.

Another experiment demonstrates the ability of the atmosphere to support a column of liquid. A glass tube, about 100 cm in length and closed at one end, is filled with mercury. The open end is covered and inverted into a reservoir of mercury, after which the cover is removed (see Figure 5.4). If care is taken so that no air enters the tube, the mercury level will fall until it is about 76 cm above the reservoir level, and a vacuum will form near the upper end of the tube. The mercury column stays up because the force exerted by the atmosphere pushing on the mercury in the outer reservoir (blue arrows) balances the gravitational force pulling on the mercury in the tube (red arrow). Mercury can move freely between tube and reservoir, so its height continuously adjusts to fluctuations in air pressure on the reservoir. The greater the pressure of the atmosphere, the higher the column of mercury. The mercury column thus functions as a **barometer**, a device for measuring atmospheric pressure.

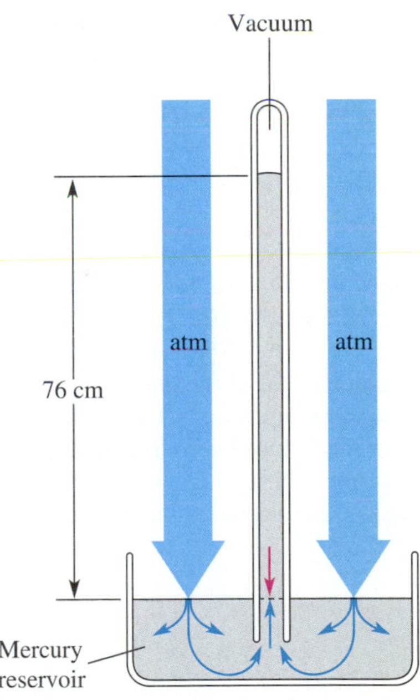

Figure 5.4

At sea level, the pressure of the atmosphere supports a column of mercury 76 cm (760 mm) high. The column stays up because the force of the atmosphere pushing the mercury up the tube (blue arrows) is exactly equal to the force of gravity pulling the mercury down (red arrow).

The atmosphere at sea level supports a column of mercury about 76 cm (760 mm) high. This pressure corresponds to 1.01×10^5 N/m^2 (see Practice Exercise 5.1). Daily pressure depends on the prevailing weather conditions, and the actual column height can vary by about 5% from the above figure, moving up with increasing pressure and down with decreasing pressure. Atmospheric pressure decreases with increasing altitude; at 10 km above sea level, for example, the atmosphere supports

DEMONSTRATION 5.1 ATMOSPHERIC PRESSURE

The white can is empty except for air. The pressures inside and outside the can are equal.

Air is pumped from the can. The outside pressure is greater than the inside pressure, so the sides of the can begin to bend inward.

The can collapses as more air is removed.

only about 21 cm of mercury. This pressure drop occurs because there is a lower concentration of gas at the higher altitude.

LEARNING HINT

The weight of the atmosphere can be estimated by multiplying the area of the earth's surface by atmospheric pressure at sea level.

Pressure is often expressed in terms of a unit called the **atmosphere** (atm). By definition, 1 atm is exactly 1.01325×10^5 N/m^2. Another convenient unit, the **torr**, is based on the observation that, at 0°C, 1 atm of pressure supports a column of mercury 760 mm high. One torr is defined to be exactly 1/760 atm; it is the pressure that will support a column of mercury that is 1 mm high at 0°C. The density of mercury, and consequently the height of a mercury column, varies slightly with temperature. Hence, 1 torr will support slightly more or less than 1 mm of mercury at temperatures other than 0°C. The variation with temperature is small, however, and in elementary work it is ignored. A barometer reading of 762 mm Hg, for example, would be reported as 762 torr. The torr is named after Evangelista Torricelli (1608–1647), who developed the first mercury barometer. To summarize:

The words *bar* and *barometer* are derived from the Greek *baros*, meaning "weight." Most desktop barometers are of the *aneroid* variety, in which the position of a needle depends on the degree to which atmospheric pressure compresses the sides of an evacuated metal box.

$$\begin{aligned} 1 \text{ atm} &= 760 \text{ torr} \\ &= 1.01325 \times 10^5 \text{ N/m}^2 \\ &= 1.01325 \text{ bar} \end{aligned}$$

since 1 bar = 10^5 N/m^2. Note that a pressure of 1 bar is almost identical to a pressure of 1 atm.

EXAMPLE 5.2

During a hurricane, the level of mercury in a barometer drops to 71.0 cm. What is the pressure in (a) atmospheres and (b) bars?

SOLUTION

(a) The barometer reading, 71.0 cm Hg = 710 mm Hg, corresponds to a pressure of 710 torr. The pressure in atmospheres is

$$710 \text{ torr} \times \frac{1 \text{ atm}}{760 \text{ torr}} = 0.934 \text{ atm}$$

(b) The relation 1 atm = 1.013 bar gives

$$0.934 \text{ atm} \times \frac{1.013 \text{ bar}}{1 \text{ atm}} = 0.946 \text{ bar}$$

The pressure at the center of the earth is believed to be 3.5 Mbar, about 3.5 million times greater than atmospheric pressure at sea level. Scientists in high-pressure research laboratories routinely produce pressures in excess of 2 Mbar. Nuclear scientists using powerful lasers have produced pressures of about 1000 Mbar, some ten times greater than the pressure at the center of the planet Jupiter.

PRACTICE EXERCISE 5.2

Convert (a) 1.04 atm to torr and (b) 750 torr to atmospheres.

PRACTICE EXERCISE 5.3

Convert a reading of 120 mm Hg into (a) torr, (b) atmospheres, and (c) bar.

The Manometer

A **manometer** (Figure 5.5) is used for measuring pressure inside a closed vessel. It consists of a U-tube partially filled with mercury that adjusts its position to compensate for unequal pressures on its two surfaces. The mercury level is lower on the side with the greater pressure.

Manometer: from the Greek *manos*, "sparse" (in reference to gaseous conditions).

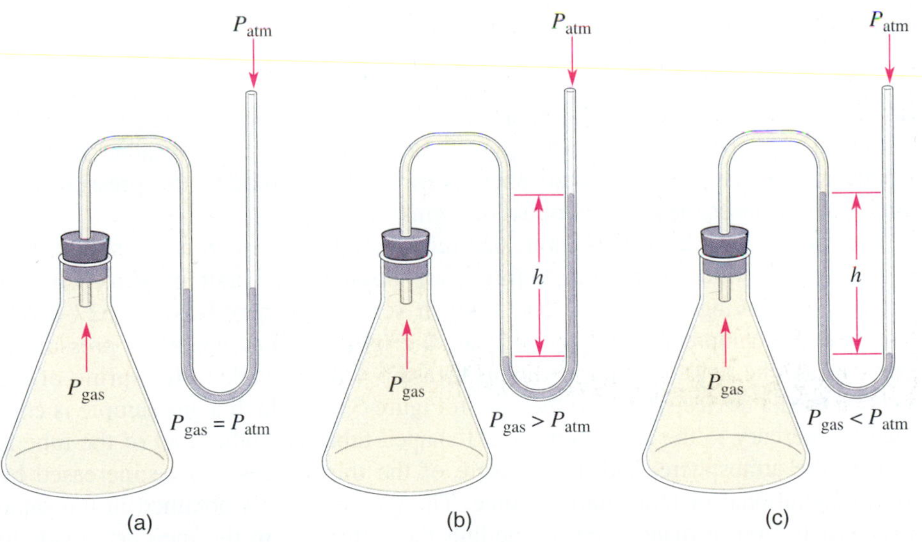

Figure 5.5

A manometer. The difference in the height of the mercury levels in mm equals the difference in torr between the gas pressure inside the flask and the atmospheric pressure outside. (a) The two mercury levels are equal, so the inside and outside pressures are also equal. (b) A lower mercury level in the inside arm shows that the gas pressure is greater than atmospheric pressure. (c) A lower level in the outside arm shows that atmospheric pressure is greater than the gas pressure.

EXAMPLE 5.3

If atmospheric pressure is 750 torr, and the difference in mercury levels is $h = 20$ mm, what is the pressure of the gas in the flask in Figure 5.5c?

SOLUTION

The atmospheric pressure is $P_{atm} = 750$ torr. This pressure is equal to the sum of the pressure in the flask (P_{gas}) and the pressure exerted by 20 mm of mercury. Hence,

$$P_{atm} = P_{gas} + 20 \text{ torr}$$
$$750 \text{ torr} = P_{gas} + 20 \text{ torr}$$
$$P_{gas} = 730 \text{ torr}$$

PRACTICE EXERCISE 5.4

The mercury level on the open side of a manometer is 100 mm higher than on the closed side (see Figure 5.5b). If atmospheric pressure is 770 torr, what is the pressure of the gas in torr?

5.3 THE GAS LAWS

Experiments have shown that the behavior of a gas depends primarily on its pressure (P), temperature (T), volume (V), and number of moles (n). These four properties, or *variables* as they are often called, are related, and a change in one of them produces changes in one or more of the others. In this section we will show that the relationship between any two of the variables, say P and V, is most conveniently obtained when the other two, T and n, are kept constant.

Boyle's Law

It is common knowledge that confined gases expand with decreasing pressure. Air bubbles, for example, become larger as they rise from high-pressure regions near the bottom of a pool to low-pressure regions near the pool surface. Weather balloons expand as they rise into the atmosphere. Conversely, increasing the pressure of a confined gas causes the gas to contract or shrink in volume.

Robert Boyle (right) with his assistant, Denis Papin.

The earliest recorded observations that relate gas pressure to volume date back to the late seventeenth century, long before it was possible to explain gaseous behavior in terms of moving molecules. The English scientist Robert Boyle (1627–1691) observed that compressed air rebounds to its original volume when the pressure is released, and he devised experiments to measure what he called the "spring of the air." He used a device similar to the one in Figure 5.6 in which a gas sample is confined by mercury to the closed end of a J-shaped tube. The other end of the tube is open to the atmosphere, and the pressure of the trapped gas can be increased by pouring additional mercury into the tube. The gas pressure is obtained in the same way as it is from a manometer, by adding the difference in the mercury levels to atmospheric pressure. The amount of gas (or in modern terms), the number of moles of gas) was constant in each of Boyle's experiments, and the readings were taken at room temperature, which was more or less constant during the experiments.

Figure 5.6 shows how the volume of a trapped gas sample decreases with increasing pressure, and Figure 5.7 shows typical data from such an experiment. The data support **Boyle's law**: *At constant temperature the volume of a fixed number of*

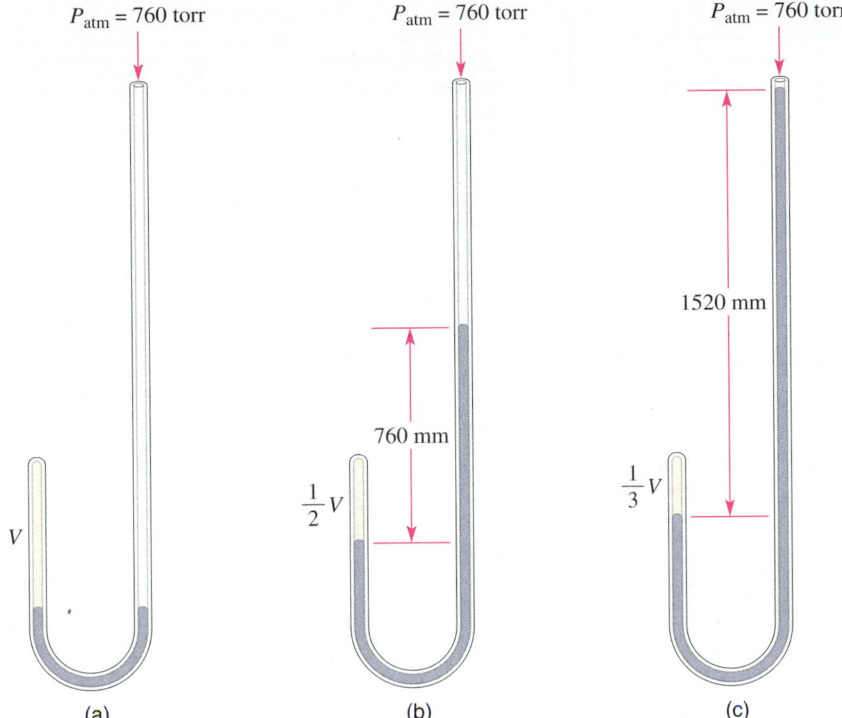

Figure 5.6

Testing Boyle's law. The pressure of the gas trapped in the closed end of the tube is obtained by adding the difference in the mercury levels to atmospheric pressure. (a) The mercury levels are equal. The trapped gas with volume *V* is at atmospheric pressure (here, 760 torr). (b) Adding mercury increases the pressure of the trapped gas to 760 torr + 760 torr, double its pressure in (a). The gas volume has decreased to $\frac{1}{2}V$. (c) Tripling the pressure of the gas decreases its volume to $\frac{1}{3}V$.

moles of gas is inversely proportional to the pressure. The phrase *inversely proportional* means that whenever the pressure is increased by a given factor, the volume will decrease by exactly the same factor; doubling the pressure will halve the volume, decreasing the pressure by a factor of 3 will triple the volume, and so forth.

EXAMPLE 5.4

A 10.0-mL sample of gas is confined in a syringe. If the pressure of the gas is tripled by pushing in the movable piston, what is the final volume? Assume that the temperature does not change.

SOLUTION

The initial volume is 10.0 mL. Increasing the pressure to three times its original value will decrease the volume to one-third of its original value. The final volume will be $\frac{1}{3}$ of 10.0 mL or 3.33 mL.

PRACTICE EXERCISE 5.5

The pressure on 15.0 mL of gas is decreased to one-fourth of its original value. Assume that the temperature remains constant and estimate the final volume.

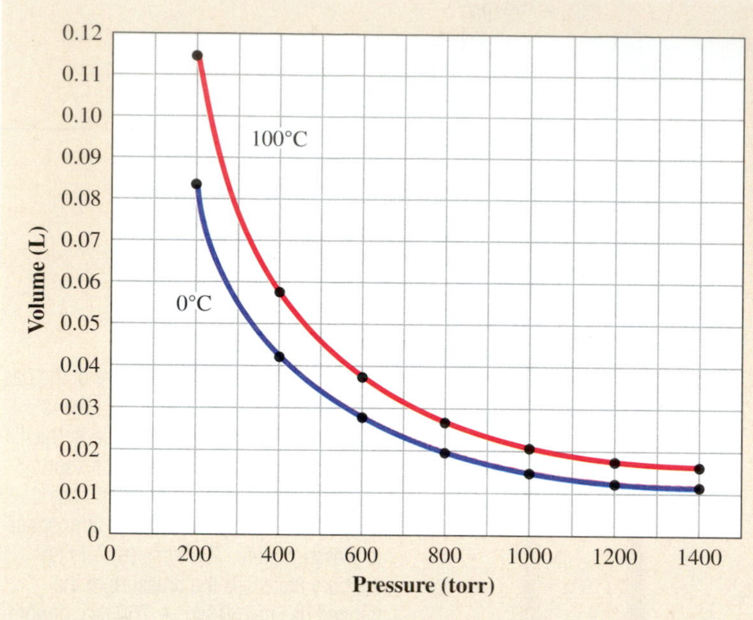

Data for Compressing 0.00100 mol of He at 100°C			
Pressure (torr)	*Pressure (atm)*	*Volume (L)*	*Pressure × Volume (L atm)*
200	0.263	0.116	0.031
400	0.526	0.058	0.031
600	0.789	0.039	0.031
800	1.05	0.029	0.030
1000	1.32	0.023	0.030
1200	1.58	0.019	0.030
1400	1.84	0.017	0.031

Data for Compressing 0.00100 mol of He at 0°C			
Pressure (torr)	*Pressure (atm)*	*Volume (L)*	*Pressure × Volume (L atm)*
200	0.263	0.085	0.022
400	0.526	0.042	0.022
600	0.789	0.028	0.022
800	1.05	0.021	0.022
1000	1.32	0.017	0.022
1200	1.58	0.014	0.022
1400	1.84	0.012	0.022

Figure 5.7

Boyle's law plots for 0.00100 mol of helium at 0°C and 100°C. Plotting volume V versus pressure P at any one temperature produces a curve of the type $PV = k_b$. The curve for 0°C lies below the curve for 100°C because k_b is smaller at 0°C than at 100°C.

EXAMPLE 5.5

A sample of gas occupies a volume of 5.2 L at 1.0 atm. If the gas is allowed to expand to 20.8 L without any change in temperature, what is its final pressure?

SOLUTION

The initial gas pressure is 1.0 atm. The volume increases by a factor of 20.8 L/5.2 L = 4.0. Hence, the gas pressure decreases by the same factor. The final pressure will be 1.0 atm/4.0 = 0.25 atm.

PRACTICE EXERCISE 5.6

The volume of a sample of gas is 25.9 mL at 155 kPa. What is the final pressure if the volume decreases to 18.5 mL with no change in temperature?

LEARNING HINT

When two quantities are inversely proportional to each other, their product is constant.

Boyle's law tells us that when the pressure is increased by some factor, the volume will decrease by the same factor. In other words, the product of the pressure and volume ($P \times V$) will remain unchanged. Boyle's law, therefore, implies that

$$P \times V = k_b \tag{5.2}$$

where k_b is a number that varies only with the number of moles of gas and the temperature. Equation 5.2 can be described graphically in the form of isothermal (constant temperature) plots of volume versus pressure. The plots for 0.00100 mol of helium at 0°C and 100°C are shown in Figure 5.7. Their shapes are similar; the only difference is that the *PV* product has a larger value at the higher temperature.

If a given quantity of gas changes isothermally from an initial pressure and volume (P_i, V_i) to a final pressure and volume (P_f, V_f), the relationship

$$P_i \times V_i = P_f \times V_f \qquad (5.3)$$

will hold because both *PV* products are equal to the same value of k_b (Equation 5.2). Equation 5.3 is often used to relate pressure and volume changes, as shown in the next example.

EXAMPLE 5.6

One mole of nitrogen occupies a volume of 22.4 L at 0°C and 1.00 atm. Estimate its volume at 0°C and a pressure of 0.550 atm.

SOLUTION

Because there is no change in temperature, we can use Equation 5.3 to calculate the final volume. Dividing both sides of the equation by P_f and rearranging gives

$$V_f = V_i \times \frac{P_i}{P_f}$$

The initial volume and pressure are $V_i = 22.4$ L and $P_i = 1.00$ atm. The final pressure is $P_f = 0.550$ atm. Substitution gives

$$V_f = 22.4 \text{ L} \times \frac{1.00 \text{ atm}}{0.550 \text{ atm}} = 40.7 \text{ L}$$

PRACTICE EXERCISE 5.7

A sample of gas occupies 28.9 mL at 25°C and 0.880 atm. What will the final pressure be if the gas is allowed to expand to 40.5 mL with no change in temperature?

In Example 5.6, Equation 5.3 was used to provide a pressure ratio or *pressure correction factor*, 1.00 atm/0.550 atm, that converts the initial volume into the final volume. In this example, the pressure factor is greater than unity (1) because reducing the pressure increases the volume. The following example shows how such factors can be devised without referring to Equation 5.3.

EXAMPLE 5.7

A sample of gas occupies 1.00 L at 720 torr. How many liters will the gas occupy if its pressure is increased to 760 torr with no change in temperature?

SOLUTION

The initial volume is $V_i = 1.00$ L. The final volume is

$$V_f = V_i \times \text{pressure factor}$$

There are two possible pressure factors: 720 torr/760 torr and 760 torr/720 torr. An increase in pressure from 720 torr to 760 torr will cause the volume to decrease, so the factor must be less than unity, or 720 torr/760 torr. Therefore,

$$V_f = 1.00 \text{ L} \times \frac{720 \text{ torr}}{760 \text{ torr}} = 0.947 \text{ L}$$

The same result could have been obtained by substituting $P_i = 720$ torr, $V_i = 1.00$ L, and $P_f = 760$ torr into Equation 5.3 and solving for V_f.

PRACTICE EXERCISE 5.8

The pressure of a sample of gas decreases from 780 torr to 760 torr. (a) Will its volume increase or decrease? (b) By what factor?

If the initial and final volumes are given, they can be used to devise a factor for calculating the final pressure, as illustrated in the next example.

EXAMPLE 5.8

A sample of gas occupies 87.5 mL at 700 torr. What will the gas pressure be if its volume decreases to 75.0 mL with no change in temperature?

SOLUTION

The pressure varies inversely with the volume. If the volume decreases from 87.5 mL to 75.0 mL, then the pressure must increase by a factor of 87.5 mL/75.0 mL. The initial pressure is $P_i = 700$ torr. The final pressure is

$$P_f = P_i \times \text{volume factor}$$

$$= 700 \text{ torr} \times \frac{87.5 \text{ mL}}{75.0 \text{ mL}} = 817 \text{ torr}$$

PRACTICE EXERCISE 5.9

The volume of a sample of gas increases from 4.0 L to 6.0 L at constant temperature. (a) Does its pressure increase or decrease? (b) By what factor?

Boyle did not know about atoms and molecules, but he probably would have been pleased by the explanation that the kinetic theory of gases now provides for his law. The pressure of a confined gas is produced by molecules bombarding the container walls (Figure 5.8), and it is equal to the total force of all the impacts divided

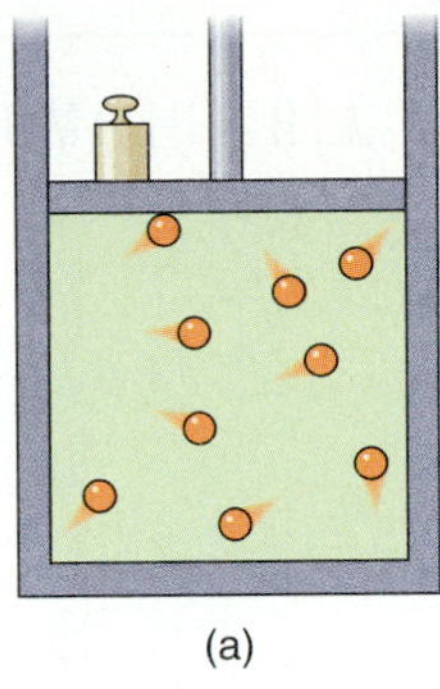

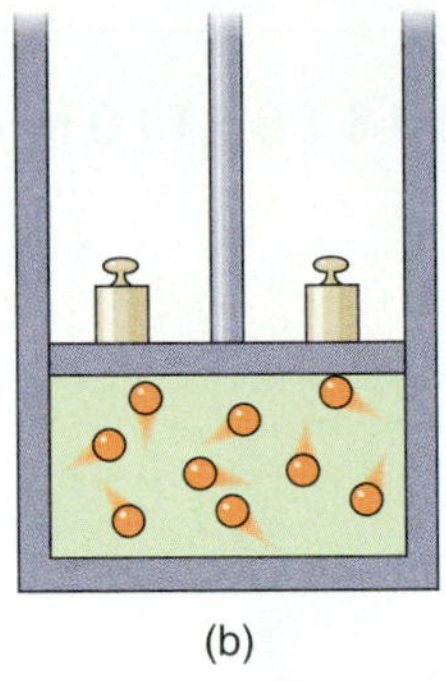

Figure 5.8

Kinetic interpretation of Boyle's law. (a) A gas is confined in a container with a movable piston. The piston is stationary because the downward force of gravity is balanced by the upward force of gas molecules hitting the piston surface.
(b) The weight of the piston is doubled. The piston will move downward until the pressure is doubled—that is, until the volume is halved and each molecule collides with the piston surface twice as often.

by the total wall area. If the volume available to the molecules is reduced by half, each molecule will, on the average, collide with the container wall twice as often, thus doubling the number of collisions per second and doubling the pressure.

Charles's Law

A heated gas will expand if given the opportunity; a cooled gas will contract. Helium-filled party balloons, for example, become smaller when exposed to cold winter air, but they return to their original size when brought back into a warm room. A simple air thermometer based on the thermal behavior of gases is shown in Demonstration 5.2. In the late 1700s the French scientist Jacques Alexandre César Charles studied the variation of gas volume with temperature. He found that if the pressure and amount (number of moles) of gas are held constant, a plot of gas volume versus Celsius temperature gives an approximately straight line. Plots for 0.0010 mol of helium at three different pressures are shown in Figure 5.9. Charles communicated his findings to Joseph-Louis Gay-Lussac, who was able to show that

Figure 5.9

Charles's law plots for 0.0010 mol of helium at three different pressures. Plotting volume versus Celsius temperature at any one pressure produces a straight line. Although the slopes increase with decreasing pressure, all Charles's law plots converge to zero volume at −273°C.

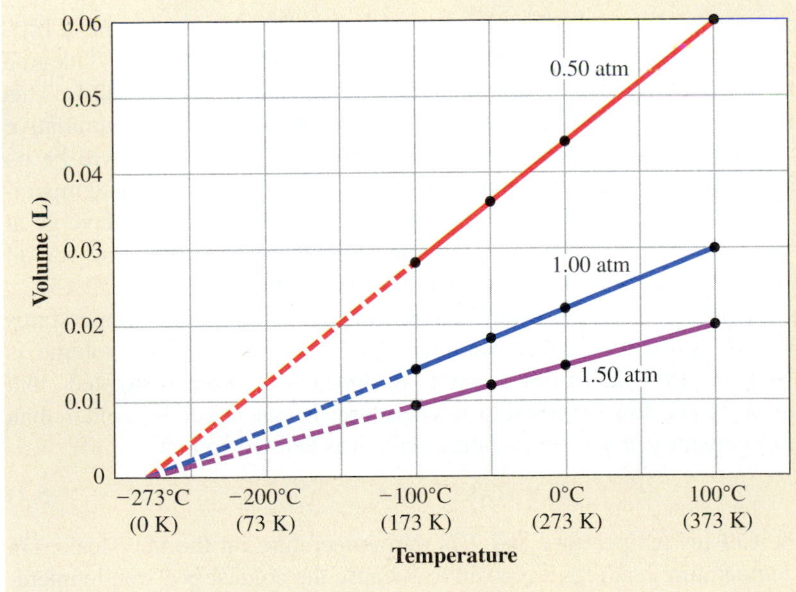

Data for warming 0.0010 mol of He at 0.50 atm			
Temperature (°C)	*Temperature (K)*	*Volume (L)*	*Volume / Temperature (L/K)*
−100	173	0.028	1.6×10^{-4}
−50	223	0.036	1.6×10^{-4}
0	273	0.045	1.6×10^{-4}
50	323	0.053	1.6×10^{-4}
100	373	0.061	1.6×10^{-4}

Data for warming 0.0010 mol of He at 1 atm			
Temperature (°C)	*Temperature (K)*	*Volume (L)*	*Volume / Temperature (L/K)*
−100	173	0.014	8.1×10^{-5}
−50	223	0.018	8.1×10^{-5}
0	273	0.022	8.1×10^{-5}
50	323	0.026	8.1×10^{-5}
100	373	0.030	8.1×10^{-5}

DEMONSTRATION 5.2 AN AIR THERMOMETER

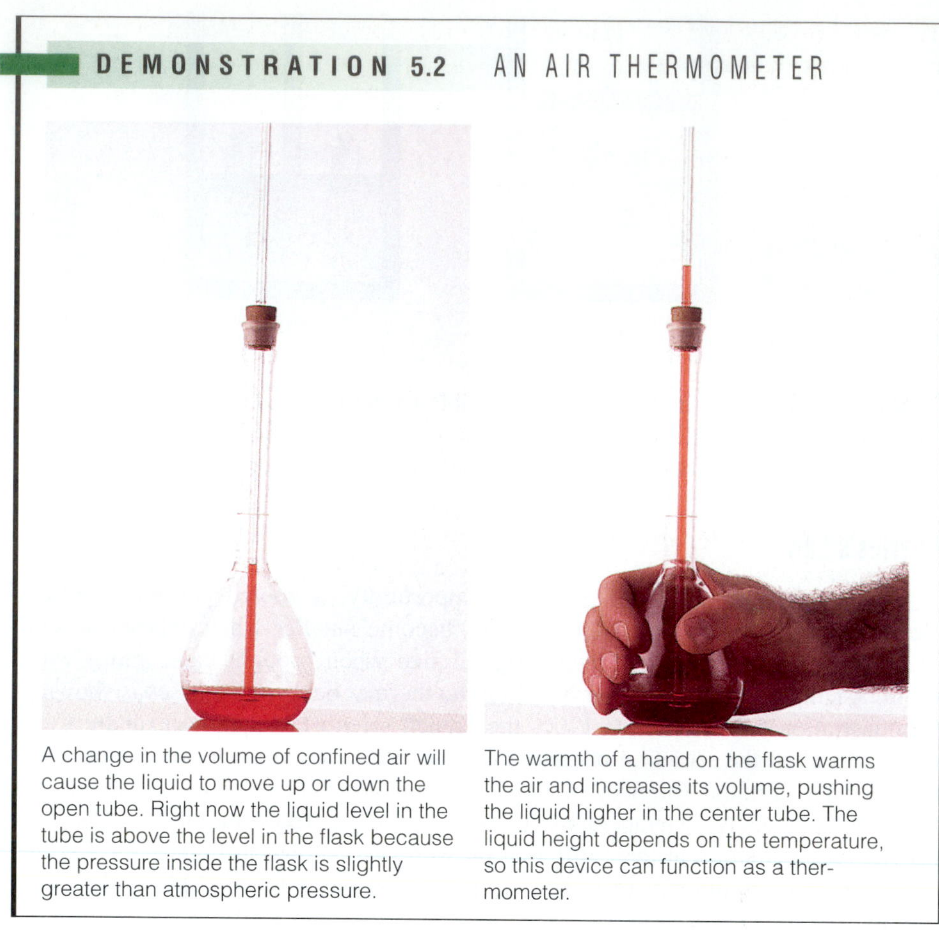

A change in the volume of confined air will cause the liquid to move up or down the open tube. Right now the liquid level in the tube is above the level in the flask because the pressure inside the flask is slightly greater than atmospheric pressure.

The warmth of a hand on the flask warms the air and increases its volume, pushing the liquid higher in the center tube. The liquid height depends on the temperature, so this device can function as a thermometer.

the volume change per degree Celsius is approximately 1/273 of the volume at 0°C. For example, if the volume of a gas sample is 22.4 L at 0°C, cooling by 1 degree will result in a volume decrease of approximately 1/273 of 22.4 L or 0.082 L. This change might seem small, but if we use this observation to predict the cumulative shrinkage as a gas is cooled indefinitely, we find that at −273°C there will be no volume left at all! We can reach the same conclusion by extrapolating the temperature–volume lines in Figure 5.9 to $V = 0$, where they are observed to converge at about −273°C. This extrapolated "zero volume" temperature is the same for all gases at low pressures; the best measurements indicate that this point is −273.15°C.

Real gases liquefy well above −273.15°C.

According to Figure 5.9, a gas would have a negative volume at temperatures below −273.15°C. Since a negative volume is absurd—even a zero volume is impossible because the molecules cannot disappear—it was suggested that −273.15°C must be the lowest possible temperature; it can never be colder than −273.15°C. Subsequently, a new temperature scale was defined so that

The Kelvin scale, originally called the *absolute temperature scale*, was introduced by William Thomson (1824–1927), who later became Lord Kelvin.

$$T = 273.15 + C \tag{5.4}$$

where C is the Celsius temperature and T is the temperature on the new scale. On this scale all temperature readings are positive because the coldest possible temperature, the **absolute zero** of temperature, corresponds to $T = 273.15 - 273.15 = 0$.

The new temperature scale is called the **Kelvin temperature scale** and readings on it are expressed in **kelvins** (K). The intervals on the Kelvin and Celsius scales are the same; that is, the size of 1 kelvin is the same as the size of 1 Celsius degree.

According to international convention, a Kelvin temperature is written without a degree sign between the last digit and the symbol K. A temperature such as 273 K is referred to as "273 kelvins," not "273 degrees Kelvin."

EXAMPLE 5.9

Express room temperature, 25°C, in kelvins.

SOLUTION

Rounding 273.15 to 273 and substituting $C = 25$ into Equation 5.4 gives

$$T = 273 + 25 = 298 \text{ K}$$

PRACTICE EXERCISE 5.10

Convert (a) −15°C into kelvins and (b) 173 K into degrees Celsius.

In Figure 5.9 each of the straight lines in the volume versus temperature plot begins where $V = 0$ L and $T = 0$ K (−273°C). Equations describing these lines have the general form

$$V = k_c \times T \tag{5.5}$$

where k_c is a number that varies only with the number of moles of gas and the pressure. (Note that k_c is not the same as k_b in Boyle's law (Equation 5.2).) Equation 5.5 implies that the volume is *directly proportional* to the Kelvin temperature. That is, when the Kelvin temperature changes by a given factor, the volume will change by the same factor; tripling the Kelvin temperature will triple the volume, halving the Kelvin temperature will halve the volume, and so forth. This relationship between volume and temperature is called **Charles's law**: *At constant pressure the volume of a fixed number of moles of gas is directly proportional to the Kelvin temperature.*

For a gas sample changing from some initial Kelvin temperature and volume (T_i, V_i) to some final Kelvin temperature and volume (T_f, V_f), Equation 5.5 tells us that

$$\frac{V_i}{T_i} = \frac{V_f}{T_f} \tag{5.6}$$

because both quotients are equal to the same value of k_c. Equation 5.6 is often used to relate temperature and volume changes, as shown in the next example.

EXAMPLE 5.10

The volume of a sample of gas is 75.0 mL at 0°C. Estimate the volume of the sample after it has been heated to 100°C with no change in pressure.

SOLUTION

First we change the Celsius temperatures to Kelvin temperatures using Equation 5.4.

LEARNING HINT

Always convert degrees Celsius to kelvins when doing gas law problems.

The initial temperature is $T_i = 273 + 0 = 273$ K. The final temperature is $T_f = 273 + 100 = 373$ K. The initial volume is $V_i = 75.0$ mL. The final volume is obtained by rearranging Equation 5.6 and substituting the appropriate data:

$$V_f = V_i \times \frac{T_f}{T_i} = 75.0 \text{ mL} \times \frac{373 \text{ K}}{273 \text{ K}} = 102 \text{ mL}$$

PRACTICE EXERCISE 5.11

The volume of a gas decreases from 1.46 L to 1.22 L with no change in pressure. If the original temperature was 20°C, what is the final temperature in (a) kelvins and (b) degrees Celsius?

The fraction 373 K/273 K in Example 5.10 is a *temperature factor*—a temperature ratio that can be used to convert an initial volume to a final volume. The answer to the example was obtained by substituting numbers in Equation 5.6, but it could also have been obtained by recognizing that an increase in temperature will increase the volume, and that the original volume must therefore be multiplied by a temperature factor greater than unity. Let us apply this "common sense" approach to the next example.

EXAMPLE 5.11

What will be the final volume if 50.0 mL of a gas is cooled from room temperature (25°C) to −15°C? Assume no change in pressure.

SOLUTION

The initial volume is $V_i = 50.0$ mL. The initial and final Kelvin temperatures are $T_i = 273 + 25 = 298$ K and $T_f = 273 - 15 = 258$ K. Cooling the gas causes its volume to decrease. For the final volume to be less than the initial volume, the temperature factor must be less than unity, or 258 K/298 K. The final volume is

$$V_f = V_i \times \text{temperature factor}$$

$$= 50.0 \text{ mL} \times \frac{258 \text{ K}}{298 \text{ K}} = 43.3 \text{ mL}$$

PRACTICE EXERCISE 5.12

A gas is warmed from − 83°C to 30°C with no change in pressure. (a) Will the volume increase or decrease? (b) By what factor?

How can we explain the thermal expansion of a gas in terms of kinetic theory? To answer this question, let us imagine warming a confined gas, as in Figure 5.10. The molecules move more rapidly as the temperature rises, so they hit the container wall more frequently and with greater force, and thus tend to increase the pressure. To maintain the original pressure, the number of collisions per second will have to

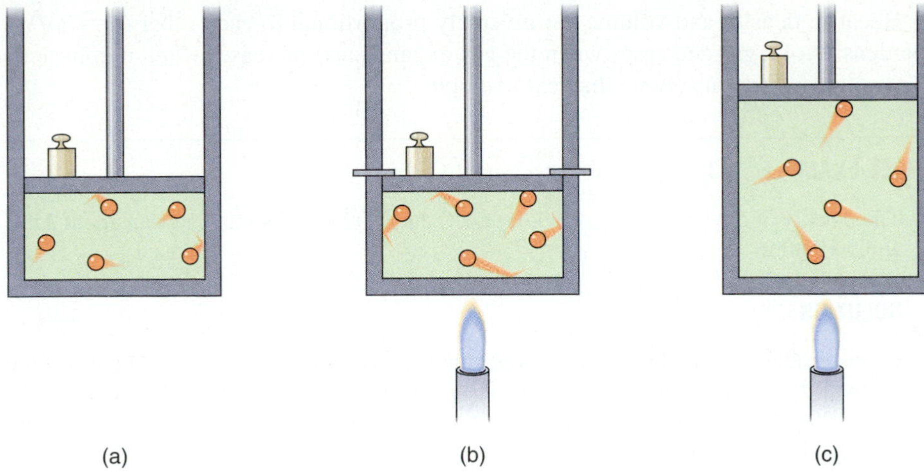

Figure 5.10
Kinetic interpretation of Charles's law. (a) The molecules of the confined gas exert an upward force on the piston equal to the downward force of gravity. (b) The gas is heated with the piston held in place. The hot gas molecules move more rapidly and collide with the piston more frequently and forcefully. (c) When released, the piston moves upward until the increase in volume reduces the number of impacts so that their total force is the same as in (a).

decrease. This can occur only if the gas expands so that the gas molecules have a greater distance to travel between collisions.

Simultaneous Changes in Temperature and Pressure

A volume change often results from temperature and pressure changes that occur at the same time. The next example shows how temperature and pressure factors can be used to sort out the individual effects.

EXAMPLE 5.12

A sample of nitrogen gas occupies a volume of 127 mL at 0°C and 740 torr. The gas is heated to 30°C and its pressure is increased to 760 torr. Calculate the final volume.

SOLUTION

The increase in temperature from 0°C = 273 K to 30°C = 303 K tends to increase the volume; the temperature factor will be greater than unity, or 303 K/273 K. The increase in pressure from 740 torr to 760 torr tends to decrease the volume; the pressure factor will be less than unity, or 740 torr/760 torr. The final volume is obtained by multiplying the initial volume, V_i = 127 mL, by both of these factors:

$$V_f = V_i \times \text{temperature factor} \times \text{pressure factor}$$

$$= 127\text{ mL} \times \frac{303\text{ K}}{273\text{ K}} \times \frac{740\text{ torr}}{760\text{ torr}} = 137\text{ mL}$$

↑ pressure factor chosen to decrease volume

↑ temperature factor chosen to increase volume

PRACTICE EXERCISE 5.13

A 55.0-mL sample of argon gas is cooled from 75°C to −10°C. At the same time, its pressure increases from 620 torr to 745 torr. Calculate the final volume.

Because density and volume are inversely proportional to each other ($d = m/V$), the density of a gas decreases when the gas expands and increases when it contracts. We use these facts in solving the next example.

EXAMPLE 5.13

The density of oxygen gas is 1.429 g/L at 0°C and 1.00 atm. Estimate its density at 25°C and 1.10 atm.

SOLUTION

In this problem we are looking for temperature and pressure factors that will convert the original density into the final density. The temperature increase from 0°C (273 K) to 25°C (298 K) tends to increase the volume and decrease the density. The temperature factor will be less than unity, or 273 K/298 K. The pressure increase from 1.00 atm to 1.10 atm tends to decrease the volume and increase the density. The pressure factor will be greater than unity, or 1.10 atm/1.00 atm. The final density is obtained by multiplying the initial density, d_i = 1.429 g/L, by both of these factors:

$$d_f = d_i \times \text{temperature factor} \times \text{pressure factor}$$

$$= 1.429\text{ g/L} \times \frac{273\text{ K}}{298\text{ K}} \times \frac{1.10\text{ atm}}{1.00\text{ atm}} = 144\text{ g/L}$$

pressure factor chosen to increase density

temperature factor chosen to increase density

PRACTICE EXERCISE 5.14

The density of an unknown gas is 0.635 g/L at 30°C and 750 torr. Calculate its density at 0°C and 1.00 atm pressure.

Avogadro's Law

If a sample of gas contains n moles and has a volume V, then the **molar volume**, or volume per mole, is V/n. The molar mass of a gas is a fixed quantity, but the molar volume will vary with temperature and pressure as the gas expands or contracts. Molar volumes can be computed from density data and molar masses; Table 5.1 lists the molar volumes for a number of gases at 0°C and 1 atm pressure. It is customary to refer to these conditions as **standard conditions** or simply **STP** (standard temperature and pressure). Volume data for numerous gases, like the data in Table 5.1, show that at standard conditions the molar volume of a gas is about 22.4 L; that is, 1 mol of any gas at 0°C and 1 atm will have a volume of about 22.4 L, 2 mol will have a volume of about 44.8 L, and so forth. Molar volumes increase with increasing temperature and decreasing pressure, but they are approximately the same for all gases at the same temperature and pressure.

Standard conditions:

0°C = 273 K

1 atm = 760 torr

EXAMPLE 5.14

Estimate the molar volume of a gas at 1.00 atm and 25°C.

SOLUTION

The molar volume of a gas is approximately 22.4 L at 1.00 atm and 0°C = 273 K. This volume will increase when the gas is warmed. The molar volume at 25°C = 298 K is obtained by multiplying 22.4 L by the temperature factor 298 K/273 K:

$$22.4 \text{ L/mol} \times \frac{298 \text{ K}}{273 \text{ K}} = 24.5 \text{ L/mol}$$

PRACTICE EXERCISE 5.15

Estimate the molar volume of a gas at 100°C and 740 torr.

TABLE 5.1 Molar Volumes of Various Gases at 0°C and 1 atm[a]

Gas	Formula	Density (g/L)	Molar Mass (g/mol)	Molar Volume (L/mol)
Carbon dioxide	CO_2	1.97694	44.010	22.262
Carbon monoxide	CO	1.25010	28.010	22.406
Dinitrogen oxide	N_2O	1.97821	44.0128	22.2488
Helium	He	0.17846	4.00260	22.429
Hydrogen	H_2	0.089873	2.0158	22.429
Hydrogen chloride	HCl	1.63915	36.461	22.244
Nitrogen	N_2	1.25046	28.0134	22.4025
Oxygen	O_2	1.42900	31.9988	22.3924

[a]The molar volume is equal to the molar mass divided by the density.

The volume of a cube about 11.1 inches on edge is 22.4 L. It is approximately equal to the volume of three regulation basketballs.

The observation that all gases have approximately the same molar volume at the same temperature and pressure is expressed by the equation

$$\frac{V}{n} = k_a \tag{5.7a}$$

or

$$V = k_a \times n \tag{5.7b}$$

where k_a varies only with the temperature and pressure. Equation 5.7 is the mathematical statement of **Avogadro's law**: *At constant temperature and pressure, the volume of a gas is directly proportional to the number of moles.* Avogadro's law tells us that it is not molecular size or mass, but only the number of moles (or the number of molecules since the mole is simply a counting unit) that determines gas volume. Thus the law implies that *equal volumes of different gases at the same temperature and pressure contain the same number of molecules;* 1 L of oxygen gas at 25°C and 1.00 atm contains the same number of molecules as 1 L of helium gas under the same conditions. This remarkable conclusion, which is called **Avogadro's hypoth-**

Amadeo Avogadro (1776–1856).

esis, was deduced and published by Avogadro in 1811, long before Avogadro's law (Equation 5.7) was formulated and named in his honor. Avogadro came to this conclusion after studying the volume changes that occur in gaseous chemical reactions.

5.4 THE IDEAL GAS LAW

In the previous section, you learned that the volume of a gas varies directly with the Kelvin temperature, directly with the number of moles, and inversely with the pressure. These three observations can be summarized in the generalized gas law

$$PV = nRT \tag{5.8}$$

Different conditions apply to each of the gas laws, so Equation 5.8 cannot be obtained by simply combining Boyle's law, Charles's law, and Avogadro's law. Methods for obtaining Equation 5.8 are described by J. D. Herron, "Derivation of the Ideal Gas Law," *J. Chem. Educ.*, **56**, 530 (1979).

where P, V, and T are the pressure, volume, and Kelvin temperature of the gas, respectively; n is the number of moles; and R is a constant that has the same value for all gases. We can show that Equation 5.8 summarizes the gas laws by rearranging it to solve for the volume:

$$V = \frac{nRT}{P} \tag{5.8a}$$

Both Boyle and Charles worked with gas samples containing a fixed number of moles (n is constant). If the temperature does not change, then the product nRT is constant and Equation 5.8a states that the volume varies inversely with pressure, in accordance with Boyle's law. On the other hand, if the pressure does not change, then the quotient nR/P is constant, and the volume varies directly with Kelvin temperature, as shown by Charles and Gay-Lussac. The assumptions in Avogadro's law are that n is variable and that both the temperature and pressure are constant. In this case, the quotient RT/P is constant and the volume varies directly with the number of moles.

In 1701 Guillaume Amontons observed that the pressure of a gas increases when the gas is heated in a container with rigid walls. The generalized gas law summarizes this observation as well. Solving Equation 5.8 for the pressure gives

$$P = \frac{nRT}{V} \tag{5.8b}$$

When a gas is confined to a container with rigid walls, its volume and number of moles will not change. Hence, the quotient nR/V will be constant and the pressure will vary directly with the Kelvin temperature. The variation of gas pressure with temperature is sometimes called **Amontons' law**: *At constant volume the pressure of a fixed number of moles of gas is directly proportional to the Kelvin temperature.*

EXAMPLE 5.15

The pressure in an oxygen tank is 10.0 atm at 0°C. What pressure will develop in the tank if it is stored in a furnace room at 45°C?

SOLUTION

The volume and number of moles of gas are fixed. The temperature increases from 0°C

= 273 K to 45°C = 318 K. The pressure will increase by a factor of 318 K/273 K:

$$P = 10.0 \text{ atm} \times \frac{318 \text{ K}}{273 \text{ K}} = 11.6 \text{ atm}$$

PRACTICE EXERCISE 5.16

The pressure in the oxygen tank in Example 5.15 drops from 10.0 atm to 9.75 atm. What is the temperature of the room in which the tank is stored?

Equation 5.8 also implies that the quotient $PV/T = nR$ is constant for a gas sample containing a fixed number of moles. If the pressure, volume, and temperature of this sample change from some initial set of values (P_i, V_i, T_i) to some final set (P_f, V_f, T_f), then the initial value of PV/T must equal its final value, that is,

$$\frac{P_i V_i}{T_i} = \frac{P_f V_f}{T_f} \tag{5.9}$$

PRACTICE EXERCISE 5.17

Refer to Example 5.12 and Practice Exercise 5.13. Solve these problems by substituting the appropriate data into Equation 5.9.

It should be emphasized that gases do not obey the gas laws rigorously. The laws are most closely obeyed by gases at low pressures and high temperatures. In later sections of this chapter, you will learn that the deviations are due largely to intermolecular forces. These forces are less effective when the molecules are far apart (low pressure) or moving rapidly (high temperature). It is convenient to define an **ideal gas** as one that obeys $PV = nRT$ under all conditions, and for this reason the generalized gas law is usually called the **ideal gas law**. There is no ideal gas in nature, although real gases are almost ideal at very low pressures. In practice, the ideal gas law is a good approximation to the behavior of real gases, and calculations based on the ideal gas law will usually be accurate to within 5% or better. For very precise work, however, the ideal gas law is not used (see Section 5.8).

Intermolecular forces are caused by attractions and repulsions between electrons and nuclei of neighboring molecules. These forces are negligibly small except when the molecules are very close to each other.

Using the Ideal Gas Law

The constant R in the ideal gas law is called the **gas constant**; it is the same for all gases. The gas constant can be estimated from the observation that 1 mol of gas occupies a volume of about 22.4 L at 0°C and 1 atm. Rearranging Equation 5.8 to solve for R and substituting $P = 1.00$ atm, $V = 22.4$ L, and $T = 0°\text{C} = 273$ K gives

$$R = \frac{PV}{nT} = \frac{1.00 \text{ atm} \times 22.4 \text{ L}}{1 \text{ mol} \times 273 \text{ K}} = 0.0821 \text{ L·atm/mol·K}$$

The ideal gas law becomes more exact at low pressures, and a more precise experimental value, $R = 0.08205784$ L·atm/mol·K, is obtained from low-pressure data.

The speed of sound in a gas is related to the value of the gas constant. The most precise value of R is obtained by measuring the speed of sound in argon gas at various pressures and extrapolating the results to zero pressure.

TABLE 5.2 Values of the Gas Constant in Various Units[a]

0.0821 L·atm/mol·K	8.3145 dm^3·kPa/mol·K
62.364 L·torr/mol·K	8.3145 J/mol·K
0.083145 L·bar/mol·K	1.9872 cal/mol·K

[a]To show that L·atm is a unit of energy, recall that 1 L = 1000 cm^3 = 10^{-3} m^3 and 1 atm = 1.01325 × 10^5 N/m^2 (Section 5.2). Hence,

$$1 \text{ L·atm} = 10^{-3}\text{ m}^3 \times 1.01325 \times 10^5\text{ N/m}^2 = 101.325\text{ N·m} = 101.325\text{ J}$$

since 1 J = 1 N·m by definition. The gas constant is

$$R = 0.0820578\ \frac{\text{L·atm}}{\text{mol·K}} \times \frac{101.325\text{ J}}{1\text{ L·atm}} = 8.31451\text{ J/mol·K}$$

$$= 1.98722\text{ cal/mol·K}$$

since 1 cal = 4.184 J by definition.

Table 5.2 shows that the units of volume × pressure, L·atm, are actually units of energy. The various values of R listed in Table 5.2 will be useful in later work.

EXAMPLE 5.16

Use the ideal gas law to estimate the volume in liters of 2.50 g of carbon dioxide at 25°C and 2.00 atm. The molar mass of carbon dioxide is 44.01 g/mol.

SOLUTION

Since the volume is required, we rearrange Equation 5.8 to give $V = nRT/P$, where T is 25°C = 298 K and P is 2.00 atm. The number of moles of carbon dioxide is n = 2.50 g × 1 mol/44.01 g = 0.0568 mol. Substituting gives

$$V = \frac{nRT}{P} = \frac{0.0568\text{ mol} \times 0.0821\text{ L·atm/mol·K} \times 298\text{ K}}{2.00\text{ atm}} = 0.695\text{ L}$$

Note carefully how the units cancel to give liters.

EXAMPLE 5.17

A gas is stored in a 50.0-L tank at 20.0 atm and 17°C. How many moles of gas are in the tank? Assume ideal behavior.

SOLUTION

Since the number of moles is required, we rearrange the ideal gas law to give $n = PV/RT$. The temperature is 17°C = 290 K. The pressure is 20.0 atm and the volume is 50.0 L. Substitution gives

$$n = \frac{PV}{RT} = \frac{20.0\text{ atm} \times 50.0\text{ L}}{0.0821\text{ L·atm/mol·K} \times 290\text{ K}} = 42.0\text{ mol}$$

PRACTICE EXERCISE 5.18

A 250-mL glass bulb contains 0.250 g of oxygen gas (O_2) at 25°C. Estimate the gas pressure in atmospheres.

Molar Mass from Gas Density

The molar mass ($\mathcal{M}$) of any substance can be found by dividing the mass (m) of a sample by the number of moles (n) in the sample: $\mathcal{M} = m/n$. If the sample is gaseous, the number of moles can be found from its volume, using the ideal gas law. A typical calculation is given in the next example.

EXAMPLE 5.18

A 250-mL sample of acetone vapor is found to weigh 0.474 g at 1.00 atm and 100°C (Figure 5.11). Estimate the molar mass of acetone.

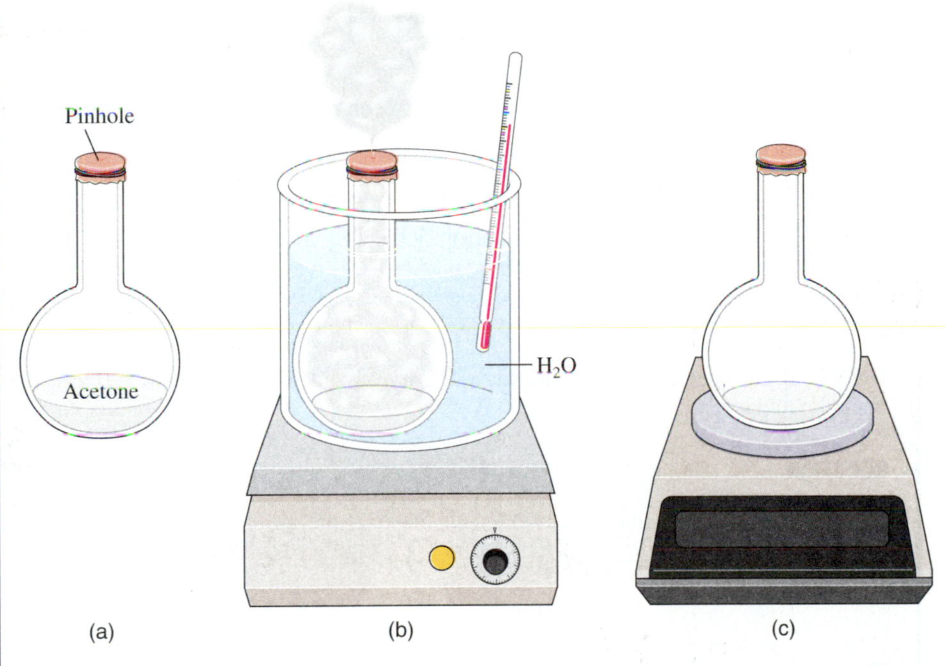

Figure 5.11

Finding the molar mass of a volatile liquid (see below). (a) A sample of acetone is placed in a weighed flask of known volume. (b) The flask is heated at a known temperature until the entire sample vaporizes and fills the flask with acetone vapor at atmospheric pressure. Excess vapor escapes through a pinhole. The number of moles of acetone is calculated from the pressure, volume, and temperature of the vapor. (c) The mass of acetone vapor is found by weighing the cooled flask plus condensed vapor and subtracting the weight of the flask.

SOLUTION

The ideal gas law is used in the form $n = PV/RT$ to find the number of moles of acetone vapor. The temperature is 100°C = 373 K and the pressure is 1.00 atm. To make the volume units consistent with those of the gas constant, 250 mL is changed to 0.250 L. Substitution gives

$$n = \frac{PV}{RT} = \frac{1.00 \text{ atm} \times 0.250 \text{ L}}{0.0821 \text{ L·atm/mol·K} \times 373 \text{ K}} = 0.00816 \text{ mol}$$

This number of moles has a mass $m = 0.474$ g. The molar mass is

$$\mathcal{M} = \frac{m}{n} = \frac{0.474 \text{ g}}{0.00816 \text{ mol}} = 58.1 \text{ g/mol}$$

PRACTICE EXERCISE 5.19

A 100-mL glass bulb contains 0.360 g of an unknown gas at 28°C and 770 torr. Estimate the molar mass of the gas.

Most molar masses are now determined by mass spectroscopy rather than by gas measurements.

A relation between molar mass and gas density can be obtained from the ideal gas law. The number of moles n is equal to $m/\mathcal{M}$, where m is the mass of the substance in grams and $\mathcal{M}$ is its molar mass in grams per mole. Substituting $m/\mathcal{M}$ for n in the gas law gives

$$PV = nRT = \frac{m}{\mathcal{M}}RT$$

which can be rearranged to

$$\mathcal{M} = \frac{m}{V}\frac{RT}{P}$$

Substituting $d = m/V$ gives the relationship between molar mass and density:

$$\mathcal{M} = d\frac{RT}{P} \qquad (5.10)$$

Example 5.19 illustrates the use of Equation 5.10.

EXAMPLE 5.19

Halothane, an inhalation anesthetic, has a density of 6.74 g/L at 65°C and 720 torr. Estimate the molar mass of halothane.

SOLUTION

To make the pressure units consistent with those of the gas constant, we convert 720 torr into atmospheres:

$$P = 720 \text{ torr} \times \frac{1 \text{ atm}}{760 \text{ torr}} = 0.947 \text{ atm}$$

The temperature T is 65°C = 338 K. The density d is 6.74 g/L. Substitution into Equation 5.10 gives the molar mass:

$$\mathcal{M} = 6.74 \text{ g/L} \times \frac{0.0821 \text{ L·atm/mol·K} \times 338 \text{ K}}{0.947 \text{ atm}} = 198 \text{ g/mol}$$

PRACTICE EXERCISE 5.20

Use Equation 5.10 to estimate the density of methane (CH_4) at 0.750 atm and 0°C.

5.5 GAS STOICHIOMETRY

Gases are usually measured by volume, so gas stoichiometry deals with the volumes of reactants and products more often than with their masses. The next few examples show that the key to these problems, as usual, is the mole.

EXAMPLE 5.20

Glucose ($C_6H_{12}O_6$) reacts with oxygen to give carbon dioxide and water vapor. How many liters of oxygen, measured at 25°C and 750 torr, are required for the oxidation of 0.500 g of glucose?

SOLUTION

We will follow the procedure given in Section 3.7 for solving stoichiometric problems. The equation for the reaction is

$$C_6H_{12}O_6(s) + 6O_2(g) \rightarrow 6CO_2(g) + 6H_2O(g)$$

The known quantity is 0.500 g of glucose and the unknown quantity is liters of oxygen. The mass of 1 mol of glucose is 180.2 g. The number of moles of oxygen is calculated as follows:

$$\text{g glucose} \rightarrow \text{mol glucose} \rightarrow \text{mol } O_2$$

$$0.500 \text{ g glucose} \times \frac{1 \text{ mol glucose}}{180.2 \text{ g glucose}} \times \frac{6 \text{ mol } O_2}{1 \text{ mol glucose}} = 1.66 \times 10^{-2} \text{ mol } O_2$$

The volume of oxygen is obtained from the ideal gas law $V = nRT/P$. The temperature is 25°C = 298 K, and the pressure is 750 torr × 1 atm/760 torr = 0.987 atm.

$$V = \frac{nRT}{P} = \frac{1.66 \times 10^{-2} \text{ mol} \times 0.0821 \text{ L·atm/mol·K} \times 298 \text{ K}}{0.987 \text{ atm}} = 0.411 \text{ L}$$

PRACTICE EXERCISE 5.21

How many liters of carbon dioxide, measured at 20°C and 1.00 atm, will be produced during the combustion of 0.250 mol of methane (CH_4)?

Stoichiometric calculations involving gases can often be simplified by using Avogadro's law, which states that at a given temperature and pressure, the volume of a gas is proportional to the number of moles. Two moles of any gas occupy twice the volume of 1 mol, 3 mol occupy three times the volume, and so forth. Avogadro's law, when applied to chemical reactions, leads to the following rule for gas volumes measured at the same temperature and pressure: *The volume ratio of the gases consumed and produced in a chemical reaction is the same as the mole ratio.* Consider, for example, the synthesis of ammonia from nitrogen and hydrogen:

$$N_2(g) + 3H_2(g) \rightarrow 2NH_3(g)$$

Two moles of ammonia will form from 1 mol of nitrogen and 3 mol of hydrogen;

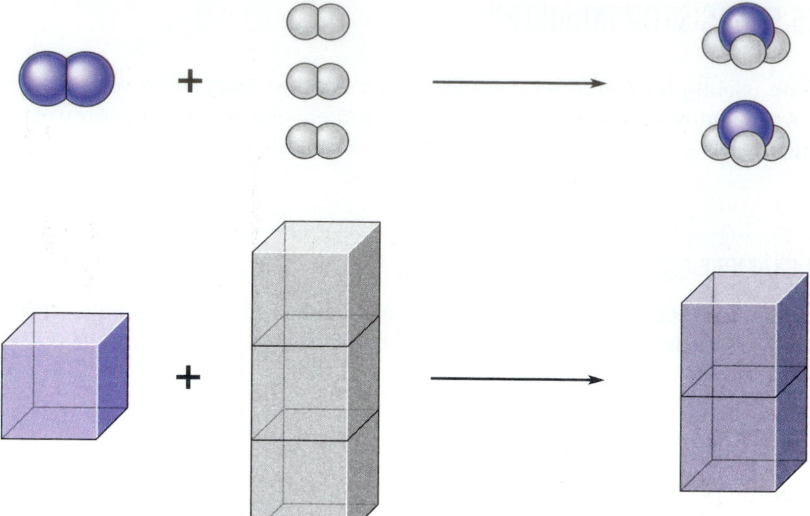

Figure 5.12
The volume ratio of the gases consumed or produced in a chemical reaction is the same as the molecule (mole) ratio. Top: one molecule of nitrogen reacts with three molecules of hydrogen to form two molecules of ammonia. Bottom: one volume of nitrogen reacts with three volumes of hydrogen to form two volumes of ammonia.

therefore, 2 L of ammonia gas will form from 1 L of nitrogen gas and 3 L of hydrogen gas (Figure 5.12). Keep in mind that this rule applies only to gaseous reactants and products at the same temperature and pressure, and it does not apply to liquids or solids.

EXAMPLE 5.21

In the synthesis of ammonia from nitrogen and hydrogen, how many liters of ammonia are produced when 12.0 L of hydrogen are consumed? Assume that the volumes are measured at the same temperature and pressure.

SOLUTION

The coefficients in the balanced equation given above tell us that 2 L of NH_3 gas will be produced from 3 L of H_2 gas. The volume of ammonia produced from 12.0 L of hydrogen is

$$12.0 \text{ L } H_2 \times \frac{2 \text{ L } NH_3}{3 \text{ L } H_2} = 8.0 \text{ L } NH_3$$

PRACTICE EXERCISE 5.22

The equation for the electrolysis of water is

$$2H_2O(l) \rightarrow 2H_2(g) + O_2(g)$$

(a) If 45.0 mL of hydrogen is formed during an electrolysis, how many milliliters of oxygen will form? (b) Is enough information given to predict the volume of water consumed?

5.6 MIXTURES OF GASES

The most familiar gas mixture is air, the ocean of gases that surrounds and sustains us. Other important gas mixtures include anesthetics, breathing mixtures for divers, fuels such as natural gas, and gases for welding. Gas molecules are relatively far apart, so the components of a gaseous mixture have little or no effect on each other. In effect, each component behaves as if it were the sole occupant of the total volume. Some of the consequences of this independent behavior are described in the following sections.

Dalton's Law of Partial Pressures

John Dalton observed in 1801 that the total pressure in a gas mixture is the sum of the pressures each gas would exert if it were alone at that volume and temperature (Figure 5.13). The pressure contributed by each gas is called a **partial pressure**, and **Dalton's law of partial pressures** states that *the total pressure of a gas mixture is equal to the sum of the partial pressures of the individual gases.* Assume, for example, that three gases, A, B, and C, are mixed in a container of volume V. Dalton's law states that the total pressure P_T is

$$P_T = P_A + P_B + P_C \qquad (5.11)$$

where P_A, P_B, and P_C are the partial pressures of each gas in the mixture.

Let us explore the implications of Equation 5.11 by using the ideal gas law and substituting $P = nRT/V$ for each of the partial pressures:

$$P_T = \frac{n_A RT}{V} + \frac{n_B RT}{V} + \frac{n_C RT}{V}$$

Factoring out the RT/V term gives

$$P_T = (n_A + n_B + n_C)\frac{RT}{V}$$

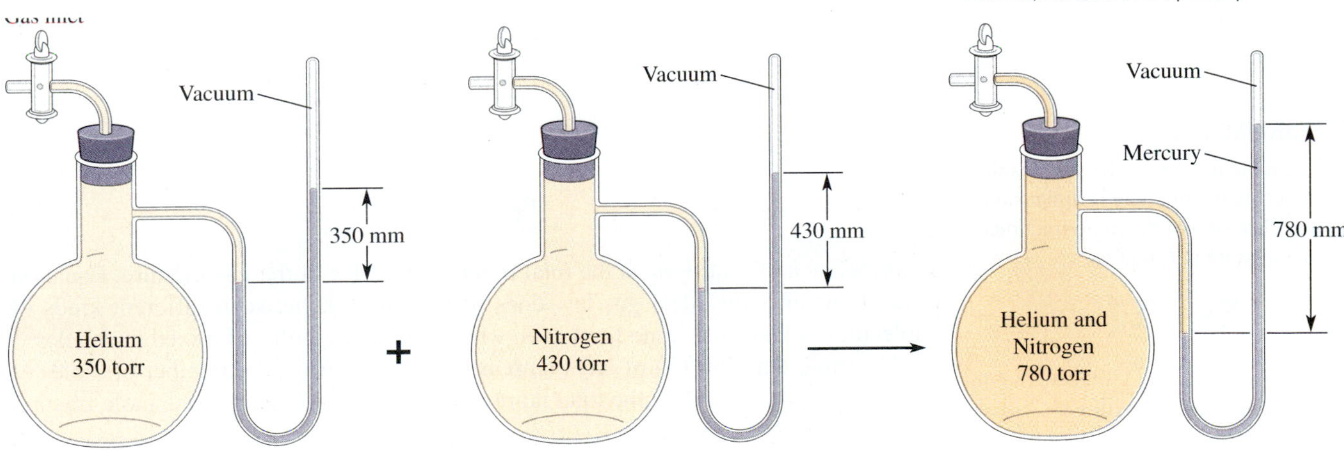

Figure 5.13
An illustration of Dalton's law of partial pressures. Each of the confined gases has the same volume and temperature. (The volume of gas in the manometer tube is negligible compared to the total volume.) When helium at 350 torr is mixed with nitrogen at 430 torr, the total pressure is 780 torr, the sum of the partial pressures.

CHEMICAL INSIGHT

"THIS MOST EXCELLENT CANOPY, THE AIR"—WILLIAM SHAKESPEARE, *HAMLET*

The atmosphere is a turbulent ocean of gases, stirred by the earth's rotation and churned by solar heat. It bathes the surface of our planet with life-giving oxygen, nitrogen, carbon dioxide, and water vapor, shields its occupants from devastating solar radiation, and supports them by its pressure. The total mass of the atmosphere has been estimated at 5.5×10^{15} tons, 99% of which is found below an altitude of 30 km.

The nature of the atmosphere at any altitude can be described in terms of its molar mass, temperature, and pressure (Figure 5.14). The molar mass of a mixture such as air is the average mass of Avogadro's number of mixed molecules. Table 5.3 shows the relative composition of clean dry air in the lower atmosphere. Between sea level and an altitude of 90 km, the atmosphere is well stirred by the weather and its composition is uniform. The molar mass is a constant 28.96 g/mol within this region. Above 90 km, many oxygen and nitrogen molecules are dissociated into less massive atoms by the intense solar radiation. The mole fractions of atomic nitrogen and oxygen increase, and the molar mass decreases to about 17 g/mol at a height of 500 km (Figure 5.14a). At still higher altitudes, there are substantial amounts of hydrogen and helium, the least massive elements, and the molar mass continues to decrease. At 1000 km, it is about 8 g/mol and, at 1500 km, it is less than 4 g/mol.

TABLE 5.3 Composition of Dry Air Near Sea Level[a]

Component		Mole Percent
Nitrogen	N_2	78.084
Oxygen	O_2	20.9476
Argon	Ar	0.934
Carbon dioxide	CO_2	0.0314
Neon	Ne	0.001818
Helium	He	0.000524
Methane	CH_4	0.0002
Krypton	Kr	0.000114
Hydrogen	H_2	0.00005
Dinitrogen oxide	N_2O	0.00005
Xenon	Xe	0.0000087

[a]The sea level atmosphere also contains small and variable amounts of ozone (O_3), sulfur dioxide (SO_2), nitrogen dioxide (NO_2), ammonia (NH_3), carbon monoxide (CO), and iodine (I_2).

Temperature forms the basis for dividing the atmosphere into the four regions shown in Figure 5.14b. The *troposphere*, a region ranging from sea level to about 11 km, is characterized by a decrease in temperature with altitude. The *stratosphere*, from 11 km to about 50 km, is a region of increasing temperature. Still higher are the *mesosphere* and the *thermosphere*, extremely rarefied regions in which the temperature decreases and then increases again. The lowest average temperature (181 K = −92°C) is found in the mesosphere at an altitude of 85 km; the highest average temperature is 1500 K (1227°C) in the thermosphere at altitudes above 400 km.

Average atmospheric pressure is 760 torr at sea level, and it falls off to 2.26×10^{-4} torr at an altitude of 100 km (Figure 5.14c). The increasing rarefication of air with height can be understood in terms of the average distance a molecule travels before it collides with another molecule. At sea level, this distance is about 1×10^{-6} cm; at 100 km, the average distance is about 3.4 cm; at 500 km, the average distance is about 2,000,000 cm (20 km)!

LEARNING HINT

The mole fraction is the same as the molecule fraction and is dimensionless. The sum of the mole fractions in a mixture is 1, that is,

$$\sum_i X_i = \sum_i \frac{n_i}{n_T} = 1$$

where the subscript i refers to the ith component in the mixture.

or

$$P_T = \frac{n_T RT}{V} \tag{5.12}$$

where $n_T = n_A + n_B + n_C$ is the total number of moles in the gas mixture. Equation 5.12 shows that the ideal gas law does not distinguish between different kinds of molecules; it takes the same form even when n refers to moles of mixed molecules.

The **mole fraction** (X) of any component in a mixture is the number of moles of the component divided by the total number of moles. For example, the mole fraction of A in the above mixture is

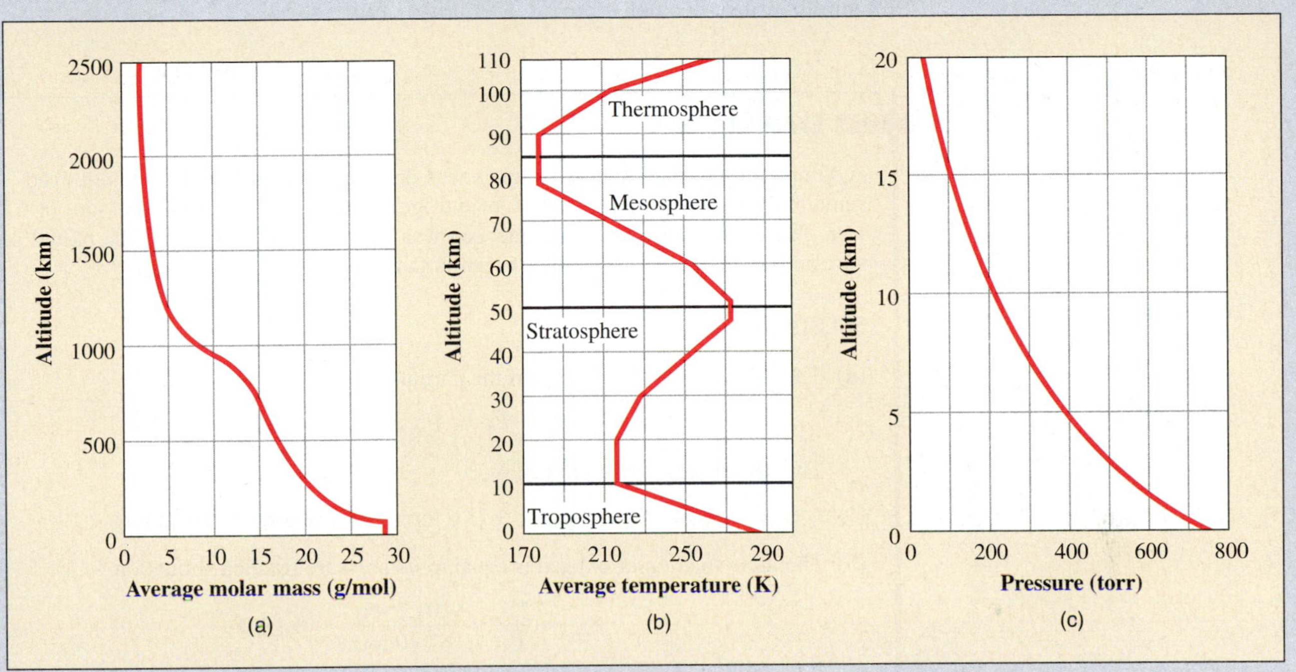

Figure 5.14

Properties of the atmosphere. (a) The average molar mass of air decreases with altitude. (b) Regions of the atmosphere are defined by temperature changes. With increasing altitude, temperature decreases in the troposphere, increases in the stratosphere, decreases in the mesosphere, and increases in the thermosphere. (c) Atmospheric pressure decreases

Thin as it is, the high-altitude atmosphere produces many colorful and interesting effects. Meteors entering from outer space collide with so many particles between 150 and 50 km that they become white-hot "falling stars," many of which completely vaporize. Above 50 km, solar radiation acts on air molecules to form ions and free electrons. This is an extremely important effect because these charged particles make radio communication possible by reflecting electromagnetic waves back to earth. Ions and other energized particles also produce the *aurora borealis* (northern lights), and *airglow*, a very faint light that is present even on the darkest nights.

$$X_A = \frac{n_A}{n_T} \tag{5.13}$$

The number of moles of A is $n_A = P_A V/RT$ and the total number of moles is $n_T = P_T V/RT$. Hence, the mole fraction of A is

$$X_A = \frac{n_A}{n_T} = \frac{P_A V/RT}{P_T V/RT}$$

or, canceling the V/RT terms,

$$X_A = \frac{P_A}{P_T} \tag{5.14}$$

Equation 5.14 shows that *the fraction of the total pressure exerted by a component in a gas mixture is equal to the mole fraction of the component.* This result is reasonable, because the pressure is caused by molecules bombarding the container walls, and we would expect the pressure contribution of each gas to be related directly to the number of molecules it provides for this activity.

EXAMPLE 5.22

A sample of air from a patient's lungs was dried and the carbon dioxide removed. The remaining air consisted principally of nitrogen and oxygen at a total pressure of 673.0 torr. The partial pressure of the nitrogen was 571.8 torr. Calculate (a) the partial pressure and (b) the mole fraction of oxygen in the sample.

SOLUTION

(a) The total pressure is the sum of the partial pressures:

$$P_T = P_{N_2} + P_{O_2}$$

The partial pressure of oxygen is

$$P_{O_2} = P_T - P_{N_2} = 673.0 \text{ torr} - 571.8 \text{ torr} = 101.2 \text{ torr}$$

(b) The mole fraction of oxygen is equal to its pressure fraction (Equation 5.14):

$$X_{O_2} = \frac{P_{O_2}}{P_T} = \frac{101.2 \text{ torr}}{673.0 \text{ torr}} = 0.1504$$

The composition of a mixture can also be expressed in **mole percent**, where mole percent = 100% × mole fraction. For example, the mole percent of oxygen in the mixture in Example 5.22 is 100% × 0.1504 = 15.04%. The mole percent of nitrogen in the mixture is 100% − 15.04% = 84.96%.

PRACTICE EXERCISE 5.23

A mixture of hydrogen gas and water vapor has a total pressure of 740 torr. If the partial pressure of water vapor is 23.8 torr, what are (a) the partial pressure, (b) the mole fraction, and (c) the mole percent of hydrogen in the mixture?

Collecting Gases Over Water

Gaseous reaction products are often collected over water as shown in Figure 5.15. The water serves as a convenient trap, but it also causes the gas to become mixed with a certain amount of water vapor. The partial pressure of a gas collected over water is obtained by subtracting the pressure of the water vapor from the total pressure of the gas mixture. The pressure of water vapor trapped above liquid water is called the **vapor pressure** of water, and it has a constant value at any given temperature. Some values are tabulated in Table 5.4. The concept of vapor pressure is explored more fully in Chapter 12.

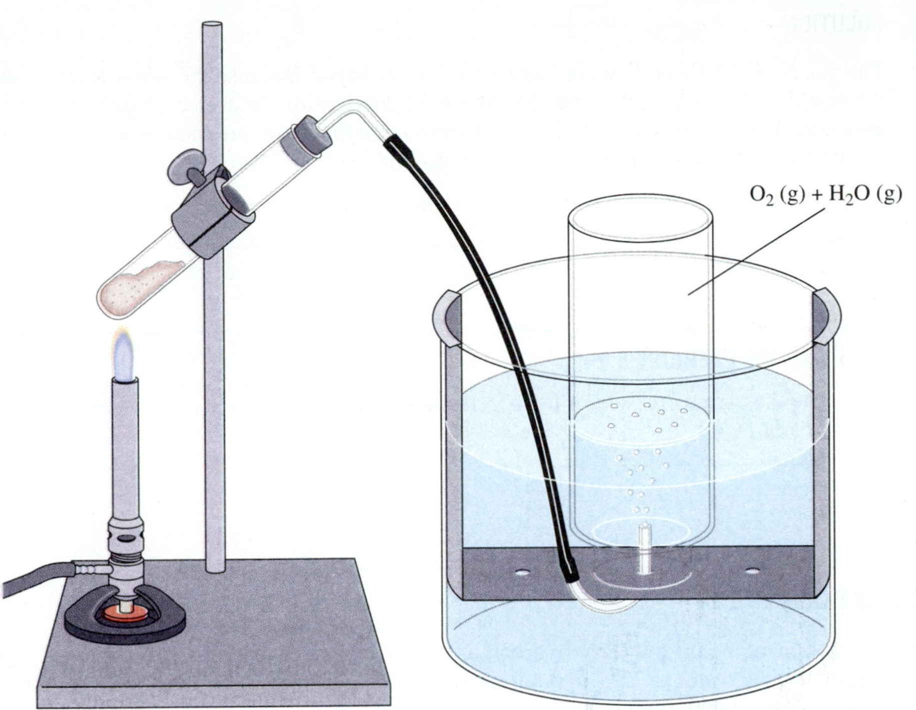

Figure 5.15

Oxygen gas produced during a chemical reaction is collected by the displacement of water. The total pressure of the confined gas equals the pressure of the oxygen plus the vapor pressure of water.

TABLE 5.4 Vapor Pressure of Water at Various Temperatures

Temperature (°C)	Vapor Pressure (torr)	Temperature (°C)	Vapor Pressure (torr)
0	4.579	30	31.824
5	6.543	35	42.175
10	9.209	40	55.324
15	12.788	45	71.88
20	17.535	50	92.51
21	18.650	60	149.4
22	19.827	70	233.7
23	21.068	80	355.1
24	22.377	90	525.8
25	23.756	100	760.0

EXAMPLE 5.23

Find the pressure of oxygen in the collection bottle in Figure 5.15 if the temperature is 20°C and atmospheric pressure is 761 torr.

SOLUTION

The gas in the bottle consists of oxygen and water vapor. Because the water level inside the bottle is the same as that outside, the total pressure of the collected gas will equal atmospheric pressure, 761 torr. Table 5.3 shows that the vapor pressure of water at 20°C is 17.5 torr. The partial pressure of oxygen is obtained from Dalton's Law:

$$P_{O_2} = P_T - P_{H_2O} = 761 \text{ torr} - 17.5 \text{ torr} = 744 \text{ torr}$$

PRACTICE EXERCISE 5.24

Calculate the partial pressure of hydrogen in a gas sample collected over water at 770 torr and 15°C.

EXAMPLE 5.24

A reaction between zinc and hydrochloric acid generated 500 mL of hydrogen gas measured over water at 22°C and a total pressure of 755 torr. How many grams of zinc were consumed in the reaction?

SOLUTION

The equation for the reaction is

$$Zn(s) + 2HCl(aq) \longrightarrow ZnCl_2(aq) + H_2(g)$$

First we calculate the number of moles of hydrogen gas, then we calculate the number of grams of zinc. The hydrogen is mixed with water vapor at 22°C and a total pressure of 755 torr. The vapor pressure of water is 19.8 torr at 22°C (Table 5.3), so the partial pressure of the hydrogen is

$$P_{H_2} = P_T - P_{H_2O}$$

$$755 \text{ torr} - 19.8 \text{ torr} = 735 \text{ torr}$$

Converting this to atmospheres gives

$$P_{H_2} = 735 \text{ torr} \times \frac{1 \text{ atm}}{760 \text{ torr}} = 0.967 \text{ atm}$$

The temperature is 22°C = 295 K and the gas volume is 500 mL = 0.500 L. Substituting these values into the ideal gas law gives the number of moles of hydrogen:

$$n_{H_2} = \frac{PV}{RT} = \frac{0.967 \text{ atm} \times 0.500 \text{ L}}{0.0821 \text{ L·atm/mol·K} \times 295 \text{ K}} = 0.0200 \text{ mol}$$

The molar mass of zinc is 65.39 g/mol; the mass of zinc consumed is calculated in the usual way:

$$\text{mol } H_2 \longrightarrow \text{mol Zn} \longrightarrow \text{g Zn}$$

$$0.0200 \text{ mol } H_2 \times \frac{1 \text{ mol Zn}}{1 \text{ mol } H_2} \times \frac{65.39 \text{ g Zn}}{1 \text{ mol Zn}} = 1.31 \text{ g Zn}$$

LEARNING HINT

The pressure and volume units used in a gas law calculation must be consistent with those of the gas constant *R*.

PRACTICE EXERCISE 5.25

When mercury(II) oxide is heated, it decomposes into liquid mercury and oxygen gas:

$$2HgO(s) \rightarrow 2Hg(l) + O_2(g)$$

How many grams of mercury(II) oxide will have decomposed if 250 mL of gas, measured at 765 torr and 21°C, is collected over water?

Graham's Law of Effusion

Gases also undergo **effusion**; that is, they can pass through small, sometimes molecular-size, openings (Figure 5.16). As a result, air slowly leaks out of balloons, basketballs, and tires, while the odors of perfume, fish, and other substances are often apparent in spite of closed bottles and sealed wrappings. The rate of escape depends on the gas—balloons filled with hydrogen or helium, for example, deflate more rapidly than balloons filled with air. In the mid-nineteenth century, Thomas Graham measured the rates at which various gases effuse through a porous plate. He found that light gases effuse more rapidly than heavy ones, and his observations are summarized in **Graham's law of effusion**: *The rates of effusion of different gases at the same temperature and pressure are inversely proportional to the square roots of their densities.* Recall that when two quantities are inversely proportional to each other, their product does not change (like P and V in Boyle's Law). Hence, for two gases, A and B, effusing under identical experimental conditions, Graham's law can be written as

$$r_A \times \sqrt{d_A} = r_B \times \sqrt{d_B} \tag{5.15}$$

where r is the *rate of effusion*, that is, the amount of gas escaping per second, and d

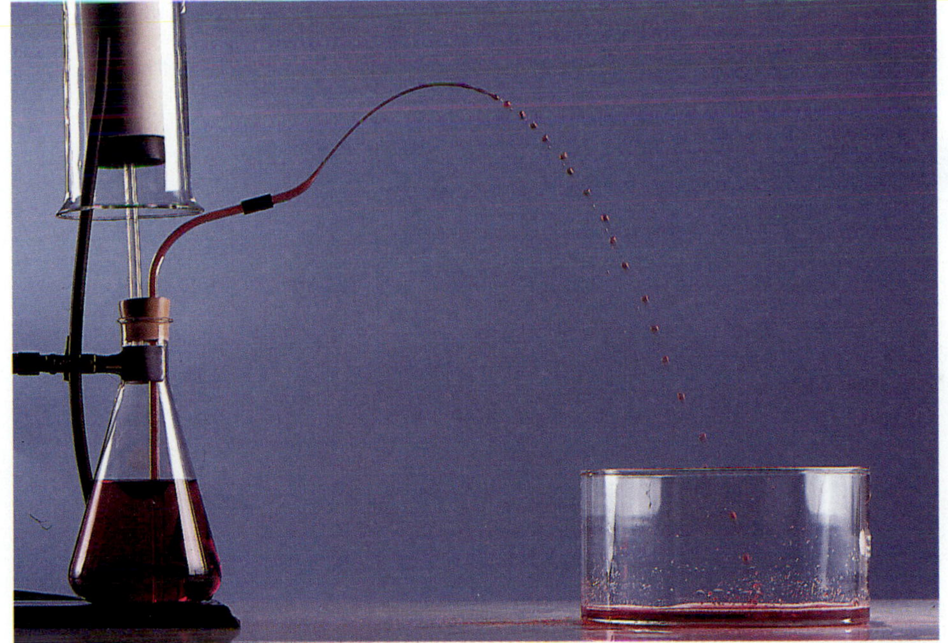

Figure 5.16

A hydrogen fountain. When the white, porous cup is bathed in hydrogen from the rubber tube in the inverted beaker, the H_2 molecules effuse into the cup more rapidly than the heavier air molecules effuse out. The pressure in the cup and attached flask increases, thus forcing liquid out the nozzle and creating a "fountain."

CHEMICAL INSIGHT

SEPARATION OF ISOTOPES BY GASEOUS EFFUSION

One practical application of differences in effusion rates is the separation of mixtures of gases whose molar masses are different but whose properties are otherwise similar. During World War II, the Manhattan Project required uranium-235 for the development of the atomic bomb. Natural uranium is principally uranium-238 with only 0.71% uranium-235. Natural uranium was converted into gaseous uranium hexafluoride (UF_6), which was then effused and recycled through thousands of porous barriers in a plant built especially for that purpose at Oak Ridge, Tennessee (Figure 5.17). (Such plants are commonly referred to as "gaseous diffusion" plants.) The gas that penetrates each barrier has a slightly higher proportion of the less massive uranium-235 isotope than the gas remaining on the other side. Passage through all of the barriers results in an enriched UF_6 with a uranium-235 content many times higher than that of natural uranium.

Gaseous effusion is still used to prepare fuel for commercial nuclear reactors, which require 2–4% uranium-235, and for submarine reactors and bombs, where the uranium-235 content must be even higher. Enrichment plants in the United States, Britain, France, and the countries of the former Soviet Union process fuel for reactors in all parts of the world. The cost of enrichment for the 30 tons of uranium fuel required each year for a standard 1000-megawatt reactor is well over 20 million dollars.

Gaseous diffusion plants are large and take up much space. The plant in Oak Ridge, Tennessee, for example, required about 4000 barriers of the type shown in Figure 5.17 and occupied an area of over 40 acres. An alternative method for separating uranium isotopes involves feeding uranium hexafluoride into gas centrifuges. A typical gas centrifuge consists of a 20-foot-tall cylinder spinning at the rate of about 1700 revolutions per second. Heavier molecules tend to move toward the cylinder walls, so the gas near the cylinder core will become enriched in the lighter uranium-235 isotope. Significant enrichment, however, requires passage of the gas through a cascade of at least 1000 centrifuges. Gas centrifuges are difficult to build, require a large source of electrical power, and frequently break down.

Figure 5.17

A gaseous diffusion plant in which the uranium-235 content of nuclear fuel is increased by effusion of UF_6 vapor. Each large cylinder contains a porous barrier through which $^{235}UF_6$ effuses slightly faster than $^{238}UF_6$. The fraction of $^{235}UF_6$ is continuously increased by passage through numerous barriers.

is the gas density. The density of an ideal gas is given by Equation 5.10 as $d = \mathcal{M}P/RT$ where $\mathcal{M}$ is the molar mass. Substituting for d in Equation 5.15 gives

$$r_A \times \sqrt{\frac{\mathcal{M}_A P}{RT}} = r_B \times \sqrt{\frac{\mathcal{M}_B P}{RT}}$$

Canceling T, P, and R leaves an alternative form of Graham's law:

$$r_A \times \sqrt{\mathcal{M}_A} = r_B \times \sqrt{\mathcal{M}_B} \qquad (5.16)$$

Equation 5.16 allows us to restate Graham's law: *The rates of effusion of different gases at the same temperature and pressure are inversely proportional to the square roots of their molar masses.* Thus, the rate of effusion decreases with increasing molar mass.

EXAMPLE 5.25

Compare the rates of effusion of molecular hydrogen and oxygen.

SOLUTION

The molar masses of hydrogen and oxygen are 2.02 and 32.0 g/mol, respectively. Substituting into Equation 5.16 gives

$$r_{H_2} \times \sqrt{2.02 \text{ g/mol}} = r_{O_2} \times \sqrt{32.0 \text{ g/mol}}$$

or

$$r_{H_2} = \sqrt{\frac{32.0}{2.02}} \times r_{O_2} = 3.98 \times r_{O_2}$$

Hydrogen should effuse 3.98 times more rapidly than oxygen.

Deviations from Graham's law occur when the gas pressure is high or when the openings are not very small. Under these conditions, collisions between molecules interfere with their free movement through the openings

PRACTICE EXERCISE 5.26

An unknown gas effuses through a small opening approximately 1.66 times more rapidly than carbon dioxide. Estimate the molar mass of the unknown gas.

Diffusion (Section 5.1), like effusion, depends on the speeds of wandering molecules. The rate at which one gas spreads into another obeys Graham's law in an approximate way; on the average, less massive gas molecules such as NH_3 are found to diffuse greater distances in a given time than more massive molecules such as HCl (see Figure 5.18).

Gas effusion: the passage of gas molecules through very small openings.

Gas diffusion: the spreading of one gas through another.

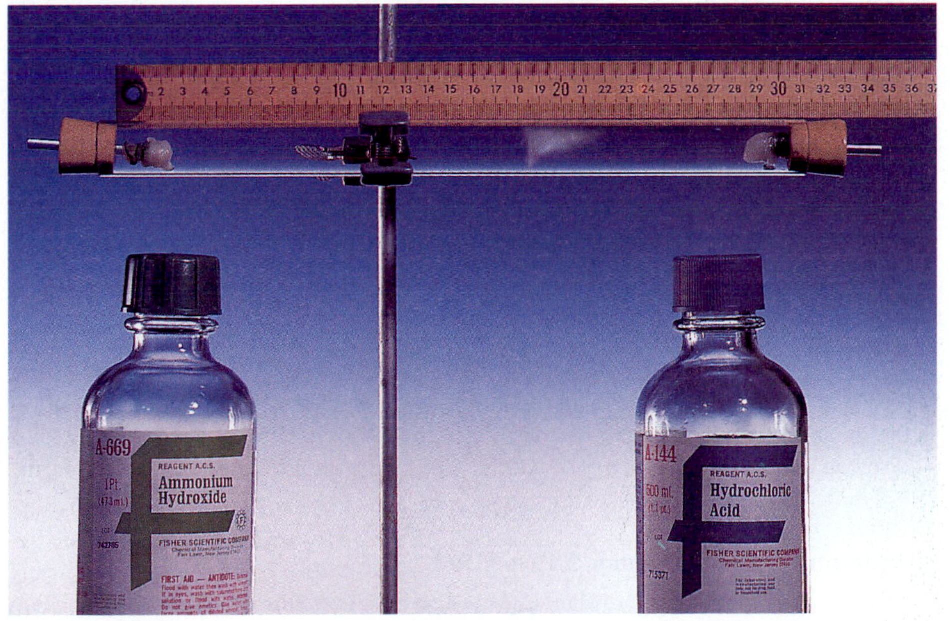

Figure 5.18

Relative rates of diffusion. Vapors of HCl and NH_3, arising from solutions on cotton swabs, diffuse from opposite ends of the glass tube. A white fog of NH_4Cl appears where the vapors meet. Observe that fog formation started near the 18.5 cm mark on the ruler, about 10 cm from the HCl swab and 16 cm from the NH_3 swab. The NH_3 molecules therefore diffuse through air 16/10, or 1.6, times faster than the HCl molecules. The fact that 1.6 is very close to the square root of the molar mass ratio, $\sqrt{36.5/17.0}$, shows that the relative rates of diffusion obey Graham's law approximately.

DIGGING DEEPER

5.7 KINETIC THEORY REVISITED

The speeds and energies referred to in this section are those of translational motion only; they do not include *rotational* (tumbling) motions or *vibrational* motions in which one part of a molecule moves back and forth relative to another part of the molecule.

Motion from one point in space to another is called **translational motion**, and the kinetic energy associated with this motion is called **translational kinetic energy**. Imagine a collection of identical gas molecules moving with different speeds in all directions. Their average kinetic energy, $\overline{\epsilon}_k$, is

$$\overline{\epsilon}_k = \tfrac{1}{2}m\overline{u^2} \qquad (5.17)$$

where m is the mass of a single molecule and $\overline{u^2}$ is the average square speed, more commonly called the **mean square speed**. (The bars over ϵ_k and u^2 indicate average quantities.) To better understand the concept of mean square speed, consider three molecules moving with speeds of 425, 450, and 600 m/s. The mean square speed is found by squaring the speed of each molecule, adding the squares, and dividing the sum by the number of molecules, in this case 3:

$$\overline{u^2} = \frac{(425 \text{ m/s})^2 + (450 \text{ m/s})^2 + (600 \text{ m/s})^2}{3} = 2.48 \times 10^5 \text{ m}^2/\text{s}^2$$

One mole of molecules contains Avogadro's number of molecules. If this number is symbolized by N_A, then the total kinetic energy E_k, of 1 mol of molecules is N_A times the average energy per molecule, that is,

$$E_k = N_A\overline{\epsilon}_k = \tfrac{1}{2}N_A m\overline{u^2}$$

LEARNING HINT

Note that the mean square speed is different from the square of the average speed which, in this case, would be 2.42×10^5 m^2/s^2. (Be sure to check this value.)

(Remember that $\overline{\epsilon}_k$ is the *average* kinetic energy of one molecule; E_k is the *total* kinetic energy of 1 mol of molecules.) The expression for E_k contains the product $N_A m$. Because N_A is the number of molecules per mole and m is the mass per molecule, their product is $\mathcal{M}$, the mass per mole (molar mass). Hence,

$$E_k = \tfrac{1}{2}\mathcal{M}\overline{u^2} \qquad (5.18)$$

Now consider the pressure exerted by confined gas molecules. The pressure is caused by molecular impacts on the container walls, and at each instant it is equal to the total force of these collisions divided by the total wall area. The total force in turn depends on the force of each impact as well as on the number of impacts per second. A mathematical analysis of this situation is complicated because the molecules are moving with different speeds in all directions, but the result of such an analysis, which we will not undertake, shows that the pressure volume product (PV) of an ideal gas is

$$PV = \tfrac{1}{3}n\mathcal{M}\overline{u^2} \qquad (5.19)$$

where n is the number of moles of gas in the container, $\mathcal{M}$ is the molar mass, and $\overline{u^2}$ is the mean square speed.

We can combine Equations 5.18 and 5.19 by rewriting Equation 5.18 in the form

$$\mathcal{M}\overline{u^2} = 2E_k$$

and substituting it into Equation 5.19 to give

$$PV = \tfrac{2}{3}nE_k \qquad (5.20)$$

Because n is the number of moles of gas and E_k is the kinetic energy per mole, the product nE_k is the total kinetic energy of the sample. Equation 5.20 leads to the conclusion that *the value of the* PV *product for an ideal gas is two-thirds of the kinetic energy of the molecules in the gas.*

Kinetic Energy and Temperature

The ideal gas law, $PV = nRT$, relates the PV product to the temperature. Equation 5.20, derived from the kinetic theory of gases, relates the PV product to the kinetic energy. Equating the two PV expressions gives

$$PV = nRT = \tfrac{2}{3}nE_k$$

which, after canceling n, can be rearranged to

$$E_k = \tfrac{3}{2}RT \qquad (5.21)$$

This important equation shows that *the kinetic energy of an ideal gas is directly proportional to the Kelvin temperature*. Furthermore, it shows that *all ideal gases have the same molar kinetic energy at the same temperature*. Equation 5.21 is consistent with the fourth assumption of the kinetic theory of gases (Section 5.1), which states that the average kinetic energy of a collection of gas molecules depends only on the temperature.

Equation 5.21 can be used to calculate the kinetic energy of a collection of gas molecules directly from its temperature, as shown in the next example.

EXAMPLE 5.26

Calculate the kinetic energy in joules of 1 mol of helium atoms at 25°C. Assume ideal gas behavior.

SOLUTION

The temperature is 25°C = 298 K. Because we are looking for energy in joules, we use $R = 8.314$ J/mol·K (Table 5.2). Substituting these values into Equation 5.21 gives

$$E_k = \tfrac{3}{2}RT = \tfrac{3}{2} \times 8.314 \text{ J/mol·K} \times 298 \text{ K} = 3.72 \times 10^3 \text{ J/mol}$$

PRACTICE EXERCISE 5.27

Calculate the kinetic energy in joules of 150 g of molecular nitrogen at 25°C. (*Hint*: First find the kinetic energy of 1 mol of nitrogen molecules.)

Root Mean Square Speeds

The mean square speed $\overline{u^2}$ of a collection of molecules can be found by equating the molar kinetic energies of Equations 5.18 and 5.21:

$$E_k = \tfrac{1}{2}\mathcal{M}\overline{u^2} = \tfrac{3}{2}RT$$

After multiplying both sides by 2, the mean square speed is obtained by rearrangement:

$$\overline{u^2} = \frac{3RT}{\mathcal{M}} \tag{5.22}$$

The **root mean square speed** u_{rms} is the square root of the mean square speed:

$$u_{\text{rms}} = \sqrt{\overline{u^2}} = \sqrt{\frac{3RT}{\mathcal{M}}} \tag{5.23}$$

The average kinetic energy of the molecules is $\frac{1}{2}m\overline{u^2}$ (Equation 5.17); hence, *the root mean square speed is the speed of a molecule possessing the average kinetic energy*.

Recall that the effusion rates of gases were observed to be inversely proportional to the square roots of their molar masses (Graham's law, Section 5.6). Since effusion rates depend on molecular speed, this observation supports the theoretical conclusion reached in Equation 5.23: *Molecular root mean square speeds are inversely proportional to the square roots of molar masses.*

EXAMPLE 5.27

The air we breathe is about 80% nitrogen. Calculate the root mean square speed of nitrogen molecules in the lungs at normal body temperature (37°C).

SOLUTION

We express all data in SI units in order to obtain the speed in meters per second. The molar mass of N_2 is 28.01 g/mol = 0.02801 kg/mol. The gas constant R is 8.314 J/mol·K = 8.314 kg·m²/s²·mol·K. (Recall that 1 J = 1 N·m and 1 N = 1 kg·m/s².) Human body temperature is 37°C = 310 K. The root mean square speed is obtained by substituting these data into Equation 5.23:

$$\begin{aligned} u_{\text{rms}} &= \sqrt{\frac{3RT}{\mathcal{M}}} \\ &= \sqrt{\frac{3 \times 8.314\ \text{kg·m}^2/\text{s}^2\text{·mol·K} \times 310\ \text{K}}{0.02801\ \text{kg/mol}}} \\ &= 525\ \text{m/s} \end{aligned}$$

Root mean square speeds of various gases are given in Table 5.5.

PRACTICE EXERCISE 5.28

Perform calculations to verify one or two of the root mean square speeds in Table 5.5.

TABLE 5.5 Root Mean Square Speeds of Various Gas Molecules at 25°C

Substance	Formula	Speed (m/s)
Ammonia	NH_3	661
Carbon dioxide	CO_2	411
Chlorine	Cl_2	324
Helium	He	1360
Hydrogen	H_2	1920
Mercury vapor	Hg	192
Methane	CH_4	681
Nitrogen	N_2	515
Oxygen	O_2	482
Uranium hexafluoride	UF_6	145
Water vapor	H_2O	642

Distribution of Speed and Energy

The molecules in a gas do not all move with the same speed. At any moment some are moving faster and others slower than average; in other words, there is a *distribution* of speeds. In 1860 James Maxwell used statistical methods to show that the speed distribution should follow a definite pattern. This pattern, when plotted as in Figure 5.19, is called a **Maxwell–Boltzmann distribution curve** in honor of Maxwell, and of Ludwig Boltzmann (1844–1906) who showed that Maxwell's distribution of speeds is a special case of a more general energy distribution. The Maxwell–Boltzmann curve changes with temperature, but at any given temperature, it shows the fraction of molecules possessing any given speed. The highest point on

LEARNING HINT

The area under each curve in Figure 5.19 may be taken as unity because it represents the sum of the fractions of all the molecules. If the total area is unity, then the fraction of molecules within some range of speeds—say, 400 and 600 m/s—is given by the area bounded by these speeds.

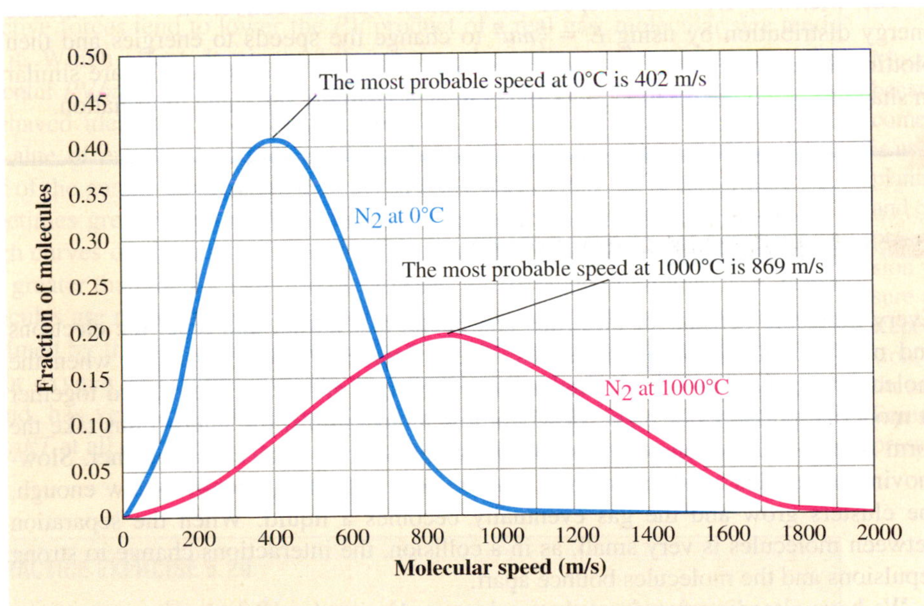

Figure 5.19
Maxwell–Boltzmann distribution curves for nitrogen molecules at 0°C and 1000°C. The highest point on the curve corresponds to the most probable speed. An increase in temperature flattens the curve and raises the most probable speed. At 1000°C, most nitrogen molecules have speeds between 500 and 1500 m/s.

KEY TERMS BY SECTION

5.1 Brownian motion
Diffusion
Kinetic theory
States of matter
Thermal motion

5.2 Atmosphere (atm)
Bar
Barometer
Manometer
Pascal (Pa)
Pressure
Torr

5.3 Absolute zero
Avogadro's hypothesis
Avogadro's law
Boyle's law
Charles's law
Kelvin (K)
Kelvin temperature scale
Molar volume
Standard conditions
STP

5.4 Amontons' law
Gas constant (R)
Ideal gas
Ideal gas law

5.6 Dalton's law of partial pressures
Effusion
Graham's law of effusion
Mole fraction (X)
Mole percent
Partial pressure
Vapor pressure

5.7*Maxwell–Boltzmann distribution curve
*Mean square speed
*Most probable speed
*Root mean square speed
*Translational motion

5.8*Equation of state
Mean free path
*Van der Waals constants
*Van der Waals equation

IMPORTANT EQUATIONS

5.1 $\text{pressure} = \dfrac{\text{force}}{\text{area over which force is exerted}}$

5.3 $P_i \times V_i = P_f \times V_f$

5.4 $T\,(\text{in K}) = 273.15 + C$

5.6 $\dfrac{V_i}{T_i} = \dfrac{V_f}{T_f}$

5.8 $PV = nRT$

5.9 $\dfrac{P_iV_i}{T_i} = \dfrac{P_fV_f}{T_f}$

5.10 $\mathcal{M} = d\dfrac{RT}{P}$

5.11 $P_T = P_A + P_B + P_C$

5.13 $X = \dfrac{n}{n_T}$

5.16 $r_A \times \sqrt{\mathcal{M}_A} = r_B \times \sqrt{\mathcal{M}_B}$

***5.21** $E_k = \frac{3}{2}RT$

***5.23** $u_{rms} = \sqrt{\dfrac{3RT}{\mathcal{M}}}$

***5.24** $\left(P + \dfrac{an^2}{V^2}\right)(V - nb) = nRT$

FINAL EXERCISES

Answers to exercises with blue numbers are given in Appendix D. *Starred exercises are based on the* Digging Deeper *sections.*

PART A. QUESTIONS AND PROBLEMS BY SECTION

Kinetic Theory (Section 5.1)

5.1 Describe some experimental evidence showing that (a) molecules are in continuous motion and (b) molecular speeds increase with increasing temperature.

5.2 How will the diffusion rate of a gas be affected by (a) increasing temperature and (b) increasing molar mass?

5.3 **(a)** State the assumptions of the kinetic theory of gases.
(b) Why is the random motion of atoms and molecules often called *thermal motion*?

5.4 Use the kinetic theory of gases to explain the following observations:
(a) Gases expand to fill their containers.
(b) Gases are easily compressed.
(c) Warm air tends to rise; cold air tends to fall.
(d) Gases can be mixed in any proportion.

Pressure (Section 5.2)

5.5 Describe at least two experiments showing that the atmosphere exerts pressure.

5.6 Use kinetic theory to explain why a gas exerts pressure in all directions.

5.7 **(a)** Is the total force exerted by the atmosphere on a basketball greater than, less than, or equal to the total force exerted on a tennis ball? Explain.
(b) Is the pressure exerted by the atmosphere on a basketball greater than, less than, or equal to the pressure exerted on a tennis ball? Explain.

5.8 Sketch the following instruments and explain how they work: (a) a mercury barometer and (b) a manometer.

5.9 The textbook states that a U.S. Roosevelt-type dime exerts a pressure of 88.5 Pa when it is lying flat on the ground. Verify this statement. The mass of the dime is 2.27 g and its diameter is 17.9 mm.

5.10 The mass of a U.S. penny is 3.11 g and its diameter is 19.0 mm. Will the pressure exerted by a penny be greater than or less than the pressure exerted by the dime in the previous problem? Perform a calculation to justify your answer.

5.11 A barometer reads 745 mm Hg. Convert this pressure into (a) atmospheres, (b) pascals, and (c) bars.

5.12 The atmospheric pressure on a certain sunny day was 1.040 bar. Convert this pressure into (a) kilopascals, (b) atmospheres, and (c) torr.

5.13 The owner's manual for an automobile recommends a tire pressure of 200 kPa. Convert this pressure into (a) atmospheres and (b) torr.

5.14 The pressure at the center of the earth is believed to be 3.5 Mbar. Convert this pressure into (a) atmospheres and (b) kilopascals.

5.15 The surface pressure on the planet Mars is 0.0060 times sea-level atmospheric pressure on earth. How many millimeters of mercury will the Martian atmosphere support?

5.16 Atmospheric pressure on the surface of the planet Venus is 90 atm. How many meters of mercury will the Venusian atmosphere support?

5.17 A manometer is attached to a flask containing gas and to a meter stick that measures the mercury levels. A nearby barometer reads 763 mm Hg.
(a) The manometer reading on the side open to the atmosphere is 5.8 cm; the reading on the flask side is 23.0 cm. What is the gas pressure in torr? (*Hint*: Sketch the apparatus.)
(b) What will the two manometer readings be if the pressure in the flask increases to 805 torr?

5.18 Find the height of a column of water that could be supported by the atmosphere on a day when atmospheric pressure is 760 torr. The densities of water and mercury are 1.00 and 13.6 g/cm^3, respectively.

The Gas Laws (Section 5.3)

5.19 List the four variables that determine the physical behavior of a gas.

5.20 Write an equation for each of the following gas laws and state the conditions under which the equation is valid:
(a) Boyle's law
(b) Charles's law
(c) Avogadro's law

5.21 How does the volume of a gas sample vary with
(a) Increasing pressure at constant temperature?
(b) Decreasing temperature at constant pressure?
(c) Increasing number of moles at constant temperature and pressure?

5.22 Explain the following observations in terms of kinetic theory:
(a) Automobile tire pressures increase during a long trip.
(b) A balloon expands when warmed.
(c) Weather balloons expand as they rise through the atmosphere.

5.23 A sample of xenon gas has a volume of 150 mL at 380 torr. Assume constant temperature and calculate the volume at the following pressures:
(a) 2.00 atm
(b) 800 torr
(c) 1.50 bar
(d) 200 kPa

5.24 A 10.0-L sample of gas exerts a pressure of 335 torr at 0°C. The sample is transferred at constant temperature to an evacuated 250-mL flask. What will be the new pressure in (a) torr, (b) atmospheres, and (c) bars?

5.25 A bubble of radius 1.5 cm forms 10 m below the surface of a pond. If the temperature of the bubble remains constant, what is its volume when it reaches the top of the pond where the pressure is 1.0 atm? The total pressure at a depth of 10 m is about 2.1 atm.

5.26 On a single graph, plot V versus $1/P$ for 0.00100 mol of helium gas at 0°C and at 100°C. Refer to Figure 5.7 for data. Each curve should have at least six points. In what way are the two plots similar? In what way are they different?

5.27 Convert the following temperatures to kelvins:
(a) normal body temperature, 37°C
(b) freezing point of water, 0°C
(c) boiling point of water, 100°C
(d) room temperature, 25°C

5.28 Convert the following temperatures to degrees Celsius:
(a) boiling point of liquid helium, 4.22 K
(b) melting point of tungsten, 3683 K
(c) surface temperature of the sun, 6273 K

(d) lowest temperature achieved in the laboratory, 2.5×10^{-10} K

5.29 A gas has a volume of 0.525 L at 25.0°C. Assume constant pressure and calculate the volume at the following temperatures:
(a) 100°C **(c)** 200 K
(b) −23.5°C **(d)** 375 K

5.30 A 750-mL sample of nitrogen is collected at 15°C and 1.00 atm. Assume constant pressure and calculate the temperature at which the sample will have a volume of (a) 2.50 L and (b) 12.5 mL. Express your answer in degrees Celsius.

5.31 The volume of an air-filled balloon is 2.5 L when its contents are at 1.00 atm and 0°C. What will be its volume at (a) 0°C and 700 torr and (b) −20°C and 1.00 atm?

5.32 A dry hydrogen sample occupies 300 mL at 25°C and 720 torr. How many milliliters will the gas occupy at (a) standard conditions and (b) 125°C and 2.50 atm?

5.33 A 1.0-L balloon contains 0.0446 mol of helium gas at 0°C and 1.00 atm.
(a) What will be the volume of a balloon containing 0.112 mol of helium under the same conditions of temperature and pressure?
(b) How many moles of helium will be in an 800-mL balloon at the same temperature and pressure?

5.34 The density of nitrogen gas is 1.250 g/L at 0°C and 1.00 atm. The molar mass of nitrogen is 28.01 g/mol. Calculate the molar volume of nitrogen under these conditions.

5.35 The density of carbon monoxide is 1.29 g/L at 1.00 atm and 0°C. Estimate the density in grams per liter at (a) 50°C and 700 torr and (b) −25°C and 1.10 atm.

5.36 The molar volume of neon gas is 22.4 L at 1.00 atm and 0°C; the molar mass of neon is 20.18 g/mol. Calculate the density of neon gas at (a) 100°C and 1.50 atm and (b) −50°C and 720 torr.

The Ideal Gas Law (Section 5.4)

5.37 **(a)** What is an ideal gas?
(b) Under what conditions do real gases behave most nearly like ideal gases?

5.38 Rearrange the ideal gas law to obtain expressions for n, P, V, and T.

5.39 Show how the following laws can be obtained from the ideal gas law:
(a) Boyle's law **(c)** Avogadro's law
(b) Charles's law **(d)** Amontons' law

5.40 How does the pressure of a confined gas vary with
(a) The number of molecules in the container?
(b) The temperature?
(c) The volume of the container?
(d) The gas density?

5.41 On a single graph, plot P versus V for 1 mol of an ideal gas at 0°C and at 137°C. Each curve should have at least six points. In what way are the two plots similar? In what way are they different?

5.42 Derive Equation 5.10, $\mathcal{M} = dRT/P$, from the ideal gas law without referring to the text.

5.43 The pressure of a confined gas is 850 torr at 0°C. Assume constant volume and calculate the pressure at the following temperatures:
(a) 45°C **(c)** 200 K
(b) −190°C **(d)** 323 K

5.44 A gas tank that is rated safe up to 35.0 atm pressure contains helium at a pressure of 20.0 atm and a temperature of 20°C. What is the maximum temperature the tank and its contents can safely withstand?

5.45 A sample of nitrogen occupying 100 L at 105 kPa and 25°C is compressed to 10.0 L at 0°C. Calculate the final pressure of the nitrogen in kPa.

5.46 The volume of a gas is 30.0 mL at −50°C and 1.80 atm. At what temperature will the gas have a pressure of 1.50 atm and a volume of 40.0 mL?

5.47 What volume in liters will 10.0 g of oxygen gas occupy at (a) 0°C and 1.00 atm and (b) −18°C and 775 torr?

5.48 A 10.0-L tank contains 80.0 g of nitrogen gas at 25°C. Calculate the nitrogen pressure in atmospheres.

5.49 An average pair of human lungs contains about 3.5 L of air after inhalation and about 3.0 L after exhalation. If the temperature is 37°C and the pressure is 1.0 atm, how many moles of gas will be present in the lungs (a) after inhalation and (b) after exhalation?

5.50 Estimate the number of moles and the number of molecules in 1.00 mL of air under the following conditions:
(a) at the beach when conditions are 1.00 atm and 22°C
(b) on the summit of Mt. Rainier at 0.60 atm and −5°C
(c) in a glass bulb evacuated to 1.5×10^{-6} torr at 20°C

5.51 The density of a gaseous element is 5.86 g/L under standard conditions.
(a) Estimate the molar mass of the element.
(b) Identify the element (consult the periodic table).

5.52 The density of dimethyl ether vapor is 0.940 g/L at 25°C and 380 torr. Estimate the molar mass of dimethyl ether.

5.53 The tank in Example 5.17 (p. 190) weighs 10.505 kg when evacuated and 10.673 kg when filled with the gas as described in the example. Calculate the molar mass of the gas.

5.54 A 1.00-L sample of a gaseous oxide weighs 2.62 g at 25°C and 1.00 atm. Estimate the molar mass of the oxide.

5.55 Estimate the density in grams per liter of the following gases:
(a) CCl_4 vapor at 20°C and 780 torr
(b) CO_2 at 37°C and 1.10 atm

5.56 Estimate the density of propane gas (C_3H_8) in grams per liter at (a) standard temperature and pressure and (b) 20°C and 750 torr.

5.57 A 1.430-g sample of a gaseous compound in a 600-mL glass bulb has a pressure of 427 torr at 70.0°C. Analysis shows that the compound contains 10.1% carbon, 0.84% hydrogen, and 89.1% chlorine. Find (a) the molar mass of the compound and (b) its formula.

5.58 "Laughing gas" is an oxide of nitrogen used as a propellant for whipped cream aerosols and also as an inhalation anesthetic and analgesic. Find the formula of laughing gas if it contains 63.65% nitrogen and has a density of 1.800 g/L at 25°C and 1 atm.

Gas Stoichiometry (Section 5.5)

5.59 How many milliliters of oxygen, measured at 0°C and 1.00 atm, are required for the complete combustion of 5.00 g of liquid methanol (CH_3OH)?

5.60 Solid mercury(II) oxide (HgO) decomposes into liquid mercury and oxygen gas when it is heated. How many grams of HgO will produce 0.100 L of oxygen measured at 25°C and 0.990 atm?

5.61 Gaseous diborane burns in oxygen according to the equation:

$$B_2H_6(g) + 3O_2(g) \rightarrow B_2O_3(s) + 3H_2O(g)$$

How many grams of B_2O_3 will form when 3.00 L of B_2H_6, measured at 20°C and 770 torr, is burned in excess oxygen?

5.62 Automobile air bags are designed to inflate rapidly in a crash. The impact closes an electrical switch and the resulting current initiates the decomposition of sodium azide:

$$2NaN_3(s) \rightarrow 2Na(s) + 3N_2(g)$$

How many grams of sodium azide will produce 50.0 L of N_2 measured at 25°C and 750 torr?

5.63 The enzyme-catalyzed reaction of urea with water is represented by the equation:

$$(NH_2)_2CO(aq) + H_2O(l) \rightarrow CO_2(g) + 2NH_3(g)$$

(a) How many milliliters of NH_3, measured at 37°C and 755 torr, will be produced when 5.00 g of urea reacts with excess water?
(b) Calculate the total volume, in milliliters, of gaseous products produced during the reaction of Part (a).

5.64 Chlorine gas has been prepared by passing HCl and O_2 gases over hot pumice stone containing a $CuCl_2$ catalyst:

$$4HCl(g) + O_2(g) \rightarrow 2Cl_2(g) + 2H_2O(g)$$

How many liters of HCl, measured at 20°C and 720 torr, are required to produce 1.00 metric ton of Cl_2 by the above process?

5.65 Nitrogen reacts with hydrogen to form ammonia according to the equation:

$$N_2(g) + 3H_2(g) \rightarrow 2NH_3(g)$$

How many liters of nitrogen are required to produce 60 L of ammonia? How many liters of hydrogen are required? Assume that all of the gases are measured at the same temperature and pressure.

5.66 Hydrogen sulfide and sulfur dioxide react as follows:

$$2H_2S(g) + SO_2(g) \rightarrow 2H_2O(l) + 3S(s)$$

(a) How many milliliters of SO_2 will react with 25.0 mL of H_2S? Assume that both volumes are measured under the same conditions.
(b) How many liters of H_2S, measured at 750 torr and 25°C, are required to produce 125 g of sulfur?

Gas Mixtures (Section 5.6)

5.67 Is the mole fraction of a component in a mixture the same as the molecule fraction? Explain.

5.68 **(a)** A mixture contains He, Ne, and Ar. Show that the sum of the mole fractions, $X_{He} + X_{Ne} + X_{Ar}$, is equal to 1.
(b) Is the relation in Part (a) always true; that is, do the mole fractions of the components in a mixture always add up to 1?

5.69 Without referring to the text, show that the pressure fraction of any component in a gaseous mixture is equal to the mole fraction of the component.

.5.70 The composition of gases is often expressed in *volume percent*, which is 100% times the volume that each isolated gas would have at the temperature and *total* pressure of the mixture, divided by the total volume of the mixture. Show that for ideal gas mixtures the volume percent is equal to the mole percent.

5.71 What is the difference between effusion and diffusion?

5.72 List the following gases in order of increasing rate of effusion: CO, SO_2, NO, NH_3, and HCl.

5.73 A gas mixture contains 1.50 mol of helium and 3.25 mol of carbon dioxide. The total pressure is 720 torr. Calculate (a) the mole fraction, (b) the mole percent, and (c) the partial pressure of each gas in the mixture.

5.74 A mixture of nitrogen and oxygen gases has a total pressure of 740 torr. The partial pressure of nitrogen is 577 torr. Calculate (a) the partial pressure of oxygen and (b) the mole fraction of each gas in the mixture.

5.75 A gas mixture contains NOCl, NO, and Cl_2 at partial pressures of 0.730, 0.270, and 0.135 atm, respectively. Calculate (a) the total pressure and (b) the mole fraction of each gas in the mixture.

5.76 The atmosphere of the planet Venus contains 96.0 mol % CO_2, 3.5 mol % N_2, and small amounts of H_2O, H_2SO_4, and HCl. The total pressure is 90 atm. Calculate the partial pressure of (a) CO_2 and (b) N_2 in the Venusian atmosphere.

5.77 A 10.0-L reaction vessel contains 0.044 mol of H_2, 0.044 mol of I_2, and 0.312 mol of HI at 458°C. Calculate (a) the partial pressure of each gas in the mixture and (b) the total pressure.

5.78 A mixture of hydrogen and oxygen has a volume of 100 mL at 27°C and 775 torr. If the gas is 40.0 mol % hydrogen, what is the partial pressure of each component?

5.79 A sample of nitrogen gas is collected over water at 25°C and a total pressure of 740 torr. Find (a) the partial pressure of nitrogen in the sample and (b) the mole percent of nitrogen.

5.80 A 350-mL sample of hydrogen gas is collected over water at 770 torr and 30°C. What volume will the hydrogen occupy if it is dried and stored at 0°C and 1.00 atm?

5.81 Calcium hydride is sometimes used as a portable source of hydrogen. Its reaction with water is

$$CaH_2(s) + 2H_2O(l) \rightarrow 2H_2(g) + Ca(OH)_2(aq)$$

How many grams of calcium hydride are required to produce 1.00 L of hydrogen gas measured over water at 24°C and a total pressure of 750 torr?

5.82 A 10.0-g mixture of potassium chlorate and potassium chloride was heated with a manganese dioxide catalyst. The potassium chlorate decomposed into oxygen and additional potassium chloride:

$$2KClO_3(s) \rightarrow 2KCl(s) + 3O_2(g)$$

The oxygen occupied a volume of 1.90 L over water at 25°C and 764 torr. Calculate the percent by weight of $KClO_3$ in the original mixture.

5.83 Which gas will effuse more rapidly under the same conditions, N_2O or O_2? How much more rapidly?

5.84 Nitrous oxide (N_2O) and diethyl ether ($C_4H_{10}O$) have been used for inhalation anesthesia. Which gas will diffuse more rapidly through the lungs? How many times more rapidly?

5.85 A given volume of neon passed through a small opening in 26.7 s. Under identical conditions, it took 38.5 s for the same volume of a hydrocarbon to pass through the opening. What is the molar mass of the hydrocarbon?

5.86 **(a)** A given volume of an unknown gas diffuses through a porous barrier in 488 s. The same volume of oxygen, under the same conditions, diffuses through the barrier in 160 s. Estimate the molar mass of the unknown gas.
(b) The unknown gas is a fluoride of tungsten. What is its formula?

Kinetic Theory Revisited (Section 5.7)

*5.87 Consider a sample of air at room temperature.
(a) Is the kinetic energy of the nitrogen molecules in the sample greater than, less than, or equal to that of the oxygen molecules? Explain.
(b) Is the root mean square speed of the nitrogen molecules greater than, less than, or equal to that of the oxygen molecules? Explain.

*5.88 What happens to the molar kinetic energy and the root mean square speed of a collection of molecules as the Kelvin temperature approaches zero?

*5.89 Sketch a typical Maxwell–Boltzmann distribution curve, locate the most probable and root mean square speeds on the curve, and state how the shape of the curve varies with temperature.

*5.90 What happens to the distribution curves of Figure 5.19 (a) as the Kelvin temperature increases without limit and (b) as the Kelvin temperature approaches zero?

*5.91 Three molecules are moving with speeds of 375, 420, and 480 m/s.
(a) Calculate the average speed, the mean square speed, and the root mean square speed.

(b) Explain why the average speed does not equal the root mean square speed.

***5.92** Ten methane molecules have the following speeds in meters per second: 680, 595, 710, 680, 690, 595, 700, 680, 680, 725. For this collection, find
(a) the root mean square speed
(b) the most probable speed
(c) the average speed
(d) the temperature

***5.93** Verify the root mean square speeds given in Table 5.5 for (a) helium and (b) uranium hexafluoride.

***5.94** (a) At what temperature do nitrogen molecules have a root mean square speed of 500 m/s?
(b) What will be the molar kinetic energy of nitrogen at the temperature found in Part (a)?

***5.95** The Kelvin temperature of a gas is doubled. Does the root mean square speed of the gas molecules increase or decrease, and by what factor?

***5.96** The volume occupied by 2.00 mol of xenon, a monatomic gas, is 20.0 L at 1.00 atm. Estimate (a) the total kinetic energy of the xenon atoms and (b) their root mean square speed.

Deviations From Ideality (Section 5.8)

5.97 Explain in terms of the kinetic molecular theory why a gas becomes less ideal when it is cold and highly compressed.

5.98 List at least four properties possessed by an ideal gas but not by a real gas.

5.99 For real gases, does the ratio PV/nRT increase or decrease with (a) increasing intermolecular attractions, (b) increasing pressure, and (c) increasing gas density?

5.100 The PV product for methane is less than that for nitrogen at 100 atm and 0°C. Which gas, methane or nitrogen, exhibits stronger intermolecular attractions?

5.101 Briefly explain the following observations: (a) real gases usually cool upon expansion and (b) a bicycle tire warms up as air is pumped into it. Would these effects occur if the gases were ideal? Explain.

***5.102** (a) Which gas in Table 5.6 exhibits the strongest intermolecular attractions? The weakest attractions?
(b) Which gas in Table 5.6 has the largest molecules? The smallest molecules?

***5.103** A 5.00-L vessel contains 1.00 mol of water vapor at 200°C. Calculate the pressure (a) assuming ideal gas behavior and (b) using the van der Waals equation and constants from Table 5.6. Which of the calculated pressures should be closer to the actual pressure?

***5.104** A 142-g sample of chlorine gas (Cl_2) is in a 2.00-L container at 25°C. Calculate the pressure from the van der Waals equation and compare it with the ideal gas pressure.

PART B. MISCELLANEOUS QUESTIONS AND PROBLEMS

5.105 Explain in terms of kinetic theory why gas molecules move around in the atmosphere instead of settling to the earth.

5.106 Part (a) of the following sketch shows two gases separated by a closed valve; Part (b) shows three gases separated by a closed valve.

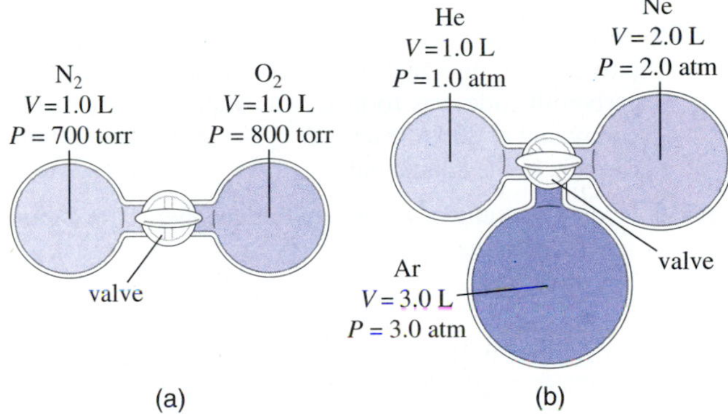

What will be the total pressure in each of these systems after the valve is opened and the system comes to equilibrium?

5.107 The density of dry air at standard conditions is 1.2929 g/L. Estimate an average molar mass for air.

5.108 A 1.5-L tank containing helium at 100 atm pressure is used to fill party balloons. If the pressure inside each balloon is 765 torr, and the average balloon volume is 1.2 L, how many balloons can be filled? Assume that the tank and the balloons remain at constant temperature.

5.109 The pressure inside a partially evacuated laboratory system was determined by compressing a 200-mL portion of its contents to a pressure of 30 torr and reading its new volume, 6.60 mL. Calculate the original pressure in torr.

5.110 If a 1.00-L sample of gas at standard conditions is expanded to 5.00 L, what temperature change would be necessary to bring the final pressure to 0.300 atm?

5.111 Use the ideal gas law to calculate the molar volume in liters of an ideal gas at standard conditions. Is your result consistent with the data reported for real gases in Table 5.1?

5.112 Refer to Figure 5.9 and show that the plot of volume versus Celsius temperature obeys the equation $V = V_0 + V_0 t/273$ where V_0 is the gas volume at 0°C and t is the Celsius temperature.

5.113 Refer to the air thermometer shown in Demonstration 5.2. The tube diameter is 6.9 mm, the volume of air in the flask is 200 mL when measured under standard conditions, and the barometric pressure is 760 torr. By how many centimeters will the liquid in the tube rise when the flask is warmed from 20 to 25°C?

5.114 The density of xylene, a compound containing only carbon and hydrogen, is 4.74 g/L at standard conditions. Combustion of a 5.00 mg sample of xylene produces water and 16.60 mg of carbon dioxide. Calculate (a) the molar mass and (b) the formula of xylene.

5.115 When 6.54 mg of an organic compound containing only carbon and hydrogen was burned, 20.5 mg of carbon dioxide was formed. The vapor density of the compound at 800 torr and 25°C is 1.215 g/L. Find the formula of the compound.

5.116 Hydrogen sulfide undergoes atmospheric oxidation according to the equation:

$$2H_2S(g) + 3O_2(g) \rightarrow 2SO_2(g) + 2H_2O(l)$$

How many milliliters of dry air are required for the oxidation of 100 mL of hydrogen sulfide? Assume that both the air and the hydrogen sulfide are measured under standard conditions. Dry air contains 20.95 mol % oxygen.

5.117 A resting astronaut exhales about 450 L of carbon dioxide, measured at 30°C and 1.00 atm, every 24 hours. The concentration of carbon dioxide in a spaceship can be kept down by cycling the air through solid lithium hydroxide, which reacts with carbon dioxide according to the equation:

$$2LiOH(s) + CO_2(g) \rightarrow Li_2CO_3(s) + H_2O(l)$$

What minimum weight of lithium hydroxide is required to remove all the carbon dioxide exhaled by an astronaut during a 6-day voyage to the moon?

5.118 In the pulmonary capillaries, 1.00 mol of hemoglobin reacts with 4.00 mol of oxygen gas to form oxyhemoglobin.

(a) Calculate the number of milliliters of oxygen, measured at 25°C and 760 torr, that combine with 1.00 g of hemoglobin. The molar mass of hemoglobin is 6.8×10^4 g/mol.

(b) The concentration of hemoglobin in blood is about 15 g per 100 mL of blood. How many milliliters of oxygen gas, measured under the conditions of Part (a), will be carried by 100 mL of blood?

5.119 Hydrochloric acid is an important component of stomach acid. One of the oldest and most familiar treatments for "acid indigestion" involves taking baking soda (sodium hydrogen carbonate), which neutralizes stomach acid according to the equation:

$$NaHCO_3(aq) + HCl(aq) \rightarrow NaCl(aq) + H_2O(l) + CO_2(g)$$

If an antacid tablet contains 520 mg of sodium in the form of sodium bicarbonate, calculate the volume of carbon dioxide, measured at 37°C and 1.00 atm, that would be produced by complete neutralization of the tablet.

5.120 A 5.00-g mixture of zinc and aluminum reacted with HCl to give 3.78 L of hydrogen measured over water at 22°C and 760 torr. The reactions are:

$$Zn(s) + 2HCl(aq) \rightarrow ZnCl_2(aq) + H_2(g)$$

and

$$2Al(s) + 6HCl(aq) \rightarrow 2AlCl_3(aq) + 3H_2(g)$$

Calculate the percent by mass of zinc in the original mixture.

5.121 Early in the nineteenth century, Gay-Lussac formulated the *law of combining volumes*, which stated that *at a given temperature and pressure, the volumes of gases consumed or produced in a chemical reaction are in the ratio of small whole numbers*. Does this law support Avogadro's hypothesis? Explain.

5.122 Ordinary dry air consists of 78.1 mol % nitrogen and 20.9 mol % oxygen. During asthmatic attacks, the bronchial tubes are constricted and only a limited amount of air can reach the alveoli of the lungs. Explain why oxygen therapy for asthma often involves using a mixture of helium and oxygen rather than nitrogen and oxygen.

5.123 Cyclopropane (C_3H_6) is an inhalation anesthetic that has been successfully used in almost every type of surgical operation. An anesthetic mixture at 1.00 atm pressure consists of 40 mol % cyclopropane, 20 mol % oxygen, and helium. Calculate (a) the partial pressure of each gas in the mixture and (b) the average molar mass of the mixture.

5.124 A mixture of dinitrogen oxide (N_2O) and oxygen is sometimes used for anesthesia. If the partial pressures of N_2O and O_2 in such a mixture are 608 and 152 torr, respectively, calculate (a) the concentration of each gas in mole percent and (b) the average molar mass of the mixture.

5.125 The combustion in air of an unknown hydrocarbon produces 3.30 mg of CO_2 and 2.05 mg of H_2O. This hydrocarbon diffuses through a given porous barrier about 2.7 times more slowly than does helium gas under identical conditions. Find the molar mass and formula of the hydrocarbon.

5.126 Consider the separation of gaseous $^{235}UF_6$ from $^{238}UF_6$ by effusion through a series of porous barriers. (See the Chemical Insight (p. 202) on separation of isotopes by gaseous effusion.)

(a) Show that passage through a single barrier will increase the amount of the lighter isotope by an enrichment factor of only 1.0043. The molar masses of U-235 and U-238 are 235.04 and 238.05 g/mol, respectively.

(b) How many diffusion stages are required for a tenfold enrichment of the lighter isotope?

5.127 **(a)** Convert 1.00 L·atm into joules.

(b) Convert $R = 0.0821$ L·atm/mol·K into J/mol·K.

***5.128** Find the root mean square speed of oxygen molecules in the lungs at normal body temperature, 37°C. Express your answer in (a) m/s and (b) miles/hour.

***5.129** For a confined gas, the frequency of wall collisions and the force per collision are each proportional to the molecular speed. Show that doubling the Kelvin temperature doubles the pressure in accordance with Charles's Law.

***5.130** Can a gas law of the form $P(V - nb) = nRT$ account for the dip that occurs in many PV/nRT versus P plots at moderate pressures? (*Hint*: Solve the above equation for PV/nRT and prepare a plot of PV/nRT versus P assuming constant temperature.)

***5.131** A 50.0-L tank contains 1.000 kg of oxygen gas at 0°C. Estimate the pressure using (a) the ideal gas law and (b) the van der Waals equation.

5.132 Use data from the graphs in Figure 5.14 to estimate the density of the atmosphere at an altitude of 10 km.

5.133 A 2.8-g sample of cerium metal is placed in a 1.0-L vessel of oxygen at 1.50 atm and 25°C and allowed to react completely according to the equation:

$$Ce + \frac{x}{2}O_2 \rightarrow CeO_x$$

At the end of the reaction, the vessel is cooled down to the initial temperature and the pressure is found to be 1.00 atm. Calculate the value of x.

5.134 A mixture of CS_2(l) and excess O_2 occupies a vessel of fixed volume at 25°C and 3.0 atm pressure. If the mixture is ignited, and the combustion products are cooled to 25°C, what will be the final pressure?

CHAPTER 6

An ammonium dichromate volcano. See Demonstration S4.5, page 1063.

THERMOCHEMISTRY

OUTLINE

PREVIEW

Lightning strikes, flame races through the forest, and early man observes an immense amount of heat and light released by burning wood, grass, and peat. About 300,000 years ago, our ancestors tamed these reactions to provide warmth and cook food; thus the cave dwellers who compared different firewoods and discovered the fuel value of peat were the first thermochemists. Eventually people found that heat promotes other useful reactions such as the reduction of ores to metals and the conversion of limestone to lime, the essential ingredient for making mortar, cement, and plaster. Chemical reactions continue to provide the energy and materials needed for civilized life. In fact, all of life, from the metabolic processes occurring in a single cell to the combined activities of millions of city dwellers, is powered by a multitude of chemical reactions that release or consume energy.

Chemical reactions are accompanied by energy changes. Some reactions release energy to the surroundings; others absorb energy. This energy usually takes the form of heat, and the study of the heat released or absorbed during chemical reactions is called **thermochemistry**. Physical changes such as mixing, melting, and crystallization often occur during chemical reactions, and the heats associated with these changes are also included within the scope of thermochemistry. Thermochemistry is part of a much broader discipline called **thermodynamics**, which deals with all forms of energy and their interconversions. All chemical processes, including the release of energy by the combustion of fuels, the storage of solar energy in plants by photosynthesis, and even human metabolism, are governed by the laws of thermodynamics.

6.1 CONSERVATION OF ENERGY

Energy can be converted from one form into another: Sunlight falling on a surface is absorbed and converted into heat energy that warms the surface; the kinetic energy of running water and the thermal energy of steam are converted into electrical energy by turbines; and the stored potential energy of fuel is converted into mechan-

LEARNING HINT

Before reading this section, you should review Section 1.7.

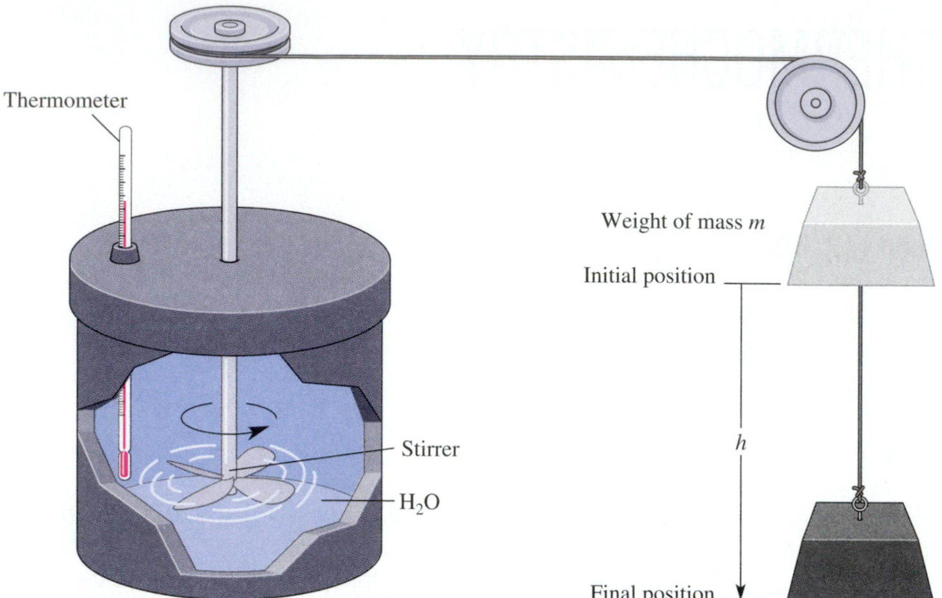

Figure 6.1
A device similar to that used by Joule to demonstrate the equivalence of mechanical and heat energy. A weight falling through a distance h causes the stirrer to rotate and the water to become warmer. The work done by the weight, mgh, is equivalent to the increase in heat energy of the water.

ical energy by automobile engines. Careful experiments have shown that energy is never lost during these transformations. For example, consider stirring some water in a container insulated from its environment (Figure 6.1). Kinetic energy is transferred from the moving stirrer to the water molecules, and the temperature of the water increases. James Prescott Joule, an English scientist who used mechanical stirrers to study this effect, was able to show that the amount of heat gained by the water sample is equivalent to the amount of work done by the stirrer. Joule's experiments, which were performed during the nineteenth century, helped provide an experimental basis for the **first law of thermodynamics**, also known as the **law of conservation of energy**: *Energy can be converted from one form into another, but it cannot be created or destroyed.* This fundamental law implies that we can account for all of the energy released or absorbed during a physical or chemical change.

The second and third laws of thermodynamics are discussed in Chapter 19.

The terms *system* and *surroundings* are often used in discussing energy transfer. A **system** is some portion of the universe arbitrarily chosen for consideration. The **surroundings** are everything not included in the system. The system could be a growing plant, and its surroundings would be the air, the ground, and everything around it. For a chemist, the system is usually a reacting substance or mixture, while the surroundings include the container, the atmosphere, the observer, and so forth. The first law of thermodynamics tells us that energy given up by the system will be absorbed by the surroundings, and vice versa.

Heat, Work, and Internal Energy

The **internal energy** E of a system is the sum of the kinetic and potential energies of its individual particles (see Figure 6.2). Unfortunately, it is impossible to evaluate the internal energy of any but the simplest systems. Chemists, however, are concerned more with changes in the internal energy than with the internal energy itself; that is, they are interested in the quantity

$$\Delta E = E_{\text{final}} - E_{\text{initial}}$$

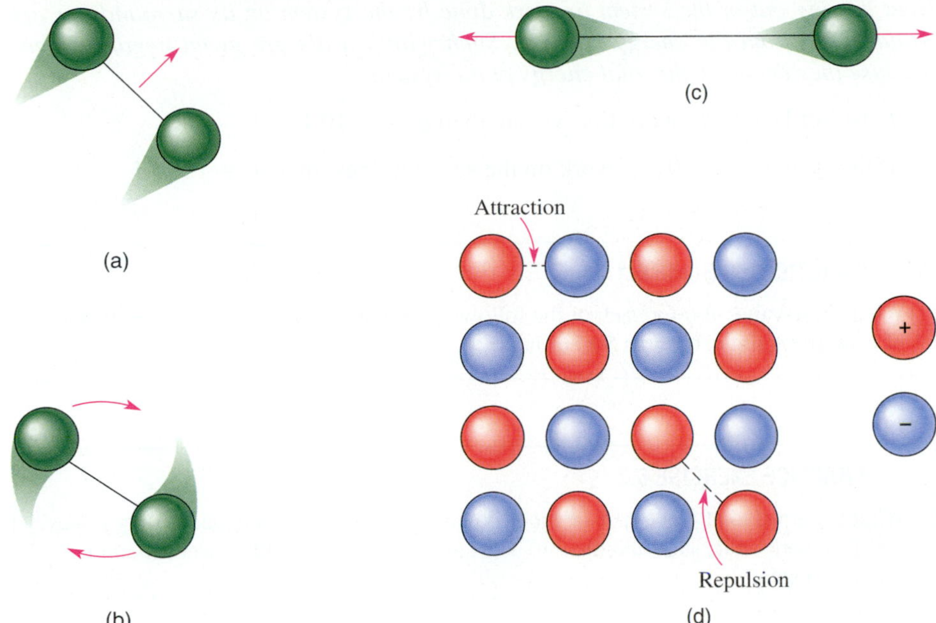

Figure 6.2

Contributions to the internal energy. Kinetic energy contributions include (a) energy of translation (motion from one location to another), (b) energy of rotation, and (c) energy of vibration. (Vibrating particles move back and forth about equilibrium positions; here they are moving away from each other.) Potential energy contributions arise from electrostatic forces between charged particles. The forces between charged ions are shown in (d).

the difference in internal energy between some initial and some final state. (The Greek letter delta Δ is often used to represent the change or difference in some quantity.) We will soon see that it is possible to determine a difference in energy without knowing the values of E_{final} and E_{initial}.

Let us review some of the ways in which a system and its surroundings can exchange energy. The flow of ***heat*** is an energy transfer that results from a temperature difference; heat always flows from warmer regions to cooler ones. Energy can also be transferred by ***work***, the action of a force through a distance. In automobile and steam engines, for example, pistons are pushed outward by the force of expanding gases; the moving pistons gain an amount of energy equal to the work done on them, and the gases lose the same amount of energy. Expansion is only one of many ways in which a system does work on its surroundings. Electrical work is performed when chemical reactions in a battery cause electrons to move through an external circuit. Magnetic substances do work when they move objects containing iron.

The first law of thermodynamics, which embodies our conviction that the energy difference ΔE can be exactly accounted for, is often given in a form that should seem familiar to anyone who has balanced a bank account. Energy can be paid into a system in two ways: by heat, q, flowing into it or by work, w, done on it. The change in the system's energy is the sum of both inputs:

$$\Delta E = q + w \tag{6.1}$$

Heat flowing into the system or work done by the surroundings on the system add to the energy account. Such additions are given positive signs because they increase the energy of the system:

- If 10 J of heat flow into the system, then $q = +10$ J.
- If the surroundings do 10 J of work on the system, then $w = +10$ J.

LEARNING HINT

Heat absorbed by a system is positive; heat released by a system is negative.

Work done on a system is positive; work done by a system is negative.

Heat flowing out of the system or work done by the system on its surroundings are withdrawals from the energy account. Such withdrawals are given negative signs because they decrease the total energy of the system:

- If 10 J of heat flow out of the system, then $q = -10$ J.
- If the system does 10 J of work on the surroundings, then $w = -10$ J.

PRACTICE EXERCISE 6.1

What is the value of q for each of the following systems? (a) A radiator loses 10 kJ of heat. (b) A liquid absorbs 5 kJ of heat.

PRACTICE EXERCISE 6.2

What is the value of w for each of the following systems? (a) An expanding gas does 500 J of work on its surroundings. (b) The surroundings do 20 kJ of work in compressing a solid.

EXAMPLE 6.1

An expanding gas does 1500 kJ of work while it absorbs 800 kJ of heat (Figure 6.3). What are (a) q, (b) w, and (c) ΔE for the gas?

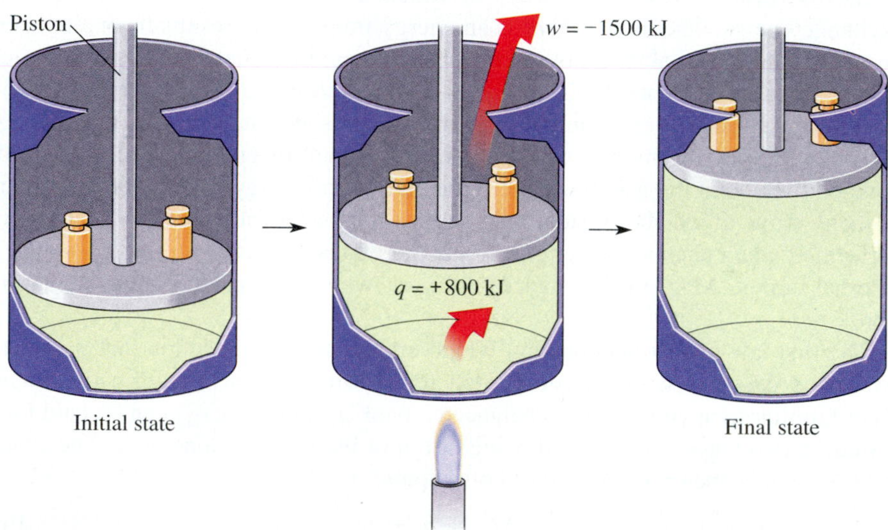

Figure 6.3 An expanding gas absorbs 800 kJ of heat from a burner flame and does 1500 kJ of work against an external pressure.

SOLUTION

(a) The gas absorbs 800 kJ of heat; $q = +800$ kJ. The sign is positive because the gas gains energy when it absorbs heat.

(b) The gas does 1500 kJ of work; $w = -1500$ kJ. The sign is negative because the gas loses energy when it does work.

(c) The change in internal energy is obtained by substituting q and w into Equation 6.1:

$$\Delta E = q + w = 800 \text{ kJ} - 1500 \text{ kJ} = -700 \text{ kJ}$$

The gas gains 800 kJ and loses 1500 kJ, for a net loss of 700 kJ.

PRACTICE EXERCISE 6.3

One gram of liquid water absorbs 2600 J of heat and is converted into steam. The steam expands and does 170 J of work on its surroundings. Find (a) q, (b) w, and (c) ΔE for the system.

The state of any system is described by properties such as pressure, volume, temperature, density, and so forth. Some or all of these properties will change whenever the state of the system changes. For example, when 1.00 L of a gas is warmed from 25°C to 100°C at constant pressure, its volume will increase to 1.25 L. The change in volume is $\Delta V = 1.25 \text{ L} - 1.00 \text{ L} = 0.25$ L, and the change in temperature is $\Delta T = 100°\text{C} - 25°\text{C} = 75°\text{C}$. These changes are simply the differences between the final and initial values—they do not depend on how the change took place, whether gradual or sudden, in one step or by stages. Any property whose change depends only on the initial and final states of the system is called a **thermodynamic property** or **state function**. Volume and temperature are state functions. Pressure, density, internal energy, and many other properties are also state functions.

Heat and work, however, are not state functions—in fact, they are not even properties of the system. The amount of heat and work required to go from some initial to some final state depends on the *pathway*—the way the change is carried out—and many different pathways can lead to the same net change. For example, a person can reach the top of a building by walking up the stairs, running up the stairs, or riding an elevator. The amount of work done by the person will be different for each pathway, but the net increase in his or her potential energy will be the same in each case. Let us reconsider the expanding gas described in Example 6.1. The loss of 700 kJ of internal energy could have occurred by other pathways, that is, by other combinations of q and w. If, for example, the gas performs less work, say 1000 kJ, while absorbing less heat, say 300 kJ, the net loss would be the same:

$$\Delta E = q + w = 300 \text{ kJ} - 1000 \text{ kJ} = -700 \text{ kJ}$$

It is not even necessary that work be done. If the gas loses 700 kJ of heat to its surroundings, and no work is done by or on the gas, then

$$\Delta E = q + w = -700 \text{ kJ} + 0 \text{ kJ} = -700 \text{ kJ}$$

6.2 HEATS OF REACTION

Figure 6.5 shows combustion occurring under three different conditions. The combustion in Figure 6.5a is an example of a **constant volume reaction**, one that takes

CHEMICAL INSIGHT

THE SOLAR ENERGY BALANCE

The earth and its inhabitants are continuously bathed in energy from the sun. In fact, more than 99.9% of the energy received by the earth is in the form of solar radiation. The sun is a massive nuclear fusion reactor radiating energy from its surface at a rate of approximately 10^{27} J/s. The earth, 150 million kilometers away, intercepts 1.73×10^{17} J/s, or less than one-billionth of the sun's vast output.

Most of the solar energy arrives in the form of visible and ultraviolet light. About 30% of this energy is immediately reflected back into space by the earth's atmosphere, clouds, and surface. The other 70%, although temporarily diverted into local activities, is eventually converted to heat and radiated into space. At present, there is an energy cycle, or solar energy balance, such that the rate at which energy is absorbed by the earth equals the rate at which it is released. If this balance did not exist, the earth would gain or lose heat, and the earth's surface would not maintain its constant average temperature.

Figure 6.4 shows some of the components of this energy cycle. Almost half (47%) of the incident solar energy heats the atmosphere, land masses, and oceans before passing back into space. This conversion of light energy into heat occurs in various ways, some involving chemical reactions. For example, oxygen in the upper atmosphere continuously absorbs light and is converted to ozone according to the equation

$$3O_2(g) \rightarrow 2O_3(g)$$

The reverse reaction, which also goes on continuously, releases the same amount of energy in the form of heat.

Another 23% of the incident solar energy is absorbed in the evaporation of water and in raising the water vapor to cloud height. The energy absorbed in evaporation is released as heat when the vapor condenses into droplets, and the potential energy of the droplets turns into the kinetic energy of rainfall, waterfalls, and rivers. Winds and ocean currents, which are driven by solar-induced temperature differences, absorb an additional small fraction, about 0.2%, of the solar energy input. A tiny amount of the kinetic energy of water and wind is temporarily diverted by people into hydroelectric and other forms of power, but ultimately this energy passes back into space as heat.

As part of the *biosphere,* the estimated 3.6×10^{14} kg of living matter dispersed in a thin layer on the earth's surface, we are particularly concerned with the very small fraction of the incident solar energy (0.02%) that is captured by photosynthesis. This reaction, which is catalyzed by *chlorophyll* (a light-capturing pigment), is used by green plants to convert carbon dioxide and water into glucose:

$$6CO_2(g) + 6H_2O(l) \rightarrow C_6H_{12}O_6(s) + 6O_2(g)$$

The photosynthesis reaction absorbs 2816 kJ of energy in the form of sunlight for each mole of glucose produced. Thus, glucose and substances derived from glucose, such as starch, cellulose, petroleum, and coal, are storehouses of chemical energy. When these substances are oxidized back into carbon dioxide and water during metabolism, combustion, and decay, their stored energy is gradually released and converted into heat.

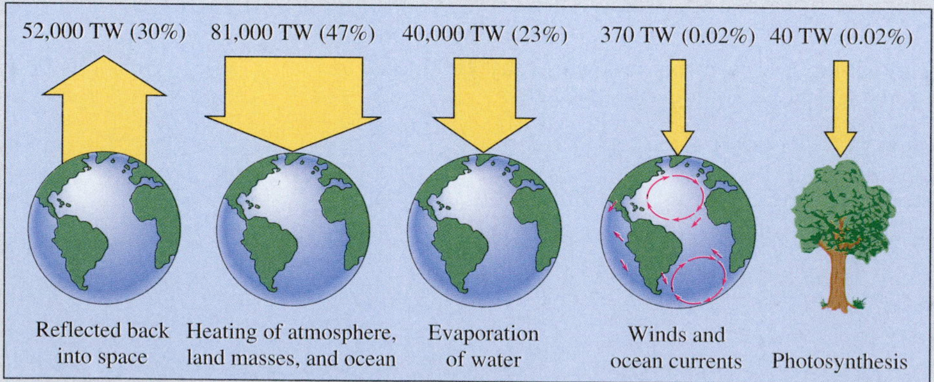

Figure 6.4
Solar energy balance. The earth intercepts 173,000 terawatts (1.73×10^{17} J/s) of the sun's energy. (One terawatt or TW is equal to 10^{12} J/s.) The upward arrow represents the fraction that is immediately reflected back into space; the downward arrows represent the fractions absorbed by the earth. The absorbed fractions eventually revert to heat energy and are returned to space.

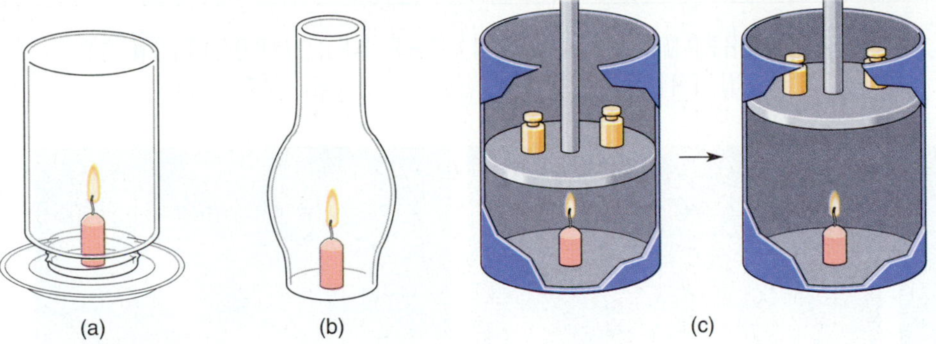

Figure 6.5
Constant volume and constant pressure reactions. (a) The burning of a candle in a sealed jar is a constant-volume reaction. (b)–(c) The burning of candles in a hurricane lamp and in a vessel with a movable piston are constant pressure reactions.

place in a sealed container with rigid walls that do not expand or contract. The pressure of the system usually changes during a constant volume reaction, and if there is an increase in the number of moles of gas, the pressure will increase significantly. A decrease in the number of moles of gas will be accompanied by a reduction in pressure.

A **constant pressure reaction** is one that occurs under constant external pressure. The reactions in an open flame (Figure 6.5b) are constant pressure reactions because the pressure of the reacting system is always equal to atmospheric pressure. A constant pressure reaction could also take place in a chamber that has flexible walls (like a living cell), or one that is equipped with a movable piston (see Figure 6.5c). The volume of the reacting system usually changes during the course of a constant pressure reaction: Escaping gases expand into the atmosphere, pistons move in or out depending on the relative volumes of the reactants and products, and so forth. Many, if not most, reactions are carried out under constant pressure conditions, and the remainder of this chapter is devoted to their study.

Enthalpy

The heat released or absorbed during a chemical reaction is called the **heat of reaction.** An **exothermic reaction**, such as the burning of natural gas or the decomposition of nitrogen triiodide (Demonstration 6.1), is one that gives off heat to its surroundings. An **endothermic reaction** absorbs heat from its surroundings; an example is the reaction of hydrated barium hydroxide with ammonium thiocyanate (Demonstration 6.2).

The heat of reaction q is given by a rearranged form of Equation 6.1:

$$q = \Delta E - w \qquad (6.2)$$

where w is the work done on the reacting system during the reaction and ΔE is the change in energy of the system.

A constant pressure reaction is usually accompanied by a volume change that automatically leads to an exchange of work with the surroundings. A gas produced during the reaction will do work on its surroundings by expanding and pushing back air molecules. Conversely, when a gas is consumed, the surroundings do work on the reacting system by compressing it to a smaller volume. Some forms of work do not involve volume changes. Examples include mechanical work done when a solution is stirred and electrical work done when a battery operates a watch or toy. In this section, we consider only work arising from volume changes.

DEMONSTRATION 6.1 AN EXOTHERMIC REACTION, THE DECOMPOSITION OF NITROGEN TRIIODIDE: $2NI_3(s) \longrightarrow N_2(g) + 3I_2(s)$

A feather is brought near a freshly made sample of nitrogen triiodide drying on a paper box. Dry NI_3 is pressure-sensitive.

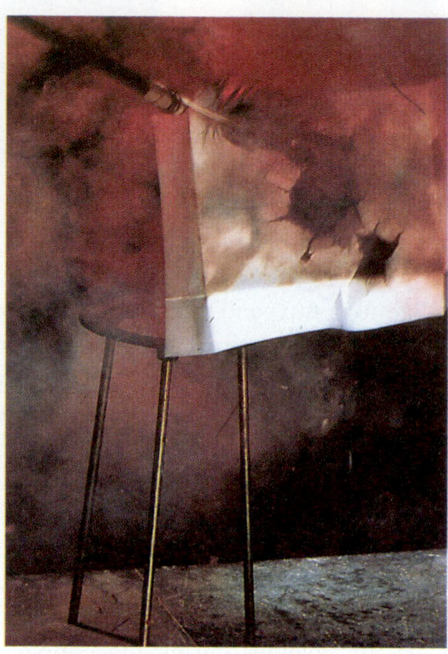

The feather's touch causes the NI_3 to explode violently. Fumes of iodine vapor are released into the surroundings.

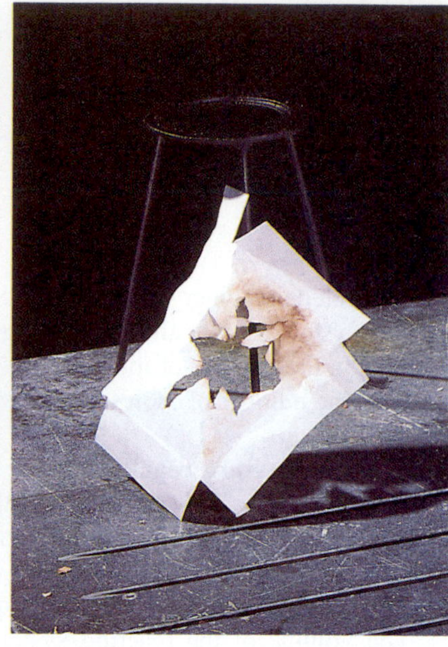

The singed and shattered box.

The work associated with the volume change of a system is called **pressure–volume work (PV work)**. Figure 6.6 shows that the PV work done on a system during a constant pressure reaction is

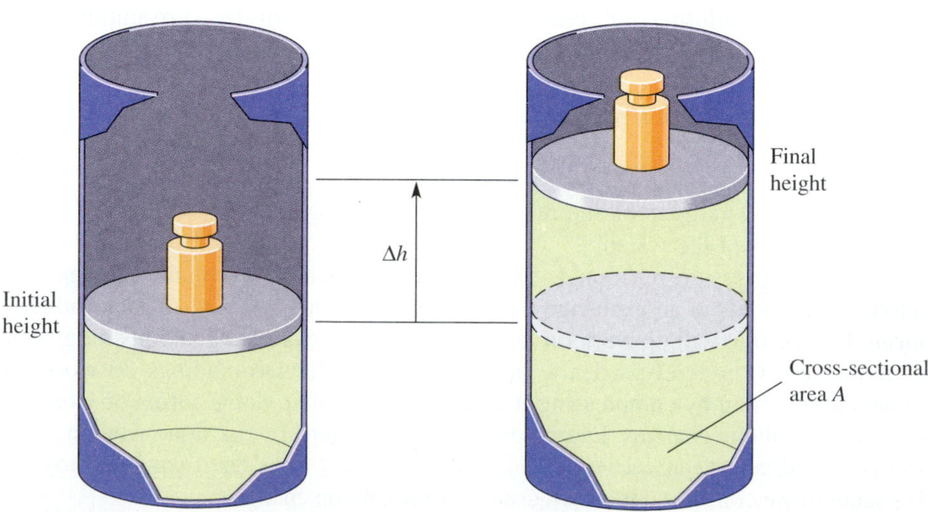

Figure 6.6

Pressure–volume work. A system is kept under constant pressure P by a piston with cross-sectional area A. If the system expands and raises the piston by the distance Δh, the volume increase is $\Delta V = A \times \Delta h$. Pressure is force per unit area, so the force exerted by the system during expansion is $P \times A$. Since work $=$ force $\times$ distance, the work done by the system on its surroundings is $P \times A \times \Delta h = P\Delta V$. The work done on the system is the negative of the work done *by* the system, or $-P\Delta V$.

DEMONSTRATION 6.2

AN ENDOTHERMIC REACTION: $Ba(OH)_2 \cdot 8H_2O(s) + 2NH_4SCN(s) \rightarrow Ba(SCN)_2(aq) + 2NH_3(g) + 10H_2O(l)$

Crystals of ammonium thiocyanate are added to crystals of barium hydroxide octahydrate in a dry flask.

A wood block is dampened with a few drops of water while the flask is swirled.

The flask is placed on the block. The endothermic reaction absorbs heat from the water, the water freezes, and the block adheres to the flask.

$$w = -P\Delta V \tag{6.3}$$

where P is the pressure of the reacting system and

$$\Delta V = V_{\text{products}} - V_{\text{reactants}}$$

is its volume change. Substituting Equation 6.3 into Equation 6.2 gives

$$q_P = \Delta E + P\Delta V \tag{6.4}$$

where the subscript P reminds us that q is now the heat of reaction at constant pressure.

For reactions involving only solids and liquids, the volume change is usually small, and $P\Delta V$ makes a negligible contribution to the heat of reaction. For reactions involving gases, however, the volume change may be substantial. Consider, for example, the decomposition of solid calcium carbonate into solid calcium oxide and gaseous CO_2:

$$CaCO_3(s) \rightarrow CaO(s) + CO_2(g)$$

The volume of this system will increase by about 22.4 L for each mole of CO_2 produced at 0°C and 1 atm. In this reaction, the $P\Delta V$ term makes a significant contribution to the heat of reaction.

It is convenient to define a new thermodynamic property called **enthalpy** (*H;* from the Greek *enthalpein,* "to heat in"). The enthalpy of a system is a state function; it is the sum of the system's internal energy E and its pressure–volume product PV:

$$H = E + PV \tag{6.5}$$

The **enthalpy change** ΔH is the enthalpy of the products minus the enthalpy of the reactants:

$$\Delta H = H_{\text{products}} - H_{\text{reactants}}$$

The enthalpy change for any reaction is also given by

$$\Delta H = \Delta E + \Delta(PV) \tag{6.6}$$

For a constant pressure reaction

$$\begin{aligned} \Delta(PV) &= PV_{\text{products}} - PV_{\text{reactants}} \\ &= P(V_{\text{products}} - V_{\text{reactants}}) \\ &= P\Delta V \end{aligned} \tag{6.7}$$

and substituting Equation 6.7 into Equation 6.6 gives

$$\Delta H = \Delta E + P\Delta V \tag{6.8}$$

A comparison of Equations 6.4 and 6.8 shows that

$$q_P = \Delta H \tag{6.9}$$

Equation 6.9 tells us that *the heat of reaction at constant pressure equals the change in enthalpy of the reacting system.*

Figure 6.7a shows an enthalpy diagram for the decomposition of calcium carbonate. The reaction is endothermic, so heat is absorbed by the system and the products have more enthalpy than the reactants. The change in enthalpy, ΔH, is positive because the system gains enthalpy. In an exothermic reaction, such as the formation of water from its elements (Figure 6.7b), heat is given off; ΔH is negative because the system loses enthalpy. Thus, *ΔH is always positive for an endothermic reaction*

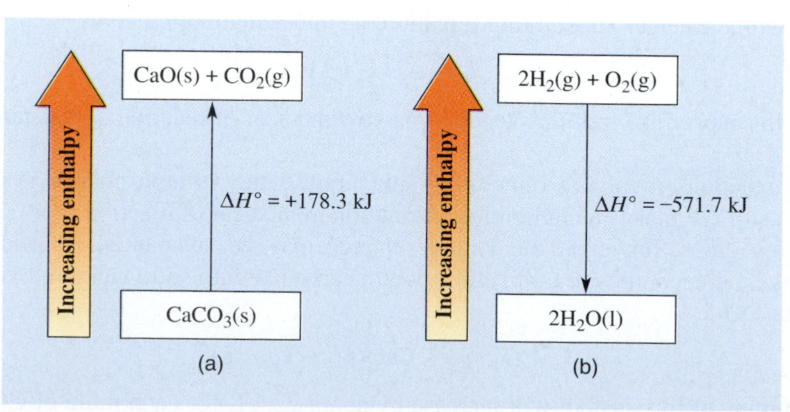

Figure 6.7
(a) An enthalpy diagram for the endothermic reaction $CaCO_3(s) \longrightarrow CaO(s) + CO_2(g)$. (b) An enthalpy diagram for the exothermic reaction $2H_2(g) + O_2(g) \longrightarrow 2H_2O(l)$.

and always negative for an exothermic reaction. Enthalpy changes are often given descriptive names such as "heat (or enthalpy) of combustion," "heat (or enthalpy) of neutralization," and "heat (or enthalpy) of formation."

Endothermic reactions: $\Delta H > 0$
Exothermic reactions: $\Delta H < 0$

PRACTICE EXERCISE 6.4

When ammonia and nitrogen dioxide form from the elements, the heats of formation are −46.1 kJ/mol and +33.2 kJ/mol, respectively. (a) Is the formation of ammonia endothermic or exothermic? (b) What about the formation of nitrogen dioxide?

PRACTICE EXERCISE 6.5

What is the sign of ΔH for the following reactions? (a) Propane gas burns in a camp stove. (b) Ammonium chloride dissolves in water and the solution becomes colder than the starting materials.

Thermochemical Equations

A **thermochemical equation** is a chemical equation that includes the enthalpy change; for example, the reaction of hydrogen with oxygen (Figure 6.7b) is represented by the thermochemical equation

$$2H_2(g, 1\ atm) + O_2(g, 1\ atm) \rightarrow 2H_2O(l) \qquad \Delta H_{298} = -571.7\ kJ$$

This equation states that 571.7 kJ of heat is given off when 2 mol of hydrogen are burned in oxygen. The initial gas pressures are specified, and the product is understood to be at a pressure of 1 atm as well. *Pressures that are not specified are generally assumed to be 1 atm.*

Note the following characteristics of thermochemical equations:

1. The subscript on ΔH gives the temperature at which the enthalpy change was measured. The hydrogen and oxygen must start at 298 K (25°C), and the products must return to that temperature for the stated ΔH value to be precisely correct; the value would be only approximate at any other temperature. *If the temperature is not specified, then it is understood to be 298 K.*

2. The enthalpy change is for the equation as written; that is, 571.7 kJ of heat is evolved when 2 mol of hydrogen is burned. Thermochemical equations are often written with fractional coefficients in order to show ΔH for 1 mol of a particular substance. For example, 1 mol of burning hydrogen evolves half as much heat as 2 mol:

$$H_2(g) + \tfrac{1}{2}O_2(g) \rightarrow H_2O(l) \qquad \Delta H = -285.8\ kJ$$

3. *The sign of ΔH changes when the equation is reversed.* One mole of liquid water absorbs 285.8 kJ when it is decomposed into its elements:

$$H_2O(l) \rightarrow H_2(g) + \tfrac{1}{2}O_2(g) \qquad \Delta H = +285.8\ kJ$$

The thermochemical equation for the decomposition of calcium carbonate into calcium oxide and carbon dioxide is

TABLE 6.2 Standard Enthalpies of Combustion at 25°C

Substance	Formula[a]	$\Delta H°$ (kJ/mol)[b]
Acetic acid	$CH_3COOH(l)$	−874.4
Acetone	$(CH_3)_2CO(l)$	−1790.4
Acetylene	$C_2H_2(g)$	−1299.6
Benzene	$C_6H_6(l)$	−3267.5
Benzoic acid	$C_6H_5COOH(s)$	−3226.9
Ethane	$C_2H_6(g)$	−1559.8
Ethanol	$C_2H_5OH(l)$	−1366.8
Ethylene	$C_2H_4(g)$	−1411.0
Glucose	$C_6H_{12}O_6(s)$	−2816
n-Hexane	$C_6H_{14}(l)$	−4163.1
Hydrogen	$H_2(g)$	−285.8
Methane	$CH_4(g)$	−890.4
Methanol	$CH_3OH(l)$	−726.5
n-Octane	$C_8H_{18}(l)$	−5450.5
Propane	$C_3H_8(g)$	−2219.9
Strychnine	$C_{21}H_{22}O_2N_2(s)$	−11237[c,d]
Sucrose	$C_{12}H_{22}O_{11}(s)$	−5640.9
2,2,4-Trimethylpentane	$C_8H_{18}(l)$	−5455.6

[a]The most stable state at 25°C is indicated with each formula.
[b]Each $\Delta H°$ value is for the combustion equation in which liquid water is a product.
[c]At 20°C.
[d]One of the combustion products of strychnine is $N_2(g)$.

Because many organic compounds are used as fuels, their heats of combustion are extensively tabulated. Values for a variety of organic compounds are given in Table 6.2.

EXAMPLE 6.3

Methane is the principal component of natural gas. (a) Use data from Table 6.2 to compare the heats given off during the combustion of 1.00 g of methane and 1.00 g of propane. (b) On a weight basis, which of these organic compounds is the more efficient fuel?

SOLUTION

(a) The molar mass of methane (CH_4) is 16.04 g/mol. Its molar heat of combustion is −890.4 kJ (Table 6.2); that is, 890.4 kJ is given off when 16.04 g of methane is burned. The heat given off per gram of methane is 890.4 kJ/16.04 g = 55.5 kJ/g.

The molar mass of propane (C_3H_8) is 44.10 g/mol, and its molar heat of combustion is −2219.9 kJ. The heat evolved per gram is 2219.9 kJ/44.10 g = 50.3 kJ/g.

(b) On a weight basis methane is the more efficient fuel because it liberates 55.5 kJ − 50.3 kJ = 5.2 kJ more heat per gram than propane.

PRACTICE EXERCISE 6.7

Liquid 2,2,4-trimethylpentane (C_8H_{18}) is used in determining the octane number of gasolines. Refer to Table 6.2 and calculate the number of kilojoules of heat given off during the combustion of 1.00 L of 2,2,4-trimethylpentane. The density of liquid 2,2,4-trimethylpentane is 0.692 g/mL.

The heat evolved per gram of fuel is called the **fuel value**. The fuel values in Table 6.3 show that when minimum mass is important, hydrogen is the fuel of choice. The main engines of the orbiter in the U.S. space shuttle, for example, are powered by hydrogen burning in oxygen. Considerations other than mass often go into the choice of a fuel. Propane (C_3H_8) and butane (C_4H_{10}) have lower fuel values than methane (see Example 6.3), but propane and butane are more easily liquefied than methane, making them more convenient for some purposes. Cylinders of "bottled gas" contain liquid propane under pressure, and disposable cigarette lighters are fueled with liquid butane.

Food is also a fuel, and average fuel values for fats, carbohydrates, and proteins are listed in Table 6.4. Fuel values in Calories per ounce or Calories per serving are often given in food labels, cookbooks, and other sources of nutritional information. A dietary Calorie (capital C) is 1 kilocalorie (4.184 kJ). An average-size adult uses from 1600 to 2400 Calories per day in normal activity.

TABLE 6.3 Fuel Values for Some Selected Fuels

Fuel	Fuel Value (kJ/g)
Hydrogen	142
Methane	56
Natural gas	53[a]
Acetylene	50
Gasoline, diesel fuel, kerosene	47[b]
Heating oil	45[b]
Carbon (graphite)	33
Coal (anthracite)	27[b]
Wood	19[b]

[a]The composition of natural gas varies with its origin. The value given is for a natural gas mixture consisting of 85% methane, 9% ethane, 3% propane, 2% nitrogen, and 1% *n*-butane.

[b]Average values for complex mixtures of variable composition.

TABLE 6.4 Average Fuel Values for Fats, Proteins, and Carbohydrates

Substance	kcal/g[a]	kJ/g
Fats	9.1	38
Proteins	4.3	18
Carbohydrates	4.1	17

[a]One kilocalorie (4.184 kJ) is equal to 1 dietary Calorie.

Enthalpies of Formation

A **standard enthalpy of formation**, ΔH_f°, is the enthalpy change accompanying the formation of 1 mol of a substance in its standard state from the most stable (lowest energy) forms of its elements in their standard states. For example, the most stable forms of mercury and chlorine at 25°C and 1 atm are the liquid and gas, respectively. When they combine to form mercury(II) chloride, 224.3 kJ of heat is given off for each mole of product formed at 1 atm pressure. Hence, the standard enthalpy of formation of $HgCl_2$ at 25°C is −224.3 kJ/mol:

$$Hg(l) + Cl_2(g) \rightarrow HgCl_2(s) \qquad \Delta H_f^\circ = -224.3 \text{ kJ}$$

As a second example, consider the formation of carbon monoxide from solid carbon and oxygen gas. Pure carbon exists as both diamond and graphite, but graphite is more stable. Hence, the standard formation equation for any carbon compound starts with graphite. The formation of 1 mol of carbon monoxide from graphite and oxygen under standard state conditions gives off 110.5 kJ of heat at 25°C:

$$C(\text{graphite}) + \tfrac{1}{2}O_2(g) \rightarrow CO(g) \qquad \Delta H_f^\circ = -110.5 \text{ kJ}$$

Other examples are the formation of hydrogen iodide, an endothermic reaction,

$$\tfrac{1}{2}H_2(g) + \tfrac{1}{2}I_2(s) \rightarrow HI(g) \qquad \Delta H_f^\circ = +26.5 \text{ kJ}$$

and the formation of liquid water

An enthalpy of formation is often called a *heat of formation;* an enthalpy of combustion is called a *heat of combustion.*

$$H_2(g) + \tfrac{1}{2}O_2(g) \rightarrow H_2O(l) \qquad \Delta H_f^\circ = -285.8 \text{ kJ}$$

Note that water forms when hydrogen burns; hence, −285.8 kJ is the heat of combustion of hydrogen gas as well as the heat of formation of water. Some standard enthalpies of formation are given in Table 6.5. A more extensive tabulation is given in Appendix B.1.

The reactants in a thermochemical formation equation must be elements in their most stable forms. Reactions that start with ions, such as

$$Na^+(g) + Cl^-(g) \rightarrow NaCl(s)$$

and reactions that start with compounds, such as

$$CaO(s) + CO_2(g) \rightarrow CaCO_3(s)$$

are not formation reactions in the thermochemical sense.

The most stable form of hydrogen at 1 atm is $H_2(g)$, so the combination of separate hydrogen atoms

$$H(g) + H(g) \rightarrow H_2(g)$$

is not a formation reaction. However, the equation

$$\tfrac{1}{2}H_2(g) \rightarrow H(g)$$

does represent a formation reaction, the formation of 1 mol of hydrogen atoms from H_2.

TABLE 6.5 Standard Enthalpies of Formation at 25°C (*A more complete listing is given in Appendix B.1*)

Substance	Formula[a]	ΔH_f° (kJ/mol)	Substance	Formula[a]	ΔH_f° (kJ/mol)
Ammonia	$NH_3(g)$	−46.1	Mercury (liquid)	$Hg(l)$	0
Ammonium chloride	$NH_4Cl(s)$	−314.4	Mercury(I) chloride	$Hg_2Cl_2(s)$	−265.2
Calcium hydroxide	$Ca(OH)_2(s)$	−986.1	Mercury(II) chloride	$HgCl_2(s)$	−224.3
Carbon (diamond)	C(diamond)	+1.9	Methane	$CH_4(g)$	−74.8
Carbon (graphite)	C(graphite)	0	Methanol	$CH_3OH(l)$	−238.7
Carbon dioxide	$CO_2(g)$	−393.5	Nitrogen dioxide	$NO_2(g)$	+33.2
Carbon disulfide	$CS_2(l)$	+89.7	*n*-Octane	$C_8H_{18}(l)$	−269.8
Carbon monoxide	$CO(g)$	−110.5	Oxygen	$O_2(g)$	0
Copper(II) nitrate	$Cu(NO_3)_2(s)$	−302.9	Ozone	$O_3(g)$	+142.7
Copper(II) oxide	$CuO(s)$	−157.3	Potassium chlorate	$KClO_3(s)$	−397.7
Ethane	$C_2H_6(g)$	−84.7	Potassium chloride	$KCl(s)$	−436.7
Ethanol	$C_2H_5OH(l)$	−277.7	Propane	$C_3H_8(g)$	−104.9
Ethylene	$C_2H_4(g)$	+52.3	Sucrose	$C_{12}H_{22}O_{11}(s)$	−2221
Glucose	$C_6H_{12}O_6(s)$	−1260	Sulfur dioxide	$SO_2(g)$	−296.8
Hydrogen chloride	$HCl(g)$	−92.3	Sulfur trioxide	$SO_3(g)$	−395.7
Hydrogen sulfide	$H_2S(g)$	−20.6	Water (gas)	$H_2O(g)$	−241.8
Mercury (gas)	$Hg(g)$	+61.3	Water (liquid)	$H_2O(l)$	−285.8

[a]The enthalpy of formation is for the state indicated in parentheses after each formula.

The definition of standard enthalpy of formation implies that *elements in their most stable form at 1 atm pressure have zero enthalpies of formation.* Less stable forms of the elements, however, have nonzero enthalpies of formation. The formation of diamond from graphite, for example, is endothermic:

$$\text{C(graphite)} \rightarrow \text{C(diamond)} \qquad \Delta H_f^\circ = +1.90\ \text{kJ}$$

So is the formation of ozone (O_3) from oxygen:

$$\tfrac{3}{2}\text{O}_2(\text{g}) \rightarrow \text{O}_3(\text{g}) \qquad \Delta H_f^\circ = +142.7\ \text{kJ}$$

EXAMPLE 6.4

The standard heat of formation of ammonia gas (NH_3) is −46.1 kJ/mol at 25°C. (a) Write a thermochemical equation for the formation of ammonia. (b) Estimate the number of kilojoules given off when 10.0 g of nitrogen react with excess hydrogen.

SOLUTION

(a) Ammonia forms from nitrogen and hydrogen, and all three substances are gases at 1 atm and 25°C. The thermochemical equation for the formation of 1 mol of NH_3 is

$$\tfrac{1}{2}\text{N}_2(\text{g}) + \tfrac{3}{2}\text{H}_2(\text{g}) \rightarrow \text{NH}_3(\text{g}) \qquad \Delta H_f^\circ = -46.1\ \text{kJ}$$

(b) The sign of the enthalpy change indicates that 46.1 kJ is given off when $\frac{1}{2}$ mol of N_2 reacts with hydrogen. The molar mass of N_2 is 28.01 g/mol. The steps in the calculation are:

$$10.0\ \text{g N}_2 \rightarrow \text{mol N}_2 \rightarrow \text{kJ}$$

They can be combined as follows:

$$10.0\ \text{g N}_2 \times \frac{1\ \text{mol N}_2}{28.01\ \text{g N}_2} \times \frac{46.1\ \text{kJ}}{0.500\ \text{mol N}_2} = 32.9\ \text{kJ}$$

The actual amount of heat evolved will depend on the conditions. If each substance is in its standard state, 32.9 kJ of heat will be given off.

PRACTICE EXERCISE 6.8

The enthalpy of formation of Fe_2O_3 (dry rust) is −824.2 kJ/mol. (a) Write a thermochemical equation for the formation of Fe_2O_3. (b) Estimate the number of kilojoules given off when 50.0 kg of iron nails turns to rust.

6.4 HESS'S LAW

A chemical reaction may occur directly in one step or indirectly in two or more consecutive steps. Carbon dioxide, for example, can be produced in one step by burning graphite in excess oxygen

$$\text{C(graphite)} + \text{O}_2(\text{g}) \rightarrow \text{CO}_2(\text{g}) \qquad \Delta H^\circ = -393.5\ \text{kJ}$$

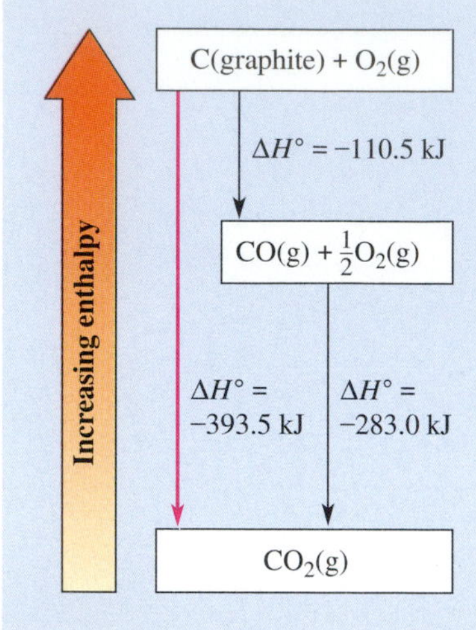

Figure 6.8

Enthalpy diagrams for the complete combustion of graphite in one step or in two steps. The total enthalpy change is the same either way, in agreement with Hess's law.

or in two steps by first burning the graphite with a limited supply of oxygen to obtain carbon monoxide

$$\text{C(graphite)} + \tfrac{1}{2}\text{O}_2\text{(g)} \rightarrow \text{CO(g)} \qquad \Delta H° = -110.5 \text{ kJ}$$

and then burning the carbon monoxide in additional oxygen

$$\text{CO(g)} + \tfrac{1}{2}\text{O}_2\text{(g)} \rightarrow \text{CO}_2\text{(g)} \qquad \Delta H° = -283.0 \text{ kJ}$$

The net change is the same in either case; 1 mol of carbon dioxide forms from 1 mol of carbon and 1 mol of oxygen. Figure 6.8 shows that the enthalpy change is also the same: −393.5 kJ in the one-step process, and −110.5 kJ − 283.0 kJ = −393.5 kJ in the two-step process. This example illustrates **Hess's law**: *The total enthalpy change for a reaction is the same whether the reaction occurs in one or several steps.* This law, which was published in 1840 by Germain Henri Hess, a Swiss chemist, is based on the fact that enthalpy is a state function, and therefore the change in enthalpy is independent of the path. In going from the same initial state, in this case graphite and oxygen, to the same final state, carbon dioxide, the enthalpy change will be the same regardless of how the reaction is carried out.

Hess's law is often used to calculate enthalpy changes that might be difficult to measure experimentally, as shown in the following example.

EXAMPLE 6.5

Use the thermochemical equations

$$\text{C(graphite)} \rightarrow \text{C(g)} \qquad \Delta H° = +716.7 \text{ kJ}$$

$$\text{C(graphite)} + \text{O}_2\text{(g)} \rightarrow \text{CO}_2\text{(g)} \qquad \Delta H° = -393.5 \text{ kJ}$$

to calculate the enthalpy change accompanying the combustion of 1 mol of carbon vapor.

SOLUTION

The combustion of graphite (Equation 3 below) can be regarded as the sum of two steps: (1) conversion of graphite to gaseous carbon and (2) combustion of the resulting carbon vapor:

1. $\text{C(graphite)} \rightarrow \text{C(g)} \qquad \Delta H_1° = +716.7 \text{ kJ}$

2. $\text{C(g)} + \text{O}_2\text{(g)} \rightarrow \text{CO}_2\text{(g)} \qquad \Delta H_2° = ?$

3. $\text{C(graphite)} + \text{O}_2\text{(g)} \rightarrow \text{CO}_2\text{(g)} \qquad \Delta H_3° = -393.5 \text{ kJ}$

Equation 3 is obtained by adding Equations 1 and 2 to give

$$\text{C(graphite)} + \cancel{\text{C(g)}} + \text{O}_2\text{(g)} \rightarrow \cancel{\text{C(g)}} + \text{CO}_2\text{(g)}$$

and canceling C(g), which appears on both sides of the sum. The enthalpy changes for Steps 1 and 3 were given in the problem. The enthalpy change for Step 2, $\Delta H_2°$, is the unknown molar enthalpy change for the combustion of carbon vapor. This enthalpy change can be found with the aid of Hess's law, which requires that

$$\Delta H_1° + \Delta H_2° = \Delta H_3°$$

Substituting the known $\Delta H°$ values gives

$$716.7\ \text{kJ} + \Delta H_2^\circ = -393.5\ \text{kJ}$$

which rearranges to

$$\Delta H_2^\circ = -393.5\ \text{kJ} - 716.7\ \text{kJ} = -1110.2\ \text{kJ}$$

The steps are diagrammed in Figure 6.9. Note that the combustion of solid graphite gives off less heat (393.5 kJ) than the combustion of carbon vapor (1110.2 kJ). The carbon atoms in the solid are closer to each other and more firmly bound than in the gas; hence, more energy is required to overcome attractive forces and less energy is released as heat.

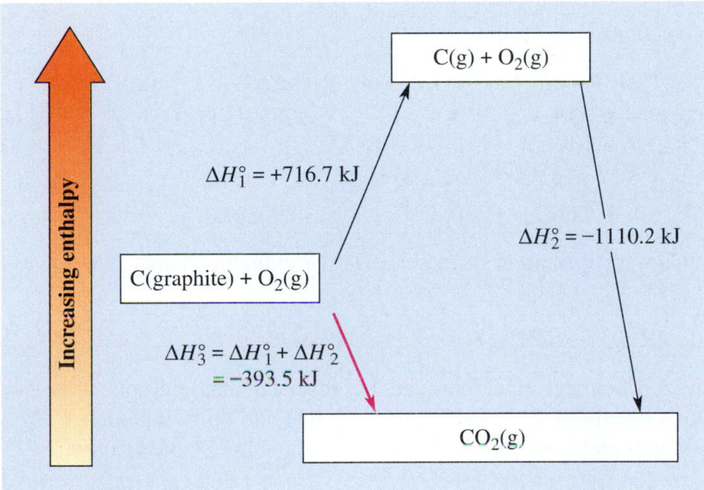

Figure 6.9
The combustion of graphite (*red* arrow) can be regarded as the sum of two steps: (1) the conversion of graphite to gaseous carbon and (2) the combustion of carbon vapor.

PRACTICE EXERCISE 6.9

Use the following enthalpy changes

$$SO_2(g) \rightarrow S(s) + O_2(g) \qquad \Delta H° = +297\ \text{kJ}$$

$$S(s) + \tfrac{3}{2}O_2(g) \rightarrow SO_3(g) \qquad \Delta H° = -396\ \text{kJ}$$

to calculate $\Delta H°$ for the reaction

$$SO_2(g) + \tfrac{1}{2}O_2(g) \rightarrow SO_3(g)$$

Recall (Section 6.2) that a thermochemical equation can be reversed or multiplied through by a factor such as $\frac{1}{2}$. Reversing an equation changes the sign of its enthalpy change, and multiplying an equation by a factor multiplies the enthalpy change by the same factor. These properties of thermochemical equations are useful in Hess's law calculations, and we use them in the next example to calculate the enthalpy of formation of methane (CH_4) from combustion data.

Methane is an important industrial compound, and its enthalpy of formation is often required for various calculations. This enthalpy cannot be directly measured because the formation reaction (carbon combining with hydrogen) occurs only when the temperature is greater than 1100°C.

EXAMPLE 6.6

The standard enthalpies of combustion of hydrogen, graphite, and methane are −285.8, −393.5, and −890.4 kJ/mol, respectively. Calculate the standard enthalpy of formation of methane.

SOLUTION

The combustion reactions are as follows:

	$\Delta H°$
1. $H_2(g) + \frac{1}{2}O_2(g) \rightarrow H_2O(l)$	−285.8 kJ
2. $C(graphite) + O_2(g) \rightarrow CO_2(g)$	−393.5 kJ
3. $CH_4(g) + 2O_2(g) \rightarrow CO_2(g) + 2H_2O(l)$	−890.4 kJ

The standard enthalpy of formation of methane is the enthalpy change for Equation 4 below.

4. $C(graphite) + 2H_2(g) \rightarrow CH_4(g)$ $\quad \Delta H_f° = ?$

The fact that thermochemical equations can be reversed and multiplied by factors allows us to change Equations 1, 2, and 3 into steps that add up to Equation 4. We will begin with the products and consider the substances in Equation 4 one at a time.

- CH_4: To obtain 1 mol of methane as a product, we write Equation 3 in reverse. The result is Equation 3a below. Note that $\Delta H°$ changes sign.
- H_2: Since 2 mol of hydrogen are needed on the reactant side, we double Equation 1 to obtain Equation 1a.
- C(graphite): Since 1 mol of graphite is needed as a reactant, we leave Equation 2 as is.

The resulting steps are

	$\Delta H°$
1a. $2H_2(g) + O_2(g) \rightarrow 2H_2O(l)$	2(−285.8 kJ)
2a. $C(graphite) + O_2(g) \rightarrow CO_2(g)$	−393.5 kJ
3a. $CO_2(g) + 2H_2O(l) \rightarrow CH_4(g) + 2O_2(g)$	+890.4 kJ

The formation reaction, Equation 4, is obtained by adding these three steps and canceling the quantities that are the same on both sides of the arrow—CO_2, $2H_2O$, and $2O_2$. (You should verify this.) The enthalpy change for Equation 4 is then obtained by adding the enthalpy changes of the three steps:

$$\Delta H_f° = 2(-285.8 \text{ kJ}) - 393.5 \text{ kJ} + 890.4 \text{ kJ} = -74.7 \text{ kJ}$$

The adjusted equations, 1a, 2a, and 3a, show methane being formed by an imaginary pathway that is equivalent to burning hydrogen and graphite in oxygen to form water and carbon dioxide (Steps 1a and 2a) and rearranging the combustion products to form methane and oxygen (Step 3a). This pathway is diagrammed in Figure 6.10.

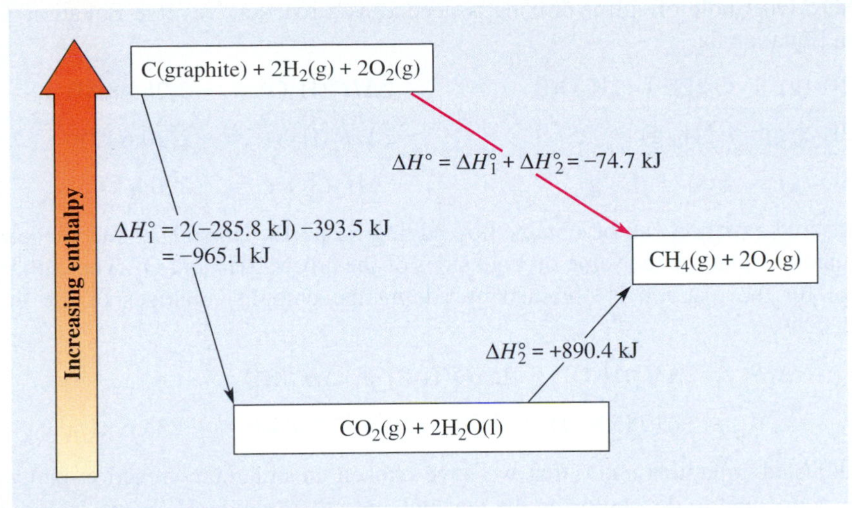

Figure 6.10

Methane is formed from its elements (*red* arrow) by an imaginary pathway equivalent to (1) burning graphite and hydrogen and (2) rearranging the combustion products to form methane.

PRACTICE EXERCISE 6.10

The molar enthalpy of combustion of acetic acid (CH_3COOH) is −874.2 kJ. Combine this with the molar enthalpies of combustion of H_2 and graphite to obtain the molar enthalpy of formation of CH_3COOH.

Enthalpy Changes from Enthalpies of Formation

Now we will use Hess's law and standard heats of formation (Table 6.5) to find the standard enthalpy change for the reaction of hydrogen sulfide with sulfur dioxide:

$$2H_2S(g) + SO_2(g) \rightarrow 3S(s) + 2H_2O(l) \qquad \Delta H^\circ = ?$$

Thermochemical equations for the formation of $H_2O(l)$, $H_2S(g)$, and $SO_2(g)$ are

1. $H_2(g) + \frac{1}{2}O_2(g) \rightarrow H_2O(l) \qquad \Delta H^\circ_f(H_2O) = -285.8\text{ kJ}$

2. $H_2(g) + S(s) \rightarrow H_2S(g) \qquad \Delta H^\circ_f(H_2S) = -\ 20.6\text{ kJ}$

3. $S(s) + O_2(g) \rightarrow SO_2(g) \qquad \Delta H^\circ_f(SO_2) = -296.8\text{ kJ}$

where the symbol $\Delta H^\circ_f(\quad)$ represents the standard heat of formation of the substance in the parentheses. Now we will use the method of Example 6.6; that is, we will convert the formation equations into steps that add up to the desired equation.

- H_2O: Two moles of water is needed on the product side; double Equation 1 to obtain Equation 1a below.

- H_2S: Two moles of hydrogen sulfide is needed on the reactant side; reverse and double Equation 2 to obtain Equation 2a below.

SO_2: One mole of sulfur dioxide is needed as a reactant; reverse Equation 3 to obtain Equation 3a.

1a. $2H_2(g) + O_2(g) \rightarrow 2H_2O(l)$ $\qquad 2\Delta H^\circ_f(H_2O) = -2(285.8 \text{ kJ})$

2a. $2H_2S(g) \rightarrow 2H_2(g) + 2S(s)$ $\qquad -2\Delta H^\circ_f(H_2S) = +2(20.6 \text{ kJ})$

3a. $SO_2(g) \rightarrow S(s) + O_2(g)$ $\qquad -\Delta H^\circ_f(SO_2) = +296.8 \text{ kJ}$

The desired equation can be obtained by adding Steps 1a, 2a, and 3a and canceling the quantities that are the same on both sides of the arrow, $2H_2$ and O_2. The enthalpy change for the reaction is obtained by adding the enthalpy changes of the three steps:

$$\begin{aligned}\Delta H^\circ &= 2\Delta H^\circ_f(H_2O) - 2\Delta H^\circ_f(H_2S) - \Delta H^\circ_f(SO_2) \\ &= -2(285.8 \text{ kJ}) + 2(20.6 \text{ kJ}) + 296.8 \text{ kJ} = -233.6 \text{ kJ}\end{aligned}$$

The adjusted equations show that we have chosen an imaginary reaction pathway that is equivalent to decomposing the reactants into their elements (Steps 2a and 3a) and reassembling those same elements into the products (Step 1a). This pathway is diagrammed in Figure 6.11.

When we examine the sum of the enthalpies, we find that we have added the heats of formation of the products (the heat of formation of sulfur is zero) and subtracted the heats of formation of the reactants. We can see this more clearly if we rewrite the sum as

$$\Delta H^\circ = 2\Delta H^\circ_f(H_2O) + 3\Delta H^\circ_f(S) - 2\Delta H^\circ_f(H_2S) - \Delta H^\circ_f(SO_2)$$

which is identical to the sum written above.

This calculation illustrates the rule that *the enthalpy change of a reaction is equal to the sum of the enthalpies of formation of the products minus the sum of the enthalpies of formation of the reactants.* In finding these sums, the molar heat of formation of each substance is multiplied by the number of moles in the thermochemical equation, as was done in the example above. This rule, which is widely used for calculating standard enthalpy changes from tabulated heats of formation, can be expressed as

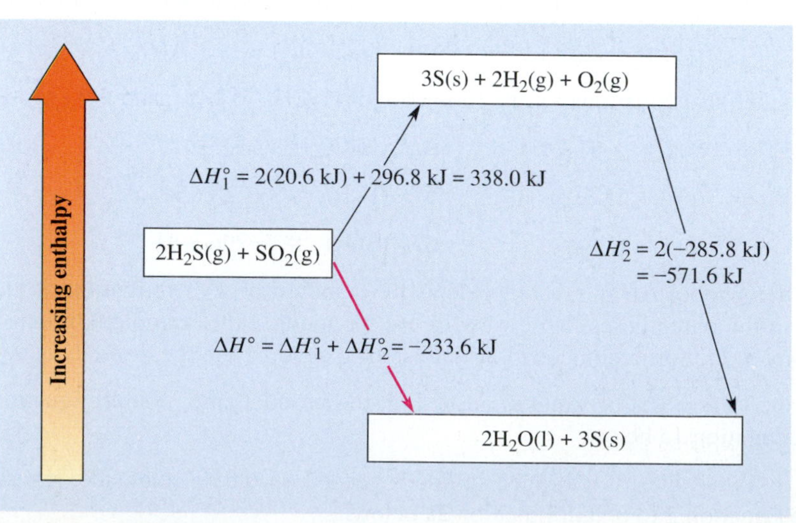

Figure 6.11

The reaction between H_2S and SO_2 (*red* arrow) is the sum of two imaginary steps: (1) decomposing the reactants into their elements and (2) reassembling the elements into the products.

$$\Delta H^\circ = \sum \left[n_p \times (\Delta H_f^\circ)_p\right] - \sum \left[n_r \times (\Delta H_f^\circ)_r\right] \tag{6.10}$$

where the Greek letter sigma Σ means "the sum of." The first sum in Equation 6.10 is the sum of $n_p \times (\Delta H_f^\circ)_p$ for each of the products; n_p is the number of moles of the product in the thermochemical equation and $(\Delta H_f^\circ)_p$ is its standard enthalpy of formation. The second sum in Equation 6.10 is the analogous sum for the reactants. Another example follows.

EXAMPLE 6.7

Use enthalpies of formation to calculate the standard enthalpy change for the photosynthetic production of glucose:

$$6CO_2(g) + 6H_2O(l) \longrightarrow C_6H_{12}O_6(s) + 6O_2(g)$$

SOLUTION

Table 6.5 gives the heats of formation of carbon dioxide, liquid water, and glucose as −393.5, −285.8, and −1260 kJ/mol, respectively. The molar heat of formation of the element oxygen is zero. Each molar heat must be multiplied by the corresponding number of moles in the balanced equation. We will keep track of the data by writing the heat of formation below each substance in the equation:

$$\begin{array}{lcccc} & 6CO_2(g) + & 6H_2O(l) \longrightarrow & C_6H_{12}O_6(s) + & 6O_2(g) \\ \Delta H_f^\circ \text{(kJ/mol)}: & -393.5 & -285.8 & -1260 & 0 \end{array}$$

The standard enthalpy change for the reaction is the sum of the heats of formation of the products minus the sum for the reactants; that is,

$$\begin{aligned} \Delta H^\circ &= 1 \text{ mol } C_6H_{12}O_6 \times \Delta H_f^\circ(C_6H_{12}O_6) + 6 \text{ mol } O_2 \times \Delta H_f^\circ(O_2) \\ &\qquad - 6 \text{ mol } CO_2 \times \Delta H_f^\circ(CO_2) - 6 \text{ mol } H_2O \times \Delta H_f^\circ(H_2O) \\ &= 1 \text{ mol} \times (-1260 \text{ kJ/mol}) + 6 \text{ mol} \times (0 \text{ kJ/mol}) \\ &\qquad - 6 \text{ mol} \times (-393.5 \text{ kJ/mol}) - 6 \text{ mol} \times (-285.8 \text{ kJ/mol}) \\ &= +2816 \text{ kJ} \end{aligned}$$

PRACTICE EXERCISE 6.11

The chromium metal used in chrome steel is obtained by heating the ore chromite ($FeCr_2O_4$) with carbon:

$$FeCr_2O_4(s) + 4C(\text{graphite}) \longrightarrow Fe(s) + 2Cr(s) + 4CO(g)$$

The ΔH° for this reaction is +988.4 kJ and the enthalpy of formation of carbon monoxide is −110.5 kJ/mol. Calculate the enthalpy of formation of $FeCr_2O_4$.

6.5 HEAT CAPACITY AND SPECIFIC HEAT

The **heat capacity** of an object is the amount of heat required to raise the temperature of the object by 1°C. The heat capacity of a large amount of material is greater than that of a small amount of the same material; it takes more heat, for example, to

The specific heat, often called the *specific heat capacity,* does not have a generally accepted symbol. The molar heat capacity is usually symbolized by C_P for a constant pressure process or C_V for a constant volume process.

warm a 2-L bottle of root beer to room temperature than it does to warm a glass of root beer. Chemists usually find it convenient to speak of heat capacity on a *per gram* or a *per mole* basis. The amount of heat required to raise the temperature of 1 g of substance by 1°C is called the **specific heat** (Table 6.6). For example, the specific heat of liquid water is 4.179 J/g·°C at 25°C; thus, when 1 g of water at 25°C absorbs 4.179 J, its temperature increases by 1°C to 26°C. The **molar heat capacity** is the amount of heat required to raise the temperature of 1 mol of a substance by 1°C. All heat capacities vary slightly with temperature.

TABLE 6.6 Specific Heats[a] in J/g·°C

Metals	
Aluminum	0.902
Copper	0.385
Gold	0.129
Iron	0.449
Lead	0.128
Mercury	0.140
Sodium	1.228
Tantalum	0.140
Water	
Ice (−2°C)	2.100
Liquid (0°C)	4.218
Liquid (4°C)	4.205
Liquid (25°C)	4.179
Liquid (100°C)	4.216
Vapor (100°C)	2.013
Other Compounds	
Ethyl ether	2.320
Ethanol (ethyl alcohol)	2.419

[a]All values are for 1 atm pressure and 25°C except where indicated.

EXAMPLE 6.8

Calculate the molar heat capacity of water at 25°C.

SOLUTION

The specific heat of water at 25°C is 4.179 J/g·°C. Its molar mass is 18.02 g. The molar heat capacity is

$$\frac{4.179\ \text{J}}{\text{g}\cdot{}^\circ\text{C}} \times \frac{18.02\ \text{g}}{1\ \text{mol}} = 75.31\ \text{J/mol}\cdot{}^\circ\text{C}$$

PRACTICE EXERCISE 6.12

The molar heat capacity of chloroform ($CHCl_3$) is 113.8 J/mol·°C. Calculate the specific heat of chloroform.

The following examples show how to calculate the amount of heat lost or gained by a substance undergoing a temperature change.

EXAMPLE 6.9

Calculate the number of joules of heat absorbed by 1.0 kg of water when it is heated from 20°C to 100°C.

SOLUTION

Although the specific heat of water varies slightly over the given temperature range, the data in Table 6.6 show that its value is constant to two significant figures, 4.2 J/g·°C. Hence, 4.2 J will increase the temperature of 1 g of water by 1°C. The number of joules required to increase the temperature of 1.0 kg (1.0×10^3 g) of water by 100°C − 20°C = 80°C is

$$1.0 \times 10^3\ \text{g} \times \frac{4.2\ \text{J}}{\text{g}\cdot{}^\circ\text{C}} \times 80^\circ\text{C} = 3.4 \times 10^5\ \text{J}$$

We can see from Example 6.9 that when a system undergoes a temperature change, the amount of heat q that enters or leaves the system is the product of the mass of the system, its specific heat, and the temperature change:

$$q = \text{mass} \times \text{specific heat} \times \Delta T \quad (6.11)$$

The temperature change ΔT is the final temperature minus the initial temperature: $\Delta T = T_{\text{final}} - T_{\text{initial}}$. When the temperature increases, ΔT is positive, heat is absorbed by the system, and q is also positive. When the temperature decreases, ΔT is negative, heat is lost by the system, and q is negative.

EXAMPLE 6.10

A 10.0-g bar of tantalum metal is heated to 99.0°C and then dropped into cold water (Figure 6.12). The final temperature of the tantalum–water mixture is 5.1°C. How much heat is lost by the tantalum? The average specific heat of tantalum over this temperature range is 0.140 J/g·°C.

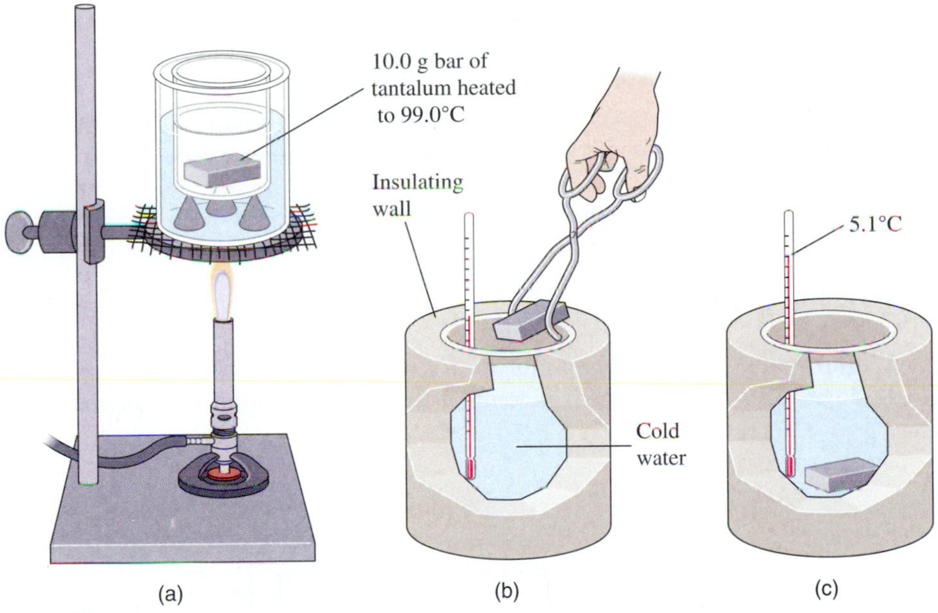

Figure 6.12

(a) A 10.0-g bar of tantalum is heated to 99.0°C. (b) The hot tantalum is dropped into cold water. (c) The final temperature of the tantalum–water mixture is 5.1°C.

SOLUTION

The mass of tantalum is 10.0 g and its temperature change is

$$\Delta T = T_{\text{final}} - T_{\text{initial}} = 5.1°\text{C} - 99.0°\text{C} = -93.9°\text{C}$$

Substitution into Equation 6.11 gives

$$q = \text{mass} \times \text{specific heat} \times \Delta T$$

$$= 10.0\ \text{g} \times \frac{0.140\ \text{J}}{\text{g}\cdot°\text{C}} \times (-93.9°\text{C}) = -131\ \text{J}$$

The negative sign shows that 131 J of heat is lost by the tantalum.

CHEMICAL INSIGHT

THE SPECIFIC HEAT OF WATER

The specific heat of liquid water, 4.2 J/g·°C, is higher than that of most other substances. Metals have especially low specific heats; the specific heat of copper, for example, is only about one-tenth that of water (Table 6.6). The high specific heat of water means that lakes, rivers, and other large bodies of water warm up more slowly than the surrounding land when the weather is hot. They act as heat reservoirs in the sense that they absorb large quantities of heat with only a small temperature rise. Conversely, in cold weather they release large quantities of heat while cooling only slightly. Thus, oceans and lakes tend to stabilize temperatures and moderate the climate.

Water plays a similar role in living organisms. The human body, which contains from 45% to 75% water by mass, can release or absorb substantial amounts of energy with virtually no change in normal body temperature. In effect, the high specific heat of water protects humans and other creatures from sudden and extreme fluctuations in external temperature. Without this protection, life would be quickly frozen or roasted into oblivion.

PRACTICE EXERCISE 6.13

A cup contains 200 g of freshly brewed coffee at 80.0°C. Estimate the temperature of the coffee after it loses 3.0 kJ of heat. (Coffee has the same specific heat as water.)

6.6 CALORIMETRY

A **calorimeter** is an instrument for measuring heats of reaction. There are several different types of calorimeter in common use, but the **bomb calorimeter** (Figure

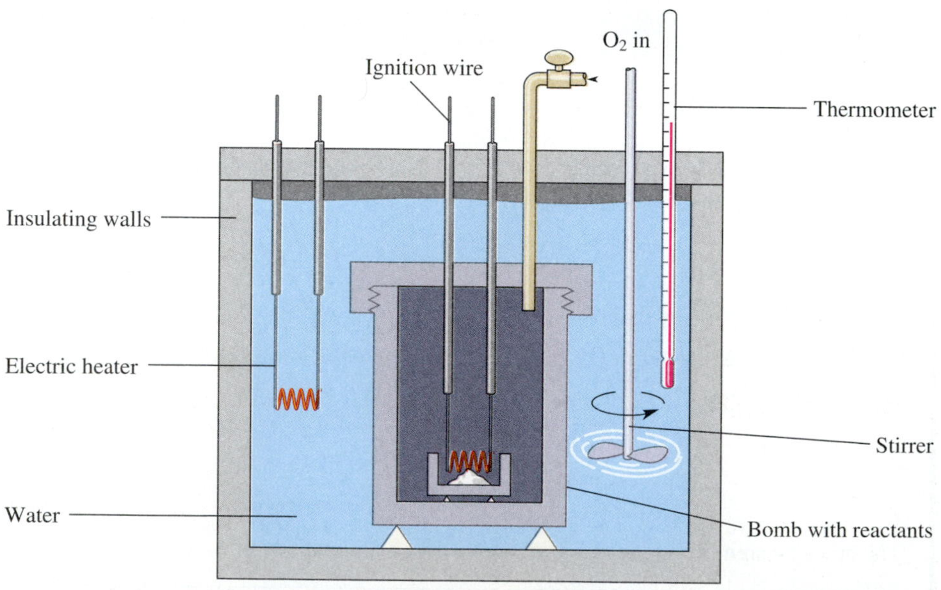

Figure 6.13

A bomb calorimeter is used to obtain heats of combustion at constant volume. The sample is placed in the inner compartment or "bomb," which has an inlet to supply oxygen under pressure and an ignition wire to start the combustion. The rise in temperature of the surrounding water is a measure of the heat released by the reaction. The calorimeter is calibrated with an electric heater that delivers a known amount of energy.

6.13) is the one most widely used for combustions and other reactions involving gases. The bomb calorimeter consists of an explosion-proof steel reaction vessel (called a *bomb*) submerged in water or some other fluid. The carefully weighed reactants are sealed inside the bomb, ignited with a spark, and allowed to react. The heat of reaction produces a temperature change in the surrounding liquid. The amount of heat given off or absorbed during the reaction is found by multiplying the calorimeter's heat capacity (the amount of energy required to raise the temperature of the calorimeter and its contents by 1°C) by the change in temperature. The heat capacity of the calorimeter is determined with great precision by heating the calorimeter electrically; the energy input is the heater wattage (1 watt = 1 joule/second) multiplied by the heating time in seconds. Example 6.11 illustrates the use of a bomb calorimeter in determining the caloric value of a food.

EXAMPLE 6.11

The fluid in a bomb calorimeter was heated for 568 s with a 15.00-watt heater. The temperature of the calorimeter increased by 2.00°C. A 0.455-g sample of sucrose (table sugar; $C_{12}H_{22}O_{11}$) was then burned in excess oxygen in the reaction vessel of the calorimeter. The temperature increased from 24.49°C to 26.25°C. Calculate (a) the calorimeter's heat capacity, (b) the heat of combustion of sucrose in kilojoules per mole, and (c) the number of dietary Calories per gram of sucrose.

SOLUTION

(a) First we will find the calorimeter's heat capacity. A 15.00-watt heater gives off 15.00 J of heat per second. The number of joules given off in 568 s is 568 s × 15.00 J/s = 8520 J. This quantity of heat produces a 2.00°C temperature rise. The heat capacity of the calorimeter is the amount of heat that would produce a temperature rise of 1°C, or 8520 J/2.00°C = 4260 J/°C.

(b) Now we can calculate the heat evolved during the combustion of 0.455 g of sucrose. The combustion causes the temperature of the calorimeter to increase by 26.25°C − 24.49°C = 1.76°C. The amount of heat given off during the combustion is

$$1.76°\text{C} \times \frac{4260\ \text{J}}{1°\text{C}} = 7.50 \times 10^3\ \text{J} = 7.50\ \text{kJ}$$

Thus 7.50 kJ of heat was given off by the combustion of 0.455 g of sucrose. The combustion of 1 mol, 342.3 g, would give off

$$\frac{342.3\ \text{g}}{1\ \text{mol}} \times \frac{7.50\ \text{kJ}}{0.455\ \text{g}} = 5.64 \times 10^3\ \text{kJ/mol}$$

Since combustions are exothermic, the heat of combustion of sucrose is -5.64×10^3 kJ/mol.

(c) One kilocalorie is 4.184 kJ. The heat given off in kilocalories per gram is

$$\frac{7.50\ \text{kJ}}{0.455\ \text{g}} \times \frac{1\ \text{kcal}}{4.184\ \text{kJ}} = 3.94\ \text{kcal/g}$$

A dietary Calorie is equal to 1 kcal; hence, the caloric content of 1 g of sucrose is 3.94 Calories.

PRACTICE EXERCISE 6.14

A 0.555-g sample of mayonnaise was burned in excess oxygen inside a bomb calorimeter with a heat capacity of 5.20 kJ/°C. The temperature rose 2.93°C. Find the number of dietary Calories in a 10.0-g serving of mayonnaise.

Heats of reaction measured in a bomb calorimeter are constant volume heats because the volume of the bomb does not change. Such heats will be enthalpy changes only if the final pressure in the bomb is the same as the initial pressure. The initial and final pressures will be approximately equal if the reaction does not involve gases or, if gases are involved, when the number of moles of gas does not change. For example, let us look at the combustion of sucrose studied in Example 6.11:

$$C_{12}H_{22}O_{11}(s) + 12O_2(g) \rightarrow 12CO_2(g) + 11H_2O(l)$$

The equation shows that 12 mol of carbon dioxide gas is produced for every 12 mol of oxygen gas consumed. (Liquid water is specified in the equation because the initial and final temperatures of the calorimeter and its contents are close to room temperature.) Because the total number of moles of gas does not change, the pressure remains constant, and the constant volume heat of combustion of sucrose determined in Example 6.11 is also an enthalpy of combustion.

When the number of moles of gas increases or decreases, as in the reaction

$$CH_4(g) + 2O_2(g) \rightarrow CO_2(g) + 2H_2O(l)$$

the pressure within the bomb increases or decreases in the same proportion, and the measured heat will not be an enthalpy change. In any event, however, the difference between a constant volume heat of reaction and an enthalpy change is small, usually less than 1%.

PRACTICE EXERCISE 6.15

Will there be a difference between the constant volume heat of reaction and the enthalpy change for the following reactions?
(a) $S(s) + O_2(g) \rightarrow SO_2(g)$
(b) $CaO(s) + H_2O(l) \rightarrow Ca(OH)_2(s)$
(c) $ZnO(s) + H_2(g) \rightarrow Zn(s) + H_2O(l)$

If a calorimeter has a loose-fitting cover or is otherwise open to the atmosphere, its contents will be at constant pressure, and the measured heat will be an enthalpy change. Figure 6.14a shows a calorimeter made from a Styrofoam cup that can be used for measuring heats of reaction in aqueous solution. Such "coffee cup" calorimeters are often used in student laboratories. Styrofoam is an insulating material that absorbs very little heat, so the solution itself will absorb virtually all the heat evolved by an exothermic reaction and will provide all the heat for an endothermic reaction.

Styrofoam does not withstand high temperatures, and it is attacked by most solvents other than water. A more durable container such as a Dewar flask (a thermoslike bottle with a wide mouth) must be used for reactions in hot or nonaqueous

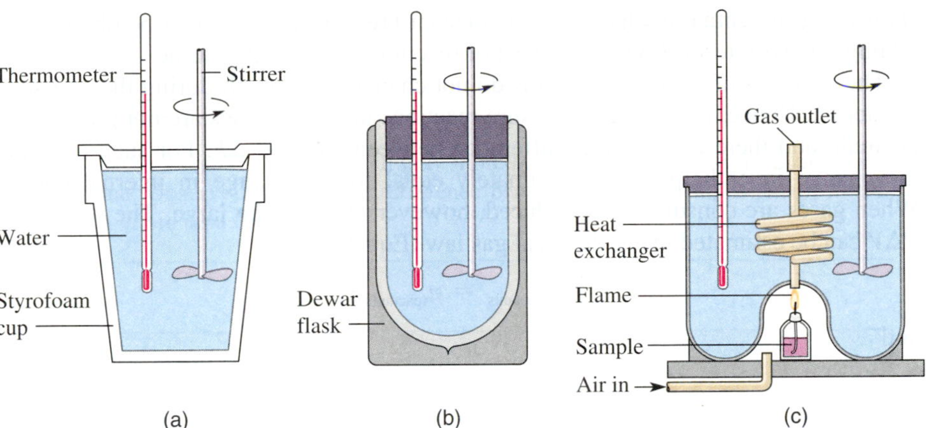

Figure 6.14
Calorimeters for obtaining enthalpy changes at atmospheric pressure. (a) A "coffee cup" calorimeter for reactions in aqueous medium. (b) A Dewar flask for reactions in hot or nonaqueous solutions. (c) A flame calorimeter for combustion reactions. As the hot gaseous products travel upward through the coil, their heat is absorbed by the surrounding water.

solutions (Figure 6.14b). Combustion reactions can be studied in open calorimeters if the hot products transfer their heat to the calorimeter fluid before escaping into the surroundings. Figure 6.14c shows a *flame calorimeter* that is often used for such reactions.

PRACTICE EXERCISE 6.16

The value of $\Delta H°$ is -52.0 kJ for the reaction

$$NH_3(aq) + HCl(aq) \rightarrow NH_4Cl(aq)$$

Estimate the increase in temperature that will occur when 100 mL of 0.50 *M* NH_3 in a Styrofoam calorimeter is neutralized by 300 mL of a solution containing excess HCl.

DIGGING DEEPER

Enthalpy Changes From Bomb Calorimeter Data

When the volume of a reacting system remains constant (as in a bomb calorimeter) and no work is done on or by the system, then $w = 0$, and Equation 6.2 becomes

$$q_V = \Delta E \qquad (6.12)$$

where the subscript V reminds us that q is now the heat of reaction at constant volume. Equation 6.12 shows that *the heat of reaction at constant volume is equal to* ΔE, the change in internal energy of the system.

A constant pressure reaction differs from a constant volume reaction in that the system exchanges PV work with its surroundings (Figure 6.6, page 230). Equation 6.8, which applies to such reactions,

$$\Delta H = \Delta E + P\Delta V \qquad (6.8)$$

KEY TERMS BY SECTION

6.1 First law of thermodynamics
Internal energy
Law of conservation of energy
State function
Surroundings
System
Thermochemistry
Thermodynamic property
Thermodynamics

6.2 Constant pressure reaction
Constant volume reaction
Endothermic reaction
Enthalpy (H)
Enthalpy change (ΔH)
Exothermic reaction
Heat of reaction
Pressure–volume work
PV work
Thermochemical equation

6.3 Fuel value
Standard enthalpy change (ΔH°)
Standard enthalpy of combustion
Standard enthalpy of formation (ΔH°_f)
Standard state

6.4 Hess's law

6.5 Heat capacity
Molar heat capacity
Specific heat

6.6 Bomb calorimeter
Calorimeter

IMPORTANT EQUATIONS

6.1 $\Delta E = q + w$

6.3 $w = -P\Delta V$

6.5 $H = E + PV$

6.8 $\Delta H = \Delta E + P\Delta V$ (for constant pressure reactions)

6.9 $q_P = \Delta H$ (for constant pressure reactions)

6.10 $\Delta H^\circ = \sum[n_p \times (\Delta H^\circ_f)_p] - \sum[n_r \times (\Delta H^\circ_f)_r]$

6.11 $q =$ mass $\times$ specific heat $\times$ ΔT

***6.12** $q_V = \Delta E$ (for constant volume reactions)

***6.13** $P\Delta V = RT\Delta n$

***6.14** $\Delta H = \Delta E + RT\Delta n$

FINAL EXERCISES

Answers to exercises with blue numbers are given in Appendix D. Starred exercises are based on the *Digging Deeper* sections.

PART A. QUESTIONS AND PROBLEMS BY SECTION

The First Law of Thermodynamics (Section 6.1)

6.1 State the first law of thermodynamics in (a) word form and (b) equation form.

6.2 What is meant by the terms *system* and *surroundings*? Make up some examples showing the difference between a system and its surroundings.

6.3 Are there ways in which a system can lose heat to its surroundings and still have an increase in energy? If so, give examples.

6.4 What is meant by the term *work*. Describe three ways in which a system and its surroundings can exchange work.

6.5 When kernels of popcorn are heated, the water inside them vaporizes and causes the kernels to explode. What are the signs of q and w for a sample of popping popcorn?

6.6 A fuel mixture of air and gasoline vapor is confined by a movable piston (see Figure 6.5c). After ignition, the gasoline burns and the piston moves upward. What are the signs of q and w for the burning fuel mixture?

6.7 What is a *state function?* List at least five state functions other than internal energy.

6.8 Why is the internal energy E, but not q or w, considered to be a state function?

6.9 Find q, w, and ΔE for the following systems:

(a) A substance gains 10.6 J of heat and does 5.3 J of work.

(b) A gas loses 25.3 J of heat while doing 12.0 J of work.

(c) A substance has 387.0 J of work done on it while absorbing 52.9 J of heat.

(d) A gas has 10.0 J of work done on it while giving off 187.5 J of heat.

6.10 For each of the following values of q and w, state whether the system (1) gained or lost heat, (2) did work or had work done on it, and (3) gained or lost internal energy.

(a) $q = +4.9$ J, $w = +7.8$ J

(b) $q = -55.8$ J, $w = -17.3$ J

(c) $q = +254$ J, $w = -13$ J
(d) $q = -12.8$ J, $w = +6.4$ J

6.11 A battery does 15 J of electrical work and gives off 2 J of heat. Find *q, w,* and ΔE for (a) the battery and (b) the surroundings.

6.12 A substance is heated by a 500-watt heater for 30.0 s. At the same time, it is compressed by a piston that does 950 J of work on it. Find *q, w,* and ΔE. (*Hint:* 1 watt = 1 J/s.)

Thermochemical Equations; Enthalpy Changes (Sections 6.2 and 6.3)

6.13 What is a *constant pressure reaction*? Give some examples of chemical reactions or other changes that normally occur under constant pressure.

6.14 What is a *constant volume reaction*? Give some examples of chemical reactions or other changes that occur at constant volume.

6.15 (a) What is pressure–volume work?
(b) How would you calculate the PV work done on a system during a constant pressure reaction?
(c) Give some examples of processes in which PV work is done.

6.16 Give three examples of work other than pressure–volume work.

6.17 (a) Under what conditions will the heat of reaction be equal to the enthalpy change?
(b) Give an example of a reacting system in which the heat of reaction is equal to the enthalpy change.

6.18 (a) Under what conditions will the heat of reaction be equal to the internal energy change?
(b) Give an example of a reacting system in which the heat of reaction is equal to the internal energy change.

6.19 Refer to the thermochemical equation on page 238 for the formation of 1 mol of liquid water.
(a) Would you expect the $\Delta H°$ value to be different if water vapor is formed instead of liquid water? Explain your answer.
(b) Explain why it is important to specify the state of each substance in a thermochemical reaction.

6.20 Refer to the thermochemical equation on page 239 for the formation of 1 mol of ozone (O_3) from oxygen.
(a) Would you expect the $\Delta H°$ value to be different if 2 mol of ozone were formed from oxygen? Explain your answer.
(b) Explain why it is important to know the specific form of the reaction equation that is associated with a $\Delta H°$ value.

6.21 Give the sign of $\Delta H°$ for each of the following changes, and state whether the change is exothermic or endothermic.
(a) Liquid bromine is vaporized.
(b) Ice melts in the sun.
(c) Carbon burns in oxygen.

6.22 Give the sign of $\Delta H°$ for each of the following changes, and state whether the change is exothermic or endothermic.
(a) Liquid mercury freezes.
(b) Water decomposes into hydrogen and oxygen.
(c) Rubbing alcohol evaporates.

6.23 Consult Table 6.2 and write a thermochemical equation for the combustion of 1 mol of (a) acetylene and (b) methanol.

6.24 Consult Table 6.2 and write a thermochemical equation for the combustion of 1 mol of (a) benzoic acid and (b) strychnine. (*Note:* N_2 is one of the combustion products of strychnine.)

6.25 Consult Table 6.5 and write a thermochemical equation for the formation of 1 mol of each of the following substances at 25°C:
(a) mercury(I) chloride (c) Hg(g)
(b) calcium hydroxide (d) $SO_3(g)$

6.26 Consult Table 6.5 and write a thermochemical equation for the formation of 1 mol of each of the following substances at 25°C:
(a) ethanol (c) $KClO_3(s)$
(b) ethylene (d) $NO_2(g)$

6.27 Which of the following equations is a standard formation equation?
(a) $Ca^{2+}(aq) + CO_3^{2-}(aq) \rightarrow CaCO_3(s)$
(b) $Ca(s) + C(graphite) + \frac{3}{2}O_2(g) \rightarrow CaCO_3(s)$
(c) $Ca(s) + C(graphite) + O_3(g) \rightarrow CaCO_3(s)$

6.28 Which of the following equations is a standard formation equation?
(a) $O_2(g) + O(g) \rightarrow O_3(g)$
(b) $\frac{3}{2}O_2(g) \rightarrow O_3(g)$
(c) $\frac{2}{3}O_3(g) \rightarrow O_2(g)$

6.29 The thermochemical equation for the combustion of carbon monoxide is

$$CO(g) + \tfrac{1}{2}O_2(g) \rightarrow CO_2(g) \qquad \Delta H° = -283.0 \text{ kJ}$$

Find $\Delta H°$ for the following reactions:
(a) $2CO(g) + O_2(g) \rightarrow 2CO_2(g)$
(b) $CO_2(g) \rightarrow CO(g) + \frac{1}{2}O_2(g)$

6.30 The thermochemical equation for the formation of 1 mol of ammonia at 25°C is

$$\tfrac{1}{2}N_2(g) + \tfrac{3}{2}H_2(g) \rightarrow NH_3(g) \qquad \Delta H° = -46.1 \text{ kJ}$$

Write a thermochemical equation and find the value of $\Delta H°$ for (a) the reaction of 1 mol of nitrogen gas

CHAPTER 7

A rainbow forms when sunlight is refracted (bent) and dispersed into an arc of spectral colors by drops of rain or mist. The analysis of such spectra has provided much information about atoms and molecules.

QUANTUM THEORY AND THE HYDROGEN ATOM

OUTLINE

PREVIEW

Light and matter are interrelated. Crystals of metallic salts tossed into a hearth fire give scarlet, emerald, and turquoise colors to the dancing flame. Various salts are used in fireworks—sodium salts for yellow starbursts, copper for blue, barium for green, and strontium for red. Electrically energized gases produce the distinctive colors of neon signs, fluorescent lamps, and the northern lights. The radiation emitted by excited atoms is different for each element, and it provides clues to their internal secrets. What are these secrets? How can atoms exist? What prevents the negative electrons from falling into the positive nucleus? We will begin to answer these questions by considering the simplest atom, hydrogen, which consists of a solitary electron moving around a proton.

The most stable arrangement of the nucleus and electrons in an atom is the one for which the total energy (kinetic plus potential) is a minimum. When an atom is exposed to heat or light or when it collides with another particle, it may absorb additional energy. An atom with extra energy is said to be *excited;* it tends to give off the energy and return to its normal state. Thus the atoms in hot, glowing objects give off various forms of radiation, including visible light, ultraviolet radiation, and infrared (heat) radiation.

Radiant energy is also emitted by excited atoms in fluorescent tubes such as mercury lamps, sodium lamps, and neon signs. The vapor in these tubes is continuously energized by an electric discharge, and the absorbed energy is given off in the form of light. The study of the absorption and emission of radiation by matter is called **spectroscopy**, and we will soon see that most of the information we have about the atom has been deduced from spectroscopic analysis.

Figure 7.6

The emission spectra of various elements. The wavelength scale is in nanometers (1 nm = 10^{-9} m).

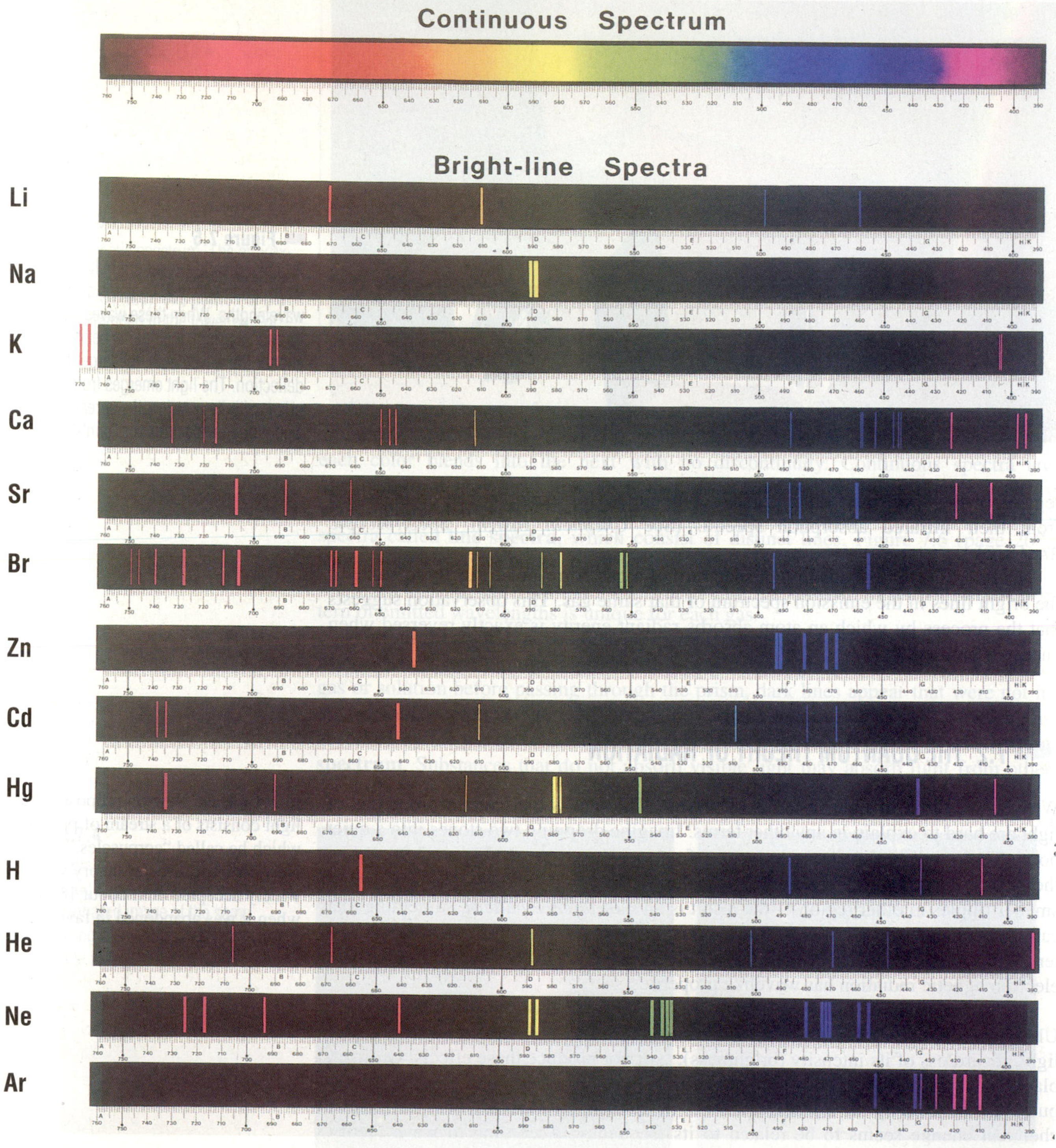

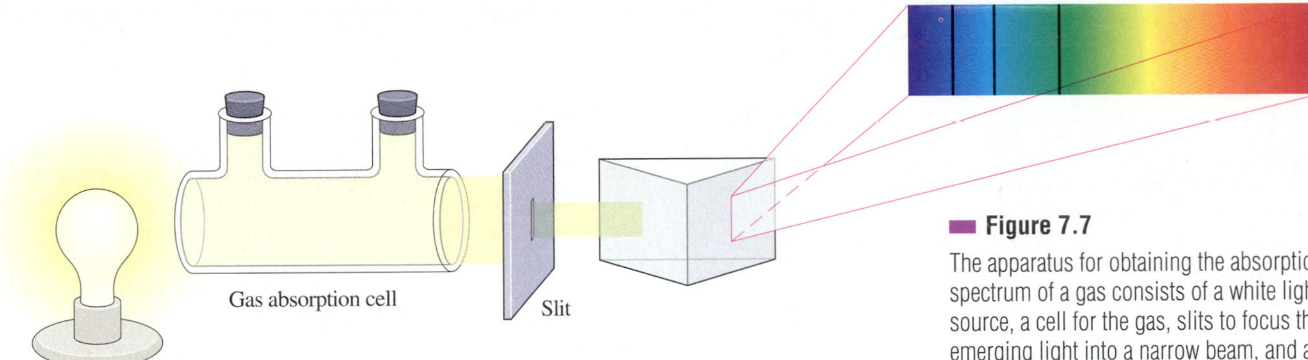

Figure 7.7

The apparatus for obtaining the absorption spectrum of a gas consists of a white light source, a cell for the gas, slits to focus the emerging light into a narrow beam, and a prism (or other device) to spread the wavelengths. The dark lines in the resulting spectrum show which wavelengths were absorbed by the gas.

light, one that cannot be predicted from wave theory.

In 1900 the German physicist Max Planck analyzed the radiation emitted by hot, glowing objects. Such objects give off a continuous spectrum of all wavelengths, and earlier investigators had found that the intensity of this emitted radiation varies with wavelength in a manner that could not be explained in terms of wave theory. To account for these experimental results, Planck proposed that radiant energy is emitted not in the continuous manner characteristic of waves, but rather in tiny bundles or packets that he called **quanta**. His proposal at first met with great skepticism, but five years later Albert Einstein extended Planck's proposition by showing that light consists of particlelike quanta, which were eventually called **photons**. (Einstein's work is described in the next section.) The energy carried by a single photon was found to be proportional to the radiation frequency ν, as expressed in **Planck's law**:

Quantum, *pl.* quanta: a quantity of something

$$E = h\nu \qquad (7.3)$$

where h, **Planck's constant**, is equal to 6.6261×10^{-34} J·s.

(a)

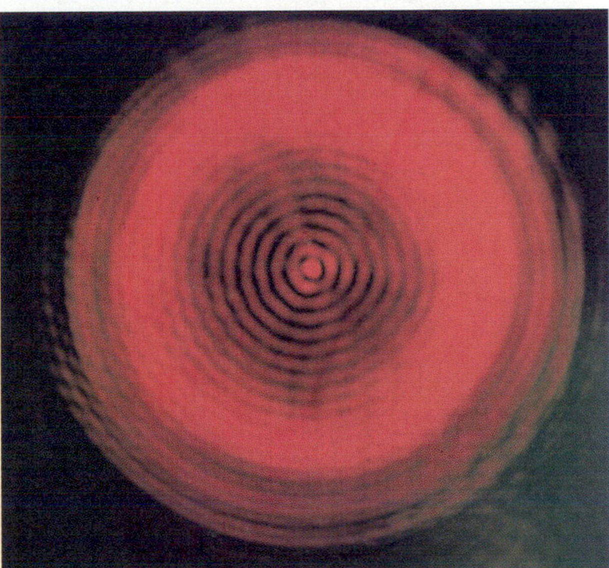

(b)

Figure 7.8

The ability to produce interference patterns is an important characteristic of waves. (a) An interference pattern produced by dropping two stones into water. (b) The interference pattern produced by light emerging from a pinhole is evidence for the wave nature of light.

CHEMICAL INSIGHT

SPECTROSCOPIC ANALYSIS

Robert W. Bunsen, a chemist, and Gustav R. Kirchhoff, a physicist, worked together in the middle of the nineteenth century to perfect the spectroscope. Because each element produces its own unique line spectrum, line spectra can be used as "fingerprints" for the elements. For example, Figure 7.9 compares a portion of the solar spectrum with the spectrum of iron vapor; the wavelengths characteristic of iron are found in the solar spectrum, thus proving that iron exists in the sun's atmosphere. The superiority of this method for identifying elements became apparent when Bunsen and Kirchhoff were able to detect sodium ion in less than 3.3×10^{-7} mg of a sodium salt, a remarkable feat for their time.

In addition to revealing the compositions of the sun and stars, spectroscopic analysis has played a major role in the discovery of new elements. Helium was discovered when the astronomers Pierre Janssen and Joseph Lockyer detected a strange line in a solar spectrum during an eclipse in 1868. (The name helium is derived from the Greek word for the sun, *helios.*) Only later was helium found on earth. Absorption spectra are especially valuable in quantitative analysis because the intensity of the absorbed line (percent of light absorbed) is a measure of the amount of absorbing material. Absorptions in the infrared and ultraviolet regions are also used to provide analytical information. The most useful instrument for this purpose is a *recording spectrophotometer,* which measures the intensity at each point in the spectrum and plots out a graph of absorption versus wavelength; examples are shown in Figure 7.10.

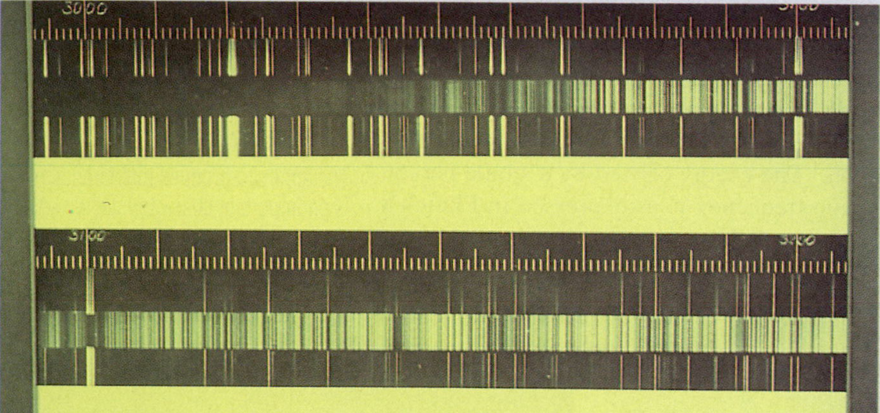

Figure 7.9

Sun and iron spectra from 3000 Å to 3200 Å (1 Å = 10^{-10} m). The dark lines in the solar spectrum (central band on each strip) are due to the absorption of light by various elements in the cooler layers of the sun's atmosphere. Some of these lines correspond exactly to the emission spectrum of iron vapor (upper and lower bands on each strip). The matching lines tell us that iron atoms are present in the sun's atmosphere.

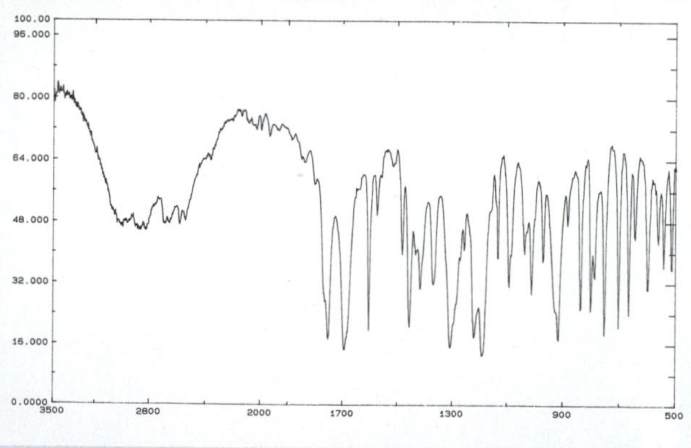

(a)

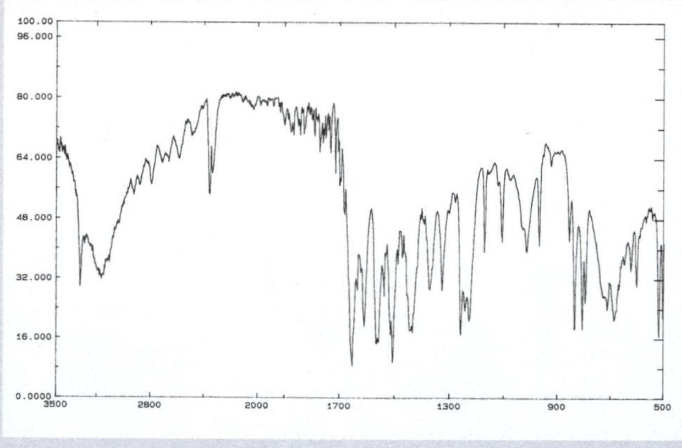

(b)

Figure 7.10

Many organic substances can be identified by their infrared absorption spectra: (a) aspirin and (b) Tylenol. These spectra show the percent of transmitted infrared light (ordinate) versus wave number in cm^{-1} (abscissa). The wave number is the reciprocal of the wavelength.

EXAMPLE 7.2

Calculate the energy in joules of (a) one photon of 580.5-nm yellow light and (b) 1 mol of these photons.

SOLUTION

(a) The frequency of the yellow light is obtained by substituting its wavelength, λ = 580.5 nm = 580.5×10^{-9} m, and the speed of light, $c = 2.9979 \times 10^8$ m/s, into Equation 7.2:

$$\nu = \frac{c}{\lambda} = \frac{2.9979 \times 10^8 \text{ m/s}}{580.5 \times 10^{-9} \text{ m}} = 5.164 \times 10^{14} \text{ s}^{-1}$$

The energy of one photon can now be obtained by substituting $h = 6.6261 \times 10^{-34}$ J·s and the frequency into Equation 7.3:

$$E = h\nu = 6.6261 \times 10^{-34} \text{ J·s} \times 5.164 \times 10^{14} \text{ s}^{-1} = 3.422 \times 10^{-19} \text{ J}$$

(b) The energy of 1 mol (Avogadro's number) of 580.5-nm photons is

$$E = \frac{3.422 \times 10^{-19} \text{ J}}{1 \text{ photon}} \times \frac{6.022 \times 10^{23} \text{ photons}}{1 \text{ mol}}$$

$$= 2.061 \times 10^5 \text{ J/mol}$$

The idea that light consists of particles should not come as a surprise. Most, if not all, things in nature are quantized. An electric current, for example, consists of a flow of electrons. The ocean consists of water molecules. Even a roll of stamps, or a roll of dimes, is quantized in the sense that it consists of individual units.

PRACTICE EXERCISE 7.2

Find the wavelength in nanometers of a photon whose energy is 1.50×10^{-18} J.

DIGGING DEEPER

The Photoelectric Effect

When light strikes the surface of a metal such as rubidium or cesium, it dislodges electrons from atoms at or near the surface (Figure 7.11). These electrons are called *photoelectrons,* and the light-induced emission of electrons from a metal surface is called the **photoelectric effect**. Figure 7.12 shows that the kinetic energy of the photoelectrons increases linearly with the frequency of the incident light and that there is a minimum frequency, ν_0, called the **threshold frequency**, below which photoelectrons will not be emitted.

Albert Einstein was able to explain the photoelectric effect in 1905 by assuming that light consists of photons, each with energy $E = h\nu$. These photons collide with the electrons in atoms near the metal surface, and if the photons have sufficient energy, they will dislodge electrons from the atoms. Photons with frequencies below the threshold frequency do not have the minimum energy W needed to eject electrons. Bombarding the surface with such photons is no more effective than trying to knock down a bowling pin by rolling marbles at it. Photons with energies in excess

In 1921 Albert Einstein was awarded the Nobel Prize for his explanation of the photoelectric effect. At the time of his work in 1905, he was a clerk in the Swiss patent office in Bern.

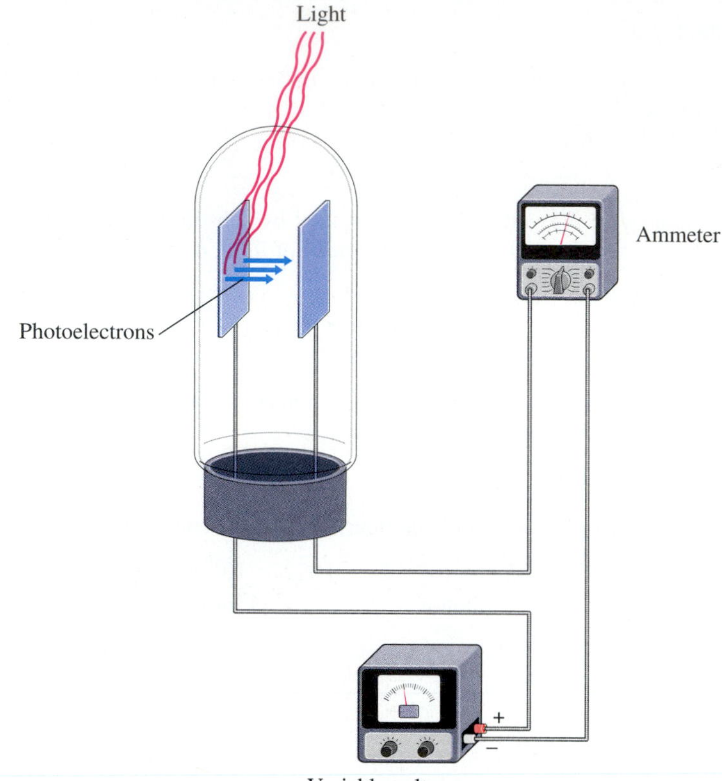

Figure 7.11

The photoelectric effect. The incident light dislodges electrons from the electrode surface on the left. The photoelectrons moving toward the opposite electrode, or grid, constitute a small current that registers on the ammeter. The voltage can be adjusted to make the grid repel the electrons and reduce the current. The kinetic energy of the photoelectrons is calculated from the voltage required to bring the current to zero.

of W not only cause electrons to be emitted, but give them some kinetic energy as well. In other words, some of the photon's energy is used to dislodge the electron, and the rest appears as kinetic energy (KE) of the photoelectron:

$$E_{\text{photon}} = h\nu_{\text{photon}} = KE_{\text{electron}} + W$$

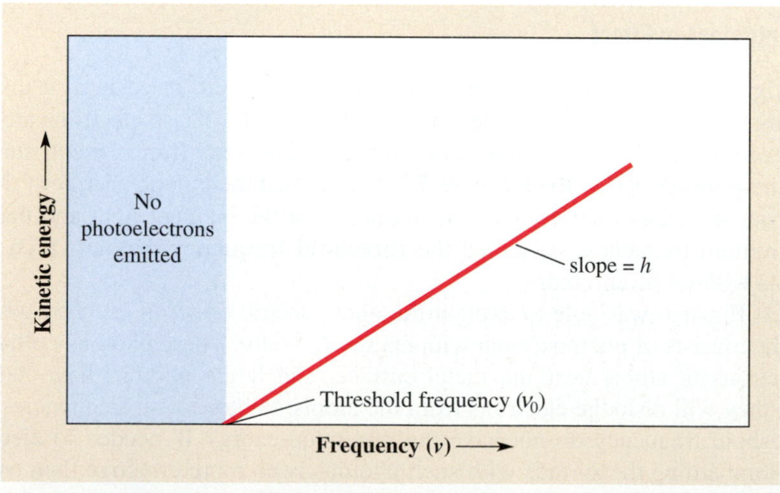

Figure 7.12

Plot of the kinetic energy of photoelectrons versus frequency of incident radiation. No electrons are ejected by radiation below a threshhold frequency ν_0, which varies from metal to metal. Above ν_0, the plot is a straight line whose slope is the same for all metals and equal to Planck's constant h.

Rearrangement gives the kinetic energy of the photoelectron as

$$KE = h\nu - W \tag{7.4}$$

Equation 7.4 predicts that the kinetic energy of a photoelectron will be a linear function of the radiation frequency ν, a prediction that is consistent with the experimental data plotted in Figure 7.12. The equation also predicts that plots of KE versus ν will have the same slope, equal to Planck's constant h, for all metals. This prediction has been verified by experimental data and provides a method for evaluating Planck's constant.

The threshold frequency ν_0 varies with the metal. Figure 7.12 shows that $KE = 0$ when $\nu = \nu_0$. Substituting these values into Equation 7.4 gives

$$0 = h\nu_0 - W$$

or

$$\nu_0 = \frac{W}{h} \tag{7.5}$$

Equation 7.5 shows that the threshold frequency increases with W, the energy required to dislodge an electron from the metal's surface. Because W varies from metal to metal, ν_0 will also vary.

EXAMPLE 7.3

It requires 7.58×10^{-19} J to dislodge an electron from a silver surface. Find the kinetic energy of a photoelectron released from silver by 150-nm ultraviolet light.

SOLUTION

The energy $h\nu$ of a 150-nm photon is found by combining Equations 7.2 and 7.3:

$$E = h\nu = \frac{hc}{\lambda}$$

Substituting $\lambda = 150 \text{ nm} = 150 \times 10^{-9}$ m, $h = 6.626 \times 10^{-34}$ J·s, and $c = 2.998 \times 10^8$ m/s gives the energy of the photon as

$$E = \frac{6.626 \times 10^{-34} \text{ J·s} \times 2.998 \times 10^8 \text{ m/s}}{150 \times 10^{-9} \text{ m}} = 1.32 \times 10^{-18} \text{ J}$$

The energy required to dislodge the electron is $W = 7.58 \times 10^{-19}$ J. The kinetic energy of the photoelectron is obtained by substituting the energy of the photon and W into Equation 7.4:

$$KE = h\nu - W = (1.32 \times 10^{-18} \text{ J}) - (7.58 \times 10^{-19} \text{ J}) = 5.6 \times 10^{-19} \text{ J}$$

PRACTICE EXERCISE 7.3

The kinetic energy of photoelectrons released from cesium with 500-nm light is 1.08×10^{-19} J. Find the threshold frequency for cesium.

7.3 THE SPECTRUM OF HYDROGEN

The visible emission spectrum of hot or "excited" hydrogen atoms (see Figure 7.13) was one of the first spectra to be studied in detail. In 1885 the mathematician Johann Balmer perceived an order in the spacing of the spectral lines and worked out an empirical formula for calculating their wavelengths. The set of visible hydrogen lines is called the **Balmer series** in his honor. Another series of lines—the **Lyman series**—was discovered in the ultraviolet region of the spectrum, and three additional series were discovered in the infrared region (Figure 7.14). Each series is named after its discoverer.

What is the significance of this spectrum? Recall that excited atoms give off energy when they return to their normal states. Each emitted photon has a frequency, and it is the wavelengths corresponding to these frequencies that we see and photograph in the emission spectrum. The appearance of some spectral lines and not others shows that excited atoms give off only certain amounts of energy (certain values of $h\nu$) during their transition from high-energy to low-energy states. The observed spectral lines can be explained by assuming that only certain energy states are possible for atoms, and that the emitted photons have energies corresponding to differences between these states. Hence, the hydrogen spectrum gave theoreticians something to work toward—a model of the atom in which energies would be limited to those that are consistent with the spectrum.

Niels Bohr (1885–1962).

The Bohr Atom

The hydrogen atom is the simplest atom; it consists of one electron moving around a positive nucleus. The emission spectrum of hydrogen remained an unexplained curiosity until 1913, when the Danish physicist Niels Bohr suggested that the electron could travel around the nucleus in one of a number of orbits similar to planetary orbits around the sun (Figure 7.15). He also assumed that when an electron absorbs energy, it jumps into a new, larger orbit farther away from the nucleus. Each orbit

Figure 7.13

The visible emission spectrum of hydrogen consists of four lines in the Balmer series.

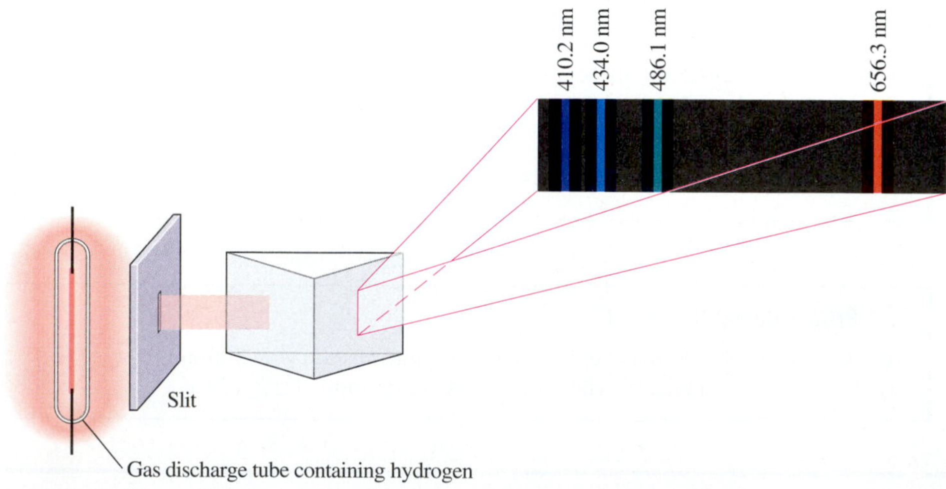

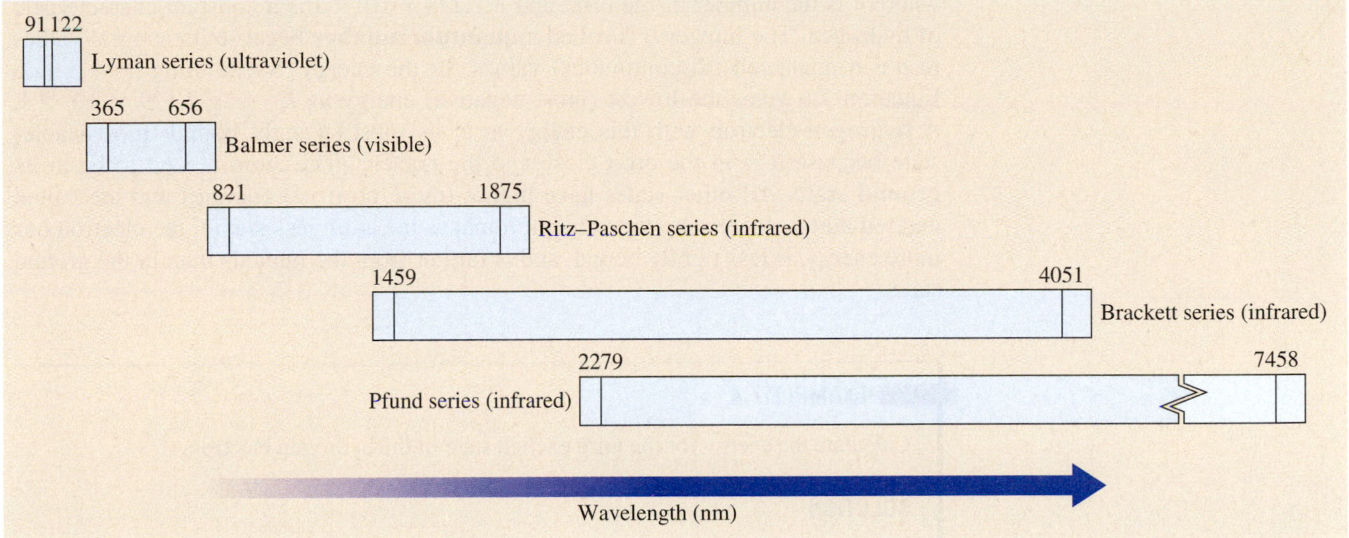

Figure 7.14

The complete emission spectrum of hydrogen contains many series of lines. The shortest and longest wavelengths of the first five series are shown here. The Lyman series, which lies entirely in the ultraviolet between 91 and 122 nm, contains the most energetic radiation.

was assigned an integral number $n = 1, 2, 3, \ldots$, where $n = 1$ represents the orbit closest to the nucleus. Using this model, Bohr was able to show that the energy of the hydrogen electron could take on the values

$$E_n = \frac{-2.179 \times 10^{-18}\ \text{J}}{n^2} \qquad (7.6)$$

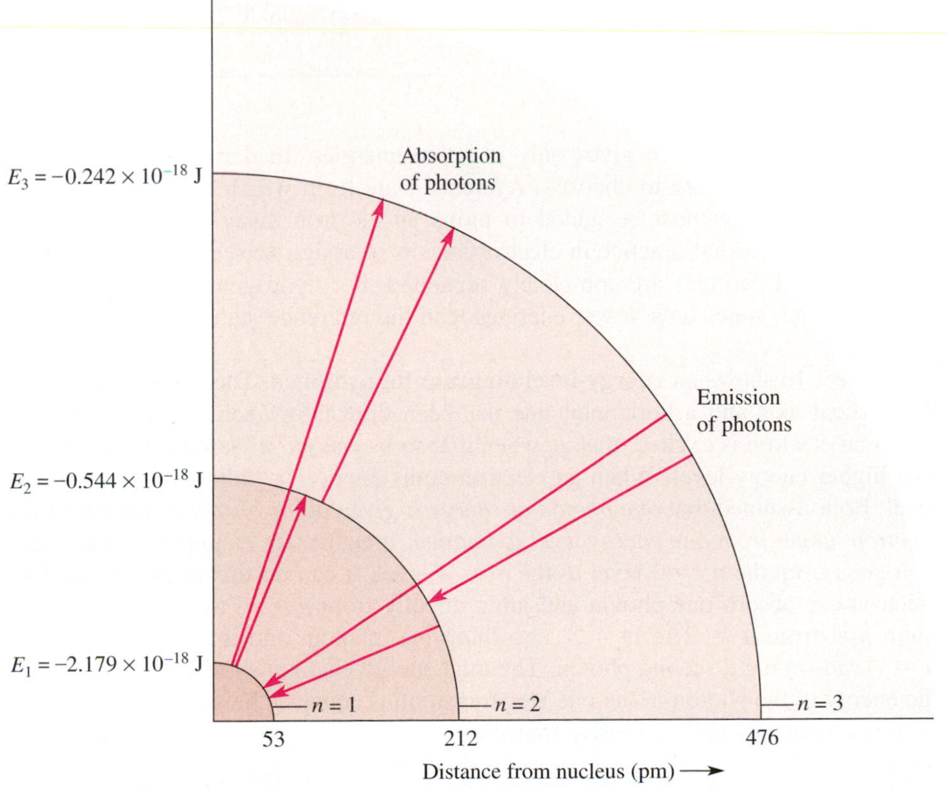

Figure 7.15

According to Bohr's model, the hydrogen electron is restricted to certain definite orbits, portions of which are shown in this diagram; the smaller the radius of the orbit, the lower the energy of the electron. An electron can move from one orbit to another by absorbing or emitting a photon with an energy equal to the energy difference between the orbits.

where n is the number of the orbit and -2.179×10^{-18} J is a constant characteristic of hydrogen. The integer n is called a **quantum number** because its integral values result in quantized (discontinuous) values for the energy. Substituting $n = 1$ into Equation 7.6 gives the lowest (most negative) energy as $E_1 = -2.179 \times 10^{-18}$ J. A hydrogen electron with this energy is in its most strongly bound (most stable) state because it is in the orbit closest to the nucleus. The atom is said to be in its **ground state**. All other states have higher (more positive) energies and are called **excited states**. Increasing the value of n makes the atom less stable; the electron has more energy, is less tightly bound, and is farther from the nucleus than in the ground state.

EXAMPLE 7.4

Calculate the energy for the third excited state of the hydrogen electron.

SOLUTION

The ground state corresponds to $n = 1$. The first, second, and third excited states have $n = 2$, 3, and 4, respectively. Substituting $n = 4$ into Equation 7.6 gives

$$E_4 = \frac{-2.179 \times 10^{-18}\ \text{J}}{4^2} = -1.362 \times 10^{-19}\ \text{J}$$

PRACTICE EXERCISE 7.4

In which state does the hydrogen electron have an energy equal to -8.716×10^{-20} J?

Note that Equation 7.6 gives only negative energies. In deriving his equation, Bohr found it necessary to choose a reference state from which to measure energy differences. Energy must be added to move an electron away from the positive nucleus, and the usual practice in electrostatics is to assign zero energy to a state in which charged particles are completely separated. For hydrogen this is the $n = \infty$ state. All other states have lower energies than the reference state and are therefore negative.

Figure 7.16 shows an **energy-level diagram** for hydrogen. The energy scale is on the vertical axis and a horizontal line has been drawn for each allowable energy. When an electron is excited, that is, when it absorbs energy, it "jumps" from a lower to a higher energy level. When an electron emits energy, it "falls back" to a lower level. Bohr assumed that *one photon of energy is given off or absorbed each time an electron moves from one energy level to another.* Imagine, for example, that an electron goes from the $n = 2$ level to the $n = 4$ level. It can do this in two ways. The electron can absorb one photon and jump directly from $n = 2$ to $n = 4$, or it can jump first from $n = 2$ to $n = 3$, absorbing one photon, and then from $n = 3$ to $n = 4$, absorbing a second photon. The total energy change is the same either way; the energy of the photon in the one-step transition is equal to the sum of the energies of the two photons in the two-step transition.

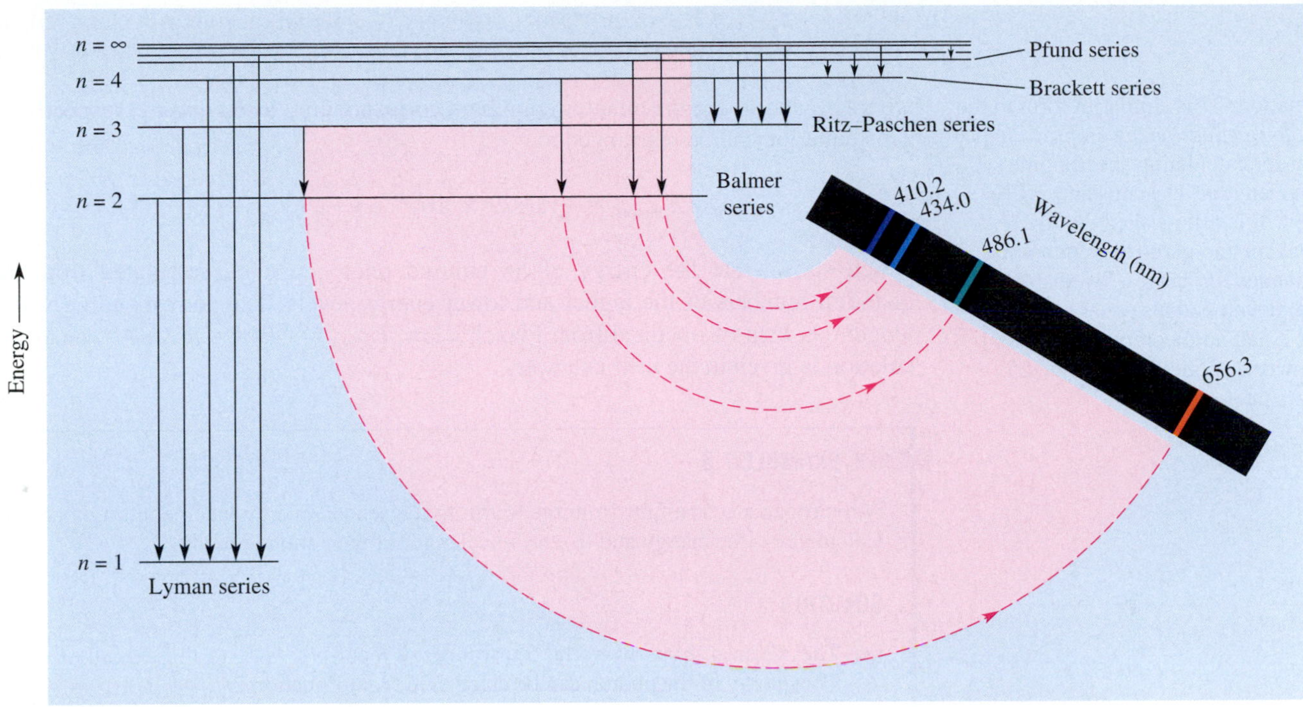

Figure 7.16

An energy-level diagram. The horizontal lines marked $n = 1, 2, \ldots, \infty$ represent the energy levels of the hydrogen electron. The vertical arrows represent possible electron transitions from higher to lower energy levels. Each transition evolves a photon and is thus the source of a spectral line; the longer the arrow, the greater the energy of the photon, and the shorter its wavelength. The curved arrows identify the transitions that give rise to the Balmer (visible) series.

EXAMPLE 7.5

Is a photon emitted or absorbed when a hydrogen electron goes from the fifth to the third level?

SOLUTION

The electron falls from the $n = 5$ level to the $n = 3$ level. The energy of the electron decreases, therefore the photon is emitted.

PRACTICE EXERCISE 7.5

Are photons emitted or absorbed during the following transitions?
(a) $n = 1 \rightarrow n = 4$ and (b) $n = 3 \rightarrow n = 2$

Bohr's real triumph, for which he was awarded the Nobel Prize in 1922, was that his model allowed him to calculate the wavelengths in the emission spectrum of hydrogen. This spectrum is caused by excited electrons falling from high energy levels to low energy levels and giving off photons in the process. The energy of each emitted photon is

$$E_{\text{photon}} = E_{\text{H}} - E_{\text{L}}$$

where E_{H} is the energy of the electron in the higher energy level and E_{L} is the energy in the lower level. Substituting Equation 7.6 into the preceding equation gives

$$E_{\text{photon}} = \frac{-2.179 \times 10^{-18}\ \text{J}}{n_{\text{H}}^2} - \frac{-2.179 \times 10^{-18}\ \text{J}}{n_{\text{L}}^2}$$

Equation 7.7 is similar in form to the *Rydberg equation*, an empirical equation for calculating spectral line wavelengths. The constant, 2.179×10^{-18} J, is often called the *Rydberg constant* and given the symbol R_{H}. Johannes Rydberg, a Swedish physicist, developed his equation in 1889, but it had no theoretical basis until the work of Niels Bohr some 25 years later.

where n_{H} and n_{L} are the quantum numbers corresponding to E_{H} and E_{L}, respectively. This equation can be rearranged to

$$E_{\text{photon}} = -2.179 \times 10^{-18}\ \text{J}\left(\frac{1}{n_{\text{H}}^2} - \frac{1}{n_{\text{L}}^2}\right) \tag{7.7}$$

which shows that the energy of an emitted photon can be calculated from the quantum numbers of the higher and lower energy levels. The corresponding wavelength can then be obtained from Planck's law, $E_{\text{photon}} = h\nu = hc/\lambda$. A typical calculation is given in the next example.

EXAMPLE 7.6

A hydrogen electron falls from the fourth to the second energy level, emitting a photon. Calculate (a) the energy and (b) the wavelength of the emitted photon.

SOLUTION

(a) The electron drops from the fourth energy level, $n_{\text{H}} = 4$, to the second, $n_{\text{L}} = 2$. The energy of the photon can be calculated from Equation 7.7:

$$E_{\text{photon}} = -2.179 \times 10^{-18}\ \text{J}\left(\frac{1}{4^2} - \frac{1}{2^2}\right) = 4.086 \times 10^{-19}\ \text{J}$$

Note that the energy calculated from Equation 7.7 is positive.

(b) The wavelength of the photon is obtained from Planck's law, $E_{\text{photon}} = hc/\lambda$, which can be rearranged to

$$\lambda = \frac{hc}{E_{\text{photon}}}$$

Substituting $E_{\text{photon}} = 4.086 \times 10^{-19}$ J, $h = 6.626 \times 10^{-34}$ J·s, and $c = 2.998 \times 10^8$ m/s gives

$$\lambda = \frac{6.626 \times 10^{-34}\ \text{J·s} \times 2.998 \times 10^8\ \text{m/s}}{4.086 \times 10^{-19}\ \text{J}}$$

$$= 4.862 \times 10^{-7}\ \text{m} = 486.2\ \text{nm}$$

Figure 7.16 shows that this corresponds to the second or blue line in the Balmer series for hydrogen.

PRACTICE EXERCISE 7.6

A hydrogen electron in the $n = 1$ energy level absorbs a 97.25-nm photon and thus makes a transition to a higher energy level. What is its energy after this transition, and which level does it occupy?

Figure 7.16 relates the hydrogen spectral lines to the electron energy levels. The arrows indicate transitions from one level to another. Each energy change is roughly

TABLE 7.1 Comparison of Calculated and Experimental Balmer Lines

Transition	Calculated Wavelength (nm)	Experimental Wavelength (nm)
$3 \rightarrow 2$	656.4	656.3
$4 \rightarrow 2$	486.2	486.1
$5 \rightarrow 2$	434.1	434.0
$6 \rightarrow 2$	410.2	410.2
$7 \rightarrow 2$	397.1	397.0
$8 \rightarrow 2$	389.0	388.9
$9 \rightarrow 2$	383.6	383.5

proportional to the length of its arrow and corresponds to a wavelength that can be calculated from Equation 7.7 and Planck's law. Transitions from $n = 2, 3, 4, \ldots$ to $n = 1$ give rise to the ultraviolet Lyman series; transitions from $n = 3, 4, 5, \ldots$ to $n = 2$ give rise to the visible Balmer series, and so forth. We know this because the calculated wavelengths of these transitions agree very closely with the experimental wavelengths. Table 7.1 compares the experimental Balmer lines with those calculated from Equation 7.7. Such precise agreement between theory and observation is remarkable.

PRACTICE EXERCISE 7.7

The Balmer lines are caused by electron transitions from higher levels to the $n = 2$ level. Calculate the wavelength in nanometers of the first Balmer line.

Limitations of the Bohr Theory

You may have noticed that we did not explain how Bohr arrived at Equation 7.6 for the hydrogen energy levels. The derivation of this formula is interesting in its own right, but it is no longer considered significant because later discoveries cannot be explained in terms of Bohr's original view of the atom. Bohr visualized a compact, well-defined electron tracing a definite path or orbit around the nucleus, much like a planet moving around the sun. The radii of Bohr's orbits were given by $r = 52.9$ pm $\times\ n^2$ (1 pm $= 10^{-12}$ m). Hence, according to Bohr, an electron can come no closer to the nucleus than 52.9 pm, a distance commonly called the **first Bohr radius** and symbolized by a_0.

This planetary model explained the hydrogen energy levels fairly well, and Bohr's concept of energy states has withstood the test of time. The electron's path, however, has turned out to be more elusive, and the concept of well-defined orbits has now been abandoned. Moreover, the Bohr model was never successfully extended to account for the observed spectra of molecules, or even atoms with more than one electron, and it finally had to give way to more powerful theories. It is only fair to point out, however, that systems consisting of three or more interacting particles (e.g., a nucleus surrounded by two electrons) are always represented by equations that have to be solved by approximation methods; that is, no solution will be absolutely perfect. This difficulty is common to all atomic theories.

7.4 MATTER WAVES

The discovery that light has both a particle nature and a wave nature created a chain of events that completely altered the chemist's way of thinking about atoms and molecules. In 1924 Louis de Broglie, a French prince, posed an interesting question: If radiation exhibits particle characteristics, should not matter exhibit wave characteristics? His belief in the symmetry of nature convinced him that particles must have both wavelike and particlelike properties. De Broglie pursued this line of reasoning in his doctoral thesis at Paris University, and was able to show on theoretical grounds that moving particles should have an associated wavelength λ given by

$$\lambda = \frac{h}{m\text{v}} \qquad (7.8)$$

where m and v are the particle's mass and speed, respectively, and h is Planck's constant. The waves associated with moving particles are called **matter waves**, and wavelengths calculated from Equation 7.8 are called **de Broglie wavelengths**.

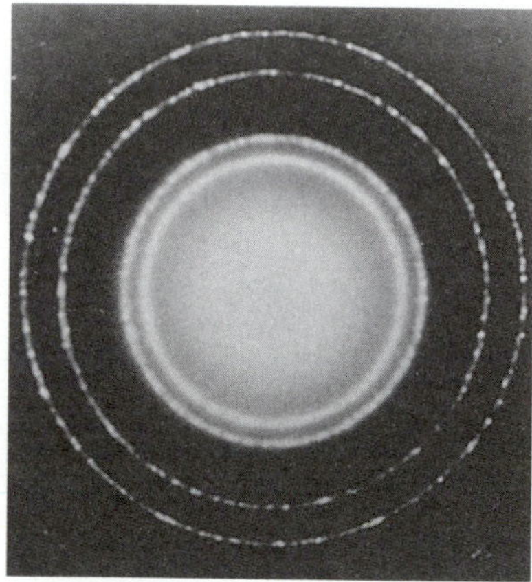

Figure 7.17

An interference pattern produced by electrons passing through aluminum foil. In effect, the foil is the equivalent of a mesh with atomic dimensions, so very short waves emerging from it would exhibit interference. This interference pattern shows that electrons can behave like waves (see Figure 7.8).

Perhaps even more astonishing than de Broglie's theoretical prediction of matter waves was their subsequent discovery by George P. Thomson in 1927. (George Thomson was the son of J. J. Thomson, who measured the charge-to-mass ratio of the electron.) The younger Thomson detected electron waves by passing streams of accelerated electrons through thin metal foils and then onto photographic plates where they formed interference patterns similar to those produced by electromagnetic radiation. A typical pattern is shown in Figure 7.17. The ability to form interference patterns is a well-known property of waves (see Figure 7.8), and Thomson's results could only mean that electrons behave like waves of light when they pass between adjacent layers of metal atoms. Similar patterns were soon obtained by beaming neutrons and protons through various crystals. In the same year that Thomson made his discovery, Clinton Davisson and Lester Germer, working at Bell Telephone Laboratories, calculated the electron wavelength from the pattern formed when electrons were reflected from a nickel surface. Their results confirmed the validity of Equation 7.8.

EXAMPLE 7.7

Calculate the de Broglie wavelength of an electron moving at a speed of 100 miles per hour (44.7 m/s).

SOLUTION

The electron mass is 9.11×10^{-31} kg. Substituting the mass, the speed, and Planck's constant, $h = 6.626 \times 10^{-34}$ J·s, into the de Broglie equation gives

$$\lambda = \frac{h}{m\text{v}} = \frac{6.626 \times 10^{-34}\ \text{J·s}}{9.11 \times 10^{-31}\ \text{kg} \times 44.7\ \text{m/s}} = 1.63 \times 10^{-5}\ \text{m}$$

The units simplify to meters because 1 J = 1 N·m = 1 kg·m^2/s^2. Note that the wave length of an electron moving at this speed is much greater than the radius of an atom (about 10^{-10} m); it is closer to the radius of a red blood cell (see Table 1.4).

PRACTICE EXERCISE 7.8

Calculate the de Broglie wavelength of a baseball moving at 100 miles per hour, and compare it with that of the electron given above. The mass of a baseball is about 140 g.

Your answer to Practice Exercise 7.8 should show that de Broglie wavelengths are extremely small for massive everyday objects such as baseballs and automobiles. Wavelengths this short cannot be detected and they do not affect our perception of ordinary objects. Subatomic particles, on the other hand, have very small masses, and their wavelengths may be very large in comparison to the size of the particles. Example 7.7 shows that an electron moving at 100 miles per hour has a wavelength of about 10^{-5} m, some 100,000 times larger than an atomic radius. We cannot ignore such waves when we try to predict the behavior of atoms and molecules.

De Broglie and his colleagues were in sharp disagreement over the interpretation of his discovery. De Broglie considered an electron wave to be as real a property of the electron as its mass, while Einstein and others rejected this view. Max Born, a German theoretical physicist at the University of Göttingen, proposed that electron waves were related to the probability of finding the electron at a particular point in space. We will come back to this interpretation in a later section.

7.5 THE UNCERTAINTY PRINCIPLE

Most of us, at one time or another, make rough visual estimates of the speed and position of moving objects. We do this almost instinctively when we play fast-moving ball games and video games. Most instruments for measuring position and speed, including the human eye, depend on electromagnetic radiation that is emitted or reflected by the moving object. For example, the speed detectors used by traffic police analyze radar waves reflected from moving vehicles, cameras use visible or infrared radiation to "freeze" the position of an object on film, and special telescopes use radio waves to locate distant galaxies.

In 1926 Werner Heisenberg, a German physicist, realized that the very act of measuring a particle's position could actually alter both its position and momentum. (The **momentum** of a particle is the product of its mass, m, and speed, v; momentum $= m \times$ v.) For example, if we measure the position of a particle with electromagnetic radiation, then we must bombard the particle with photons. The impact of these photons will cause the particle's position and speed to change, and this effect can be appreciable if the particle is very small. The situation can be compared to bombarding an object with ping-pong balls. If the object is large and heavy, like a car or a baseball, the lighter balls will have virtually no effect on its position or speed. However, if the object is small and light, like a feather or another ping-pong ball, the bombardment will change both its position and speed. Photons bouncing off planets, off the walls of a room, or off the pages of this textbook, have

CHEMICAL INSIGHT

HOW TO PHOTOGRAPH AN ATOM

Atoms and molecules cannot be photographed with visible light because their dimensions are smaller than visible wavelengths. The wave engulfs the particle and all detail is lost. The wavelengths of x-rays and gamma rays are short enough, but they cannot be focused to give clear pictures. Electron waves, however, offer another possibility. Equation 7.8 suggests that the wavelength of an electron beam can be adjusted by controlling its speed; greater speed produces a smaller wavelength. Furthermore, electron beams can be focused with electric and magnetic fields to produce clear, sharp pictures similar to those on a television screen. Present-day *electron microscopes* routinely use the short wavelengths associated with very fast electrons to reveal details of biological and chemical structures that are invisible under ordinary optical microscopes. Many molecules and atoms have now been photographed. The electron micrograph in Figure 7.18, for example, shows an individual gold atom.

A more recent technique for imaging both surface atoms and the electron bonds that hold them in place makes use of an instrument called the *scanning tunneling microscope* (STM). This device uses a needlelike probe whose tip may be as narrow as a single atom. A current of electrons flows between the probe and the surface to be studied. As the probe passes over the surface atoms and bonds, the current tends to change. These changes are sensed by a computer, and the vertical position of the probe is adjusted to maintain a constant current, that is, a constant tip-to-surface distance. The probe thus moves up and down as it scans the surface, and a computer-generated plot of this motion produces images such as the one shown in Figure 7.19.

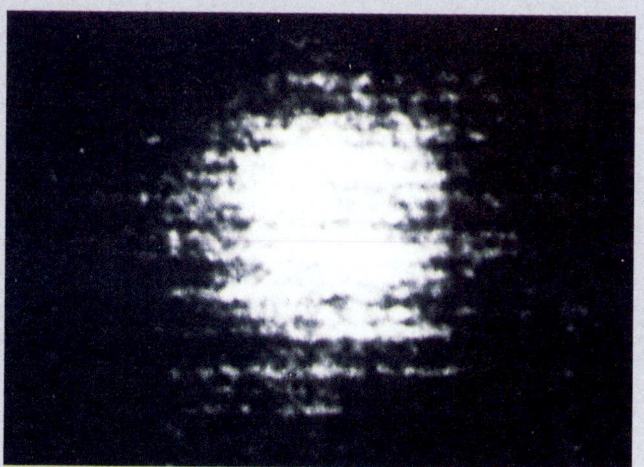

Figure 7.18 Electron micrograph of a single gold atom.

Figure 7.19 Scanning tunneling micrograph of the surface of a silicon crystal.

no noticeable effect, but photons bouncing off an electron will drastically alter the electron's course (Figure 7.22). In fact, we have already seen how photon bombardment can cause the hydrogen electron to jump from one energy level to another.

The **Heisenberg uncertainty principle** states that *it is impossible to make simultaneous and exact measurements of both the position and momentum of a particle.* Thus there will always be some uncertainty in these quantities. Heisenberg was also able to show that the uncertainties in position and momentum are related; the more precisely the position is known, the greater the uncertainty in the momentum, and

The tip of an STM can be adapted to pick up and move individual atoms from one site on a crystal surface to another (Figure 7.20). In 1990 IBM scientists used this technique to spell their corporate logo "IBM" by positioning 35 xenon atoms on a nickel surface. The ability to manipulate individual atoms and molecules ushers in the era of *molecular nanotechnology* that may someday allow people to build nanometer-size electronic components and other devices. (The length of four xenon atoms is about 1 nm.)

Closely related to the STM is the *atomic force microscope* (AFM; See Figure 7.21). The AFM is a mechanical device containing a sharp tip (stylus) attached to a spring. The spring moves up and down as the tip scans the atoms on the sample surface. The motion of the spring is measured by a sensing device, and a display system maps out the atomic landscape. The AFM has been used to image ions on a sodium chloride crystal surface, individual carbon atoms in graphite, and a number of other atomic structures.

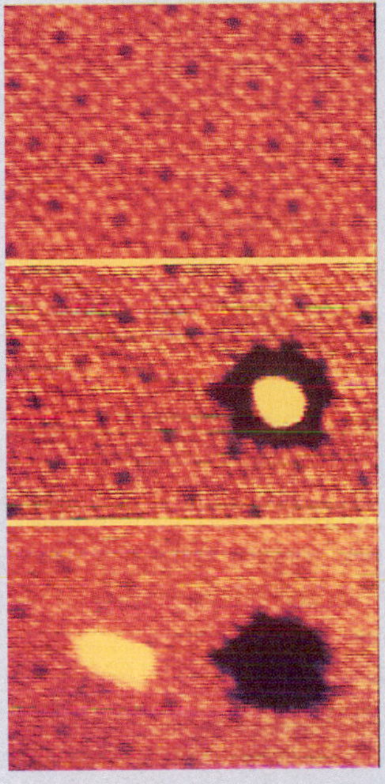

Figure 7.20

Silicone atoms are moved with the tip of a scanning tunneling microscope. Top image: Atoms on a silicon surface. Middle image: A mound of silicon atoms that were in the now-empty black region. Lower image: The mound has been moved from one region of the silicon surfact to another.

Figure 7.21

An atomic force microscope.

vice versa. It is important to understand that the uncertainties do not arise from poor or defective measuring instruments—they are inherent in the measuring process itself and cannot be eliminated by improvements in technique. The importance of the uncertainty principle is not only that it puts a limit on the precision of our measurements, but that it also describes an inexactness in what can be known about the subatomic universe. This inexactness means that subatomic particles are not entirely subject to the same physical laws or even to the "common sense" logic of cause and effect that seems to rule the everyday world.

To come to terms with uncertainty and to incorporate it into our concept of the

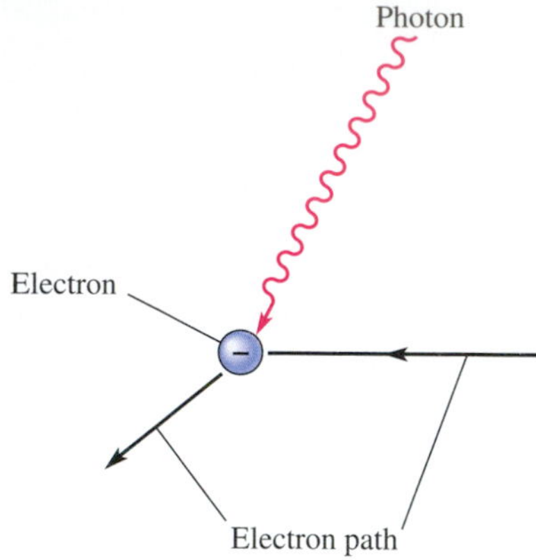

Figure 7.22
The collision of a photon with an electron changes the path of the electron.

atom, we must first accept the fact that we can never exactly measure the location or the speed of an electron. Electrons are not like cars or spaceships or planets whose paths can be plotted. Without exact position and momentum values, we cannot predict a precise path or orbit for an electron or for any other subatomic particle. Hence, there is no way of experimentally verifying Bohr's assumption that the electron moves in a well-defined path around the nucleus. It is for this reason that the Bohr model could not be developed in greater detail and was ultimately abandoned in favor of a less precise model. If our knowledge of the atom cannot be based on certainty, then what is left? What knowledge can we hope for?

The answer is knowledge based on probabilities, the same sort of knowledge with which we conduct most of our daily affairs. Farmers, for example, follow schedules based on the most probable temperature and rainfall patterns for their section of the country. Automobile insurance companies base their annual rates on the average number of accidents that occur each year. Probabilities will not predict single events such as one rainstorm or one accident, but farmers and insurance companies have no other choice. In the long run, they will be successful only if they act on the basis of the most probable conditions. In the remaining sections of this chapter, we will find that, even though we cannot predict the exact position of an electron in a hydrogen atom, we can predict its most probable location and behavior.

7.6 WAVE MECHANICS

The discovery of matter waves and the uncertainties predicted by Heisenberg called for an entirely new approach to the behavior of very small particles. Planck, Einstein, and Bohr had already shown that matter absorbs and emits energy in the form of tiny photons or quanta, and their work became part of a new branch of science called **quantum mechanics**, which applies the concept of quantization to the structure and behavior of atoms and molecules.

Quantum mechanics entered a new phase in 1926 when Erwin Schrödinger (Figure 7.23), a Viennese physicist working at the University of Zurich, used the concept of matter waves to develop a theory called **wave mechanics**. Wave mechanics is a version of quantum mechanics that recognizes, as does Bohr's older and simpler version, that only certain states are available to an electron in an atom or molecule. Each state is described by a mathematical expression called a **wave function**. Wave functions are more complicated than the expressions in Bohr's theory because they deal with the wave nature of the electron, and with the fact that a wave can persist only if it is in a state where it does not destructively interfere with itself (see Figure 7.24). Wave functions are also consistent with the Heisenberg uncertainty principle; in place of the precise radii of the Bohr orbits, wave functions allow only the calculation of probabilities, such as the most probable distance of the electron from the nucleus, and the probability of finding the electron in some specified region of space. In the remainder of this chapter, we consider wave functions for the hydrogen atom.

Figure 7.23
Erwin Schrödinger (1887–1961) formulated a wave equation that has become one of the chemist's most powerful theoretical tools. The equation was developed by adapting equations that describe ordinary waves to small particle behavior. When Schrödinger's equation is applied to a system of an electron plus a proton, its solutions are the orbitals of the hydrogen atom.

Electron Orbitals and Quantum Numbers

Wave functions are also called **orbitals**, a term that should not be confused with the word *orbit:* an orbital is a mathematical expression, an orbit is a curved path.

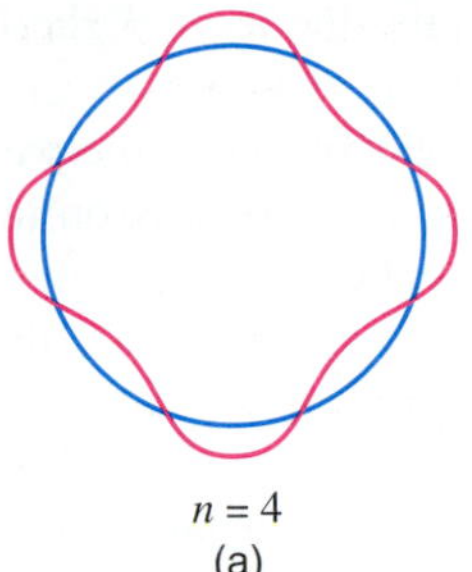

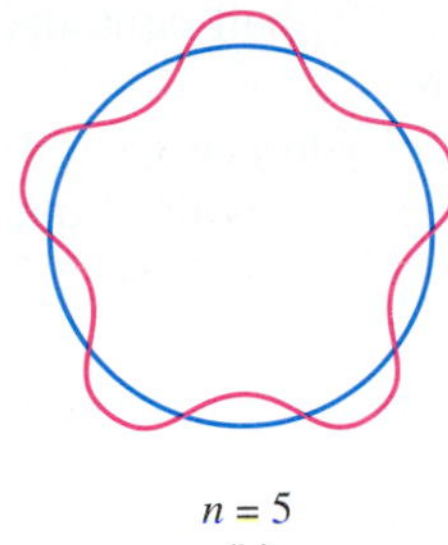

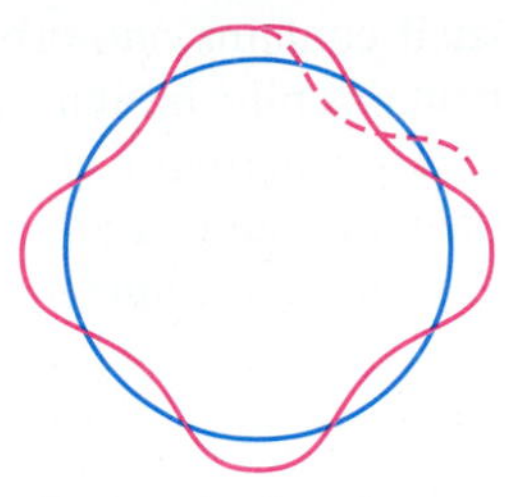

Figure 7.24
Various waves confined to one plane and moving in a given circular path. Waves (a) and (b) will persist because in each case an integral number, *n,* of wavelengths fit around the circle. The resulting wave, which does not destructively interfere with itself, is called a *standing wave.* The wave in (c) interferes with itself destructively and thus cannot last. Although three-dimensional waves like those of the hydrogen electron are much more complicated than the waves shown here, the same principle holds: Allowed states of the electron, also called *stationary states,* are those in which the wave will not destructively interfere with itself.

Because each state is associated with a unique wave function, the terms *electron state* and *orbital* are often used interchangeably.

In Bohr's model of hydrogen, each state (or orbit) was identified by a single quantum number, *n*. Wave mechanics is more complex—*each state (or orbital) is identified by the following quantum numbers: the principal quantum number, n; the azimuthal quantum number, l; and the magnetic quantum number,* m_l. Table 7.2 summarizes the quantum numbers for 30 different orbitals of hydrogen.

The **principal quantum number** *n* is identical to the quantum number discovered by Bohr, and it can have any integral value starting with $n = 1$. Inspection of Table 7.2 shows that more than one orbital can have the same value of *n*. For example, the table shows four orbitals with $n = 2$. One of these orbitals has a zero value for both *l* and m_l. The other three have an *l* value of 1, and m_l values of 1, 0, and −1. Table 7.2 also includes 9 orbitals for which $n = 3$, and 16 orbitals for which $n = 4$. A set of orbitals with the same principal quantum number is called a **shell**, and *the total number of orbitals in a shell is always equal to* n^2. The $n = 1$

TABLE 7.2 Values for the Quantum Numbers *n, l,* and m_l from $n = 1$ through $n = 4$[a]

n	*l*	Orbital Designation	m_l	Number of Orbitals
1	0	1*s*	0	1
2	0	2*s*	0	1
	1	2*p*	1, 0, −1	3
3	0	3*s*	0	1
	1	3*p*	1, 0, −1	3
	2	3*d*	2, 1, 0, −1, −2	5
4	0	4*s*	0	1
	1	4*p*	1, 0, −1	3
	2	4*d*	2, 1, 0, −1, −2	5
	3	4*f*	3, 2, 1, 0, −1, −2, −3	7

[a]NOTE:
1. For a given *n*, *l* can take on the values 0, 1, 2, 3, . . . , $(n - 1)$.
2. For a given *l*, m_l can take on the values 0, ±1, ±2, ±3, . . . , ±*l*.

shell contains one orbital, the $n = 2$ shell contains four orbitals, the $n = 3$ shell contains nine orbitals, and so forth.

The energy of an electron in a hydrogen orbital is determined by the principal quantum number and the Bohr energy formula (Equation 7.6). Hence, an electron in any one of the nine $n = 3$ orbitals listed in Table 7.2 would have the same energy:

$$E_3 = \frac{-2.179 \times 10^{-18}\ \text{J}}{3^2} = -2.421 \times 10^{-19}\ \text{J}$$

It should not come as a surprise that wave mechanics gives the same energies for the hydrogen atom as does the Bohr theory. These energies agree with experiment, so any successful theoretical model of the hydrogen atom must predict their values.

The **azimuthal quantum number** l determines the shape of the orbital. (Orbital shapes are discussed in Section 7.7.) The azimuthal quantum number is a secondary quantum number whose allowed values depend on the value assigned to n. Table 7.2 shows that for a given value of n, l has integral values ranging from 0 to $(n - 1)$. For example, an $n = 1$ orbital has only one value of the azimuthal quantum number, $l = 0$. For $n = 2$, there are orbitals with $l = 0$ and with $l = 1$. The orbitals in the $n = 3$ shell have l equal to 0, 1, or 2.

Historically, the letters *s*, *p*, *d*, and *f* were derived from the names of various spectral series arising from transitions between these orbitals: *s* from *sharp*, *p* from *principal*, *d* from *diffuse*, and *f* from *fundamental*.

An orbital is usually described by giving the n value followed by a letter symbol for l. The symbols used for l are

l:	0	1	2	3	4	5	
symbol:	*s*	*p*	*d*	*f*	*g*	*h*	(alphabetical from here on)

For example, the $n = 1$, $l = 0$ orbital is called 1*s* and the $n = 3$, $l = 1$ orbital is called 3*p*. Other examples are given in Table 7.2. Within a shell, all orbitals with a given l value are referred to as a **subshell**. Table 7.2 shows that there are three orbitals in any *p* subshell, five orbitals in any *d* subshell, and seven orbitals in any *f* subshell.

The **magnetic quantum number** m_l determines the orientation of the orbital in space and also the energy of its electron in a magnetic field. (Orbital orientations will be discussed more fully in Section 7.7.) The values for this number depend on the value of l. For a given l value, the magnetic quantum number may have positive or negative integral values ranging from $-l$ to $+l$, including 0. The magnetic quantum number also determines the number of orbitals in a subshell. The ***s* orbitals** (those for which $l = 0$) all have $m_l = 0$; no other m_l value is possible, so each *s* subshell contains only one orbital. The ***p* orbitals** come in subshells of three because when $l = 1$, there are three possible m_l values: $m_l = -1$, $m_l = 0$, and $m_l = 1$. Inspection of Table 7.2 shows that *the number of orbitals in a subshell is* $2l + 1$, *where* l *is the azimuthal quantum number.*

EXAMPLE 7.8

How many 4*d* orbitals are possible for a hydrogen atom?

SOLUTION

For a *d* subshell, $l = 2$ and $2l + 1 = 5$. Every *d* subshell will contain five orbitals regardless of its n value. These orbitals correspond to the five possible m_l values: 2, 1, 0, −1, −2.

PRACTICE EXERCISE 7.9

How many orbitals are in the 5*f* subshell?

EXAMPLE 7.9

Is there a 3*f* orbital?

SOLUTION

If n is 3, the allowed l values are 0, 1, and 2. Hence, the $n = 3$ shell contains only s ($l = 0$), p ($l = 1$), and d ($l = 2$) orbitals. There are no 3*f* orbitals.

PRACTICE EXERCISE 7.10

Is there (a) a 1*p* orbital? (b) a 2*p* orbital?

The Spinning Electron

Experiments have shown that the line spectrum of an element changes when its atoms are placed in a magnetic field (see *Digging Deeper: Evidence for Electron Spin*). In 1925 George Uhlenbeck and Samuel Goudsmit, two young graduate students at the University of Leiden in Holland, explained the effect of magnetic fields on line spectra by proposing that an electron behaves like a tiny magnet with its own north and south pole (see Figure 7.25). Moving charges are known to generate magnetic fields, and Uhlenbeck and Goudsmit accounted for the electron's magnetism by assuming that the electron spins on its axis as it moves about the nucleus in much the same way that the earth spins on its axis as it revolves around the sun. Experiments also show that the magnetic field of a spinning electron can be oriented in an external magnetic field in only two ways, as illustrated in Figure 7.26—there are no intermediate possibilities. Thus electron spin is quantized and an electron will be in one of two **spin states**. Each spin state is described by a quantum number called the **spin quantum number** m_s; its possible values are $+\frac{1}{2}$ and $-\frac{1}{2}$. The half-integral values arise from the mathematics used to describe the electron's magnetism, but the important point to remember is that these values designate equal but opposite spin effects.

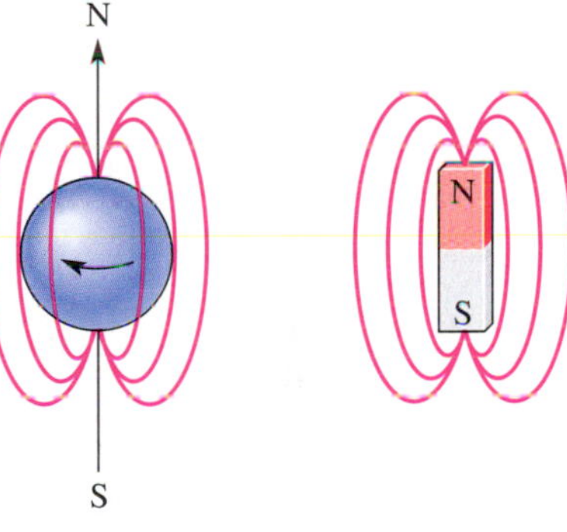

Figure 7.25

A "spinning" electron behaves like a tiny magnet with a north and south pole. (The magnetic field is represented by the curved lines.) The direction of the spin determines the direction of the magnetic field; reversing the spin reverses the field.

The model of an electron that spins on its axis like a child's top is an oversimplification that does not take into account the wave nature of the electron. We use the term *spin* as a convenience, although we know that the actual origin of the electron's magnetism is more complicated. A theoretical explanation for the electron's magnetism was given in 1928 by the English physicist and mathematician Paul Dirac, who modified Schrödinger's theory to take into account Einstein's theory of relativity. Dirac and Schrödinger were awarded a Nobel Prize in 1933 for their achievements.

Electron spin and the existence of two spin states means that four quantum numbers are required to completely specify each state of the hydrogen atom. Three of these quantum numbers, n, l, and m_l, identify the orbital governing the electron's

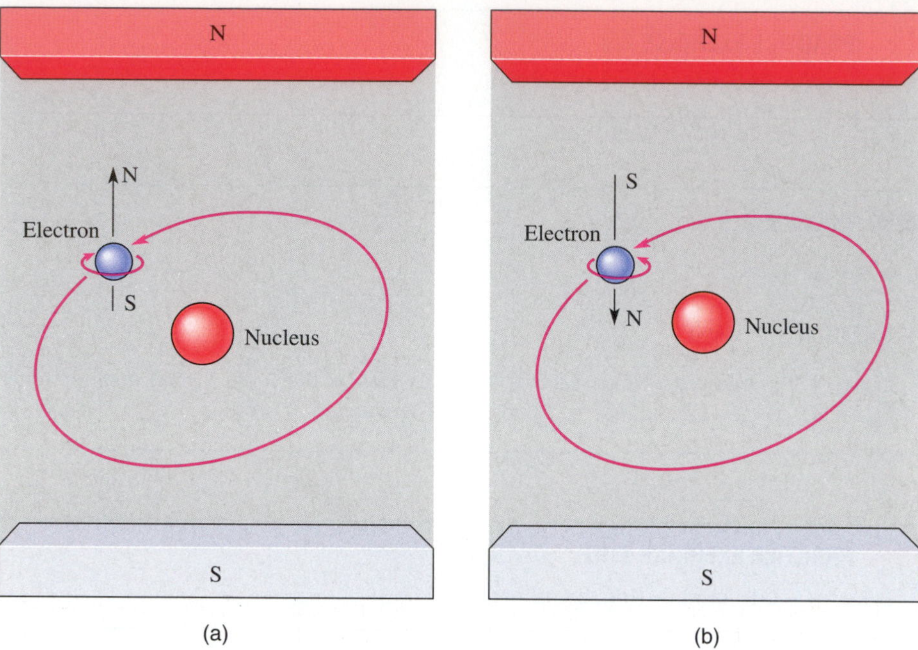

Figure 7.26
The magnetic field of a spinning electron can be oriented in an external magnetic field in only two ways: (a) fields aligned (north pointing to north) or (b) fields opposed (north pointing to south). The electron has lower energy when its field opposes the external field as in (b) than when the fields are aligned as in (a).

motion. The fourth, m_s, describes the spin state of the electron in the orbital. The importance of electron spin will become apparent when we deal with many-electron atoms in the next chapter.

EXAMPLE 7.10

How many different states are possible for the hydrogen electron if it occupies an orbital in the $n = 2$ shell?

SOLUTION

For $n = 2$ there are four orbitals (one 2*s* orbital and three 2*p* orbitals). Two spin states are possible for each of these orbitals, so the total number of states for the electron is $4 \times 2 = 8$. These states are listed in Table 7.3.

PRACTICE EXERCISE 7.11

How many states are possible for a 3*d* electron?

DIGGING DEEPER

EVIDENCE FOR ELECTRON SPIN

The existence of two spin orientations for the electron is consistent with experimental work done by the German physicists Otto Stern and Walter Gerlach in 1922.

TABLE 7.3 Electron States and Their Quantum Numbers for the $n = 1$ and $n = 2$ Shells

n	l	Orbital Designation	m_l	m_s	Number of States
1	0	1*s*	0	$+\frac{1}{2}, -\frac{1}{2}$	2
2	0	2*s*	0	$+\frac{1}{2}, -\frac{1}{2}$	2
	1	2*p*	−1	$+\frac{1}{2}, -\frac{1}{2}$	2
	1	2*p*	0	$+\frac{1}{2}, -\frac{1}{2}$	2
	1	2*p*	+1	$+\frac{1}{2}, -\frac{1}{2}$	2

LEARNING HINT

The total number of orbitals in a shell is n^2; the total number of states, including spin, is $2n^2$.

Because each state is uniquely described by three orbital quantum numbers and a spin quantum number, it is sometimes referred to as a *spin-orbital.*

They observed that a beam of silver ions passing through a nonuniform magnetic field splits into two beams (Figure 7.27). (The same effect was later observed with a beam of hydrogen atoms.) The atomic number of silver is 47, and a silver atom resembles a hydrogen atom in that it possesses a single electron moving around a positive core; in this case, the core consists of a nucleus and 46 inner electrons. The net magnetic field of the core is zero, so the magnetic field of the atom is simply that of its lone outer electron. In effect, each atom behaves like a tiny magnet that will be deflected by the external magnetic field. The extent of the deflection, that is, the degree to which the path changes, depends on the orientation of the electron's magnetic field with respect to the external field. If all orientations are possible, then a beam of many atoms should spread out into a wide band during passage through the field. If only two orientations are possible (e.g., those shown in Figure 7.26), then the beam should split into two narrow beams that bend in opposite directions. The experimental results obtained by Stern and Gerlach show clearly that only two ori-

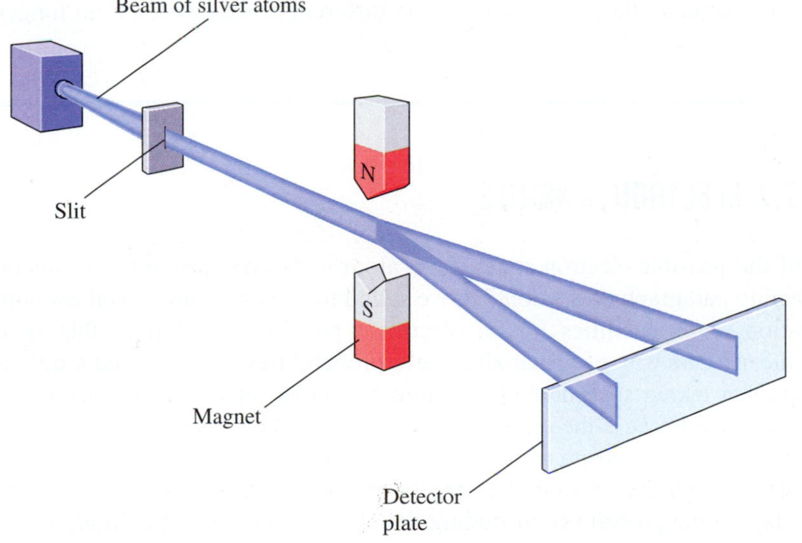

Figure 7.27

The Stern–Gerlach experiment. When a silver atom travels through a nonuniform magnetic field generated by differently shaped poles, its path will bend in a direction that depends on the spin orientation of its one unpaired electron. The splitting of the beam into two equally intense beams that are deflected in opposite directions shows that there are two, and only two, orientations for the electron spin.

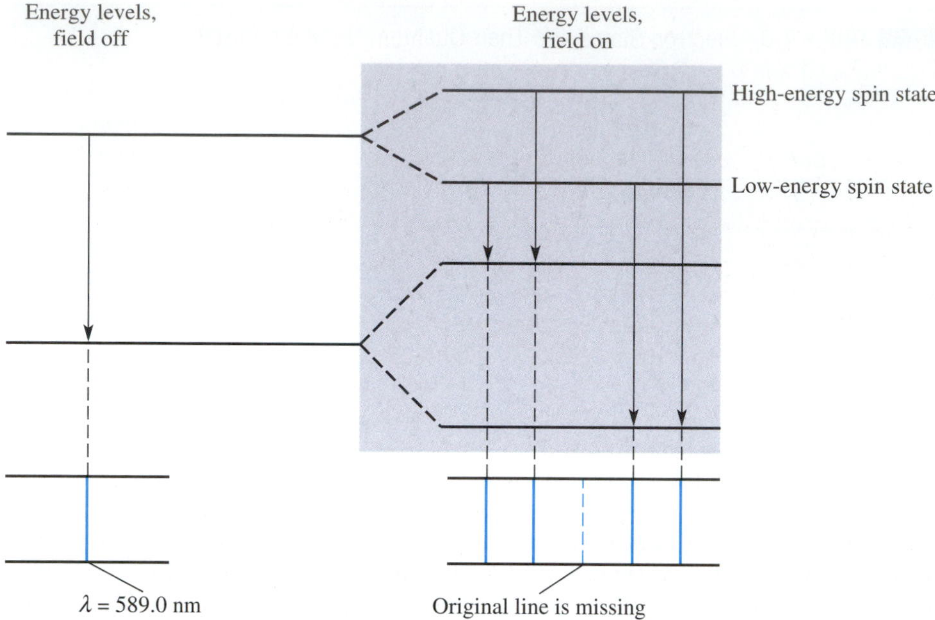

Figure 7.28

The splitting of spectral lines in a magnetic field is known as the Zeeman effect. In the upper left are the energy levels of two orbitals in the absence of an external magnetic field. The arrow represents a transition from the higher level to the lower; underneath is the single spectral line that would result from the transition. In the upper right are each of the original levels split into two by a magnetic field. The arrows show four possible transitions; the spectrum underneath shows that the original single line has become four lines, two with higher wavelengths and two with lower wavelengths than the original line.

entations are possible, thus confirming the existence of two and only two electron spin states.

The electron's magnetism is also indicated by spectral line effects. For example, when spectra are taken in a magnetic field, each of the spectral lines is broken up into a set of three or more lines (Figure 7.28). This phenomenon was first observed by the Dutch physicist Pieter Zeeman in 1896, and it is now called the **Zeeman effect**. Uhlenbeck and Goudsmit (see p. 287) explained this multiplicity of lines by proposing that when the external magnetic field is turned on, the magnetic field of the electron can take one or the other of the orientations shown in Figure 7.26. The spin state in Figure 7.26a has a slightly higher energy than the spin state in Figure 7.26b, and transitions from the different spin states account for the Zeeman effect. When the magnetic field is turned off, the different orientations are no longer possible, and the Zeeman line splitting vanishes.

7.7 ELECTRON DENSITIES

Each of the possible electron states in an atom is described by a wave function. In the opening paragraph of Section 7.6, we stated that a wave function allows only the calculation of probabilities for an electron's position. What does this statement really mean? One way of visualizing these probabilities is to imagine a tiny atomic photographer taking snapshots of the atom at regular intervals. The electron shows up as a dot, but because the electron moves rapidly, the dot is in a different place in each snapshot. A composite of many snapshots for the 1*s* orbital of hydrogen would resemble Figure 7.29a. A high dot density indicates a region that the electron visits frequently, so the probability of finding the electron there is high. In regions where dots are sparse, the probability of finding the electron is low. The crowded dots near the center (i.e., near the nucleus) in Figure 7.29a indicate that the electron visits each small region near the nucleus more often than it visits similar-size regions farther out.

The cloud-of-dots cross section is one way to picture an orbital; in fact, the cloud metaphor is often used for the electron itself. The fast-moving electron covers all locations in such a short time that we refer to it as an **electron cloud**.

Now we will relate the electron cloud to the wave function. A wave function has a definite numerical value at each point in space; the symbol for this value is the lowercase Greek letter ψ (psi, pronounced "sigh"). In 1926 Max Born proposed that *the probability of finding an electron in a small volume of space centered about some point is proportional to the value of* ψ^2 *at that point.* According to this proposal, the dot density varies directly with ψ^2. A high dot density corresponds to a large value of ψ^2; a low dot density corresponds to a small value. A graph of ψ^2 versus the distance r from the nucleus is given in Figure 7.29b for the 1*s* orbital of hydrogen. Comparison of this figure with Figure 7.29a shows that both ψ^2 and the dot density decrease with increasing r. Because the square of the wave function varies directly with the density of the electron cloud, ψ^2 is called the **electron probability density** or simply the **electron density**.

Figure 7.29a shows that in the 1*s* orbital the variation of electron density is the same in all directions; the density depends only on the distance from the nucleus and is a maximum at the nucleus. Such an orbital is *spherically symmetrical:* Points with the same electron density lie on the surface of a sphere. All *s* orbitals are spherically symmetrical, but their electron density plots are not all alike. Figure 7.30 shows that the 2*s* and 3*s* plots, unlike the 1*s* plot, have radial distances at which the electron density is zero. A point or surface at which the electron density is zero is called a **node**. The electron density of a 2*s* orbital is high at the nucleus, decreases to zero at the nodal radius, builds up beyond that, and then decreases to zero again. A 3*s* orbital has two nodal radii. Electron densities for *p* orbitals are graphed in Figure 7.31. These orbitals are not spherically symmetrical; each one has a **nodal plane** (a plane of zero electron density) containing the nucleus.

Orbitals can also be represented by **balloon pictures** like those shown in Figure

(a)

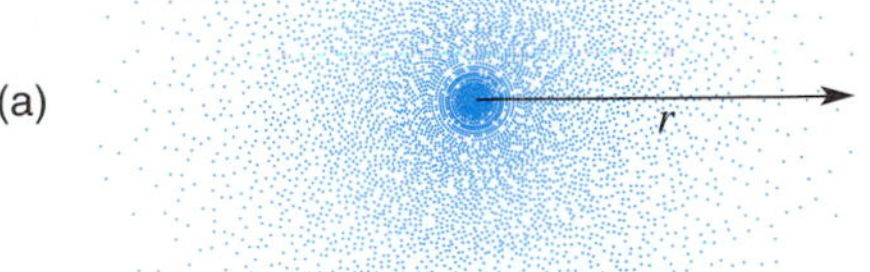

(b)

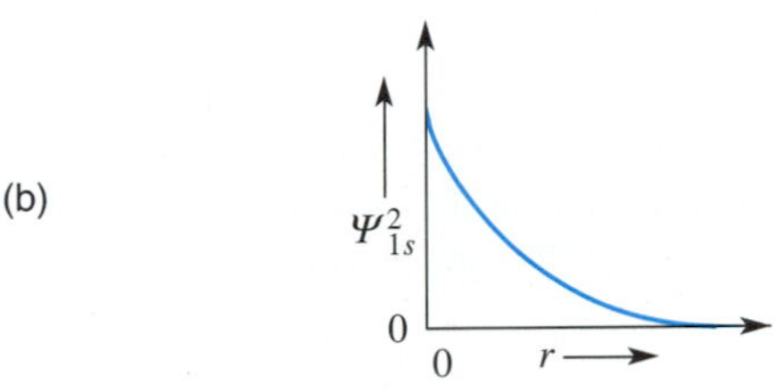

Figure 7.29

Electron probability density in the 1*s* orbital. (a) A cross section of the hydrogen atom in which electron density is represented by dot density. (b) A plot of electron probability density (ψ^2) versus distance (r) from the nucleus. As r increases, the density decreases, approaching zero but never quite reaching it.

The square of the wave function for the 1*s* orbital of hydrogen is

$$\psi^2_{1s} = \frac{1}{\pi}\left(\frac{1}{a_0}\right)^3 e^{-2r/a_0}$$

where r is the radial distance from the nucleus, a_0 is the first Bohr radius (p. 279), and e is the base of natural logarithms.

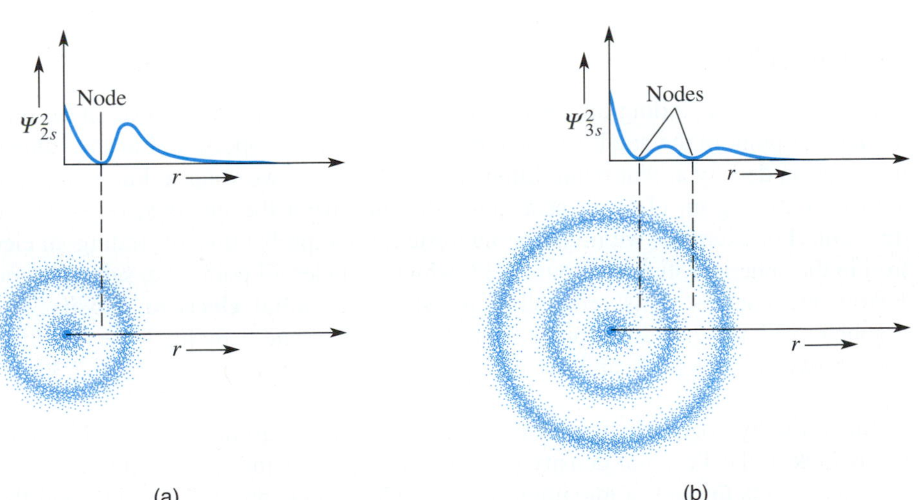

Figure 7.30

Electron density in the 2*s* and 3*s* orbitals. (a) The cross-sectional dot representation for the 2*s* orbital shows two spherical regions of high electron density separated by a spherical shell (nodal shell) of zero density. A plot of electron density versus r (upper figure) shows two peaks with the node in between. (b) The 3*s* orbital has two nodal spheres of zero density separating three spherical regions of high density.

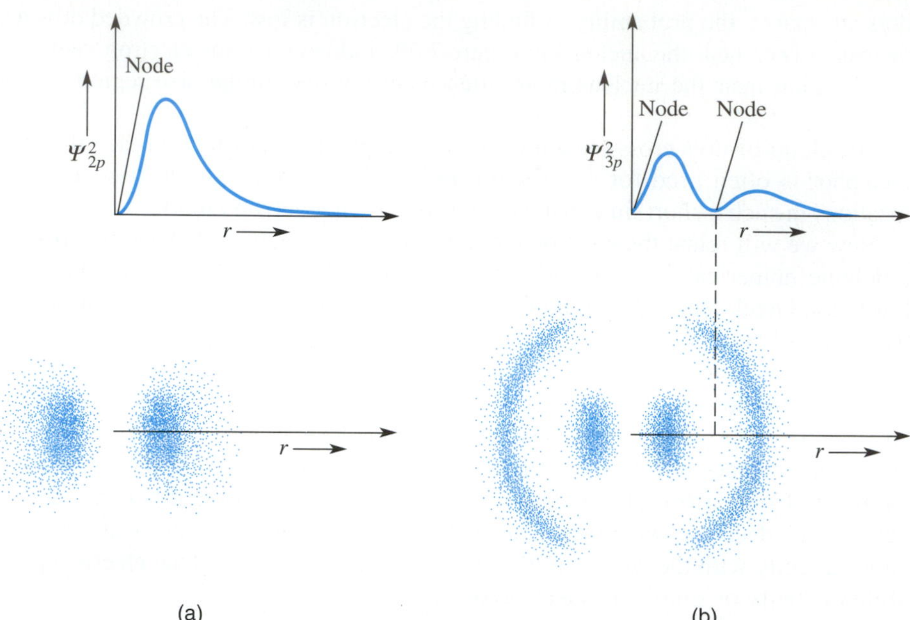

Figure 7.31

Electron density in the $2p$ and $3p$ orbitals. (a) The $2p$ orbital consists of two lobes of density separated by a nodal plane that contains the nucleus. The upper figure is a plot of electron density versus r along the axis of one lobe. (b) The $3p$ orbital also consists of two lobes separated by a nodal plane through the nucleus. Another nodal surface, this one spherical, intersects both lobes.

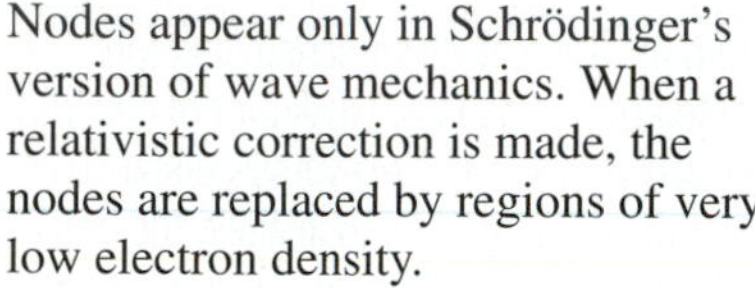
Nodes appear only in Schrödinger's version of wave mechanics. When a relativistic correction is made, the nodes are replaced by regions of very low electron density.

7.32. A balloon picture depicts a surface of constant electron density chosen so that there is a large probability (say, 90%) of finding the electron within its boundary. The balloon picture for any s orbital is a sphere. The pictures for the three p orbitals are shaped somewhat like dumbbells because the orbitals are divided into two lobes by their nodal planes. Most d orbitals have four lobes of density and two nodal planes, and f orbitals are even more elaborate. Note that p and d orbitals, unlike s orbitals, have electron densities of zero at the nucleus.

Orbitals that differ only in their m_l values, such as a set of p orbitals or a set of d orbitals, will have the same energy and often the same shape, but they are turned in different directions. Figure 7.32b shows, for example, that the three p orbitals are oriented at right angles to each other. It is customary to distinguish the p orbitals by the symbols p_x, p_y, and p_z. Spatial orientations for the five d orbitals are shown in Figure 7.32c.

Radial Densities

The probability of finding an electron in some specific small region of space (e.g., around the point in the small segment of Figure 7.33) is proportional to the value of the electron density at that point. Suppose, however, that we want to know the probability of finding an electron at a given distance from the nucleus, regardless of direction. For example, we might be interested in the probability of finding an electron in the spherical shell of Figure 7.33, which includes all points the same distance r from the nucleus. This probability is called the **radial electron density**; it is graphed in Figure 7.34 for orbitals from the $n = 1$, 2, and 3 shells. Note the difference between these plots and the electron density plots of Figures 7.29, 7.30, and 7.31.

Since the hydrogen electron is normally found in the ground state or $1s$ orbital, let us look at the $1s$ radial density plot in some detail. It increases from zero at the nucleus, passes through a maximum at $r = 52.9$ pm (1 pm $= 10^{-12}$ m), and then

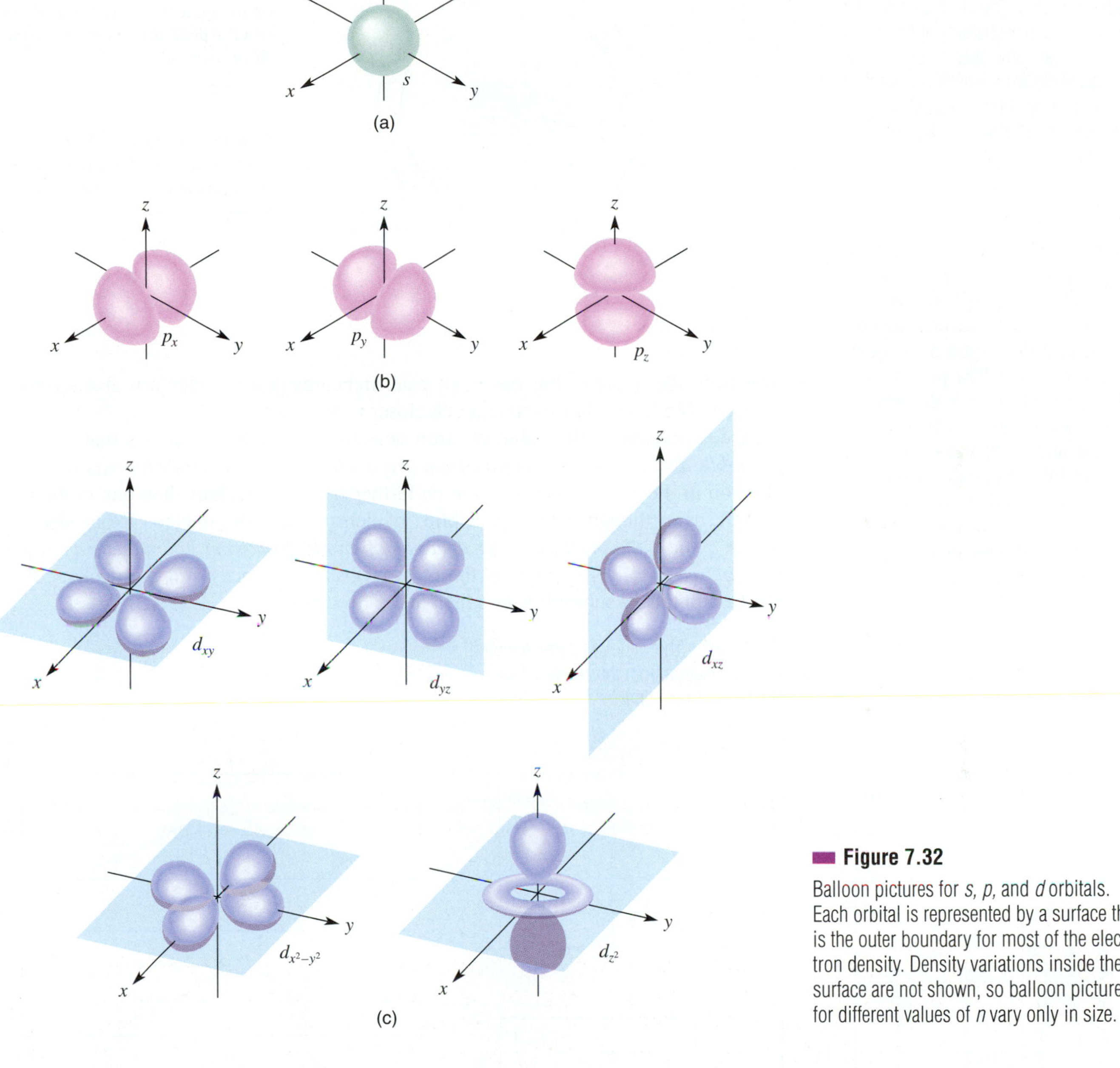

Figure 7.32

Balloon pictures for *s*, *p*, and *d* orbitals. Each orbital is represented by a surface that is the outer boundary for most of the electron density. Density variations inside the surface are not shown, so balloon pictures for different values of *n* vary only in size.

approaches zero at large values of *r*. The **most probable distance** from the nucleus—that is, the distance at which the electron spends the greatest fraction of time—is 52.9 pm. It is interesting to note that this distance corresponds to the radius calculated by Bohr for his lowest energy orbit. The radial electron density of the 2*s* plot, which corresponds to an excited state for hydrogen, has a large peak at 280 pm and a smaller peak at 42 pm. Between the peaks is a node where the probability drops to zero. The large peak corresponds to the most probable distance of a 2*s* elec-

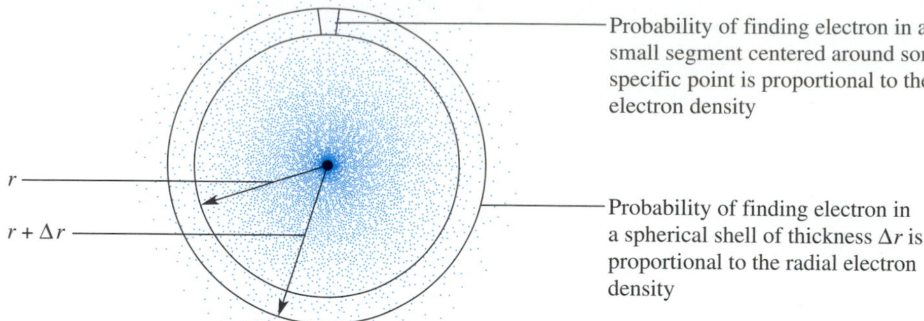

Figure 7.33 Comparison of electron density and radial electron density. The electron density is a measure of the probability of finding the electron near some specific point in space. The radial electron density is a measure of the probability of finding an electron in a spherical shell between radius r and radius $r + \Delta r$.

For those of you who are math buffs, the volume of the spherical shell in Figure 7.33 is approximately $4\pi r^2 \Delta r$, where $4\pi r^2$ is the surface area of the sphere of radius r and Δr is the thickness of the shell. The probability of finding the electron in the spherical shell is the product of the shell volume and ψ^2, the electron density or probability per unit volume. Hence, the radial electron density is approximately $4\pi r^2 \psi^2 \Delta r$. Because of the r^2 term, a radial density plot is always zero at the origin where $r = 0$.

tron from the nucleus, but the small peak indicates that the electron also spends a considerable fraction of its time much closer to the nucleus.

A comparison of the radial electron densities for hydrogen shows that *the most probable distance of an electron from the nucleus increases with increasing n.* An electron in the $n = 3$ shell tends to be farther from the nucleus than one in the $n = 2$ shell, even though it has small blips of density near the nucleus. Note also that *within a shell, the most probable distance of an electron from the nucleus decreases with increasing l.* For example, the most probable distance from the nucleus of an electron in the 3*d* subshell is less than in the 3*p* subshell.

Figure 7.34 Plots of radial electron density versus r. The value of r corresponding to the largest peak is the most probable distance of the electron from the nucleus, that is, the distance at which the electron spends the greatest fraction of time.

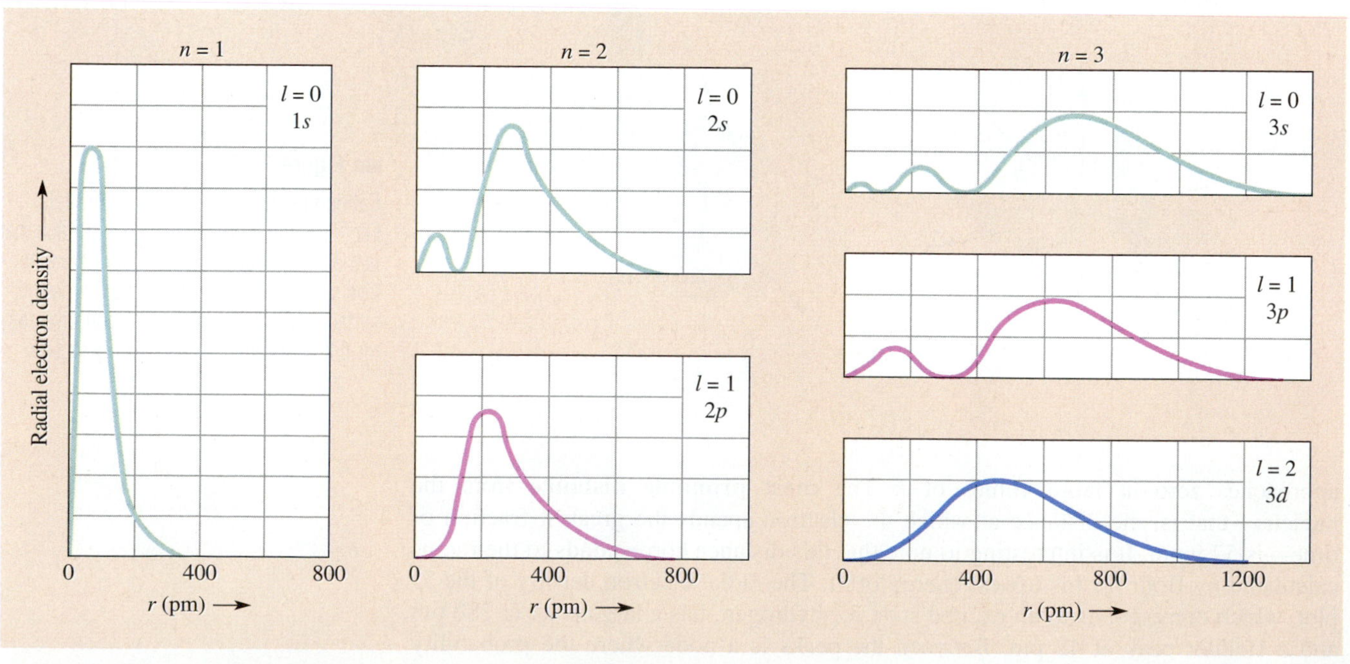

CHAPTER REVIEW

Starred entries are based on the *Digging Deeper* sections.

LEARNING OBJECTIVES BY SECTION

7.1 1. Calculate the frequency of electromagnetic radiation from its wavelength, and vice versa.
2. List the regions of the electromagnetic spectrum in order of increasing frequency or decreasing wavelength.
3. List the colors of the visible spectrum in order of increasing frequency or decreasing wavelength.
4. Distinguish between a continuous spectrum and a line spectrum.
5. Distinguish between an emission spectrum and an absorption spectrum.

7.2 1. Calculate the frequency, wavelength, or photon energy, given one of the quantities.
*2. Explain how the value of Planck's constant can be obtained from a plot of photoelectron kinetic energy versus photon frequency.
*3. Calculate W from ν_0, and vice versa.
*4. Use W (or ν_0) and ν to calculate the kinetic energy of a photoelectron.

7.3 1. Calculate excited and ground state energies for the hydrogen electron, given the quantum number n.
2. Calculate the energy, frequency and wavelength of a photon emitted or absorbed during an electron transition in the hydrogen atom.
3. Identify the electron transitions that give rise to the Lyman and Balmer series of spectral lines.

7.4 1. Calculate the de Broglie wavelength associated with a beam of moving particles.
2. State how the de Broglie wavelength varies with increasing particle mass and speed.

7.5 1. State the uncertainty principle, and explain why we cannot predict precise paths or orbits for very small particles such as electrons.

7.6 1. Write symbols for the orbital and spin quantum numbers, and give their range of values.
2. State the number of orbitals in each shell and in each subshell.
3. State the total number of states available to an electron in each shell or subshell.
*4. Describe some of the evidence for electron spin.

7.7 1. Sketch balloon pictures for *s*, *p*, and *d* orbitals.
2. Distinguish between p_x, p_y, and p_z orbitals.
3. Distinguish between electron density and radial electron density.
4. State how the most probable distance of an electron from the nucleus varies with increasing n and, within a shell, with increasing l.

KEY TERMS BY SECTION

7.1 Absorption spectrum
Continuous emission spectrum
Electromagnetic radiation
Electromagnetic spectrum
Emission spectrum
Infrared radiation
Line spectrum
Spectroscope
Spectroscopy
Spectrum
Speed of light
Ultraviolet radiation
Wave
Wave frequency (ν)
Wavelength (λ)
White light

7.2 *Photoelectric effect
Photon
Planck's constant (h)
Planck's law
Quantum
*Threshold frequency (ν_0)

7.3 Balmer series
Energy-level diagram
Excited state
First Bohr radius (a_0)
Ground state
Lyman series
Quantum number

7.4 De Broglie wavelength
Matter wave

7.5 Heisenberg uncertainty principle
Momentum

7.6 Azimuthal quantum number (l)
d orbital
Magnetic quantum number (m_l)
Orbital
p orbital
Principal quantum number (n)
Shell
s orbital
Spin quantum number (m_s)
Spin state
*Stern–Gerlach experiment
Subshell
Wave function
*Zeeman effect

7.7 Balloon picture
Electron cloud
Electron density (ψ^2)
Most probable distance (of an electron)
Nodal plane
Node
Radial electron density

IMPORTANT EQUATIONS

7.2 $\lambda\nu = c$

7.3 $E = h\nu$

***7.4** $KE = h\nu - W$

***7.5** $\nu_0 = \frac{W}{h}$

7.6 $E_n = \frac{-2.179 \times 10^{-18}\text{ J}}{n^2}$

7.7 $E_{\text{photon}} = -2.179 \times 10^{-18}\text{ J}\left(\frac{1}{n_H^2} - \frac{1}{n_L^2}\right)$

7.8 $\lambda = \frac{h}{mv}$

FINAL EXERCISES

Answers to exercises with blue numbers are given in Appendix D. Starred exercises are based on the *Digging Deeper* sections.

PART A. QUESTIONS AND PROBLEMS BY SECTION

Electromagnetic Radiation and Spectra (Section 7.1)

7.1 List the following spectral regions in order of increasing frequency: gamma rays, infrared, x-rays, radio, ultraviolet, microwave, visible.

7.2 List the following spectral regions in order of increasing wavelength: infrared, yellow light, ultraviolet, blue light, red light, green light.

7.3 Which region of the electromagnetic spectrum has
(a) the shortest wavelength?
(b) the lowest frequency?
(c) the highest energy?

7.4 List the colors in the visible spectrum in order of increasing frequency. Which color has the longest wavelength? The highest energy?

7.5 Which region of the electromagnetic spectrum contains "heat waves"? Are heat waves more or less energetic than visible light waves?

7.6 Which region of the electromagnetic spectrum contains "radar waves"? Does radar have a longer or shorter wavelength than ultraviolet rays?

7.7 **(a)** What is an emission spectrum? What is the difference between a continuous emission spectrum and a line spectrum?
(b) Diagram the apparatus for obtaining a visible emission spectrum.

7.8 **(a)** What is an absorption spectrum?
(b) Diagram the apparatus for obtaining a visible absorption spectrum.

7.9 The planet Uranus, which is 1.69×10^{12} miles from Earth, was visited by Voyager 2 in January 1986. How many hours did it take for a radio signal from Voyager 2 to reach Earth?

7.10 At the point of closest approach, it takes 2.42 s for microwaves to travel from the earth to the moon and back again. Calculate the earth–moon distance in kilometers. (This method has actually been used to determine the earth–moon distance.)

7.11 Calculate (a) the frequency of 650-nm light and (b) the wavelength in nanometers of 6.00×10^{16} Hz radiation. In which spectral regions would you expect to find these radiations? (Refer to Figure 7.3.)

7.12 Calculate (a) the wavelength in nanometers of 7.5×10^{15} Hz radiation and (b) the frequency of 7.40-μm radiation. In which regions of the spectrum would you look for these radiations? (Refer to Figure 7.3.)

7.13 The emission spectrum of sodium contains a closely spaced pair of lines at 589.0 nm and 589.6 nm.
(a) Calculate the frequency of each of these lines.
(b) In which spectral region will these lines appear?

7.14 The radiation emitted by a helium–neon laser has a frequency of 4.74×10^{14} Hz.
(a) Calculate the wavelength in nanometers of this radiation.
(b) In which spectral region will this radiation appear?

Quantum Theory of Radiation (Section 7.2)

7.15 Describe some of the evidence suggesting that radiation is quantized.

7.16 Describe some of the evidence suggesting that radiation consists of waves.

7.17 How does the energy of a photon vary with (a) increasing wavelength and (b) increasing frequency?

7.18 How do the frequency and wavelength of electromagnetic radiation vary with increasing photon energy?

7.19 For each of the electromagnetic waves in the following table, ν or λ is given. Calculate the remaining quantities and fill each empty space in the table. Be sure to specify units.

Range in Which Wave Is Found	Frequency ν	Wavelength λ	E_{photon}
Ultraviolet	1×10^{16} Hz	?	?
Radio	1×10^{6} Hz	?	?
Visible (green)	?	500 nm	?

7.20 For each of the electromagnetic waves in the following table, λ or E_{photon} is given. Calculate the remaining quantities and fill each empty space in the table. Be sure to specify units.

Range in Which Wave Is Found	Frequency ν	Wavelength λ	E_{photon}
Infrared	?	6 μm	?
Visible (red)	?	7000 Å[a]	?
?	?	?	6×10^{-17} J

[a]One angstrom unit (Å) is 10^{-10} m.

7.21 The sodium vapor lamps used for street lighting emit an intense spectral line at 589.6 nm.
(a) Calculate the frequency and energy of an emitted photon.
(b) Explain why sodium vapor lamps cast an orange-yellow glow on the objects they illuminate.

7.22 Mercury vapor lamps emit intense spectral lines at 546.1, 453.8, and 253.7 nm.
(a) Calculate the corresponding photon energies.
(b) In what region of the spectrum will each of these lines appear?

7.23 Visible radiation with wavelengths 460 and 610 nm is emitted when lithium is excited by a flame or electric discharge. For each of these wavelengths, calculate (a) the frequency, (b) the photon energy in joules, and (c) the energy in kilojoules of 1.00 mol of photons.

7.24 A radio-frequency transmitter used in nuclear magnetic resonance experiments operates at 60 MHz. For this frequency, calculate (a) the wavelength, (b) the photon energy in joules, and (c) the energy in kilojoules of 1.00 mol of photons.

7.25 How many photons are required to provide 1.00 J of each of the following forms of radiation?
(a) 285.2-nm ultraviolet
(b) 501.5-nm green light
(c) 45-μm infrared

7.26 How many photons are required to provide
(a) 1.00 kJ of 8.4-GHz shortwave radiation?
(b) 0.500 mJ of 1.00×10^{-15} m gamma rays?
(c) 1.00 J of radiation from the helium–neon laser of Exercise 7.14?

The Photoelectric Effect (Section 7.2)

***7.27** What is the photoelectric effect? What role did the photoelectric effect play in the development of the quantum theory of radiation?

***7.28** How are photoelectron data used to obtain the value of Planck's constant?

***7.29** The minimum energy required to dislodge electrons from potassium metal is 3.69×10^{-19} J. Will photoelectrons be produced when visible radiation of 600-nm wavelength shines on a clean potassium surface?

***7.30** The longest wavelength radiation that will dislodge electrons from a gold surface is 257 nm.
(a) Will photoelectrons be ejected from gold by visible light? Explain.
(b) Calculate the threshold frequency and W for a gold surface.

***7.31** Refer to Exercise 7.29 and calculate the kinetic energy of an electron ejected from a potassium surface by 400-nm radiation.

***7.32** Refer to Exercise 7.30 and calculate the kinetic energy of an electron ejected from a gold surface by 240-nm radiation.

The Hydrogen Spectrum and the Bohr Atom (Section 7.3)

7.33 In which region of the spectrum is the Lyman series? The Balmer series?

7.34 Which lines of the Balmer series are in the visible region of the spectrum, and which are in the ultraviolet? (Refer to Table 7.1.)

7.35 Write Bohr's formula for the hydrogen energy levels.
(a) Why is the integer n called a *quantum number*?
(b) What is the lowest allowed energy?
(c) What value of n gives the energy of Part (b)?

7.36 Write Bohr's formula for the orbital radii of hydrogen.
(a) Express the first Bohr radius in picometers and in angstrom units.
(b) What value of n gives the radius of Part (a)?

7.37 What does it mean when we say that the hydrogen atom is in the *ground state*? What value of n corresponds to the ground state?

7.38 What does it mean when we say that the hydrogen atom is in an *excited state*? What value of n corresponds to the first excited state? The second excited state?

7.39 A hydrogen electron drops directly from the $n = 4$ level to the $n = 1$ level. How many photons are emitted? Explain.

7.40 Diagram all the possible paths by which a hydrogen electron can drop from the $n = 4$ level to the $n = 1$ level. How many photons will be emitted for each path?

7.41 Calculate the hydrogen energy associated with each level from $n = 1$ to $n = 6$. Use your results and some graph paper to prepare an accurately scaled energy-level diagram for hydrogen.

7.42 Refer to Bohr's energy formula and your energy-level diagram from the previous exercise.
(a) As n increases without limit, what happens to the Bohr energy levels in terms of energy? Of spacing?
(b) What is the highest possible value for the electron's energy? Add this value to your energy-level diagram.

7.43 Calculate the Bohr radius for each level from $n = 1$ to $n = 6$. Use your results, some graph paper, and a compass to prepare an accurately scaled diagram similar to Figure 7.15.

7.44 Refer to Bohr's radius formula and your graph from the previous exercise.
(a) As n increases without limit, what happens to the Bohr orbits in terms of distance from the nucleus? What happens to the distance between neighboring orbits?
(b) What happens to the Bohr radius as the energy approaches its maximum value?

7.45 Use Bohr's energy formula to calculate the energy of an electron in (a) the $n = 4$ state and (b) the second excited state of the hydrogen atom.

7.46 The energy of a hydrogen electron is -4.45×10^{-20} J. What level does it occupy?

7.47 A hydrogen electron makes a transition from $n = 2$ to $n = 3$.
(a) Is a photon emitted or absorbed during this transition?
(b) Calculate the energy, frequency, and wavelength of the photon.
(c) In what region of the spectrum would you look for the photon?

7.48 A hydrogen electron makes a transition from $n = 4$ to $n = 3$.
(a) Is a photon emitted or absorbed during this transition?
(b) Calculate the energy, frequency, and wavelength of the photon.
(c) In what region of the spectrum would you look for the photon?

7.49 Calculate the energy difference between the $n = 2$ and the $n = 1$ levels of the hydrogen electron. Explain why all of the Lyman lines are in the ultraviolet region of the spectrum.

7.50 Calculate the wavelengths of the first four Lyman lines.

7.51 After a certain hydrogen atom emits a 3740-nm photon, the energy of its electron is -8.716×10^{-20} J.
(a) What was the electron's energy before photon emission, and what level did it occupy?
(b) In which spectral series would you look for the emission line?

7.52 The emission spectrum of a certain element contains a green line whose wavelength is 500 nm. If the initial energy of the electron making the transition was -7.0×10^{-19} J, what was its final energy? Can this element be hydrogen? Why or why not?

Matter Waves (Section 7.4)

7.53 **(a)** Write the de Broglie formula for the wavelength associated with a moving particle.
(b) How does the particle's wavelength vary with increasing speed? With increasing particle mass?

7.54 A beam of electrons and a beam of protons are moving with the same speed. Which has the longer de Broglie wavelength?

7.55 Describe some experiments that provide evidence for the existence of matter waves.

7.56 Why are we generally not aware of matter waves in everyday life? Why must these waves be taken into consideration when we deal with subatomic particles?

7.57 Calculate and compare the de Broglie wavelengths associated with (a) an electron, (b) a proton, and (c) a 0.40-kg piece of space debris, each moving with a speed of 50 m/s. (See Table 2.1, p. 37, for the particle masses.)

7.58 Electron waves and neutron waves are used to examine structures that are too small to be seen with the longer waves of visible light. Calculate the speed needed for (a) electrons to have 0.10-nm waves and (b) neutrons to have 0.30-nm waves.

The Uncertainty Principle (Section 7.5)

7.59 State the uncertainty principle, and explain how it affects our ability to predict the behavior of very small particles such as electrons.

7.60 Explain why the uncertainty principle does not hinder us in determining planetary orbits, rocket trajectories, and distances between heavenly bodies.

7.61 What effect do photons of visible light have on the position and speed of particles that we observe with this light? Will the effect be greater or less for a subatomic particle than for a uranium atom? Explain.

7.62 Do you think that future measuring techniques might eventually be precise enough to eliminate the uncertainties described by the uncertainty principle? Explain.

7.63 Why did the uncertainty principle eventually lead scientists to abandon Bohr's theory?

7.64 One consequence of the uncertainty principle is that we cannot determine precise paths and positions for the electrons in atoms and molecules. Can any predictions be made, and on what basis?

Wave Mechanics and Hydrogen Orbitals (Section 7.6)

7.65 Which aspects of the Bohr model are retained in wave mechanics? Which aspects are abandoned?

7.66 What is the difference between a Bohr *orbit* and a Schrödinger *orbital*?

7.67 How many quantum numbers are needed to specify an orbital? List these numbers by name and symbol, and state the restrictions on their values.

7.68 What property is described by each of the orbital quantum numbers?

7.69 **(a)** Which quantum number is shared by all the orbitals in a given shell?
(b) Which quantum numbers are shared by the orbitals in a given subshell?

7.70 **(a)** Which quantum numbers do all *s* orbitals have in common?
(b) Which quantum numbers do all *d* orbitals have in common?

7.71 Which quantum numbers are shared by all the orbitals in the 3*d* subshell? In the 4*f* subshell?

7.72 **(a)** Use orbital notation to identify each of the subshells in the $n = 4$ shell.
(b) What property do hydrogen electrons in any $n = 4$ orbital have in common?

7.73 Give the number of orbitals in
(a) a *p* subshell
(b) a subshell with $l = 2$
(c) the lowest energy shell ($n = 1$)
(d) the seventh shell

7.74 How many orbitals are in
(a) the 3*d* subshell?
(b) the $n = 4$ shell?
(c) the 4*f* subshell?

7.75 Which of the following sets of quantum numbers are allowed, and which are not? State what is wrong with each set that is not allowed.
(a) $n = 2, l = 1, m_l = 0$
(b) $n = 2, l = 2, m_l = 2$
(c) $n = 2, l = 1, m_l = -1$
(d) $n = 2, l = 1, m_l = -2$

7.76 Which of the following sets of quantum numbers are allowed, and which are not? State what is wrong with each set that is not allowed.
(a) $n = 0, l = 0, m_l = 0$
(b) $n = 1, l = 0, m_l = 0$
(c) $n = 6, l = 4, m_l = -5$
(d) $n = 6, l = 5, m_l = -4$

7.77 Which of the following symbols represent possible orbitals, and which do not? Explain what is wrong with each incorrect symbol.
(a) 2*s* **(c)** 2*d*
(b) 3*p* **(d)** 2*f*

7.78 Which of the following symbols represent possible orbitals, and which do not? Explain what is wrong with each incorrect symbol.
(a) 1*p* **(c)** 3*d*
(b) 8*s* **(d)** 3*g*

7.79 How many orbitals are in
(a) the $n = 3$ shell? **(c)** the 4*d* subshell?
(b) the 3*d* subshell? **(d)** the 5*g* subshell?

7.80 Expand Table 7.2 to include all the orbitals for the $n = 5$ shell.

7.81 What property of the electron is explained on the basis of spin? Which quantum number is associated with this property? What are its allowed values?

7.82 Explain the following statement: Each state of the hydrogen atom is completely described by four quantum numbers.

7.83 Give the four quantum numbers for each of the two possible electron states in a 1*s* orbital.

7.84 How many different electron states are possible for an electron in the $n = 3$ shell? Give the quantum numbers for each of these states.

***7.85** Describe the Stern–Gerlach experiment, and explain how it confirmed the existence of two spin states for the electron.

***7.86** Describe the Zeeman effect, and explain it in terms of electron spin.

Electron Density (Section 7.7)

7.87 What is the general shape of (a) all *s* orbitals and (b) all *p* orbitals?

7.88 **(a)** Explain, with the help of drawings, the difference between a p_x, a p_y, and a p_z orbital.
(b) Do all d orbitals have the same shape?

7.89 Draw neat balloon pictures showing the shapes and spatial orientations of the following orbitals: (a) s, (b) p_x, and (c) d_{xy}. Be sure to include the x and y axes.

7.90 For each of the following pairs of orbitals, draw balloon pictures that show the difference between them:
(a) $1s$ and $2s$
(b) $2p_x$ and $2p_z$
(c) $3d_{xy}$ and $3d_{x^2-y^2}$

7.91 Figures 7.29 through 7.32 show various ways of representing the spatial characteristics of atomic orbitals. Which of these diagrams, if any, represent the path of the electron? Explain your answer.

7.92 Sketch a balloon picture for a $2p_y$ orbital. Are the following statements true or false for an electron in this orbital? Explain.
(a) The electron is always located somewhere on the surface of your sketch.
(b) The electron is always located somewhere within the surface of your sketch.
(c) The energy of the electron will change as the radial distance of the electron from the nucleus changes.

7.93 Explain in your own words what the phrase *electron density* means. Where is the electron density greatest for an s electron? For a $2p$ electron?

7.94 Explain what is meant by the phrase *radial electron density.* Where does the radial electron density have its maximum value for a $2s$ electron? For a $2p_x$ electron?

7.95 Which of the orbitals in Exercise 7.89 have one or more nodal planes? Describe the location of these planes.

7.96 Which of the orbitals in Figures 7.29 through 7.31 have a nodal surface other than a plane?

7.97 Estimate, with the help of a ruler, the most probable electron distance for each of the orbitals in Figure 7.34. How does this distance vary with increasing n? With increasing l within a shell?

7.98 What is the most probable distance of a $1s$ electron from the nucleus? What relation does this value bear to the first Bohr radius?

PART B. MISCELLANEOUS QUESTIONS AND PROBLEMS

7.99 Why does the gas in a fluorescent tube emit light?

7.100 It is often said that electromagnetic radiation has a dual nature. What does this mean?

7.101 Sunlight coming through a glass window will not produce a sunburn because glass absorbs ultraviolet radiation. What ultimately happens to the energy of the absorbed photons?

7.102 The solar spectrum contains several dark lines called *Fraunhofer lines.* Do these lines represent an absorption spectrum or an emission spectrum? Explain.

7.103 Measurements at corneal surfaces have shown that the human eye can detect 3.15×10^{-17} J of 510-nm radiation. How many photons does this correspond to?

7.104 One mole of photons is called an *einstein.* Calculate the energy in joules of 1.00 einstein of 460-nm light.

7.105 A microwave source emits radiation with a wavelength of 12 cm. How many photons of this radiation are required to raise the temperature of 250 mL of water (about 1 cup) from 25°C to 95°C? Assume that the density of water is 1.00 g/mL and its specific heat is 4.18 J/g·°C.

7.106 The absorption of a 784.5-nm photon dissociates an I_2 molecule into atoms. Calculate the energy required to dissociate 1.00 mol of iodine molecules into iodine atoms.

7.107 One mole of Br_2 molecules is dissociated into atoms by the absorption of 193.9 kJ. If each molecule dissociates by absorbing a single photon, calculate (a) the minimum energy and (b) the maximum wavelength of the photon. What spectral region will provide light of this wavelength?

***7.108** It requires 7.4×10^{-19} J to dislodge an electron from a silver surface. Could silver be used in making a light-sensitive photoelectron tube (an electric eye)? Why or why not? What properties must a metal surface have to be useful in such a tube?

***7.109** When the intensity of light shining on a metal surface is increased, the number of photoelectrons is also increased. The kinetic energy of the photoelectrons, however, does not change. Explain these observations in terms of the quantum theory of radiation.

7.110 Which two adjacent energy levels of hydrogen have the greatest spacing between them; that is, which adjacent levels are farthest apart in energy?

7.111 Use Planck's law and Equation 7.7 to find a general expression for $1/\lambda$, where $1/\lambda$ is the reciprocal of the wavelength of a photon emitted during the transition from n_H to n_L. (This expression is called the *Rydberg equation.*)

7.112 Calculate the amount of energy required to remove an electron from a hydrogen atom, that is, to raise the electron from the ground state to $n = \infty$. (This energy is called the *ionization energy*; it is discussed more fully in Chapter 8.)

7.113 One hydrogen atom has an electron in the $n = 2$ level, and a second hydrogen atom has an electron in the $n = 3$ level.
(a) Calculate the energy required to remove the electron from each atom.
(b) Which electron is more tightly bound?
(c) Which atom is more easily ionized?

7.114 The Humphreys series, which involves transitions from higher levels to $n = 6$, is another series in the emission spectrum of atomic hydrogen.
(a) Calculate the wavelength of the first Humphreys line.
(b) In what spectral region should one look for the Humphreys series?

7.115 Use the Rydberg equation from Exercise 7.111 to calculate the wavelength of the first Humphreys line. (Your answer should agree with the answer calculated in Exercise 7.114.)

7.116 Bohr's energy formula is a special case of a more general formula that applies to any system consisting of a nucleus and one electron: $E_n = -Z^2 \times 2.179 \times 10^{-18}$ J/n^2, where Z is the atomic number of the nucleus. Calculate the following:
(a) the ground-state energy of a helium ion, He^+
(b) the wavelength of radiation that would cause the electron to jump from the ground state to the $n = 2$ level

7.117 Two of the allowed energy levels for the electron in He^+ are -21.8×10^{-19} J and -9.68×10^{-19} J. A student claims that one of the green lines in the helium spectrum is due to a transition between these levels. Is the student correct? Explain. (Refer to Exercise 7.116.)

7.118 The Li^{2+} emission spectrum is similar to that of atomic hydrogen. Will the lines in each spectral series of Li^{2+} be shifted to shorter or longer wavelengths than in the corresponding series for hydrogen? Explain. (Refer to Exercise 7.116.)

7.119 A beam of electrons has a de Broglie wavelength of 1.50×10^{-4} m. Calculate the speed of the electron beam in meters per second.

***7.120** The energy required to eject an electron from a surface atom in calcium metal is 4.60×10^{-19} J. If calcium is irradiated with 400-nm photons, what is the de Broglie wavelength of the resulting photoelectron beam?

7.121 **(a)** What is meant by the phrase *standing wave*? *Stationary state*?
(b) Why are the allowed states for hydrogen sometimes referred to as stationary states?

7.122 The radius of a Bohr orbit is given by 52.9 pm $\times\ n^2$, where n is the quantum number of the orbit. Calculate the speed at which the wavelength of the electron is exactly equal to the circumference of the first Bohr orbit.

7.123 Consider the following electronic transitions in the hydrogen atom: $5p \longrightarrow 1s$, $5p \longrightarrow 2s$, and $5p \longrightarrow 3s$. Which of these transitions
(a) emits the longer wavelength photon?
(b) gives a visible spectral line?
(c) gives a line in the Lyman series?

7.124 Is the $1s$ electron in He^+ more or less tightly bound than the $1s$ electron in the hydrogen atom? Explain. (Refer to Exercise 7.116.)

7.125 Figure 7.29 shows that the electron density of a $1s$ orbital is a maximum at the nucleus. Refer to Figure 7.34, and explain why the radial electron density for this orbital is zero at the nucleus and a maximum at some distance from the nucleus.

7.126 Refer to the expression given on page 291 for the $1s$ electron density. On a piece of graph paper, plot the electron density as a function of r. For what value of r is the electron density a maximum? A minimum?

7.127 Refer to the expression given on page 294 for the radial electron density. On a piece of graph paper, plot the radial electron density for the $1s$ orbital as a function of r. Assume that $\Delta r = 1$. For what value of r is the radial electron density a maximum? A minimum?

CHAPTER 8

The periodic table, one of the most fertile constructions in all of science, is based on the discovery that the properties of the elements are periodic functions of their atomic numbers.

MANY-ELECTRON ATOMS AND THE PERIODIC TABLE

OUTLINE

PREVIEW

The hydrogen atom has 1 electron moving about its nucleus; an atom of fermium has 100 electrons. Other atoms have anywhere from 2 to 109 electrons. How are the electrons in these atoms arranged? Many electrons, all drawn toward the nucleus and all repelling each other, could be a prescription for chaos. Yet each atom turns out to have an orderly hierarchy with a separate state for each electron and a firm set of "no trespassing" rules. We will explore the systematic way in which electrons are distributed among the orbitals of an atom, and we will discover that these arrangements lead directly to the periodic table.

A **many-electron atom** is one that contains two or more electrons; in other words, it is any atom other than hydrogen. The electrons in a many-electron atom repel each other and, at the same time, they are attracted by the nucleus. These interactions are complex, and the orbitals of many-electron atoms, though similar in some respects to those of hydrogen, are nevertheless different. The equations of quantum mechanics tell us that the orbitals of many-electron atoms have the same quantum numbers, n, l, and m_l, and the same *s, p, d,* and *f* shapes as hydrogen orbitals. However, these orbitals differ from those of hydrogen in their electron energies and in the most probable distances of the electron from the nucleus. We will explore these similarities and differences in the next few sections, and then we will show how the theory of many-electron atoms provides us with a basis for understanding the periodic table.

8.1 ELECTRON ENERGIES IN MANY-ELECTRON ATOMS

The Bohr energy formula

$$E_n = \frac{-2.179 \times 10^{-18}\ \text{J}}{n^2}$$

shows that the energy of a hydrogen electron varies only with the principal quantum number n of the orbital it occupies. A hydrogen electron in a $3d$ orbital, for example, has the same energy as an electron in a $3p$ or $3s$ orbital, as illustrated in Figure 8.1a. In many-electron atoms, however, the Bohr formula does not apply, and the energy level diagram for each element is unique. Electron energies in many-electron atoms depend strongly on n, but they also depend to a lesser extent on the azimuthal quantum number l. Figure 8.1b illustrates that, *in a many-electron atom, the energy of an electron in a given shell increases with the l value of its orbital.* An electron in the d orbital of a given shell is slightly higher in energy than an electron in the p orbital, and a p electron is slightly higher in energy than an s electron. Figure 8.1 also shows that *the energy of an electron in a given orbital decreases with increasing atomic number.* An electron in the $1s$ orbital of potassium ($Z = 19$), for example, has a lower energy than an electron in the $1s$ orbital of lithium ($Z = 3$). The greater attractive force exerted by the more positive potassium nucleus pulls the electron closer to the nucleus, binding it more tightly, and lowering its energy.

PRACTICE EXERCISE 8.1

Which electron has a lower energy: (a) a $1s$ electron in a hydrogen atom or a $1s$ electron in a carbon atom; (b) a $3s$ electron in a chlorine atom or a $3p$ electron in the same atom?

The Shielding Effect

To understand why the l quantum number affects the energy, consider an atom with electrons occupying $1s$, $2s$, and $2p$ orbitals. The $1s$ electrons are close to the nucleus, and they exert a repulsive effect that partially protects or shields the outer $2s$ and $2p$ electrons from the attractive force of the nucleus. This is an example of the **shielding effect**, a decrease in the nuclear attraction for an electron caused by the presence of other electrons in underlying orbitals. As a result of the shielding effect,

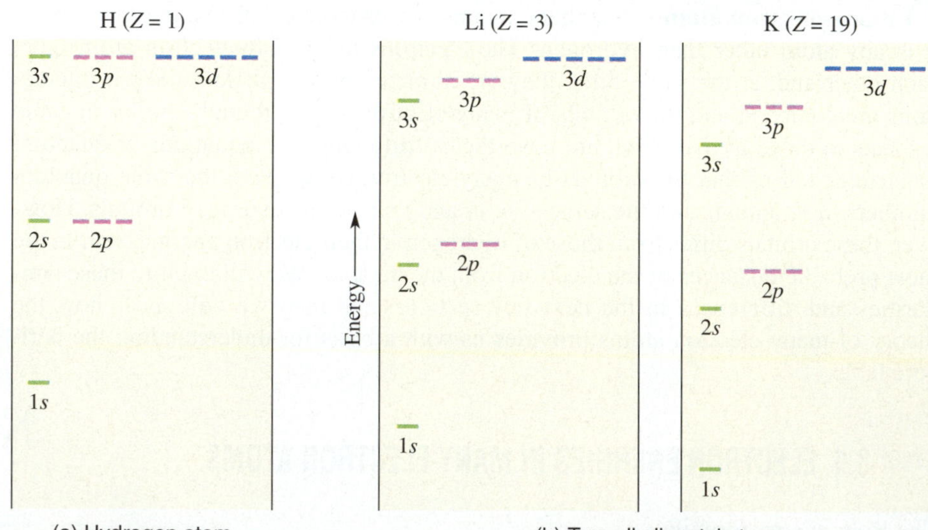

Figure 8.1
Relative orbital energies (not drawn to scale). (a) When only one electron is present, all orbitals with a given n (such as $3s$, $3p$, and $3d$) have the same energy. (b) When two or more electrons are present, the energies vary with both n and l so that $2p > 2s$ and $3d > 3p > 3s$. The dependence on l increases with the number of electrons.

the outer electrons, in this case the 2*s* and 2*p* electrons, are attracted not by the actual nuclear charge *Z,* but by some positive charge that is less than the nuclear charge. This lesser charge, called the **effective nuclear charge**, decreases with the number of shielding electrons. The 1*s* electrons are barely shielded, and they experience virtually the full force of the nuclear attraction. For these electrons, the effective charge is almost identical to the nuclear charge. The 2*s* and 2*p* electrons are shielded by the 1*s* electrons. The effective charge is smaller than the nuclear charge, so the $n = 2$ electrons experience a weaker attractive force. Electrons in the $n = 3$ shell would experience an even weaker force because they are shielded by electrons from two underlying shells.

Electrons in different orbitals within the same shell (e.g., the 2*s* and 2*p* electrons) experience different degrees of shielding because their radial density distributions (Figure 7.34, p. 294) are not the same. The 2*s* radial density distribution in Figure 7.34 has a minor peak near the nucleus, showing that a 2*s* electron spends a significant fraction of its time close to the nucleus, where it is not well shielded. The 2*p* orbital has very little density near the nucleus, so a 2*p* electron is shielded more completely than a 2*s* electron. A 2*s* electron, therefore, experiences a greater effective nuclear charge, is more tightly bound, and has a lower energy than a 2*p* electron. *The greater the effective nuclear charge experienced by an electron, the lower its energy.*

Figure 7.34 also shows that, within a shell, the number of minor peaks of density near the nucleus decreases with increasing *l;* the 3*s* ($l = 0$) orbital, for example, has two such peaks; the 3*p* ($l = 1$) orbital has one minor peak; and the 3*d* ($l = 2$) orbital has no peaks near the nucleus. Thus electrons with high *l* values tend to have less of their density very close to the nucleus and are therefore more effectively shielded than electrons with low *l* values. Within the $n = 4$ shell, for example, *f* electrons will be less tightly bound and have higher energies than *d* electrons, *d* electrons will have higher energies than *p* electrons, and *p* electrons will have higher energies than *s* electrons.

8.2 THE PAULI EXCLUSION PRINCIPLE

A hydrogen atom is in its ground state (lowest energy state) when its electron is in the 1*s* orbital. In the ground state of a many-electron atom, we might suppose that all of the electrons would be in the 1*s* orbital, but this is not the case. Instead, the electrons are distributed over the various orbitals—some will occupy the 1*s* orbital, others the 2*s,* still others the 2*p,* and so forth. An important rule regarding electron distributions was deduced from spectral data: *Only two electrons can occupy any one orbital, and these electrons must have opposite spins.* This rule, originally stated in somewhat different form by the Viennese physicist Wolfgang Pauli in 1925, is called the **Pauli exclusion principle**. Consider an orbital occupied by two electrons. Both electrons have the same orbital quantum numbers *n, l,* and m_l, but since they have opposite spins, one electron has a spin quantum number $m_s = +\frac{1}{2}$ and the other electron has $m_s = -\frac{1}{2}$. We already know that electrons in different orbitals have different orbital quantum numbers. The fact that electrons in the same orbital have different spin quantum numbers leads to an alternative statement of the exclusion principle: *No two electrons in an atom can have the same values for all four quantum numbers.*

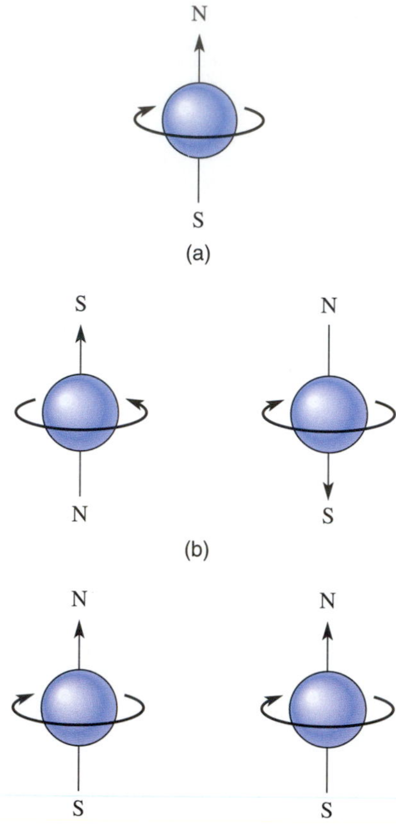

Figure 8.2

(a) A spinning electron behaves like a tiny magnet. (b) Paired electrons have opposite (antiparallel) spins and opposing magnetic fields; they can occupy the same atomic orbital. (c) Unpaired electrons have parallel spins and parallel magnetic fields; they cannot occupy the same orbital.

Recall our simplified model in which a spinning electron behaves like a magnet with a north and south pole (see Section 7.6 and Figure 8.2a). A single electron in an orbital is said to be **unpaired**. Two electrons in the same orbital must have opposite (antiparallel) spins, and their magnetic fields will be oriented as illustrated in Figure 8.2b. These electrons are **paired**. If two electrons have parallel spins (Figure 8.2c), the exclusion principle tells us that they cannot be in the same orbital. The importance of the exclusion principle should not be underestimated. The severe restriction it places on orbital occupancy requires the electrons in many-electron atoms to be spread out over many orbitals, and we will soon see that the resulting electron distributions are largely responsible for the physical and chemical properties of the elements.

8.3 ELECTRON CONFIGURATIONS

With orbital occupancy limited to two electrons, it becomes apparent that more than one orbital will have to be occupied when there are three or more electrons in an atom. The distribution of electrons in the orbitals of a many-electron atom is called an **electron configuration**. We are most interested in the **ground-state configuration**, the configuration that provides the lowest energy for the atom as a whole.

The Aufbau Procedure

To find the ground-state configuration of an atom, we must know which orbitals fill first. The **orbital filling order** has been determined by analysis of atomic spectra, and for the first 36 elements, hydrogen to krypton (atomic numbers $Z = 1$ through $Z = 36$), the orbital filling order is

$$\underrightarrow{1s \quad 2s \quad 2p \quad 3s \quad 3p \quad 4s \quad 3d \quad 4p}$$

The arrow shows that for these elements the $1s$ orbital fills first, then the $2s$ orbital, then the $2p$ orbital, and so forth.

One way of remembering the orbital filling order for the first 36 elements is to use the "**$n + l$ rule**." This rule states that (a) orbitals fill in order of increasing $(n + l)$ value and (b) when two orbitals have the same $(n + l)$ value, the orbital with the lower n value fills first. The rule predicts, for example, that the $4s$ orbital $(n + l = 4 + 0 = 4)$ will fill before the $3d$ orbital $(n + l = 3 + 2 = 5)$.

PRACTICE EXERCISE 8.2

Use the $n + l$ rule to confirm the filling order given above.

The number of electrons in a neutral atom is equal to its atomic number Z. The ground-state configuration for the atom is obtained by placing these electrons into the various orbitals one at a time according to the above order, with the understanding that no orbital can hold more than two electrons. The simplest atom is hydrogen with one electron ($Z = 1$). Its ground-state configuration is

H (Z = 1) $1s^1$ or [↑] 1s

where the superscript in the symbol $1s^1$ means that one electron occupies the 1*s* orbital. The representation with the vertical arrow, called an **orbital diagram**, shows that the lone electron is unpaired.

The configuration for helium with two electrons is

He (Z = 2) $1s^2$ or [↑↓] 1s

showing that two electrons occupy the 1*s* orbital. The opposing arrows in the orbital diagram indicate different spins and emphasize that the electrons are paired. In helium the 1*s* orbital is full—it cannot accommodate any more electrons. Note that helium, a noble gas, is the second and last element in the first horizontal row of the periodic table (see inside front cover).

The symbols $1s^1$ and $1s^2$ are read as "one s one" and "one s two," not "one s to the first power" nor "one s square."

The 2*s* orbital is the next one to fill. The configurations of lithium and beryllium, the first two elements in the second row of the periodic table, are

Li (Z = 3) $1s^2 2s^1$ or [↑↓] 1s [↑] 2s

Be (Z = 4) $1s^2 2s^2$ or [↑↓] 1s [↑↓] 2s

These configurations contain the $1s^2$ configuration of helium and are said to possess a **noble gas core**. They are often written in abbreviated form as

Li (Z = 3) $[He]2s^1$

Be (Z = 4) $[He]2s^2$

where $[He] = 1s^2$ represents the helium core.

The 2*s* orbital is now full, so the fifth electron of boron will be unpaired in one of the three 2*p* orbitals:

B (Z = 5) $[He]2s^2 2p^1$ or [↑↓] 1s [↑↓] 2s [↑][ ][ ] 2p

This electron could be in any one of the three *p* orbitals; it makes no difference because the energy of the atom will be the same regardless of which *p* orbital is occupied.

This method for obtaining ground-state configurations is known as the **aufbau procedure** (*aufbau* means "building-up" in German) because the configuration for each atom is obtained by building on the configuration of an atom with one less electron.

Hund's Rule

The next element, carbon (Z = 6), has the configuration $[He]2s^2 2p^2$. Are the two *p* electrons in the same *p* orbital or different *p* orbitals? We can imagine three possible arrangements:

Friedrich Hund, a German physicist who specialized in quantum mechanics, used spectroscopic data to show that electrons in a subshell tend to remain unpaired. His results are summarized in **Hund's rule:** *Electrons occupy the orbitals of a subshell singly and with parallel spins until each orbital has one electron.* Single occupancy is preferred because it allows the electrons to minimize their mutual repulsion by staying farther apart. Pairing will occur only when there is more than one electron for each orbital in the subshell. The ground-state configuration of carbon obeys Hund's rule:

C ($Z = 6$) [He]$2s^22p^2$ or $1s$ $2s$ $2p$

as do the configurations of elements 7 through 10:

N ($Z = 7$) [He]$2s^22p^3$ or $1s$ $2s$ $2p$

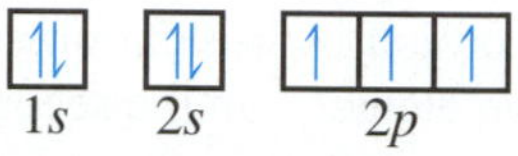

O ($Z = 8$) [He]$2s^22p^4$ or $1s$ $2s$ $2p$

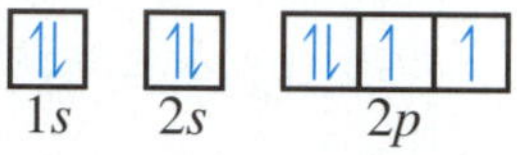

F ($Z = 9$) [He]$2s^22p^5$ or $1s$ $2s$ $2p$

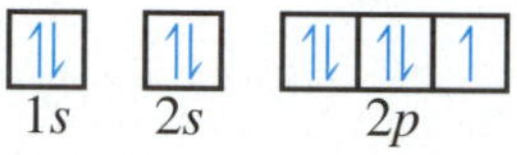

Ne ($Z = 10$) [He]$2s^22p^6$ or $1s$ $2s$ $2p$

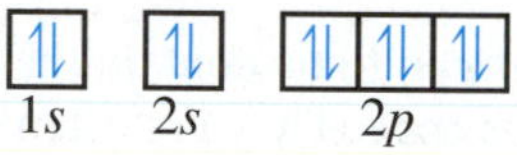

With neon, the $2p$ orbitals are full and the $n = 2$ shell is completely occupied. Note that neon, another noble gas, is the last element in the second row of the periodic table.

The configurations for the third-row elements, sodium through argon, are similar to those of lithium through neon, except that electrons occupy the $3s$ and $3p$ orbitals. These elements possess the neon core. The configuration of sodium, for example, is

$$\text{Na}\ (Z = 11) \qquad [\text{Ne}]3s^1$$

where [Ne] $= 1s^22s^22p^6$ represents the 10-electron configuration of neon. Argon, the last element in the third row of the periodic table, has the configuration

$$\text{Ar}\ (Z = 18) \qquad [\text{Ne}]3s^23p^6$$

PRACTICE EXERCISE 8.3

Write the ground-state configurations for the elements magnesium ($Z = 12$) through chlorine ($Z = 17$).

The orbital filling order given at the beginning of this section shows that the $4s$ orbital is next to fill after the $3p$ orbital. Hence, the ground-state configurations for potassium and calcium, the first two elements in the fourth row of the periodic table, are

$$\text{K } (Z = 19) \qquad [\text{Ar}]4s^1$$

$$\text{Ca } (Z = 20) \qquad [\text{Ar}]4s^2$$

where $[\text{Ar}] = 1s^2 2s^2 2p^6 3s^2 3p^6$ represents the 18-electron noble gas core of argon. With scandium, element 21, the $3d$ orbitals begin to fill. The configurations of scandium and iron, for example, are

$$\text{Sc } (Z = 21) \qquad [\text{Ar}]4s^2 3d^1$$

$$\text{Fe } (Z = 26) \qquad [\text{Ar}]4s^2 3d^6$$

Configurations are often written with the orbitals in order of increasing n. The configurations for scandium and iron may also be written as

$$\text{Sc } (Z = 21) \qquad [\text{Ar}]3d^1 4s^2$$

$$\text{Fe } (Z = 26) \qquad [\text{Ar}]3d^6 4s^2$$

EXAMPLE 8.1

The atomic number of selenium is 34. (a) Write its ground-state electron configuration. (b) Give an orbital diagram for its outer three subshells.

SOLUTION

(a) Selenium has 34 electrons. The atom will have the 18-electron core of argon and 16 additional electrons. The argon core ends with $3p^6$. According to the orbital filling order, 2 of the 16 additional electrons will occupy the $4s$ orbital, 10 will occupy the $3d$ orbitals, and 4 will occupy the $4p$ orbitals:

the argon core contains 18 electrons

$$\text{Se } (Z = 34) \qquad [\text{Ar}]4s^2 3d^{10} 4p^4$$

the remaining 16 electrons fill the $4s$ and $3d$ orbitals and partially fill the $4p$ orbitals

Orbitals beyond $4p$ remain empty. If preferred, the configuration for selenium may also be written in order of increasing n as

$$\text{Se } (Z = 34) \qquad [\text{Ar}]3d^{10} 4s^2 4p^4$$

or in expanded form as

$$\text{Se } (Z = 34) \qquad 1s^2 2s^2 2p^6 3s^2 3p^6 3d^{10} 4s^2 4p^4$$

(b) The outer three subshells are $4s^2 3d^{10} 4p^4$. Hund's rule applies to the partially filled $4p$ subshell, so two of the four electrons are unpaired. The orbital diagram is

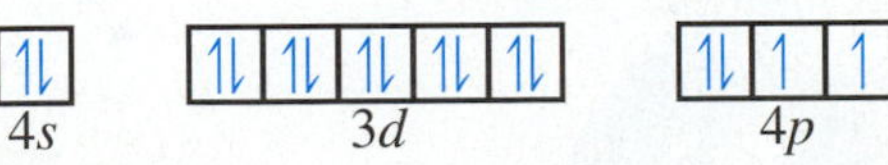

PRACTICE EXERCISE 8.4

(a) Write the electron configuration for vanadium, V ($Z = 23$). (b) Give an orbital diagram for the subshells outside the argon core.

Now you should be able to use the aufbau procedure, the Pauli exclusion principle, and Hund's rule to predict the ground-state configuration for any of the first 36 elements. Try writing configurations for all the fourth-row elements and compare your configurations with the actual configurations given in Table 8.1. You will find that only two elements, chromium and copper, do not follow the rules. The actual configurations of chromium and copper are

$$\text{Cr } (Z = 24) \qquad [\text{Ar}]4s^1 3d^5$$

$$\text{Cu } (Z = 29) \qquad [\text{Ar}]4s^1 3d^{10}$$

The aufbau procedure predicts $4s^2 3d^4$ for chromium and $4s^2 3d^9$ for copper; however, half-filled d subshells (such as d^5) and full d subshells (such as d^{10}) have especially low energies, so that chromium and copper gain stability by losing electrons from the $4s$ orbitals to the $3d$ orbitals. In every case, *the actual ground-state configuration will be the one that gives a minimum energy to the atom as a whole.*

8.4 THE PERIODIC TABLE

In 1869 Dmitri Mendeleev, a professor at St. Petersburg University, presented a paper to the Russian Chemical Society in which he stated that "the elements, if arranged according to their atomic weights, show a distinct periodicity of their properties." The paper included a periodic table arranged so that horizontal rows contained elements with similar chemical properties (see Figure 8.3). Somewhat later in 1869, Julius Lothar Meyer, a German chemist, published a similar table and similar conclusions. Meyer's table was obtained independently and without knowledge of Mendeleev's table, and it was based more on the periodicity of physical properties (e.g., atomic volumes) than on chemical properties. Devising a periodic table in 1869 required a great deal of chemical insight and ingenuity, and the conclusions reached by Mendeleev and Meyer, though not immediately accepted, were truly

(a)

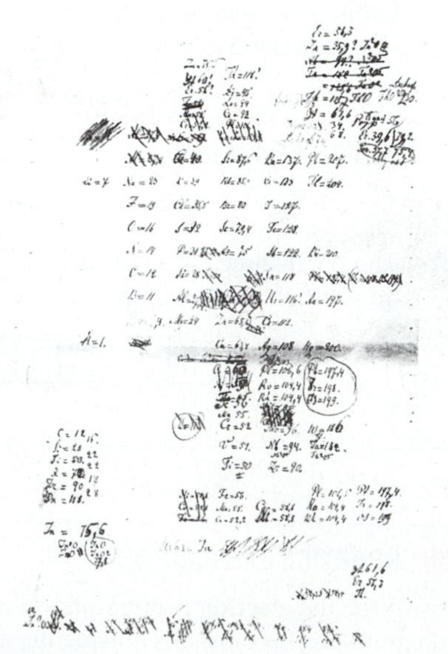

(b)

Figure 8.3
(a) Dmitri Ivanovitch Mendeleev (1834–1907). (b) One of the sketches of the first periodic system by Mendeleev, 1869.

TABLE 8.1 The Ground-State Electron Configurations of the Elements

Atomic Number	Element	Electron Configuration	Atomic Number	Element	Electron Configuration	Atomic Number	Element	Electron Configuration
1	H	$1s^1$	38	Sr	[Kr]$5s^2$	75	Re	[Xe]$6s^24f^{14}5d^5$
2	He	$1s^2$	39	Y	[Kr]$5s^24d^1$	76	Os	[Xe]$6s^24f^{14}5d^6$
3	Li	[He]$2s^1$	40	Zr	[Kr]$5s^24d^2$	77	Ir	[Xe]$6s^24f^{14}5d^7$
4	Be	[He]$2s^2$	41	Nb[a]	[Kr]$5s^14d^4$	78	Pt[a]	[Xe]$6s^14f^{14}5d^9$
5	B	[He]$2s^22p^1$	42	Mo[a]	[Kr]$5s^14d^5$	79	Au[a]	[Xe]$6s^14f^{14}5d^{10}$
6	C	[He]$2s^22p^2$	43	Tc	[Kr]$5s^24d^5$	80	Hg	[Xe]$6s^24f^{14}5d^{10}$
7	N	[He]$2s^22p^3$	44	Ru[a]	[Kr]$5s^14d^7$	81	Tl	[Xe]$6s^24f^{14}5d^{10}6p^1$
8	O	[He]$2s^22p^4$	45	Rh[a]	[Kr]$5s^14d^8$	82	Pb	[Xe]$6s^24f^{14}5d^{10}6p^2$
9	F	[He]$2s^22p^5$	46	Pd[a]	[Kr]$4d^{10}$	83	Bi	[Xe]$6s^24f^{14}5d^{10}6p^3$
10	Ne	[He]$2s^22p^6$	47	Ag[a]	[Kr]$5s^14d^{10}$	84	Po	[Xe]$6s^24f^{14}5d^{10}6p^4$
11	Na	[Ne]$3s^1$	48	Cd	[Kr]$5s^24d^{10}$	85	At	[Xe]$6s^24f^{14}5d^{10}6p^5$
12	Mg	[Ne]$3s^2$	49	In	[Kr]$5s^24d^{10}5p^1$	86	Rn	[Xe]$6s^24f^{14}5d^{10}6p^6$
13	Al	[Ne]$3s^23p^1$	50	Sn	[Kr]$5s^24d^{10}5p^2$	87	Fr	[Rn]$7s^1$
14	Si	[Ne]$3s^23p^2$	51	Sb	[Kr]$5s^24d^{10}5p^3$	88	Ra	[Rn]$7s^2$
15	P	[Ne]$3s^23p^3$	52	Te	[Kr]$5s^24d^{10}5p^4$	89	Ac	[Rn]$7s^26d^1$
16	S	[Ne]$3s^23p^4$	53	I	[Kr]$5s^24d^{10}5p^5$	90	Th[a]	[Rn]$7s^26d^2$
17	Cl	[Ne]$3s^23p^5$	54	Xe	[Kr]$5s^24d^{10}5p^6$	91	Pa	[Rn]$7s^25f^26d^1$
18	Ar	[Ne]$3s^23p^6$	55	Cs	[Xe]$6s^1$	92	U	[Rn]$7s^25f^36d^1$
19	K	[Ar]$4s^1$	56	Ba	[Xe]$6s^2$	93	Np	[Rn]$7s^25f^46d^1$
20	Ca	[Ar]$4s^2$	57	La	[Xe]$6s^25d^1$	94	Pu[a]	[Rn]$7s^25f^6$
21	Sc	[Ar]$4s^23d^1$	58	Ce	[Xe]$6s^24f^15d^1$	95	Am[a]	[Rn]$7s^25f^7$
22	Ti	[Ar]$4s^23d^2$	59	Pr[a]	[Xe]$6s^24f^3$	96	Cm	[Rn]$7s^25f^76d^1$
23	V	[Ar]$4s^23d^3$	60	Nd[a]	[Xe]$6s^24f^4$	97	Bk[a]	[Rn]$7s^25f^9$
24	Cr[a]	[Ar]$4s^13d^5$	61	Pm[a]	[Xe]$6s^24f^5$	98	Cf[a]	[Rn]$7s^25f^{10}$
25	Mn	[Ar]$4s^23d^5$	62	Sm[a]	[Xe]$6s^24f^6$	99	Es[a]	[Rn]$7s^25f^{11}$
26	Fe	[Ar]$4s^23d^6$	63	Eu[a]	[Xe]$6s^24f^7$	100	Fm[a]	[Rn]$7s^25f^{12}$
27	Co	[Ar]$4s^23d^7$	64	Gd	[Xe]$6s^24f^75d^1$	101	Md[a]	[Rn]$7s^25f^{13}$
28	Ni	[Ar]$4s^23d^8$	65	Tb[a]	[Xe]$6s^24f^9$	102	No[a]	[Rn]$7s^25f^{14}$
29	Cu[a]	[Ar]$4s^13d^{10}$	66	Dy[a]	[Xe]$6s^24f^{10}$	103	Lr	[Rn]$7s^25f^{14}6d^1$
30	Zn	[Ar]$4s^23d^{10}$	67	Ho[a]	[Xe]$6s^24f^{11}$	104	Unq	[Rn]$7s^25f^{14}6d^2$
31	Ga	[Ar]$4s^23d^{10}4p^1$	68	Er[a]	[Xe]$6s^24f^{12}$	105	Unp	[Rn]$7s^25f^{14}6d^3$
32	Ge	[Ar]$4s^23d^{10}4p^2$	69	Tm[a]	[Xe]$6s^24f^{13}$	106	Unh	[Rn]$7s^25f^{14}6d^4$
33	As	[Ar]$4s^23d^{10}4p^3$	70	Yb[a]	[Xe]$6s^24f^{14}$	107	Ns	[Rn]$7s^25f^{14}6d^5$
34	Se	[Ar]$4s^23d^{10}4p^4$	71	Lu	[Xe]$6s^24f^{14}5d^1$	108	Hs	[Rn]$7s^25f^{14}6d^6$
35	Br	[Ar]$4s^23d^{10}4p^5$	72	Hf	[Xe]$6s^24f^{14}5d^2$	109	Mt	[Rn]$7s^25f^{14}6d^7$
36	Kr	[Ar]$4s^23d^{10}4p^6$	73	Ta	[Xe]$6s^24f^{14}5d^3$			
37	Rb	[Kr]$5s^1$	74	W	[Xe]$6s^24f^{14}5d^4$			

Noble gas cores:

[He] = $1s^2$ [Kr] = [Ar]$4s^23d^{10}4p^6$

[Ne] = [He]$2s^22p^6$ [Xe] = [Kr]$5s^24d^{10}5p^6$

[Ar] = [Ne]$3s^23p^6$ [Rn] = [Xe]$6s^24f^{14}5d^{10}6p^6$

[a]Configurations of these elements differ from those predicted by the methods given in the text.

remarkable. At the time of their work, many elements had not been discovered and one group of elements, the noble gases, was completely missing. The ordering in the early tables was based on atomic weights rather than atomic numbers because electrons had not yet been identified and atomic numbers were unknown.

Since 1869 the periodic table has been extended, the missing elements have been discovered, and a theoretical basis for the table in terms of atomic numbers and electron configurations has been developed. A modern periodic table with atomic symbols, atomic numbers, and atomic weights is shown inside the front cover of this book. The periodic table in Figure 8.4 shows the outermost electrons, those most

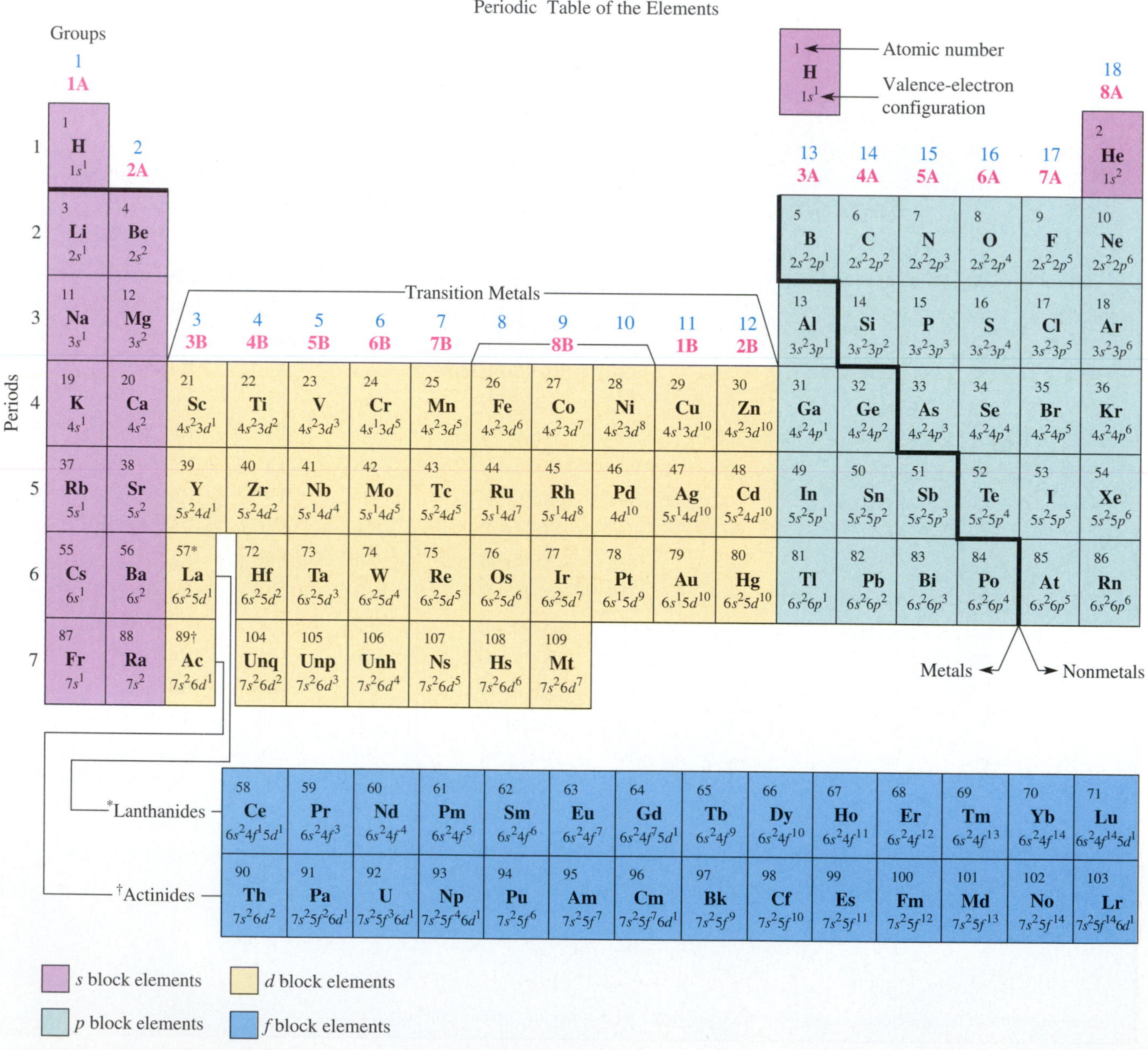

Figure 8.4

A modern periodic table showing atomic numbers, atomic symbols, and valence-electron configurations. In this text we use standard American notation, which assigns the letter A to representative groups and the letter B to transition groups. The International Union of Pure and Applied Chemistry (IUPAC) has recommended that the *s*, *p*, and *d* block groups be numbered consecutively from 1 to 18.

TABLE 8.2 The Representative Groups and Their Names

Group	Name
Group 1A (except hydrogen)	Alkali metals
Group 2A	Alkaline earth metals
Group 3A	Boron group
Group 4A	Carbon group
Group 5A	Nitrogen group
Group 6A	Oxygen group
Group 7A	Halogens
Group 8A	Noble gases

likely to participate in chemical reactions. These electrons are called *valence electrons.* (Valence electrons will be discussed more fully in the next section.)

Let us examine Figure 8.4 in more detail. The vertical columns, called **groups**, are families of elements with similar chemical properties. Each column is headed by a number and letter that identify the group.

The horizontal rows are called **periods**. If we look at the electron configurations from one atom to the next, we see that the $n = 1$ shell fills across the *first period* with hydrogen (H, $1s^1$) and helium (He, $1s^2$). The $n = 2$ shell fills as we move across the *second period* from lithium (Li, $[\text{He}]2s^1$) to neon (Ne, $[\text{He}]2s^22p^6$). The $3s$ and $3p$ orbitals fill as we move across the *third period* from sodium (Na, $[\text{Ne}]3s^1$) to argon. (The apparent gaps after beryllium and magnesium have no significance; they simply make room on the page for subsequent longer periods.) The second and third periods each contain eight elements because the s and p subshells can hold a total of eight electrons. Note that *each period begins with the addition of an* s *electron to an unoccupied shell and ends with the formation of a noble gas atom.*

Let's continue to examine the table. The *fourth period* begins with electrons filling the $4s$ orbitals in potassium and calcium. Then electrons fill the $3d$ subshell beginning with scandium (Sc, $[\text{Ar}]4s^23d^1$) and ending with zinc (Zn, $[\text{Ar}]4s^23d^{10}$). Finally, the $4p$ subshell fills from gallium through krypton. The $4s$, $3d$, and $4p$ subshells hold a total of 18 electrons; hence, there are 18 elements in the fourth period.

Before we look at the elements past krypton, note that the periodic table in Figure 8.4 is divided into *s, p, d,* and *f* blocks. Groups 1A and 2A (and helium) constitute the ***s* block** because each member adds an *s* electron to the previous configuration. Groups 3A through 8A (excluding helium) form the ***p* block** because their configurations result from the addition of *p* electrons. The elements in the *s* and *p* blocks are called **representative elements**. The eight vertical groups, or A groups, that contain these elements are called **representative groups** or **main groups**. All of the elements in the first three periods and all of the nonmetallic elements are found in the main groups. Some of the groups have special names; for example, Group 7A contains the *halogens* and Group 8A contains the *noble gases* (see Table 8.2).

The ***d* block** and the ***f* block** consist of elements in which the last electron enters a *d* and an *f* orbital, respectively. These elements are all metals; they lie between Groups 2A and 3A, and they are known collectively as **transition elements**. The *f* block elements are often called **inner transition elements**. The *d* block numbers and letters used in this book run from 3B through 8B (8B consists of three groups), followed by 1B and 2B. (An alternative numbering system, the IUPAC system, is shown in Figure 8.4.) Transition groups are often identified by the name of the first member of the group; for example, the copper group (1B) consists of copper, silver, and gold; the zinc group (2B) consists of zinc, cadmium, and mercury.

The first transition series contains 10 metals beginning with scandium and ending with zinc; the second transition series begins with yttrium and ends with cadmium. The 14 metals of the first inner transition series, cerium ($Z = 58$) through lutetium ($Z = 71$), are called **lanthanides** because they immediately follow the element lanthanum, which they greatly resemble. Lanthanides are sometimes called **rare earth elements** because they were found in oxide ores called *earths* and because they were once thought to be very rare. The second inner transition series, thorium ($Z = 90$) through lawrencium ($Z = 103$), follows the element actinium, so these elements are called **actinides**. Only the first three actinides, thorium, protactinium, and uranium, occur naturally; the others have been synthesized by nuclear chemists. Lanthanides and actinides are usually written below the main table in order to save space.

Now let us look at the filling of the electron subshells in the fifth, sixth, and seventh periods. Figure 8.4 shows that the *fifth period* begins with rubidium (Rb, $[Kr]5s^1$) and ends with xenon. The $5s$, $4d$, and $5p$ subshells fill across this period. The fifth period, like the fourth period, contains 18 elements. The *sixth period* begins with electrons filling the $6s$ orbitals in cesium and barium. Lanthanum follows barium with one electron in a $5d$ orbital (La, $[Xe]6s^25d^1$). Then the $4f$ subshell fills, beginning with cerium (Ce, $[Xe]6s^24f^15d^1$) and extending for 14 elements to lutetium (Lu, $[Xe]6s^24f^{14}5d^1$). The $5d$ and $6p$ subshells then fill until the sixth period ends with radon. There are 32 elements in the sixth period because the $6s$, $4f$, $5d$, and $6p$ subshells hold a total of 32 electrons. The *seventh period* starts with francium, but it is incomplete.

Valence Electrons

The original periodic table was a clever device to group elements with similar properties together. Today the periodic table has even greater significance because it provides a link between chemical properties and electron configuration. Figure 8.4 shows that, with few exceptions, *elements from the same group have similar outer electron configurations.* All Group 1A elements, for example, have an ns^1 outer configuration, where n is the period in which the element is found. Group 2A elements have an ns^2 outer configuration, and the noble gases (Group 8A) have the configuration ns^2np^6. (Helium, a noble gas with the configuration $1s^2$, is an exception.) We now know that the group similarities discovered by early chemists are due to similarities in electron configuration.

The word *valence* is derived from the Latin *valentia* meaning "strength" or "capacity." The term is used by chemists in a number of ways, but in its broadest sense it refers to the capacity of an atom to combine with other atoms.

The electrons most likely to participate in an atom's chemical behavior come from the outer electron shells and are called **valence electrons**. The shells containing these electrons are called **valence shells**. Inspection of Figure 8.4 shows that the shell with the largest value of n is always a valence shell. Valence electrons also come from the $n-1$ shell for transition (d block) elements and the $n-2$ shell for inner transition (f block) elements. Comparison of the configurations in Figure 8.4 with those in Table 8.1 shows that the valence-shell configuration may or may not be identical to the configuration of electrons outside the noble gas core. For example, tin ($Z = 50$) contains 14 electrons ($5s^24d^{10}5p^2$) outside the krypton core, but only 4 of these ($5s^25p^2$) are valence electrons.

If we look at the representative elements (A-group elements) in Figure 8.4, we will see that, *with the single exception of helium, the number of valence electrons in an atom of a representative element is the same as its group number.* For example, lithium in Group 1A has one valence electron, aluminum in Group 3A has three valence electrons, and phosphorus in Group 5A has five valence electrons.

EXAMPLE 8.2

State the number of valence electrons and give the valence-shell configuration for an atom of (a) oxygen and (b) bromine.

SOLUTION

(a) Oxygen ($Z = 8$), a Group 6A element, has six valence electrons. The electron configuration of oxygen is $[He]2s^22p^4$. The valence shell is the $n = 2$ shell. Its configuration is $2s^22p^4$.

(b) Bromine ($Z = 35$), a Group 7A element, has seven valence electrons. The electron configuration of bromine is $[Ar]4s^23d^{10}4p^5$. The valence shell is the outermost $n = 4$ shell. Its configuration is $4s^24p^5$.

PRACTICE EXERCISE 8.5

How many valence electrons are in an atom of (a) iodine, (b) magnesium, and (c) helium?

Electron Configurations from the Periodic Table

The division into *s, p, d,* and *f* blocks makes it easy for us to find the ground-state configurations of most elements from their positions in the periodic table. Each period ends with one of the noble gases of Group 8A, so that an atom in the next period consists of the preceding noble gas core plus the additional electrons accumulated across its period. Inspection of Figure 8.4 shows, for example, that calcium ($Z = 20$) is in the *s* block of the fourth period. Calcium therefore has an argon core (18 electrons) plus two additional electrons; its configuration is $[Ar]4s^2$. The periodic table can be used to find the electron configuration of any element, but it is especially convenient for elements beyond krypton, as shown in the next example.

EXAMPLE 8.3

Use the periodic table inside the front cover to write the electron configuration for tin, element 50.

SOLUTION

Tin is in the fifth period, Group 4A. The preceding noble gas is krypton in the fourth period, so the tin configuration starts with [Kr]. Traveling from krypton to tin takes us 14 spaces across the fifth period, adding 2 electrons in the *s* block, 10 electrons in the *d* block, and finally 2 *p* electrons. The configuration of tin is $[Kr]5s^2 4d^{10} 5p^2$.

PRACTICE EXERCISE 8.6

Use the periodic table inside the front cover to write the electron configuration for lead, element 82. (*Hint:* Don't forget that some *f* block elements fall between elements 57 and 72.)

An electron configuration predicted from the periodic table does not always agree with the actual electron configuration. Table 8.1 shows that predicted configurations are identical to actual configurations for the representative elements, but only approximate for most of the *f* block elements and some of the *d* block elements. The irregularities are due to small energy differences among the electrons in *d* or *f* orbitals, and there is no simple way of predicting them all.

Electron Configurations of Monatomic Ions

The electron configuration of a positive ion can be found from that of its parent atom: *When an atom forms a positive ion, its electrons are removed in order of decreasing n; electrons with the same n value are removed in order of decreasing l.* For example, let's look at the formation of an iron(II) ion (Fe^{2+}). The configuration of iron is obtained from the periodic table:

$$\text{Fe}\ (Z = 26) \qquad [\text{Ar}]4s^2 3d^6$$

During ionization the 4*s* electrons (highest *n* value) are lost first and the configuration of the iron(II) ion is

$$\text{Fe}^{2+} \qquad [\text{Ar}]3d^6$$

Note that the order in which electrons are removed is not always the reverse of the filling order. For example, 4*s* electrons are lost before 3*d* electrons even though 4*s* orbitals fill before 3*d* orbitals.

Now consider the indium(I) ion (In^+). The configuration of indium is

$$\text{In}\ (Z = 49) \qquad [\text{Kr}]5s^2 4d^{10} 5p^1$$

The 5*s* and the 5*p* electrons have the same *n* value. During ionization, the 5*p* electron (highest *l* value) is lost first, and the configuration of the indium(I) ion is

$$\text{In}^+ \qquad [\text{Kr}]5s^2 4d^{10}$$

EXAMPLE 8.4

Write the ground-state electron configurations of K^+, Mn^{2+}, and Fe^{3+}.

SOLUTION

First we use the periodic table to write configurations for the neutral atoms:

$K\ (Z = 19)$ $[Ar]4s^1$

$Mn\ (Z = 25)$ $[Ar]4s^23d^5$

$Fe\ (Z = 26)$ $[Ar]4s^23d^6$

Electrons with the highest n are lost first. Potassium will lose one $4s$ electron, manganese will lose two $4s$ electrons, and iron will lose two $4s$ electrons and one $3d$ electron. The ion configurations are

K^+ $[Ar]$

Mn^{2+} $[Ar]3d^5$

Fe^{3+} $[Ar]3d^5$

Observe that the iron(III) and manganese(II) ions have the same configuration.

PRACTICE EXERCISE 8.7

Write the ground-state electron configuration of (a) Ba^{2+} and (b) Pb^{2+}.

The ground-state configuration of a negative monatomic ion is the same as that of an atom with the same number of electrons. Moreover, a stable negative ion will have the configuration of the noble gas at the end of its period. This is illustrated in the next example.

EXAMPLE 8.5

Write the ground-state configuration of the oxide ion (O^{2-}).

SOLUTION

The atomic number of oxygen is 8, so the neutral atom has 8 electrons. The oxide ion has 2 additional electrons, or 10 in all. Its configuration is identical to that of neon, $1s^22s^22p^6$, or simply [Ne].

PRACTICE EXERCISE 8.8

Write the electron configuration of the iodide ion (I^-).

8.5 ATOMIC PROPERTIES

Each element is unique because its atoms are different from those of other elements. The properties of atoms are determined in large part by their atomic numbers and electron configurations, and they exhibit periodic variations and trends that can be traced across a period and down a group. These trends make it possible for a chemist to predict much of an element's behavior simply from its position in the periodic table.

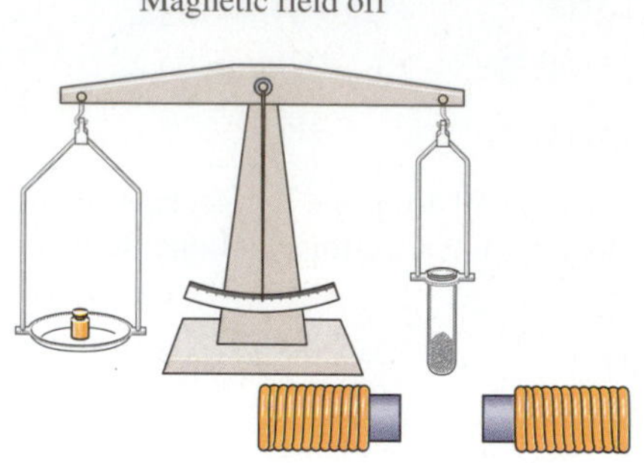

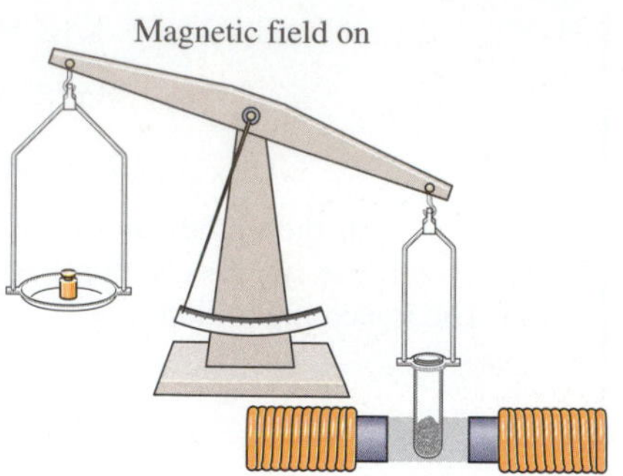

Figure 8.5

Diagram of a Gouy balance for measuring paramagnetism. On the left, the tube containing a paramagnetic sample is balanced by known weights with the magnetic field off. On the right, the magnetic field exerts an additional downward force on the paramagnetic substance. The added weights needed to restore balance are a measure of the paramagnetism.

Paramagnetism

Two electrons that share an orbital have opposing spins, and their magnetic fields cancel as shown in Figure 8.2. The field of an unpaired electron is not canceled, so atoms and ions containing unpaired electrons can be drawn into an external magnetic field (see Figure 8.5). Such atoms are said to be **paramagnetic**. Atoms and ions with no unpaired electrons are **diamagnetic**, a term that essentially denotes the absence of paramagnetism.

The degree of paramagnetism increases with the number of unpaired electrons. A carbon atom ([He]$2s^2 2p^2$) with two unpaired *p* electrons shows a higher degree of paramagnetism than a lithium atom ([He]$2s^1$) with one unpaired electron. The response of atoms and ions to an external magnetic field provides data from which the number of unpaired electrons can be calculated; this information helps in determining their electron configurations.

EXAMPLE 8.6

Compare the atoms of cobalt and copper with respect to paramagnetism.

SOLUTION

The electron configuration for cobalt ($Z = 27$) is [Ar]$4s^2 3d^7$. For copper ($Z = 29$), the configuration is [Ar]$4s^1 3d^{10}$. The cobalt atom has three unpaired electrons in the $3d$ orbitals

and exhibits a greater degree of paramagnetism than the copper atom, which has only one unpaired electron in the $4s$ orbital.

PRACTICE EXERCISE 8.9

Are the following atoms paramagnetic or are they diamagnetic: (a) calcium, (b) sodium, and (c) sulfur?

CHEMICAL INSIGHT

FERROMAGNETISM

Most of us are familiar with the magnets that operate toys, hold cupboard doors shut, and pull scrap iron out of a junkpile. We are also familiar with magnetic compass needles that point north. These magnets are made of iron, and their magnetic action is due to *ferromagnetism,* an effect that is many times stronger than paramagnetism.

Paramagnetism is a property of individual atoms. Paramagnetic measurements are made on samples of independent particles, such as vaporized atoms or ions in a solution. The magnetic properties of a collection of atoms in close contact with each other may be quite different from the magnetic properties of isolated atoms. Solid copper, for example, is diamagnetic even though its atoms are paramagnetic. Neighboring copper atoms in the solid interact so that their magnetic fields cancel and there is no net magnetic field. The atoms of a ferromagnetic solid such as iron behave differently—they align their individual magnetic fields so that they reinforce each other and produce magnetic regions within the solid called *domains* (Figure 8.6b). The various domains tend to be oriented randomly with respect to each other (Figure 8.6c), but an external magnetic field will cause them to line up as in Figure 8.6d. The domains remain aligned after the external field is removed, and the iron becomes a permanent magnet (see Figure 8.7).

Ferromagnetism is exhibited to a lesser degree by cobalt and nickel, the elements immediately following iron in the periodic table. Certain alloys and oxides are also ferromagnetic. Strong permanent magnets, for example, are often made from *alnico,* an alloy of aluminum, nickel, cobalt, and iron. Magnetic recording tapes are coated with ferromagnetic oxides such as chromium dioxide (CrO_2) or the mixed oxide of iron and cobalt ($Fe_2O_3 \cdot CoO$).

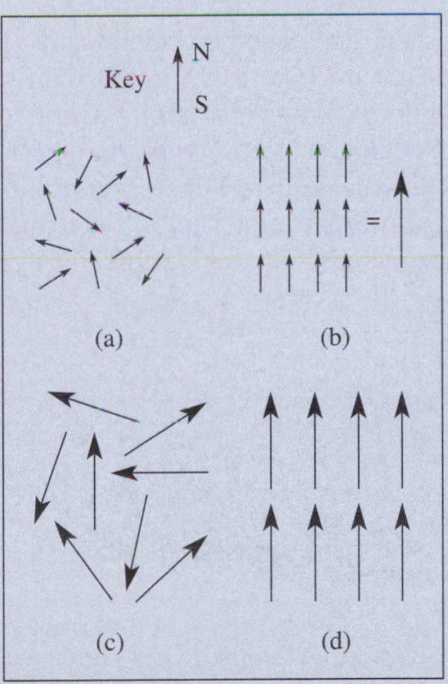

Figure 8.6

Ferromagnetism. (a) Neighboring paramagnetic atoms in random alignment. (Small arrow = magnetic field of a single atom.) (b) In iron and other ferromagnetic metals, groups of paramagnetic atoms tend to align themselves into magnetic domains. (Heavy arrow = net magnetic field of a domain.) (c) Usually the domains are randomly aligned, so the net magnetic field of the iron is zero. (d) An external magnetic field causes the domains to line up, and the iron becomes a magnet.

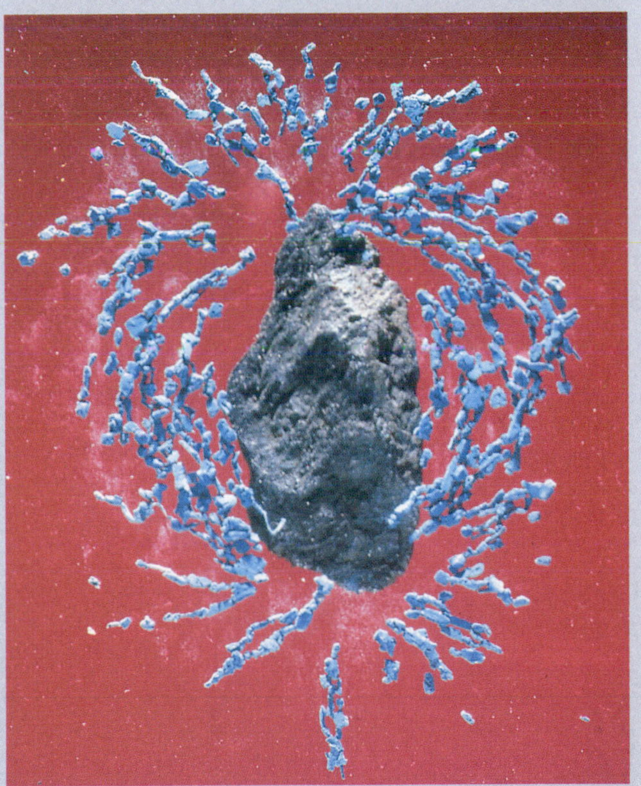

Figure 8.7

Iron filings arrange themselves along the magnetic lines of force surrounding a piece of magnetite (Fe_3O_4). (The filings are painted blue.)

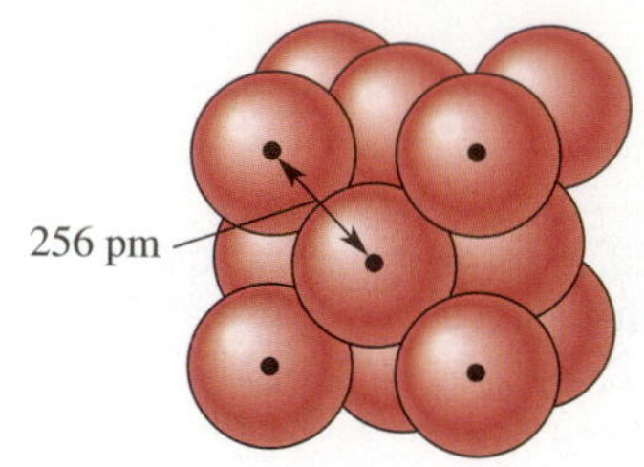

$$\text{Atomic radius} = \frac{256\text{ pm}}{2} = 128\text{ pm}$$

Copper

(a)

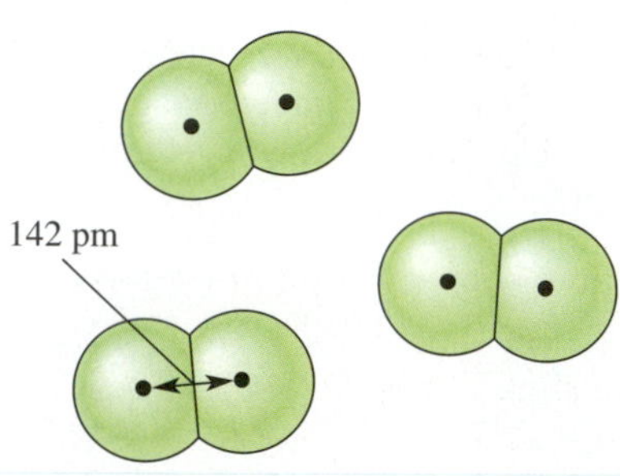

$$\text{Atomic radius} = \frac{142\text{ pm}}{2} = 71\text{ pm}$$

Fluorine (F_2)

(b)

Figure 8.8

An atomic radius is one-half the average distance between centers of identical neighboring atoms. (a) In copper metal, where the atoms form a cubic lattice, each copper atom has a radius of 128 pm. (b) The radius of a fluorine atom in F_2 is 71 pm.

Atomic Size

How big is a carbon atom or an atom of gold? These questions are not easy to answer because most of an atom's volume is occupied by an insubstantial electron cloud with no boundary to separate inside from outside. Yet in many respects atoms behave like ordinary objects, bouncing off each other as if boundaries did exist. The apparent boundary that prevents one atom from merging with another is an electrostatic one. Atoms are drawn toward each other by the attraction of each electron cloud for the other nucleus, but at the same time they are pushed apart by the repulsion between electrons and between nuclei. An **atomic radius** is one-half the average distance between centers of identical neighboring atoms, as shown in Figure 8.8. The atomic radius varies with the environment of the atom; it decreases as the atom becomes more firmly bound. For example, when lithium metal is vaporized, a few of the liberated atoms combine to form Li_2 molecules; the atoms in isolated Li_2 molecules have smaller radii than the atoms in solid lithium metal.

Internuclear distances can be measured to the nearest picometer (1 pm = 10^{-12} m). As shown in Figure 8.8a, the average distance between two copper nuclei in copper metal is 256 pm, so the radius of a copper atom is one-half of this distance or 128 pm. Values for atomic radii are given in Figure 8.9 and plotted in Figure 8.10.

Figures 8.9 and 8.10 show that, with some exceptions, *atomic radii increase from top to bottom within a group.* The radii of Group 1A atoms, for example, increase from 152 pm for lithium to 265 pm for cesium. This trend is to be expected because each successive period starts a new shell, and each new shell has its peak electron density at a greater distance from the nucleus. Figures 8.9 and 8.10 also show (again with some exceptions) that *atomic radii decrease from left to right across a period.* This decrease in size occurs because the added electrons fill vacancies in an already existing shell. The nuclear charge (atomic number) increases from left to right across the period, and because electrons in the same shell do not shield each other effectively from the nuclear force, the increased attractive force brings all the electrons closer to the nucleus.

EXAMPLE 8.7

Refer only to the periodic table inside the front cover, and list the following atoms in order of increasing atomic radius: Mg, K, and Ca.

SOLUTION

Ca (Group 2A) lies to the right of K (Group 1A) in the fourth period, so Ca is smaller than K. Mg lies just above Ca in Group 2A; therefore, Mg is smaller than Ca. Mg is the smallest and K is the largest: Mg $<$ Ca $<$ K.

PRACTICE EXERCISE 8.10

Refer only to the periodic table inside the front cover, and predict which atom in each of the following pairs has the larger atomic radius: (a) bromine or iodine; (b) silicon or phosphorus.

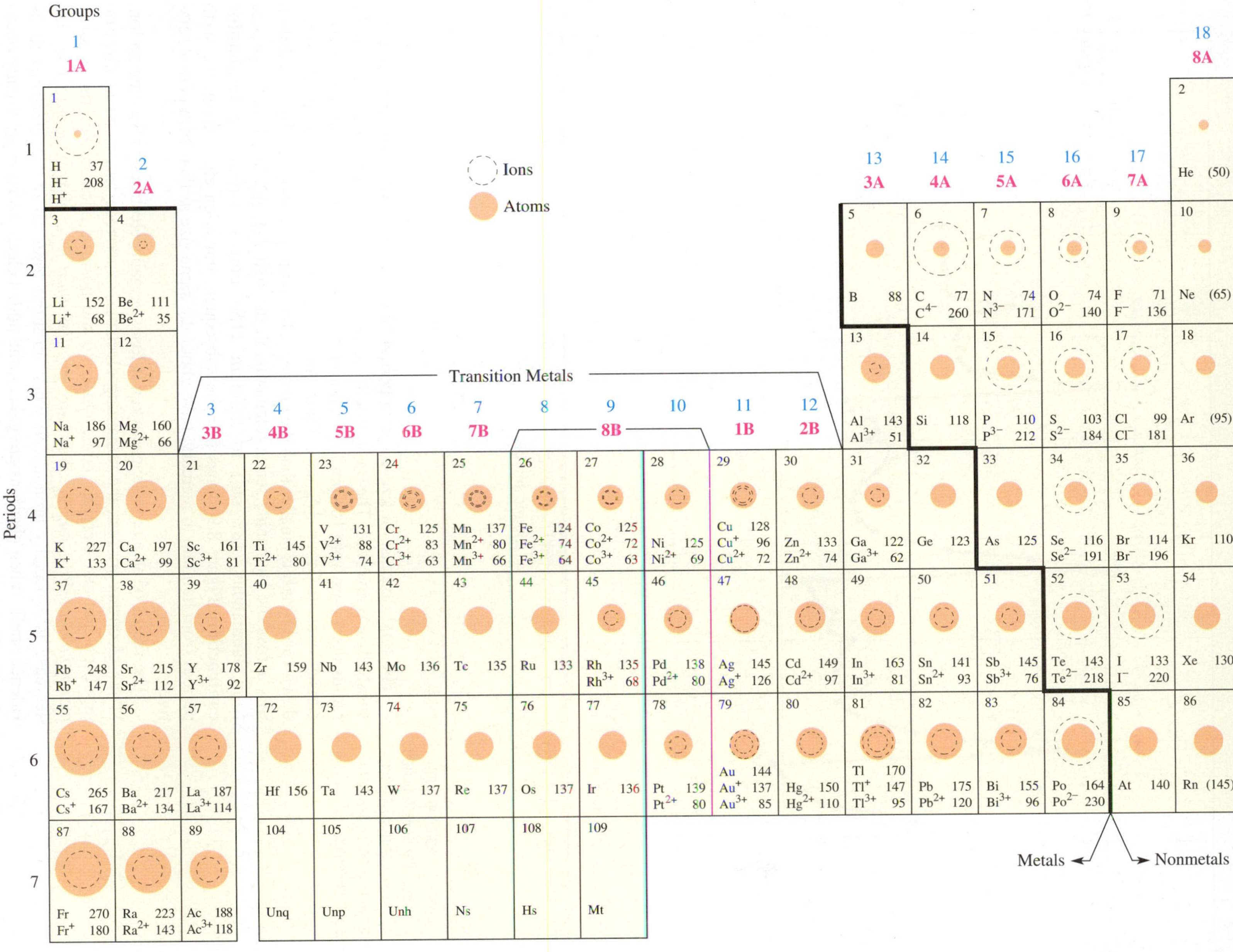

Figure 8.9

Atomic and ionic radii, with values in picometers (1 m = 10^{12} pm). (The values in parentheses are estimated.) With some exceptions, radii increase from top to bottom within a group and decrease from left to right across a period. Positive ions are always smaller than their parent atoms, and negative ions are always larger.

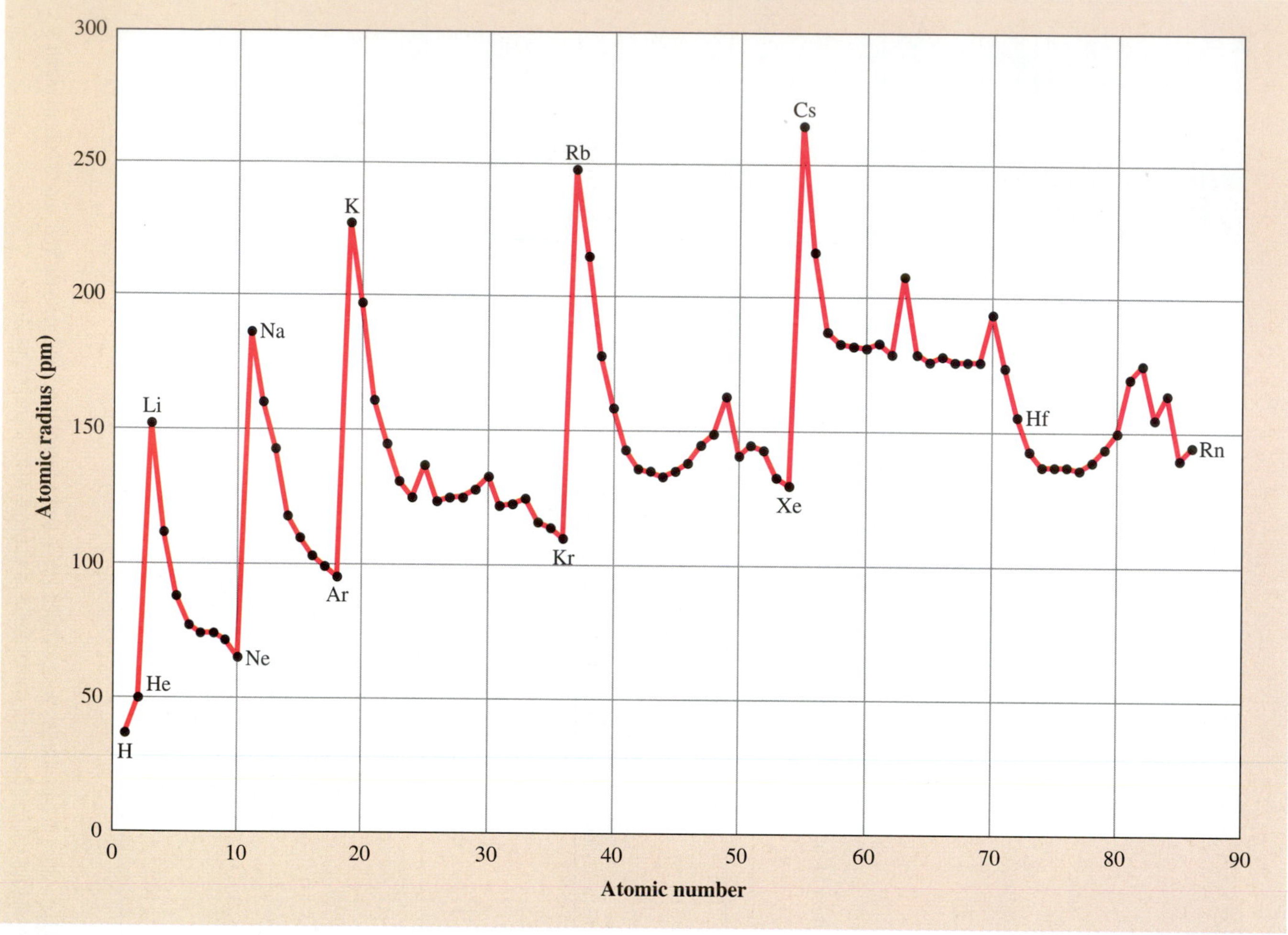

Figure 8.10

Plot of atomic radius versus atomic number for the first 86 elements. In each period, the Group 1A atom has the largest atomic radius, and the Group 8A noble gas atom has the smallest atomic radius.

The total decrease in size across a period can be appreciable, especially if *d* or *f* subshells are involved. The 3*d* subshell fills across the first transition series, and the radius decreases from 161 pm for scandium to 133 pm for zinc. The decrease in size carries over to Group 3A, so gallium (in one of the few exceptions to the usual downward trend) is smaller than the element aluminum just above it. The 4*f* subshell fills across the lanthanide series (elements 58 through 71), and the decrease in size from lanthanum (187 pm) to hafnium (156 pm) is called the **lanthanide contraction**. Figure 8.9 shows that the subsequent transition elements in the sixth period, from hafnium to mercury, have about the same size as their fifth period relatives.

The radius of an atom changes when it loses or gains electrons to become an ion. When electrons are lost, the remaining electrons experience less electron–electron repulsion and can therefore come closer to the positive nucleus. For this reason, *monatomic positive ions (cations) are smaller than their parent atoms.* The size decrease is especially pronounced when an atom loses all of the electrons in its valence shell. When electrons are gained, the added electrons enter an already occupied shell and mutual repulsion tends to push all the electrons apart. Consequently, *monatomic negative ions (anions) are larger than the parent atoms.* Some ionic radii are reported in Figure 8.9.

PRACTICE EXERCISE 8.11

Which will have the larger radius: (a) Ba or Ba^{2+}; (b) S or S^{2-}?

Ionization Energy

The energy required to remove an electron from a gaseous ground-state atom or ion is called the **ionization energy** (IE). The first ionization energy is the amount of energy needed to remove the most loosely bound electron from the valence shell of the neutral atom. The second ionization energy is the amount needed to remove the second electron after the first is gone. Ionization energies are usually expressed in kilojoules per mole of atoms, and some values are given in Figure 8.11 and Table 8.3. For example, the first and second ionization energies of magnesium are

$$Mg(g) \rightarrow Mg^{+}(g) + e^{-} \qquad \text{first IE} = +737.7 \text{ kJ}$$

$$Mg^{+}(g) \rightarrow Mg^{2+}(g) + e^{-} \qquad \text{second IE} = +1450.7 \text{ kJ}$$

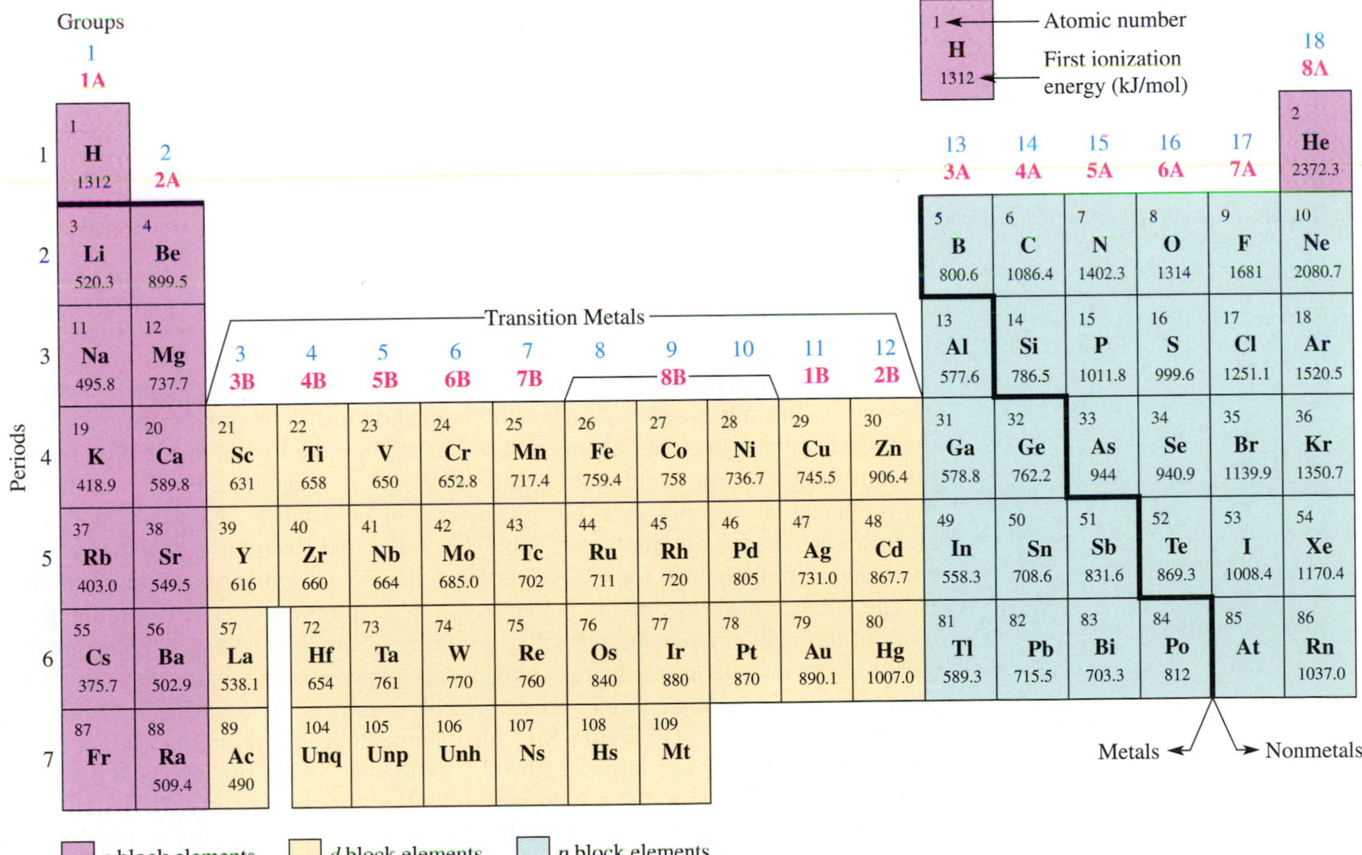

Figure 8.11
First ionization energies of the elements in kilojoules per mole of atoms. With some exceptions, ionization energies decrease toward the bottom of a group and increase from left to right across a period.

TABLE 8.3 Ionization Energies of the First 20 Elements

		Ionization Energies (kJ/mol)									
Z	*Element*	*First*	*Second*	*Third*	*Fourth*	*Fifth*	*Sixth*	*Seventh*	*Eighth*	*Ninth*	*Tenth*
1	H	1312									
2	He	2372	5250								
3	Li	520	7298	11,815							
4	Be	900	1757	14,849	21,007						
5	B	801	2427	3660	25,026	32,827					
6	C	1086	2353	4621	6223	37,830	47,277				
7	N	1402	2856	4578	7475	9445	53,266	64,360			
8	O	1314	3388	5300	7469	10,990	13,326	71,335	84,078		
9	F	1681	3374	6050	8408	11,023	15,164	17,868	92,038	106,434	
10	Ne	2081	3952	6122	9370	12,178	15,238	19,999	23,069	115,379	131,431
11	Na	496	4512	6912	9544	13,353	16,610	20,115	25,490	28,934	141,363
12	Mg	738	1451	7733	10,540	13,628	17,995	21,704	25,656	31,643	25,462
13	Al	578	1817	2745	11,578	14,831	18,378	23,295	27,459	31,861	38,457
14	Si	787	1577	3232	4356	16,091	19,785	23,786	29,252	33,877	38,733
15	P	1012	1903	2912	4957	6274	21,269	25,397	29,854	35,867	40,959
16	S	1000	2251	3361	4564	7013	8496	27,106	31,670	36,578	43,138
17	Cl	1251	2297	3822	5158	6540	9362	11,018	33,605	38,598	43,962
18	Ar	1521	2666	3931	5771	7238	8781	11,995	13,842	40,760	46,187
19	K	419	3051	4411	5877	7976	9649	11,343	14,942	16,964	48,576
20	Ca	590	1145	4912	6474	8144	10,496	12,320	14,207	18,192	20,385

NOTE: The ionization energies of valence-shell electrons are in color. Note that electrons below the valence shell have extremely high ionization energies.

The removal of electrons from an atom is an endothermic process, so ionization energies are always positive. The second ionization energy is greater than the first, the third is greater than the second, and so on, because the loss of an electron leaves the remaining electrons held more tightly by the positive nucleus. Each atom has as many ionization energies as it has electrons, but only those for the valence-shell electrons are chemically important. Table 8.3 shows that electrons below the valence shell have extremely high ionization energies; they do not ordinarily leave the atom.

The valence electrons of the larger atoms in a group are easier to remove than those of the smaller atoms because they are farther from the nucleus and, despite the fact that the nucleus has a larger charge, they experience a weaker net attractive force. Hence, *ionization energies tend to decrease toward the bottom of a group,* as illustrated in Figure 8.11. Figure 8.11 also shows that *ionization energies tend to increase from left to right across a period.* Moving across a period brings a decrease in atomic radius along with an increase in nuclear charge. Because the outer electrons are closer to the nucleus and held more strongly, more energy is required to remove them from the atom.

A graph of first ionization energy versus atomic number (see Figure 8.12) shows

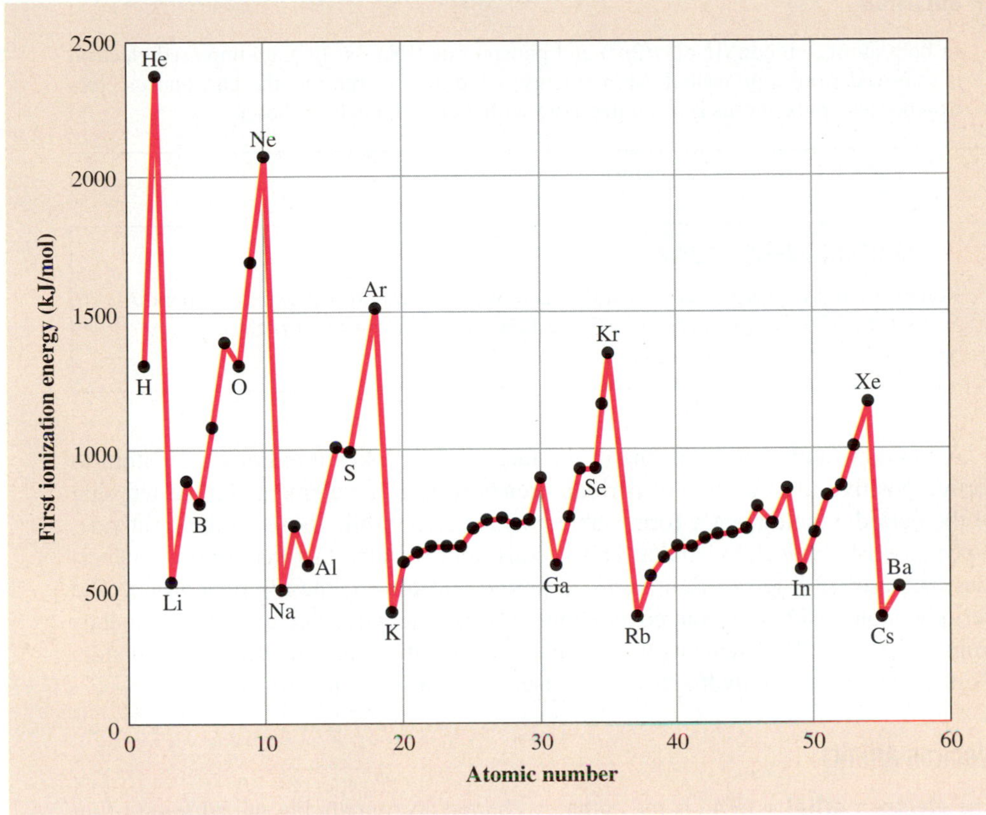

Figure 8.12

Plot of first ionization energy versus atomic number for the first 56 elements. In each period, the Group 1A atom has the lowest ionization energy, and the Group 8A noble gas atom has the highest ionization energy.

that each period starts with a low ionization energy in Group 1A and ends with a high ionization energy for the noble gas terminating the period. There are a few irregularities, some of which are easily explained. Boron, for example, has a lower ionization energy than beryllium. The lone *p* electron in the valence shell of a boron atom ($2s^22p^1$) experiences a smaller effective charge and is less tightly bound than either of the two electrons in the valence shell of a beryllium atom ($2s^2$); the lone *p* electron is easier to remove than the *s* electrons, so boron has a lower ionization energy.

LEARNING HINT

Review the discussion of the shielding effect given in Section 8.1. The shielding effect and effective charge can be estimated from the ionization energy. (See Final Exercise 8.111.)

Another irregularity occurs at oxygen, which has a lower ionization energy than nitrogen. Oxygen has a $2s^22p^4$ valence shell configuration containing two paired *p* electrons and two unpaired *p* electrons. Nitrogen has a $2s^22p^3$ configuration with three unpaired *p* electrons.

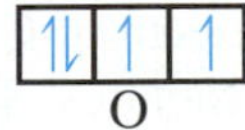

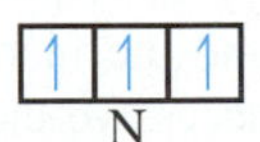

Electron–electron repulsion causes paired electrons in a subshell to have slightly higher energies than unpaired electrons, so it takes less energy to remove a paired $2p$ electron from oxygen than an unpaired $2p$ electron from nitrogen.

EXAMPLE 8.8

Explain why the first ionization energy of phosphorus is greater than that of silicon.

SOLUTION

In both atoms, silicon ([Ne]$3s^23p^2$) and phosphorus ([Ne]$3s^23p^3$), an unpaired electron is removed from a $3p$ orbital. More energy is required to remove the phosphorus electron because phosphorus is a smaller atom with a greater nuclear charge.

PRACTICE EXERCISE 8.12

Refer only to the periodic table inside the front cover and predict which element in each of the following pairs has the higher first ionization energy: (a) chlorine or bromine; (b) sodium or magnesium.

A low ionization energy is one of the factors that makes it easier for an atom to form a positive ion. A study of periodic trends shows that elements on the left side of the periodic table readily form stable positive ions, while those on the right side, especially the upper right, do not. This is consistent with the fact that ionization energies increase from left to right across a period and toward the top of a group. The periodic table inside the front cover shows a heavy stepwise line that divides metals from nonmetals. The elements that readily form positive ions are on the metal side. Nonmetals, except for hydrogen, do not usually form stable positive ions.

Electron Affinity

The **electron affinity** (EA) is the enthalpy change accompanying the addition of an electron to a gaseous ground-state atom or ion. The electron affinity of chlorine, for example, is −349 kJ/mol:

$$\text{Cl(g)} + \text{e}^- \rightarrow \text{Cl}^-\text{(g)} \qquad \Delta H^\circ_{\text{EA}} = -349 \text{ kJ}$$

The negative sign means that 349 kJ of heat is evolved during the formation of 1 mol of chloride ions. (Some authors reverse the sign for electron affinity, so watch out!) The electron affinities listed in Figure 8.13 and plotted in Figure 8.14 show that the variation of electron affinity is periodic, but there are many irregularities. With the notable exceptions of the Group 2A elements and the noble gases, most atoms evolve at least a small amount of energy when they acquire one electron. The greatest amounts of energy are evolved by the relatively small halogen atoms of Group 7A.

The gain of a second or third electron by a negative gaseous ion is always endothermic because energy is required to overcome the electrostatic repulsion between the negative ion and the incoming electron. Consider the formation of a gaseous O^{2-} ion by two successive electron captures. The first electron capture is exothermic and evolves 141 kJ/mol; the second electron capture is endothermic and absorbs 880 kJ/mol:

$$\text{O(g)} + \text{e}^- \rightarrow \text{O}^-\text{(g)} \qquad \Delta H^\circ_{\text{EA}} = -141 \text{ kJ}$$

$$\text{O}^-\text{(g)} + \text{e}^- \rightarrow \text{O}^{2-}\text{(g)} \qquad \Delta H^\circ_{\text{EA}} = +880 \text{ kJ}$$

The net reaction is endothermic:

$$\text{O(g)} + 2\text{e}^- \rightarrow \text{O}^{2-}\text{(g)} \qquad \Delta H^\circ = +639 \text{ kJ}$$

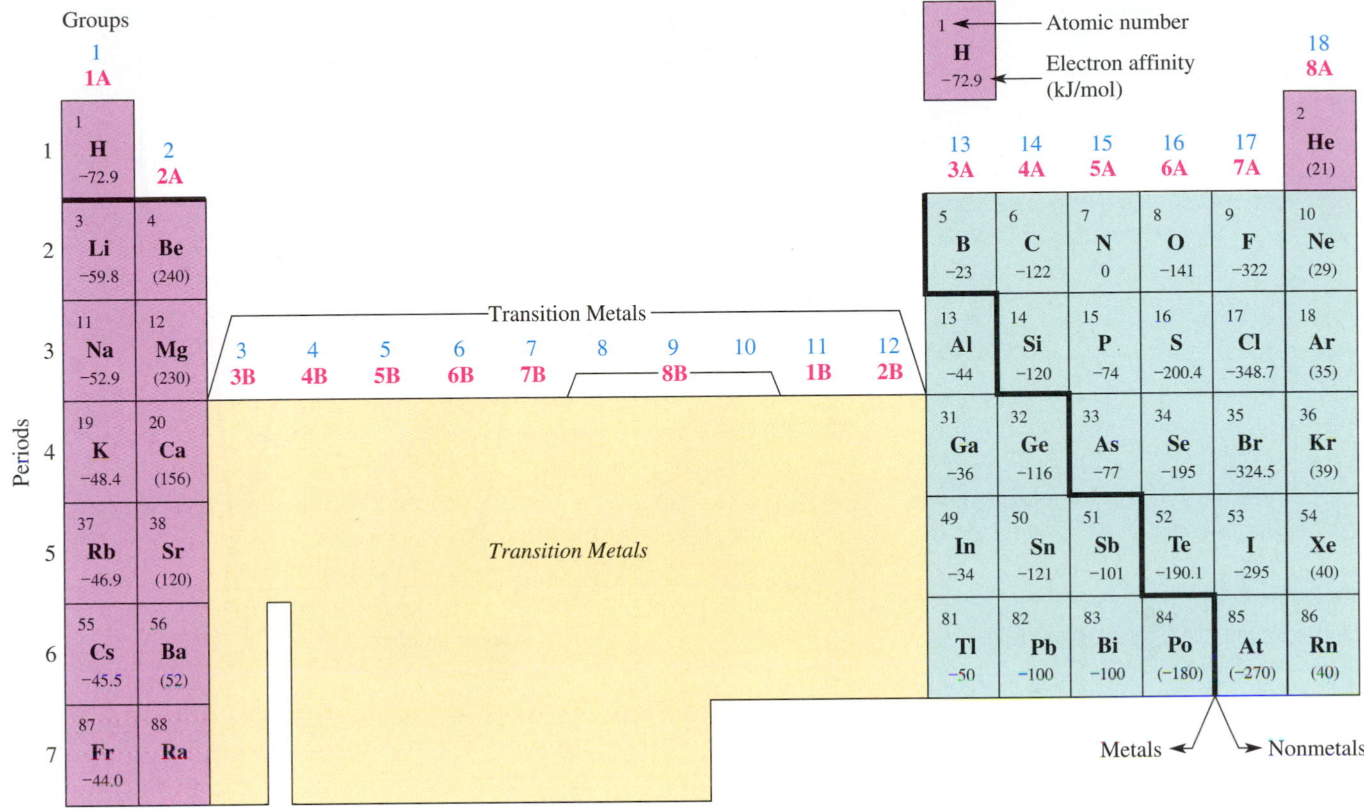

Figure 8.13

Electron affinities of the representative elements in kilojoules per mole of atoms. (A value in parentheses is estimated.) Most atoms evolve energy when they acquire electrons; hence, most atoms have negative electron affinities.

8.6 DESCRIPTIVE CHEMISTRY AND THE PERIODIC TABLE

The periodic tables published by Mendeleev and Meyer in 1869 were based on a wealth of facts about the 63 elements known at that time. Their densities and colors, their reactions, the formulas of their compounds, and the acidity or basicity of their oxides were among the many properties that were found to have periodicity with respect to atomic weight. Now we know that the periodic table also reflects a periodicity in the electron configurations of the atoms making up the elements. Because the chemical properties of an element are determined primarily by the valence-shell configurations of its atoms, we should be able to use the periodic table to predict some of the experimental behavior that produced the table in the first place.

Traditionally, the periodic table is divided by a stepwise line into metallic and nonmetallic elements. An examination of Figure 8.15 shows that half of the representative elements are **metals** and half are **nonmetals**. The transition elements, however, are all metals, so 87 of the 109 known elements are metals. Metals possess a characteristic shine or luster, are good conductors of electricity and heat, and are usually *malleable* and *ductile*. (A malleable substance is one that can be shaped by hammering or pounding; a ductile substance is one that can be drawn into wires.)

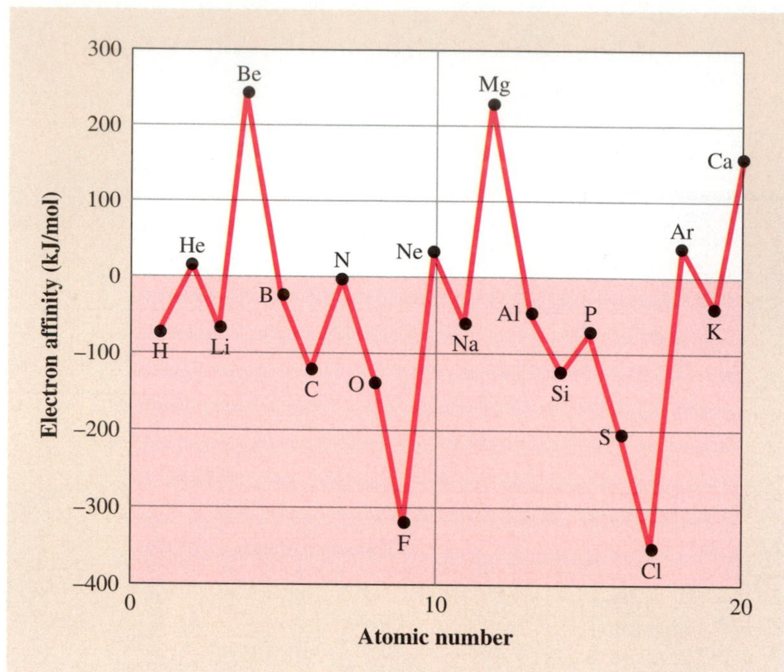

Figure 8.14

Plot of electron affinity versus atomic number for the first 20 elements.

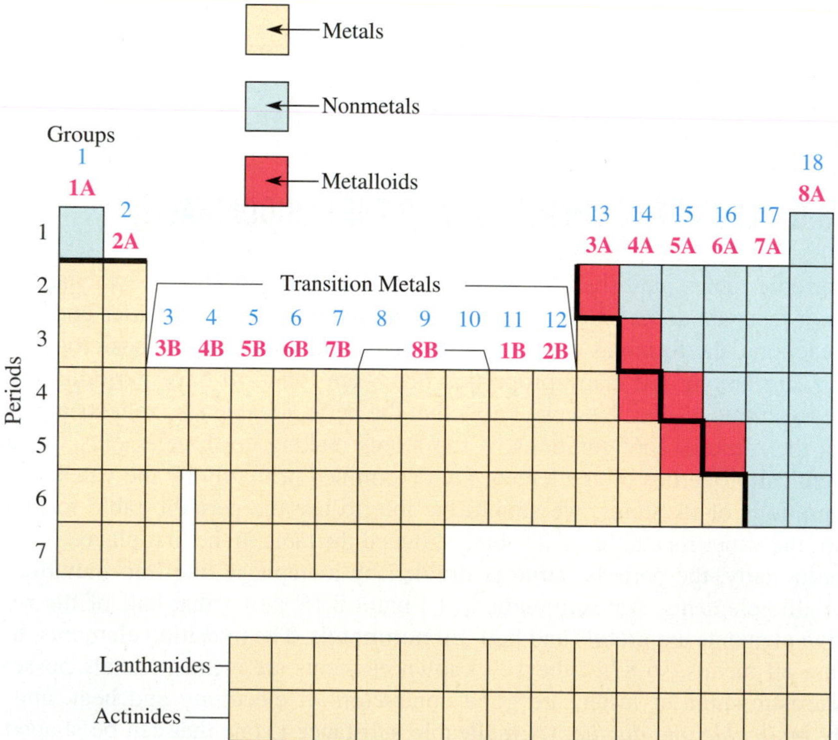

Figure 8.15

Periodic table showing the metallic elements, the nonmetallic elements, and the metalloids.

TABLE 8.4 Some Properties of Metallic and Nonmetallic Elements

Metallic Elements
Possess a characteristic luster
Good conductors of heat and electricity
Usually malleable and ductile
Atoms tend to lose electrons

Nonmetallic Elements
Not lustrous
Poor conductors of heat and electricity; good insulators
Solids tend to be brittle
Atoms tend to gain electrons

Nonmetals do not usually possess these properties; they are not lustrous, are poor conductors of electricity and heat, and their solid forms tend to be brittle. Some of the properties of metals and nonmetals are compared in Table 8.4. The change from metallic to nonmetallic character is not as sharp as the stepwise line in the periodic table might suggest. Many of the elements immediately adjacent to the line have some metallic and some nonmetallic characteristics. Antimony, for example, has a bluish white metallic luster, but it is extremely brittle. Such elements are sometimes called **semimetals** or **metalloids**.

You will soon learn that most of the physical and chemical properties of metals are associated with loosely held valence electrons. Atoms of metals have lower ionization energies than atoms of nonmetals, and they have less tendency to accept electrons. Metal atoms always lose electrons to some extent when they combine with nonmetal atoms. In fact, the most active metal atoms lose their valence electrons altogether and form positive ions.

Much of the descriptive chemistry of the elements and their compounds is woven into the main chapters of this book. The chemistry of the most important elements and periodic groups, however, is systematically described in four survey chapters. These surveys have been positioned to reinforce the concepts of the theoretical chapters immediately preceding them, but you may actually read them any time after the chapter that precedes each one has been covered. Survey 1, which immediately follows this chapter, discusses hydrogen and the alkali metals (Group 1A), the alkaline earth metals (Group 2A), and the noble gases (Group 8A). Survey 2, which describes oxygen, nitrogen, and the halogens (Group 7A), uses some of the concepts developed in Chapter 10 (Chemical Bonding Theory) and Chapter 11 (Oxidation–Reduction Reactions). Survey 3 follows Chapter 18 (Solubility and Complex Ion Equilibria) and describes the properties of boron, carbon, phosphorus, sulfur, and some of the more important semimetals. Finally, Survey 4, which follows Chapter 21 (Metals and Coordination Chemistry), describes the important metals in the *p* and *d* blocks of the periodic table.

CHAPTER REVIEW

LEARNING OBJECTIVES BY SECTION

8.1 1. List the similarities and differences between the orbitals of a many-electron atom and those of hydrogen.
2. List the orbitals within a shell in order of increasing electron energy.
3. State how the energy of an electron in a given orbital varies with increasing atomic number.
4. Describe the shielding effect, and explain why the energy of an electron increases with increasing *l* within a shell.

8.2 1. State the Pauli exclusion principle.

8.3 1. State the orbital filling order for the first 36 elements.
2. Write ground-state electron configurations in both abbreviated and expanded forms for the first 36 elements.
3. Write orbital diagrams for the first 36 elements.

8.4 1. Identify the *s, p, d,* and *f* blocks in the periodic table.
2. Given a periodic table, identify the representative elements, the transition elements, the lanthanides, and the actinides.
3. Give the name of each of the representative groups.
4. Explain why the elements in a periodic group exhibit similar chemical properties.
5. Use group numbers to state the number of valence electrons in a representative element.
6. Use the periodic table to write ground-state electron configurations of atoms and monatomic ions.

8.5 1. Use electron configurations to predict whether atoms and ions are paramagnetic or diamagnetic.
2. Use periodic trends to compare atomic and ionic radii.
3. Use periodic trends to compare ionization energies of atoms and ions.
4. Use periodic trends to compare electron affinities of atoms.
5. Explain why the second electron affinity for an atom is always positive.

8.6 1. Distinguish between metals and nonmetals on the basis of their physical and chemical properties.

KEY TERMS BY SECTION

8.1 Effective nuclear charge
Many-electron atom
Shielding effect

8.2 Paired electrons
Pauli exclusion principle
Unpaired electron

8.3 Aufbau procedure
Electron configuration
Ground-state configuration
Hund's rule
$n + l$ rule
Noble gas core
Orbital diagram

8.4 Actinides
d block elements
f block elements
Group (in the periodic table)
Inner transition elements
Lanthanides
Main group elements
p block elements
Period (in the periodic table)
Rare earth elements
Representative elements
s block elements
Transition elements
Valence electrons
Valence shell

8.5 Atomic radius
Diamagnetism
Electron affinity (EA)
Ionization energy (IE)
Lanthanide contraction
Paramagnetism

8.6 Metal
Metalloid
Nonmetal
Semimetal

FINAL EXERCISES

Answers to exercises with blue numbers are given in Appendix D.

PART A. QUESTIONS AND PROBLEMS BY SECTION

Orbitals in Many-Electron Atoms (Section 8.1)

8.1 Describe two ways in which the orbitals of many-electron atoms resemble hydrogen orbitals and two ways in which they differ from hydrogen orbitals.

8.2 Give at least two reasons why the orbitals in a many-electron atom cannot be exactly like those of hydrogen.

8.3 List the following electrons in order of increasing energy: 3*s*, 3*d*, and 3*p*.

8.4 List the orbitals in the $n = 4$ shell in order of increasing electron energy.

8.5 In which orbital does an electron have the lower energy, the 2*p* orbital of barium or the 2*p* orbital of strontium? In which orbital is the electron more tightly bound to the atom?

8.6 Which electron has the higher energy, a 2*s* electron in calcium or a 2*s* electron in iodine? Which electron is more tightly bound to the atom?

8.7 **(a)** Describe the shielding effect.
(b) Explain in terms of radial electron densities why the energy of an electron increases with increasing *l* within a shell.

8.8 **(a)** What is meant by the phrase *effective nuclear charge?* How is the effective nuclear charge related to the shielding effect?
(b) Is the effective charge experienced by a 3*p* electron greater than or less than the effective charge experienced by a 3*d* electron in the same atom? Explain.

8.9 In which orbital of aluminum, 3*s* or 3*p*, will an electron be more tightly bound? Explain your answer in terms of the shielding effect and the effective nuclear charge experienced by the electron.

8.10 In which orbital of boron, 2*s* or 2*p*, will an electron have a lower energy? Explain your answer in terms of the shielding effect and the effective nuclear charge experienced by the electron.

Pauli Exclusion Principle (Section 8.2)

8.11 State two versions of the Pauli exclusion principle.

8.12 What is the difference between paired electrons and unpaired electrons? Between parallel spins and antiparallel spins?

8.13 What is the maximum number of electrons that can occupy each of the following shells?
(a) $n = 2$
(b) $n = 4$
(c) $n = n$

8.14 How many electrons are in
(a) a filled *p* subshell? **(c)** a half-filled *f* subshell?
(b) a filled *d* subshell? **(d)** a filled $n = 3$ shell?

8.15 Give the four quantum numbers for each electron in (a) a full 2*s* subshell of a sodium atom and (b) a full 2*p* subshell of a chlorine atom.

8.16 How many electrons in a given atom can have the following quantum numbers?
(a) $n = 4, l = 1$ **(c)** $n = 2, m_s = +\frac{1}{2}$
(b) $n = 3$ **(d)** $n = 4, l = 2, m_l = -1$

Electron Configurations (Section 8.3)

8.17 What is a ground-state electron configuration? Write the ground-state configurations of oxygen and hydrogen.

8.18 What is an excited-state electron configuration? Write an excited-state configuration for oxygen and for hydrogen.

8.19 State the orbital filling order for the first 36 elements.

8.20 State Hund's rule. Draw an orbital diagram for the nitrogen atom, and explain how it illustrates Hund's rule.

8.21 What is the aufbau, or "building-up," procedure? Describe how the aufbau procedure is used in writing the ground-state electron configuration of the carbon atom.

8.22 What is a *noble gas core*? Write symbols for the noble gas cores of fluorine and iron.

8.23 Write expanded and abbreviated ground-state electron configurations for each of the following atoms:
(a) beryllium **(d)** vanadium
(b) phosphorus **(e)** selenium
(c) calcium **(f)** fluorine

Compare your configurations with those in Table 8.1.

8.24 Write expanded and abbreviated ground-state electron configurations for each of the following atoms:
(a) nitrogen **(d)** nickel
(b) chlorine **(e)** arsenic
(c) gallium **(f)** iron

Do your configurations agree with those in Table 8.1?

8.25 Draw an orbital diagram for each of the atoms in Exercise 8.23. How many unpaired electrons are in each atom?

8.26 Draw an orbital diagram for each of the atoms in Exercise 8.24. How many unpaired electrons are in each atom?

8.27 Use the $n + l$ rule to write electron configurations for chromium and copper. Do these configurations agree with those in Table 8.1?

8.28 Use the $n + l$ rule to write electron configurations for manganese and zinc. Do these configurations agree with those in Table 8.1?

The Periodic Table (Section 8.4)

8.29 What does the term *periodic* mean? Which of the following exhibit some degree of periodicity?
(a) days of the week
(b) your age
(c) atomic numbers
(d) valence-shell electron configurations

8.30 Why did Meyer and Mendeleev base their periodic tables on atomic weights rather than atomic numbers? What kinds of properties did they find to be periodic?

8.31 Sketch the general form of the periodic table without looking at the text.
(a) How many columns do you put in the *s* block? The *d* block? The *p* block?
(b) In which row does the *d* block start?
(c) How many spaces do you put in the lanthanide series?

8.32 Sketch (or imagine) the general form of the periodic table without looking at the text.
(a) Which period is the shortest?
(b) Which period is the longest?
(c) Between which two elements does the stepwise line dividing metals from nonmetals begin? Between which two elements does it end?

8.33 Which subshells fill from left to right across (a) the second period, (b) the third period, and (c) the fourth period?

8.34 Which subshells fill from left to right across (a) the fifth period, (b) the sixth period, and (c) the seventh period?

8.35 In which blocks of the periodic table are the following elements found?
(a) representative elements **(d)** metals
(b) transition elements **(e)** nonmetals
(c) inner transition elements **(f)** rare earth elements

8.36 In which block of the periodic table are the following elements found?
(a) halogens **(e)** coinage metals (copper group)
(b) alkaline earth metals **(f)** iron triad (Fe, Co, and Ni)
(c) actinides
(d) helium

8.37 Refer to the periodic table, and identify the noble gas core for each of the following atoms:
(a) Be **(d)** Al
(b) Fe **(e)** U
(c) Sb **(f)** Ra

8.38 Refer to the periodic table, and identify the noble gas core for each of the following atoms:
(a) Cu **(d)** Gd
(b) Se **(e)** Pt
(c) Au **(f)** Ag

8.39 Use the periodic table to write abbreviated ground-state configurations for the atoms in Exercise 8.37. Compare your configurations with those in Table 8.1. Which, if any, of these elements show irregularities?

8.40 Use the periodic table to write abbreviated ground-state configurations for the atoms in Exercise 8.38. Check your configurations with those in Table 8.1. Which, if any, of these elements show irregularities?

8.41 Use the periodic table to write abbreviated ground-state configurations for the following atoms:
(a) Cr **(d)** I
(b) Sm **(e)** Hg
(c) Mo **(f)** Ba

Check your configurations with those in Table 8.1. Which, if any, of these elements show irregularities?

8.42 Use the periodic table to identify the following elements, and write their abbreviated ground-state electron configurations:
(a) $Z = 17$ **(d)** $Z = 28$
(b) $Z = 39$ **(e)** $Z = 77$
(c) $Z = 98$ **(f)** $Z = 109$

Do your configurations agree with those in Table 8.1?

8.43 Draw orbital diagrams for all partially filled subshells in the atoms of Exercise 8.37.

8.44 Write an abbreviated electron configuration for zirconium, and draw an orbital diagram for each partially filled subshell.

8.45 Refer to Table 8.1, and draw orbital diagrams for the 4*s* and 3*d* orbitals of each element in the first transition series. Which of these elements show irregularities?

8.46 Refer to Table 8.1, and draw orbital diagrams for the 5*s* and 4*d* orbitals of each element in the second transition series. Are the irregularities in this series the same or different from those in the first transition series (see Exercise 8.45)?

8.47 How many valence electrons does each of the following atoms have?
(a) Sr **(c)** Cs
(b) In **(d)** Si

8.48 How many valence electrons does each of the following atoms have?
(a) I **(c)** Bi
(b) Se **(d)** Pb

8.49 Use the periodic table to write the valence-shell configuration of each of the following representative elements:

(a) Na (c) Te
(b) Ca (d) Br

8.50 Use the periodic table to write the valence-shell configuration of each of the following representative elements:
(a) Ga (c) Ar
(b) P (d) Sn

8.51 Write the valence-shell configuration for each of the following groups:
(a) Halogens (d) Alkaline earth metals
(b) Alkali metals (e) Group 4A
(c) Group 2A (f) Group 2B

8.52 Write the valence-shell configuration for each of the following groups:
(a) Oxygen group (d) Group 3A
(b) Noble gases (e) Group 5A
(c) Group 1A (f) Group 1B

8.53 Write an abbreviated electron configuration for each of the following ions:
(a) Zn^{2+} (e) V^{2+}
(b) Co^{3+} (f) F^-
(c) Al^{3+} (g) S^{2-}
(d) Ni^{2+} (h) Ag^+

8.54 Write an abbreviated electron configuration for each of the following ions:
(a) Ca^{2+} (e) Fe^{2+}
(b) N^{3-} (f) Na^+
(c) Cr^{3+} (g) I^-
(d) O^{2-} (h) Hg^{2+}

Atomic Properties; Periodic Trends (Section 8.5)

8.55 Briefly explain why paramagnetic substances are drawn into external magnetic fields while diamagnetic substances are not.

8.56 In which periodic groups are paramagnetic atoms found? Are the majority of elements paramagnetic or diamagnetic?

8.57 Given the fact that an atom has no outer surface, how do we justify the concept of an atomic radius?

8.58 How is the radius of an atom defined? Does the radius of a given atom always have the same value? Explain.

8.59 Which element has the largest atomic radius? The smallest?

8.60 Do atomic radii tend to increase or decrease (a) from top to bottom in a periodic group and (b) from left to right across a period? Account for these trends on the basis of atomic structure and electron configuration.

8.61 Which element has the greatest first ionization energy? The smallest?

8.62 Do ionization energies tend to increase or decrease (a) from top to bottom in a periodic group and (b) from left to right across a period? Account for these trends on the basis of atomic structure and electron configuration.

8.63 (a) Why are ionization energies always positive?
(b) Why is the second ionization energy of an atom always greater than its first ionization energy?

8.64 Which is more favorable for the formation of positive ions, high ionization energies or low? Explain.

8.65 Lithium is next to beryllium in the periodic table. Table 8.3 shows that the first ionization energy of lithium is less than that of beryllium. Lithium's second ionization energy, however, is substantially greater than that of beryllium. Account for these facts.

8.66 Explain why the first ionization energies of B, O, Al, and S are less than the first ionization energies of the elements immediately preceding them in the periodic table, contrary to the general periodic trend.

8.67 Explain why the electron affinity value is a measure of the tendency of an atom to gain an electron.

8.68 Which element has the most positive electron affinity? The most negative electron affinity?

8.69 Explain why the second electron affinity of an atom is always positive.

8.70 (a) Which periodic groups have positive electron affinities?
(b) Which group has the lowest (most negative) electron affinities?

8.71 Which of the first 10 elements have paramagnetic atoms?

8.72 (a) Which of the fourth-period elements have paramagnetic atoms?
(b) Which have diamagnetic atoms?

8.73 How many unpaired electrons does each of the following atoms have?
(a) N (c) Fe
(b) Zn (d) Ag

8.74 How many unpaired electrons does each of the following atoms have?
(a) Cl (c) As
(b) Cr (d) Ti

8.75 Which ions in Exercise 8.53 are paramagnetic?

8.76 Which ions in Exercise 8.54 are diamagnetic?

8.77 Write the ground-state configurations of Cu, Cu^+, and Cu^{2+}. (*Caution:* Copper is an exception to the usual filling rules.) Compare the paramagnetism of these species.

8.78 Consult Table 8.1, and find the element whose atom should exhibit the highest degree of paramagnetism.

8.79 Use the periodic table to predict which atom in each pair has the larger atomic radius:
(a) S or Cl (b) Ca or Ba (c) Si or S (d) Mg or Na (e) N or C (f) Br or I

8.80 Use the periodic table to predict which atom in each pair has the smaller atomic radius:
(a) Ar or Xe (b) Ba or Cl (c) Al or S (d) K or As (e) Cl or I (f) C or Si

8.81 Use the periodic table to predict the order of increasing atomic radius in each of following sets:
(a) I, Cl, and Br
(b) C, F, and Si
(c) K, Mg, and Ca

8.82 Use the periodic table to predict the order of decreasing atomic radius in each of the following sets:
(a) Rb, Kr, and K
(b) O, F, and Mg
(c) Se, S, and As

8.83 Which species should have the larger radius?
(a) K or K^+ (b) Cl or Cl^- (c) Fe^{2+} or Fe^{3+} (d) P or P^{3-} (e) S^{2-} or O^{2-} (f) I^- or Cl^-

8.84 Which species should have the smaller radius?
(a) Mg or Mg^{2+} (b) O or O^{2-} (c) H^+ or H^- (d) Au^+ or Au^{3+} (e) Se^{2-} or S^{2-} (f) N^{3-} or F^-

8.85 Use the periodic table to predict which atom in each pair has the greater first ionization energy:
(a) K or Ca (b) F or C (c) S or Se (d) Ba or Mg (e) Al or P (f) N or F

8.86 Use the periodic table to predict which atom in each pair has the smaller first ionization energy:
(a) P or Ga (b) K or Cl (c) Sr or S (d) O or Si (e) Mg or Rb (f) N or Br

8.87 Refer to Exercise 8.83. Which species in each pair should have the greater first ionization energy?

8.88 Refer to Exercise 8.84. Which species in each pair should have the smaller first ionization energy?

8.89 Use the periodic table to predict the order of increasing ionization energy for each set of atoms in Exercise 8.81.

8.90 Use the periodic table to predict the order of decreasing ionization energy for each set of atoms in Exercise 8.82.

Metals and Nonmetals (Section 8.6)

8.91 How many of the representative elements are metals? How many are nonmetals?

8.92 How many of the *d* and *f* block elements are metals? How many are nonmetals?

8.93 How many of the elements are metals? How many are nonmetals?

8.94 What is a semimetal (metalloid)? How many elements are semimetals? Name them.

8.95 List some of the physical properties by which a metal can be distinguished from a nonmetal.

8.96 Compare metal and nonmetal atoms with respect to the following properties:
(a) ionization energies
(b) tendency to give up electrons
(c) tendency to accept electrons

PART B. MISCELLANEOUS QUESTIONS AND PROBLEMS

8.97 A lithium atom consists of a nucleus and three electrons. Keep track of the electrons by labeling them 1, 2, and 3, and count the number of different electrostatic attractions and repulsions that are operating within this atom.

8.98 Assume that no two electrons in an atom have the same values for all four quantum numbers. Show how this assumption leads to the conclusion that orbital occupancy is limited to two electrons of opposite spin.

8.99 Write abbreviated electron configurations for elements 57 through 72 using (a) the periodic table and (b) the $n + l$ rule. Compare your configurations with those in Table 8.1. Which method gives better results for these elements?

8.100 Explain why the elements in a group exhibit similar properties.

8.101 (a) Refer to Figure 8.11, and calculate the longest wavelength radiation that will ionize gaseous sodium atoms.
(b) In what spectral region will radiation of this wavelength be found?

8.102 Bohr's formula for electron energies in hydrogen is a special case of the following general formula that applies to any atom or ion with only one electron: $E_n = -Z^2 \times 2.179 \times 10^{-18}\ \text{J}/n^2$. Do the ionization energies of one-electron ions increase or decrease with increasing atomic number Z? Explain. Confirm your prediction by calculating ionization energies for the ions He^+ and Li^{2+}.

8.103 What is the lanthanide contraction? Explain it on the basis of periodic trends.

8.104 Explain in terms of electron configurations and atomic properties why the lanthanides (elements 58 through 71) all exhibit similar behavior.

8.105 Explain why the noble gases have positive electron affinities.

8.106 Use the periodic table to compare the atoms in the following pairs with respect to paramagnetism, atomic radius, first ionization energy, and electron affinity: (a) Na and S, (b) N and Sb.

8.107 It is often said that "metallic character" increases from top to bottom within a group and decreases from left to right across a period. Explain these trends in terms of atomic properties.

8.108 Which member of Group 6A should have
- **(a)** the highest first ionization energy?
- **(b)** the most negative electron affinity?
- **(c)** the smallest atomic radius?
- **(d)** the least tendency to form positive ions?
- **(e)** the least tendency to form negative ions?
- **(f)** the most metallic character?

8.109 Potassium and bromine are in the fourth period. Compare these elements with respect to the following properties:
- **(a)** atomic size
- **(b)** first ionization energy
- **(c)** electron affinity
- **(d)** metallic character
- **(e)** tendency to give up electrons
- **(f)** tendency to accept electrons

8.110 Look up the melting and boiling points of the first 36 elements in the *Handbook of Chemistry and Physics*, and plot them as a function of atomic number. Are these properties periodic?

8.111 The effective nuclear charge (Z^*) experienced by an electron in a many-electron atom is given by $Z^* = Z - \sigma$ where Z is the actual nuclear charge and σ, called the shielding constant, is a measure of the effect of electron repulsions. The energy of an electron in a many-electron atom can be approximated by a modified Bohr formula:

$$E_n = -(Z^*)^2 \times 2.179 \times 10^{-18}\ \text{J}/n^2.$$

Use the ionization energies in Table 8.3 to calculate effective charges and shielding constants for the $1s$ and $2s$ electrons in lithium. Do your results make sense? Why or why not? (*Note:* You might want to try Exercise 8.102 before attempting this one.)

The principal ingredient of seashells is calcium carbonate ($CaCO_3$).

GROUPS 1A, 2A, AND 8A

OUTLINE

PREVIEW

Each element has its own niche in the periodic table; it shares the family characteristics conferred on it by the valence-shell configuration, and it has unique properties that are largely determined by its size. Family resemblances are most consistent among the representative elements, especially in the first two and the last two groups. Groups like these, in which most of the members' behavior can be predicted on the basis of periodic properties and group trends, are said to be "well behaved." In this chapter we look at Groups 1A, 2A, and 8A, three of the best behaved groups.

Each element belongs to a group or "family" of elements in the periodic table. However, like individuals in a human family, each element has properties that distinguish it from its relatives and neighbors. When we study a group of elements, we will often single out the more important members for special attention. A thorough acquaintance with an element includes knowledge of its occurrence, preparation, properties, and uses, as well as the occurrence, preparation, properties, and uses of its principal compounds. This information is not just a catalog of facts; it forms part of the much larger pattern that makes up the entire periodic mosaic of elemental properties.

S1.1 HYDROGEN

Hydrogen, which forms more compounds than any other element, is placed in Group 1A because its atoms have a single *s* valence electron. Hydrogen atoms can give up their electrons to form positive $H^+(aq)$ ions; they can share electrons as in the nonionic compounds H_2O, CH_4, and NH_3; and they can capture electrons to form negative ***hydride ions*** (H^-). The other Group 1A elements differ from hydrogen in that their atoms do not readily share electrons or form negative ions. The unique behavior of hydrogen is largely due to the small size, high ionization energy, and electron affinity of its atoms.

TABLE S1.1 Elemental Composition of the Sun and the Universe[a]

Element	Sun	Universe
Hydrogen	92.5	90.87
Helium	7.3	9.08
All others	0.2	0.05

[a]In percent by atoms.

TABLE S1.2 Properties of Hydrogen

Atomic Properties	
Abundance in earth's crust and waters	8.7 g/kg
Electron configuration	$1s^1$
Atomic radius	37 pm
Ionization energy	1312 kJ/mol
Physical Properties	
Density, STP	0.0899 g/L
Melting point	−259°C
Boiling point	−253°C

Occurrence and Preparation of Hydrogen

Hydrogen is by far the most abundant element in the universe, with helium a distant second (Table S1.1). It is believed that hydrogen and helium nuclei formed even before matter coalesced into stars and galaxies, and that they served as the raw material from which all other elements were later synthesized. (The origin of the elements is discussed in Section 22.8). In spite of its prevalence in the universe, hydrogen is much less abundant on earth, and hydrogen atoms provide less than 1% of the total mass of the earth's crust, water, and atmosphere (Table S1.2).

The usual elemental form of hydrogen (H_2) is a colorless and odorless gas at room temperature. Atomic hydrogen is found in the upper atmosphere, above 3500 km. Hydrogen atoms are lighter than any other atoms, and many of them escape from the earth's atmosphere. Earth's *escape velocity* is 11.2 km/s; this is the minimum velocity an object must have in order to escape from the earth's surface. We have seen from Graham's law (Equation 5.16, p. 201) that less massive gas molecules travel faster on the average than more massive molecules at the same temperature. Therefore, light atoms such as hydrogen have a better chance of attaining the escape velocity and wandering off into space. Nearly all of the hydrogen atoms that remain on earth are combined with heavier elements in molecules such as water.

Petroleum, natural gas, and other fossil fuels are composed largely of **hydrocarbons**, compounds that contain only hydrogen and carbon (Figure S1.1). The number of different hydrocarbons runs into hundreds of thousands because carbon atoms have the ability to join together in an almost endless variety of chains and rings. The simplest hydrocarbon is methane (CH_4), the main component of natural gas. Other gaseous hydrocarbons are ethane (C_2H_6), propane (C_3H_8), and acetylene (C_2H_2). Larger hydrocarbon molecules are found in liquid mixtures such as gasoline, kerosene, and fuel oil, and in solids such as paraffin.

Hydrogen is an important industrial gas and is produced in a number of ways. Most hydrogen is obtained by the reactions of methane and other small-molecule hydrocarbons with steam at 800–1000°C in the presence of finely divided nickel. (The nickel acts as a *catalyst*, a substance that is not used up in the reaction but whose presence makes the reaction go faster.)

Figure S1.1
Structural formulas and ball-and-stick models for some hydrocarbon molecules.

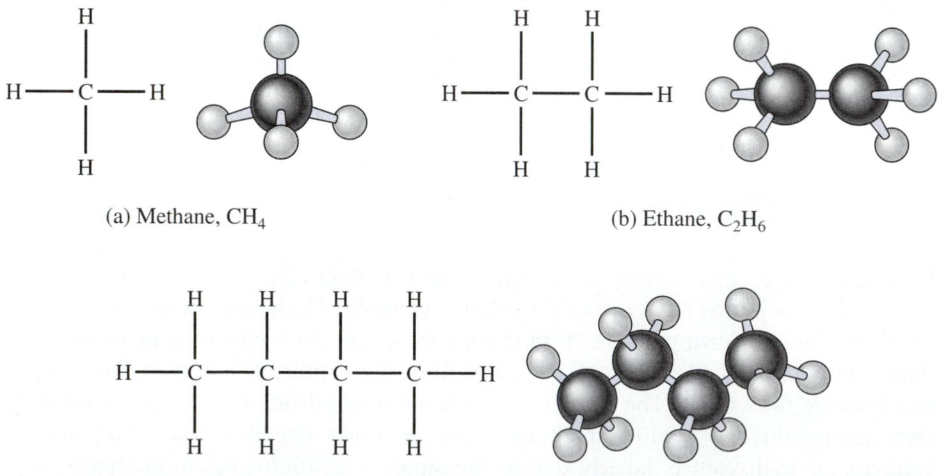

(a) Methane, CH_4

(b) Ethane, C_2H_6

(c) *n*-Butane, C_4H_{10}

$$CH_4(g) + H_2O(g) \xrightarrow[\text{Ni}]{900^\circ\text{C}} CO(g) + 3H_2(g)$$

A mixture of CO and H_2 is called *synthesis gas* because it is often used as the starting material for the synthesis of various organic compounds. If pure hydrogen is desired, the synthesis gas is transferred to a second reaction vessel where the carbon monoxide reacts with water vapor in the presence of iron oxide:

$$CO(g) + H_2O(g) \xrightarrow[\text{iron oxide}]{450^\circ\text{C}} CO_2(g) + H_2(g)$$

The resulting mixture is then passed through water, which dissolves carbon dioxide, but not hydrogen.

An older process consists of passing steam over red-hot coal or coke:

$$\underset{\text{coke or coal}}{C(s)} + H_2O(g) \xrightarrow{\text{about } 1000^\circ\text{C}} CO(g) + H_2(g)$$

The equimolar mixture of carbon monoxide and hydrogen, called *water gas*, was once widely used as an industrial fuel. Hydrogen is also obtained during oil refining when large hydrocarbon molecules are *cracked* (split into smaller fragments).

Very pure hydrogen can be prepared by the electrolysis of water (see Figure 3.1, p. 84). Small amounts of hydrogen are sometimes prepared for laboratory use by allowing an active metal to displace hydrogen from an acid solution (see Figure 3.5, p. 102) or from the reaction of metal hydrides with water (see below).

Reactions and Compounds of Hydrogen

Hydrogen forms covalent compounds with all nonmetals except the noble gases. Many hydrogen compounds, such as water, ammonia, and hydrogen chloride, are already familiar from earlier chapters. Water, hydrogen's principal compound, covers 70% of the earth's surface; in fact, earth is sometimes referred to as the "water planet." The ions found in acidic and basic solutions, $H^+(aq)$ and $OH^-(aq)$, each contain hydrogen and are central to the aqueous chemistry of acids and bases.

Hydrogen reacts with the metals of Groups 1A and 2A to form solid ionic compounds called *hydrides* (see Sections S1.2 and S1.3). Examples are lithium hydride (LiH) and calcium hydride (CaH_2). The hydride ion (H^-) is unstable in the presence of water and reacts instantly to form hydrogen gas and hydroxide ions:

$$\underset{\substack{\text{in solid}\\\text{hydride}}}{H^-} + H_2O(l) \rightarrow H_2(g) + OH^-(aq)$$

PRACTICE EXERCISE S1.1

Write the electron configuration of the hydride ion.

Industry uses about 48 billion cubic feet (1.4×10^{12} L) of hydrogen annually in metallurgy and chemical synthesis. Most of this is used in the **Haber process**, which combines hydrogen with nitrogen from air to make ammonia (NH_3):

$$N_2(g) + 3H_2(g) \xrightarrow[\text{catalyst}]{\substack{500 \text{ atm} \\ 450°C}} 2NH_3(g)$$

The synthesis is carried out at high pressure (500 atm) and temperature (450°C) in the presence of a catalyst. Ammonia is used to make fertilizers and other industrial chemicals; its preparation is described in greater detail in Chapter 15.

Two other important industrial chemicals synthesized from hydrogen are hydrogen chloride (HCl) and methanol (CH_3OH, methyl alcohol):

$$H_2(g) + Cl_2(g) \rightarrow 2HCl(g)$$

$$2H_2(g) + CO(g) \rightarrow CH_3OH(g)$$

The latter reaction is carried out at high temperature and pressure in the presence of a catalyst consisting of aluminum and chromium oxides.

Because hydrogen bonds so strongly to oxygen, it is a good reducing agent for preparing metals from their oxides. Hydrogen reductions, however, are expensive and are used only when less costly reducing agents such as carbon do not yield the desired purity of product. For example, tungsten (W), the metal used for light bulb filaments, is prepared in very pure form by heating its oxide with hydrogen:

$$WO_3(s) + 3H_2(g) \rightarrow W(s) + 3H_2O(g)$$

Hydrogen gas is also used as a fuel in torches for cutting and welding, and liquid hydrogen is used as a rocket fuel. For every gram of hydrogen that burns in air to form water vapor, 120 kJ is released—more energy per gram than any of the hydrocarbon fuels. Combustion of hydrogen is efficient, clean, and nonpolluting. Some futurists hope for an eventual *hydrogen economy* in which the world's energy problems will be solved by using solar energy to convert water into its elements, thus providing a virtually inexhaustible supply of hydrogen. Unfortunately, a cost-effective method for solar water decomposition has not yet been developed.

S1.2 GROUP 1A: THE ALKALI METALS

Lithium, sodium, potassium, rubidium, cesium, and francium are the Group 1A metals, also called the **alkali metals**. Sodium and potassium are the most abundant and best-known members of the group. Since ancient times, the ashes of wood and plants have served as a source of the strongly basic sodium and potassium compounds used for making soap and glass. The word *alkali*, a synonym for "strong base," is derived from the Arabic word *al-qaliy*, meaning "ashes of saltwort," a seaside plant.

Occurrence, Preparation, and Uses of the Metals

Sodium and potassium are extensively distributed throughout the earth's crust and biosphere. Lithium, rubidium, and cesium are also widely distributed, but are less abundant. Francium is a rare, radioactive element that we will not consider further; according to the *CRC Handbook of Chemistry and Physics*, there is probably less than 28 g of francium in the earth's crust at any time. The alkali metals occur in nature only in compounds, never as free metals. Their compounds are ionic and generally soluble, and are abundant in oceans, salt lakes, and brine wells. A liter of seawater, for example, contains about 0.47 mol of sodium ion. Alkali metal salts are also found in underground salt caves and, where the climate is dry, in surface deposits called *salt flats* (Figure S1.2).

Some of the properties of the alkali metals are summarized in Table S1.3. The atoms of an alkali metal are larger than those of the other elements in its period (see Figure 8.9, p.321), and the metals have remarkably low densities. Lithium has the lowest density of any metal; its density is half that of water. The alkali metals are soft, have low melting points, and have no structural uses.

Recall that each Group 1A metal has the lowest ionization energy of any element in its period (see Figure 8.11, p. 323). The alkali metals easily donate their valence electrons to other atoms, forming compounds in which they appear as 1+ ions. Because they are so reactive, the alkali metals are stored under oil to prevent them from reacting with oxygen and moisture (Figure S1.3). Rubidium and cesium, the most reactive of all the metals, are usually kept in sealed glass ampoules and are rarely seen or used.

The alkali metals can be obtained by passing an electric current through their molten chlorides or hydroxides. Sodium metal, which is the least expensive and most useful, is prepared from sodium chloride:

$$2NaCl(l) \xrightarrow{\text{electrolysis}} 2Na(l) + Cl_2(g)$$

Chlorine gas is an important coproduct. Preparation of an active metal requires a great deal of energy; it takes 411 kJ to decompose 1 mol of sodium chloride into sodium metal and chlorine gas.

Sodium, like all metals, is a good conductor of heat. It has a low melting point (98°C) and is used as a liquid heat exchanger in some nuclear reactors. The sodium vapor lamps used for streetlights are electric discharge tubes in which energized atoms of sodium vapor emit yellow 589-nm radiation. Sodium vapor lamps provide more visible light per joule of energy consumed than conventional bulbs because nearly all of their energy input is re-emitted in the form of visible photons. A standard light bulb, in contrast, has a tungsten filament that radiates most of its energy as unwanted heat.

(a)

(b)

Figure S1.2

Natural forms of sodium chloride. (a) Salt flats at Lake Hart, South Australia. (b) Crystals of the mineral halite.

TABLE S1.3 Properties of the Group 1A (Alkali) Metals

Element	Abundance[a] (g/kg)	Configuration	Atomic Radius (pm)	Ionic Radius (pm)	Ionization Energy (kJ/mol)	Density[b] (g/cm^3)	Melting Point (°C)	Boiling Point (°C)
Li	0.02	$[He]2s^1$	152	68	520	0.53	180	1347
Na	24	$[Ne]3s^1$	186	97	496	0.97	98	883
K	24	$[Ar]4s^1$	227	133	419	0.86	64	774
Rb	0.09	$[Kr]5s^1$	248	147	403	1.53	39	688
Cs	0.003	$[Xe]6s^1$	265	167	376	1.87	28	678
Fr	—	$[Rn]7s^1$	—	180	—	—	27	677

[a]In grams per kilogram of the earth's crust.
[b]At 20°C.

(a)

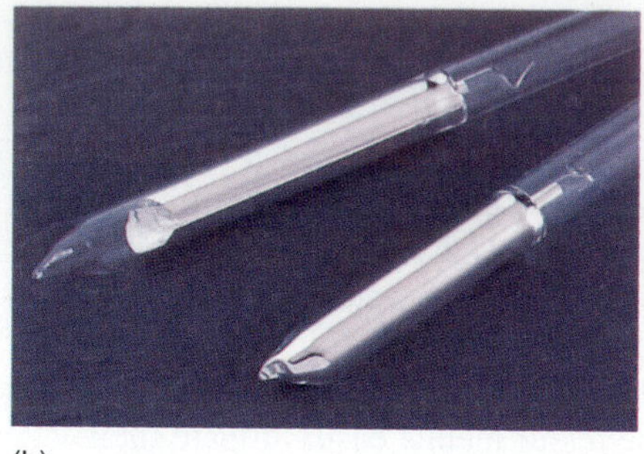
(b)

(c)

Figure S1.3

The alkali metals. (a) Lithium and sodium protected by oil. Lithium (density 0.53 g/cm^3) floats in the oil. Both metals are coated with oxidation products. (b) Potassium and rubidium sealed in glass ampoules. (c) Cesium in a glass ampoule. Note that cesium has a golden cast.

The ionization energies of rubidium and cesium are very low, and the energy of visible light is sufficient to eject electrons from their solid surfaces. These metals are used in light meters, electric eyes, and other photoelectric devices.

Reactions of the Alkali Metals

The alkali metals have low ionization energies and will react vigorously with any substance that will accept their valence electrons. Some of these reactions are summarized below and in Table S1.4.

With Water. Alkali metals displace hydrogen from water, leaving OH^- ions in solution. Any alkali metal can be substituted for sodium in the following equation:

$$2Na(s) + 2H_2O(l) \rightarrow H_2(g) + \underset{\text{sodium hydroxide}}{2NaOH(aq)} \qquad (= Na^+ + OH^-)$$

TABLE S1.4 Reactions of the Alkali Metals

With Water (very rapid and exothermic):

$2M(s) + 2H_2O(l) \rightarrow 2MOH(aq) + H_2(g)$ (M = any alkali metal)

With Halogens (very rapid and exothermic):

$2M(s) + X_2 \rightarrow 2MX(s)$ (M = any alkali metal; X = F, Cl, Br, I)

With Hydrogen:

$2M(s) + H_2(g) \xrightarrow{\text{heat}} 2MH(s)$ (M = any alkali metal)

With Oxygen (very rapid and exothermic):

$4Li(s) + O_2(g) \rightarrow 2Li_2O(s)$

$2Na(s) + O_2(g) \rightarrow Na_2O_2(s)$ (also K)

$K(s) + O_2(g) \rightarrow KO_2(s)$ (also Rb and Cs)

With Nitrogen:

$6Li(s) + N_2(g) \xrightarrow{\text{heat}} 2Li_3N(s)$ (Li only)

DEMONSTRATION S1.1 THE REACTIONS OF ALKALI METALS WITH WATER

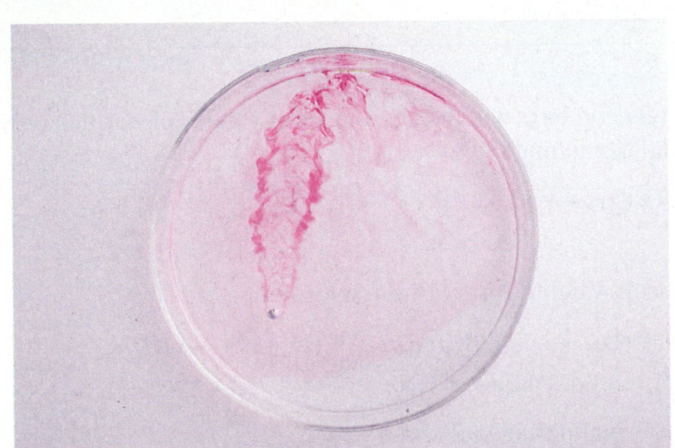

Sodium reacts with water to form hydrogen gas and aqueous sodium hydroxide. The heat of reaction converts the remaining sodium into a molten ball that scoots across the water on a cushion of hydrogen gas. The sodium hydroxide trail turns dissolved phenolphthalein red.

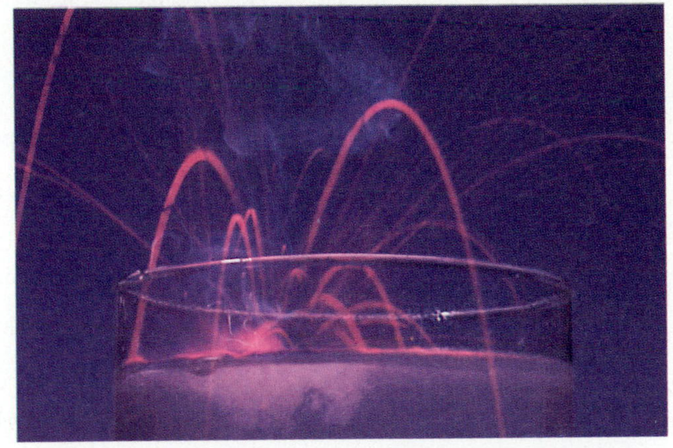

The reaction of potassium with water is much more violent than that of sodium. The evolved heat is intense enough to produce burning sparks of potassium and ignite the hydrogen gas.

The heat produced by the reaction of sodium with water converts the remaining metal into a molten sphere that is buoyed up by a cushion of hydrogen and skids about on the water surface until it is consumed (see Demonstration S1.1). The reaction of potassium and water generates such intense heat that the hydrogen gas ignites and envelops the molten potassium ball in a sheath of violet flame. One practical application of this reaction is in the use of sodium wire to remove traces of moisture from nonaqueous solvents such as ether. The hydrogen gas bubbles out of the liquid, while sodium hydroxide, which is insoluble in ether, falls to the bottom.

With Halogens. The alkali metals react with fluorine, chlorine, bromine, and iodine to form white crystalline salts. These reactions are rapid and violent, and evolve heat and light (see Demonstration 3.2, p. 100). The reaction of potassium with chlorine is an example:

$$2K(s) + Cl_2(g) \rightarrow \underset{\text{potassium chloride}}{2KCl(s)} \qquad (= K^+ + Cl^-)$$

With Hydrogen. When heated with hydrogen, the alkali metals form white crystalline salts containing the hydride ion (H^-):

$$2Na(s) + H_2(g) \rightarrow \underset{\text{sodium hydride}}{2NaH(s)} \qquad (= Na^+ + H^-)$$

As we mentioned in Section S1.1, hydrides react with water to form H_2 and OH^-:

$$NaH(s) + H_2O(l) \rightarrow NaOH(aq) + H_2(g)$$

PRACTICE EXERCISE S1.2

Write equations for the reactions of (a) potassium with water, (b) sodium with bromine, and (c) lithium with hydrogen.

With Oxygen. Alkali metals combine with oxygen to form oxides of various types. Lithium forms a compound containing the ***oxide ion*** (O^{2-}):

$$4Li(s) + O_2(g) \rightarrow \underset{\text{lithium oxide}}{2Li_2O(s)} \qquad (= 2Li^+ + O^{2-})$$

Lithium oxide combines with water to form lithium hydroxide

$$Li_2O(s) + H_2O(l) \rightarrow \underset{\text{lithium hydroxide}}{2LiOH(aq)} \qquad (= Li^+ + OH^-)$$

and with carbon dioxide to form lithium carbonate

$$Li_2O(s) + CO_2(g) \rightarrow \underset{\text{lithium carbonate}}{Li_2CO_3(s)} \qquad (= 2Li^+ + CO_3^{2-})$$

Note that only the oxide ion changes in the last two reactions.

Other alkali metals form oxide ions only when the oxygen supply is limited. With abundant oxygen they yield the ***peroxide ion*** (O_2^{2-}) and the ***superoxide ion*** (O_2^-) instead. Large anions such as O_2^{2-} and O_2^- tend to be more stable in the company of large cations such as K^+ and Rb^+, therefore, *the larger the Group 1A ion, the more oxygen it combines with.* Sodium forms sodium peroxide Na_2O_2; potassium forms the peroxide K_2O_2 and the superoxide KO_2; and rubidium and cesium, the largest Group 1A ions, form mainly superoxides. Equations for some of these reactions are as follows:

$$2Na(s) + O_2(g) \rightarrow \underset{\text{sodium peroxide}}{Na_2O_2(s)} \qquad (= 2Na^+ + O_2^{2-})$$

$$Rb(s) + O_2(g) \rightarrow \underset{\text{rubidium superoxide}}{RbO_2(s)} \qquad (= Rb^+ + O_2^-)$$

Peroxides and superoxides have a number of uses, and their chemistry is discussed in the second survey chapter.

With Nitrogen. Lithium is the only alkali metal that reacts with nitrogen. It forms lithium nitride (Li_3N), an ionic compound containing the ***nitride ion*** (N^{3-}):

$$6Li(s) + N_2(g) \rightarrow \underset{\text{lithium nitride}}{2Li_3N(s)} \qquad (= 3Li^+ + N^{3-})$$

The nitride ion combines with water to form ammonia gas and hydroxide ions:

$$\underset{\text{in solid nitride}}{N^{3-}} + 3H_2O(l) \rightarrow NH_3(g) + 3OH^-(aq)$$

$$Li_3N(s) + 3H_2O(l) \rightarrow NH_3(g) + 3LiOH(aq) \qquad (= Li^+ + OH^-)$$

Alkali Metal Compounds

Most alkali metal compounds are ionic and soluble in water. Lithium salts tend to be somewhat less soluble than other alkali metal salts, and a few lithium salts such as LiF and Li_2CO_3 are sparingly soluble. Salts with small ions often have limited solubilities, and Li^+ has the smallest ionic radius in Group 1A. Some of the important alkali metal compounds and their uses are described below.

Alkali Metal Cations. Alkali metal cations have no color in aqueous solution, and they undergo very few reactions; in fact, Na^+ and K^+ are the chemist's favorite spectator ions. Their compounds supply anions such as hydroxide, carbonate, and phosphate in aqueous solution, while the Na^+ and K^+ ions do not interfere with the reactions for which these negative ions are intended.

Rubidium (from the Latin *rubidus,* red) and cesium (from the Latin *caesius*, blue) were the first two elements to be discovered with a spectroscope. Each of them is named after the most prominent line in its emission spectrum.

Sodium and potassium ions have important biochemical roles: Na^+ is the principal cation in extracellular fluids such as blood plasma, while K^+ predominates in intracellular fluid. Life depends on maintaining these ions in proper balance. Potassium is an important plant nutrient, and fertilizers usually provide potassium ion in the form of KCl, KNO_3, or K_2SO_4.

Alkali metal ions exhibit characteristic colors when their compounds are vaporized in a flame. Electrons are excited by the flame to higher energy states and then give off visible photons as they return to lower energy states (see Figure 7.6, p. 268). Flame tests are often used to show whether Group 1A elements are present in a compound (Figure S1.4). Compounds containing lithium, sodium, and potassium salts give red, yellow, and violet flame colors, respectively. The yellow flame that appears when glass is heated for bending or fire-polishing is produced by vaporized sodium compounds. Rubidium ions impart a lavender color to a flame, and cesium ions color a flame purple.

Figure S1.4

Flame tests for alkali metals. The colors are produced by heating small crystals of the metal chlorides or nitrates in an almost colorless flame. (a) Lithium. (b) Sodium. (c) Potassium. Rubidium and cesium produce flame colors similar to that of potassium.

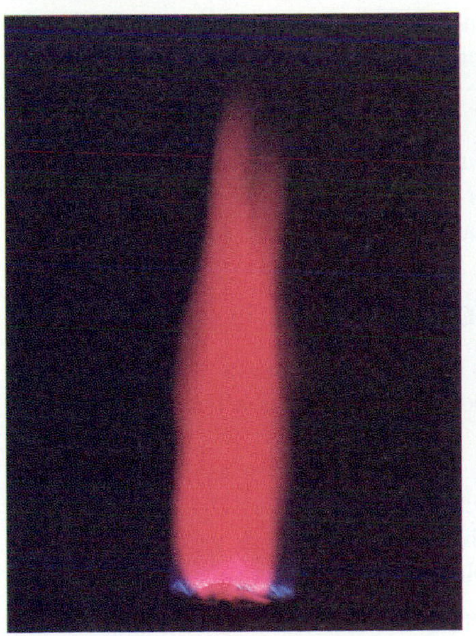
(a)

(b)

(c)

Alkali Metal Chlorides. ***Sodium chloride*** (NaCl, *table salt*) is the most abundant alkali metal compound. It is easily melted, dissolved, and purified, and it serves as the starting point for the manufacture of many other compounds, including sodium hydroxide, sodium carbonate, and hydrochloric acid. Sodium ions and chloride ions are both essential in the diet. Animals crave salt and humans often tend to ingest more than is good for them. Salt was a precious commodity in earlier days, one of the few items a rural family could not produce on its own. Part of a Roman soldier's pay consisted of a salt ration or *salarium*, which is the origin of the word *salary*. ***Potassium chloride*** (KCl) is often used as a salt substitute for persons restricted to a low-sodium diet.

Alkali Metal Hydroxides. ***Sodium hydroxide*** (NaOH, *caustic soda*) is among the top 10 industrial chemicals manufactured in the United States; most of it is produced by the electrolysis of brine (concentrated NaCl solution):

$$2NaCl(aq) + 2H_2O(l) \xrightarrow{\text{electrolysis}} 2NaOH(aq) + H_2(g) + Cl_2(g)$$

Hydrogen and chlorine are important coproducts.

Sodium hydroxide is a strong, soluble base. It is used in the manufacture of plastics and paper, in petroleum refining to neutralize sulfuric acid and other acids, and in the manufacture of soap from fat (Chapter 23). ***Potassium hydroxide*** (KOH) is similar in behavior to sodium hydroxide, but more expensive. *Lye* and other concentrated alkaline solutions for cleaning ovens and clogged drains contain sodium and potassium hydroxides. These solutions will gradually attack glass, so they are usually stored in plastic bottles. Strong alkalis attack the proteins in skin, hair, silk, and wool, and should be handled with extreme caution.

Carbonates and Hydrogen Carbonates. ***Sodium carbonate*** (Na_2CO_3), also known as *soda ash* or *washing soda*, is an important industrial chemical. Its principal source in the United States is the mineral *trona*, $Na_2CO_3 \cdot NaHCO_3 \cdot 2H_2O$, which was deposited millions of years ago in what is now a dried lake bed in Wyoming. On heating, the sodium hydrogen carbonate decomposes into sodium carbonate:

$$2NaHCO_3(s) \rightarrow Na_2CO_3(s) + H_2O(g) + CO_2(g)$$

Most of the sodium carbonate supply is used in making glass, a process that is described in the third survey chapter. Other uses are based on the ability of carbonate ions to neutralize acids (Section 4.2) and to form a basic solution by reacting with water:

$$CO_3^{2-}(aq) + H_2O(l) \rightarrow HCO_3^{-}(aq) + OH^{-}(aq)$$

Washing soda is used in dishwater detergents and in other heavy-duty detergents, the types that are not "kind to your hands." Sodium carbonate is a harsh reagent that should be handled with care.

Sodium hydrogen carbonate ($NaHCO_3$), commonly known as *baking soda* or *bicarbonate of soda*, is prepared from salt and limestone ($CaCO_3$) by the Solvay process (Survey Chapter 3). It is a mildly basic reagent that is used to neutralize acids when a strong base would not be suitable:

$$NaHCO_3(s) + HCl(aq) \rightarrow NaCl(aq) + H_2O(l) + CO_2(g)$$

and it is an ingredient of many commercial antacids sold for the relief of "acid indigestion." In baking it is used as a leavening agent that reacts with acidic components such as the lactic acid ($HC_3H_5O_3$) in sour milk or with potassium hydrogen tartrate ($KHC_4H_4O_6$), *cream of tartar*) in baking powder. Carbon dioxide gas is evolved and retained by the sticky dough. The gas expands when heated, causing the dough to rise. The escaping gas causes the baked product to have a porous structure.

Lithium carbonate (Li_2CO_3) is used to control the symptoms of manic depression and enables many sufferers of this illness to lead normal lives. The mechanism by which lithium ions affect the nervous system is still being studied. Evidence indicates that lithium ions exert a regulating effect on the transmission of impulses from one nerve cell to another and thus help to compensate for hereditary defects in the body's own control system.

S1.3 GROUP 2A: THE ALKALINE EARTH METALS

The **alkaline earth metals** are beryllium, magnesium, calcium, strontium, barium, and radium. This group acquired its name from the alkaline character of its hydroxides and from the low solubilities of magnesia (MgO) and lime (CaO), which led early chemists to classify these oxides as "earths." Group 2A metals resemble Group 1A metals in many ways, but they also exhibit differences that are consistent with their smaller ionic radii and higher ionization energies. As you study these elements, it will help to keep the Group 1A elements in mind for comparison. Some properties of the alkaline earth metals are summarized in Table S1.5.

The Metals and Their Uses

Group 2A atoms are heavier and smaller than those of Group 1A; consequently, an alkaline earth metal is harder and denser and has a higher melting point than the neighboring alkali metal in its period. You can verify these differences by comparing the data in Tables S1.3 and S1.5.

The alkaline earth metals are almost as reactive as the Group 1A metals, and they

TABLE S1.5 Properties of the Group 2A (Alkaline Earth) Metals

Element	Abundance[a] (g/kg)	Configuration	Atomic Radius (pm)	Ionic Radius (pm)	Ionization Energy (kJ/mol) 1st	Ionization Energy (kJ/mol) 2nd	Density[b] (g/cm^3)	Melting Point (°C)	Boiling Point (°C)
Be	0.028	$[He]2s^2$	111	35	899	1757	1.85	1278	2970
Mg	20	$[Ne]3s^2$	160	66	738	1451	1.74	649	1090
Ca	42	$[Ar]4s^2$	197	99	590	1145	1.55	839	1484
Sr	0.375	$[Kr]5s^2$	215	112	550	1064	2.54	768	1384
Ba	0.425	$[Xe]6s^2$	217	134	503	965	3.5	725	1640
Ra	—	$[Rn]7s^2$	—	143	509	979	5	700	1737

[a]In grams per kilogram of the earth's crust.
[b]At 20°C.

Figure S1.5
Beryl ($3BeO \cdot Al_2O_3 \cdot 6SiO_2$).

occur in nature only as 2+ ions. The principal source of beryllium is the mineral *beryl*, $3BeO \cdot Al_2O_3 \cdot 6SiO_2$ (Figure S1.5). The gemstones *aquamarine*, whose color is due to the presence of Fe^{3+} ions, and *emerald*, which contains Cr^{3+} ions, are precious forms of beryl. Most crystals of beryl, however, are imperfect and either colorless or brown. Magnesium, the eighth most abundant element in the earth's crust, occurs in a number of insoluble minerals such as the carbonate ($MgCO_3$) and in soluble salts such as $MgCl_2$ and $MgSO_4$ in salt deposits, brine wells, and seawater. Calcium, the fifth most abundant element, is an essential ingredient of bones, teeth, coral, and seashells. It occurs as *limestone* ($CaCO_3$), *gypsum* ($CaSO_4 \cdot 2H_2O$), and *fluorite* (CaF_2), and it is also found in most silicate and phosphate rocks. Marble and chalk are mainly calcium carbonate (see Figure S1.6). The mineral *dolomite*, $CaCO_3 \cdot MgCO_3$, is an important source of both magnesium and calcium. Strontium and barium occur in various sulfate and carbonate minerals.

All of the Group 2A metals except beryllium are prepared by first converting their natural compounds to chlorides and then electrolyzing the molten chloride:

$$MgCl_2(l) \xrightarrow{\text{electrolysis}} Mg(l) + Cl_2(g)$$

Beryllium is prepared by the reaction of beryllium fluoride with magnesium metal.

Beryllium and all its compounds are extremely toxic to humans.

Beryllium and magnesium, the two lightest members of the group, need no special protection from air and moisture—not because they are nonreactive, but because their surfaces develop adherent oxide films that protect the remaining metal. Beryllium is transparent to x-rays and is used for windows in x-ray tubes. Beryllium is also used to slow down neutrons in nuclear reactors. Magnesium, which burns with a bright white light (see Demonstration S1.2), is used in flash bulbs and flares. Beryllium and magnesium are strong enough to be used in lightweight structural alloys. Ladders are often made of magnesium alloyed with aluminum, another strong, light metal.

Calcium, strontium, and barium are more like the alkali metals in reactivity and lack of mechanical strength. All three are attacked by air and moisture. Calcium metal reacts slowly and is usually kept in an ordinary, closed container, but strontium and barium metals are stored under oil. Calcium, strontium, and barium do not have many uses as metals. Radium is a rare radioactive metal whose compounds have been used in the treatment of cancer. Its use for this purpose, however, has

(a)

(b)

Figure S1.6
Some calcium minerals. (a) Fluorite (CaF_2). Although pure calcium fluoride is colorless, the natural mineral exhibits a variety of colors such as pink, amber, blue, and violet. (b) Deposits of natural chalk ($CaCO_3$).

DEMONSTRATION S1.2 THE COMBUSTION OF MAGNESIUM: $2Mg(s) + O_2(g) \rightarrow 2MgO(s)$

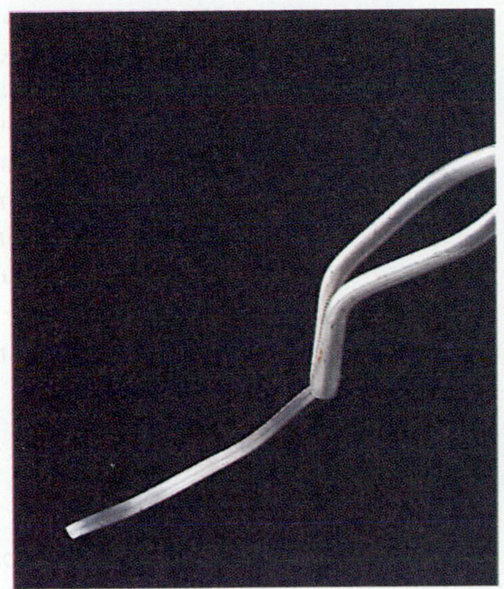

A strip of magnesium metal held with a set of tongs.

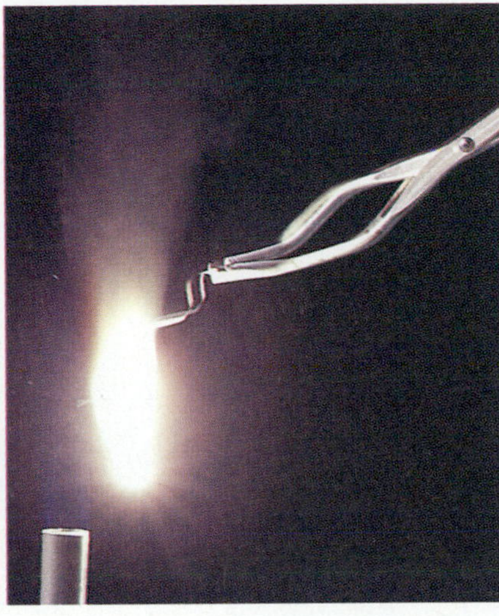

After ignition, the magnesium burns with a blinding white light.

The white ash of magnesium oxide from the burning of several magnesium strips.

declined with the development of radioactive sources that can be administered directly to diseased tissues while causing little damage to healthy tissue.

Reactions of the Alkaline Earth Metals

Alkaline earth metal atoms have an s^2 valence-shell configuration and they readily form 2+ ions and ionic compounds. Calcium, strontium, and barium are almost as active as the alkali metals and undergo many of the same reactions. Magnesium is only slightly less active, but its reactivity is often masked by the presence of its oxide film. Beryllium, which has the smallest atomic radius and highest ionization energy in the group, is less active than the other alkaline earth metals and forms fewer ionic compounds. The reactions of the alkaline earth metals are summarized in Table S1.6 and described below.

With Water. Calcium, strontium, and barium react with water at ordinary temperatures to form hydrogen gas and aqueous hydroxides (see Demonstration S1.3):

$$Ba(s) + 2H_2O(l) \rightarrow \underset{\text{barium hydroxide}}{Ba(OH)_2(aq)} + H_2(g)$$

Magnesium will react with hot water, but beryllium does not react with water at all.

With Halogens. All members of Group 2A react vigorously to form ionic halides, as illustrated by the reaction of magnesium and chlorine:

$$Mg(s) + Cl_2(g) \rightarrow \underset{\text{magnesium chloride}}{MgCl_2(s)}$$

TABLE S1.6 Reactions of the Alkaline Earth Metals

Reaction	Conditions
With Water:	
$M(s) + 2H_2O(l) \rightarrow M(OH)_2(aq) + H_2(g)$	(M = Mg, Ca, Sr, Ba; Mg requires heat; vigorous with Ca, Sr, Ba)
With Halogens (very rapid and exothermic):	
$M(s) + X_2 \rightarrow MX_2(s)$	(M = any Group 2A metal; X = F, Cl, Br, I)
With Hydrogen:	
$M(s) + H_2(g) \xrightarrow{\text{heat}} MH_2(s)$	(M = Ca, Sr, Ba)
With Oxygen:	
$2M(s) + O_2(g) \rightarrow 2MO(s)$	(M = any Group 2A metal; Be and Mg require heat)
$Ba(s) + O_2(g) \xrightarrow{\text{heat}} BaO_2(s)$	(Ba only)
With Nitrogen:	
$3M(s) + N_2(g) \xrightarrow{\text{heat}} M_3N_2(s)$	(M = any Group 2A metal)

With Hydrogen and Nitrogen. Calcium, strontium, and barium react with hydrogen at high temperature to form ionic hydrides:

$$Ca(s) + H_2(g) \xrightarrow{\text{heat}} \underset{\text{calcium hydride}}{CaH_2(s)}$$

Beryllium and magnesium do not react directly with hydrogen.

All of the Group 2A metals form nitrides when heated in nitrogen:

$$3Mg(s) + N_2(g) \rightarrow \underset{\text{magnesium nitride}}{Mg_3N_2(s)}$$

(Recall that, in Group 1A, only lithium reacts with nitrogen.) Nitrides and hydrides are decomposed by water. Nitrides form ammonia gas:

$$Ca_3N_2(s) + 6H_2O(l) \rightarrow 2NH_3(g) + 3Ca(OH)_2(aq)$$

while hydrides liberate hydrogen:

$$CaH_2(s) + 2H_2O(l) \rightarrow 2H_2(g) + Ca(OH)_2(aq)$$

Calcium hydride is sometimes used as a portable source of hydrogen. Controlled addition of water to the solid releases the hydrogen gas as needed.

PRACTICE EXERCISE S1.3

Write equations for the reactions of (a) calcium with water, (b) barium with chlorine, (c) strontium with hydrogen, and (d) calcium with nitrogen.

DEMONSTRATION S1.3 THE REACTIONS OF CALCIUM AND MAGNESIUM WITH WATER

Calcium reacts with water to form hydrogen gas and aqueous calcium hydroxide.

Magnesium strips are placed in cold water that contains a few drops of phenolphthalein. No change is visible.

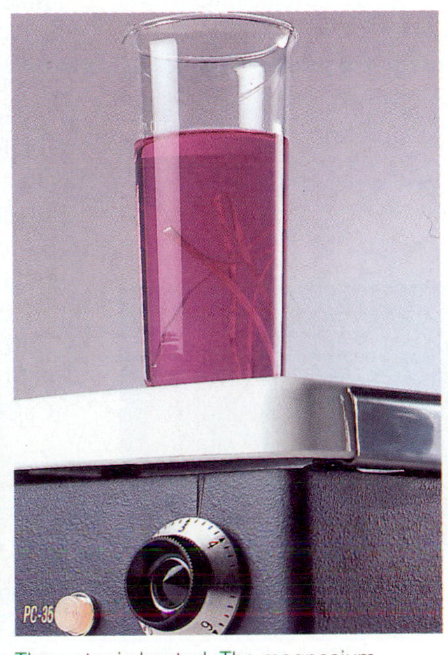

The water is heated. The magnesium reacts slowly to form bubbles of hydrogen gas and aqueous magnesium hydroxide. The hydroxide ions cause the phenolphthalein to turn pink.

With Oxygen. Calcium, strontium, and barium react with oxygen at ordinary temperatures to form compounds containing the oxide ion. Beryllium and magnesium react more reluctantly, but both will combine with oxygen if ignited:

$$2Mg(s) + O_2(g) \rightarrow \underset{\text{magnesium oxide}}{2MgO(s)}$$

The oxide ion in calcium, strontium, and barium oxides reacts with water to form the hydroxide ion:

$$CaO(s) + H_2O(l) \rightarrow \underset{\text{calcium hydroxide}}{Ca(OH)_2(aq)}$$

Beryllium oxide and magnesium oxide are sparingly soluble and do not react directly with water.

Barium has larger atoms than the elements above it in Group 2A and, if heated persistently with excess oxygen, it will form a peroxide:

$$Ba(s) + O_2(g) \rightarrow \underset{\text{barium peroxide}}{BaO_2(s)}$$

The radius of Ba^{2+} is similar to that of K^+ in Group 1A (see Figure 8.9, p. 321), which also forms a peroxide.

Alkaline Earth Metal Compounds

The alkaline earth metals form fewer soluble compounds than their neighbors in Group 1A. Group 2A halides and nitrates are soluble, but the hydroxides and many other salts are not. Some alkaline earth metal compounds and their uses are listed in Table S1.7.

Group 2A Ions. Because Group 2A compounds tend to be less soluble than Group 1A compounds, the Group 2A ions undergo more precipitation reactions than those of Group 1A. The ions Ca^{2+} and Mg^{2+} are often present in natural waters along with HCO_3^-, SO_4^{2-}, and other anions; water containing these ions is said to be "hard." Hard water forms a precipitate with soap (soap scum), which makes laundering difficult and more expensive. Furthermore, when hard water is heated, it forms a scale of $CaCO_3$ and $CaSO_4$:

$$Ca^{2+}(aq) + HCO_3^-(aq) \rightarrow CaCO_3(s) + CO_2(g) + H_2O(l)$$

$$Ca^{2+}(aq) + SO_4^{2-}(aq) \rightarrow CaSO_4(s)$$

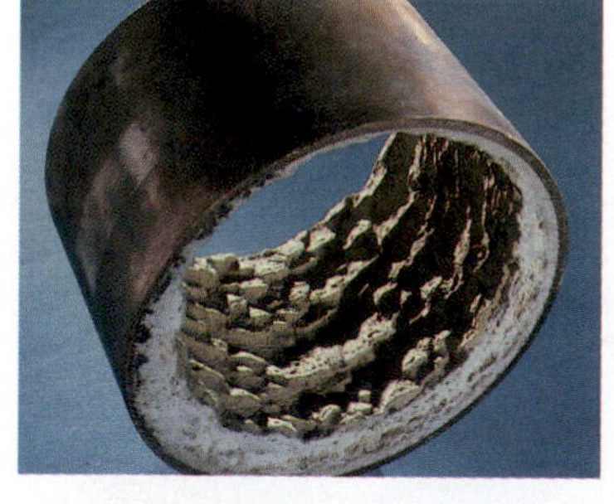

Figure S1.7
Scale deposited by hard water in a hot-water pipe.

($CaSO_4$ is less soluble in hot water than in cold water.) The resulting scale can clog steam irons, boiler pipes, and hot-water heating systems (Figure S1.7).

Calcium and magnesium ions are essential in the body. Calcium ions play a role in heart action and muscle contraction; magnesium ions are necessary for the functioning of a number of enzymes.

TABLE S1.7 Some Alkaline Earth Metal Compounds and Their Uses

Compound	Formula	Common Name and Familiar Uses
Magnesium hydroxide	$Mg(OH)_2$	Milk of magnesia is a suspension of $Mg(OH)_2$ in water
Magnesium oxide	MgO	Magnesia; used to make firebricks
Magnesium sulfate	$MgSO_4$	$MgSO_4 \cdot 7H_2O$ is Epsom salts; a laxative
Calcium carbonate	$CaCO_3$	The principal ingredient in limestone, marble, coral, and shells
Calcium oxide	CaO	Lime; used in mortar, cement, and plaster
Calcium hydroxide	$Ca(OH)_2$	Slaked lime; an aqueous suspension is used as whitewash
Calcium chloride	$CaCl_2$	A desiccant (absorber of moisture); also used to melt ice on roads
Calcium sulfate	$CaSO_4$	$CaSO_4 \cdot 2H_2O$ is gypsum; used in blackboard chalk, in wallboard, and to condition soil; $CaSO_4 \cdot \frac{1}{2}H_2O$ is plaster of Paris; used to make molds
Barium sulfate	$BaSO_4$	Opaque to x-rays, used in x-ray examination of the gastrointestinal tract.

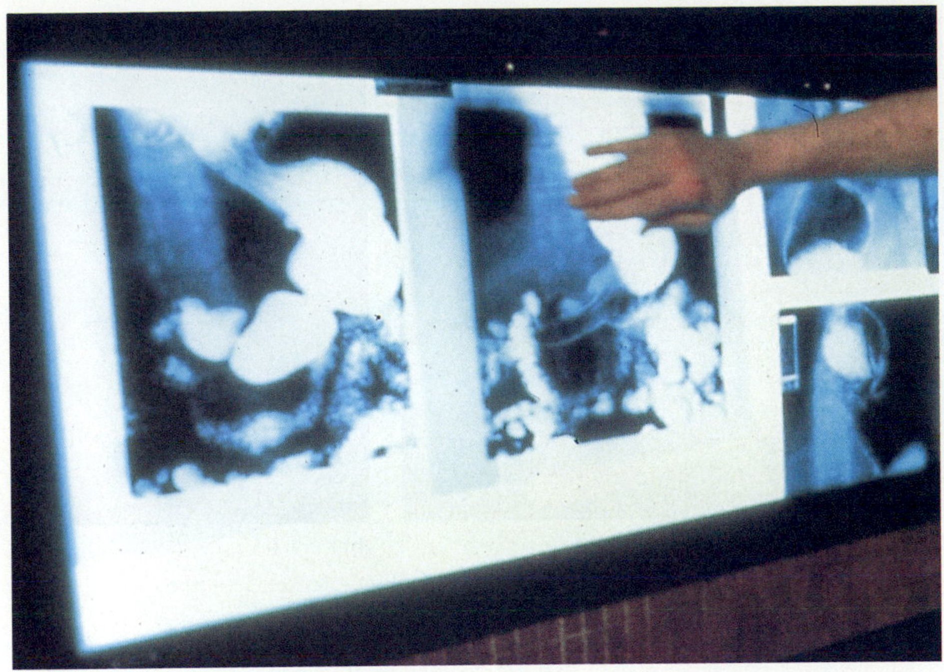

Figure S1.8

Abdominal x-rays taken after the patient has drunk a barium sulfate suspension. The gastrointestinal tract appears white because barium sulfate is opaque to x-rays.

Barium ions are opaque to x-rays, and a suspension of barium sulfate is given internally to outline the gastrointestinal tract in x-ray examinations (Figure S1.8). Although barium ions are poisonous, $BaSO_4$ itself is harmless because it is sparingly soluble and very few ions are released in the body.

The flame colors of calcium, strontium, and barium salts are orange-red, scarlet, and light green, respectively (Figure S1.9). Strontium, which has a particularly vivid color, is used in fireworks and red signal flares. Beryllium and magnesium compounds do not give colored flame tests.

Figure S1.9

Some Group 2A flame tests. (a) Calcium. (b) Strontium. (c) Barium.

(a)

(b)

Figure S1.10

Various forms of calcium carbonate. (a) Dogtooth calcite. (b) A starfish skeleton, coral, seashells, and bird's eggs.

The Lime Family. Much of the earth's calcium is found in the form of ***calcium carbonate*** ($CaCO_3$), the principal ingredient of minerals such as limestone, marble, calcite, and natural chalk, and of animal products such as shells and coral (Figure S1.10). Limestone, fourth on the list of the most widely used natural inorganic materials in the United States, is exceeded in annual tonnage only by air, water, and sodium chloride.

Marble and limestone are common building materials. Granular limestone is used as a slow-working neutralizer for acid soil:

$$CaCO_3(s) + \underset{\uparrow\text{from acidic soil ingredients}}{H^+(aq)} \rightarrow Ca^{2+}(aq) + HCO_3^-(aq)$$

Most industrial calcium carbonate, however, is converted to ***calcium oxide*** (CaO), commonly known as *lime* or *quicklime,* and to ***calcium hydroxide*** ($Ca(OH)_2$), also known as *slaked lime*. Lime is prepared by heating crushed shells or crushed limestone in a lime kiln like the one shown in Figure S1.11:

$$CaCO_3(s) \xrightarrow{\text{heat}} CaO(s) + CO_2(g)$$

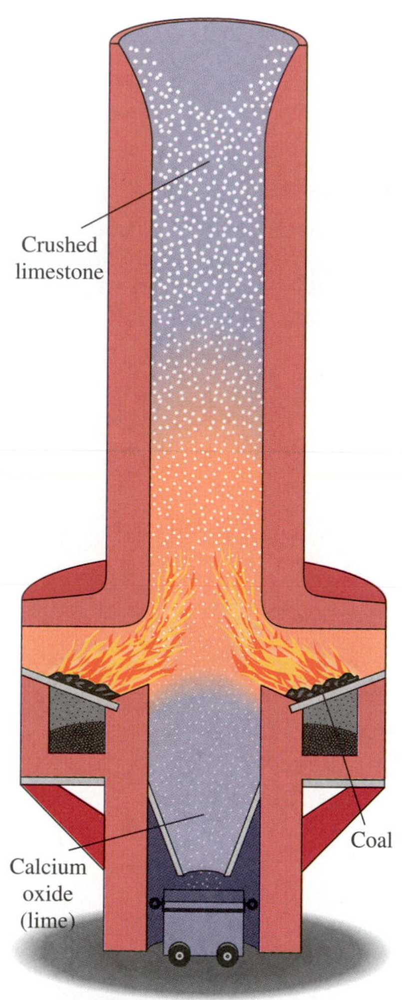

Figure S1.11

A lime kiln. Crushed limestone ($CaCO_3$) is added through the hopper at the top. The heat of burning coal converts the limestone into lime (CaO): $CaCO_3(s) \rightarrow CaO(s) + CO_2(g)$. Lime is removed at the bottom and carbon dioxide escapes at the top.

Lime absorbs atmospheric moisture rapidly and exothermically to form calcium hydroxide:

$$CaO(s) + H_2O(g) \rightarrow Ca(OH)_2(s) \qquad \Delta H^\circ = -109.2 \text{ kJ}$$

The heat generated by this reaction has been known to char wood and start fires.

For centuries the major use of lime has been in the preparation of mortar, cement, and plaster; it was used to build the Egyptian pyramids over 4000 years ago. Fresh mortar is a paste of lime, sand, and water that sets to a solid as the lime absorbs water. The solid becomes progressively harder as a result of slow neutralization, partly by atmospheric carbon dioxide

$$CaO(s) + CO_2(g) \rightarrow CaCO_3(s)$$

$$Ca(OH)_2(s) + CO_2(g) \rightarrow CaCO_3(s) + H_2O(l)$$

and partly by ingredients in sand. The reaction with sand is quite complex; it can be represented by the simplified equation

$$\underset{\text{silicon dioxide in sand}}{CaO(s) + SiO_2(s)} \rightarrow \underset{\text{calcium silicate}}{CaSiO_3(s)}$$

Calcium Sulfate. *Gypsum*, $CaSO_4{\cdot}2H_2O$, is a naturally occurring mineral that is used to make wallboard and other plaster-containing materials. Powdered gypsum is first converted to *plaster of Paris* by heating it to about 100°C:

$$\underset{\text{gypsum}}{CaSO_4{\cdot}2H_2O(s)} \rightarrow \underset{\text{plaster of Paris}}{CaSO_4{\cdot}\tfrac{1}{2}H_2O(s)} + \tfrac{3}{2}H_2O(g)$$

Moistened plaster of Paris sets to rock-hard gypsum by reabsorbing the water. Because plaster of Paris expands while setting, it forms a sharp impression of the details of its container, and thus it is used by artists, jewelry makers, and dentists for making castings. The earliest known description of a plaster of Paris cast for fractured bones was found in a Persian pharmacy text of A.D. 975.

Modern *alabaster* (a soft stone used for carvings) is a translucent form of gypsum. The ancient building material called "alabaster" is actually a form of calcium carbonate.

S1.4 GROUP 8A: THE NOBLE GASES

The Group 8A elements, or **noble gases**, are helium, neon, argon, krypton, xenon, and radon. Group 8A is the only group in which all the elements are gaseous and in which they all normally exist as independent atoms. Helium has a $1s^2$ electron configuration, and the other Group 8A atoms have s^2p^6 valence-shell configurations. These configurations, which are associated with high ionization energies and positive electron affinities, have little tendency to interact with other electron clouds and, consequently, the noble gases are not very reactive. The use of the term *noble* for nonreactive elements goes back to the ancient Greek philosophers' belief that unchanging entities are complete in themselves and therefore noble. The elements in Group 8A have also been called *inert gases* and *rare gases*, but neither name is entirely suitable in the light of present knowledge.

Occurrence and Preparation of the Noble Gases

The invisible and inconspicuous noble gases were the last group of elements to be discovered. The gases are colorless, odorless, and tasteless, and liquefy only at very low temperatures (Table S1.8). Although helium is present in natural gas—as much as 6% in some deposits—the first hint of its existence came from unidentified lines seen in the solar spectrum obtained during the solar eclipse of 1868. Some astronomers believed they had found a new element, which they named helium after the Greek word for sun, *helios*, but many scientists remained skeptical. In 1895 helium was discovered on earth. Argon, which makes up almost 1% of the atmosphere, had been identified a year earlier in 1894, and the other gases were found in the atmosphere shortly afterward.

TABLE S1.8 Properties of the Noble Gases

Element	Concentration in Atmosphere at Sea Level (mol %)	Valence-Shell Configuration	Atomic Radius[a] (pm)	Ionization Energy (kJ/mol)	Density[b] (g/L)	Melting Point (°C)	Boiling Point[c] (°C)
He	5.24×10^{-4}	$1s^2$	(50)	2372	0.179	—[d]	−268.9
Ne	1.82×10^{-3}	$2s^22p^6$	(65)	2081	0.900	−248.7	−246.0
Ar	0.939	$3s^23p^6$	(95)	1521	1.784	−189.2	−185.7
Kr	1.14×10^{-4}	$4s^24p^6$	110	1351	3.733	−156.6	−152.3
Xe	8.7×10^{-6}	$5s^25p^6$	130	1170	5.887	−111.9	−107.1
Rn	1×10^{-19}	$6s^26p^6$	(145)	1037	9.73	−71	−61.8

[a]Values in parentheses are estimated.

[b]At 0°C and 1 atm.

[c]Note that the liquid exists only within a very narrow temperature range.

[d]At 1 atm pressure, liquid helium does not freeze even at the lowest measured temperature. The freezing point at 26 atm is below −272.2°C.

Helium is the second most abundant element in the universe, but it is not abundant on earth. Helium, like hydrogen, escapes earth's gravity, and most of the helium now present on earth was produced underground. Radioactive minerals constantly emit alpha particles (helium nuclei), which capture electrons and become helium atoms. Some of this helium remains trapped within the rock, some seeps into gas deposits, and some reaches the atmosphere and is lost.

Most of the earth's original argon, which consisted largely of argon-36 and argon-37, has also escaped into the atmosphere. The argon now present on earth is almost all argon-40, which is produced by the decay of radioactive potassium-40. The loss of the lighter argon isotopes accounts for an apparent anomaly in the periodic table where argon, despite a lower atomic number, has a greater chemical atomic weight than potassium.

Radon is produced by the radioactive decay of radium. Radon itself soon decays to polonium by emitting alpha particles; in other words, all the radon in the universe is temporary, in transit from radium to polonium. It has been estimated that the earth contains 1 g of buried radon for every square mile of surface area. Some of this radon seeps to the surface and may enter a building through cracks in the cellar floor and through openings for gas and water pipes. Recent studies have shown that some houses built on uranium-bearing geological formations have higher radon levels than those found in uranium mines. Such levels are a cause for concern because radon and its decay product polonium can enter the lungs, damage the cells, and eventually cause lung cancer. An inexpensive radon detector utilizing activated carbon (charcoal with a porous structure and extensive internal surface area) can be used to absorb some of the indoor radon in order to assess its level.

All of the noble gases except radon can be obtained from liquid air by *fractional distillation*, that is, by raising the temperature of liquid air gradually so that its components boil off one by one. Radon, the rarest of the noble gases, is obtained from the gas that collects above solutions of radium salts.

PRACTICE EXERCISE S1.4

Dry air consists largely of nitrogen (boiling point (bp), −195.8°C) and oxygen (bp, −183.0°C), with small amounts of carbon dioxide (bp, −78.5°C), methane (bp, −164°C), and each of the noble gases. (a) Consult Table S1.8 and state the temperature to which air must be cooled in order to liquefy all of its components. (b) In what order will the components boil off when liquefied air is warmed?

The percentage of helium is higher in some natural gas deposits than in air, and most of the helium used in industry is obtained from these deposits. Helium is an important industrial chemical, but one whose total supply is limited. About 800 million cubic feet of helium are used in the United States or exported each year. The reserves in natural gas fields have been estimated at about 250 billion cubic feet. A highly rarefied but abundant supply of helium exists in the atmosphere between 1100 and 3500 km above the earth, but so far there is no economical method for collecting it.

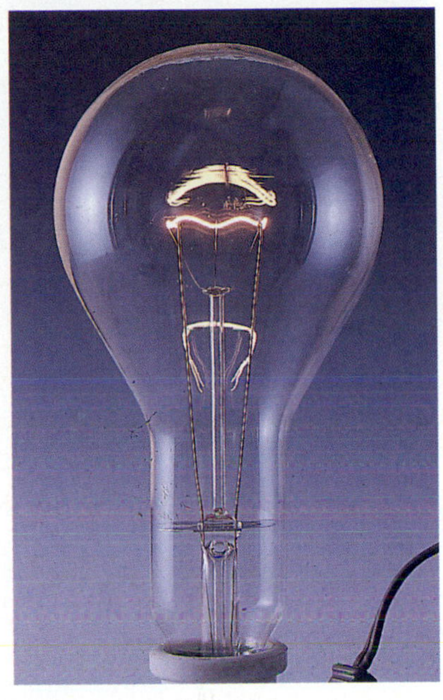

The hot tungsten filament in an incandescent light bulb will not corrode when the bulb is filled with argon gas.

Properties and Uses of the Noble Gases

Most uses for the noble gases are based on their inertness. Argon and krypton, for example, are used to fill electric light bulbs because, unlike oxygen, they will not attack and corrode the hot tungsten filament.

Almost one-quarter of all helium production is used to provide a protective (nonreactive) atmosphere for welding and other high-temperature work. Although helium gas is more dense than hydrogen, it is preferred for balloons and blimps because it is not flammable. Helium is less soluble in blood than nitrogen, so it is used instead of nitrogen in breathing mixtures for divers; the elimination of nitrogen from such mixtures helps to prevent the "bends," a painful and sometimes lethal condition that occurs when bubbles of nitrogen gas collect in joints and blood vessels during decompression (see *Chemical Insight: Decompression Illness*, p. 596). Breathing helium causes the voice to become high-pitched. The rise in pitch occurs because the speed of sound in helium is greater than in air. In fact, helium transmits sound more rapidly than any gas except hydrogen.

Helium has the lowest boiling point of any element. Liquid helium in equilibrium with its vapor maintains a temperature of 4.2 K; it is thus invaluable for low-temperature experiments. Liquid helium is also used to maintain the extremely low temperatures required for certain metals to become *superconducting*, that is, to lose their electrical resistance so that current will flow indefinitely with almost no loss of power.

A class of crystalline compounds that will superconduct at temperatures up to 133 K was discovered in 1993. The commercial development of these compounds, which are composed of mercury, barium, calcium, copper, and oxygen, would allow liquid nitrogen (bp, 77 K) to replace liquid helium as the coolant.

When liquid helium is cooled below 2.18 K, it undergoes an abrupt change of phase and enters a *superfluid* state in which it loses all viscosity (resistance to flow). The superfluid helium seems to defy gravity by spontaneously climbing the inner walls of its container, flowing down the outer walls, and dripping from the underside. This property, as far as we know, is not shared by any other substance.

Radon is an alpha particle emitter that has been used in the treatment of malignant tumors. Small glass tubes filled with radon gas are surgically implanted in the center of the tumor and later removed. Radiation from the radon destroys the cancerous cells, thus slowing the growth of the tumor. A recent addition to the tech-

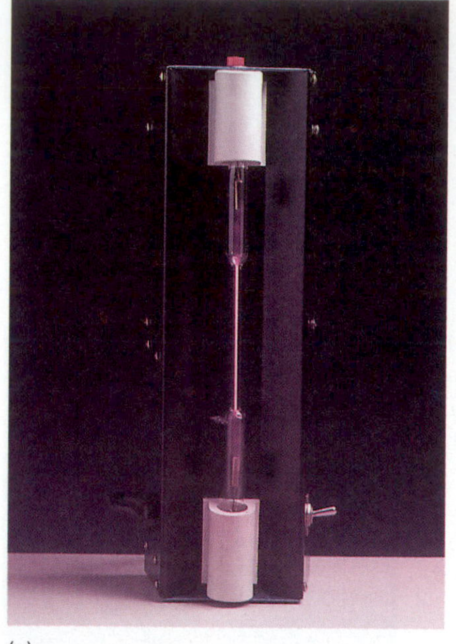

(a)

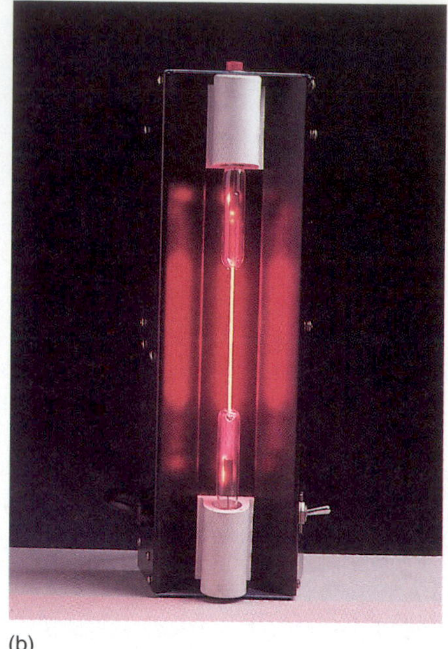

(b)

(c)

Figure S1.12

Colors emitted by gases in discharge tubes. (a) Helium. (b) Neon. (c) Mixtures of gases produce a variety of colors in "neon" display signs.

niques of earthquake prediction is based on increased seepage of underground radon from rock fissures just prior to an earthquake.

Characteristic colors are emitted when electric currents pass through tubes containing noble gases at low pressure (Figure S1.12). Neon gives the familiar red-orange color seen in neon advertising signs. Other colors are obtained by passing currents through mixtures of neon, argon, helium, and various other gases.

DIGGING DEEPER

Noble Gas Compounds

For decades after their discovery, the inertness of the noble gases was barely questioned. In 1933 Linus Pauling suggested that the ionization energy of xenon might be low enough to allow it to combine with elements like fluorine that have high electron affinities, but attempts at that time to react xenon with fluorine were frustrated by the fact that neither element was in good supply. In 1962 Neil Bartlett and N. K. Jha, of the University of British Columbia, startled chemists by reporting the synthesis of $XePtF_6$, a feat that was accomplished simply by mixing xenon and platinum hexafluoride (PtF_6) in a sealed glass tube at room temperature. They subsequently produced XeF_4 (xenon tetrafluoride) by heating the $XePtF_6$ (Figure S1.13). Other workers soon synthesized XeF_2, XeF_4, and XeF_6 directly from the elements, and they found that all of these compounds were much more stable than expected. Oxygen compounds such as $XeOF_4$ and the highly explosive XeO_3 were obtained by allowing the fluorides to react with water:

The first ionization energy of O_2 ($O_2 \longrightarrow O_2^+ + e^-$; IE = 1175 kJ/mol) is about the same as the first ionization energy of Xe ($Xe \longrightarrow Xe^+ + e^-$; IE = 1170 kJ/mol). This fact, coupled with the known reaction between O_2 and PtF_6 to form O_2PtF_6, led Bartlett to suspect that xenon would also react with PtF_6.

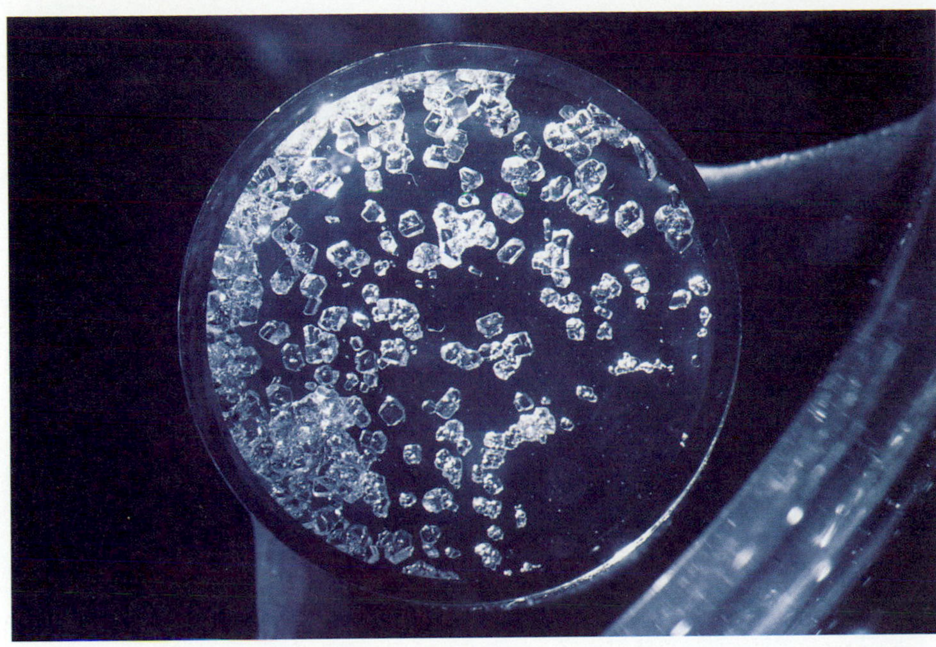

Figure S1.13
Crystals of xenon tetrafluoride (XeF_4).

$$XeF_6(s) + 3H_2O(l) \rightarrow XeO_3(aq) + 6HF(aq)$$

Compounds containing Xe—N and Xe—C bonds have been reported, but they are less stable than compounds containing Xe—O and Xe—F bonds. A number of krypton compounds, including KrF_2 and the ionic $KrF^+Sb_2F_{11}^-$, have been prepared, and at least one ionic radon fluoride compound has been reported. Helium, neon, and argon still have no known chemistry, and the krypton compounds are less stable than the xenon compounds. The periodic implications are clear. Only the largest noble gas atoms have outer electrons sufficiently removed from the nucleus to interact significantly with other atoms, and only fluorine, oxygen, and nitrogen atoms are sufficiently small and electron-greedy to convert this interaction into a stable arrangement.

CHAPTER REVIEW

Starred entries are based on the *Digging Deeper* sections.

LEARNING OBJECTIVES BY SECTION

S1.1 1. Describe the occurrence, properties, preparation, and uses of hydrogen.

2. Write equations for the reactions of hydrogen with oxygen, nitrogen, and halogens.

S1.2 1. List the alkali metals, and describe their occurrence, preparation, properties, and uses.
2. Write equations for the reactions of alkali metals with water, oxygen, halogens, hydrogen, and nitrogen.
3. Write equations for the reactions of alkali metal oxides, nitrides, and hydrides with water.
4. State the formulas, names, common names, and uses for the more important alkali metal compounds.

S1.3 1. List the alkaline earth metals, and describe their occurrence, preparation, properties, and uses.
2. Compare Group 1A and Group 2A metals with respect to atomic radius, ionization energy, hardness, density, and melting point.
3. Write equations for the reactions of Group 2A metals with water, oxygen, hydrogen, halogens, and nitrogen.
4. Write equations for the reactions of Group 2A oxides, nitrides, and hydrides with water.
5. State the formulas, names, common names, and uses for the more important Group 2A compounds.
6. Write equations for the preparation of calcium oxide and calcium hydroxide from limestone.

S1.4 1. List the noble gases, and describe their occurrence, preparation, properties, and uses.
*2. Write formulas for some noble gas compounds, and relate their formation to group trends in atomic properties.

KEY TERMS BY SECTION

S1.1 Haber process
Hydrocarbon
Hydride ion

S1.2 Alkali metal
Nitride ion
Peroxide ion
Superoxide ion

S1.3 Alkaline earth metal

S1.4 Noble gas

FINAL EXERCISES

Answers to exercises with blue numbers are given in Appendix D. Starred exercises are based on the *Digging Deeper* sections.

PART A. QUESTIONS AND PROBLEMS BY SECTION

Hydrogen (Section S1.1)

S1.1 Write a formula for each of the following species:
(a) a hydrogen atom **(c)** a hydrogen ion
(b) a hydrogen molecule **(d)** a hydride ion

S1.2 If the hydrogen atom loses its valence electron, what is left? Give two names for this particle.

S1.3 **(a)** Name the two monatomic ions of hydrogen and write their formulas.
(b) Which ion is larger?
(c) Which ion is more common in our environment?

S1.4 **(a)** Which of the two monatomic ions of hydrogen is found in aqueous solution?
(b) Which ion reacts with water, and what products are formed?

S1.5 Explain why the earth's atmosphere contains so little hydrogen. Would you expect to find a higher proportion of hydrogen on the planet Jupiter? Explain.

S1.6 Which compound of hydrogen is the most abundant on earth? List some other naturally occurring compounds that contain hydrogen.

S1.7 **(a)** What is *synthesis gas*?
(b) What is *water gas*? Is water gas a form of synthesis gas?
(c) From what natural materials can synthesis gas be prepared?
(d) What is synthesis gas used for?

S1.8 List three ways of preparing pure hydrogen.

S1.9 List the principal uses of hydrogen, and explain why hydrogen is especially suitable for these purposes.

S1.10 Discuss the advantages and disadvantages of hydrogen as a fuel from practical, economic, and environmental points of view.

S1.11 Write an equation for the preparation of
(a) water gas
(b) synthesis gas from methane
(c) pure hydrogen from methane and steam (two equations)

S1.12 Write an equation for each of the following reactions:
(a) electrolysis of water
(b) reaction of zinc metal with sulfuric acid
(c) combustion of water gas (two equations)

S1.13 Write an equation for the reaction of hydrogen with (a) nitrogen, (b) chlorine, and (c) oxygen. Name the product of each reaction.

S1.14 Write an equation for each of the following reactions:
- **(a)** combustion of hydrogen
- **(b)** hydrogen with bromine
- **(c)** hydride ion with water

S1.15 Write an equation for the industrial synthesis of
- **(a)** ammonia (Haber process)
- **(b)** methanol
- **(c)** hydrogen chloride

S1.16 Write an equation for the reduction by hydrogen of
- **(a)** WO_3 to tungsten metal
- **(b)** Fe_2O_3 to iron metal
- **(c)** MoO_3 to molybdenum metal

Alkali and Alkaline Earth Metals (Sections S1.2 and S1.3)

S1.17 Compare Na and Mg with respect to
- **(a)** electron configuration
- **(b)** atomic radius
- **(c)** first ionization energy
- **(d)** second ionization energy
- **(e)** formula of cation

S1.18 Compare K and Ca with respect to
- **(a)** electron configuration
- **(b)** atomic radius
- **(c)** first ionization energy
- **(d)** second ionization energy
- **(e)** formula of cation

S1.19 Compare K and Cs with respect to
- **(a)** valence-shell configuration
- **(b)** atomic radius
- **(c)** first ionization energy
- **(d)** ionic radius

S1.20 Compare Mg and Ba with respect to
- **(a)** valence-shell configuration
- **(b)** atomic radius
- **(c)** first ionization energy
- **(d)** ionic radius

S1.21 Compare adjacent alkali and alkaline earth metals with respect to
- **(a)** reactivity
- **(b)** hardness
- **(c)** density
- **(d)** melting point

Explain any differences in terms of atomic properties.

S1.22 Explain why the Group 1A and Group 2A metals are not found free in nature. How is each of the metals stored?

S1.23 A crusty piece of sodium does not look like a metal. What is the crust composed of? How could you demonstrate the metallic properties of the sodium sample?

S1.24 Explain why beryllium and magnesium are inert to air and moisture under ordinary conditions.

S1.25 State the principal mineral sources of
- **(a)** sodium
- **(b)** magnesium
- **(c)** beryllium
- **(d)** calcium

S1.26 Explain the role of solubility in the natural occurrence of Group 1A and Group 2A compounds.

S1.27 List some industrial uses for the following metals:
- **(a)** sodium
- **(b)** beryllium
- **(c)** magnesium
- **(d)** cesium

S1.28 State the biochemical role or roles of each of the following elements:
- **(a)** sodium
- **(b)** potassium
- **(c)** magnesium
- **(d)** calcium
- **(e)** lithium

S1.29
- **(a)** Which of the Group 1A and 2A ions can be identified by a flame test?
- **(b)** Explain the origin of the flame colors.
- **(c)** What flame colors are produced by the nitrates of (1) sodium, (2) strontium, (3) calcium, (4) potassium, and (5) barium?

S1.30 What is a sodium vapor lamp? Why are such lamps preferred over ordinary lamps for street lighting?

S1.31 Give the names, formulas, and uses of five alkali metal compounds.

S1.32 Give the names, formulas, and uses of five alkaline earth metal compounds.

S1.33 Identify the principal ingredient(s) of the following substances:
- **(a)** caustic soda
- **(b)** soda ash
- **(c)** baking soda
- **(d)** washing soda
- **(e)** bicarbonate of soda
- **(f)** lye

S1.34 Identify the principal ingredient(s) of the following substances:
- **(a)** lime
- **(b)** limestone
- **(c)** marble
- **(d)** gypsum
- **(e)** chalk
- **(f)** seashells
- **(g)** quicklime
- **(h)** slaked lime

S1.35 Write an equation for the electrolysis of (a) molten potassium chloride and (b) molten calcium chloride. [*Hint*: Calcium deposits as a solid in Part (b).]

S1.36 Pure beryllium is made by reacting beryllium fluoride with magnesium. Write the equation.

S1.37 Write equations for the reactions, if any, of each alkali metal with the following substances, and name the products:
- **(a)** hydrogen
- **(b)** nitrogen
- **(c)** excess oxygen
- **(d)** chlorine
- **(e)** water

Which of the above reactions require special conditions?

S1.38 Write equations for the reactions, if any, of each alkaline earth metal with the following substances, and name the products:
(a) hydrogen
(b) nitrogen
(c) excess oxygen
(d) chlorine
(e) water
Which of the above reactions require special conditions?

S1.39 Write equations for the reactions of the following substances with water, and name all the products:
(a) potassium
(b) calcium oxide
(c) sodium hydride
(d) lithium nitride

S1.40 Write equations for the reactions of the following substances with water, and name all the products:
(a) calcium
(b) lithium oxide
(c) calcium hydride
(d) magnesium nitride

S1.41 Write equations for the reactions between the following elements, and name the compound formed in each case:
(a) $Ba + Br_2$
(b) $K + O_2$
(c) $Li + N_2$
(d) $Sr + H_2$

S1.42 Write equations for the reactions between the following elements, and name the compound formed in each case:
(a) $Rb + I_2$
(b) $Sr + O_2$
(c) $Mg + N_2$
(d) $K + H_2$

S1.43 Write an equation for each of the following reactions:
(a) the electrolysis of aqueous potassium chloride
(b) the thermal decomposition of sodium hydrogen carbonate
(c) sodium carbonate with hydrochloric acid
(d) sodium hydrogen carbonate with sulfuric acid

S1.44 Write the equation for the reaction that occurs when
(a) excess stomach acid is treated with sodium bicarbonate
(b) limestone is converted to lime
(c) lime is converted to slaked lime
(d) mortar sets
(e) acid soil is neutralized by pulverized limestone
(f) plaster of Paris sets

Noble Gases (Section S1.4)

S1.45 Suggest reasons why the noble gases remained undiscovered for so long.

S1.46 Explain (a) why most earthly helium is found underground and (b) why so little radon exists.

S1.47 Which of the noble gases are continually synthesized by nuclear reactions on earth?

S1.48 (a) Why is the chemical atomic weight of argon greater than that of potassium?
(b) Can we assume that chemical atomic weights determined on earth will be the same elsewhere in the universe? Which types of elements might show the greatest discrepancies?

S1.49 List the principal uses of helium, and explain why helium is especially suitable for these purposes.

S1.50 Explain the use of helium in breathing mixtures for divers.

S1.51 Give at least one use for
(a) neon
(b) krypton
(c) argon
(d) radon

S1.52 Suggest reasons why argon and krypton are better than helium, nitrogen, and air for filling light bulbs.

S1.53 (a) Write the valence-shell configuration of each of the noble gases.
(b) Explain why helium, which has two electrons in its outer shell, is placed in Group 8A rather than in Group 2A.

***S1.54** Write the formula for each of the following xenon compounds:
(a) xenon difluoride
(b) xenon tetrafluoride
(c) xenon hexafluoride
(d) xenon trioxide
(e) xenon hexafluoroplatinate (the first xenon compound)

PART B. MISCELLANEOUS QUESTIONS AND PROBLEMS

S1.55 It is often said that hydrogen does not truly belong to any periodic group. Discuss and explain this statement on the basis of atomic structure and data from Section 8.5.

S1.56 Write the equation for the preparation of synthesis gas from ethane, C_2H_6. How does this mixture differ from the synthesis gas prepared from methane?

S1.57 Refer to Table 6.5, and calculate the fuel value per mole of (a) hydrogen, (b) methane, and (c) water gas. (*Hint*: Remember that 1 mol of water gas contains $\frac{1}{2}$ mol of each of its components.)

S1.58 Which of the gases in Exercise S1.57 has (a) the highest and (b) the lowest fuel value per gram.

S1.59 (a) Consult Table 6.5, and calculate the amount of heat absorbed during the conversion of 1.00 kg of carbon to water gas.
(b) Calculate and compare the heat energy released by burning 1.00 kg of carbon directly with the net amount of energy released by converting 1.00 kg of carbon to water gas and then burning the gas.
(c) What is the principal advantage of converting coke to water gas?

S1.60 The annual consumption of hydrogen by industry is about 48 billion cubic feet measured at standard conditions. Express the annual consumption of hydrogen in (a) moles and (b) metric tons.

S1.61 Why are Group 2A compounds less soluble than Group 1A compounds?

S1.62 Do you think that Group 1A and Group 2A compounds are easily decomposed? Why or why not? Explain on the basis of atomic properties.

S1.63 The reaction of potassium with water is more spectacular than the reaction of sodium or lithium because it is the most rapid reaction of the three. The lithium reaction is the slowest. Consider the nature of the atoms and the solid metals, and suggest a reason for the differences in reaction rate.

S1.64 Nineteenth-century chemists observed what they called a "diagonal relationship" between elements of the second and third period—Li resembles Mg in some of its properties, Be resembles Al, and so forth. The resemblances between Li and Mg include the following: Li (unlike other Group 1A elements) and Mg form nitrides; Li and Mg form oxides rather than peroxides or superoxides; and Li salts, like those of Mg, tend to be less soluble than other alkali metal salts. How can the resemblance between Li and Mg be explained on the basis of effective nuclear charge and/or other atomic properties?

***S1.65** Offer an explanation for the apparent inability of helium, neon, and argon to react with fluorine.

S1.66 Argon and oxygen have similar boiling points, so argon obtained from the fractional distillation of liquid air contains a large proportion of oxygen. The following procedure is used to obtain pure argon from this mixture: (1) Excess hydrogen is added and the mixture is ignited. (2) The mixture from (1) is passed over hot copper oxide (CuO). (3) The mixture from (2) is passed over a drying agent. Explain the purpose of each step and write equations for the reactions in (1) and (2).

S1.67 The density of dry air is 1.185 g/L at 25°C and 1.00 atm.

(a) Estimate the densities of hydrogen, helium, argon, and krypton under the same conditions.

(b) Would a balloon float in air if it were filled with argon? With krypton?

(c) Would you expect the ratio of argon to N_2 to increase or decrease with increasing altitude? What about the ratio of krypton to N_2?

S1.68 The ejection of an electron from a cesium surface requires 3.43×10^{-19} J of energy.

(a) Calculate the maximum wavelength to which a photoelectric sensor containing metallic cesium will respond.

(b) Will the sensor be affected by visible radiation? Explain.

S1.69 Calcium hydride has been used as a portable source of hydrogen. Assuming 100% efficiency, how many liters of water could be heated from 25°C to 100°C by burning the hydrogen obtained from the reaction of 100.0 g of calcium hydride with water?

S1.70 A 1.00-g sample of a solid consisting of calcium hydride and inert impurities is dissolved in 200.0 mL of water. A 25.0-mL sample of the resulting solution was titrated with 0.200 *M* HCl and found to require 29.4 mL of the acid. Calculate the percentage of calcium hydride in the solid.

S1.71 When 1.000 mol of magnesium was burned in air, the resulting mixture of MgO and Mg_3N_2 was found to weigh 38.20 g. What fraction of the product was MgO?

S1.72 A 0.340-g sample of lithium was burned in dry air to form a mixture of lithium oxide and lithium nitride. The resulting mixture was then allowed to react with water to form 0.500 L of solution. A 25.0-mL sample of the solution was titrated to a neutral endpoint with 26.0 mL of 0.110 *M* HCl. What fraction of the product was Li_3N? (*Hint*: Write equations for the reactions of the oxide and nitride with water. Assume the ammonia remains in solution.)

CHAPTER 9

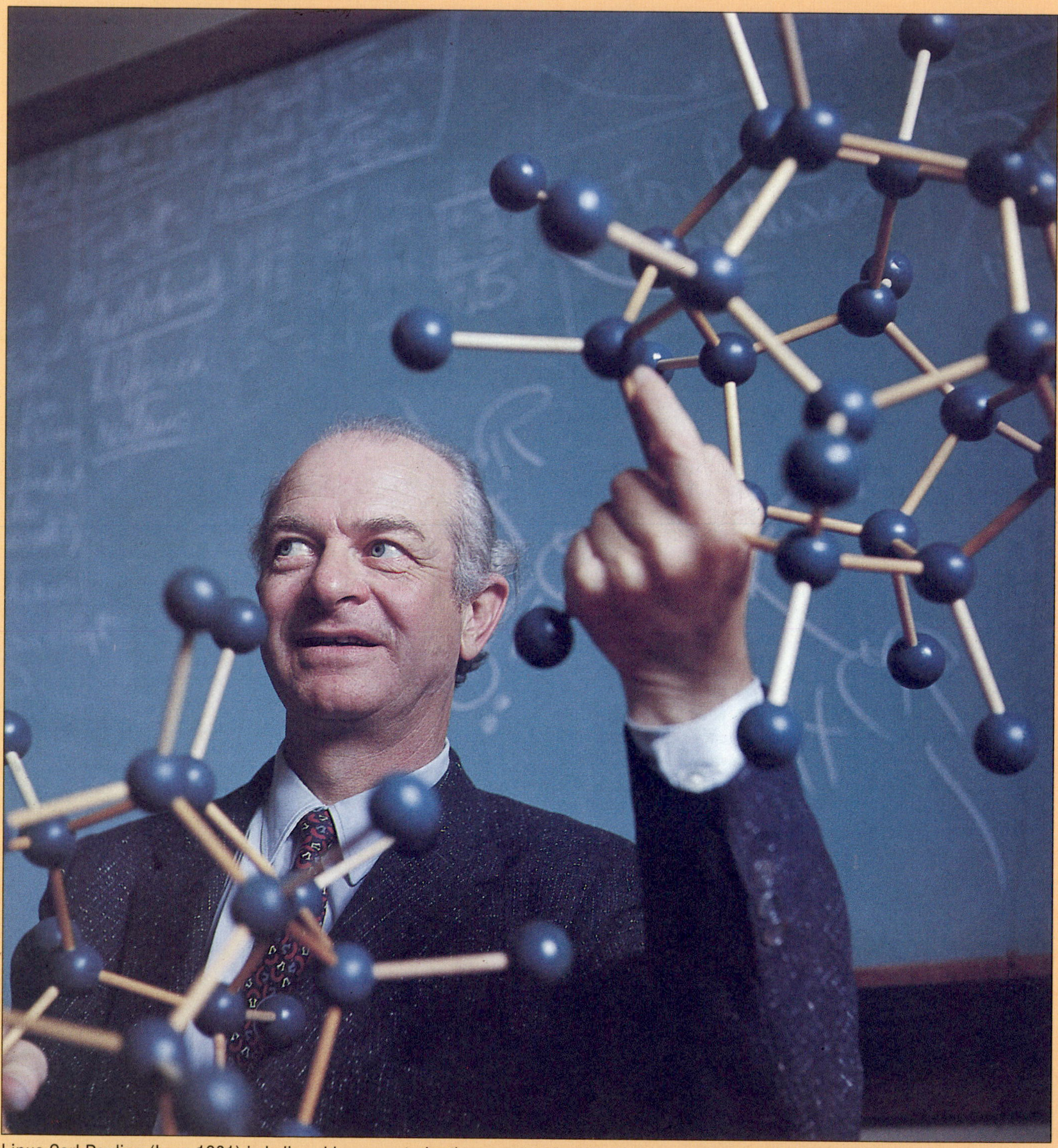

Linus Carl Pauling (born 1901) is believed by many to be the greatest chemist of the twentieth century. His epoch-making book, *The Nature of the Chemical Bond,* and earlier papers bearing the same general title, were among the first to consider the chemical bond in terms of quantum theory.

THE CHEMICAL BOND

OUTLINE

PREVIEW

Very few atoms remain single in our earthly environment; most form chemical bonds with their neighbors. Bonding more than anything else determines the nature of a substance, whether it is a tenuous gas, an impervious solid, or a conductive metal. The shapes of molecules—linear, planar, or three-dimensional—are also determined by bonding. In his famous book titled *Valence,* Charles Alfred Coulson pointed out that a theory of chemical bonding must answer three questions: (1) Why do atoms combine to form compounds? (2) Why do they combine in definite proportions? (3) Why do molecules have their characteristic shapes? In this chapter we will discover that the answers to these questions lie with the valence electrons. To explore the realm of bonding is to venture into complicated terrain, and in this first expedition, we will establish some general principles that will serve as a base for later, more penetrating, investigations.

The term **chemical bond** describes the strong link between the atoms in a molecule or a crystal, as distinguished from the weaker forces that attract one molecule to another. The energies required to break most chemical bonds lie between 100 and 1000 kJ per mole of bonds and are much greater than the energies required to overcome attractions between molecules. In nearly all stable substances, elements and compounds alike, each atom is chemically bonded to at least one other atom. Only the noble gases of Group 8A exist normally as isolated atoms.

The energies required to separate the atoms in some metals, for example, potassium and mercury, are less than 100 kJ/mol.

There are three principal types of chemical bonds: ionic, covalent, and metallic. An **ionic bond** is the electrostatic attraction between oppositely charged ions. Ionic bonds form when electrons are transferred from one atom to another, usually from a metal to a nonmetal. Recall from Chapter 2 that an ionic solid such as sodium chloride (NaCl) consists of positive and negative ions arranged in a regular pattern called an **ionic lattice**, in which each ion is surrounded by, and attracted to, neighbors of opposite charge (see Figure 2.19, p. 63). There are no molecules in an ionic solid.

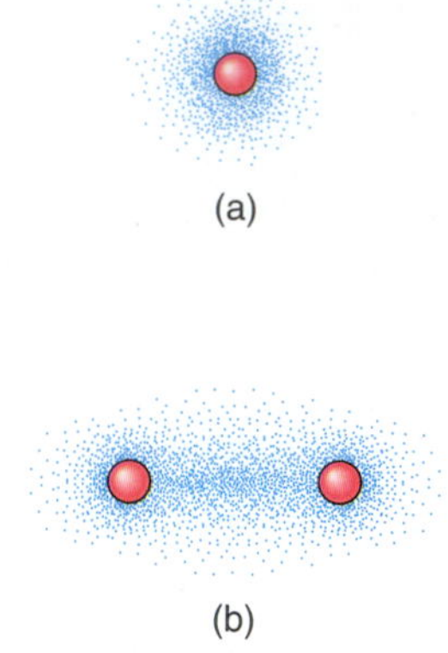

Figure 9.1

Dot representations of electron density in (a) the hydrogen atom and (b) the hydrogen molecule. The two protons in H_2 are bonded together by the high concentration of electron density between them.

Electron transfer is unlikely between identical atoms or atoms with similar electron-attracting tendencies; such atoms are held together by either covalent bonds or metallic bonds. The simplest **covalent bond**, also called an *electron-pair bond,* is a pair of electrons shared between two atoms and attracted by both nuclei. More complex covalent bonds consist of two or even three pairs of electrons.

We can describe covalent bonds in terms of electron-density patterns. For example, consider the covalent bond in H_2, where each hydrogen atom contributes one electron to the shared pair. Recall that the electron density in an isolated hydrogen atom is spherically symmetrical about the nucleus (Figure 9.1a). When two hydrogen atoms bond to form H_2, the electron-density patterns change, as shown in Figure 9.1b. Formation of the covalent bond is accompanied by an increase of electron density between the two hydrogen nuclei; in other words, the shared electrons spend most of their time in this region. The negative electron density between the positive nuclei results in an electrostatic attraction that holds the molecule together and gives the covalent bond its strength.

Atoms in a metal are held together by **metallic bonding**, in which the valence electrons are not confined to individual atoms or to individual bonds but flow freely through the metal. Metallic structure (Figure 9.2) is sometimes described as a lattice of positive ions bathed by a sea of mobile electrons; these electrons provide metals with their conductivity and other distinctive properties.

9.1 LEWIS SYMBOLS

Atoms of representative elements (those in Groups 1A through 8A) use only their valence (outer shell) electrons in the formation of chemical bonds. Recall from Section 8.4 (p. 315) that the number of valence electrons in a main group atom is equal to its group number. (Helium, with two valence electrons, is the only exception to this rule.) The **Lewis symbol** for an atom consists of the atomic symbol surrounded by a number of dots equal to the number of valence electrons (Figure 9.3). Lewis symbols for the second-period elements are given in Table 9.1. Because the atoms in any one group all have the same number of valence electrons, their symbols will all have the same number of dots. The Lewis symbols for Group 1A atoms, for example, all have one dot.

Figure 9.2

Dot representation of electron density in a metal. The positive metal ions are arrayed in a regular pattern, while the valence electrons move freely among them. The electron density is greatest around each ion.

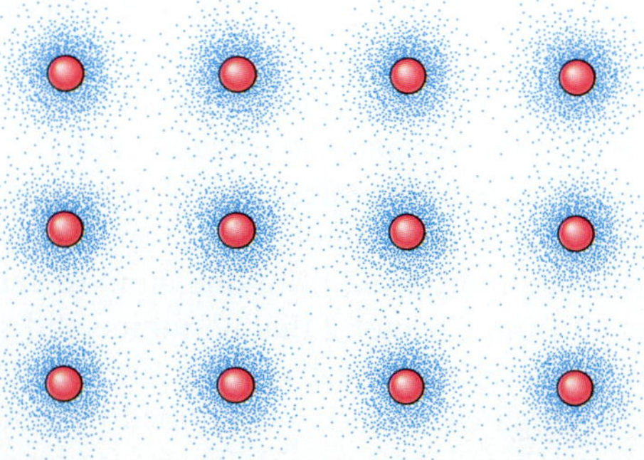

(a)

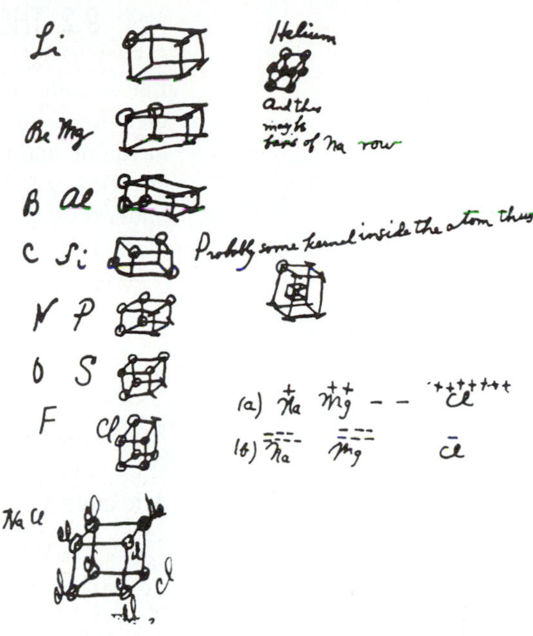

(b)

Figure 9.3

(a) Gilbert Newton Lewis (1875–1946) developed many of the bonding concepts discussed in this chapter. In 1916 he proposed the theory that covalent bonding is due to shared electron pairs. (b) Lewis's original sketch of the octet rule.

TABLE 9.1 Lewis Symbols for the Second-Period Elements[a]

Group	1A	2A	3A	4A	5A	6A	7A	8A
Symbol	Li·	·Be·	·Ḃ·	·Ċ·	·N̈· (one dot below)	·Ö· (two dots below)	:F̈· (two dots below)	:N̈e: (two dots below)

[a]Each dot represents a valence electron; hence, the number of dots around any symbol is equal to the group number. For Groups 1A through 4A, the dots are arranged so that there is no more than one dot on each side of the symbol. For Groups 5A through 8A, the dots are paired, one at a time, until there are four pairs around the Group 8A symbol.

EXAMPLE 9.1

Write the Lewis symbol for selenium, Se.

SOLUTION

Selenium is in Group 6A of the periodic table, so the selenium atom has six valence electrons. The Lewis symbol is

PRACTICE EXERCISE 9.1

Write Lewis symbols for (a) phosphorus, (b) bromine, and (c) silicon.

9.2 THE IONIC BOND

Let us compare the electron configurations of some main group atoms with those of their ions. We will start with the metallic elements of Groups 1A and 2A, which have one and two valence electrons, respectively. These atoms have low ionization energies (see Table 8.3, p. 324), and their valence electrons are completely removed under ordinary reaction conditions. Group 1A atoms form 1+ cations:

LEARNING HINT

Some important monatomic ions are shown in Figure 2.22 on page 67. The periodic trends will become more apparent if you refer to this table.

$$\underset{[Ne]3s^1}{Na\cdot} \rightarrow \underset{[Ne]}{Na^+} + e^-$$

$$\underset{[Ar]4s^1}{K\cdot} \rightarrow \underset{[Ar]}{K^+} + e^-$$

Atoms and ions with the same electron configuration are said to be **isoelectronic**. Each Group 1A ion is isoelectronic with the noble gas atom preceding it in the periodic table. Group 2A atoms form 2+ ions that also have noble gas configurations:

$$\underset{[Ar]4s^2}{\cdot Ca\cdot} \rightarrow \underset{[Ar]}{Ca^{2+}} + 2e^-$$

$$\underset{[Xe]6s^2}{\cdot Ba\cdot} \rightarrow \underset{[Xe]}{Ba^{2+}} + 2e^-$$

A positive monatomic ion of an alkali (Group 1A) or alkaline earth (Group 2A) metal is isoelectronic with the noble gas atom preceding its period.

Metal atoms from Groups 3A, 4A, and 5A would have to lose three or more electrons in order to become isoelectronic with a noble gas; the large ionization energy required to remove this number of electrons is not available under ordinary reaction conditions, so most cations from these groups do not have noble gas configurations. Aluminum in Group 3A is a notable exception. The aluminum ion has the electron configuration of neon:

$$\underset{[Ne]3s^23p^1}{\cdot\dot{Al}\cdot} \rightarrow \underset{[Ne]}{Al^{3+}} + 3e^-$$

The atoms of nonmetals form negative ions by gaining just enough electrons to fill the s and p orbitals in their valence shells. Thus, they acquire the electron configurations of the noble gas atoms at the ends of their periods. The halogen atoms of Group 7A each gain one electron to form a halide ion. Bromine atoms, for example, form bromide ions:

$$\underset{[Ar]4s^23d^{10}4p^5}{:\ddot{\underset{\cdot\cdot}{Br}}\cdot} + e^- \rightarrow \underset{[Kr]}{:\ddot{\underset{\cdot\cdot}{Br}}:^-}$$

Group 6A atoms gain two electrons; for example, oxygen forms the oxide ion:

$$\underset{[He]2s^22p^4}{\cdot\ddot{\underset{\cdot\cdot}{O}}\cdot} + 2e^- \rightarrow \underset{[Ne]}{:\ddot{\underset{\cdot\cdot}{O}}:^{2-}}$$

In general, *a negative monatomic ion is isoelectronic with the noble gas atom at the end of its period.* Another example is nitrogen (Group 5A), which forms the nitride ion:

$$\cdot\ddot{\text{N}}\cdot + 3e^- \rightarrow :\ddot{\underset{\cdot\cdot}{\text{N}}}:^{3-}$$

$$[\text{He}]2s^22p^3 \qquad [\text{Ne}]$$

The nonmetals of Groups 3A and 4A do not normally form negative ions.

EXAMPLE 9.2

Write formulas for (a) the strontium ion and (b) the hydride ion.

SOLUTION

(a) Strontium is a Group 2A metal. Its valence shell contains two electrons, both of which are lost. The formula of the strontium ion is Sr^{2+}.

(b) The electron configuration of hydrogen is $1s^1$. To make the negative hydride ion (remember, the *ide* ending denotes a negative ion), one electron must be added to complete the shell and produce the helium configuration. The formula of the hydride ion is H^-.

PRACTICE EXERCISE 9.2

Write formulas for the monatomic ions formed by (a) magnesium, (b) iodine, and (c) sulfur.

Noble gases, because of their high positive electon affinities and high ionization energies, are stable and quite unreactive. Therefore, it is not surprising to find noble gas configurations duplicated in many stable ions. Except for the helium atom with a $1s^2$ configuration, each noble gas atom has an s^2p^6 outer shell configuration of eight electrons. This configuration is often referred to as an *octet.*

In many binary ionic compounds (ionic compounds consisting of only two elements), each ion is isoelectronic with a noble gas atom; for example, in NaCl, Na^+ has the octet of neon and Cl^- the octet of argon:

$$\underset{[\text{Ne}]}{\text{Na}^+} \qquad \underset{[\text{Ar}]}{\text{Cl}^-}$$

The statement that *bonded atoms have noble gas configurations* is known as the **octet rule**. This rule is frequently but not invariably obeyed by ions in ionic compounds. It is always obeyed when ionic compounds are formed from the metals of Groups 1A or 2A and the nonmetals of Groups 5A, 6A, or 7A. Calcium and oxygen ions, for example, possess noble gas configurations in calcium oxide (CaO):

$$\underset{[\text{Ar}]}{\text{Ca}^{2+}} \qquad \underset{[\text{Ne}]}{\text{O}^{2-}}$$

as do the ions in lithium nitride (Li_3N):

$$\underset{[\text{He}]}{\text{Li}^+} \qquad \underset{[\text{Ne}]}{\text{N}^{3-}}$$

Unlike the main group ions described above, the ions of transition metals and post-transition metals (metals to the right of the transition groups in the periodic table) do not usually have noble gas configurations and thus do not usually obey the octet rule. The atoms of these metals would have to lose large numbers of electrons to achieve noble gas configurations, and the ionization energy is usually not available. The heavier post-transition metal atoms tend to lose only *p* electrons from the valence shell. The most important examples are

$$\underset{[Kr]5s^2 4d^{10} 5p^2}{Sn} \rightarrow \underset{[Kr]5s^2 4d^{10}}{Sn^{2+}} + 2e^-$$

$$\underset{[Xe]6s^2 4f^{14} 5d^{10} 6p^2}{Pb} \rightarrow \underset{[Xe]6s^2 4f^{14} 5d^{10}}{Pb^{2+}} + 2e^-$$

$$\underset{[Xe]6s^2 4f^{14} 5d^{10} 6p^3}{Bi} \rightarrow \underset{[Xe]6s^2 4f^{14} 5d^{10}}{Bi^{3+}} + 3e^-$$

A transition metal atom will lose its outer *s* electrons and perhaps one or two inner shell *d* electrons as well:

$$\underset{[Ar]4s^2 3d^8}{Ni} \rightarrow \underset{[Ar]3d^8}{Ni^{2+}} + 2e^-$$

$$\underset{[Ar]4s^1 3d^5}{Cr} \rightarrow \underset{[Ar]3d^3}{Cr^{3+}} + 3e^-$$

Many transition metals form more than one ion. Iron, for example, forms iron(II) and iron(III) ions:

$$\underset{[Ar]4s^2 3d^6}{Fe} \rightarrow \underset{[Ar]3d^6}{Fe^{2+}} + 2e^-$$

$$\underset{[Ar]4s^2 3d^6}{Fe} \rightarrow \underset{[Ar]3d^5}{Fe^{3+}} + 3e^-$$

There are no easy rules for predicting the charges on transition metal cations—the important ions simply have to be memorized.

PRACTICE EXERCISE 9.3

Which of the ions in the following binary compounds obey the octet rule and which do not: (a) lithium oxide (Li_2O), (b) lead chloride ($PbCl_2$), and (c) zinc bromide ($ZnBr_2$)?

The Strength of Ionic Bonding: Lattice Energy

It is true for compounds as well as for atoms that the most stable arrangement of particles is the one that provides the lowest possible energy. In an ionic solid, the ions arrange themselves so that attractive forces between unlike charges are maximized and repulsive forces between like charges are minimized. In sodium chloride, for example, each ion is surrounded by six ions of opposite charge (see Figure 2.19, p. 63), and there are no like charges next to each other. One measure of the

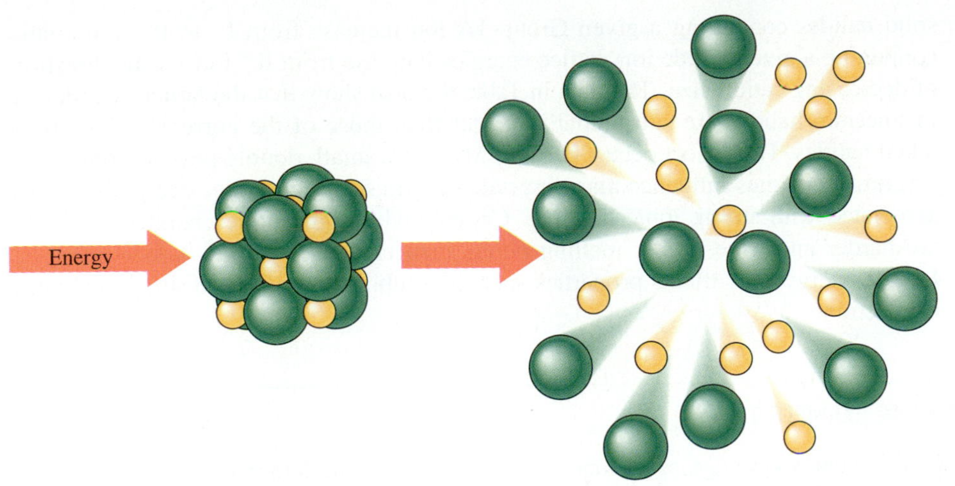

Figure 9.4
Lattice energy is the energy that would be required to break up 1 mol of a crystalline compound into independent gaseous ions.

strength of the bonding in an ionic crystal is the **lattice energy**, the energy required to break up 1 mol of the crystal into independent gaseous ions (see Figure 9.4). The lattice energy of sodium chloride, for example, is 786 kJ/mol:

$$NaCl(s) \rightarrow Na^+(g) + Cl^-(g) \qquad \Delta H^\circ = +786 \text{ kJ}$$

The 786 kJ is needed to separate the ions in 1 mol of NaCl and disperse them to an "infinite distance," that is, to distances so great that none of the ions interact. Conversely, if the widely separated ions come together to form 1 mol of crystalline NaCl, 786 kJ of energy will be released.

Some chemists define *lattice energy* as the energy *released* when 1 mol of an ionic solid forms from its separated gaseous ions. According to this definition, the lattice energy of sodium chloride would be $\Delta H^\circ = -786$ kJ/mol.

The electrostatic force F between a positive ion of charge q_+ and a negative ion of charge q_- is directly proportional to the product of the charges and inversely proportional to the square of the distance r between the ion centers

$$F = k\frac{q_+q_-}{r^2} \qquad (9.1)$$

where k, the proportionality constant, is equal to 8.988×10^9 N·m²/C². Each ion in a crystal is surrounded by ions of opposite charge; the greater the attractive force between these ions, the greater the energy required to separate them. Hence, Equation 9.1 implies that *the lattice energy of a crystal increases with increasing ionic charge and with decreasing ionic radii.* These trends are confirmed by the data in Table 9.2. Halide ions decrease in size from I^- to F^-, and the lattice energies of

TABLE 9.2 Lattice Energies

Compound	Lattice energy (kJ/mol)	Compound	Lattice Energy (kJ/mol)	Compound	Lattice Energy (kJ/mol)	Compound	Lattice Energy (kJ/mol)
LiF	1036	NaF	923	KF	821	MgF_2	2957
LiCl	853	NaCl	786	KCl	715	$MgCl_2$	2526
LiBr	807	NaBr	747	KBr	682	$MgBr_2$	2440
LiI	757	NaI	704	KI	649	MgI_2	2327

solid halides containing a given Group 1A ion increase from I^- to F^-. For solids containing a given halide ion, lattice energies increase from K^+ to Li^+, the direction of decreasing cation size. The data in Table 9.2 also show that the lattice energies of magnesium halides are substantially greater than those of the corresponding alkali metal halides. This effect occurs because Mg^{2+} is a small, doubly positive ion.

Lattice energies influence many crystal properties. A high lattice energy is always associated with strong ionic bonding. Crystals with high lattice energies generally are harder and have higher melting points than crystals with low lattice energies. Lattice energy also affects properties such as solubility, response to stress, and rates of reaction.

EXAMPLE 9.3

Use your knowledge of periodic trends to compare the lattice energies and melting points of lithium fluoride (LiF) and cesium chloride (CsCl).

SOLUTION

The ions in CsCl and LiF are all singly charged, so any difference in lattice energies will be the result of size differences. We know that Li^+ (second period) is smaller than Cs^+ (sixth period) and F^- (second period) is smaller than Cl^- (third period). Because it has smaller ions, LiF should have a higher lattice energy and a higher melting point than CsCl.

The lattice energies of LiF and CsCl are 1036 and 659 kJ/mol, respectively; LiF melts at 845°C, CsCl at 645°C.

PRACTICE EXERCISE 9.4

Which would you expect to have the higher lattice energy: (a) MgF_2 or $CaCl_2$; (b) MgF_2 or MgO; (c) MgF_2 or AlF_3?

DIGGING DEEPER

How Lattice Energies Are Determined

As illustrated in Figure 9.5, the formation of an ionic solid from its elements can be thought of as occurring directly in one step or indirectly in five distinct steps. Hess's law (Section 6.4) tells us that the enthalpy change for the direct formation of sodium chloride from sodium metal and chlorine gas

$$\text{Na(s)} + \tfrac{1}{2}\text{Cl}_2\text{(g)} \rightarrow \text{NaCl(s)} \qquad \Delta H^\circ = -411 \text{ kJ}$$

can be equated to the sum of the enthalpy changes of the five steps in the indirect reaction. These steps are as follows:

1. Sodium atoms sublime (evaporate) from the solid metal. Energy is absorbed; the enthalpy of sublimation is 108 kJ per mole of sodium atoms:

$$\text{Na(s)} \rightarrow \text{Na(g)} \qquad \Delta H_1^\circ = +108 \text{ kJ}$$

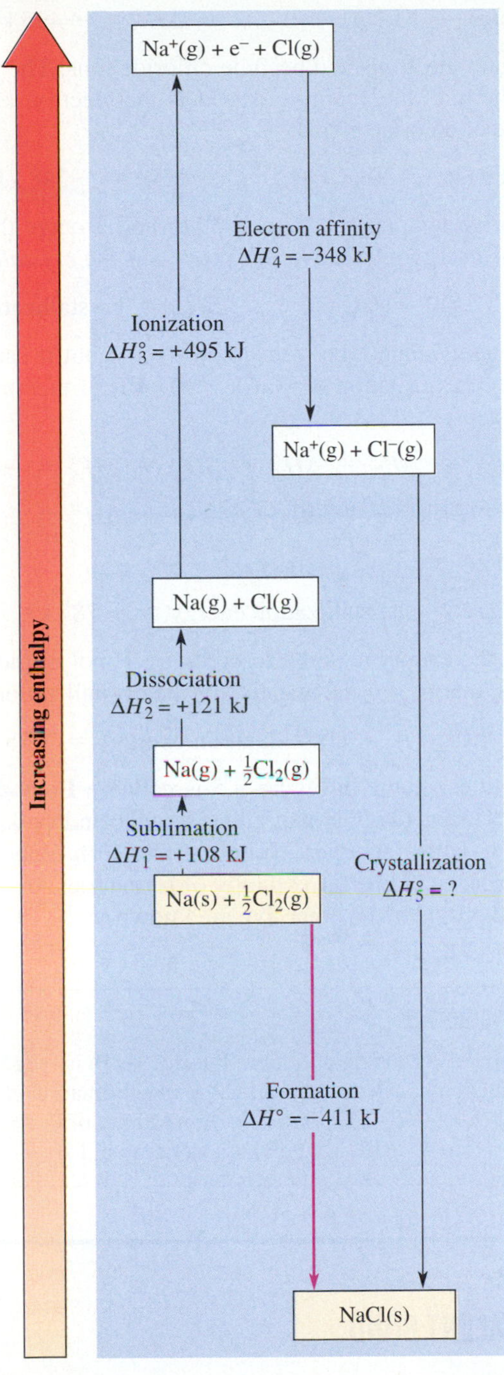

Figure 9.5

Diagram of the Born–Haber cycle for the formation of 1 mol of crystalline sodium chloride from its elements. The direct formation reaction is indicated by the red arrow. The black arrows show an imaginary five-step pathway for the same reaction. The sum of the $\Delta H°$ values for the five steps equals the overall $\Delta H°$ of formation, −411 kJ.

2. Chlorine molecules dissociate into atoms. Energy is absorbed; the enthalpy of dissociation is 121 kJ per mole of chlorine atoms:

$$\tfrac{1}{2}Cl_2(g) \rightarrow Cl(g) \qquad \Delta H_2^\circ = +121 \text{ kJ}$$

3. Sodium atoms lose electrons to become sodium ions. Energy is absorbed; the enthalpy of ionization is the ionization energy, 495 kJ per mole of sodium atoms:

LEARNING HINT

One mole of chlorine atoms is produced from $\frac{1}{2}$ mol of chlorine molecules.

$$Na(g) \rightarrow Na^+(g) + e^- \qquad \Delta H_3^\circ = +495 \text{ kJ}$$

4. Chlorine atoms gain electrons and become chloride ions. The formation of chloride ions is exothermic; the energy released is the electron affinity of chlorine, −348 kJ per mole of chlorine atoms:

$$Cl(g) + e^- \rightarrow Cl^-(g) \qquad \Delta H_4^\circ = -348 \text{ kJ}$$

5. Sodium and chloride ions approach each other and arrange themselves into the crystal lattice. The energy released in this process is the *crystallization energy*:

$$Na^+(g) + Cl^-(g) \rightarrow NaCl(s) \qquad \Delta H_5^\circ = \text{crystallization energy}$$

The crystallization energy cannot be measured directly, but it can be calculated by equating the enthalpy of formation of NaCl, −411 kJ, to the sum of the enthalpy changes for the five steps:

$$\Delta H_1^\circ + \Delta H_2^\circ + \Delta H_3^\circ + \Delta H_4^\circ + \Delta H_5^\circ = -411 \text{ kJ}$$

$$108 \text{ kJ} + 121 \text{ kJ} + 495 \text{ kJ} - 348 \text{ kJ} + \Delta H_5^\circ = -411 \text{ kJ}$$

Solving this equation gives

$$\Delta H_5^\circ = \text{crystallization energy} = -787 \text{ kJ}$$

The lattice energy is the energy required to break up 1 mol of the crystal into independent gaseous ions; that is, it is the negative of the crystallization energy. Hence,

$$\text{lattice energy} = -(\text{crystallization energy}) = +787 \text{ kJ}$$

The cycle of reactions shown in Figure 9.5 is called a **Born–Haber cycle** after Max Born and Fritz Haber, two German chemists who independently devised this method for calculating lattice energies. Note that the Born–Haber cycle involves a total of six energy terms including the enthalpy of formation. Any one of these energies can be calculated provided the other five are known.

PRACTICE EXERCISE 9.5

The standard enthalpy of formation of lithium fluoride (LiF) is −612.1 kJ/mol. Use the following information to calculate the enthalpy of sublimation of lithium metal: ionization energy of Li = +520.3 kJ/mol, electron affinity of F = −322 kJ/mol, dissociation energy of F_2 = +157 kJ/mol, lattice energy of LiF = +1036 kJ/mol.

9.3 THE COVALENT BOND

Most ionic compounds consist of metal atoms that have lost electrons and nonmetal atoms that have gained electrons. The vast majority of compounds, however, contain only nonmetal atoms, between which there is little likelihood of electron transfer. The atoms in these substances are held together by *covalent bonds,* that is, by electron pairs shared between the bonded atoms. Simple molecules such as H_2, O_2, N_2, H_2O, and NH_3 are covalent, and so are more complex substances such as sugar, rubber, Teflon, and silk. Silicates, which make up most of the rocks in the earth's crust, contain covalent networks of oxygen and silicon atoms. Living matter consists of even more intricate molecules containing covalently bonded atoms of carbon, hydrogen, oxygen, nitrogen, sulfur, and phosphorus. Metals, especially the transition and post-transition metals, also form covalent bonds, as in the chromate ion (CrO_4^{2-}) and in nonionic halides such as $SnCl_4$.

There are only 22 nonmetals in the periodic table, yet over 90% of the known compounds consist solely of these elements.

Lewis Structures

Covalently bonded atoms often obey the octet rule. In the hydrogen molecule (H_2), the two atoms share electrons to make one electron-pair bond:

$$\mathrm{H\cdot + \cdot H \longrightarrow H:H}$$

shared electron pair

Each hydrogen atom thus acquires the two-electron configuration of helium. In the water molecule, the oxygen atom shares one valence electron with each of the two hydrogen atoms:

$$\mathrm{H\cdot + \cdot\ddot{\underset{\cdot\cdot}{O}}\cdot + \cdot H \longrightarrow H:\ddot{\underset{\cdot\cdot}{O}}:H}$$

shared electron pairs

The oxygen atom has an octet (four electron pairs) around it, and each hydrogen atom has a "duet" (one electron pair). Observe that two of the electron pairs on the oxygen atom are not shared. Unshared pairs of valence electrons are often referred to as **lone pairs** or **nonbonding pairs**; the oxygen atom in H_2O has two lone pairs.

A covalent molecule can be represented by a **Lewis structure** in which shared electron pairs are shown as dashes between the bonded atoms and unshared valence electrons are shown as dots. The Lewis structure of hydrogen is

$$\mathrm{H{-}H}$$

and the Lewis structure of water is

$$\mathrm{H{-}\ddot{\underset{\cdot\cdot}{O}}{-}H}$$

Lewis structures do not usually show molecular geometry. Water is a bent molecule with an H—O—H angle of about 105°, so a more accurate representation is

$$\mathrm{H\diagup\ddot{\underset{\cdot\cdot}{O}}\diagdown H}$$

Each dash in a Lewis structure represents a shared electron pair, so there are two electrons in the structure of hydrogen and eight electrons in the structure of water. All of the electrons in a Lewis structure are valence electrons. The eight electrons in the structure of water come from three atoms: one electron from the valence shell of each hydrogen atom and six electrons from the valence shell of the oxygen atom. *The total number of electrons in the Lewis structure of a molecule is the sum of the valence electrons of the individual atoms.*

The Lewis structure of H_2O shows each hydrogen atom bonded to oxygen by a **single bond**, that is, by one shared pair of electrons. Some molecules contain **double bonds** consisting of two shared pairs of electrons. An example is ethylene (C_2H_4):

$$\begin{matrix}\mathrm{H} & & & \mathrm{H}\\ & \diagdown & \diagup & \\ & \mathrm{C{=}C} & & \\ & \diagup & \diagdown & \\ \mathrm{H} & & & \mathrm{H}\end{matrix}$$

The systematic names for ethylene and acetylene are *ethene* and *ethyne,* respectively.

Acetylene (C_2H_2) is an example of a molecule with a **triple bond** consisting of three shared pairs:

$$\mathrm{H{-}C{\equiv}C{-}H}$$

These structures satisfy the octet rule; each hydrogen atom shares two electrons and each carbon atom shares eight electrons.

The atoms in some covalent molecules do not obey the octet rule, but for now, we will not consider such cases. The octet rule is obeyed in each of the following molecules:

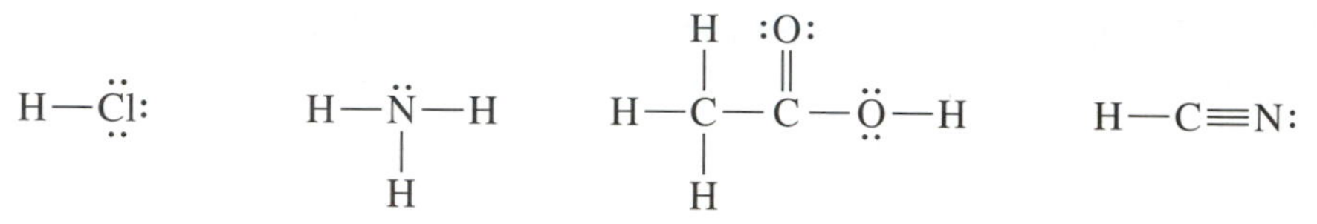

$$\mathrm{H{-}\ddot{\underset{..}{Cl}}{:}} \qquad \mathrm{H{-}\ddot{N}(H){-}H} \qquad \mathrm{H{-}CH_2{-}C({=}\ddot{O}){-}\ddot{O}{-}H} \qquad \mathrm{H{-}C{\equiv}N{:}}$$

Hydrogen chloride — Ammonia — Acetic acid — Hydrogen cyanide

The Strength of Covalent Bonds: Bond Energies and Bond Lengths

Bonded atoms are in constant motion. They vibrate back and forth under the influence of electrostatic forces in much the same way as do two objects connected by a spring (Figure 9.6). When the atoms move away from each other, attractive forces act to pull them back, and when they move toward each other, repulsive forces push them away again. A **bond length** is the equilibrium distance between the centers of two bonded atoms. It is the distance at which the attractive and repulsive forces are in balance and the net force on the atoms is zero. (If the bond were a real spring, the bond length would be the length of the spring when it is neither stretched nor compressed.) Some covalent bond lengths are reported in Table 9.3.

The strength of a covalent bond can be measured in terms of the energy required to break it. The average energy required to break a chemical bond in an isolated molecule is called the *bond dissociation energy* (D) or simply the **bond energy**. Bond energies are expressed in kilojoules per mole of bonds, and some values are given in Table 9.3. For diatomic molecules such as H_2, the bond energy is simply the energy needed to dissociate 1 mol of gaseous molecules into gaseous atoms. The H—H bond energy, for example, is 436 kJ/mol at 1 atm pressure:

$$\mathrm{H_2(g) \rightarrow 2H(g)} \qquad D(\mathrm{H{-}H}) = \Delta H^\circ = +436\ \mathrm{kJ}$$

The formation of 1 mol of H_2 from its atoms is the reverse of this reaction; it is

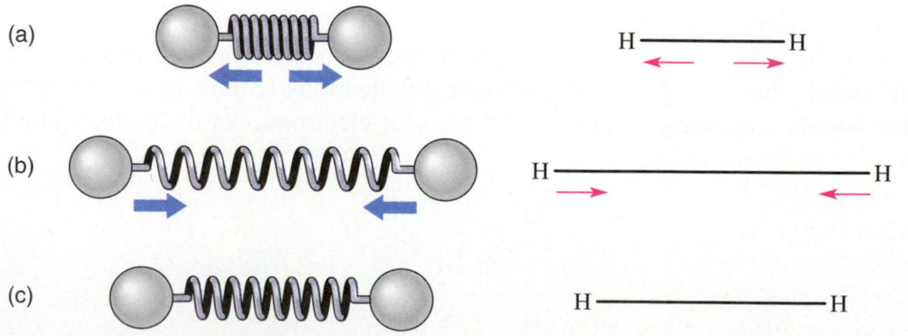

Figure 9.6

A covalent bond is like a spring; both tend to vibrate (periodically lengthen and shorten) around some equilibrium length. (a) When the bond is compressed, repulsive forces tend to lengthen it. (b) When the bond is stretched, attractive forces tend to shorten it. (c) Attractive and repulsive forces are balanced at the equilibrium bond length.

(a) H—H

(b) H—H

(c) H—H

TABLE 9.3 Bond Energies (kJ/mol) and Bond Lengths (pm)

Single Bonds[a,b]

	H	C	N	O	S	F	Cl	Br	I
H	435.9 74	413 109	391 101	463 96	366 134	568.1 92	432.0 127	366.2 141	298.3 161
C		348 154	308 147	360 143	272 182	488 135	330 177	288 194	216 214
N			170 145	204 140	— —	286 136	201 175	— —	— —
O				145 148	— —	193 142	220 —	204 —	204 —
S					268 205	287 156	258 207	217 227	— —
F						158.0 142	255.1 163	284.7 176	281.5 191
Cl							243.4 199	218.9 214	210.7 232
Br								192.9 228	177.9 —
I									151.2 267

Multiple bonds[c]

Bond	*Energy*	*Length*	*Bond*	*Energy*	*Length*
C—C	348	154	C—O	360	143
C=C	614	134	C=O	802	120
C≡C	839	120	C≡O	1077	113
C—N	308	147	N≡N	945.4	110
C=N	618	—	O=O	498.3	121
C≡N	890	116			

[a]The upper value of each entry is the bond energy in kilojoules per mole; the lower value is the bond length in picometers. For example, the bond energy of H—F is 568.1 kJ/mol and its bond length is 92 pm. Values for the F—H bond are the same as for the H—F bond and are not repeated in the table.

[b]Bond energies for diatomic molecules were obtained directly from thermodynamic data and are given to four significant figures. The other bond energies, which are average values, are given to three significant figures.

[c]Some single bonds are included for comparison.

accompanied by the release of 436 kJ of energy. *Bond breaking is endothermic; bond formation is exothermic.*

Each bond in a polyatomic molecule has its own dissociation energy. Imagine dissociating an H_2O molecule, one atom at a time. The two O—H bonds are equally strong in the original water molecule, but after one bond is broken, the remaining bond is somewhat easier to break. In other words, it takes more energy to remove the first hydrogen atom from water than the second:

$$H_2O(g) \rightarrow H(g) + OH(g) \qquad \Delta H^\circ = +499 \text{ kJ}$$

$$OH(g) \rightarrow H(g) + O(g) \qquad \Delta H^\circ = +428 \text{ kJ}$$

The O—H bond energy in water is defined as the average of the two successive dissociation energies, that is,

$$D(\text{O—H}) = \frac{499 \text{ kJ/mol} + 428 \text{ kJ/mol}}{2} = 464 \text{ kJ/mol}$$

The N—H bond energy in ammonia (NH_3) and the C—Cl bond energy in carbon tetrachloride (CCl_4) are defined in a similar way. Bond energies remain approximately constant from one compound to the next, and some of the bond energies in Table 9.3 are values averaged from a number of different molecules.

The data in Table 9.3 show that *bonds between small atoms tend to be stronger and shorter than bonds between large atoms.* For example, let us compare the Cl—Cl and I—I bonds. Chlorine is above iodine in the periodic table, and chlorine atoms are smaller than iodine atoms. The bond energy of Cl_2 (243.4 kJ/mol) is greater than that of I_2 (151.2 kJ/mol) and the stronger Cl—Cl bond is shorter than the I—I bond. (This rule has some exceptions. The bond in F_2, for example, is weaker than the longer bond in Cl_2.) The data in Table 9.3 also illustrate that *bonds between a given pair of atoms become stronger and shorter with increasing multiple-bond character*; a triple bond is stronger than a double bond, and a double bond is stronger than a single bond.

EXAMPLE 9.4

Which molecule, acetylene (HC≡CH) or ethylene (H_2C═CH_2), should have the shorter carbon–carbon bond? The higher carbon–carbon bond energy?

SOLUTION

Acetylene has more multiple-bond character than ethylene. The triple bond in acetylene should be shorter and should have a higher bond energy than the double bond in ethylene. The data in Table 9.3 confirm that a triple bond between two carbon atoms is shorter and stronger than a double bond.

PRACTICE EXERCISE 9.6

Bromine is below chlorine in the periodic table. Which molecule, HCl or HBr, should have (a) the higher bond energy and (b) the longer bond length? Do your predictions agree with the data in Table 9.3?

Enthalpies of Reaction from Bond Energies

We can use bond energies to estimate enthalpies of reaction by treating bond breaking and bond formation as reaction steps and adding their energies to get the enthalpy change for the reaction. Keep in mind, however, that many bond energies are average values, and their use will give only approximate enthalpy changes. Reaction enthalpies calculated from heats of formation (Section 6.4) are usually more accurate.

EXAMPLE 9.5

Use a table of bond energies to find $\Delta H°$ for the reaction

$$H_2(g) + Br_2(g) \longrightarrow 2HBr(g)$$

SOLUTION

The molar bond energies reported in Table 9.3 are $D(\text{H—H}) = +435.9$ kJ, $D(\text{Br—Br}) = +192.9$ kJ and $D(\text{H—Br}) = +366.2$ kJ. We can think of the reaction as occurring in a series of steps in which reactant bonds are first broken and then product bonds are formed. The enthalpy change for each step is the bond energy, which is positive when a bond is broken and negative when a bond is formed. The steps are

Step 1: Breaking 1 mol of H—H bonds:

$$H_2(g) \longrightarrow 2H(g) \qquad \Delta H° = D(\text{H—H}) = +435.9 \text{ kJ}$$

Step 2: Breaking 1 mol of Br—Br bonds:

$$Br_2(g) \longrightarrow 2Br(g) \qquad \Delta H° = D(\text{Br—Br}) = +192.9 \text{ kJ}$$

Step 3: Forming 2 mol of H—Br bonds:

$$2H(g) + 2Br(g) \longrightarrow 2HBr(g) \qquad \Delta H° = -2 \times D(\text{H—Br})$$

$$= -2 \times 366.2 = -732.4 \text{ kJ}$$

The sum of these three steps is the net reaction, and its enthalpy change is the sum of the enthalpy changes for the three steps:

$$\Delta H° = 435.9 \text{ kJ} + 192.9 \text{ kJ} - 732.4 \text{ kJ} = -103.6 \text{ kJ}$$

The value of an enthalpy change obtained from bond energies will agree with the experimental value when the reactants and products are diatomic molecules. For example, the experimental value of $\Delta H°$ for the reaction in Example 9.5 is -103.7 kJ, which is virtually identical to the calculated value of -103.6 kJ. The agreement is good because the bond energies for diatomic molecules are obtained directly from thermodynamic data and are not average values. For a calculation involving polyatomic molecules, average bond energies are used, and the calculated $\Delta H°$ value will be a rough estimate of the experimental $\Delta H°$ value. The following example illustrates this point.

EXAMPLE 9.6

Estimate $\Delta H°$ for the hydrogenation of ethylene:

```
H       H                  H   H
 \     /                   |   |
  C=C      + H2(g) ⟶ H— C — C —H
 /     \                   |   |
H       H                  H   H
```

SOLUTION

The steps include breaking a C=C bond, breaking an H—H bond, forming a C—C bond, and forming two C—H bonds. (The four C—H bonds in ethylene also appear in the product and can be ignored.) Molar bond energies from Table 9.3 are $D(\text{C=C}) = +614$ kJ, $D(\text{H—H}) = +436$ kJ, $D(\text{C—C}) = +348$ kJ, and $D(\text{C—H}) = +413$ kJ. The steps can be summarized as follows:

Step		$\Delta H°$
1. Breaking 1 mol of C=C bonds:		+614 kJ
2. Breaking 1 mol of H—H bonds:		+436 kJ
3. Forming 1 mol of C—C bonds:		−348 kJ
4. Forming 2 mol of C—H bonds:	2 × (−413 kJ) =	−826 kJ
	$\Delta H°$ = sum =	−124 kJ

The estimated enthalpy change is −124 kJ per mole of ethylene. This value differs from the experimental value, −136.9 kJ/mol, by about 10%.

PRACTICE EXERCISE 9.7

Use a table of bond energies to estimate $\Delta H°$ for the formation of hydrazine (N_2H_4) from its elements:

$$2H_2(g) + :\text{N}\equiv\text{N}:(g) \longrightarrow \begin{matrix} \text{H} & & \text{H} \\ & \ddot{}\text{N—N}\ddot{} & \\ \text{H} & & \text{H} \end{matrix}$$

9.4 DRAWING LEWIS STRUCTURES

Drawing a Lewis structure is a valuable prelude to studying the bonding and behavior of any molecule or polyatomic ion, and it is important to learn how to do it. The principal steps are as follows:

1. *Find the total number of electrons in the Lewis structure by adding up the number of valence electrons for each atom in the molecule or ion; add one extra electron for each negative charge; subtract one electron for each positive charge.* (Remember, for a representative element, the number of valence electrons is the same as its group number.) The electrons are usually in pairs (bond pairs and lone pairs), so it will simplify matters if you divide the total number of electrons by 2 to get the number of electron pairs.

2. *Draw a skeleton structure of the molecule or ion by arranging the atoms and putting one single bond (one pair of electrons) between atoms that are bonded together.* You will have to find out or guess which atoms are bonded to each other. If you have no other information, pick the most symmetrical arrangement of the atoms; in most cases, this will be the correct arrangement.

It also helps to remember that a hydrogen atom has only one electron and can therefore be bonded to only one atom.

3. *Distribute the remaining electrons so as to satisfy the octet rule as closely as possible.* Keep in mind that these electrons may form lone pairs, double bonds, or triple bonds.

EXAMPLE 9.7

Draw the Lewis structure for ammonia (NH_3).

SOLUTION

We will follow the steps listed above.

Step 1: *Add the valence electrons.* Nitrogen is in Group 5A and has five valence electrons; hydrogen is in Group 1A and has one valence electron. The total number of electrons in the Lewis structure is

$$\begin{array}{lrl} \text{N:} & & 5 \\ \text{3 H:} & 3 \times 1 = & \underline{3} \\ & & 8 \text{ electrons or 4 electron pairs} \end{array}$$

Step 2: *Draw a skeleton structure for the molecule.* Each of the hydrogen atoms is bonded to the nitrogen atom:

```
H—N—H
   |
   H
```

Any other arrangement would have hydrogen atoms bonded to two or more atoms, in violation of the rule that hydrogen forms only one bond. Note that one single bond (one electron pair) has been placed between each of the bonded atoms.

Step 3: *Distribute the remaining electrons according to the octet rule.* The three single bonds use three pairs of electrons, so one more pair must be added to the structure. Each hydrogen atom already has a duet of electrons, but the nitrogen atom does not have an octet. The remaining pair of electrons becomes a lone pair on the nitrogen atom:

```
    ..
H—N—H
   |
   H
```

LEARNING HINTS

Remember that hydrogen acquires the two-electron configuration of helium.

The H atom in NH_3 is a *terminal atom,* one that is bonded to only one other atom. The N atom is a *central atom,* one that is bonded to two or more other atoms.

PRACTICE EXERCISE 9.8

Draw a Lewis structure for methane (CH_4).

EXAMPLE 9.8

Draw a Lewis structure for carbon dioxide (CO_2).

SOLUTION

Step 1: Carbon is in Group 4A and oxygen is in Group 6A. The number of valence electrons is

$$\begin{array}{rl} \text{C:} & \quad\quad\quad\; 4 \\ 2\,\text{O:} & 2 \times 6 = \underline{12} \\ & \quad\quad\quad\; 16 \text{ electrons or 8 electron pairs} \end{array}$$

Step 2: Possible skeleton structures are

O—C—O or C—O—O or O—O (ring with C: each O bonded to C)

We will choose the most symmetrical arrangement:

O—C—O

Step 3: The single bonds in the O—C—O skeleton use up two of the eight electron pairs. Six pairs remain to be added. To satisfy the octet rule, each oxygen atom needs three more pairs and the carbon atom needs two more pairs. There seems to be a shortage: six pairs are available, eight pairs are needed. As in everything else, the fairest way to deal with a shortage is to increase the amount of sharing, in this case, with multiple bonds. One arrangement that satisfies the octet rule has a double bond between the carbon atom and each oxygen atom:

$$\ddot{\underset{\cdot\cdot}{\text{O}}}\!=\!\text{C}\!=\!\ddot{\underset{\cdot\cdot}{\text{O}}}$$

Another arrangement that satisfies the octet rule is

$$:\ddot{\underset{\cdot\cdot}{\text{O}}}\!-\!\text{C}\!\equiv\!\text{O}:$$

(Remember that the symmetry suggestion in Step 2 refers to atoms, not electrons.) You will find that it is often possible to draw more than one structure that satisfies the octet rule. We will discuss alternative structures in more detail in the next section.

LEARNING HINT

The following atoms often form multiple bonds: C, O, N, S, and P.

PRACTICE EXERCISE 9.9

Draw a Lewis structure for sulfur dioxide (SO_2).

EXAMPLE 9.9

Draw a Lewis structure for hydrogen cyanide (HCN). (*Note:* The atoms are in the order given in the formula.)

SOLUTION

Step 1: Hydrogen is in Group 1A, carbon is in Group 4A, and nitrogen is in Group 5A, so there are $1 + 4 + 5 = 10$ valence electrons or 5 electron pairs.

Step 2: The skeleton structure is

H—C—N

Step 3: The single bonds use up two of the five electron pairs. The hydrogen atom already has two electrons. The carbon atom needs two more pairs to satisfy the octet rule, and the nitrogen atom needs three more pairs. Again there seems to be a shortage; only three electron pairs are available. Electrons must be shared; in this case, a triple bond will satisfy the octet rule:

$$H—C≡N:$$

PRACTICE EXERCISE 9.10

Draw a Lewis structure for HNC, a molecule that has been detected in interstellar gas clouds. Assume that the atoms are in the order given in the formula.

The following rule will help you decide how the atoms are arranged in oxygen-containing acids such as HNO_3 and anions such as NO_3^-: *In acids such as HNO_3, $HClO_4$, and H_2SO_4, and in their anions, the oxygen atoms are usually bonded to the central atom (N, Cl, and S in the examples given), and the hydrogen atoms are usually bonded to oxygen atoms.*

EXAMPLE 9.10

Draw a Lewis structure for the hydrogen sulfate ion (HSO_4^-) in which each atom satisfies the octet rule.

SOLUTION

Step 1: Hydrogen is in Group 1A, and sulfur and oxygen are in Group 6A. Remember that one electron must be added for the negative charge on the ion.

H:		1
S:		6
4O:	$4 \times 6 =$	24
charge:		1
		32 electrons or 16 electron pairs

Step 2: According to the above rule, the oxygen atoms are bonded to the sulfur atom, and the hydrogen atom is bonded to one of the oxygen atoms. The structure is

$$\left[\begin{array}{c} \quad\quad\quad O \\ \quad\quad\quad | \\ H—O—S—O \\ \quad\quad\quad | \\ \quad\quad\quad O \end{array}\right]^-$$

Step 3: The single bonds use five electron pairs, giving the hydrogen atom its one bond and the sulfur atom its octet. Eleven more pairs of electrons must be added; these are exactly enough to complete the oxygen octets:

$$\left[\begin{array}{c} \quad\quad\quad :\ddot{O}: \\ \quad\quad\quad | \\ H—\ddot{\underset{..}{O}}—S—\ddot{\underset{..}{O}}: \\ \quad\quad\quad | \\ \quad\quad\quad :\underset{..}{O}: \end{array}\right]^-$$

LEARNING HINT

Another, and perhaps preferred, structure for HSO_4^- is

$$\left[\begin{array}{c} \quad\quad\quad :\ddot{O}: \\ \quad\quad\quad \| \\ H—\ddot{\underset{..}{O}}—S—\ddot{\underset{..}{O}}: \\ \quad\quad\quad \| \\ \quad\quad\quad :\underset{..}{O}: \end{array}\right]^-$$

where the sulfur atom has an *expanded octet* (more than eight electrons). We will discuss such structures in Section 9.6 and explain why they are preferred by many chemists.

PRACTICE EXERCISE 9.11

Draw Lewis structures that satisfy the octet rule for (a) carbonic acid (H_2CO_3) and (b) the hydrogen carbonate ion (HCO_3^-).

Considering the elusive nature of an electron—its high speed and its uncertain location—it might seem surprising that so much importance is attached to diagrams that show electrons as immovable dots. You should regard a Lewis structure as you would any useful diagram; it is simple, it leaves out a lot of detail, and it provides information up to a point. You will have many occasions to draw more Lewis structures, and you will soon discover how helpful they can be.

Formal Charge

If you count the electrons around each atom in a Lewis structure and assign half of the shared electrons to each bonded atom, you will find that most atoms have their original number of valence electrons. Consider the HCN molecule of Example 9.9:

$$\mathrm{H{-}C{\equiv}N{:}}$$

The number of valence electrons assigned to each atom is

H: half of 1 shared pair = 1 electron

C: half of 4 shared pairs = 4 electrons

N: half of 3 shared pairs (3 electrons) plus 1 lone pair (2 electrons) = 5 electrons

According to this way of assigning electrons, each atom in HCN has the same number of valence electrons that it had as an independent atom.

Sometimes the number of electrons assigned to an atom differs from the original number of valence electrons. Consider the following Lewis structure for sulfur dioxide:

$$\mathrm{:\ddot{\underset{..}{O}}{-}\ddot{S}{=}\ddot{\underset{..}{O}}}$$

Oxygen and sulfur are in Group 6A, so the independent atoms each have six valence electrons. In the structure given above, the left-hand oxygen atom is assigned three lone pairs (six electrons) plus one-half of the shared pair (one electron) for a total of seven electrons, one more than its original number. The sulfur atom is assigned one lone pair (two electrons) plus one-half of three shared pairs (three electrons) for a total of five electrons, one less than its original number. These atoms are said to carry a **formal charge**. The formal charge on the left-hand oxygen atom with one extra electron is -1; the formal charge on the sulfur atom with one less electron is $+1$.

The formal charge on any atom in a Lewis structure is equal to the original number of valence electrons minus the number of electrons assigned to it in the structure. All unshared electrons and half of the bond electrons are counted in the assignment. The formal charge is therefore:

$$\begin{matrix}\text{formal}\\\text{charge}\end{matrix} = \begin{pmatrix}\text{number of}\\\text{valence electrons}\end{pmatrix} - \begin{pmatrix}\text{number of}\\\text{unshared electrons}\end{pmatrix} - \frac{1}{2}\begin{pmatrix}\text{number of}\\\text{bond electrons}\end{pmatrix} \quad (9.2)$$

EXAMPLE 9.11

The following Lewis structure for sulfuric acid (H_2SO_4) resembles the structure of the hydrogen sulfate ion in Example 9.10, but it has one more hydrogen atom:

```
          ..
         :O:
          |
     ..        ..
  H—O—S—O—H
     ..   |    ..
         :O:
          ..
```

Find the formal charge on each atom.

SOLUTION

(a) Each hydrogen atom contributes one valence electron, has no unshared electrons, and forms one single bond. Hence,

$$\text{formal charge on each H} = 1 - 0 - \tfrac{1}{2}(2) = 0$$

(b) Oxygen is in Group 6A, and each oxygen atom contributes six valence electrons to the structure. The left and right oxygen atoms (those bonded to hydrogen) each have four unshared electrons, and there are four electrons in the two single bonds. The formal charge on each of these atoms is

$$\text{formal charge on left and right O} = 6 - 4 - \tfrac{1}{2}(4) = 0$$

The upper and lower oxygen atoms each have six unshared electrons, and there are two electrons in the single bond. The formal charge on each of these atoms is

$$\text{formal charge on upper and lower O} = 6 - 6 - \tfrac{1}{2}(2) = -1$$

(c) Sulfur is also in Group 6A, and the sulfur atom contributes six valence electrons. It has no unshared electrons, but there are eight electrons in the four single bonds. Hence,

$$\text{formal charge on sulfur} = 6 - 0 - \tfrac{1}{2}(8) = +2$$

PRACTICE EXERCISE 9.12

Find the formal charge on each atom in (a) $\ddot{\underset{..}{O}}{=}C{=}\ddot{\underset{..}{O}}$, (b) H_3PO_4, (c) $NH_4{}^+$, and (d) OH^-.

Formal charges may be included in Lewis structures. In H_2SO_4, for example, the formal charges may be shown in circles as follows:

```
          ..
         :O: (-1)
          |  (+2)
     ..        ..
  H—O—S—O—H
     ..   |    ..
         :O: (-1)
          ..
```

Observe that the sum of the formal charges in H_2SO_4 is zero. *Formal charges in a neutral molecule always add up to zero; formal charges in an ion add up to the charge on the ion.*

Later in this chapter you will learn that the electrons in a covalent bond are not always shared equally by the bonded atoms—they may spend more time near one nucleus than the other. Hence, formal charges may not indicate actual centers of charge within the molecule. Nevertheless, you will soon find the concept of formal charge to be very useful.

9.5 RESONANCE STRUCTURES

LEARNING HINT

Although it is not necessary, Lewis structures are sometimes drawn to show the molecule's shape. The NO_3^- ion is planar, with the oxygen atoms at the corners of an imaginary equilateral triangle, as indicated in the structures to the left.

Two or more Lewis structures, each with the same arrangement of atoms and the same number of electron pairs, can often be drawn for one molecule or ion. For example, the following structures for the nitrate ion show the double bond in three possible positions:

$$\left[\begin{array}{c} :\text{O}: \\ \| \\ :\ddot{\text{O}}-\text{N}-\ddot{\text{O}}: \end{array}\right]^- \quad \left[\begin{array}{c} :\ddot{\text{O}}: \\ | \\ \ddot{\text{O}}=\text{N}-\ddot{\text{O}}: \end{array}\right]^- \quad \left[\begin{array}{c} :\ddot{\text{O}}: \\ | \\ :\ddot{\text{O}}-\text{N}=\ddot{\text{O}} \end{array}\right]^-$$

Since doubly bonded atoms are more tightly bound to each other than are singly bonded atoms, any one of these structures by itself would suggest that one of the bonds in the nitrate ion is shorter and stronger than the other bonds. Experimental evidence, however, shows that the nitrate ion is completely symmetrical—all of the bonds are equally strong, and each bond length is 124 pm. Apparently, the three bonds share the double-bond character equally. The double bond in the nitrate ion is an example of a **delocalized bond**, one in which certain bonding electrons are spread out over several atoms and help to bond all of them. Delocalization of a bond cannot be shown in a single Lewis structure, so we imagine the molecule or ion as a composite of several different Lewis structures. These different structures are called **resonance structures** or **contributing structures**, and the molecule or ion is referred to as a **resonance hybrid**.

Resonance structures for a given molecule or ion differ only in the positions of their electron pairs, as illustrated in the above structures for NO_3^- and also in the following example.

EXAMPLE 9.12

The molecules of ozone (O_3) are V-shaped. Draw resonance structures for O_3 and comment on the expected bond lengths.

SOLUTION

Oxygen is in Group 6A. The three atoms each contribute 6 electrons for a total of 18 electrons or 9 electron pairs. The single bonds use two electron pairs:

$$\text{O}-\text{O}-\text{O}$$

Seven pairs are left to complete the octets. The end oxygen atoms need three more pairs each, and the center oxygen atom needs two more pairs, or eight pairs in all. Since only seven electron pairs are available, one pair will have to be shared. The double bond can go to either end atom:

$$\ddot{\text{O}}{=}\ddot{\text{O}}{-}\ddot{\text{O}}\qquad \ddot{\text{O}}{-}\ddot{\text{O}}{=}\ddot{\text{O}}$$

These two resonance structures differ only in the position of the double bond, and we can assume they will contribute equally to the resonance hybrid. The oxygen–oxygen bonds in the actual molecule should be identical and intermediate in length between single and double bonds. This prediction agrees with experimental data, which shows that each of the bonds in ozone is 128 pm long.

PRACTICE EXERCISE 9.13

Write resonance structures for CO_3^{2-}. Comment on the expected bond lengths in this ion.

Double-headed arrows are often used to show bond delocalization over two or more positions. The ozone molecule, for example, can be represented by

$$\ddot{\text{O}}{=}\ddot{\text{O}}{-}\ddot{\text{O}} \longleftrightarrow \ddot{\text{O}}{-}\ddot{\text{O}}{=}\ddot{\text{O}}$$

The double-headed arrow does not mean that the molecule changes back and forth from one structure to the other. It simply means that the actual structure of the molecule is intermediate to those represented by the contributing structures.

Resonance Structures and Formal Charge

All the resonance structures that represent a given molecule must have the same arrangement of atoms and the same number of paired electrons. However, all structures that meet these criteria do not necessarily represent equally stable arrangements. *The actual molecule, that is, the resonance hybrid, is more stable (has a lower energy) than any of the contributing structures.* Hence, the most stable resonance structure makes the greatest contribution to the hybrid; in other words, it is the structure that most nearly represents the actual molecule. Less stable structures can sometimes be ignored altogether. Comparative stabilities can often be evaluated on the basis of formal charge, using the following rules:

1. The most stable structure has the least formal charge.
2. Structures in which adjacent atoms have formal charges of the same sign are especially unstable.

For example, consider the Lewis structure for carbon dioxide from Example 9.8:

$$\ddot{\text{O}}{=}\text{C}{=}\ddot{\text{O}}$$

The formal charge on each atom in this structure is zero (be sure to verify this). Two other structures that satisfy the octet rule and have the same atomic arrangement are

$$\overset{-1}{:\ddot{\text{O}}}{-}\text{C}{\equiv}\overset{+1}{\text{O}}: \quad \text{and} \quad \overset{+1}{:\text{O}}{\equiv}\text{C}{-}\overset{-1}{\ddot{\text{O}}}:$$

The formal charges on the oxygen atoms indicate that these arrangements are less stable than the symmetrical one, so their contributions to the resonance hybrid are generally disregarded.

EXAMPLE 9.13

Draw resonance structures for nitric acid (HNO_3), and state which structure (if any) makes the greater contribution to the hybrid.

SOLUTION

Hydrogen contributes 1 electron, nitrogen contributes 5 electrons, and three oxygen atoms contribute 18 electrons for a total of 24 electrons or 12 electron pairs. The atoms are arranged according to the rule for oxygen-containing acids; that is, the oxygen atoms are bonded to the nitrogen atom, and the hydrogen atom is bonded to one of the oxygen atoms:

```
          O
         /
H—O—N
         \
          O
```

The single bonds use four pairs of electrons, leaving eight pairs to complete the octets. Three resonance structures can be draw:

```
           :O:                        :Ö: (-1)                    :Ö: (-1)
      (+1) //                    (+1) /                  (+1) (+1)/
H—Ö—N                       H—Ö—N                    H—Ö=N
   ..     \                     ..    \\                          \
          :Ö: (-1)                    :O:                         :Ö: (-1)
     I                            II                         III
```

(Be sure to verify the formal charges.) Structure III has more formal charge on its atoms than structure I or II, and it has positive charges on adjacent atoms as well. It is the least stable of the three structures. The HNO_3 molecule is principally a resonance hybrid of structures I and II.

PRACTICE EXERCISE 9.14

"Laughing gas" is dinitrogen oxide (N_2O). Its molecule has a terminal oxygen atom. (a) Draw three resonance structures for N_2O. (b) Which structure contributes least to the hybrid?

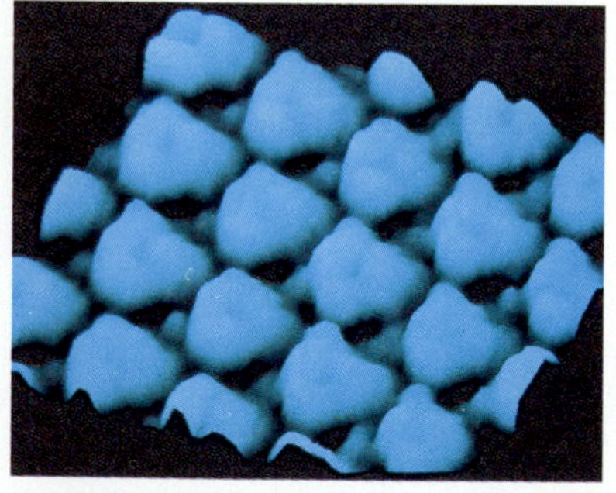

Figure 9.7

This image, generated by a scanning tunneling microscope (see p. 282), is one of the first pictures showing how atoms are arranged in individual benzene molecules. The molecules are lined up in rows, and each ring-shaped cluster is a single benzene molecule.

The Benzene Molecule

The resonance approach is one way of explaining the bonding in benzene (C_6H_6; see Figure 9.7), an important industrial chemical. The conventional Lewis structure for benzene is drawn with alternating single and double bonds so that each carbon has an octet. Two structures can be drawn with the double bonds in different positions:

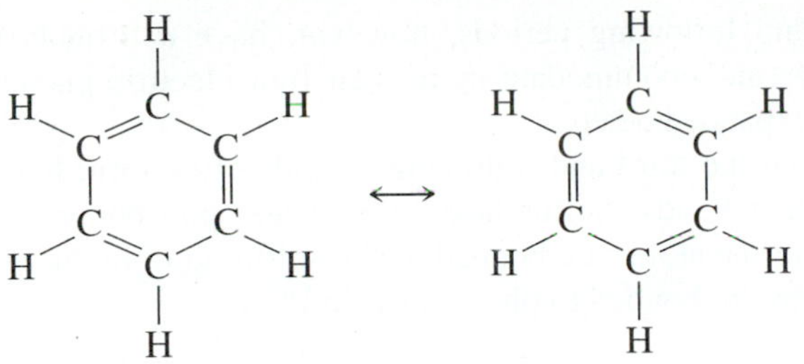

These cyclic structures are called *Kekulé structures* after the German chemist August Kekulé who first suggested them in 1865. They are often abbreviated as

In his memoirs, Kekulé reported that his insight into the benzene structure came during a dream in which he saw a whirling serpent catch its own tail.

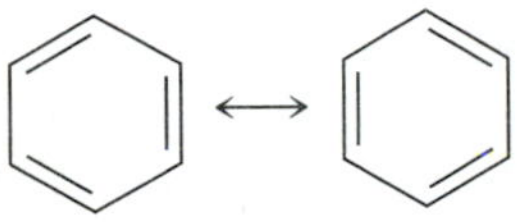

where each corner of the hexagon is understood to represent one carbon atom and its associated hydrogen atom.

Double bonds are shorter than single bonds, so the individual Kekulé structures would suggest that the molecule is lopsided, with alternating long and short bonds. Experimental evidence, however, shows that the benzene molecule is a regular hexagon with each carbon–carbon bond 140 pm long. The equal bond lengths can be explained by assuming that the molecule is a hybrid to which the two Kekulé structures make equal contributions. The carbon–carbon bonds are identical; each bond consists of one electron pair plus an equal share of the three remaining electron pairs.

At least half of all known organic compounds have benzenelike rings in their structures. Benzene is so important that it is usually represented by a distinctive symbol, a hexagon with a circle inside it to indicate the three delocalized pairs of electrons:

9.6 EXCEPTIONS TO THE OCTET RULE

The octet rule is not always obeyed by the atoms in a molecule. Some molecules have atoms with **expanded octets** (more than eight electrons) and some have atoms with incomplete octets (fewer than eight electrons). Molecules with an odd number of valence electrons—NO_2 is an example—can never obey the octet rule because at least one electron will be unpaired.

More Than Four Electron Pairs

The second-period atoms, Li through Ne, have only four orbitals in the valence shell, one 2*s* orbital and three 2*p* orbitals. These atoms cannot accommodate more than four electron pairs and can never have more than eight electrons in a Lewis structure. Consequently, *second-period atoms never exhibit octet expansion.* Atoms from the third and following periods, however, have *d* orbitals in their valence shells; thus, they can accommodate more than four electron pairs, and such atoms frequently have expanded octets.

Sulfur hexafluoride, an unusual substance that is widely used as an insulating gas in high-voltage generators, is not attacked by water even at 500°C. The six fluorine atoms surround the sulfur atom on all sides, making SF_6 one of a number of compounds that owe their volatility and/or inertness to the unbroken protection of an electron cloud provided by terminal fluorine atoms. Other examples are UF_6 (p. 202) and Teflon (p. 516).

Nonmetals from the third and following periods often form five, six, seven, and even eight covalent bonds. Sulfur hexafluoride and phosphorus pentachloride are examples. Six fluorine atoms are bonded to the sulfur atom in the SF_6 molecule, and five chlorine atoms are bonded to phosphorus in PCl_5:

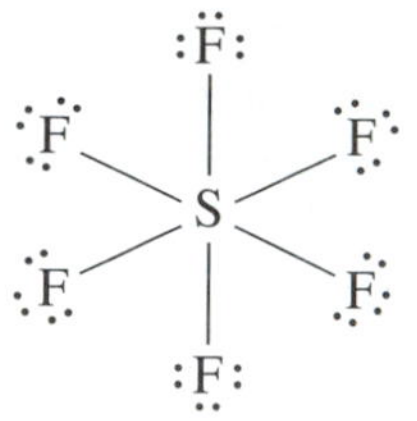

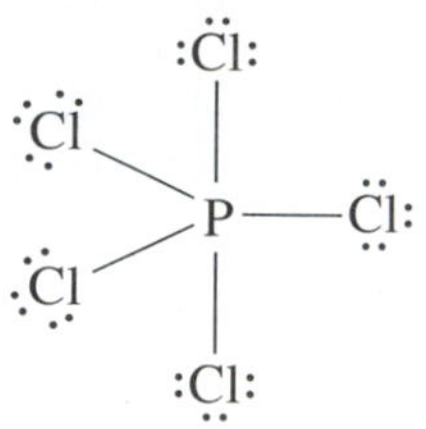

:F: :F: :F: S :F: :F: :F: :Cl: :Cl: P :Cl: :Cl: :Cl:

The following examples illustrate how formal charge can help in deciding which octets should be expanded.

EXAMPLE 9.14

Halogen atoms can combine with each other to form a number of *interhalogen* compounds such as BrF_3, ICl_3, and IF_7, in which the heavier halogen atom is in the center with the lighter atoms attached to it. Draw a Lewis structure for iodine trichloride (ICl_3).

SOLUTION

Each halogen atom has seven valence electrons. The total for the four atoms is $4 \times 7 = 28$ electrons or 14 pairs. The single-bonded framework with iodine in the center uses three electron pairs:

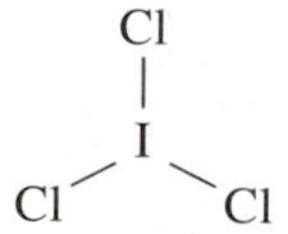

Cl — I (Cl)(Cl)

Eleven pairs are left, but only 10 pairs are needed to complete the octets. The additional pair of electrons is given to the iodine atom:

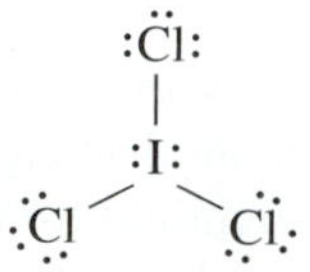

:Cl: — :I: (:Cl:)(:Cl:)

The extra electron pair was placed on iodine rather than chlorine to avoid formal charge.

LEARNING HINT

Terminal atoms do not usually exhibit octet expansion.

PRACTICE EXERCISE 9.15

Draw Lewis structures for (a) BrF_3 and (b) IF_7.

The structure drawn in Example 9.11 for H_2SO_4 obeys the octet rule. Sulfur, however, is a third-period atom, and its octet can be expanded to give a structure without formal charge. Compare the following H_2SO_4 structures:

```
      :Ö: (-1)                        :O:
       | (+2)                          ||
H—Ö—S—Ö—H                    H—Ö—S—Ö—H
       |                               ||
      :Ö: (-1)                        :O:
```

follows octet rule; formal charges present — sulfur octet expanded; no formal charges

Formal charge was eliminated in the right-hand structure by allowing the positively charged sulfur atom to share additional electron pairs with the oxygen atoms that have excess negative charge.

We know that the formal charges in the left-hand structure are not real, because such large charge differences could not exist within a small molecule. The structure on the right indicates a completely even distribution of charge, one that may not be entirely real either. Although the H_2SO_4 molecule can be regarded as a resonance hybrid of both structures, the larger contribution comes from the structure with the expanded octet. This structure is more consistent with the known stability of the sulfuric acid molecule than the left-hand structure is, because it eliminates formal charge and at the same time provides stronger bonding between the sulfur and oxygen atoms.

PRACTICE EXERCISE 9.16

(a) Write a Lewis structure that satisfies the octet rule for $HClO_3$. Calculate the formal charge on each atom in your structure. (b) Write another Lewis structure without formal charge for $HClO_3$. Which structure makes the larger contribution to the resonance hybrid?

Lewis, who died in 1946, would perhaps have been amazed to learn that some of the noble gases are able to form covalent bonds by expanding their valence shells to accommodate more electrons than the four pairs they already have. The first xenon compound, $XePtF_6$, was synthesized in 1962, and the fluorides, XeF_2, XeF_4, and XeF_6, followed shortly afterward. (The synthesis of noble gas compounds is described in Section S1.4 of Survey 1.)

EXAMPLE 9.15

Draw a Lewis structure for xenon tetrafluoride (XeF_4).

SOLUTION

Xenon contributes 8 electrons and four fluorine atoms contribute $4 \times 7 = 28$ electrons, for a total of 36 electrons or 18 electron pairs. The most symmetrical arrangement is one with all the fluorine atoms bonded to a central xenon atom:

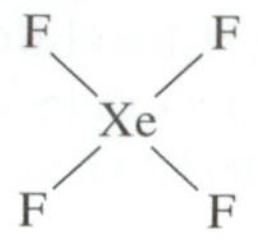

F F
Xe
F F

The single bonds use 4 electron pairs, and 14 pairs remain. Twelve pairs are used to complete the octets on the fluorine atoms. Fluorine, a second-period element, cannot exhibit octet expansion; hence, the remaining two pairs are assigned to xenon:

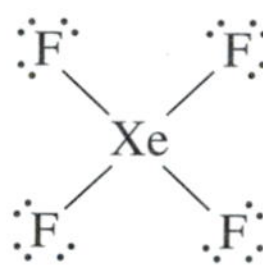

PRACTICE EXERCISE 9.17

Draw a Lewis structure for xenon difluoride (XeF_2).

Fewer Than Four Electron Pairs

Beryllium, boron, and aluminum, from Groups 2A and 3A, form a variety of covalent compounds. The atoms of these elements, however, cannot complete their octets without acquiring a negative formal charge. Boron, for example, has three valence electrons and often forms three bonds. The Lewis structure for boron trifluoride (BF_3) is usually drawn with only six electrons on the boron atom (structure I):

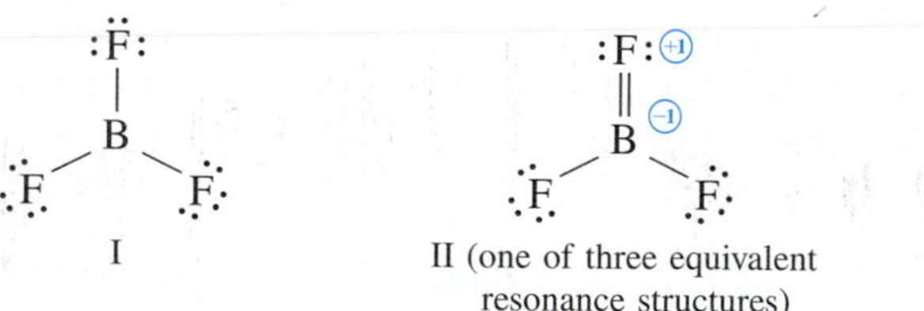

I

II (one of three equivalent resonance structures)

Structure I has no formal charge on any of its atoms. Resonance structures that obey the octet rule (such as structure II) can be drawn, but they will have a formal charge. Although structure II contributes to the hybrid, it is not as important as structure I.

PRACTICE EXERCISE 9.18

Draw Lewis structures without formal charge for the covalent molecules (a) $BeCl_2$, (b) BCl_3, and (c) $SnCl_2$.

Free Radicals

A **free radical** is any atom, molecule, or ion that contains at least one unpaired electron. Free radicals have been detected in flames, in comets, in stellar atmospheres, and in outer space. Most free radicals are unstable and have only a fleeting existence under ordinary conditions. A few, however, are well-known molecules. Nitrogen dioxide (NO_2), nitrogen monoxide (NO), and chlorine dioxide (ClO_2) each have one unpaired electron, and oxygen (O_2) has two unpaired electrons. Free radicals are often colored: Nitrogen dioxide is reddish brown, chlorine dioxide is yellow, and liquid oxygen is pale blue.

EXAMPLE 9.16

Nitrogen dioxide (NO_2) is a brown toxic gas produced by certain reactions of nitric acid. Draw its Lewis structure.

SOLUTION

Nitrogen is in Group 5A and oxygen is in Group 6A. Nitrogen contributes 5 electrons and oxygen contributes $2 \times 6 = 12$ electrons for a total of 17 electrons, or 8 pairs plus 1 odd electron. The symmetrical arrangement

$$\mathrm{O{-}N{-}O}$$

uses up two pairs of electrons. Six electron pairs and one odd electron are left. The resonance forms that most nearly conform to the octet rule are

$$\mathrm{\ddot{\underset{..}{O}}{=}\ddot{N}{-}\ddot{\underset{..}{O}}\cdot \leftrightarrow \cdot\ddot{\underset{..}{O}}{-}\ddot{N}{=}\ddot{\underset{..}{O}} \leftrightarrow \ddot{O}{=}\dot{N}{-}\ddot{\underset{..}{O}}: \leftrightarrow :\ddot{\underset{..}{O}}{-}\dot{N}{=}\ddot{\underset{..}{O}}}$$

Two of these structures have formal charges. Can you find them?

PRACTICE EXERCISE 9.19

Draw three resonance structures without formal charge for chlorine dioxide (ClO_2). (*Hint:* Chlorine exhibits octet expansion.)

Nearly all stable molecules have even numbers of electrons, and in most of them the electrons are neatly paired. The common oxygen molecule (O_2) is a surprising exception. It exhibits a degree of paramagnetism (Section 8.5) consistent with two unpaired electrons per molecule and is therefore a free radical (Figure 9.9, page 396) as well as an outstanding example of a species for which good Lewis structures cannot be written. A structure that obeys the octet rule (structure I)

$$\underset{\text{I}}{\mathrm{\ddot{\underset{..}{O}}{=}\ddot{\underset{..}{O}}}} \qquad \underset{\text{II}}{\mathrm{\cdot\ddot{\underset{..}{O}}{-}\ddot{\underset{..}{O}}\cdot}}$$

does not have unpaired electrons. A structure with two unpaired electrons (structure II) has only a single bond, yet the stability and strength of the actual O_2 bond are much greater than would be expected from a single bond. The bonding in this interesting molecule is explored more fully in Chapter 10.

CHEMICAL INSIGHT

FREE RADICALS AND THE OZONE LAYER

Because of their unpaired electrons, most free radicals tend to be very reactive. A number of free radicals have been detected in outer space, where the average density of matter is less than one atom per cubic centimeter. Many of these radicals are uncombined atoms that could not exist in the crowded atmosphere of earth; for example, 50–90% of interstellar hydrogen is in the form of atoms. Table 9.4 shows some of the molecules that have been identified in interstellar space. The free radicals on this list include familiar species such as nitrogen monoxide (NO) as well as exotic species such as C_4H, which no earthbound chemist has ever been able to synthesize.

Somewhat closer to home are the free radicals of earth's upper atmosphere. They are produced when chemical bonds are disrupted by the intense solar radiation, a process known as *photodissociation.* At altitudes above 130 km, most oxygen is in atomic form and water vapor has been largely photodissociated as well:

$$O_2(g) \xrightarrow{h\nu} 2O(g)$$

$$H_2O(g) \xrightarrow{h\nu} H(g) + OH(g)$$

$$OH(g) \xrightarrow{h\nu} H(g) + O(g)$$

where $h\nu$ is the energy of a photon.

Still closer to earth, in the stratosphere, we find a mixture of molecules, atoms, and free radicals undergoing various cycles of photodissociation and recombination. The reactions of ozone (O_3) concern us the most. The photodissociation of ozone into oxygen atoms is caused by ultraviolet solar radiation with wavelengths ranging from about 200 to 310 nm:

$$O_3(g) \xrightarrow{200\text{–}310\text{ nm}} O_2(g) + O(g)$$

Radiation at wavelengths less than 242 nm dissociates oxygen molecules as well:

$$O_2(g) \xrightarrow{<242\text{ nm}} O(g) + O(g)$$

TABLE 9.4 Some Free Radicals Detected in Interstellar Space

Name	Lewis Structure
Cyanogen	·C≡N·
Methylidyne	·C—H
Hydroxyl	·O—H
Nitrogen monosulfide	·N=S
Nitrogen monoxide	·N=O
Ethynyl	H—C≡C·
Formyl	H—C=O
Cyanoethynyl	:N≡C—C≡C·
Butadiynyl	H—C≡C—C≡C·

Ozone is regenerated when oxygen atoms and oxygen molecules combine:

$$O(g) + O_2(g) \rightarrow O_3(g)$$

A balance between the dissociation and regeneration of ozone causes the ozone concentration to peak at altitudes between 25 and 30 km, where its average concentration is about 10 ppm (10 parts per million). This region is called the *ozone layer.* The ozone layer absorbs most of the sun's ultraviolet radiation, thus protecting life on earth from its deleterious effects.

Ozone reacts readily with a number of free radicals in the stratosphere. In the following cycle, the R· symbolizes any free radical:

$$R\cdot(g) + O_3(g) \rightarrow RO\cdot(g) + O_2(g)$$

$$RO\cdot(g) + O(g) \rightarrow R\cdot(g) + O_2(g)$$

9.7 POLAR AND NONPOLAR COVALENT BONDS

A Lewis structure places each bonding electron pair exactly between two atoms, which seems to imply that the electrons are shared equally. Experiments show, how-

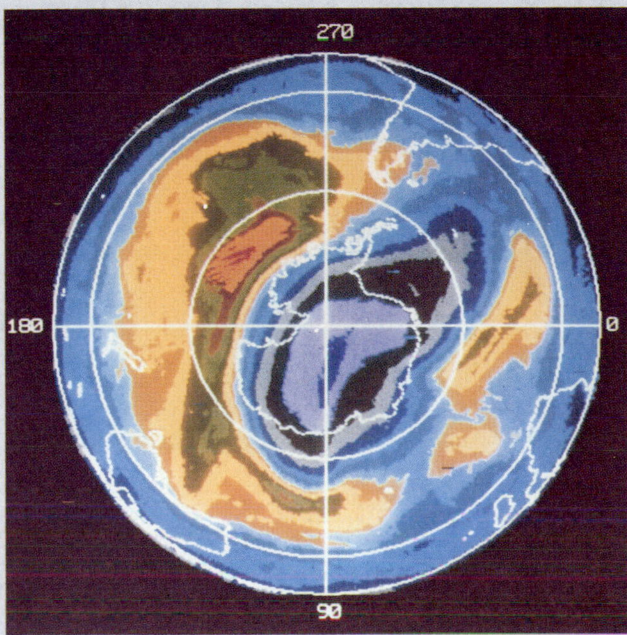

Figure 9.8

Antarctic ozone hole. This map, which was computer-generated from data obtained by a Total Ozone Mapping Spectrometer (TOMS) aboard a Nimbus-7 satellite, shows ozone concentrations above the South Pole (center of map) and Antarctica (outlined in white). The central region of very low total ozone (dark and light violet) covers an area about the size of the United States.

These reactions not only use up ozone from the ozone layer, but they also regenerate the free radical, which can then attack more ozone. Naturally occurring free radicals such as O and OH do not cause any problems, but free radicals introduced by human activities could lower the ozone concentration to levels that permit a substantial increase in the amount of ultraviolet radiation reaching the surface of the earth. This radiation could cause genetic damage to animals and plants, increase the incidence of diseases such as skin cancer, and raise the average surface temperature of the earth to a point where agricultural and climatic problems occur.

In recent years, a severe drop in stratospheric ozone (an "ozone hole") has been observed over Antarctica each spring. Evidence indicates that much of the ozone loss is caused by chlorine atoms produced by the photodissociation of chlorofluorocarbons (CFCs), which have been widely used in refrigerants, rigid and flexible foam packing, and spray can propellants. The first photodissociation step of dichlorodifluoromethane (CF_2Cl_2), a typical CFC, is

$$CF_2Cl_2 \xrightarrow{195\text{–}220\text{ nm}} CF_2Cl\cdot + Cl\cdot$$

Once produced, the Cl atoms can destroy ozone by means of the cycle described above. (Substitute Cl for R in the equations.) In addition, a number of other cycles involving Cl and ClO are believed to contribute to the depletion of ozone. The potential for depleting the ozone layer has alarmed both scientists and politicians, and recently, 87 nations have agreed to gradually phase out the production of CFCs.

In the lowest regions of the atmosphere, we find the free radicals NO and NO_2, which are produced during high-temperature combustions, particularly in automobile engines. Nitrogen oxides are toxic in themselves, but they also raise the levels of atomic oxygen and ozone by the following reactions:

$$NO_2(g) \xrightarrow{\text{sunlight, 400 nm}} NO(g) + O(g)$$

$$O(g) + O_2(g) \rightarrow O_3(g)$$

Ozone, desirable as it may be in the ozone layer, is an unwelcome component in the lower atmosphere. It can result in the death of small organisms and can cause health problems for larger ones. Atomic oxygen and ozone attack rubber and plastics and combine with hydrocarbons to form still more free radicals that may be damaging to health and the environment.

ever, that most covalent bonds are **polar**; that is, they have an uneven charge distribution with a greater electron density near one nucleus. An example of a polar molecule is HCl. The chlorine atom attracts electrons more strongly than hydrogen does, and the resulting displacement of electron density (Figure 9.10) causes the chlorine

Figure 9.9

Liquid oxygen, poured over the poles of an electromagnet, adheres to the poles and fills the gap between them. The O_2 molecule has two unpaired electrons and is paramagnetic.

end of the molecule to be more negative than the hydrogen end. The charge difference is small—much less than the charge on a single electron. (Only an ionic bond can have a charge difference of one electron or more.) A partial charge difference (less than one electron) is indicated by $\delta+$ and $\delta-$ in the Lewis structure:

$$\overset{\delta+}{\text{H}}-\overset{\delta-}{\ddot{\underset{\cdot\cdot}{\text{Cl}}}}\text{:}$$

(δ is the lowercase Greek delta.)

Nonpolar covalent bonds are relatively rare; two atoms will share electron density equally only if they are identical in all respects. Therefore, the only completely nonpolar covalent bonds are found in elements such as H_2, O_2, and Cl_2 and in the center bonds of symmetrical molecules such as hydrogen peroxide H_2O_2 (HO—OH) and ethane ($H_3C—CH_3$).

Dipole Moments

An object whose ends carry equal but opposite charges is called a **dipole**, and any polar molecule (e.g., HCl) is a molecular dipole (see Demonstration 9.1). A molec-

Figure 9.10

(a) A nonpolar covalent bond. The electron density of the bond electrons is equally distributed between the two atoms. (b) A polar covalent bond. The bond electrons spend more time near the more electronegative atom.

DEMONSTRATION 9.1 POLAR AND NONPOLAR MOLECULES

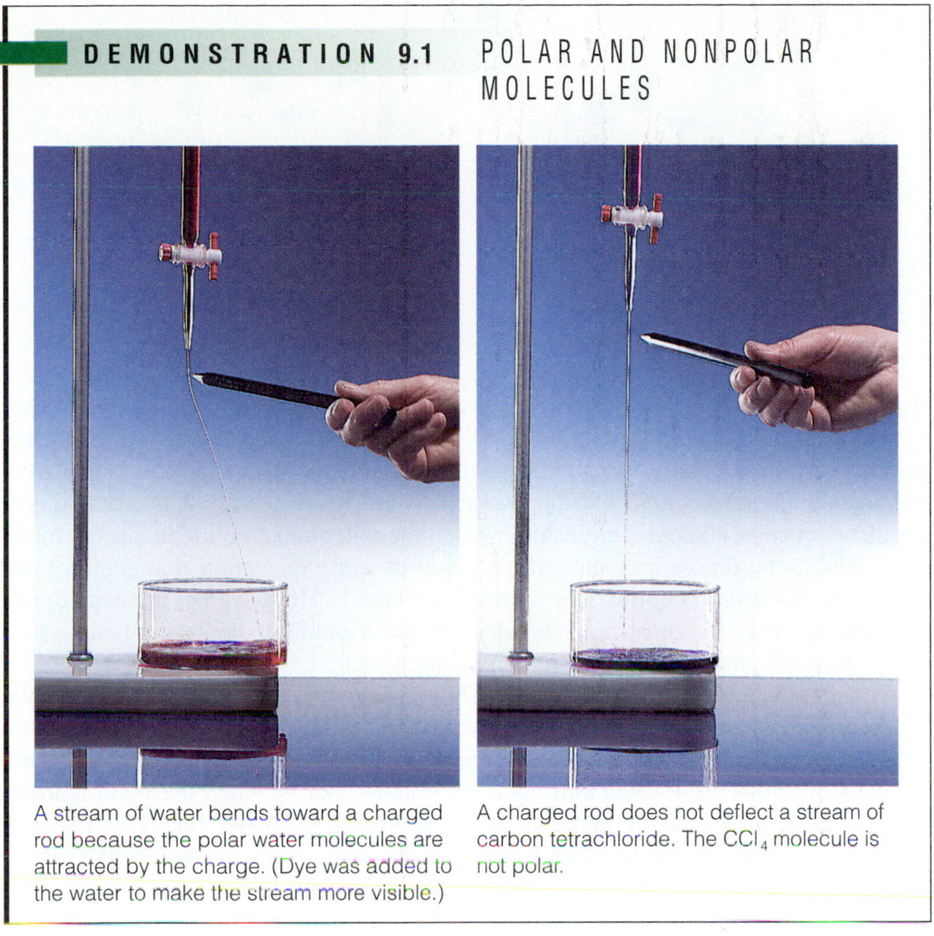

A stream of water bends toward a charged rod because the polar water molecules are attracted by the charge. (Dye was added to the water to make the stream more visible.)

A charged rod does not deflect a stream of carbon tetrachloride. The CCl_4 molecule is not polar.

ular dipole is represented by a crossed arrow directed toward the negative atom:

$$\overset{\delta+}{\text{H}}-\overset{\delta-}{\text{Cl}}$$
$$\longmapsto$$

The crossed or + end of the arrow lies near the positive atom. The **dipole moment**, whose symbol is the lowercase Greek letter mu (μ), is a measure of the polarity; it is defined as the magnitude q of the separated charges multiplied by the distance d between their centers.

$$\longleftarrow d \longrightarrow$$
$$(+q \qquad\qquad -q)$$
$$\mu = q \times d$$

The SI unit for dipole moment is the coulomb meter, but a much smaller unit, the **debye** (pronounced deh-BYE), is commonly used for molecules. One debye (D) equals 3.33×10^{-30} C·m. The debye is named after Peter Joseph William Debye (1884–1966), a Dutch-American physical chemist noted for his pioneering work on molecular structure.

Dipole moments are measured by placing the molecules between two charged plates as shown in Figure 9.11. Polar molecules tend to orient themselves with their

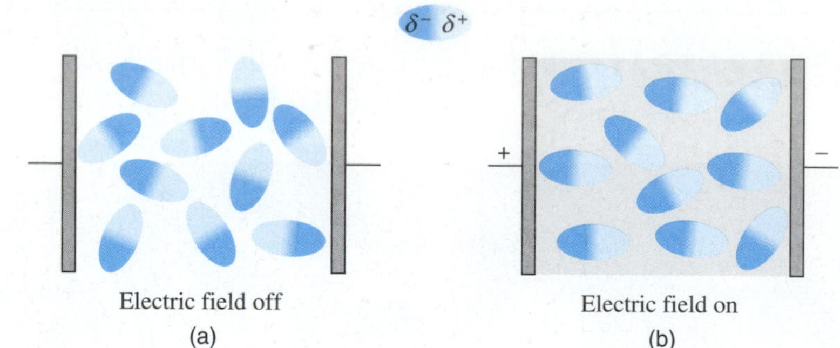

Figure 9.11

Polar molecules in an electric field. (a) When no field is present, the molecules are randomly oriented. (b) The molecules tend to align themselves in an electric field, but thermal motion preserves some of the randomness.

negative ends pointing toward the positive plate and their positive ends toward the negative plate. This alignment allows the plates to hold a greater charge than if the molecules were randomly oriented, and the dipole moment can be calculated from the magnitude of this effect. Table 9.5 gives the experimentally measured dipole moments of some diatomic molecules. When the two atoms are identical, the molecule exhibits no dipole moment, which confirms our expectation that such a bond would be nonpolar. Of the four hydrogen halides, HF has the greatest dipole moment and thus the most polar bond; HI has the smallest dipole moment, which shows that its bonding electrons are more equally shared.

The dipole-moment data in Table 9.5 show that some polar bonds are more polar than others. Greater polarity results when the bonding electron density is more concentrated near one of the two atoms, so the ultimate polarity should occur when both bonding electrons are located entirely on one atom, that is, when the bond is ionic. Thus it is best to think of nonpolar covalent bonds and ionic bonds as opposite extremes of a broad spectrum of bonding behavior (see Figure 9.12). Polarity is sometimes referred to as *ionic character*; For example, a slightly polar bond can be described as "covalent with some ionic character." Other bonds (see Figure 9.12c) are said to be "ionic with some covalent character."

How then do we decide whether a given substance should be called ionic or covalent? We might look at its behavior; a truly ionic compound does not form separate molecules, so it cannot melt and evaporate at the low temperatures typical of many covalent compounds. However, many covalent substances do not form molecules either, so their melting and boiling points are also very high (diamond and silicon dioxide are examples). The most infallible criterion for an ionic compound is that it

TABLE 9.5 Molecular Dipole Moments

Molecule	Dipole Moment (D)[a]	Molecule	Dipole Moment (D)[a]
H_2	0	HF	1.82
F_2	0	HCl	1.08
Cl_2	0	HBr	0.82
Br_2	0	HI	0.44
I_2	0		

[a] 1 debye (D) = 3.33×10^{-30} C·m.

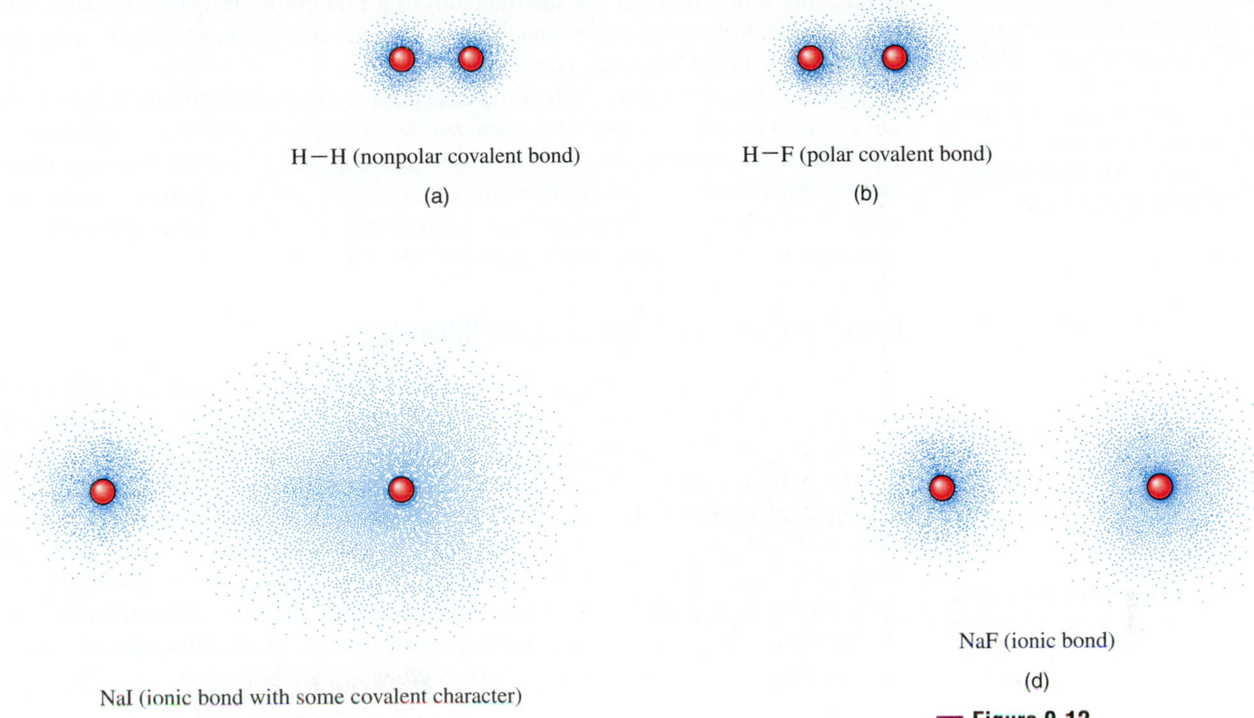

Figure 9.12
Electron-density diagrams that show the progression from covalent to ionic bonding. (a) The nonpolar covalent bond in H_2. (b) The polar covalent bond in HF. (c) The ionic bond in NaI has some covalent character due to polarization of the large I^- ion by neighboring Na^+ ions. (d) The ionic bond in NaF has very little covalent character.

conducts electricity when molten, thus showing that its ions can exist independently of the ionic lattice. The fact that some molten ionic compounds conduct electricity better than others is evidence that ionic compounds have varying degrees of "covalent character." Many metal halides such as $AlCl_3$, $BeCl_2$, $SnCl_2$, and SbI_3 are more covalent than ionic in the pure state. Compounds of this type have lower melting points than ionic halides such as NaF and $CaCl_2$, and in the molten state they are poorer conductors of electricity.

Electronegativity

Electronegativity is the term used to describe the general tendency of an atom to attract shared electrons. Bond polarity is caused by the electronegativity difference between the bonded atoms; the more electronegative atom draws the electron density toward itself—away from the less electronegative atom. Linus Pauling, one of the few persons ever to win two Nobel Prizes (Figure 9.13), was the first to provide a quantitative basis for the concept of electronegativity. He noted that polar bonds are generally stronger and have higher bond energies than nonpolar bonds composed of the parent atoms. For example, the bond energy for H—F is greater than the bond energy of either H—H or F—F (you can verify this from the data in Table 9.3). Pauling attributed the enhanced strength of a polar bond to the electrostatic attraction between the opposite partial charges on the bonded atoms. His analysis of bond energies led him to devise an **electronegativity scale** in which relative values were assigned to each element (see Figure 9.14). These dimensionless numbers

The electronegativity (X) of an element is related to both the ionization energy (IE) and the electron affinity (EA) of the element. R. S. Mulliken proposed in 1934 that the electronegativity is proportional to the difference between the ionization energy and the electron affinity energy, that is

$$X = k(\text{IE} - \text{EA})$$

where k is a proportionality constant. Because of the difficulty in obtaining electron affinities, Mulliken's proposal has had, at least until recently, limited value.

LEARNING HINT

Remember that fluorine is the most electronegative element and oxygen is the second most electronegative element.

Figure 9.13

Linus Pauling and his wife, Ava, in Stockholm examining the Nobel Peace Prize. Pauling's pioneer work has covered many fields of chemistry, ranging from quantum mechanics to molecular biology. In 1954 he received the Nobel Prize in chemistry for his work on protein structures; in 1962 he received the Nobel Peace Prize for his campaign against nuclear weapons testing.

range from a minimum of 0.7 for francium to a maximum of 4.0 for fluorine. Other electronegativity scales have since been devised, but Pauling's scale, with some modification, is still in general use.

Figure 9.14 shows that, with some exceptions, *electronegativities increase from left to right across a period and decrease down a group.* Nonmetals are generally more electronegative than metals. Electronegativities vary in much the same manner as ionization energies and electron affinities. Highly electronegative atoms tend to retain their electrons and have high ionization energies. They also acquire new electrons easily and have large exothermic electron affinities.

Bond Polarities and Electronegativity Differences

The electronegativity difference between two atoms tells us something about the kind of bond they will form. Atoms with identical electronegativities will share electrons equally and form a nonpolar bond. If their electronegativities are different, the sharing will be unequal and the bond will be polar. The data in Table 9.6 show that for similar molecules such as the hydrogen halides, the dipole moment increases with increasing electronegativity difference between the bonded atoms. In general, *the greater the electronegativity difference between two bonded atoms, the more polar the bond.* Unequal electron sharing reaches its extreme when the more electronegative atom takes over the bonding electrons entirely so that ions are formed. The following rules of thumb work fairly well for most bonds:

- When the electronegativity difference is zero, as in H_2, O_2, and N_2, electrons are shared equally and the bond is nonpolar.
- When the electronegativity difference is between 0 and 1.7, the bond is polar, but predominantly covalent.
- When the electronegativity difference is greater than 1.7, the bond is predominantly ionic.

These electronegativity rules are not infallible. For example, the electronegativity difference in lithium hydride (LiH) is less than 1.7, but the bond is ionic. Lithium hydride is a crystalline compound that melts at 688°C and conducts electricity when molten. The electronegativity difference in hydrogen fluoride (HF) is 1.8, but its

TABLE 9.6 Molecular Dipole Moments and Electronegativity Differences

Substance	Dipole Moment[a] (D)	Electronegativity Difference[b]
HF	1.82	1.8
HCl	1.08	1.0
HBr	0.82	0.8
HI	0.44	0.4

[a]1 debye (D) = 3.33×10^{-30} C·m.

[b]The electronegativity difference of the atoms in a given bond can be obtained from Figure 9.14 by subtracting the smaller electronegativity from the larger electronegativity.

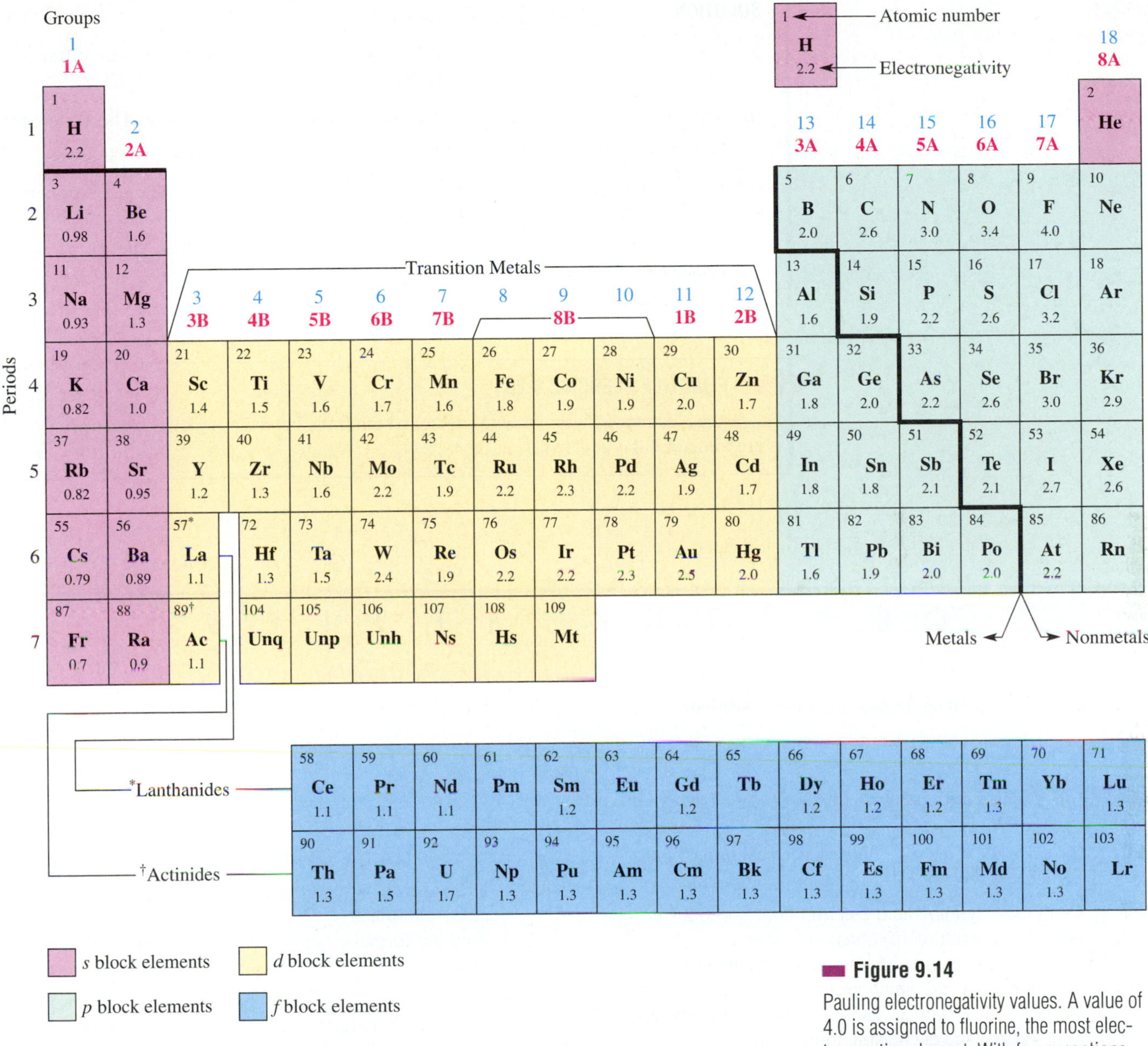

Figure 9.14

Pauling electronegativity values. A value of 4.0 is assigned to fluorine, the most electronegative element. With few exceptions, electronegativities increase from left to right across a period and decrease down a group.

bond is polar covalent. Hydrogen fluoride is a liquid at room temperature and a poor conductor of electricity.

EXAMPLE 9.17

Use the electronegativity scale in Figure 9.14 to predict the nature of the bonds in (a) NaCl, (b) HI, and (c) H_2O.

SOLUTION

(a) The electronegativities of Na and Cl are 0.9 and 3.2, respectively. The difference ($3.2 - 0.9 = 2.3$) exceeds 1.7, so NaCl is predominantly ionic: $Na^{+}Cl^{-}$.

(b) The electronegativities of H and I are 2.2 and 2.7, respectively. The difference ($2.7 - 2.2 = 0.5$) indicates a polar covalent bond: $\overset{\delta+}{H}—\overset{\delta-}{I}$

(c) The electronegativity of oxygen is 3.4. The electronegativity difference between O and H is $3.4 - 2.2 = 1.2$. The difference is less than 1.7, so each O—H bond is polar covalent: $\overset{\delta-}{O}—\overset{\delta+}{H}$. The O—H bond is more polar than the H—I bond because its electronegativity difference is greater.

PRACTICE EXERCISE 9.20

Use the electronegativity scale to predict whether the bonds in (a) BCl_3 and (b) RbCl are ionic or polar covalent.

CHAPTER REVIEW

Starred entries are based on the *Digging Deeper* sections.

LEARNING OBJECTIVES BY SECTION

9.1 1. Describe the three principal types of chemical bonds.
2. Write Lewis symbols for the main group atoms.

9.2 1. Use Lewis symbols and electron configurations to represent the formation of monatomic ions.
2. Know which groups of ions generally have noble gas configurations and which groups do not.
3. Describe how lattice energy varies with increasing ionic charge and decreasing ionic radii.
4. List some properties of crystals associated with high lattice energy.
*5. Diagram the Born–Haber cycle for the formation of an ionic solid, and use it to calculate lattice energies from experimental data.
*6. Given five of the six enthalpy changes in the Born–Haber cycle, calculate the missing enthalpy change.

9.3 1. Distinguish between single, double, and triple bonds in Lewis structures.
2. State how bond energies vary with multiple-bond character and with atom size.
3. Use bond energy data to estimate reaction enthalpies.

9.4 1. Draw Lewis structures for covalently bonded molecules and ions.
2. Find the formal charges on each atom in a Lewis structure.

9.5 1. Draw resonance structures for molecules and ions.
2. Use formal charge to determine which resonance structures make the greatest contribution to the resonance hybrid.
3. Draw the important resonance structures of benzene.

9.6 1. State which elements can exhibit octet expansion, and draw Lewis structures for molecules containing these elements.
2. State which elements are likely to have incomplete octets, and draw Lewis structures for molecules containing these elements.
3. Give examples of free radicals, and draw their Lewis structures.
4. State some of the properties of free radicals.

9.7 1. Give examples of polar and nonpolar molecules.
2. Use crossed arrows to represent dipoles in diatomic molecules.
3. State how electronegativities vary across a period and down a group.
4. Use electronegativities to predict whether a bond will be ionic, polar covalent, or nonpolar.

KEY TERMS BY SECTION

9.1 Chemical bond
Covalent bond
Ionic bond
Ionic lattice
Lewis symbol
Metallic bonding

9.2* Born–Haber cycle
Isoelectronic
Lattice energy
Octet rule

9.3 Bond energy (D)
Bond length
Double bond
Lewis structure
Lone pair (of electrons)
Single bond
Triple bond

9.4 Formal charge

9.5 Delocalized bond
Resonance hybrid
Resonance structures

9.6 Expanded octet
Free radical

9.7 Debye (D)
Dipole
Dipole moment (μ)
Electronegativity
Nonpolar bond
Pauling electronegativity scale
Polar covalent bond

IMPORTANT EQUATIONS

9.1 $F = k\dfrac{q_+q_-}{r^2}$

9.2 $\text{Formal charge} = \left(\text{number of valence electrons}\right) - \left(\text{number of unshared electrons}\right) - \dfrac{1}{2}\left(\text{number of bond electrons}\right)$

FINAL EXERCISES

Answers to exercises with blue numbers are given in Appendix D. Starred exercises are based on the *Digging Deeper* sections.

PART A. QUESTIONS AND PROBLEMS BY SECTION

Bond Types: Lewis Symbols (Introduction; Section 9.1)

9.1 Describe the difference between the following bonds:
(a) an ionic bond and a metallic bond
(b) a covalent bond and an ionic bond

9.2 Explain why molecular compounds are always covalent, but covalent compounds are not always molecular. Give examples.

9.3 Ammonium nitrate (NH_4NO_3) contains both ionic and covalent bonds. Identify them.

9.4 Identify the types of chemical bonds that exist in K_2SO_4, an ionic solid.

9.5 Give a Lewis symbol for each of the following atoms:
(a) C **(b)** As **(c)** N **(d)** K **(e)** Al **(f)** O **(g)** Mg **(h)** Cl

9.6 Give a Lewis symbol for each of the following atoms:
(a) S **(b)** Ra **(c)** B **(d)** I **(e)** Sn **(f)** Kr **(g)** Bi **(h)** Sb

The Ionic Bond; Lattice Energies (Section 9.2)

9.7 Explain why monatomic ions do not normally have charges greater than 3+ or 3−.

9.8 Would you expect silicon, a Group 4A element, to form monatomic ions? Explain your answer.

9.9 Explain each of the following observations:

(a) Sodium does not normally form an Na^{2+} ion.
(b) Chlorine does not form a Cl^{2-} ion.

9.10 Explain why magnesium does not normally form an Mg^{+} ion.

9.11 State the octet rule. Why is it called the *octet rule*?

9.12 Which groups of monatomic ions generally obey the octet rule? Which groups do not?

9.13 How do the lattice energies of crystals vary with (a) increasing ionic charge and (b) increasing ionic radius?

9.14 List some properties of crystals associated with high lattice energy.

***9.15** Diagram the Born–Haber cycle for Li_2O.

***9.16** Diagram the Born–Haber cycle for Mg_3N_2.

9.17 Write formulas for the monatomic ions formed by
(a) Se **(d)** O
(b) Rb **(e)** P
(c) F **(f)** Ba

9.18 Write formulas for the monatomic ions formed by
(a) Al **(d)** Cs
(b) Br **(e)** Ra
(c) N **(f)** Ga

9.19 Write the formula for at least one ionic compound in which all ions are isoelectronic with the krypton atom.

9.20 Write the formula for at least one ionic compound in which all ions are isoelectronic with the xenon atom.

9.21 Write formulas for the monatomic ions formed by (a) Sn and (b) Fe (two ions).

9.22 Which of the ions in the following binary compounds obey the octet rule, and which do not: (a) silver fluoride (AgF); (b) barium nitride (Ba_3N_2)?

9.23 State which ionic compound in each of the following pairs should have the higher lattice energy, and explain why: (a) $Mg(NO_3)_2$ or $Be(NO_3)_2$; (b) NaOH or $Ca(OH)_2$.

9.24 State which ionic compound in each of the following pairs should have the higher melting point, and explain why: (a) KCl or $CaCl_2$; (b) KCl or KI.

***9.25** Estimate the lattice energy of calcium chloride. Use data from the ionization energy and electron affinity tables in Chapter 8 and the following information: the enthalpy of sublimation of calcium is +178.2 kJ/mol; the enthalpy of dissociation of Cl_2 is +243.4 kJ/mol; the enthalpy of formation of $CaCl_2$ is −795.8 kJ/mol.

***9.26** (a) Estimate the lattice energy of cesium iodide. Use data from the ionization energy and electron affinity tables in Chapter 8 and the following information: the enthalpy of formation of CsI is −346.6 kJ/mol; the enthalpy of sublimation of Cs is +76.1 kJ/mol; the enthalpy of formation of I(g) is +106.8 kJ/mol. (b) Compare the lattice energy of CsI with those given for other alkali metal iodides in Table 9.2, and account for any trend you observe.

The Covalent Bond; Bond Energies and Bond Lengths (Section 9.3)

9.27 What is a *Lewis structure*?

9.28 What is the difference between a single bond, a double bond, and a triple bond? How many electrons are in each of these bonds?

9.29 Define *bond length.* Explain why a covalent bond is sometimes compared to a spring.

9.30 Define *bond energy.* Does the formation of a bond absorb or release energy? Explain.

9.31 How do bond energies and bond lengths vary with (a) the size of the bonded atoms and (b) the multiplicity of the bond?

9.32 Rank the C—C, C═C, and C≡C bonds in order of (a) increasing bond length and (b) increasing bond strength.

9.33 Explain why small atoms tend to form stronger bonds than larger atoms.

9.34 Explain why multiple bonds tend to be shorter than single bonds.

9.35 Explain why the bond energy in Cl_2 is greater than the bond energy in I_2.

9.36 Explain why ICl has a greater bond energy than I_2.

9.37 Draw Lewis structures for (a) water, (b) acetylene, and (c) ammonia. Identify each bond as a single bond, double bond, or triple bond.

9.38 Draw Lewis structures for (a) ethylene, (b) acetic acid, and (c) hydrogen chloride. How many lone pairs are in each of these molecules?

9.39 Which molecule, NH_3 or PH_3, should have the stronger bond? The shorter bond?

9.40 Which molecule, acetylene or ethylene, should have the stronger bond? The shorter bond?

9.41 Sulfur is above selenium in Group 6A. Which bond, H—S or H—Se, should have (a) the greater bond energy and (b) the longer bond length?

9.42 Consult the periodic table, and rank the C═C, C═O, and C═S bonds in expected order of (a) increasing bond energy and (b) increasing bond length.

9.43 Use bond energy data to calculate $\Delta H°$ for the following reaction:

$$2HCl(g) + F_2(g) \longrightarrow 2HF(g) + Cl_2(g)$$

9.44 Use bond energy data to estimate the enthalpy of formation of HOCl.

9.45 The H—H and H—F bond energies are 435.9 and 568.1 kJ/mol, respectively. The enthalpy of formation of gaseous HF is −268.6 kJ/mol. Use this information to calculate the F—F bond energy.

9.46 The enthalpy of formation of gaseous ammonia (NH_3) is −46.1 kJ/mol. The N≡N and H—H bond energies are 945.4 and 435.9 kJ/mol, respectively. Use this information to determine the N—H bond energy in ammonia.

9.47 Use bond energy data to estimate the enthalpy change for the formation of methyl chloride from methane and chlorine:

$$H-\underset{\underset{H}{|}}{\overset{\overset{H}{|}}{C}}-H + Cl_2(g) \longrightarrow H-\underset{\underset{H}{|}}{\overset{\overset{H}{|}}{C}}-Cl + HCl(g)$$

9.48 Hydrogen and carbon monoxide are produced by the reaction of methane with steam:

$$CH_4(g) + H_2O(g) \longrightarrow 3H_2(g) + :C{\equiv}O:(g) \qquad \Delta H^\circ = +206.1 \text{ kJ}$$

Use C—H, O—H, and H—H bond energies from Table 9.3 to calculate the C≡O bond energy. How does your value compare with the C≡O bond energy reported in Table 9.3?

Drawing Lewis Structures; Formal Charge (Section 9.4)

9.49 Why is hydrogen never found as a central atom in a Lewis structure?

9.50 Can fluorine be a central atom in a Lewis structure? Explain.

9.51 What is the sum of the formal charges in the Lewis structure of a neutral molecule? Of an ion?

9.52 Do the formal charges in a Lewis structure correspond to actual charges in the molecule? Explain your answer.

9.53 Draw Lewis structures that obey the octet rule for the following molecules:
(a) hydrogen sulfide, H_2S
(b) nitrogen triiodide, NI_3
(c) silane, SiH_4
(d) ethane, C_2H_6
(e) formaldehyde, H_2CO (both H atoms are bonded to carbon)

9.54 Draw Lewis structures that obey the octet rule for the following molecules:
(a) nitrogen, N_2
(b) hydrazine, N_2H_4 ($H_2N—NH_2$)
(c) hydrogen peroxide, H_2O_2 (HO—OH)
(d) cyanogen, C_2N_2 (NC—CN)
(e) sulfur, S_8 (a ring of eight sulfur atoms)

9.55 Draw Lewis structures that obey the octet rule for the following ions:
(a) hydronium ion, H_3O^+
(b) hydroxide ion, OH^-
(c) hypochlorite ion, OCl^-
(d) ammonium ion, NH_4^+

9.56 Draw Lewis structures that obey the octet rule for the following species:
(a) ozone, O_3
(b) nitrite ion, NO_2^-
(c) carbonate ion, CO_3^{2-}
(d) carbon disulfide, CS_2

9.57 Find the formal charges, if any, in the Lewis structures of (a) Exercise 9.53 and (b) Exercise 9.54.

9.58 Find the formal charges, if any, in the Lewis structures of (a) Exercise 9.55 and (b) Exercise 9.56.

Resonance Structures (Section 9.5)

9.59 Explain the concept of a resonance hybrid. In what way, if at all, do the contributing structures represent the actual molecule?

9.60 In what way do the resonance structures for a given molecule differ from each other? In what ways must they be similar?

9.61 What is a *delocalized bond*? How are delocalized bonds represented by resonance structures?

9.62 What criteria are used in deciding which resonance structures make the greatest contribution to a resonance hybrid?

9.63 Draw resonance structures for the following species:
(a) nitrite ion, NO_2^-
(b) dinitrogen tetroxide, N_2O_4 ($O_2N—NO_2$)
(c) hydrazoic acid, HN_3 (HNNN)
(d) sulfur dioxide, SO_2

9.64 Draw resonance structures for the following species:
(a) cyanamide, H_2NCN
(b) nitrous acid, HNO_2
(c) carbon monoxide, CO
(d) acetate ion, CH_3COO^-

9.65 Find the formal charges in the resonance structures of Exercise 9.63. Which structure, if any, makes the greatest contribution to the hybrid?

9.66 Find the formal charges in the resonance structures of Exercise 9.64. Which structure, if any, makes the greatest contribution to the hybrid?

9.67 Draw the Kekulé structures for benzene.

9.68 Draw and explain the abbreviated structures for benzene.

Exceptions to the Octet Rule (Section 9.6)

9.69 Is the best Lewis structure invariably the one that follows the octet rule? Explain your answer.

9.70 What is meant by the phrase *octet expansion*? For which elements are expanded octets possible? Which elements are frequently given an expanded octet in a Lewis structure?

9.71 What is meant by the phrase *incomplete octet*? Which elements usually exhibit incomplete octets in a Lewis structure?

9.72 What is a *free radical*? List some of the properties of free radicals.

9.73 Draw Lewis structures for the following species:
(a) sulfur tetrafluoride, SF_4
(b) hexafluoroantimonate ion, SbF_6^-

9.74 Draw Lewis structures for the following species:
(a) xenon hexafluoride, XeF_6
(b) dichloroiodide ion, ICl_2^-

9.75 Draw Lewis structures without formal charge for the following molecules:
(a) sulfur trioxide, SO_3
(b) perbromic acid, $HBrO_4$
(c) thionyl chloride, $SOCl_2$
(d) sulfuryl chloride, SO_2Cl_2

9.76 Draw Lewis structures without formal charge for the following molecules:
(a) phosphorus oxychloride, $POCl_3$
(b) chlorine trifluoride, ClF_3
(c) bromine pentafluoride, BrF_5
(d) chloric acid, $HClO_3$

9.77 **(a)** Draw Lewis structures that obey the octet rule for SO_2 and for O_3. Compare the two.
(b) Write a structure for SO_2 that has no formal charge. Can a similar structure be written for O_3? Explain.

9.78 Draw a Lewis structure for $BeCl_2$ that obeys the octet rule. Draw another structure for $BeCl_2$ that does not have formal charge. Which structure contributes most to the resonance hybrid?

9.79 Draw Lewis structures with incomplete octets for the following species:
(a) aluminum chloride, $AlCl_3$
(b) dilithium, Li_2
(c) dichlorothallium ion, $TlCl_2^+$

9.80 Draw Lewis structures without formal charge for the following molecules:
(a) boron trichloride, BCl_3
(b) beryllium fluoride, BeF_2
(c) beryllium hydride, BeH_2

9.81 Draw Lewis structures for the following free radicals: (a) CH_3 and (b) ClO_4.

9.82 The following free radicals have been detected in flames. Draw a Lewis structure for each one.
(a) CH
(b) C_2
(c) OH
(d) HCO

Polar Molecules; Dipole Moments (Section 9.7)

9.83 Distinguish between the following bond types:
(a) polar and nonpolar covalent bonds
(b) polar covalent bonds and ionic bonds

9.84 Which physical property is the most important for determining whether a compound is ionic?

9.85 Define and give units for the dipole moment. Explain how molecular dipole moments are measured.

9.86 What is a polar molecule? Explain why the dipole moment of a diatomic molecule indicates the degree of polarity in the bond.

9.87 What is meant by "ionic character" in a covalent bond? Give examples of covalent bonds with considerable ionic character.

9.88 What is meant by "covalent character" in ionic bonding? Give examples of ionic compounds that have considerable covalent character.

Electronegativity (Section 9.7)

9.89 Define *electronegativity*. What is the difference between electronegativity and electron affinity?

9.90 How do the electronegativities of the elements vary (a) from left to right across a period and (b) from the bottom to the top of a group?

9.91 Which element has (a) the highest electronegativity, (b) the second highest electronegativity, and (c) the lowest electronegativity?

9.92 Use the electronegativity table to find
(a) the most electronegative metal
(b) the least electronegative nonmetal
(c) all nonmetals that are less electronegative than hydrogen
(d) all metals that are more electronegative than hydrogen

9.93 Consult the periodic table inside the front cover, and pick the more electronegative element from each of the following pairs:
(a) Cl or Br
(b) Si or O
(c) N or O
(d) Ga or S

9.94 Consult the periodic table inside the front cover, and pick the more electronegative element from each of the following pairs:
(a) P or S
(b) Li or Rb
(c) As or Br
(d) Ba or I

9.95 Use what you know about periodic trends to predict the polarity of the following bonds: (a) N—O, (b) Br—Cl, (c) S—H.

9.96 Use your knowledge of periodic trends to predict the polarity of the following bonds: (a) B—N, (b) F—Br, (c) P—Cl.

9.97 Use the table of electronegativities to predict whether the following compounds are predominantly ionic or predominantly covalent:
(a) beryllium chloride, $BeCl_2$
(b) boron trifluoride, BF_3
(c) potassium fluoride, KF

9.98 Use the table of electronegativities to predict whether the following compounds are predominantly ionic or predominantly covalent:
(a) cesium bromide, CsBr
(b) aluminum fluoride, AlF_3
(c) hydrogen bromide, HBr

9.99 Draw Lewis structures for the following molecules, and indicate the polarity of each bond by a crossed arrow:
(a) hydrogen bromide, HBr
(b) boric acid, H_3BO_3
(c) ammonia, NH_3

9.100 Draw Lewis structures for the following molecules, and indicate the polarity of each bond by a crossed arrow:
(a) chlorine trifluoride, ClF_3
(b) methanol, CH_3OH
(c) sulfur dioxide, SO_2

PART B. MISCELLANEOUS QUESTIONS AND PROBLEMS

9.101 A *refractory* is a ceramic material that is not easily melted or broken down by heat. Which substance, MgO or $MgCl_2$, would you expect to be a better refractory? Explain your answer in terms of lattice energies.

9.102 Why is it difficult to predict the charges on transition metal and post-transition metal ions?

9.103 The enthalpy of vaporization of liquid bromine is +30.9 kJ/mol. Use this information and data from Table 9.3 to estimate the standard enthalpy of formation of hydrogen bromide. (Liquid bromine is the most stable form of bromine at room temperature.)

9.104 The equation for the formation of diselenium dichloride (Cl—Se—Se—Cl) is

$$2Se(s) + Cl_2(g) \rightarrow Se_2Cl_2(g)$$

Use the following information to estimate its enthalpy of formation. The Cl—Cl, Se—Cl, and Se—Se bond energies are 243, 246, and 175 kJ/mol, respectively. The enthalpy of sublimation of solid selenium is +227 kJ/mol.

9.105 Determine the C—H bond energy in methane from the following data: the enthalpy of formation of methane, CH_4, is −74.8 kJ/mol; the H—H bond energy is +435.9 kJ/mol; the enthalpy of sublimation of graphite (carbon) is +716.7 kJ/mol.

9.106 Draw a Lewis structure for carbon monoxide (CO) that obeys the octet rule. Note that it has formal charge. Draw another structure that does not have formal charge. Evidence indicates that the oxygen end of CO is slightly positive despite the fact that oxygen is more electronegative than carbon. Is this evidence consistent with the above resonance structures?

9.107 The chlorobenzene molecule, C_6H_5Cl, is a benzene molecule with a chlorine atom substituted for one of the hydrogens. Draw the Kekulé structures for chlorobenzene.

9.108 Nitrogen forms nitrogen dioxide (NO_2), nitrogen monoxide (NO, a pollutant emitted by internal combustion engines), and dinitrogen oxide (N_2O, an anesthetic known as "laughing gas"). Draw plausible Lewis structures for each oxide. (The N_2O arrangement is N—N—O.) Which, if any, of these oxides are free radicals?

9.109 Boron trifluoride (BF_3) reacts with ammonia (NH_3) to form F_3BNH_3, a molecule containing a B—N bond. Draw Lewis structures for the reactants and for the product. Do the formal charges on boron and nitrogen change during the reaction?

9.110 Two molecules of nitrogen dioxide (NO_2) react to form N_2O_4, a molecule containing an N—N bond. Draw Lewis structures for the reactant and the product.

9.111 Passage of electric discharges through gases at low pressure has produced the following free radicals and free radical ions:
(a) CN
(b) NH
(c) N_2^+
(d) CO^+
(e) CO_2^+
Draw a Lewis structure for each of these species.

9.112 **(a)** Which ion, NO_3^- or NO_2^-, would you expect to have the shorter N—O bonds? (*Hint:* Draw resonance structures for each ion.)

(b) Consider the resonance structures of HNO_3 (see Example 9.13). How should the N—O bond lengths compare with each other? How should they compare with the bond lengths in the NO_3^- ion?

9.113 The ozone molecule (O_3) contains three oxygen atoms, yet it has a dipole moment of 0.53 D. Try to account for this by drawing resonance structures for the molecule. Is the ozone molecule linear or bent?

9.114 The internuclear separation in HCl is 128 pm.

(a) Calculate the dipole moment that HCl would exhibit if it were 100% ionic (H^+Cl^-). The electron charge is 1.602×10^{-19} C.

(b) The experimental dipole moment of HCl is 1.08 D. Estimate the percent ionic character of HCl.

9.115 Robert S. Mulliken proposed that the electronegativity (X) of an atom is proportional to the ionization energy (IE) minus the electron affinity energy (EA); that is, $X = k(\text{IE} - \text{EA})$, where k is a proportionality constant. What is the rationale of Mulliken's proposal; that is, why does electronegativity depend on both ionization energy and electron affinity? Your explanation should account for the minus sign in Mulliken's formula.

9.116 Use Mulliken's formula (see Exercise 9.115) and data from the ionization energy and electron affinity tables in Chapter 8 to calculate values of (IE − EA) for the halogens. Find the best value of k that will give electronegativity values comparable to those of Pauling.

CHEMISTS AT WORK

THE CHEMICAL TECHNICIAN

The success of an experiment or experimental procedure is frequently a function of the skills of the operator; such operators are often experienced chemical technicians. These professionals are really experimental artists who design and execute procedures for obtaining meaningful, relevant, and reproducible data. Diane Johnson, one of these professionals, works in Central Research at Pfizer, Inc., a major manufacturer of pharmaceuticals, in Groton, Connecticut.

Because of her strong interest in applied science, Diane elected in her sophomore year in high school to pursue a program in science and math at Coventry's Vocational School. "After high school, I chose to further my chemical technician education at the Community College of Rhode Island," she says. Shortly after graduation in 1975, she was hired at Pfizer with the title "Technician." During her tenure she has had five promotions and now holds the position "Scientist." (For reference: at Pfizer, most "Scientists" have, at a minimum, a master's degree and 5 years experience.)

The central research facility of any major corporation is where ideas are born and tested for practicality. In a pharmaceutical company, the ideas are directly related to the development, verification, and documentation of the efficacy of a compound or preparation for the relief or cure of some disease or physical condition. Diane's first position was in the drug metabolism department. "During my first few years," she says, "I was constantly supervised, but I was also learning the necessary techniques and procedures. However, as time passed, I was given more responsibility and worked more independently." Work in the metabolism area included developing and testing assays for the new product, doing toxicological studies, and gathering data so that approval for human experimentation could be obtained.

DIANE JOHNSON

In addition to her work-related experience, Diane has continually strengthened her background by taking courses both in house and at the Avery Point campus of the University of Connecticut. In part, her success is due to her personal career philosophy. "Be inquisitive! For every project, experiment, procedure, or assignment, ask questions. Go to the library. Read the manual. Learn all the aspects of the task. Learn and understand it well enough to teach it."

Her career has been rich in experiences and accomplishments. For 6 years, she worked on the research team responsible for bringing the drug Feldene, an antiinflamatory, to the market. The Federal Drug Administration examined the data that she collected on the product and was impressed with its excellent quality. Diane has also been part of teams that have brought drugs as diverse as antibiotics and antihypertensives to the market. "Seeing rewards from research are not easy," she states. "You must learn to deal with a large number of failures as compared to a small number of successes. It is quite rewarding seeing a product on which you have worked finally go to market." She has had the opportunity to teach what she has learned by presenting papers at scientific meetings and authoring other papers.

Her career has changed and grown from being an experimentalist to designing the experiments that generate the required data. In 1993, her responsibilities included training new professionals who join the staff of the laboratory, supervising as many as three other technicians, and taking the lead for the design of the studies that will be used on a potential new product. "Now I spend less than 20% of my time doing laboratory work."

"One of Diane's favorite sayings is Pfizer's motto: 'Science for the world's well being.' She looks at her work as more than just a job. Every aspect of her career is ultimately devoted to bettering the health of human beings."[1]

1. Susanne Cabral, *25th Anniversary Edition,* Community College of Rhode Island, Spring 1989.

CHAPTER 10

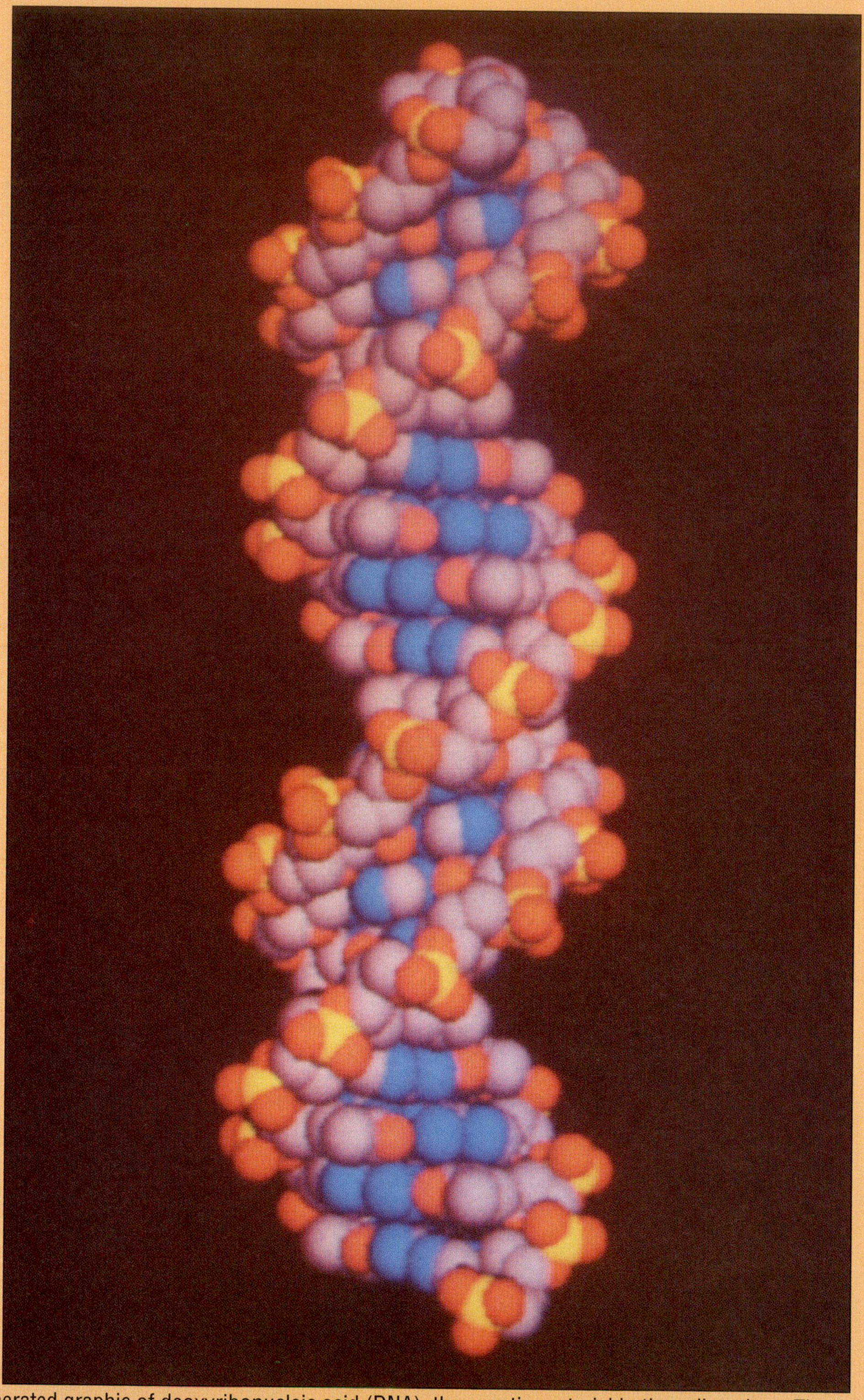

A computer-generated graphic of deoxyribonucleic acid (DNA), the genetic material in the cell nucleus. The double helix structure of DNA is maintained by covalent bonds and hydrogen bonds. (Hydrogen bonds are dicussed in Chapter 12.)

MOLECULAR GEOMETRY AND CHEMICAL BONDING THEORY

OUTLINE

PREVIEW

A molecule is a group of atoms that stays together and retains its general shape even while it twists, stretches, rolls, and travels through space. We know that shared electrons keep molecules from flying apart, but we have not described the density patterns associated with shared electrons, nor have we really explained how covalent bonds form in the first place. We have treated bonding only in terms of rules involving octets and noble gas configurations—and these rules have many exceptions. Electrons in molecules, like electrons in atoms, are governed by the laws of quantum mechanics, and chemical bonds must be described in terms of these laws and their restrictions. This description will ultimately lead to a deeper and more satisfying explanation of molecular properties and geometries than any explanation offered thus far.

Molecules have many different shapes; some are linear, others are flat (planar), and still others have complex three-dimensional structures. The shape of a molecule determines many of its properties. The compounds *n*-butane and isobutane, for example, both have the formula C_4H_{10}, but they have different physical and chemical properties because their molecules have different shapes:

$CH_3 — CH_2 — CH_2 — CH_3$

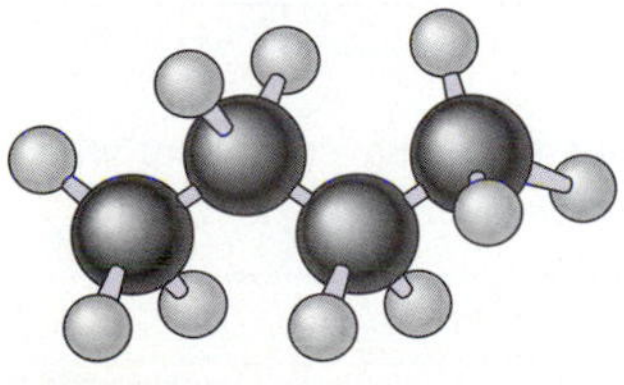

n-Butane
Melting point: −138°C
boiling point: −0.5°C

$$\begin{array}{c} CH_3 \\ | \\ CH_3 — CH — CH_3 \end{array}$$

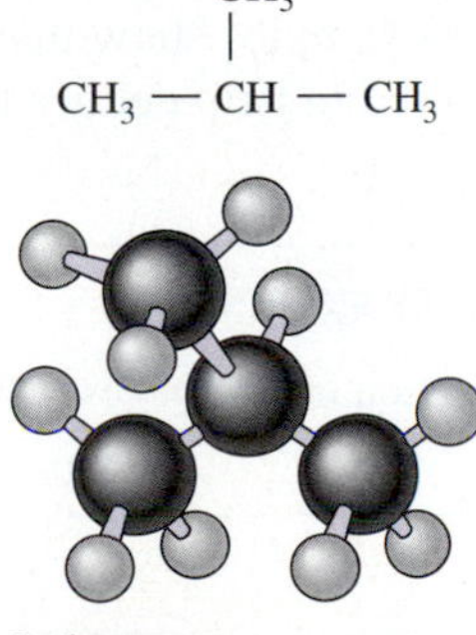

Isobutane
Melting point: −159°C
boiling point: −12°C

Enzymes, the molecules that control reactions within the body, operate with miraculous efficiency because their contours are compatible with those of the molecules whose reactions they promote. The efficacy of drugs, antibiotics, perfumes, and flavoring agents also depends on molecular shape.

The shape of a molecule is described in terms of bond lengths and bond angles. A **bond length** was defined in Section 9.3 as the equilibrium distance between the centers of two bonded atoms; some bond lengths are tabulated in Table 9.3 (p. 377). A **bond angle** is the equilibrium angle between the lines that join two atoms to a third atom (see Figure 10.1). Computerized treatment of x-ray and spectral data provides precise information about molecular geometries; many bond lengths have been determined to within 0.1 pm and many bond angles to within 0.01°.

10.1 DIPOLE MOMENTS AND MOLECULAR GEOMETRY

Most molecules have more than one bond, and each bond has its own dipole moment. The net dipole moment of such a molecule depends on the magnitudes of the individual bond moments and also on the geometry of the molecule. Carbon dioxide, for example, is observed to have no molecular dipole moment, although other evidence indicates that each of its two bonds has a dipole moment of 2.3 D. The bond moments must therefore be oriented so that they cancel each other:

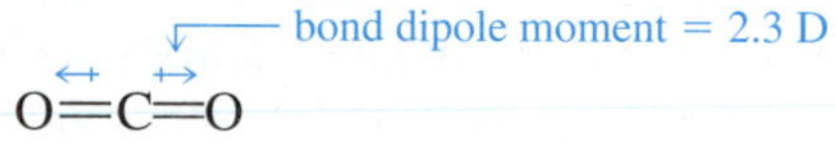

bond dipole moment = 2.3 D

O=C=O

molecular dipole moment = 0

In other words, CO_2 is linear with the two C=O bonds pointing in exactly opposite directions.

Water, on the other hand, has an observed dipole moment of 1.85 D and two bond dipole moments of 1.5 D each:

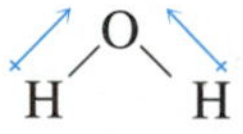

each bond dipole moment = 1.5 D

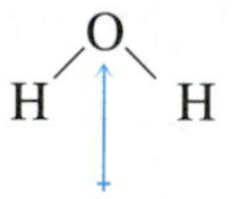

molecular dipole moment = 1.85 D

The two bond moments do not cancel because the water molecule is not linear. Instead, they partially reinforce each other to give a resultant dipole moment of 1.85 D. The oxygen atom carries a partial negative charge because oxygen is more electronegative than hydrogen.

For those of you who are familiar with vectors, the resultant dipole moment of water is the vector sum of the two bond moments.

The above examples show that the observed value of a molecular dipole moment can often be used to help deduce the molecule's shape.

PRACTICE EXERCISE 10.1

The dipole moment of OF_2 is 0.297 D. Suggest a geometry consistent with this value.

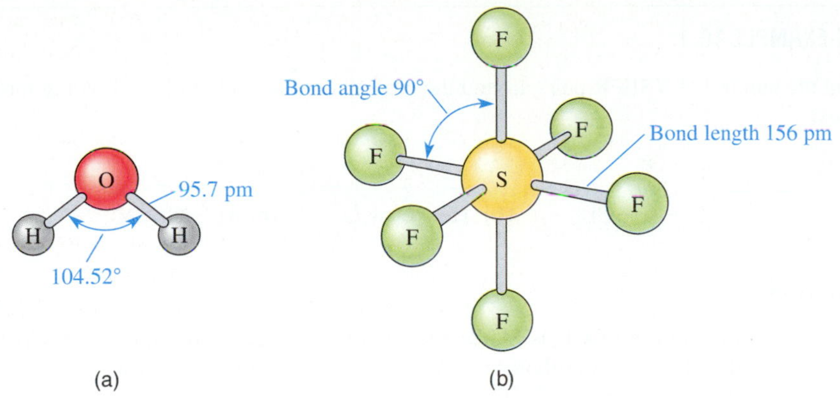

Figure 10.1
Bond lengths and bond angles in (a) H_2O and (b) SF_6.

10.2 THE VSEPR MODEL

The shape of a molecule can be determined by a number of experimental techniques, including measurement of its dipole moment. Even without such information, however, we can often roughly predict the shape of a molecule by assuming that *electron pairs tend to stay as far from each other as possible.* This assumption, when applied to valence shell electron pairs, is the basis for the Valence Shell Electron Pair Repulsion model, or simply **VSEPR** (pronounced "vesper").

Initially, we will apply this model to molecules with one **central atom**, that is, one atom to which two or more other atoms are bonded. For example, the oxygen and sulfur atoms in Figure 10.1 are central atoms; the oxygen atom is bonded to two hydrogen atoms, the sulfur atom to six fluorine atoms. Each of the bonds consists of an electron pair that spends most of its time between the bonded atoms; if these pairs are as far from each other as possible, then the bonds will be separated by the largest possible angle. In other words, bond angles are determined by the arrangement of the electron pairs around the central atom, and this arrangement, in turn, is determined by the number of electron pairs.

The first step in the VSEPR procedure for predicting the geometry of a molecule or ion is to write the Lewis structure and count the number of electron pairs around the central atom. We are interested only in **VSEPR electron pairs**, that is, electron pairs that determine bond angles. The number of VSEPR pairs is not always the same as the total number of pairs, and they are counted according to the following rules:

1. Count each bond—single, double, or triple—as one VSEPR pair.
2. Count each lone (nonbonding) pair as one VSEPR pair.

Extra electrons in multiple bonds are not counted because they play only a secondary role in determining molecular shape. Lone pairs are counted because, according to the VSEPR model, they repel neighboring pairs and thus influence the positions of bond pairs. With few exceptions, the number of VSEPR electron pairs ranges from two to six.

LEARNING HINTS

Each bond represents a region of high electron density between two nuclei, and each lone pair represents a region of high electron density near a nucleus. Hence, the number of VSEPR pairs is simply the total number of such regions.

Some chemists refer to VSEPR pairs as *structural pairs.*

EXAMPLE 10.1

Find the number of VSEPR pairs around the central atom in each of the following molecules:

(a) :C̤̈l—S̤(=Ö:)—C̤̈l: (b) H—C(H)=Ö̤ (c) H—C≡N:

SOLUTION

(a) The central atom in $SOCl_2$ is S. The two single bonds each contribute one VSEPR pair, the double bond contributes one VSEPR pair, and there is one lone pair around sulfur; hence, there are four VSEPR pairs.

(b) The central atom in CH_2O is C. The single bonds each contribute one VSEPR pair, the double bond contributes one VSEPR pair, and there are no lone pairs around carbon; hence, there are three VSEPR pairs.

(c) The central atom in HCN is C. The single and triple bond each contribute one VSEPR pair, for a total of two VSEPR pairs. There are no lone pairs around the carbon atom.

PRACTICE EXERCISE 10.2

How many VSEPR pairs surround (a) carbon in the CO_2 molecule and (b) nitrogen in the NH_3 molecule?

The second step in the VSEPR procedure is to arrange the VSEPR pairs around the central atom so that they are as far apart as possible, thereby minimizing the repulsions between them. For example, if the VSEPR count is two, the two pairs would be on opposite sides of the central atom, 180° apart. The arrangement would be linear as in Figure 10.2a. If there are three VSEPR pairs, they would be 120° apart at the corners of an equilateral triangle (Figure 10.2b). Counts of four or more would lead to the three-dimensional arrangements illustrated in Figures 10.2c–e. We will consider these arrangements one at a time.

Two VSEPR Pairs. Two VSEPR electron pairs tend to be on opposite sides of the central atom. If these are bonding pairs, the bonds will form a 180° angle and the molecule will be linear. Such molecules are classified as AX_2 molecules, where A represents the central atom (Figure 10.3a); examples include carbon dioxide, hydrogen cyanide (HCN), and beryllium chloride ($BeCl_2$):

Ö̤=C=Ö̤ (180°) H—C≡N: (180°) :C̤̈l—Be—C̤̈l: (180°)

Three VSEPR Pairs. Three VSEPR electron pairs tend to arrange themselves at the corners of an equilateral triangle. If all three are bond pairs, as in boron trifluoride, the molecule is classified as AX_3 (Figure 10.3b). The atoms all lie in the same plane

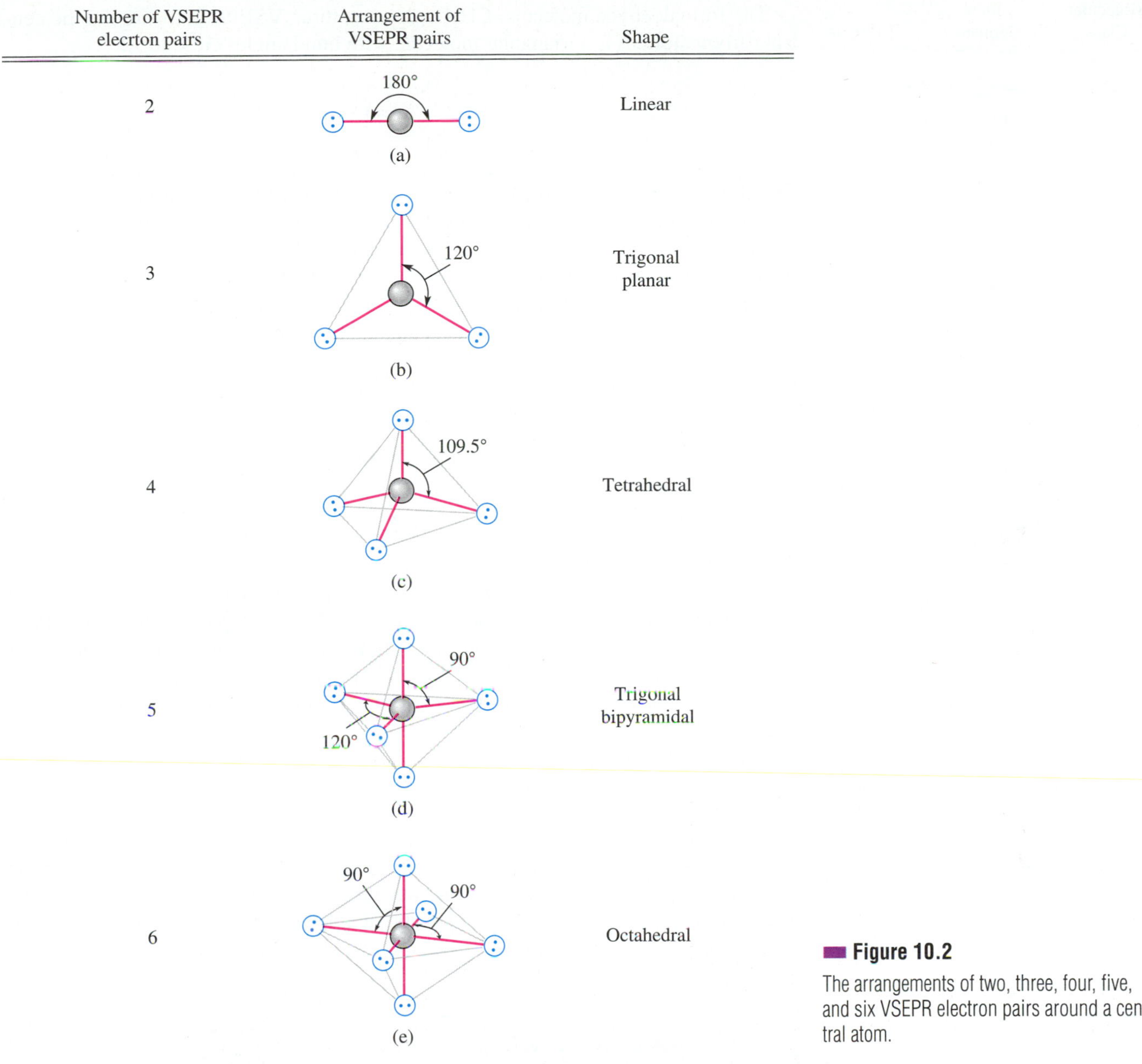

Number of VSEPR elecrton pairs
Arrangement of VSEPR pairs
Shape
2
180°
(a)
Linear
3
120°
(b)
Trigonal planar
4
109.5°
(c)
Tetrahedral
5
90°
120°
(d)
Trigonal bipyramidal
6
90°
90°
(e)
Octahedral

Figure 10.2
The arrangements of two, three, four, five, and six VSEPR electron pairs around a central atom.

and the bonds make angles of 120°, an arrangement called **trigonal planar**. The nitrate ion (NO_3^-) and the carbonate ion (CO_3^{2-}) are also trigonal planar.

PRACTICE EXERCISE 10.3

Draw the (a) nitrate and (b) carbonate ions. Recall that the bonds in these ions are identical because of resonance.

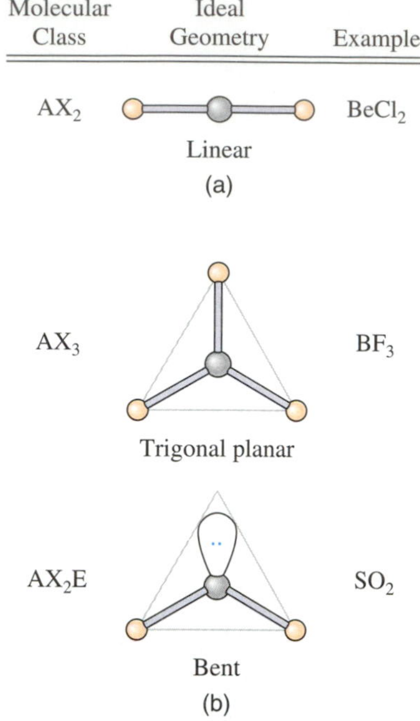

Figure 10.3

Idealized geometries for molecules with one central atom and (a) 2 or (b) 3 VSEPR pairs of electrons. The molecules in a class (left column) have the same general formula in which A is the central atom, X is an atom bonded to A, and E is a lone pair on A. To visualize the shape of AX_2E, consider the lone pair to be present but "invisible."

The formaldehyde molecule, CH_2O, also has three VSEPR pairs around the central carbon atom. It is a triangular molecule with bond angles close to 120°:

118° H₂C=O 121°

The slight distortion from the trigonal planar angle of 120° occurs because the bonds are not completely identical; they involve different atoms and have different multiplicities (one is a double bond and the others are single bonds). *Angles that include a multiple bond, such as the H—C=O angle in formaldehyde, are usually somewhat larger than angles between single bonds.* Multiple bonds contain more electrons than single bonds, and their greater repulsive force widens the bond angle.

Sulfur dioxide (SO_2) also has a VSEPR count of three; one of the VSEPR pairs is a lone pair of electrons on the sulfur atom:

O=S=O 119.5°

The three VSEPR pairs adopt a trigonal planar arrangement, but only two of them are bond pairs. The two bonds are identical and are directed toward two corners of an imaginary triangle with sulfur at its center. The lone pair occupies the third corner of the triangle. There is no atom bonded by the lone pair, so the sulfur dioxide molecule is bent, or V-shaped. This type of molecule is classified as AX_2E in Figure 10.3b. The E indicates that the third position is "empty"; that is, it is not occupied by an atom.

Experimental data show the sulfur dioxide angle to be 119.5°, slightly less than the trigonal planar value of 120°. *A lone pair on the vertex atom of an angle acts to compress the angle,* an effect called **lone-pair repulsion**. This compression can be explained in terms of electron-density patterns. The electron (or charge) density of a bond pair is drawn out between two nuclei. The electron density of a lone pair, on the other hand, is concentrated on one nucleus. The greater charge density of the lone pair repels the lower charge density of adjacent bond pairs and, in effect, pushes the bond pairs closer together.

PRACTICE EXERCISE 10.4

Draw the ozone molecule, O_3.

Four VSEPR Pairs. Four VSEPR electron pairs will experience minimum repulsion if they are oriented toward the corners of a **regular tetrahedron**, as in the AX_4 molecule in Figure 10.4. A tetrahedron has four triangular faces; in a regular tetrahedron, the faces are identical equilateral triangles. Lines from the corners of a regular tetrahedron to its center meet at angles of 109.5°, the **tetrahedral angle**. More molecular geometry is based on the tetrahedron than on any other shape. Methane (CH_4), for example, is a tetrahedral molecule with four bonds coming from the central carbon atom:

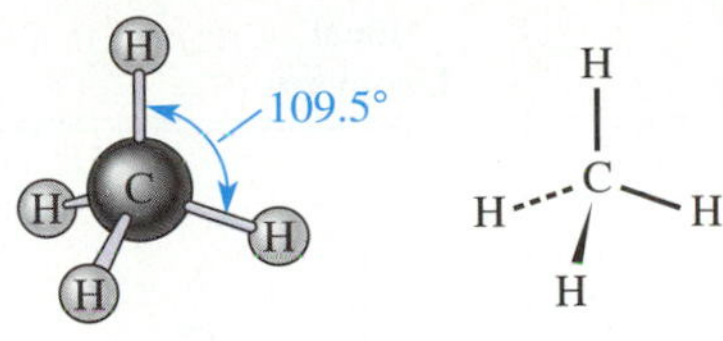

The bonds point toward the corners of the tetrahedron, carbon is at its center, and each bond angle is 109.5°. (*Note:* In the above diagram, the solid lines represent bonds that lie in the plane of the paper. The triangular wedge represents a bond that comes out of the plane of the paper toward the reader, and the dashed line represents a bond that falls below the plane of the paper away from the reader.) Other tetrahedral species are the ammonium ion (NH_4^+), the sulfate ion (SO_4^{2-}), and silicon tetrafluoride (SiF_4).

PRACTICE EXERCISE 10.5

Draw (a) the NH_4^+ ion, (b) the SO_4^{2-} ion, and (c) the SiF_4 molecule.

The ammonia molecule, NH_3, has three bonds and one lone pair tetrahedrally arranged around the central nitrogen atom. It has the AX_3E structure shown in Figure 10.4. The three bonds point toward three corners of the tetrahedron. There is no atom at the fourth corner, so the molecule is shaped like a shallow pyramid formed by the nitrogen atom at the apex and three hydrogen atoms in the base:

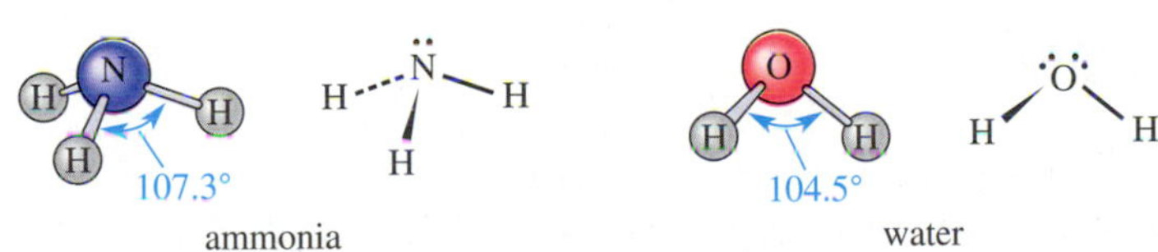

The H—N—H bond angle, 107.3°, is slightly less than the tetrahedral value of 109.5° because of lone-pair repulsion.

The water molecule has two bonds and two lone pairs tetrahedrally arranged around the oxygen atom (AX_2E_2 in Figure 10.4). Two corners of the tetrahedron are unoccupied by atoms, so all we see is the V-shape made by the central oxygen atom and the two hydrogen atoms. The bond angle, 104.5°, is quite a bit less than the tetrahedral value of 109.5°. A comparison of the angles in H_2O and NH_3 shows that the two lone pairs in water exert more repulsion than the one lone pair in ammonia.

PRACTICE EXERCISE 10.6

Draw (a) the phosphine (PH_3) molecule and (b) the hydrogen sulfide (H_2S) molecule. The bond angles in phosphine and hydrogen sulfide are 93° and 92°, respectively.

LEARNING HINT

Practice drawing various three-dimensional molecules.

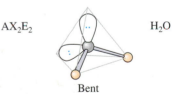

Molecular Class	Ideal Geometry	Example
AX_4	Tetrahedral	CH_4
AX_3E	Trigonal pyramidal	NH_3
AX_2E_2	Bent	H_2O

Figure 10.4

Idealized geometries for molecules in which the central atom has 4 VSEPR pairs of electrons. Keep in mind that lone pairs are "invisible" when you visualize the molecular shape.

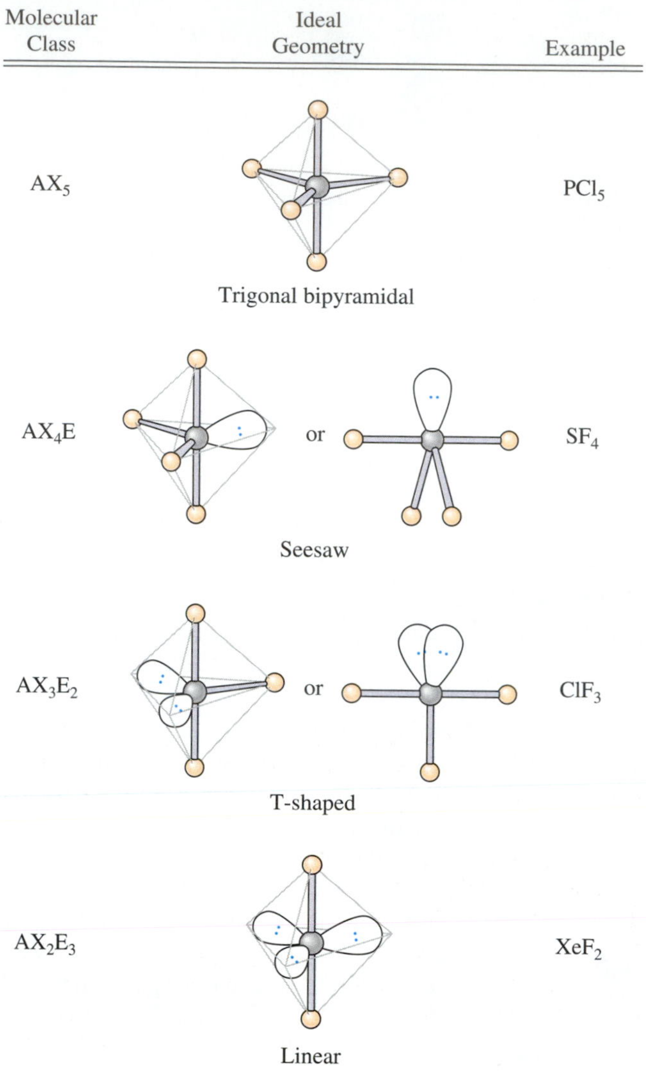

Molecular Class	Ideal Geometry	Example
AX_5	Trigonal bipyramidal	PCl_5
AX_4E	or Seesaw	SF_4
AX_3E_2	or T-shaped	ClF_3
AX_2E_3	Linear	XeF_2

Figure 10.5

Idealized geometries for molecules in which the central atom has 5 VSEPR pairs of electrons. The AX_4E and AX_3E_2 arrangements are each drawn twice, from different perspectives.

Five VSEPR Pairs. Five VSEPR electron pairs are much less common than two, three, or four VSEPR pairs, because five pairs can be accommodated only by certain elements in the third and following periods (Section 9.6). The five electron pairs are located at the corners of a **trigonal bipyramid** (AX_5 in Figure 10.5), a structure that is equivalent to two pyramids sharing a common triangular face. The three bonds in the center plane, called *equatorial bonds,* make 120° angles with each other and 90° angles with the two *axial bonds.* Some examples of trigonal bipyramidal molecules are PF_5, PCl_5, and AsF_5.

Lone pairs tend to occupy positions in the equatorial plane where the 120° angles give them room to spread out. The ideal geometry of molecules containing one lone pair (AX_4E), two lone pairs (AX_3E_2), and three lone pairs (AX_2E_3) is shown in Figure 10.5. Examples of such molecules are SF_4, ClF_3, and XeF_2, respectively. In

SF_4 and ClF_3, lone-pair repulsion compresses the angles between the axial and equatorial bonds to less than their ideal value of 90°:

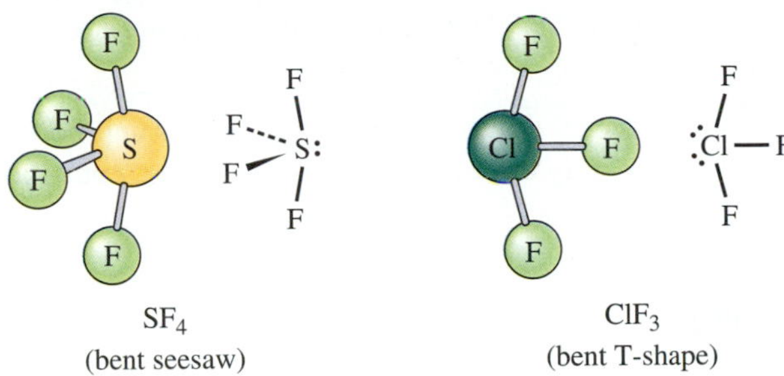

PRACTICE EXERCISE 10.7

Draw the molecules of (a) AsF_5, (b) $TeCl_4$, and (c) BrF_3. Your sketches should show the effects, if any, of lone-pair repulsion. (*Hint:* Lone pairs occupy equatorial positions.)

Six VSEPR Pairs. Six VSEPR electron pairs minimize repulsions by occupying the six corners of a **regular octahedron** (AX_6 in Figure 10.6). This shape, which is second in importance only to the tetrahedron, can be visualized as two square pyramids sharing their square base. Lines drawn from the six corners to the center of an octahedron meet at right angles, so the octahedral bond angle is 90°. Sulfur hexafluoride (SF_6) is octahedral with sulfur at the center and the six fluorine atoms at the corners. The BrF_5 molecule (AX_5E in Figure 10.6) has five bonds and one lone pair arranged octahedrally around the bromine atom. One octahedral position is unoccupied, so the remaining atoms form a square pyramid that is slightly distorted in the actual molecule by lone-pair repulsion. The XeF_4 molecule has four bonds and two lone pairs (AX_4E_2 in Figure 10.6). Two positions are unoccupied, and the remaining atoms form a square around the xenon atom.

Molecular Class	Ideal Geometry	Example
AX_6	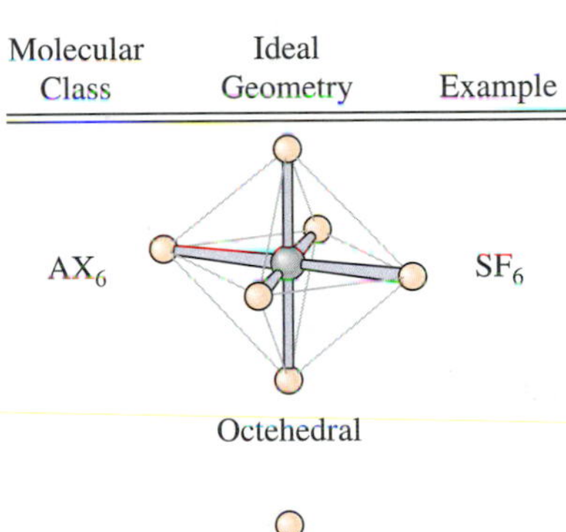 Octehedral	SF_6
AX_5E	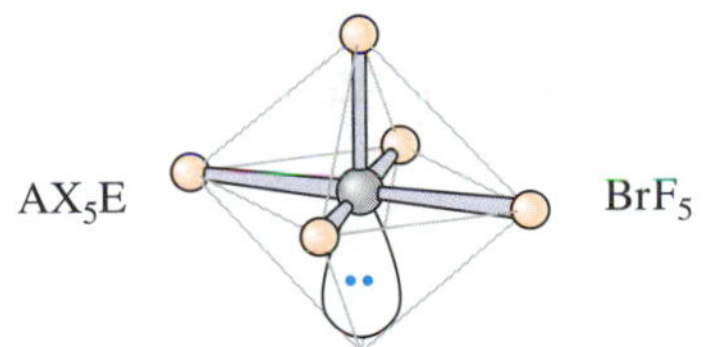 Square pyramidal	BrF_5
AX_4E_2	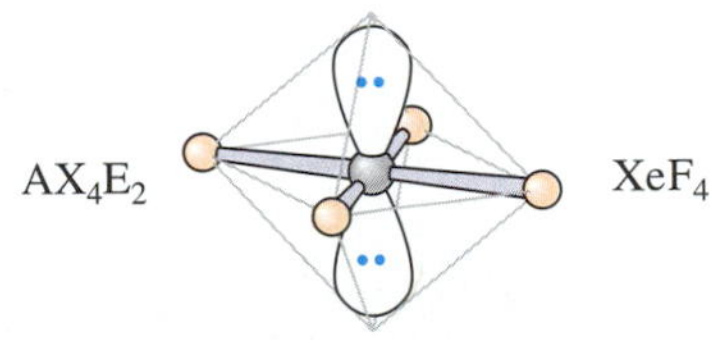 Square planar	XeF_4

Figure 10.6

Idealized geometries for molecules in which the central atom has 6 VSEPR pairs of electrons.

PRACTICE EXERCISE 10.8

Draw the (a) BrF_5 molecule and (b) the ICl_4^- ion. Your sketches should show the effects, if any, of lone-pair repulsion.

Predicting Molecular Shapes

The VSEPR model can be used to predict the shape of a molecule from its formula. The steps are as follows:

- ***Step 1:*** Write the Lewis structure.
- ***Step 2:*** Count the number of VSEPR pairs around the central atom. Be sure to use the VSEPR counting rules.

Step 3: Pick the arrangement of VSEPR pairs (linear, tetrahedral, etc.) that minimizes repulsion. The five possible arrangements are shown in Figure 10.2.

Step 4: Using the arrangement obtained in Step 3, and ignoring lone pairs, determine the shape of the molecule. The most common shapes are summarized in Figures 10.3 through 10.6.

Step 5: If you are interested in the fine points, note which bond angles would be expanded by bond multiplicity or compressed by lone-pair repulsion.

EXAMPLE 10.2

Discuss the geometry of the thionyl fluoride (SOF_2) molecule. Sulfur is the central atom.

SOLUTION

We will follow the steps listed above.

Step 1: The sulfur and oxygen atoms each contribute 6 valence electrons and each fluorine atom contributes 7 electrons for a total of 26 electrons or 13 pairs. The Lewis structure is

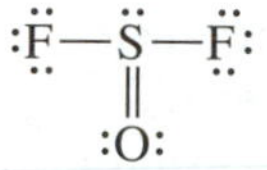

:F̤̈—S̈—F̤̈:
‖
:O:

LEARNING HINT

A resonance structure containing only single bonds can be drawn for SOF_2. Draw this structure, and show that it has the same number of VSEPR pairs as the structure in Example 10.2.

Step 2: Each of the three bonds contributes one VSEPR pair, and there is one lone pair around the sulfur atom; hence, there are four VSEPR pairs.

Step 3: Four VSEPR pairs are tetrahedrally arranged around the sulfur atom.

Step 4: The three bonds point toward the corners of a tetrahedron whose center is the sulfur atom. The fourth corner of the tetrahedron is missing, leaving a pyramid with sulfur at the apex:

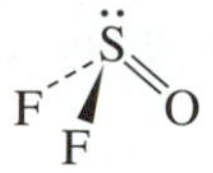

S̈
F F O

Step 5: The bond angles in an ideal tetrahedral arrangement would be 109.5°. In this case, however, lone-pair repulsion should compress the bond angles to less than this value. We can also expect the two F—S═O angles with their double bonds to be larger than the F—S—F angle.

(*Note:* These predictions agree very well with the known structure of SOF_2. The molecule is pyramidal with sulfur at the apex. The F—S═O angle is 106.8° and the F—S—F angle is 92.8°.)

PRACTICE EXERCISE 10.9

Predict the shape and bond angle in the nitrite ion (NO_2^-).

Predicting Molecular Polarity

The net dipole moment of a molecule with more than one bond depends on the individual bond dipole moments and on the geometry of the molecule. If the geometry of a molecule and the polarities of its bonds are known, then the polarity of the molecule can be predicted, as illustrated in the next example.

EXAMPLE 10.3

Both sulfur dioxide (SO_2) and boron trichloride (BCl_3) are flat molecules. Sulfur dioxide is bent or V-shaped and BCl_3 is trigonal planar with its bonds pointing to the corners of an equilateral triangle. (Be sure to verify these shapes.) Use this information and electronegativity differences to predict whether these molecules are polar.

SOLUTION

(a) SO_2 is a bent molecule. Since oxygen is more electronegative than sulfur, the molecule will be polar with the oxygen ends negative.

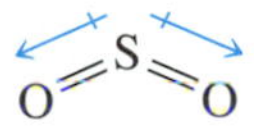

bond dipole movements

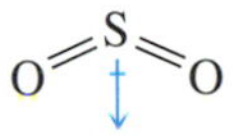

molecular dipole moment

(b) BCl_3 is a trigonal planar molecule. Chlorine is more electronegative than boron, so the bond moments are not zero. The three B—Cl bonds are symmetrically arranged around the boron atom, and the resultant of the two B—Cl bond moments on the left cancels out the B—Cl bond moment on the right:

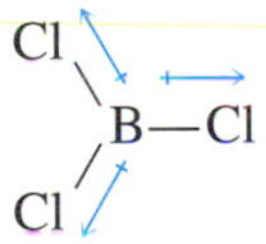

Hence, the molecule is nonpolar.

PRACTICE EXERCISE 10.10

Use crossed arrows to indicate the bond dipoles and the net dipole in the nitrogen trifluoride (NF_3) molecule.

The predictions made in the preceding example can be made for any of the molecular geometries in Figures 10.3 through 10.6. An examination of these geometries shows that when identical A—X bonds are symmetrically arranged around the central atom, as in BF_3 or CH_4, then the bond dipoles will cancel and the molecule will be nonpolar (zero dipole moment). If the A—X bonds are not symmetrically arranged around the central atom, as in SO_2 and NH_3, then the molecule will usually be polar (nonzero dipole moment). In general, *symmetrical AX_n molecules are nonpolar; nonsymmetrical AX_n molecules are usually polar.*

PRACTICE EXERCISE 10.11

Consult Figures 10.3 through 10.6 and predict whether the following molecules are polar or nonpolar: (a) BrF_5, (b) SF_6, (c) PCl_5, (d) ClF_3, and (e) XeF_4.

10.3 THE SHAPES OF SOME HYDROCARBON MOLECULES

The purpose of this section is to introduce the structural features of some organic compounds. A more detailed introduction to organic chemistry is given in Chapter 23.

Nowhere is bonding and molecular geometry more important than in the study of compounds containing carbon, which constitute over 90% of all compounds that are known today. The number and variety of carbon compounds are due to a fortunate combination of factors. The first one is the stable carbon–carbon bond, which enables carbon atoms to form long chains containing hundreds of atoms. Then there are carbon's four valence electrons, which allow it to form branched chains and to attach other elements onto the chain. A third factor is the small size of the carbon atom, which facilitates the formation of multiple bonds and makes possible such compounds as ethylene ($H_2C{=}CH_2$), acetylene ($HC{\equiv}CH$), and benzene (C_6H_6).

Alkanes

A **hydrocarbon** is a compound containing only carbon and hydrogen atoms. An **alkane** is a hydrocarbon whose molecules contain only single bonds. The three simplest alkanes are *methane* (CH_4), *ethane* (C_2H_6), and *propane* (C_3H_8) (Figure 10.7):

$$\begin{array}{ccccc} & & H & & \\ & & | & & \\ H & - & C & - & H \\ & & | & & \\ & & H & & \end{array} \qquad \begin{array}{ccccccc} & & H & & H & & \\ & & | & & | & & \\ H & - & C & - & C & - & H \\ & & | & & | & & \\ & & H & & H & & \end{array} \qquad \begin{array}{ccccccccc} & & H & & H & & H & & \\ & & | & & | & & | & & \\ H & - & C & - & C & - & C & - & H \\ & & | & & | & & | & & \\ & & H & & H & & H & & \end{array}$$

methane (CH_4) ethane (CH_3CH_3) propane ($CH_3CH_2CH_3$)

The bond angles in methane (an AX_4 type of molecule) have the tetrahedral value of 109.5°. The ethane molecule has two central carbon atoms with tetrahedrally oriented bonds. Each half of the molecule can rotate independently around the C—C

Figure 10.7

Models of hydrocarbon molecules (carbon, black; hydrogen, white). (a) Methane (CH_4). (b) Ethane (C_2H_6); each half of the molecule can rotate independently about the single C—C bond. (c) Propane (C_3H_8) has a V-shaped arrangement of carbon atoms. Observe that all of the bonds in these molecules are tetrahedrally oriented.

(a)

(b)

(c)

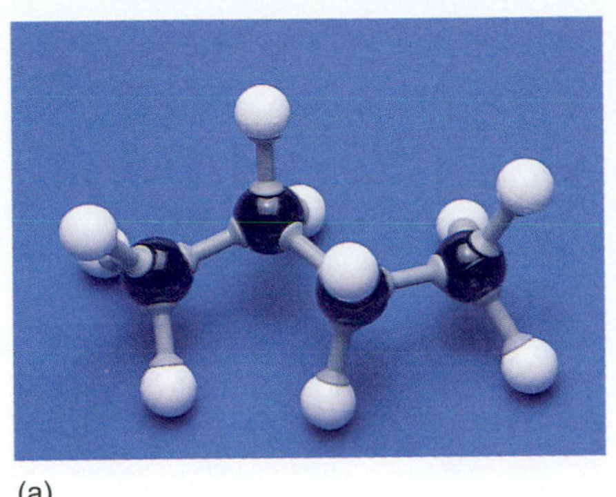
(a)

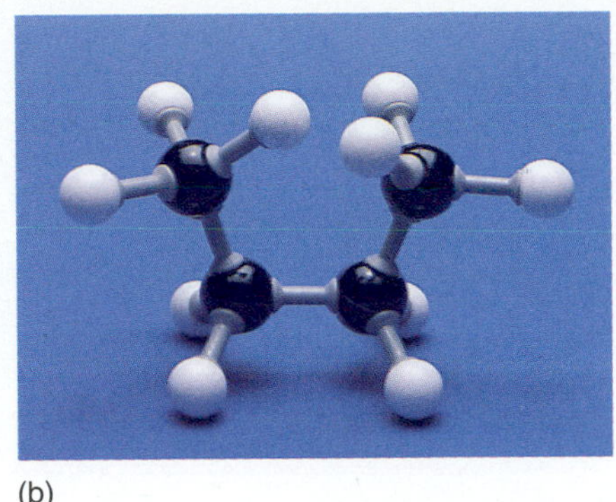
(b)

(c)

Figure 10.8
Structural isomers of butane (C_4H_{10}). (a) and (b) Two conformations of the *n*-butane isomer. (c) Isobutane.

bond at the center. The propane molecule has a chain of three carbon atoms and a V-shaped conformation as shown in Figure 10.7c. (**Conformations** are the shapes a molecule can achieve by twisting, bending, or rotating its parts, but without breaking bonds.) Chains with more than three carbon atoms adopt puckered conformations that change continuously as different sections rotate around the single bonds (Figure 10.8a and b).

Structural Isomers

The formula C_4H_{10} could represent either the continuous-chain alkane called "normal" *butane* (*n*-butane), or the branched molecule that is informally called *isobutane*. These molecules have different shapes (compare Figures 10.8a and b with Figure 10.8c) and different physical properties:

$$CH_3-CH_2-CH_2-CH_3 \qquad \begin{array}{c} CH_3 \\ | \\ CH_3-CH-CH_3 \end{array}$$

n-butane isobutane

The two butanes are **isomers**, compounds with the same composition but a different arrangement of atoms. More specifically, *n*-butane and isobutane are **structural isomers**, that is, isomers in which the sequence of atoms or bonds is different. In *n*-butane, the four carbon atoms form a continuous chain; in isobutane, the chain is branched.

There are three isomeric five-carbon alkanes, C_5H_{12} (Figure 10.9):

$$CH_3-CH_2-CH_2-CH_2-CH_3 \qquad \begin{array}{c} CH_3 \\ | \\ CH_3-CH-CH_2-CH_3 \end{array} \qquad \begin{array}{c} CH_3 \\ | \\ CH_3-C-CH_3 \\ | \\ CH_3 \end{array}$$

n-pentane isopentane neopentane

One isomer can be converted into another only by breaking bonds and forming new ones. Rotating part of a molecule around a single bond or viewing a molecule from a different angle does not make a new isomer. For example, isopentane could be drawn in many ways, such as

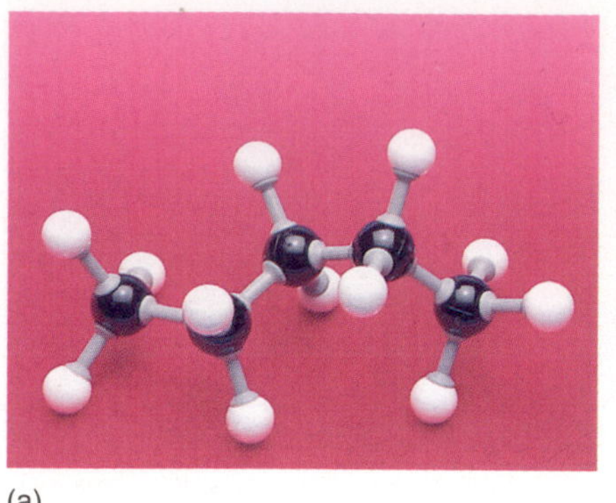

(a)

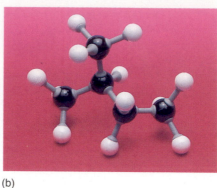

(b)

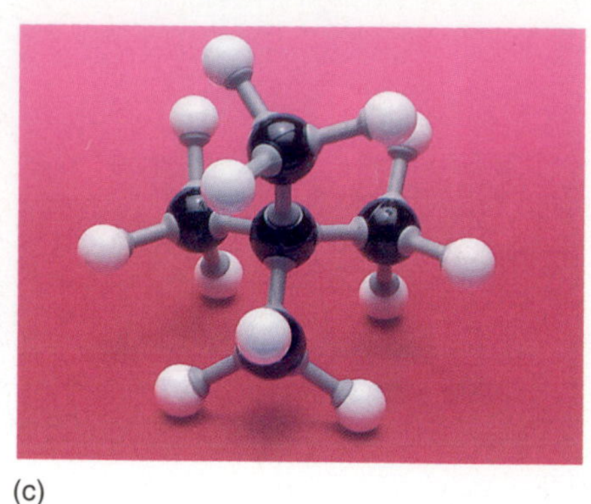

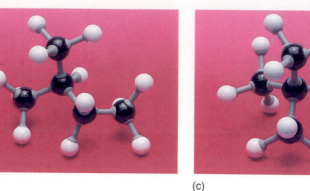

(c)

Figure 10.9

The three isomers of pentane (C_5H_{12}). (a) *n*-Pentane. (b) Isopentane. (c) Neopentane.

$$\begin{array}{ccccccc} & & CH_3 & & & & \\ & & | & & & & \\ CH_3 & — & CH & — & CH_2 & — & CH_3 \end{array} \qquad \begin{array}{ccccccc} CH_3 & — & CH & — & CH_2 & — & CH_3 \\ & & | & & & & \\ & & CH_3 & & & & \end{array}$$

$$\begin{array}{ccccc} & & CH_3 & & \\ & & | & & \\ CH_3 & — & CH & — & CH_2 \\ & & & & | \\ & & & & CH_3 \end{array} \qquad \begin{array}{ccc} CH_2 & — & CH_3 \\ | & & \\ CH & — & CH_3 \\ | & & \\ CH_3 & & \end{array}$$

All of these structures have four carbons in a continuous chain, and all represent the same isomer.

PRACTICE EXERCISE 10.12

Which of the following C_6H_{14} structures represent different isomers? Which, if any, are the same?

$$\text{(a)}\ \begin{array}{ccccccccc} CH_3 & — & CH & — & CH_2 & — & CH_2 & — & CH_3 \\ & & | & & & & & & \\ & & CH_3 & & & & & & \end{array} \qquad \text{(b)}\ \begin{array}{ccccccc} & & CH_3 & & & & CH_3 \\ & & | & & & & | \\ CH_3 & — & CH & — & CH_2 & — & CH_2 \end{array}$$

$$\text{(c)}\ \begin{array}{ccccccc} & & & & CH_3 & & \\ & & & & | & & \\ CH_3 & — & CH_2 & — & CH & — & CH_2 \\ & & & & & & | \\ & & & & & & CH_3 \end{array}$$

Ethene is produced in small quantities by citrus and tomato plants, where it serves as a ripening agent. Larger quantities of ethene are used commercially to ripen citrus fruit after it has been picked. It is also an important starting material for the manufacture of alcohol and plastics.

Alkenes and Alkynes

An **alkene** is a hydrocarbon whose molecules contain a carbon–carbon double bond. The simplest alkene is *ethene* ($CH_2{=}CH_2$), often called *ethylene,* with only two carbon atoms (Figure 10.10a).

(a)

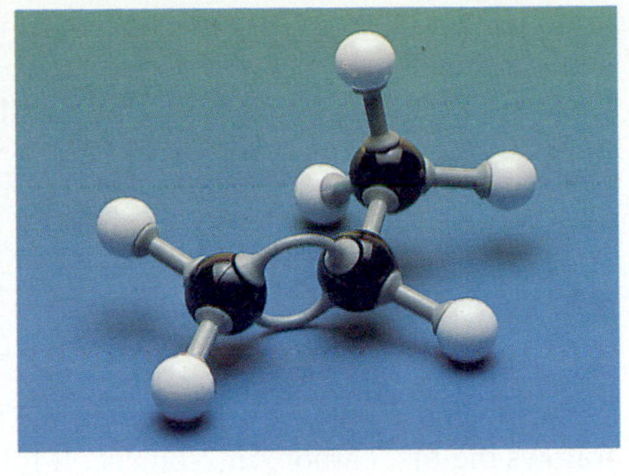

(b)

Figure 10.10

(a) Ethene (C_2H_4), a rigid planar molecule. (b) Propene ($CH_2{=}CH{-}CH_3$). The double bond causes most of the molecule to be planar and rigid; only the CH_3 group rotates freely.

```
H       H          H       H
 \     /            \     /
  C=C                C=C
 /     \            /     \
H       H          H       CH3
ethene (ethylene)  propene (propylene)
```

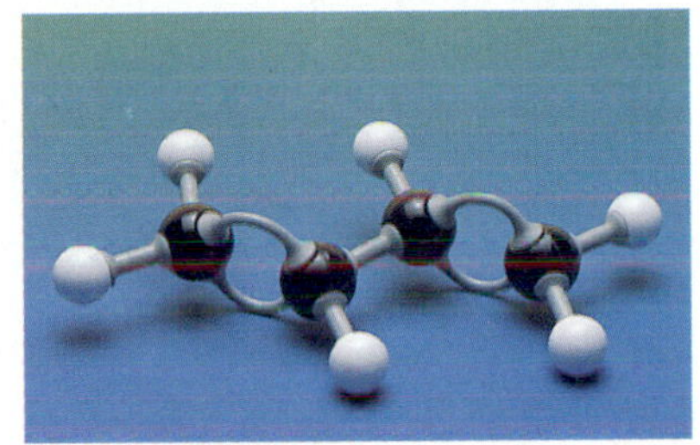

Figure 10.11

Butadiene ($CH_2{=}CH{-}CH{=}CH_2$), a molecule with two double bonds.

Each carbon in ethene has three VSEPR pairs and a trigonal planar arrangement of bonds. All angles in ethene are approximately 120°. The C═C bond is stronger than a single bond, and the two halves of the molecule cannot rotate around it. As a result, the entire ethene molecule lies in one plane.

Propene ($CH_2{=}CH{-}CH_3$), also called *propylene,* has a three-carbon chain with one carbon–carbon double bond (Figure 10.10b). The double-bonded portion of the propene molecule is flat and rigid, but the CH_3 group is tetrahedral and free to rotate around the single bond.

Butadiene ($CH_2{=}CH{-}CH{=}CH_2$), a hydrocarbon used to make synthetic rubber, contains two double bonds (Figure 10.11). *β-Carotene,* a yellow pigment found in carrots and green leaves, is a natural product with many double bonds (Figure 10.12). The double bonds in butadiene and β-carotene are *conjugated*; that is, they alternate with single bonds.

The **alkynes** consist of molecules containing a carbon–carbon triple bond. The simplest hydrocarbon with a C≡C bond is *ethyne* (CH≡CH), which is commonly called *acetylene* (Figure 10.13a). *Propyne* ($CH{\equiv}C{-}CH_3$) has three carbon atoms and one triple bond (Figure 10.13b).

$$H{-}C{\equiv}C{-}H \qquad H{-}C{\equiv}C{-}CH_3$$

acetylene (ethyne) propyne

The number of VSEPR pairs on a triple-bonded carbon atom is two, and the linear

Figure 10.12

β-Carotene. In this type of notation, a carbon chain is represented by a zigzag line in which each corner represents one carbon atom with its bonded hydrogen atoms, if any. Double bonds are shown as double lines. This carbon chain has alternating single and double bonds.

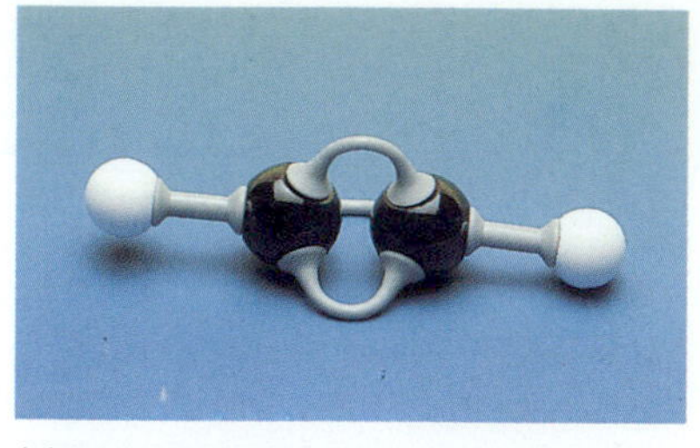

(a)

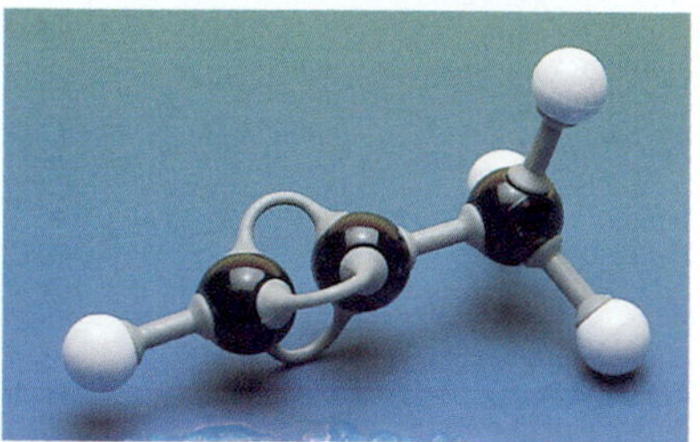

(b)

Figure 10.13

(a) Acetylene (C_2H_2), a linear molecule. (b) Propyne ($CH{\equiv}C{-}CH_3$). The three carbons lie in a straight line.

arrangement of the bonds means that in any alkyne, at least four atoms will be in a straight line.

A multiple bond increases the possibilities for isomerism. Butene (C_4H_8) has three structural isomers:

$$CH_2{=}CH{-}CH_2{-}CH_3 \qquad CH_3{-}CH{=}CH{-}CH_3 \qquad CH_2{=}\overset{\overset{\displaystyle CH_3}{|}}{C}{-}CH_3$$

1-butene, 2-butene, isobutene

1-Butene and 2-butene differ only in the position of the double bond: in 1-butene, it follows the first carbon; in 2-butene, it follows the second carbon.

cis–trans Isomers

Cis–trans isomerism arises whenever each carbon in a double bond has two different substituents. The ***cis* isomer** will have similar substituents on the same side of the double bond, and the ***trans* isomer** will have them on opposite sides. In 2-butene, each double-bonded carbon is joined to a hydrogen and a methyl group. The isomers are (see Figure 10.14):

$$\begin{array}{ccc} H_3C & & CH_3 \\ & \diagdown\ \ \diagup & \\ & C{=}C & \\ & \diagup\ \ \diagdown & \\ H & & H \end{array} \qquad \begin{array}{ccc} H_3C & & H \\ & \diagdown\ \ \diagup & \\ & C{=}C & \\ & \diagup\ \ \diagdown & \\ H & & CH_3 \end{array}$$

cis-2-butene *trans*-2-butene

The rigid double bond prevents either isomer from changing into the other.

EXAMPLE 10.4

Which of the following can have *cis–trans* isomers? Draw them.

(a) 1-butene **(b)** $CH_3{-}CH_2{-}CH{=}\underset{\underset{\displaystyle CH_2{-}CH_3}{|}}{C}{-}CH_3$

SOLUTION

(a) The formula for 1-butene is $CH_2{=}CH{-}CH_2{-}CH_3$; the first double-bonded carbon does not have two different substituents, so there are no *cis–trans* isomers.

(b) Each double-bonded carbon has two different substituents. The isomers are

$$\begin{array}{ccc} CH_3{-}CH_2 & & CH_3 \\ & \diagdown\ \ \diagup & \\ & C{=}C & \\ & \diagup\ \ \diagdown & \\ H & & CH_2{-}CH_3 \end{array} \qquad \begin{array}{ccc} H & & CH_3 \\ & \diagdown\ \ \diagup & \\ & C{=}C & \\ & \diagup\ \ \diagdown & \\ CH_3{-}CH_2 & & CH_2{-}CH_3 \end{array}$$

trans *cis*

(a)

(b)

Figure 10.14

(a) *cis*-2-Butene (both CH_3 groups are on one side of the double bond; both hydrogen atoms are on the other) and (b) *trans*-2-butene.

PRACTICE EXERCISE 10.13

Does this compound have *cis–trans* isomers?

$$\begin{array}{c} \qquad\qquad CH_2-CH_3 \\ \qquad\quad | \\ CH_3-CH=C-CH_2-CH_3 \end{array}$$

Because *cis* and *trans* isomers have different shapes, they differ in physical properties; any chemical properties that depend on shape will also differ. Rubber, for example, consists of long hydrocarbon molecules with double bonds in the *cis* configuration. Similar molecules with the *trans* configuration are not rubbery. Unsaturated fats such as those found in fish and vegetable oils contain double bonds with the *cis* configuration, and the enzymes that help the body metabolize fat are shaped to handle only *cis* isomers.

Aromatic Hydrocarbons

Aromatic hydrocarbons are those that contain benzene rings. If you look at the Lewis structures given for benzene in Section 9.5, you will see that each carbon has three VSEPR pairs and thus a planar arrangement of bonds at 120° angles. As a result, the entire benzene ring is planar and rigid. Recall also that the benzene ring is usually drawn as a hexagon with an inner circle representing the delocalized bonds. Each corner of the hexagon represents a carbon atom whose substituent is a hydrogen atom unless otherwise specified:

LEARNING HINT

Review the discussion of the benzene molecule given in Section 9.5.

benzene (C_6H_6)

Some of the compounds that contain benzene rings were first found in fragrant substances such as balsam and almond oil, so the term **aromatic** came to be used for

TABLE 10.1 Some Aromatic Compounds with Fused Rings

Name	Structural Formula	Source and Properties
Naphthalene		The most abundant constituent of coal tar; used in moth balls.
Anthracene		Found in coal tar; used in the manufacture of dyes.
Phenanthrene		Found in coal tar; a carcinogen.
3,4-Benzopyrene		Found in cigarette smoke; a carcinogen.

all members of this group. Petroleum and coal tar are the chief sources of aromatic compounds today. Benzene itself was discovered in 1825 by Michael Faraday in an oily film deposited by coal gas.

Table 10.1 shows some aromatic molecules that contain fused benzene rings. The structure of napthalene (Figure 10.15b) is typical in that each carbon atom, except for those joining the rings, is bonded to a hydrogen atom.

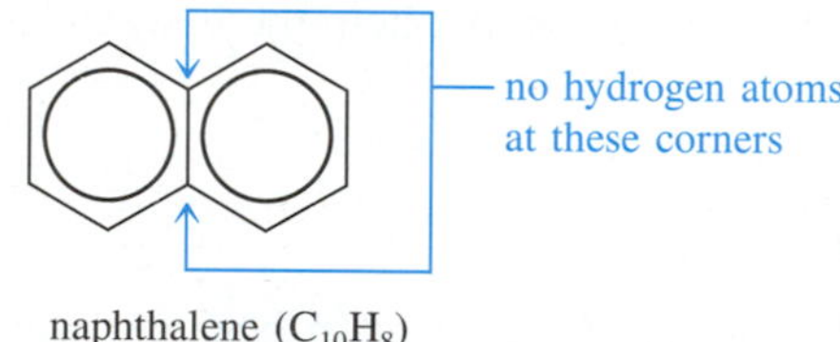

naphthalene ($C_{10}H_8$)

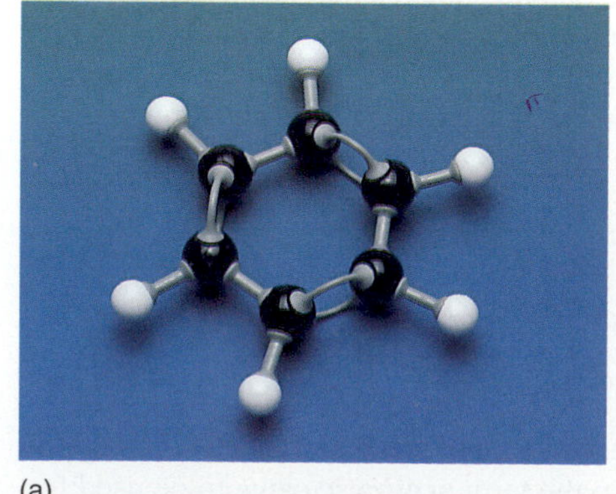

(a)

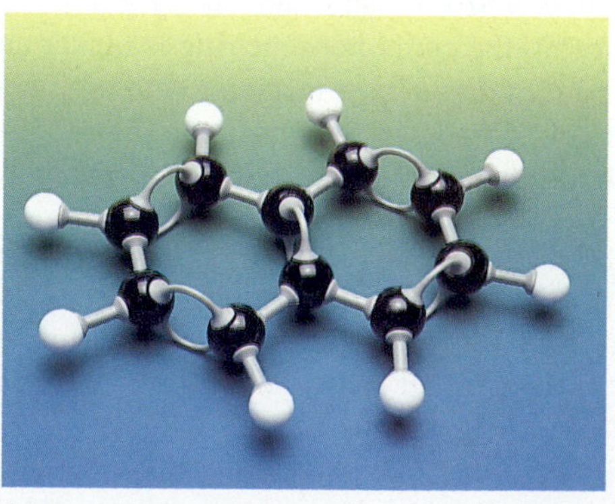

(b)

Figure 10.15
(a) Benzene (C_6H_6) and (b) naphthalene ($C_{10}H_8$). (Due to the limitations of these models, the aromatic rings are represented with localized rather than delocalized bonds.)

Other groups may replace hydrogen atoms to give benzene derivatives. *Toluene* (Figure 10.16) is a widely used solvent, and *chlorobenzene* is used in the synthesis of many useful compounds:

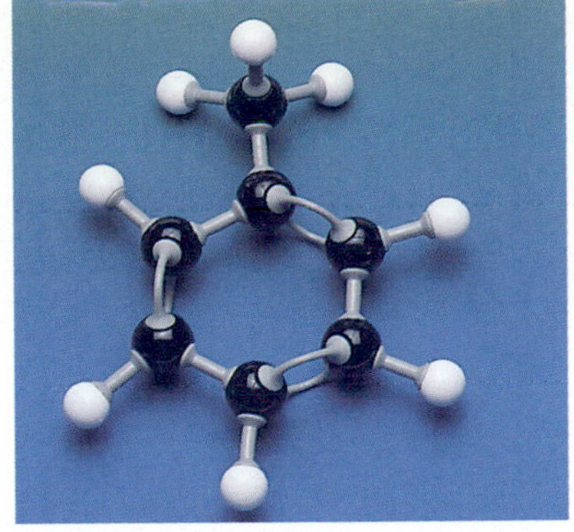

Figure 10.16
Toluene ($C_6H_5CH_3$).

10.4 VALENCE BOND THEORY

Lewis structures and VSEPR are useful for predicting molecular geometry, but they do not provide detailed information about chemical bonds. They do not tell us why covalent bonds form, nor do they describe what happens to the atomic orbitals during the process of bond formation. Furthermore, the original basis for Lewis structures, the octet rule, has many glaring exceptions; in the vapor state, for example, we find unusual species such as dilithium (Li_2), with only two valence electrons, and the hydrogen molecule ion (H_2^+), which is held together by a single electron.

Two theories, both based on quantum mechanics, are used to describe the formation of covalent bonds: the **valence bond theory**, which is discussed in this section, and the **molecular orbital theory**, which is discussed in Section 10.5.

The valence bond model of a molecule assumes that each bond is **localized**; that is, each bond connects just two atoms, as in the Lewis structure. The simplest version of valence bond theory assumes that atomic orbitals do not change during the formation of a bond. Instead, they merely **overlap** (share a common region of space) as the atoms approach each other. Figure 10.17a shows the overlap of the two 1*s* hydrogen orbitals in the hydrogen molecule. Electron density is higher in the overlap region than anywhere else, and the buildup of negative electron charge between the positive nuclei provides the electrostatic attraction that holds the atoms together. The energy of the system (Figure 10.17b) reaches a minimum when the nuclei are 74 pm apart. This distance corresponds to the equilibrium bond length and most stable state of the hydrogen molecule. If the nuclei approach more closely than 74 pm, repulsion between them will cause the energy to rise sharply, and they will move apart again.

Figure 10.18a shows the bond produced when two iodine atoms ($5s^25p^5$) approach each other to form the iodine molecule. The singly occupied 5*p* orbitals overlap, and there is a buildup of electron density between the nuclei. In the formation of the H—F bond in hydrogen fluoride (Figure 10.18b), the half-filled hydrogen 1*s* orbital overlaps the half-filled 2*p* orbital of fluorine ($2s^22p^5$).

A single bond consists of two electrons of opposite spin; hence, *the two atomic orbitals that give rise to a single bond can have no more than two electrons in total.* In the examples of Figures 10.17 and 10.18, each atom contributes one electron to the bond, but later we will see examples where both electrons come from the same atom. When two atomic orbitals overlap to build up electron density along the axis between two nuclei, the resulting localized bond is called a **sigma bond** (σ).

Sigma, σ, is the lowercase Greek "s."

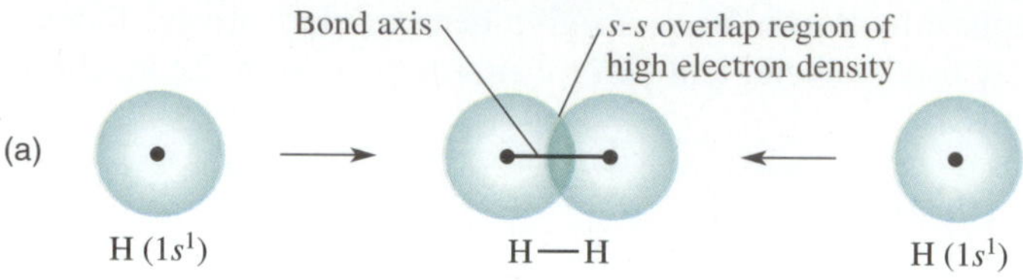

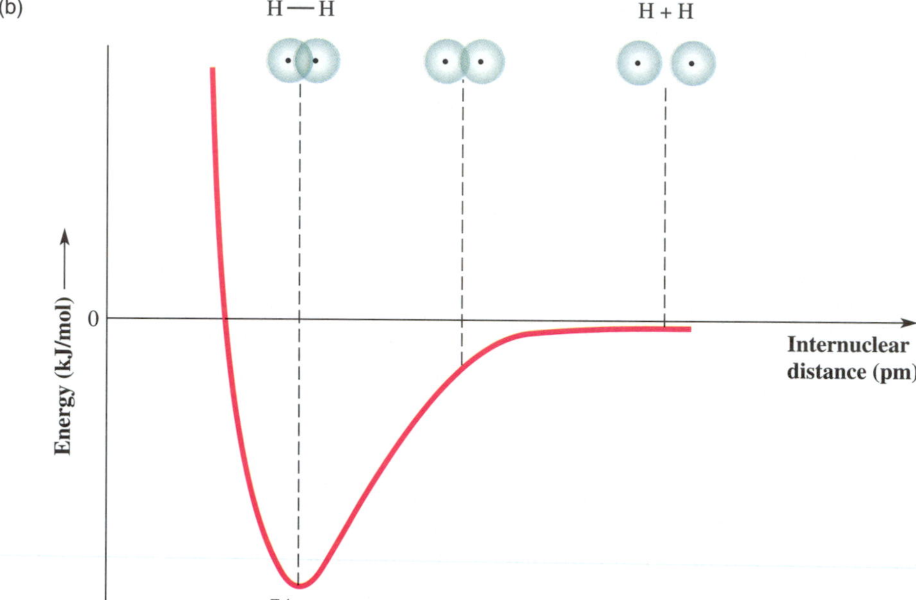

Figure 10.17

(a) The formation of sigma bonds by overlapping *s* orbitals in H_2. (b) Potential energy versus internuclear distance for two hydrogen atoms. The potential energy is zero when the atoms are completely separated. As the distance between the atoms decreases, the potential energy also decreases, reaching a minimum at 74 pm, the equilibrium bond length. At distances closer than 74 pm, repulsive forces become predominant and cause the potential energy to rise.

EXAMPLE 10.5

Identify the overlapping orbitals that form the single bond in BrCl.

SOLUTION

Chlorine ($3s^23p^5$) has one half-filled $3p$ orbital; bromine ($4s^24p^5$) has one half-filled $4p$ orbital. These orbitals will overlap to form a sigma bond.

PRACTICE EXERCISE 10.14

Sketch the sigma bonds in the following molecules: (a) Cl_2 and (b) HCl.

Orbital Hybridization and Electron Promotion

Let us use valence bond theory to explain the geometry of beryllium chloride ($BeCl_2$), a linear molecule with two identical Be—Cl bonds. The valence shell of each chlorine atom ($3s^23p^5$) contains a half-filled *p* orbital that can readily overlap

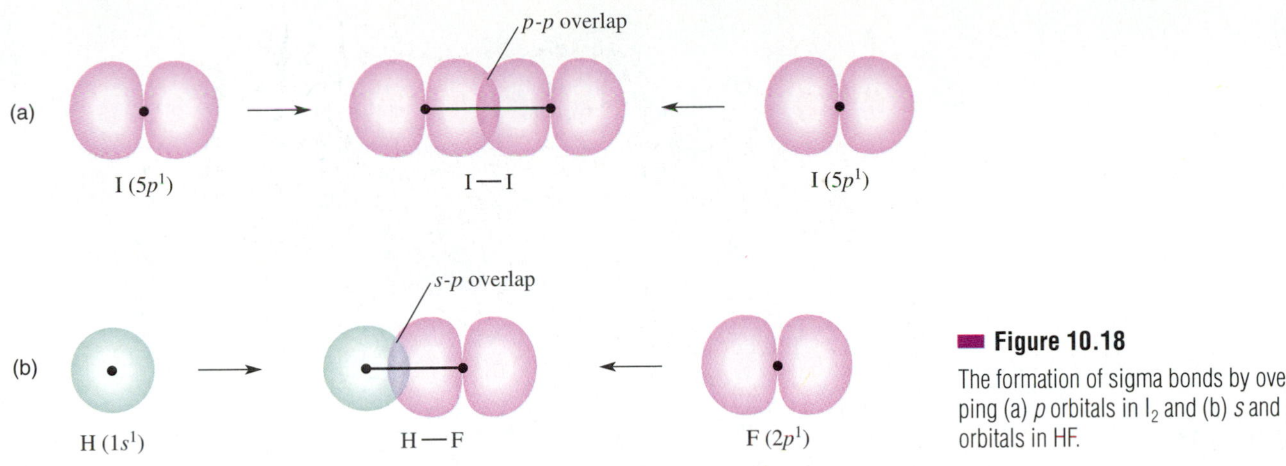

Figure 10.18
The formation of sigma bonds by overlapping (a) *p* orbitals in I_2 and (b) *s* and *p* orbitals in HF.

an orbital from beryllium. However, beryllium has the configuration $1s^22s^2$, and its full 2*s* orbital cannot form a sigma bond with a half-filled *p* orbital of chlorine. (Remember that overlapping orbitals can have a total of no more than two electrons.) The conversion of a ground-state beryllium atom into an atom ready to form two bonds separated by a 180° bond angle requires adding two more concepts to the valence bond picture: (1) orbital hybridization and (2) electron promotion.

Beryllium is more electronegative than the other metals of Group 2A, so its compounds exhibit a higher degree of covalence. Above 750°C, beryllium chloride vapor consists entirely of covalent molecules

:C̤̈l—Be—C̤̈l:

Hybridization. **Hybridization** is an imaginary mixing process in which the orbitals of an atom rearrange to form new atomic orbitals called **hybrid orbitals**. Hybridization can occur in various ways, but we are interested in the formation of hybrid orbitals whose electron densities are concentrated along the major bond axes. In $BeCl_2$, there are two identical bonds coming from the central beryllium atom. We can obtain two identical atomic orbitals and the 180° bond angle by mixing the 2*s* orbital of beryllium with one of its 2*p* orbitals to form two ***sp* hybrid orbitals** that are half *s* and half *p* in character. The designation *sp* indicates that one *s* and one *p* orbital were used for the mixture. The new hybrid atomic orbitals have energies that are identical to one another and intermediate between *s* and *p*:

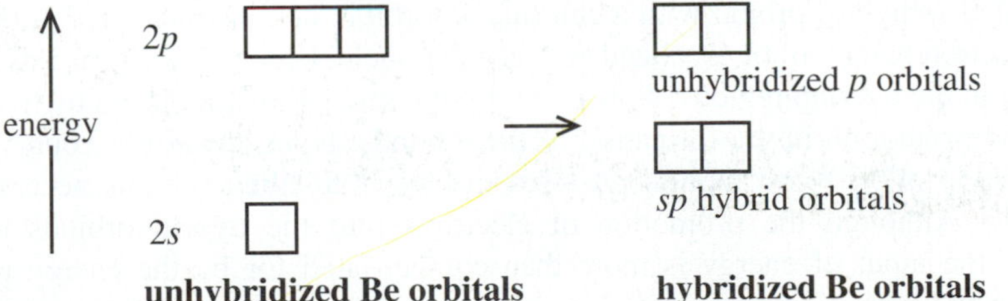

The two *sp* hybrid orbitals have the same shape (Figure 10.19); a major lobe contains most of the electron density and a small lobe contains a minor amount of electron density. Their only difference is that the major lobes point in opposite directions, 180° apart.

Promotion. Electron **promotion** is the shift of an electron from its orbital to an orbital that is slightly higher in energy. In beryllium, the promotion of electrons

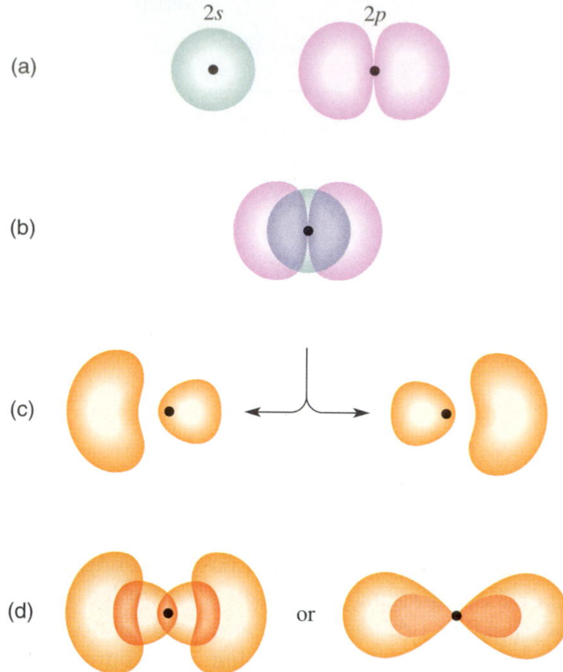

Figure 10.19

Formation of *sp* hybrid orbitals on a central atom. (a) Separated *s* and *p* orbitals. (b) The *s* and *p* orbitals on one central atom. (c) The shapes of two separated *sp* hybrid orbitals. (d) The pair of *sp* hybrid orbitals on the central atom. (*Note:* The diagram on the right in (d) is drawn with narrowed lobes for easier visualization.)

from the 2*s* ground-state orbital to the *sp* hybrid orbitals provides the two half-filled orbitals needed for two bonds:

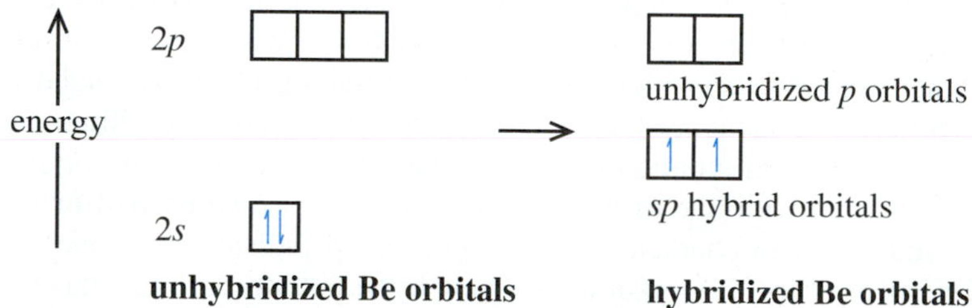

The sigma bonds in beryllium chloride can now be formed by overlapping each half-filled *sp* hybrid orbital with a chlorine 3*p* orbital, as illustrated in Figure 10.20. The electron pairs will be localized between the nuclei because electron densities are highest in the overlap regions. Note carefully the separate roles played by hybridization and promotion in the formation of these bonds: Hybridization accounts for the bond angle of 180°; promotion provides the two half-filled orbitals necessary for bonding. Although the promotion of electrons into the hybrid orbitals requires energy, the input of energy is more than compensated for by the energy released during bond formation.

PRACTICE EXERCISE 10.15

Some chemists have proposed that BeH_2 molecules can exist in the vapor state. Sketch the sigma bonds in the BeH_2 molecule.

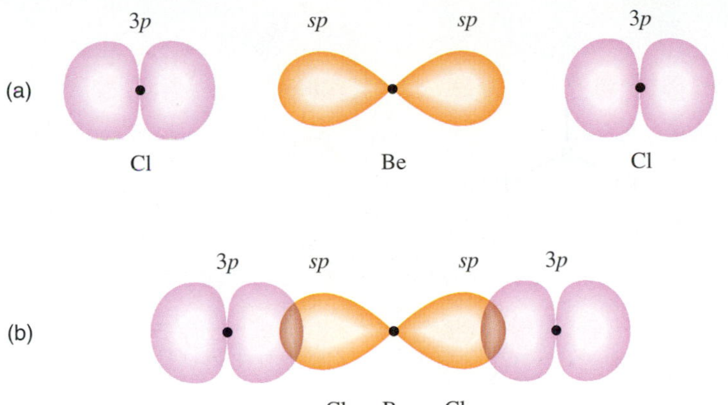

Figure 10.20

The valence bond model of the $BeCl_2$ molecule. (a) The four orbitals that will overlap when the three atoms come together. (The minor lobes of the two *sp* hybrid orbitals are omitted for easier visualization.) Each orbital contains one electron. (b) Each *sp* hybrid orbital overlaps a chlorine 3*p* orbital to form a sigma bond in which two electrons share the overlap region.

In working with the concepts of valence bond theory—orbital hybridization, electron promotion, and orbital overlap—keep in mind that these are not separate stages that electrons and orbitals actually pass through. There is no evidence, for example, to suggest that any electron ever occupied anything resembling a hybrid orbital on an independent atom. Hybridization, promotion, and overlap are treated as independent steps simply to give us a better understanding of the transition from isolated atoms to chemical bonds.

sp^2 Hybridization

The BF_3 molecule is trigonal planar with 120° angles between three identical bonds. The valence shell of boron ($2s^22p^1$) has only one half-filled orbital. To achieve trigonal planar geometry, the *s* orbital must be mixed with two *p* orbitals to give three **sp^2 hybrid orbitals** (Figure 10.21). Each of the hybrid orbitals contains one electron.

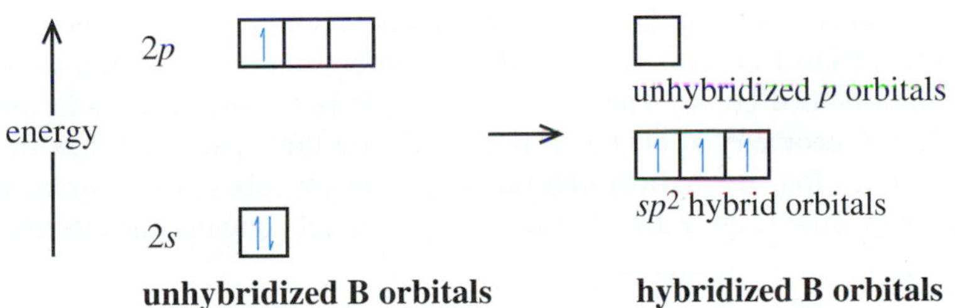

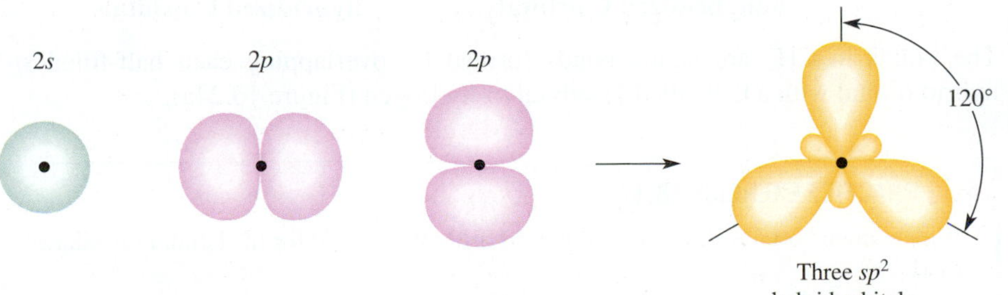

Figure 10.21

The formation of sp^2 hybrid orbitals. Left of the arrow: the *s* orbital and two *p* orbitals. Right of the arrow: the three sp^2 orbitals obtained by hybridization. The three orbital axes lie in the same plane and meet at 120° angles.

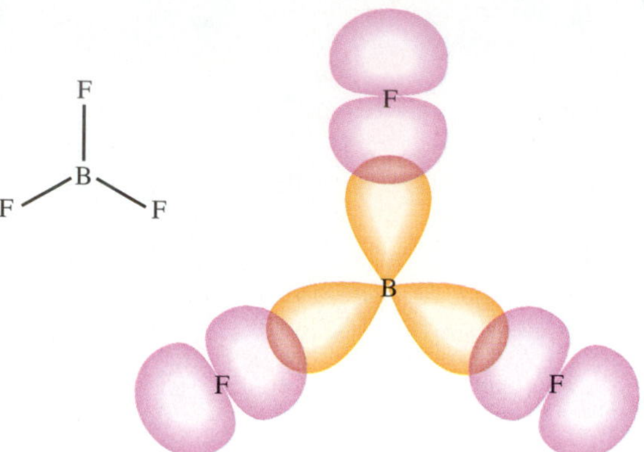

Figure 10.22

The valence bond model of the BF_3 molecule. Each of the three half-filled sp^2 hybrid orbitals (the minor lobes are omitted for easier visualization) overlaps one half-filled $2p$ orbital of a fluorine atom.

Figure 10.21 shows that the sp^2 hybrid orbitals on boron are oriented toward the corners of an equilateral triangle. Each hybrid orbital forms a sigma bond by overlapping a *p* orbital on a fluorine atom (Figure 10.22). The unhybridized $2p$ orbital of boron is perpendicular to the plane of the molecule and is not involved in the overlap. Note once again how hybridization accounts for the bond angles, while electron promotion and rearrangement provide the half-filled orbitals necessary for bonding.

PRACTICE EXERCISE 10.16

Sketch the sigma bonds in AlH_3, a covalent molecule that has been detected in mixtures of aluminum vapor and hydrogen.

sp^3 Hybridization

Carbon and silicon are the key elements in the biological and mineral worlds, respectively. In most of their compounds, the carbon and silicon atoms have four tetrahedrally oriented bonds. The simplest example is the methane molecule, CH_4. The tetrahedral geometry is obtained by hybridizing the *s* and *p* orbitals of carbon ($2s^22p^2$) to form four sp^3 **hybrid orbitals** whose major lobes point toward the corners of a tetrahedron (Figure 10.23). Each hybrid orbital contains one electron:

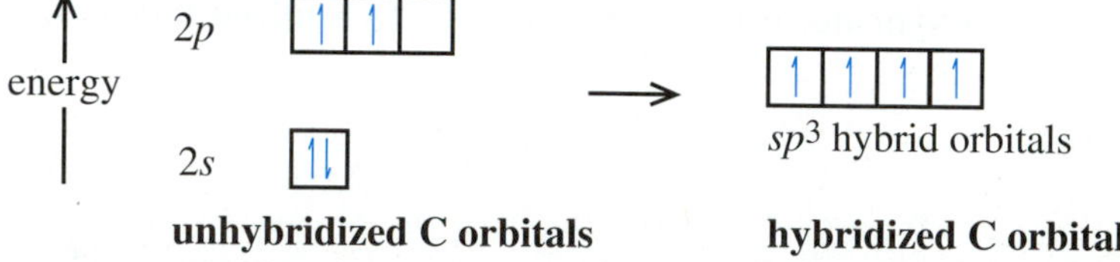

The bonds in CH_4 are sigma bonds formed by overlapping each half-filled sp^3 hybrid orbital with a half-filled $1s$ orbital of hydrogen (Figure 10.24a).

PRACTICE EXERCISE 10.17

Which atomic orbitals form the sp^3 hybrid orbitals required for the bonding in silane (SiH_4)?

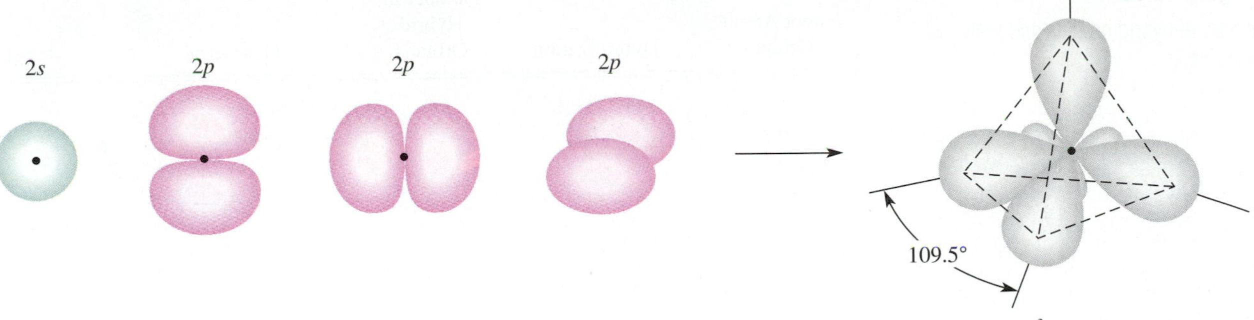

Figure 10.23

The formation of sp^3 hybrid orbitals. Left of the arrow: the *s* orbital and the three mutually perpendicular *p* orbitals. Right of the arrow: the four sp^3 hybrid orbitals. The major lobes of the hybrid orbitals point to the corners of a regular tetrahedron and make 109.5° angles with each other.

Hybridization provides orbitals for lone pairs as well as for bonding electrons. Consider the ammonia molecule, NH_3, in which four electron pairs are arranged tetrahedrally around the central nitrogen atom. The valence shell of nitrogen ($2s^22p^3$) contains five electrons, and the tetrahedral geometry is accounted for by sp^3 hybridization:

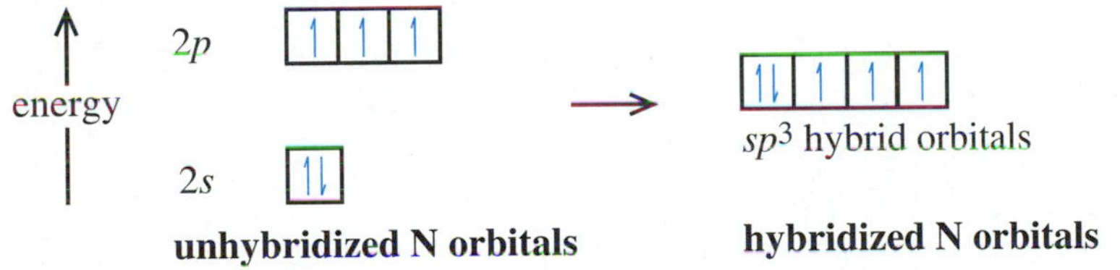

One sp^3 hybrid orbital contains the lone pair and the other three form bonds to hydrogen (Figure 10.24b). The bonding in the water molecule is similar; oxygen is sp^3 hybridized with two lone pairs occupying hybrid orbitals (Figure 10.24c).

A summary of the various hybrid orbitals and their geometries is given in Figure 10.25. The theoretical basis for the VSEPR counting procedure (Section 10.2) becomes apparent when Figures 10.2 and 10.25 are compared. The number of VSEPR electron pairs around a central atom is equal to the number of hybrid

Figure 10.24

The valence bond models of methane, ammonia, and water. In each case, there is sp^3 hybridization on the central atom. (a) In CH_4, all four hybrid orbitals form sigma bonds; the molecule is tetrahedral. (b) In NH_3, three of the hybrid orbitals form sigma bonds and one contains a lone pair; the molecule is shaped like a pyramid. (Remember that lone pairs are invisible.) (c) In H_2O, two of the hybrid orbitals form sigma bonds and two contain lone pairs; the molecule is V-shaped.

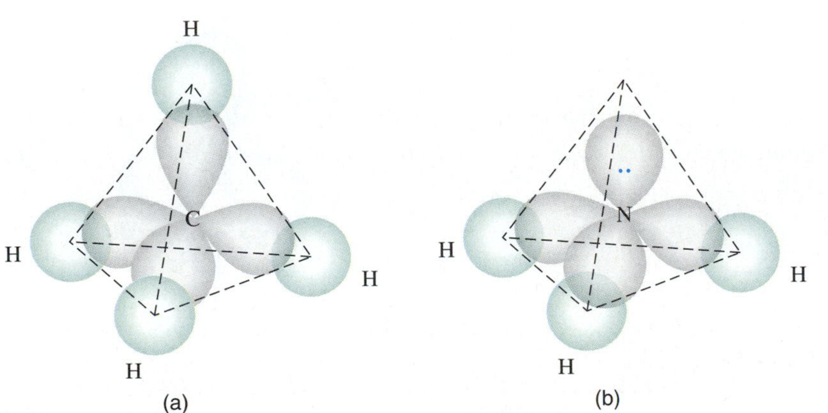

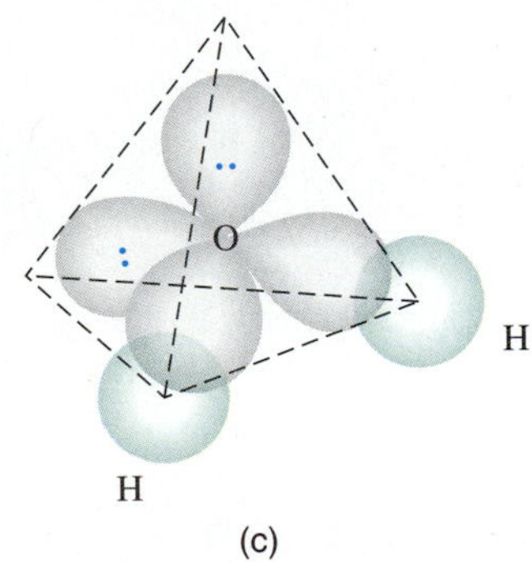

Figure 10.25
Summary of hybrid orbitals and their geometries.

Parent Atomic Orbitals	Hybridization	Number of Hybrid Orbitals	Geometry	Shape
$s + p$	sp	2	 180°	Linear
$s + p + p$	sp^2	3	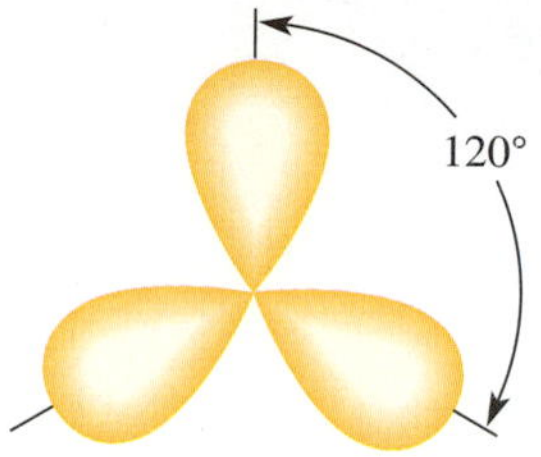 120°	Trigonal planar
$s + p + p + p$	sp^3	4	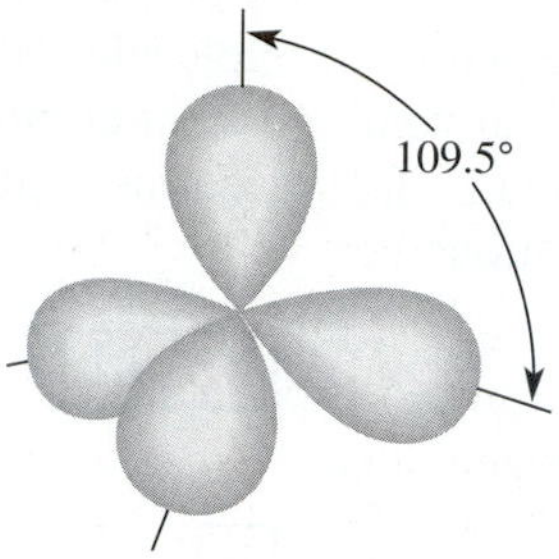 109.5°	Tetrahedral
$s + p + p + p + d$	sp^3d	5	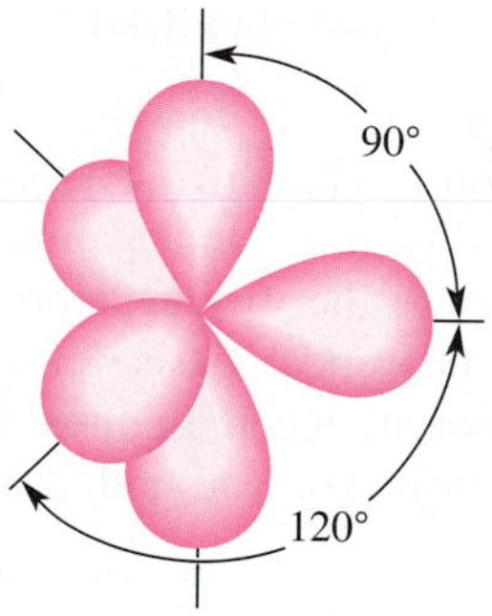 90°, 120°	Trigonal bipyramidal
$s + p + p + p + d + d$	sp^3d^2	6	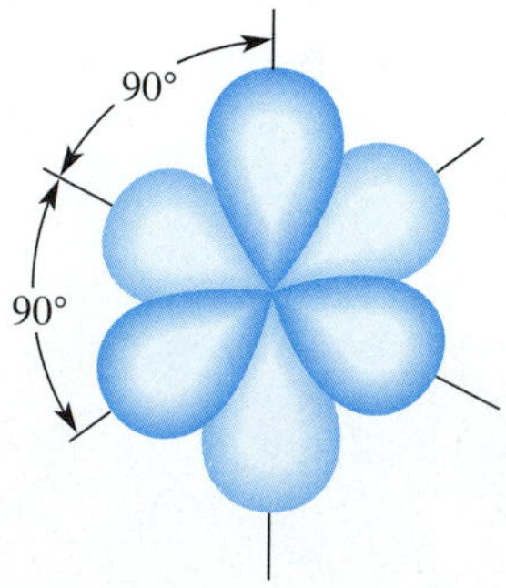 90°, 90°	Octahedral

orbitals for the atom. This observation implies that *each VSEPR pair around a central atom is either a lone pair in a hybrid orbital or a bond pair in a sigma bond.*

Hybridization Involving *d* Orbitals

When the total number of bonds and lone pairs on a particular atom exceeds four, orbitals from the *d* subshell enter the hybridization scheme. Phosphorus, for example, uses the five electrons in its valence shell ($3s^23p^3$) to form the five bonds in PCl_5. The five electrons occupy hybrid orbitals formed from the 3*s* orbital, the three 3*p* orbitals, and one of the 3*d* orbitals:

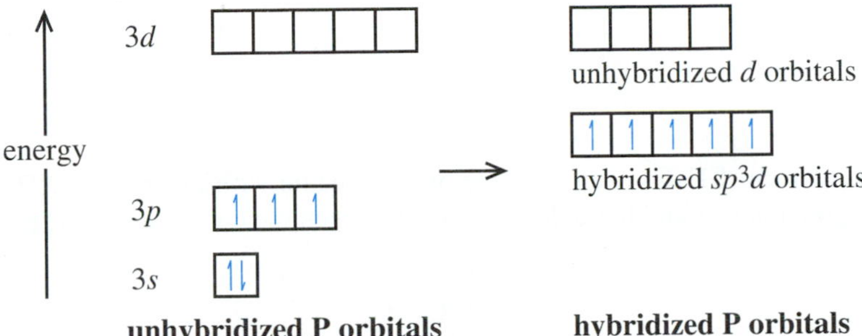

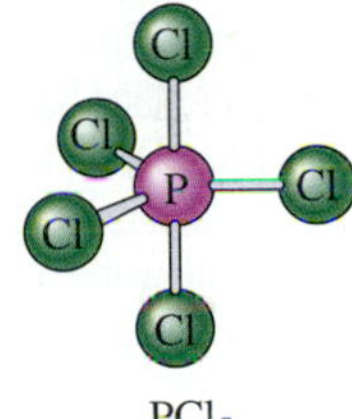

PCl_5

Figure 10.25 shows that the five sp^3d orbitals are oriented toward the corners of a trigonal bipyramid. Each bond in PCl_5 is a sigma bond formed by the overlap of a half-filled phosphorus sp^3d orbital with a half-filled *p* orbital of chlorine. The XeF_2 and ClF_3 molecules also have sp^3d hybridization around the central atom; each of these molecules has lone pairs in some of the hybrid orbitals.

Figure 10.25 also shows that hybridization of an *s* orbital, three *p* orbitals, and two *d* orbitals gives a set of sp^3d^2 hybrid orbitals oriented toward the corners of an octahedron. Sulfur ($3s^23p^4$) with six electrons in its valence shell, is sp^3d^2 hybridized in the SF_6 molecule. Other species with this type of hybridization are SiF_6^{2-}, BrF_5, and XeF_4.

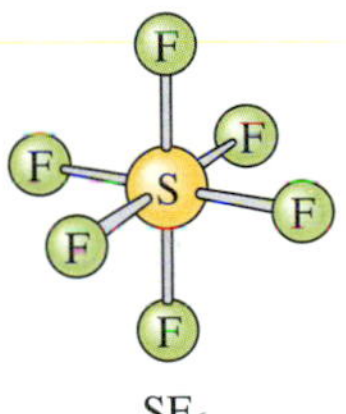

SF_6

Multiple Bonds

Thus far we have considered only sigma bonds, bonds in which the maximum electron density lies between the nuclei along a line through the atomic centers. A single bond is always a sigma bond, and molecular geometry is determined by the orientation of these bonds. A multiple bond consists of one sigma bond plus one or more **pi bonds** (π) formed by the overlap of parallel *p* orbitals (See Figure 10.26). All of the electron density in a pi bond is found in lobes above and below the line connecting the atomic centers.

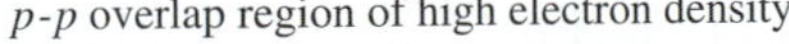

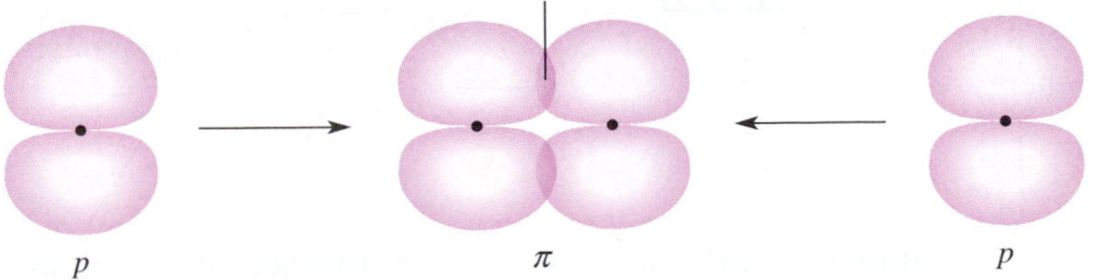

Figure 10.26
Formation of a pi bond. Two half-filled *p* orbitals perpendicular to the internuclear axis may overlap sideways to form a pi bond. A pi bond has lobes of electron density above and below the internuclear axis, and zero density along the axis.

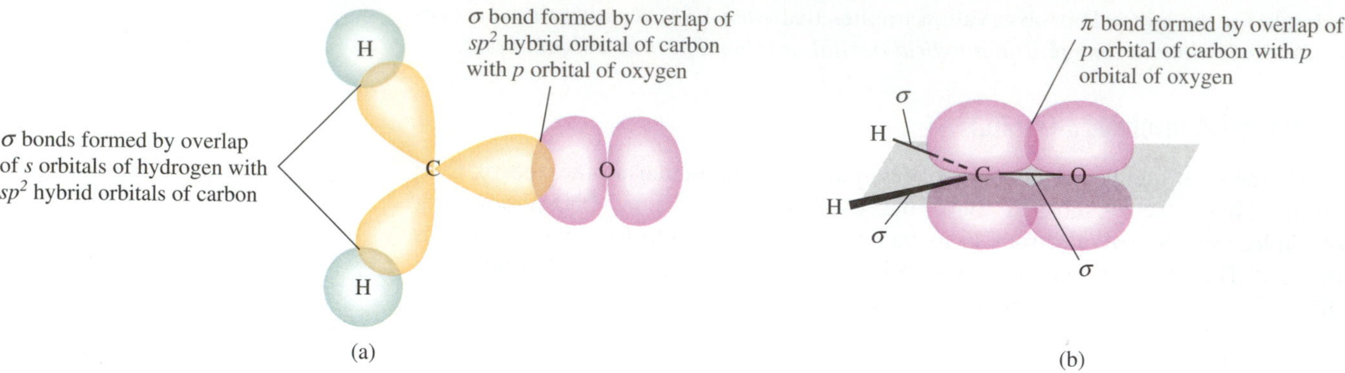

Figure 10.27

Valence bond representation of formaldehyde. (a) The three sigma bonds formed by the carbon sp^2 hybrid orbitals, the hydrogen 1*s* orbitals, and a half-filled oxygen 2*p* orbital. (b) The pi bond formed from the unhybridized carbon 2*p* orbital and the other half-filled oxygen 2*p* orbital. In this image, the molecular plane has been rotated so that it is perpendicular to the plane of the paper; the pi bond has lobes of electron density above and below the molecular plane

Formaldehyde (CH_2O) is an example of a molecule with one pi bond. The molecule is trigonal planar and has a double bond between the carbon and oxygen atoms:

H
118° (C=Ö:
H 121°

The carbon atom uses sp^2 hybrid orbitals to form the three sigma bonds:

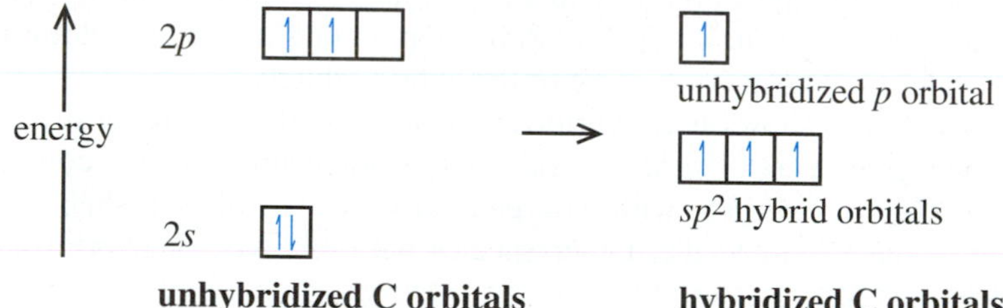

Figure 10.27 shows that two of the hybrid orbitals overlap the 1*s* orbitals of the two hydrogen atoms, and the third overlaps a half-filled 2*p* orbital of oxygen ($2s^22p^4$). The unhybridized 2*p* orbital of carbon and the other half-filled 2*p* orbital of oxygen are perpendicular to the plane of the sigma bonds. These orbitals form a pi bond with lobes above and below the molecular plane. The C=O bond in formaldehyde is typical in that *a double bond consists of one sigma bond plus one pi bond.*

The hydrogen cyanide molecule

H—C≡N:

is a linear molecule with a triple bond. The carbon atom has two *sp* hybrid orbitals and two unhybridized *p* orbitals, each containing one electron:

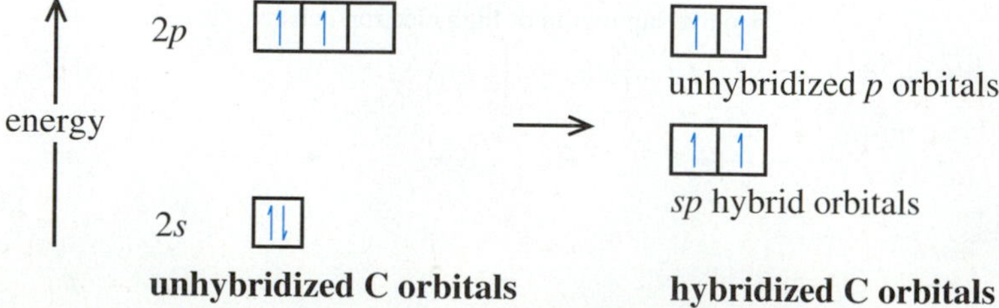

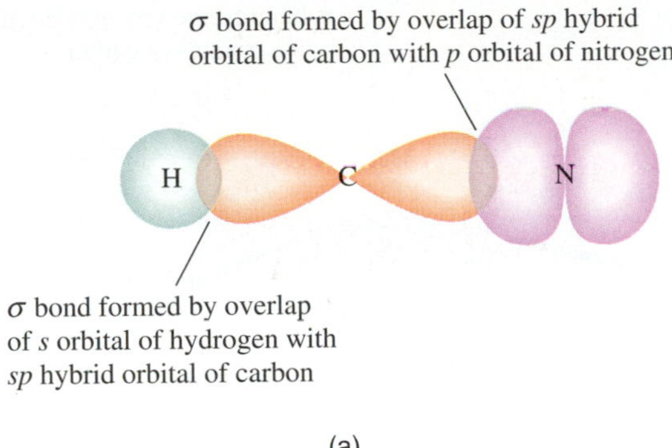

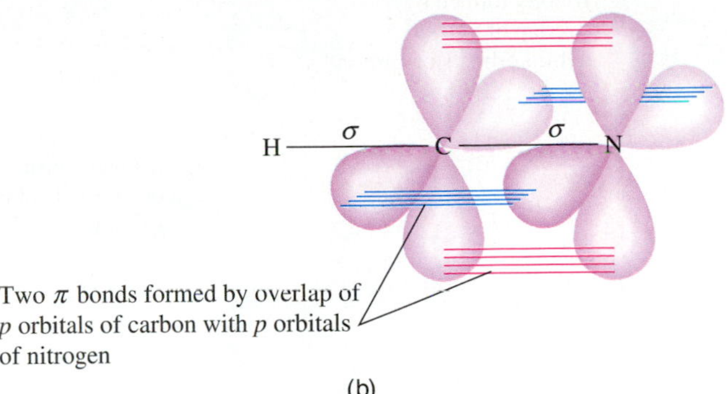

Figure 10.28

Valence bond representation of hydrogen cyanide. (a) The two sigma bonds formed by the overlap of orbitals lying along the molecular axis. (b) The half-filled *p* orbitals on carbon and nitrogen overlap to form two pi bonds; one in the plane of the paper and one perpendicular to it. The orbital lobes have been elongated for the sake of clarity.

One *sp* hybrid orbital overlaps the hydrogen 1*s* orbital, and the other overlaps one of the three half-filled *p* orbitals in nitrogen ($2s^22p^3$) to form the two sigma bonds that establish the linear geometry of the molecule (Figure 10.28). The two unhybridized *p* orbitals on the carbon atom are perpendicular to the molecular axis and to each other, as are the two remaining *p* orbitals of nitrogen. These orbitals overlap to form two pi bonds. *A triple bond consists of one sigma bond plus two pi bonds.*

Molecules with Two or More Central Atoms

A molecule of ethane (C_2H_6) has two central carbon atoms with tetrahedrally oriented bonds about each carbon (p. 422):

```
    H   H
    |   |
H — C — C — H
    |   |
    H   H
```

Thus according to valence bond theory, we would assume that each carbon forms four sp^3 hybrid orbitals. Three of these orbitals overlap the *s* orbitals of three hydrogen atoms, and the fourth orbital overlaps an sp^3 orbital from the other carbon atom. The four sigma bonds are oriented as shown in Figure 10.29, and each half of the molecule can rotate independently around the central sigma bond.

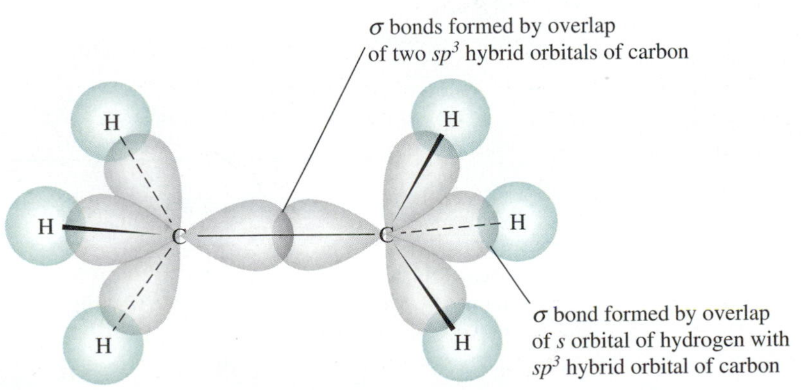

Figure 10.29

The bonding in ethane (C_2H_6). Each carbon atom has four bonds, so we assume sp^3 hybridization. All bonds in ethane are sigma bonds.

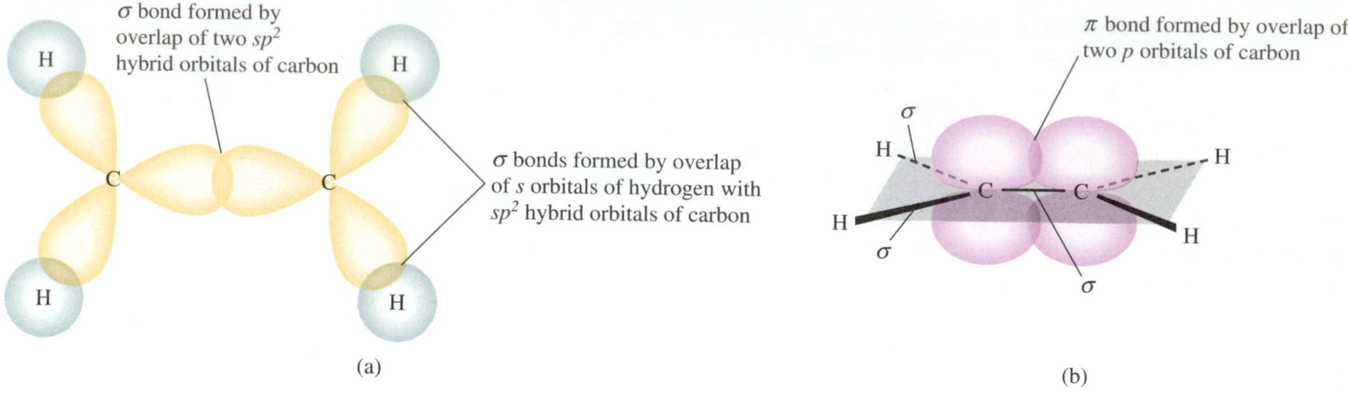

Figure 10.30

The bonding in ethene. (a) The sigma bonds formed by the overlap of sp^2 hybrid orbitals with hydrogen orbitals and with each other. (b) The unhybridized p orbitals on the carbon atoms overlap to form a pi bond with lobes of electron density above and below the molecular plane. (Here the molecular plane is perpendicular to the plane of the paper.)

In ethene (C_2H_4), there is a trigonal planar arrangement of bonds about each carbon (p. 425):

```
H       H
 \     /
  C=C
 /     \
H       H
```

To achieve this arrangement, we assume that each carbon is sp^2 hybridized. The carbon atoms are joined by a double bond that consists of a sigma bond formed from overlapping sp^2 hybrid orbitals and a pi bond formed from the parallel overlap of unhybridized p orbitals perpendicular to the molecular plane (Figure 10.30). The two halves of the molecule cannot rotate independently without breaking the pi bond; thus the entire molecule is rigid and planar.

In acetylene (C_2H_2), each carbon is sp hybridized with a linear configuration:

$$H—C≡C—H$$

The two halves of the molecule are aligned by the overlapping sp orbitals so that all four atoms lie in a straight line (Figure 10.31). Each carbon atom has two unhy-

Figure 10.31

The bonding in acetylene. (a) The sigma bonds formed by the overlap of sp hybrid orbitals with hydrogen orbitals and with each other. (b) The unhybridized p orbitals on the carbon atoms overlap to form two pi bonds, one in the plane of the paper and one perpendicular to it. The orbital lobes have been elongated for the sake of clarity.

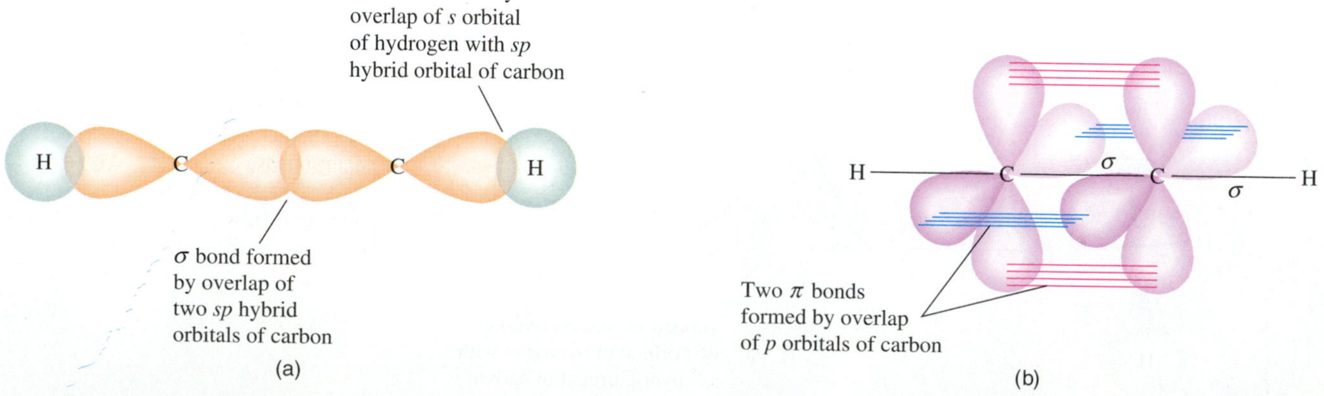

bridized p orbitals perpendicular to the axis of the molecule and to each other. The two sets of half-filled p orbitals overlap to form two pi bonds.

As these examples show, we can predict the geometry of a molecule with two or more central atoms by first considering each central atom separately. The number of VSEPR pairs around each of these atoms allows us to predict its bond angles and type of hybridization. The bonding between central atoms then allows us to predict the geometry of the entire molecule; single bonds allow parts of the molecule to rotate (as in ethane), while a double bond produces a rigid planar region (as in ethene).

EXAMPLE 10.6

(a) Which type of hybridization is exhibited by each of the central atoms in acetic acid (CH_3COOH)? (b) What are the approximate C—C═O, H—C—C, and C—O—H bond angles?

SOLUTION

(a) The Lewis structure for acetic acid is

```
      H  :O:
      |   ‖    ..
  H — C — C — O — H
      |        ..
      H
```

The left-hand carbon atom with four VSEPR electron pairs forms four sigma bonds; they will be tetrahedrally oriented and the hybridization is sp^3. The middle carbon atom with three VSEPR pairs forms three sigma bonds; they are trigonal planar and the hybridization is sp^2. The right-hand oxygen atom with four VSEPR pairs forms two sigma bonds and has two lone pairs; its hybridization is sp^3.

(b) The bond angles around each central atom are consistent with its hybridization (Figure 10.25). The central carbon in angle C—C═O is sp^2 hybridized; the angle is approximately 120°. The central carbon in angle H—C—H is sp^3 hybridized and the angle is approximately 109.5°. The central oxygen in angle C—O—H is also sp^3 hybridized and the angle is approximately 109.5°. The actual angles are somewhat different from their ideal values because of the effects of lone-pair repulsion and the presence of a double bond (Figure 10.32).

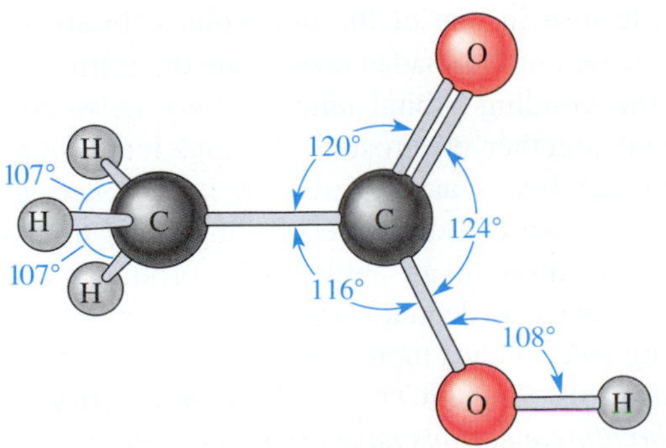

Figure 10.32
Geometry of the acetic acid molecule (see Example 10.6). The right-hand carbon is sp^2 hybridized, and the three sigma bonds around this carbon lie in a plane. The left-hand carbon is sp^3 hybridized, and its four bonds have a tetrahedral orientation.

PRACTICE EXERCISE 10.18

The structural formula for methylamine is

```
    H   H
    |   |
H—N—C—H
   ..   |
        H
```

Give approximate values for the bond angles H—N—H, N—C—H, and H—C—H.

10.5 MOLECULAR ORBITAL THEORY FOR DIATOMIC MOLECULES

In its simplest form, valence bond theory assumes that atomic orbitals do not change when atoms form a covalent bond; the atomic orbitals simply overlap, and electrons tend to stay in the overlap region where the electron density is highest. **Molecular orbital theory**, on the other hand, assumes that when atoms come together, their orbitals not only overlap but are simultaneously transformed into new orbitals with different shapes, different energies, and different distributions of electron density. These new orbitals, called **molecular orbitals**, play the same role for molecules that atomic orbitals play for atoms. Atomic orbitals are the allowed states for an electron moving in the field of one nucleus; molecular orbitals are the allowed states for an electron moving in the field of several nuclei. Electrons occupy molecular orbitals according to the same rules developed for atomic orbitals: The aufbau procedure is followed, and the Pauli exclusion principle and Hund's rule are obeyed.

Molecular Orbitals for Hydrogen and Helium

According to molecular orbital theory, *the total number of orbitals does not change when atomic orbitals combine to form molecular orbitals*; two atomic orbitals will always form two molecular orbitals. Figure 10.33 shows the molecular orbitals formed by the combination of two 1*s* atomic orbitals and the relative energies that an electron would have in each orbital. The lower energy molecular orbital is called a **bonding orbital**; the higher energy orbital is an **antibonding orbital**. Diagrams of molecular orbitals are similar to those of atomic orbitals—they show a surface of constant electron density chosen so that the electron has a high probability of being inside it. Thus an electron in one of the molecular orbitals of Figure 10.33a will spend most of its time within the shaded area of the diagram.

An electron in the bonding orbital tends to stay between the positive nuclei, thereby holding them together electrostatically and increasing the stability of the molecule. Increased stability is associated with lower energy, so the energy of a bonding electron is less than that of an electron in one of the parent atomic orbitals (Figure 10.33b). An electron in an antibonding orbital spends most of its time beyond the nuclei; it tends to decrease molecular stability by drawing the nuclei apart. An antibonding electron has more energy than an electron in one of the parent orbitals. *Bonding electrons have lower energies than electrons in the parent atomic orbitals, while antibonding electrons have higher energies.*

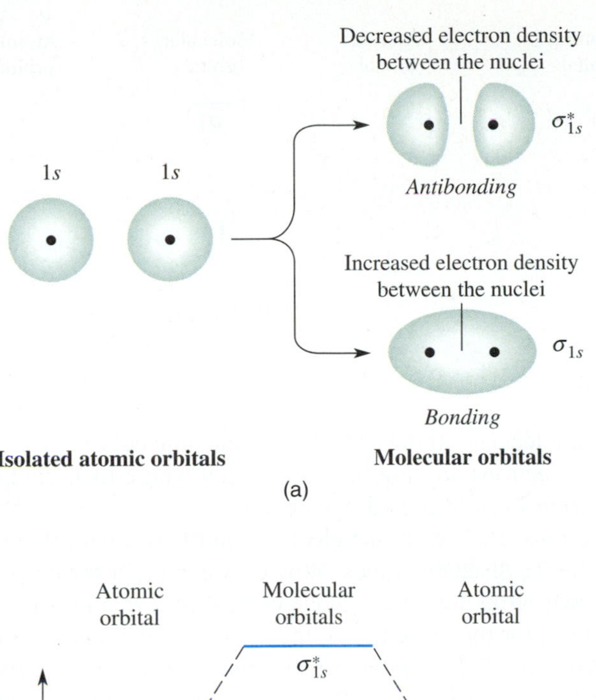

Figure 10.33

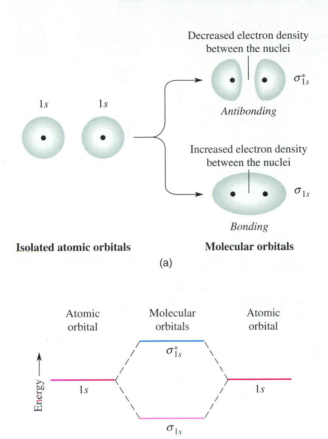

Molecular orbitals formed by the $1s$ orbitals of two identical atoms. (a) The atomic orbitals form two molecular orbitals. The σ_{1s} bonding orbital has its maximum electron density between the nuclei. The σ^*_{1s} antibonding orbital has a node of zero electron density between the nuclei. (b) Relative orbital energies. The energy of an electron in the σ_{1s} orbital is less than the energy of an electron in the $1s$ atomic orbital. The energy of an electron in the σ^*_{1s} orbital is greater than the energy of an electron in the $1s$ orbital.

The bonding orbital in Figure 10.33 is characterized by a buildup of electron density between the nuclei and along the line of centers connecting the nuclei. An orbital whose major density lies along this line of centers is called a **sigma orbital**. The sigma bonding orbital in Figure 10.33 is given the symbol σ_{1s} ("sigma one s") to indicate that it is formed from $1s$ atomic orbitals. The symbol for an antibonding orbital has a superscript star; the antibonding orbital in Figure 10.33 is written σ^*_{1s} ("sigma one s star"). A σ^* antibonding orbital has reduced electron density between the nuclei and increased density along the line of centers beyond the nuclei. A nodal plane of zero electron density bisects the line midway between atomic centers. Loosely speaking, we say that the σ_{1s} orbital forms when electron density "flows" into the space between the nuclei and that the σ^*_{1s} orbital forms when electron density "flows" out beyond the nuclei.

LEARNING HINT

A sigma orbital is not the same as a sigma bond. A sigma bond is a shared pair of electrons; a sigma orbital is a molecular orbital that may or may not be occupied by electrons.

Now let us look at the hydrogen molecule, H_2, in terms of molecular orbital theory. The molecule has two electrons, one from each hydrogen atom. These electrons will be in molecular orbitals, and the ground-state electron configuration for H_2 is determined by the same rules that were used for many-electron atoms. Because σ_{1s} is the lowest energy molecular orbital and because each molecular orbital can hold two electrons (the Pauli principle), both electrons in hydrogen are placed in the σ_{1s} bonding orbital, as shown in the orbital occupancy diagram of Figure 10.34a. The ground-state configuration is written as $(\sigma_{1s})^2$, which means that two electrons

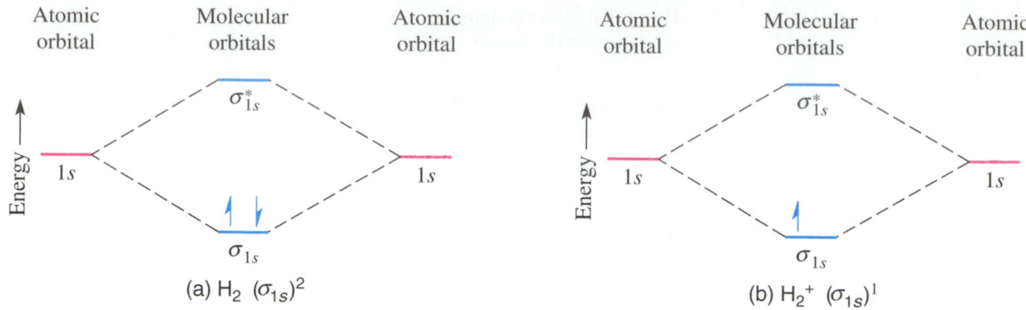

Figure 10.34

Molecular orbital occupancy diagrams for (a) H_2 and (b) H_2^+. The electrons occupy the lowest energy orbitals available. The H_2 molecule, with two bonding electrons, is more stable than the H_2^+ ion, which has only one bonding electron.

occupy the σ_{1s} molecular orbital. The hydrogen molecule is stable (i.e., it exists) because the two electrons are in a bonding orbital where their energy is lower than that of two electrons in isolated hydrogen atoms.

The diatomic ions, H_2^+, with one electron, and He_2^+, with three electrons, have been identified in gas discharge tubes. How do we explain bonds with odd numbers of electrons? Lewis structures and valence bond theory require shared electron pairs and cannot account for the existence of these ions. Molecular orbital theory, on the other hand, predicts that H_2^+ is stable because it has one electron in a bonding orbital (Figure 10.34b); its ground-state configuration is $(\sigma_{1s})^1$. The He_2^+ ion has three electrons. Two of them are in the σ_{1s} orbital, and because this orbital is full, the third electron is in the next higher energy molecular orbital, σ^*_{1s} (Figure 10.35a). The ground-state configuration is $(\sigma_{1s})^2(\sigma^*_{1s})^1$. The He_2^+ ion is stable because its formation results in a net decrease in energy; that is, the total energy of the two bonding electrons and one antibonding electron is lower than that of three electrons in 1*s* atomic orbitals. *A molecule is stable with respect to its atoms whenever the number of bonding electrons is greater than the number of antibonding electrons.*

LEARNING HINT

A molecule or polyatomic ion is stable with respect to its atoms if its energy is lower than the total energy of the independent atoms. The lower the energy, the more stable the species. The existence of a stable molecule or ion is confirmed by measuring one or more of its properties.

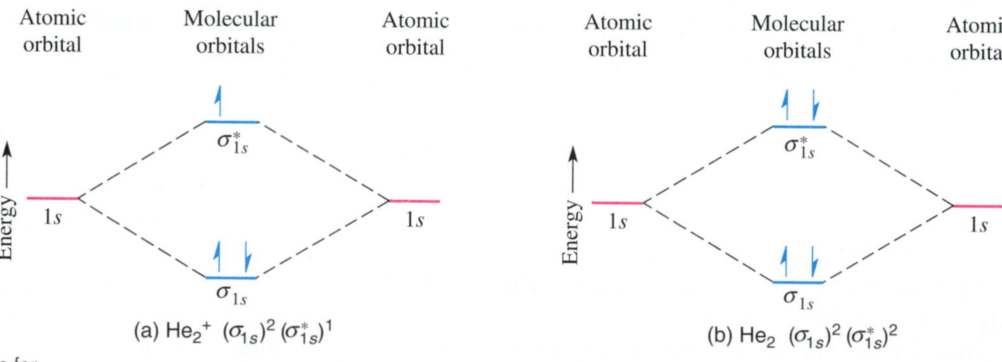

Figure 10.35

Molecular orbital occupancy diagrams for (a) He_2^+ and (b) He_2. The He_2^+ ion can exist because it has more bonding than antibonding electrons; its fleeting presence has been detected in discharge tubes. The He_2 molecule would have as many antibonding as bonding electrons; it should not exist and has never been detected.

EXAMPLE 10.7

Write ground-state molecular orbital configurations for (a) H_2^- and (b) He_2. Do you expect either of these species to be stable with respect to its atoms? Explain.

SOLUTION

(a) The H_2^- ion has three electrons. Its ground-state configuration would be $(\sigma_{1s})^2(\sigma_{1s}^*)^1$, the same as the ground-state configuration of He_2^+. There are more bonding than antibonding electrons, so the ion should be stable with respect to its atoms.

(b) The atomic number of helium is 2; therefore, He_2 would have four electrons. The ground-state configuration would be $(\sigma_{1s})^2(\sigma_{1s}^*)^2$ (Figure 10.35b). The number of bonding electrons is equal to the number of antibonding electrons, so there is no net decrease in energy. The total energy of the electrons in the molecular orbitals would not be less than the energy of the electrons in two separated atoms. The molecule would not be stable with respect to its atoms.

Bond Order

We have seen that a molecule cannot be stable unless it has more bonding electrons than antibonding electrons. One measure of stability is the **bond order**, which is defined as

$$\text{Bond order} = \frac{1}{2}\left(\begin{matrix}\text{number of}\\ \text{bonding electrons}\end{matrix} - \begin{matrix}\text{number of}\\ \text{antibonding electrons}\end{matrix}\right) \qquad (10.1)$$

The excess of bonding electrons is divided by two in order to be consistent with the traditional definition of a covalent bond as a shared electron pair.

EXAMPLE 10.8

Calculate the bond order for (a) H_2 and (b) He_2^+.

SOLUTION

(a) The ground-state configuration for H_2 is $(\sigma_{1s})^2$. It has two bonding electrons and no antibonding electrons. The bond order is $\frac{1}{2}(2 - 0) = 1$.

(b) The configuration for He_2^+ is $(\sigma_{1s})^2(\sigma_{1s}^*)^1$. It has two bonding electrons and one antibonding electron. The bond order is $\frac{1}{2}(2 - 1) = \frac{1}{2}$.

PRACTICE EXERCISE 10.19

Calculate the bond order for (a) H_2^+, (b) H_2^-, and (c) He_2.

A stable molecule has an excess of bonding electrons; hence, *molecules with bond orders of $\frac{1}{2}$ or greater will be stable with respect to their atoms; molecules with bond orders of 0 will be unstable.* Furthermore, for comparable species, a greater bond order usually indicates a stronger and shorter bond. The data in Table 10.2 show that stability, measured in terms of increasing bond dissociation energy and decreasing bond length, tends to increase with bond order. For example, H_2, with a bond order of 1, has a stronger and shorter bond than H_2^+, with a bond order of $\frac{1}{2}$.

TABLE 10.2 Ground-State Molecular Orbital Configurations and Bond Properties of First-Period Homonuclear Diatomic Ions and Molecules

Species	Molecular Orbital Configuration	Bond Order	Bond Energy (kJ/mol)	Bond Length (pm)
H_2^+	$(\sigma_{1s})^1$	$\frac{1}{2}$	260	106
H_2	$(\sigma_{1s})^2$	1	436	74
He_2^+	$(\sigma_{1s})^2(\sigma_{1s}^*)^1$	$\frac{1}{2}$	322	108
He_2	$(\sigma_{1s})^2(\sigma_{1s}^*)^2$	0	unstable	

Molecular Orbitals for Homonuclear Diatomic Molecules

Recall that the three *p* orbitals are distinguished by the subscripts *x, y,* and *z* because they are oriented at right angles to each other (Figure 7.32, p. 293).

We have examined the molecular orbitals formed from atomic orbitals in the $n = 1$ shell; now we look at molecular orbitals formed from atomic orbitals in the $n = 2$ shell. These molecular orbitals accommodate the valence electrons of homonuclear (identical nuclei) diatomic molecules and ions (such as N_2 and O_2^+) formed from second-period atoms. The $n = 2$ shell has four atomic orbitals: $2s$, $2p_x$, $2p_y$, and $2p_z$; hence, the two atoms in a homonuclear molecule have eight atomic orbitals from which eight molecular orbitals can form (Figure 10.36).

The $2s$ orbitals combine to form a σ_{2s} bonding orbital and a σ_{2s}^* antibonding orbital (Figure 10.36a) with relative energies and shapes similar to those of the $1s$ combinations in Figure 10.33. Keep in mind, however, that just as $2s$ electrons are more energetic than $1s$ electrons, electrons in σ_{2s} orbitals will be more energetic than electrons in σ_{1s} orbitals.

The $2p$ atomic orbitals can combine in two ways. If we designate the line between atomic centers as the *z* direction, then the $2p_z$ orbitals will merge along the line of centers to form the bonding σ_{2p_z} and the antibonding $\sigma_{2p_z}^*$ orbitals (Figure 10.36b). The bonding orbital is characterized by a buildup of electron density between the nuclei. The antibonding orbital has a nodal plane between the nuclei and high concentrations of electron density beyond the nuclei. An electron in the bonding orbital will spend most of its time between the nuclei and consequently will have a lower energy than an antibonding electron.

LEARNING HINTS

A preferred direction in an atom or molecule, for example, the internuclear axis in H_2, is usually designated as the *z* axis.

A pi orbital is not the same as a pi bond. A pi bond is a shared pair of electrons. A pi orbital is a molecular orbital that may or may not be occupied by electrons.

Molecular orbitals are designated by Greek letters σ, π, and δ, which are analogous to the *s, p,* and *d* of atomic orbitals.

The $2p_y$ orbitals are perpendicular to the line between atomic centers and overlap in parallel as shown in Figure 10.36c. They form bonding and antibonding molecular orbitals with lobes of electron density above and below the internuclear axis. These orbitals are called **pi orbitals** (π); the bonding orbital is π_{2p_y} and the antibonding orbital is $\pi_{2p_y}^*$. The $2p_x$ orbitals also overlap in parallel and form π_{2p_x} and $\pi_{2p_x}^*$ molecular orbitals (Figure 10.36d). These orbitals are perpendicular to the orbitals in the π_y set, but they are identical in all other respects. Electrons in either of the two bonding pi orbitals would have the same energy. Electrons in the antibonding pi orbitals would also have identical energies, but the energies would be higher than those of electrons in the bonding pi orbitals.

Electron Configurations for Homonuclear Diatomic Molecules

Lithium, the first element in the second period, has a $1s^22s^1$ configuration and forms Li_2 molecules in the vapor state. The relative energies of atomic and molecular

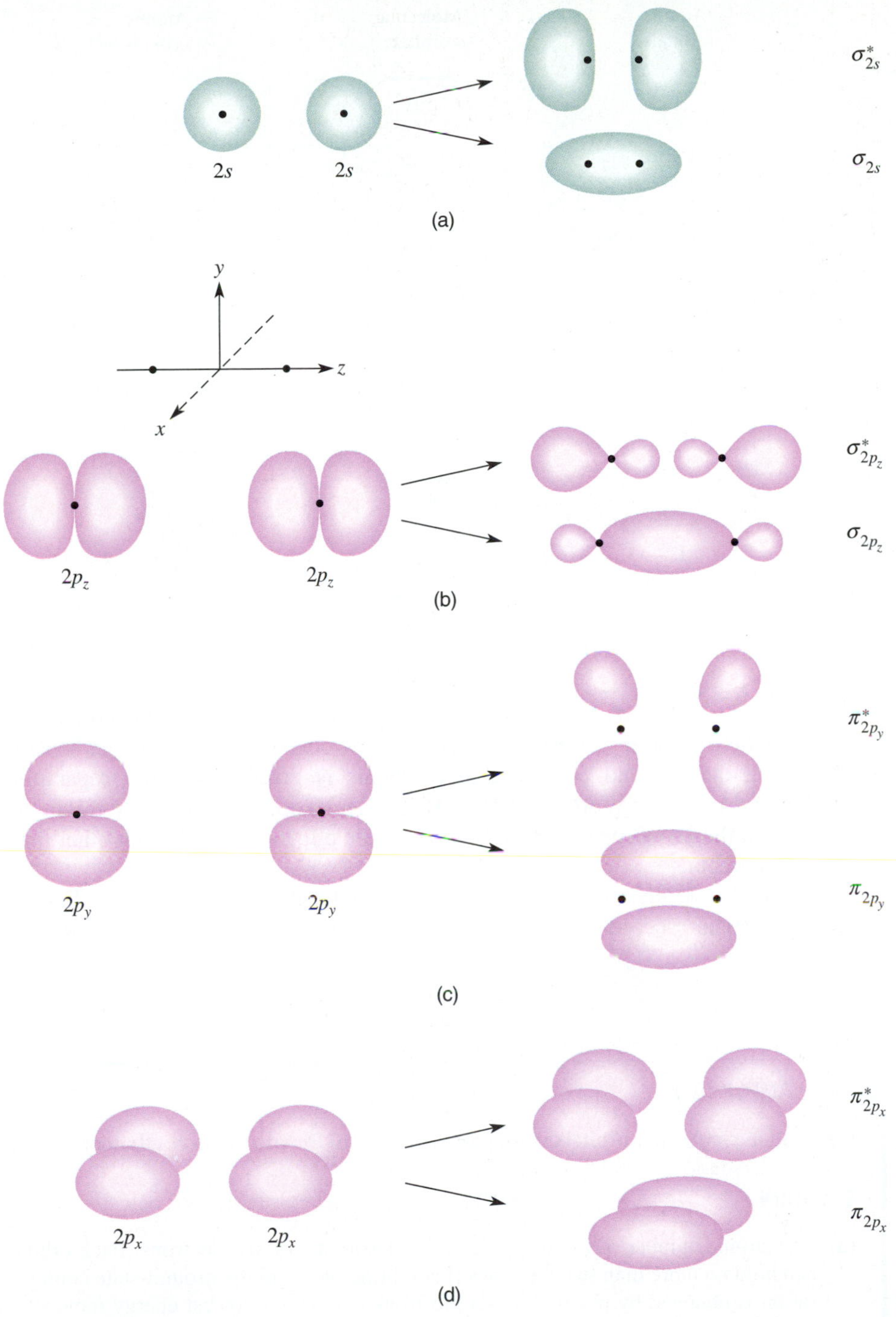

Figure 10.36

Molecular orbitals formed by the eight $n = 2$ atomic orbitals of two identical atoms. Each pair of atomic orbitals (left) overlaps and forms two molecular orbitals (right). The molecular orbitals are (a) σ_{2s} and σ^*_{2s}, (b) σ_{2p_z} and $\sigma^*_{2p_z}$, (c) π_{2p_y} and $\pi^*_{2p_y}$, and (d) π_{2p_x} and $\pi^*_{2p_x}$.

orbitals for lithium have been experimentally established and are shown in Figure 10.37. This diagram illustrates a number of features of molecular orbital theory. First, we observe that molecular orbital energies depend primarily on the energies of the original atomic orbitals—lower energy atomic orbitals give rise to lower energy

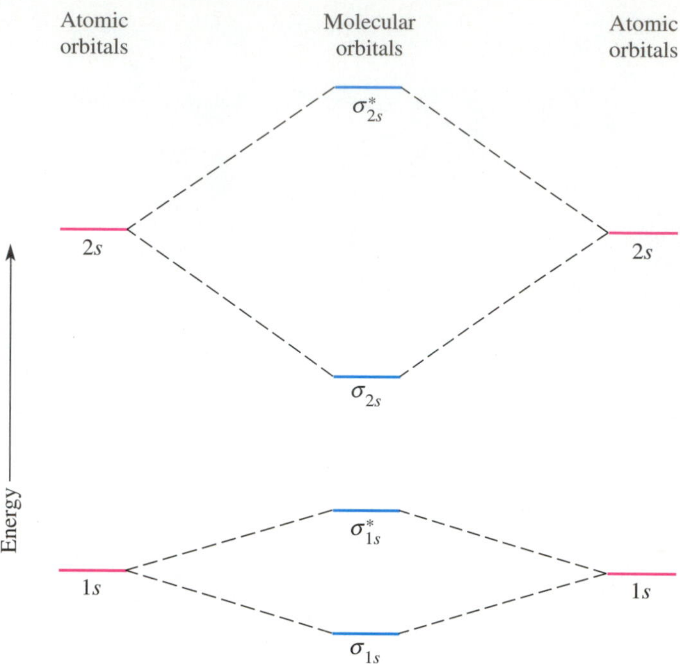

Figure 10.37

Energy-level diagram for molecular orbitals formed from 1*s* and 2*s* atomic orbitals.

molecular orbitals. The σ_{1s} molecular orbital is lowest in energy, followed by σ^*_{1s}. Then come the σ_{2s} and σ^*_{2s} molecular orbitals. Now observe that the energy separation between the σ_{2s} and σ^*_{2s} molecular orbitals is greater than the separation between the σ_{1s} and σ^*_{1s} orbitals. The greater separation indicates, as might be expected, that the 2*s* valence electrons interact more effectively with each other than do the core electrons.

The ground-state electron configuration of a molecule is obtained by filling its molecular orbitals according to the usual rules, as illustrated in the following example.

EXAMPLE 10.9

(a) Write the molecular orbital configuration of Li_2. (b) Determine its bond order.

SOLUTION

(a) The atomic number of lithium is 3, so Li_2 has a total of six electrons. Each orbital can hold no more than two electrons (Pauli principle), and the ground-state configuration is obtained by placing the six electrons in the three lowest energy molecular orbitals of Figure 10.37 as follows:

$$(\sigma_{1s})^2(\sigma^*_{1s})^2(\sigma_{2s})^2$$

(b) There are four bonding and two antibonding electrons. Substitution into Equation 10.1 gives the bond order as $\frac{1}{2}(4 - 2) = 1$.

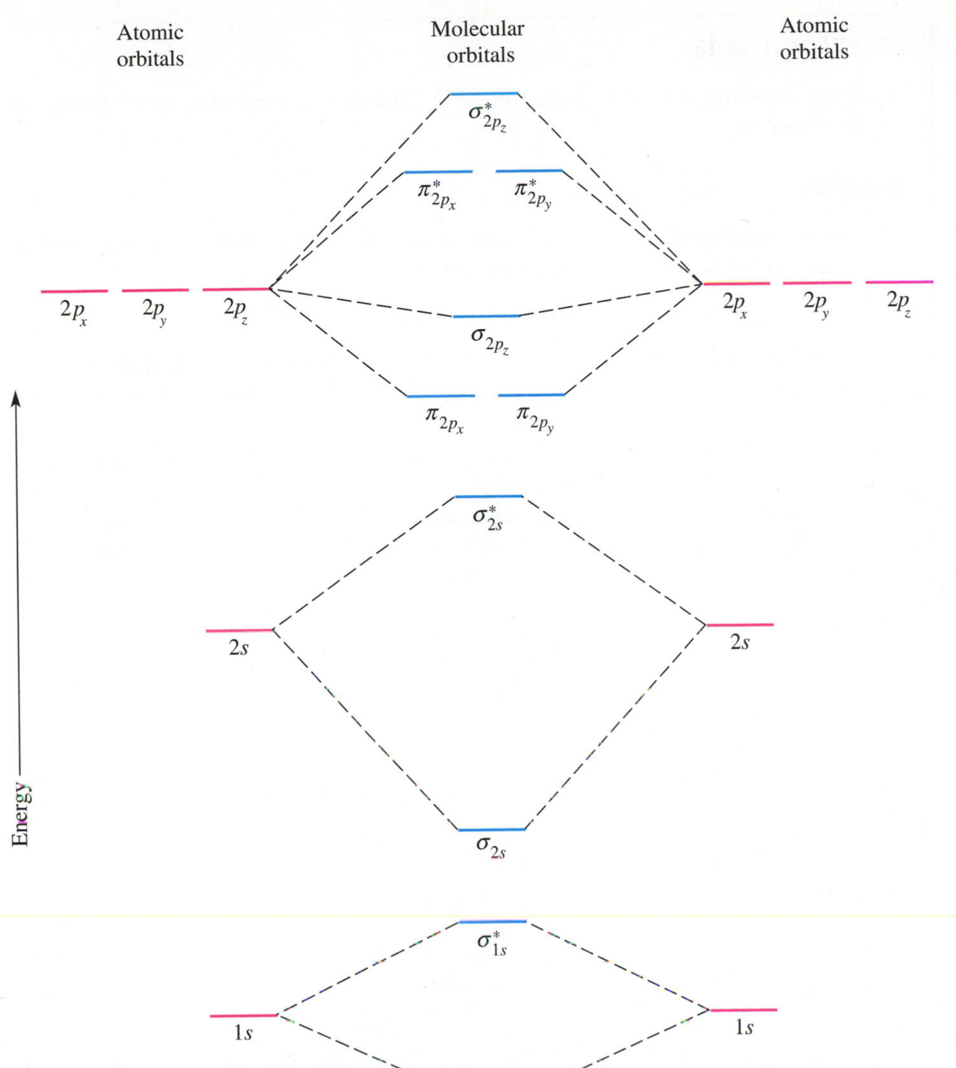

Figure 10.38
An approximate energy-level diagram for molecular orbitals formed from 1*s*, 2*s*, and 2*p* atomic orbitals. The actual order of the orbitals varies from left to right across the period, and more detailed diagrams for O_2 and F_2 would show the σ_{2p_z} orbital below the π_{2p_x} and π_{2p_y} orbitals.

PRACTICE EXERCISE 10.20

(a) Refer to the energy-level diagram in Figure 10.37, and write the ground-state configuration for Be_2. (b) Predict whether this molecule will exist.

The valence electrons of Li and Be are in the 2*s* orbital, so the 2*p* orbitals are not included in the energy-level diagram of Figure 10.37. The remaining second-period elements (B through Ne) have both 2*s* and 2*p* valence electrons. Their energy-level diagrams vary from one element to the next, but the approximate diagram given in Figure 10.38 can be used to predict the behavior of their diatomic molecules and ions.

EXAMPLE 10.10

Write the molecular orbital configuration of O_2, and draw an orbital occupancy diagram for the molecule.

SOLUTION

The atomic number of oxygen is 8, so there are 16 electrons in O_2. The ground-state configuration can be obtained from Figure 10.38:

$$(\sigma_{1s})^2(\sigma^*_{1s})^2(\sigma_{2s})^2(\sigma^*_{2s})^2(\pi_{2p_x})^2(\pi_{2p_y})^2(\sigma_{2p_z})^2(\pi^*_{2p_x})^1(\pi^*_{2p_y})^1$$

The orbital occupancy diagram for O_2 is given in Figure 10.39. Note that the π^* electrons occupy separate orbitals in accordance with Hund's rule. Thus each oxygen molecule is paramagnetic with two unpaired electrons.

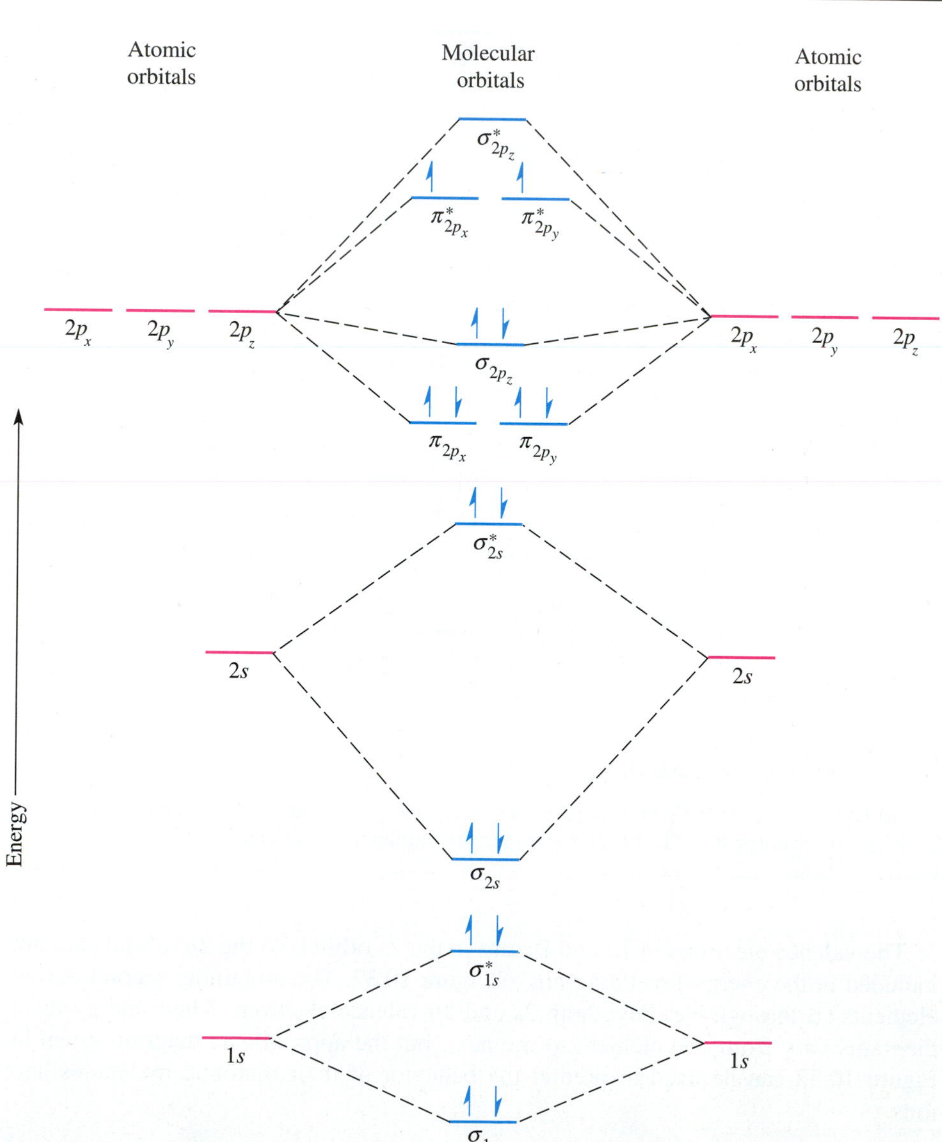

Figure 10.39 Molecular orbital occupancy diagram for O_2. The two π^* orbitals contain single electrons with parallel spins (Hund's rule). This diagram is approximate (see caption to Figure 10.38). σ_{2p_z} orbital in O_2 is actually slightly lower in energy than the π_{2p_x} and π_{2p_y} orbitals.

PRACTICE EXERCISE 10.21

(a) Write the molecular orbital configuration for N_2. (b) Will N_2 be paramagnetic?

The prediction of two unpaired electrons in the oxygen molecule was one of the early triumphs of molecular orbital theory. Oxygen was known to be paramagnetic (see Section 9.5 and Figure 9.9, p. 396), but the necessary unpaired electrons could not be incorporated into a satisfactory Lewis structure or valence bond theory. Oxygen is a *diradical,* that is, a molecule with two unpaired electrons. Most free radicals are extremely reactive and exist only fleetingly. Oxygen is a notable exception.

EXAMPLE 10.11

(a) Compare the bond orders of O_2 and N_2. (b) Which of these molecules should have the higher bond energy? The shorter bond length?

SOLUTION

(a) O_2 has 10 bonding electrons and 6 antibonding electrons (Example 10.10); the bond order is $\frac{1}{2}(10 - 6) = 2$. N_2 has 10 bonding electrons and 4 antibonding electrons (Practice Exercise 10.21); the bond order is $\frac{1}{2}(10 - 4) = 3$.

(b) N_2 has a higher bond order, so we would expect nitrogen to have a higher bond energy and a shorter bond length than oxygen. This conclusion agrees with the experimental data in Table 10.3.

TABLE 10.3 Ground-State Molecular Orbital Configurations and Bond Properties of Second-Period Homonuclear Diatomic Ions and Molecules

Species	Molecular Orbital Configuration	Bond Order	Bond Energy (kJ/mol)	Bond Length (pm)
Li_2	$(\sigma_{1s})^2(\sigma^*_{1s})^2(\sigma_{2s})^2$	1	103	267
Be_2	$(\sigma_{1s})^2(\sigma^*_{1s})^2(\sigma_{2s})^2(\sigma^*_{2s})^2$	0	unstable	
B_2	$(\sigma_{1s})^2(\sigma^*_{1s})^2(\sigma_{2s})^2(\sigma^*_{2s})^2(\pi_{2p})^2$	1	295	159
C_2	$(\sigma_{1s})^2(\sigma^*_{1s})^2(\sigma_{2s})^2(\sigma^*_{2s})^2(\pi_{2p})^4$	2	602	124
N_2	$(\sigma_{1s})^2(\sigma^*_{1s})^2(\sigma_{2s})^2(\sigma^*_{2s})^2(\pi_{2p})^4(\sigma_{2p_z})^2$	3	945	110
O_2	$(\sigma_{1s})^2(\sigma^*_{1s})^2(\sigma_{2s})^2(\sigma^*_{2s})^2(\pi_{2p})^4(\sigma_{2p_z})^2(\pi^*_{2p})^2$	2	498	121
F_2	$(\sigma_{1s})^2(\sigma^*_{1s})^2(\sigma_{2s})^2(\sigma^*_{2s})^2(\pi_{2p})^4(\sigma_{2p_z})^2(\pi^*_{2p})^4$	1	158	142
Ne_2	$(\sigma_{1s})^2(\sigma^*_{1s})^2(\sigma_{2s})^2(\sigma^*_{2s})^2(\pi_{2p})^4(\sigma_{2p_z})^2(\pi^*_{2p})^4(\sigma^*_{2p_z})^2$	0	unstable	

PRACTICE EXERCISE 10.22

Compare O_2^{2-} and O_2^- with respect to (a) paramagnetism, (b) bond energy, and (c) bond length.

Heteronuclear Diatomic Molecules

The molecular orbitals for homonuclear molecules described in the last section were obtained by combining identical atomic orbitals, 1*s* with 1*s*, 2*p* with 2*p*, and so forth. The parent atomic orbitals in heteronuclear (different nuclei) molecules may have different energies. Consequently, the energy-level diagrams and electron-density patterns for heteronuclear molecular orbitals will differ from those of homonuclear molecular orbitals. Recall (Section 8.1) that the greater the charge on an atomic nucleus, the lower the energy associated with each of its atomic orbitals. A 2*p* electron on oxygen, for example, has a lower energy than a 2*p* electron on carbon. The difference in energy is relatively small for atoms that lie close together in the same period, and the molecular orbital configuration for a heteronuclear molecule composed of such atoms can be obtained from the approximate diagram given in Figure 10.38.

EXAMPLE 10.12

(a) Write the molecular orbital configuration of the CO molecule and (b) give its bond order.

LEARNING HINT

The number of electrons in a complete molecular orbital configuration is the sum of the atomic numbers, not the sum of the group numbers.

SOLUTION

(a) CO contains 14 electrons, 6 from carbon and 8 from oxygen. Carbon and oxygen are in the same period, so the ground-state configuration of CO can be obtained from Figure 10.38:

$$(\sigma_{1s})^2(\sigma^*_{1s})^2(\sigma_{2s})^2(\sigma^*_{2s})^2(\pi_{2p_x})^2(\pi_{2p_y})^2(\sigma_{2p_z})^2$$

(b) CO has 10 bonding electrons and 4 antibonding electrons. Substitution into Equation 10.1 gives the bond order as $\frac{1}{2}(10 - 4) = 3$.

PRACTICE EXERCISE 10.23

(a) Write the electron configuration for HeH^+. (b) Will this ion be stable?

The configuration and bond order of CO are the same as those of N_2 (Practice Exercise 10.21), but there are differences in energy and electron density that do not appear in the configurations. Figure 10.40 shows that the molecular orbitals of CO are asymmetrical. The bonding orbitals have more electron density near oxygen, while the antibonding orbitals have more density near carbon.

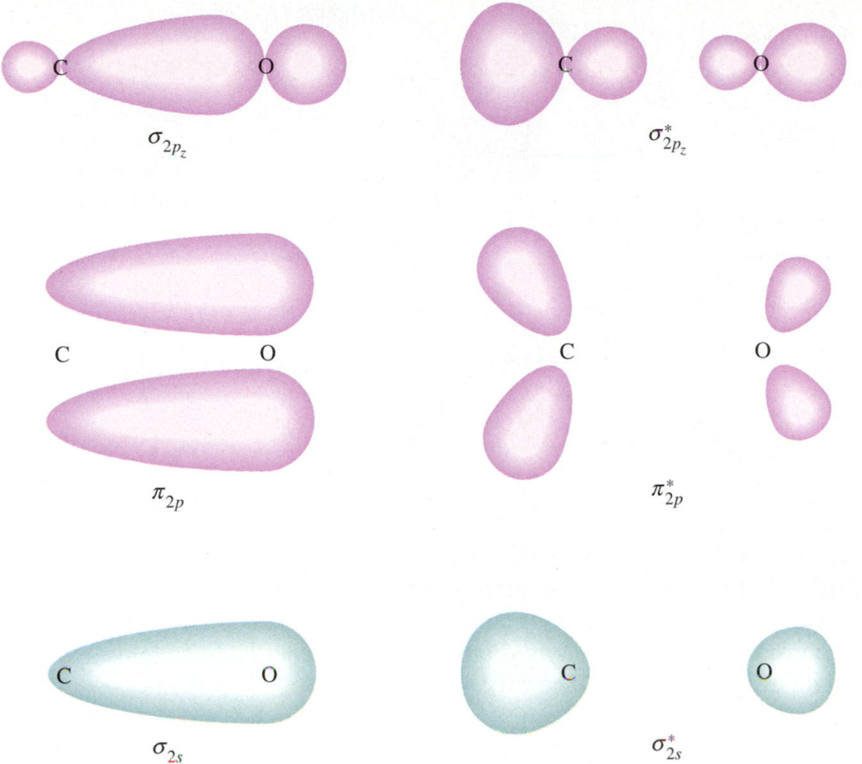

Figure 10.40

Approximate shapes of molecular orbitals in the carbon monoxide molecule. Each orbital is an asymmetrical version of a similar orbital in Figure 10.36. Observe that the bonding orbitals have more electron density near the oxygen nucleus, while the antibonding orbitals have more density near the carbon nucleus.

10.6 A COMBINED MODEL FOR POLYATOMIC MOLECULES

Valence bond theory assumes that each pair of bonding electrons spends most of its time in the overlap region between two atoms, while molecular orbital theory assumes that all of the electrons in a molecule are distributed over bonding and antibonding orbitals. Because sigma valence bonds are easy to visualize and because hybrid orbitals lend themselves to the VSEPR method for finding molecular shapes, valence bond theory is usually more convenient than molecular orbital theory for describing the single bonds of polyatomic ("many-atom") molecules. The behavior of pi electrons, however, is often easier to explain in terms of π and π^* molecular orbitals. Hence, in practice, chemists often treat the sigma and pi electrons of a molecule as separate and independent electron systems.

This dual approach to chemical bonding can be used to explain the properties and structure of the benzene (C_6H_6) molecule, a flat hexagon with bond angles of 120°. There is no single Lewis structure that can account for the geometry of benzene, and in Section 9.5 we explained its geometry by assuming that benzene is a resonance hybrid of two Kekulé structures with alternating single and double bonds:

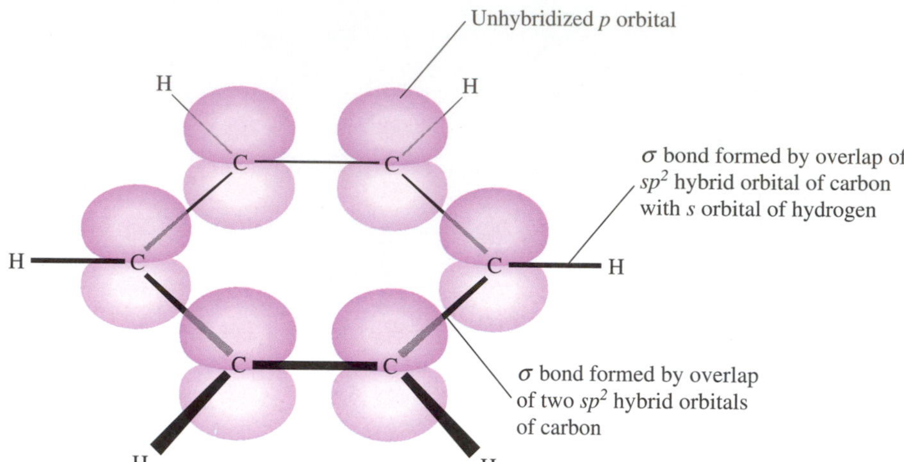

Figure 10.41

The sigma bonding framework in the hexagonal planar benzene molecule is represented by lines and wedges. Each carbon has a half-filled *p* orbital perpendicular to the plane of the molecule.

Valence bond theory assumes that each carbon atom ($2s^2 2p^2$) in benzene is sp^2 hybridized. The hybrid orbitals point to the corners of an equilateral triangle, and each carbon atom forms three trigonal planar sigma bonds by overlapping its sp^2 hybrid orbitals with the sp^2 hybrid orbitals of two other carbon atoms and with the 1*s* orbital of hydrogen (see Figure 10.41).

In addition to the hybrid orbitals, each carbon has a half-filled *p* orbital perpendicular to the plane of the molecule. These orbitals are treated in terms of molecular orbital theory. All six *p* orbitals combine to form three bonding and three antibonding molecular orbitals with lobes above and below the molecular plane (Figure 10.42). The six *p* electrons occupy the three bonding orbitals.

Notice how this approach to the structure of benzene eliminates the need for resonance structures. We no longer have to imagine different structures for one unchanging benzene molecule. The combined model gives us a single structure with 24 electrons in sigma valence bonds and 6 electrons in pi molecular orbitals.

Molecular orbitals in which electron densities are spread out over three or more atoms are said to be **delocalized**. The pi orbitals of benzene are examples. The nitrate ion (NO_3^-) and the carbonate ion (CO_3^{2-}) also have delocalized pi molecular orbitals. The nitrate ion is planar with three identical bonds at 120° angles—this geometry was explained in terms of three resonance structures in Section 9.5. Our combined model, however, requires only one structure. The nitrogen atom is sp^2 hybridized, and each hybrid orbital overlaps with a *p* orbital on oxygen to form a sigma valence bond (Figure 10.43a). The nitrogen atom has an unhybridized *p* orbital perpendicular to the plane of the ion, as does each of the three oxygen atoms. The four parallel *p* orbitals overlap to form four pi molecular orbitals. The lowest energy bonding orbital is delocalized over all four atoms (see Figure 10.43b).

10.7 THE CHEMICAL BOND: SOME FINAL THOUGHTS

At this point, you might wonder which is the correct way to look at a molecule. If the molecular orbital treatment is correct, is the valence bond treatment wrong? The answer is that each representation—Lewis structure, VSEPR configuration, valence bond, or molecular orbital—is a model that is useful in a particular context.

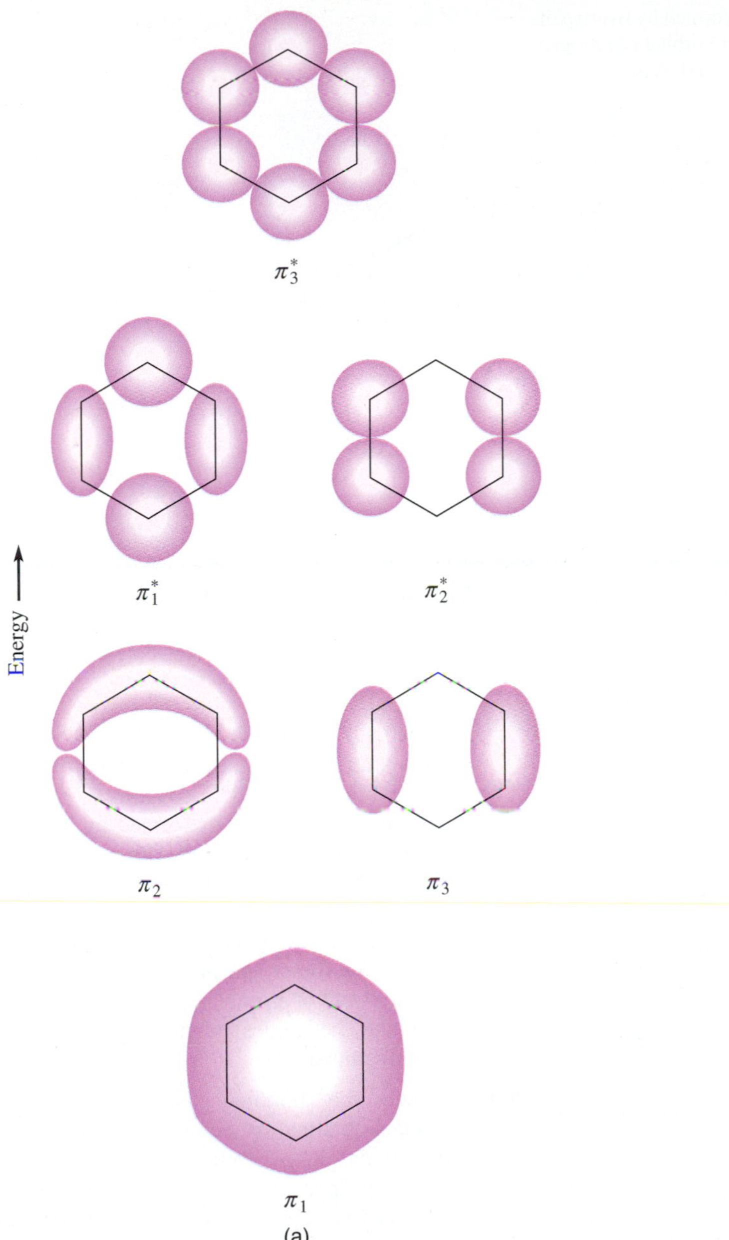

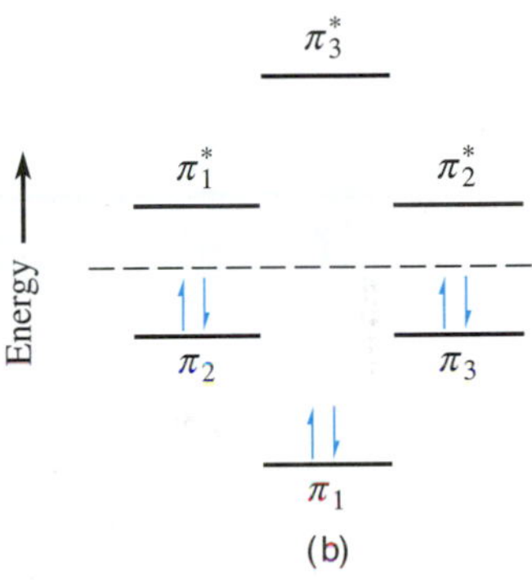

Figure 10.42

(a) Top view of the six pi molecular orbitals formed by the six half-filled *p* orbitals in the benzene molecule. The ring of carbon atoms lies in the plane of the paper. The orbital lobes that lie above the plane are shown; for each one, we must imagine an opposite lobe that lies below the plane. (b) Molecular orbital occupancy diagram. The six electrons from the six *p* orbitals occupy the three bonding molecular orbitals. The dashed line represents the energy of the parent *p* orbitals.

For some purposes, we prefer a simple model just as we might sometimes prefer a simple map, because we want broad outlines without too much detail. Lewis structures give the number of bonds; VSEPR gives the geometry. Valence bond theory, with its localized sigma and pi bonds, considers separate parts of the molecule but does not consider the molecule as a whole; it explains why electron pairs are shared between two atoms but fails to account for bonds that are delocalized over more than two atoms. Although molecular orbital theory is much more complex than valence bond theory, it provides the most complete bonding model. It takes the entire molecule into account; it explains the bonding in all species; and it provides more detailed information about molecular structures, energies, and properties than any of the other models.

Figure 10.43

Bonding in the nitrate ion. (a) The framework of sigma bonds (lines and wedges) is formed from the overlap of nitrogen sp^2 hybrid orbitals with oxygen p orbitals. Each atom also has a p orbital that is perpendicular to the molecular plane. (b) The four perpendicular p orbitals form four pi orbitals with lobes of electron density above and below the molecular plane. The lowest energy bonding orbital is shown here.

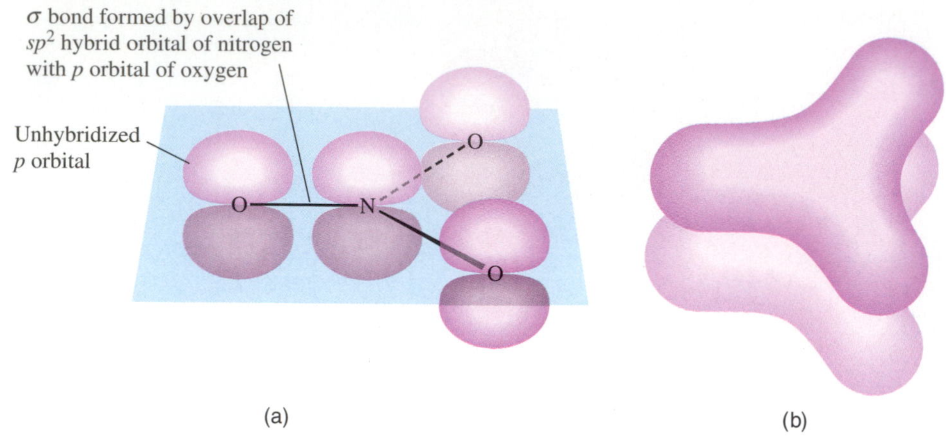

CHAPTER REVIEW

LEARNING OBJECTIVES BY SECTION

10.1 1. Use dipole moments to predict the shapes of simple molecules.

10.2 1. Determine the number of VSEPR electron pairs around a central atom.
2. Draw linear, trigonal planar, tetrahedral, trigonal bipyramidal, and octahedral molecules.
3. Explain deviations from ideal geometry in terms of lone-pair repulsion and multiple-bond character, and draw molecules exhibiting such deviations.
4. Use VSEPR to predict bond angles, molecular shapes, and polarities.

10.3 1. Write structural formulas for simple alkanes, alkenes, and alkynes.
2. Explain why there is free rotation around carbon–carbon single bonds, but not around carbon–carbon double bonds.
3. Draw and name the structural isomers of butane, pentane, and butene.
4. State whether a given alkene could have *cis–trans* isomers, and draw the isomers.
5. Write the structural formulas of benzene, napthalene, and anthracene.

10.4 1. Describe the formation of a covalent bond in terms of valence bond theory.
2. Sketch the ways s and p orbitals overlap to form sigma bonds.
3. State the roles of orbital hybridization and electron promotion in valence bond theory.
4. Sketch an s orbital, a p orbital, and the general shape of a hybrid orbital.
5. Sketch the sigma bonds formed by overlap of hybrid orbitals with s orbitals and with p orbitals.
6. Sketch the sigma bonds and lone pairs in molecules such as CH_4, NH_3, and H_2O.
7. Sketch the formation of a pi bond from two p orbitals.
8. Sketch the sigma and pi bonds in molecules with two or more central atoms.
9. Estimate the angles in molecules with two or more central atoms.

10.5 1. Sketch the bonding and antibonding molecular orbitals obtained by combining s orbitals, p_x orbitals, p_y orbitals, and p_z orbitals.
2. Draw an approximate energy-level diagram showing the relative energies of the first 10 homonuclear diatomic molecular orbitals and their parent atomic orbitals.
3. Write ground-state molecular orbital configurations, and draw orbital occupancy diagrams for first- and second-period homonuclear diatomic molecules.
4. Use molecular orbital configurations to predict whether homonuclear diatomic molecules are stable and whether they are paramagnetic.
5. Calculate bond orders from molecular orbital configurations.
6. Use bond orders to compare relative bond energies and bond lengths for similar species.

7. Write molecular orbital configurations for diatomic molecules containing different atoms from the same period.

10.6 1. Use valence bond theory to describe the sigma-bonded framework of a polyatomic molecule and molecular orbital theory to describe its pi electron system.

2. Give examples showing how modern bonding theory eliminates the need for resonance structures.

KEY TERMS BY SECTION

10.1 Bond angle

10.2 Central atom
Lone-pair repulsion
Octahedron
Tetrahedral angle
Tetrahedron
Trigonal bipyramid
Trigonal planar
VSEPR
VSEPR electron pair

10.3 Alkane
Alkene
Alkyne
Aromatic hydrocarbon
Cis–trans isomers
Conformation
Hydrocarbon
Isomers
Structural isomers

10.4 Hybridization
Hybrid orbital
Localized bond
Overlap (of orbitals)
Pi bond
Promotion (of electrons)
Sigma bond
Valence bond theory

10.5 Antibonding orbital
Bonding orbital
Bond order
Molecular orbital
Molecular orbital theory
Pi orbital
Sigma orbital

10.6 Delocalized orbital

IMPORTANT EQUATION

10.1 $$\text{Bond order} = \frac{1}{2}\left(\begin{array}{c}\text{number of}\\\text{bonding electrons}\end{array} - \begin{array}{c}\text{number of}\\\text{antibonding electrons}\end{array}\right)$$

FINAL EXERCISES

Answers to exercises with blue numbers are given in Appendix D.

PART A. QUESTIONS AND PROBLEMS BY SECTION

Dipole Moments and Molecular Geometry (Introduction; Section 10.1)

10.1 What is a *bond angle*? Illustrate your definition with a drawing.

10.2 Draw a Lewis structure for (a) CH_4 and (b) H_2SO_4. How many bond angles are in each of these structures? Identify them.

10.3 Distinguish between a *bond dipole moment* and a *molecular dipole moment*. Under what circumstances will the two moments be identical?

10.4 A triatomic molecule AB_2 has polar A—B bonds. Sketch a geometry in which the molecule has (a) a net dipole moment and (b) no dipole moment.

10.5 The dipole moment of H_2S is 0.97 D.
(a) Is the molecule linear or bent? Explain your reasoning.
(b) Use crossed arrows to indicate the direction of the bond dipoles and the net dipole in the H_2S molecule.

10.6 The dipole moment of NH_3 is 1.47 D.
(a) Draw a three-dimensional shape consistent with this value.
(b) Use crossed arrows to indicate the direction of the bond dipoles and the net dipole in the NH_3 molecule.

10.7 The dipole moment of BF_3 is zero. Does this fact tell us anything about the shape of the molecule? Explain.

10.8 Carbon tetrachloride (CCl_4) has polar bonds, but its net dipole moment is zero. Draw two shapes consistent with this information.

The VSEPR Model (Section 10.2)

10.9 What does "VSEPR" stand for? State the basic assumptions of VSEPR theory.

10.10 (a) What is a *VSEPR electron pair*?
(b) Under what circumstances will the number of VSEPR pairs around a central atom be less than the total number of pairs?
(c) Can the number of VSEPR pairs around a central atom ever be greater than the total number of pairs? Explain.

10.11 Explain why a lone pair of electrons tends to compress the bond angles on its atom.

10.12 Explain why bond angles that include multiple bonds tend to be larger than angles formed from single bonds.

10.13 Sketch and give one example of a molecule with each of the following shapes:
(a) linear
(b) bent
(c) trigonal planar
(d) tetrahedral
(e) square planar
(f) trigonal pyramidal

10.14 Sketch and give one example of a molecule with each of the following shapes:
(a) square pyramidal
(b) seesaw
(c) T-shape
(d) trigonal bipyramidal
(e) octahedral

10.15 Use Lewis structures and VSEPR to predict the shapes and bond angles of the following molecules:
(a) $BeCl_2$
(b) BCl_3
(c) $SiCl_4$
(d) PCl_5
(e) PCl_3
(f) SF_4

10.16 Use Lewis structures and VSEPR to predict the shapes and bond angles of the following molecules.
(a) H_2S
(b) CO_2
(c) IF_5
(d) OF_2
(e) $SnCl_2$
(f) N_2O (N is the central atom)

10.17 Which of the molecules in Exercise 10.15 have a dipole moment?

10.18 Which of the molecules in Exercise 10.16 have a dipole moment?

10.19 Use Lewis structures and VSEPR to predict the shapes and bond angles of the following species:
(a) $TeCl_4$
(b) BF_4^-
(c) BrF_6^+
(d) ICl_2^-
(e) $SbCl_6^-$
(f) SbF_5

10.20 Use Lewis structures and VSEPR to predict the shapes and bond angles of the following species. (The central atom is underlined in some of the formulas.)
(a) $\underline{Xe}OF_2$
(b) $\underline{P}OCl_3$
(c) O_3
(d) $\underline{Xe}OF_4$
(e) $SnCl_3^-$
(f) $\underline{S}O_2Cl_2$

10.21 Use Lewis structures and VSEPR to predict the shapes and bond angles of the following ions:
(a) PO_4^{3-}
(b) SO_4^{2-}
(c) IF_4^+
(d) CO_3^{2-}

10.22 Use Lewis structures and VSEPR to predict the shapes and bond angles of the following ions:
(a) ClO_2^-
(b) ClO_3^-
(c) ClO_4^-
(d) ICl_4^-

10.23 Compare the shapes and bond angles of the following species:
(a) H_2O and H_3O^+
(b) NO_2^+ and NO_2^-
(c) SO_3 and SO_3^{2-}

10.24 Compare the shapes and bond angles of the following species:
(a) PH_3 and PH_4^+
(b) NO_2^- and NO_2
(c) $\underline{S}OCl_2$ and $\underline{S}O_2Cl_2$ (central atom underlined)

Hydrocarbons (Section 10.3)

10.25 List three factors that allow carbon atoms to form millions of different compounds.

10.26 Silicon, like carbon, has four valence electrons. Give at least one reason why you would expect silicon to form fewer compounds than carbon.

10.27 What is the difference between an *alkane,* an *alkene,* and an *alkyne*?

10.28 Write structural formulas for (a) ethane, (b) ethene, and (c) acetylene (ethyne).

10.29 Explain why the different *conformations* of a molecule are not *isomers.*

10.30 Explain why *cis–trans* isomers are not structural isomers.

10.31 Describe the geometry of the bonds around the carbon atoms in an alkane such as propane. What value do all the bond angles have?

10.32 *n*-Butane is said to have a straight chain of four carbon atoms. Is the chain really straight? Explain.

10.33 Describe the effect that a double bond has on the geometry of a hydrocarbon molecule.

10.34 Describe the effect that a triple bond has on the geometry of a hydrocarbon molecule.

10.35 Draw and name the structural isomers of (a) butane and (b) butene.

10.36 Draw and name the structural isomers of pentane.

10.37 Draw the five structural isomers of hexane (C_6H_{14}), a six-carbon alkane. (Make sure they are all different.)

10.38 Draw the nine structural isomers of heptane (C_7H_{14}), a seven-carbon alkane. (Make sure they are all different.)

10.39 Write structural formulas for the *cis–trans* isomers, if any, of the following alkenes:
(a) $CH_3—CH{=}CH—CH_2—CH_3$
(b) $CH_2{=}CH—CH_2—CH_2—CH_3$

10.40 Write structural formulas for the *cis–trans* isomers, if any, of the following alkenes:
(a) $CH_3—CH{=}CH_2$
(b) $CH_3—CH{=}C(CH_3)—CH_2—CH_3$

10.41 Write the structural formulas of (a) benzene, (b) napthalene, and (c) anthracene.

10.42 Refer to the structure of 3,4-benzopyrene in Table 10.1 and state which of the carbon atoms are not bonded to hydrogen atoms.

Valence Bond Theory (Section 10.4)

10.43 What is a *sigma bond*? Sketch the sigma bond formed by the overlap of two *s* orbitals.

10.44 Sketch the sigma bond formed by the overlap of (a) two *p* orbitals and (b) an *s* and a *p* orbital.

10.45 Explain the role of hybrid orbitals in valence bond theory.

10.46 Explain the role of electron promotion in valence bond theory.

10.47 Sketch (a) an *s* orbital, (b) a *p* orbital, and (c) the general shape of a hybrid orbital formed from *s* and *p* orbitals.

10.48 Sketch the general shape of a sigma bond formed by the overlap of a hybrid orbital with (a) an *s* orbital and (b) a *p* orbital.

10.49 State the ideal bond angles and type of geometry associated with each of the following orbitals:
(a) sp
(b) sp^2
(c) sp^3

10.50 State the ideal bond angles and type of geometry associated with each of the following orbitals:
(a) sp^3d
(b) sp^3d^2

10.51 What is a *pi bond*? Sketch the general shape of a pi bond formed from two *p* orbitals.

10.52 Explain why there is free rotation around single bonds such as the C—C bond in ethane but not around double bonds such as the C═C bond in ethene.

10.53 Sketch the sigma and pi bonds in the N_2 molecule.

10.54 Sketch the sigma bonds in (a) HF and (b) F_2.

10.55 Which type of hybridization is exhibited by the central atom of each of the molecules in (a) Exercise 10.15 and (b) Exercise 10.16?

10.56 Which type of hybridization is exhibited by the central atom of each of the species in (a) Exercise 10.19 and (b) Exercise 10.20?

10.57 Draw the following molecules and give the type of hybridization around each central atom:
(a) NF_3
(b) CCl_4
(c) HCN

10.58 Draw the following molecules and give the type of hybridization around each central atom:
(a) $COCl_2$ (C is the central atom)
(b) XeF_4
(c) BrF_3

10.59 **(a)** Sketch the sigma and pi bonds for each of the molecules in Exercise 10.57.
(b) Which central atoms in Part (a) have lone pairs? Which orbitals do they occupy?

10.60 **(a)** Sketch the sigma and pi bonds for each of the molecules in Exercise 10.58.
(b) Which central atoms in Part (a) have lone pairs? Which orbitals do they occupy?

10.61 Explain in terms of the availability of *d* orbitals why SF_6 exists but OF_6 does not.

10.62 Explain in terms of the availability of *d* orbitals why PCl_5 exists but NCl_5 does not.

10.63 **(a)** Sketch the sigma and pi bonds in ethane, ethene, and acetylene.
(b) Which of these molecules should have the greatest carbon–carbon bond energy? The shortest carbon–carbon bond length?

10.64 Explain in terms of valence bond theory why the ethene molecule is planar.

10.65 Write structural formulas for the following molecules, determine the hybridization of each central atom, and estimate all bond angles:
(a) hydrogen peroxide, H_2O_2 (HO—OH)
(b) ethanol, CH_3CH_2OH
(c) dimethylamine, $(CH_3)_2NH$
(d) propene, C_3H_6 ($CH_2{=}CH—CH_3$)
(e) hydrazine, N_2H_4 ($H_2N—NH_2$)

10.66 Write structural formulas for the following molecules, determine the hybridization of each central atom, and estimate all bond angles:
(a) acetone, $(CH_3)_2CO$ (contains a C═O bond)
(b) formic acid, HCOOH

(c) *n*-butane, C_4H_{10} ($CH_3—CH_2—CH_2—CH_3$)
(d) acetaldehyde, CH_3CHO (contains a C═O bond)

Molecular Orbital Theory (Section 10.5)

10.67 Sketch the bonding and antibonding molecular orbitals obtained by combining (a) two *s* orbitals and (b) two p_z orbitals.

10.68 Sketch the bonding and antibonding molecular orbitals obtained by combining (a) two p_x orbitals and (b) two p_y orbitals.

10.69 How do bonding and antibonding sigma orbitals differ with respect to (a) electron energy and (b) distribution of electron density?

10.70 How do bonding and antibonding pi orbitals differ with respect to (a) electron energy and (b) distribution of electron density?

10.71 What is the difference between a sigma orbital and a sigma bond?

10.72 What is the difference between a pi orbital and a pi bond?

10.73 Why do bonding electrons tend to stabilize molecules while antibonding electrons tend to destabilize them?

10.74 Explain in terms of bonding and antibonding electrons why some diatomic species are stable with respect to their atoms and others are not.

10.75 Draw an approximate energy-level diagram showing the relative energies of the first 10 homonuclear diatomic molecular orbitals and their parent atomic orbitals.

10.76 What familiar rules are followed in assigning electrons to the available molecular orbitals?

10.77 (a) Write molecular orbital configurations and calculate bond orders for B_2, C_2, F_2, and Ne_2.
(b) Will these molecules exist? Explain.
(c) Will they be paramagnetic? Explain.

10.78 (a) Write molecular orbital configurations and calculate bond orders for F_2^+ and Ne_2^+.
(b) Will these ions exist?
(c) Will they be paramagnetic?
(d) Which ion would you expect to have the longer bond length? The greater bond energy?

10.79 (a) Make a table of molecular orbital configurations and bond orders for O_2, O_2^{2-}, O_2^-, and O_2^+.
(b) Which of the species in Part (a) is most stable with respect to its atoms? Least stable?
(c) Which of the species in Part (a) is paramagnetic?

10.80 List the oxygen species of Exercise 10.79 in order of (a) increasing bond energy and (b) increasing bond length.

10.81 (a) Write a molecular orbital configuration and calculate the bond order for N_2^+, a free radical.
(b) Are any of the molecules in Exercise 10.77 free radicals? Which?

10.82 Write molecular orbital configurations and calculate bond orders for the following free radicals: (a) CN, (b) CO^+, and (c) BO.

10.83 The He_2^+ and HeH^+ ions are stable with respect to their atoms. Do you think that HeH and HeH^- will also be stable? Justify your answer in terms of molecular orbital theory.

10.84 (a) Write molecular orbital configurations, draw orbital occupancy diagrams, and give bond orders for NO and NO^+.
(b) Which of these species should have the greater bond energy? The longer bond length?

10.85 Write molecular orbital configurations and calculate bond orders for (a) CN^- and (b) BN.

10.86 Use molecular orbital theory to list the following species in order of increasing stability: CN, CN^-, and CN^+.

Delocalized Pi Orbitals and the Combined Model (Sections 10.6 and 10.7)

10.87 Describe the difference between the valence bond and molecular orbital theories of chemical bonding. Which theory provides the simplest explanation of Lewis structures with localized bonds? Which theory best accounts for resonance structures with delocalized bonds?

10.88 What is the *combined model* for polyatomic molecules? Why do chemists often use valence bond theory to describe the geometry of a molecule and molecular orbital theory to describe its delocalized pi electron system?

10.89 Sketch the sigma-bond framework and pi bonding and antibonding orbitals in (a) ethene (C_2H_4) and (b) formaldehyde (H_2CO).

10.90 The bonding in formic acid (HCOOH) can be described in terms of the combined model for polyatomic molecules. Sketch the sigma-bond framework and pi molecular orbitals for this molecule.

10.91 Butadiene (CH_2═CH—CH═CH_2) has a delocalized pi electron system.
(a) Sketch the sigma-bond framework for butadiene and show any unhybridized *p* orbitals.
(b) How many orbitals are in the delocalized pi system? Sketch the lowest energy pi orbital.
(c) How many electrons are in each pi orbital?

10.92 The nitrite ion (NO_2^-) has a delocalized pi electron system.

(a) Sketch the sigma-bond framework for the ion, and show the location of the *p* orbitals that contribute to the delocalized system.
(b) Sketch the lowest energy pi orbital.

10.93 Describe the following species in terms of localized sigma bonds and a delocalized pi electron system:
(a) the V-shaped ozone molecule, O_3
(b) the planar carbonate ion, CO_3^{2-}

10.94 Describe the following molecules in terms of localized sigma bonds and a delocalized pi electron system: (a) SO_2 and (b) SO_3. Sketch the lowest energy pi orbital.

PART B. MISCELLANEOUS QUESTIONS AND PROBLEMS

10.95 Explain how a dipole moment measurement could distinguish between the two isomers of dichlorobenzene shown below. (The isomer on the right is an ingredient of mothballs and moth crystals.)

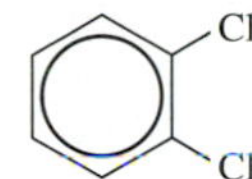

Cl — C_6H_4 — Cl

orthodichlorobenzene paradichlorobenzene

10.96 The dipole moment of water is 1.85 D. The O—H bond length is 96 pm and the H—O—H bond angle is 104.5°. Calculate the individual O—H bond moments. (*Hint:* This problem requires a knowledge of vectors.)

10.97 Describe some of the factors that cause actual bond angles to deviate somewhat from the ideal VSEPR bond angles given in Figures 10.3 through 10.6. Give examples of molecules in which you would expect the actual bond angles to be the same as the ideal angles.

10.98 Explain why the XeF_2 molecule is linear while the OF_2 molecule is bent.

10.99 The IF_7 molecule has five equatorial and two axial fluorine atoms.
(a) Do you think that this geometry could have been predicted using the basic assumptions of VSEPR?
(b) What is the value of the equatorial F—I—F angle?

10.100 Consider the structure of sulfuryl chloride, SO_2Cl_2 (Exercise 10.20f).
(a) Will the bond angles be exactly equal? Why or why not?
(b) Which angle is largest? Which smallest?

10.101 Cyclohexane (C_6H_{12}) is an alkane in which the six carbons form a ring.
(a) Write its structural formula.
(b) Will the ring be puckered or planar? Explain.

10.102 The structural formula for butadiene is $CH_2{=}CH{-}CH{=}CH_2$.
(a) What are the bond angles around each of the four carbon atoms?
(b) Write out the structural formula so that it indicates approximately correct bond angles.

10.103 Discuss the question, Do hybrid orbitals exist?

10.104 (a) Sketch the sigma and pi valence bonds in the linear CO_2 molecule.
(b) Are the pi bonds oriented in the same plane or in different planes with respect to each other?

10.105 The basic structural unit of proteins is the *peptide linkage*. The sigma-bond framework for this linkage is

```
             H
    |        |   |
  —C—C——N—C—
    |  |     |
       O
```

where all six atoms lie in the same plane.
(a) Assume that the nitrogen atom and the carbon atom bonded to oxygen are sp^2 hybridized, and explain why there is no rotation around the central C—N bond.
(b) What are the expected bond angles in the peptide linkage?

10.106 A Lewis structure for formamide, $HCONH_2$, is

```
     H  :O:
     |   ||
  H—N—C—H
     ··
```

Each atom in this structure satisfies the octet rule, and there is no formal charge. The observed bond angles are H—N—H = 119° and N—C=O = 123.6°.
(a) Draw a resonance structure for formamide with formal charges on the nitrogen and oxygen atoms.
(b) Which of the two resonance structures contributes more to the hybrid?

10.107 The S_8 molecule consists of eight sulfur atoms in a ring.
(a) Draw a Lewis structure for the molecule.
(b) A flat eight-sided ring would have internal angles of 135°. The experimental value of the S—S—S angle is 108°, which indicates that the sulfur atom is sp^3 hybridized. Describe and sketch the shape of the S_8 molecule.

10.108 Are hybrid orbitals atomic orbitals or molecular orbitals? Explain your answer.

10.109 Use molecular orbital theory to predict which member of the following pairs has the greater first ionization energy: (a) H_2 or H; (b) O_2 or O.

10.110 Use molecular orbital theory to explain why the bond length in F_2 is longer than the bond lengths in C_2, N_2, and O_2.

10.111 **(a)** Describe the allene molecule ($H_2C{=}C{=}CH_2$) in terms of localized sigma bonds and a delocalized pi electron system.
(b) Sketch the molecule and explain why the two sets of hydrogen atoms must lie in different planes.

10.112 **(a)** Describe the formamide molecule (Exercise 10.106) in terms of localized sigma bonds and a delocalized pi electron system.
(b) Over which nuclei would the pi orbitals be delocalized?

CHEMISTS AT WORK

THE CHEMIST AND THE ARTS

Imagine that, as part of your laboratory duties, you're standing with a scalpel in your hand ready to take a "sample" from a painting by Rembrandt. This is not an unusual experience for Dr. Suzanne Lomax, an organic chemist at the National Gallery of Art in Washington, D.C., one of the world's greatest museums. She is one of the relatively small number of chemists who work to conserve and preserve objects of art.

During her senior year at the University of Maryland and at the suggestion of a graduate student, she elected to participate in an undergraduate research project. That project developed into a study of the photochemistry of organic compounds containing nitrogen, which was her doctoral research, also at the University of Maryland. After postdoctoral study at Northwestern University, she accepted a position with the Environmental Protection Agency (EPA) in Washington where she evaluated, along with a chemical engineer and an economist, the safety of potential new products and their manufacturing processes. Since the EPA position was entirely a desk job and she wanted to get back into laboratory work, she applied for a position as organic chemist at the National Gallery in 1986 and was selected from a pool of approximately 60 applicants.

Before joining the museum staff, she had never had any formal education or training in art. "I had never even taken a course in art history," she says, "but was always interested in art and visited museums whenever I could. Since accepting this position, I have taken several courses in art appreciation and art history." Her education in art continues because the conservators at the museum provide a continuous array of learning opportunities through seminars and other activities.

The general process and experience of research chemistry are an important foundation in her current position. During her doctoral research, she learned how to manipulate and analyze very small samples by a variety of instrumental techniques and now uses many of them when working with microgram-sized samples that originate from works of art. Included in her repertoire are various chromatographic techniques, infrared and UV/visible spectroscopy, optical microscopy, and x-ray diffraction and fluorescence. Typical problems might involve any of the varied components of a particular painting, for example: What is the nature of the pigment—organic, inorganic? What kind of treatment did the canvas have before the paint was applied? Did the artist add a filler or some other material? How and what kind of surface coating was applied to the painting? When was the coating applied?

Typically, one of the conservators might have a problem with a given painting. What kind of varnish was used? In trying to remove a coating from a painting, some of the pigment seems to be coming off. Are the various layers on the painting from the same period of time as the original work? Have additives been used to tone down a pigment or a varnish? Dr. Lomax then attempts to answer these questions based on chemical investigations. She says that the work is always a challenge because each new project is unique, and a unique solution, if it can be found, is required. Even when a solution is found, there is the challenge of explaining the results and their ramifications for the preservation of the object to nonchemists.

The chemistry laboratory at the National Gallery has nine staff members: the lab director, six chemists, a biochemist, and a botanist who work directly with conservators. They also pursue research on developing new methods of analysis for very small-sized samples and new products that can be used in conservation.

Dr. Lomax says she feels a sense of accomplishment when leading a tour through the museum and seeing the paintings on which she has worked: her contribution to the preservation of the nation's and the world's cultural heritage.

DR. SUSAN LOMAX

CHAPTER 11

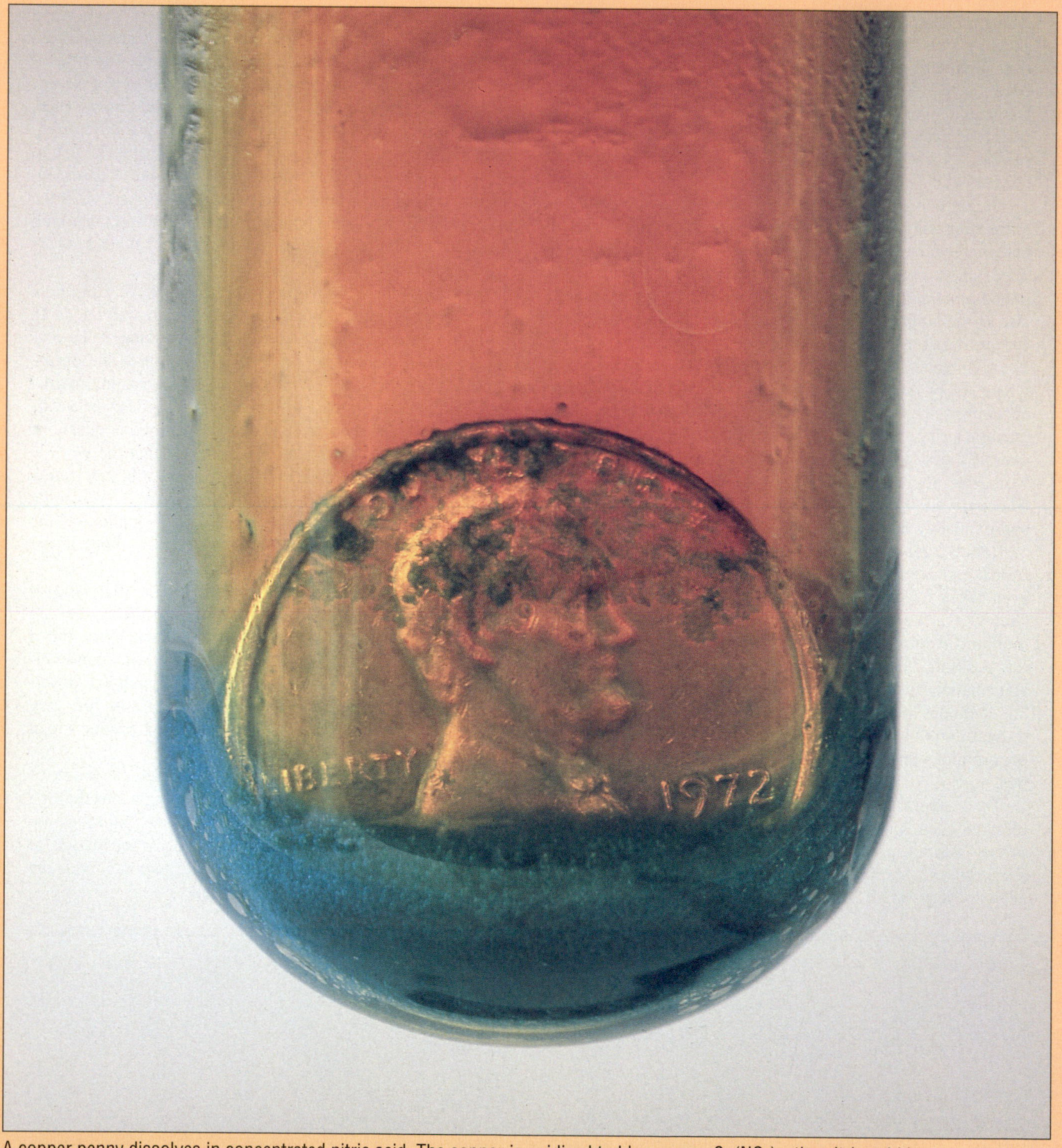

A copper penny dissolves in concentrated nitric acid. The copper is oxidized to blue copper $Cu(NO_3)_2$; the nitric acid is reduced to brown nitrogen dioxide gas (NO_2).

OXIDATION–REDUCTION REACTIONS

OUTLINE

PREVIEW

Photosynthesis, metabolism, corrosion, decay, and combustion are among the many natural processes that involve the transfer of electrons from one substance to another. Electron transfer also occurs when batteries are used to provide power, when bleaches are used to remove stains, and when antiseptics are used on wounds. The importance of electron transfer, more commonly called oxidation–reduction, cannot be overstated. The making of bread, cheese, wine, beer, metals, dyes, drugs, and fertilizers all involve oxidation–reduction reactions.

When sodium metal is warmed with chlorine gas (see Demonstration 3.2, p. 100), the metal bursts into flame and solid sodium chloride forms:

$$2Na(s) + Cl_2(g) \rightarrow 2NaCl(s)$$

Sodium chloride consists of positive sodium ions and negative chloride ions; therefore, electrons are transferred from sodium atoms to chlorine molecules during this reaction. The reverse reaction occurs when an electric current is passed through molten sodium chloride; the chloride ions lose electrons and the sodium ions regain them:

$$2NaCl(l) \xrightarrow{\text{electrolysis}} 2Na(l) + Cl_2(g)$$

Electron transfer occurs in all reactions in which elements are consumed or produced, and it occurs in many other reactions as well. In this chapter we will study electron transfer reactions in general, and later on, in Chapter 20, we will show how such reactions are used to produce useful work.

11.1 OXIDATION–REDUCTION REACTIONS

A reaction in which electrons are transferred from one substance to another is called an **oxidation–reduction** or **redox reaction**. The reaction between sodium and chlorine is one example; another is the spontaneous displacement reaction that occurs

DEMONSTRATION 11.1 A SILVER TREE

A copper tree stands beside a beaker of colorless silver nitrate solution.

Copper displaces silver from the silver nitrate solution. The copper atoms are oxidized to Cu^{2+} ions, which color the solution blue. Aqueous Ag^{+} ions are reduced to metallic silver, which deposits on the copper branches.

when a piece of copper metal is dipped into silver nitrate solution (see Demonstration 11.1). Filaments of metallic silver form on the surface of the copper, showing that silver ions (Ag^{+}) gain electrons to become neutral silver atoms. At the same time, the original colorless solution acquires the characteristic blue color of hydrated Cu^{2+} ions that form when neutral copper atoms lose electrons. The equation for this displacement reaction is

$$Cu(s) + 2Ag^{+}(aq) \rightarrow Cu^{2+}(aq) + 2Ag(s)$$

The loss of electrons by an element or compound is called **oxidation**—copper atoms are oxidized in this reaction. The gain of electrons is called **reduction**—the silver ions are reduced. Oxidation and reduction always occur together because one reactant must gain the electrons that the other reactant loses.

The loss and gain of electrons can be shown more clearly by breaking the equation into two **half-reactions**, one that shows copper atoms giving up electrons, and one that shows silver ions accepting electrons:

$$Cu(s) \rightarrow Cu^{2}(aq) + 2e^{-} \qquad \text{(oxidation)}$$

$$Ag^{+}(aq) + e^{-} \rightarrow Ag(s) \qquad \text{(reduction)}$$

The symbol e^- represents a single electron with its negative charge. Since the number of electrons gained must be the same as the number lost, two silver ions will be reduced for every copper atom that is oxidized. We represent this by doubling the reduction equation. The overall equation is obtained by adding

$$Cu(s) \rightarrow Cu^{2+}(aq) + 2e^-$$

$$2Ag^+(aq) + 2e^- \rightarrow 2Ag(s)$$

to give:

$$Cu(s) + 2Ag^+(aq) + \cancel{2e^-} \rightarrow Cu^{2+}(aq) + \cancel{2e^-} + 2Ag(s)$$

Note that the electrons on each side cancel.

An **oxidizing agent** is a substance that takes electrons from another substance, causing the other substance to be oxidized. In our example, silver nitrate is the oxidizing agent because it provides the silver ions that take electrons from copper. A **reducing agent** is a substance that gives electrons to another substance, causing it to be reduced. Copper is the reducing agent because it gives electrons to the silver ions.

EXAMPLE 11.1

The equation for the formation of sodium chloride from its elements is

$$2Na(s) + Cl_2(g) \rightarrow 2NaCl(s)$$

(a) Write equations for the half-reactions and (b) identify the oxidizing and reducing agents.

SOLUTION

(a) Sodium metal consists of sodium atoms; chlorine gas consists of chlorine molecules. The product, sodium chloride, consists of sodium ions (Na^+) and chloride ions (Cl^-). Thus sodium is oxidized to Na^+ and chlorine is reduced to Cl^-. The half-reactions are

$$Na(s) \rightarrow Na^+ + e^- \qquad \text{(oxidation)}$$

$$Cl_2(g) + 2e^- \rightarrow 2Cl^- \qquad \text{(reduction)}$$

(b) Chlorine accepts electrons and is the oxidizing agent. Sodium gives up electrons and is the reducing agent.

PRACTICE EXERCISE 11.1

Zinc metal displaces hydrogen from aqueous solutions of acids according to the equation:

$$Zn(s) + 2H^+(aq) \rightarrow Zn^{2+}(aq) + H_2(g)$$

(a) Write equations for the half-reactions and (b) identify the oxidizing and reducing agents.

PRACTICE EXERCISE 11.2

(a) Is the oxidizing agent oxidized or reduced in a redox reaction? (b) What happens to the reducing agent?

The oxidation–reduction concept can also be applied to reactions that do not involve ions. Recall (Section 9.7) that covalent bonds between different atoms have some degree of polarity due to electronegativity differences. For example, the chlorine atom in the hydrogen chloride molecule ($\overset{\delta+}{H}-\overset{\delta-}{Cl}$) draws electron density away from the hydrogen atom. The formation of HCl

$$H_2(g) + Cl_2(g) \rightarrow 2HCl(g)$$

can therefore be regarded as a transfer of electron density from the less electronegative hydrogen atoms to the more electronegative chlorine atoms. In this sense, the reaction is an oxidation–reduction reaction in which hydrogen is oxidized and chlorine is reduced.

LEARNING HINT

This is a good time to review Section 3.6 on oxidation and reduction.

The term *oxidation* is derived from *oxygen*, the most abundant and ubiquitous oxidizing agent in our environment. Because oxygen is more electronegative than all other elements except fluorine, it oxidizes everything it combines with (except fluorine). Reactants that gain oxygen are oxidized; reactants that lose oxygen are reduced.

11.2 OXIDATION NUMBERS

Oxidation–reduction reactions are easily recognized when monatomic ions are consumed or produced, as in the formation of sodium chloride from its elements or the reaction of copper metal with silver nitrate solution. The redox reactions of covalent substances, however, are harder to identify. How can we tell, for example, whether the formation of ammonia

$$N_2(g) + 3H_2(g) \rightarrow 2NH_3(g)$$

is a redox reaction? Furthermore, if it is a redox reaction, how do we identify the oxidizing and reducing agents? The concept of oxidation numbers will help us answer these questions.

The **oxidation number** or **oxidation state** of a bonded atom is the charge it would have if the electrons in each bond were given to the more electronegative atom. In ionic compounds, the more electronegative element already has full possession of the bonding electrons. For example, since chlorine is much more electronegative than calcium, the calcium atom in $CaCl_2$ has lost two electrons to form Ca^{2+} and has an oxidation number of $+2$. Each chlorine atom has gained one electron to form Cl^- and has an oxidation number of -1. There are two chloride ions for every calcium ion, so the oxidation numbers in $CaCl_2$ add up to zero. This example illustrates that *the oxidation number of a monatomic ion is the same as its charge*.

Charges are written as $2+$ or $1-$, but oxidation numbers are written as $+2$ or -1.

In a covalent compound, oxidation numbers are obtained by assigning the electrons in each covalent bond to the more electronegative atom, as if the bonds were ionic. In water, for example, the bond electrons are assigned to oxygen because oxygen is more electronegative than hydrogen:

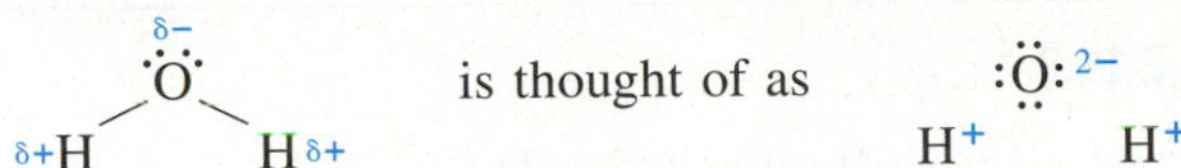

The oxidation numbers of oxygen and each hydrogen are −2 and +1, respectively, and they add up to zero for the neutral molecule. Note carefully that we are not dealing with real charges; the actual charge on each atom is the result of bond polarity and is much less than the assigned charge.

Oxidation numbers are often used to classify compounds and to keep track of electrons in redox reactions. The procedure used for finding the oxidation number of each atom in H_2O—writing a Lewis structure and assigning bond electrons to the more electronegative atom—is often inconvenient and time-consuming for other species. For this reason, the following set of rules has been devised for defining oxidation numbers without reference to Lewis structures. Use the rules in the order given. Each rule takes precedence over the rules that follow, so you can stop as soon as you have the value you are looking for.

1. *The oxidation number of an atom in a free element is zero, regardless of the formula.* For example, the oxidation numbers of oxygen in O_2 or O_3, phosphorus in P_4, and mercury in Hg are zero.

2. *The sum of the oxidation numbers in a neutral molecule is zero; the sum for an ion is equal to the charge on the ion.* For example, the sum of the oxidation numbers is 0 in H_2O, +1 in NH_4^+, and −3 in PO_4^{3-}. This rule implies that *the oxidation number of a monatomic ion is equal to the charge on the ion*. For example, the oxidation numbers of Al^{3+} and S^{2-} are +3 and −2, respectively.

3. *Hydrogen and the alkali metals (Group 1A) have an oxidation number of +1 in their compounds; the alkaline earth metals (Group 2A) have an oxidation number of +2.* An exception to this rule occurs in metal hydrides such as CaH_2. The oxidation number of hydrogen in the hydride ion (H^-) is, according to Rule 2, −1. (Remember that Rule 2 has precedence over Rule 3).

4. *The most electronegative element, fluorine, always has an oxidation number of −1 in its compounds. The second-most electronegative element, oxygen, has an oxidation number of −2 except when it is bonded to fluorine or to itself.* This rule and the preceding rules require that the oxidation number of oxygen be +2 in OF_2, −1 in peroxides such as H_2O_2, and $-\frac{1}{2}$ in the superoxide ion O_2^-.

5. *The remaining halogens (Cl, Br, and I) have an oxidation number of −1 except when bonded to oxygen or to a more electronegative halogen.* The oxidation number of chlorine, for example, is −1 in PCl_3, and +1 in Cl_2O and ClF.

6. *Unless Rules 1–5 require otherwise, atoms from Groups 5A and 6A are assigned the oxidation numbers they would have in their monatomic negative ions (i.e., −3 for Group 5A elements and −2 for Group 6A elements).* For example, the oxidation number of sulfur in CS_2 is −2 and the oxidation number of nitrogen in HCN is −3.

LEARNING HINT

Rules 1–6 occasionally give oxidation numbers that differ from those obtained by assigning electrons in a Lewis structure. It is important that oxidation numbers be consistent; in this text we will always use values obtained from these rules.

EXAMPLE 11.2

Find the oxidation number of sulfur in Na_2SO_4.

SOLUTION

The sum of the oxidation numbers is zero (Rule 2). If we let x represent the oxidation number of an element, then

$$2x_{Na} + x_S + 4x_O = 0$$

Substituting +1 for x_{Na} (Rule 3) and −2 for x_O (Rule 4) gives

$$2(+1) + x_S + 4(-2) = 0$$

and

$$x_S = +6$$

PRACTICE EXERCISE 11.3

Find the oxidation number of each atom in $COCl_2$. The chlorine atoms are bonded to carbon, not to oxygen.

EXAMPLE 11.3

Find the oxidation number of nitrogen in NH_4^+.

SOLUTION

The sum of the oxidation numbers is +1 (Rule 2). Hence,

$$x_N + 4x_H = +1$$

Substituting +1 for x_H (Rule 3) gives

$$x_N + 4(+1) = +1$$

and

$$x_N = -3$$

PRACTICE EXERCISE 11.4

Determine the oxidation number of each atom in (a) NO_3^-, (b) NO_2^-, and (c) CN^-.

A crystal of Fe_3O_4 contains Fe^{2+} and Fe^{3+} in a 1 to 2 ratio and is sometimes written as $FeO{\cdot}Fe_2O_3$. The structure of Co_3O_4 is analogous to that of Fe_3O_4. The compound Pb_3O_4 has a different structure and is sometimes written as $2PbO{\cdot}PbO_2$.

When a formula contains two or more differently bonded atoms of the same element, the oxidation number obtained from the above rules will be an average for the atoms and may not even be an integer. For example, each metal atom in the oxides Fe_3O_4, Co_3O_4, and Pb_3O_4 has an oxidation number of $+\frac{8}{3}$ according to our rules. Such fractional oxidation numbers work just as well as whole numbers for the purposes described in this chapter.

PRACTICE EXERCISE 11.5

What is the oxidation number of sulfur in sodium dithionate, $Na_2S_4O_6$?

Oxidation Numbers and the Periodic Table

The principal oxidation states of the elements are shown in Figure 11.1. Observe that the oxidation numbers of main group elements do not exceed the group number, which is the number of electrons in the valence shell. Sulfur, for example, is in Group 6A, so up to six of its electrons can be used in bonding; +6 is its highest oxidation state. Some elements, notably oxygen and fluorine, never exhibit the maximum oxidation state for their group. Observe also that metals have only positive

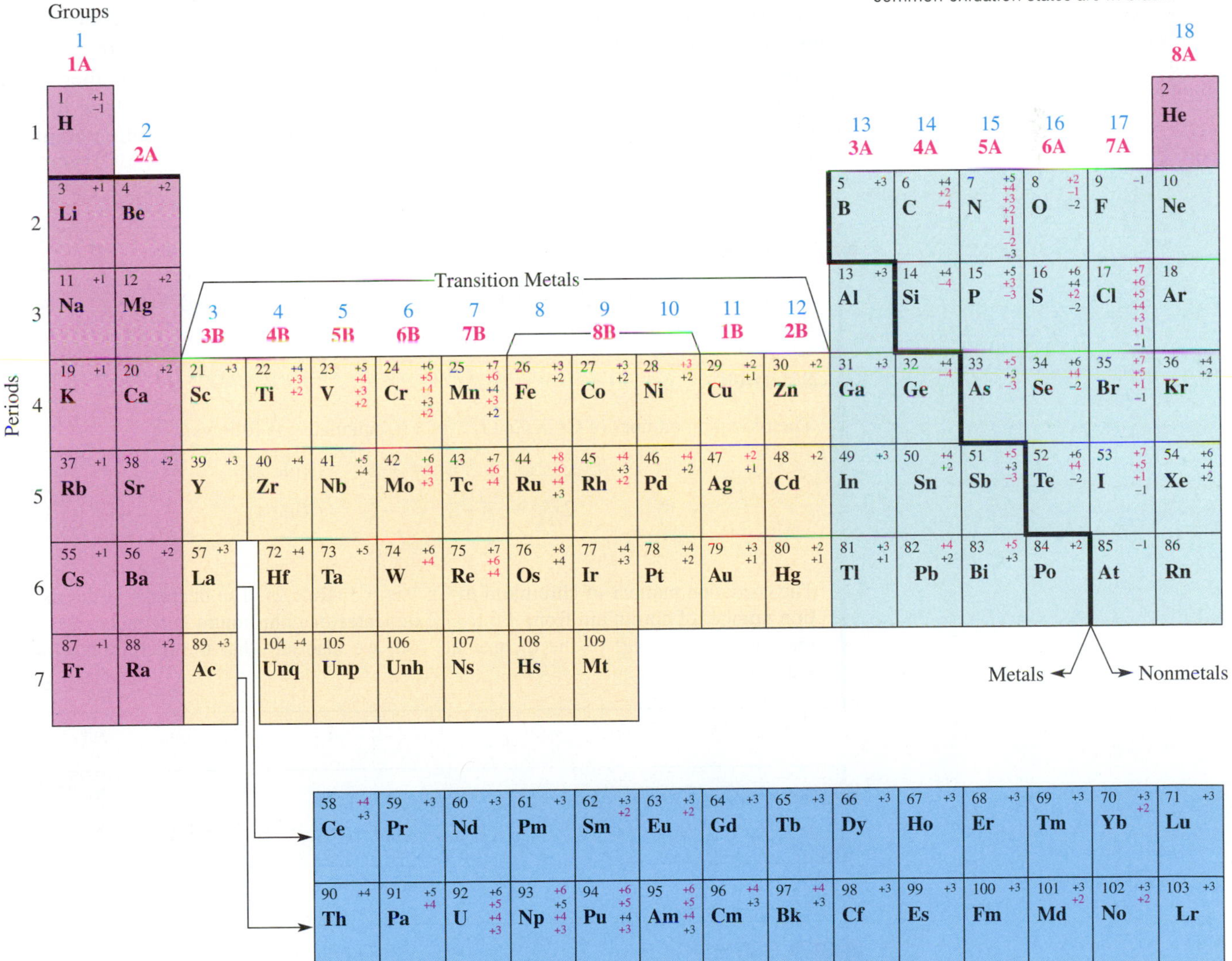

Figure 11.1

Oxidation states of the elements. The more common oxidation states are in black.

oxidation states, while nonmetals have both positive and negative states. The maximum number of electrons a nonmetal can gain is eight minus the group number, which is the number of electrons needed to complete its octet; its lowest (most negative) oxidation state is the negative of this number or the group number minus eight. Nitrogen, for example, is in Group 5A and its lowest oxidation state is $5 - 8 = -3$. Oxidation states lower than -4 have not been observed.

PRACTICE EXERCISE 11.6

List the following acids in order of the increasing oxidation number of bromine: $HBrO_3$, HBr, $HBrO_4$, and $HBrO_2$.

Using Oxidation Numbers

Oxidation: increase in oxidation number.
Reduction: decrease in oxidation number.

Oxidation is characterized by a loss of electrons, reduction by a gain of electrons. Thus the oxidation number of at least one atom increases (becomes more positive) during an oxidation half-reaction, while the oxidation number of another atom decreases during the reduction half-reaction. The following examples show that an oxidation–reduction reaction can be recognized by changes in the oxidation numbers of the reacting substances.

EXAMPLE 11.4

Ethyl alcohol converts dichromate ion ($Cr_2O_7^{2-}$) to chromium(III) ion (Cr^{3+}). Is the dichromate ion oxidized or reduced in this reaction?

SOLUTION

The oxidation number of Cr in $Cr_2O_7^{2-}$ is +6, calculated as follows:

$$2x_{Cr} + 7x_O = -2 \qquad \text{(Rule 2)}$$

$$2x_{Cr} + 7(-2) = -2 \qquad \text{(Rule 4)}$$

$$x_{Cr} = +6$$

The oxidation number of chromium in Cr^{3+} is +3 (Rule 2). The decrease in the oxidation number of chromium from +6 to +3 indicates that chromium is reduced. Because the chromium is part of the dichromate ion, we also say that the dichromate ion is reduced.

PRACTICE EXERCISE 11.7

Is PCl_3 oxidized or reduced when it is converted to PCl_5?

EXAMPLE 11.5

Verify that the reaction

$$2H_2SO_4(aq) + Cu(s) \rightarrow CuSO_4(aq) + SO_2(g) + 2H_2O(l)$$

is an oxidation–reduction reaction.

SOLUTION

Copper is in the form of Cu atoms on the reactant side and Cu^{2+} ions on the product side. ($CuSO_4$ consists of Cu^{2+} and SO_4^{2-} ions.) Hence, the oxidation number of copper changes from 0 (Rule 1) to +2 (Rule 2), and copper is oxidized. The oxidation number of sulfur in H_2SO_4 on the reactant side is +6. On the product side, the oxidation number of S is +6 in SO_4^{2-} and +4 in SO_2. Therefore, some of the sulfuric acid is reduced.

PRACTICE EXERCISE 11.8

Identify the substances that are oxidized and reduced in the reaction

$$2FeCl_3(aq) + 3H_2S(aq) \rightarrow 2FeS(s) + S(s) + 6HCl(aq)$$

Many important reactions fall into the oxidation–reduction category. Oxidizing agents are used in water purification and waste treatment; as bleaching agents, antiseptics, analytical reagents, and reagents for chemical synthesis; and for many other purposes. The selection of a good oxidizing agent usually involves a number of practical considerations, but one characteristic is essential: *An oxidizing agent must contain an element in one of its higher oxidation states.* An element that is in its lowest oxidation state cannot take electrons from other substances and cannot act as an oxidizing agent.

The primary oxidizing agent in earth's natural environment is O_2, a molecule in which the oxygen atoms have an oxidation number of 0. During the formation of oxides and most other oxygen-containing compounds, each oxygen atom gains enough electrons to complete its octet. Hence, its oxidation number changes from 0 to −2. Oxidizing agents more powerful than oxygen include oxygen-rich acids and anions such as HNO_3 and ClO_4^-. Each of these species contains a central atom in a highly positive oxidation state.

Reducing agents are used in metallurgy (see Section 3.6), in photography, and as reagents in analytical chemistry and chemical synthesis. *A reducing agent must contain an element in one of its lower oxidation states.* An element that is in its highest oxidation state cannot lose electrons to form a still higher state. Elemental carbon, hydrogen, and active metals are good reducing agents because they readily give up electrons and form compounds in which their oxidation states are positive.

Hydrogen peroxide (H_2O_2) is an example of a compound that can function either as an oxidizing agent or as a reducing agent. The oxygen atoms in H_2O_2 have an oxidation number of −1. Depending on the nature of the other reactants, the oxidation number can rise to 0 with O_2 as a principal product, or fall to −2, in which case water or an oxide or hydroxide will be a principal product.

EXAMPLE 11.6

What is the oxidation number of S in (a) hydrogen sulfide (H_2S) and (b) sulfuric acid

(H_2SO_4)? Which of these substances can be oxidized? Reduced? (*Note:* Assume that there is no change in the oxidation states of hydrogen and oxygen.)

SOLUTION

(a) The oxidation number of S in H_2S is -2. Examination of Figure 11.1 shows that -2 is the lowest oxidation state of S. Hence, hydrogen sulfide can be oxidized, but not reduced.

(b) The oxidation number of S in H_2SO_4 is $+6$. This number corresponds to the highest oxidation state of sulfur (Figure 11.1), so sulfuric acid cannot be oxidized. It can, however, be reduced.

PRACTICE EXERCISE 11.9

Can sulfur trioxide (SO_3) be used as an oxidizing agent? A reducing agent? How about sulfur dioxide (SO_2)?

11.3 WRITING AND BALANCING REDOX EQUATIONS

Many oxidation–reduction equations are simple and easy to balance, but there are others that defy the normal efforts of balancing by inspection. These difficulties can be overcome by using half-reactions and oxidation numbers. In the **half-reaction method** of balancing oxidation–reduction reactions, equations for the half-reactions are written and added together in a way that equalizes electron gain and loss. This method was used in Section 11.1 for the reaction between copper metal and silver nitrate solution. Now we apply it to more complicated redox reactions.

Balancing Redox Equations in Acidic Solution

Consider the reaction between concentrated hydrochloric acid and potassium permanganate, a compound consisting of K^+ and MnO_4^- ions. Chlorine gas evolves, water forms, and the intense purple color of the permanganate ion fades as it changes into the pale pink manganese(II) ion (Mn^{2+}) (Figure 11.2). The potassium ions are spectator ions; that is, they do not participate in the reaction. The following equation, incomplete and unbalanced, summarizes these observations:

$$KMnO_4(aq) + HCl(aq) \rightarrow Mn^{2+}(aq) + Cl_2(g) + K^+(aq) + H_2O(l)$$

This equation is difficult to balance by inspection (try it!), but the half-reaction method provides a systematic approach. The steps in the method are as follows:

LEARNING HINTS

When balancing an equation by the half-reaction method, never add molecules or ions that are not in the reaction mixture.

Recall (Section 4.2) that a net ionic equation does not include spectator ions.

Step 1: *Write an unbalanced net ionic equation.* First, write down the actual molecules and dissolved ions known to be present; then eliminate all spectator ions. In this reaction, $KMnO_4(aq)$ is ionic, $HCl(aq)$ is ionic, $Cl_2(g)$ is covalent, and $H_2O(l)$ is covalent:

$$K^+(aq) + MnO_4^-(aq) + H^+(aq) + Cl^-(aq) \rightarrow Mn^{2+}(aq) + Cl_2(g) + K^+(aq) + H_2O(l)$$

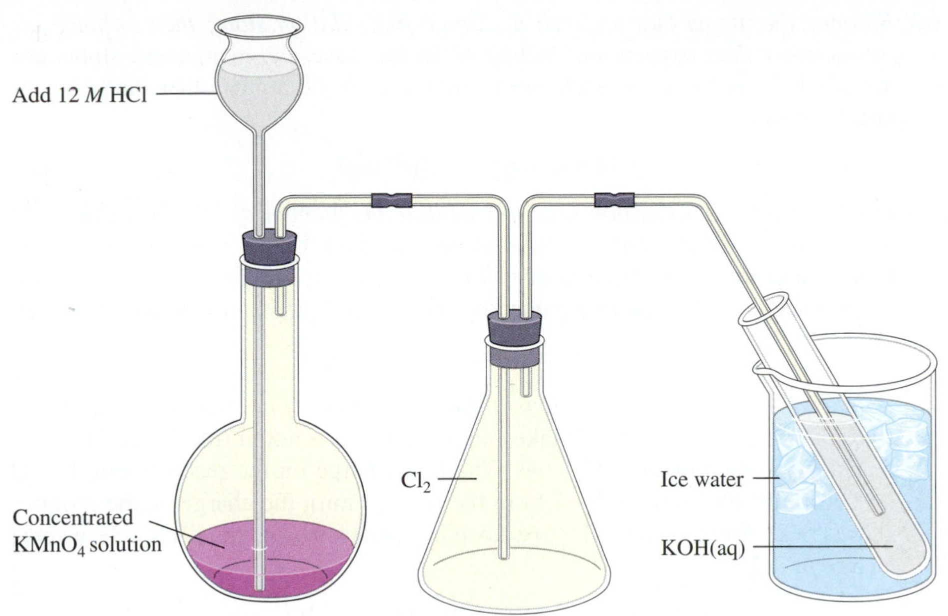

Figure 11.2
The reaction between potassium permanganate ($KMnO_4$) and hydrochloric acid (HCl) produces chlorine gas. The chlorine gas collects in the center flask. Excess chlorine is absorbed by a cold solution of potassium hydroxide, which converts it to potassium chloride and potassium hypochlorite: $Cl_2(g) + 2KOH(aq) \rightarrow KCl(aq) + KOCl(aq) + H_2O(l)$.

The potassium ions are spectator ions. They are eliminated, and the unbalanced net ionic equation is

$$MnO_4^-(aq) + H^+(aq) + Cl^-(aq) \rightarrow Mn^{2+}(aq) + Cl_2(g) + H_2O(l)$$

Step 2: *Find the substances that are oxidized and reduced, and write unbalanced equations for the half-reactions.* Look for changes in oxidation numbers to identify the oxidized and reduced substances. In this reaction, the oxidation number of chlorine increases from -1 in Cl^- to 0 in Cl_2; hence, chloride ion is oxidized:

$$Cl^-(aq) \rightarrow Cl_2(g) \qquad \text{(oxidation)}$$

The oxidation number of manganese decreases from $+7$ in MnO_4^- to $+2$ in Mn^{2+}, hence, permanganate ion is reduced:

$$MnO_4^-(aq) \rightarrow Mn^{2+}(aq) \qquad \text{(reduction)}$$

Step 3: *Balance the half-reaction equations with respect to atoms and with respect to charge.* First, consider the oxidation half-reaction. The equation is simple enough to be balanced by inspection. The atoms are balanced by putting a 2 in front of the Cl^- ion:

$$2Cl^-(aq) \rightarrow Cl_2(g)$$

The charge is balanced by adding two electrons to the product side, thus giving each side a charge of -2:

$$2Cl^-(aq) \rightarrow Cl_2(g) + 2e^-$$

Now consider the reduction half-reaction. This reaction is more complicated than the previous one because it contains both manganese and oxygen atoms. There are a number of ways to balance an equation involving oxygen, but the following procedure works well in both acidic and basic solutions:

(a) *Balance the atoms that undergo a change in oxidation state; then balance all atoms other than oxygen and hydrogen.* In this case, the manganese atoms are already balanced (one on each side), and there are no atoms other than oxygen and hydrogen:

$$MnO_4^-(aq) \rightarrow Mn^{2+}(aq)$$

(b) *Use the change in oxidation state to calculate the number of electrons gained or lost by the reactants, and add them to the equation.* In our example, the oxidation state of manganese falls from $+7$ on the reactant side to $+2$ on the product side; hence, five electrons are gained by the reactants and added to the equation:

$$MnO_4^-(aq) + 5e^- \rightarrow Mn^{2+}(aq)$$

(c) *Balance the charge using H^+ ions if the solution is acidic and OH^- ions if the solution is basic.* Our reaction takes place in acidic solution (recall that HCl is a reactant), so we must use H^+ ions. The total charge on the reactant side is -6 (-5 from the electrons and -1 from the MnO_4^- ion); the charge on the product side is $+2$. Adding eight H^+ ions to the reactant side gives each side a total charge of $+2$:

$$MnO_4^-(aq) + 8H^+(aq) + 5e^- \rightarrow Mn^{2+}(aq)$$

(d) *Finally, balance the oxygen and hydrogen atoms by adding H_2O molecules to the equation.* The reactant side has eight hydrogen atoms and four oxygen atoms. They can be balanced by adding four H_2O molecules to the product side:

$$MnO_4^-(aq) + 8H^+(aq) + 5e^- \rightarrow Mn^{2+}(aq) + 4H_2O(l)$$

Step 4: *Multiply the half-reaction equations by factors that equalize the numbers of electrons lost and gained.* The total number of electrons produced by oxidation must equal the total number consumed by reduction. In our example, two of the reduction half-reactions must be accompanied by five of the oxidation half-reactions, for a total of 10 electrons transferred:

$$2 \times [MnO_4^-(aq) + 8H^+(aq) + 5e^- \rightarrow Mn^{2+}(aq) + 4H_2O(l)]$$

$$5 \times [2Cl^-(aq) \rightarrow Cl_2(g) + 2e^-]$$

Step 5: *Add the half-reaction equations from Step 4 to obtain the balanced net ionic equation.* The electrons will cancel:

$$2MnO_4^-(aq) + 16H^+(aq) + 10Cl^-(aq) \rightarrow 2Mn^{2+}(aq) + 8H_2O(l) + 5Cl_2(g)$$

The net ionic equation should be checked to verify that both sides have the same numbers of atoms and the same net charge.

Step 6: A net ionic equation can be converted to the so-called molecular form by adding enough ions to complete the neutral formulas. In our example, adding two K^+ and six Cl^- ions to each side gives

$$2KMnO_4(aq) + 16HCl(aq) \rightarrow 2MnCl_2(aq) + 8H_2O(l) + 5Cl_2(g) + 2KCl(aq)$$

EXAMPLE 11.7

Solid copper(I) oxide (Cu_2O) is oxidized to $CuCl_4^{2-}$ by aqua regia, a mixture of nitric and hydrochloric acids. Write the balanced equation for the oxidation half-reaction.

SOLUTION

Equations for half-reactions are obtained from Steps 2 and 3 of the above procedure.

Step 2: The unbalanced equation for the half-reaction is

$$Cu_2O(s) \rightarrow CuCl_4^{2-}(aq)$$

Note that copper(I) oxide is not written in ionic form; it is a sparingly soluble solid and forms relatively few dissolved ions.

Step 3: The unbalanced equation contains copper, chlorine, and oxygen atoms. We will balance it by using the four substeps given previously for half-reactions involving oxygen.

(a) The copper atoms change their oxidation state. They are balanced by putting a 2 in front of the $CuCl_4^{2-}$ ion:

$$Cu_2O(s) \rightarrow 2CuCl_4^{2-}(aq)$$

The chlorine atoms are balanced by putting eight Cl^- ions on the reactant side (aqueous HCl is the source of the Cl^- ions):

$$Cu_2O(s) + 8Cl^-(aq) \rightarrow 2CuCl_4^{2-}(aq)$$

(b) The oxidation state of each copper atom increases from +1 to +2. There are two copper atoms, so the total change in oxidation state is +2. Hence, two electrons are lost by the reactants and added to the product side:

$$Cu_2O(s) + 8Cl^-(aq) \rightarrow 2CuCl_4^{2-}(aq) + 2e^-$$

(c) The charge on the reactant side is −8; the total charge on the product side is −6 (−2 from the electrons and −4 from the two $CuCl_4^{2-}$ ions). The solution is acidic, so the charge is balanced by adding two H^+ ions to the reactant side:

$$Cu_2O(s) + 8Cl^-(aq) + 2H^+(aq) \rightarrow 2CuCl_4^{2-}(aq) + 2e^-$$

The total charge on each side is now −6.

(d) The hydrogen and oxygen atoms are balanced by adding one H_2O molecule to the product side:

$$Cu_2O(s) + 8Cl^-(aq) + 2H^+(aq) \rightarrow 2CuCl_4^{2-}(aq) + H_2O(l) + 2e^-$$

LEARNING HINT

When using the method of half-reactions, keep in mind that sparingly soluble solids and covalent molecules are not written in ionic form.

LEARNING HINT

Always begin Step 3 by balancing the atoms that change their oxidation states.

LEARNING HINT

The number of electrons lost or gained by the reactants is equal to the total change in oxidation number.

Increase in oxidation number = electrons lost by reactants

Decrease in oxidation number = electrons gained by reactants

PRACTICE EXERCISE 11.10

Write a half-reaction equation for the reduction of aqueous iodate ion (IO_3^-) to solid iodine (I_2) in acidic solution.

PRACTICE EXERCISE 11.11

Silver metal dissolves in nitric acid with the formation of silver ions. Gaseous NO is evolved. Write (a) the balanced net ionic equation and (b) the molecular equation.

Balancing Redox Equations in Basic Solution

The method for balancing redox equations in basic solution is essentially the same as in acidic solution—the only difference is that OH^- ions are used instead of H^+ ions.

EXAMPLE 11.8

When zinc metal dissolves in aqueous base, zinc is oxidized to zincate ion ($Zn(OH)_4^{2-}$), and water is reduced to hydrogen gas. Write the net ionic equation for the reaction.

SOLUTION

We will follow the steps given above.

Step 1: Write an unbalanced net ionic equation:

$$Zn(s) + H_2O(l) \rightarrow Zn(OH)_4^{2-}(aq) + H_2(g)$$

Step 2: Write unbalanced equations for the half-reactions. Zinc is oxidized to zincate ion and water is reduced to hydrogen gas:

$$Zn(s) \rightarrow Zn(OH)_4^{2-}(aq) \qquad \text{(oxidation)}$$

$$H_2O(l) \rightarrow H_2(g) \qquad \text{(reduction)}$$

Step 3: The equation for the oxidation half-reaction can be balanced by inspection. The solution is basic, and the atoms are balanced by adding four OH^- ions to the reactant side:

$$Zn(s) + 4OH^-(aq) \rightarrow Zn(OH)_4^{2-}(aq)$$

The charge is balanced by adding two electrons to the product side, thus giving each side a charge of -4:

$$Zn(s) + 4OH^-(aq) \rightarrow Zn(OH)_4^{2-}(aq) + 2e^-$$

The reduction equation is balanced by using the four substeps given for half-reactions involving oxygen.

(a) The hydrogen atoms are already balanced (two on each side):

$$H_2O(l) \rightarrow H_2(g)$$

(b) The oxidation state of each hydrogen atom falls from $+1$ in H_2O to 0 in H_2. There are two hydrogen atoms, so two electrons are gained by the reactants:

$$H_2O(l) + 2e^- \rightarrow H_2(g)$$

(c) The total charge on the reactant side is -2; the charge on the product side is 0. The solution is basic, so the charge is balanced by adding two OH^- ions to the product side:

$$H_2O(l) + 2e^- \rightarrow H_2(g) + 2OH^-(aq)$$

Each side now has a charge of -2.

(d) The hydrogen and oxygen atoms are balanced by adding one H_2O molecule to the reactant side:

$$2H_2O(l) + 2e^- \rightarrow H_2(g) + 2OH^-(aq)$$

Step 4: Equalize the numbers of electrons lost and gained. This step is unnecessary in this example because the two half-reaction equations already have the same number of electrons:

$$Zn(s) + 4OH^-(aq) \rightarrow Zn(OH)_4^{2-}(aq) + 2e^-$$

$$2H_2O(l) + 2e^- \rightarrow H_2(g) + 2OH^-(aq)$$

Step 5: Add the half-reaction equations to obtain the balanced net ionic equation:

$$Zn(s) + 4OH^-(aq) + 2H_2O(l) \rightarrow Zn(OH)_4^{2-}(aq) + H_2(g) + 2OH^-(aq)$$

The equation can be simplified by subtracting two OH^- ions from each side:

$$Zn(s) + 2OH^-(aq) + 2H_2O(l) \rightarrow Zn(OH)_4^{2-}(aq) + H_2(g)$$

PRACTICE EXERCISE 11.12

Sodium dithionite ($Na_2S_2O_4$) is used in analytical chemistry as a strong reducing agent. Write the half-reaction equation for the oxidation of the dithionite ion ($S_2O_4^{2-}$) to sulfite ion (SO_3^{2-}) in basic solution.

PRACTICE EXERCISE 11.13

Write the net ionic equation for the reaction in basic solution of chromate ion (CrO_4^{2-}) with I^-. The products are Cr^{3+} and IO_3^-.

Disproportionation Reactions

A **disproportionation reaction** is a redox reaction in which one substance is both oxidized and reduced. Each of the half-reactions starts with the same reactant, as illustrated in the next example.

EXAMPLE 11.9

Chlorine forms chloride and chlorate ions (ClO_3^-) in hot aqueous base. Write the net ionic equation.

SOLUTION

Step 1: The unbalanced equation is

$$Cl_2(g) \rightarrow Cl^-(aq) + ClO_3^-(aq)$$

This is a disproportionation reaction because the oxidation state of chlorine changes from 0 in Cl_2 to -1 in Cl^- and $+5$ in ClO_3^-.

Step 2: Unbalanced equations for the half-reactions are

$$Cl_2(g) \rightarrow Cl^-(aq) \qquad \text{(reduction)}$$

$$Cl_2(g) \rightarrow ClO_3^-(aq) \qquad \text{(oxidation)}$$

Step 3: The balanced half-reaction equations are

$$Cl_2(g) + 2e^- \rightarrow 2Cl^-(aq)$$

$$Cl_2(g) + 12OH^-(aq) \rightarrow 2ClO_3^-(aq) + 6H_2O(l) + 10e^-$$

(Be sure to verify these equations.)

Step 4: To equalize the electrons, the reduction half-reaction is multiplied by 5; the oxidation half-reaction is left as is:

$$5 \times [Cl_2(g) + 2e^- \rightarrow 2Cl^-(aq)]$$

$$Cl_2(g) + 12OH^-(aq) \rightarrow 2ClO_3^-(aq) + 6H_2O(l) + 10e^-$$

Step 5: Adding the half-reaction equations gives the net ionic equation:

$$6Cl_2(g) + 12OH^-(aq) \rightarrow 10Cl^-(aq) + 2ClO_3^-(aq) + 6H_2O(l)$$

Dividing by 2 simplifies the coefficients:

$$3Cl_2(g) + 6OH^-(aq) \rightarrow 5Cl^-(aq) + ClO_3^-(aq) + 3H_2O(l)$$

PRACTICE EXERCISE 11.14

Manganate ion (MnO_4^{2-}) disproportionates in acidic solution to permanganate ion (MnO_4^-) and solid MnO_2. Write the net ionic equation.

DIGGING DEEPER

The Oxidation Number Method for Balancing Redox Equations

In a redox reaction, the number of electrons gained by the oxidizing agent is equal to the number lost by the reducing agent. The total change in oxidation number is zero. The **oxidation number method** for balancing oxidation–reduction equations equalizes the oxidation number gain and loss before balancing the other atoms. Half-reactions are not written.

Consider the oxidation of platinum by a mixture of nitric and hydrochloric acids. The principal products are nitrogen dioxide gas and chloroplatinic acid (H_2PtCl_6). The steps in the oxidation number method are:

- ***Step 1:*** *Write the unbalanced equation:*

$$Pt(s) + HNO_3(aq) + HCl(aq) \rightarrow NO_2(g) + H_2PtCl_6(aq)$$

- ***Step 2:*** *Identify the oxidation number changes.* In this reaction, the oxidation number of platinum increases from 0 in Pt to +4 in H_2PtCl_6. The oxidation number of nitrogen decreases from +5 in HNO_3 to +4 in NO_2.

- ***Step 3:*** *Determine the molar ratio of the substances oxidized and reduced.* Pt atoms are oxidized from 0 to +4; N atoms are reduced from +5 to +4. Because the total change in oxidation number is zero, four N atoms must be reduced for every Pt atom oxidized. Hence, there must be 4 mol of HNO_3 for each mole of Pt on the reactant side of the equation and 4 mol of NO_2 for each mole of H_2PtCl_6 on the product side:

$$Pt(s) + 4HNO_3(aq) + HCl(aq) \rightarrow H_2PtCl_6(aq) + 4NO_2(g)$$

Note that up to this point we have balanced only the nitrogen and platinum atoms.

Step 4: *Balance the remaining atoms and charges (if any) by inspection, but do not tamper with the ratios established in Step 3.* Balancing the chlorine atoms gives:

$$Pt(s) + 4HNO_3(aq) + 6HCl(aq) \rightarrow H_2PtCl_6(aq) + 4NO_2(g)$$

The shortage of oxygen and hydrogen atoms on the product side can be corrected by adding four H_2O molecules:

$$Pt(s) + 4HNO_3(aq) + 6HCl(aq) \rightarrow H_2PtCl_6(aq) + 4NO_2(g) + 4H_2O(l)$$

The equation is now balanced.

PRACTICE EXERCISE 11.15

Balance the equations in Examples 11.8 and 11.9 by the oxidation number method.

The oxidation number method is quick, especially in the hands of an expert. Furthermore, it can be used to balance redox reactions that do not occur in aqueous solutions and also reactions that do not involve ions. You should practice the half-reaction method first because it will help you build up a stock of half-reactions useful for future work. After you learn this method, you can then add the oxidation number method to your collection of skills.

11.4 REDOX STOICHIOMETRY

Stoichiometry uses the mole ratios in an equation to answer questions of *How much*? This is also true of redox stoichiometry, but here the calculations are often based on net ionic equations rather than molecular equations. Example 11.10 deals with a *redox titration* in which the concentration of a potassium permanganate solution is determined by titrating it against oxalic acid. Deep purple permanganate ion is reduced to pale pink Mn^{2+} ion during the titration (see Demonstration 11.2). As long as any oxalic acid remains in the flask, each added drop of potassium permanganate loses its color; when all the oxalic acid has been consumed, the first excess drop of potassium permanganate will retain its deep color, indicating the endpoint of the titration. In effect, the potassium permanganate acts as its own indicator.

Titrations were described in Section 4.6. In a redox titration, an oxidation–reduction reaction is used to obtain quantitative information about one of the reactants.

EXAMPLE 11.10

A 23.30-mL sample of a $KMnO_4$ solution is decolorized by 0.1111 g of oxalic acid ($H_2C_2O_4$). The products are Mn^{2+} and CO_2 gas. Calculate the molarity of the $KMnO_4$ solution.

SOLUTION The first step in any stoichiometric problem is to write a balanced equation. The net ionic equation for this reaction is

$$2MnO_4^-(aq) + 6H^+(aq) + 5H_2C_2O_4(aq) \rightarrow 2Mn^{2+}(aq) + 10CO_2(g) + 8H_2O(l)$$

(Use the steps in Section 11.3 to verify the balancing of this equation.) Note that 2 mol of MnO_4^- (or 2 mol of $KMnO_4$) react with 5 mol of $H_2C_2O_4$.

The problem is solved in the usual way:

$$0.1111 \text{ g } H_2C_2O_4 \rightarrow \text{mol } H_2C_2O_4 \rightarrow \text{mol } KMnO_4 \rightarrow \text{molarity of } KMnO_4$$

The molar mass of oxalic acid is 90.04 g. Hence,

$$0.1111 \text{ g } H_2C_2O_4 \times \frac{1 \text{ mol } H_2C_2O_4}{90.04 \text{ g } H_2C_2O_4} \times \frac{2 \text{ mol } KMnO_4}{5 \text{ mol } H_2C_2O_4} = 4.936 \times 10^{-4} \text{ mol } KMnO_4$$

Thus 23.30 mL (0.02330 L) of solution contains 4.936×10^{-4} mol $KMnO_4$. The molarity of the $KMnO_4$ solution is

$$M = \frac{4.936 \times 10^{-4} \text{ mol}}{0.02330 \text{ L}} = 0.02118 \text{ mol/L}$$

DEMONSTRATION 11.2

STANDARDIZING A POTASSIUM PERMANGANATE SOLUTION WITH OXALIC ACID

A $KMnO_4$ solution of unknown concentration is added from a buret to a flask containing a weighed amount of oxalic acid ($H_2C_2O_4$) dissolved in water. Each drop of purple $KMnO_4$ solution is immediately decolorized by the oxalic acid.

Finally, a drop of $KMnO_4$ solution retains its color, showing that the oxalic acid has been used up. The $KMnO_4$ concentration can be calculated from the mass of oxalic acid and the volume of solution delivered by the buret (see Example 11.10).

PRACTICE EXERCISE 11.16

The oxidation of Fe^{2+} ions by permanganate in acidic solution is accompanied by the formation of Fe^{3+} ions and Mn^{2+} ions. (a) What is the mole ratio of MnO_4^- to Fe^{2+} in this reaction? (b) How many moles of $FeSO_4$ would be oxidized by 100.0 mL of 0.02118 *M* $KMnO_4$ solution?

CHAPTER REVIEW

Starred entries are based on the *Digging Deeper* sections.

LEARNING OBJECTIVES BY SECTION

11.1 1. Write equations for the half-reactions that occur during oxidation–reduction reactions involving ions.
2. Identify the oxidizing and reducing agents in an oxidation–reduction reaction.

11.2 1. Calculate the oxidation number of each atom in a molecule or ion.
2. State the maximum and minimum oxidation numbers associated with each group of representative elements.
3. Use oxidation numbers to determine whether a reaction is a redox reaction.
4. Use oxidation numbers to predict whether a substance can be an oxidizing agent, a reducing agent, or both.

11.3 1. Balance redox equations by the half-reaction method.
2. Balance equations for disproportionation reactions.
*3. Balance redox equations by the oxidation number method.

11.4 1. Perform stoichiometric calculations that involve redox reactions.

KEY TERMS BY SECTION

11.1 Half-reaction
Oxidation
Oxidation–reduction reaction
Oxidizing agent
Redox reaction
Reducing agent
Reduction

11.2 Oxidation number
Oxidation state

11.3 Disproportionation reaction
Half-reaction method
*Oxidation number method

FINAL EXERCISES

Answers to exercises with blue numbers are given in Appendix D. Starred exercises are based on the *Digging Deeper* sections.

PART A. QUESTIONS AND PROBLEMS BY SECTION

Oxidation–Reduction Reactions (Section 11.1)

11.1 Define *oxidation* in terms of electron transfer. Use your definition to explain why substances that combine with oxygen are oxidized.

11.2 Define *reduction* in terms of electron transfer. Use your definition to explain why substances that lose oxygen are reduced.

11.3 (a) Why do oxidation and reduction always occur together?
(b) What substance is reduced when iron forms rust (an oxide of iron)?

11.4 (a) Define *oxidizing agent* and *reducing agent*, and state what happens to each one during the course of an oxidation–reduction reaction.

(b) Identify the oxidizing and reducing agents in the rusting of iron.

11.5 What is a *half-reaction*? What species always appears in a half-reaction equation but never appears in a net equation?

11.6 (a) Are electrons placed on the reactant side or on the product side of an oxidation half-reaction equation? A reduction half-reaction equation?
(b) What must be done to half-reaction equations before they can be added to give a net equation?

11.7 For each of the following reactions, write the half-reaction equations and the net ionic equation. (*Hint*: The activity series, Table 4.6, p. 144, may be helpful.)
(a) magnesium metal plus hydrochloric acid
(b) aluminum metal plus aqueous silver nitrate
(c) $Zn(s) + Br_2(aq) \rightarrow ZnBr_2(aq)$

11.8 For each of the following reactions, write the half-reaction equations and the net ionic equation. (*Hint*: The activity series, Table 4.6, p. 144, may be helpful.)
(a) lead metal plus aqueous copper sulfate
(b) $2Na(s) + 2H_2O(l) \rightarrow 2NaOH(aq) + H_2(g)$
(c) $Cl_2(g) + 2KBr(aq) \rightarrow Br_2(l) + 2KCl(aq)$

11.9 Identify the oxidizing and reducing agents for each of the reactions in Exercise 11.7. State which substance is oxidized and which is reduced.

11.10 Identify the oxidizing and reducing agents for each of the reactions in Exercise 11.8. State which substance is oxidized and which is reduced.

11.11 Write an equation for the formation of ammonia (NH_3) from its elements. State which element is oxidized and which is reduced.

11.12 Write an equation for the electrolysis of water. State which element is oxidized and which is reduced.

Oxidation Numbers (Section 11.2)

11.13 (a) Explain why the maximum oxidation number of a main group element does not exceed its group number.
(b) Which main group elements do not exhibit the maximum oxidation number of their group?

11.14 (a) How would you find the lowest (most negative) oxidation state of a nonmetallic element from its group number?
(b) Is there any nonmetal that does not exhibit the lowest oxidation state of its group?

11.15 Explain how oxidation numbers can be used to identify potential oxidizing agents.

11.16 Explain how oxidation numbers can be used to identify potential reducing agents.

11.17 Why can substances such as H_2O_2 and SO_2 function as oxidizing agents and as reducing agents?

11.18 Give examples of elements that have both positive and negative oxidation states. Where are such elements located in the periodic table?

11.19 List the following nitrogen oxides in order of increasing oxidation number of nitrogen: N_2O_4, N_2O, NO_2, NO, N_2O_3, and N_2O_5.

11.20 List the following acids in order of increasing oxidation number of chlorine: $HClO_3$, $HOCl$, $HClO_4$, and $HClO_2$.

11.21 Give the oxidation number of each element in the following compounds:
(a) $XeOF_2$
(b) H_3BO_3
(c) KIF_4
(d) BrF_3
(e) $KSbF_6$
(f) $H_4I_2O_9$
(g) $Na_2Sn(OH)_6$
(h) $KCrO_3Cl$
(i) $(NH_4)_2Cr_2O_7$
(j) $(CH_3)_2S$

11.22 Give the oxidation number of each element in the following ions:
(a) AlO_2^-
(b) $Sn(OH)_4^{2-}$
(c) PH_4^+
(d) $CuCl_4^{2-}$
(e) NH_3OH^+
(f) $CuOH^+$
(g) $TaOF_6^{3-}$
(h) CoF_6^{3-}
(i) $MoO_3(OH)^-$
(j) CN^-

11.23 Give the oxidation number of oxygen in each of the following molecules and ions:
(a) hydrogen peroxide, H_2O_2
(b) barium peroxide, BaO_2
(c) superoxide ion, O_2^-
(d) oxygen difluoride, OF_2
(e) dioxygen difluoride, O_2F_2

11.24 Give the oxidation number of sulfur in each of the following molecules and ions:
(a) H_2SO_4
(b) S_8
(c) H_2S
(d) thiosulfate ion, $S_2O_3^{2-}$
(e) tetrasulfur tetranitride, S_4N_4

11.25 Give the oxidation number of carbon in each of the following molecules and ions:
(a) CO_2
(b) CH_4
(c) methanol, CH_3OH
(d) CCl_4
(e) formate ion, $HCOO^-$
(f) buckminsterfullerene, C_{60}

11.26 Give the oxidation number of nitrogen in each of the following molecules and ions:

(a) HNO_3
(b) NH_3
(c) Mg_3N_2
(d) hydroxylamine, NH_2OH
(e) NO_2^-
(f) NO_2

11.27 Which of the following reactions are redox reactions? For each redox reaction, identify the oxidizing and reducing agents.
(a) $N_2O_5(g) + H_2O(l) \longrightarrow 2HNO_3(aq)$
(b) $3NO_2(g) + H_2O(l) \longrightarrow 2HNO_3(aq) + NO(g)$
(c) $2SO_2(g) + O_2(g) \longrightarrow 2SO_3(g)$
(d) $3K_2S(aq) + K_2Cr_2O_7(aq) + 7H_2O(l) \longrightarrow 3S(s) + 2KCr(OH)_4(aq) + 6KOH(aq)$

11.28 Which of the following reactions are redox reactions? For each redox reaction, identify the oxidizing and reducing agents.
(a) $2O_3(g) \longrightarrow 3O_2(g)$
(b) $4NH_3(g) + 5O_2(g) \longrightarrow 4NO(g) + 6H_2O(g)$
(c) $CH_3OH(g) \longrightarrow CH_2O(g) + H_2(g)$
(d) $2PbCl_2(aq) + K_2Cr_2O_7(aq) + H_2O(l) \longrightarrow 2PbCrO_4(s) + 2KCl(aq) + 2HCl(aq)$

11.29 Use the periodic table inside the front cover to predict (a) the highest and (b) the lowest (most negative) oxidation states for arsenic.

11.30 Use the periodic table inside the front cover to predict (a) the highest and (b) the lowest (most negative) oxidation states for selenium.

11.31 Use oxidation numbers and Figure 11.1 to predict which of the following species can be oxidized:
(a) HNO_3
(b) H_2Se
(c) MnO_2
(d) Co^{2+}
(e) HI
(f) H_3PO_3
(g) CrO_4^{2-}
(h) $HClO_4$

11.32 Use oxidation numbers and Figure 11.1 to predict which of the following species can be reduced:
(a) CH_4
(b) F_2
(c) HNO_2
(d) CO_2
(e) H_2SO_3
(f) Bi_2S_3
(g) CO
(h) MnO_4^-

11.33 Which of the species in Exercise 11.31 can be used (a) only as an oxidizing agent, (b) only as a reducing agent, and (c) both as an oxidizing agent and a reducing agent?

11.34 Which of the species in Exercise 11.32 can be used (a) only as an oxidizing agent, (b) only as a reducing agent, and (c) both as an oxidizing agent and a reducing agent?

Balancing Redox Equations (Section 11.3)

11.35 List the steps for balancing redox reactions by the *half-reaction method.*

***11.36** List the steps for balancing redox reactions by the *oxidation number method.*

11.37 Use the half-reaction method to balance the following equations for reactions in acidic solution:
(a) $MnO_4^-(aq) + HSO_3^-(aq) \longrightarrow Mn^{2+}(aq) + HSO_4^-(aq)$
(b) $CuS(s) + NO_3^-(aq) \longrightarrow Cu^{2+}(aq) + S(s) + NO(g)$
(c) $Fe(NO_3)_3(aq) + H_2S(aq) \longrightarrow FeS(s) + S(s) + HNO_3(aq)$
(d) $HNO_3(aq) + H_2S(aq) \longrightarrow S(s) + NO(g)$
(e) $Pb(s) + PbO_2(s) + H_2SO_4(aq) \longrightarrow PbSO_4(s)$
(f) $MnO_2(s) + PbO_2(s) + HNO_3(aq) \longrightarrow HMnO_4(aq) + Pb(NO_3)_2(aq) + H_2O(l)$
(g) $ClO^-(aq) + I_2(s) \longrightarrow Cl^-(aq) + IO_3^-(aq)$
(h) $Br_2(aq) + SO_2(g) \longrightarrow Br^-(aq) + SO_4^{2-}(aq)$

11.38 Use the half-reaction method to balance the following equations for reactions in acidic solution:
(a) $P_4(s) + HNO_3(aq) \longrightarrow H_3PO_4(aq) + NO(g)$
(b) $HgCl_2(aq) + SnCl_2(aq) \longrightarrow Hg_2Cl_2(s) + SnCl_6^{2-}(aq)$
(c) $Cr_2O_7^{2-}(aq) + CH_3OH(aq) \longrightarrow Cr^{3+}(aq) + CH_2O(aq)$
(d) $H_2SO_4(aq) + H_2S(aq) \longrightarrow S(s)$
(e) $HNO_3(aq) + Fe(s) \longrightarrow Fe^{2+}(aq) + NH_4^+(aq)$
(f) $HNO_3(aq) + Ag(s) \longrightarrow Ag^+(aq) + NO(g)$
(g) $Ru(s) + HCl(aq) + HNO_3(aq) \longrightarrow RuCl_6^{3-}(aq) + NO_2(g)$
(h) $Ta(s) + HF(aq) + HNO_3(aq) \longrightarrow H_2TaOF_5(aq) + NO(g)$

11.39 Use the half-reaction method to balance the following equations for reactions in basic solution:
(a) $Al(s) + OH^-(aq) \longrightarrow Al(OH)_4^-(aq) + H_2(g)$
(b) $MnO_4^-(aq) + S^{2-}(aq) \longrightarrow MnO_2(s) + S(s)$
(c) $Zn(s) + NO_3^-(aq) + OH^-(aq) \longrightarrow NH_3(aq) + Zn(OH)_4^{2-}(aq)$
(d) $Pb(OH)_3^-(aq) + OCl^-(aq) \longrightarrow PbO_2(s) + Cl^-(aq)$

11.40 Use the half-reaction method to balance the following equations for reactions in basic solution:
(a) $MnO_4^-(aq) + C_2O_4^{2-}(aq) \longrightarrow MnO_2(s) + CO_3^{2-}(aq)$
(b) $Bi(OH)_3(s) + Sn(OH)_3^-(aq) \longrightarrow Bi(s) + Sn(OH)_6^{2-}(aq)$
(c) $MnO_4^-(aq) + SO_3^{2-}(aq) \longrightarrow MnO_2(s) + SO_4^{2-}(aq)$
(d) $Cu(NH_3)_4^{2+}(aq) + S_2O_4^{2-}(aq) \longrightarrow Cu(s) + SO_3^{2-}(aq) + NH_3(aq)$

11.41 Hydrogen peroxide can function as either an oxidizing agent or a reducing agent. Write half-reaction equations for
(a) H_2O_2 reduced to H_2O in acidic solution
(b) H_2O_2 oxidized to O_2 in acidic solution

11.42 Hydrogen peroxide can function as either an oxidizing agent or a reducing agent. Write half-reaction equations for
(a) H_2O_2 reduced to OH^- in basic solution
(b) H_2O_2 oxidized to O_2 in basic solution

11.43 Use the half-reactions from Exercise 11.41 to write balanced equations for the following reactions in acidic solution:
(a) $Cr_2O_7^{2-}(aq) + H_2O_2(aq) \rightarrow Cr^{3+}(aq) + O_2(g)$
(b) $H_2S(aq) + H_2O_2(aq) \rightarrow S(s)$
(c) $MnO_4^-(aq) + H_2O_2(aq) \rightarrow Mn^{2+}(aq) + O_2(g)$
(d) $Sn^{2+}(aq) + H_2O_2(aq) + HCl(aq) \rightarrow SnCl_6^{2-}(aq) + H_2O(l)$
(e) $HAsO_2(aq) + H_2O_2(aq) \rightarrow H_3AsO_4(aq)$

11.44 Use the half-reactions from Exercise 11.42 to write balanced equations for the following reactions in basic solution:
(a) $CrO_2^-(aq) + H_2O_2(aq) \rightarrow CrO_4^{2-}(aq)$
(b) $Fe(OH)_2(s) + H_2O_2(aq) \rightarrow Fe(OH)_3(s)$
(c) $Mn(OH)_2(s) + H_2O_2(aq) \rightarrow MnO_2(s)$
(d) $Co(NH_3)_6^{3+}(aq) + H_2O_2(aq) \rightarrow Co(NH_3)_6^{2+}(aq) + O_2(g)$
(e) $IO^-(aq) + H_2O_2(aq) \rightarrow I^-(aq) + O_2(g)$

11.45 Use the half-reaction method to balance the following disproportionation reactions:
(a) $NO_2(g) + H_2O(l) \rightarrow HNO_3(aq) + NO(g)$
(b) $ClO_2(g) \rightarrow ClO_2^-(aq) + ClO_3^-(aq)$ (basic solution)

11.46 Use the half-reaction method to balance the following disproportionation reactions that occur in basic solution:
(a) $Se(s) \rightarrow Se^{2-}(aq) + SeO_3^{2-}(aq)$
(b) $P_4(s) + OH^-(aq) \rightarrow PH_3(g) + H_2PO_2^-(aq)$

11.47 The following reaction takes place in acidic solution:

$$HAsO_3^{2-}(aq) + BrO_3^-(aq) \rightarrow Br^-(aq) + H_3AsO_4(aq)$$

(a) Write the balanced net ionic equation.
(b) Write the molecular equation. Assume that the reactants are Na_2HAsO_3, $KBrO_3$, and HCl.

11.48 Write a balanced molecular equation for the reaction of sodium bismuthate ($NaBiO_3$) with manganese(II) nitrate ($Mn(NO_3)_2$) in a solution acidified with HNO_3. Sodium bismuthate is a sparingly soluble solid; the products are $Bi(NO_3)_3$ and $NaMnO_4$.

***11.49** Balance the following equations by the oxidation number method:
(a) $Mn(OH)_2(s) + O_2(g) \rightarrow Mn(OH)_3(s)$ (basic solution)
(b) $Ti^{3+}(aq) + RuCl_5^{2-}(aq) \rightarrow Ru(s) + TiO^{2+}(aq)$ (basic solution)
(c) $H_2S(g) + SO_2(g) \rightarrow S(s) + H_2O(g)$
(d) $FeSO_4(aq) + K_2Cr_2O_7(aq) + H_2SO_4(aq) \rightarrow Fe_2(SO_4)_3(aq) + Cr_2(SO_4)_3(aq) + KHSO_4(aq) + H_2O(l)$

***11.50** Balance the following equations by the oxidation number method:
(a) $N_2H_4(aq) + Cu(OH)_2(s) \rightarrow Cu(s) + N_2(g)$
(b) $Cr(OH)_2(s) \rightarrow Cr_2O_3(s) + H_2O(g) + H_2(g)$
(c) $HNO_3(g) \rightarrow NO_2(g) + O_2(g) + H_2O(g)$
(d) $KCN(aq) + K_2CrO_4(aq) + H_2O(l) \rightarrow KCr(OH)_4(aq) + KCNO(aq) + KOH(aq)$

Redox Stoichiometry (Section 11.4)

11.51 The oxidation of copper metal by nitric acid is accompanied by the formation of copper(II) ions and NO gas. How many grams of copper must react with excess nitric acid to produce 500 mL of NO gas measured at 20°C and 1.00 atm?

11.52 How many grams of Cr_2O_3 will be reduced to chromium metal by 1.00 mol of H_2?

11.53 The reaction between sodium oxalate ($Na_2C_2O_4$) and an acidified solution of potassium permanganate ($KMnO_4$) produces CO_2 and manganese(II) ions (Mn^{2+}). How many grams of sodium oxalate will be oxidized by 0.500 g of potassium permanganate?

11.54 Sodium bismuthate ($NaBiO_3$) oxidizes $MnCl_2$ to $NaMnO_4$ and is itself reduced to $Bi(OH)_3$. How many grams of $MnCl_2$ will be oxidized by 10.0 g of $NaBiO_3$?

11.55 Iodine reacts with thiosulfate ion ($S_2O_3^{2-}$) in acidic solution to form iodide ion and tetrathionate ion ($S_4O_6^{2-}$). Calculate the volume in milliliters of 0.100 M $Na_2S_2O_3$ needed to react with 7.50 g of I_2.

11.56 The reaction between potassium dichromate and tin(II) chloride in hydrochloric acid solution forms chromium(III) ions and $SnCl_6^{2-}$ ions. Calculate the volume in milliliters of 0.250 *M* $K_2Cr_2O_7$ needed to react with 3.50 g of $SnCl_2$.

11.57 How many grams of iron(II) sulfate can be oxidized in acidic solution by 25.00 mL of 0.600 *M* potassium permanganate? The unbalanced equation is

$$Fe^{2+}(aq) + MnO_4^-(aq) \rightarrow Mn^{2+}(aq) + Fe^{3+}(aq)$$

11.58 How many grams of bismuth metal will form when excess bismuth hydroxide ($Bi(OH)_3$) is treated with 25.0 mL of 0.100 *M* sodium stannite (Na_2SnO_2)? The sodium stannite is oxidized to sodium stannate (Na_2SnO_3).

11.59 Refer to Exercise 11.53. How many grams of CO_2 will be produced when 1.500 g of $Na_2C_2O_4$ reacts with 30.0 mL of 0.100 *M* $KMnO_4$?

11.60 How many grams of MnO_2 will form when 0.500 g of sodium sulfite (Na_2SO_3) is added to 50.0 mL of 0.0500 *M* $KMnO_4$ in the presence of a base? The sulfite ions are oxidized to sulfate ions.

PART B. MISCELLANEOUS QUESTIONS AND PROBLEMS

11.61 Does the oxidation number of an element in a compound ever correspond to an actual charge? Give examples to illustrate your answer.

11.62 Discuss the relationship between the definitions given for oxidation and reduction in Section 3.6 and those given in this chapter. Explain how the two sets of definitions are related. Which definitions are more comprehensive?

11.63 **(a)** It is often said that "oxidation numbers are arbitrary and not fundamental—only changes in oxidation number have significance." Is this statement true or false? Explain your answer.

(b) Suppose you try to balance redox equations and solve redox stoichiometry problems using oxidation numbers that are all two units higher than the accepted values. What, if anything, would this do to your results? Is your answer consistent with the statement in Part (a)? Explain.

11.64 It is generally assumed that an organic molecule is oxidized if it acquires more oxygen atoms. Test this assumption by comparing the oxidation states of carbon in ethane (CH_3CH_3), ethanol (CH_3CH_2OH), acetaldehyde (CH_3CHO), and acetic acid (CH_3COOH). Is it enough just to look at the number of oxygen atoms? What else should be considered?

11.65 A strong oxidizing agent known as *Caro's acid* has the formula H_2SO_5. Keeping in mind that sulfur is in periodic Group 6A, show that Caro's acid must contain peroxide oxygen. How many O—O bonds must there be in each molecule of H_2SO_5?

11.66 Some gold ores contain no more than 10 g of gold per ton. The gold is extracted from such ores by the *cyanide leaching* process, which consists of crushing the ore, mixing it with a basic solution containing CN^- ions, and aerating the mixture for several days.

(a) In this process, the gold is oxidized to $Au(CN)_2^-$ and the oxygen is reduced to OH^-. Write the net ionic equation for the reaction.

(b) Gold metal is recovered from the $Au(CN)_2^{2-}$ solution by treating it with powdered zinc. The zinc is oxidized to $Zn(CN)_4^{2-}$. Write the net ionic equation for the recovery reaction.

11.67 Salts of the cyanate ion (CNO^-) can be prepared by the reaction of cyanide ion (CN^-) with $Cu(NH_3)_2^{2+}$ in basic solution. The unbalanced equation is:

$$Cu(NH_3)_4^{2+}(aq) + CN^-(aq) \rightarrow CuCN(s) + CNO^-(aq) + NH_3(aq)$$

Complete and balance the equation.

11.68 Ce^{4+} is reduced to Ce^{3+} by $Fe(CN)_6^{4-}$ in basic solution. The unbalanced equation is

$$Ce^{4+}(aq) + Fe(CN)_6^{4-}(aq) \rightarrow Ce(OH)_3(s) + Fe(OH)_3(s) + NO_3^-(aq) + CO_3^{2-}(aq)$$

Complete and balance the equation.

***11.69** A little-used method for preparing alkali cyanides consists of heating a ferrocyanide (a compound containing the $Fe(CN)_6^{4-}$ ion) with a carbonate. If potassium salts are used, the unbalanced equation is

$$K_4Fe(CN)_6(s) + K_2CO_3(s) \rightarrow KCN(s) + KOCN(s) + Fe(s) + CO_2(g)$$

Balance this equation by the oxidation number method.

11.70 When solid KI is treated with a concentrated solution of sulfuric acid, I_2 and H_2S are formed. How many grams of I_2 will form if 3.00 g of H_2SO_4 is reduced to H_2S?

11.71 The reaction between iodine and sodium thiosulfate ($Na_2S_2O_3$) in basic solution produces I^- and $S_4O_6^{2-}$.

(a) Write the net ionic equation for the reaction.

(b) Calculate the number of grams of iodine that will dissolve in 50.0 mL of 0.362 *M* $Na_2S_2O_3$.

11.72 Hydrogen peroxide reduces cerium(IV) to cerium(III) in acid solution. The unbalanced equation is

$$Ce^{4+}(aq) + H_2O_2(aq) \rightarrow Ce^{3+}(aq) + O_2(g)$$

Calculate the molarity of a hydrogen peroxide solution if 10.0 mL of the solution reduces 40.0 mL of 0.500 *M* cerium(IV) solution.

11.73 Sulfur dioxide and hydrogen sulfide gases react on contact to form solid sulfur and water vapor.

(a) If 1.00 mol of SO_2 is mixed with 1.00 mol of H_2S, how many grams of sulfur are formed?

(b) What gases, and how many moles of each, will be in the vapor that remains after the reaction of Part (a)?

11.74 Zinc reacts with nitric acid to form Zn^{2+} and NH_4^+.

(a) Calculate the number of moles of ammonium ion produced when 1.00 g of zinc dissolves in 150 mL of 6.00 *M* HNO_3.

(b) Calculate the final molarities of the ammonium and nitrate ions in Part (a). Assume that there is no change in the solution volume.

11.75 Most iron(II) salts in aqueous solution are gradually oxidized to Fe(III) by dissolved oxygen. The oxygen is reduced to H_2O. What volume of oxygen, measured in milliliters at standard conditions, is absorbed by an acidic solution in which 5.0 g of $FeCl_2$ is converted to $FeCl_3$?

11.76 A 1.00-mol sample of Cl_2 disproportionates to Cl^- and ClO_3^- in hot potassium hydroxide solution. How many grams each of KCl and $KClO_3$ will be present in the residue when the solution is evaporated?

11.77 In acidic solution, permanganate ion (MnO_4^-) converts cobaltinitrite ion ($Co(NO_2)_6^{3-}$) to Co^{2+} and NO_3^-, and is itself reduced to Mn^{2+}.

(a) Write the balanced net ionic equation for this reaction. (*Hint*: Keep in mind that the ratio of Co to N is fixed.)

(b) The reaction in Part (a) has been used to determine the mass of potassium after it is precipitated as $K_2Na[Co(NO_2)_6]$. If one such precipitate decolorized 55.5 mL of 0.0250 *M* $KMnO_4$, how many moles of $K_2Na[Co(NO_2)_6]$ and how many milligrams of potassium did it contain?

11.78 Chlorine dioxide, ClO_2, a gas used in industry to bleach wood pulp and other materials, can be prepared by reacting sodium chlorate with hydrochloric acid; Cl_2 gas is a coproduct. A student used the half-reaction method to obtain the following net ionic equation:

$$2ClO_3^-(aq) + 2Cl^-(aq) + 4H^+(aq) \rightarrow 2ClO_2(g) + Cl_2(g) + 2H_2O(l)$$

Another student balanced the equation by trial and error and obtained a net ionic equation with different molar ratios:

$$4ClO_3^-(aq) + 12Cl^-(aq) + 16H^+(aq) \rightarrow 2ClO_2(g) + 7Cl_2(g) + 8H_2O(l)$$

How would you explain the results obtained by the above students? Your explanation should discuss the possibility of balancing the equation with still other coefficients that provide different molar ratios. (The following reference may be helpful: Carlos A. L. Filgueiras, *J. Chem. Educ.* 1992, **69**, 276–277.)

CHEMISTS AT WORK

THE DEVELOPMENT CHEMIST

Sometime today you used or will use a product that contains a surfactant. Surfactants are large molecules in which one end is a hydrocarbon and nonpolar, and the other end is ionic and polar. Since these compounds exhibit both nonpolar and polar characteristics, they have unique properties that function in cleaning (detergency), foaming, wetting, emulsifying, solubilizing, and dispersing. Surfactants are important components in a variety of products such as detergents, shampoos, lotions, toothpaste, dish cleaners, and cosmetics. Brian Frank, a 1989 graduate of Illinois State University, is a development chemist working with surfactants at the Stepan Company in Northfield, Illinois, a leading merchant-producer of the compounds.

Brian's interest in chemistry was first prompted by an enthusiastic and dynamic high school chemistry teacher. "My interest in chemistry continued during my freshman year in college," he says. "I learned that there were many opportunities based on the study of chemistry." As part of his collegiate program, he spent three semesters as a co-op student at a Fortune 500 company learning about quality assurance, polymer synthesis, and "scaling-up" reactions. Because of these experiences, he learned how industry works, how to interact with supervisors and colleagues, how to accept responsibility, and how to be accountable for results. During on-campus terms, Brian pursued an undergraduate research project on photoreduction of biological molecules sponsored by another Fortune 500 company; the results were reported at a symposium at the corporate headquarters.

A development chemist attempts to find profitable new uses for a company's existing or potential products; to use the company's products to meet specifications required by customers; to produce new products to meet the needs of, or to make improvements for, customers or markets; and/or to demonstrate commercial importance of new products or technologies. The development chemist begins where the fundamental research scientist ends. These are the responsibilities that Brian has in his job.

BRIAN FRANK

Some kind of surfactant is required in emulsion polymerizations, and since Brian had some experience in that technology, that's where his career began. His first responsibility was to evaluate the effectiveness and performance of various surfactants, existing and new, for the production of macromolecules (very large polymeric molecules); he did this for customers and for potential customers. His current assignment is to evaluate surfactants (and formulations that include surfactants) for use in cleaning products for the laundry, dishes, and hard surfaces. In this new position, he works with customers, both in his lab and in their labs, to determine whether formulations meet the needs and specifications required; new materials and formulations are developed and tested with the goal of increasing the performance or improving some property of a consumer product.

In addition to the general skills of a scientist, Brian also uses many chemical techniques and principles in his work. When new compounds are developed, one part of their characterization is the determination of their structure, and techniques such as NMR (Nuclear Magnetic Resonance Spectroscopy) are typically used. The properties of new compounds are frequently explained in terms of acid-base theory and/or colloid behavior.

Brian is an active member of the Education Committee of the Chemical Industrial Council of Illinois, an organization of volunteers from approximately 120 companies in the Chicago area that delivers services to secondary schools, teachers, and students in the Chicago area. He has participated in their Saturday workshops for teachers, and has been a mentor for a student from one of the local high schools. He has also produced a resource manual for teachers in which educational services available from the Council companies are described. Since 1992, he has been the chairman of the Council's annual spring Career Conference. Over 3200 high school students attend the event to learn about careers in chemistry. "I really enjoy these activities," he says. "This is one way I keep from becoming one-dimensional in my career development."

SURVEY OF THE ELEMENTS 2

Lightning over Los Angeles, California. Atmospheric nitrogen is converted to nitric (HNO_3) and nitrous (HNO_2) acids during thunderstorms. Rainfall deposits the acids on the ground, where they are utilized by plants.

OXYGEN, NITROGEN, AND THE HALOGENS

OUTLINE

PREVIEW

Fires and explosions are oxidation–reduction reactions, and so are the quiet changes of metabolism and growth. Among the principal actors in these chemical dramas are the atmospheric elements nitrogen and oxygen, whose redox reactions cause them to circulate continuously through the earth's crust and biosphere and back into the atmosphere again. The halogens, a family of elements whose chemistry is rich in redox reactions, also play important roles in the environment and in the chemistry of modern technology.

In this survey we will study nitrogen, oxygen, and the halogens (Group 7A). These important elements have rich and distinctive chemistries, and most of their reactions fall into the oxidation–reduction category. Nitrogen, oxygen, and fluorine are the lightest members of Groups 5A, 6A, and 7A, respectively, and like other second-period elements, each has unique properties that are not shared by other members of its group. The atoms of these elements form shorter and stronger bonds than larger atoms from the third and following periods. In addition to single bonds, oxygen forms double bonds, and nitrogen forms both double and triple bonds. The pi bonds of second-period atoms such as oxygen and nitrogen are especially strong because parallel *p* orbitals can undergo extensive overlap when the atoms are small.

S2.1 OXYGEN

Oxygen is the most reactive element in Group 6A, the only one that is a gas under ordinary conditions, and the only one that has no important positive oxidation states. Some of the atomic properties of oxygen are summarized in Table S2.1.

Oxygen is the most abundant element in the earth's crust, which contains 62.6% oxygen by atom count and 46.6% oxygen by mass. Oxygen compounds are major constituents of oceans, rocks, and living things. Elemental oxygen (O_2) constitutes

TABLE S2.1 Atomic Properties of Oxygen

Electron configuration	$[He]2s^22p^4$
Electronegativity	3.4
First ionization energy	1314 kJ/mol
Electron affinity	−141 kJ/mol
Atomic radius	74 pm
Ionic radius, O^{2-}	140 pm

20.9% of the atmosphere by molecule count, where it is second in abundance only to nitrogen.

Most of the oxygen in the atmosphere is produced by *photosynthesis*, a process in which green plant pigments called *chlorophylls* utilize solar energy to convert carbon dioxide and water into glucose ($C_6H_{12}O_6$):

$$6CO_2(g) + 6H_2O(l) \xrightarrow{h\nu} C_6H_{12}O_6(aq) + 6O_2(g)$$

A smaller amount of oxygen is produced in the upper atmosphere by the photolysis (light-induced decomposition) of water molecules:

$$2H_2O(g) \xrightarrow{h\nu} 2H_2(g) + O_2(g)$$

Oxygen is removed from the atmosphere and converted back into CO_2 and H_2O by the metabolism, combustion, and decay of organic compounds. It is estimated that the entire atmospheric oxygen supply is used and renewed every 2000 years.

Preparation, Properties, and Uses of Oxygen

Ordinary oxygen (O_2) is a colorless, odorless gas that dissolves slightly in water and condenses to a blue liquid at −183°C (Table S2.2). Commercial amounts of pure oxygen are obtained by the fractional distillation of liquid air. Limited amounts of oxygen can be prepared in the laboratory by the electrolysis of water or by the

TABLE S2.2 Physical Properties of Oxygen and Ozone

Property	O_2	O_3
Melting point	−218.4°C	−192.7°C
Boiling point	−183.0°C	−111.9°C
Density at STP	1.429 g/L	2.144 g/L
Solubility in 100 mL H_2O at 0°C[a]	4.89 cm^3	49 cm^3
Color	Colorless gas Pale blue liquid Pale blue solid	Bluish gas Blue-black liquid Violet-black solid

[a]Gas volumes measured at standard conditions.

decomposition of potassium chlorate with a manganese dioxide catalyst and heat:

$$2KClO_3(s) \xrightarrow{MnO_2(s)} 2KCl(s) + 3O_2(g)$$

Oxygen is evolved in a number of other decomposition reactions. The oxides of mercury and silver, for example, decompose when heated (see Demonstration 3.3, p. 101):

$$2HgO(s) \rightarrow 2Hg(l) + O_2(g)$$

$$2Ag_2O(s) \rightarrow 4Ag(s) + O_2(g)$$

A redox reaction occurs when ionic peroxides (compounds containing the O_2^{2-} ion) are heated or treated with water. For example, when BaO_2 is heated, the O_2^{2-} ion (which contains oxygen in a -1 oxidation state) disproportionates to form the oxide ion (O^{2-}) and elemental oxygen (zero oxidation state):

$$2BaO_2(s) \xrightarrow{>600°C} 2BaO(s) + O_2(g)$$

A similar reaction occurs when sodium peroxide is treated with water:

$$2Na_2O_2(s) + 2H_2O(l) \longrightarrow O_2(g) + 4NaOH(aq)$$

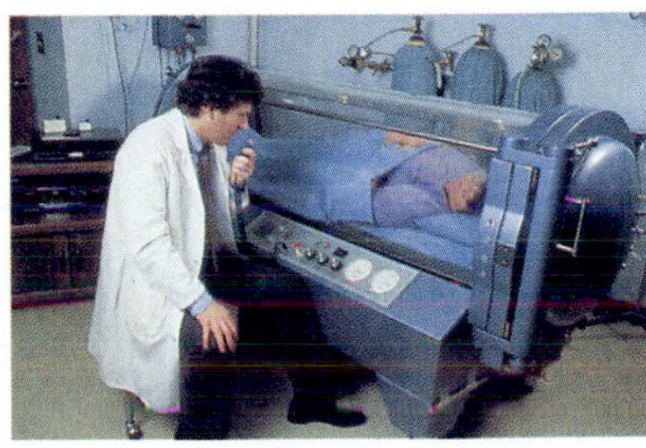

A patient is treated in a hyperbaric oxygen chamber.

Most industrial oxygen is used in the preparation of steel, where it serves to oxidize and remove excess carbon and certain impurities. Oxygen is also used to synthesize oxygen-containing compounds and to produce the high-temperature flames needed for metal fabrication (Figure S2.1). Liquid oxygen is used as an oxidizing agent for rocket fuels such as liquid hydrogen and hydrazine (N_2H_4); the expanding gaseous combustion products provide the thrust that lifts the rocket. Oxygen is used medically in life-support systems and in the *hyperbaric oxygen chamber*, where patients breathe 100% oxygen at a pressure of 2 atm, high enough to saturate their body fluids. Conditions that respond to hyperbaric oxygen treatment include carbon monoxide poisoning, gas gangrene, and bone infections. Gangrene and bone infections are often caused by organisms that cannot survive prolonged contact with oxygen.

Oxygen is the second-most electronegative element, and all elements except fluorine surrender electrons to some degree when they combine with oxygen. Thus, the term *oxidation*, which means "combining with oxygen," has also acquired the larger meaning of "losing electrons." Oxygen forms bonds with every element except helium, neon, and argon. Oxygen reacts slowly under room conditions, but at a high temperature or with a catalyst, it combines directly and exothermically with most

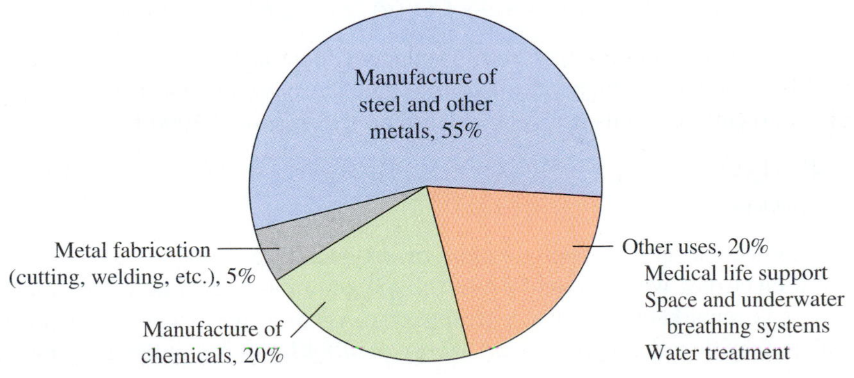

Figure S2.1

The commercial uses of oxygen.

DEMONSTRATION S2.1 STEEL WOOL BURNING IN OXYGEN

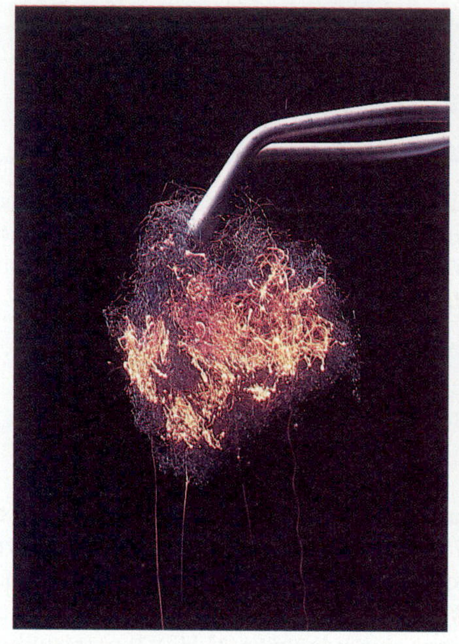

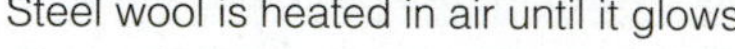

Steel wool is heated in air until it glows.

An oxygen atmosphere causes the hot steel wool to burst into flame. The principal product under these conditions is Fe_3O_4.

elements (Demonstration S2.1):

$$3Fe(s) + 2O_2(g) \rightarrow Fe_3O_4(s) \qquad \Delta H^\circ = -1118 \text{ kJ}$$

$$Zn(s) + \tfrac{1}{2}O_2(g) \rightarrow ZnO(s) \qquad \Delta H^\circ = -348 \text{ kJ}$$

$$S(s) + O_2(g) \rightarrow SO_2(g) \qquad \Delta H^\circ = -297 \text{ kJ}$$

Oxygen also combines with most compounds that are not already fully oxidized:

$$SO_2(g) + \tfrac{1}{2}O_2(g) \xrightarrow{Pt(s)} SO_3(g) \qquad \Delta H^\circ = -99 \text{ kJ}$$

$$CO(g) + \tfrac{1}{2}O_2(g) \longrightarrow CO_2(g) \qquad \Delta H^\circ = -283 \text{ kJ}$$

Using a welding torch.

Organic compounds burn in oxygen to produce carbon dioxide and water. The combustion of acetylene is highly exothermic, and the oxyacetylene torches used for welding and cutting metals generate temperatures in excess of 3000°C:

$$\underset{\text{acetylene}}{2C_2H_2(g)} + 5O_2(g) \rightarrow 4CO_2(g) + 2H_2O(g) \qquad \Delta H^\circ = -2511 \text{ kJ}$$

The molecular orbital configuration of oxygen (see Table S2.3) shows two unpaired electrons, which should cause the oxygen molecule to be highly reactive. In fact, it is remarkable that the reactions of oxygen do not get out of hand more often than they do—the high O_2 bond energy of 498 kJ/mol is a moderating factor.

CHEMICAL INSIGHT

OXYGEN, THE BREATH OF LIFE

Life's need for oxygen is paradoxical. Most organisms need oxygen, yet oxygen is lethal to cells that cannot cope with its reactivity. Initially, the earth's atmosphere contained no free oxygen, and the earth's early organisms were *anaerobes* (life-forms that do not tolerate oxygen). Anaerobes derive their energy not from oxidation but from the fermentation of organic compounds—the same type of reaction that produces ethyl alcohol from glucose. One mole of glucose releases 210 kJ of energy when it is fermented to ethyl alcohol:

$$C_6H_{12}O_6(aq) \xrightarrow{\text{enzymes}} \underset{\text{ethyl alcohol}}{2C_2H_5OH(aq)} + 2CO_2(g) \qquad \Delta H^\circ = -210 \text{ kJ}$$

Eventually, some plants evolved the ability to photosynthesize glucose from carbon dioxide and water, releasing oxygen as a by-product—a disastrous form of air pollution from an anaerobe's point of view. Photosynthetic plants changed the lower atmosphere from a reducing blanket of methane, ammonia, and nitrogen into a harsh oxidizing medium containing 20% oxygen. Some anaerobes learned to survive by avoiding air altogether. The most successful organisms, however, developed enzymes that use oxygen to oxidize glucose:

$$C_6H_{12}O_6(aq) + 6O_2(g) \xrightarrow{\text{enzymes}} 6CO_2(g) + 6H_2O(g) \qquad \Delta H^\circ = -2816 \text{ kJ}$$

The complete oxidation of glucose releases almost 14 times more energy than the fermentation of glucose, thus making it possible for larger and more energy-intensive species to survive.

An average person requires about 200 mL of oxygen (measured at STP) per minute while resting, and up to 39 times that volume during periods of activity. The alert human brain, which constitutes only about 2% of the body's mass, consumes over 20% of the body's resting oxygen requirement.

TABLE S2.3 Bond Data for Various Dioxygen Species

Name	Formula	Molecular Orbital Configuration of Valence Electrons	Bond Energy (kJ/mol)	Bond Length (pm)
Oxygen	O_2	$(\sigma_{2s})^2 (\sigma^*_{2s})^2 (\pi_{2p})^4 (\sigma_{2p})^2 (\pi^*_{2p})^2$	498	121
Superoxide ion	O_2^-	$(\sigma_{2s})^2 (\sigma^*_{2s})^2 (\pi_{2p})^4 (\sigma_{2p})^2 (\pi^*_{2p})^3$	393	126
Peroxide ion	O_2^{2-}	$(\sigma_{2s})^2 (\sigma^*_{2s})^2 (\pi_{2p})^4 (\sigma_{2p})^2 (\pi^*_{2p})^4$	—	149
Dioxygenyl ion[a]	O_2^+	$(\sigma_{2s})^2 (\sigma^*_{2s})^2 (\pi_{2p})^4 (\sigma_{2p})^2 (\pi^*_{2p})^1$	646	112

[a]Dioxygenyl ion is a prevalent form of oxygen in the ionosphere, the region of the atmosphere that lies between 50 and 200 km above the earth.

Ozone

Ozone (O_3) is an elemental form of oxygen that consists of triatomic molecules, and like oxygen, it is a gas at room temperature. Oxygen and ozone are **allotropes**—different forms of an element that exist in the same physical state. Resonance structures for the V-shaped ozone molecule

$$\ddot{O}{=}\ddot{O}{-}\ddot{O}: \qquad :\ddot{O}{-}\ddot{O}{=}\ddot{O}$$

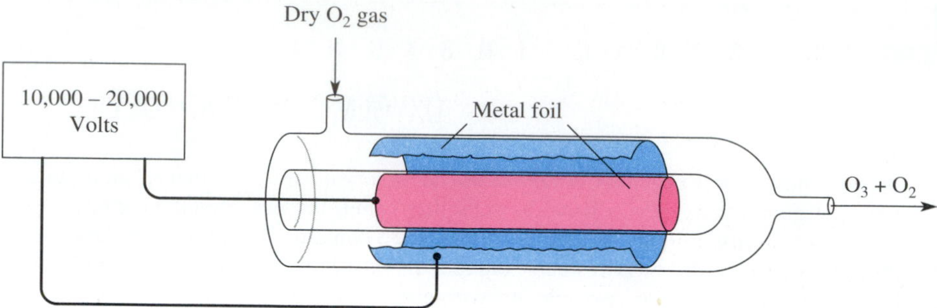

Figure S2.2

Schematic diagram of an ozonizer. An electric discharge between the inner and outer metal foils converts some of the oxygen that passes through the ozonizer to ozone.

suggest that one pair of bonding electrons is in a delocalized pi orbital shared by all three atoms. Both bond lengths are 128 pm and the bond angle is 117°, somewhat less than the trigonal planar value of 120°. The properties of ozone and O_2 are compared in Table S2.2.

Ozone is a pale blue gas that condenses to a deep blue liquid at $-112°C$. In the ozone layer of the upper atmosphere, ozone absorbs much of the sun's ultraviolet radiation and partially protects life on earth from mutations, cancer, sunburn, and other afflictions that can be induced by ultraviolet light (see *Chemical Insight: Free Radicals and the Ozone Layer* in Chapter 9). Direct contact with ozone is deleterious, however, and in the lower atmosphere ozone is a pollutant. Concentrations above 0.1 ppm by volume are considered unsafe for prolonged exposure.

Ozone is generated artificially by passing oxygen or air through an *ozonizer* (Figure S2.2) where it is subjected to an electrical discharge:

$$3O_2(g) \rightarrow 2O_3(g) \qquad \Delta H° = +285.4 \text{ kJ}$$

The distinctive odor of ozone is sometimes noticeable near sparking electrical equipment. The presence of ozone can be verified by passing air through a solution of potassium iodide and starch. Ozone, but not O_2, will oxidize iodide ion to iodine. The iodine forms a deep blue complex with starch (Demonstration S2.2):

$$\underset{\text{colorless}}{2I^-(aq)} + O_3(g) + H_2O(l) \rightarrow 2OH^-(aq) + \underset{\substack{\text{blue color}\\\text{with starch}}}{I_2(aq)} + O_2(g)$$

PRACTICE EXERCISE S2.1

Write the half-reaction equations for the reaction of aqueous iodide ion with ozone.

Ozone is a more vigorous oxidizing agent than O_2, and its uses are based on this property. Ozonizers were once used in restaurants and hospitals to destroy airborne germs, but this use has been discontinued because ozone can be harmful to humans as well as to bacteria. Ozone is increasingly favored as a replacement for chlorine in the treatment of public water supplies. Like chlorine, ozone is capable of destroying bacteria and oxidizing organic wastes. Although ozone is more expensive than chlorine, it does not affect the taste of the water and it produces no harmful chlorinated by-products.

DEMONSTRATION S2.2

TESTING FOR OZONE WITH POTASSIUM IODIDE:
$2I^-(aq) + O_3(g) + H_2O(l) \longrightarrow 2OH^-(aq) + I_2(aq) + O_2(g)$

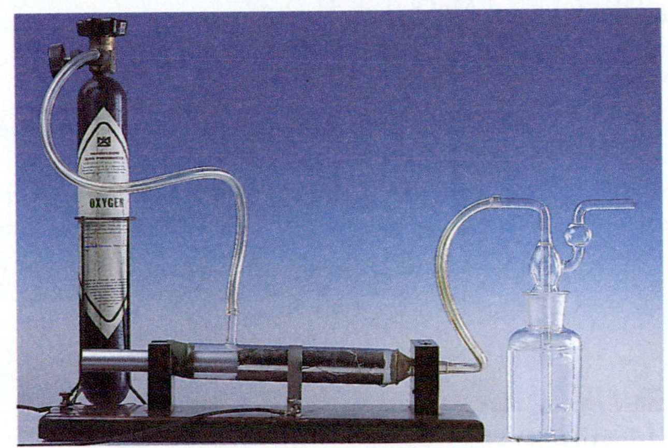

The ozonizer is turned off. Oxygen bubbles into a colorless solution of potassium iodide and starch.

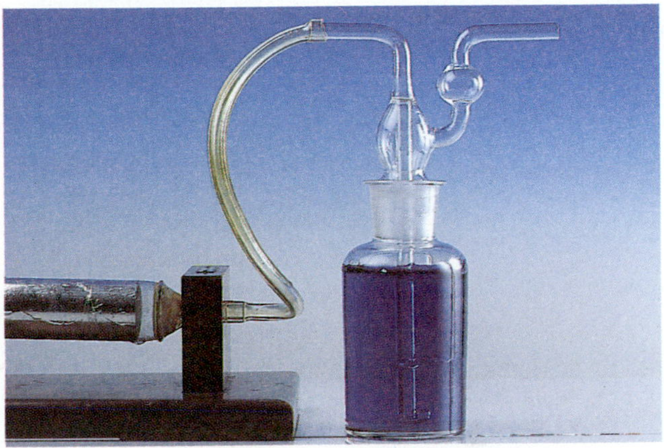

The ozonizer is turned on. Ozone oxidizes iodide ion to iodine, which reacts with starch to form a blue compound.

Oxides

Most oxides contain oxygen in the −2 oxidation state. These "normal" oxides are discussed in this section. (Peroxides and superoxides, which contain oxygen in more positive oxidation states, are discussed in the next section.) The bonding in normal oxides ranges from ionic in the oxides of very active metals ($CaO = Ca^{2+} + O^{2-}$) to covalent in nonmetal oxides such as SiO_2, CO_2, and SO_2. The degree of covalent character increases with the electronegativity of the atom bonded to oxygen, that is, from lower left to upper right in the periodic table. Oxides containing small metal atoms in the +2 or +3 oxidation state generally have significant covalent character; examples include BeO, Al_2O_3, and transition metal oxides such as Cr_2O_3.

Reactions of Metal Oxides. Most metal oxides are **basic oxides**; that is, they react with water to form ionic hydroxides and with aqueous acids to form water. The oxides of the Group 1A metals and the oxides of calcium, strontium, and barium in Group 2A are soluble and react directly with water. The oxide ion is the actual reactant:

$$O^{2-} + H_2O(l) \rightarrow 2OH^-(aq)$$

$$Na_2O(s) + H_2O(l) \rightarrow 2NaOH(aq) \qquad (= Na^+(aq) + OH^-(aq))$$

$$CaO(s) + H_2O(l) \rightarrow Ca(OH)_2(aq) \qquad (= Ca^{2+}(aq) + 2OH^-(aq))$$

Other metal oxides such as MgO and Al_2O_3 are sparingly soluble and react with water only to an insignificant extent. Their hydroxides can be prepared by metathesis

$$MgCl_2(aq) + 2NaOH(aq) \rightarrow Mg(OH)_2(s) + 2NaCl(aq)$$
$$Al_2(SO_4)_3(aq) + 6NaOH(aq) \rightarrow 2Al(OH)_3(s) + 3Na_2SO_4(aq)$$

and the oxide–hydroxide relationship can be shown by heating the hydroxide to eliminate the water:

$$Mg(OH)_2(s) \xrightarrow{\text{heat}} MgO(s) + H_2O(g)$$
$$2Al(OH)_3(s) \xrightarrow{\text{heat}} Al_2O_3(s) + 3H_2O(g)$$

Note that the oxidation state of a metal does not change when its oxide is converted to the hydroxide, or vice versa.

The reaction of a basic oxide with aqueous acid is similar to that of the corresponding hydroxide. In each case, $H^+(aq)$ is a reactant, and the metal ion and water are products:

$$MgO(s) + 2H^+(aq) \rightarrow Mg^{2+}(aq) + H_2O(l)$$
$$Mg(OH)_2(s) + 2H^+(aq) \rightarrow Mg^{2+}(aq) + 2H_2O(l)$$

$$Fe_2O_3(s) + 6H^+(aq) \rightarrow 2Fe^{3+}(aq) + 3H_2O(l)$$
$$Fe(OH)_3(s) + 3H^+(aq) \rightarrow Fe^{3+}(aq) + 3H_2O(l)$$

The oxidation state of the metal remains unchanged during these reactions.

The oxides of very active metals react with carbon dioxide to form carbonates. The oxide ion is the reactant:

$$O^{2-} + CO_2(g) \rightarrow CO_3^{2-}$$
$$Li_2O(s) + CO_2(g) \rightarrow Li_2CO_3(s)$$
$$CaO(s) + CO_2(g) \rightarrow CaCO_3(s)$$

Acidic and basic oxides are often referred to as acidic and basic *anhydrides*. Anhydride means "without water."

Reactions of Nonmetal Oxides. Most nonmetal oxides are **acidic oxides**; that is, they react with water to form acids and with aqueous bases to form water. Some examples are

$$P_4O_{10}(s) + 6H_2O(l) \rightarrow \underset{\text{phosphoric acid}}{4H_3PO_4(aq)}$$

$$SO_2(g) + H_2O(l) \rightarrow \underset{\text{ions of "sulfurous acid"}}{H^+(aq) + HSO_3^-(aq)}$$

(Sulfurous acid would be H_2SO_3, but as mentioned on p. 142, there is no evidence that this molecule actually exists.)

The reaction of an acidic oxide with aqueous base is similar to that of the corresponding acid. Both reactants produce water and the same anion:

$$CO_2(g) + 2OH^-(aq) \rightarrow \underset{\text{carbonate ion}}{CO_3^{2-}(aq)} + H_2O(l)$$

$$H_2CO_3(aq) + 2OH^-(aq) \rightarrow CO_3^{2-}(aq) + 2H_2O(l)$$

The oxidation state of the nonmetal does not change during these reactions.

PRACTICE EXERCISE S2.2

Identify the products of the reaction of $SO_3(g)$ with (a) $H_2O(l)$ and (b) $NaOH(aq)$.

Peroxides and Superoxides

Peroxides are compounds that contain O—O bonds. The best-known example is hydrogen peroxide (H_2O_2, Figure S2.3), a covalent compound. Peroxides of active metals such as Na_2O_2 and BaO_2 contain the peroxide ion (O_2^{2-}). Superoxides contain the O_2^- ion; KO_2 is an example.

$$H-\ddot{\underset{..}{O}}-\ddot{\underset{..}{O}}-H$$

hydrogen peroxide, H_2O_2

Hydrogen Peroxide. Small amounts of aqueous hydrogen peroxide can be prepared by treating barium peroxide with cold dilute sulfuric acid:

$$BaO_2(s) + H_2SO_4(aq) \xrightarrow{0°C} BaSO_4(s) + H_2O_2(aq)$$

At room temperature, pure hydrogen peroxide is a colorless, syruplike liquid; it freezes at $-0.41°C$ and boils at $150.2°C$. Peroxide oxygen has an oxidation state of -1, intermediate between the zero state of oxygen in O_2 and the usual -2 state in other compounds; hence, hydrogen peroxide is readily subject to disproportionation. As a result, the pure liquid can decompose violently and exothermically into water and oxygen:

$$2H_2O_2(l) \rightarrow 2H_2O(l) + O_2(g) \qquad \Delta H° = -196 \text{ kJ}$$

This reaction is catalyzed by light and by numerous substances, including dust particles, blood, and metal ions such as Cu^{2+} and Fe^{3+}. Hydrogen peroxide is usually supplied in aqueous solution, and its decomposition is inhibited by storage in dark bottles with preservatives that bind and inactivate metal ions.

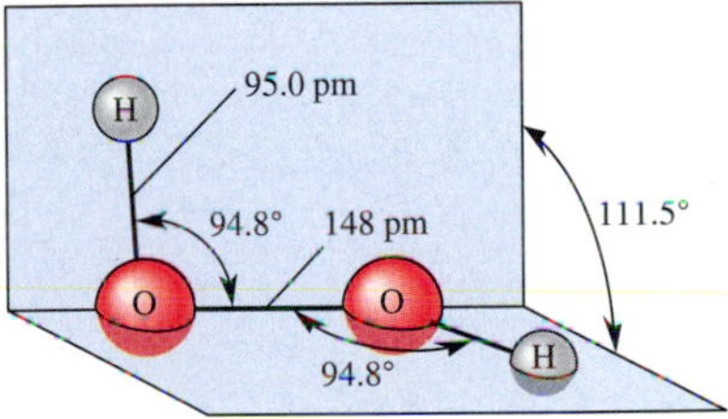

Figure S2.3
Structure of the H_2O_2 molecule in the gas phase. Two lone pairs of electrons on each oxygen atom compress the bond angles to 94.8°, well below the tetrahedral value of 109.5°. The planes containing the O—H bonds make an angle of 111.5°.

Because of the intermediate oxidation state of its oxygen, hydrogen peroxide can function as either an oxidizing agent or a reducing agent. It is a strong oxidizing agent that is reduced to H_2O in acidic solution and to OH^- in basic solution:

$$H_2SO_3(aq) + H_2O_2(aq) \rightarrow H_2SO_4(aq) + H_2O(l)$$

$$2I^-(aq) + H_2O_2(aq) \rightarrow I_2(aq) + 2OH^-(aq)$$

Hydrogen peroxide is a weak reducing agent. In acidic solutions, it is oxidized to oxygen:

$$8H^+(aq) + Cr_2O_7^{2-}(aq) + 3H_2O_2(aq) \rightarrow 2Cr^{3+}(aq) + 7H_2O(l) + 3O_2(g)$$

PRACTICE EXERCISE S2.3

Write half-reaction equations for (a) the reduction of H_2O_2 in acidic solution, (b) the reduction of H_2O_2 in basic solution, and (c) the oxidation of H_2O_2 in acidic solution.

Most commercial uses of hydrogen peroxide are based on its strong oxidizing ability. A 30% aqueous solution of H_2O_2 is used in water purification, in bleaching wood pulp, and in chemical synthesis. A 3% solution used as a mild antiseptic and hair bleach is available for household use.

A 30% solution of hydrogen peroxide must be handled with great caution. It will rapidly oxidize many substances (including skin) and may decompose explosively when heated.

DEMONSTRATION S2.3 BLEACHING LEAD SULFIDE WITH HYDROGEN PEROXIDE: $PbS(s) + 4H_2O_2(aq) \rightarrow PbSO_4(s) + 4H_2O(l)$

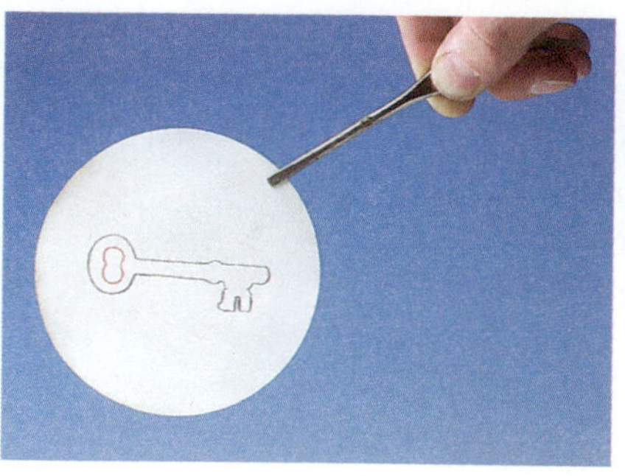

A drawing on paper impregnated with colorless lead nitrate.

The paper is immersed in sodium sulfide solution, which converts lead nitrate to black lead sulfide.

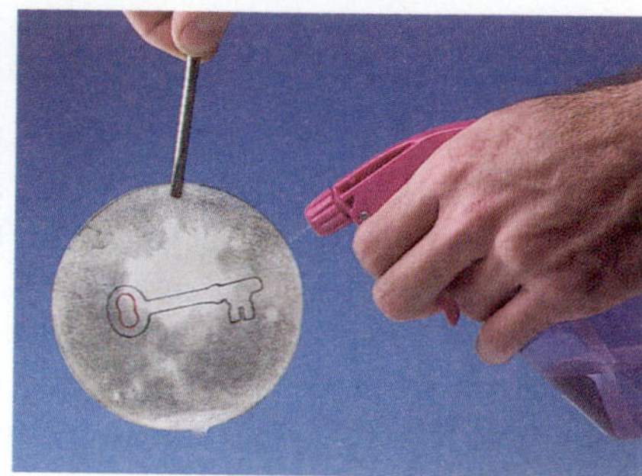

A spray of 3% H_2O_2 oxidizes the PbS to white $PbSO_4$, and the drawing becomes visible again.

Hydrogen peroxide has also been used in the restoration of old paintings. Many paints contain "white lead," a pigment that is gradually converted to black lead sulfide (PbS) as a result of exposure to airborne sulfides such as H_2S:

$$\underset{\text{white lead}}{PbCO_3{\cdot}Pb(OH)_2(s)} + 2H_2S(g) \longrightarrow \underset{\text{black}}{2PbS(s)} + CO_2(g) + 3H_2O(g)$$

Washing with aqueous hydrogen peroxide oxidizes the black lead sulfide to white lead sulfate (Demonstration S2.3):

$$\underset{\text{black}}{PbS(s)} + 4H_2O_2(aq) \longrightarrow \underset{\text{white}}{PbSO_4(s)} + 4H_2O(l)$$

Ionic Peroxides and Superoxides. Peroxides and superoxides are formed by the direct combination of active metals with oxygen, as described in Sections S1.2 and S1.3. The important peroxides are Na_2O_2 and BaO_2, which are used industrially as bleaching agents for textiles and paper.

The molecular orbital configurations in Table S2.3 show that peroxide and superoxide ions have more antibonding electrons than O_2 molecules. Therefore, these ions have lower bond energies and are more powerful oxidizing agents than O_2. Both ions form in the body when O_2 is reduced during metabolism, but a series of enzymatic reactions quickly disproportionates them into oxygen and water before they can do any harm:

$$2O_2^-(aq) + 2H^+(aq) \xrightarrow{\text{enzymes}} H_2O_2(aq) + O_2(g)$$

$$2H_2O_2(aq) \xrightarrow{\text{enzymes}} 2H_2O(l) + O_2(g)$$

Some reactions of oxide, peroxide, and superoxide anions are compared in Table S2.4. *Potassium superoxide* is used in emergency breathing masks and in other

TABLE S2.4 Reactions of Oxygen Anions

Oxide (O^{2-})	
With water	$O^{2-} + H_2O(l) \rightarrow 2OH^-(aq)$
With CO_2	$O^{2-} + CO_2(g) \rightarrow CO_3^{2-}$
With acid	$O^{2-} + 2H^+(aq) \rightarrow H_2O(l)$
Peroxide (O_2^{2-})	
With water	$2O_2^{2-} + 2H_2O(l) \rightarrow 4OH^-(aq) + O_2(g)$
With CO_2	$2O_2^{2-} + 2CO_2(g) \rightarrow 2CO_3^{2-} + O_2(g)$
With acid	$2O_2^{2-} + 4H^+(aq) \rightarrow 2H_2O(l) + O_2(g)$
Superoxide (O_2^-)	
With water	$4O_2^- + 2H_2O(l) \rightarrow 4OH^-(aq) + 3O_2(g)$
With CO_2	$4O_2^- + 2CO_2(g) \rightarrow 2CO_3^{2-} + 3O_2(g)$
With acid	$4O_2^- + 4H^+(aq) \rightarrow 2H_2O(l) + 3O_2(g)$

NOTE: All oxygen anions react with water to form hydroxide ions, with carbon dioxide to form carbonate ions, and with aqueous acid to form water. Peroxides and superoxides form O_2 in addition to the other products.

closed breathing systems because of its ability to absorb carbon dioxide while releasing oxygen:

$$4KO_2(s) + 2CO_2(g) \rightarrow 2K_2CO_3(s) + 3O_2(g)$$

S2.2 NITROGEN

Nitrogen is the lightest member of Group 5A and the only gaseous element in the group. It exhibits every oxidation state from -3 to $+5$ and has a rich redox chemistry probably unequaled by any other element. Nitrogen is a key element in a great variety of compounds ranging from simple ionic salts to the giant molecules of protein and DNA. Some atomic properties of nitrogen are summarized in Table S2.5.

TABLE S2.5 Atomic Properties of Nitrogen

Electron configuration	$[He]2s^22p^3$
Electronegativity	3.0
First ionization energy	1402 kJ/mol
Electron affinity	0
Atomic radius	74 pm
Ionic radius, N^{3-}	171 pm

Preparation, Properties, and Uses of Nitrogen

Elemental nitrogen (N_2) is a colorless, odorless gas (Table S2.6). The triple bond in N_2 is one of the strongest bonds known (see Table 9.3, p. 377) and N_2 molecules are extremely stable. The atmosphere contains about 78% N_2 by molecule count and is the ultimate repository of elemental nitrogen. Nitrogen is continuously removed from the atmosphere by lightning, bacterial activity, and industrial reactions; it is continuously returned by the combustion, metabolism, and decay of organic nitrogen compounds. (The nitrogen cycle is shown in Figure S2.4.) Most inorganic nitrogen compounds are water soluble; thus the solid portion of the earth's crust contains very few nitrogenous minerals, although deposits of KNO_3 and $NaNO_3$ exist in certain arid regions.

TABLE S2.6 Physical Properties of Nitrogen (N_2)

Melting point	−209.86°C
Boiling point	−195.8°C
Density at 1 atm and 0°C	1.25046 g/L
Solubility in 100 mL H_2O at 0°C	2.33 cm^3 measured at STP
Color	Colorless in all states

Figure S2.4
The nitrogen cycle. Atmospheric nitrogen is fixed (converted into compounds) by lightning, by industry (Haber process), and by nitrogen-fixing bacteria. Nitrogen is released into the atmosphere by the action of various microorganisms on proteins and other nitrogen compounds.

Pure nitrogen is obtained from liquid air by fractional distillation. Since nitrogen is unreactive at ordinary temperatures, it is used as an inexpensive and inert atmosphere for protecting reactive chemicals and for processing food. Liquid nitrogen at −196°C is used for freezing food and as an environment for low-temperature reactions. The largest use of industrial nitrogen is in the manufacture of ammonia. The uses of nitrogen are summarized in Table S2.7.

Liquid helium (b.p. −268.9°C) is now used to cool superconducting devices to the temperatures at which they offer no resistance to the flow of electric current. The development of "high-temperature" superconducting compounds (Section S1.4) may eventually allow liquid nitrogen (b.p. −195.8°C) to replace the more expensive liquid helium for this purpose.

Nitrogen fixation (the formation of compounds from atmospheric nitrogen) is accomplished by nature in two principal ways. One type of fixation occurs during thunderstorms when lightning shatters the bonds of nitrogen and oxygen molecules in its path. Nitrogen monoxide (NO) is one of the products of the recombining atoms. The overall equation is

$$N_2(g) + O_2(g) \rightarrow 2NO(g)$$

Nitrogen monoxide then reacts with additional oxygen to form nitrogen dioxide:

$$2NO(g) + O_2(g) \rightarrow 2NO_2(g)$$

which then reacts with moisture to form nitric and nitrous acids:

$$2NO_2(g) + H_2O(l) \xrightarrow{\text{cold, dilute}} \underset{\text{nitric acid}}{HNO_3(aq)} + \underset{\text{nitrous acid}}{HNO_2(aq)}$$

Rainfall deposits the acids on the ground, where they are utilized by plants. An estimated 7.6 million metric tons (17 billion pounds) of nitrogen is fixed annually by lightning.

TABLE S2.7 Uses of Elemental Nitrogen

Major Use
Production of ammonia
Other Uses
As a blanketing atmosphere in:
Petroleum refining
Storing air-sensitive substances
Protecting air-sensitive reaction mixtures
Electronic components manufacture
Metals treating and processing
As a purging atmosphere for:
Drying reagents and other materials
As a freezing and cooling agent (liquid nitrogen) for:
Freezing food
Preserving biological tissues
Liquefying gaseous reaction products
Low-temperature reaction environments
In the manufacture of products such as:
Hydrazine (N_2H_4) for use as a rocket fuel
Solid covalent nitrides, such as BN, for cutting and drilling

An even larger quantity of nitrogen—about 54 million metric tons—is absorbed by bacteria that live in nodules on the roots of certain plants such as beans, peas, and alfalfa. This process is nature's second way of fixing nitrogen (see *Chemical Insight: Nature's Quiet Fixers*).

Ammonia and Ammonium Salts

TABLE S2.8 Properties of Ammonia (NH_3)

Melting point	−77.7°C
Boiling point	−33.3°C
Density at STP	0.7714 g/L
Solubility in 100 mL H_2O at 0°C	89.9 g
Odor	Pungent
Color	Colorless

Ammonia (NH_3) and ammonium salts (salts containing the NH_4^+ ion) contain nitrogen in the −3 oxidation state. Ammonia is a colorless gas that is pungent in odor and extremely soluble in water (Table S2.8). The saturated aqueous ammonia on the laboratory shelf is about 15 *M*. Ammonia solutions are weakly basic and are sometimes labeled "ammonium hydroxide":

$$NH_3(aq) + H_2O(l) \rightleftharpoons NH_4^+(aq) + OH^-(aq)$$

The efficacy of ammonia as a household cleaner is due to the ability of hydroxide ion to react with films of grease.

Ammonia reacts with acids to form ammonium salts:

$$NH_3(aq) + H^+(aq) \rightarrow NH_4^+(aq)$$

$$2NH_3(aq) + H_2SO_4(aq) \rightarrow (NH_4)_2SO_4(aq)$$

Compounds containing ammonium ion give off ammonia gas when they are gently heated with aqueous base:

$$NH_4^+(aq) + OH^-(aq) \xrightarrow{\text{heat}} NH_3(g) + H_2O(l)$$

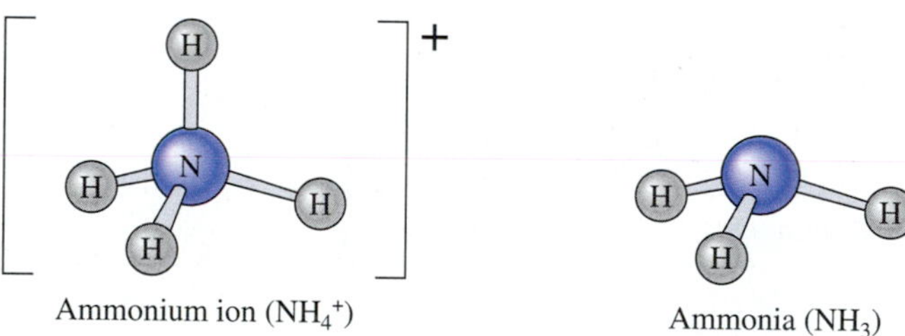

Ammonium ion (NH_4^+) Ammonia (NH_3)

This reaction is the reverse of the ionization of aqueous ammonia (see above reaction of NH_3 with H_2O). Because ammonia can be easily recognized by its odor, the reaction serves as a simple laboratory test for ammonium ions.

Ammonia ranks among the top three or four leading industrial chemicals. It is used for preparing fertilizers, fibers and plastics (e.g., nylons and acrylics), and explosives (see Figure S2.6). Ammonia, either in solution or as a pure liquid under pressure, is now the most widely used synthetic nitrogen fertilizer (Figure S2.7). Water-soluble fertilizers made from ammonia include *ammonium sulfate* ($(NH_4)_2SO_4$), *ammonium nitrate* (NH_4NO_3), the *ammonium phosphates* ($NH_4H_2PO_4$ and $(NH_4)_2HPO_4$), and *urea* (H_2NCONH_2).

The huge production of nitrogenous fertilizers and the resulting agricultural "green revolution" were made possible by the nitrogen-fixation process developed in 1909 by the German chemist Fritz Haber. The success of the **Haber process** is principally due to special iron oxide catalysts; without a catalyst, the reaction would be too slow to be practical:

CHEMICAL INSIGHT

NATURE'S QUIET FIXERS

The Haber process is the chemist's way of fixing nitrogen. Industrial production of ammonia and ammonium compounds, however, is dwarfed by the accomplishments of a host of nitrogen-fixing organisms that inhabit soil, ocean water, desert sand, and even the guts of animals. These algae and bacteria quietly convert large quantities of atmospheric nitrogen to ammonium ion under normal outdoor conditions—a feat that we are just beginning to understand and cannot as yet duplicate on any practical scale. It is estimated that about three times more nitrogen is fixed by microorganisms than by all of industry.

Some nitrogen-fixing bacteria inhabit nodules on the roots of plants such as beans, clover, and alfalfa (Figure S2.5). The ammonium ion produced in these nodules is absorbed by the plant, which then converts it to protein and other essential biomolecules. *Denitrifying bacteria* accomplish the reverse process. They inhabit decaying plant and animal tissue, where they help release nitrogen and return it to the atmosphere.

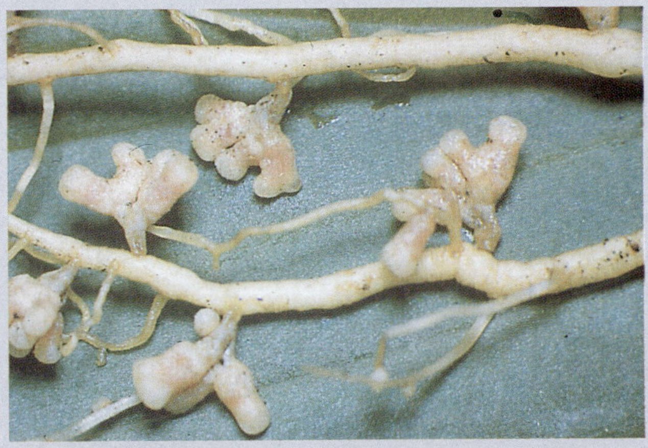

Figure S2.5
Root nodules that contain nitrogen-fixing bacteria.

Genetic engineers are working to modify other plants so that their roots, like those of beans and alfalfa, will harbor nitrogen-fixing bacteria. If they are successful, major crops such as wheat, corn, and barley might someday be self-fertilizing, and farmlands could be harvested continuously without dependence on commercial fertilizers. Regions that now depend on other areas for their food supplies could become self-sufficient, and water supplies would no longer be threatened by pollution from fertilizer runoff.

Chemists are also at work trying to understand and mimic natural nitrogen fixation. The nitrogen-fixing enzymes in some organisms have been found to contain atoms of iron and molybdenum, and molybdenum compounds have now been synthesized that will bind N_2 in such a way that it can be reduced to ammonia at room temperature. Special expensive reducing agents are needed, however, and economical nitrogen fixation at ordinary temperatures and pressures is not yet in sight.

$$N_2(g) + 3H_2(g) \xrightarrow[\text{catalyst}]{400°\text{C, 250 atm}} 2NH_3(g)$$

The efficiency of the reaction is further enhanced by elevated pressure and temperature. The Haber process is discussed more fully in Chapter 15.

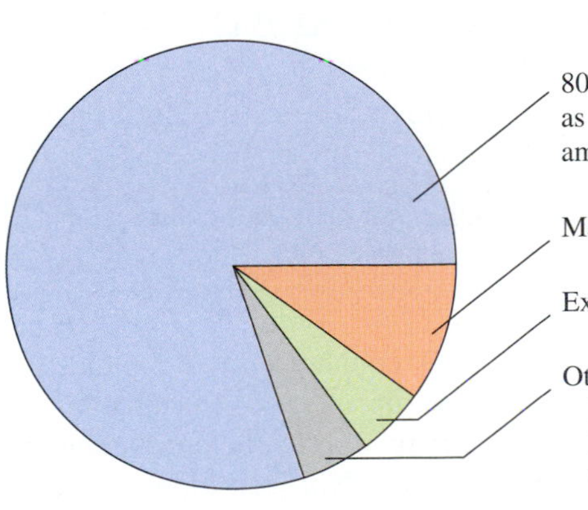

Figure S2.6
The commercial uses of ammonia.

Figure S2.7
Fertilizing with liquid ammonia. Plants convert ammonia into protein.

Oxides, Oxo Acids, and Oxo Anions of Nitrogen

The positive oxidation states of nitrogen are represented by oxides and oxo acids, as shown in Table S2.9. An **oxo acid** is an acid in which the central atom is bonded to oxygen; HNO_3 is an example. The anion of an oxo acid is called an **oxo anion**.

Oxidation State +1. The only important compound in which nitrogen exhibits a +1 oxidation state is the colorless gas *dinitrogen oxide* (N_2O), also known as *nitrous oxide*. N_2O is used as a general anesthetic in dentistry; it is also called "laughing gas" because inhalation may produce exhilaration and laughter. N_2O is soluble in cream and serves as the propellant in cans of whipped cream.

Dinitrogen oxide is prepared by gently heating ammonium nitrate:

$$NH_4NO_3(s) \xrightarrow{\text{heat}} N_2O(g) + 2H_2O(g)$$

Although N_2O is not reactive at room temperature, it decomposes explosively when heated:

$$N_2O(g) \rightarrow N_2(g) + \tfrac{1}{2}O_2(g) \qquad \Delta H^\circ = -82.05 \text{ kJ}$$

PRACTICE EXERCISE S2.4

Use VSEPR to predict whether the N_2O molecule is linear or bent. The sequence of atoms is N—N—O.

Oxidation States +2 and +4. *Nitrogen monoxide* (NO) contains nitrogen in the +2 state. It is a colorless gas produced by the combination of nitrogen and oxygen during lightning and other electric discharges and during high-temperature combustions such as those that take place in automobile engines. Nitrogen monoxide oxi-

TABLE S2.9 Oxidation States of Nitrogen

Oxidation State	Hydride or Oxide	Other Molecules and Ions of Importance
−3	NH_3	NH_4^+ N^{3-} (nitride ion) CN^- (cyanide ion)
−2	N_2H_4 (hydrazine)	
−1		NH_2OH (hydroxylamine)
0		N_2
+1	N_2O	
+2	NO	
+3	N_2O_3[a]	HNO_2 (nitrous acid) NO_2^- (nitrite ion)
+4	NO_2	N_2O_4
+5	N_2O_5	HNO_3 (nitric acid) NO_3^- (nitrate ion)

[a]Not stable above −21°C.

dizes in air to *nitrogen dioxide*, a toxic brown gas with a choking odor, which contains nitrogen in the +4 state:

$$N_2(g) + O_2(g) \xrightarrow[\text{}]{\text{heat or electric discharge}} 2NO(g)$$

$$\underset{\text{colorless}}{2NO(g)} + O_2(g) \longrightarrow \underset{\text{brown}}{2NO_2(g)}$$

The brown NO_2 quickly establishes an equilibrium with *dinitrogen tetroxide*, N_2O_4 ($O_2N—NO_2$), a colorless gas in which nitrogen has the same +4 oxidation state as in NO_2:

$$\underset{\text{brown}}{2NO_2(g)} \rightleftharpoons \underset{\text{colorless}}{N_2O_4(g)}$$

NO_2 predominates in the equilibrium mixture at high temperatures, N_2O_4 at low temperatures (see Figure 15.7 on page 724).

Nitrogen monoxide is synthesized by various cells in the body. Recent research has shown that this substance plays a role in the relaxation of blood vessels, thus helping to regulate blood pressure. Other studies have shown that NO is involved in the mechanism by which white blood cells kill tumor cells and bacteria that cause disease. Nitrogen dioxide has no known role in the human body. Both NO and NO_2 are free radicals (Section 9.6); each contains an odd number of electrons and is paramagnetic. They are longer lasting than most free radicals, and their persistence in automobile exhaust and in the atmosphere contributes to the problems of smog and acid rain (see *Chemical Insight: Air Pollution and Photochemical Smog*).

Oxidation State +3. *Nitrous acid* (HNO_2) and *nitrites* (compounds containing the NO_2^- ion) contain nitrogen in the +3 oxidation state. Nitrites can be made by dissolving NO and NO_2 in basic solution:

$$NO(g) + NO_2(g) + 2OH^-(aq) \rightarrow 2NO_2^-(aq) + H_2O(l)$$

or by treating sodium nitrate ($NaNO_3$) with a reducing agent such as coke:

$$NaNO_3(l) + C(s) \xrightarrow{\text{heat}} NaNO_2(l) + CO(g)$$

Sodium nitrite is used as an additive in cured meats, where it inhibits the growth of deadly botulinus bacteria. It has been suggested that the nitrites might produce harmful concentrations of carcinogenic nitrosamines (carbon compounds containing the N—N=O group) during cooking or during the process of digestion. Studies so far seem to indicate that this is not a significant problem.

PRACTICE EXERCISE S2.5

Use VSEPR to predict whether the nitrite ion is linear or bent.

The action of an acid on a nitrite produces pale blue nitrous acid, HNO_2 (H—O—N=O), an unstable weak acid that exists only in cold solution:

$$H^+(aq) + NO_2^-(aq) \rightarrow HNO_2(aq)$$

$$2HCl(aq) + Ca(NO_2)_2(aq) \rightarrow 2HNO_2(aq) + CaCl_2(aq)$$

At room temperature, nitrous acid gradually disproportionates into nitric acid and NO gas:

$$3HNO_2(aq) \rightarrow HNO_3(aq) + 2NO(g) + H_2O(l)$$

Nitrites can be oxidized to nitrates (NO_3^-) or reduced to NO (See Demonstration S2.4).

PRACTICE EXERCISE S2.6

Write half-reaction equations for (a) the reduction of HNO_2 in acidic solution and (b) the oxidation of NO_2^- in basic solution.

Oxidation State +5. *Dinitrogen pentoxide* (N_2O_5), *nitric acid* (HNO_3), and *nitrates* contain nitrogen in the +5 oxidation state. At room temperature, dinitrogen pentoxide is a colorless, crystalline solid composed of nitronium ions (NO_2^+) and nitrate ions (NO_3^-). The solid sublimes (vaporizes) at 32.5°C, and the unstable vapor decomposes into NO_2 and O_2:

$$2N_2O_5(g) \rightarrow 2NO_2(g) + O_2(g)$$

CHEMICAL INSIGHT

AIR POLLUTION AND PHOTOCHEMICAL SMOG

Photochemical smog is the yellow-brown haze that at one time or another blankets almost every modern industrial city, irritating the lungs and blighting the vegetation. Engine exhaust contains the *primary pollutants* that cause smog—the nitrogen oxides NO and NO_2, known collectively as NO_x, and unsaturated hydrocarbons (hydrocarbons with C=C double bonds). The brown color of photochemical smog is mainly due to NO_2 (see Figure S2.8).

The reactions of nitrogen oxides in the atmosphere resemble a ball game in which an oxygen atom is tossed back and forth:

Figure S2.8
Photochemical smog over Mexico City.

$$2NO(g) + O_2(g) \longrightarrow 2NO_2(g)$$

$$NO_2(g) \xrightarrow[400\text{ nm}]{\text{sunlight}} NO(g) + O(g)$$

$$O(g) + O_2(g) \longrightarrow O_3(g)$$

$$O_3(g) + NO(g) \longrightarrow NO_2(g) + O_2(g)$$

This shuffle adds another undesirable pollutant, ozone, to the NO and NO_2 already present. Ozone attacks the unsaturated hydrocarbons from engine exhaust, forming a variety of free radicals including the peroxyacetyl radical:

$$CH_3—\overset{\overset{:\ddot{O}:}{\|}}{C}—\ddot{\underset{..}{O}}—\ddot{\underset{..}{O}}\cdot$$

Nitrogen dioxide, which is also a free radical, reacts with peroxyacetyls to form products such as

$$CH_3—\overset{\overset{:\ddot{O}:}{\|}}{C}—\ddot{\underset{..}{O}}—\ddot{\underset{..}{O}}—NO_2$$

↑ peroxide bond

which belong to a family of compounds called peroxyacylnitrates or PAN. (Other peroxyacylnitrates have different hydrocarbon groups in place of the CH_3.) Ozone and PAN, the *secondary pollutants*, are even more noxious than the primary ones. Photochemical smog is worst when the polluted air is trapped by a temperature inversion (cool air beneath warm air). The inversion not only slows the dispersal of polluted air, it also gives the secondary pollutants more time to form.

PRACTICE EXERCISE S2.7

Use VSEPR to predict the shape of (a) the nitronium ion and (b) the nitrate ion.

Nitric acid is an important industrial chemical. Small batches of pure nitric acid can be prepared by heating nitrates with concentrated sulfuric acid:

$$NaNO_3(s) + H_2SO_4(18\ M) \xrightarrow{\text{heat}} HNO_3(g) + NaHSO_4(aq)$$

The vapor condenses to a colorless liquid at 83°C (Figure S2.9).

DEMONSTRATION S2.4 REDOX REACTIONS OF THE NITRITE ION, NO_2^-

Acidified solutions of potassium iodide (colorless) and potassium permanganate (purple). The watch glasses contain sodium nitrite crystals.

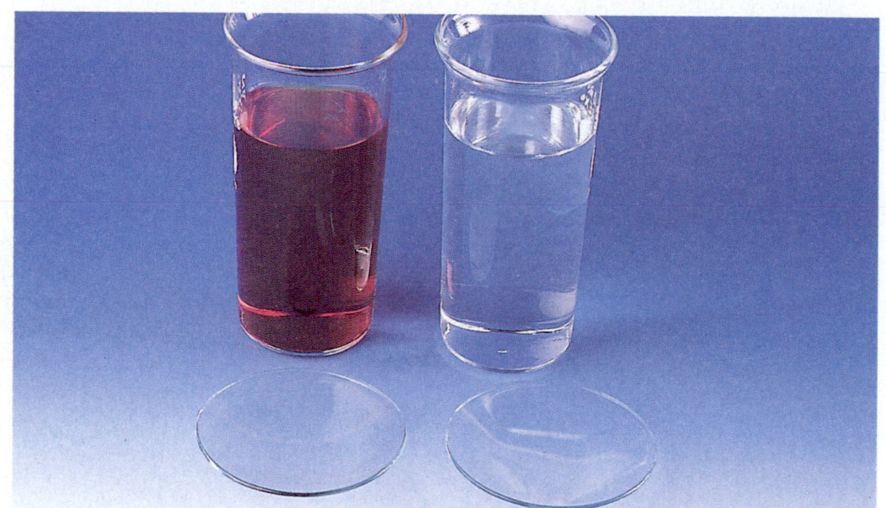

Sodium nitrite is added to the solutions. *Left:* I^- is oxidized to I_2 (brown solution); NO_2^- is reduced to NO(g). *Right:* MnO_4^- is reduced to Mn^{2+} (almost colorless); NO_2^- is oxidized to NO_3^-.

Most of the nitric acid used in industry comes in the form of approximately 16 *M* aqueous solution. It is produced from ammonia by the three-step **Ostwald process**:

Step 1: Ammonia is burned in excess oxygen over a platinum catalyst to form NO:

$$4NH_3(g) + 5O_2(g) \xrightarrow[900°C]{Pt} 4NO(g) + 6H_2O(g)$$

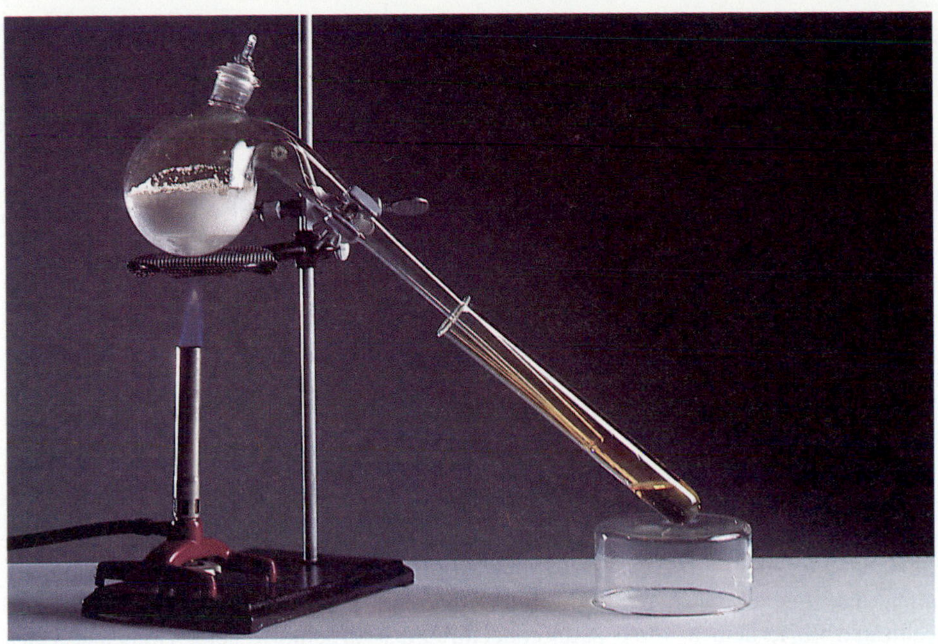

Figure S2.9

Laboratory preparation of nitric acid. A mixture of sodium nitrate and concentrated sulfuric acid is warmed in a retort. Pure HNO_3 boils out of the mixture and condenses in the test tube. (The yellow color is caused by a small amount of NO_2 that forms during the reaction.)

Step 2: Additional air is added to cool the mixture and oxidize NO to NO_2:

$$2NO(g) + O_2(g) \xrightarrow{\text{cool}} 2NO_2(g)$$

Step 3: The NO_2 gas is bubbled into warm water where it disproportionates to nitric acid and NO. The NO is recycled to Step 2:

$$3NO_2(g) + H_2O(l) \rightarrow 2HNO_3(aq) + \underset{\text{recycled}}{NO(g)}$$

The yearly production of nitric acid exceeds 14 billion pounds. The largest portion of this production is used to synthesize ammonium nitrate, a water-soluble fertilizer. Large quantities are also used to make plastics, drugs, and explosives such as trinitrotoluene (TNT) and nitroglycerine (see *Chemical Insight: Fireworks, Rockets, and Explosives* in the next section).

PRACTICE EXERCISE S2.8

The yellow color often seen in concentrated solutions of nitric acid is due to dissolved NO_2, which is produced along with O_2 by the slow, light-induced decomposition of HNO_3. Use the method of half-reactions to obtain an equation for the reaction. (*Hint:* H_2O is oxidized to O_2.)

Nitric acid is both a strong acid and a strong oxidizing agent. Concentrated (16 *M*) nitric acid is the strongest oxidizing agent readily available in aqueous solution. The reduction products are usually a mixture of NO_2, NO, and NH_4^+, but some N_2O and N_2 may also form. The acid concentration, the temperature, and the strength of the reducing agent determine which product predominates. Selected examples of nitric acid oxidation are given below.

Figure S2.10

The action of nitric acid on copper. *Left:* Copper does not react with 1 *M* HNO_3. *Center:* Copper oxidizes slowly in 6 *M* HNO_3; the gas is colorless NO. *Right:* Copper oxidizes rapidly in 12 *M* HNO_3; the gas is brown NO_2.

Nonmetals. Concentrated HNO_3 oxidizes the nonmetals carbon, phosphorus, sulfur, and iodine to CO_2, H_3PO_4, H_2SO_4, and HIO_3. The principal reduction product is NO_2:

$$S(s) + 6HNO_3(16\,M) \rightarrow H_2SO_4(aq) + 6NO_2(g) + 2H_2O(l)$$

Moderately Active Metals. Ammonium ion is the principal reduction product when moderately active metals such as zinc and manganese are oxidized by dilute (6 *M*) nitric acid:

$$4Zn(s) + 10HNO_3(6\,M) \rightarrow NH_4NO_3(aq) + 4Zn(NO_3)_2(aq) + 3H_2O(l)$$

Less Active Metals. Copper and many metals less active than hydrogen reduce nitric acid to either NO or NO_2, depending on the acid concentration (Figure S2.10):

$$3Cu(s) + 8HNO_3(6\,M) \rightarrow 3Cu(NO_3)_2(aq) + 2NO(g) + 4H_2O(l)$$

$$Cu(s) + 4HNO_3(12\,M) \rightarrow Cu(NO_3)_2(aq) + 2NO_2(g) + 2H_2O(l)$$

Noble Metals. Gold, platinum, rhodium, and iridium are not attacked by nitric acid alone, but they all dissolve to some degree in *aqua regia*, a mixture of one part concentrated nitric acid to three parts concentrated hydrochloric acid. Gold and platinum dissolve especially well in this mixture. Chloride ion helps the metal to dissolve by incorporating the oxidized atoms into stable complexes such as $AuCl_4^-$ and $PtCl_6^{2-}$. The net ionic equation for the dissolving of gold in aqua regia is

Aqua regia ("royal water") was given its name by the alchemists who discovered its ability to dissolve gold, a "noble" metal.

$$Au(s) + NO_3^-(aq) + 4H^+(aq) + 4Cl^-(aq) \rightarrow AuCl_4^-(aq) + NO(g) + 2H_2O(l)$$

PRACTICE EXERCISE S2.9

Write half-reaction equations for the reaction of (a) carbon with 16 *M* nitric acid and (b) copper with 6 *M* HNO_3.

Nitrates are salts of nitric acid. They are soluble and are, therefore, good vehicles for introducing metal ions into solution. Because of their solubility, natural deposits of solid nitrates are not abundant. *Potassium nitrate* (KNO_3, saltpeter) formed by decaying organic matter was an essential ingredient of old-fashioned gunpowder (saltpeter, sulfur, and charcoal). Before the Haber process was developed, *sodium nitrate* from the Chilean deserts (Chile saltpeter) was the principal source of inorganic nitrogen for fertilizer. *Ammonium nitrate* is especially rich in nitrogen. Like all nitrates, it is potentially explosive in large quantities, especially if tightly packed and poorly ventilated. The decomposition products at moderate temperatures (200–260°C) are dinitrogen oxide and water:

$$NH_4NO_3(s) \rightarrow N_2O(g) + 2H_2O(g) \qquad \Delta H° = -36.0\ kJ/mol$$

At the high temperatures produced by most explosions, the N_2O decomposes into N_2 and O_2.

Other Nitrogen Compounds

Nitrides. Nitrides contain nitrogen in the −3 oxidation state. *Ionic nitrides* are white crystalline solids that contain the N^{3-} ion. They are formed by direct combination of nitrogen with lithium, aluminum, and most Group 2A metals (Sections S1.2 and S1.3). The nitride ion reacts with water to form ammonia:

$$N^{3-} + 3H_2O(l) \rightarrow NH_3(g) + 3OH^-(aq)$$

$$Mg_3N_2(s) + 6H_2O(l) \rightarrow 2NH_3(g) + 3Mg(OH)_2(s)$$

Cyanides. The *cyanide ion* (CN^-), which has the same number of electrons as the N_2 molecule, is a stable ion with a short, strong triple bond. It contains nitrogen in the -3 oxidation state. Ionic cyanides are toxic, and so is the corresponding weak acid HCN. An ionic cyanide in contact with acid releases lethal HCN vapor:

Cyanide salts should NEVER be acidified. Solutions that contain cyanide salts must always be kept basic.

$$HCl(aq) + NaCN(aq) \rightarrow NaCl(aq) + HCN(g)$$

Hydrazine. *Hydrazine* (N_2H_4) contains nitrogen in the -2 oxidation state. It is a colorless liquid that is synthesized by oxidizing ammonia with sodium hypochlorite (NaOCl):

H—N̈—N̈—H
  | |
  H   H

hydrazine (N_2H_4)

$$2NH_3(l) + OCl^-(aq) \rightarrow N_2H_4(aq) + Cl^-(aq) + H_2O(l)$$

Hydrazine fumes are toxic; therefore, household bleaches and cleansers containing hypochlorite or "chlorine" should not be mixed with household ammonia.

Hydrazine and *methylhydrazines* (hydrazine with CH_3 groups substituted for hydrogen atoms) have been used as rocket fuels. They are readily oxidized to nitrogen by both liquid oxygen and hydrogen peroxide:

$$N_2H_4(l) + O_2(l) \rightarrow N_2(g) + 2H_2O(g)$$

S2.3 THE HALOGENS

The elements of Group 7A—fluorine, chlorine, bromine, iodine, and astatine—are called the **halogens**, a name that comes from a Greek word meaning "salt-former." Indeed, the most abundant halogen compound is sodium chloride, table salt. The halogens have many properties in common, and their periodic trends (see Tables S2.10 and S2.11) are consistent. Fluorine, the lightest member of the group, is the

TABLE S2.10 Abundances and Atomic Properties of the Halogens

Element	Abundance on Earth		Electron Configuration	Electro-Negativity	First Ionization Energy (kJ/mol)	Electron Affinity (kJ/mol)	Atomic Radius (pm)	Ionic Radius (X^-, pm)
	Crust (g/kg)	*Oceans (mg/L)*						
F	0.625	1.3	$[He]2s^22p^5$	4.0	1681	−322	71	136
Cl	0.13	19,000	$[Ne]3s^23p^5$	3.2	1251	−349	99	181
Br	0.0025	65	$[Ar]3d^{10}4s^24p^5$	3.0	1140	−325	114	195
I	0.0005	0.05	$[Kr]4d^{10}5s^25p^5$	2.7	1008	−295	133	216

TABLE S2.11 Physical Properties of the Halogens

Formula	Melting Point (°C)	Boiling Point (°C)	Solubility in H_2O at 25°C (mol/L)	State and Color at 1 atm and 25°C
F_2	−220	−188	reacts	Pale yellow gas
Cl_2	−101	−35	0.2141	Yellow-green gas
Br_2	−7	59	0.0013	Red-brown fuming liquid
I_2	113	184	0.092	Gray crystals subliming to violet vapor

most electronegative of all elements; its oxidation state in compounds is always −1. The other halogens exhibit oxidation states ranging from −1 to +7.

Astatine is the heaviest Group 7A element and was the last to be discovered. Its name is derived from the Greek word *astatos*, meaning "unstable." Astatine is produced in the stepwise radioactive decay of certain uranium and thorium isotopes and is itself radioactive; most of its atoms decay within a few hours of their formation. It has been estimated that the earth's entire astatine content at any given moment is probably less than 30 g. Tests on small amounts of artificially produced astatine show that it strongly resembles iodine. The known chemistry of astatine is not important and we will not consider it further.

Three halogens are known to be essential in the body. The chloride ion (Cl^-) is the principal anion in body fluids, where its presence helps to maintain fluid volume and internal cell pressure. Iodide ion (I^-) is needed in small amounts for thyroid hormones, such as thyroxine (Figure S2.11). Traces of fluoride ion (F^-) are required for the development of strong teeth and bones and may play a role in promoting general growth. The consumption of small amounts of fluoride ion has been shown to lower the incidence of tooth decay.

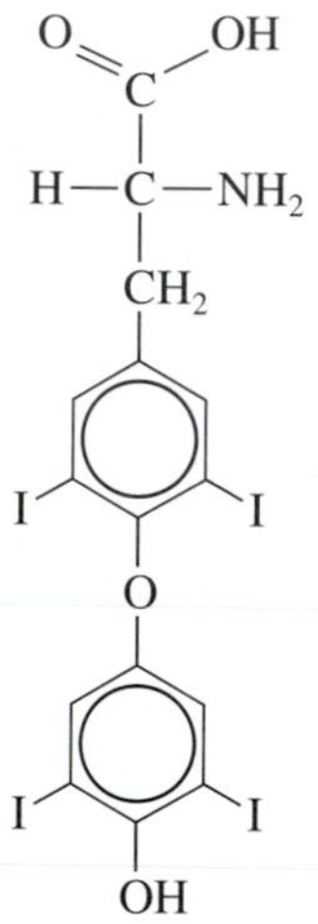

O OH C H—C—NH$_2$ CH$_2$ I I O I I OH

Figure S2.11
Thyroxine is an iodine-containing hormone that is essential for growth and metabolism.

Preparation, Properties, and Uses of the Halogens

Elemental halogens exist as the diatomic molecules F_2, Cl_2, Br_2, and I_2. At room temperature, fluorine is a pale yellow gas, chlorine a greenish yellow gas, bromine a fuming red liquid, and iodine a gray solid that sublimes to a violet vapor (Figure S2.12).

The natural sources of these colorful reactive elements are colorless, unreactive salts containing the halide ions F^-, Cl^-, Br^-, and I^-. The relatively small radius of the fluoride ion makes for large lattice energies, so fluorides tend to be less soluble than other halides. They occur in a number of solid minerals such as *fluorspar* (CaF_2), *cryolite* (Na_3AlF_6), and *fluorapatite* ($Ca_5(PO_4)_3F$). Chlorides, bromides, and iodides are found in seawater and in the brine wells and salt deposits that are the remains of ancient inland oceans. A liter of seawater contains about 0.54 mol chloride ion and smaller quantities of iodide and bromide ions. Iodides tend to concentrate in seaweed and some shellfish—iodine was first discovered in the ashes of seaweed. Iodates, salts containing IO_3^-, occur as impurities in South American sodium nitrate deposits.

A *brine* is a concentrated salt (NaCl) solution.

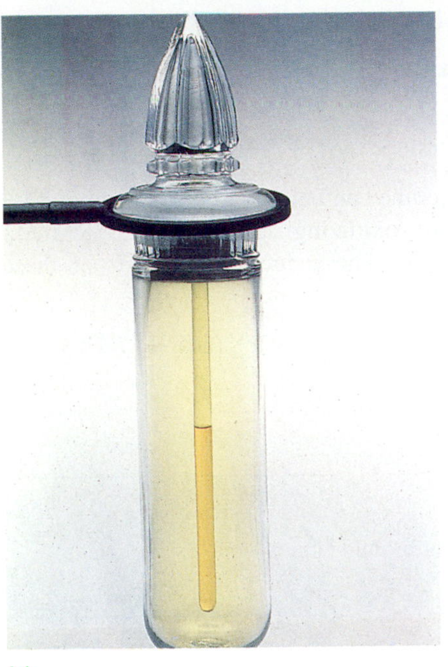

(a)

(b)

(c)

Figure S2.12

The halogens. (a) Chlorine gas (outer cylinder) and liquid chlorine under pressure (inside tube). (b) Bromine vapor above a small amount of dark red liquid bromine. (c) Iodine crystals hanging from a glass surface in violet iodine vapor.

Free halogens are prepared by oxidizing halide salts. Fluorine is prepared by electrolysis, because chemical agents are not strong enough to oxidize the fluoride ion. The first step is to heat fluorspar (CaF_2) with concentrated sulfuric acid:

$$CaF_2(s) + H_2SO_4(18\,M) \xrightarrow{\text{heat}} 2HF(g) + CaSO_4(s)$$

The hydrogen fluoride is then dissolved in molten potassium fluoride, and the hot solution is electrolyzed:

$$2HF\text{ (in KF)} \xrightarrow{\text{electrolysis}} H_2(g) + F_2(g)$$

Fluorine is a powerful oxidizing agent that reacts with glass, metal, most minerals, and most plastics, the materials from which containers are generally made. Vessels for preparing and storing fluorine are constructed from special steel alloys; the initial reaction of fluorine with the alloy produces a thin adherent coating of metal fluoride that protects the remaining metal from further attack.

Electrolysis is also the most economical method for preparing chlorine from aqueous sodium chloride:

$$2NaCl(aq) + 2H_2O(l) \xrightarrow{\text{electrolysis}} 2NaOH(aq) + Cl_2(g) + H_2(g)$$

To prevent NaOH and Cl_2 from reacting with each other, the products are not allowed to mix.

Bromine is obtained commercially by oxidizing the bromides in seawater with chlorine:

$$\underset{\text{in seawater}}{2Br^-(aq)} + Cl_2(g) \rightarrow Br_2(l) + 2Cl^-(aq)$$

The bromine is swept out of solution by a stream of air.

Iodine is obtained from brine wells and seaweed in a similar fashion

$$2I^-(aq) + Cl_2(g) \rightarrow I_2(s) + 2Cl^-(aq)$$

and it can also be obtained by reducing naturally occurring iodates with sodium hydrogen sulfite ($NaHSO_3$):

$$2IO_3^-(aq) + 5HSO_3^-(aq) \rightarrow I_2(s) + 5SO_4^{2-}(aq) + 3H^+(aq) + H_2O(l)$$

Small amounts of chlorine, bromine, and iodine can be prepared in the laboratory by treating their acidified halides with strong oxidizing agents such as $KMnO_4$, $K_2Cr_2O_7$, and MnO_2 (see Figure 11.2, p. 475). The reduction products of these reagents are Mn^{2+} and Cr^{3+}. A sample equation is

$$6Br^-(aq) + Cr_2O_7^{2-}(aq) + 14H^+(aq) \rightarrow 3Br_2(l) + 2Cr^{3+}(aq) + 7H_2O(l)$$

PRACTICE EXERCISE S2.10

Write half-reaction equations for the reaction of bromide ion with potassium dichromate.

a small portion of a Teflon molecule

The refurbished Statue of Liberty has an inner coating of inert Teflon that prevents corrosion-causing contact between the copper lady and her inner steel framework.

Many uses of the halogens are based on their oxidizing ability. *Chlorine water* (an aqueous chlorine solution) is an industrial and household bleaching agent, and chlorine gas has been used as a poison gas during war. The halogens are bactericides; chlorine is used to sterilize public water supplies, and iodine crystals are used for the emergency treatment of personal drinking water. *Bromine chloride* (BrCl) is also used to disinfect water. *Tincture of iodine* (iodine dissolved in alcohol) was once a widely used first-aid antiseptic; alcohol, however, has a painful sting, and iodine is somewhat damaging to healthy tissues as well as to bacteria. Safer, more pleasant, and less expensive antiseptics are now in use.

The halogens, especially chlorine and bromine, are used in the synthesis of thousands of industrial compounds. Important chlorine products include *chloroform* ($CHCl_3$), an anesthetic, *carbon tetrachloride* (CCl_4), a solvent for fats and oils, and *trichloroethylene* ($Cl_2C{=}CHCl$) and *tetrachloroethylene* ($Cl_2C{=}CCl_2$), dry-cleaning solvents.

Large-scale fluorine technology developed as a consequence of the need for UF_6 in preparing the World War II atomic bomb (see *Chemical Insight: Separation of Isotopes by Gaseous Effusion* in Chapter 5, p. 202). Since then many new fluorine compounds have been synthesized and put to use. *Fluorocarbons* (compounds containing only carbon and fluorine) have unusual and useful properties. The most familiar example is polytetrafluoroethylene, a heat-resistant, water-repellent, chemically inert plastic material better known as Teflon. Teflon is used to make nonstick coatings for kitchenware, fibers for rainwear fabric, and tubing and stopcocks for the laboratory. Teflon is a *polymer*; that is, its long molecules are constructed of many repeating units.

Another product of the postwar fluorine technology is the group of *chlorofluorocarbons* (CFCs) known as Freons. Freons are used as refrigerants, and they were once used as spray can propellants. The latter use has been abandoned in the United States and some other countries because volatile CFCs pose a threat to the ozone layer (see *Chemical Insight: Free Radicals and the Ozone Layer*, p. 394).

Certain liquid fluorocarbons have a capacity for dissolving and transporting oxygen and carbon dioxide that equals or surpasses that of natural blood. An aqueous emulsion containing perfluorodecalin and perfluorotripropylamine has been used for transfusions when compatible blood was not available or acceptable to the patient.

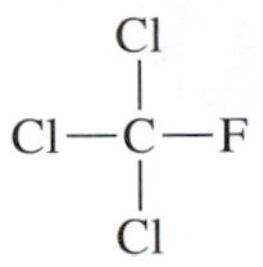

Freon-11 (CCl_3F)

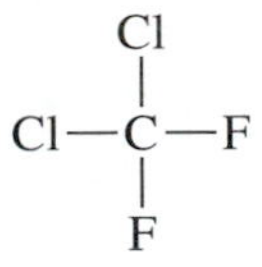

Freon-12 (CCl_2F_2)

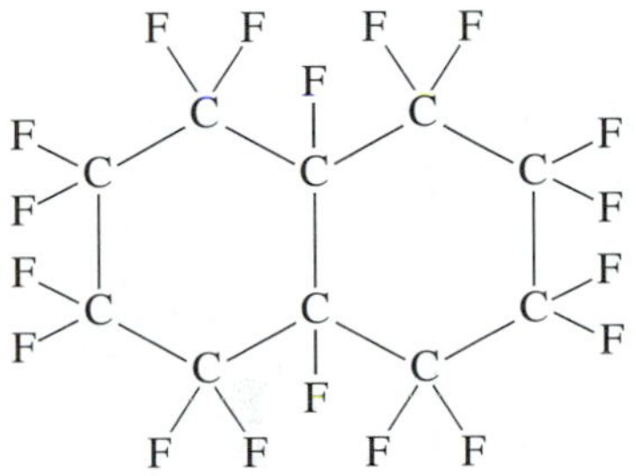

perfluorodecalin ($C_{10}F_{18}$)

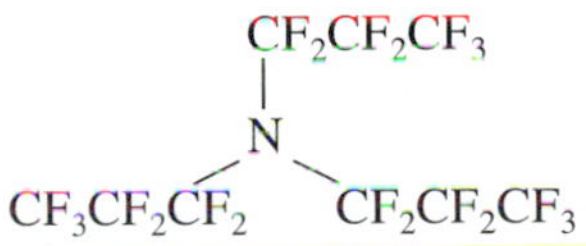

$N(CF_2CF_2CF_3)_3$

perfluorotripropylamine

Reactions of the Halogens

Halogen atoms are one electron short of a noble gas configuration (Table S2.10). Thus, they are "electron-hungry" and strive to fill the vacancy by forming halide ions or by sharing electrons with other nonmetal atoms. Fluorine, which is the most reactive element in the group, forms the strongest, shortest bonds. Iodine is the least reactive and forms the weakest bonds. Some reactions are given below.

With Metals. Halogens combine with most metals to form metal halides. Even unreactive "noble" metals such as gold and platinum succumb to attack by chlorine:

$$Zn(s) + Br_2(l) \rightarrow ZnBr_2(s)$$

$$2Au(s) + 3Cl_2(g) \xrightarrow{150°C} 2AuCl_3(s)$$

With Hydrogen. Halogens combine with hydrogen to form colorless gaseous *hydrogen halides*:

$$H_2(g) + Cl_2(g) \rightarrow 2HCl(g)$$

With Water. Chlorine, bromine, and iodine dissolve in water to a limited extent. The solutions are weakly acidic due to reactions such as

$$Cl_2(aq) + H_2O(l) \rightleftharpoons H^+(aq) + Cl^-(aq) + HOCl(aq)$$

Most of the acidity of the solution comes from the HCl (or HBr or HI), which ionizes completely. *Hypochlorous acid* (HOCl) is a weak acid, as are HOBr and HOI.

Fluorine displaces oxygen and other nonmetals from compounds—these reactions are often violent. Water, glass, and asbestos all burn brightly in a fluorine atmosphere. The principal reaction with water is

$$2F_2(g) + 2H_2O(l) \rightarrow 4HF(g) + O_2(g)$$

Some ozone, hydrogen peroxide, and oxygen fluorides are also found among the products.

With Oxygen. Although oxygen reacts directly with most other elements, it does not combine directly with halogens. Halogen oxides such as OF_2, Cl_2O_7, and BrO_2 are formed by indirect methods and tend to be unstable.

With Halide Ions. The oxidizing strength of the halogens decreases from fluorine to iodine. Chlorine, for example, can oxidize both bromide and iodide ions in aqueous solution (Demonstration S2.5):

$$Cl_2(aq) + 2Br^-(aq) \rightarrow 2Cl^-(aq) + Br_2(aq)$$

$$Cl_2(aq) + 2I^-(aq) \rightarrow 2Cl^-(aq) + I_2(aq)$$

but the reverse reactions do not occur. Bromine oxidizes aqueous iodide ions

DEMONSTRATION S2.5 **DISPLACEMENT OF IODIDE ION BY CHLORINE: $Cl_2(aq) + 2I^-(aq) \rightarrow 2Cl^-(aq) + I_2(aq)$**

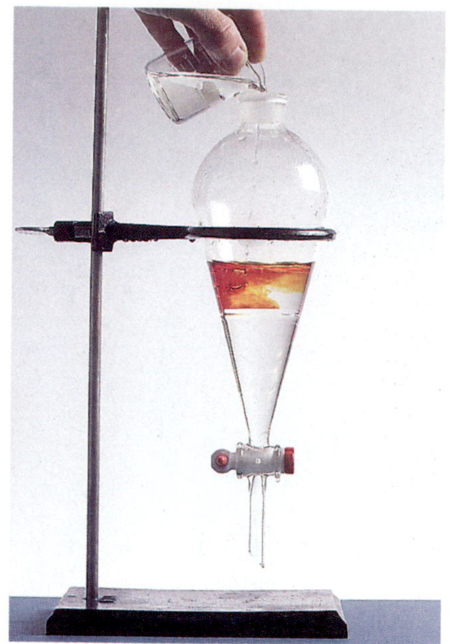

The two immiscible, colorless liquids are aqueous sodium iodide solution (top layer) and carbon tetrachloride (bottom layer). Chlorine water is added to the aqueous layer. The iodide ions are oxidized to iodine, which looks brown when dissolved in water.

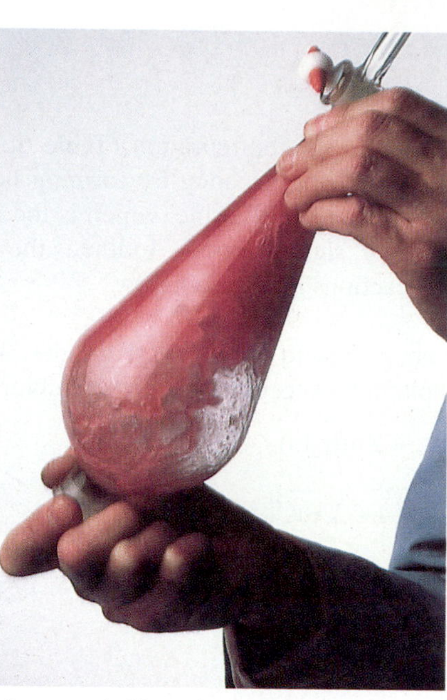

Vigorous shaking mixes the layers and allows CCl_4 to extract iodine from the aqueous layer.

When the layers separate, we see the violet color that is characteristic of iodine dissolved in CCl_4.

$$Br_2(aq) + 2I^-(aq) \rightarrow 2Br^-(aq) + I_2(aq)$$

but iodine does not oxidize bromide ion. Note that *each halogen displaces lower members of the group from aqueous solutions of their halides.* (This rule does not apply to fluorine, which reacts with water and does not exist in aqueous solution.) These displacement reactions are the basis for the preparation of bromine and iodine from seawater (see p. 515).

With Other Halogens. A compound composed of two different halogens is called an **interhalogen compound**. Some of these compounds are formed by direct combination:

$$Cl_2(g) + F_2(g) \rightarrow 2ClF(g)$$

$$(l) + 3F_2(g) \rightarrow 2BrF_3(l)$$

Others are formed by the reaction of halogens with ionic halides:

$$3F_2(g) + KCl(s) \rightarrow ClF_5(g) + KF(s)$$

$$4F_2(g) + KI(s) \rightarrow IF_7(g) + KF(s)$$

Interhalogen formulas are of the type XY, XY_3, XY_5, or XY_7, where X is the larger and less electronegative halogen atom. All interhalogen compounds are strong oxidizing agents; many are unstable.

PRACTICE EXERCISE S2.11

Use VSEPR to predict the shapes of the following interhalogen compounds: (a) IF_3 and (b) BrF_5. (*Hint*: Recall that nonmetals from the third and following periods often exhibit octet expansion.)

Hydrogen Halides and Their Salts

Hydrogen halides are produced by treating ionic halides with a concentrated nonvolatile acid. The hydrogen halide evolves as a gas. Sulfuric acid is the acid of choice for producing HF and HCl because it is readily available and inexpensive:

$$CaF_2(s) + H_2SO_4(18\ M) \rightarrow CaSO_4(s) + 2HF(g)$$

$$NaCl(s) + H_2SO_4(18\ M) \rightarrow NaHSO_4(s) + HCl(g)$$

Phosphoric acid is used to prepare HBr and HI

$$NaI(s) + H_3PO_4(15\ M) \rightarrow HI(g) + NaH_2PO_4(aq)$$

because sulfuric acid would oxidize HBr and HI to Br_2 and I_2:

$$2HI(aq) + H_2SO_4(18\ M) \rightarrow I_2(s) + SO_2(g) + 2H_2O(l)$$

The hydrogen halides are acidic in aqueous solution: HF is a weak acid while HCl, HBr, and HI are strong acids:

$$HF\ (aq) \rightleftharpoons H^+(aq) + F^-(aq)$$

$$HCl(aq) \rightarrow H^+(aq) + Cl^-(aq)$$

Hydrochloric acid is an important industrial chemical, used whenever a strong nonoxidizing acid is needed.

All hydrogen halides are toxic and irritating, but *hydrogen fluoride* is exceptionally lethal and destructive. Hydrogen fluoride converts many oxygen compounds to fluorides, and it disintegrates the silicates in glass and ceramic containers:

$$SiO_2(s) + 4HF(g) \rightarrow SiF_4(g) + 2H_2O(g)$$

$$CaSiO_3(s) + 6HF(g) \rightarrow SiF_4(g) + CaF_2(s) + 3H_2O(g)$$

Hydrogen fluoride vapor is used to etch designs on glass. The areas that are not to be etched are protected with wax. Frosted electric light bulbs and engraved glassware such as thermometers and burets are often manufactured in this manner.

Ordinary table salt (NaCl) is the best known halogen compound. "Iodized" salt contains about 0.01% NaI. Iodide in the diet helps to prevent goiter, an enlargement of the thyroid gland resulting from iodine deficiency. *Sodium bromide* and *potassium bromide* are mild sedatives that have been used in headache powders.

Iodide ion (I^-) is easily oxidized to I_2, and a solution of potassium iodide and starch is used to detect the presence of a strong oxidizing agent such as ozone (see

CHEMICAL INSIGHT

FIREWORKS, ROCKETS, AND EXPLOSIVES

The first intentional chemical explosion was an undocumented and perhaps prehistoric event. It is known, however, that "black powder" (gunpowder) made from potassium nitrate (saltpeter), charcoal, and sulfur was used in China for launching aerial fireworks as early as the ninth century. It is still used all over the world for this purpose.

An explosive mixture has two essential components, an oxidizing agent or *oxidizer* and a reducing agent or *fuel*. Oxidizers can contain chlorate ions, perchlorate ions, nitrate ions, or nitro ($-NO_2$) groups. A simple but effective oxidizer is liquid oxygen (LOX). Fuels can contain ammonium ions or other compounds with N—H or C—H bonds, or they can be oxidizable elements such as carbon, sulfur, and hydrogen. Carbon and sulfur are the fuels in black powder; the oxidizer is potassium nitrate. The ratio by weight of the components used to make black powder—15 parts charcoal, 10 parts sulfur, and 75 parts KNO_3—has remained unchanged for hundreds of years.

The explosion itself is an exothermic reaction that is usually complex and produces a variety of gaseous products. When black powder explodes, large volumes of carbon dioxide, nitrogen, and carbon monoxide are suddenly produced. The explosive force is a result of the rapid expansion of these hot gaseous reaction products. If the main purpose of the explosion is to produce light, the fuel may include combustible metals that burn with a bright flame, such as zinc, aluminum, magnesium, or titanium. The flash powder used at rock concerts is a mixture of potassium perchlorate and powdered magnesium.

Fireworks for aerial display have a primary rocket stage that propels them into the air with an explosion of black powder. Then, as shown in Figure S2.13, a multiple-stage fuse successively ignites other charges that produce the sound and light effects. Burning magnesium and aluminum provide white light, and compounds containing Sr^{2+}, Cu^{2+}, Na^+, and Ba^{2+} provide red, blue, yellow, and green colors, respectively. Blue flames ($CuCl_2$) and purple and violet flames (a mixture of $CuCl_2$ and $SrCl_2$) are the most difficult to achieve, and pyrotechnicians (those who make fireworks) often judge each other by the richness of these colors.

A high reaction temperature may produce an excess of white light that washes out the color, so the oxidizer for a color stage should have a relatively low ignition temperature. Chlorates provide the lowest temperatures and the brightest displays; chlorate mixtures are hazardous, however, and have been known to explode even at room temperature. Potassium perchlorate is safer and almost as effective.

The type of reaction that explodes fireworks will also propel rockets into space. Instead of a sudden explosion, however, a rocket needs a continuous regulated thrust, which is achieved by the controlled mixing of oxidizer and fuel.

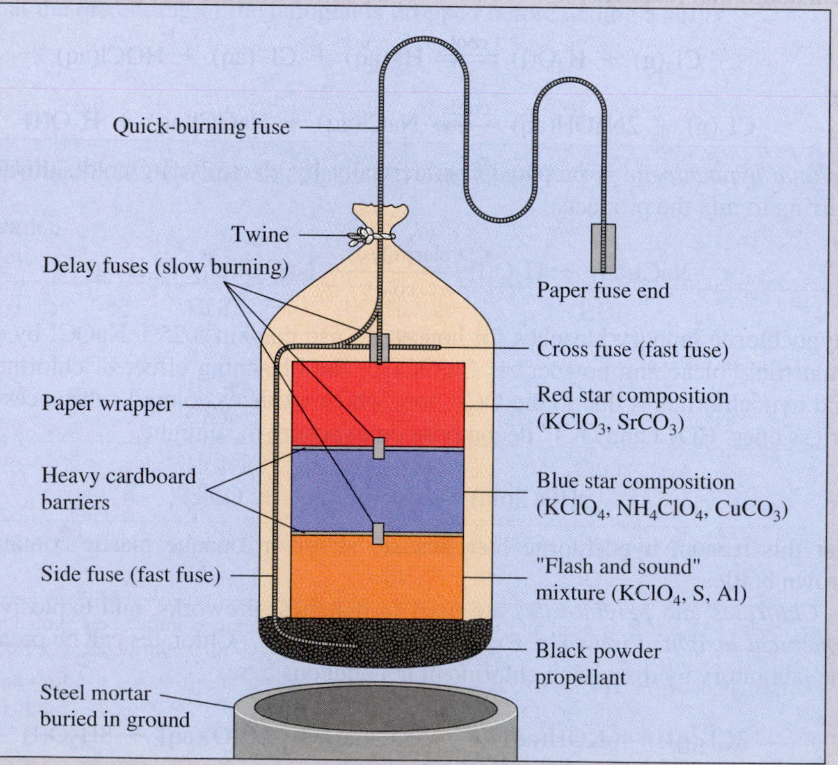

Figure S2.13

Diagram of a shell used for aerial fireworks. The shell, about two to eight inches in diameter, is lowered into the buried steel mortar. The quick-burning fuse ignites the upper slow-burning delay fuse and the black powder propellant. The gases produced by the burning propellant lift the shell several hundred feet into the air. Once airborne, the upper delay fuse ignites the upper charge, producing a burst of red stars and igniting the middle delay fuse. Then there is a burst of blue stars and the lower delay fuse is ignited. The final "boom" occurs when the "flash and sound" mixture ignites.

Liquid dimethylhydrazine ($(CH_3)_2N—NH_2$) fueled the lunar excursion module of the Apollo space vehicles. The oxidizer was liquid dinitrogen tetroxide (N_2O_4). The liquids react on contact:

$$(CH_3)_2NNH_2(l) + 2N_2O_4(l) \rightarrow 3N_2(g) + 4H_2O(g) + 2CO_2(g) \qquad \Delta H° = -1764.7 \text{ kJ}$$

Booster rockets for the early space shuttles used a solid propellant containing ammonium perchlorate (NH_4ClO_4). More than 1.5 million pounds of this chemical were consumed with each shuttle launching. When ammonium perchlorate is ignited at 200°C, the perchlorate ion oxidizes the ammonium ion:

$$2NH_4ClO_4(s) \rightarrow N_2(g) + Cl_2(g) + 2O_2(g) + 4H_2O(g) \qquad \Delta H° = -376.7 \text{ kJ}$$

The solid-fuel components of newer booster rockets contain a mixture of perchlorates, aluminum powder, and iron oxide. The shuttle engine, which takes over once the shuttle is in space, is powered by liquid hydrogen burning in liquid oxygen, the fuel–oxidizer mixture that provides the greatest thrust per unit mass. At liftoff, a typical shuttle carries more than 385,000 gallons of liquid hydrogen and more than 140,000 gallons of liquid oxygen (Figure S2.14).

Explosives used for blasting and demolition usually contain compounds such as ammonium perchlorate and ammonium nitrate, which combine the oxidizing and reducing components in a single substance. Ammonium nitrate may become explosive at temperatures as low as 170°C. Below 300°C, its decomposition products are N_2O and H_2O:

$$NH_4NO_3(s) \rightarrow N_2O(g) + 2H_2O(g) \qquad \Delta H° = -36.0 \text{ kJ}$$

An even more exothermic decomposition occurs at temperatures in excess of 300°C:

$$2NH_4NO_3(s) \rightarrow 2N_2(g) + O_2(g) + 4H_2O(g) \qquad \Delta H° = -236.2 \text{ kJ}$$

Storing ammonium nitrate in warm areas without adequate ventilation has resulted in disastrous explosions. In 1947, an ammonium nitrate explosion in Texas City, Texas claimed 576 lives.

Many explosives contain N—O bonds. *Nitroglycerin* is made by treating glycerol with nitric acid:

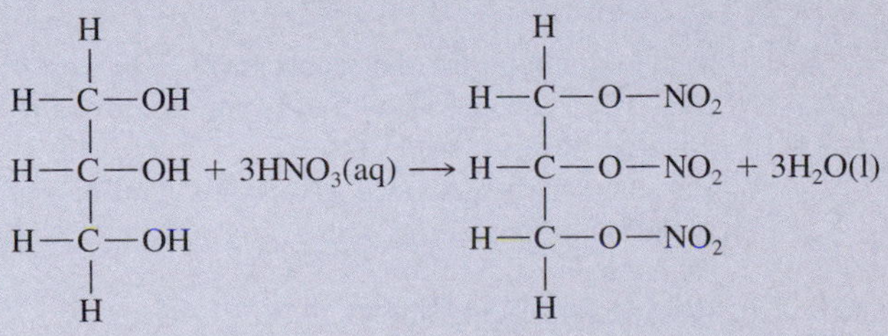

glycerol (glycerin) trinitroglycerol (nitroglycerin)

Figure S2.14

Liftoff of the space shuttle Atlantis. The booster rockets, each of which is about 149 feet long and 12 feet in diameter, are attached to the external fuel tank (red).

The decomposition of nitroglycerin produces only gases—N_2, H_2, O_2, and CO_2. Pure nitroglycerin is a hazardous and unpredictable substance that may be set off by the least shock. Dynamite, which is 75% nitroglycerin absorbed into 25% diatomaceous earth (a mixture of porous silicates), is less sensitive.

A *high explosive* is one that remains stable until it receives a strong shock from a lesser explosive, or *detonator. Trinitrotoluene* (TNT) is an example:

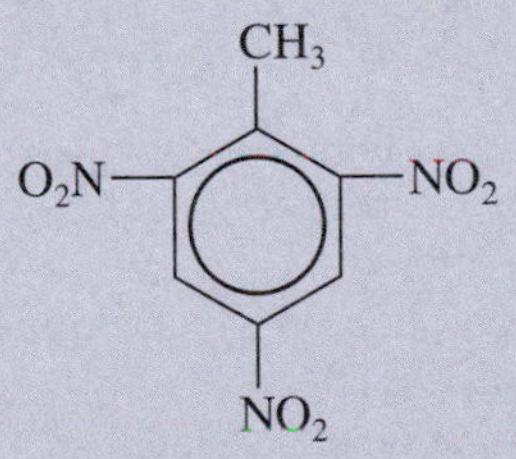

trinitrotoluene (TNT)

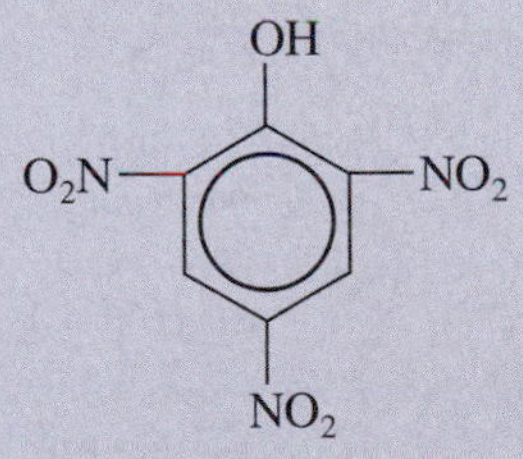

trinitrophenol (picric acid)

TNT has become the standard for measuring explosive power. A 1 megaton bomb, for example, has the explosive equivalent of 1 million tons of TNT. The greatest explosion caused by humans prior to the atomic bomb occurred in Halifax, Nova Scotia, when two ships, one carrying munitions, collided in the harbor. The resulting mile-high explosion of picric acid (trinitrophenol), TNT, and guncotton (cellulose nitrate) claimed 1600 lives and shattered windows 100 miles away.

Chlorates are prepared commercially by electrolyzing hot, concentrated chloride solutions and stirring to mix the products:

$$KCl(aq) + 3H_2O(l) \xrightarrow[\text{hot}]{\text{electrolysis}} KClO_3(aq) + 3H_2(g)$$

Prolonged electrolysis oxidizes chlorate ion to perchlorate ion. The half-reaction equation is:

$$ClO_3^-(aq) + H_2O(l) \rightarrow ClO_4^-(aq) + 2H^+(aq) + 2e^-$$

A perchlorate can also be obtained by gently heating a pure chlorate in the absence of a catalyst:

$$4KClO_3(s) \xrightarrow{\text{heat}} 3KClO_4(s) + KCl(s)$$

Strong heating in the presence of a catalyst will liberate oxygen:

$$2KClO_3(s) \xrightarrow[\text{heat}]{MnO_2} 2KCl(s) + 3O_2(g)$$

PRACTICE EXERCISE S2.15

Use VSEPR to predict the shapes of the following ions: (a) ClO_2^-, (b) ClO_3^-, and (c) ClO_4^-.

CHAPTER REVIEW

LEARNING OBJECTIVES BY SECTION

S2.1
1. Using equations whenever possible, describe how oxygen is produced in nature, in industry, and in the laboratory.
2. Write equations for the preparation of oxygen from $KClO_3$, HgO, BaO_2, Na_2O_2, and H_2O.
3. Write an equation for the formation of O_3 from O_2.
4. Compare the properties of oxygen and ozone, and list the principal uses of each allotrope.
5. Give examples of basic oxides, and write equations for their reactions with water and $H^+(aq)$.
6. Give examples of acidic oxides, and write equations for their reactions with water and $OH^-(aq)$.
7. Describe the structure, properties, and uses of hydrogen peroxide.
8. Write equations for the reaction of BaO_2 with cold H_2SO_4, the decomposition of H_2O_2, and the oxidation and reduction half-reactions of H_2O_2.
9. Write equations for the reactions of oxide, peroxide, and superoxide ions with water, $H^+(aq)$, and CO_2.
10. List the principal uses of peroxides and superoxides.

S2.2
1. Describe the occurrence, preparation, properties, and uses of nitrogen.
2. Write equations for nitrogen fixation in the atmosphere and by the Haber process.
3. Describe the properties and uses of ammonia and its salts.
4. Write equations for the reaction of NH_3 with water, NH_3 with aqueous acids, and ammonium salts with aqueous bases.
5. List the principal oxides, oxo acids, and oxo anions of nitrogen in order of increasing nitrogen oxidation number.
6. Describe the preparation and uses of N_2O.

7. Write equations for the formation in the atmosphere of NO, NO_2, and N_2O_4.
8. Write equations for the formation of nitrites and nitrous acid, and list their uses.
9. Using equations, describe the Ostwald process for the synthesis of nitric acid.
10. Write half-reaction equations for the reduction of HNO_3 to NO_2, NO, N_2O, N_2, and NH_4^+.
11. List the principal uses of nitric acid and nitrates.
12. Write equations for the reaction of nitrides with water, cyanides with aqueous acids, and hydrazine with oxygen.

S2.3
1. Describe the occurrence, preparation (including equations), physical properties, and principal uses of the halogens.
2. Write equations for the reactions of the halogens with metals, hydrogen, and water.
3. List the halogens in order of increasing oxidizing strength, and write equations for halogen–halide displacement reactions.
4. Write equations for the preparation of hydrogen halides from their salts.
5. List the principal uses of the hydrogen halides and their salts.
6. Given the name of a halogen oxo acid or oxo anion, write its formula; write the name when the formula is given.
7. List the halogen oxo acids in order of increasing oxidation number, increasing acid strength, and increasing ability to act as an oxidizing agent.
8. Write equations for the preparation of hypochlorites, chlorates, and perchlorates, and list their principal uses.

KEY TERMS BY SECTION

S2.1 Acidic oxide
Allotropes
Basic oxide
Ostwald process
Oxo acid
Oxo anion

S2.2 Haber process
Nitrogen fixation

S2.3 Interhalogen compound

FINAL EXERCISES

Answers to exercises with blue numbers are given in Appendix D.

PART A. QUESTIONS AND PROBLEMS BY SECTION

Oxygen (Section S2.1)

S2.1 Discuss the abundance and occurrence of oxygen in (a) the atmosphere, (b) the lithosphere (the solid outer layers of the earth), (c) the hydrosphere (the natural waters of the earth), and (d) the biosphere.

S2.2 List at least three processes that remove oxygen from the atmosphere. By what processes is atmospheric oxygen renewed?

S2.3 **(a)** What role does ozone play in the upper atmosphere?
(b) Is the presence of ozone in the lower atmosphere beneficial or harmful to living things? Explain your answer.

S2.4 **(a)** Describe and give an equation for the generation of ozone.
(b) Explain why the odor of ozone is sometimes detected near sparking electrical equipment.

S2.5 Describe a test for ozone and write the equation.

S2.6 **(a)** Define the term *allotrope.*
(b) Explain why oxygen and ozone are considered to be allotropes.
(c) Are graphite and diamond allotropes? Why or why not?
(d) Are solid sodium and sodium vapor allotropes? Why or why not?

S2.7 **(a)** Give three examples of basic oxides. What type of element forms a basic oxide?
(b) Give three examples of acidic oxides. What type of element forms an acidic oxide?
(c) How can one determine in the laboratory whether a given oxide is acidic or basic?

S2.8 Selenium, a nonmetallic element of Group 6A, forms the oxides SeO_2 and SeO_3. Would you expect these oxides to be ionic or covalent? Acidic or basic?

S2.9 List some uses of hydrogen peroxide that are based on its oxidizing properties.

S2.10 Explain in terms of bond energies why hydrogen peroxide is a stronger oxidizing agent than oxygen.

S2.11 Diagram the hydrogen peroxide molecule, and discuss the bond angles in terms of VSEPR.

S2.12 **(a)** Write Lewis structures for the ozone molecule.
(b) Does the molecule contain delocalized electrons, and if so, where are they located?
(c) Explain why the molecule is bent.

S2.13 Give the molecular orbital configuration of O_2. Discuss the relationship between this configuration and the properties of O_2.

S2.14 Give the molecular orbital configurations of peroxide and superoxide ions. Explain why these ions are more reactive than O_2.

S2.15 Write an equation for the formation of O_2 by each of the following reactions:
(a) electrolysis of water
(b) decomposition of potassium chlorate
(c) decomposition of hydrogen peroxide
(d) photosynthesis

S2.16 Write an equation for the formation of oxygen from each of the following compounds:
(a) HgO **(c)** Na_2O_2
(b) BaO_2 **(d)** Ag_2O

S2.17 Write an equation for the reaction of oxygen with each of the following substances:
(a) iron **(d)** phosphorus (P_4)
(b) zinc **(e)** sulfur dioxide
(c) sulfur **(f)** acetylene (C_2H_2)

S2.18 Write an equation for the reaction of oxygen with each of the following substances:
(a) magnesium **(d)** methane
(b) aluminum **(e)** carbon monoxide
(c) carbon **(f)** ethanol

S2.19 Write an equation for the dehydration of each of the following hydroxides:
(a) $Mg(OH)_2$ **(c)** $Fe(OH)_2$
(b) LiOH **(d)** $Fe(OH)_3$
Remember that oxidation numbers do not change during dehydration.

S2.20 Write an equation for the reaction of each of the following oxides with water:
(a) SO_2 **(d)** N_2O_5
(b) SO_3 **(e)** P_4O_{10}
(c) CO_2 **(f)** P_4O_6
Remember that oxidation numbers do not change during hydration.

S2.21 Write equations for the reactions of (a) BaO and (b) Na_2O with water.

S2.22 Write equations for the reactions of (a) CaO and (b) Li_2O with CO_2.

S2.23 Write equations for the reactions of the following oxides with NaOH(aq) and with $Ca(OH)_2$(aq):
(a) SO_2
(b) SO_3
(c) CO_2

S2.24 Write equations for the reactions of the following oxides with HCl(aq) and with H_2SO_4(aq):
(a) CaO
(b) Li_2O
(c) Fe_2O_3

S2.25 Describe, with the aid of a balanced equation, a method for preparing hydrogen peroxide in the laboratory.

S2.26 Write the equation for the decomposition of hydrogen peroxide. What characteristics of this reaction make it potentially explosive?

S2.27 Write equations for the reactions of oxide, peroxide, and superoxide ions with (a) water, (b) H^+(aq), and (c) CO_2.

S2.28 Explain, with an equation, how potassium superoxide functions in a closed breathing system.

S2.29 Write an equation for each of the following reactions:
(a) hydrogen peroxide with lead sulfide
(b) sodium peroxide with water
(c) sodium peroxide with carbon dioxide
(d) potassium superoxide with carbon dioxide

S2.30 Write an equation for each of the following reactions:
(a) the liberation of oxygen by heated barium peroxide
(b) barium peroxide with cold sulfuric acid
(c) potassium superoxide with hydrochloric acid
(d) the oxidation of $Fe(OH)_2$ to $Fe(OH)_3$ by H_2O_2 in basic solution

Nitrogen (Section S2.2)

S2.31 **(a)** Discuss the occurrence of nitrogen in the atmosphere, the lithosphere (the rocky part of the earth), and the biosphere.
(b) Explain why very few nitrogen compounds are found in the lithosphere.

S2.32 **(a)** Describe two natural processes that remove nitrogen from the atmosphere.
(b) How is nitrogen restored to the atmosphere?

S2.33 (a) How is nitrogen obtained industrially?
(b) List some uses for elemental nitrogen.

S2.34 List the principal uses of the following nitrogen compounds:
(a) NH_3
(b) N_2O
(c) $NaNO_2$
(d) HNO_3
(e) NH_4NO_3
(f) N_2H_4

S2.35 When a sample of nitrogen is subjected to an electric discharge, it becomes reactive and combines with many elements such as sulfur and sodium to which it would ordinarily be inert. Explain why this happens.

S2.36 Give the molecular orbital configuration for N_2, and explain why the molecule is unreactive under ordinary conditions.

S2.37 Write an equation for the reaction of nitrogen with (a) hydrogen, (b) lithium, and (c) magnesium.

S2.38 Write equations for nitrogen fixation as it occurs in the atmosphere.

S2.39 (a) Describe and give the equation for the Haber process.
(b) Why is this process important?

S2.40 Write an equation for each of the following reactions:
(a) ammonia with water
(b) ammonia with aqueous HCl
(c) solid NH_4Cl heated with concentrated sodium hydroxide

S2.41 Illustrate each oxidation state of nitrogen by writing the formula of a substance in which it has that oxidation state.

S2.42 List (a) the oxides, (b) the oxo acids, and (c) the oxo anions of nitrogen in order of increasing oxidation number.

S2.43 Write equations for reactions that produce the following oxides:
(a) N_2O
(b) NO
(c) NO_2
(d) N_2O_4

S2.44 Write an equation for each of the following reactions:
(a) oxidation of nitrogen monoxide in air
(b) equilibrium between nitrogen dioxide and dinitrogen tetroxide
(c) nitrogen dioxide with water

S2.45 Write equations for (a) the preparation of nitrous acid and (b) its decomposition in aqueous solution.

S2.46 Write an equation for the preparation of sodium nitrite from sodium nitrate.

S2.47 (a) Describe and write an equation for each step in the industrial production of nitric acid (Ostwald process).
(b) Which step would you expect to be the most difficult and expensive?

S2.48 Pure HNO_3 can be dehydrated to N_2O_5 by treating it with solid P_4O_{10}. The other product is phosphoric acid. Write a balanced equation for the reaction.

S2.49 Write an equation for the half-reaction in which HNO_3 is reduced to
(a) NO_2
(b) NO
(c) N_2O
(d) N_2
(e) NH_4^+

S2.50 Write equations for the half-reactions that occur when (a) zinc reacts with 6 *M* nitric acid and (b) copper reacts with 12 *M* nitric acid.

S2.51 Use the method of half-reactions to write equations for the oxidation of the following elements by concentrated nitric acid:
(a) phosphorus
(b) sulfur
(c) iodine

S2.52 Write equations for the half-reactions that occur when aqua regia dissolves (a) gold and (b) platinum.

S2.53 Write an equation for each of the following reactions:
(a) sodium carbonate with nitric acid
(b) magnesium hydroxide with nitric acid
(c) dinitrogen pentoxide with water
(d) lithium nitride with water
(e) sodium cyanide with hydrochloric acid
(f) synthesis of hydrazine

S2.54 Write an equation for each of the following reactions:
(a) nitride ion with water
(b) cyanide ion with $H^+(aq)$
(c) hydrazine with O_2
(d) liquid hydrazine with hydrogen peroxide
(e) dimethylhydrazine with liquid oxygen
(f) the light-induced decomposition of HNO_3

S2.55 Draw a Lewis structure for each of the following molecules:
(a) nitrogen monoxide
(b) dinitrogen oxide
(c) nitrogen dioxide
(d) dinitrogen trioxide
(e) dinitrogen tetroxide
(f) dinitrogen pentoxide

S2.56 Draw a Lewis structure for each of the following species:
(a) nitrous acid, HNO_2
(b) nitric acid, HNO_3
(c) hydroxylamine, NH_2OH
(d) hydrazine, N_2H_4
(e) cyanide ion, CN^-
(f) amide ion, NH_2^-

S2.57 Use VSEPR to sketch the shape of each species in (a) Exercise S2.55 and (b) Exercise S2.56.

S2.58 Which of the nitrogen oxides in Exercise S2.55 would be attracted into a magnetic field? Justify your answer.

The Halogens (Section S2.3)

S2.59 (a) In what form do the halogens occur naturally?
(b) Give the names and formulas of at least five naturally occurring halogen compounds.
(c) Which halogen compound is most abundant?

S2.60 State which halogens are essential in the body, and explain the function of each.

S2.61 Compare the physical properties of the halogens, and show how they exemplify group trends.

S2.62 Which halogen is the strongest oxidizing agent? The weakest? Explain your answers in terms of atomic properties and group trends.

S2.63 List the principal uses of fluorine and chlorine.

S2.64 List the principal uses of the following halogen compounds:
(a) $Ca(OCl)_2$ (d) NaOCl
(b) HCl (e) $KClO_3$
(c) ClO_2 (f) $KClO_4$

S2.65 Name each of the following compounds:
(a) HOI (d) HIO_3
(b) $KClO_3$ (e) $NaBrO_4$
(c) $HClO_4$ (f) $Mg(OCl)_2$

S2.66 Write a formula for each of the following compounds:
(a) calcium hypochlorite (d) hypobromous acid
(b) barium iodate (e) potassium perchlorate
(c) periodic acid (f) chlorous acid

S2.67 Write an equation for the commercial preparation of each of the following halogens:
(a) fluorine (c) bromine
(b) chlorine (d) iodine

S2.68 Write equations for the oxidation and reduction half-reactions when (a) HCl reacts with $KMnO_4$ and (b) an acidified solution of sodium iodide reacts with manganese dioxide.

S2.69 Write an equation for the reaction of each halogen with the following substances:
(a) sodium (c) hydrogen
(b) zinc (d) water

S2.70 Predict the principal products when the following substances react with fluorine:
(a) methane
(b) carbon dioxide
(c) boron trioxide (B_2O_3)

S2.71 (a) Will chlorine gas react with sodium iodide solution?
(b) Will bromine react with potassium chloride solution?
Write an equation for each reaction that occurs.

S2.72 Dilute aqueous solutions of bromine and iodine are both faintly yellowish. Describe at least one simple laboratory test that would distinguish between them.

S2.73 Write an equation for the commercial preparation of each of the following hydrogen halides:
(a) HF (c) HBr
(b) HCl (d) HI

S2.74 Write an equation for each of the following reactions:
(a) $CaCl_2$ with concentrated H_2SO_4
(b) NaF with concentrated H_2SO_4
(c) NaBr with concentrated H_3PO_4
(d) KI with concentrated H_3PO_4

S2.75 Write an equation for the reaction of HF with (a) magnesium, (b) water, and (c) silicon dioxide.

S2.76 Write an equation for the reaction of (a) HBr and (b) HI with concentrated sulfuric acid. Why is phosphoric acid, and not sulfuric acid, used for preparing HBr and HI?

S2.77 List the oxo acids of bromine in order of (a) increasing bromine oxidation number, (b) increasing acid strength, and (c) increasing ability to act as an oxidizing agent.

S2.78 Write the formula for the oxo acid that would result from the hydration of (a) Cl_2O_7 and (b) I_2O_5.

S2.79 Write an equation for each of the following reactions:
(a) iodine with potassium hydroxide solution
(b) decomposition of hypochlorous acid
(c) catalyzed decomposition of potassium chlorate

S2.80 The following substances can be prepared by electrolyzing aqueous chloride solutions:
(a) chlorine (c) potassium chlorate
(b) sodium hypochlorite (d) potassium perchlorate
Write an equation for the electrolysis in each case, and state the conditions under which the electrolysis is carried out.

S2.81 Draw a Lewis structure, and sketch the shape of each of the following species:
(a) $HClO_3$ (d) I_3^-
(b) IO_2^- (e) ClO_2
(c) HOI (f) ClO_4^-

S2.82 Draw a Lewis structure, and sketch the shape of each of the following interhalogen molecules and ions:
(a) ClF_3 (d) ICl_2^-
(b) IF_5 (e) BrF_4^-
(c) BrCl (f) IF_4^+

PART B. MISCELLANEOUS QUESTIONS AND PROBLEMS

S2.83 Astatine, the radioactive member of the halogen family, is difficult to study, and less is known about it than about any other member of the group.
- **(a)** On the basis of periodic trends what would you predict for the physical state and color of astatine?
- **(b)** How would you expect At_2 to react with metals and with water?
- **(c)** Should At_2 react with potassium iodide solution? Explain.

S2.84 Chemists have observed that the stability of a polyatomic ion is enhanced when the other ion in the ionic compound is of similar size. Would you say that this applies to the peroxides and superoxides of Group 1A and 2A? Justify your answer.

S2.85 When fluorine was first prepared in 1886, it was stored in vessels fashioned from the mineral fluorspar, CaF_2. Fluorine is now kept in special steel containers. Explain why fluorine does not destroy (a) fluorspar and (b) the specially formulated steel.

S2.86 **(a)** What factors make a compound potentially explosive?
- **(b)** Why are nitrogen compounds so often explosive?

S2.87 Give the chemical names of the substances or mixtures that have the following common names:

(a) laughing gas	**(f)** fluorspar
(b) tincture of iodine	**(g)** Teflon
(c) aqua regia	**(h)** Freon-12
(d) saltpeter	**(i)** bleaching powder
(e) Chile saltpeter	

S2.88 Perbromates cannot be prepared by the electrolytic processes that produce perchlorates and periodates, and for some time it was believed that they could not exist. The BrO_4^- ion has now been prepared by oxidizing aqueous bromate ion with xenon difluoride. The other products are xenon and hydrogen fluoride. Use the method of half-reactions to obtain the balanced equation for this reaction.

S2.89 Bromine is prepared commercially by passing chlorine into brine from natural brine wells.
- **(a)** If the brine has a density of 1.1 g/mL and contains 4000 ppm by mass of bromine in the form of bromide ion, how many liters of the brine would be needed to produce 1.00 kg of bromine?
- **(b)** What volume of chlorine measured at STP would be needed? Assume that the process is 100% efficient.

S2.90 A sodium chloride solution is 25% sodium chloride by mass and has a density of 1.2 g/mL. How many liters of the solution must be electrolyzed to produce 1.00 kg of chlorine? Assume that the electrolysis is 80% efficient.

S2.91 How many kilograms of sodium fluoride are needed each day to fluoridate the water supply for a community of 1,500,000 persons? Assume a fluoride ion concentration of 1.0 ppm by mass, an average water use of 100 gallons per day per person, and a water density of 1.00 g/mL.

S2.92 Which of the following monatomic ions cannot exist in aqueous solution?

(a) H^+	**(d)** Cl^-
(b) H^-	**(e)** O^{2-}
(c) F^-	**(f)** N^{3-}

Write an equation for the reaction, if any, between each ion and water.

S2.93 Calculate the molarity of the following solutions:
- **(a)** concentrated aqueous ammonia, 25% NH_3 by mass, density 0.95 g/mL
- **(b)** antiseptic hydrogen peroxide, 3.0% H_2O_2 by mass, density 1.0 g/mL
- **(c)** isotonic sodium chloride for injection, 0.90% NaCl by mass, density 1.0 g/mL

S2.94 A 4.15-L sample of dry air, measured at 25°C and 1.00 atm, was passed through an ozonizer and then through a solution of potassium iodide, where 75.1 mg of I_2 was formed. Assume that air is 20.9 mol % O_2, and calculate the percent yield in the ozonizer reaction:

$$3O_2(g) \rightarrow 2O_3(g)$$

S2.95 A solution contains 6.0% H_2O_2 by mass and has a density of 1.00 g/mL. (a) Calculate the volume of oxygen produced at 25°C and 1.00 atm when all of the hydrogen peroxide in 1.00 L of the solution decomposes to water and oxygen. (b) An older name for the solution was "20-volume H_2O_2." How accurate was this designation?

S2.96 *Chemical and Engineering News*, a publication of the American Chemical Society, reported that 38.99 billion pounds of oxygen were produced in the United States in 1990. How much oxygen was produced in (a) metric tons and (b) liters measured at STP? (*Hint:* 1 metric ton = 1000 kg.)

S2.97 Liquid hydrazine (N_2H_4) has been used with hydrogen peroxide as a rocket propellant.
- **(a)** Write the equation for the reaction of these compounds to produce nitrogen gas and water vapor.
- **(b)** Refer to Appendix B.1, and calculate the heat released per mole of hydrazine. The standard enthalpies of formation of $N_2H_4(l)$ and $H_2O_2(l)$ are +50.63 kJ/mol and −187.78 kJ/mol, respectively.

S2.98 The following reaction occurs during the high-temperature explosive decomposition of ammonium nitrate:

$$2NH_4NO_3(s) \rightarrow N_2(g) + 4H_2O(g) + O_2(g)$$

(a) Refer to Appendix B.1, and calculate the standard enthalpy change for this reaction.

(b) How many liters of gas, measured at 800°C and 1.00 atm, would be produced by the explosion of 1.00 kg of ammonium nitrate?

S2.99 Ammonium perchlorate decomposes according to the equation

$$2NH_4ClO_4(s) \rightarrow N_2(g) + Cl_2(g) + 2O_2(g) + 4H_2O(g)$$

The early space shuttles consumed more than 1.5 million pounds of ammonium perchlorate with each launching. Refer to Appendix B.1, and calculate the quantity of heat released when 1.5 million pounds of ammonium perchlorate decompose under standard conditions. The standard enthalpy of formation of NH_4ClO_4 is -295.3 kJ/mol.

S2.100 The Winkler method for determining the concentration of dissolved oxygen in water consists of the following steps: (1) A measured volume of water reacts with excess Mn^{2+} in basic solution. The oxygen is reduced to OH^- and Mn^{2+} is oxidized to $MnO_2(s)$. (2) The solid MnO_2 is treated with excess I^- in acidic solution to give I_2 and Mn^{2+}. (3) Starch is added to give a dark blue color with the iodine, and the solution is titrated with a standard thiosulfate solution until the blue color disappears. The I_2 is reduced to I^- and the $S_2O_3^{2-}$ is oxidized to $S_4O_6^{2-}$.

(a) Write a balanced equation for each step in the Winkler method.

(b) A 1.00-L sample of water required 10.4 mL of 0.360 *M* $Na_2S_2O_3$ when analyzed by the Winkler method. Calculate the concentration of oxygen in the sample. Express your answer in milligrams of oxygen per milliliter of water.

CHEMISTS AT WORK

THE CHEMIST IN BUSINESS

Modern corporations, large and small, employ chemists in many diverse functions, such as sales and marketing, personnel, safety, and public relations. The skills and fundamental knowledge that comprise a scientific background are important in understanding and representing both the nature of products and the operation of a company. Many chemists pursue such careers in what is typically termed *the business area*, either directly after graduation or after several years of experience.

When Anne Clark entered college at the University of Wisconsin—Eau Claire, she knew that she wanted to major in a science but wasn't sure which one. She decided to be a chemistry major because the professor in her general chemistry course stimulated her interest. "The faculty went out of their way to make classes interesting," she reminisces. Specifically, Anne elected the major Chemistry with Business Emphasis, a curriculum composed of chemistry and business courses. For Anne, this preparation led to a career in the sciences with creative outlets not solely based in the laboratory.

After graduation in December 1986, she began her career at Vista Chemical Company of Houston, Texas. An international company, Vista's two divisions are the Surfactants and Specialties Division, including surfactants ("wetting agents," see Chapter 12), solvents, and alumina (Al_2O_3), and the Olefins and Vinyl Division, including polyvinyl chloride (PVC) and vinyl chloride monomer. Her initial positions were in the company's Business Career Development Program. "This is not just an ordinary entry-type training program," she states. "These are jobs that must be done but which also provide an opportunity to learn about the company and its products." During her tenure in this program, a marketing manager served as Anne's mentor, an individual with whom Anne could discuss, in a confidential manner, her personal and professional development.

ANNE CLARK

Anne's first assignment was in supply and transportation. Her specific responsibilities were in the planning and execution of efficient and cost-effective manners of product delivery. In this regard, she functioned as a logistics analyst, planning the route and method of transportation to be used while always being aware of the nature of the product.

Her next assignment was as a customer service representative where she could deal directly with customers and their specific problems. Duties increased rapidly, from working with a few customers from a small geographical area to representing one product nationally. Customer service was followed by a special project for the Vice-President of Manufacturing in which she determined that the established sales territories for surfactants were effective and efficient.

Anne then spent over 2 years in sales, first in Chicago where she represented PVC, both pure polymer and compounded materials, in a six-state territory in the upper Midwest. The PVC was used by other manufacturers to make products such as pipes, siding, and wall coverings. After an internal realignment of the PVC business, she returned to Houston and represented only the compounded materials, also in a six-state area. "I liked the work in sales. I liked meeting new people; I liked finding new business; I liked the challenge involved in sales; I did not like all the travel."

In 1991, she became Product Safety Specialist in the Environmental Department where she is in charge of working with Vista's nine plants on environmental, safety, and regulatory issues. In this position, she certifies that all sites are complying with state, federal, and sometimes international rules and regulations related to the transportation of chemicals and that they stay current with rule changes. She also writes Material Safety Data Sheets for the products of the company and conducts internal audits to determine whether the company is complying with EPA and OSHA regulations. She says, "I like this position because now I'm using much more chemistry."

CHAPTER 12

Water droplets and gas bubbles trapped in amber.

LIQUIDS, SOLIDS, AND INTERMOLECULAR FORCES

OUTLINE

PREVIEW

The study of the three states of matter engages our sense of the material world perhaps more than any other topic: Gas—the hiss of steam or the scent of lilacs; liquid—the sound and swirl of flowing water; solid—the elegance of a gemstone or a freshly fallen snowflake. What forces hold atoms together in the liquid and solid states? How do these forces give rise to the flow of liquids and the symmetry of crystals? To answer these questions, we must consider the collective behavior of closely packed atoms, ions, and molecules.

Ordinary matter exists in three states—solid, liquid, and gas. Solids have the most orderly arrangement of molecules, liquids are less orderly, and gases are the most disordered of all. Molecular motion increases with increasing temperature, so warming favors the formation of more chaotic states (Figure 12.1). Cooling slows the molecules down so that intermolecular attractions can restore order. Increasing the pressure also increases order by bringing molecules nearer to each other. At any given temperature and pressure, the most stable state of a substance represents a compromise between disorder arising from random molecular motion and order imposed by attractive forces.

Matter is composed of atoms, molecules, or ions. It is convenient to refer to these particles collectively as "molecules."

The molecules of a gas are relatively far apart, so intermolecular attractions are weak. Gas molecules are essentially independent of each other and tend to wander off in all directions; hence, gases expand to fill their containers and do not have fixed shapes or volumes. The molecules of a liquid or solid, on the other hand, are close together, and the attractive forces between them are fairly strong. Liquids and solids are not easily compressed, they do not expand like gases, and they have definite volumes. The intermolecular forces in a liquid are strong enough to prevent it from expanding, but they are not strong enough to prevent its molecules from flowing past each other. Hence, despite its fixed volume, a liquid will take the shape of its container. Solids have definite shapes as well as fixed volumes. The molecules of a solid vibrate (move back and forth) about fixed positions, but the forces

Less familiar states of matter include the *liquid crystal state*, the *supercritical fluid state*, and the *plasma state*. These states will be discussed in this and later chapters.

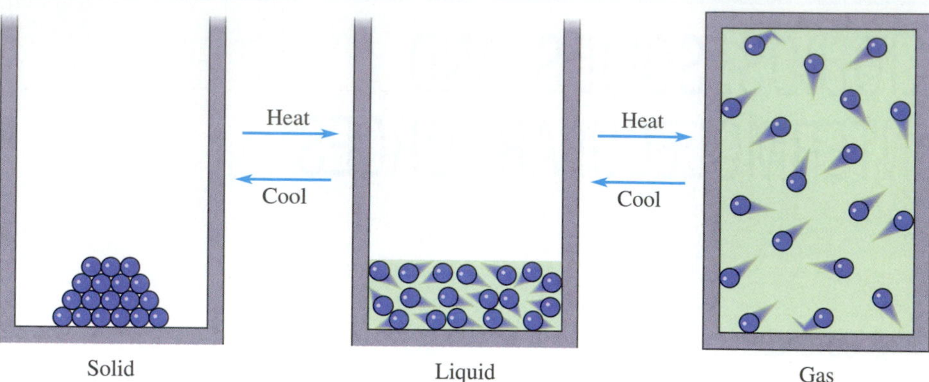

Figure 12.1
Heating, which increases molecular motion, converts the orderly solid to the partially disordered liquid and finally to the completely chaotic gas. Cooling, which decreases molecular motion, produces the reverse sequence of events.

between them are strong enough to keep the structure rigid. Some properties of gases, liquids, and solids are compared in Table 12.1.

12.1 PHASE CHANGES

A change in state is called a **phase change** (Figure 12.2). The following phase changes are endothermic: **melting** (solid to liquid), **vaporization** (liquid to gas), and **sublimation** (solid to gas). The reverse changes, **freezing** (liquid to solid), **condensation** (gas to liquid), and **deposition** (gas to solid), are exothermic. The phase changes of water from ice to liquid to steam and back are the ones we know best,

TABLE 12.1 Properties of Gases, Liquids, and Solids

Gas	Liquid	Solid
Takes the shape and volume of its container	Has a fixed volume but takes the shape of its container	Has a fixed shape and volume
Highly compressible	Only slightly compressible	Least compressible of all
Densities variable but always low	Densities much higher than those of gases	Densities comparable to liquid densities
Free space very large compared to volume of molecules	Very little free space between molecules	Usually less free space than in liquids
Molecules move at random everywhere; rapid diffusion	Molecules are able to flow past each other; slow diffusion	Molecules held in fixed positions; very slow diffusion
Molecular motion predominates; attractive forces play a minor role	Molecular motion and attractive forces are both important	Bond forces and intermolecular forces predominate; molecular motion plays a minor role

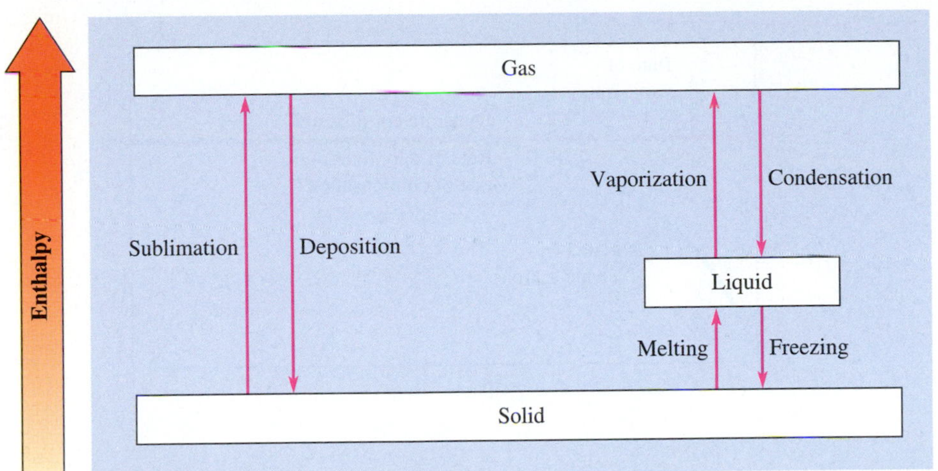

Figure 12.2
An enthalpy diagram showing some phase changes.

but many others are also familiar. For example, vaporization and sublimation allow us to detect vapors from perfumes and mothballs, and melting causes electrical fuses to blow when their metal strips overheat.

Vapor Pressure of Liquids and Solids

Imagine a liquid confined as in Figure 12.3. Its molecules move randomly, and the more energetic ones near the liquid surface overcome the intermolecular attractions and vaporize into the space above the liquid. The vapor molecules also move randomly, and some of them will strike the liquid surface and be recaptured. As more molecules vaporize, more will return from gas to liquid, and soon the rate of condensation becomes equal to the rate of vaporization. When the two rates are equal, the concentration of molecules in the vapor no longer changes with time. The liquid and vapor are now in a state of **dynamic equilibrium** (Figure 12.4), a state in which no net change occurs because opposing processes are taking place at the same rate.

LEARNING HINT
Review the characteristics and previous examples of dynamic equilibrium given in Section 4.1.

The pressure exerted by a vapor in equilibrium with its liquid is called the **vapor pressure** of the liquid. One way of measuring vapor pressure is illustrated in Figure 12.5, and Table 12.2 lists the vapor pressures of several liquids at 25°C. (You have

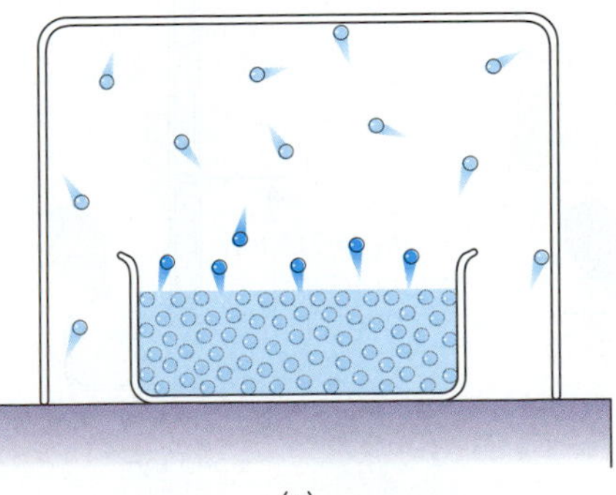

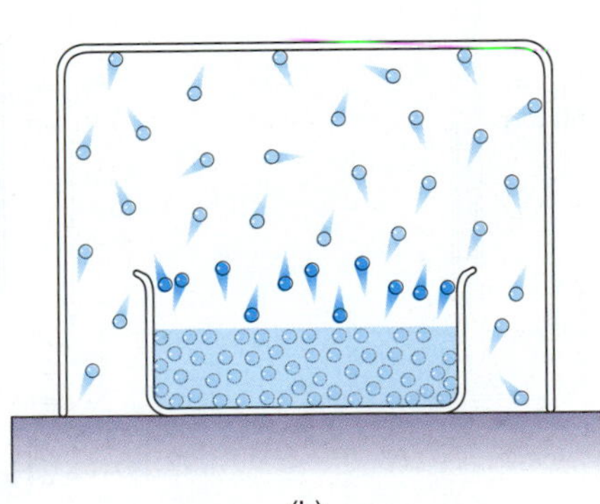

Figure 12.3
Evaporation and condensation in a closed system. (a) When a liquid begins to evaporate, the rate at which molecules escape from the liquid surface is greater than their rate of return. (b) As the vapor concentration increases, the rate of return will increase until it equals the rate of escape.

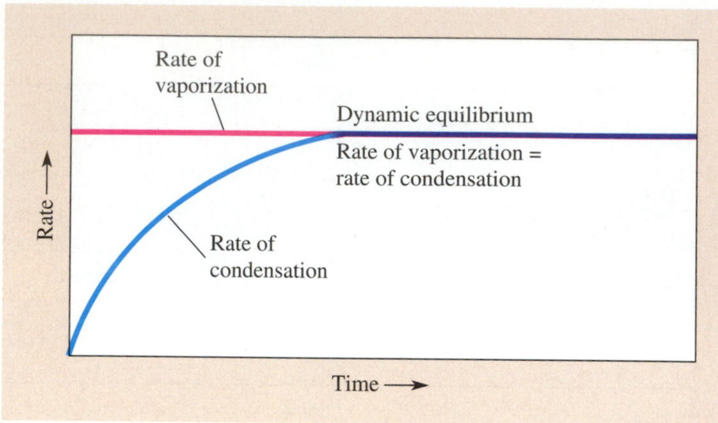

Figure 12.4

Rates of vaporization and condensation versus time in a closed system. As more vapor forms, the condensation rate increases until it equals the rate of vaporization. When the two rates are equal, the liquid and vapor are in equilibrium.

already used the vapor pressure of water in partial pressure calculations; see Table 5.4, p. 199.)

LEARNING HINT

The vapor pressure of a liquid is independent of the size or shape of its container.

Vapor pressure is a rough indicator of intermolecular attraction. Strong attractive forces between molecules cause a liquid to have a lower vapor pressure and to be less volatile (less easily vaporized) than a liquid with weak intermolecular forces. Figure 12.6 shows that vapor pressure increases with increasing temperature. This increase is a result of increased molecular speed. Fast molecules are more likely to escape from the liquid, and once in the vapor, they are less likely to return. The result is a greater concentration of vapor molecules at equilibrium and a greater vapor pressure.

Solids also have vapor pressures. We can smell the vapors of solids such as menthol, camphor, mothballs, and iodine. A volatile solid like frozen CO_2 (Dry Ice) sublimes rapidly when warmed to room temperature because it has a high vapor pressure. Nonvolatile solids such as sodium chloride have negligible vapor pressures and do not sublime to any appreciable extent.

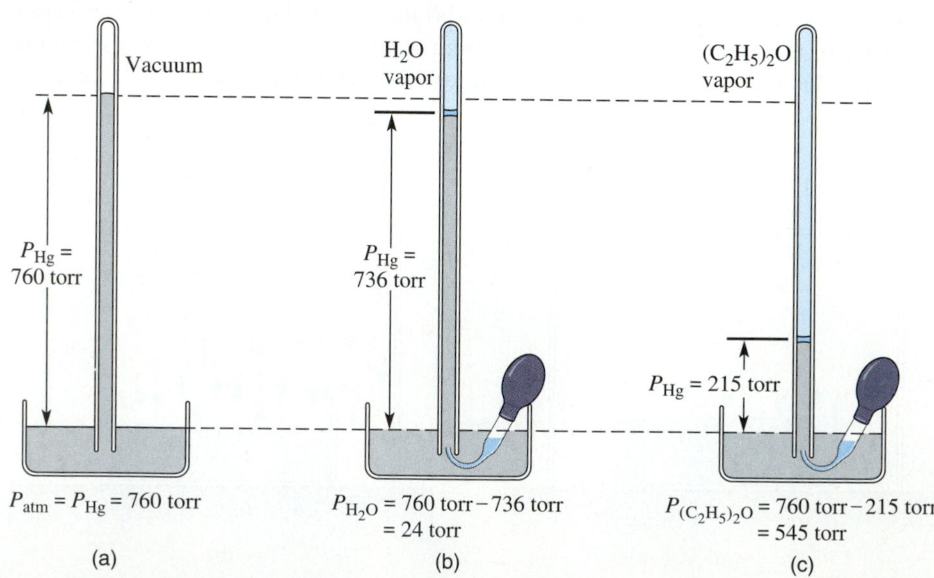

Figure 12.5

Measuring the vapor pressure of a liquid. (a) A 760-mm column of mercury is supported by atmospheric pressure. (b) Water injected into the column rises to the top where some of it evaporates; its vapor pressure pushes the mercury column down by 24 mm. Hence, the vapor pressure of water is 24 torr. (c) The vapor pressure of diethyl ether ($(C_2H_5)_2O$) is 545 torr at 25°C.

TABLE 12.2 Vapor Pressures[a] of Various Liquids at 25°C

Substance	Formula	Vapor Pressure (torr)	Intermolecular Forces
Carbon disulfide	CS_2	336	Weak
Carbon tetrachloride	CCl_4	115	Moderate
Diethyl ether	$C_2H_5OC_2H_5$	545	Weak
Ethanol	C_2H_5OH	54	Strong
Methanol	CH_3OH	122	Strong
Water	H_2O	23.8	Very strong

[a]Substances with strong intermolecular attractions tend to be less volatile and have lower vapor pressures than those with weak attractions.

Boiling

Now let us consider what happens when a liquid is heated in an open container. The vapor pressure of the liquid will increase as the temperature rises and will eventually become equal to the external pressure. When this happens, bubbles of vapor begin to form throughout the liquid. The bubbles rise and the liquid boils. The **boiling point** of a liquid is the temperature at which its vapor pressure is equal to the external pressure. The **normal boiling point** is the boiling point under an external pressure of 1 atm; it is the approximate temperature at which a liquid will boil in an open container at sea level. The normal boiling point of water is 100°C; normal boiling points of several other substances are listed in Table 12.3.

If the external pressure increases, the vapor pressure required for boiling also increases, and boiling occurs at a higher temperature. Similarly, a lower external pressure lowers the boiling point (see Demonstration 12.1). At 1 mile above sea

Underwater explorers in submarines are subject to pressures in excess of 1 atm. If they heat water or coffee to boiling, they may find it too hot to drink.

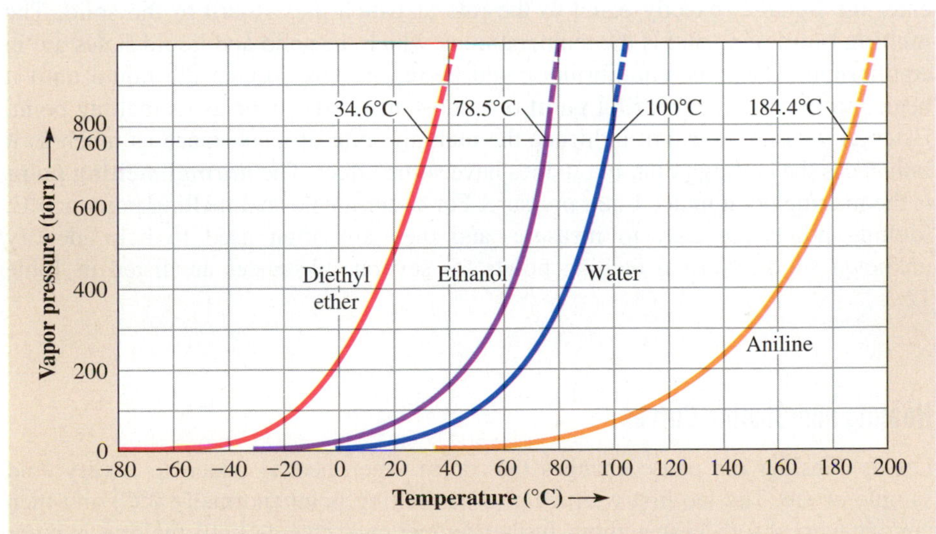

Figure 12.6

Plots of vapor pressure versus temperature for various liquids. The normal boiling point of each liquid is the temperature at which its vapor pressure equals 760 torr. The normal boiling point of diethyl ether, for example, is 34.6°C.

TABLE 12.3 Some Thermal Properties

Substance	Normal Melting Point (°C)	Normal Boiling Point (°C)	Specific Heat at 25°C (J/g·°C)	Molar Heat of Fusion (kJ/mol)[a]	Molar Heat of Vaporization (kJ/mol)[b]
Aluminum	658.6	2327	0.902	10.7	283
Copper	1083.1	2595	0.385	13.0	304
Diethyl ether	−116.3	34.51	2.320	7.27	26.0
Ethanol	−114.5	78.5	2.419	5.02	38.6
Gold	1063.01	2660	0.129	12.7	310
Iron	1535	2735	0.449	14.9	354
Lead	327.5	1750	0.128	5.12	178
Mercury	−38.87	356.58	0.140	2.33	58.5
Water	0.0	100.0	4.179	6.01	40.67

[a]At the normal melting point.
[b]At the normal boiling point.

level, for example, atmospheric pressure drops to about 610 torr and water boils at about 94°C. The curves in Figure 12.6, which show vapor pressure as a function of temperature, are also boiling point curves. For example, the vapor pressure of ethanol is 100 torr at 35°C. Hence, if the pressure above a sample of ethanol is reduced from 760 torr (1 atm) to 100 torr, the boiling point of ethanol will be lowered from its normal value of 78.5°C to 35°C.

It takes longer to cook a "three-minute egg" in Denver than in New York City.

Solid–Liquid Equilibrium

At 0°C and 1 atm pressure, ice and liquid water are in equilibrium; in other words, both states are stable under these conditions, and the rate at which water molecules enter the liquid is exactly equal to the rate at which they return to the solid. The **melting point** of a solid is the temperature at which the solid and liquid states are in equilibrium. The same equilibrium could be reached by cooling the liquid until it begins to freeze—the **freezing point** of a substance is the same as its melting point. External pressure does not influence the melting point of a substance as much as it influences the boiling point, but it does have some effect. The **normal melting point** is the melting point under 1 atm pressure. For water the normal melting point is 0°C. Melting points are easy to measure, and they are often used to help identify unknown solids. Normal melting points for several substances are listed in Table 12.3.

Heating and Cooling Curves

Let us consider the phase changes that occur when heat is added to a very cold sample of ice. The ice first warms up to its melting point (normally 0°C) and then continues to absorb heat without further temperature change until melting is com-

DEMONSTRATION 12.1 BOILING UNDER REDUCED PRESSURE

We usually heat water to make it boil, but we could lower the pressure instead. The vacuum pump removes air from the flask, reducing the pressure on the aqueous solution inside.

The solution boils when the pressure in the flask is less than or equal to the vapor pressure of the solution (about 24 torr).

plete. Continued heating of the liquid causes the temperature to rise further. When the boiling point (normally 100°C) is reached, the liquid vaporizes without temperature change until the entire sample is converted into steam. Additional heating of the steam causes its temperature to rise above 100°C.

A plot of temperature versus heat absorbed is called a **heating curve**. Figure 12.7 shows the heating curve for 1 mol of water. The upward slanting portions of the

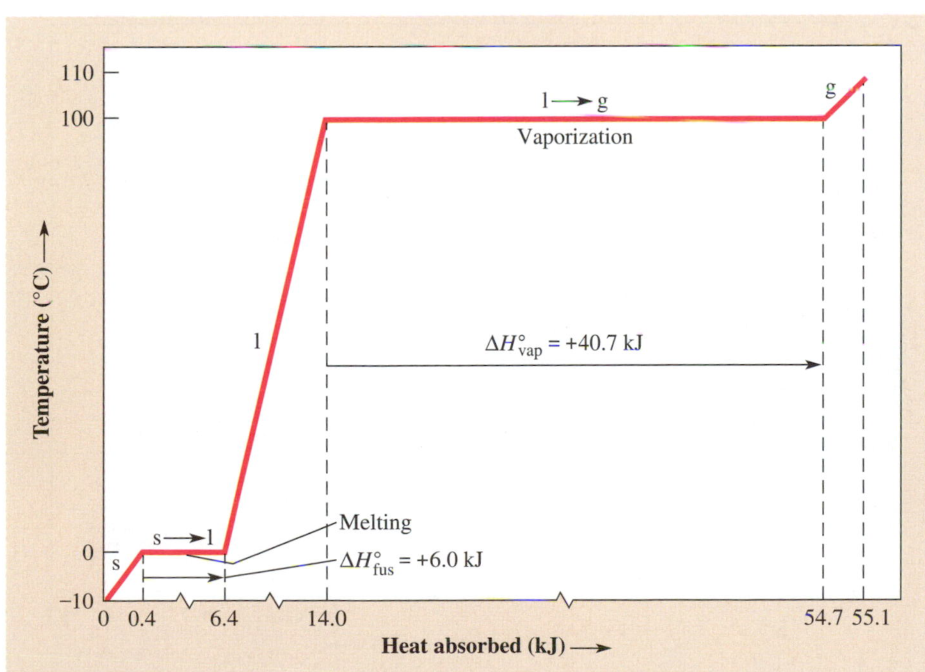

Figure 12.7

This heating curve for 1 mol of water at 1 atm starts with ice at −10°C and ends with steam at 110°C. The upward slanting portions represent warming of the solid (s), liquid (l), and gaseous (g) phases. The horizontal portions correspond to constant temperature phase changes.

curve represent phase warming. The molecules of water absorb heat energy, move more rapidly, and the temperature goes up. The horizontal portions of the curve correspond to phase changes. During a phase change, the added energy disrupts the ordered arrangement of a solid or vaporizes the molecules of a liquid. Heat is converted to potential energy as the molecules move apart. The temperature remains constant because there is no change in kinetic energy; the average speed of the molecules remains the same. When the phase change is complete, continued addition of energy produces more rapid motion. The kinetic energy of the molecules increases, and there is a corresponding increase in temperature.

A **cooling curve**, that is, a plot of temperature versus heat withdrawn, is essentially the reverse of a heating curve. It shows how each phase cools, condenses to the next phase, and then cools again as heat is removed from the system. An experimentally determined cooling curve like the one in Figure 12.8 usually shows a temperature dip just before the appearance of solid crystals. The dip indicates the existence of a **supercooled liquid** at temperatures below the normal freezing point. This metastable condition exists because randomly moving molecules cannot instantaneously organize themselves into the regular pattern required by the solid. For a solid to grow, there must be "seed" crystals to provide sites for molecules to occupy. Once a few molecules manage to establish the correct pattern, others quickly add to it. The solid then grows rapidly and the temperature rises back to the freezing point. Although supercooled water is very unstable, carefully conducted experiments have produced liquid water at temperatures as low as −40°C. Supercooling can be minimized by active stirring or by suspended particles—dust or other material—that provide centers around which crystals will form. Liquids with large complex molecules are especially difficult to crystallize and may remain in the supercooled state for long periods of time at temperatures well below their freezing points.

A supercooled liquid is in many ways analogous to the supersaturated solution discussed in Section 4.1.

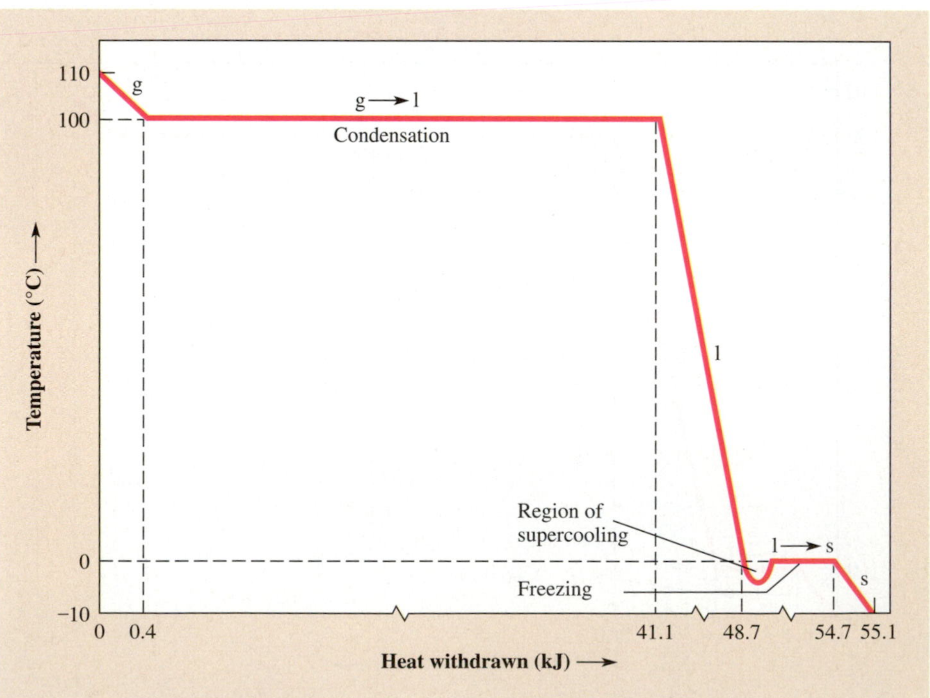

Figure 12.8

A cooling curve for 1 mol of water. The cooling curve is the reverse of the heating curve (Figure 12.7) except in the region of supercooling.

The quantity of heat needed to melt 1 mol of a solid is called the **molar heat of fusion** (ΔH°_{fus}). The quantity of heat needed to vaporize 1 mol of a liquid is the **molar heat of vaporization** (ΔH°_{vap}). These heats are measured at constant pressure; thus they are enthalpy changes. Some values are given in Table 12.3 and represented graphically in the heating curve of Figure 12.7. Large heats of fusion and vaporization indicate strong intermolecular attractions.

Fusion: a synonym for melting.

EXAMPLE 12.1

How many kilojoules of heat are absorbed by 100.0 g of aluminum when it melts at its normal melting point?

SOLUTION

The molar mass of aluminum is 26.98 g/mol. The molar heat of fusion is reported in Table 12.3 as +10.7 kJ/mol. First we will convert 100.0 g of aluminum to moles, then we will find the heat q in kilojoules:

$$q = 100.0 \text{ g Al} \times \frac{1 \text{ mol Al}}{26.98 \text{ g Al}} \times \frac{10.7 \text{ kJ}}{1 \text{ mol Al}} = 39.7 \text{ kJ}$$

PRACTICE EXERCISE 12.1

How many kilojoules of heat are evolved when 10.0 g of steam at 100°C condenses to liquid water at 100°C?

DIGGING DEEPER

THE CLAUSIUS–CLAPEYRON EQUATION

The vapor pressure–boiling point curves of Figure 12.6 all have a similar shape. If you are a math buff, their steadily increasing upward slopes might remind you of an exponential curve, the same sort of curve that describes a bank account growing at compound interest or the exploding world population. The general equation for a vapor pressure curve is

$$P = Ae^{-\Delta H^\circ_{vap}/RT} \tag{12.1}$$

where P is the vapor pressure at the Kelvin temperature T, R is the gas constant, ΔH°_{vap} is the molar heat of vaporization, and A is a proportionality constant. Both A and ΔH°_{vap} are different for each liquid. Taking the natural logarithm (logarithm to the base e) of both sides gives

$$\ln P = -\frac{\Delta H^\circ_{vap}}{RT} + B \tag{12.2}$$

where ln P is the natural logarithm of P, and $B = \ln A$ is another constant. (For a

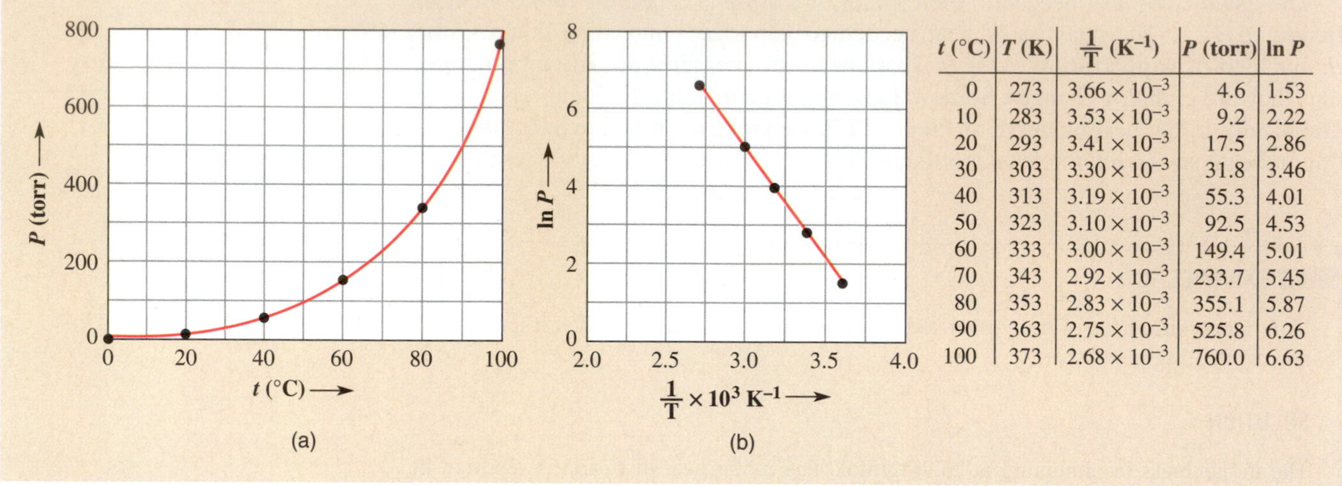

t (°C)	T (K)	$\frac{1}{T}$ (K^{-1})	P (torr)	ln P
0	273	3.66×10^{-3}	4.6	1.53
10	283	3.53×10^{-3}	9.2	2.22
20	293	3.41×10^{-3}	17.5	2.86
30	303	3.30×10^{-3}	31.8	3.46
40	313	3.19×10^{-3}	55.3	4.01
50	323	3.10×10^{-3}	92.5	4.53
60	333	3.00×10^{-3}	149.4	5.01
70	343	2.92×10^{-3}	233.7	5.45
80	353	2.83×10^{-3}	355.1	5.87
90	363	2.75×10^{-3}	525.8	6.26
100	373	2.68×10^{-3}	760.0	6.63

Figure 12.9

(a) A plot of vapor pressure (P) versus Celsius temperature (t) for water. (b) The natural logarithm of P versus the reciprocal of Kelvin temperature ($1/T$) is a straight line with slope $= -\Delta H^\circ_{vap}/R$ (see Equation 12.2).

review of logarithms, both natural and to the base 10, see Appendix A.2.) Over short temperature ranges, ΔH°_{vap} is approximately constant, and a plot of ln P versus $1/T$ is a straight line of slope equal to $-\Delta H^\circ_{vap}/R$ and intercept B (Figure 12.9). Equation 12.2 is known as the **Clausius–Clapeyron equation** after Emile Clapeyron, who first deduced it in 1834, and Rudolf Clausius, who modified it in 1850. The Clausius–Clapeyron equation can be used to estimate vapor pressures at various temperatures, boiling points at various pressures, and unknown heats of vaporization.

EXAMPLE 12.2

Estimate the temperature of boiling water in a pressure cooker set for 2 atm (the approximate pressure when the regulator is set for "15 pounds"). The ΔH°_{vap} for water is +40.7 kJ/mol.

SOLUTION

First we must find the value of the constant B for water. We know that the vapor pressure P of water is 1.00 atm at 100°C = 373 K. The gas constant R is 8.314 J/mol·K (Table 5.2, p. 190). The heat of vaporization of water is 40.7 kJ/mol = 40.7×10^3 J/mol. This information is inserted into Equation 12.2:

$$\ln 1.00 = \frac{-40.7 \times 10^3 \text{ J/mol}}{8.314 \text{ J/mol·K} \times 373 \text{ K}} + B$$

Rearranging and simplifying the fraction gives

$$B = \ln 1.00 + 13.1$$
$$= 13.1$$

since ln 1.00 = 0.00. Now we can use the value of B and Equation 12.2 to find the temperature T that corresponds to P = 2.00 atm. Substituting the appropriate data into Equation 12.2 gives

$$\ln 2.00 = \frac{-40.7 \times 10^3 \text{ J/mol}}{8.314 \text{ J/mol·K} \times T} + 13.1$$

LEARNING HINT

Use the LN button on your calculator. Remember that the logarithm of a number has no units.

Solving for T gives

$$T = 395 \text{ K} = 122°\text{C}$$

An alternative form of the Clausius–Clapeyron equation, which can be used for problems like Example 12.2, is obtained by letting P_1 and P_2 be the vapor pressures at T_1 and T_2, respectively. Then

$$\ln P_1 = -\frac{\Delta H^\circ_{vap}}{RT_1} + B \quad (12.2a)$$

$$\ln P_2 = -\frac{\Delta H^\circ_{vap}}{RT_2} + B \quad (12.2b)$$

Since B does not depend on temperature, we can subtract Equation 12.2b from Equation 12.2a to get

$$\ln P_1 - \ln P_2 = -\frac{\Delta H^\circ_{vap}}{RT_1} + \frac{\Delta H^\circ_{vap}}{RT_2}$$

This equation can be rearranged to

$$\ln\left(\frac{P_1}{P_2}\right) = \frac{\Delta H^\circ_{vap}}{R}\left(\frac{1}{T_2} - \frac{1}{T_1}\right) \quad (12.3)$$

The advantage of using this form of the Clausius–Clapeyron equation is that you do not have to find the intercept B. (Try using Equation 12.3 to solve Example 12.2.)

Dissolved compounds are often recovered from solution by boiling off (distilling) the solvent. A delicate compound, however, might decompose if the distillation is carried out at the normal boiling temperature of the solution. To prevent decomposition, the external pressure can be lowered so that boiling occurs at a lower temperature where the compound is stable. The reduced pressure required to distill the solution can be estimated from the Clausius–Clapeyron equation or from graphs similar to those in Figure 12.9.

PRACTICE EXERCISE 12.2

A solution contains a small amount of a heat-sensitive organic compound dissolved in ethanol. To what value must the pressure be reduced to distill ethanol from the solution at 25°C? The normal boiling point of ethanol is 78.5°C, and its heat of vaporization is +38.6 kJ/mol.

12.2 PHASE DIAGRAMS

A **phase diagram** is a graph that shows the most stable state of a substance at any given temperature and pressure. Figure 12.10 is a phase diagram for water at moderate pressures. Each point on the graph represents a different temperature and pressure, and each of the colored regions is labeled to show the most stable state of

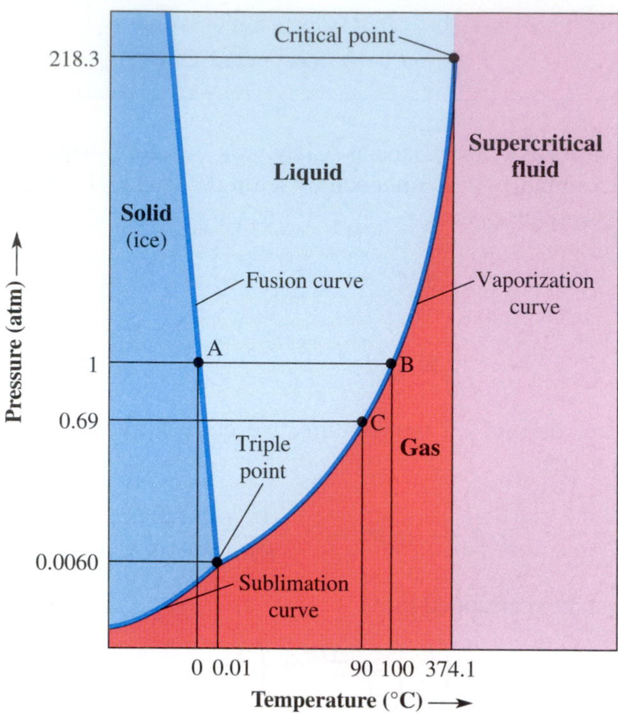

Figure 12.10

Phase diagram for water at moderate pressures (not drawn to scale). Ice is the most stable form of water at temperatures and pressures within the dark blue region, liquid is the most stable form within the light blue region, and gas within the red region. The vaporization, fusion, and sublimation curves show the temperatures and pressures under which two states can coexist in equilibrium. All three states can coexist only at the triple point.

water at temperatures and pressures within that region. The diagram shows, for example, that ice is the most stable form of water at −5°C and 1 atm and that the liquid form of water is unstable at pressures below 0.0060 atm.

PRACTICE EXERCISE 12.3

Use Figure 12.10 to identify the most stable form of water at (a) 50°C and 0.9 atm and (b) 0°C and 1.1 atm.

The solid blue lines separating the regions of the phase diagram show the conditions under which two states can coexist in equilibrium. The **fusion curve** shows the temperatures and pressures at which ice will be in equilibrium with liquid, the **sublimation curve** describes the conditions for ice–vapor equilibrium, and the **vaporization curve** describes the conditions for liquid–vapor equilibrium.

Let us use the phase diagram to trace the fate of a piece of ice as it is heated under a constant pressure of one atmosphere. The process can be followed by moving from left to right along the 1-atm line in Figure 12.10. The ice will warm up until its temperature reaches 0°C (point A on the fusion curve). This temperature is the normal melting point of ice. After the ice melts, continued heating will raise the temperature of the liquid to its normal boiling point, 100°C (point B on the vaporization curve). The liquid will vaporize, and after vaporization is complete, further heating will raise the temperature of the gaseous water.

The effect of pressure can also be traced on a phase diagram. Consider, for example, the effect of increasing the pressure on a sample of gaseous water kept at a constant temperature of 100°C. The pressure change can be followed by moving vertically upward along the 100°C line in Figure 12.10. The diagram shows that the sample will condense to liquid water when the pressure reaches 1 atm.

When considering the effect of pressure, it helps to remember that substances tend to shrink when compressed, so an increase in pressure produces a decrease in volume. Figure 12.10 shows that at temperatures below 374.1°C, an increase in pressure will eventually cause any sample of gaseous water to condense into the more compact liquid or solid state. The phase diagram also shows that an increase in pressure at 0°C can cause ice to melt. The reason is that liquid water at 0°C has a greater density than ice, so a given mass of liquid water occupies a smaller volume than the same mass of ice. This property of water is unusual; most substances become less dense and occupy a greater volume when they melt.

EXAMPLE 12.3

The vapor pressure of liquid water at 90°C is 0.69 atm. (a) Locate this point on the phase diagram. (b) What is the most stable state of water at 90°C and 0.5 atm? (c) What will happen if the pressure is increased from 0.5 atm to 0.9 atm while the temperature remains at 90°C?

SOLUTION

(a) The vapor pressure of a liquid is the pressure of the vapor in equilibrium with the liquid. Hence, 90°C and 0.69 atm is a point on the vaporization curve (point C on Figure 12.10).

(b) The phase diagram shows that the gas phase is the most stable state of water at 90°C and 0.5 atm.

(c) Increasing the pressure with no change in temperature is equivalent to moving up a vertical line in the phase diagram. As the pressure increases from 0.5 atm to 0.9 atm at 90°C, water will condense to the liquid state.

PRACTICE EXERCISE 12.4

A sample of water is at 1.1 atm and 0°C. What states will it pass through (a) if the pressure is gradually decreased with no change in temperature and (b) if the temperature is gradually increased with no change in pressure?

The effect of pressure on the boiling point of water is summarized by the vaporization curve in Figure 12.10. This curve, which is identical to the curve given for water in Figure 12.6, slants to the right as it rises, showing that the temperature of the liquid–vapor equilibrium increases with increasing pressure. The shape of the curve illustrates what you learned in the last section; namely, water boils at a higher temperature when the pressure is increased. The fusion curve shows the effect of pressure on the melting point. The curve for water slants upward to the left, showing that the melting point of ice decreases with increasing pressure. This effect occurs because, as discussed above, liquid water is more closely packed (denser) than ice

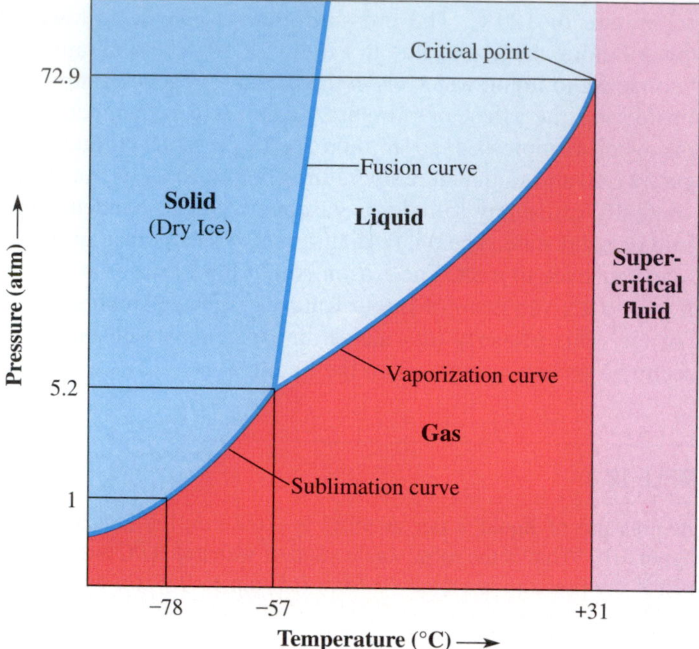

Figure 12.11

Phase diagram for carbon dioxide at moderate pressures (not drawn to scale). Liquid CO_2 is unstable at pressures below 5.2 atm, so the only phase changes observed under ordinary conditions are sublimation and deposition.

under ordinary conditions. The sublimation curve describes the effect of pressure on the **sublimation temperature**, that is, the temperature at which the solid and gaseous states are in equilibrium.

PRACTICE EXERCISE 12.5

Figure 12.11 is a phase diagram for carbon dioxide. Does the melting point of carbon dioxide increase or decrease as the pressure increases? How about the sublimation temperature?

The phase diagram for water contains only one point at which all three states coexist in equilibrium. This point is the **triple point** where the vaporization, fusion, and sublimation curves meet. For water the triple point occurs at 0.0060 atm (4.58 torr) and 0.01°C. The phase diagram also shows that the liquid form of water is unstable at pressures below 0.0060 atm. Hence, ice will sublime when it is warmed at pressures below the triple point pressure.

EXAMPLE 12.4

Refer to Figure 12.11. (a) What phase change occurs when frozen carbon dioxide is warmed under 1 atm pressure? (b) What is the minimum pressure needed for liquid carbon dioxide to be stable? (c) From the slope of the fusion curve, deduce whether the density of frozen carbon dioxide is greater or less than the density of liquid carbon dioxide.

SOLUTION

(a) At $P = 1$ atm, a horizontal line drawn across the diagram in the direction of increasing temperature shows that carbon dioxide goes directly from the solid state to the vapor state; that is, it sublimes.

(b) The lowest point in the liquid area is the triple point at 5.2 atm. The liquid form of carbon dioxide is not stable below the triple point pressure.

(c) The fusion curve is slanted toward the right. A vertical line drawn in the direction of increasing pressure from any point on the fusion curve takes carbon dioxide into the solid state; that is, an increase in pressure will cause liquid carbon dioxide to solidify. The solid carbon dioxide must therefore be more compact than the liquid and have a greater density.

PRACTICE EXERCISE 12.6

The atmosphere on Mars has a surface pressure of about 0.007 atm and a temperature ranging from -143°C to 27°C. It consists mostly of carbon dioxide gas (95%), but some water vapor is present. Can (a) liquid carbon dioxide and (b) liquid water exist on the Martian surface?

Critical Temperature

The vaporization curve in the phase diagram for water comes to an abrupt end at 374.1°C and 218.3 atm. This point, called the **critical point**, is indicated in Figure 12.10. The temperature and pressure at the critical point are referred to as the **critical temperature** and **critical pressure**. To understand the significance of the critical point, consider the following experiment. A sample of water occupies a sealed tube like the one shown in Figure 12.12. The liquid and vapor states are distinct from each other, and the *meniscus* (the curved boundary between the liquid and the vapor) is clearly visible. If the tube is slowly heated, the vapor pressure above the liquid will gradually increase, and the state of the system will follow the upward path of the vaporization curve in Figure 12.10. (*CAUTION*: Heating liquids in sealed

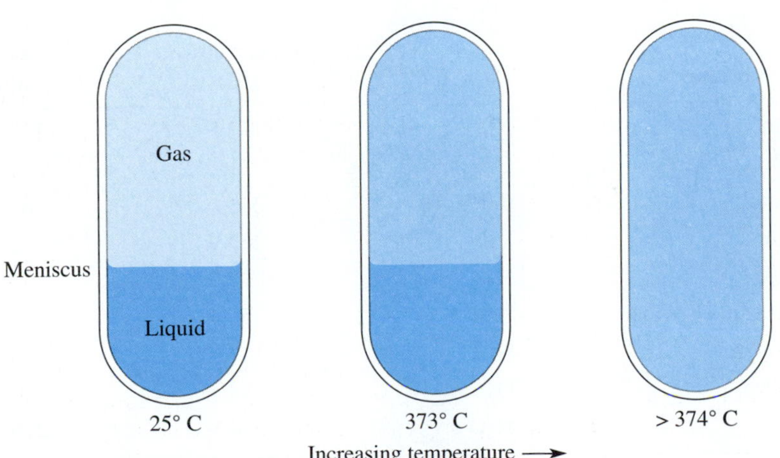

Figure 12.12

Behavior at the critical point. A sealed tube is filled to 32% of its volume with liquid water and then heated. As long as the temperature remains below 374°C, some of the liquid vaporizes, the remaining liquid expands, and the position of the meniscus does not change. At 374°C, the density of the vapor becomes equal to the density of the liquid, and the meniscus vanishes. Above this temperature, the system remains homogeneous at all pressures.

CHEMICAL INSIGHT

USING THE SUBLIMATION OF ICE TO PRESERVE AND RESTORE OBJECTS

The fact that ice sublimes at pressures below 4.58 torr has been put to use in a number of interesting ways. One application is the preservation of various products by *freeze drying*. The material to be freeze-dried—freshly prepared food, for example—is rapidly frozen and placed into a vacuum chamber. The pressure in the chamber is then reduced to less than 1 torr by a vacuum pump. The ice crystals in the food sublime and leave behind a dehydrated product that can be reconstituted by the addition of water. Freeze-drying does not seriously impair the flavor or nutritional value of the food, and it is used commercially to preserve meat and fish and to manufacture instant coffee and tea. This technique produces lightweight products with long shelf lives that are ideal for backpackers and space travelers. Nonfood products that are preserved by freeze-drying include blood plasma, vaccines, and various antibiotics.

Freeze-drying has recently been adapted to the salvaging of waterlogged books and documents. For many years, libraries and book collectors had to throw away rare books and manuscripts that were damaged by floods. If wet books are frozen and stored before deterioration sets in, they can later be freeze-dried in a vacuum chamber. The books emerge dry and brittle, but they can be rehydrated by exposure to moist air.

LEARNING HINT

You are probably wondering why the container is filled to 32% of its volume with liquid water. The *total* density of water (gas + liquid) in the tube remains unchanged throughout the experiment. Since the density of water at the critical point is 0.32 g/mL, this must also be the density at the beginning of the experiment. A container that is 32% full will have 32 mL of water (32 g) for every 100 mL of total volume. The total density will be 32 g/100 mL = 0.32 g/mL.

tubes is very dangerous because of the large pressures involved.) Some of the liquid vaporizes, causing the density of the vapor to increase. At the same time, the liquid expands, so its density decreases. At the critical point, the density of the vapor becomes equal to the density of the remaining liquid. (The common density is 0.32 g/mL.) The meniscus vanishes and the system becomes homogeneous. If the temperature is above 374.1°C, the meniscus will not reappear, no matter how high the pressure. The critical temperature can be considered a cutoff point beyond which the kinetic energy of the molecules is large enough to prevent intermolecular attractions from producing an ordinary liquid state. Hence, a substance with strong intermolecular attractions will have a higher critical temperature than one with weak attractions. Some critical temperatures and pressures are given in Table 12.4.

TABLE 12.4 Some Critical Temperatures and Pressures

Substance	Formula	Critical Temperature (°C)	Critical Pressure (atm)
Ammonia	NH_3	132.5	112.5
Carbon dioxide	CO_2	31	72.9
Ethanol	C_2H_5OH	243	63
Helium	He	−267.9	2.26
Hydrogen	H_2	−239.9	12.8
Hydrogen chloride	HCl	51.4	82.1
Nitrogen	N_2	−147	33.5
Oxygen	O_2	−118.4	50.1
Water	H_2O	374.1	218.3

A substance with a temperature above its critical temperature is referred to as a **supercritical fluid** (see Figures 12.10 and 12.11). A supercritical fluid has gaslike properties when its pressure is less than the critical pressure. For example, supercritical CO_2 behaves like a gas when its pressure is less than 72.9 atm. Above the critical temperature and pressure, where the distinction between gas and liquid disappears (recall that the vaporization curve ends at the critical point), supercritical fluids have unusual properties that are just beginning to be understood and utilized in industry. Such supercritical fluids have been used as solvents; supercritical CO_2, for example, is used to dissolve and remove caffeine from coffee beans. For further applications of supercritical fluid technology, see *Chemical Insight: Supercritical Fluids—Solvents of the Future* in Chapter 13.)

12.3 INTERMOLECULAR FORCES

An **intermolecular attractive force** is any force that causes otherwise independent atoms and molecules to cluster together. Intermolecular forces are responsible for phase changes such as condensation and solidification, and these forces must be overcome during changes such as melting and vaporization. Intermolecular attractive forces are often referred to as **van der Waals forces** in honor of the Dutch physicist Johannes van der Waals, who recognized in 1873 that such forces cause real gases to deviate from ideal behavior (Section 5.8).

Intermolecular forces should not be confused with chemical bonding forces. Covalent bonds and ionic bonds are strong; intermolecular forces are weak in comparison. For example, the average energy required to break the covalent O—H bond in water is 463 kJ/mol. The separation of water molecules from each other, however, requires only the heat of vaporization, 40.7 kJ/mol.

Dipole–Dipole Forces

Many molecules have permanent dipoles (Chapters 9 and 10) that give rise to intermolecular forces. The positive ends of polar molecules will be attracted to the negative ends of other molecules, but the similarly charged ends will repel each other. Although the molecules are in constant motion and their orientations keep changing, the polar molecules will, on average, be oriented so that oppositely charged ends are closer to each other than ends with like charges (Figure 12.13). The result is a net

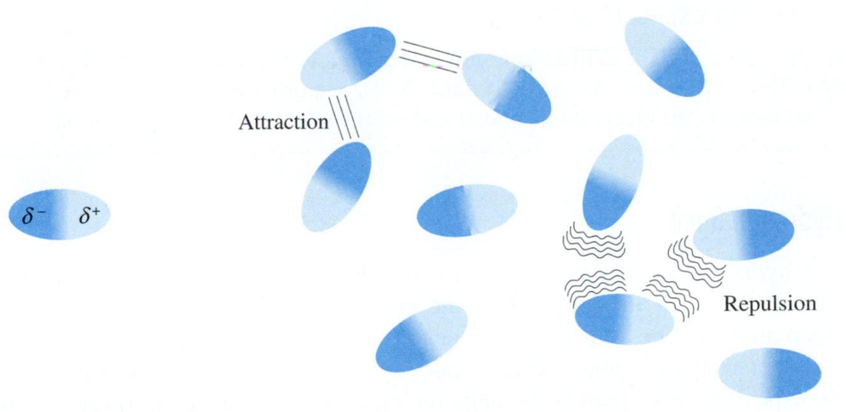

Figure 12.13
Dipole–dipole interactions. Nearby polar molecules orient themselves so that oppositely charged ends are closer to each other, on average, than similarly charged ends. The net force is one of attraction.

attractive force, called a **dipole–dipole force** (or **dipole–dipole interaction**), that holds the molecules together. The strength of the dipole–dipole interaction increases with the polarity of the participating molecules.

The force between two ions is inversely proportional to d^2, where d is the distance between charge centers. Doubling the distance decreases the force by a factor of $2^2 = 4$. The force between two dipoles is inversely proportional to d^4. Doubling the distance decreases the force by a factor of $2^4 = 16$. Dipole–dipole forces are much weaker than the forces between ions.

The force that draws two dipoles together is much weaker than the force between a pair of ions, partly because the charge differences within a polar molecule are much smaller than ionic charges, and partly because the attraction between oppositely charged dipole ends is counteracted to some extent by the repulsion between similarly charged ends. Attractions in a liquid are further reduced by random thermal motion, which causes many dipoles to be unfavorably aligned. Because of this combination of attraction and repulsion, dipole–dipole forces are *short range* in nature, effective only for molecules that are relatively close to each other.

The effect of dipole–dipole forces becomes apparent when we compare the properties of polar substances with those of similar nonpolar substances. The molecules of bromine (Br_2) and iodine monochloride (ICl), for example, are isolectronic (i.e., they have the same numbers of electrons). Bromine molecules are nonpolar, and bromine boils at 59°C. Iodine monochloride, which consists of polar ICl molecules, boils at 97°C—almost 40 degrees higher. The higher boiling point of iodine monochloride shows that its intermolecular attractions are stronger than those of bromine. Any substance with strong dipole–dipole forces has to absorb substantially more energy in order to melt and boil; it will have higher melting and boiling points and greater heats of fusion and vaporization than comparable nonpolar substances.

EXAMPLE 12.5

Nitrogen (N_2) and carbon monoxide (CO) are isoelectronic. Which substance should have the higher melting point?

SOLUTION

Carbon monoxide is slightly polar; nitrogen is not. Hence, dipole–dipole attractions in carbon monoxide should cause it to have a higher melting point than nitrogen. This prediction agrees with experimental data, which show that the normal melting points of carbon monoxide and nitrogen are -199 and -210°C, respectively.

PRACTICE EXERCISE 12.7

The phosphine (PH_3) and hydrogen sulfide (H_2S) molecules each contain 18 electrons. Their dipole moments are 0.58 and 0.97 D, respectively. Which substance should have (a) the higher boiling point and (b) the greater heat of vaporization?

The Hydrogen Bond

When a hydrogen atom is bonded to a small, highly electronegative atom such as nitrogen, oxygen, or fluorine, it is at the positive end of a very polar bond. This hydrogen atom will be attracted to the negative ends of nearby polar molecules and may interact with their lone electron pairs as well. The dipole–dipole interactions will be especially strong when the negative ends are also nitrogen, oxygen, or fluorine atoms. The intermolecular attraction that results when a hydrogen atom comes between two small, highly electronegative atoms is called a **hydrogen bond**. The three atoms involved in a hydrogen bond tend to have a linear arrangement, as shown in the examples of Figure 12.14. The hydrogen bond is the strongest intermolecular attractive force. From 13 to 40 kJ/mol are required to disrupt hydrogen bonds, compared to 1–25 kJ/mol for other types of dipole interactions.

Effective hydrogen bonding occurs when the electronegative atoms are nitrogen, oxygen, and fluorine. A hydrogen atom between larger atoms such as chlorine or iodine, or between less electronegative atoms such as sulfur, gives rise to an intermolecular attraction that does not differ significantly from ordinary dipole–dipole attraction.

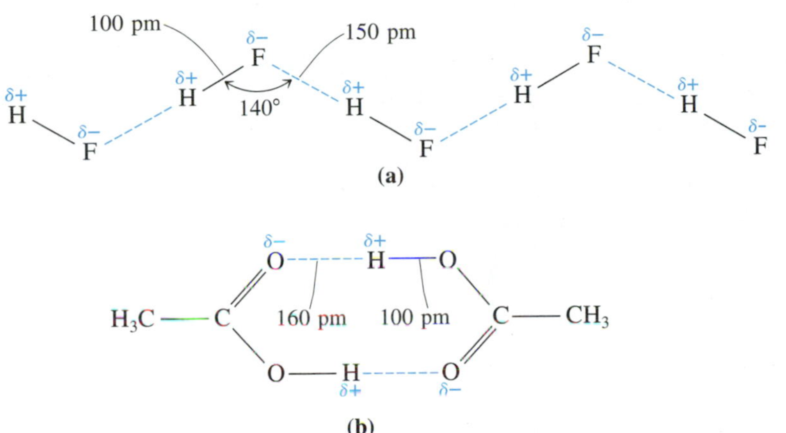

Figure 12.14
(a) Hydrogen bonding links hydrogen fluoride molecules in a zigzag chain. The chains in solid HF contain many molecules; in liquid HF, the chains are shorter. (b) Pure liquid acetic acid contains a high proportion of dimers, or double molecules, held together by hydrogen bonding between the COOH groups.

EXAMPLE 12.6

Would you expect to find a significant degree of hydrogen bonding between molecules of (a) NH_3, (b) CH_3OH, and (c) HBr?

SOLUTION

(a) The NH_3 molecule is polar with nitrogen at its negative end (Figure 12.15a). The hydrogen atom in the N—H bond will be attracted to a nitrogen atom from a nearby NH_3 molecule and should form a strong hydrogen bond.

(b) The CH_3OH molecule contains three C—H bonds and an O—H bond (Figure 12.15b). The electronegativity difference between carbon and hydrogen is small (Figure 9.14, p. 401), so the C—H bonds are not strongly polar. The O—H bond, however, is strongly polar with oxygen at the negative end. The hydrogen atom in an O—H bond will be attracted to the oxygen atom from a neighboring CH_3OH molecule and should form a strong hydrogen bond.

(c) HBr is a polar molecule,

$$\overset{\delta+}{\text{H}}-\overset{\delta-}{\ddot{\underset{\cdot\cdot}{\text{Br}}}}\!:$$

and we can expect the hydrogen end of one molecule to be attracted to the bromine end of another. Bromine, however, is a relatively large third-period atom. Even though its electronegativity is similar to that of nitrogen, its size prevents the formation of a strong hydrogen bond. Intermolecular attractions in HBr will not significantly differ from ordinary dipole–dipole interactions.

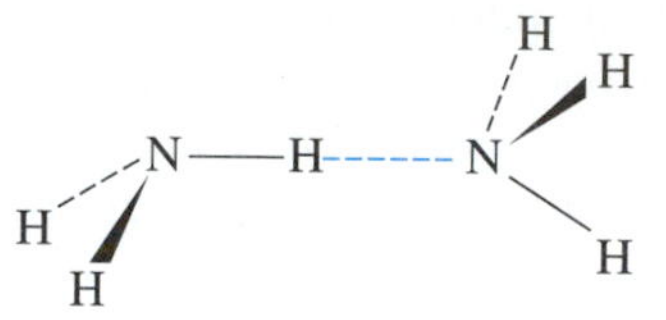

(a)

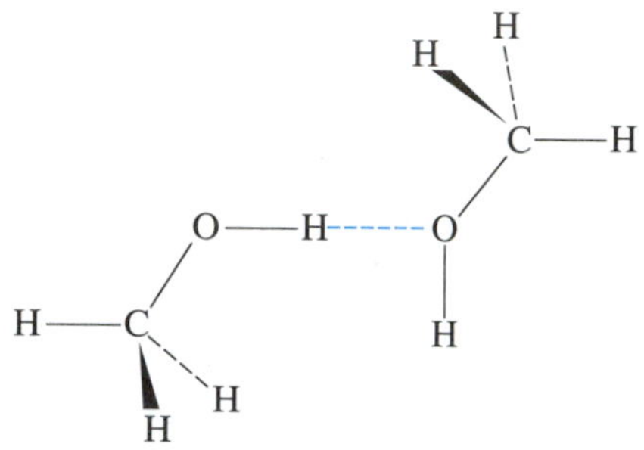

(b)

Figure 12.15

Hydrogen bonding in (a) ammonia and (b) methanol.

PRACTICE EXERCISE 12.8

Would you expect to find significant hydrogen bonding between molecules of (a) PH_3, (b) CH_3F, and (c) CH_3NH_2?

Hydrogen bonds have a pronounced influence on physical properties. Substantial energy is required to break these bonds, and thus hydrogen-bonded substances have low volatilities, high heats of vaporization, and high specific heats. The data in Figure 12.16 for hydrogen compounds of Groups 5A, 6A, and 7A show that the boiling points of HF, H_2O, and NH_3 are much higher than might be predicted on the basis of periodic trends. These compounds have polar molecules that undergo extensive hydrogen bonding in the liquid state. The boiling point of water, for example, is about 160 degrees higher than the boiling point of H_2S, a less polar compound in which hydrogen bonding does not play a significant role.

Hydrogen bonding also helps to maintain the structure that causes ice to be less dense than liquid water. Ice has an open structure in which each oxygen atom forms two hydrogen bonds in addition to its own two covalent bonds. The four bonds are tetrahedrally arranged around the oxygen atom as shown in Figure 12.17. When ice melts, many of the hydrogen bonds break; the open structure collapses, the volume becomes less, and the density increases. The lower density of ice plays a crucial role in the survival of life. Because ice floats on the surfaces of lakes and rivers, organisms are able to winter on the lake bottoms in contact with their food supply.

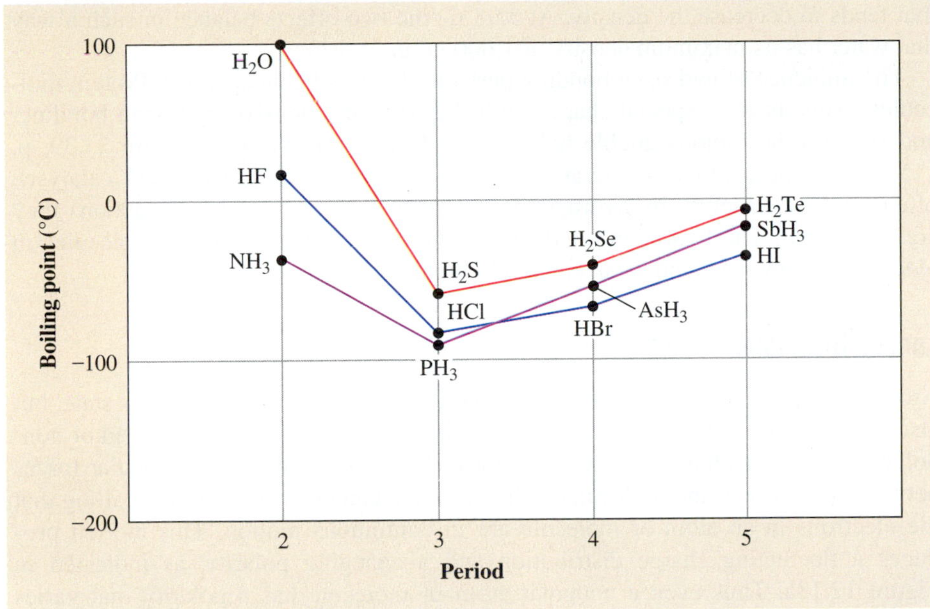

Figure 12.16

Boiling points of the binary hydrogen compounds of Groups 5A, 6A, and 7A. The boiling points of H_2O, HF, and NH_3 are unusually high because of hydrogen bonding.

Some hydrogen bonding persists in liquid water and produces temporary regions of local order as bonds continually break and form. Heating liquid water has two opposing effects on its density. Heat energy breaks some of the hydrogen bonds, which tends to reduce the empty space and thus increase the density. At the same time, increased thermal motion causes the water molecules to move apart, an effect

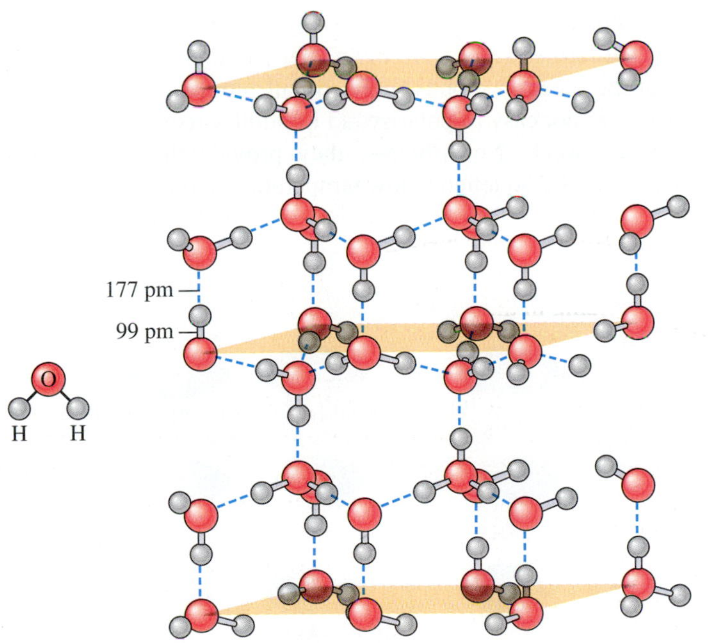

Figure 12.17

The structure of ice. Each oxygen atom is tetrahedrally bonded to four hydrogen atoms. Its two shorter bonds are the covalent bonds within the H_2O molecule; its two longer bonds are hydrogen bonds to adjacent molecules.

that tends to decrease the density. At 3.98°C, the two effects balance in such a way that water has its maximum density of 1.000 g/cm^3.

The influence of hydrogen bonding pervades the biochemical world. Protein molecules maintain their special shapes with the help of extensive hydrogen bonding, and so does the famous double helix of the DNA molecule (see Figure 23.39, p. 1157). Enzymes, which are large molecules that serve as biochemical catalysts, often attach themselves by hydrogen bonds to the molecules whose reactions they are promoting. The hydrogen bonds are strong enough to hold these molecules in place while they react and weak enough to break when the reaction is over.

London Dispersion Forces

Atomic substances such as argon and krypton exist not only in the gaseous state, but also in the liquid and solid states. The same is true for substances composed of nonpolar molecules such as Cl_2 and O_2. What is the nature of the intermolecular forces between such atoms and molecules? This question can be answered by recalling that the electrons in an atom or molecule are in continuous motion. This motion produces a fluctuating charge distribution and a changing polarity, as indicated in Figure 12.18a. Thus even a nonpolar atom or molecule has a polarity that varies from instant to instant, although it averages out to zero over time. An atom or molecule that is temporarily polarized is said to have an **induced dipole**.

The fluctuating charge distributions in an atom or molecule tend to polarize neighboring atoms and molecules (Figure 12.18b). These induced dipoles are temporary. They come and go, but many will be present at any one instant, and their interaction results in a net attractive force, called a **London dispersion force**. London forces, like dipole–dipole forces, are effective only at very short distances.

In 1928, Fritz London, a German physicist, was able to show that dispersion forces are inversely proportional to d^7; hence, doubling the distance d between two induced dipoles decreases the force by a factor of $2^7 = 128$.

The strength of the London force depends on the magnitude of the induced dipole and this, in turn, depends on how easily the electron cloud is polarized. It is generally true that *large atoms and molecules are more easily polarized than small atoms and molecules*. For example, I_2 is more easily polarized than F_2 because the outer electrons in I_2 are farther away from the nuclei and their clouds are more subject to distortion than the electron clouds in F_2. Small atoms and molecules with tightly bound electrons are not easy to polarize, so London forces in substances like helium and hydrogen are weak. Nevertheless, they provide the slight "stickiness" that allows these gases to be liquefied at low temperature and high pressure.

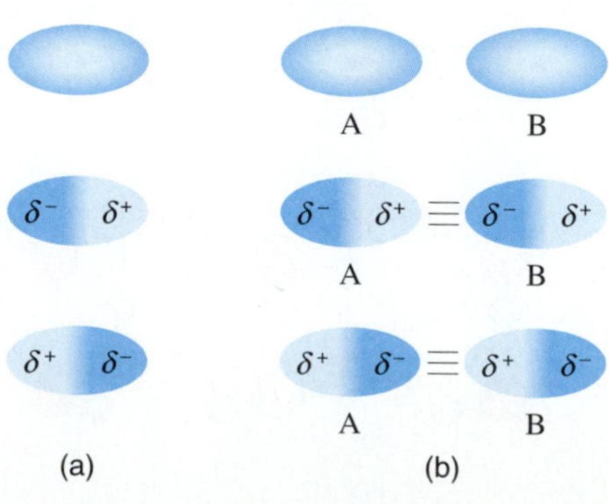

Figure 12.18
London dispersion forces. (a) Three of the many charge distributions that can arise from the random motion of electrons in a nonpolar molecule. (b) The fluctuating charge distributions in molecule A polarize the electron cloud in molecule B. The result is a net attractive force.

TABLE 12.5 Normal Melting and Boiling Points of Related Nonpolar Substances

Substance	Molar Mass (g/mol)	Melting Point (°C)	Boiling Point (°C)
He	4.0	−272	−269
Ne	20.2	−249	−246
Ar	39.9	−189	−186
Kr	83.8	−157	−152
Xe	131.3	−112	−107
F_2	38.0	−220	−188
Cl_2	71.0	−101	−35
Br_2	159.8	−7	59
I_2	253.8	114	184
CF_4	88.0	−150	−129
CCl_4	153.8	−23	77
CBr_4	331.6	90	190

London forces increase dramatically with molecular size, and when other effects are absent, *substances consisting of large atoms and molecules will have higher melting and boiling points than comparable substances consisting of small atoms and molecules.* This rule implies that melting and boiling points should increase toward the bottom of a periodic group. The data in Table 12.5 confirm this rule for the noble gases, the halogens, and some nonpolar carbon tetrahalides. Large atoms and molecules are more massive than small atoms and molecules, so it is tempting to conclude that the increases in melting and boiling points are caused by increased mass. It should be kept in mind, however, that intermolecular forces are electrostatic in origin and that their apparent dependence on mass is coincidental.

EXAMPLE 12.7

Which element should have the higher melting point, O_2 or Cl_2?

SOLUTION

Oxygen and chlorine are in the second and third periods, respectively. Both molecules are nonpolar, but Cl_2 has more electrons and is larger than O_2. The chlorine molecule is more easily polarized, and stronger London dispersion forces should cause chlorine to have a higher melting point than oxygen. This prediction agrees with experimental data, which show that the normal melting points of Cl_2 and O_2 are −101°C and −218.4°C, respectively.

EXAMPLE 12.8

Propane ($CH_3CH_2CH_3$) and butane ($CH_3CH_2CH_2CH_3$) consist of nonpolar molecules. Which should have the higher melting and boiling points?

SOLUTION

The London forces in butane, with its longer carbon chain, should be stronger than those in propane. Hence, butane should have the higher melting and boiling points. These predictions also agree with experiment. The normal melting points of propane and butane are −189.7 and −138.4°C, respectively; the normal boiling points are −42.1 and −0.5°C. (*Note*: Melting point predictions are not as reliable as boiling point predictions. The normal melting point of ethane (CH_3CH_3), for example, is −183.3°C, higher than that of propane.)

PRACTICE EXERCISE 12.9

Which substance in each pair should have the higher boiling point? (a) I_2 or Br_2; (b) CH_4 or SiH_4; (c) CH_4 or CH_3CH_3.

12.4 THE LIQUID STATE

The liquid state exhibits neither the complete chaos of a gas nor the neat order of a solid. Like gases, liquids flow freely and take the shape of whatever container they happen to be in. Like solids, liquids maintain constant volume and do not expand to fill their containers. The fact that liquids are generally more compressible and have higher vapor pressures than solids suggests that molecules in a liquid are less tightly bound than in a solid. Although liquid molecules have no definite arrangement, x-ray diffraction studies indicate a certain amount of *short-range order* in which small groups of molecules form orderly clusters. These orderly domains, however, are continually breaking apart and reforming, so that disorder prevails over the liquid as a whole. As might be expected, the patches of order shrink when the temperature is raised and the molecules move more rapidly. The order and disorder of the liquid state make it impossible to develop a simple ideal model for liquids that is comparable to the ideal gas model for gases. There is, however, a wealth of experimental data on liquid properties, some of which are described below.

X-ray diffraction (see Section 12.6) is a method usually used to find the positions of atoms and molecules in a solid lattice.

Viscosity

The **viscosity** of a liquid is a measure of its resistance to flow. Molasses, glycerin, motor oil, and other liquids that pour slowly are said to be viscous. Freely flowing liquids like water and gasoline have low viscosities. The viscosity of a liquid can be calculated from its rate of flow through a thin tube (see Figure 12.19) or from the speed attained by small steel balls falling through it. Viscosities of some common liquids are listed in Table 12.6.

Viscous liquids pour slowly because of the strong intermolecular attractions between their molecules. Liquids with polar molecules tend to be more viscous than

TABLE 12.6 Viscosities and Surface Tensions for Some Liquids at 20°C

Liquid	Viscosity (cP)[a]	Surface Tension (N/m)[b]
Benzene	0.652	0.0289
Carbon tetrachloride	0.969	0.0270
Castor oil	986	—
Chloroform	0.58	0.0271
Diethyl ether	0.233	0.0170
Ethanol	1.200	0.0228
Glycerol (glycerin)	1490	0.0634
Mercury	1.554	0.436
Olive oil	84	—
Water (at 20°C)	1.002	0.0728
(at 40°C)	0.653	0.0696
(at 60°C)	0.467	0.0662
(at 80°C)	0.355	0.0626
(at 100°C)	0.282	0.0589

[a]The SI unit of viscosity is the $N \cdot s/m^2$. More commonly used units are the poise (P) where $1\ P = 0.1\ N \cdot s/m^2$ and the centipoise (cP).

[b]Surface tension is the energy (work) required to produce a unit change in surface area. The unit of surface tension is work/area $= N \cdot m/m^2 = N/m$.

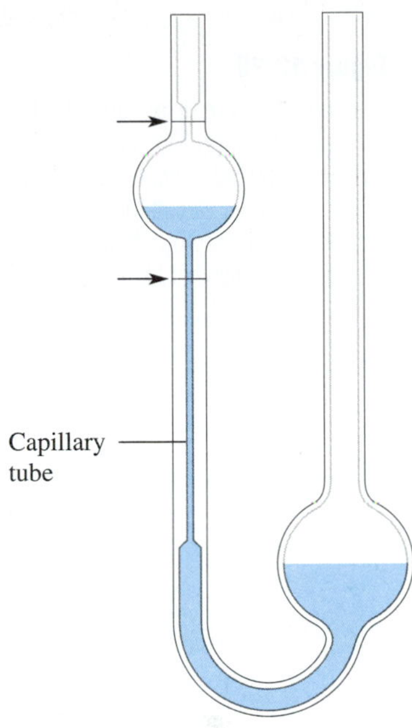

Figure 12.19

An Ostwald viscometer. The viscosity of a liquid is proportional to the time it takes for the volume between the arrows to flow through the capillary tube into the lower reservoir.

others, and strongly hydrogen-bonded liquids such as glycerol tend to be very viscous:

$$\begin{array}{c} \text{H} \\ | \\ \text{H—C—}\ddot{\text{O}}\text{—H} \\ | \\ \text{H—C—}\ddot{\text{O}}\text{—H} \\ | \\ \text{H—C—}\ddot{\text{O}}\text{—H} \\ | \\ \text{H} \end{array}$$

glycerol

Molecular size and shape also affect viscosity; substances with large, irregularly shaped molecules often have high viscosities. Liquid viscosities decrease with increasing temperature because the molecules move faster, have more free space, and have less short-range order. Motor oil, for example, becomes "thinner" (less viscous) as an automobile engine warms up.

Surface Tension

A molecule beneath the surface of a liquid is attracted by its neighbors in all directions as shown in Figure 12.20a. A surface molecule (Figure 12.20b), on the other

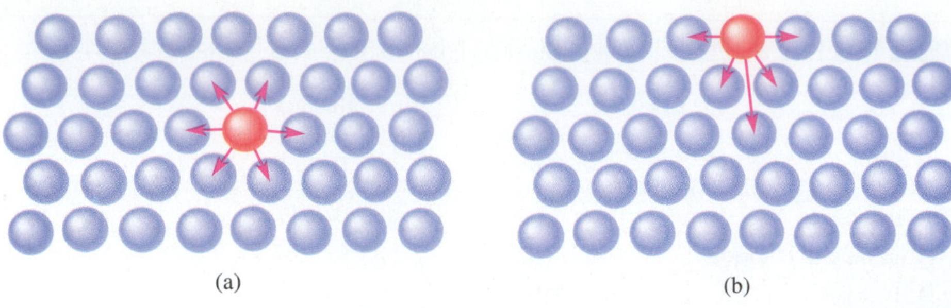

Figure 12.20

Unbalanced forces are responsible for liquid surface tension. (a) Molecules below the surface are pulled almost evenly in all directions. (b) Molecules at the surface experience a net inward pull that makes the surface behave somewhat like an elastic skin.

hand, has no molecules above it to counteract the downward pull of those below, and the unbalanced forces provide the surface with special properties not found in the rest of the liquid. These forces cause an apparent "tightening" of the liquid surface so that the outer layer of molecules behaves somewhat like an elastic skin. This effect allows water striders and other insects to walk on water and is the basis for such tricks as floating a needle on water. Like any elastic skin, the surface resists stretching and makes itself as small as possible. The tendency for a liquid to minimize its surface area causes liquid droplets to be spherical (Figure 12.21). (A sphere provides the smallest surface area for a given volume.) Various devices have been perfected for stretching or compressing oil slicks, soap films, and other thin surfaces, and **surface tension** is defined as the energy required to "stretch" the surface of a liquid by some unit amount. Some surface tension values are listed in Table 12.6. As might be expected, high surface tension is associated with strong intermolecular forces. Like liquid viscosity, liquid surface tension decreases with increasing temperature.

Figure 12.21

Water droplets on a spider web. The tendency for water to minimize its surface area causes water droplets to be spherical.

CHEMICAL INSIGHT

HOW TO MAKE LIQUIDS WETTER

The intermolecular attraction between like molecules is called *cohesion*. Strong cohesive forces result in viscous liquids and high surface tensions. The intermolecular attraction between unlike molecules is called *adhesion*. When a liquid such as water or paint wets (spreads out on) a surface, the liquid is said to adhere to the surface. Liquids with high surface tensions (strong cohesive forces) are poor *wetting agents* and, like mercury, they tend to form droplets when spilled on table tops or on the floor. The wetting ability of a liquid can be increased by adding surface active agents (*surfactants*) that lower the surface tension. Soap and household detergents are examples of surfactants. They lower the surface tension of water so that it can "wet" and dissolve grease. (The mechanism of detergent action is discussed in Section 13.6.) Bile, a bitter fluid secreted by the liver and stored in the gallbladder, also contains surfactants (bile salts) that aid digestion by helping water break up fats and oils in the intestinal tract.

Surfactants also play a significant role in maintaining the stability of pulmonary alveoli, the small sacs in the lungs that fill with air during respiration. A surface tension of about 0.050 N/m has been reported for the mucous fluid lining the alveolar walls. This tension would be high enough to collapse the small alveolar sacs and prevent them from holding air. The alveolar walls, however, secrete surfactants that reduce the surface tension to about 0.003 N/m and stabilize the alveoli against collapse. Premature babies lacking these surfactants have difficulty breathing, and there is a demonstrable lack of surfactants in lung extracts from infants dying of respiratory distress syndrome (hyaline membrane disease) compared to lung extracts from infants dying of nonpulmonary causes.

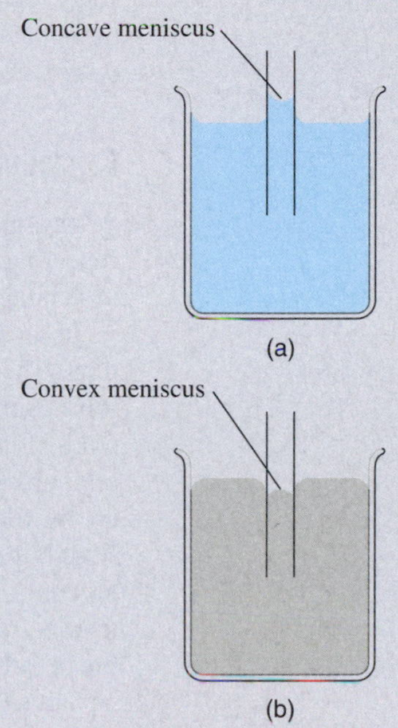

Figure 12.22

(a) Water in a capillary tube rises above the outside water level and has a concave meniscus. (b) Mercury in a capillary tube falls below the outside mercury level and has a convex meniscus.

Adhesion and cohesion are responsible for the spontaneous rise or fall of liquids in narrow tubes (Figure 12.22), an effect known as *capillary action*. (The word "capillary" is derived from the Latin *capillus* for "hair"; very thin tubes are often called capillary tubes.) If a glass capillary tube, or even a straw, is placed in water, the water rises as shown in Figure 12.22a. Adhesive forces between glass and water molecules are stronger than the cohesive forces in water, so the water creeps up and forms a thin film on the inner walls of the tube. Additional water is pulled up as surface tension acts to reduce the surface area of the film. The water continues to rise until the upward force—that is, the adhesive force between water and glass—is balanced by the downward force of gravity. The two forces combine to produce a concave meniscus. This effect is reversed when cohesive forces within the liquid are greater than the liquid–glass attraction. For example, in a glass capillary tube containing mercury (Figure 12.22b), the mercury level falls and the mercury meniscus is convex. The magnitude of the capillary effect increases with increasing surface tension and decreasing capillary width.

Tubules of animal and vegetable material are similar to glass in that they exert a great attraction on water and other polar molecules. Fabrics and paper towels have capillaries that draw in water and hold it, which accounts for their ability to soak up liquid. Fabrics can be made water-repellent by permeating the material with silicones (Survey Chapter 3) or other substances that repel water. Such substances reduce capillary action while preserving the porosity of the fabric.

12.5 THE SOLID STATE

In the solid state, attractive forces predominate. They confine atoms, molecules, and ions to fixed positions, and they limit molecular motion to a mere jiggling in place. A solid can be held together by the forces of chemical bonding—ionic, covalent, or metallic—or by intermolecular forces. The physical properties associated with each type of bonding are discussed in this section and summarized in Table 12.7.

Crystalline Solids

A **crystalline solid** consists of atoms, ions, or molecules arranged in a regular, repeating pattern called a **crystal lattice**. Crystalline solids can be classified as ionic, molecular, network covalent, or metallic.

In an **ionic solid** the particles that make up the lattice are positive and negative ions. The sodium chloride lattice shown in Figure 2.19 (p. 63) is an example. Each ion is surrounded by neighbors of opposite charge and there are no separate molecules. Ions are not easily moved from their lattice sites, so that typical ionic solids are hard and rigid, with high melting points. In spite of their hardness, ionic crystals are brittle. They shatter easily and cannot be shaped by bending or hammering. Figure 12.23 shows how a deformation created by bending or hammering pushes layers of ions away from their oppositely charged neighbors and brings them closer to ions of like charge. The increase in electrostatic repulsion causes the crystal to break along the displaced plane. An ionic crystal does not conduct electricity because the ions are not free to move in the lattice. When the crystal melts or dissolves in water, however, the ions become mobile and the resulting liquid is a good conductor of electricity.

TABLE 12.7 Types of Crystalline Solids

Type	Attractive Forces	Properties	Examples
Ionic	Electrostatic forces between positive and negative ions	Hard and brittle High melting points; low volatility High heats of fusion and vaporization Poor conductors of heat and electricity except when molten	NaCl, MgO, $CaCO_3$
Molecular	Dipole–dipole forces Hydrogen bonds Dispersion forces	Soft to hard and brittle Low melting points; volatile Low heats of fusion and vaporization Poor conductors of heat and electricity (good insulators)	H_2O, CO_2 (Dry Ice), $C_{12}H_{22}O_{11}$ (sucrose), O_2, N_2, I_2
Network covalent	Covalent bonds	Very hard and brittle Extremely high melting points; very low volatility Very high heats of fusion and vaporization Poor conductors of heat and electricity (good insulators)	SiO_2 (quartz), SiC (carborundum)
Metallic	Metallic bond	Very soft to hard; often malleable and ductile Very low to high melting points; low volatility Variable heats of fusion and vaporization Good conductors of heat and electricity Metallic luster	Fe, Al, Hg, and other metals

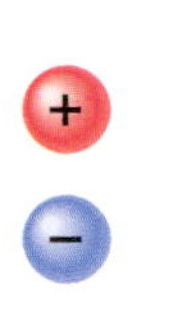

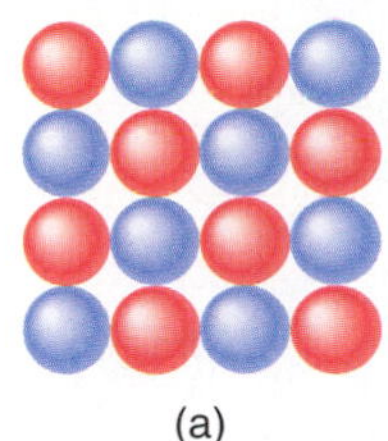

(a)

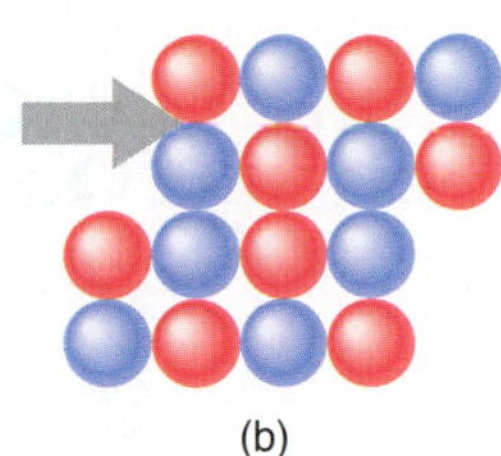

(b)

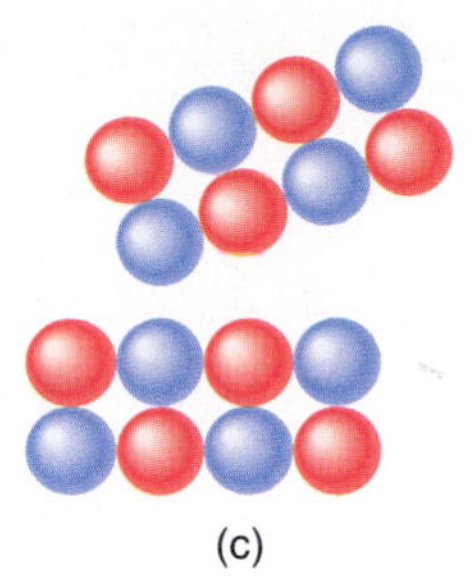

(c)

Figure 12.23

An ionic crystal is brittle, not malleable. (a) Each ion is surrounded by ions of opposite charge. (b) Any deformation tends to bring similarly charged ions closer together, thus increasing repulsive forces within the crystal. (c) Repulsive forces cause the crystal to split along the plane of deformation.

Molecular solids contain molecules held in place by intermolecular forces. Examples of molecular solids include ice (Figure 12.17), frozen carbon dioxide (Dry Ice, Figure 12.24), sugar, and paraffin (a mixture of large hydrocarbon molecules). Because intermolecular forces are weak, molecular solids often have low melting points, high vapor pressures, and little mechanical strength. They are poor conductors of electricity.

Network covalent solids such as diamond and graphite (see Figure 2.18, p. 62), carborundum (SiC, Figure 12.25), and quartz (SiO_2, Figure 2.20, p. 63) contain no molecules. Each atom in the network is covalently bonded to its neighbors, which are in turn bonded to other neighbors. The continuous bonding causes most network solids to be very hard and nonvolatile, and to have high melting points. Their crystals resist deformations that distort the natural bond angles and, like ionic crystals, they can be shattered but not bent. With some exceptions (notably graphite), they are poor conductors of electricity.

Atoms in a **metallic solid** are held together by metallic bonding of the type depicted in Figure 9.2 (p. 366). The metal atoms occupy regular lattice sites. The valence electrons, however, are not localized on individual atoms or even in individual bonds, but flow freely throughout the crystal. Metallic bonding imparts certain properties not found in other types of solids. The loose electrons are efficient

Figure 12.24

Crystal structure of frozen carbon dioxide. The molecules are held in the lattice by weak intermolecular forces.

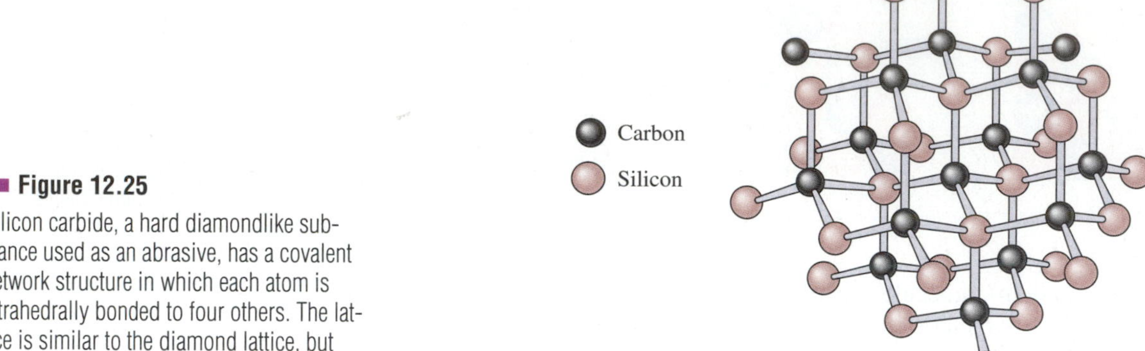

Figure 12.25
Silicon carbide, a hard diamondlike substance used as an abrasive, has a covalent network structure in which each atom is tetrahedrally bonded to four others. The lattice is similar to the diamond lattice, but with alternating silicon and carbon atoms.

carriers of electric charge and kinetic energy, so metals are outstandingly good conductors of electricity and heat. Mobile electrons also produce the typical sheen or **metallic luster** characteristic of metals. They do this by absorbing photons and then re-emitting visible wavelengths in all directions. Most metals are **malleable** (they can be shaped by hammering) and **ductile** (they can be drawn into wires). Metals have this ability to acquire new shapes because there are no individual bonds to resist deformation and the atoms easily slide from one lattice site to another.

Amorphous Solids

Amorphous: from the Greek word *amorphos* meaning "without form."

Many solids, including glass, rubber, plastics, and even hard candy, are **amorphous**; that is, their atoms and molecules are not arranged in a regular lattice (Figure 12.26). An amorphous solid forms when the molecules are unable to achieve an ordered arrangement, either because they are too complex in shape, because they were frozen too rapidly, or because impurities are present that would not fit into the reg-

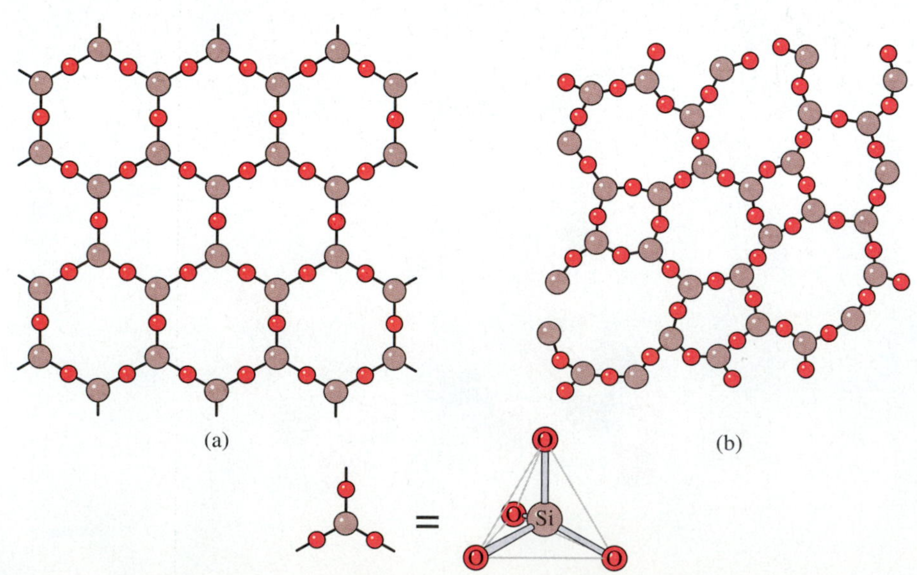

Figure 12.26
Two-dimensional representation of a silicate lattice. In the actual lattice, each silicon atom is tetrahedrally bonded to four oxygen atoms. (a) In a crystalline silicate, the silicon and oxygen atoms form a regular lattice with equal Si—O bond lengths. (b) An amorphous (noncrystalline) silicate has an irregular structure with random bond lengths.

CHEMICAL INSIGHT

LIQUID CRYSTALS

An intermediate state of matter with astonishing, entertaining, and sometimes useful properties is exhibited by certain organic substances. Compounds with long, rigid, rodlike molecules often pass through a *liquid crystal* phase in the transition from liquid to solid. One such compound is *p*-azoxyanisole:

$$CH_3-\ddot{\underset{\cdot\cdot}{O}}-C_6H_4-\underset{}{\overset{:\ddot{O}:}{\overset{|}{N}}}=\ddot{N}-C_6H_4-\ddot{\underset{\cdot\cdot}{O}}-CH_3$$

It is now recognized that liquid crystals are neither liquids nor crystals but represent a different and distinct state of matter. They pour like liquids, but x-ray diffraction shows that their molecules are in a semiordered state intermediate between liquid disorder and crystalline order. Thousands of organic and biochemical molecules are known to have at least one liquid crystal phase, and some of them have several.

Figure 12.27 illustrates the classification of liquid crystals according to their molecular arrangement. The *nematic* (Greek for "threadlike") liquid crystal phase has the least order and is most like a true liquid. The long axes of its molecules are parallel, but otherwise the molecules are free to slide or roll. A *smectic* (Greek for "soaplike") phase has more order in that its molecules are arranged in layers. Their long axes are parallel to each other and perpendicular to the plane of the layer. The layers can slide past each other, but an individual molecule can only twirl about its long axis while remaining within its layer. Because its structure is more rigid, the smectic phase is more viscous and opaque than the nematic phase. In a *cholesteric* (Greek for "hard, solid") phase, the molecules are confined to layers with their long axes parallel to each other and to the plane of the layer. The axis alignment in each layer is rotated by an angle from that of the layer below, giving rise to a long-range spiral order.

Liquid crystals have a number of unusual optical and electrical properties that lend themselves to a variety of uses. It has been discovered, for example, that an electric field pulls the polar molecules of a nematic phase out of alignment and causes the substance to become opaque. Transparency returns when the electrical signal is removed. This effect is utilized in watches, electronic calculators, and other liquid crystal display (LCD) instruments in which electronically generated information is displayed in black numbers against a cloudy background. (Such instruments should not be confused with light-emitting diode (LED) instruments, which use semiconductors to create their own light and show red numbers against a black background. The LCD displays, unlike LED displays, cannot be seen in the dark.)

A cholesteric liquid crystal may undergo a series of color changes as the layer thickness and the angle of rotation between layers change with temperature. Such liquid crystals can be used as thermometers within narrow temperature ranges of about 3°C. Cholesteric liquid crystals chosen for the right temperature range are used in adhesive-backed indicator tapes to monitor body temperature or to spot areas of overheating in mechanical systems. They are also used in novelty items such as the "mood rings" of the 1970s, whose color varied with the temperature of the skin. Of more importance is the discovery of liquid crystal character in some biochemical fluids. An increased understanding of this type of order may help elucidate some of life's processes.

Figure 12.27

Types of liquid crystal structure: (a) nematic, (b) smectic, (c) cholesteric. (d) A special effects photograph showing superimposed images of a liquid crystal display.

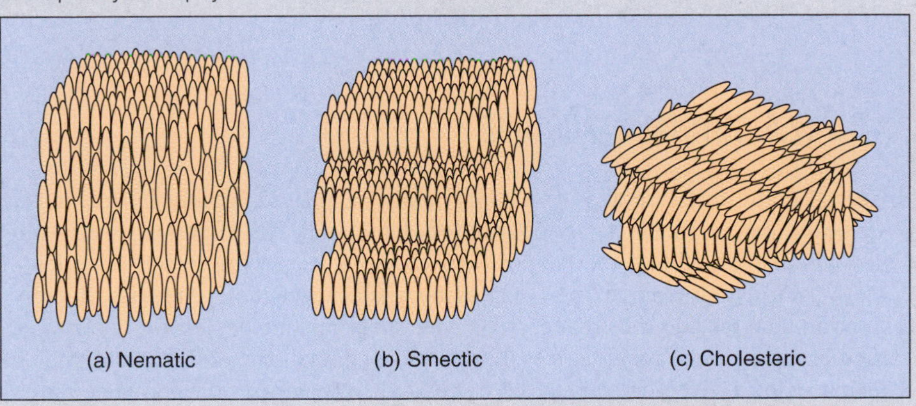

(d)

(a)

(b)

Figure 12.28
(a) Quartz crystals of all sizes, even broken crystals, have planar faces and constant angles that reflect the underlying regularity of its atomic array. (b) Glass is an amorphous solid; its broken surfaces are curved and irregular.

ular lattice. Synonyms for amorphous solid are **vitreous solid, glassy solid**, or simply **glass**. Some chemists reserve the term *solid* solely for crystalline materials and prefer to define an amorphous solid as a supercooled liquid of very high viscosity.

LEARNING HINT
Use a magnifying glass to examine some salt from a salt shaker. You will see the flat crystal faces.

Crystalline and amorphous materials can be distinguished from each other by their physical properties. A glassy solid can have any shape and will take the form of the mold in which it solidifies. A crystal, on the other hand, has planar surfaces and angles that reflect the internal arrangement of its particles. The pieces of a shattered crystal also have planar surfaces, but the surfaces of a broken glassy solid are curved and irregular (Figure 12.28). Crystalline solids have sharp melting points. Their molecules are uniformly spaced so the forces holding them in place are uniform, and all the bonds begin to break at the same temperature. In an amorphous solid, however, the intermolecular distances vary, so some molecules become free to move at a lower temperature than others. An amorphous solid softens gradually over a range of temperature and has no sharp melting point.

12.6 CRYSTAL STRUCTURE

Crystal enthusiasts are able to identify many substances simply by looking at the shapes of their crystals. **X-ray diffraction**, which consists of passing x-rays through crystals and examining the pattern formed by the emerging beam, provides much more information than can be obtained by visual inspection. Analysis of x-ray diffraction patterns confirms that crystals contain atoms, molecules, or ions arranged in regular three-dimensional lattices. Furthermore, the lattice geometry and interatomic distances of a given crystal can be calculated from measurements made on its dif-

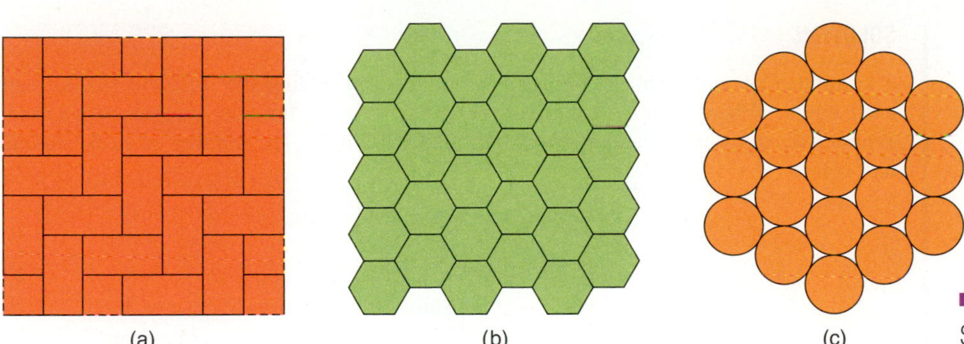

Figure 12.29

Some two-dimensional arrays. (a) A patio of rectangular bricks. A dotted line indicates half of a brick. (b) Hexagonal tiles cover a floor without leaving gaps. (c) Circular tiles leave gaps.

fraction pattern (see *Digging Deeper: How X-ray Diffraction Works* at the end of this section).

Unit Cells

Just as the basic unit of a sheet of identical postage stamps is one stamp, the basic unit of a crystal lattice is the **unit cell**, a three-dimensional arrangement of atoms or ions that repeats itself throughout the lattice. A unit cell has one atom or ion centered at each corner, and there may be atoms or ions in the faces and in the interior of the cell as well. The shape of a crystal is determined by the shape of its unit cell. For example, a crystal built of cubic unit cells will have right angles between its faces. Unit cells are limited to shapes that fill all the available space when they are stacked together. To visualize space-filling in two dimensions, think about tiling a floor. The tiles can be rectangles, triangles, or hexagons, because these shapes will fit snugly together (Figure 12.29). They cannot be pentagons or circles, which would leave gaps in the floor. Example 12.9 is a problem involving floor tiles; it is similar to some problems you will consider shortly in three dimensions.

EXAMPLE 12.9

A number of square tiles (we will call them "unit cells") are placed on the floor to form a pattern of circles (see Figure 12.30). How many circles are in each unit cell?

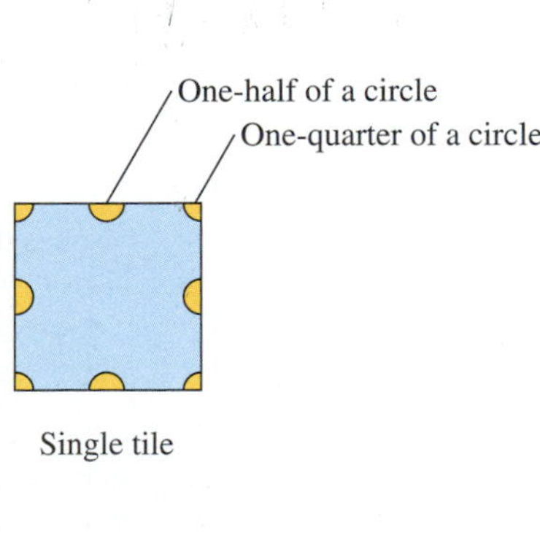

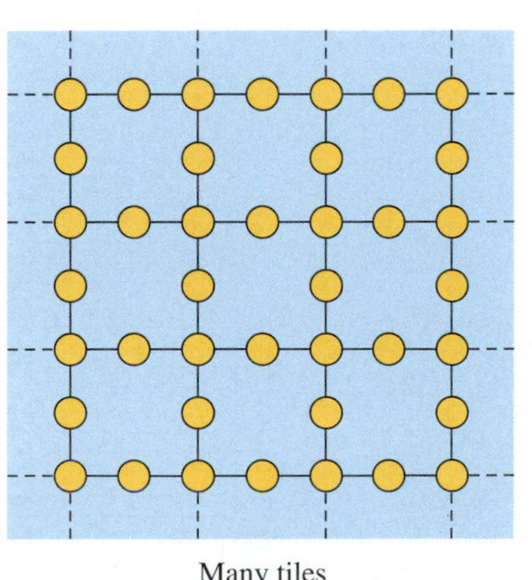

Figure 12.30

A single tile and a pattern formed by many tiles. Each tile contains parts of eight different circles; the circles are completed when the tiles are laid together.

SOLUTION

The number of circles in the unit cell is obtained by adding the different fractions of each circle. Where four cell corners meet, each corner has one-quarter of a circle. Where two edges meet, each edge contains half of a circle. The number of circles in one unit cell is

$$\begin{aligned} 4 \text{ corner circles} \times \tfrac{1}{4} &= 1 \text{ circle} \\ 4 \text{ edge circles} \times \tfrac{1}{2} &= \underline{2 \text{ circles}} \\ \text{Sum} &= 3 \text{ circles} \end{aligned}$$

Thus, each unit cell contains the equivalent of three full circles.

Most bricks and building blocks are shaped like rectangular boxes or cubes, and these are common shapes for unit cells as well. In visualizing the arrangement of atoms in a unit cell, it helps to think of the atoms as spheres. We begin by examining three cubic lattices with different arrangements of identical spheres.

A **simple cubic** arrangement of atoms has eight identical spheres at the corners of a cube (Figure 12.31a). When the atoms are stacked in a simple cubic lattice (Figure 12.31b), each one will be shared by eight cubes, and the resulting unit cell will contain one-eighth of each corner atom (Figure 12.31c). A **body-centered cubic** arrangement of nine atoms is shown in Figure 12.32a. This arrangement is equivalent to a simple cubic arrangement that has been expanded to fit another atom into the center. The unit cell for a body-centered cubic lattice is drawn in Figure 12.32b. Once again, each corner atom is shared by eight cubes. The center atom, however, belongs entirely to the unit cell. A **face-centered cubic** arrangement of 14 atoms is shown in Figure 12.33a. The center of the cube is empty, but each of the six faces contains one atom. Figure 12.33b shows the unit cell for a face-centered cubic lattice. Each corner sphere is shared by eight cubes, and each face sphere is shared by two cubes.

Finding the number of atoms in the unit cell of a cubic lattice is analogous to solving the tile problem of Example 12.9.

(a)

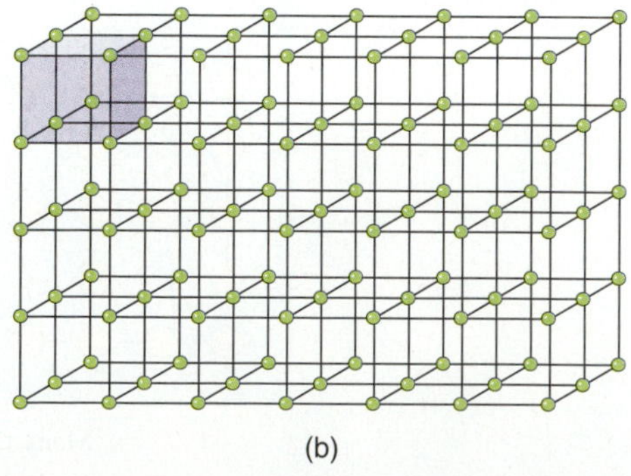

(b)

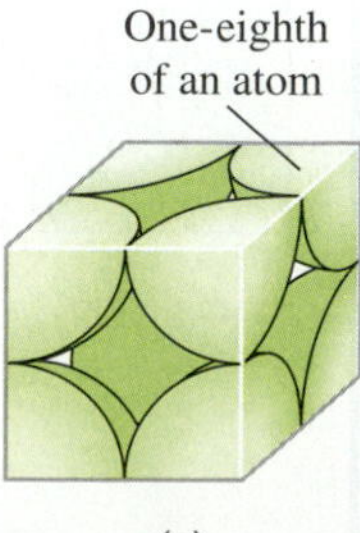

(c)

Figure 12.31

(a) Eight atoms packed in a simple cubic arrangement. (b) A lattice formed of simple cubic unit cells, one of which is shaded. Each atom is shared by eight unit cells. (c) The simple cubic unit cell showing one-eighth of an atom at each corner. When the unit cells are stacked together as in (b), each corner will have a complete atom.

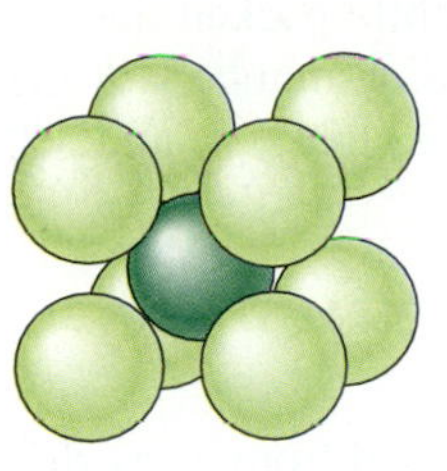

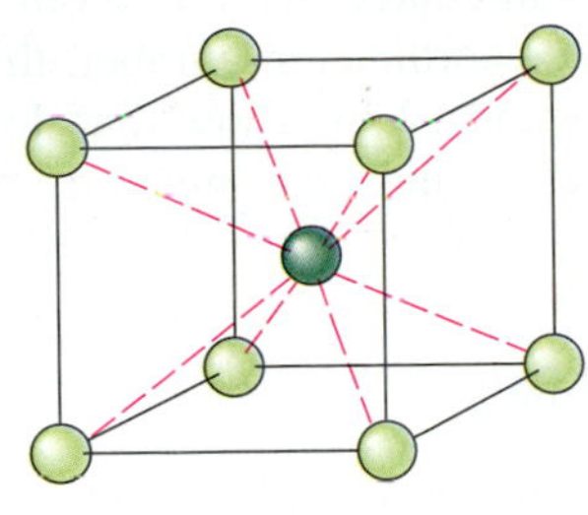

(a) (b)

Figure 12.32

(a) A body-centered cubic arrangement of nine atoms. (b) *Left:* A body-centered cube showing only atomic centers. Dotted lines connect atoms that touch. *Right:* A body-centered cubic unit cell with a complete atom at the center and one-eighth of an atom at each corner. When these cells are stacked together to make a lattice, each corner will have a complete atom. (*Note:* All of the atoms in a body-centered arrangement are identical; the center atom has been shaded differently to distinguish it from the corner atoms.)

EXAMPLE 12.10

How many atoms are in the unit cell of (a) a simple cubic unit lattice and (b) a face-centered cubic unit lattice?

SOLUTION

(a) A simple cubic unit cell contains one-eighth of each of the eight corner atoms, or $8 \times \frac{1}{8} = 1$ atom per unit cell.

(b) A face-centered unit cell contains one-eighth of each corner atom and one-half of each face atom:

$$\begin{aligned} 8 \text{ corner atoms} \times \tfrac{1}{8} &= 1 \text{ atom} \\ 6 \text{ face atoms} \times \tfrac{1}{2} &= \underline{3 \text{ atoms}} \\ \text{Sum} &= 4 \text{ atoms per unit cell} \end{aligned}$$

PRACTICE EXERCISE 12.10

How many atoms are in the unit cell of a body-centered cubic lattice?

Figure 12.33

(a) A face-centered cubic arrangement of 14 atoms. Eight atoms occupy the eight corners of the cube, and six atoms occupy the six faces. (b) *Left:* A face-centered cube showing only atomic centers. Dotted lines connect atoms that touch. *Right:* A unit cell showing one-eighth of an atom at each corner and one-half of an atom in each face. (c) The face atom shared by two unit cells (shown in red) touches 12 others, four corner atoms and eight face atoms. Each atom in this lattice has a coordination number of 12. (*Note:* All of the atoms in a face-centered arrangement are identical; the face atoms have been shaded differently to distinguish them from the corner atoms.)

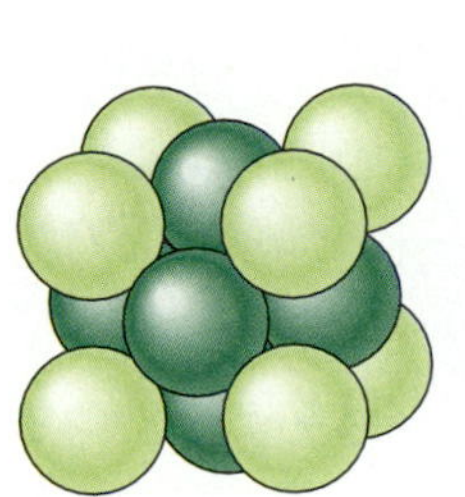

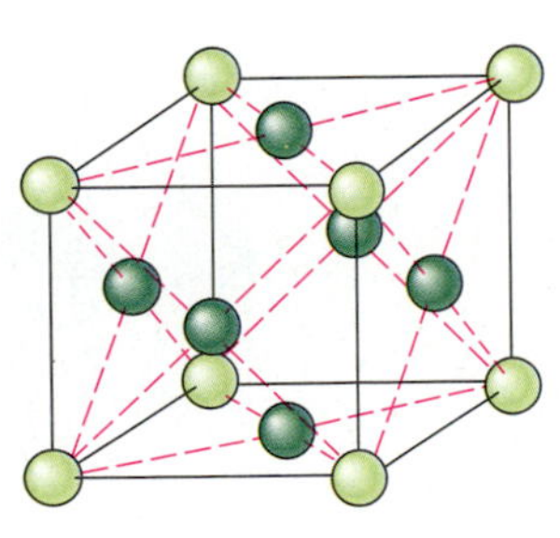

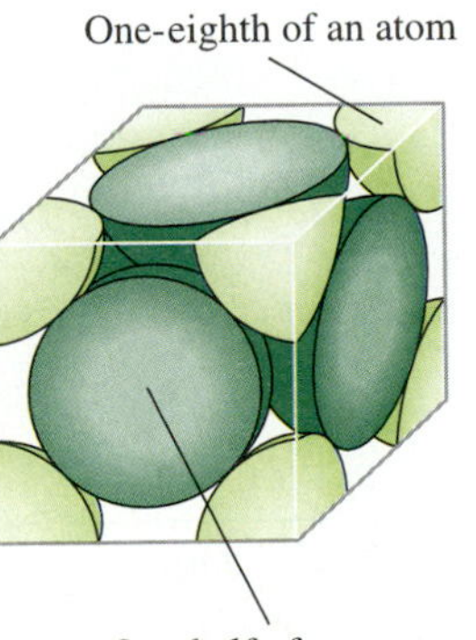

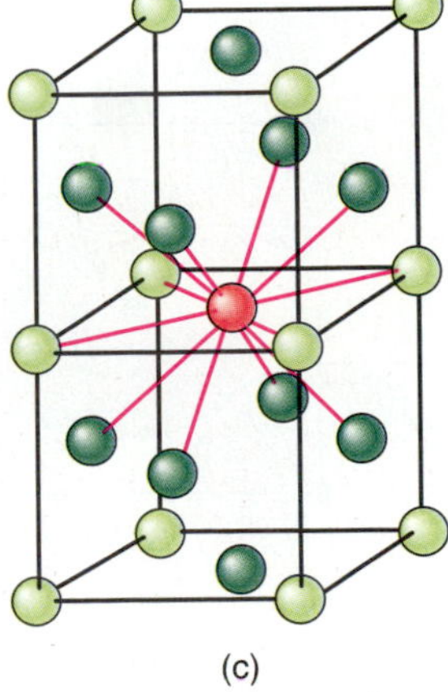

(a) (b) (c)

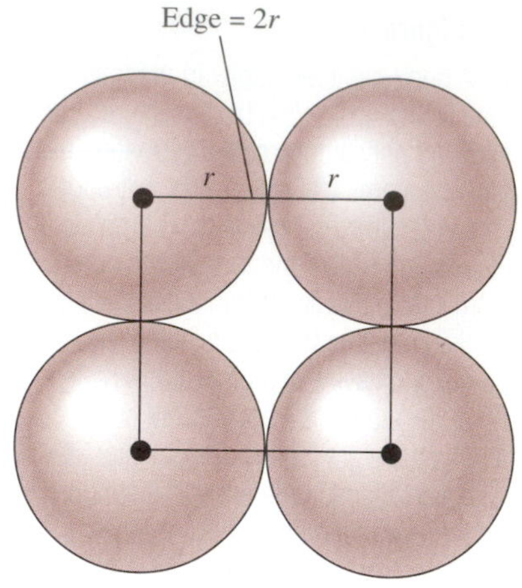

Figure 12.34

Four atoms form the face of a simple cubic unit cell. The edge of the unit cell equals two atomic radii ($2r$).

The number of neighbors in contact with an atom or ion is called its **coordination number**. The higher the coordination number, the more tightly packed the lattice. Figures 12.31b, 12.32b, and 12.33c show that the coordination numbers of the atoms in simple, body-centered, and face-centered cubic lattices are 6, 8, and 12, respectively.

Atomic Radii from Unit Cell Data

The lattice type and unit cell dimensions of a crystal can be obtained from x-ray diffraction measurements. If we assume that the atoms making up the unit cells are identical spheres, then the edge of the unit cell is related to the atomic radius r. Figure 12.34 shows that each corner atom in a simple cubic unit cell touches two other corner atoms and that the unit cell edge is twice the atomic radius:

$$\text{edge of a simple cubic unit cell} = 2r \tag{12.4}$$

In face-centered and body-centered unit cells, the corner atoms do not touch each other (Figures 12.32 and 12.33), and each unit cell edge is greater than two atomic radii. In Figures 12.35 and 12.36 it is shown that

$$\text{edge of a face-centered cubic unit cell} = 2\sqrt{2}r \tag{12.5}$$

$$\text{edge of a body-centered cubic unit cell} = \frac{4r}{\sqrt{3}} \tag{12.6}$$

The edge lengths and other important properties of cubic unit cells are summarized in Table 12.8.

Equations 12.4, 12.5, and 12.6 can be used to estimate atomic radii from unit cell data.

Figure 12.35

(a) Five atoms form the face of a face-centered cubic unit cell. The length of the face diagonal is four atomic radii ($4r$). (b) The two cell edges form a right triangle with the face diagonal as the hypotenuse. The Pythagorean theorem requires that $(\text{edge})^2 + (\text{edge})^2 = (4r)^2$. Hence, the edge of the unit cell equals $2\sqrt{2}r$.

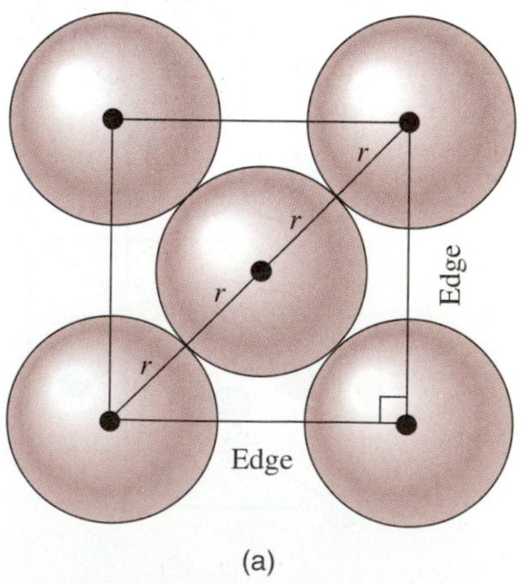

(a)

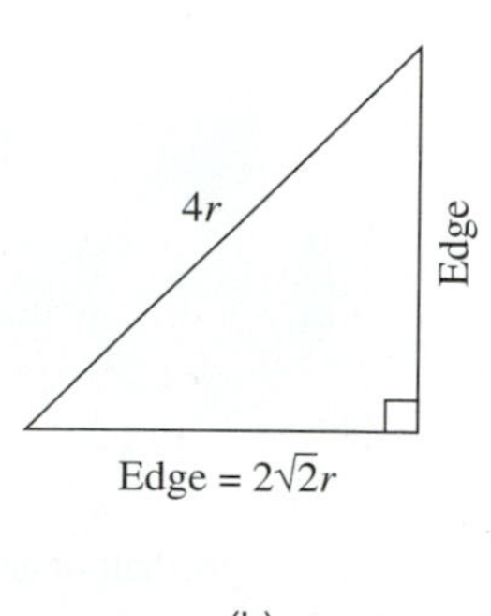

(b)

Figure 12.36

Three atoms touch along the body diagonal b of a body-centered cubic unit cell. The length of this diagonal is $4r$. The face diagonal f and one of the cell edges form a right triangle with the body diagonal as the hypotenuse. The Pythagorean theorem requires that $(f)^2 + (\text{edge})^2 = (4r)^2$. The Pythagorean theorem also requires that $(f)^2 = (\text{edge})^2 + (\text{edge})^2$. Substitution gives $3(\text{edge})^2 = (4r)^2$. Hence, the edge of the unit cell equals $4r/\sqrt{3}$.

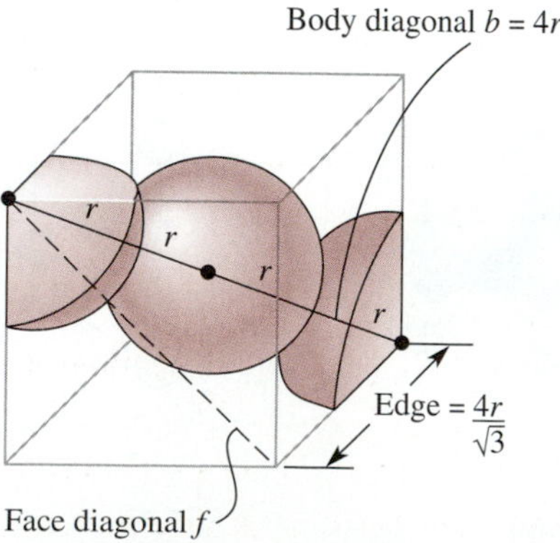

TABLE 12.8 Properties of Cubic Unit Cells

Type of Cell	Atoms per Cell	Coordination Number	Percent Occupied Space	Edge Length[a]
Simple	1	6	52.36	$2r$
Body-centered	2	8	68.02	$\frac{4r}{\sqrt{3}}$
Face-centered	4	12	74.05	$2\sqrt{2}r$

[a]The edge length is expressed in terms of the radius r of the atoms making up the cells.

EXAMPLE 12.11

Iron crystallizes in a body-centered cubic lattice with a unit cell edge of 286 pm. Estimate the radius of the iron atom.

SOLUTION

Equation 12.6 applies to body-centered unit cells. The atomic radius is obtained by rearrangement and substitution as follows:

$$r = \frac{\sqrt{3}}{4} \times \text{edge} = \frac{\sqrt{3}}{4} \times 286 \text{ pm} = 124 \text{ pm}$$

The crystal radius calculated here for iron is the same as the radius reported in Figure 8.9 (page 321).

PRACTICE EXERCISE 12.11

Silver crystallizes in a face-centered cubic lattice with a unit cell edge of 408 pm. Estimate the radius of the silver atom.

Crystal Structure Calculations

Avogadro's number has been estimated by several different methods, but the most precise determination of this important constant is based on the analysis of crystal structure data.

EXAMPLE 12.12

X-ray diffraction measurements show that copper crystallizes in a face-centered cubic lattice with a unit cell edge of 361.50 pm. The density of copper is 8.9340 g/cm^3 and its molar mass is 63.546 g. Use this information to determine Avogadro's number.

SOLUTION

Avogadro's number is the number of atoms in 1 mol (63.546 g) of copper. To find this

PRACTICE EXERCISE 12.13

Calculate the percentage of occupied space in (a) a body-centered cubic lattice of atoms and (b) a face-centered cubic lattice.

The body-centered cubic lattice of Figure 12.32 has a more efficient packing arrangement than the simple cubic lattice. Each atom in the body-centered lattice touches eight neighbors rather than six, and the percentage of occupied space is 68.0%, compared to 52.4% for the simple cubic array (Table 12.8). Fourteen metals, including iron, chromium, and all of the alkali metals, crystallize in body-centered cubic lattices.

An even more efficient packing arrangement, one that packs the maximum number of oranges, atoms, or other spherical objects into a given volume, is called *close-packing*. To pack as many atoms or spheres as possible in a given volume, you would probably start with a bottom layer in which each sphere is circled by six others, as shown in Figure 12.37a. There will be a small hollow in the surface of the layer wherever three spheres meet. The next layer will resemble the bottom layer, but it will be placed so that each of its spheres settles into one of the hollows of the first layer. Observe that only three spheres will fit into the six hollows of the first layer; only half of the hollows can be filled.

There are now two ways of positioning the third layer of spheres, shown in Figures 12.37 and 12.38. Figure 12.37b shows the third layer placed directly over the spheres in the first layer; similarly, the fourth layer is directly over the second, and so forth, in an alternating ABABAB arrangement. Note that the first three layers of

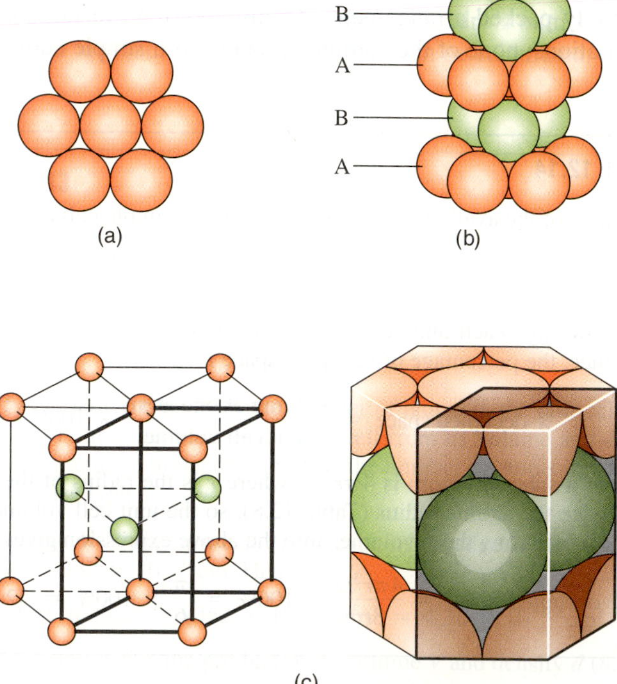

Figure 12.37

Hexagonal close-packing. (a) Six spheres surrounding a seventh. Note the hollows wherever three spheres meet. (b) The spheres of the second layer B rest in the hollows of the first layer A. Only half of the hollows are occupied. The spheres of the third layer rest in the second-layer hollows that are directly above spheres in the first layer. Successive layers repeat the alternating ABAB pattern. (c) *Left:* The centers of the ABA layers in (b) form a hexagonal prism consisting of three unit cells. One unit cell is outlined with heavy lines. *Right:* The hexagonal prism showing parts of shared spheres at corners and faces.

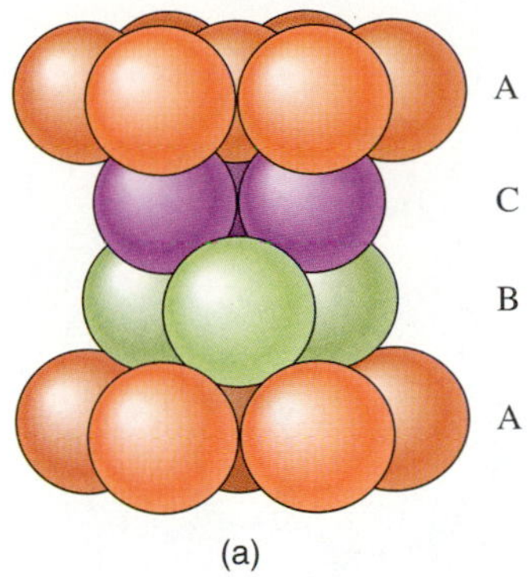

(a)

(b)

Figure 12.38

Cubic close-packing. (a) Each sphere is surrounded by six others in layer A. Spheres of the second layer B rest in the hollows of the first layer A. Spheres of the third layer C rest in hollows that are directly above unoccupied hollows of the first layer. Successive layers repeat the three-layer ABCABC sequence; the fourth layer of spheres is directly above the first layer, and so on. (b) *Left:* Selected parts of four layers containing a total of 14 spheres. *Right:* Viewed from a different angle, the arrangement shown on the left is seen to be a face-centered cube. Thus the cubic close-packed lattice is also a face-centered cubic lattice.

this arrangement (Figure 12.37c) form a hexagonal prism consisting of three unit cells. This form of close-packing is called **hexagonal close-packing**.

The other close-packing scheme, called **cubic close-packing**, is shown in Figure 12.38. The first two layers are placed as in Figure 12.37. The third layer is placed in hollows that are directly above the unoccupied hollows of the first layer. The spheres of the fourth layer are then placed directly over the spheres of the first layer to repeat the pattern, which can be represented as ABCABCABC. This three-layer pattern looks complicated at first, but it turns out to be unexpectedly simple. Viewed from another angle (Figure 12.38b), the array can be recognized as the familiar face-centered cubic lattice.

The packing of atoms or spheres in a crystal lattice is not too different from the packing of oranges in a box. In both forms of close-packing, each atom is surrounded by and touching 12 others, a hexagon of six atoms in the same plane plus a triangle of three atoms above the plane and another triangle of three below the plane. Thus the hexagonal and face-centered cubic lattices have a coordination number of 12. The percentage of occupied space in each of these lattices is 74.0% (Table 12.8). They are more tightly packed than the simple or body-centered cubic lattices and are the most efficient ways of packing atoms.

Because the atoms in a close-packed lattice must be of identical size, these lattices are limited to the elements. Helium crystallizes in a hexagonal lattice; the other noble gases crystallize in face-centered cubic lattices. About 25 metals, including magnesium, zinc, and cobalt, crystallize in the hexagonal lattice, and 14 metals, including aluminum, copper, nickel, lead, and the precious metals, platinum, silver, and gold, crystallize in the face-centered lattice. Metals with face-centered cubic lattices are the most malleable because this arrangement provides more planes of slippage along which layers can glide past each other.

A close-packed pyramid of cannon balls. Oranges at the fruit stand are often stacked in hexagonal close-packed pyramids so that they will not slip.

Ionic Lattices

An ionic compound contains positive and negative ions that may be quite different in size. The lattice of an ionic compound must allow each ion to touch as many oppositely charged ions as possible while preserving a reasonable distance between ions of the same charge. In the familiar alternating lattice of sodium chloride, shown once again in Figure 12.39, each sodium ion touches six chloride ions and each chloride ion touches six sodium ions; every ion has a coordination number of six. Pretend for a moment that the sodium ions are absent and look only at the chloride

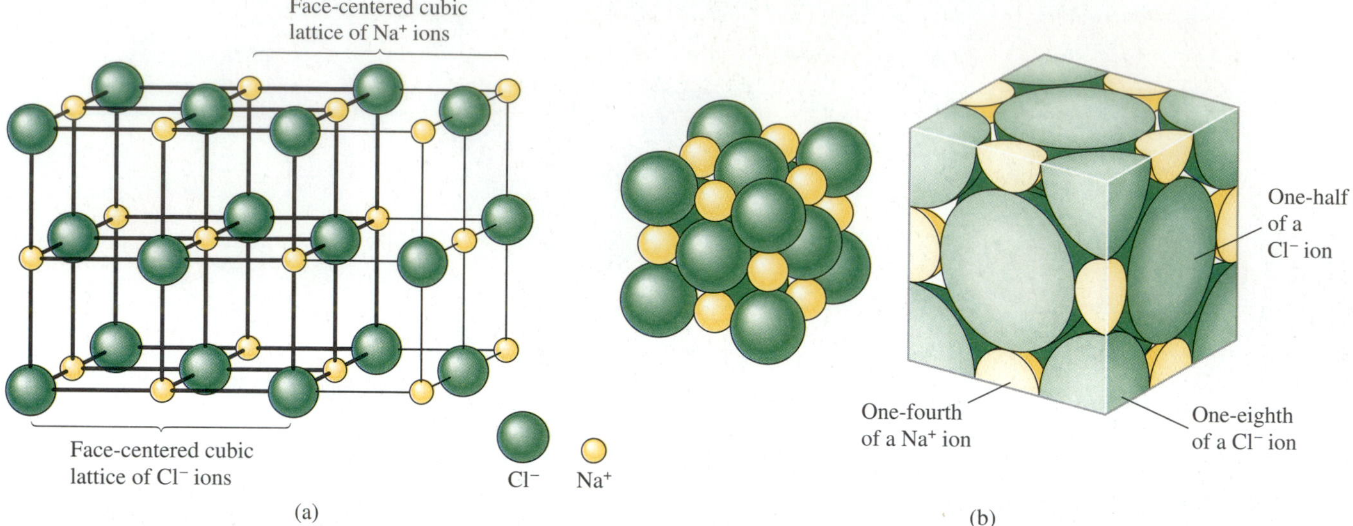

Figure 12.39

The sodium chloride lattice. (a) The lattice of alternating sodium and chloride ions can be regarded as two interlocking face-centered cubic lattices—one of sodium ions and one of chloride ions. Each ion is in contact with six ions of the opposite charge. (b) *Left:* The NaCl lattice is shown as a face-centered cube of chloride ions with sodium ions in all the edges. *Right:* A unit cell of sodium chloride.

ions in Figure 12.39a; you will see that they form a face-centered cubic lattice. The sodium ions also form a face-centered cubic lattice, and the complete array is the superposition of the two separate lattices. The unit cell is the face-centered cube shown in Figure 12.39b. Silver chloride, calcium and magnesium oxides, and most alkali metal halides crystallize in the sodium chloride arrangement, often referred to as the **rock salt structure**.

The type of lattice adopted by an ionic compound depends on several factors, one of which is the ratio of positive ions to negative ions in the formula. For example, the calcium fluoride (CaF_2) lattice, which contains two fluoride ions for each calcium ion, must be different from the sodium chloride lattice, which contains one chloride ion for each sodium ion. The type of lattice also depends on the relative sizes of the ions. Cesium ions (Cs^+) are about the same size as chloride ions, and in the cesium chloride (CsCl) lattice each ion has a coordination number of eight (Figure 12.40). In the sodium chloride lattice, the coordination number of each ion is only six. The coordination number in sodium chloride is smaller than in cesium chloride because sodium ions are smaller than cesium ions and fewer chloride ions can fit around them.

Figure 12.40

The cesium chloride lattice. The unit cell is a cube with one ion (either Cs^+ or Cl^-) at the center and one-eighth of the other ion at each of the corners.

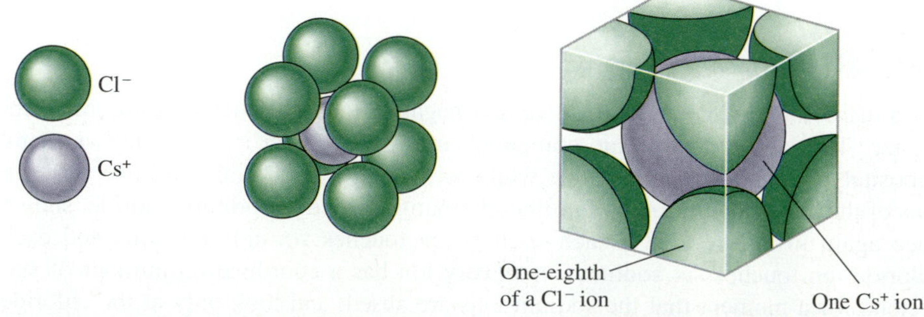

EXAMPLE 12.15

The unit cell of a cesium chloride lattice is a cube with chloride ions at the corners and a cesium ion in the center (Figure 12.40). How many (a) chloride ions and (b) cesium ions are in the unit cell?

SOLUTION

(a) There are eight corner chloride ions, so the number of chloride ions per cell is $8 \times \frac{1}{8} = 1$.

(b) The cesium ion is in the center of the cell; hence, there is one cesium ion per cell.

PRACTICE EXERCISE 12.14

Refer to Figure 12.39, and determine the number of (a) chloride ions and (b) sodium ions in the unit cell of NaCl. (*Hint*: Don't forget the center sodium ion.)

Lattice Imperfections

All of the crystals discussed thus far have been ideal crystals with perfect unit cells in perfect alignment. Real crystals contain numerous imperfections or **lattice defects** such as missing atoms or ions, dislocations (misplaced atoms or ions), and impurities. It has been calculated that an ordinary sodium chloride crystal at room temperature has about 50 million vacant lattice sites per cubic centimeter, to say nothing of other imperfections. Lattice defects influence many properties of crystalline substances. For example, metals with many dislocated atoms tend to be brittle and mechanically weak. A sheet of metal will break when it is bent back and forth many times because bending increases the number of dislocations. The number of lattice defects is also increased by radiation, as evidenced by the premature deterioration of piping and other metal components in nuclear reactors. Defects also increase the susceptibility of crystals to chemical attack; metals corrode more rapidly at joints and other areas where mechanical stress has produced lattice irregularities.

DIGGING DEEPER

How X-ray Diffraction Works

When a beam of light passes through two adjacent pinholes, it forms a pattern of alternating bright and dark regions similar to the patterns in Figure 7.8 (p. 269). Such patterns, called **diffraction patterns**, occur whenever the distance between the pinholes is comparable to the wavelength of the light. The bright regions appear where light waves reinforce each other by arriving in phase, an effect called *constructive interference*. The intervening dark areas occur where light waves arrive out of phase and cancel each other (*destructive interference*). Similarly, x-rays passing

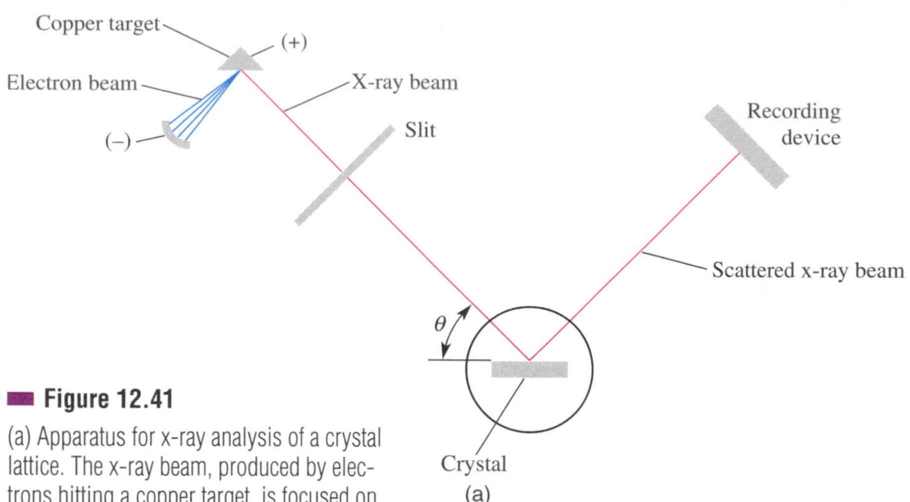

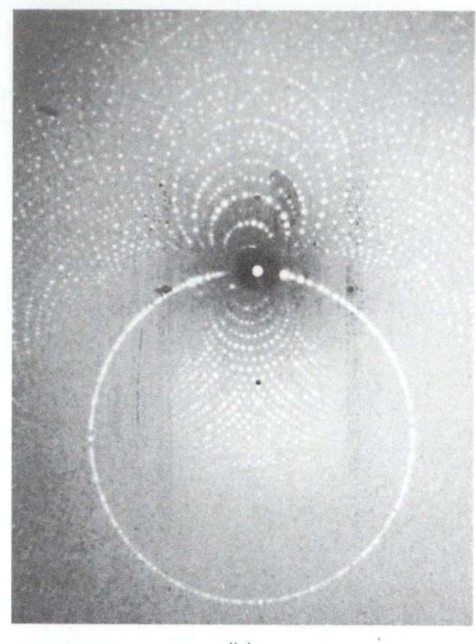

Figure 12.41

(a) Apparatus for x-ray analysis of a crystal lattice. The x-ray beam, produced by electrons hitting a copper target, is focused on the crystal as it is rotated through successive angles of incidence (θ). Scattering occurs at angles that satisfy the Bragg equation (Equation 12.7). (b) Laue diffraction pattern of a crystal of lysozome (an enzyme) from hen egg white taken at CHESS, the synchrotron facility at Cornell University.

through a crystal will show a diffraction pattern because the spaces between atoms are comparable to x-ray wavelengths. The first x-ray diffraction pattern was obtained in 1912 after Max von Laue, a German physicist, predicted that crystals should give this effect. The pattern was obtained by directing a beam of x-rays at a mounted crystal (see Figure 12.41), and the emerging beam was recorded on photographic film. In other methods for obtaining diffraction patterns, the x-rays are beamed at rotating crystals or powdered samples, and the angle at which each diffraction line appears is recorded.

In 1913 the father-and-son team of William Henry Bragg and William Lawrence Bragg worked out the mathematical method for obtaining interatomic distances from x-ray diffraction patterns. Figure 12.42 shows x-rays of wavelength λ scattered from successive atomic layers separated by some distance d. The paths that the two rays travel differ in length by an amount equal to $2d \sin \theta$, where θ is the scattering angle. (The symbols λ and θ are the small Greek letters lambda and theta, respectively.) For the waves to enter and leave in phase, the extra distance traveled by the second ray must be an integral number of wavelengths; that is, it must be exactly one, two, three, or more wavelengths. Thus the beams will emerge in phase only if they satisfy the **Bragg equation**:

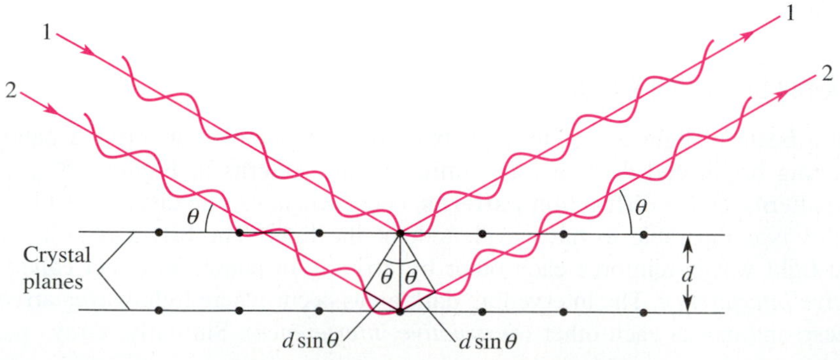

Figure 12.42

The Bragg equation. Rays 1 and 2 are part of a wavefront of x-rays entering the crystal at angle θ. Ray 1 is scattered from the upper layer of atoms and ray 2 from the lower layer. Note that the difference in path length between these rays is $2d \sin \theta$, where d is the distance between the crystal planes. For the waves to enter and leave in phase, the extra distance traveled by ray 2 must be an integral number of wavelengths; that is, $2d \sin \theta$ must equal $n\lambda$, where n is an integer and λ is the wavelength of the incident radiation.

$$n\lambda = 2d \sin \theta \tag{12.7}$$

where n is an integer. The Bragg equation implies that for a given wavelength λ and interatomic spacing d, reinforcement will occur at angles whose sine is $\lambda/2d$, $2\lambda/2d$, $3\lambda/2d$, and so forth. Many of these scattered beams will not be seen, however, because their intensity drops off sharply with increasing n. The most intense scattering, called *first-order-scattering*, occurs when $n = 1$.

EXAMPLE 12.16

The first-order scattering angle is 5.97° when 0.05869-nm x-rays strike a sodium chloride crystal. How far apart are the layers of ions responsible for this effect?

SOLUTION

In first-order scattering, $n = 1$. The wavelength λ of the incident radiation is 0.05869 nm and the diffraction angle θ is 5.97°. Substituting $n = 1$ into the Bragg equation and solving for d gives

$$d = \frac{\lambda}{2 \sin \theta} = \frac{0.05869 \text{ nm}}{2 \sin 5.97°} = 0.282 \text{ nm} = 282 \text{ pm}$$

PRACTICE EXERCISE 12.15

Refer to Example 12.16. At what angle should one look for second-order scattering ($n = 2$)?

CHAPTER REVIEW

Starred entries are based on the *Digging Deeper* sections.

LEARNING OBJECTIVES BY SECTION

12.1
1. Compare the properties of the three states of matter, and relate these properties to molecular motion and attractive forces.
2. Give names for commonly encountered phase changes.
3. Describe the molecular process by which a liquid or solid achieves its equilibrium vapor pressure.
4. State how the vapor pressure of a liquid varies with temperature and with intermolecular attractive forces.
5. Describe the relation between a liquid's vapor pressure and its boiling point.
6. Distinguish between the boiling point of a liquid and its normal boiling point, and describe how the boiling point varies with external pressure.
7. Describe, in molecular terms, the equilibrium that exists at the melting point.
8. Distinguish between the melting point of a solid and its normal melting point.

9. Given the appropriate data, sketch heating and cooling curves for a substance.
10. Calculate the amount of heat absorbed or evolved during a phase change.
*11. Given the appropriate data, use the Clausius–Clapeyron equation to estimate vapor pressures at various temperatures, boiling points at various pressures, and heats of vaporization.

12.2 1. Sketch the phase diagram for water, and label the various regions and curves.
2. Locate the normal melting point, normal boiling point, triple point, and critical point on a phase diagram.
3. Use a phase diagram to identify the most stable form of a substance at a given temperature and pressure, and describe what happens to a sample of the substance as its temperature or pressure changes.
4. Use the phase diagram to predict which of two states has the greater density.
5. Use fusion curves to determine the variation of melting point with pressure.
6. State how critical temperatures vary with increasing intermolecular attractions.

12.3 1. Distinguish between intermolecular attractive forces and chemical bonding forces.
2. Describe the origin of dipole–dipole forces, and predict the effect of such forces on melting points, boiling points, and other thermal properties.
3. Describe the hydrogen bond, and identify molecules in which hydrogen bonding might occur.
4. State some of the anomalous properties of water attributable to hydrogen bonding.
5. Describe the origin of London dispersion forces, and predict the effect of such forces on melting points, boiling points, and other thermal properties.

12.4 1. Explain what is meant by the statement that liquids exhibit short-range order.
2. State how the viscosity and surface tension of liquids vary with increasing intermolecular attractions and increasing temperature.
3. Explain why liquids tend to minimize their surface area.

12.5 1. Describe the bonding in ionic, molecular, network covalent, and metallic solids, and list the physical properties associated with each type of solid.
2. Give examples of each type of crystalline solid.
3. Distinguish between crystalline and amorphous solids in terms of physical properties and in terms of molecular arrangement.
4. Give examples of commonly encountered amorphous solids.

12.6 1. Draw simple cubic, body-centered cubic, and face-centered cubic unit cells.
2. State the number of atoms per unit cell and the coordination number for each of the three cubic lattices.
3. Use unit cell data to estimate atomic radii.
4. Calculate the missing quantity given four of the following: type of lattice, edge length, crystal density, molar mass, and Avogadro's number.
5. Calculate the percentage of occupied space in each of the three cubic lattices.
6. Draw the common atomic-packing arrangements for crystalline elements.
7. Draw the rock salt structure and its unit cell.
8. Describe some lattice imperfections, and state how they affect the properties of crystals.
*9. Derive the Bragg equation, and use it to calculate interatomic spacings from measured diffraction angles.

KEY TERMS BY SECTION

12.1 Boiling point
*Clausius–Clapeyron equation
Cooling curve
Dynamic equilibrium
Freezing point
Heating curve
Melting point
Molar heat of fusion
Molar heat of vaporization
Normal boiling point
Normal melting point
Phase change
Sublimation
Supercooled liquid
Vapor pressure
Vaporization

12.2 Critical point
Critical pressure
Critical temperature
Fusion curve
Phase diagram
Sublimation curve
Sublimation temperature
Supercritical fluid
Triple point
Vaporization curve

12.3 Dipole–dipole force
Hydrogen bond
Induced dipole
Intermolecular attractive force
London dispersion force
Van der Waals force

12.4 Surface tension
Viscosity

12.5 Amorphous solid
Crystal lattice
Crystalline solid
Ductility
Glass (glassy solid)
Ionic solid
Malleability
Metallic luster
Metallic solid
Molecular solid
Network covalent solid
Vitreous solid

12.6 Body-centered cubic unit cell
*Bragg equation
Coordination number
Cubic close-packing
*Diffraction pattern
Face-centered cubic unit cell
Hexagonal close-packing
Lattice defect
Rock salt structure
Simple cubic unit cell
Unit cell
X-ray diffraction

IMPORTANT EQUATIONS

***12.1** $P = Ae^{-\Delta H^\circ_{vap}/RT}$

***12.2** $\ln P = -\dfrac{\Delta H^\circ_{vap}}{RT} + B$

***12.3** $\ln\left(\dfrac{P_1}{P_2}\right) = \dfrac{\Delta H^\circ_{vap}}{R}\left(\dfrac{1}{T_2} - \dfrac{1}{T_1}\right)$

12.4 Edge of a simple cubic unit cell $= 2r$

12.5 Edge of a face-centered cubic unit cell $= 2\sqrt{2}r$

12.6 Edge of a body-centered cubic unit cell $= \dfrac{4r}{\sqrt{3}}$

***12.7** $n\lambda = 2d \sin\theta$

FINAL EXERCISES

Answers to exercises with blue numbers are given in Appendix D. Starred exercises are based on the *Digging Deeper* sections.

PART A. QUESTIONS AND PROBLEMS BY SECTION

States of Matter; Phase Changes (Introduction; Section 12.1)

12.1 List the three states of matter in order of (a) increasing intermolecular attractions and (b) increasing molecular disorder.

12.2 Do liquids have a fixed shape? A fixed volume? What about gases?

12.3 Explain why the formation of solids and liquids is favored by (a) high pressures and (b) low temperatures.

12.4 Will a substance with strong intermolecular attractions have
(a) a high or a low vapor pressure?
(b) a high or a low boiling point?
(c) a high or a low heat of vaporization?

12.5 What does the phrase *dynamic equilibrium* mean? Give examples of systems in dynamic equilibrium.

12.6 Explain why the vapor pressure of a solid at its melting point must be equal to the vapor pressure of the liquid.

12.7 What effect does increasing temperature have on the vapor pressure of a liquid?

12.8 What effect does increasing pressure have on the boiling point of a liquid?

12.9 Will the boiling point of water be higher or lower than 100°C on top of Mt. Everest? Explain.

12.10 Use Figure 12.6 to estimate the boiling point of diethyl ether at 500 torr.

12.11 Explain why a burn from steam at 100°C is more severe than a burn from the same mass of water at 100°C.

12.12 Which would be more effective in cooling a drink—an ice cube at 0°C or an equal mass of liquid water at 0°C? Explain.

12.13 Explain why the temperature remains constant during a phase change such as melting or vaporization.

12.14 Supercooled liquids are often compared with supersaturated solutions. In what ways are these systems analogous?

12.15 Use data from Table 12.3 to sketch a heating curve for 1 mol of mercury over the temperature range of −50 to 500°C.

12.16 Use data from Table 12.3 to sketch a cooling curve for 1 mol of ethanol (C_2H_5OH). Your curve should extend from vapor at 95°C to solid at −125°C. Show the temperatures of all phase changes, indicate molar heats of fusion and vaporization on the plateaus, and show approximate molar heat capacities along the slanted lines.

12.17 How many kilojoules are required to (a) melt 1.00 kg of ice at 0°C and (b) vaporize 1.00 kg of liquid water at 100°C? Assume that atmospheric pressure is 1.00 atm.

12.18 How many kilojoules are required to melt a 1.00-cm-diameter aluminum marble originally at 25°C? The density of aluminum is 2.70 g/cm^3. Refer to Table 12.3 for additional data.

12.19 How much heat is required to convert 1.00 mol of ice at 0°C and 1.00 atm to steam at 100°C and 1.00 atm? Give your answer in (a) kilojoules and (b) kilocalories.

12.20 A 500-g iron bar is heated to 600°C and plunged into 100 g of water at 25°C. Calculate the number of grams of water that will vaporize by the time equilibrium is reached. Assume that no heat is lost to the surroundings and that the average specific heat of iron is 0.45 J/g·°C over the temperature range involved.

***12.21** Use the Clausius–Clapeyron equation to estimate the boiling point of water on top of Mt. Everest, where atmospheric pressure is about 0.28 atm.

***12.22** Use the Clausius–Clapeyron equation and any two points on the diethyl ether vapor pressure curve (Figure 12.6) to estimate the heat of vaporization of diethyl ether.

Phase Diagrams (Section 12.2)

12.23 The phase diagram of carbon dioxide is given in Figure 12.11. What is the most stable state of carbon dioxide at (a) 1 atm and 25°C and (b) 75 atm and 35°C?

12.24 Refer to Figure 12.11, and describe the phase changes that occur when carbon dioxide at 10 atm is warmed from −78° to 100°C.

12.25 Solid benzene is more dense than liquid benzene. Would you expect the melting point of benzene to increase or decrease with increasing pressure? Explain.

12.26 Explain in terms of densities why the sublimation curve for water slants to the right as it rises.

12.27 What is a *triple point* on a phase diagram? What will happen to a sample of ice if it is warmed from 0°C to 25°C at a pressure just below the triple point pressure?

12.28 What is the *critical point* on a phase diagram? Refer to Figure 12.12, and explain why the meniscus vanishes at the critical point.

12.29 Some chemists reserve the term *vapor* for a gaseous substance below its critical temperature. Can a vapor be liquefied by pressure alone? By cooling alone? Explain your answers.

12.30 The critical temperatures of hydrogen, oxygen, nitrogen, and carbon monoxide are −240, −118, −147, and −140°C, respectively.

(a) List these gases in order of increasing intermolecular attractions.

(b) A research student working at room temperature attempted to liquefy each of the gases by subjecting them to very high pressures. Did this attempt succeed or fail? Explain.

12.31 Use the following information to construct a rough phase diagram for hydrogen: normal melting point, −259.34°C; normal boiling point, −252.87°C; triple point, −259.35°C and 0.0695 atm; critical temperature, −240.17°C; critical pressure, 12.76 atm.

12.32 Use the following information to construct a rough phase diagram for oxygen: normal melting point, −218.4°C; normal boiling point, −182.96°C; triple point, −218.80°C and 1.138 torr; critical temperature, −118.57°C; critical pressure, 49.77 atm.

12.33 Is the density of solid hydrogen greater or less than the density of liquid hydrogen? Refer to the phase diagram of Exercise 12.31.

12.34 Is the density of solid oxygen greater or less than the density of liquid oxygen? Refer to the phase diagram of Exercise 12.32.

12.35 A low-temperature phase diagram of helium (Figure 12.43) shows the existence of two liquid states labeled Liquid I and Liquid II.

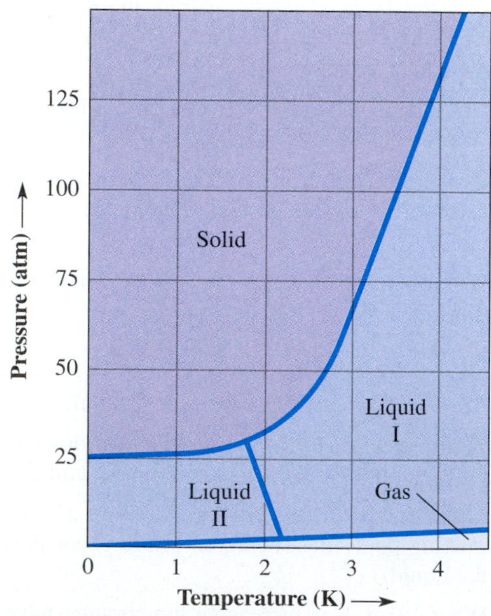

Figure 12.43
Low-temperature phase diagram for helium.

(a) What is the maximum number of states that can coexist in equilibrium? Identify them.
(b) Is it possible to solidify helium under 1 atm pressure?
(c) Does the melting point of the solid increase or decrease with increasing pressure?

12.36 The phase diagram of sulfur (Figure 12.44) shows the existence of two crystalline forms, rhombic and monoclinic.
(a) How many triple points does sulfur have?
(b) Describe what happens when rhombic sulfur at 60°C and 1 atm is heated at constant pressure to 160°C.

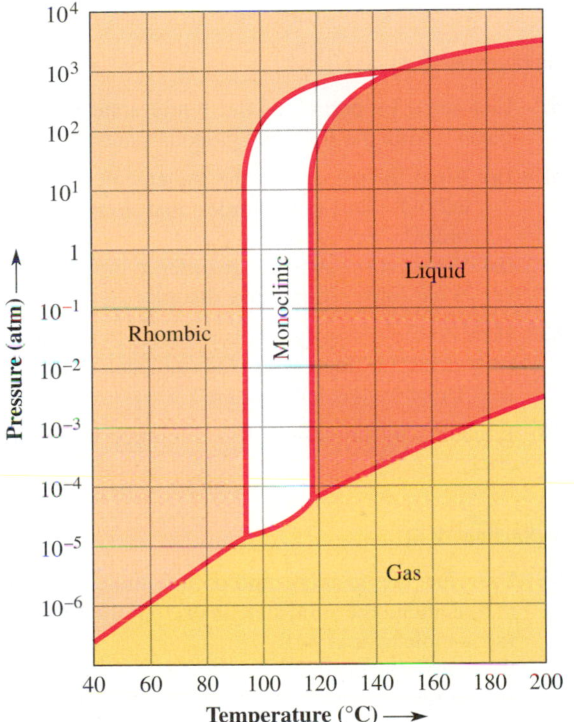

Figure 12.44
Phase diagram for sulfur.

12.37 At a temperature of 2 K, an increase in pressure converts liquid helium II into liquid helium I (Figure 12.43). Which form of helium has the greater density?

12.38 Refer to Figure 12.44, and predict which form of sulfur, rhombic or monoclinic, has the greater density at 100°C.

12.39 Figure 12.45 is the phase diagram of carbon.
(a) What is the stable form of carbon at ordinary temperatures and pressures? At 100,000 atm and 1700°C?

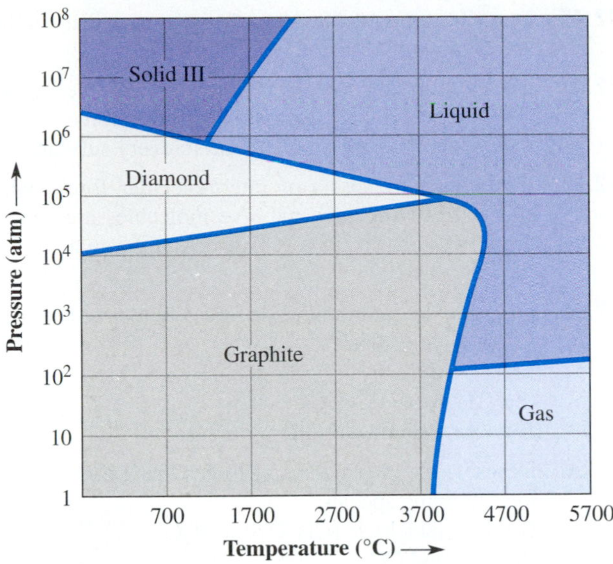

Figure 12.45
Phase diagram for carbon.

(b) Estimate the normal sublimation temperature of graphite, that is, the temperature at which solid graphite has a vapor pressure of 1 atm.

12.40 Use Figure 12.45 to determine whether the density of diamond is greater or less than that of graphite at the same temperature.

Intermolecular Forces; Hydrogen Bonding (Section 12.3)

12.41 Distinguish between intermolecular forces and chemical bonding forces. Which type of force is stronger? Justify your answer by comparing some typical bond energies (Table 9.3, p. 377) with molar heats of vaporization (Table 12.3).

12.42 (a) Explain, with the aid of a diagram, the origin of dipole–dipole forces.
(b) Ion–dipole forces are a type of interaction not discussed in this chapter. Explain, with the aid of a diagram, how such forces might originate.

12.43 Define and give examples of a *hydrogen bond*. Explain why we should be careful not to use the term "hydrogen bond" for bonds like those in H_2 and H_2S.

12.44 (a) Which atoms participate in the formation of hydrogen bonds?
(b) The electronegativity of chlorine is greater than that of nitrogen. Why doesn't chlorine form hydrogen bonds?

12.45 What is the arrangement of atoms in a hydrogen bond? Suggest a reason for this arrangement.

12.46 Describe some of the properties of water that can be attributed to hydrogen bonding.

12.47 (a) Describe the difference between a dipole–dipole interaction and a London dispersion force.
(b) Which interaction is present in every substance?

12.48 Explain on the basis of intermolecular forces why substances with large massive molecules are generally less volatile than substances with small molecules.

12.49 Describe the intermolecular forces that exist between the following species:
(a) I_2 and CCl_4
(b) CH_3Br and CH_3Cl
(c) CH_3OH and CH_3NH_2

12.50 Describe the intermolecular forces that exist between the following species:
(a) Formaldehyde (CH_2O) and H_2O
(b) CS_2 and CCl_4
(c) BrF and BrCl

12.51 In which of the following substances would you expect to find a significant degree of hydrogen bonding? Diagram the hydrogen bond in each case.
(a) HF
(b) HI
(c) CH_2O

12.52 In which of the following substances would you expect to find a significant degree of hydrogen bonding? Diagram the hydrogen bond in each case.
(a) CH_3COOH
(b) C_2H_5OH
(c) C_2H_5SH

12.53 The normal boiling points of N_2, O_2, and NO are −196, −183, and −152°C, respectively. Explain why the boiling point of NO is substantially higher than the boiling points of nitrogen and oxygen.

12.54 Which would you expect to have the higher heat of vaporization, C_2H_5OH or $CH_3CH_2CH_3$? Explain your answer.

12.55 The molar mass of methanol (CH_3OH) is very close to that of ethane (CH_3CH_3). Which of these substances will have (a) the higher boiling point and (b) the higher heat of vaporization? Explain your answers.

12.56 Which substance would you expect to have the higher boiling point, carbon dioxide (CO_2) or formaldehyde (CH_2O)? Explain your answer.

12.57 Which substance would you expect to have the higher vapor pressure, CH_3OCH_3 or CH_3CH_2OH? Explain your answer.

12.58 Arrange the following elements in order of increasing London dispersion forces: Xe, Ar, and Kr.

12.59 Which substance in each pair would you expect to have the higher boiling point? Explain your answer in each case.
(a) Ar or Xe
(b) HF or HBr
(c) CH_3CH_2OH or $HOCH_2CH_2OH$

12.60 Which substance in each pair would you expect to have the higher boiling point? Explain your answer in each case.
(a) PH_3 or AsH_3
(b) C_2H_5SH or C_2H_5OH
(c) Na or K

Liquids (Section 12.4)

12.61 List some properties of the liquid state that distinguish it from the solid and gaseous states.

12.62 (a) What is meant by the statement that liquids exhibit short-range order and long-range disorder?
(b) What causes short-range order in liquids?
(c) What causes long-range disorder?

12.63 How do the viscosity and surface tension of a liquid vary with (a) increasing intermolecular attractions and (b) increasing temperature?

12.64 Explain why liquids tend to minimize their surface area.

Solids (Section 12.5)

12.65 Describe the bonding and/or intermolecular forces that maintain the solid state in each of the following:
(a) a molecular crystal
(b) a metallic crystal
(c) an ionic crystal
(d) a network covalent crystal
(e) an amorphous solid

12.66 Prepare a table comparing ionic, molecular, network covalent, and metallic solids with respect to the following physical properties:
(a) hardness and fragility
(b) malleability and ductility
(c) melting and boiling points
(d) heats of fusion and vaporization
(e) ability to conduct electricity
(f) ability to conduct heat

12.67 Give two examples of (a) a molecular crystal, (b) a metallic crystal, (c) an ionic crystal, (d) a network covalent crystal, and (e) an amorphous solid. For each of the substances you choose as examples, list two or three properties that reflect the nature of its structure and bonding.

12.68 Describe three properties that distinguish amorphous solids from crystalline solids. Give five examples of amorphous solids.

12.69 Identify each of the following solids as metallic, ionic, molecular, or network covalent, and defend your choice on the basis of physical properties:
(a) mothballs
(b) sand
(c) ice
(d) table salt
(e) a toaster filament

12.70 Identify each of the following solids as crystalline or amorphous, and defend your choice on the basis of physical properties:
(a) sugar
(b) rubber
(c) polyethylene
(d) aluminum
(e) tar
(f) potassium sulfate
(g) diamond

12.71 What type of solid (ionic, molecular, metallic, or network covalent) is each of the following?
(a) PrI_3, melts at 733°C, boils at 1377°C; liquid conducts electricity
(b) Osmium hexafluoride (OsF_6), melts at 32°C, boils at 46°C; liquid does not conduct electricity
(c) Borazon (BN), sublimes at about 3000°C

12.72 What type of solid (ionic, molecular, metallic, or network covalent) is each of the following?
(a) Polonium (Po), melts at 254°C, boils at 962°C; solid conducts electricity
(b) $SiBr_4$, melts at 5.4°C, boils at 154°C; liquid does not conduct electricity
(c) Silicon (Si), melts at 1410°C, boils at 2355°C; solid does not conduct electricity

Unit Cells; Crystal Structure Calculations (Section 12.6)

12.73 Identify the unit cell in (a) a sheet of postage stamps, (b) a chessboard pattern, and (c) a pyramidal stack of grapefruit.

12.74 Use identical spheres (marbles, styrofoam balls, etc.) and whatever is necessary to hold them together (glue, toothpicks, etc.) to construct (a) a simple cubic lattice, (b) a body-centered cubic lattice, and (c) a face-centered cubic lattice. Keep adding unit cells until you are sure you understand the geometry of the lattice.

12.75 Draw the following unit cells:
(a) simple cubic
(b) body-centered cubic
(c) face-centered cubic

12.76 For each of the cubic lattices, list (a) the number of atoms per unit cell, (b) the coordination number, and (c) the percentage of occupied space. Which cubic lattice has the most efficient packing arrangement? The least efficient packing arrangement?

12.77 Using pennies or other coins, convince yourself that only six identical circles fit neatly around a seventh.

12.78 Using marbles or other identical spheres, convince yourself that only six spheres can surround and touch a center sphere in one layer.

12.79 Use identical spheres as in Exercise 12.74 to construct (a) a hexagonal close-packed lattice and (b) a cubic close-packed lattice. Keep adding spheres until you are sure you understand the geometry of each lattice.

12.80 Use different size spheres (see Exercise 12.74) to construct (a) the sodium chloride lattice (Figure 12.39) and (b) the cesium chloride lattice (Figure 12.40). Keep adding unit cells until you are sure you understand the geometry of each lattice.

12.81 List three types of lattice defects.

12.82 Explain why metals with many dislocated atoms tend to be brittle and mechanically weak.

***12.83** Does the first-order scattering angle for a given wavelength increase or decrease as the distance between atomic layers increases?

***12.84** Would you expect to get a diffraction effect by passing 500-nm visible radiation through a crystalline solid? Explain.

12.85 Gold crystallizes as a face-centered cubic lattice with a unit cell edge of 407.86 pm. Calculate (a) the radius of a gold atom in picometers and (b) the density of gold in grams per cubic centimeter.

12.86 Chromium crystallizes in a body-centered cubic lattice. The density of chromium is 7.2 g/cm^3 and its molar mass is 52.00 g. Calculate (a) the unit cell edge and (b) the radius of a chromium atom. Express your answers in picometers.

12.87 Use the data in Example 12.11 to calculate the density of iron. Compare your result with the observed density of 7.87 g/cm^3.

12.88 Use the data in Practice Exercise 12.11 (p. 569) to calculate the density of silver.

12.89 Lead crystallizes in a cubic lattice with a unit cell edge of 495.05 pm. The density of lead is 11.3 g/cm^3.
(a) Calculate the number of atoms per unit cell and identify the lattice type.
(b) Calculate the radius in picometers of a lead atom.

12.90 Aluminum crystallizes in a face-centered cubic lattice. The radius of an aluminum atom is 142.8 pm. Calculate (a) the unit cell edge in picometers and (b) the mass of the unit cell in grams.

12.91 (a) Draw the unit cell for zinc, a metal that crystallizes in a hexagonal close-packed lattice.
(b) How many zinc atoms are in the unit cell?
(c) What is the coordination number of each zinc atom?

12.92 (a) Draw the unit cell for argon, an element that crystallizes in a cubic close-packed lattice.
(b) How many argon atoms are in the unit cell?
(c) What is the coordination number of each argon atom?

12.93 Draw the unit cell for sodium chloride. What is the coordination number of (a) the sodium ion and (b) the chloride ion?

12.94 Titanium(I) chloride (TiCl) crystallizes in a cesium chloride lattice (Figure 12.40).
(a) Draw the unit cell of TiCl.
(b) What is the coordination number of the Ti^+ ion? The Cl^- ion?

***12.95** A strontium sample is irradiated with 0.154-nm x-rays. The first-order scattering angle is 14.7°. How far apart are the atomic layers responsible for this effect?

***12.96** The distance between layers in a KCl lattice is 318 pm. What will be the first-order scattering angle produced by these layers when a potassium chloride crystal is irradiated with 0.225-nm x-rays?

PART B. MISCELLANEOUS QUESTIONS AND PROBLEMS

12.97 Explain the following observations:
(a) Evaporating part of a liquid sample leaves the remaining liquid at a lower temperature.
(b) Evaporation of water from the skin removes excess heat and helps to maintain constant body temperature.
(c) The temperature of a tree on hot days is usually several degrees cooler than its immediate environment.

12.98 Ethyl chloride, a volatile substance with a normal boiling point of 12.3°C, has been used as a local anesthetic for minor surgery. The liquid, which is initially under pressure, is simply sprayed on tissue where an incision is to be made. Suggest how ethyl chloride might anesthetize the tissue.

12.99 Explain why perspiration cools a body more effectively on a dry day than on a damp day.

12.100 A sample of liquid water initially at 25°C and 1.00 atm is treated as follows: (1) The sample is compressed at constant temperature to a pressure greater than the critical pressure (218.3 atm); (2) The compressed sample is warmed at constant pressure to a temperature greater than the critical temperature (374.1°C); (3) The pressure of the warm sample is reduced at constant temperature to 1.00 atm.
(a) What is the final state of the water?
(b) Were any phase boundaries crossed during the above steps? (Refer to the phase diagram for water, Figure 12.10.)
(c) Devise a series of steps that would convert water vapor at 40°C and 5 torr to liquid water at 25°C and 1.00 atm without crossing a phase boundary. Discuss what happens to the water during each step.

12.101 Which gaseous fuel will be the first to liquefy in very cold weather, butane (C_4H_{10}), methane (CH_4), or propane (C_3H_8)?

12.102 The reciprocal of the viscosity is called the *fluidity*. List the substances of Table 12.6 in order of increasing fluidity.

12.103 Explain the following observations:
(a) The viscosity of blood plasma is greater than that of water.
(b) Mercury forms spherical droplets when spilled on the floor.
(c) The meniscus of water is concave, but the meniscus of mercury is convex.

12.104 Explain how a substance with cubic unit cells can form thin flat crystals.

12.105 Without referring to the text, derive Equations 12.4, 12.5, and 12.6.

***12.106** Derive the Bragg equation without referring to the text.

12.107 A 50.0-g sample of ice at 0°C and a 50.0-g sample of copper at 25°C are dropped into 250 g of liquid water at 75°C. Calculate the final equilibrium temperature. Refer to Table 12.3 for pertinent data.

12.108 Refer to the experiment described in Figure 12.12 and the Learning Hint on page 548. What will happen to the meniscus if the sealed tube initially contains more than 32% of its volume in water? Less than 32% of its volume?

12.109 Diamond crystallizes as a cubic unit cell with an edge length of 356.7 pm. The density of diamond is 3.513 g/cm³ at 25°C.
(a) How many carbon atoms are in the unit cell of diamond?

(b) Draw the unit cell, keeping in mind that each carbon atom forms four covalent bonds.

(c) Is the unit cell one of those listed in Table 12.8?

12.110 Refer to Figure 12.37 and calculate the percentage of occupied space in a hexagonal close-packed lattice.

12.111 X-ray diffraction measurements show that the centers of the Na^+ and Cl^- ions in a sodium chloride crystal are 282 pm apart. Refer to Figure 12.39 and estimate the density of NaCl. Compare your result with the observed density of 2.17 g/cm^3.

12.112 Zinc sulfide (ZnS) often crystallizes in the *zinc blende* structure. The unit cell of zinc blende (Figure 12.46) is a face-centered cube of sulfide ions with four zinc ions inside it.

(a) Show that there are equal numbers of Zn^{2+} and S^{2-} ions in the unit cell.

(b) What is the coordination number of each Zn^{2+} ion? Each S^{2-} ion?

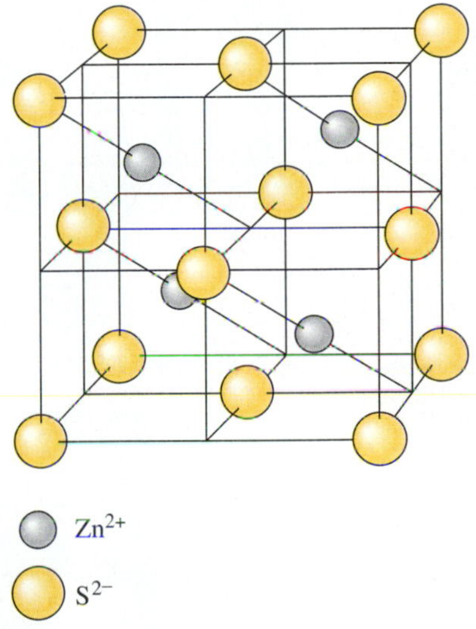

Figure 12.46
Zinc blende (ZnS) unit cell.

12.113 The density of calcium oxide (CaO) is 3.35 g/cm^3. Diffraction data shows that calcium oxide has a cubic unit cell whose edge is 480 pm. What is the lattice type? (*Hint*: Determine how many CaO formula units are contained in one unit cell.)

***12.114** The distance between layers in a graphite lattice is 335 pm.

(a) What will be the first- and second-order scattering angles produced by these layers when graphite is irradiated with 0.154-nm x-rays?

(b) What will be the angles if 0.225-nm x-rays are used?

CHAPTER 13

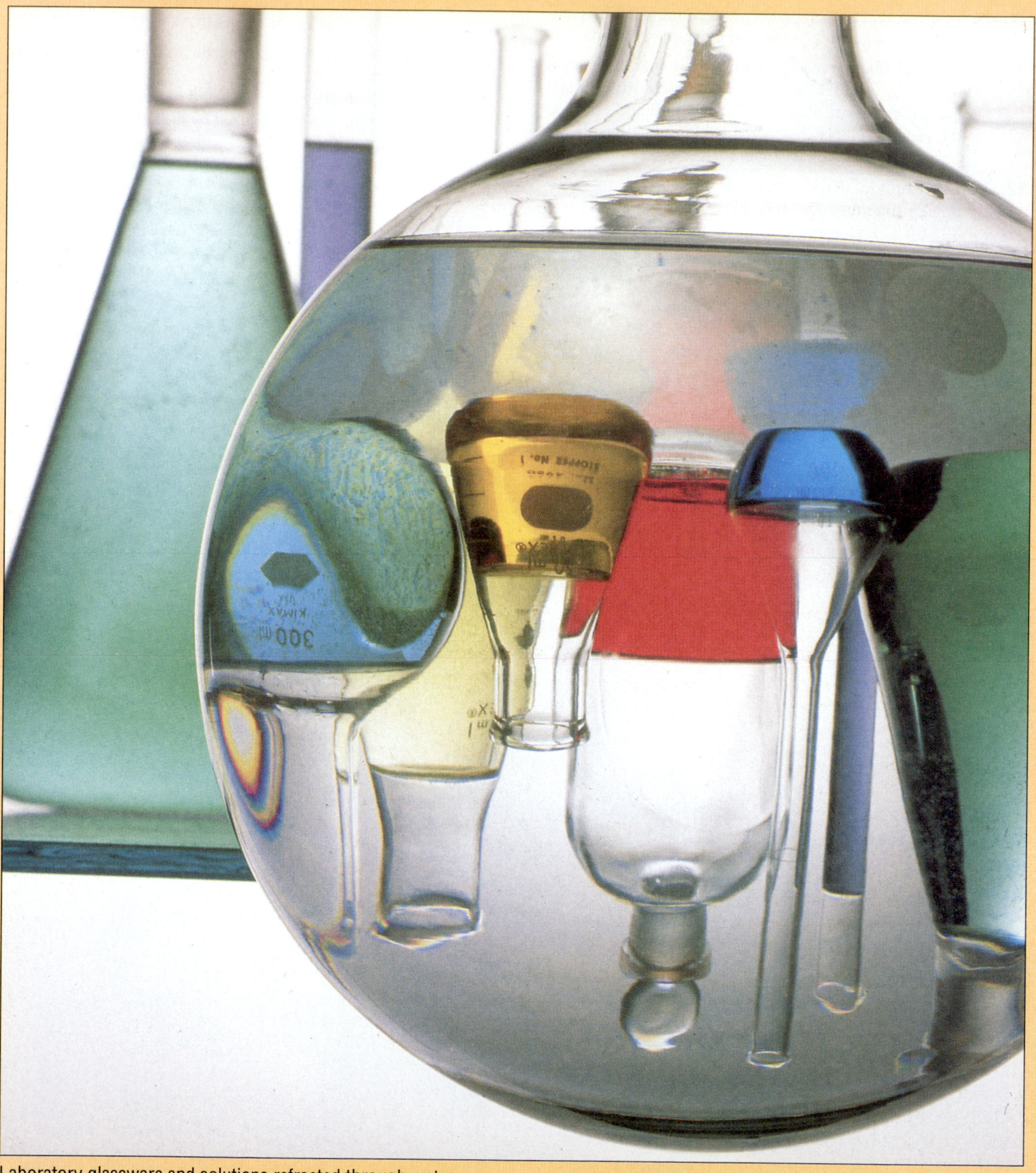

Laboratory glassware and solutions refracted through water.

SOLUTIONS

OUTLINE

PREVIEW

The importance of solutions cannot be overstated; they transport dissolved substances, and they provide an environment in which dissolved molecules can meet and react. Living organisms use solutions to carry nutrients and oxygen to their cells and to carry away waste products. Oceans and other natural waters contain untold numbers of dissolved substances, and much of the chemical work of the earth's crust and biosphere is done in these natural solutions or at their boundaries.

The mixtures known as colloids are also important. Their dispersed particles are somewhat larger than those in solutions but still small enough to have unique properties not exhibited by coarser aggregates. Colloids include such diverse substances as milk, clouds, germs, and smoke. Protoplasm, the basic material of all living cells, is also colloidal.

A **solution** is a homogeneous mixture of two or more substances. In this chapter we will examine solutions on the molecular level, with particular emphasis on how solutions form and on how their properties vary with composition. Our primary interest will be liquid solutions that consist of one or more **solutes** dissolved in some **solvent**.

LEARNING HINT

Review solution terminology (Section 4.1) and concentration units such as molarity (Section 4.4) and mole fraction (Section 5.6).

13.1 SOLUBILITY

When one substance dissolves in another, there is usually a limit to the amount that will dissolve. For example, only 35.7 g of sodium chloride will dissolve in 100 mL of water at 0°C. The **solubility** of a substance is the amount of the substance that will saturate a given amount of solvent at a given temperature. Solubilities vary widely: large quantities of sugar will dissolve in water, aspirin is much less soluble, and oil barely mixes with water at all. In this section we study some of the factors that influence solubility and solution formation.

Solubility and Intermolecular Forces

Moving particles tend to mix with each other. A drop of ink or dye spreads through water, and natural gas mixes rapidly with air when the pilot light on a stove goes out. Taken alone, the tendency to mix would cause all substances to dissolve completely in each other. But intermolecular attractions and bonding forces must also be considered. Strong attractive forces between solute particles tend to keep the particles together and reduce their solubility. On the other hand, attractive forces between solute and solvent molecules make dissolving easier and help to keep particles in solution. The solubility of one substance in another depends to a great extent on these forces and to a much lesser extent on conditions such as temperature and pressure.

The effectiveness of anesthetic gases, such as dinitrogen oxide and diethyl ether, depends on their ability to dissolve in fatty tissue.

Substances will dissolve in each other readily when the intermolecular forces in the solution are similar to those in the pure components. This fact is reflected in practical sayings such as "like dissolves like" and "oil and water don't mix." Nonpolar substances such as benzene, turpentine, oils, and grease mix well with each other, but they do not dissolve strongly polar or ionic substances such as water or sodium chloride. A polar or ionic solute needs a polar solvent.

Water is the principal solvent on our planet, and it is certainly the solvent used most frequently by the beginning chemist. Water molecules are strongly associated with each other through hydrogen bonding (Section 12.3), and compounds that form hydrogen bonds tend to be soluble in water. Small hydrogen-bonded molecules such as methanol (CH_3OH) and ethanol (C_2H_5OH) are completely miscible with water (Figure 13.1a). Compounds such as glucose (Figure 13.1b), which consist of larger organic molecules with many OH groups, are also very soluble.

Water is often referred to as an *ionizing solvent* because of its ability to stabilize ions in solution. Water dissolves ionic compounds by orienting its polar molecules around each ion, as illustrated in Figure 13.2. The surrounding layers of water, called *hydration layers,* partially shield the ions from each other, helping them break away from the crystal lattice and preventing them from coming together once they

Figure 13.1

(a) *Upper:* Hydrogen bonding in methanol. *Lower:* Hydrogen bonding between methanol and water. (b) A hydrogen bond between glucose and water. The glucose molecule has five OH groups that can form hydrogen bonds.

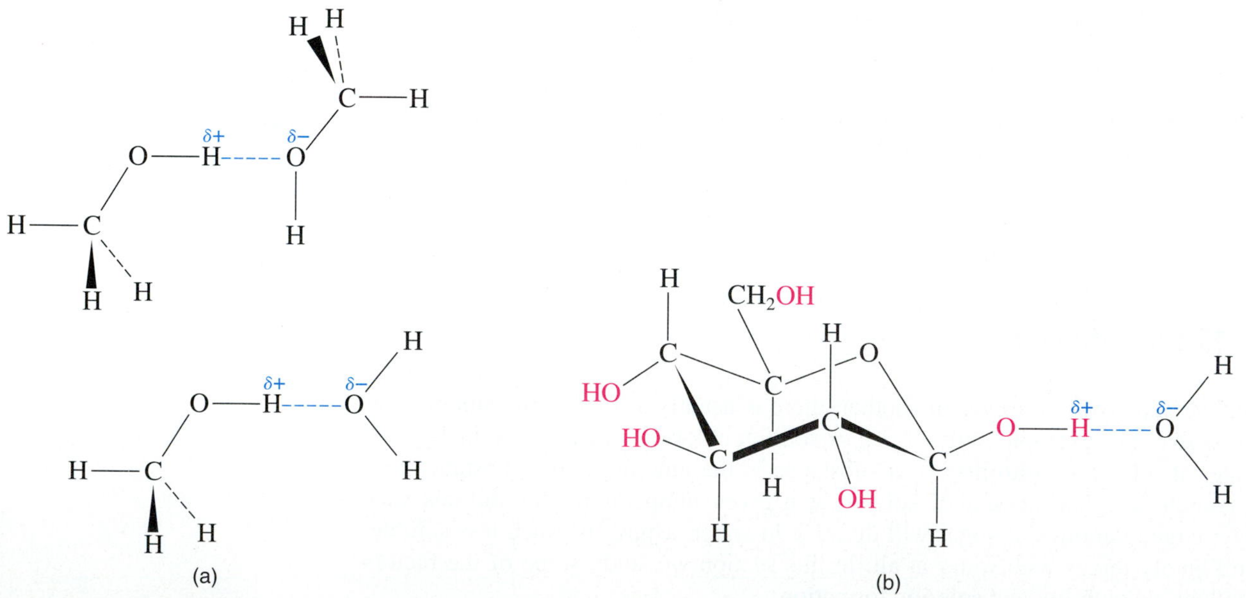

(a)

(b)

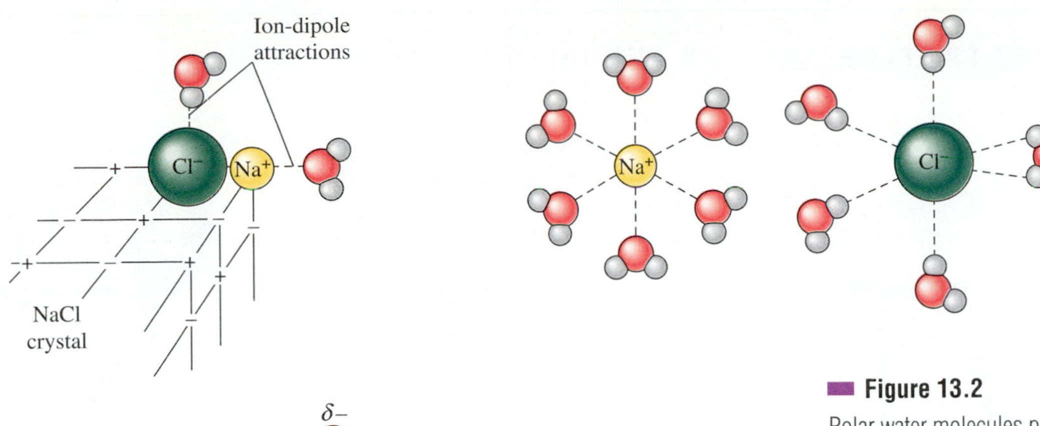

Figure 13.2
Polar water molecules promote the dissolving of ionic compounds by providing ion–dipole attractions to replace ion–ion attractions in the crystal. The water molecules surround the dissolved ions, thus shielding them and partially reducing their attractions for each other.

are in solution. The attraction between an ion and a polar molecule, called an **ion–dipole attraction**, is strongest when the ionic charge is high and the ionic radius is small. Small, highly charged positive ions often hold water molecules so firmly that they form hydrated species of definite composition, such as $Mg(H_2O)_6{}^{2+}$ and $Al(H_2O)_6{}^{3+}$.

Some gases react with water to form hydrated ions. Hydrogen chloride, for example, ionizes completely into $H^+(aq)$ and $Cl^-(aq)$ ions:

$$HCl(g) \rightarrow H^+(aq) + Cl^-(aq)$$

The hydration layers around the positive and negative ions inhibit their recombination. The solubility of hydrogen chloride in water is very high; commercially available concentrated hydrochloric acid contains about 12 mol of HCl per liter. Ammonia, carbon dioxide, and chlorine gases also react with water to form equilibrium mixtures of molecules and ions (see Demonstration 13.1):

$$NH_3(aq) + H_2O(l) \rightleftharpoons NH_4{}^+(aq) + OH^-(aq)$$

$$CO_2(g) + H_2O(l) \rightleftharpoons H^+(aq) + HCO_3{}^-(aq)$$

$$Cl_2(aq) + H_2O(l) \rightleftharpoons H^+(aq) + Cl^-(aq) + HOCl(aq)$$

Gases that do not react with water are not very soluble; H_2, N_2, O_2, and CH_4 (methane) are examples.

DIGGING DEEPER

Salting-In and Salting-out

The solubility of a sparingly soluble ionic compound in water is increased by the presence of other salts that do not have an ion in common with the compound. Silver chloride (AgCl), for example, is more soluble in potassium nitrate solution than in pure water. The K^+ ions tend to surround the negative chloride ions, while the $NO_3{}^-$ ions surround the positive silver ions. In effect, each ion is surrounded by an "ionic

DEMONSTRATION 13.1 AN AMMONIA FOUNTAIN

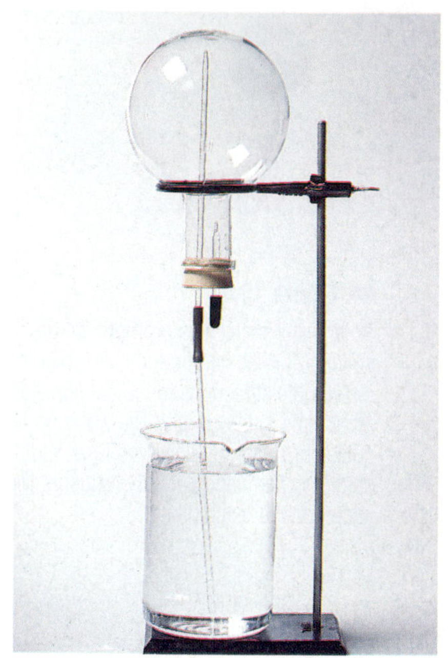

The inverted flask contains dry ammonia gas. The long tube leads into water that contains phenolphthalein, an indicator that turns red in basic solution. The medicine dropper contains a small amount of water.

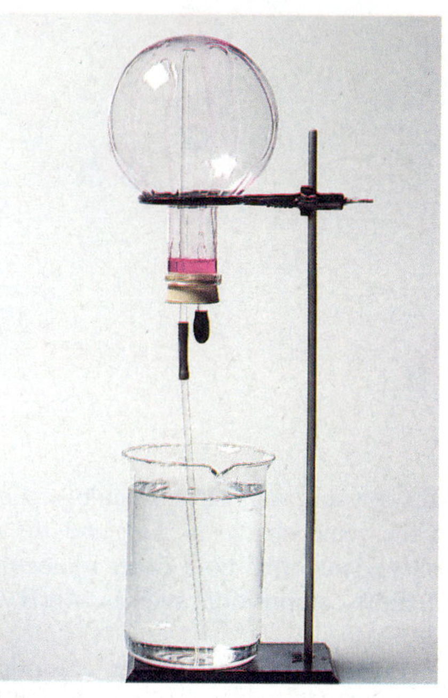

Ammonia is very soluble in water; a squirt of water from the dropper dissolves the ammonia in the flask. Water rushes up to fill the resulting partial vacuum.

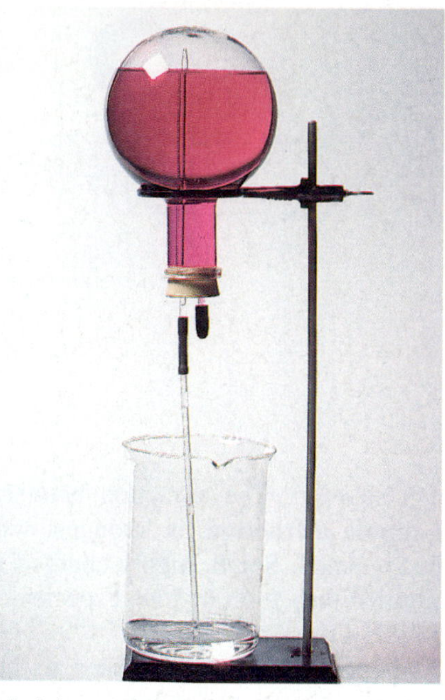

All of the ammonia has dissolved and only a little air remains.

atmosphere" of the opposite charge. The ionic atmosphere helps bring the silver and chloride ions into solution, thus increasing the solubility of silver chloride. Adding one salt to enhance the solubility of another is called **salting-in**.

An opposite effect, called **salting-out**, occurs with solutes of low polarity. It has been observed, for example, that oxygen, nitrogen, and other gases that do not react with water are less soluble in salt water than in pure water. When oxygen gas comes in contact with pure water, some of the oxygen molecules become hydrated and dissolve. In a solution of sodium chloride, however, the polar water molecules preferentially cluster around the dissolved sodium and chloride ions. Fewer water molecules are available to hydrate oxygen molecules, so less oxygen enters the solution.

Heats of Solution

The formation of a solution is often accompanied by a temperature change. When ammonium chloride (NH_4Cl) dissolves in water, heat is absorbed from the surroundings and the container becomes cold to the touch. Sulfuric acid, on the other hand,

CHEMICAL INSIGHT

SUPERCRITICAL FLUIDS—SOLVENTS FOR THE FUTURE

Do oil and water mix? Most people will say "No," but they are wrong. Oil and water do mix when the water is a supercritical fluid near its critical temperature and pressure (Figure 12.10, p. 544). In this state, water is miscible in all proportions with hydrogen and oxygen, with methane and other alkanes, and it dissolves many substances that would be only sparingly soluble under more familiar conditions.

How can mere water become such a different solvent from the one we know? Under familiar conditions, liquid water has a density close to 1 g/mL. The density of supercritical water depends on its temperature and pressure, and can vary from very low (gaslike) values to about 2 g/mL. (At the critical point, 374°C and 218 atm, the density of water is 0.32 g/mL.) The ability of supercritical water to form solvation layers around molecules and ions increases with increasing density. Hence, the solubility of a substance in supercritical water can be controlled by adjusting the temperature and pressure of the mixture. Given the proper conditions, supercritical water will dissolve polar and nonpolar substances as well as ionic compounds.

Supercritical water is already employed in some industrial processes (Figure 13.3), and further potential uses are being actively explored. For example, organic compounds, when dissolved with oxygen or hydrogen peroxide in supercritical water, are smoothly and completely oxidized to water and carbon dioxide. Hazardous wastes, sewage sludge, and paper mill waste have been converted to harmless waste products in prototype *supercritical water reactors* without the residues of soot and tar that often accompany conventional incineration. These reactors operate at substantially lower temperatures than incinerators, so there is a fuel savings as well. Oxidation in supercritical water is so efficient that the U.S. Department of Defense is planning a pilot plant to use this process for the destruction of chemical weapons, explosives, and obsolete rocket fuels.

Because the critical point of carbon dioxide (31°C, 72.9 atm) is fairly easy to reach, supercritical CO_2 is already widely used as a medium for *supercritical fluid extraction.* One example is in the manufacture of decaffeinated coffee; supercritical CO_2 near its critical point will penetrate the coffee beans and dissolve out (extract) the caffeine, while leaving behind the substances that make coffee enjoyable. Techniques are also being explored that use supercritical CO_2 to extract cholesterol from milk, to extract fat from meat, and to deodorize cooking oil.

Figure 13.3 These ultrapure quartz bricks, grown from silicon dioxide dissolved in supercritical water, will be cut with diamond saws and polished into wafers.

gives off heat when it dissolves, and produces a rise in temperature. The **heat of solution** ($\Delta H^\circ_{\text{solution}}$) is the enthalpy change that accompanies the dissolving of 1 mol of solute in a given solvent. The value of $\Delta H^\circ_{\text{solution}}$ is positive when heat is absorbed from the surroundings (endothermic mixing) and negative when heat is evolved (exothermic mixing).

A heat of solution depends on concentration. For example, 1 mol of sulfuric acid dissolving in 100 g of water evolves 73.6 kJ of heat, but the same amount of sulfuric acid dissolving in 1000 g of water evolves 78.4 kJ. A typical heat of solution becomes more negative (more heat is evolved) with increasing dilution, but it eventually levels off and approaches a constant value as the solution becomes more and

TABLE 13.1 Molar Heats of Solution at Infinite Dilution

Substance[a]	$\Delta H^\circ_{solution}$ (kJ/mol)[b]
$AlCl_3(s)$	−373.63
$CO_2(g)$	− 19.41
HCl(g)	− 74.84
HF(g)	− 51.5
$HNO_3(l)$	− 33.28
$H_2SO_4(l)$	− 95.28
KCl(s)	17.22
$KNO_3(s)$	34.89
LiCl(s)	− 37.03
$LiNO_3(s)$	− 2.51
NaCl(s)	3.88
$NaNO_3(s)$	20.50
NaOH(s)	− 44.51
$NH_3(g)$	− 30.50
$NH_4Cl(s)$	14.77

[a]The initial state of each substance is indicated in parentheses after the formula.

[b]In water at 25°C. A negative sign signifies an exothermic mixing process.

more dilute. This limiting value is called the heat of solution at "infinite dilution." For H_2SO_4, the heat of solution at infinite dilution is −95.28 kJ/mol. Some heats of solution are given in Table 13.1.

Heats of solution can be better understood if we imagine the solution process

$$\text{pure solvent} + \text{pure solute} \longrightarrow \text{solution}$$

to be a composite of three steps (Figure 13.4):

1. *Solute particles separate from each other.* The separation of solute particles is *endothermic* because energy is required to overcome the forces holding the solute particles together. This step can be ignored for gases because their molecules are already widely separated.
2. *Some of the solvent molecules move apart to make room for the incoming solute particles.* This step is also *endothermic* because energy is required to overcome the forces holding the solvent molecules together. The amount of energy required for the separation of solvent molecules, however, is usually small compared to the energies of the other two steps.
3. *Solvent molecules surround the solute particles.* Each solute particle is surrounded by a layer or "cage" of solvent molecules that is attracted to it by intermolecular forces. This step, which is *exothermic*, is referred to as **solvation**, and the energy released is called the **solvation energy**. The terms **hydration** and **hydration energy** are often used when water is the solvent.

The heat of solution, $\Delta H^\circ_{solution}$, is the sum of the enthalpies associated with these three steps, two endothermic (positive ΔH°) and one exothermic (negative ΔH°). Figure 13.5 illustrates how the heat of solution can be positive or negative, depending on the magnitude of these terms. The heat of solution of a gas is usually negative because Step 1 can be ignored and the energy absorbed in Step 2 is negligible compared to that released in Step 3.

Temperature and Solubility

Most solids are more soluble in hot water than in cold (Figure 13.6). The solubility of sucrose, for example, increases greatly with temperature. The solubility of sodium chloride also increases, but only slightly. A few solids (e.g., calcium hydroxide, cerium sulfate, and some lithium salts) become less soluble at higher

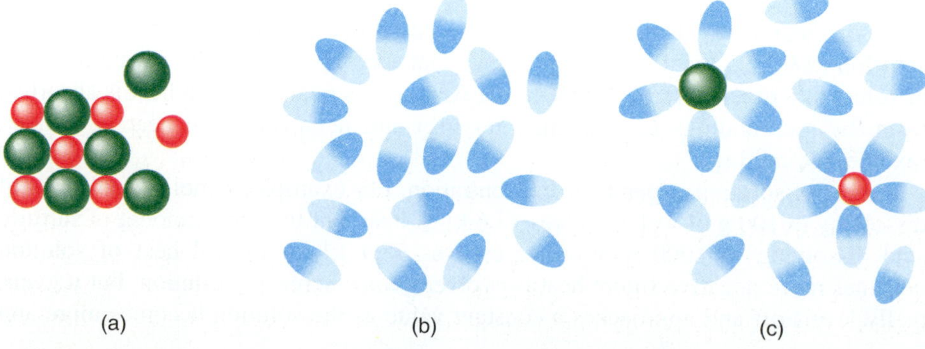

Figure 13.4
Solution formation can be visualized in terms of three steps. (a) Solute particles separate from each other. (b) Solvent molecules move apart to make room for the solute. (c) Solvent molecules surround the solute particles. Steps (a) and (b) absorb energy; step (c) releases energy.

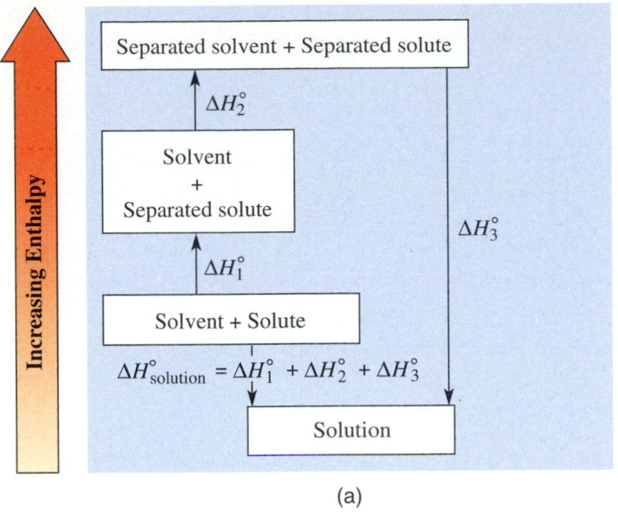

(a)

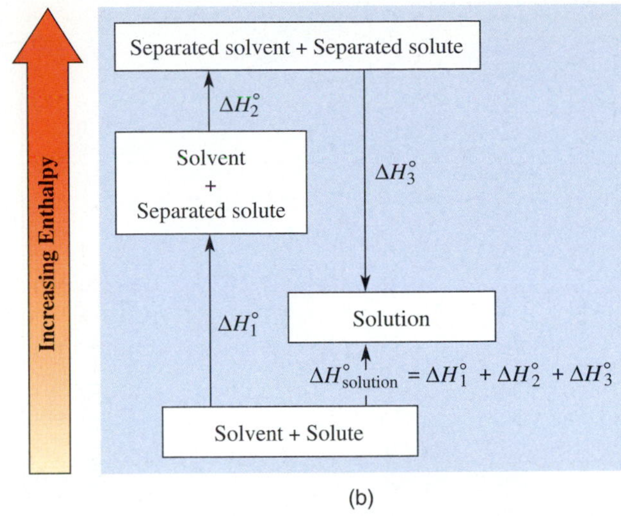

(b)

Figure 13.5

Enthalpy diagrams for the solution process. (a) Starting with "solvent + solute", the total heat absorbed in Steps 1 and 2 is less than the heat evolved in the solvation Step 3, so solution formation is exothermic. (b) The total heat absorbed in Steps 1 and 2 is greater than the heat evolved in Step 3, so solution formation is endothermic.

temperatures. To understand why temperature affects solubility, we must consider the enthalpy change that occurs when a small amount of the solid enters a saturated or nearly saturated solution. If this enthalpy change is positive, the solid will absorb heat when it enters the solution. More heat is available for absorption at higher temperatures, so more of the solid will dissolve, and its solubility will increase with temperature. On the other hand, if the solid gives off heat when it enters the solution, it will then absorb heat when it precipitates. Precipitation will be favored by high temperatures, and the solubility of the solid will decrease with temperature. *Note carefully that the heat of solution at the saturation point (not the overall heat of solution) determines the temperature effect.* When NaOH dissolves in nearly pure water, the solution process is strongly exothermic. When it dissolves in an almost saturated solution, however, the process is endothermic. The solubility of NaOH increases with temperature even though the overall process is exothermic.

The solubility in water of nearly all gases, unlike that of most solids, decreases with increasing temperature. (In other solvents, gas solubilities often increase with temperature.) Recall from the preceding section that gases usually give off heat when they dissolve in water. Hence, an increase in temperature will favor the reverse process, which absorbs heat and tends to drive the gas out of solution. An open bottle of soda left standing at room temperature will retain less carbon dioxide than an open bottle stored in the refrigerator.

Gas Solubility and Partial Pressure

The solubility of a liquid or solid does not change very much with pressure. The solubility of a gas, on the other hand, increases with the partial pressure of the gas above the solution surface. This effect has long been used in the beverage industry, where carbonated beverages such as soda and beer are bottled under carbon dioxide pressures of up to 4 atm. It is also used in oxygen therapy for carbon monoxide poisoning, for some disorders of the circulatory system, and for certain infections such as gas gangrene; the use of high-pressure oxygen chambers (hyperbaric chambers)

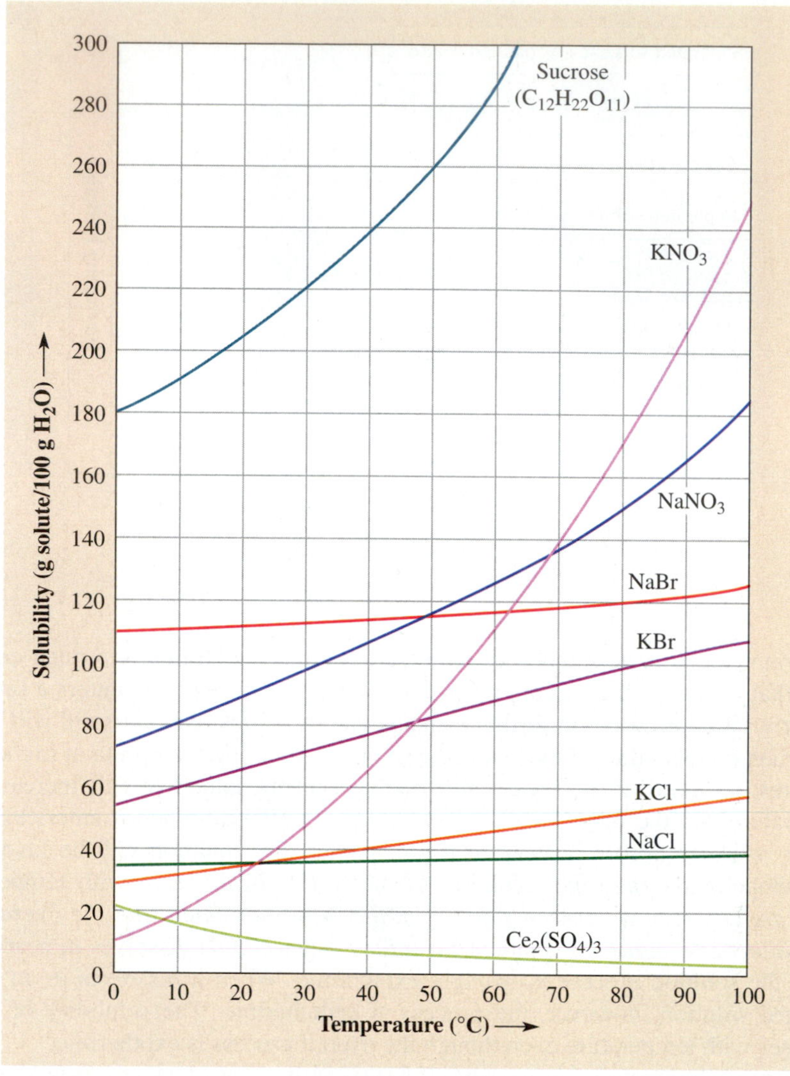

Figure 13.6

Solubility versus temperature plots for a number of solutes in water.

to treat these and other conditions is based on the increased solubility of oxygen in blood at elevated oxygen pressures.

We can better understand the effect of pressure on gas solubility by looking at Figure 13.7. When a solution is saturated with gas, the rate at which gas molecules enter the solution equals the rate at which they leave. An increase in the gas pressure above the solution disturbs the equilibrium by causing more gas molecules to collide with the surface each second. The solubility of the gas increases because the solution absorbs some of these molecules as it settles into a new equilibrium. William Henry, an English chemist, observed in 1803 that *the solubility of a sparingly soluble gas is directly proportional to the pressure of the gas above the solution*. This observation is now known as **Henry's law**. Very soluble gases such as ammonia and hydrogen chloride also become more soluble as their partial pressures increase, but not in a way that can be predicted from Henry's law.

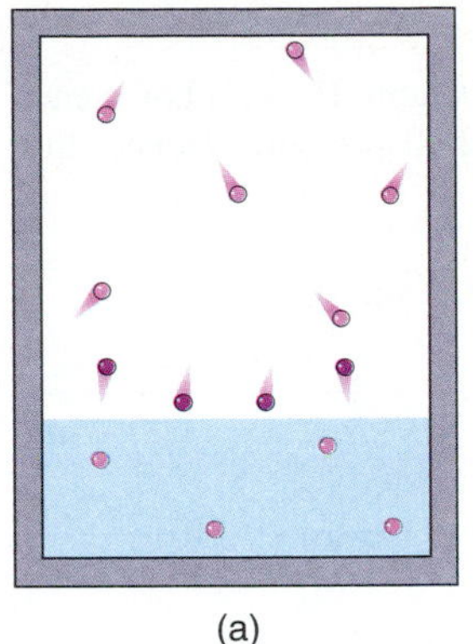
(a)

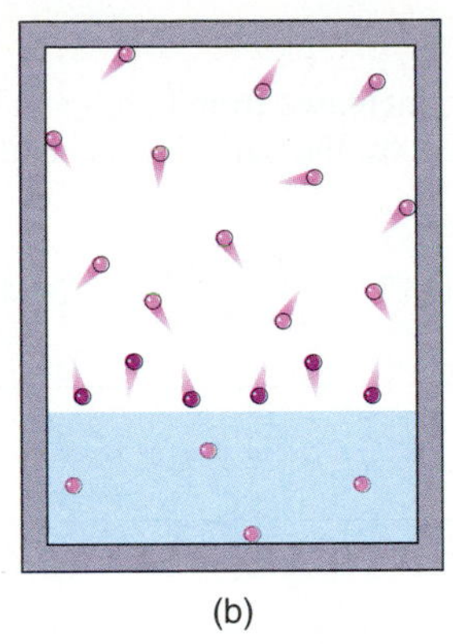
(b)

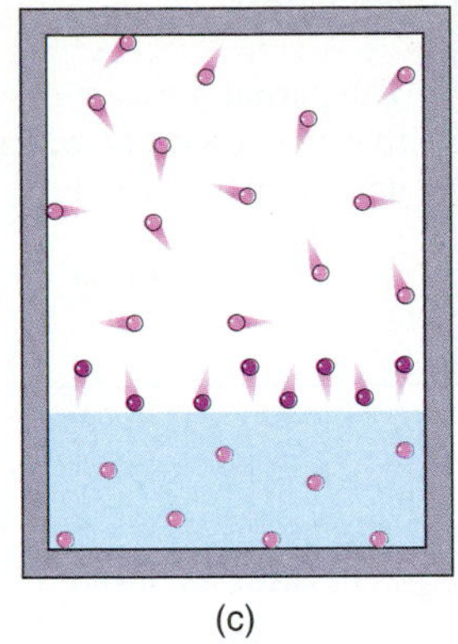
(c)

Figure 13.7
The solubility of a gas increases with the partial pressure of the gas above the solution. (a) The solution is in equilibrium with undissolved gas; equal numbers of molecules enter and leave the solution. (b) An increase in partial pressure increases the number of gas molecules entering the solution, so the system is temporarily out of equilibrium. (c) The concentration of dissolved molecules increases until a new equilibrium is established.

CHEMICAL INSIGHT

DECOMPRESSION ILLNESS

A medically significant effect of increased gas solubility under pressure is the painful and occasionally fatal condition known as *decompression illness*, or "the bends," which is caused by a sudden reduction in atmospheric pressure. Symptoms can appear, for example, after a rapid ascent in a nonpressurized airplane to a height above 25,000 feet, or after a sudden return to a normal atmosphere from the compressed atmosphere used by divers. Some of the nitrogen in air dissolves in body fluids and a sudden reduction in pressure makes the nitrogen less soluble, causing bubbles of the gas to form in the tissues, bloodstream, and lung capillaries. These bubbles exert pressure, clog small blood vessels, and result in symptoms ranging from mild itching to total paralysis.

Decompression illness can be prevented by allowing decompression to occur slowly over a longer period of time in a decompression chamber. Gradual decompression allows the dissolved nitrogen to escape from the body fluids gradually with no bubble formation. Decompression illness can also be minimized by using a breathing mixture in which helium has been substituted for nitrogen. Helium is less soluble than nitrogen, so less gas will be evolved from body fluids during decompression.

Technicians operate a decompression chamber.

EXAMPLE 13.1

The solubility of oxygen in water at 25°C and 0.968 atm is 1.23×10^{-3} mol O_2 per liter of solution. Estimate the molar solubility if the partial pressure of oxygen above the solution is increased to 2.00 atm with no change in temperature.

SOLUTION

The partial pressure of oxygen increases from 0.968 atm to 2.00 atm. Henry's law states that the solubility will increase by the same factor as the partial pressure. Hence, the new solubility will be

$$1.23 \times 10^{-3} \text{ mol/L} \times \frac{2.00 \text{ atm}}{0.968 \text{ atm}} = 2.54 \times 10^{-3} \text{ mol/L}$$

PRACTICE EXERCISE 13.1

The solubility of nitrogen in water is 1.05×10^{-4} mol/L at 0.993 atm and 0°C. Estimate the solubility of nitrogen at 0°C if its partial pressure above a solution is 0.100 atm.

13.2 IDEAL SOLUTIONS AND REAL SOLUTIONS

The properties of a solution reflect to some extent the properties of its individual components. Sugar water, for example, has the liquidity of water, the sweet taste of sugar, and a density intermediate between the two. It is not possible, however, to make exact predictions about the properties of a solution because the components interact in various ways, and each molecule undergoes subtle changes in behavior when it dissolves and becomes surrounded by other molecules. A component may even alter its nature completely; covalent HCl molecules, for example, are converted to $H^+(aq)$ and $Cl^-(aq)$ ions in aqueous solution.

Ideal Solutions and Raoult's Law

An **ideal solution** is an imaginary solution in which the component molecules are subject to forces that are identical to those they would experience in the pure state. The ideal solution, like the ideal gas, does not exist in nature. Nevertheless, the concept of an ideal solution is useful because the properties of such a solution can be predicted from the properties of its components. Real solutions are not ideal, but to the extent that real solutions resemble ideal solutions, their properties can be predicted as well.

The different molecules in an ideal solution attract each other neither more nor less than they attract their own kind. The molecules would not pull closer together nor spread out more than they would in the pure state, and the ease with which they move about and evaporate would be unchanged. Therefore, two properties of an ideal solution would be:

1. The volume of an ideal solution is the sum of its pure component volumes; there is no expansion or contraction on mixing.
2. The heat of solution is zero; mixing is neither exothermic nor endothermic.

The vapor pressure above an ideal solution can be predicted from the vapor pressures of its pure components. Figure 13.8a depicts a pure liquid A in equilibrium

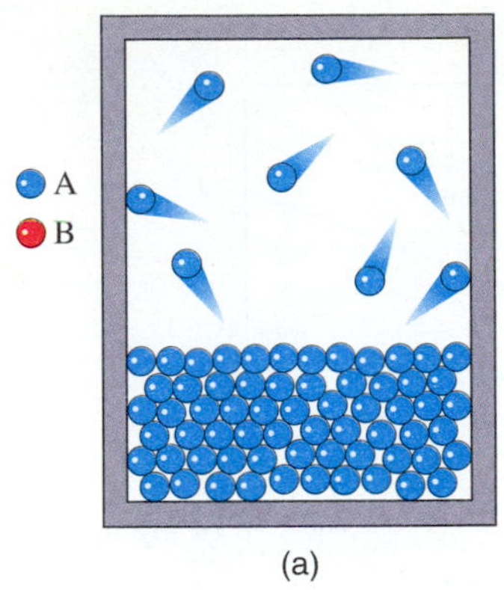

(a)

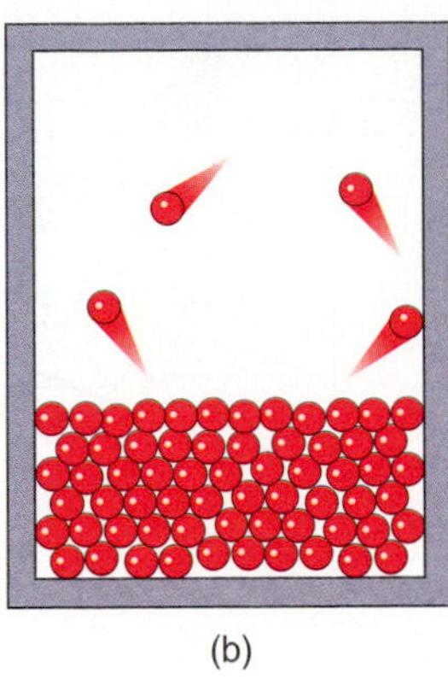

(b)

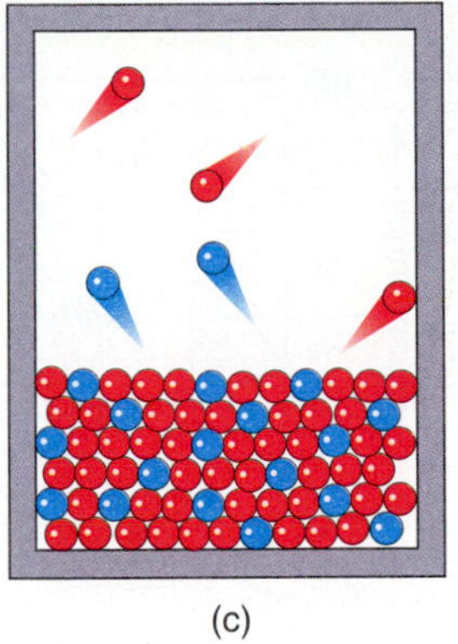

(c)

Figure 13.8

Raoult's law. (a) Pure liquid A in equilibrium with its vapor; the vapor pressure is P_A°. (b) Pure liquid B in equilibrium with its vapor; the vapor pressure is P_B° (B is less volatile than A). (c) An ideal solution in which one-fourth of the molecules are A's ($X_A = \frac{1}{4}$; $X_B = \frac{3}{4}$). The vapor pressure of A is $\frac{1}{4}P_A^\circ$; the vapor pressure of B is $\frac{3}{4}P_B^\circ$.

with its vapor. At the equilibrium vapor pressure, which we will call P_A°, the rate at which A's molecules escape from the surface is exactly equal to their rate of return. Figure 13.8b depicts a less volatile liquid B with an equilibrium vapor pressure P_B°. Now consider the ideal solution in Figure 13.8c, in which the mole fraction of A is $\frac{1}{4}$ and the mole fraction of B is $\frac{3}{4}$; in other words, one-fourth of the molecules are A and three-fourths are B. Because the solution is ideal, each molecule behaves exactly as it does in the pure state, and any difference in properties is due entirely to the difference in the number of molecules. The surface of the solution contains one-fourth as many A molecules as are present in pure A. Hence, the rate of escape of A molecules will be one-fourth of the rate in pure A, and the vapor pressure of A above the solution, P_A, will be one-fourth of P_A°. Similarly, the vapor pressure of component B above the solution will be three-fourths of P_B°, where P_B° is the vapor pressure above pure B. Since the mole fraction is equal to the molecule fraction, we can write

Recall (Section 5.6) that the mole fraction of a component in a solution is the number of moles of the component divided by the total number of moles.

$$P = XP^\circ \tag{13.1}$$

where P is the vapor pressure of a component above the ideal solution, X is the mole fraction of the component in the solution, and P° is the vapor pressure of the component in the pure state. Equation 13.1 states that *the vapor pressure of a component above an ideal solution is proportional to the mole fraction of the component in the solution.* This relationship was discovered in 1886 by the French chemist François-Marie Raoult and is known as **Raoult's law**. Graphs of Raoult's law for each component of a two-component solution are given in Figure 13.9.

Two additional properties of ideal solutions can now be stated:

3. Each component of an ideal solution obeys Raoult's law (Equation 13.1).

4. The vapor pressure above the solution is also ideal, so the total vapor pressure is given by Dalton's law of partial pressures (Equation 5.11, p. 195); that is, the total vapor pressure is equal to the sum of the individual vapor pressures:

$$P_{\text{total}} = P_A + P_B \tag{13.2}$$

where P_A and P_B are the individual vapor pressures of components A and B, respectively. Substituting Equation 13.1 into Equation 13.2 gives

$$\begin{aligned} P_{\text{total}} &= P_A + P_B \\ &= X_A P_A^\circ + X_B P_B^\circ \end{aligned} \tag{13.3}$$

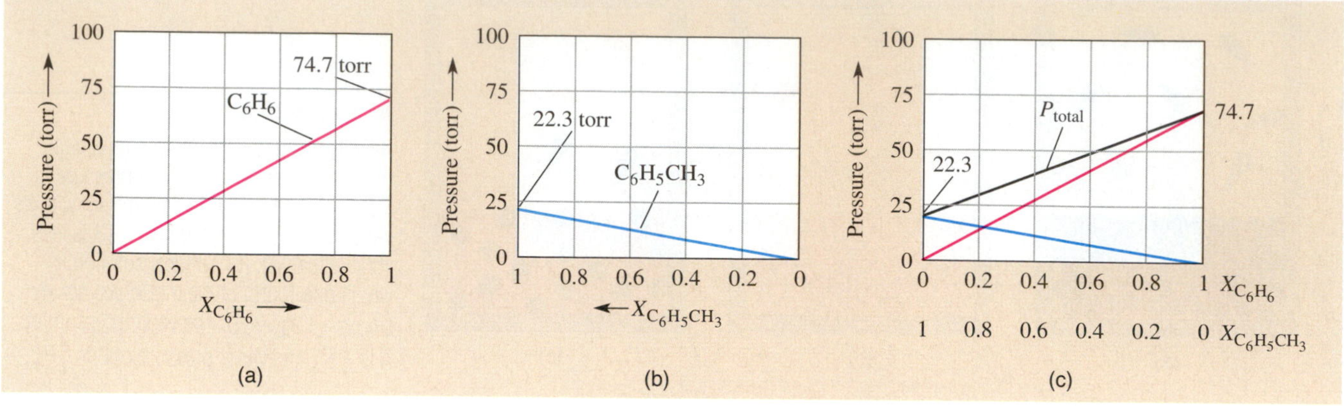

Figure 13.9

Vapor pressure versus composition for solutions of benzene (C_6H_6) and toluene ($C_6H_5CH_3$) at 20°C. (a) Vapor pressure of benzene versus mole fraction of benzene. (b) Vapor pressure of toluene versus mole fraction of toluene. (c) The benzene and toluene plots of (a) and (b) are superimposed. The total vapor pressure at any mole fraction, P_{total}, is equal to the sum of the individual vapor pressures.

A graph of Equation 13.3 is given in Figure 13.9c.

EXAMPLE 13.2

A nearly ideal solution forms when 1.00 mol of benzene (C_6H_6) is mixed with 3.00 mol of toluene ($C_6H_5CH_3$). At 20°C, the vapor pressure of pure benzene is 74.7 torr, while that of pure toluene is 22.3 torr. Estimate the partial pressure of each component and the total vapor pressure above the solution.

SOLUTION

The mole fractions are $X_{benzene} = 1.00/4.00 = 0.250$ and $X_{toluene} = 3.00/4.00 = 0.750$. The vapor pressures of the pure liquids are $P^\circ_{benzene} = 74.7$ torr and $P^\circ_{toluene} = 22.3$ torr. The partial pressure of each component is obtained from Equation 13.1:

$$P_{benzene} = X_{benzene}\, P^\circ_{benzene}$$

$$= 0.250 \times 74.7 \text{ torr} = 18.7 \text{ torr}$$

$$P_{toluene} = X_{toluene} P^\circ_{toluene}$$

$$= 0.750 \times 22.3 \text{ torr} = 16.7 \text{ torr}$$

The total pressure is obtained by adding the two partial pressures (Equation 13.2):

$$P_{total} = P_{benzene} + P_{toluene}$$

$$= 18.7 \text{ torr} + 16.7 \text{ torr} = 35.4 \text{ torr}$$

PRACTICE EXERCISE 13.2

At 20°C, the vapor pressures of pure methanol (CH_3OH) and pure ethanol (C_2H_5OH) are 88.7 torr and 44.5 torr, respectively. An almost ideal solution is prepared by mixing 1.00 mol of methanol with 2.00 mol of ethanol. Estimate the partial pressure of each component and the total vapor pressure above the solution.

Real Solutions

Solutions tend to be nearly ideal when the solute and solvent molecules are similar in size and polarity. Mixtures of hexane (C_6H_{14}) and heptane (C_7H_{16}) are almost ideal, as are mixtures of ethyl bromide (C_2H_5Br) and ethyl iodide (C_2H_5I). The forces experienced by molecules in each of these solutions are very similar to the forces experienced by the molecules in the pure state. Deviations from ideal behavior occur when the component molecules experience forces different from those in the pure state.

All **real solutions** deviate to some extent from Raoult's law. If, for example, the attractive force between the molecules of components A and B is greater than the A—A force or the B—B force, each molecule will be less likely to escape from the solution than from the pure substance. Thus the vapor pressure will be lower than the ideal value predicted by Equation 13.1, and the graph of vapor pressure versus mole fraction will show a **negative deviation** from Raoult's law, as illustrated in Figure 13.10 for solutions of acetone (CH_3COCH_3) and chloroform ($CHCl_3$). In these solutions, the deviations are caused by strong dipole–dipole interactions between acetone and chloroform molecules (Figure 13.10b). These bond-forming interactions give off energy; hence, negative heats of solution often accompany negative deviations from Raoult's Law.

When the attractive force between the component molecules is less than the A—A force or the B—B force, there will be a **positive deviation** from Raoult's law. This type of behavior is illustrated in Figure 13.11 for solutions of methanol (methyl alcohol, CH_3OH) and carbon tetrachloride (CCl_4). In pure methanol, the molecules are extensively hydrogen-bonded (Figure 13.1a on page 588). The addition of nonpolar carbon tetrachloride molecules disrupts many of the hydrogen bonds and does not replace them with other strong attractions. The net loss of attractive force means that the molecules are more likely to evaporate from the solution than from the pure liquids, and the vapor pressure above the solution will be higher than the ideal vapor pressure. Breaking bonds is an endothermic process, so positive heats of solution often accompany positive deviations from Raoult's law.

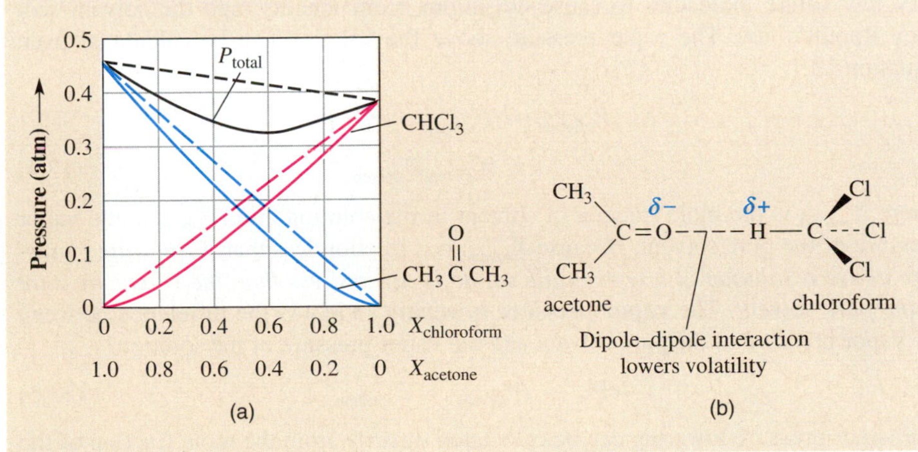

Figure 13.10

(a) Solutions of chloroform and acetone show negative deviations from Raoult's law. The solid curves are observed vapor pressures; the dashed lines are ideal (Raoult's law) vapor pressures. (b) The negative deviations are caused by dipole–dipole interactions that lower the tendency of the dissolved molecules to escape from the solution.

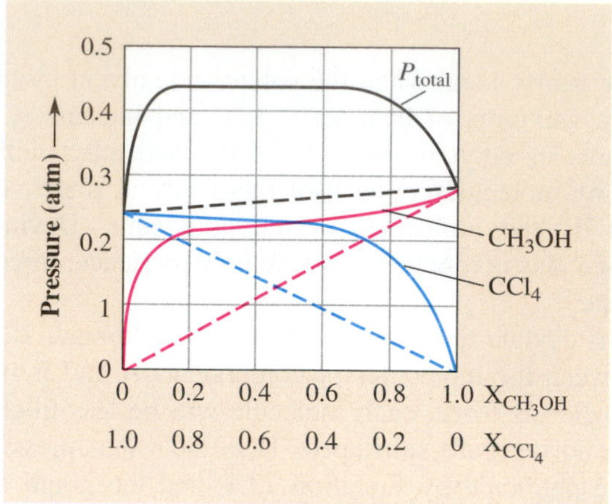

Figure 13.11

Solutions of methanol and carbon tetrachloride show positive deviations from Raoult's law. The solid curves are observed vapor pressures; the dashed lines are Raoult's law vapor pressures.

13.3 COLLIGATIVE PROPERTIES

Certain solution properties depend on the concentration of solute particles and not on their nature. These properties are called **colligative properties**. For a given solvent, they will be the same for different solutes. The principal colligative properties are vapor pressure lowering, boiling point elevation, freezing point depression, and osmotic pressure. We will study each of these properties in turn, and we will find that the magnitude of each effect can be calculated when the solution is dilute, that is, when the solute is present in such low concentration that it has little effect on the solvent. Our initial discussion of colligative properties will be confined to solutions in which the solute is nonionic and nonvolatile (not easily vaporized).

Vapor Pressure Lowering

When a nonvolatile solute is dissolved in a volatile solvent, the vapor pressure above the solution is entirely due to the solvent. If the solution is dilute, there will be very few solute molecules to cause deviations from ideality, and the solvent will obey Raoult's law. The vapor pressure above the solution can be calculated from Equation 13.1:

$$P_{\text{solution}} = P_{\text{solvent}}$$

$$= X_{\text{solvent}} P^\circ_{\text{solvent}} \qquad (13.4)$$

where X_{solvent} is the mole fraction of solvent in the solution and P°_{solvent} is the vapor pressure of the pure solvent. Because X_{solvent} is a fraction less than 1, *the vapor pressure above a solution of a nonvolatile solute is always less than the vapor pressure of the pure solvent.* The **vapor pressure lowering** (VPL) is the difference between the vapor pressure of the pure solvent and the vapor pressure of the solution:

Many chemists use ΔP instead of VPL to symbolize the vapor pressure lowering.

$$\text{VPL} = P^\circ_{\text{solvent}} - P_{\text{solution}} \qquad (13.5)$$

The vapor pressure lowering can be calculated directly from the mole fraction of the solute. Substituting Equation 13.4 into Equation 13.5 gives

$$\begin{aligned} \text{VPL} &= P^\circ_{\text{solvent}} - X_{\text{solvent}} P^\circ_{\text{solvent}} \\ &= (1 - X_{\text{solvent}}) P^\circ_{\text{solvent}} \\ &= X_{\text{solute}} P^\circ_{\text{solvent}} \end{aligned} \tag{13.6}$$

since $1 - X_{\text{solvent}} = X_{\text{solute}}$. (Recall that the sum of the mole fractions is 1.) Equation 13.6 shows that the vapor pressure lowering is proportional to the mole fraction of solute; vapor pressure lowering is a colligative property that increases with increasing solute concentration.

EXAMPLE 13.3

The vapor pressure of pure water at 25°C is 23.8 torr. Estimate the vapor pressure lowering when 50.0 g of glucose ($C_6H_{12}O_6$) is dissolved in 100 g of water at 25°C.

SOLUTION

The vapor pressure lowering is calculated from Equation 13.6, but first we must find the mole fraction of glucose. The molar masses of glucose and water are 180.2 and 18.02 g/mol, respectively. The number of moles of each component is

$$n_{\text{glucose}} = 50.0 \text{ g} \times \frac{1 \text{ mol}}{180.2 \text{ g}} = 0.277 \text{ mol}$$

$$n_{\text{water}} = 100 \text{ g} \times \frac{1 \text{ mol}}{18.02 \text{ g}} = 5.55 \text{ mol}$$

The mole fraction of glucose is

$$X_{\text{glucose}} = \frac{\text{number of moles of glucose}}{\text{number of moles of glucose} + \text{number of moles of water}}$$

$$= \frac{0.277 \text{ mol}}{0.277 \text{ mol} + 5.55 \text{ mol}} = 0.0475$$

Substituting $X_{\text{glucose}} = 0.0475$ and $P^\circ_{\text{water}} = 23.8$ torr into Equation 13.6 gives the vapor pressure lowering:

$$\text{VPL} = X_{\text{glucose}} P^\circ_{\text{water}} = 0.0475 \times 23.8 \text{ torr} = 1.13 \text{ torr}$$

PRACTICE EXERCISE 13.3

At 25°C, the vapor pressure above an aqueous solution of sucrose is 22.0 torr. Estimate the mole fraction of sucrose in this solution.

The answer to Example 13.3 could also have been obtained by using Raoult's law (Equation 13.1) to calculate the vapor pressure of water above the solution and then subtracting the result from the vapor pressure of pure water. Regardless of how the problem is solved, the answer will be accurate only to the extent that the solvent obeys Raoult's law. The calculated results will become less accurate with increasing solute concentration because as the solute concentration increases, deviations from ideality also increase.

DEMONSTRATION 13.2 SOLVENT MIGRATION INTO A CONCENTRATED SOLUTION

The liquids in the closed system are pure water and copper sulfate solution (blue).

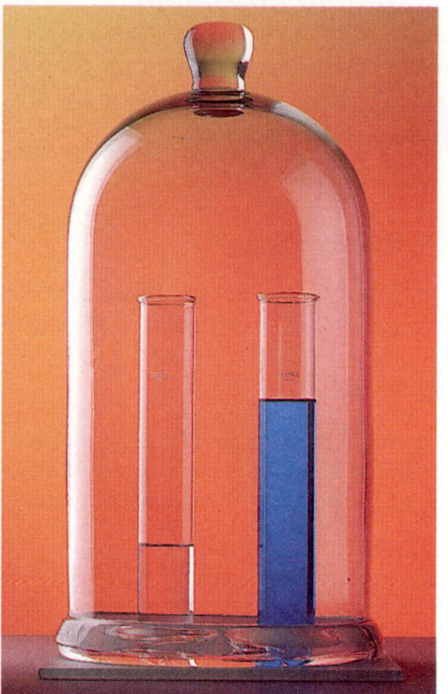

As time passes, the solution level gradually rises and the water level drops, showing that water vapor condenses into the solution and pure water evaporates to replace it. The process will continue until all of the water has evaporated.

One of the more interesting vapor pressure effects is the migration of solvent molecules from a less concentrated solution to a more concentrated solution via the vapor phase (Demonstration 13.2). The vapor pressure of the solvent above a pure liquid or a dilute solution is greater than that above a concentrated solution. If containers of both solutions are present in a closed system, the solvent will evaporate from the dilute solution, and solvent vapor will condense into the concentrated solution. This migration will continue until both solutions achieve the same concentration and the same vapor pressure. If one container holds pure solvent rather than a dilute solution, the pure solvent will completely disappear.

Boiling Point Elevation

Boiling water will stop boiling when salt or sugar is added to it. If heating is continued, the solution will soon boil again, but at a higher temperature. This rise in boiling point occurs when any nonvolatile solute is dissolved in a liquid solvent; it is a consequence of the vapor pressure lowering produced by the solute molecules. A liquid boils when its vapor pressure equals the external or atmospheric pressure. If

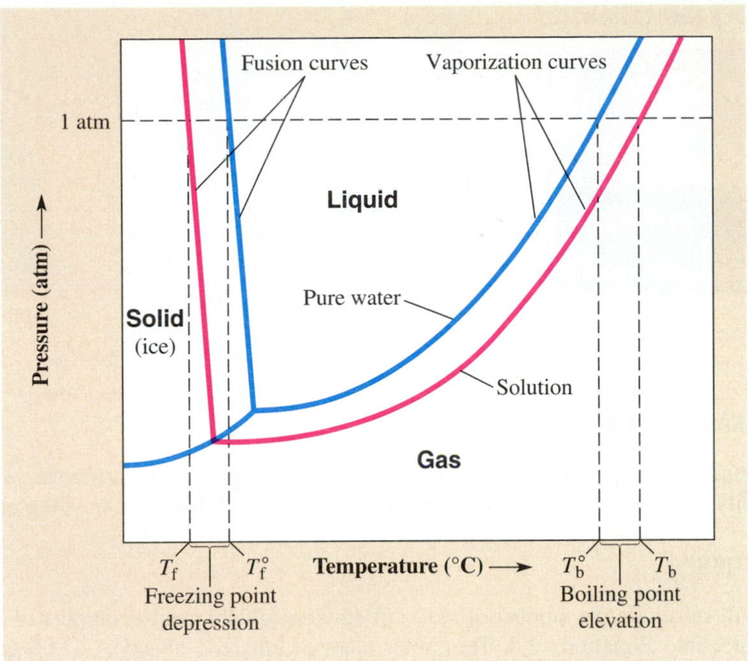

Figure 13.12
The phase diagram for an aqueous solution of a nonvolatile solute (red lines) superimposed on the phase diagram for pure water (blue lines). The boiling point of the solution, T_b, is greater than that of pure water, and the freezing point, T_f, is less than that of pure water.

the vapor pressure is lowered by the presence of a solute, the solution must be heated to a higher temperature to achieve the vapor pressure required for boiling. *A solution containing a nonvolatile solute boils at a higher temperature than the pure solvent.*

The same conclusion could have been reached by studying Figure 13.12, which shows a phase diagram of an aqueous solution containing a nonvolatile solute superimposed on the phase diagram of water. The vaporization curve for the solution lies below that of water and, at any given temperature, the vapor pressure of the solution is less than that of pure water. Observe that the boiling point of the solution, T_b, is greater than that of the solvent, T_b°, as discussed in the preceding paragraph.

LEARNING HINT
When a solution containing a nonvolatile solute boils, the vapor consists only of the solvent.

The **boiling point elevation** (BPE) is the difference between the boiling point of the solution and the boiling point of the solvent; that is,

$$\text{BPE} = T_b - T_b^\circ \tag{13.7}$$

The boiling point elevation increases with increasing concentration of solute, but in calculating its effect we will need to use a new concentration unit called **molality** (m). Molality is the number of moles of solute per kilogram of solvent:

The boiling point elevation (BPE) is often symbolized by ΔT_b.

$$\text{molality} = \frac{\text{number of moles of solute}}{\text{number of kilograms of solvent}} \tag{13.8}$$

A 2 m (2 molal) aqueous sodium chloride solution, for example, can be made by dissolving 2 mol of NaCl in 1 kg of water (Figure 13.13). Molality (m) should not be confused with molarity (M). Molality has kilograms of solvent in the denominator; molarity has liters of solution. Molarity depends on volume and will vary with temperature as the solution expands or contracts. Molality values do not involve volume and are temperature independent.

Figure 13.13

A 2.0 molal NaCl solution is made by dissolving 2 mol of NaCl (116.9 g) in 1 kg of water.

EXAMPLE 13.4

Ethylene glycol ($C_2H_6O_2$) is the main component of automobile antifreeze. What is the molality of a solution containing 200 g of ethylene glycol dissolved in 900 g of water?

SOLUTION

We will substitute the number of moles of ethylene glycol and the number of kilograms of water into Equation 13.8. The molar mass of ethylene glycol is 62.09 g/mol. The number of moles of ethylene glycol is

$$n = 200\text{ g} \times \frac{1\text{ mol}}{62.09\text{ g}} = 3.22\text{ mol}$$

This number of moles is dissolved in 0.900 kg of water, so the molality is

$$\text{molality} = \frac{\text{number of moles of ethylene glycol}}{\text{number of kilograms of water}}$$

$$= \frac{3.22\text{ mol}}{0.900\text{ kg}} = 3.58\text{ mol/kg}$$

The solution can also be described as "3.58 molal" or simply "3.58 *m*."

LEARNING HINT

The important chemical concentration units are summarized below. Be sure you understand how each one differs from the others. (*M* signifies molarity; *X* signifies mole fraction; *m* signifies molality.)

$$M = \frac{\text{moles of solute}}{\text{liters of solution}}$$

$$X = \frac{\text{moles of solute}}{\text{total number of moles}}$$

$$m = \frac{\text{moles of solute}}{\text{kilograms of solvent}}$$

PRACTICE EXERCISE 13.4

Find the molality of ethanol in a solution containing 6.00 g of ethanol (C_2H_5OH) dissolved in 75.0 g of water.

For dilute solutions of nonionic solutes, the magnitude of the boiling point elevation (Equation 13.7) is proportional to the molality of the solute; that is,

$$\text{BPE} = T_b - T_b^\circ = K_b m \tag{13.9}$$

The proportionality constant K_b is called the **boiling point elevation constant**; it varies from solvent to solvent. For water, $K_b = 0.512°\text{C}\cdot\text{kg/mol}$; values for other solvents are given in Table 13.2. (The units of K_b must be °C·kg/mol in order for the product of K_b and molality to have units of degrees Celsius.) Equation 13.9 shows that boiling point elevation is a colligative effect; it depends only on the concentration of the solute, not on its identity.

TABLE 13.2 Normal Boiling and Freezing Point Data for Various Solvents

Solvent	Boiling Point (°C)	K_b (°C·kg/mol)	Freezing Point (°C)	K_f (°C·kg/mol)
Acetic acid	117.9	3.07	16.6	3.90
Benzene	80.1	2.53	5.53	4.90
Carbon tetrachloride	76.6	5.03	− 23.0	31.8
Chloroform	61.7	3.63	− 63.5	4.68
Ethanol	78.5	1.22	−117.30	1.99
Water	100.00	0.512	0.00	1.86

EXAMPLE 13.5

The normal boiling point of water is 100.00°C. Estimate the boiling point of the ethylene glycol antifreeze solution in Example 13.4.

SOLUTION

The boiling point elevation constant of water is 0.512°C·kg/mol. The molality of the ethylene glycol solution in Example 13.4 is 3.58 mol/kg. Substitution into Equation 13.9 gives

$$\text{BPE} = K_b m = \frac{0.512°\text{C·kg}}{1 \text{ mol}} \times \frac{3.58 \text{ mol}}{1 \text{ kg}} = 1.83°\text{C}$$

The boiling point of the solution, T_b, is obtained from Equation 13.7 by adding the boiling point elevation to the boiling point of water:

$$T_b = T_b^\circ + \text{BPE} = 100.00°\text{C} + 1.83°\text{C} = 101.83°\text{C}$$

The calculated result is valid only to the extent that Equation 13.9 is valid. The discrepancy between the calculated and actual boiling points of a solution increases with increasing solute concentration.

EXAMPLE 13.6

The boiling point of pure carbon disulfide is 46.30°C, and its boiling point elevation constant K_b is 2.34°C·kg/mol. A solution of sulfur in liquid carbon disulfide boils at 46.71°C. Estimate the molality of sulfur in this solution.

SOLUTION

The boiling point elevation is obtained by subtracting the boiling point of pure carbon disulfide from the boiling point of the solution (Equation 13.7):

$$\text{BPE} = 46.71°\text{C} - 46.30°\text{C} = 0.41°\text{C}$$

The molality is obtained by substituting the boiling point elevation and K_b into Equation 13.9:

$$m = \frac{\text{BPE}}{K_b} = \frac{0.41°\text{C}}{2.34°\text{C}\cdot\text{kg/mol}} = 0.18 \text{ mol/kg}$$

PRACTICE EXERCISE 13.5

A solution contains 1.25 g of biphenyl, $(C_6H_5)_2$, dissolved in 50.0 g of benzene. The boiling point of the solution is 0.411°C higher than that of pure benzene. Estimate K_b for benzene.

Freezing Point Depression

Antifreeze is put into the cooling system of an automobile to prevent the coolant from freezing at 0°C. Other substances such as salt and urea are spread on roads and sidewalks to prevent the formation of ice in cold weather. These practices are based on the fact that *a solution freezes at a lower temperature than the pure solvent.*

LEARNING HINT

The freezing point of a solution such as salt water is defined as the temperature at which the solution is in equilibrium with solid solvent.

Let us see how the solute achieves this effect. When a solution begins to freeze, the solid that forms is almost always pure solvent. Pure ice, for example, crystallizes out of a solution of salt water. Solute particles, because of their different size and shape, do not fit into the crystal lattice of the frozen solvent. They tend to concentrate in the remaining liquid, and they interfere with the freezing process by getting in the way of solvent molecules looking for lattice sites. This interference causes the solution to freeze at a lower temperature than the pure solvent. The more concentrated the solute, the greater the interference and the lower the freezing point of the solution.

The phase diagrams in Figure 13.12 show that the fusion curve of a solution is shifted toward lower temperatures at all pressures; thus the freezing point of the solution, T_f, is less than that of pure water, T_f°. The **freezing point depression** (FPD) is the difference between the freezing point of the pure solvent and the freezing point of the solution:

The freezing point depression (FPD) is often symbolized by ΔT_f.

$$\text{FPD} = T_f^\circ - T_f \tag{13.10}$$

Note that both the freezing point depression (Equation 13.10) and the boiling point elevation (Equation 13.7) are defined as positive; the smaller temperature is subtracted from the larger temperature.

For dilute solutions of nonionic solutes, the equation relating freezing point depression to concentration is analogous to Equation 13.9 for boiling point elevation:

$$\text{FPD} = T_f^\circ - T_f = K_f m \tag{13.11}$$

Here K_f is the **freezing point depression constant** and m is the molality of the solute. The value of K_f for water is 1.86°C·kg/mol; other values for K_f are given in Table 13.2. Equation 13.11 shows that freezing point depression is a colligative effect.

EXAMPLE 13.7

The normal freezing point of water is 0.00°C. Estimate the freezing point of the ethylene glycol antifreeze solution in Example 13.4.

SOLUTION

The freezing point depression constant of water is 1.86°C·kg/mol. The molality of the ethylene glycol solution in Example 13.4 is 3.58 mol/kg. Substituting into Equation 13.11 gives

$$\text{FPD} = K_f m = \frac{1.86^\circ\text{C}\cdot\text{kg}}{1\ \text{mol}} \times \frac{3.58\ \text{mol}}{1\ \text{kg}} = 6.66^\circ\text{C}$$

The freezing point of the solution, T_f, is obtained from Equation 13.10 by subtracting the freezing point depression from the freezing point of pure water:

$$T_f = T_f^\circ - \text{FPD} = 0.00^\circ\text{C} - 6.66^\circ\text{C} = -6.66^\circ\text{C}$$

PRACTICE EXERCISE 13.6

A solution of an organic compound in benzene freezes at 3.41°C. Use data from Table 13.2 to estimate the molality of the compound in this solution.

A dissolved solute will lower the freezing point of a solution and also raise its boiling point. Both of these effects are useful. The antifreeze solution that protects the engine coolant from freezing also protects it from boiling in hot weather. The best antifreeze solutions contain about equal volumes of ethylene glycol and water. The concentration of ethylene glycol in such solutions is about 18 *m*, which unfortunately is too high for Equations 13.9 and 13.11 to be valid. The actual freezing and boiling points of concentrated solutions must be determined experimentally.

Osmotic Pressure

A living cell has an outer membrane that separates its contents from the solution surrounding it. This cell wall is a type of **semipermeable membrane**, one that allows only selected materials to pass from one side to the other. Inert semipermeable membranes (the membrane of a living cell is not inert) can be prepared in the laboratory from animal tissues, from organic substances such as cellophane or cellulose acetate, and even from certain inorganic materials. An inert membrane allows small molecules like water to pass through while serving as a barrier to larger molecules.

Consider a solution of sugar water separated from a sample of pure water by a membrane that is permeable to water but not to sugar (see Figure 13.14 and Demonstration 13.3). Molecules collide continually and randomly with both sides of the membrane. Sugar and water molecules collide with the solution side; only water molecules collide with the solvent side. Because more water molecules hit the solvent side than the solution side in any time period, there is a net flow of water from the pure solvent into the solution. This flow is an example of **osmosis**, the net movement of solvent molecules through a semipermeable membrane from a pure solvent or a dilute solution into a more concentrated solution. The solution becomes diluted and, if it is confined as shown in Figure 13.14, its level will rise. The level will con-

Osmosis: from the Greek *osmos*, meaning "action of pushing."

CHEMICAL INSIGHT

OSMOTIC PRESSURE IN ACTION

Osmotic pressure plays an important role in biological systems. The membranes of living cells selectively allow the passage of water, nutrients, wastes, and assorted ions. Human blood, which consists of blood cells surrounded by plasma (Figure 13.15), exhibits an average osmotic pressure of about 7.7 atm at normal body temperature (37°C). If for some reason the plasma is diluted, that is, if its osmotic pressure falls below 7.7 atm, water will flow into the blood cells, causing them to expand and possibly rupture, a phenomenon known as *hemolysis*. (Note carefully that water flows from a dilute solution to a more concentrated one; from a solution with a low osmotic pressure to one with a high osmotic pressure.) Conversely, if the plasma becomes concentrated, there will be a loss of cell fluid into the plasma. This loss, which is accompanied by cell shrinkage, is called *crenation*. The maintenance of the osmotic balance between cells and plasma is essential to life and health.

Solutions are said to be *isotonic* if they exhibit the same osmotic pressure. For example, a 0.16 *M* sodium chloride solution ("physiological saline solution") is isotonic with normal red blood cells. A red blood cell placed in this solution will undergo neither hemolysis nor crenation. Solutions of medicine or nutrients to be administered intravenously are usually made isotonic with blood by dissolving salts in them.

Osmosis is also part of the mechanism by which water rises from the roots to the leaves of trees and other plants. Tremendous quantities of water evaporate each day from the leaves of a tree, and the fluids in the leaf tend to become concentrated. This effect is counteracted by the osmotic flow of water from the soil into the roots, trunk, branches, stems, and ultimately the leaves.

Osmosis causes prunes to swell when soaked in water, cucumbers to turn into pickles when placed in brine, and cut flowers to stay fresh longer when their stems are kept in water. The presence of salt in preserved meat, and sugar in jelly and jam, helps to destroy potentially dangerous bacteria. The bacteria cells shrink and ultimately die because of water loss to the more concentrated surrounding fluids.

Reverse osmosis takes place when a concentrated solution inside a semipermeable membrane is subjected to a pressure greater than its osmotic pressure. The excess pressure forces solvent from the concentrated solution into the dilute solution outside the membrane. This procedure has been used to desalinate seawater (Figure 13.16). Seawater contains a large quantity of dissolved material and exhibits an osmotic pressure of about 26 atm. If seawater is placed on one side of a suitable membrane and subjected to pressures in excess of 26 atm, fairly pure water can be obtained. A second pass through the membrane will purify the water even further. One of the difficulties with reverse osmosis is that the high pressure required for an appreciable flow of reclaimed water—about 40 to 100 atm—tends to deteriorate the membrane. A number of relatively small reverse osmosis plants have been installed on various Caribbean islands, and a 15,000,000 gal/day plant has recently been built in Saudi Arabia. According to the International Desalination Association, about 39% of the world's desalinated water is currently obtained by reverse osmosis; the rest is obtained by distillation.

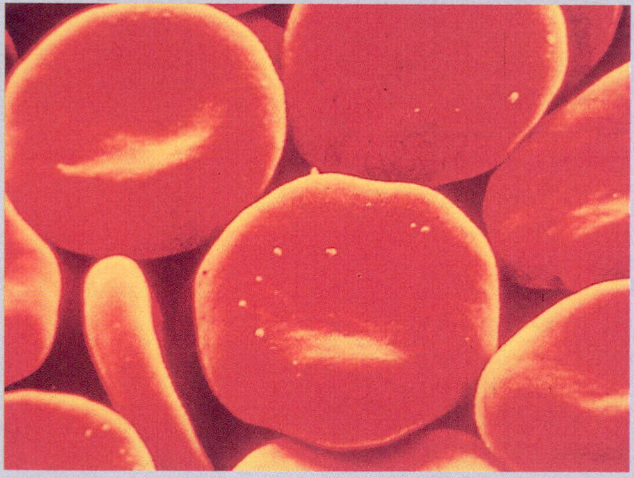

Figure 13.15
Electron micrograph of red blood cells.

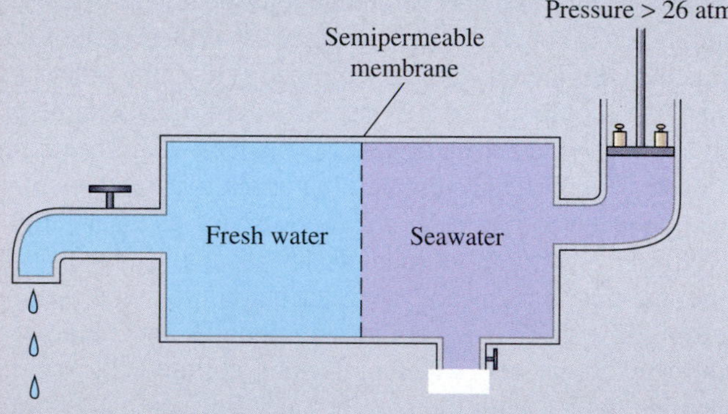

Figure 13.16
Reverse osmosis can be used to desalinate seawater. A pressure greater than the osmotic pressure of seawater (26 atm) forces pure water through the semipermeable membrane.

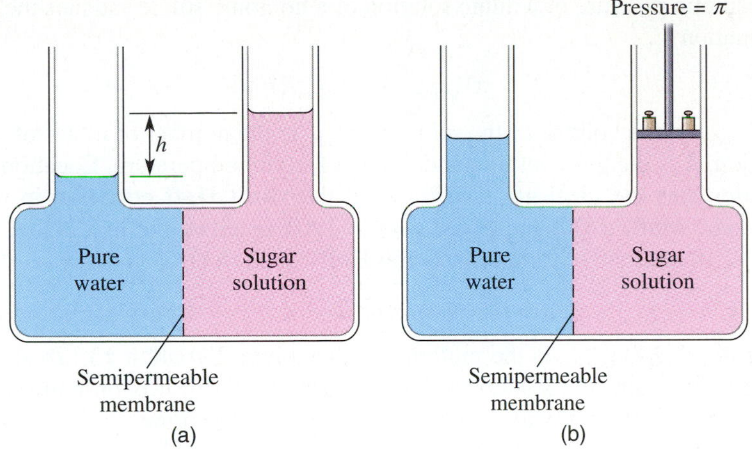

Figure 13.14

Osmosis. (a) A net flow of water into the sugar solution causes the solution level to rise to some equilibrium height *h*. At equilibrium, the pressure exerted by the higher solution level sends water molecules back through the membrane at a rate equal to the forward rate of flow. (b) The osmotic pressure π of the solution is the pressure required to stop the flow of water. It is equal to the equilibrium pressure exerted by the higher solution level in (a).

tinue to rise on the solution side until the added pressure due to the higher level sends water molecules back through the membrane at a rate equal to their forward flow. The pressure required to stop the net flow of solvent into the solution is called the **osmotic pressure** (π) of the solution.

DEMONSTRATION 13.3 OSMOSIS

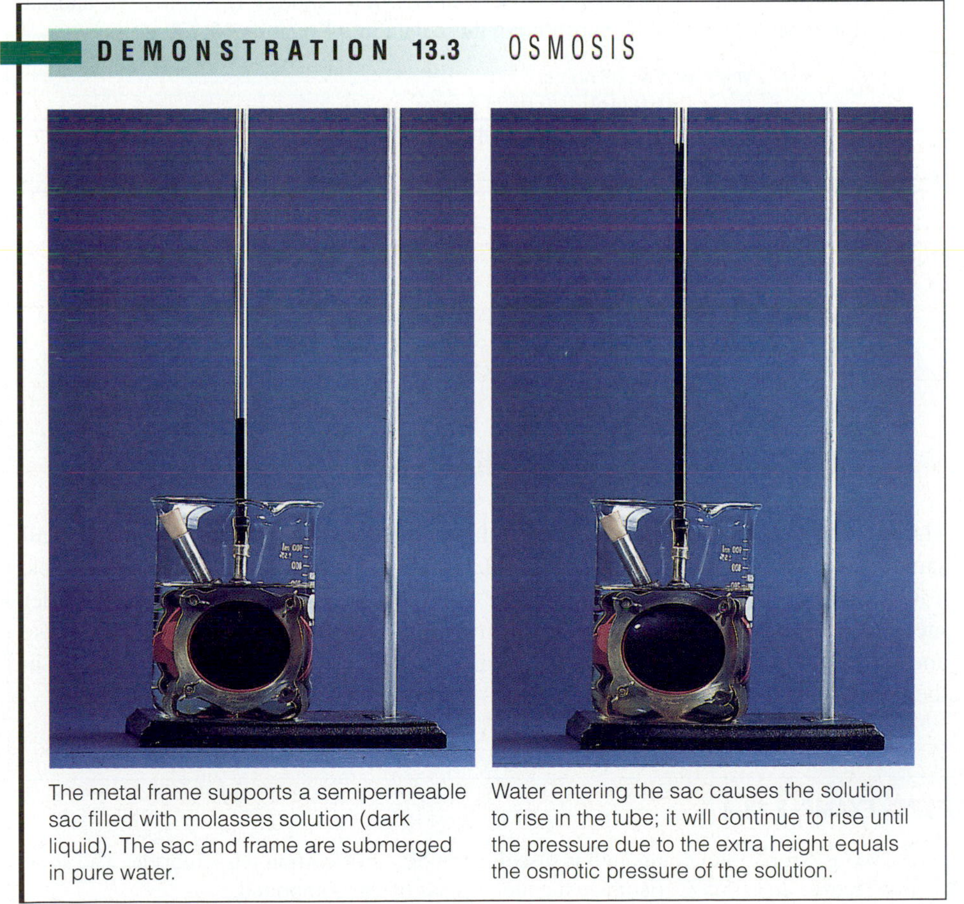

The metal frame supports a semipermeable sac filled with molasses solution (dark liquid). The sac and frame are submerged in pure water.

Water entering the sac causes the solution to rise in the tube; it will continue to rise until the pressure due to the extra height equals the osmotic pressure of the solution.

The osmotic pressure of a dilute solution of a nonionic solute satisfies the approximate equation

$$\pi V_{\text{solution}} = n_{\text{solute}}RT \tag{13.12a}$$

where V_{solution} is the volume of the solution, n_{solute} is the number of moles of solute in the solution, R is the gas constant, and T is the Kelvin temperature. Equation 13.12a, which resembles the ideal gas law, is called the **van't Hoff equation** in honor of Jacobus van't Hoff, a Dutch chemist who in 1901 received the first Nobel Prize in chemistry. An alternative form of the van't Hoff equation is

$$\pi = MRT \tag{13.12b}$$

where $M = n_{\text{solute}}/V_{\text{solution}}$ is the molarity of the solute. Equation 13.12b shows that the osmotic pressure of a solution is a colligative property; it is proportional to the molarity of the solute molecules, but independent of their nature.

EXAMPLE 13.8

Estimate the osmotic pressure at 25°C of a 0.010 *M* protein solution.

SOLUTION

The molarity of the solution is $M = 0.010$ mol/L, $R = 0.0821$ L·atm/mol·K, and the Kelvin temperature is 298 K. Substituting these data into Equation 13.12b gives

$$\begin{aligned}\pi &= MRT \\ &= \frac{0.010\ \text{mol}}{1\ \text{L}} \times \frac{0.0821\ \text{L}\cdot\text{atm}}{\text{mol}\cdot\text{K}} \times 298\ \text{K} = 0.24\ \text{atm}\end{aligned}$$

PRACTICE EXERCISE 13.7

Estimate the osmotic pressure at 25°C of a 1.00 *M* solution of any nonvolatile and nonionic solute.

13.4 MOLAR MASSES FROM COLLIGATIVE PROPERTIES

The modern method for determining molar masses requires vaporization of small samples in a mass spectrometer (Section 2.3). Substances with very large molecules, however, may be difficult to vaporize or may decompose when heated. An older method of determining molar masses from colligative effects is still useful for these and other molecules. The procedure consists of four steps, which are illustrated in the next example.

EXAMPLE 13.9

A 20.0-g sample of a compound is dissolved in 500 g of carbon tetrachloride. The solution freezes at −35.5°C. Estimate the molar mass of the compound.

SOLUTION

Step 1: *Determine the magnitude of the colligative effect.* In this case, the colligative effect is the freezing point depression. The normal freezing point of carbon tetrachloride is −23.0°C, and its freezing point depression constant is 31.8°C·kg/mol (Table 13.2). The freezing point depression is obtained from Equation 13.10 by subtracting the freezing point of the solution, $T_f = -35.5°C$, from the normal freezing point of carbon tetrachloride, T_f°:

$$\text{FPD} = T_f^\circ - T_f = -23.0°\text{C} - (-35.5°\text{C}) = 12.5°\text{C}$$

Step 2: *Use the colligative effect to calculate the concentration of the solute.* The molality of the compound is obtained by substituting the freezing point depression and $K_f = 31.8°\text{C·kg/mol}$ into Equation 13.11:

$$m = \frac{\text{FPD}}{K_f} = \frac{12.5°\text{C}}{31.8°\text{C·kg/mol}} = 0.393 \text{ mol/kg}$$

Step 3: *Use the concentration from Step 2 to calculate the number of moles of solute.* The number of moles of compound dissolved in 500 g (0.500 kg) of solvent is obtained from the molality:

$$n = 0.500 \text{ kg solvent} \times \frac{0.393 \text{ mol}}{1 \text{ kg solvent}} = 0.196 \text{ mol}$$

Step 4: *Divide the number of grams of solute by the number of moles to obtain the molar mass.* We have just found that there is 0.196 mol of the compound in 500 g of solvent. The problem stated that there are 20.0 g of compound in 500 g of the solvent. Hence, 0.196 mol is equivalent to 20.0 g. The molar mass is

$$\mathcal{M} = \frac{20.0 \text{ g}}{0.196 \text{ mol}} = 102 \text{ g/mol}$$

LEARNING HINT

Beginning students sometimes misinterpret the units of molality; remember that mol/kg refers to moles of **solute** per kilogram of **solvent**.

EXAMPLE 13.10

The sulfur solution in Example 13.6 was prepared by dissolving 2.26 g of sulfur in 50.0 g of carbon disulfide. Estimate (a) the molar mass and (b) the formula of the dissolved sulfur molecules.

SOLUTION

(a) ***Steps 1 and 2:*** The molality of the sulfur solution was estimated in Example 13.6 to be 0.18 mol sulfur/kg solvent.

Step 3: The number of moles of sulfur in 50.0 g (0.050 kg) of solvent is

$$n = 0.050 \text{ kg solvent} \times \frac{0.18 \text{ mol sulfur}}{1 \text{ kg solvent}}$$
$$= 0.0090 \text{ mol sulfur}$$

Step 4: The problem states that there are 2.26 g of sulfur in 50.0 g of solvent. Step 3 shows that there is 0.0090 mol of sulfur in 50.0 g of solvent. Hence, the molar mass of sulfur is

$$\mathcal{M} = \frac{2.26 \text{ g}}{0.0090 \text{ mol}} = 250 \text{ g/mol}$$

(b) The molar mass of the dissolved sulfur is 250 g/mol. The molar mass of sulfur atoms is 32.07 g/mol. The number of atoms per molecule is 250/32.07 = 7.8, which rounds off to 8. The formula of the sulfur molecules is S_8.

PRACTICE EXERCISE 13.8

Cholesterol, the main constituent of gallstones, is found in every cell of the body. The boiling point of a solution containing 1.00 g of cholesterol dissolved in 25.0 g of chloroform is 0.376°C higher than the boiling point of pure chloroform. Use data from Table 13.2 to estimate the molar mass of cholesterol.

Freezing points are often easier to measure than boiling points; moreover, the freezing point depression in a given solution is usually greater than the boiling point elevation (compare the K_f and K_b values in Table 13.2). For these reasons, freezing point data usually provide more accurate molar mass values than boiling point data. Neither method, however, can be used for polymers, proteins, and other molecules with very large molar masses. When the molar mass is large, the molar concentration of the dissolved solute will be small. The freezing point depression and boiling point elevation caused by the solute will also be small—too small to measure. Osmotic pressures, however, are relatively large and easy to measure (see Demonstration 13.3), even for solutes with large molar masses. The next example shows how molar masses are determined from osmotic pressure measurements.

EXAMPLE 13.11

An aqueous solution contains 20.0 g of hemoglobin per liter. If the osmotic pressure of this solution is 5.6 torr at 27°C, estimate the molar mass of hemoglobin.

SOLUTION

Step 1: The osmotic pressure of the solution is 5.6 torr × 1 atm/760 torr = 7.4×10^{-3} atm.

Steps 2 and 3: The number of moles of hemoglobin in 1 L of solution is obtained by substituting $\pi = 7.4 \times 10^{-3}$ atm, $V = 1.00$ L, $R = 0.0821$ L·atm/mol·K, and $T = 300$ K into Equation 13.12a:

$$n = \frac{\pi V_{\text{solution}}}{RT}$$

$$= \frac{7.4 \times 10^{-3} \text{ atm} \times 1.00 \text{ L}}{0.0821 \text{ L·atm/mol·K} \times 300 \text{ K}} = 3.0 \times 10^{-4} \text{ mol}$$

The earliest value for the molecular weight of hemoglobin was obtained from an osmotic pressure measurement.

Step 4: The mass of 3.0×10^{-4} mol of hemoglobin is 20.0 g. The molar mass is

$$\mathcal{M} = \frac{20.0 \text{ g}}{3.0 \times 10^{-4} \text{ mol}} = 6.7 \times 10^{4} \text{ g/mol}$$

PRACTICE EXERCISE 13.9

An aqueous solution contains 28.8 mg of an unknown protein per milliliter of solution. The osmotic pressure of the solution is 12.0 torr at 25°C. Estimate the molar mass of the protein.

13.5 COLLIGATIVE PROPERTIES OF IONIC SOLUTIONS

We have seen that the magnitude of a colligative effect depends on the concentration of solute particles. If 1 mol of solute dissociates into more than 1 mol of dissolved particles (as occurs when NaCl dissolves in water), then we would expect the effect on vapor pressure, boiling point, freezing point, and osmotic pressure to be greater than that calculated on the basis of 1 mol.

EXAMPLE 13.12

Compare the expected freezing point depression of (a) a 0.500 *m* NaCl solution and (b) a 0.500 *m* K_2SO_4 solution with that of a 0.500 *m* aqueous solution of a nonionized solute such as glycerol. Assume complete dissociation of both electrolytes.

SOLUTION

The freezing point depression caused by the nonionized solute is obtained by substituting its molality (m = 0.500 mol/kg) and the freezing point depression constant for water (K_f = 1.86°C·kg/mol) into Equation 13.11:

$$\text{FPD} = K_f m = \frac{1.86^\circ\text{C}\cdot\text{kg}}{1\text{ mol}} \times \frac{0.500\text{ mol}}{1\text{ kg}} = 0.930^\circ\text{C}$$

(a) NaCl dissociates according to the equation

$$NaCl(aq) \longrightarrow Na^+(aq) + Cl^-(aq)$$

Hence, 0.500 mol of NaCl will produce 2 × 0.500 mol of ions. Each colligative effect in a 0.500 *m* NaCl solution should be twice that in a 0.500 *m* solution of glycerol. The expected freezing point depression is 2 × 0.930°C = 1.86°C.

(b) K_2SO_4 dissociates according to the equation

$$K_2SO_4(aq) \longrightarrow 2K^+(aq) + SO_4^{2-}(aq)$$

The 0.500 mol of K_2SO_4 will produce 3 × 0.500 mol of ions. Each colligative effect in a 0.500 *m* K_2SO_4 solution should be three times that in a 0.500 *m* solution of glycerol. The expected freezing point depression is 3 × 0.930°C = 2.79°C.

PRACTICE EXERCISE 13.10

Confirm the calculated freezing point depressions listed in Tables 13.3 and 13.4.

The colligative effects in ionic solutions are always smaller than the effects cal-

TABLE 13.3 Observed Freezing Point Depressions for 1.00 *m* Solutions of Various Ionic Solutes in Water Compared to the Freezing Point Depressions Calculated on the Basis of Complete Dissociation

Solute	Calculated FPD (°C)	Observed FPD (°C)
NaCl	3.72	3.37
$MgSO_4$	3.72	2.03
H_2SO_4	5.58	4.04
$ZnCl_2$	5.58	5.21

TABLE 13.4 Observed Freezing Point Depressions for Various Concentrations of Aqueous $MgSO_4$ Compared to the Freezing Point Depressions Calculated on the Basis of Complete Dissociation

Concentration (*m*)	Calculated FPD (°C)	Observed FPD (°C)	*i* Factor
1.000	3.72	2.03	1.09
0.100	0.372	0.225	1.21
0.0100	0.0372	0.0285	1.53
0.00100	0.00372	0.00339	1.82

Figure 13.17

In an ionic solution, each ion tends to be surrounded by an "atmosphere" of oppositely charged ions.

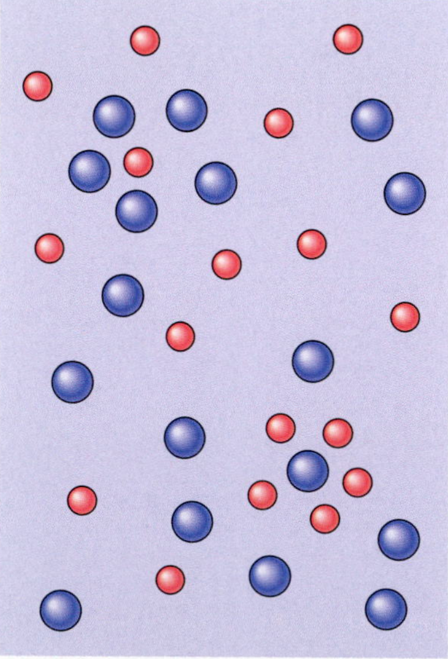

culated on the basis of complete dissociation. Table 13.3 compares the observed and calculated freezing point depressions for various 1.00 *m* ionic solutions. Table 13.4 makes a similar comparison for $MgSO_4$ solutions of different concentrations. Notice that *the observed colligative effect for an ionic solution is less than the calculated effect, but it approaches the calculated effect as the solution becomes more dilute.*

There are various reasons for the increase in colligative effects with increasing dilution. Weak electrolytes such as ammonia and acetic acid ionize only partially in concentrated solutions:

$$NH_3(aq) + H_2O(l) \rightleftharpoons NH_4^+(aq) + OH^-(aq)$$

$$CH_3COOH(aq) \rightleftharpoons H^+(aq) + CH_3COO^-(aq)$$

The fraction of molecules ionized, however, increases with increasing dilution, and the greater number of ions produces a larger colligative effect. Strong electrolytes such as HCl and $MgSO_4$, on the other hand, are always completely dissociated in aqueous solution. In concentrated solutions, however, each ion is loosely associated with one or more ions of opposite charge (Figure 13.17). This "ionic atmosphere" creates the effect of fewer ions and decreases the magnitude of any colligative effect. With increasing dilution, the ionic atmosphere thins out, and the observed colligative effects approach their expected values.

The degree to which ions behave independently also depends on their charge. The interactions between doubly charged Mg^{2+} and SO_4^{2-} ions are greater than the interactions between singly charged ions such as Na^+ and Cl^-. The data in Table 13.3

show that the difference between the calculated and observed freezing point depressions for a 1.00 *m* $MgSO_4$ solution is almost five times the difference for a 1.00 *m* NaCl solution.

DIGGING DEEPER

The van't Hoff *i* Factor

The effective number of moles of ions produced by 1 mol of solute is often expressed in terms of the **van't Hoff factor** (*i*):

$$i = \frac{\text{observed colligative effect}}{\text{calculated effect assuming no dissociation}} \tag{13.13}$$

EXAMPLE 13.13

Refer to Table 13.4 and calculate the van't Hoff factor for a 1.00 *m* $MgSO_4$ solution.

SOLUTION

Table 13.4 shows that the observed freezing point depression for 1.00 *m* $MgSO_4$ is 2.03°C. If we assume no dissociation, the calculated effect for a 1.00 *m* solution would be 1.86°C. Substituting these values into Equation 13.13 gives

$$i = \frac{2.03°\text{C}}{1.86°\text{C}} = 1.09$$

PRACTICE EXERCISE 13.11

Verify the other *i* factors reported in Table 13.4.

The *i* factor of 1.09 calculated in Example 13.13 means that the $MgSO_4$ solution behaves as if 1 mol of solute produces 1.09 mol of ions. Since $MgSO_4$ is a strong electrolyte, we would expect 1 mol to produce exactly 2 mol of ions. The strong ionic atmosphere that exists in a 1.00 *m* solution, however, reduces the effective number of moles from 2 to 1.09. Note that the *i* factors reported in Table 13.4 approach the expected value of 2 as the solution becomes more dilute, that is, as the ionic atmosphere becomes thinner.

An *i* factor obtained from one colligative effect in a given solution can be used to calculate other colligative effects for the same solution, as illustrated in the following example.

EXAMPLE 13.14

Estimate the boiling point of a 0.100 *m* $MgSO_4$ solution.

SOLUTION

The i factor for 0.100 m $MgSO_4$ is 1.21 (Table 13.4). If there were no dissociation, the boiling point elevation for a 0.100 m aqueous solution would be 0.0512°C. (Be sure to verify this number.) Substituting the data into Equation 13.13, gives

$$\text{observed BPE} = i \times \text{calculated BPE assuming no dissociation}$$
$$= 1.21 \times 0.0512°\text{C} = 0.0620°\text{C}$$

The boiling point of the solution will be 100.00°C + 0.0620°C = 100.06°C.

PRACTICE EXERCISE 13.12

Use data from Table 13.3 to estimate the boiling point of a 1.00 m H_2SO_4 solution.

13.6 COLLOIDS

Colloid: from the Greek *kolla*, meaning "glue."

A true solution is a homogeneous mixture containing dissolved particles of molecular or ionic size. The particles do not settle out on standing, and their dimensions are usually less than 1000 pm. A **suspension**, on the other hand, is a heterogeneous mixture containing dispersed particles with dimensions exceeding 100,000 pm. A mixture of mud and water is a suspension, as is the mixture of products formed during a precipitation reaction. Unlike the particles in a solution, the particles in a solid–liquid suspension may settle out on standing, and they can also be separated from the liquid by ordinary filtration. Between these two extremes lie **colloids** or **colloidal dispersions**, which are heterogeneous mixtures containing dispersed particles with dimensions greater than 1000 pm, but with at least one dimension smaller than 100,000 pm. Colloidal particles can be aggregates of atoms, ions, or molecules, or they can be giant molecules. Many everyday mixtures such as milk, butter, cheese, jelly, shaving cream, fog, smoke, and gems are colloidal. The importance of this class of substances should not be underestimated—even the protoplasm in living cells is colloidal.

Colloids resemble true solutions in that the particles are invisible to the naked eye, do not settle out, and pass through ordinary filters. Unlike the particles in a true solution, colloidal particles are retained by special membranes with very small pores (Figure 13.18). Colloidal particles also scatter light so that the path of a light beam passing through the mixture is visible from the side (Figure 13.19). This phenomenon, called the **Tyndall effect**, is familiar to anyone who has been in a dusty movie theater and observed the beam of light passing from the projector to the screen. True solutions such as clean air and salt water do not give a Tyndall effect.

Colloids can consist of solids dispersed in liquids, liquids dispersed in gases, gases dispersed in solids, and so forth (Table 13.5). All combinations of physical states are possible for a colloid except gas dispersed in gas (a mixture of gases is always a true solution). Special names are given to different types of colloids. **Sols** and **gels** each contain solids dispersed in a liquid. A sol is a liquid, while in a gel the

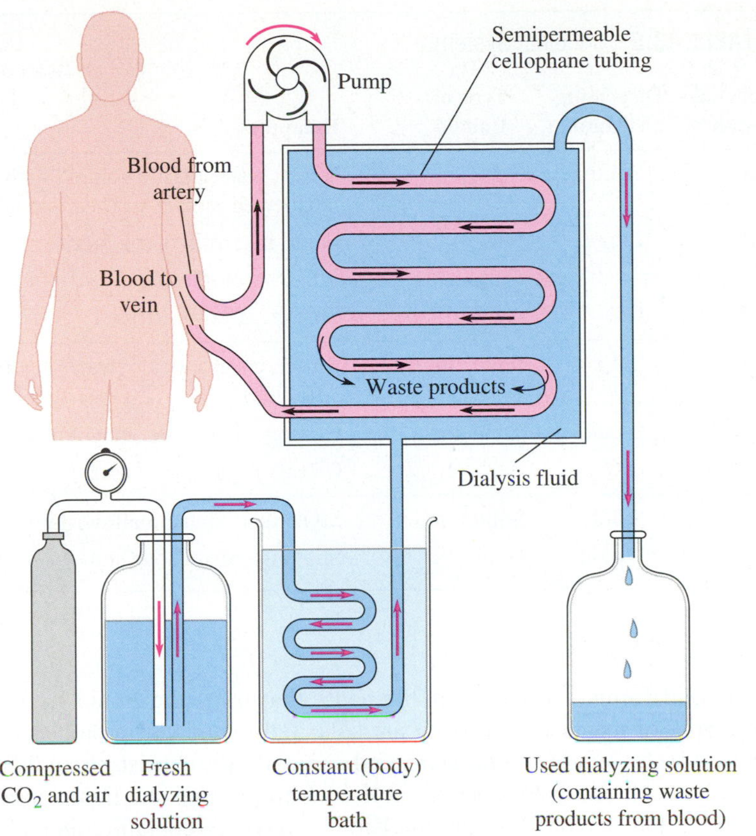

Figure 13.18

Dialysis is the selective passage of water, ions, and small molecules through special membranes that retain larger molecules and colloidal particles. A hemodialysis machine functions like a kidney; both remove waste products such as urea (NH_2CONH_2, a small molecule) from the blood by dialysis through membranes that retain proteins and blood cells. The dialyzing solution contains the same concentration of essential ions as the blood, so no ions are lost or gained.

Figure 13.19

The Tyndall effect. The light beam passes from left to right through a purple gold sol (a colloid), a blue copper sulfate solution, and colloidal iron(III) hydroxide. The light path can be seen in both colloids, but not in the copper sulfate solution.

TABLE 13.5 Colloidal Systems

Dispersed Substance	Dispersing Medium	Type of Colloid	Examples
Solid	Solid	Solid sol	Gems, glass, plastic colored with dispersed metals, iron-carbon alloys
Solid	Liquid	Sol	Paint, ink, milk of magnesia
		Gel	Jell-O, protoplasm
Solid	Gas	Aerosol	Smoke, airborne viruses
Liquid	Solid	Solid emulsion	Butter, cheese, opals (water in silicon dioxide)
Liquid	Liquid	Emulsion	Homogenized milk, mayonnaise
Liquid	Gas	Aerosol	Clouds, fog, aerosol sprays
Gas	Solid	Solid foam	Styrofoam, marshmallow, pumice stone
Gas	Liquid	Foam	Shaving cream, soap suds, beer foam

dispersed solid forms a network of filaments that traps the liquid in a semirigid structure. Milk of magnesia and paint are sols; Jell-O and cell protoplasm are gels. **Solid sols**, which include certain gems and some alloys, consist of solids dispersed in other solids. An **emulsion** contains colloidal droplets of one liquid suspended in another; milk and mayonnaise are emulsions. **Aerosols** contain colloidal particles dispersed in gas. The particles can be liquid, as in fog and aerosol sprays, or solid, as in smoke. Gases dispersed in liquids or solids are called **foams**. Whipped cream is a liquid foam and Styrofoam is a solid foam.

Hydrophilic and Hydrophobic Colloids

Colloidal particles in water are either hydrophilic (water-seeking) or hydrophobic (water-rejecting). **Hydrophilic colloids** contain polar molecules arranged so that the polar groups can form hydrogen bonds with water molecules near the surface of the colloidal particle (Figure 13.20). Such molecules, which include proteins and carbohydrates such as starch, have a natural affinity for water and disperse slowly but spontaneously when put into water. The water that is hydrogen-bonded to the colloidal particles stabilizes the colloid by preventing the particles from coagulating (coming together).

Hydrophobic colloids contain particles that have very little affinity for water. Examples include aqueous dispersions of metals such as gold (gold sols), of nonmetals such as sulfur, and of sparingly soluble solids such as silver chloride. The particles in a hydrophobic colloid often have densities greater than water, and one might expect them to settle out in the course of time. Such colloids, however, are surprisingly permanent. Sols containing gold have been known to last for over a hundred years. Colloidal sulfur, which forms in reactions such as

$$2H_2S(aq) + H_2SO_3(aq) \rightarrow 3S(s) + 3H_2O(l)$$

is also quite stable. The persistence of hydrophobic colloids is due to the tendency

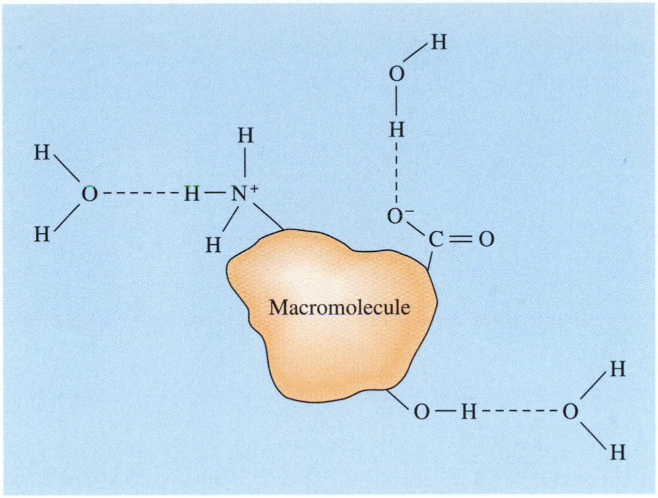

Figure 13.20

The polar groups protruding from a hydrophilic macromolecule (a very large molecule) form hydrogen bonds with surrounding water molecules.

of their particles to selectively adsorb dissolved positive or negative ions on their surfaces, as illustrated in Figure 13.21. The adsorbed ions provide all the particles with charges of the same sign. The colloidal particles do not coagulate because mutual repulsion keeps them away from each other.

Hydrophobic colloids often become unstable and precipitate in the presence of dissolved salts (Figure 13.22). Clay sols, for example, are coagulated by seawater. The salts in the water provide ions that neutralize the charges on the surface of the colloidal particles. River deltas, the triangular deposits of mud at the mouths of rivers, form in this manner. River water contains negatively charged colloidal particles of clay and silt. At the river mouth, where river water and seawater mix, the colloidal particles are neutralized by Na^+ and Mg^{2+} ions from seawater. The clay and silt gradually settle out and form the delta (Figure 13.23). The most efficient coagulants are salts containing highly charged ions such as Al^{3+} and Fe^{3+}.

The adsorption of charge is a major factor in the stabilization of many colloids. Smoke, for example, is an aerosol consisting of charged carbon and dust particles dispersed in air. It can be removed from smokestack gases by the Cottrell precipitator, in which hot gases pass over charged plates that neutralize the smoke particles

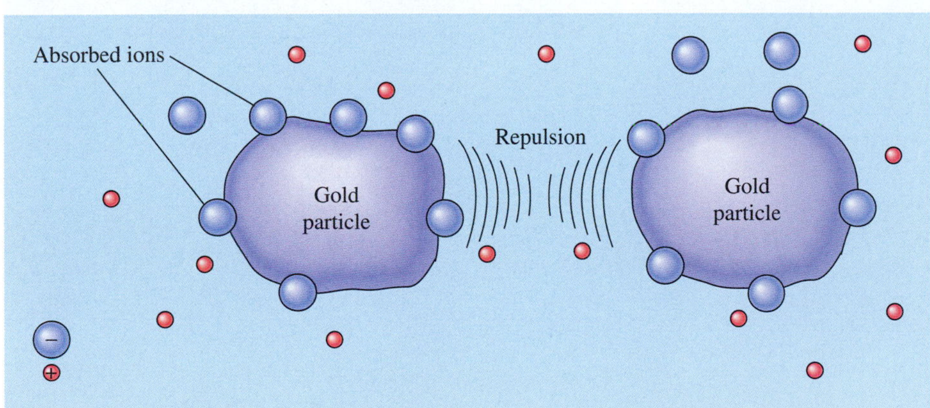

Figure 13.21

Hydrophobic gold particles selectively adsorb negative ions. The adsorbed ions cause the particles to repel each other and prevent them from settling out.

Figure 13.22

Both cylinders originally contained a colloidal suspension of arsenic (III) sulfide (As_2S_3). A few milliliters of aluminum sulfate solution was added to the cylinder on the left. The dissolved aluminum and sulfate ions destabilized the colloid, which then settled out as a yellow precipitate. The clear solution above the precipitate does not show a Tyndall effect.

and cause them to settle out (Figure 13.24). The soot and dust can then be disposed of or treated for the recovery of valuable impurities.

Some hydrophobic colloids are stabilized as emulsions (liquid suspended in liquid) by the adsorption of **emulsifying agents** on their surfaces. An emulsifying agent consists of fairly long molecules with one polar end and one nonpolar end. Soaps and detergents (Figure 13.25) are typical emulsifying agents whose polar ends usually contain oxygen atoms and whose nonpolar ends consist mainly of carbon and hydrogen atoms. The effect of a detergent on an otherwise insoluble oil stain is shown in Figure 13.28. Each minute oil droplet in the stain becomes surrounded by detergent molecules. The nonpolar ends of these molecules dissolve in the oil drop and the protruding polar ends form hydrogen bonds with the surrounding water. The hydrogen bonds keep the oil drops apart and make it easier for them to be washed away.

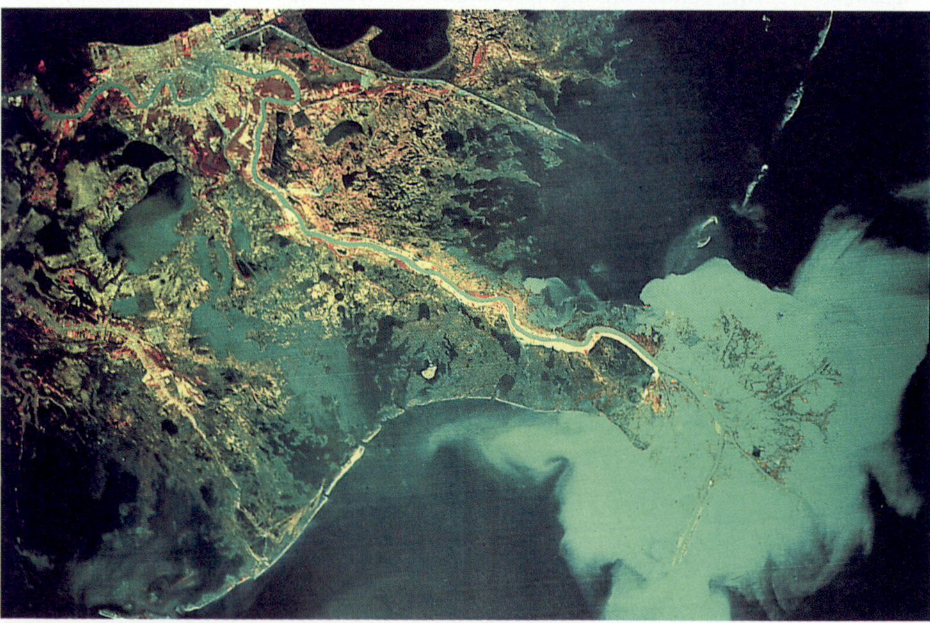

Figure 13.23

Mississippi river delta. This color-enhanced satellite image shows the Mississippi River flowing past New Orleans (upper center) into the Gulf of Mexico. The cloudy light-green region is where fresh water mixes with salt water.

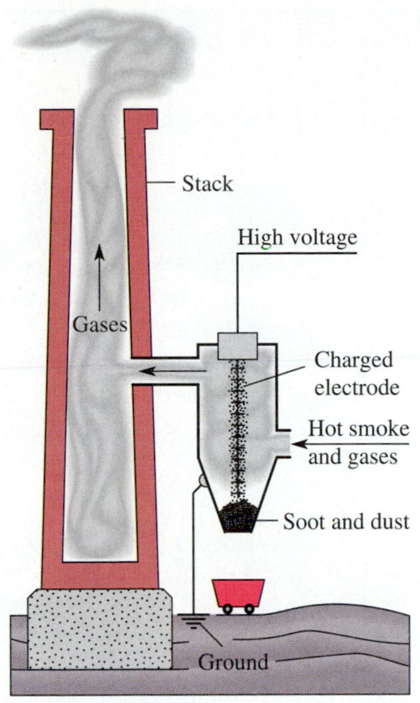

Figure 13.24
A Cottrell precipitator. Before entering the smokestack, soot-laden gases pass over charged plates that neutralize charged smoke particles and cause them to settle out.

Emulsifying agents are also used to stabilize a wide variety of foods. Mayonnaise, for example, is an emulsion of salad oil in vinegar or lemon juice; the emulsifying agent is raw egg yolk that is beaten into the mixture. Salad dressings, ice cream, soft drinks made with oil flavorings, and even bread contain emulsifying agents. You will find them listed as additives on the package label—look for names such as lecithin, polysorbate, carrageenan, and sorbitan monostearate.

Figure 13.25
Soaps and detergents consist of long molecules with hydrophilic heads and hydrophobic tails. (a) Sodium stearate, a typical soap. The hydrophobic tail is a hydrocarbon chain and the hydrophilic head (COO^-) bears a negative charge. (b) Sodium lauryl sulfate, a synthetic detergent. (c) A symbol that is often used to represent a detergent anion.

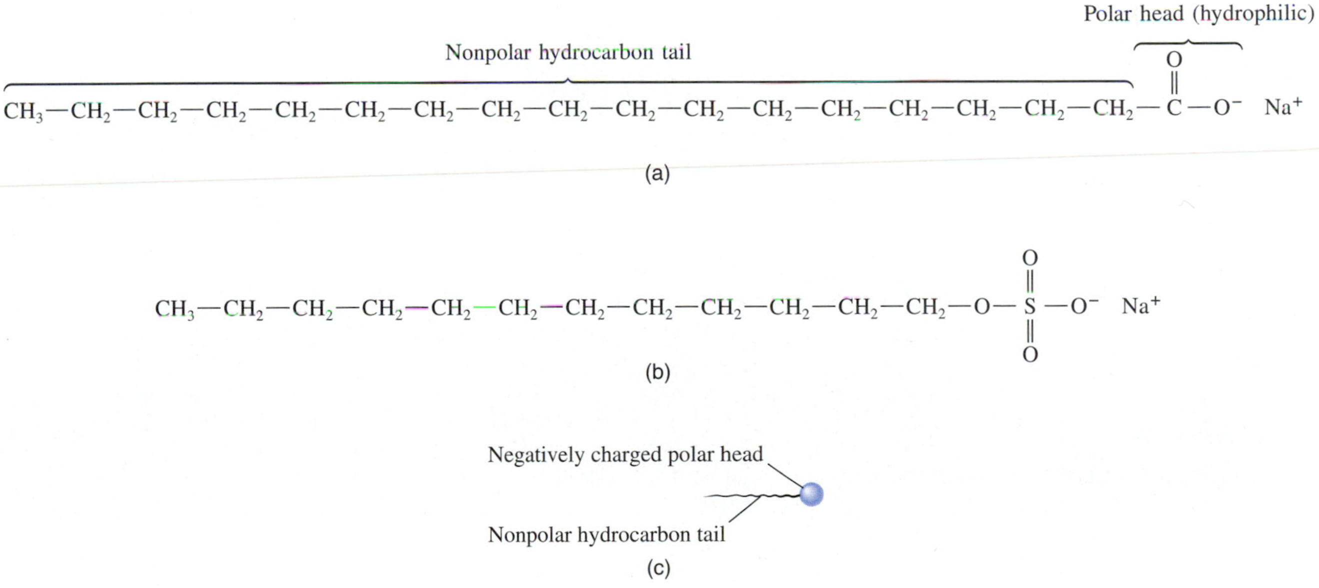

Nonpolar hydrocarbon tail — Polar head (hydrophilic)

$$CH_3-CH_2-CH_2-CH_2-CH_2-CH_2-CH_2-CH_2-CH_2-CH_2-CH_2-CH_2-CH_2-CH_2-CH_2-CH_2-CH_2-\overset{\displaystyle O}{\overset{\|}{C}}-O^- \quad Na^+$$

(a)

$$CH_3-CH_2-CH_2-CH_2-CH_2-CH_2-CH_2-CH_2-CH_2-CH_2-CH_2-CH_2-O-\underset{\underset{\displaystyle O}{\|}}{\overset{\overset{\displaystyle O}{\|}}{S}}-O^- \quad Na^+$$

(b)

Negatively charged polar head

Nonpolar hydrocarbon tail

(c)

CHEMICAL INSIGHT

SICKLE-CELL ANEMIA, A MOLECULAR DISEASE

Sickle-cell anemia is a "molecular disease" caused by a hereditary defect in hemoglobin, the protein molecule that carries oxygen in red blood cells. Normal hemoglobin molecules are complex, three-dimensional structures consisting of long chains folded over each other. The chains are made up of a number of repeating units, one of which is the glutamate unit (Figure 13.26a). Each glutamate unit contains a polar side group that protrudes from the hemoglobin chain. The hydrophilic polar groups form hydrogen bonds with water and thus help to keep normal hemoglobin dispersed as a gel within the red cells. Sickle-cell hemoglobin is similar to normal hemoglobin, except that valine units replace the glutamate units (Figure 13.26b). Valine differs from glutamate in that it possesses a nonpolar side group of carbon and hydrogen atoms. Such hydrocarbon groups are hydrophobic and cannot form hydrogen bonds with water. Thus sickle-cell hemoglobin is less soluble than normal hemoglobin. Deoxygenated hemoglobin (hemoglobin that has released its oxygen to the tissues) is especially insoluble and precipitates as elongated fibrous crystals within the red cells.

A red blood cell contains many hemoglobin molecules and normally has the biconcave disk shape shown in the electron micrograph of Figure 13.15 (page 608). Red cells that contain elongated hemoglobin crystals are deformed and take on a crescent or "sickle" shape (Figure 13.27). Sickled red cells do not flow as readily as normal cells. They tend to bunch up in small blood vessels and impede the flow of blood to various body tissues and organs. Symptoms of sickle-cell disease include severe abdominal and leg pains, leg ulcers, swollen joints, and jaundice. In the past, sickle-cell anemia was considered to be a fatal disease, but good medical care has substantially prolonged the lives of its victims. Despite intensive research efforts, however, there is still no cure for the disease.

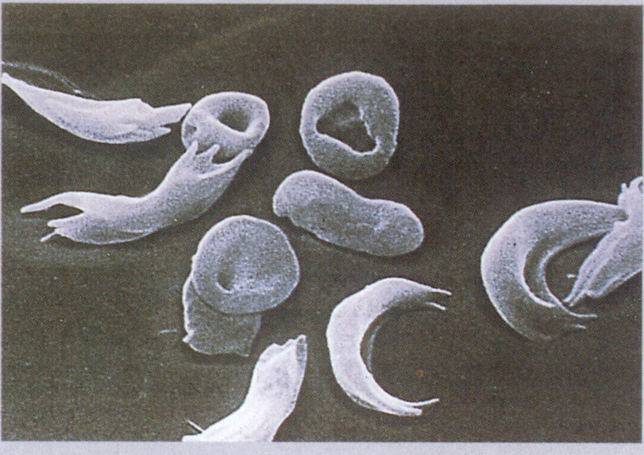

Figure 13.27

Normal red blood cells (biconcave disks) and red cells that show sickling.

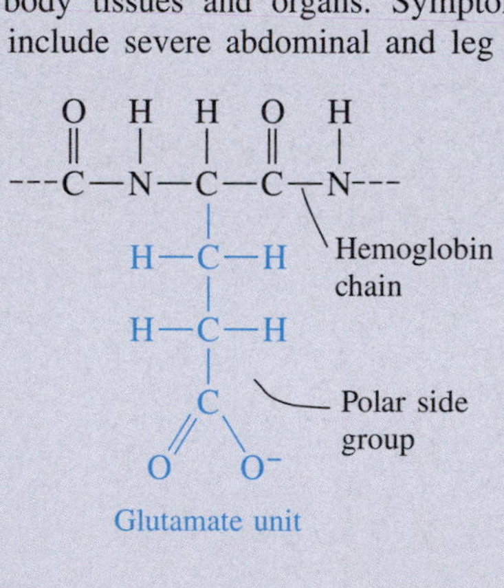

Figure 13.26

(a) The hydrophilic glutamate unit of a normal hemoglobin molecule. (b) In sickle-cell hemoglobin, the glutamate unit is replaced by a hydrophobic valine unit.

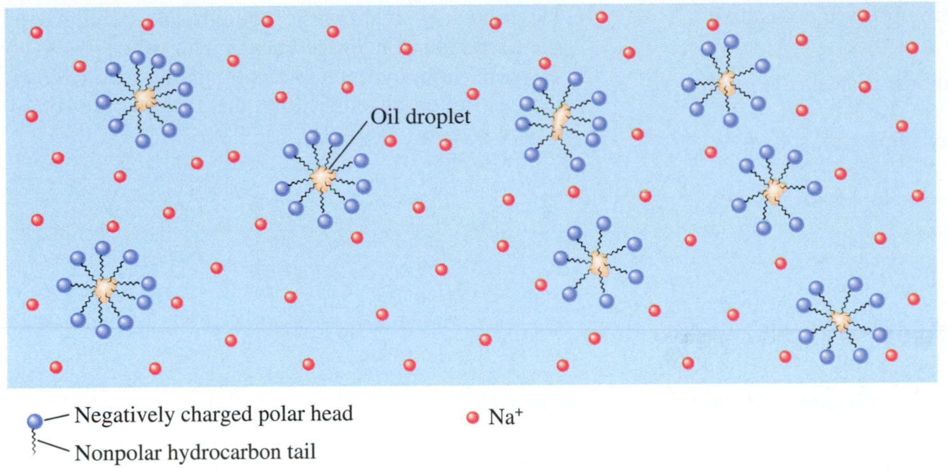

Figure 13.28
Emulsifying action. The nonpolar tails of detergent molecules dissolve in oil droplets while the polar heads protrude in the surrounding water. The oily particle thus formed is called a *micelle*. The polar heads form hydrogen bonds with water, making it easy for the micelles to be washed away.

CHAPTER REVIEW

Starred entries are based on the *Digging Deeper* sections.

LEARNING OBJECTIVES BY SECTION

13.1 1. Describe the effect of solute–solute and solute–solvent attractions on solution formation, and explain why "like dissolves like."
2. Describe how polar water molecules dissolve hydrogen-bonded substances such as C_2H_5OH, ionic solids such as NaCl, and gases such as HCl.
*3. Explain the salting-in and salting-out effects.
4. List each step of the solution process, and state whether the step is endothermic or exothermic.
5. Explain why the heat of solution of a gas in water is usually negative.
6. Explain the effect of temperature on the solubilities of solids and gases.
7. State and explain the effect of increasing pressure on gas solubilities.
8. Use Henry's law to calculate the variation of gas solubility with pressure.

13.2 1. List four properties of ideal solutions, and explain why an ideal solution would have these properties.
2. Calculate the partial pressure of each component and the total vapor pressure above an ideal solution.
3. Explain why some solutions are almost ideal while other solutions deviate substantially from ideality.
4. Explain the causes of positive and negative deviations from Raoult's law.

13.3 1. Describe the following colligative effects and explain why they occur: vapor pressure lowering, boiling point elevation, freezing point depression, and osmotic pressure.
2. Given two of the three quantities, calculate vapor pressure lowering, mole fraction of solute, or vapor pressure of solvent.
3. Calculate the molality of a solution from the masses of solute and solvent.
4. Given two of the three quantities, calculate the boiling point elevation of a solution, molality of solute, or K_b for the solvent.
5. Given two of the three quantities, calculate the freezing point depression of a solution, molality of solute, or K_f for the solvent.
6. Calculate the osmotic pressure of a solution from its molarity, or vice versa.

13.4 1. Calculate molar masses from colligative effects and concentration data.

13.5 1. Calculate the colligative effects of an ionic solute assuming complete dissociation.
2. Explain why the observed colligative effects of an ionic solute are less than the calculated effects.
3. State how the observed colligative effects of an ionic solute vary with increasing ionic charge and increasing concentration of ions.

*4. Calculate van't Hoff factors from observed colligative effects, and vice versa.

13.6 1. Compare the properties of solutions, suspensions, and colloidal dispersions.
2. Distinguish between, and give examples of, sols, gels, emulsions, aerosols, and foams.
3. Distinguish between, and give examples of, hydrophilic and hydrophobic colloids.
4. Describe how hydrophilic colloids are stabilized by the formation of hydrogen bonds with water.
5. Describe how hydrophobic colloids are stabilized by the adsorption of ions and by the adsorption of emulsifying agents.

KEY TERMS BY SECTION

13.1 Heat of solution ($\Delta H^\circ_{\text{solution}}$)
Henry's Law
Hydration
Hydration energy
Ion–dipole attraction
*Salting-in
*Salting-out
Solvation
Solvation energy

13.2 Ideal solution
Negative deviation (from Raoult's law)
Positive deviation (from Raoult's law)
Raoult's law
Real solution

13.3 Boiling point elevation (BPE)
Boiling point elevation constant (K_b)
Colligative property
Freezing point depression (FPD)
Freezing point depression constant (K_f)
Molality (m)
Osmosis
Osmotic pressure (π)
Semipermeable membrane
Van't Hoff equation
Vapor pressure lowering (VPL)

13.5* Van't Hoff factor (i)

13.6 Aerosol
Colloid
Colloidal dispersion
Emulsifying agent
Emulsion
Foam
Gel
Hydrophilic colloid
Hydrophobic colloid
Sol
Solid sol
Suspension
Tyndall effect

IMPORTANT EQUATIONS

13.1 $P = XP^\circ$

13.5 $\text{VPL} = P^\circ_{\text{solvent}} - P_{\text{solution}}$

13.6 $\text{VPL} = X_{\text{solute}}P^\circ_{\text{solvent}}$

13.7 $\text{BPE} = T_b - T_b^\circ$

13.8 $\text{Molality} = \dfrac{\text{number of moles of solute}}{\text{number of kilograms of solvent}}$

13.9 $\text{BPE} = K_b m$

13.10 $\text{FPD} = T_f^\circ - T_f$

13.11 $\text{FPD} = K_f m$

13.12a $\pi V_{\text{solution}} = n_{\text{solute}}RT$

13.12b $\pi = MRT$

***13.13** $i = \dfrac{\text{observed colligative effect}}{\text{calculated effect assuming no dissociation}}$

FINAL EXERCISES

Answers to exercises with blue numbers are given in Appendix D. Starred exercises are based on the *Digging Deeper* sections.

PART A. QUESTIONS AND PROBLEMS BY SECTION

Solubilities (Section 13.1)

13.1 The solubility of one substance in another is a compromise between intermolecular attractions and the natural tendency of the substances to mix. Explain and discuss this statement.

13.2 Describe the three steps of the solution process. State whether each step is endothermic or exothermic.

13.3 Why is water referred to as an "ionizing solvent"? How does water stabilize the existence of ions in solution?

13.4 Explain why "like dissolves like" and why "oil and water don't mix."

13.5 Describe how water would dissolve each of the following substances:
(a) ethanol, C_2H_5OH
(b) hydrogen bromide gas, HBr
(c) potassium nitrate, KNO_3

13.6 Describe how water would dissolve each of the following substances:
(a) calcium chloride, $CaCl_2$
(b) sulfur dioxide gas, SO_2
(c) oxygen gas, O_2

13.7 Why do most solids become more soluble with increasing temperature?

13.8 Why do the solubilities of most gases in water decrease with increasing temperature?

13.9 Explain why the heat of solution of a gas in water is usually negative.

13.10 What effect does (a) increasing temperature and (b) increasing pressure have on the solubility of a gas in water?

13.11 Explain why the solubilities of solids and liquids are not appreciably affected by pressure.

13.12 Explain why soda water bubbles when the cap is removed from the bottle.

***13.13** Account for the fact that nitrogen is slightly less soluble in blood plasma than in pure water.

***13.14** Would you expect barium sulfate, a sparingly soluble solid, to be more soluble or less soluble in seawater than in pure water? Explain your answer.

13.15 The solubility of CO_2 in water is 3.1×10^{-2} mol/L at 25°C and a CO_2 partial pressure of 1.00 atm. Calculate the molar concentration of CO_2 in seltzer water bottled at 25°C and a CO_2 partial pressure of 4.0 atm.

13.16 The solubilities of O_2 and N_2 in water at 0°C are 2.2 and 1.03 mmol/L, respectively, when each gas is at a partial pressure of 1.00 atm. Assume that air contains 21% O_2 and 79% N_2 by molecule count, and compute the millimolar concentrations of each gas in water that is saturated with air at a total pressure of 1.00 atm.

Ideal Solutions and Real Solutions (Section 13.2)

13.17 List four properties of an ideal solution, and explain why such a solution would have these properties.

13.18 Explain in terms of molecular interactions why aqueous solutions are not ideal.

13.19 State the factors that cause (a) negative deviations and (b) positive deviations from Raoult's law.

13.20 Would you expect a positive or a negative deviation from Raoult's law if the mixing process is exothermic? Endothermic?

13.21 At 19°C, the vapor pressures of methanol (CH_3OH) and ethanol (C_2H_5OH) are 89 and 40 torr, respectively. Assume ideal behavior, and sketch Raoult's law graphs similar to those in Figure 13.9.

13.22 At 27°C, the vapor pressures of carbon tetrachloride (CCl_4) and chloroform ($CHCl_3$) are 124 and 214 torr, respectively. Assume ideal behavior and prepare a Raoult's law graph similar to Figure 13.9c for carbon tetrachloride–chloroform solutions.

13.23 A solution contains 100 g of ethyl acetate ($CH_3COOC_2H_5$) and 100 g of carbon disulfide (CS_2) at 27°C.
(a) Assume ideal behavior, and calculate the partial pressure of each component above the solution. The vapor pressures of ethyl acetate and carbon disulfide are 100 and 390 torr, respectively, at 27°C.
(b) Calculate the total pressure above the solution.

13.24 A solution contains 30.0 g of chloroform ($CHCl_3$) dissolved in 70.0 g of carbon tetrachloride (CCl_4) at 27°C.
(a) Assume ideal behavior, and calculate the partial pressure of each component above the solution. Vapor pressure data are given in Exercise 13.22.
(b) Calculate the total pressure above the solution.

13.25 The vapor pressure of pure water is 23.76 torr at 25°C. The vapor pressure of a dilute aqueous solution of urea, a nonvolatile, nonionic solute, is 22.50 torr at the same temperature. Assume that the solvent obeys Raoult's law, and calculate the mole fraction of each component in the solution.

13.26 Using vapor pressure data from Exercise 13.21, and assuming ideality, calculate the mole fraction of each component in a solution of methanol and ethanol whose equilibrium vapor pressure is 60 torr at 19°C.

13.27 The vapor pressure of pure ethanol (C_2H_5OH) is 43.5 torr at 20°C. The mole fraction of ethanol in an aqueous solution is 0.282 and the vapor pressure of ethanol above the solution is 23.4 torr. Does the solution exhibit a deviation from Raoult's law? Is the deviation positive or negative?

13.28 The vapor pressure of acetone (CH_3COCH_3) is 230 torr at 25°C.
(a) Assume ideal behavior, and calculate the partial pressure of acetone above an aqueous solution

containing 0.63 mol of acetone and 0.87 mol of water at 25°C.

(b) The actual vapor pressure of acetone above the solution of Part (a) is 160 torr. Does the solution exhibit a deviation from Raoult's law? Is the deviation positive or negative?

13.29 At 20°C, benzene has a higher vapor pressure and is more volatile than toluene. Calculate the mole fraction of benzene in the vapor above the solution of Example 13.2 (p. 598), and show that it is greater than the mole fraction of benzene in the solution. (This problem illustrates that the vapor above a nearly ideal solution is always richer than the solution in the more volatile component.)

13.30 Refer to Exercise 13.24, and calculate the mole fraction of each component in the vapor above the solution.

Molality (Section 13.3)

13.31 What is the difference between *molality, molarity,* and *mole fraction*?

13.32 Molarity is temperature dependent, but molality is not. Explain why.

13.33 Calculate the molality of a solution containing 0.615 g of dichlorobenzene ($C_6H_4Cl_2$) dissolved in 21.8 g of benzene.

13.34 Calculate the molality of an aqueous solution containing 36.00% aluminum chloride ($AlCl_3$) by mass.

13.35 A 1.50 *m* $CaCl_2$ solution is prepared by dissolving 1.50 mol of $CaCl_2$ in 1000 g of water. What is the total mass of the solution?

13.36 How would you make 500 g of a 1.50 *m* calcium chloride solution?

13.37 A solution of sulfuric acid contains 10.0 g of H_2SO_4 for every 90.0 g of water. Its density is 1.07 g/mL. Calculate (a) the molarity, (b) the mole fraction, and (c) the molality of sulfuric acid in the solution.

13.38 A solution of ammonia contains 24.0% ammonia by mass and has a density of 0.910 g/mL. Calculate (a) the mole fraction, (b) the molality, and (c) the molarity of the ammonia.

13.39 A 2.45 *M* H_2SO_4 solution has a density of 1.15 g/mL. Calculate the molality of this solution.

13.40 The density of a 2.38 *m* aqueous glucose solution is 1.125 g/mL at 20°C. What is the molarity of this solution?

Colligative Properties (Section 13.3)

13.41 In terms of events on the molecular level, explain the cause of (a) vapor pressure depression and (b) boiling point elevation.

13.42 In terms of events on the molecular level, explain the cause of (a) freezing point depression and (b) osmotic pressure.

13.43 For a given solvent, how will each of the colligative effects vary with (a) increasing mass of solute and (b) increasing molar mass of solute?

13.44 In Demonstration 13.2, why do solvent molecules evaporate from the pure solvent and condense into the concentrated solution?

13.45 Which aqueous solution, 0.10 *m* glucose or 0.010 *m* urea, will have (a) the higher boiling point and (b) the higher freezing point?

13.46 Estimate (a) the boiling point and (b) the freezing point of the dichlorobenzene solution in Exercise 13.33. Refer to Table 13.2 for appropriate data.

13.47 Use data from Table 13.2 to estimate the freezing point of the following solutions:

(a) 3.00 g of carbon tetrachloride (CCl_4) in 190 g of benzene

(b) an aqueous 1.51 *m* methanol (CH_3OH) solution

13.48 Use data from Table 13.2 to estimate the boiling point of the following solutions:

(a) 1.00 g of naphthalene ($C_{10}H_8$) in 25.0 g of carbon tetrachloride

(b) 0.0025 mol of benzoic acid (C_6H_5COOH) in 200 g of water

13.49 A solution contains 15.0 g of mannitol ($C_6H_{14}O_6$) dissolved in 500 g of water at 40°C. Estimate (a) the vapor pressure, (b) the boiling point, and (c) the freezing point of the solution. The vapor pressure of water at 40°C is 55.3 torr.

13.50 A 0.10-mol sample of urea ($(NH_2)_2CO$) is dissolved in 100 g of water at 25°C. Estimate (a) the vapor pressure, (b) the boiling point, and (c) the freezing point of the solution. The vapor pressure of water at 25°C is 23.8 torr.

13.51 The osmotic pressure of blood is 7.7 atm at 37°C. Estimate the freezing point of blood plasma. (*Hint*: The molality of a dilute aqueous solution is approximately equal to its molarity.)

13.52 The osmotic pressure of blood is 7.7 atm at 37°C. How would you prepare 0.500 L of a glucose solution that is isotonic with (has the same osmotic pressure as) blood? The formula for glucose is $C_6H_{12}O_6$.

Molar Masses from Colligative Effects (Section 13.4)

13.53 Which solvent, chloroform or ethanol, would you use for a more accurate molar mass determination based on boiling point elevation data? Explain your answer.

13.54 Explain why osmotic pressure is the best colligative effect to use when determining the molar masses of proteins and other large molecules.

13.55 The normal freezing point of cyclohexane is 6.5°C. A solution containing 5.00 g of naphthalene ($C_{10}H_8$) in 100 g of cyclohexane freezes at −1.4°C. A solution containing 2.00 g of an unknown compound in 100 g of cyclohexane freezes at 3.4°C.
- **(a)** Calculate the freezing point depression constant of cyclohexane.
- **(b)** Calculate the molar mass of the unknown compound.

13.56 Paramethadione, a drug that has been used as an anticonvulsant, is freely soluble in benzene. The freezing point of a solution containing 0.286 g of paramethadione in 25.0 g of benzene is 5.17°C. Calculate the molar mass of paramethadione.

13.57 Reserpine is a natural product that has been used as a tranquilizer. The boiling point of a solution containing 3.00 g of reserpine in 20.0 g of chloroform is 0.893°C higher than the boiling point of pure chloroform. Calculate the molar mass of reserpine.

13.58 A solution containing 3.00 g of an unknown compound dissolved in 60.0 g of water boils at 100.28°C. Calculate the molar mass of the unknown compound.

13.59 A solution contains 25.0 g of a nonvolatile organic compound dissolved in 250 g of benzene. The vapor pressure of the solution is 6.35 torr less than the vapor pressure of pure benzene. If the vapor pressure of pure benzene is 74.7 torr, calculate the molar mass of the compound.

13.60 An aqueous solution contains 0.0120 g of bovine insulin per milliliter. The osmotic pressure of this solution is 38.9 torr at 28°C. Calculate the molar mass of bovine insulin.

Colligative Properties of Ionic Solutions (Section 13.5)

13.61 Compare seawater and pure water with respect to the following properties:
- **(a)** boiling point
- **(b)** freezing point
- **(c)** vapor pressure
- **(d)** osmotic pressure

13.62 Sodium chloride and calcium chloride are used for melting ice and snow on roads and sidewalks. Why are these substances preferred over covalent substances such as sucrose and urea?

13.63 **(a)** Why are the observed colligative effects of an ionic solute always smaller than the effects calculated on the basis of complete dissociation?
(b) Why do the observed effects approach the calculated values when the solution becomes dilute?

13.64 How do the observed colligative effects of an ionic solute vary with (a) increasing ionic charge and (b) increasing concentration of ions?

13.65 Which solution will have the lower freezing point, 1.0 *m* $CaCl_2$ or 1.0 *m* NaCl? Explain your answer.

13.66 List the following aqueous solutions in order of increasing boiling point: 0.1 *m* $MgCl_2$ (ionic), 0.1 *m* urea (nonionic), 0.2 *m* glucose (nonionic).

13.67 Assuming complete dissociation, estimate the freezing and boiling points of a 0.100 *m* $Al_2(SO_4)_3$ solution. Will the actual freezing and boiling points be greater than, less than, or the same as your estimated values? Explain your answer.

13.68 Assuming complete dissociation, estimate the freezing and boiling points of (a) the H_2SO_4 solution in Exercise 13.39 and (b) the $AlCl_3$ solution in Exercise 13.34.

13.69 A seawater sample freezes at −2.00°C. Calculate (a) the total molality and (b) the boiling point of the sample.

13.70 A 1.00 *m* aqueous solution of $ZnCl_2$ freezes at −5.21°C. Estimate the boiling point of the solution.

***13.71** The observed freezing point depression of a 1.00 *m* H_2SO_4 solution is 4.04°C. Calculate the van't Hoff *i* factor.

***13.72** The van't Hoff *i* factor for a 0.0100 *m* $MgSO_4$ solution is 1.53. Calculate the boiling point of the solution.

***13.73** The van't Hoff *i* factor for a 0.100 *m* NaCl solution is 1.87. Calculate (a) the freezing point, (b) the boiling point, and (c) the osmotic pressure of the solution at 25°C. Assume that the density of the solution is 1.00 g/mL.

***13.74** A 0.100 *m* aqueous solution of phosphoric acid freezes at −0.230°C. Calculate the van't Hoff *i* factor, and use it to estimate the boiling point of the solution.

Colloids (Section 13.6)

13.75 Distinguish between solutions, suspensions, and colloidal dispersions on the basis of particle size. Give three examples of each.

13.76 Compare the properties of solutions, suspensions, and colloidal dispersions.

13.77 What is the Tyndall effect and what causes it? Will suspensions give a Tyndall effect?

13.78 How can one experimentally distinguish between a colloidal dispersion and a solution?

13.79 Give at least two examples of the following colloidal dispersions:
(a) sol
(b) gel
(c) emulsion
(d) aerosol
(e) foam

13.80 Give at least two examples of the following colloidal dispersions:
(a) solid sol
(b) liquid foam
(c) solid foam
(d) aerosols containing liquid colloidal particles
(e) aerosols containing solid colloidal particles

13.81 What is a *hydrophilic sol*? Give two examples of this type of sol.

13.82 What is a *hydrophobic sol*? Give two examples of this type of sol.

13.83 Explain why aqueous dispersions of hydrophilic colloids are stable.

13.84 Explain why hydrophobic sols are not stable in ionic solutions.

13.85 Discuss the role of adsorbed ions in preventing colloidal particles from settling out of air or aqueous solutions.

13.86 Give some examples of emulsifying agents, and describe how they stabilize hydrophobic colloids in aqueous solutions.

PART B. MISCELLANEOUS QUESTIONS AND PROBLEMS

13.87 Explain why the solubility of sodium hydroxide increases with increasing temperature even though the overall solution process is exothermic.

13.88 Is the following statement true or false? Explain. The freezing point of a solution such as salt water is the temperature at which the vapor pressure of the solution is equal to the vapor pressure of the solid solvent.

13.89 Briefly explain each of the following observations:
(a) Salads made with salt and vinegar wilt within short periods of time.
(b) People get thirsty after eating salty foods.
(c) It is sometimes possible to freshen wilted flowers by transferring them to pure water.
(d) Many chefs prepare oil and vinegar salads by first tossing the vegetables with oil and then adding the vinegar.

13.90 **(a)** Explain why the molality of a dilute aqueous solution is approximately equal to its molarity.
(b) Would this equality exist in nonaqueous solutions? Explain.

13.91 What difficulties might arise if we tried to use molalities in acid–base titrations? Molarities in dealing with freezing and boiling point changes?

13.92 *Association* is the opposite of dissociation. For example, acetic acid (CH_3COOH) associates to form species such as $(CH_3COOH)_2$ in concentrated aqueous solution. What effect will association have on colligative properties?

13.93 The solubility of potassium nitrate is 13.3 g/100 g of water at 0°C and 247 g/100 g of water at 100°C.
(a) Estimate the minimum volume of water needed to dissolve 50.0 g of potassium nitrate at 100°C.
(b) How many grams of potassium nitrate will recrystallize if the solution in Part (a) is cooled to 0°C?

13.94 Two or more soluble substances can be partially separated by taking advantage of the solubility decrease on cooling. At 25°C, the solubilities of the amino acids glycine and alanine are 24.99 g/100 g water and 16.72 g/100 g water, respectively. At 75°C, the solubilities increase to 54.39 g/100 g water and 31.89 g/100 g water, respectively. Suggest a method for separating a mixture containing 25.0 g of each amino acid. Which amino acid component and how much of it will be obtained in a pure state? (This method is called *fractional crystallization*.)

13.95 A benzene–toluene solution, containing benzene at a mole fraction of 0.451, boils at 94.8°C under an external pressure of 1.00 atm. The vapor pressures of pure benzene and pure toluene are 1108 and 474 torr, respectively, at this temperature.
(a) Assume ideal behavior, and calculate the composition of the vapor above the solution.
(b) If the vapor of Part (a) is condensed to a liquid, what would be its vapor pressure at 94.8°C?

13.96 The density of the hemoglobin solution in Example 13.11 (p. 612) is 1.00 g/mL. Estimate (a) the boiling point, (b) the freezing point, and (c) the vapor pressure of the solution. The vapor pressure of pure water at 27°C is 26.7 torr.

13.97 **(a)** Estimate the number of grams of ethylene glycol ($C_2H_6O_2$) that must be added to 1.0 L of water to prevent freezing at −25°F.
(b) Estimate the boiling point of the solution in Part (a).
(c) Do you think that your answers to Parts (a) and (b) will agree with experiment? Explain why or why not.

13.98 Use the results of Exercise 13.16 to estimate the freezing point of water saturated with air at 0°C.

13.99 An osmotic pressure experiment is set up as in Figure 13.14. Pure water is on one side of the membrane and a 0.010 *M* protein solution is on the other side. Estimate the maximum height that can be attained by the protein solution at 25°C. (*Hint*: Assume that the density of mercury is 13.6 times greater than the density of the solution.)

13.100 Two beakers are placed next to each other as in Demonstration 13.2. One beaker contains 2.00 g of sucrose ($C_{12}H_{22}O_{11}$) dissolved in 30.0 g of water. The other contains 2.00 g of vitamin C ($C_6H_8O_6$) dissolved in 30.0 g of water. The temperature is 30°C.

(a) Which solution will become more concentrated, and which will become more dilute?

(b) If the temperature is kept constant, estimate the mole fraction of each solute at equilibrium.

13.101 The male hormone testosterone contains 79.12% C, 9.79% H, and 11.09% O. A solution containing 0.363 g of the hormone in 5.00 g of benzene freezes at 4.27°C.

(a) Use data from Table 13.2 to obtain an approximate molar mass for testosterone.

(b) Use your result from Part (a) to find the formula and precise molar mass of testosterone.

13.102 **(a)** Calculate the freezing point of a 0.0100 *m* aqueous acetic acid solution assuming that acetic acid does not ionize.

(b) The actual freezing point of a 0.0100 *m* acetic acid solution has been reported as −0.0195°C. What fraction of the acetic acid molecules are ionized? The ionization equation is

$$CH_3COOH(aq) \rightleftharpoons H^+(aq) + CH_3COO^-(aq)$$

CHAPTER 14

A mixture of hydrogen and oxygen gases reacts on the surface of a glowing platinum wire. The platinum is a contact catalyst for this reaction.

CHEMICAL KINETICS

OUTLINE

PREVIEW

Iron rusts slowly but wood burns quickly—different reactions occur at different rates. The rate of any one reaction also depends on conditions. For example, sugar oxidizes rapidly in a flame, slowly in a living cell, and not at all on a pantry shelf. Understanding reaction rates often helps us to control them. We put food in the refrigerator to retard spoilage, and we put catalytic converters on cars to speed up the conversion of pollutants into harmless gases. When chemicals are synthesized in industry, the reaction conditions are carefully chosen to produce the maximum yield of product in the shortest period of time.

The study of motion is called **kinetics**, from the Greek word *kinesis,* meaning "movement." **Chemical kinetics** is the study of the rates of chemical reactions. Chemical kinetics is also concerned with the atomic rearrangements that occur during chemical reactions. A knowledge of these rearrangements or "mechanisms" helps us to understand why reactions are fast or slow and may help us find methods for controlling their rates.

14.1 REACTION RATES AND THE FACTORS THAT INFLUENCE THEM

Rate is change per unit time. The rate of motion of a moving vehicle, for example, is expressed in miles per hour or kilometers per hour and is obtained by dividing the distance traveled by the elapsed time; rate of motion = distance/time. A **chemical reaction rate** is expressed as a change in the amount or concentration of some reactant or product per unit time. Time can be measured in seconds, minutes, hours, days, or years, depending on the speed of the reaction. If a reaction rate is expressed in terms of the amount of substance, the rate units will be moles/second, liters/minute, or some other units giving amount per unit time. If a rate is expressed in terms of a concentration such as molarity or partial pressure, the units will be molarity/second (mol/L·s), torr/minute, and so forth.

The following factors influence the rates of chemical reactions:

1. the nature of the reactants
2. the reactant concentrations and state of subdivision
3. the temperature
4. the presence or absence of a catalyst

We will briefly discuss each of these factors in preparation for more detailed discussions later in the chapter.

Nature of the Reactants

Some chemical reactions are completed in seconds while others go on for years. Each reaction has its own rate under a given set of conditions, and there is no complete set of rules for predicting whether that rate will be fast or slow. The following observations, however, are useful for making limited predictions.

1. *Ions in solution tend to react quickly.* Most ionic precipitates seem to form instantly, and most aqueous acid–base neutralizations are complete within microseconds or less. Ionic reactions are rapid because of the high mobility of dissolved ions and the electrostatic attractions between them.
2. *Covalent molecules, especially large ones, tend to react slowly.* Solutions of organic reactants may require hours or days to yield an appreciable amount of product.
3. *Molecules with strong covalent bonds tend to react more slowly than molecules with weak bonds.* Elemental oxygen and nitrogen, for example, persist in the atmosphere because O_2 and N_2 have high bond dissociation energies. The chlorine molecule, on the other hand, has a relatively low bond energy, so Cl_2 usually reacts quickly.

The dissociation energies of N_2, O_2, and Cl_2 are 945, 498, and 243 kJ/mol, respectively.

Concentration and Subdivision

Molecules and ions must collide in order to react with each other. In discussing the effect of collisions, it is useful to distinguish between homogeneous and heterogeneous reactions. A **homogeneous reaction** is one in which all substances are dissolved in a single solution. The molecules, especially those in liquid or gaseous solutions, are free to move around and collide with each other. Increasing the concentration of one of the reactants will increase the frequency with which its molecules encounter others, and this will often lead to an increased reaction rate.

This grain elevator was destroyed by a dust explosion.

A **heterogeneous reaction** is one in which the reactants are in different states and the reaction occurs at the surface or *interface* where the two states meet. Examples of heterogeneous reactions include coal burning in air and zinc metal reacting with hydrochloric acid. The rate of a heterogeneous reaction can be increased by subdividing the condensed phase to provide more surface area. Lumps of coal, for example, burn slowly, but coal dust suspended in air may burn fast enough to produce an explosion (see Demonstration 14.1). The concentration of reactants in the less condensed state also affects the rate of a heterogeneous reaction. Zinc metal

DEMONSTRATION 14.1 A DUST EXPLOSION

A compact pile of lycopodium powder (spores of certain mosses) burns sluggishly.

A similar sample of lycopodium powder in a blowpipe.

The blown powder burns instantly with a bright flash.

reacts more rapidly with a concentrated solution of hydrochloric acid than with a dilute solution (Figure 14.1). A cigarette that glows in air (20% oxygen) will burst into flame when the oxygen supply is enriched.

Temperature

All reaction rates increase with increasing temperature. The effect of temperature is well known; we heat food to cook it, and we cool food to delay its spoiling. Before the days of air-conditioned darkrooms, photographers observed that the time needed to develop a film was approximately halved whenever the temperature increased by 10°C. From this observation came the "photographers' rule of thumb": *Reaction rates approximately double for each 10-degree rise in Celsius temperature.* A reaction following this rule would go twice as fast at 30°C as at 20°C, four times as fast at 40°C, and eight times as fast at 50°C. Many reactions show temperature effects of about this magnitude, but many others do not, so any prediction based on this rule is, at best, a rough estimate.

Figure 14.1

The effect of concentration on reaction rate. Zinc reacts visibly faster (more hydrogen bubbles) with 3*M* hydrochloric acid (right) than with 1.5 *M* HCl (left).

EXAMPLE 14.1

Water boils at about 120°C in a pressure cooker set at 15 lb/in^2 above atmospheric pressure. Compare the approximate cooking times for vegetables boiled in the pressure cooker and vegetables boiled in an open pot of water at 100°C.

SOLUTION

The temperature in the pressure cooker is 20°C higher than in the open pot. This increase is equivalent to two temperature rises of 10 degrees each. The rate of cooking is therefore doubled and redoubled. The vegetables in the pressure cooker will cook $2 \times 2 = 4$ times faster than in the open pot; the cooking time should decrease by a factor of 4.

PRACTICE EXERCISE 14.1

The growth rate of a certain colony of bacteria obeys the photographers' rule of thumb. Compare the growth rate of this colony at 35°C with the rate at 5°C.

Catalysts

A **catalyst** is a substance that increases the rate of a chemical reaction without itself being consumed in the reaction. Catalysts are widely used in industry to speed up reactions that would otherwise be too slow to be practical, and they are often used to carry out processes at lower and more economical temperatures.

In **homogeneous catalysis**, the catalyst is in solution with the reactants. Gaseous nitrogen monoxide, for example, speeds up the gas phase oxidation of sulfur dioxide to sulfur trioxide; the NO does not appear in the balanced equation, although it may be written over the arrow:

$$2SO_2(g) + O_2(g) \xrightarrow{NO(g)} 2SO_3(g)$$

In **heterogeneous catalysis**, the catalyst provides a surface on which the reaction takes place. Such catalysts are called **contact catalysts**. Platinum powder is a contact catalyst that accelerates the combination of gaseous hydrogen and oxygen to form water:

$$2H_2(g) + O_2(g) \xrightarrow{Pt(s)} 2H_2O(g)$$

Without either a catalyst or a spark to initiate the reaction, a mixture of hydrogen and oxygen gases could remain unchanged indefinitely.

A catalyst that works for one reaction may or may not work for another reaction. Powdered platinum has broad catalytic activity; it accelerates numerous reactions that involve hydrogen, oxygen, or gaseous hydrocarbons. **Enzymes**, the catalysts for biological reactions, have narrow catalytic activity; most of the thousands of enzymes present in living cells promote just one step each in the interlocking complex of biochemical processes.

14.2 MEASURING REACTION RATES

To determine a reaction rate, we must measure the changing amount of some reactant or product. Various techniques can be used, depending on the substance being measured. A volume of gas released during a reaction, for example, can be monitored using a gas buret (Figure 14.2). Concentration changes can be measured by analyzing small samples withdrawn from a reacting mixture at fixed time intervals. Properties related to concentration can often be monitored without disturbing the reacting system. A color change, for example, can be followed with a spectrophotometer, an instrument that measures the fraction of light absorbed at a chosen wavelength. Changes in ionic concentration can be followed by measuring electrical conductance. These methods are suitable for reactions that proceed at moderate rates; very fast reactions—those that are complete within milliseconds or less—are more difficult to study and usually present a challenge to the investigator's ingenuity.

Consider the decomposition of nitrogen dioxide gas into nitrogen monoxide and oxygen:

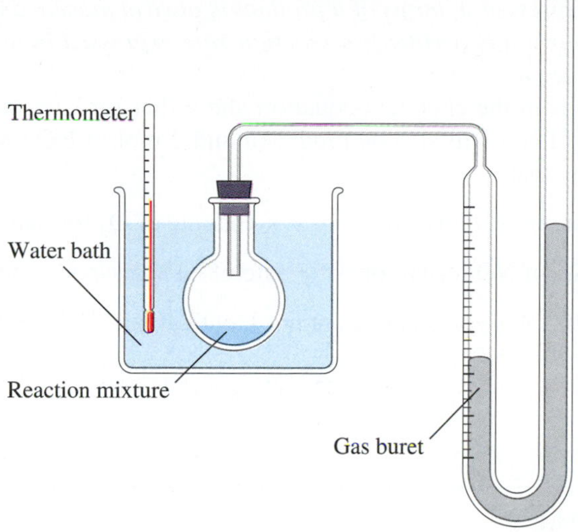

Figure 14.2
A reaction vessel with a gas buret for monitoring the volume of evolved gas. The reaction rate is proportional to the increase in volume of the gas divided by the elapsed time.

$$\underset{\text{reddish brown}}{2NO_2(g)} \rightarrow \underset{\text{colorless}}{2NO(g) + O_2(g)}$$

The rate of this reaction can be found by measuring the rate at which NO is formed. It can also be found by measuring the rate of formation of O_2 or the rate of disappearance of NO_2.

Imagine that we measure the concentration of NO in the reaction vessel. Later, when more NO has been produced, we measure it again. The increase in the concentration of NO is

$$\Delta[NO] = \text{final concentration of NO} - \text{initial concentration of NO}$$

The symbol [NO] represents the concentration of NO in moles per liter. (Brackets are frequently used to represent the molarity of a substance.) The time interval over which this change in concentration occurs is

$$\Delta t = \text{final time} - \text{initial time}$$

The average reaction rate can be expressed as

$$\text{rate of NO formation} = \frac{\Delta[NO]}{\Delta t}$$

The expression $\Delta[NO]/\Delta t$ gives the rate at which the NO concentration increases over the time interval. The reaction rate can also be expressed in terms of the rate of formation of oxygen:

$$\text{rate of } O_2 \text{ formation} = \frac{\Delta[O_2]}{\Delta t}$$

Note that the rates $\Delta[NO]/\Delta t$ and $\Delta[O_2]/\Delta t$ are positive because $\Delta[NO]$, $\Delta[O_2]$, and Δt are positive.

The rate at which NO_2 disappears is $\Delta[NO_2]/\Delta t$, but this quantity is negative because $\Delta[NO_2]$ is negative. A reaction rate is inherently positive, so when we write the reaction rate in terms of NO_2, we express it as $-\Delta[NO_2]/\Delta t$, that is

$$\text{reaction rate} = -\text{rate of } NO_2 \text{ disappearance} = -\frac{\Delta[NO_2]}{\Delta t}$$

A reaction rate expressed in terms of a product is always preceded by a positive sign (which may or may not be written); a reaction rate expressed in terms of a reactant is preceded by a negative sign.

The coefficients in the chemical equation show that for each mole of O_2 that is produced, 2 mol of NO will also be produced and 2 mol of NO_2 will be consumed. These ratios imply that

$$\text{rate of NO formation} = 2 \times \text{the rate of } O_2 \text{ formation}$$

$$\text{rate of NO formation} = -\text{rate of } NO_2 \text{ disappearance}$$

In other words, the rates given above are not identical but are related as follows:

$$\frac{\Delta[\mathrm{NO}]}{\Delta t} = 2\frac{\Delta[\mathrm{O_2}]}{\Delta t} = -\frac{\Delta[\mathrm{NO_2}]}{\Delta t}$$

EXAMPLE 14.2

One step in the commercial synthesis of nitric acid (the Ostwald process) consists of passing nitrogen dioxide through water to produce aqueous nitric acid and nitrogen monoxide gas:

$$3NO_2(g) + H_2O(l) \longrightarrow 2HNO_3(aq) + NO(g)$$

The nitrogen dioxide is consumed at the rate of 0.30 mol/L·s. Express the reaction rate in terms of the concentration of HNO_3.

SOLUTION

The coefficients in the equation show that 2 mol of HNO_3 are produced for every 3 mol of NO_2 consumed. Hence,

$$\begin{aligned}\text{reaction rate} &= \frac{\Delta[\mathrm{HNO_3}]}{\Delta t} \\ &= \frac{0.30 \text{ mol } \mathrm{NO_2}}{\mathrm{L \cdot s}} \times \frac{2 \text{ mol } \mathrm{HNO_3}}{3 \text{ mol } \mathrm{NO_2}} \\ &= 0.20 \text{ mol } \mathrm{HNO_3/L \cdot s}\end{aligned}$$

PRACTICE EXERCISE 14.2

Find $\Delta[\mathrm{NO}]/\Delta t$ for the reaction in Example 14.2.

DIGGING DEEPER

Average and Instantaneous Reaction Rates

A typical kinetics experiment yields a series of concentration and time readings that must be converted into rates. In the nitrogen dioxide decomposition described above, the reactant NO_2 is reddish brown and the products NO and O_2 are colorless;

hence, the color of the reacting mixture will fade as the reaction proceeds. The instantaneous concentration of NO_2 in the reaction mixture can be determined from the intensity of its color. The results of one experiment are given in Figure 14.3.

The **average reaction rate** over a time interval is equal to the change in concentration divided by the change in time. Consider the interval between 20 and 40 seconds, during which the concentration of NO_2 drops from 0.0116 mol/L to 0.0082 mol/L. The change in concentration over this interval is

$$\Delta[NO_2] = 0.0082 \text{ mol/L} - 0.0116 \text{ mol/L} = -0.0034 \text{ mol/L}$$

The time interval is

$$\Delta t = 40 \text{ s} - 20 \text{ s} = 20 \text{ s}$$

Thus the average reaction rate between 20 and 40 seconds is

Figure 14.3

Plot of NO_2 concentration versus time for the reaction $2NO_2(g) \longrightarrow 2NO(g) + O_2(g)$ at 354°C. The average reaction rate between 20 and 40 s is the negative of the slope of the straight line connecting the two points.

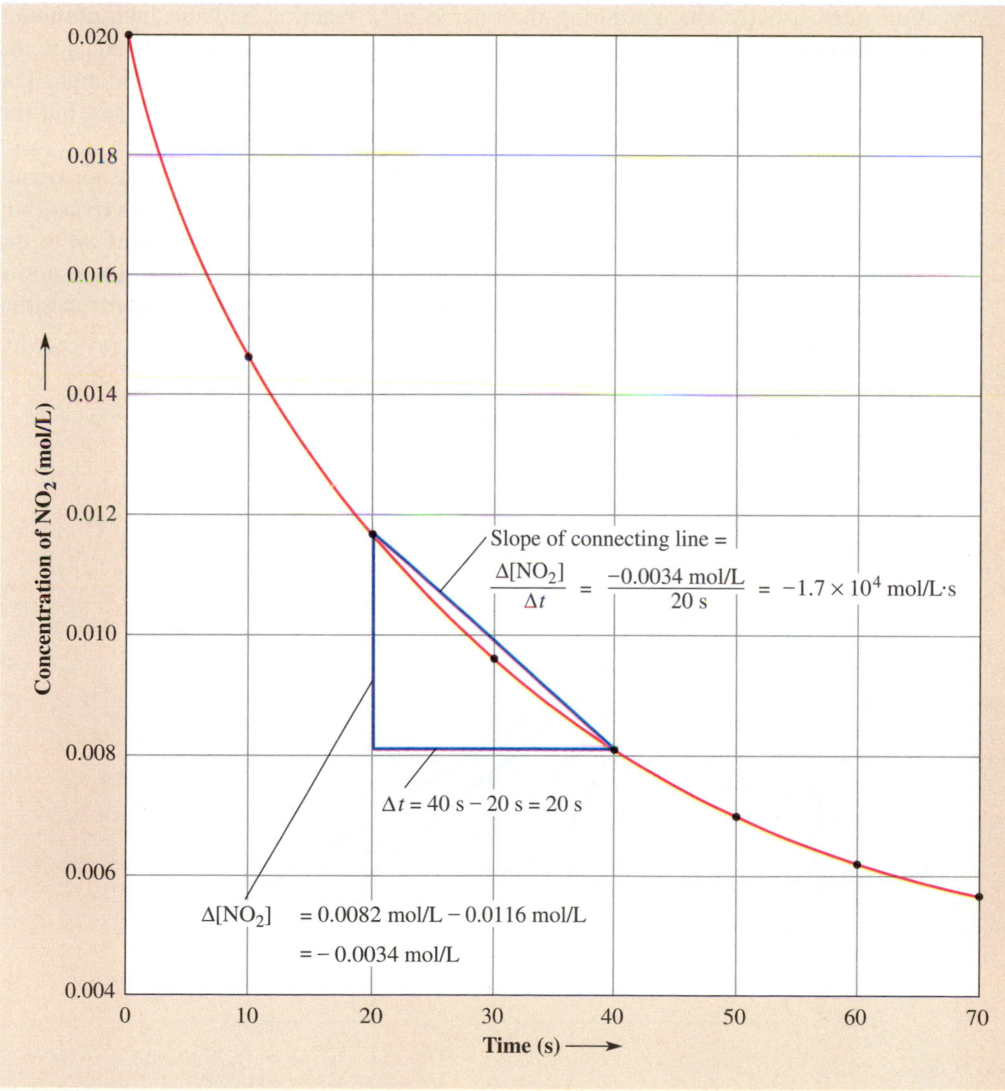

time (s)	$[NO_2]$ (mol/L)
0.0	0.0200
10.0	0.0147
20.0	0.0116
30.0	0.0096
40.0	0.0082
50.0	0.0071
60.0	0.0063
70.0	0.0057

LEARNING HINT

Review Appendix A.3. The equation for a straight line is $y = mx + b$, where m is the slope of the line and b is its intercept with the y axis. The slope $m = \Delta y/\Delta x$ is positive if y increases with increasing x and negative if y decreases with increasing x.

$$\text{average rate} = -\frac{\Delta[NO_2]}{\Delta t} = -\frac{(-0.0034 \text{ mol/L})}{20 \text{ s}} = 1.7 \times 10^{-4} \text{ mol/L·s}$$

The negative sign is used because NO_2 is a reactant and its concentration is decreasing. Inspection of Figure 14.3 shows that the average reaction rate between 20 and 40 s is the negative of the slope of the straight line connecting the points (t = 20 s, $[NO_2]$ = 0.0116 mol/L) and (t = 40 s, $[NO_2]$ = 0.0082 mol/L).

PRACTICE EXERCISE 14.3

Refer to Figure 14.3 and show that the average reaction rate between 50 and 70 s is 7.0×10^{-5} mol/L·s.

Observe that the average rate between 50 and 70 s is less than the rate between 20 and 40 s—the reaction rate decreases as the concentration of NO_2 decreases. Reaction rates usually change during the course of a reaction and the **instantaneous reaction rate**, that is, the rate at some particular instant of time, is rarely equal to the average rate. Imagine, as an analogy, a car accelerating from 40 mph to 50 mph. The instantaneous speed during the period of acceleration constantly increases, but the average speed during the interval is 45 mph.

Instantaneous reaction rates can be obtained from graphs of concentration versus time. Figure 14.4 shows that when the time interval between two measurements is very small, the straight line connecting the measurements is almost identical to the tangent line to the curve. In Figure 14.4, P_1 is a point corresponding to some initial concentration c_1 at time t_1. The points P_2, P_3, and P_4 correspond to later measure-

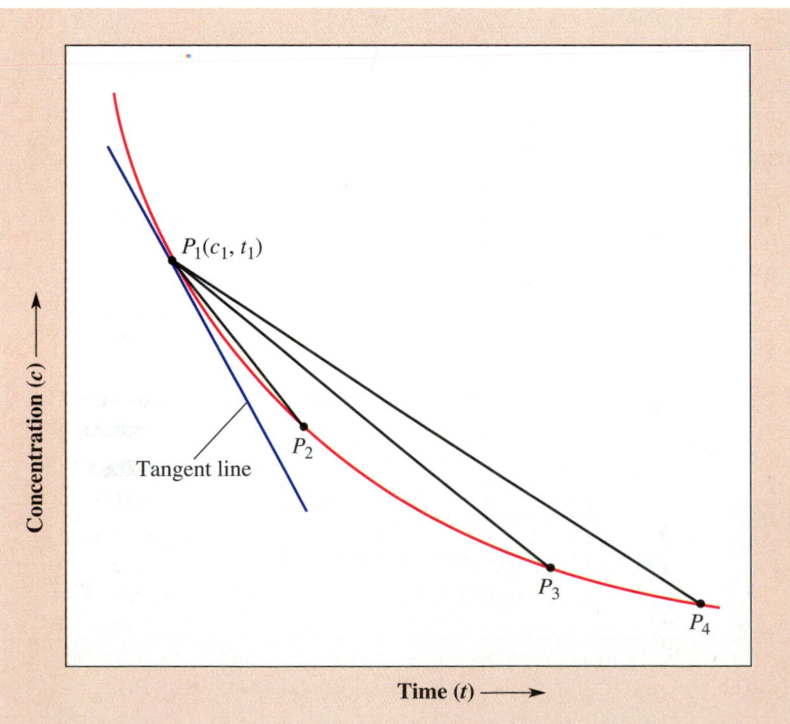

Figure 14.4

When the time interval between two measurements becomes small, the straight line connecting the measurements approaches the tangent line to the curve. The point P_1 corresponds to some initial measurement; the points P_2, P_3, and P_4 correspond to measurements made at longer time intervals.

ments made at longer and longer time intervals. The sketch shows that as the time interval decreases, the measured points approach P_1 and the connecting lines approach the tangent to the curve at P_1. Thus the instantaneous rate at P_1 is equal to the negative of the slope of the tangent line at P_1. The use of tangent lines for determining instantaneous reaction rates is illustrated in the following example.

EXAMPLE 14.3

Use Figure 14.5 to estimate the instantaneous reaction rate for the NO_2 decomposition at $t = 20$ s.

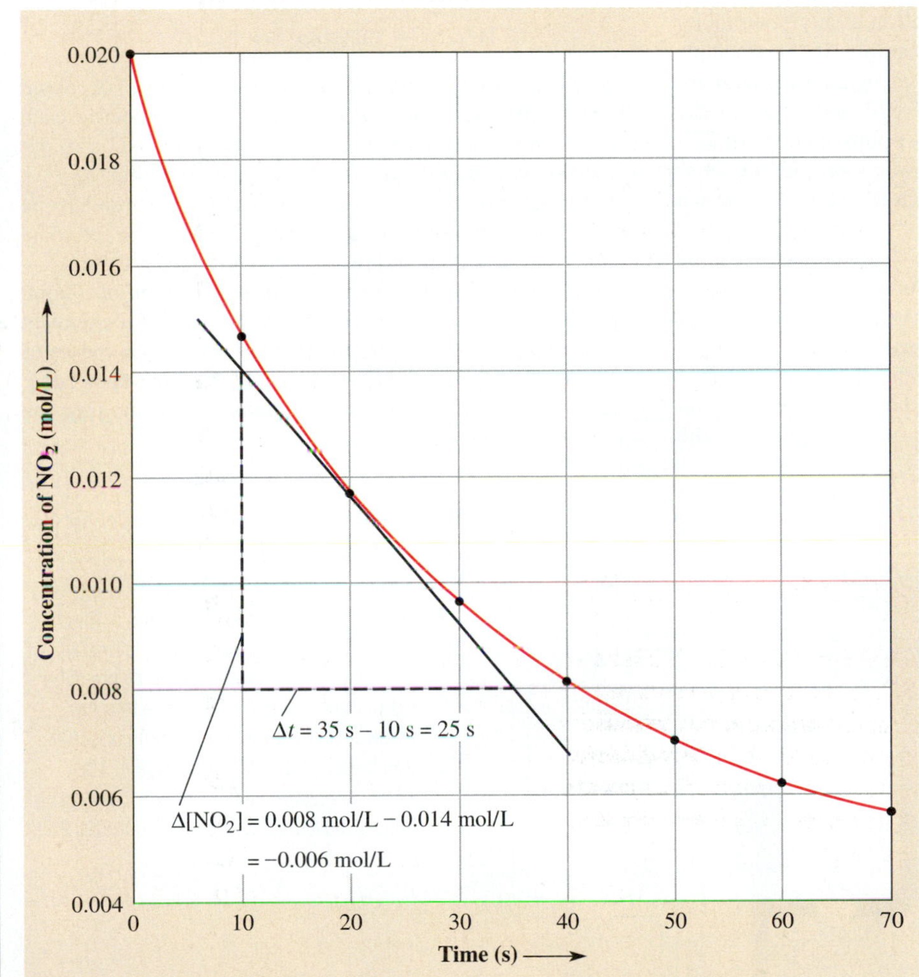

Figure 14.5
The NO_2 concentration versus time curve of Figure 14.3 with a tangent line drawn at $t = 20$ s.

LEARNING HINT
The instantaneous reaction rate is equal to the negative slope of the tangent line only when the concentration decreases with time. When the concentration increases, the rate is equal to the positive slope of the tangent line. Try drawing some sketches to convince yourself of this.

SOLUTION

Figure 14.5 shows the tangent line to the $[NO_2]$ versus time curve at $t = 20$ s. A right triangle has been drawn with part of the tangent line as the hypotenuse. The legs of the

CHEMICAL INSIGHT

HOW FAST IS FAST?

Reactions that are over within milliseconds (10^{-3} s) can be investigated with *flow techniques* (see Figure 14.6). The reactants are mixed at the entrance of a transparent tube. The mixture is forced through the tube at a known and constant speed so that each point along the tube corresponds to a known reaction time. The concentration of any reactant or product changes as it flows along the tube, but because the rate of flow is constant, the concentration at any point along the tube will remain constant. For example, if the reactants have a violet color and the products are green, a stationary display of violet shading into green will appear along the tube. Color measurements at various points and the time it takes for the mixture to reach those points provide the concentration versus time data from which the reaction rate can be determined.

Processes that we tend to regard as "instantaneous" can be investigated by *relaxation methods.* For example, to measure the rate of ionization of acetic acid (CH_3COOH), an aqueous solution of the acid is subjected to a sudden pulse of energy that raises it to a temperature where the fraction of ionized molecules would normally be higher than at the lower initial temperature. During the microseconds or less required for the system to "relax," that is, to reach the degree of ionization appropriate to the new temperature, the changing ionic concentrations are monitored by an oscillograph trace of the solution's electrical conductivity. Other techniques have extended rate determinations into the nanosecond (10^{-9} s) and picosecond (10^{-12} s) ranges.

One of the fastest known reactions is

$$H^+(aq) + OH^-(aq) \longrightarrow H_2O(l)$$

It has been found that water forms at a rate of about 10^{11} mol/s in a solution that is 1 molar with respect to each ion. One mole of water will form in about 10^{-11} s, or, to put it another way, most ions find their partners within 10^{-11} s. How far down on the time scale can kinetic measurements be made? Each vibration of a chemical bond—the periodic lengthening and contracting of the internuclear distance—takes about 10^{-13} s. The formation or dissociation of a bond requires at least the time of one vibration, so that a chemical change will not occur more rapidly than this. Laser pulses of a few hundred femtoseconds (1 fs $=$ 10^{-15} s) are now being used to probe certain gaseous reactions and to monitor the actual bond-breaking process.

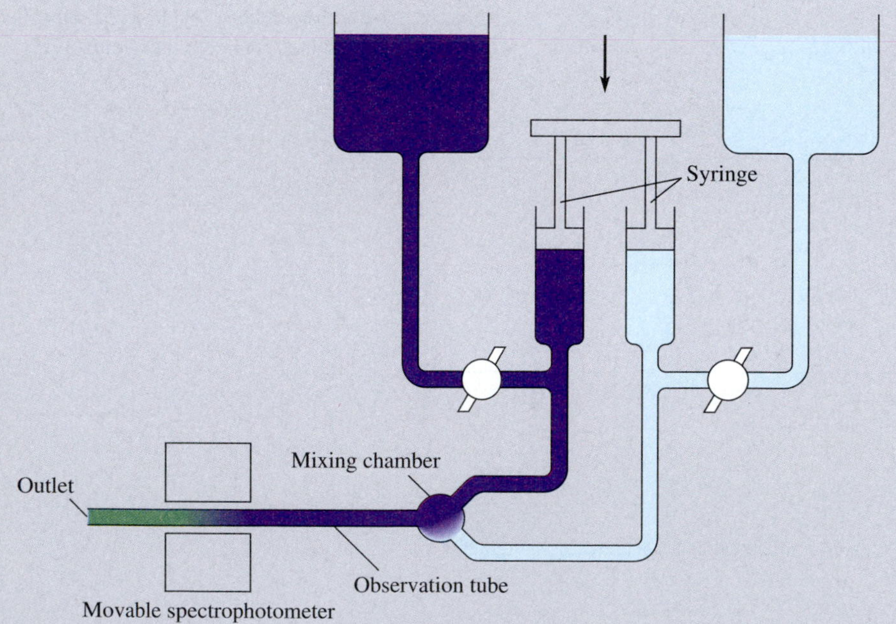

Figure 14.6
A flow apparatus for measuring the rate of a fast reaction. The double syringe is filled with reactants (one violet, one colorless) from the two reservoirs above. Depressing the pistons sends the reactants into the mixing chamber and through the observation tube at a constant known rate. The extent of reaction at a given point along the tube is measured by the spectrophotometer.

triangle are $\Delta[NO_2] = -0.006$ mol/L and $\Delta t = 25$ s. The slope of the tangent line is

$$\text{slope} = \frac{\Delta[NO_2]}{\Delta t} = \frac{-0.006 \text{ mol/L}}{25 \text{ s}} = -2 \times 10^{-4} \text{ mol/L·s}$$

Because concentration is decreasing with time, the instantaneous reaction rate is the negative of the slope, or 2×10^{-4} mol/L·s.

PRACTICE EXERCISE 14.4

Estimate the instantaneous rate for the NO_2 decomposition at 40 s.

14.3 RATE LAWS

Most chemical reaction rates vary with reactant concentrations; some rates vary with product concentrations as well. The relation between concentration and rate must be determined by experiment. One way to determine this relation, called the *method of initial reaction rates,* is to run the reaction several times using different starting concentrations each time and keeping the temperature and other conditions constant. The initial reaction rate is measured for each run.

We will illustrate this method for the reaction

$$2NO(g) + 2H_2(g) \rightarrow N_2(g) + 2H_2O(g)$$

Table 14.1 summarizes initial rates and starting reactant concentrations for three different runs at 866°C. The effect of a given reactant on the rate is determined by comparing runs that differ only in the concentration of that reactant. Compare Run 1 with Run 2. The NO concentration is the same in both runs, but the concentration of hydrogen, $[H_2]$, is four times greater in Run 2 than in Run 1. The rate in Run 2 is also four times greater than in Run 1 ($0.246/0.062 = 4$). The reaction rate changes by the same factor as the hydrogen concentration; therefore, the rate is proportional to $[H_2]$. Now compare Runs 1 and 3, in which the hydrogen concentration is con-

LEARNING HINT

The initial rate of a reaction is the instantaneous rate at the beginning of a run. It is determined from the slope of the tangent line to the concentration vs. time graph at time = 0. See *Digging Deeper:* Average and Instantaneous Reaction Rates.

TABLE 14.1 Dependence of Initial Rate on Initial Reactant Concentration in the Reaction $2NO(g) + 2H_2(g) \rightarrow N_2(g) + 2H_2O(g)$ at 866°C

Run	Reactant Concentrations (mol/L)		Rate of Decrease in [NO] (mol/L·s)
	[NO]	$[H_2]$	
1	0.10	0.010	0.062
2	0.10	0.040	0.246
3	0.30	0.010	0.558

> **LEARNING HINT**
> If $y = x^2$, then tripling x will increase y by a factor of 9. For example, when $x = 2$, $y = 4$; when $x = 6$, $y = 36$.

stant. Here a threefold increase in the NO concentration results in a ninefold rate increase (0.558/0.062 = 9). Thus the rate is proportional to $[NO]^2$, the square of the NO concentration. Combining both concentration effects gives

$$\text{rate} = k[H_2][NO]^2$$

where k is a proportionality constant. Any expression that relates reaction rate and concentration is called a **rate law** or **rate equation**. Note that the exponent of [NO] in the above rate law is the same as the coefficient of NO in the reaction equation, but the exponent of $[H_2]$ differs from the coefficient of H_2. *The exponents in a rate law are usually different from the coefficients in the chemical equation and consequently cannot be predicted from the equation.*

EXAMPLE 14.4

The following initial rates were obtained for the reaction

$$2NO(g) + O_2(g) \rightarrow 2NO_2(g)$$

Run	[NO] (mol/L)	$[O_2]$ (mol/L)	Rate (mol/L·s)
1	0.0240	0.0350	0.143
2	0.0150	0.0350	0.0559
3	0.0240	0.0450	0.184

Assume that the rate law is of the form

$$\text{rate} = k[NO]^m[O_2]^n$$

and find (a) m and (b) n.

SOLUTION

(a) The concentration of O_2 is the same in Runs 1 and 2. Substituting the concentrations of NO and O_2 into the rate law and dividing the larger rate (Run 1) by the smaller rate (Run 2) gives:

$$\frac{\text{rate 1}}{\text{rate 2}} = \frac{0.143 \text{ mol/L·s}}{0.0559 \text{ mol/L·s}} = \frac{\cancel{k}(0.0240)^m \cancel{(0.0350)^n}}{\cancel{k}(0.0150)^m \cancel{(0.0350)^n}} = \left(\frac{0.0240}{0.0150}\right)^m$$

or

$$2.56 = (1.60)^m$$

Taking the logarithm of each side (see Appendix A.2) gives

$$\log 2.56 = m \log 1.60$$

and

$$m = \frac{\log 2.56}{\log 1.60} = 2$$

> **LEARNING HINT**
> Instead of using logarithms, you might first try substituting integers for m. If one of these integers satisfies the equation, fine. If m is not an integer, then you must use logarithms.

(b) The concentration of NO is the same in Runs 1 and 3. Dividing the larger rate (Run 3) by the smaller rate (Run 1) gives:

$$\frac{\text{rate 3}}{\text{rate 1}} = \frac{0.184 \text{ mol/L·s}}{0.143 \text{ mol/L·s}} = \frac{\cancel{k(0.0240)^m}(0.0450)^n}{\cancel{k(0.0240)^m}(0.0350)^n} = \left(\frac{0.0450}{0.0350}\right)^n$$

or

$$1.29 = (1.29)^n$$

and n must equal 1. Substituting $m = 2$ and $n = 1$ into the rate law gives

$$\text{rate} = k[NO]^2[O_2]$$

PRACTICE EXERCISE 14.5

The following rates were obtained for the reaction

$$2I^-(aq) + S_2O_8^{2-}(aq) \rightarrow I_2(aq) + 2SO_4^{2-}(aq)$$

at 25°C.

Run	$[I^-]$	$[S_2O_8^{2-}]$	Rate (mol/L·s)
1	0.15	0.45	2.6×10^{-4}
2	0.15	0.25	1.4×10^{-4}
3	0.50	0.45	8.6×10^{-4}

Find the rate law for the reaction.

Reaction Order

The **order** of a reaction is the sum of the exponents in its rate law. The reaction in Example 14.4 is third order because the sum of its exponents is $2 + 1 = 3$. (Any unwritten exponent such as the exponent of $[O_2]$ is assumed to be 1.) Reaction order can also be defined with respect to a single substance; this reaction is second order with respect to NO and first order with respect to O_2. A reactant whose concentration does not affect the rate is not included in the rate law; in effect, the concentration of such a reactant has an exponent of zero, and the reaction is said to be *zero order* with respect to that substance.

LEARNING HINT

Recall that any quantity raised to the zero power equals 1; $x^0 = 1$.

EXAMPLE 14.5

The rate law for the reaction

$$NO_2(g) + CO(g) \rightarrow NO(g) + CO_2(g)$$

at 200°C is

$$\text{rate} = k[NO_2]^2$$

State (a) the order of this reaction with respect to each reactant and (b) the overall reaction order.

SOLUTION

(a) The reaction is second order with respect to NO_2 because the exponent of $[NO_2]$ is 2. The rate does not depend on [CO], so the exponent of [CO] is understood to be zero. The reaction is zero order with respect to CO.

(b) The sum of the exponents is $2 + 0 = 2$. The reaction is second order overall.

PRACTICE EXERCISE 14.6

The decomposition of dinitrogen pentoxide

$$2N_2O_5(g) \longrightarrow 4NO_2(g) + O_2(g)$$

is first order with respect to N_2O_5. Write the rate law.

Many rate laws are a simple product of concentration factors and have the general form

$$\text{rate} = k[A]^m[B]^n[C]^p \ldots \tag{14.1}$$

The exponents m, n, and p are often positive integers, but fractional and negative exponents may also appear. For example, the conversion of ozone to oxygen

$$2O_3(g) \longrightarrow 3O_2(g)$$

obeys the rate law

$$\text{rate} = k\frac{[O_3]^2}{[O_2]} = k[O_3]^2[O_2]^{-1}$$

This law tells us that the rate at which ozone is converted to oxygen decreases as the concentration of oxygen increases.

A catalyst will appear in the rate law even though it does not appear in the balanced equation. The decomposition of hydrogen peroxide, for example, is accelerated by iodide ion:

$$2H_2O_2(aq) \xrightarrow{I^-(aq)} 2H_2O(l) + O_2(g)$$

The rate law shows that the speed of the reaction is proportional to the concentration of iodide ion:

$$\text{rate} = k[H_2O_2][I^-]$$

PRACTICE EXERCISE 14.7

The acid-catalyzed decomposition of thiosulfate in the presence of dissolved oxygen

$$2S_2O_3^{2-}(aq) + O_2(aq) \xrightarrow{H^+(aq)} 2SO_4^{2-}(aq) + 2S(s)$$

is first order in thiosulfate ion, zero order in $O_2(aq)$, and first order in $H^+(aq)$. Write the rate law.

Rate Constants

The constant k in the rate law is called a **rate constant**. The value of k depends on the nature of the reactants and on the temperature, but it does not vary with the reactant or product concentrations. A reaction with a large rate constant is inherently faster than one with a small rate constant. The increase of reaction rates with temperature reflects the fact that *rate constants increase with increasing temperature.* The next example shows how a rate constant is determined from experimental data.

EXAMPLE 14.6

Use data from Table 14.1 to calculate the rate constant for the reaction

$$2NO(g) + 2H_2(g) \rightarrow N_2(g) + 2H_2O(g)$$

at 866°C.

SOLUTION

We have already used the data in Table 14.1 to show that

$$\text{rate} = k[H_2][NO]^2$$

Any set of data in the table can be substituted into this rate law to find k. Using data from the second run gives

$$0.246 \text{ mol/L·s} = k(0.040 \text{ mol/L})(0.10 \text{ mol/L})^2$$

Hence $$k = 6.2 \times 10^2 \text{ L}^2/\text{mol}^2\text{·s}$$

PRACTICE EXERCISE 14.8

The reaction A + 2B → C is zero order in A and first order in B. The rate is 0.012 mol/L·s when [A] is 0.020 *M* and [B] is 0.60 *M*. Determine k (include units), and find the reaction rate when A and B are each 0.010 *M*.

The preceding example and practice problem show that the units of the rate constant depend on the exponents in the rate law. The rate constant of a first-order reaction, for example, will have different units than the rate constant of a zero-, second-, or third-order reaction.

14.4 FIRST- AND SECOND-ORDER REACTIONS

LEARNING HINT

A knowledge of logarithms and linear (straight-line) equations is required for this section. A review of logarithms, both natural and to the base 10, is given in Appendix A.2. Linear equations are reviewed in Appendix A.3.

In the usual first-order reaction, the rate is proportional to the first power of a single reactant concentration. For example, the decomposition of nitrogen pentoxide

$$2N_2O_5(g) \rightarrow 4NO_2(g) + O_2(g)$$

obeys the rate law

$$\text{rate} = k[N_2O_5]$$

This rate law follows the general form

$$\text{rate} = k[A] \tag{14.2}$$

where [A] is the concentration of some reactant and k is a rate constant.

In a first-order reaction, the concentration of A decreases with time according to the equation

$$[A] = [A]_0 e^{-kt} \tag{14.3}$$

where $[A]_0$ is the concentration of A at the beginning of the experiment (i.e., at time

LEARNING HINT

With many calculators, the value of e can be obtained by using the e^x button. Set $x = 1$, push the button, and you will find that $e = 2.7183$. (With some calculators, you have to use the INV and ln x buttons to evaluate e^x. Read your instruction manual.)

$t = 0$), [A] is the concentration remaining at some later time t, and e is the base of natural logarithms. Note that when $t = 0$, $e^{-kt} = e^0 = 1$, and $[A] = [A]_0$. (If you are familiar with calculus, read the *Digging Deeper* on Integrated Rate Laws for First- and Second-Order Reactions, later in this section, for a derivation of Equation 14.3.)

For the N_2O_5 decomposition, A is N_2O_5 and Equation 14.3 becomes

$$[N_2O_5] = [N_2O_5]_0 e^{-kt}$$

Concentration versus time data for the decomposition of N_2O_5 at 85°C are plotted in Figure 14.7a. Note how the concentration of N_2O_5 decreases rapidly in the early stages of the reaction and then more slowly as the N_2O_5 becomes depleted. All first-order decompositions give curves similar to one in Figure 14.7a.

Another form of Equation 14.3 can be obtained by rearranging it to

$$\frac{[A]}{[A]_0} = e^{-kt}$$

where $[A]/[A]_0$ is the fraction of A remaining at time t. Taking the natural logarithm of both sides gives

$$\ln \frac{[A]}{[A]_0} = -kt \tag{14.4}$$

since $\ln e^{-kt} = -kt \ln e = -kt$. (Recall that $\ln e = 1$. You can verify this with your calculator.)

Equation 14.4 can be used to determine whether a reaction is first order. Since $\ln (A/B) = \ln A - \ln B$, Equation 14.4 can be written as

$$\ln \frac{[A]}{[A]_0} = \ln [A] - \ln [A]_0 = -kt$$

or

$$\ln [A] = -kt + \ln [A]_0 \tag{14.5}$$

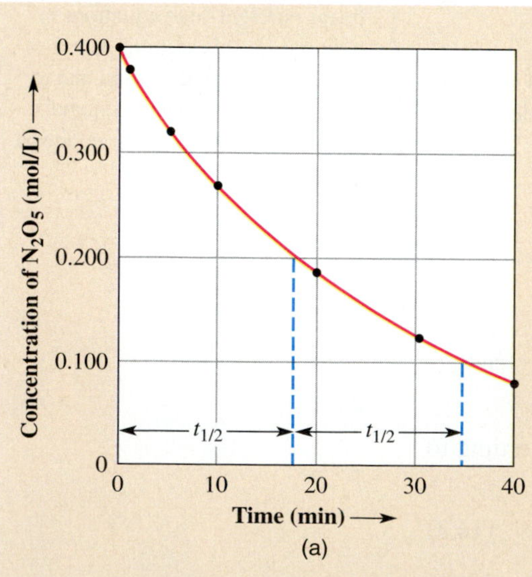

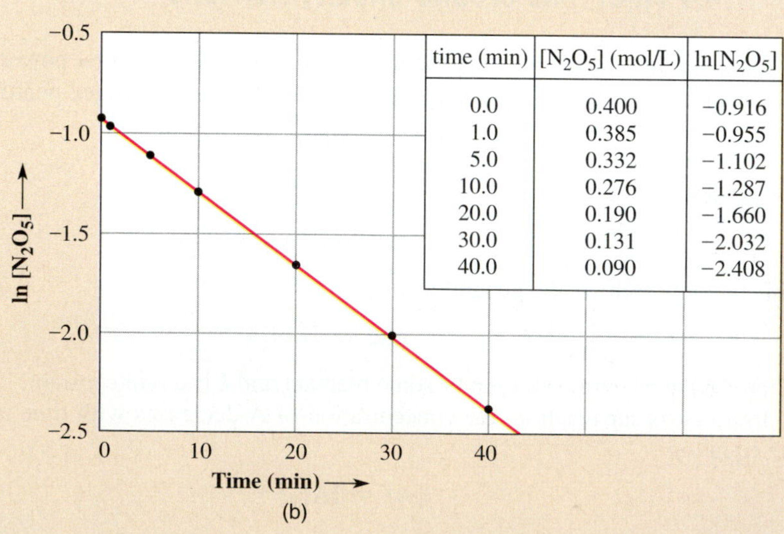

time (min)	$[N_2O_5]$ (mol/L)	$\ln[N_2O_5]$
0.0	0.400	−0.916
1.0	0.385	−0.955
5.0	0.332	−1.102
10.0	0.276	−1.287
20.0	0.190	−1.660
30.0	0.131	−2.032
40.0	0.090	−2.408

Figure 14.7

First-order reaction kinetics. (a) A plot of N_2O_5 concentration versus time for the first-order reaction $2N_2O_5(g) \longrightarrow 4NO_2(g) + O_2(g)$ at 85°C. The half-life periods, two of which are marked off along the time coordinate, are constant. (b) The plot of $\ln [N_2O_5]$ versus time is a straight line with slope $= -k$.

Equation 14.5 is of the form $y = mx + b$

$$\begin{array}{ccccc} \ln [A] & = & -kt & + & \ln [A]_0 \\ \uparrow & & \uparrow\uparrow & & \uparrow \\ y & = & mx & + & b \end{array}$$

which describes a straight line (ln [A] and t correspond to the variables y and x, respectively). The slope of the line, $m = -k$, is the negative of the rate constant; its intercept (the value of y at $t = 0$) is $b = \ln [A]_0$. Equation 14.5 tells us that *a graph of the logarithm of concentration versus time will be a straight line for a first-order reaction.* For the N_2O_5 decomposition, Equation 14.5 becomes

$$\ln [N_2O_5] = -kt + \ln [N_2O_5]_0$$

and the corresponding straight-line graph is plotted in Figure 14.7b. Such a graph provides an excellent test for first-order kinetics, because any deviation from a straight line would indicate that the reaction is not first order.

Because $m = -k$, the rate constant for a first-order reaction can be obtained from the slope of the graph of ln [A] versus time. It can also be estimated using Equation 14.4 and any two concentration values, as illustrated in the next example.

LEARNING HINT

If you prefer to use logarithms to the base 10, symbolized by "log," then simply substitute "2.303 log" wherever "ln" occurs. (see Appendix A.2).

EXAMPLE 14.7

Use Equation 14.4 and data from Figure 14.7 to find the rate constant for the decomposition of N_2O_5 at 85°C.

SOLUTION

The data in Figure 14.7 show that the initial concentration of dinitrogen pentoxide is $[N_2O_5]_0 = 0.400$ mol/L. After $t = 40$ min, the concentration falls to $[N_2O_5] = 0.090$ mol/L. (Any other time interval on the graph could have been chosen to solve this problem.) Substituting these values into Equation 14.4 gives

$$\ln \frac{[N_2O_5]}{[N_2O_5]_0} = -kt$$

$$\ln \frac{0.090 \text{ mol/L}}{0.400 \text{ mol/L}} = -k \times 40 \text{ min}$$

Rearranging and solving for k gives

$$-k \times 40 \text{ min} = \ln \frac{0.090}{0.400} = -1.49$$

and

$$k = \frac{1.49}{40 \text{ min}} = 0.037 \text{ min}^{-1}$$

LEARNING HINTS

Most calculators have an ln or ln x button. To find ln x, enter x in the display, push the ln or ln x button, and read the answer.

The number of decimal places in the logarithm of a number is equal to the number of significant figures in the number. The number obtained by dividing 0.090 by 0.400 has two significant figures. Hence, its logarithm, -1.49, will have two decimal places. See Appendix A.2 for a more detailed explanation of this rule.

CAUTION: The left side of Equation 14.4 is ln ([A]/[A]$_0$), the natural logarithm of a quotient. Do not make the mistake of dividing ln [A] by [A]$_0$.

PRACTICE EXERCISE 14.9

Use the concentration of N_2O_5 at $t = 1$ min and recalculate the value of k. Compare your answer with the value obtained in Example 14.7. (*Note:* The average rate constant calculated from several sets of data will be more reliable than the rate constant calculated from any one set of data.)

PRACTICE EXERCISE 14.10

Use the data given in Figure 14.7b to determine the slope *m* of the straight line. Then use the slope to find the rate constant *k*. Compare your answer with those obtained in Example 14.7 and Practice Exercise 14.9.

Equation 14.4 can also be used to predict reactant concentrations at any time during the reaction, as illustrated in the next example.

EXAMPLE 14.8

The conversion of cyclobutane (C_4H_8) to ethylene (C_2H_4)

$$C_4H_8(g) \rightarrow 2C_2H_4(g)$$

is a first-order reaction with $k = 0.0277\ \text{min}^{-1}$ at 449°C. If 0.400 mol of cyclobutane is placed in a 1.00-L reaction vessel at this temperature, what will its concentration be after 20.0 min?

H H / \ H—C—C—H | | H—C—C—H \ / H H

cyclobutane, C_4H_8

SOLUTION

Substitute C_4H_8 for A in Equation 14.4:

$$\ln \frac{[C_4H_8]}{[C_4H_8]_0} = -kt$$

The time $t = 20.0$ min, $k = 0.0277\ \text{min}^{-1}$, and the original concentration of cyclobutane is $[C_4H_8]_0 = 0.400$ mol/L. Hence,

$$\ln \frac{[C_4H_8]}{0.400\ \text{mol/L}} = -0.0277\ \text{min}^{-1} \times 20.0\ \text{min} = -0.554$$

This equation is solved for $[C_4H_8]$ by finding the antilog of -0.554:

$$\frac{[C_4H_8]}{0.400\ \text{mol/L}} = e^{-0.554} = 0.575$$

and

$$[C_4H_8] = 0.575 \times 0.400\ \text{mol/L} = 0.230\ \text{mol/L}$$

LEARNING HINT

If $\ln x = y$, then $x = e^y$.

The next example shows how Equation 14.4 can be used to calculate the time required for some fraction of the reactant to disappear in a first-order reaction.

EXAMPLE 14.9

Refer to Example 14.8, and determine how long it will take for 65.0% of the original cyclobutane sample to be consumed.

SOLUTION

If 65.0% of the cyclobutane disappears, then 35.0% remains; the fraction remaining is $[C_4H_8]/[C_4H_8]_0 = 0.350$. This fraction and $k = 0.0277\ \text{min}^{-1}$ are substituted into Equation 14.4:

$$\ln \frac{[C_4H_8]}{[C_4H_8]_0} = -kt$$

$$\ln 0.350 = -0.0277\ \text{min}^{-1} \times t$$

$$-1.050 = -0.0277\ \text{min}^{-1} \times t$$

and

$$t = \frac{1.050}{0.0277\ \text{min}^{-1}} = 37.9\ \text{min}$$

PRACTICE EXERCISE 14.11

The rate constant for the decomposition of N_2O_5 is 0.037 min^{-1} at 85°C (Example 14.7). What fraction of an N_2O_5 sample will remain after 45 min at this temperature?

PRACTICE EXERCISE 14.12

The acid-catalyzed conversion of sucrose to glucose and fructose in dilute solution is a first-order reaction. After 139 min, the concentration of sucrose decreases to two-thirds of its original value. Determine the rate constant k.

Half-Life

The **half-life** ($t_{1/2}$) of a reactant is the time required for its concentration to decrease to one-half of its initial value. Half-life periods for the N_2O_5 decomposition are marked off in Figure 14.7a. During the first half-life period of about 18.5 min, the concentration of N_2O_5 drops from its initial value of 0.400 mol/L to 0.200 mol/L. During the second half-life period, which is also 18.5 min, the concentration halves again, dropping from 0.200 mol/L to 0.100 mol/L.

Half-life periods are not unique to first-order reactions. Every naturally occurring process has a half-life. The human body, for example, constantly renews its red blood cells, and there is a half-life period during which half of the cells are replaced. The time it takes for half of a population to die or for half of a tree's leaves to fall in the autumn is also a half-life period. In every reaction the quantity of some reactant is reduced to half of its original value in the first half-life period. During the second half-life period, half of what remains disappears, so that only one-quarter of the original amount is left. Half of that amount is lost during the third half-life period, leaving one-eighth of the original amount, and so on.

A useful relation between the half-life and the rate constant of first-order reaction is obtained by substituting the half-life into Equation 14.4. At $t = t_{1/2}$, the concentration [A] equals $\frac{1}{2}[A]_0$. Equation 14.4 becomes

$$\ln \frac{[A]}{[A]_0} = -kt$$

$$\ln \frac{\frac{1}{2}\cancel{[A]_0}}{\cancel{[A]_0}} = \ln\left(\frac{1}{2}\right) = -kt_{1/2}$$

or

$$kt_{1/2} = -\ln\left(\frac{1}{2}\right)$$

The relation ln (A/B) = −ln (B/A) gives

$$kt_{1/2} = \ln 2 \tag{14.6}$$

Equation 14.6, which is valid only for first-order reactions, shows that the half-life of such a reaction is inversely proportional to the rate constant; a reaction with a large rate constant has a short half-life, and vice-versa. Note that the half-lives of first-order reactions are independent of concentration and will not change as the reaction proceeds.

EXAMPLE 14.10

The rate constant for the conversion of cyclobutane to ethylene at 449° is 0.0277 min^{-1} (see Example 14.8). Find the half-life of this reaction.

SOLUTION

Substituting $k = 0.0277\ \text{min}^{-1}$ into Equation 14.6 gives

$$t_{1/2} = \frac{\ln 2}{k} = \frac{0.693}{0.0277\ \text{min}^{-1}} = 25.0\ \text{min}$$

$$\begin{array}{c} \text{H}\quad\text{H} \\ |\quad\ | \\ \text{H}-\text{C}-\text{C}-\text{H} \\ \diagdown\ \diagup \\ \text{O} \end{array}$$

ethylene oxide, $(CH_2)_2O$

PRACTICE EXERCISE 14.13

The decomposition of ethylene oxide

$$(CH_2)_2O(g) \rightarrow CH_4(g) + O_2(g)$$

is first order with a half-life of 56.3 min at 415°C. Calculate the rate constant.

Radioactive decay (Chapter 22) follows first-order kinetics. The half-life of a radioactive isotope is one of its most important characteristics. Uranium-238, for example, has a half-life of 4.51×10^9 years, while the short-lived sodium-26 isotope has a half-life of only 1.0 second.

EXAMPLE 14.11

The half-life of francium-220 is 27.5 s. What fraction of atoms in a francium sample will decay in 1.00 min?

SOLUTION

We will first find $[Fr]/[Fr]_0$, the fraction of francium atoms remaining after 1.00 min. This fraction can be obtained from Equation 14.4 provided the rate constant k is known. The rate constant is found by substituting $t_{1/2} = 27.5$ s into Equation 14.6:

$$k = \frac{\ln 2}{t_{1/2}} = \frac{0.693}{27.5\ \text{s}} = 0.0252\ \text{s}^{-1}$$

Substituting A = Fr, $k = 0.0252\ \text{s}^{-1}$, and $t = 1.00\ \text{min} = 60.0\ \text{s}$ into Equation 14.4 gives

$$\ln \frac{[\text{Fr}]}{[\text{Fr}]_0} = -0.0252\ \text{s}^{-1} \times 60.0\ \text{s} = -1.51$$

The fraction that remains is obtained by finding the antilog of -1.51:

$$\frac{[\text{Fr}]}{[\text{Fr}]_0} = e^{-1.51} = 0.22$$

The fraction of francium that has decayed is 1 minus the fraction that remains; that is, $1 - 0.22 = 0.78$.

PRACTICE EXERCISE 14.14

Find the time required for 90% of a francium sample to decay.

Second-Order Reactions

The simplest type of second-order reaction is one in which the rate depends on the square of a single reactant concentration:

$$\text{rate} = k[\text{A}]^2 \qquad (14.7)$$

For example, the decomposition of nitrogen dioxide at high temperatures

$$2NO_2(g) \rightarrow 2NO(g) + O_2(g)$$

obeys the rate law

$$\text{rate} = k[NO_2]^2$$

For second-order reactions of this type, the concentration of A varies with time according to the equation

$$\frac{1}{[\text{A}]} = kt + \frac{1}{[\text{A}]_0} \qquad (14.8)$$

where $[\text{A}]_0$ is the initial concentration of A. (The derivation of Equation 14.8 is discussed in the next section.)

PRACTICE EXERCISE 14.15

At 383°C, the rate constant for the decomposition of nitrogen dioxide is k = 10.1 L/mol·s. What will the NO_2 concentration be after 30.0 s if the initial concentration is 0.0300 mol/L?

According to Equation 14.8, a graph of 1/[A] versus t for a second-order reaction will be a straight line whose slope is k. (The 1/[A] and t correspond to the variables y and x, respectively.) An example of such a graph is given in Figure 14.8 for the reaction

$$2NOCl(g) \rightarrow 2NO(g) + Cl_2(g)$$

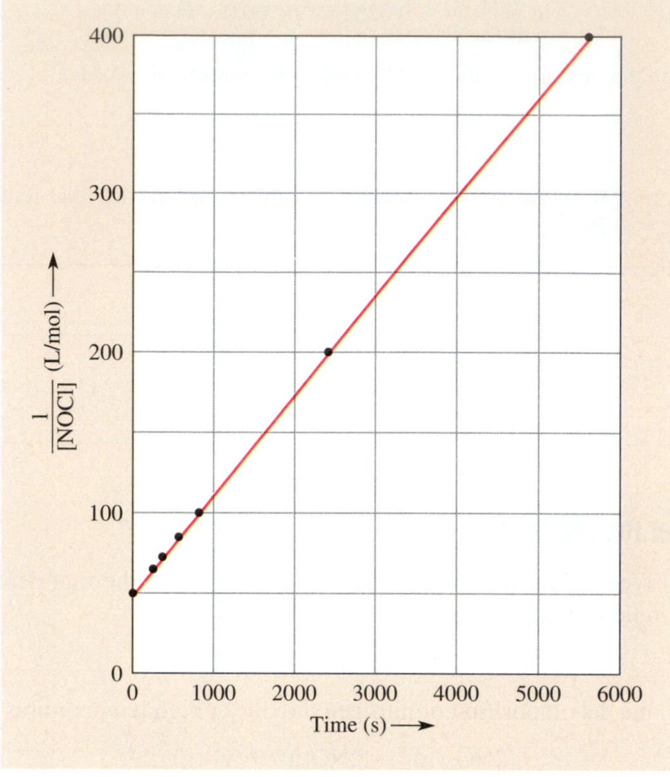

Figure 14.8

Second-order reaction kinetics. A plot of $1/[NOCl]$ versus time yields a straight line for the second-order reaction $2NOCl(g) \longrightarrow 2NO(g) + Cl_2(g)$ at 200°C.

for which

$$\text{rate} = k[NOCl]^2$$

For a first-order reaction, a graph of ln [A] versus t is a straight line; for a second-order reaction, a graph of $1/[A]$ versus t is a straight line.

The linearity of a graph of $1/[A]$ versus time is often used as a test for second-order kinetics.

Second-order half-lives are not particularly useful because they vary with concentration. Substituting $t = t_{1/2}$ and $[A] = \frac{1}{2}[A]_0$ into Equation 14.8 and solving for $t_{1/2}$ will show that for second-order reactions

$$t_{1/2} = \frac{1}{k[A]_0} \tag{14.9}$$

(You should derive Equation 14.9.) Each successive half-life period begins with a smaller concentration of the reactant A. Because this concentration appears in the denominator of Equation 14.9, second-order half-life periods become progressively longer as the reaction proceeds. The half-life periods marked off in Figure 14.9 show this type of increase.

PRACTICE EXERCISE 14.16

Refer to Practice Exercise 14.15, and calculate the first and second half-life periods for the decomposition of the nitrogen dioxide sample.

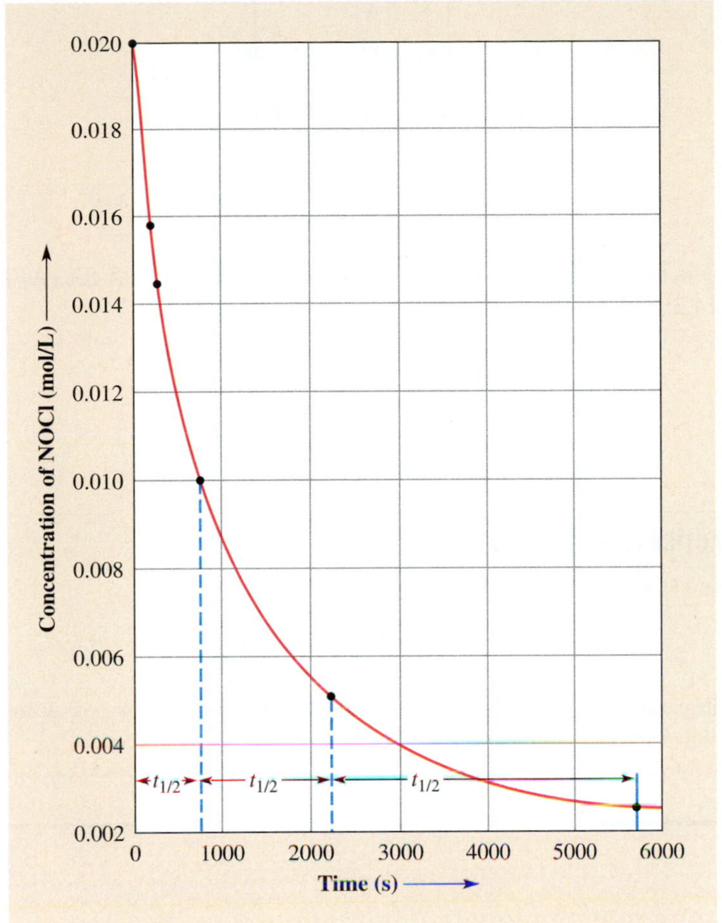

Figure 14.9

A plot of NOCl concentration versus time for the second-order reaction $2NOCl(g) \longrightarrow 2NO(g) + Cl_2(g)$ at 200°C. The half-life intervals become increasingly longer as the reaction proceeds.

DIGGING DEEPER

Integrated Rate Laws for First- and Second-Order Reactions

Equations 14.3 and 14.8 are examples of **integrated rate laws**, laws that describe how concentration varies with time. You can use these equations without deriving them, but you might like to see how they are obtained.

The general form of a first-order rate law using differential notation is

$$\text{rate} = -\frac{d[\text{A}]}{dt} = k[\text{A}]$$

where [A] is the concentration of the reactant being consumed and $-d[\text{A}]/dt$ is its time rate of change. Rearranging gives

$$\frac{d[\text{A}]}{[\text{A}]} = -k\,dt$$

Integrating between time 0 and time t gives Equation 14.4:

$$\int_{[A]_0}^{[A]} \frac{d[A]}{[A]} = -k \int_0^t dt$$

or

$$\ln [A] - \ln [A]_0 = -k(t - 0)$$

or

$$\ln \frac{[A]}{[A]_0} = -kt \qquad (14.4)$$

where $[A]_0$ is the concentration at $t = 0$. Solving Equation 14.4 for $[A]/[A]_0$ gives Equation 14.3:

$$\frac{[A]}{[A]_0} = e^{-kt}$$

or

$$[A] = [A]_0 e^{-kt} \qquad (14.3)$$

PRACTICE EXERCISE 14.17

The differential form of Equation 14.7 (p. 651) is

$$\text{rate} = -\frac{d[A]}{dt} = k[A]^2$$

Derive Equation 14.8 by rearranging Equation 14.7 to $d[A]/[A]^2 = -k\,dt$ and integrating between time 0 and time t.

14.5 REACTION MECHANISMS

A chemical equation identifies the substances present at the beginning and at the end of a reaction, but it does not describe the actual process that takes place during the reaction. In general, we envision this process in terms of a **reaction mechanism**, or series of consecutive steps leading from reactants to products, and we try to figure out from the available data what those steps might be. For example, the following two-step mechanism might be proposed for the decomposition of NO_2 into NO and O_2 at high temperatures:

Step 1: $NO_2(g) + NO_2(g) \rightarrow NO_2(g) + NO(g) + O(g)$

Step 2: $O(g) + NO_2(g) \rightarrow O_2(g) + NO(g)$

Each step in a mechanism represents a specific event and is called an **elementary step**. The proposed mechanism for the decomposition of NO_2 has two elementary steps. In Step 1, a collision between two NO_2 molecules breaks one bond, splitting one NO_2 molecule into NO and an oxygen atom. Observe that the remaining NO_2 molecule is not canceled out of the equation for the elementary step; it has a role and must be shown. In Step 2, the oxygen atom collides with another NO_2 molecule to form O_2 and a second molecule of NO. The oxygen atom is an **intermediate**, a substance produced in one step of the mechanism and used up in another. Many intermediates are highly reactive species such as single atoms or molecular fragments. The sum of the steps in the above mechanism, obtained by adding up the species on each side and canceling those that appear on both sides, is the balanced equation for the reaction:

$$\textit{Steps 1 + 2:} \quad 2NO_2(g) \rightarrow 2NO(g) + O_2(g)$$

In effect, a reaction mechanism is a pathway for change, and we will find that even the simplest change can be slow if the pathway is difficult.

The **molecularity** of an elementary step is the number of molecules or atoms that participate in that step. A **unimolecular step** is one in which a single molecule breaks down or rearranges itself into another species. Examples are the dissociation of ethane into two methyl radicals:

$$C_2H_6(g) \rightarrow 2CH_3(g)$$

and the arrangement of methyl isocyanide into acetonitrile:

$$CH_3NC(g) \rightarrow CH_3CN(g)$$

The majority of elementary steps are **bimolecular**; that is, they involve the collision of two molecules or atoms. An example is the reaction of nitrogen monoxide with ozone:

$$NO(g) + O_3(g) \rightarrow NO_2(g) + O_2(g)$$

A **termolecular step** is one in which three molecules or atoms simultaneously collide. The probability of three particles coming together at the same time is much less than that of two coming together; hence, termolecular steps are relatively rare compared to bimolecular steps, and reactions involving them tend to be slow. It has been proposed that the reaction

$$2NO(g) + Cl_2(g) \rightarrow 2NOCl(g)$$

occurs in one termolecular step.

The rate of an elementary step is proportional to the number of collisions between the molecules or atoms participating in the step, and the number of collisions in turn is proportional to the concentration of each species. The rate of the bimolecular step $A + B \rightarrow C$, for example, is proportional to the concentration of A and to the concentration of B:

$$\text{rate} = k[A][B]$$

A different step, $B + B \rightarrow C$, would have a rate proportional to $[B] \times [B]$ or $[B]^2$:

$$\text{rate} = k[B]^2$$

The exponents in the rate law of an elementary step are identical to the coefficients in the equation for the step. Table 14.2 lists rate laws for various elementary steps.

EXAMPLE 14.12

The mechanisms for some gaseous reactions include a step in which two atoms combine on colliding with a third species (Figure 14.10). The third molecule or atom carries off some energy, so the two combining atoms cannot fly apart again. Argon gas, for example, promotes the recombination of dissociated iodine atoms in the elementary step

$$I(g) + I(g) + Ar(g) \longrightarrow I_2(g) + Ar(g)$$

State the molecularity of this step and write its rate law.

SOLUTION

The step is termolecular because it involves the simultaneous collision of three atoms. The rate law is

$$\text{rate} = k[I][I][Ar] = k[I]^2[Ar]$$

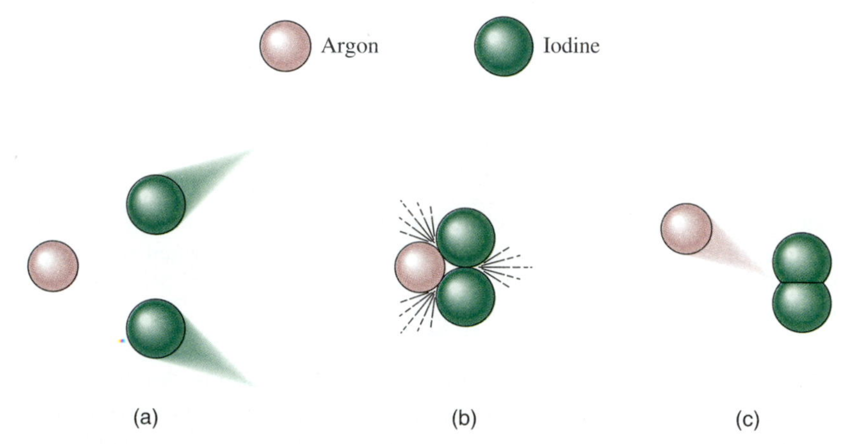

Figure 14.10

A termolecular reaction step in which an argon atom and two iodine atoms simultaneously collide. The argon atom absorbs excess energy, thus allowing an I_2 molecule to form.

PRACTICE EXERCISE 14.18

Write rate laws for the following elementary steps:

(a) $N_2O(g) + O(g) \longrightarrow N_2(g) + O_2(g)$

(b) $Br_2(g) \longrightarrow 2Br(g)$

TABLE 14.2 Rate Laws for Elementary Steps

Step	Molecularity	Law
$A \longrightarrow$ products	1	rate $= k[A]$
$A + A \longrightarrow$ products	2	rate $= k[A]^2$
$A + B \longrightarrow$ products	2	rate $= k[A][B]$
$2A + B \longrightarrow$ products	3	rate $= k[A]^2[B]$
$A + B + C \longrightarrow$ products	3	rate $= k[A][B][C]$

The rate law for an elementary step can be deduced from the equation for the step because the equation includes every participating molecule or atom. Remember, however, that the rate law for a complete reaction cannot be deduced from its equation. The reason for this is that the complete reaction is the sum of several elementary steps, and its overall equation does not always include every participating species. Atoms or molecules produced in one step may be consumed in a later step and will thus cancel when the steps are added.

Devising Reaction Mechanisms

Observing a chemical reaction in a flask or test tube is like watching a magic show; finding a reaction mechanism is like deducing what takes place behind the scenes. A proposed mechanism must meet two basic requirements:

1. It must account for the observed products.
2. It must explain the experimental rate law.

The second requirement can often be satisfied by assuming that one step in the mechanism is so much slower than the others that it alone determines how fast the reaction can go. This slow step, called the **rate-determining step** or "bottleneck step," is analogous to a traffic bottleneck in that molecules from previous steps pile up and wait to get through. *The rate law for the overall reaction is determined by the rate law for the bottleneck step.*

Let us devise a mechanism for the iodide-catalyzed decomposition of hydrogen peroxide (Demonstration 14.2):

$$2H_2O_2(aq) \xrightarrow{I^-(aq)} 2H_2O(l) + O_2(g)$$

for which the experimental rate law is:

$$\text{rate} = k[H_2O_2][I^-]$$

The iodide catalyst is not consumed; the number of moles of iodide ion is the same at the end of the reaction as at the beginning. This tells us that the mechanism must have at least two steps, one in which the iodide ion is a reactant and one in which it is a product. A simple assumption would be that the first step involves a reaction between iodide ion and hydrogen peroxide

$$\textbf{\textit{Step 1:}} \quad H_2O_2(aq) + I^-(aq) \rightarrow \text{products}$$

A number of different products might result from this encounter, but some are more probable than others. Let us assume that two well-known substances, H_2O and OI^- (hypoiodite ion), are formed. With these products, the complete step would be

$$\textbf{\textit{Step 1:}} \quad H_2O_2(aq) + I^-(aq) \rightarrow H_2O(aq) + OI^-(aq)$$

The rate for this bimolecular step would be proportional to the concentration of each reactant:

$$\text{rate of Step 1} = k[H_2O_2][I^-]$$

Note that this rate law is identical to the experimental rate law. Thus we can explain the observed rate law by assuming that Step 1 is the bottleneck step. Steps that come after Step 1 will not affect the rate; therefore, any number of subsequent steps could

DEMONSTRATION 14.2 CATALYZED DECOMPOSITION OF HYDROGEN PEROXIDE: $2H_2O_2(aq) \rightarrow 2H_2O(l) + O_2(g)$

The 3% hydrogen peroxide solution in the beaker is fairly stable, until a few crystals of potassium iodide are added.

Catalyzed by iodide ion, the decomposition proceeds rapidly and there is a vigorous evolution of oxygen gas. (The brown color is due to a side reaction in which some I^- is oxidized to I_2.)

be postulated as long as they are reasonable and account for the products. The simplest possibility is a two-step mechanism in which the second step is

$$\textit{Step 2:} \quad H_2O_2(aq) + OI^-(aq) \rightarrow H_2O(l) + O_2(g) + I^-(aq)$$

Step 2 regenerates the catalyst by consuming the intermediate OI^- formed in Step 1. The equations for Steps 1 and 2 add up to the overall reaction equation.

EXAMPLE 14.13

The experimental rate law for the reaction

$$2ICl(g) + H_2(g) \rightarrow 2HCl(g) + I_2(g)$$

is rate $= k[ICl][H_2]$. Devise a two-step mechanism with a slow first step.

SOLUTION

A slow first step starting with

$$ICl(g) + H_2(g) \rightarrow \text{products}$$

would account for the observed rate law. It is reasonable to assume that the known molecules HI and HCl are products:

$$\textit{Step 1:}\quad ICl(g) + H_2(g) \longrightarrow HI(g) + HCl(g)$$

A second step to complete the mechanism brings in the second ICl molecule and eliminates the intermediate HI:

$$\textit{Step 2:}\quad HI(g) + ICl(g) \longrightarrow HCl(g) + I_2(g)$$

You can verify that these steps add up to the overall equation.

PRACTICE EXERCISE 14.19

The experimental rate law for the reaction

$$2NO_2(g) + F_2(g) \longrightarrow 2NO_2F(g)$$

is rate $= k[NO_2][F_2]$. Devise a two-step mechanism with a slow first step.

It is often possible to devise more than one mechanism that meets the basic requirements. Although experimental evidence such as the spectroscopic detection of certain intermediates in the reaction mixture may sometimes help us to choose a probable mechanism, it is rarely possible to prove any mechanism beyond the shadow of a doubt. The best we can do is devise a mechanism consistent with all available data.

DIGGING DEEPER

The Steady-State Approximation

Most mechanisms are not worked out as easily as those in the previous examples. Fitting mechanisms to observed rate laws is a skill that often requires considerable expertise, especially when the slow step does not come first. Consider the conversion of ozone to oxygen:

$$2O_3(g) \longrightarrow 3O_2(g)$$

a reaction that occurs extensively in the stratosphere. The observed rate law is

$$\text{rate} = \frac{k[O_3]^2}{[O_2]} \qquad (14.10)$$

and the proposed mechanism consists of the following steps

$$\textit{Step 1:}\quad O_3(g) + M(g) \xrightarrow{k_1} O_2(g) + O(g) + M(g) \qquad \text{fast}$$

$$\textit{Step 2:}\quad O_2(g) + O(g) + M(g) \xrightarrow{k_{-1}} O_3(g) + M(g) \qquad \text{fast}$$

$$\textit{Step 3:}\quad O(g) + O_3(g) \xrightarrow{k_3} 2O_2(g) \qquad \text{slow}$$

where k_1, k_{-1}, and k_3 are rate constants corresponding to Steps 1, 2, and 3, respectively. (Observe that Step 2 is the reverse of Step 1. The rate constant for a reverse reaction is usually written with a negative subscript—in this case, k_{-1} for the reverse of the reaction with rate constant k_1.) The reaction is initiated by a fast first step in which an ozone molecule collides with some inert particle M and breaks up into an O_2 molecule and an oxygen atom. The particle M may be another molecule, a grain of dust, or even an ice crystal. This step generates a continuous supply of oxygen atoms. Most of the oxygen atoms quickly recombine with O_2 in the fast Step 2, which is the exact reverse of Step 1; however, a few will encounter O_3 molecules and form O_2 in the slow Step 3. The conversion of ozone to oxygen, $2O_3(g) \rightarrow 3O_2(g)$, is the sum of Steps 1 and 3. Note that this mechanism differs from our previous mechanisms in that the overall equation is not the sum of all the steps in the mechanism.

The rate of Step 3, the rate-determining step, is proportional to the concentrations of oxygen atoms and ozone:

$$\text{rate} = k_3[\text{O}][\text{O}_3] \tag{14.11}$$

This equation, which includes oxygen atoms, bears little resemblance to the observed rate law (Equation 14.10), which contains only the initial reactant and the final product. We could make a better comparison of the two rate laws if we eliminate [O] from Equation 14.11. We can do this by expressing [O] in terms of $[O_3]$ and/or $[O_2]$.

Atomic oxygen is an intermediate; it is formed in Step 1 and consumed in Steps 2 and 3. An intermediate is fairly reactive, and it is reasonable to assume that at some point during the reaction it will be in a **steady state**; that is, it will be used up as fast as it is produced. In the **steady-state approximation**, we assume that *the concentration of an intermediate in a reacting mixture does not change; its rate of formation is equal to its rate of consumption.* For the ozone mechanism, the steady-state approximation requires that the rate of formation of oxygen atoms in Step 1 be equal to their rate of consumption in Steps 2 and 3. The rate of formation is

$$\text{rate of formation} = \text{rate of Step 1} = k_1[\text{O}_3][\text{M}]$$

The rate of consumption is equal to the sum of the rates of Steps 2 and 3, but since Step 3 is much slower than Step 2, we can ignore the rate of Step 3 and write:

$$\text{rate of consumption} = \text{rate of Step 2} = k_{-1}[\text{O}_2][\text{O}][\text{M}]$$

Equating these rates gives:

$$\text{rate of formation} = \text{rate of consumption}$$

$$k_1[\text{O}_3][\text{M}] = k_{-1}[\text{O}_2][\text{O}][\text{M}]$$

Canceling [M] and solving for [O] gives the steady-state concentration of atomic oxygen:

$$[\text{O}] = \frac{k_1[\text{O}_3]}{k_{-1}[\text{O}_2]}$$

If we substitute this expression for [O] into Equation 14.11, the rate law for the bottleneck step becomes:

$$\text{rate} = k_3[\text{O}][\text{O}_3] = k_3\frac{k_1[\text{O}_3]}{k_{-1}[\text{O}_2]}[\text{O}_3] = \frac{k_3k_1[\text{O}_3]^2}{k_{-1}[\text{O}_2]} = k\frac{[\text{O}_3]^2}{[\text{O}_2]}$$

where $k = k_3k_1/k_{-1}$. This equation is identical to Equation 14.10, the experimental rate law. Hence, the proposed mechanism satisfies our minimum requirements: It accounts for the products and it explains the observed rate law.

PRACTICE EXERCISE 14.20

A reaction that contributes to smog formation is the combination of oxygen with NO from automobile exhaust:

$$2NO(g) + O_2(g) \rightarrow 2NO_2(g)$$

Show that the observed rate law

$$\text{rate} = k[NO]^2[O_2]$$

is consistent with the following mechanism:

$$NO(g) + O_2(g) \underset{k_{-1}}{\overset{k_1}{\rightleftharpoons}} NO_3(g) \qquad \text{fast}$$

$$NO_3(g) + NO_2(g) \xrightarrow{k_2} 2NO_2(g) \qquad \text{slow}$$

Note: Two steps that are the reverse of each other are often combined as in this Practice Exercise. The double arrow indicates that the reaction is reversible (i.e., it can go either way).

14.6 FREE RADICAL CHAIN REACTIONS

A **chain reaction** is one in which an initial step leads to a succession of repeating steps that continue indefinitely. The mechanisms of chain reactions are both interesting and important. An example of such a reaction is the formation of hydrogen bromide from its elements:

$$H_2(g) + Br_2(g) \rightarrow 2HBr(g)$$

The initial step, called the **chain initiation step**, occurs when a bromine molecule absorbs a photon of visible light (indicated by the symbol $h\nu$) and dissociates into atoms.

chain initiation step:

$$Br_2(g) \xrightarrow{h\nu} Br\cdot(g) + Br\cdot(g)$$

The chain itself consists of the following steps, called **chain propagation steps**, which add up to the overall reaction and continuously produce the product.

chain propagation steps:

$$Br\cdot(g) + H_2(g) \rightarrow HBr(g) + H\cdot(g)$$

$$H\cdot(g) + Br_2(g) \rightarrow HBr(g) + Br\cdot(g)$$

The chain propagation steps can repeat indefinitely as long as there are H_2 and Br_2 molecules for the free atoms to collide with.

The initiation step of a chain reaction is typically an endothermic dissociation that produces *free radicals* (reactive species with unpaired electrons). A free radical can be a single atom such as H· or Br· or ·O·, or a group of atoms such as $\cdot CH_3$. (The unpaired electrons in a free radical are often indicated by dots next to the formula.) Each propagation step uses up one free radical and forms another free radical, so the reaction is *self-propagating;* that is, it keeps itself going. Chain initiation may be an infrequent event, but self-propagation allows a limited number of initiation steps to produce an extensive reaction.

The propagation cycle is repeated until the reactants are depleted or until the chain is halted by a **chain termination step**, which removes free radicals from circulation.

chain termination steps:

$$Br\cdot(g) + Br\cdot(g) \rightarrow Br_2(g)$$

$$H\cdot(g) + H\cdot(g) \rightarrow H_2(g)$$

$$H\cdot(g) + Br\cdot(g) \rightarrow HBr(g)$$

Chain reactions are common in gaseous systems and have been identified in the combustion of fuels, in explosions, in the reactions of halogen vapors, in the formation of smog, and in the atmospheric ozone layer. The destruction of ozone in the ozone layer over Antarctica (see *Chemical Insight:* Free Radicals and the Ozone Layer, p. 394) is believed to involve a chain mechanism. The initiation step is the sunlight-induced dissociation of CFCs (chlorofluorocarbon compounds such as CF_2Cl_2) into chlorine atoms:

$$CF_2Cl_2(g) \xrightarrow{h\nu} CF_2Cl\cdot(g) + Cl\cdot(g)$$

The chlorine atoms have an odd number of electrons and thus function as free radicals. They convert ozone into oxygen by the following chain propagation steps:

$$2 \times [Cl\cdot(g) + O_3(g) \longrightarrow ClO\cdot(g) + O_2(g)]$$

$$ClO\cdot(g) + ClO\cdot(g) + M(g) \longrightarrow ClOOCl(g) + M(g)$$

$$ClOOCl(g) \xrightarrow{h\nu} Cl\cdot(g) + ClOO\cdot(g)$$

$$ClOO\cdot(g) + M(g) \longrightarrow Cl\cdot(g) + O_2(g) + M(g)$$

where M is a neutral molecule such as N_2 or O_2. The sum of the four propagation steps is

$$2O_3(g) \rightarrow 3O_2(g)$$

Chlorine atoms produced in the third and fourth propagation steps begin the cycle over again.

The ozone depletion process is complicated by the fact that for most of the year, chlorine atoms are bound in relatively inactive "reservoir species" such as hydrogen chloride (HCl) and chlorine nitrate ($ClONO_2$). During the Antarctic winter, however, stratospheric ice particles provide a surface on which the "reservoir species" break down into chlorine atoms:

$$ClONO_2(g) + HCl(s) \xrightarrow{\text{ice}} HNO_3(s) + Cl_2(g)$$

$$Cl_2(g) \xrightarrow{h\nu} 2Cl\cdot(g)$$

A stratospheric ozone hole (severe ozone depletion) forms each year during the early Antarctic spring (October) because the chlorine atom concentration is a maximum at this time. A smaller and less intense ozone hole is expected to appear over Arctic regions sometime in the not too distant future.

14.7 ACTIVATION ENERGY

Consider two molecules on a collision course. Will they react or will they rebound unchanged? The answer depends on the nature of the molecules, their orientation, and the energy of their impact. A reaction will take place only if the molecules collide with sufficient energy to disrupt some of the bonds. The minimum energy required for a reaction to occur is called the **activation energy** (E_a). Because different molecules have different bond strengths, the activation energy varies from one reaction to another.

When you strike a match, the motion of your hand provides the activation energy for ignition.

Imagine that a collision between a hydrogen molecule and a bromine atom results in the elementary step

$$H_2(g) + Br(g) \rightarrow H(g) + HBr(g)$$

The H—H bond stretches, weakens, and eventually breaks. The H—Br bond, long and weak at first, becomes shorter and stronger as the reaction proceeds:

H—H bond is breaking

H—H + Br ⟶ H--H-----Br ⟶ H---H---Br

reactants

H—Br bond is forming

—— Increasing potential energy ⟶

Recall (Chapter 9) that bond breaking is endothermic and bond formation is exothermic. The lengthening H—H bond initially absorbs more energy than is evolved by the newly forming H—Br bond, and the potential energy of the three-atom complex H--H----Br increases. Energy must come from somewhere, so the increase in potential energy is accompanied by a decrease in the kinetic energy of the reacting molecules. At some point, the potential energy reaches a maximum; it then begins to decline as the configuration develops into H----H--Br and finally into H + HBr.

H—H bond is breaking

H---H---Br ⟶ H-----H--Br ⟶ H + H—Br

H—Br bond is forming

products

—— Decreasing potential energy ⟶

As the potential energy of the complex decreases, the kinetic energy increases and the product molecules fly apart.

Figure 14.11 is a **potential energy profile** that illustrates the potential energy changes that occur during the transition from reactants to products. The state of the system at the highest point on the profile is called the **transition state**, and the configuration of atoms in the transition state is called the **activated complex**. The reaction cannot be completed unless the reacting system passes through the transition state. The activation energy (78.5 kJ/mol for this reaction) is the difference between the potential energy of the activated complex and the initial potential energy of the reactants. It takes 78.5 kJ to convert 1 mol of H_2 and 1 mol of Br into 1 mol of activated complex. The reaction will occur only if the reactant molecules have the acti-

Recall (p. 250) that $\Delta H° = \Delta E°$ for reactions in which the number of moles of gas does not change.

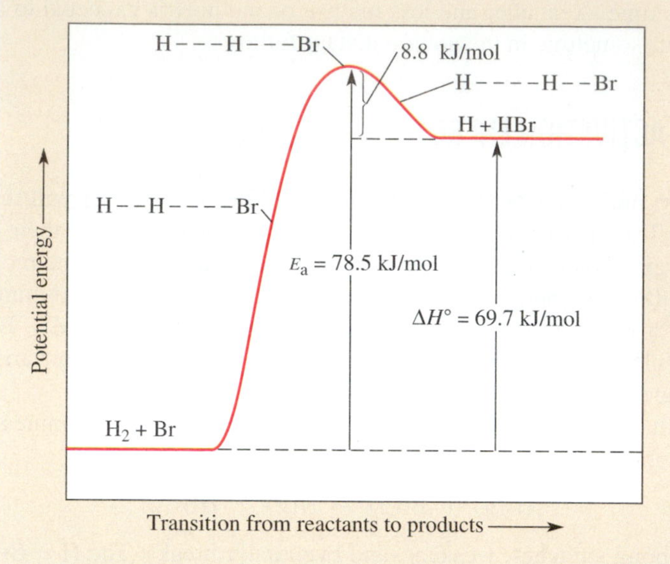

Figure 14.11

Potential energy profile for the elementary step: $H_2(g) + Br(g) \longrightarrow H(g) + HBr(g)$. For the reaction to occur, the reactants must overcome the activation energy barrier, $E_a = 78.5$ kJ/mol. The net change in potential energy at constant pressure, 69.7 kJ/mol, is $\Delta H°$ for the reaction. The reverse reaction follows the pathway from right to left, with an activation energy of 8.8 kJ/mol.

vation energy and are oriented in such a way that they form the activated complex when they collide. If the molecules do not have the activation energy, they will simply bounce apart.

Figure 14.11 shows that the activation energy is, in effect, a potential energy barrier that must be surmounted in order for the molecules to react. The reverse reaction

$$HBr(g) + H(g) \longrightarrow H_2(g) + Br(g)$$

also requires an activation energy. The reverse reaction can be traced on Figure 14.11 by moving from right to left (backwards) along the transition pathway. Note that both the forward and reverse reactions pass through the same activated complex. Because HBr and H are 69.7 kJ higher in energy than H_2 and Br, they need less energy to achieve the transition state. The activation energy for the reverse reaction is only 78.5 kJ/mol − 69.7 kJ/mol = 8.8 kJ/mol.

Figure 14.11 also shows that $\Delta H°$, the enthalpy change for the forward reaction, is equal to the activation energy for the forward reaction minus the activation energy for the reverse reaction:

$$\Delta H° = E_a(\text{forward}) - E_a(\text{reverse}) \qquad (14.12)$$

EXAMPLE 14.14

The conversion of gaseous *cis*-2-butene into gaseous *trans*-2-butene

$$\begin{matrix} H_3C & & CH_3 \\ & C{=}C & \\ H & & H \end{matrix} \longrightarrow \begin{matrix} H_3C & & H \\ & C{=}C & \\ H & & CH_3 \end{matrix}$$

occurs in one step with $\Delta H° = -4.4$ kJ/mol and $E_a = 218$ kJ/mol. (a) Draw a potential energy profile for this reaction. (b) Calculate E_a for the reverse reaction.

SOLUTION

(a) The activation energy is 218 kJ/mol; hence, the activated complex is 218 kJ higher in energy than the reactants. The enthalpy change is negative, so the reaction is exothermic and the products are 4.4 kJ lower in energy than the reactants. The potential energy profile is sketched in Figure 14.12.

(b) The activation energy for the reverse reaction is obtained by substituting $\Delta H° =$ -4.4 kJ/mol and E_a(forward) $=$ 218 kJ/mol into Equation 14.12:

$$-4.4 \text{ kJ} = 218 \text{ kJ} - E_a(\text{reverse})$$

and

$$E_a(\text{reverse}) = 218 \text{ kJ} + 4.4 \text{ kJ} = 222 \text{ kJ}$$

Note that this answer could also have been obtained by inspecting Figure 14.12.

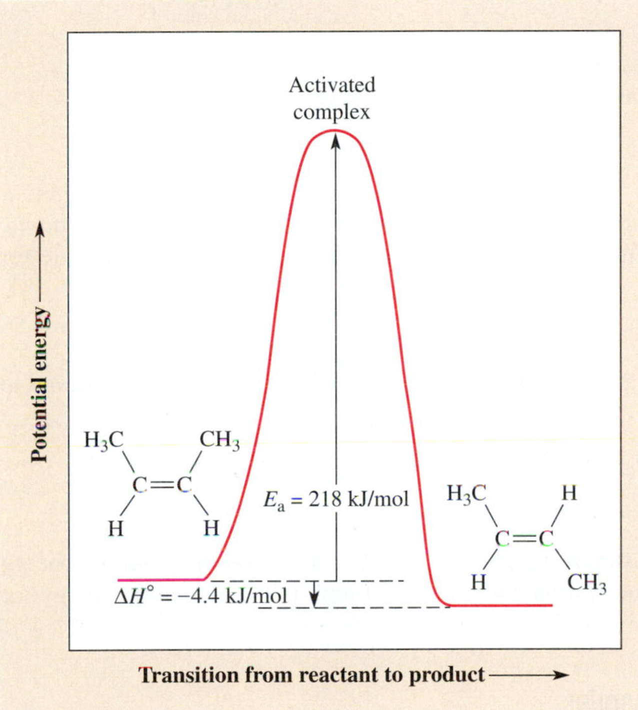

Figure 14.12
The potential energy profile for the conversion of *cis*-2-butene to *trans*-2-butene (see Example 14.14).

The high activation energy for this reaction consists largely of energy needed to break the pi bond that hinders free rotation in *cis*-2-butene.

PRACTICE EXERCISE 14.21

The forward activation energy for an elementary step is 42 kJ/mol and the reverse activation energy is 32 kJ/mol. Calculate $\Delta H°$ for the step, and state whether it is exothermic or endothermic.

Each step in a mechanism has its own activation energy. At a given temperature, there are fewer high-energy collisions than low-energy collisions, so *the step that is*

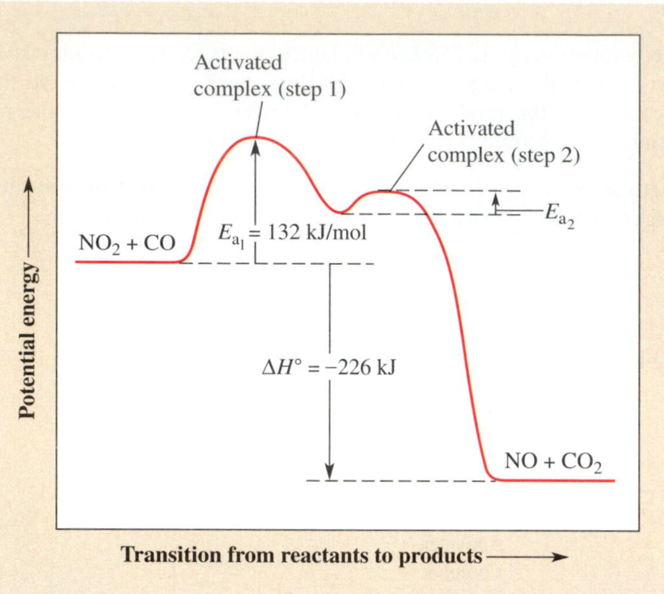

Figure 14.13
The potential energy profile for the reaction $NO_2(g) + CO(g) \longrightarrow NO(g) + CO_2(g)$. Each step in the two-step mechanism has its own activation energy.

highest on the potential energy profile is less likely to occur and will usually be the slow or rate-determining step. Figure 14.13 shows the potential energy profile for the reaction

$$NO_2(g) + CO(g) \longrightarrow NO(g) + CO_2(g) \qquad \Delta H° = -226 \text{ kJ}$$

At low temperatures, the rate is second order in NO_2. The mechanism is believed to consist of two steps:

$$\textit{\textbf{Step 1:}} \quad NO_2(g) + NO_2(g) \longrightarrow NO_3(g) + NO(g) \qquad \text{slow}$$

$$\textit{\textbf{Step 2:}} \quad NO_3(g) + CO(g) \longrightarrow NO_2(g) + CO_2(g) \qquad \text{fast}$$

Observe that the slow first step is higher than the second step on the potential energy profile. The experimental activation energy, 132 kJ/mol, is the activation energy for the slow first step.

Determining Activation Energies

Svante Arrhenius, a Swedish chemist who developed the concept of ionization in aqueous solution, also studied the effect of temperature on rate constants. In 1889 he found that k increases exponentially with temperature (Figure 14.14) according to the equation

$$k = Ae^{-E_a/RT} \tag{14.13}$$

In this equation, now called the **Arrhenius equation**, k is the rate constant, R is the gas constant, and T is the Kelvin temperature; the activation energy E_a and the pre-exponential factor A are constants for a given reaction.

Equation 14.13 can be converted to the straight line form, $y = mx + b$, by taking the natural logarithm of both sides:

$$\ln k = \ln e^{-E_a/RT} + \ln A$$

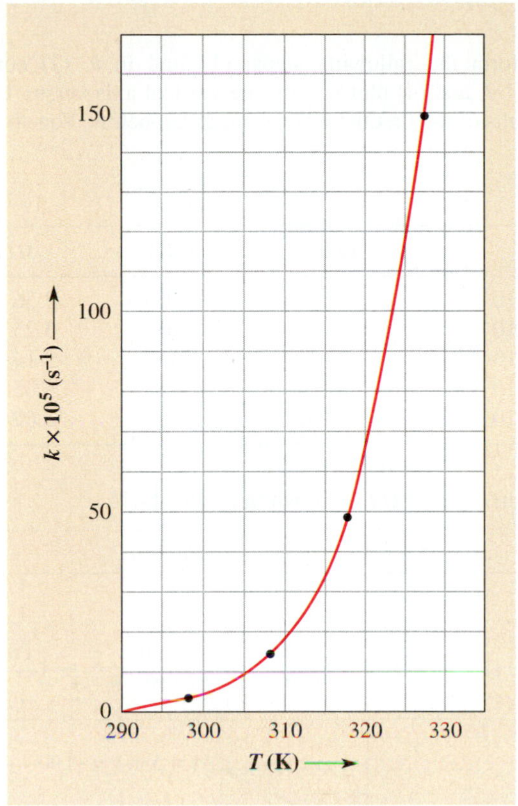

Figure 14.14

A plot of rate constant versus Kelvin temperature for the decomposition of $N_2O_5(g)$ is an exponential curve. The plotted values are given in Example 14.15.

or

$$\underset{y}{\ln k} = -\underset{m}{\frac{E_a}{R}} \times \underset{x}{\frac{1}{T}} + \underset{b}{\ln A} \qquad (14.14)$$

A plot of ln k versus $1/T$ yields a straight line; its slope is $-E_a/R$ and its intercept on the vertical axis is ln A. The following example shows how such a plot can be used to evaluate activation energies.

EXAMPLE 14.15

(a) Prepare a graph of ln k versus $1/T$ from the following data for the decomposition of gaseous N_2O_5, and (b) use your graph to determine E_a for this reaction.

T(°C)	k(s^{-1})
25	3.46×10^{-5}
35	1.35×10^{-4}
45	4.98×10^{-4}
55	1.50×10^{-3}
100	1.43×10^{-1}

SOLUTION

(a) We will perform the following steps: (1) find ln k, (2) convert T to kelvins, (3) calculate $1/T$, and (4) plot ln k on the vertical axis versus $1/T$ on the horizontal axis. The results of the first three steps are given below. You should verify each calculation.

k	ln k	T(K)	$1/T$(K^{-1})
3.46×10^{-5}	-10.272	298	3.36×10^{-3}
1.35×10^{-4}	-8.910	308	3.25×10^{-3}
4.98×10^{-4}	-7.605	318	3.14×10^{-3}
1.50×10^{-3}	-6.502	328	3.05×10^{-3}
1.43×10^{-1}	-1.945	373	2.68×10^{-3}

The graph of ln k versus $1/T$ is shown in Figure 14.15.

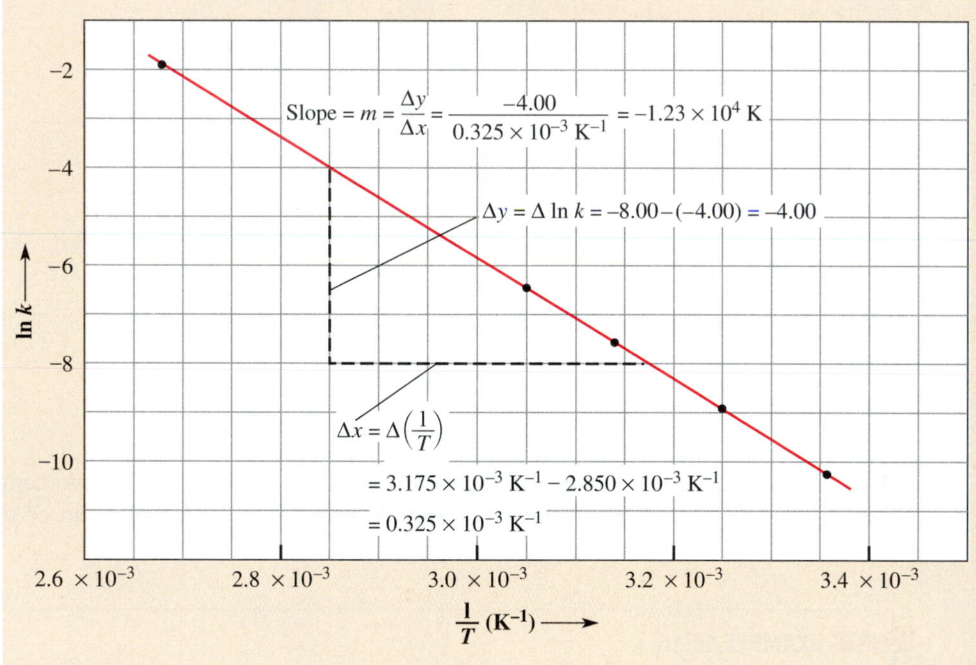

Figure 14.15
A plot of ln k versus $1/T$ for the decomposition of N_2O_5 is a straight line with slope $= -E_a/R$. The plotted values are given in Example 14.15.

(b) The slope of the straight line in Figure 14.15 is

$$m = \frac{-4.00}{0.325 \times 10^{-3}\ \text{K}^{-1}} = -1.23 \times 10^4\ \text{K}$$

The slope m is also equal to $-E_a/R$ (Equation 14.14). To get the value of E_a in joules, we use $R = 8.314$ J/mol·K (Table 5.2, p. 190):

$$-\frac{E_a}{R} = -\frac{E_a}{8.314 \text{ J/mol·K}} = -1.23 \times 10^4 \text{ K}$$

and $$E_a = 1.02 \times 10^5 \text{ J/mol} = 102 \text{ kJ/mol}$$

For most chemical reactions, a graph of ln k versus $1/T$ is a straight line, as predicted by Equation 14.14. (Explosions and some enzyme-catalyzed reactions are exceptions to this rule.) It has been suggested that any process with an activation energy in excess of 21 kJ/mol involves bond breaking and is probably chemical in nature. Some unusual examples of processes whose rates increase exponentially with temperature are the chirping of crickets and the flashing of fireflies. When the logarithm of a cricket's chirp rate or the logarithm of the number of firefly flashes per second is plotted against $1/T$, the resulting graph is a straight line. The mechanisms for the production of chirps and flashes are not completely understood, but the activation energy in each case is about 51 kJ/mol. This value is consistent with a chemical mechanism.

An activation energy can also be estimated from rate constants at two different temperatures without making a graph. Equation 14.14 is rearranged to give

$$\ln A = \ln k + \frac{E_a}{RT}$$

Let k_1 and k_2 be the rate constants at the Kelvin temperatures T_1 and T_2. Since A is a constant for the reaction, ln A will be the same at both temperatures. Hence,

$$\ln k_1 + \frac{E_a}{RT_1} = \ln k_2 + \frac{E_a}{RT_2}$$

Collecting terms gives

$$\ln k_1 - \ln k_2 = \frac{E_a}{RT_2} - \frac{E_a}{RT_1}$$

which can be simplified to

$$\ln \frac{k_1}{k_2} = \frac{E_a}{R}\left(\frac{1}{T_2} - \frac{1}{T_1}\right) \qquad (14.15)$$

LEARNING HINT

It is better to learn how to derive Equation 14.15 than to try to memorize it.

The following example illustrates the use of this equation to calculate E_a.

EXAMPLE 14.16

The rate constant for the reaction

$$2NO_2(g) \longrightarrow 2NO(g) + O_2(g)$$

is 0.522 L/mol·s at 592 K and 1.70 L/mol·s at 627 K. Calculate E_a.

SOLUTION

Let k_1 and T_1 be 0.522 L/mol·s and 592 K, respectively; let k_2 and T_2 be 1.70 L/mol·s and 627 K, respectively; R is 8.314 J/mol·K. Substituting these values into Equation 14.15 gives:

$$\ln \frac{0.522}{1.70} = \frac{E_a}{8.314 \text{ J/mol·K}}\left(\frac{1}{627 \text{ K}} - \frac{1}{592 \text{ K}}\right)$$

Solving gives $E_a = 1.0 \times 10^5$ J/mol $= 1.0 \times 10^2$ kJ/mol. (*Note:* When using this method, keep in mind that a value calculated from two points is usually less accurate than one calculated from a line drawn through several points.)

PRACTICE EXERCISE 14.22

The activation energy for a reaction is 18.7 kJ/mol. The rate constant for the reaction is 0.0400 s^{-1} at 0°C. Calculate the rate constant at 25°C.

DIGGING DEEPER

The Collision Theory of Reaction Rates

The rate of a bimolecular elementary step

$$A + B \longrightarrow \text{products}$$

depends on the concentrations of A and B and also on the rate constant k:

$$\text{rate} = k[A][B]$$

If k is large, the step is inherently fast; if k is small, the step will be slow at all concentrations. In addition to accounting for the effect of concentration, a theory of reaction rates should give us some insight into why one rate constant is different from another.

A bimolecular reaction will not occur unless three conditions are satisfied:

1. Molecules A and B must collide with each other.
2. The colliding molecules must have the activation energy required for the reaction.
3. The colliding molecules must be oriented so as to bring together the atoms that will form new bonds.

The **collision theory of reaction rates** assumes that the rate of a bimolecular reaction is equal to the total number of bimolecular collisions per second multiplied by the fraction of collisions having the necessary energy and by the fraction having a suitable orientation, that is,

$$\text{rate} = Z \times f_a \times p \tag{14.16}$$

- p: fraction of collisions with suitable orientation
- f_a: fraction of collisions having the activation energy
- Z: total number of bimolecular collisions per second

Each fraction has a value between 0 and 1, so the number of *reactive collisions* (i.e., the number of collisions that result in chemical change) will be less than the total number of collisions. Any change that increases one or more of the factors in Equation 14.16 will also increase the reaction rate. We will consider each of these factors in turn.

Collisions Per Second. The number of bimolecular collisions that occur each second in a given sample is called the **collision frequency** (Z). Under ordinary conditions, the collision frequency is enormous. Calculations show, for example, that there are about 8×10^{31} bimolecular collisions per second in a liter of oxygen gas at standard conditions. This number is essentially the same for all gases under the same conditions. The number of molecules in 1 L of gas at 0°C and 1 atm is:

$$1\ \text{L} \times \frac{1\ \text{mol}}{22.4\ \text{L}} \times \frac{6.022 \times 10^{23}\ \text{molecules}}{1\ \text{mol}} = 3 \times 10^{22}\ \text{molecules}$$

Hence, the number of collisions per second, 8×10^{31}, is almost 3 billion times greater than the number of molecules present! If every collision resulted in a reaction, everything would be over in less than one-billionth of a second. The fact that most gaseous reactions are not nearly this fast leads us to conclude that most collisions leave the molecules unchanged.

The collision frequency Z increases with increasing molecular concentration and increasing molecular speed. In Section 5.7 we showed that the average speed of a collection of molecules is proportional to the square root of the Kelvin temperature. A more detailed analysis shows that *the frequency of bimolecular collisions in a gas is proportional to the concentration of each of the colliding molecules and to* $\sqrt{T}$.

EXAMPLE 14.17

By what factor will the number of collisions in a sample of gas increase if the temperature is raised from 25°C to 35°C?

SOLUTION

The corresponding Kelvin temperatures are 298 K and 308 K. The collision frequencies are proportional to the square roots of these temperatures:

$$\frac{\text{collision frequency at 308 K}}{\text{collision frequency at 298 K}} = \sqrt{\frac{308\ \text{K}}{298\ \text{K}}} = 1.02$$

The collision frequency will increase by a factor of 1.02.

PRACTICE EXERCISE 14.23

By what factor will the frequency of collisions between A and B increase if the concentration of A is tripled while the concentration of B is halved?

Recall (Section 14.1) that a 10-degree rise in temperature causes many reaction rates to double. However, Example 14.17 shows that only a very small part of this increase is due to an increase in collision frequency. This calculation, coupled with our earlier conclusion that most colliding molecules bounce off each other unchanged, provides strong support for Equation 14.16, that is, for our belief that bimolecular reaction rates depend on factors other than the total number of collisions.

Activation Energy. The number of collisions with energies greater than or equal to the activation energy depends on the magnitude of the activation energy and also on the temperature. The shaded areas under the energy distribution curves of Figure 14.16 represent the number of collisions that have energies in excess of E_a. Figure 14.16a shows that this number decreases with increasing E_a; Figure 14.16b shows that it increases with increasing temperature.

A statistical analysis of molecular energies leads to the conclusion that the fraction, f_a, of collisions with energies greater than or equal to E_a is given by

$$f_a = e^{-E_a/RT} \tag{14.17}$$

where R is the gas constant, T is the Kelvin temperature, and e is the base of natural logarithms. The appearance of f_a in both the theoretical expression for reaction rates (Equation 14.16) and the experimental Arrhenius equation for rate constants (Equation 14.13) gives additional support to the collision theory of reaction rates.

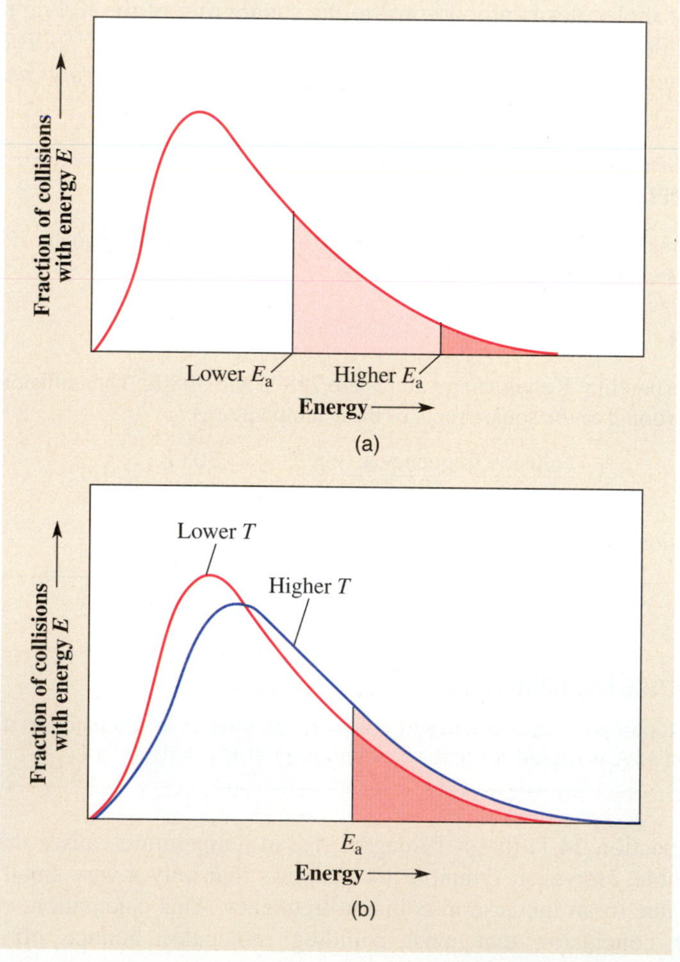

Figure 14.16

Energy distribution curves for colliding molecules. The area under the curve between any two energies represents the fraction of collisions in that energy range. (a) The number of collisions with energies above E_a is lower when the value of E_a is higher. (b) Energy distribution curves at two temperatures. The shift to a higher temperature increases the number of collisions with energies above E_a.

EXAMPLE 14.18

Calculate the fraction of collisions with energies in excess of 50.0 kJ/mol at (a) 25°C and (b) 35°C.

SOLUTION

According to Equation 14.17, each fraction is $e^{-E_a/RT}$, where E_a is 50.0 kJ/mol = 50,000 J/mol and $R = 8.314$ J/mol·K.

(a) Substituting E_a, R, and $T = 298$ K into Equation 14.17 gives

$$f_a = e^{-(50{,}000\ \text{J/mol})/(8.314\ \text{J/mol·K})(298\ \text{K})}$$
$$= 1.72 \times 10^{-9}$$

(b) Substituting E_a, R, and $T = 308$ K into Equation 14.17 gives $f_a = 3.31 \times 10^{-9}$.

Example 14.18 shows that increasing the temperature from 25°C to 35°C almost doubles the number of collisions with energies equal to or greater than 50.0 kJ/mol. This increase would double the reaction rate given by Equation 14.16 and thus account for the photographers' rule of thumb (Section 14.1). Keep in mind, however, that the effect of temperature on f_a depends on the magnitude of E_a; thus the photographers' rule will not always be obeyed.

PRACTICE EXERCISE 14.24

A reaction has an activation energy of 10.0 kJ/mol. By what factor will the reaction rate increase if the temperature is raised from 25°C to 35°C? (*Hint:* Ignore the effect of temperature on Z.)

Orientation Factor. For many reactions, the experimental rate is much less than the product of the collision frequency Z and the fraction of collisions possessing the activation energy f_a. As mentioned earlier, a collision must also bring together the atoms that will form the new bonds. Consider the elementary step

$$NO(g) + O_3(g) \rightarrow NO_2(g) + O_2(g)$$

Various orientations for colliding NO and O_3 molecules are shown in Figure 14.17, where we see that relatively few collisions would put the nitrogen atom in a position to capture an oxygen atom from one end of the O_3 molecule. The **orientation factor** (p in Equation 14.16) is the fraction of collisions that have the geometry necessary for a reaction to take place. This fraction can be quite small. It is estimated to be about 0.017 for the reaction between NO and O_3; in other words, only 1.7% of the NO–O_3 collisions are oriented so that the change could occur. The percentage is even smaller when larger molecules are involved.

By way of contrast, consider the elementary step

$$H(g) + H(g) \rightarrow H_2(g)$$

The hydrogen atoms are symmetrical, so any collision angle can result in reaction.

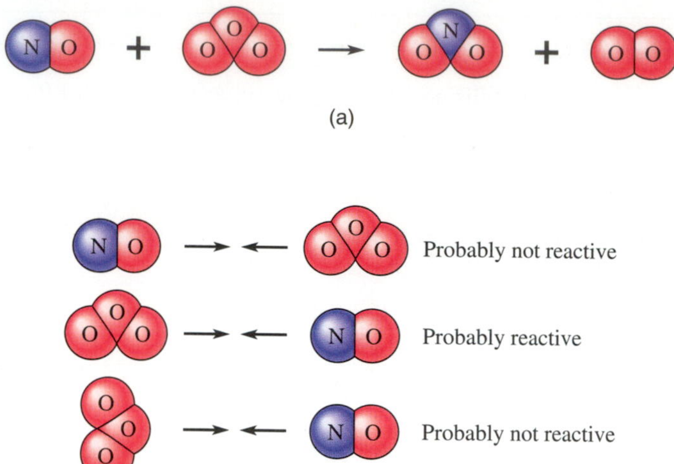

Figure 14.17

(a) NO reacts with O_3 to form NO_2 and O_2. (b) Reactive and nonreactive collisions between NO and O_3 molecules. In the reactive collision, two of the oxygen atoms in ozone remain bonded while the third forms a bond with the nitrogen atom.

The percentage of collisions with a satisfactory orientation is 100%, and the orientation factor p is equal to 1.

14.8 MORE ABOUT CATALYSTS

Let us consider the effect of a catalyst on the decomposition of hydrogen peroxide in aqueous solution:

$$2H_2O_2(aq) \rightarrow 2H_2O(l) + O_2(g)$$

In the absence of a catalyst, this reaction proceeds slowly with an activation energy of 75.3 kJ/mol H_2O_2. When iodide ion is present, the reaction proceeds by a two-step mechanism with an activation energy of only 56.5 kJ/mol.

$$\text{slow: } H_2O_2(aq) + I^-(aq) \rightarrow H_2O(l) + OI^-(aq) \qquad E_a = 56.5 \text{ kJ}$$

$$\text{fast: } H_2O_2(aq) + OI^-(aq) \rightarrow H_2O(l) + O_2(g) + I^-(aq)$$

(This mechanism was discussed in Section 14.5.) Figure 14.18 compares the potential energy profiles for the catalyzed and uncatalyzed reactions. Even though the first step of the catalyzed reaction is slow, it will occur more rapidly than the uncatalyzed reaction because of its lower activation energy.

The decomposition of hydrogen peroxide in the presence of iodide ion is an example of **homogeneous catalysis** (Section 14.1). It is similar to all catalyzed reactions in that the catalyst first combines with a reactant to form an activated complex whose potential energy is less than that of the complex formed in the absence of the catalyst. A subsequent step of the mechanism then regenerates the catalyst. In effect, *a catalyst provides an alternative reaction mechanism with a lower activation energy than the uncatalyzed reaction.*

In **heterogeneous catalysis** (Section 14.1), a contact catalyst provides a surface on which the reaction takes place. The reactant molecules diffuse from the sur-

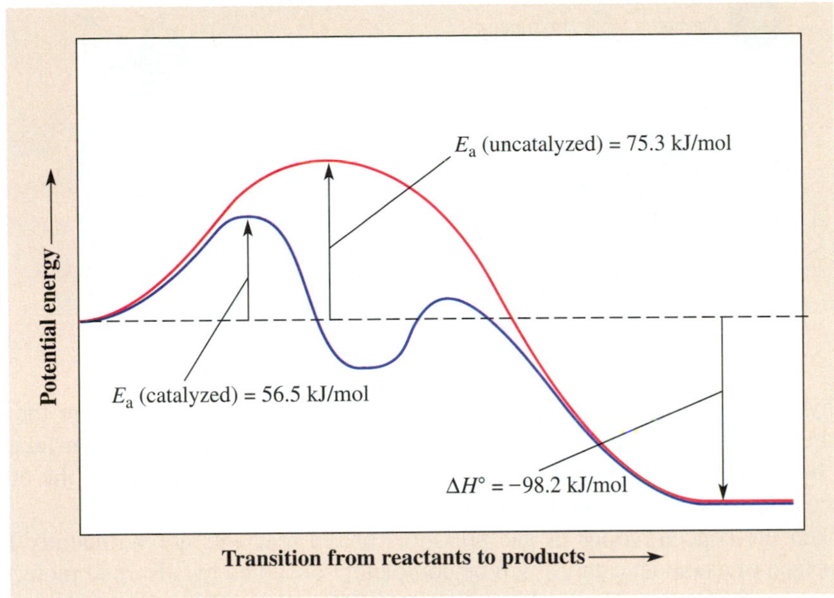

Figure 14.18
Potential energy profiles for the catalyzed and uncatalyzed decomposition of hydrogen peroxide. The uncatalyzed reaction has an activation energy of 75.3 kJ/mol. The two-step iodide-catalyzed reaction has an activation energy of 56.5 kJ/mol for the rate-determining step.

rounding liquid or gas to the surface where they are adsorbed. Weak chemical bonds form between the surface atoms and the adsorbed molecules; hence, the process is called **chemisorption**. When, for example, hydrogen gas is chemisorbed onto platinum metal, the H—H bond is stretched and virtually broken as the hydrogen atoms interact with platinum atoms (Figure 14.19). Although 435 kJ would ordinarily be needed to dissociate 1 mol of H_2 molecules into free atoms, the platinum surface makes the hydrogen atoms available to other reactants at a much lower energy cost.

An example of heterogeneous catalysis is the hydrogenation of ethene (C_2H_4) to form ethane (C_2H_6) on a finely divided nickel or platinum surface (Figure 14.20):

$$\underset{\text{ethene}}{H_2C{=}CH_2} + H_2 \xrightarrow{\text{Ni or Pt}} \underset{\text{ethane}}{H{-}CH_2{-}CH_2{-}H}$$

74 pm
Platinum surface
158 pm
(a)

(b)

Figure 14.19
(a) Chemisorption of hydrogen onto a platinum catalyst. The distance between hydrogen nuclei stretches from the normal bond length of 74 pm to 158 pm, which is the internuclear distance in platinum. (b) A mixture of hydrogen and oxygen gases reacts on the surface of a glowing platinum wire.

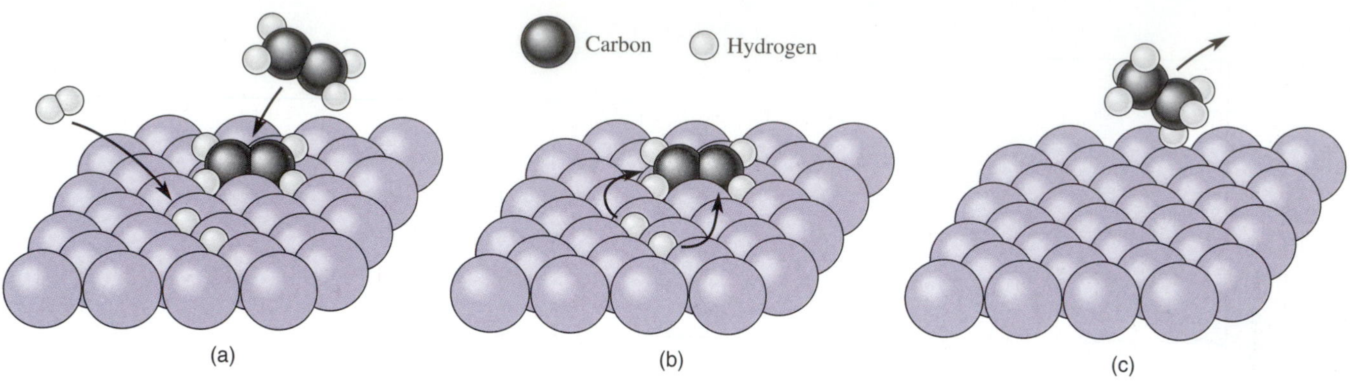

Figure 14.20

The reaction of C_2H_4 with H_2 on a platinum surface: $C_2H_4(g) + H_2(g) \longrightarrow C_2H_6(g)$. (a) The H_2 and C_2H_4 molecules diffuse to and are adsorbed on a platinum surface. (b) The loosely bound hydrogen atoms migrate across the platinum surface and form bonds with the adsorbed carbon atoms. (c) The ethane molecule diffuses away from the surface.

The hydrogen and ethane molecules diffuse to the catalytic surface, where they are adsorbed. The loosely bound hydrogen atoms then migrate across the surface and form bonds with the adsorbed carbon atoms. When the reaction is over, the ethane molecules diffuse away from the sruface.

When the concentrations or gas pressures of the reactants are sufficiently high, the surface of a contact catalyst will be completely occupied by adsorbed molecules. Since only adsorbed molecules react, and since all of the surface sites are occupied, the rate of reaction will level off to a constant value. Recall (p. 643) that when the reaction rate is constant, that is, when the rate is independent of concentration, the reaction is said to be zero order. For example, the decomposition of ammonia on a tungsten surface

$$2NH_3(g) \xrightarrow{W} N_2(g) + 3H_2(g)$$

is zero order at 856°C and an ammonia partial pressure of 200 torr. The rate law for the reaction is

$$\text{rate} = k[NH_3]^0 = k$$

Under these conditions, the concentration of ammonia decreases with time according to the equation

$$[NH_3] = [NH_3]_0 - kt$$

where t is the time and $[NH_3]_0$ is the initial concentration of ammonia, that is, the concentration at $t = 0$. (If you are interested in the derivation of this expression, try Final Exercise 14.111.)

The most effective contact catalysts are fine powders with a large area of spongy, convoluted surface on which chemisorption can readily occur. Contact catalysts eventually lose their effectiveness because they accumulate impurities that bond strongly to the surface; the inactivated catalyst is said to be **poisoned**. A poisoned catalyst can be regenerated by heating it to drive off the adsorbed impurities or by some chemical treatment that removes the impurities.

Contact catalysts composed of metals and metal oxides are used in thousands of industrial reactions. A catalyst composed largely of solid vanadium pentoxide (V_2O_5) is used to accelerate the oxidation of SO_2 to SO_3 in the contact process for preparing sulfuric acid. In the Haber process, ammonia forms from its elements on the surface of a catalyst composed of finely divided iron mixed with small amounts of various oxides (Al_2O_3, MgO, CaO, and K_2O). Platinum catalysts are used in the petroleum industry for cracking and reforming hydrocarbon molecules and in the

CHEMICAL INSIGHT

ENZYMES ARE BIOLOGICAL CATALYSTS

Biological reactions are controlled by thousands of enzymes that entice molecules along the reaction pathways that provide the organism with energy, growth, and response. Enzymes are highly specific; each one acts on a single substance (or a group of structurally similar substances) called the *substrate.* The names of most enzymes are derived from the substrate and end in *-ase;* for example, sucrase catalyzes the conversion of sucrose to glucose and fructose, and cellulase catalyzes the breakdown of cellulose to glucose.

The profound effect of enzymes on activation energy and reaction rate is illustrated by the action of urease on the decomposition of urea, a compound formed during human metabolism. Rate versus temperature studies show that the activation energy for the decomposition

$$CO(NH_2)_2(aq) + H^+(aq) + 2H_2O(l) \rightarrow 2NH_4^+(aq) + HCO_3^-(aq)$$

is 137 kJ/mol without a catalyst and only 37.0 kJ/mol when urease is present. The rate of the catalyzed reaction is more than 3×10^{17} times greater than the rate without the catalyst.

Enzymes are large molecules with molar masses ranging from 5000 grams to more than 1 million grams. Studies of enzyme structures indicate that each enzyme molecule has one or more *active sites* that attach the substrate in a way that promotes its subsequent reaction (Figure 14.21). The active site is a small crevice or hollow that adapts to the shape of the substrate molecule. An active site is also equipped with polar atoms and atoms that can form hydrogen bonds, all precisely located to attract corresponding atoms on the substrate and facilitate the formation of a low-energy transition state. When the substrate concentration is high enough to occupy all the active sites, the reaction goes at its maximum rate. Adding more substrate at this point cannot make it any faster; the reaction is now zero order with respect to substrate.

The original concept of geometrical and chemical fit between enzyme and substrate was first proposed in 1906 and is called the *lock-and-key model* (Figure 14.21a). This model assumes that the enzyme and substrate are rigid and that the substrate fits into the enzyme much as a key fits into a lock. More recent evidence indicates, however, that the active site is flexible. It adapts to the shape of the substrate more like a glove than a rigid lock (Figure 14.21b). This modified model of enzyme–substrate interaction is called the *induced-fit model.*

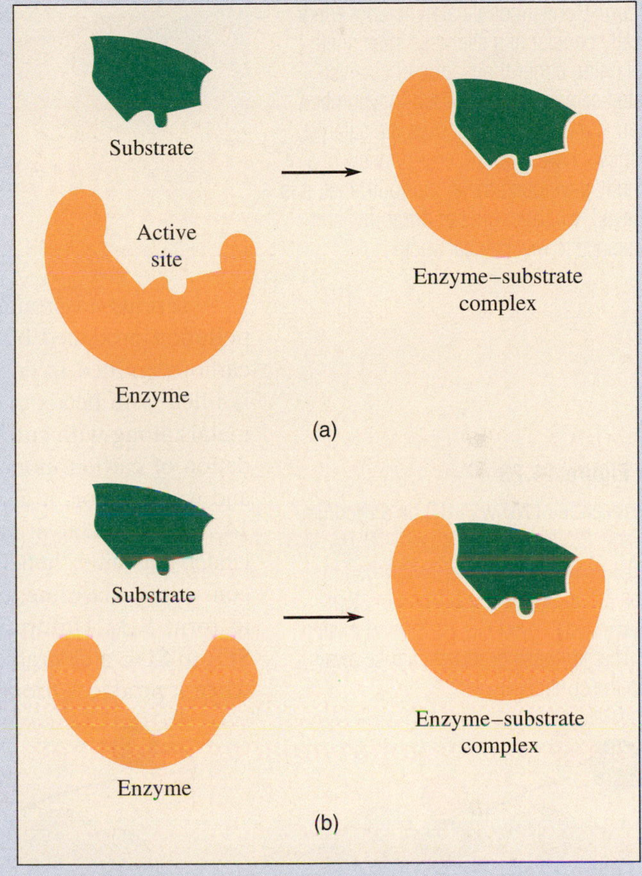

Figure 14.21
Models of enzyme action. (a) *Lock-and-key model:* A rigid substrate molecule exactly fits into a rigid active site. (b) *Induced-fit model:* The active site adapts to the shape of the substrate molecule.

food industry for hydrogenating unsaturated fats. (Unsaturated fats such as corn oil and safflower oil contain carbon–carbon double bonds and are usually liquid. Eliminating some of the double bonds by adding hydrogen results in solid products such as margarine and solid shortening.)

Figure 14.22

Bottom: Conventional catalytic converters usually consist of a stainless steel shell containing either thousands of catalyst-coated ceramic pellets or a honeycombed ceramic monolith that carries the catalyst, as shown here. *Top:* An end-on view of a recently developed metal monolith that, it is believed, will offer greater durability and resistance to mechanical shock.

A serious environmental problem associated with the automobile age is the air pollution produced by carbon monoxide, nitrogen monoxide, and unburned hydrocarbons emitted in engine exhaust. The catalytic converter required on modern cars is filled with beads of metallic elements such as platinum, palladium, and rhodium metals along with some metal oxides (Figure 14.22). This mixture catalyzes the oxidation of carbon monoxide and unburned hydrocarbons to harmless carbon dioxide and water vapor; it also promotes the decomposition of NO into N_2 and O_2 (Figure 14.23). The catalyst is poisoned by tetraethyl lead, $Pb(C_2H_5)_4$, an additive used in leaded gasoline; hence, only unleaded gasoline can be used in cars with catalytic converters. Gasoline, like all fossil fuels, contains some sulfur compounds that burn to form SO_2. Unfortunately, the platinum catalyst also promotes the oxidation of SO_2 to SO_3, a desirable reaction in a sulfuric acid plant, but an undesirable reaction when it produces droplets of sulfuric acid in city and country air.

Figure 14.23

The reaction of NO with CO on a rhodium surface: $2NO(g) + 2CO(g) \xrightarrow{Rh} N_2(g) + 2CO_2(g)$. The reactants are adsorbed in steps (a) and (b), N_2 is formed in steps (c) and (d), and CO_2 is formed in step (e). The harmless products diffuse away from the surface.

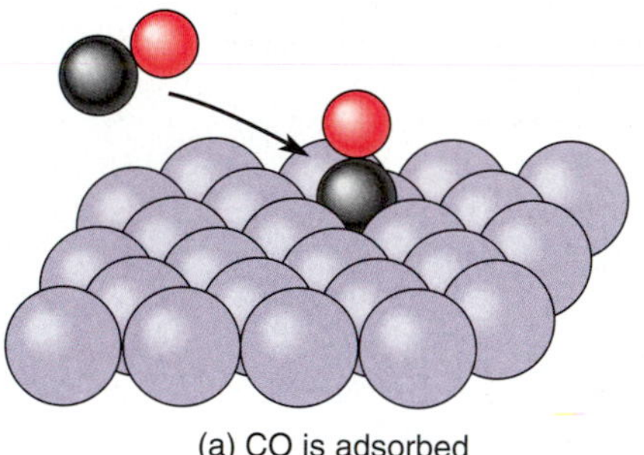

(a) CO is adsorbed

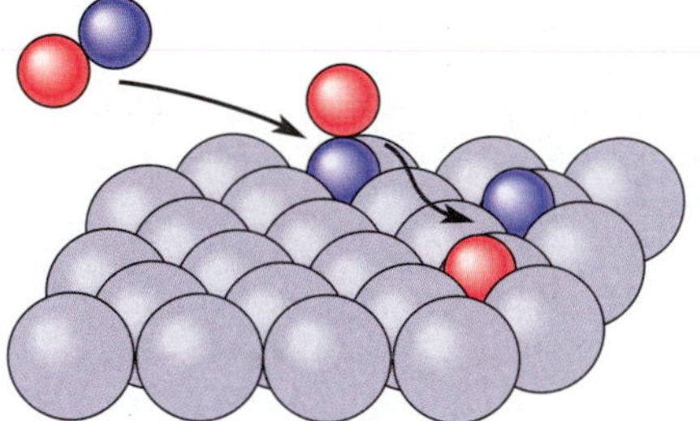

(b) NO is adsorbed and some of it dissociates into atoms

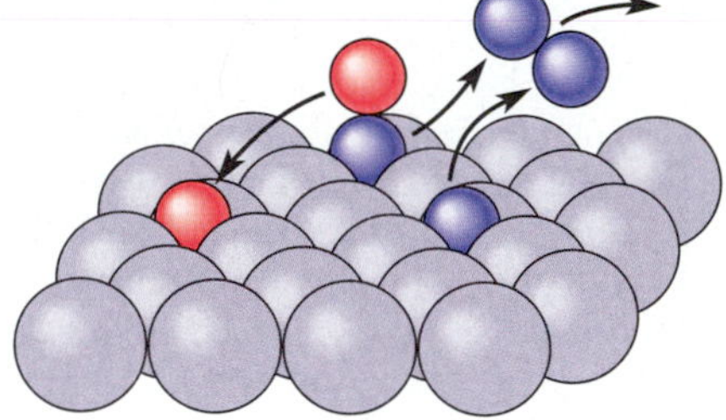

(c) Adsorbed N and NO form N_2 and adsorbed O

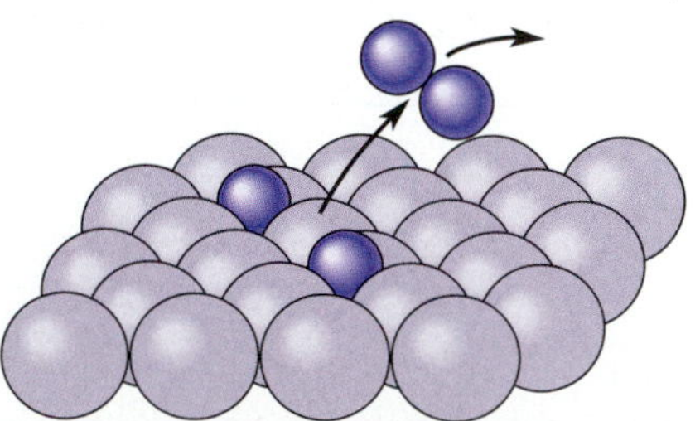

(d) Adsorbed N forms N_2

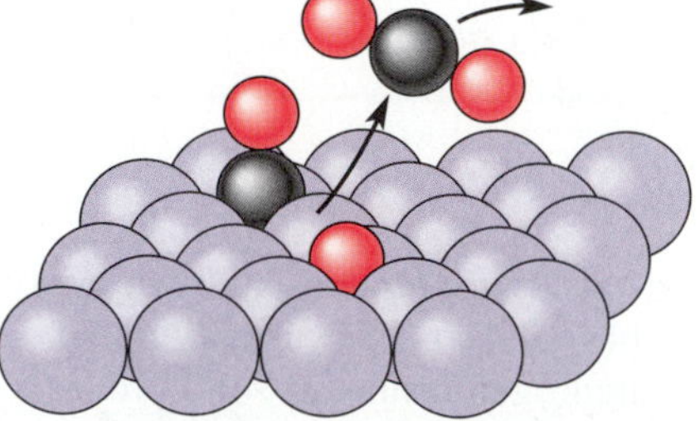

(e) Adsorbed CO and O form CO_2

CHAPTER REVIEW

Starred entries are based on the *Digging Deeper* sections.

LEARNING OBJECTIVES BY SECTION

14.1 1. Give several examples of the units used to describe reaction rates.
2. List the factors that influence reaction rates.
3. Distinguish between homogeneous and heterogeneous reactions.
4. Use the photographers' rule of thumb to estimate the effect of a temperature change on a reaction rate.
5. Distinguish between homogeneous and heterogeneous catalysis.

14.2 1. Use the balanced equation for a reaction to relate rates of change of products and reactants.
*2. Obtain average reaction rates and instantaneous reaction rates from concentration versus time data.

14.3 1. Use the method of initial reaction rates to find the rate law and rate constant for a reaction.
2. Determine the reaction order from the rate law.
3. Use the rate law to calculate the reaction rate for any given set of concentrations.

14.4 1. Use concentration versus time data to identify first-order reactions and to calculate their rate constants.
2. Given three of the four quantities, [A], $[A]_0$, k, and t, for a first-order reaction, calculate the missing quantity.
3. Calculate the half-life of a first-order reaction from its rate constant, and vice versa.
4. Use concentration versus time data to identify second-order reactions and to calculate their rate constants.
5. Given three of the four quantities, [A], $[A]_0$, k, and t, for a second-order reaction, calculate the missing quantity.
6. Calculate half-life periods for second-order reactions.
*7. Derive the integrated rate laws for first- and second-order reactions.

14.5 1. State the molecularity and write the rate law for any given elementary step.
2. State the requirements for a reaction mechanism.
3. Propose mechanisms for reactions in which the first step is rate determining.
*4. Use the steady-state approximation to find rate laws for reactions in which steps other than the first are rate determining.

14.6 1. Identify the initiation, propagation, and termination steps in a chain mechanism.
2. Describe the mechanism by which CFCs deplete the ozone layer.

14.7 1. Use values of E_a and $\Delta H°$ to sketch the potential energy profile for an elementary step.
2. Given two of the three quantities, E_a(forward), E_a(reverse), and $\Delta H°$, calculate the missing quantity.
3. Use activation energies to identify the slowest step in a mechanism.
4. Use rate constant versus temperature data to calculate activation energies.
5. Use the activation energy and the rate constant measured at one temperature to calculate the rate constant at another temperature.
*6. State how the collision frequency varies with concentration and with temperature.
*7. State how the fraction of reactive collisions varies with temperature and with activation energy.
*8. Use the activation energy to calculate the effect of changing temperature on the rate of a reaction.
*9. State how the orientation factor varies with increasing molecular complexity.

14.8 1. Explain why a reaction goes faster in the presence of a catalyst.
2. Describe the mechanism by which platinum catalyzes the reactions of hydrogen and certain other gases.
3. Explain why many heterogeneous catalysis reactions are zero-order.

KEY TERMS BY SECTION

14.1 Catalyst
Chemical kinetics
Contact catalyst
Enzyme
Heterogeneous catalysis
Heterogeneous reaction
Homogeneous catalysis
Homogeneous reaction
Reaction rate

14.2 *Average reaction rate
*Instantaneous reaction rate

14.3 First-order reaction
Order (of a reaction)
Rate constant
Rate equation
Rate law
Zero-order reaction

14.4 Half-life ($t_{1/2}$)
*Integrated rate law

14.5 Bimolecular step
Elementary step
Intermediate
Molecularity (of a step)
Rate-determining step
Reaction mechanism
*Steady state
*Steady-state approximation
Termolecular step
Unimolecular step

14.6 Chain initiation step
Chain propagation step
Chain reaction
Chain termination step

14.7 Activated complex
Activation energy (E_a)
Arrhenius equation
*Collision frequency (Z)
*Orientation factor (p)
Potential energy profile
Transition state

14.8 Chemisorption
Poisoning (of a catalyst)

IMPORTANT EQUATIONS

14.1 $\text{rate} = k[\text{A}]^m[\text{B}]^n[\text{C}]^p \ldots$

14.2 $\text{rate} = k[\text{A}]$

14.3 $[\text{A}] = [\text{A}]_0 e^{-kt}$

14.4 $\ln\dfrac{[\text{A}]}{[\text{A}]_0} = -kt$

14.5 $\ln[\text{A}] = -kt + \ln[\text{A}]_0$

14.6 $kt_{1/2} = \ln 2$

14.7 $\text{rate} = k[\text{A}]^2$

14.8 $\dfrac{1}{[\text{A}]} = kt + \dfrac{1}{[\text{A}]_0}$

14.9 $t_{1/2} = \dfrac{1}{k[\text{A}]_0}$

14.12 $\Delta H^\circ = E_a(\text{forward}) - E_a(\text{reverse})$

14.13 $k = Ae^{-E_a/RT}$

14.14 $\ln k = -\dfrac{E_a}{RT} + \ln A$

14.15 $\ln\dfrac{k_1}{k_2} = \dfrac{E_a}{R}\left(\dfrac{1}{T_2} - \dfrac{1}{T_1}\right)$

***14.16** $\text{rate} = Z \times f_a \times p$

***14.17** $f_a = e^{-E_a/RT}$

FINAL EXERCISES

Answers to exercises with blue numbers are given in Appendix D. Starred exercises are based on the *Digging Deeper* sections.

PART A. QUESTIONS AND PROBLEMS BY SECTION

Factors that Influence Reaction Rates (Section 14.1)

14.1 List four factors that influence the rate of a reaction.

14.2 Which of the factors affecting reaction rates are involved in (a) coal dust explosions, (b) cold storage, and (c) vacuum packing to prevent spoilage?

14.3 What simple adjustments would you make in conditions to maximize the rate of the following reactions?
(a) burning wood
(b) rusting iron
(c) combining NO and O_2 gases

14.4 What simple adjustments would you make in conditions to maximize the rate of the following reactions?
(a) absorption of CO_2 gas by solid CaO to form solid $CaCO_3$
(b) oxidation of solid sodium sulfite to sodium sulfate by atmospheric oxygen
(c) decomposition of solid NH_4Cl into HCl and NH_3 gases

14.5 Explain why reactions between ions in solution tend to be rapid.

14.6 Explain why increasing surface area increases the rate of a heterogeneous reaction.

14.7 Distinguish between heterogeneous and homogeneous catalysis. What is a contact catalyst?

14.8 Which type of catalysis is illustrated by the following observations?
(a) The gaseous decomposition of acetaldehyde at 500°C is accelerated by iodine vapor.
(b) The decomposition of N_2O gas at 900°C occurs more rapidly in the presence of gold metal than in its absence.

14.9 The body temperature of a hibernating animal drops from 35.5°C to 15.5°C. By what factor would you expect the animal's oxygen consumption to decrease? Assume that metabolic reaction rates approximately follow the "photographers' rule of thumb".

14.10 Water boils at about 110°C inside a pressure cooker set at the "low" setting. What cooking time would you allow in the pressure cooker for a stew that would normally take $1\frac{1}{2}$ hours to cook in an open pot?

Measuring Reaction Rates (Section 14.2)

14.11 List at least three physical properties that might be used to monitor the rate of a reaction—try to think of some properties that were not mentioned in the text.

***14.12** (a) Distinguish between an average reaction rate and an instantaneous reaction rate.
(b) Why are these rates seldom equal to each other?
(c) Under what circumstances will they be equal?

***14.13** (a) How would you use concentration versus time data to calculate the average rate of a reaction during some specified time interval?
(b) How would you determine the instantaneous rate at some given time?

***14.14** How would you use concentration versus time data to determine the initial rate of a reaction?

14.15 One step in the commercial production of nitric acid by the Haber–Ostwald process is the oxidation of ammonia gas:

$$4NH_3(g) + 5O_2(g) \rightarrow 4NO(g) + 6H_2O(g)$$

Let $\Delta[NO]/\Delta t$ represent the rate of formation of NO. Express the rate of formation of H_2O and the rates of disappearance of NH_3 and O_2 in terms of $\Delta[NO]/\Delta t$.

14.16 Consider the reaction

$$2CO(g) + O_2(g) \rightarrow 2CO_2(g)$$

Express the rates of change of O_2 and CO_2 in terms of $\Delta[CO]/\Delta t$.

14.17 Hydrogen peroxide decomposes according to the equation

$$2H_2O_2(aq) \rightarrow 2H_2O(l) + O_2(g)$$

At 27°C and 1.00 atm, a 25.0-mL sample of aqueous hydrogen peroxide decomposes at a rate that produces a 10.0 mL/s increase in the oxygen volume above the solution. Assume ideal gas behavior and convert this rate into (a) moles of oxygen produced per second and (b) change in molarity of H_2O_2 per second.

14.18 Paraldehyde ($(CH_3CHO)_3$) dissociates into acetaldehyde (CH_3CHO) at elevated temperatures:

$$(CH_3CHO)_3(g) \rightarrow 3CH_3CHO(g)$$

The reaction rate can be monitored by measuring the rate at which the total pressure changes. Derive an expression relating the change in total pressure to the change in the number of moles of paraldehyde. Assume that the volume and temperature of the reacting system do not change.

***14.19** Sucrose hydrolyzes (reacts with water) in acidic solution to form glucose and fructose. The rate of the reaction can be monitored by optical techniques, and the following molarity versus time data were obtained for one run at 25°C:

t (min)	[sucrose]
0	0.584
20.0	0.549
60.0	0.483
100	0.427
160	0.350
220	0.291
440	0.146
660	0.073

Prepare a graph of concentration versus time, and use it to estimate (a) the initial reaction rate and (b) the rate at 440 min.

***14.20** The following data for the hydrolysis of $(CH_3)_3CBr$ in a solvent consisting of 10% water and 90% acetone were obtained at 25°C:

t (h)	$[(CH_3)_3CBr]$
0	0.1039
3.15	0.0896
4.10	0.0859
6.20	0.0776
8.20	0.0701
10.0	0.0639
26.0	0.0270

Prepare a graph of concentration versus time, and use it to estimate (a) the initial reaction rate and (b) the rate at 14.0 h.

***14.21** Use data from Exercise 14.19 to estimate the average rate of hydrolysis of sucrose for the first 100 min.

***14.22** Use data from Exercise 14.20 to estimate the average rate of hydrolysis of $(CH_3)_3CBr$ for the first 10.0 h.

Rate Laws (Section 14.3)

14.23 What is a *rate law*? Can the exponents in a rate law be deduced from a balanced equation?

14.24 Briefly describe how rate laws are obtained by the *method of initial reaction rates.*

14.25 What is meant by the *order* of a reaction?

14.26 What does it mean when we say that a reaction is *zero order* with respect to a reactant?

14.27 What is a *rate constant*? How, in general, are rate constants determined?

14.28 How is the value of a rate constant affected by (a) reactant concentrations, (b) product concentrations, (c) temperature, and (d) pressure?

14.29 If concentration and time are expressed in moles per liter and seconds, respectively, what will be the units of the rate constant for each of the following reactions?
(a) a first-order reaction
(b) a second-order reaction

14.30 If concentration and time are expressed in moles per liter and minutes, respectively, what will be the units of the rate constant for each of the following reactions?
(a) a third-order reaction
(b) a zero-order reaction

14.31 A study of the ammonia synthesis at 450°C

$$N_2(g) + 3H_2(g) \longrightarrow 2NH_3(g)$$

shows that the rate increases by a factor of 9 when the partial pressure of hydrogen is tripled and the nitrogen pressure is kept constant. The rate increases by a factor of 3 when the partial pressure of nitrogen is tripled and the hydrogen pressure is kept constant. Write the rate law for the synthesis.

14.32 The gas-phase decomposition of acetaldehyde (CH_3CHO) produces methane (CH_4) and carbon monoxide:

$$CH_3CHO(g) \longrightarrow CH_4(g) + CO(g)$$

The order of the decomposition is 1.5 with respect to acetaldehyde.
(a) Write the rate law for the decomposition.
(b) By what factor will the reaction rate increase if the partial pressure of acetaldehyde is doubled?

14.33 The following initial rates were obtained for the reaction

$$A + B \longrightarrow \text{products}$$

Run	[A]	[B]	rate (mol/L·s)
1	0.10	0.60	0.012
2	0.30	0.60	0.108
3	0.10	0.30	0.006

(a) Write the experimental rate law and state the order of the reaction.
(b) Calculate the rate constant (include units).
(c) What is the rate of this reaction when A and B are each 0.010 *M*?

14.34 Some initial rates and concentration data for the reaction

$$NH_4^+(aq) + HNO_2(aq) \longrightarrow N_2(g) + 2H_2O(l) + H^+(aq)$$

are summarized below:

$[HNO_2]$	$[NH_4^+]$	rate (mol/L·s)
0.0092	0.098	34.9×10^{-8}
0.0092	0.047	16.6×10^{-8}
0.0498	0.196	335.0×10^{-8}
0.0227	0.196	156.0×10^{-8}

(a) Write the rate law for the reaction.
(b) Calculate the rate constant (include units).
(c) Calculate the rate of N_2 production in (mol/L·s) when each reactant concentration is 0.040 *M.*

14.35 The following initial rates were obtained at 620 K for the reaction

$$2C_2F_4(g) \longrightarrow C_4F_8(g)$$

mmol C_2F_4/L	rate (mmol C_2F_4/L·s)
10.14	6.39×10^{-12}
9.31	5.38×10^{-12}
6.88	2.94×10^{-12}

(a) Write the experimental rate law and state the order of the reaction.
(b) Evaluate the rate constant (include units).

(c) What will the reaction rate be when the concentration of C_2F_4 is 0.200 mmol/L?

14.36 The following initial rates were obtained for the reaction of mercury(II) chloride with oxalate ion at 20°C:

$$2HgCl_2(aq) + C_2O_4^{2-}(aq) \rightarrow 2Cl^-(aq) + 2CO_2(g) + Hg_2Cl_2(s)$$

$[HgCl_2]$	$[C_2O_4^{2-}]$	rate (mol/L·min)
0.15	0.20	9.2×10^{-5}
0.15	0.32	2.4×10^{-4}
0.25	0.32	4.0×10^{-4}

(a) Write the experimental rate law and state the order of the reaction.
(b) Evaluate the rate constant (include units).
(c) Suggest at least one way that this rate might have been monitored.

14.37 The rate of decomposition of N_2O_4

$$N_2O_4(g) \rightarrow 2NO_2(g)$$

depends on the concentration of nitrogen gas in the reaction mixture. Initial rates and concentration data for the reaction have been reported as follows:

$[N_2O_4]$	$[N_2]$	Rate (mol/L·s)
1.80×10^{-4}	0.0225	6.93
2.25×10^{-4}	0.0200	7.70
2.00×10^{-4}	0.0225	7.69

(a) Write the rate law for the reaction.
(b) Calculate the rate constant (include units).
(c) What is the order of the reaction with respect to N_2O_4? With respect to N_2?

14.38 The combination of iodine atoms in the presence of argon is represented by the equation:

$$2I(g) + Ar(g) \rightarrow I_2(g) + Ar(g)$$

Initial rates and concentration data for the reaction have been reported as follows:

[I]	[Ar]	Rate (mol/L·s)
2.50×10^{-5}	3.00×10^{-3}	1.63×10^{-2}
3.25×10^{-5}	3.25×10^{-3}	2.99×10^{-2}
3.75×10^{-5}	3.00×10^{-3}	3.67×10^{-2}

(a) Write the rate law for the reaction.
(b) Calculate the rate constant (include units).
(c) What is the order of the reaction with respect to I? With respect to Ar?

First- and Second-Order Reactions (Section 14.4)

14.39 What is a *first-order reaction*? How would you use concentration versus time data to identify a first-order reaction?

14.40 How would you use concentration versus time data to distinguish between a first-order reaction and a second-order reaction?

14.41 **(a)** What is meant by *half-life*?
(b) How does the half-life of a first-order reaction differ from the half-life of a second-order reaction?

14.42 Derive the relation $kt_{1/2} = \ln 2$ from Equation 14.4 without referring to the text.

14.43 Refer to the data for Exercise 14.19.
(a) Show graphically that the hydrolysis of sucrose follows first-order kinetics.
(b) Evaluate the rate constant at 25°C.
(c) How many minutes would it take to hydrolyze 20% of a sample of sucrose at 25°C?

14.44 Refer to the data for Exercise 14.20.
(a) Show graphically that the hydrolysis of $(CH_3)_3CBr$ follows first-order kinetics.
(b) Evaluate the rate constant at 25°C.
(c) How many hours would it take to hydrolyze 80% of a sample of $(CH_3)_3CBr$ at 25°C?

14.45 The decomposition

$$2N_2O_5(g) \rightarrow 4NO_2(g) + O_2(g)$$

is a first-order reaction with $k = 4.98 \times 10^{-4}\ s^{-1}$ at 45°C.
(a) If 1.00 mol of N_2O_5 is placed in a container at 45°C, how many moles will remain after 1.00 h?
(b) If the N_2O_5 pressure is originally 0.500 atm and the volume is constant, what will be the total gas pressure after 1.00 h?

14.46 A study of the reaction

$$SO_2Cl_2(g) \rightarrow SO_2(g) + Cl_2(g)$$

at 593 K shows that it is first order and that 10.0% of the SO_2Cl_2 decomposes in 80.0 min.
(a) Calculate k for the reaction at 593 K.
(b) How many minutes will it take for a 5.00 mmol sample of SO_2Cl_2 to decompose to 3.50 mmol?

14.47 Refer to Exercise 14.45.
(a) Calculate the half-life for the N_2O_5 decomposition.
(b) How many minutes will it take for 15.0% of an N_2O_5 sample to decompose?

14.48 Refer to Exercise 14.46.
(a) Calculate the half-life for the SO_2Cl_2 decomposition.
(b) What fraction of an SO_2Cl_2 sample will remain after 125 min?

14.49 Refer to Exercise 14.43, and calculate the half-life for the acid-catalyzed hydrolysis of sucrose.

14.50 Refer to Exercise 14.44, and calculate the half-life for the hydrolysis of $(CH_3)_3CBr$.

14.51 The half-life of a first-order decomposition, A $\rightarrow$ products, is 32.0 min. How many minutes would it take for the concentration of A to drop from 0.256 *M* to 0.0160 *M*.

14.52 Refer to Exercise 14.51. How many minutes would it take for the concentration of A to drop by one-fourth, that is, to 75% of its original value?

14.53 The half-life of a second-order decomposition, A $\rightarrow$ products, is 32.0 min when the initial concentration of A is 0.256 *M*. How many minutes would it take for the concentration of A to drop from 0.256 *M* to 0.0160 *M*? Compare this answer with the answer you obtained in Exercise 14.51.

14.54 Refer to Exercise 14.53.
(a) Calculate the second half-life of A, that is, the time required for the concentration to drop to one-fourth of 0.256 *M*.
(b) Calculate the third half-life of A.

14.55 The following concentration versus time data have been reported for the conversion of ammonium cyanate to urea:

$$NH_4CNO(aq) \rightarrow (NH_2)_2CO(aq)$$

t (min)	$[NH_4CNO]$
0	0.1000
45.0	0.0808
72.0	0.0716
107.0	0.0638
230.0	0.0463

(a) Show that this reaction is second order.
(b) Find the value of the rate constant.
(c) How many minutes would it take for the concentration of product to reach 0.0750 *M*?

14.56 Ethyl acetate $(CH_3COOC_2H_5)$ reacts with hydroxide ion in aqueous solution according to the equation

$$CH_3COOC_2H_5(aq) + OH^-(aq) \rightarrow CH_3COO^-(aq) + C_2H_5OH(aq)$$

The reaction is known to be second order. Use the following concentration versus time data to (a) determine the rate constant, (b) calculate the number of seconds required for one-half the original amount of ethyl acetate to react, and (c) calculate the time required for 75% of the ethyl acetate to react.

time (s)	*M*(of each reactant)
0	0.01000
60.0	0.00917
120.0	0.00840
180.0	0.00775
240.0	0.00724
300.0	0.00675

Reaction Mechanisms (Section 14.5)

14.57 **(a)** What is a *reaction mechanism*?
(b) State the requirements that must be satisfied by any mechanism.

14.58 **(a)** What is an *elementary step*?
(b) Explain why the exponents in the rate law of an elementary step are identical to the coefficients in the equation for the step.
(c) Why doesn't a similar relation hold for a reaction that is not an elementary step?

14.59 Distinguish between *reaction order* and *molecularity.*

14.60 Why are termolecular steps (steps with molecularity of three) relatively uncommon?

14.61 What is the difference between a *catalyst* and an *intermediate* in a reaction mechanism?

***14.62** What is the *steady-state approximation*? What is the purpose of this approximation?

14.63 Write a rate law for each of the following elementary steps:
(a) $Cl_2(g) \rightarrow Cl(g) + Cl(g)$
(b) $Cl(g) + Cl(g) \rightarrow Cl_2(g)$

14.64 Write a rate law for each of the following elementary steps:
(a) $Cl(g) + Cl(g) + N_2(g) \rightarrow Cl_2(g) + N_2(g)$
(b) $Br_2(g) + H(g) \rightarrow HBr(g) + Br(g)$

14.65 The reaction

$$2ClO_2(g) + F_2(g) \longrightarrow 2FClO_2(g)$$

is first order in ClO_2 and first order in F_2. Write the rate law, and devise a two-step mechanism consistent with the rate law.

14.66 The rate law for the reaction

$$2Fe^{2+}(aq) + Cl_2(aq) \longrightarrow 2Cl^-(aq) + 2Fe^{3+}(aq)$$

is rate $= k[Fe^{2+}][Cl_2]$. Devise a possible mechanism.

Steady State Approximation (Section 14.5)

***14.67** Hydrogen iodide forms from its elements according to the equation

$$H_2(g) + I_2(g) \longrightarrow 2HI(g)$$

The principal mechanism at moderate temperatures is believed to consist of the following steps:

$$I_2(g) \underset{k_{-1}}{\overset{k_1}{\rightleftharpoons}} 2I(g) \qquad \text{fast}$$

$$H_2(g) + 2I(g) \overset{k_2}{\longrightarrow} 2HI(g) \qquad \text{slow}$$

Derive the rate law for this reaction.

***14.68** The rate law for the reaction

$$2NO(g) + Cl_2(g) \longrightarrow 2NOCl(g)$$

is rate $= k[NO]^2[Cl_2]$. Show that the following mechanism is consistent with this rate law:

$$2NO(g) \underset{k_{-1}}{\overset{k_1}{\rightleftharpoons}} N_2O_2(g) \qquad \text{fast}$$

$$N_2O_2(g) + Cl_2(g) \overset{k_2}{\longrightarrow} 2NOCl(g) \qquad \text{slow}$$

***14.69** The following mechanism has been proposed for the reaction between NO and O_2 gases:

$$2NO(g) \underset{k_{-1}}{\overset{k_1}{\rightleftharpoons}} N_2O_2(g) \qquad \text{fast}$$

$$N_2O_2(g) + O_2(g) \overset{k_2}{\longrightarrow} 2NO_2(g) \qquad \text{slow}$$

(a) Write the overall equation and identify any intermediates that are present.
(b) Use the steady-state approximation to derive the rate law for the reaction.

***14.70** Consider the following mechanism for an enzyme (E) acting on a substrate (S) to form a product (P):

$$E + S \underset{k_{-1}}{\overset{k_1}{\rightleftharpoons}} ES \qquad \text{fast}$$

$$ES \overset{k_2}{\longrightarrow} E + P \qquad \text{slow}$$

(a) Write an equation for the overall reaction.
(b) Identify the catalyst and any intermediates in the above mechanism.
(c) Use the steady-state approximation to determine the rate law for the reaction.

Chain Reactions (Section 14.6)

14.71 What is the difference between a *chain initiation step*, a *chain propagation step*, and a *chain termination step*?

14.72 **(a)** Describe, with equations, the mechanism by which CFCs deplete the ozone layer over Antarctica. Identify the chain initiation and chain propagation steps.
(b) Write an equation for the overall reaction. Which steps add up to this reaction?
(c) Explain, with equations, why the ozone depletion is most severe during the early Antarctic spring.

14.73 The chlorination of methane

$$CH_4(g) + Cl_2(g) \longrightarrow CH_3Cl(g) + HCl(g)$$

proceeds by a free-radical chain mechanism. The initiation step is the dissociation of a chlorine molecule that has been energized by light or heat.
(a) Write equations for the chain initiation step, the chain propagation steps, and three possible chain termination steps. (*Hint:* HCl is formed in the first propagation step.)
(b) Which steps add up to the overall reaction?

***14.74** Refer to the chain mechanism given for the formation of hydrogen bromide in Section 14.6. The chain initiation step and the second chain propagation step are fast; the first propagation step is slow. Use the steady-state approximation to derive the rate law for this mechanism. Assume that the chain termination steps do not contribute to the mechanism.

Activation Energy (Section 14.7)

14.75 How is the rate of an elementary step affected by the magnitude of its activation energy?

14.76 Does the activation energy of an elementary step vary with temperature? Explain your answer.

14.77 Sketch a potential energy profile for an endothermic bimolecular step.

14.78 Sketch a potential energy profile for an exothermic unimolecular step.

14.79 The reaction of nitrogen dioxide with oxygen

$$NO_2(g) + O_2(g) \longrightarrow NO(g) + O_3(g)$$

is endothermic with $\Delta H° = +199$ kJ and $E_a =$ 209 kJ/mol. Sketch a potential energy profile for this reaction. Your diagram should show the forward and reverse activation energies as well as $\Delta H°$.

14.80 The hydrolysis of sucrose to glucose and fructose in acid solution has an activation energy of 108 kJ/mol. The enthalpy change for the reaction is −20.1 kJ/mol.

(a) Calculate the activation energy for the reverse reaction

$$\text{glucose} + \text{fructose} \longrightarrow \text{sucrose}$$

(b) Sketch the potential energy profile for the hydrolysis of sucrose. Your diagram should show the forward and reverse activation energies as well as $\Delta H°$.

14.81 The reaction A + B → D is catalyzed by C in a two-step mechanism:

Step 1: $A + C \longrightarrow X$

Step 2: $X + B \longrightarrow D + C$

The activation energies are 40 kJ/mol for Step 1 and 100 kJ/mol for Step 2. The enthalpy changes are −20 kJ for Step 1 and −315 kJ for Step 2.

(a) Sketch the potential energy profile for this reaction.

(b) Which step is rate determining?

(c) Calculate the value of $\Delta H°$ for the overall reaction.

14.82 Consider the following mechanism for the catalyzed decomposition of acetaldehyde (CH_3CHO) vapor:

$$CH_3CHO(g) + I_2(g) \longrightarrow CH_3I(g) + HI(g) + CO(g) \qquad \text{slow}$$

$$CH_3I(g) + HI(g) \longrightarrow CH_4(g) + I_2(g) \qquad \text{fast}$$

The activation energy for the reaction is 108 kJ/mol and $\Delta H°$ is −19.1 kJ.

(a) To which step, first or second, does the activation energy apply?

(b) Sketch a potential energy profile for the two-step mechanism.

Determining Activation Energies (Section 14.7)

14.83 Write the Arrhenius equation. How does the rate constant of a reaction vary with (a) increasing E_a and (b) increasing T?

14.84 Derive Equation 14.15 from the Arrhenius equation without referring to the text.

14.85 Calculate the activation energy for a reaction whose rate doubles between 25 and 35°C.

14.86 Calculate the activation energy for a reaction whose rate triples between 25 and 35°C.

14.87 The rate constant for a first-order reaction is 2.20×10^{-5} min^{-1} at 457.6 K and 3.07×10^{-3} min^{-1} at 510.1 K.

(a) Calculate the activation energy for this reaction.

(b) Calculate the preexponential factor A.

(c) Calculate the rate constant at 475.0 K.

14.88 The rate constant for a second-order reaction is 0.105 L/mol·s at 759 K and 0.343 L/mol·s at 791 K.

(a) Calculate the activation energy for this reaction.

(b) Calculate the preexponential factor A.

(c) Calculate the rate constant at 780 K.

14.89 The following data were obtained for the reaction

$$2C_2H_4(g) \longrightarrow C_4H_8(g)$$

T (°C)	k (L/mol·s)
25	1.6×10^{-20}
35	1.3×10^{-19}
45	8.7×10^{-19}
55	5.4×10^{-18}
65	3.0×10^{-17}

Prepare a graph of ln k versus $1/T$, and use it to determine A and E_a for the reaction.

14.90 The following data were obtained for the decomposition of N_2O_5 dissolved in carbon tetrachloride:

T (°C)	k (s^{-1})
25	4.1×10^{-5}
35	1.5×10^{-4}
45	6.2×10^{-4}
55	2.1×10^{-3}

Prepare a graph of ln k versus $1/T$, and use it to determine E_a. (*Note:* E_a for this reaction will not be the same as E_a obtained for the gas phase decomposition in Example 14.15.)

Collision Theory of Reaction Rates (Section 14.7)

***14.91** State the principal assumption underlying the collision theory of reaction rates.

***14.92** List the factors that determine the rate of a bimolecular reaction.

***14.93** How does the bimolecular collision frequency of a gas vary with (a) increasing temperature and (b) increasing concentration?

***14.94** Some reactions involving free radicals or charged particles apparently have zero activation energy. How will the rate of such a reaction respond to a change in temperature? (*Hint:* Don't forget the collision frequency.)

***14.95** How does the orientation factor vary with increasing molecular complexity?

***14.96** The fraction of collisions with an orientation suitable for reaction is quite low for the elementary step

$$NO_2(g) + NO_2(g) \longrightarrow NO_3(g) + NO(g)$$

Sketch various orientations for the colliding molecules. Which orientations do you think would result in a reaction?

***14.97** A gaseous reaction with $E_a = 85$ kJ/mol is carried out first at 273 K and then at 373 K.

(a) By what factor will the number of bimolecular collisions increase?

(b) By what factor does $e^{-E_a/RT}$ increase?

(c) By what factor will the reaction rate increase?

***14.98** The Br—Br bond energy is 193 kJ/mol. What is the probability that a collision will result in the dissociation of a Br_2 molecule at (a) 300 K and (b) 3000 K? (*Hint:* Assume that $p = 1$ and that the probability of a dissociation is the same as the fraction of reactive collisions.)

Catalysis (Section 14.8)

14.99 Explain why a reaction goes faster in the presence of a catalyst.

14.100 What effect does a catalyst have on each of the following?

(a) reaction mechanism
(b) energy of activation
(c) forward reaction rate
(d) reverse reaction rate
(e) enthalpy of reaction

14.101 Briefly describe the general mechanism of *heterogeneous catalysis.*

14.102 Explain why many heterogeneous catalysis reactions are zero order.

14.103 The activation energy for the decomposition of ammonia

$$NH_3(g) \longrightarrow \tfrac{1}{2}N_2(g) + \tfrac{3}{2}H_2(g)$$

is 335 kJ/mol for the uncatalyzed reaction and 165 kJ/mol when the reaction is catalyzed by powdered tungsten. The enthalpy change for the reaction is +46 kJ/mol. Calculate the molar activation energy for the formation of ammonia from its elements when the reaction is (a) catalyzed and (b) uncatalyzed.

14.104 The activation energy for the decomposition of urea is reduced from 137 kJ/mol to 37.0 kJ/mol by the enzyme urease. Show that the enzyme increases the reaction rate by a factor of about 3×10^{17} at 298 K.

PART B. MISCELLANEOUS QUESTIONS AND PROBLEMS

14.105 Explain why nitrogen (N_2), which is ordinarily unreactive, reacts quickly with many substances when it is subjected to an electrical discharge.

14.106 The equation for the enzyme catalyzed decomposition of glucose is

$$C_6H_{12}O_6(s) \longrightarrow 3CO_2(g) + 3CH_4(g)$$

The reaction rate can be followed by measuring the total pressure of the evolving gas. Derive an expression that relates the change in pressure to the change in the number of moles of glucose. Assume that the volume and temperature of the evolving gas remain constant.

14.107 The reaction

$$H_2(g) + I_2(g) \longrightarrow 2HI(g)$$

is first order with respect to both hydrogen and iodine at 400°C. The rate constant is 2.42×10^{-2} L/mol·s. Calculate the reaction rate when $[H_2] = 0.104$ *M* and $[I_2] = 0.0100$ *M*.

14.108 The half-life of radon-222, a radioactive isotope, is 3.823 days.

(a) Calculate the rate constant *k*.

(b) What fraction of a radon-222 sample will remain after 1 week?

14.109 Acetone (CH_3COCH_3) decomposes at high temperatures to give carbon monoxide and a mixture of hydrocarbons. The decomposition is first order with a half-life of 81 s at 600°C. How many seconds would it take for the partial pressure of acetone to drop from 0.51 atm to 0.45 atm at 600°C?

14.110 The following data were obtained for the decomposition of nitrosyl chloride at 100°C:

$$2NOCl(g) \longrightarrow 2NO(g) + Cl_2(g)$$

t (s)	0	10	20	30
[NOCl]	0.50	0.36	0.28	0.23

(a) Determine the order of this reaction and find the rate constant at 100°C.
(b) What change would there be in the initial rate if the volume of the container were doubled without adding more reactant?

*14.111 The reaction, A $\longrightarrow$ products, is zero order; that is, the rate is constant and independent of concentration.
(a) Derive the integrated rate law for this reaction.
(b) Show that a plot of [A] versus time is linear.
(c) Derive the half-life expression for this reaction.

*14.112 How would you use concentration versus time data to distinguish between a zero-order and a first-order reaction?

*14.113 The rate law for the reaction, A + B $\longrightarrow$ products, is rate = k[A][B]. Derive the integrated rate law.

14.114 The rate constant for the elementary step

$$H^+(aq) + OH^-(aq) \longrightarrow H_2O(l)$$

is 1.3×10^{11} L/mol·s at 25°C. If a strong acid and base are mixed in concentrations such that both H^+ and OH^- are initially 1.0 M, how many seconds would it take for each of their concentrations to be reduced to 10^{-7} M? (*Hint:* Show that the rate law is of the form, rate = $k[A]^2$.)

14.115 Can a third-order reaction have a mechanism consisting only of bimolecular steps? Explain.

*14.116 The rate law for the reaction

$$H_2O_2(aq) + 2H^+(aq) + 2Br^-(aq) \longrightarrow 2H_2O(l) + Br_2(aq)$$

is rate = $k[H_2O_2][H^+][Br^-]$. Show that the following mechanism, in which Steps 1 and 2 are fast, and Step 3 is rate determining, could account for the reaction products and the observed rate law:

Step 1: $H_2O_2(aq) + H^+(aq) \longrightarrow H_3O_2^+(aq)$

Step 2: $H_3O_2^+(aq) \longrightarrow H^+(aq) + H_2O_2(aq)$

Step 3: $H_3O_2^+(aq) + Br^-(aq) \longrightarrow H_2O(l) + HOBr(aq)$

Step 4: $HOBr(aq) + Br^-(aq) \longrightarrow Br_2(aq) + OH^-(aq)$

Step 5: $H^+(aq) + OH^-(aq) \longrightarrow H_2O(l)$

*14.117 The following mechanism has been proposed for the reaction between Cl_2 and $CHCl_3$ gases:

$$Cl_2(g) \underset{k_{-1}}{\overset{k_1}{\rightleftharpoons}} 2Cl(g) \qquad \text{fast}$$

$$CHCl_3(g) + Cl(g) \xrightarrow{k_2} HCl(g) + CCl_3(g) \qquad \text{slow}$$

$$CCl_3(g) + Cl(g) \xrightarrow{k_3} CCl_4(g) \qquad \text{fast}$$

(a) Write the overall equation and identify any intermediates that are present.
(b) Use the steady-state approximation to determine the rate law for the reaction.

14.118 Refer to the *Chemical Insight* on smog (p. 394), and account for the observation that the ozone component of smog is at its maximum on sunny days.

14.119 Egg albumin solutions were heated at 65.0 and 70.2°C to denature the protein. (Denaturation is a chemical change that makes albumin insoluble.) The weights of soluble protein remaining per milliliter of solution were reported at various time intervals as follows:

65.0°C		70.2°C	
t (h)	mg protein/mL	t (min)	mg protein/mL
0	7.77	0	6.75
5	7.37	30	6.13
10	7.00	80	5.25
20	6.32	120	4.68
30	5.72	200	3.72
40	5.22	300	2.76

(a) Find the rate constant at each temperature.
(b) Determine the activation energy for egg albumin denaturation.

14.120 The uncatalyzed decomposition of hydrogen peroxide is first order with an activation energy of 75.3 kJ/mol and a half-life of 6.0 hours at 40°C. How many hours would it take for a solution of H_2O_2 to lose half of its strength at 25°C?

14.121 The half-life of a certain first-order decomposition is 20.0 s at 25°C and 1.0 s at 50°C. Calculate the activation energy and the preexponential factor A for this reaction.

*14.122 The Cl—Cl bond energy is 243 kJ/mol. Estimate the temperature at which one collision per mole of collisions would involve enough energy to dissociate a Cl_2 molecule.

***14.123** Consider the reaction

$$2NO_2(g) \rightarrow 2NO(g) + O_2(g)$$

At a certain temperature, the rate at which NO_2 decreases is 4.8×10^{-6} mol/L·min. The frequency of collisions having the activation energy for the rate-determining step is estimated to be 1.0×10^{15} collisions/mL·s. Estimate the orientation factor *p*. (*Hint:* Convert mol/L·min to collisions/mL·s.)

14.124 The decomposition of hydrogen peroxide

$$2H_2O_2(aq) \rightarrow 2H_2O(l) + O_2(g)$$

has an activation energy of 75.3 kJ/mol when no catalyst is present. The activation energy is 56.5 kJ/mol when the decomposition is catalyzed by I^- and 8.4 kJ/mol when it is catalyzed by the enzyme catalase. A solution of H_2O_2 evolves 1.00 L of O_2 in 10.0 s when catalase is added at 25°C.

(a) How long would it take for 1.00 L of O_2 to evolve if I^- had been added instead of catalase?

(b) How long would it take if no catalyst had been added?

14.125 Finely divided platinum adsorbs both H_2 and O_2. The adsorbed atoms can react either with each other or with molecules from the surroundings.

(a) Suggest several possible mechanisms for the platinum-catalyzed reaction of hydrogen with oxygen to form water.

(b) Platinum adsorbs oxygen more strongly than hydrogen. Which of your mechanisms would you expect to predominate? Would the mechanism depend on concentration? Explain your answer.

14.126 Refer to Exercise 14.73. The Cl—Cl bond energy is 243 kJ/mol.

(a) What minimum energy must a photon have to dissociate a chlorine molecule?

(b) What spectral region would provide such photons?

CHAPTER 15

The effect of temperature on the equilibrium $Co(H_2O)_6^{2+}$ (aq) + $4Cl^-$ (aq) $\rightleftharpoons$ $CoCl_4^{2-}$(aq) + $6H_2O$(l). Warming the equilibrium mixture converts pink $Co(H_2O)_6^{2+}$ ions to blue $CoCl_4^{2-}$ ions.

CHEMICAL EQUILIBRIUM

OUTLINE

PREVIEW

Imagine a jar of water open to the atmosphere. The water will gradually evaporate and disappear. Now imagine a closed jar of water. The water will remain in the jar for years. What is the difference between these two jars? In both of them, water molecules are escaping from and returning to the surface of the liquid. In the open jar, the rate of escape is greater than the rate of return and the water evaporates. In the closed jar, the rate of escape equals the rate of return, so the total amount of liquid remains unchanged.

On the molecular level, chemical and physical changes invariably consist of two processes, one undoing the work of the other. When one process occurs more rapidly than the other, we see something happening, and we say that a change is taking place. When both processes occur at the same rate, nothing seems to happen, and we say that the system is at equilibrium. Systems in equilibrium are sometimes referred to as "dead"—a run-down automobile battery, for example, is called a "dead" battery. Yet this is not death as we normally think of it. The molecules remain in constant motion, and the two opposing reactions continue with full and unabated vigor. In Chapter 14 we studied the approach to equilibrium. In this chapter we will study the equilibrium state itself—and we will find that systems in equilibrium are very much "alive."

Dynamic equilibrium is a state in which no net change takes place because two opposing processes are occurring at the same rate. You have already studied equilibrium systems such as a solid or liquid in equilibrium with its vapor (Section 12.1) and a solid in equilibrium with a saturated solution (Section 4.1). There are two requirements for equilibrium—a **reversible process** (one that goes backward as well as forward) and a **closed system** (a system from which substances cannot escape). True equilibria are rarely found in nature because few real systems are entirely closed. In an open container, the products tend to disperse and the reverse reaction cannot take place. Even though many systems do not reach equilibrium, there are no exceptions to the rule that *all chemical and physical changes tend toward an equilibrium state.* In this chapter we will study chemical equilibrium and the mixtures produced by reversible chemical change.

15.1 REVERSIBLE REACTIONS AND THE EQUILIBRIUM STATE

When N_2O_4 and NO_2 gases are mixed, two reactions take place, one undoing the work of the other:

$$N_2O_4(g) \rightarrow 2NO_2(g)$$

$$2NO_2(g) \rightarrow N_2O_4(g)$$

This reaction is an example of a **reversible reaction**, one that can proceed in both the forward and backward directions. It is customary to represent reversibility by a double arrow:

$$N_2O_4(g) \rightleftharpoons 2NO_2(g)$$

In principle, all chemical reactions are reversible as long as the products remain in contact with the reactants.

Nitrogen dioxide (NO_2) is a reddish brown gas with a very irritating odor. Dinitrogen tetroxide (N_2O_4) is colorless with a normal boiling point of 21°C.

Let us consider an experiment that involves a mixture of N_2O_4 and NO_2. A 0.0250-mol sample of cold liquid N_2O_4 is placed in a 1-L flask, which is then sealed and warmed to 25°C. The N_2O_4 vaporizes and some of the vapor decomposes to reddish brown NO_2. The color intensifies as more NO_2 forms, but after a short period of time there is no further change in color. Chemical analyses from this time on show that both gases are present in the flask and that their concentrations remain constant. The N_2O_4 concentration is 0.0202 mol/L, down from its original value of 0.0250 mol/L, and the NO_2 concentration, which was originally zero, is 0.00966 mol/L. The data for this experiment and two other similar ones are summarized in Table 15.1 and diagrammed in Figure 15.1. Note that the final concentrations are different in each experiment.

Figure 15.2 shows how the forward and reverse reaction rates in the first experiment vary as the reaction proceeds. The rate of the forward reaction decreases as the N_2O_4 disappears. The rate of the reverse reaction increases from an initial value of

TABLE 15.1 Three Sets of Initial and Equilibrium Concentrations for the $N_2O_4(g) \rightleftharpoons 2NO_2(g)$ Reaction at 25°C

Experiment	Initial Molarity		Equilibrium Molarity		K_c^a
	$[N_2O_4]$	$[NO_2]$	$[N_2O_4]$	$[NO_2]$	
1	0.0250	0	0.0202	0.00966	0.00462
2	0.0125	0.0125	0.0146	0.00823	0.00464
3	0	0.0250	0.00923	0.00654	0.00463

[a] K_c is the equilibrium value of $[NO_2]^2/[N_2O_4]$.

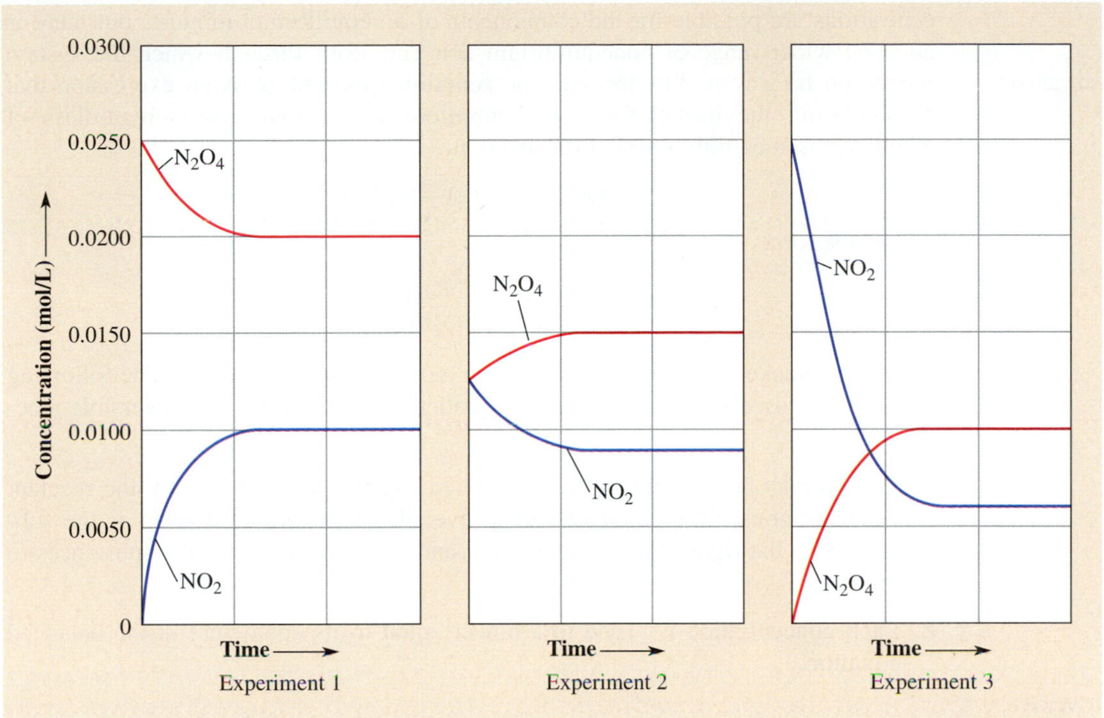

Figure 15.1
Concentrations versus time in gaseous mixtures of NO_2 and N_2O_4. The concentrations level off to equilibrium values that are different in each mixture. *Experiment 1:* Only N_2O_4 is present initially. *Experiment 2:* The initial mixture contains equimolar concentrations of NO_2 and N_2O_4. *Experiment 3:* Only NO_2 is present initially.

zero as NO_2 forms. When the two rates become equal, that is, when NO_2 recombines as fast as it forms, the color of the mixture and the concentration of each species stop changing, and the system is at equilibrium.

The forward and reverse reactions do not stop at equilibrium—they continue, but at exactly the same rate. *All chemical equilibria have the following characteristics in common: (1) Two opposing reactions occur at the same rate, and (2) the concentration of each substance does not change with time.*

15.2 REACTION QUOTIENTS AND EQUILIBRIUM CONSTANTS

Figure 15.1 shows that the final composition of an N_2O_4–NO_2 equilibrium mixture varies with the initial concentrations of the reacting substances. Many different con-

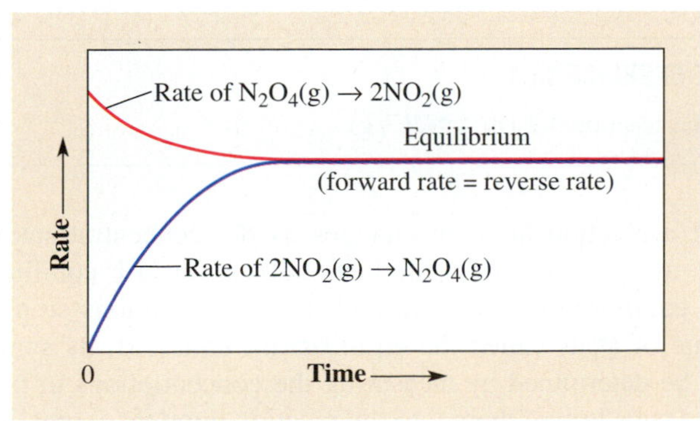

Figure 15.2
Reaction rates versus time in a mixture that initially contains only N_2O_4. The forward rate (N_2O_4 decomposition) decreases while the reverse rate increases. Equilibrium is reached when the two rates become equal.

Reaction quotients are also called **mass action expressions**.

centrations are possible for the components of an equilibrium mixture, but there is an even wider range of nonequilibrium concentrations through which the system moves on its way to equilibrium. The **reaction quotient** Q_c is an expression that combines all the instantaneous concentrations in a given system regardless of whether it is at equilibrium. For the reaction

$$2NO(g) + O_2(g) \rightleftharpoons 2NO_2(g)$$

Q_c has the form:

$$Q_c = \frac{[NO_2]^2}{[NO]^2[O_2]}$$

where the brackets signify concentration in moles per liter (molarity). The following rules can be used to write the reaction quotient expression for any reversible reaction:

1. The product concentrations are multiplied together and divided by the reactant concentrations. (The "products" of a reversible reaction are defined as the substances to the right of the double arrow and the "reactants" are the substances to the left.)
2. Each concentration is raised to a power equal to its coefficient in the balanced equation.

EXAMPLE 15.1

Use the above rules to write Q_c for the following reactions:

(a) $H_2(g) + I_2(g) \rightleftharpoons 2HI(g)$

(b) $4HCl(g) + O_2(g) \rightleftharpoons 2H_2O(g) + 2Cl_2(g)$

CAUTION: Beginning students sometimes make the mistake of adding rather than multiplying concentrations in the numerator and denominator of the reaction quotient.

SOLUTION

(a) $Q_c = \frac{[HI]^2}{[H_2][I_2]}$

(b) $Q_c = \frac{[H_2O]^2[Cl_2]^2}{[HCl]^4[O_2]}$

PRACTICE EXERCISE 15.1

Write Q_c for the reaction: $Cl_2(g) \rightleftharpoons 2Cl(g)$

The value of a reaction quotient changes as the concentrations of the various species change and the reacting mixture moves toward equilibrium. It stops changing when equilibrium is reached and the concentrations stop changing. The equilibrium value of Q_c is called the **equilibrium constant**; its symbol is K_c. The value of K_c can be determined by measuring the concentrations in the final equilibrium mixture and substituting them into the reaction quotient expression.

EXAMPLE 15.2

Find K_c for the reaction

$$N_2O_4(g) \rightleftharpoons 2NO_2(g)$$

using data from the first experiment in Table 15.1.

SOLUTION

The reaction quotient is

$$Q_c = \frac{[NO_2]^2}{[N_2O_4]}$$

The equilibrium constant K_c is the value of Q_c at equilibrium, that is,

$$K_c = (Q_c)_{eq} = \frac{[NO_2]^2_{eq}}{[N_2O_4]_{eq}}$$

The equilibrium concentrations for the first experiment in Table 15.1 are $[N_2O_4]_{eq}$ = 0.0202 mol/L and $[NO_2]_{eq}$ = 0.00966 mol/L. Substituting these concentrations into the K_c expression gives

$$K_c = \frac{(0.00966)^2}{0.0202} = 4.62 \times 10^{-3}$$

LEARNING HINT

Experimental equilibrium constants and reaction quotients are sometimes written with units, but we will follow the more common practice of omitting units from such expressions. In advanced work (Chapter 19), equilibrium constants are obtained from thermodynamic data. Such constants, called *thermodynamic equilibrium constants,* have no units.

PRACTICE EXERCISE 15.2

Verify the equilibrium constants given for the second and third experiments in Table 15.1.

We often need to find the concentration of each species in an equilibrium mixture. Equilibrium concentrations can be obtained in three steps:

- ***Step 1:*** Use the given data to find the initial (i.e., the original) concentration of each species.
- ***Step 2:*** Use the balanced equation to find the change in each concentration as the system comes to equilibrium.
- ***Step 3:*** Find the equilibrium concentration of each species by adding the results of Steps 1 and 2.

These steps are illustrated in the next example.

EXAMPLE 15.3

When phosgene, a poisonous gas that has been used as a chemical warfare agent, is heated, it decomposes into carbon monoxide and chlorine:

$$COCl_2(g) \rightleftharpoons CO(g) + Cl_2(g)$$

When 2.00 mol of phosgene is put into an empty 1.00-L flask at 395°C and allowed to come to equilibrium, the final mixture contains 0.0398 mol of chlorine. Find K_c.

SOLUTION

We must find the equilibrium concentration of each species.

Step 1: The initial concentration of $COCl_2$ is 2.00 mol/1.00 L = 2.00 mol/L. There is no Cl_2 or CO initially, so their original concentrations are zero. These concentrations are recorded in the concentration summary given below.

Step 2: The equation shows that for each mole of Cl_2 that forms, 1 mol of CO will also form, and 1 mol of $COCl_2$ will be used up. If 0.0398 mol/L of Cl_2 forms, then 0.0398 mol/L of CO will form and 0.0398 mol/L of $COCl_2$ will be used up. These changes are also recorded in the concentration summary.

Step 3: The equilibrium concentrations are obtained by adding the changes in Step 2 to the initial concentrations in Step 1. They are recorded in the final line of the concentration summary.

LEARNING HINT

Start each equilibrium problem with a balanced equation, and fill in the concentration summary as you go along.

CONCENTRATION SUMMARY

Equation:	$COCl_2(g)$	$\rightleftharpoons$	$CO(g)$	+	$Cl_2(g)$
Initial concentrations (mol/L):	2.00		0		0
Concentration changes (mol/L):	−0.0398		+0.0398		+0.0398
Equilibrium concentrations (mol/L):	1.96		0.0398		0.0398

The value of K_c is obtained by substituting the equilibrium concentrations into the Q_c expression:

$$K_c = (Q_c)_{eq} = \frac{[CO]_{eq}[Cl_2]_{eq}}{[COCl_2]_{eq}} = \frac{(0.0398)(0.0398)}{1.96} = 8.08 \times 10^{-4}$$

EXAMPLE 15.4

When 1.000 mol of hydrogen and 1.000 mol of iodine are introduced into a 5.000-L reaction vessel at 458°C, the iodine vaporizes and reacts with the hydrogen to form hydrogen iodide:

$$H_2(g) + I_2(g) \rightleftharpoons 2HI(g)$$

The concentration of I_2 at equilibrium, as measured by the intensity of its violet color, is 0.0440 mol/L. Calculate K_c for the reaction.

SOLUTION

We will use a three-step concentration summary to find the equilibrium concentration of each species, just as we did in Example 15.3.

Step 1: The initial concentrations of H_2 and I_2 are each 1.000 mol/5.000 L = 0.2000 mol/L. The initial concentration of HI is zero.

Step 2: The I_2 concentration decreases by 0.2000 mol/L − 0.0440 mol/L = 0.1560 mol/L. The equation shows that 1 mol of H_2 will be used up and 2 mol of HI will form for each mole of I_2 that is consumed. If the concentration of I_2 decreases by

0.1560 mol/L, then the H_2 concentration must also decrease by 0.1560 mol/L, and the HI concentration must increase by 2×0.1560 mol/L $= 0.3120$ mol/L.

Step 3: The equilibrium concentrations are obtained by adding the changes in Step 2 to the initial concentrations in Step 1. The results are recorded in the concentration summary given below.

CONCENTRATION SUMMARY

Equation:	$H_2(g)$	+	$I_2(g)$	$\rightleftharpoons$	$2HI(g)$
Initial concentrations (mol/L):	0.2000		0.2000		0
Concentration changes (mol/L):	−0.1560		−0.1560		+0.3120
Equilibrium concentrations (mol/L):	0.0440		0.0440		0.3120

Substituting the equilibrium concentrations into the Q_c expression gives us K_c:

$$K_c = (Q_c)_{eq} = \frac{[HI]^2_{eq}}{[H_2]_{eq}[I_2]_{eq}} = \frac{(0.3120)^2}{(0.0440)(0.0440)} = 50.3$$

PRACTICE EXERCISE 15.3

When 1.00 mol of H_2O and 1.00 mol of CO are introduced into an empty 1.00-L reaction vessel at 959 K and allowed to come to equilibrium, the equilibrium mixture contains 0.422 mol of water vapor. Find K_c at 959 K for the reaction

$$H_2O(g) + CO(g) \rightleftharpoons H_2(g) + CO_2(g)$$

The Law of Chemical Equilibrium

The three experiments summarized in Table 15.1 begin with different concentrations of N_2O_4 and NO_2. The equilibrium concentrations are also different, but the value obtained for the equilibrium constant (about 0.00463) is essentially the same in each experiment. Additional experiments would show that no matter what starting concentrations are chosen, the equilibrium value of Q_c at 25°C is always about 0.00463. Other reversible reactions exhibit similar behavior. Equilibrium constants change with temperature and vary from one reaction to another, but for any given reaction at a given temperature, the equilibrium constant is independent of the starting concentrations. This observation is summarized in the **law of chemical equilibrium**: *The equilibrium constant is independent of the initial concentrations and varies only with the temperature.*

The law of chemical equilibrium is sometimes referred to as the **law of mass action**. It was first derived in 1864 by the Norwegian chemists Cato Maximilian Guldberg and Peter Waage.

All systems move toward equilibrium, so the reaction quotient Q_c will keep changing until it reaches its equilibrium value. By comparing the instantaneous value of Q_c with K_c for a reacting mixture, we can determine whether the mixture is at equilibrium. Furthermore, if it is not at equilibrium, we can predict the direction of the reaction. Three possibilities exist:

1. *If Q_c is less than K_c*, then Q_c will increase as the system moves toward equilib-

rium. Because Q_c contains product concentrations divided by reactant concentrations, the product concentrations must increase and the reactant concentrations must decrease. Thus the reaction will move in the forward direction.

2. *If Q_c equals K_c*, then the system is at equilibrium.
3. *If Q_c is greater than K_c*, then Q_c will decrease as the system moves toward equilibrium. The product concentrations will decrease, the reactant concentrations will increase, and the reaction will move in the reverse direction.

The use of these rules is illustrated in the following example.

EXAMPLE 15.5

The equilibrium constant for the reaction

$$2NOCl(g) \rightleftharpoons 2NO(g) + Cl_2(g)$$

is 4.4×10^{-4} at 500°C. In which direction will the reaction move when the concentration of each substance in the reacting mixture is 0.100 mol/L?

SOLUTION

Substituting 0.100 mol/L for each concentration in the reaction quotient gives:

$$Q_c = \frac{[NO]^2[Cl_2]}{[NOCl]^2} = \frac{(0.100)^2(0.100)}{(0.100)^2} = 0.100$$

The equilibrium constant is 4.4×10^{-4}. Because Q_c is greater than K_c, Q_c will decrease with time. The reaction will move in the reverse direction. Some of the NO and Cl_2 will be used up, and more NOCl will form.

PRACTICE EXERCISE 15.4

The value of K_c for the reaction

$$I_2(g) + Cl_2(g) \rightleftharpoons 2ICl(g)$$

is 9.1. A 1.00-L reaction vessel contains 0.0025 mol ICl and 0.10 mol each of I_2 and Cl_2. Does the reaction move in the forward or reverse direction?

DIGGING DEEPER

The Kinetic Basis for the Law of Chemical Equilibrium

The law of chemical equilibrium states that the equilibrium constant for a given reaction varies only with temperature and is independent of the initial concentrations. Although experimental data support this law, they do not explain why the law should be true. In this section we show that the law of chemical equilibrium can be derived from the kinetic relationships that exist in the equilibrium state.

Let us look again at the reversible reaction

$$N_2O_4(g) \rightleftharpoons 2NO_2(g)$$

In this system the forward and reverse reactions consist of one elementary step each. Recall (Section 14.5) that we can write the rate law for an elementary step by using the coefficients in the chemical equation as exponents of the concentrations. Hence, the forward and reverse rate expressions are

$$\text{forward rate} = k_f[N_2O_4] \qquad \text{reverse rate} = k_r[NO_2]^2$$

where k_f and k_r are the forward and reverse rate constants, respectively. Equilibrium is established when the two rates are equal, that is, when

$$k_f[N_2O_4]_{eq} = k_r[NO_2]^2_{eq}$$

This equation can be rearranged to

$$\frac{k_f}{k_r} = \frac{[NO_2]^2_{eq}}{[N_2O_4]_{eq}} = K_c$$

Hence, for the N_2O_4–NO_2 reaction, the value of K_c equals the forward rate constant divided by the reverse rate constant. Because the rate constants vary only with temperature, K_c will also vary only with temperature, in agreement with the law of chemical equilibrium.

Now let us see if a similar relation holds for the reversible reaction

$$2O_3(g) \rightleftharpoons 3O_2(g)$$

for which the mechanism consists of two reversible elementary steps:

$$\textit{Step 1:} \quad O_3(g) \rightleftharpoons O_2(g) + O(g)$$

$$\textit{Step 2:} \quad O(g) + O_3(g) \rightleftharpoons 2O_2(g)$$

When the system is in equilibrium, each elementary step will also be in equilibrium. Equating equilibrium reaction rates for the first step gives

$$k_{f_1}[O_3]_{eq} = k_{r_1}[O_2]_{eq}[O]_{eq}$$

which can be rearranged to

$$\frac{k_{f_1}}{k_{r_1}} = \frac{[O_2]_{eq}[O]_{eq}}{[O_3]_{eq}} = K_{c_1}$$

where K_{c_1} is the equilibrium constant for the first step.

Equating rates for the second step gives

$$k_{f_2}[O]_{eq}[O_3]_{eq} = k_{r_2}[O_2]^2_{eq}$$

and

$$\frac{k_{f_2}}{k_{r_2}} = \frac{[O_2]^2_{eq}}{[O]_{eq}[O_3]_{eq}} = K_{c_2}$$

where K_{c_2} is the equilibrium constant for the second step.

Multiplying K_{c_1} by K_{c_2} gives:

$$K_{c_1} \times K_{c_2} = \frac{k_{f_1}}{k_{r_1}} \times \frac{k_{f_2}}{k_{r_2}}$$

$$= \frac{[O_2]_{eq}[O]_{eq}}{[O_3]_{eq}} \times \frac{[O_2]^2_{eq}}{[O]_{eq}[O_3]_{eq}}$$

$$= \frac{[O_2]^3_{eq}}{[O_3]^2_{eq}} = K_c$$

The first and last lines of this multiplication show that the product of the forward rate constants divided by the product of the reverse rate constants is equal to K_c, the equilibrium constant for the overall reaction. Because rate constants vary only with temperature, the equilibrium constant will vary only with temperature, in accord with the law of chemical equilibrium. A similar result would be obtained in the analysis of any multistep reaction.

This analysis also shows that $K_c = K_{c_1} \times K_{c_2}$. *The equilibrium constant for a multistep reaction is equal to the product of the equilibrium constants for each of the steps.* We will come back to this important rule later in this chapter.

15.3 CALCULATIONS USING K_c

If the initial concentrations in a reaction mixture are known, the equilibrium constant can be used to calculate the equilibrium concentrations.

EXAMPLE 15.6

The value of K_c for the reaction

$$H_2(g) + I_2(g) \rightleftharpoons 2HI(g)$$

is 50.3 at 458°C (see Example 15.4). The concentrations of H_2, I_2, and HI in a certain mixture are 0.0100 mol/L, 0.0100 mol/L, and 0.0200 mol/L, respectively. What concentrations can be expected at equilibrium?

SOLUTION

First we determine whether the reaction is moving in the forward or reverse direction. The initial value of Q_c is obtained by substituting the given concentrations into the reaction quotient expression:

$$Q_c = \frac{[HI]^2}{[H_2][I_2]} = \frac{(0.0200)^2}{(0.0100)(0.0100)} = 4.00$$

The value of Q_c is less than its equilibrium value of 50.3, so the reaction is moving in the forward direction. The concentrations of H_2 and I_2 will decrease; the concentration of HI will increase.

Now we can find the equilibrium concentrations.

Step 1: The initial concentrations of H_2, I_2, and HI are recorded as usual in the concentration summary below.

Step 2: The equation for the reaction shows that 1 mol of H_2 reacts with 1 mol of I_2; therefore, the decrease in the H_2 concentration will equal the decrease in the I_2 concentration. At this point, we do not know what the concentration changes are, so we will let x be the decrease in the concentrations of H_2 and I_2. The equation also shows that 2 mol of HI form for each mole of H_2 and I_2 consumed; therefore, the concentration of HI will increase by $2x$.

Step 3: The equilibrium concentrations are obtained by adding the changes in Step 2 to the initial concentrations in Step 1. The results are recorded in the concentration summary.

CONCENTRATION SUMMARY

Equation:	$H_2(g)$	+	$I_2(g)$	$\rightleftharpoons$	$2HI(g)$
Initial concentrations (mol/L):	0.0100		0.0100		0.0200
Concentration changes (mol/L):	$-x$		$-x$		$+2x$
Equilibrium concentrations (mol/L):	$0.0100 - x$		$0.0100 - x$		$0.0200 + 2x$

Substituting the equilibrium concentrations and $K_c = 50.3$ into the equilibrium constant expression gives

$$K_c = \frac{[HI]^2_{eq}}{[H_2]_{eq}[I_2]_{eq}}$$

$$50.3 = \frac{(0.0200 + 2x)^2}{(0.0100 - x)(0.0100 - x)} = \left(\frac{0.0200 + 2x}{0.0100 - x}\right)^2$$

The equation can be simplified by taking the square root of both sides:

$$7.09 = \frac{0.0200 + 2x}{0.0100 - x}$$

Cross multiplying and solving for x gives

$$7.09(0.0100 - x) = 0.0200 + 2x$$
$$9.09x = 0.0509$$
$$x = 5.60 \times 10^{-3}$$

The equilibrium concentrations are

$$[H_2] = [I_2] = 0.0100 - x = 0.0100 \text{ mol/L} - 0.0056 \text{ mol/L} = 0.0044 \text{ mol/L}$$

$$[HI] = 0.0200 + 2x = 0.0200 \text{ mol/L} + 2(0.0056) \text{ mol/L} = 0.0312 \text{ mol/L}$$

LEARNING HINT

The square roots of 50.3 are $+7.09$ and -7.09. The negative root is not used because it would lead to a negative value for the equilibrium HI concentration, which is not physically possible.

EXAMPLE 15.7

The value of K_c for the reaction

$$N_2O_4(g) \rightleftharpoons 2NO_2(g)$$

is 4.63×10^{-3} at 25°C (see Table 15.1). What will the concentration of each gas be if 0.100 mol of N_2O_4 comes to equilibrium in a 5.00-L flask at 25°C?

SOLUTION

There is no NO_2 initially, so the reaction can only move in the forward direction. (The initial value of Q_c is zero.) The concentration of NO_2 increases while that of N_2O_4 decreases.

Step 1: The initial concentration of N_2O_4 is 0.100 mol/5.00 L = 0.0200 mol/L. The initial concentration of NO_2 is zero.

Step 2: Assume that x moles per liter of N_2O_4 decompose by the time equilibrium is reached. The equation shows that $2x$ moles per liter of NO_2 will form.

Step 3: The equilibrium concentrations are obtained in the usual manner and recorded in the concentration summary.

CONCENTRATION SUMMARY

Equation:	$N_2O_4(g)$	$\rightleftharpoons$	$2NO_2(g)$
Initial concentrations (mol/L):	0.0200		0
Concentration changes (mol/L):	$-x$		$+2x$
Equilibrium concentrations (mol/L):	$0.0200 - x$		$2x$

The expression for the equilibrium constant is

$$K_c = \frac{[NO_2]^2}{[N_2O_4]}$$

(The use of K_c implies that equilibrium concentrations are involved, so from now on we will omit the subscript "eq" from the concentrations in K_c expressions.) Substituting the equilibrium concentrations and $K_c = 4.63 \times 10^{-3}$ into the above expression gives

$$4.63 \times 10^{-3} = \frac{(2x)^2}{0.0200 - x}$$

This equation cannot be simplified by taking the square root of both sides, as we did in Example 15.6, but it can be changed into a more familiar form by multiplying both sides by $(0.0200 - x)$ and rearranging to

$$4x^2 + 4.63 \times 10^{-3}x - 9.26 \times 10^{-5} = 0$$

We now have a quadratic equation of the general form $ax^2 + bx + c = 0$, which can be solved using the quadratic formula (see Appendix A.4):

$$x = \frac{-b \pm \sqrt{b^2 - 4ac}}{2a}$$

In our case $a = 4$, $b = 4.63 \times 10^{-3}$, and $c = -9.26 \times 10^{-5}$. Hence,

$$x = \frac{-4.63 \times 10^{-3} \pm \sqrt{(4.63 \times 10^{-3})^2 - 4(4)(-9.26 \times 10^{-5})}}{2 \times 4}$$

$$= 4.27 \times 10^{-3}$$

This solution was obtained by using the positive sign before the square root. The negative sign would have given a negative value for x and, since $[NO_2] = 2x$, a negative equilibrium concentration for NO_2. A negative concentration has no meaning, so the negative root must be rejected. The equilibrium concentrations are

$$[NO_2] = 2x = 2 \times 0.00427 \text{ mol/L} = 0.00854 \text{ mol/L}$$

$$[N_2O_4] = 0.0200 - x = 0.0200 \text{ mol/L} - 0.00427 \text{ mol/L} = 0.0157 \text{ mol/L}$$

LEARNING HINT

The concentration summary for Example 15.7 shows that x must be greater than zero but less than 0.0200. Always check to make sure that your answer falls within the expected range.

PRACTICE EXERCISE 15.5

The equilibrium constant for

$$Br_2(g) \rightleftharpoons 2Br(g)$$

is $K_c = 2.64 \times 10^{-4}$ at 1400 K. What will the concentration of bromine atoms be if 0.500 mol of Br_2 comes to equilibrium in a 0.500-L reaction vessel at this temperature?

Approximate Calculations

The magnitude of an equilibrium constant is an indication of how far the reaction will proceed before reaching equilibrium. A very large constant such as $K_c = 1.0 \times 10^{15}$ shows that the reaction goes virtually to completion; that is, the equilibrium mixture will contain mostly product and very little reactant. A small constant such as $K_c = 5.0 \times 10^{-10}$ implies that the equilibrium mixture will contain mostly reactant and very little product. These predictions may help to simplify equilibrium calculations, as illustrated in the next example.

EXAMPLE 15.8

The value of K_c for the reaction

$$2H_2S(g) \rightleftharpoons 2H_2(g) + S_2(g)$$

is 4.20×10^{-6} at 830°C. What concentration of S_2 can be expected when 0.0700 mol of H_2S comes to equilibrium at 830°C in an otherwise empty 1.00-L vessel?

Hydrogen sulfide (H_2S) is responsible for the odor of rotten eggs.

SOLUTION

Since only hydrogen sulfide is present initially, the reaction must proceed in the forward direction. We will set up the usual concentration summary.

Sulfur vapor contains S_8, S_6, S_4, and S_2 molecules. At very high temperatures, S_2 predominates.

Step 1: The initial concentration of H_2S is 0.0700 mol/L. The initial concentrations of H_2 and S_2 are zero.

Step 2: The equation shows that 2 mol of H_2S decompose and 2 mol of H_2 form for every mole of S_2 that forms. If we let x be the increase in the S_2 concentration at equilibrium, then the H_2S concentration will decrease by $2x$ and the H_2 concentration will increase by $2x$.

Step 3: The equilibrium concentrations are obtained by adding the results of Steps 1 and 2.

CONCENTRATION SUMMARY

Equation:	$2H_2S(g)$	$\rightleftharpoons$	$2H_2(g)$	+	$S_2(g)$
Initial concentrations (mol/L):	0.0700		0		0
Concentration changes (mol/L):	$-2x$		$+2x$		$+x$
Equilibrium concentrations (mol/L):	$0.0700 - 2x$		$2x$		x

Substituting the equilibrium concentrations and $K_c = 4.20 \times 10^{-6}$ into the equilibrium constant expression gives

$$K_c = \frac{[H_2]^2[S_2]}{[H_2S]^2}$$

$$4.20 \times 10^{-6} = \frac{(2x)^2(x)}{(0.0700 - 2x)^2}$$

This equation contains an x^3 term in the numerator, and although it can be solved, the mathematics is quite complicated. Fortunately, the magnitude of K_c allows us to make a simplifying approximation. The small value of the equilibrium constant indicates that very little product will be present at equilibrium. Hence, $2x$, the equilibrium concentration of H_2, will be small compared to 0.0700, the initial concentration of H_2S. Thus we can simplify the calculation by assuming that $0.0700 - 2x$ is approximately equal to 0.0700. Replacing $0.0700 - 2x$ with 0.0700 in the denominator of the preceding equation gives

$$4.20 \times 10^{-6} = \frac{(2x)^2(x)}{(0.0700)^2}$$

Cross multiplying and rearranging gives

$$x^3 = 5.14 \times 10^{-9}$$

and taking the cube root gives

$$x = [S_2] = 1.73 \times 10^{-3} \text{ mol/L}$$

LEARNING HINT

To find the cube root of a number, use the y^x button on your calculator. The cube root of 5.14×10^{-9} is obtained by setting $y = 5.14 \times 10^{-9}$ and $x = \frac{1}{3} = 0.3333$. Similarly, the nth root of any number can be found by setting $x = 1/n$.

Example 15.8 illustrates the fact that very small quantities may be ignored when they are added to or subtracted from much larger quantities. This type of approximation reduces the labor required to solve certain equations, and we will use it whenever possible in this and later chapters. Equilibrium constants are not very precise, and a 5% error is considered acceptable for most equilibrium calculations. A useful rule of thumb for approximations is as follows: *If the neglected quantity turns out to be less than 5% of the retained quantity, then the approximation is valid.* In Example 15.8 the retained quantity is 0.0700 and 5% of the retained quantity is $0.05 \times 0.0700 = 3.50 \times 10^{-3}$. The neglected quantity is $2x$, or $2 \times 1.73 \times 10^{-3} = 3.46 \times 10^{-3}$. The neglected quantity is somewhat less than 5% of the retained quantity, so the approximation is valid.

PRACTICE EXERCISE 15.6

The value of K_c for the reaction

$$2CO_2(g) \rightleftharpoons 2CO(g) + O_2(g)$$

is 4.50×10^{-23} at 745°C. Estimate the concentration of oxygen that will develop when 0.100 mol of CO_2 is allowed to come to equilibrium in a 2.00-L container at 745°C.

DIGGING DEEPER

The Method of Successive Approximations

The answer to Example 15.8 was barely acceptable according to the 5% rule, and the approximation would have failed if the initial concentration of H_2S had been slightly lower than 0.0700 mol/L. (Convince yourself of this by trying to solve Example 15.8 using an initial H_2S concentration of 0.0500 mol/L.) When a simple approximation does not work, or when we want a more accurate answer, we can often use the **method of successive approximations** to obtain the answer. We will illustrate this method by improving our answer to the cubic equation in Example 15.8:

$$4.20 \times 10^{-6} = \frac{(2x)^2(x)}{(0.0700 - 2x)^2} = \frac{4x^3}{(0.0700 - 2x)^2}$$

First Approximation. The first approximation is identical to the one we used in Example 15.8; that is, we will assume that x in the denominator is zero and drop the $2x$ term. The equation becomes

$$4.20 \times 10^{-6} = \frac{4x^3}{(0.0700)^2}$$

Solving this equation for x (see Example 15.8 for details) gives

$$x = 1.73 \times 10^{-3}$$

Second Approximation. The value of x obtained in the first approximation is substituted into the denominator of the original equation, and the equation is solved again:

$$4.20 \times 10^{-6} = \frac{4x^3}{[0.0700 - (2 \times 1.73 \times 10^{-3})]^2}$$

and $$x = 1.67 \times 10^{-3}$$

Third Approximation. The procedure is repeated. Substitute x from the second approximation into the denominator of the original equation, and solve again:

$$4.20 \times 10^{-6} = \frac{4x^3}{[0.0700 - (2 \times 1.67 \times 10^{-3})]^2}$$

and $$x = 1.67 \times 10^{-3}$$

The answers from the second and third approximations are identical to three significant figures, so we can stop here. (Convince yourself that another approximation would give the same answer.) In general, *the approximations are continued until the last two answers agree to the desired degree of precision.* Note that this improved answer is about 3.6% higher than the approximate answer (1.73×10^{-3}) obtained in Example 15.8.

The method of successive approximations can also be used as an alternative to the quadratic formula for solving quadratic equations. Consider, for example, the quadratic equation in Example 15.7:

$$4.63 \times 10^{-3} = \frac{(2x)^2}{0.0200 - x}$$

As in Example 15.8, the first approximation is made by substituting $x = 0$ into the denominator. It is left as an exercise for you to show that agreement to three significant figures will be obtained after five iterations. The calculated values of x should be 4.81×10^{-3}, 4.19×10^{-3}, 4.28×10^{-3}, 4.27×10^{-3}, and 4.27×10^{-3}.

In equilibrium calculations, the method of successive approximations usually works when the equilibrium constant and the anticipated value of x are small, as in the above examples. In such cases, *the first approximation is made by substituting $x = 0$ into terms that are added to or subtracted from constants.* As another example, consider the following equation:

$$1.7 \times 10^{-5} = x(0.040 + 2x)^2$$

The first approximation is made by assuming that $(0.040 + 2x) = 0.040$. After eight iterations, the calculated values of x will be 0.011, 0.0044, 0.0071, 0.0058, 0.0064, 0.0061, 0.0062, and 0.0062. Hence, the final answer is $x = 0.0062$. (Be sure to verify these numbers.)

PRACTICE EXERCISE 15.7

Use the method of successive approximations to solve the following equation for x:

$$8.00 \times 10^{-2} = \frac{x(1.00 + 4x)^4}{0.0500 - x}$$

(*Hint:* Begin by assuming that $(1.00 + 4x) = 1.00$ and $(0.0500 - x) = 0.0500$.)

The method of successive approximations does not always converge to an answer, especially when the value of the equilibrium constant is large. Consider for example, the equation

$$19.95 = \frac{x}{(0.0100 - x)(2.00 - 2x)^2}$$

where the conditions of the problem require that x lie between 0 and 0.0100. If we start by assuming that $x = 0$, that is, $(0.0100 - x) = 0.0100$ and $(2.00 - 2x) = 2.00$, then the calculated values of x after three iterations will be 0.798, -2.57, and 2.62×10^3. The calculated values do not converge; they alternate between positive and negative numbers, and rapidly increase in magnitude. In such cases, the equation must be solved by more sophisticated approximation techniques or by trial and error.

15.4 EQUILIBRIA INVOLVING SOLVENTS

Reaction quotients and equilibrium constants have so many practical uses that we try to keep them as simple as possible. One way to simplify such expressions is to eliminate concentrations that remain essentially constant. Consider the following reaction:

$$\underset{\text{ethyl acetate}}{CH_3COOC_2H_5(aq)} + H_2O(l) \rightleftharpoons \underset{\text{acetic acid}}{CH_3COOH(aq)} + \underset{\text{ethanol}}{C_2H_5OH(aq)}$$

for which the complete equilibrium constant expression is

$$K'_c = \frac{[CH_3COOH][C_2H_5OH]}{[CH_3COOC_2H_5][H_2O]}$$

When this reaction is carried out in aqueous solution, water is both reactant and solvent. In a dilute solution, the change in the amount of water is negligible compared to its total amount, so the concentration of water is effectively constant. Multiplying K'_c by the concentration of water gives

$$K'_c[H_2O] = \frac{[CH_3COOH][C_2H_5OH]}{[CH_3COOC_2H_5]} = K_c$$

where K_c is constant as long as water is the solvent and present in large excess. Thus K_c is a practical equilibrium constant under these conditions. *The concentration of solvent is omitted from expressions for Q_c and K_c when a reaction occurs in dilute solution.*

EXAMPLE 15.9

Write K_c expressions for the following reactions which occur in dilute aqueous solution:

(a) $NH_3(aq) + H_2O(l) \rightleftharpoons NH_4^+(aq) + OH^-(aq)$

(b) $Al^{3+}(aq) + H_2O(l) \rightleftharpoons Al(OH)^{2+}(aq) + H^+(aq)$

SOLUTION

The concentration of water is not included in the equilibrium constant expressions. Hence,

(a) $K_c = \dfrac{[NH_4^+][OH^-]}{[NH_3]}$

(b) $K_c = \dfrac{[Al(OH)^{2+}][H^+]}{[Al^{3+}]}$

PRACTICE EXERCISE 15.8

Write K_c expressions for the following reactions which occur in dilute aqueous solution:

(a) $CN^-(aq) + H_2O(l) \rightleftharpoons HCN(aq) + OH^-(aq)$

(b) $Be^{2+}(aq) + H_2O(l) \rightleftharpoons Be(OH)^+(aq) + H^+(aq)$

EXAMPLE 15.10

For the reaction of ethyl acetate with water (see the equation at the top of this page), K_c is 0.25 at room temperature. Calculate the final concentration of ethanol when a 1.00 *M* solution of ethyl acetate comes to equilibrium.

SOLUTION

We will set up the usual concentration summary to find the equilibrium concentrations.

Step 1: The initial concentration of ethyl acetate is 1.00 mol/L; the initial concentrations of acetic acid and ethanol are zero.

Step 2: Initially, Q_c is zero, so the reaction can only move in the forward direction. Let x be the concentration of ethanol at equilibrium. The equation shows that the changes in the ethyl acetate and acetic acid concentrations will be $-x$ and $+x$, respectively.

Step 3: The equilibrium concentrations are obtained by adding the results of Steps 1 and 2.

CONCENTRATION SUMMARY

Equation:	$CH_3COOC_2H_5(aq)$ + $H_2O(l)$ ⇌	$CH_3COOH(aq)$ +	$C_2H_5OH(aq)$
Initial concentrations (mol/L):	1.00	0	0
Concentration changes (mol/L):	$-x$	$+x$	$+x$
Equilibrium concentrations (mol/L):	$1.00 - x$	x	x

(The concentration of water is ignored because it is not included in the equilibrium constant expression.) Substituting the equilibrium concentrations and $K_c = 0.25$ into the K_c expression gives

$$K_c = \frac{[CH_3COOH][C_2H_5OH]}{[CH_3COOC_2H_5]}$$

$$0.25 = \frac{(x)(x)}{(1.00 - x)}$$

which is a quadratic equation. If we make the approximation that $1.00 - x = 1.00$, the equation becomes $x^2 = 0.25$, and $x = 0.50$. This value of x, however, does not satisfy the 5% rule for approximate calculations; 5% of 1.00 is 0.0500, and x, the neglected quantity, is 10 times this value. Hence, we must solve the quadratic equation exactly. Multiplying both sides of the equation by $(1.00 - x)$ and rearranging gives

$$x^2 + 0.25x - 0.25 = 0$$

The value of x can now be obtained from the quadratic formula with $a = 1$, $b = 0.25$, and $c = -0.25$:

$$x = \frac{-0.25 \pm \sqrt{(0.25)^2 - 4(1)(-0.25)}}{2}$$

Inspection of the concentration summary shows that x will be between 0 and 1.0; hence, the positive sign is chosen for the square root term. Solving gives $x = [C_2H_5OH] = 0.39$ mol/L.

PRACTICE EXERCISE 15.9

For the reaction of ammonia with water

$$NH_3(aq) + H_2O(l) \rightleftharpoons NH_4^+(aq) + OH^-(aq)$$

K_c is 1.77×10^{-5} at 25°C. Calculate the concentration of OH^- ions when a 0.10 *M* solution of ammonia comes to equilibrium.

15.5 GAS PHASE EQUILIBRIUM: Q_p AND K_p

Molarity is not always the most convenient unit to use in equilibrium calculations. For gas phase reactions, it is often more practical to express concentrations in terms of partial pressures. The partial pressure of a gas in a mixture, say gas A, is given by the ideal gas law as

$$P_A = \frac{n_A RT}{V} = [A]RT \tag{15.1}$$

where $[A] = n_A/V$ is the concentration of A in moles per liter, R is the gas constant, and T is the Kelvin temperature. Equation 15.1 shows that, at a given temperature, the partial pressure of a gas is proportional to its molarity.

If partial pressures are used, the reaction quotient and equilibrium constant are represented by Q_p and K_p, respectively. Unless otherwise specified, it will be understood that the partial pressures are in atmospheres. For example, Q_p for the reaction

$$H_2(g) + I_2(g) \rightleftharpoons 2HI(g)$$

is

$$Q_p = \frac{P_{HI}^2}{P_{H_2}P_{I_2}}$$

where P_{HI}, P_{H_2}, and P_{I_2} are the partial pressures of HI, H_2, and I_2, respectively. Calculations using partial pressures and K_p are analogous to those using molarities and K_c.

EXAMPLE 15.11

Phosphorus pentachloride vapor partially decomposes into phosphorus trichloride and chlorine:

$$PCl_5(g) \rightleftharpoons PCl_3(g) + Cl_2(g)$$

Some PCl_5, initially at 0.200 atm, is allowed to come to equilibrium with its decomposition products at a fixed temperature and constant volume. The total pressure at equilibrium is found to be 0.345 atm. Calculate K_p.

SOLUTION

We will set up a pressure summary similar to the concentration summaries used in previous problems.

Step 1: The initial pressure of PCl_5 is 0.200 atm. The initial pressures of PCl_3 and Cl_2 are zero.

Step 2: The reaction moves in the forward direction. The partial pressure of each gas is proportional to its molarity, so we can use the coefficients in the reaction equation to relate the pressure changes. The equation shows that 1 mol of PCl_5 decomposes into 1 mol each of PCl_3 and Cl_2. If the pressure of PCl_3 increases by x, then the pressure of Cl_2 will also increase by x, and the pressure of PCl_5 will decrease by x.

Step 3: The equilibrium pressures are obtained by adding the pressure changes in Step 2 to the initial pressures in Step 1.

PRESSURE SUMMARY

Equation:	$PCl_5(g)$	$\rightleftharpoons$	$PCl_3(g)$	+	$Cl_2(g)$
Initial pressures (atm):	0.200		0		0
Pressure changes (atm):	$-x$		$+x$		$+x$
Equilibrium pressures (atm):	$0.200 - x$		x		x

The total pressure at equilibrium, 0.345 atm, is the sum of the partial pressures. Therefore,

$$0.345 \text{ atm} = P_{PCl_5} + P_{PCl_3} + P_{Cl_2}$$

$$= 0.200 \text{ atm} - x + x + x = 0.200 \text{ atm} + x$$

and

$$x = 0.145 \text{ atm}$$

The equilibrium pressures are

$$P_{PCl_5} = 0.200 - x = 0.200 \text{ atm} - 0.145 \text{ atm} = 0.055 \text{ atm}$$

$$P_{PCl_3} = x = 0.145 \text{ atm}$$

$$P_{Cl_2} = x = 0.145 \text{ atm}$$

Substituting these values into the K_p expression gives

$$K_p = \frac{P_{PCl_3}P_{Cl_2}}{P_{PCl_5}} = \frac{0.145 \times 0.145}{0.055} = 0.38$$

PRACTICE EXERCISE 15.10

When a sample of pure NOCl at 1.000 atm and 250°C comes to equilibrium with its decomposition products, the pressure increases to 1.135 atm. The equation is

$$2NOCl(g) \rightleftharpoons 2NO(g) + Cl_2(g)$$

Calculate (a) the partial pressure of each gas in the equilibrium mixture and (b) K_p.

The next example shows how a K_p value can be used to calculate an equilibrium pressure.

EXAMPLE 15.12

The value of K_p for the decomposition of sulfur trioxide

$$2SO_3(g) \rightleftharpoons 2SO_2(g) + O_2(g)$$

is 1.79×10^{-5} at 350°C. If SO_3, initially at 1.00 atm pressure, comes to equilibrium in an otherwise empty tank at 350°C, what will the final pressure be?

SOLUTION

The final pressure is the sum of the partial pressures of each gas in the equilibrium mixture. We begin by preparing a pressure summary.

Step 1: The initial pressure of SO_3 is 1.00 atm. There is no SO_2 or O_2 present at the start, so their initial pressures are zero.

Step 2: The reaction moves in the forward direction. Let us assume that the pressure of O_2 increases by x. The equation coefficients show that the SO_2 pressure will increase by $2x$ and the SO_3 pressure will decrease by $2x$.

Step 3: The equilibrium pressures are obtained in the usual manner and summarized below.

PRESSURE SUMMARY

Equation:	$2SO_3(g)$	$\rightleftharpoons$	$2SO_2(g)$	+	$O_2(g)$
Initial pressures (atm):	1.00		0		0
Pressure changes (atm):	$-2x$		$+2x$		$+x$
Equilibrium pressures (atm):	$1.00 - 2x$		$2x$		x

Substituting $K_p = 1.79 \times 10^{-5}$ and the equilibrium pressures into the K_p expression gives a cubic equation:

$$K_p = \frac{P_{SO_2}^2 P_{O_2}}{P_{SO_3}^2}$$

$$1.79 \times 10^{-5} = \frac{(2x)^2(x)}{(1.00 - 2x)^2}$$

The value of K_p is small, so only a small amount of SO_3 will decompose and $2x$ should be small relative to 1.00. Approximating $(1.00 - 2x)$ by 1.00 gives

$$1.79 \times 10^{-5} = \frac{(2x)^2(x)}{(1.00)^2} = 4x^3$$

and $x = 0.0165$ atm. The approximation is justified because $2x = 2 \times 0.0165$ atm $=$ 0.0330 atm, which is less than 5% of 1.00 atm.

The equilibrium pressures (see pressure summary) are calculated using the approximate value of x:

$$P_{SO_3} = 1.00 - 2x = 1.00 \text{ atm} - 0.0330 \text{ atm} = 0.97 \text{ atm}$$

$$P_{SO_2} = 2x = 0.0330 \text{ atm}$$

$$P_{O_2} = x = 0.0165 \text{ atm}$$

The total pressure is the sum of the partial pressures

$$P_{total} = 0.97 \text{ atm} + 0.0330 \text{ atm} + 0.0165 \text{ atm} = 1.02 \text{ atm}$$

PRACTICE EXERCISE 15.11

The value of K_p for the reaction between acetylene and hydrogen

$$C_2H_2(g) + 3H_2(g) \rightleftharpoons 2CH_4(g)$$

is 2.41×10^{-4} at 1725°C. A mixture of C_2H_2 and H_2, each initially at 1.00 atm pressure, comes to equilibrium in a closed container. Estimate the partial pressure of CH_4 in the equilibrium mixture.

The Relationship Between K_p and K_c

Equation 15.1, which shows that the partial pressure of a gas is proportional to its molarity, allows us to derive a relationship between K_p and K_c. Consider, for example, the equilibrium

$$N_2(g) + 3H_2(g) \rightleftharpoons 2NH_3(g)$$

The K_p expression is

$$K_p = \frac{P^2_{NH_3}}{P_{N_2}P^3_{H_2}}$$

From Equation 15.1, we know that $P_{NH_3} = [NH_3]RT$, $P_{N_2} = [N_2]RT$, and $P_{H_2} = [H_2]RT$. Substituting these values into K_p gives

$$K_p = \frac{[NH_3]^2(RT)^2}{[N_2](RT)[H_2]^3(RT)^3} = \frac{[NH_3]^2}{[N_2][H_2]^3} \times \frac{1}{(RT)^2} = K_c(RT)^{-2}$$

Observe that the final exponent of RT, -2, equals the number of moles of products minus the number of moles of reactants in the balanced equation. Applying this procedure to any gaseous reaction yields the general formula

$$K_p = K_c(RT)^{\Delta n} \tag{15.2}$$

where Δn is numerically equal to the change in the number of moles of gas in the forward equation. In the formation of ammonia, there is a decrease of 2 mol of gas, so Δn is equal to -2. The use of Equation 15.2 is illustrated in the following examples.

EXAMPLE 15.13

For the reaction

$$N_2O_4(g) \rightleftharpoons 2NO_2(g)$$

K_c is 4.63×10^{-3} at 25°C. Find K_p.

SOLUTION

For this reaction, $\Delta n = 2 - 1 = 1$ and Equation 15.2 becomes $K_p = K_cRT$. Substituting $K_c = 4.63 \times 10^{-3}$, $R = 0.0821$ L·atm/mol·K, and $T = 298$ K gives

$$K_p = K_cRT = 4.63 \times 10^{-3} \times 0.0821 \times 298 = 0.113$$

(Note that you do not have to keep track of the units because units are not used for equilibrium constants. Remember, however, that temperature must be changed to kelvins.)

LEARNING HINT

Because K_p is defined in terms of atmospheres and K_c in terms of moles per liter, the value of R to use in calculations involving K_p and K_c is 0.0821 L·atm/mol·K.

EXAMPLE 15.14

At 491°C, the value of K_p for the reaction

$$H_2(g) + I_2(g) \rightleftharpoons 2HI(g)$$

is 45.6. What is the value of K_c?

SOLUTION

For this reaction, $\Delta n = 2 - 2 = 0$. Hence, $K_c = K_p = 45.6$.

PRACTICE EXERCISE 15.12

Refer to Example 15.12 for data on the SO_3–SO_2 equilibrium at 350°C and find (a) Δn and (b) K_c.

15.6 FORMS OF THE REACTION QUOTIENT

The form of a reaction quotient or equilibrium constant depends on how the equation for the reaction is written. Consider the following equations, one of which is the reverse of the other:

1. $2CO(g) + O_2(g) \rightleftharpoons 2CO_2(g) \qquad K_{p_1} = \dfrac{P^2_{CO_2}}{P^2_{CO}P_{O_2}}$

2. $2CO_2(g) \rightleftharpoons 2CO(g) + O_2(g) \qquad K_{p_2} = \dfrac{P^2_{CO}P_{O_2}}{P^2_{CO_2}}$

Note that K_{p_2} is the reciprocal of K_{p_1}. The same relationship is found in the expressions for K_c and the reaction quotient. This example illustrates the following general rule: *When the equation for a reaction is reversed, the new equilibrium constant is the reciprocal of the original equilibrium constant.*

An equation is sometimes written with fractional coefficients in order to show just 1 mol of product. For example, the equation for the synthesis of ammonia can be written in the following ways:

Remember that you can use either K_c or K_p for reactions that involve gases.

1. $N_2(g) + 3H_2(g) \rightleftharpoons 2NH_3(g) \qquad K_{c_1} = \dfrac{[NH_3]^2}{[N_2][H_2]^3}$

2. $\frac{1}{2}N_2(g) + \frac{3}{2}H_2(g) \rightleftharpoons NH_3(g) \qquad K_{c_2} = \dfrac{[NH_3]}{[N_2]^{1/2}[H_2]^{3/2}}$

Note that K_{c_2} is the square root of K_{c_1}; that is, K_{c_2} equals K_{c_1} raised to the one-half power. This example illustrates a second general rule: *When an equation is multiplied by a factor, the new equilibrium constant is the original equilibrium constant raised to a power equal to that factor.*

The use of these rules is illustrated in the next example.

EXAMPLE 15.15

The equilibrium constant K_p for the reaction

$$2SO_2(g) + O_2(g) \rightleftharpoons 2SO_3(g)$$

is 9.00×10^2 at 530°C. Find K_p for

(a) $2SO_3(g) \rightleftharpoons 2SO_2(g) + O_2(g)$

(b) $SO_2(g) + \frac{1}{2}O_2(g) \rightleftharpoons SO_3(g)$

SOLUTION

(a) The equation for (a) is obtained by reversing the original equation, so its K_p is the reciprocal of the original equilibrium constant:

$$K_p = \frac{1}{900} = 1.11 \times 10^{-3}$$

(b) The equation for (b) is obtained by multiplying the original equation by $\frac{1}{2}$, so its K_p is the square root of the original equilibrium constant:

$$K_p = (900)^{1/2} = 30.0$$

LEARNING HINT

Any of the three values of K_p in Example 15.15 could be used in an equilibrium problem involving SO_2, SO_3, and O_2. Although the constants are numerically different, each will give the same answer when it is used with its form of the equation.

PRACTICE EXERCISE 15.13

Consider the equilibrium

$$2H_2O(g) \rightleftharpoons 2H_2(g) + O_2(g)$$

for which $K_p = 7.6 \times 10^{-16}$ at 1000°C. Find K_p for

(a) $H_2O(g) \rightleftharpoons H_2(g) + \frac{1}{2}O_2(g)$

(b) $\frac{1}{2}O_2(g) + H_2(g) \rightleftharpoons H_2O(g)$

When a stepwise reaction comes to equilibrium, each of the steps will also be in equilibrium. Consider, for example, the two-step reaction:

Step 1: $2NO(g) + O_2(g) \rightleftharpoons 2NO_2(g)$ $\qquad K_1 = \dfrac{P^2_{NO_2}}{P^2_{NO}P_{O_2}}$

Step 2: $2NO_2(g) \rightleftharpoons N_2O_4(g)$ $\qquad K_2 = \dfrac{P_{N_2O_4}}{P^2_{NO_2}}$

The sum of Steps (1) and (2) is

$$2NO(g) + O_2(g) \rightleftharpoons N_2O_4(g)$$

for which

$$K_{sum} = \frac{P_{N_2O_4}}{P^2_{NO}P_{O_2}}$$

Observe that $K_{sum} = K_1 \times K_2$. *The equilibrium constant for a stepwise reaction is the product of the equilibrium constants for the steps.* This relation is true both for values of K_c and for values of K_p.

EXAMPLE 15.16

The following equilibrium constants have been determined at 25°C:

$$H_2C_2O_4(aq) \rightleftharpoons H^+(aq) + HC_2O_4^-(aq) \qquad K_1 = 5.90 \times 10^{-2}$$

$$HC_2O_4^-(aq) \rightleftharpoons H^+(aq) + C_2O_4^{2-}(aq) \qquad K_2 = 6.40 \times 10^{-5}$$

Find K for

$$H_2C_2O_4(aq) \rightleftharpoons 2H^+(aq) + C_2O_4^{2-}(aq)$$

SOLUTION

The final equation is the sum of the first two equations. Hence,

$$K = K_1 \times K_2 = (5.90 \times 10^{-2})(6.40 \times 10^{-5}) = 3.78 \times 10^{-6}$$

PRACTICE EXERCISE 15.14

Find K_p for the reaction

$$2NO(g) + O_2(g) \rightleftharpoons 2NO_2(g)$$

from the following K_p values:

$$N_2(g) + O_2(g) \rightleftharpoons 2NO(g) \qquad K_p = 4.27 \times 10^{-31}$$

$$N_2(g) + 2O_2(g) \rightleftharpoons 2NO_2(g) \qquad K_p = 6.90 \times 10^{-19}$$

15.7 HETEROGENEOUS EQUILIBRIA

So far we have considered only **homogeneous equilibria**, in which all the reacting substances are in the same solution. Now we will look at **heterogeneous equilibria**, in which the opposing reactions occur at the surface of a solid or liquid.

Consider, for example, a hot gaseous mixture of CO and CO_2 in equilibrium with solid graphite:

$$C(graphite) + CO_2(g) \rightleftharpoons 2CO(g)$$

As Figure 15.3 shows, the opposing reactions occur in a layer of atoms at the graphite surface, where gas molecules mingle with carbon atoms. The gas molecules are free to enter and leave the surface layer, and the gas concentrations will continue

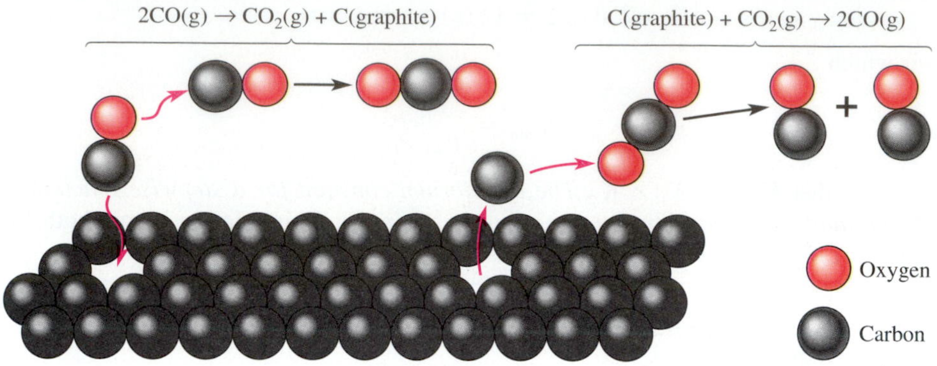

$2CO(g) \rightarrow CO_2(g) + C(\text{graphite})$

$C(\text{graphite}) + CO_2(g) \rightarrow 2CO(g)$

Oxygen

Carbon

Figure 15.3

Heterogeneous equilibrium. The reaction $C(\text{graphite}) + CO_2(g) \rightleftharpoons 2CO(g)$ takes place on the surface of solid carbon, where the concentration of carbon atoms is constant.

to change until the system reaches equilibrium. The concentration of carbon atoms, on the other hand, is constant. (The concentration of a pure solid such as graphite is proportional to its density, and the density will not change unless the temperature changes.) Adding more graphite (or breaking up the pieces of graphite already there) will increase the surface area and the number of surface atoms, but the concentration of these atoms remains the same. The reaction quotient combines only those concentrations that vary with time, so the reaction quotient for this equilibrium includes CO and CO_2 but not graphite:

$$Q_c = \frac{[CO]^2}{[CO_2]}$$

or in terms of partial pressures

$$Q_p = \frac{P_{CO}^2}{P_{CO_2}}$$

The omission of graphite from these expressions illustrates the general rule that *pure solids and pure liquids are omitted from the Q and K expressions for heterogeneous reactions.*

EXAMPLE 15.17

Write K_p expressions for

(a) $CaCO_3(s) \rightleftharpoons CaO(s) + CO_2(g)$

(b) $2H_2O(l) \rightleftharpoons 2H_2(g) + O_2(g)$

Write the K_c expression for

(c) $AgCl(s) \rightleftharpoons Ag^+(aq) + Cl^-(aq)$

SOLUTION

All of these reactions are heterogeneous, so the solid or liquid is omitted from the equilibrium constant expression.

(a) Both $CaCO_3$ and CaO are solids; $K_p = P_{CO_2}$.

(b) H_2O is a liquid; $K_p = P^2_{H_2}P_{O_2}$.

(c) AgCl is a solid; $K_c = [Ag^+][Cl^-]$.

PRACTICE EXERCISE 15.15

Write K_c expressions for

(a) $NO(g) + \frac{1}{2}Br_2(l) \rightleftharpoons NOBr(g)$

(b) $Sb^{3+}(aq) + Cl^-(aq) + H_2O(l) \rightleftharpoons SbOCl(s) + 2H^+(aq)$

EXAMPLE 15.18

The value of K_p for the reaction

$$NH_4Cl(s) \rightleftharpoons NH_3(g) + HCl(g)$$

is 0.25 at 340°C. What pressure will develop if excess ammonium chloride is introduced into a closed container that already contains NH_3 at 0.20 atm and 340°C?

SOLUTION

The final pressure will be the sum of the equilibrium pressures of NH_3 and HCl.

Step 1: The initial pressure of NH_3 is 0.20 atm; the initial pressure of HCl is zero.

Step 2: The balanced equation shows that NH_3 and HCl gases are produced in equal molar amounts. If the pressure of ammonia increases by x, the pressure of HCl will also increase by x.

Step 3: The equilibrium pressures will be $P_{NH_3} = 0.20 + x$ and $P_{HCl} = 0 + x = x$.

PRESSURE SUMMARY

Equation:	$NH_4Cl(s)$	$\rightleftharpoons$	$NH_3(g)$	+	$HCl(g)$
Initial pressures (atm):			0.20		0
Pressure changes (atm):			$+x$		$+x$
Equilibrium pressures (atm):			$0.20 + x$		x

NH_4Cl is not included in the pressure summary because it is a solid and its concentration does not change. Substituting the equilibrium pressures and $K_p = 0.25$ into the equilibrium constant expression gives

$$K_p = P_{NH_3}P_{HCl}$$

$$0.25 = (0.20 + x)(x)$$

This quadratic equation can be rearranged to

$$x^2 + 0.20x - 0.25 = 0$$

The quadratic formula with $a = 1$, $b = 0.20$, and $c = -0.25$ gives

$$x = \frac{-0.20 \pm \sqrt{(0.20)^2 - 4(1)(-0.25)}}{2}$$

Inspection of the pressure summary shows that x, the equilibrium pressure of HCl, must be positive. Choosing the positive sign for the square root gives $x = 0.41$ atm. The equilibrium pressures are

$$P_{HCl} = x = 0.41 \text{ atm}$$

$$P_{NH_3} = 0.20 + x = 0.20 \text{ atm} + 0.41 \text{ atm} = 0.61 \text{ atm}$$

The total pressure is

$$P_{total} = P_{NH_3} + P_{HCl} = 0.61 \text{ atm} + 0.41 \text{ atm} = 1.02 \text{ atm}$$

PRACTICE EXERCISE 15.16

The equilibrium constant K_p for the reaction

$$C(graphite) + S_2(g) \rightleftharpoons CS_2(g)$$

is 5.60 at 1010°C. Find the partial pressure of each gas when carbon disulfide, initially at 1.00 atm, comes to equilibrium with its decomposition products at 1010°C.

15.8 LE CHÂTELIER'S PRINCIPLE

A system in equilibrium will remain in equilibrium if it is not disturbed. The slightest change in conditions or concentrations, however, will upset the delicate balance between the opposing processes and will cause the system to move toward a new equilibrium state. The effects of such changes can be predicted with the aid of **Le Châtelier's principle**: *When a system in equilibrium is disturbed, the system adjusts to a new equilibrium in a way that partially counteracts the disturbance.* We will see how this principle predicts the response of a system at equilibrium to changes in concentration, volume, pressure, and temperature.

Henri Louis Le Châtelier (1850–1936) was a French industrial chemist known for his work on the structure of alloys. His name is pronounced "luh shat′elyáy."

Concentration Changes

When potassium thiocyanate (KSCN) is added to a solution containing Fe^{3+} ions, the solution turns red (see Demonstration 15.1). A simplified equation for the reaction is

$$Fe^{3+}(aq) + SCN^-(aq) \rightleftharpoons \underset{\text{red}}{FeSCN^{2+}(aq)}$$

If a salt containing Fe^{3+} is added to the equilibrium mixture, the red color deepens, indicating the formation of a new equilibrium mixture containing more $FeSCN^{2+}$ ions. This response is what we would expect from Le Châtelier's principle—disturbing the system by adding more Fe^{3+} promotes the reaction that uses up Fe^{3+}. We can also interpret this observation in terms of the reaction quotient

$$Q_c = \frac{[FeSCN^{2+}]}{[Fe^{3+}][SCN^-]}$$

CHEMICAL INSIGHT

DRYING AGENTS AND LABORATORY DESICCATORS

Many salt crystals have definite proportions of water in their lattices. Hydrated salts of this type include $Na_2CO_3 \cdot 10H_2O$ (washing soda), $CuSO_4 \cdot 5H_2O$ (copper sulfate), and $Na_2B_4O_7 \cdot 10H_2O$ (borax). These salts can lose their *water of crystallization* to form anhydrous solids, that is, solids without water. The loss of water is reversible, and in a closed container the hydrated and anhydrous solids exist in equilibrium with water vapor. The calcium chloride equilibrium is an important example:

$$\underset{\text{hydrated solid}}{CaCl_2 \cdot H_2O(s)} \rightleftharpoons \underset{\text{anhydrous solid}}{CaCl_2(s)} + H_2O(g)$$

The reaction quotient for this heterogeneous equilibrium is

$$Q_p = P_{H_2O}$$

where P_{H_2O} is the vapor pressure of water above the solid. The value of K_p is only 0.14 torr at 25°C. The small value of K_p means that Q_p will be greater than K_p in all but the driest of atmospheres. (A water vapor pressure of 0.14 torr corresponds to less than 0.6% relative humidity at 25°C.) Thus if anhydrous calcium chloride is exposed to the atmosphere, it will absorb water until all of it is hydrated or until the relative humidity drops to 0.6%. This property (along with the fact that calcium chloride is relatively inexpensive) explains why anhydrous calcium chloride is widely used as a *desiccant* (drying agent) on subway track beds and in laboratory *desiccators* (Figure 15.4). Other substances that are used as drying agents include anhydrous $CaSO_4$ (Drierite), phosphorous pentoxide (P_4O_{10}), and silica gel, a partially hydrated form of SiO_2. Many hydrated salts can be converted to the anhydrous form by heating, and some spent desiccants can be rejuvenated by baking them in a hot oven for several hours.

Figure 15.4
A common type of laboratory desiccator. The desiccant, in this case anhydrous $CaCl_2$, is put in the bottom compartment while the materials to be kept dry are stored on the rack above. (The thick glass walls make the contents appear blurred.)

Before the salt is added, the mixture is in equilibrium and Q_c is equal to K_c. Increasing the concentration of Fe^{3+} increases the denominator of the reaction quotient, so Q_c is momentarily less than K_c. For Q_c to increase back to its equilibrium value, some of the added Fe^{3+} must react with SCN^- to form additional $FeSCN^{2+}$. Figure 15.5 shows how the concentration of each ion varies with time. Note that not all of the added Fe^{3+} ions are consumed.

The addition of more KSCN also causes the red color of the solution to deepen. This change is in accordance with Le Châtelier's principle, and its mechanism is similar to the one described for the addition of Fe^{3+}. The increased concentration of

DEMONSTRATION 15.1

AN EXAMPLE OF LE CHÂTELIER'S PRINCIPLE: $Fe^{3+}(aq) + SCN^{-}(aq) \rightleftharpoons FeSCN^{2+}(aq)$

The four beakers contain identical solutions in which Fe^{3+} (colorless), SCN^{-} (colorless), and $FeSCN^{2+}$ (red) are in equilibrium. The watch glasses from left to right contain crystals of $FeCl_3$, NaSCN, and $Na_2C_2O_4$.

The crystals are added to the beakers. Dissolved $FeCl_3$ and NaSCN intensify the red color; $Na_2C_2O_4$ lightens it.

SCN^{-} ions causes a temporary decrease in the reaction quotient. Some of the SCN^{-} ions combine with Fe^{3+} to form more $FeSCN^{2+}$ ions, and Q_c increases until it again equals K_c.

What happens when Fe^{3+} ions are removed from the solution? We can find out by adding sodium oxalate ($Na_2C_2O_4$), a substance that combines with Fe^{3+}:

$$Fe^{3+}(aq) + 3C_2O_4^{2-}(aq) \rightarrow Fe(C_2O_4)_3^{3-}(aq)$$

The removal of Fe^{3+} causes Q_c to become momentarily greater than K_c, so the reac-

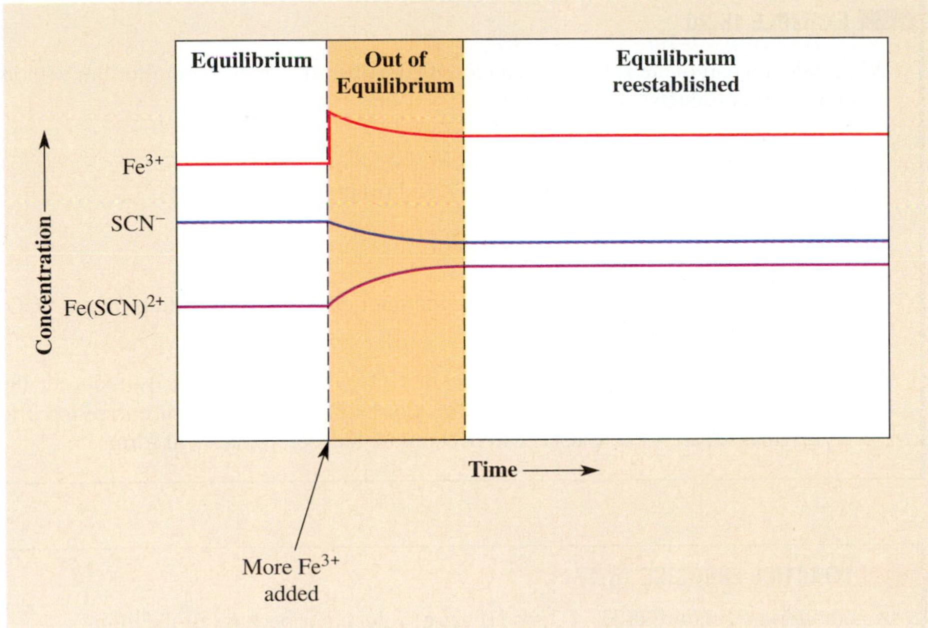

Figure 15.5
Effect of additional Fe^{3+} on the $Fe^{3+}(aq) + SCN^-(aq) \rightleftharpoons FeSCN^{2+}(aq)$ equilibrium. *Left:* A set of equilibrium concentrations. *Tan region:* The concentrations change in a way that uses up added Fe^{3+}. *Right:* The new equilibrium concentrations.

tion adjusts by moving in the reverse direction. The red color fades because of the dissociation of $FeSCN^{2+}$ and levels off to a lighter shade in the new equilibrium mixture. Le Châtelier's principle is once again satisfied—removal of Fe^{3+} ions is partially counteracted by the formation of a new equilibrium mixture in which some Fe^{3+} ions are restored.

These examples illustrate that *equilibrium mixtures adjust to concentration changes by forming new equilibrium mixtures in which the changes are partially counteracted.*

EXAMPLE 15.19

The following system is at equilibrium in a closed container:

$$2SO_2(g) + O_2(g) \rightleftharpoons 2SO_3(g)$$

Predict the effects of (a) adding some SO_3 to the container and (b) decreasing the partial pressure of O_2 by removing some of it.

SOLUTION

(a) The concentration of SO_3 has increased, so some of the added SO_3 will be consumed. The reaction will move in the reverse direction, and a new equilibrium mixture containing more SO_2 and more O_2 will form.

(b) Decreasing the partial pressure of oxygen is equivalent to decreasing its concentration (recall that the partial pressure of a gas is proportional to its molarity). The reaction will move in the reverse direction to restore some of the O_2. The new equilibrium mixture will contain more SO_2 and less SO_3.

EXAMPLE 15.20

When calcium carbonate is heated in a closed container, it comes to equilibrium with its decomposition products:

$$CaCO_3(s) \rightleftharpoons CaO(s) + CO_2(g)$$

Predict the effects of (a) adding solid $CaCO_3$ and (b) removing CO_2.

SOLUTION

(a) The solids are not included in the reaction quotient, therefore adding more $CaCO_3$ will not affect the equilibrium.

(b) The concentration and partial pressure of CO_2 will be momentarily reduced; the reaction will proceed in the forward direction until the CO_2 pressure is restored to its original value. Some $CaCO_3$ will decompose and some CaO will form.

PRACTICE EXERCISE 15.17

An equilibrium mixture of H_2, I_2, and HI has a violet color because of the iodine vapor:

$$H_2(g) + I_2(g) \rightleftharpoons 2HI(g)$$

Will this color become more or less intense if (a) more HI is added to the mixture and (b) the partial pressure of hydrogen is reduced?

Figure 15.6

A heterogeneous equilibrium: $NH_4Cl(s) \rightleftharpoons NH_3(g) + HCl(g)$. When the system is compressed by the addition of weights to the piston, the gases will react to form more NH_4Cl. When weights are removed from the piston, the system expands and NH_4Cl decomposes.

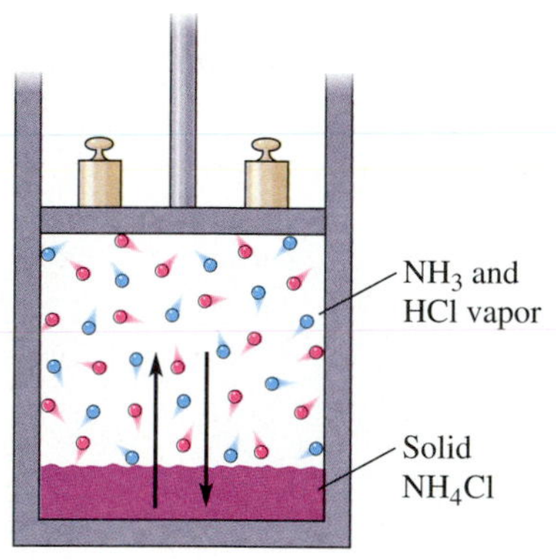

Volume Changes

Solid ammonium chloride placed in a closed container will come into equilibrium with its decomposition products (Figure 15.6):

$$NH_4Cl(s) \rightleftharpoons NH_3(g) + HCl(g)$$

The reaction quotient

$$Q_p = P_{NH_3} \times P_{HCl}$$

equals K_p as long as the mixture is not disturbed. If the mixture is suddenly compressed to a smaller volume, the partial pressures of NH_3 and HCl increase, and Q_p becomes momentarily greater than K_p. The reaction moves in the reverse direction. The gases react to form solid NH_4Cl, the value of Q_p falls, and a new equilibrium mixture is formed. An increase in the container volume has the opposite effect—the partial pressure of each gas decreases and Q_p becomes less than K_p. The reaction moves in the forward direction, ammonium chloride decomposes to produce more gas, and Q_p will increase until equilibrium is again restored.

Compression: Reducing the volume of a system by increasing the external pressure.

These experiments show that compressing the reaction mixture results in the formation of solid NH_4Cl, the substance with the smallest volume. (Recall that, mole for mole, liquid and solid volumes are much smaller than gas volumes. One mole of liquid water, for example, occupies only 18 mL at 0°C and 1 atm; 1 mol of water vapor occupies about 22.4 L = 22,400 mL.) Increasing the volume of the reaction mixture has the opposite effect—the substances with the greatest volume (NH_3 and

HCl gases) form. These observations are consistent with Le Châtelier's principle: *Compression of an equilibrium mixture favors the formation of substances that occupy less volume; increasing the volume of the mixture favors the formation of substances that occupy more volume.*

EXAMPLE 15.21

Predict the effect of compression on the following equilibrium systems:

(a) $2H_2(g) + O_2(g) \rightleftharpoons 2H_2O(g)$

(b) $H_2(g) + Br_2(l) \rightleftharpoons 2HBr(g)$

(c) $H_2(g) + Cl_2(g) \rightleftharpoons 2HCl(g)$

SOLUTION

(a) Gas volumes are proportional to the numbers of moles, so 2 mol of water vapor occupy a smaller volume than 2 mol of hydrogen and 1 mol of oxygen. The reaction will move in the forward direction to form a new equilibrium mixture containing more H_2O and less H_2 and O_2.

(b) The volume of liquid bromine is negligible; therefore, 1 mol of hydrogen and 1 mol of liquid bromine occupy a smaller volume than 2 mol of hydrogen bromide gas. The reaction will move in the reverse direction. The new equilibrium mixture will contain more H_2 and Br_2 and less HBr.

(c) There are 2 mol of gas on each side of the equation—reactants and products occupy the same volume. Compression will have no appreciable effect on the equilibrium.

LEARNING HINT

When gases are involved in an equilibrium, compression favors the direction that produces fewer moles of gas; expansion favors the direction that produces more moles of gas.

EXAMPLE 15.22

When ammonia is synthesized from its elements

$$N_2(g) + 3H_2(g) \rightleftharpoons 2NH_3(g)$$

should a high or low pressure be used to obtain the best yield?

SOLUTION

Two moles of ammonia occupy a smaller volume than 1 mol of nitrogen and 3 mol of hydrogen. Higher pressure reduces the volume and favors a larger proportion of ammonia in the equilibrium mixture.

PRACTICE EXERCISE 15.18

NO_2 forms an equilibrium mixture with N_2O_4:

$$N_2O_4(g) \rightleftharpoons 2NO_2(g)$$

What effect will an increase in volume have on the relative amounts of N_2O_4 and NO_2 in the mixture?

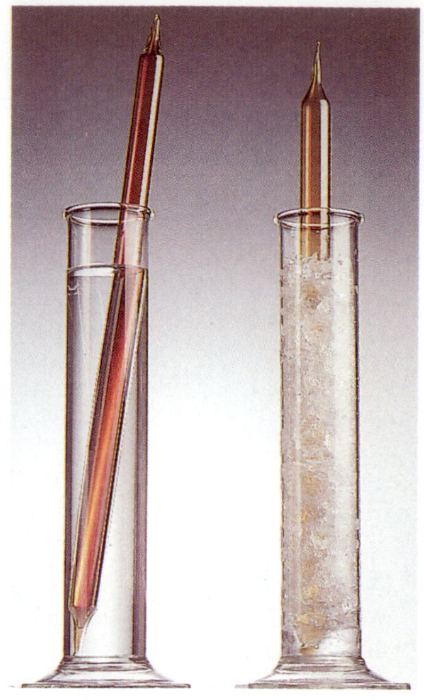

Figure 15.7

The effect of temperature on the equilibrium $N_2O_4(g) \rightleftharpoons 2NO_2(g)$. The sealed tubes, which contain brown NO_2 gas in equilibrium with colorless N_2O_4 gas, were initially the same color. The tube on the left, which is immersed in hot water, has turned darker; heat favors the formation of NO_2. The tube on the right, which is immersed in ice water, has become lighter; cold favors the formation of N_2O_4. From these observations we conclude that the forward reaction is endothermic.

Temperature Changes

Consider the formation of ammonia from its elements

$$N_2(g) + 3H_2(g) \rightleftharpoons 2NH_3(g) \qquad \Delta H^\circ = -92.2 \text{ kJ}$$

The forward reaction evolves heat; the reverse reaction absorbs heat. At equilibrium, the net heat flow is zero because the forward and reverse reactions are occurring at the same rate. The equilibrium constant has been measured at various temperatures and is found to decrease with increasing temperature. For example, K_p is 6.8×10^5 at 25°C, but only 1.7×10^{-4} at 400°C. Because K_p is of the form *product pressures/reactant pressures,* the decrease in K_p indicates a shift away from product formation. Thus increasing the temperature causes the reaction to move in the reverse or endothermic direction. An equilibrium mixture at a high temperature will contain more H_2 and N_2 and less NH_3 than one at a low temperature.

The response to temperature change could have been predicted from Le Châtelier's principle. The addition of heat to raise the temperature is counteracted when the reaction moves in the endothermic direction, the direction that absorbs some of the added heat. *Increasing the temperature of an equilibrium system favors the endothermic reaction; decreasing the temperature favors the exothermic reaction* (see Figure 15.7).

EXAMPLE 15.23

Consider the equilibrium

$$N_2(g) + O_2(g) \rightleftharpoons 2NO(g) \qquad \Delta H^\circ = +180.5 \text{ kJ}$$

What effect will an increase in temperature have on (a) the amount of NO in the equilibrium mixture and (b) K_p?

SOLUTION

(a) The forward reaction is endothermic. Increasing the temperature will cause the reaction to move in the forward direction, thus increasing the amount of NO.

(b) The partial pressure of NO will increase; the partial pressures of N_2 and O_2 will decrease. Hence, K_p will increase.

EXAMPLE 15.24

The value of K_p for the reaction

$$PCl_5(g) \rightleftharpoons PCl_3(g) + Cl_2(g)$$

increases with increasing temperature. Is the forward reaction exothermic or endothermic?

SOLUTION

The increase in K_p shows that an increase in temperature favors the formation of products. The forward reaction is favored by heating, so it must be endothermic.

CHEMICAL INSIGHT

KINETICS AND EQUILIBRIUM IN THE FORMATION OF NITROGEN OXIDE POLLUTANTS

Air, which contains nitrogen and oxygen, has the potential for forming nitrogen monoxide (NO):

$$N_2(g) + O_2(g) \rightleftharpoons 2NO(g) \qquad \Delta H^\circ = +180.5 \text{ kJ}$$

At room temperature, K_p is very small (4.1×10^{-31} at 25°C), and virtually no nitrogen monoxide forms. The reaction is endothermic, however, so K_p increases with temperature, and significant amounts of NO will form at high temperatures. Some NO always forms, for example, during the combustion of fossil fuels. Once in the atmosphere, NO is further oxidized by oxygen to NO_2, and both oxides play a role in the formation of photochemical smog and acid rain. (See *CHEMICAL INSIGHT: Air Pollution and Photochemical Smog,* p. 509.) High temperatures are produced in automobile and diesel engines as well as in industrial furnaces; for this reason, nitrogen oxide pollutants reach their highest levels over heavily populated industrial areas. The photochemical smog that occurs over Los Angeles is a much publicized consequence of this type of pollution.

At this point you may be thinking, "If high temperatures favor the formation of NO, then low temperatures favor its decomposition, so why doesn't NO decompose into its elements after it cools down?" The answer to this question lies in kinetics—cooling slows the reaction rate. Even though decomposition is favored, the rate of decomposition is so slow that nitrogen oxides tend to remain once they are formed. An appropriate catalyst in the cooler regions of an exhaust system will speed up the oxide decomposition, and catalytic converters are installed on automobiles for this purpose. Engineers are also studying ways to reduce NO at its source by developing cooler running engines and fuel mixtures that contain water. Water vapor has a greater heat capacity than air, and the inclusion of water vapor in an air–fuel mixture maintains the excess air—that is, air not used in the combustion—at a lower temperature.

PRACTICE EXERCISE 15.19

Consider the equilibrium

$$H_2(g) + F_2(g) \rightleftharpoons 2HF(g) \qquad \Delta H^\circ = -542 \text{ kJ}$$

What effect will an increase in temperature have on (a) the amount of each substance in the equilibrium mixture and (b) K_p?

Nonparticipating Substances and Catalysts

A substance that does not react with any of the participants in a chemical equilibrium will not alter the concentrations that appear in the reaction quotient; hence, such substances have no effect on the equilibrium.

EXAMPLE 15.25

The following equilibrium exists in a closed vessel at an elevated temperature:

$$COCl_2(g) \rightleftharpoons CO(g) + Cl_2(g)$$

What will happen to the equilibrium mixture if the total pressure in the vessel is increased by the addition of argon, an inert gas?

SOLUTION

The addition of argon at constant volume will not change the partial pressures of the other gases; hence, it will not affect the equilibrium.

Although a catalyst participates in a reaction, it has no effect on the composition of the equilibrium mixture. Recall that a catalyst speeds up a reaction by providing an alternative reaction path with a lower activation energy. The reverse reaction simply retraces the forward path, so its activation energy is lowered as well, and the forward and reverse rates increase by exactly the same factor. Hence, a catalyst will decrease the time it takes for a system to reach equilibrium, but it will not change the equilibrium constant or the position of equilibrium.

DIGGING DEEPER

The Constancy of *K* and the van't Hoff Equation

Although the law of chemical equilibrium states that equilibrium constants vary only with temperature, many K_c and K_p values have been found to change slightly with concentration and total pressure as well. At 400°C, for example, the K_p value for

$$N_2(g) + 3H_2(g) \rightleftharpoons 2NH_3(g)$$

varies from 4.34×10^{-5} at 10 atm total pressure to 4.76×10^{-5} at 50 atm. An equilibrium constant based on concentration or pressure is constant only to the extent that the system is ideal—that is, to the extent that intermolecular forces can be ignored and that molecules are randomly dispersed. In systems that are nearly ideal, such as dilute solutions and gases at low pressure, the effect of intermolecular forces is minimal and K values are virtually constant. At high concentrations and pressures, however, intermolecular forces prevent particles from being independent, and K values deviate from constancy. Recall that a similar situation exists with the various laws governing gas behavior (Section 5.8).

Aside from the effect of intermolecular forces at high concentrations, temperature is the only factor that influences the magnitude of an equilibrium constant. The variation in the value of K over a small temperature range is given by the formula

$$\ln K = -\frac{\Delta H°}{RT} + B \qquad (15.3)$$

where ln K is the natural logarithm of K, $\Delta H°$ is the enthalpy change for the forward reaction, R is the gas constant, T is the Kelvin temperature, and B is a constant that depends on the reaction. (Notice that Equation 15.3 is similar in form to Equation 12.2 (p. 541), the Clausius–Clapeyron equation.) The value of $\Delta H°$ does not change very much with temperature, and a plot of ln K versus $1/T$ will be a straight line with a slope equal to $-\Delta H°/R$ and an intercept equal to B. A typical plot is given in Figure 15.8 for the reaction

$$N_2(g) + O_2(g) \rightleftharpoons 2NO(g)$$

If K_1 and K_2 are the equilibrium constants at temperatures T_1 and T_2, respectively,

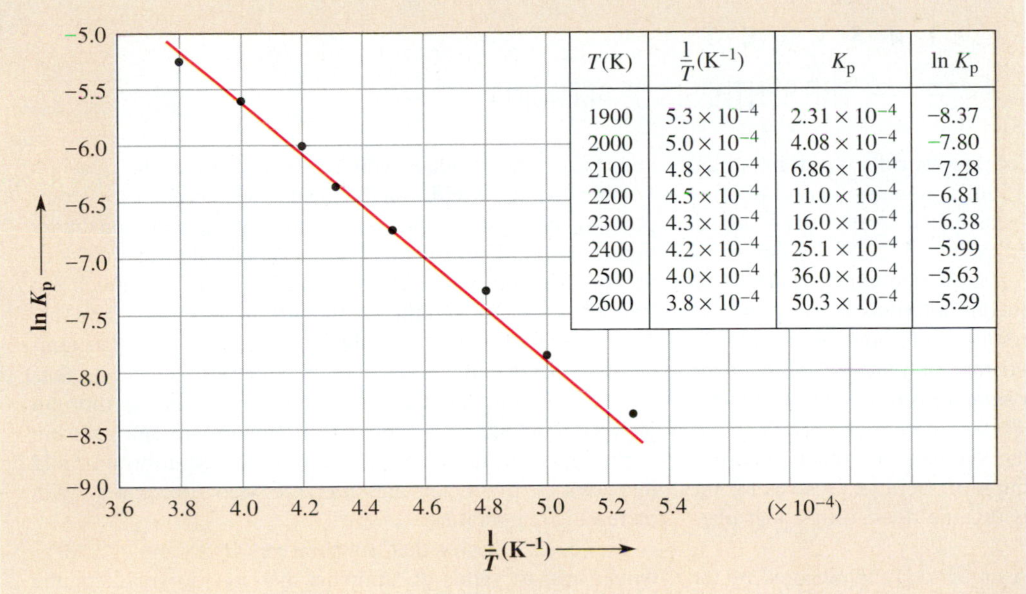

T(K)	$\frac{1}{T}$ (K^{-1})	K_p	ln K_p
1900	5.3×10^{-4}	2.31×10^{-4}	−8.37
2000	5.0×10^{-4}	4.08×10^{-4}	−7.80
2100	4.8×10^{-4}	6.86×10^{-4}	−7.28
2200	4.5×10^{-4}	11.0×10^{-4}	−6.81
2300	4.3×10^{-4}	16.0×10^{-4}	−6.38
2400	4.2×10^{-4}	25.1×10^{-4}	−5.99
2500	4.0×10^{-4}	36.0×10^{-4}	−5.63
2600	3.8×10^{-4}	50.3×10^{-4}	−5.29

Figure 15.8

A plot of ln K_p versus $1/T$ for the $N_2(g) + O_2(g) \rightleftharpoons 2NO(g)$ equilibrium is linear.

then

$$B = \ln K_1 + \frac{\Delta H^\circ}{RT_1} = \ln K_2 + \frac{\Delta H^\circ}{RT_2}$$

This equation can be rearranged to

$$\ln \frac{K_2}{K_1} = -\frac{\Delta H^\circ}{R}\left(\frac{1}{T_2} - \frac{1}{T_1}\right) \tag{15.4}$$

Equation 15.4, which can be used to estimate equilibrium constants at various temperatures, was derived by the Dutch chemist Jacobus van't Hoff and is called the **van't Hoff equation**.

EXAMPLE 15.26

At 2000 K, $K_p = 4.08 \times 10^{-4}$ for the reaction

$$N_2(g) + O_2(g) \rightleftharpoons 2NO(g)$$

If ΔH° for the forward reaction is +181 kJ/mol, estimate K_p at 2200 K.

SOLUTION

Let K_2 be the unknown equilibrium constant; then $T_2 = 2200$ K, and K_1 and T_1 are 4.08×10^{-4} and 2000 K, respectively. Substituting these data and $R = 8.314 \times 10^{-3}$ kJ/mol·K into Equation 15.4 gives

$$\ln \frac{K_2}{4.08 \times 10^{-4}} = \frac{-181 \text{ kJ/mol}}{8.314 \times 10^{-3} \text{ kJ/mol·K}}\left(\frac{1}{2200 \text{ K}} - \frac{1}{2000 \text{ K}}\right)$$

Solving this equation for the unknown equilibrium constant gives $K_2 = 1.10 \times 10^{-3}$.

CHEMICAL INSIGHT

THE SYNTHESIS OF AMMONIA

Ammonia (NH_3) is one of the world's most important industrial chemicals. About 15 million metric tons (34 billion pounds) are produced each year in the United States alone. Because of its large nitrogen content (82.3% N), ammonia makes an excellent fertilizer. It is easy to apply—pure liquid ammonia or concentrated aqueous solutions of ammonia can be sprayed directly into the soil—and about one-quarter of the annual production is used for this purpose. Most of the remainder is converted into solid fertilizers such as ammonium phosphate ($(NH_4)_3PO_4$) and urea ($(NH_2)_2CO$). Ammonia is also used in the synthesis of other important industrial chemicals, including nitric acid, explosives for the mining industry, and various organic dyes, drugs, and plastics.

Most of the ammonia consumed today is prepared by the direct reaction of nitrogen with hydrogen:

$$N_2(g) + 3H_2(g) \rightleftharpoons 2NH_3(g) \qquad \Delta H° = -92.2 \text{ kJ}$$

Nitrogen for the synthesis is obtained by the fractional distillation of liquid air, and hydrogen is produced by allowing steam to react with methane, the major component of natural gas:

$$CH_4(g) + H_2O(g) \rightleftharpoons CO(g) + 3H_2(g)$$

This reaction is carried out at temperatures ranging from 600°C to 1000°C, using finely divided nickel as a catalyst.

The industrial synthesis of ammonia illustrates how the *rate* of a chemical reaction must be balanced against its *equilibrium yield* to maximize output and minimize cost. The value of K_p for the synthesis of ammonia is quite large at room temperature ($K_p = 6.9 \times 10^5$ at 25°C), so an equilibrium mixture will contain relatively large amounts of ammonia. The product volume is less than the reactant volume, so the yield can be made even greater by compressing the system, that is, by running the reaction under high pressure. Unfortunately, the reaction rate, regardless of pressure, is very slow at 25°C, and it would take much too long to produce a significant quantity of ammonia. The rate can be increased by raising the temperature, but higher temperature favors the reverse or endothermic reaction, thus reducing both K_p and the equilibrium yield. The equilibrium yield of ammonia as a function of temperature and pressure is presented in Table 15.2. Note how the equilibrium yield increases with increasing pressure and decreases with increasing temperature.

Table 15.2 shows that temperatures in excess of 600°C would lead to yields of ammonia that are too small for the

TABLE 15.2 Equilibrium Yield of Ammonia in Mole Percent as a Function of Temperature and Total Pressure[a]

T (°C)	Pressure (atm)					
	1	100	200	300	600	1000
200	15	82	86	90	95	98
300	2	52	63	71	84	93
400	0.4	25	36	47	65	80
500	0.1	11	18	26	42	57
600	0.05	4	8	14	23	31

[a]For each entry, the initial mixture contained 3 mol of hydrogen per mole of nitrogen.

CHAPTER REVIEW

Starred entries are based on the *Digging Deeper* sections.

LEARNING OBJECTIVES BY SECTION

15.1 1. State the characteristics common to all chemical equilibria.

2. Explain why the concentration of each substance in an equilibrium mixture does not change with time.

process to be economically feasible. High temperatures have the further disadvantage of being expensive to maintain. Modern ammonia plants strike a balance between reaction rate and yield by using temperatures around 400°C and pressures of about 250 atm. The rate is increased by an iron oxide catalyst that contains small amounts of potassium and aluminum oxides. A schematic diagram of the ammonia synthesis is shown in Figure 15.9. Note that the reactants and products never actually come to equilibrium. The reaction mixture contains about 20% ammonia after passing over the catalyst, and the ammonia is separated from the reactants by refrigeration and liquefaction (ammonia liquefies at −33.4°C). The remaining nitrogen and hydrogen are then recycled.

The ammonia synthesis, or *Haber process,* was developed by the German chemist Fritz Haber and put into operation in 1913. Ammonia can be converted to nitric acid and, during World War I, the Haber process allowed Germany to base its explosives industry on ammonia rather than on imported Chilean nitrates. The greatest significance of the Haber process, however, lies in the fact that it opened the way for massive production of nitrogen fertilizers. Haber received a Nobel Prize in 1918 for his achievement.

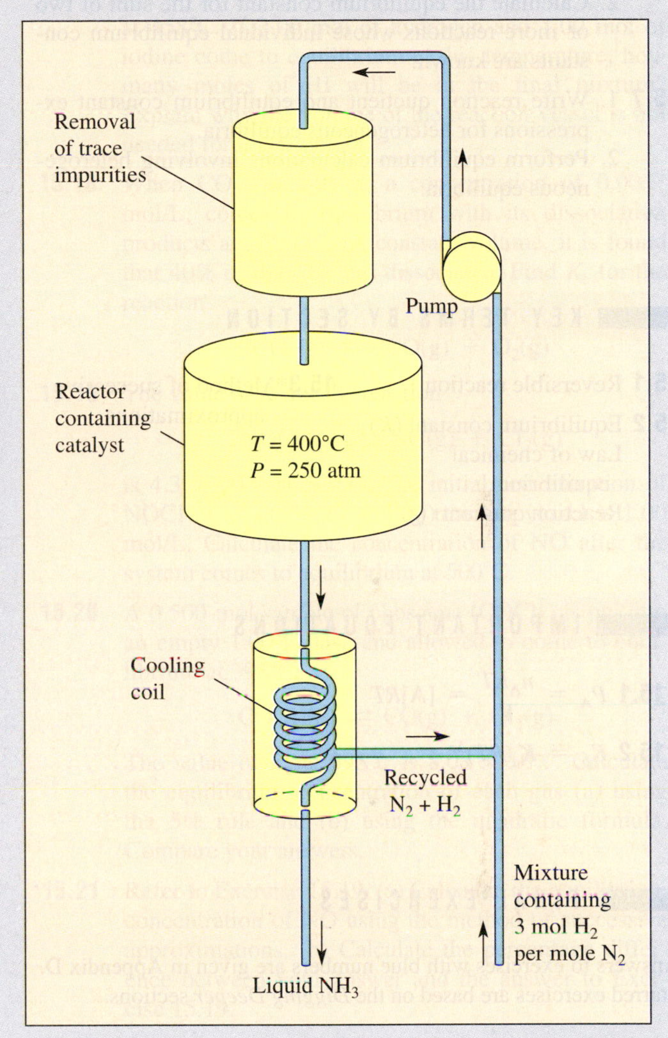

Figure 15.9
The Haber process for the synthesis of ammonia. The reaction takes place in the center cylinder, where the highly compressed hydrogen and nitrogen gases come in contact with a catalyst. Ammonia (boiling point, −33°C) is removed from the mixture by condensation, and the unreacted gases are recycled.

15.2 1. Write reaction quotients for reversible reactions.
2. Use equilibrium concentrations to calculate the value of an equilibrium constant.
3. State the law of chemical equilibrium.
4. Use Q_c and K_c to predict whether a reaction is moving in the forward or reverse direction.
*5. Derive the law of chemical equilibrium from the kinetic relations that exist in the equilibrium state.

15.3 1. Use K_c and the initial concentrations of a mixture to calculate its equilibrium composition.
2. Use simplifying approximations to solve equilibrium problems, and check to see if the approximations are valid.
*3. Use the method of successive approximations to solve equilibrium problems.

15.4 1. Write Q_c and K_c expressions, and solve equilibrium problems for reactions involving a solvent.

15.5 1. Write Q_p and K_p expressions.
2. Use partial pressures in gas phase equilibrium calculations.
3. Convert K_p to K_c, and vice versa.

CH_3OH if CO and H_2, each initially at 0.100 mol/L, come to equilibrium at 250°C.

Equilibria Involving Solvents (Section 15.4)

15.25 Write a K_c expression for each of the following aqueous equilibria:
(a) $Cr_2O_7^{2-}(aq) + 2OH^-(aq) \rightleftharpoons 2CrO_4^{2-}(aq) + H_2O(l)$
(b) $CO_3^{2-}(aq) + H_2O(l) \rightleftharpoons HCO_3^-(aq) + OH^-(aq)$
(c) $H_2O(l) \rightleftharpoons H^+(aq) + OH^-(aq)$

15.26 Explain why the solvent is omitted from Q_c and K_c expressions for reactions in dilute solution.

15.27 At 25°C, $K_c = 3.70 \times 10^{-4}$ for the reaction of methylamine with water:

$$CH_3NH_2(aq) + H_2O(l) \rightleftharpoons CH_3NH_3^+(aq) + OH^-(aq)$$

Find the concentration of $CH_3NH_3^+$ ions when a 0.250 *M* solution of methylamine comes to equilibrium.

15.28 The equilibrium

$$H_2O(l) \rightleftharpoons H^+(aq) + OH^-(aq)$$

occurs in all aqueous solutions. The value of K_c for this reaction is 1.00×10^{-14} at room temperature. If the equilibrium concentration of hydroxide ions is 0.50 *M*, what is the concentration of hydrogen ions?

Gas Phase Equilibrium: Q_p and K_p (Section 15.5)

15.29 For what type of reaction do we use a K_p? What units are used in formulating a K_p?

15.30 Are the numerical values of K_p and K_c for a given equation identical (a) in all cases, (b) in some cases, or (c) never? Explain your answer.

15.31 Without referring to the text, derive the relation between the pressure of a gas and its molar concentration.

15.32 Without referring to the text, derive the relation between K_p and K_c for a given reaction.

15.33 Write a Q_p expression for each of the following gas phase reactions:
(a) $2O_3(g) \rightleftharpoons 3O_2(g)$
(b) $C_2H_2(g) + H_2(g) \rightleftharpoons C_2H_4(g)$

15.34 Write a K_p expression for each of the following gas phase reactions:
(a) $CO(g) + 2H_2(g) \rightleftharpoons CH_3OH(g)$
(b) $S_2(g) \rightleftharpoons 2S(g)$

15.35 A sample of oxygen has a pressure of 770 torr at 25°C. Calculate its molarity.

15.36 The atmosphere contains about 3.8×10^{-4} mol of argon gas per liter at 25°C. Find the partial pressure of atmospheric argon at this temperature. Express your answer in atmospheres.

15.37 The value of K_p for the reaction

$$Cl_2(g) \rightleftharpoons 2Cl(g)$$

increases from 4.8×10^{-16} at 600 K to 0.570 at 2000 K. Does the fraction of chlorine molecules dissociated increase or decrease with increasing temperature?

15.38 The equilibrium

$$2NO(g) \rightleftharpoons N_2(g) + O_2(g)$$

has a K_p value of 309 at 2400°C. Find the total pressure if NO, initially at 1.00 atm, comes to equilibrium in a closed container. (*Hint:* The answer can be obtained without an equilibrium calculation.)

15.39 Refer to Exercise 15.13. Calculate (a) the equilibrium partial pressure of each gas in the flask and (b) K_p for the reaction.

15.40 Refer to Exercise 15.14. Calculate (a) the equilibrium partial pressure of each gas in the flask and (b) K_p for the reaction.

15.41 A K_p value of 0.113 has been reported at 25°C for the reaction

$$N_2O_4(g) \rightleftharpoons 2NO_2(g)$$

Find the partial pressure of each gas if N_2O_4, initially at 0.500 atm, is allowed to come to equilibrium in a closed flask at 25°C.

15.42 At 25°C, $K_p = 1.08$ for the equilibrium

$$H_2(g) + Cl_2(g) \rightleftharpoons 2HCl(g)$$

The three gases, each at a partial pressure of 1.00 atm, are introduced into a reaction vessel. Determine (a) the direction in which the reaction moves and (b) the equilibrium partial pressure of each gas.

15.43 For the reaction

$$Br_2(g) \rightleftharpoons 2Br(g)$$

K_p is 0.255 at 1800 K. A sample of Br_2 that is initially at 2.00 atm comes to equilibrium at 1800 K. Find (a) the equilibrium pressure of each gas and (b) the fraction of Br_2 molecules dissociated.

15.44 One volume of nitrogen is mixed with three volumes of hydrogen in a reaction vessel at 400°C and a total pressure of 10.0 atm. The equilibrium mixture contains 3.85 mol % ammonia. Calculate K_p for

$$N_2(g) + 3H_2(g) \rightleftharpoons 2NH_3(g)$$

***15.45** Solve Example 15.12 (page 711) using the method of successive approximations.

***15.46** Solve Practice Exercise 15.11 (page 712) using the method of successive approximations.

15.47 Express K_c in terms of K_p, R, and T for each of the reactions in Exercise 15.33.

15.48 Express K_c in terms of K_p, R, and T for each of the reactions in Exercise 15.34.

15.49 The value of K_c for the equilibrium

$$CO(g) + \tfrac{1}{2}O_2(g) \rightleftharpoons CO_2(g)$$

is 2.5×10^{15} at 500°C. Find K_p.

15.50 The value of K_p for the reaction

$$2SO_3(g) \rightleftharpoons 2SO_2(g) + O_2(g)$$

is 1.79×10^{-5} at 350°C. Find K_c.

15.51 Calculate K_p for the reactions in (a) Exercise 15.19 and (b) Exercise 15.24.

15.52 Calculate K_c for the reactions in (a) Exercise 15.42 and (b) Exercise 15.43.

Forms of the Reaction Quotient (Section 15.6)

15.53 The value of K_p for

$$O_2(g) \rightleftharpoons 2O(g)$$

is 5.2×10^{-7} at 1725°C. Find K_p for the following equilibria:
(a) $2O(g) \rightleftharpoons O_2(g)$
(b) $O(g) \rightleftharpoons \tfrac{1}{2}O_2(g)$

15.54 Consider the equilibrium

$$N_2O_4(g) \rightleftharpoons 2NO_2(g)$$

for which K_c is 0.90 at 120°C. Find K_c for the following equilibria:
(a) $\tfrac{1}{2}N_2O_4(g) \rightleftharpoons NO_2(g)$
(b) $2NO_2(g) \rightleftharpoons N_2O_4(g)$

15.55 Find K_c for

$$CrO_4^{2-}(aq) + H_2O(l) \rightleftharpoons HCrO_4^-(aq) + OH^-(aq)$$

from the following equilibrium constants:

$$H_2O(l) \rightleftharpoons H^+(aq) + OH^-(aq) \quad K_c = 1.0 \times 10^{-14}$$

$$HCrO_4^-(aq) \rightleftharpoons H^+(aq) + CrO_4^{2-}(aq) \quad K_c = 3.1 \times 10^{-7}$$

15.56 The following equilibrium constants were determined at 1400 K:

$$H_2(g) + Br_2(g) \rightleftharpoons 2HBr(g) \quad K_p = 6.7 \times 10^4$$

$$H_2(g) \rightleftharpoons 2H(g) \quad K_p = 2.96 \times 10^{-11}$$

$$Br_2(g) \rightleftharpoons 2Br(g) \quad K_p = 3.03 \times 10^{-2}$$

Find K_p for

$$HBr(g) \rightleftharpoons H(g) + Br(g)$$

Heterogeneous Equilibria (Section 15.7)

15.57 Distinguish between homogeneous and heterogeneous equilibria. Illustrate each type of equilibrium with one or two examples.

15.58 Why are substances in the condensed state omitted from equilibrium constant expressions?

15.59 Write a K_p expression for each of the following equilibria:
(a) $C(graphite) + H_2O(g) \rightleftharpoons H_2(g) + CO(g)$
(b) $4H_2(g) + Fe_3O_4(s) \rightleftharpoons 4H_2O(g) + 3Fe(s)$
(c) $Zn(s) + Cl_2(g) \rightleftharpoons ZnCl_2(s)$

15.60 Write a K_p expression for each of the following equilibria:
(a) $2HgO(s) \rightleftharpoons 2Hg(l) + O_2(g)$
(b) $H_2O(l) \rightleftharpoons H_2O(g)$
(c) $CaO(s) + SO_2(g) \rightleftharpoons CaSO_3(s)$

15.61 Write a K_c expression for each of the equilibria in Exercise 15.59.

15.62 Write a K_c expression for each of the equilibria in Exercise 15.60.

15.63 Write a K_c expression for each of the following equilibria:
(a) $Bi_2S_3(s) \rightleftharpoons 2Bi^{3+}(aq) + 3S^{2-}(aq)$
(b) $Br_2(l) + H_2O(l) \rightleftharpoons H^+(aq) + Br^-(aq) + HOBr(aq)$

15.64 Write a K_c expression for each of the following equilibria:
(a) $I_2(s) \rightleftharpoons I_2(aq)$
(b) $AgBr(s) + 2NH_3(aq) \rightleftharpoons Ag(NH_3)_2^+(aq) + Br^-(aq)$

15.65 Ammonium carbamate (NH_2COONH_4) decomposes into ammonia and carbon dioxide at 30°C. The reaction is

$$NH_2COONH_4(s) \rightleftharpoons 2NH_3(g) + CO_2(g)$$

with $K_p = 2.87 \times 10^5$. What total pressure will develop if solid ammonium carbamate is placed into an empty flask and allowed to come to equilibrium at 30°C?

15.66 At 850°C, K_p is 14.1 for the reaction

$$C(graphite) + CO_2(g) \rightleftharpoons 2CO(g)$$

A reaction vessel at 850°C initially contains CO_2 at 0.500 atm, CO at 1.00 atm, and graphite. Determine (a) the direction in which the reaction moves and (b) the equilibrium partial pressure of each gas.

15.67 Solid NH_4HS is placed in a closed flask at 25°C. The solid dissociates according to the equation

$$NH_4HS(s) \rightleftharpoons NH_3(g) + H_2S(g)$$

and the total pressure increases by 0.66 atm. Find K_p.

15.68 Refer to the NH_4HS equilibrium in Exercise 15.67. If solid NH_4HS is placed into a flask containing H_2S gas at 0.750 atm and 25°C, determine (a) the total pressure at equilibrium and (b) the mole percent of ammonia in the equilibrium mixture.

15.69 Given the following K_p values at 25°C:

$$C(graphite) + \tfrac{1}{2}O_2(g) \rightleftharpoons CO(g) \qquad K_p = 1.13 \times 10^{24}$$

$$C(graphite) + O_2(g) \rightleftharpoons CO_2(g) \qquad K_p = 1.25 \times 10^{69}$$

Find K_p for

$$2CO(g) + O_2(g) \rightleftharpoons 2CO_2(g)$$

15.70 Refer to Exercise 15.69 and calculate K_c for each equilibrium involving graphite.

Le Châtelier's Principle (Section 15.8)

15.71 In your own words, give a general rule for predicting the effect on an equilibrium mixture of (a) concentration changes, (b) volume changes, and (c) temperature changes.

15.72 Consider the general reaction

$$A + B \rightleftharpoons C$$

(a) Use Le Châtelier's principle to predict the effect of (1) increasing concentration of A, (2) decreasing concentration of B, and (3) increasing concentration of C on the equilibrium mixture.

(b) Write the reaction quotient for the above reaction, and use it to explain the effects predicted in Part (a).

15.73 How will the color of the equilibrium mixture

$$\underset{\text{orange}}{Cr_2O_7^{2-}(aq)} + 2OH^-(aq) \rightleftharpoons \underset{\text{yellow}}{2CrO_4^{2-}(aq)} + H_2O(l)$$

be affected by the addition of (a) sodium hydroxide and (b) hydrochloric acid?

15.74 How will the color of the equilibrium mixture

$$\underset{\text{pink}}{Co(H_2O)_6^{2+}(aq)} + 4Cl^-(aq) \rightleftharpoons \underset{\text{blue}}{CoCl_4^{2-}(aq)} + 6H_2O(l) \qquad \Delta H° > 0$$

be affected by (a) the addition of HCl and (b) increasing the temperature?

15.75 For the equilibrium

$$N_2O_4(\text{colorless gas}) \rightleftharpoons 2NO_2\ (\text{reddish-brown gas})$$

the value of K_c increases from 0.90 at 120°C to 3.2 at 150°C. Assume constant volume, and predict the effect of each of the following on the color of the equilibrium mixture:

(a) increasing temperature
(b) increasing concentration of N_2O_4
(c) increasing partial pressure of NO_2
(d) addition of helium gas

15.76 Carbon monoxide reduces nickel oxide to nickel metal:

$$NiO(s) + CO(g) \rightleftharpoons Ni(s) + CO_2(g)$$

The value of K_p for this reduction falls from 4540 at 936 K to 1580 at 1125 K. Predict the effect of each of the following on the equilibrium yield of nickel metal:

(a) increasing the total pressure by compression
(b) increasing the total pressure by addition of helium gas
(c) increasing the temperature
(d) increasing the partial pressure of CO_2
(e) increasing the partial pressure of CO
(f) increasing the mass of NiO

15.77 Refer to the N_2O_4–NO_2 equilibrium in Exercise 15.75. Is the forward reaction exothermic or endothermic? Explain.

15.78 Refer to the NiO–Ni equilibrium in Exercise 15.76. Is the forward reaction exothermic or endothermic? Explain.

15.79 The conversion of sulfur dioxide to sulfur trioxide

$$2SO_2(g) + O_2(g) \rightleftharpoons 2SO_3(g) \qquad \Delta H° = -197\text{ kJ}$$

is an important step in the industrial production of sulfuric acid. How will the equilibrium yield of sulfur trioxide be affected by (a) compression and (b) increasing temperature?

15.80 Methanol (CH_3OH) is an important industrial chemical prepared by the reaction of carbon monoxide with hydrogen:

$$CO(g) + 2H_2(g) \rightleftharpoons CH_3OH(g) \qquad \Delta H° = -90.7\text{ kJ}$$

Suggest four ways in which the equilibrium yield of methanol might be increased.

15.81 Consider the equilibrium

$$I_2(g) + Cl_2(g) \rightleftharpoons 2ICl(g) \qquad \Delta H^\circ = -26.9 \text{ kJ}$$

How will the equilibrium yield of ICl be affected by (a) increasing temperature, (b) increasing partial pressure of Cl_2, (c) compression, and (d) adding a catalyst?

15.82 Consider the equilibrium

$$2H_2S(g) \rightleftharpoons 2H_2(g) + S_2(g)$$

(a) How will the equilibrium yield of S_2 be affected by (1) the addition of hydrogen, (2) the addition of H_2S, and (3) compression.

(b) Use reaction quotients to explain the effects predicted in Part (a).

***15.83** Refer to Exercise 15.75. Estimate ΔH° for the N_2O_4–NO_2 equilibrium.

***15.84** Refer to Exercise 15.76. Estimate ΔH° for the NiO–Ni equilibrium.

***15.85** At 25°C, K_p for

$$N_2(g) + 3H_2(g) \rightleftharpoons 2NH_3(g)$$

is 6.9×10^{-5} and $\Delta H^\circ = -92.2$ kJ. Estimate K_p at 100°C.

***15.86** Refer to Exercise 15.76. Estimate K_p for the NiO–Ni equilibrium at 1025 K.

PART B. MISCELLANEOUS QUESTIONS AND PROBLEMS

15.87 The following equilibrium constants have been reported in aqueous solution at 25°C:

$$Ag(NH_3)_2^+ \rightleftharpoons Ag^+(aq) + 2NH_3(aq) \qquad K_c = 6.8 \times 10^{-8}$$

$$Ag(S_2O_3)_2^{3-}(aq) \rightleftharpoons Ag^+(aq) + 2S_2O_3^{3-}(aq) \qquad K_c = 6.0 \times 10^{-14}$$

(a) Which ion, $Ag(NH_3)_2^+$ or $Ag(S_2O_3)_2^{3-}$, is more stable in aqueous solution?

(b) Which has the greater concentration of free Ag^+ ion, a solution that contains 1.0 *M* $Ag(NH_3)_2^+$ or one that contains 1.0 *M* $Ag(S_2O_3)_2^{3-}$?

15.88 A mixture containing 0.250 mol $SbCl_3$, 0.250 mol Cl_2, and 0.500 mol $SbCl_5$ comes to equilibrium in a 5.00-L reaction vessel at 648°C. The reaction is

$$SbCl_3(g) + Cl_2(g) \rightleftharpoons SbCl_5(g)$$

with $K_c = 40.0$. Determine (a) the direction in which the reaction moves and (b) the equilibrium concentration of each gas.

15.89 Consider the equilibrium

$$CH_3COOH(aq) \rightleftharpoons H^+(aq) + CH_3COO^-(aq)$$

for which $K_c = 1.76 \times 10^{-5}$ at 25°C. A solution is initially 0.50 *M* in CH_3COOH, 0.50 *M* in CH_3COO^-, and 1.0×10^{-7} *M* in H^+.

(a) In which direction will the reaction move?

(b) Find the equilibrium concentration of each species.

15.90 At 25°C, $K_p = 113$ for the reaction

$$2NO(g) + Br_2(g) \rightleftharpoons 2NOBr(g)$$

The Br_2 gas pressure in a reaction vessel is maintained at a constant value of 0.282 atm by having some liquid bromine present. Some NOBr is introduced at an initial partial pressure of 0.150 atm and allowed to come to equilibrium. What are the equilibrium pressures of NO and NOBr?

15.91 Hydrogen can be prepared from carbon monoxide and water at elevated temperatures:

$$CO(g) + H_2O(g) \rightleftharpoons CO_2(g) + H_2(g)$$

For this reaction, $K_p = 0.227$ at 1725°C. Equal volumes of carbon monoxide and water vapor are mixed in a reaction vessel at 1725°C and 25.0 atm total pressure. Calculate the mole percent of H_2 in the equilibrium mixture.

***15.92** For the reaction

$$C_2H_2(g) + 3H_2(g) \rightleftharpoons 2CH_4(g)$$

K_p is 2.41×10^{-4} at 1725°C. Estimate the partial pressure of each gas if CH_4, initially at 0.500 atm, comes to equilibrium at this temperature.

15.93 Consider the reaction

$$2ICl(g) \rightleftharpoons I_2(g) + Cl_2(g)$$

for which $K_c = 0.110$. A 2.00-L vessel initially contains 0.500 mol each of the three gases. Calculate the equilibrium concentration of ICl.

***15.94** The equilibrium constant K_p for the synthesis of ammonia

$$N_2(g) + 3H_2(g) \rightleftharpoons 2NH_3(g)$$

is 5.2×10^{-2} at 250°C. Find the partial pressure of NH_3 if nitrogen and hydrogen, each initially at 1.00 atm partial pressure, come to equilibrium in a reaction vessel maintained at 250°C.

15.95 The vapor pressure of water at 20°C is 17.5 torr. Find K_p for

$$H_2O(l) \rightleftharpoons H_2O(g)$$

15.96 At 25°C, K_p is 6.07×10^5 for

$$H_2(g) + S(s) \rightleftharpoons H_2S(g)$$

A mixture of solid sulfur and H_2S gas at 1.00 atm is placed in an otherwise empty vessel and allowed to come to equilibrium at 25°C. Determine the partial pressure of hydrogen in the final mixture.

15.97 The dissociation of carbonyl chloride is endothermic:

$$COCl_2(g) \rightleftharpoons CO(g) + Cl_2 \qquad \Delta H° = +112\text{ kJ}$$

How will the fraction of carbonyl chloride dissociated be affected by (a) increasing temperature, (b) compression, and (c) increasing partial pressure of Cl_2?

***15.98** Compare the van't Hoff equation (Equation 15.3) with the Clausius–Clapeyron vapor pressure equation (Equation 12.2). Keep in mind that the vapor pressure of a liquid is constant at any given temperature, and explain why the vapor pressure equation is a special case of the van't Hoff equation. (*Hint:* See Exercise 15.95.)

***15.99** Refer to the equilibrium in Exercise 15.93. A 2.00-L reaction vessel contains 1.00 mol of I_2, 0.500 mol of Cl_2, and no ICl. Calculate the equilibrium concentration of ICl.

CHEMISTS AT WORK

THE LIAISON

If you examine the responsibilities and day-to-day tasks of some chemists, it would appear that these people are not really chemists at all. Yet they provide unique and vital services in many varied endeavors by using their scientific knowledge, skills, problem-solving abilities, and experiences. John Higuchi, a 1989 graduate of Hope College, is one of these chemists.

After John's sophomore year in high school, he participated in a summer program at the University of Utah, studying stoichiometry and inorganic chemistry. His experiences in this program led first to a part-time job in high school and then to summer jobs at the Center for Engineering Design at the University in college. He worked first at simple tasks such as maintaining laboratories and determining viscosities of materials, then at more responsible activities such as research into the design and operation of a system of delivering medications transdermally via electrical current, and finally in the management of the manufacture of these devices.

"In college, life became complicated," he says. "Because I had AP [Advanced Placement*] credit, I had the opportunity to explore a variety of courses. A political science professor suggested that there was a great deal that I could learn from a semester's internship in Washington, and I accepted the invitation. That semester really opened my eyes." In Washington, John had two internships. The first was in the College Education Office of the American Association for the Advancement of Science, and the second was with the U.S. Senate Committee on Labor and Human Resources. His duties included developing a questionnaire on decision-making processes in colleges and universities, general office work, and observing processes related to his work. "I learned how federal policy decisions to spend large amounts of money are made," John says, "and that scientists play a major role in informing the public about issues that affect our everyday lives." Senator Orrin Hatch of Utah told him that it was important for people with science backgrounds to be involved in policy decisions because of the increasing complexity of many issues. After his semester in Washington, John's perspective about career options changed because he realized there was much more to do in and with science.

JOHN W. HIGUCHI

After graduation, his first two positions were with public policy groups in Washington. In one, John had to describe health-care issues related to children and young people; in the other, he studied how county governments were dealing with health-care problems. During his nearly 2 years' tenure at these organizations, he sharpened his skills in oral and written communication, research, building consensus, and politics.

In 1991, John joined the Office of Government Relations and Science Policy at the American Chemical Society (ACS). He worked as a liaison with both the membership and leadership of the ACS in communicating to Congress and various agencies issues of importance to scientific professionals. In 1992, he became manager of the ACS Student Affiliate Office where he is responsible for assisting in the delivery of services and programs to undergraduates across the United States. These include publishing the magazine *In Chemistry,* various newsletters, and marketing materials; arranging programs for students at national and regional ACS meetings; representing student needs to the society and to the profession; and being the liaison between students and the profession.

"I am grateful for the many mentors during my career," he says. "One thing that I have learned is that you must stretch yourself. For example, your first major presentation is the one that is most intimidating. But after you succeed once, the next one is easier. Everyone should take the time to explore options because you never know how successful you can be."

*A program elected in high school wherein certain courses can earn college credit.

CHAPTER 16

Katsura Palace Garden, Japan. The survival of a healthy ecological system is strongly dependent on the pH of soil and water. The azaleas in this garden require a somewhat acidic soil for best growth.

ACIDS AND BASES

OUTLINE

PREVIEW

Acids and bases have been used since ancient times in medicine, cookery, cleaning, and technology. Today's most widely used industrial chemical is sulfuric acid (H_2SO_4); the top 10 chemicals also include phosphoric acid (H_3PO_4) and three bases—ammonia (NH_3), lime (CaO), and sodium hydroxide (NaOH). We use vinegar in cooking, and we enjoy the acid taste of oranges, soda pop, and pickles. Household and heavy-duty cleansers contain basic (alkaline) substances such as ammonia, strong soap, sodium carbonate, and lye.

Acids and bases have long been recognized by their behavior toward other substances (Table 16.1). In 1884 Svante Arrhenius formally defined an **acid** to be any substance that produces $H^+(aq)$ ions in aqueous solution, a **base** to be any substance that produces $OH^-(aq)$ ions, and **neutralization** to be the reaction of $H^+(aq)$ ions with $OH^-(aq)$ ions to form water. The Arrhenius definitions describe most aspects of acid–base behavior in aqueous solutions, but they are not broad enough to include anhydrous neutralizations such as

$$NH_3(g) + HCl(g) \rightarrow NH_4Cl(s) \quad (= NH_4^+ + Cl^-)$$

This reaction (see Figure 16.1), which produces the same salt as the reaction between aqueous solutions of NH_3 and HCl, illustrates the need for acid–base definitions that apply to nonaqueous as well as aqueous systems.

LEARNING HINT
Before studying this chapter, review the discussion of acids and bases in Section 4.2.

16.1 BRØNSTED–LOWRY ACIDS AND BASES

Johannes Brønsted, a Danish chemist, and Thomas Lowry, an English chemist, each concluded independently that the transfer of a proton (H^+, a hydrogen ion) is the central process in an acid–base reaction. They published the following definitions in 1923: An **acid** is a substance that donates protons, a **base** is a substance that accepts

TABLE 16.1 Properties of Acids and Bases

Acids	Bases
Neutralize bases	Neutralize acids
Taste sour	Taste bitter
Burn the skin if strong	Feel soapy; burn skin if strong
Turn litmus red[a]	Turn litmus blue
Furnish $H^+(aq)$	Furnish $OH^-(aq)$
Dissolve active metals	Dissolve certain metals[b]
Dissolve carbonates	React with fats to form soaps[c]

[a]Litmus and other colored acid–base indicators are discussed in Chapter 17.
[b]The solubility of metals in bases is discussed in Chapter 18.
[c]The reaction of strong bases with fat is discussed in Chapter 23.

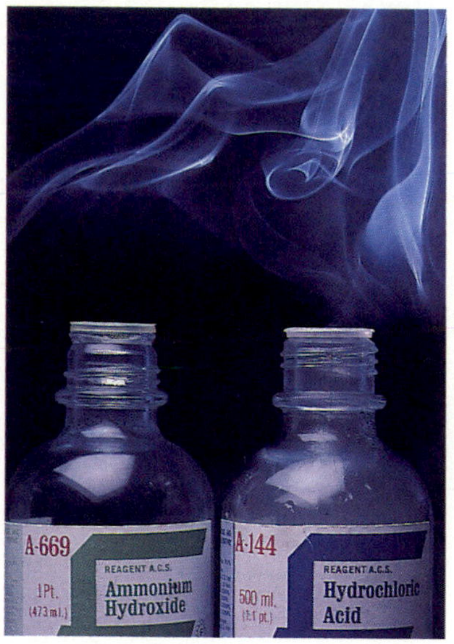

Figure 16.1
Ammonia and hydrogen chloride vapors rising from their concentrated solutions form a fog of ammonium chloride crystals.

protons, and **neutralization** is proton transfer from acid to base. Consider the reaction between NH_3 and HCl gases in terms of these definitions. When the reaction equation is written with Lewis structures

$$\underset{\text{base}}{\mathrm{H{-}\overset{H}{\underset{H}{N}}{:}}} + \underset{\text{acid}}{\mathrm{H{-}\ddot{\underset{..}{Cl}}{:}}} \longrightarrow \left[\mathrm{H{-}\overset{H}{\underset{H}{N}}{-}H}\right]^+ + \mathrm{:\ddot{\underset{..}{Cl}}{:}^-}$$

it shows that a proton is transferred from the HCl molecule to the NH_3 molecule. Hence, HCl is a Brønsted acid, NH_3 is a Brønsted base, and the reaction is a neutralization. Another example of proton transfer is the reaction of hydrogen cyanide gas with solid sodium hydroxide. The products are sodium cyanide and water vapor:

$$HCN(g) + NaOH(s) \rightarrow NaCN(s) + H_2O(g)$$

The Na^+ ion does not change; it is a spectator ion present in both solids. The net ionic equation

$$\underset{\text{acid}}{\mathrm{H{-}CN}} + \underset{\text{base}}{\mathrm{:\ddot{\underset{..}{O}}{-}H^-}} \longrightarrow CN^- + H_2O(g)$$

shows that a proton is transferred from HCN, the Brønsted acid, to OH^-, the Brønsted base. (NaOH is also called a base because it supplies OH^- ions.)

A Brønsted acid is a proton donor; hence, *any Brønsted acid must contain at least one hydrogen atom.* Familiar acids such as HCl, H_2SO_4, and CH_3COOH are Brønsted acids. Ions and molecules that donate protons to strong bases are also Brønsted acids. Some examples are ammonium ion (NH_4^+), hydrogen carbonate ion (HCO_3^-), and water:

$$\left[\begin{array}{c} \mathrm{H} \\ | \\ \mathrm{H{-}N{-}H} \\ | \\ \mathrm{H} \end{array}\right]^+ \qquad \left[\mathrm{H{-}\ddot{\underset{..}{O}}{-}\overset{\overset{:O:}{\|}}{C}{-}\ddot{\underset{..}{O}}:}\right]^- \qquad \begin{array}{c}\mathrm{:\ddot{O}{-}H} \\ | \\ \mathrm{H}\end{array}$$

ammonium ion — hydrogen carbonate ion — water

A Brønsted base accepts protons. The best known Brønsted base is the hydroxide ion found in hydroxide bases such as NaOH and $Ca(OH)_2$, but many other species such as ammonia (NH_3), cyanide ion (CN^-), and acetate ion (CH_3COO^-) also accept protons:

$$\begin{array}{c} \mathrm{H} \\ | \\ \mathrm{H{-}\underset{..}{N}{-}H} \end{array} \qquad [:\mathrm{C{\equiv}N}:]^- \qquad \left[\mathrm{CH_3{-}\overset{\overset{:O:}{\|}}{C}{-}\ddot{\underset{..}{O}}:}\right]^-$$

ammonia — cyanide ion — acetate ion

Every Brønsted base has at least one lone pair of electrons that is used to form a bond with the captured proton.

LEARNING HINT

Draw Lewis structures—they can help you to identify acidic hydrogen atoms and lone electron pairs.

Ammonia and various amines make up an important group of Brønsted bases. An ***amine*** is a weak base related to ammonia by the substitution of a hydrocarbon group for one of the hydrogen atoms. Examples are methylamine (CH_3NH_2) and aniline ($C_6H_5NH_2$):

$$\begin{array}{c} \mathrm{H{-}\ddot{N}{-}H} \\ | \\ \mathrm{H} \end{array} \qquad \begin{array}{c} \mathrm{H{-}\ddot{N}{-}CH_3} \\ | \\ \mathrm{H} \end{array} \qquad \begin{array}{c} \mathrm{H{-}\ddot{N}{-}C_6H_5} \\ | \\ \mathrm{H} \end{array}$$

ammonia — methylamine — aniline

Like ammonia, amines have a lone pair of electrons on the nitrogen atom. Their reactions with water and other Brønsted acids are similar to those of ammonia:

$$\underset{\text{base}}{\begin{array}{c} \mathrm{H} \\ | \\ \mathrm{H{-}N{:}} \\ | \\ \mathrm{H} \end{array}}(g) + \underset{\text{acid}}{\mathrm{:\ddot{O}{-}H}}\ (l) \rightleftharpoons NH_4^+(aq) + OH^-(aq)$$

$$\underset{\text{base}}{\begin{array}{c} \mathrm{H} \\ | \\ \mathrm{CH_3{-}N{:}} \\ | \\ \mathrm{H} \end{array}}(l) + \underset{\text{acid}}{\mathrm{:\ddot{O}{-}H}}\ (l) \rightleftharpoons CH_3NH_3^+(aq) + OH^-(aq)$$

The Hydronium Ion

Experimental evidence shows that the small, positively charged proton is hydrated

Mass spectrometer studies have shown that in the gas phase, protonated water clusters contain hydronium ions surrounded by cagelike structures of 20 water molecules held together by hydrogen bonds.

in aqueous solution; it combines with a water molecule to form a **hydronium ion** (H_3O^+) (Figure 16.2). This trigonal pyramidal ion, which contains three covalent bonds, also exists in the gas phase and in the crystal lattice of ionic solids such as $H_3O^+ClO_4^-$ and $H_3O^+NO_3^-$. In solution, the hydronium ion is further hydrated; that is, it is hydrogen bonded to several water molecules. Most chemists use the symbols $H^+(aq)$ and $H_3O^+(aq)$ interchangeably to represent the proton in solution. The ionization of HCl, for example, is written as either

$$HCl(aq) \rightarrow H^+(aq) + Cl^-(aq)$$

or

$$HCl(aq) + H_2O(l) \rightarrow H_3O^+(aq) + Cl^-(aq)$$

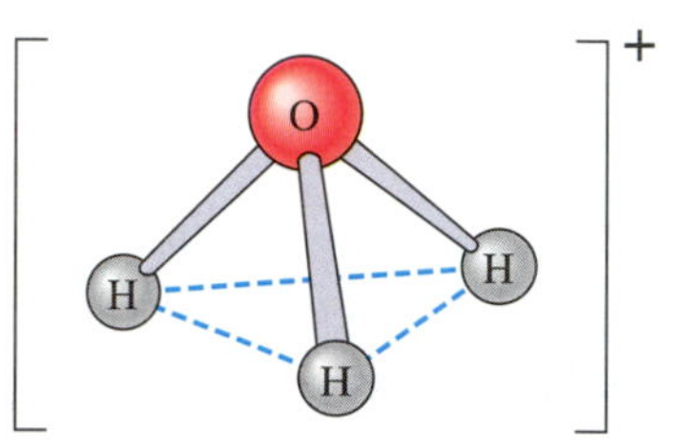

Figure 16.2

The combination of a hydrogen ion (proton) and a water molecule produces the pyramidal hydronium ion, H_3O^+.

The latter equation has the advantage of showing that the ionization of HCl in water is actually a proton transfer reaction with water acting as the Brønsted base. We will use the hydronium ion formula whenever it is important to emphasize the role played by water molecules in acid–base chemistry.

PRACTICE EXERCISE 16.1

Rewrite the following equations using hydronium ion notation:

(a) $CH_3COOH(aq) \rightleftharpoons H^+(aq) + CH_3COO^-(aq)$

(b) $H^+(aq) + OH^-(aq) \rightleftharpoons H_2O(l)$

Amphiprotic Molecules and Ions

Many molecules and ions, such as H_2O, HCO_3^-, and NH_3, contain both hydrogen atoms and lone electron pairs:

```
 ..                      :O:                     ..
:O—H                 [    ||      ]−          H—N—H
 |                   [ H—O—C—O:   ]              |
 H                   [   ..  ..   ]              H
                          ..
water              hydrogen carbonate ion      ammonia
```

Amphi-: from the Greek *ampho,* meaning "both." Amphiprotism is a special case of a more general phenomenon known as "amphoterism" (Chapter 18).

Such species are **amphiprotic**—they can accept or donate protons. An amphiprotic substance acts as a base toward stronger acids and as an acid toward stronger bases. We have already seen how water accepts protons from acids such as acetic acid:

$$\underset{\text{acid}}{CH_3COOH(aq)} + \underset{\text{base}}{H_2O(l)} \rightleftharpoons H_3O^+(aq) + CH_3COO^-(aq)$$

and gives up protons to bases such as ammonia

$$\underset{\text{base}}{NH_3(aq)} + \underset{\text{acid}}{H_2O(l)} \rightleftharpoons NH_4^+(aq) + OH^-(aq)$$

The hydrogen carbonate ion (HCO_3^-) reacts with strong bases to produce carbonate ion

$$\underset{\text{acid}}{HCO_3^-(aq)} + \underset{\text{base}}{OH^-(aq)} \rightarrow CO_3^{2-}(aq) + H_2O(l)$$

and with strong acids to liberate CO_2

$$\underset{\text{base}}{HCO_3^-(aq)} + \underset{\text{acid}}{HCl(aq)} \longrightarrow [H_2CO_3](aq) + Cl^-(aq)$$
$$[H_2CO_3] \longrightarrow H_2O(l) + CO_2(g)$$

The bracketed formula for carbonic acid indicates that it is an unstable molecule.

In some cases, the acidic or basic character of an amphiprotic substance seems to predominate. Ammonia, for example, accepts protons from water, so it is more familiar as a base (see the equation given above). A very strong base such as sodium hydride (NaH), however, will remove protons from ammonia, leaving the amide ion (NH_2^-):

$$\underset{\text{acid}}{NH_3(l)} + \underset{\text{base}}{H^-} \rightarrow \underset{\text{amide ion}}{NH_2^-} + H_2(g)$$

Acid and base strengths are discussed in the next section.

Since the hydride ion reacts with water (Section S1.1), this reaction must be carried out in anhydrous liquid ammonia.

Amphiprotic substances also undergo **autoionization** (self-ionization) reactions in which a proton is transferred from one molecule of the substance to another molecule of the same substance. Liquid water, for example, always contains small concentrations of hydronium and hydroxide ions:

$$\underset{\text{base}}{H_2O(l)} + \underset{\text{acid}}{H_2O(l)} \rightleftharpoons H_3O^+(aq) + OH^-(aq)$$

Pure water, although essentially covalent, is a feeble conductor of electricity because of the presence of these ions. Many liquids are amphiprotic and undergo similar self-ionizations. The autoionization reactions of anhydrous liquid ammonia and liquid ethanol are as follows:

Ammonia is a gas at room temperature but condenses to a liquid at −33°C.

$$\underset{\text{base}}{NH_3(l)} + \underset{\text{acid}}{NH_3(l)} \rightleftharpoons NH_4^+ + NH_2^-$$

$$\underset{\text{base}}{C_2H_5{-}\ddot{O}{-}H\ (l)} + \underset{\text{acid}}{C_2H_5{-}\ddot{O}{-}H\ (l)} \rightleftharpoons C_2H_5OH_2^+ + C_2H_5O^-$$

PRACTICE EXERCISE 16.2

Write an equation for the autoionization reaction that occurs in liquid methylamine (CH_3NH_2).

CHEMICAL INSIGHT

HOW HYDROGEN AND HYDROXIDE IONS TRAVEL THROUGH WATER

In aqueous solution, all ions are hydrated to some extent. Layers of associated water molecules surround the ions so that they all have approximately the same size, and most ions tend to migrate through water with about the same rate of speed. Experiments show, however, that $H^+(aq)$ and $OH^-(aq)$ seem to move many times faster than other ions. Size differences are not the cause: $H^+(aq)$, for example, has about the same effective diameter as $K^+(aq)$, but $H^+(aq)$ moves almost five times more rapidly than $K^+(aq)$. These unusually high speeds play an important role in biological systems and are believed to be the result of protons "jumping" from one water molecule to another. Figure 16.3 shows that when a proton travels through a solution, there is a sort of relay race in which each successive water molecule accepts one proton and gives up another. The same effect occurs in the solid state—protons travel about 50 times faster in ice than in liquid water. Random molecular motion in a liquid tends to interfere with proton jumps, and the increased mobility of protons in ice is believed to be a consequence of its more rigid hydrogen-bonded structure.

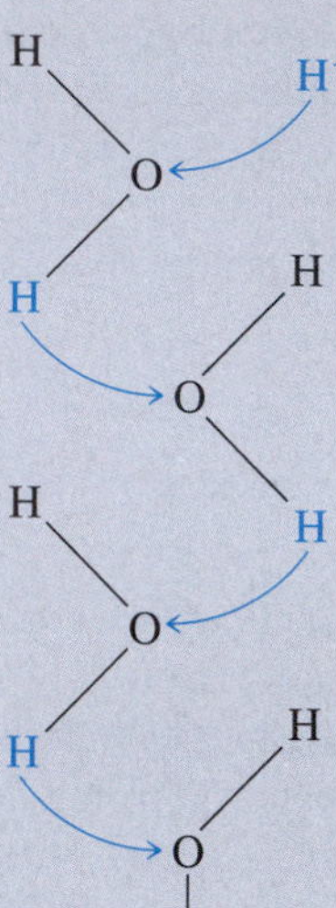

Figure 16.3
The apparent speed with which a proton travels through water is the result of rapid proton transfers from one water molecule to another.

16.2 CONJUGATE ACID–BASE PAIRS

When a Brønsted acid loses a proton, it forms a Brønsted base. For example, HCN forms CN^-:

$$\underset{\text{acid}}{HCN} \underset{+H^+}{\overset{-H^+}{\rightleftharpoons}} \underset{\text{base}}{CN^-}$$

Conjugate means "coupled" or "joined together" (as in a pair).

Conversely, when a Brønsted base accepts a proton, it becomes an acid. We call HCN and CN^- a **conjugate acid–base pair**, an acid and a base whose formulas differ by a single proton: HCN is the acid; CN^- is the base. Other conjugate acid–base pairs are NH_4^+ and NH_3, H_2O and OH^-, and HSO_4^- and SO_4^{2-}. More examples are given in Table 16.2.

PRACTICE EXERCISE 16.3

Write formulas for (a) the conjugate acid of HPO_4^{2-} and (b) the conjugate base of OH^-.

TABLE 16.2 Conjugate Acid–Base Pairs

	Acids		Bases		
Strongest (ACID STRENGTH ↑)				Weakest (BASE STRENGTH ↓)	
Strong Acids (Stronger than H_3O^+)	Perchloric acid	$HClO_4$	ClO_4^-	Perchlorate ion	Very Weak Bases (Weaker than H_2O)
	Nitric acid	HNO_3	NO_3^-	Nitrate ion	
	Hydrochloric acid	HCl	Cl^-	Chloride ion	
	Sulfuric acid	H_2SO_4	HSO_4^-	Hydrogen sulfate ion	
	Hydronium ion	H_3O^+	H_2O	Water	
Weak Acids	Hydrogen sulfate ion	HSO_4^-	SO_4^{2-}	Sulfate ion	Weak Bases
	Phosphoric acid	H_3PO_4	$H_2PO_4^-$	Dihydrogen phosphate ion	
	Formic acid	$HCOOH$	$HCOO^-$	Formate ion	
	Nitrous acid	HNO_2	NO_2^-	Nitrite ion	
	Hydrofluoric acid	HF	F^-	Fluoride ion	
	Acetic acid	CH_3COOH	CH_3COO^-	Acetate ion	
	Carbonic acid	H_2CO_3	HCO_3^-	Hydrogen carbonate ion	
	Hydrogen sulfide	H_2S	HS^-	Hydrogen sulfide ion	
	Dihydrogen phosphate ion	$H_2PO_4^-$	HPO_4^{2-}	Hydrogen phosphate ion	
	Ammonium ion	NH_4^+	NH_3	Ammonia	
	Hydrogen cyanide	HCN	CN^-	Cyanide ion	
	Hydrogen carbonate ion	HCO_3^-	CO_3^{2-}	Carbonate ion	
	Methylammonium ion	$CH_3NH_3^+$	CH_3NH_2	Methylamine	
	Hydrogen phosphate ion	HPO_4^{2-}	PO_4^{3-}	Phosphate ion	
	Water	H_2O	OH^-	Hydroxide ion	
Very Weak Acids (Weaker than H_2O)	Hydrogen sulfide ion	HS^-	S^{2-}	Sulfide ion	Strong Bases (Stronger than OH^-)
	Ammonia	NH_3	NH_2^-	Amide ion	
	Hydrogen	H_2	H^-	Hydride ion	
	Hydroxide ion	OH^-	O^{2-}	Oxide ion	
	Methane	CH_4	CH_3^-	Methide ion	
Weakest				Strongest	

In the reaction between NH_3 and HCl

$$\underset{\text{base}_1}{NH_3(g)} + \underset{\text{acid}_2}{HCl(g)} \rightarrow \underset{\text{acid}_1}{NH_4^+} + \underset{\text{base}_2}{Cl^-}$$

NH_4^+ and NH_3 are a conjugate acid–base pair designated by the subscript 1. The base, NH_3, accepts a proton and forms its conjugate acid, NH_4^+. Another conjugate acid–base pair, HCl and Cl^-, is designated by the subscript 2. The acid, HCl, gives up a proton and forms its conjugate base, Cl^-.

EXAMPLE 16.1

Identify the conjugate acid–base pairs in the reaction between hydrogen bromide and methylamine:

$$CH_3NH_2(l) + HBr(g) \rightarrow CH_3NH_3Br(s) \quad (= CH_3NH_3^+ + Br^-)$$

SOLUTION

HBr gives up a proton and is an acid; its conjugate base is Br^-. CH_3NH_2 accepts a proton and is a base; its conjugate acid is $CH_3NH_3^+$.

PRACTICE EXERCISE 16.4

Identify the conjugate acid–base pairs in the following reaction:

$$HC_2O_4^-(aq) + HS^-(aq) \rightarrow C_2O_4^{2-}(aq) + H_2S(g)$$

Acid and Base Strengths

Acid strength refers to the ability of an acid to relinquish protons; a strong acid gives up protons more easily than a weak acid. **Base strength** refers to the ability of a base to accept protons; a strong base accepts protons more readily than a weak base. The apparent strength of an acid (or base) depends to some extent on the base (or acid) with which it reacts. Formic acid, for example, transfers a limited number of protons to water and appears to be a weak acid in aqueous solution:

$$\underset{\text{formic acid}}{HCOOH(aq)} + \underset{\substack{\text{weak}\\\text{base}}}{H_2O(l)} \rightleftharpoons HCOO^-(aq) + H_3O^+(aq)$$

But with a stronger base such as sodium hydroxide, formic acid donates all of its protons and appears to be a strong acid:

$$\underset{\text{formic acid}}{HCOOH(aq)} + \underset{\substack{\text{strong}\\\text{base}}}{OH^-(aq)} \rightarrow HCOO^-(aq) + H_2O(l)$$

The strengths of two acids can be compared only by testing the acids with the same base. The comparison is usually made in aqueous solution where the base is water. A strong acid will ionize to a greater extent and produce more hydronium ions per mole than a weak acid. We know, for example, that formic acid is stronger than acetic acid because the concentration of hydronium ions in 0.1 *M* formic acid is almost four times greater than in 0.1 *M* acetic acid.

If an acid is strong and easily loses protons, then its conjugate base does not readily accept protons. Hence, the strength of an acid is inversely related to the strength of its conjugate base. An examination of Table 16.2 shows that *the stronger the acid, the weaker the conjugate base; the stronger the base, the weaker the conjugate acid.* Formic acid is a stronger acid than acetic acid, so formate ion ($HCOO^-$) is a weaker base than acetate ion (CH_3COO^-).

Acids above hydronium ion in Table 16.2 are traditionally referred to as **strong acids**. They are almost completely ionized in aqueous solution. Nitric acid, for example, forms hydronium and nitrate ions:

$$HNO_3(l) + H_2O(l) \rightarrow H_3O^+(aq) + NO_3^-(aq)$$

This reaction goes virtually to completion, so that $H_3O^+(aq)$ becomes the only active acid remaining in the solution. All strong acids react completely with water to form $H_3O^+(aq)$; hence, *hydronium ion is the strongest acid that can exist in an aqueous solution.* Water is said to exert a **leveling effect** on strong acids because they are completely ionized and thus appear to be of equal strength.

Water also exerts a leveling effect on **strong bases**, those below hydroxide ion in Table 16.2. Strong bases appear to be equally strong in aqueous solution because they take protons from water, leaving hydroxide ion as the only active base. Hydride ions, for example, take protons from water to form hydrogen gas and hydroxide ions:

$$H^- + H_2O(l) \rightarrow H_2(g) + OH^-(aq)$$

Such reactions show that *hydroxide ion is the strongest base that can exist in aqueous solution.*

PRACTICE EXERCISE 16.5

Write equations for (a) the ionization of perchloric acid in water and (b) the reaction of amide ion with water.

DIGGING DEEPER

Solvents Other Than Water

All amphiprotic solvents exert a leveling effect. Liquid ammonia, for example, self-ionizes according to the equation

$$2NH_3(l) \rightleftharpoons NH_4^+ + NH_2^-$$

If liquid ammonia is used as a solvent, all acids stronger than ammonium ion (NH_4^+) will appear to have the same strength, and so will all bases stronger than amide ion (NH_2^-). *The strongest acid that can exist in any amphiprotic solvent is the conjugate acid of the solvent; the strongest base that can exist is the conjugate base of the solvent.*

EXAMPLE 16.2

Identify the strongest acid and the strongest base than can exist in liquid methanol (CH_3OH).

SOLUTION

Liquid methanol self-ionizes according to the equation:

$$2CH_3OH(l) \rightleftharpoons CH_3OH_2^+ + CH_3O^-$$

The strongest acid that can exist in this liquid is $CH_3OH_2^+$; the strongest base is CH_3O^-.

PRACTICE EXERCISE 16.6

Identify (a) the strongest base that can exist in liquid hydrogen sulfide (H_2S) and (b) the strongest acid that can exist in liquid methylamine (CH_3NH_2).

A solvent that can distinguish differences in acid and base strength is called a **differentiating solvent**. Water is a differentiating solvent for the weak acids and weak bases that lie between hydronium ion and hydroxide ion in Table 16.2. For example, equimolar aqueous solutions of formic acid, nitrous acid, acetic acid, and hydrocyanic acid contain different concentrations of $H_3O^+(aq)$, thus showing that these acids differ in their ability to transfer protons to water. Acids stronger than $H_3O^+(aq)$ are completely ionized in aqueous solution, so water cannot be used to differentiate between them. Neither can water be used to differentiate between bases stronger than $OH^-(aq)$.

Very strong acids can be differentiated by dissolving them in each other. The weaker one acts as a base and accepts protons from the stronger one. Perchloric acid is placed at the top of Table 16.2 because it *protonates* (gives protons to) all other acids in the table:

$H_2NO_3^+$

$$HClO_4(l) + HNO_3(l) \rightleftharpoons H_2NO_3^+ + ClO_4^-$$

$$HClO_4(l) + HCl(g) \rightleftharpoons H_2Cl^+ + ClO_4^-$$

H_2Cl^+

Strong acids can also be differentiated by using a solvent that is not as easily protonated as water. Liquid acetic acid, for example, is protonated by both HCl and H_2SO_4:

$$HCl(l) + CH_3COOH(l) \rightleftharpoons CH_3COOH_2^+ + Cl^-$$

$$H_2SO_4(l) + CH_3COOH(l) \rightleftharpoons CH_3COOH_2^+ + HSO_4^-$$

Acetic acid is more extensively protonated by HCl than by H_2SO_4; thus HCl is the stronger acid.

$CH_3COOH_2^+$

16.3 OTHER PROTON TRANSFER REACTIONS

Proton transfer reactions are as prevalent and diverse as electron transfer reactions (Chapter 11) and are equally important. The following types of proton transfer reactions are all acid–base reactions in the Brønsted–Lowry sense. They will be considered in more detail in Chapter 17.

1. *The ionization of weak acids and weak bases in water:* The weak acids and weak bases in the central portion of Table 16.2 are only partially ionized in aqueous solution. Each ionization produces two conjugate acid–base pairs that are in equilibrium with each other. The ionizations of HCN, a weak acid, and NH_3, a weak base, are examples:

$$HCN(aq) + H_2O(l) \rightleftharpoons H_3O^+(aq) + CN^-(aq)$$

$$NH_3(aq) + H_2O(l) \rightleftharpoons NH_4^+(aq) + OH^-(aq)$$

2. *The reaction of strong acids with hydroxide bases in aqueous solution:* Strong acids, and hydroxide bases such as NaOH or $Ca(OH)_2$, are completely ionized in water. Solutions of strong acids contain hydronium ions while solutions of hydroxide bases contain hydroxide ions. The neutralization reaction is always

$$H_3O^+(aq) + OH^-(aq) \rightarrow 2H_2O(l)$$

In this reaction, the hydronium ion gives up a proton to the hydroxide ion:

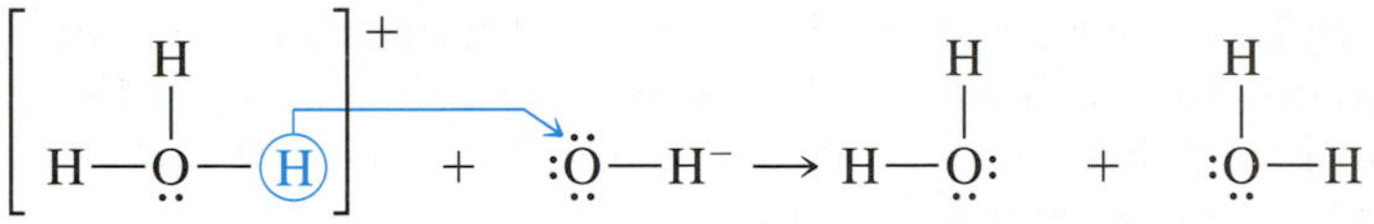

$$[H_3O]^+ + {:}\ddot{O}{-}H^- \rightarrow H{-}\ddot{O}{:}(H) + {:}\ddot{O}{-}H(H)$$

3. *The reactions of weak acids with hydroxide bases in aqueous solution:* The weak acid gives protons directly to the OH^- ions of hydroxide bases. The reaction of acetic acid with any hydroxide base is an example:

$$CH_3COOH(aq) + OH^-(aq) \longrightarrow CH_3COO^-(aq) + H_2O(l)$$

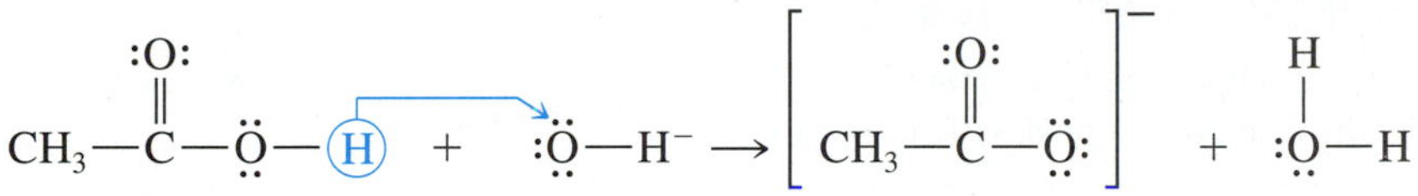

$$CH_3{-}C({=}O){-}\ddot{O}{-}H + {:}\ddot{O}{-}H^- \longrightarrow [CH_3{-}C({=}O){-}\ddot{O}{:}]^- + {:}\ddot{O}{-}H(H)$$

4. *The reactions of weak bases with strong acids in aqueous solution:* The weak base accepts protons directly from the H_3O^+ ions produced by the strong acid. The reaction of ammonia with any strong acid is an example:

$$NH_3(aq) + H_3O^+(aq) \longrightarrow NH_4^+(aq) + H_2O(l)$$

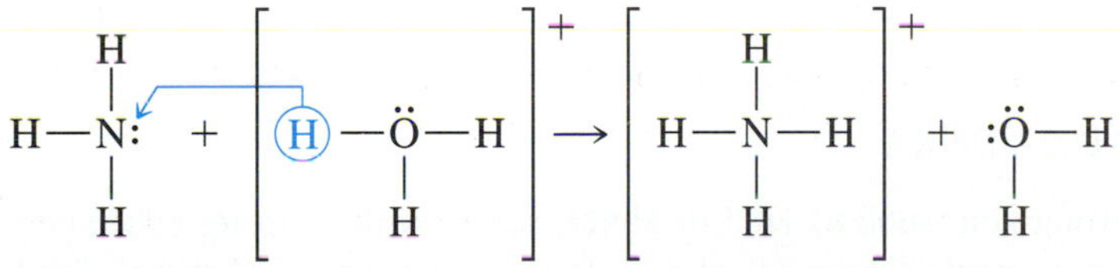

$$H{-}\ddot{N}(H)_2 + [H{-}\ddot{O}{-}H(H)]^+ \longrightarrow [H{-}N(H)_2{-}H]^+ + {:}\ddot{O}{-}H(H)$$

PRACTICE EXERCISE 16.7

Write net ionic equations for the following reactions in aqueous solution: (a) HF with NaOH and (b) NaCN with HCl.

5. *The reactions of weak acids with weak bases in aqueous solution:* The weak base accepts protons directly from the weak acid. The reaction of acetic acid with cyanide ion is an example:

$$CH_3COOH(aq) + CN^-(aq) \rightleftharpoons CH_3COO^-(aq) + HCN(aq)$$

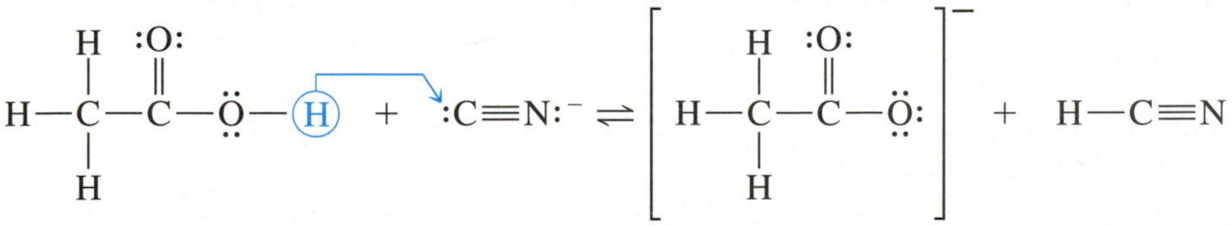

$$H{-}C(H)_2{-}C({=}O){-}\ddot{O}{-}H + {:}C{\equiv}N{:}^- \rightleftharpoons [H{-}C(H)_2{-}C({=}O){-}\ddot{O}{:}]^- + H{-}C{\equiv}N$$

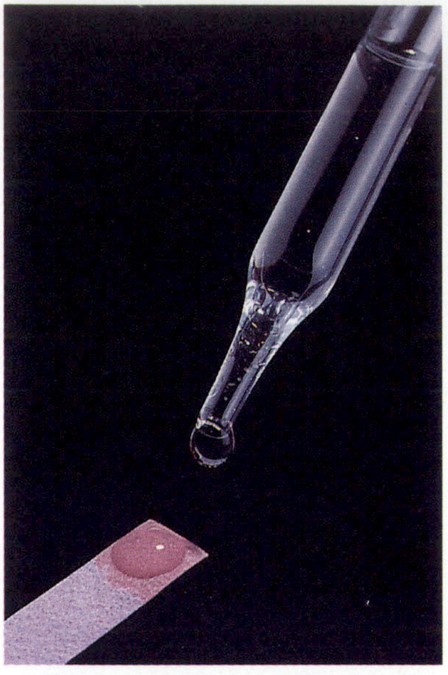

Figure 16.4

A drop of $Al_2(SO_4)_3$ solution turns blue litmus red, showing that the solution is acidic.

PRACTICE EXERCISE 16.8

Write net ionic equations for the following reactions in aqueous solution: (a) HF with NaCN and (b) NH_3 with CH_3COOH.

6. The reactions of *hydrated ions in aqueous solution:* Certain metal ions behave as acids when they are in aqueous solution. Dissolved salts of Al^{3+}, Cu^{2+}, and Fe^{3+}, for example, exhibit all of the traditional acid properties; they turn litmus red, dissolve iron nails, and release CO_2 from carbonates (Figure 16.4). In each case, the acid is a hydrated cation such as $Cu(H_2O)_4{}^{2+}$ or $Fe(H_2O)_6{}^{3+}$ that transfers protons to water molecules:

$$Fe(H_2O)_6{}^{3+}(aq) + H_2O(l) \rightleftharpoons Fe(H_2O)_5(OH)^{2+}(aq) + H_3O^+(aq)$$

We will discuss the acidity of hydrated metal ions in more detail in Section 16.5.

PRACTICE EXERCISE 16.9

Aqueous copper sulfate contains $Cu(H_2O)_4{}^{2+}$ ions. Write an equation that accounts for the acidity of copper sulfate solutions.

7. *Nonaqueous proton transfer reactions:* The gas phase reaction between NH_3 and HCl (see p. 739) is an example of a proton transfer reaction that occurs in the absence of water. Another example is the reaction between liquid aniline ($C_6H_5NH_2$)and hydrogen bromide gas:

$$C_6H_5NH_2(l) + HBr(g) \rightarrow C_6H_5NH_3Br(s) \quad (= C_6H_5NH_3{}^+ + Br^-)$$

Heats of Neutralization

When a strong acid such as HCl or HNO_3 reacts with a strong base such as NaOH, the enthalpy of neutralization is always found to be −55.8 kJ per mole of $H_3O^+(aq)$ or $OH^-(aq)$ neutralized. This observation supports our assumption that all strong acid–strong base neutralizations have the same net ionic equation, namely:

$$H_3O^+(aq) + OH^-(aq) \rightarrow 2H_2O(l) \qquad \Delta H^\circ = -55.8 \text{ kJ}$$

The other ions present during such neutralizations are simply spectator ions.

The enthalpies of neutralization of weak acids and weak bases are all different, showing that each such neutralization is unique. For example, the enthalpy of neutralization of acetic acid by a hydroxide base

$$CH_3COOH(aq) + OH^-(aq) \rightarrow CH_3COO^-(aq) + H_2O(l) \qquad \Delta H^\circ = -56.1 \text{ kJ}$$

is the sum of the enthalpy changes for the following steps:

$$CH_3COOH(aq) + H_2O(l) \rightarrow CH_3COO^-(aq) + H_3O^+(aq) \qquad \Delta H^\circ = -0.3 \text{ kJ}$$

$$H_3O^+(aq) + OH^-(aq) \rightarrow 2H_2O(l) \qquad \Delta H^\circ = -55.8 \text{ kJ}$$

If an acid other than acetic acid is neutralized, then the enthalpy change for the first step would be different.

Polyprotic Acids

A **polyprotic acid** has more than one acidic hydrogen atom per molecule. Sulfuric acid, for example, has two such hydrogen atoms and is a **diprotic acid**. Its ionization occurs in two steps:

$$H_2SO_4(aq) + H_2O(l) \rightarrow H_3O^+(aq) + HSO_4^-(aq)$$

$$HSO_4^-(aq) + H_2O(l) \rightleftharpoons H_3O^+(aq) + SO_4^{2-}(aq)$$

Note that H_2SO_4 reacts completely with water and is thus a strong acid. The HSO_4^- ion, however, is a weak acid.

Carbon dioxide, a normal component of the atmosphere and an essential participant in the biological cycle, dissolves in water to form small concentrations of diprotic carbonic acid (H_2CO_3):

$$CO_2(g) + H_2O(l) \rightleftharpoons H_2CO_3(aq)$$

Carbonic acid reacts with water to form HCO_3^- and CO_3^{2-}:

$$H_2CO_3(aq) + H_2O(l) \rightleftharpoons H_3O^+(aq) + HCO_3^-(aq)$$

$$HCO_3^-(aq) + H_2O(l) \rightleftharpoons H_3O^+(aq) + CO_3^{2-}(aq)$$

The slightly sour taste of seltzer and other carbonated beverages is due to the hydronium ions produced during these ionizations.

PRACTICE EXERCISE 16.10

Phosphoric acid (H_3PO_4) is a weak triprotic acid (three acidic hydrogen atoms). Write equations for its stepwise ionization in water.

Some hydrated cations behave like polyprotic acids. We have seen, for example, that $Fe(H_2O)_6^{3+}$ gives up a proton to water:

$$Fe(H_2O)_6^{3+}(aq) + H_2O(l) \rightleftharpoons Fe(H_2O)_5(OH)^{2+}(aq) + H_3O^+(aq)$$

A stronger base can remove up to three protons:

$$Fe(H_2O)_6^{3+}(aq) + 3OH^-(aq) \rightarrow Fe(H_2O)_3(OH)_3(s) + 3H_2O(l)$$

Each proton is removed from a different water molecule, so three waters of hydration are converted to three OH^- groups that remain attached to the Fe^{3+} ion. $Fe(H_2O)_3(OH)_3$ is the hydrated form of $Fe(OH)_3$, an insoluble hydroxide. Water of hydration is usually not included in an equation unless there is some reason to emphasize its role; hence, the preceding equation is usually written as

The number of water molecules that remain attached to a hydroxide precipitate will vary with the age of the precipitate and with other conditions.

$$Fe^{3+}(aq) + 3OH^-(aq) \rightarrow Fe(OH)_3(s)$$

PRACTICE EXERCISE 16.11

Write an equation for the precipitation of zinc hydroxide from a solution containing $Zn(H_2O)_4^{2+}$ and hydroxide ions.

16.4 ACIDIC, BASIC, AND NEUTRAL SOLUTIONS

In every sample of water, some water molecules take protons from other water molecules. The autoionization of water

$$2H_2O(l) \rightleftharpoons H_3O^+(aq) + OH^-(aq)$$

which may also be written as

$$H_2O(l) \rightleftharpoons H^+(aq) + OH^-(aq)$$

provides small concentrations of hydrogen and hydroxide ions in any solution that contains water. The equilibrium constant for this reaction is called the **water constant**, K_w, and its value at 25°C is

$$K_w = [H^+][OH^-] = 1.00 \times 10^{-14} \quad (16.1)$$

The water constant K_w increases from 1.14×10^{-15} at 0°C to 5.35×10^{-14} at 50°C.

where the bracketed quantities, as usual, represent molar concentrations. The concentration of H_2O is omitted from the K_w expression because its concentration is virtually constant in all dilute solutions (Section 15.4).

In pure water, the $H^+(aq)$ and $OH^-(aq)$ concentrations are equal. Because their product is 1.00×10^{-14}, each concentration is

$$[H^+] = [OH^-] = 1.00 \times 10^{-7}\,M$$

Pure water is said to be **neutral**, and any aqueous solution in which the hydrogen and hydroxide ion concentrations are equal is a **neutral solution**. Sugar water is an example of a neutral solution.

When an acid such as HCl or CH_3COOH is dissolved in water, the $H^+(aq)$ concentration increases. The resulting solution, in which the concentration of $H^+(aq)$ is greater than 1.00×10^{-7} *M*, is an **acidic solution**. The extra $H^+(aq)$ shifts the water equilibrium back toward molecular H_2O in accord with Le Châtelier's principle. Fewer water molecules are ionized, so the $OH^-(aq)$ concentration is less than 1.00×10^{-7} *M*. The product of the two concentrations, however, still satisfies Equation 16.1.

A **basic solution**, formed by dissolving a base such as NaOH or NH_3 in water, has an $OH^-(aq)$ concentration greater than 1.00×10^{-7} *M*. The addition of base also shifts the water equilibrium back, so that in basic solutions the $H^+(aq)$ concentration is less than 1.00×10^{-7} *M*.

The conditions that exist in acidic, basic, and neutral solutions are summarized below:

- Acidic solutions: $[H^+] > 1.00 \times 10^{-7}\,M$; $[OH^-] < 1.00 \times 10^{-7}\,M$
- Neutral solutions: $[H^+] = [OH^-] = 1.00 \times 10^{-7}\,M$
- Basic solutions: $[H^+] < 1.00 \times 10^{-7}\,M$; $[OH^-] > 1.00 \times 10^{-7}\,M$
- All solutions: $[H^+][OH^-] = 1.00 \times 10^{-14}$

The fact that in acidic and basic solutions the water equilibrium is shifted in the direction of fewer ions means that *the self-ionization of water is a maximum in neutral solutions.*

The value of K_w can be used to calculate the concentration of $H^+(aq)$ or $OH^-(aq)$ if one of the concentrations is known.

LEARNING HINTS

The symbol $>$ means "greater than"; $<$ means "less than."

Note the distinction between an acid and an acidic solution. An acid is a proton donor; an acidic solution is an aqueous solution in which $[H^+] > [OH^-]$. Similarly, a base is a proton acceptor; a basic solution is an aqueous solution in which $[OH^-] > [H^+]$.

EXAMPLE 16.3

Calculate the hydrogen and hydroxide ion concentrations at 25°C in (a) 0.25 *M* HCl and (b) 0.50 *M* NaOH.

SOLUTION

(a) HCl is a strong acid that ionizes completely in water; therefore, 0.25 mol of dissolved HCl provides 0.25 mol of $H^+(aq)$. The self-ionization of water contributes relatively few hydrogen ions (less than 10^{-7} mol/L) and does not add significantly to the $H^+(aq)$ provided by the acid. Hence, $[H^+] = 0.25\ M$. The hydroxide ions come only from the water. Their concentration is obtained from Equation 16.1:

$$[OH^-] = \frac{K_w}{[H^+]} = \frac{1.00 \times 10^{-14}}{0.25} = 4.0 \times 10^{-14}\ M$$

(b) NaOH is an ionic compound; 0.50 mol of NaOH furnishes 0.50 mol of $OH^-(aq)$. The hydroxide ion furnished by the water is negligible compared to that furnished by the NaOH, so $[OH^-] = 0.50\ M$. The $H^+(aq)$ ions come only from the water. Their concentration is

$$[H^+] = \frac{K_w}{[OH^-]} = \frac{1.00 \times 10^{-14}}{0.50} = 2.0 \times 10^{-14}\ M$$

Observe that both hydrogen and hydroxide ions are present in the acidic solution of (a) as well as in the basic solution of (b).

PRACTICE EXERCISE 16.12

Calculate the hydrogen and hydroxide ion concentrations in $1.0 \times 10^{-3}\ M\ Ca(OH)_2$.

The pH Scale

The acidity of an aqueous solution is often expressed in terms of **pH**, where pH is defined as the negative logarithm of the hydrogen ion concentration:

$$pH = -\log [H^+] \qquad (16.2)$$

(The term "log" means "logarithm to the base 10." A review of logarithms is given in Appendix A.2.)

EXAMPLE 16.4

Calculate the pH of (a) a neutral solution, (b) an acidic solution in which $[H^+] = 0.15$ *M*, and (c) a basic solution in which $[OH^-] = 2 \times 10^{-3}\ M$.

SOLUTION

(a) In a neutral solution $[H^+] = 1.0 \times 10^{-7}\ M$. Hence,

$$pH = -\log (1.0 \times 10^{-7}) = -(-7.00) = 7.00$$

The pH has no units because a logarithm is dimensionless.

LEARNING HINTS

Use the log x button on your electronic calculator.

Keep in mind that the number of decimal places in the logarithm of a number is equal to the number of significant figures in the number. The number 0.15 has two significant figures; hence, its logarithm will have two decimal places.

(b) $[H^+] = 0.15\ M$. Hence,

$$pH = -\log 0.15 = -(-0.82) = 0.82$$

(c) We must first find $[H^+]$. Substituting $[OH^-] = 2 \times 10^{-3}\ M$ into Equation 16.1 gives

$$[H^+] = \frac{K_w}{[OH^-]} = \frac{1.00 \times 10^{-14}}{2 \times 10^{-3}} = 5 \times 10^{-12}\ M$$

The pH can now be obtained from Equation 16.2:

$$pH = -\log [H^+] = -\log (5 \times 10^{-12}) = -(-11.3) = 11.3$$

PRACTICE EXERCISE 16.13

Calculate the pH of a solution in which the hydrogen ion concentration is $1.5 \times 10^{-2}\ M$.

Acidic solutions: $pH < 7.0$
Neutral solutions: $pH = 7.0$
Basic solutions: $pH > 7.0$

Because there is a negative sign in Equation 16.2, the pH will decrease as the hydrogen ion concentration increases; in other words, *the more acidic the solution, the lower its pH; the more basic the solution, the higher its pH.* The pH of a neutral solution at 25°C is 7.0. An acidic solution will have a pH less than 7.0; a basic solution will have a pH greater than 7.0.

Because the pH scale is logarithmic, *a change of one pH unit corresponds to a tenfold change in hydrogen ion concentration.* For example, the hydrogen ion concentration in a solution with a pH of 5.0 is 1000 times greater than in a solution with a pH of 8. Approximate pH values for various aqueous mixtures are given in Table 16.3. Figure 16.5 shows a pH meter that is used to measure solution pH.

PRACTICE EXERCISE 16.14

Refer to Table 16.3, and state whether seawater is acidic, basic, or neutral.

TABLE 16.3 The Approximate pH of Various Aqueous Mixtures

Human Biological Fluids	pH	Other Fluids	pH
Bile	6.8–7.0	Lemon juice	2.2–2.4
Blood plasma	7.3–7.5	Cow's milk	6.3–6.6
Gastric juice	1.0–3.0	Seawater	8 (average)
Milk	6.6–7.6	Soft drinks	2.0–4.0
Pancreatic juice	7.8–8.0	Tomato juice	4.0–4.4
Saliva	6.5–7.5	Vinegar	2.4–3.4
Spinal fluid	7.3–7.5	Drinking water	6.5–8.0
Urine	4.8–8.4	Wine	2.8–3.8

Figure 16.5
A laboratory pH meter.

The pH notation, which was originally devised in 1909 by the Danish biochemist Soren P. Sørensen, has since been extended to other exponential quantities, and it is especially convenient for numbers that are very small. Hydroxide concentrations, for example, can be expressed in terms of **pOH**, where

$$\text{pOH} = -\log [\text{OH}^-] \qquad (16.3)$$

The "p" in pH and pOH originally meant "power" (German *potenz*), as in exponential power.

EXAMPLE 16.5

Use the results of Example 16.3a to find the pH and pOH of a 0.25 *M* solution of HCl.

SOLUTION

The H^+ concentration of the solution in Example 16.3a is 0.25 *M*; the OH^- concentration is 4.0×10^{-14} *M*. Hence,

$$\text{pH} = -\log [\text{H}^+] = -\log 0.25 = 0.60$$

$$\text{pOH} = -\log [\text{OH}^-] = -\log (4.0 \times 10^{-14}) = 13.40$$

PRACTICE EXERCISE 16.15

Use the results of Example 16.3b to calculate the pH and pOH of a 0.50 *M* NaOH solution.

In Example 16.5, notice that the sum of the pH and the pOH is 14.00. The value of this sum is not a coincidence. At 25°C, it is always true that

$$\text{pH} + \text{pOH} = 14.000 \tag{16.4}$$

Equation 16.4 can be obtained from Equation 16.1 as follows:

$$[H^+][OH^-] = K_w = 1.00 \times 10^{-14}$$

Taking the logarithm of both sides gives

$$\log [H^+] + \log [OH^-] = \log (1.00 \times 10^{-14}) = -14.000$$

Multiplying through by -1 gives

$$-\log [H^+] - \log [OH^-] = 14.000$$

Equation 16.4 is obtained by substituting $\text{pH} = -\log [H^+]$ and $\text{pOH} = -\log [OH^-]$ into the above expression.

EXAMPLE 16.6

A sample of lemon juice has a pH of 2.3. Calculate its pOH.

SOLUTION

The pH is 2.3. Hence,

$$\text{pOH} = 14.00 - \text{pH} = 14.00 - 2.3 = 11.7$$

PRACTICE EXERCISE 16.16

What is the pOH of a neutral solution?

Equations 16.1 through 16.4 relate the hydrogen ion concentration, the hydroxide ion concentration, the pH, and the pOH. If one of these quantities is known, the other three can be calculated.

EXAMPLE 16.7

The pH of a dilute baking soda ($NaHCO_3$) solution is 8.70. What is the hydrogen ion concentration?

SOLUTION

We will use Equation 16.2, which relates pH and hydrogen ion concentration. Substituting pH = 8.70 gives

$$\text{pH} = 8.70 = -\log [H^+]$$

or

$$\log [H^+] = -8.70$$

$$[H^+] = \text{antilog} (-8.70) = 10^{-8.70} = 2.0 \times 10^{-9}\ M.$$

LEARNING HINTS

If $\log x = y$, then $x = \text{antilog } y = 10^y$.

An expression such as $10^{-8.70}$ can be evaluated using the 10^x, the y^x, or the INV and LOG buttons on your calculator. For this example, $x = -8.70$ and $y = 10$.

The answer has two significant figures because its logarithm, −8.70, has two digits after the decimal place.

PRACTICE EXERCISE 16.17

The pH of a certain fruit drink is 4.3. Find the hydrogen ion concentration.

EXAMPLE 16.8

The pOH of a soap solution is 4.0. Find (a) the pH, (b) the hydrogen ion concentration, and (c) the hydroxide ion concentration.

SOLUTION

(a) The pOH = 4.0. From Equation 16.4,

$$\text{pH} = 14.00 - \text{pOH} = 14.00 - 4.0 = 10.0$$

(b) The concentration of hydrogen ions is found by substituting pH = 10.0 into Equation 16.2:

$$\text{pH} = 10.0 = -\log\,[\text{H}^+]$$

$$\log\,[\text{H}^+] = -10.0$$

$$[\text{H}^+] = \text{antilog}\,(-10.0) = 10^{-10.0} = 1 \times 10^{-10}\,M$$

(c) The concentration of hydroxide ions is found by substituting $[\text{H}^+] = 1 \times 10^{-10}$ M into Equation 16.1:

$$[\text{OH}^-] = \frac{K_w}{[\text{H}^+]} = \frac{1.00 \times 10^{-14}}{1 \times 10^{-10}} = 1 \times 10^{-4}\,M$$

(The hydroxide ion concentration can also be obtained by substituting pOH = 4.0 into Equation 16.3. Try it!)

PRACTICE EXERCISE 16.18

The soap in Example 16.8 would not be "kind to your hands"; it is sufficiently alkaline to be harsh on the skin. A somewhat milder soap forms a solution with a pH of 8.0. Show that the $OH^-(aq)$ concentration in this solution is 100 times smaller than in the solution of pH 10.0.

16.5 MOLECULAR STRUCTURE AND ACID STRENGTH

Can the formula of an acid or base give us any idea of its strength? Often it does, and a satisfying number of predictions can be based on simple Lewis structures. We need only look for structural features that make it easy for an acid to lose a proton or for a base to gain a proton.

Ionic Charge

Since a proton has a positive charge, it can part more readily from a positively charged ion such as NH_4^+ than from a neutral molecule such as NH_3; NH_4^+ is therefore a stronger acid than NH_3. The SO_4^{2-} ion with a double negative charge accepts protons more readily and is a stronger base than the singly charged HSO_4^- ion. These examples show that *for species differing only in the number of protons, acid strength increases with increasing positive charge; base strength increases with increasing negative charge.*

EXAMPLE 16.9

List phosphoric acid (H_3PO_4) and its ions in order of (a) increasing acidity and (b) increasing basicity.

SOLUTION

(a) Acid strength increases with increasing positive charge; hence,

$$HPO_4^{2-} < H_2PO_4^- < H_3PO_4$$

——increasing acid strength——→

The PO_4^{3-} ion is not an acid.

(b) Base strength increases with increasing negative charge; hence,

$$H_3PO_4 < H_2PO_4^- < HPO_4^{2-} < PO_4^{3-}$$

————increasing base strength————→

The ionization of a polyprotic acid such as H_3PO_4 (Section 16.3) occurs in distinct steps because each successive anion is more negative and thus a substantially weaker acid than the one preceding it.

PRACTICE EXERCISE 16.19

(a) Which is the stronger acid, H_2S or HS^-? (b) Which is the stronger base, OH^- or O^{2-}?

Oxo Acids

Another term for *oxo acid* is *oxyacid.* The oxo acids of nitrogen and the halogens were discussed in Chapter S2.

Oxo acids are acids that contain hydrogen, oxygen, and at least one other element. Some examples are shown in Table 16.4. The acidic hydrogen in each of these acids is at the positive end of a polar O—H bond

$$\overset{\delta-}{X-\ddot{\underset{..}{O}}}-\overset{\delta+}{H}$$

↑ acidic hydrogen

and is therefore vulnerable to capture by the electron pair of a sufficiently strong

TABLE 16.4 Some Oxo Acids

Acetic acid	CH_3COOH	$H-\underset{H}{\overset{H}{C}}-\overset{:O:}{\overset{\Vert}{C}}-\ddot{O}-H$
Nitric acid	HNO_3	$H-\ddot{O}-\overset{:O:}{\overset{\Vert}{N}}-\ddot{O}:$
Perchloric acid	$HClO_4$	$H-\ddot{O}-\underset{:O:}{\overset{:O:}{Cl}}=\ddot{O}$
Phosphoric acid	H_3PO_4	$H-\ddot{O}-\underset{:O:-H}{\overset{:O:}{P}}-\ddot{O}-H$
Sulfuric acid	H_2SO_4	$H-\ddot{O}-\underset{:O:}{\overset{:O:}{S}}-\ddot{O}-H$

The Lewis structures given for the oxo acids in this section have minimum formal charge on their atoms. Other resonance structures are possible (see Chapter 9).

base. The conjugate base, that is, the negative ion left after loss of the proton, is called an **oxo anion**. Hypochlorous acid (HOCl), for example, surrenders some of its protons to water to form $H_3O^+(aq)$ and the oxo anion OCl^-:

$$H-\underset{H}{\ddot{O}}: + \overset{\delta+}{H}-\overset{\delta-}{\ddot{O}}-\ddot{Cl}: \rightleftharpoons \left[H-\underset{H}{\ddot{O}}-H\right]^+ + :\ddot{O}-\ddot{Cl}:^-$$

Oxo acid — Oxo anion

The principal factor affecting the acidity of an oxo acid is the polarity of the O—H bond; *the more polar the O—H bond, the stronger the oxo acid.* The polarity of a bond is determined mainly by the difference in electronegativities of its two atoms, but it is also modified by other atoms in the molecule that exert a pull on the electron cloud. Atoms that influence the O—H bond polarity are as follows:

LEARNING HINT

Recall that oxygen is the second most electronegative element.

1. *An electronegative central atom:* The central atom is the one to which the OH group is attached. In chloric acid ($HClO_3$), the central atom is Cl; in iodic acid (HIO_3), the central atom is I. Chlorine is more electronegative than iodine and therefore more effective in drawing electron density away from the hydrogen atom. Thus the H—O bond is more polar in $HClO_3$ than in HIO_3

$$H-\ddot{O}-\overset{:O:}{\overset{\Vert}{\ddot{Cl}}}=\ddot{O} \qquad H-\ddot{O}-\overset{:O:}{\overset{\Vert}{\ddot{I}}}=\ddot{O}$$

more polar; more electronegative — less polar; less electronegative

and $HClO_3$ is a stronger acid than HIO_3. *For oxo acids with the same number of*

CHEMICAL INSIGHT

ACID RAIN

"Gentle rain from heaven" (William Shakespeare, *The Merchant of Venice*) has been regarded throughout the ages as one of nature's mildest substances. Rainwater has a reputation for being kind to skin, hair, and delicate fabrics, and in the past it was often collected in household rain barrels because of its superiority to mineral-laden well water. Unfortunately, this reputation is changing. Rainfall in many parts of the globe has become increasingly acidic, showering the landscape with dilute solutions of nitric and sulfuric acids and occasionally exhibiting a pH comparable to that of strong vinegar. Acid rain, which dissolves archaeological treasures, contributes to the corrosion of buildings and other materials, poisons fish, and blights vegetation (Figure 16.6), is caused by volatile acidic oxides generated by smelting ores and burning fossil fuels. The damage caused by acid rain is complex and not confined to the effects of hydrogen ion alone. Acid droplets, for example, dissolve aluminum, cadmium, lead, and copper from particles emitted by industrial smokestacks, thus contaminating groundwater with toxic ions. Acid rain also dissolves and washes away desirable nutrients in the soil, to the detriment of crops and other vegetation.

Carbon dioxide, a component of the atmosphere since the early ages of biological evolution, does not contribute to the acid rain problem. Its solubility is limited, and H_2CO_3 is a weak acid. The pH of rain in equilibrium with atmospheric CO_2 is about 5.6, a value that is considered normal. The carbon dioxide content of the atmosphere is increasing because of an increase in the combustion of wood and fossil fuels, and this increase presents a problem, but not one of acidity (see *Chemical Insight: The Greenhouse Effect,* p. 884).

The principal culprits in the production of acid rain are large quantities of sulfur dioxide (SO_2) and the nitrogen oxides NO and NO_2, which are collectively referred to as

(a)

(b)

Figure 16.6

Damage caused by acid rain in the northeastern United States: (a) a limestone statue and (b) trees on a mountain slope.

oxygen atoms, acid strength increases with increasing central atom electronegativity. This rule is illustrated in Table 16.5 for some oxo acids from Groups 5A, 6A, and 7A.

2. *Other electronegative atoms attached to the central atom:* When atoms such as F, Cl, and O are attached to the central atom, they help it to withdraw electron density from the O—H group. Many oxo acids contain oxygen atoms in addition to those in the O—H groups. Sulfuric acid (H_2SO_4) and sulfurous acid (H_2SO_3) each have two O—H groups, but sulfuric acid with two extra oxygen atoms is strong, while sulfurous acid with one extra oxygen atom is weak:

NO_x. In the presence of sunlight, oxygen, and water vapor, these molecules are further oxidized and hydrated to produce the strong acids H_2SO_4 and HNO_3. The total acidity of rain in the northeastern United States, that is, the total concentration of strong and weak acids, has increased about sixfold since 1930. The hydrogen ion concentration, however, has increased by a factor of about 25. Thus, a greater fraction of the acidity is now due to strong acids such as HNO_3 and H_2SO_4.

The average pH of rain is currently around 4.2, down from its normal value of 5.6 (Figure 16.7). Clouds and fog, which tend to concentrate acids, have an even lower average pH. A fog pH as low as 1.7 has been recorded. Dry particles such as dust and soot also absorb acidic pollutants; hence, acid damage occurs even when rainfall is scanty. Some of this acidity is neutralized when groundwater is absorbed in regions where limestone (calcium carbonate) is the predominant rock

$$CaCO_3(s) + H^+(aq) \rightarrow Ca^{2+}(aq) + HCO_3^-(aq)$$

thus giving a certain amount of protection to vegetation and aquatic life in those areas. Unfortunately, much groundwater runs off into lakes and streams before neutralization is complete. In areas where granite (silicate) rock predominates, as in Scandinavia and the Adirondack Mountains of New York, neutralization does not occur. In these regions acid deposition, both dry and wet, is producing tremendous economic, ecological, and esthetic damage.

"Liming" (air-dropping large quantities of ground limestone) has been used to reduce the acidity and to restore life in a number of dying lakes in Sweden and in New York. This treatment, which does not protect forests, buildings, or statuary, is expensive and has to be repeated every year. It is at best a stopgap measure.

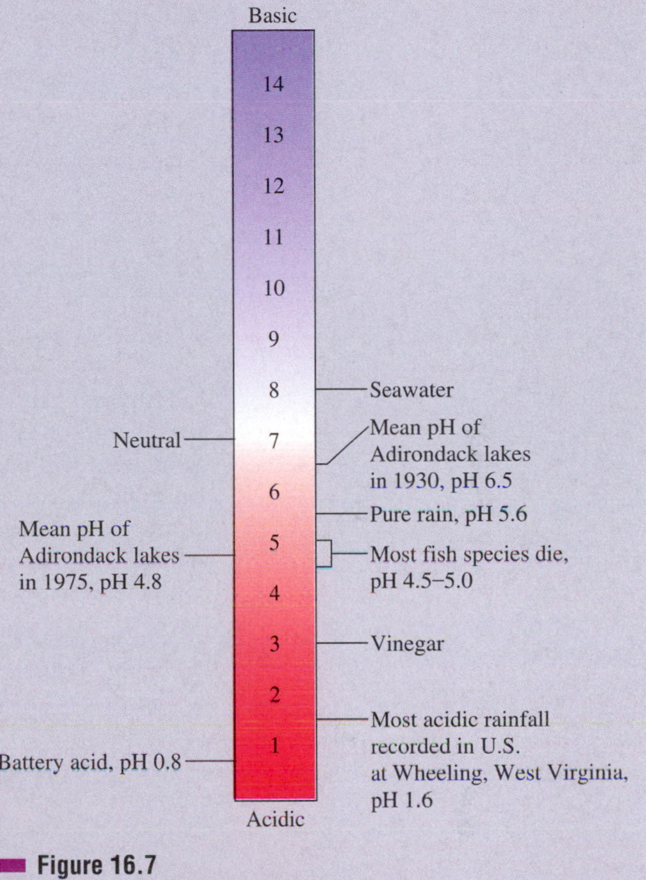

Figure 16.7
The pH of rainwater and other aqueous solutions.

```
          :O:                        :O:
           ‖                          ‖
   H—Ö—S—Ö—H                  H—Ö—S—Ö—H
     ¨   ‖  ¨                    ¨  ¨  ¨
          :O:
sulfuric acid, strong        sulfurous acid, weak
```

In general, *the strength of oxo acids increases with the number of oxygen atoms bonded to the central atom.* Table 16.6 shows the effect of additional oxygen atoms on the strength of the chlorine oxo acids.

TABLE 16.5 Dependence of Oxo Acid Strength on Central Atom Electronegativity[a]

Periodic Group: 5A	6A	7A	
H_3PO_4	H_2SO_4	$HClO_4$	**Decreasing Acid Strength** ↓
H_3AsO_4	H_2SeO_4	$HBrO_4$	
H_3SbO_4	H_2TeO_4	HIO_4	
Increasing Acid Strength ⟶			

[a]For oxo acids with the same number of oxygen atoms, acid strength increases with increasing electronegativity of the central atom. Hence, with a few exceptions, acid strength increases from left to right across a period and decreases from top to bottom in a group.

Extra oxygen atoms also make the oxo anions larger and less basic. Because the charge on a large oxo anion is spread out or *delocalized* over a greater volume, it will attract protons less effectively than the more concentrated charge on a small oxo anion with the same central atom. The charge on SO_4^{2-}, for example, is more delocalized than on SO_3^{2-}, and the charge on ClO_4^- is more delocalized than on ClO^-. In each case, the more delocalized charge is associated with the weaker base.

3. *Electronegative atoms elsewhere in the molecule:* Even remote atoms can exert a surprising influence. Compare acetic acid with the related chloroacetic acid:

```
      H  :O:                   H  :O:
      |   ‖                  ..    |   ‖     ..
  H—C—C—O—H         :Cl—C—C—O—H
      |       ..            ..   |       ..
      H                         H
```

acetic acid, weaker   chloroacetic acid, stronger

TABLE 16.6 Relative Strengths of the Oxo Acids of Chlorine[a]

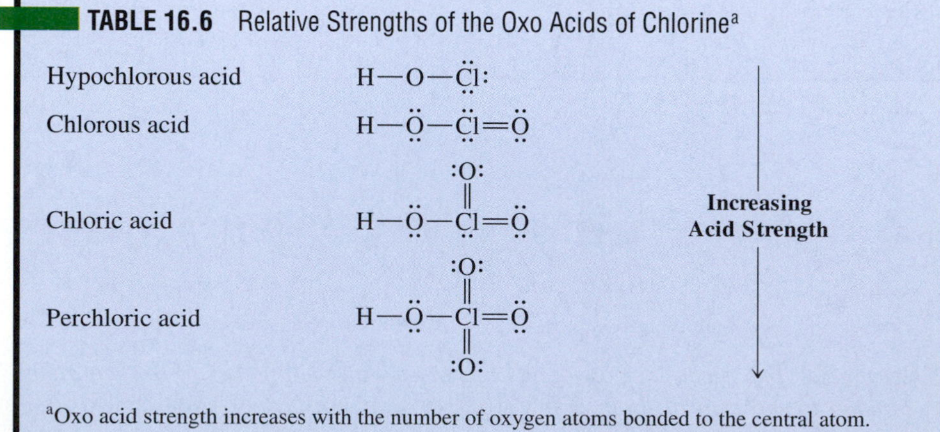

[a]Oxo acid strength increases with the number of oxygen atoms bonded to the central atom.

Replacing one of the nonacidic hydrogen atoms with the more electronegative chlorine atom draws electron density away from the COOH group, thus increasing the O—H bond polarity and making chloroacetic acid stronger than acetic acid.

EXAMPLE 16.10

(a) Which is the stronger acid, H_3PO_4 or H_3AsO_4?

(b) Which is the weaker base, BrO_2^- or BrO_3^-?

SOLUTION

(a) P is above As in Group 5A; hence, P atoms are more electronegative than As atoms. The O—H bond is more polar in H_3PO_4 than in H_3AsO_4, so H_3PO_4 is the stronger acid.

(b) The charge on BrO_3^- is more delocalized and less intense than on the smaller BrO_2^- ion. Therefore, BrO_3^- has less attraction for protons and is the weaker base.

LEARNING HINT

Draw Lewis structures—they will help you predict acid strengths.

PRACTICE EXERCISE 16.20

(a) Which is the stronger acid, $CH_2ClCOOH$ or $CHCl_2COOH$? (b) Which is the stronger base, OI^- or OCl^-?

Binary Acids

A **binary acid** is one that contains hydrogen and only one other element; HCl and H_2S are examples of binary acids. As Table 16.7 shows, *the strength of binary acids increases from left to right across a period and from top to bottom within a group.* The increase across a period is best explained in terms of electronegativity, which also increases from left to right across the period. The O—H bond in water, for

TABLE 16.7 Relative Strengths of Binary Acids[a]

Periodic Group			
5A	6A	7A	
NH_3	H_2O	HF	Increasing Acid Strength ↓
PH_3	H_2S	HCl	
AsH_3	H_2Se	HBr	
SbH_3	H_2Te	HI	

—— Increasing Acid Strength ⟶

[a]The strength of a binary acid increases from left to right across a period and from top to bottom in a group.

example, is more polar than the N—H bond in ammonia, so H_2O is a better proton donor than NH_3.

Within a group, the effect of anion size and charge delocalization takes precedence over polarity in determining the strengths of binary acids. Hydrogen fluoride, for example, is a weaker acid than hydrogen chloride, even though the H—F bond is more polar than the H—Cl bond. Recall (Figure 8.9, p. 321) that the fluoride ion (F^-) is smaller than the chloride ion (Cl^-). The extra electron on the F^- ion is crowded into a small volume that is shared with seven other electrons in the second valence shell. Because this concentrated charge has a very high affinity for protons, F^- is a relatively strong base and HF is a correspondingly weak acid. The extra electron in Cl^-, on the other hand, is accommodated in the spacious third valence shell, and its charge is spread over a relatively large volume. Chloride ion has little affinity for protons; it is a relatively poor base and HCl is a correspondingly strong acid. Hydrogen iodide produces the largest anion and is the strongest of the binary halogen acids.

PRACTICE EXERCISE 16.21

(a) Which is the stronger acid, HCl or HBr? (b) Which is the stronger base, HS^- or OH^-?

Hydrated Cations

The acidity of hydrated cations is illustrated by reactions such as

$$Al(H_2O)_6^{3+}(aq) + H_2O(l) \rightleftharpoons Al(H_2O)_5(OH)^{2+}(aq) + H_3O^+(aq)$$

$$Zn(H_2O)_4^{2+}(aq) + H_2O(l) \rightleftharpoons Zn(H_2O)_3(OH)^{+}(aq) + H_3O^+(aq)$$

These cations are acidic because the positive central ion attracts the electron clouds of the bonded water molecules and thus increases the O—H bond polarity (Figure 16.8). The polarizing influence of a positive ion increases with the intensity of its charge; therefore, *hydrated cations that form from small, highly charged positive ions tend to be more acidic than those that form from larger, less highly charged ions.* Hydrated trivalent cations are the most acidic; for example, $Fe(H_2O)_6^{3+}$ is almost as strong as phosphoric acid, and $Al(H_2O)_6^{3+}$ is similar in strength to acetic

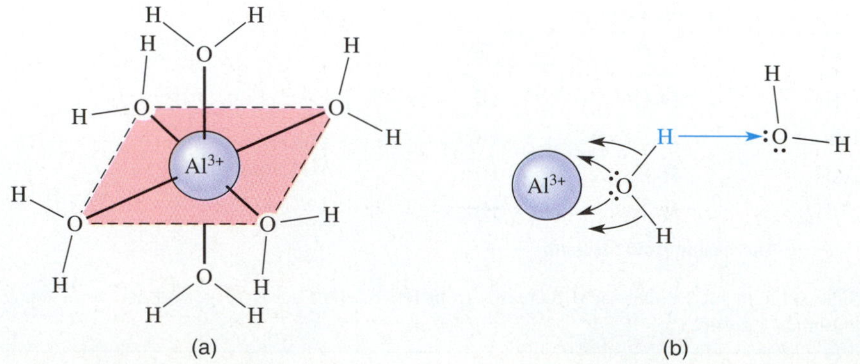

Figure 16.8

(a) Six water molecules surround Al^{3+} in the hydrated aluminum ion. (b) The positive charge of Al^{3+} attracts the electron cloud of a bonded water molecule, thus increasing its ability to donate a proton to another water molecule.

(a)

(b)

acid. Hydrated ions that form from small divalent cations such as Be^{2+}, or from transition metal and post-transition metal ions such as Fe^{2+}, Cu^{2+}, and Pb^{2+}, are also acidic, but less so than trivalent cations. Iron(II) salts, for example, are less acidic than iron(III) salts. Hydrated monovalent cations such as Ag^+ and Na^+ are not very acidic at all, nor are the hydrated ions that form from alkaline earth metal ions other than Be^{2+}.

Because each water of hydration is a potential proton donor, hydrated cations behave like polyprotic acids. Each proton lost from a hydrated cation decreases its positive charge and makes further proton removal more difficult.

PRACTICE EXERCISE 16.22

Which is the stronger acid, $Cr(H_2O)_6{}^{3+}$ or $Ni(H_2O)_6{}^{2+}$?

16.6 LEWIS ACIDS AND BASES

Gilbert N. Lewis, who originated the theory of the electron pair bond and for whom Lewis structures are named, devised a system of acid–base concepts that is even more comprehensive than the Brønsted–Lowry system. He published the following definitions in 1923: A **base** is a substance that donates electron pairs, an **acid** is a substance that accepts electron pairs, and **neutralization** is the sharing of an electron pair between an acid and a base.

A Lewis base must have a lone electron pair; therefore, all Lewis bases are Brønsted bases. Lewis acids, on the other hand, include many species that are not included in the Brønsted–Lowry definition. A Lewis acid is any molecule or ion that has an affinity for an extra pair of electrons; Lewis acids include all of the Brønsted acids, the proton itself, all positive ions, and any molecule with a positive region that can accommodate an additional electron pair. Lewis acids are often referred to as **electrophiles**, seekers of electrons. Lewis bases are called **nucleophiles**, seekers of a positive nucleus.

The Lewis acid–base definitions are especially useful for reactions in solvents other than water.

Neutralization occurs when the Lewis acid and base share the electron pair provided by the base, forming a new covalent bond that holds the two together. Lewis acid–base neutralizations are of the form

$$E + :Nu \rightarrow E\text{—}Nu$$

where E is the electrophile (Lewis acid) and :Nu is the nucleophile (Lewis base). An example is the reaction of hydrogen and hydroxide ions to form water:

$$\underset{\substack{\text{Lewis acid}\\\text{electrophile}}}{H^+(aq)} + \underset{\substack{\text{Lewis base}\\\text{nucleophile}}}{:\ddot{\underset{..}{O}}\text{—}H^-(aq)} \longrightarrow :\underset{\underset{H}{|}}{\ddot{O}}\text{—}H(l)$$

(new covalent bond)

Another example of Lewis acid–base neutralization is the reaction of ammonia with boron trifluoride. Here the electron-deficient boron atom is attracted to the lone pair of electrons on the nitrogen atom:

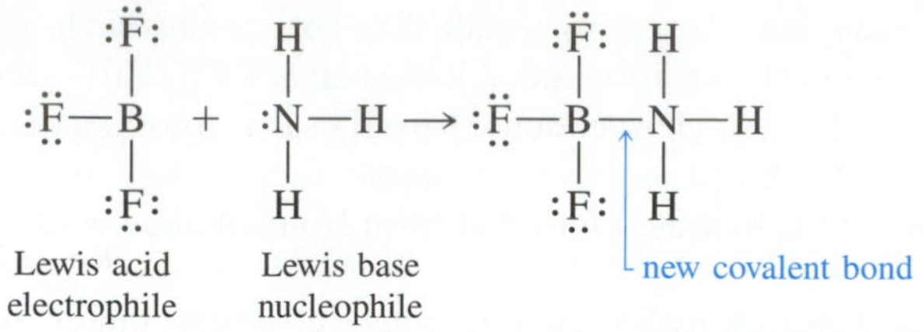

$$BF_3 + :NH_3 \longrightarrow F_3B{-}NH_3$$

Lewis acid electrophile — Lewis base nucleophile — new covalent bond

Lewis acid–base reactions often involve the breaking of bonds. In the reaction between NH_3 and HCl, for example, the H—Cl bond is broken as the new N—H bond forms:

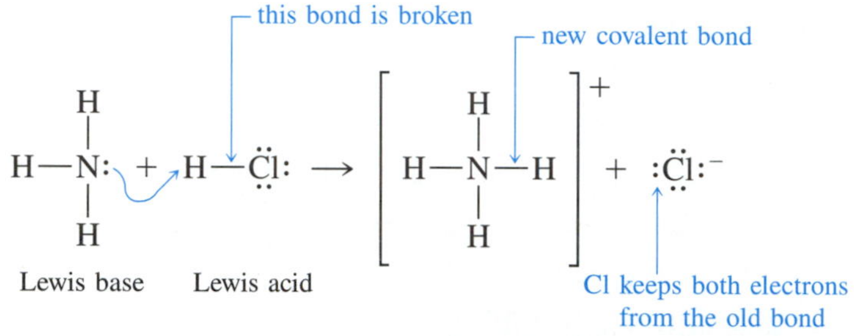

$$H_3N: + H{-}\ddot{Cl}: \longrightarrow [H{-}NH_3]^+ + :\ddot{Cl}:^-$$

Lewis base — Lewis acid — this bond is broken — new covalent bond — Cl keeps both electrons from the old bond

The Lewis definitions provide the best basis for describing the acidic behavior of carbon dioxide. It is well known that carbon dioxide neutralizes basic oxides and hydroxides even when they are not dissolved in water. An example is the reaction between solid calcium oxide and carbon dioxide gas:

$$CaO(s) + CO_2(g) \rightarrow CaCO_3(s)$$

The Arrhenius and Brønsted–Lowry definitions cannot be used to describe this neutralization because the reaction does not involve protons. In terms of the Lewis definitions, however, carbon dioxide is an acid whose positive center is the carbon atom

$$\overset{\delta-}{\ddot{O}}{=}\overset{\delta+}{C}{=}\overset{\delta-}{\ddot{O}}$$

The mechanism by which CO_2 neutralizes calcium oxide is shown below. The oxide ion is the base, the calcium ion is a spectator ion, and carbonate ion is the product.

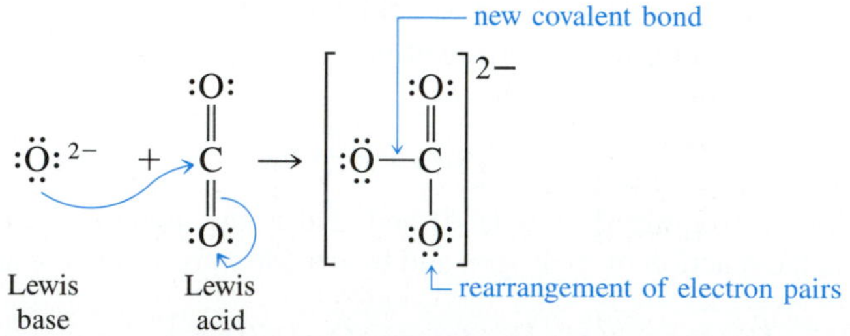

$$:\ddot{O}:^{2-} + O{=}C{=}O \longrightarrow [\ddot{O}{-}C(=O){-}O]^{2-}$$

Lewis base — Lewis acid — new covalent bond — rearrangement of electron pairs

The formation of the new bond forces a pair of electrons from one of the double bonds to relocate in order to preserve carbon's octet. Without this rearrangement, the association of the acid with the base would not be permanent, and the reaction would not take place. Internal electron rearrangements of this type often occur during Lewis acid–base reactions.

EXAMPLE 16.11

In the following reaction, which reactant is the Lewis acid and which is the Lewis base?

$$NH_4^+(aq) + S^{2-}(aq) \rightarrow NH_3(aq) + HS^-(aq)$$

SOLUTION

The equation, written with Lewis structures,

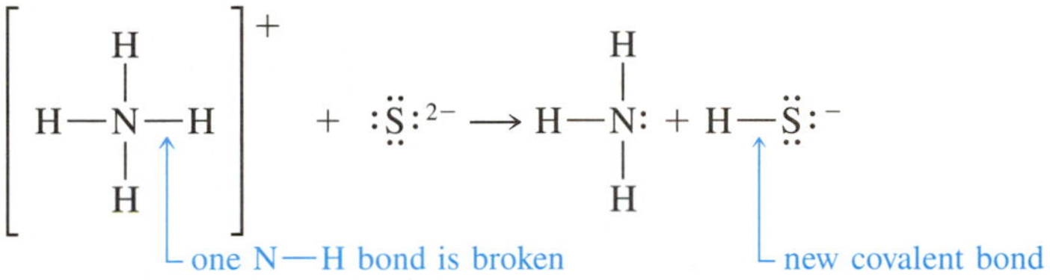

$$\left[\begin{array}{c} H \\ | \\ H-N-H \\ | \\ H \end{array}\right]^+ + :\ddot{\underset{..}{S}}:^{2-} \rightarrow H-\underset{H}{\overset{H}{N}}: + H-\ddot{\underset{..}{S}}:^-$$

one N—H bond is broken; new covalent bond

shows that the ammonium ion provides a proton that accepts an electron pair on the sulfide ion. Hence, NH_4^+ is the Lewis acid and S^{2-} is the Lewis base.

LEARNING HINT

Draw Lewis structures—they will help you to identify Lewis acids and bases.

PRACTICE EXERCISE 16.23

Identify the Lewis acid and base in the following reaction:

$$AlCl_3 + Cl^- \rightarrow AlCl_4^-$$

A Comparison of Acid–Base Theories

The principal features of the Arrhenius, Brønsted–Lowry, and Lewis acid–base systems are summarized in Table 16.8. Each system has its advantages. The Arrhenius system is the most limited, but its very simplicity makes it easy to use. The Brønsted–Lowry approach is more comprehensive and provides more insight into the reaction process. Chemists prefer the Brønsted–Lowry system whenever protons are involved, especially for reactions in water and in other amphiprotic solvents such as liquid ammonia. The Lewis system is the most comprehensive of all; it is favored by organic chemists who work mostly with nonaqueous mixtures.

TABLE 16.8 Summary of Arrhenius, Brønsted–Lowry, and Lewis Acid–Base Definitions

Type	Acid	Base	Neutralization
Arrhenius	Produces $H^+(aq)$ in solution	Produces $OH^-(aq)$ in water	$H^+(aq) + OH^-(aq) \rightarrow H_2O(l)$
Brønsted–Lowry	Proton donor	Proton acceptor	$HA + B \rightarrow A^- + HB^+$
Lewis	Electrophile	Nucleophile	$E + :Nu \rightarrow E:Nu$

16.78 Which base in each of the following pairs is stronger?
(a) HCO_3^- or CO_3^{2-}
(b) CH_3NH_2 or CH_3NH^-
(c) HS^- or S^{2-}

16.79 Explain the following observations in terms of molecular structure:
(a) HCl is a weaker acid than HBr.
(b) $HClO_3$ is a stronger acid than $HBrO_3$.
(c) HNO_2 is a weaker acid than HNO_3.
(d) $Al(H_2O)_6^{3+}$ is a stronger acid than $Zn(H_2O)_4^{2+}$.

16.80 Explain the following observations in terms of molecular structure:
(a) NH_3 is a weaker acid than H_2O.
(b) H_5IO_6 is a stronger acid than H_6TeO_6.
(c) $As(OH)_3$ is a stronger acid than $Sb(OH)_3$.
(d) HS^- is a stronger acid than OH^-.

16.81 Use the periodic table to pick the stronger acid from each of the following pairs. Explain each choice.
(a) HBr or HI
(b) H_3PO_4 or H_3AsO_4
(c) H_2O or H_2S
(d) HS^- or HCl
(e) $Cr(H_2O)_6^{3+}$ or $Cr(H_2O)_6^{2+}$
(f) $HClO_3$ or $HClO_4$
(g) $CH_3CHClCOOH$ or $CH_3CHBrCOOH$

16.82 Use the periodic table to pick the stronger base from each of the following pairs. Explain each choice.
(a) S^{2-} or O^{2-}
(b) Cl^- or Br^-
(c) ClO_2^- or ClO_3^-
(d) NH_3 or NH_2OH
(e) SO_3^{2-} or SeO_3^{2-}
(f) CO_3^{2-} or HCO_3^-
(g) $H_2PO_4^-$ or $H_2SbO_4^-$

Lewis Acids and Bases (Section 16.6)

16.83 State the Lewis definitions of acid, base, and neutralization, and show how they differ from the Brønsted–Lowry definitions.

16.84 What advantages do the Lewis acid–base definitions have over the Brønsted–Lowry definitions? What disadvantages?

16.85 Why is a Lewis acid called an electrophile? What is the corresponding term for a Lewis base?

16.86 What happens on the molecular level when a Lewis acid reacts with a Lewis base?

16.87 What molecular feature do Lewis acids have in common? List five Lewis acids that are not Brønsted acids.

16.88 What molecular feature do Lewis bases have in common? Are there any Lewis bases that are not Brønsted bases? Explain your answer.

16.89 Diagram each of the following reactions using Lewis structures. Identify the Lewis acid and Lewis base in each reaction.
(a) $BF_3 + F^- \rightarrow BF_4^-$
(b) $CH_3NH_2(l) + HCl(g) \rightarrow CH_3NH_3Cl(s) \quad (= CH_3NH_3^+ + Cl^-)$

16.90 Diagram each of the following reactions using Lewis structures. Identify the Lewis acid and Lewis base in each reaction.
(a) $Ag^+(aq) + 2NH_3(aq) \rightarrow Ag(NH_3)_2^+(aq)$
(b) $CO_2(aq) + OH^-(aq) \rightarrow HCO_3^-(aq)$

16.91 Interpret the reaction between methylamine and hydrochloric acid

$$CH_3NH_2(aq) + HCl(aq) \rightarrow CH_3NH_3Cl(aq)$$

in terms of (a) an Arrhenius neutralization, (b) a Brønsted–Lowry proton transfer, and (c) a Lewis acid–base reaction. Which interpretation seems simplest?

16.92 A sodium hydroxide solution can be partially neutralized by bubbling carbon dioxide gas through the solution. Write equations and interpret the reaction in terms of (a) an Arrhenius neutralization, (b) a Brønsted–Lowry proton transfer, and (c) a Lewis acid–base reaction. Which interpretation seems simplest?

PART B. MISCELLANEOUS QUESTIONS AND PROBLEMS

16.93 Is it possible to write an exact formula for a hydrated proton? Explain.

16.94 Explain how the nature of water influences our concept of acid and base strength. Which of the weak acids listed in Table 16.2 would a chemist consider strong on a strange, cold planet where the oceans, lakes, and rivers consist of liquid ammonia?

16.95 Write equations for the following neutralizations. Add water if it is needed to balance the equation.
(a) $CaO + H_3PO_4$
(b) $NaOH + P_4O_{10}$
(c) $CaO + SO_2$
(d) $MgO + HNO_3$
(e) $Ba(OH)_2 + SO_2$
(f) $Cr_2O_3 + H_2SO_4$
(g) $KOH + CO_2$

16.96 Coral, like eggshells, is mostly calcium carbonate. Explain why it would not be advisable to use vinegar to clean coral.

16.97 A solution is prepared by mixing 50.0 mL of 0.500 *M* NaOH with 50.0 mL of 0.500 *M* $Ca(OH)_2$. Calculate the pH of the solution.

16.98 The pH of a gastric juice sample is 1.1. How many grams of $Mg(OH)_2$ (administered as milk of magnesia) would it take to raise the pH of a 100.0-mL sample to 1.9? (Gastric juice contains the strong acid HCl.)

16.99 Compare the following over-the-counter antacid tablets with respect to the number of moles of $H^+(aq)$ each can neutralize: Brand A, 300 mg $Al(OH)_3$, 200 mg $Mg(OH)_2$; Brand B, 600 mg $Al(OH)_3$; Brand C, 500 mg $CaCO_3$.

16.100 The acidity of gastric juice has been determined by titrating the contents of the stomach with 0.1 *M* NaOH using dimethylaminoazobenzene as an indicator. For most people, 100 mL of stomach contents requires about 27 mL of 0.1 *M* NaOH. Estimate the number of moles of $H^+(aq)$ per liter of "stomach acid."

16.101 Calculate the number of milliliters of 0.100 *M* HNO_3 that will neutralize 1.00 g of $Mg(OH)_2$.

16.102 A 2.00-g sample of impure solid NaOH is neutralized by 60.0 mL of 0.500 *M* HCl. Calculate the percentage of NaOH in the sample.

16.103 An antacid tablet weighing 0.800 g is neutralized by 100 mL of 0.150 *M* HCl. The active ingredient in the tablet is $Al(OH)_3$. Calculate (a) the number of grams of $Al(OH)_3$ in the tablet and (b) the percentage of $Al(OH)_3$ in the tablet.

16.104 The nitrogen content of organic samples can be determined by the *Kjeldahl method.* Bound nitrogen is first converted into ammonium sulfate by warming the sample with concentrated sulfuric acid. Addition of concentrated base then liberates ammonia

$$NH_4^+(aq) + OH^-(aq) \longrightarrow NH_3(g) + H_2O(l)$$

which is distilled into a standard acid solution.

(a) The ammonia produced from a 2.00-mL pathological spinal fluid sample is distilled into 5.00 mL of 0.1010 *M* HCl. The excess acid is titrated with 2.43 mL of 0.1112 *M* NaOH. How many millimoles of ammonia were produced?

(b) If all of the nitrogen in the sample came from protein, calculate the number of grams of protein per 100 mL of spinal fluid. Assume that 6.25 g of protein contains 1.00 g of nitrogen.

CHAPTER 17

Exercise alters body chemistry. Blood and other body fluids contain buffers (Section 17.7) that maintain a constant pH in spite of these changes.

ACID–BASE EQUILIBRIA IN AQUEOUS SOLUTIONS

OUTLINE

PREVIEW

Earth is sometimes called the "water planet" because most of its surface is covered with water. Aqueous chemistry dominates the biosphere, and many natural reactions depend on the concentration of hydrogen ions in the reaction mixture. The local pH determines which plants will flourish in our gardens, whether sediments will deposit in river beds, and even whether our teeth will decay. Many functions of the body are restricted to narrow ranges of hydrogen ion concentration, and physicians must work to restore the hydrogen ion balance when the normal controlling mechanisms go awry. Most of these mechanisms involve aqueous equilibria between Brønsted acids and their conjugate bases.

Weak acids and weak bases ionize in water to produce equilibrium mixtures containing ions and their parent molecules. Whenever pH is crucial to function, as it is in living organisms, we find it controlled by such equilibria. Therefore, it is important to be able to relate the pH of a solution to the concentration of weak acid or base and to be able to calculate one from the other. Such calculations are done with the aid of **ionization constants**, which are the equilibrium constants for weak electrolytes in aqueous solution.

17.1 ACID IONIZATION CONSTANTS

The equilibrium constant for the ionization of a weak acid in water is called an **acid ionization constant** (**K_a**). For example, acetic acid ionizes in water:

$$CH_3COOH(aq) \rightleftharpoons H^+(aq) + CH_3COO^-(aq)$$

An expanded table of acid ionization constants is given in Appendix B.2.

TABLE 17.1 Ionization Constants[a] of Weak Acids at 25°C

Acid	Formula	K_a	pK_a
Acetic acid	CH_3COOH	1.76×10^{-5}	4.754
Aluminum ion	$Al(H_2O)_6^{3+}$	1.4×10^{-5}	4.85
Arsenic acid	H_3AsO_4	2.5×10^{-4}	3.60
	$H_2AsO_4^-$	5.6×10^{-8}	7.25
	$HAsO_4^{2-}$	3×10^{-13}	12.5
Ascorbic acid[b]	$H_2C_6H_6O_6$	7.9×10^{-5}	4.10
	$HC_6H_6O_6^-$	1.6×10^{-12}	11.80
Barbituric acid	$HC_4H_3N_2O_3$	9.8×10^{-5}	4.01
Carbonic acid	$CO_2 + H_2O$[c]	4.30×10^{-7}	6.367
	HCO_3^-	5.61×10^{-11}	10.251
Chloroacetic acid	$CH_2ClCOOH$	1.40×10^{-3}	2.85
Diethylbarbituric acid	$HC_8H_{11}N_2O_3$	3.7×10^{-8}	7.43
Formic acid	$HCOOH$	1.9×10^{-4}	3.72
Hydrocyanic acid	HCN	4.93×10^{-10}	9.307
Hydrofluoric acid	HF	3.53×10^{-4}	3.452
Hydrogen sulfide	H_2S	1.02×10^{-7}	6.991
	HS^-[d]	$\sim 1 \times 10^{-19}$	~19.0
Hypobromous acid	$HOBr$	2.06×10^{-9}	8.686
Hypochlorous acid	$HOCl$	2.8×10^{-8}	7.55
Hypoiodous acid	HOI	2.3×10^{-11}	10.64
Iron(III) ion	$Fe(H_2O)_6^{3+}$	7.9×10^{-3}	2.10
Malonic acid	$H_2C_3H_2O_4$	1.49×10^{-3}	2.827
	$HC_3H_2O_4^-$	2.03×10^{-6}	5.693
Nicotinic acid	$HC_6H_4NO_2$	1.4×10^{-5}	4.85
Nitrous acid	HNO_2	6.0×10^{-4}	3.22
Oxalic acid	$H_2C_2O_4$	5.90×10^{-2}	1.229
	$HC_2O_4^-$	6.40×10^{-5}	4.194
Phosphoric acid	H_3PO_4	7.52×10^{-3}	2.124
	$H_2PO_4^-$	6.23×10^{-8}	7.206
	HPO_4^{2-}	4.5×10^{-13}	12.35
o-Phthalic acid	$H_2C_8H_4O_4$	1.3×10^{-3}	2.89
	$HC_8H_4O_4^-$	3.1×10^{-6}	5.51
Sulfurous acid	H_2SO_3	1.71×10^{-2}	1.767
	HSO_3^-	6.0×10^{-8}	7.22
Sulfuric acid	H_2SO_4	Strong	
	HSO_4^-	1.20×10^{-2}	1.92
Trichloroacetic acid	CCl_3COOH	2×10^{-1}	0.7
Veronal (see Diethylbarbituric acid)			

[a]The first ionization constant of a polyprotic acid such as H_3PO_4 is usually symbolized by K_{a_1}, the second by K_{a_2}, and so forth.

[b]Ascorbic acid is the chemical name for vitamin C.

[c]H_2CO_3 is unstable; the first ionization constant is for the reaction $CO_2(aq) + H_2O(l) \rightleftharpoons H^+(aq) + HCO_3^-(aq)$.

[d]The value of K_a for HS^- is controversial. The reported value was taken from R. J. Meyers, *J. of Chem. Ed.*, **63**, 687 (1986).

and its ionization constant at 25°C is

$$K_a = \frac{[H^+][CH_3COO^-]}{[CH_3COOH]} = 1.76 \times 10^{-5}$$

Ionization constants are often expressed in terms of the p-notation used for pH and pOH, that is,

Acid and base ionization constants are often called *dissociation constants.*

$$pK = -\log K \tag{17.1}$$

The pK_a of acetic acid at 25°C is

$$pK_a = -\log(1.76 \times 10^{-5}) = 4.754$$

The K_a and pK_a values of some weak acids are listed in Table 17.1 and Appendix B.2. *The larger the ionization constant, the smaller the* pK_a *value and the stronger the acid.* For example, nitrous acid with $K_a = 6.0 \times 10^{-4}$ ($pK_a = 3.22$) is a stronger acid than benzoic acid with $K_a = 6.5 \times 10^{-5}$ ($pK_a = 4.19$).

PRACTICE EXERCISE 17.1

The pK_a values of lactic and hydrofluoric acids are 3.86 and 3.45, respectively, at 25°C. Which is the stronger acid?

Calculations involving ionization constants are similar to the equilibrium calculations described in Chapter 15. The next example shows how K_a values can be determined experimentally.

EXAMPLE 17.1

A 0.0500 *M* acetic acid solution is tested with a pH meter at 25°C and found to have a pH of 3.03. Use this information to calculate K_a for acetic acid.

SOLUTION

We will set up a concentration summary based on the equation given above for the ionization of acetic acid. We will use the pH value to deduce the equilibrium concentration of each species in the solution, then we will calculate K_a by substituting these concentrations into the K_a expression.

Step 1: The initial concentration of acetic acid is 0.0500 *M*; the initial concentrations of CH_3COO^- and H^+ are zero. (Hydrogen ions produced by the self-ionization of water are ignored because, as shown in Section 16.4, their concentration is less than 10^{-7} mol/L. These ions would be of concern only if the acid were extremely dilute or extremely weak.)

Step 2: The equilibrium concentration of $H^+(aq)$ is calculated from the pH of the solution:

$$pH = 3.03 = -\log[H^+]$$

$$[H^+] = \text{antilog}(-3.03) = 10^{-3.03} = 9.3 \times 10^{-4}\,M$$

The equation for the ionization shows that the formation of 9.3×10^{-4} mol of H^+ will be accompanied by the formation of 9.3×10^{-4} mol of CH_3COO^- and the loss of 9.3×10^{-4} mol of CH_3COOH.

Step 3: The equilibrium concentrations are found by adding the changes in Step 2 to the initial concentrations in Step 1.

CONCENTRATION SUMMARY

Equation:	$CH_3COOH(aq)$	$\rightleftharpoons H^+(aq)$	$+ CH_3COO^-(aq)$
Initial concentrations (mol/L):	0.0500	0	0
Concentration changes (mol/L):	-9.3×10^{-4}	$+9.3 \times 10^{-4}$	$+9.3 \times 10^{-4}$
Equilibrium concentrations (mol/L):	0.0491	9.3×10^{-4}	9.3×10^{-4}

Substituting the equilibrium concentrations into the K_a expression gives

$$K_a = \frac{[H^+][CH_3COO^-]}{[CH_3COOH]} = \frac{(9.3 \times 10^{-4})^2}{0.0491} = 1.8 \times 10^{-5}$$

PRACTICE EXERCISE 17.2

The equation for the ionization of formic acid is

$$HCOOH(aq) \rightleftharpoons H^+(aq) + HCOO^-(aq)$$

The pH of a 0.200 *M* formic acid solution is 2.23. Calculate K_a and check your value with that reported in Table 17.1.

The next example shows how a tabulated K_a value can be used to calculate the pH of a solution of known molarity.

EXAMPLE 17.2

Calculate the pH of 0.10 *M* hypochlorous acid (HOCl). Use the K_a value from Table 17.1.

SOLUTION

The equation for the ionization of hypochlorous acid is shown in the concentration summary. To calculate the pH, we must find the concentration of $H^+(aq)$ in the equilibrium mixture. If x is the concentration of $H^+(aq)$, then the equilibrium concentrations of OCl^- and HOCl will be x and $0.10 - x$, respectively.

CONCENTRATION SUMMARY

Equation:	$HOCl(aq)$	$\rightleftharpoons H^+(aq)$	$+ OCl^-(aq)$
Initial concentrations (mol/L):	0.10	0	0
Concentration changes (mol/L):	$-x$	$+x$	$+x$
Equilibrium concentrations (mol/L):	$0.10 - x$	x	x

Substituting the equilibrium concentrations and $K_a = 2.8 \times 10^{-8}$ from Table 17.1 into the K_a expression gives

$$K_a = \frac{[H^+][OCl^-]}{[HOCl]}$$

$$2.8 \times 10^{-8} = \frac{x^2}{0.10 - x}$$

An exact solution for x can be obtained using the quadratic formula. However, because K_a is small, only a small amount of HOCl will ionize. The value of x should be small relative to 0.10, and we can approximate the solution by substituting 0.10 for $0.10 - x$:

$$2.8 \times 10^{-8} = \frac{x^2}{0.10}$$

and

$$x = 5.3 \times 10^{-5}\,M = [H^+]$$

Recall (Section 15.3) that an approximation is acceptable in an equilibrium calculation if the neglected quantity turns out to be less than 5% of the retained quantity. This criterion is justified because tabulated equilibrium constants generally have a precision of two significant figures or less. The calculated value of x, 5.3×10^{-5}, is less than 5% of 0.10, so the approximation is valid. The pH is

$$pH = -\log [H^+] = -\log (5.3 \times 10^{-5}) = 4.28$$

PRACTICE EXERCISE 17.3

Calculate the concentrations of hydrogen and cyanide ions in a 0.050 M HCN solution.

Although approximations are appropriate in most ionic equilibrium problems, we must nevertheless watch for cases in which K_a is unusually large or the molarity unusually small. The following example is one in which the approximation does not work.

LEARNING HINT

According to the 5% criterion, the initial molarity must be at least 400 times larger than K_a for the approximation to be valid.

EXAMPLE 17.3

Calculate the hydrogen ion concentration and pH of a 0.10 M solution of dichloroacetic acid ($CHCl_2COOH$; $K_a = 3.3 \times 10^{-2}$).

SOLUTION

The concentration summary is similar to that of Example 17.2.

CONCENTRATION SUMMARY

Equation:	$CHCl_2COOH(aq) \rightleftharpoons$	$H^+(aq)$ +	$CHCl_2COO^-(aq)$
Initial concentrations (mol/L):	0.10	0	0
Concentration changes (mol/L):	$-x$	$+x$	$+x$
Equilibrium concentrations (mol/L):	$0.10 - x$	x	x

Substituting the equilibrium concentrations into the K_a expression gives

$$K_a = \frac{[H^+][CHCl_2COO^-]}{[CHCl_2COOH]}$$

$$3.3 \times 10^{-2} = \frac{x^2}{0.10 - x}$$

LEARNING HINT

Inspection of the concentration summary shows that $0 < x < 0.10$

If we assume that $0.10 - x = 0.10$, then $x^2 = (0.10)(3.3 \times 10^{-2})$, and $x = 0.057\ M$. This value of x is more than 50% of 0.10, so the approximation is not valid. The correct value of x is obtained by rearranging the equation into quadratic form

$$x^2 + 3.3 \times 10^{-2}x - 3.3 \times 10^{-3} = 0$$

and using the quadratic formula (Appendix A.4):

$$x = \frac{-3.3 \times 10^{-2} \pm \sqrt{(3.3 \times 10^{-2})^2 - (4)(1)(-3.3 \times 10^{-3})}}{(2)(1)}$$

The negative root is rejected because it would give a negative hydrogen ion concentration. The positive root gives $x = 0.043\ M$. The pH is

$$\begin{aligned} pH &= -\log [H^+] \\ &= -\log 0.043 \\ &= 1.37 \end{aligned}$$

The following practice exercise illustrates that *for a given acid, approximate calculations become less accurate as the initial molarity of the acid decreases.*

PRACTICE EXERCISE 17.4

Calculate the hydrogen ion concentration in a 0.100 M acetic acid solution first with and then without an approximation. Repeat the calculation for a $1.00 \times 10^{-4}\ M$ solution.

17.2 BASE IONIZATION CONSTANTS

The equilibrium constant for the reaction of a weak base with water is called a **base ionization constant (K_b)**. Ammonia is a weak base

$$NH_3(aq) + H_2O(l) \rightleftharpoons NH_4^+(aq) + OH^-(aq)$$

with $K_b = 1.77 \times 10^{-5}$ at 25°C

$$K_b = \frac{[NH_4^+][OH^-]}{[NH_3]} = 1.77 \times 10^{-5}$$

Water is omitted from the K_b expression because it is the solvent (Section 15.4). Other base ionization constants and pK_b values are given in Table 17.2 and Appendix B.3.

Additional base ionization constants are given in Appendix B.3.

TABLE 17.2 Ionization Constants of Weak Bases at 25°C

Base	Formula	K_b	pK_b
Ammonia	NH_3	1.77×10^{-5}	4.752
Aniline	$C_6H_5NH_2$	4.3×10^{-10}	9.37
Methylamine	CH_3NH_2	3.70×10^{-4}	3.432
Ethylamine	$C_2H_5NH_2$	6.41×10^{-4}	3.193[a]
Trimethylamine	$(CH_3)_3N$	6.3×10^{-5}	4.20

[a]At 20°C.

PRACTICE EXERCISE 17.5

The pK_b values of ammonia and trimethylamine are 4.75 and 4.19, respectively, at 25°C. Which is the stronger base?

Calculations involving weak bases are similar to those for weak acids.

EXAMPLE 17.4

Calculate (a) the OH^- concentration and (b) the pH of a 0.075 *M* ammonia solution.

SOLUTION

Our calculation will be based on the equation given above for the ionization of NH_3. We will ignore hydroxide ions produced by the self-ionization of water because, as shown in Section 16.4, their concentration is less than 10^{-7} mol/L. First we use K_b to calculate the OH^- concentration; then we calculate the pH.

(a) If x is the equilibrium concentration of hydroxide ion, the concentration of NH_4^+ is also x, and the concentration of NH_3 is $0.075 - x$.

CONCENTRATION SUMMARY

Equation:	$NH_3(aq) + H_2O(l) \rightleftharpoons$		$NH_4^+(aq)$	$+\ OH^-(aq)$
Initial concentrations (mol/L):	0.075		0	0
Concentration changes (mol/L):	$-x$		$+x$	$+x$
Equilibrium concentrations (mol/L):	$0.075 - x$		x	x

LEARNING HINT

In Example 17.4, as in the previous examples, we ignored the ionization of water because its effect on the pH is insignificant compared to that of the solute.

Substituting the equilibrium concentrations into the K_b expression (see the beginning of this section) gives

$$1.77 \times 10^{-5} = \frac{x^2}{0.075 - x}$$

Making the usual approximation, $0.075 - x = 0.075$, we get

$$1.77 \times 10^{-5} = \frac{x^2}{0.075}$$

and

$$x = 1.2 \times 10^{-3}\,M = [OH^-]$$

The value of x is less than 5% of 0.075, so the approximation is valid.

(b) The pOH of the solution is

$$pOH = -\log [OH^-] = -\log (1.2 \times 10^{-3}) = 2.92$$

The pH is

$$pH = 14.00 - pOH = 14.00 - 2.92 = 11.08$$

PRACTICE EXERCISE 17.6

Calculate the pH of a 0.15 *M* aqueous solution of aniline ($C_6H_5NH_2$).

The Relation Between K_a and K_b of a Conjugate Acid–Base Pair

The fact that a strong acid has a weak conjugate base prompts us to look for some inverse relationship between the ionization constants of the members of the conjugate pair. Consider an acid HA and its conjugate base A^-. The acid protonates water with a strength measured by its K_a:

$$HA(aq) \rightleftharpoons H^+(aq) + A^-(aq)$$

$$K_a = \frac{[H^+][A^-]}{[HA]}$$

The base accepts protons from water with a strength measured by its K_b:

$$A^-(aq) + H_2O(l) \rightleftharpoons HA(aq) + OH^-(aq)$$

$$K_b = \frac{[HA][OH^-]}{[A^-]}$$

Multiplying the two expressions and canceling common concentrations gives

$$K_aK_b = \frac{[H^+]\cancel{[A^-]}}{\cancel{[HA]}} \times \frac{\cancel{[HA]}[OH^-]}{\cancel{[A^-]}} = [H^+][OH^-] = K_w$$

The relationship

$$K_aK_b = K_w = 1.00 \times 10^{-14} \qquad (17.2)$$

applies to any conjugate acid–base pair in aqueous solution at 25°C. If K_a for the acid is known, then K_b can be calculated for its conjugate base, and vice versa. Handbooks customarily make use of this fact by listing only one ionization constant for each conjugate pair.

LEARNING HINT

One can also show that

$$pK_a + pK_b = 14.00$$

Try Exercise 17.8.

EXAMPLE 17.5

For acetic acid, K_a is 1.76×10^{-5}. Calculate K_b for the reaction of acetate ion with water:

$$CH_3COO^-(aq) + H_2O(l) \rightleftharpoons CH_3COOH(aq) + OH^-(aq)$$

SOLUTION

Acetate ion is the conjugate base of acetic acid. Substituting $K_a = 1.76 \times 10^{-5}$ and $K_w = 1.00 \times 10^{-14}$ into Equation 17.2 gives

$$K_b = \frac{K_w}{K_a} = \frac{1.00 \times 10^{-14}}{1.76 \times 10^{-5}} = 5.68 \times 10^{-10}$$

PRACTICE EXERCISE 17.7

The value of K_b for NH_3 is 1.77×10^{-5}. Calculate K_a for the ammonium ion:

$$NH_4^+(aq) \rightleftharpoons H^+(aq) + NH_3(aq)$$

17.3 PERCENT IONIZATION

Ionic equilibria, like all other equilibria, obey Le Châtelier's principle. If the solution is diluted, or if the temperature changes, or if ions are added or removed, the system responds by shifting its composition in the direction that counteracts the change. Weak acids such as HOCl

$$HOCl(aq) + H_2O(l) \rightleftharpoons H_3O^+(aq) + OCl^-(aq)$$

and weak bases such as NH_3

$$NH_3(aq) + H_2O(l) \rightleftharpoons NH_4^+(aq) + OH^-(aq)$$

are no exception; they respond to dilution (addition of water) by undergoing further ionization to increase the number of product particles. The extent to which an acid or base ionizes is given in terms of the **percent ionization**, where

$$\begin{aligned} \%\text{ ionization} &= 100\% \times \text{fraction of molecules ionized} \\ &= 100\% \times \frac{\text{moles of acid or base ionized per liter}}{\text{initial moles of acid or base per liter}} \qquad (17.3) \end{aligned}$$

EXAMPLE 17.6

Calculate the percent of HOCl molecules ionized in (a) a 0.10 *M* solution and (b) a 0.010 *M* solution. Use the results of Example 17.2 for Part (a).

SOLUTION

(a) A 0.10 *M* HOCl solution produces an equilibrium hydrogen ion concentration of

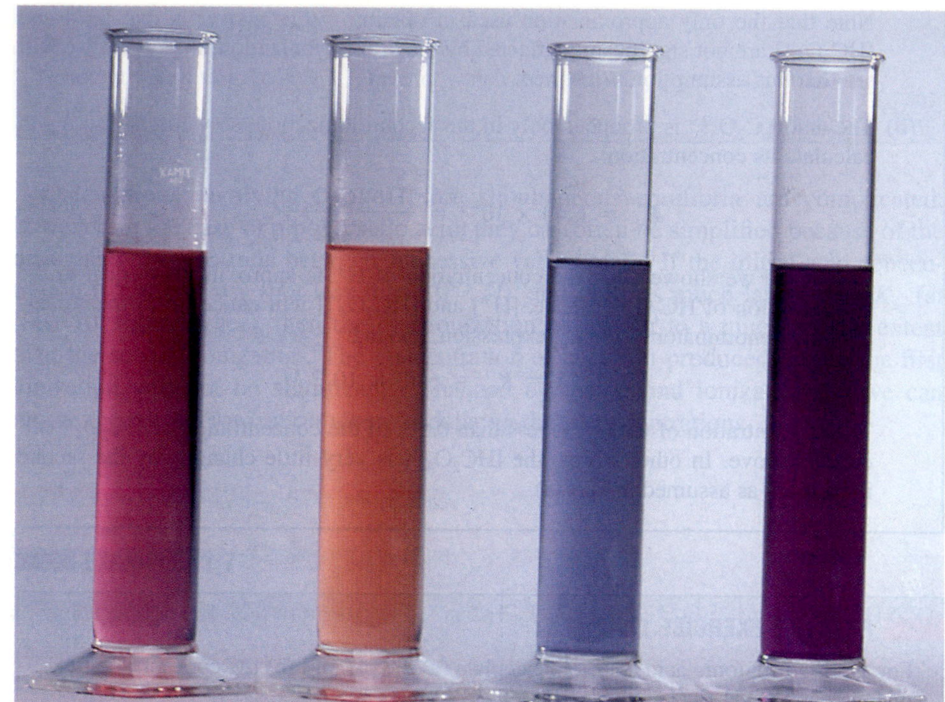

Figure 17.2

Universal indicator added to aqueous solutions of (from left to right) H_3PO_4, NaH_2PO_4, Na_2HPO_4, and Na_3PO_4. Universal indicator changes colors continuously from red (at pH 2), through orange, yellow, yellow-green (at pH 7), green, blue, to purple (at pH 12). The colors show that H_3PO_4 is the most acidic, $H_2PO_4^-$ is less acidic, HPO_4^{2-} is slightly basic, and PO_4^{3-} is very basic.

$$CH_3COO^-(aq) + H_2O(l) \rightleftharpoons CH_3COOH(aq) + OH^-(aq)$$

Because acetic acid is also produced in this reaction, the solution has a vinegary smell. Carbonates, cyanides, and other salts of weak acids react in a similar way:

$$CO_3^{2-}(aq) + H_2O(l) \rightleftharpoons HCO_3^-(aq) + OH^-(aq)$$

$$CN^-(aq) + H_2O(l) \rightleftharpoons HCN(aq) + OH^-(aq)$$

Moist potassium cyanide gives off the odor of HCN, a toxic gas that smells like bitter almonds. In general, *the weaker the acid, the more basic its anion.* Hydrogen cyanide is a weaker acid than acetic acid (Table 16.2, p. 745), therefore a 0.1 *M* solution of potassium cyanide will be more basic than a 0.1 *M* solution of sodium acetate.

The base constant for the reaction of acetate ion with water is

$$K_b = \frac{[CH_3COOH][OH^-]}{[CH_3COO^-]} = 5.68 \times 10^{-10}$$

LEARNING HINT

Recall that $K_aK_b = K_w$ (Equation 17.2). Hence, if K_a is large, K_b must be small, and vice versa.

Recall that this value of K_b was calculated from K_a for acetic acid (Example 17.5). The reaction of any basic anion with water is governed by a K_b expression analogous to that for the acetate ion.

PRACTICE EXERCISE 17.12

Write the equation and K_b expression for the reaction of sulfide ion (S^{2-}) with water.

Cations

In principle, every hydrated cation is a Brønsted acid because acidic hydrogen atoms are present in its waters of hydration. You have already learned (Section 16.5) that small, highly charged, metal cations such as $Al(H_2O)_6^{3+}$ are acidic with respect to water. Not *all* hydrated ions are acidic, however. Recall that monovalent metal cations and divalent alkaline earth metal cations (except for beryllium) are essentially neutral toward water.

LEARNING HINT

This is a good time to review the reactions of hydrated ions (Sections 16.3 and 16.5).

Ammonium ion (NH_4^+) and protonated amines such as methylammonium ion ($CH_3NH_3^+$) are also acidic in water. The acidity of ammonium chloride solution, for example, is due to the reaction

$$NH_4^+(aq) + H_2O(l) \rightleftharpoons NH_3(aq) + H_3O^+(aq)$$

or

$$NH_4^+(aq) \rightleftharpoons NH_3(aq) + H^+(aq)$$

The reaction of a salt with water is often called **hydrolysis**, a term that can be used whenever water is a reactant. The K_a and K_b values of ions are sometimes referred to as **hydrolysis constants**.

Because ammonia is also formed, moist ammonium salts always have the pungent smell of ammonia.

The acid ionization constant for ammonium ion is

$$K_a = \frac{[NH_3][H^+]}{[NH_4^+]} = 5.65 \times 10^{-10}$$

This value of K_a was calculated from K_b for NH_3 in Practice Exercise 17.7.

PRACTICE EXERCISE 17.13

Write the equation and K_a expression for the reaction of methylammonium ion ($CH_3NH_3^+$) with water.

The pH of Salt Solutions

The acidity or basicity of a salt solution can be predicted from the behavior of the ions in solution, as will be seen in the next example.

EXAMPLE 17.8

Would you expect solutions of the following salts to be acidic, basic, or neutral: (a) sodium fluoride, (b) potassium perchlorate, and (c) methylammonium chloride (CH_3NH_3Cl)?

SOLUTION

(a) A solution of sodium fluoride contains $Na^+(aq)$ and $F^-(aq)$. The hydrated sodium ion does not donate or accept protons; it will not react with water, and it has no effect on the pH of the solution. The fluoride ion is the anion of the weak acid HF. It is a weak base that will accept protons from water:

$$F^-(aq) + H_2O(l) \rightleftharpoons HF(aq) + OH^-(aq)$$

The solution will be basic.

LEARNING HINT

If you are not sure whether an acid is weak or strong, look for it in Table 16.2 (p. 745) or in a table of ionization constants.

(b) A solution of potassium perchlorate contains $K^+(aq)$ and $ClO_4^-(aq)$. The hydrated potassium ion is monovalent and will not react with water. The perchlorate ion is the anion of a strong acid ($HClO_4$); it will not react with water. The solution will be neutral.

(c) A solution of methylammonium chloride contains $CH_3NH_3^+(aq)$ and $Cl^-(aq)$. The chloride ion is the anion of a strong acid and will not react with water. Methylammonium ion, a cation derived from the weak base methylamine (CH_3NH_2), will form hydrogen ions:

$$CH_3NH_3^+(aq) \rightleftharpoons CH_3NH_2(aq) + H^+(aq)$$

The solution will be acidic.

EXAMPLE 17.9

Will a solution of ammonium nitrite (NH_4NO_2) be acidic, basic, or neutral?

SOLUTION

An ammonium nitrite solution contains $NH_4^+(aq)$ and $NO_2^-(aq)$. The NH_4^+ tends to make the solution acidic:

$$NH_4^+(aq) \rightleftharpoons NH_3(aq) + H^+(aq)$$

NO_2^- is the anion of nitrous acid (HNO_2), a weak acid. It tends to make the solution basic:

$$NO_2^-(aq) + H_2O(l) \rightleftharpoons HNO_2(aq) + OH^-(aq)$$

The net effect depends on which ion has the larger K value.

The value of K_a for the ammonium ion is 5.65×10^{-10} (Practice Exercise 17.7). The K_b value for the nitrite ion can be calculated from K_a for its conjugate acid, HNO_2. Using Equation 17.2 and data from Table 17.1,

$$K_b = \frac{K_w}{K_a} = \frac{1.00 \times 10^{-14}}{6.0 \times 10^{-4}} = 1.7 \times 10^{-11}$$

The K_a value for NH_4^+ is slightly greater than the K_b value for NO_2^-; hence, the acid strength of NH_4^+ is slightly greater than the base strength of NO_2^-. The solution will be slightly acidic.

PRACTICE EXERCISE 17.14

Predict whether solutions of the following salts will be acidic, basic, or neutral: (a) Na_3PO_4, (b) $Cu(NO_3)_2$, (c) $LiNO_3$, and (d) $(CH_3)_3NHF$.

The pH of a salt solution can be calculated from the initial molarity and the appropriate K_a or K_b value, as illustrated in the next example.

EXAMPLE 17.10

Calculate the pH of a 0.10 *M* NH_4Cl solution. The value of K_a for NH_4^+ is 5.65×10^{-10}.

SOLUTION

Ammonium chloride provides the ions $NH_4^+(aq)$ and $Cl^-(aq)$. The chloride ion does not react with water and does not affect the pH of the solution. The ammonium ion is acidic and will form hydrogen ions. Our concentration summary will be based on this reaction.

CONCENTRATION SUMMARY

Equation:	$NH_4^+(aq) \rightleftharpoons$	$NH_3(aq)$ +	$H^+(aq)$
Initial concentrations (mol/L):	0.10	0	0
Concentration changes (mol/L):	$-x$	$+x$	$+x$
Equilibrium concentrations (mol/L):	$0.10 - x$	x	x

If x is the equilibrium H^+ concentration, then the equilibrium concentrations of NH_3 and NH_4^+ are x and $0.10 - x$, respectively. Approximating $0.10 - x$ by 0.10 and substituting the equilibrium concentrations into the K_a expression gives

$$K_a = \frac{[NH_3][H^+]}{[NH_4^+]}$$

$$5.65 \times 10^{-10} = \frac{x^2}{0.10}$$

and

$$x = 7.5 \times 10^{-6}\,M = [H^+]$$

The pH of the solution is

$$pH = -\log[H^+] = -\log(7.5 \times 10^{-6}) = 5.12$$

LEARNING HINT

In Example 17.10 we ignored the ionization of water because K_b is much greater than K_w. When the value of K_a or K_b is close to that of K_w, the ions coming from water must be taken into account, and the calculations become more complicated.

The ions from water must also be included when the solution is very dilute, as in $10^{-7}\,M\,NH_4^+$.

PRACTICE EXERCISE 17.15

Calculate the pH of 0.200 *M* barium acetate, $Ba(CH_3COO)_2$ $(= Ba^{2+} + 2\,CH_3COO^-)$.

Amphiprotic Ions

An **amphiprotic ion** such as HS^- or HPO_4^{2-} can accept or donate protons. The reactions of HS^- with water are

$$HS^-(aq) \rightleftharpoons H^+(aq) + S^{2-}(aq) \qquad K_a = 1 \times 10^{-19}$$

$$HS^-(aq) + H_2O(l) \rightleftharpoons H_2S(aq) + OH^-(aq) \qquad K_b = 1.0 \times 10^{-7}$$

The K_b value is greater than the K_a value, showing that HS^- is stronger as a base than as an acid. A solution containing HS^- ions will be basic because the second reaction will occur to a greater extent than the first reaction.

Other amphiprotic ions give rise to acidic solutions. A solution of sodium hydrogen sulfate, for example, is acidic. The predominant reaction, that is, the one with the larger K value, is

$$HSO_4^-(aq) \rightleftharpoons H^+(aq) + SO_4^{2-}(aq)$$

PRACTICE EXERCISE 17.16

Use data from Table 17.1 to predict whether a solution of $NaHCO_3$ will be acidic or basic. Write an equation for the reaction that predominates in aqueous solution. (*Hint*: Calculate K_b for HCO_3^-; then compare the values of K_a and K_b.)

17.6 THE COMMON ION EFFECT

Consider the equilibrium that exists in an aqueous solution of acetic acid:

$$CH_3COOH(aq) \rightleftharpoons CH_3COO^-(aq) + H^+(aq)$$

Adding more CH_3COO^- ions shifts the equilibrium to the left.

If we dissolve sodium acetate in the solution (Demonstration 17.1a), Le Châtelier's principle is obeyed and the equilibrium shifts to the left. A new equilibrium mixture containing fewer hydrogen ions and more acetic acid molecules is formed. This response is an example of the **common ion effect**, a shift in equilibrium caused by the addition of an ion common to the equilibrium mixture. Acetic acid is a weak electrolyte, and this illustration of the common ion effect shows that *the ionization of a weak electrolyte is repressed by the addition of a solute having an ion in common with the electrolyte.*

The repression of the ionization of ammonia by an ammonium salt is another example of the common ion effect (Demonstration 17.1b):

$$NH_3(aq) + H_2O(l) \rightleftharpoons NH_4^+(aq) + OH^-(aq)$$

Adding more NH_4^+ ions shifts the equilibrium to the left.

Still another example is the repression of the self-ionization of water by the addition of acid or base (Section 16.4):

$$H_2O(l) \rightleftharpoons H^+(aq) + OH^-(aq)$$

Adding H^+ or OH^- ions shifts the equilibrium to the left.

PRACTICE EXERCISE 17.17

What salt might you add to repress the ionization of (a) benzoic acid (C_6H_5COOH) and (b) aniline ($C_6H_5NH_2$)? Would the pH in each case be raised or lowered by the added salt?

Most problems involving the common ion effect are equilibrium problems in which two of the concentrations are known.

DEMONSTRATION 17.1 THE COMMON ION EFFECT

(a) The beakers originally contained 0.1 *M* acetic acid (CH_3COOH) plus a few drops of methyl red (red below pH 4.4, yellow above pH 6.2). Sodium acetate added to the beaker on the right raised the pH by repressing the ionization of acetic acid.

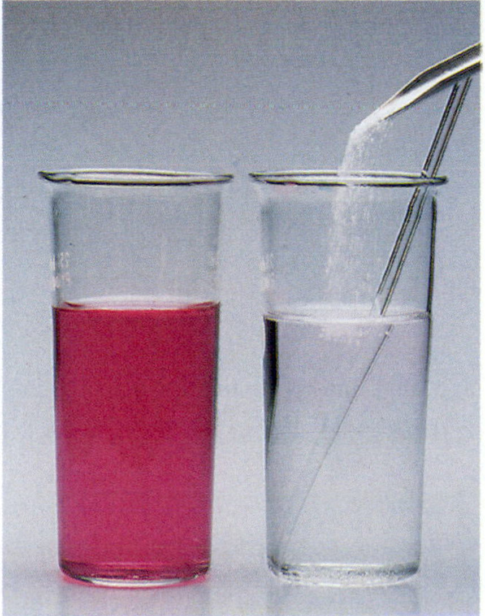

(b) Both beakers originally contained 0.1 *M* ammonia (NH_3) plus a few drops of phenolpthalein (red above pH 10, colorless below pH 8.2). Ammonium chloride added to the beaker on the right lowered the pH by repressing the ionization of ammonia.

EXAMPLE 17.11

Calculate the pH of a 0.050 *M* acetic acid solution that is also 0.10 *M* in sodium acetate.

SOLUTION

The concentration summary is based on the equation for the equilibrium between acetic acid and acetate ions. Sodium acetate is a salt and is therefore completely ionized; a 0.10 *M* solution of sodium acetate is 0.10 *M* in acetate ions. The initial concentrations of acetic acid and acetate ions are 0.050 and 0.10 *M*, respectively.

CONCENTRATION SUMMARY

Equation:	$CH_3COOH(aq)$ $\rightleftharpoons$	$CH_3COO^-(aq)$ +	$H^+(aq)$
Initial concentrations (mol/L):	0.050	0.10	0
Concentration changes (mol/L):	$-x$	$+x$	$+x$
Concentration changes (mol/L):	$-x$	$+x$	$+x$
Equilibrium concentrations (mol/L):	$0.050 - x$	$0.10 + x$	x
Approximate concentrations (mol/L):	0.050	0.10	x

The partial ionization of acetic acid provides x moles per liter of H^+ and an additional x moles per liter of acetate ion. The magnitude of x is small because of the common ion effect, so the calculation can be simplified by assuming that $0.050 - x = 0.050$ and $0.10 + x = 0.10$. These approximations have been included in the concentration summary. Substituting the approximate concentrations and $K_a = 1.76 \times 10^{-5}$ (Table 17.1) into the K_a expression gives

$$K_a = \frac{[CH_3COO^-][H^+]}{[CH_3COOH]}$$

$$1.76 \times 10^{-5} = \frac{(0.10)(x)}{0.050}$$

and

$$x = 8.8 \times 10^{-6}\,M = [H^+]$$

Note that our approximation satisfies the 5% criterion; the calculated value of x is less than 5% of 0.050 M. The pH is

$$\begin{aligned} pH &= -\log [H^+] \\ &= -\log (8.8 \times 10^{-6}) \\ &= 5.06 \end{aligned}$$

PRACTICE EXERCISE 17.18

Calculate the OH^- concentration in a solution that is 0.200 M in methylamine (CH_3NH_2) and 0.200 M in methylammonium chloride (CH_3NH_3Cl).

Compare the hydrogen ion concentration in the 0.050 M acetic acid solution of Example 17.1 with that of the 0.050 M acetic acid solution in Example 17.11. Observe that the additional acetate ions reduce the concentration of H^+ from $9.3 \times 10^{-4}\,M$ to $8.8 \times 10^{-6}\,M$, more than a hundredfold reduction. Adding more sodium acetate would reduce the H^+ concentration even further. This observation suggests that the common ion effect can be used to adjust the concentration of the other ion, in this case H^+. We will develop this idea in Section 17.7.

Two salts can also exhibit a common ion effect. In a solution of NaH_2PO_4 and Na_2HPO_4, for example, the HPO_4^{2-} ions will repress the ionization of $H_2PO_4^-$ ions:

$$H_2PO_4^-(aq) \rightleftharpoons H^+(aq) + HPO_4^{2-}(aq)$$

The HPO_4^{2-} ions from Na_2HPO_4 shift the equilibrium to the left.

A similar effect occurs in solutions containing HCO_3^- and CO_3^{2-} ions, HPO_4^{2-} and PO_4^{3-} ions, and so forth.

EXAMPLE 17.12

Find the pH of a solution containing 0.20 mol of NaH_2PO_4 and 0.10 mol of Na_2HPO_4 per liter.

SOLUTION

The equilibrium that concerns us is

$$H_2PO_4^-(aq) \rightleftharpoons H^+(aq) + HPO_4^{2-}(aq)$$

which is the second ionization step of phosphoric acid and for which $K_a = 6.23 \times 10^{-8}$ (Table 17.1). (*Note*: The ions $H_2PO_4^-$ and HPO_4^{2-} are amphiprotic, so H_3PO_4 and PO_4^{3-} are also present in the equilibrium mixture. The equilibrium constants for the reactions that produce H_3PO_4 and PO_4^{3-}, however, are much smaller than 6.23×10^{-8}, so these reactions can be ignored relative to the above reaction.)

CONCENTRATION SUMMARY

Equation:	$H_2PO_4^-(aq) \rightleftharpoons$	$H^+(aq)$ +	$HPO_4^{2-}(aq)$
Initial concentrations (mol/L):	0.20	0	0.10
Concentration changes (mol/L):	$-x$	$+x$	$+x$
Equilibrium concentrations (mol/L):	$0.20 - x$	x	$0.10 + x$
Approximate concentrations (mol/L):	0.20	x	0.10

The initial concentrations of $H_2PO_4^-$ and HPO_4^{2-} are 0.20 and 0.10 M, respectively. Let x be the equilibrium hydrogen ion concentration. The value of x will be small because of the common ion effect, so we will assume that $0.20 - x = 0.20$ and $0.10 + x = 0.10$. Substituting the approximate concentrations and $K_a = 6.23 \times 10^{-8}$ into the K_a expression gives

$$K_a = \frac{[H^+][HPO_4^{2-}]}{[H_2PO_4^-]}$$

$$6.23 \times 10^{-8} = \frac{(x)(0.10)}{0.20}$$

and

$$x = 1.2 \times 10^{-7}\ M = [H^+]$$

The pH is calculated in the usual way:

$$pH = -\log [H^+] = -\log (1.2 \times 10^{-7}) = 6.92$$

PRACTICE EXERCISE 17.19

Refer to Table 17.1, and calculate the pH of a solution that is 0.10 M in potassium malonate ($K_2C_3H_2O_4$) and 0.10 M in potassium hydrogen malonate ($KHC_3H_2O_4$).

Mixtures of Strong and Weak Acids or Bases

Hydrogen ion is the common ion in a solution containing more than one acid. If the solution contains substantial amounts of both a strong acid and a weak acid, the H^+ provided by the complete ionization of the strong acid will repress the ionization of the weak acid. The ionization of acetic acid, for example, is repressed by the addition of HCl:

$$CH_3COOH(aq) \rightleftharpoons CH_3COO^-(aq) + H^+(aq)$$

The H^+ ions coming from HCl shift the equilibrium to the left.

Because very little of the weak acid is ionized, *virtually all of the hydrogen ions in a mixture of a strong acid and a weak acid come from the strong acid.* The pH of such a solution will be almost identical to that of a solution containing the strong acid alone.

PRACTICE EXERCISE 17.20

The acid component of gastric juice is HCl. The gastric juice of a certain patient has a volume of approximately 100 mL and a pH of 1.20. What is the pH of the gastric juice after the patient ingests a 325-mg aspirin tablet? Aspirin is acetylsalicylic acid ($HC_{16}H_{11}O_6$, $K_a = 3.5 \times 10^{-4}$).

A similar situation exists in mixtures of strong and weak bases. The ionization of NH_3, for example, is repressed by the addition of NaOH, a strong base:

$$NH_3(aq) + H_2O(l) \rightleftharpoons NH_4{}^+(aq) + OH^-(aq)$$

The OH^- ions coming from NaOH shift the equilibrium to the left.

PRACTICE EXERCISE 17.21

Calculate the pH of a solution prepared by dissolving 0.010 mol of NaOH in 250 mL of 0.50 M NH_3.

17.7 BUFFER SOLUTIONS

Imagine a solution that contains appreciable amounts of both a weak acid and its conjugate base; for example, a solution that has 0.10 mol of acetic acid and 0.10 mol of sodium acetate per liter. The equilibrium between acetate ions and molecular acetic acid

$$CH_3COOH(aq) \rightleftharpoons CH_3COO^-(aq) + H^+(aq)$$

Both species are present in substantial concentration.

protects the solution against attempts to change its pH. If a small quantity of strong acid is added to the solution, the additional hydrogen ions will be consumed by acetate ions

$$H^+(aq) + CH_3COO^-(aq) \rightarrow CH_3COOH(aq)$$

Similarly, added hydroxide ions will be consumed by molecular acetic acid

$$CH_3COOH(aq) + OH^-(aq) \rightarrow CH_3COO^-(aq) + H_2O(l)$$

This solution is an example of a **buffer solution**, a solution that resists pH changes.

Buffer solutions are used in the laboratory whenever a constant pH is required for running a reaction or for making consistent measurements. Even more important is the role played by buffers in maintaining the pH of body fluids within the limits compatible with life (see *Chemical Insight: Buffers in the Blood* later in this section).

A *mechanical buffer* is a device for absorbing or lessening the shock of an impact, for example, the springs and shock absorbers of a car. A *chemical buffer* protects a solution against abrupt pH changes that would result from the addition of an acid or base.

Buffer solutions can be prepared from weak acids or weak bases and their salts. The acetic acid–sodium acetate buffer was made from a weak acid and its salt. An ammonia–ammonium chloride buffer containing NH_3 and its conjugate acid NH_4^+ can be made by dissolving NH_4Cl in aqueous ammonia:

$$NH_3(aq) + H_2O(l) \rightleftharpoons NH_4^+(aq) + OH^-(aq)$$

Ammonia counteracts the addition of acid

$$NH_3(aq) + H^+(aq) \rightarrow NH_4^+(aq)$$

and ammonium ions counteract the addition of base

$$NH_4^+(aq) + OH^-(aq) \rightarrow NH_3(aq) + H_2O(l)$$

A buffer solution can also be made from two salts that provide a conjugate acid–base pair. A solution containing NaH_2PO_4 and Na_2HPO_4 is a phosphate buffer that depends on the equilibrium

$$H_2PO_4^-(aq) \rightleftharpoons HPO_4^{2-}(aq) + H^+(aq)$$

to neutralize acid by converting HPO_4^{2-} to $H_2PO_4^-$ and to neutralize base by converting $H_2PO_4^-$ to HPO_4^{2-}.

LEARNING HINT

Strong acids and strong bases cannot form buffer solutions because their reactions with water are virtually complete, even in the presence of a common ion.

In effect, a buffer solution is a common ion system that maintains a steady $H^+(aq)$ concentration by shifting the equilibrium between a conjugate acid–base pair. The pH of a buffer system can be calculated by substituting the acid and base concentrations into the K_a expression, as in Examples 17.11 and 17.12. An equivalent but more direct method involves rearranging the K_a expression to yield pH directly. This method is shown below for an acetic acid–sodium acetate buffer. The K_a expression for this system is

$$K_a = \frac{[H^+][CH_3COO^-]}{[CH_3COOH]}$$

The hydrogen ion concentration is obtained by rearrangement

$$[H^+] = K_a \times \frac{[CH_3COOH]}{[CH_3COO^-]}$$

and the pH is found by taking the negative logarithm of both sides

$$pH = -\log [H^+] = -\log K_a - \log \frac{[CH_3COOH]}{[CH_3COO^-]}$$

We then substitute the relation $pK_a = -\log K_a$ and use the identity $\log (A/B) = -\log (B/A)$ to give

$$pH = pK_a + \log \frac{[CH_3COO^-]}{[CH_3COOH]}$$

A similar derivation for any buffer system consisting of a conjugate acid–base pair leads to the general equation

$$pH = pK_a + \log \frac{[\text{base}]}{[\text{acid}]} \tag{17.4}$$

where [acid] and [base] are the molar concentrations of the acid and conjugate base, respectively, and K_a is the acid ionization constant. Equation 17.4 is known as the **Henderson–Hasselbalch equation**. It is widely used for routine buffer calculations in biology and biochemistry laboratories.

The concentration ratio, [base]/[acid], is called the **buffer ratio**. Equation 17.4 shows that the pH of a buffer solution depends directly on the buffer ratio; it increases when the ratio increases and decreases when the ratio decreases. Observe that only the ratio of base to acid is important, not the concentrations themselves. In most buffer calculations, numbers of moles can be substituted for concentrations because the mole ratio in a given solution is the same as the concentration ratio.

EXAMPLE 17.13

Use Equation 17.4 to calculate the pH of a solution that is 0.40 *M* in NH_3 and 0.30 *M* in NH_4Cl.

SOLUTION

NH_4^+ is the acid and NH_3 is the base. Equation 17.4 becomes

$$pH = pK_a + \log \frac{[NH_3]}{[NH_4^+]}$$

The value of K_a for NH_4^+ is 5.65×10^{-10} (Practice Exercise 17.7); hence,

$$pK_a = -\log K_a = -\log(5.65 \times 10^{-10}) = 9.248$$

The approximate concentrations of NH_3 and NH_4^+ are 0.40 and 0.30 *M*, respectively. (*Note*: Some NH_3 will react with water to form additional NH_4^+, but the extent of this reaction will be small because ammonium ion is already present.) Substituting the pK_a and approximate concentrations into the above equation gives the pH:

$$pH = 9.248 + \log \frac{0.40}{0.30} = 9.37$$

LEARNING HINT

When the K_a value is much smaller than the concentrations, as it is for most buffers, the initial acid and base concentrations can be substituted for equilibrium concentrations in the Henderson–Hasselbalch equation. Changes in the initial concentrations are routinely neglected in buffer calculations.

PRACTICE EXERCISE 17.22

The pH of a formic acid–sodium formate buffer is 3.95. Calculate the buffer ratio. The K_a value for formic acid is 1.9×10^{-4}.

Even though buffer solutions resist changes in pH, slight changes do occur when acid or base is added. Example 17.14 compares the effects of adding the same quantity of strong acid to buffered and unbuffered solutions of the same pH.

EXAMPLE 17.14

A buffer solution has a volume of 100 mL and contains 0.0100 mol each of sodium benzoate and benzoic acid (C_6H_5COOH, $pK_a = 4.190$). The pH of this buffer is 4.19, the same as the pH of an unbuffered 6.46×10^{-5} *M* HCl solution. (You should verify the pH of each solution.) Calculate the pH changes that occur when 0.10 mL (about two drops) of 10 *M* HCl is added to (a) 100 mL of the unbuffered acid and (b) the buffer solution.

SOLUTION

The number of moles of HCl added to each solution is

$$0.10 \text{ mL} \times \frac{1 \text{ L}}{1000 \text{ mL}} \times \frac{10 \text{ mol HCl}}{1 \text{ L}} = 0.0010 \text{ mol HCl}$$

(a) The volume of the unbuffered solution (100 mL) is not significantly changed by the addition of 0.10 mL of HCl (100 mL + 0.10 mL = 100 mL = 0.100 L). The additional 0.0010 mol of HCl ionizes completely, so the H^+ concentration increases by 0.0010 mol/0.100 L = 0.010 *M*. The new H^+ concentration is

$$6.46 \times 10^{-5}\,M + 0.010\,M = 0.010\,M$$

The new pH is

$$\text{pH} = -\log\,[\text{H}^+] = -\log 0.010 = 2.0$$

The pH changes from 4.19 to 2.0, a decrease of 2.2 units.

(b) Since HCl is a strong acid, it will react completely with the basic component of the buffer; 0.0010 mol HCl will convert 0.0010 mol of $C_6H_5COO^-$ to 0.0010 mol of C_6H_5COOH:

$$\underset{-0.0010\text{ mol}}{C_6H_5COO^-(aq)} + H^+(aq) \longrightarrow \underset{+0.0010\text{ mol}}{C_6H_5COOH(aq)}$$

The new quantities of benzoate ion and benzoic acid are:

$$C_6H_5COO^-\text{: } 0.0100 \text{ mol} - 0.0010 \text{ mol} = 0.0090 \text{ mol}$$

$$C_6H_5COOH\text{: } 0.0100 \text{ mol} + 0.0010 \text{ mol} = 0.0110 \text{ mol}$$

The new pH is calculated from Equation 17.4:

$$\text{pH} = pK_a + \log \frac{[C_6H_5COO^-]}{[C_6H_5COOH]}$$

$$= 4.190 + \log \frac{0.0090}{0.0110} = 4.10$$

(The mole ratio, which is the same as the concentration ratio, was used in this calculation.) The pH drops from 4.19 to 4.10, a change of only 0.09 units. Note that the pH change in the unbuffered solution of Part (a) is almost 25 times greater than in the buffered solution. Buffers really do work! (See Demonstration 17.2.)

PRACTICE EXERCISE 17.23

Calculate the final pH after 0.0400 g (1.00×10^{-3} mol) of solid NaOH is dissolved in the buffer of Example 17.14.

DEMONSTRATION 17.2 EFFECTIVENESS OF A BUFFER

Both solutions, unbuffered HCl on the left and an acetic acid–acetate buffer on the right, have a pH of 5.0 and contain a few drops of thymol blue indicator (orange-yellow below pH 8.0, purple-blue above pH 9.6). The graduated cylinders each contain 25.0 mL of 0.1 *M* NaOH

1.0 mL of the base was added to each solution. The purple-blue color of the unbuffered solution shows that its pH is 9.6 or above, a change of at least 4.6 units. The buffered solution shows no color change.

Addition of the remaining 24.0 mL of base to the buffered solution produces no change in indicator color.

Preparing Buffer Solutions

The number of moles of strong acid or strong base needed to change the pH of 1 liter of buffer by one unit is called the **buffer capacity**. The larger the buffer capacity, the more resistant the buffer is to changes in pH. The capacity of a given buffer is determined by the concentrations of acid and base and also by the buffer ratio. For example, if the buffer solution in Example 17.14 contained 1.00 mol of each component rather than 0.0100 mol of each, its buffer ratio would be the same, and its initial pH would still be 4.19. The new, more concentrated buffer, however, would be able to neutralize more acid or base than the original buffer and would thus be more resistant to pH change. *For a given buffer ratio, the capacity of a buffer system increases with increasing concentrations of conjugate acid and base.*

How is the buffer capacity affected by the buffer ratio? In Example 17.14 we found that the addition of 0.0010 mol HCl changed the base/acid ratio from 0.0100/0.0100 = 1.00 to 0.0090/0.0110 = 0.82, a decrease of 18%. If the buffer had originally contained 0.0020 mol of base and 0.0180 mol of acid, the same amount of HCl would have changed the ratio from 0.0020/0.0180 = 0.11 to 0.0010/0.0190 = 0.053, a decrease of more than 50%. These calculations show that a buffer with a ratio of 1.0 is less sensitive to the addition of strong acid than a buffer with a ratio of 0.11. Further calculations of this sort would show that the buffer ratio and, consequently, the pH are most resistant to change when the ratio is exactly 1, that is, when the concentrations of acid and base are equal. When the buffer ratio is 1, Equation 17.4 becomes

$$\mathrm{pH} = \mathrm{p}K_\mathrm{a} + \log\frac{[\mathrm{base}]}{[\mathrm{acid}]} = \mathrm{p}K_\mathrm{a} + \log 1 = \mathrm{p}K_\mathrm{a}$$

since log 1 = 0. In other words, *the most effective buffer for maintaining a given pH is one whose $\mathrm{p}K_\mathrm{a}$ is exactly equal to the pH.* Since there are not enough acids to match every pH with a $\mathrm{p}K_\mathrm{a}$, chemists must often settle for buffers with $\mathrm{p}K_\mathrm{a}$ values that are close to the desired pH. In practice, a buffer is usually acceptable if the $\mathrm{p}K_\mathrm{a}$ of the acid is within one unit of the desired pH.

Most efficient buffer: $\mathrm{p}K_\mathrm{a} = \mathrm{pH}$
Acceptable buffer: $\mathrm{p}K_\mathrm{a} = \mathrm{pH} \pm 1$

EXAMPLE 17.15

Use the table of acid ionization constants to choose a buffer system for pH 7.00. Identify the substances to be used.

SOLUTION

We will look in Table 17.1 for acids with $\mathrm{p}K_\mathrm{a}$ values within one unit of the desired pH. Several candidates present themselves: aqueous CO_2 with $\mathrm{p}K_{\mathrm{a}_1} = 6.37$, H_2S with $\mathrm{p}K_{\mathrm{a}_1} = 6.99$, HOCl with $\mathrm{p}K_\mathrm{a} = 7.55$, veronal with $\mathrm{p}K_\mathrm{a} = 7.43$, and phosphoric acid with $\mathrm{p}K_{\mathrm{a}_2} = 7.21$. Practical considerations eliminate carbonic acid (unstable), H_2S (smelly), and HOCl (unstable). Veronal would be satisfactory, but it may not be readily available. Let us choose a phosphate buffer that utilizes the second ionization of H_3PO_4:

Veronal and its sodium salts have been used clinically since 1903 as sedatives and hypnotics.

$$H_2PO_4^-(aq) \rightleftharpoons H^+(aq) + HPO_4^{2-}(aq) \qquad \mathrm{p}K_\mathrm{a} = 7.21$$

The buffer would be made from soluble salts containing $H_2PO_4^-$ and HPO_4^{2-}. Sodium salts are usually the least expensive.

PRACTICE EXERCISE 17.24

What substances could you choose to prepare a buffer solution of pH 3.52?

Example 17.15 shows that the pK_a of a buffer system does not usually equal the desired pH. The next example shows how the buffer ratio is adjusted to make up the difference.

EXAMPLE 17.16

How many moles of solid Na_2HPO_4 must be added to 250 mL of 0.25 *M* NaH_2PO_4 to make a buffer solution of pH = 7.00?

SOLUTION

The pK_a for the ionization of $H_2PO_4^-$ is 7.21 (Example 17.15). The pH of the buffer solution must satisfy the Henderson–Hasselbalch equation, that is,

$$\mathrm{pH} = \mathrm{p}K_a + \log \frac{[\mathrm{HPO_4^{2-}}]}{[\mathrm{H_2PO_4^-}]}$$

$$7.00 = 7.21 + \log \frac{[\mathrm{HPO_4^{2-}}]}{[\mathrm{H_2PO_4^-}]}$$

Rearrangement gives

$$\log \frac{[\mathrm{HPO_4^{2-}}]}{[\mathrm{H_2PO_4^-}]} = 7.00 - 7.21 = -0.21$$

and

$$\frac{[\mathrm{HPO_4^{2-}}]}{[\mathrm{H_2PO_4^-}]} = \text{antilog}\,(-0.21) = 10^{-0.21} = 0.62$$

The concentration of $H_2PO_4^-$ is 0.25 *M*. Hence, the concentration of HPO_4^{2-} must be

$$[\mathrm{HPO_4^{2-}}] = 0.62 \times [\mathrm{H_2PO_4^-}] = 0.62 \times 0.25\,M = 0.16\,M$$

The number of moles of Na_2HPO_4 that must be added to 250 mL of the solution is

$$0.250\ \mathrm{L} \times \frac{0.16\ \mathrm{mol\ Na_2HPO_4}}{1\ \mathrm{L}} = 0.040\ \mathrm{mol\ Na_2HPO_4}$$

PRACTICE EXERCISE 17.25

How many moles of solid sodium formate must be added to 500 mL of 0.10 *M* formic acid to make a buffer of pH = 3.52?

CHEMICAL INSIGHT

BUFFERS IN THE BLOOD

The pH of normal arterial blood fluctuates only slightly around its average value of 7.40. A deviation as small as 0.2 pH units will seriously disturb blood chemistry, while a deviation of 0.4 units is usually fatal. The principal buffer systems that maintain blood pH are acidic and basic proteins, the most important of which is hemoglobin (HHb):

$$HHb(aq) \rightleftharpoons H^+(aq) + Hb^-(aq)$$

The pK_a values for these proteins are approximately equal to the blood pH of 7.4, so in addition to being the most abundant buffer systems, they are also the most effective.

Another buffer system involved in the regulation of blood pH is the carbon dioxide–bicarbonate buffer

$$CO_2(aq) + H_2O(l) \rightleftharpoons H^+(aq) + HCO_3^-(aq)$$

for which $pK_a = 6.10$ at normal body temperature. At pH 7.4, the $[HCO_3^-]/[CO_2]$ ratio in blood plasma is about 20/1; the bicarbonate concentration is some 20 times greater than the concentration of dissolved CO_2. This ratio manages to persist despite all of the concentration changes that result from eating, exercise, stress, illness, and so forth.

The pH level of blood is under constant assault by the production of carbon dioxide during cell metabolism. An increase in blood CO_2 would shift the carbon dioxide equilibrium (see above), producing more H^+ and lowering the blood pH; this potentially serious condition is called *acidosis*. To prevent acidosis, the body rids itself of excess CO_2 through a mechanism that involves both hemoglobin molecules and bicarbonate ions.

Hemoglobin molecules pick up oxygen in the lungs and release it in the cells. Bicarbonate ions carry carbon dioxide from the cells and release it in the lungs. The detailed mechanism for gas transport in the blood is complex, but the overall picture is not difficult to understand. Let us start in the lungs, where the circulating blood contains both hemoglobin and dissolved bicarbonate ions. Here oxygen is in good supply, and hemoglobin is converted to its oxygenated form, HbO_2^-:

$$HHb(aq) + O_2(aq) \rightarrow H^+(aq) + HbO_2^-(aq)$$

(This equation is very much simplified. One hemoglobin molecule actually carries up to four molecules of oxygen.) The hydrogen ions released during the formation of HbO_2^- combine with the dissolved bicarbonate ions to form carbonic acid, which then becomes CO_2:

$$H^+(aq) + HCO_3^-(aq) \rightarrow H_2CO_3(aq) \rightarrow H_2O(l) + CO_2(g)$$

The carbon dioxide gas is exhaled.

The oxygenated hemoglobin (HbO_2^-) is carried by the blood to the cells of the body, where it provides the oxygen necessary for cell metabolism:

$$HbO_2^-(aq) + H^+(aq) \rightarrow HHb(aq) + O_2(aq)$$

During metabolism, glucose and other cell nutrients are converted to carbon dioxide and water. This CO_2 enters the bloodstream, where it forms carbonic acid and, ultimately, bicarbonate ions:

$$CO_2(aq) + H_2O(l) \rightarrow H_2CO_3(aq) \rightarrow H^+(aq) + HCO_3^-(aq)$$

The deoxygenated hemoglobin and the bicarbonate ions then travel back to the lungs, where the cycle starts over again.

These equations show that the reactions that occur in the cells are, in effect, the reverse of those that occur in the lungs. The conversion of CO_2 to HCO_3^- in the bloodstream and back again to CO_2 in the lungs is the body's major mechanism for disposing of excess CO_2.

17.8 A CLOSER LOOK AT ACID–BASE TITRATIONS

In a titration, a solution of one reactant is gradually added to a measured amount of another reactant until some color change or meter reading signals an endpoint. The endpoint should coincide as closely as possible with the **equivalence point**, that is, the point at which neither reactant is in excess. The equivalence point for the addition of sodium hydroxide to hydrochloric acid

$$NaOH(aq) + HCl(aq) \rightarrow NaCl(aq) + H_2O(l)$$

LEARNING HINT

Review the material on titrations, endpoints, and indicators (Section 4.6).

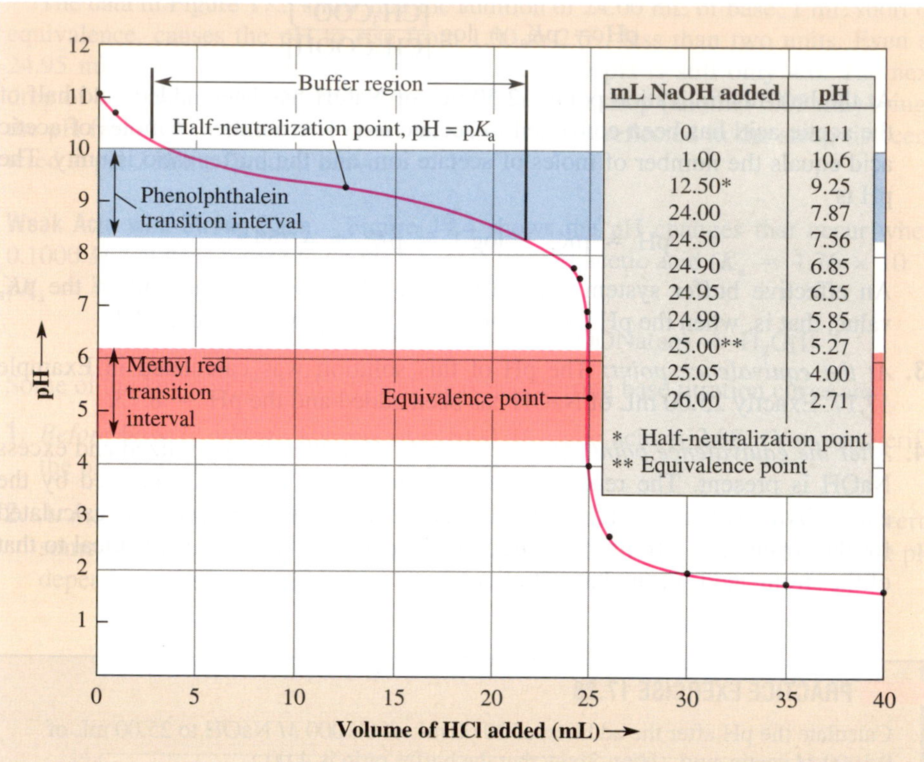

mL NaOH added	pH
0	11.1
1.00	10.6
12.50*	9.25
24.00	7.87
24.50	7.56
24.90	6.85
24.95	6.55
24.99	5.85
25.00**	5.27
25.05	4.00
26.00	2.71

Figure 17.5

A weak base–strong acid titration curve, showing how the pH decreases as 0.1000 *M* HCl is added to 25.00 mL of 0.1000 *M* NH_3. The equivalence point pH is 5.27. The steep portion of the curve includes the transition interval of methyl red but not of phenolphthalein.

More About Indicators

Most indicators are complex organic molecules that change color over a characteristic pH range called a **transition interval** (see Figure 17.6 and Table 17.3). Litmus, the active ingredient in litmus test paper, is extracted from lichen, and many other indicators are also obtained from vegetable material. Phenolphthalein, on the other hand, is a synthetic product with medicinal as well as chemical uses.

An indicator is a weak acid or base whose two conjugate forms differ in color (Figure 17.7). If we represent the acid form by HIn, and the base form by In^-, the equilibrium in aqueous solution is

$$\underset{\text{acid color}}{HIn\,(aq)} \rightleftharpoons H^+(aq) + \underset{\text{base color}}{In^-(aq)}$$

The color changes can be explained in terms of Le Châtelier's principle. In an acidic solution, the high concentration of hydrogen ions shifts the equilibrium to the left to form more HIn and a deeper acid color. Conversely, the high concentration of hydroxide ions in basic solutions shifts the equilibrium toward In^- and a deeper base color.

The Henderson–Hasselbalch equation (Equation 17.4)

$$pH = pK_a + \log \frac{[In^-]}{[HIn]}$$

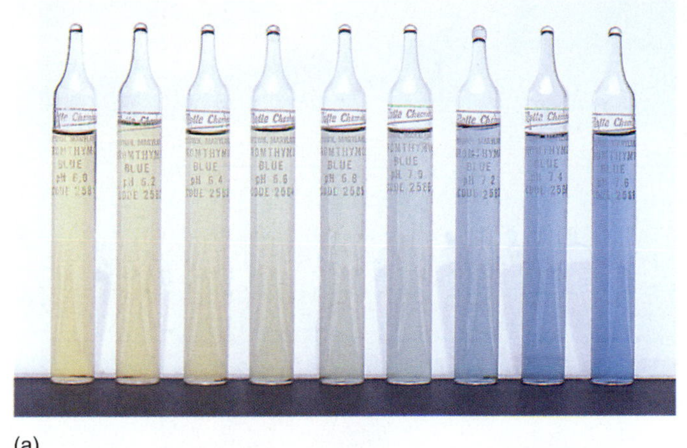
(a)

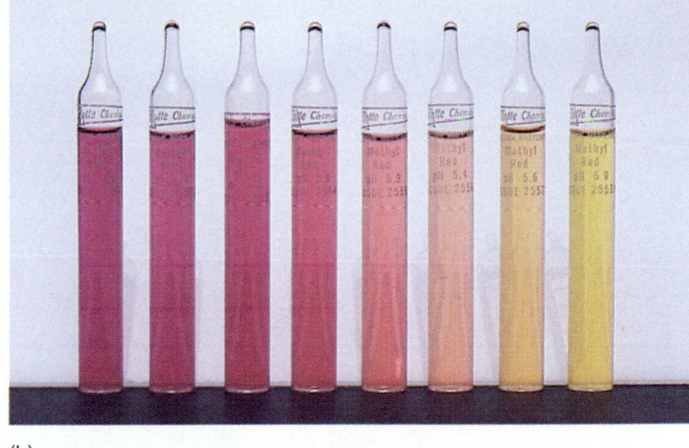
(b)

Figure 17.6

Color changes within the transition interval of (a) bromthymol blue (yellow below pH 6.0, blue above pH 7.6) and (b) methyl red (red below pH 4.4, yellow above pH 6.2).

shows that when [HIn] is equal to $[In^-]$, the pH of the solution will equal the pK_a of the indicator. At this pH, the color of the solution will be a half-and-half blend of the acid and base colors. The human eye is not sensitive enough to detect this point exactly, but it can identify the interval over which the color change takes place. The width of the transition interval varies somewhat with the hue and intensity of the indicator, but as shown in Table 17.3, the typical range is about two pH units. If we assume that $[HIn] = [In^-]$ at the midpoint of this range, it follows that the transition interval of a typical indicator extends from pH = pK_a − 1 to pH = pK_a + 1.

The titration curve may provide guidance in choosing an appropriate indicator for

TABLE 17.3 Transition Intervals of Selected Acid–Base Indicators

Indicator	Acid color	Transition Interval	Base Color
Methyl violet	Yellow	0.0–1.6	Blue
Thymol blue (acid range)	Red	1.2–2.8	Yellow
Bromphenol blue	Yellow	3.0–4.6	Blue
Congo red	Blue	3.0–5.0	Red
Methyl orange	Red	3.1–4.4	Yellow
Bromcresol green	Yellow	3.8–5.4	Blue
Methyl red	Red	4.4–6.2	Yellow
Litmus	Red	4.5–8.3	Blue
Bromcresol purple	Yellow	5.2–6.8	Purple
Bromthymol blue	Yellow	6.0–7.6	Blue
Phenol red	Yellow	6.8–8.4	Red
Thymol blue (basic range)	Yellow	8.0–9.6	Blue
Phenolphthalein	Colorless	8.2–10.0	Red
Thymolphthalein	Colorless	9.3–10.5	Blue
Alizarin yellow R	Yellow	10.0–12.1	Red

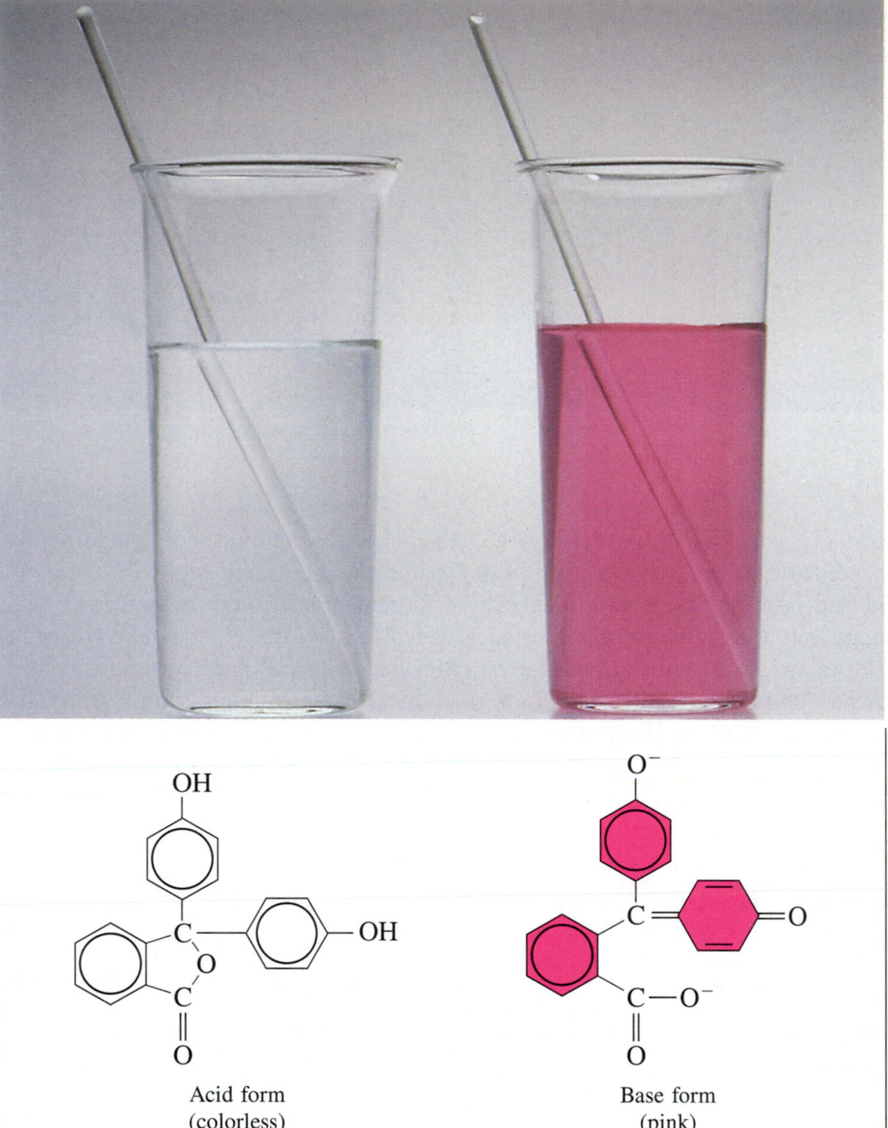

Figure 17.7

Phenolphthalein. *Left*: Acid form (colorless). *Right*: Base form (red). One advantage of phenolphthalein over other indicators is that for most people, the change from colorless to red is easier to see than other color changes.

a given titration. The steep pH rise in the strong acid–strong base titration of Figure 17.3 means that any indicator whose transition interval falls between pH 4 and pH 10 will change color within about one drop of the equivalence point. Methyl red and phenolphthalein are commonly used for strong acid–strong base titrations. Their transition intervals are superimposed on Figure 17.3.

The shorter rise in weak acid–strong base titration curves (Figure 17.4) limits the choice of indicators for these titrations. For example, the superimposed transition intervals in Figure 17.4 show that methyl red is not a good indicator for the CH_3COOH–NaOH titration, but phenolphthalein is. A good rule of thumb for acid–base titrations is to *choose an indicator whose transition interval includes the pH at the equivalence point.* Methyl red would be a suitable indicator for the HCl–NH_3 titration (Figure 17.5) since its transition interval includes the equivalence point pH.

LEARNING HINT

Only a few drops of indicator are required to produce color changes visible to the eye. Hence, the amount of HIn can be neglected in calculating the pH of the solution.

PRACTICE EXERCISE 17.30

The titration of 0.050 *M* aniline with HCl has an equivalence point pH of 3.1. Refer to Table 17.3, and pick an indicator suitable for the titration.

CHAPTER REVIEW

LEARNING OBJECTIVES BY SECTION

17.1 1. Write the K_a expression for a weak acid.
2. Use K_a and pK_a values to compare acid strengths.
3. Solve problems relating K_a, equilibrium concentrations, pH, and the initial molarity of a weak acid.

17.2 1. Write the K_b expression for a weak base.
2. Use K_b and pK_b values to compare base strengths.
3. Solve problems relating K_b, equilibrium concentrations, pH, and the initial molarity of a weak base.
4. Given K_a, calculate K_b for the conjugate base, and vice versa.

17.3 Use the ionization constant and initial molarity to calculate the percent ionization of a weak acid or weak base.

17.4 1. Write the K_a expression for each step in the ionization of a polyprotic acid.
2. Calculate the concentration of each ion in a polyprotic acid solution.

17.5 1. State whether a given ion is acidic, basic, or neutral with respect to water.
2. Predict whether the solution of a given salt will be acidic, basic, or neutral.
3. Calculate the pH of a salt solution.

17.6 1. Calculate the pH of a solution containing a weak acid or a weak base and its salt.
2. Calculate the pH of (a) a solution containing both a weak and a strong acid and (b) a solution containing both a weak and a strong base.

17.7 1. Explain, with equations, how a buffer solution resists changes in pH.
2. Solve problems that relate pH to the composition of a buffer solution.
3. Calculate the pH changes produced by the addition of small amounts of strong acid or strong base to a buffer solution.
4. Use a table of ionization constants to choose an appropriate buffer system for a given pH.

17.8 1. Calculate the equivalence point pH for a given titration.
2. Sketch and compare titration curves for (a) a strong acid titrated with a strong base, (b) a weak acid titrated with a strong base, and (c) a weak base titrated with a strong acid.
3. Use a titration curve to choose a suitable indicator for the titration.

KEY TERMS BY SECTION

17.1 Acid ionization constant (K_a)
pK_a

17.2 Base ionization constant (K_b)
pK_b

17.3 Percent ionization

17.5 Hydrolysis
Hydrolysis constant

17.6 Common ion effect

17.7 Buffer capacity
Buffer ratio
Buffer solution
Henderson–Hasselbalch equation

17.8 Equivalence point
Titration curve
Half-neutralization point
Transition interval (of an indicator)

IMPORTANT EQUATIONS

17.1 $pK = -\log K$

17.2 $K_aK_b = K_w$

17.3 % ionization

$= 100\% \times$ fraction of molecules ionized

$$= 100\% \times \frac{\text{moles of acid or base ionized per liter}}{\text{initial moles of acid or base per liter}}$$

17.4 $\text{pH} = \text{p}K_a + \log \frac{[\text{base}]}{[\text{acid}]}$

FINAL EXERCISES

Answers to exercises with blue numbers are given in Appendix D.

PART A. QUESTIONS AND PROBLEMS BY SECTION

Ionization Constants (Sections 17.1 and 17.2)

17.1 Look up hypoiodous acid and trichloroacetic acid in Table 17.1.
(a) Write the K_a expression for each acid.
(b) Which is the stronger acid?

17.2 Look up nicotinic acid and nitrous acid in Table 17.1.
(a) Write the K_a expression for each acid.
(b) Which is the stronger acid?

17.3 The pK_b of ammonia varies from 4.86 at 0°C to 4.72 at 50°C. Does ammonia become a stronger base or a weaker base as the temperature goes up?

17.4 The pK_a of formic acid is 3.72 at 25°C and 3.77 at 5°C. What happens to the pH of a formic acid solution as it cools from 25°C to 5°C?

17.5 The pH of a certain acid solution is 4.0. What are the H^+ and OH^- concentrations in this solution? Explain why the autoionization of water can usually be ignored when calculating the pH of a solution.

17.6 Under what circumstances must the autoionization of water be taken into account when calculating the pH of a solution?

17.7 Write the K_a and K_b expressions for NH_4^+ and NH_3, respectively, and use them to illustrate that $K_aK_b = K_w$.

17.8 Use Equation 17.2 to show that pK_a + pK_b = 14.00 for any conjugate acid–base pair.

17.9 Para-aminobenzoic acid ($C_6H_4NH_2COOH$), popularly known as PABA, is a vitamin used in sunscreen preparations. Its pK_a is 4.92. Calculate K_a.

17.10 Malic acid ($H_2C_4H_4O_5$) is a diprotic acid that is found in apples and other fruits. Its first and second ionization constants are $K_{a_1} = 3.9 \times 10^{-4}$ and $K_{a_2} = 7.8 \times 10^{-6}$. Calculate p$K_{a_1}$ and pK_{a_2}.

17.11 The K_b value for ethylamine ($C_2H_5NH_2$) is 6.41×10^{-4}. Calculate K_a for the ethylammonium ion ($C_2H_5NH_3^+$).

17.12 Consult Table 17.1 and calculate K_b for the following ions:
(a) OCl^- **(c)** CN^-
(b) $C_6H_5COO^-$ **(d)** $Al(H_2O)_5(OH)^{2+}$

17.13 The pH of 0.100 *M* propionic acid (CH_3CH_2COOH) is 2.96 at 25°C. Calculate (a) the hydrogen ion concentration and (b) K_a for propionic acid.

17.14 The pH of a 0.100 *M* hydroxylamine (NH_2OH) solution is 9.51 at 25°C. Calculate (a) the hydroxide ion concentration and (b) K_b for hydroxylamine. The reaction of hydroxylamine with water is as follows:

$$NH_2OH(aq) + H_2O(l) \rightleftharpoons NH_3OH^+(aq) + OH^-(aq)$$

17.15 **(a)** Calculate the pH of a 0.180 *M* lactic acid solution. Can the usual approximation be used in this calculation?
(b) Calculate the pH of a 0.0180 *M* lactic acid solution. Can the usual approximation be used? Explain your answer.

17.16 Calculate the pH of (a) 0.0800 *M* formic acid and (b) 0.0100 *M* formic acid. (*Note:* Be sure to check the validity of any approximations you make.)

Percent Ionization (Section 17.3)

17.17 Do the following quantities increase, decrease, or remain the same when a solution of acetic acid is diluted?
(a) $[CH_3COOH]$ **(d)** $[OH^-]$
(b) $[CH_3COO^-]$ **(e)** K_a
(c) pH **(f)** percent ionization

17.18 Do the following quantities increase, decrease, or remain the same when a solution of ethylamine is diluted?
(a) $[C_2H_5NH_2]$
(b) $[C_2H_5NH_3^+]$
(c) pH
(d) $[OH^-]$
(e) K_b
(f) percent ionization

17.19 Calculate the percent ionization in (a) 0.10 *M* lactic acid and (b) 0.25 *M* methylamine.

17.20 Calculate the percent ionization in (a) 0.15 *M* HF and (b) 0.050 *M* aniline.

17.21 A 0.0010 *M* solution of formic acid is 34% ionized. Use this information to calculate (a) the pH of the solution and (b) K_a for formic acid.

17.22 The pH of a 0.10 *M* hydrazine solution is 10.61. Write the equation for the reaction of hydrazine (H_2NNH_2) with water, and calculate (a) the percent ionization and (b) K_b.

17.23 Calculate (a) the pH and (b) the fraction of HCN ionized in a 0.250 *M* solution of HCN.

17.24 Calculate (a) the pH and (b) the fraction of CH_3NH_2 ionized in a 0.175 *M* solution of CH_3NH_2.

Polyprotic Acids (Section 17.4)

17.25 Explain why each step in the ionization of a polyprotic acid has a smaller K_a than the preceding step.

17.26 Explain why the second ionization step can often be neglected in calculating the pH of a polyprotic acid solution.

17.27 Calculate the concentrations of H^+, HSO_3^-, and SO_3^{2-} in a 0.150 *M* H_2SO_3 solution.

17.28 Calculate (a) the pH and (b) the concentrations of $H_2PO_4^-$, HPO_4^{2-}, and PO_4^{3-} in a 0.200 *M* phosphoric acid solution.

17.29 The concentration of dissolved CO_2 in a solution saturated with the gas at 25°C is 0.033 *M*. Use K_a values from Table 17.1 to calculate (a) the pH of the solution and (b) the concentrations of HCO_3^- and CO_3^{2-}.

17.30 **(a)** Use K_a of the hydrated aluminum ion ($Al(H_2O)_6^{3+}$) to estimate the pH of a 0.100 *M* $Al_2(SO_4)_3$ solution.
(b) Assume that K_{a_2} is approximately 10^5 times less than K_{a_1}, and estimate the concentration of $Al(H_2O)_4(OH)_2^+$ in the solution of Part (a).

Acidic and Basic Salts (Section 17.5)

17.31 Which types of ions are neutral with respect to water? Give three examples of each type.

17.32 Which types of ions are basic with respect to water? Give three examples of each type, and write equations for their reactions with water.

17.33 Which types of ions are likely to be acidic in solution? Give three examples of each type, and write equations for their reactions with water.

17.34 State whether the following ions are acidic, basic, or neutral in aqueous solution.
(a) K^+
(b) Fe^{3+}
(c) barbiturate ion
(d) ClO_4^-

17.35 Write a formula for each of the following salts, and predict whether it will be acidic, basic, or neutral in aqueous solution:
(a) aluminum nitrate
(b) potassium hypobromite
(c) calcium hyrogen carbonate
(d) ammonium perchlorate

17.36 Write a formula for each of the following salts, and predict whether it will be acidic, basic, or neutral in aqueous solution:
(a) sodium carbonate
(b) calcium iodide
(c) potassium dihydrogenphosphate
(d) iron(III) chloride

17.37 Use information from Table 17.1 to predict whether a solution of potassium cyanide will be more or less basic than an equimolar solution of potassium fluoride. Write the equation and K_b expression for the reaction of each anion with water.

17.38 Use information from Table 17.2 to predict whether a solution of ammonium chloride will be more or less acidic than an equimolar solution of methylammonium chloride. Write the equation and K_a expression for the reaction of each cation with water.

17.39 Use information from Table 17.1 to predict whether a solution of aluminum chloride will be more or less acidic than an equimolar solution of iron(III) chloride. Write the equation and K_a expression for the reaction of each cation with water.

17.40 Use information from Tables 17.1 and 17.2 to predict whether the following salts will be acidic or basic in solution:
(a) ammonium cyanide
(b) methylammonium fluoride
(c) iron(III) acetate
Might any of these salts be almost neutral? Explain your answer.

17.41 Calculate (a) K_a of the ammonium ion and (b) the pH of a 0.15 *M* NH_4Cl solution.

17.42 Calculate (a) K_b of the benzoate ion and (b) the pH of a 0.050 *M* calcium benzoate solution.

17.43 Calculate the pH of a 0.075 *M* KCN solution.

17.44 Calculate the pH of a 0.10 *M* ethylammonium chloride solution.

17.45 Calculate the pH of a 0.500 *M* NaOBr solution.

17.46 Calculate the pH of a 0.325 *M* solution of sodium formate (HCOONa).

17.47 Sodium phosphate (Na_3PO_4), also called "trisodium phosphate" or "TSP," is an ingredient in certain heavy-duty cleaning preparations. It helps to cut grease by forming a basic solution. Calculate the pH of a 0.50 *M* solution of Na_3PO_4.

17.48 How many moles of sodium acetate should be dissolved in 1.00 L of water to give a solution of pH 9.00?

The Common Ion Effect (Section 17.6)

17.49 If methylammonium chloride (CH_3NH_3Cl) is added to an aqueous solution of methylamine, will the following quantities increase, decrease, or remain the same?
(a) $[OH^-]$
(b) $[H^+]$
(c) $[CH_3NH_2]$
(d) pH
(e) pOH
(f) pK_b

17.50 If sodium lactate ($NaC_3H_5O_3$) is added to a solution of lactic acid, will the following quantities increase, decrease, or remain the same?
(a) $[OH^-]$
(b) $[H^+]$
(c) $[HC_3H_5O_3]$
(d) pH
(e) pOH
(f) pK_a

17.51 Calculate the pH of a solution formed by adding 10.0 g of sodium formate (HCOONa) to 1.00 L of 0.0800 *M* formic acid. Compare your answer with the pH found in Exercise 17.16 for 0.0800 *M* formic acid alone.

17.52 **(a)** Calculate the pH of a 0.100 *M* benzoic acid solution.
(b) How many moles of sodium benzoate must be added to 1.00 L of 0.100 *M* benzoic acid to bring the pH up to 4.00?

17.53 Calculate the pH of a solution that is 0.050 *M* in $NaHCO_3$ and 0.050 *M* in Na_2CO_3.

17.54 Calculate the pH at which one-third of the aluminum ions in 0.300 *M* aluminum sulfate will be in the form of $Al(H_2O)_5(OH)^{2+}(aq)$.

17.55 Calculate the concentrations of H^+ and ascorbate ion ($C_6H_6O_6^{2-}$) in (a) 0.10 *M* ascorbic acid and (b) a mixture that is 0.10 *M* in ascorbic acid and 0.010 *M* in HCl.

17.56 Calculate the concentrations of OH^- and anilinium ion ($C_6H_5NH_3^+$) in (a) 0.255 *M* aniline and (b) a mixture that is 0.255 *M* in aniline and 0.100 *M* in NaOH.

Buffers (Section 17.7)

17.57 What is a *buffer system*? How do the following buffer systems resist changes in pH?
(a) HCOOH–HCOONa
(b) Na_2CO_3–$NaHCO_3$

17.58 Explain, using equations, how the following buffer systems resist changes in pH:
(a) benzoic acid–sodium benzoate
(b) CH_3NH_2–CH_3NH_3Cl

17.59 Without referring to the text, derive the Henderson–Hasselbalch equation for the equilibrium

$$HA(aq) \rightleftharpoons H^+(aq) + A^-(aq)$$

17.60 Derive the Henderson–Hasselbalch equation for the equilibrium

$$NH_3(aq) + H_2O(l) \rightleftharpoons NH_4^+(aq) + OH^-(aq)$$

17.61 What is a *buffer ratio*? When a buffer solution is diluted, what changes, if any, occur in (a) the buffer ratio and (b) the pH? Explain your answers.

17.62 What can be said about the composition of a buffer solution whose pH is equal to the pK_a of the acid component?

17.63 Describe how the following quantities must be related in order to have the most effective buffer system:
(a) H^+ and K_a
(b) pH and pK_a
(c) OH^- and K_b
(d) pOH and pK_b
(e) the concentrations of conjugate acid and conjugate base

17.64 Can a buffer system be prepared from strong acids and their salts? Strong bases and their salts? Explain your answer.

17.65 Refer to Tables 17.1 and 17.2 and choose a buffer system for pH 4.50. Give the names and formulas of the substances needed, and describe how you would go about making 500 mL of this buffer in the laboratory.

17.66 Refer to Tables 17.1 and 17.2, and describe how you would prepare 250 mL of a pH 9.50 buffer.

17.67 Barbituric acid and veronal (diethylbarbituric acid) and their salts are often used to make buffers in biology and biochemistry laboratories. Find the pH of (a) a buffer containing equal molar quantities of barbituric acid and sodium barbiturate and (b) a buffer containing twice as many moles of barbituric acid as sodium barbiturate.

17.68 Calculate the pH of a solution containing NaH_2PO_4 and Na_2HPO_4 in a three-to-one molar ratio.

17.69 How many grams of solid ammonium chloride must be dissolved in 1.00 L of 0.200 *M* ammonia to make a buffer of pH 8.50?

17.70 How many mL of 12.0 *M* HCl must be added to 500 mL of 0.50 *M* sodium acetate to make a buffer of pH 4.75?

17.71 **(a)** How many grams of solid sodium formate (HCOONa) must be added to 1.00 L of 0.100 *M* formic acid to make a buffer of pH 3.52?
(b) What will the pH be if the buffer of Part (a) is diluted with an equal volume of water?

17.72 A 25.0-mL sample of 0.100 *M* acetic acid is mixed with 10.0 mL of 0.100 *M* sodium hydroxide. Calculate the pH of the resulting buffer.

17.73 Calculate the pH change that occurs when 1.00 mL of 10.0 *M* HCl is added to (a) 100 mL of an unbuffered 1.76×10^{-5} *M* HCl solution and (b) 100 mL of a buffer solution that contains 0.100 mol acetic acid and 0.100 mol sodium acetate.

17.74 How many mL of 10.0 *M* NaOH are needed to produce a pH increase of one unit in (a) 1.00 L of a buffer consisting of 0.100 *M* acetic acid and 0.100 *M* sodium acetate and (b) 1.00 L of 1.76×10^{-5} *M* HCl? (Note that the solutions in Parts (a) and (b) have the same pH.)

17.75 **(a)** Calculate the pH of a solution containing 0.10 *M* NH_3 and 0.25 *M* NH_4Cl.
(b) Calculate the pH after 1.00 L of the solution in Part (a) absorbs 500 mL of HCl gas measured at 1.00 atm and 25°C.

17.76 Lactic acid and sodium lactate are to be used to make 500 mL of a pH 3.00 buffer.
(a) If 0.015 mol of lactic acid is used, how many moles of sodium lactate must be added?
(b) How many grams of solid sodium hydroxide would have to be dissolved in the buffer to raise its pH from 3.00 to 3.30?

Acid–Base Titration (Section 17.8)

17.77 Distinguish between the endpoint of an acid–base titration and the equivalence point. Do the two points always coincide?

17.78 Use Le Châtelier's principle to explain the color changes exhibited by an acid–base indicator.

17.79 Explain why only a small amount of indicator is used in an acid–base titration.

17.80 What is meant by the *transition interval* of an indicator? How would you choose an indicator for a given acid–base titration?

17.81 Calculate the equivalence point pH when 25.0 mL of 0.150 *M* HCN is titrated with 0.100 *M* NaOH.

17.82 What is the pH when half of a sample of (a) 0.200 *M* HCl and (b) 0.200 *M* benzoic acid has been neutralized with 0.200 *M* NaOH?

17.83 Sketch a titration curve for 50.0 mL of 0.200 *M* NaOH titrated with 0.100 *M* HCl. Calculate each of the following pH values and include them in your sketch:
(a) initial pH
(b) equivalence point pH
(c) pH at half-neutralization
(d) pH 1.0 mL short of the equivalence point
(e) pH 1.0 mL past the equivalence point

17.84 Sketch a titration curve for 25.0 mL of 0.200 *M* benzoic acid titrated with 0.100 *M* NaOH. Calculate at least five points including the initial pH, the equivalence point pH, and the pH at half-neutralization.

17.85 Consult Table 17.3, and choose an indicator for the titration in (a) Exercise 17.83 and (b) Exercise 17.84.

17.86 The K_a value of lysergic acid ($C_{15}H_{15}N_2COOH$) is 6.3×10^{-4}. What indicator could be used in titrating a 0.100 *M* lysergic acid solution with 0.100 *M* NaOH?

17.87 Choose an indicator for each of the following titrations:
(a) HNO_3 versus $Ca(OH)_2$
(b) 0.100 *M* benzoic acid versus 0.100 *M* KOH
(c) 1.00 *M* HCl versus 0.500 *M* NH_3

17.88 Use Table 17.3 to estimate the pK_a value of each of the following indicators:
(a) congo red
(b) methyl orange
(c) phenolphthalein
(d) methyl red

PART B. MISCELLANEOUS QUESTIONS AND PROBLEMS

17.89 Use the K_a value from Table 17.1 to calculate $[H^+]$ and the percent ionization in 0.50, 0.10, 0.050, and 0.010 *M* benzoic acid. Plot a rough curve of percent ionization versus concentration, and comment on the general trend.

17.90 Calculate the pH of 1.0×10^{-8} *M* HCl. (*Hint:* The self-ionization of water is not negligible.)

17.91 Consider the reaction:

$$HA + B \rightleftharpoons HB^+ + A^-$$

where HA and B are a weak acid and a weak base, with ionization constants K_a and K_b, respectively. Write the equations for the reactions of HA with water and B with water, and use them to show that the equilibrium constant for the above reaction is

$$K_{eq} = K_aK_b/K_w.$$

17.92 Write the equation for the reaction of formic acid with aniline, and use the formula derived in Exercise 17.91 to calculate its equilibrium constant.

17.93 Will the following quantities increase, decrease, or remain the same when solid $NaNO_2$ is stirred into a 0.10 *M* solution of nitrous acid, HNO_2?
(a) pH
(b) $[OH^-]$
(c) fraction of HNO_2 ionized
(d) K_a

17.94 Aspirin (acetylsalicylic acid, $C_8H_7O_2COOH$) is a weak acid with $K_a = 3.3 \times 10^{-4}$. Calculate the percent of aspirin ionized when one five-grain tablet (325 mg) is dissolved in a 100-mL sample of gastric juice. Assume that the gastric juice has a pH of 1.20 and that its hydrogen ions are supplied by HCl. (This example shows that aspirin tablets do not appreciably increase stomach acidity, although they may produce other undesirable effects.)

17.95 What fraction of the ions in a dissolved iron(III) salt will be in the form of $Fe(H_2O)_5(OH)^{2+}$ at pH = 3.00?

17.96 **(a)** Rank the following solutions in order of increasing pH: 0.10 *M* KF, 0.10 *M* $CaCl_2$, and 0.10 *M* CH_3NH_3Br.
(b) Write an equation for the reaction, if any, of each ion with water.

17.97 Calculate the pH of a 0.250 *M* KF solution.

17.98 **(a)** Which should be the stronger base, HCO_3^- or CO_3^{2-}?
(b) Calculate the pH of a 0.050 *M* Na_2CO_3 solution.

17.99 Acid indigestion occurs when the pH of the stomach's contents falls below its normal range of 1.0–3.0. How many grams of $NaHCO_3$ should be dissolved in a 200-mL glass of water to raise the pH of 200 mL of gastric juice from 0.80 to 1.20? The active acid in gastric juice is HCl.

17.100 Many laboratories keep a set of stock buffers with pH values ranging from 3.0 to 11.0. Consult the tables of ionization constants, and state what compounds you might use to make each of the following buffers:
(a) pH 3.0
(b) pH 4.0
(c) pH 5.0
(d) pH 6.0
(e) pH 7.0
(f) pH 8.0
(g) pH 9.0
(h) pH 10.0
(i) pH 11.0

17.101 Consider the phthalic acid–potassium hydrogen phthalate buffer system:

$$H_2C_8H_4O_4(aq) \rightleftharpoons H^+(aq) + HC_8H_4O_4^-(aq) \qquad pK_a = 2.89$$

(a) A buffer system has a pH = 2.75 and is 0.200 *M* in phthalic acid. Calculate the concentration of $HC_8H_4O_4^-$.
(b) Calculate the pH after 0.0100 mol of solid NaOH is stirred into 200 mL of the buffer system in Part (a). (*Hint:* Assume that the addition of NaOH does not change the volume of the solution.)

17.102 The transport of CO_2 in blood and the control of blood pH involves the carbonic acid–bicarbonate buffer system

$$CO_2(aq) + H_2O(l) \rightleftharpoons H^+(aq) + HCO_3^-(aq)$$

for which $pK_a = 6.10$ at 37°C.
(a) Calculate the buffer ratio, $[HCO_3^-]/[CO_2]$, for a normal blood pH of 7.4.
(b) Does more of the blood CO_2 exist as dissolved CO_2 or as HCO_3^-?
(c) A sample of plasma from arterial blood has a pH of 7.40 and a total carbonate concentration (i.e., $[CO_2] + [HCO_3^-]$) of 1.2 mmol/L. Calculate the individual concentrations of CO_2 and HCO_3^-.

17.103 An important inorganic buffer system in human physiology is the $H_2PO_4^-/HPO_4^{2-}$ system for which pK_a = 7.21. If the pH of a urine sample is 5.6 and its total phosphate concentration (i.e., $[H_2PO_4^-] + [HPO_4^{2-}]$) is 46.6 mmol/L, calculate the concentration of each phosphate species in the sample.

17.104 Hemoglobin (HHb) and oxyhemoglobin ($HHbO_2$) are weak acids:

$$HHb(aq) \rightleftharpoons H^+(aq) + Hb^-(aq) \qquad pK_a = 7.93$$

$$HHbO_2(aq) \rightleftharpoons H^+(aq) + HbO_2^- \qquad pK_a = 6.68$$

(a) Does the pH rise or fall when hemoglobin is oxygenated to oxyhemoglobin?
(b) A 2.50×10^{-3} *M* solution of oxyhemoglobin comes to equilibrium at the blood pH of 7.40. Calculate the equilibrium concentrations of $HHbO_2$ and HbO_2^-.

17.105 One way of making a buffer is to partially neutralize a weak acid with a strong base. The base is added until enough acid has been converted to its conjugate base to give the desired buffer ratio. How many moles of NaOH must be added to 1.00 L of 1.00 *M* veronal to make a buffer of pH 8.00?

17.106 What volume of 1.00 *M* NaOH would have to be added to 1.00 L of 0.0100 *M* nicotinic acid to make a buffer of pH 5.00?

17.107 THAM hydrochloride, ($(CH_2OH)_3CNH_3Cl$) is used to make buffers for clinical chemistry. The equilibrium is

$$(CH_2OH)_3CNH_3^+(aq) \rightleftharpoons (CH_2OH)_3CNH_2(aq) + H^+(aq) \qquad pK_a = 8.08$$

How many moles of potassium hydroxide must be added to 1.00 L of 0.100 *M* THAM hydrochloride to make a buffer of pH 7.80?

17.108 Calculate the pH of 1.000 L of 1.000 *M* HCl after the addition of the following quantities of NaOH: (1) 0.0 g, (2) 4.0 g, (3) 20.0 g, (4) 36.0 g, (5) 39.0 g, (6) 39.9 g, and (7) 40.0 g (1 mol). Prepare a graph of pH versus moles of NaOH added, and answer the following questions:

(a) What was the total pH change after the addition of 1 mol of NaOH?

(b) What fraction of the total pH change was achieved by adding $\frac{1}{2}$ mol of NaOH?

(c) Approximately what fraction of a mole of NaOH is needed to achieve one-half of the total pH change?

17.109 Under what circumstances will the equivalence point pH in an acid–base titration be (a) greater than 7.0, (b) equal to 7.0, and (c) less than 7.0?

17.110 Suppose methyl orange is used as an indicator for the titrations plotted in Figures 17.3 and 17.4. Will the endpoint of each titration be appreciably different from the equivalence point? If so, by how many milliliters will the titration be in error?

17.111 Phenolphthalein is used as an indicator when 25.00 mL of a veronal solution is titrated with 0.0700 *M* NaOH. If 23.45 mL of NaOH is required to reach the endpoint, what was the original concentration of veronal?

17.112 Malonic acid ($H_2C_3H_2O_4$) is a diprotic acid. Sketch a titration curve for 25.00 mL of a 0.1000 *M* malonic acid solution titrated with 0.1000 *M* NaOH. Refer to Table 17.1 for K_a values. (*Note:* When a diprotic acid has been half-neutralized, $[H^+] = \sqrt{K_{a_1} \times K_{a_2}}$ and $pH = \frac{1}{2}(pK_{a_1} + pK_{a_2})$.

17.113 Water contains 0.033 *M* CO_2 when it is in equilibrium with CO_2 gas at 25°C and 1 atm partial pressure.

(a) Estimate the concentration of CO_2 in water in equilibrium with air at 25°C and 1 atm. Air contains approximately 0.030 mol % CO_2.

(b) Estimate the HCO_3^- and CO_3^{2-} concentrations in this solution.

***17.114** Calculate the pH of 1.00×10^{-7} *M* NH_3. (*Hint:* Set up two simultaneous equations, one for the ionization of ammonia and one for the autoionization of water. Assume that $[OH^-] = x + y$, where x is the contribution from NH_3 and y is the contribution from H_2O. Eliminate x from one of the equations and solve for y using the method of successive approximations.)

17.115 Derive the relation given in the note to Example 17.112.

CHAPTER 18

Mineral terraces at Mammoth Hot Springs in Wyoming's Yellowstone Park. These deposits form when warm, underground mineral-laden water comes to the surface and cools.

SOLUBILITY AND COMPLEX ION EQUILIBRIA

OUTLINE

PREVIEW

A transparent crystal lying motionless in a clear solution seems to be a picture of rest; on the molecular level, however, it is a scene of activity and ferment. Ions in the crystal break away and float off in envelopes of water molecules. Ions in solution are caught up by the crystal and become part of the solid again. This game of dissolving and precipitating is responsible for many natural phenomena. For example, the ocean above the Bahama Banks, where the water is rich in calcium ions and dissolved carbon dioxide, is often milky with minute calcium carbonate crystals. Near Carlsbad, New Mexico, the underground dissolving and redeposition of calcium carbonate has hollowed out vast caves draped and studded with strange limestone shapes. Less spectacular, but of more immediate importance, are the slow formation of kidney stones from calcium oxalate and the pollution of lake beds with heavy-metal salts.

Consider some calcium oxalate crystals lying at the bottom of a saturated solution of calcium oxalate. The solid is in equilibrium with its dissolved ions:

$$CaC_2O_4(s) \rightleftharpoons Ca^{2+}(aq) + C_2O_4{}^{2-}(aq)$$

Like any equilibrium system, a saturated solution of calcium oxalate obeys Le Châtelier's principle and is governed by an equilibrium constant. Increasing the concentration of either ion favors precipitation; decreasing the concentration of either ion will cause solid calcium oxalate to dissolve.

18.1 THE SOLUBILITY PRODUCT

A solid in equilibrium with dissolved ions is a heterogeneous equilibrium similar to those studied in Chapter 15. Other examples of such equilibria are

$$AgCl(s) \rightleftharpoons Ag^+(aq) + Cl^-(aq)$$
$$PbI_2(s) \rightleftharpoons Pb^{2+}(aq) + 2I^-(aq)$$
$$Li_2CO_3(s) \rightleftharpoons 2Li^+(aq) + CO_3^{2-}(aq)$$

The equilibrium constant for a solid in equilibrium with dissolved ions is called a **solubility product constant** (K_{sp}). The solubility product constants for the above equilibria at 25°C are

$$AgCl: \quad K_{sp} = [Ag^+][Cl^-] = 1.8 \times 10^{-10}$$
$$PbI_2: \quad K_{sp} = [Pb^{2+}][I^-]^2 = 7.9 \times 10^{-9}$$
$$Li_2CO_3: \quad K_{sp} = [Li^+]^2[CO_3^{2-}] = 1.7 \times 10^{-3}$$

The values of some other solubility product constants are listed in Table 18.1.

Note that K_{sp} expressions, like other heterogeneous equilibrium constants, do not include solids. Note also that the exponents in a K_{sp} expression are the same as the formula subscripts—a K_{sp} expression can thus be written from the formula alone.

EXAMPLE 18.1

Write the K_{sp} expression for (a) calcium phosphate and (b) mercury(I) chloride.

SOLUTION

(a) Calcium phosphate ($Ca_3(PO_4)_2$) contains three Ca^{2+} ions for every two PO_4^{3-} ions. The solubility product constant is

$$K_{sp} = [Ca^{2+}]^3[PO_4^{3-}]^2$$

(b) Mercury(I) chloride (Hg_2Cl_2), contains one Hg_2^{2+} ion for every two Cl^- ions. The solubility product constant is

$$K_{sp} = [Hg_2^{2+}][Cl^-]^2$$

PRACTICE EXERCISE 18.1

Write the K_{sp} expressions for (a) CaC_2O_4 and (b) Bi_2S_3.

The substances listed in Table 18.1 are sparingly soluble salts whose ionic concentrations are small even in a saturated solution. The use of K_{sp} expressions is limited to sparingly soluble salts because equilibrium constants are reliable only when the ionic concentrations are low enough for interionic attractions to be ignored.

For salts having the same formula type (e.g., AB, AB_2, *or* A_2B*), the lower the* K_{sp}*, the lower the molar solubility of the salt.* Silver chromate (Ag_2CrO_4, an A_2B type), with a K_{sp} of 1.2×10^{-12}, is much less soluble than lithium carbonate (Li_2CO_3), which has a K_{sp} of 1.7×10^{-3}. When salts have different formula types, their relative molar solubilities cannot be predicted by simply comparing K_{sp} values. For example, AgCl is less soluble than Ag_2CrO_4, even though its K_{sp} (1.8×10^{-10}) is more than 100 times greater than the K_{sp} of Ag_2CrO_4.

TABLE 18.1 Solubility Product Constants at 25°C

Precipitate	Color	K_{sp}
Acetates		
CH_3COOAg	White	2.5×10^{-3}
Bromides		
AgBr	Cream	5.0×10^{-13}
Hg_2Br_2	Yellow	1.3×10^{-22}
$PbBr_2$	White	2.1×10^{-6}
Carbonates		
Ag_2CO_3	Yellow	8.1×10^{-12}
$BaCO_3$	White	5.0×10^{-9}
$CaCO_3$	White	4.5×10^{-9}
Li_2CO_3	White	1.7×10^{-3}
$MgCO_3$	White	1.6×10^{-6}
$SrCO_3$	White	9.3×10^{-10}
$PbCO_3$	White	7.4×10^{-14}
Chlorides		
AgCl	White	1.8×10^{-10}
CuCl	White	1.7×10^{-7}
Hg_2Cl_2	White	1.3×10^{-18}
$PbCl_2$	White	1.7×10^{-5}
Chromates		
Ag_2CrO_4	Rust red	1.2×10^{-12}
$BaCrO_4$	Yellow	1.2×10^{-10}
$CaCrO_4$	Yellow	7.1×10^{-4}
$PbCrO_4$	Yellow	2.8×10^{-13}
$SrCrO_4$	Yellow	3.6×10^{-5}
Fluorides		
BaF_2	White	1.3×10^{-6}
CaF_2	White	3.9×10^{-11}
[illegible]	[illegible]	3.7×10^{-8}
	White	3.6×10^{-8}
		7.9×10^{-10}
Hydroxides		
$Al(OH)_3$	White	3×10^{-34}
$Ba(OH)_2$	White	5×10^{-3}

Precipitate	Color	K_{sp}
Hydroxides, continued		
$Ca(OH)_2$	White	6.5×10^{-6}
$Cr(OH)_3$	Gray-green	6.3×10^{-31}
$Cu(OH)_2$	Pale blue	1.6×10^{-19}
$Fe(OH)_2$	Pale green	7.9×10^{-16}
$Fe(OH)_3$	Rust red	1.6×10^{-39}
$Mg(OH)_2$	White	7.1×10^{-12}
$Mn(OH)_2$	Pink	6×10^{-14}
$Pb(OH)_2$	White	1.2×10^{-15}
$Sr(OH)_2$	White	3.2×10^{-4}
$Zn(OH)_2$	White	4.5×10^{-17}
Iodides		
AgI	Yellow	8.3×10^{-17}
Hg_2I_2	Yellow	4.5×10^{-29}
PbI_2	Yellow	7.9×10^{-9}
Oxalates		
$g_2C_2O_4$	White	1.1×10^{-11}
aC_2O_4	White	1.6×10^{-7}
aC_2O_4	White	1.3×10^{-9}
MgC_2O_4	White	8.6×10^{-5}
PbC_2O_4	White	8.3×10^{-12}
SrC_2O_4	White	5.6×10^{-8}
Phosphates		
PO_4	Yellow	1.8×10^{-18}
$(PO_4)_2$	White	6×10^{-39}
$(PO_4)_2$	White	1.3×10^{-32}
$(PO_4)_2$	White	1×10^{-31}
Sulfates		
$_2SO_4$	White	$1.4 \times$ [illegible]
aSO_4	White	[illegible] 10^{-10}
aSO_4	White	[illegible] $\times 10^{-5}$
$PbSO_4$	White	[illegible]$.3 \times 10^{-7}$
$SrSO_4$	White	3.2×10^{-7}

PRACTICE EXERCISE 18.2

Which would have the larger concentratio[illegible], a saturated solution of lead bromide or a saturated solution of lead [illegible] [illegible]er to Table 18.1 for K_{sp} values.

If the dissolved salt consists [illegible] do not react with water or some other species in the solution, then th[illegible] can be calculated from the solubility of the salt, and vice versa.

EXAMPLE 18.2

The *Handbook of Chemistry and Physics* reports the solubility of BaF_2 as 1.2 g/L at 25°C. Assume that the hydrolysis of fluoride ions is negligible, and calculate the K_{sp} of barium fluoride.

SOLUTION

The K_{sp} of BaF_2 is

$$K_{sp} = [Ba^{2+}][F^-]^2$$

We will evaluate the K_{sp} by (1) converting grams of BaF_2 per liter to moles per liter, (2) finding the molarity of each ion, and (3) substituting the molarities into the K_{sp} expression.

Step 1: The molar mass of barium fluoride is 175.3 g/mol; the molar solubility is obtained as follows:

$$\frac{1.2 \text{ g BaF}_2}{1 \text{ L}} \times \frac{1 \text{ mol BaF}_2}{175.3 \text{ g BaF}_2} = 6.8 \times 10^{-3} \text{ mol BaF}_2/\text{L}$$

Step 2: One mole of BaF_2 provides 1 mol of Ba^{2+} ions and 2 mol of F^- ions. Therefore, the ionic concentrations in 6.8×10^{-3} *M* BaF_2 are

$$[Ba^{2+}] = 6.8 \times 10^{-3} \text{ mol/L}$$
$$[F^-] = 2 \times 6.8 \times 10^{-3} \text{ mol/L}$$
$$= 1.4 \times 10^{-2} \text{ mol/L}$$

Step 3: Substituting the concentrations into the K_{sp} expression gives

$$K_{sp} = [Ba^{2+}][F^-]^2 = (6.8 \times 10^{-3})(1.4 \times 10^{-2})^2$$
$$= 1.3 \times 10^{-6}$$

PR… CE EXERCISE 18.3

The solubil… ver chloride is $8.9 \times$ …0^{-4} g/L at 10°C. Calc…
at this tempera…

EXAMPLE 18.3

The K_{sp} of lead bromide is 2.1 … of solu… 5°C (Table 18.1). Calculate the solubility of $PbBr_2$ in grams per 100 mL of solu…

SOLUTION

The concentration summary given below is … equation for the dissolving of $PbBr_2$. Initially there are no lead ions or brom… solution. Let x be the molar solubility of lead bromide. The equation shows … $PbBr_2$ produces x mol of Pb^{2+} and $2x$ mol of Br^-.

CONCENTRATION SUMMARY

	$PbBr_2(s) \rightleftharpoons$	$Pb^{2+}(aq)$	$+ \ 2Br^-(aq)$
Equation:	$PbBr_2(s) \rightleftharpoons$	$Pb^{2+}(aq)$	$+ \ 2Br^-(aq)$
Initial concentrations (mol/L):		0	0
Concentration changes (mol/L):		$+x$	$+2x$
Equilibrium concentrations (mol/L):		x	$2x$

Substituting the equilibrium concentrations into the K_{sp} expression gives

$$K_{sp} = [Pb^{2+}][Br^-]^2$$

$$2.1 \times 10^{-6} = (x)(2x)^2 = 4x^3$$

and

$$x = 8.1 \times 10^{-3} \text{ mol/L} = [Pb^{2+}]$$

The equilibrium concentration of Pb^{2+}, 8.1×10^{-3} mol/L, is also the molar solubility of $PbBr_2$. The molar mass of $PbBr_2$ is 367.0 g/mol; hence, its solubility in grams per liter is

$$\frac{8.1 \times 10^{-3} \text{ mol PbBr}_2}{1 \text{ L}} \times \frac{367.0 \text{ g PbBr}_2}{1 \text{ mol PbBr}_2} = 3.0 \text{ g PbBr}_2\text{/L}$$

which is equivalent to 0.30 g per 100 mL. (Solubilities are often tabulated in grams per 100 milliliters.)

LEARNING HINT

The cube root of a number y is the number raised to the $\frac{1}{3}$ power, $y^{1/3}$. Use the y^x or the x^y button on your electronic calculator.

PRACTICE EXERCISE 18.4

The K_{sp} of Ag_2SO_4 is 1.4×10^{-5}. How many grams of silver sulfate will remain undissolved when 3.00 g of finely divided silver sulfate is shaken with 500 mL of water?

The Common Ion Effect

Consider the effect of adding bromide ions to a saturated solution of lead bromide (Figure 18.1). The solubility equilibrium

$$PbBr_2(s) \rightleftharpoons Pb^{2+}(aq) + 2Br^-(aq)$$

will shift to use up some of the added Br^- ions, and some $PbBr_2$ will precipitate. This example is another illustration of the common ion effect covered in Section 17.6.

EXAMPLE 18.4

How many moles of lead bromide can dissolve in 1.00 L of 0.10 *M* calcium bromide solution? The K_{sp} of $PbBr_2$ is 2.1×10^{-6}.

SOLUTION The initial Pb^{2+} concentration is zero. One mole of $CaBr_2$ supplies 2 mol of Br^-, so the initial Br^- concentration is twice the calcium bromide concentration, or 0.20 *M*. The concentration summary is based on the equation for dissolving $PbBr_2$.

PRACTICE EXERCISE 18.6

A 50-mL portion of 0.050 *M* $Ca(NO_3)_2$ solution is poured into 50 mL of 0.050 *M* K_2CrO_4. Will $CaCrO_4$ precipitate?

The next example shows how the concentration of a dissolved ion can be controlled by a precipitation reaction.

EXAMPLE 18.6

The concentration of silver ion in a certain reagent should not exceed 1.0×10^{-7} *M*. What concentration of HCl in the reagent would guarantee this? The K_{sp} of AgCl is 1.8×10^{-10}.

SOLUTION

The HCl provides Cl^- ions, which tend to precipitate Ag^+ ions as silver chloride. We must find the concentration of HCl that will lower the silver ion concentration to 1.0×10^{-7} *M* or less. Substituting $[Ag^+] = 1.0 \times 10^{-7}$ *M* into the K_{sp} expression and solving for the chloride concentration gives

$$K_{sp} = [Ag^+][Cl^-] = 1.8 \times 10^{-10}$$

$$(1.0 \times 10^{-7})[Cl^-] = 1.8 \times 10^{-10}$$

$$[Cl^-] = 1.8 \times 10^{-3}\,M$$

If the HCl, which is completely ionized, is maintained at 1.8×10^{-3} *M* or greater, the concentration of silver ion will not exceed 1.0×10^{-7} *M*.

PRACTICE EXERCISE 18.7

The concentration of lead ion can be controlled by the addition of chloride ion. Calculate the minimum concentration of HCl required to maintain $[Pb^{2+}]$ at 1.0×10^{-3} *M* or less.

Fractional Precipitation

Suppose we want to extract the calcium ions from a sample of dolomite, a mineral composed of calcium and magnesium carbonates. We would first dissolve the sample in HCl to obtain a solution containing magnesium and calcium ions. Then we would add a solution of sodium sulfate to precipitate calcium sulfate and leave the magnesium ions in solution. This separation is easy because magnesium sulfate is very soluble while calcium sulfate is sparingly soluble ($K_{sp} = 2.4 \times 10^{-5}$). Separations are more difficult when both ions form similar insoluble compounds so that any reagent that precipitates one tends to precipitate both. Even in these cases, separation can often be achieved by taking advantage of slight differences in solubility, as illustrated in the next example.

EXAMPLE 18.7

A solution contains 0.10 M Ba^{2+} and 0.10 M Ca^{2+}. What concentration of CrO_4^{2-} will precipitate the maximum amount of $BaCrO_4$ without precipitating any $CaCrO_4$? The K_{sp} values are as follows: $BaCrO_4$, 1.2×10^{-10}; $CaCrO_4$, 7.1×10^{-4}.

SOLUTION

$BaCrO_4$ has the same formula type as $CaCrO_4$; hence, $BaCrO_4$, with the smaller K_{sp}, is less soluble than $CaCrO_4$. Because of the solubility difference, it should be possible to add enough chromate ion (usually in the form of dissolved K_2CrO_4) to precipitate nearly all the Ba^{2+} as $BaCrO_4$ without building up the chromate concentration to a level that would precipitate $CaCrO_4$. The optimum chromate concentration is the one that just saturates the solution with respect to $CaCrO_4$; its value is determined by substituting $[Ca^{2+}] = 0.10\ M$ into the K_{sp} expression for $CaCrO_4$:

$$K_{sp} = [Ca^{2+}][CrO_4^{2-}]$$

$$7.1 \times 10^{-4} = (0.10)[CrO_4^{2-}]$$

$$[CrO_4^{2-}] = 7.1 \times 10^{-3}\ M$$

A chromate concentration larger than $7.1 \times 10^{-3}\ M$ will precipitate both calcium and barium chromates; a smaller concentration will precipitate only barium chromate. The maximum separation will occur when the chromate concentration reaches $7.1 \times 10^{-3}\ M$.

PRACTICE EXERCISE 18.8

A solution is 0.10 M in Mg^{2+} and 0.10 M in Ca^{2+}. Calculate the concentration of NaOH required to precipitate the maximum amount of $Mg(OH)_2$ without precipitating any $Ca(OH)_2$. Use K_{sp} values from Table 18.1.

The use of a precipitating agent to precipitate one ion selectively and leave others in solution is called **fractional precipitation**. One way to judge the efficiency of the separation in Example 18.7 is to calculate the concentration of Ba^{2+} remaining in solution at the point of maximum separation. For a good separation, this concentration should be small relative to the starting concentration.

EXAMPLE 18.8

Refer to Example 18.7, and calculate the concentration of Ba^{2+} remaining in solution at the point of maximum separation.

SOLUTION

Maximum separation occurs when $[CrO_4^{2-}] = 7.1 \times 10^{-3}\ M$. At this concentration, the solution contains some barium ions in equilibrium with the barium chromate precipitate and all of the calcium ions. The concentration of barium ions is found by substituting the chromate concentration into the K_{sp} expression of $BaCrO_4$:

$$K_{sp} = [Ba^{2+}][CrO_4^{2-}]$$

$$1.2 \times 10^{-10} = [Ba^{2+}](7.1 \times 10^{-3})$$

and

$$[Ba^{2+}] = 1.7 \times 10^{-8}\ M$$

Note that the concentration of barium ions has decreased from its original value of 0.10 M to 1.7×10^{-8} M. Most of the barium ions are in the precipitate (less than $2 \times 10^{-5}\%$ of them remain in solution); all of the calcium ions are in the solution. This separation would be good enough for most purposes.

PRACTICE EXERCISE 18.9

Refer to Practice Exercise 18.8. Calculate the concentration of Mg^{2+} remaining in the final solution. Do you think that NaOH is a good reagent for separating magnesium and calcium ions?

18.3 SPARINGLY SOLUBLE HYDROXIDES

Imagine adding sodium hydroxide to a solution of an iron(II) salt such as $FeSO_4$. Because $Fe(OH)_2$ is sparingly soluble, the following equilibrium is quickly established:

$$Fe^{2+}(aq) + 2OH^-(aq) \rightleftharpoons Fe(OH)_2(s)$$

Increasing the OH^- concentration shifts the equilibrium toward solid iron(II) hydroxide. Acidifying the solution reduces the hydroxide concentration and shifts the equilibrium back toward dissolved Fe^{2+}. *Sparingly soluble hydroxides tend to precipitate in basic solution and dissolve in acidic solution.*

EXAMPLE 18.9

Dissolved oxygen gradually converts $Fe^{2+}(aq)$ to $Fe^{3+}(aq)$ and $Fe(OH)_3(s)$.

Find the concentration of Fe^{2+} that can remain in an oxygen-free solution at pH 7.00. The K_{sp} of $Fe(OH)_2$ is 7.9×10^{-16}.

SOLUTION

The H^+ and OH^- concentrations are each 1.0×10^{-7} M at pH 7.00. Substituting $[OH^-] = 1.0 \times 10^{-7}\ M$ into the K_{sp} expression of $Fe(OH)_2$ gives

$$K_{sp} = [Fe^{2+}][OH^-]^2$$

$$7.9 \times 10^{-16} = [Fe^{2+}](1.0 \times 10^{-7})^2$$

and

$$[Fe^{2+}] = 0.079\ M$$

PRACTICE EXERCISE 11.2

To what value must the pH be lowered if a solution is to contain 0.10 M Al^{3+}? The K_{sp} of $Al(OH)_3$ is 3×10^{-34}.

Hydroxide precipitation is often controlled by buffering the solution to keep the OH^- concentration below a certain level.

EXAMPLE 18.10

The pH of an ammonia solution can be controlled by the addition of an ammonium salt. Calculate the minimum NH_4^+ concentration needed to prevent precipitation of $Mg(OH)_2$ from a solution that is 0.10 *M* in Mg^{2+} and 0.075 *M* in NH_3. The K_{sp} value of $Mg(OH)_2$ is 7.1×10^{-12}; the K_b value of NH_3 is 1.8×10^{-5}.

SOLUTION

First we calculate the hydroxide ion concentration in a saturated solution of $Mg(OH)_2$. The precipitation of $Mg(OH)_2$ will not occur if the actual hydroxide concentration is less than or equal to this value. Then we calculate the concentration of NH_4^+ ions needed to maintain the hydroxide concentration at this level.

Step 1: The concentration of hydroxide ions in the saturated solution is obtained by substituting $[Mg^{2+}] = 0.10\ M$ into the K_{sp} expression of $Mg(OH)_2$:

$$K_{sp} = [Mg^{2+}][OH^-]^2 = 7.1 \times 10^{-12}$$

$$(0.10)[OH^-]^2 = 7.1 \times 10^{-12}$$

and

$$[OH^-] = 8.4 \times 10^{-6}\ M$$

Step 2: The hydroxide ions come from the ionization of ammonia:

$$NH_3(aq) + H_2O(l) \rightleftharpoons NH_4^+(aq) + OH^-(aq)$$

The concentration of ammonium ions required to maintain the level of hydroxide ions at $8.4 \times 10^{-6}\ M$ is obtained by substituting $[OH^-] = 8.4 \times 10^{-6}\ M$ and $[NH_3] = 0.075\ M$ into the K_b expression of ammonia

$$K_b = \frac{[NH_4^+][OH^-]}{[NH_3]}$$

$$1.8 \times 10^{-5} = \frac{[NH_4^+](8.4 \times 10^{-6})}{(0.075)}$$

$$[NH_4^+] = 0.16\ M$$

Magnesium hydroxide will not precipitate if the concentration of ammonium ions is greater than or equal to 0.16 *M*.

PRACTICE EXERCISE 11.2

If the concentration of Mg^{2+} in Example 18.10 is increased to 0.20 *M* and the NH_3 concentration is reduced to 0.050 *M*, what concentration of ammonium ion is needed to prevent precipitation of magnesium hydroxide?

Amphoteric Hydroxides

Hydroxides that have the ability to neutralize both acids and bases are said to be **amphoteric**. Chromium(III) hydroxide and zinc hydroxide, for example, dissolve in

Figure 18.2

The reactions of an amphoteric hydroxide. Initially all three beakers contained a suspension of chromium(III) hydroxide, $Cr(OH)_3$. The middle beaker remains untreated. The hydroxide on the left was dissolved by adding concentrated HCl(aq) to form violet-colored Cr^{3+}(aq). The hydroxide on the right was dissolved by adding concentrated NaOH to form green $Cr(OH)_4^-$(aq).

strong acids to form salts in which the metal is the positive ion (Figure 18.2):

$$Cr(OH)_3(s) + 3H^+(aq) \rightarrow Cr^{3+}(aq) + 3H_2O(l)$$

$$Zn(OH)_2(s) + 2H^+(aq) \rightarrow Zn^{2+}(aq) + 2H_2O(l)$$

These hydroxides also dissolve in strong bases to produce salts in which the metal is part of a negative ion:

$$Cr(OH)_3(s) + OH^-(aq) \rightarrow Cr(OH)_4^-(aq)$$

$$Zn(OH)_2(s) + 2OH^-(aq) \rightarrow Zn(OH)_4^{2-}(aq)$$

An amphoteric hydroxide behaves the way it does because it is amphiprotic (see Section 16.1). For example, the composition of wet zinc hydroxide is more nearly represented by the formula $Zn(OH)_2(H_2O)_2$. The OH groups can accept protons; the waters of hydration can donate protons. The reactions of zinc hydroxide with acid and strong base can also be written as follows:

$$Zn(OH)_2(H_2O)_2(s) + 2H^+(aq) \rightarrow Zn(H_2O)_4^{2+}(aq)$$

$$Zn(OH)_2(H_2O)_2(s) + 2OH^-(aq) \rightarrow Zn(OH)_4^{2-}(aq) + 2H_2O(l)$$

Table 18.2 gives the formulas of the most common amphoteric hydroxides and their associated ions.

EXAMPLE 18.11

Refer to Table 18.2, and write both ionic and molecular equations for the reactions that occur when lead(II) hydroxide dissolves in (a) aqueous HNO_3 and (b) aqueous KOH.

SOLUTION

(a) $Pb(OH)_2$ reacts with HNO_3 to form a salt containing Pb^{2+} ions. The net ionic and molecular equations are

$$Pb(OH)_2(s) + 2H^+(aq) \longrightarrow Pb^{2+}(aq) + 2H_2O(l)$$

$$Pb(OH)_2(s) + 2HNO_3(aq) \longrightarrow Pb(NO_3)_2(aq) + 2H_2O(l)$$

(b) $Pb(OH)_2$ reacts with strong base to form the plumbite ion, $Pb(OH)_4^{2-}$ (Table 18.2). The net ionic and molecular equations are

$$Pb(OH)_2(s) + 2OH^-(aq) \longrightarrow Pb(OH)_4^{2-}(aq)$$

$$Pb(OH)_2(s) + 2KOH(aq) \longrightarrow \underset{\text{potassium plumbite}}{K_2Pb(OH)_4(aq)}$$

PRACTICE EXERCISE 18.12

Which of the following mixed precipitates can be separated by the addition of strong (6 *M*) NaOH solution: (a) $Ca(OH)_2$ and $Fe(OH)_2$, (b) $Be(OH)_2$ and $Ca(OH)_2$, and (c) $Pb(OH)_2$ and $Zn(OH)_2$?

An oxide has the same acidic or basic character as the corresponding hydroxide. Accordingly, the dehydration of an amphoteric hydroxide produces an amphoteric oxide that will also dissolve in acids and in strong base. Compare, for example, the

TABLE 18.2 Some Amphoteric Hydroxides and the Ions They Form in Acidic and Basic Solutions

Hydroxide	Acidic Solution	Basic Solution
$Al(OH)_3$ Aluminum hydroxide	Al^{3+} Aluminum ion	$Al(OH)_4^-$ Aluminate ion
$Sb(OH)_3$ Antimony(III) hydroxide	Sb^{3+} Antimony(III) ion	$Sb(OH)_4^-$ Antimonite ion
$Be(OH)_2$ Beryllium hydroxide	Be^{2+} Beryllium ion	$Be(OH)_4^{2-}$ Beryllate ion
$Cr(OH)_3$ Chromium(III) hydroxide	Cr^{3+} Chromium(III) ion	$Cr(OH)_4^-$ Chromite ion
$Pb(OH)_2$ Lead(II) hydroxide	Pb^{2+} Lead(II) ion	$Pb(OH)_4^{2-}$ Plumbite ion
$Sn(OH)_2$ Tin(II) hydroxide	Sn^{2+} Tin(II) ion	$Sn(OH)_4^{2-}$ Stannite ion
$Zn(OH)_2$ Zinc hydroxide	Zn^{2+} Zinc ion	$Zn(OH)_4^{2-}$ Zincate ion

CHEMICAL INSIGHT

DENTAL HEALTH, PH, AND FLUORIDATION

The enamel that makes up the outer layer of teeth is the hardest substance in the body. About 98% of the 2-mm-thick enamel consists of hydroxyapatite ($Ca_{10}(PO_4)_6(OH)_2$, a sparingly soluble substance that is also found in bone) with small and varying amounts of Mg^{2+}, CO_3^{2-}, Cl^-, and F^- incorporated into the crystal structure. The remaining 2% is mostly protein. Underneath the enamel is dentin, a softer material containing about 70% hydroxyapatite and 30% organic material.

About 95% of the population is afflicted to some degree with *caries* (tooth decay), a disease in which portions of the teeth become demineralized and cavities form. The demineralization process can be represented in simplified form as an equilibrium between solid hydroxyapatite and its dissolved ions:

$$Ca_{10}(PO_4)_6(OH)_2(s) \underset{\text{mineralization}}{\overset{\text{demineralization}}{\rightleftharpoons}} 10Ca^{2+}(aq) + 6PO_4^{3-}(aq) + 2OH^-(aq)$$

Saliva has an average pH of about 6.8 and is almost neutral. Very little hydroxyapatite will dissolve at this pH, so conditions in the "normal" mouth favor mineralization. An acidic environment, however, promotes demineralization by removing both OH^- and PO_4^{3-} ions:

$$H^+(aq) + OH^-(aq) \rightarrow H_2O(l)$$

$$H^+(aq) + PO_4^{3-}(aq) \rightarrow HPO_4^{2-}(aq)$$

Demineralization becomes rapid below pH 6. Acid also tends to protonate and decompose the protein component of tooth enamel.

Acid is generated in *plaque*, a gelatinous mass consisting mostly of bacteria. Plaque adheres to the teeth because it contains *glucans*, a sticky substance manufactured by the bacteria. The bacteria live on energy obtained by decomposing sugars into various products including lactic acid ($CH_3CHOHCOOH$). The more sugars that come their way, the more the bacteria multiply, and the more lactic acid they produce.

Saliva fights the bacterial depredations by (1) replenishing the PO_4^{3-} and Ca^{2+} ions, (2) providing a compound called *lactoferrin* that removes iron needed by the bacteria, and (3) raising the pH by means of various buffers. The buffering action of saliva is due to bicarbonate ion and also to a number of polypeptides (compounds that contain both COOH and NH_2 groups). These amphiprotic components allow the saliva to quickly counteract pH changes caused by most normal dietary ingredients. Saliva cannot, however, easily counteract the effect of large quantities of sugar. Experiments have shown that one rinse with a glucose mouthwash sends the oral pH from its normal value of 6.8 to 5.5 within 2 minutes. The pH then stays at 5.5 or below for about 20 minutes during which time the teeth are actively decaying. After 20 minutes, the pH begins to rise, and it returns to 6.8 in about 40 minutes (Figure 18.3). Demineralization is effectively stopped above pH 6.0.

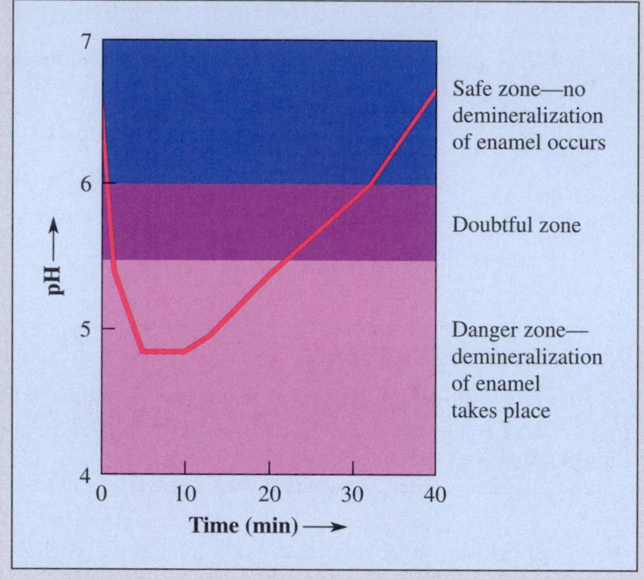

Figure 18.3

The pH of dental plaque versus time after rinsing the mouth with a glucose solution. The danger range for cavity formation—pH 5.5 and below—lasts for about 20 minutes.

The natural fluoride content of drinking water in the United States ranges from 0.05 ppm to 8 ppm. Surveys have shown that children develop fewer cavities in regions where the water contains at least 1 ppm fluoride ion than in regions where the water contains 0.1 ppm fluoride or less. The various mechanisms by which fluoride exerts this protective effect are still under study. It is known, however, that fluoride ions can replace some of the hydroxide ions in hydroxyapatite to give fluorapatite:

$$Ca_{10}(PO_4)_6(OH)_2(s) + 2F^-(aq) \rightarrow \underset{\text{fluorapatite}}{Ca_{10}(PO_4)_6F_2(s)} + 2OH^-(aq)$$

Fluorapatite is harder and more dense than hydroxyapatite, and the fluoride ion is less easily protonated by acid than the hydroxide ion. In lactic acid solution, fluorapatite is approximately 100 times less soluble than hydroxyapatite.

following reactions of zinc oxide with those given on page 832 for zinc hydroxide:

$$ZnO(s) + 2H^+(aq) \rightarrow Zn^{2+}(aq) + H_2O(l)$$

$$ZnO(s) + 2OH^-(aq) + H_2O(l) \rightarrow Zn(OH)_4^{2-}(aq)$$

PRACTICE EXERCISE 18.13

Write net ionic equations for the reactions of $Al_2O_3(s)$ with (a) aqueous HNO_3 and (b) aqueous NaOH.

18.4 SOLUBILITY AND pH

As we discovered in Section 17.5, most salts of weak acids have basic anions. Barium fluoride, for example, is a sparingly soluble salt of the weak acid HF. Its solubility equilibrium is

$$BaF_2(s) \rightleftharpoons Ba^{2+}(aq) + 2F^-(aq)$$

Adding hydrochloric acid to a barium fluoride solution reduces the concentration of fluoride ions

$$H^+(aq) + F^-(aq) \rightarrow HF(aq)$$

and shifts the barium fluoride equilibrium forward in the direction of increased solubility. The overall reaction can be represented by the net ionic equation

$$BaF_2(s) + 2H^+(aq) \rightarrow Ba^{2+}(aq) + 2HF(aq)$$

or by the molecular equation

$$BaF_2(s) + 2HCl(aq) \rightarrow BaCl_2(aq) + 2HF(aq)$$

As this example illustrates, *sparingly soluble salts of weak acids become more soluble in acidic solution*. The hydrogen ions can be provided by any strong acid that does not contain another precipitating ion; for example, hydrochloric acid dissolves BaF_2, but sulfuric acid converts BaF_2 into sparingly soluble $BaSO_4$:

$$BaF_2(s) + H_2SO_4(aq) \rightarrow BaSO_4(s) + 2HF(aq)$$

EXAMPLE 18.12

Would you expect the addition of nitric acid to increase the solubility of (a) AgBr and (b) CdS? Write a molecular equation for any reaction that occurs.

SOLUTION

(a) AgBr is the salt of the strong acid HBr. Since Br^- is not a basic anion, it will not accept protons in aqueous solution. The solubility of AgBr should not increase with the addition of nitric acid.

(b) CdS is the salt of hydrogen sulfide (H_2S), a weak acid. The sulfide ions are easily protonated, so CdS will dissolve in acid, producing H_2S (Figure 18.4).

$$CdS(s) + 2HNO_3(aq) \rightarrow Cd(NO_3)_2(aq) + H_2S(aq)$$

Hydrogen sulfide is a toxic gas with the odor of rotten eggs.

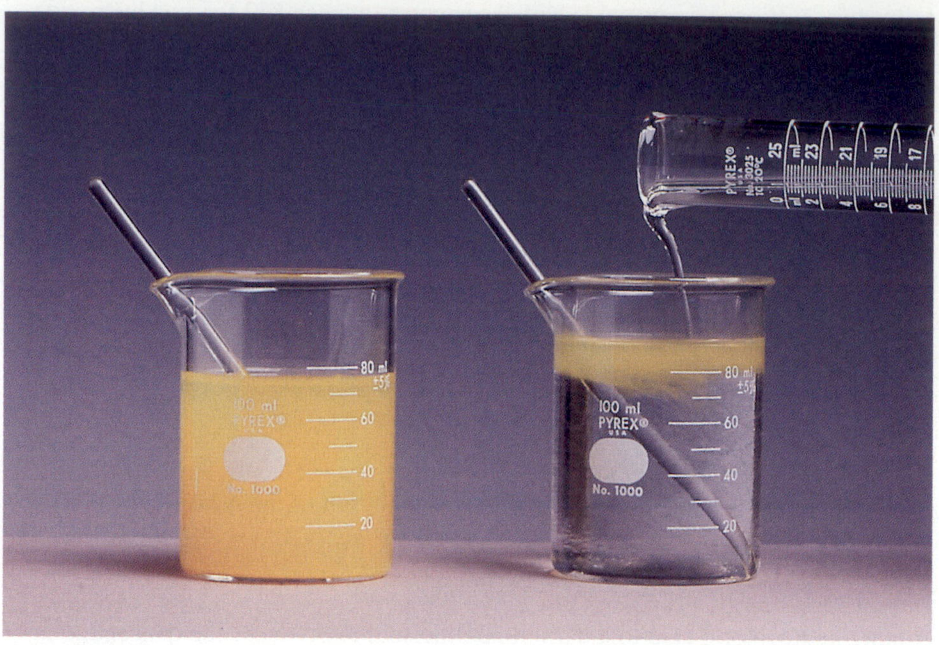

Figure 18.4

Both beakers contain freshly precipitated yellow cadmium sulfide (CdS). The addition of concentrated hydrochloric acid dissolves the precipitate on the right.

PRACTICE EXERCISE 18.14

Which of the following lead salts will dissolve in aqueous nitric acid: (a) $PbCO_3$, (b) PbI_2, (c) $PbCl_2$, and (d) PbF_2? Write a molecular equation for each reaction that takes place.

pH Control of Sulfide Precipitations

Hydrogen sulfide (H_2S) is an analytical reagent often used for the identification of metal ions in solution. Hydrogen sulfide gas can be passed directly into a solution of metal ions (Demonstration 18.1), or it can be generated by heating dissolved thioacetamide (CH_3CSNH_2):

$$CH_3CSNH_2(aq) + 2H_2O(l) \rightarrow NH_4^+(aq) + CH_3COO^-(aq) + H_2S(aq)$$

The following equilibria occur in a solution of H_2S:

$$H_2S(aq) \rightleftharpoons H^+(aq) + HS^-(aq) \qquad K_{a_1} = 1.02 \times 10^{-7}$$

$$HS^-(aq) \rightleftharpoons H^+(aq) + S^{2-}(aq) \qquad K_{a_2} = 1 \times 10^{-19}$$

The extremely low value of K_{a_2} shows that there is virtually no sulfide ion in an aqueous solution of H_2S; that is, the second equilibrium can be ignored. The addition of acid to an H_2S solution favors the reverse reactions, and at pH = 5 or less, virtually all of the sulfide will be in the form of H_2S (aq). (You can verify this state-

LEARNING HINT

The sulfide ion is a strong base that reacts almost completely with water:

$$S^{2-}(aq) + H_2O(l) \rightleftharpoons HS^-(aq) + OH^-(aq) \qquad K_b = 1 \times 10^5$$

The concentration of S^{2-} is negligible compared to the HS^- concentration, even at high pH values. Try to verify this statement by using the above equilibrium constant to show that $[HS^-] = 10^9 \times [S^{2-}]$ at pH = 10.

DEMONSTRATION 18.1 SOME METAL SULFIDES

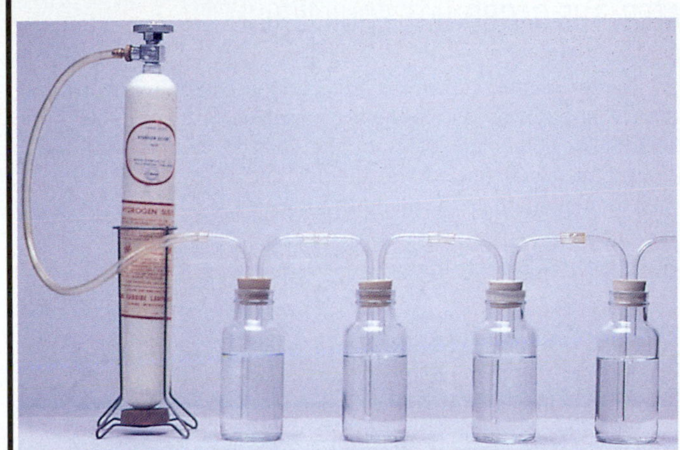

From left to right, the four solutions contain nitrates of zinc, antimony(III), lead, and cadmium.

Hydrogen sulfide gas bubbling through the solutions produces precipitates of ZnS (white), Sb_2S_3 (red), PbS (black), and CdS (yellow).

ment by using the first equation to show that $[H_2S] = 100 \times [HS^-]$ at pH = 5.)

We are often concerned with the pH at which a given metal sulfide will precipitate. Under acidic conditions, the precipitation is governed by equilibria like the one shown below for FeS:

$$FeS(s) + 2H^+(aq) \rightleftharpoons Fe^{2+}(aq) + H_2S(aq)$$

$$K_{spa} = \frac{[Fe^{2+}][H_2S]}{[H^+]^2} = 6 \times 10^2$$

The symbol K_{spa} ("solubility product in acid solution") has been given to the equilibrium constants for this type of reaction. Additional K_{spa} values are given in Table 18.3.

The sulfides of most metals, except for those in Groups 1A and 2A, are sparingly soluble. The equation given for FeS illustrates that sulfides, like all sparingly soluble salts of weak acids, become more soluble as the solution becomes more acidic.

When a continuous supply of H_2S gas is used as a precipitating agent, it saturates the solution to give a constant 0.10 *M* H_2S concentration at room temperature. This fact helps to simplify the calculations in problems such as the following:

TABLE 18.3 K_{spa} Values for Metal Sulfides[a,b]

Sulfide	K_{spa}
MnS	3×10^{10}
FeS	6×10^{2}
ZnS	3×10^{-2}
SnS	1×10^{-5}
CdS	8×10^{-7}
PbS	3×10^{-7}
CuS	6×10^{-16}
Ag_2S	6×10^{-30}
HgS	2×10^{-32}

[a]For $MS(s) + 2H^+(aq) \rightleftharpoons M^{2+}(aq) + H_2S(g)$

[b]Data taken from R. J. Meyers, *J. of Chem. Ed.*, **63**, 687 (1986).

EXAMPLE 18.13

Find the maximum concentration of Zn^{2+} that can remain unprecipitated in a 0.50 *M* HCl solution that is saturated with H_2S.

SOLUTION

The reaction is

CHEMICAL INSIGHT

CLAM SHELLS, CAVES, AND HARD WATER: THE CARBONATE EQUILIBRIUM

Ocean water contains about 420 mg of calcium (1.0×10^{-2} mol) and 29 mg of carbon (2.4×10^{-2} mol) per liter. Calcium exists as Ca^{2+}, and most of the carbon is in the form of HCO_3^-, with smaller quantities existing as aqueous CO_2 and CO_3^{2-}. Several natural processes that occur in the oceans and fresh waters of the earth depend on four simultaneous equilibria involving these species:

1. $CO_2(g) \rightleftharpoons CO_2(aq)$
2. $CO_2(aq) + H_2O(l) \rightleftharpoons H^+(aq) + HCO_3^-(aq)$
3. $HCO_3^-(aq) \rightleftharpoons H^+(aq) + CO_3^{2-}(aq)$
4. $Ca^{2+}(aq) + CO_3^{2-}(aq) \rightleftharpoons CaCO_3(s)$

Lakes and oceans absorb carbon dioxide from the air to an extent that depends on the pH, the temperature, and the CO_2 partial pressure. A low H^+ concentration (high pH) shifts the second and third equilibria (see above) in the forward direction. More CO_2 dissolves and the concentration of CO_3^{2-} increases. The increased carbonate concentration increases the likelihood that $CaCO_3$ will precipitate.

The precipitation of calcium carbonate is assisted by numerous organisms. Marine algae, for example, concentrate calcium ions from the water in order to line their cells with calcium carbonate. Marine animals such as coral, clams, and oysters form their shells and skeletons in coastal waters where the pH may be as high as 9.

Some calcium carbonate precipitation occurs without the intervention of living organisms, especially in regions where cold CO_2-rich water rises to the surface and is warmed. Warming drives out carbon dioxide and converts dissolved calcium bicarbonate to calcium carbonate:

$$Ca^{2+}(aq) + 2HCO_3^-(aq) \xrightarrow{\text{warm}} CO_2(g) + H_2O(l) + CaCO_3(s)$$

Calcium carbonate comes out of solution in the form of minute crystals that eventually sink to the ocean floor, where they mix with shells and other sediment.

The earth's crust is constantly rising and falling. After millions of years, the massive quantities of calcium carbonate formed in the ocean appear on land as dolomite mountains and limestone deposits. The formation of calcium carbonate is then slowly reversed as rainwater containing dissolved carbon dioxide seeps into limestone deposits and reconverts some of the CO_3^{2-} to HCO_3^-:

$$CaCO_3(s) + H_2O(l) + CO_2(aq) \rightarrow Ca^{2+}(aq) + 2HCO_3^-(aq)$$

The rocks dissolve and limestone caves are formed. Because of this reaction, water flowing through limestone regions usually contains a substantial concentration of dissolved $Ca(HCO_3)_2$. When such water evaporates or is warmed, the reaction is reversed once again and calcium carbonate is

$$ZnS(s) + 2H^+(aq) \rightleftharpoons Zn^{2+}(aq) + H_2S(aq)$$

and the K_{spa} from Table 18.3 is

$$\frac{[Zn^{2+}][H_2S]}{[H^+]^2} = K_{spa} = 3 \times 10^{-2}$$

The strong acid HCl supplies virtually all of the hydrogen ions. Substituting $[H^+] = 0.50\ M$ and $[H_2S] = 0.10\ M$ into the K_{spa} expression gives

$$\frac{[Zn^{2+}](0.10)}{(0.50)^2} = 3 \times 10^{-2}$$

and

$$[Zn^{2+}] = 0.08\ M$$

A zinc ion concentration in excess of 0.08 M will shift the equilibrium in the reverse direction, causing ZnS to precipitate.

redeposited as stalactites, stalagmites, and other cave formations (Figure 18.5).

Water containing salts of Ca^{2+}, Mg^{2+}, or Fe^{2+} is called *hard water*. Slightly hard water contains about 20 mg of these ions per liter, and very hard water may contain 10 times this amount. Hard water has several disadvantages: (1) the dissolved ions react with soap to form precipitates ("soap scum"); they also inactivate many of the detergents used instead of soap; (2) when heated, the dissolved salts precipitate to form deposits that clog boilers, hot water pipes, and tea kettles (see Figure S1.7, p. 352); and (3) the taste is unpleasant to those who are not used to it. When the dissolved salts are present in the form of bicarbonates, as they are in water that comes from limestone-rich regions, the hardness is said to be "temporary" because heating drives off the CO_2 and reprecipitates the carbonate. Water with temporary hardness can be "softened" by boiling and discarding the solid residue. More practical methods are available for softening large water supplies. The *lime–soda process* is a method in which hydrated lime ($Ca(OH)_2$) and sodium carbonate (Na_2CO_3) are added in the correct proportion to precipitate all of the Ca^{2+} as $CaCO_3$; Mg^{2+} and Fe^{2+} also precipitate as carbonates and/or hydroxides. More modern methods use *ion-exchange resins*—solids composed of organic macromolecules that adsorb Ca^{2+} and Mg^{2+} and release other ions, usually Na^+, in their place.

Figure 18.5

A lake in a limestone cave, Luray Caverns, Virginia. The stalactites (hanging) and stalagmites (standing) consist primarily of calcium carbonate.

PRACTICE EXERCISE 18.15

(a) Calculate the pH that would allow 1.0×10^{-8} M Cd^{2+} ion to remain in a solution that is saturated with H_2S. (b) Which way would you adjust the pH to reduce the dissolved cadmium ion concentration even further?

The identification of metal ions in solution is part of the branch of chemistry known as **quantitative analysis**. Sulfides are important in qualitative analysis because they exhibit a wide range of solubilities—some dissolve completely in acid, others are sparingly soluble at any pH. They thus lend themselves to selective precipitation, as illustrated in the next example.

EXAMPLE 18.14

Calculate the pH that will achieve the maximum separation of Zn^{2+} and Cd^{2+} when a

$$K_f = \frac{[Ag(NH_3)_2^+]}{[Ag^+][NH_3]^2} = K_{f_1} \times K_{f_2}$$
$$= (2.2 \times 10^3)(7.2 \times 10^3) = 1.6 \times 10^7$$

The larger the overall formation constant, the more stable is the complex ion with respect to its constituents; $Ag(CN)_2^-$ with $K_f = 1 \times 10^{21}$ is much more stable than $Ag(NH_3)_2^+$ with $K_f = 1.6 \times 10^7$. Some formation constants are given in Table 18.4. The reciprocal of the formation constant is called a **dissociation constant** (K_d); the more stable the ion, the smaller the dissociation constant.

PRACTICE EXERCISE 18.17

Refer to Table 18.4 and determine which complex ion, $Cd(NH_3)_4^{2+}$, $Fe(SCN)^{2+}$, or HgI_4^{2-}, is the most stable. Which is the least stable?

Calculations Involving Complex Ions

When the concentration of complexing agent in a solution is low relative to the concentration of metal ions, many of the metal ions will not receive their full complement of ligands, and partially complexed species such as $Ag(NH_3)^+$ and $Cu(NH_3)_2^{2+}$ will predominate in the equilibrium mixture. When the complexing agent is present in large excess, most of the metal ions will be fully complexed, and the concentrations of intermediate species will be negligible; in this case, calculations are fairly simple.

EXAMPLE 18.15

Estimate the equilibrium Ag^+ concentration when 0.010 mol of $AgNO_3$ is dissolved in 1.00 L of 2.00 *M* NH_3. The K_f of $Ag(NH_3)_2^+$ is 1.6×10^7.

SOLUTION

Because K_f is large and because ammonia is present in large excess, most of the silver will be in the form of $Ag(NH_3)_2^+$. The simplest way to solve this problem is to assume that all of the silver is initially in the form of the complex, which then breaks down to form x mol of silver ion at equilibrium. That is, we will assume that $[Ag(NH_3)_2^+]$ is initially 0.010 *M*. Because each mole of complex contains 2 mol of ammonia, the formation of the complex reduces the concentration of NH_3 by 0.020 *M*. Hence, $[NH_3]$ is initially 2.00 *M* − 0.020 *M* = 1.98 *M*.

CONCENTRATION SUMMARY

Equation:	$Ag(NH_3)_2^+(aq)$ ⇌	$Ag^+(aq)$ +	$2NH_3(aq)$
Initial concentrations (mol/L):	0.010	0	1.98
Concentration changes (mol/L):	$-x$	$+x$	$+2x$
Equilibrium concentrations (mol/L):	$0.010 - x$	x	$1.98 + 2x$
Approximate concentrations (mol/L):	0.010	x	1.98

LEARNING HINT

Another and perhaps more natural way of solving Example 18.15 is to assume that the initial concentrations are $[Ag^+] = 0.010\ M$, $[NH_3] = 2.00\ M$, and $[Ag(NH_3)_2^+] = 0$. At equilibrium, the concentrations will be $[Ag^+] = 0.010 - x$, $[NH_3] = 2.00 - 2x$, and $[Ag(NH_3)_2^+] = x$, where x is now the equilibrium concentration of the complex. This approach, however, leads to a cubic equation that cannot be simplified because x is large relative to 0.010.

Because x, the concentration of free Ag^+, is small, we make the usual approximations; that is, $0.010 - x = 0.010$ and $1.98 + 2x = 1.98$. The equation is written as a dissociation, so we will use the K_d expression of $Ag(NH_3)_2{}^+$. Substituting

$$K_d = \frac{1}{K_f} = \frac{1}{1.6 \times 10^7} = 6.2 \times 10^{-8}$$

and the approximate concentrations into the K_d expression gives

$$K_d = \frac{[Ag^+][NH_3]^2}{[Ag(NH_3)_2{}^+]}$$

$$6.2 \times 10^{-8} = \frac{(x)(1.98)^2}{(0.010)}$$

and $$x = 1.6 \times 10^{-10}\, M$$

The silver ion concentration at equilibrium will be $1.6 \times 10^{-10}\, M$.

PRACTICE EXERCISE 18.18

Calculate the concentration of uncomplexed Zn^{2+} ion when 0.10 mol of $Zn(NO_3)_2$ is dissolved in 1.00 L of 2.0 *M* NaOH. The K_d of $Zn(OH)_4{}^{2-}$ is 4.5×10^{-17}.

EXAMPLE 18.16

Will AgCl precipitate if the solution in Example 18.15 is made 1.0 *M* in NaCl? The K_{sp} of AgCl is $1.8 \times 10^{-10.}$

SOLUTION

The Ag^+ concentration is $1.6 \times 10^{-10}\, M$. If the Cl^- concentration is 1.0 *M*, the ion product will be

$$Q = [Ag^+][Cl^-] = (1.6 \times 10^{-10})(1.0)$$
$$= 1.6 \times 10^{-10}$$

The ion product is less than the K_{sp}, so there will be no precipitate.

PRACTICE EXERCISE 18.19

Will AgBr precipitate if the solution in Example 18.15 is made 1.0 *M* in NaBr? The K_{sp} of AgBr is 5.0×10^{-13}.

When a complexing agent is used to dissolve a sparingly soluble solid, the overall reaction can be thought of as a stepwise process. For example, photographers' "hypo" (sodium thiosulfate, $Na_2S_2O_3$) dissolves unexposed silver bromide during film processing in a reaction that involves the following equilibria:

TABLE 18.4 Formation Constants[a] for Complex Ions at 25°C

Complex Ion	K_f
Ammonia complexes	
$Ag(NH_3)_2{}^+$	1.6×10^7
$Cd(NH_3)_4{}^{2+}$	1×10^7
$Cu(NH_3)_4{}^{2+}$	1.1×10^{13}
$Ni(NH_3)_6{}^{2+}$	5.5×10^8
$Zn(NH_3)_4{}^{2+}$	2.9×10^9
Cyanide complexes	
$Ag(CN)_2{}^-$	1×10^{21}
$Au(CN)_2{}^-$	2×10^{38}
$Cd(CN)_4{}^{2-}$	1.2×10^{17}
$Cu(CN)_2{}^-$	1.0×10^{24}
$Cu(CN)_3{}^{2-}$	2×10^{27}
$Fe(CN)_6{}^{4-}$	1.0×10^{24}
$Fe(CN)_6{}^{3-}$	1.0×10^{31}
$Ni(CN)_4{}^{2-}$	1×10^{22}
$Hg(CN)_4{}^{2-}$	2×10^{41}
$Zn(CN)_4{}^{2-}$	6×10^{16}
Halide complexes	
$AgCl_2{}^-$	2.5×10^5
$AlF_6{}^{3-}$	6.7×10^{19}
$HgCl_4{}^{2-}$	1.2×10^{15}
$HgI_4{}^{2-}$	1.9×10^{30}
Hydroxide complexes	
$Al(OH)_4{}^-$	7.7×10^{33}
$Zn(OH)_4{}^{2-}$	2.2×10^{16}
Other complexes	
$Ag(S_2O_3)_2{}^{3-}$	1.7×10^{13}
$Fe(C_2O_4)_3{}^{3-}$	3×10^{20}
$Fe(SCN)^{2+}$	1.2×10^2
$Zn(EDTA)^{2-}$	3.8×10^{16}

[a]The dissociation constant (K_d) is the reciprocal of the formation constant. Formation constants are also called **stability constants** (K_{stab}); dissociation constants are called **instability constants** (K_{instab}).

CHEMICAL INSIGHT

WAX REMOVAL, RUST REMOVAL, AND CHELATING AGENTS

In recent decades, hundreds of new commercial and medical uses have been found for complexing agents, and more uses are constantly being developed. To take an example close to home, complexing agents have made possible the formulation of floor waxes that can be removed easily with ammonia cleansers. These liquid waxes contain the zinc–ammonia complex ($Zn(NH_3)_4^{2+}$). After the wax is applied, the ammonia evaporates and releases zinc ions

$$Zn(NH_3)_4^{2+}(aq) \rightarrow Zn^{2+}(aq) + 4NH_3(g)$$

that remain behind and become incorporated into the structure of the dried wax film. Ammoniacal cleansers reverse the reaction. They recomplex the zinc ion, thus disintegrating the film and allowing it to be removed easily. Other more powerful wax removers contain even stronger complexing agents than ammonia.

Rust is partially dehydrated iron(III) hydroxide. Rust stains can be removed with preparations containing oxalate ion ($C_2O_4^{2-}$):

$$Fe^{3+}(aq) + 3C_2O_4^{2-}(aq) \rightarrow Fe(C_2O_4)_3^{3-}(aq)$$

Oxalate ions have the structure

$$\begin{array}{c} :\ddot{O}{:}\ \ :\ddot{O}{:} \\ \| \quad\ \ \| \\ ^{-}{:}\ddot{\underset{..}{O}}-C-C-\ddot{\underset{..}{O}}{:}^{-} \end{array}$$

They bend around and bond the Fe^{3+} ion at two sites, holding it as if between two fingers (Figure 18.7). A ligand that occupies two or more sites in a complex ion is called a *chelating agent* because of its pincer or clawlike action. (Greek *chelos* means "claw.") Complexes formed with chelating agents tend to be very stable.

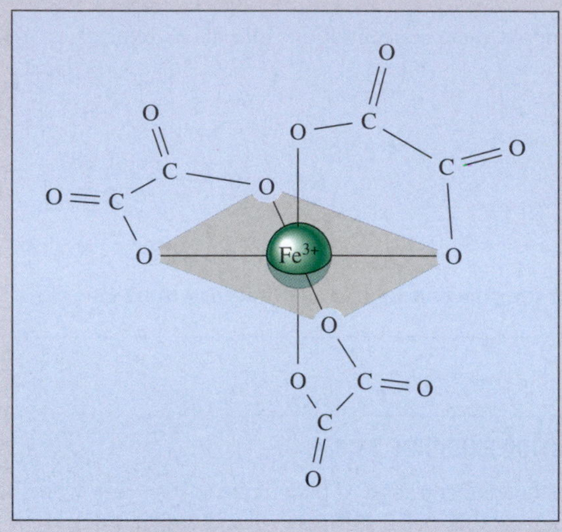

Figure 18.7
Each of three oxalate ions binds the iron(III) ion at two sites to form the $Fe(C_2O_4)_3^{3-}$ complex ion.

Ethylenediaminetetraacetate ion (EDTA) is a chelating agent with a multitude of uses. Figure 18.8 shows how one EDTA ion can surround a central metal ion and bond it in six positions. The EDTA ion, usually in the form of the disodium calcium salt ($Na_2CaEDTA$), is added to bottled salad dressings, canned beans, lotions, perfumes, and other products to complex and inactivate all traces of heavy-metal ions such as copper, iron, and manganese. These ions, if free, would catalyze the oxidation reactions that produce rancidity,

1. $AgBr(s) \rightleftharpoons Ag^+(aq) + Br^-(aq)$ $\qquad K_1 = K_{sp}$

2. $Ag^+(aq) + 2S_2O_3^{2-}(aq) \rightleftharpoons Ag(S_2O_3)_2^{3-}(aq)$ $\qquad K_2 = K_f$

The net reaction is the sum of Steps 1 and 2

$$AgBr(s) + 2S_2O_3^{2-}(aq) \rightleftharpoons Ag(S_2O_3)_2^{3-}(aq) + Br^-(aq)$$

and the combined equilibrium constant is

$$\frac{[Ag(S_2O_3)_2^{3-}][Br^-]}{[S_2O_3^{2-}]^2} = K_1 \times K_2 = K_{sp} \times K_f$$

These steps do not take partially complexed species into account, so the use of a combined equilibrium constant in a complex ion calculation is valid only when the

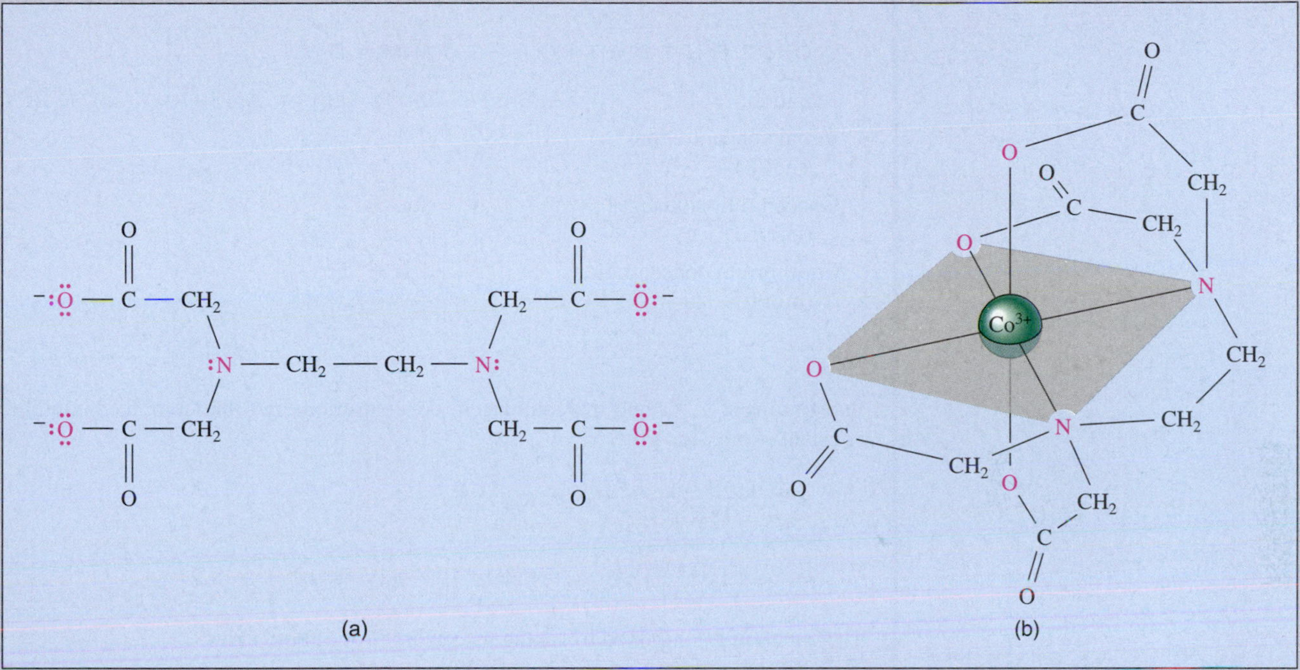

(a)

(b)

Figure 18.8

(a) The ethylenediaminetetraacetate ion (EDTA). The ion has a total charge of 4−. The six atoms that may bind EDTA to a metal ion are shown in red. (b) The cobalt(III)–EDTA complex. One EDTA ligand completely encloses the metal ion, binding it at six sites.

discoloration, and deterioration of fragrance. EDTA is also used as a chelating agent in numerous cleaning products such as wax removers that do not contain ammonia and bathtub cleansers that complex the calcium ion responsible for soap film.

EDTA has also been used as an antidote for heavy-metal poisoning. It has the disadvantage, however, of being nonselective, and it may remove essential ions from the body along with the heavy-metal ions. Special chelating agents have been developed that will selectively remove toxic ions such as Pb^{2+} or Cu^{2+} from the bloodstream. Heavy-metal complexes are discussed further in Chapter 21.

concentrations of such species are negligible, that is, when there is a large excess of complexing agent.

EXAMPLE 18.17

Use the equations given above to calculate the molar solubility of AgBr in a 2.0 *M* $Na_2S_2O_3$ solution. The K_{sp} of AgBr $= 5.0 \times 10^{-13}$; the K_f of $Ag(S_2O_3)_2^{3-} = 1.7 \times 10^{13}$.

SOLUTION

Let x be the molar solubility of AgBr. The concentration summary shows that the concentrations of $Ag(S_2O_3)_2^{3-}$ and Br^- each increase by x. The concentration of $S_2O_3^{2-}$ decreases by $2x$ because 1 mol of Ag^+ is complexed by 2 mol of thiosulfate ion.

CONCENTRATION SUMMARY

Equation:	$AgBr(s)$ +	$2S_2O_3^{2-}(aq)$ ⇌	$Ag(S_2O_3)_2^{3-}(aq)$ +	$Br^-(aq)$
Initial concentrations (mol/L):		2.0	0	0
Concentration changes (mol/L):		$-2x$	$+x$	$+x$
Equilibrium concentrations (mol/L):		$2.0 - 2x$	x	x

Substituting K_{sp}, K_f, and the equilibrium concentrations into the combined equilibrium constant expression gives

$$\frac{[Ag(S_2O_3)_2^{3-}][Br^-]}{[S_2O_3^{2-}]^2} = K_{sp} \times K_f$$

$$\frac{(x)(x)}{(2.0 - 2x)^2} = (5.0 \times 10^{-13})(1.7 \times 10^{13}) = 8.5$$

The equation is simplified by taking the square root of both sides:

$$\frac{x}{2.0 - 2x} = 2.9$$

and

$$x = 0.85\ M$$

The solubility of AgBr in the $Na_2S_2O_3$ solution is 0.85 M. (Note that the square root of 8.5 is +2.9 or −2.9. The negative root is not used because it would lead to a negative value for the concentration of $S_2O_3^{2-}$.)

PRACTICE EXERCISE 18.20

Calculate the molar solubility of AgCl in 2.0 M aqueous ammonia. The K_{sp} of AgCl $= 1.8 \times 10^{-10}$; the K_f of $Ag(NH_3)_2^+ = 1.6 \times 10^7$.

CHAPTER REVIEW

LEARNING OBJECTIVES BY SECTION

18.1
1. Write K_{sp} expressions for sparingly soluble solids.
2. Calculate K_{sp} values from solubilities, and vice versa.
3. Calculate the solubility of a sparingly soluble solid in a solution containing a common ion.

18.2 1. Perform calculations to predict whether the mixing of two solutions will produce a precipitate.
2. Calculate the concentration of an ion required to maintain a given concentration of some other ion.
3. Calculate the concentration of a precipitating agent that will result in the maximum separation of two ions.
4. Calculate the concentration of ions remaining in solution after the addition of a precipitating agent.

18.3 1. Calculate the OH^- concentration needed to precipitate a sparingly soluble hydroxide.
2. Calculate the pH below which a sparingly soluble hydroxide will not precipitate.
3. Calculate the composition of a buffer that will prevent a sparingly soluble hydroxide from precipitating.
4. Write equations for the reaction of amphoteric hydroxides and their corresponding oxides with strong acids and strong hydroxide bases.

18.4 1. Predict whether the solubility of a given salt will be affected by pH.
2. Write equations for the reactions that occur when salts of weak acids dissolve in strong acids.
3. Calculate the concentration of metal ion that will remain unprecipitated in an H_2S solution at a given pH.
4. Calculate the pH at which two ions can be separated using H_2S as the precipitating agent.

18.5 1. Give examples of complex ions and their uses.
2. Use formation and/or dissociation constants to predict which of two complex ions is more stable.
3. Calculate the concentration of uncomplexed metal ion in the presence of excess complexing agent.
4. Calculate the molar solubility of a salt in excess complexing agent.

KEY TERMS BY SECTION

18.1 Solubility product constant (K_{sp})

18.2 Fractional precipitation
Ion product

18.3 Amphoteric hydroxide

18.5 Aquo-complex
Complex ion
Dissociation constant (K_d; of a complex ion)
Formation constant (K_f)
Ligand

FINAL EXERCISES

Answers to exercises with blue numbers are given in Appendix D.

PART A. QUESTIONS AND PROBLEMS BY SECTION

Solubility Product Constants (Section 18.1)

18.1 Explain why solubility product constants are used only for sparingly soluble solids.

18.2 Under what circumstances will the use of molar solubilities lead to incorrect K_{sp} values? Give examples to illustrate your answer.

18.3 Describe the *common ion effect* as it applies to solubility equilibria.

18.4 Will $CaCO_3$ be more soluble or less soluble in calcium chloride solution than in pure water? Explain your answer in terms of the common ion effect.

18.5 Write a K_{sp} expression for each of the following compounds:
(a) lead sulfate
(b) antimony(III) sulfate
(c) iron(III) hydroxide

18.6 Write a K_{sp} expression for each of the following compounds:
(a) lead bromide
(b) silver phosphate
(c) mercury(I) iodide

18.7 Calculate K_{sp} values from the following solubility data:
(a) cobalt iodate ($Co(IO_3)_2$); solubility = 0.011 *M* at 18°C
(b) barium thiosulfate (BaS_2O_3); solubility = 0.008 *M* at 20°C

18.52 Explain why PbF_2 becomes more soluble in acid but PbI_2 does not.

18.53 Which of the following sparingly soluble copper(I) compounds should be more soluble in acidic solution than in neutral solution?
(a) CuCl **(c)** CuCN
(b) CuI **(d)** Cu_2S

18.54 Which of the following sparingly soluble lead compounds should be more soluble in acidic solution than in neutral solution?
(a) $PbCl_2$ **(c)** $PbCO_3$
(b) PbF_2 **(d)** $PbSO_4$

18.55 A white precipitate is either $BaSO_4$ or $BaSO_3$. What reagent could be added to identify it?

18.56 A white precipitate is either $SrSO_4$ or $SrCO_3$. What reagent might you add to find out which it is?

18.57 Write an equation for the reaction that occurs when each of the following solids dissolves in excess HCl:
(a) CaC_2O_4
(b) $CaCO_3$
(c) MnS

18.58 Gaseous H_2S is sometimes prepared by treating solid FeS with concentrated HCl. Write an equation for the reaction.

18.59 Calculate the molar solubility of $CaCO_3$ at pH 9.00. (*Hints*: (1) Only the first protonation of CO_3^{2-} needs to be considered at this pH. (2) Show that $[CO_3^{2-}] + [HCO_3^-] = x$, where x is the molar solubility of $CaCO_3$; then find the $[HCO_3^-]/[CO_3^{2-}]$ ratio.)

18.60 Calculate the number of moles of $CaSO_4$ that will dissolve in 1.00 L of HCl solution maintained at pH 1.00. The value of K_{a_2} for $H_2SO_4 = 1.2 \times 10^{-2}$. (*Hint*: Show that $[SO_4^{2-}] + [HSO_4^-] = x$, where x is the molar solubility of $CaSO_4$; then find the $[HSO_4^-]/[SO_4^{2-}]$ ratio.)

Sulfide Precipitations (Section 18.4)

18.61 Do sparingly soluble sulfides become more or less soluble with increasing pH? Explain your answer.

18.62 Write a general equation for the reaction that occurs when a sparingly soluble sulfide dissolves in acid.

18.63 What is a K_{spa}? Write the K_{spa} expression for MnS.

18.64 How does a K_{spa} expression for a sulfide such as ZnS differ from its K_{sp} expression?

18.65 What is the maximum concentration of Cd^{2+} that can remain unprecipitated in a 0.75 *M* HCl solution that is saturated with H_2S?

18.66 Calculate the molar solubility of SnS in a solution buffered to pH = 3.5 and saturated with H_2S.

18.67 Will a precipitate form when a solution containing 0.10 *M* $CuCl_2$, 0.10 *M* $MnCl_2$, and 0.10 *M* HCl is saturated with H_2S? If so, identify the precipitate.

18.68 Refer to Exercise 18.67. Is it possible to make the HCl so concentrated that no precipitate will appear? Use calculations to justify your answer.

18.69 A solution containing 0.0010 *M* Pb^{2+} and 0.0010 *M* Sn^{2+} is saturated with H_2S.
(a) Calculate the $H^+(aq)$ concentration needed to achieve maximum separation of Pb^{2+} and Sn^{2+}.
(b) Calculate the concentrations of Pb^{2+} and Sn^{2+} remaining in solution at the point of maximum separation.

18.70 A solution containing 0.10 *M* $ZnSO_4$ and 0.10 *M* $MnSO_4$ is saturated with H_2S.
(a) Calculate the pH that will achieve maximum separation of Zn^{2+} and Mn^{2+}.
(b) What percent of each ion remains in solution at the pH calculated in Part (a)? Do you think that the separation is effective? Why or why not?

18.71 In Example 18.14 (page 839) it was found that saturated H_2S at a pH = 0.2 would leave 0.10 *M* Zn^{2+} in solution while precipitating CdS. What would happen if (a) the pH is less than 0.2 and (b) the pH is greater than 0.2? Which of these pH changes is more likely to spoil the separation?

18.72 Refer to Exercise 18.70. What percent of each ion will remain in solution if the pH is lowered from the calculated value to pH = 3.5? Do you think that the separation at this pH is effective? Why or why not?

Complex Ions (Section 18.5)

18.73 What is the difference between a *stepwise formation constant* and an *overall formation constant*? Which constant do we usually use in calculations involving complex ions?

18.74 What is the relation between the *formation constant* and the *dissociation constant* of a complex ion? How do these constants vary with the stability of the ion?

18.75 Starting with equal concentrations of Ag^+ and complexing agent, which of the following complex ions would form in the greatest concentration: $Ag(NH_3)_2^+$, $Ag(CN)_2^-$, or $Ag(S_2O_3)_2^{3-}$?

18.76 Which complex, $Zn(NH_3)_4^{2+}$ or $Zn(OH)_4^{2-}$, is more stable with respect to dissociation in aqueous solution? Given equal concentrations, which complex will provide a greater concentration of free Zn^{2+} ion?

18.77 The formation constant of AlF_6^{3-} is 6.7×10^{19}. Calculate K_d of the ion.

18.78 The dissociation constant of $Cu(NH_3)_4^{2+}$ is 9.1×10^{-14}. Calculate K_f of the ion.

18.79 Each of the following reactions produces a complex ion. Write the equation.
(a) solid tin(II) hydroxide plus concentrated NaOH
(b) solid silver acetate plus aqueous NH_3
(c) solid copper(I) oxide plus aqueous KCN

18.80 Each of the following reactions produces a complex ion. Write the equation.
(a) lead metal plus concentrated NaOH
(b) aluminum metal plus concentrated KOH
(*Hint*: These reactions are redox reactions in which the metal is oxidized to a complex ion and water is reduced to H_2.)

18.81 Calculate the concentration of Fe^{3+} in a solution that contains 0.25 *M* $Fe(SCN)^{2+}$ and 2.0 *M* SCN^-.

18.82 What will be the concentration of Ni^{2+} in a solution that is 1.50 *M* in CN^- and 0.100 *M* in $Ni(CN)_4^{2-}$?

18.83 Calculate the concentration of Cu^{2+} in a solution prepared by dissolving 0.150 mol of $CuSO_4$ in 1.00 L of 3.00 *M* NH_3.

18.84 A solution is made by dissolving 1.2 mol of NaCN in 1.0 L of 0.050 *M* $Ni(NO_3)_2$. Calculate the equilibrium concentration of Ni^{2+}.

18.85 Calculate the equilibrium concentrations of Cl^- and Hg^{2+} after 0.010 mol of $Hg(NO_3)_2$ has been dissolved in 1.0 L of 0.20 *M* NaCl.

18.86 Calculate the equilibrium concentrations of Cd^{2+} and $Cd(CN)_4^{2-}$ after 0.100 mol of $Cd(NO_3)_2$ has been dissolved in 1.00 L of 3.00 *M* NaCN.

18.87 The red complex $Fe(SCN)^{2+}$, which is visible to the eye at concentrations down to 1×10^{-4} *M*, is often used to detect small concentrations of Fe^{3+} ion. What is the lowest concentration of Fe^{3+} that will become visible if a solution is made 0.010 *M* in KSCN?

18.88 **(a)** What color change is observed when 1.00 mol of gaseous NH_3 is dissolved in 1.00 L of 0.100 *M* $CuSO_4$?
(b) Calculate the concentrations of $Cu(NH_3)_4^{2+}$ and Cu^{2+} in the solution of Part (a).

18.89 A solution is made by dissolving 1.0 g of $ZnCl_2$ in 500 mL of 0.50 *M* NH_3. Estimate the equilibrium concentrations of $Zn(NH_3)_4^{2+}$, Zn^{2+}, and NH_3.

18.90 Calculate the OH^- concentration in the solution described in Exercise 18.89. Is it possible that $Zn(OH)_2$ might precipitate? Explain your answer.

18.91 The formation constant of $AgCl_2^-$ is 2.5×10^5. Calculate the molar solubility of AgCl in 1.0 *M* HCl solution.

18.92 Calculate the molar solubility of $Zn(OH)_2$ in 1.0 *M* NaOH.

18.93 Sodium thiosulfate ($Na_2S_2O_3$), known to photographers as "hypo," is used to dissolve unexposed silver halides out of film after the exposed silver halide grains have been reduced to a silver image. Calculate the concentration of $S_2O_3^{2-}$ needed to dissolve 20 g of AgBr per liter of solution.

18.94 Calculate the concentration of ethylenediaminetetraacetate ion (EDTA) that will dissolve 0.50 mol of $Zn(OH)_2$ per liter.

PART B. MISCELLANEOUS QUESTIONS AND PROBLEMS

18.95 The solubility of $Ca(OH)_2$ is 0.185 g/100 mL at 0°C. Calculate the K_{sp} at this temperature. Refer to Table 18.1, and state whether the solubility of $Ca(OH)_2$ increases or decreases with increasing temperature.

18.96 **(a)** Calculate the molar solubility of $Ca(OH)_2$ in (1) pure unbuffered water and (2) a solution that is buffered at pH 11.00. Assume a temperature of 25°C.
(b) Explain why $Ca(OH)_2$ is more soluble in the higher pH buffer than in pure water. Is this result an exception to the common ion effect?

18.97 Calculate the molar solubility of $Fe(OH)_3$ in water. The K_{sp} of $Fe(OH)_3$ is 1.6×10^{-39} at 25°C.

18.98 Calculate the number of moles of iron(II) hydroxide that will dissolve in (a) 500 mL of water, (b) 500 mL of 0.10 *M* NaOH, and (c) 500 mL of 0.10 *M* $FeSO_4$. The K_{sp} of $Fe(OH)_2$ is 7.9×10^{-16} at 25°C.

18.99 The Environmental Protection Agency recommends that the concentration of lead ion in drinking water not exceed 1.0×10^{-7} *M*. Consult Table 18.1 to find reagents that could reduce Pb^{2+} to this level. What concentration of reagent would be needed?

18.100 Seawater contains 1300 mg Mg^{2+}, 420 mg Ca^{2+}, and 14 mg Sr^{2+} per liter. Consult Table 18.1, and offer one explanation (with calculations) for the fact that marine animals form shells and other structures from $CaCO_3$ rather than $MgCO_3$ or $SrCO_3$.

18.101 A 500-mL portion of saturated silver acetate (CH_3COOAg) was mixed with 500 mL of 2.0 *M* Na_2S. The resulting black Ag_2S precipitate was filtered, dried, and found to weigh 3.00 g. Use these data to calculate the K_{sp} of CH_3COOAg.

18.102 Because barium sulfate is opaque to x-rays, patients are given a barium sulfate suspension to drink prior to x-ray examination of the gastrointestinal tract. The suspension contains very little free Ba^{2+}(aq) and is therefore harmless. Calculate the number of moles and the number of grams of dissolved Ba^{2+} in a "barium meal" consisting of finely divided $BaSO_4$ suspended in 250 mL of H_2O.

18.103 A urine specimen contains calcium oxalate crystals. If the concentration of dissolved Ca^{2+} ions in the specimen is 11.8 mg/dL, calculate the molarity of dissolved oxalate ions.

18.104 A *Mohr titration* is a procedure for finding the amount of chloride ion in a solution by titrating it with aqueous $AgNO_3$:

$$Ag^+(aq) + Cl^-(aq) \rightarrow AgCl(s)$$

(a) Calculate the Ag^+ concentration at the equivalence point, that is, at the point where the number of moles of added silver ion is equal to the number of moles of chloride ion initially in the solution.

(b) K_2CrO_4 is used as an indicator because excess $AgNO_3$ will form a red precipitate of Ag_2CrO_4. If the CrO_4^{2-} concentration is 0.0020 *M*, at what Ag^+ concentration will the red precipitate form?

(c) Explain why Ag_2CrO_4 does not form before the equivalence point.

18.105 Lead chloride can be leached out of a mixed $PbCl_2$–AgCl precipitate with boiling water.

(a) How many millimoles of AgCl and how many millimoles of $PbCl_2$ will precipitate if 5.0 mL of a solution containing 0.10 *M* $AgNO_3$ and 0.10 *M* $Pb(NO_3)_2$ is made 0.50 *M* in Cl^-?

(b) The mixed precipitate of Part (a) is filtered, washed, and boiled with 10 mL of water. How many millimoles of each precipitate will dissolve? The K_{sp} values at 100°C are as follows: AgCl, 2.2×10^{-10}; $PbCl_2$, 2.1×10^{-4}.

18.106 Seawater contains an average of 1.30 g of Mg^{2+} per liter. The initial step in the Dow process for obtaining magnesium metal from seawater consists of treating the seawater with lime (CaO) to precipitate $Mg(OH)_2$ at a pH of about 11.3.

(a) Write equations for the reaction of CaO with water and for the reaction of the resulting solution with Mg^{2+}.

(b) What percent of the Mg^{2+} is not removed from seawater at this pH?

(c) What mass of magnesium can ideally be obtained from 1000 L of seawater using this method?

18.107 Calcium oxalate kidney stones are sometimes associated with a diet containing a large proportion of oxalate-rich foods such as spinach and rhubarb.

(a) The pH of urine may vary between 4.8 and 8.0. Does a high pH favor or hinder the formation of kidney stones? Explain your answer.

(b) A normal adult excretes from 1.25 mmol to 3.75 mmol of calcium ion daily in a volume of urine that ranges from 600 mL to 1500 mL. Find the range of daily oxalate consumption that might put a person at risk for calcium oxalate stones. Assume an average urine pH of 7.0, and assume that all of the dietary oxalate enters the urine as $C_2O_4^{2-}$.

18.108 **(a)** Use K_{a_1} and K_{a_2} for H_2CO_3 to estimate the $[HCO_3^-]/[H_2CO_3]$ and $[CO_3^{2-}]/[HCO_3^-]$ ratios in a solution containing dissolved carbon dioxide at (1) pH 7, (2) pH 8 (the average pH of seawater), and (3) pH 9 (the pH of many coastal waters).

(b) Estimate the CO_3^{2-} molarity at each pH if the total concentration of CO_2, HCO_3^-, and CO_3^{2-} is 2.3×10^{-3} *M*.

(c) Should $CaCO_3$ precipitate at the pH values in Part (a) if the solutions contain 420 mg/L of calcium ion? (*Note*: Seawater is a very complex mixture, and any predictions based on simple K_{sp} calculations are highly approximate.)

18.109 Calculate the molar solubility of BaF_2 in a solution buffered at pH 2.46. (*Hint*: Show that $[F^-] + [HF] = 2x$, where x is the molar solubility of BaF_2; then find the $[F^-]/[HF]$ ratio.)

18.110 Calculate the pH of a buffer that will dissolve 0.50 mol of silver acetate per liter.

18.111 Calculate the concentration of H^+ required to prepare a solution containing 0.010 mol of dissolved magnesium oxalate per liter. (*Hint*: Oxalic acid is a diprotic acid.)

18.112 **(a)** Calculate the $C_2O_4^{2-}$ concentration in a solution containing 0.20 *M* oxalic acid and 0.0050 *M* HCl. (*Hint*: Oxalic acid is a diprotic acid.)

(b) Would the solution in Part (a) form a precipitate with 0.010 *M* $Mg(NO_3)_2$? With 0.010 *M* $Ca(NO_3)_2$?

18.113 Calculate the number of moles of KCN that must be added to 250 mL of 0.0750 *M* $AgNO_3$ to reduce the concentration of Ag^+ to 1.0×10^{-10} *M*.

18.114 **(a)** Calculate the molar solubilities of AgCl, AgBr, and AgI in 1.0 *M* NH_3. For which of these compounds is aqueous NH_3 a fairly good solvent?

(b) Calculate the molar solubilities of AgBr and AgI in 1.0 *M* $Na_2S_2O_3$. Do your results suggest why $Na_2S_2O_3$ rather than NH_3 is used in the darkroom to dissolve silver halides out of photographic film?

(c) Refer to Table 18.4. Is there a more effective complexing agent than NH_3 or $S_2O_3^{2-}$ for dissolving AgI?

CHEMISTS AT WORK

THE CHEMIST WHO TEACHES

"I think like a high school kid because I never graduated from high school mentally," Lee Marek says with an obvious smile in his voice. Lee is a chemist who teaches at Naperville North High School, which has an enrollment of approximately 2500 students and which is located in suburban Chicago. "I always wanted to be a teacher," he reminisces, "but friends and family counseled that there would be greater financial rewards in another field." After completing the program for a bachelor's degree in chemical engineering in 1968 at the University of Illinois, he went to work for a petroleum company. "It was a job that did not suit my personality. So I left after a year and found a job teaching in a junior high school."

When Lee started teaching, there was a critical shortage of teachers and he was able to teach while pursuing his certification. He muses, "It was funny. I had been teaching for 5 years before I did my student teaching." Since he started teaching, he has accumulated over 95 hours of credit in chemistry, physics, and math courses and has earned two master's degrees in the teaching of science. "You have to be a saint to teach in junior high," he says. "The kids are . . . well . . . highly exothermic . . . effervescent. They really like the hands-on learning." After 5 years at the junior high level, he accepted a teaching position in high school. Now his usual teaching load is five sections of chemistry courses divided between regular and advanced placement sections.

Lee claims that he follows a fairly standard curriculum but spices it with lots of experiments, demonstrations, and experiences. For example, to describe careers in chemistry and the nature of the collegiate experience, he arranged for a group of alumni—a surgeon, pharmacist, two engineers, and two Ph.D. chemists—to make presentations to the current classes. Across the years, he has had students who participated in the International Olympiad Camp and the Chicago Area ACS Contest and who were in the final 40 for Westinghouse Scholarships.

LEE MAREK

His teaching is enriched by the constant experiences that he undertakes in developing new and exciting ways to present science in the classroom and by sharing that learning with other teachers. For 6 years, he directed a summer program for high school teachers that was sponsored jointly by the National Science Foundation and the Fermi Laboratories in Batavia, Illinois. The program's purpose was to provide teachers with a better background and to generate more excitement for teaching. One outcome of this program was the formation of an "Alliance" of chemistry teachers in the Chicago area; it is a mechanism for them to share ideas, concerns, and successes about their teaching.

"One evening in 1985 in the back of a station wagon on the way to an Alliance meeting, several of us mused that there had to be better and more interesting ways to do demonstrations in classrooms. We needed something different—something off-the-wall," he said. This was the beginning of "Weird Science," developed by Lee and Dewayne Lieneman, Bob Lewis, Bill West, and Mike Offut, all teachers in suburban Chicago. Weird Science is simply a dramatic and maybe somewhat theatrical way to illustrate scientific principles. For example, Boyle's Law is illustrated with the "potato gun." A piece of potato is wedged in a cylinder and another piece of potato is forced into the other end of the tube. As the gas between pieces is compressed, it creates enough force to send pieces of potato spraying from the tube.

Weird Science has been a tremendous success. The group has developed a packet of materials to share with other teachers and made presentations to over 200 groups in the United States and Canada. They have also appeared on the David Letterman Show on NBC-TV and recorded four half-hour TV shows for another major network. In addition, Lee and others in the group have received well-deserved awards and recognition for their work.

His advice for potential teachers is simple, "Keep a positive mental attitude."

SURVEY OF THE ELEMENTS 3

Erupting volcano and lava flows, Kilauea Rift, Hawaii. Volcanic emissions are rich in compounds of sulfur.

THE REMAINING NONMETALS

OUTLINE

PREVIEW

Silicon provides the backbone for the earth's rocky crust, and carbon is the structural element of life. Phosphorus is the element that enables cells to use energy and store information. Sulfur provides us with sulfuric acid, a most versatile industrial chemical. Boron forms novel compounds that bring a new dimension to our concept of covalent bonding. These elements are discussed in this chapter along with the other nonmetals in their groups.

The differences between metals and nonmetals have been recognized since ancient times. Metals are shiny substances that can be hammered and shaped; nonmetals lack these properties—they are generally nonlustrous, poor conductors of heat and electricity, and devoid of mechanical strength. Of the 92 elements that occur naturally on the face of the earth, only 22 are considered to be nonmetals.

The stepwise line in the periodic table separates metallic elements from nonmetallic elements. Along the line, however, lie certain elements that have intermediate properties. These elements, which include boron, silicon, germanium, arsenic, antimony, and tellurium, are often called **metalloids** or **semimetals**. Most semimetals exhibit nonmetallic chemical behavior but have physical properties that border on the metallic.

Some nonmetallic elements and several of the semimetals are *semiconductors*—they exhibit a special type of weak electrical conductivity different from that of a metal. The nature of semiconductors is described more fully in Section 21.3.

The nonmetallic elements have an outstanding ability to form covalent bonds with each other; as a result, the 22 nonmetals have a rich chemistry, forming more compounds than all of the metals put together.

S3.1 GROUP 6A: SULFUR, SELENIUM, AND TELLURIUM

The sixth periodic group contains earth's most abundant element, oxygen, and four that are less abundant—sulfur, selenium, tellurium, and polonium. The metallic character of the elements increases going down the group, in accordance with the usual periodic trend. Oxygen, sulfur, and selenium are nonmetals; tellurium is generally regarded as a semimetal; and the rare, radioactive element polonium is a metal. Selenium and tellurium are semiconductors.

Group 6A atoms have s^2p^4 configurations; other atomic properties are summarized in Table S3.1. Scanning the group from top to bottom, we note the decreases in ionization energy and electronegativity that accompany increasing metallic character and the increase in atomic size that leads to longer weaker bonds. We also notice an "irregularity" that is present in other groups as well: The magnitudes of the electron affinity and homonuclear bond energy are less for the second-period element (oxygen) than for the third-period element (sulfur). The oxygen atom, although small and reactive, has a small valence shell with no *d* orbitals. An added electron will experience a great deal of repulsion from other electrons within the crowded shell. There is much less repulsion in the more spacious valence shell of sulfur, so the S—S bond, despite the sulfur atom's larger size, is stronger than the O—O bond.

Oxygen, the lightest element in Group 6A, differs from the heavier ones in another important way. An oxygen atom has only two vacancies in its valence shell and usually forms only two bonds. The atoms of sulfur, selenium, and tellurium, though less reactive than those of oxygen, can form as many as six bonds by involving empty *d* orbitals in their valence shells.

Occurrence, Preparation, Properties, and Uses of the Elements

The strong flavors of plants such as mustard, garlic, and onion are due to organic sulfur compounds.

Sulfur constitutes less than 0.1% of the earth's crust (Figure S3.1). Metal sulfides such as *pyrite* (FeS_2) contain much of this sulfur; lesser amounts are found in the mineral *anhydrite* ($CaSO_4$) and other sulfates. Sulfur is a constituent of protein and is thus present in every living cell. The earth also contains vast deposits of elemental sulfur lying on domes of salt, often at depths of several hundred feet. This sulfur is believed to have been produced by the action of bacteria on $CaSO_4$ in marshy areas at the edge of vanished oceans. The **Frasch process**, a method in which sulfur is forced to the surface by hot water and air (Figure S3.2), is used to mine sulfur from subsurface deposits along the Gulf coast of Texas and Louisiana. Sulfur obtained by the Frasch process is usually 99.5–99.9% pure.

TABLE S3.1 Properties of Group 6A Atoms

Element	Atomic Radius (pm)	First Ionization Energy (kJ/mol)	Electron Affinity (kJ/mol)	Electronegativity	Homonuclear Single Bond Energy (kJ/mol)
O	74	1314	−141	3.4	142[a]
S	103	1000	−200	2.6	266
Se	116	941	−195	2.6	172
Te	143	869	−190	2.1	126
Po	—	812	−180	2.0	—

[a]In H_2O_2

Figure S3.1
Sulfur deposits within a volcanic crater on Mt. Papandayan, Java. The sulfur is produced by the reaction of H_2S and SO_2 gases:

$$2H_2S(g) + SO_2(g) \longrightarrow 3S(s) + 2H_2O(l).$$

Sulfur compounds, especially hydrogen sulfide, are being removed to an ever-increasing extent from natural gas and petroleum to make these fuels less polluting, and these compounds provide yet another source of sulfur. Some of the hydrogen sulfide produced in the desulfurization of fuel is burned in air to produce sulfur dioxide:

$$2H_2S(g) + 3O_2(g) \rightarrow 2SO_2(g) + 2H_2O(g)$$

The sulfur dioxide and remaining hydrogen sulfide are converted to sulfur by passing the hot gases over a Fe_2O_3 or Al_2O_3 catalyst:

$$SO_2(g) + 2H_2S(g) \xrightarrow[300^\circ C]{\text{catalyst}} 3S(l) + 2H_2O(g)$$

Sulfur is an attractive, bright yellow solid that usually comes in the form of powder ("flowers of sulfur") or in easily broken sticks of "roll sulfur." Table S3.2 lists the physical properties of sulfur's allotropic forms. *Rhombic sulfur* (Figure S3.3a) is the stable allotrope at room temperature. *Monoclinic sulfur* (Figure S3.3b), which is stable above 95.5°C, appears as slender needles when liquid sulfur solidifies at the melting point; it will remain in a metastable state for some time after cooling. Rhombic and monoclinic sulfur consist of ring-shaped S_8 molecules in different crystal lattices (Figure S3.4).

When sulfur melts, the S_8 rings in the hot liquid rupture to form open chains that randomly link with each other (Figure S3.5). Continued heating causes the chains to grow; they become tangled and the liquid becomes dark and viscous (see Demonstration S3.1). Above 190°C, the chains begin to break apart and the liquid again becomes mobile. Sudden cooling of this liquid produces an amorphous (noncrystalline) mass known as *plastic sulfur*, a metastable form in which the chains form

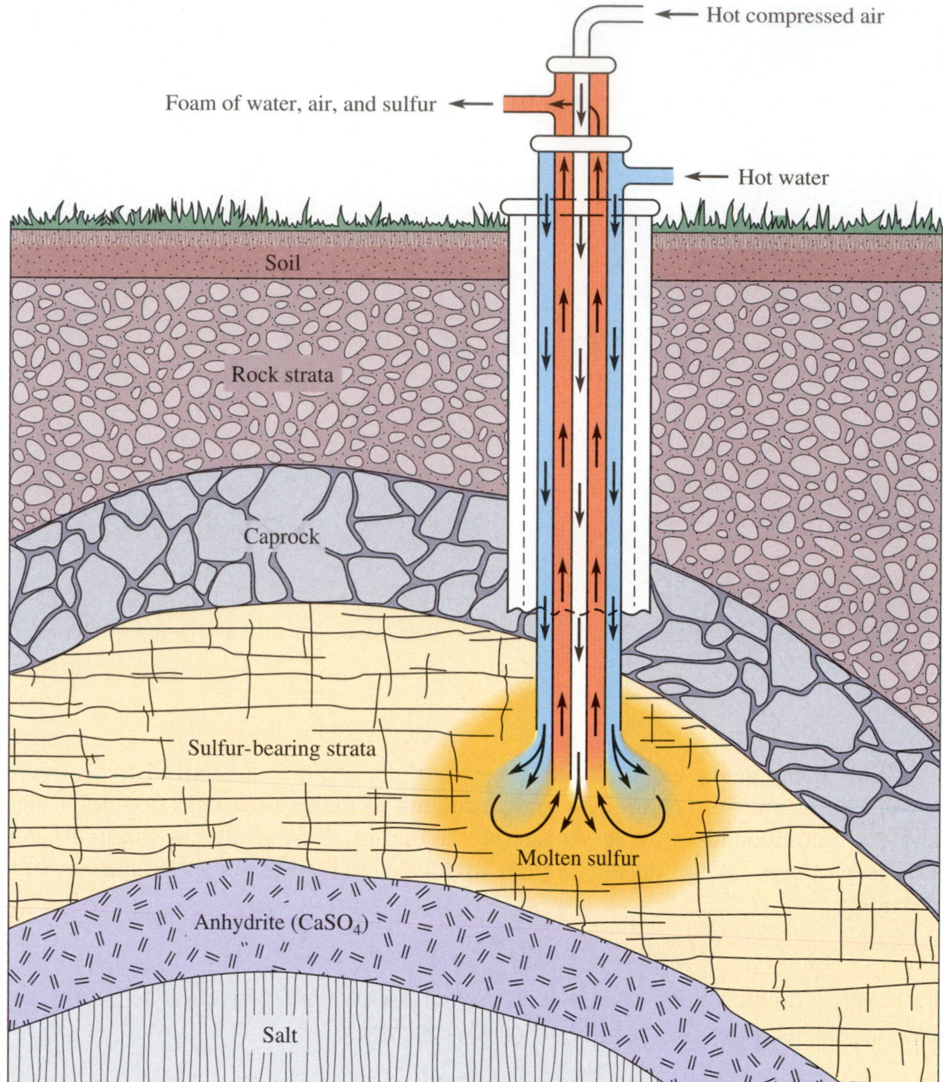

Figure S3.2

The Frasch process. Three concentric pipes are inserted through a drill hole into the sulfur-bearing strata, which may be as much as 700 m below the surface. Compressed superheated water (170°C under 7 atm) is sent down the outer pipe, and compressed air is sent down the center pipe. A froth of water, air, and molten sulfur rises through the middle pipe.

spirals that cause the mass to become stretchable and rubbery. Over a period of time, plastic sulfur becomes crystalline because its atoms migrate and reform S_8 rings.

Sulfur vapor at the boiling point contains S_8, S_6, and S_2 molecules. Above 1000°C, the vapor is blue and consists entirely of S_2 molecules; S_2 is a paramagnetic molecule whose electron configuration resembles that of O_2 (Section 10.5).

PRACTICE EXERCISE S3.1

Consult the phase diagram for sulfur in Figure 12.44 (p. 581) and estimate the temperature at which rhombic and monoclinic sulfur are in equilibrium under 1 atm pressure.

TABLE S3.2 Physical Properties of Sulfur

Property	Rhombic sulfur	Monoclinic sulfur
Density (g/cm^3)	2.07	1.96
Melting point (°C)	112.8[a]	119.0
Boiling point (°C)	444.7	444.7
Solubility (both forms)	Insoluble in water Soluble in CS_2 and CCl_4 Slightly soluble in benzene	

[a]Rhombic sulfur slowly transforms into monoclinic sulfur at 95.5°C and 1 atm. If rhombic sulfur is heated rapidly, it will pass the transition point without change and will melt at 112.8°C.

(a)

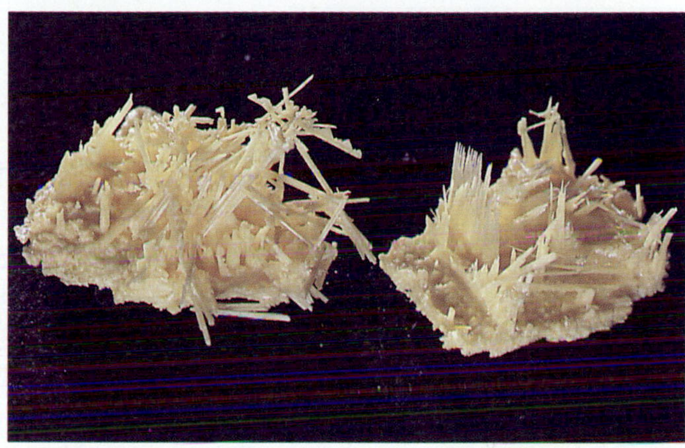

(b)

Figure S3.3

Forms of sulfur: (a) rhombic and (b) monoclinic.

Sulfur is the alchemists' "brimstone" (stone that burns); its combustion produces a blue flame and the choking, acrid fumes of SO_2:

$$S(s) + O_2(g) \rightarrow SO_2(g)$$

Sulfur combines directly with all elements except gold, platinum, and the noble gases. Table S3.3 summarizes the principal reactions of sulfur, and Table S3.4 lists representative compounds of sulfur's principal oxidation states.

Selenium and tellurium are rare and occur principally as selenides and tellurides that are mixed with the sulfide ores of copper, silver, lead, and iron. The most stable allotropes of selenium and tellurium are nonmolecular, although selenium also forms a less stable allotrope that contains Se_8 molecules. Selenium and tellurium, like sulfur, burn in air to form dioxides, and their principal oxidation states are $+6$, $+4$, and -2. Selenium dioxide is an ingredient in the red glass that was once used for automobile tail lights. Selenium and tellurium and their compounds are highly toxic; nevertheless, a trace amount of selenium (50–200 micrograms/day) is essential in the diet. Tellurium has no known biological function.

Figure S3.4

The sulfur molecule is a puckered ring of eight sulfur atoms.

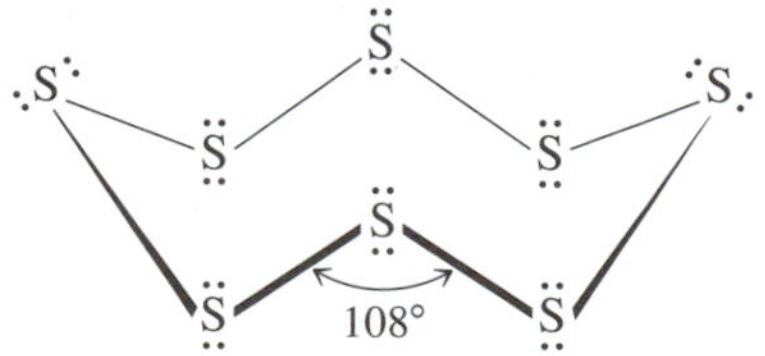

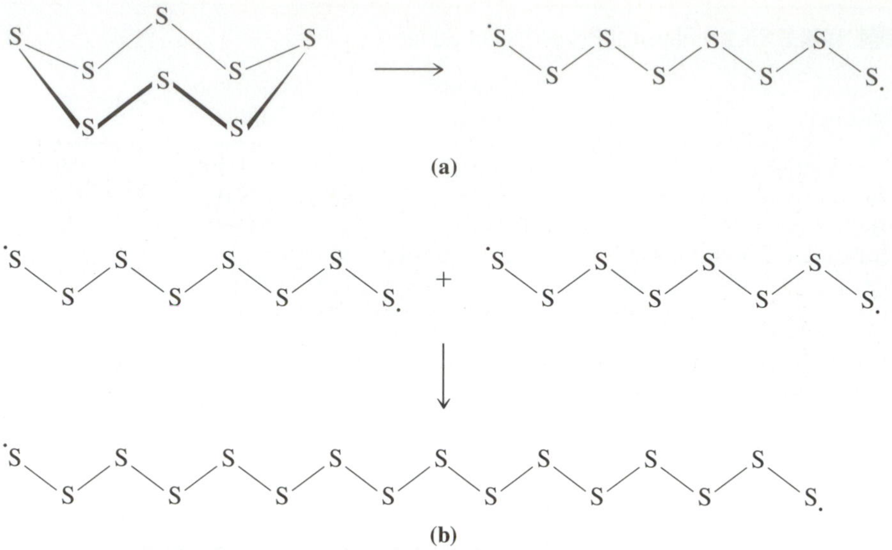

Figure S3.5
Rings and chains in molten sulfur.
(a) Thermal energy causes some S_8 rings to break. (b) The reactive open ends of the chains can link together.

TABLE S3.3 Some Reactions of Sulfur[a]

With Oxygen:

$S(s) + O_2(g) \xrightarrow{\text{heat}} SO_2(g)$

With Hydrogen:

$S(s) + H_2(g) \xrightarrow{\text{heat}} H_2S(g)$

With Halogens:

$S(s) + Cl_2(g) \rightarrow SCl_2(l)$

$2S(l) + Cl_2(g) \xrightarrow{\text{heat}} S_2Cl_2(l)$

$S(s) + 3F_2(g) \longrightarrow SF_6(g)$

With Metals:

$S(s) + Cu(s) \longrightarrow CuS(s)$

$S(s) + 2Ag(s) \longrightarrow Ag_2S(s)$

[a]Elemental sulfur is usually represented in equations by S.

The +6 Oxidation State: Sulfuric Acid and Sulfates

Most sulfur is used to make ***sulfuric acid***, the most abundantly produced industrial chemical. Sulfuric acid is also made from the SO_2 that is recovered in the treatment of sulfide ores and fossil fuels. The conversion of sulfur to sulfuric acid

$$S(s) \xrightarrow{O_2} SO_2(g) \xrightarrow{O_2} SO_3(g) \xrightarrow{H_2O} H_2SO_4(aq)$$

occurs spontaneously, although very slowly, in moist air (see *Chemical Insight: Acid Rain*, p. 760). The **contact process**, the method by which sulfuric acid is prepared from SO_2 in industry, consists of three steps:

TABLE S3.4 Principal Oxidation States of Sulfur

Oxidation State	Hydride or Oxide	Other Molecules and Ions of Importance
−2	H_2S	S^{2-}
0		S_8
+2		SCl_2
+4	SO_2	H_2SO_3 (sulfurous acid) HSO_3^- (hydrogen sulfite ion) SO_3^{2-} (sulfite ion)
+6	SO_3	H_2SO_4 (sulfuric acid) HSO_4^- (hydrogen sulfate ion) SO_4^{2-} (sulfate ion) SF_6

DEMONSTRATION S3.1 HEATING SULFUR

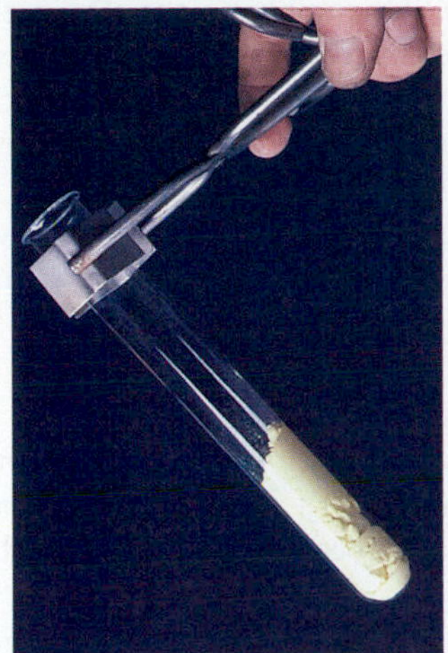
Powdered rhombic sulfur is heated and begins to melt.

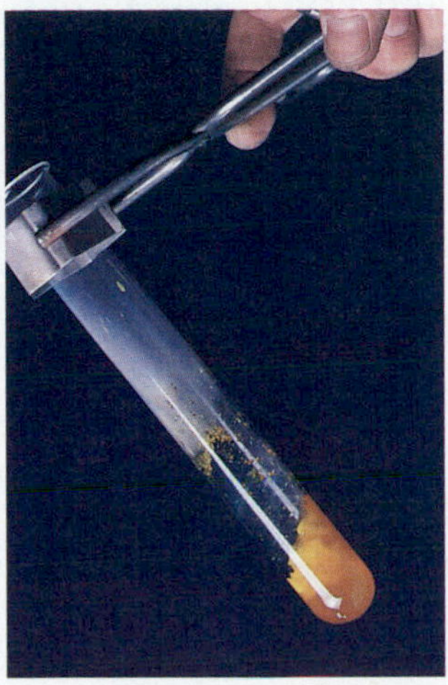
The sulfur becomes completely molten at 112.8°C.

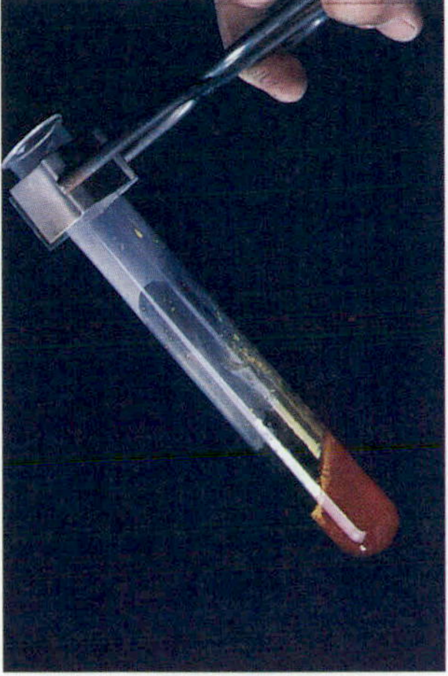
The color of the liquid sulfur deepens as the temperature rises.

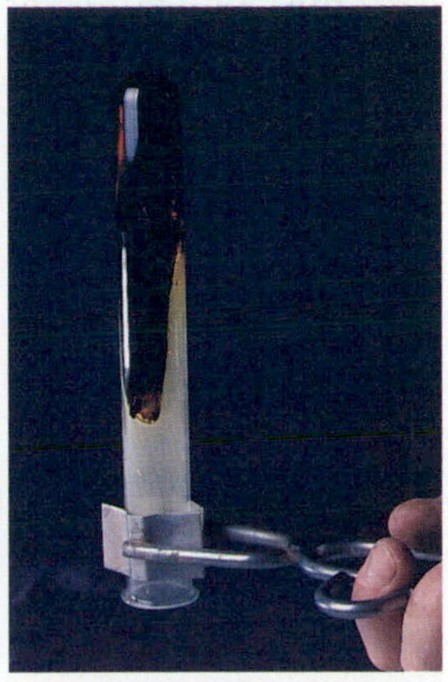
The liquid reaches its maximum viscosity at 190°C; it will hardly pour.

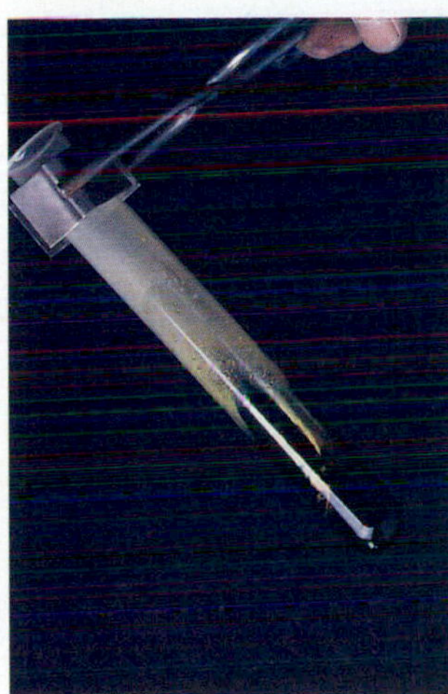
The very dark liquid becomes mobile again above 190°C.

The hot liquid is quickly poured into cold water.

The sulfur that has suddenly solidified does not resemble the original rhombic form. It has an amorphous structure and is known as *plastic sulfur.*

Plastic sulfur is dark and elastic.

Step 1: $2SO_2(g) + O_2(g) \xrightarrow[400°C]{V_2O_5 \text{ catalyst}} 2SO_3(g)$

Step 2: $SO_3(g) + H_2SO_4(aq, 98\%) \rightarrow \underset{\text{pyrosulfuric acid}}{H_2S_2O_7}$

Step 3: $H_2S_2O_7(l) + H_2O(l) \rightarrow 2H_2SO_4(aq, 98\%)$

$$\mathrm{H{-}\ddot{O}{-}S(=O)_2{-}\ddot{O}{-}S(=O)_2{-}\ddot{O}{-}H}$$

pyrosulfuric acid ($H_2S_2O_7$)

The rate of conversion of SO_2 to SO_3 (Step 1) is slow, but it is increased by heating the reaction mixture to 400°C in the presence of a contact catalyst, usually vanadium pentoxide (V_2O_5). The hydration of SO_3 is complicated by the tendency of water and SO_3 to form a dense fog of H_2SO_4 droplets. Water and SO_3 combine more efficiently if the SO_3 is first absorbed by concentrated H_2SO_4 (Step 2) to form *pyrosulfuric acid* (also called "fuming sulfuric acid" or "oleum"). Pyrosulfuric acid reacts quickly with water to form H_2SO_4 (Step 3). The end product is 98% commercial sulfuric acid, which has a density of 1.8 g/mL and boils at 317°C.

Sulfuric acid is a versatile reagent that can function as a strong acid, as an oxidizing agent, and as a dehydrating agent. Some of its uses are described in Figure S3.6 and in the following paragraphs.

As a Strong Acid. About 70% of the sulfuric acid produced in the United States is used by the fertilizer industry to convert ammonia to ammonium sulfate and to convert naturally occurring $Ca_3(PO_4)_2$ to $CaHPO_4$ and $Ca(H_2PO_4)_2$. The hydrogen phosphate salts are more soluble than $Ca_3(PO_4)_2$ and are absorbed more quickly by the roots of plants.

Sulfuric acid is used to dissolve metal oxides from iron and steel surfaces before coating them with zinc (galvanizing) or coating them with tin to make "tin" cans.

PRACTICE EXERCISE S3.2

Write the equation for the reaction of iron(III) oxide with sulfuric acid.

Concentrated sulfuric acid is used to prepare more volatile acids such as HCl and HNO_3 from their salts. Heating the reaction mixture drives off the lower boiling acid (see Figure S2.9, p. 511):

$$\underset{\text{(bp 338°C)}}{H_2SO_4(98\%)} + NaNO_3(s) \xrightarrow{\text{heat}} NaHSO_4(s) + \underset{\text{(bp 83°C)}}{HNO_3(g)}$$

$$H_2SO_4(98\%) + NaCl(s) \longrightarrow NaHSO_4(s) + HCl(g)$$

As an Oxidizing Agent. Concentrated H_2SO_4 oxidizes most metals and many nonmetals as well. The usual reduction product is $SO_2(g)$ (top of next page):

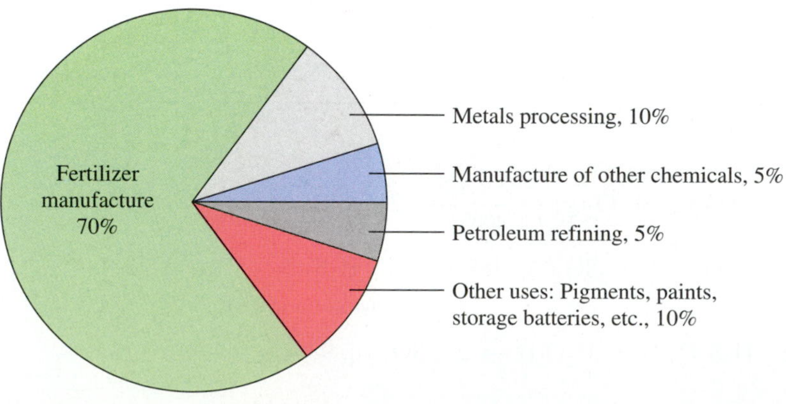

Figure S3.6
The uses of sulfuric acid.

$$\underset{\text{concentrated}}{Cu(s) + 2H_2SO_4(aq)} \longrightarrow CuSO_4(aq) + SO_2(g) + 2H_2O(l)$$

$$\underset{\text{concentrated}}{C(s) + 2H_2SO_4(aq)} \xrightarrow{\text{heat}} CO_2(g) + 2SO_2(g) + 2H_2O(g)$$

Stronger reducing agents may reduce H_2SO_4 to solid sulfur or to H_2S gas.

PRACTICE EXERCISE S3.3

Sodium iodide reduces concentrated sulfuric acid to H_2S. The iodide ion is oxidized to iodine. (a) Write the half-reaction equation for the reduction of H_2SO_4 to H_2S. (b) Write the net ionic equation for the reaction of H_2SO_4 with I^-.

As a Dehydrating Agent. Concentrated sulfuric acid combines readily with water, evolving large quantities of heat, and forming hydrates with the general formula $H_2SO_4{\cdot}nH_2O$, where nH_2O represents from one to eight water molecules. When concentrated H_2SO_4 is diluted, the acid must always be added slowly to the water so that the heat will be dissipated as the dense acid sinks. Water added to concentrated acid would float, evaporate, and spatter. Because of its affinity for water, concentrated sulfuric acid is sometimes used to absorb water vapor from air and from other gaseous mixtures. The concentrated acid will even remove the elements of water from molecules that do not contain water as such; for example, it removes hydrogen and oxygen from carbohydrates such as sugar and starch and converts them to carbon, as shown in Demonstration S3.2:

The chemists' rule is "Acid into water, never water into acid."

$$\underset{\text{sucrose}}{C_{12}H_{22}O_{11}(s)} + 11H_2SO_4(98\%) \rightarrow 12C(s) + 11H_2SO_4{\cdot}H_2O(l)$$

Concentrated H_2SO_4 also chars paper, fabric, and skin.

Natural sulfate minerals include $CaSO_4{\cdot}2H_2O$ (*gypsum* and *alabaster*) and $MgSO_4{\cdot}7H_2O$ (*epsomite*). Some of the uses for Group 1A and 2A sulfates were given in Sections S1.2 and S1.3 of the first survey chapter. Certain ions, most notably Ba^{2+}, are identified by the formation of sulfate precipitates:

$$Ba^{2+}(aq) + SO_4^{2-}(aq) \rightarrow BaSO_4(s)$$

The +4 Oxidation State: Sulfur Dioxide and Sulfites

Sulfur dioxide (SO_2) contains sulfur in the +4 oxidation state. When sulfur dioxide dissolves in water, it is generally assumed that *sulfurous acid* (H_2SO_3), a weak diprotic acid, forms:

$$SO_2(g) + H_2O(l) \rightarrow H_2SO_3(aq)$$

(Molecules of H_2SO_3 are unstable and have never been isolated.) The reaction of SO_2 or H_2SO_3 with limited base produces a hydrogen sulfite; excess base produces a sulfite:

$$H_2SO_3(aq) + OH^-(aq) \rightarrow \underset{\text{hydrogen sulfite ion}}{HSO_3^-(aq)} + H_2O(l)$$

DEMONSTRATION S3.2

THE DEHYDRATING ACTION OF SULFURIC ACID ON SUGAR

Concentrated sulfuric acid is added to table sugar (sucrose, $C_{12}H_{22}O_{11}$).

Sulfuric acid draws the elements of water from sucrose molecules, leaving black carbon. Intense heat is evolved; vapor bubbles turn the carbon into a rising foam.

Reaction is almost over.

$$\underset{\text{}}{HSO_3^-(aq)} + OH^-(aq) \rightarrow \underset{\text{sulfite ion}}{SO_3^{2-}(aq)} + H_2O(l)$$

LEARNING HINT

The reaction of acids with sulfite salts is dicussed on page 142.

Sulfites can be recognized by the acrid SO_2 odor that they produce on contact with acid. Acid protonates the SO_3^{2-} ion, and the resulting H_2SO_3 decomposes to SO_2 and H_2O:

$$BaSO_3(s) + 2H^+(aq) \rightarrow Ba^{2+}(aq) + SO_2(g) + H_2O(l)$$

Sulfur dioxide reacts with carbonates to form sulfites, which are oxidized in air to sulfates:

$$CaCO_3(s) + SO_2(g) \rightarrow CaSO_3(s) + CO_2(g)$$

$$2CaSO_3(s) + O_2(g) \rightarrow 2CaSO_4(s)$$

These reactions occur when marble and limestone statues, which consist of $CaCO_3$, are attacked by the SO_2 in acid rain and polluted air. The resulting $CaSO_4$ does not fit into the $CaCO_3$ lattice, so the marble crumbles away. On the other hand, the reaction between SO_2 and carbonates is useful for "scrubbing" smokestack gases. The SO_2 is removed from these gases when they are passed through a limestone–water slurry.

DEMONSTRATION S3.3

THE REACTION BETWEEN H_2S AND SO_2: $2H_2S(g) + SO_2(g) \rightarrow 3S(s) + 2H_2O(l)$

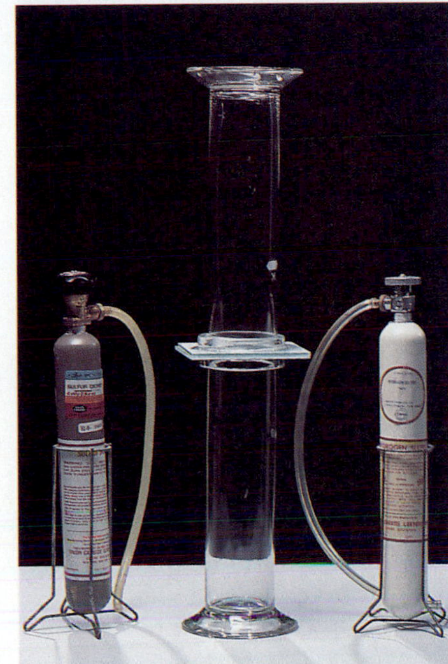

H_2S gas and SO_2 gas occupy two damp cylinders separated by a glass plate.

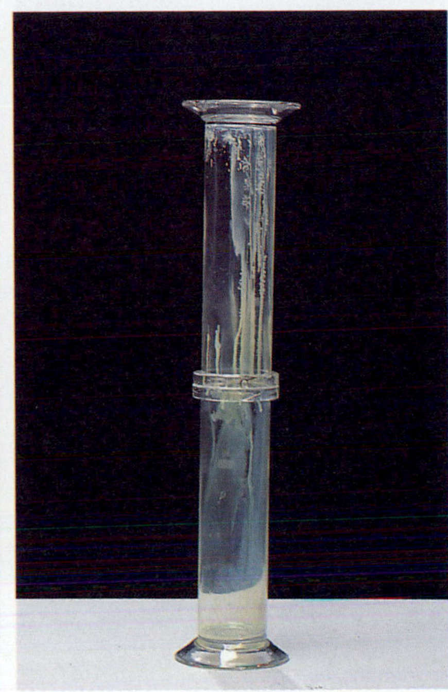

The glass plate is removed and the gases react: H_2S is oxidized; SO_2 is reduced. Since moisture is a catalyst for this reaction, the sulfur appears as a coating on the wet glass.

The +4 oxidation state is an intermediate one for sulfur; therefore, sulfites and SO_2 can be either oxidized or reduced (Demonstration S3.3). They function best as reducing agents; in this capacity, they are used as bleaching agents for fabrics and dried fruit (Demonstration S3.4). Sulfites are also used as food preservatives and as fungicides because they are toxic to many microorganisms. Their principal application, however, is in the pulp and paper industry where they are used to dissolve the lignin that holds bundles of wood fibers together.

The Thiosulfate Ion

Sulfur forms a number of different anions that are structurally related to SO_4^{2-}. One of these is the thiosulfate ion, $S_2O_3^{2-}$. The *thio* prefix indicates that an oxygen atom in the parent ion, SO_4^{2-} in this case, has been replaced by a sulfur atom.

A thiosulfate is formed by heating a sulfite solution with sulfur:

$$SO_3^{2-}(aq) + S(s) \rightarrow S_2O_3^{2-}(aq)$$

thiosulfate ion ($S_2O_3^{2-}$)

DEMONSTRATION S3.4

THE BLEACHING ACTION OF SULFUR DIOXIDE

An African violet in full color.

The violet after exposure to SO_2 gas. The SO_2 reduces the violet pigments to colorless compounds.

Acid reverses this reaction, decomposing the thiosulfate to SO_2 and free sulfur:

$$S_2O_3^{2-}(aq) + 2H^+(aq) \rightarrow S(s) + \underset{\text{unstable}}{H_2SO_3(aq)}$$

$$H_2SO_3(aq) \rightarrow SO_2(g) + H_2O(l)$$

Strong oxidizing agents convert thiosulfate to sulfate:

$$8MnO_4^-(aq) + 5S_2O_3^{2-}(aq) + 14H^+(aq) \rightarrow 8Mn^{2+}(aq) + 10SO_4^{2-}(aq) + 7H_2O(l)$$

Weaker oxidizing agents convert it to the *tetrathionate* ion, $S_4O_6^{2-}$:

$$2S_2O_3^{2-}(aq) + I_2(s) \rightarrow S_4O_6^{2-}(aq) + 2I^-(aq)$$

This reaction can be used to decolorize iodine stains because the thiosulfate ion and its products are colorless (Demonstration S3.5).

Sodium thiosulfate ($Na_2S_2O_3$) is known to photographers as "hypo"; it is used as a "fixer" to dissolve unexposed silver bromide out of photographic film. The thiosulfate complexes the silver ion:

DEMONSTRATION S3.5

REMOVING AN IODINE STAIN WITH SODIUM THIOSULFATE: $2S_2O_3^{2-}(aq) + I_2(s) \rightarrow 2I^-(aq) + S_4O_6^{2-}(aq)$

A brown iodine stain on filter paper.

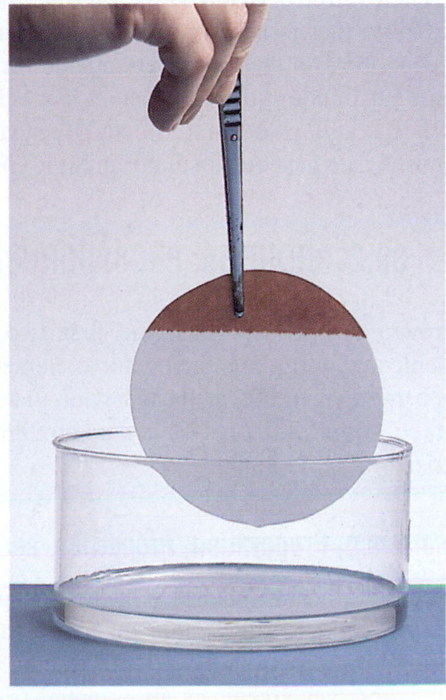

A $Na_2S_2O_3$ solution bleaches the stain by reducing I_2 to colorless I^-.

$$AgBr(s) + 2S_2O_3^{2-}(aq) \rightarrow Ag(S_2O_3)_2^{3-}(aq) + Br^-(aq)$$

The −2 Oxidation State: Sulfides

Hydrogen sulfide (H_2S) is a toxic gas with the unforgettable smell of rotten eggs. Protein contains sulfur, and H_2S is produced whenever protein decays in the absence of air. Concentrations of H_2S above 13 ppm are unsafe to breathe, but since the odor becomes unbearable at concentrations well below this, H_2S claims few victims. The weak acidic properties of H_2S and its use as a precipitating agent for metal ions were discussed in Chapters 17 and 18.

Various *polysulfide* ions can be formed by treating sulfur with a hot solution of a sulfide such as Na_2S. The reaction that forms disulfide ion is:

$$S^{2-}(aq) + S(s) \rightarrow S_2^{2-}(aq)$$

More sulfur atoms can then add on to produce chains up to six sulfur atoms long.

a polysulfide ion (S_4^{2-})

Sulfur Halides

Sulfur forms a number of different compounds with fluorine and chlorine. Chlorine and sulfur combine at room temperature in the presence of a catalyst to give SCl_2, a dark red liquid. At higher temperatures, chlorine oxidizes molten sulfur to give S_2Cl_2, an orange-yellow liquid with an offensive smell. Sulfur chlorides are used as solvents for sulfur in the vulcanization of rubber, a process in which rubber is combined with sulfur to increase its hardness and elasticity. All sulfur chlorides are decomposed by water.

When sulfur is oxidized by fluorine, the principal product is SF_6, the only Group 6A halide that is not attacked by moisture. The SF_6 molecule is unreactive partly because +6 is a stable oxidation state for sulfur and partly because the six fluorine atoms surrounding the sulfur atom leave little opening for attack by another reagent. Sulfur hexafluoride is not affected by steam even at 500°C; because of its inertness, it is used as a gaseous insulator in high-voltage equipment.

S3.2 GROUP 5A: PHOSPHORUS, ARSENIC, AND ANTIMONY

Nitrogen and phosphorus, the first two elements in Group 5A, are nonmetals. Arsenic and antimony, the next two elements, are semimetals; antimony, however, is much more metallic in its behavior than arsenic. The last element, bismuth, is a metal. Group 5A atoms have s^2p^3 configurations; other atomic properties are summarized in Table S3.5.

Occurrence, Preparation, Properties, and Uses of the Elements

Phosphorus occurs naturally in the phosphate minerals of rocks, bones, and teeth, and in the biochemical molecules that make up genes and cell membranes. Arsenic and antimony are relatively rare. Arsenic compounds have achieved some notoriety as poisons, yet arsenic is an essential trace element in the diet. Antimony has no known biological function.

Elemental phosphorus is obtained from *phosphate rock*, a mineral that contains various phosphorus compounds, including *fluorapatite, chlorapatite,* and *hydroxyapatite*. The formula for fluorapatite is $3Ca_3(PO_4)_2 \cdot CaF_2$ (i.e., $Ca_3(PO_4)_2$ and CaF_2 in a 3:1 mole ratio); in the chlor- and hydroxy- versions, F is replaced by Cl and OH, respectively. In the preparation of phosphorus, the phosphate rock is crushed and

TABLE S3.5 Properties of Group 5A Atoms

Element	Atomic Radius (pm)	First Ionization Energy (kJ/mol)	Electron Affinity (kJ/mol)	Electronegativity	Homonuclear Single Bond Energy (kJ/mol)
N	74	1402	0	3.0	163
P	110	1012	−74	2.2	200
As	125	944	−77	2.2	150
Sb	145	832	−101	2.1	120
Bi	155	703	−100	2.0	—

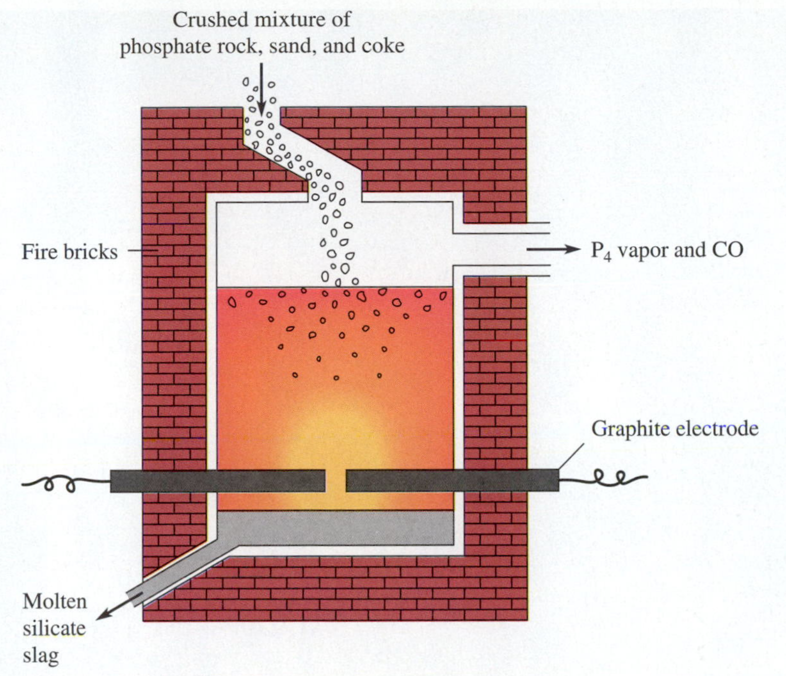

Figure S3.7

The industrial preparation of phosphorus. The furnace is heated to 1450°C by an electric current that passes between two large carbon electrodes. The charge of crushed phosphate rock, sand, and coke is fed in at the top. Vapors of P_4 and CO emerge from the upper outlet. A water spray condenses the phosphorus vapor to liquid, which is tapped off and allowed to solidify.

then heated with sand and coke to 1450°C in an electric furnace (Figure S3.7). Silicon displaces the phosphorus from calcium phosphate

$$2Ca_3(PO_4)_2(l) + \underset{\text{sand}}{6SiO_2(l)} \rightarrow P_4O_{10}(g) + 6CaSiO_3(l)$$

and carbon reduces it to the element

$$P_4O_{10}(g) + \underset{\text{coke}}{10C(s)} \rightarrow P_4(g) + 10CO(g)$$

The product condenses to *white phosphorus* (P_4), a waxy solid that must be stored under water to prevent it from reacting with oxygen.

The P_4 molecule is a tetrahedron in which each atom forms three bonds while retaining a lone pair of electrons (Figure S3.8). According to the VSEPR model (Section 10.2), the bond angles around an atom with three single bonds and one lone pair should be close to 109.5°. The bond angles in the P_4 molecule, however, are only 60°, and the resulting distortion causes the molecule to be very reactive. A lump of P_4 will ignite spontaneously in air at 31°C, while dissolved P_4 will catch fire at lower temperatures. Even below its ignition point, white phosphorus undergoes a continuous slow oxidation that evolves energy in the form of light (Figure S3.9). The resulting glow, which is visible in the dark, is the *phosphorescence* for which this element is named. White phosphorus vapor attacks bones and other tissues and causes burns that do not readily heal.

White phosphorus gradually changes to the most stable form of phosphorus, a mulberry-colored solid called *red phosphorus* (Figure S3.10). This allotrope can also be prepared by heating white phosphorus in the absence of air. Red phosphorus has a variable covalent network consisting of up to six different lattice types; its extended structure causes it to be less volatile, less soluble, and less reactive than

Figure S3.8

The P_4 molecule.

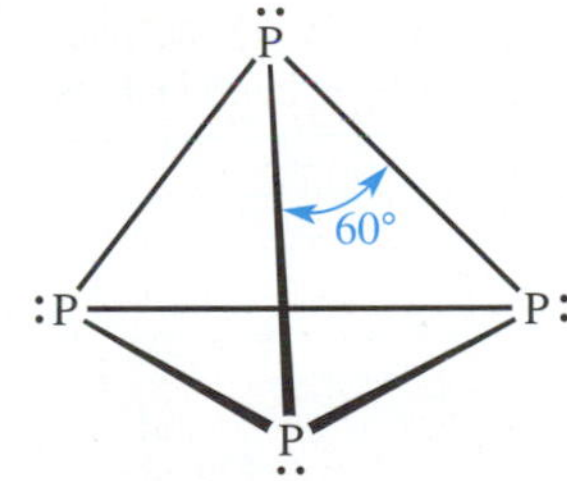

Figure S3.9

"The Alchemist," a painting by Joseph Wright, depicts an alchemist's fascination with phosphorescence.

white phosphorus (Table S3.6). Black, violet, and scarlet allotropes of phosphorus have also been prepared.

Phosphorus reacts with oxygen, sulfur, the halogens, and many metals (Table

Figure S3.10

Allotropic forms of phosphorus. *Left:* Red phosphorus. *Right:* White phosphorus stored under water to protect it from air.

TABLE S3.6 Properties of White and Red Phosphorus

Property	White Phosphorus	Red Phosphorus
Density (g/cm^3)	1.82	2.34
Melting point (°C)	44.1	Sublimes at 417
Boiling point (°C)	280	—
Spontaneous ignition temperature (°C)	31	260
Solubility (g/100 mL)		
Water	0.0003	Sparingly soluble
Carbon disulfide	125	Sparingly soluble
Ethyl alcohol	0.3	Sparingly soluble
Benzene	2.9	Sparingly soluble

S3.7). White phosphorus has been used in weapons, and some phosphorus is used in fireworks, rodent poisons, and matches. Most of its industrial output, however, is used to synthesize phosphorus compounds (Figure S3.11).

Arsenic and antimony occur in nature as sulfides, usually mixed with other sulfide minerals (Figure S3.12). The stable form of each element is a nonmolecular solid with a somewhat metallic appearance. Less stable allotropes that contain As_4 and Sb_4 molecules have also been prepared. Both arsenic and antimony are used in the manufacture of various alloys.

Antimony is the only semimetal to form positive ions. Salts such as $Sb_2(SO_4)_3$ contain the Sb^{3+} ion, and SbOCl contains the *antimonyl* ion, SbO^+.

TABLE S3.7 Some Reactions of Phosphorus

With Oxygen:

$$P_4(s) + 3O_2(g) \xrightarrow{\text{limited oxygen}} P_4O_6(s)$$

$$P_4(s) + 5O_2(g) \xrightarrow{\text{excess oxygen}} P_4O_{10}(s)$$

With Other Nonmetals:

$$P_4(s) + 6X_2 \xrightarrow{\text{limited halogen}} 4PX_3 \qquad (X = \text{any halogen})$$

$$P_4(s) + 10X_2 \xrightarrow{\text{excess halogen}} 4PX_5 \qquad (X = F, Cl, Br)$$

$$P_4(l) + 3S(l) \xrightarrow{\text{heat}} P_4S_3(s) \qquad (\text{and other sulfides such as } P_4S_5 \text{ and } P_4S_7)$$

With Group 1A and Group 2A Metals:

$$12Na(s) + P_4(s) \xrightarrow{\text{heat}} 4Na_3P(s)$$

$$6Ca(s) + P_4(s) \xrightarrow{\text{heat}} 2Ca_3P_2(s)$$

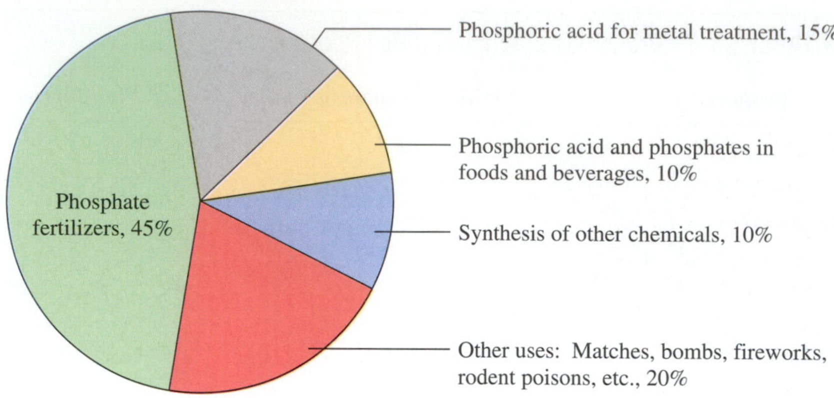

Figure S3.11
The uses of phosphorus.

Oxides

When phosphorus is burned in a limited supply of oxygen, the product is P_4O_6, a white solid that has a low melting point and contains phosphorus in the +3 oxidation state. The P_4O_6 molecule can be visualized as a P_4 molecule expanded to accommodate an oxygen atom between every two phosphorus atoms (Figure S3.13a). In excess oxygen, phosphorus burns to form P_4O_{10}, a white solid. The P_4O_{10} molecule resembles the P_4O_6 molecule but has an additional oxygen bonded to each of the four phosphorus atoms (Figure S3.13b). P_4O_{10}, which contains phosphorus in the +5 state, is sometimes called *phosphorus pentoxide* after its empirical formula; it is also called *phosphoric anhydride* because it is an effective drying agent that avidly absorbs water to form phosphoric acid:

$$P_4O_{10}(s) + 6H_2O(l \text{ or } g) \rightarrow 4H_3PO_4(l)$$

PRACTICE EXERCISE S3.4

Use VSEPR to predict the P—O—P angles in (a) P_4O_6 and (b) P_4O_{10}. Compare your predicted angles with the observed angles given in Figure S3.13.

Figure S3.12
Some arsenic and antimony ores: (a) native arsenic, (b) orpiment (As_2S_3), and (c) black stibnite (Sb_2S_3).

(a)

(b)

(c)

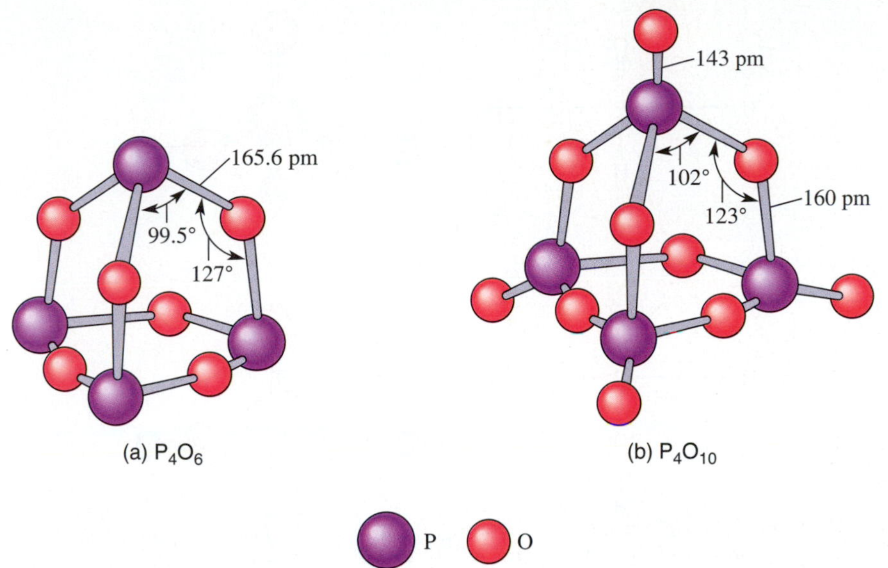

Figure S3.13
The structure of (a) P_4O_6 and (b) P_4O_{10}.

The principal oxidation states of the Group 5A elements are +5, +3, and −3 (Table S3.8). The +5 state predominates in compounds of phosphorus, the +3 state in compounds of arsenic and antimony. Arsenic and antimony, for example, burn in limited or excess oxygen to form As_4O_6 and Sb_4O_6. The +5 oxides of these elements do not form during the combustion, but they can be obtained by oxidizing the lower oxides with nitric acid. The behavior of the Group 5A elements illustrates a characteristic trend in main periodic groups: *Lower positive oxidation states become relatively more stable toward the bottom of a group.*

The famous "arsenic" of murder mysteries is arsenic trioxide, As_4O_6, which is fatal in doses as low as 0.1 g.

Oxo Acids and Anions of Phosphorus

The principal oxo acids of phosphorus are listed in Table S3.8. *Phosphorous acid* (H_3PO_3; spelled with an *-ous*) contains phosphorus in the +3 state. It is formed by the hydration of P_4O_6:

$$P_4O_6(s) + 6H_2O(l) \rightarrow 4H_3PO_3(aq)$$

TABLE S3.8 Some Oxidation States of Phosphorus

Oxidation State	Representative Compounds
+5	P_4O_{10} (phosphorus pentoxide[a])
	H_3PO_4 (orthophosphoric acid)
	PCl_5 (phosphorus pentachloride)
+3	P_4O_6 (phosphorus trioxide[a])
	H_3PO_3 (phosphorous acid)
	PCl_3 (phosphorus trichloride)
−3	Ca_3P_2 (calcium phosphide)
	PH_3 (phosphine)

[a]Phosphorus pentoxide and phosphorus trioxide are named after their empirical formulas.

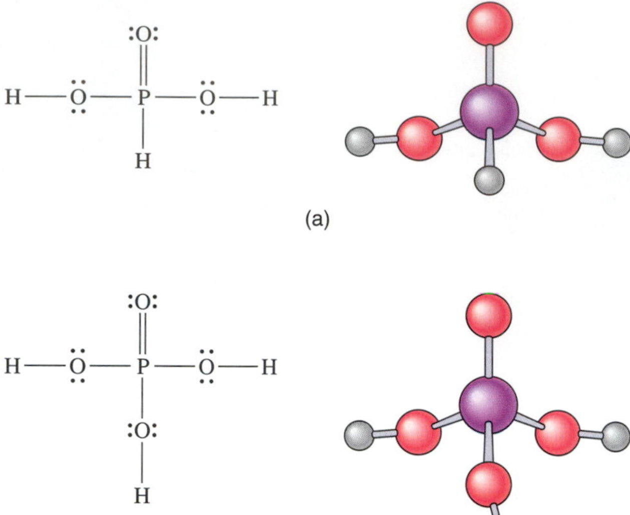

Figure S3.14
The structure of (a) phosphorous acid (H_3PO_3), a diprotic acid, and (b) orthophosphoric acid (H_3PO_4), a triprotic acid.

H_3PO_3 is a diprotic acid because one of its three hydrogen atoms is bonded directly to the phosphorus atom and is not acidic (Figure S3.14a).

The +5 oxidation state is by far the most stable state for phosphorus and the only one we will consider in detail. ***Orthophosphoric acid*** (Figure S3.14b), often simply called phosphoric acid, can be prepared by treating phosphate rock with sulfuric acid:

$$Ca_3(PO_4)_2(s) + 3H_2SO_4(aq) \rightarrow 2H_3PO_4(85\%) + 3CaSO_4(s)$$

This reaction proceeds in the forward direction because the calcium sulfate precipitates and is removed from the solution. Very pure phosphoric acid can be prepared by hydrating P_4O_{10} (the equation is given in the section on oxides). H_3PO_4 is a moderately weak triprotic acid and a poor oxidizing agent. Phosphoric acid and its salts are used in buffer solutions; the acid–base properties of H_3PO_4, $H_2PO_4^-$, HPO_4^{2-}, and PO_4^{3-} were discussed in Chapters 17 and 18.

About 85% of the industrial output of phosphoric acid is used in the fertilizer industry. The phosphorus in fertilizer comes from phosphate rock that has been treated to make its phosphates more soluble. A mixture of $Ca(H_2PO_4)_2$ and $CaSO_4$ is called *superphosphate*. It is produced by reacting phosphate rock with sulfuric acid:

$$Ca_3(PO_4)_2(s) + 2H_2SO_4(aq, 98\%) \rightarrow \underset{\text{superphosphate}}{Ca(H_2PO_4)_2(s)} + 2CaSO_4(s)$$

Triple superphosphate, which is obtained from phosphate rock and phosphoric acid, has a higher phosphate content:

$$Ca_3(PO_4)_2(s) + 4H_3PO_4(aq) \rightarrow \underset{\text{triple superphosphate}}{3Ca(H_2PO_4)_2(s)}$$

Sodium phosphate (Na_3PO_4), known commercially as *trisodium phosphate* or TSP, dissolves in water to give a pH of about 12. It is used as a specialized cleansing agent, degreaser, and paint remover.

PRACTICE EXERCISE S3.5

Write the equation for the preparation of trisodium phosphate from sodium hydroxide and phosphoric acid.

A number of polymeric phosphates can be prepared by removing the elements of water from orthophosphoric acid (H_3PO_4) or from its sodium salts. The first dehydration product is *pyrophosphoric acid* ($H_4P_2O_7$):

$$
\begin{array}{c}
:\!O\!: \\ \| \\ H-\ddot{O}-P-\ddot{O}-H \\ | \\ :\!O\!: \\ | \\ H
\end{array}
+
\begin{array}{c}
:\!O\!: \\ \| \\ H-\ddot{O}-P-\ddot{O}-H \\ | \\ :\!O\!: \\ | \\ H
\end{array}
\xrightarrow{215^\circ}
H_2O +
\begin{array}{c}
:\!O\!: \quad\quad :\!O\!: \\ \| \quad\quad\quad \| \\ H-\ddot{O}-P-\ddot{O}-P-\ddot{O}-H \\ | \quad\quad\quad | \\ :\!O\!: \quad\quad :\!O\!: \\ | \quad\quad\quad | \\ H \quad\quad\quad H
\end{array}
$$

pyrophosphoric acid

The structure of the pyrophosphate ion is shown in Figure S3.15a. Additional dehydration produces long-chain ***polyphosphates*** such as the *triphosphate* ion ($P_3O_{10}{}^{5-}$) shown in Figure S3.15b. More drastic dehydration produces a mixture of *trimetaphosphates* and *hexametaphosphates*, in which the phosphorus and oxygen atoms are joined in rings (Figure S3.15c).

Figure S3.15 Polymeric oxo anions of phosphorus.

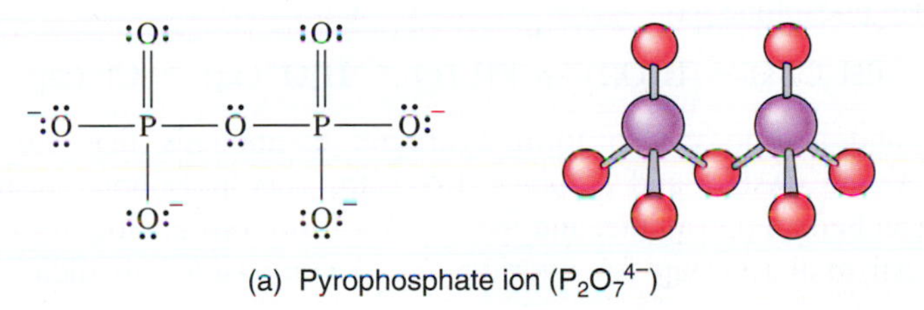

(a) Pyrophosphate ion ($P_2O_7{}^{4-}$)

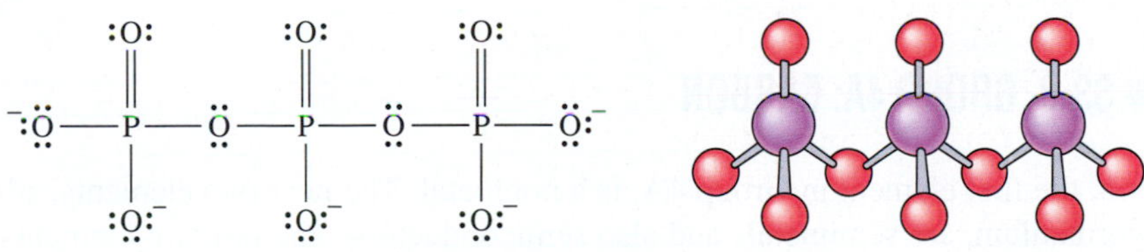

(b) Triphosphate ion ($P_3O_{10}{}^{5-}$)

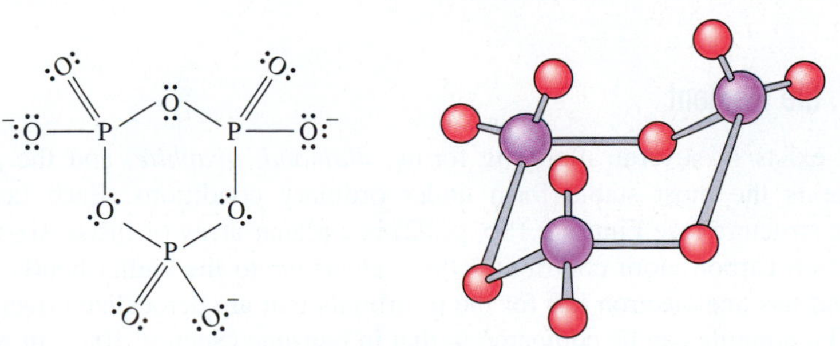

(c) Trimetaphosphate ion ($P_3O_9{}^{3-}$)

Sodium polyphosphates and metaphosphates were once widely used in laundry detergents because they *sequester* (envelop and hold) the Ca^{2+} and Mg^{2+} ions that produce soap scum. The use of phosphates in household laundry detergents has been largely discontinued because the runoff is harmful to the local water. Other sequestering agents are used instead. For a further description of the effect of phosphorus effluent on lakes see *Chemical Insight: Phosphorus, the Limiting Element* (p. 114).

The −3 Oxidation State: Phosphides and Phosphine

Phosphides are binary compounds of phosphorus with a less electronegative element. The ionic phosphides are salts of active metals and contain the phosphide ion (P^{3-}). Calcium phosphide can be prepared by direct combination of the elements:

$$6Ca(s) + P_4(s) \xrightarrow{\text{heat}} 2Ca_3P_2(s)$$

Ionic phosphides have properties similar to those of ionic nitrides (see p. 512). They react with water to give basic solutions and *phosphine* (PH_3), a toxic, colorless gas:

$$Ca_3P_2(s) + 6H_2O(l) \rightarrow 2PH_3(g) + 3Ca(OH)_2(aq)$$

The phosphine molecule, like the ammonia molecule, is pyramidal with a lone pair of electrons at the phosphorus apex. PH_3, however, is a much weaker base than NH_3. The addition of a proton to PH_3 produces the *phosphonium* ion ($PH_4{}^+$), but phosphonium salts, unlike ammonium salts, are not very stable. They react with water to form phosphine gas:

$$PH_4Cl(s) + H_2O(l) \rightarrow PH_3(g) + H_3O^+(aq) + Cl^-(aq)$$

Arsenic and antimony also form hydrogen compounds that are analogs of ammonia. *Arsine* (AsH_3) and *stibine* (SbH_3) are very poisonous gases; they are much less stable than phosphine, and they do not form ions analogous to the ammonium ion. All of the Group 5A hydrides are decomposed into their elements by heating.

S3.3 GROUP 4A: CARBON

Carbon, the first element in Group 4A, is a nonmetal. The next two elements, silicon and germanium, are semimetals and also semiconductors. The last two elements, tin and lead, are metals. Group 4A atoms have s^2p^2 configurations; other atomic properties are summarized in Table S3.9.

Carbon, the Element

Carbon exists in several allotropic forms, *diamond, graphite*, and the *fullerenes*. Graphite is the most stable form under ordinary conditions. Each layer in the graphite structure (see Figure 2.18b, p. 62) is a planar array of fused, six-membered rings. Each carbon atom contributes three electrons to the sigma bonds within the plane and has one electron left for the pi orbitals that are delocalized over the entire layer. The bonding can be compared to that in benzene (Section 10.6), in which each carbon contributes an electron to pi molecular orbitals above and below the plane of

TABLE S3.9 Properties of Group 4A Atoms

Element	Atomic Radius (pm)	First Ionization Energy (kJ/mol)	Electron Affinity (kJ/mol)	Electronegativity	Homonuclear Single Bond Energy (kJ/mol)
C	77	1086	−122	2.6	348
Si	118	787	−120	1.9	226
Ge	123	762	−116	2.0	188
Sn	141	709	−121	1.8	151
Pb	175	716	−100	1.9	—

the molecule. The carbon–carbon distance within a layer (142 pm) is similar to the bond length in benzene. The distance between adjacent layers (335 pm) is too long for significant orbital overlap; the layers are bound to each other mainly by van der Waals forces.

The unusual structure of graphite provides it with some unique properties. Graphite is the only naturally occurring nonmetallic substance that conducts electricity; its electrons travel easily through the delocalized pi orbitals in each layer, so that its conductivity is greatest in the directions that are parallel to the layers. Graphite's conductivity increases with pressure; hence, graphite is used in microphones and telephone receivers to pick up the fluctuating pressure of sound waves.

Graphite absorbs air and moisture between its layers, and this helps the layers slide past each other. Powdered graphite is used as a lubricant. Graphite compressed with clay and wax is pencil "lead"; the graphite layers rub off as marks on the paper. Graphite loses its lubricating quality when it is heated in a vacuum to drive out the cushioning air.

Soot, ashes, and carbon black are finely divided forms of carbon known collectively as **amorphous carbon**. *Carbon black* is a very pure carbon powder produced by the controlled oxidation of hydrocarbons. The atoms in amorphous carbon do not form a consistent lattice; such order as they do have, however, resembles that of graphite. Amorphous carbon contains many incompletely bonded carbon atoms that attract other molecules, so it is an effective adsorbent. *Activated charcoal*, charcoal that has been steam heated and then dried to remove foreign matter, is used in filters to purify air, water, beverages, syrups, and oil (Figure S3.16).

Diamonds (density = 3.53 g/cm^3) are denser than graphite (density = 2.51 g/cm^3) and do not conduct electricity. Their outstanding characteristics are hardness, which enables them to scratch every other known substance; a high refractive index, which makes them sparkle; and rarity, which makes them valuable. Each atom in the *diamond lattice* is singly bonded to four tetrahedrally placed neighbors (see Figure 2.18a, p. 62). Diamond becomes stable with respect to graphite only under extremely high pressure (see the phase diagram of carbon in Figure 12.45, p. 581); in other words, all the diamonds we ever see are in a metastable state.

Diamond surpasses most metals in its ability to conduct heat, and at the same time, it is one of nature's best electrical insulators. This combination of properties is unusual since, in most substances, thermal conductivity and electrical conductivity go together. These characteristics, along with hardness and inertness (resistance to

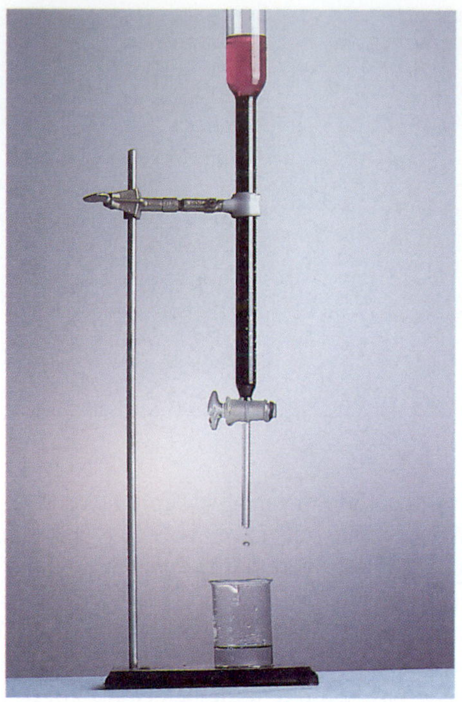

Figure S3.16

Activated charcoal as a purifying agent. Water containing a red dye (top) trickles through the charcoal (black solid in tube). The charcoal adsorbs the dye, and the purified water emerges below.

chemical attack), make diamond a highly prized material for industrial use. Since 1954, diamonds of industrial grade have been synthesized from graphite in a molten metal catalyst at 1500 K and pressures of at least 50,000 atm (Figure S3.17). More recently, a method has been developed for the slow deposition of diamond films from a mixture of methane and hydrogen under ordinary pressure. The mixture is first strongly heated to decompose methane into carbon and hydrogen, and then irradiated to generate excess atomic hydrogen. As the solid carbon slowly deposits, the hydrogen atoms react with any carbon–carbon double bonds that may exist, thus preventing the formation of a graphite lattice.

Figure S3.17

(a) Native diamond in a rock matrix.
(b) Industrial diamonds.

(a)

(b)

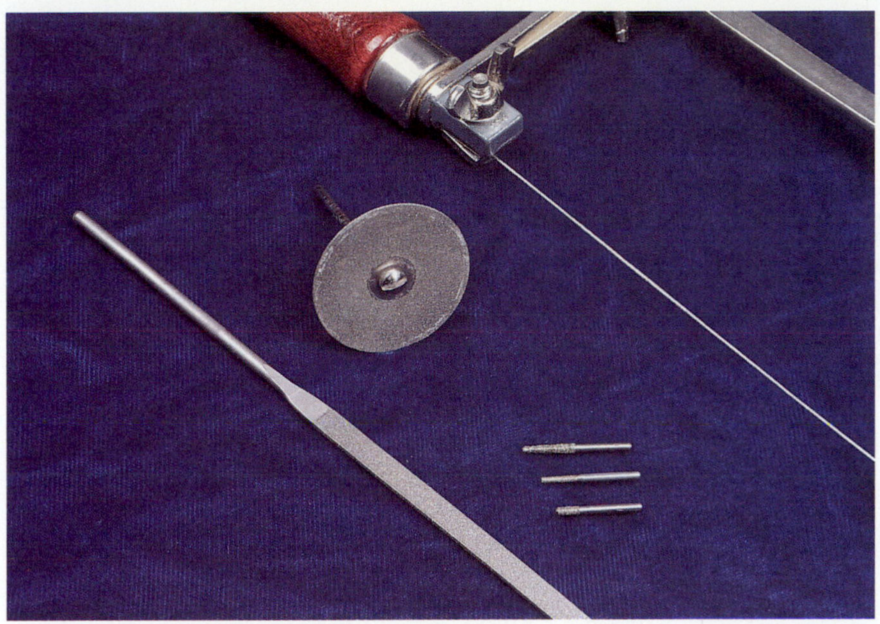

Figure S3.18
Diamond-coated cutting tools: saw, rotary slitting disc, three rotary bits, and file.

Diamond films, like other forms of diamond, are transparent to visible light, ultraviolet radiation, x-rays, and most of the infrared spectrum. Such films can be used as scratch-proof coatings on scientific glassware (such as x-ray windows) and optical lenses (e.g., eyeglass lenses) with no loss in optical transmission. Diamond films can also be used to make abrasion-resistant industrial tools such as knives and saws that never need sharpening (Figure S3.18). Isotopically pure carbon-12 diamond films are expected to find uses in the design of more powerful and faster computers, fiber-optic communications cables, improved laser cutting devices, and more accurate laser measurement devices.

The fullerenes, unlike diamond and graphite, consist of discrete molecules. *Buckminsterfullerene* (C_{60}), the principal member of this group, was detected in 1985 in the mass spectrum of laser-vaporized graphite and prepared in bulk in 1990 (see *Chemical Insight: Buckyballs and Other Fullerenes*). C_{60} has a cagelike structure shaped like a soccer ball (Figure S3.19a). The nearly spherical surface consists of 12 pentagons, 20 hexagons, and 60 vertices. A carbon atom is located at each vertex. Buckminsterfullerene molecules are often called "buckyballs" for short.

Buckminsterfullerene and related fullerenes are very stable and have a number of potential uses. For example, research scientists hope to use the hollow carbon cages as catalytic surfaces and also to carry small molecules and metal ions (Figure S3.19c). It is hoped that these stable molecules may one day be used to transport drugs and medication to specific body sites. The remarkably impervious and nearly spherical nature of fullerene molecules suggests that they may ultimately provide a new class of lubricants. Chemists have recently prepared "teflon" buckyballs ($C_{60}F_{60}$), which may have exceptional lubricating properties.

Scientists at AT&T Bell Laboratories have produced a "buckide" salt, K_3C_{60}, consisting of a face-centered cubic structure of buckyballs with potassium ions filling the spaces between the buckyballs. K_3C_{60} behaves like a metal at room temperature and becomes superconducting (no resistance to electrical flow) below 18 K.

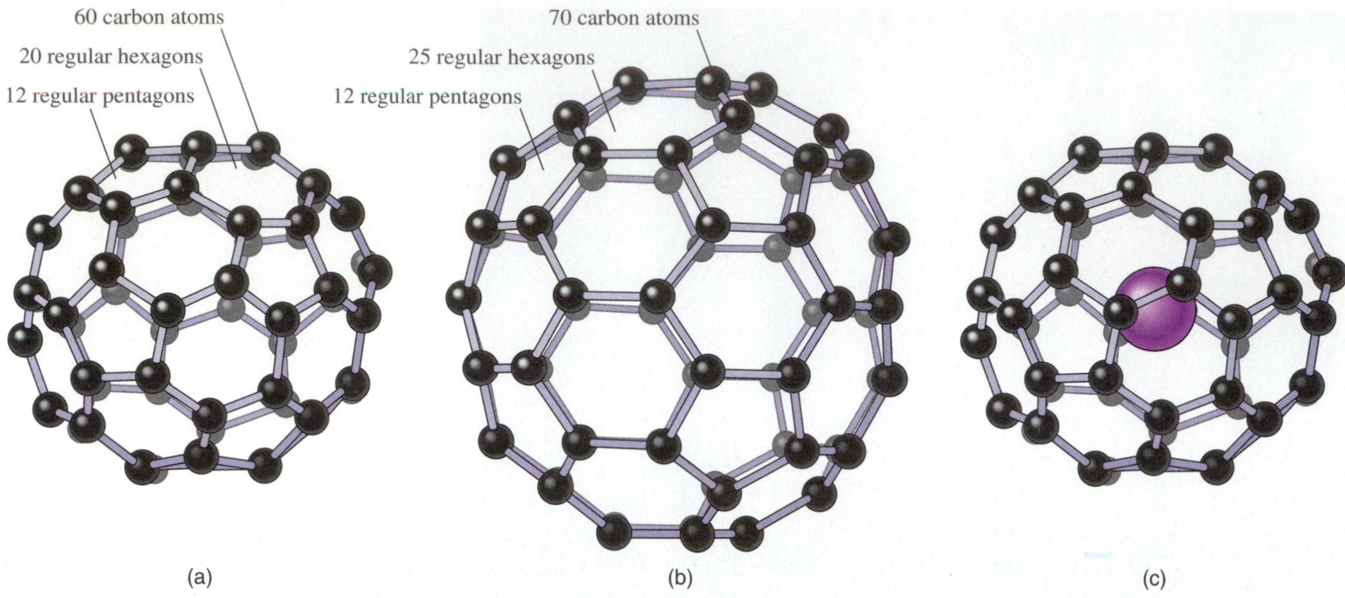

Figure S3.19

Fullerene molecules. The term *fullerene* is used collectively for all carbon cages that contain 12 pentagons and some number of hexagons. (a) The buckminsterfullerene (C_{60}) molecule, with 12 pentagons and 20 hexagons, has a soccer-ball structure with a carbon atom at each of the 60 vertices. (b) The C_{70} molecule, with 12 pentagons and 25 hexagons, has been described as "somewhat egg-shaped" or "reminiscent of a rugby ball." (c) A lanthanum atom encapsulated in C_{60}.

C_{60} films doped with rubidium and thallium also become superconducting at temperatures below 42.5 K. The nature and chemical behavior of the fullerenes are just being explored, so we can only speculate impatiently about the interesting discoveries still to come.

Carbon is an important reducing agent. The reduction of metal ores with coke was described in Section 3.6. Some other reactions of elemental carbon are summarized in Table S3.10 (see also Demonstration S3.6).

TABLE S3.10 Some Reactions of Carbon

With Oxygen:

$$2C(s) + O_2(g) \xrightarrow[\text{or } T > 500°C]{\text{limited oxygen}} 2CO(g)$$

$$C(s) + O_2(g) \xrightarrow[\text{or } T < 500°C]{\text{exces oxygen}} CO_2(g)$$

With Oxides:

$$C(s) + \text{metal oxide} \xrightarrow{\text{heat}} \text{metal} + CO(g)$$

(metal oxide = Fe_2O_3, ZnO, SnO_2, and so forth)

$$C(s) + H_2O(g) \xrightarrow{1000°C} H_2(g) + CO(g)$$

With Other Nonmetals:

$$C(s) + 2S(s) \xrightarrow{800°C} CS_2(g)$$

$$C(s) + 2F_2(g) \xrightarrow{25°C} CF_4(g)$$

CHEMICAL INSIGHT

BUCKYBALLS AND OTHER FULLERENES

In 1985 chemists from Rice University (Richard Smalley, Robert Curl, and their students James Heath and Sean O'Brien) and the University of Sussex in England (Harold Kroto) detected hitherto unrecognized carbon molecules—C_{60}, C_{70}, and others—in the mass spectrum of products that form when graphite is vaporized with laser pulses in a helium atmosphere. The C_{60} molecule seemed to be unusually stable, and the scientists proposed that the 60 carbon atoms were arranged in a hollow cagelike structure similar to that of a soccer ball (Figure S3.19a). This arrangement, formally known as a truncated icosahedron, is the most symmetrical and strain-free of all possible closed three-dimensional shells. Because this same arrangement was adapted by the engineer R. Buckminster Fuller for the construction of his geodesic domes, the C_{60} molecule quickly acquired the name *buckminsterfullerene*, or "buckyball" for short. A somewhat more complicated geodesic dome structure was also proposed for the C_{70} molecule (Figure S3.19b).

Research scientists from the University of Arizona (Donald Huffman and his student Lowell Lamb) and the Max Planck Institute for Nuclear Physics in Heidelberg (Wolfgang Krätschmer and his student Konstantinos Fostiropoulos) soon discovered that fullerene molecules were present in smoke samples prepared by striking an arc between graphite electrodes in a helium atmosphere (Figure S3.20). They discovered that these molecules dissolve in benzene to form a wine-red solution and can thus be separated from other components of the specially prepared soot.

Evaporation of the benzene solvent left a yellow solid composed of the cagelike fullerenes. About 75% of the solid contained C_{60} molecules, 23% contained C_{70} molecules, and the remaining 2% contained fullerene molecules with still higher molar masses. The scientists at the University of Arizona and the Max Planck Institute made chemical history—they were the first to prepare, observe, and handle bulk quantities of a new allotrope of carbon.

Since 1990 small amounts of fullerenes have been found in nature, in sooty candle flames, and in certain carbon-rich rocks. Fullerenes are now synthesized in bulk, and hardly a day passes without some interesting development in fullerene research.

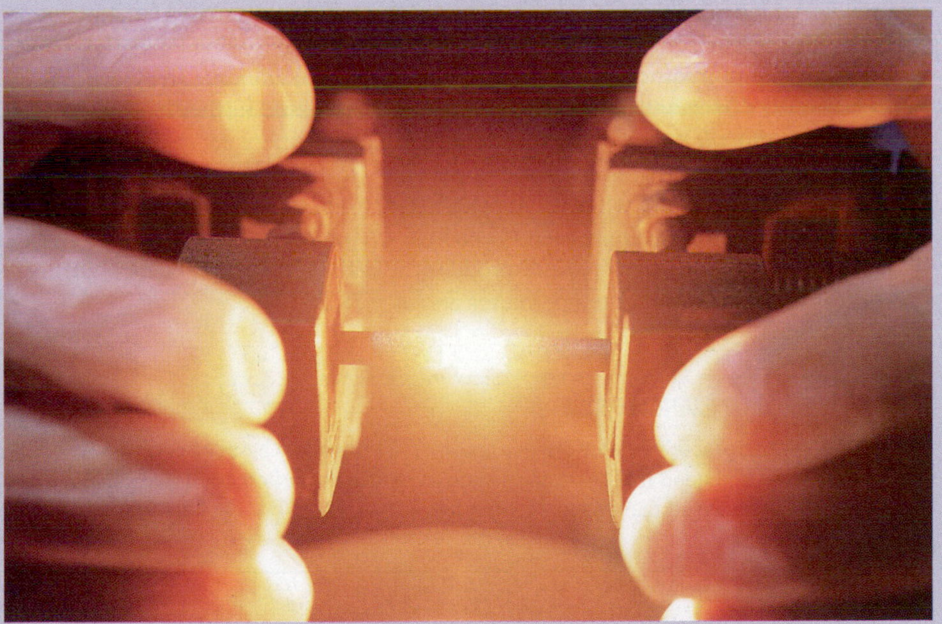

Figure S3.20
Macroscopic quantities of buckminsterfullerene (C_{60}) are produced when an electric arc bridges the gap between two graphite electrodes in a helium atmosphere. In this image, gloved hands hold the electrode clamps.

DEMONSTRATION S3.6 BURNING A DIAMOND: $C(diamond) + O_2(g) \rightarrow CO_2(g)$

A small diamond is inserted into a quartz tube.

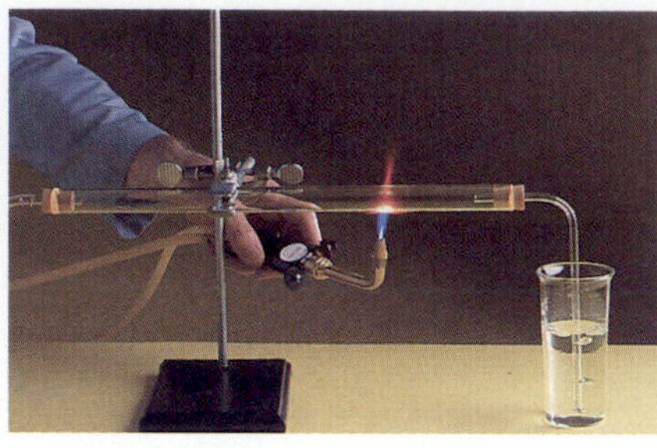

A stream of oxygen is passed over the diamond and into a beaker of $Ca(OH)_2$ solution.

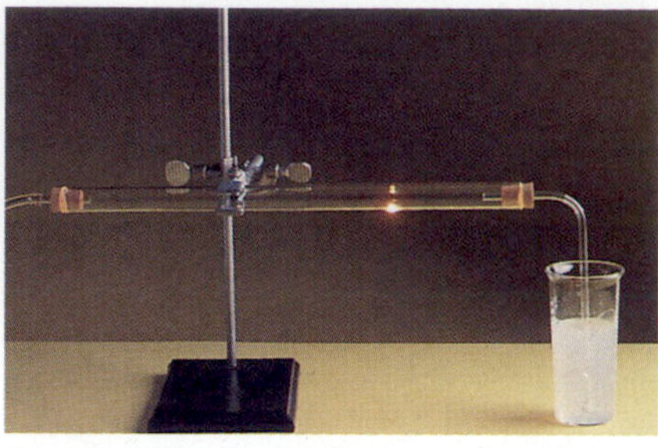

The diamond burns brightly in the oxygen. The combustion product, CO_2 gas, reacts with the $Ca(OH)_2$ solution to form a milky white $CaCO_3$ precipitate: $CO_2(g) + Ca(OH)_2(aq) \longrightarrow CaCO_3(s) + H_2O(l)$.

Carbon Dioxide and Carbonates

The compounds of carbon are classified somewhat arbitrarily into *organic compounds*, which include the hydrocarbons and their derivatives, and *inorganic compounds*, which are chiefly carbon dioxide and carbonates. No distinct boundary exists between these divisions, and their chemistry is intertwined. You have already met a number of hydrocarbons and are familiar with the bonding in ethane, ethylene, acetylene, and benzene. Organic compounds are discussed further in Chapter 23.

$\ddot{\underset{\cdot\cdot}{O}}=C=\ddot{\underset{\cdot\cdot}{O}}$

carbon dioxide (CO_2)

Carbon dioxide is a colorless, odorless gas at ordinary temperatures and pressures. It is somewhat soluble in water, harmless to breathe at concentrations below 1%, and useful in many ways (Figure S3.21). Carbon dioxide is heavier than air and can smother fires. Leavening agents generate CO_2 bubbles that cause dough to rise, and CO_2 dissolved under pressure produces sparkling beverages. The solid form of CO_2 (Dry Ice) sublimes at temperatures above $-78.5°C$.

PRACTICE EXERCISE S3.6

The old "soda–acid" type of fire extinguisher contained a vessel of concentrated sulfuric acid suspended in a solution of sodium hydrogen carbonate. Inverting the extinguisher caused the reagents to mix and form CO_2. Write the equation for this reaction.

The photosynthetic conversion of carbon dioxide and water to glucose and starch was described in Section S2.1. This reaction, which requires sunlight and green plants, provides the biosphere with fuel and substance, and the atmosphere with oxygen. The acid–base properties of CO_2, its hydrated form H_2CO_3, and the ions HCO_3^- and CO_3^{2-} were discussed in Chapters 16 and 17.

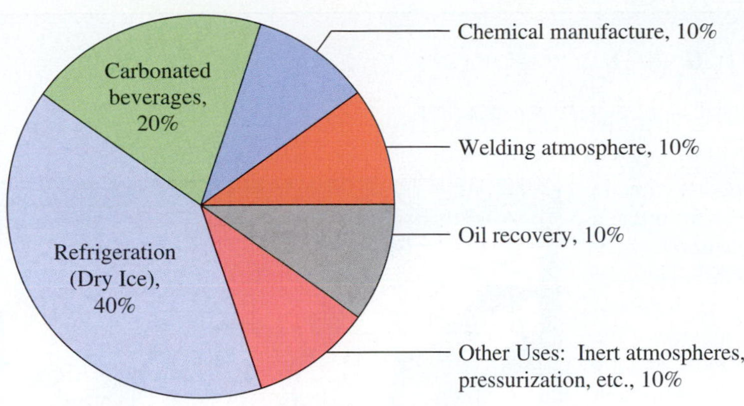

Figure S3.21
Industrial uses of carbon dioxide.

Sodium carbonate (Na_2CO_3) ranks among the leading industrial chemicals produced in North America. Its preparation from the mineral *trona* was described in Section S1.2. *Sodium hydrogen carbonate* ($NaHCO_3$) can be prepared from salt water and limestone ($CaCO_3$) by the **Solvay process**. Crushed limestone is first decomposed to CaO and CO_2:

Step 1: $CaCO_3(s) \xrightarrow{\text{heat}} CaO(s) + CO_2(g)$

Then the CO_2 is converted into sparingly soluble $NaHCO_3$ by passing it into concentrated salt water that has been made basic with ammonia:

Step 2: $CO_2(g) + NaCl(aq) + H_2O(l) + NH_3(aq) \rightarrow NaHCO_3(s) + NH_4Cl(aq)$

The solubility of $NaHCO_3$ in pure water is about 70 g/L at 0°C. It precipitates in Step 2 because the concentrated solution contains a large excess of $Na^+(aq)$ and is kept very cold.

$NaHCO_3$ is separated from the solution by filtration. Ammonia is the most expensive reactant in the process, but most of the ammonia can be recovered by adding CaO from Step 1 to the NH_4Cl solution from Step 2:

Step 3: $CaO(s) + 2NH_4Cl(aq) \rightarrow CaCl_2(aq) + H_2O(l) + 2NH_3(g)$

The recovered ammonia is then recycled and $CaCl_2$ is a coproduct.

Carbon Monoxide

Carbon monoxide is produced when carbon compounds are oxidized in limited oxygen or at temperatures above 500°C:

$$C(s) + CO_2(g) \rightleftharpoons 2CO(g) \qquad K_p = 4.8 \times 10^4 \text{ at } 2000°C$$

The molecules of CO and N_2 are isoelectronic and have the same molecular orbital configurations, but CO has a bond energy greater than that of any other chemical bond (See Table 9.3, p. 377). The CO molecule uses the lone pair of electrons on its carbon atom to help it form complexes with transition metal atoms such as iron and nickel. For example, when carbon monoxide is inhaled, it enters the bloodstream and binds to the iron atoms in hemoglobin so that they can no longer bind with oxygen. A carbon monoxide victim looks flushed and rosy because the hemoglobin–CO complex is cherry red. Carbon monoxide poisoning causes dizziness, drowsiness, and eventually death. The only antidote is prompt breathing of pure oxygen to displace the carbon monoxide.

$:C\equiv O:$

carbon monoxide (CO)

$:N\equiv N:$

nitrogen (N_2)

CHEMICAL INSIGHT

THE GREENHOUSE EFFECT

The carbon dioxide that is absorbed by photosynthetic plants is released back to the atmosphere when organic matter decays or burns. Until recently, the natural carbon cycle maintained an atmospheric CO_2 concentration between 200 and 300 parts per million (ppm). Within the last century, however, humans have overloaded the CO_2 output side of the cycle by increasing the combustion of fossil fuels and by clearing areas where forests once absorbed CO_2. Although some of the excess CO_2 is absorbed by the ocean, much of it remains in the atmosphere. The average global concentration of CO_2 is now about 356 ppm, and it is increasing at the rate of about 1.5 ppm per year.

The average temperature at the earth's surface represents a balance between energy income and outgo. The energy of absorbed sunlight goes through various transformations and is eventually transformed into heat. Heat radiation lies in the infrared region of the spectrum; therefore, earth's outward radiation consists largely of infrared photons. Certain atmospheric components such as CO_2, H_2O, O_3, the synthetic chlorofluorocarbons (CFCs), and the trace gases N_2O and CH_4 absorb and reemit these photons, thus returning some of the radiated heat back to the earth. Such gases are called "greenhouse gases" because they produce the same effect as a greenhouse, which traps heat within its transparent walls.

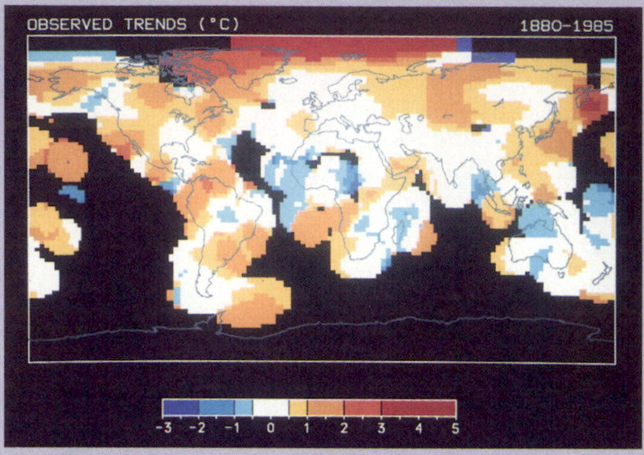

Figure S3.22

A computer-generated diagram showing temperature trends from 1880 to 1985. The numbers on the color key indicate degrees above and below the average temperature for 1951–1980 (shown in white). The temperature rise is caused by "greenhouse gases" that return heat radiation back to the earth.

Even a greenhouse can become too hot for the plants that are growing in it. The average global air temperature has increased by 0.5°C in the past century, and it is now feared that increasing concentrations of greenhouse gases in the atmosphere may raise earth's mean temperature by as much as 4–5°C in the next century (Figure S3.22). An increase of this magnitude would melt glaciers, cause ocean water to expand, and probably result in a 2- to 3-foot rise in sea level. Coastal areas would be flooded, rainfall patterns would be disrupted, and many ecosystems would be thrown out of balance.

The warming of the atmosphere cannot be reversed, but it can be slowed down and possibly leveled off. Necessary steps include the preservation of forest areas, especially large rain forests, and energy conservation. Converting to fuels with lower carbon-to-hydrogen ratios can also be of some help. For example, for a given amount of energy, natural gas (CH_4) releases less than half the CO_2 that coal does. Hydrogen releases no CO_2 at all.

Enough carbon monoxide is pumped into the air by natural processes and by human activity to double its atmospheric concentration every few years. Fortunately, the doubling does not occur, thanks to at least 16 varieties of soil fungi and various types of bacteria that metabolize excess CO and convert it to CO_2—another example of the interdependence of living species.

Carbides and Cyanides

A **carbide** is a binary compound of carbon with a less electronegative element. Active metals form *ionic carbides*. Some ionic carbides such as Be_2C and Al_4C_3 contain the carbide ion (C^{4-}). They react with water to form methane:

$$Be_2C(s) + 4H_2O(l) \rightarrow 2Be(OH)_2(s) + CH_4(g)$$

$$Al_4C_3(s) + 12H_2O(l) \rightarrow 4Al(OH)_3(s) + 3CH_4(g)$$

Other ionic carbides contain the *acetylide* ion (C_2^{2-}). The best known of these is calcium carbide, which can be produced by heating lime (CaO) with coke:

$$CaO(s) + 3C(s) \xrightarrow{2000^\circ C} CaC_2(s) + CO(g)$$

Acetylides react with water to release acetylene, which can be used as a fuel:

$$CaC_2(s) + 2H_2O(l) \rightarrow Ca(OH)_2(s) + C_2H_2(g)$$

Less active metals form **interstitial carbides**, in which the small carbon atoms occupy vacancies in the metal's crystal lattice and have a hardening effect on the metal. Steel contains carbon in the form of small grains of iron carbide (Fe_3C). Tungsten carbide (WC) and titanium carbide (TiC) retain their hardness at high temperatures and are used for lathes and cutting tools.

Nonmetals and semimetals form *covalent carbides* such as boron carbide (B_4C) and silicon carbide (SiC). Silicon carbide, also called *carborundum*, crystallizes in a diamond lattice and is almost as hard as diamond. Silicon carbide remains stable up to its decomposition temperature of 2700°C. It is used in resistance elements for electric furnaces and in solid-state electronic devices that must withstand high temperatures.

Carbon forms a number of compounds with sulfur and nitrogen. Carbon disulfide (CS_2) is a valuable solvent in spite of its volatility, flammability, and toxicity. Hydrogen cyanide (HCN) is a weak acid. The toxicity of HCN and cyanides is well known. Cyanide salts must be disposed of carefully, because any contact with acid will liberate hydrogen cyanide into the air:

$$NaCN(s) + H^+(aq) \rightarrow Na^+(aq) + HCN(g)$$

Cyanide, either ingested or inhaled, combines with and disables the enzyme *cytochrome oxidase*, which is essential to a living cell's ability to use oxygen. As little as 10^{-8} *M* CN^- will inhibit this enzyme completely.

PRACTICE EXERCISE S3.7

Write Lewis structures for (a) C_2^{2-}, (b) CS_2, and (c) HCN.

S3.4 GROUP 4A: SILICON AND GERMANIUM

Silicon comprises 27.7% of the earth's crust and provides the structure of most minerals and rocks. It is also an essential trace element that is needed for proper bone growth in higher animals. The natural sources of silicon are the silicate minerals, which contain silicon bonded to oxygen. Germanium, by contrast, is a rare element that tends to bond to sulfur; it is found as a sulfide mixed with the sulfide ores of arsenic, copper, and zinc.

Elemental silicon was once thought to be of little interest, but the development of computers and other semiconducting devices has turned its production into an essential industry (Figure S3.23). Crude silicon is obtained by reducing silicon dioxide with coke in an electric furnace:

$$SiO_2(s) + 2C(s) \xrightarrow{3000^\circ C} Si(l) + 2CO(g)$$

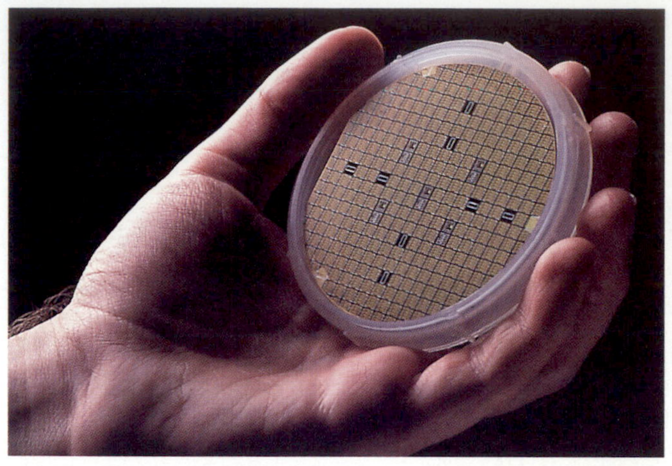

(a)

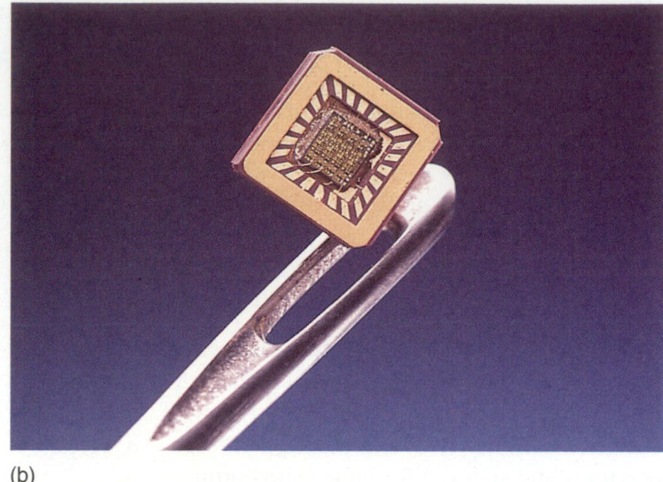

(b)

Figure S3.23

(a) The silicon base for the integrated circuits on this wafer was prepared from a crystal that grew from an ultrapure silicon melt. (b) A circuit chip cut from a wafer similar to the one in (a) is small enough to fit through the eye of a needle.

Ultrapure silicon is obtained by **zone refining** (Figure S3.24), a method that was developed to supply extremely pure material (less than $10^{-9}\%$ impurities by atom count) for use in semiconductors. A bar of the substance to be purified is slowly passed through a ring-shaped heater so that successive regions melt and resolidify. Impurities tend to stay in the melt and are swept along with the melted region as it travels from one end of the bar to the other. Successive passes through the heater

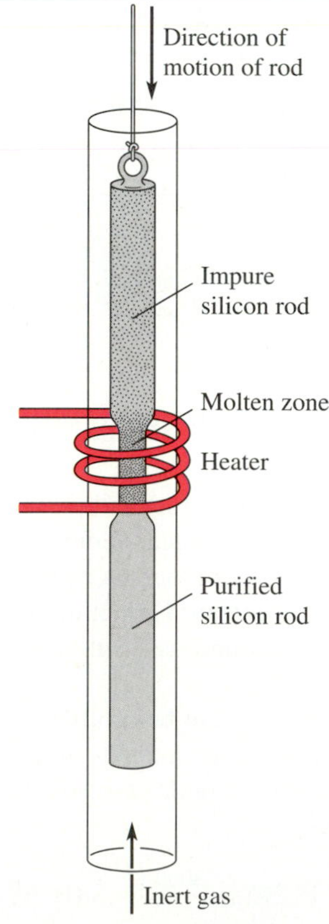

Figure S3.24

Zone refining. A rod of impure silicon or germanium is slowly lowered through a circular heater. The impurities accumulate in the molten portion and are swept to the top of the bar.

make the leading end of the bar progressively purer, while the end with accumulated impurities is eventually cut off and discarded.

Silicon and germanium crystallize in a diamond lattice. Their atoms, however, are larger than carbon atoms, so the bonds are weaker and their crystals are not as hard as diamond. Silicon and germanium do not have graphitelike forms because their atoms are too large to form strong pi bonds.

Silica and Silicates

Very few single covalent bonds are as strong as the one between silicon and oxygen, and this bond strength helps to explain the stability and prevalence of silicate minerals. (Figure S3.25). Every silicon atom in a silicate is surrounded tetrahedrally by four oxygen atoms. ***Orthosilicates*** contain isolated SiO_4^{4-} ions (Figure S3.26a). Sodium silicate (Na_4SiO_4) is the only common silicate that is soluble in water; its saturated solution is known as *water glass*. A crystal of another soluble salt placed in water glass will quickly sprout branches of insoluble silicates that grow upward in the dense solution (Demonstration S3.7). Other orthosilicates include the gemstone *zircon* ($ZrSiO_4$) and the yellowish green mineral *olivine*. Olivine is the principal mineral of the earth's mantle, the thick layer that lies below the crust. The formula for olivine is written as $(Mg,Fe)_2SiO_4$ to show that its Mg/Fe ratio is variable.

Most silicate minerals have extended anions in which SiO_4 tetrahedra are joined by shared oxygen atoms. The *pyrosilicate* anion ($Si_2O_7^{6-}$) contains two tetrahedra (Figure S3.26b). Longer silicate chains contain multiple SiO_3^{2-} units (Figure

Figure S3.25
Natural silicates: (a) jadeite, (b) asbestos, (c) talc, and (d) mica.

DEMONSTRATION S3.7 A CHEMICAL GARDEN

The beaker contains water glass (concentrated sodium silicate solution) on a layer of sand. The crystals on the watch glasses are, clockwise from bottom left: $FeCl_3$, $Co(NO_3)_2$, $Ni(NO_3)_2$, $Fe_2(SO_4)_3$, $Mn(NO_3)_2$, $Cu(NO_3)_2$, and $Zn(NO_3)_2$.

Crystals dropped into the silicate solution appear to grow stems and branches. Actually the soluble salts are being converted to insoluble silicates that "precipitate" upward because they are less dense than the solution in which they form.

S3.26c). The mineral *jadeite* (Figure S3.25a) is a chain silicate with the empirical formula $NaAl(SiO_3)_2$.

Other silicate anions contain tetrahedra joined in rings, double chains, sheets, or three-dimensional networks (Table S3.11). *Beryl* (see Figure S1.5, p. 348) contains rings of $Si_6O_{18}^{12-}$ ions in which each tetrahedral unit shares two oxygen atoms with other units. *Amphiboles* are minerals with double silicate chains in which the

Figure S3.26
Structures of silicate anions.

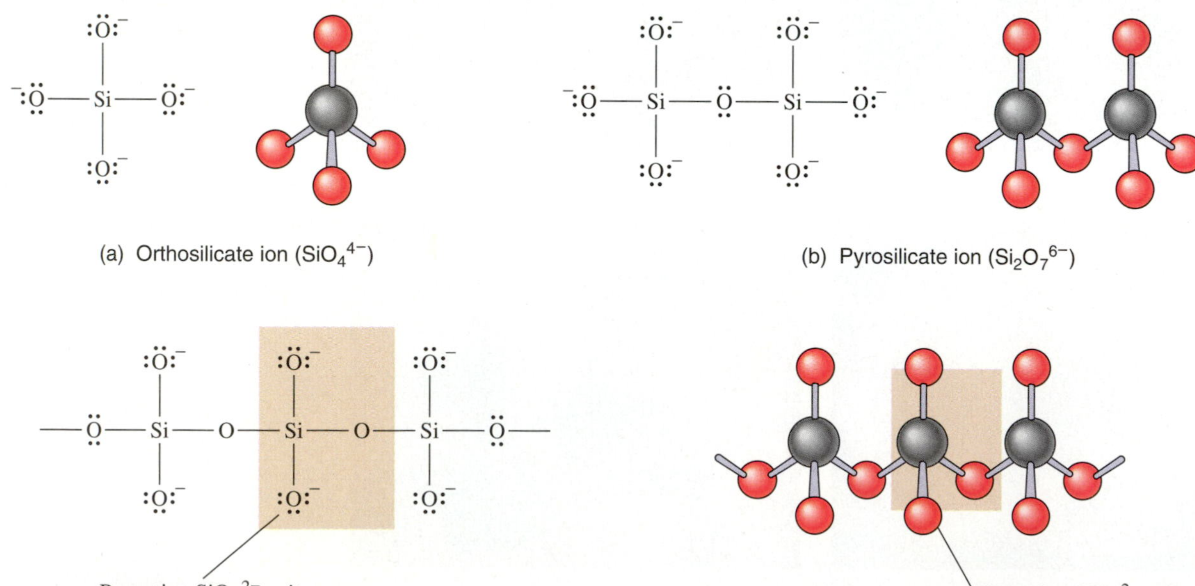

(a) Orthosilicate ion (SiO_4^{4-})

(b) Pyrosilicate ion ($Si_2O_7^{6-}$)

(c) $(SiO_3^{2-})_n$ chain

TABLE S3.11 Types of Silicate Minerals

Structural Type	Representation	Formula of Anion	Examples
Orthosilicates (no oxygen atoms shared)		SiO_4^{4-}	Zircon, $ZrSiO_4$ Olivine, $(Mg,Fe)_2SiO_4$
Pyrosilicates (two SiO_4 tetrahedra share one oxygen atom)	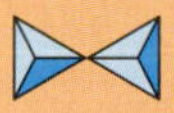	$Si_2O_7^{6-}$	Akermanite, $Ca_2MgSi_2O_7$
Pyroxenes (single chains of SiO_4 tetrahedra; each tetrahedron shares two oxygen atoms)	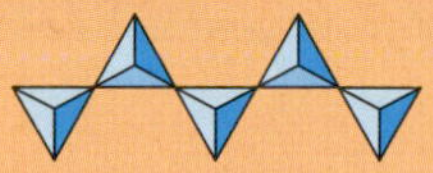	$(SiO_3^{2-})_n$	Jadeite, $NaAl(SiO_3)_2$ Diopside, $CaMg(SiO_3)_2$
Rings (each SiO_4 tetrahedron shares two oxygen atoms)	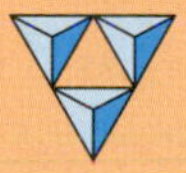 	$Si_3O_9^{6-}$ $Si_6O_{18}^{12-}$	Benitoite, $BaTiSi_3O_9$ Beryl, $Al_2Be_3Si_6O_{18}$
Amphiboles (double chains of SiO_4 tetrahedra; the tetrahedra alternately share two and three oxygen atoms)	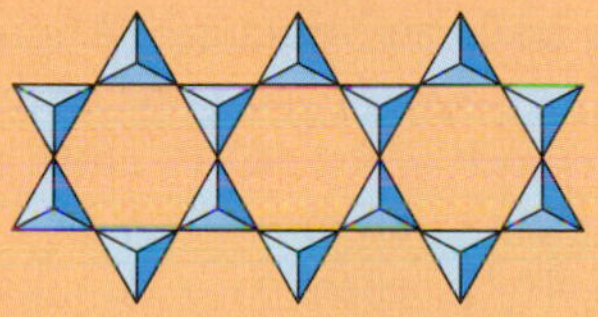	$(Si_4O_{11}^{6-})_n$	Tremolite, $Ca_2Mg_5(Si_4O_{11})_2(OH)_2$ (a form of asbestos)
Sheets (each SiO_4 tetrahedron shares three oxygen atoms)	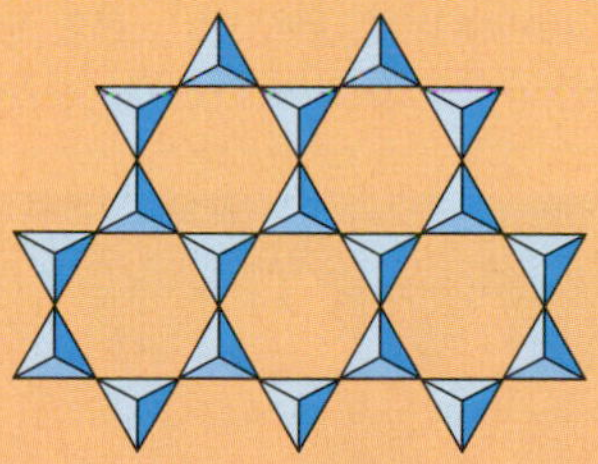	$(Si_2O_5^{2-})_n$	Talc, $Mg_2(Si_2O_5)_2 \cdot Mg(OH)_2$ Kaolinite, $Al_2Si_2O_5(OH)_4$
Three-dimensional networks (each SiO_4 tetrahedron shares four oxygen atoms)	See Figure 2.20 on page 63.	Nonionic	Quartz, SiO_2

Key

represents Si O^- O^- O^- O^-

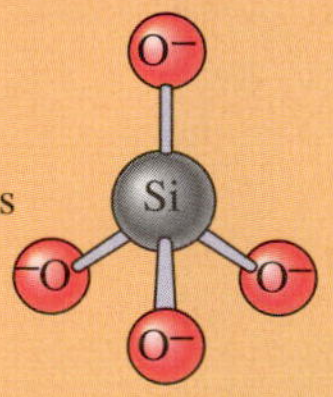

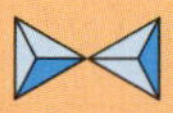

represents Si Si O^- O^- O^- O O^- O^- O^-

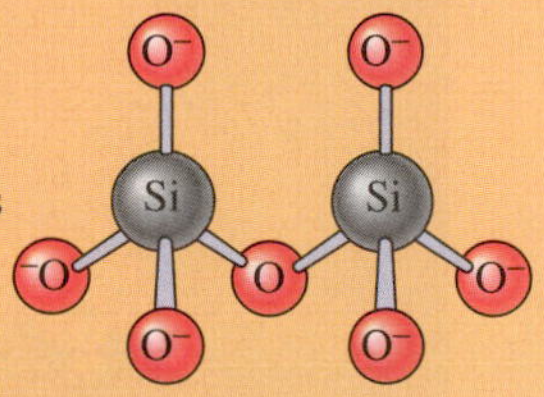

Figure S3.27
Feldspar.

Small asbestos fibers from inhaled asbestos dust can lodge in the lungs and cause serious lesions that may ultimately become cancerous.

repeating unit is $Si_4O_{11}^{6-}$; the fibrous minerals known as *asbestos* are of this type (Figure S3.25b). Flaky minerals such as *talc* and *mica* (Figure S3.25c and d) contain sheetlike structures in which the repeating unit is $Si_2O_5^{2-}$. *Kaolinite*, a clay used in pottery, also has a sheetlike lattice.

Silicon dioxide, also called ***silica***, has a nonionic three-dimensional framework with the empirical formula SiO_2 (see Figure 2.20, p. 63). Every oxygen atom is shared by two SiO_4 tetrahedra. *Quartz* is a crystalline form of silica, and *flint* is a hard amorphous form. Silicas are the principal ingredients of sand. *Silica gel* is a partially hydrated silica that is used for absorbing moisture and various vapors.

Aluminosilicates are minerals in which aluminum atoms have replaced some of the silicon atoms in a silicate structure. (Aluminum and silicon atoms are similar in size.) *Feldspars* (Figure S3.27) and some of the micas are aluminosilicates. *Granite*, the principal igneous rock of the earth's crust, consists of mica, quartz, and feldspar.

Zeolites are metal aluminosilicates that have the general formula $M_{2/n}\cdot(AlO_2)_2\cdot xSiO_2\cdot yH_2O$, where M is an alkali or alkaline earth metal and n is its oxidation number. They crystallize in an open structure with cavities that can absorb small molecules such as H_2O, N_2, and C_2H_6 (Figure S3.28). Synthetic zeolites, sometimes called "molecular sieves," are designed for selective absorption of

Figure S3.28
Model of a synthetic zeolite or "molecular sieve." Some small atoms, molecules, and ions are reversibly absorbed during passage through the zeolite channels.

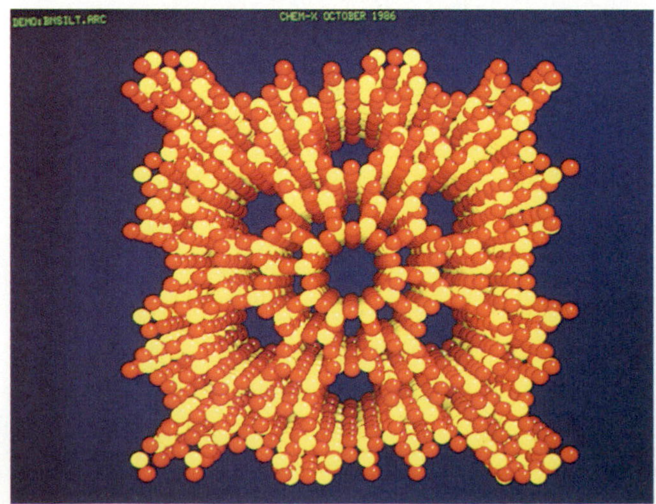

CHEMICAL INSIGHT

GLASS, CLAY, AND CERAMICS

Silica and silicates do not usually recover their original crystalline structures when they are melted and resolidified. Instead, they form glasses in which the SiO_4 tetrahedra are joined in an irregular structure. One example is *obsidian*, a shiny dark natural glass that forms during volcanic action (Figure S3.29).

Figure S3.29
Snowflake obsidian.

As early as 3000 B.C., the Egyptians had discovered that sand mixed with limestone ($CaCO_3$) and soda ash (Na_2CO_3) would melt at a lower temperature than sand alone. (The melting point of pure silica sand is 1700°C.) The heating process drives off CO_2, and the cooled product is an amorphous mixture of calcium and sodium silicates known as *soda–lime glass*. Most glass made today is still of this type and has the approximate composition $Na_2O{\cdot}CaO{\cdot}6SiO_2$. The Na^+ and Ca^{2+} ions are irregularly scattered among the randomly attached SiO_4 units.

The chemical composition of glass can be adjusted to serve various special purposes. Glass made entirely of silica, for example, is resistant to chemical attack and useful for certain laboratory vessels. A *borosilicate glass* such as Pyrex has a high proportion of B_2O_3. Borosilicate glasses are used for ordinary laboratory glassware and cookware because they have low coefficients of thermal expansion ("ovenproof") and are resistant to shock. *Flint glass* contains lead monoxide (PbO). It has a high refractive index and is used for lenses and decorative cut glass. Other special glasses have been formulated for ultraviolet lamps, signal lights, and so forth.

The manufacture of clay pottery probably predates the manufacture of glass. Clays are finely divided sheet silicates and aluminosilicates that contain OH groups. *Kaolinite*, found in the kaolin clays from which fine pottery is made, has the empirical formula $Al_2Si_2O_5(OH)_4$. The silicate sheets adsorb water to produce the familiar cohesive, wet, slippery clay that can be molded and shaped. The firing of clay in a kiln first drives out the adsorbed water and then locks the silicate sheets together by eliminating water from adjacent OH groups. As the temperature rises, various reactions and phase changes produce a hard, sturdy structure. The glaze that is applied to pottery in the kiln is essentially a glass that gives the surface an attractive finish and makes it impervious to water. Decorative glazes contain transition metal ions to impart color.

chosen ions or molecules. They are used industrially as catalysts and drying agents. Certain zeolites are used as *ion exchangers* in water-softening devices, where they absorb Ca^{2+} and Mg^{2+} ions from the water and release Na^+ or H^+ ions in their place. The zeolite is regenerated by running concentrated NaCl solution through it until the exchange has been reversed.

Some Synthetic Silicon Compounds

A **silicone** contains silicon–oxygen chains with hydrocarbon side groups (Figure S3.30). Silicones combine the flexibility of a chain structure, the water-repelling properties of hydrocarbon groups, and the thermal stability of silicon–oxygen bonds. In *cross-linked* silicones, the chains are connected by Si—O—Si bridges that confer a certain degree of rigidity.

The properties of silicones vary with the structure, and a great variety of silicones have been synthesized. Liquid silicones are used in brake fluids and as water repel-

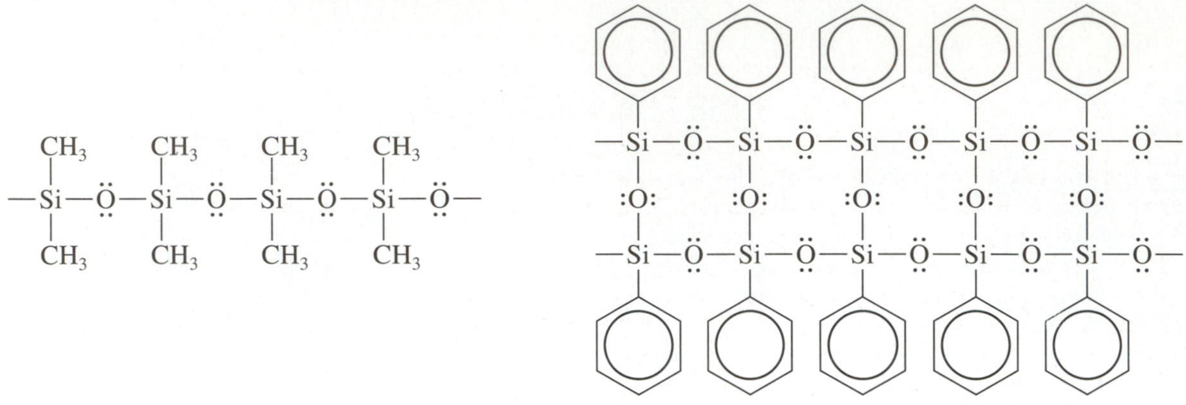

(a) Methyl silicone

(b) A cross-linke phenyl silicone

Figure S3.30

Some silicone structures. (The attached C_6H_5 ring is called a phenyl group.)

lants. Greaselike silicones are good lubricants and are used in cosmetic products such as lipstick and suntan lotion. Elastic silicones ("Silly Putty" is an example) can be substituted for rubber under severe conditions because, unlike rubber, they do not deteriorate at high temperatures nor lose elasticity at low temperatures. They are used in adhesives, refrigerator door gaskets, and chewing gum. Rigid silicones are used as high-temperature insulators.

Silanes are silicon–hydrogen compounds that are analogous to hydrocarbons, but they are much less stable. The simplest silane is SiH_4. The ability to *catenate* (form chains) is more limited in silicon than in carbon; the upper limit seems to be reached in hexasilane (Si_6H_{14}). All of the silanes are hydrolyzed by water and ignite spontaneously in air:

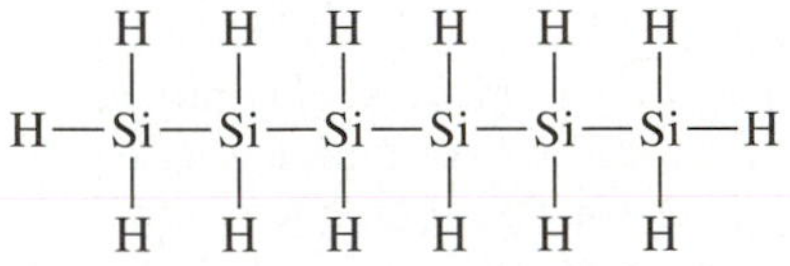

hexasilane (Si_6H_{14})

$$SiH_4(g) + 2O_2(g) \rightarrow SiO_2(s) + 2H_2O(l)$$

Silicon nitride (Si_3N_4) is an extremely stable, hard, nonmolecular compound with a high melting point. Ball bearings made of silicon nitride are twice as hard but only half as heavy as steel, and they can be used at much higher temperatures than steel.

S3.5 GROUP 3A: BORON

Boron, the only nonmetal in Group 3A, is a semimetal that is also a semiconductor. Elemental boron is a hard, lustrous, and brittle black solid whose crystal lattice is composed of B_{12} clusters interconnected by covalent bonds. The atoms in each B_{12} group occupy the corners of a regular *icosahedron*, a 12-cornered figure with 20 triangular sides (Figure S3.31).

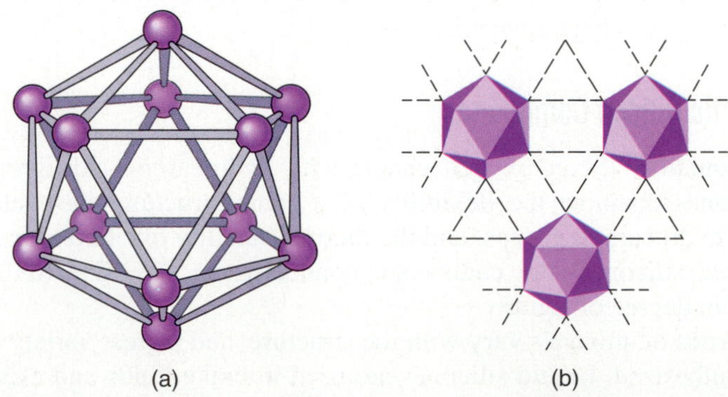

Figure S3.31

The structure of crystalline boron. (a) The basic structural unit is an icosahedron (a figure with 20 triangular faces and 12 corners) with a boron atom at each corner. (b) Packing of the B_{12} units in red alpha-rhombohedral boron, one of three boron allotropes. The icosahedra are held together by covalent bonds.

CHEMICAL INSIGHT

THE WHOLE EARTH

The principal layers of our planet are the outer crust, the inner core, and the mantle that lies in between (Figure S3.32). The part familiar to us is the crust, which occupies less than 1% of the total volume and consists largely of silicates and aluminosilicates. Table 2.2 (p. 45) shows the elemental composition of the crust; note that oxygen and silicon alone comprise almost three-fourths of the total. The average density of granite, the outer crust's most abundant rock, is about 2.6 g/cm^3, slightly less than the average density of the crust as a whole.

The mantle occupies about 85% of the earth's volume. It has an average density of 4.5 g/cm^3, and 90% of its atoms are silicon, oxygen, magnesium, and iron. Olivine is the most abundant mineral in the upper mantle, which is solid except for pockets of molten rock (magma).

The transition zone and lower mantle contain structures in which each silicon atom is surrounded by six oxygen atoms instead of the usual four; these structures form only under high pressure. One such form of silica, called *stishovite*, was first synthesized in the laboratory and subsequently discovered in meteor craters where the meteor impact once produced a high temperature and pressure.

The earth's core, with an average density of 11 g/cm^3, is the simplest region chemically. It consists entirely of elements—an estimated 85% iron, 7% nickel, and the rest mostly silicon, cobalt, and possibly some hydrogen. Studies of the seismic waves produced by earthquakes and explosions have led geologists to conclude that the outer core is molten and the inner core is solid.

Our conclusions about the nature of the core and lower mantle are based entirely on observations made on or near the earth's surface. Although millions of visitors to the Smithsonian National Air and Space Museum in Washington, D.C. have touched a rock from the moon, no one has yet touched a sample of the earth's deep interior.

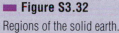

Figure S3.32

Regions of the solid earth.

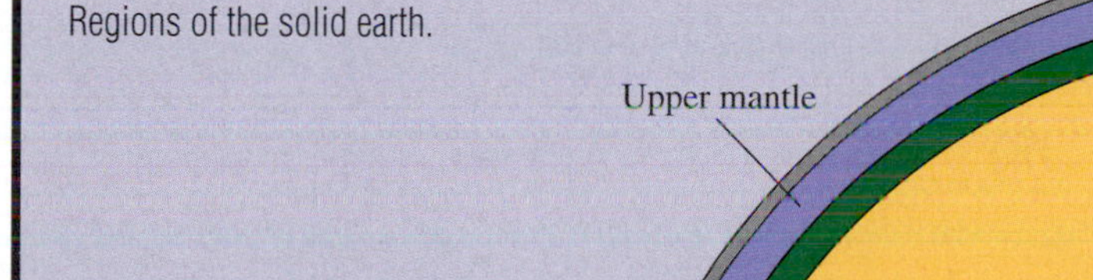

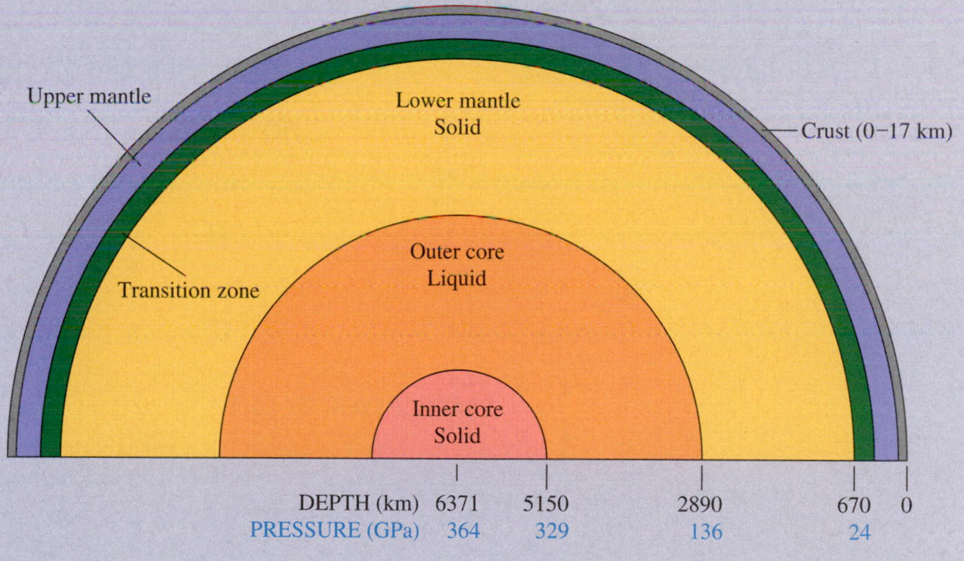

In many ways, boron resembles silicon more than it resembles the other members of its own group. Boron, like silicon, is naturally associated with oxygen. Some of its minerals are shown in Figure S3.33. The most common is *tourmaline*, an aluminosilicate that contains about 10% boron. The commercial source of boron is *borax*, a white mineral found in dried lake beds in desert areas of the Middle East and American Southwest. The formula of borax (hydrated sodium tetraborate) is often written $Na_2B_4O_7{\cdot}10H_2O$, although its structure is more accurately represented as $Na_2B_4O_5(OH)_4{\cdot}8H_2O$. The tetraborate ion is shown in Figure S3.34. Borax precipi-

(a)

(b)

(c)

Figure S3.33
(a) Boron, (b) tourmaline, and (c) borax.

tates the Ca^{2+} and Mg^{2+} of hard water and is used in the laundry as a water softener. Its aqueous solutions are mildly basic:

$$B_4O_7^{2-}(aq) + 7H_2O(l) \rightleftharpoons 4H_3BO_3(aq) + 2OH^-(aq)$$

It has recently been determined that boron is an essential nutrient. Trace amounts of boron compounds are necessary in the human diet.

Boric acid (H_3BO_3) is used as a gentle antiseptic in eyewash preparations. Some boric acid occurs naturally, but it can also be prepared by treating borax with strong acid (the equilibrium in the above reaction shifts to the right). The slippery white platelike crystals of boric acid are composed of trigonal planar molecules that are connected through hydrogen bonding (Figure S3.35). The formula for boric acid is often written as $B(OH)_3$; it is not a Brønsted acid but behaves toward water as a Lewis acid, displacing protons to form a faintly acidic solution containing $B(OH)_4^-$ ions:

$$B(OH)_3(aq) + H_2O(l) \rightleftharpoons B(OH)_4^-(aq) + H^+(aq)$$

electron-deficient center → B (bonded to HO, OH, OH) + :O(H)H ⇌ [B(HO)(OH)(HO)(OH)]$^-$ + H^+

trigonal planar molecule — tetrahedral anion

PRACTICE EXERCISE S3.8

What is the formal charge on the boron atom in $B(OH)_4^-$?

Figure S3.34
Structure of the tetraborate ion, $B_4O_5(OH)_4^{2-}$. Two of the boron atoms have trigonal planar bonds to three oxygen atoms; the other two boron atoms are tetrahedrally bonded to four oxygen atoms.

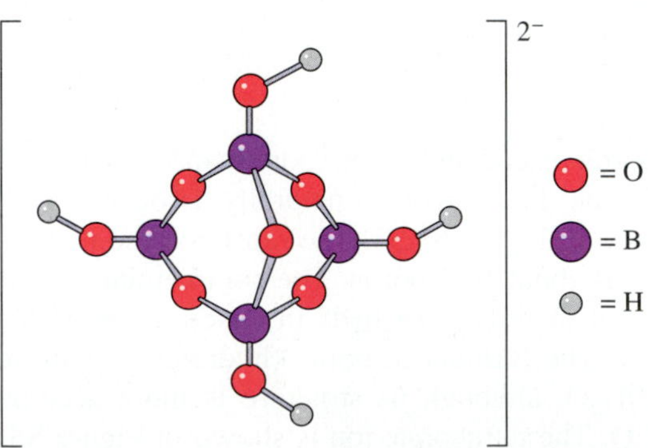

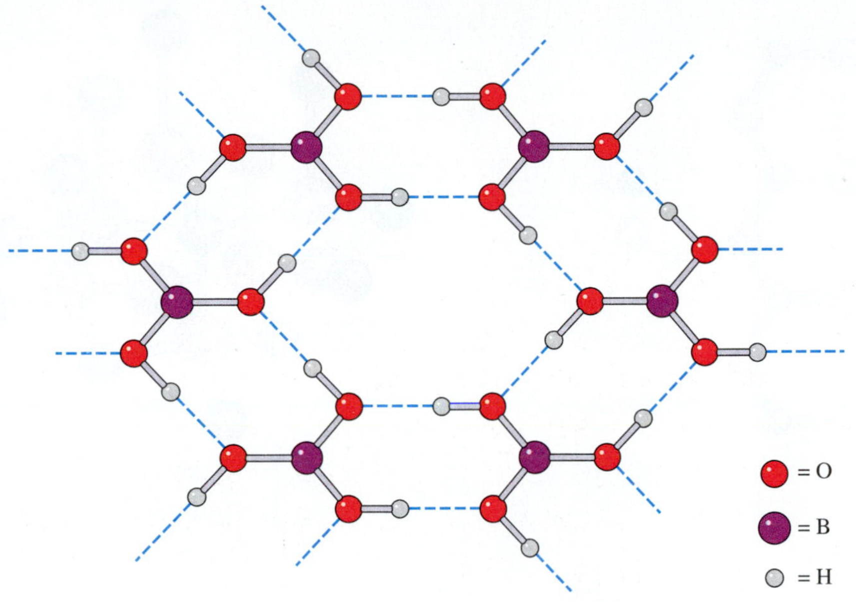

Figure S3.35
Solid boric acid is composed of $B(OH)_3$ units joined by hydrogen bonds. This planar configuration is reflected in the delicate platelike form of the crystals.

When heated, boric acid dehydrates to the oxide B_2O_3, which is the starting material for preparing boron and boron compounds:

$$2H_3BO_3(s) \xrightarrow{\text{heat}} B_2O_3(s) + 3H_2O(g)$$

Impure boron is prepared by reducing B_2O_3 with magnesium at high temperature:

$$B_2O_3(s) + 3Mg(s) \xrightarrow{\text{heat}} 2B(s) + 3MgO(s)$$

Synthetic boron compounds include the trihalides, which have trigonal planar structures similar to $B(OH)_3$ and are strong Lewis acids. The most stable trihalide is BF_3. It is used as an acid catalyst in a number of organic reactions. Very pure boron is obtained by the high-temperature reduction of BCl_3 with hydrogen:

:F:
B
:F: :F:
boron trifluoride (BF_3)

$$2BCl_3(s) + 3H_2(g) \xrightarrow{\text{heat}} 2B(s) + 6HCl(g)$$

Boron nitride (BN) has the same average number of electrons per atom as carbon. It forms a graphitelike lattice in which B and N atoms alternate (Figure S3.36a), and it is a slippery, white, lubricating solid. Like graphite, boron nitride can be converted under high pressure (50,000 atm) to a denser form with a diamond lattice. This crystalline form, called *borazon*, is as hard as diamond (Figure S3.36b).

Boranes

Boron's most exotic compounds are the boron hydrides, or **boranes**, of which the simplest is diborane (B_2H_6). Diborane can be synthesized by reacting sodium borohydride ($NaBH_4$, an important reducing agent) with iodine:

$$2NaBH_4(s) + I_2(s) \rightarrow B_2H_6(g) + 2NaI(s) + H_2(g)$$

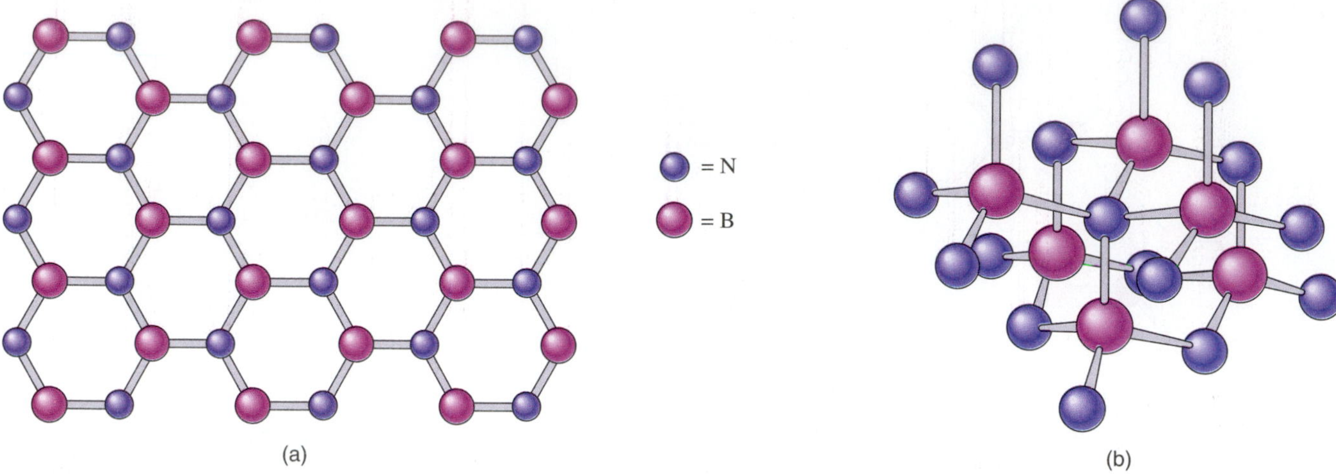

Figure S3.36

Crystal structures of boron nitride. (a) The more common form of boron nitride has a graphitelike layer structure with alternating boron and nitrogen atoms. (b) In the diamondlike material called *borazon*, each atom forms four tetrahedral bonds.

The boranes are not noted for stability; many are spontaneously flammable in air, and all react with water to form boric acid and hydrogen:

$$B_2H_6(g) + 3O_2(g) \rightarrow 2H_3BO_3(s)$$

$$B_2H_6(g) + 6H_2O(l) \rightarrow 2H_3BO_3(s) + 6H_2(g)$$

The boranes exhibit an unusual type of multicentered bonding (see Figure S3.37a). The two boron atoms in the B_2H_6 molecule are connected by covalent *hydrogen bridges*. At first glance, the bridging hydrogen atoms appear to violate the usual rule of only one bond per hydrogen atom; however, a count of valence electrons shows that the two B—H—B bridges have only four electrons between them. (There are 12 valence electrons in the B_2H_6 molecule, and 8 of these are used in the four B—H single bonds.) Each bridge, therefore, has one pair of electrons that holds three atoms together; one bond, in effect, doing the work of two. The hydrogen atom is thus in the middle of a single *three-center bond*, in which the bonding pair of electrons occupies an orbital formed by the overlap of the hydrogen 1*s* orbital with an sp^3 hybrid orbital from each boron atom (Figure S3.37b).

Figure S3.37

(a) Structure of the diborane (B_2H_6) molecule. The four outer hydrogen atoms lie in a plane that is perpendicular to the plane of the paper; the other atoms lie in the plane of the paper. (b) Bonding in the diborane molecule. On each boron atom, two sp^3 hybrid orbitals (not shown) are used for bonds with the outer hydrogen atoms. The other two hybrid orbitals overlap the *s* orbitals of hydrogen to form the three-center B—H—B bonds.

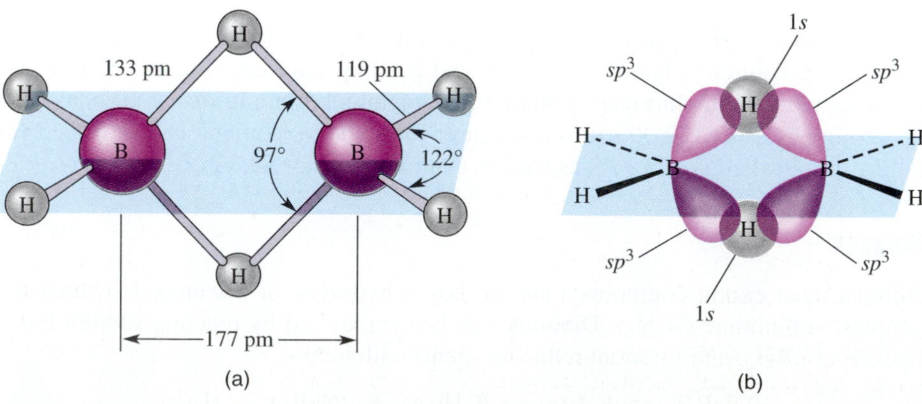

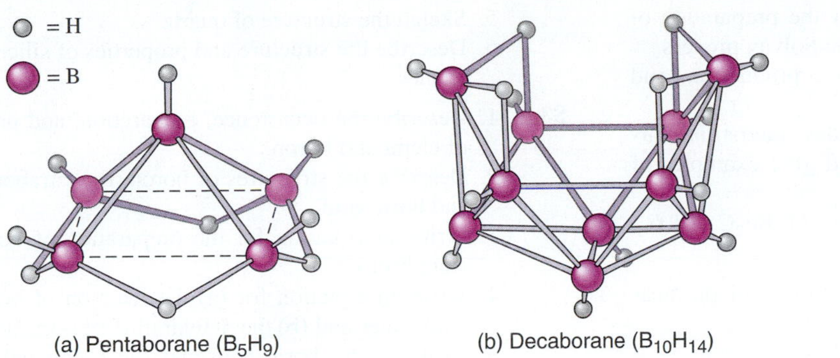

Figure S3.38
Some higher boranes.

The numerous boranes include compounds like B_4H_{10}, B_5H_9, $B_{10}H_{14}$, and $B_{12}H_{12}^{2-}$. Their structures would provide a field day for a geometry buff. The $B_{12}H_{12}^{2-}$ anion, for example, is a complete closed icosahedron, and a number of the other species can be regarded as icosahedral fragments (Figure S3.38). The study of boranes has led to the development of new aspects of molecular orbital theory and to the synthesis of many novel compounds.

CHAPTER REVIEW

LEARNING OBJECTIVES BY SECTION

S3.1
1. Describe the occurrence, preparation, and properties of elemental sulfur.
2. Describe the molecular structure of the various forms of solid, liquid, and gaseous sulfur.
3. Write equations for the reactions of sulfur with oxygen, chlorine, fluorine, and metals.
4. Describe the contact process for the preparation of sulfuric acid; include the appropriate equations.
5. Write equations for reactions in which sulfuric acid is an acid, an oxidizing agent, and a dehydrating agent.
6. State the principal uses of sulfuric acid.
7. Write equations for the preparation of SO_2 and its reactions with $H_2O(l)$, $OH^-(aq)$, and $CaCO_3(s)$.
8. Write equations for the preparation of thiosulfate ion and its reactions with $H^+(aq)$, $MnO_4^-(aq)$, $I_2(s)$, and AgBr(s).
9. Describe the properties and uses of hydrogen sulfide.
10. Write formulas for some polysulfide ions, and write equations for their formation.

S3.2
1. Describe the preparation of elemental phosphorus; include the appropriate equations.
2. Compare the properties of red and white phosphorus.
3. Write equations for the reactions of phosphorus with oxygen, the halogens, sulfur, and active metals.
4. Sketch the structures of P_4, P_4O_6, and P_4O_{10}.
5. Write equations for the reactions of P_4O_6 and P_4O_{10} with water.
6. Write equations for the preparation of phosphorous acid, orthophosphoric acid (two equations), and pyrophosphoric acid.
7. Write Lewis structures for phosphorous acid, orthophosphoric acid, and pyrophosphoric acid.
8. Write equations for the preparation of superphosphate and triple superphosphate.
9. Write Lewis structures for the pyrophosphate ion, the triphosphate ion, and the metaphosphate ion.
10. Write equations for the preparation of ionic phosphides and their reaction with water.
11. List the three principal oxidation states of the Group 5A elements, and illustrate each state with one phosphorus, one arsenic, and one antimony compound.

S3.3
1. Compare the structure and properties of graphite, diamond, and buckminsterfullerene.
2. Write equations for the reactions of carbon with oxygen, metal oxides, sulfur, and fluorine.
3. Describe the properties and uses of carbon dioxide.

4. Describe and write equations for the preparation of sodium hydrogen carbonate by the Solvay process.
5. Describe the molecular structure, production, and toxicity of carbon monoxide.
6. Distinguish between ionic carbides, interstitial carbides, and covalent carbides, and give examples of each.
7. Write equations for the reactions of Be_2C, Al_4C_3, and CaC_2 with water.

S3.4 1. Describe the occurrence, preparation, and purification of silicon.
2. Describe the structure and properties of silicon.
3. Write the structures of the orthosilicate and pyrosilicate ions.
4. Sketch the structures of silicate anions containing tetrahedra linked in rings, double chains, and sheets.
5. Sketch the structure of quartz.
6. Describe the structure and properties of silicones and silanes.

S3.5 1. Describe the occurrence, preparation, and properties of elemental boron.
2. Describe the structures of boron, the tetraborate ion, and boric acid.
3. Write an equation for the preparation of boric acid from borax.
4. Write an equation for (a) the reaction of boric acid with water and (b) the dehydration of boric acid.
5. Explain why boric acid and the boron halides are Lewis acids.
6. Sketch the diborane molecule, and discuss its bonding.

KEY TERMS BY SECTION

S3.1 Contact process
Frasch process
Metalloid
Semimetal

S3.2 Phosphide

S3.3 Amorphous carbon
Carbide
Interstitial carbide
Solvay process

S3.4 Aluminosilicate
Silane
Silicone
Zone refining

S3.5 Borane

FINAL EXERCISES

Answers to exercises with blue numbers are given in Appendix D.

PART A. QUESTIONS AND PROBLEMS BY SECTION

Sulfur and Group 6A (Section S3.1)

S3.1 List the Group 6A elements. Which of the Group 6A elements are semimetals? Which are semiconductors?

S3.2 Give specific examples of physical and chemical behavior in Group 6A to illustrate the trend toward increasing metallic character going down a group.

S3.3 Describe and diagram the Frasch process for mining sulfur.

S3.4 Describe, with equations, the process by which sulfur is recovered from fuels.

S3.5 Describe the occurrence of sulfur in the earth's crust.

S3.6 What is the principal use of sulfur?

S3.7 **(a)** Describe the changes that occur when solid sulfur is gradually heated from room temperature to 1000°C and then cooled abruptly.
(b) Explain, with sketches, what happens to the sulfur molecules during the changes in Part (a).

S3.8 Name and describe the solid allotropes of sulfur. Which allotrope is stable at 25°C and 1 atm?

S3.9 Describe the industrial production of sulfuric acid. Write the equation and state the conditions for each step.

S3.10 **(a)** Write equations for the preparation of HCl and HNO_3 by the action of concentrated sulfuric acid on their salts.
(b) Diagram an apparatus that could be used if the prepared acid is to be collected in the form of an aqueous solution.

S3.11 Draw the Lewis structure for the S_8 molecule. Use VSEPR to predict the approximate bond angles. Explain why the ring is puckered.

S3.12 **(a)** Under what conditions are S_2 molecules found?
(b) Write the electron configuration of S_2, and explain why the molecule is paramagnetic.

S3.13 Draw a Lewis structure for each of the following species:
(a) sulfur dioxide (c) sulfurous acid
(b) sulfur trioxide (d) sulfate ion

S3.14 Draw a Lewis structure for each of the following species:
(a) sulfuric acid (c) sulfite ion
(b) pyrosulfuric acid (d) thiosulfate ion

S3.15 Draw a Lewis structure for (a) SF_6, (b) S_2Cl_2 (Cl—S—S—Cl), and (c) SCl_2.

S3.16 Draw a Lewis structure for (a) the peroxydisulfate ion ($S_2O_8^{2-}$) and (b) the tetrathionate ion ($S_4O_6^{2-}$). (*Hint*: The $S_2O_8^{2-}$ ion contains an O—O linkage in the center; the $S_4O_6^{2-}$ ion contains a chain of four sulfur atoms.)

S3.17 On the basis of electron configuration, suggest a reason why sulfur forms SF_6 while oxygen forms OF_2.

S3.18 On the basis of electron configuration, suggest a reason why SF_4 is reactive while SF_6 is not.

S3.19 Write an equation for the reaction of sulfur with each of the following elements:
(a) oxygen
(b) hydrogen
(c) fluorine

S3.20 Write an equation for the reaction of sulfur with each of the following elements:
(a) copper
(b) chlorine
(c) silver

S3.21 Find the oxidation number of sulfur in (a) $S_2O_3^{2-}$, (b) $S_2O_6^{2-}$, and (c) $S_4O_6^{2-}$.

S3.22 Use the method of half-reactions to obtain a balanced equation for each of the following oxidations by hot, concentrated sulfuric acid:
(a) copper to copper sulfate
(b) iodine to iodate ion
(c) carbon to carbon dioxide

S3.23 Complete and balance the following equations:
(a) $Na_2SO_3(aq) + HCl(aq) \rightarrow$
(b) $S_2O_3^{2-}(aq) + I_2(s) \rightarrow$
(c) $S^{2-}(aq) + CrO_4^{2-}(aq) \xrightarrow{\text{base}} Cr(OH)_4^-(aq) + S(s)$
(d) $Na_2SO_3(aq) + S(s) \xrightarrow{\text{heat}}$

S3.24 Complete and balance the following equations:
(a) $AgBr(s) + Na_2S_2O_3(aq) \longrightarrow$
(b) $S_2O_3^{2-}(aq) + MnO_4^-(aq) \xrightarrow{\text{acid}}$
(c) $H_2S(aq) + HNO_3(aq, conc) \longrightarrow H_2SO_4(aq) + NO_2(g)$
(d) $K_2S(aq) + S(s) \xrightarrow{\text{heat}}$

S3.25 Write an equation for the reaction of sulfur dioxide gas with (a) liquid water, (b) aqueous hydroxide ions, and (c) solid calcium carbonate.

S3.26 Write an equation for the reaction of aqueous thiosulfate ion with each of the following species:
(a) $H^+(aq)$ (c) $I_2(s)$
(b) $MnO_4^-(aq)$ (d) $AgBr(s)$

S3.27 When treated with concentrated H_2SO_4, a wood chip turns black, a blue crystal of $CuSO_4{\cdot}5H_2O$ turns white, a white crystal of sodium bromide turns brown, and a piece of copper dissolves. Explain each reaction and write equations where possible.

S3.28 Two unknown solid sodium salts gave off colorless gases when treated with hot, concentrated H_2SO_4. The gas from the first salt was passed into a silver nitrate solution, where it produced a white precipitate. The gas from the second salt was passed into a tube containing moist copper filings, which reacted to produce a brown gas. Identify each salt, and write an equation for each of the reactions described above.

S3.29 (a) Explain why a sulfur atom can take the place of an oxygen atom in a number of molecules.
(b) How should the replacement of oxygen by sulfur affect the strength of an oxo acid?

S3.30 Use periodic trends (see Table S3.1) to predict (a) whether CuS or CuTe is the more ionic compound, (b) whether H_2S or H_2Te is the stronger acid, and (c) whether H_2SO_3 or H_2TeO_3 is the stronger acid.

Phosphorus and Group 5A (Section S3.2)

S3.31 List the Group 5A elements. Which of the Group 5A elements are semimetals?

S3.32 Which of the Group 5A elements are known to have some biological function?

S3.33 Describe the industrial preparation of phosphorus. Include the equations and a sketch of the apparatus.

S3.34 Compare white and red phosphorus with respect to
(a) molecular structure (c) chemical properties
(b) physical properties (d) toxicity
How are the differences in properties related to the difference in molecular structure?

S3.35 Describe some commercial uses for phosphorus.

S3.36 Describe some commercial uses for sodium polyphosphates and sodium metaphosphate.

S3.37 Sketch the P_4 molecule, the P_4O_6 molecule, and the P_4O_{10} molecule. How are these structures related?

S3.38 Draw a Lewis structure for (a) phosphorous acid, (b) orthophosphoric acid, and (c) pyrophosphoric acid. What is the oxidation state of phosphorus in each of these acids?

S3.39 H_3PO_4 is a triprotic acid, but H_3PO_3 is not. Explain this observation in terms of Lewis structures.

S3.40 Write a Lewis structure for (a) the pyrophosphate ion, (b) the triphosphate ion, and (c) the trimetaphosphate ion.

S3.41 Write an equation for the reaction of phosphorus with each of the following elements:
(a) limited oxygen **(c)** sulfur
(b) excess oxygen **(d)** sodium

S3.42 Write an equation for the reaction of phosphorus with each of the following elements:
(a) limited chlorine **(c)** calcium
(b) excess fluorine

S3.43 Write an equation for the reaction of water with (a) P_4O_6 and (b) P_4O_{10}.

S3.44 P_4O_{10} will dehydrate pure nitric acid to N_2O_5. Write the equation for this reaction.

S3.45 Write equations for two ways of preparing orthophosphoric acid. Which method gives a purer product? Why?

S3.46 Write an equation for the preparation of (a) phosphorous acid and (b) pyrophosphoric acid.

S3.47 Write an equation for each of the following reactions:
(a) the hydrolysis of PCl_5 to form phosphoric and hydrochloric acids
(b) the hydrolysis of $AsCl_3$ to form $As(OH)_3$ and HCl

S3.48 Write an equation for each of the following reactions:
(a) the combustion of P_4S_3, a compound used in match heads
(b) the reaction of calcium phosphide with water

S3.49 Write an equation for the reaction of sulfuric acid with calcium phosphate to produce (a) calcium hydrogen phosphate and (b) calcium dihydrogen phosphate. Why do these products make better fertilizers than the original phosphate rock?

S3.50 Write an equation for the preparation of (a) superphosphate and (b) triple superphosphate. Which of these fertilizers has the higher phosphate content?

S3.51 H_3AsO_4 is a stronger oxidizing agent than H_3PO_4. What periodic trend is illustrated by this observation? Explain your answer.

S3.52 **(a)** Rank the acids HNO_3, H_3PO_4, and H_3AsO_4 in order of increasing strength. What atomic property is important in determining the strengths of these acids?
(b) Which is the stronger base, NH_3 or AsH_3? How do you know?

Group 4A: Carbon (Section S3.3)

S3.53 List the Group 4A elements. Which of the Group 4A elements are semimetals? Which are semiconductors?

S3.54 Show how the atomic properties of the Group 4A elements (see Table S3.9) illustrate the trend toward increasing metallic character going down a periodic group.

S3.55 Sketch the graphite lattice. List some of the properties of graphite, and explain them in terms of lattice structure.

S3.56 Describe and sketch the diamond lattice. Compare the physical properties of diamond and graphite, and explain the differences in terms of lattice structure.

S3.57 Describe (a) the structural characteristics of fullerenes and (b) the structure of buckminsterfullerene, C_{60}.

S3.58 How are diamond films prepared? List some uses for such films.

S3.59 Describe, with equations, the preparation of $NaHCO_3$ by the Solvay process.

S3.60 Write an equation for the conversion of $NaHCO_3$ into Na_2CO_3.

S3.61 Describe at least three commercial uses of CO_2.

S3.62 Explain why carbon monoxide is toxic. How is carbon monoxide poisoning treated?

S3.63 What is a *carbide*? Give some uses for at least three carbides.

S3.64 Distinguish between ionic, interstitial, and covalent carbides. Give examples of each.

S3.65 Draw a Lewis structure for
(a) carbon dioxide **(c)** hydrogen cyanide
(b) carbonic acid, H_2CO_3 **(d)** carbon disulfide

S3.66 Draw a Lewis structure for (a) N_2, (b) CO, (c) CN^-, and (d) C_2^{2-}. What features do these structures have in common? What properties, if any, do the ions and molecules have in common?

S3.67 Write an equation for the combustion of carbon to form (a) CO and (b) CO_2. What conditions favor the formation of CO?

S3.68 Write an equation for the reaction of carbon with (a) Fe_2O_3, (b) sulfur, and (c) fluorine.

S3.69 Explain why cyanide solutions must be kept basic. Write an equation for the acidification of aqueous potassium cyanide.

S3.70 When carbon is heated with 16 *M* HNO_3, the products include CO_2 and NO_2 gases. Use the method of half-reactions to obtain the balanced equation.

S3.71 What is an *acetylide*? Write an equation for (a) the formation of CaC_2 and (b) its reaction with water.

S3.72 Write an equation for the reaction of water with (a) Be_2C and (b) Al_4C_3.

Group 4A: Silicon (Section S3.4)

S3.73 Describe and compare the occurrence and abundance of silicon and germanium.

S3.74 Describe the preparation and purification of elemental silicon.

S3.75 (a) How do the crystal lattices of silicon and germanium resemble the diamond lattice of carbon? How do they differ?
(b) Explain why silicon and germanium do not form graphite-type lattices.

S3.76 Describe the zone-refining method for obtaining ultra-pure semiconductor materials.

S3.77 What is the difference between a *silicate* and an *aluminosilicate*? Give some examples of aluminosilicates.

S3.78 What is a *zeolite*? What are zeolites used for?

S3.79 Draw a Lewis structure for (a) the orthosilicate ion and (b) the pyrosilicate ion. What is "water glass"?

S3.80 Refer to Table S3.11, and sketch the silicate ion found in (a) olivine, (b) asbestos, and (c) talc.

S3.81 Draw the structural formula of methyl silicone. What factors contribute to the stability of silicones? Give some uses for silicones.

S3.82 Draw the Lewis structure of (a) SiH_4 and (b) Si_3H_8. Write an equation for the combustion of each of these silanes.

Boron (Section S3.5)

S3.83 Describe the occurrence, preparation, and properties of elemental boron.

S3.84 List some of the uses for (a) boric acid and (b) borax.

S3.85 Describe the structure of (a) boron, (b) the tetraborate ion, and (c) boric acid.

S3.86 (a) Sketch the graphite and diamond lattices of boron nitride (BN).
(b) What conditions are needed to convert boron nitride to borazon, and why?

S3.87 Sketch the diborane molecule and discuss its bonding.

S3.88 How does a hydrogen bridge bond differ from a hydrogen bond? What types of compounds possess hydrogen bridge bonds?

S3.89 Write an equation and use structural formulas to account for the acidity of boric acid. Is H_3BO_3 a Brønsted acid?

S3.90 Draw the Lewis structure for BF_3 and explain why it is a Lewis acid.

S3.91 Write an equation for each of the following reactions:
(a) the preparation of boric acid from borax
(b) the reaction of boric acid with water
(c) the dehydration of boric acid

S3.92 Write an equation for each of the following reactions:
(a) the combustion of diborane in air
(b) the conversion of diborane, on heating, to B_5H_9 and hydrogen
(c) the reaction of B_5H_9 with chlorine to produce BCl_3 and HCl

PART B. MISCELLANEOUS QUESTIONS AND PROBLEMS

S3.93 Identify the following substances by formula and chemical name:
(a) brimstone
(b) epsom salts
(c) gypsum
(d) oil of vitriol
(e) oleum
(f) phosphoric anhydride
(g) silica gel
(h) talc
(i) borax

S3.94 Suggest a reason why carbon forms both carbon monoxide and carbon dioxide while silicon and germanium form only dioxides.

S3.95 Describe the environmental effects that led to the banning of phosphates in laundry detergents. Why doesn't phosphate rock have a harmful effect?

S3.96 Describe test tube reactions that could distinguish between
(a) solid samples of calcium sulfate and calcium sulfite
(b) solutions of sodium sulfite and sodium thiosulfate

S3.97 The formation of a lead sulfate precipitate can be used to detect the presence of lead ion in solution.
(a) Perform a calculation to show that $PbSO_4$ will form when 1.0 mL of 0.10 *M* Na_2SO_4 is added to 10 mL of 0.10 *M* $Pb(NO_3)_2$ solution.
(b) Calculate the concentration of $Pb^{2+}(aq)$ in the final equilibrium mixture.

S3.98 Sulfur dioxide emissions in the flue gases of burning coal can be reduced by mixing the crushed coal with finely ground limestone.
(a) Write the equation for the conversion of sulfur, oxygen, and $CaCO_3$ to $CaSO_4$. What is the other product, and what happens to it?
(b) Find the theoretical quantity of limestone needed to react with the sulfur in 1.0 metric ton of coal containing 3.5% sulfur by mass.

S3.99 Among the various methods for removing SO_2 from smokestack gases are (a) passing the gas through limewater (a suspension of calcium hydroxide) and (b) passing the gas over magnesium oxide. Write an equation for each reaction.

S3.100 The reaction between H_2S and SO_2 gases forms solid sulfur.
- **(a)** Write a balanced equation.
- **(b)** This reaction proceeds only when moisture is present. Can you suggest why?

S3.101 The Si—Si, Si—H, and Si—O bond energies are 222, 318, and 452 kJ/mol, respectively. Refer to Table 9.3 (p. 377) for additional bond energies and estimate $\Delta H°$ for the combustion of 1.00 mol of Si_3H_8. Compare your answer with $\Delta H°$ for the combustion of 1.00 mol of propane (C_3H_8). Why is propane used for fuel while the corresponding silane is not?

S3.102 The standard state of phosphorus was chosen to be white phosphorus (P_4) because this allotrope has a well-defined structure. The standard enthalpies of formation at 298 K are −17.6 kJ/mol for P(red) and −2984.0 kJ/mol for $P_4O_{10}(s)$. Calculate the standard molar enthalpy of combustion in excess oxygen of (a) white phosphorus and (b) red phosphorus.

S3.103
- **(a)** Consider the molecular orbitals formed by the valence-shell *s* and *p* orbitals in S_2, and give the number of electrons in each. Compare this configuration with the molecular orbital configuration of O_2.
- **(b)** What effect does the size of the sulfur atom have on the overlapping of *p* orbitals to form pi molecular orbitals? Suggest a reason why S_2 is not the stable form of sulfur at room temperature.

S3.104 B_2 molecules have been found in boron vapor. Write the molecular orbital configuration of B_2. Do you expect B_2 to be paramagnetic? Explain your answer.

S3.105 A certain gaseous borane is found to have a density of 1.09 g/L at 25°C and 0.500 atm.
- **(a)** Estimate its molar mass.
- **(b)** Devise a reasonable molecular formula for this borane. (*Hint*: The number of hydrogen atoms per boron atom cannot exceed three.)

S3.106 A 24.1-mL sample of an unknown $KMnO_4$ solution is decolorized by a solution that contains 3.16 g $Na_2S_2O_3$. In this reaction, the permanganate is reduced to Mn^{2+}, and the thiosulfate is oxidized to tetrathionate ion ($S_4O_6^{2-}$). Calculate the molarity of the $KMnO_4$ solution.

S3.107 A starch indicator and 25.00 mL of 0.1000 *M* NaI are added to 25.00 mL of an acidified $KMnO_4$ solution. The permanganate oxidizes some of the iodide ion to I_2, which then produces a blue color with the starch:

$$2MnO_4^-(aq) + 10I^-(aq) + 16H^+(aq) \rightarrow 2Mn^{2+}(aq) + 5I_2(aq) + 8H_2O(l)$$

The mixture is titrated with 0.1010 *M* $Na_2S_2O_3$, and the blue color disappears after the addition of 29.70 mL of the thiosulfate solution. Calculate the molarity of the original $KMnO_4$ solution.

S3.108 The three numbers on a bag of fertilizer represent, respectively, the percentage of nitrogen as N, the percentage of phosphorus as P_2O_5, and the percentage of potassium as K_2O. How many pounds of nitrogen, phosphorus, and potassium are in a 25-pound bag of 20-10-5 fertilizer?

CHEMISTS AT WORK

THE MANAGER

Many chemists find that as they gain work experience they spend less time in the laboratory and undertake more responsibilities as managers. That is exactly the path Leon Clark's career has taken at the Upjohn Company in Kalamazoo, Michigan.

When Leon entered Grambling State University on a partial football scholarship and a grant from the chemistry department, he already had an avid interest in science and math. Even though his first love was math, because of the grant he decided to pursue chemistry. During his time at Grambling, Leon's interest in analytical chemistry deepened, especially through his work as an assistant for an instrumental analysis course and also through his participation in a research project. "I really got turned on to chemistry because of these experiences," he declares. "They taught me how chemistry is applied in real situations." During the summers of his junior and senior years, Leon further developed his interests in chemistry by participating in the Summer Minority Internship Program at the Bell Laboratories in Murray Hill, New Jersey. Upon graduation, he received a fellowship from Bell Labs and went on to graduate school at Iowa State University where he pursued a thesis-based master's degree in analytical chemistry.

In March 1978, he went to work for the Upjohn Company, a major pharmaceutical concern. Some of his early assignments in research and development related to the behavior of potential medications in the body. One project involved developing tests to determine the rate of dissolution of compounds; another focused on developing and evaluating the testing of absorption rates of various formulations in animals. In time he earned the title of "Group Leader" and responsibilities for scheduling work in the lab, implementing new assays, and trouble shooting testing on new drugs and potential drug candidates were added to his bench-work.

Leon was eventually invited to accept a "Training Supervisor Position," which at Upjohn is a supervised trial period to determine whether a candidate has the skills and abilities to become a manager and also whether the candidate is interested in pursuing a management career. In this program, Leon supervised the work of seven people in a lab where the shelf-life of various drugs and formulations is determined. He enjoyed this experience and decided to embark on a career in management. In addition to on-the-job training, Leon has also taken a variety of management courses offered both by the company or by private vendors.

Since 1985, Leon has been Manager of Assay Services and Spectroscopy, where he supervises the work of 13 people. Initially, his responsibilities involved supervising the Quality Controls Lab, which is involved in routine testing of various formulations using a variety of spectroscopic techniques. Over time his responsibilities expanded to encompass the identification of impurities with TLC (Thin Layer Chromatography), all the testing related to the hair-restoring product Rogaine, and all the assay methods development for the Atomic Absorption Lab. "One of the hardest things that I had to learn was how to delegate work," he says. "It is difficult to give up the hands-on work, but I enjoy working with people and helping them to obtain their professional goals. I don't evaluate people based solely on their academic degree, but also on their abilities and accomplishments."

Leon's accomplishments and responsibilities in the sciences are not founded solely on his work in the laboratory. He is also responsible for minority recruitment at a group of colleges in the South, including his alma mater. To attract students to the sciences, he and 25 other people from the company present annual demonstrations to over 600 students in area elementary and secondary schools. In 1992, Leon developed a program where secondary school teachers spend a week in the summer at Upjohn observing chemists at work. Leon's philosophy? "Many people helped me along the way and I want to return the favors."

LEON CLARK

CHAPTER 19

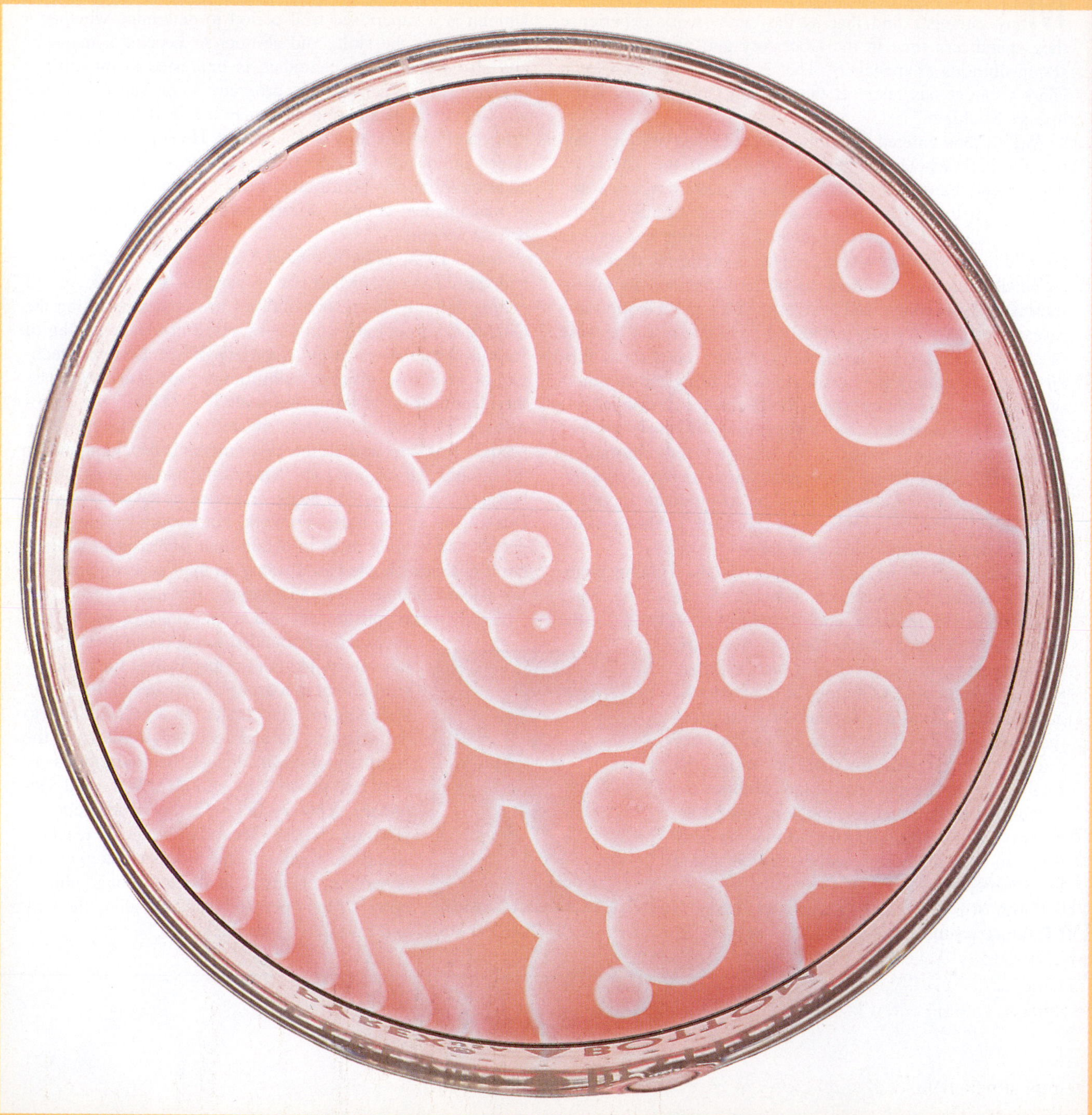

The Belousov-Zhabotinskii reaction. These ripple patterns, which formed from an initially homogeneous mixture containing a visual indicator, show that temporary domains of order can arise spontaneously out of disorder. The reacting mixture oscillates in composition and thus in the color of the added indicator. Oscillating chemical reactions occur in some industrial processes and also in biochemical systems.

FREE ENERGY, ENTROPY, AND THE SECOND LAW OF THERMODYNAMICS

OUTLINE

PREVIEW

Many events occur spontaneously. Other events occur only when there is some sort of intervening action. For example, a clock runs down by itself (spontaneously), but must be rewound manually. We can often anticipate the direction of spontaneous change—we know that ice melts in warm water, that leaves scatter in the wind, and that living things grow, mature, and die. Experience gives us a sense of the forward direction of events, and we do not expect tomorrow to become yesterday. This feeling for the way things proceed does not have its basis in energy alone. The first law of thermodynamics tells us that tomorrow's energy will be the same as yesterday's energy, that none will be or has been lost. As far as total energy is concerned, events might just as well roll backward as forward. Why don't they? In answering this question, we may discover the common thread that makes the passage of time a one-way trip.

The reactions that we have studied—neutralizations, precipitations, and so forth—take place because chemical systems always move toward a state of equilibrium. Chemical reactions are reversible; that is, the forward and reverse reactions occur simultaneously. When the system is not at equilibrium, one of the opposing reactions occurs at a faster rate than the other. For example, when silver nitrate is added to aqueous sodium chloride

$$Ag^+(aq) + Cl^-(aq) \rightleftharpoons AgCl(s)$$

the rate at which solid silver chloride forms is initially greater than the rate at which it dissolves. The faster reaction (in this case, precipitation) is the reaction we observe, so we call it the *spontaneous reaction* and say that there is a *net driving force* in the forward direction. At equilibrium, the rates of the forward and reverse reactions are equal, so the net driving force is zero.

A reversible reaction is somewhat analogous to a tug of war between two children. If one child is stronger than the other, a net driving force exists, and the rope will move in the direction of the stronger child. If the children are evenly matched and exert equal forces on the rope, there is no net driving force, and the rope will be at equilibrium even though the exertions of each child may cause momentary shifts in one direction or the other. Another example of an equilibrium system is ice at 0°C floating in liquid water at 0°C. The ice is continuously melting and the water is continuously freezing. The two states coexist because the net driving force is zero. The situation is quite different when ice at 0°C is dropped into water at 70°C. The temperature difference provides a strong net driving force that causes all of the ice to melt.

19.1 FREE ENERGY AND USEFUL WORK

LEARNING HINT

Review the concepts of heat, work, internal energy, and enthalpy in Sections 6.1 and 6.2.

A spontaneous reaction can do work in its drive toward equilibrium. We have already studied the pressure–volume (PV) work done by gaseous products that expand and push back their surroundings (see Figure 6.6, p. 230). In this chapter we are concerned with **useful work**, a term used to describe work other than PV work. (This definition does not mean that PV work is useless; in fact, the work done by steam and automobile engines is all PV work. For most of the reactions that will concern us, however, PV work simply pushes back the atmosphere and is not used for any useful purpose.) The most familiar form of useful work is the electrical work performed by redox reactions in batteries (Chapter 20), but chemical reactions can also be used to provide mechanical, magnetic, osmotic, and photochemical work.

In Chapter 15 we learned how to predict the direction of a spontaneous reaction by comparing the reaction quotient Q with the equilibrium constant K. Now let us look for another property that will tell us how much useful work can be obtained from a chemical system during its passage to equilibrium. We already know that in mechanical systems the potential energy decreases during a spontaneous change. Water, for example, has more potential energy when it is in an elevated position—thus water runs downhill spontaneously (Figure 19.1a). As the water flows, its potential energy changes into kinetic energy that can be used to turn wheels or to do other forms of useful work. The maximum useful work that can be obtained from a sample of water is equal to the decrease in its potential energy; that is, the maximum useful work is equal to the difference between its potential energy on top of the hill and its potential energy at the bottom of the hill.

Chemical systems possess a property analogous to potential energy; it is called the **free energy** (G), and it decreases as the reaction proceeds. It reaches its lowest value at equilibrium, when the reaction quotient is equal to the equilibrium constant (Figure 19.1b). The Gibbs free energy is defined so that its change at constant temperature and pressure

LEARNING HINT

In this chapter we are discussing the work done *by* a chemically reacting system. Keep in mind that w, the work done *on* a system, is the negative of the work done by the system (Section 6.1).

$$\Delta G = G(\text{products}) - G(\text{reactants}) \qquad (19.1)$$

is equal in magnitude but opposite in sign to the maximum amount of useful work that can be obtained from the reaction:

$$\Delta G = -\text{maximum useful work} \qquad (19.2)$$

The negative sign is included in Equation 19.2 because work done on the surround-

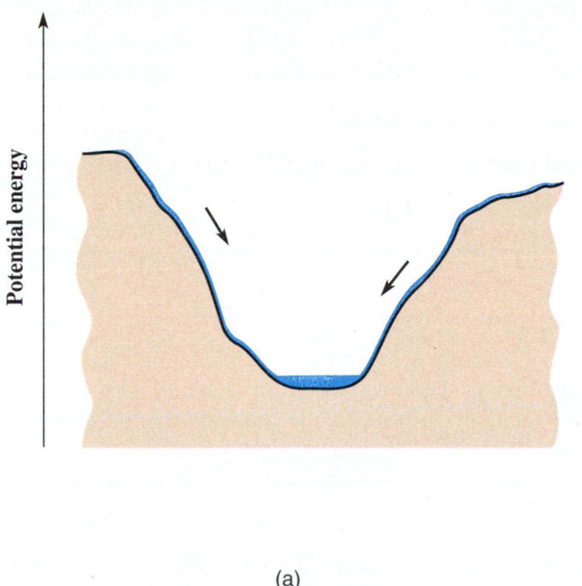

(a)

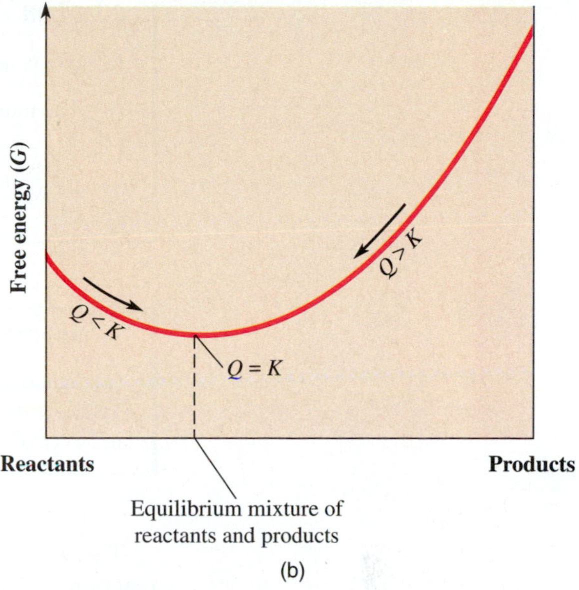

(b)

Figure 19.1

Potential energy and free energy. (a) Water flows spontaneously to a lower level, in the direction of lower potential energy. (b) A chemical system proceeds spontaneously toward equilibrium, in the direction of lower free energy, as shown in this sketch of free energy versus composition.

ings by a spontaneously reacting system is positive while the free-energy change of the system is negative (if G decreases, then $\Delta G < 0$).

Every spontaneous reaction has the potential for doing useful work, and Equation 19.2 states that, for such reactions, ΔG is always negative. The further the system is from equilibrium, the more work the reaction can do, and the more negative the value of ΔG.

As the reacting mixture approaches equilibrium, its potential for doing useful work decreases, so ΔG becomes less negative and increases toward zero. At equilibrium, the system can no longer do useful work, and ΔG is zero. Thus the sign of ΔG can be used to predict whether a reaction will be spontaneous. Consider, for example, a general reaction in which A and B undergo the reversible change

The concept of free energy was formulated by Josiah Willard Gibbs (1839–1903), a mathematical physicist at Yale University.

$$A \rightleftharpoons B$$

- If ΔG is negative ($\Delta G < 0$), the forward reaction is spontaneous; that is, $A \rightarrow B$.
- If $\Delta G = 0$, the system is at equilibrium.
- If ΔG is positive ($\Delta G > 0$), the reverse reaction is spontaneous; that is, $B \rightarrow A$.

EXAMPLE 19.1

For the reaction

$$CO(g, 1\ atm) + \tfrac{1}{2}O_2(g, 1\ atm) \rightleftharpoons CO_2(g, 1\ atm)$$

ΔG is -257.2 kJ at 25°C. (a) Is the forward or reverse reaction spontaneous at this temperature? (b) What is the maximum amount of useful work that can be obtained from the conversion of 1 mol of CO to CO_2?

SOLUTION

(a) ΔG is negative for the reaction as written, so the forward reaction is spontaneous.

(b) The maximum work is calculated from Equation 19.2:

$$\text{maximum useful work} = -\Delta G = -(-257.2 \text{ kJ})$$
$$= 257.2 \text{ kJ}$$

PRACTICE EXERCISE 19.1

(a) What is the value of ΔG when 1 mol of liquid water at its normal boiling point of 100°C forms steam at 100°C? (b) Could any useful work be obtained from the change in Part (a)?

The first law of thermodynamics requires that the total energy change of a given reaction be constant. Hence, the energy change will be the same whether or not the reacting system does useful work. The potential energy lost by a reacting system appears as heat or is used by the system to do work on its surroundings. If some of the energy is used to do work, then less energy will appear as heat. For example, the enthalpy change for the combustion of 1 mol of carbon monoxide is −283.0 kJ. If no useful work is done during the combustion, then 283 kJ of heat will be evolved when 1 mol of carbon monoxide is burned. On the other hand, if the maximum amount of useful work is done (see Example 19.1), then only 283.0 kJ − 257.2 kJ = 25.8 kJ of heat will be evolved.

PRACTICE EXERCISE 19.2

For the combustion of methane

$$CH_4(g) + 2O_2(g) \rightarrow CO_2(g) + 2H_2O(g)$$

$\Delta H°$ is −890.4 kJ/mol at 25°C and ΔG is −817.9 kJ/mol. (a) What is the maximum useful work that can be obtained from the combustion of 2.00 mol of methane at 25°C? (b) How many kilojoules of heat would be evolved if this much work were actually done?

It is important to remember that the free-energy change (see Equation 19.1) depends only on the products and reactants. Free energy, like enthalpy, is a state function (Section 6.1)—it does not depend on the pathway or mechanism by which the reactants are converted into products; therefore, it provides no information about the reaction speed. Similarly, when a sample of water runs down a hill, the potential energy change is the same regardless of whether the water trickles or cascades down the hill. A reaction with a negative ΔG may proceed so slowly that its rate is hardly observable on a human time scale. The conversion of diamonds to graphite, for example, is a spontaneous reaction with $\Delta G = -2.9$ kJ/mol at 25°C and 1 atm. The reaction is so slow, however, that we feel quite secure in buying diamonds and handing them down from one generation to the next.

19.2 FREE-ENERGY CHANGE AND CONCENTRATION

Free-energy terminology is similar to the terminology used for enthalpy. A **standard free-energy change** ($\Delta G°$) is the free energy of reaction when the pressure of each gas is 1 atmosphere and the concentration of each dissolved substance is 1 molar; that is, it is the free-energy change when reactants in their standard states are converted to products in their standard states. The standard-state conventions were defined in Section 6.3 and are summarized for convenience in Table 19.1.

LEARNING HINT

Review Sections 6.3 and 6.4 on standard enthalpy changes and Hess's law.

The net driving force and even the direction of a reaction can be changed by altering the concentrations of reactants and products. The instantaneous value of ΔG varies with the instantaneous value of the reaction quotient Q according to the equation

$$\Delta G = \Delta G° + RT \ln Q \tag{19.3}$$

where $\Delta G°$ is the standard free-energy change, $R = 8.314$ J/mol·K is the gas constant, T is the Kelvin temperature, and ln Q is the natural logarithm of the reaction quotient. (The derivation of Equation 19.3 requires a detailed knowledge of the second law of thermodynamics and is beyond the scope of this book.) If the gases each have a pressure equal to 1 atm and the dissolved substances are each 1 M, then $Q = 1$, $\ln Q = 0$, and $\Delta G = \Delta G°$. Because Q contains product concentrations divided by reactant concentrations, the value of Q will increase as the product concentrations (or pressures) increase. The term $RT \ln Q$ will also increase, ΔG will rise (become less negative or more positive), and the net driving force of the forward reaction will decrease. In other words, *the driving force of the forward reaction decreases as the products accumulate*. An increase in reactant concentrations has the opposite effect: The values of Q and $RT \ln Q$ will decrease, ΔG will become lower (less positive or more negative), and the forward reaction will have a greater driving force. In effect, the $RT \ln Q$ term converts $\Delta G°$, the standard free-energy change, into ΔG, the free-energy change at conditions other than standard-state conditions.

$\Delta G°$ and the Equilibrium Constant

At equilibrium, the reaction quotient is equal to the equilibrium constant, and the free-energy change is zero. Substituting $Q = K$ and $\Delta G = 0$ into Equation 19.3

TABLE 19.1 Thermodynamic Standard States[a]

Substance	Standard State
Solid	Pure solid at 1 atm
Liquid	Pure liquid at 1 atm
Gas	Ideal gas at 1 atm
Solute	Ideal solution at 1 M concentration[b]

[a]The International Union of Pure and Applied Chemistry (IUPAC) has recommended that 1 bar rather than 1 atm (1.01325 bar) be used to define the standard state. This change will not significantly alter thermodynamic values.

[b]One molar is approximate. The precise standard state for a solute is the ideal solution at unit molality.

gives

$$\Delta G = \Delta G^\circ + RT \ln Q$$
$$0 = \Delta G^\circ + RT \ln K$$

and

$$\Delta G^\circ = -RT \ln K \quad (19.4)$$

Equation 19.4 allows us to calculate the standard free-energy change of a reaction from the equilibrium constant.

EXAMPLE 19.2

Calculate ΔG° for the reaction

$$N_2(g) + 3H_2(g) \rightleftharpoons 2NH_3(g)$$

where $K_p = 5.80 \times 10^5$ at 25°C.

SOLUTION

Substituting $K_p = 5.80 \times 10^5$. $T = 298$ K, and $R = 8.314$ J/mol·K into Equation 19.4 gives

$$\begin{aligned}\Delta G^\circ &= -RT \ln K_p \\ &= -8.314 \text{ J/mol·K} \times 298 \text{ K} \times \ln (5.80 \times 10^5) \\ &= -32{,}900 \text{ J/mol} = -32.9 \text{ kJ/mol}\end{aligned}$$

Observe that the units for ΔG° are kilojoules per mole. At this point you may ask, −32.9 kilojoules per mole of what? In this case, *mole* refers to a *mole of reactions*, that is, to an Avogadro number of reactions as represented by the chemical equation. The equation for the formation of ammonia tells us that 1 mol of N_2, 3 mol of H_2, and 2 mol of NH_3 take part in 1 mol of reaction. Hence, −32.9 kJ/mol can also be written as −32.9 kJ/1 mol N_2, −32.9 kJ/3 mol H_2, or −32.9 kJ/2 mol NH_3.

PRACTICE EXERCISE 19.3

The value of K_p for the reaction

$$H_2(g) + I_2(g) \rightleftharpoons 2HI(g)$$

is 50.3 at 458°C. Calculate ΔG° for the reaction at this temperature.

A **standard free energy of formation** (ΔG°_f) is the standard free-energy change for the formation of 1 mol of substance from its elements in their most stable forms. The standard free energy of formation of ammonia at 25°C, for example, is half of the free-energy change calculated in Example 19.2:

$$\Delta G^\circ_f(NH_3) = \frac{-32.9 \text{ kJ}}{2 \text{ mol } NH_3} = -16.4 \text{ kJ/mol } NH_3$$

Some standard free energies of formation at 25°C are given in Table 19.2. Note that ΔG°_f is zero for the most stable form of an element under standard-state conditions.

TABLE 19.2 Selected Thermodynamic Data[a] at 25°C *(A More Complete Listing is Given in Appendix B.1)*

Substance	ΔH°_f (kJ/mol)	ΔG°_f (kJ/mol)	S° (J/mol·K)
Ag(s)	0	0	42.55
AgCl(s)	−127.07	−109.79	96.2
Ag_2O(s)	−31.05	−11.20	121.3
Al(s)	0	0	28.33
Al_2O_3(s)	−1675.7	−1582.3	50.92
Br_2(l)	0	0	152.23
Br_2(g)	30.91	3.11	245.46
BrCl(g)	14.64	−0.98	240.10
C(g)	716.68	671.26	158.10
C(diamond)	1.895	2.900	2.377
C(graphite)	0	0	5.740
CH_4(g)	−74.81	−50.72	186.26
C_2H_2(g)	226.73	209.20	200.94
C_2H_4(g)	52.26	68.15	219.56
C_2H_6(g)	−84.68	−32.82	229.60
C_6H_6(l)	49.04	124.42	172.80
CO(g)	−110.525	−137.168	197.67
CO_2(g)	−393.509	−394.359	213.74
CS_2(l)	89.70	65.27	151.34
Ca(s)	0	0	41.42
$CaCO_3$(s, calcite)	−1206.92	−1128.79	92.9
CaO(s)	−635.09	−604.03	39.75
$CaSO_4$(s)	−1434.11	−1321.79	106.7
Cl_2(g)	0	0	223.07
Cu(s)	0	0	33.15
CuO(s)	−157.3	−129.7	42.63
F_2(g)	0	0	202.78
Fe(s)	0	0	27.28
Fe_2O_3(s)	−824.2	−742.2	87.40
H_2(g)	0	0	130.684
HCl(g)	−92.31	−95.30	186.91
HBr(g)	−36.40	−53.45	198.70
HF(g)	−271.1	−273.2	173.80
HI(g)	26.48	1.70	206.59
HNO_3(l)	−174.10	−80.71	155.60
H_2O(g)	−241.818	−228.572	188.825
H_2O(l)	−285.83	−237.129	69.91
H_2S(g)	−20.63	−33.56	205.79
H_2SO_4(l)	−813.99	−690.00	156.90
Hg(g)	61.317	31.820	174.96
Hg(l)	0	0	76.02
$HgCl_2$(s)	−224.3	−178.6	146.0
Hg_2Cl_2(s)	−265.22	−210.745	192.5
HgO(s, red)	−90.83	−58.539	70.29
I_2(g)	62.438	19.327	260.69
I_2(s)	0	0	116.135
KCl(s)	−436.75	−409.14	82.59
Mg(s)	0	0	32.68
MgO(s)	−601.70	−569.43	26.94
N_2(g)	0	0	191.61
NH_3(g)	−46.11	−16.45	192.45
NO(g)	90.25	86.55	210.76
NO_2(g)	33.18	51.31	240.06
N_2O(g)	82.05	104.20	219.85
N_2O_4(g)	9.16	97.89	304.29
NOCl(g)	51.71	66.08	261.69
NaCl(s)	−411.15	−384.14	72.13
$NaHSO_4$(s)	−1125.5	−992.8	113.0
NaI(s)	−287.78	−286.06	98.53
Na_2SO_4(s)	−1387.08	−1270.16	149.58
O_2(g)	0	0	205.138
O_3(g)	142.7	163.2	238.93
P_4(g)	58.91	24.44	279.98
PCl_3(g)	−287.0	−267.8	311.78
PCl_5(g)	−374.9	−305.0	364.58
PbS(s)	−100.4	−98.7	91.2
S(s, rhombic)	0	0	31.80
SO_2(g)	−296.83	−300.19	248.22
SO_3(g)	−395.72	−371.06	256.76
SiO_2(s, quartz)	−910.94	−856.64	41.84
Zn(g)	130.73	95.15	160.98
Zn(s)	0	0	41.63
ZnO(s)	−348.28	−318.30	43.64
ZnS(s, sphalerite)	−205.98	−201.29	57.7

[a]Data are based on a standard-state pressure of 1 bar, the standard pressure recommended by IUPAC (see Table 19.1). Because 1 atm = 1.01325 bar, calculations based on these data will not significantly differ from those based on a standard pressure of 1 atm, as shown by the following conversion equations:

$\Delta H_f(1\text{ atm}) = \Delta H_f(1\text{ bar})$ $S(1\text{ atm}) = S(1\text{ bar})$ (for solids and liquids)

$\Delta G_f(1\text{ atm}) = \Delta G_f(1\text{ bar}) + \Delta n \times 0.03263\text{ kJ/mol}$ $S(1\text{ atm}) = S(1\text{ bar}) - 0.1094\text{ J/mol·K}$ (for gases)

where Δn is the change in the number of moles of gas in the formation equation.

The standard free-energy change of a reaction can be calculated from free energies of formation according to the following rule: *The free-energy change of a reaction is equal to the sum of the free energies of formation of the products minus the sum of the free energies of formation of the reactants.* That is,

$$\Delta G^\circ = \sum[n_p(\Delta G_f^\circ)_p] - \sum[n_r(\Delta G_f^\circ)_r] \qquad (19.5)$$

where the symbols "p" and "r" refer to products and reactants, respectively, and n is the number of moles of each reactant or product. Note that this rule is analogous to the rule developed in Section 6.4 for calculating enthalpy changes; Equation 19.5 is analogous to Equation 6.10 (p. 245).

EXAMPLE 19.3

Very pure samples of certain metals can be prepared by reducing their purified oxides with hydrogen. Consult Table 19.2 to determine whether H_2 at 1 atm and 25°C would spontaneously reduce (a) CuO to Cu and (b) Al_2O_3 to Al.

SOLUTION

(a) The standard free energies of formation are −129.7 kJ/mol for CuO(s), −237.1 kJ/mol for H_2O(l), and zero for the elements. The copper oxide reduction is

$$\begin{array}{lcccc} & CuO(s) + & H_2(g) \longrightarrow & Cu(s) + & H_2O(l) \\ \Delta G_f^\circ \text{ (kJ/mol):} & -129.7 & 0 & 0 & -237.1 \end{array}$$

The value of ΔG° for the reaction is obtained by substituting the number of moles and the standard free energy of formation of each reactant and product into Equation 19.5; that is,

$$\begin{aligned}\Delta G^\circ &= 1 \text{ mol Cu} \times \Delta G_f^\circ(\text{Cu}) + 1 \text{ mol } H_2O \times \Delta G_f^\circ(H_2O) \\ &\qquad - 1 \text{ mol CuO} \times \Delta G_f^\circ(\text{CuO}) - 1 \text{ mol } H_2 \times \Delta G_f^\circ(H_2) \\ &= 1 \text{ mol} \times (0 \text{ kJ/mol}) + 1 \text{ mol} \times (-237.1 \text{ kJ/mol}) \\ &\qquad - 1 \text{ mol} \times (-129.7 \text{ kJ/mol}) - 1 \text{ mol} \times (0 \text{ kJ/mol}) \\ &= -107.4 \text{ kJ}\end{aligned}$$

LEARNING HINT

ΔG_f° and ΔH_f° are zero for the most stable form of an element in its standard state.

The reduction of copper oxide with hydrogen at 1 atm and 25°C is accompanied by a negative free-energy change, so it is spontaneous in the thermodynamic sense. Thus the reaction is possible, although we do not know whether it occurs fast enough to be practical.

(b) The standard free energy of formation of Al_2O_3(s) is −1582.3 kJ/mol. The reduction equation is

$$\begin{array}{lcccc} & Al_2O_3(s) + & 3H_2(g) \longrightarrow & 2Al(s) + & 3H_2O(l) \\ \Delta G_f^\circ \text{ (kJ/mol):} & -1582.3 & 0 & 0 & -237.1 \end{array}$$

The standard free-energy change for the reaction is

$$\begin{aligned}\Delta G^\circ &= 2 \text{ mol} \times (0 \text{ kJ/mol}) + 3 \text{ mol} \times (-237.1 \text{ kJ/mol}) \\ &\qquad - 1 \text{ mol} \times (-1582.3 \text{ kJ/mol}) - 3 \text{ mol} \times (0 \text{ kJ/mol}) \\ &= +871.0 \text{ kJ}\end{aligned}$$

The value of $\Delta G°$ is positive for the reduction of aluminum oxide, so this reaction will not occur under the stated conditions.

PRACTICE EXERCISE 19.4

(a) Refer to Table 19.2, and calculate $\Delta G°$ for the conversion of 1 mol of liquid water to water vapor at 25°C. (b) State what happens when liquid water is in contact with water vapor at 1 atm pressure and 25°C.

Tabulated free energies of formation are often used for calculating equilibrium constants, as illustrated in the next example. An equilibrium constant calculated from thermodynamic data is called a **thermodynamic equilibrium constant**.

EXAMPLE 19.4

Refer to Table 19.2, and calculate K_p for the reaction

$$N_2O_4(g) \rightleftharpoons 2NO_2(g)$$

at 25°C.

SOLUTION

We will first calculate $\Delta G°$, then we will use Equation 19.4 to calculate K_p. The free energies of formation of NO_2 and N_2O_4 are +51.31 kJ/mol and +97.89 kJ/mol, respectively. The standard free-energy change per mole of reaction is obtained from Equation 19.5:

$$\Delta G° = 2 \text{ mol } NO_2 \times \Delta G^\circ_f(NO_2) - 1 \text{ mol } N_2O_4 \times \Delta G^\circ_f(N_2O_4)$$
$$= 2 \text{ mol} \times (51.31 \text{ kJ/mol}) - 1 \text{ mol} \times (97.89 \text{ kJ/mol}) = 4.73 \text{ kJ}$$

Substituting $\Delta G° = 4.73$ kJ/mol $= 4730$ J/mol, $R = 8.314$ J/mol·K, and $T = 298$ K into Equation 19.4 gives

$$\Delta G° = -RT \ln K_p$$
$$4730 \text{ J/mol} = -8.314 \text{ J/mol·K} \times 298 \text{ K} \times \ln K_p$$

Hence,
$$\ln K_p = -1.91$$

and
$$K_p = \text{antilog}(-1.91) = e^{-1.91} = 0.15$$

LEARNING HINT

The free-energy change calculated in Example 19.4 is actually the free-energy change per "mole of reaction," as explained in Example 19.2. Hence, the answer can also be expressed as $\Delta G° = 4.73$ kJ/mol. You will find it convenient to use kilojoules per mole when working with Equations 19.3 and 19.4.

PRACTICE EXERCISE 19.5

Calculate $\Delta G°$ and K_p for the reaction

$$2NOCl(g) \rightleftharpoons 2NO(g) + Cl_2(g)$$

at 25°C.

Calculating ΔG for Nonstandard Conditions

The standard free-energy change ΔG° applies only to a mixture of reactants and products in which each gas is at 1 atmosphere partial pressure and each dissolved substance is 1 molar. The free-energy change at other pressures and concentrations can be calculated from Equation 19.3.

EXAMPLE 19.5

The value of ΔG° is $+4.73$ kJ/mol for the reaction

$$N_2O_4(g) \rightleftharpoons 2NO_2(g)$$

at 25°C (Example 19.4). (a) Calculate ΔG when the partial pressures of N_2O_4 and NO_2 are 0.700 atm and 0.300 atm, respectively. (b) Which direction, forward or reverse, is spontaneous under these conditions?

SOLUTION

(a) Before we can use Equation 19.3 for calculating ΔG, we must find Q_p by substituting the partial pressures into the reaction quotient expression:

$$Q_p = \frac{P^2_{NO_2}}{P_{N_2O_4}} = \frac{(0.300)^2}{0.700} = 0.129$$

Substituting $\Delta G^\circ = +4.73$ kJ/mol, $R = 8.314 \times 10^{-3}$ kJ/mol·K, $T = 298$ K, and $Q_p = 0.129$ into Equation 19.3 gives

$$\begin{aligned}\Delta G &= \Delta G^\circ + RT \ln Q_p \\ &= 4.73 \text{ kJ/mol} + 8.314 \times 10^{-3} \text{ kJ/mol·K} \times 298 \text{ K} \times \ln 0.129 \\ &= -0.34 \text{ kJ/mol}\end{aligned}$$

(b) The value of ΔG is negative, so the forward reaction is spontaneous under these conditions.

PRACTICE EXERCISE 19.6

Liquid water is in equilibrium with steam at 1.00 atm and 100°C; in other words, $\Delta G^\circ = 0$ for

$$H_2O(l) \rightleftharpoons H_2O(g)$$

at the normal boiling point. (a) Calculate ΔG for the vaporization of 1 mol of liquid water in contact with steam at 2.00 atm and 100°C. (b) Which process, condensation or vaporization, is thermodynamically favored when liquid water is in contact with pressurized steam at 100°C?

19.3 ENTROPY AND THE SECOND LAW OF THERMODYNAMICS

What do spontaneous changes have in common with each other? Let us examine a few changes that at first glance appear to be quite different from each other. Imagine

(a)

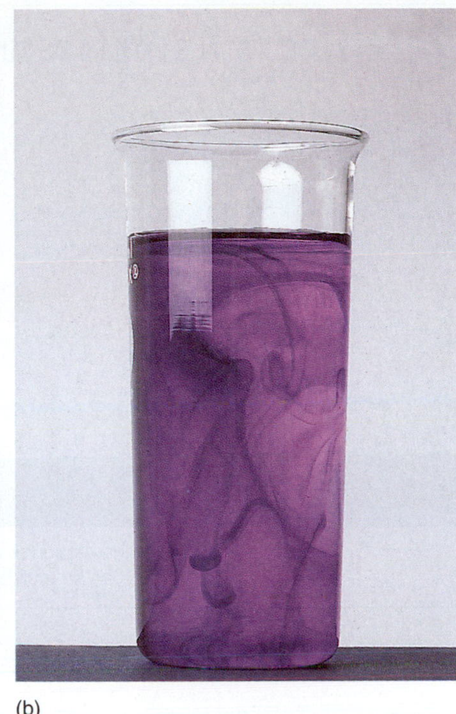
(b)

Figure 19.2
(a) Dye is released from a pipet at the bottom of a beaker of water. (b) The molecules of dye diffuse spontaneously through the water.

a small drop of dye added to a large volume of water (Figure 19.2). The molecules of dye will diffuse spontaneously through the water. At equilibrium, the concentration of dye will be the same everywhere in the solution, and the color will be uniform. Now imagine a second experiment in which a bar of hot metal is dropped into cold water as in Figure 19.3. Heat will flow spontaneously from the metal to the water—cooling the metal and warming the water—until at equilibrium both the metal and the water have the same temperature. Finally, consider two flasks, each containing a different gas, joined together as in Demonstration 19.1. When the stop-

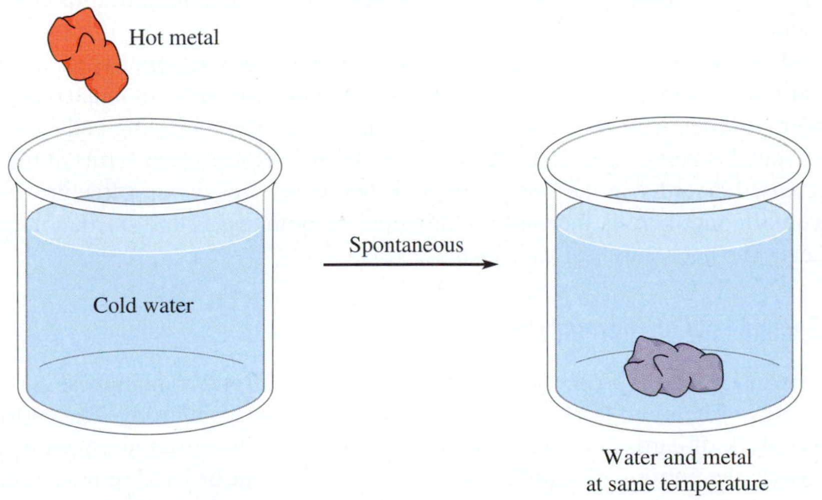

Figure 19.3
Heat flows from hot metal to cold water until both have the same temperature. This spontaneous process produces a more random distribution of energy.

DEMONSTRATION 19.1 MIXING IS A SPONTANEOUS PROCESS

Two flasks containing different gases.

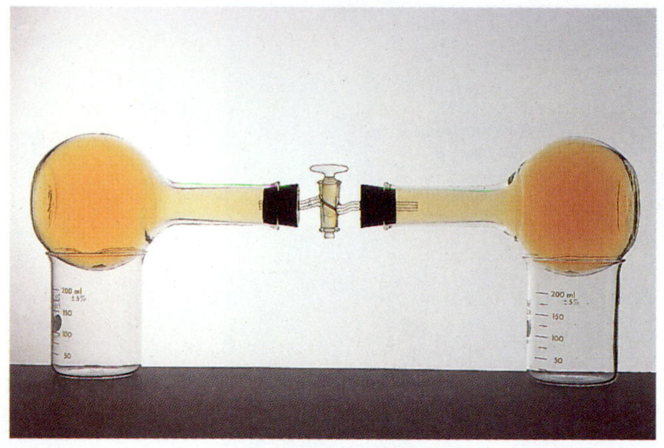

When the stopcock is opened, the gases begin to mix and continue to do so until both flasks have a uniform color.

cock between the flasks is opened, the gases will mix spontaneously by flowing from one flask into the other. At equilibrium, the gases will be uniformly distributed between the two flasks.

These three spontaneous processes are quite different, but they all have a common feature: *Each process results in a more random, that is, a more disordered, distribution of matter or energy*. The dye that was originally localized in one small drop diffused throughout the water; the heat energy that was originally localized in the metal bar dispersed throughout the bar–water mixture; and the gases that were originally localized in separate flasks dispersed into a larger volume. These examples illustrate that *in every spontaneous change, the system or the surroundings (or both) enter a more disordered state*. Similarly, disorder increases when an ordered deck of cards is thrown into the air, when combustion products from a fire scatter throughout the atmosphere, and when a hurricane leaves behind a trail of chaos and destruction.

The Latin poet Lucretius (99–55 B.C.) recognized the increasing disorder associated with spontaneous processes in his poem *de Rerum Natura*: "For we see that anything can be more speedily disintegrated than put together again. Hence, what the long day of time, the bygone eternity, has already shaken and loosened to fragments could never in the residue of time be reconstructed." (Lucretius, *The Nature of the Universe*, translated by Ronald E. Latham, Harmondsworth: Penguin Books, 1951.)

We must take *both* the system and its surroundings into account when we search for increased disorder. It sometimes appears that order can arise spontaneously from disorder, as when ice crystals form from liquid water or when living cells grow in a chaotic environment. This order, however, is deceptive; it always forms at the price of increased disorder in the surroundings. When water freezes or a tree grows, heat is given off, and there is increased motion and randomness in the surrounding environment. The net result is always increased disorder.

The Second Law of Thermodynamics

The term *entropy* is derived from the Greek *en trope*, meaning "in change."

The thermodynamic property that measures disorder and randomness is called **entropy** (S). The more disordered the system, the greater its entropy. The entropy of a mixed deck of cards is greater than that of a deck that has been arranged in a specific order; the entropy of liquid water is greater than that of ice. We have seen that spontaneous changes produce disorder, either in the system, in the surroundings, or in both. Because the system and its surroundings make up the universe, our observations about increasing disorder can be summarized as follows: *All spontaneous changes are accompanied by an increase in the entropy of the universe*. This statement is known as the **second law of thermodynamics**. Rudolf Clausius, the German physicist who introduced the concept of entropy in the mid-nineteenth century, summarized the first two laws of thermodynamics in their most compact form: "Die Energie der Welt is konstant; die Entropie der Welt strebt einem Maximum zu." (The energy of the universe is constant; the entropy of the universe tends towards a maximum.)

An entropy change (ΔS) is defined as the final entropy minus the initial entropy

$$\Delta S = S_{\text{final}} - S_{\text{initial}}$$

The entropy change in the universe is the sum of the entropy changes in the system and in the surroundings. Because the entropy of the universe always increases during a spontaneous process, this sum is always positive, that is,

$$\Delta S_{\text{univ}} = \Delta S_{\text{sys}} + \Delta S_{\text{surr}} > 0 \tag{19.6}$$

Remember that the only requirement of the second law of thermodynamics is an increase in the *total* entropy (ΔS_{univ}) during a spontaneous process. The entropy of the system alone can increase or decrease. The same is true for the entropy of the surroundings. If the entropy of a system decreases, as it does when water freezes, then the entropy of the surroundings must increase enough for the total entropy change to be positive.

The entropy of a changing system generally increases with physical changes such as expansion, warming, melting, evaporating, or mixing, all of which reduce orderliness and increase chaos (Figure 19.4). Conversely, compression, cooling, freezing, condensing, and separating are processes that decrease the entropy of the system. Entropy also increases during chemical changes that evolve gases or increase the number of moles of particles.

EXAMPLE 19.6

Does the entropy of the system increase or decrease during the following changes? (a) Sugar is stirred into a cup of coffee; (b) A student sorts the contents of a desk drawer; (c) $NH_4Cl(s) \longrightarrow NH_3(g) + HCl(g)$; (d) $NaCl(aq) + AgNO_3(aq) \longrightarrow NaNO_3(aq) + AgCl(s)$

SOLUTION

(a) A mixture of sugar and coffee is more disordered than the separated components; the entropy increases.

(b) An organized drawer has less disorder than a cluttered drawer; the entropy decreases.

(c) Two moles of gas are more disordered than 1 mol of solid; the entropy increases.

(d) Two moles of products, one of which is a solid, have less disorder than 2 mol of mixed reactants in solution; the entropy decreases.

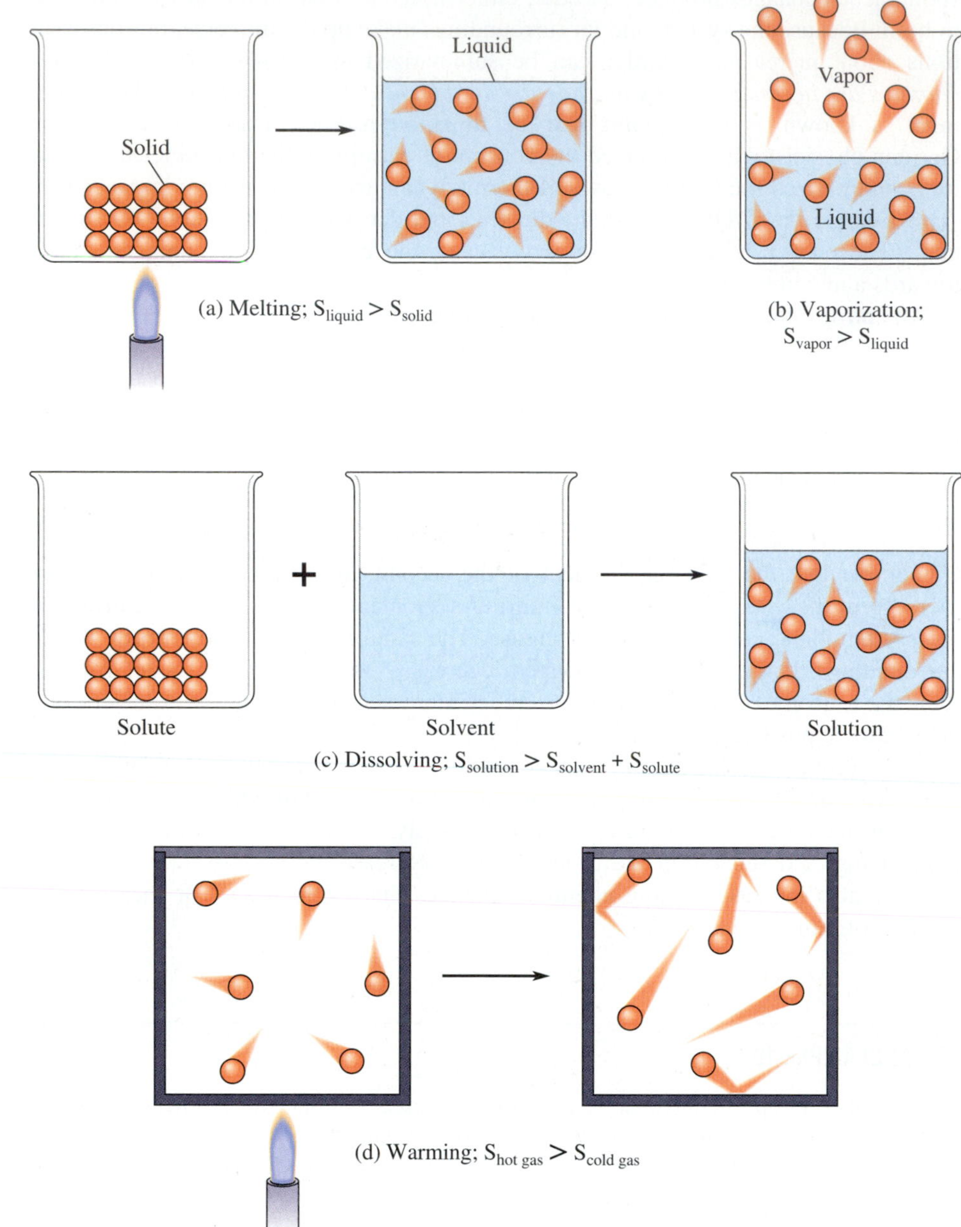

Figure 19.4

Some changes that increase the entropy of the system. (a) Melting, (b) vaporization, (c) mixing, and (d) warming.

PRACTICE EXERCISE 19.7

Does the entropy of the system increase or decrease during the following processes? (a) A TV dinner is defrosted, (b) $2NO_2(g) \rightarrow N_2O_4(g)$, (c) $H_2(g) + Br_2(l) \rightarrow 2HBr(g)$

19.4 ENTROPY CALCULATIONS

At constant temperature and pressure, the free-energy change (ΔG), the enthalpy change (ΔH), and the entropy change (ΔS) of a reacting system are related by the equation

$$\Delta G = \Delta H - T\Delta S \qquad (19.7)$$

where T is the Kelvin temperature. When the reacting system is at equilibrium, $\Delta G = 0$, and Equation 19.7 can be rearranged to

$$\Delta S = \frac{\Delta H}{T} \qquad (19.8)$$

Entropy, like enthalpy and free energy, is a state function. A derivation of Equation 19.7 is given in the *Digging Deeper* section entitled "The Relation Between ΔG of a System and ΔS of the Universe."

Equation 19.8 is a special case of the more general equation

$$\Delta S = \frac{q_{rev}}{T} \qquad (19.9)$$

where q_{rev} is the amount of heat absorbed by the system when the reaction is carried out under **reversible conditions**, that is, under conditions where the driving force of the forward reaction is only infinitesimally greater than the driving force of the reverse reaction. These conditions are said to be "reversible" because the reaction is virtually at equilibrium and its direction can easily be reversed by slight changes in the reactant or product concentrations.

Equations 19.7 and 19.8 are remarkable in that they allow us to calculate the entropy change—the change in disorder—for a reacting system, as illustrated in the following examples.

EXAMPLE 19.7

Calculate the entropy change when 500 g of ice melts at 0°C and 1 atm. The heat of fusion of ice under these conditions is 6.01 kJ/mol (Table 12.3, p. 538).

SOLUTION

The melting of ice at 0°C and 1 atm is an equilibrium process, therefore Equation 19.8 applies. The change in enthalpy is

$$\Delta H^\circ = 500\text{ g} \times \frac{1\text{ mol}}{18.02\text{ g}} \times \frac{6.01\text{ kJ}}{1\text{ mol}} = 167\text{ kJ}$$

The superscript on the ΔH is used because the melting takes place at 1 atm pressure (standard-state conditions). Substituting $\Delta H^\circ = 167$ kJ and $T = 273$ K into Equation 19.8 gives

$$\Delta S^\circ = \frac{\Delta H^\circ}{T} = \frac{167\text{ kJ}}{273\text{ K}} = 0.612\text{ kJ/K} = 612\text{ J/K}$$

Note that the units for entropy and entropy change are energy per kelvin, J/K.

CHEMICAL INSIGHT

ENTROPY AND PROBABILITY

Why do spontaneous changes create disorder? Let us begin to answer this question by considering the possible results of flipping three coins (Figure 19.5). There is one ordered arrangement with all heads, one with all tails, and six disordered arrangements with mixed heads and tails. The probability of obtaining an ordered arrangement of heads or tails is only two out of eight, so a disordered arrangement is much more likely to occur. Random arrangements for ten coins are even more numerous than for three coins, but there is still only one way to get each orderly arrangement of 10 heads or 10 tails. The larger the number of coins, the more certain we can be that flipping them will result in a disordered arrangement. The disordered (high-entropy) outcome of coin flipping is simply a consequence of the fact that there are more possibilities for disorder than for order.

These concepts can be extended to molecular systems. In a container holding three nitrogen molecules and three oxygen molecules, there is a small but finite probability that all of the nitrogen molecules will be in one-half of the container and all of the oxygen molecules in the other half (Figures 19.6a and 19.6d). There are many more arrangements in which one or two oxygen molecules are exchanged with nitrogen molecules (Figures 19.6b and 19.6c). If more molecules are added to the container, the number of possible arrangements rises astronomically, and the probability of a sorted arrangement becomes correspondingly minute. The mixing of two gases is spontaneous because the disordered arrangements of a mixture are more numerous and therefore more likely to occur than the ordered arrangements in which the gases are separated. In a system that contains measurable amounts of two gases, even as little as 1 picomole (6×10^{11} molecules) of each, the probability of spontaneous ordering is so remote as to be virtually impossible. Once gases are mixed, they inevitably stay that way.

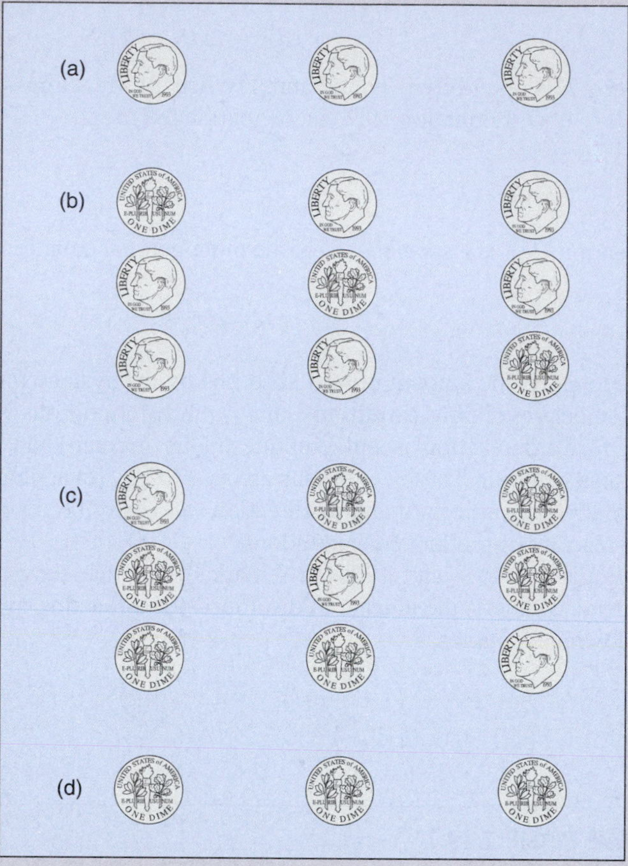

Figure 19.5

The eight possible results of flipping three coins. Each result is equally probable. (a) The only way to get three heads. (b) The three ways to get two heads and one tail. (c) The three ways to get one head and two tails. (d) The only way to get three tails. Six of the eight ways give mixtures of heads and tails; therefore, a mixed arrangement (higher entropy) is three times more probable than a uniform arrangement (lower entropy).

PRACTICE EXERCISE 19.8

Calculate the entropy change when 1 mol of liquid water at 100°C and 1 atm is converted to steam at 100°C and 1 atm. The heat of vaporization of water under these conditions is 40.7 kJ/mol.

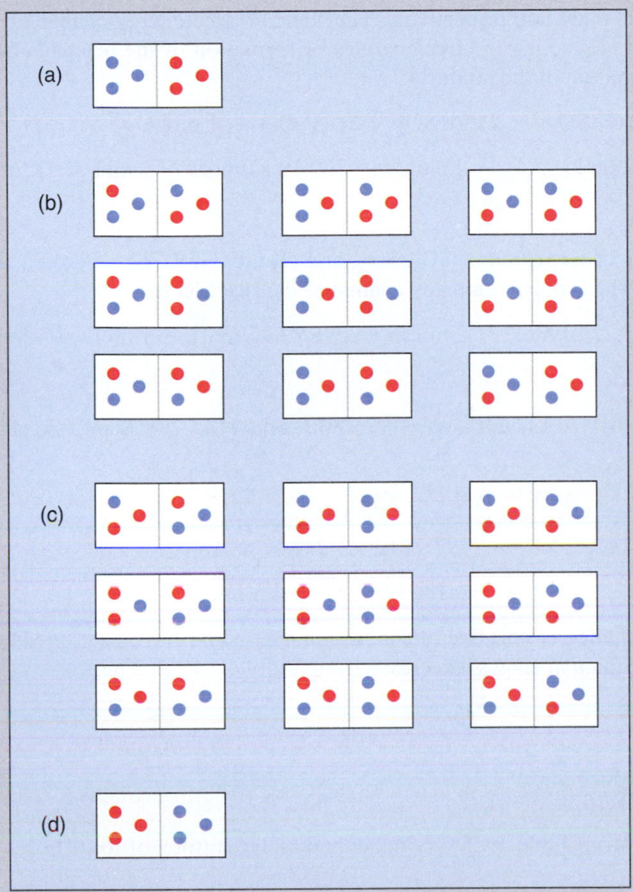

Figure 19.6

Three nitrogen (blue) and three oxygen (red) molecules in a box. (a) The three nitrogen molecules are on one side of the box and the three oxygen molecules are on the other side, an improbable arrangement. (b) The nine ways in which one nitrogen molecule could exchange places with one oxygen molecule to give a more probable arrangement. (c) The nine ways in which two nitrogen molecules could exchange places with two oxygen molecules. The arrangements in (b) and (c) are equally probable. (d) The one way that three nitrogen molecules could exchange places with three oxygen molecules. The arrangements in (a) and (d) are equally probable.

Mixing is not the only cause of disorder. Pure substances exhibit disorder because of the numerous ways that the energy of the system can be divided among its individual molecules. A disordered energy distribution in which the molecules have different amounts of energy is more likely to occur than a distribution in which all molecules have the same energy.

Disorder also resides in the way each individual molecule allots its kinetic energy between rotational, vibrational, and translational motion. Because complex molecules have more ways of stretching, bending, and twisting than simple molecules, disorder tends to increase with increasing molecular complexity.

In 1896 Ludwig Boltzmann showed that the entropy (S) of a system is given by

$$S = k \ln W$$

where $\ln W$ is the natural logarithm of the number of possible arrangements of the particles composing the system and k is a proportionality constant. Boltzmann made many contributions to physics and chemistry (see the Maxwell–Boltzmann distribution curves on p. 207), but the above equation relating entropy and probability was one of his greatest. The equation is inscribed on his tombstone in Vienna.

EXAMPLE 19.8

Use Equation 19.7 and data from Table 19.2 to calculate $\Delta S°$ at 25°C for the reaction

$$2SO_2(g) + O_2(g) \rightarrow 2SO_3(g)$$

SOLUTION

First we will use data from Table 19.2 to calculate ΔG° and ΔH°. Then we will use Equation 19.7 to calculate ΔS°. (The reaction is not at equilibrium, so Equation 19.8 does not apply.) The standard free energies of formation of SO_3, SO_2, and O_2 are -371.06, -300.19, and 0 kJ/mol, respectively. The standard free-energy change of the reaction is obtained by subtracting the free energies of formation of the reactants from the free energies of formation of the products:

$$\begin{aligned}\Delta G^\circ &= 2 \text{ mol } SO_3 \times \Delta G^\circ_f(SO_3) - 2 \text{ mol } SO_2 \times \Delta G^\circ_f(SO_2) - 1 \text{ mol } O_2 \times \Delta G^\circ_f(O_2)\\ &= 2 \text{ mol} \times (-371.06 \text{ kJ/mol}) - 2 \text{ mol} \times (-300.19 \text{ kJ/mol}) - 1 \text{ mol} \times 0 \text{ kJ/mol}\\ &= -141.74 \text{ kJ}\end{aligned}$$

The standard enthalpies of formation of SO_3, SO_2, and O_2 are -395.72, -296.83, and 0 kJ/mol, respectively. The standard enthalpy change of the reaction is

$$\begin{aligned}\Delta H^\circ &= 2 \text{ mol} \times (-395.72 \text{ kJ/mol}) - 2 \text{ mol} \times (-296.83 \text{ kJ/mol}) - 1 \text{ mol} \times 0 \text{ kJ/mol}\\ &= -197.78 \text{ kJ}\end{aligned}$$

Substituting $\Delta G^\circ = -141.74$ kJ, $\Delta H^\circ = -197.78$ kJ, and $T = 298$ K into Equation 19.7 gives

$$\Delta G^\circ = \Delta H^\circ - T\Delta S^\circ$$

$$-141.74 \text{ kJ} = -197.78 \text{ kJ} - 298 \text{ K} \times \Delta S^\circ$$

and

$$\Delta S^\circ = -0.188 \text{ kJ/K} = -188 \text{ J/K}$$

Note that the entropy change is negative, a conclusion that could have been predicted from the chemical equation that shows a decrease from 3 mol of gas to 2 mol.

PRACTICE EXERCISE 19.9

Use ΔH°_f and ΔG°_f data from Table 19.2 to calculate the molar entropy of formation, ΔS°_f, of NH_3 at 25°C.

Absolute Entropies

Every substance has a certain degree of disorder that arises from the motion of its particles. Even a "perfect" crystal (one with no defects) would be somewhat disordered because of molecular motion. The amount of disorder in such a crystal would decrease when the temperature is lowered and should become very small as the temperature approaches 0 K. This conclusion is supported by low-temperature experiments that show that the molar entropies of pure crystalline substances decrease toward a common value as the temperature falls. The experiments do not provide enough information to determine the value of the common entropy, but it is reasonable to assign it a value of zero. One version of the **third law of thermodynamics** states that *the entropy of a pure, perfectly crystalline element or compound may be taken as zero at 0 K.*

Low-temperature research has produced temperatures in the vicinity of 2.5×10^{-10} K. Analysis of these experiments leads to the interesting conclusion that absolute zero can be approached, but never attained. This conclusion is summarized in another version of the third law of thermodynamics: *It is impossible to reduce the temperature of any system to absolute zero in a finite number of steps.*

Assigning the value of zero to entropies at 0 K provides a starting point for calculating the entropies of pure substances at other temperatures. Imagine a pure substance, for example, N_2, being warmed from 0 K and 1 atm to 298 K and 1 atm. The crystalline nitrogen absorbs heat, melts, and eventually vaporizes as the temperature rises:

$$N_2(s, 0\text{ K}, 1\text{ atm}) \rightarrow N_2(g, 298\text{ K}, 1\text{ atm})$$

The change in entropy for this process is

$$\Delta S = S_{298\text{ K}} - S_{0\text{ K}} = S_{298\text{ K}}$$

since $S_{0\text{ K}}$ (the entropy at 0 K) is zero for a pure, perfectly crystalline sample of N_2. In other words, the measured value of ΔS is equal to the entropy of nitrogen at 298 K. (The actual determination of ΔS involves adding the entropy changes for each of the transitions that occur when nitrogen is heated from 0 K to 298 K. The details of this procedure are described in any physical chemistry textbook.) Entropies determined in this way are called **absolute entropies** or **third-law entropies**.

The third-law entropy of 1 mol of substance at 1 atm pressure is called a standard molar entropy and is symbolized by $S°$. For N_2, the **standard molar entropy** at 25°C is $S° = +191.6$ J/mol·K. Table 19.2 lists some additional standard molar entropies at 25°C.

The **standard entropy change** ($\Delta S°$) of a reaction can be found by subtracting the entropies of the reactants from the entropies of the products, as illustrated in the next example.

EXAMPLE 19.9

Use entropy data from Table 19.2 to calculate the standard entropy change for the combustion of 1 mol of acetylene (C_2H_2) at 25°C.

SOLUTION

The equation and $S°$ values from the table are

$$\begin{array}{lcccc} & C_2H_2(g) \; + & \tfrac{5}{2}O_2(g) \; \rightarrow & 2CO_2(g) \; + & H_2O(l) \\ S°\ (\text{J/mol·K}): & 200.94 & 205.14 & 213.74 & 69.91 \end{array}$$

The standard entropy change for the reaction is the sum of the entropies for the products minus the sum for the reactants, that is,

$$\begin{aligned} \Delta S° &= 2\text{ mol } CO_2 \times S°(CO_2) + 1\text{ mol } H_2O \times S°(H_2O) \\ &\qquad - 1\text{ mol } C_2H_2 \times S°(C_2H_2) - \tfrac{5}{2}\text{mol } O_2 \times S°(O_2) \\ &= 2\text{ mol} \times (213.74\text{ J/mol·K}) + 1\text{ mol} \times (69.91\text{ J/mol·K}) \\ &\qquad - 1\text{ mol} \times (200.94\text{ J/mol·K}) - \tfrac{5}{2}\text{mol} \times (205.14\text{ J/mol·K}) \\ &= -216.40\text{ J/K} \end{aligned}$$

An alternative way of obtaining the standard entropy change would be to calculate $\Delta G°$ and $\Delta H°$ from the data in Table 19.2 and then use Equation 19.7 to calculate $\Delta S°$, as we did in Example 19.8. (Try it!) Note that $\Delta S°$ is negative, a result that is consistent with the conversion of 3.5 mol of gas into 2 mol of gas and 1 mol of liquid.

PRACTICE EXERCISE 19.10

Calculate the standard entropy change for the conversion of 1 mol of diamond to graphite at 25°C. Which lattice is more ordered, diamond or graphite?

DIGGING DEEPER

The Relation Between ΔG of a System and ΔS of the Universe

The ultimate criterion for any spontaneous change is embodied in the second law of thermodynamics—the entropy of the universe must increase. There is another criterion for a spontaneous change at constant temperature and pressure—the free energy of the system must decrease. Thus the quantities ΔG_{sys} and ΔS_{univ} must be related in such a way that one will be negative when the other is positive. To find the relation between ΔG_{sys} and ΔS_{univ}, we will first derive Equation 19.7, the equation relating ΔG of the system to ΔS of the system.

The energy change of a reacting system is given by Equation 6.1:

$$\Delta E = q + w$$

where q is the heat absorbed by the system and w is the work done on the system. Imagine that the reaction is carried out at constant temperature and pressure and in such a way that maximum work is obtained from the system. The latter restriction implies that the reaction is carried out under reversible conditions, that is, under conditions such that the driving forces of the forward and reverse reactions differ by only an infinitesimal amount. (Two tennis players, for example, will do maximum work when they are evenly matched.)

The total work done *by* the reacting system is equal to the sum of the expansion work ($P\Delta V$) and the maximum useful work ($-\Delta G$). The work done *on* the system, w, is the negative of this sum:

$$w = \Delta G - P\Delta V$$

Because the reaction is proceeding under reversible conditions, the heat absorbed by the reacting system is given by rearranging Equation 19.9:

$$q = q_{rev} = T\Delta S$$

Substituting these expressions for q and w into the expression for ΔE gives

$$\Delta E = q + w$$

$$\Delta E = T\Delta S + \Delta G - P\Delta V$$

Rearrangement gives

$$\Delta G = \Delta E + P\Delta V - T\Delta S$$

Because the pressure is constant, $\Delta E + P\Delta V = \Delta H$ (Equation 6.8). Substituting this expression for ΔH into the above equation for ΔG gives Equation 19.7:

$$\Delta G = \Delta H - T\Delta S \qquad (19.7)$$

Let us now find the relation between ΔG_{sys} and ΔS_{univ}. We begin by rewriting Equation 19.7 to emphasize that it applies to the system:

$$\Delta G_{sys} = \Delta H_{sys} - T\Delta S_{sys} \tag{19.7a}$$

This expression for ΔG contains ΔS_{sys}. Can we show that it contains ΔS_{univ}?

Since the surroundings absorb heat when the system evolves heat, and vice versa, the heat absorbed by the surroundings (q_{surr}) is equal in magnitude but opposite in sign to the heat absorbed by the system (q_{sys}):

$$q_{surr} = -q_{sys}$$

The system is at constant temperature and pressure, so

$$q_{sys} = \Delta H_{sys}$$

Hence,

$$q_{surr} = -\Delta H_{sys}$$

The change is carried out under reversible conditions, so Equation 19.9 applies and

$$\Delta S_{surr} = \frac{q_{surr}}{T} = \frac{-\Delta H_{sys}}{T}$$

Rearrangement gives

$$\Delta H_{sys} = -T\Delta S_{surr}$$

and substitution into Equation 19.7a gives

$$\Delta G_{sys} = -T\Delta S_{surr} - T\Delta S_{sys} = -T(\Delta S_{surr} + \Delta S_{sys})$$

But $\Delta S_{surr} + \Delta S_{sys} = \Delta S_{univ}$, therefore

$$\Delta G_{sys} = -T\Delta S_{univ} \tag{19.10}$$

Equation 19.10 relates the free-energy change of the reacting system to the entropy change of the universe. It shows that the two requirements for a spontaneous reaction, negative ΔG_{sys} and positive ΔS_{univ}, are essentially equivalent for changes occurring at constant temperature and pressure.

19.5 THE VARIATION OF $\Delta G°$ WITH TEMPERATURE

The enthalpy and entropy changes of a reaction vary only slightly with temperature. For example, the standard enthalpy of formation of gaseous water is −241.8 kJ/mol at 25°C and −242.6 kJ/mol at 100°C, a difference of less than 0.5%. The values of ΔG and $\Delta G°$, however, depend strongly on temperature (Equations 19.3 and 19.4). Because enthalpy and entropy changes are approximately constant, we can write Equation 19.7 in terms of standard changes

$$\Delta G° = \Delta H° - T\Delta S° \tag{19.7}$$

and use it to estimate the value of $\Delta G°$ at various temperatures.

CHEMICAL INSIGHT

TIME'S ARROW AND COUPLED REACTIONS

The second law of thermodynamics states that all spontaneous reactions are accompanied by an increase in the disorder of the universe. Thus useful energy—energy that could be converted into work—is gradually being dissipated into unavailable energy in the form of random molecular motion. Every spontaneous process is a movement of the universe toward its equilibrium state which, according to thermodynamic theory, would be a state of maximum disorder at some uniform temperature. This rather bleak picture has been aptly referred to as the "heat death of the universe." The astronomer and physicist Sir Arthur Eddington suggested in 1927 that the dissipation of energy and increase of disorder is the physical basis of our concept of time. The forward passage of time is the direction of increasing entropy: "Entropy is time's arrow."

We cannot stop the universe from rolling toward equilibrium, but we can arrange for its descent to follow pathways that are helpful to us. A river, for example, flows downhill, losing potential energy and generating disorder in the form of random kinetic energy and heat. We cannot prevent its ultimate descent, but we can divert the water so that it raises the level of a reservoir, thus saving some of the potential energy for use at a later time. In other words, we can *couple* two systems such as a river and a reservoir, using the decreasing potential energy of one to raise the potential energy of the other.

Chemical reactions can also be coupled. The decreasing free energy of a spontaneous change can be stored in substances whose formation would not ordinarily be spontaneous. Living organisms have a remarkable ability to couple reactions in order to make the most of the free energy that passes through. The principal biological free-energy reservoir is the ATP–ADP system (Figure 19.7). When adenosine diphosphate (ADP) molecules add phosphate units to produce adenosine triphosphate (ATP) molecules, they store 31 kJ/mol of free energy that can be released when the reaction is reversed:

$$ADP^{3-}(aq) + HPO_4^{2-}(aq) + H^+(aq) \rightarrow ATP^{4-}(aq) + H_2O(l) \qquad \Delta G^\circ = +31 \text{ kJ}$$

In a living cell, this free energy is supplied by the oxidation of glucose to CO_2 and H_2O:

$$C_6H_{12}O_6(s) + 6O_2(g) \rightarrow 6CO_2(g) + 6H_2O(l) \qquad \Delta G^\circ = -2870 \text{ kJ}$$

The oxidation of 1 mol of glucose releases 2870 kJ of free energy. Some of this energy is used to convert 36 mol of ADP to ATP:

$$36ADP^{3-}(aq) + 36HPO_4^{2-}(aq) + 36H^+(aq) \rightarrow 36ATP^{4-}(aq) + 36H_2O(l)$$

$$\Delta G^\circ = 36 \times 31 \text{ kJ} = +1100 \text{ kJ}$$

Hence, 1100 kJ, or 38%, of the energy released by 1 mol of glucose is stored in ATP. The ATP later releases this energy as needed. Some of the energy is used for muscle action, and some is used to synthesize highly organized biomolecules and structural elements. In this way, living organisms are able to exist as temporary islands of low entropy in an increasingly disordered environment. Eventually, the stored energy is released, and the universe obtains its full entropy increase.

EXAMPLE 19.10

Consider the conversion of ozone to oxygen:

$$2O_3(g) \rightleftharpoons 3O_2(g)$$

Use data from Table 19.2 to estimate ΔG° at 150°C.

SOLUTION

First we calculate the standard enthalpy and entropy changes at 25°C. The standard molar enthalpies of formation of oxygen and ozone are 0 kJ and +142.7 kJ, respectively. Hence,

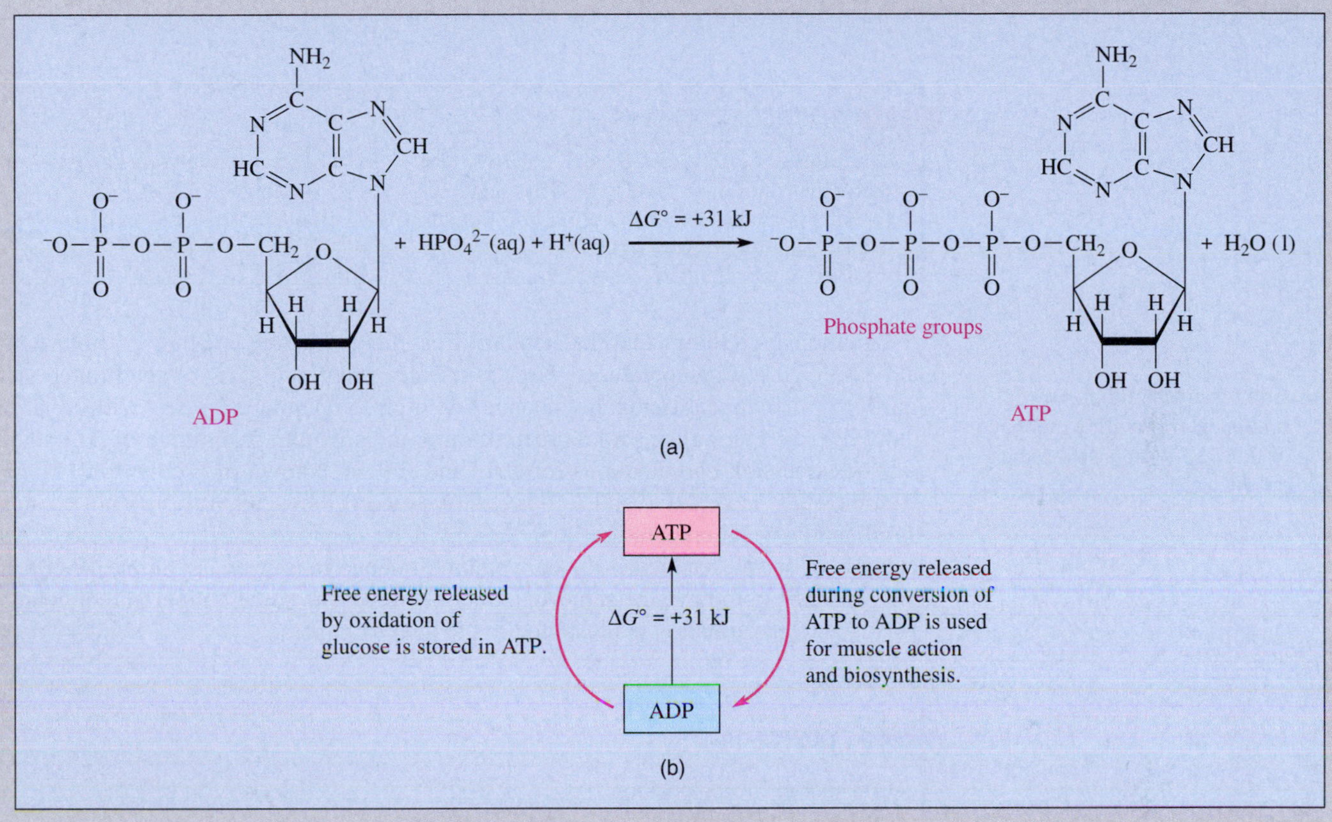

Figure 19.7

(a) Free energy is stored when an ADP molecule gains a phosphate group to form ATP. (b) Free-energy diagram for the interconversion of ADP and ATP. The oxidation of glucose provides energy to drive this reaction in the nonspontaneous direction, ADP $\longrightarrow$ ATP. The spontaneous reaction ATP $\longrightarrow$ ADP then releases the free energy as needed for biosynthesis and other cell activities.

$$\Delta H° = 3 \text{ mol } O_2 \times \Delta H_f°(O_2) - 2 \text{ mol } O_3 \times \Delta H_f°(O_3)$$

$$= 3 \text{ mol} \times (0 \text{ kJ/mol}) - 2 \text{ mol} \times (142.7 \text{ kJ/mol}) = -285.4 \text{ kJ}$$

The standard entropies of oxygen and ozone are 205.14 J/mol·K and 238.93 J/mol·K, respectively. Hence,

$$\Delta S° = 3 \text{ mol } O_2 \times S°(O_2) - 2 \text{ mol } O_3 \times S°(O_3)$$

$$= 3 \text{ mol} \times (205.14 \text{ J/mol·K}) - 2 \text{ mol} \times (238.93 \text{ J/mol·K})$$

$$= 137.56 \text{ J/K} = 0.13756 \text{ kJ/K}$$

We will assume that the values of $\Delta H°$ and $\Delta S°$ do not change significantly between 25°C and 150°C (423 K). Substituting these values and $T = 423$ K into Equation 19.7 gives

$$\Delta G^\circ = \Delta H^\circ - T\Delta S^\circ$$
$$= -285.4 \text{ kJ} - 423 \text{ K} \times 0.13756 \text{ kJ/K} = -343.6 \text{ kJ}$$

The standard free-energy change is approximately −343.6 kJ per mole of reaction at 150°C.

PRACTICE EXERCISE 19.11

The values of ΔH° and ΔG° are −2816 kJ and −2870 kJ, respectively, for the oxidation of 1 mol of glucose to $H_2O(l)$ and $CO_2(g)$ at 25°C. Estimate ΔG° at 37°C (normal body temperature). (*Hint*: Use Equation 19.7 to find ΔS° at 25°C, and then use it again to find ΔG° at 37°C.)

A reaction in which the forward direction is determined by the magnitude of the $T\Delta S$ term is said to be *entropy-driven.*

Equation 19.7 shows that the standard free-energy change depends on both ΔH° and $T\Delta S^\circ$. At low temperatures, $T\Delta S^\circ$ is usually small, and ΔG° is determined primarily by the standard enthalpy change. At high temperatures, $T\Delta S^\circ$ will be large and will play a significant role in determining the sign and magnitude of ΔG°. The four possible sign combinations for ΔH° and ΔS° are shown in Figure 19.8. If, for example, ΔH° is negative and ΔS° is positive, and if these quantities do not change significantly with temperature, then ΔG° will be negative at all temperatures. An example of such a reaction is the conversion of ozone to oxygen (Example 19.10). If ΔH° and ΔS° both have the same sign, then there will be some temperature at which $\Delta G^\circ = 0$. This situation is illustrated in the next example.

EXAMPLE 19.11

Consider the decomposition

$$2HgO(s) \longrightarrow 2Hg(g) + O_2(g)$$

for which $\Delta H^\circ = +304.3$ kJ and $\Delta S^\circ = +0.414$ kJ/K at 25°C. Estimate the temperature at which $\Delta G^\circ = 0$.

SOLUTION We will assume that ΔH° and ΔS° do not change with temperature. Substituting their values and $\Delta G^\circ = 0$ into Equation 19.7 gives

$$\Delta G^\circ = \Delta H^\circ - T\Delta S^\circ$$
$$0 \text{ kJ} = +304.3 \text{ kJ} - T \times 0.414 \text{ kJ/K}$$

and
$$T = 735 \text{ K} = 462^\circ\text{C}$$

The calculation in Example 19.11 shows that the standard free-energy change (ΔG°) of the HgO decomposition will be positive at temperatures below 462°C and negative at temperatures above 462°C. In other words, if each reactant and product is in its standard state, spontaneous decomposition will occur when the temperature is greater than 462°C. When the substances are not in their standard states, as is usually the case, then ΔG must be used for determining the temperature at which the reaction becomes spontaneous.

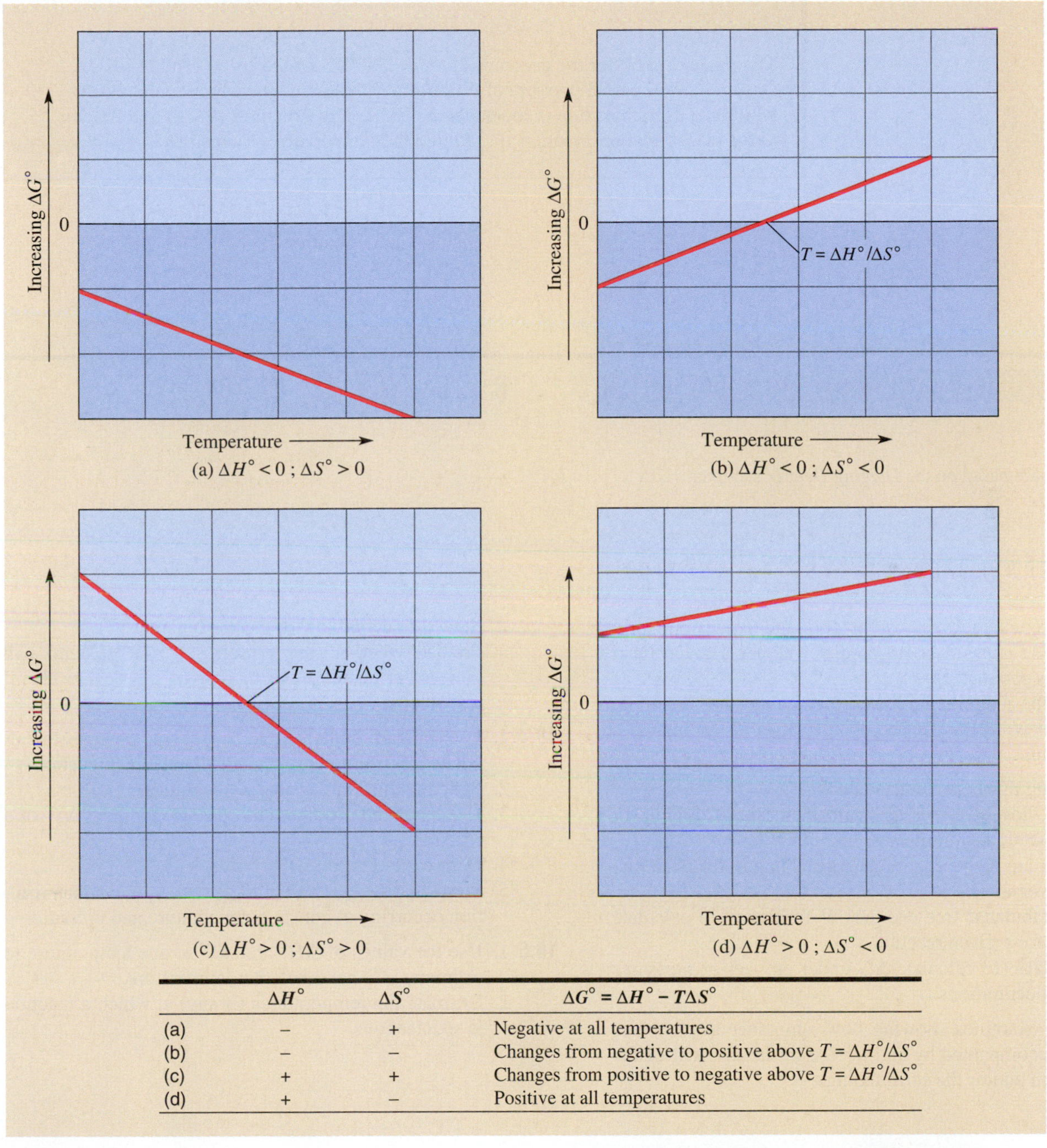

	$\Delta H°$	$\Delta S°$	$\Delta G° = \Delta H° - T\Delta S°$
(a)	–	+	Negative at all temperatures
(b)	–	–	Changes from negative to positive above $T = \Delta H°/\Delta S°$
(c)	+	+	Changes from positive to negative above $T = \Delta H°/\Delta S°$
(d)	+	–	Positive at all temperatures

Figure 19.8

Effect of temperature on $\Delta G°$. The direction of the change depends on the signs of $\Delta H°$ and $\Delta S°$, as summarized in the table found beneath the graphs.

PRACTICE EXERCISE 19.12

Calcium oxide (lime) is prepared by heating shells, limestone, and other forms of $CaCO_3$ in a lime kiln (see Figure S1.11, p. 354):

$$CaCO_3(s) \rightarrow CaO(s) + CO_2(g)$$

The value of $\Delta H°$ for the decomposition is $+178.3$ kJ; $\Delta S°$ is $+0.1606$ kJ/K. Assume a CO_2 partial pressure of 0.10 atm, and estimate the temperature above which the decomposition is spontaneous. (*Hint:* Use Equation 19.7 to find $\Delta G°$ in terms of T, then use Equation 19.3 to find the temperature at which $\Delta G = 0$.)

CHAPTER REVIEW

Starred entries are based on the *Digging Deeper* sections.

LEARNING OBJECTIVES BY SECTION

19.1 1. Distinguish between useful work and expansion work.
2. Use the sign of ΔG to predict whether a given reaction is spontaneous.
3. Use the magnitude of ΔG to calculate the maximum useful work that can be obtained from a spontaneous reaction.

19.2 1. Distinguish between ΔG and $\Delta G°$.
2. State how ΔG varies with increasing reactant or product concentrations.
3. Calculate standard free-energy changes from K, and vice versa.
4. Use tabulated free energies of formation to calculate $\Delta G°$ for a given reaction.
5. Use $\Delta G°$ to calculate ΔG under any set of pressures or concentrations.

19.3 1. Give examples showing how spontaneous processes are accompanied by an increase in the disorder of the system and/or the surroundings.
2. State the second law of thermodynamics.
3. Predict whether the entropy of the system will increase or decrease during a given physical or chemical change.

19.4 1. Use standard enthalpy and free-energy changes to calculate $\Delta S°$ for a phase change or chemical reaction.
2. State the third law of thermodynamics.
3. Use tabulated absolute entropies to calculate $\Delta S°$ of a reaction.
*4. Derive the relation $\Delta G = \Delta H - T\Delta S$.
*5. Show that $\Delta G_{sys} = -T\Delta S_{univ}$ for a spontaneous reaction occurring at constant temperature and pressure.

19.5 1. Use the values of $\Delta H°$ and $\Delta S°$ at one temperature to estimate $\Delta G°$ at some other temperature.
2. Estimate the temperatures (if any) at which a reaction is spontaneous.

KEY TERMS BY SECTION

19.1 Free energy (G)
Useful work
Thermodynamic equilibrium constant

19.2 Standard free-energy change ($\Delta G°$)
Standard free energy of formation ($\Delta G_f°$)

19.3 Entropy
Second law of thermodynamics

19.4 Absolute entropy
Reversible conditions
Standard entropy change ($\Delta S°$)
Standard molar entropy ($S°$)
Third-law entropy
Third law of thermodynamics

IMPORTANT EQUATIONS

19.1 $\Delta G = G(\text{products}) - G(\text{reactants})$

19.2 $\Delta G = -\text{maximum useful work}$

19.3 $\Delta G = \Delta G^\circ + RT \ln Q$

19.4 $\Delta G^\circ = -RT \ln K$

19.5 $\Delta G^\circ = \sum[n_p(\Delta G_f^\circ)_p] - \sum[n_r(\Delta G_f^\circ)_r]$

19.6 $\Delta S_{univ} = \Delta S_{sys} + \Delta S_{surr} > 0$

19.7 $\Delta G = \Delta H - T\Delta S$

19.8 $\Delta S = \dfrac{\Delta H}{T}$

19.9 $\Delta S = \dfrac{q_{rev}}{T}$

***19.10** $\Delta G_{sys} = -T\Delta S_{univ}$

FINAL EXERCISES

Answers to exercises with blue numbers are given in Appendix D. Starred exercises are based on the *Digging Deeper* sections.

PART A. QUESTIONS AND PROBLEMS BY SECTION

Free Energy and Useful Work (Section 19.1)

19.1 Distinguish between *pressure–volume work* and *useful work*. Give some examples of useful work done by chemical reactions.

19.2 Explain what is meant by *maximum useful work*. Is the work obtained from a chemical reaction usually the maximum work? Why or why not?

19.3 What is the relationship between the useful work done during a reaction and the quantity of heat evolved?

19.4 Can useful work ever be obtained from an endothermic reaction? Explain your answer.

19.5 The following reactions occur at constant temperature and pressure. Give the sign of ΔG for each of them.
(a) a reaction at equilibrium
(b) a spontaneous reaction in the forward direction
(c) a spontaneous reaction in the reverse direction

19.6 Give the sign of the free-energy change for each of the following processes:
(a) Ice melts at 0°C and 1 atm.
(b) Ice melts at 10°C and 1 atm.
(c) Hydrogen and iodine gases react to form hydrogen iodide.
(d) Nitrogen and hydrogen gases are in equilibrium with ammonia gas.

19.7 The standard enthalpy of formation of liquid water is −285.8 kJ/mol; the standard free energy of formation is −237.1 kJ/mol.
(a) What is the maximum amount of useful electrical work that can be obtained from a fuel cell in which 1.00 mol of oxygen reacts with 2.00 mol of hydrogen. Express your answer in kilojoules.
(b) How many kilojoules of heat would be evolved if the maximum work is done during the reaction of Part (a)?

19.8 The standard free-energy change for the complete combustion of CH_4 is −817.9 kJ/mol; the standard enthalpy of combustion of CH_4 is −890.4 kJ/mol. How many kilojoules of heat would be evolved when 1.00 mol of methane is burned in a fuel cell that has an efficiency of 20% (i.e., the cell produces only 20% of the maximum useful work)?

Free-Energy Change and Concentration (Section 19.2)

19.9 Distinguish between ΔG and ΔG° for a reaction. Write a formula relating the two quantities.

19.10 In thermodynamics, how is the standard state defined for (a) a solid, (b) a liquid, (c) a gas, and (d) a solute? What problems would arise if the concept of standard state had not been developed?

19.11 How does ΔG vary with (a) increasing reactant concentrations and (b) increasing product concentrations?

19.12 **(a)** When will ΔG for a reaction equal ΔG°?
(b) When will ΔG equal zero?
(c) When will ΔG° equal zero?

19.13 Use data from Table 19.2 to determine which of the following nitrogen compounds are stable with respect to their elements at 25°C under standard-state conditions:
(a) NH_3
(b) HNO_3
(c) N_2O
(d) NO
(e) NO_2

Which compound is most stable? Which is least stable?

19.14 Benzene (C_6H_6) is not very reactive at ordinary temperatures and is commonly used as an organic solvent. Consult Table 19.2, and state whether benzene is stable with respect to its elements under standard-state conditions. Would you say the inertness of benzene is caused by thermodynamic or by kinetic factors?

19.15 Assume that each substance is in its standard state, and use data from Table 19.2 to find the spontaneous direction of the following reactions at 25°C:
(a) $2SO_2(g) + O_2(g) \rightleftharpoons 2SO_3(g)$
(b) $Hg_2Cl_2(s) \rightleftharpoons HgCl_2(s) + Hg(l)$

19.16 Assume that each substance is in its standard state, and use data from Table 19.2 to find the spontaneous direction of the following reactions at 25°C:
(a) $2H_2S(g) + 3O_2(g) \rightleftharpoons 2H_2O(l) + 2SO_2(g)$
(b) $Br_2(l) + Cl_2(g) \rightleftharpoons 2BrCl(g)$

19.17 Hearing aids and other small devices are often powered by mercury cells that utilize the reaction

$$Zn(s) + HgO(s) \longrightarrow ZnO(s) + Hg(l)$$

Assume that each substance is in its standard state, and calculate the maximum electrical work that could be obtained when 100 mg of HgO reacts with excess zinc at 25°C.

19.18 Assume that each substance is in its standard state, and calculate the maximum useful work (1) per mole and (2) per gram that could be obtained by completely oxidizing each of the following at 25°C:
(a) hydrogen
(b) methane
(c) acetylene (C_2H_2)
(d) graphite
Which substance would provide the most work per mole? Per gram?

19.19 Calculate ΔG at 25°C for the reaction

$$2SO_2(g) + O_2(g) \rightleftharpoons 2SO_3(g)$$

when the partial pressures of SO_2, O_2, and SO_3 are 1.0×10^{-4}, 0.20, and 0.10 atm, respectively. Which direction, forward or reverse, is spontaneous under these conditions?

19.20 Calculate ΔG at 25°C for the reaction

$$Br_2(l) + Cl_2(g) \rightleftharpoons 2BrCl(g)$$

when the partial pressures of Cl_2 and BrCl are 1.0×10^{-3} and 0.10 atm, respectively. Which direction, forward or reverse, is spontaneous under these conditions?

19.21 Calculate ΔG at 25°C for the reaction

$$H_2(g) + Cl_2(g) \rightleftharpoons 2HCl(g)$$

when each gas is at a partial pressure of 0.50 atm. Is the reacting system closer to equilibrium under these conditions or under standard-state conditions? Explain.

19.22 Calculate ΔG at 25°C for the reaction

$$NH_4Cl(s) \rightleftharpoons NH_3(g) + HCl(g)$$

when (a) the partial pressures of NH_3 and HCl are 0.25 and 0.40 atm, respectively, and (b) the partial pressure of each gas is 0.30 atm. Under which set of pressures will the reacting system be closer to equilibrium?

ΔG° and the Equilibrium Constant (Section 19.2)

19.23 Derive Equation 19.4 from Equation 19.3 without referring to the textbook.

19.24 **(a)** What is the value of K for a reaction for which $\Delta G^\circ = 0$?
(b) What can be said about K when ΔG° is negative? When ΔG° is positive?

19.25 At 500°C, $K_p = 2.7 \times 10^{-2}$ for the reaction

$$2NOCl(g) \rightleftharpoons 2NO(g) + Cl_2(g)$$

(a) Calculate ΔG° at 500°C
(b) Calculate ΔG at 500°C when each gas is at a partial pressure of 0.50 atm.
(c) Which direction, forward or reverse, is spontaneous under the conditions of Part (b)?

19.26 At 25°C, $K_p = 98.4$ for the reaction

$$2NO(g) + Br_2(g) \rightleftharpoons 2NOBr(g)$$

(a) Calculate ΔG° at 25°C.
(b) Calculate ΔG at 25°C when the partial pressures of NO, Br_2, and NOBr are 0.20, 0.50, and 0.30 atm, respectively.
(c) Is the reacting system closer to equilibrium under the conditions of Part (b) or under standard-state conditions?

19.27 Refer to Table 19.2, and calculate K_p for each of the following reactions at 25°C:
(a) $N_2(g) + O_2(g) \rightleftharpoons 2NO(g)$
(b) $3O_2(g) \rightleftharpoons 2O_3(g)$
(c) $C_2H_4(g) + H_2(g) \rightleftharpoons C_2H_6(g)$

19.28 Refer to Table 19.2, and calculate K_p for each of the following reactions at 25°C:
(a) $CaO(s) + SO_2(g) + \frac{1}{2}O_2(g) \rightleftharpoons CaSO_4(s)$
(b) $H_2(g) + Br_2(l) \rightleftharpoons 2HBr(g)$
(c) $4HCl(g) + O_2(g) \rightleftharpoons 2H_2O(g) + 2Cl_2(g)$

19.29 **(a)** The vapor pressure of water at 20°C is 17.5 torr. Calculate K_p and ΔG° for the reaction

$$H_2O(l) \rightleftharpoons H_2O(g)$$

at 20°C.

(b) What will be the values of K_p and $\Delta G°$ at the normal boiling point of water?

19.30 Consult Table 19.2, and calculate (a) $\Delta G°$ for the evaporation of 1.00 mol of water at 25°C and (b) the vapor pressure of water at 25°C.

19.31 Refer to Tables 17.1, 17.2, and 18.4 (pp. 776, 781, and 843, respectively), and calculate $\Delta G°$ for each of the following reactions at 25°C:
(a) $CH_3COOH(aq) \rightleftharpoons CH_3COO^-(aq) + H^+(aq)$
(b) $Fe^{3+}(aq) + SCN^-(aq) \rightleftharpoons Fe(SCN)^{2+}(aq)$
(c) $NH_3(aq) + H_2O(l) \rightleftharpoons NH_4^+(aq) + OH^-(aq)$

19.32 Refer to Tables 17.1, 18.1, and 18.4 (pp. 776, 823, and 843, respectively), and calculate $\Delta G°$ for each of the following reactions at 25°C:
(a) $H_2S(aq) \rightleftharpoons H^+(aq) + HS^-(aq)$
(b) $Zn(NH_3)_4^{2+}(aq) \rightleftharpoons Zn^{2+}(aq) + 4NH_3(aq)$
(c) $Ag^+(aq) + Cl^-(aq) \rightleftharpoons AgCl(s)$

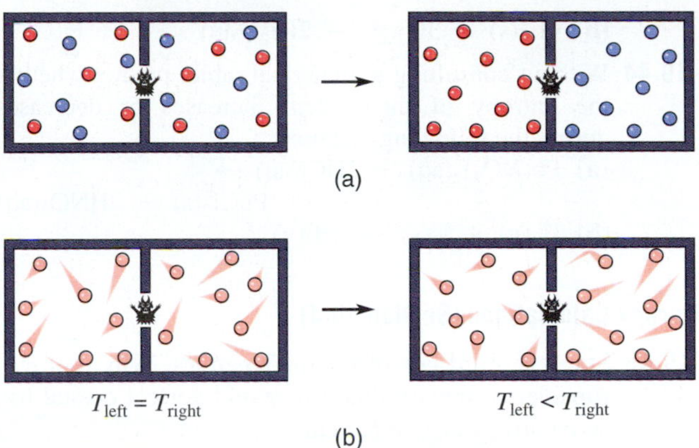

Figure 19.9
Maxwell's demon sorting out molecules. (a) The demon separates the gases by allowing only nitrogen molecules (blue) into the right chamber and oxygen molecules (red) into the left chamber. (b) The demon creates a temperature difference by allowing only faster than average molecules into the right chamber and slower than average molecules into the left chamber.

Entropy and the Second Law of Thermodynamics (Section 19.3)

19.33 State the second law of thermodynamics, and use some everyday examples to illustrate what it means.

19.34 How would you explain the concept of *entropy* to someone who is not a science student? Your explanation should include everyday examples to illustrate what you mean.

19.35 Maxwell's demon (perhaps a microrobot) stands at the opening in the partition of Figure 19.9a. The demon separates the gases by allowing only nitrogen molecules to move into the right chamber and only oxygen molecules to move into the left chamber. Molecules moving in the wrong direction are sent back. Does the entropy of the system increase or decrease?

19.36 A single gas at uniform temperature is separated into two portions by a partition with an opening (Figure 19.9b). Maxwell's demon turns up once again and creates a temperature difference by allowing only faster-than-average molecules to pass through the partition from left to right and slower-than-average molecules to pass from right to left. Does the entropy of the system increase or decrease?

19.37 Explain why the entropy of a gas increases when (a) it is warmed at constant volume and (b) it expands at constant temperature.

19.38 Explain why a substance with large molecules has a higher molar entropy than one with small molecules.

19.39 What is the sign of the entropy change during the formation of a precipitate? Explain your answer.

19.40 Discuss the thermodynamic basis for our confidence that air will not spontaneously separate into its components.

19.41 For each of the following occurrences, give the sign of the entropy change in the system and in the universe:
(a) An ordered deck of cards falls and scatters on the ground.
(b) The scattered cards are picked up and rearranged into an ordered deck.
(c) Molten iron crystallizes.
(d) A baby grows into an adult.
(e) A living organism dies.

19.42 Does the entropy of the system increase, decrease, or remain the same during each of the following changes?
(a) Dry Ice sublimes at its normal transition temperature of −78.5°C to form CO_2 vapor at 1 atm.
(b) The CO_2 vapor from Part (a) expands into an evacuated container to four times its original volume.
(c) The vapor from Part (b) is warmed at constant volume to 25°C.
(d) The warmed vapor from Part (c) is allowed to mix with an equal volume of air at 25°C.
(e) The gaseous mixture from Part (d) is passed over solid CaO, which absorbs the CO_2 to form solid $CaCO_3$.

19.43 Without consulting a numerical table, predict whether the entropy of the system increases or decreases during the following reactions:
(a) $2KClO_3(s) \rightarrow 2KCl(s) + 3O_2(g)$

(b) $4Fe(s) + 3O_2(g) \rightarrow 2Fe_2O_3(s)$

19.44 Without consulting a numerical table, predict whether the entropy of the system increases or decreases during the following reactions:
(a) $Pb(NO_3)_2(aq) + 2HCl(aq) \rightarrow PbCl_2(s) + 2HNO_3(aq)$
(b) $H_2(g) + I_2(s) \rightarrow 2HI(g)$

Entropy Calculations (Section 19.4)

19.45 State the third law of thermodynamics. Why is it reasonable to suppose that a pure and perfect crystal has zero entropy at zero Kelvin?

19.46 Which of the following, if uncontaminated and free of defects, would have zero entropy at zero Kelvin?
(a) sodium chloride
(b) vitamin C (ascorbic acid)
(c) pyrex glass
(d) tin
(e) pewter

***19.47** Derive Equation 19.7 without referring to the text.

***19.48** Derive Equation 19.10 without referring to the text.

***19.49** For a given change within a system, what is the relationship between q for the system and q for the surroundings?

***19.50** Discuss the relationship between q for the system and ΔS for the system. What factors other than the magnitude of q have to be considered?

19.51 The heat of fusion of benzene (C_6H_6) is 126 J/g. Calculate the entropy change when 1.00 mol of solid benzene is converted to liquid at its normal melting point of 5.4°C.

19.52 Calculate the entropy change when 250 g of lead melts at its normal melting point of 328°C. The heat of fusion of lead is 5.12 kJ/mol.

19.53 Calculate the entropy change when 45.0 g of ethanol (C_2H_5OH) vaporizes at its normal boiling point of 78.5°C. The heat of vaporization of ethanol is 38.6 kJ/mol.

19.54 Calculate the entropy change when a 3.00-g sample of methanol vapor (CH_3OH) condenses to liquid at its normal boiling point of 65.0°C. The heat of vaporization of methanol is 37.6 kJ/mol.

19.55 Use $\Delta G°$ and $\Delta H°$ values from Table 19.2 to calculate $\Delta S°$ for each of the reactions in Exercise 19.27.

19.56 Use $\Delta G°$ and $\Delta H°$ values from Table 19.2 to calculate $\Delta S°$ for each of the reactions in Exercise 19.28.

19.57 Use $\Delta G°$ and $\Delta H°$ values from Table 19.2 to calculate standard molar entropies of formation at 25°C for (a) CO_2 and (b) Hg(g).

19.58 Use $\Delta G°$ and $\Delta H°$ values from Table 19.2 to calculate standard molar entropies of formation at 25°C for (a) N_2O and (b) CH_4.

19.59 Use entropy values from Table 19.2 to calculate $\Delta S°$ for each of the following reactions at 25°C:
(a) $2NaI(s) + Cl_2(g) \rightarrow 2NaCl(s) + I_2(s)$
(b) $N_2(g) + 3H_2(g) \rightarrow 2NH_3(g)$

19.60 Use entropy values from Table 19.2 to calculate $\Delta S°$ for each of the following reactions at 25°C:
(a) $NaCl(s) + NaHSO_4(s) \rightarrow Na_2SO_4(s) + HCl(g)$
(b) $SO_2(g) + H_2O(l) + \frac{1}{2}O_2(g) \rightarrow H_2SO_4(l)$

19.61 Use entropy values from Table 19.2 to verify each of the $\Delta S°$ values calculated in Exercise 19.55.

19.62 Use entropy values from Table 19.2 to verify each of the $\Delta S°$ values calculated in Exercise 19.56.

$\Delta G°$ and Temperature (Section 19.5)

19.63 Will the sign of $\Delta G°$ change with temperature if
(a) $\Delta H°$ and $\Delta S°$ are positive?
(b) $\Delta H°$ and $\Delta S°$ are negative?
(c) $\Delta H°$ and $\Delta S°$ have opposite signs?

19.64 Use ΔH and ΔS factors to explain why all decompositions, endothermic or exothermic, are thermodynamically favored by heating.

19.65 Use data from Table 19.2 to estimate $\Delta G°$ values at 300°C for each of the reactions in Exercise 19.27.

19.66 Use data from Table 19.2 to estimate $\Delta G°$ values at 250°C for each of the reactions in Exercise 19.28.

19.67 Estimate the temperature at which N_2O_4 and NO_2, each at 1.00 atm partial pressure, will be in equilibrium. The reaction is

$$N_2O_4(g) \rightleftharpoons 2NO_2(g)$$

19.68 Assume that each substance is in its standard state and determine (1) which of the following reactions are spontaneous at 25°C and (2) which can be reversed by raising the temperature:
(a) $2NO_2(g) \rightarrow 2NO(g) + O_2(g)$
(b) $Fe_2O_3(s) + 2Al(s) \rightarrow Al_2O_3(s) + 2Fe(s)$
(c) $C(gr) \rightarrow C(d)$
(d) $C_2H_2(g) + H_2(g) \rightarrow C_2H_4(g)$

19.69 Use tabulated data to determine whether silver oxide (Ag_2O) is thermodynamically stable with respect to its elements at (a) 25°C and (b) 180°C. Assume that the partial pressure of oxygen is 1.00 atm.

19.70 The partial pressure of oxygen in the atmosphere is about 0.2 atm. Will a sample of Ag_2O exposed to the atmosphere be thermodynamically stable with respect to its elements at (a) 25°C and (b) 180°C?

PART B. MISCELLANEOUS QUESTIONS AND PROBLEMS

19.71 What is the difference between the Gibbs free energy G and the internal energy E?

19.72 At 25°C, sunlight slowly decomposes silver bromide into silver metal and bromine vapor. Account for this observation, given the fact that $\Delta G°$ is positive for the decomposition.

19.73 (a) How does the performance of useful work affect the heat evolved during a reaction?
(b) What fraction of the evolved energy can ideally be obtained as useful work when 1.00 mol of steam condenses to liquid water at 100°C and 1.00 atm?

19.74 Can a state of zero entropy ever be observed? Explain.

19.75 Imagine that you have four molecules distributed over two flasks. How many ways are there to have (a) four molecules in one flask and none in the other, (b) three molecules in one flask and one in the other, and (c) two molecules in one flask and two in the other? Which of these arrangements is most likely to occur? Which arrangement will have the highest entropy?

19.76 Do the efforts of Maxwell's demon in Exercises 19.35 and 19.36 violate the second law of thermodynamics? Explain your answer.

19.77 Use thermodynamic data to explain why diamonds cannot be made from pure graphite at 1 atm pressure.

19.78 The ore sphalerite (ZnS) can be reduced to zinc vapor by a high-temperature process consisting of two steps:

Step 1: $2ZnS(s) + 3O_2(g) \rightarrow 2ZnO(s) + 2SO_2(g)$

Step 2: $ZnO(s) + C(gr) \rightarrow Zn(g) + CO(g)$

Zinc metal is then obtained by cooling and condensing the vapor. Assume that each substance is in its standard state, and use data from Table 19.2 to estimate the temperature above which the first two steps would be spontaneous.

19.79 The source of energy for muscle action and biosynthesis is the hydrolysis of adenosine triphosphate (ATP) into adenosine diphosphate (ADP):

$$ATP^{4-}(aq) + H_2O(l) \rightleftharpoons ADP^{3-}(aq) + HPO_4^{2-}(aq) + H^+(aq)$$

(a) The standard free-energy change for the forward reaction is −31 kJ at 25°C. Calculate the equilibrium constant.
(b) In human red blood cells, the concentrations of ATP^{4-}, ADP^{3-}, and HPO_4^{2-} are 2.25 mM, 0.25 mM, and 1.65 mM, respectively. Estimate ΔG, and use it to show which reaction, forward or reverse, is capable of providing useful work. Assume that the pH is 7.00 and the temperature is 25°C.

19.80 The value of $\Delta G°$ for the complete oxidation of glucose is −2870 kJ/mol. The biological oxidation of glucose occurs in a series of steps that also involves the conversion of 36 mol of ADP to ATP. Refer to Exercise 19.79, and show that 38% of the free energy released by the glucose is stored in ATP.

***19.81** Use Equations 19.4 and 19.7 to derive the van't Hoff equation (p. 727).

19.82 (a) The value of K_p for the reaction

$$H_2(g) + I_2(g) \rightleftharpoons 2HI(g)$$

is 50.3 at 458°C. Calculate $\Delta G°$ at this temperature.
(b) The value of $\Delta H°$ for the reaction in Part (a) is 51.9 kJ. Estimate K_p at 400°C. (*Note*: K_p can be obtained from the value of $\Delta G°$ at 400°C or from the van't Hoff equation (p. 727).)

19.83 Some industrial chlorine was once prepared by the catalytic oxidation of hydrogen chloride at 400°C:

$$4HCl(g) + O_2(g) \rightarrow 2Cl_2(g) + 2H_2O(g)$$

(a) Calculate $\Delta G°$ for this reaction (1) at 25°C and (2) at 400°C.
(b) Calculate K_p for this reaction (1) at 25°C and (2) at 400°C. (*Note*: The value of K_p at 400°C can be obtained from $\Delta G°$ at 400°C or from the van't Hoff equation (p. 727).
(c) Why was the industrial process carried out at 400°C rather than at 25°C?

19.84 The vapor pressure of benzene is 45 torr at 15°C and its normal boiling temperature is 80.1°C. Estimate $\Delta H°_{vap}$ for benzene.

C H A P T E R 20

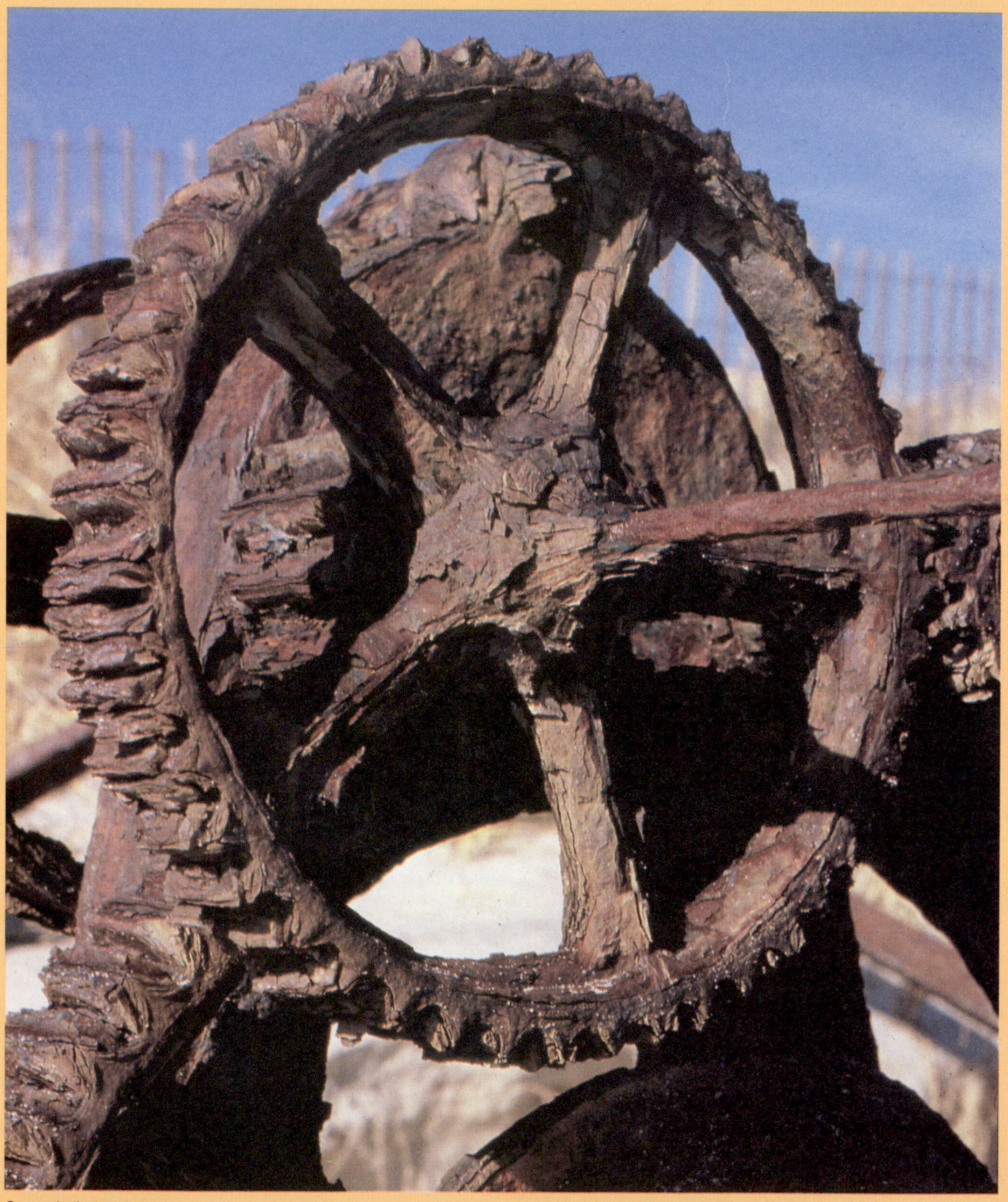

Corroded metal gears at the seashore. Corrosion proceeds by an electrochemical mechanism.

ELECTROCHEMISTRY

OUTLINE

PREVIEW

In 1786 the Italian physician Luigi Galvani observed that a frog's leg twitches when it is touched with two different metals. In attempting to explain this observation, he incorrectly concluded that the source of electricity was the animal tissue. Alessandro Volta, an Italian physicist, disputed this conclusion, and the resulting controversy led to the discovery that electric currents could be produced by chemical reactions. In 1800 Volta used this discovery to invent the first chemical battery.

All chemistry is electrical in the sense that it involves the behavior of electrons and other charged particles; the term **electrochemistry**, however, is generally reserved for processes that convert chemical energy into electrical energy, or vice versa.

One branch of electrochemistry deals with **galvanic cells** such as batteries and fuel cells. (Galvanic cells are also called **voltaic cells**.) These devices use spontaneous oxidation–reduction reactions to produce electric currents and do electrical work. A galvanic cell is constructed so that electrons flow from one terminal to the other when the terminals are connected by an external circuit. The driving force that impels the electrons through the circuit is generated by the negative free-energy change of the spontaneous reaction.

The terms *galvanic* and *voltaic* are derived from the names of Luigi Galvani (1737–1798) and Alessandro Volta (1745–1827).

Another branch of electrochemistry deals with **electrolysis**, a process in which an electric current is used to bring about an oxidation–reduction reaction by removing electrons from one reactant and giving them to another. Electrolysis is used to decompose stable compounds such as water and salt, to electroplate metals, and to produce high-energy substances such as sodium, fluorine, peroxy-compounds, and perchlorates.

20.1 GALVANIC CELLS

When zinc metal is immersed in copper sulfate solution (Demonstration 20.1), some zinc dissolves and some copper metal forms on the surface of the remaining zinc.

DEMONSTRATION 20.1 ZINC METAL AND COPPER SULFATE SOLUTION: $Zn(s) + Cu^{2+}(aq) \rightarrow Cu(s) + Zn^{2+}(aq)$

A zinc rod is put into copper sulfate solution.

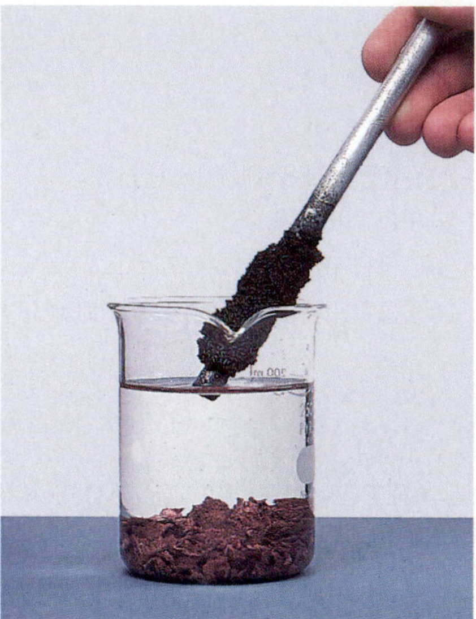

Some hours later. The blue color of $Cu^{2+}(aq)$ has disappeared. Copper metal has appeared; some clings to the zinc rod.

The copper metal and the pitted remains of the zinc rod.

The net reaction

$$Cu^{2+}(aq) + Zn(s) \rightarrow Cu(s) + Zn^{2+}(aq)$$

is a redox reaction in which electrons are transferred from zinc atoms to copper ions. The half-reaction equations are

$$Zn(s) \rightarrow Zn^{2+}(aq) + 2e^- \qquad \text{oxidation}$$

$$Cu^{2+}(aq) + 2e^- \rightarrow Cu(s) \qquad \text{reduction}$$

LEARNING HINT

Review redox equations and half-reactions in Chapter 11.

The solution is in contact with the metal; hence, the electrons flow directly from the zinc atoms on the rod's surface to copper ions that randomly move near the surface. The energy released by this spontaneous redox reaction can do no useful work; it all appears as heat.

If the half-reactions could be made to occur in separate compartments and the electrons were transferred through a wire from one compartment to the other, then we would have a galvanic cell in which the energy of the flowing electrons could be used to do useful work, for example, running a motor, lighting a bulb, or operating a radio. A diagram of such a cell, similar to one constructed by John Frederick Daniell in 1836, is shown in Figure 20.1, and a photograph of it is shown in Figure 20.2a. One compartment of the Daniell cell contains a zinc electrode immersed in a solution of zinc sulfate ($ZnSO_4$); the other compartment contains a copper electrode in a solution of copper sulfate ($CuSO_4$). When the two electrodes are connected by a wire, the zinc atoms give up electrons, forming positive zinc ions:

Electrode: a conducting substance, usually a metal, through which electrons enter or leave a conducting medium.

$$Zn(s) \rightarrow Zn^{2+}(aq) + 2e^-$$

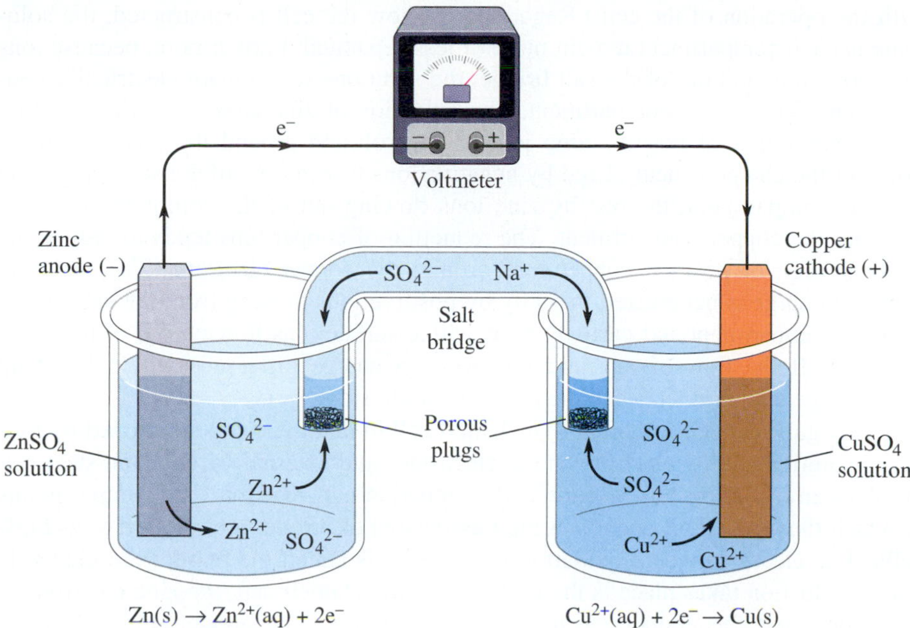

Figure 20.1

Diagram of the Daniell cell. The cell reaction is as follows:

$$Zn(s) + Cu^{2+}(aq) \longrightarrow Cu(s) + Zn^{2+}(aq)$$

The zinc ions repel each other and enter the solution. The negative electrons also repel each other and travel through the wire to the copper electrode, where they are accepted by copper ions from the surrounding solution:

$$Cu^{2+}(aq) + 2e^- \rightarrow Cu(s)$$

The resulting copper atoms "plate out" (deposit as copper metal) on the surface of the copper electrode. The Daniell cell reaction is identical to the reaction that occurs when zinc metal is dipped directly into copper sulfate solution—zinc atoms are oxidized and copper ions are reduced—but the cell is constructed so that the electrons pass through an external circuit where they can do useful work.

The solutions in the two halves of a galvanic cell must be connected to complete the circuit, but they must not mix. The connection can be made through a salt bridge as in Figures 20.1 and 20.2a, through a porous partition as in Figure 20.2b, or by direct contact as in Figure 20.2c. (A **salt bridge** is an inverted U-tube filled with an electrolyte solution, such as Na_2SO_4 or KNO_3, chosen so that it does not interfere

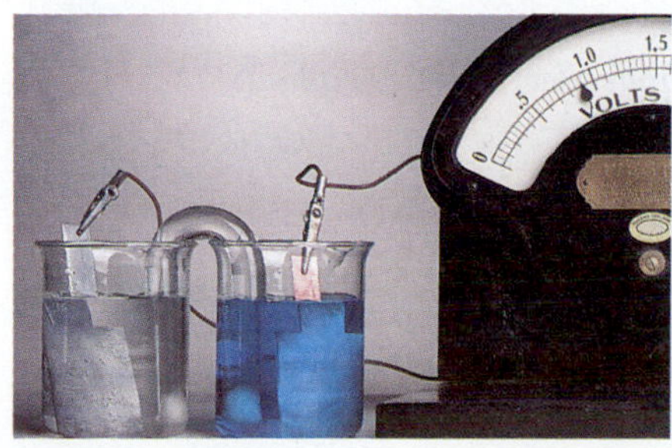

(a)

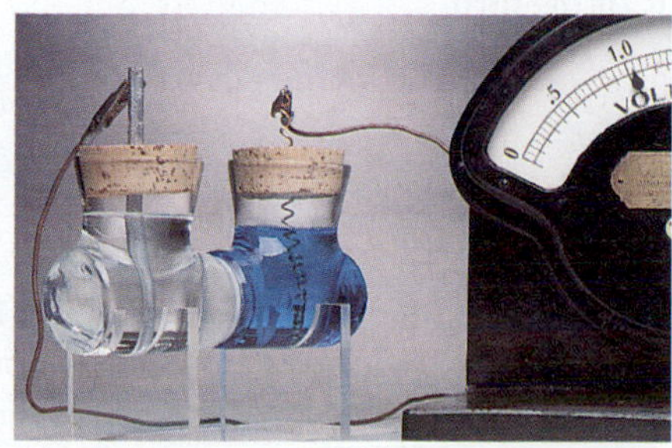

(b)

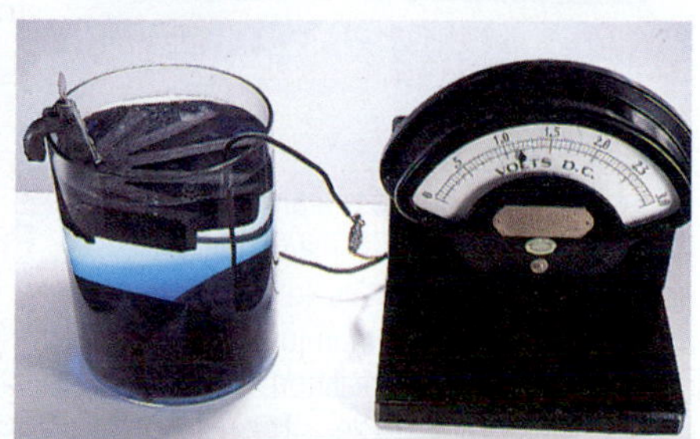

(c)

Figure 20.2

Three versions of the Daniell cell. The two electrolytes are prevented from mixing in (a) by a salt bridge filled with Na_2SO_4 solution, in (b) by a porous glass partition, and in (c) by gravity. The gravity cell is made by putting a saturated $CuSO_4$ solution into the beaker and floating a less dense $ZnSO_4$ solution on top of it. The zinc electrode (top) is of the "crowfoot" type. The copper electrode (bottom) consists of metal strips connected to an insulated wire.

with the operation of the cell.) Regardless of how the cell is constructed, the solutions in each compartment remain more or less separated. Furthermore, because ions can move into and out of the salt bridge, the solutions will remain electrically neutral. Consider the zinc compartment. The oxidation of zinc atoms tends to build up the concentration of positive zinc ions in the solution around the zinc electrode. Some of the charge is neutralized by negative ions flowing out of the salt bridge and into the compartment, the rest by zinc ions flowing out of the compartment. Now consider the copper compartment. The reduction of copper ions tends to reduce the concentration of positive copper ions around the copper electrode. The resulting negative charge is neutralized partially by positive ions flowing from the salt bridge into the compartment and partially by negative sulfate ions flowing out of the compartment. If the flow of ions did not occur, a charge difference would build up between the compartments and the reaction would stop.

Anode—oxidation

Cathode—reduction

Every galvanic cell is similar to a Daniell cell in that the charge is carried by ions in the solution and by electrons in the electrodes and external wires. Each such cell involves an oxidation half-reaction and a reduction half-reaction. The compartments in which these reactions occur and their associated electrodes are referred to as **half-cells**. The electrode where oxidation takes place is called the **anode**; the electrode where reduction takes place is the **cathode**. In the Daniell cell, the zinc electrode is the anode and the copper electrode is the cathode. The anode sends electrons through the connecting wire; it is the electron source or negative terminal of the cell. The cathode receives electrons, so it is the positive terminal. *When the terminals of a galvanic cell are connected by an external circuit, there will always be a flow of electrons from anode to cathode through the circuit.*

Cell Notation

A system of **cell notation** allows us to describe a galvanic cell without drawing a diagram. For example, a Daniell cell in which the electrolyte concentrations are each 1 molar is represented as

$$\mathrm{Zn(s)} \mid \mathrm{Zn^{2+}}(1\ M) \parallel \mathrm{Cu^{2+}}(1\ M) \mid \mathrm{Cu(s)}$$

The anode is written on the left, the cathode on the right, and concentrations and other data are given in parentheses. The vertical lines indicate phase boundaries, and the double vertical line indicates a salt bridge or porous partition. The cell notation for the Daniell cell tells us that the zinc anode is dipped into a 1 *M* solution of zinc ions, the copper cathode is dipped into a 1 *M* solution of copper ions, and the two half-cells are separated by a salt bridge or porous partition. (If the two half-cells were in direct contact as in Figure 20.2c, the double vertical line would be replaced by a single line.) The notation for any galvanic cell begins with the anode on the left and is of the form

$$\textit{anode} \mid \textit{anode electrolyte} \parallel \textit{cathode electrolyte} \mid \textit{cathode}$$

Figure 20.3

A lemon cell. A voltage will always develop between different metals in a conducting solution. These electrodes are an iron nail and a copper penny; lemon juice is the acidic electrolyte. The reaction at the nail is $Fe(s) \longrightarrow Fe^{2+}(aq) + 2e^-$. The reaction at the penny is $2H^+(aq) + 2e^- \longrightarrow H_2(g)$.

Practical Considerations

Every redox reaction is the sum of an oxidation half-reaction and a reduction half-reaction, so every redox system could, in principle, serve as the basis for a galvanic cell (Figure 20.3). Some reactions, however, are more easily adapted to this purpose than others. The following examples illustrate some of the practical considerations.

EXAMPLE 20.1

Diagram and give the cell notation for a galvanic cell based on the reaction

$$Fe(s) + Hg_2^{2+}(aq) \rightarrow Fe^{2+}(aq) + 2Hg(l)$$

Assume that the solution concentrations are 0.10 *M*.

SOLUTION

The half-reactions are

$$Fe(s) \rightarrow Fe^{2+}(aq) + 2e^- \qquad \text{anode, oxidation}$$

$$Hg_2^{2+}(aq) + 2e^- \rightarrow 2Hg(l) \qquad \text{cathode, reduction}$$

The mercury, being liquid, lies on the floor of the cathode half-cell with a wire from the external circuit dipping into it (Figure 20.4). The wire must be a metal that is not attacked by mercury or by the solution—platinum would be a good choice. The cell notation is

$$Fe(s) \mid Fe^{2+}(0.10\ M) \parallel Hg_2^{2+}(0.10\ M) \mid Hg(l)$$

The platinum wire is not included in the cell notation because its only function is to bring electrons to the mercury electrode.

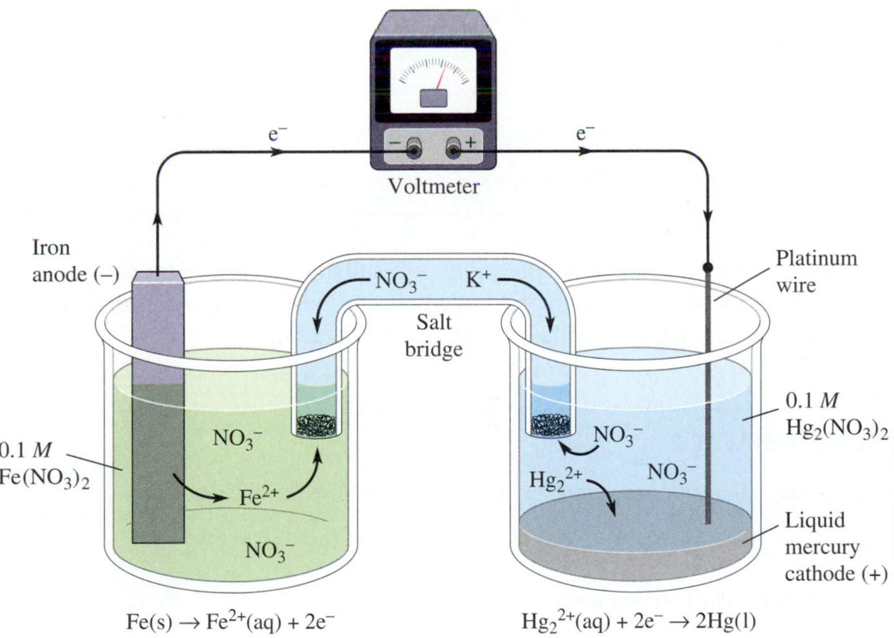

Figure 20.4
Schematic diagram of the cell

$$Fe(s) \mid Fe^{2+}(0.10\ M) \parallel Hg_2^{2+}(0.10\ M) \mid Hg(l)$$

The electrolytes are 0.10 *M* $Fe(NO_3)_2$ and 0.10 *M* $Hg_2(NO_3)_2$. The salt bridge contains KNO_3.

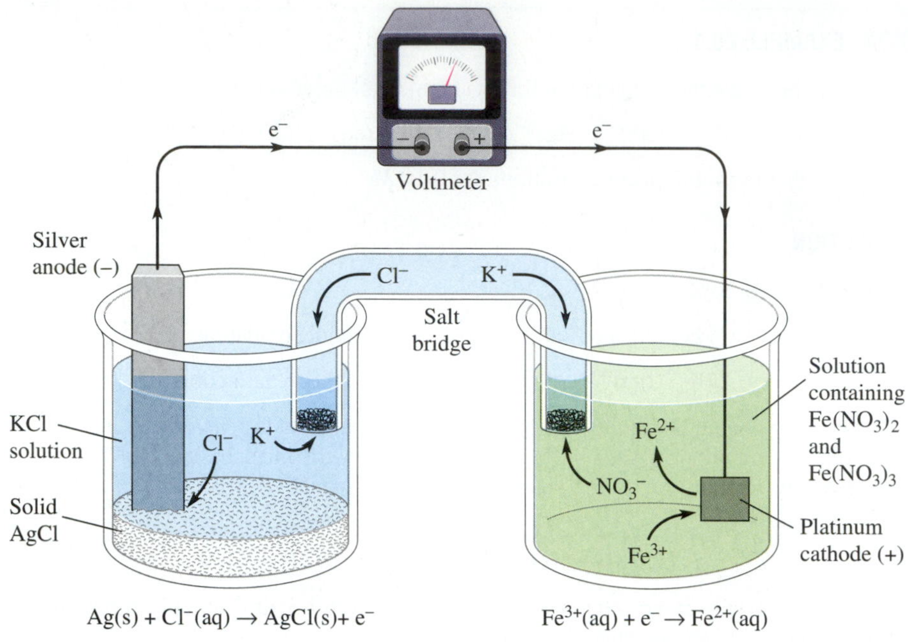

Figure 20.5 Schematic diagram of the cell

$$Ag(s) \mid AgCl(s) \mid Cl^-(aq) \parallel Fe^{3+}(aq), Fe^{2+}(aq) \mid Pt(s)$$

The anode electrolyte is KCl(aq); the cathode electrolyte is a solution containing $Fe(NO_3)_3$ and $Fe(NO_3)_2$. The salt bridge contains KCl.

PRACTICE EXERCISE 20.1

Diagram the cell

$$Al(s) \mid Al_2(SO_4)_3(aq) \parallel AgNO_3(aq) \mid Ag(s)$$

and write equations for the electrode reactions and the complete cell reaction.

When the half-reaction does not involve a suitable substance for an electrode, a solid inert electrode such as graphite or platinum is used to provide the surface on which the reaction occurs, as illustrated in the next example.

EXAMPLE 20.2

Give the cell notation and diagram a galvanic cell based on the reaction

$$Fe^{3+}(aq) + Ag(s) + Cl^-(aq) \rightarrow Fe^{2+}(aq) + AgCl(s)$$

SOLUTION The half-reactions are

$$Ag(s) + Cl^-(aq) \rightarrow AgCl(s) + e^- \qquad \text{anode, oxidation}$$
$$Fe^{3+}(aq) + e^- \rightarrow Fe^{2+}(aq) \qquad \text{cathode, reduction}$$

The cell is diagrammed in Figure 20.5. In the anode half-cell, silver metal is the electrode, aqueous KCl is the electrolyte, and solid AgCl is in contact with both. The cathode reaction involves only dissolved ions, so an inert substance such as platinum is used as the electrode. The cell notation is

$$Ag(s) \mid AgCl(s) \mid Cl^-(aq) \parallel Fe^{3+}(aq), Fe^{2+}(aq) \mid Pt(s)$$

Two phase-boundary lines are used in the anode half-cell notation because two different solids are in contact with the solution in the anode compartment. The cathode half-cell contains two dissolved ions that participate in the reaction, Fe^{3+} and Fe^{2+}; such ions are conventionally written in the order in which they appear in the reaction equation. Platinum is included in the cell notation because, in contrast to its role as a connecting wire in Example 20.1, it is now the electrode on whose surface the cathode half-reaction occurs.

PRACTICE EXERCISE 20.2

Give the cell notation for a galvanic cell based on the reaction

$$Zn(s) + Br_2(l) \rightarrow ZnBr_2(aq)$$

(*Hint*: Bromine attacks most metals but does not attack graphite.)

Standard Cells

A **standard cell** is one in which all substances—products as well as reactants—are in their standard states. A standard cell based on the reaction between zinc metal and copper sulfate will have electrodes of pure zinc and pure copper dipped into 1 molar solutions of zinc sulfate and copper sulfate, respectively. Each half of the standard cell is called a **standard half-cell** or **standard electrode**.

Recall (Table 19.1, p. 909) that the standard state for a solute is the ideal solution at unit molality. Unit molarity is a convenient approximation.

EXAMPLE 20.3

Use cell notation to describe a standard cell based on the reaction

$$5Fe^{2+}(aq) + MnO_4^-(aq) + 8H^+(aq) \rightarrow 5Fe^{3+}(aq) + Mn^{2+}(aq) + 4H_2O(l)$$

Assume that the half-reactions occur at platinum electrodes.

SOLUTION

The half-reactions are

$$Fe^{2+}(aq) \rightarrow Fe^{3+}(aq) + e^- \qquad \text{anode, oxidation}$$

$$MnO_4^-(aq) + 8H^+(aq) + 5e^- \rightarrow Mn^{2+}(aq) + 4H_2O(l) \qquad \text{cathode, reduction}$$

In a standard cell, all participating ions will be 1 molar. The oxidation of Fe^{2+} is the anode reaction and the reduction of MnO_4^- is the cathode reaction. Both half-cells contain platinum electrodes. The cell notation is

$$Pt(s) \mid Fe^{2+}(1\,M), Fe^{3+}(1\,M) \parallel MnO_4^-(1\,M), H^+(1\,M), Mn^{2+}(1\,M) \mid Pt(s)$$

PRACTICE EXERCISE 20.3

Refer to Practice Exercise 20.2. Use cell notation to describe a standard cell based on the reaction between zinc and bromine.

20.2 ELECTRICAL WORK AND FREE ENERGY

When electrons move through a wire, they encounter resistance in the form of other particles that block their paths. The driving force that allows the electrons to overcome this resistance and move around the circuit is called an **electromotive force** (EMF). The electromotive force in a galvanic cell comes from the redox reaction that supplies electrons to one electrode and removes them from the other. Electrons on the negative electrode repel each other and have more potential energy than electrons on the positive electrode. This *potential difference* causes electrons to flow through the external circuit from the negative terminal to the positive terminal. The energy of the moving electrons is used to overcome resistance in the external circuit, and it can also be harnessed to do various forms of electrical work. Energy that overcomes resistance appears as heat or as light emitted by a glowing filament (as in light bulbs). Electrical work is used for operating meters, starting cars, running watches, radios, and computers, electrolyzing chemicals, and so forth.

The terms *electromotive force, voltage,* and *potential difference* are often used interchangeably even though they represent somewhat different concepts.

The SI unit of charge is the coulomb (C). Electromotive force is expressed in terms of the potential difference that causes one coulomb of charge to move. It is measured in **volts** (V), where one volt is one joule per coulomb (1 V = 1 J/C). For example, one coulomb of charge at the negative terminal of a 12-volt battery would have 12 joules more potential energy than one coulomb at the positive terminal. Hence, 12 joules is the maximum work that can be done when one coulomb moves from the negative terminal to the positive terminal. The maximum electrical work that can be done during the transfer of Q coulombs is

$$\text{maximum electrical work (J)} = Q\ (\text{C}) \times E\ (\text{J/C}) \tag{20.1}$$

where E is the electromotive force of the cell in joules per coulomb, or volts. The actual electrical work will always be less than the theoretical maximum because some energy is inevitably lost as heat.

LEARNING HINT

Electric current is the rate of flow of charge. It is expressed in units of coulombs per second (C/s). The concept of current will be developed more fully in Section 20.8.

EXAMPLE 20.4

Calculate the maximum electrical work that can be done by a 12-volt automobile battery supplying a current of 10 C/s for 30 seconds.

SOLUTION

Substituting $Q = 10\ \text{C/s} \times 30\ \text{s} = 300\ \text{C}$ and $E = 12\ \text{V} = 12\ \text{J/C}$ into Equation 20.1 gives

$$\text{maximum electrical work} = 300\ \text{C} \times 12\ \text{J/C} = 3600\ \text{J}$$

PRACTICE EXERCISE 20.4

The mercury cell used in electric watches and some hearing aids has an EMF of about 1.35 V. Compute the maximum electrical work that can be done when a current of 1.00×10^{-7} C/s is drawn from the cell for 10.0 min.

The electromotive force of a galvanic cell depends on the cell reaction and is not affected by the volume of electrolyte or by the size of the electrodes. Fresh flashlight batteries, for example, all have an EMF of about 1.5 volts regardless of their size (AAA, AA, C, or D). Large batteries, however, last longer and deliver more current than small batteries. The greater supply of reactant in the larger battery will transfer more charge before it is used up, and the larger electrodes will allow more charge to be transferred each second. Neither of these factors, however, affects the electromotive force that sends each electron along its way.

ΔG and Cell Voltage

A spontaneous reaction at constant temperature and pressure is accompanied by a decrease in free energy (Chapter 19). Furthermore, the free-energy change is the negative of the maximum useful work that can be obtained from the reaction (Equation 19.2, p. 906). In galvanic cells, useful work is electrical work, so

$$\Delta G = -\text{maximum electrical work} \tag{20.2}$$

Combining Equations 20.1 and 20.2 gives

$$\Delta G = -QE \tag{20.3}$$

where ΔG is the free-energy change of the cell reaction, E is the cell voltage, and Q is the charge transferred during the reaction. It is often convenient to express Q in terms of the number of moles of electrons transferred. The charge on 1 mol of electrons, 96,485 C, is called the **faraday** and is represented by F ($1\ F = 96{,}485$ C/mol e^-). If n moles of electrons are transferred, then the number of coulombs is

Michael Faraday (1791–1867) discovered that the number of moles of substance consumed or produced at an electrode is directly proportional to the number of coulombs passing through the cell.

$$Q = n\,(\text{mol } e^-) \times F\,(\text{C/mol } e^-) \tag{20.4}$$

and Equation 20.3 becomes

$$\Delta G = -nFE \tag{20.5a}$$

For a reaction in a standard cell, the free-energy change is $\Delta G°$, the cell voltage is $E°$, and Equation 20.5a becomes

$$\Delta G° = -nFE° \tag{20.5b}$$

The chemical reaction that drives a galvanic cell is spontaneous, so the associated free-energy change must be negative. The negative sign in Equation 20.5a means that *the voltage of a galvanic cell is always positive.* The more positive the voltage, the more negative ΔG will be, and the greater the driving force of the cell reaction. The best measurements of cell voltage are obtained with a *potentiometer,* a device that records the maximum voltage because it draws virtually no current (Figure 20.6). Ordinary voltmeters are not used for precise measurements because they give readings somewhat less than the maximum.

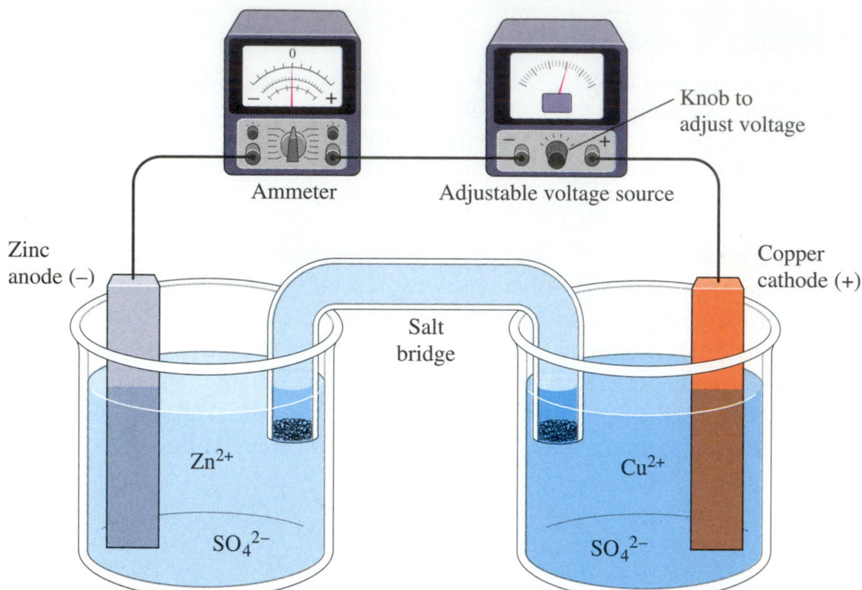

Figure 20.6

A potentiometer measures the EMF of a cell under a no-current condition. When the variable voltage source is adjusted so the current through the ammeter registers zero, the voltage reading will be equal and opposite to the EMF of the cell.

EXAMPLE 20.5

A standard cell using the reaction

$$Zn(s) + Cu^{2+}(aq) \longrightarrow Zn^{2+}(aq) + Cu(s)$$

gives a potentiometer reading of 1.10 V. Calculate $\Delta G°$ for this reaction.

SOLUTION

The oxidation of 1 mol of Zn atoms by Cu^{2+} ions requires the transfer of 2 mol of electrons. Substituting $n = 2$ mol e^-, $F = 96{,}485$ C/mol e^-, and $E° = 1.10\text{ V} = 1.10\text{ J/C}$ into Equation 20.5b gives

$$\begin{aligned}\Delta G° &= -nFE° \\ &= -2\text{ mol e}^- \times \frac{96{,}485\text{ C}}{1\text{ mol e}^-} \times \frac{1.10\text{ J}}{1\text{ C}} \\ &= -2.12 \times 10^5\text{ J} = -212\text{ kJ}\end{aligned}$$

PRACTICE EXERCISE 20.5

The standard free-energy change of the reaction

$$2Al(s) + 3Cu^{2+}(aq) \longrightarrow 3Cu(s) + 2Al^{3+}(aq)$$

is -1166 kJ. What would the potentiometer reading be if this reaction were carried out in a standard cell?

20.3 STANDARD ELECTRODE POTENTIALS

The total free-energy change for a redox reaction is the sum of the contributions from each half-reaction. The standard free-energy change ΔG° is therefore

$$\Delta G^\circ = \Delta G^\circ_{\text{oxid}} + \Delta G^\circ_{\text{red}}$$

where $\Delta G^\circ_{\text{oxid}}$ and $\Delta G^\circ_{\text{red}}$ are contributions from the oxidation and reduction half-reactions, respectively. If the reaction takes place in a galvanic cell, we can substitute $-nFE^\circ$ for ΔG° (Equation 20.5) to obtain

$$-nFE^\circ = -nFE^\circ_{\text{oxid}} - nFE^\circ_{\text{red}}$$

Dividing through by $-nF$ gives

$$E^\circ = E^\circ_{\text{oxid}} + E^\circ_{\text{red}} \qquad (20.6)$$

where E°_{oxid}, the **standard oxidation potential**, is the contribution of the oxidation half-reaction to the total voltage E°, and E°_{red}, the **standard reduction potential**, is the contribution of the reduction half-reaction. Equation 20.6 shows that *the total voltage of a galvanic cell is equal to the sum of the oxidation and reduction potentials.* The potentials E°_{oxid} and E°_{red} are often called **half-cell potentials**.

A half-cell potential is determined from the voltage obtained when the half-cell is coupled to the **standard hydrogen electrode** (Figure 20.7). This electrode, which is accepted by international agreement as the reference electrode for determining half-cell potentials, consists of platinum metal dipped into a 1 *M* solution of hydrochloric acid. Hydrogen gas at a partial pressure of 1 atm is bubbled through the acid solution surrounding the platinum. The half-reaction for the standard hydrogen electrode is

$$2H^+(1\,M) + 2e^- \rightleftharpoons H_2(g,\ 1\ \text{atm})$$

Figure 20.7a shows the standard hydrogen electrode coupled to a standard copper electrode. Hydrogen is oxidized, copper ions are reduced, and electrons flow through the external circuit from the hydrogen electrode to the copper electrode. The observed voltage is 0.34 V and the reactions are

The standard state for $H^+(aq)$ is a 1 *m* (1 molal) ideal solution (Table 19.1, p. 909); a 1 *M* (1 molar) solution is an approximation.

$$\text{anode:}\quad H_2(g,\ 1\ \text{atm}) \rightarrow 2H^+(1\,M) + 2e^- \qquad E^\circ_{\text{oxid}} = ?$$

$$\text{cathode:}\quad Cu^{2+}(1\,M) + 2e^- \rightarrow Cu(s) \qquad E^\circ_{\text{red}} = ?$$

$$\text{sum:}\quad H_2(g,\ 1\ \text{atm}) + Cu^{2+}(1\,M) \rightarrow 2H^+(1\,M) + Cu(s) \qquad E^\circ = 0.34\ \text{V}$$

The half-cell potentials of the hydrogen and copper electrodes, E°_{oxid} and E°_{red}, cannot be measured individually because the electrodes work together as a pair. Equation 20.6, however, shows that their sum is equal to 0.34 V, the total cell voltage:

$$E^\circ = E^\circ_{\text{oxid}} + E^\circ_{\text{red}} = 0.34\ \text{V}$$

If we arbitrarily define the oxidation potential of the standard hydrogen electrode to be zero ($E^\circ_{\text{oxid}} = 0$ V), then the *relative* standard reduction potential of the copper electrode will be $E^\circ_{\text{red}} = 0.34$ V. In other words, we can write:

$$Cu^{2+}(1\,M) + 2e^- \rightleftharpoons Cu(s) \qquad E^\circ_{\text{red}} = +0.34\ \text{V}$$

The positive sign of the reduction potential shows that the copper electrode in this cell is positive; that is, it accepts electrons from the standard hydrogen electrode.

Figure 20.7b shows the standard hydrogen electrode coupled to a standard zinc electrode. Zinc is oxidized, hydrogen ions are reduced, and electrons flow from the zinc electrode to the hydrogen electrode. The observed voltage is 0.76 V and the reactions are

anode: $Zn(s) \longrightarrow Zn^{2+}(1\ M) + 2e^-$ $\qquad E^\circ_{oxid} = ?$

cathode: $2H^+(1\ M) + 2e^- \longrightarrow H_2(g, 1\ atm)$ $\qquad E^\circ_{red} = ?$

sum: $Zn(s) + 2H^+(1\ M) \longrightarrow Zn^{2+}(1\ M) + H_2(g, 1\ atm)$ $\qquad E^\circ = 0.76\ V$

The half-cell potentials of the zinc and hydrogen electrodes are E°_{oxid} and E°_{red}, respectively. (Note the different roles played by the hydrogen electrode. In this cell it is the cathode; in the previous cell it was the anode.) The total cell voltage is 0.76 V, so

$$E^\circ = E^\circ_{oxid} + E^\circ_{red} = 0.76\ V$$

LEARNING HINT

By convention, a value of zero is arbitrarily assigned to the potential of the standard hydrogen electrode:

$$2H^+(1\ M) + 2e^- \rightleftharpoons H_2(g, 1\ atm)$$
$$E^\circ_{red} = E^\circ_{oxid} = 0$$

If we arbitrarily define the reduction potential of the standard hydrogen electrode to be zero ($E^\circ_{red} = 0$ V), then the relative standard oxidation potential of the zinc electrode will be $E^\circ_{oxid} = 0.76$ V; that is,

$$Zn(s) \rightleftharpoons Zn^{2+}(1\ M) + 2e^- \qquad E^\circ_{oxid} = 0.76\ V$$

Reversing the reaction so that it becomes a reduction gives

$$Zn^{2+}(1\ M) + 2e^- \rightleftharpoons Zn(s) \qquad E^\circ_{red} = -0.76\ V$$

The reduction potential is the negative of the oxidation potential because ΔG° is equal to $-nFE^\circ$, and ΔG° changes sign when the reaction equation is reversed. We can use the negative sign of the reduction potential as a reminder that the zinc electrode in Figure 20.7b is negative; that is, it sends electrons to the standard hydrogen electrode.

Standard reduction potentials of other half-reactions can be obtained by coupling the half-reaction to either the standard hydrogen electrode or to some other half-cell whose reduction potential is known. A list of standard reduction potentials is given in Table 20.1. Note that each half-reaction in the table is written as a reduction and followed by the value of E°_{red}. It is useful to remember that *half-cells with positive reduction potentials accept electrons from the hydrogen electrode; half-cells with negative reduction potentials send electrons to the hydrogen electrode.*

PRACTICE EXERCISE 20.6

The standard reduction potential for the calcium electrode is −2.869 V. What is the standard oxidation potential?

Using a Table of Reduction Potentials

LEARNING HINT

Oxidation—loss of electrons

Reduction—gain of electrons

A great deal of chemistry can be learned by studying a table of reduction potentials. A high reduction potential (large positive value) implies a strong tendency to accept electrons, and substances at the top of Table 20.1 tend to be strong oxidizing agents. Fluorine (F_2) has the highest reduction potential in this table and is the strongest oxidizing agent shown. Fluorine is difficult to prepare and work with because it vigorously removes electrons from most other substances; in nature it occurs only as the

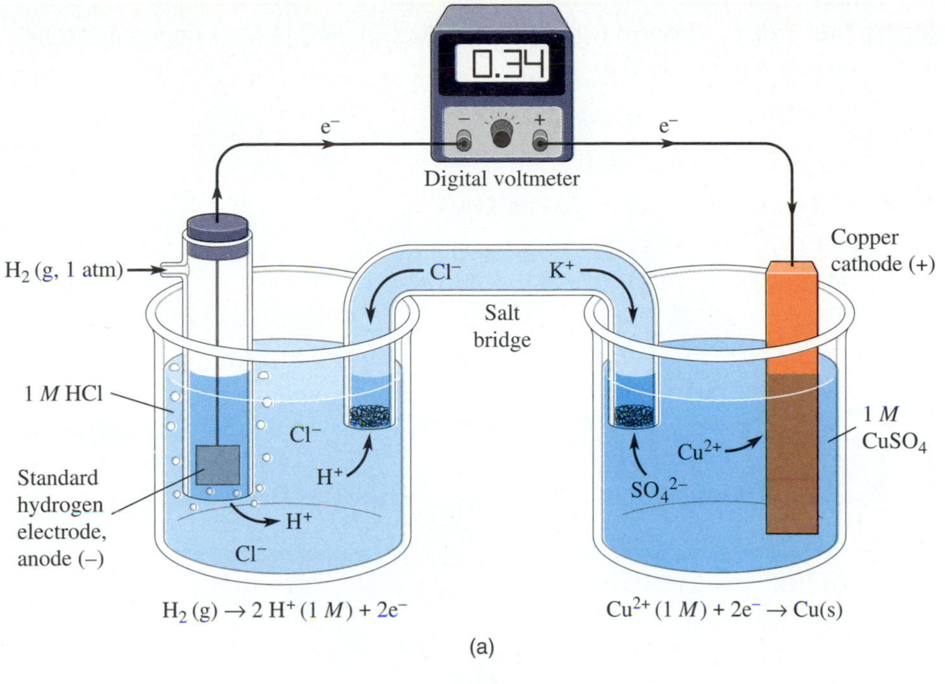

(a)

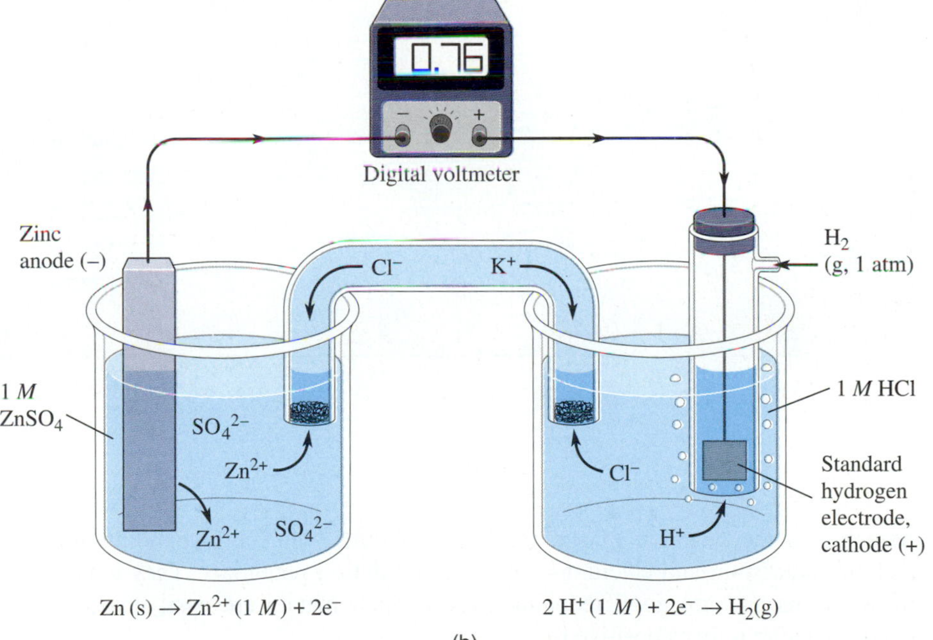

(b)

Figure 20.7

The standard hydrogen electrode. (a) When coupled to a standard copper electrode, the standard hydrogen electrode is the anode; H_2 is oxidized. (b) When coupled to a standard zinc electrode, the standard hydrogen electrode is the cathode; H^+ is reduced. (*Note*: A digital voltmeter draws a negligible amount of current and is often used instead of a potentiometer.)

fluoride ion (F^-). Chlorine, hydrogen peroxide, permanganate ion, dichromate ion, and other strong oxidizing agents also have high reduction potentials.

Cations such as Na^+, K^+, and Ca^{2+} are found near the bottom of Table 20.1. They have low reduction potentials and rarely accept electrons. Since the oxidation

The recent recommendation by IUPAC to change the standard-state pressure from 1 atm to 1 bar will change $E°$ values by less than 0.5 millivolt, a change that is smaller than the uncertainty of most EMF measurements.

TABLE 20.1 Standard Reduction Potentials[a] at 25°C (A More Complete Listing is Given in Appendix B.6)

Half-Reaction	$E°$ (V)
$F_2(g) + 2e^- \rightleftharpoons 2F^-(aq)$	2.889
$H_2O_2(aq) + 2H^+(aq) + 2e^- \rightleftharpoons 2H_2O(l)$	1.763
$Au^+(aq) + e^- \rightleftharpoons Au(s)$	1.691
$MnO_4^-(aq) + 8H^+(aq) + 5e^- \rightleftharpoons Mn^{2+}(aq) + 4H_2O(l)$	1.512
$Cl_2(g) + 2e^- \rightleftharpoons 2Cl^-(aq)$	1.360
$Cr_2O_7^{2-}(aq) + 14H^+(aq) + 6e^- \rightleftharpoons 2Cr^{3+}(aq) + 7H_2O(l)$	1.33
$MnO_2(s) + 4H^+(aq) + 2e^- \rightleftharpoons Mn^{2+}(aq) + 2H_2O(l)$	1.229
$O_2(g) + 4H^+(aq) + 4e^- \rightleftharpoons 2H_2O(l)$	1.229
$Br_2(l) + 2e^- \rightleftharpoons 2Br^-(aq)$	1.078
$NO_3^-(aq) + 4H^+(aq) + 3e^- \rightleftharpoons NO(g) + 2H_2O(l)$	0.964
$Hg^{2+}(aq) + 2e^- \rightleftharpoons Hg(l)$	0.852
$Ag^+(aq) + e^- \rightleftharpoons Ag(s)$	0.799
$Hg_2^{2+}(aq) + 2e^- \rightleftharpoons 2Hg(l)$	0.796
$Fe^{3+}(aq) + e^- \rightleftharpoons Fe^{2+}(aq)$	0.769
$O_2(g) + 2H^+(aq) + 2e^- \rightleftharpoons H_2O_2(aq)$	0.695
$I_2(s) + 2e^- \rightleftharpoons 2I^-(aq)$	0.535
$Cu^+(aq) + e^- \rightleftharpoons Cu(s)$	0.518
$O_2(g) + 2H_2O(l) + 4e^- \rightleftharpoons 4OH^-(aq)$	0.401
$Cu^{2+}(aq) + 2e^- \rightleftharpoons Cu(s)$	0.339
$Hg_2Cl_2(s) + 2e^- \rightleftharpoons 2Hg(l) + 2Cl^-(aq)$	0.268
$AgCl(s) + e^- \rightleftharpoons Ag(s) + Cl^-(aq)$	0.222
$S(s) + 2H^+(aq) + 2e^- \rightleftharpoons H_2S(aq)$	0.144
$AgBr(s) + e^- \rightleftharpoons Ag(s) + Br^-(aq)$	0.0732
$2H^+(aq) + 2e^- \rightleftharpoons H_2(g)$	0.000

↑ Stronger Oxidizing Agents (upward) — Stronger Reducing Agents ↓ (downward)

potential is the negative of the reduction potential, the corresponding metals—sodium, potassium and calcium—have high oxidation potentials. They tend to give up electrons and are good reducing agents. Such metals are easily oxidized and occur in nature only as positive ions.

EXAMPLE 20.6

Which should be the stronger oxidizing agent, an acidified solution of hydrogen peroxide or an acidified solution of potassium dichromate? Assume 1 *M* concentrations.

SOLUTION The standard reduction potentials are:

Half-Reaction	$E°$ (V)
$2H^+(aq) + 2e^- \rightleftharpoons H_2(g)$	0.000
$Pb^{2+}(aq) + 2e^- \rightleftharpoons Pb(s)$	−0.127
$Sn^{2+}(aq) + 2e^- \rightleftharpoons Sn(s)$	−0.14
$AgI(s) + e^- \rightleftharpoons Ag(s) + I^-(aq)$	−0.152
$Ni^{2+}(aq) + 2e^- \rightleftharpoons Ni(s)$	−0.236
$Co^{2+}(aq) + 2e^- \rightleftharpoons Co(s)$	−0.282
$Ag(CN)_2^-(aq) + e^- \rightleftharpoons Ag(s) + 2CN^-(aq)$	−0.31
$Cd^{2+}(aq) + 2e^- \rightleftharpoons Cd(s)$	−0.402
$Fe^{2+}(aq) + 2e^- \rightleftharpoons Fe(s)$	−0.409
$Cr^{3+}(aq) + 3e^- \rightleftharpoons Cr(s)$	−0.74
$Zn^{2+}(aq) + 2e^- \rightleftharpoons Zn(s)$	−0.762
$2H_2O(l) + 2e^- \rightleftharpoons H_2(g) + 2OH^-(aq)$	−0.828
$SO_4^{2-}(aq) + H_2O(l) + 2e^- \rightleftharpoons SO_3^{2-}(aq) + 2OH^-(aq)$	−0.936
$Zn(NH_3)_4^{2+}(aq) + 2e^- \rightleftharpoons Zn(s) + 4NH_3(aq)$	−1.015
$Mn^{2+}(aq) + 2e^- \rightleftharpoons Mn(s)$	−1.182
$Al^{3+}(aq) + 3e^- \rightleftharpoons Al(s)$	−1.68
$Mg^{2+}(aq) + 2e^- \rightleftharpoons Mg(s)$	−2.357
$Na^+(aq) + e^- \rightleftharpoons Na(s)$	−2.714
$Ca^{2+}(aq) + 2e^- \rightleftharpoons Ca(s)$	−2.869
$Ba^{2+}(aq) + 2e^- \rightleftharpoons Ba(s)$	−2.906
$K^+(aq) + e^- \rightleftharpoons K(s)$	−2.936
$Li^+(aq) + e^- \rightleftharpoons Li(s)$	−3.040

Stronger Oxidizing Agents ↑

Stronger Reducing Agents ↓

[a]Each substance is in its standard state; that is, approximately 1 *M* for dissolved species and 1 atm for gases (see Table 19.1, p. 909).

$$H_2O_2(aq) + 2H^+(aq) + 2e^- \rightleftharpoons 2H_2O(l) \qquad E° = 1.76\ V$$

$$Cr_2O_7^{2-}(aq) + 14H^+(aq) + 6e^- \rightleftharpoons 2Cr^{3+}(aq) + 7H_2O(l) \qquad E° = 1.33V$$

An acidified solution of hydrogen peroxide has a more positive reduction potential and thus a greater tendency to accept electrons than an acidified solution of potassium dichromate; the peroxide solution will be the stronger oxidizing agent.

EXAMPLE 20.7

Which metal, potassium or lithium, is the stronger reducing agent in an aqueous medium?

SOLUTION

The oxidation potentials of potassium and lithium are obtained by reversing the half-reactions in Table 20.1 and changing the signs of the $E°$ values:

$$Li(s) \rightleftharpoons Li^+(aq) + e^- \qquad E°_{oxid} = 3.040\ V$$

$$K(s) \rightleftharpoons K^+(aq) + e^- \qquad E°_{oxid} = 2.936\ V$$

Lithium has a higher (more positive) oxidation potential and a greater tendency to give up electrons than potassium; lithium will be the stronger reducing agent.

PRACTICE EXERCISE 20.7

(a) Which ion is more easily oxidized in aqueous solution, Cl^- or I^-? (b) Which ion is more easily reduced in aqueous solution, Cr^{3+} or Co^{2+}?

The more positive the reduction potential, the greater the tendency to accept electrons. As a result, *when two half-reactions are coupled, the reaction with the higher (more positive) reduction potential will proceed as a reduction; the other will proceed as an oxidation.*

EXAMPLE 20.8

Use the table of reduction potentials to predict whether bromine will oxidize manganese metal. If it will, write an equation for the reaction.

SOLUTION

Table 20.1 provides the following reduction potentials:

$$Br_2(l) + 2e^- \rightleftharpoons 2Br^-(aq) \qquad E° = 1.08\ V$$

$$Mn^{2+}(aq) + 2e^- \rightleftharpoons Mn(s) \qquad E° = -1.18\ V$$

The bromine half-reaction has the more positive reduction potential; it will go forward as a reduction and the manganese half-reaction will go in reverse as an oxidation. (The direction of each half-reaction is indicated by the curved arrow drawn between the equations.) Hence, bromine will oxidize manganese metal. The overall reaction will be the sum of the oxidation and reduction half-reactions

reduction: $Br_2(l) + 2e^- \rightarrow 2Br^-(aq)$

oxidation: $Mn(s) \rightarrow Mn^{2+}(aq) + 2e^-$

sum: $Br_2(l) + Mn(s) \rightarrow 2Br^-(aq) + Mn^{2+}(aq)$

EXAMPLE 20.9

Use Table 20.1 to find reagents that might be used to produce chlorine from sodium chloride.

SOLUTION

The following half-reaction, which involves both Cl_2 and Cl^-, is listed in Table 20.1:

$$Cl_2(g) + 2e^- \rightleftharpoons 2Cl^-(aq) \qquad E° = 1.36\ V$$

We must find a reagent that will force this half-reaction to go in reverse, that is, to proceed as an oxidation. We will look in the top part of Table 20.1 for an oxidizing agent with a reduction potential greater than 1.36 V. Fluorine (F_2) with a reduction potential of 2.89 V, would be the most efficient oxidizing agent, but working with fluorine is dangerous and impractical. An acidified solution of permanganate ion (MnO_4^-), with a reduction potential of 1.51 V, would be a better choice.

PRACTICE EXERCISE 20.8

Will aqueous nitric acid oxidize (a) bromide ions and (b) iodide ions? (*Hint*: Aqueous nitric acid contains $H^+(aq)$ and $NO_3^-(aq)$.)

Remember that predictions based on standard potentials are valid only for aqueous solutions in which each substance is in its standard state. In Section 20.4 we will learn to estimate the effect of changing concentrations, and we will find that such effects are usually small.

Calculating Standard Cell Voltages and Free-Energy Changes

Recall (Equation 20.6) that the voltage of a galvanic cell is the sum of the half-cell voltages. This is illustrated in the next example.

EXAMPLE 20.10

A galvanic cell is based on the reaction of iron(II) ions with bromine. (a) Write the equation for the reaction. (b) Calculate $E°$ for the cell.

SOLUTION

(a) The reduction potentials from Table 20.1 are

$$Br_2(l) + 2e^- \rightleftharpoons 2Br^-(aq) \qquad E° = 1.08\ V$$

$$Fe^{3+}(aq) + e^- \rightleftharpoons Fe^{2+}(aq) \qquad E° = 0.77\ V$$

The bromine half-reaction has the more positive reduction potential, so it will go forward as a reduction, and the iron half-reaction will go in reverse as an oxidation. The half-reactions are

$$\text{reduction:}\quad Br_2(l) + 2e^- \rightarrow 2Br^-(aq) \qquad E°_{red} = 1.08\ V$$

$$\text{oxidation:}\quad 2Fe^{2+}(aq) \rightarrow 2Fe^{3+}(aq) + 2e^- \qquad E°_{oxid} = -0.77\ V$$

The oxidation half-reaction is multiplied by 2 because two Fe^{2+} ions are oxidized for every Br_2 molecule reduced. *Note carefully that the oxidation potential stays the same when the equation is doubled—voltage is the number of joules per coulomb, and energy per coulomb is independent of the amount of the reaction.* The cell reaction is the sum of the adjusted half-reactions:

$$Br_2(l) + 2Fe^{2+}(aq) \longrightarrow 2Br^-(aq) + 2Fe^{3+}(aq)$$

(b) The cell voltage is obtained by substituting $E^\circ_{oxid} = -0.77$ V and $E^\circ_{red} = 1.08$ V into Equation 20.6:

$$E^\circ = E^\circ_{oxid} + E^\circ_{red} = -0.77 \text{ V} + 1.08 \text{ V} = 0.31 \text{ V}$$

PRACTICE EXERCISE 20.9

A galvanic cell is based on the following reaction:

$$2MnO_4^-(aq) + 10I^-(aq) + 16H^+(aq) \longrightarrow 2Mn^{2+}(aq) + 5I_2(s) + 8H_2O(l)$$

(a) Write the half-reactions. (b) Calculate E° for the cell.

If E° for a redox reaction is known, then the standard free-energy change can be calculated from Equation 20.5.

EXAMPLE 20.11

Use data from Table 20.1 to (a) predict whether the following reaction is spontaneous and (b) calculate ΔG°:

$$Cd^{2+}(1\,M) + 2Ag(s) \longrightarrow Cd(s) + 2Ag^+(1\,M)$$

SOLUTION

(a) The reaction is the sum of the following half-reactions:

$$Cd^{2+}(1\,M) + 2e^- \longrightarrow Cd(s) \qquad E^\circ_{red} = -0.40 \text{ V}$$

$$2Ag(s) \longrightarrow 2Ag^+(1\,M) + 2e^- \qquad E^\circ_{oxid} = -E^\circ_{red} = -0.80 \text{ V}$$

The total voltage is

$$E^\circ = -0.80 \text{ V} - 0.40 \text{ V} = -1.20 \text{ V}$$

Since E° is negative, ΔG° is positive and the reaction is not spontaneous under standard-state conditions.

LEARNING HINT

Remember that a spontaneous redox reaction is associated with a positive E and a negative ΔG.

(b) The value of ΔG° can be obtained from Equation 20.5. Two moles of electrons are transferred for each mole of Cd^{2+} reduced; $n = 2$ mol e^-. $E^\circ = -1.20$ V $= -1.20$ J/C. Substituting these quantities and $F = 96{,}485$ C/mol e^- into Equation 20.5 gives

$$\begin{aligned}\Delta G^\circ &= -nFE^\circ \\ &= -2 \text{ mol e} \times \frac{96{,}485 \text{ C}}{1 \text{ mol e}^-} \times \frac{-1.20 \text{ J}}{1 \text{ C}} \\ &= 2.32 \times 10^5 \text{ J} = 232 \text{ kJ}\end{aligned}$$

Observe that ΔG° is positive, as predicted in Part (a).

PRACTICE EXERCISE 20.10

Use data from Table 20.1 to calculate the standard free-energy change of the following reaction:

$$Cu(s) + 2Ag^{+}(1\ M) \rightarrow Cu^{2+}(1\ M) + 2Ag(s)$$

20.4 EMF AND CONCENTRATION

The concentration of each dissolved reactant and product changes during the operation of a galvanic cell. These changes are accompanied by a decrease in the voltage of the cell and a corresponding decrease in the ability of the cell to do electrical work.

The free-energy change of a cell reaction is related to the cell voltage E by Equation 20.5a:

$$\Delta G = -nFE \tag{20.5a}$$

The free-energy change is also related to the reaction quotient Q by Equation 19.3:

$$\Delta G = \Delta G^\circ + RT \ln Q \tag{19.3}$$

The dependence of cell voltage on concentration is found by substituting Equation 20.5 into Equation 19.3:

$$-nFE = -nFE^\circ + RT \ln Q$$

Rearrangement gives

$$E = E^\circ - \frac{RT}{nF} \ln Q \tag{20.7a}$$

In a standard cell the concentrations and partial pressures are all unity; thus $Q = 1$, $\ln Q = 0$, and $E = E^\circ$. When a cell is in use, the reactants form products; hence, Q and $\ln Q$ increase in magnitude. (Recall that Q contains product concentrations divided by reactant concentrations.) Equation 20.7a shows that the cell voltage will decrease accordingly.

Equation 20.7a is called the **Nernst equation** in honor of Walther Hermann Nernst (1864–1941), a German physical chemist who, in addition to his work in electrochemistry, helped develop the third law of thermodynamics. A convenient form of the Nernst equation for use at room temperature is obtained by substituting $R = 8.314$ J/mol·K, $F = 96{,}485$ C/mol e^-, $T = 298.15$ K, and $\ln Q = 2.303 \log Q$ into Equation 20.7a. The result is

$$E = E^\circ - \frac{0.0592}{n} \log Q \tag{20.7b}$$

If n is expressed as moles of electrons per mole of reaction (mol e^-/mol), then the units associated with RT/nF or $0.0592/n$ will be volts. (Be sure to convince yourself of this statement.)

LEARNING HINT

Do not confuse the reaction quotient Q with the symbol Q used for coulombs earlier in this chapter.

If you prefer to work with natural logarithms, another form of the Nernst equation is

$$E = E^\circ - \frac{0.0257}{n} \ln Q$$

EXAMPLE 20.12

For the Daniell cell illustrated in Figure 20.2, $E°$ is 1.10 V. Calculate the voltage when the concentration of zinc sulfate is 1.2 *M* and the concentration of copper sulfate is 1.2×10^{-3} *M*.

SOLUTION

The cell reaction and standard voltage are

$$Cu^{2+}(aq) + Zn(s) \longrightarrow Zn^{2+}(aq) + Cu(s) \qquad E° = 1.10\ V$$

The reaction quotient is

$$Q = \frac{[Zn^{2+}]}{[Cu^{2+}]} = \frac{1.2}{1.2 \times 10^{-3}} = 1.0 \times 10^3$$

(Recall that solids do not appear in the reaction quotient.) Two moles of electrons are transferred per mole of Cu^{2+} reduced, therefore $n = 2$. Substituting Q, n, and $E° =$ 1.10 V into the Nernst equation gives

$$E = 1.10\ V - \frac{0.0592\ V}{2} \times \log(1.0 \times 10^3) = 1.01\ V$$

PRACTICE EXERCISE 20.11

Refer to Example 20.10. Calculate the cell voltage when the concentrations of Fe^{2+} and Fe^{3+} are each 0.10 *M* and the concentration of Br^- is 0.050 *M*.

The voltage calculated for the Daniell cell in Example 20.12 is less than the standard-state voltage of 1.10 V. We could have predicted this without a calculation. The equation for the cell reaction shows that an increase in the concentration of zinc ions and a decrease in the concentration of copper ions will lower the driving force of the reaction, thus decreasing the cell voltage.

The Nernst equation can also be applied to half-cell potentials, as illustrated in the next example.

EXAMPLE 20.13

Calculate the reduction potential of a half-cell consisting of a zinc electrode in 5.0×10^{-4} *M* zinc sulfate solution.

SOLUTION

The half-reaction and standard reduction potential are

$$Zn^{2+}(aq) + 2e^- \rightleftharpoons Zn(s) \qquad E° = -0.76\ V$$

Before making the calculation, let us try to predict the direction of the voltage change. The concentration of Zn^{2+} is less than 1 *M*; hence, the forward reaction will have a smaller driving force than the reaction in a standard zinc half-cell. The reduction potential should be less (more negative) than −0.76 V. We can verify our prediction by substituting $n = 2$, $E° = -0.76$ V, and $Q = 1/[Zn^{2+}] = 1/(5.0 \times 10^{-4}) = 2.0 \times 10^3$ into the Nernst equation:

$$E = -0.76\text{ V} - \frac{0.0592\text{ V}}{2} \times \log(2.0 \times 10^3) = -0.86\text{ V}$$

PRACTICE EXERCISE 20.12

Calculate the reduction potential of a hydrogen electrode dipped into a solution of pH 7.00. Assume that the hydrogen gas bubbling around the electrode remains at 1 atm pressure.

$E°$ and the Equilibrium Constant

When a galvanic cell produces a large voltage, the reaction occurring in the cell is far from equilibrium, and the free-energy change is negative. With continued use, the cell reaction proceeds toward equilibrium; the voltage of the cell decreases and so does its ability to produce useful electrical work. At equilibrium, when the cell is "dead," there is no more driving force and both E and ΔG are zero.

The standard free-energy change of a cell reaction is given by Equation 20.5b:

$$\Delta G° = -nFE° \tag{20.5b}$$

The standard free-energy change is also related to the equilibrium constant by Equation 19.4:

$$\Delta G° = -RT \ln K \tag{19.4}$$

Combining these equations gives

$$-nFE° = -RT \ln K$$

and

$$E° = \frac{RT}{nF} \ln K \tag{20.8a}$$

LEARNING HINT

Equation 20.8a can also be obtained by substituting the equilibrium values $E = 0$ and $Q = K$ into the Nernst equation (Equation 20.7a).

Equation 20.8a, like the Nernst equation, has a convenient room temperature version:

$$E° = \frac{0.0592}{n} \log K \tag{20.8b}$$

This equation allows us to calculate the equilibrium constant for any reaction whose standard voltage is known; that is, for any reaction that can be written as the sum of two redox half-reactions. EMF measurements have good precision, and electrochemical methods are often used when accurate equilibrium constants are needed.

CHEMICAL INSIGHT

ANALYTICAL GALVANIC CELLS

The sensitivity of galvanic cell voltage to concentration can be put to use measuring concentrations of specific ions in solution. An analytical galvanic cell contains a *reference electrode* with a known half-cell potential coupled to an *ion-sensitive electrode* whose half-cell potential varies with the concentration of the ion to which the electrode is sensitive.

The analytical galvanic cell most commonly encountered in chemistry laboratories is the pH meter (Figure 20.8). The reference electrode in this cell is a saturated *calomel electrode*. Calomel is mercury(I) chloride, the electrolyte is saturated KCl, and the electrode reaction is

$$Hg_2Cl_2(s) + 2e^- \rightleftharpoons 2Hg(l) + 2Cl^-(aq)$$

The half-cell potential of the calomel electrode remains constant during a pH measurement because KCl, the only dissolved substance, is in a saturated solution, where its concentration does not change.

The ion-sensitive electrode in the pH meter is called a *glass electrode*. It consists of a Ag/AgCl electrode surrounded by 1 *M* HCl, all sealed inside a thin glass membrane. The driving force of the half-reaction

$$AgCl(s) + e^- \rightleftharpoons Ag(s) + Cl^-(1\ M)$$

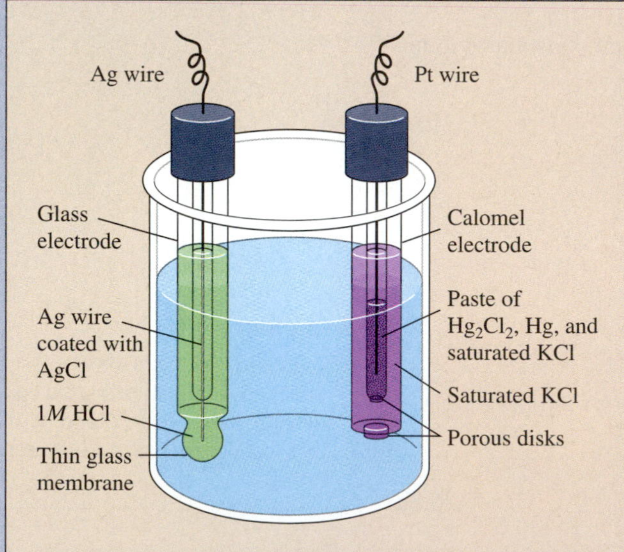

Figure 20.8
Schematic diagram of a pH meter.

is influenced by a potential difference across the glass membrane separating the 1 *M* HCl solution from the unknown solution. This potential difference, called a *membrane potential*, is caused by hydrogen ions adsorbed on the inner and outer glass surfaces of the electrode. The degree of adsorption and thus the magnitude of the potential difference vary with the concentration of H^+ in the unknown solution. The half-cell potential of the glass electrode depends on the H^+ concentration according to the equation:

$$E(\text{glass electrode}) = E' + 0.0592\text{ V} \times \log [H^+]$$
$$= E' - 0.0592\text{ V} \times \text{pH}$$

where E' is a constant that is characteristic of the electrode. Because pH is the only variable that contributes to the pH meter voltage, the meter dial can be calibrated to read pH directly.

The glass–calomel pH meter is extremely versatile and can be used to determine pH under a wide variety of conditions. Miniature electrode systems have been developed for finding the pH of small amounts (a drop or less) of solution. A pH meter is even available for observing gastric acidity—it is enclosed, along with a battery and radio frequency transmitter, in a small polyethylene capsule that can be swallowed. An antenna wrapped around the waist picks up the signal.

Electrodes similar to the glass electrode are used for measuring the concentrations of ions other than H^+. Their development has depended on finding membranes—special glasses, cellophane, Teflon, polyethylene, and inorganic crystals—that will selectively adsorb the ions to be tested. Electrodes are available to measure cations such as Na^+, K^+, Ca^{2+}, Mg^{2+}, Ag^+, and $NH_4{}^+$, and molecular species such as dissolved oxygen and urea.

A *probe* is a galvanic sensor that contains both the concentration-sensitive electrode and the reference electrode in a single compact device. Miniaturization methods have produced needlelike microprobes for monitoring ion and gas concentrations in living tissue. Some electrodes are so small that they can even be inserted into individual living cells without disrupting their normal activity. For example, microsensors with a tip diameter of about 0.5 μm can be implanted into single cells and used to detect NO levels (NO is an important bioregulator; see p. 507) as low as 10^{-2} attomol (10^{-20} mol). Still smaller electrodes, called *nanodes*, have tip diameters of about 2000 pm—the width of about seven platinum atoms.

EXAMPLE 20.14

For the Daniell cell, $E°$ is 1.10 V at 25°C. Calculate the equilibrium constant for the reaction

$$Cu^{2+}(aq) + Zn(s) \rightleftharpoons Zn^{2+}(aq) + Cu(s)$$

SOLUTION Two moles of electrons are transferred per mole of reaction; hence $n = 2$. Substituting $E° = 1.10$ V and $n = 2$ into Equation 20.8b gives

$$1.10\text{ V} = \frac{0.0592\text{ V}}{2}\log K$$

Hence $\log K = 37.2$

and $K = \text{antilog } 37.2 = 10^{37.2} = 2 \times 10^{37}$

LEARNING HINT

Remember to use common logarithms (logarithms to the base 10) if you work with Equations 20.7b and 20.8b.

PRACTICE EXERCISE 20.13

The equilibrium constant for the reaction

$$2Cu^{+}(aq) \rightleftharpoons Cu(s) + Cu^{2+}(aq)$$

is $K = 1.1 \times 10^6$ at 25°C. Find (a) $E°$ and (b) $\Delta G°$. (*Hint*: Use oxidation number changes to find n.)

As the Daniell cell reaction illustrates, any reaction with a substantial $E°$ will also have a large equilibrium constant. In such reactions, the conversion of reactants to products is virtually complete at equilibrium. EMF measurements can also be used to determine very small equilibrium constants, even equilibrium constants for nonredox reactions, as demonstrated in the next example.

EXAMPLE 20.15

Use data from Table 20.1 to calculate the solubility product constant for AgCl.

SOLUTION The K_{sp} is the equilibrium constant for the reaction

$$AgCl(s) \rightleftharpoons Ag^{+}(aq) + Cl^{-}(aq)$$

This is not a redox reaction, but it can be written as the sum of two half-reactions whose potentials can be found in Table 20.1:

$$Ag(s) \rightarrow Ag^{+}(aq) + e^{-} \qquad E°_{oxid} = -E°_{red} = -0.799\text{ V}$$

$$AgCl(s) + e^{-} \rightarrow Ag(s) + Cl^{-}(aq) \qquad E°_{red} = 0.222\text{ V}$$

The sum of the half-cell voltages is

$$E° = E°_{oxid} + E°_{red} = -0.799\text{ V} + 0.222\text{ V} = -0.577\text{ V}$$

The negative sign of $E°$ shows that AgCl will not spontaneously dissolve under standard-state conditions. (When the concentrations of Ag^{+} and Cl^{-} are each 1 *M*, a precipitate of AgCl will form.) Nevertheless, we can still use $E°$ to calculate a value for the solubility product constant. Substituting $n = 1$ (each half-reaction involves 1 mol of electrons) and $E° = -0.577$ V into Equation 20.8b gives

$$-0.577\text{ V} = \frac{0.0592\text{ V}}{1} \times \log K_{sp}$$

Hence $\log K_{sp} = -9.75$

and $K_{sp} = \text{antilog}(-9.75) = 10^{-9.75} = 1.8 \times 10^{-10}$

PRACTICE EXERCISE 20.14

Use $E°$ values from Table 20.1 to find K for the reaction

$$Zn^{2+}(aq) + 4NH_3(aq) \rightleftharpoons Zn(NH_3)_4^{2+}(aq)$$

20.5 BATTERIES AND FUEL CELLS

A **battery** is a package of one or more galvanic cells used to produce and store electrical energy. The principal advantages offered by batteries are convenience and portability rather than efficient use of energy. An ordinary flashlight battery, for example, provides less energy than is required to manufacture it.

Some batteries can be recharged by using an outside power source to reverse the electrode reactions. In theory, if E is the voltage of the battery, the minimum voltage required for recharging is also E. In practice, however, the actual voltage required for recharging is always greater than the theoretical minimum. The difference between the actual voltage and the theoretical minimum is called **overvoltage**. Its magnitude depends not only on the cell reaction and concentrations, but also on factors such as cell construction, impurities on the electrode surface, and lack of mixing in the electrolyte. These factors affect the rates of ion transport to and from the electrodes and also the rates of electron transfer at the electrode surfaces; thus overvoltage is more a matter of kinetics than of thermodynamics. Because of overvoltage, the energy output of a rechargeable battery will always be less than the energy required to charge it.

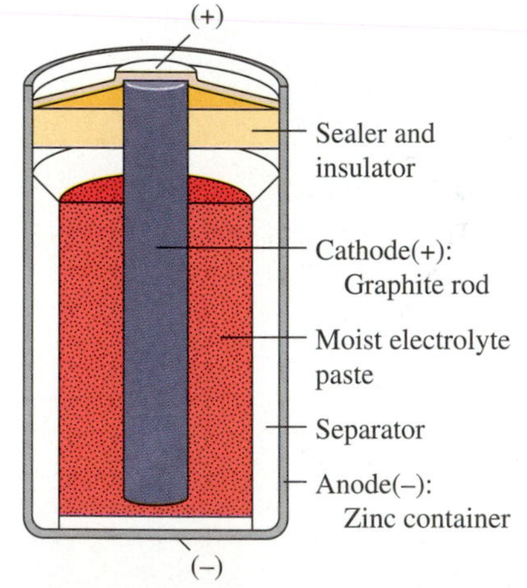

Figure 20.9

The Leclanché dry cell or ordinary "flashlight battery." The anode is a zinc container covered with an insulating wrapper; the cathode is the carbon rod at the center. The electrolyte paste contains $ZnCl_2$, NH_4Cl, MnO_2, starch, graphite, and water. A fresh dry cell delivers about 1.5 V.

Zinc Batteries

The original "flashlight battery" (Figure 20.9 and Demonstration 20.2) was developed in 1865 by Georges Leclanché. It is often called a *dry cell* because the electrolyte is incorporated into a nonflowing paste. The zinc anode, which is covered with an insulating wrapper, forms the container. The cathode is an inert graphite rod surrounded by a mixture of manganese dioxide (the oxidizing agent), zinc and ammonium chlorides, carbon and starch powders, and enough water to make a paste. The following equations give a simplified version of the electrode reactions:

$$Zn(s) \rightarrow Zn^{2+}(aq) + 2e^- \qquad \text{anode}$$

$$2MnO_2(s) + 2NH_4^+(aq) + 2e^- \rightarrow Mn_2O_3(s) + 2NH_3(aq) + H_2O(l) \qquad \text{cathode}$$

The Zn^{2+} produced at the anode combines with the NH_3 produced at the cathode to form $Zn(NH_3)_4^{2+}$, a soluble complex ion. Excess ammonia may form when large currents are drawn from the battery. The gaseous ammonia surrounds the cathode,

DEMONSTRATION 20.2 MAKING A ZINC-CARBON BATTERY

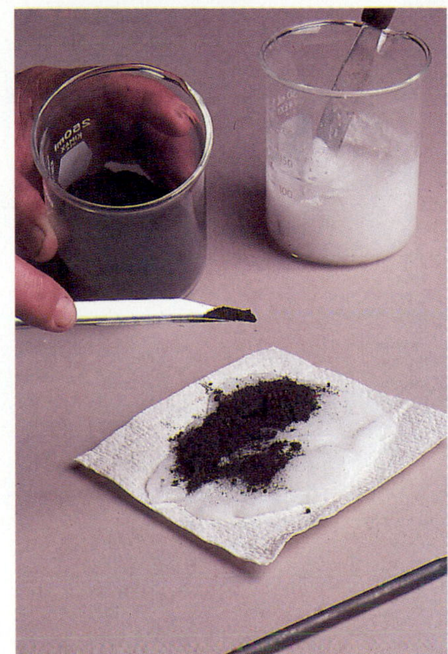

Manganese dioxide (MnO_2) is added to a moist paste containing zinc chloride ($ZnCl_2$), ammonium chloride (NH_4Cl), and starch. Carbon powder will also be added.

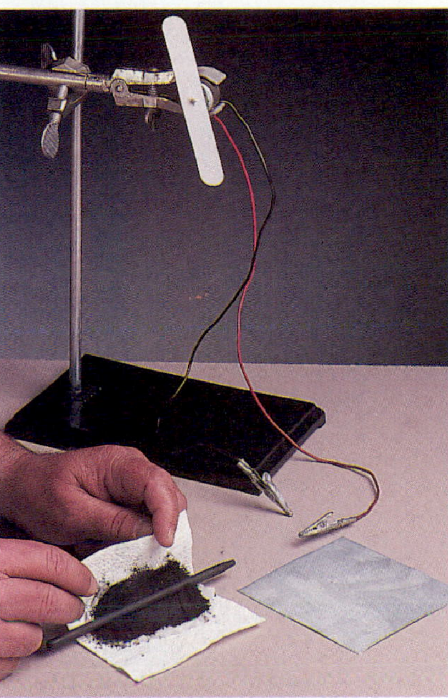

A graphite rod is placed in the electrolyte mixture. The sheet of zinc metal (foreground) and the mixture will be wrapped around the rod.

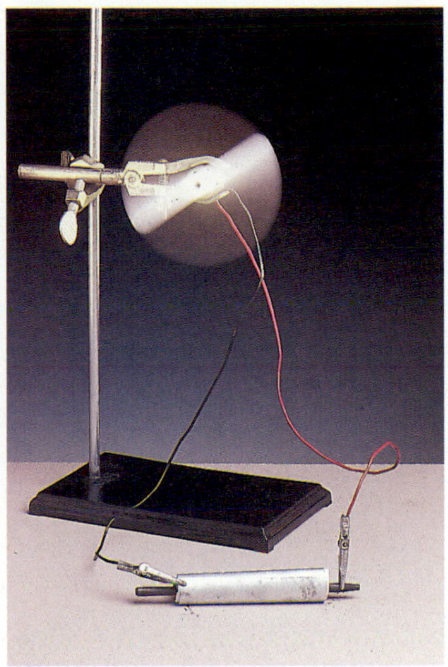

The battery in operation. The zinc metal is the anode; the graphite rod is the cathode.

insulating it from the electrolyte, and the voltage drops. The battery will "bounce back" after a period of rest because the ammonia eventually diffuses into the paste and reacts with the zinc ions. A fresh dry cell delivers about 1.5 volts and cannot be recharged. Higher voltages are obtained by connecting several dry cells in series; that is, by connecting the anode of one to the cathode of another. A 9-volt battery, for example, contains six dry cells packaged together.

The *alkaline dry cell* is more expensive than the Leclanché cell, but it is also more efficient. Again, zinc is the anode and manganese dioxide the oxidizing agent (Figure 20.10). The electrolyte is 40% KOH saturated with zinc oxide (ZnO). The electrode reactions in the alkaline cell are

$$Zn(s) + 2OH^-(aq) \rightarrow ZnO(s) + H_2O(l) + 2e^- \qquad \text{anode}$$

$$2MnO_2(s) + H_2O(l) + 2e^- \rightarrow Mn_2O_3(s) + 2OH^-(aq) \qquad \text{cathode}$$

The alkaline cell has a longer shelf life and a more reliable operating life than the Leclanché cell, partially because of better construction and purity of starting materials, partially because the alkaline paste causes less corrosion than the acidic ammonium chloride paste, and partially because it produces no gaseous NH_3.

The *mercury cell*, developed in 1942, is another zinc dry cell devised for use in small appliances such as watches (Figure 20.11). It delivers about 1.35 volts and has

LEARNING HINT

The voltage of a dry cell, like that of any other cell, is governed by the Nernst equation. In the *Leclanché dry cell*, the product concentrations increase with time and the battery voltage drops with use. In the *alkaline dry cell* and *mercury cell*, most of the substances are solids and liquids that do not appear in the reaction quotients (Q). The OH^- concentration is constant because it is consumed and produced at the same rate. Thus the reaction quotients do not change as the reactions proceed, and the cell voltages remain more or less constant.

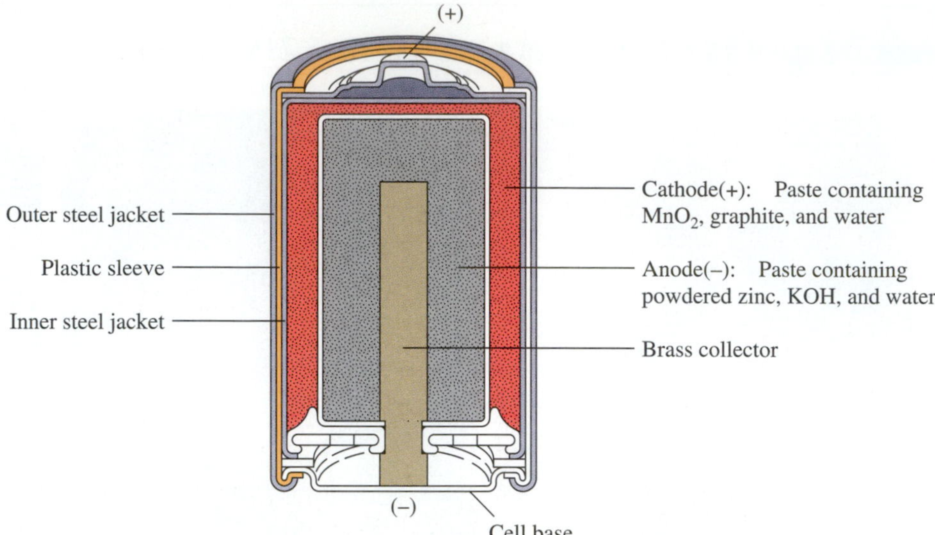

Figure 20.10

The alkaline dry cell. The anode is a paste of zinc, KOH, and water, which donates electrons to the cell base via a brass collector. The cathode is a paste of MnO_2, graphite, and water, which takes electrons from the inner steel case. A plastic sleeve separates the inner steel case from the outer steel jacket.

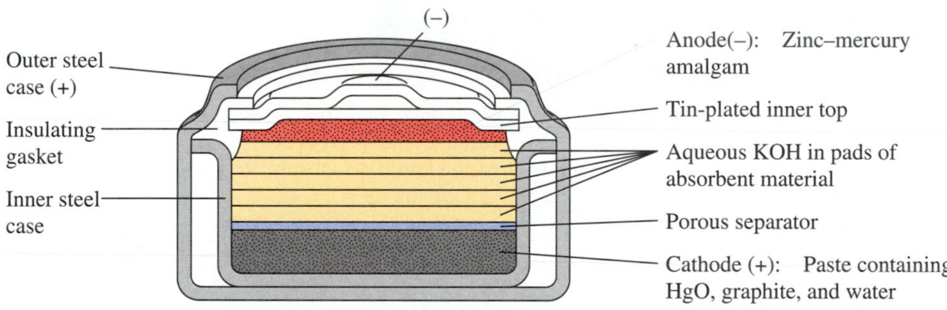

Figure 20.11

A mercury battery. A zinc–mercury amalgam is the anode; the cathode is a paste of HgO, graphite, and water. Mercury batteries, some of which are smaller than a pencil eraser, deliver about 1.35 V.

the advantage of maintaining a fairly constant voltage during its lifetime. In spite of its high initial expense, it is an efficient choice for hearing aids and other devices where reliability is of the utmost importance. The electrode reactions are

$$Zn(s) + 2OH^-(aq) \rightarrow ZnO(s) + H_2O(l) + 2e^- \qquad \text{anode}$$

$$HgO(s) + H_2O(l) + 2e^- \rightarrow Hg(l) + 2OH^-(aq) \qquad \text{cathode}$$

Storage Batteries

The automobile or *lead storage battery* is the best-known example of a battery that can be recharged by applying an outside voltage to reverse the electrode reactions. Each cell consists of alternating sheets of lead and lead dioxide (Figure 20.12). The lead sheets are connected to form the negative electrode, and the lead dioxide sheets are connected to form the positive electrode. The large electrode areas allow a high rate of electron transfer so that more current can be drawn. The electrolyte, commonly called "battery acid," is a 30% solution of sulfuric acid with a density of 1.28 g/mL. The spontaneous discharging reactions are

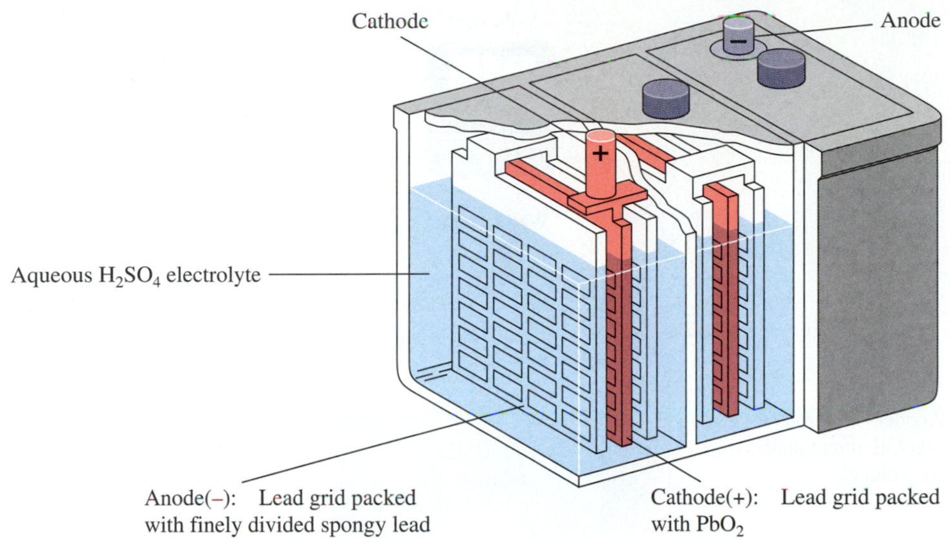

Figure 20.12

A lead storage cell. The interconnected lead sheets form the anode; the sheets filled with lead dioxide form the cathode. The electrolyte is sulfuric acid. A fully charged cell delivers about 2.0 V. This 6-V battery consists of three cells connected in series (connections not shown).

$$Pb(s) + HSO_4^-(aq) \rightarrow PbSO_4(s) + H^+(aq) + 2e^- \qquad \text{anode}$$

$$PbO_2(s) + HSO_4^-(aq) + 3H^+(aq) + 2e^- \rightarrow PbSO_4(s) + 2H_2O(l) \qquad \text{cathode}$$

A lead–acid cell in good condition provides about 2 volts. Higher voltages are obtained by connecting several cells in series; the usual 12-volt car battery consists of six such cells. The cell equations show that discharging depletes the sulfuric acid electrolyte and deposits solid lead sulfate on both electrodes. The electrolyte density can be measured by a hydrometer; if it is below 1.20 g/mL, the battery should be recharged.

Recharging reverses the above reactions; the outside power source sends electrons into the lead plates and withdraws them from the lead dioxide plates. The lead sulfate adhering to each plate is reconverted to the original electrode material, and sulfuric acid is also regenerated.

More than 2 billion batteries are sold each year in the United States alone. These batteries contain toxic and carcinogenic metals, and they should be collected and recycled rather than dispersed in trash, landfills, and incinerators.

No rechargeable cell can compete with the lead storage battery for low cost and availability of materials. Other more expensive cells, however, are more compact, have longer lives, and are more easily recharged. The *nickel–cadmium cell*, for example, is used in hand calculators, electronic flash units, and other small battery-operated devices. The nickel–cadmium cell has electrodes of cadmium and NiO(OH), and the electrolyte is a 20% aqueous KOH solution. The discharging reactions are

$$Cd(s) + 2OH^-(aq) \rightarrow Cd(OH)_2(s) + 2e^- \qquad \text{anode}$$

$$NiO(OH)(s) + H_2O(l) + e^- \rightarrow Ni(OH)_2(s) + OH^-(aq) \qquad \text{cathode}$$

As in the lead storage battery, the solid discharge products remain on the electrodes, so that the reactions are easily reversed by the recharging voltage.

Fuel Cells

A **fuel cell** is a galvanic cell in which reactants are continuously fed in and products are continuously removed. Fuel cells differ from batteries in that they do not store

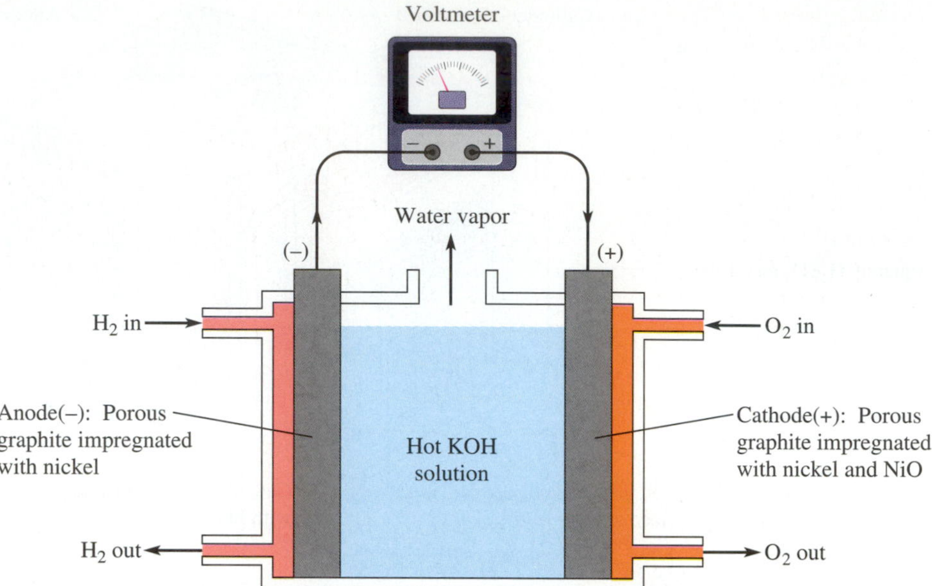

Figure 20.13

One type of fuel cell. Porous carbon electrodes are impregnated with catalysts and saturated with an electrolyte of hot potassium hydroxide. Hydrogen and oxygen gases circulate beside the porous plates; electrode reactions occur within the pores. The operating temperature is about 140°C; the maximum voltage is about 0.9 V.

electrical energy. The imperatives of the space program have brought fuel cell research to limited fruition. When the Apollo astronauts traveled to the moon, electricity and pure water for their spacecraft were provided by a fuel cell that used hydrogen as the fuel and oxygen as the oxidizing agent (Figure 20.13):

$$H_2(g) + 2OH^-(aq) \rightarrow 2H_2O(l) + 2e^- \quad \text{anode}$$

$$O_2(g) + 2H_2O(l) + 4e^- \rightarrow 4OH^-(aq) \quad \text{cathode}$$

$$2H_2(g) + O_2(g) \rightarrow 2H_2O(l) \quad \text{net}$$

This cell was able to produce 3 kilowatts (3000 J/s) for about 14 days without accumulating objectionable by-products.

The chemical energy of a fuel such as coal or natural gas is usually converted into electrical energy in two stages. First the fuel is burned to provide heat; then the heat energy is used to generate electricity. The total efficiency of the conventional process is 40% or less. The conversion is much more efficient in a fuel cell because electricity is produced in a single step. In H_2/O_2 fuel cells, for example, about 70% of the available chemical energy is transformed into electrical energy. Gas companies are currently trying to develop small fuel cells that will generate electricity by the oxidation of natural gas; the only waste products would be carbon dioxide and water. The environmental impact of these cells would be minimal, and they might one day be used to provide electricity for apartment complexes and small businesses. A 4.8-megawatt gas-fueled experimental fuel cell system is currently in operation in Tokyo.

20.6 CORROSION

Corrosion is a natural galvanic process in which metals become pitted and eaten away by slow oxidation. We can see the effects of corrosion all around us in the form of rusted iron, pitted aluminum, tarnished statues and jewelry, and so forth (Figure 20.14). Much of the annual iron production in this country is used to replace products lost by rusting, and the yearly economic loss due to all corrosion is probably in excess of 200 billion dollars.

Figure 20.14
A rusted chain.

The corrosion of iron and steel produces the familiar red rust that is a hydrated form of Fe_2O_3. The reaction proceeds via a galvanic mechanism in which the electrolyte consists of water containing dissolved impurities. Corrosion begins when regions of iron covered by water become anodic and discharge Fe^{2+} ions into the water (Figure 20.15):

$$Fe(s) \rightarrow Fe^{2+}(aq) + 2e^-$$

The electrons travel to some other region on the metal surface, where a variety of cathode reactions can occur. In an acidic medium, atmospheric oxygen is quickly reduced to water:

$$4H^+(aq) + O_2(g) + 4e^- \rightarrow 2H_2O(l) \qquad E^\circ_{red} = 1.23\ V$$

A less vigorous reduction occurs at a higher pH:

$$O_2(g) + 2H_2O(l) + 4e^- \rightarrow 4OH^-(aq) \qquad E^\circ_{red} = 0.40\ V$$

Corrosion requires both oxygen and water.

Once the minigalvanic system is established, the two half-reactions promote each other, and the anode region under the water develops into a corroded pit. Rust forms when the Fe^{2+} ions are oxidized by air to Fe^{3+} ions:

$$4Fe^{2+}(aq) + O_2(g) + 4H^+(aq) \rightarrow 4Fe^{3+}(aq) + 2H_2O(l)$$

$$2Fe^{3+}(aq) + (3 + x)H_2O(l) \rightarrow Fe_2O_3{\cdot}xH_2O(s) + 6H^+(aq)$$

The x in the second equation indicates that rust (hydrated iron(III) oxide) has a variable water content.

Salt accelerates the corrosion process by making the water more conductive. Because hydrogen ions promote the step in which oxygen is reduced, acid rain and the acid water from strip mines are extremely corrosive. Hydroxide ions, on the other hand, inhibit oxygen reduction, and the corrosion of iron is almost nil in solutions of high pH (Figure 20.16). Soapy water, for example, is slightly alkaline, so steel wool stored under soapy water lasts much longer than steel wool in an open dish.

One obvious way to prevent corrosion is to cover the metal with a *protective coating* such as oil, porcelain, paint, or some less active metal. Tin cans, for example, are actually steel cans plated with a thin layer of the less active metal tin.

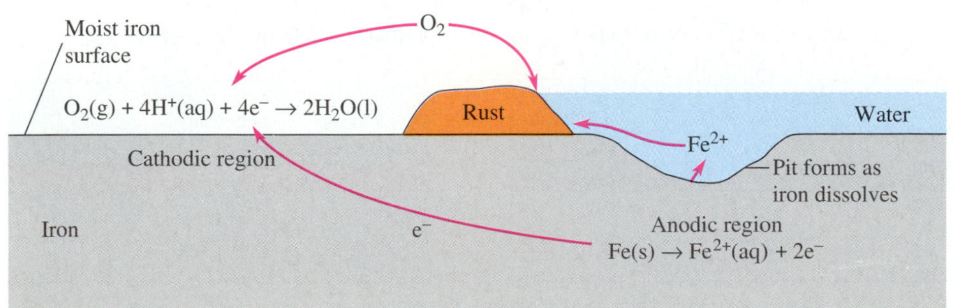

Figure 20.15
Electrochemical mechanism of corrosion. A water-covered region of iron becomes anodic and sends electrons through the iron to the moist iron surface, where oxygen is reduced to water. The entire galvanic process is accelerated by dissolved electrolytes such as salt, which transport the charge.

Figure 20.16

An iron nail rusts rapidly in tap water (left), which contains dissolved oxygen and has a pH between 5 and 6. The nail does not rust in oxygen-free water (center). The nail rusts very slowly in a basic solution (right). The latter solution is faintly pink because of added phenolphthalein indicator.

Protective coatings are good only as long as the coating remains impervious at all points.

Corrosion inhibitors (certain phosphates, chromates, and organic compounds) are often added to circulating solutions such as automobile radiator coolants to retard corrosion within the system. These substances are adsorbed onto the iron surface, where they slow the rate of electron transfer. Corrosion inhibitors such as red lead (Pb_3O_4) and zinc chromate ($ZnCrO_4$) are also added to certain protective paints.

A more effective method for the prevention of corrosion is shown in Demonstration 20.3. This method, known as **cathodic protection**, involves putting a more active metal such as zinc or magnesium in contact with the metal to be protected. **Galvanized iron**, for example, is iron coated with zinc. The zinc coating has a higher oxidation potential than iron, so if a minigalvanic system starts up in a scratched area, the more active zinc will become the anode and dissolve (Figure 20.17). All of the iron will become the cathode, and the reduction half-reaction will take place on its surface. Thus the zinc will corrode and the iron will be protected. The same principle is used on a larger scale when zinc plates are bolted to the steel

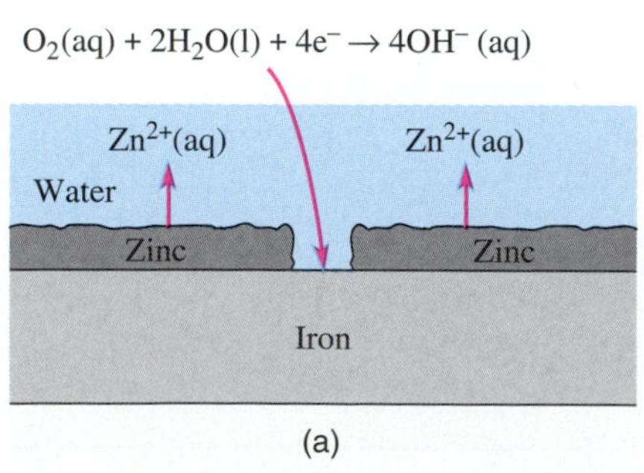

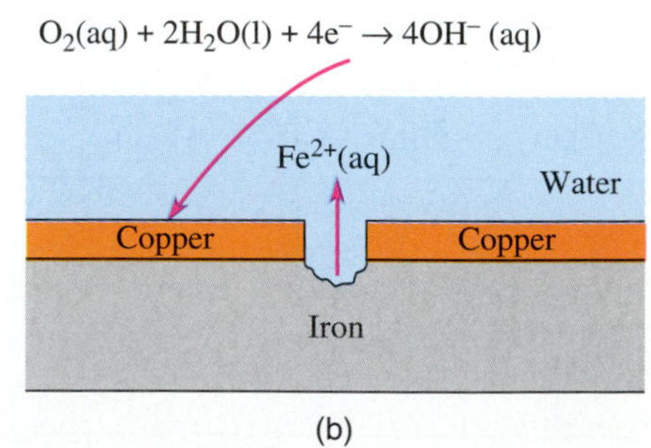

Figure 20.17

A layer of zinc protects iron from corrosion; a layer of copper does not. (a) Even a broken layer of zinc will provide iron with cathodic protection. Zinc is more easily oxidized and corrodes instead of the iron. (b) A layer of copper may hasten the corrosion of iron. At every break in the copper layer, the copper becomes the cathode while the iron oxidizes.

DEMONSTRATION 20.3 CATHODIC PROTECTION

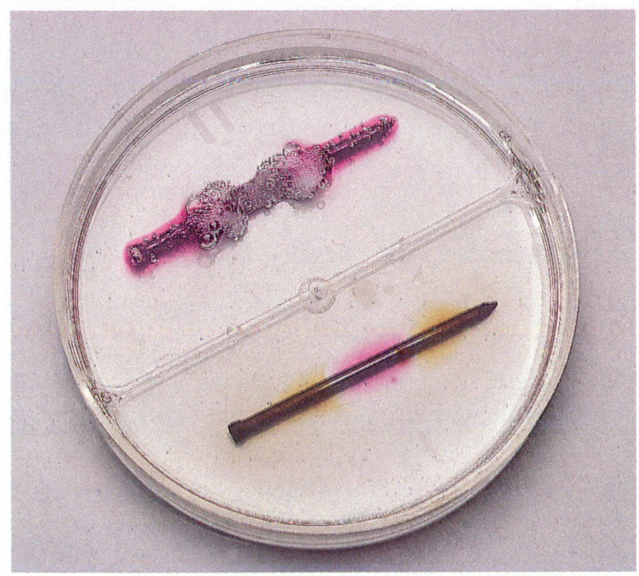

An unwrapped nail and a nail partially wrapped in magnesium wire were placed in a gel containing a few drops of phenolphthalein. Both ends of the unwrapped nail are anodic and begin to oxidize (shown by the brown iron oxide color). The center of the nail is cathodic and does not corrode; the reaction $O_2(g) + 2H_2O(l) + 4e^- \longrightarrow 4OH^-(aq)$ occurs on its surface. The pink color of the indicator is due to hydroxide ions.

In the wrapped nail, the magnesium wire is the anode: $Mg(s) \longrightarrow Mg^{2+}(aq) + 2e^-$. The entire nail is cathodic and protected from corrosion; no brown oxides are visible. The cathode reaction, $2H_2O(l) + 2e^- \longrightarrow H_2(g) + 2OH^-(aq)$, produces hydroxide ions and bubbles of hydrogen.

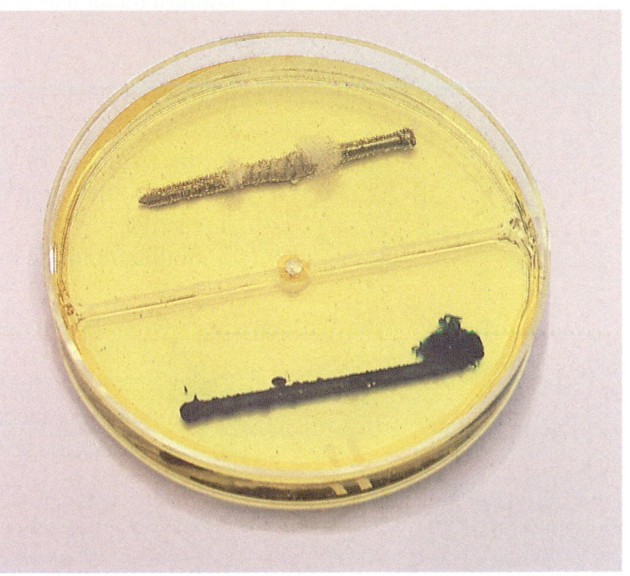

An unwrapped nail (bottom) and a nail partially wrapped in magnesium wire (top) were placed in a gel containing some potassium ferricyanide. The unwrapped nail oxidizes to Fe^{2+} ions that convert the amber ferricyanide to a Prussian blue precipitate that contains iron in both the +2 and +3 oxidation states. The wrapped nail is cathodically protected and is still shiny.

hulls of ships and when underground storage tanks and pipes are connected by wires to buried magnesium blocks (Figure 20.18). The magnesium block is a "sacrificial anode" that slowly oxidizes and sends electrons to the iron tank. The iron tank becomes the protected cathode on which the reduction half-reaction occurs. Although the magnesium block will eventually have to be replaced, the iron tank will remain intact.

Figure 20.18

Zinc wire placed along the Alaskan oil pipeline provides cathodic protection for the pipeline.

20.7 ELECTROLYSIS

Just as a pump can be used to force water uphill, an outside voltage source can be used to reverse a spontaneous redox reaction. The use of an electric current to bring about an oxidation–reduction reaction is called **electrolysis**. Electrolysis is an important method for preparing certain elements and high-energy compounds that are difficult or even impossible to prepare by other chemical methods.

The Electrolysis of Molten Sodium Chloride

The electrolysis of molten sodium chloride (melting point, 801°C) produces liquid sodium metal and chlorine gas:

$$2NaCl(l) \rightarrow 2Na(l) + Cl_2(g)$$

The electrolysis is carried out in an **electrolytic cell** consisting of two electrodes immersed in the hot molten electrolyte (Figure 20.19). The electrodes are connected to the terminals of a battery or other voltage source that sends electrons into one electrode, making it negative, and draws electrons out of the other electrode, making it positive.

Sodium ions are reduced to sodium metal at the negative electrode:

$$Na^+ + e^- \rightarrow Na(l) \qquad \text{reduction}$$

(e^- provided by the negative electrode)

This electrode is the cathode because, by definition, the cathode is the electrode at which reduction takes place. Chloride ions are oxidized to chlorine gas at the positive electrode (the anode):

$$2Cl^- \rightarrow Cl_2(g) + 2e^- \qquad \text{oxidation}$$

($2e^-$ given to the positive electrode)

Two reductions occur for every oxidation, and the complete cell reaction is

$$2Na^+ + 2Cl^- \rightarrow 2Na(l) + Cl_2(g)$$

or

$$2NaCl(l) \rightarrow 2Na(l) + Cl_2(g)$$

Note that the half-reactions in this cell are the reverse of those that occur when sodium metal reacts spontaneously with chlorine gas. Note also that the electrode

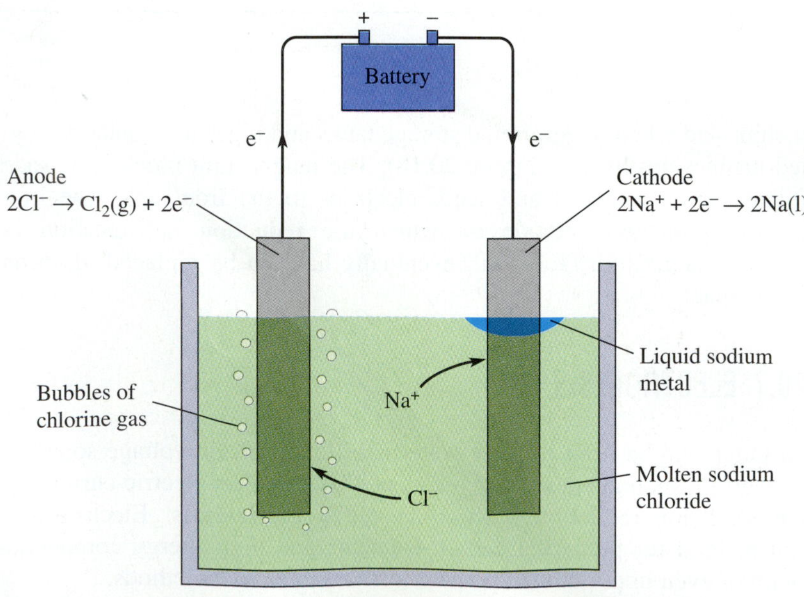

Figure 20.19
Electrolysis of molten sodium chloride. Inert electrodes are used. Chlorine gas is produced at the anode, liquid sodium at the cathode.

polarities—the anode is positive, the cathode is negative—are the reverse of those in a galvanic cell. *Electrolytic cells differ from galvanic cells in that the anode is positive and the cathode is negative.*

	Cathode	Anode
Galvanic cell	+	−
Electrolytic cell	−	+

Production of Sodium and Other Active Metals

The compounds of very active metals are difficult to decompose, so electrolytic reduction, in spite of its expense, is the most economical method for obtaining such metals. Electrolysis can also be used to prepare less active metals when very pure samples are needed. Molten metal chlorides are usually used as electrolytes because they generally have lower melting points than other readily available salts. Sodium, lithium, magnesium, calcium, and barium are among the metals produced by electrolysis.

PRACTICE EXERCISE 20.15

Write equations for the electrode reactions that occur during the electrolysis of molten magnesium chloride ($MgCl_2$).

Over 20,000 metric tons of sodium are produced in the United States each year for use in sodium vapor lamps, as a heat-transfer medium in some nuclear reactors, and as a strong reducing agent. A **Downs cell** (Figure 20.20) is used for the electrolysis. The cathode is a hollow iron cylinder that surrounds the electrolyte. A graphite rod is used as the anode because a metal electrode would be attacked by the chlorine gas. The electrolyte is sodium chloride with enough calcium chloride added to reduce the melting point to about 600°C, thus allowing the cell to operate more economically. Calcium ions are less easily reduced than sodium ions and remain in the

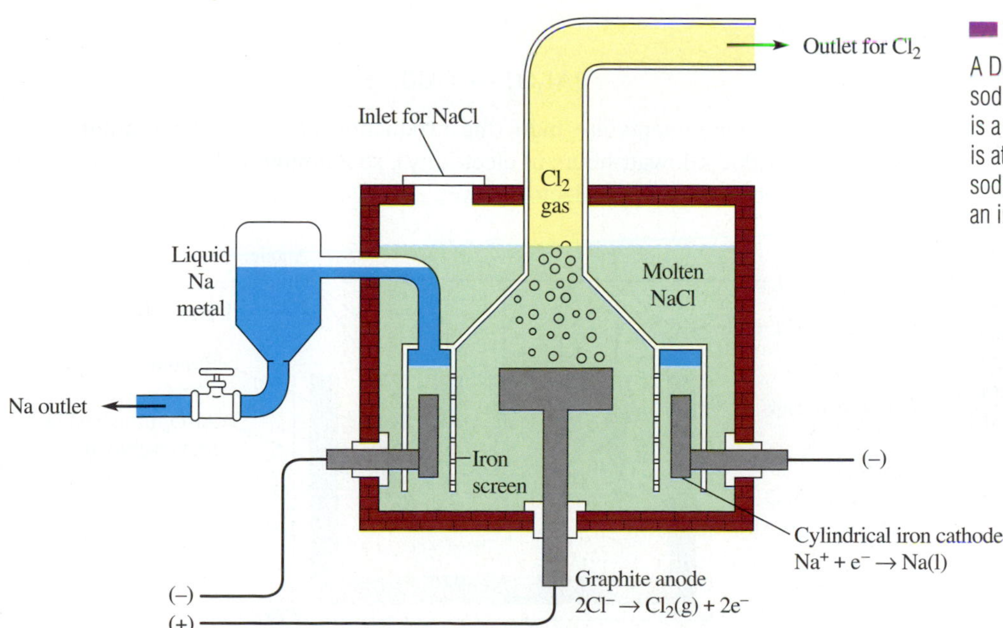

Figure 20.20
A Downs cell used for the production of sodium from sodium chloride. The cathode is a hollow iron cylinder; a graphite anode is at its center. The chlorine gas and molten sodium metal are prevented from mixing by an iron screen.

melt. Liquid sodium metal (melting point, 98°C) rises to the top of the melt and is continuously tapped off. Air and moisture are excluded, and an iron screen protects the sodium from the chlorine. The chlorine gas is a useful coproduct.

Electrolytic Production of Aluminum

Although aluminum is the earth's third most abundant element, the pure metal was unknown until 1827. In that year the German chemist Friedrich Wöhler produced a small, expensive sample by reducing aluminum chloride with potassium vapor. Attempts to produce aluminum by the electrolysis of molten aluminum chloride failed because the molten chloride is a poor conductor of electricity. An alternative electrolytic process was developed in 1886 by two students, each aged 22, each working independently and unknown to the other. One student was Charles Martin Hall, a chemistry major at Oberlin College; the other was Paul Héroult in France.

Hall and Héroult tried to electrolyze naturally occurring aluminum compounds. Molten cryolite (Na_3AlF_6; melting point, 1000°C) proved resistant to reduction. Aluminum oxide (Al_2O_3; melting point, 2000°C), abundantly available in the ore bauxite, had a melting point that was too high. The young men discovered, however, that aluminum oxide dissolves in cryolite to form an ionic solution that can be electrolyzed at moderate temperatures. The **Hall–Héroult process** uses a 5% mixture of Al_2O_3 in cryolite with a melting point of about 950°C. (Recent modifications of the process use somewhat different mixtures with even lower melting temperatures.) Liquid aluminum metal (melting point, 660°C) lies at the bottom of the electrolytic cell (Figure 20.21). The electrode reactions are

$$Al^{3+} + 3e^- \rightarrow Al(l) \qquad \text{cathode, reduction}$$

$$2O^{2-} \rightarrow O_2(g) + 4e^- \qquad \text{anode, oxidation}$$

About 4.5% of the electricity used in the United States is used in aluminum production.

Four reductions occur for every three oxidations, and the complete cell reaction is

$$4Al^{3+} + 6O^{2-} \rightarrow 4Al(l) + 3O_2(g)$$

or

$$2Al_2O_3 \rightarrow 4Al(l) + 3O_2(g)$$

The electricity requirements are high (the production of 1 ton of aluminum consumes about 17,000 kilowatt-hours of electricity), so aluminum plants are often built near abundant sources of hydroelectric power.

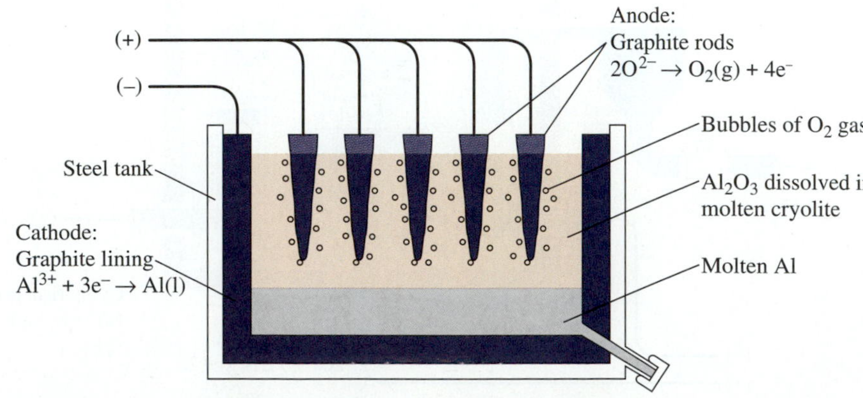

Figure 20.21
A Hall–Héroult cell for the electrolytic production of aluminum from aluminum oxide in cryolite. The anode consists of graphite rods; the cathode is a steel tank lined with graphite. The graphite rods are consumed by the liberated oxygen and must be replaced frequently.

The Electrolysis of Water

When water is electrolyzed between inert electrodes as in Figure 20.22 (see also Figure 3.1, p. 84), the half-reactions are

$$2H_2O(l) \rightarrow 4H^+(aq) + O_2(g) + 4e^- \qquad \text{anode, oxidation}$$

$$2H_2O(l) + 2e^- \rightarrow H_2(g) + 2OH^-(aq) \qquad \text{cathode, reduction}$$

Multiplying the cathode equation by 2 to balance the electrons, adding the adjusted equations, and canceling the four water molecules formed by combining the hydrogen and hydroxide ions ($4H^+(aq) + 4OH^-(aq) \rightarrow 4H_2O(l)$) gives the net equation

$$2H_2O(l) \rightarrow 2H_2(g) + O_2(g)$$

Pure water is a feeble conductor of electricity, so a small amount of ionic solute is added to carry the current. The added ions must be more resistant to oxidation and reduction than water is; sodium sulfate is often used.

What voltage is needed to electrolyze water? Ideally, an electrolysis needs only enough voltage to oppose the EMF of the spontaneous reverse reaction. The value of $\Delta G°$ for the formation of 2 mol of water under standard-state conditions is -474 kJ. The minimum voltage is obtained by substituting $\Delta G°$ and $n = 4$ mol e$^-$ into Equation 20.5b:

$$E° = -\frac{\Delta G°}{nF} = -\frac{-474{,}000\ \text{J}}{(4\ \text{mol e}^-)(96{,}485\ \text{C/mol e}^-)}$$

$$= 1.23\ \text{J/C} = 1.23\ \text{V}$$

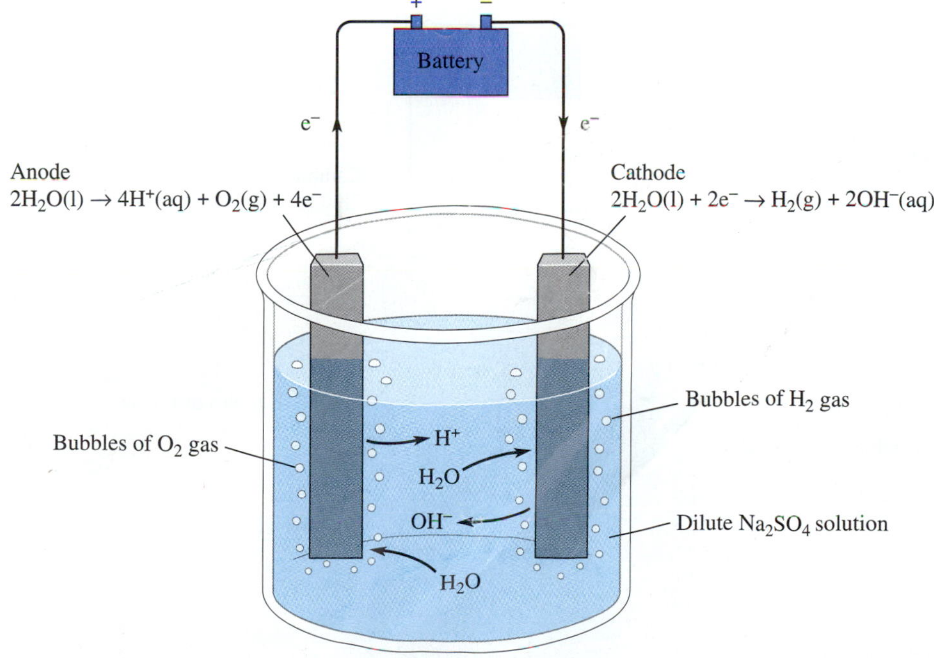

Figure 20.22
Electrolysis of water. Hydrogen gas and hydroxide ions are produced at the cathode, oxygen gas and hydrogen ions at the anode. A small amount of sodium sulfate is added to the water to help carry the current.

In practice, quite a bit more than the ideal minimum voltage is required for any electrolysis, and more than 1.23 V is needed to electrolyze water (see the discussion of overvoltage at the beginning of Section 20.5).

The Electrolysis of Brine

The electrolysis of aqueous ionic solutions is often complicated by the presence of water. The electrolysis of brine (concentrated salt water) illustrates this point. Chlorine gas forms at the anode just as it does when molten sodium chloride is electrolyzed:

$$2Cl^-(aq) \rightarrow Cl_2(g) + 2e^- \qquad \text{anode, oxidation}$$

At the cathode, however, hydrogen is produced, and hydroxide ions accumulate in the nearby solution (Figure 20.23 and Demonstration 20.4). Water molecules, not sodium ions, are reduced at this electrode:

$$2H_2O(l) + 2e^- \rightarrow H_2(g) + 2OH^-(aq) \qquad \text{cathode, reduction}$$

The net ionic equation is the sum of the half-reactions:

$$2Cl^-(aq) + 2H_2O(l) \rightarrow Cl_2(g) + H_2(g) + 2OH^-(aq)$$

The molecular equation is obtained by adding the spectator ions (in this case, two sodium ions) to each side of the net ionic equation:

$$2NaCl(aq) + 2H_2O(l) \rightarrow Cl_2(g) + H_2(g) + 2NaOH(aq)$$

The electrolysis of brine is complicated because two half-reactions can occur at each electrode. The anode possibilities are

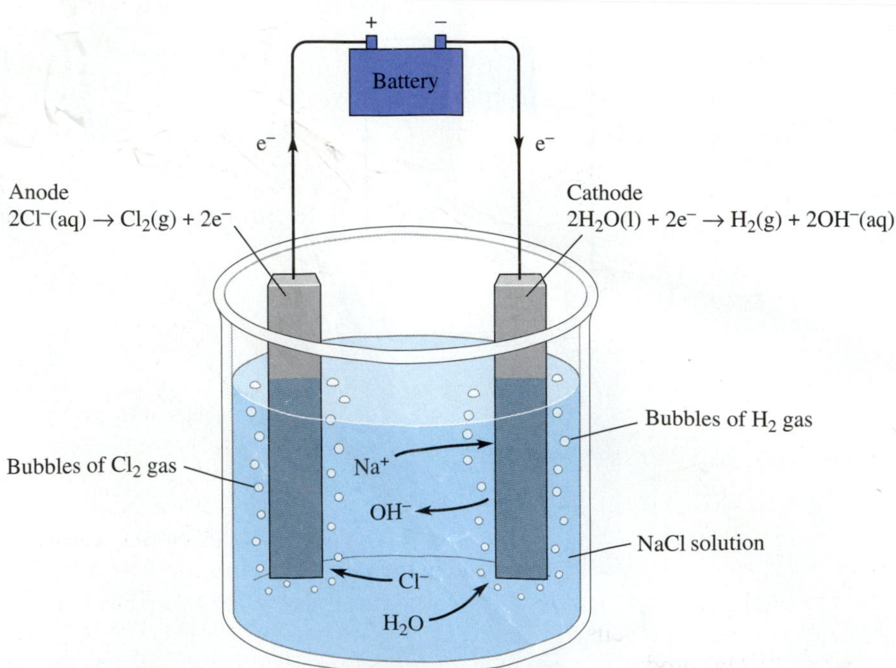

Figure 20.23
Electrolysis of brine. Water is reduced to hydrogen gas and hydroxide ions at the cathode; chloride ions are oxidized to chlorine gas at the anode.

DEMONSTRATION 20.4 **ELECTROLYSIS OF AQUEOUS SODIUM IODIDE: $2NaI(aq) + 2H_2O(l) \rightarrow I_2(aq) + H_2(g) + 2NaOH(aq)$**

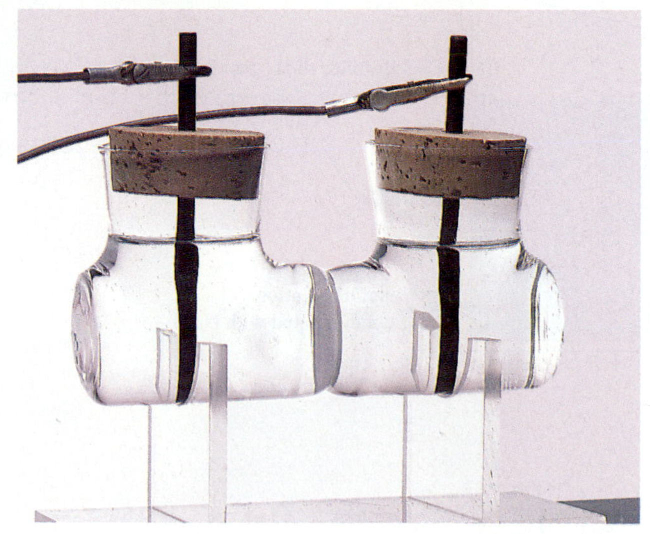

The electrolytic cell is fitted with two carbon electrodes and filled with a solution of sodium iodide containing a few drops of phenolphthalein. A glass frit in the center inhibits mixing.

Water is reduced at the cathode, producing hydrogen gas and hydroxide ions, which turn the phenolphthalein red. Iodide ions are oxidized at the anode to iodine, which produces a brown color in the anode compartment.

$$2Cl^-(aq) \rightarrow Cl_2(g) + 2e^- \qquad E^\circ_{oxid} = -1.36 \text{ V}$$

$$2H_2O(l) \rightarrow 4H^+(aq) + O_2(g) + 4e^- \qquad E^\circ_{oxid} = -1.23 \text{ V}$$

The standard oxidation potentials of these two half-reactions are comparable, but the actual oxidation potentials will depend on the concentrations of the ions and the partial pressures of the gases. When the chloride concentration is high, as in a concentrated solution of sodium chloride, the oxidation of chloride ion is favored and chlorine is produced. When there are relatively few chloride ions, as in a dilute solution of sodium chloride, the oxidation of water is favored and oxygen is produced.

The possible cathode reactions are

$$Na^+(aq) + e^- \rightarrow Na(s) \qquad E^\circ_{red} = -2.71 \text{ V}$$

$$2H_2O(l) + 2e^- \rightarrow H_2(g) + 2OH^-(aq) \qquad E^\circ_{red} = -0.83 \text{ V}$$

The reduction potential of water is much higher (more positive) than that of sodium ion, so the reduction of water is the favored reaction at all concentrations of NaCl.

The electrolysis of brine is the major commercial source of both chlorine and sodium hydroxide (NaOH). Sodium hydroxide is used industrially in petroleum refining, in the manufacture of textiles, and in the production of soap. It is one of the top 10 industrial chemicals, and over 12 million tons of it are produced each year.

A number of different cells are used for the electrolysis of brine. The *diaphragm cell* (Figure 20.24a) produces a sodium hydroxide solution that is contaminated with

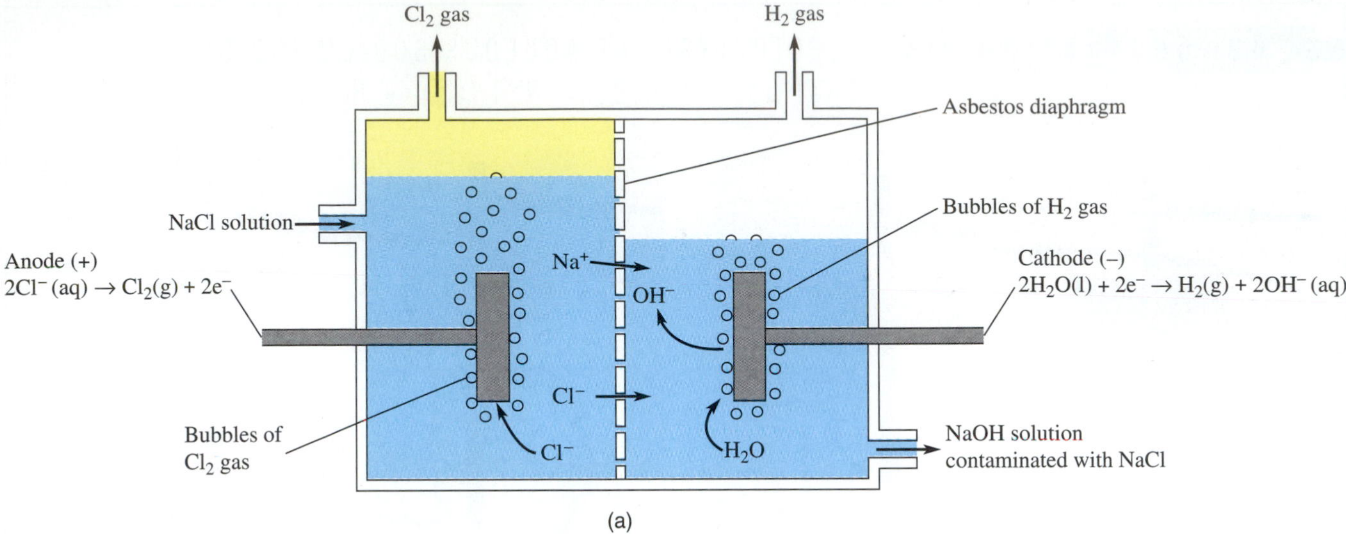

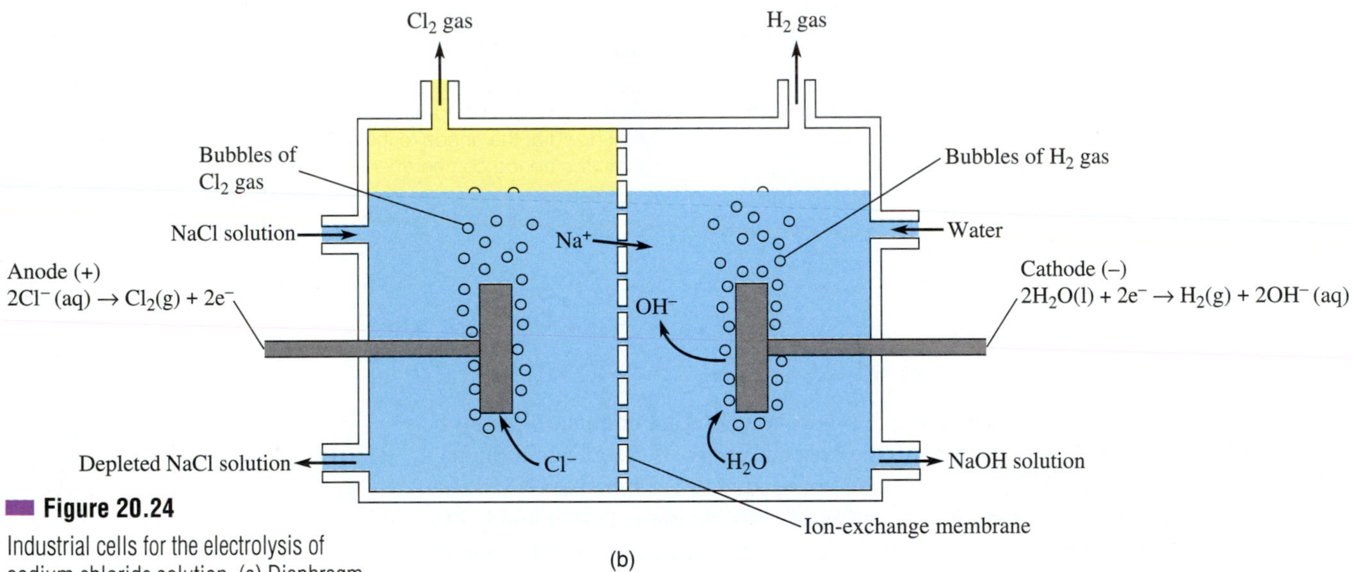

Figure 20.24

Industrial cells for the electrolysis of sodium chloride solution. (a) Diaphragm cell. The wet asbestos diaphragm prevents contact between the hydrogen and chlorine gases. Although dissolved ions can move through the diaphragm, the higher pressure on the anode side (caused by the higher solution level) inhibits the migration of OH^- ions toward the anode and improves the yield of NaOH obtained from the cathode compartment. (b) Membrane cell. An ion exchange membrane allows only sodium ions to pass through. A better yield of sodium hydroxide is obtained because chloride ions remain in the anode compartment.

unreacted sodium chloride. The *membrane cell* (Figure 20.24b) produces a pure solution of sodium hydroxide.

Electroplating

Electroplating is a process whereby one metal is electrolytically deposited or "plated out" on the surface of another metal. Electroplating is often done to protect the base metal from corrosion or simply to give it a more pleasant or richer appearance. The electrolyte contains the plating metal in the form of dissolved ions. The anode is often (but not always) made of the plating metal, while the cathode is the object being plated (Demonstration 20.5).

DEMONSTRATION 20.5 ELECTROPLATING CHROMIUM

The electrodes are two strips of copper. The electrolyte is chromium (VI) oxide (CrO_3) in concentrated sulfuric acid.

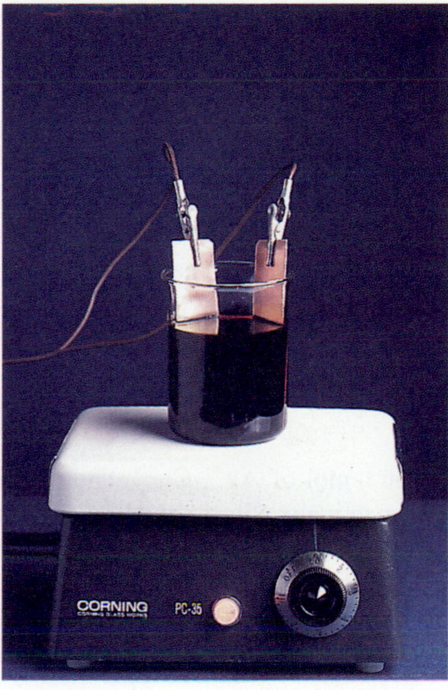

The current causes copper to dissolve from the anode strip: $Cu(s) \longrightarrow Cu^{2+}(aq) + 2e^-$. Chromium metal plates out on the cathode strip: $CrO_3(aq) + 6H^+(aq) + 6e^- \longrightarrow Cr(s) + 3H_2O(l)$.

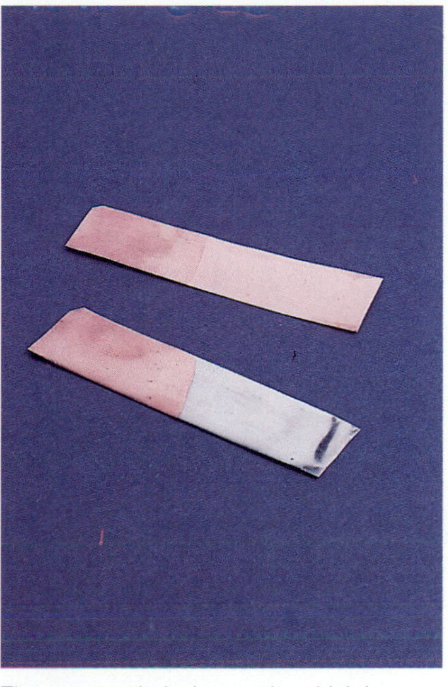

The upper strip is the anode, which lost copper; the lower strip is the chromium-plated cathode.

Copper can be plated out from an acidified copper sulfate solution using a copper anode. The anode gradually dissolves as current passes through the cell

$$Cu(s) \rightarrow Cu^{2+}(aq) + 2e^- \qquad \text{anode, oxidation}$$

while the metal object at the cathode acquires a copper coating

$$Cu^{2+}(aq) + 2e^- \rightarrow Cu(s) \qquad \text{cathode, reduction}$$

Other substances are often added to the electrolyte to increase conductivity or to improve the quality of the plate. Potassium cyanide (KCN) is usually added to alkaline copper-plating solutions because copper metal plates out more slowly and evenly when the copper is in the form of the complex ion $Cu(CN)_4{}^{3-}$. The electrode reactions are

$$Cu(s) + 4CN^-(aq) \rightarrow Cu(CN)_4{}^{3-}(aq) + e^- \qquad \text{anode, oxidation}$$

$$Cu(CN)_4{}^{3-}(aq) + e^- \rightarrow Cu(s) + 4CN^-(aq) \qquad \text{cathode, reduction}$$

Silver and gold, which form the complexes $Ag(CN)_2{}^-$ and $Au(CN)_2{}^-$, are also plated from alkaline cyanide solutions. The electrode reactions for silver are similar to those for copper:

$$Ag(s) + 2CN^-(aq) \rightarrow Ag(CN)_2^-(aq) + e^- \qquad \text{anode, oxidation}$$
$$Ag(CN)_2^-(aq) + e^- \rightarrow Ag(s) + 2CN^-(aq) \qquad \text{cathode, reduction}$$

20.8 ELECTROCHEMICAL STOICHIOMETRY

The equation for an electrode reaction relates the number of moles of substance consumed or produced during the reaction to the quantity of charge that passes through the cell. For example, the half-reactions

$$Ag^+(aq) + e^- \rightarrow Ag(s)$$
$$Cu^{2+}(aq) + 2e^- \rightarrow Cu(s)$$
$$AuCl_4^-(aq) + 3e^- \rightarrow Au(s) + 4Cl^-(aq)$$

show that 1 mol of Ag^+, Cu^{2+}, and $AuCl_4^-$ ions pick up 1, 2, and 3 mol of electrons, respectively. Since the charge carried by 1 mol of electrons is 1 faraday or 96,485 C, passage of 96,485 C through an electrolysis cell will deposit 1 mol of silver, 2 × 96,485 C will deposit 1 mol of copper, and 3 × 96,485 C will deposit 1 mol of gold.

In 1832 Michael Faraday laid the foundation for electrochemical stoichiometry with the publication of his experimentally determined laws of electrolysis. Faraday's laws state that *the mass of a substance liberated during electrolysis is proportional to the quantity of electric charge that passed through the cell and to the equivalent weight of the substance.* (In modern terms, the "equivalent weight" of a substance is the mass liberated by the passage of one mole of electrons. In 1832 electrons had not yet been discovered.)

EXAMPLE 20.16

How many grams of copper will be deposited when 5000 C pass through a $CuSO_4$ solution?

SOLUTION

The equation given above for the electrode reaction shows that 1 mol (63.55 g) of copper metal is deposited by 2 mol of electrons or 2 × 96,485 C. The steps in the calculation are similar to those of previous stoichiometry calculations:

$$5000\text{ C} \rightarrow \text{mol e}^- \rightarrow \text{mol Cu} \rightarrow \text{g Cu}$$

$$5000\text{ C} \times \frac{1\text{ mol e}^-}{96{,}485\text{ C}} \times \frac{1\text{ mol Cu}}{2\text{ mol e}^-} \times \frac{63.55\text{ g Cu}}{1\text{ mol Cu}} = 1.65\text{ g Cu}$$

PRACTICE EXERCISE 20.16

How many grams of gold will be deposited when 5000 C pass through a solution containing $AuCl_4^-$ ions?

The deposition of silver metal is so reliable that a small electrolytic cell consisting of a silver electrode dipping into a solution containing silver ions can be used as a **coulometer** for measuring the total number of coulombs that pass through the circuit. The number of coulombs is calculated from the mass of silver deposited, as illustrated in the next example.

EXAMPLE 20.17

A silver coulometer is connected in series with an operating electrolytic cell. If the mass of the silver electrode increases by 2.75 g, how many coulombs have passed through the cell?

SOLUTION

The number of coulombs that pass through the cell is the same as the number of coulombs that pass through the coulometer. The electrode reaction

$$Ag^+(aq) + e^- \longrightarrow Ag(s)$$

shows that 1 mol (107.9 g) of silver is deposited by 1 mol of electrons or 96,485 C. The steps in the calculation are

$$2.75 \text{ g Ag} \longrightarrow \text{mol Ag} \longrightarrow \text{mol e}^- \longrightarrow \text{C}$$

$$2.75 \text{ g Ag} \times \frac{1 \text{ mol Ag}}{107.9 \text{ g Ag}} \times \frac{1 \text{ mol e}^-}{1 \text{ mol Ag}} \times \frac{96{,}485 \text{ C}}{1 \text{ mol e}^-} = 2.46 \times 10^3 \text{ C}$$

PRACTICE EXERCISE 20.17

The zinc electrode in a Daniell cell weighs 350 mg. If the cell operates until all of the zinc is dissolved, how many coulombs will have passed through the circuit?

Electric current is the rate of flow of charge. The SI unit for current is the **ampere** (A). One ampere is 1 coulomb per second (1 A = 1 C/s). The current I in amperes is given by the equation

$$I = \frac{Q}{t} \qquad (20.9)$$

where Q is the charge in coulombs and t is the time in seconds.

EXAMPLE 20.18

A miniature battery supplies a continuous current of 5.56 microamperes (μA). How many coulombs pass through the circuit in 1.00 hour?

SOLUTION

The current $I = 5.56\ \mu\text{A} = 5.56 \times 10^{-6}\ \text{A} = 5.56 \times 10^{-6}$ C/s. The time t = 1.00 h = 3600 s. Substituting I and t into Equation 20.9 gives

$$Q = I \times t = 5.56 \times 10^{-6} \text{ C/s} \times 3600 \text{ s} = 2.00 \times 10^{-2} \text{ C}$$

PRACTICE EXERCISE 20.18

An electroplating process draws a current of 0.556 A. How many coulombs will pass through the electrolyte in 25.0 min?

EXAMPLE 20.19

How many hours would it take for a 3.50-A current to electroplate 129 g of gold? The electrode reaction is

$$Au(CN)_2^-(aq) + e^- \longrightarrow Au(s) + 2CN^-(aq)$$

SOLUTION

The current is 3.50 A = 3.50 C/s. The electrode equation shows that 1 mol of electrons (or 96,485 C) will deposit 1 mol (197.0 g) of gold. The steps in the calculation are

$$129 \text{ g Au} \longrightarrow \text{mol Au} \longrightarrow \text{mol e}^- \longrightarrow \text{C} \longrightarrow t$$

$$129 \text{ g Au} \times \frac{1 \text{ mol Au}}{197.0 \text{ g Au}} \times \frac{1 \text{ mol e}^-}{1 \text{ mol Au}} \times \frac{96{,}485 \text{ C}}{1 \text{ mol e}^-} \times \frac{1 \text{ s}}{3.50 \text{ C}} \times \frac{1 \text{ h}}{3600 \text{ s}} = 5.01 \text{ h}$$

PRACTICE EXERCISE 20.19

How many liters of chlorine, measured at 1.00 atm and 0°C, are produced per minute when molten sodium chloride is electrolyzed by a 15.0-A current?

CHAPTER REVIEW

LEARNING OBJECTIVES BY SECTION

20.1 1. Identify the anode, the cathode, the direction of electron flow, and the direction of ion flow in a given galvanic cell.
2. Describe galvanic cells using cell notation.
3. Given the cell reaction or the cell notation for a galvanic cell, sketch a diagram of the cell.
4. Distinguish between standard and nonstandard cells and half-cells.

20.2 1. Use the cell voltage to calculate the maximum electrical work that can be obtained from the transfer of a given charge.
2. Calculate the free-energy change of a cell reaction from the cell voltage, and vice versa.
3. Explain why the voltage of a galvanic cell must be positive.

20.3 1. Diagram the standard hydrogen electrode, and explain how it is used to find half-cell potentials.
2. Convert standard reduction potentials into standard oxidation potentials, and vice versa.
3. Use standard reduction potentials to compare the strengths of oxidizing and reducing agents and to predict whether a given redox reaction is spontaneous.
4. Use standard reduction potentials to calculate $E°$ and $\Delta G°$ for a given redox reaction.

20.4 1. Calculate cell voltages and half-cell potentials at concentrations other than standard-state concentrations.
2. Calculate $\Delta G°$, $E°$, or K, given one of the three quantities.

20.5 1. Describe the construction and operation of zinc batteries, storage batteries, and the H_2/O_2 fuel cell.
2. Describe the difference between a fuel cell and a battery.

20.6 1. Describe the galvanic mechanism by which iron forms rust.
2. List three ways of preventing corrosion, and explain how they work.

20.7 1. Diagram and write electrode reactions for the electrolytic production of sodium and aluminum.
2. Diagram and write electrode reactions for the electrolysis of water.
3. Diagram and write electrode reactions for the electrolysis of brine in the diaphragm cell and in the membrane cell.
4. Diagram an electroplating cell, and write equations for the electrodeposition of a metal from a solution of its salt.
5. Use ΔG to calculate the minimum voltage required for an electrolysis.

20.8 1. Solve problems relating current, time, and quantity of substance consumed or produced in an electrolytic cell.

KEY TERMS BY SECTION

20.1 Anode
Cathode
Cell notation
Electrochemistry
Galvanic cell
Half-cell
Salt bridge
Standard cell
Standard electrode
Standard half-cell
Voltaic cell

20.2 Electromotive force (EMF)
Faraday (F)
Volt (V)

20.3 Half-cell potential
Standard hydrogen electrode
Standard oxidation potential (E°_{oxid})
Standard reduction potential (E°_{red})

20.4 Nernst equation

20.5 Battery
Fuel cell
Overvoltage

20.6 Cathodic protection
Corrosion
Corrosion inhibitor
Galvanized iron

20.7 Downs cell
Electrolysis
Electrolytic cell
Electroplating
Hall–Héroult process

20.8 Ampere (A)
Coulometer
Electric current (I)

IMPORTANT EQUATIONS

20.1 maximum electrical work $= QE$

20.2 $\Delta G = -$maximum electrical work

20.3 $\Delta G = -QE$

20.4 $Q = nF$

20.5a $\Delta G = -nFE$

20.5b $\Delta G^\circ = -nFE^\circ$

20.6 $E^\circ = E^\circ_{oxid} + E^\circ_{red}$

20.7a $E = E^\circ - \dfrac{RT}{nF}\ln Q$

20.7b $E = E^\circ - \dfrac{0.0592}{n}\log Q$

20.8a $E^\circ = \dfrac{RT}{nF}\ln K$

20.8b $E^\circ = \dfrac{0.0592}{n}\log K$

20.9 $I = \dfrac{Q}{t}$

FINAL EXERCISES

Answers to exercises with blue numbers are given in Appendix D.

PART A. QUESTIONS AND PROBLEMS BY SECTION

Galvanic Cells (Section 20.1)

20.1 Is the anode positive or negative in a galvanic cell? What is the charge on the cathode? What is the direction of electron flow in the external circuit of a galvanic cell?

20.2 What is the direction of anion flow in a galvanic cell? What is the direction of cation flow?

20.3 Describe the function of a salt bridge. What would happen if the half-cell compartments of a galvanic cell were not connected?

20.4 Explain why a salt bridge is not necessary in (a) the gravity cell in Figure 20.2c, (b) the lead storage battery in Figure 20.12, and (c) the fuel cell in Figure 20.13.

20.5 Draw diagrams for galvanic cells based on the following reactions. Label the anode and cathode, and write the electrode reactions. Indicate electrode materials, electrolyte compositions, and directions of electron flow and ion migration.
(a) $Fe(s) + Sn^{2+}(aq) \longrightarrow Fe^{2+}(aq) + Sn(s)$
(b) $H_2(g) + Br_2(l) \longrightarrow 2HBr(aq)$
(c) $2KI(aq) + Cl_2(g) \longrightarrow 2KCl(aq) + I_2(s)$

20.6 Draw diagrams for galvanic cells based on the following reactions. Label the anode and cathode, and write the electrode reactions. Indicate electrode materials, electrolyte compositions, and directions of electron flow and ion migration.
(a) $Cr_2O_7^{2-}(aq) + 2Cr(s) + 14H^+(aq) \longrightarrow 4Cr^{3+}(aq) + 7H_2O(l)$
(b) $Cd(s) + Cu^{2+}(aq) \longrightarrow Cd^{2+}(aq) + Cu(s)$
(c) $4Fe^{2+}(aq) + O_2(g) + 4H^+(aq) \longrightarrow 4Fe^{3+}(aq) + 2H_2O(l)$

20.7 Write the cell notation for each of the cells in Exercise 20.5.

20.8 Write the cell notation for each of the cells in Exercise 20.6.

20.9 Write the cell notation for standard cells based on the following reactions:
(a) the oxidation of silver by nitric acid (assume that nitric acid is reduced to NO)
(b) $Zn(s) + 2Ag(NH_3)_2^+(aq) \longrightarrow 2Ag(s) + Zn(NH_3)_4^{2+}(aq)$

20.10 Write the cell notation for standard cells based on the following reactions:
(a) $2H_2(g) + O_2(g) \longrightarrow 2H_2O(l)$
(b) $3Ni(s) + 2MnO_4^-(aq) + 4H_2O(l) \longrightarrow 3Ni^{2+}(aq) + 2MnO_2(s) + 8OH^-(aq)$

Electrical Work and Free Energy (Section 20.2)

20.11 What are the SI units for (a) charge and (b) voltage?

20.12 Which of the following quantities depend on cell size?
(a) voltage
(b) current
(c) free-energy change per mole of reaction
(d) work done during the cell's lifetime

20.13 Explain how redox half-reactions give rise to a potential difference within a galvanic cell.

20.14 Explain why the work actually done by a galvanic cell is invariably less than the theoretical maximum.

20.15 An almost dead Daniell cell supplies only 0.35 V. Calculate ΔG for the cell reaction.

20.16 The standard free energies of formation of $Hg^{2+}(aq)$ and $Hg_2^{2+}(aq)$ are +164.4 and +153.5 kJ/mol, respectively. Calculate $E°$ for a galvanic cell that utilizes the reaction

$$Hg^{2+}(aq) + Hg(l) \longrightarrow Hg_2^{2+}(aq)$$

20.17 The value of $E°$ for the cell described in Example 20.2 is 0.547 V. Calculate the standard free energy per mole of Fe^{3+} reduced.

20.18 The standard free-energy change for the combustion of hydrogen

$$H_2(g) + \tfrac{1}{2}O_2(g) \longrightarrow H_2O(l)$$

is −237.13 kJ.
(a) Calculate $E°$ for a fuel cell utilizing this reaction.
(b) What is the maximum electrical work that could be obtained from a cell fueled with 1.00 kg of hydrogen? Assume that each substance is in its standard state.

Standard Reduction Potentials (Section 20.3)

20.19 Sketch the standard hydrogen electrode, and explain how it is used to determine reduction potentials.

20.20 Briefly describe how you would determine the standard reduction potential of (a) silver and (b) aluminum. Include a diagram and cell notation for each determination.

20.21 A half-cell is coupled with the standard hydrogen electrode. Explain why the standard reduction potential of the half-cell has the same sign as the polarity of its electrode.

20.22 Refer to Table 20.1, and find standard reduction potentials for the Cl_2/Cl^- chlorine and Cr^{3+}/Cr chromium half-cells. What are the standard oxidation potentials for these half-cells?

20.23 Ni^{2+} has a higher (more positive) reduction potential than Cd^{2+}.
(a) Which ion is more easily reduced to the metal?
(b) Which metal, Ni or Cd, is more easily oxidized?

20.24 Refer to Table 20.1, and predict which metal in each of the following pairs is the stronger reducing agent:
(a) sodium or potassium
(b) magnesium or barium

20.25 Which is the stronger oxidizing agent, chlorine water or an acidified solution of potassium permanganate?

20.26 Refer to Table 20.1, and list the halogens in order of decreasing strength as oxidizing agents.

20.27 You need to oxidize Mn^{2+} to MnO_2, and you have the following reagents on hand: $KMnO_4$, H_2S, $Hg(NO_3)_2$, and $FeCl_3$. Which ones are worth trying? Explain your answer.

20.28 Suggest reagents for the following reactions:
(a) oxidizing NaI to I_2
(b) reducing $MnCl_2$ to Mn
(c) reducing Fe^{3+} to Fe^{2+}
(d) oxidizing Cu to Cu^{2+}
(e) reducing $KMnO_4$ to Mn^{2+}

20.29 Use the table of reduction potentials to predict the results of the following experiments. Write an equation for each reaction that occurs.
(a) Zinc and copper metals are individually dipped into 1 *M* HCl.
(b) Silver and gold metals are individually dipped into nitric acid solution.
(c) Chlorine gas is bubbled into a solution containing fluoride ions.

20.30 Use the table of reduction potentials to predict the results of the following experiments. Write an equation for each reaction that occurs.
(a) Chlorine gas is bubbled into a solution containing bromide ions.
(b) $FeSO_4$ is added to an acidified solution of potassium dichromate.
(c) Iodine, silver nitrate, and iron(II) chloride are each treated with aqueous hydrogen peroxide in the presence of 1 *M* acid. (*Note*: Hydrogen peroxide (H_2O_2) can be either oxidized or reduced. Both potentials appear in Table 20.1.)

20.31 Calculate $E°$ and $\Delta G°$ for each of the galvanic cells diagrammed in Exercise 20.5.

20.32 Calculate $E°$ and $\Delta G°$ for each of the galvanic cells diagrammed in Exercise 20.6.

20.33 Calculate $E°$ and $\Delta G°$ for each spontaneous reaction in Exercise 20.29.

20.34 Calculate $E°$ and $\Delta G°$ for each spontaneous reaction in Exercise 20.30.

EMF and Concentration (Section 20.4)

20.35 Explain why the voltage of a cell decreases as the reaction proceeds. At what point does the cell voltage reach zero?

20.36 When a galvanic cell has been allowed to run until it no longer produces any voltage, what can be said about (a) the final half-cell *E* values and (b) the final electrolyte concentrations?

20.37 Derive the Nernst equation without referring to the text.

20.38 Derive the relationship between $E°$ and K_{eq} without referring to the text.

20.39 A standard Cu/Cu^{2+} half-cell is coupled with a similar half-cell in which the Cu^{2+} is 0.10 *M*. Which half-cell is the anode? Explain your answer.

20.40 Diagram, label, and write the half-reaction equations for a galvanic cell that might be used to determine the copper ion concentration in an unknown solution.

20.41 Calculate *E* for each of the following cells:
(a) a standard silver half-cell coupled with a copper electrode dipping into a 1.5 *M* $CuSO_4$ solution
(b) a cadmium electrode in a 1.2 *M* cadmium nitrate solution coupled with a cadmium electrode in 0.0030 *M* cadmium nitrate

20.42 (a) Calculate $E°$ for a galvanic cell based on the reaction

$$2MnO_4^-(aq) + 5HNO_2(aq) + H^+(aq) \rightarrow 2Mn^{2+}(aq) + 5NO_3^-(aq) + 3H_2O(l)$$

(b) Use the Nernst equation to calculate *E* for a cell in which the pH is 2.00, Mn^{2+} and NO_3^- are each 0.50 *M*, and MnO_4^- and HNO_2 are each 0.0010 *M*.

20.43 Calculate a single electrode potential for each of the following half-cells:
(a) $Cr_2O_7^{2-}(0.30\ M) + 14H^+(1\ M) + 6e^- \rightarrow 2Cr^{3+}(0.50\ M) + 7H_2O(l)$
(b) $O_2(g,\ 1\ atm) + 4H^+(aq,\ pH = 7.0) + 4e^- \rightarrow 2H_2O(l)$

20.44 Calculate a single electrode potential for each of the following half-cells:
(a) $Fe^{3+}(0.50\ M) + e^- \rightarrow Fe^{2+}(0.50\ M)$
(b) $Fe^{3+}(0.50\ M) + e^- \rightarrow Fe^{2+}(0.0050\ M)$

20.45 A galvanic cell is composed of two MnO_4^-/Mn^{2+} half-cells, one with an electrolyte pH = 0.0 and the other with an electrolyte pH = 2.0. All other concentrations are 1 *M*. Calculate the EMF of this cell.

20.46 Find the EMF of a cell composed of a standard hydrogen electrode coupled with a hydrogen electrode in which the electrolyte pH is 5.0.

20.47 A standard Ni/Ni^{2+} half-cell is coupled with a Ni/Ni^{2+} half-cell in which the Ni^{2+} concentration is unknown. If the standard half-cell is the cathode and the cell EMF is 0.054 V, find the unknown Ni^{2+} concentration.

20.48 A standard hydrogen electrode is coupled with a Cu/Cu^{2+} half-cell in which the Cu^{2+} concentration is unknown. The result is a galvanic cell in which the hydrogen electrode is the anode and the cell EMF is 0.0185 V. Find the unknown Cu^{2+} concentration.

20.49 Use reduction potentials to compute an equilibrium constant for each of the following reactions at 25°C:
(a) $O_2(g) + 4I^-(aq) + 4H^+(aq) \rightleftharpoons 2I_2(s) + 2H_2O(l)$
(b) $Zn(NH_3)_4^{2+}(aq) \rightleftharpoons Zn^{2+}(aq) + 4NH_3(aq)$

20.50 Use reduction potentials to compute an equilibrium constant for each of the following reactions at 25°C:
(a) $Ag(s) + \frac{1}{2}I_2(s) \rightleftharpoons AgI(s)$
(b) $H_2O(l) \rightleftharpoons H^+(aq) + OH^-(aq)$

20.51 Use reduction potentials to calculate the K_{sp} of mercury(I) chloride (Hg_2Cl_2).

20.52 The K_{sp} of AgBr is 5.3×10^{-13}. Calculate $E°$ for

$$AgBr(s) + e^- \rightleftharpoons Ag(s) + Br^-(aq)$$

Compare your result with the reduction potential reported in Table 20.1.

20.53 Copper metal is dipped into 0.010 *M* silver nitrate solution. Calculate the concentrations of Cu^{2+} and Ag^+ in the final equilibrium solution.

20.54 A 1.00-g piece of cobalt is placed in 10 mL of 0.100 *M* $Ni(NO_3)_2$. Calculate the concentrations of Co^{2+} and Ni^{2+} at equilibrium.

Batteries and Fuel Cells (Section 20.5)

20.55 Sketch a diagram for (a) the Leclanché dry cell and (b) the lead storage battery. Briefly describe the processes that occur during the discharge of these cells.

20.56 What advantages do alkaline dry cells have over Leclanché dry cells? What disadvantages do they have?

20.57 What is *overvoltage*? Why does it take more energy to recharge a battery than can be obtained from the battery?

20.58 Write equations for the reactions that occur when the lead storage battery is recharged. Explain why it is relatively easy to recharge this battery but very difficult to recharge a Leclanché dry cell.

20.59 Briefly explain the following observations:
(a) The electrolyte in a discharged lead storage battery freezes at a higher temperature than in a fully charged battery.
(b) Shaking and sudden shock shorten the life of a lead storage battery.

20.60 Explain why the voltage of a Leclanché dry cell drops with use, but the voltage of an alkaline dry cell remains more or less constant.

20.61 What is a *fuel cell*? How do fuel cells differ from batteries and the other types of galvanic cells described in this chapter?

20.62 **(a)** Diagram and give equations for the H_2/O_2 fuel cell.
(b) What advantage does this cell have over first burning the hydrogen and then converting the heat energy into electrical energy?

Corrosion (Section 20.6)

20.63 Why is corrosion considered to be a galvanic process?

20.64 Describe the mechanism (include equations) by which an iron railing corrodes in acid rain.

20.65 Explain why corrosion requires both oxygen and water.

20.66 What effect will each of the following have on the rate at which an iron nail corrodes? Explain your answer in each case.
(a) immersing the nail in an acidic solution
(b) immersing the nail in a basic solution
(c) immersing the nail in a solution of salt water
(d) drying the nail
(e) placing the nail in a nitrogen atmosphere

20.67 List three ways of protecting an iron object from corrosion.

20.68 What is a *corrosion inhibitor*? How do corrosion inhibitors work?

20.69 What is *galvanized iron*? How does galvanizing an iron pail protect it from corrosion?

20.70 Very old homes often contain galvanized iron pipe in their plumbing systems. Would you advise for or against replacing a section of damaged pipe with copper tubing? Explain.

20.71 What is *cathodic protection*? Suggest a method for cathodically protecting an underground oil storage tank. Explain why your method should work.

20.72 Tin cans are actually steel cans plated with a protective layer of tin. Does tin plating also provide cathodic protection for the underlying steel? Why do damaged tin cans rust more rapidly than unplated steel?

Electrolysis (Section 20.7)

20.73 Is the anode positive or negative in an electrolytic cell? What is the charge on the cathode? How do these charges compare with those in a galvanic cell?

20.74 What is the direction of anion flow in an electrolytic cell? What is the direction of cation flow? How do these directions of flow compare with those in a galvanic cell?

20.75 Explain why sodium chloride must be molten to be electrolyzed.

20.76 What happens to the electrical energy consumed in the electrolysis of molten sodium chloride? Can we get this energy back? How?

20.77 Write the electrode equations and the net equation for the electrolysis of (a) molten KCl and (b) molten $CaCl_2$.

20.78 Explain why electrolytic reduction rather than chemical reduction is often used to obtain active metals from their compounds.

20.79 Explain why water must be excluded in the electrolytic preparation of an active metal such as potassium. What will happen if water gets into the cell?

20.80 Describe the electroplating process. Why is a complexing agent sometimes used?

20.81 Diagram the Downs cell. Identify the anode and cathode and give the electrode reactions.

20.82 Diagram the Hall–Héroult cell. Identify the anode and cathode, and give the electrode reactions.

20.83 Write the electrode equations and the net equation for the electrolysis of (a) aqueous potassium chloride and (b) aqueous calcium chloride. Explain why potassium and calcium are not produced during these electrolyses.

20.84 Diagram and write the electrode reactions for the following cells used for the electrolysis of brine: (a) the diaphragm cell; (b) the membrane cell. List some of the advantages and disadvantages of each cell.

20.85 The electrolysis of a sodium sulfate solution produces hydrogen and oxygen gases. Diagram the electrolytic cell, write the electrode reactions, and show the direction in which each ion moves. What would happen if sodium sulfate were not present?

20.86 An aqueous solution of copper(II) sulfate is electrolyzed between inert electrodes. Copper metal plates out at one of the electrodes while oxygen gas and hydrogen ions form at the other electrode.

(a) Write the half-reaction equations and the net ionic equation for the electrolysis.

(b) Which electrode is the anode and which is the cathode?

Electrochemical Stoichiometry (Section 20.8)

20.87 Consider the electrode reaction

$$2H_2O(l) + 2e^- \longrightarrow H_2(g) + 2OH^-(aq)$$

(a) How many grams of water are consumed by the passage of 15,000 C?

(b) How many liters of hydrogen, measured at 25°C and 1.00 atm, are produced by the passage of 1.25 F?

20.88 Consider the electrode reaction

$$2H_2O(l) \longrightarrow O_2(g) + 4H^+(aq) + 4e^-$$

(a) How many moles of water are consumed by the passage of 1.75 F?

(b) How many liters of oxygen, measured at STP, will be produced by the passage of 180,000 C?

20.89 How many grams of lead will be deposited from a solution of Pb^{2+} ions by a 0.15-A current flowing for 1.00 h?

20.90 How many hours would it take for a current of 25.0 A to produce 1.00 kg of aluminum in the Hall–Héroult cell?

20.91 **(a)** How many grams of aluminum are produced when a 75.0-A current passes through a Hall–Héroult cell for 24.0 h?

(b) How many liters of oxygen, measured at 25°C and 800 torr, will be produced along with the aluminum in Part (a)?

20.92 **(a)** How many hours would it take for an 80-mA current to liberate 3.0 mmol of oxygen in the electrolysis of water?

(b) How many millimoles of hydrogen would be liberated along with the oxygen?

20.93 A current of 1.10 A passing through a copper sulfate solution plates out 0.650 g of copper in 30.0 minutes. Assume that the copper ion is Cu^{n+}, and use the data to show that $n = 2$.

20.94 Passage of a 50.0-mA current through a solution of a nickel compound for 6.00 h resulted in the deposition of 0.328 g of nickel metal. Use this information to calculate the oxidation state of nickel in the compound.

20.95 How many faradays are required to deposit 1.00 mol of gold if the electrolyte contains (a) $AuCl_4^-$ and (b) $Au(CN)_2^-$?

20.96 How many coulombs are required for the deposition of 1.00 mol of chromium from an acidified solution of CrO_4^{2-}?

20.97 A cell with a copper cathode in aqueous $CuSO_4$ and a cell with a nickel cathode in aqueous $NiSO_4$ are connected in series with a silver coulometer. A current passes through the circuit, and the mass of the silver cathode increases by 0.156 g. Calculate the change in mass of (a) the copper electrode and (b) the nickel electrode.

20.98 A current passing through a large vat of molten calcium chloride produces 1.00 kg of calcium every 10.0 days.

(a) Calculate the current in amperes.
(b) How many liters of chlorine gas, measured at 30°C and 710 torr, will be produced during the same time period?

PART B. MISCELLANEOUS QUESTIONS AND PROBLEMS

20.99 Suppose the standard copper electrode had been chosen as the official reference half-cell rather than the standard hydrogen electrode.
(a) What would be the standard reduction potential for hydrogen?
(b) What effect would there be on the standard reduction potentials for other half-cells?
(c) Which, if any, of the negative reduction potentials in Table 20.1 would become positive? Which, if any, of the positive reduction potentials would become negative?

20.100 Use reduction potentials to account for the following observations:
(a) Copper pennies dissolve in nitric acid but not in hydrochloric acid.
(b) Iron nails placed in a bottle of aqueous $FeSO_4$ inhibit the oxidation of Fe^{2+} to Fe^{3+}.
(c) Cu^+ disproportionates in aqueous solution according to the equation

$$2Cu^+(aq) \rightarrow Cu^{2+}(aq) + Cu(s)$$

20.101 Use Table 20.1 to find combinations of half-cells that could be used to construct a galvanic cell with an EMF of about 1.5 V. Which combinations would be most practical?

20.102 Refer to the derivation of Equation 20.6, and perform a similar derivation to show that the total voltage will equal the sum of the individual voltages when two or more cells are connected in series.

20.103 What is the maximum work that could be done by a 1.5-V battery that supplies 30 mA of current for 2.0 s?

20.104 (a) Find $E°$ and K for the reaction

$$Ag(CN)_2^-(aq) \rightleftharpoons Ag^+(aq) + 2CN^-(aq)$$

(b) Will the EMF increase or decrease if the concentration of CN^- is made greater than 1 M? Explain your answer.

20.105 The equilibrium constant for the reaction

$$ClO_4^-(aq) + ClO_2^-(aq) \rightleftharpoons 2ClO_3^-(aq)$$

is 1.9×10^4 at 25°C. Find (a) $E°$ and (b) $\Delta G°$. (*Hint*: Use oxidation-number changes to find n.)

20.106 Standard reduction potentials for the nickel and cadmium half-cells of a nickel–cadmium battery are 0.49 and −0.82 V, respectively. The aqueous electrolyte contains 20% KOH by mass and has a density of 1.20 g/mL. What will be the voltage of the nickel–cadmium battery?

20.107 (a) Diagram a galvanic cell that uses the reaction

$$2MnO_4^-(aq) + 3SO_3^{2-}(aq) + H_2O(l) \rightarrow 2MnO_2(s) + 3SO_4^{2-}(aq) + 2OH^-(aq)$$

Write the anode and cathode half-reactions, and indicate the directions of electron flow and ion migration.
(b) Calculate $\Delta G°$ and K for the cell reaction.
(c) Calculate E for the cell when all of the dissolved substances are 0.10 M.

20.108 Acidic foods such as citrus juice and tomatoes used to be sold in cans with lead-soldered seams. When these cans were opened, air mixed in and some of the lead solder dissolved. Occasionally, the level of lead in the food became high enough to constitute a health hazard. Look over Table 20.1, and suggest a set of half-reactions to explain why the solder dissolves. Explain in terms of ΔG why refrigeration will not prevent the process from occurring.

20.109 Tin metal is usually added to solutions of Sn^{2+} to prevent air oxidation.
(a) Find the equilibrium constant for

$$2Sn^{2+}(aq) \rightleftharpoons Sn^{4+}(aq) + Sn(s)$$

(b) Starting with 1.0 M Sn^{2+}, calculate the fraction of Sn^{2+} that will be converted to Sn^{4+} at equilibrium.

20.110 The Downs cell electrolyte is a mixture of sodium chloride and calcium chloride. What is the function of the calcium chloride? Why isn't calcium metal produced along with sodium metal?

20.111 The mercury cell described in Section 20.5 has an EMF of about 1.35 V. If a current of 0.10 μA is drawn for 1.00 h, calculate (a) the number of coulombs that pass through the external circuit, (b) the mass in micrograms of mercury that forms, and (c) the maximum work in joules that can be done by the battery.

20.112 A fuel cell generates electricity by the oxidation of natural gas at 1200°. The electrolyte contains a molten carbonate, and the equations are

anode: $CH_4(g) + 4CO_3^{2-} \rightarrow 2H_2O(g) + 5CO_2(g) + 8e^-$

cathode: $2CO_2(g) + O_2(g) + 4e^- \rightarrow 2CO_3^{2-}$

(a) Write the net equation for the reaction.
(b) Assume that each gas is at 1.00 atm pressure, and estimate $\Delta G°$ for the reaction at 1200°C. (*Hint*: Remember that $\Delta H°$ and $\Delta S°$ do not vary with temperature.)

(c) Estimate the voltage that would be produced by this fuel cell if it were run under standard-state conditions.

20.113 An ordinary flashlight battery has an EMF of about 1.5 V. A current of 0.50 A is drawn for 5.0 minutes.

(a) How many grams of the zinc electrode are consumed?

(b) How many moles of ammonia are produced by the cathode reaction?

(c) Find the power output of the battery. (Power is measured in watts, where 1 watt = 1 J/s).

20.114 One hundred brass spheres, each 5.0 cm in diameter, are to be plated with silver to a thickness of 0.020 cm.

(a) How many grams of silver are required to plate the spheres? The density of silver is 10.5 g/cm^3.

(b) What minimum current must be used to complete the plating job in 5.0 h? Assume that the electroplating solution contains $Ag(CN)_2^-$.

(c) If the current efficiency—the percent of current that actually plates silver—is 95%, what actual current is needed to complete the plating job in 5.0 h?

20.115 The passage of 1.00 F through a molten mixture of NaF and HF produces 12.8 L of fluorine gas measured at 40°C and 1.00 atm. Assume that the formula for fluorine is F_x, and use this information to show that $x = 2$.

20.116 A current of 4.00 A is passed through a $CuSO_4$ solution for 6.25 minutes. The mass of copper deposited is 0.494 g.

(a) Calculate the value of the Faraday, that is, the charge on 1 mol of electrons.

(b) The magnitude of the charge on one electron is 1.6022×10^{-19} C (Table 2.1). Use your answer from Part (a) to estimate the value of the Avogadro number (N_A).

CHAPTER 21

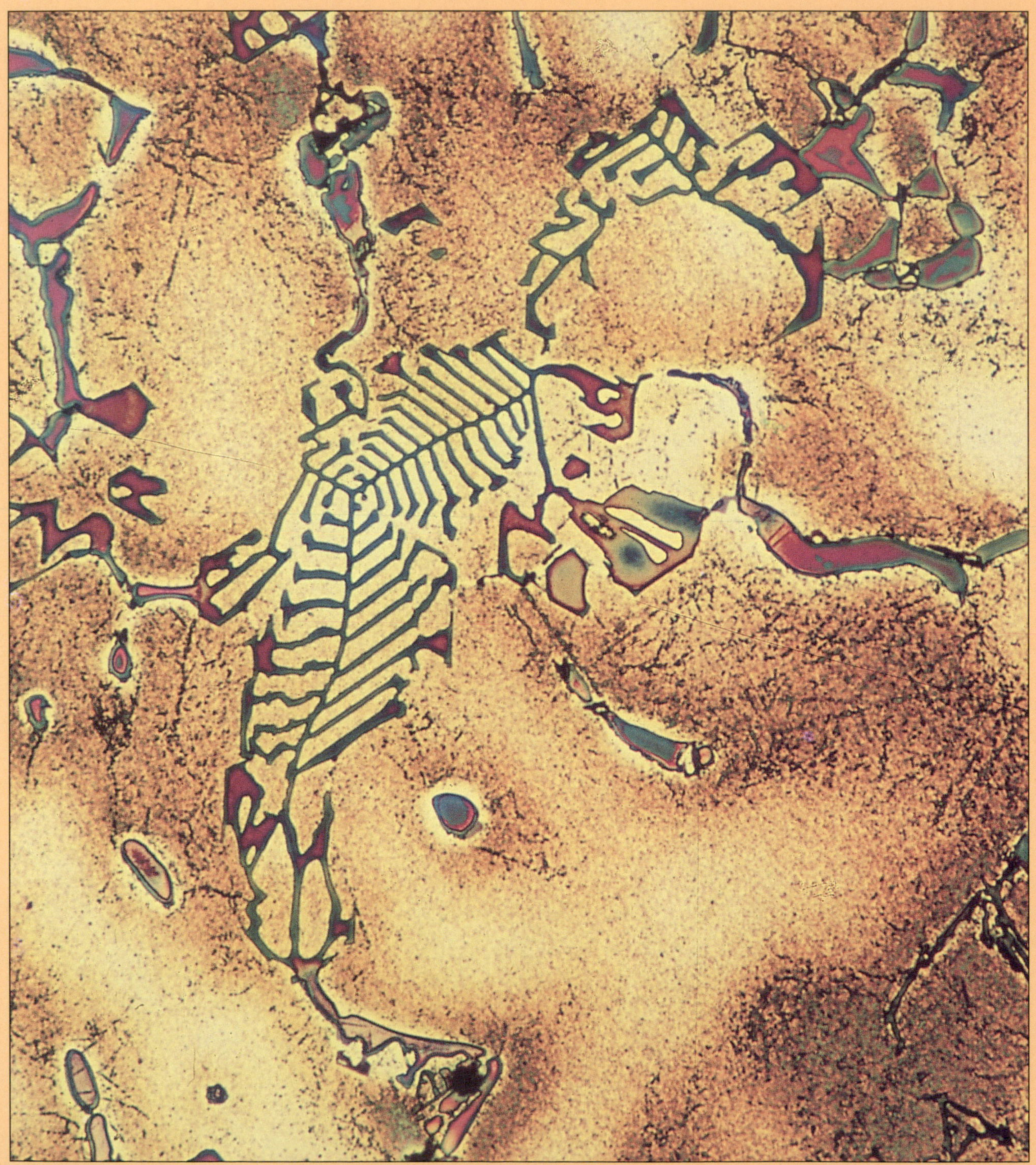

Micrograph of aluminum alloyed with copper, iron, manganese, and silicon.

METALS AND COORDINATION CHEMISTRY

OUTLINE

PREVIEW

The word *metal* brings to mind steel girders, tools, coins, and jewelry; we think of metallic objects as hard, shiny things that clank or jingle. In their compounds, however, metals are a quiet part of everything—they are dissolved in the ocean, they are part of the chlorophyll of green plants, and they flow with the blood in our veins. Iron compounds carry oxygen to our cells, zinc compounds are involved in taste and smell, and other metal compounds regulate our heartbeats and our moods. Small metal atoms centered in large biomolecules recycle the earth's supply of nitrogen and oxygen, literally keeping the world alive. The role played by metal atoms in nature is fascinating and important, and new uses for metals are discovered almost every day.

The metallic elements exhibit a wide range of physical and chemical properties. Most metals are hard, high-melting solids that are *malleable* (can be shaped by hammering) and *ductile* (can be drawn into wires). There are some exceptions. For example, calcium and bismuth are brittle and break into pieces when hammered, sodium and potassium are soft, and mercury is a liquid at room temperature. All metals, however, are good conductors of heat and electricity, and their freshly cut surfaces exhibit a typical metallic luster. Chemically, metals range from those that are extremely reactive, like sodium, to those that are almost inert, like gold. Metals are generally less electronegative than nonmetals, and the metal atom is the positive partner when it is bonded to a nonmetal atom. Most metals form positive ions; nonmetals, with the exception of hydrogen, ordinarily form negative ions.

21.1 PHYSICAL PROPERTIES OF METALS

The physical properties of metals, which include *luster, conductivity, compactness,* and *deformability*, are related to the nature of the metallic bond in which loosely held valence electrons move freely through the lattice.

LEARNING HINT

Review the discussion of metallic solids and metallic bonding in Section 12.5.

All metals are lustrous, and most are described as silvery, yet each one has a distinctive appearance. Silver, for example, is whitish. Tungsten is dark. Gold and copper have the most pronounced colors, but cesium is distinctly yellowish and manganese has a pinkish brown cast.

Metals are good conductors of heat and electricity because their mobile valence electrons transport kinetic energy and charge. The electrical conductivity of a metal decreases somewhat as the temperature rises because the increased motion of atoms in the lattice impedes the electron flow. Metallic conduction is also reduced by impurities.

Electrical conductivities have their peak values in the coinage group (copper, silver, and gold); copper and silver are the best conductors, followed closely by gold. Aluminum is also very conductive, but it is less suited for electrical wiring because its low density makes it bulky and its chemical activity makes it more subject to deterioration than less active metals such as copper.

The electrons in a nonmetal are confined to separate molecules or to localized bonds within a covalent network; therefore, most nonmetals cannot conduct electricity. Seven elements—boron, silicon, germanium, arsenic, selenium, antimony, and tellurium—are **semiconductors** whose electrical conductivities are intermediate between those of metals and nonmetals. Electric charge in a semiconductor is carried only by exceptionally energetic electrons. More electrons become activated at higher temperatures; hence, semiconduction, unlike metallic conduction, increases with increasing temperature.

Metals are dense and compact because each atom in a metal has many close neighbors, ranging from 8 in a body-centered cubic lattice to 12 in a close-packed arrangement. This efficient use of space distinguishes a metallic lattice from a more open covalent lattice, in which each atom's nearest neighbors are limited by the number of covalent bonds formed by the atom and rarely exceed 4. The exact density of a metal depends on atomic mass, atomic radius, and lattice type. Lithium is the lightest metal (0.53 g/cm^3), and osmium is the heaviest (22.5 g/cm^3). Lead, supposedly synonymous with heaviness, has a density of 11.35 g/cm^3, while liquid mercury, named after a winged god, is even denser (13.6 g/cm^3). The density of iron is a "mere" 7.9 g/cm^3.

Deformability, a result of nondirectional bonding, varies from metal to metal. The deformability of gold is the most phenomenal; gold can be hammered into translucent foil a mere 100 nm thick, and 1 g of gold can be drawn into a virtually invisible wire 3000 m long. It is not surprising that gold and other highly malleable metals have cubic close-packed structures; recall from Section 12.6 that this type of lattice provides many planes in which layers of atoms can slip past each other.

The hardness, melting point, and boiling point of a metal depend on the strength of the metallic bond, which decreases with increasing atomic size and increases with the number of valence electrons. Group 1A metals, with large atoms and only one valence electron per atom, have low melting points and are soft. Group 2A metals, with smaller atoms and two valence electrons per atom, have higher melting points and are much harder than the Group 1A metals. Most structural metals, that is, metals used for building and machinery, are transition elements in which underlying *d* electrons contribute to the bond strength.

The strength of the metallic bond can be measured in terms of the enthalpy of sublimation, that is, the energy needed to separate a mole of metal into gaseous atoms. Figure 21.1 shows that the enthalpy of sublimation peaks at vanadium (V, $4s^2 3d^3$) in the fourth period; niobium (Nb, $5s^2 4d^3$) in the fifth period; and tung-

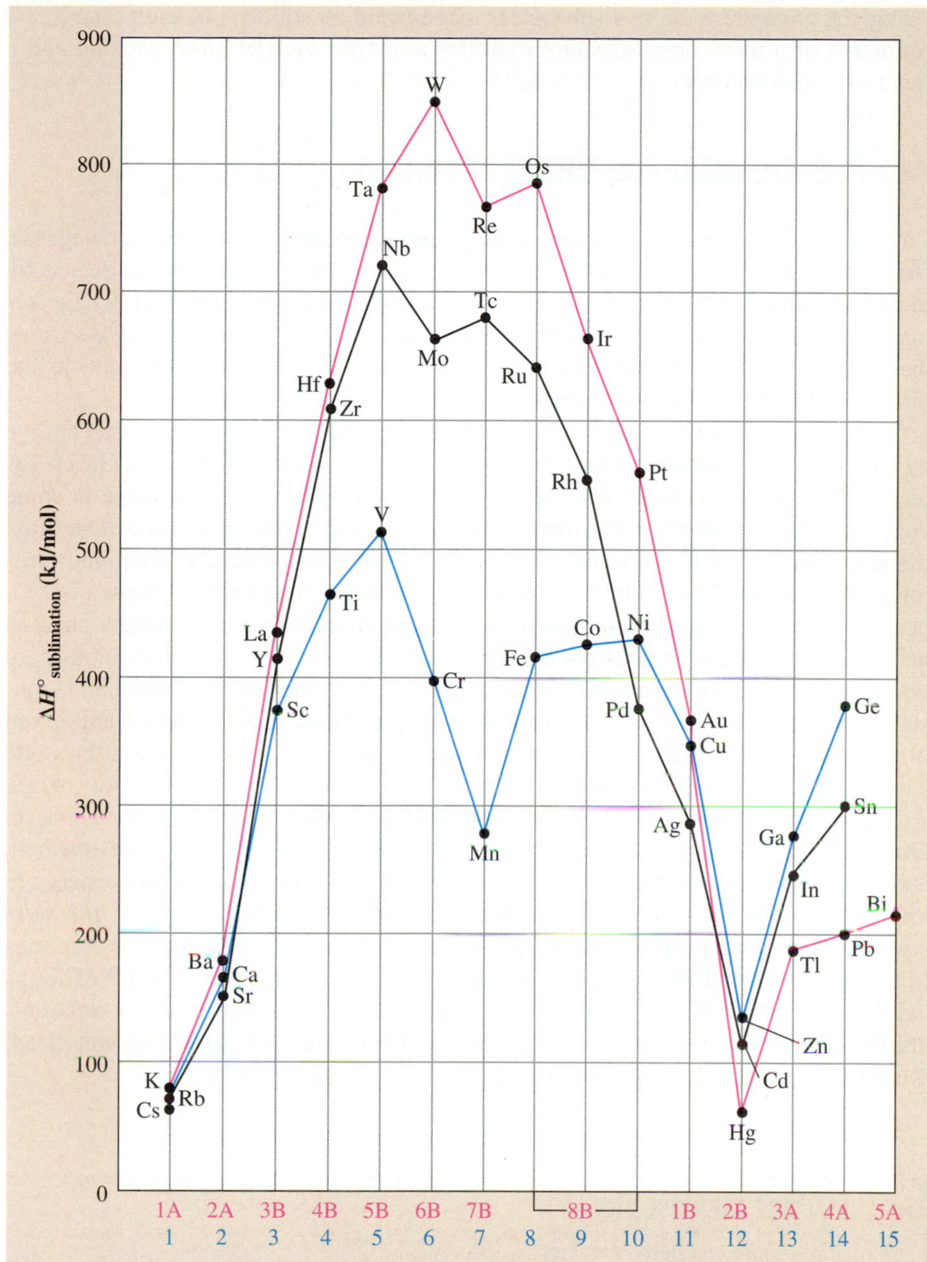

Figure 21.1
Enthalpies of sublimation for fourth-period (blue), fifth-period (black), and sixth-period (red) metals. The lanthanides are not included.

sten (W, $6s^2 5d^4$) in the sixth period. The melting and boiling points in each period exhibit a similar rise and fall, and they peak within one or two atomic numbers of the above metals. Tungsten, the metal used in light bulb filaments, has the highest melting (3410°C) and boiling (5660°C) points of all the metallic elements.

Metals to the right of the transition series in the periodic table have characteristics that suggest a trend toward covalence. Melting points and conductivities drop. Although tin is classified as a metal, below 18°C it slowly converts to an allotropic form with a covalent lattice and nonmetallic properties. Antimony and bismuth in

Most substances become denser when they freeze. Water, antimony, and bismuth are among the exceptions.

Group 5A are among the few substances that expand on solidifying from a melt; the formation of a more open structure indicates a distinct lack of close packing and a trend toward covalence.

21.2 CHEMICAL PROPERTIES OF METALS

Chemically, metals are distinguished from nonmetals by the absence of negative oxidation states. Two other chemical properties that are shared to some degree by most metals are (1) the ability to form positive ions and ionic bonds and (2) the formation of basic oxides. These two properties become less pronounced as we move diagonally across the periodic table from the lower left to the upper right, in the direction of higher ionization energies and smaller atoms.

The most straightforward periodic trends are exhibited by the representative metals. Group 1A metals almost always form ionic bonds, as do most Group 2A metals. Beryllium and the metals of Groups 3A, 4A, and 5A often engage in some degree of electron sharing. The trend toward covalence across a period can be illustrated by comparing the physical properties of sodium bromide and aluminum bromide. Solid sodium bromide is an ionic solid. It has a high melting point (747°C) because each ion in the crystal lattice is attracted equally by its oppositely charged neighbors on all sides; it does not associate with one neighbor more than another, so the lattice is not easily disrupted. Solid aluminum bromide, on the other hand, consists of Al_2Br_6 molecules. The intermolecular attractions in a crystal of aluminum bromide are weaker than the attractions in an ionic crystal, and aluminum bromide has a relatively low melting point of 97.5°C.

Transition and post-transition metals (those that follow the transition metals in Groups 3A through 6A) can draw bonding electrons from more than one subshell, and most of them can form more than one positive oxidation state. It takes energy to remove electrons; thus, *for a given element, the higher the oxidation state, the more covalent the bonding*. For example, tin and lead in Group 4A (s^2p^2) lose p electrons to form the Sn^{2+} and Pb^{2+} ions found in ionic salts such as $Sn(NO_3)_2$ and $Pb(ClO_4)_2$. On the other hand, their +4 oxidation states, which involve both s and p electrons, are found only in covalent substances such as PbO_2 and $Pb(C_2H_5)_4$ (tetraethyllead, the gasoline additive), and in anions such as $SnCl_6^{2-}$ and $Pb(OH)_6^{2-}$.

PRACTICE EXERCISE 21.1

Which has more covalent character, VCl_2 or VCl_3?

Activity: The Energetics of Cation Formation

Metals tend to form positive oxidation states; that is, they act as reducing agents and become oxidized. A strongly reducing metal is said to be *active*. The metal's oxidation potential, which is the negative of its reduction potential, is a measure of its activity in an aqueous environment. According to the $E°$ values in Table 20.1 (p. 950), Groups 1A and 2A contain the most active metals, and activity falls off in a somewhat irregular fashion from left to right across the periodic table. Most main-group metals are more active than hydrogen; only bismuth and polonium do not displace hydrogen from 1 *M* H^+(aq).

The activity trends are irregular because the conversion of a solid metal into aqueous ions is a complex process in which the metal atoms must (1) separate from the solid, (2) lose electrons, and (3) become hydrated. These steps can be represented by the thermochemical equations

1. $M(s) \rightarrow M(g) \qquad \Delta H^\circ_{sub}$

2. $M(g) \rightarrow M^+(g) + e^- \qquad \Delta H^\circ_{ion}$

3. $M^+(g) \rightarrow M^+(aq) \qquad \Delta H^\circ_{hyd}$

where M symbolizes a Group 1A metal and ΔH°_{sub}, ΔH°_{ion}, and ΔH°_{hyd} are the enthalpies of sublimation, ionization, and hydration, respectively. Let us see if we can estimate the standard oxidation potential of a metal from what we know about the enthalpy and entropy changes associated with the above three steps.

1. The energy needed to separate individual atoms from the solid, that is, the sublimation energy, is generally small compared to the energies of the other steps, but it plays an important role nevertheless. A large atom with one valence electron escapes from its lattice more easily than a small atom with two or more valence electrons. Group 1A metals are therefore more easily vaporized than Group 2A metals, and the metals near the bottom of a representative group are more easily vaporized than the metals at the top. Some enthalpies of sublimation are listed in Table 21.1.
2. A Group 1A atom loses electrons easily, so it has a low first-ionization energy. A Group 2A atom has a higher first-ionization energy and needs still more energy to remove the second valence electron.
3. If a positive ion is not stabilized by its environment, it will quickly gain electrons and lose its charge. In a solid, the ions are stabilized by the attraction of oppositely charged neighbors. In aqueous solution, they are stabilized by ion–dipole interactions with surrounding water molecules. These interactions are accompanied by the release of large amounts of hydration energy. Ion–dipole forces are strongest when the ionic charge is high and the ionic radius is small, so more energy is evolved in the hydration of small ions such as Mg^{2+} and Al^{3+} than in the hydration of large, singly charged ions such as Na^+. Within a group, the energy of hydration is greatest for the small ions at the top. The enthalpies of hydration reported in Table 21.1 are consistent with these conclusions.

LEARNING HINT

Review the discussion of ion–dipole interactions and the formation of hydrated ions in Section 13.1.

The total enthalpy change for the oxidation of a metal in aqueous solution, ΔH°_{oxid}, is the sum of the enthalpy changes for each step:

$$M(s) \rightarrow M^+(aq) + e^- \qquad \Delta H^\circ_{oxid} = \Delta H^\circ_{sub} + \Delta H^\circ_{ion} + \Delta H^\circ_{hyd}$$

Values for this sum are given in Table 21.1.

The standard free-energy change for the oxidation, ΔG°_{oxid}, is obtained by subtracting the entropy term, $T\Delta S^\circ$, from the enthalpy change (Equation 19.7):

$$\Delta G^\circ_{oxid} = \Delta H^\circ_{oxid} - T\Delta S^\circ_{oxid} \qquad (19.7)$$

TABLE 21.1 Thermodynamic Data for Steps in the Formation of Aqueous Metal Ions[a]

Group 1A Metal Ions

Element	*Atomic Radius (pm)*	ΔH°_{sub}	ΔH°_{ion}	ΔH°_{hyd}[b]	ΔH°_{oxid}	$T\Delta S^\circ_{oxid}$[c]	ΔG°_{oxid}	E°_{oxid} *(J/C)*
Li	152	159	526	−963	−278	15	−293	3.04
Na	186	107	502	−849	−240	22	−262	2.72
K	227	89	425	−766	−252	31	−283	2.93
Rb	248	81	409	−741	−251	33	−284	2.94
Cs	265	76	382	−716	−258	34	−292	3.03

Third-Period Ions Na^+, Mg^{2+}, and Al^{3+}

Element	*Atomic Radius (pm)*	ΔH°_{sub}	ΔH°_{ion}	ΔH°_{hyd}[b]	ΔH°_{oxid}	$T\Delta S^\circ_{oxid}$[c]	ΔG°_{oxid}	E°_{oxid} *(J/C)*
Na	186	107	502	−849	−240	22	−262	2.72
Mg	160	148	2201	−2816	−467	−12	−455	2.36
Al	143	326	5157	−6014	−531	−46	−485	1.68

[a]The ΔH°, $T\Delta S^\circ$, and ΔG° values are in kilojoules per mole.

[b]The standard enthalpy of hydration is based on the convention that the enthalpy, free energy, and entropy of formation for $H^+(aq)$ are zero. The basis for this convention is (1) the formation process for an ion in solution is defined in terms of the reaction $M(s) + H^+(aq) \longrightarrow M^+(aq) + \frac{1}{2}H_2(g)$ and (2) E° and ΔG° are defined to be zero for the standard hydrogen electrode.

[c]Evaluated at $T = 298$ K.

The oxidation potential of the metal, E°_{oxid}, is calculated from ΔG°_{oxid} using Equation 20.5:

$$\Delta G^\circ_{oxid} = -nFE^\circ_{oxid} \qquad (20.5)$$

The experimental oxidation potentials are the negative of the reduction potentials listed in Table 20.1 (p. 950). The estimated oxidation potentials in Table 21.1 agree well with these values.

The data in Table 21.1 show that the metals of Groups 1A and 2A all end up with rather similar activities (similar values of E°_{oxid}). Note that lithium, at the top of Group 1A, is a slightly stronger reducing agent than cesium, at the bottom of the group. On the basis of ionization energies alone, we might expect cesium to be more active than lithium, but the relatively large hydration energy released by the small lithium ion causes lithium to have a somewhat higher oxidation potential. Note also that aluminum (in Group 3A) is only slightly less active than the metals of Groups 1A and 2A. An aluminum atom must lose three electrons and has a very high ionization energy. Nevertheless, the oxidation potential remains high because of the large hydration energy released by the small, highly charged aluminum ion.

Metallic Oxides and Hydroxides

Most elements combine with oxygen, so one way to compare the chemistry of different elements is to compare the nature of their oxides and hydroxides. With the exception of beryllium, the metals of Groups 1A and 2A form oxides and hydroxides that are ionic. The O^{2-} and OH^- ions are strong proton acceptors; hence, compounds like CaO and $Ca(OH)_2$ are strong bases. As we move up and to the right in the periodic table, we find the metal–oxygen bonds losing some of their ionic character and becoming more covalent. For example, the oxygen atoms in $Be(OH)_2$ and $Al(OH)_3$ share electrons with the metal atoms; thus, they retain less electron density and are less able to attract protons. Covalence in a metal–oxygen bond makes the oxygen less basic:

ionic bond
$$M^+ \quad :\ddot{\underset{..}{O}}-H^-$$
strong proton acceptor

covalent bond
$$M-\ddot{\underset{..}{O}}-H$$
weak proton acceptor

The decrease in the basicity of metal oxides from left to right across a period and from the bottom to the top of a group manifests itself in the appearance of amphoteric oxides. Recall (Section 18.3) that an amphoteric oxide can dissolve in both acidic and basic solutions. The representative metals whose oxides are amphoteric form a diagonal swath across the periodic table (Table 21.2). Aluminum oxide is a typical amphoteric oxide; its reactions with acid and base are

$$Al_2O_3(s) + 6H^+(aq) \rightarrow 2Al^{3+}(aq) + 3H_2O(l)$$

$$Al_2O_3(s) + 2OH^-(aq) + 3H_2O(l) \rightarrow 2Al(OH)_4^-(aq)$$

When a metal forms more than one oxide, the oxide in which the metal has the lower oxidation state will be more ionic and more basic. The oxide in which the

TABLE 21.2 Acid–Base Character of Main-Group Metallic Oxides

Group 1A	Group 2A	Group 3A	Group 4A	Group 5A
Li_2O[a] Basic	BeO Amphoteric			
Na_2O[a] Basic	MgO Basic	Al_2O_3 Amphoteric		
K_2O[a] Basic	CaO[a] Basic	Ga_2O_3 Amphoteric		
Rb_2O[a] Basic	SrO[a] Basic	In_2O_3 Basic	SnO[b] Amphoteric	
Cs_2O[a] Basic	BaO[a] Basic	Tl_2O_3 Basic	PbO[b] Amphoteric	Bi_2O_3[b] Basic

[a]Dissolves in water to form hydroxide ions.

[b]Higher oxides of these elements are acidic.

TABLE 21.3 Oxides and Hydroxides of Chromium

Oxidation State	Oxide or Hydroxide	Character
+2[a]	$Cr(OH)_2$ Yellow brown	Basic
+3	Cr_2O_3 Gray-green	Amphoteric
+6	CrO_3 Scarlet	Acidic

[a]The +2 oxidation state of chromium is less stable than the +3 and +6 states.

metal has the higher oxidation state will be more covalent and less basic—sometimes even acidic. The effect of oxidation state is illustrated in Table 21.3 for the oxides of chromium.

PRACTICE EXERCISE 21.2

Which of the following substances should be more basic (a) Ga_2O_3 or In_2O_3; (b) $Fe(OH)_2$ or $Fe(OH)_3$?

Special Properties of the Transition Metals

The transition elements are found in the B groups that lie between Groups 2A and 3A in the periodic table. Each of their electron configurations is completed by adding an electron to an underlying *d* or *f* subshell. Some chemists confine the term *transition element* to elements that have partially filled *d* orbitals in at least one observed oxidation state. This definition would exclude the final *d*-block members, Zn, Cd, and Hg (Group 2B), and as one might expect, this is the group that most resembles the representative elements.

The periodic trends among transition elements are less regular than those in the representative groups, and some trends are even reversed. For example, the most active representative metals (those with the highest oxidation potentials) are usually found at the bottom of their groups, while the most active transition metals are found at the top—Ni and Cu in the fourth period have higher oxidation potentials than Pt and Au in the sixth period.

The following properties are characteristics of transition metals:

1. *Small compact atoms*: Refer to Figure 8.9 (p. 321) and note the relatively small size of transition-metal atoms. Note also that the atomic size decreases across a period, goes through a minimum, and then increases as the atoms become less metallic. Some transition-metal properties that are associated with small atomic size are hardness, high density, and high melting and boiling points.

2. *Paramagnetism*: Most transition-metal atoms and many of their compounds are

TABLE 21.4 Common Oxidation States of Fourth-Period Transition Metals

Sc	Ti	V	Cr	Mn	Fe	Co	Ni	Cu	Zn
								+1	
	+2	+2	+2	+2	+2	+2	+2	+2	+2
+3	+3	+3	+3	+3	+3	+3			
	+4	+4		+4					
		+5							
			+6	+6					
				+7					

paramagnetic. Recall from Section 8.5 that paramagnetism shows the presence of unpaired electrons.

3. *Variable oxidation states*: The higher oxidation states involve underlying *d* electrons and *d* orbitals. The principal oxidation states of the fourth-period transition metals are shown in Table 21.4.
4. *Stable complexes*: Transition metals form a greater number of complex ions and molecules than do representative metals, and their complexes tend to be more stable. The chemistry of the heavier transition elements is almost entirely the chemistry of complexes.
5. *Colored compounds*: Among the brightly colored species are oxides, oxo ions, sulfides, and numerous complex ions.
6. *Catalytic activity*: Many industrial catalysts contain transition metals, and transition-metal atoms are found in a number of enzymes. Much of the catalytic activity can be attributed to the ability of these atoms to shift from one oxidation state to another and to their ability to form complexes that have a temporary existence.

21.3 BAND THEORY OF CONDUCTORS, SEMICONDUCTORS, AND INSULATORS

Each electron in the sea of electrons that constitutes the metallic bond must occupy an energy state or orbital. The Pauli exclusion principle imposes the additional requirement that each orbital can hold no more than two electrons. Consider the molecular orbitals formed from overlapping the 2*s* orbitals of two lithium atoms, as shown in Figure 21.2a. The valence electrons, one from each atom, pair up in the low-energy bonding orbital. The antibonding orbital, which is separated from the bonding orbital by a wide **energy gap**, is empty.

LEARNING HINT
Review the discussion of molecular orbitals in Section 10.5.

Figure 21.2b shows the 5 bonding and 5 antibonding orbitals formed by 10 lithium atoms. Each molecular orbital represents a slightly different way of combining the atomic orbitals, and the energies of the molecular orbitals are spread out so that the energy gap between the bonding and the antibonding orbitals is smaller

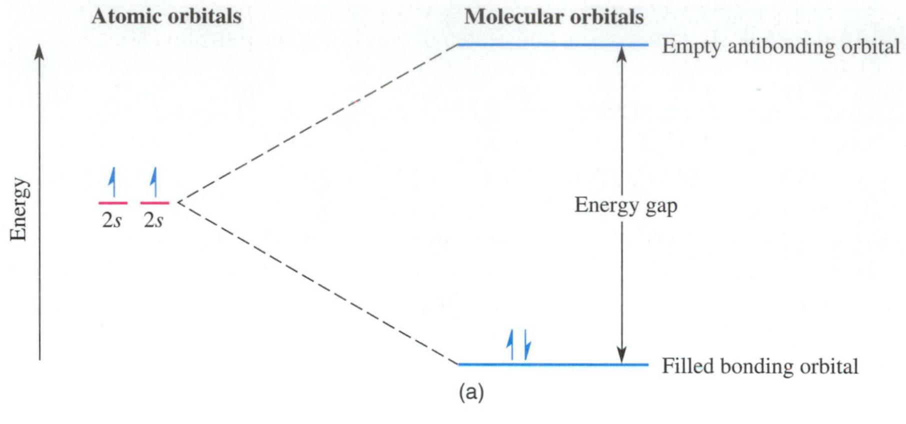

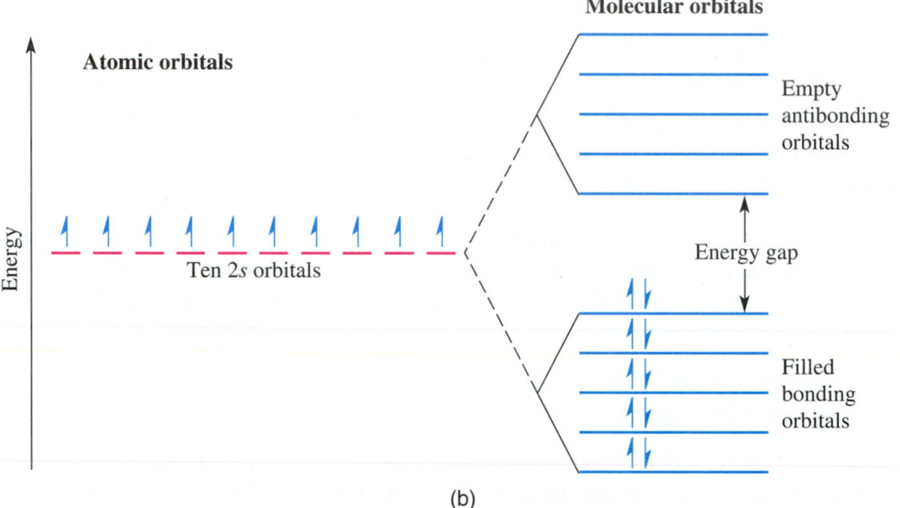

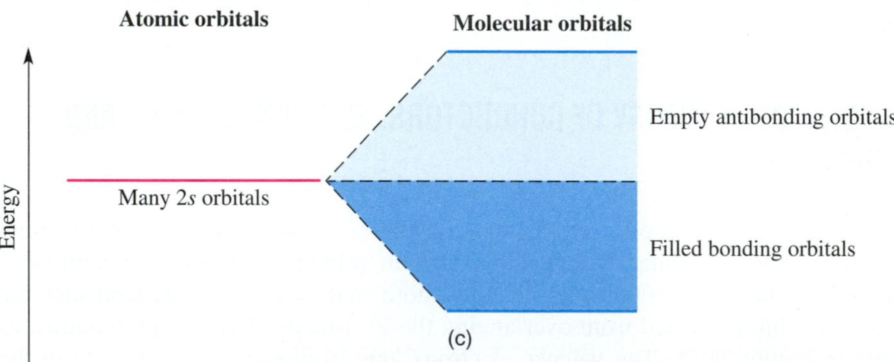

Figure 21.2

Band formation in lithium metal. (a) The two molecular orbitals formed by the *s* orbitals of two lithium atoms. (b) The 10 lithium atoms form 5 bonding and 5 antibonding orbitals. The energy gap is smaller than that for two atoms. (c) A large number of lithium atoms forms a band of molecular orbitals with virtually no energy gap.

than it was when there were only two lithium atoms. The 10 valence electrons occupy the 5 bonding orbitals in accordance with the Pauli principle.

The greater the number of lithium atoms, the more spread out the molecular orbitals and the smaller the energy gap. In a crystal composed of N lithium atoms, where N is a large number, the N molecular orbitals are spaced so closely that the energy gap between the bonding and antibonding orbitals virtually disappears (Figure 21.2c).

A collection of closely spaced molecular orbitals is called a **band**. The energy required for a transition from one orbital to another within a band is vanishingly small and, since the lithium band contains N vacancies as well as N electrons, it is easy for electrons to move from filled orbitals into empty orbitals. Thus, the vacancies in the band give electrons mobility and allow them to carry current.

A Group 2A metal such as magnesium has two valence electrons per atom. In a crystal composed of N magnesium atoms, the N molecular orbitals formed from the $3s$ atomic orbitals will be completely filled by the $2N$ valence electrons. Metallic conduction always involves electrons moving within partially filled bands. But if the band formed by the s orbitals is full, then how do the electrons move? The answer is shown in Figure 21.3. Another band, formed from $3p$ atomic orbitals, overlaps the s band. There are no $3p$ electrons in magnesium, so the empty p band provides the vacancies needed for metallic conduction.

Many properties of metals can be explained in terms of band theory. For example, metallic conduction is lowered by the presence of impurities and/or warming of the metal. Electrons in partially filled bands flow most efficiently past atoms that occupy regular lattice positions. The decrease in conduction occurs because impurities and increased thermal motion disrupt the crystal lattice. Metals also have high thermal conductivities because mobile electrons can transport kinetic energy from one end of the metal to the other. Band theory also accounts for metallic luster. Since energy levels in the band are virtually continuous, all wavelengths of light can be absorbed by electrons moving to higher orbitals in the band.

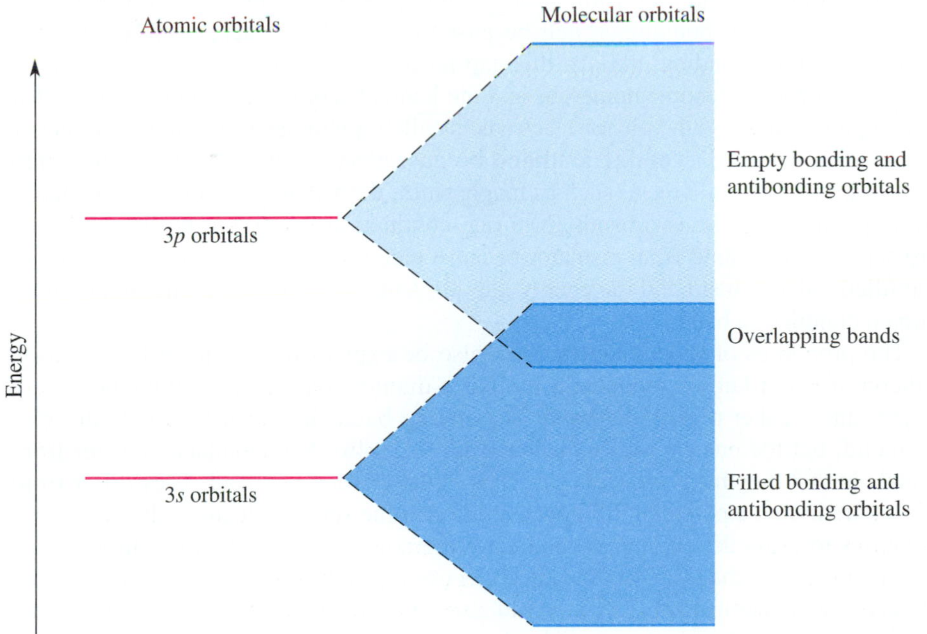

Figure 21.3

Band formation in magnesium metal. The band formed from empty $3p$ atomic orbitals overlaps the full s band. The empty p band provides the vacancies needed for metallic conduction.

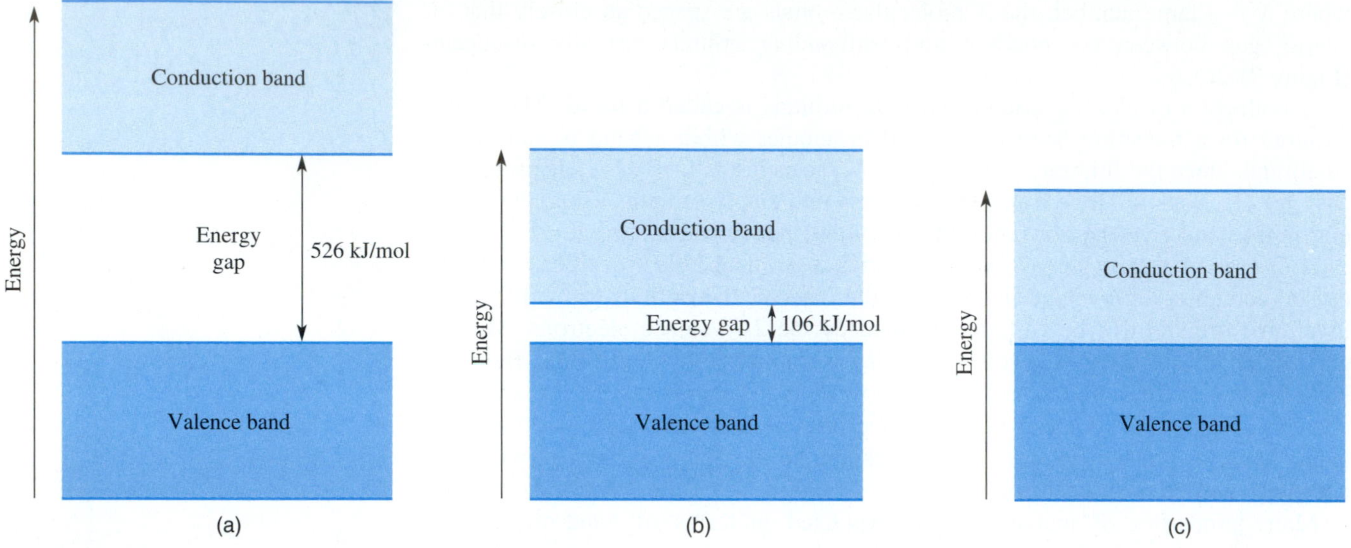

Figure 21.4

(a) Diamond, an insulator, has a 526 kJ/mol energy gap between the valence band and the empty conduction band. (b) Silicon, a semiconductor, has a 106 kJ/mol energy gap between the valence and conduction bands. (c) Magnesium, a metallic conductor, has no gap between the valence and conduction bands.

The luster consists of the continuous spectrum of light emitted as the electrons return to lower energy orbitals.

The difference between a metallic conductor like magnesium and an insulator like diamond is illustrated in Figure 21.4. The carbon atoms in diamond are smaller than metal atoms, and they form stronger bonds. Strong bonding means that most electrons stay in the bonding orbitals because there is a large energy gap between the bonding and antibonding orbitals; this gap is too large to disappear even when the molecular orbitals become numerous and the bands become continuous. A band that is completely filled with valence electrons is called a **valence band**; a higher energy vacant band is called a **conduction band** because electrons would have to move into it in order to flow. In a metal such as magnesium, the valence and conduction bands overlap (in effect, the two bands form one conduction band), but in diamond they are separate. Diamond is an insulator because electrons have no room to migrate in the filled valence band and the energy gap prevents them from moving to the unoccupied conduction band.

The properties of semiconductors can also be explained in terms of band theory. Silicon, the best-known semiconductor, has a diamondlike lattice, but the bonds are longer and weaker than in diamond. The energy bands in silicon resemble those in diamond, but the energy gap is smaller (Figure 21.4b). Semiconductors differ from metals in that they have very few electrons in the conduction band. They differ from insulators in that a potential difference applied to the semiconductor will cause more electrons to cross the energy gap and enter the conduction band. A rise in temperature will increase the fraction of excited electrons and thus increase the conductivity. (Recall that a rise in temperature would *decrease* the conductivity of a metal.) The electrons in a semiconductor may also be energized by radiant energy falling on the semiconductor surface.

CHEMICAL INSIGHT

SEMICONDUCTORS

Semiconduction can be enhanced by increasing the number of electrons in the conduction band or by creating vacancies in the valence band. These effects can be produced by the addition of carefully selected impurity atoms to the lattice, a process called *doping*. The dopant atoms must be few in number (about 100 ppm) so as not to disrupt the basic lattice, and each atom must bring the "wrong" number of electrons into the covalent network. For example, silicon (from Group 4A) can be doped with phosphorus or arsenic atoms from Group 5A or with boron or gallium atoms from Group 3A.

The dopant atoms give rise to new molecular orbitals that lie between the valence and conduction bands and act as stepping stones within the energy gap. Phosphorus atoms, for example, have five valence electrons, not four; the new orbitals created by the introduction of phosphorus atoms into silicon will contain the extra phosphorus electrons (Figure 21.5a). An applied voltage will excite these electrons into the conduction band, where they will enhance the conductivity.

A dopant from a higher periodic group (such as phosphorus or arsenic in silicon) is called a *donor impurity* because it provides or donates additional electrons to the lattice. A semiconductor with a donor impurity is called an *n-type semiconductor* where *n* implies an excess of *negatively* charged electrons.

A dopant from a lower periodic group is called an *acceptor impurity*. Boron has only three valence electrons, and the new orbitals created by the presence of boron in the silicon lattice will accept electrons from the valence band (Figure 21.5b). The resulting vacancies, or *positive holes* as they are often called, turn the valence band into a conduction band. The current is carried by electrons jumping from one positive hole to the other. The positive holes seem to migrate and, for this reason, a semiconductor with an acceptor impurity is called a *p-type semiconductor*.

Semiconductors are used to control the flow of electrons in radios, television sets, calculators, personal computers, and other electronic devices. Figure 21.6 shows a solar battery (photovoltaic cell) consisting of adjacent *p*-type and *n*-type silicon semiconductors. When light strikes the *p*-type silicon, it causes electrons to jump from the valence band to the conduction band. The electrons then cross the *p–n* junction and enter the conduction band of the *n*-type silicon. From there they flow through the external circuit and eventually return to the *p*-type silicon.

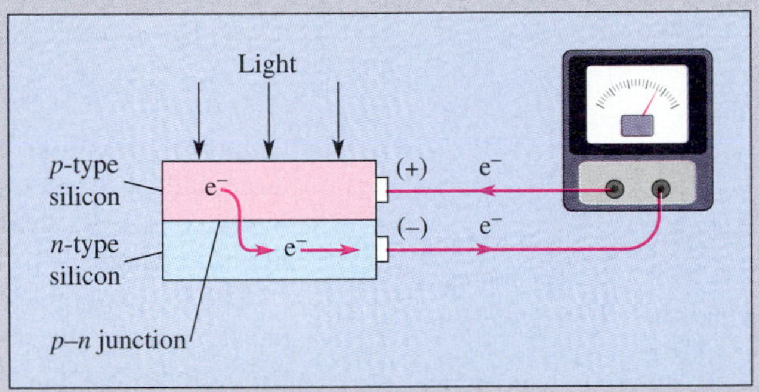

Figure 21.6
A solar battery. When energized by light, electrons flow through the external circuit from the *n*-type semiconductor to the *p*-type semiconductor.

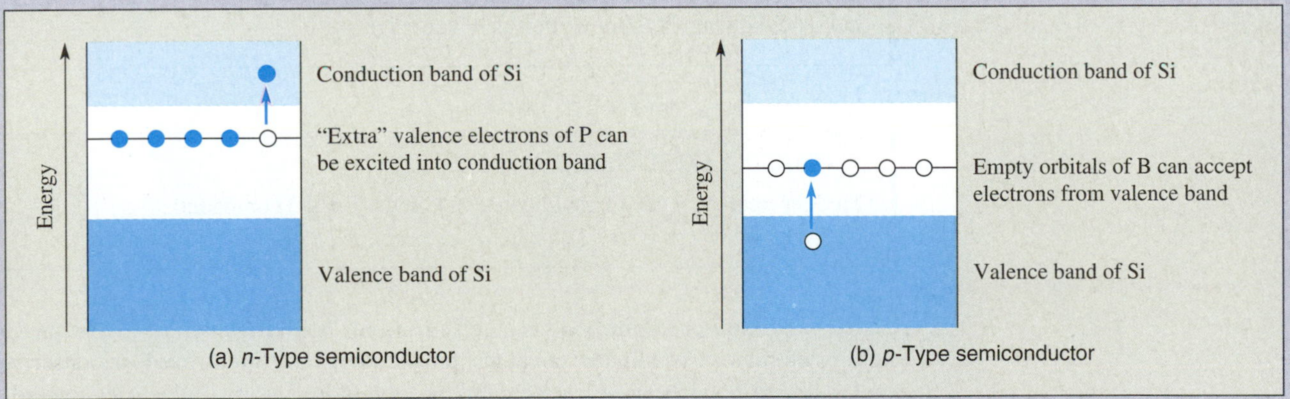

Figure 21.5 The molecular orbitals formed by dopant atoms lie between the valence and conduction bands of a semiconductor. (a) Silicon doped with a Group 5A element such as phosphorus is an *n*-type semiconductor. (b) Silicon doped with a Group 3A element such as boron is a *p*-type semiconductor.

LEARNING HINT

Review the introduction to complex ion equilibria in Section 18.5.

21.4 METAL COMPLEXES

A **metal complex** is a molecule or ion that consists of a central metal atom bonded to one or more groups called **ligands**. Examples include complex ions such as $Ag(NH_3)_2^+$ and neutral molecules such as $Ni(CO)_4$. Every ligand in the complex has a lone pair of electrons that is directed toward the metal atom. For example,

$$Ag(NH_3)_2^+ = [H_3N\!:\!Ag\!:\!NH_3]^+$$

Complex formation is especially characteristic of transition metals and governs a large part of their colorful and varied chemistry. Although main-group metals form fewer complexes than transition metals, every metal ion can be complexed by some ligand or another.

The IUPAC system of nomenclature places all complexes in brackets. Examples of this notation are $[Cu(NH_3)_4]^{2+}$ and $[Ni(CO)_4]$. In practice, the brackets are often omitted, especially when they are also used to indicate concentration.

The formula of a complex is written with the central metal atom first, followed by the ligands. In the formula of a compound, a complex component is distinguished from other components by brackets. For example, $[Cu(NH_3)_4]SO_4$ is the formula of a salt containing the $Cu(NH_3)_4^{2+}$ complex and the SO_4^{2-} ion. The charge on a complex is the sum of the metal and ligand charges.

EXAMPLE 21.1

(a) Identify the complex ion and ligands in the compound $K_3[Fe(CN)_5CO]$. (b) Find the oxidation number of the metal atom in the complex ion.

SOLUTION

(a) The complex ion is the bracketed portion, $[Fe(CN)_5CO]^{3-}$. (We know the charge is 3− because the three K^+ ions have a total charge of 3+.) The ligands are cyanide ion (CN^-) and carbon monoxide (CO).

(b) The sum of the oxidation numbers in the complex is −3. Let x_{Fe} be the oxidation number of iron. Each CN^- has a charge of 1− and CO is neutral, so

$$x_{Fe} + 5 \times (-1) + 0 = -3$$

and

$$x_{Fe} = +2$$

The oxidation number of iron is +2. (Recall that ionic charges are written as 2+ or 1−, but oxidation numbers are written as +2 or −1.)

PRACTICE EXERCISE 21.3

Find the charge on the nitrosyl ligand, NO, in the Co(III) compound $Na_2[Co(CN)_5NO]$.

A **coordinate covalent** bond is an electron-pair bond in which one atom provides both electrons. Coordinate covalent terminology is often used to describe a metal–ligand bond because each ligand orients its lone pair of electrons toward the metal atom. The ligands are said to *coordinate* the central metal atom. The ligand atom with the lone electron pair is called the **coordinating atom**. Compounds that contain complexes are called **coordination compounds**. A number of ligands and their coordinating atoms are shown in Table 21.5.

TABLE 21.5 Some Ligands and their Coordinating Atoms[a]

Ligand	Name	Ligand	Name
H_2O	Aqua	NO_2^-	Nitro
NH_3	Ammine	ONO^-	Nitrito
F^-	Fluoro	OH^-	Hydroxo
Cl^-	Chloro	SO_4^{2-} (OSO_3^{2-})	Sulfato
Br^-	Bromo	$S_2O_3^{2-}$ (SSO_3^{2-})	Thiosulfato
I^-	Iodo	CO	Carbonyl
CN^-	Cyano	NO	Nitrosyl
SCN^-	Thiocyanato	CO_3^{2-} (OCO_2)	Carbonato
NCS^-	Isothiocyanato	C_6H_5N (⌬N:)	Pyridine (py)

[a]The coordinating atoms are in color.

The nature of many coordination compounds has been deduced from conductivity measurements in solution, which give the number of moles of ions in 1 mol of compound, and from precipitation stoichiometry. For example, conductivity measurements show that 1 mol of $CrCl_3 \cdot 5NH_3$ forms 3 mol of ions when it dissolves. Furthermore, 1 mol of the dissolved compound reacts with aqueous $AgNO_3$ to form 2 mol of solid AgCl. The latter observation suggests that there are 2 mol of Cl^- ion per mole of compound. The formula $[Cr(NH_3)_5Cl]Cl_2$ is consistent with these experimental results:

$$[Cr(NH_3)_5Cl]Cl_2(s) \rightarrow Cr(NH_3)_5Cl^{2+}(aq) + 2Cl^-(aq)$$

The formulas of some platinum coordination compounds are given in Table 21.6.

PRACTICE EXERCISE 21.4

One mole of $CoCl_3 \cdot 4NH_3$ contains 2 mol of ions and reacts with aqueous silver nitrate to form 1 mol of AgCl. Write a formula consistent with these observations.

TABLE 21.6 Formulas of Platinum(IV) Coordination Compounds Deduced from Conductivity Data and Precipitation Stoichiometry

Composition	Ions per Mole of Complex[a]	Moles of Cl^- per Mole of Complex[b]	Formula
$PtCl_4 \cdot 6NH_3$	5	4	$[Pt(NH_3)_6]Cl_4$
$PtCl_4 \cdot 5NH_3$	4	3	$[Pt(NH_3)_5Cl]Cl_3$
$PtCl_4 \cdot 4NH_3$	3	2	$[Pt(NH_3)_4Cl_2]Cl_2$
$PtCl_4 \cdot 3NH_3$	2	1	$[Pt(NH_3)_3Cl_3]Cl$
$PtCl_4 \cdot 2NH_3$	0	0	$[Pt(NH_3)_2Cl_4]$

[a]From solution conductivity. [b]From precipitation stoichiometry.

The Geometry of Complexes

The number of ligand atoms in contact with the central metal atom is called the **coordination number** of the atom. Coordination numbers from 2 through 12 have been observed; 6 is the most common, but 4 and 2 also occur frequently. Two coordinating atoms usually adopt a linear configuration, with one coordinating atom on each side of the central atom. Four coordinating atoms will occupy either the corners of a tetrahedron or the corners of a square; in either case, the metal ion is in the center. Six coordinating atoms usually occupy the corners of an octahedron. Examples of these arrangements are shown in Figure 21.7.

Coordination number	Shape	Geometry	Example
2	Linear	L—(M)—L	$Ag(NH_3)_2^+$
4	Tetrahedral	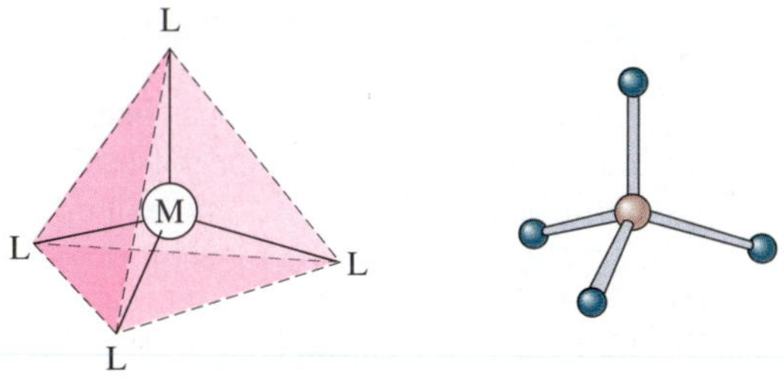	$Zn(NH_3)_4^{2+}$
4	Square planar	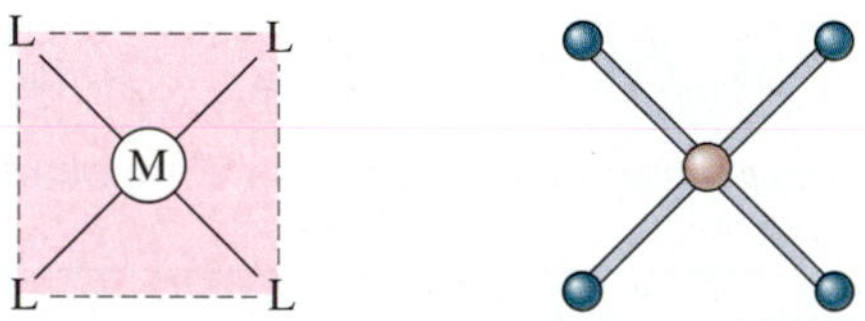	$Pt(NH_3)_4^{2+}$
6	Octahedral	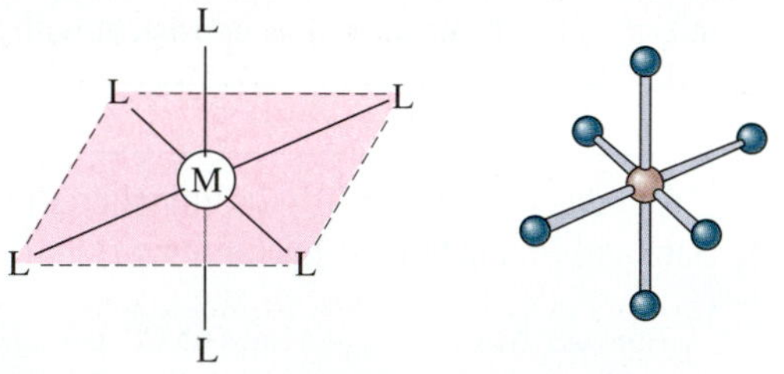	$Co(NH_3)_6^{3+}$

Figure 21.7

The most common structures for metal complexes. M represents a metal atom or ion while L represents a ligand. The M–L lines represent metal–ligand bonds.

EXAMPLE 21.2

Prussian blue, a pigment that is used in great quantity to make ink, carbon paper, and other products, is a salt containing Fe^{3+} and ferrocyanide ion. The latter ion is a cyanide complex of Fe^{2+} with a coordination number of 6. (a) Write the formula of the ferrocyanide ion and sketch its configuration. (b) Write the formula for Prussian blue.

SOLUTION

(a) The total charge on a complex formed from Fe^{2+} and six CN^- ligands is 4−. The formula is $Fe(CN)_6{}^{4-}$. The cyanide ion coordinates through carbon (Table 21.5). The configuration is octahedral with the coordinating atom of each ligand facing the central metal atom:

$$\left[\begin{array}{c} \text{CN} \\ \text{NC} \quad \text{CN} \\ \text{Fe} \\ \text{NC} \quad \text{CN} \\ \text{CN} \end{array}\right]^{4-}$$

(b) Four Fe^{3+} ions will combine with three $Fe(CN)_6{}^{4-}$ ions. The formula for Prussian blue is $Fe_4[Fe(CN)_6]_3$.

Some coordination complexes contain more than one metal atom. An example is the golden-yellow compound $Mn_2(CO)_{10}$:

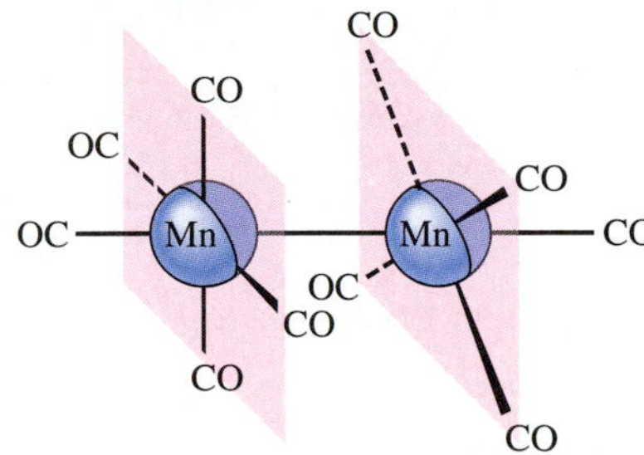

Compounds such as $Mn_2(CO)_{10}$ or $Fe(CO)_5$, which contain bonds between metal and carbon atoms, are called *organometallic compounds*.

PRACTICE EXERCISE 21.5

Sketch the following complexes: (a) $Ni(CO)_4$ (tetrahedral) and (b) $Ni(CN)_4{}^{2-}$ (square planar).

A ligand is said to be **monodentate**, **bidentate**, **tridentate**, and so forth, according to whether it has one, two, three, or more coordinating atoms. The ligands in Table 21.5 are monodentate; some polydentate ligands are shown in Figure 21.8. A **polydentate ligand** that binds a metal atom in two or more positions, as if with a pincers, is called a **chelating agent** (from the Greek *chelos*, meaning "claw"). Ethylenediamine (Figure 21.9a) is a bidentate ligand that coordinates through two nitrogen atoms. Platinum(II) has a coordination number of 4, and Figure 21.11b shows how two ethylenediamine ligands can fully complex this atom in a square planar arrangement. The most effective chelating agents form five- or six-membered rings with the metal atom. Five- and six-membered rings are large enough to permit normal bond angles of 109° to 120° within the ligand backbone, and small enough for both ends of the ligand to be bonded to the same metal atom. The ethylenediamine–platinum(II) complex in Figure 21.9 contains two five-membered rings.

The word *dentate* is derived from the Latin *dens, dent-*, meaning "tooth."

EXAMPLE 21.3

For the neutral complex $[Fe(acac)_2Cl]$, find (a) the oxidation number of iron and (b) the coordination number of iron. Consult Figure 21.8.

SOLUTION

(a) The chloride ligand has a 1− charge; *acac* is the acetylacetonate ion, a bidentate ligand that also has a 1− charge. The ligands provide a total charge of 3−. The complex is neutral; therefore, the oxidation number of iron is +3.

(b) The chloride coordinates one position, and each of the two *acacs* coordinates two positions, to give a coordination number of five for the iron in this complex.

Bidentate:

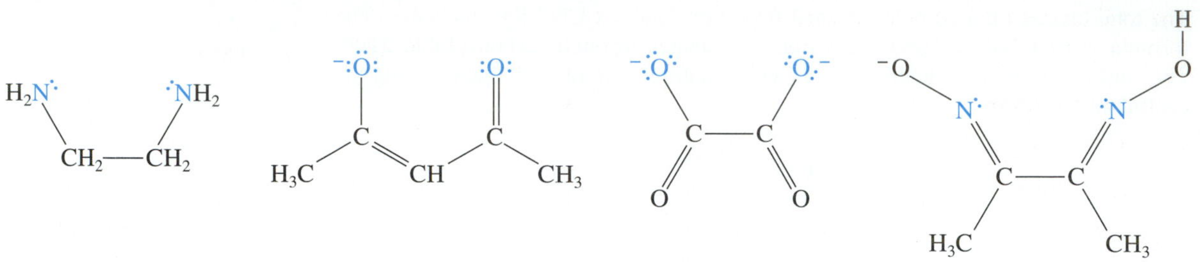

ethylenediamine (en)

acetylacetonato (acac)

oxalato (ox)

dimethylglyoxime (dmg)

Tridentate:

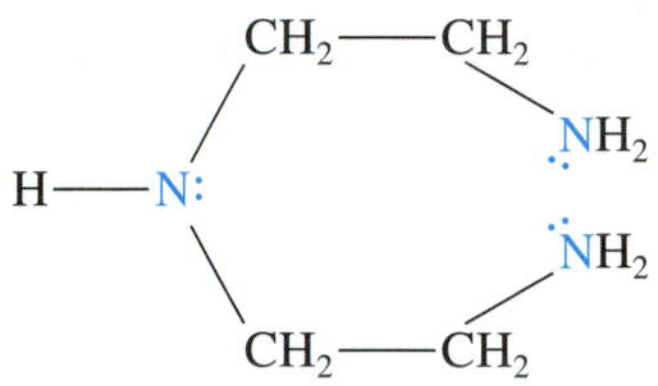

diethylenetriamine (dien)

Hexadentate:

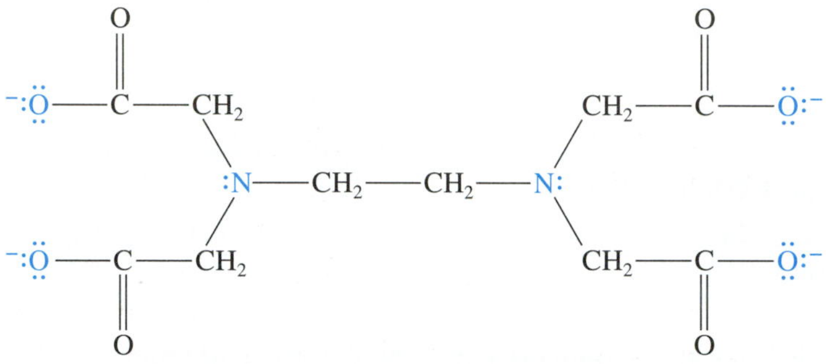

ethylenediaminetetraacetato (EDTA)

Figure 21.8 (above)

Some polydentate ligands. The coordinating atoms and their electron pairs are in blue.

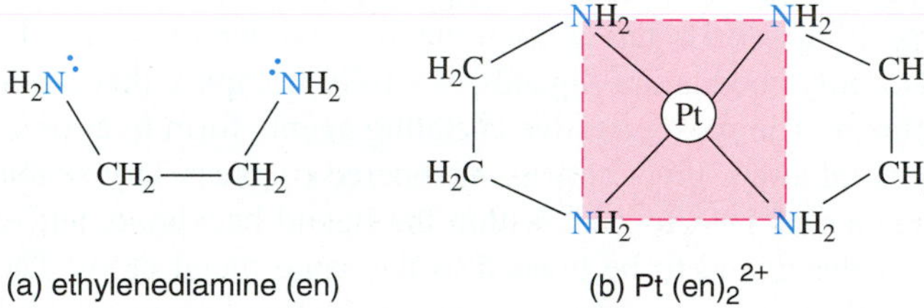

Figure 21.9 (right)

(a) Ethylenediamine. (b) Two ethylenediamine ligands fully complex Pt(II) in a square planar arrangement.

PRACTICE EXERCISE 21.6

How many ethylenediamine ligands are needed to occupy all the sites of an octahedral complex?

DIGGING DEEPER

Naming Complexes

Complexes are so numerous that chemists have developed a precise system of nomenclature for them. The following system is recommended by IUPAC, the International Union of Pure and Applied Chemistry.

1. Ligands are named according to the following rules:

(a) Anionic ligands are given names ending in *-o*.

Cl^- chloro — SO_4^{2-} sulfato

(b) Neutral ligands keep their own names.

$H_2NCH_2CH_2NH_2$ ethylenediamine — C_5H_5N pyridine

(c) A few ligands have special names.

H_2O aqua — NH_3 ammine

Note that *ammine* meaning NH_3 is spelled with two *m*'s, while *amine* meaning an NH_2 group is spelled with one *m*. The names of some common ligands are given in Table 21.5 (p. 1001) and Figure 21.8.

2. The name of a complex combines the names of the ligands and the central metal atom according to the following rules:

(a) The ligands are named first in alphabetical order.

(b) The prefixes *di-, tri-, tetra-, penta-, hexa-,* and so forth, are used with inorganic ligands to indicate the number of each.

diammine = two NH_3 ligands — tetrachloro = four Cl^- ligands

(c) The prefixes *bis-* for 2, *tris-* for 3, and *tetrakis-* for 4 are used with organic ligands. The ligand name is put in parentheses. (These prefixes are used to avoid confusion with organic names that may already contain *di-, tri-,* and so forth.)

bis(ethylenediamine) = two ethylenediamine ligands

(d) Prefixes are disregarded in alphabetizing the ligands.

3. The metal name follows the ligand names. The oxidation number in Roman numerals is included if needed.

(a) In a positive or neutral complex, the metal is given its usual name.

(b) In a negative (anionic) complex, the suffix *-ate* is added to the root of the metal name. The Latin root is used for metals whose symbols have a Latin derivation. Some examples are

platinate	negative complexes of platinum (Pt)
argentate	negative complexes of silver (Ag)
plumbate	negative complexes of lead (Pb)

4. The name of a complex is written without spaces, and the name of a charged complex is followed by the word *ion*. Some examples are:

$Ag(NH_3)_2^+$	diamminesilver ion
$[Co(NH_3)_3(NO_2)_3]$	triamminetrinitrocobalt(III)
$Cr(dien)_2^{3+}$	bis(diethylenetriamine)chromium(III) ion
$ZnCl_4^{2-}$	tetrachlorozincate ion
$Fe(H_2O)_6^{3+}$	hexaaquairon(III) ion
$Fe(CN)_6^{3-}$	hexacyanoferrate(III) ion
$Fe(CN)_6^{4-}$	hexacyanoferrate(II) ion

Chemists do not always adhere to rigorous official nomenclature. For example, $Ag(NH_3)_2^+$ is often simply called the silver–ammonia complex; $Fe(CN)_6^{4-}$ is also known as the *ferrocyanide* ion; and $Fe(CN)_6^{3-}$ is the *ferricyanide* ion.

CHEMICAL INSIGHT

CHELATING AGENTS AS ANTIDOTES

A number of chelating agents are used as antidotes for poisoning by heavy metals such as lead, mercury, and arsenic. One antidote is EDTA (Figure 21.8), a polydentate ligand with six coordinating atoms. Victims of lead poisoning, usually children, are given intravenous injections of $Na_2CaEDTA$. This salt combines with lead to form a soluble EDTA complex (Figure 21.10) that is ultimately eliminated in the urine. EDTA has also been used to rid the body of radioactive metals such as plutonium. Calcium ions are included in the reagent to help maintain the body's supply of calcium. If the tetrasodium salt (Na_4EDTA) were used instead of $Na_2CaEDTA$, the body would lose calcium ions along with heavy-metal ions.

Many chelating agents are nonspecific; that is, they remove essential ions as well as heavy-metal ions. Specific chelating agents can be designed by tailoring the geometry of the chelating agent to the size and coordination number of the ion to be removed. Copper(II) ions, for example, tend to form square planar complexes, so chemists synthesized a compound called glycylglycyl-L-histidine-N-methyl amide, which has four binding sites that fall naturally into a square the right size for a Cu^{2+} ion (Figure 21.11). It is hoped that this agent, which binds copper(II) exclusively, can be used in the treatment of Wilson's disease, a genetic disorder in which the body is unable to rid itself of excess copper ions.

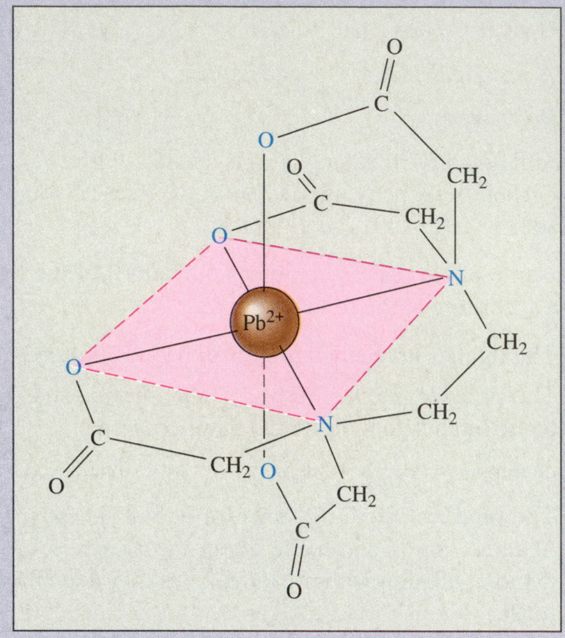

Figure 21.10
$Pb(EDTA)^{2-}$, the ethylenediaminetetraacetato complex of Pb^{2+}.

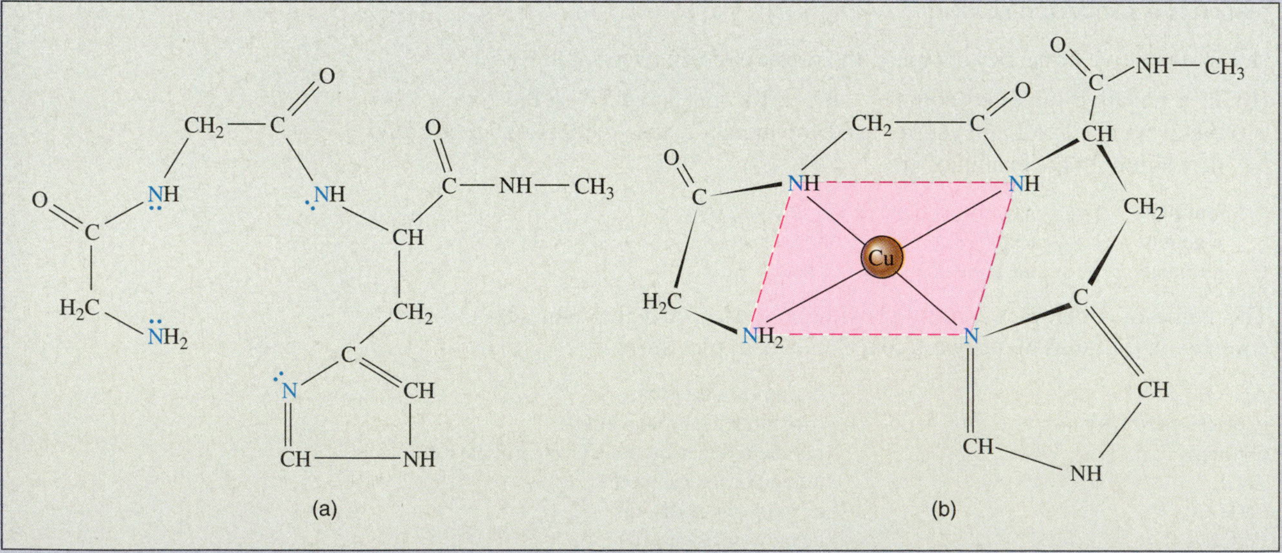

Figure 21.11
(a) Glycylglycyl-L-histidine-N-methyl amide, a specific chelating agent for copper(II) ions. (b) The structure of the complex, showing the square planar arrangement of ligand atoms around the copper ion.

EXAMPLE 21.4

Name (a) $[Pt(py)_2(NO_2)_2]$ and (b) $Au(CN)_2^{2-}$.

SOLUTION

(a) In alphabetical order, the ligands are nitro and pyridine. The prefix *bis* is used with the pyridine. The complex is neutral; therefore, the name of the metal is unchanged. The two NO_2^- ligands provide a charge of 2−, and the other ligands are neutral; therefore, the platinum has an oxidation number of +2. The name is dinitrobis(pyridine)platinum(II).

(b) The two cyanide ions are *dicyano*. The complex is a negative ion, so the Latin root *aur-* followed by *-ate* is used for the gold. The gold is in the +1 oxidation state. The name is dicyanoaurate(I) ion.

EXAMPLE 21.5

Write the formula of diaquabis(oxalato)chromate(III) ion.

SOLUTION

Bis(oxalato) means two oxalate ions; *diaqua* stands for two water molecules. The ionic charges are 4− for the two oxalates and 3+ for the chromium(III), for a total of 1−. The formula is $Cr(C_2O_4)_2(H_2O)_2^-$. The optional abbreviation *ox* may be used for oxalate, in which case the formula is written $Cr(ox)_2(H_2O)_2^-$.

PRACTICE EXERCISE 21.7

(a) Give the official name of $Na_3[Co(NO_2)_6]$. (b) Write the formula for sodium ethylenediaminetetraacetatoplumbate(II).

21.5 ISOMERISM IN METAL COMPLEXES

Recall that isomers are compounds with the same composition but different arrangements of atoms. In **structural isomers** the atoms are joined in a different sequence, so that each isomer has a different set of bonds. **Stereoisomers** have the same bonds, but as we shall soon see, they differ in the orientation of the bonds with respect to each other. Coordination compounds can exhibit both structural isomerism and stereoisomerism.

LEARNING HINT

Review the discussion of isomerism in Section 10.3.

Structural Isomerism

One type of structural isomerism found in some ionic coordination compounds is **ionization isomerism**, in which each isomer has a different ion outside the complex. This type of isomerism is illustrated by the compounds shown at the top of page 1010.

CHEMICAL INSIGHT

HEMOGLOBIN AND OTHER PORPHYRIN COMPLEXES

The porphyrins are flat tetradentate molecules containing various organic groups attached to a central porphin structure (Figure 21.12). Metalloporphyrin complexes include chlorophyll (Figure 21.13), which contains magnesium, and heme (Figure 21.14), which contains iron in the +2 oxidation state. Vitamin B-12, a porphyrin complex that is necessary for the growth of red blood cells, contains cobalt.

Oxyhemoglobin is a complex protein molecule that incorporates four units of heme. The Fe(II) atom in heme has six coordination positions, four of which are occupied by the porphyrin nitrogens. A nitrogen from the protein occupies the fifth coordination position of the Fe(II) atom, and oxygen takes up the sixth position (Figure 21.15). As oxyhemoglobin travels through the bloodstream, it releases oxygen to the cells and, as shown in Figure 21.15, the iron atom moves out of the heme plane.

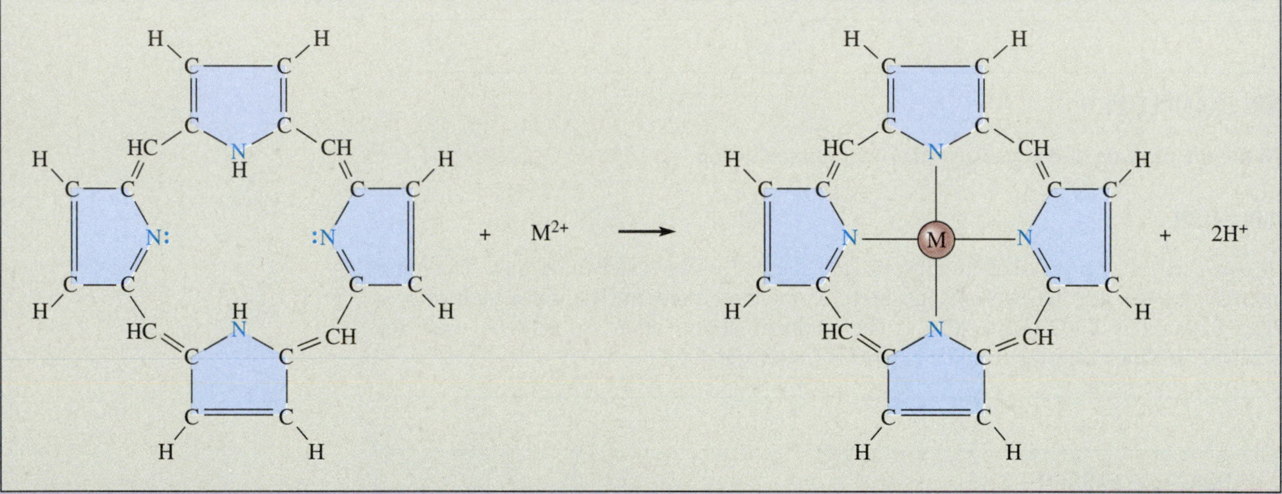

$+ \; M^{2+} \longrightarrow$ $+ \; 2H^+$

Figure 21.12 (above)
The porphin molecule is a tetradentate chelating agent.

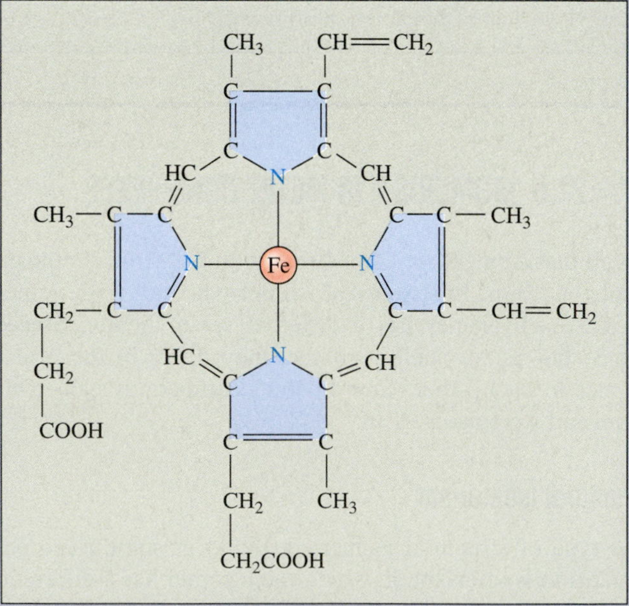

Figure 21.14 (right)
Heme, which is part of the hemoglobin molecule, is a porphyrin complex of iron(II).

Substances other than oxygen can preempt the sixth position on the iron. Inhalation of automobile exhaust fumes or cigarette smoke results in the formation of carboxyhemoglobin, in which carbon monoxide occupies the spot usually reserved for oxygen. The CO–hemoglobin linkage is more than 200 times stronger than the O_2–hemoglobin linkage, so body cells will not receive their normal supply of oxygen if a significant concentration of carbon monoxide is inhaled. This condition is especially serious when the blood supply to the heart and other organs is already reduced by deposits of fat and cholesterol in the blood vessels. The problem is compounded for cigarette smokers, who inhale nicotine along with carbon monoxide. Nicotine is a potent drug—70 mg can kill an average person within several minutes—that accelerates the cardiac rate, thus creating a demand for more oxygen than is normally required.

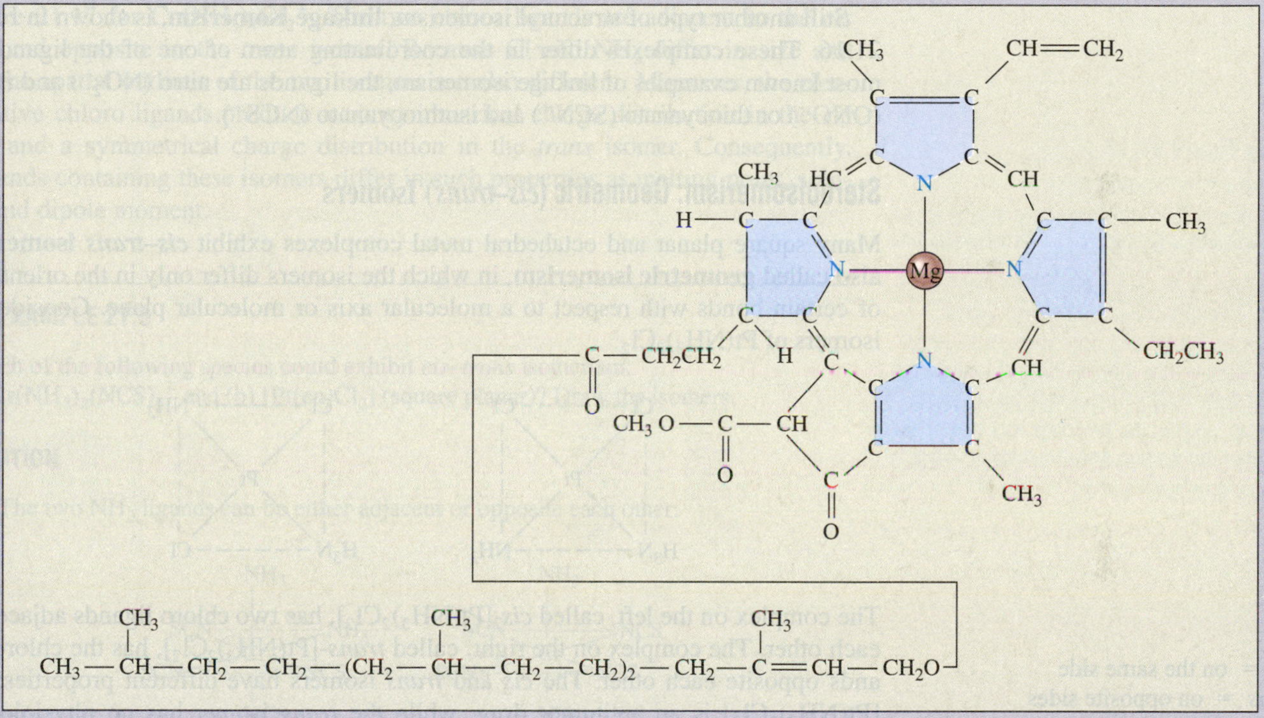

Figure 21.13 (above)

Chlorophyll, a porphyrin complex of magnesium(II), is the catalytic agent responsible for photosynthesis.

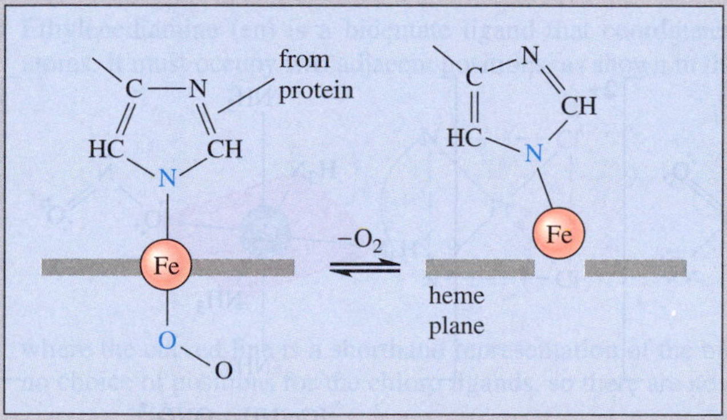

Figure 21.15 (left)

In oxyhemoglobin, a nitrogen atom from the protein portion of the molecule coordinates the fifth position on each iron(II) atom and an oxygen molecule coordinates the sixth position. When oxyhemoglobin releases oxygen to the cells, the iron atom moves out of the heme plane.

CHEMICAL INSIGHT

GEMSTONES

Red rubies and green emeralds owe their colors to electron transitions between t_{2g} and e_g^* orbitals (Figure 21.29). Rubies consist basically of corundum (Al_2O_3), and emeralds consist of beryl ($Be_3Al_2(SiO_3)_6$); in both gems, a few Al^{3+} ions have been replaced by Cr^{3+}. Each Al^{3+} and Cr^{3+} ion is surrounded octahedrally by six oxygen atoms, and the interaction of these atoms with Cr^{3+} (a d^3 ion) produces an orbital splitting energy that lies in the visible range.

In ruby, the splitting is such that electron transitions within Cr^{3+} absorb photons from the yellow-green regions of the spectrum. The red and blue wavelengths transmitted through the crystal give rubies a deep red color tinged with purple. In emerald (beryl), each oxygen atom is bonded to a silicon atom as well as to a metal atom, so its bonds with the chromium ions are weaker than in ruby. Thus the orbital splitting is smaller in emerald than in ruby, and emerald absorbs photons from the longer wavelength red region of the spectrum. Emeralds transmit green and some blue wavelengths and vary in color from light to deep green.

Many other gemstones owe their colors to transition-metal ions. Aquamarine, garnet, and topaz contain iron(III), blue sapphires contain cobalt(II) or vanadium(III), turquoise and malachite contain copper(II), and amethyst contains manganese(II).

(a)

(b)

Figure 21.29 (a) Uncut rubies in a matrix of corundum and a synthetic ruby produced by fusing and tinting alumina. (b) Beryl and emerald; emerald is a transparent variety of beryl.

those absorbed by the original complex; thus the new ligand produces a smaller splitting energy and must have a weaker field than the original ligand.

The order of ligand field strengths is nearly always the same for different metal ions; NCS^-, for example, is usually weaker than NH_3, and CN^- is usually stronger than NH_3. A list of ligands in order of field strength (Table 21.8) is called a **spectrochemical series** because it is based on the observed spectra of transition-metal complexes.

PRACTICE EXERCISE 21.10

Which complex will absorb higher frequency light, $Ni(NH_3)_6^{2+}$ or $Ni(en)_3^{2+}$?

Figure 21.30

Deducing ligand field strengths. The cations in these compounds differ only in their sixth ligand; from left to right they are $[Co(NH_3)_6)]^{3+}$ (yellow), $[Co(NH_3)_5H_2O]^{3+}$ (orange-red), $[Co(NH_3)_5Cl]^{2+}$ (rose-red), and $[Co(NH_3)_5Br]^{2+}$ (deep red). This sequence of colors increases in photon wavelength, as do the corresponding absorbed colors (see the color wheel in Figure 21.28). The splitting energy decreases with increasing wavelength, so the ligands—NH_3, H_2O, Cl^-, and Br^-—decrease in field strength from NH_3 to Br^-.

Spin Pairing and Paramagnetism

Three electrons will occupy the three lower t_{2g} orbitals in an octahedral complex. Where will a fourth electron go? If the electron occupies a higher energy e_g^* orbital, it must absorb an amount of energy equal to the orbital splitting energy; if it pairs with another electron in the lower energy t_{2g} orbital, it must absorb enough energy to overcome the mutual repulsion of two electrons in one orbital. An electron will enter a higher energy orbital when it is easier to do so, that is, when the splitting energy is less than the pairing energy. Pair formation, on the other hand, will occur when the pairing energy is less than the splitting energy.

These two possibilities are illustrated in Figure 21.31, which shows how the six d electrons of Fe^{2+} are distributed among the t_{2g} and e_g^* orbitals in $Fe(H_2O)_6^{2+}$ and in $Fe(CN)_6^{4-}$. Since H_2O is a weak-field ligand, the splitting energy in $Fe(H_2O)_6^{2+}$ is small. The six electrons are distributed over all five orbitals. Pairing is minimized

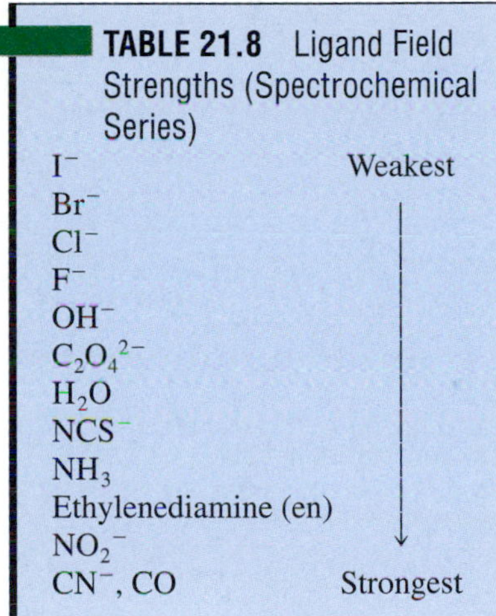

TABLE 21.8 Ligand Field Strengths (Spectrochemical Series)

Ligand	
I^-	Weakest
Br^-	↓
Cl^-	
F^-	
OH^-	
$C_2O_4^{2-}$	
H_2O	
NCS^-	
NH_3	
Ethylenediamine (en)	
NO_2^-	
CN^-, CO	Strongest

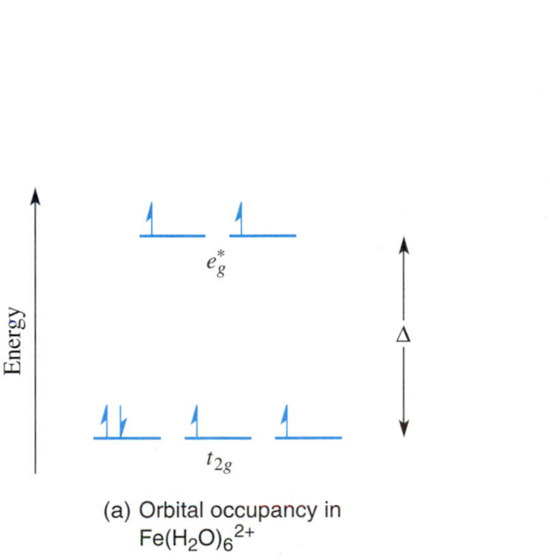

(a) Orbital occupancy in $Fe(H_2O)_6^{2+}$

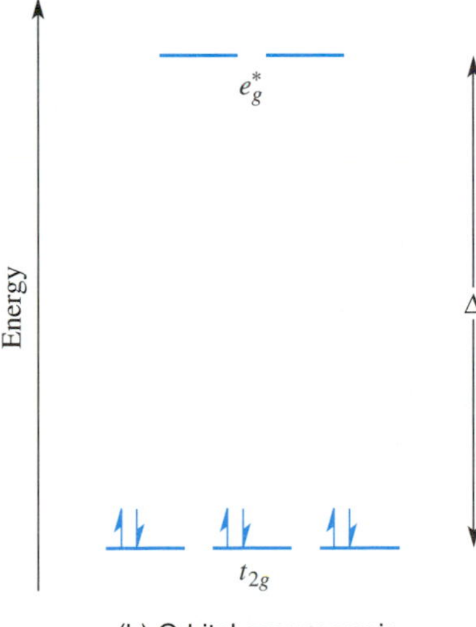

(b) Orbital occupancy in $Fe(CN)_6^{4-}$

Figure 21.31

The orbital splitting and orbital occupancy in high- and low-spin complexes of iron(II), a d^6 ion. (a) The weak ligands in $Fe(H_2O)_6^{2+}$ provide a splitting energy that is smaller than the pairing energy. The electrons remain unpaired and the complex is high spin. (b) The strong ligands in $Fe(CN)_6^{4-}$ provide a splitting energy that is greater than the pairing energy. The electrons pair up and the complex is low spin.

because it takes less energy to occupy higher e_g^* orbitals than to form pairs in lower t_{2g} orbitals. The CN^- ion, unlike H_2O, is a strong-field ligand. The splitting energy in $Fe(CN)_6{}^{4-}$ is high, and all six electrons pair up in the lower t_{2g} orbitals. $Fe(H_2O)_6{}^{2+}$, with the same number of parallel spins and the same paramagnetism as an isolated Fe^{2+} ion, is called a **high-spin complex**. $Fe(CN)_6{}^{4-}$, in which all the electrons are paired, is a **low-spin complex**.

EXAMPLE 21.9

The paramagnetism of $Mn(CN)_5(NO)^{3-}$ is lower than that of $Mn(H_2O)_6{}^{2+}$. Diagram the orbital configuration of Mn^{2+} in each complex.

SOLUTION

$Mn(H_2O)_6{}^{2+}$, the more paramagnetic complex, is the high-spin complex; $Mn(CN)_5(NO)^{3-}$ is the low-spin complex. Mn^{2+} has five *d* electrons, so the configurations will be identical to those shown in Figure 21.32.

PRACTICE EXERCISE 21.11

An octahedral iron complex has one unpaired electron. What is the oxidation state of the iron? (*Hint*: Study the high- and low-spin configurations of Fe^{2+} and Fe^{3+}.)

Tetrahedral Complexes

The four ligands in a tetrahedral complex may be visualized as approaching the central atom from alternate corners of a cube (Figure 21.33). Because of their location at the cube corners, the ligand orbitals do not approach any of the *p* or *d* orbital

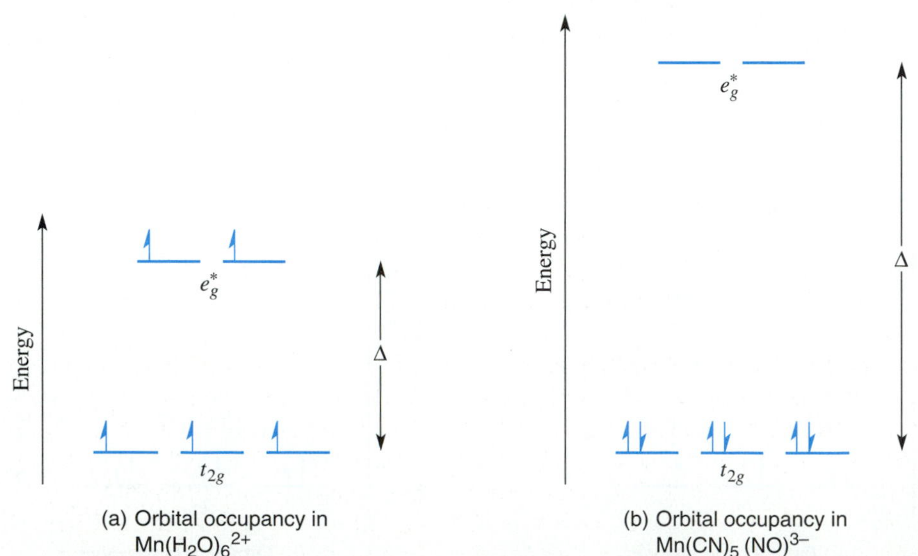

Figure 21.32
The orbital splitting and orbital occupancy in (a) $Mn(H_2O)_6{}^{2+}$, a high-spin complex, and (b) $Mn(CN)_5(NO)^{3-}$, a low-spin complex.

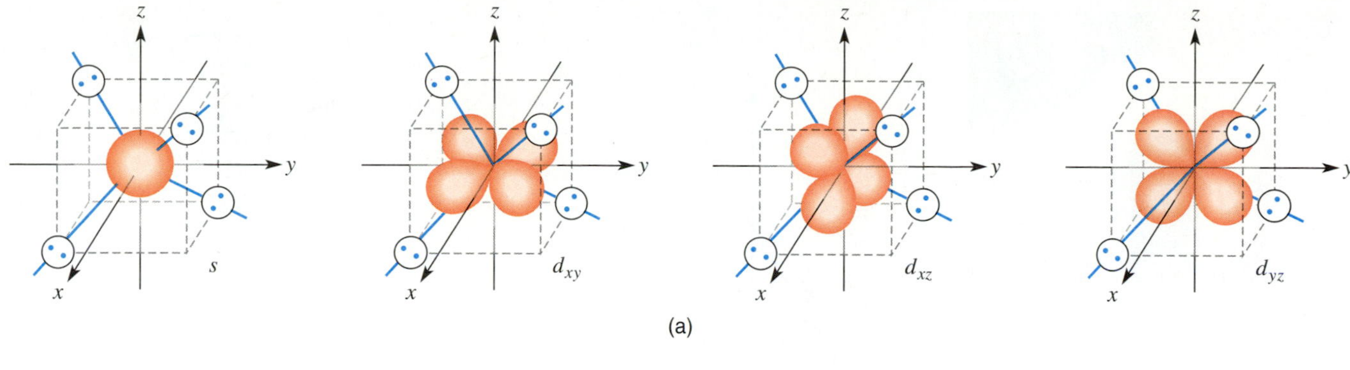

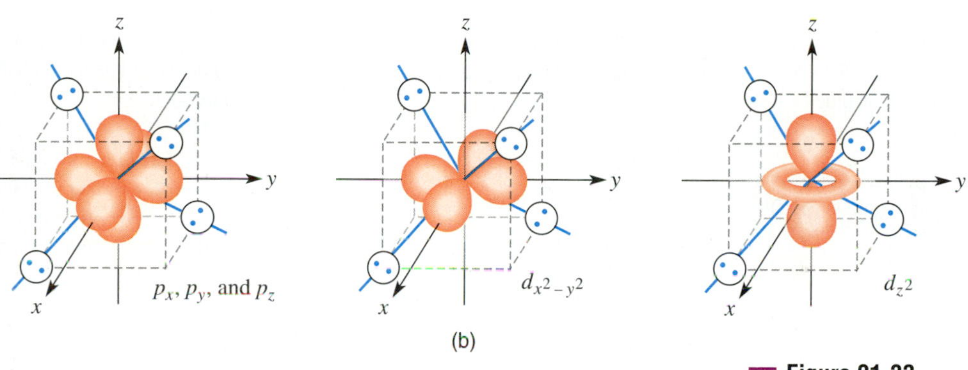

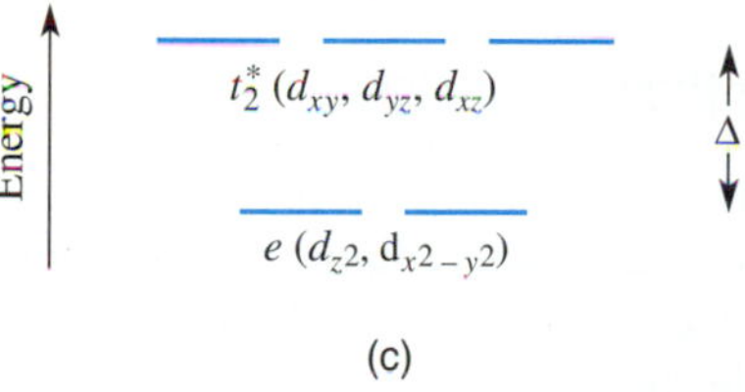

Figure 21.33

Formation of a tetrahedral complex. The ligands, symbolized by lone electron pairs at the corners of a cube, are tetrahedrally oriented between the *x, y,* and *z* axes. (a) The ligand orbitals interact with the metal orbitals oriented between the axes and pointing toward the cube edges (*s*, d_{xy}, d_{xz}, and d_{yz}). (b) The ligand orbitals do not interact with the metal orbitals lying along the axes (*p*, $d_{x^2-y^2}$, and d_{z^2}). (c) The *e* and t_2^* orbital sets are reversed from those in an octahedral field, and the value of Δ is smaller than in most octahedral complexes. (Note that the *g* subscript is not used for tetrahedral complexes.)

lobes head on. They will interact weakly with the metal orbitals closest to the corners of the cube, that is, with the orbitals oriented between the axes and pointing toward the cube edges—the *s*, d_{xy}, d_{xz}, and d_{yz} orbitals (Figure 21.33a). There will be virtually no overlap between the ligand orbitals and the metal orbitals oriented along the axes and pointing toward the cube faces—the *p*, $d_{x^2-y^2}$, and d_{z^2} orbitals (Figure 21.33b).

The effect of this interaction on the orbital splitting pattern is shown in Figure 21.33c. Observe that the lower energy set (*e*) consists of the nonbonding $d_{x^2-y^2}$ and d_{z^2} orbitals, while the higher energy set (t_2^*) consists of antibonding orbitals formed by ligand overlap with the metal d_{xy}, d_{xz}, and d_{yz} orbitals. The orbital splitting pattern in tetrahedral complexes is just the reverse of that in octahedral complexes.

The ligand–metal bonding in tetrahedral complexes is weak, and orbital splitting is small. Hence, all tetrahedral complexes are high spin. Many tetrahedral com-

Figure 21.34

Concentrated HCl turns a pink cobalt(II) chloride solution blue. The added chloride ions shift the equilibrium from the pink octahedral aquo complex to the blue tetrahedral chloro complex:

$$Co(H_2O)_6^{2+}(aq, pink) + 4Cl^-(aq) \rightleftharpoons CoCl_4^{2-}(aq, blue) + 6H_2O(l)$$

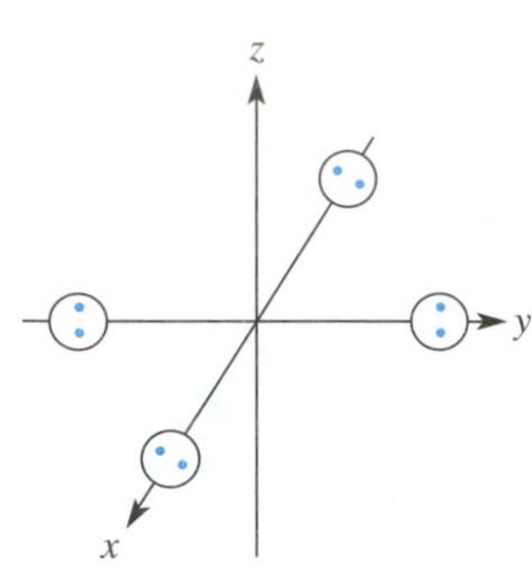

Figure 21.35

A square planar complex has ligands oriented along the *x* and *y* axes; there are no ligands on the *z* axis.

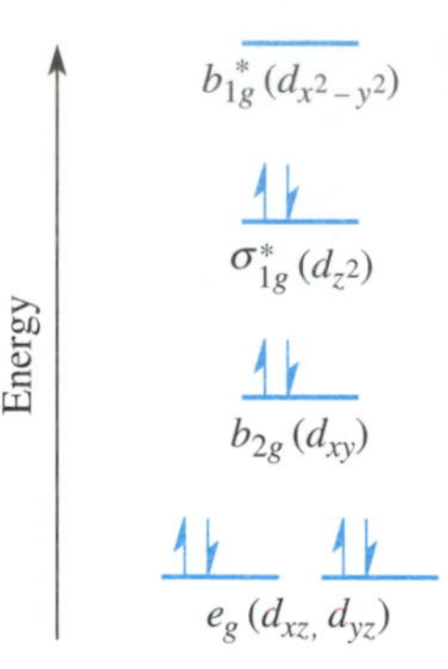

Figure 21.36

The four sets of orbitals that accommodate the metal *d* electrons in a square planar complex and their occupancy in a d^8 complex such as $Pt(CN)_4^{2-}$. The e_g, b_{2g}, σ^*_{1g}, and the b^*_{1g} symbols come from the mathematics used in molecular orbital theory. The closely related orbitals of the free metal atom are given in parentheses. (The relative energies of the four low-lying orbitals depend on the degree of pi bonding between the metal atom and ligand.)

plexes are intensely colored. The Co^{2+} ion, for example, forms a bright blue $CoCl_4^{2-}$ complex (Figure 21.34).

Figure 21.37

Some nickel complexes. From left to right: green $Ni(H_2O)_6^{2+}(aq)$, blue $Ni(NH_3)_6^{2+}$, and a suspension of red $Ni(dmg)_2(s)$.

Square Planar Complexes

A square planar complex has ligands centered along the *x* and *y* axes; there are no ligands on the *z* axis (Figure 21.35). Most square planar complexes are formed by d^8 metals (see Table 21.7, p. 1016) coordinating with strong ligands; $Ni(CN)_4^{2-}$ is an example. In such cases, the bonding orbitals are very low in energy, while the energies of the antibonding orbitals, including those that form from the metal *d* orbitals, are correspondingly high. Without going into details, the important thing to note about the square planar arrangement is that the parent *d* orbitals are split into four sets of molecular orbitals, as shown in Figure 21.36. The eight *d* electrons of a d^8 metal ion such as Pt^{2+} occupy the four lower energy orbitals in a spin-paired arrangement, leaving one very high energy orbital (derived from the $d_{x^2-y^2}$ metal orbital) vacant.

Ni^{2+}, a d^8 ion, forms square planar complexes with strong ligands such as CN^- and dimethylglyoxime (dmg) and octahedral complexes with weak ligands such as NH_3 and H_2O (Figure 21.37). The square planar complex is favored for strong ligands because it takes less energy to pair all eight electrons in four square planar orbitals than to put two of them into high-energy e_g^* orbitals in an octahedral complex. Paladium(II) and platinum(II), which also have eight *d* electrons, form square planar complexes almost exclusively.

CHAPTER REVIEW

Starred entries are based on the *Digging Deeper* sections.

LEARNING OBJECTIVES BY SECTION

21.1 1. List the physical properties common to most metals.
2. Distinguish between metals, nonmetals, and semiconductors on the basis of electrical conductivity.
3. Describe how metallic bond strength is measured.
4. List the factors that influence metallic bond strength.
5. State how the hardness, melting points, and boiling points of metals vary across a period and down a group.

21.2 1. List the chemical properties associated with metallic character.
2. State how the ionic versus covalent character of representative metal bonds varies with the periodic position and oxidation state of the metal.
3. List each step in the formation of an aqueous cation, and describe the periodic trends in energy for each step.
4. Describe how oxidation potentials are calculated from thermodynamic data.
5. State how the acid–base character of a metal oxide varies with the periodic position and oxidation state of the metal.
6. List the special properties of the transition metals.

21.3 1. Diagram the formation of the valence and conduction bands in an atomic solid.
2. Distinguish between metallic conductors, insulators, and semiconductors in terms of band theory.

21.4 1. Use conductivity data and precipitation stoichiometry to determine the formula of a complex.
2. Identify the complex in a coordination compound, and determine the oxidation number of its metal atom.
3. Sketch the geometries of two-, four-, and six-coordinate complexes.
4. Give examples of monodentate and polydentate ligands, and identify their coordinating atoms.
*5. Given the formula of a complex, write its IUPAC name, and vice versa.

21.5 1. List and give examples of three types of structural isomerism exhibited by coordination compounds.
2. State whether a given complex could have geometric isomers, and sketch the *cis* and *trans* forms.
3. Describe how two enantiomers differ from each other in structure and properties.
4. State whether a given complex could exhibit optical isomerism, and sketch the enantiomers.

21.6 *1. Describe the electrostatic model of metal–ligand bonding.
*2. List some properties of transition-metal complexes that are not explained by the electrostatic model.
*3. Draw an energy-level diagram for the orbitals that are occupied by the metal d electrons in an octahedral complex, and show the configurations adopted by one, two, and three d electrons.
*4. Use observed colors to predict relative ligand field strengths.
*5. Use the spectrochemical series to predict color shifts when one ligand is substituted for another in a given metal complex.
*6. Diagram the high- and low-spin configurations for octahedral d^4, d^5, d^6, and d^7 complexes.
*7. Use the observed paramagnetism of a complex to determine whether it has a high- or low-spin configuration.
*8. Draw energy-level diagrams for the orbitals that are occupied by the metal d electrons in tetrahedral and square planar complexes.
*9. Explain why many d^8 complexes are square planar.

KEY TERMS BY SECTION

21.1 Semiconductor

21.2 Activity (of a metal)

21.3 Band
Band theory
Conduction band
Energy gap
Valence band

21.4 Chelating agent
Coordinate covalent bond
Coordinating atom
Coordination compound
Coordination number
Ligand
Metal complex

Monodentate ligand
Polydentate ligand

21.5 Achiral
Chiral
cis–trans Isomerism
Coordination isomerism
Dextrorotatory
Enantiomers
Geometric isomerism
Ionization isomerism
Isomers
Levorotatory
Linkage isomerism
Optical activity
Optical isomers
Plane of symmetry
Racemic mixture
Stereoisomers
Structural isomers

21.6 *Electrostatic model
*High-spin complex
*Low-spin complex
*Orbital splitting energy (Δ)
*Spectrochemical series
*Strong-field ligand
*Weak-field ligand

FINAL EXERCISES

Answers to exercises with blue numbers are given in Appendix D. Starred exercises are based on the *Digging Deeper* sections.

PART A. QUESTIONS AND PROBLEMS BY SECTION

Physical Properties of Metals (Section 21.1)

21.1 What physical properties are common to all metals? What other physical properties are found to some degree in most metals? Give examples of metals that have unusual properties.

21.2 Metals are said to be *lustrous*. Why do we not immediately see the luster of lead, calcium, and sodium? How can their luster be made evident?

21.3 What is meant by *metallic conduction*? Explain why the conductivity of a metal decreases with temperature.

21.4 What is meant by *semiconduction*? Explain why semiconductivity increases with temperature. Which elements are semiconducting?

21.5 Explain in terms of atomic structure and bonding why Group 1A metals are softer and have lower melting and boiling points than their adjacent neighbors in Group 2A.

21.6 Consult a table of atomic radii and explain why (a) the melting points of representative metals decrease from top to bottom in a group and (b) the melting points in many transition groups increase from top to bottom.

21.7 How would you expect melting points to vary across the lanthanide series? Explain your answer.

21.8 Explain why a metallic lattice is more compact than a covalent lattice. What does the expansion of bismuth on freezing indicate about its bonding?

Chemical Properties of Metals (Section 21.2)

21.9 What chemical property is shared by all metals? What other chemical properties are found to some degree in most metals?

21.10 Use periodic trends in atomic properties to explain why the stepwise line separating metals from nonmetals slants from the upper left to the lower right in the periodic table.

21.11 Describe the three steps in the conversion of a solid metal to positive ions in aqueous solution. Which steps are exothermic and which are endothermic?

21.12 Describe how the energy of each step in Exercise 21.11 varies going down a group and across a period. Explain these trends in terms of atomic properties.

21.13 Where in the periodic table would you look for metals with multiple oxidation states? Why aren't there any metals with multiple oxidation states in Group 1A?

21.14 Explain why the metals of Group 2A do not have multiple oxidation states.

21.15 The ionic character of the bonds formed by a given metal decreases as the oxidation state increases. Explain why.

21.16 Explain in terms of atomic structure and bonding why the compounds of Li, Be, and Al have more covalent character than the compounds of other metals in their groups.

21.17 Thallium forms both Tl(I) and Tl(III) compounds.
- **(a)** Write formulas for thallium(I) hydroxide and thallium(III) hydroxide. One of these hydroxides is a strong and soluble base. Predict which one.
- **(b)** Write formulas for Tl(I) and Tl(III) chlorides. One chloride melts at 430°C and the other melts at 25°C. Predict which is which.

21.18
- **(a)** $Ca(OH)_2$ and $Zn(OH)_2$ both react with HCl. Which reaction involves independent OH^- ions? Write the equations.
- **(b)** Write equations for the reactions that occur when the acid solutions of Part (a) are neutralized with NaOH and then treated with excess concentrated NaOH.
- **(c)** Explain how the differences between the two hydroxides in Part (a) are related to atomic size,

electronegativity, and the nature of the metal–oxygen bond.

21.19 Are most of the representative metals above hydrogen or below hydrogen in the activity series? How about the transition metals?

21.20 Which of the transition metals in Table 20.1 (p. 950) will be oxidized by 1 *M* $H^+(aq)$?

21.21 What atomic properties are characteristic of transition-metal atoms? Why do some chemists not include Zn, Cd, and Hg with the transition elements?

21.22 List some physical and chemical properties that are especially characteristic of transition metals.

21.23 Calculate $\Delta G°$ and $\Delta S°$ for the following reaction at 25°C:

$$Mg(s) \rightarrow Mg^{2+}(aq) + 2e^- \qquad \Delta H° = -467 \text{ kJ}$$

(*Hint*: Look up the standard reduction potential of Mg.)

21.24 The enthalpy of sublimation of calcium is 178 kJ/mol. The enthalpy of hydration of Ca^{2+} is −2469 kJ/mol. Consult Tables 8.3 and 20.1 (pp. 324 and 950, respectively) and calculate $\Delta H°$, $\Delta G°$, and $\Delta S°$ for the following reaction at 25°C:

$$Ca(s) \rightarrow Ca^{2+}(aq) + 2e^-$$

Band Theory (Section 21.3)

21.25 Explain how the application of molecular orbital theory to atomic solids leads to the concept of energy bands. Use this model to explain the differences between metallic conductors, insulators, and semiconductors.

21.26 Diagram the formation of the conduction band in sodium metal. Describe the electron occupancy of the band, and explain why sodium is a metallic conductor.

21.27 **(a)** Diagram the formation of the conduction and valence bands in magnesium.
(b) Describe the electron occupancy of the bands, and explain why *p* orbitals have to be included in order to explain the conductivity of magnesium.

21.28 Sketch the energy bands in silicon, and explain how they differ from those in diamond. Explain why diamond is an insulator and silicon is not.

Metal Complexes and Coordination Compounds (Section 21.4)

21.29 Conductivity measurements indicate that 1 mol of $CrCl_3 \cdot 4NH_3$ dissociates into 2 mol of ions in solution. Write (a) the formula of the coordination compound and (b) an equation for the dissociation.

21.30 A 0.10 *M* solution of each of the following compounds is prepared: $K_4[Fe(CN)_6]$, $[Co(NH_3)_6]Cl_3$, and $[Pt(NH_3)_2Cl_4]$. Rank these solutions in order of increasing conductivity.

21.31 Two coordination compounds A and B both have a composition that can be represented as $PtBr(NO_3)\cdot 3NH_3$. Aqueous silver nitrate added to a solution of each compound produces a cream-colored AgBr precipitate in A but no precipitate in B. Write (a) formulas for A and B and (b) an equation for the reaction of A with silver nitrate.

21.32 Write an equation for the reaction, if any, of the following compounds with aqueous silver nitrate: $[Co(NH_3)_6]Cl_3$, $[Pt(NH_3)_2Cl_2F_2]$, and $[Pt(NH_3)_4(NO_2)_2]Cl_2$.

21.33 Write formulas for octahedral complexes formed from the following metals and ligands.
(a) Fe^{3+} and chloride ion
(b) Fe^{3+} and oxalate ion

21.34 Write formulas for octahedral complexes formed from the following metals and ligands.
(a) Ni^{2+} and diethylenetriamine
(b) Ir(III) and equimolar amounts of NH_3 and Cl^-

21.35 Give the oxidation state of the central metal atom in each of the following coordination compounds:
(a) $[Co(NH_3)_6]Cl_3$ **(c)** $K_4[Fe(CN)_6]$
(b) $K_3[FeF_6]$ **(d)** $K_3[Fe(CN)_6]$

21.36 Give the oxidation state of the central metal atom in each of the following coordination compounds:
(a) $Na[AlCl_4]$ **(c)** $[Pt(NH_3)_2Cl_4]$
(b) $K_2[CuCl_4]$ **(d)** $[V(H_2O)_6]SO_4$

21.37 Sketch a structure for each of the following complex ions:
(a) $PdCl_4^{2-}$ (square planar)
(b) $Be(OH)_4^{2-}$ (tetrahedral)
(c) $AgCl_2^-$
(d) $Co(NH_3)_5(CO_3)^+$

21.38 Sketch a structure for each of the following complex ions:
(a) $Cu(NH_3)_4^{2+}$ (square planar)
(b) VCl_4^- (tetrahedral)
(c) $Cr(NCS)_6^{3-}$
(d) $V(CN)_6^{4-}$

21.39 Sketch a structure for each of the following complex ions:
(a) $Pt(en)_2^{2+}$
(b) $Co(en)_2(H_2O)Cl^+$
(c) $Ni(NH_3)_3(dien)^{2+}$

21.40 Sketch a structure for each of the following complex ions:

(a) $Co(en)_3^{3+}$
(b) $Al(C_2O_4)_3^{3-}$
(c) $Fe(NH_3)Cl(en)_2^{2+}$

Nomenclature (Section 21.4)

***21.41** Write formulas for the following complexes:
(a) hexaaquanickel(II) ion
(b) tetraamminecopper(II) ion
(c) triamminetrichloroiridium(III)
(d) tetraamminedichlorocobalt(II)
(e) hexaaquairon(III) ion

***21.42** Write formulas for the following complex ions:
(a) hexacyanoferrate(III) ion
(b) tetrachloroaluminate(III) ion
(c) diamminesilver(I) ion
(d) tetrachloroplatinate(II) ion
(e) trioxalatovanadate(III) ion

***21.43** Name the complexes in (a) Exercise 21.35, (b) Exercise 21.37, and (c) Exercise 21.39.

***21.44** Name the complexes in (a) Exercise 21.36, (b) Exercise 21.38, and (c) Exercise 21.40.

Isomers (Section 21.5)

21.45 What is *ionization isomerism*? Give an example of ionization isomerism other than the one in the textbook.

21.46 What is *coordination isomerism*? Give an example of coordination isomerism other than the one in the textbook.

21.47 What is *linkage isomerism*? Give an example of linkage isomerism other than the one in the textbook.

21.48 What is the difference between *structural isomers* and *stereoisomers*?

21.49 What is the difference between *geometric isomers* and *enantiomers*? Why are enantiomers often called *optical isomers*?

21.50 What is a *chiral molecule*? Is BF_3 chiral? How do you know?

21.51 Draw the *cis–trans* isomers of (a) $[Pt(H_2O)_2Br_2]$ and (b) $Fe(H_2O)_4(OH)_2^+$.

21.52 Draw the *cis–trans* isomers of (a) $Cr(en)_2(NCS)_2^{2+}$ and (b) $Ir(C_2O_4)_2Cl_2^{3-}$.

21.53 Draw all the isomers of the following complexes. Label the *cis* and *trans* forms of any geometric isomers, and identify all enantiomer pairs.
(a) $[Co(NH_3)_3Cl_2Br]$
(b) $[Pt(NH_3)_2Cl_2F_2]$
(c) $Rh(H_2O)_4Cl_2^+$
(d) $[Pt(NH_3)_2BrF]$

21.54 Draw all the isomers of the following complexes. Label the *cis* and *trans* forms of any geometric isomers, and identify all enantiomer pairs.
(a) $[Cr(NH_3)_3Cl_3]$
(b) $[Ni(H_2O)_4(OH)_2]$
(c) $Co(en)(H_2O)_2Cl_2^+$
(d) $Rh(en)_3^{3+}$

21.55 Use sketches to show that (a) the square planar complex $[Pt(NH_3)_2BrCl]$ has only two isomers and (b) the square planar complex $[Pt(py)(NH_3)BrCl]$ has three isomers.

21.56 State whether each of the following pairs are *cis–trans* isomers, optical isomers, or identical. All have the formula $Co(en)_2Cl_2^+$.

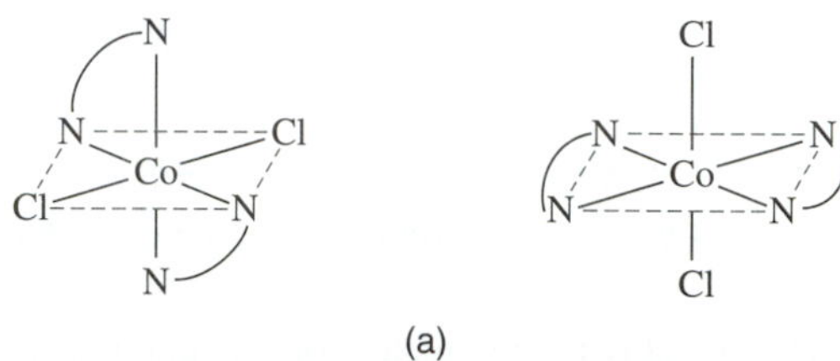
(a)

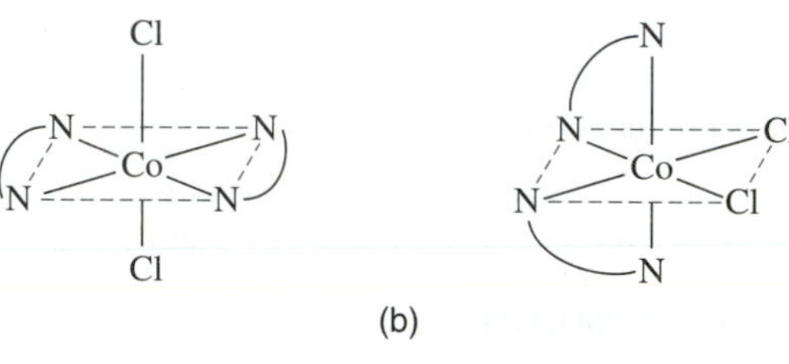
(b)

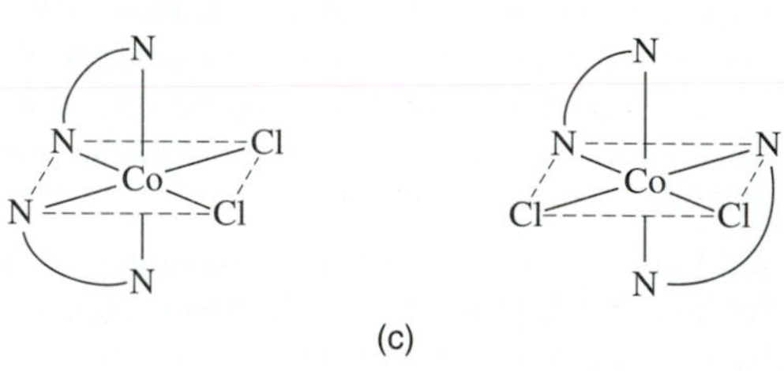
(c)

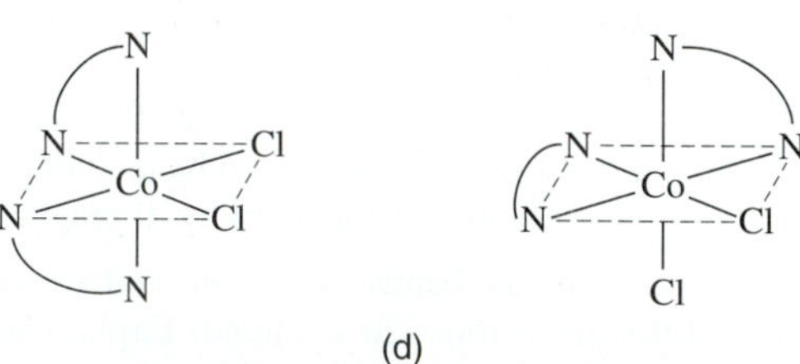
(d)

Bonding in Metal Complexes (Section 21.6)

***21.57** Briefly describe the electrostatic model of metal–ligand bonding. State the strong points and shortcomings of this model. For what type of complexes does it work best?

***21.58** Describe some properties of transition-metal complexes that are not explained by the electrostatic model.

***21.59** Sketch the five *d* orbitals, and show how each one is oriented in relation to the ligands in an octahedral complex.

***21.60** Sketch the five *d* orbitals, and show how each one is oriented in relation to the ligands in (a) a tetrahedral complex and (b) a square planar complex.

***21.61** Sketch an energy-level diagram of the orbitals that accommodate the metal *d* electrons in an octahedral complex. Explain what is meant by *orbital splitting*.

***21.62** Sketch an energy-level diagram of the orbitals that accommodate the metal *d* electrons in a tetrahedral complex.

***21.63** Sketch an energy-level diagram of the orbitals that accommodate the metal *d* electrons in a square planar complex.

***21.64** Under what circumstances will a d^8 ion such as Ni^{2+} form an octahedral complex? A square planar complex? Your discussion should include an orbital occupancy diagram for each complex.

***21.65** According to the electrostatic model, which complex ion in each of the following pairs would be expected to have the larger formation constant K_f?
(a) $Be(H_2O)_4^{2+}$ or $Cu(H_2O)_4^{2+}$
(b) $Fe(H_2O)_6^{2+}$ or $Fe(H_2O)_6^{3+}$

***21.66** According to the electrostatic model, which complex ion in each of the following pairs would be expected to have the larger formation constant K_f?
(a) $Ni(H_2O)_6^{2+}$ or $Al(H_2O)_6^{3+}$
(b) $Zn(NH_3)_4^{2+}$ or $Cd(NH_3)_4^{2+}$

***21.67** Which complex ion should absorb light of a shorter wavelength, $Cr(H_2O)_6^{3+}$ or $CrCl_6^{3-}$?

***21.68** Zinc, unlike the elements immediately preceding it in the periodic table, forms colorless complexes. Explain why this behavior is to be expected.

***21.69** On the basis of the following colors, list NH_3, H_2O, Cl^-, and NCS^- in order of increasing ligand strength. Compare your list with the spectrochemical series in Table 21.8.

$Co(NH_3)_6^{3+}$	orange-yellow
$Co(NH_3)_5(H_2O)^{3+}$	red
$Co(NH_3)_5Cl^{2+}$	purple
$Co(NH_3)_5(NCS)^{2+}$	orange

***21.70** A solution containing $Ni(NH_3)_6^{2+}$ is blue. Your laboratory partner has a blue-green solution that she claims contains $Ni(en)_3^{2+}$. Should you believe her? Explain.

***21.71** Draw an orbital occupancy diagram for each of the following high-spin complexes:
(a) $Ni(NH_3)_6^{2+}$ **(c)** $Fe(H_2O)_6^{2+}$
(b) $Co(H_2O)_6^{2+}$ **(d)** $Fe(H_2O)_6^{3+}$

***21.72** Draw an orbital occupancy diagram for each of the following low-spin complexes:
(a) $Co(en)_3^{3+}$ **(c)** $Fe(CN)_6^{3-}$
(b) $Fe(CN)_6^{4-}$ **(d)** $Pt(NH_3)_4Cl_2^{2+}$

***21.73** How many unpaired electrons are in each of the following complexes? (*Note*: All except (a) have low-spin configurations.)
(a) $Ni(NH_3)_6^{2+}$ **(d)** $Pt(NH_3)_3Cl^+$
(b) $Pt(NH_3)_3Cl_3^+$ **(e)** $K_4[Mn(CN)_6]$
(c) $MoCl_6^{3-}$ **(f)** $K_3[RuCl_6]$

21.74 Draw an orbital occupancy diagram for d^9 square planar complexes such as $Cu(H_2O)_4^{2+}$ and $Cu(NH_3)_4^{2+}$. Which molecular orbital has a vacancy?

***21.75** The conversion of $Mn(H_2O)_6^{3+}$ to $Mn(CN)_6^{3-}$ is accompanied by a reduction in paramagnetism. Draw an orbital occupancy diagram for each ion.

***21.76** $Co(NH_3)_6^{3+}$ is not paramagnetic. Can we use this fact to deduce whether $Co(CN)_6^{3-}$ is paramagnetic? Can we deduce whether $Co(H_2O)_6^{3+}$ is paramagnetic? (*Hint*: See Table 21.8.)

PART B. MISCELLANEOUS QUESTIONS AND PROBLEMS

21.77 A glance at the divisions in the periodic table shows that metals far outnumber nonmetals. Now count only the main-group metals and nonmetals, and comment on their relative numbers.

21.78 Look at the elemental abundances in Table 2.2 (p. 45). What is the approximate ratio of metal to nonmetal atoms in the earth's crust?

21.79 Gold can be hammered into a foil only 100 nm thick. Consult Figure 8.9 (p. 321), and calculate the number of layers of atoms in the foil assuming (a) a simple cubic array of atoms and (b) a cubic close-packed array of atoms (the usual structure for gold).

21.80 Explain the metallic conduction of aluminum in terms of band theory.

21.81 The energy gap in pure silicon is 106 kJ/mol. Find the lowest radiation frequency that would excite an electron across the gap. Explain why visible light increases the conductivity of silicon.

***21.82** Novelty "weather predictors" often have a clown's nose or a donkey's tail that turns blue in dry air and pink in moist air. The blue compound is tetraaquacobalt(II) chloride and the pink compound is hexaaquacobalt(II) chloride. Write formulas for each of these compounds, and write an equation for the conversion of one complex into the other.

21.83 Describe a precipitation test that would distinguish between $[Co(NH_3)_5Br]SO_4$ and $[Co(NH_3)_5SO_4]Br$.

21.84 In making wine, undesirable iron(III) salts are sometimes removed by adding a controlled amount of potassium ferrocyanide. How does this work? Write the equation.

21.85 A brick-red compound and a purple compound each have the composition $CoCl_3 \cdot 5NH_3 \cdot H_2O$. Excess silver nitrate solution added to 100 mL of a 0.037 *M* solution of the purple compound precipitated 1.06 g of AgCl. Excess silver nitrate added to 100 mL of a 0.10 *M* solution of the brick-red compound precipitated 4.3 g of AgCl. Calculate the number of moles of ions per mole of each compound. Write a tentative formula for each compound showing the complex ion in brackets.

21.86 $[Pt(py)(NH_3)(NO_2)(Cl)(Br)(I)]$ is one of the very few octahedral complexes that have been prepared with six different ligands. It has been shown that there are 15 geometric isomers, each with 2 enantiomers, corresponding to this formula. Draw as many as you can. Be careful to avoid duplication.

21.87 Can a tetrahedral complex such as $Be(OH)_2Cl_2^{2-}$ have *cis* and *trans* isomers? Explain.

***21.88** Which aspect of the interaction between metal and ligand is neglected in the simple electrostatic model of complex formation but included in the molecular orbital model?

***21.89** Chloride ions, when present, often replace some of the water molecules in a hydrated cation. Explain why aqueous solutions of $CrCl_3$ often appear green even though $Cr(H_2O)_6^{3+}$ is violet.

***21.90** Hydrogen peroxide alone cannot oxidize a simple aqueous Co(II) salt to Co(III), but a mixture of hydrogen peroxide and ammonia can. (a) Draw an orbital occupancy diagram for the high- and low-spin configurations of Co^{2+}, and (b) use them to explain why formation of the ammonia complex promotes the oxidation. (*Hint*: Consider the effect of removing one electron on the energy of each configuration.)

***21.91** For Co^{3+} the pairing energy is 251 kJ/mol, and the orbital splitting energies produced by F^- and NH_3 are 155 and 275 kJ/mol, respectively. Draw orbital occupancy diagrams for CoF_6^{3-} and $Co(NH_3)_6^{3+}$. Compare the paramagnetism of the two complexes.

***21.92** The ligand H_2O does not cause electrons to pair in either Fe^{2+} or Fe^{3+}. The ligand CN^- causes pairing in both.

(a) Use this information to deduce orbital occupancy diagrams for (1) $Fe(H_2O)_6^{2+}$, (2) $Fe(H_2O)_6^{3+}$, (3) $Fe(CN)_6^{4-}$, and (4) $Fe(CN)_6^{3-}$.

(b) Use the configurations of Part (a) to explain why $Fe(H_2O)_6^{2+}$ will oxidize in air to $Fe(H_2O)_6^{3+}$.

(c) Use the configurations of Part (a) to explain how a strong ligand would stabilize the +2 oxidation state of iron. (*Hint*: Recall that configurations are especially stable when sets of orbitals with the same energy are filled or half-filled.)

CHEMISTS AT WORK

THE FORENSIC CHEMIST

Debra Minton always wanted to be a forensic scientist. When she graduated with a bachelor's degree in chemistry from Eastern Illinois University in 1975, positions in the field were not available. She therefore elected to begin her career in an industrial firm. However, she never lost sight of her original goal, and in 1987 she accepted a position with the Illinois State Police Bureau of Forensic Sciences. Now Debra is one of 200 scientists who work in seven laboratories scattered across Illinois and who provide a range of services to local law enforcement agencies.

Deb attended a 1-year training program at the Southern Illinois Forensic Science Centre, operated by the state, as part of the conditions of her employment. Individuals with degrees in either chemistry or biology are typically recruited for positions at the centre where they pursue specializations for 1–3 years. Areas of specialization include chemistry, latent fingerprints, serology (body fluids and DNA identification), firearms and tool marks, trace chemistry, documents, polygraphs, and toxicology. The training encompasses lectures, laboratory work, exams, testifying in court, record keeping, and preserving the integrity of evidence.

As a chemist in the system, her principal responsibilities are in the area of controlled substances, and during her tenure, she has been qualified as an expert in drug identification. She receives a sample from a local agency and then identifies it as marijuana, heroin, cocaine, LSD, and so on. In a typical month, she performs between 80 and 90 analyses. Sometimes an analysis may take as little as 20 minutes because she is able to identify a material by its appearance and then make a positive identification by appropriate procedures. Other times the identification may take several days. She says, "You must be careful about the quality assurance of your results. The bottom line is that you are determining someone's future, and mistakes can't be made."

In 1993, because of the increasing case load and number of samples to analyze, Deb undertook a pilot project to attempt to increase efficiency without sacrificing the quality of the results. She showed how an "autosampler," a type of robotic apparatus, can be used to deliver samples and record their UV (ultraviolet) and mass spectra. The apparatus can analyze up to 100 samples overnight. This procedure was recommended to other laboratories in the system.

DEBRA MINTON

There are five major skills or attributes that are necessary for the typical forensic scientist. The first is good laboratory skills; this includes the ability to observe, correlate, and analyze and the ability to use scientific instrumentation and computers effectively. The second is meticulous record keeping. This is essential since scientists never know when they will be called on to testify about a sample. Since the analyst may analyze 600–800 samples a year and it may be as much as several months to several years before testimony is required, very accurate records are needed. Good communication skills are also important since the analyst must explain the results of sophisticated scientific techniques to jury members, attorneys, judges, and other professionals who have never studied much science. Good people skills are necessary since the forensic scientist deals not only with law enforcement professionals, but also with defendants and attorneys often in adversarial situations. Finally, forensic scientists must possess confidence in their abilities to obtain totally accurate data; someone insecure in his or her abilities will find court appearances very uncomfortable.

On average, Deb is required to appear in court about 10 times per year. In the beginning, she did not relish court appearances. "It is sometimes difficult to face someone who is accused. It is easier when the process is impersonal."

She is very enthusiastic about her work and is proud to explain her job to people. "I feel as if I'm really doing something worthwhile for society since I'm a contributor to fighting the war on drugs. Also, I feel that I'm doing something for my children since I'm helping, in a small way, to keep drugs off the streets."

SURVEY OF THE ELEMENTS 4

Gold, silver, and platinum.

METALS AND METALLURGY

OUTLINE

PREVIEW

Seven metals were known to the ancient Romans: gold, silver, copper, iron, lead, tin, and mercury. Seventy metals have now been found in the earth's crust, and others have been created in nuclear reactions. The metallic elements stretch across the periodic table from lithium to polonium, occupying the entire *s* block (except for hydrogen and helium), the entire *d* and *f* blocks, and part of the *p* block. Each metal in the table is unique. Almost every one has some special use or function that was never dreamed of by our ancestors. In this chapter we describe the more important metals of the *p* and *d* blocks.

The earliest known metal objects date back to before 9000 B.C. They were made of copper, silver, and gold—elements that could be found in the native or uncombined state. Tin and lead were discovered later in minerals that could be reduced by the hot charcoal of wood fires, perhaps the same fires that were used to harden pottery. The Bronze Age (bronze contains copper and tin) preceded the Iron Age because iron is much more difficult to obtain than copper and tin; its oxides can be reduced only in very hot furnaces. In fact, the first samples of iron were found in meteorites (Figure S4.1) and the Egyptians referred to iron as "metal from the sky."

An **alloy** is a mixture of metals; some alloys contain nonmetals as well. Nearly every metal product in everyday use is an alloy of some kind. (The pure copper used for electrical wiring is a major exception.) Pure metals tend to be relatively soft, and much of the alloying is done to increase hardness and resistance to wear. Alloys are designed for corrosion resistance, for high tensile strength, for toughness, or for the ability to retain a sharp edge. Alloys with other special properties have been developed for magnets, fuses, heating filaments, and numerous other devices. Steels are **ferrous alloys**—their principal ingredient is iron. Bronze, brass, and pewter are **nonferrous alloys**: they do not contain iron. Some important nonferrous alloys are listed in Table S4.1.

Figure S4.1

An iron–nickel meteorite. Its composition, about 75% iron and 25% nickel, is believed to resemble that of the earth's core.

TABLE S4.1 Some Nonferrous Alloys

Alloy	Composition (wt %)
Antimonial tin solder (for use with drinking-water pipes)	95% Sn, 5% Sb
Brass (red)	90% Cu, 10% Zn
Brass (yellow)	67% Cu, 33% Zn
Bronze	90% Cu, 10% Sn (variable composition)
Dental alloy (to be mixed with mercury)	65% Ag, 21% Sn, 12% Cu, 2% Zn
German silver	64% Cu, 24% Zn, 12% Ni
Gold (14 karat)	58% Au; 12–28% Cu; 14–30% Ag
Pewter	75% Sn, 25% Pb (variable composition)
Solder (soft; melting point, 192°C)	67% Pb, 33% Sn
Sterling silver	92.5% Ag, 7.5% Cu
Wood's metal (melting point, 70°C; used in fuses)	50% Bi, 25% Pb, 12.5% Sn, 12.5% Cd

S4.1 METALLURGY

Molybdenite (MoS_2), an ore of molybdenum.

Metallurgy—the science of extracting metals from their ores, refining them, and making alloys—developed by trial and error over a long period of time. Metallic elements are found in **minerals**, which are naturally occurring inorganic substances with a definite crystal structure and a definite range of composition. More than 2000 minerals have been identified in the earth's crust; rocks are aggregates of minerals in varying proportions.

An **ore** is a rock or mineral from which some element can be extracted with profit; in this section we are concerned with the ores of metals. **Native ores** contain metals that are not chemically combined, usually noble metals such as gold or platinum. Most other metal ores contain oxides, sulfides, halides, or carbonates (see Table S4.2). Silicate minerals are stubbornly difficult to decompose and, although they comprise 87% of the earth's crust, they are not used as ores if other compounds are available.

The three principal steps that are carried out to obtain a metal from its ore are concentrating the ore, extracting the metal, and refining (purifying) the crude metal.

LEARNING HINT

Review the metallurgy of zinc and tin in Section 3.6 and the metallurgy of sodium and aluminum in Section 20.7.

Concentrating the Ore

Concentrating an ore increases the fraction of metal-bearing mineral by eliminating most of the accompanying rock and soil, called **gangue**. The ore must first be finely divided, by crushing and grinding if necessary; then the gangue can be removed in a number of ways. The old forty-niner panned for gold by swishing sandy ore in running water. This technique was a *gravity separation*; dense gold particles settled to

TABLE S4.2 Some Common Ores

Native Metals		Sulfides	
Bismuth	Bi	Argentite	Ag_2S
Copper	Cu	Chalcocite	Cu_2S
Gold	Au	Cinnabar	HgS
Iridium	Ir	Galena	PbS
Osmium	Os	Millerite	NiS
Palladium	Pd	Sphalerite	ZnS
Platinum	Pt		
Silver	Ag		
Oxides		**Carbonates**	
Bauxite	Al_2O_3	Calcite	$CaCO_3$
Cassiterite	SnO_2	Cerussite	$PbCO_3$
Hematite	Fe_2O_3	Dolomite	$CaCO_3 \cdot MgCO_3$
Magnetite	Fe_3O_4	Siderite	$FeCO_3$
Pyrolusite	MnO_2	Smithsonite	$ZnCO_3$
Rutile	TiO_2	Rhodocrosite	$MnCO_3$

Other ores include chromates ($PbCrO_4$), halides (NaCl, CaF_2), hydroxides ($Mg(OH)_2$), phosphates ($Ca_3(PO_4)_2$), and sulfates ($BaSO_4$).

the bottom while less dense particles of sand and soil were swept away. *Magnetic separation* can be used for ferromagnetic minerals; for example, particles of magnetite (Fe_3O_4) will adhere to a rotating magnetic drum, leaving nonmagnetic particles behind. Sulfide ores are concentrated by **flotation**. The finely ground ore is mixed with oil, which wets sulfides but not silicates. The mixture is then agitated with a solution of detergent and frothing agent. The detergent molecules surround the oily particles and carry them up in the foam (Demonstration S4.1).

The **Bayer process** for obtaining pure aluminum oxide from bauxite is an example of a chemical separation. Bauxite consists principally of alumina ($Al_2O_3 \cdot xH_2O$) with oxides of iron and silicon as impurities. The crushed ore is treated with hot 30% aqueous sodium hydroxide, and pressure is applied to prevent boiling. The amphoteric aluminum oxide dissolves:

$$Al_2O_3(s) + 2OH^-(aq, 30\%) + 3H_2O(l) \xrightarrow{150\text{–}220°C} 2Al(OH)_4^-(aq)$$

Iron oxide and other basic oxides are not affected by the base, and the silicate impurities are converted to insoluble aluminosilicates (Section S3.4). The solution is filtered, cooled, and diluted to reduce the OH^- concentration. Aluminum hydroxide then precipitates and the anhydrous oxide is obtained by heating:

$$Al(OH)_4^-(aq) \xrightarrow{\text{dilution}} Al(OH)_3(s) + OH^-(aq)$$

$$2Al(OH)_3(s) \xrightarrow{\text{heat}} Al_2O_3(s) + 3H_2O(g)$$

DEMONSTRATION S4.1 CONCENTRATING A SULFIDE ORE BY FLOTATION

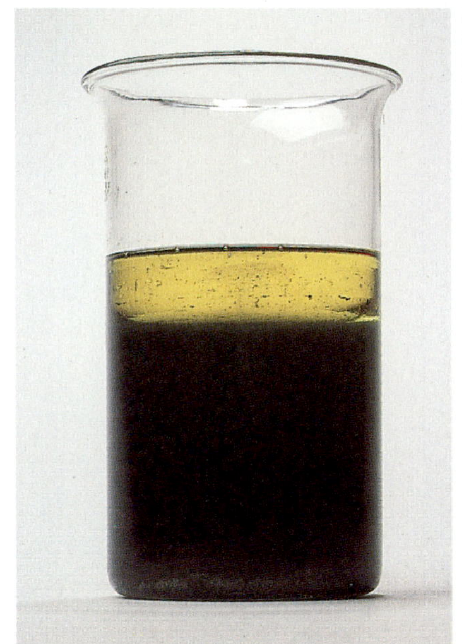

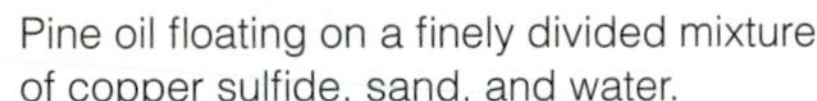

Pine oil floating on a finely divided mixture of copper sulfide, sand, and water.

The mixture is stirred vigorously.

When the oil and water separate, the black copper sulfide is found in the upper oil layer, the sand in the water layer.

Obtaining the Metal: Pretreatment

Pretreatment converts a mineral to a form that is easy to reduce. Sulfides, for example, are converted to reducible oxides by heating in dry air, a process called **roasting**:

$$2PbS(s) + 3O_2(g) \xrightarrow{\text{heat}} 2PbO(s) + 2SO_2(g)$$

PRACTICE EXERCISE S4.1

Write the equation for the reaction that occurs when nickel(III) sulfide is roasted to nickel(II) oxide.

If a metal is to be obtained by electrolytic reduction, its oxide, hydroxide, or carbonate is often treated with HCl to convert it to a chloride. Chlorides are preferred for electrolytic reductions because they usually melt at low temperatures to form conductive liquids. (Aluminum chloride is an important exception; the molten chloride consists of covalent molecules and is a poor conductor of electricity.) This pretreatment method is used in the **Dow process** for obtaining magnesium metal from seawater. Seawater contains 1.3 g of magnesium ions per liter. These ions are first precipitated as magnesium hydroxide and then converted to magnesium chloride.

The hydroxide ions are provided by lime from shells that abound near the ocean. The equations are

$$\underset{\text{shells}}{CaCO_3(s)} \xrightarrow{\text{heat}} \underset{\text{lime}}{CaO(s)} + CO_2(g)$$

$$CaO(s) + H_2O(l) \longrightarrow \underset{\text{saturated solution}}{Ca(OH)_2(aq)}$$

$$\underset{\text{seawater}}{Mg^{2+}(aq)} + Ca(OH)_2(aq) \longrightarrow Mg(OH)_2(s) + Ca^{2+}(aq)$$

Recall that the solubility of $Mg(OH)_2$ is quite a bit lower than that of $Ca(OH)_2$ (see Table 18.1, p. 823).

The magnesium hydroxide is filtered and converted to magnesium chloride with HCl:

$$Mg(OH)_2(s) + 2HCl(aq) \rightarrow MgCl_2(aq) + 2H_2O(l)$$

Pure $MgCl_2$ for electrolysis is recovered by evaporating the water.

Obtaining the Metal: Reduction

Electrolytic reduction, as described in Section 20.7, is the most economical method for obtaining active metals such as lithium, sodium, magnesium, calcium, and aluminum. On the other hand, potassium, rubidium, and cesium are usually produced by chemical reduction. Electrolysis is inconvenient for these metals because they are extremely volatile at the temperature of the chloride melt and their vapors attack the carbon electrodes. Potassium is obtained by reducing molten KCl with liquid sodium:

$$KCl(l) + Na(l) \rightarrow NaCl(l) + K(g)$$

The forward reaction is promoted by continuous removal of the potassium vapor.

Smelting produces a molten metal by high-temperature chemical reduction. Carbon is the most common reducing agent. The ancients smelted ores with charcoal, which is carbon obtained by charring wood. Today's smelters use *coke*, carbon obtained by heating coal. Copper, tin, lead, zinc, manganese, cobalt, nickel, and iron are obtained by reducing their oxides with coke at high temperatures:

$$PbO(s) + C(s) \xrightarrow{\text{heat}} Pb(l) + CO(g)$$

A **blast furnace** for smelting iron ore is shown in Figure S4.2. Smaller blast furnaces are used for lead and manganese. The furnace is charged from the top with a mixture of crushed ore, coke, and limestone; blasts of heated air or oxygen are blown through the furnace from below. The oxidation of carbon to carbon monoxide produces additional heat, and the gaseous carbon monoxide reduces the ore to molten metal in a series of steps. The overall reaction for the reduction of hematite ore by carbon monoxide is

$$Fe_2O_3(s) + 3CO(g) \rightarrow 2Fe(l) + 3CO_2(g)$$

Some oxide is reduced by solid carbon as well:

$$Fe_2O_3(s) + 3C(s) \rightarrow 2Fe(l) + 3CO(g)$$

(a)

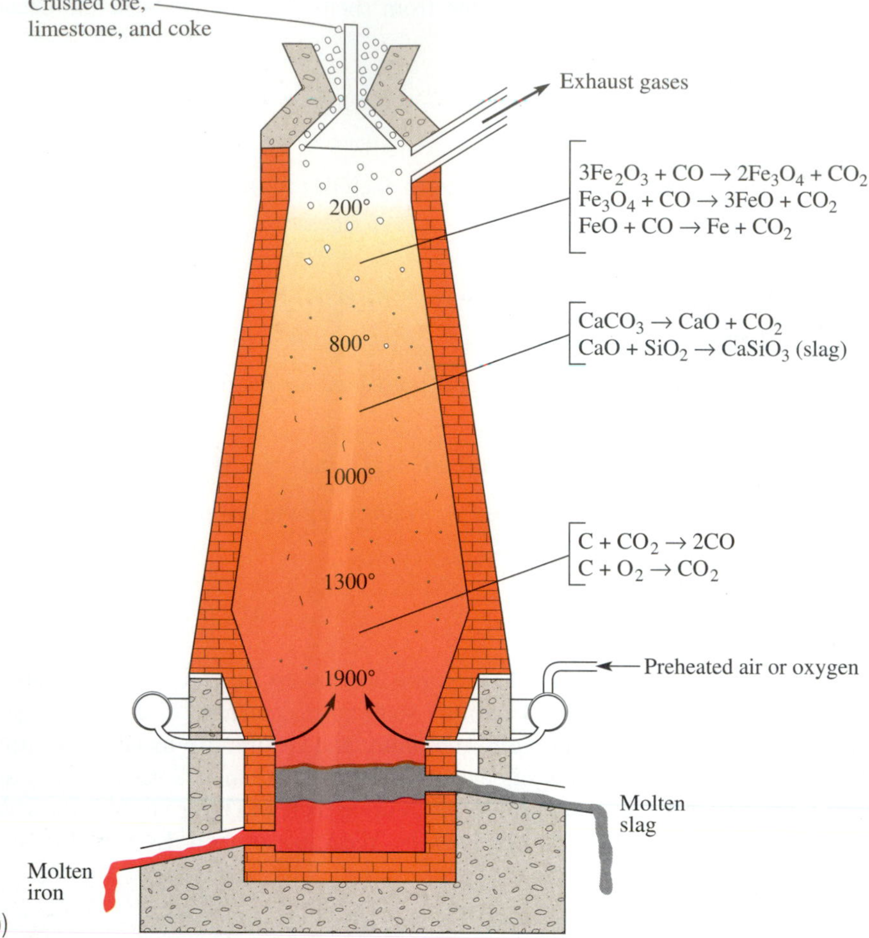

(b)

Figure S4.2

(a) Pouring molten iron for machine tools. (b) A blast furnace for the production of iron. The charge of ore, coke, and limestone is added at the top of the hot furnace; compressed air is blown in at the bottom. The temperature is highest near the bottom of the furnace where the molten iron and slag are tapped off.

The limestone is converted to calcium oxide, which removes silica from the ore by forming a mixture of molten calcium silicates called **slag**:

$$CaCO_3(s) \xrightarrow{\text{heat}} CaO(s) + CO_2(g)$$

$$CaO(s) + SiO_2(s) \longrightarrow CaSiO_3(l)$$

The molten iron and slag trickle through the charge and collect in layers at the bottom. The silicate slag floats on top of the more dense molten metal.

Coke converts certain metals such as titanium and vanadium to carbides, so it cannot be used to reduce their ores. Alternative reducing agents include hydrogen, aluminum, magnesium, calcium, and sodium. Hydrogen reduction produces the purest metal because the other product, H_2O, is volatile. Pure tungsten is prepared in this way:

$$WO_3(s) + 3H_2(g) \rightarrow W(s) + 3H_2O(g)$$

Reduction with aluminum, called **aluminothermy**, evolves a great deal of heat and produces molten products:

The formation of Al_2O_3 is exothermic; its standard heat of formation is -1676 kJ/mol at 25°C.

$$Cr_2O_3(s) + 2Al(s) \rightarrow 2Cr(l) + Al_2O_3(l)$$

(See Demonstration 3.4, p. 103.)

Chemical means are sometimes used to remove native metals from ores that are too finely divided for mechanical separation. In the **cyanide leaching process**, a crushed gold ore is aerated for several days in a solution of sodium cyanide. Cyanide acts as a complexing agent and stabilizes the oxidized gold:

$$4Au(s) + 8CN^-(aq) + O_2(g) + 2H_2O(l) \rightarrow 4Au(CN)_2^-(aq) + 4OH^-(aq)$$

Powdered zinc is then used as a reducing agent to displace gold from the complex:

$$2Au(CN)_2^-(aq) + Zn(s) \rightarrow 2Au(s) + Zn(CN)_4^{2-}(aq)$$

Silver can also be recovered in this fashion. The reactions and equations are analogous to those for gold.

PRACTICE EXERCISE S4.2

Identify the oxidizing and reducing agents when gold ore is leached with sodium cyanide solution.

Refining the Crude Metal

Numerous techniques are available for separating the impurities from a crude metal. Physical methods include distillation, differential dissolving, and zone refining (see Figure S3.24, p. 886). Chemical methods are also used. For example, hot air is used to oxidize and remove excess carbon and other impurities from crude molten iron (see Section S4.5).

Electrorefining, an elegant method for producing a very pure product, is used to obtain the 99.9% pure copper needed for electrical wiring. Some tin, lead, nickel, chromium, zinc, and other metals are also purified in this way. The anode, a bar of crude copper, and the cathode, a thin sheet of pure copper, are immersed in an aqueous solution of copper sulfate (Figure S4.3). The voltage applied to the electrodes is just high enough to oxidize the copper atoms in the anode:

$$Cu(s) \rightarrow Cu^{2+}(aq) + 2e^- \qquad \text{(at anode)}$$

The copper ions enter the solution and travel to the cathode, where they are reduced to pure copper metal:

$$Cu^{2+}(aq) + 2e^- \rightarrow Cu(s) \qquad \text{(at cathode)}$$

Impurities more active than copper also dissolve from the anode, but because their reduction potentials are lower than that of copper, they are not reduced at the cathode. They therefore remain in the electrolyte solution. Less active impurities are not oxidized and stay behind in the anode. As the anode disintegrates, these impurities gradually fall to the bottom and form **anode sludge**, a muddy deposit that is an important source of less active metals. The sludge from a copper anode yields significant quantities of the precious metals gold, silver, platinum, and palladium.

S4.2 METALS OF THE *p* BLOCK

The *p*-block metals include aluminum and the post-transition metals of Groups 3A through 6A. The most important metals in these groups are aluminum in Group 3A, tin and lead in Group 4A, and bismuth in Group 5A.

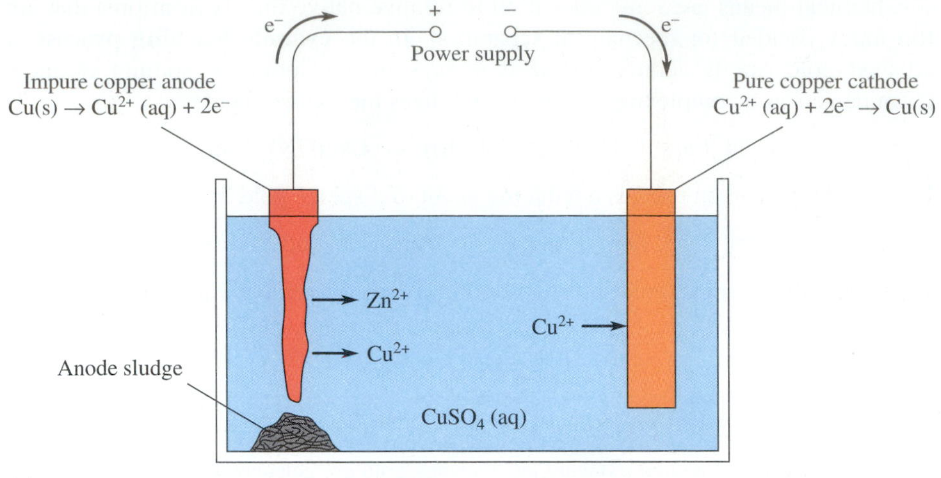

(a)

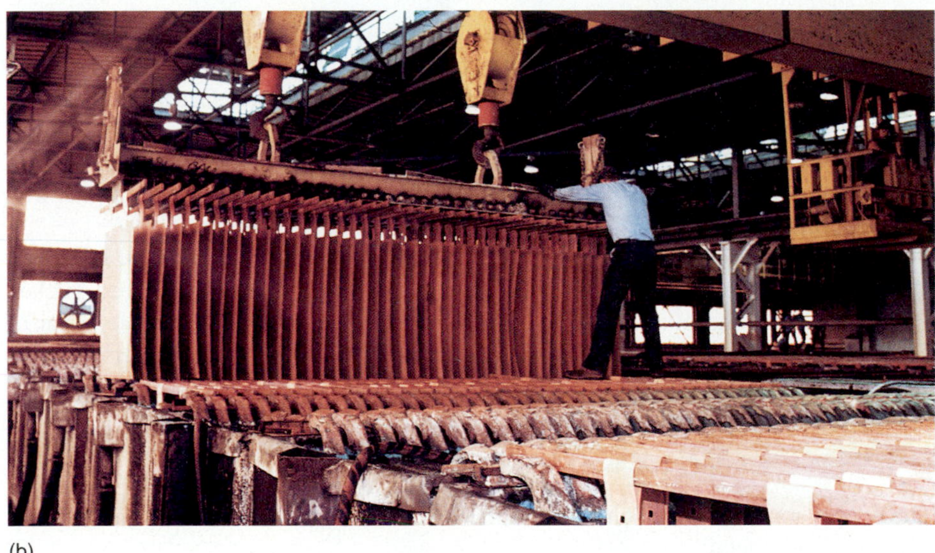

(b)

Figure S4.3

Electrorefining copper. (a) The anode, which consists of impure copper, gradually disintegrates as the copper and impurities more active than copper (e.g., Zn) are oxidized. Less active impurities fall to the bottom as anode sludge. At the pure copper cathode, copper ions are reduced to copper metal; the more active impurities remain in solution. (b) Sheets of very pure copper are lowered between bars of crude copper in a commercial electrorefining cell.

Group 3A: Aluminum

Aluminum, the third most abundant element in the earth's crust, is widely distributed in aluminosilicates such as feldspars, micas, and clays. Oddly enough, this widespread element has no known physiological function, although most other light elements do. Very little aluminum is found in the normal human brain, but significant amounts have been found in the cortex and hypothalamus regions of the brains of Alzheimer's victims. The role of aluminum in this disease is not yet understood.

Aluminum metal is obtained electrolytically from bauxite by the Hall–Héroult process (Section 20.7). Aluminum is conductive and lustrous, and with proper alloying, it is excellent for lightweight construction.

Aluminum is the only *p*-block metal that has no *d* electrons; it resembles the metals of Groups 1A and 2A in that it is active and exhibits a single oxidation state in all its compounds. The activity of aluminum, like that of beryllium and magne-

DEMONSTRATION S4.2 THE REACTIVITY OF AMALGAMATED ALUMINUM

A portion of an aluminum sheet is rubbed with moistened mercury(II) chloride. The mercury compound is reduced to mercury metal, which penetrates the protective aluminum oxide layer and forms an amalgam with aluminum.

The amalgamated aluminum reacts with oxygen and moisture in the air to form whiskers of $Al(OH)_3$. The untreated aluminum does not react.

sium, is often masked by an invisible oxide film that under ordinary conditions protects it from air and water. This film of Al_2O_3 can be thickened by **anodizing**, a process in which water containing a little sulfuric acid is electrolyzed with the aluminum as the anode. The electrode reaction is

$$2Al(s) + 3H_2O(l) \rightarrow Al_2O_3(s) + 6H^+(aq) + 6e^-$$

The anodized surface appears to be colored because of interference between light rays reflected from the oxide surface and those reflected from the underlying aluminum (Figure S4.4). Dyes absorbed from the electrolyte may add to the color.

Aluminum's defense against chemical attack can be breached by rubbing its surface with mercury(II) chloride to form an *amalgam* (a solution of metal in mercury). The amalgamated surface does not retain an oxide film, so it quickly reacts with oxygen and moisture in the air (Demonstration S4.2).

(a)

(b)

Figure S4.4

(a) Aluminum blocks. (b) Civic Bank in Los Angeles. The red sculpture and the metallic portions of the building facade are made of anodized aluminum.

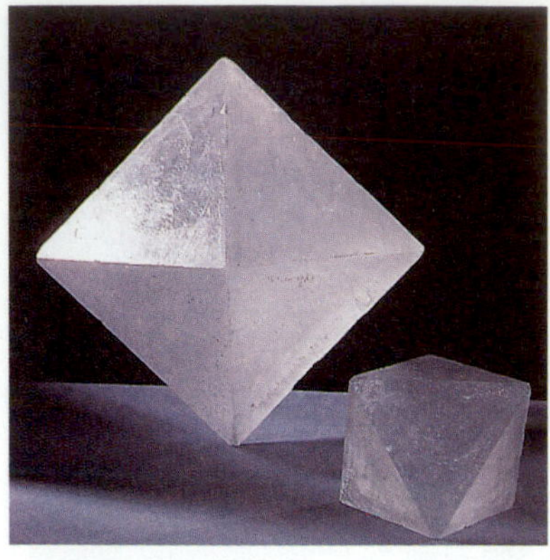

Figure S4.5

Octahedral crystals of potassium aluminum sulfate ($KAl(SO_4)_2 \cdot 12H_2O$).

The amphoteric oxide film can also be dissolved by strong bases and acids, which then oxidize the aluminum, evolving hydrogen gas. The reaction with strong base produces aluminate ion:

$$Al_2O_3(s) + 2OH^-(aq, conc) + 3H_2O(l) \rightarrow 2Al(OH)_4^-(aq)$$

$$2Al(s) + 2OH^-(aq, conc) + 6H_2O(l) \rightarrow 2Al(OH)_4^-(aq) + 3H_2(g)$$

Some commercial drain cleaners contain sodium hydroxide crystals and bits of aluminum. When this mixture is added to a clogged drain, some of the sodium hydroxide attacks grease and protein, and some reacts with the aluminum to release hydrogen gas. The escaping bubbles of hydrogen help to relieve the obstruction in the drain.

The reaction of the oxide film and aluminum with acid produces aluminum ion:

$$Al_2O_3(s) + 6H^+(aq) \rightarrow 2Al^{3+}(aq) + 3H_2O(l)$$

$$2Al(s) + 6H^+(aq) \rightarrow 2Al^{3+}(aq) + 3H_2(g)$$

Recall (Chapter 16) that the aluminum ion in solution is hydrated; its formula is actually $Al(H_2O)_6^{3+}$. It is acidic and will protonate any basic anion such as S^{2-} or CO_3^{2-}. In the following reaction, the hydrated aluminum ions donate protons to the carbonate ion, and the resulting carbonic acid decomposes into water and CO_2 gas.

$$2Al(H_2O)_6^{3+}(aq) + 3CO_3^{2-}(aq) \rightarrow 2Al(OH)_3(H_2O)_3(s) + 3H_2O(l) + 3CO_2(g)$$

PRACTICE EXERCISE S4.3

Predict the products that will form when solutions of $Al(NO_3)_3$ and Na_2S are mixed.

An *astringent* causes tissues such as sweat glands and capillaries to contract.

A number of uses for aluminum salts are based on the acidic character of $Al^{3+}(aq)$ and on its astringent effect on tissues. Sodium aluminum sulfate ($NaAl(SO_4)_2$) is the slow-acting acid ingredient in "double action" baking powder. (The action of baking powder is explained on p. 347). Aluminum chlorhydrate ($Al_2(OH)_5Cl \cdot 2H_2O$) and aluminum sulfate ($Al_2(SO_4)_3 \cdot 18H_2O$) are astringents used in antiperspirants and styptic pencils. Potassium aluminum sulfate ($KAl(SO_4)_2 \cdot 12H_2O$), often called *alum*, is noted for its distinctive octahedral crystals (Figure S4.5). It is also an inexpensive source of $Al^{3+}(aq)$ for the treatment of water supplies. Dissolved in water and neutralized with lime, alum forms a gelatinous aluminum hydroxide precipitate that absorbs impurities.

Aluminum's small atom gives it a greater capacity for covalence than is found in most other active metals. Anhydrous aluminum halides become increasingly covalent as the halide ion becomes larger and more polarizable. Solid aluminum fluoride is an ionic solid in which Al^{3+} ions are octahedrally surrounded by F^- ions; it sublimes at 1291°C and 1 atm without melting. Solid aluminum chloride is also ionic, but its much lower sublimation temperature (183°C) shows that it has more covalent character than the fluoride. In the gaseous and liquid states, aluminum chloride, bromide, and iodide consist of molecules with the general formula Al_2X_6 (Figure S4.6). These molecules also exist in solid aluminum bromide and iodide. At high temperatures, the covalent halides dissociate into trigonal planar AlX_3 molecules:

See page 990 for a comparison of the bonding in sodium bromide and aluminum bromide.

$$Al_2Cl_6(g) \xrightarrow{750°C} 2AlCl_3(g)$$

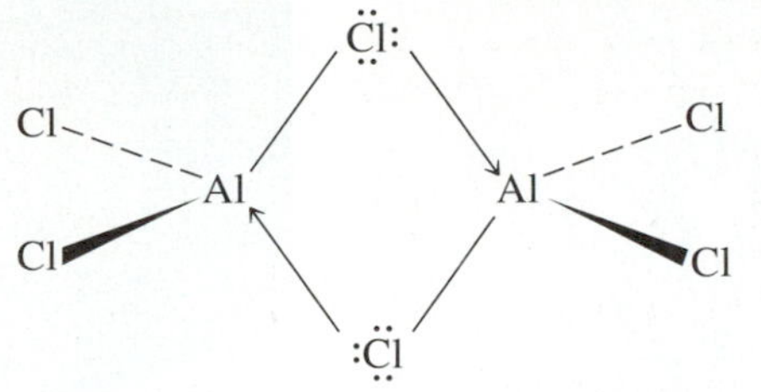

Figure S4.6

The structure of Al_2Cl_6. The joining of two $AlCl_3$ units to form an Al_2Cl_6 molecule allows each aluminum atom to achieve a tetrahedral configuration and a full octet of electrons.

Group 3A: Gallium, Indium, and Thallium

The properties of the Group 3A metals are compared in Table S4.3. Gallium, indium, and thallium are much less abundant than aluminum and less widely used. Gallium will melt on a hot day and has the distinction of being liquid over a wider temperature range than any other known substance. All three metals are moderately active, but as shown by their $E°$ values, they are much less active than aluminum.

The tendency toward covalence decreases with increasing atomic size. This trend is reflected in the behavior of the Group 3A oxides: Al_2O_3 and Ga_2O_3 are amphoteric, but In_2O_3 and Tl_2O_3 are basic. Gallium forms a molecular chloride (Ga_2Cl_6), but the anhydrous chlorides of indium and thallium are ionic.

As usual, the lower oxidation state becomes more important toward the bottom of the group. Compounds of aluminum(I), gallium(I), and indium(I) are unstable. Thallium, on the other hand, forms numerous stable compounds containing the thallium(I) ion. Tl^+ has about the same radius as K^+, and its hydroxide is soluble like that of K^+. Thallium compounds are extremely toxic; Tl_2SO_4 has been used in rat poisons and in ant and roach baits.

The Metals of Groups 4A and 5A

Tin and lead are the only metals in Group 4A; bismuth is the only metal in Group 5A. Even though tin and lead are not abundant, they are found in concentrated and workable ores and have been known since ancient times. The principal tin ore is *cassiterite* (SnO_2; see Figure 3.6, p. 105). The principal lead ore is *galena* (PbS; see Figure S4.7), which can be converted to PbO by roasting. Bismuth, a less-known metal, also occurs in sulfide ores. All three metals are obtained by reducing their oxides with carbon. Bismuth is also obtained from the anode sludge produced during the electrorefining of copper.

TABLE S4.3 Properties of the Group 3A Metals (s^2p^1)

Element	Melting Point (°C)	Boiling Point (°C)	Density (g/cm³)	$E°$ for Reduction of Listed Ion (V)	Important Oxidation States	Nature of Oxide
Al	660.4	2467	2.70	-1.68 (Al^{3+})	+3	Amphoteric (Al_2O_3)
Ga	29.8	2403	5.9	-0.53 (Ga^{3+})	+3	Amphoteric (Ga_2O_3)
In	156.6	2080	7.3	-0.34 (In^{3+})	+3	Basic (In_2O_3)
Tl	303.5	1457	11.8	-0.34 (Tl^+)	+1, +3	Basic (Tl_2O, Tl_2O_3)

(a)

(b)

(c)

Figure S4.7

Some lead ores. (a) Galena (PbS) crystals in a rock. (b) Cerussite ($PbCO_3$). (c) Crocoite ($PbCrO_4$).

Physical Properties. Some properties of tin, lead, and bismuth are summarized in Table S4.4 (see also Figure S4.8). Tin is hard, shiny, and springy—it is the best metal for organ pipes. It is also nontoxic and safe for use in food containers (tin cans). A trace of tin is essential in the diet. Lead is bluish-white, shiny, soft, malleable, and very toxic. It is a poor conductor of electricity and very resistant to corrosion. Lead is used in storage batteries, cable sheathing, and ammunition. Bismuth has a reddish cast, is brittle, and is the poorest conductor of all metals. All three metals have low melting points, and their mixtures melt at even lower temperatures. Alloys of tin and lead are used for type metal and solder, and alloys containing bismuth are used in fuses and automatic sprinklers (see Table S4.1). (Tin cans soldered with lead-based alloys may become unsafe for food after air is admitted; see Exercise 20.108.)

The familiar metallic form of tin is called *white tin*. Tin also forms a nonmetallic allotrope called *gray tin* that is stable below 13.2°C and has a covalent diamond lattice like that of carbon, silicon, and germanium. The covalent lattice is less dense than the metallic lattice, so tin cooled to a low temperature will expand and crumble as the white form slowly converts to the gray form. This change is called "tin disease"; it has been known to affect organ pipes and other tin objects when they are subjected to prolonged cold.

Reactions of the Metals and Their Oxides. Tin, lead, and bismuth are the least active of the representative metals. Tin forms both Sn(II) and Sn(IV) compounds; all four of its outermost electrons are readily involved in bond formation. When tin combines directly with nonmetals, it usually forms Sn(IV) compounds such as SnO_2, $SnCl_4$, and SnS_2.

In base, SnO_2 dissolves to give *stannate* ions ($Sn(OH)_6^{2-}$):

$$SnO_2(s) + 2OH^-(aq) + 2H_2O(l) \rightarrow Sn(OH)_6^{2-}(aq)$$

TABLE S4.4 Properties of Tin, Lead, and Bismuth

Element	Electron Configuration	Melting Point (°C)	Boiling Point (°C)	Density (g/cm³)	$E°$ for Reduction of Listed Ion (V)	Important Oxidation States	Nature of Oxide
Sn	s^2p^2	231.9	2270	7.28 (white) 5.75 (gray)	−0.136 (Sn^{2+})	+2, +4	Amphoteric (SnO, SnO_2)
Pb	s^2p^2	327.4	1620	11.29	−0.127 (Pb^{2+})	+2, +4	Amphoteric (PbO)
Bi	s^2p^3	271	1560	9.80	+0.32 (BiO^+)	+3	Basic (Bi_2O_3)

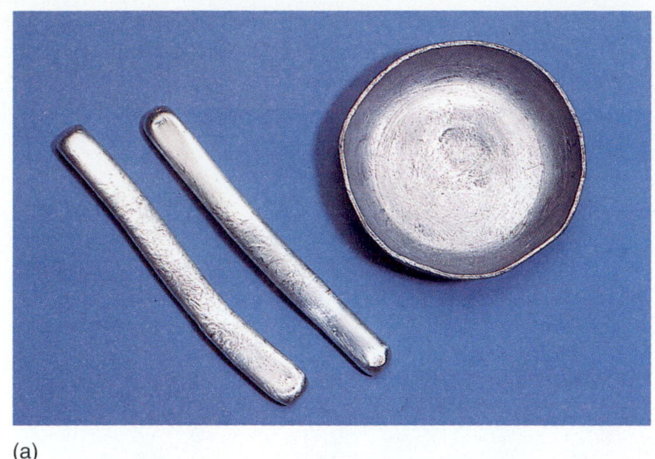
(a)

(b)

Figure S4.8
The metals of Groups 4A and 5A.
(a) Tin rods and a lead dish.
(b) Bismuth.

In acid, SnO_2 forms anionic complexes such as $SnCl_6^{2-}$:

$$SnO_2(s) + 6HCl(aq) \rightarrow 2H^+(aq) + SnCl_6^{2-}(aq) + 2H_2O(l)$$

The Sn^{4+} ion probably does not exist as an independent species.

The lower oxide of tin, SnO, is amphoteric, dissolving in acids to give Sn^{2+}

$$SnO(s) + 2H^+(aq) \rightarrow Sn^{2+}(aq) + H_2O(l)$$

and in bases to give *stannite* ion ($Sn(OH)_3^-$)

$$SnO(s) + OH^-(aq) + H_2O(l) \rightarrow Sn(OH)_3^-(aq)$$

An important Sn(II) compound, tin(II) fluoride, also called *stannous fluoride*, is used in fluoride toothpastes (see *Chemical Insight: Dental Health, pH, and Fluoridation*, p. 834). Compounds containing tin in the +2 oxidation state are good reducing agents, especially in alkaline solution.

PRACTICE EXERCISE S4.4

Refer to Appendix B.6 and state whether 1 *M* aqueous $Sn(NO_3)_2$ will reduce (a) Hg^{2+} to Hg and (b) AgBr to Ag.

Lead, which is below tin in Group 4A, differs from tin in that it favors its lower oxidation state—only its two outermost *p* electrons are readily involved in bond formation. For example, lead combines with oxygen to form PbO, with sulfur to form PbS, and with chlorine and fluorine to form $PbCl_2$ and PbF_2. PbO is a yellow pigment that has been used in pottery glazes. (Such glazes should not be used to decorate pottery intended for food storage or handling. See *Chemical Insight: Lead, Mercury, and Cadmium Poisoning*, p. 1048.) PbO and $Pb(OH)_2$ are amphoteric; they dissolve in acids to give lead(II) ion and in base to give *plumbite* ion ($Pb(OH)_3^-$). The equations are analogous to those given above for SnO.

The +2 state of lead is more stable than the +4 state, so Pb(IV) compounds are good oxidizing agents and relatively rare. Lead(IV) oxide can be prepared by oxidizing plumbite ion with hypochlorite ion in basic solution:

$$Pb(OH)_3^-(aq) + OCl^-(aq) \rightarrow PbO_2(s) + OH^-(aq) + Cl^-(aq) + H_2O(l)$$

Figure S4.9

The oxides of lead. Lower right: litharge (PbO); lower left: red lead (Pb_3O_4); top: lead dioxide (PbO_2).

LEARNING HINT

The sixth-period metals, Tl, Pb, and Bi, all tend to avoid using their two 6*s* electrons in compound formation. This phenomenon, called the *inert pair effect*, occurs because the energy required to involve the 6*s* electrons in bonding is more than the energy that would be released during the formation of two additional bonds.

PbO_2 is analogous to SnO_2 in that it reacts with strong base to form plumbate ion ($Pb(OH)_6^{2-}$). The role of PbO_2 in the lead storage battery was described in Section 20.5. Pb_3O_4 is a mixed oxide of lead that is used as a red pigment ("red lead") in rust-inhibiting paints (Figure S4.9).

Bismuth, a Group 5A metal, favors the +3 oxidation state; that is, only the three outermost *p* electrons are readily involved in bond formation. Bismuth reacts with oxygen, sulfur, and halogens (X_2) to form Bi_2O_3, Bi_2S_3, and BiX_3, respectively. Bi_2O_3 is a basic oxide that reacts with acid to give insoluble compounds such as BiOCl, which contain the *bismuthyl* ion BiO^+. Bismuthyl salts are used in various preparations to relieve gastric and intestinal irritation. Pepto-Bismol, for example, contains bismuthyl salicylate, $BiO(OOCC_6H_4OH)$. At a very low pH, the BiO^+ ion is converted to Bi^{3+}:

$$BiOCl(s) + 2HCl(aq) \rightarrow Bi^{3+}(aq) + 3Cl^-(aq) + H_2O(l)$$

S4.3 GROUP 2B: ZINC, CADMIUM, AND MERCURY

Although zinc, cadmium, and mercury are found in the *d* block, they behave in many ways like representative elements. The underlying *d* subshells in their $d^{10}s^2$ configurations are filled and are not used in bonding, so they resemble the Group 2A elements in their same periods (calcium, strontium, and barium). The atoms of the zinc group, however, have greater nuclear charges and smaller radii than the corresponding Group 2A atoms. They also have higher ionization energies, are less active, and exhibit a greater degree of covalence. Like other transition elements, zinc, cadmium, and mercury form numerous complexes, and their activity decreases with increasing size, mercury being the least reactive.

(a)

(b)

Figure S4.10

(a) A broken bar of zinc metal. (b) A 1.7-kg (3.8-lb) bronze block (density 7.8 g/cm^3) floats in liquid mercury (density 13.6 g/cm^3).

The Metals

Zinc is a fairly active, blue-white metal (Figure S4.10). It is brittle at room temperature but becomes malleable at 100–150°C. Cadmium is similar in appearance to zinc, but it is less active and much softer. Pure cadmium can be cut with a knife. Mercury is a silver-colored liquid; it is the only metal that is liquid at room temperature. Some properties of these metals are summarized in Table S4.5.

Zinc, cadmium, and mercury are obtained from sulfide ores (Figure S4.11). The preparation of zinc was described in Section 3.6 (see also Exercise 19.78, p. 935). *Sphalerite* (ZnS) is roasted to form zinc oxide, which is then reduced with coke. Because cadmium sulfide is usually mixed in with zinc sulfide, cadmium forms along with the zinc. The two metals are separated by fractional distillation. Mercury is prepared by roasting bright red *cinnabar* (HgS) in air:

$$HgS(s) + O_2(g) \rightarrow SO_2(g) + Hg(g)$$

The mercury vapor distills from the furnace and is condensed to the liquid.

PRACTICE EXERCISE S4.5

What temperature is required to separate zinc and cadmium by distillation? Which metal will volatize?

TABLE S4.5 Properties of the Group 2B Metals ($d^{10}s^2$)

Element	Melting Point (°C)	Boiling Point (°C)	Density (g/cm^3)	$E°$ for Reduction of Listed Ion (V)	Important Oxidation States	Nature of Oxide
Zn	419.5	907	7.14	−0.762 (Zn^{2+})	+2	Amphoteric (ZnO)
Cd	320.9	767	8.64	−0.402 (Cd^{2+})	+2	Basic (CdO)
Hg	−38.9	357	13.59	+0.852 (Hg^{2+})	+1, +2	Basic (HgO)

(a)

(b)

(c)

Figure S4.11

Some ores of Group 2B metals. (a) Sphalerite (ZnS) on a quartz matrix. (b) Smithsonite ($ZnCO_3$), named after the founder of the Smithsonian Institution in Washington, D.C. (c) Cinnabar (HgS).

Zinc is an abundant metal with many uses; it is used in alloys (Table S4.1), to galvanize iron (Section 20.6) and as the anode in flashlight batteries (Section 20.5). Trace amounts of zinc are necessary for growth and metabolism. Cadmium is similar to zinc in many respects. It is used in alloys, and like zinc, it is used as a protective coating for iron and steel. Cadmium is also used in the control rods of nuclear reactors.

CHEMICAL INSIGHT

LEAD, MERCURY, AND CADMIUM POISONING

The heavy elements barium, cadmium, mercury, thallium, and lead are highly toxic, and so are two of the lighter metals, beryllium and nickel. Lead, mercury, and cadmium are the most widespread of these metals, and they present the greatest health problems.

Lead interferes with metabolism and attacks the nervous system. The ancient Greeks knew that illness could result from drinking acidic fermented beverages (beer, wine, or cider) made in vessels coated with lead-based glazes or pigments such as chrome yellow ($PbCrO_4$). Some researchers have suggested that wine kept in lead-glazed pottery contributed to the deterioration of the Roman ruling class and hastened the fall of the Roman Empire. Skeletons from Herculaneum, which, like Pompeii, was buried by the eruption of Mount Vesuvius in 79 A.D., have been found to contain toxic levels of lead.

Incidents of lead poisoning have been reported in many industries where workers handle lead or lead products. Clinical symptoms of lead poisoning begin at a level of about 30 μg of lead per 100 mL of blood, but levels as low as 10 μg per 100 mL have been associated with hyperactivity and learning disorders in children. Such lead levels are not uncommon in urban areas, where they have been linked to the use of leaded gasoline. Over 200,000 tons of lead a year were once sent into the atmosphere by automobile exhaust. Fortunately, this pollution has been sharply reduced by the changeover to unleaded gas, and lead levels in blood have declined accordingly (Figure S4.12).

Children become victims of lead poisoning when they eat or chew flakes of lead-based paint or breathe its dust. Although the pigment *white lead* ($Pb(CO_3)_2(OH)_2$) has been banned for use in interior house paint for many years, many older buildings still contain peeling layers of lead-based paint. Colored package labels, food wrappers, and even "spitballs" are also potential sources of lead. The "manufacture" of just a few spitballs by chewing colored paper from newspapers, comics, or magazines might well result in the ingestion of more than 300 μg of lead.

Mercury is a nerve poison that causes muscular tremors, loss of coordination, and eventual insanity. Mercury metal is more volatile than most metals; it has a relatively high vapor pressure and the vapor is readily absorbed through the skin and the respiratory tract. Containers of mercury should be securely covered, and mercury spills should be promptly and carefully cleaned up. *Hatters' disease*, the affliction caricatured by the famous Mad Hatter of *Alice in Wonderland*, was once common among hat makers who used $HgCl_2$ to convert fur into felt. More recent cases of mercury poisoning have resulted from eating seed grain contaminated with mercury fungicides and from eating contaminated fish.

A potential problem exists in Latin American and Caribbean communities where mercury is often sold by small local stores called botanicas. Consumers believe that the silvery liquid has the power to drive away evil spirits, and they carry it around in packets, spread it on their floors at home, and place it in candles. The resulting mercury vapor levels

Mercury is the rarest and most distinctive of the three metals. Its unique combination of liquidity and inertness make it irreplaceable in devices such as thermometers, barometers, and mercury switches. Mercury does not wet glass as most other liquids do; its meniscus in a glass tube is convex rather than concave. (See *Chemical Insight: How to Make Liquids Wetter*, p. 559.) Mercury amalgamates (dissolves) nearly all metals except iron and platinum. Dental amalgam is a solution of mercury with silver, tin, and other metals (Table S4.1); it hardens into a stable alloy and expands slightly to fill cracks and make a tight fit.

Never let mercury touch gold or silver jewelry.

Reactions and Compounds

The atoms of zinc, cadmium, and mercury use two valence electrons in bonding, and the +2 oxidation state is characteristic of all three metals. Mercury also forms a +1 state in mercury(I) compounds such as Hg_2Cl_2, in which each atom devotes one of its valence electrons to a covalent Hg—Hg bond. Most mercury(I) salts are insoluble.

$Hg_2^{2+} = [Hg{:}Hg]^{2+}$

may be a source of danger, especially to small children who sleep and play on the floor.

Until recently, industrial waste containing mercury was discharged directly into open waters on the assumption that the mercury would sink to the bottom and do no harm. It is now known that microorganisms convert the discarded mercury into dimethylmercury ($Hg(CH_3)_2$), which is readily absorbed and retained by fatty tissue. Mercury compounds become more concentrated as they move upward in the aquatic food chain, reaching their peak concentrations in fish and shellfish.

Cadmium is considered by many to be the most dangerous heavy-metal pollutant because is it widespread and poorly controlled. Five milligrams of cadmium per kilogram of body weight is lethal to animals; lower levels have been implicated in high blood pressure, in diseases of the liver, heart, and kidney, and in some forms of cancer. The average kidney cadmium level in the general population is already between one-third and one-half of the estimated toxic level.

Cadmium is found in zinc ores, and airborne emissions from zinc smelters are a major source of cadmium contamination in air, water, soil, and all crops, including tobacco. The average normal diet now contains somewhat less than 1 μg of cadmium per day. Smoking one package of cigarettes will increase the body's cadmium content by as much as 1.5 μg. The effect is cumulative, so that the body's cadmium level at any time in life correlates closely with the total number of cigarettes smoked.

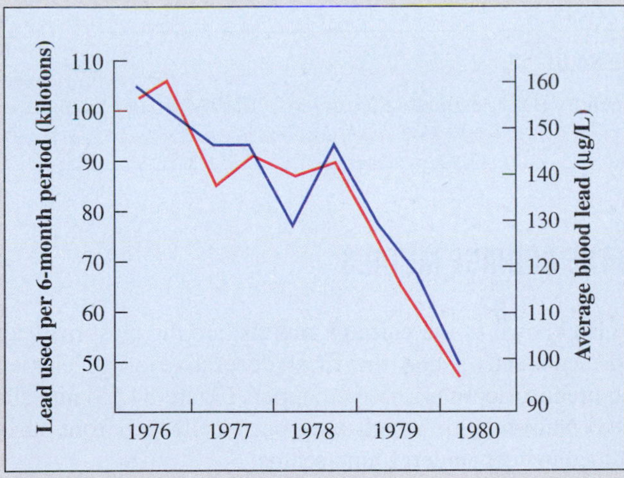

Figure S4.12

Lead level in blood (red line) versus lead used in gasoline (blue line). (This graph was compiled from EPA data.)

Under suitable conditions, all three metals form oxides, halides, and sulfides such as HgO, $CdCl_2$, and ZnS. Zinc and cadmium displace hydrogen from solutions containing 1 M H^+(aq); mercury does not. All three metals are attacked by oxidizing acids such as HNO_3 and H_2SO_4. Concentrated nitric acid oxidizes mercury to either Hg^{2+} or Hg_2^{2+}:

$$3Hg(l) + 8HNO_3(aq) \xrightarrow{\text{excess acid}} 3Hg(NO_3)_2(aq) + 2NO(g) + 4H_2O(l)$$

$$6Hg(l) + 8HNO_3(aq) \xrightarrow{\text{excess mercury}} 3Hg_2(NO_3)_2(aq) + 2NO(g) + 4H_2O(l)$$

Hot, concentrated sulfuric acid oxidizes mercury to $HgSO_4$. The reduction product is SO_2.

Zinc oxide and zinc hydroxide are amphoteric. Their reactions with acids and concentrated NaOH were discussed in Section 18.3. Cadmium and mercury oxides are basic; they are dissolved by acids but not by NaOH. Zinc and cadmium oxides dissolve in aqueous ammonia to form the ammine complexes $Zn(NH_3)_4^{2+}$ and $Cd(NH_3)_4^{2+}$:

$$ZnO(s) + 4NH_3(aq) + H_2O(l) \rightarrow Zn(NH_3)_4^{2+}(aq) + 2OH^-(aq)$$

HgO does not form an ammine complex.

The compounds of zinc are essentially ionic. ZnO is a white pigment. ZnS is used in television screens because it emits light when struck by a beam of electrons. $ZnSO_4$ and ZnO are astringents and antiseptics used in treating skin disorders such as acne and poison ivy. The active ingredient in calamine lotion, for example, is ZnO.

Cadmium compounds exhibit more covalence than those of zinc; $CdCl_2$, for example, ionizes only partially in aqueous solution. Most mercury compounds have a high degree of covalent character, and only the nitrates and perchlorates are both soluble and ionic; $HgCl_2$ and other Hg(II) compounds form very few ions, even in solution. Mercury(I) chloride (Hg_2Cl_2), called *calomel*, is a white solid used in the standard calomel electrode (Section 20.4).

Figure S4.13
Native copper.

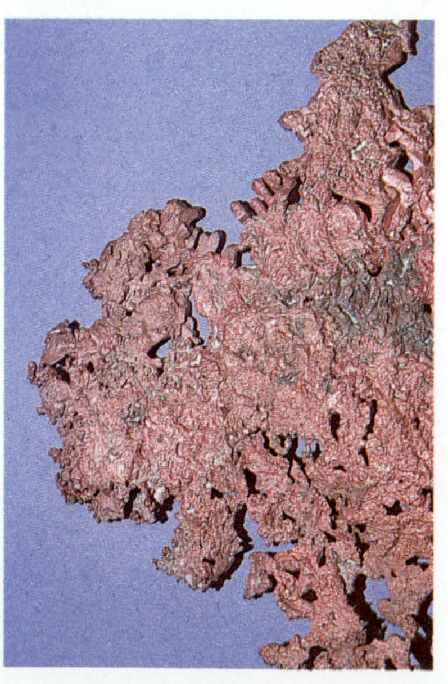

PRACTICE EXERCISE S4.6

Use the K_{sp} value in Appendix B.4 to estimate the molar solubility of Hg_2Cl_2 in water.

S4.4 GROUP 1B: THE COINAGE METALS

Copper, silver, and gold, also known as the **coinage metals**, are the most malleable, the most ductile, the most conductive, and the most decorative of all elements. Prized and collected since prehistoric times, native copper (Figure S4.13) and silver have virtually disappeared. Native gold, which was once collected from surface deposits, is now mostly mined in deep underground seams.

Most of the copper currently produced is obtained from low-grade sulfide ores that contain less than 1% copper. First the ore is concentrated to about 30% copper by flotation (Section S4.1). Then the impurities, especially FeS, are converted to a silicate slag by roasting and smelting the ore with SiO_2. The slag, which floats on

Figure S4.14

Molten copper being poured from a transport ladle into a blister mold.

top of molten copper sulfide, is poured off. The remaining copper sulfide, called *copper matte*, is reduced to copper metal by blowing air through it:

$$Cu_2S(l) + O_2(g) \rightarrow 2Cu(l) + SO_2(g)$$

The 98–99% pure product has a blistered appearance due to gas bubbles and is known as *blister copper* (Figure S4.14). Blister copper can be used for plumbing, decorative purposes, and other applications where purity is not important. For electrical use, however, it must be refined by electrolysis (Section S4.1).

The anode sludge formed during the electrorefining of copper contains significant amounts of both gold and silver. These metals are leached out with cyanide solution and reprecipitated with zinc, as described in Section S4.1. The resulting mixture of silver and gold is separated with concentrated sulfuric acid, which dissolves silver but not gold.

Silver is used in dental and other alloys (Table S4.1), printed circuits, electrical contacts, and batteries. Gold is used for coating the electrical contacts in electronic devices. Both silver and gold are used in coins and jewelry; gold has been used as an international currency standard.

Reactions and Compounds

Some properties of copper, silver, and gold are summarized in Table S4.6. The usual oxidation state of copper is +2. The best-known copper(II) compound, the soluble, blue hydrate $CuSO_4{\cdot}5H_2O$, is used in copper plating and as a fungicide and algicide. Insoluble copper(II) compounds include light blue $Cu(OH)_2$ and black CuS. Very low concentrations of $Cu^{2+}(aq)$ can be detected by adding ammonia to form the intensely colored $Cu(NH_3)_4{}^{2+}$ complex (see Figure 18.6a, p. 841). The Cu^+ ion is rarely found in solution because it tends to disproportionate to Cu^{2+} and Cu:

Copper is an essential trace nutrient. The human body seems to have no need of gold or silver.

$$2Cu^+(aq) \rightarrow Cu^{2+}(aq) + Cu(s)$$

The Cu(I) state is stable in sparingly soluble compounds such as black Cu_2S, white CuI, and in complexes such as colorless $Cu(CN)_2{}^-$.

TABLE S4.6 Properties of the Group 1B Metals ($d^{10}s^1$)

Element	Melting Point (°C)	Boiling Point (°C)	Density (g/cm³)	$E°$ for Reduction of Listed Ion (V)	Important Oxidation States	Nature of Oxide
Cu	1083	2595	8.96	$+0.339$ (Cu^{2+})	+1, +2	Basic (Cu_2O, CuO)
Ag	960.8	2210	10.5	$+0.799$ (Ag^+)	+1	Basic (Ag_2O)
Au	1063	2970	19.3	$+1.69$ (Au^+)	+1, +3	Unstable (Au_2O, Au_2O_3)

PRACTICE EXERCISE S4.7

Use data from Appendix B.6 to calculate $\Delta G°$ for the disproportionation of Cu^+ ions.

Very finely divided metals often appear black.

The usual oxidation state of silver is +1. Silver nitrate ($AgNO_3$), a soluble salt, is used in silver plating, in the preparation of silver mirrors, and as a disinfectant, astringent, and caustic. Silver is one of the many heavy transition metals that precipitates an oxide rather than a hydroxide:

$$\underset{}{2Ag^+(aq)} + 2OH^-(aq) \rightarrow \underset{\text{black}}{Ag_2O(s)} + H_2O(l)$$

Silver ion is identified in solution by the formation of a white AgCl precipitate that dissolves in aqueous ammonia to form colorless $Ag(NH_3)_2{}^+$ ions:

$$Ag^+(aq) + Cl^-(aq) \rightarrow AgCl(s)$$

$$AgCl(s) + 2NH_3(aq) \rightarrow Ag(NH_3)_2{}^+(aq) + Cl^-(aq)$$

Silver salts tend to darken in the light due to the formation of finely divided elemental silver (see Demonstration S4.3):

$$\underset{\text{white}}{2AgCl(s)} \xrightarrow{h\nu} \underset{\text{black}}{2Ag(s)} + Cl_2(g)$$

The effect of light on silver salts is the basis of photography. Photographic film contains grains of silver bromide that become "sensitized," or more susceptible to decomposition, when they absorb small quantities of light energy. The *developing solution* contains a mild reducing agent that rapidly reduces the sensitized grains to black grains of silver metal. The resulting image is blackest where the light exposure was strongest. Unreduced silver bromide is dissolved out of the film with a *fixing solution* that contains sodium thiosulfate ("hypo"); the thiosulfate ion complexes the silver ion:

$$AgBr(s) + 2S_2O_3{}^{2-}(aq) \rightarrow Ag(S_2O_3)_2{}^{3-}(aq) + Br^-(aq)$$

Color film also contains silver halides that are sensitized by light. A complex developing process produces a colored image from layers of dyes in the film.

Copper, silver, and gold are attacked by chlorine, forming $CuCl_2$, AgCl, and $AuCl_3$, respectively. Copper and silver, but not gold, are tarnished by the natural sul-

DEMONSTRATION S4.3 LIGHT-INDUCED DECOMPOSITION OF A SILVER HALIDE—THE PHOTOGRAPHIC EFFECT

A key is placed on freshly prepared silver bromide crystals.

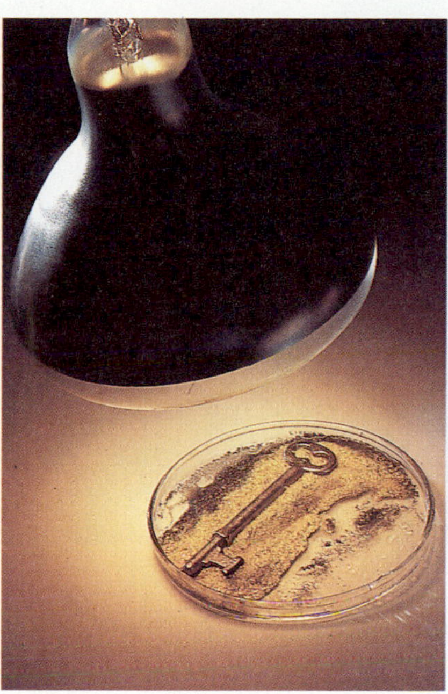

A bright light shines for several minutes.

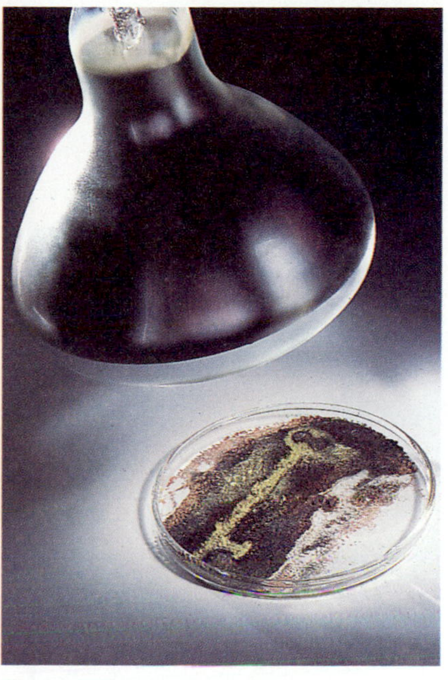

The exposed silver bromide has partially decomposed to dark elemental silver. The key has been removed, but its image remains on the crystals.

fides in eggs and other food, as well as by sulfur compounds in the air:

$$4Ag(s) + 2H_2S(g) + O_2(g) \rightarrow \underset{\text{silver tarnish}}{2Ag_2S(s)} + 2H_2O(g)$$

The green coating or *patina* that develops on copper, brass, and bronze surfaces exposed to moist air is usually $CuCO_3{\cdot}Cu(OH)_2$ (Figure S4.15). High SO_2 concentrations (which sometimes exist in New York Harbor near the copper-clad Statue of Liberty) may convert some of the coating to $CuSO_4{\cdot}2Cu(OH)_2$, which tends to flake off and allow further erosion.

The coinage metals have positive reduction potentials and do not displace hydrogen from acids. Copper, the most active of the three metals, will dissolve gradually in HCl if the solution is well aerated; in this case, the oxidizing agent is oxygen and the low pH enhances the reaction:

$$2Cu(s) + 4H^+(aq) + O_2(g) \rightarrow 2Cu^{2+}(aq) + 2H_2O(l)$$

Copper and silver are oxidized to Cu^{2+} and Ag^+ by either HNO_3 or hot, concentrated H_2SO_4:

$$2Ag(s) + 4H^+(aq) + SO_4^{2-}(aq) \rightarrow 2Ag^+(aq) + SO_2(g) + 2H_2O(l)$$

Figure S4.15

These copper ornaments are covered with a patina of mixed carbonate and hydroxide.

PRACTICE EXERCISE S4.8

Copper metal dissolves in concentrated nitric acid with the evolution of NO_2 gas (see Figure S2.10, p. 512). Write a net ionic equation for the reaction.

Gold does not dissolve in either HNO_3 or H_2SO_4, but it is attacked by certain mixtures that contain complexing agents. One is the air–cyanide mixture used in gold metallurgy (p. 1039), and another is ***aqua regia***, a mixture of concentrated HCl and HNO_3 in which the oxidizing power of HNO_3 is assisted by the formation of the $AuCl_4^-$ complex ion:

$$Au(s) + 4H^+(aq) + NO_3^-(aq) + 4Cl^-(aq) \rightarrow AuCl_4^-(aq) + NO(g) + 2H_2O(l)$$

Chlorauric acid ($HAuCl_4$) is used in gold plating, photography, and the manufacture of ruby glass.

S4.5 THE IRON TRIAD

In the periodic table, iron, cobalt, and nickel make up a horizontal trio of elements called the **iron triad**. They are similar in a number of ways. Iron is ferromagnetic, as are cobalt and nickel to a lesser degree (See *Chemical Insight: Ferromagnetism*, p. 319). Iron is the fourth most abundant element in the earth's crust. Cobalt and nickel are less abundant, but nickel has been found in meteorites and is believed to be a major component of the earth's core. The body needs a supply of iron to make hemoglobin and other essential biomolecules, and it also needs small amounts of cobalt and nickel. Each molecule of vitamin B_{12} has a cobalt atom at its center, and nickel is a constituent of several enzymes.

CHEMICAL INSIGHT

GOLD

Gold is the only bright yellow metal, and one of the few metals that does not rust, corrode, or tarnish (Figure S4.16). Since early times gold has been associated with the sun; the alchemists' symbol for gold was a circular sunburst. The first known coinage was established in 700 B.C. in Lydia using coins of *electrum*, a native alloy of gold that contains about 25% silver. The purity of gold is measured in carats; 24 carats is 100% pure, 12 carats is 50% pure, and so forth.

Early metal workers fashioned gold into filigree and gold leaf and decorated cloth with interwoven gold threads. No other metal is so malleable and ductile; gold leaf can be hammered so thin that it becomes translucent and appears blue or green by transmitted light. Gold is an excellent reflector of infrared radiation, and gold-plated heat shields have been used to protect astronauts walking in space. The electrical conductivity of gold and its chemical inertness make it ideal for coating electrical contacts, especially in computers.

A colloidal suspension of gold can be obtained by reducing a solution containing $AuCl_4^-$ with a mild reducing agent such as tin. Gold colloids exhibit a wide range of color (see Figure 13.19, p. 617), from blue (larger particles) through ruby red to orange (smaller particles). Medieval workers used colloidal gold to make the famous ruby glass seen in cathedral windows. A colloidal dispersion of radioactive gold, gold-198, $t_{1/2} = 2.7$ days, is given by injection in the treatment of certain cancers; the gold atoms are chemically unreactive and do not harm the body. Nonradioactive gold compounds such as sodium dithiosulfatoaurate(III) ($NaAu(S_2O_3)_2$) are used in the treatment of rheumatoid arthritis.

It has been estimated that all the gold ever collected by man could be fitted into a cube 60 feet on edge. Although gold is widely distributed throughout the rocks of the earth, there are few concentrated deposits. Seawater contains approximately 4×10^{-6} mg of gold per liter, but no one has yet devised a profitable method for mining the ocean's gold.

Figure S4.16 Gold bars.

At least one species of algae adsorbs gold ions such as $AuCl_4^-$ and reduces them to minute crystals of elemental gold. Scanning electron micrographs show that finely divided gold particles from Alaskan sites consist of filaments arranged in a lacelike structure. It has been suggested that these filaments were produced by budding bacteria (bacteria that grow by sending out shoots) whose metabolism depended on accumulating and removing soluble forms of gold from water and soil.

Sources and Metallurgy

The principal iron ores are *hematite* (Fe_2O_3), *magnetite* (Fe_3O_4), and *siderite* ($FeCO_3$). (See Figure S4.17 and Figure 3.6a, p. 105). *Iron pyrite* (FeS_2) is plentiful, but its high sulfur content makes it impractical to use as an ore (Figure S4.18). *Taconite ores*, which contain iron oxides mixed with silica, were once considered too difficult to work with. Now that the better ores have been depleted, the taconites are also being used.

Figure S4.17 Hematite kidney ore (Fe_2O_3).

Iron ores are reduced in a blast furnace like the one shown in Figure S4.2. The molten iron is drawn off at the bottom of the furnace and allowed to solidify in the

Figure S4.18

Iron pyrite (FeS_2). The deceptively golden look of this mineral is the source of its common name, "fool's gold."

form of ingots called "pigs." *Pig iron* contains about 4% carbon, 2% silicon, and lesser amounts of manganese, phosphorus, and sulfur. *Cast iron* is obtained by pouring molten pig iron into molds and allowing it to solidify. Because of its impurities, cast iron tends to be very hard and brittle, and it cannot be used where it will be subjected to high stress. Iron pipes, steam and hot water radiators, and automobile engine blocks are usually made out of cast iron.

Steels are ferrous alloys containing controlled amounts of carbon (0.1–1.5%). Different metals are usually added to impart some desired mix of hardness, flexibility, corrosion resistance, tensile strength, and other special properties (Table S4.7). Most of the steel made in the United States is produced by the **basic oxygen process** in which molten pig iron and scrap iron are poured into large barrel-like furnaces (Figure S4.19). Streams of powdered limestone and oxygen gas are blown into the molten mixture. The impurities—excess carbon, manganese, phosphorous, and silicon—are oxidized. Some oxides escape as gases, and some react with CaO (a decomposition product of limestone) to form a slag. Sample reactions are

$$P_4O_{10}(l) + 6CaO(s) \rightarrow 2Ca_3(PO_4)_2(l)$$

$$MnO(s) + SiO_2(s) \rightarrow MnSiO_3(l)$$

The contents of the hot, molten mixture are monitored spectroscopically, and other metals are added at intervals to produce whatever type of steel is desired. The basic oxygen process is very efficient; 100 tons of steel can be produced in less than 30 minutes.

It has been said that the first makers of steel planted the seeds of industrial society. The earliest known steel artifacts were found in burial caves in Jordan. They had been made of steel containing about 0.5% carbon and have been dated between 1200 and 1500 B.C.

The physical properties of a steel can be modified by heat treatment. Some of the carbon in steel is present as iron carbide (Fe_3C), a substance also known as *cementite*:

$$3Fe(s) + C(graphite) \rightleftharpoons Fe_3C(s)$$

The forward reaction is endothermic, so raising the temperature favors the formation of cementite. When the steel is heated to very high temperatures (red heat) and instantly cooled (a process called **quenching**), most of the carbon remains in the form of cementite, and the steel is hard and brittle. On the other hand, moderate heating (250–300°C) followed by slow cooling (a process called **tempering**) gives the reaction time to move in the reverse direction. The result is a stronger and more flexible steel containing minute crystals of graphite. The graphite crystals impart a

TABLE S4.7 Some Alloy Steels

Alloy	Composition[a]	Characteristics
Alnico I	20% Ni, 12% Al, 5% Co	Magnetic
Duriron	14.5% Si, 0.35% Mn	Wear resistance, acid resistance
Invar	36% Ni	Low expansion coefficient
Nickel-chrome	3% Ni, 1.5% Cr	Very hard but flexible, high tensile strength
18-8 Stainless	18% Cr, 8% Ni	Corrosion resistance
Tungsten	5% W	Heat resistance

[a]All steels contain iron and up to 1.5% carbon.

gray color to the tempered steel. It is also possible to prepare steels that are hard on the surface but strong and flexible in the interior.

Cobalt and nickel were sources of frustration to early metal workers, who gave them the unflattering names they still carry. Cobalt ores (a *kobold* is a deceitful impish spirit) looked like iron ores but yielded no iron, while the reddish ore *kupfer-nickel* ("devil's copper") is a nickel arsenide (NiAs) that failed to yield the copper it seemed to promise. Both metals are now obtained by roasting their sulfide ores and reducing the resulting oxides. Cobalt and nickel are used in steels and other alloys and as catalysts. A wear-resistant copper–nickel alloy forms the outer layers of the U.S. nickel coin. Cobalt and nickel are ferromagnetic and are used in magnetic alloys (Table S4.7).

Figure S4.19
Molten iron being poured into a basic oxygen furnace. A typical furnace is loaded with 200 tons of pig iron, 100 tons of scrap iron, and 20 tons of limestone.

Reactions and Compounds

Some of the properties of iron, cobalt, and nickel are summarized in Table S4.8. The finely divided metals burn in oxygen; iron forms Fe_2O_3 and/or Fe_3O_4, depending on the temperature, cobalt forms Co_3O_4, and nickel forms NiO. Iron corrodes in damp air to give rust, a hydrated form of Fe_2O_3 (Section 20.6). The red pigment Fe_2O_3 is an important industrial compound used in paints, rubber, and ceramics. It is also an ingredient of the rouge used by jewelers for polishing glass and gems, and it is used in magnetic tapes.

When heated with carbon monoxide, nickel forms *nickel carbonyl* ($Ni(CO)_4$), an extremely toxic gas:

$$Ni(s) + 4CO(g) \rightleftharpoons Ni(CO)_4(g)$$

Iron and cobalt form a number of different carbonyls. Metal carbonyls are unusual in that the metal has an oxidation state of zero.

The most common oxidation states of iron and cobalt are +2 and +3. The two states occur together in the oxides Fe_3O_4 ($= FeO{\cdot}Fe_2O_3$) and Co_3O_4 ($= CoO{\cdot}Co_2O_3$).

Iron, cobalt, and nickel are moderately active metals that displace hydrogen from acids:

$$Fe(s) + 2H^+(aq) \rightarrow Fe^{2+}(aq) + H_2(g)$$

The aqueous Fe^{2+} ion is pale green. $FeSO_4$ is used in electroplating and as a nutritional supplement. Its hydroxide is the sparingly soluble green $Fe(OH)_2$, which is basic. Aqueous Fe^{2+} is easily oxidized to Fe^{3+} by oxygen from the air:

$$4Fe^{2+}(aq) + O_2(g) + 4H^+(aq) \rightarrow 4Fe^{3+}(aq) + 2H_2O(l)$$

TABLE S4.8 Properties of Iron, Cobalt, and Nickel

Element	Electron Configuration	Melting Point (°C)	Boiling Point (°C)	Density (g/cm³)	$E°$ for Reduction of Listed Ion (V)	Important Oxidation States	Nature of Oxide
Fe	d^6s^2	1535	2750	7.87	−0.409 (Fe^{2+})	+2, +3	Basic (FeO, Fe_2O_3, Fe_3O_4)
Co	d^7s^2	1495	2870	8.9	−0.282 (Co^{2+})	+2, +3	Basic (CoO, Co_2O_3, Co_3O_4)
Ni	d^8s^2	1453	2732	8.90	−0.236 (Ni^{2+})	+2	Basic (NiO)

Figure S4.20

Iron nails placed in a $FeSO_4$ solution counteract the oxidation of Fe^{2+} to Fe^{3+} by dissolved O_2.

The oxidation can be counteracted by keeping iron nails in the iron(II) solution (Figure S4.20):

$$Fe(s) + 2Fe^{3+}(aq) \rightarrow 3Fe^{2+}(aq)$$

The aqueous Fe^{3+} ion, actually $Fe(H_2O)_6{}^{3+}$, is faintly violet, almost colorless; however, aqueous Fe^{3+} salts often have a light amber color due to complexes such as $Fe(H_2O)_5(OH)^{2+}$ and $Fe(H_2O)_5Cl^{2+}$. The sparingly soluble red-brown $Fe(OH)_3$ is a basic hydroxide. Sulfide ion reduces Fe^{3+} to an insoluble mixture of sulfur and black FeS:

$$2Fe^{3+}(aq) + 3S^{2-}(aq) \rightarrow \underset{\text{black}}{2FeS(s)} + S(s)$$

Fe^{3+} can be identified in solution by adding potassium thiocyanate (KSCN) to form deep red $Fe(SCN)^{2+}(aq)$, or by adding potassium ferrocyanide ($K_4Fe(CN)_6$) to form a dark suspension of Prussian blue ($Fe_4[Fe(CN)_6]_3$) (see Figure 18.6b, p. 841). Prussian blue is a pigment used in inks.

The stable oxidation state of cobalt is +2 when it is not complexed by any ligand stronger than water. Sparingly soluble Co(II) compounds include pink $Co(OH)_2$ and black CoS. The blue tetrahedral complex $Co(H_2O)_4{}^{2+}$ can exist in equilibrium with pink $Co(H_2O)_6{}^{2+}$; the prevalent form depends on the relative concentrations of H_2O and Cl^-. The reversible reaction

$$\underset{\text{blue}}{[Co(H_2O)_4]Cl_2(s)} + 2H_2O(g) \underset{\text{low humidity}}{\overset{\text{high humidity}}{\rightleftharpoons}} \underset{\text{pink}}{[Co(H_2O)_6]Cl_2(s)}$$

has been used somewhat unreliably in weather-predicting gadgets (see Demonstration S4.4).

When cobalt is complexed by ligands stronger than H_2O, its +3 state becomes more stable than the +2 state. If ammonia is added to a Co^{2+} salt, the presence of air alone is sufficient to oxidize it to the amber complex $Co(NH_3)_6{}^{3+}$ (Figure S4.21):

$$4Co^{2+}(aq) + 24NH_3(aq) + O_2(g) + 4H^+(aq) \rightarrow \underset{\text{amber}}{4Co(NH_3)_6{}^{3+}(aq)} + 2H_2O(l)$$

A reaction that is often used to identify Co^{2+} in solution makes use of KNO_2 as both an oxidizing agent and a complexing agent to form the yellow precipitate $K_3[Co(NO_2)_6]$:

$$Co^{2+}(aq) + 7KNO_2(aq) + 2H^+(aq) \rightarrow \underset{\text{yellow}}{K_3[Co(NO_2)_6](s)} + 4K^+(aq) + NO(g) + H_2O(l)$$

The most common oxidation state of nickel is +2. The aqueous Ni^{2+} ion is green; it reacts with base to produce green $Ni(OH)_2$ and with sulfide ion to form black NiS. The $Ni(OH)_2$ does not dissolve in excess NaOH, but it does dissolve in ammonia to produce a blue complex, $Ni(NH_3)_6{}^{2+}$:

$$\underset{\text{green}}{Ni(OH)_2(s)} + 6NH_3(aq) \rightarrow \underset{\text{blue}}{Ni(NH_3)_6{}^{2+}(aq)} + 2OH^-(aq)$$

Ni^{2+} is identified by the insoluble red complex it forms with dimethylglyoxime (see Figure 21.37, p. 1024).

DEMONSTRATION S4.4 THE $Co(H_2O)_4^{2+}$–$Co(H_2O)_6^{2+}$ EQUILIBRIUM

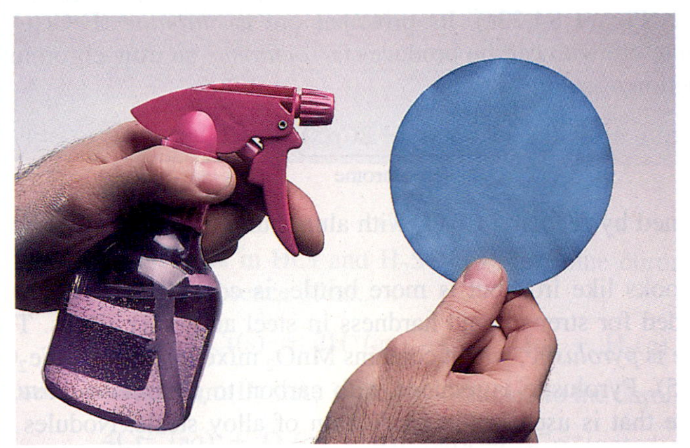

A piece of filter paper soaked in a pink solution of $CoCl_2$ turns blue when it is dried.

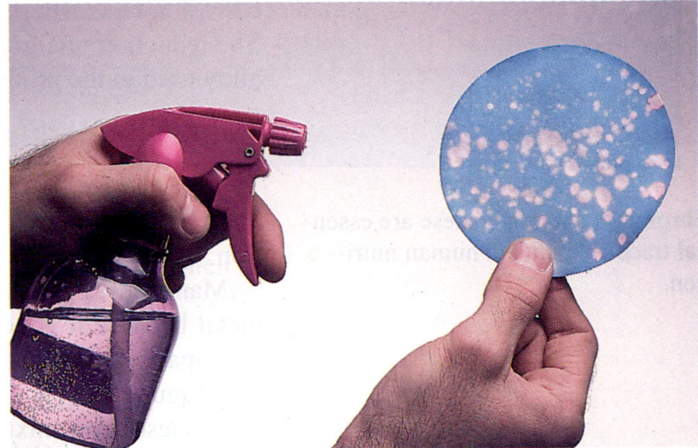

A spray of water restores the pink color.

The nickel +4 state is rare and strongly oxidizing. The use of NiO_2 in the nickel–cadmium cell is described in Section 20.5.

PRACTICE EXERCISE S4.9

How many 3*d* electrons are present on the central atom in (a) $CoCl_4^{2-}$, (b) $Co(NO_2)_6^{3-}$, and (c) $Ni(NH_3)_6^{2+}$?

Figure S4.21

The formation of cobalt(III) complexes from a cobalt(II) salt. Left beaker: the starting solution, aqueous cobalt(II) chloride ($CoCl_2$), contains the pink $Co(H_2O)_6^{2+}$ ion. Middle beaker: addition of aqueous ammonia produces an amber solution that contains the $Co(NH_3)_6^{3+}$ ion. Right beaker: addition of an acidified potassium nitrite (KNO_2) solution produces a yellow precipitate of $K_3Co(NO_2)_6$.

S4.62 Explain how the properties of a steel are affected by (a) quenching and (b) tempering. What is *cementite*?

S4.63 Describe an analytical test for identifying each of the following ions: (a) $Fe^{3+}(aq)$, (b) $Co^{2+}(aq)$, and (c) $Ni^{2+}(aq)$.

S4.64 Name and give a commercial use of at least one important compound of (a) iron, (b) cobalt, and (c) nickel.

S4.65 Write equations for the reactions of iron, cobalt, and nickel with (a) oxygen and (b) HCl.

S4.66 Write the electron configuration of each of the following species:
- **(a)** Fe, Fe^{2+}, and Fe^{3+}
- **(b)** Co, Co^{2+}, and Co^{3+}
- **(c)** Ni and Ni^{2+}

S4.67 **(a)** Explain why an aqueous solution of Fe^{3+} is acidic.
- **(b)** Which should be more acidic, $Fe^{3+}(aq)$ or $Fe^{2+}(aq)$? Explain your answer.

S4.68 Write equations for the reactions of $Fe^{2+}(aq)$, $Co^{2+}(aq)$, and $Ni^{2+}(aq)$ with (a) NaOH, (b) NH_3, and (c) S^{2-}.

S4.69 Which is a better reducing agent, Fe or Ni?

S4.70 Which ion in each of the following pairs is a better oxidizing agent?
- **(a)** $Fe^{2+}(aq)$ or $Fe^{3+}(aq)$
- **(b)** $Fe^{2+}(aq)$ or $Ni^{2+}(aq)$

S4.71 Write the half-reaction equations and the net ionic equation for the oxidation of Fe^{2+} to Fe^{3+} in aqueous solutions exposed to oxygen.

S4.72 Why are iron nails often placed in bottles of freshly prepared $FeSO_4$ solution? Explain with the aid of an equation.

Chromium and Manganese (Section S4.6)

S4.73 State (a) the principal sources and (b) the principal uses of chromium and manganese.

S4.74 Name and give a commercial use for at least one important compound of (a) chromium and (b) manganese.

S4.75 Chromium is said to resemble aluminum in many ways.
- **(a)** List some of the chemical and physical properties that are similar in these elements.
- **(b)** List their principal differences.

S4.76 Draw a Lewis structure for each of the following ions: (a) $CrO_4{}^{2-}$, (b) $Cr_2O_7{}^{2-}$, and (c) $MnO_4{}^{-}$.

S4.77 Write equations for the reduction of (a) chromite ore, (b) pyrolusite ore, and (c) Cr_2O_3. What is *ferrochrome*?

S4.78 Write equations for the reactions of Cr and Mn with aqueous HCl.

S4.79 Write equations for the reactions of aqueous Cr^{3+} and Mn^{2+} with NaOH.

S4.80 Write an equation describing the chromate–dichromate equilibrium. What is the effect of pH on this equilibrium?

PART B. MISCELLANEOUS QUESTIONS AND PROBLEMS

S4.81 Mercury metal forms when mercury(II) sulfide is roasted in air at 500°C. Zinc oxide forms when zinc sulfide is roasted under the same conditions.
- **(a)** Consult Appendix B.1 and explain why the metal is obtained from one sulfide but not from the other.
- **(b)** What should form, silver metal or silver oxide, when silver sulfide is roasted at 500°C?

S4.82 Seawater (d = 1.024 g/mL) contains 0.13% by weight of Mg^{2+}. Assume that 1.00 metric ton (1000 kg) of magnesium is to be recovered by the Dow process and that the process is 70% efficient.
- **(a)** What volume of seawater is needed?
- **(b)** What mass of shells is required if the average shell is 97% $CaCO_3$?
- **(c)** What volume of concentrated (12 *M*) HCl is required?
- **(d)** How many coulombs of electricity are required if the electrolytic cell is 80% efficient?
- **(e)** Natural gas (CH_4) is burned in a mixture of air plus the chlorine from the electrolysis. The products are CO and HCl, which is recycled. Write the equation for the combustion.

S4.83 **(a)** Consult Appendix B.1 and calculate $\Delta H°$ for the reduction of 1.00 mol of Fe_2O_3 with 2.00 mol of Al (thermite reaction).
- **(b)** Assume that the reactants in Part (a) start at 25°C and that all of the evolved heat is absorbed in raising the temperature of the products. What will be the final temperature? The molar heat capacities of Fe and Al_2O_3 are 68 J/mol·K and 124 J/mol·K, respectively.

S4.84 Consult Appendix B.4 and calculate the fraction of Mg^{2+} remaining in solution when seawater (d = 1.024 g/mL, 0.13% Mg by mass) is mixed with an equal volume of saturated $Ca(OH)_2$. The solubility of $Ca(OH)_2$ is 1.85 g/L.

S4.85 **(a)** Using data from the tables in Appendix B, calculate the concentration of dissolved Cu^+ in (1) saturated CuCl and (2) 1.0 *M* $Na[Cu(CN)_2]$.

(b) Use data from Appendix B to show that the Cu^+ ion will not disproportionate appreciably in the solutions of Part (a).

S4.86 A 1.00-g sample of gold can be drawn into a wire 3000 m long. Calculate the diameter of this wire in millimeters.

S4.87 Sulfur dioxide emissions from copper smelters are a serious environmental problem.

(a) What mass of SO_2 would be evolved during the production of 1000 metric tons of copper from copper ore that is 50% CuS and 50% Cu_2S?

(b) If the SO_2 from Part (a) is converted to H_2SO_4, what volume of marble statuary ($CaCO_3$, d = 2.93 g/cm^3) could it disintegrate?

S4.88 Pure chromium is prepared from chromite by the following steps: (1) the ore is fused with solid sodium carbonate in the presence of air, yielding sodium chromate, iron(III) oxide, and carbon dioxide gas; (2) the sodium chromate is dissolved and converted to sodium dichromate by the addition of acid; (3) the sodium dichromate is crystallized, dried, and reduced to chromium(III) oxide with carbon; (4) the chromium(III) oxide is reduced to chromium metal with aluminum. Write the equation for each step.

S4.89 Draw the Lewis structure for Hg_2^{2+}.

S4.90 Give the color and at least one commercial use for each of the following pigments: PbO, Pb_3O_4, Cr_2O_3, $KFe[Fe(CN)_6]$, and MnO_2.

S4.91 Consult Appendix B.1 and calculate $\Delta G°$ for the reduction of Fe_2O_3 with CO at (a) 25°C and (b) 1000°C. Comment on your results as they apply to the smelting of iron.

S4.92 Barium chloride is obtained by heating the ore *barites* ($BaSO_4$) with calcium chloride and carbon:

$$BaSO_4(s) + 2C(s) + CaCl_2(s) \longrightarrow BaCl_2(s) + CaS(s) + 2CO_2(g)$$

$BaCl_2$ is leached out of the final mixture with water at 80°C.

(a) Calculate the volume of water needed to dissolve the $BaCl_2$ produced from 1.00 kg of $BaSO_4$. The solubilities of $BaCl_2$ and CaS in water at 80°C are 53 g/100 mL and 0.048 g/100 mL, respectively.

(b) What fraction of the CaS produced in Part (a) will be leached out along with the $BaCl_2$?

S4.93 Calculate K_c for the following disproportionation reaction:

$$Fe(s) + 2Fe^{3+}(aq) \rightleftharpoons 3Fe^{2+}(aq)$$

CHAPTER 22

A charged particle passing through a cloud chamber ionizes gas molecules along its path. Water vapor condenses on the ionized molecules leaving visible tracks, as shown in this false-color photograph.

NUCLEAR CHEMISTRY

OUTLINE

PREVIEW

The nuclei of elements other than hydrogen are assembled in hot stars and eventually hurled into space. Some of these nuclei are stable; they make up the elements and compounds that are the building blocks of our world. Other nuclei are unstable; they radiate energy and small particles, and we say that they are *radioactive*. The study of radioactive nuclei has had almost unbelievable consequences. It has shown us how to achieve the alchemists' dream of changing one element into another, and it has taught us how to make new elements. In fact, it has carried us to the point where we may soon be able to obtain heat, light, and matter from the same reactions that fuel the stars.

In 1896 Antoine Henri Becquerel, a professor at the École Polytechnic in Paris, discovered that crystals of potassium uranyl sulfate (a uranium salt) will produce images on unused, well-wrapped photographic plates. After an intensive investigation, Becquerel concluded that uranium atoms were emitting some sort of penetrating radiation, and he called this phenomenon **radioactivity**. Other radioactive elements were soon discovered. In the mineral pitchblende (Figure 22.1), a complex ore whose principal constituent is uranium oxide (U_3O_8), Marie Curie, a student of Becquerel, found a new radioactive element that she named "polonium" in honor of her native Poland. A few months later, in 1898, Mme. Curie and her husband, Pierre Curie, isolated an even more intensely radioactive element that they named "radium." In 1903 the Curies and Becquerel shared a Nobel Prize for their discoveries.

22.1 NATURAL RADIOACTIVITY

When the emissions from radioactive minerals are passed between charged plates (see Figure 22.2), they separate into positively charged particles called **alpha particles** (α), negatively charged particles called **beta particles** (β), and a form of elec-

(a)

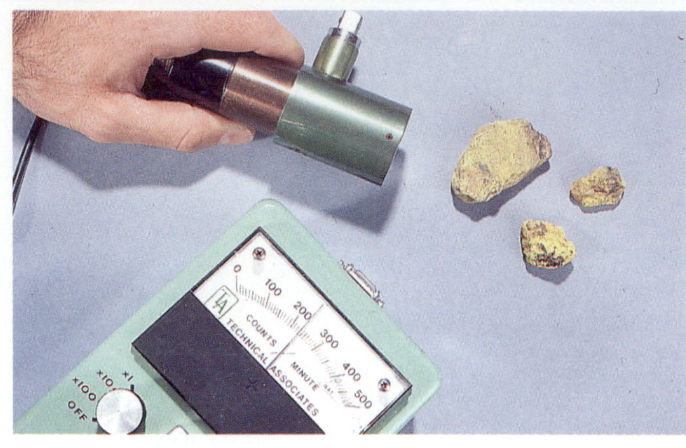

(b)

Figure 22.1

Some uranium ores. (a) Pitchblende. (b) The radiation from carnotite ($K_2O \cdot 2U_2O_3 \cdot V_2O_5 \cdot 3H_2O$) is measured with a Geiger counter.

tromagnetic radiation called **gamma rays** (γ). All three forms of radiation are extremely energetic. Further study shows that alpha particles are the nuclei of helium atoms (two protons and two neutrons) and that beta particles are electrons moving at very high speeds. The properties of these particles are summarized in Table 2.1 (p. 37). Gamma rays consist of photons with very high frequencies (about 10^{20} s^{-1}).

The names alpha, beta, and gamma were given to the three types of radioactive emanation by Ernest Rutherford and his associate Frederick Soddy. Rutherford went on to identify the alpha particle as a He^{2+} ion, and subsequently used it in the gold foil experiment (Section 2.2).

The radiation emitted by a radioactive element is the same whether the element is free or combined in a compound, and it is not affected by conditions such as temperature or pressure. These facts suggest that radioactivity is a property of the atomic nucleus. It is now known that radioactive elements have unstable nuclei that spontaneously decay into more stable nuclei by emitting energetic particles and photons. For example, a radium atom (element 88) becomes a radon atom (element 86) by emitting an alpha particle from its nucleus. The radon atom is also unstable and emits another alpha particle to become polonium (element 84), and after several more emissions, it becomes a stable atom of lead (element 82). Nuclei that decay spontaneously are called **radionuclides** or **radioisotopes**.

The sequence of steps by which a radioisotope is transformed into other radioisotopes and finally into a stable atom is called a **decay series**. The elements beyond bismuth (element 83) are all radioactive, and each of them belongs to one of three decay series: the *uranium series*, in which uranium-238 decays to lead-206; the *actinium series*, in which uranium-235 decays to lead-207; or the *thorium series*, in which thorium-232 decays to lead-204. The complete uranium series is shown in Figure 22.3. A uranium ore such as pitchblende contains some of each decay product along with the uranium. The ejected alpha particles acquire electrons to

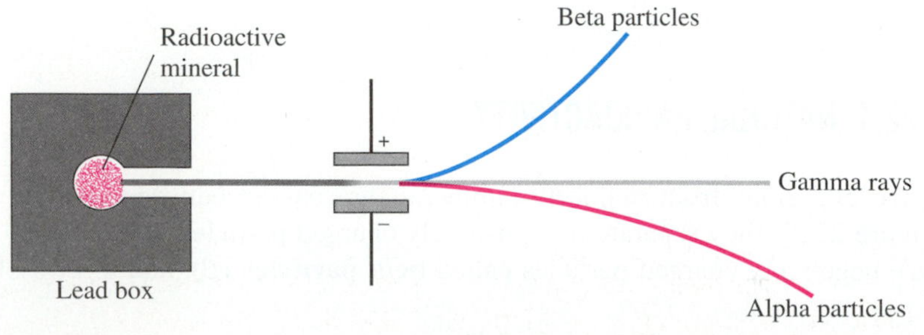

Figure 22.2

The radiation from the various atomic species in a radioactive mineral separates into three beams when it passes between charged plates. The negatively charged beta particles are less massive and exhibit a sharper deflection than the positive alpha particles. The gamma rays are not affected by electric charge.

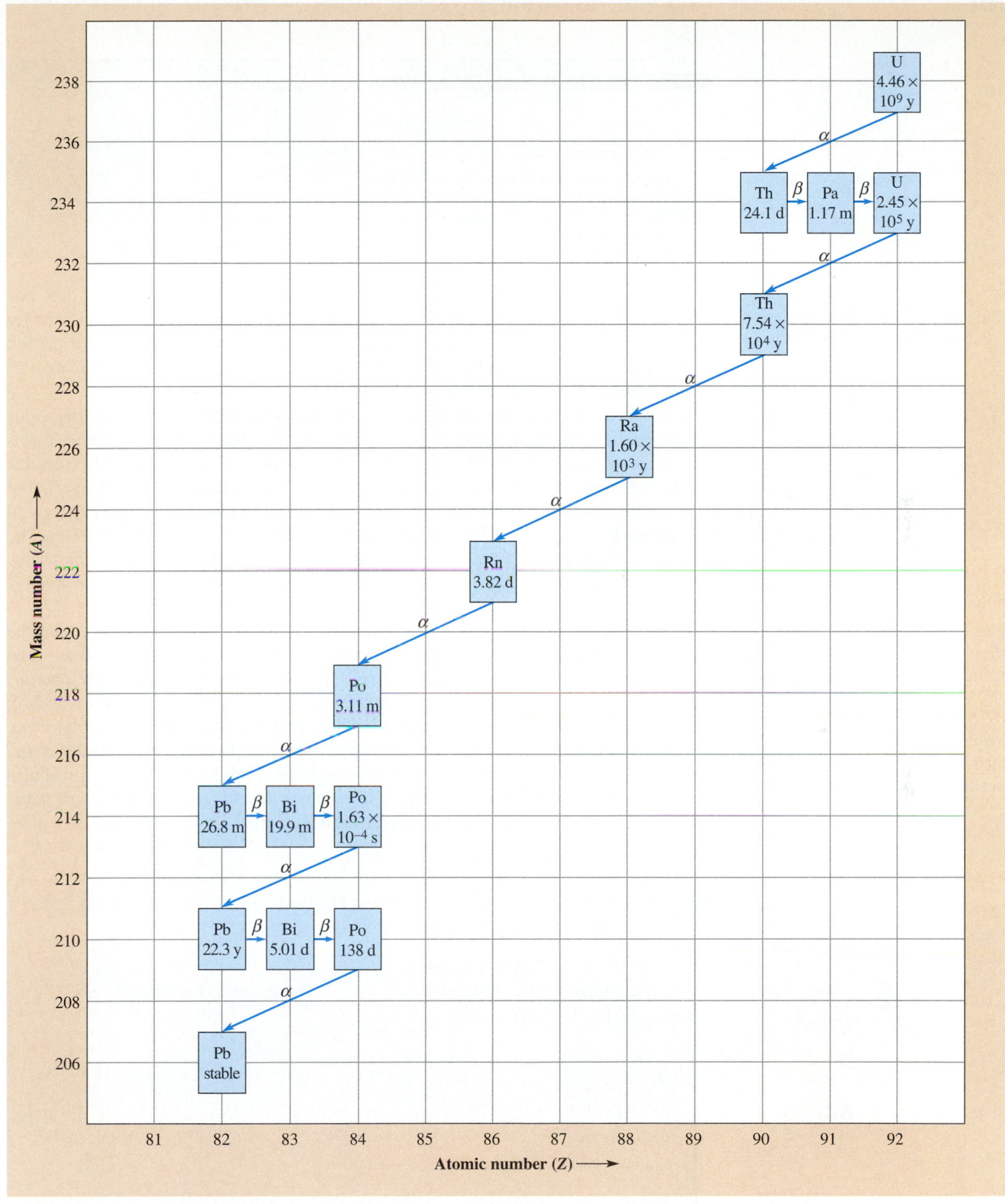

Figure 22.3 The uranium-238 decay series. The nuclei and their half-lives (y = years, d = days, m = minutes, s = seconds) are located on a graph of mass number versus atomic number. Emission of an alpha particle reduces the mass number by 4 and the atomic number by 2. Emission of a beta particle raises the atomic number by 1 but has no effect on the mass number.

TABLE 22.1 Some Energy Units Used in Nuclear Chemistry

1 electron volt (eV)[a] = 1.602×10^{-19} J
1 megaelectron volt (MeV) = 10^6 eV = 1.602×10^{-13} J
1 atomic mass unit (u) = 931.5 MeV

[a]One eV is equal to the energy acquired by an electron when it is accelerated by a potential difference of 1 V.

The hazards from radon that seeps into buildings is described on page 356.

become helium gas, much of which remains trapped in the ore. Radon, another gaseous decay product trapped in rocks, is now being monitored to help predict earthquakes. Increased radon emission near a geological fault may be indicative of subterranean movement that causes the trapped gas to escape.

Radioactive emissions are a form of **ionizing radiation**; that is, the emissions have enough energy to ionize and fragment particles along their paths. Other forms of ionizing radiation include x-rays, electron beams, and high-energy laser beams. Ionizing radiation is hazardous to living organisms because it can damage living cells and cause genetic defects. The energy of ionizing radiation is usually expressed in **megaelectron volts** (MeV), where 1 MeV = 1.602×10^{-13} J (Table 22.1). The energy of a typical alpha particle is about 5 MeV, while typical beta particles have energies between 0.05 and 1 MeV. Gamma-ray photons have energies of about 1 MeV.

Because most alpha particles are stopped by skin and clothing, alpha particles outside the body are generally harmless. However, alpha particles generated inside the body are extremely destructive because of their relatively large mass and high kinetic energy.

The average distance that radiation will travel in a given medium is called the **penetrating ability** of the radiation. Penetrating abilities depend on the nature of the medium as well as on the energy, mass, and charge of the particles or photons involved (Table 22.2). Alpha particles can be stopped by thin layers of cardboard, clothing, and skin, while beta particles will penetrate several millimeters into flesh. Although alpha particles are more energetic than beta particles, their greater mass, lower speed, and higher charge make them more likely to be absorbed by molecules along their path. Gamma rays and x-rays carry no charge; therefore, they have greater penetrating abilities than alpha or beta particles. Gamma rays will pass

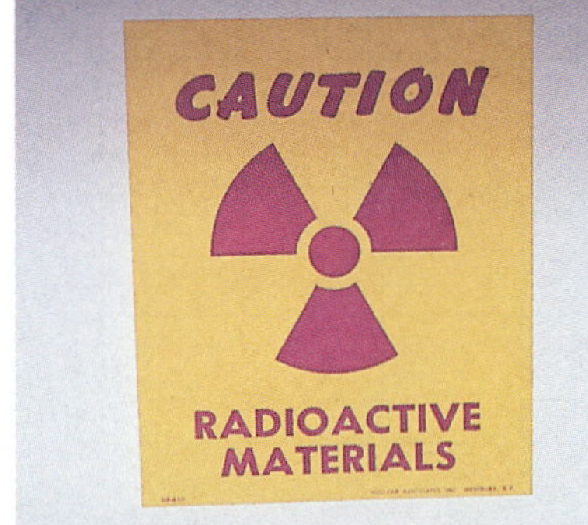

This international caution symbol alerts viewers to the presence of radiation sources.

TABLE 22.2 Energy and Penetrating Ability of Alpha, Beta, and Gamma Radiation from Typical Radioactive Sources

Radiation Type	Approximate Energy (MeV)	Approximate Penetration (cm)		
		Dry Air	*Living Tissue*	*Lead*
Alpha particles	5.0	4	0.005	0
Beta particles	0.05 to 1.0	6 to 300	0.006 to 0.40	0.0005 to 0.01
Gamma photons	1.0	40,000[a]	50[a]	3[a]

SOURCE: J. B. Little, *New England J. Medicine*, Vol. 275, pp. 929–936 (1966).
[a]These thicknesses reduce the intensity of the gamma-ray beam to 10% of its initial intensity.

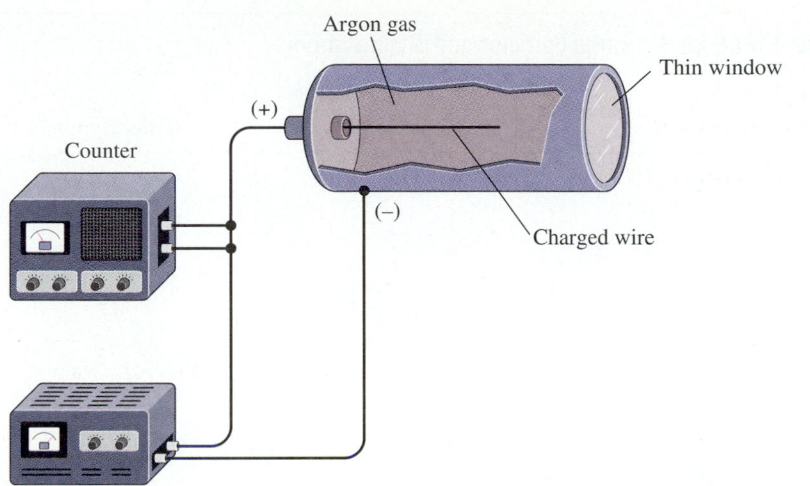

Figure 22.4
Diagram of a Geiger–Müller counter. The wire running through the center of the tube is positively charged with respect to the tube wall. Each photon or particle that passes through the thin window ionizes the argon atoms in its path. Free electrons rush toward the positive wire, colliding with and ionizing additional atoms along the way. The result is a momentary surge of current that is registered on a meter or converted into an audible click or flash of light.

through the human body, through several centimeters of a metal such as aluminum, and even through several feet of concrete. They are more dangerous because their passage leaves longer trails of broken molecules and ionized atoms.

Measuring Radiation

Various devices are used to detect radiation and measure its intensity. Radioactivity was first discovered through its effect on photographic film, and people who work with x-rays and other forms of radiation wear film badges to monitor their exposure. Individual particles and photons can be detected with **ionization counters** such as the *Geiger–Müller counter* (Figure 22.4), in which each energetic photon or particle causes a momentary surge of current that either registers on a meter or is converted into an audible click.

A *scintillation counter* (Figure 22.5) contains a *phosphor*, a substance that emits a flash of light when a particle or photon strikes it. Sodium iodide, zinc sulfide, and thallium iodide are among the phosphors used for this purpose. The intensity of the flash is proportional to the energy of the absorbed radiation; thus, the scintillation counter not only counts particles and photons but measures their energies as well. Since the source of radioactivity can often be identified by the values of the energies it emits, a scintillation counter is the instrument of choice for many purposes.

Scintillation: from the Latin *scintilla*, meaning a "spark."

22.2 EQUATIONS FOR NUCLEAR REACTIONS

A **nuclear reaction** is one in which a nucleus changes. The equation for a nuclear reaction includes the mass number and atomic number of every participating nucleus and particle. The symbols for some of the particles commonly encountered in nuclear reactions are given in Table 22.3. Nuclear equations obey the following rules:

1. *The sum of the mass numbers does not change*. Protons may sometimes change into neutrons, and vice versa, but the total number of nucleons (protons plus neutrons) remains the same.

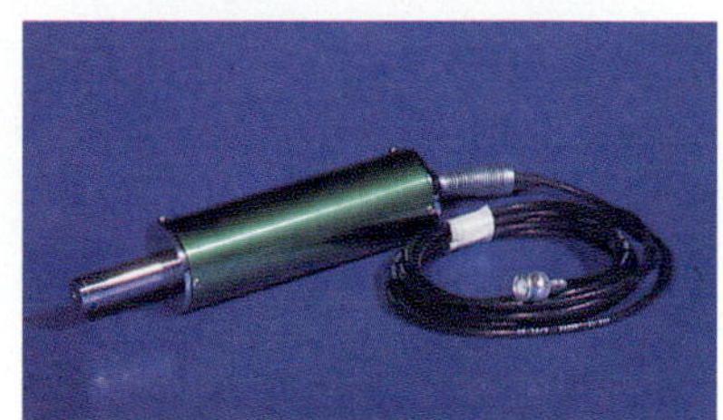

Figure 22.5
A scintillation counter.

LEARNING HINT

Review isotopic notation on page 45.

TABLE 22.3 Some Particles and Their Symbols

Name of Particle	Symbol Preferred for Equations	Other Symbols
Alpha particle or helium nucleus	${}^{4}_{2}\mathrm{He}$	α
Beta particle or electron	${}^{0}_{-1}\mathrm{e}$	β^-
Deuteron	${}^{2}_{1}\mathrm{H}$	D
Gamma photon	γ	—
Neutron	${}^{1}_{0}\mathrm{n}$	n
Positron	${}^{0}_{1}\mathrm{e}$	β^+
Proton	${}^{1}_{1}\mathrm{H}$	p

2. *The sum of the atomic numbers does not change.* Atomic numbers represent charges, and the total charge does not change during a nuclear reaction.

LEARNING HINT

The charge on a nuclear particle is given by the atomic number subscript; it is not written as a right superscript as in chemical equations.

For example, the equation for the alpha decay of uranium-238 into thorium-234 is

$$ {}^{238}_{92}\mathrm{U} \rightarrow {}^{4}_{2}\mathrm{He} + {}^{234}_{90}\mathrm{Th} $$

On each side of the equation, the sum of the mass numbers is 238 and the sum of the atomic numbers is 92. The alpha particle (a helium nucleus) contains two protons and two neutrons; hence, *emission of an alpha particle reduces the mass number of a nucleus by four units and the atomic number by two units.*

The decay of thorium-234 into protactinium-234 by beta-particle emission is represented by the equation

$$ {}^{234}_{90}\mathrm{Th} \rightarrow {}^{0}_{-1}\mathrm{e} + {}^{234}_{91}\mathrm{Pa} $$

The beta particle is an electron that has been produced in a nuclear decay reaction. Like all electrons, it has a mass number of zero and an effective atomic number of -1. *Emission of a beta particle raises the atomic number of a nucleus by one unit and leaves the mass number unchanged.* In effect, one of the neutrons in the nucleus has decayed into a proton and an electron. Similar behavior has been observed for neutrons outside the nucleus.

Most nuclear transformations are accompanied by the emission of one or more gamma photons. A photon does not affect the mass number or the atomic number, and its inclusion in the equation is optional.

EXAMPLE 22.1

Technetium-99, the most stable isotope of element 43, decays by beta emission. Write the nuclear equation.

SOLUTION

The incomplete equation is

$$ {}^{99}_{43}\mathrm{Tc} \rightarrow {}^{0}_{-1}\mathrm{e} + \mathrm{X} $$

where X represents the product nucleus. The atomic number of X must be 44 and its mass number must be 99:

$$^{99}_{43}\text{Tc} \longrightarrow {}^{0}_{-1}\text{e} + {}^{99}_{44}\text{X}$$

The atomic number determines the identity of the element. Atomic number 44 belongs to ruthenium (see periodic table). The complete equation is

$$^{99}_{43}\text{Tc} \longrightarrow {}^{0}_{-1}\text{e} + {}^{99}_{44}\text{Ru}$$

Neutrons become protons during beta-particle emission; the atomic number of the nucleus increases by 1.

Some nuclear reactions involve particles called **positrons**. A positron has the same mass as an electron, but it has a positive charge. Thus, it has a zero mass number and an effective atomic number of +1 (Table 22.3). *Emission of a positron decreases the atomic number of a nucleus by one unit and leaves the mass number unchanged.*

EXAMPLE 22.2

Write the nuclear equation for the decay of phosphorus-30 by positron emission.

SOLUTION

The atomic number of phosphorus is 15. The positron symbol is ${}^{0}_{1}\text{e}$ (Table 22.3), so the incomplete equation is

$$^{30}_{15}\text{P} \longrightarrow {}^{0}_{1}\text{e} + {}^{30}_{14}\text{X}$$

Element number 14 is silicon. The complete equation is

$$^{30}_{15}\text{P} \longrightarrow {}^{0}_{1}\text{e} + {}^{30}_{14}\text{Si}$$

Protons become neutrons during positron emission; the atomic number of the nucleus decreases by 1.

PRACTICE EXERCISE 22.1

(a) What are the mass number, atomic number, and symbol of the nucleus that remains after mercury-187 emits an alpha particle? (b) Complete the following nuclear equation by adding missing symbols, mass numbers, and atomic numbers:

$$^{28}\text{Al} \longrightarrow {}^{28}\text{Si} +$$

22.3 NUCLEAR BOMBARDMENT REACTIONS

Changing one element into another is called **transmutation**, a term that goes back to the alchemists who hoped to transmute base metals like lead and iron into gold. Transmutation can now be accomplished by nuclear bombardment, in which streams of energetic particles are directed at target nuclei. The first nuclear transformation to be carried out in a laboratory was accomplished by Ernest Rutherford in 1919, when he directed alpha particles into a nitrogen sample. The particles that scored direct hits on the nitrogen nuclei produced oxygen-17 nuclei and protons:

$$^{14}_{7}\text{N} + {}^{4}_{2}\text{He} \longrightarrow {}^{17}_{8}\text{O} + {}^{1}_{1}\text{H}$$

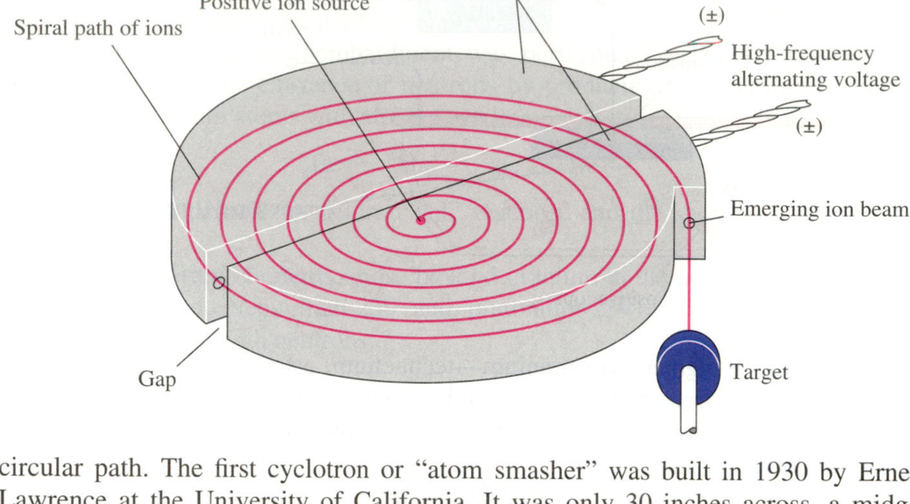

Figure 22.7

Diagram of a cyclotron. An ion introduced at the center is subjected to an alternating voltage that sends it back and forth between the two hollow, D-shaped electrodes. An ever-widening spiral path is imparted to the ion's motion by magnetic poles above and below the electrodes. The speed of the ion increases with the radius of curvature until the ion emerges on a course that leads to the target.

circular path. The first cyclotron or "atom smasher" was built in 1930 by Ernest Lawrence at the University of California. It was only 30 inches across, a midget compared to the larger ones in operation today. The *Bevatron*, built in the 1950s at the University of California at Berkeley, can produce protons with energies up to one gigaelectron volt (1 GeV $= 10^9$ eV). The *Tevatron* at the Fermi National Accelerator Laboratory in Batavia, Illinois, now accelerates protons to energies just short of one teraelectron volt (1 TeV $= 10^{12}$ eV).

A **linear accelerator** contains a series of straight tubes in which particles are subjected to electric fields that impart tremendous energies and speeds (Figure 22.8). The alternating field in each tube is timed to attract a charged particle as it enters the tube and to repel the particle as it leaves for the next tube. Linear accelerators vary from a few hundred feet to several miles in length. The longer accelerators handle larger projectiles and give them higher energies. The *Bevelac* at Berkeley combines the Heavy-Ion Linear Accelerator (HILAC) with the Bevatron to provide a two-stage acceleration of large nuclei such as iron, argon, and uranium. Speeds up to 95% of the speed of light and energies up to 2.1 GeV per nucleon have been obtained. An accelerator proposed for construction in Texas, called the Superconducting Super Collider (SSC), would be the largest and most costly scientific instrument in the world. It is designed to accelerate protons, first in a linear accelerator, and then in three progressively larger circular accelerators, to energies of about 20 TeV.

The collisions of very energetic particles are expected to provide deep insight into the fundamental nature of matter, and they may have practical uses as well. For example, the carefully targeted impact of heavy superenergetic nuclei may provide more effective antitumor radiation therapy and have fewer side effects than the x-rays and gamma rays now used.

(a)

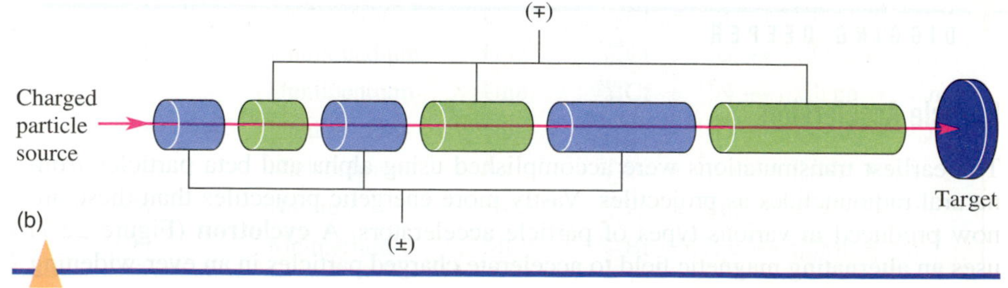

(b)

Figure 22.8

(a) The Heavy-Ion Linear Accelerator (SUPER-HILAC) at Lawrence Berkeley Laboratory, University of California, is used to impart energies up to 8.5 MeV per nucleon to ions ranging from carbon to uranium. (b) Diagram of a linear accelerator. The alternating electric field in each tube is timed to attract a charged particle as it enters the tube and to repel the particle as it leaves for the next tube.

22.4 ACTIVITY AND HALF-LIFE

LEARNING HINT

Review first-order reactions and half-lives in Section 14.4.

The **activity** of a radioactive sample is the number of atoms that disintegrate per unit time. The SI unit of activity is the **becquerel** (Bq), where 1 becquerel is 1 disintegration per second (dps). Activity can also be expressed in **curies** (Ci), an older non-SI unit defined such that 1 curie is equal to the disintegration rate of 1 g of radium (3.7×10^{10} dps). Thus, 1 Ci $= 3.7 \times 10^{10}$ Bq.

Radioactive decay obeys a first-order rate law; that is, the activity at each instant is proportional to the number of radioactive atoms in the sample

$$\text{activity} = kN$$

where N is the number of the radionuclides (or some quantity proportional to their number) and k is the rate constant. If N_0 is the number of radionuclides at time $t = 0$, then the fraction of atoms remaining at time t is N/N_0. This fraction is given by Equation 14.4:

$$\ln \frac{N}{N_0} = -kt \tag{22.1}$$

The half-life ($t_{1/2}$) of a radioisotope is related to the rate constant by Equation 14.6:

$$kt_{1/2} = \ln 2 = 0.693 \tag{22.2}$$

Half-lives range from fractions of a second to billions of years (Table 22.5) and are not affected by chemical combination or by conditions such as temperature and pressure.

EXAMPLE 22.3

On a visit to the United States in 1921, Madame Marie Curie (Figure 22.9) was presented with 1.00 g of pure radium as a gift from the women of America. Estimate how much of that radium, which was later given to the Curie Institute of France, still existed in 1990. The half-life of the principal isotope of radium, radium-226, is 1.60×10^3 y.

SOLUTION

Equation 22.1 relates the remaining fraction to the elapsed time t. The value of t is 1990 y $-$ 1921 y $=$ 69 y. The rate constant is obtained from the half-life using Equation 22.2:

$$k = \frac{0.693}{t_{1/2}} = \frac{0.693}{1.60 \times 10^3 \text{ y}} = 4.33 \times 10^{-4} \text{ y}^{-1}$$

Substituting k and $t = 69$ y into Equation 22.1 gives

$$\ln \frac{N}{N_0} = -4.33 \times 10^{-4} \text{ y}^{-1} \times 69 \text{ y} = -0.030$$

and

$$\frac{N}{N_0} = e^{-0.030} = 0.970$$

Hence, 97.0% or 0.970 g of the radium present in 1921 still existed in 1990.

Figure 22.9

Mme. Marie Sklodowska Curie (1867–1934) is one of only four scientists to be awarded two Nobel Prizes. Mme. Curie, her husband Pierre, and Antoine Henri Becquerel shared the Nobel Prize for physics in 1903. After her husband's death in 1906, Marie continued their work, and in 1911 she received the Nobel Prize in chemistry.

TABLE 22.5 Half-Lives of Some Naturally Occurring Radionuclides

Nuclide	Half-Life[a]	Principal Decay Mode
Primordial Nuclides[b]:		
Neodymium-144	2.1×10^{15} y	Alpha
Potassium-40	1.26×10^{9} y	Electron capture
Rubidium-87	4.88×10^{10} y	Beta
Thorium-232	1.41×10^{10} y	Alpha
Uranium-235	7.04×10^{8} y	Alpha
Uranium-238	4.46×10^{9} y	Alpha
Nuclides Continuously Formed by Cosmic-Ray Bombardment:		
Carbon-14	5730 y	Beta
Tritium (hydrogen-3)	12.32 y	Beta
Nuclides Formed as Intermediates in Decay Series:		
Astatine-218	1.6 s	Alpha
Polonium-210	138.4 d	Alpha
Polonium-214	1.63×10^{-4} s	Alpha
Protactinium-234	6.70 h	Beta
Radon-222	3.823 d	Alpha
Radium-226	1.599×10^{3} y	Alpha
Thorium-234	24.10 d	Beta

[a]y = years, d = days, h = hours, s = seconds.
[b]*Primordial*: existing from the beginning.

PRACTICE EXERCISE 22.3

The activity of a sample containing beta-emitting phosphorus-32 declined in 48.0 hours to 90.8% of its initial value. Calculate the half-life of phosphorus-32.

Radioisotope Dating

Radioisotope dating uses the concentration of a radionuclide to establish the age of an ancient object. **Radiocarbon dating** makes use of the decay of carbon-14, an isotope that is continuously formed in the atmosphere by the action of cosmic rays on nitrogen. **Cosmic rays** are energetic particles that originate in the stars and travel through space; they consist mostly of electrons, protons, and nuclei of low atomic mass. Their interactions with molecules in the upper atmosphere add neutrons to the particle shower. The collision of a neutron with a nitrogen nucleus produces carbon-14:

$$^{14}_{7}\text{N} + ^{1}_{0}\text{n} \rightarrow ^{14}_{6}\text{C} + ^{1}_{1}\text{H}$$

Figure 22.10

Pages from the Dead Sea Scrolls, Book of Isaiah. Radiocarbon dating has been used to show that these scrolls were written or copied between 100 B.C. and 50 A.D. They are almost 1000 years older than any other biblical manuscript.

The carbon-14 atoms are incorporated into atmospheric carbon dioxide, which is absorbed by plants during photosynthesis and then absorbed by animals that feed on the plants. In this way, the carbon that recycles through the biosphere and atmosphere maintains a fairly constant ratio of one carbon-14 atom for every 10^{12} carbon-12 atoms. Carbon-14 is a beta emitter ($t_{1/2} = 5730$ y)

$$^{14}_{6}\text{C} \rightarrow {}^{0}_{-1}\text{e} + {}^{14}_{7}\text{N}$$

and each gram of carbon in living tissue has a constant beta-ray activity of 15.3 disintegrations per minute (15.3 dpm). After a plant or animal dies, it stops exchanging carbon with its surroundings and its carbon-14 decays without being replenished. The disintegration rate of the remaining carbon-14 then becomes a clock from which we can read the time elapsed since the organism was alive. Materials suitable for radio-carbon dating include wood, cloth, bones, and shells. The radiocarbon method is most reliable when it is used to date objects ranging in age from 1000 to 50,000 years; fortunately, this is the time interval of greatest interest to archaeologists (Figure 22.10).

EXAMPLE 22.4

A piece of charcoal from a prehistoric campsite is found to emit 3.85 beta particles per minute per gram of carbon. Estimate the age of the campsite. Assume that each gram of carbon in living tissue has a disintegration rate of 15.3 dpm, and use 5730 y for the half-life of carbon-14.

SOLUTION

We will first find the fraction of carbon atoms remaining, then we will use Equation 22.1 to solve for t, the time elapsed since the charred wood was part of a living tree. Originally, each gram of carbon emitted 15.3 beta particles per minute. Because of its decreased carbon-14 content, each gram now emits 3.85 beta particles per minute. The number of carbon-14 atoms is proportional to the disintegration rate; therefore, the fraction of carbon-14 remaining is

$$\frac{N}{N_0} = \frac{^{14}\text{C now}}{^{14}\text{C at } t = 0} = \frac{3.85 \text{ dpm}}{15.3 \text{ dpm}} = 0.252$$

The disintegration rates of carbon from the yearly rings of bristlecone pine (the oldest living trees) show that the $^{14}C/^{12}C$ ratio in living material has varied somewhat over the past 8000 years. The variation is attributed to fluctuations in cosmic-ray activity. Very precise carbon dating is adjusted for these known variations.

The half-life of carbon-14 is 5730 y. Substituting

$$k = \frac{\ln 2}{t_{1/2}} = \frac{\ln 2}{5730 \text{ y}} = 1.21 \times 10^{-4} \text{ y}^{-1}$$

and $N/N_0 = 0.252$ into Equation 22.1 gives

$$\ln \frac{N}{N_0} = -kt$$

$$\ln 0.252 = -1.21 \times 10^{-4} \text{ y}^{-1} \times t$$

and $$t = 1.14 \times 10^4 \text{ y} = 11{,}400 \text{ y}$$

PRACTICE EXERCISE 22.4

The decay rate of the carbon in an old bone is one-eighth that of the carbon in a living tree. How old is the bone?

Dating methods suitable for even older objects make use of the decay rates of uranium-238, thorium-232, rubidium-87, and potassium-40 (Table 22.5). Uranium-238 has a half-life of 4.46×10^9 years and decays in a series of steps to lead-206. The $^{206}Pb/^{238}U$ ratio can be used to estimate the age of a uranium-bearing rock. (A rock's age is the time that has elapsed since the rock solidified and there was no further mixing of its components.)

EXAMPLE 22.5

A moon rock contains lead-206 atoms and uranium-238 atoms in a 1.00-to-1.09 ratio. The absence of other lead isotopes in the rock indicates that all of the lead-206 was produced by uranium decay. Estimate the age of the rock. The half-life of uranium is 4.46×10^9 y.

SOLUTION

The 1.00-to-1.09 ratio means that 100 lead-206 atoms are present for every 109 uranium-238 atoms. The 100 lead atoms were originally uranium; in other words, for every 109 uranium atoms now present, there were 109 + 100 = 209 uranium atoms at the time the rock solidified. The fraction of uranium-238 remaining in the rock is therefore

$$\frac{N}{N_0} = \frac{109}{209} = 0.522$$

Substituting this fraction and

$$k = \frac{\ln 2}{t_{1/2}} = \frac{\ln 2}{4.46 \times 10^9 \text{ y}} = 1.55 \times 10^{-10} \text{ y}^{-1}$$

into Equation 22.1 gives

$$\ln 0.522 = -1.55 \times 10^{-10}\ \text{y}^{-1} \times t$$

and
$$t = 4.19 \times 10^{9}\ \text{y}$$

This moon rock must have solidified soon after the solar system formed about 4.5 billion (4.5×10^9) years ago.

PRACTICE EXERCISE 22.5

Rubidium-87 has a half-life of 4.88×10^{11} y and decays by beta emission to strontium-87. Estimate the age of a rock if it contains 56 strontium-87 atoms for every 10,000 rubidium-87 atoms.

DIGGING DEEPER

Dosage

The extent of radiation damage and the magnitude of any therapeutic effect both depend on how much radiation is actually absorbed. Radiation dosage is usually expressed in **rads** (**r**adiation **a**bsorbed **d**ose), where 1 rad is the amount of radiation that results in the absorption of 10^{-2} J of energy per kilogram of irradiated material. A radiation dose of 500 rads, for example, means that each kilogram of irradiated material absorbs $500 \times 10^{-2} = 5.00$ J of energy. The SI unit of dosage is the **gray** (Gy) named for the British radiologist Harold Gray. One gray is the amount of radiation that results in the absorption of 1 joule of energy per kilogram of irradiated material; 1 Gy = 100 rads.

Experiments have shown that the magnitude of a biological effect varies with the type of radiation as well as with the dose. A 1-rad dose of alpha radiation, for example, generally produces 10 times the effect of a 1-rad dose of beta or gamma radiation because the larger, doubly charged alpha particles are more likely to collide with and fragment tissue molecules than are beta particles or gamma rays. Each type of radiation can be assigned a value indicating its **relative biological effectiveness** (RBE). X-rays, gamma rays, and beta particles have about the same biological effectiveness and are assigned RBE values of unity (1). Alpha particles, protons, and neutrons are assigned the value 10; a 1-rad dose of these particles will produce 10 times the biological effect of a 1-rad dose of the other forms of radiation.

The **rem** (**r**adiation **e**quivalent for **m**an) is a unit for radiation dosage that takes into account the actual dose and the biological effectiveness of the radiation involved. The number of rems is obtained by multiplying the dose in rads by the RBE value for the radiation:

$$\text{dose (in rems)} = \text{dose (in rads)} \times \text{RBE} \tag{22.3}$$

Neutrons, for example, have an RBE value of 10. Hence, a 50-rad dose of neutrons is equivalent to $50 \times 10 = 500$ rems. This dose will have the same biological effect as a 500-rad dose of gamma rays, which also supplies $500 \times 1 = 500$ rems.

CHEMICAL INSIGHT

RADIATION IN BIOLOGY AND MEDICINE

Radiation damage to living organisms results from the ability of energetic particles and photons to ionize atoms and break chemical bonds. The irradiation of water, for example, produces ions and free radicals such as H·, ·OH, and H_2O^+, which then interact to form substances such as H_2 and H_2O_2.

The biological effects of radiation damage are classified as *genetic* or *somatic*. Genetic damage affects the genes and can be passed along to future generations. Somatic damage, which cannot be inherited, consists of molecular changes that interfere with body functions such as cell growth and cell division. Impaired production of red blood cells and antibodies is an example of somatic damage; this disorder results in anemia and increased risk of infection.

Radiation can also have beneficial aspects. Controlled doses of radiation are used for diagnostic and therapeutic purposes. At one time, thin tubes containing radium salts (or their decay product radon gas) were surgically implanted in malignant tumors. The gamma radiation emanating from the tube destroyed the tumor tissue. Today such tumors are often treated nonsurgically using external sources of radiation such as x-rays or gamma radiation from cobalt-60.

Radiopharmaceuticals are radioactive preparations that can be ingested or injected. Their use in therapy or diagnosis is based on their tendency to concentrate in specific areas of the body. When iodine-131, a beta emitter, is administered as sodium iodide in drinking water, almost all of it will find its way to the thyroid. The rate of iodine-131 uptake, determined with a Geiger counter or other scanning device, indicates whether the thyroid is functioning properly (Figure 22.11). Thyroid malignancies can be given direct continuous radiation treatment by including radioactive iodine in the diet. Another diagnostic tool is a solution of sodium chloride containing sodium-24, which can be injected into the bloodstream for the study of blood circulation. The beta particles emitted by the sodium-24 are followed with a scintillation counter as the blood travels through the body, and impaired circulation to any region is immediately detected.

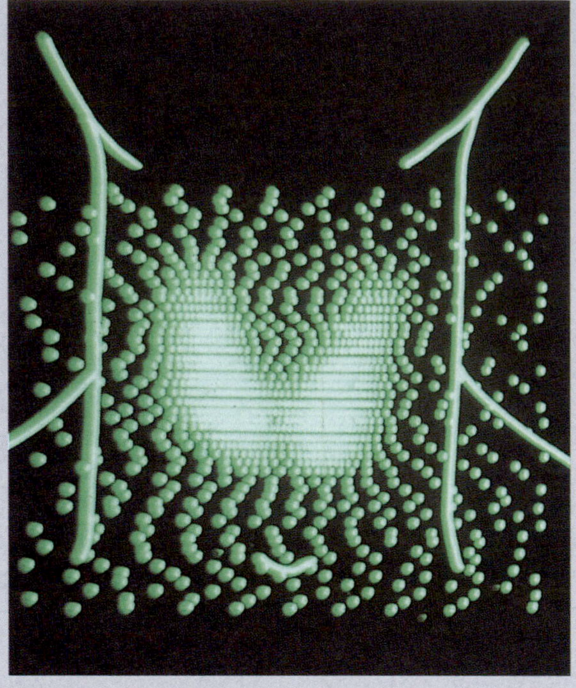

Figure 22.11

Iodine-131 scan of a normal thyroid section. An enlarged or otherwise diseased thyroid would show a different pattern of dots.

Gamma-ray emitters are especially favored for diagnosis because very little of their radiation is stopped in the body. The gamma emitters thallium-201 and technetium-99m are used for external heart-scanning studies. (The "m" in technetium-99m signifies that this radioisotope is in an excited, or "metastable" state.) A thallium-201 compound injected into the bloodstream will concentrate in normal heart muscle but will not remain in damaged tissue. A photograph with a nuclear scintillation camera allows the physician to locate the

Radiation exposure limits are normally expressed in rems. Acceptable limits, as currently set by the U.S. Environmental Protection Agency, are 5 rems per year for radiation workers and 0.5 rem per year for the general population. Average exposure from natural sources is about 0.15 rem per year. It has been estimated that a whole-body single dose of 500 rems (not 500 rems per year) will kill half of the exposed population within 30 days of exposure.

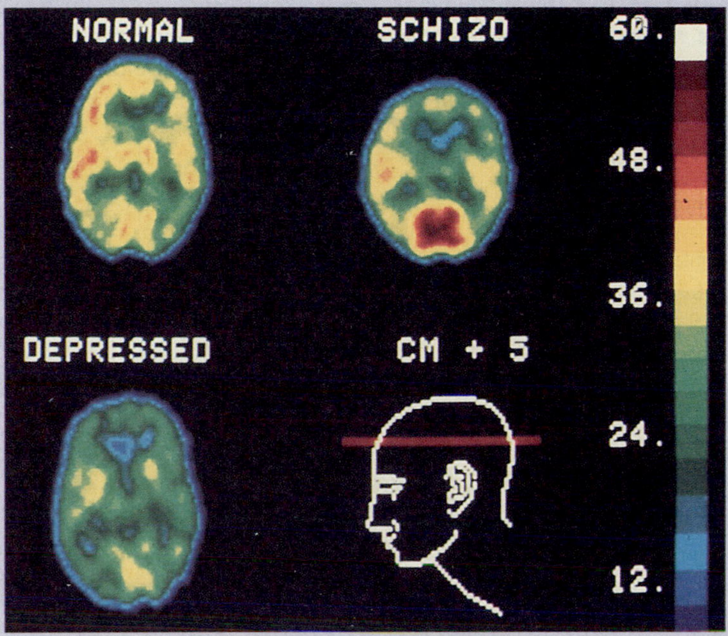

Figure 22.12

PET scans of brain cross-sections from normal, schizophrenic, and depressed patients. (Each scan was taken at the level of the red line in the lower right image.) Brain activity, as measured by glucose consumption, is lowest in the blue regions and highest in the red regions.

damaged areas. Technetium-99m, on the other hand, deposits in damaged heart cells; it will also concentrate in a brain tumor so that the tumor can be located with a brain scan. Technetium-99m is probably the most widely used radioisotope in medicine today; it is a decay product of molybdenum-99.

Reagents can be "labeled" or "tagged" with radioactive atoms to follow the course of a reaction. Such *radiotracers* are often used to investigate biochemical mechanisms. The vitamin B_{12} molecule, for example, contains cobalt, and vitamin B_{12} tagged with radiocobalt has been used to study the absorption of the vitamin from the gastrointestinal tract. Compounds tagged with iron-59 and iron-55 have been used to study the metabolism of iron. Metabolic intermediates can be isolated, and the amount of radiotracer in each can be determined by means of a counter. Carbon dioxide labeled with carbon-14 enabled the biochemist Melvin Calvin to work out the process by which plants photosynthesize carbohydrate from carbon dioxide and water.

Positron emission tomography (PET) is a radioisotope scanning technique in which compounds tagged with short-lived positron emitters are injected into the body. An example is glucose tagged with carbon-11 (half-life, 20.3 minutes). The radioisotope circulates through the body, and the positrons emitted in the heart, brain or some other organ are monitored by a PET detector. A computer uses this information to construct an image (called a PET scan) of the organ that is being examined (Figure 22.12). PET scans have been used to study the effects of drugs on cancers, to measure damage in victims of stroke or heart attack, and to study chemical changes that occur during epileptic seizures. PET scans of the brain have been particularly useful. They have shown, for example, that certain regions of the brain in manic-depressive patients use more glucose during the manic phase than the same regions in normal individuals. Schizophrenic patients, on the other hand, have been shown to use less glucose than normal individuals.

22.5 NUCLEAR STABILITY

The atomic nucleus is truly remarkable in that so many nucleons (protons and neutrons) are able to occupy such a minute volume in spite of their mutual electrostatic repulsion. Nuclear particles are held together by a powerful attractive force that we do not directly experience in our everyday world. This nuclear force is many times stronger than the electrostatic force, but it is effective only over distances that do not

If two protons are 10^{-15} m apart, the nuclear force is about 100 times stronger than the electrostatic force. If the protons are slightly farther apart (10^{-14} m) the nuclear force is about 10 times weaker than the electrostatic force.

exceed 10^{-15} m, the approximate diameter of a nucleus. Even though the nuclear force is strong, the number of protons that can coexist in a stable nucleus is limited. Elements with atomic numbers above 83 (bismuth) are radioactive, elements above 92 (uranium) are virtually absent in nature, and elements above 109 have not yet been synthesized.

Nuclear Energy Levels

A graph of elemental abundance versus atomic number (Figure 22.13) shows that hydrogen and helium are the most abundant elements in the universe. The graph has a zigzag appearance because elements with even atomic numbers are more abundant than their odd-numbered neighbors. Of the 10 most abundant elements in the earth's crust, only 3—aluminum, sodium, and potassium—have odd atomic numbers. Each of these odd-numbered elements, however, has an even number of neutrons in its most stable isotope. Of the 264 nuclear species known to be stable, a disproportionate number have an even number of neutrons, an even number of protons, or both, as shown in Table 22.6.

It is now known that nucleons have a property analogous to electron spin and a corresponding tendency to pair up with each other. This pairing accounts for the prevalence of even numbers. It is also known that nucleons occupy discrete energy levels within the nucleus, in much the same way that electrons occupy energy levels outside the nucleus. Abundance data and nuclear bombardment data show that the most stable nuclei have certain **magic numbers** of either protons or neutrons,

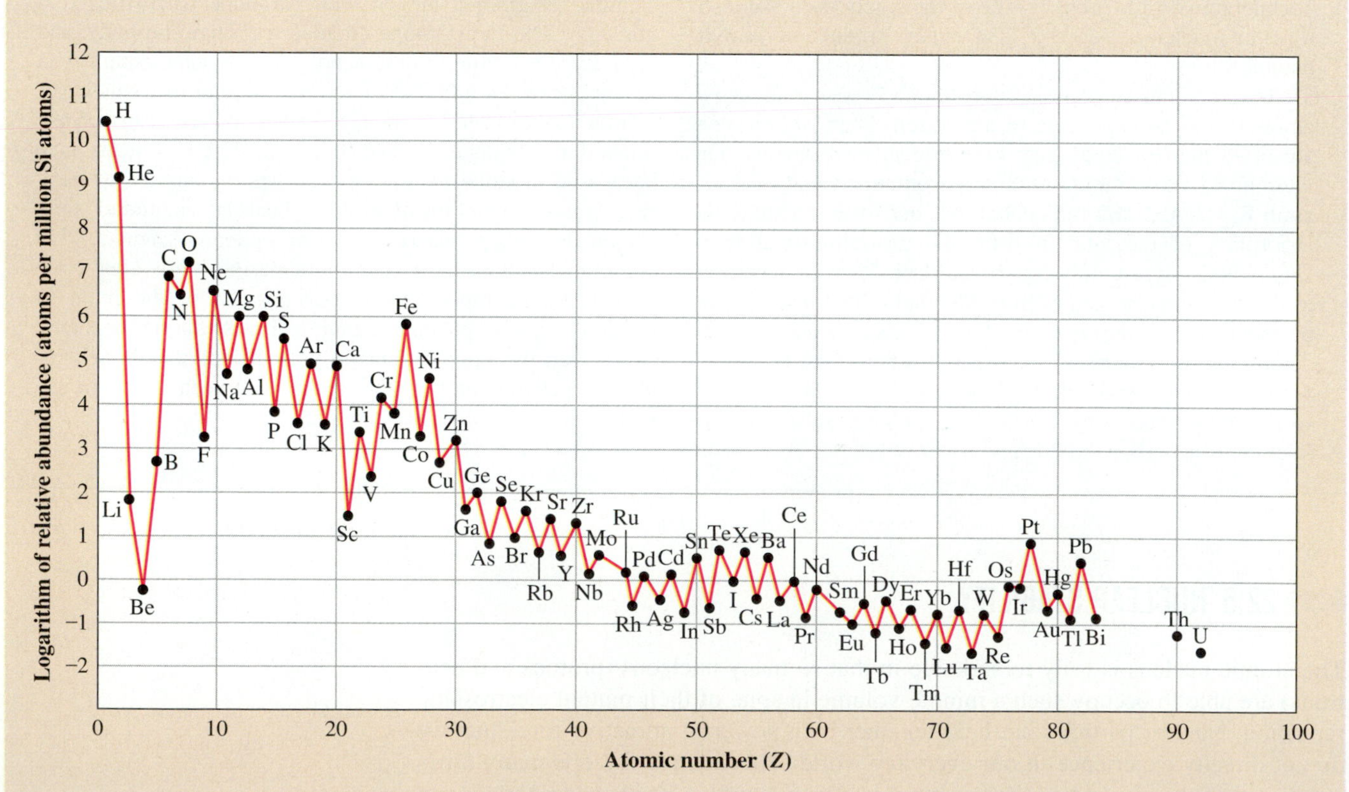

Figure 22.13

Relative abundances of the elements in the universe. The vertical scale is logarithmic. Silicon, because of its prevalence and abundance on earth, is taken as the standard and assigned a relative abundance of six (log 10^6 = 6).

TABLE 22.6 Odd and Even Numbers of Protons and Neutrons in Stable Nuclei

Number of Protons	Number of Neutrons	Number of Stable Nuclides	Examples
Even	Even	157	${}^{4}_{2}He$, ${}^{12}_{6}C$, ${}^{16}_{8}O$
Even	Odd	52	${}^{9}_{4}Be$, ${}^{25}_{12}Mg$, ${}^{87}_{38}Sr$
Odd	Even	50	${}^{7}_{3}Li$, ${}^{19}_{9}F$, ${}^{23}_{11}Na$
Odd	Odd	5	${}^{6}_{3}Li$, ${}^{10}_{5}B$, ${}^{14}_{7}N$

namely, 2, 8, 20, 28, 40, 50, or 82 (for neutrons the numbers 126 and 184 are also "magic"). These numbers correspond to filled nuclear shells, which, like filled electron shells, are more stable than partially filled shells. Nuclei such as ${}^{4}_{2}He$, ${}^{16}_{8}O$, ${}^{40}_{20}Ca$, and ${}^{208}_{82}Pb$ are "doubly magic"; they have magic numbers of both protons and neutrons and are especially stable.

The existence of nuclear energy levels accounts for the gamma-ray photons emitted during radioactive decay. The emission of these photons is analogous to the emission of visible photons during electronic transitions from high to low energy levels. A nucleus produced during a decay reaction is usually in an excited state, and its subsequent downward transitions toward the ground state then release discrete quantities of energy. For example, three gamma-ray frequencies are observed when calcium-47 nuclei decay into scandium-47 nuclei and beta particles. The origin of these photons is illustrated in Figure 22.14, which shows that the new scandium nucleus may be in one of two different excited states. A scandium nucleus in the higher excited state can fall directly to the ground state by emitting one gamma photon, or it can fall in two stages, emitting two gamma photons with different frequencies. Both processes take place, so three different photons are produced. Note also that the kinetic energy of the emitted beta particle depends on the energy level of the excited scandium nucleus: the higher the energy level of the nucleus, the lower the kinetic energy of the beta particle.

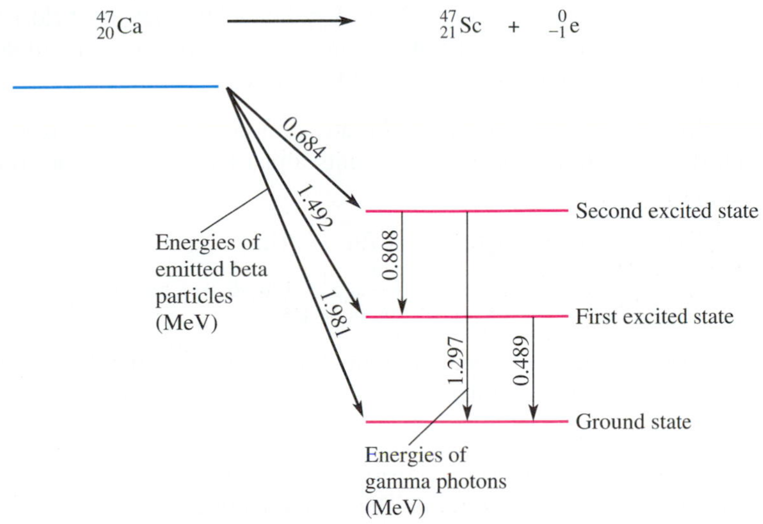

Figure 22.14

Beta particles with three different energies are emitted by calcium-47 as it decays to scandium-47. A lower energy beta particle leaves more energy in the scandium nucleus; the excited nucleus eventually reaches the ground state by emitting gamma photons. As this diagram shows, three different photons can be produced by the excited states of scandium-47.

CHEMICAL INSIGHT

NEUTRON ACTIVATION ANALYSIS

A Rembrandt painting, an ancient pottery sherd, a rag from the scene of a crime, hair from an exhumed corpse—these are examples of objects that have been analyzed for certain elements by *neutron activation analysis*. One great advantage of this method is that it leaves the analyzed substances essentially unchanged; a painting, for example, would not be ruined. The material to be examined is bombarded with neutrons that raise some of its nuclei to metastable energy states. A *metastable state* is an excited state that does not immediately decay. The activated nuclei eventually return to the ground state, emitting their extra energy in the form of gamma photons. Each element has a characteristic set of metastable states with characteristic decay times and characteristic decay energies. A scintillation counter is used to monitor the intensities and energies of the gamma rays emitted by the irradiated object over a period of days, and the results show not only which elements are present but their relative amounts as well.

Neutron activation analysis sometimes detects forgeries by revealing the presence of modern pigments in paintings that had been attributed to old masters, but it also may reveal a master's painting underneath an inferior one. Applied to ancient pottery, neutron activation analysis can be used to determine where the potter's clays and pigments were obtained, thus providing clues to the history of that civilization. Neutron activation analysis even has a place in forensic medicine. Analysis of hair from the corpse of Napoleon has revealed an arsenic concentration that is consistent with gradual arsenic poisoning.

The Stability Band

When a graph is made of the number of neutrons in a nucleus versus the number of protons, the stable nuclei are found to lie within a narrow region called the **stability band** (see Figure 22.15). Stability apparently requires the neutron/proton ratio (n/p ratio) to be about 1.0 for the lightest elements and to increase up to about 1.5 for the heaviest elements. The need for excess neutrons to accompany protons in heavy nuclei suggests that neutrons play an important role in counteracting proton–proton repulsions.

Radionuclides have n/p ratios that fall outside the stability band, and when they decay, they tend to form stable species that lie within the band. The most common decay modes are summarized in Table 22.7. A radionuclide that lies below the stability band (i.e., to the right of the band) has an excess of protons and will decay by a mode that increases the n/p ratio. The possible modes are:

1. *Alpha emission*: The decay series of Figure 22.3 contains several examples of alpha emission. Radium, for example, emits alpha particles as it decays to the noble gas radon:

$$^{226}_{88}Ra \rightarrow {}^{222}_{86}Rn + {}^{4}_{2}He$$

$^{226}_{88}Ra$: n/p = 138/88 = 1.57; $^{222}_{86}Rn$: n/p = 136/86 = 1.58

Alpha emission is the most common decay mode for radionuclides heavier than lead (atomic number 82). Elements below atomic number 66 do not usually emit alpha particles.

2. *Positron emission:* The loss of one positive charge converts a nuclear proton into a neutron, thus raising the n/p ratio. This decay mode is fairly common among

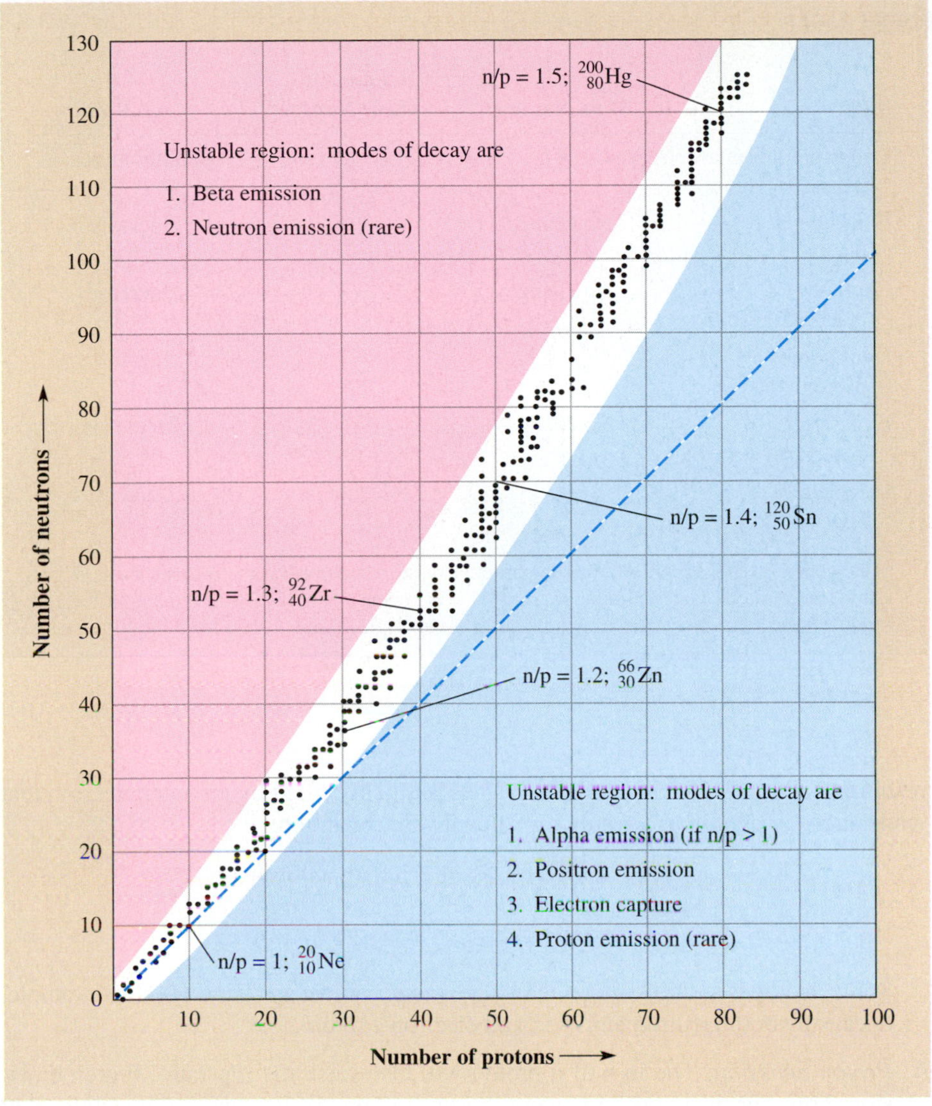

Figure 22.15

The stability band. On a plot of number of neutrons versus number of protons, the known stable nuclei occupy a narrow band in which the neutron/proton ratio rises from about 1 to 1.5 with increasing atomic number. The dashed line represents a neutron/proton ratio of 1. A nucleus outside the band will decay by a mechanism that brings its neutron/proton ratio into the stable region.

The heaviest known stable nucleus is $^{209}_{83}Bi$.

the lighter radioisotopes:

$$^{31}_{16}S \rightarrow ^{31}_{15}P + ^{0}_{1}e$$

n/p = 15/16 = 0.94 (for $^{31}_{16}S$); n/p = 16/15 = 1.07 (for $^{31}_{15}P$)

The positron and the electron are *antiparticles*—on contact they disappear in a burst of energy:

$$^{0}_{-1}e + ^{0}_{1}e \rightarrow 2 \text{ gamma-ray photons}$$

Most of the positrons emitted during nuclear reactions are annihilated by electrons within 1 nanosecond (10^{-9} s) of ejection.

3. *Electron capture*: A 1*s* electron captured from the innermost extranuclear shell increases the n/p ratio by converting one nuclear proton into a neutron. The

TABLE 22.7 Modes of Radioactive Decay

Mode	Mass Number	Change in Atomic Number	n/p Ratio
Emission of alpha particle (^4_2He)	−4	−2	Usually increases[a]
Emission of positron (^0_1e)	No change	−1	Increases
Capture of electron ($^0_{-1}e$)	No change	−1	Increases
Emission of proton (^1_1H)	−1	−1	Increases
Emission of beta particle ($^0_{-1}e$)	No change	+1	Decreases
Emission of neutron (^1_0n)	−1	No change	Decreases

[a]The change in the n/p ratio for alpha-particle emission depends on the original value of the ratio. The ratio will increase if the original value is greater than one, it will remain unchanged if the original value is one, and it will decrease if the original value is less than one.

resulting vacancy in the inner electron shell is quickly filled by an outer electron that emits an x-ray photon as a result of its downward transition:

$$^{60}_{30}Zn + {}^{0}_{-1}e \rightarrow {}^{60}_{29}Cu + \text{x-ray photon}$$

($^{0}_{-1}e$: 1*s* electron; $^{60}_{29}Cu$: increased n/p ratio)

Some isotopes exhibit more than one mode of decay. Zinc-60, for example, decays by both positron emission and electron capture.

4. *Proton emission*: The loss of a proton will also raise the n/p ratio. Proton emission, however, is a very rare event; boron-9 is one of the few radionuclides that decay this way:

$$^{9}_{5}B \rightarrow {}^{8}_{4}Be + {}^{1}_{1}H$$

($^{8}_{4}Be$: increased n/p ratio)

A radionuclide that lies above the stability band (i.e., to the left of the band) has an excess of neutrons and will decay by a mode that decreases the n/p ratio. The possible modes are:

1. *Beta emission*: Beta emission is by far the most common mode for decreasing the n/p ratio. The loss of a beta particle converts one neutron into a proton. Carbon-14, the isotope of carbon used in radiodating, decays by beta emission to produce nitrogen atoms:

$$^{14}_{6}C \rightarrow {}^{14}_{7}N + {}^{0}_{-1}e$$

($^{14}_{6}C$: n/p = 8/6 = 1.33; $^{14}_{7}N$: n/p = 7/7 = 1)

2. *Neutron emission*: This type of decay, though rare, has been observed in a few synthetic radionuclides. An example is the decay of iodine-137:

$$^{137}_{53}I \rightarrow ^{136}_{53}I + ^{1}_{0}n$$

$n/p = 84/53 = 1.58$ (for $^{137}_{53}I$); $n/p = 83/53 = 1.57$ (for $^{136}_{53}I$)

Iodine-137 atoms can also decay by beta emission.

EXAMPLE 22.6

What decay mode would you predict for tin-109?

SOLUTION

The atomic number of tin is 50. Tin-109, with 50 protons and 59 neutrons, is just below the stability band in Figure 22.15, so it will decay by a mode that raises the n/p ratio. Alpha emission is not common for elements with atomic numbers below 66, and proton emission rarely occurs. The most likely possibilities are positron emission and electron capture:

$$^{109}_{50}Sn \rightarrow ^{109}_{49}In + ^{0}_{1}e$$

$$^{109}_{50}Sn + ^{0}_{-1}e \rightarrow ^{109}_{49}In$$

(It has been shown that tin-109 decays by both of these modes.)

PRACTICE EXERCISE 22.6

Predict the decay mode of rubidium-95.

22.6 MASS–ENERGY CONVERSIONS

The products of a nuclear decay reaction always have a lower mass than the reactants. We can explain this observation in terms of the modern view of mass as a concentrated form of energy. An unstable nucleus has a higher energy and therefore a greater mass than its stable decay products. With each spontaneous nuclear transformation, a small quantity of mass is transformed into kinetic or radiant energy, and the total mass therefore decreases. Consider the emission of an alpha particle by the americium-241 nucleus:

$$^{241}_{95}Am \rightarrow ^{4}_{2}He + ^{237}_{93}Np$$

The nuclear masses of the species in the reaction (see Table 22.8) are

Reactant mass:	^{241}Am	241.0047 u
Product masses:	^{237}Np	236.9972 u
	^{4}He	4.0015 u
Total product mass:		240.9987 u

TABLE 22.8 Selected Particle and Nuclear Masses in Atomic Mass Units[a]

Particle	Mass
Alpha particle (helium-4 nucleus)	4.00150
Electron	0.00054858
Hydrogen-2 (deuteron)	2.01355
Hydrogen-3 (tritium nucleus)	3.01550
Neutron	1.00866
Proton	1.00728

Nucleus	Nuclear Mass[b]
Americium-241	241.00471
Argon-40	39.95251
Beryllium-9	9.00999
Boron-10	10.01019
Boron-11	11.00656
Carbon-12	11.99671
Helium-3	3.01493
Helium-4	4.00150
Hydrogen-2	2.01355
Hydrogen-3	3.01550
Iron-56	55.92068
Iron-57	56.92113
Lithium-6	6.01348
Lithium-7	7.01436
Neptunium-237	236.99715
Nickel-58	57.91999
Oxygen-16	15.99053
Potassium-40	39.95358
Silicon-28	27.96925
Thorium-234	233.99422
Plutonium-239	239.00059
Uranium-235	234.99346
Uranium-238	238.00032

[a] $1 \text{ g} = 6.0221 \times 10^{23}$ u.

[b] The nuclear mass is the isotopic mass minus the total mass of the extranuclear electrons.

The law of conservation of mass does not apply to nuclear reactions.

The vanished mass is 241.0047 u − 240.9987 u = 0.0060 u. This amount may seem inconsequential, but it is 0.0025% of the original reactant mass. For every kilogram of americium-241 that decays, there is a conversion of 0.025 g of mass into energy.

The energy equivalent of a mass m is given by Albert Einstein's **mass–energy equation** (Figure 22.16):

$$E = mc^2 \tag{22.4}$$

where E is the energy evolved in the nuclear transformation, m is the mass lost, and c is the speed of light in a vacuum, 2.9979×10^8 m/s. The energy will be in joules if the mass is in kilograms (recall that $1\text{ J} = 1\text{ N}\cdot\text{m} = 1\text{ kg}\cdot\text{m}^2/\text{s}^2$). The energy equivalent of 1 atomic mass unit (u) is obtained from Equation 22.4 as follows:

$$\begin{aligned} \text{E} &= 1\text{ u} \times \frac{1\text{ g}}{6.0221 \times 10^{23}\text{ u}} \times \frac{1\text{ kg}}{1000\text{ g}} \times (2.9979 \times 10^8\text{ m/s})^2 \\ &= 1.4924 \times 10^{-10}\text{ J} \end{aligned}$$

This amount of energy is very small, and nuclear chemists prefer the electron volt and megaelectron volt (Table 22.1) as energy units for individual particles. The conversion factor $1\text{ MeV} = 1.6022 \times 10^{-13}$ J gives the energy equivalent of 1 u as

$$1\text{ u} = 1.4924 \times 10^{-10}\text{ J} \times \frac{1\text{ MeV}}{1.6022 \times 10^{-13}\text{ J}} = 931.47\text{ MeV}$$

Figure 22.16
Albert Einstein (1879–1955). Einstein's famous equation, $E = mc^2$, describes the equivalence of mass and energy.

EXAMPLE 22.7

As we have seen, the decay of an americium-241 atom to neptunium-237 is accompanied by a mass decrease of 0.0060 u. Calculate the released energy in (a) joules and (b) megaelectron volts.

SOLUTION

The energy equivalent of 1 u is $1.492 \times 10^{-10}\text{ J} = 931.5$ MeV; therefore,

(a) $0.0060\text{ u} \times 1.492 \times 10^{-10}\text{ J/u} = 9.0 \times 10^{-13}\text{ J}$

(b) $0.0060\text{ u} \times 931.5\text{ MeV/u} = 5.6\text{ MeV}$

The calculations in Example 22.7 show that 9.0×10^{-13} J is released during the decay of one americium atom. The energy released during the decay of 1 mol of americium atoms would be 6.022×10^{23} times greater, or 5.4×10^{11} J. This amount of energy is equivalent to more than 150,000 kilowatt hours of electrical energy.

PRACTICE EXERCISE 22.7

The excited technetium-99m isotope used in medical diagnosis emits a 96.5-keV (keV = kiloelectron volt) gamma photon when it decays to its ground state. (The m in technetium-99m stands for "metastable." A metastable state is a long-lived excited state.) What mass does the isotope lose during this transition? Give your answer in atomic mass units.

Einstein's equation is universally true. Any system, nuclear or chemical, will exhibit a mass change when its energy increases or decreases. The law of conservation of mass (Section 3.2) is valid for ordinary chemical reactions only because the mass changes that occur during endothermic and exothermic reactions are such minute fractions of the total mass that they cannot be measured directly.

Nuclear Binding Energy

The energy needed to decompose a nucleus into protons and neutrons is called the **binding energy**. This energy, which is analogous to chemical bond energy, is required because an assembled nucleus is more stable than its parts. The amount of energy required to decompose the nucleus is the same as the amount of energy released when the nucleus is formed. Because the energy of the nucleus is less than that of its independent protons and neutrons, its mass must also be less. The difference in mass, called the **mass defect**, can be converted into the binding energy using either Equation 22.4 or the relation 1 u = 931.5 MeV. Consider, for example, the decomposition of an alpha particle into two protons and two neutrons:

$$^{4}_{2}\text{He} \rightarrow 2\,^{1}_{1}\text{H} + 2\,^{1}_{0}\text{n}$$

The mass defect is found by subtracting the mass of the alpha particle from the sum of the masses of the separated protons and neutrons. The total mass of the separate nucleons is obtained from Table 22.8:

$$\begin{aligned} 2 \text{ protons} \times 1.00728 \text{ u/proton} &= 2.01456 \text{ u} \\ 2 \text{ neutrons} \times 1.00866 \text{ u/neutron} &= \underline{2.01732 \text{ u}} \\ \text{Total} &= 4.03188 \text{ u} \end{aligned}$$

The mass defect is

$$\begin{aligned} \text{mass defect} &= \text{mass of separate nucleons} - \text{mass of alpha particle} \\ &= 4.03188 \text{ u} - 4.00150 \text{ u} \\ &= 0.03038 \text{ u} \end{aligned}$$

The binding energy is the energy equivalent of the mass defect, or

$$0.03038 \text{ u} \times 931.5 \text{ MeV/u} = 28.30 \text{ MeV}$$

When comparing different nuclei, it is customary to consider the *binding energy per nucleon*. The alpha particle has four nucleons, so the binding energy per nucleon is one-fourth of the total binding energy or 7.075 MeV.

PRACTICE EXERCISE 22.8

The nickel-58 nucleus contains 28 protons and 30 neutrons. Use data from Table 22.8 to calculate (a) the mass defect in atomic mass units, (b) the binding energy in megaelectron volts, and (c) the binding energy per nucleon.

When binding energies per nucleon are plotted against mass number as in Figure 22.17, they show an increase with mass number for lighter nuclei, a peak value at iron-56 (the most abundant iron isotope), then a gradual decrease with mass number for heavier nuclei. The most stable nuclei lie in the vicinity of iron-56. This graph also shows that energy will be released when less stable reactants from the extreme

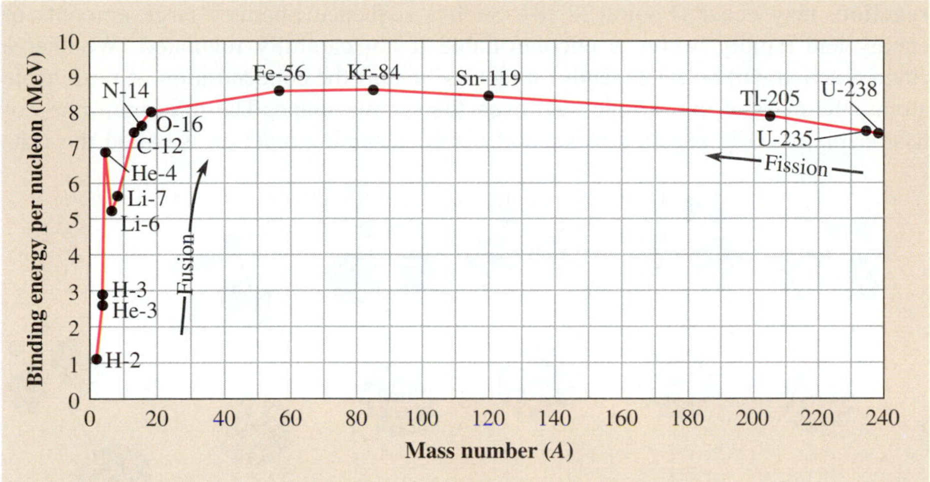

Figure 22.17
A plot of binding energy per nucleon versus mass number reaches a maximum at iron-56. The most stable nuclei lie in the vicinity of iron-56; energy is released when lighter nuclei fuse together and when heavier nuclei undergo fission.

regions of the curve are converted to more stable products that lie near the center of the curve. Very light nuclei will evolve energy when they combine to form heavier nuclei, a process called **fusion**, and very heavy nuclei will evolve energy when they split into lighter nuclei, a process called **fission**. Note that the increase in binding energy for the light elements is much steeper than the decrease for heavy elements. This observation suggests that fusion reactions should give out more energy than fission reactions. Both fission and fusion reactions are discussed more fully in the following sections.

22.7 FISSION

The nuclear binding energy curve (Figure 22.17) shows that fission of a heavy nucleus into two or more lighter nuclei is accompanied by the release of energy. Fortunately, most naturally occurring nuclei do not readily undergo fission reactions. Uranium-235 is an exception. The fission of uranium-235 is initiated by the absorption of stray neutrons from the atmosphere. The uranium-235 nucleus forms an unstable uranium-236 nucleus, which then disintegrates into two nuclei with masses in the neighborhood of 140 and 90 u. Many decay modes are possible, and more than one neutron is given off during each disintegration:

$$
{}^{235}_{92}\text{U} + {}^{1}_{0}\text{n} \longrightarrow
\begin{cases}
{}^{141}_{56}\text{Ba} + {}^{92}_{36}\text{Kr} + 3\,{}^{1}_{0}\text{n} \\
{}^{144}_{54}\text{Xe} + {}^{90}_{38}\text{Sr} + 2\,{}^{1}_{0}\text{n} \\
{}^{144}_{55}\text{Cs} + {}^{90}_{37}\text{Rb} + 2\,{}^{1}_{0}\text{n} \\
\text{other combinations of nuclides}
\end{cases}
$$

Of the nuclei that can undergo fission, only uranium-235 and plutonium-239 are currently of practical importance.

A fission event produces more neutrons than it consumes; on the average, 2.4 neutrons are produced for each atom of uranium-235. These neutrons can initiate further fission reactions, which produce still more neutrons, which initiate additional fission reactions, and so on. If conditions are such that each fission triggers one or more others, a self-sustaining sequence of fission reactions, called a **nuclear chain**

reaction, may occur (Figure 22.18). Such a sequence liberates large amounts of energy and rapidly becomes uncontrollable if not carefully regulated. Whether a chain reaction will occur depends on factors such as the concentration of fissionable atoms, the mass of the sample, and even the sample shape. The minimum mass of fissionable material required to support a self-sustaining reaction is called the **crit-**

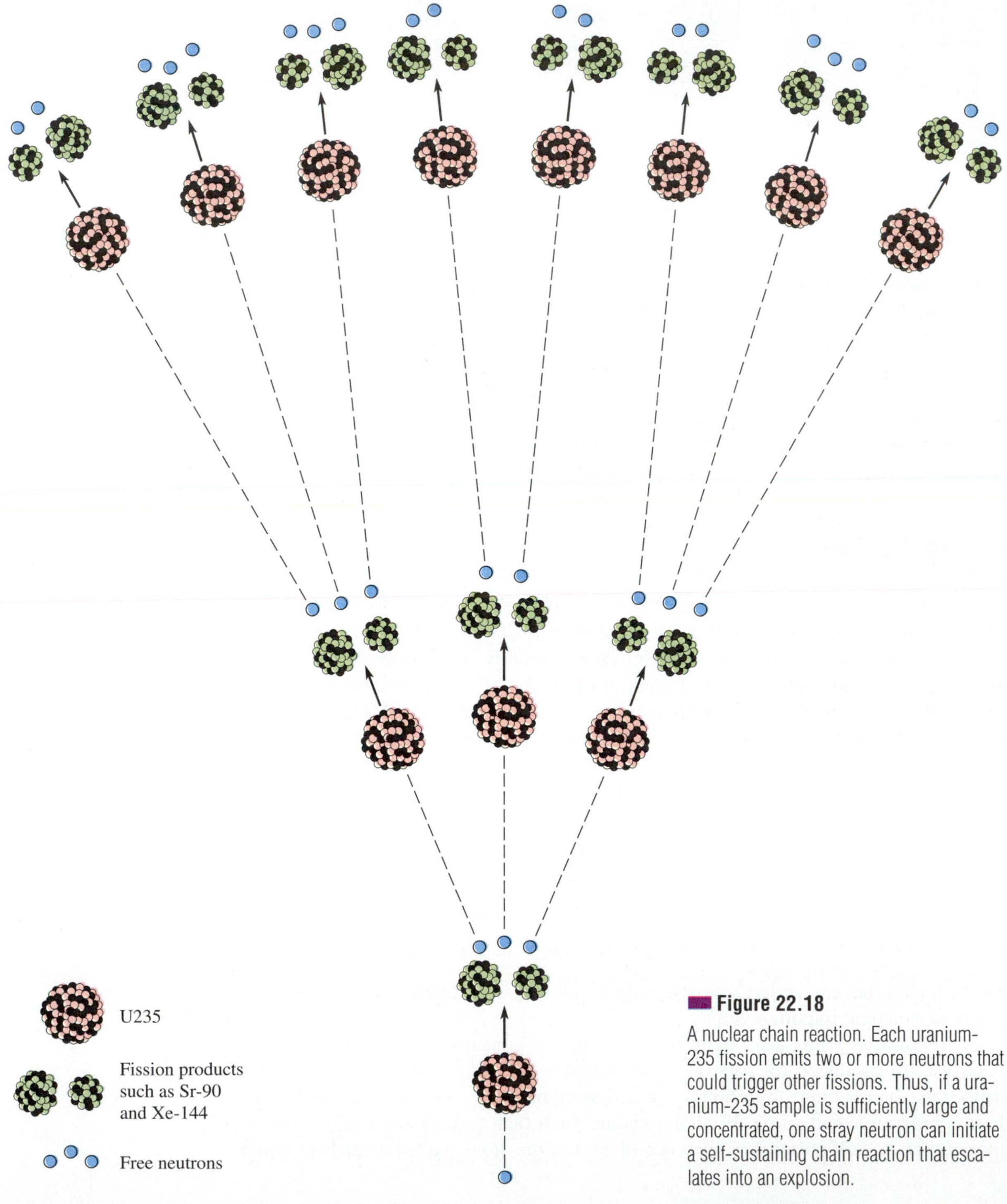

Figure 22.18

A nuclear chain reaction. Each uranium-235 fission emits two or more neutrons that could trigger other fissions. Thus, if a uranium-235 sample is sufficiently large and concentrated, one stray neutron can initiate a self-sustaining chain reaction that escalates into an explosion.

ical mass. When the mass of a sample is subcritical, that is, when it is less than the critical value, most of the neutrons generated by fission will escape into the surroundings before they strike a nucleus. When the mass exceeds its critical value, most neutrons will strike other uranium-235 nuclei, and the resulting nuclear chain reaction will escalate into an explosion. In fact, an explosion is inevitable once the requisite mass is assembled and held together for a long enough time to allow the chain reaction to occur. The first atomic bombs contained subcritical masses of uranium-235 or plutonium-239 (Figure 22.19). Just prior to detonation, a chemical explosive was used to force the masses together so that the nuclear chain reaction and explosion could occur. The explosion energies of these first bombs ranged from 0.01 to 0.02 megatons, where 1 **megaton** is the energy equivalent of 1 million tons of TNT (Figure 22.20).

The critical mass of a spherical uranium-235 sample is less than 1 kg.

Nuclear Reactors

The energy released by the fission of 1 g of uranium-235 is equivalent to that produced by the combustion of 13.7 barrels of crude oil or 2700 kg of coal; hence, fission is regarded by many as an attractive energy source. There are more than 100 nuclear reactors in the United States, and nuclear power plants are now a major source of electrical power. The nuclear fuel used for controlled fission in a power plant usually consists of UO_2 pellets prepared from uranium containing about 3% uranium-235. This concentration is far below the 90% concentration required for an explosion, but there are nevertheless a number of serious problems associated with its use. These problems include the cost and difficulty of obtaining enriched uranium fuel (ordinary uranium contains only 0.7% uranium-235), the reprocessing of spent fuel, and the storage of the highly radioactive waste products. Some of the fission products produced in nuclear reactors have long half-lives and will remain active for thousands of years. Another problem is the possibility of disaster due to overheating, explosions, or earthquakes, which could rupture the reactor shielding and release substantial amounts of radiation and radioactive material into the environment.

Even if these difficulties are dealt with, uranium-235 is a rare isotope and we can eventually expect shortages. One solution may be to manufacture more nuclear fuel in a **breeder reactor**, which uses neutrons from the uranium-235 fission to convert uranium-238 into fissionable plutonium-239:

$$^{238}_{92}U + ^{1}_{0}n \rightarrow ^{239}_{92}U$$

$$^{239}_{92}U \xrightarrow{t_{1/2} = 24 \text{ min}} ^{239}_{93}Np + ^{0}_{-1}e$$

$$^{239}_{93}Np \xrightarrow{t_{1/2} = 2.3 \text{ days}} ^{239}_{94}Pu + ^{0}_{-1}e$$

Each uranium-235 fission produces two or more neutrons, enough to sustain the fission reaction itself and produce plutonium at the same time. Despite the potential for conserving scarce nuclear fuel, the large-scale development of breeder reactors has been delayed for a number of reasons. Plutonium is an extremely potent poison and carcinogen; inhalation of only a few micrograms can be fatal. Plutonium fuel, which contains about 70% plutonium-239, is more concentrated than uranium fuel and the handling, storage, and transportation of large quantities present major safety problems. Furthermore, plutonium-239 is used in atomic bombs and other nuclear weapons, and it is feared that extraordinary security precautions would have to be taken to prevent its theft by terrorist groups.

Figure 22.19

Schematic diagram of a fission bomb. A chemical explosive drives subcritical uranium-235 masses together to produce a supercritical mass, which explodes. (a) A *gun-type* bomb. (b) In an *implosion* bomb, the chemical explosive pushes the wedges of plutonium-239 toward the center of a sphere.

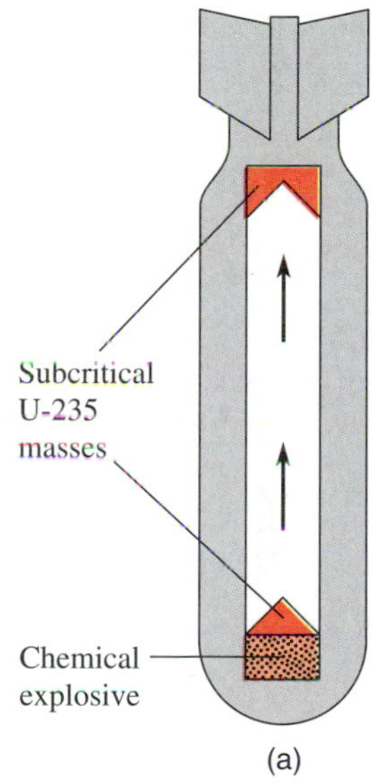

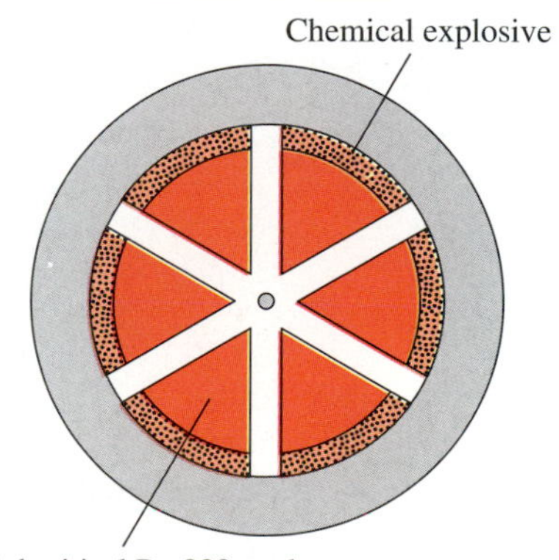

(a) 0.006 seconds (b) 0.053 seconds

(c) 2.0 seconds (d) 5.0 seconds

Figure 22.20

The first atomic bomb was exploded over Trinity Site, New Mexico, at 5:29:45 A.M., mountain war time, July 16, 1945. The flash of light from this explosion marked the dawn of the atomic age.

How a Reactor Works

Enriched uranium fuel is made by converting *yellow cake* (UO_2; prepared from uranium ores) to UF_6, and then passing the UF_6 through gas centrifuges or porous effusion barriers (see *Chemical Insight: Separation of Isotopes by Gaseous Effusion*, p. 202).

The heart of a nuclear power plant is a **reactor core** like the one shown in Figure 22.21. The **fuel rods** usually contain uranium dioxide (UO_2) in which the uranium-235 has been enriched from its natural abundance of 0.7% to about 3%. The operation of a nuclear reactor requires that the rate of the fission process and the speed of the released neutrons be kept under rigid control. The fission rate is regulated with a set of **control rods** containing neutron-absorbing atoms such as cadmium or, more commonly, boron:

$$^{10}_{5}\text{B} + ^{1}_{0}\text{n} \rightarrow ^{7}_{3}\text{Li} + ^{4}_{2}\text{He}$$

Ideally, there should be just enough free neutrons to keep the reaction self-sustaining; too few neutrons will allow the reaction to die out, and too many neutrons may cause overheating and meltdown of the reactor core and shield. The fission rate is increased by raising the control rods out of the reactor core and decreased by lowering them into the core. Full lowering of the control rods stops the reaction and shuts down the reactor.

Neutron speed is controlled with the aid of a **moderator**, a substance whose molecules slow the neutrons by successive collisions until they have the most favorable energy for initiating uranium-235 fission. If the neutrons are moving too fast, they will react with uranium-238 to produce unwanted neptunium-239 and plutonium-239 (Section 22.3). Graphite, light water (H_2O), and heavy water (D_2O) are used as moderators in different types of reactors. Graphite has the disadvantage of being combustible. The Chernobyl disaster in Ukraine in 1986 occurred when the graphite in a Soviet nuclear reactor caught fire. The reactor did not have an adequate containment dome, and the accompanying release of radioactive debris into the atmosphere

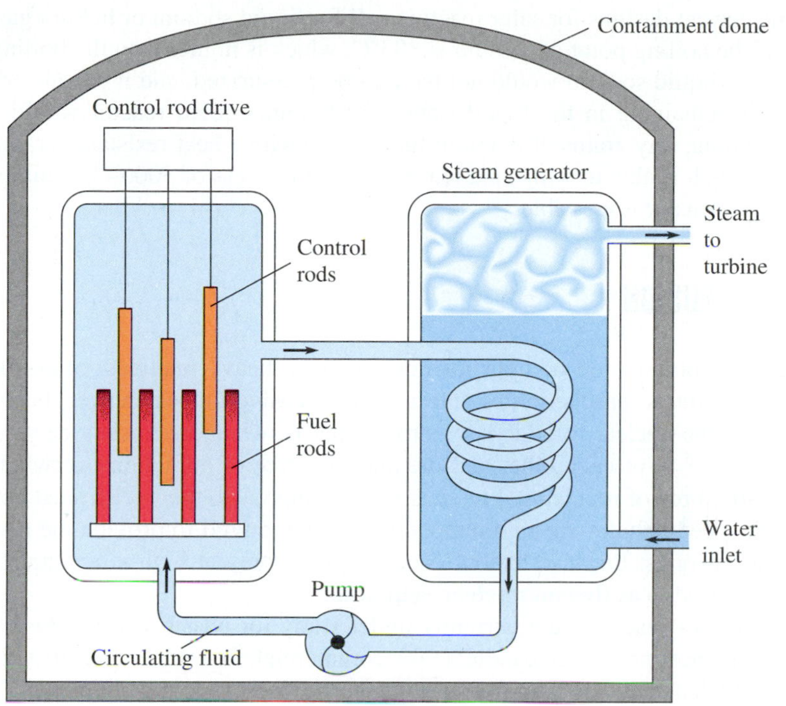

Figure 22.21

(a) Schematic diagram of a common type of nuclear power plant. The core consists of fuel rods containing uranium-235 and movable control rods that regulate the chain reaction by absorbing neutrons. The circulating fluid serves both as a moderator and as a heat exchanger that carries heat to the steam generator. The steam drives a turbine that produces electricity. (b) Loading of fuel rods into a reactor core; the rods are long, thin, metal tubes filled with enriched UO_2 pellets.

necessitated the evacuation of thousands of people from their homes and resulted in widespread pollution of farmlands and livestock. There were 31 deaths, mostly from radiation poisoning, and tens of thousands of people are expected to develop radiation-induced cancers over the next 50 years.

Water is not combustible and has the additional advantage of being able to carry off the heat generated in a reactor. Most reactors in the United States use light water as both a moderator and coolant. The water is usually under pressure to prevent it from boiling. Pumps circulate the water through the coils of a steam generator as shown in Figure 22.21, and the resulting steam is used to drive turbines that generate electricity. In 1979 a combination of a water pump failure and human error resulted in the overheating and partial meltdown of fuel rods at the Three Mile Island nuclear reactor in Pennsylvania. Some radioactivity was released into the atmosphere, but fortunately, the built-in safety systems were able to prevent a potentially disastrous explosion.

Nuclear reactors using light water (H_2O) as a moderator require fuel made from enriched uranium (3% uranium-235) because many of the neutrons in such a reactor combine with hydrogen nuclei instead of uranium atoms:

$$^{1}_{1}\text{H} + ^{1}_{0}\text{n} \rightarrow ^{2}_{1}\text{H}$$

Heavy water (D_2O) is a more efficient moderator because neutrons do not readily combine with deuterium nuclei. A heavy-water reactor is thus able to run on less expensive fuel prepared from natural uranium (0.7% uranium-235). Heavy water, however, is expensive in its own right—energy is needed to produce it. Heavy-water reactors are used in some research laboratories and in Canada, where inexpensive hydroelectric power keeps down production costs. The CANDU nuclear power plant in eastern Canada uses heavy water as a moderator and light water to convey heat to the turbines.

Some current designs for safer reactors call for liquid sodium or helium gas as the coolant. The boiling point of sodium is 883°C, which is higher than the boiling point of water, so liquid sodium would not have to be pressurized, and it can absorb more heat while remaining in the liquid state. The helium-cooled reactor would have a core containing tiny grains of uranium fuel coated with a heat-resistant ceramic. The ceramic, which is able to withstand temperatures in excess of 3000°C, would make a meltdown virtually impossible.

22.8 FUSION

Fusion, the combination of light nuclides to form heavy nuclides, offers the best hope for the future production of safe and inexpensive nuclear power. Fusion will occur when the nuclear force is very strong, that is, when the nuclear centers come to within 10^{-15} m of each other and the nuclear force of attraction overwhelms the electrostatic force of repulsion. For such a close approach, the nuclei must collide at very high speeds; thus very high temperatures are required to provide the necessary activation energies. Because fusion reactions occur only at high temperatures, they are also referred to as **thermonuclear reactions**.

A number of reactions are currently under study for possible use in *fusion reactors*, but the most promising candidate for commercial, first-generation fusion power plants is the deuterium–tritium fusion reaction:

$$^{2}_{1}\text{H} + ^{3}_{1}\text{H} \rightarrow ^{4}_{2}\text{He} + ^{1}_{0}\text{n} + 17.6 \text{ MeV}$$

This reaction requires an activation energy of 10 keV (0.010 MeV), but it releases 17.6 MeV of energy.

Tritium is a scarce material, but when the deuterium–tritium fusion is under way, additional tritium could be generated by allowing fusion-produced neutrons to escape into a lithium-bearing blanket surrounding the fusion chamber (Figure 22.22):

$$^{7}_{3}\text{Li} + ^{1}_{0}\text{n} \rightarrow ^{3}_{1}\text{H} + ^{4}_{2}\text{He} + ^{1}_{0}\text{n}$$

$$^{6}_{3}\text{Li} + ^{1}_{0}\text{n} \rightarrow ^{3}_{1}\text{H} + ^{4}_{2}\text{He}$$

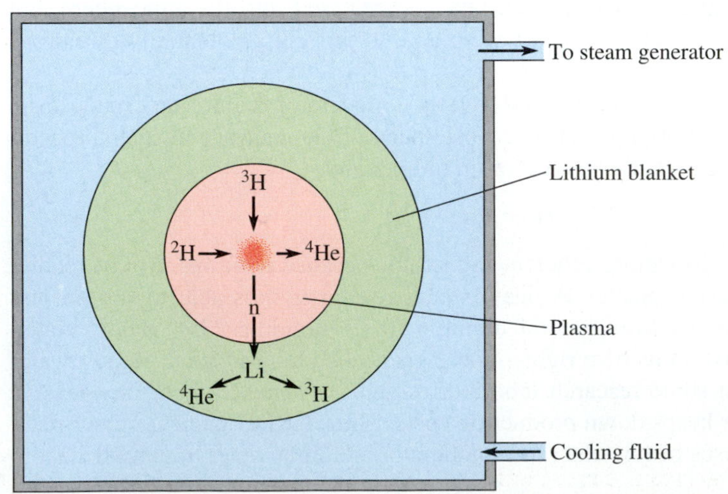

Figure 22.22
A projected scheme for obtaining power from deuterium–tritium fusion. The fusion reaction occurs in the central chamber. Neutrons formed during the reaction enter the lithium blanket surrounding the chamber, where they produce additional tritium fuel. The cooling fluid carries the heat of reaction to an energy conversion system such as a steam generator.

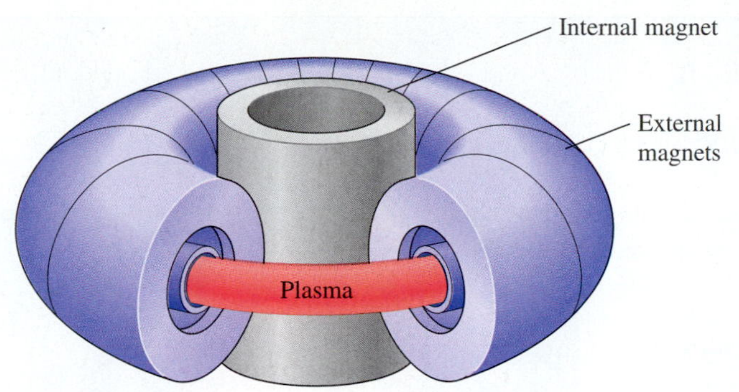

Figure 22.23

Confining plasma in a "magnetic bottle." A *tokamak* is a reactor that uses two magnetic fields, one internal and one external, to force charged plasma particles to remain within a doughnut-shaped region.

Deuterium can be obtained from seawater, which contains about one deuterium atom for every 6700 ordinary hydrogen atoms. One cubic meter of seawater will provide about 34 g of deuterium, which, if converted to helium-4 by the deuterium–tritium fusion, would liberate about 2.9×10^{13} J of energy. The amount of energy released per gram of deuterium is about 10 times more than the energy released during the fission of 1 g of uranium-235. Also, unlike fission reactions, the fusion reaction does not produce long-lived radioactive waste products. Its products are relatively harmless, and tritium leakage would be the principal hazard. Tritium, a beta emitter with a half-life of 12.3 years, is one of the less toxic radioisotopes.

The high activation energies required to initiate fusion reactions are the major obstacle to the development of controlled fusion power. The required temperatures are almost unbelievable—2×10^8 K for the deuterium–tritium reaction. At this temperature, matter exists only as a **plasma**, a gas composed of separate electrons and positive nuclei. A number of methods, including the use of particle accelerators, magnetic shock waves, and laser beams, have been explored for producing plasmas, and plasma temperatures in the vicinity of 6×10^9 K—more than enough to ignite the deuterium–tritium fusion—have now been achieved. Another, more complicated, problem is how to confine the hot plasma. All ordinary containers would vaporize at plasma temperatures. It is known that magnetic fields restrict the motion of charged particles in space, and some progress has been made in confining plasmas by means of "magnetic bottles" (Figure 22.23).

Another technique for generating fusion energy involves the laser implosion (inward explosion) of minute, plastic-coated glass spheres containing deuterium and tritium under high pressure (Figure 22.24). As the plastic vaporizes in the focused laser beam, it emits a shock wave that implodes the sphere and compresses its contents to achieve the pressure and temperature needed for fusion. A series of such implosions could release a steady supply of fusion energy.

Sad to say, the most successful use of fusion so far has been in weapons of destruction. The hydrogen bomb is a thermonuclear weapon in which fusion is initiated by the energy of a fission bomb. A two-stage fission–fusion bomb contains lithium-6 deuteride ($^{6}Li^{2}H$) and a fission bomb. Gamma rays from the fission blast provide the activation energy required to initiate the fusion of lithium and deuterium:

$$^{6}_{3}\text{Li} + {}^{1}_{0}\text{n} \rightarrow {}^{4}_{2}\text{He} + {}^{3}_{1}\text{H} + 4.8 \text{ MeV}$$

$$^{2}_{1}\text{H} + {}^{3}_{1}\text{H} \rightarrow {}^{4}_{2}\text{He} + {}^{1}_{0}\text{n} + 17.6 \text{ MeV}$$

$$\text{Overall:} \quad {}^{6}_{3}\text{Li} + {}^{2}_{1}\text{H} \rightarrow 2\,{}^{4}_{2}\text{He} + 22.4 \text{ MeV}$$

CHEMICAL INSIGHT

THE ORIGIN OF THE ELEMENTS

Nuclear fusion is the source of all elements beyond hydrogen and the primal source of most of our energy as well. As mentioned in the introduction to Chapter 1, the universe is expanding, and the observed rate of expansion indicates that it all began about 15 billion years ago with a "Big Bang." According to present theory, about 3 minutes after the Big Bang, the temperature had cooled to about 1 billion degrees and the universe consisted of a plasma suitable for fusion. Protons began to fuse with neutrons to form nuclei of deuterium, helium-3, and helium-4, and within a few minutes the universe had achieved what is still its general isotopic composition, about 91% hydrogen and 9% helium-4.

The universe continued to expand and cool, and when the temperature reached 4000 K—some 500,000 years after the Big Bang—the nuclei combined with electrons to form hydrogen and helium atoms. The heavier elements had not yet appeared, and the universe was filled with giant clouds of hydrogen and helium gas. These clouds were not homogeneous; rather, like fog in the night, some regions were thicker with gas than others. Gravitational attraction by the more massive, thicker regions attracted atoms away from the thinner regions. After a while, some regions became dense

Figure 22.25
The Horsehead Nebula is a cloud of interstellar dust that blocks light from the stars behind it. The heavier elements in this and similar clouds were produced by fusion reactions in the stars.

Figure 22.24
The NOVA fusion laser at the Lawrence Livermore National Laboratory in California is one of the world's most powerful lasers. This photograph shows a small fusion reaction created by firing blue laser light at a 1-mm-diameter pellet containing deuterium and tritium.

The first thermonuclear test explosion, on October 31, 1952, caused the Pacific island of Elugelab to vanish in a blinding flash of light.

A few grams of tritium inside the bomb helps to produce a more efficient reaction. Recently developed thermonuclear bombs are reputed to have TNT equivalents of up to 60 megatons—in comparison, the total energy equivalent of all the bombs exploded in the Second World War was only 6 megatons.

enough to coalesce into galaxies and into stars. Material condensing to form a star is accelerated and heated by gravity to such an extent that its center reverts to plasma and nuclear fusion begins once again. Our galaxy, the Milky Way, is a collection of stars shaped somewhat like a pancake with a bulge in its center, and our sun is a typical star located near the edge of the pancake.

Fusion reactions in the stars continue to produce helium as well as heavier elements (Figure 22.25). The fusion of protons into helium nuclei is believed to be the principal source of energy for stars similar in mass to our sun, where internal temperatures are about 10^7 K. One mechanism for stellar helium production is the *proton–proton chain*:

$$2({}^{1}_{1}H + {}^{1}_{1}H \rightarrow {}^{2}_{1}H + {}^{0}_{1}e)$$

positron

$$2({}^{2}_{1}H + {}^{1}_{1}H \rightarrow {}^{3}_{2}He)$$

$${}^{3}_{2}He + {}^{3}_{2}He \rightarrow {}^{4}_{2}He + 2{}^{1}_{1}H$$

The net fusion reaction,

$$4{}^{1}_{1}H \rightarrow {}^{4}_{2}He + 2{}^{0}_{1}e$$

releases 26.7 MeV per atom of helium-4.

Virtually all the elements heavier than helium, up to and including iron, are produced in successively deeper and hotter layers of massive stars, especially in very hot "second-generation stars." Second-generation stars are new stars formed from material ejected by "living" stars and debris left by "dying" stars—old interstellar matter that already contains a great variety of different nuclei. (The sun is a second-generation star, but not massive or hot enough to create any but the lightest elements.) Carbon-12, for example, is produced at 10^8 K by the *triple alpha reaction*, a fusion of three helium nuclei:

$$3{}^{4}_{2}He \rightarrow {}^{12}_{6}C$$

Oxygen forms at about 8×10^8 K:

$${}^{12}_{6}C + {}^{4}_{2}He \rightarrow {}^{16}_{8}O$$

Nuclei heavier than iron, and lighter nuclei as well, are formed during the terminal explosions of dying giant stars (supernovas). These elements are flung into space by the shock waves that accompany the explosion. Radiation received in December 1987 from the "nearby" (170,000 light years away) Supernova 1987A (see Figure 1.2, p. 3) indicated extensive production of iron, cobalt, and nickel.

CHAPTER REVIEW

Starred entries are based on the *Digging Deeper* sections.

LEARNING OBJECTIVES BY SECTION

22.1 1. State the nature of alpha, beta, and gamma radiation, and describe how each form of radiation behaves in an electric field.
2. Compare alpha, beta, and gamma radiation in terms of average energy and penetrating ability.
3. Distinguish between ionization counters and scintillation counters.

22.2 1. Describe the atomic and mass number changes that accompany the emission of alpha particles, beta particles, and positrons.
2. Write balanced equations for nuclear reactions involving alpha particles, beta particles, and positrons.

22.3 1. Write equations for nuclear bombardment reactions.
*2. Describe and diagram the operation of a cyclotron and a linear accelerator.

22.4 1. Use first-order kinetics in calculations relating half-life, elapsed time, and the remaining fraction of radioactive atoms.
2. Calculate the age of a carbon-containing artifact from its carbon-14 beta activity.
3. Calculate the age of an object from the isotope ratio of a radionuclide and its decay product.
*4. Use RBE factors to calculate the biologically effective radiation dose in rems from the actual dose in rads, and vice versa.

22.5 1. Describe the role of magic numbers in accounting for the stability of a nucleus.
2. Describe the source of gamma rays in nuclear decay reactions.
3. Use the stability band and the neutron/proton ratio to predict possible decay modes for a radionuclide.

22.6 1. Use nuclear masses to calculate the energy released during a nuclear reaction.
2. Use nuclear masses to calculate the mass defect, the binding energy, and the binding energy per nucleon of any nucleus.
3. Sketch the binding energy curve, and describe how the binding energy per nucleon varies with mass number.
4. Explain why both fission and fusion reactions release energy.

22.7 1. List some fissionable nuclei and write typical fission equations.
2. Explain how a fission reaction becomes self-sustaining.
3. Compare the advantages and disadvantages of ordinary fission reactors and breeder reactors.
4. Diagram a reactor core, and state the function of the fuel rods, the control rods, and the moderator.

22.8 1. Write the equation for the deuterium–tritium fusion reaction.
2. Describe the potential advantages of fusion power and the difficulties associated with its development.

KEY TERMS BY SECTION

22.1 Alpha particle (α or $^4_2$He)
Beta particle (β or $_{-1}^{0}$e)
Decay series
Gamma ray (γ)
Ionization counter
Ionizing radiation
Megaelectron volt (MeV)
Penetrating ability
Radioactivity
Radioisotope
Radionuclide

22.2 Nuclear reaction
Positron

22.3 *Cyclotron
*Linear accelerator
Transmutation
Transuranium element

22.4 Activity
Becquerel (Bq)
Cosmic rays
Curie (Ci)
*Gray (Gy)
*Rad
Radiocarbon dating
Radioisotope dating
*Relative biological effectiveness (RBE)
*Rem

22.5 Electron capture
Magic number
Stability band

22.6 Binding energy
Fission
Fusion
Mass defect
Mass–energy equation

22.7 Breeder reactor
Control rods
Critical mass
Fuel rods
Megaton (explosive unit)
Moderator
Nuclear chain reaction
Reactor core

22.8 Plasma
Thermonuclear reaction

IMPORTANT EQUATIONS

22.1 $\ln \frac{N}{N_0} = -kt$

22.2 $kt_{1/2} = \ln 2$

***22.3** dose (in rems) = dose (in rads) × RBE

22.4 $E = mc^2$

FINAL EXERCISES

Answers to exercises with blue numbers are given in Appendix D. Starred exercises are based on the *Digging Deeper* sections.

PART A. QUESTIONS AND PROBLEMS BY SECTION

Natural Radioactivity (Section 22.1)

22.1 What are *alpha particles*, *beta particles*, and *gamma rays*? What happens to the atomic number and the mass number of the nucleus when each of these radiations is emitted?

22.2 Diagram the paths taken by alpha particles, beta particles, and gamma rays when they pass between charged plates. What do these paths reveal about the nature of each type of radiation?

22.3 Explain why beta particles are deflected more than alpha particles by the same electric field.

22.4 What is *ionizing radiation*? List some types of ionizing radiation. In what ways does ionizing radiation damage living cells?

22.5 Compare alpha, beta, and gamma radiation in terms of (a) energy and (b) penetrating ability. Give reasons for the differences in penetrating ability.

22.6 How is the penetrating ability of a particle affected by the density of the medium through which it travels?

22.7 Diagram a Geiger–Müller counter and explain how it works.

22.8 Explain how a scintillation counter works. What advantage does a scintillation counter have over a Geiger–Müller counter?

Nuclear Equations (Section 22.2)

22.9 Write an equation for each of the following radioactive decay reactions:
(a) alpha emission by thorium-228
(b) beta emission by aluminum-28
(c) electron capture by rubidium-82
(d) positron emission by carbon-11

22.10 Write an equation for each of the following radioactive decay reactions:
(a) beta emission by iodine-129
(b) positron emission by rubidium-93
(c) alpha emission by actinium-227
(d) neutron emission by krypton-90

22.11 Refer to Figure 22.3, and write an equation for each step of the uranium decay series.

22.12 Francium-223, a beta emitter with a half-life of 22 minutes, was discovered in 1939 among the decay products of actinium. Write an equation for (a) the decay of actinium-227 into francium-223 and (b) the beta decay of francium-223.

Nuclear Bombardment Reactions (Section 22.3)

***22.13** With the aid of a diagram, explain how a particle is accelerated in a cyclotron.

***22.14** With the aid of a diagram, explain how a particle is accelerated in a linear accelerator.

22.15 Complete the following equations by adding missing symbols, mass numbers, and atomic numbers:
(a) $^{10}B + {}^{4}He \rightarrow ? + {}^{1}n$
(b) $? + {}^{1}n \rightarrow {}^{9}Li$
(c) $^{238}U + {}^{1}n \rightarrow ? + {}_{-1}^{0}e$

22.16 Complete the following equations by adding missing symbols, mass numbers, and atomic numbers:
(a) ^{98}Mo + deuterium → ? + a neutron
(b) ? + a proton → ^{21}Mg + 3 neutrons
(c) $? + {}^{12}C \rightarrow {}^{149}Tb + 4{}^{1}n$

22.17 Use the "in, out" shorthand notation to describe each reaction in Exercise 22.15.

22.18 Use the "in, out" shorthand notation to describe each reaction in Exercise 22.16.

22.19 An isotope of neptunium forms when uranium-238 is bombarded with deuterium nuclei. This isotope decays by beta emission into plutonium-238. Write the sequence of equations for the production of plutonium-238 from uranium-238.

22.20 Actinium, which exists only in trace amounts in nature, is prepared in milligram batches by bombarding radium-227 with neutrons. Neutron capture is followed by beta decay. Write the equations for the synthesis of actinium.

22.21 The halogen astatine, missing in nature, was prepared in 1940 by bombarding bismuth with alpha particles. Write the equation for the conversion of bismuth-209 to astatine-211.

22.22 About 60% of astatine-211 decays by electron capture and 40% decays by alpha emission. Write an equation for each mode of decay.

22.23 The americium-241 used in smoke detectors is produced by intense neutron bombardment of plutonium-239. Suggest a sequence of nuclear reactions for this synthesis, and write an equation for each step.

22.24 Write an equation for (a) the formation of carbon-14 in the atmosphere and (b) the beta decay of carbon-14.

Decay Rates and Radioisotope Dating (Section 22.4)

22.25 Can radiocarbon dating be used to find the age of an arrowhead made of flint (a kind of stone)? Explain.

22.26 In Example 22.5 (p. 1084) the half-life of uranium-238 was used to calculate the age of a moon rock. Explain why the half-lives of the other steps in the uranium decay series (Figure 22.3) were ignored.

22.27 The half-life of argon-41, a beta emitter, is 1.83 h. How many minutes will it take for the activity of an argon-41 sample to fall to 75.0% of its initial value?

22.28 It takes 28.3 days for the activity of sulfur-35, a beta emitter, to drop by 20.0%. Calculate the half-life of sulfur-35.

22.29 Strontium-90 ($t_{1/2}$ = 28.1 y) is a nuclear fission product that tends to concentrate in bones and milk.
(a) If strontium-90 is accidently released into the environment, what fraction will remain after 1 year? After 10 years?
(b) What fraction of the strontium-90 released in the atomic bomb explosions of 1945 would still be remaining in the year 2000?

22.30 Calculate the fraction of the original carbon-14 activity remaining in a carbon sample after (a) 5000 years, (b) 10,000 years, (c) 25,000 years, and (d) 50,000 years. Explain why radiocarbon dating cannot be effectively used for dating objects more than 50,000 years old.

22.31 The uranium-238 decay series begins with uranium-238 and ends with lead-206 (Figure 22.3). Estimate the age of a rock containing 85 g of lead-206 for every 100 g of uranium-238.

22.32 A wooden bowl from an ancient tomb is found to have a carbon-14 activity of 11.0 dpm per gram of carbon. Estimate the age of the bowl.

22.33 A sealed bottle of scotch whiskey was found in a sunken aircraft. The bottle contained 2.26×10^5 tritium atoms per mole of hydrogen atoms. Fresh water contains 6.02×10^5 tritium atoms per mole of hydrogen atoms. Estimate the age of the whiskey.

22.34 The decay of rubidium-87 to strontium-87 has been used for radiodating. The oldest and best-preserved skeleton of any erect human ancestor—the "Lucy" skeleton—was found in sedimentary rock in Hadar, Ethiopia. The rock was analyzed by mass spectrometry and found to contain 4.97 strontium-87 atoms for every 100,000 rubidium-87 atoms. How old is the skeleton?

Dosage (Section 22.4)

***22.35** Which radiation unit is used by the U.S. Environmental Protection Agency in setting acceptable exposure for the general population? What is the average exposure from natural sources of radiation? Identify some of these natural sources.

***22.36** Explain why the relative biological effectiveness of beta particles and gamma rays is less than that of (a) alpha particles and (b) neutrons.

22.37 A blood sample contains 5.0 pCi of sodium-24 per milliliter. How many sodium-24 atoms disintegrate per minute in a 10-mL sample?

22.38 A 20.0-mCi dose of technetium-99m ($t_{1/2} = 6.0$ h) is injected into a patient. Assuming that none of the isotope is eliminated by biological means, how many millicuries will remain after (a) 1 day, (b) 2 days, and (c) 1 week?

22.39 Colloidal suspensions of gold-198, a beta emitter with $t_{1/2} = 2.69$ days, have been used for the treatment of certain types of cancer and also for liver-scanning studies. If 100 mCi of this isotope is administered and none is lost by biological elimination, estimate (a) the number of days it will take for the activity to fall to one-tenth of its original value and (b) the activity remaining after 2 weeks.

***22.40** Half of an exposed population will die within 30 days from a 500-rem whole-body radiation dose. How many rads of each of the following radiations will provide this dose?

(a) alpha particles
(b) gamma rays
(c) x-rays
(d) neutrons
(e) beta particles
(f) protons

Nuclear Stability (Section 22.5)

22.41 Is the force that keeps nucleons in the nucleus an electrostatic force? A gravitational force? What are the limitations of the nuclear force?

22.42 What are "magic numbers"? How are magic numbers interpreted in terms of nuclear structure?

22.43 Nuclear species with an even number of nucleons are more numerous than those with an odd number. How is this phenomenon explained?

22.44 What is the source of the gamma rays emitted in nuclear decay reactions?

22.45 Which decay modes result in an increased neutron/proton ratio? Illustrate each mode with a nuclear equation.

22.46 Which decay modes result in a decreased neutron/proton ratio? Illustrate each mode with a nuclear equation.

22.47 Which decay mode would you predict for strontium-90? Explain your answer.

22.48 Which decay mode would you predict for chlorine-38? Explain your answer.

22.49 The stable isotopes of oxygen are oxygen-16, oxygen-17, and oxygen-18. Suggest possible modes of decay for the following radioactive isotopes of oxygen:

(a) oxygen-13
(b) oxygen-14
(c) oxygen-15
(d) oxygen-19
(e) oxygen-20

(*Hint*: Electron capture is not involved.)

22.50 Suggest possible decay modes for the following radionuclides:

(a) argon-39
(b) sulfur-30
(c) sulfur-37
(d) manganese-56
(e) zinc-69m
(f) zinc-69
(g) helium-6

22.51 Titanium-45 decays by both positron emission and electron capture.

(a) Write the decay equations.
(b) Calculate the neutron/proton ratio for titanium-45 and for the decay product.
(c) Is titanium-45 above or below the stability band?

22.52 Sulfur-35 decays by beta emission.

(a) Write the decay equation.
(b) Calculate the neutron/proton ratio for sulfur-35

and for the decay product.
(c) Is sulfur-35 above or below the stability band?

Mass-Energy Conversions (Section 22.6)

22.53 What is meant by *binding energy*? Which nuclide has the highest binding energy per nucleon?

22.54 Refer to the binding energy curve (Figure 22.17), and explain why fusion reactions should give off more energy than fission reactions.

22.55 Find the energy equivalent in kJ/mol of (a) a 1.0-GeV particle produced in the Bevatron and (b) a 1.0-TeV particle produced in the Tevatron.

22.56 How many kilograms of liquid water at 25°C could be converted into steam at 100°C by the energy equivalent of 1.00 g of mass?

22.57 The overall reaction for the stellar production of helium is

$$4\,^{1}_{1}H \longrightarrow \,^{4}_{2}He + 2\,^{0}_{1}e$$

Use data from Table 22.8 to calculate the energy released in (a) joules per helium atom and (b) joules per gram of helium.

22.58 Plutonium-239 decays by alpha emission:

$$^{239}_{94}Pu \longrightarrow \,^{235}_{92}U + \,^{4}_{2}He$$

Use data from Table 22.8 to calculate the energy released in (a) megaelectron volts per atom of plutonium and (b) joules per gram of plutonium.

22.59 Write the equation for the decay of uranium-238 by alpha emission, and use data from Table 22.8 to calculate the energy released. Give your answer in (a) megaelectron volts per atom and (b) joules per gram.

22.60 How many kilojoules of energy are evolved when 1.00 mol of helium-4 is converted into oxygen-16? The equation is:

$$4\,^{4}_{2}He \longrightarrow \,^{16}_{8}O$$

22.61 Refer to Table 22.8, and calculate the average binding energy per nucleon for (a) beryllium-9, (b) boron-10, and (c) silicon-28. List these isotopes in order of increasing stability.

22.62 Perform calculations to show which nuclide, iron-56 or iron-57, is more stable.

Fission and Fusion (Sections 22.7 and 22.8)

22.63 Describe and explain the conditions needed for a self-sustaining fission reaction.

22.64 What is meant by the phrase *critical mass*.

22.65 Describe the operation of a uranium-235 fission reactor. Your description should include a sketch of the reactor core.

22.66 Explain the function of (a) the control rods and (b) the moderator in a fission reactor.

22.67 Explain why an atomic bomb is required to detonate a hydrogen bomb.

22.68 Explain why fusion reactions require high activation energies.

22.69 Summarize the advantages and disadvantages of generating nuclear power with (a) fission reactors and (b) fusion reactors. Which type of reactor should have less impact on the environment? Explain.

22.70 Describe the major obstacle in the way of achieving controlled self-sustaining fusion reactions. What approaches are being made to this problem?

22.71 (a) Write equations for the sequence of nuclear reactions in which uranium-238 is converted to plutonium-239 in a breeder reactor.
(b) Describe the advantages and disadvantages of breeder reactors.

22.72 It has been suggested that fissionable uranium-233 could be bred from thorium-232 by a three-step mechanism consisting of neutron capture followed by two beta emissions. Write a nuclear equation for each step.

22.73 A uranium-235 atom captures a neutron and undergoes fission. Write a nuclear equation for the reaction that yields the following fission products. (*Note*: Neutrons are also produced in each of these reactions.)
(a) ^{90}Rb and ^{144}Cs
(b) ^{90}Sr and ^{143}Xe
(c) ^{90}Sr and ^{144}Xe
(d) ^{97}Y and ^{137}I
(e) ^{137}Te and ^{97}Zr

22.74 Write a nuclear equation for the fission of plutonium-239 into cerium-144 and krypton-85. Neutrons are the only other fission product.

22.75 The fission of one uranium-235 nucleus releases about 200 MeV of energy. The explosion of 1 g of TNT releases about 2.8 kJ of energy. Estimate the number of tons of TNT equivalent in energy to 1 g of uranium-235 undergoing fission.

22.76 The activation energy for the deuterium–tritium fusion reaction is about 10 KeV for each helium-4 nucleus produced. Express this energy in kilojoules per mole.

22.77 Refer to Table 22.8, and calculate the energy released by the deuterium–tritium fusion reaction. Give your answer in (a) megaelectron volts per atom of product and (b) kilojoules per mole of product.

22.78 (a) How many kilojoules of energy are released when 1.00 g of deuterium fuses with tritium to form helium-4?

(b) Refer to Exercise 22.75, and estimate the number of tons of TNT that must be exploded to release the same amount of energy as in Part (a).

PART B. MISCELLANEOUS QUESTIONS AND PROBLEMS

22.79 The majority of artificial radionuclides are now synthesized by neutron bombardment. Why did Rutherford use alpha particles for the first transmutation?

22.80 Complete the following "in-out" shorthand notations, and write a nuclear equation for each reaction:

(a) $^{10}B(n, \alpha)$
(b) $(n, p)^{9}Li$
(c) $^{238}U(n, 2\beta)$
(d) $^{98}Mo(D, n)$
(e) $(p, 3n)^{21}Mg$
(f) $(^{12}C, 4n)^{149}Tb$

22.81 What practical use is made of the following synthetic radioisotopes?

(a) iodine-131
(b) thallium-201
(c) technetium-99m
(d) sodium-24
(e) cobalt-60
(f) plutonium-238
(g) americium-241

22.82 Explain how each of the following natural radioisotopes has proved to be useful:

(a) radium-226
(b) carbon-14
(c) potassium-40
(d) uranium-238
(e) radon-222

22.83 A uranium-235 nucleus goes through a sequence of decay reactions and ultimately forms lead-207. How many alpha particles and how many beta particles are emitted during the sequence?

22.84 In 1937 technetium was discovered in a molybdenum sample that had been bombarded with deuterium nuclei. In 1961 traces of technetium-99 were found in the uranium ore pitchblende. Technetium is not part of any natural decay series, and it was suggested that the technetium in pitchblende resulted from uranium-235 fission with subsequent beta decay of mass-99 nuclei. Write nuclear equations showing how the technetium-99 may have formed.

22.85 When copper-68 decays by beta emission, it gives off beta particles having energies of 3.5, 2.7, and 2.3 MeV.

(a) Sketch an energy-level diagram similar to Figure 22.14 for the decay of copper-68.
(b) How many different gamma ray photons will be emitted? What will be their energies in megaelectron volts?

22.86 Suggest and describe a radioisotope dilution procedure for determining (a) the volume of a large body of water and (b) the blood volume of an animal. How would you choose a radioisotope for each of these determinations?

22.87 The combustion of 12.00 g of carbon liberates 39.35 kJ of heat.

(a) Calculate the change in mass that accompanies this reaction.
(b) Explain why the law of conservation of mass, as stated in Section 3.2, is valid for chemical reactions.

22.88 If only natural decay takes place, how many years will it take for earth's supply of uranium-235 to be reduced by 25.0%?

22.89 A thermonuclear weapon is based on the fusion of lithium-6 with deuterium:

$$^{6}Li + {}^{2}H \rightarrow 2\,{}^{4}He$$

How much energy, in megaelectron volts per atom of lithium, is released by this reaction?

22.90 The mass of the sun is estimated to be 1.99×10^{33} g. The earth receives about 173,000 terawatts of energy from the sun (1 terawatt = 10^{12} watts = 10^{12} J/s). The total energy radiated by the sun is about 50 billion times greater than the energy received by the earth. Estimate (a) the sun's mass loss per second and (b) the percent of the sun's mass that is converted to energy every 100 years.

22.91 The fission of a uranium-235 atom releases about 200 MeV of energy. The text states that the energy released per gram of deuterium in the deuterium–tritium fusion is about 10 times the energy released per gram of uranium-235. Verify this statement.

22.92 The half-life of sodium-24 is 15.0 hours. A sodium chloride solution containing 4.0 μCi of sodium-24 is injected into an animal to study blood circulation. Calculate (a) the fraction of the radionuclide remaining and (b) its activity in microcuries after 2 hours.

22.93 Each gram of carbon in living tissue has a beta ray activity of 15.3 dpm. Express this activity in microcuries per gram.

22.94 Radioisotopes are eliminated from the body by biological means as well as by radioactive decay. The rate of biological elimination often follows first-order kinetics. Iron-59 has a radioactive half-life of 45.1 days and a biological half-life in red blood cells of 60 days. What fraction of the initial iron-59 activity will remain in a sample of red blood cells after 1 week?

22.95 About 89.3% of a potassium-40 sample decays into argon-40 by electron capture; the remaining 10.7% forms calcium-40 by beta emission. The half-life of potassium-40 is 1.26×10^{9} years. A rock brought back by a space traveler is found to contain potassium-40 and trapped argon-40 in a 1.5-to-1.0 molar ratio. Estimate the age of the rock. (*Hint:* $N = {}^{40}K$; $N_0 = {}^{40}K + {}^{40}Ca + {}^{40}Ar$.)

CHAPTER 23

Many scientists believe that the chemicals of life were originally synthesized by solar energy and lightning.

ORGANIC CHEMISTRY AND THE CHEMICALS OF LIFE

OUTLINE

PREVIEW

At one time, *organic chemistry* dealt only with substances that were produced by living organisms, presumably with the aid of a "vital force" that could not be duplicated in the laboratory. The compounds of life are indeed unusual in that they contain a large proportion of carbon, an element that is not abundant in the mineral world. In 1828, however, Friedrich Wöhler demonstrated that urea, an organic compound produced in living cells, could also be produced by heating the inorganic compounds potassium cyanate (KOCN) and ammonium chloride:

$$KOCN + NH_4Cl \longrightarrow \underset{\text{urea}}{H_2N{-}\overset{\overset{\displaystyle O}{\|}}{C}{-}NH_2} + KCl$$

Thus the need for a "vital force" disappeared, along with the old distinction between what is organic and what is not. The term *organic* has persisted, however, and now includes all carbon compounds except for a few that are of mineral origin.

Several million carbon compounds have been isolated and characterized, and new ones are synthesized or extracted from natural substances every day. This activity in organic chemistry persists in part because the versatile carbon atom is a challenge to a chemist's creative urge, but mainly because organic compounds are the stuff of life and the stuff we live with. Organic substances include plastics, fuels, food, drugs, hormones, viruses, and genes.

LEARNING HINT

Review Section 10.3: *The Shapes of Some Hydrocarbon Molecules.*

In Chapter 10 you were introduced to the bonding and structure of various hydrocarbon molecules. In this chapter you will learn more about the properties of hydrocarbons and about other types of organic compounds that are related to them.

23.1 SATURATED HYDROCARBONS: ALKANES

The **alkanes**—methane, ethane, and so forth—are called **saturated hydrocarbons** because every carbon atom is bonded to four other atoms, the maximum number for a second-period atom with four valence electrons.

Alkanes are all closely related. In our imagination, we can "make" an ethane molecule from methane by adding a CH_2 group:

```
                                 ┌──────────── CH2 group ────────────┐
                                 ↓                                   ↓
        H                    H   H                       H   H   H
        |                    |   |                       |   |   |
     H—C—H               H—C—C—H                 H—C—C—C—H
        |                    |   |                       |   |   |
        H                    H   H                       H   H   H
```

methane (CH_4) ethane (CH_3CH_3) propane ($CH_3CH_2CH_3$)

We can make propane from ethane by adding another CH_2 group; the process can be repeated until the entire class of alkanes has been constructed by successive CH_2 increments. The carbon atoms may form a continuous chain (often called a *straight chain* although it is far from straight in the geometrical sense), or they may form branches. Straight or branched, the *general formula* for all alkanes is C_nH_{2n+2}, where n is the number of carbon atoms. The alkanes are an example of a **homologous series**, a series of compounds that differ by regular increments and have a common general formula. Table 23.1 lists the first 10 straight-chain members of the alkane series and their boiling points. Observe that alkanes with four or fewer carbon atoms are gases at room temperature. Alkanes with 18 carbon atoms or more are solid. Paraffin wax is a mixture of solid alkanes.

An alkane molecule with one hydrogen atom removed is called an **alkyl group**; for example, CH_3 is the *methyl* group and C_2H_5 is the *ethyl* group. An alkyl group is not a complete molecule, but it can be part of a molecule; for example, ethyl alcohol (C_2H_5OH) is an ethyl group plus OH. A hydrocarbon group is usually represented in a general formula by "R." For example, ROH represents any hydrocarbon group attached to OH.

Alkane Nomenclature

Identifying several million carbon compounds by name is difficult enough, and without a good system of nomenclature it would be impossible. As it is, the IUPAC organic nomenclature section in the yearly *Handbook of Chemistry and Physics* generally runs to more than 50 pages. Fortunately, only a few rules are needed to name most alkanes:

1. An open chain hydrocarbon takes its name from its longest continuous carbon chain. The italicized prefixes in Table 23.1 (*meth-*, *eth-*, *prop-*, *but-*, *pent-*, etc.) indicate the number of carbon atoms in this chain.

TABLE 23.1 The First 10 Straight-Chain Alkanes

Name[a]	Molecular Formula	Condensed Structural Formula	Boiling Point[b] (°C)	Related Alkyl Group
*Meth*ane	CH_4	CH_4	−162	Methyl CH_3—
*Eth*ane	C_2H_6	CH_3CH_3	−88.5	Ethyl C_2H_5—
*n-Prop*ane	C_3H_8	$CH_3CH_2CH_3$	−42	Propyl C_3H_7—
*n-But*ane	C_4H_{10}	$CH_3CH_2CH_2CH_3$	−0.5	Butyl C_4H_9—
*n-Pent*ane	C_5H_{12}	$CH_3(CH_2)_3CH_3$	36	Pentyl C_5H_{11}—
*n-Hex*ane	C_6H_{14}	$CH_3(CH_2)_4CH_3$	69	Hexyl C_6H_{13}—
*n-Hept*ane	C_7H_{16}	$CH_3(CH_2)_5CH_3$	98	Heptyl C_7H_{15}—
*n-Oct*ane	C_8H_{18}	$CH_3(CH_2)_6CH_3$	126	Octyl C_8H_{17}—
*n-Non*ane	C_9H_{20}	$CH_3(CH_2)_7CH_3$	151	Nonyl C_9H_{19}—
*n-Dec*ane	$C_{10}H_{22}$	$CH_3(CH_2)_8CH_3$	174	Decyl $C_{10}H_{21}$—

[a]The prefix *n* stands for "normal," or straight chain.
[b]Boiling points and melting points are often used to identify organic compounds.

2. Atoms (other than hydrogen) and groups of atoms that are attached to the straight chain are called *substituents*. The name of each substituent is included as a prefix, preceded by *di-*, *tri-*, *tetra-*, and so forth if there are more than one of the group. The entire prefix is preceded by numbers that locate the substituents on the chain:

$$CH_3-\underset{}{\overset{\overset{\displaystyle CH_3}{|}}{CH}}-CH_2-CH_3$$

2-methylbutane

$$CH_3-\overset{\overset{\displaystyle CH_3}{|}}{CH}-\overset{\overset{\displaystyle CH_3}{|}}{CH}-CH_3$$

2,3-dimethylbutane

Observe that multiple-position numbers are separated by commas and that the last number is followed by a dash.

3. Different substituents on the same carbon chain are given in alphabetical order. The carbon atoms in the chain are numbered from the end that gives a lower number to the first substituent on the chain, or if numbering from either end gives the same number to a first substituent, then numbering proceeds from the end that gives a lower number to the second substituent. Examples are

$$\underset{①}{CH_3}-\underset{②}{\overset{\overset{\displaystyle CH_3}{|}}{CH}}-\underset{③}{CH_2}-\underset{④}{CH_3}$$

2-methylbutane
(*not* 3-methylbutane)

$$\underset{①}{CH_3}-\underset{②}{\overset{\overset{\displaystyle CH_3}{|}}{CH}}-\underset{③}{\overset{\overset{\overset{\displaystyle CH_3}{|}}{\displaystyle CH_2}}{\overset{|}{CH}}}-\underset{④}{CH_2}-\underset{⑤}{\overset{\overset{\displaystyle CH_3}{|}}{CH}}-\underset{⑥}{CH_3}$$

3-ethyl-2,5-dimethylhexane
(The six-membered carbon chain is numbered from the left, putting substituents on carbons 2, 3, and 5. Numbering from the right would have put substituents on carbons 2, 4, and 5.)

EXAMPLE 23.1

Write the structural formula for 2-methyl-4-propyloctane.

SOLUTION

An octane has a continuous chain of eight carbon atoms. A methyl ($-CH_3$) group is attached to the second carbon and a propyl ($-CH_2CH_2CH_3$) group to the fourth carbon. The structural formula is

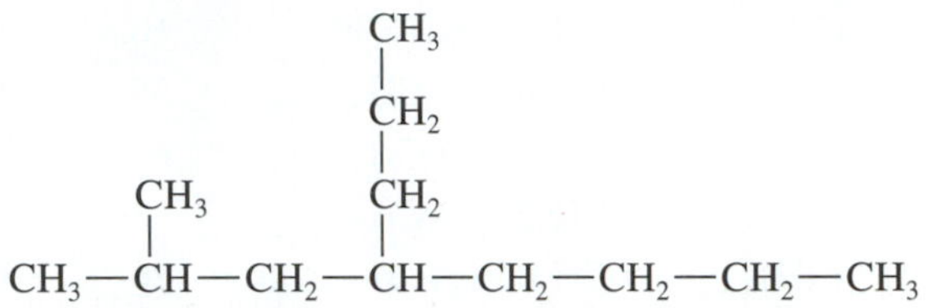

$$\begin{array}{ccccccccccccccc}
 & & & & & & CH_3 & & & & & & & & \\
 & & & & & & | & & & & & & & & \\
 & & & & & & CH_2 & & & & & & & & \\
 & & & & & & | & & & & & & & & \\
 & & CH_3 & & & & CH_2 & & & & & & & & \\
 & & | & & & & | & & & & & & & & \\
CH_3 & - & CH & - & CH_2 & - & CH & - & CH_2 & - & CH_2 & - & CH_2 & - & CH_3
\end{array}$$

EXAMPLE 23.2

Give the IUPAC name for

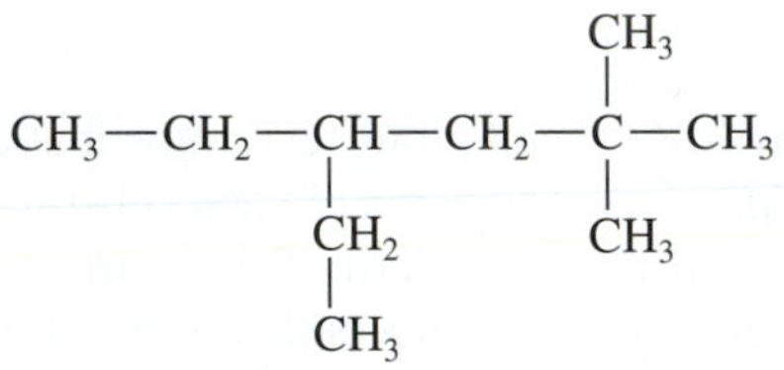

$$\begin{array}{ccccccccccc}
 & & & & & & & & CH_3 & & \\
 & & & & & & & & | & & \\
CH_3 & - & CH_2 & - & CH & - & CH_2 & - & C & - & CH_3 \\
 & & & & | & & & & | & & \\
 & & & & CH_2 & & & & CH_3 & & \\
 & & & & | & & & & & & \\
 & & & & CH_3 & & & & & &
\end{array}$$

SOLUTION

The molecule is a hexane because its longest continuous chain contains six carbon atoms. Although there are a number of ways (six, to be exact) to count six continuous carbon atoms in this molecule, a correctly chosen chain is one in which two methyl groups are on the second carbon and an ethyl group is on the fourth carbon:

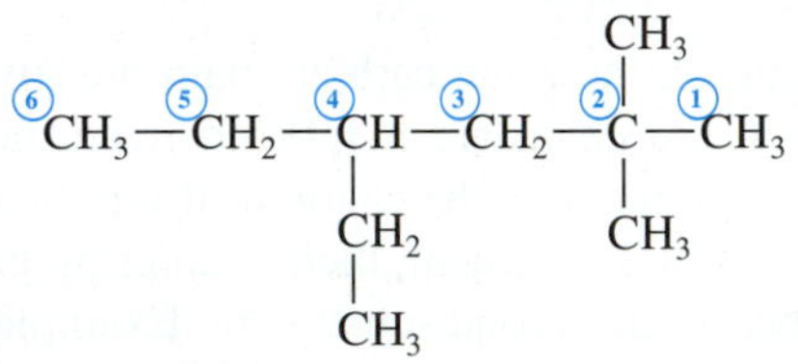

$$\begin{array}{ccccccccccc}
 & & & & & & & & CH_3 & & \\
 & & & & & & & & | & & \\
\overset{6}{CH_3} & - & \overset{5}{CH_2} & - & \overset{4}{CH} & - & \overset{3}{CH_2} & - & \overset{2}{C} & - & \overset{1}{CH_3} \\
 & & & & | & & & & | & & \\
 & & & & CH_2 & & & & CH_3 & & \\
 & & & & | & & & & & & \\
 & & & & CH_3 & & & & & &
\end{array}$$

The name is 4-ethyl-2,2-dimethylhexane

PRACTICE EXERCISE 23.1

Name the molecule

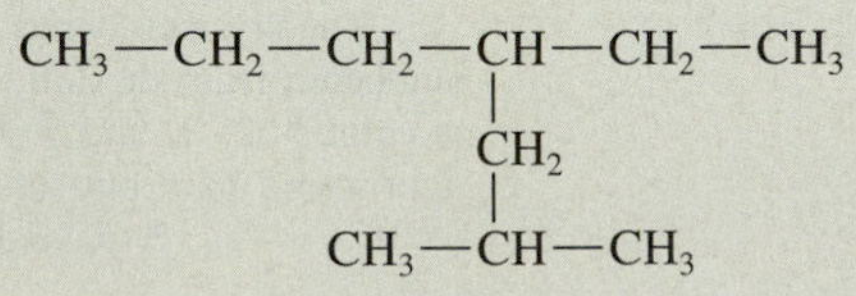

$$\begin{array}{ccccccccccc}
CH_3 & - & CH_2 & - & CH_2 & - & CH & - & CH_2 & - & CH_3 \\
 & & & & & & | & & & & \\
 & & & & & & CH_2 & & & & \\
 & & & & & & | & & & & \\
 & & & & CH_3 & - & CH & - & CH_3 & &
\end{array}$$

Cyclic Alkanes

Cycloalkanes contain rings of carbon atoms. A one-ring cycloalkane has two fewer hydrogen atoms than the corresponding open-chain alkane; it has the general formula C_nH_{2n}.

cyclopropane cyclobutane cyclohexane

Cyclopropane is a potent anesthetic that allows the patient to remain conscious. *Cyclohexane* is an important starting material for chemical syntheses; the manufacture of nylon alone uses more than 2 billion pounds each year.

The structural formula of a carbon ring is often drawn as a polygon in which each corner represents a carbon atom plus its correct number of bonded hydrogen atoms:

cyclopropane C_3H_6 cyclobutane C_4H_8 cyclopentane C_5H_{10} 1,2-dimethylcyclohexane $C_6H_{10}(CH_3)_2$

The bond angles in cyclopropane and cyclobutane are much smaller than the tetrahedral value of 109.5°, and these molecules are less stable than larger cycloalkanes, which can adjust their bond angles by adopting a puckered conformation.

The two conformations that provide tetrahedral angles in a six-member ring are called the **boat** and the **chair** (Figure 23.1). Cyclohexane moves rather easily between chair and boat conformations, and the shape of this molecule by itself is not particularly important. As part of a larger molecule, however, a six-membered ring may be locked into a single conformation that determines the shape and function of the entire molecule. The biologically important group of substances called **steroids,**

Figure 23.1

Conformations of cyclohexane (C_6H_{12}): (a) boat and (b) chair. The bent hexagons are common symbols for these conformations.

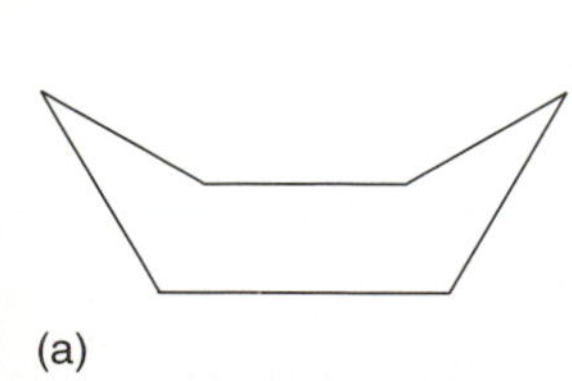
(a)
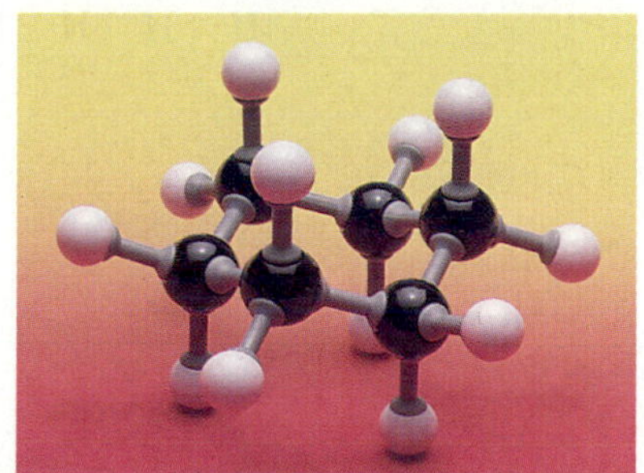

(b)

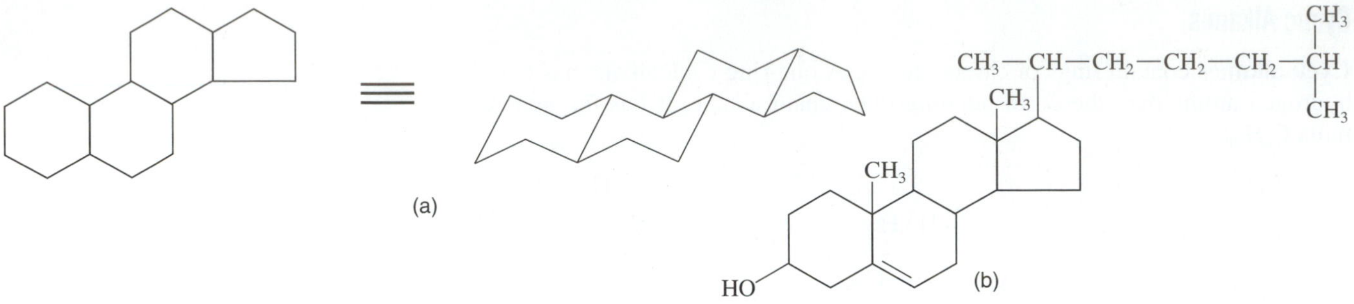

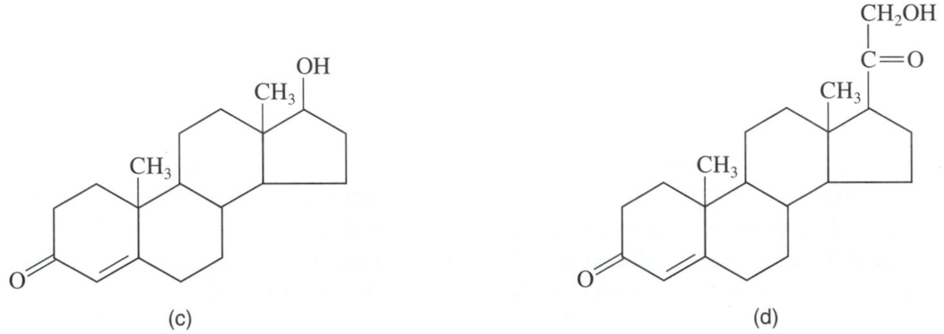

Figure 23.2

Some steroids. (a) The basic steroid skeleton consists of four fused rings. The three six-membered rings have the chair conformation. (b) Cholesterol, a component of all animal tissue; about 3–5 g are synthesized daily in the human liver. (c) Testosterone, a male sex hormone. (d) Progesterone, a pregnancy hormone.

Biological molecules generally have very specific geometries that enable them to fulfill their functions within the living cell.

which includes cholesterol and the sex hormones, has carbon skeletons composed of four fused rings. Three of these are six-membered rings in the chair configuration; the fourth ring has five carbons. All steroid molecules have the basic shape shown in Figure 23.2.

Reactions of Alkanes

All hydrocarbons will undergo combustion with the release of energy; they are our principal fuels. At moderate temperatures, however, hydrocarbon molecules may undergo less drastic changes that leave the carbon skeleton intact; these reactions are the ones that chemists find most interesting.

LEARNING HINT

Fuel values for various hydrocarbons and hydrocarbon mixtures are compared in Example 6.3 and Table 6.3 (p. 237).

Saturated hydrocarbons react with very few substances at room temperature. Halogens will attack a hydrocarbon in the presence of light or a catalyst and replace some of the hydrogen atoms with halogen atoms; this type of reaction is called a **substitution reaction**. Demonstration 23.1 shows the reaction of *n*-hexane with bromine:

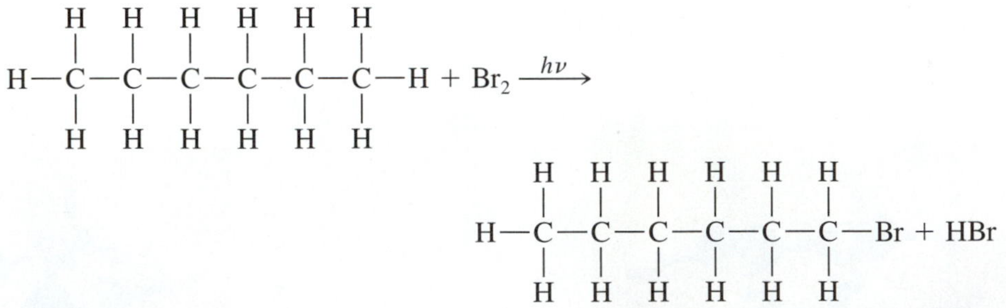

$$\mathrm{H{-}CH_2{-}CH_2{-}CH_2{-}CH_2{-}CH_2{-}CH_2{-}H} + Br_2 \xrightarrow{h\nu} \mathrm{H{-}CH_2{-}CH_2{-}CH_2{-}CH_2{-}CH_2{-}CH_2{-}Br} + HBr$$

1-bromohexane
(one of many possible substitution products)

DEMONSTRATION 23.1 THE REACTION OF BROMINE WITH A SATURATED HYDROCARBON: $C_6H_{14} + Br_2 \rightarrow C_6H_{13}Br + HBr$

A solution of bromine (red-brown) being added to hexane (colorless). The persistence of the brown color shows that bromine has not reacted.

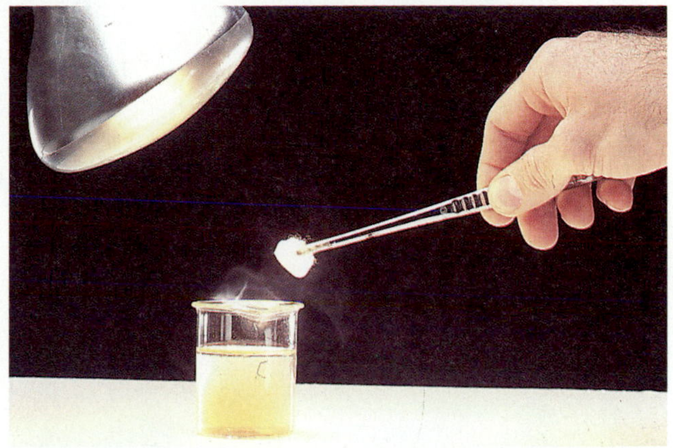

Bright light activates the reaction and the bromine color disappears. One product is HBr gas, which is detected by the white fog of $NH_4Br(s)$ that forms when HBr reacts with ammonia gas (from aqueous NH_3 on the cotton ball).

Halogen substitution proceeds by a free-radical chain mechanism (Section 14.6). The function of the photons (represented by $h\nu$ in the equation) is to dissociate Br_2 into the Br atoms that initiate the reaction. A mixture of products results, partly because substitution may occur at different points along the carbon chain and partly because some molecules may acquire two or more bromine atoms in successive steps.

23.2 UNSATURATED HYDROCARBONS: ALKENES AND ALKYNES

An **alkene**, as you may recall from Chapter 10, is a hydrocarbon whose molecules contain a carbon–carbon double bond. The simplest alkene is *ethene* ($CH_2{=}CH_2$) with one double bond. Adding successive CH_2 units to ethene produces the homologous alkene series whose members have one double bond and the general formula C_nH_{2n}. Each member of the alkene series has two fewer hydrogen atoms than the corresponding alkane.

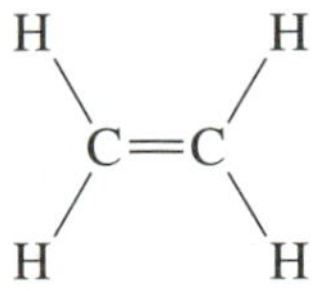

Ethene and propene are often called by their common names, *ethylene* and *propylene*.

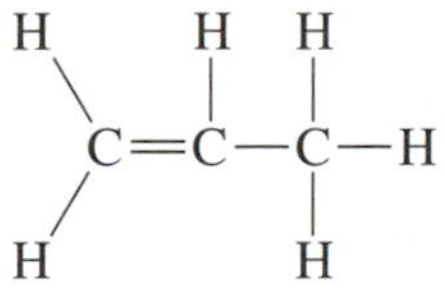

```
H       H        H     H   H
 \     /          \    |   |
  C==C             C==C--C--H
 /     \          /        |
H       H        H         H
```

ethene
(ethylene)

propene
(propylene)

The names of the alkenes end in *-ene*. The position of the double bond is indicated by a number; for example, in 1-butene a double bond follows the first carbon atom in a four-carbon chain, while in 2-butene the double bond follows the second carbon atom.

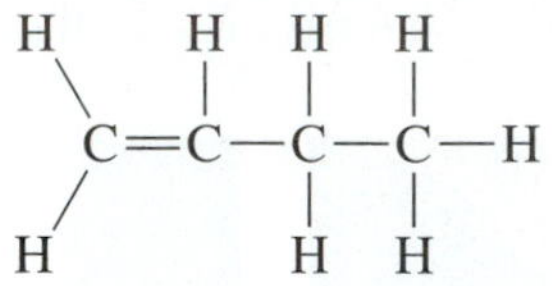

```
H       H  H  H              H  H  H  H
 \      |  |  |              |  |  |  |
  C=C---C--C--H           H--C--C=C--C--H
 /         |  |              |        |
H          H  H              H        H
      1-butene                  2-butene
```

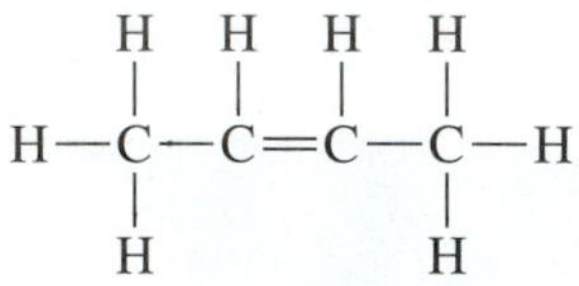

Dienes contain two double bonds; their names end in *-diene*. An example is 1,3-hexadiene:

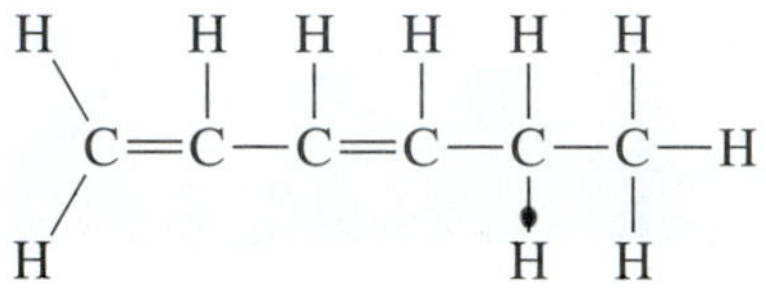

```
H      H  H  H  H  H
 \     |  |  |  |  |
  C=C--C=C--C--C--H
 /             |  |
H              H  H
      1,3-hexadiene
```

Cyclic alkenes and dienes also exist. An example is cyclohexene, which may be written in expanded or polygon form:

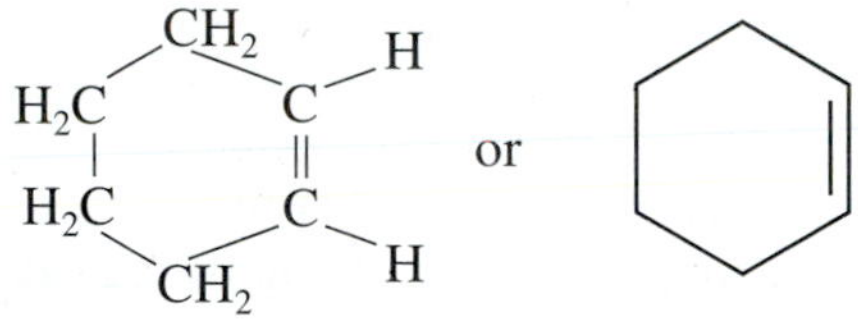

```
        CH2      H
      /     \   /
  H2C         C
   |          ||        or   [hexagon with one double bond]
  H2C         C
      \     /   \
        CH2      H
```

cyclohexene (C_6H_{10})

Hydrocarbons with triple bonds have names that end in *-yne*. The **alkyne** series contains hydrocarbons with one triple bond and the general formula C_nH_{2n-2}. The simplest examples are *ethyne* (more commonly called *acetylene*) and propyne:

$$H-C\equiv C-H \qquad\qquad H-C\equiv C-CH_3$$

ethyne (acetylene) propyne

Alkenes and alkynes are said to be **unsaturated** because they contain fewer than the maximum number of hydrogen atoms and can thus undergo various reactions that are not possible for alkanes.

Addition Reactions

Unsaturated hydrocarbons are generally more reactive than saturated hydrocarbons because each multiple bond provides a reactive site that can add various molecules. Some of these **addition reactions** are as follows.

Hydrogenation. Hydrogen addition converts an unsaturated hydrocarbon into a saturated one. A catalyst such as finely divided platinum is needed to help disrupt the strong H—H bond of the hydrogen molecule:

$$H_2C{=}CH_2 + H_2 \xrightarrow{Pt} H{-}CH_2{-}CH_2{-}H$$

$$H{-}C{\equiv}C{-}H + 2H_2 \xrightarrow{Pt} H{-}CH_2{-}CH_2{-}H$$

Halogen Addition. These rapid reactions do not require light or a catalyst (see Demonstration 23.2):

$$H_2C{=}CH_2 + Br_2 \longrightarrow H{-}CHBr{-}CHBr{-}H$$

1,2-dibromoethane (Figure 23.3)

$$H{-}C{\equiv}C{-}H + 2Br_2 \longrightarrow H{-}CBr_2{-}CBr_2{-}H$$

1,1,2,2-tetrabromoethane

Figure 23.3
1,2-Dibromoethane (CH_2BrCH_2Br). The two bromine atoms (green) are on different carbon atoms.

DEMONSTRATION 23.2 THE REACTION OF BROMINE WITH AN UNSATURATED HYDROCARBON: $C_6H_{12} + Br_2 \rightarrow C_6H_{12}Br_2$

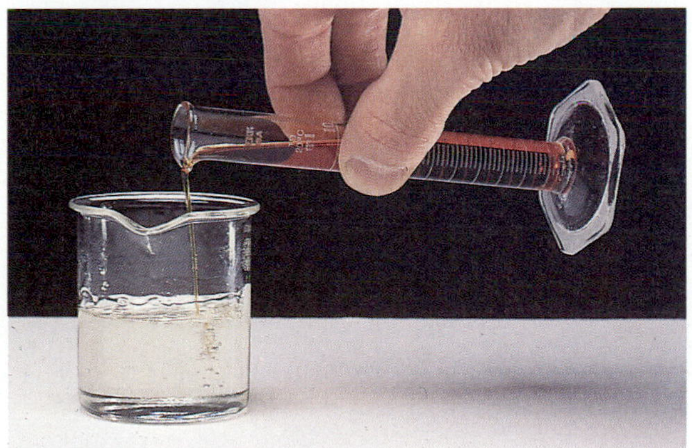

Bromine added to 1-hexene loses its red-brown color immediately. No light is needed for the reaction and no gas is evolved. (Compare with Demonstration 23.1.)

The remainder of the bromine solution is added. The last drops react as quickly as the first.

PRACTICE EXERCISE 23.2

Give the name and structural formula of the product resulting from the addition of chlorine to 2-pentene.

Hydrogen Halide Addition. HCl and HBr add rapidly to double and triple bonds:

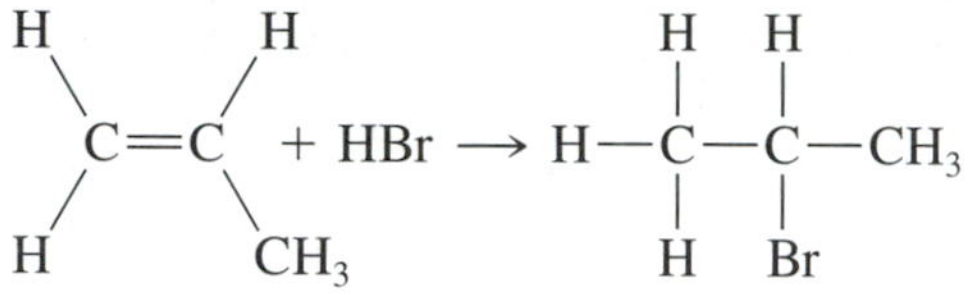

$$\underset{\text{propene}}{\mathrm{H_2C{=}CH{-}CH_3}} + \mathrm{HBr} \longrightarrow \underset{\text{2-bromopropane}}{\mathrm{H{-}CH_2{-}CHBr{-}CH_3}}$$

Note that in the above reaction, the hydrogen atom adds to the left-hand carbon atom in the propene structure. In 1870 the Russian chemist Vladimir Markovnikov, who was studying hydrogen halide addition, published his observation that the hydrogen adds to the carbon that already has more hydrogen ("the rich get richer"). **Markovnikov's rule** has since been generalized to cover the addition of other polar molecules, and its updated form states: *When an unsymmetrical reactant adds to a double bond, the more positive part will add to the carbon that has more hydrogen atoms.* The unsymmetrical HBr molecule is polar, with hydrogen more positive than bromine. Hence, the hydrogen atom adds to the CH_2 carbon of propene, and the bromine atom adds to the center carbon.

EXAMPLE 23.3

Water will add to small alkenes if acid is present. Predict the product of adding H_2O to 2-methylpropene.

SOLUTION

2-Methylpropene is

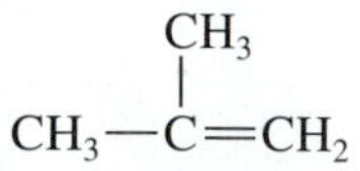

$$\mathrm{CH_3{-}C(CH_3){=}CH_2}$$

The water molecule is polar:

$$\overset{\delta+}{\mathrm{H}}{-}\overset{\delta-}{\mathrm{OH}}$$

The hydrogen atom, being more positive than the OH, will add to the CH_2 carbon and OH will add to the center carbon, which has no hydrogen. The product is

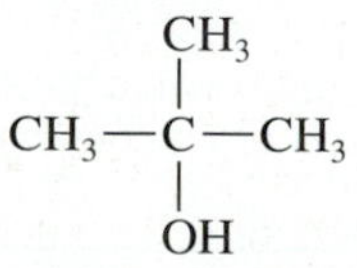

$$\mathrm{CH_3{-}C(CH_3)(OH){-}CH_3}$$

PRACTICE EXERCISE 23.3

Name and write the structural formula of the principal product formed when HCl adds to butadiene ($CH_2{=}CH{-}CH{=}CH_2$).

Polymerization

A **polymer** (from the Greek, *poly* + *meros* = "many parts") is a large molecule formed from many small molecules, which are called **monomers**. In the presence of a suitable catalyst, small molecules with double bonds will polymerize to produce long, saturated chains. Ethylene, for example, forms *polyethylene:*

$$n\left[\begin{array}{ccc} H & & H \\ & \diagdown\ \diagup & \\ & C{=}C & \\ & \diagup\ \diagdown & \\ H & & H \end{array}\right] \longrightarrow \begin{array}{ccccccccccc} & H & H & H & H & H & H & H & H & H & H \\ & | & | & | & | & | & | & | & | & | & | \\ - & C- & C- & C- & C- & C- & C- & C- & C- & C- & C- \\ & | & | & | & | & | & | & | & | & | & | \\ & H & H & H & H & H & H & H & H & H & H \end{array}$$

This unit is repeated n times.

The polymer chain can also be represented as

$$\left[\begin{array}{c} H\quad H \\ |\quad\ | \\ -C-C- \\ |\quad\ | \\ H\quad H \end{array}\right]_n$$

A polyethylene sample consists of many tangled chains of different lengths. Polymers formed from ethylene and its derivatives are called **addition polymers** because the monomers simply add—there is no product other than the polymer.

Substituents on the double-bond carbons will appear as substituents on the chain. Propene (propylene), for example, polymerizes to give *polypropylene:*

$$n\left[\begin{array}{ccc} H & & H \\ & \diagdown\ \diagup & \\ & C{=}C & \\ & \diagup\ \diagdown & \\ CH_3 & & H \end{array}\right] \longrightarrow \begin{array}{ccccccccc} & H & H & H & H & H & H & H & H \\ & | & | & | & | & | & | & | & | \\ - & C- & C- & C- & C- & C- & C- & C- & C- \\ & | & | & | & | & | & | & | & | \\ & CH_3 & H & CH_3 & H & CH_3 & H & CH_3 & H \end{array}$$

or

$$\left[\begin{array}{c} H\quad H \\ |\quad\ | \\ -C-C- \\ |\quad\ | \\ CH_3\ H \end{array}\right]_n$$

Other addition polymers are listed in Table 23.2.

Double-bond polymerization proceeds by a free-radical chain reaction mechanism. The initiator that starts the chains is a compound that decomposes when mildly heated into free hydrocarbon radicals such as phenyl ($C_6H_5\cdot$), which begin and terminate each chain. The free radical ($R\cdot$) binds to a monomer and creates a

TABLE 23.2 Some Addition Polymers

Monomer	Monomer Name	Polymer	Polymer Name
$CH_2{=}CH_2$	Ethylene	$[-CH_2-CH_2-]_n$	Polyethylene
$CH_2{=}CH(Cl)$	Vinyl chloride	$[-CH_2-CH(Cl)-]_n$	Polyvinylchloride (PVC, vinyl)
$CF_2{=}CF_2$	Tetrafluoroethylene	$[-CF_2-CF_2-]_n$	Teflon
$CH_2{=}CH(CN)$	Acrylonitrile	$[-CH_2-CH(CN)-]_n$	Orlon, Acrilan
$CH_2{=}C(CH_3)-C(=O)-O-CH_3$	Methyl methacrylate	$[-CH_2-C(CH_3)(C(=O)-O-CH_3)-]_n$	Lucite Plexiglas Acrylic

CHEMICAL INSIGHT

ELASTOMERS

Rubbery polymers are called *elastomers*. The natural rubber obtained from latex (a milky juice secreted by rubber trees) is a polymer of 2-methylbutadiene, commonly called *isoprene*. The polyisoprene chains in natural rubber vary in molar mass from 10^5 to 10^6 g.

$$CH_2{=}C(CH_3)-CH{=}CH_2 \quad \text{isoprene}$$

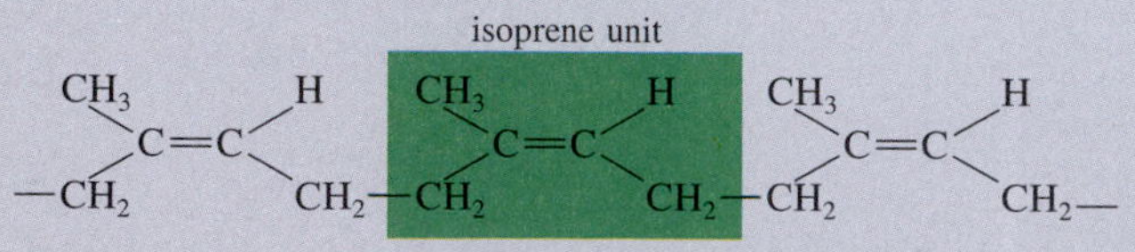

isoprene unit

$$-CH_2-C(CH_3){=}CH-CH_2-CH_2-C(CH_3){=}CH-CH_2-CH_2-C(CH_3){=}CH-CH_2-$$

polyisoprene (rubber)

The monomer has two double bonds, and each unit of the polymer retains one. The double bonds in rubber have the *cis* configuration with the $-CH_2-CH_2-$ groups on the same side of each double bond; this configuration makes the chain elastic. A few tropical plants secrete the *trans* isoprene polymer, which is called *gutta percha*. Gutta percha is pliable but not elastic; it has been used as an insulator and in dentistry as a space filler.

Since isoprene is butadiene with a methyl group added, it was natural to try various other butadiene derivatives for making synthetic elastomers. *Neoprene*, a polymer of 2-chlorobutadiene, is not dissolved by liquid hydrocarbons as rubber is, so it is useful for stoppers and for automobile engine belts. *SBR rubber*, a polymer that combines butadiene and styrene ($C_6H_5-CH{=}CH_2$), is used for automobile tires.

Pure rubber is sticky and soft, but its elasticity and hardness can be improved by the process of **vulcanization**, which was discovered by Charles Goodyear in 1839. When rubber is heated with sulfur, the cyclic sulfur molecules open up and form bridges of sulfur atoms between adjacent polymer chains. These links keep the chains from sliding past each other and help the rubber to regain its original shape after stretching.

longer free radical, which then attacks another monomer:

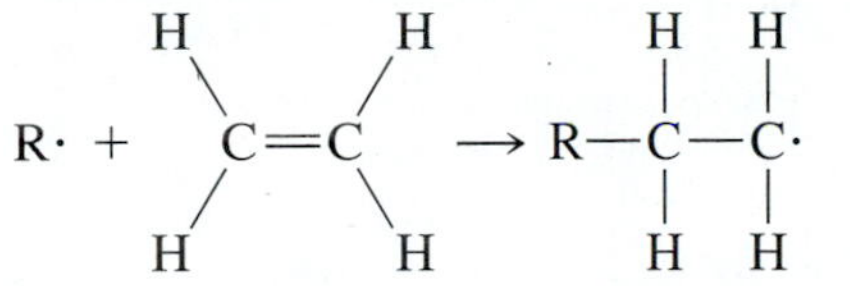

$$\begin{array}{ccccc}
 & \text{H} \quad\quad \text{H} & & \text{H} \quad \text{H} \\
 & \diagdown \quad\;\; \diagup & & | \quad\;\; | \\
\text{R}\cdot + & \text{C}{=}\text{C} & \longrightarrow & \text{R}{-}\text{C}{-}\text{C}\cdot \\
 & \diagup \quad\;\; \diagdown & & | \quad\;\; | \\
 & \text{H} \quad\quad \text{H} & & \text{H} \quad \text{H}
\end{array}$$

monomer — the start of a free-radical chain

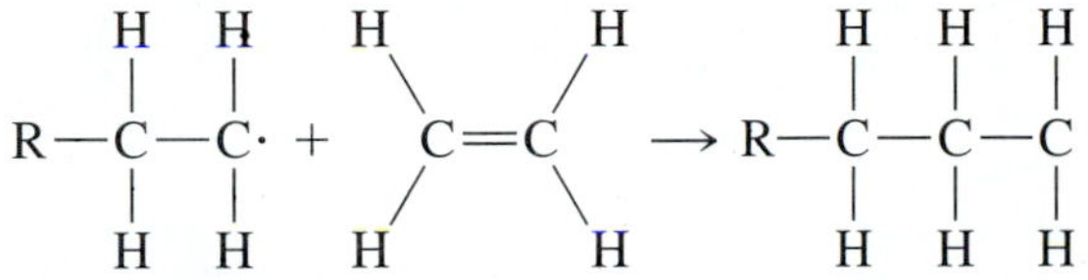

$$\begin{array}{ccccc}
\text{H} \quad \text{H} & & \text{H} \quad\quad \text{H} & & \text{H} \quad \text{H} \quad \text{H} \\
| \quad\;\; | & & \diagdown \quad\;\; \diagup & & | \quad\;\; | \quad\;\; | \\
\text{R}{-}\text{C}{-}\text{C}\cdot & + & \text{C}{=}\text{C} & \longrightarrow & \text{R}{-}\text{C}{-}\text{C}{-}\text{C} \\
| \quad\;\; | & & \diagup \quad\;\; \diagdown & & | \quad\;\; | \quad\;\; | \\
\text{H} \quad \text{H} & & \text{H} \quad\quad \text{H} & & \text{H} \quad \text{H} \quad \text{H}
\end{array}$$

A chain usually adds thirty or forty thousand units before its growth is terminated by combination with another free radical.

23.3 AROMATIC HYDROCARBONS

Aromatic compounds include benzene and compounds that contain at least one benzene ring (Section 10.3). The positions of two similar substituents on a benzene ring may be described as *o*- (*ortho*, adjacent), *m*- (*meta*, separated by one carbon), or *p*- (*para*, opposite), as illustrated by the three dimethylbenzenes, commonly called xylenes:

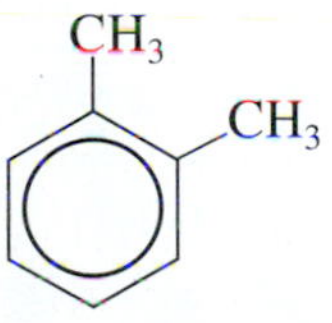

CH_3 CH_3 CH_3 CH_3 CH_3 CH_3

o-xylene
1,2-dimethylbenzene

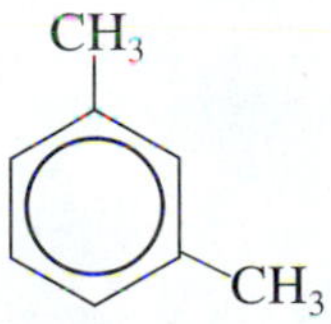

m-xylene
1,3-dimethylbenzene

p-xylene
1,4-dimethylbenzene

The xylene used to clean microscope lenses is a mixture of the three isomers.

The C_6H_5 group is called **phenyl**:

$CH_3{-}CH{-}CH_2{-}CH_3$

phenyl

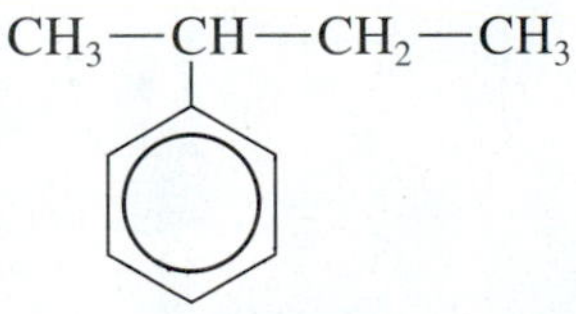

2-phenylbutane

Aromatic molecules do not generally participate in addition reactions. However, with high pressure and a catalyst, benzene will add hydrogen to form cyclohexane:

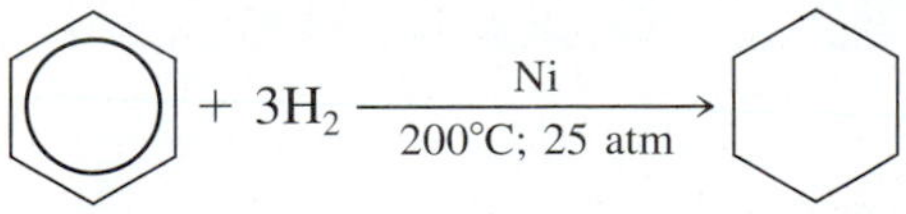

$$\text{benzene} + 3H_2 \xrightarrow[200^\circ\text{C};\ 25\ \text{atm}]{\text{Ni}} \text{cyclohexane}$$

benzene

cyclohexane

CHEMICAL INSIGHT

FOSSIL FUELS

Coal, petroleum, and natural gas are believed to have formed over millions of years from buried organic matter that has slowly lost the elements of water, a process called *carbonization*. Peat, a low-grade fuel found in bogs, contains vegetable matter in an early carbonization stage. Fossil fuels provide more than fuel; they also feed the vast *petrochemical* industry, which converts alkanes into alkenes, aromatics into substituted aromatics, and these compounds into plastics, fertilizers, pesticides, dyes, adhesives, drugs, cosmetics, perfumes, and other materials that have become essential to our civilization.

Coal contains a higher proportion of carbon than the other fossil fuels (Figure 23.4). When coal is *pyrolyzed* (heated in the absence of air), the products are *coke*, which is mainly carbon; *coal tar*, a rich source of aromatic compounds; and *coal gas*, which is principally H_2, CH_4, and CO.

Natural gas in the western hemisphere contains 60–90% methane and varying amounts of ethane, propane, butane, pentane, and helium; the natural gas found in Europe is nearly all methane. Natural gas is a clean-burning fuel, and it is also the principal source of industrial hydrogen, which is prepared by reacting methane with steam (see Section S1.1).

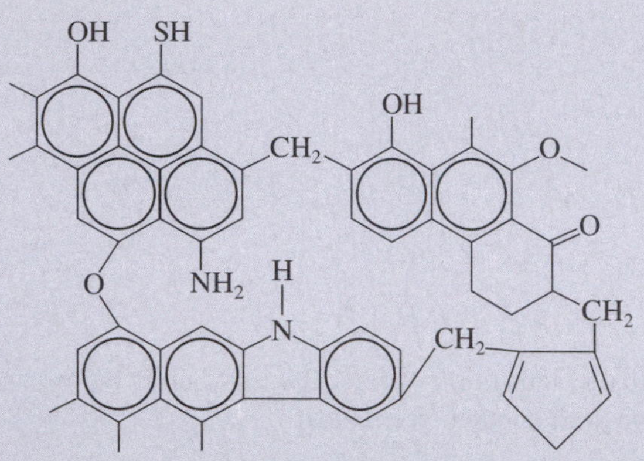

Figure 23.4

A model showing some of the structures found in coal. Note that hydrogen, oxygen, nitrogen, and sulfur atoms are present in addition to carbon. (This image shows some of the many possible ways that these structures can be linked together.)

Figure 23.5

An oil refinery at Port Arthur, Texas.

TABLE 23.3 The Major Fractions of Petroleum

Fraction	Number of Carbon Atoms[a]	Boiling Range[a](°C)	Uses
Gas	1–4	−162–30	Fuel gas; starting material for plastics manufacture
Petroleum ether	5–6	30–60	Solvent, gasoline additives
Gasoline	5–12	40–200	Gasoline
Kerosene	11–16	175–275	Diesel fuel, jet fuel, heating oil
Heating oil	15–18	275–375	Industrial heating
Lubricating oil	17–24	Over 350	Lubricants
Paraffin	20 and up	Solid residue	Candles, toiletries, wax paper
Asphalt	30 and up	Solid residue	Road surfacing

[a]The exact ranges may vary.

Petroleum is a dark, oily liquid composed mainly of alkanes, along with some aromatics and some molecules that contain oxygen, nitrogen, and sulfur. Petroleum is distilled at oil refineries (Figure 23.5) into the fractions listed in Table 23.3. Various reactions are used to augment and improve the gasoline fraction. *Cracking* breaks larger molecules into smaller ones with the help of catalysts and a high temperature:

$$C_{18}H_{38}(g) \rightarrow \underset{\text{an alkane}}{C_9H_{20}(g)} + \underset{\text{an alkene}}{C_9H_{18}(g)}$$

Other catalytic processes combine small hydrocarbons into larger ones:

$$\underset{\substack{\text{methylpropane}\\(\text{``isobutane''})}}{CH_3-\overset{\overset{\large CH_3}{|}}{CH}-CH_3} + \underset{\text{1-butene}}{CH_2{=}CH-CH_2-CH_3} \rightarrow \underset{\substack{\text{2,2,4-trimethylpentane}\\(\text{isooctane})}}{CH_3-\overset{\overset{\large CH_3}{|}}{\underset{\underset{\large CH_3}{|}}{C}}-CH_2-\overset{\overset{\large CH_3}{|}}{C}-CH_3}$$

Branched hydrocarbons and aromatic hydrocarbons burn more smoothly in an automobile engine than do the straight-chain hydrocarbons, which tend to knock. The *octane number* of a gasoline is a measure of its relative antiknock tendency. Isooctane (2,2,4-trimethylpentane) is taken as the standard for smoothness and assigned an octane number of 100. The straight-chain compound *n*-heptane, which knocks badly, is given an octane number of zero. The octane number of a gasoline is equal to the volume percent of isooctane in a mixture of isooctane and *n*-heptane with the same antiknock properties as the gasoline.

Most reactions of aromatic molecules are substitution reactions that preserve the planar ring and the delocalized pi bonds that are the source of its stability. Substitution for one or more hydrogen atoms may occur in the presence of certain acidic catalysts. Reactions that introduce halogen atoms and nitro groups are starting points for the synthesis of thousands of useful benzene derivatives:

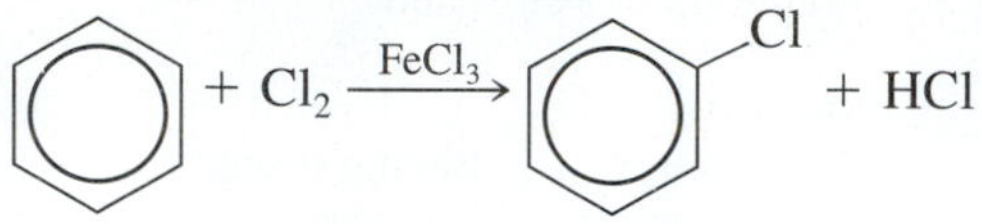

$$\text{benzene} + Cl_2 \xrightarrow{FeCl_3} \text{chlorobenzene} + HCl$$

$$\xrightarrow{H_2SO_4} \text{(2,4,6-trinitrotoluene: } CH_3 \text{ at C1; } O_2N, NO_2, NO_2 \text{ on ring)} + 3H_2O$$

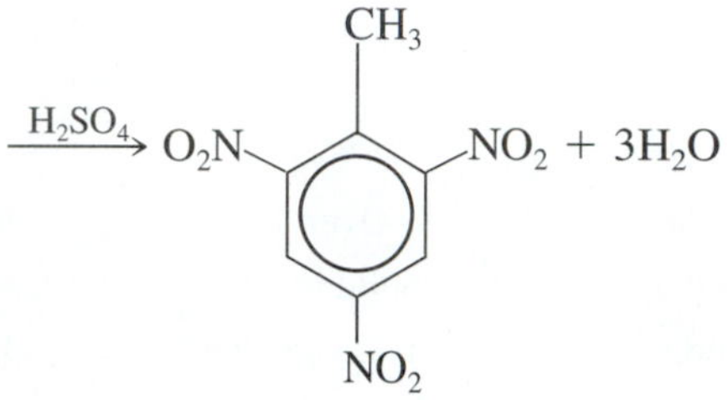

The carbon atoms in the benzene ring of trinitrotoluene are numbered so that the methyl group is attached to carbon 1.

Reactions with $AlCl_3$ as the catalyst are used to introduce alkyl groups:

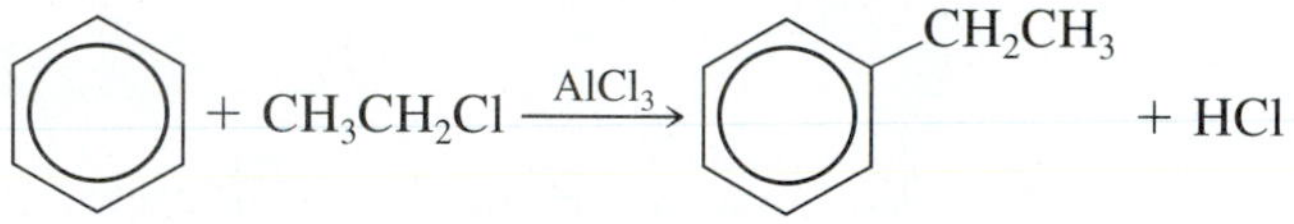

$$\text{benzene} + CH_3CH_2Cl \xrightarrow{AlCl_3} \text{ethylbenzene } (C_6H_5CH_2CH_3) + HCl$$

23.4 FUNCTIONAL GROUPS

The behavior of organic molecules is dominated by **functional groups**, which are combinations of atoms that act as reactive sites. Some functional groups are listed in Table 23.4. Compounds with the same functional group will have many properties in common; for example, molecules containing the COOH group will neutralize bases. A functional group such as a double bond, a halogen atom, or a group containing oxygen, nitrogen, or sulfur may convert a rather dull hydrocarbon into a lively molecule ready to participate in numerous reactions.

Many of the functional groups listed in Table 23.4 contain oxygen. An oxygen atom inserted into a hydrocarbon molecule can have either two single bonds (—O—) or one double bond (=O), so it can link two atoms or be doubly bonded to one atom. We will consider some of these possibilities in the remainder of this section.

Ethers

An **ether** contains an oxygen atom linking two hydrocarbon groups (Figure 23.6):

$R_1—O—R_2$	$CH_3—O—CH_3$	$CH_3CH_2—O—CH_2CH_3$
ether (general formula)	dimethyl ether	diethyl ether

TABLE 23.4 Classification of Organic Compounds by Functional Groups

Class	Functional Group	General Formula	Example
Alcohol	—OH	R—OH	CH_3OH Methanol
Aldehyde	H above —C=O (H bonded to C)	R—C(H)=O	HCHO Formaldehyde
Alkene	>C=C<		$CH_2=CH_2$ Ethene
Alkyne	—C≡C—		CH≡CH Acetylene
Amine (primary)	$—NH_2$	$R—NH_2$	CH_3NH_2 Methylamine
Carboxylic acid	—C(=O)—OH	R—C(=O)—OH	HCOOH Formic acid
Ether	—C(—)(—)—O—C(—)(—)—	R—O—R′	CH_3OCH_3 Dimethyl ether
Halide	Halogen atom, X	R—X	CH_3Cl Methyl chloride
Ketone	>C=O	R—C(=O)—R′	CH_3COCH_3 Acetone

(a)

(b)

Figure 23.6
(a) Dimethyl ether (CH_3OCH_3) and (b) diethyl ether ($C_2H_5OC_2H_5$). (Red spheres represent oxygen atoms.)

Diethyl ether is highly volatile and flammable. Partly for this reason, its use as an anesthetic has been discontinued in a number of countries.

The best-known ether is the anesthetic, *diethyl ether*. The ether linkage, C—O—C, is not particularly reactive; for this reason, diethyl ether is a widely used solvent for organic reactions.

The ether linkage is found in a variety of cyclic and noncyclic compounds. *Dioxane* ($C_4H_4O_2$) is a solvent for cellulose. *Dioxin* ($C_{12}H_4Cl_4O_2$) consists of two partially chlorinated benzene rings connected by two ether linkages. It is a contaminant in industrial effluents, incinerator smoke, and certain herbicides.

Dioxin first came to public notice as a toxic contaminant in *Agent Orange*, a defoliant used by the U.S. armed forces in the Vietnam War.

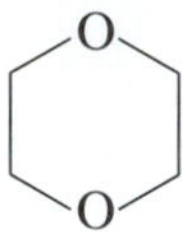

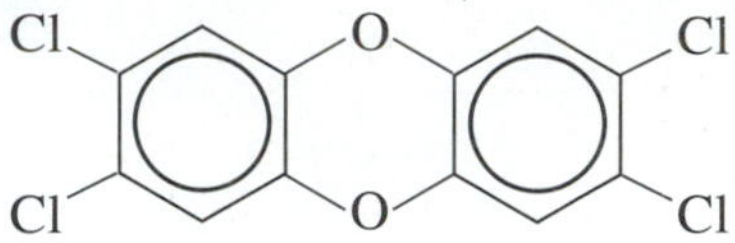

2,3,7,8-tetrachlorodibenzo-*p*-dioxin
(TCDD; dioxin)

Alcohols and Phenols

An alcohol contains an OH group covalently bonded to a hydrocarbon group; the general formula is R—OH. The IUPAC name for an alcohol contains the hydrocarbon root plus the suffix *-ol:*

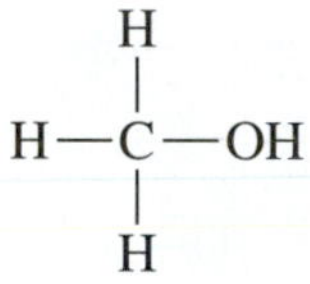

$$\begin{array}{ccc} & H & \\ & | & \\ H- & C & -OH \\ & | & \\ & H & \end{array} \qquad\qquad \begin{array}{cccc} & H & H & \\ & | & | & \\ H- & C- & C & -OH \\ & | & | & \\ & H & H & \end{array}$$

methanol (methyl alcohol) ethanol (ethyl alcohol)

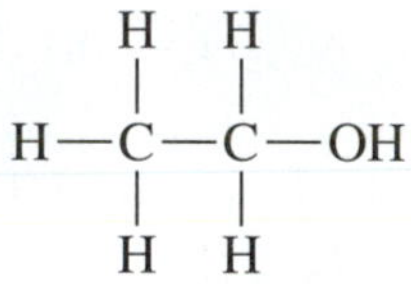

Methanol and ethanol are called **primary alcohols** because the OH group is on an end carbon. The OH group in a **secondary alcohol** is attached to a carbon that bonds two other carbon atoms. The OH group in a **tertiary alcohol** is attached to a carbon atom that bonds three other carbon atoms:

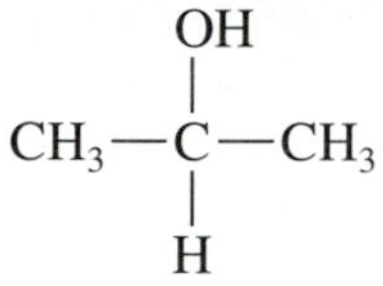

$$\begin{array}{ccc} & OH & \\ & | & \\ CH_3- & C & -CH_3 \\ & | & \\ & H & \end{array} \qquad\qquad \begin{array}{ccc} & OH & \\ & | & \\ CH_3- & C & -CH_3 \\ & | & \\ & CH_3 & \end{array}$$

2-propanol (isopropyl alcohol, a secondary alcohol) 2-methyl-2-propanol (*tert*-butyl alcohol, a tertiary alcohol)

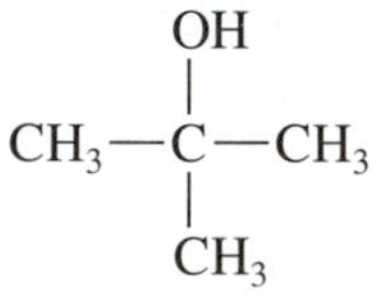

In IUPAC nomenclature, the OH position is indicated by a number; for example, in 2-propanol the OH is attached to carbon 2 in the chain.

EXAMPLE 23.4

Write the structural formula for 3-pentanol, and state whether it is a primary, secondary, or tertiary alcohol.

SOLUTION

The *pent-* means five carbons in a continuous chain; the *3-* and *-ol* indicate an OH group on the third carbon. The structure of this secondary alcohol is

$$\begin{array}{ccccccccccc} & & H & & H & & OH & & H & & H & & \\ & & | & & | & & | & & | & & | & & \\ H & - & C & - & C & - & C & - & C & - & C & - & H \\ & & | & & | & & | & & | & & | & & \\ & & H & & H & & H & & H & & H & & \end{array}$$

Methanol (Figure 23.7a), also known as *wood alcohol*, was originally obtained by distilling wood chips in the absence of air. It is now made by the catalytic combination of carbon monoxide and hydrogen:

$$CO(g) + 2H_2(g) \xrightarrow[\text{heat, pressure}]{\text{catalyst}} CH_3OH(g)$$

Methanol is used as a solvent for shellac, as a fuel that gives a small safe flame in spirit lamps, and as a starting material for the preparation of many other organic compounds. If ingested, methanol causes blindness, paralysis, and death.

Ethanol (Figure 23.7b) is also known as *grain alcohol*. Ethanol is obtained by the catalytic hydration of ethylene (a product of oil refining):

$$CH_2{=}CH_2 + H_2O \xrightarrow{\text{catalyst}} CH_3CH_2OH$$

and also from the fermentation of carbohydrates:

$$\underset{\text{glucose}}{C_6H_{12}O_6} \xrightarrow{\text{enzymes}} \underset{\text{ethanol}}{2CH_3CH_2OH} + 2CO_2(g)$$

Ethanol is used as a reagent in chemical synthesis and also as a fuel, solvent, antiseptic, and beverage ingredient. A mixture of ethyl alcohol and water is often used to dissolve medicinal ingredients that are not soluble in water alone; these preparations are called *tinctures*. An example is *tincture of iodine*. *Isopropyl alcohol* (Figure 23.7c), also called *rubbing alcohol*, is a secondary alcohol with antiseptic and cooling properties similar to those of ethanol. *tert-Butyl alcohol* (see bottom of page 1130) is used as a solvent in numerous products such as perfumes and paint thinner.

When ingested, ethyl alcohol depresses the central nervous system, impairs coordination, and over a period of time may irreversibly damage the liver, heart, and brain. In many states, a blood alcohol level of 100 mg/100 mL has been defined as medical/legal evidence of intoxication; a level of 400–500 mg/100 mL may result in death.

Figure 23.7

(a) Methanol, CH_3OH (methyl alcohol), (b) ethanol, CH_3CH_2OH (ethyl alcohol) and (c) isopropanol, $CH_3CHOHCH_3$ (isopropyl alcohol).

(a)

(b)

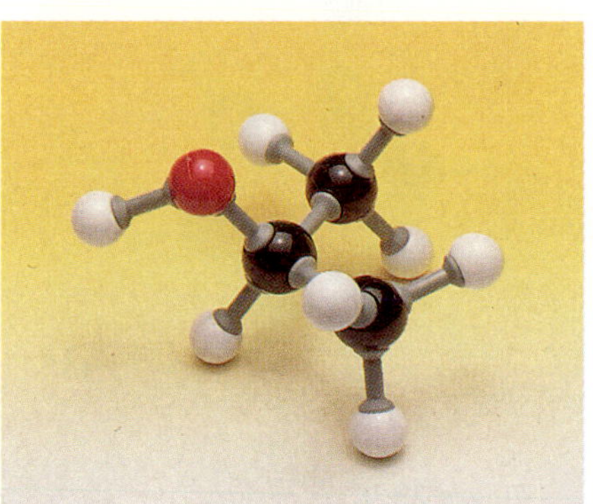

(c)

(a)

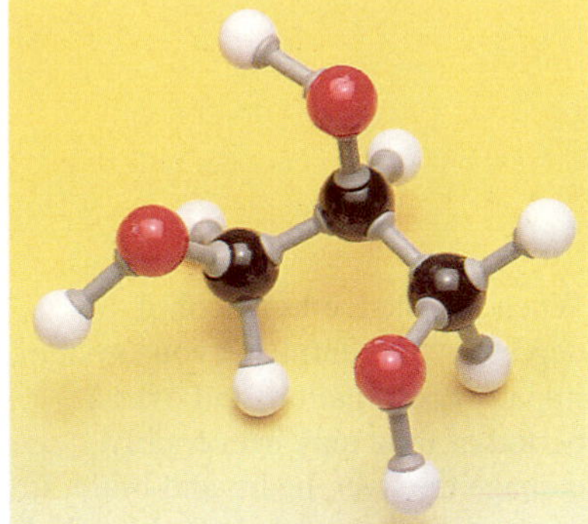

(b)

Figure 23.8

Alcohols with two or more OH groups. (a) Ethylene glycol (CH_2OHCH_2OH). (b) Glycerol ($CH_2OHCHOHCH_2OH$).

Hydrogen bonding (see Figure 13.1, p. 588) accounts for the relatively high boiling points of methanol and ethanol and also for their miscibility with water in all proportions. Methyl alcohol, which mixes with hydrocarbons and burns when they do, is useful as a gasoline antifreeze because it ties up small amounts of water that might otherwise form ice crystals in the gas line. Hydrogen bonding is less effective in alcohols with large hydrocarbon groups, so that butanol (four carbon atoms) mixes with water to only a limited extent, and higher alcohols are even less miscible.

Polyhydroxy alcohols contain two or more OH groups, each on a different carbon atom. They have strong hydrogen-bonding tendencies and a high affinity for water. *Ethylene glycol* (Figure 23.8a) is the usual automobile antifreeze, and it is also a solvent for ballpoint ink. *Glycerol* (Figure 23.8b), which is also called *glycerin*, is used as an emollient (an agent that soothes and softens the skin) in lotions or wherever its ability to attract and hold water is of value. *Nitroglycerin*, the explosive ingredient in dynamite, is made by treating glycerol with nitric acid.

$$\begin{array}{ccc}
\begin{array}{c} \mathrm{H} \\ | \\ \mathrm{H{-}C{-}OH} \\ | \\ \mathrm{H{-}C{-}OH} \\ | \\ \mathrm{H} \end{array} &
\begin{array}{c} \mathrm{H} \\ | \\ \mathrm{H{-}C{-}OH} \\ | \\ \mathrm{H{-}C{-}OH} \\ | \\ \mathrm{H{-}C{-}OH} \\ | \\ \mathrm{H} \end{array} &
\begin{array}{c} \mathrm{H} \\ | \\ \mathrm{H{-}C{-}O{-}NO_2} \\ | \\ \mathrm{H{-}C{-}O{-}NO_2} \\ | \\ \mathrm{H{-}C{-}O{-}NO_2} \\ | \\ \mathrm{H} \end{array} \\
\text{ethylene glycol} & \text{glycerol (glycerin)} & \text{glyceryl trinitrate (nitroglycerin)}
\end{array}$$

A **phenol** contains an OH group attached to an aromatic ring. The simplest phenol (C_6H_5OH), called simply *phenol*, is the disinfectant also known as *carbolic acid* (Figure 23.9). Other phenols, such as *p-cresol*, are also powerful disinfectants, fungicides, and fumigants.

[Structures: phenol — benzene ring with OH; *p*-cresol — benzene ring with OH and para CH_3]

phenol   *p*-cresol

Figure 23.9

Phenol (C_6H_5OH).

Phenols tend to be acidic because the aromatic ring withdraws electron density from the O—H bond. Phenol itself has a K_a of 1.3×10^{-10}; it is slightly more acidic than water and will neutralize bases in aqueous solution.

Aldehydes and Ketones

Aldehydes and ketones both contain the **carbonyl group**, $>C{=}O$. When an oxygen atom replaces two hydrogens on an end carbon atom, an **aldehyde** is formed; when it replaces two hydrogens on an interior carbon, a **ketone** is formed.

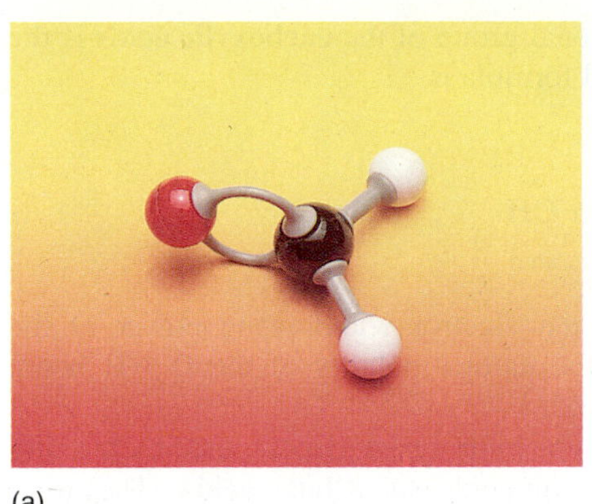

(a)

(b)

Figure 23.10

(a) Formaldehyde ($\mathrm{H{-}\overset{\overset{\textstyle O}{\|}}{C}{-}H}$).

(b) Acetaldehyde ($\mathrm{CH_3{-}\overset{\overset{\textstyle O}{\|}}{C}{-}H}$).

The general formulas are

$$\underset{\text{aldehyde}}{\mathrm{R{-}\overset{\overset{\textstyle H}{|}}{C}{=}O}} \qquad \underset{\text{ketone}}{\mathrm{R{-}\overset{\overset{\textstyle O}{\|}}{C}{-}R'}}$$

where R and R′ represent different hydrocarbon groups.

The simplest aldehyde is *formaldehyde* (CH_2O, Figure 23.10a). A 40% aqueous formaldehyde solution known as *formalin* is used as a disinfectant, a preservative for biological specimens, and an embalming fluid. Acetaldehyde (Figure 23.10b) is used in the production of acetic acid. The simplest ketone is *acetone* (CH_3COCH_3, Figure 23.11), a widely used solvent that mixes with water as well as organic liquids. It also dissolves lacquers such as nail polish, plastic objects including some eyeglass frames, and many synthetic fabrics.

A number of *pheromones* (sex attractants) contain carbonyl groups. Some examples are shown in Figure 23.12.

Figure 23.11

Acetone ($\mathrm{CH_3{-}\overset{\overset{\textstyle O}{\|}}{C}{-}CH_3}$).

PRACTICE EXERCISE 23.4

Identify each of the following compounds as an ether, alcohol, phenol, aldehyde, or ketone:

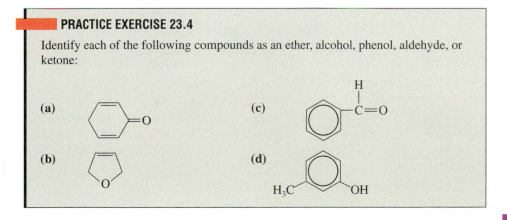

(a) =O

(c) H, C=O

(b) O

(d) H_3C, OH

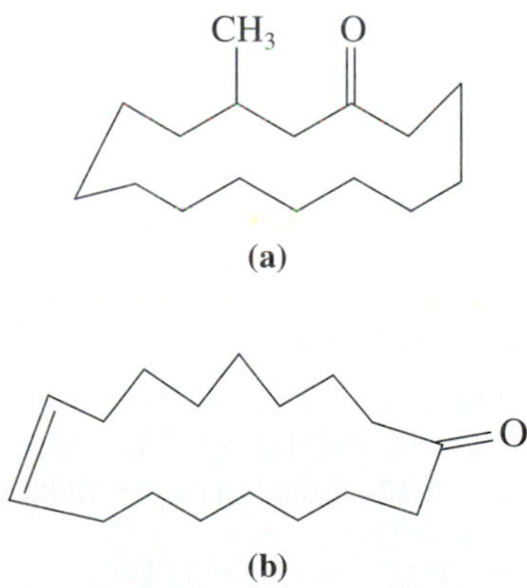

Figure 23.12

Two pheromones that are cyclic ketones. (a) Muskone ($C_{16}H_{30}O$), a musk deer pheromone. (b) Civetone ($C_{17}H_{30}O$), a pheromone of the civet cat.

Carboxylic Acids

Formic acid (HCOOH; Figure 23.13a) is the acid derived from methane; its rarely used IUPAC name is *methanoic acid. Acetic acid* (CH_3COOH, *ethanoic acid*; Figure

(a)

(b)

Figure 23.13
(a) Formic acid (HCOOH).
(b) Acetic acid (CH_3COOH).

23.13b) is an ethane derivative. The functional group of the **carboxylic acids** is the **carboxyl group** (COOH), and their general formula is

$$\begin{array}{c} \text{O} \\ \| \\ \text{R—C—OH} \end{array}$$

carboxylic acid

The carboxyl group owes its acidity to the electron-withdrawing power of the double-bonded oxygen, which enhances the natural polarity of the O—H bond. Some carboxylic acids are listed in Table 23.5.

Under certain conditions, fats and oil can be hydrolyzed (decomposed by reacting with water) to produce glycerol and long chain acids called **fatty acids**. (The structure of fats is described in Section 23.5.) Many natural fatty acids contain a single carboxyl group at the end of a straight chain of 14, 16, or 18 carbon atoms. Unsaturated fatty acids have one or more double bonds in the carbon chain, and these double bonds nearly always have a *cis* configuration (Figure 23.14). Some fatty acids are listed in Table 23.6.

A salt of a fatty acid is a **soap**. Sodium and potassium soaps are soluble, and their anions function in solution as **detergents** because the COO^- head is hydrophilic and the long hydrocarbon tail is hydrophobic (see Section 13.6). Calcium and magnesium soaps are insoluble; they form the "soap scum" that precipitates when soluble soaps encounter the Ca^{2+} and Mg^{2+} ions that abound in hard water.

Oxidation and Reduction

Arranging methane and its oxygen derivatives in order of increasing ratio of oxygen to hydrogen atoms gives

$$\underset{\text{methane}}{CH_4} \rightarrow \underset{\text{methanol}}{CH_3OH} \rightarrow \underset{\text{formaldehyde}}{CH_2O} \rightarrow \underset{\text{formic acid}}{HCOOH} \rightarrow \underset{\text{carbon dioxide}}{CO_2}$$

The derivatives of ethane can be arranged similarly:

$$\underset{\text{ethane}}{CH_3CH_3} \rightarrow \underset{\text{ethanol}}{CH_3CH_2OH} \rightarrow \underset{\text{acetaldehyde}}{CH_3CHO} \rightarrow \underset{\text{acetic acid}}{CH_3COOH} \rightarrow \underset{\text{carbon dioxide}}{2CO_2}$$

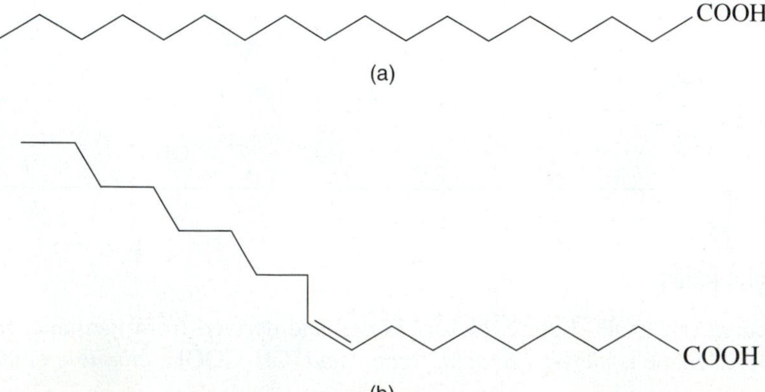

Figure 23.14
Saturated and unsaturated fatty acids. (a) Stearic acid ($CH_3(CH_2)_{16}COOH$) is a saturated fatty acid with a flexible hydrocarbon chain. (b) Oleic acid ($CH_3(CH_2)_7CH{=}CH(CH_2)_7COOH$) is a monounsaturated fatty acid with a rigid bend at the site of the double bond. (Refer to Figure 10.12, p. 425, for an explanation of the zigzag notation.)

TABLE 23.5 Some Organic Acids

Name	Formula	Source and Uses
Acetic acid	CH_3COOH	A component of vinegar; widely used in organic synthesis.
Benzoic acid	C_6H_5COOH (COOH on a benzene ring)	Found in berries; used in antifungal ointments.
Butyric acid	$CH_3(CH_2)_2COOH$	Responsible for the smell of rancid butter.
Formic acid	$HCOOH$	Originally obtained from ants (Latin *formicus*); dangerously caustic to skin; used as a reducing agent.
Oxalic acid	$HOOC{-}COOH$	Found in low concentrations in fruits and vegetables—larger amounts are toxic; bleaches wood; removes rust stains.
Propionic acid	CH_3CH_2COOH	Mold inhibitor; flavoring agent.
Salicylic acid	benzene ring with COOH and adjacent OH	Analgesic and antipyretic (fever reducer) found in willow bark; used to manufacture aspirin.

TABLE 23.6 Some Fatty Acids

Name	Formula	Source
Saturated Fatty Acids		
Lauric acid	$CH_3(CH_2)_{10}COOH$	Coconut oil
Myristic acid	$CH_3(CH_2)_{12}COOH$	Nutmeg butter Milk fat
Palmitic acid	$CH_3(CH_2)_{14}COOH$	Palm oil
Stearic acid	$CH_3(CH_2)_{16}COOH$	Animal fat
Unsaturated Fatty Acids		
Linoleic acid	$CH_3(CH_2)_4CH{=}CHCH_2CH{=}CH(CH_2)_7COOH$	Linseed oil Vegetable oils
Linolenic acid	$CH_3CH_2CH{=}CHCH_2CH{=}CHCH_2CH{=}CH(CH_2)_7COOH$	Linseed oil
Oleic acid	$CH_3(CH_2)_7CH{=}CH(CH_2)_7COOH$	Olive oil

This order is also the order of increasing oxidation state of carbon (you can verify this by calculating the oxidation number of carbon in each molecule). Each molecule is an oxidation product of the molecules to its left, and a reduction product of the molecules to its right. Thus alcohols, aldehydes, ketones, and carboxylic acids may be viewed as arising from the controlled oxidation of parent hydrocarbons. The complete oxidation of a hydrocarbon produces carbon dioxide and water.

In organic and biochemical reactions, oxidation usually consists of acquiring oxygen atoms or giving up hydrogen atoms, while reduction is just the reverse (see Demonstration 23.3). Oxidizing and reducing agents are often represented by the general symbols [O] and [H], respectively. The oxidation of isopropyl alcohol to acetone, for example, can be written as

Recall (Figure S4.25, p. 1063) the "Breathalyzer" test in which ethanol is oxidized to acetaldehyde while reducing orange $Cr_2O_7^{2-}$ to Cr^{3+}.

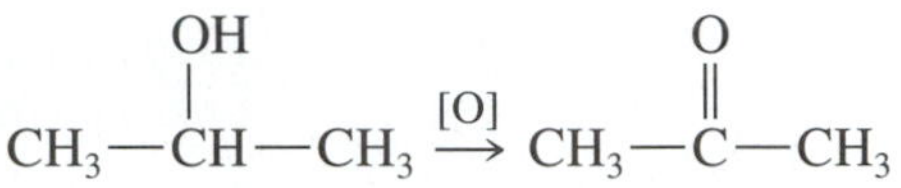

$$\underset{\text{isopropyl alcohol}}{CH_3-\overset{\overset{\large OH}{|}}{C}H-CH_3} \xrightarrow{[O]} \underset{\text{acetone}}{CH_3-\overset{\overset{\large O}{\|}}{C}-CH_3}$$

where [O] represents any oxidizing agent that will do the job. Chromium trioxide is one that works well in this particular case.

The reverse reaction can be represented as

When a reactant is written over the arrow, you are not expected to balance the equation.

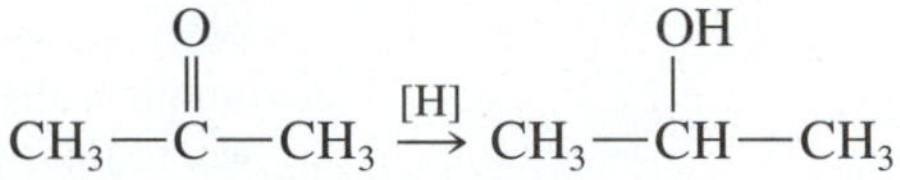

$$CH_3-\overset{\overset{\large O}{\|}}{C}-CH_3 \xrightarrow{[H]} CH_3-\overset{\overset{\large OH}{|}}{C}H-CH_3$$

The [H] stands for a reducing agent. One possible reducing agent is lithium aluminum hydride ($LiAlH_4$), which contains hydrogen in its -1 oxidation state.

Glucose ($C_6H_{12}O_6$), the simple sugar that serves as the principal source of energy for most living organisms (see *Chemical Insight: Time's Arrow and Coupled Reactions*, p. 926), is produced in plants by the photosynthetic reduction of carbon dioxide:

$$6CO_2(g) + 6H_2O(l) \xrightarrow[\text{chlorophyll}]{\text{sunlight}} \underset{\text{glucose}}{C_6H_{12}O_6(s)} + 6O_2(g) \qquad \Delta H^\circ = +2{,}816\text{ kJ}$$

The glucose molecule usually exists as a six-membered ring (Section 23.8). However, it spends a small fraction of its time in the form of an open-chain polyhydroxy aldehyde:

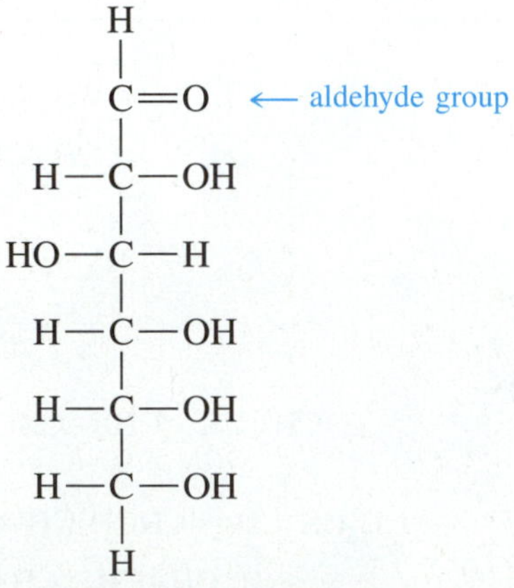

```
        H
        |
        C=O   ← aldehyde group
        |
    H—C—OH
        |
   HO—C—H
        |
    H—C—OH
        |
    H—C—OH
        |
    H—C—OH
        |
        H
     glucose
```

DEMONSTRATION 23.3 PRODUCTION OF A SILVER MIRROR

Formaldehyde is added to a flask that contains a solution of silver nitrate dissolved in aqueous ammonia.

After covering the neck opening with a plastic film, the contents of the flask are mixed.

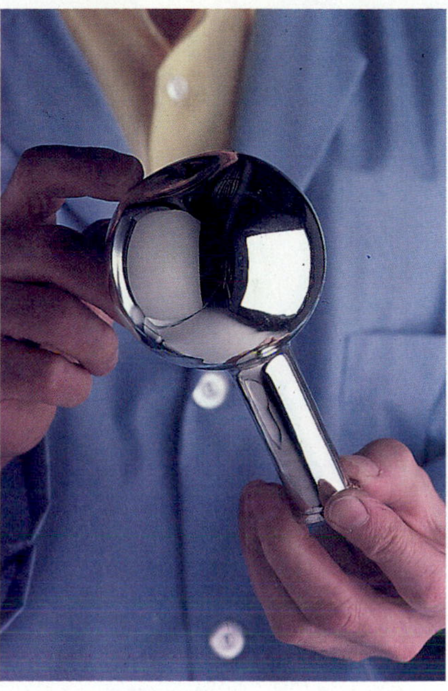

The silver ion is reduced to silver, which forms a mirror on the walls of the flask. Formaldehyde is oxidized to formic acid.

Glucose, like all aldehydes, is a reducing agent; it can be oxidized to an organic acid. This property is the basis for tests that determine glucose concentrations in samples of blood and urine.

PRACTICE EXERCISE 23.5

The oxidation of glucose in the body takes place in a series of enzyme-catalyzed steps. Early steps split the glucose molecule into two molecules of glyceraldehyde, which are then oxidized to pyruvic acid and finally to CO_2:

$$
\begin{array}{ccc}
 & \mathrm{H} & \\
 & | & \\
 & \mathrm{C{=}O} & \\
 & | & \\
\mathrm{H-} & \mathrm{C} & \mathrm{-OH} \\
 & | & \\
\mathrm{H-} & \mathrm{C} & \mathrm{-OH} \\
 & | & \\
 & \mathrm{H} &
\end{array}
\qquad\qquad
\begin{array}{ccc}
 & \mathrm{O} & \\
 & \| & \\
 & \mathrm{C} & \mathrm{-OH} \\
 & | & \\
 & \mathrm{C{=}O} & \\
 & | & \\
\mathrm{H-} & \mathrm{C} & \mathrm{-H} \\
 & | & \\
 & \mathrm{H} &
\end{array}
$$

glyceraldehyde pyruvic acid

Find the ratio of oxygen to hydrogen atoms in (a) glucose, (b) glyceraldehyde, and (c) pyruvic acid. Which parts of the biochemical sequence

$$\text{glucose} \longrightarrow \text{glyceraldehyde} \longrightarrow \text{pyruvic acid} \longrightarrow \text{carbon dioxide}$$

are oxidation?

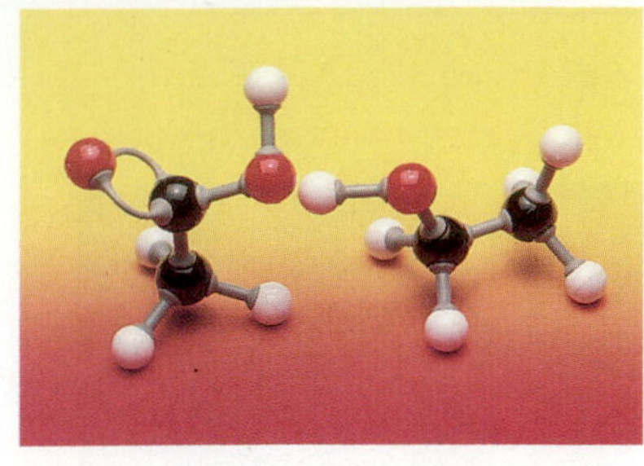

(a)

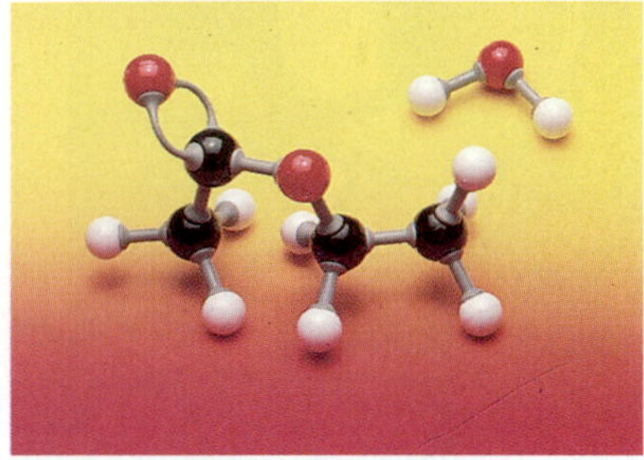

(b)

Figure 23.15

The formation of an ester. (a) *Left*: a molecule of acetic acid (CH_3COOH). *Right*: a molecule of ethyl alcohol (CH_3CH_2OH). (b) The two molecules join at the site of their OH groups, losing a water molecule (upper right) in the process. The product is a molecule of ethyl acetate ($CH_3COOCH_2CH_3$).

23.5 THE ESTER LINKAGE

An organic acid (functional group, —COOH) and an alcohol (functional group, —OH), may combine with the loss of a water molecule to form a larger molecule called an **ester**. This reaction is an example of a **condensation**, a combination reaction that eliminates a small molecule (see Figure 23.15). The OH in the eliminated H_2O comes from the acid, the H from the alcohol:

$$\underset{\text{acetic acid}}{CH_3-\overset{\displaystyle O}{\overset{\|}{C}}-OH} + \underset{\text{ethyl alcohol}}{H-O-CH_2CH_3} \longrightarrow \underset{\text{ethyl acetate}}{CH_3-\overset{\displaystyle O}{\overset{\|}{C}}-O-CH_2CH_3} + H_2O$$

The group

$$-\overset{\displaystyle O}{\overset{\|}{C}}-O-$$

is called the **ester linkage**. The general formula for an ester is

$$R-\overset{\displaystyle O}{\overset{\|}{C}}-O-R'$$

where R and R′ are the hydrocarbon groups from the acid and alcohol, respectively. Esters are named after their hypothetical "parents," first the alcohol group, and then the acid with the suffix *-ate* (e.g., *ethyl acetate*).

Naturally occurring esters contribute to the pleasant odors of fruits and flowers. Similar synthetic esters are often used in perfumes and artificial flavorings. Other esters are found in fats and waxes (Table 23.7).

Aspirin (Figure 23.16), which is used for the relief of pain and to reduce inflammation and fever, is an ester of acetic acid and *salicylic acid* (the salicylic acid acts as the alcohol in this case):

$$\underset{\text{salicylic acid}}{C_6H_4(COOH)-OH} + \underset{\text{acetic acid}}{HO-\overset{\displaystyle O}{\overset{\|}{C}}-CH_3} \longrightarrow \underset{\text{aspirin}}{C_6H_4(COOH)-\underbrace{O-\overset{\displaystyle O}{\overset{\|}{C}}}_{\text{ester linkage}}-CH_3} + H_2O$$

Under certain conditions an ester will hydrolyze (react with water) releasing the acid and alcohol. Aspirin hydrolyzes in the body to release salicylic acid, which is the active pain reliever.

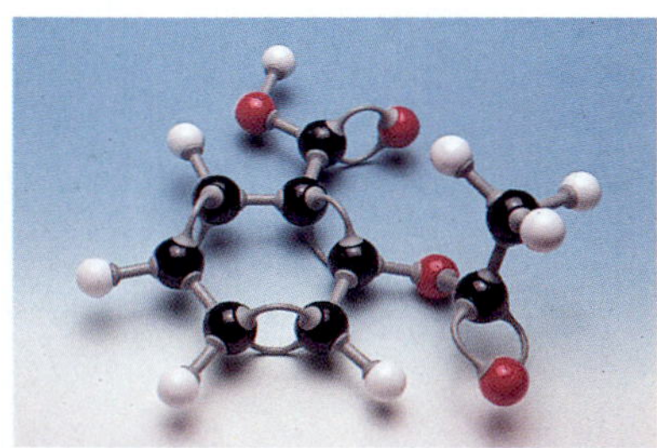

Figure 23.16

Aspirin.

PRACTICE EXERCISE 23.6

Write the equation for the reaction of 1-butanol with propionic acid (a three-carbon acid). Name the ester formed.

TABLE 23.7 Some Esters

Name	Formula	Source
Ethyl butyrate	$CH_3(CH_2)_2-C(=O)-O-CH_2CH_3$	Pineapple aroma
Glycerol tristearate (stearin)	$H_2C-O-C(=O)-(CH_2)_{16}CH_3$ $HC-O-C(=O)-(CH_2)_{16}CH_3$ $H_2C-O-C(=O)-(CH_2)_{16}CH_3$	Major ingredient in beef fat
Methyl salicylate	$C_6H_4(OH)-C(=O)-O-CH_3$	Oil of wintergreen
Myricyl palmitate (myricin)	$CH_3(CH_2)_{14}-C(=O)-O-(CH_2)_{30}CH_3$	Major ingredient in beeswax
Octyl acetate	$CH_3-C(=O)-O-(CH_2)_7CH_3$	Orange aroma

Polyesters

A dicarboxylic acid (two COOH groups) and a dihydroxy alcohol (two OH groups) can polymerize by condensation to form a **polyester** chain with alternating units. One well-known polyester is synthesized from terephthalic acid (*p*-phthalic acid) and ethylene glycol:

$$\underset{\text{terephthalic acid}}{HO-C(=O)-C_6H_4-C(=O)-OH} + \underset{\text{ethylene glycol}}{HO-CH_2CH_2-OH} \longrightarrow$$

$$HO-C(=O)-C_6H_4-C(=O)-O-CH_2CH_2-OH + H_2O$$

Another alcohol can condense here. (arrow to HO at left)

Another acid can condense here. (arrow to OH at right)

Figure 23.17
This photograph shows one of the mylar-covered foot pads of a lunar module that has landed on the moon. The shadow of the foot pad darkens the powdery lunar surface; the footprints are those of astronauts who walked on the surface.

The condensation of many acid and alcohol molecules in alternating sequence can be represented as

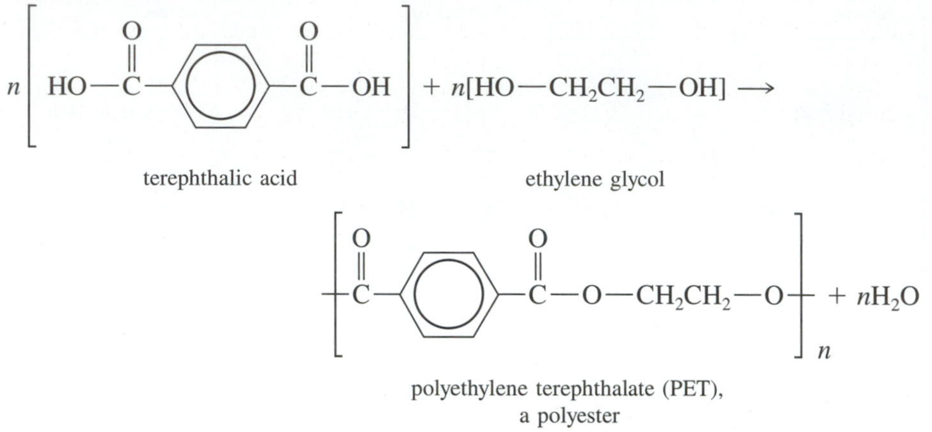

$$n\left[\mathrm{HO{-}\overset{\overset{\Large O}{\|}}{C}{-}C_6H_4{-}\overset{\overset{\Large O}{\|}}{C}{-}OH}\right] + n[\mathrm{HO{-}CH_2CH_2{-}OH}] \longrightarrow$$

terephthalic acid — ethylene glycol

$$\left[\mathrm{\overset{\overset{\Large O}{\|}}{C}{-}C_6H_4{-}\overset{\overset{\Large O}{\|}}{C}{-}O{-}CH_2CH_2{-}O}\right]_n + n\mathrm{H_2O}$$

polyethylene terephthalate (PET), a polyester

This polymer can be spun into fibers of Dacron or rolled into sheets of Mylar film (Figure 23.17).

Polyesters are widely used in clothing—check your labels.

Fats and Oils

Triglycerides are esters of fatty acids (Figure 23.14) and glycerol (Figure 23.8). The general formula for a triglyceride is

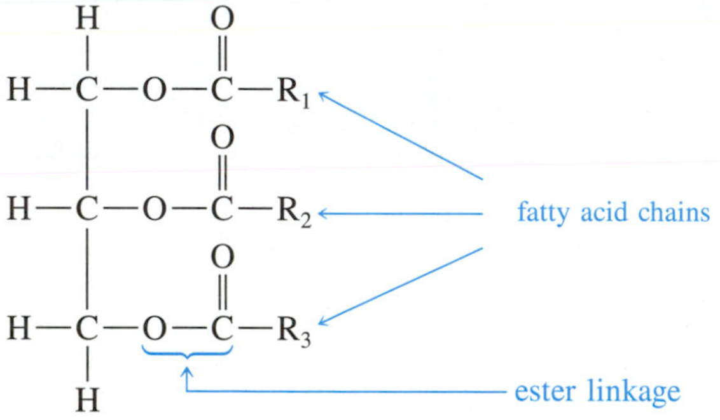

$$\begin{array}{l}\mathrm{H{-}\overset{\overset{\large H}{|}}{C}{-}O{-}\overset{\overset{\large O}{\|}}{C}{-}R_1}\\ \quad\ |\\ \mathrm{H{-}C{-}O{-}\overset{\overset{\large O}{\|}}{C}{-}R_2}\\ \quad\ |\\ \mathrm{H{-}\underset{\underset{\large H}{|}}{C}{-}O{-}\overset{\overset{\large O}{\|}}{C}{-}R_3}\end{array}$$

fatty acid chains

ester linkage

Evidence indicates that saturated fat promotes the deposition of solids (plaque) in the arteries. A severe blockage can result in heart attack or stroke. It is currently recommended that total dietary fat should be no more than 30% of the calories consumed; saturated fat should be no more than 10% of the calories consumed.

The triglyceride esters that form from saturated fatty acids are solid at room temperature and are commonly called **fats**. *Stearin*, or glyceryl tristearate (Figure 23.18), is a principal ingredient of animal fat; it is the ester of glycerol and stearic acid ($C_{17}H_{35}COOH$).

Triglyceride esters of unsaturated fatty acids tend to be liquid at room temperature; they are called **oils**. For example, 85% of corn oil is unsaturated fat. The inflex-

Figure 23.18
Stearin (glyceryl tristearate), a saturated fat.

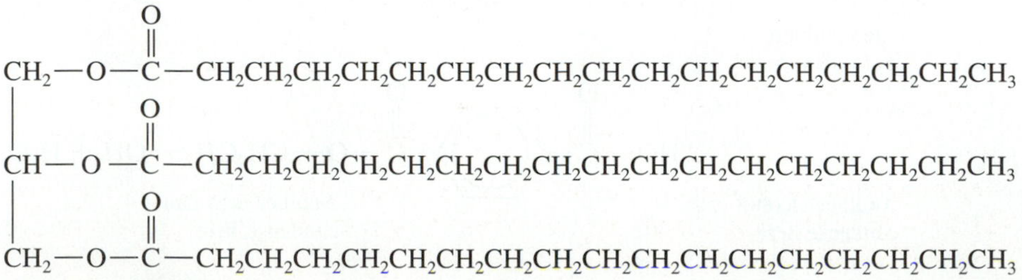

$$\begin{array}{l}\mathrm{CH_2{-}O{-}\overset{\overset{\large O}{\|}}{C}{-}CH_2CH_2CH_2CH_2CH_2CH_2CH_2CH_2CH_2CH_2CH_2CH_2CH_2CH_2CH_2CH_2CH_3}\\ \ |\\ \mathrm{CH{-}O{-}\overset{\overset{\large O}{\|}}{C}{-}CH_2CH_2CH_2CH_2CH_2CH_2CH_2CH_2CH_2CH_2CH_2CH_2CH_2CH_2CH_2CH_2CH_3}\\ \ |\\ \mathrm{CH_2{-}O{-}\overset{\overset{\large O}{\|}}{C}{-}CH_2CH_2CH_2CH_2CH_2CH_2CH_2CH_2CH_2CH_2CH_2CH_2CH_2CH_2CH_2CH_2CH_3}\end{array}$$

ible bends produced by the *cis* double-bond configuration (see Figure 23.14) make the unsaturated triglyceride less compact and thus give it a lower melting point. Oils are sometimes *hydrogenated* to make them more solid by saturating some or all of the double bonds. Oleomargarine often contains hydrogenated cottonseed or soybean oil.

Fats provide the body with concentrated stored energy. The oxidation of 1 g of fat will yield about 9 kcal, compared to about 4 kcal per gram for carbohydrates and proteins. Moreover, fats are hydrophobic and store compactly without absorbing water.

EXAMPLE 23.5

Would you expect a triglyceride made from oleic acid, $C_{17}H_{33}COOH$, to be solid or liquid at room temperature?

SOLUTION

A saturated 17-carbon alkyl group would have the formula $C_{17}H_{35}$. The hydrocarbon groups in oleic acid have only 33 hydrogen atoms and must therefore be unsaturated. The triglyceride would be liquid.

The hydrolysis of an ester in the presence of base produces the alcohol and a salt of the acid; this process, called **saponification**, is often used for making soap from fat:

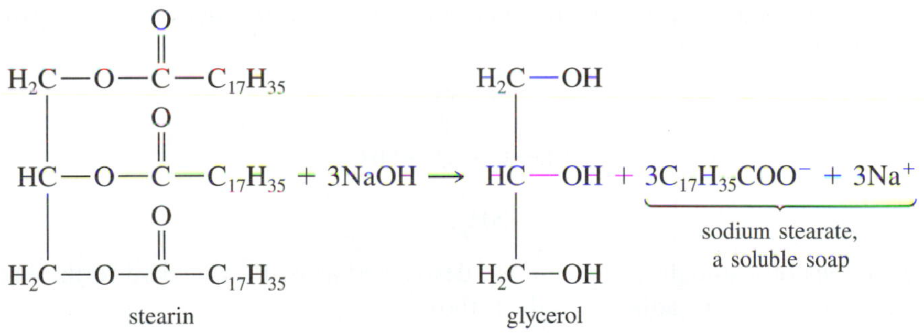

$$\begin{array}{l} H_2C{-}O{-}\overset{\overset{\displaystyle O}{\|}}{C}{-}C_{17}H_{35} \\ \;| \\ HC{-}O{-}\overset{\overset{\displaystyle O}{\|}}{C}{-}C_{17}H_{35} \\ \;| \\ H_2C{-}O{-}\overset{\overset{\displaystyle O}{\|}}{C}{-}C_{17}H_{35} \end{array} + 3NaOH \longrightarrow \begin{array}{l} H_2C{-}OH \\ \;| \\ HC{-}OH \\ \;| \\ H_2C{-}OH \end{array} + \underbrace{3C_{17}H_{35}COO^- + 3Na^+}_{\text{sodium stearate, a soluble soap}}$$

stearin — glycerol

23.6 AMINES

Amines can be regarded as derivatives of ammonia in which one, two, or all three of the hydrogen atoms have been replaced by hydrocarbon groups to form **primary**, **secondary**, and **tertiary** amines, respectively (Figure 23.19).

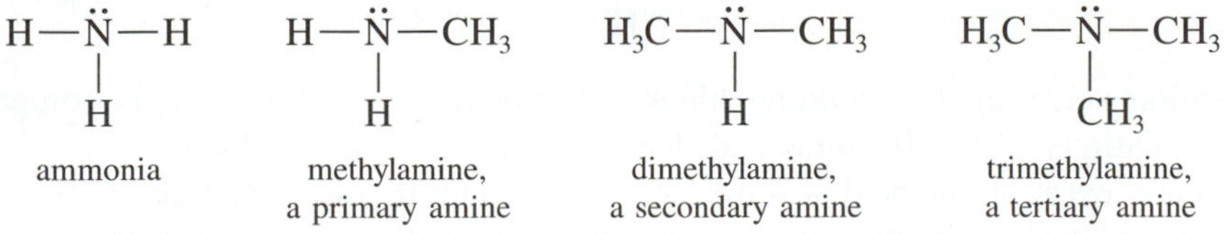

$$\begin{array}{cccc} H{-}\ddot{N}{-}H & H{-}\ddot{N}{-}CH_3 & H_3C{-}\ddot{N}{-}CH_3 & H_3C{-}\ddot{N}{-}CH_3 \\ | & | & | & | \\ H & H & H & CH_3 \end{array}$$

ammonia; methylamine, a primary amine; dimethylamine, a secondary amine; trimethylamine, a tertiary amine

The functional group in a primary amine is the **amino group**, NH_2. *Aniline* (Figure 23.20), one of the best-known primary amines, is used in the manufacture of dyes, pharmaceuticals, varnishes, perfumes, and shoe polish.

(a)

(b)

Figure 23.19

Primary and secondary amines. (a) Methylamine (CH_3NH_2) is a primary amine. (b) Dimethylamine ($(CH_3)_2NH$) is a secondary amine. (Blue spheres represent nitrogen atoms.)

Figure 23.20

Aniline ($C_6H_5NH_2$) is a primary amine.

Amines, like ammonia, are bases; the nitrogen atom retains its lone pair of electrons and its ability to attract protons. The neutralization of an amine by a strong acid produces an amine salt (the reaction is analogous to the production of an ammonium salt by the neutralization of ammonia):

$$CH_3NH_2 + HBr \rightarrow \underset{\text{methylammonium bromide}}{CH_3NH_3Br} \qquad (= CH_3NH_3^+ + Br^-)$$

Amino Acids and Chirality

An **amino acid** is a carboxylic acid with an amino group on the carbon chain. It has the properties of an acid and of an amine, and some special properties of its own as well. *Glycine* is the simplest amino acid:

$$\begin{array}{c} \quad\ \ \text{H}\ \ \ \text{O} \\ \quad\ \ |\ \ \ \ \| \\ \text{H—C—C—OH} \\ |\quad\quad \\ \ \ \text{NH}_2\quad\quad \end{array}$$

The carbons of a straight-chain acid are designated as α, β, γ, and so forth, where the α-carbon is the one adjacent to the carboxyl group:

$$—\overset{\gamma}{C}H_2—\overset{\beta}{C}H_2—\overset{\alpha}{C}H_2—COOH$$

The general formula of an **α-amino acid** is

$$\begin{array}{c} \text{H} \\ | \\ \text{R—C—COOH} \\ | \\ \text{NH}_2 \end{array}$$

α-Amino acids are the building blocks of protein. Natural protein is composed almost entirely of the 20 amino acids listed in Table 23.8 (page 1144).

LEARNING HINT

Review the discussion of enantiomers in Section 21.5.

Nearly every amino acid is chiral; that is, it cannot be superimposed on its mirror image, and each of its molecules is therefore one of two optical isomers, or enantiomers (see Section 21.5). *A carbon atom with four different substituents is always chiral*, as shown in Figure 23.21. A chiral carbon atom is also called an **asymmetric**

carbon and may be labeled with an asterisk:

$$\begin{array}{c} \text{H} \\ | \\ \text{CH}_3\text{—C*—COOH} \\ | \\ \text{OH} \end{array} \qquad \begin{array}{c} \text{H} \\ | \\ \text{CH}_3\text{—C*—COOH} \\ | \\ \text{NH}_2 \end{array}$$

lactic acid — alanine (an amino acid, Figure 23.22)

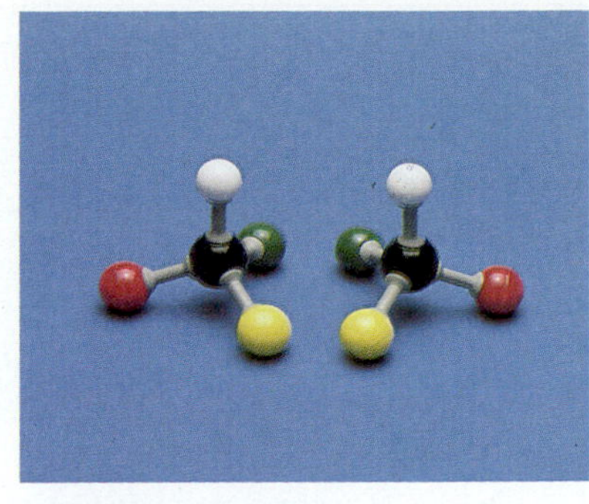

Figure 23.21

A carbon atom with four different substituents is chiral. These two molecules, which are mirror images of each other, are alike but not identical.

EXAMPLE 23.6

Do the following molecules have enantiomers? Identify any asymmetric carbon atoms.

$$\textbf{(a)}\ \begin{array}{c} \text{Cl} \\ | \\ \text{CH}_3\text{—CH}_2\text{—C—CH}_3 \\ | \\ \text{Br} \end{array} \qquad \textbf{(b)}\ \begin{array}{c} \text{H}\quad\text{Cl} \\ |\quad\ | \\ \text{CH}_3\text{—C—C—CH}_2\text{—CH}_3 \\ |\quad\ | \\ \text{OH}\ \text{Cl} \end{array}$$

SOLUTION

(a) The third carbon from the left is bonded to four different groups, CH_3CH_2, CH_3, Cl, and Br. This carbon atom is asymmetric, so the molecule has enantiomers.

(b) The second carbon from the left is bonded to H, OH, CH_3, and $CCl_2CH_2CH_3$. This carbon is also asymmetric, so the molecule has enantiomers.

Tartaric acid HOOC—CHOH—CHOH—COOH, contains two asymmetric carbon atoms (Can you identify them?), was the first compound found to exist in mirror-image forms. The crystals of tartaric acid are also mirror images. In 1848 Louis Pasteur isolated the two enantiomers by sorting crystals of their sodium ammonium salts with tweezers under a hand lens.

PRACTICE EXERCISE 23.7

Which of the following chlorinated propanols are chiral? Identify the asymmetric carbon atoms.

(a) $CH_2Cl—CH_2—CH_2OH$ (b) $CH_3—CHCl—CH_2OH$

We might think that chirality is a trivial property, since the resemblance between two enantiomers seems overwhelming compared to their subtle difference. This difference, however, is profoundly important. Many compounds of biological significance are chiral, but usually only one enantiomer is produced in nature and fulfills a biochemical function. The enantiomers of α-amino acids are designated D and L:

$$\begin{array}{c} \text{R} \\ | \\ \text{HOOC}\blacktriangleleft\text{C}\blacktriangleright\text{NH}_2 \\ \vdots \\ \text{H} \end{array} \qquad \begin{array}{c} \text{R} \\ | \\ \text{H}_2\text{N}\blacktriangleleft\text{C}\blacktriangleright\text{COOH} \\ \vdots \\ \text{H} \end{array}$$

D-amino acid — L-amino acid

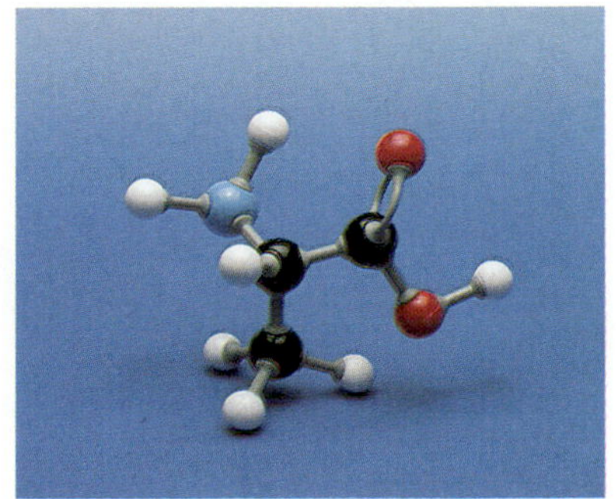

Figure 23.22

L-Alanine (CH_3CHNH_2COOH), an amino acid.

Proteins contain only the L-amino acids. The enantiomers of lactic acid provide another example:

TABLE 23.8 The 20 Amino Acids Found in Proteins

Essential[a]

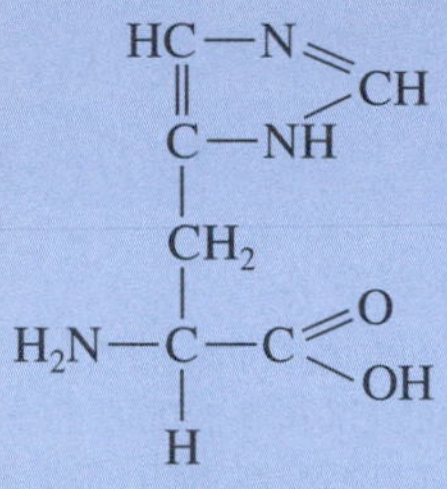

histidine (His)	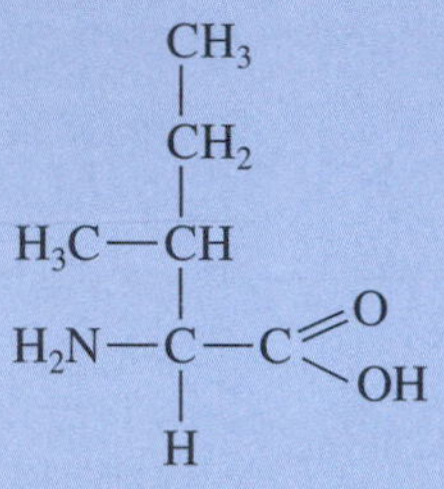 isoleucine (Ile)	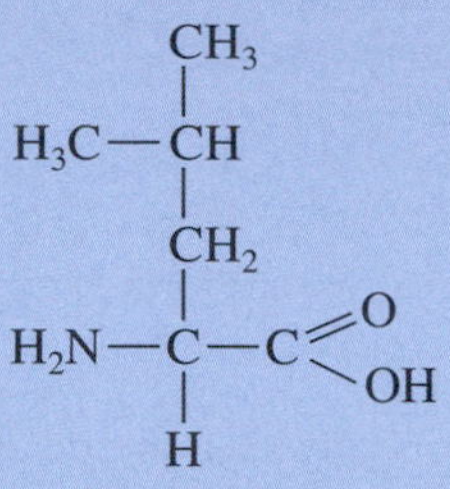 leucine (Leu)	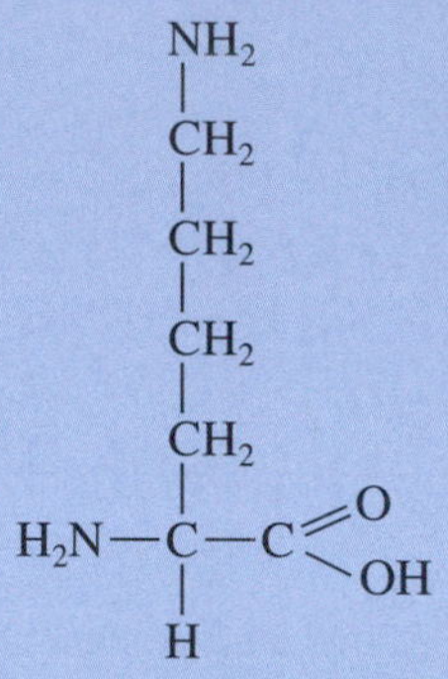 lysine (Lys)	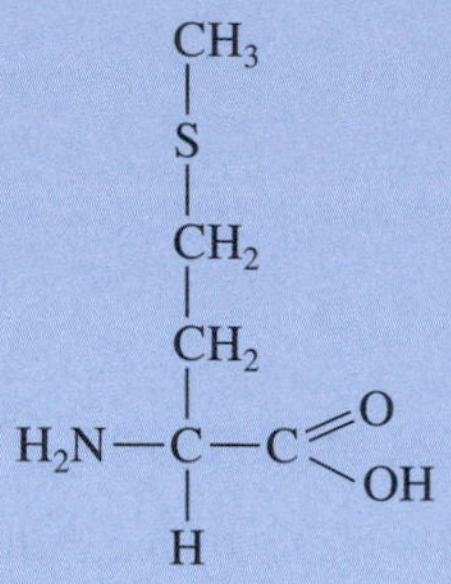 methionine (Met)
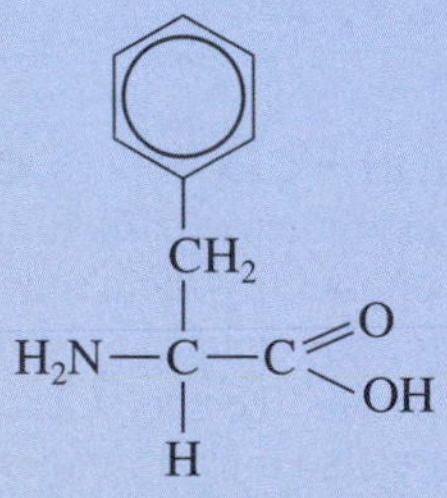 phenylalanine (Phe)	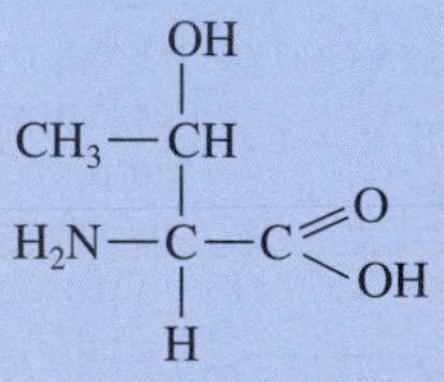 threonine (Thr)	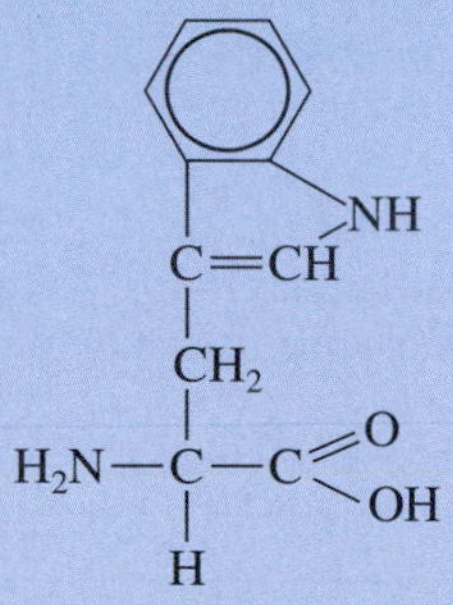 tryptophan (Trp)	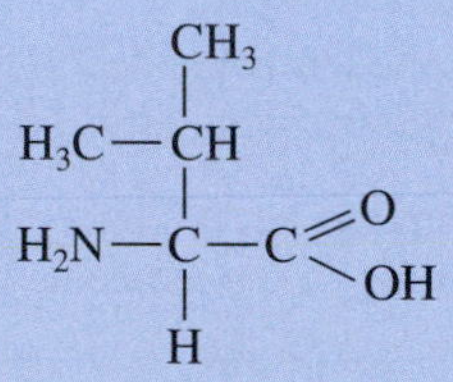 valine (Val)	

[a]Essential amino acids are not synthesized by the body; they must be supplied in the diet.

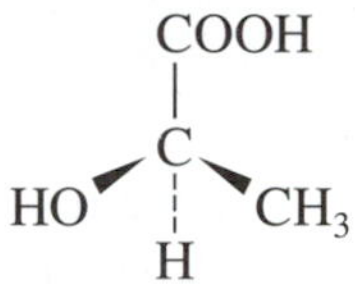

D-lactic acid

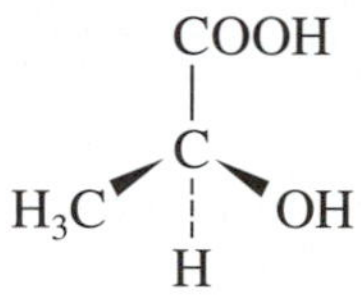

L-lactic acid

Only the L-lactic acid is produced when muscles contract. Still another example is the common sugar dextrose (D-glucose), which is readily metabolized by animals while its enantiomer (L-glucose) is not. It almost appears as though Alice (in *Through the Looking Glass* by Lewis Carroll) knew what she was talking about when she said, "Perhaps looking-glass milk isn't good to drink."

Many drugs on the market today are mixtures of enantiomers. One enantiomer produces the desired effect while the other enantiomer is inactive or, even worse, produces undesirable side effects. A case in point is *thalidomide*, a sedative prescribed for morning sickness in the 1950s. One enantiomer was indeed a sedative, but the other, which was mixed in with the first, caused hundreds of children to be born with severe birth defects.

LEARNING HINT

The symbols D and L do **not** stand for dextrorotatory and levorotatory. They refer instead to the arrangement or configuration of atoms in the molecule. Some D compounds are dextrorotatory and others are levorotatory.

TABLE 23.8 The 20 Amino Acids Found in Proteins (*Continued from previous page*)

Nonessential

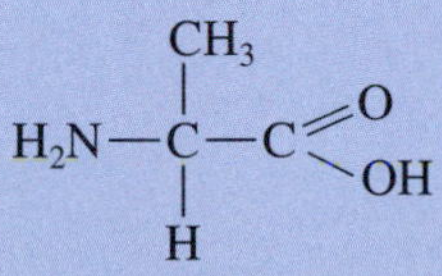

alanine (Ala)

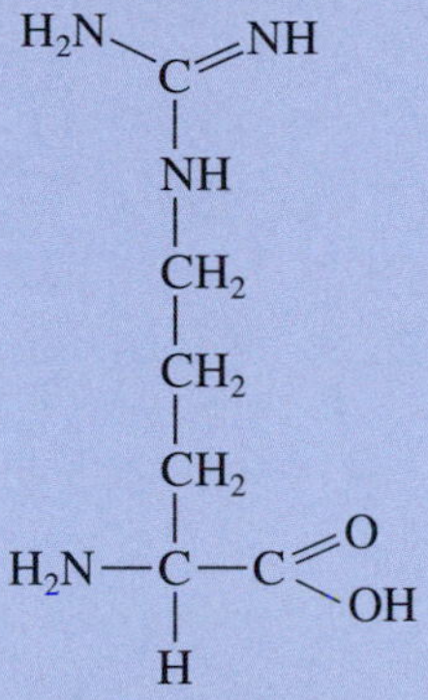

arginine (Arg)

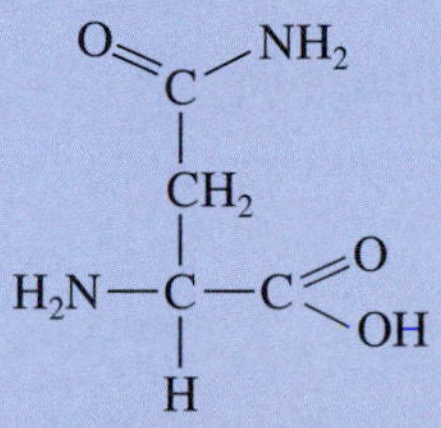

asparagine (Asn)

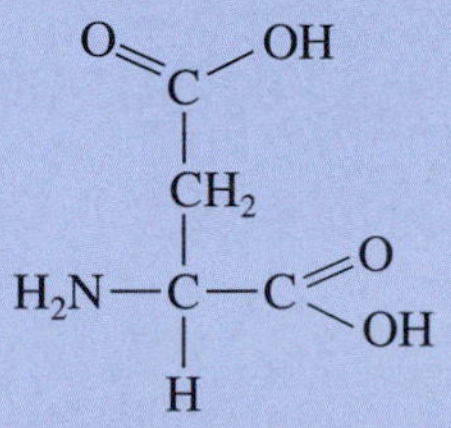

aspartic acid (Asp)

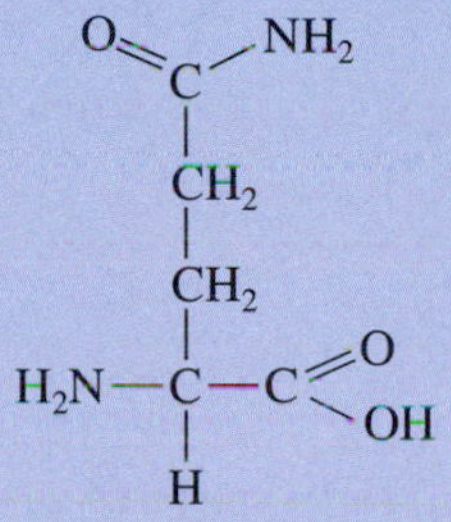

glutamine (Gln)

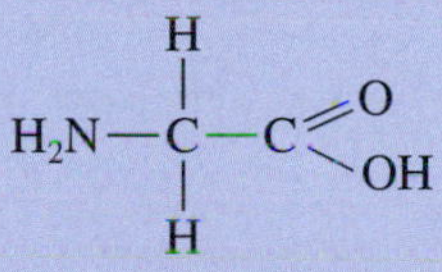

glycine (Gly)

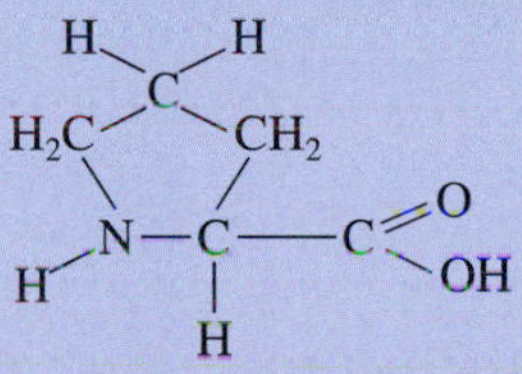

proline (Pro)

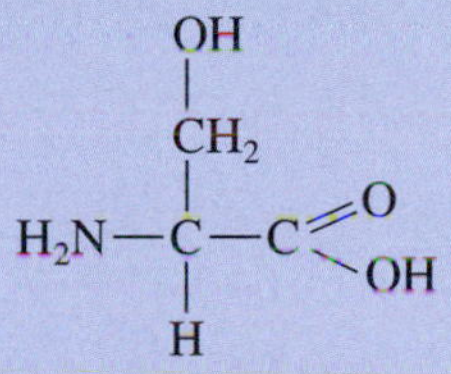

serine (Ser)

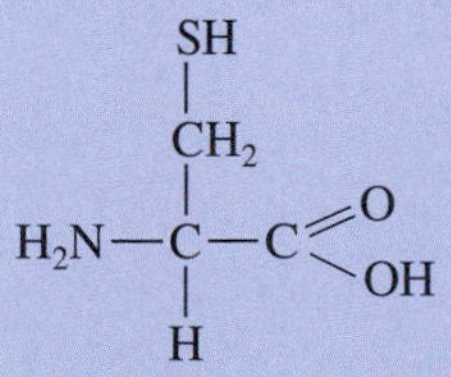

cysteine (Cys)

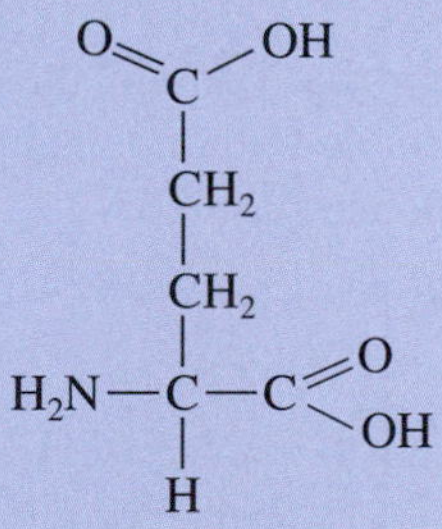

glutamic acid (Glu)

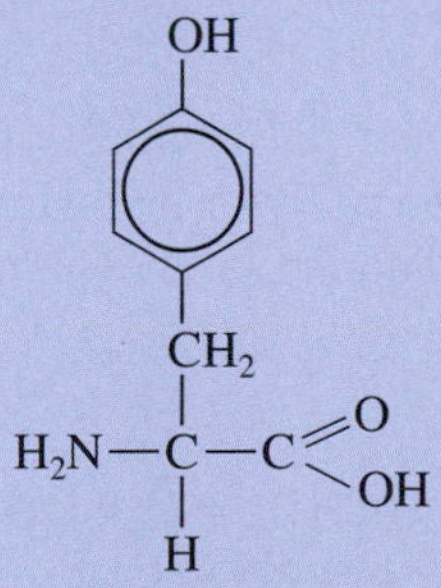

tyrosine (Tyr)

Figure 23.23
Acetamide (CH_3CONH_2).

23.7 THE AMIDE LINKAGE

An **amide** is a product of condensation between a COOH group and a NH_2 group. Acetic acid and ammonia, for example, condense to form *acetamide* (Figure 23.23):

$$\underset{\text{acetic acid}}{CH_3-\overset{\overset{\displaystyle O}{\|}}{C}-OH} + \underset{\text{ammonia}}{H-\overset{\overset{\displaystyle H}{|}}{N}-H} \longrightarrow \underset{\text{acetamide}}{CH_3-\overset{\overset{\displaystyle O}{\|}}{C}-\overset{\overset{\displaystyle H}{|}}{N}-H} + H_2O$$

The **amide linkage** is

$$-\overset{\overset{\displaystyle O}{\|}}{C}-\underset{\underset{\displaystyle H}{|}}{N}-$$

A number of amides have analgesic properties. *Acetaminophen*, the pain-relieving ingredient in Tylenol, can be formed from acetic acid and *p*-aminophenol (Figure 23.24):

$$\underset{\text{acetic acid}}{CH_3-\overset{\overset{\displaystyle O}{\|}}{C}-OH} + \underset{p\text{-aminophenol}}{H-\underset{\underset{\displaystyle H}{|}}{N}-C_6H_4-OH} \longrightarrow$$

$$\underset{\substack{p\text{-acetamidophenol}\\ \text{(acetaminophen)}}}{CH_3-\overbrace{\overset{\overset{\displaystyle O}{\|}}{C}-\underset{\underset{\displaystyle H}{|}}{N}}^{\text{amide linkage}}-C_6H_4-OH} + H_2O$$

Polyamides

The nylons are **polyamides**, polymers that contain the amide linkage. Most nylon used in fabric is Nylon-66, which is made by condensing a six-carbon diacid with a six-carbon diamine, as shown in Demonstration 23.4:

$$n\left[HO-\overset{\overset{\displaystyle O}{\|}}{C}-(CH_2)_4-\overset{\overset{\displaystyle O}{\|}}{C}-OH\right] + n\left[H-\underset{\underset{\displaystyle H}{|}}{N}-(CH_2)_6-\underset{\underset{\displaystyle H}{|}}{N}-H\right] \longrightarrow$$

adipic acid; 1,6-diaminohexane

$$\underset{\text{Nylon-66}}{\left[\overset{\overset{\displaystyle O}{\|}}{C}-(CH_2)_4-\overbrace{\overset{\overset{\displaystyle O}{\|}}{C}-\underset{\underset{\displaystyle H}{|}}{N}}^{\text{amide linkage}}-(CH_2)_6-\underset{\underset{\displaystyle H}{|}}{N}\right]_n} + 2nH_2O$$

DEMONSTRATION 23.4

MAKING NYLON, A POLYAMIDE

The reactants occupy separate immiscible layers. The lower layer contains adipyl chloride ($ClCO(CH_2)_4COCl$) in CCl_4; the upper layer is 1,6-diaminohexane in water. The reactants combine at the interface; the product is Nylon-66, a form of nylon. (Adipyl chloride differs from adipic acid (see text) in that a chlorine atom has replaced each OH group.)

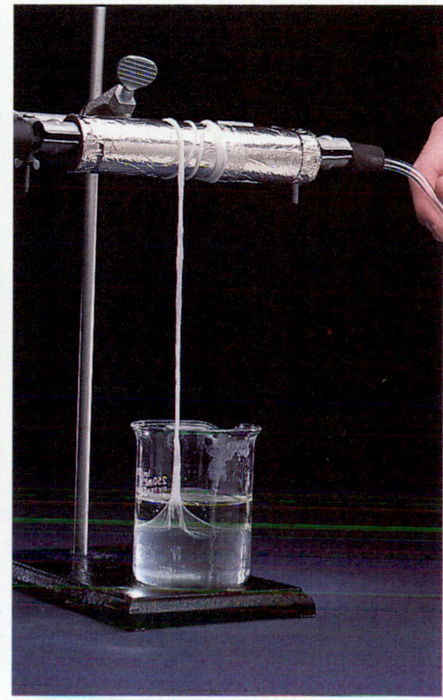

The rotating cylinder continuously pulls a nylon strand from the interface. The strand will continue to form until one reactant is used up.

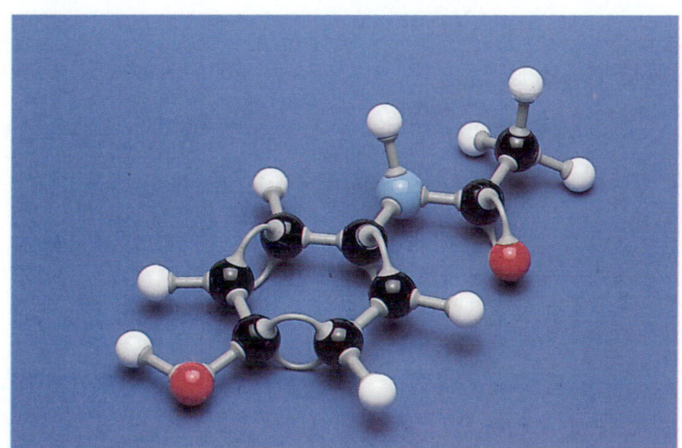

Figure 23.24
Acetaminophen (Tylenol™) is a phenol and an amide.

The two monomers, adipic acid and diaminohexane, are made from cyclohexane, a refinery product. About 80% of the synthetic fabrics that we wear and use are made from petroleum.

Polypeptides

Amino acids can combine with each other through their amine and carboxyl groups. The amide linkage between amino acids is often called a **peptide linkage**:

$$\underset{\text{alanine, an amino acid}}{\mathrm{H{-}N(H){-}C(H)(CH_3){-}C(=O){-}OH}} + \underset{\text{glycine, an amino acid}}{\mathrm{H{-}N(H){-}C(H)(H){-}C(=O){-}OH}} \longrightarrow \underset{\text{alanyl glycine, a dipeptide}}{\mathrm{H{-}N(H){-}C(H)(CH_3){-}\overbrace{C(=O){-}N(H)}^{\text{peptide linkage}}{-}C(H)(H){-}C(=O){-}OH}} + H_2O$$

A **dipeptide** contains two **amino acid residues** (the term used for units in a peptide chain). A **polypeptide** consists of many residues. The general formula for a polypeptide chain formed from n α-amino acids is

$$\mathrm{H{-}N(H){-}C(H)(R_1){-}C(=O){-}\underbrace{N(H){-}C(H)(R_2){-}C(=O)}_{\text{one residue}}{-}N(H){-}C(H)(R_3){-}C(=O){-}\cdots{-}N(H){-}C(H)(R_n){-}C(=O){-}OH}$$

where the R's represent the side groups of the different amino acids.

The formula of a polypeptide is conventionally written with the free amino group on the left and the free carboxyl group on the right. It may also be described by giving the amino acid sequence in terms of the three-letter symbols given in Table 23.8.

EXAMPLE 23.7

Write the structural formula for the polypeptide Val-Phe-Leu.

SOLUTION

The backbone will contain three $\mathrm{{-}N(H){-}C(H)(R){-}C(=O){-}}$ residues:

$$\mathrm{H{-}N(H){-}C(H)(R_1){-}C(=O){-}N(H){-}C(H)(R_2){-}C(=O){-}N(H){-}C(H)(R_3){-}C(=O){-}OH}$$

From left to right, the three R groups will be those of valine (Val), phenylalanine (Phe), and leucine (Leu). Table 23.8 gives the formulas:

```
             valine      phenylalanine     leucine
                                              CH3
                              (C6H5)          |
              CH3               |         H3C—CH
              |                 |             |
         H3C—CH  O            CH2 O           CH2 O
              |  ||            |  ||           |  ||
        H—N—C—C—N—C—C—N—C—C—OH
           |  |        |  |        |  |
           H  H        H  H        H  H
```

PRACTICE EXERCISE 23.8

One of the body's natural opiates that regulate the reception of painful stimuli by the brain is the polypeptide Met-enkephalin, Tyr-Gly-Gly-Phe-Met. Write its structural formula.

DIGGING DEEPER

Protein Structure

A **protein** is a natural polypeptide chain with anywhere from 50 to 2000 amino acid residues, a definite composition, and a definite conformation. Some protein molecules consist of two or more polypeptide chains in a well-defined spatial relationship. The body contains over 100,000 proteins that serve different functions. Enzymes, hormones, and antibodies are proteins. Skin, cartilage, and contractile material such as muscle fiber are also composed of protein.

The **primary structure** of a protein is simply the sequence of amino acid residues. The first primary structure ever to be determined was that of bovine insulin, a small protein with 51 amino acid residues (Figure 23.25).

Protein chains are not stretched out as Figure 21.25 might suggest. An actual chain has a three-dimensional structure, or conformation, arising from the interaction of various groups in the chain with each other and with the aqueous medium. A polypeptide chain invariably folds and twists into a conformation that maximizes the attractions and achieves the lowest energy.

In many proteins, these interactions give rise to a **secondary structure** consisting of a regular repeating conformation that is held in place by hydrogen bonding between the N—H and C=O groups along the backbone of the polypeptide chain. Two common repeating structures are the **α-helix** and the **β-pleated sheet**. The α-helix, shown in Figure 23.26, is a spiral resembling a coiled spring. The helical structure tends to be elastic; it is found in the protein fibers of hair and wool. The pleated sheet consists of a side-by-side array of extended polypeptide chains with hydrogen bonding between adjacent chains (Figure 23.27). This arrangement, which allows very little stretching, is found in the fibers of silk.

Many proteins also have a **tertiary structure** consisting of folds and bends that are not part of a repeating secondary structure. The principal types of interaction that

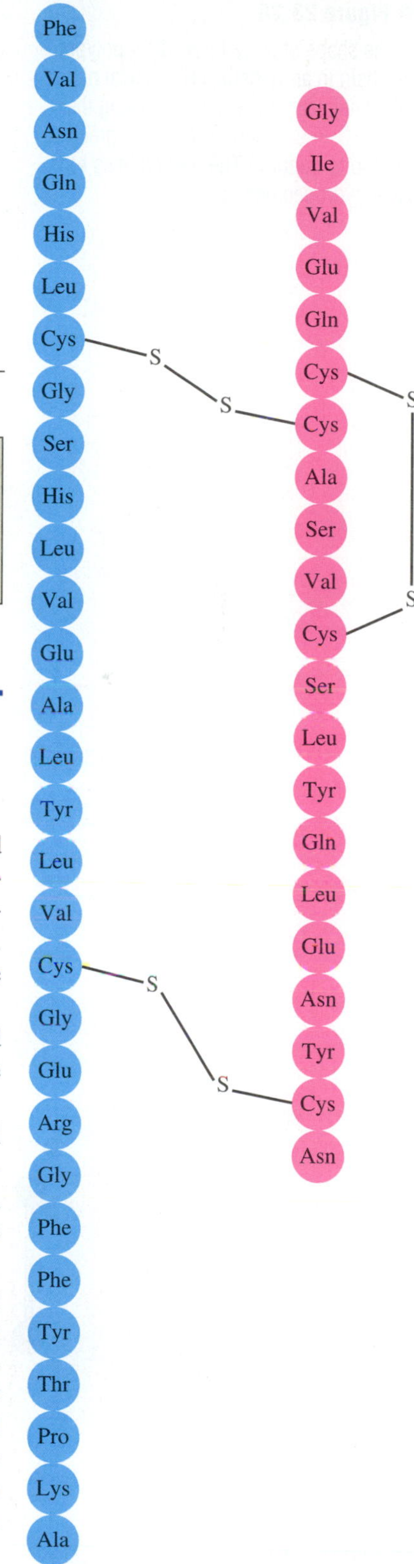

Figure 23.25
The primary structure of bovine insulin.

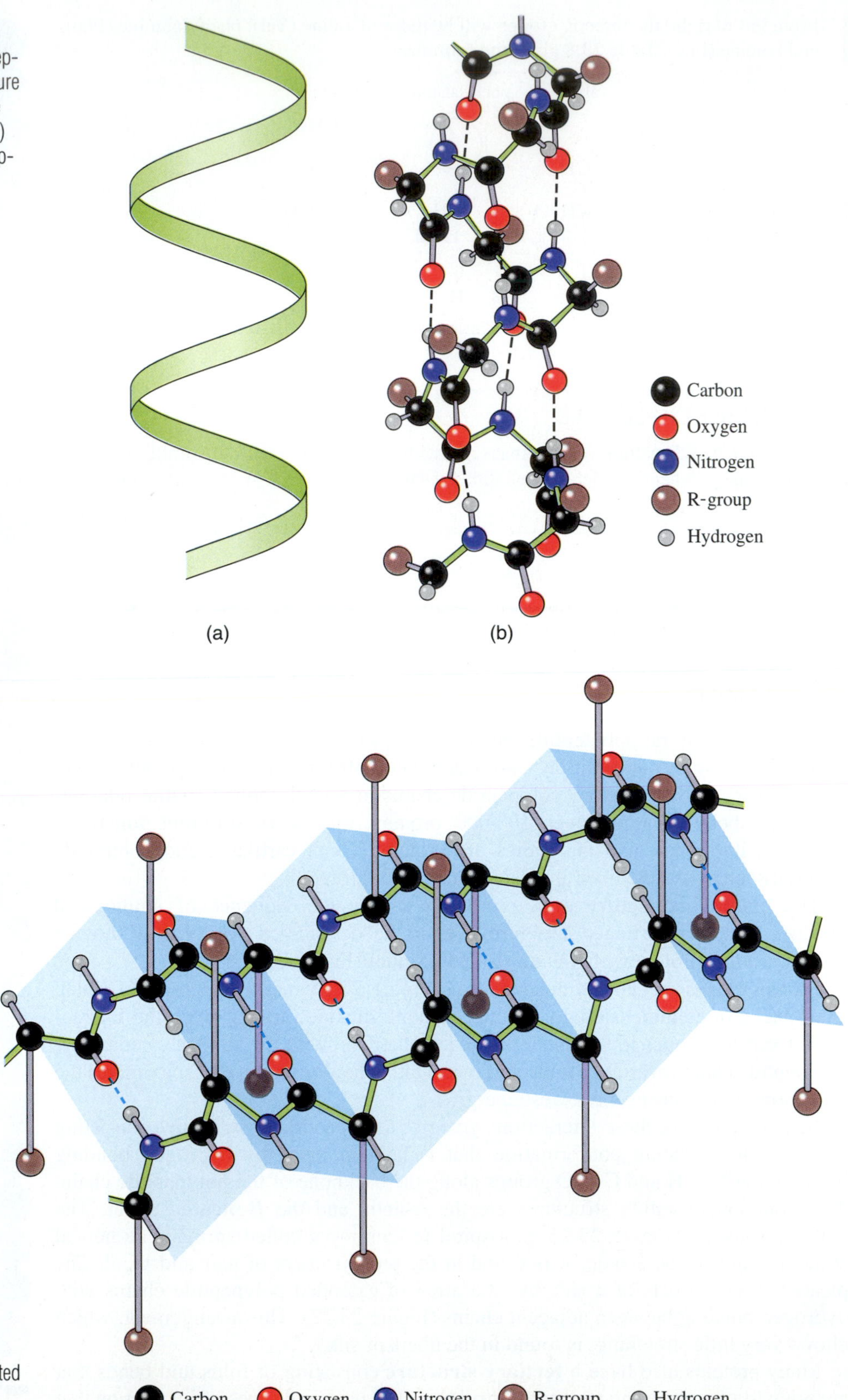

Figure 23.26

(a) The shape of an α-helix. (b) A polypeptide chain in an α-helix. (The helical nature of the chain can be seen by following the —N—C—C— bonds (colored green) from top to bottom.) The dashed lines represent hydrogen bonds.

Figure 23.27

Adjacent polypeptide chains in a β-pleated sheet.

maintain the tertiary structure are illustrated in Figure 23.28. The tertiary structure is important in **globular proteins**, in which the polypeptide chain folds into a ball-like shape, which appears to be irregular but is nevertheless well defined. Secondary and tertiary structures can be present together; myoglobin, for example, is a globular protein with both helical segments and nonrepeating portions.

Some proteins combine two or more polypeptide chains, each with its own secondary and tertiary structure. The spatial relationships among these chains, called the **quaternary structure**, is maintained by hydrogen bonding, ionic interactions, and hydrophobic interactions. Figure 23.29 shows the quaternary structure of hemoglobin, a protein with four myoglobin subunits.

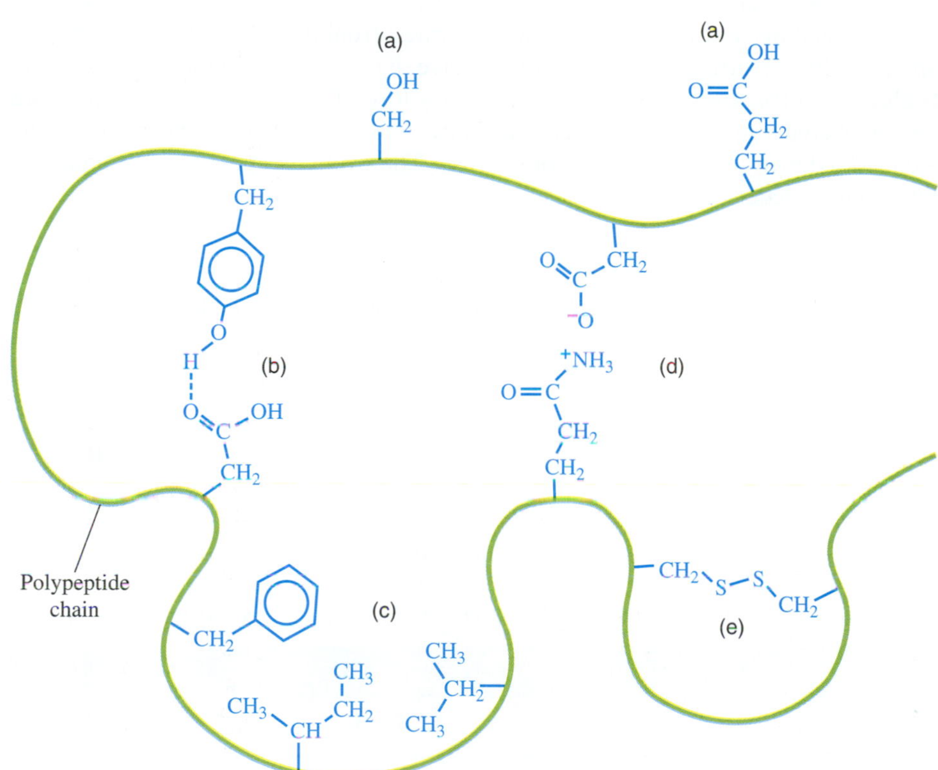

Figure 23.28

Interactions that maintain tertiary protein structure. (a) Hydrophilic groups in contact with water. (b) Hydrogen bonding. (c) Interaction between hydrophobic groups. (d) Ionic interaction. (e) Disulfide bridges.

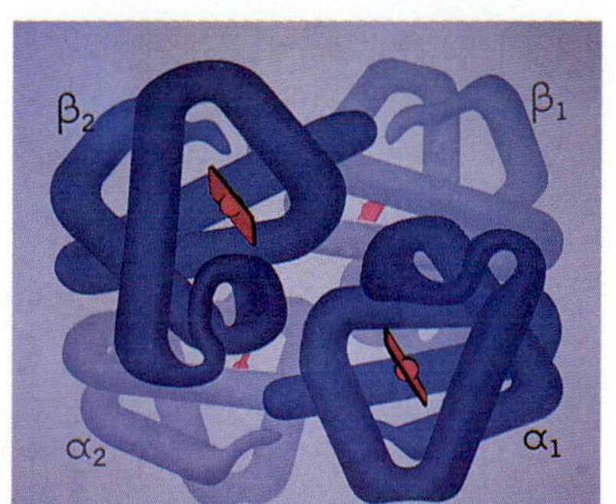

Figure 23.29

Top view of hemoglobin. The molecule is composed of four myoglobin subunits (α_1, α_2, β_1, β_2) in a definite spatial relationship called the *quaternary structure.* The four small red structures are the heme groups. (See *Chemical Insight: Hemoglobin and Other Porphyrin Complexes*, p. 1008.)

23.8 CARBOHYDRATES

Carbohydrates include the **monosaccharides** (simple sugars such as glucose) and monosaccharide polymers such as starch and cellulose. Monosaccharides are polyhydroxy aldehydes and ketones with the general formula $(CH_2O)_n$, where *n* ranges from 3 to 7. The names of most saccharides end in *-ose*. The monosaccharides found in food are mostly **hexoses**, which have six carbon atoms and the formula $C_6H_{12}O_6$; examples are *glucose*, which is an **aldose** (aldehyde sugar), and *fructose*, which is a **ketose** (ketone sugar). *Ribose*, a **pentose** (five carbons), is a component of nucleic acids such as RNA (Section 23.9).

A $C_6H_{12}O_6$ aldose has four asymmetric carbon atoms and is therefore one of 16 possible isomers made up of 8 enantiomeric pairs. Some of these are shown in Figure 23.30. Each member of a pair is classified as D or L according to the configuration around the asymmetric carbon atom farthest from the >C=O group. Nearly all naturally occurring monosaccharides have the D configuration, for example, D-glucose, D-fructose, D-mannose, and D-galactose. Their preferred structures are six-membered rings that can open and close with a slight rearrangement of the atoms, as shown in Figure 23.31 for glucose. The cyclic isomers are distinguished by the Greek letters alpha and beta.

Figure 23.30 Four naturally occurring hexoses. All D-hexoses have the same configuration around the carbon atom next to the CH_2OH group. In this type of notation, the carbon atoms are rotated so that the C—H and C—OH bonds point toward the viewer, as indicated in the left diagram.

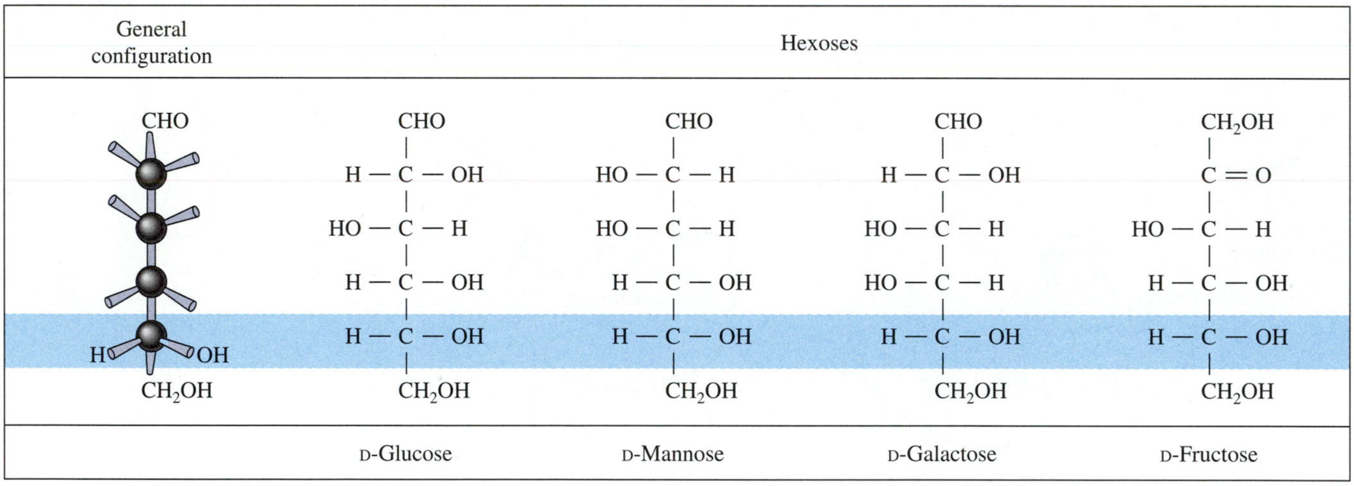

General configuration	Hexoses			
CHO	CHO	CHO	CHO	CH_2OH
	H — C — OH	HO — C — H	H — C — OH	C = O
	HO — C — H	HO — C — H	HO — C — H	HO — C — H
	H — C — OH	H — C — OH	HO — C — H	H — C — OH
H OH	H — C — OH	H — C — OH	H — C — OH	H — C — OH
CH_2OH	CH_2OH	CH_2OH	CH_2OH	CH_2OH
	D-Glucose	D-Mannose	D-Galactose	D-Fructose

Figure 23.31

Equilibrium between the open and cyclic forms of glucose. The orientation of the H and OH bonds on carbon 1 are reversed in the α- and β-cyclic forms.

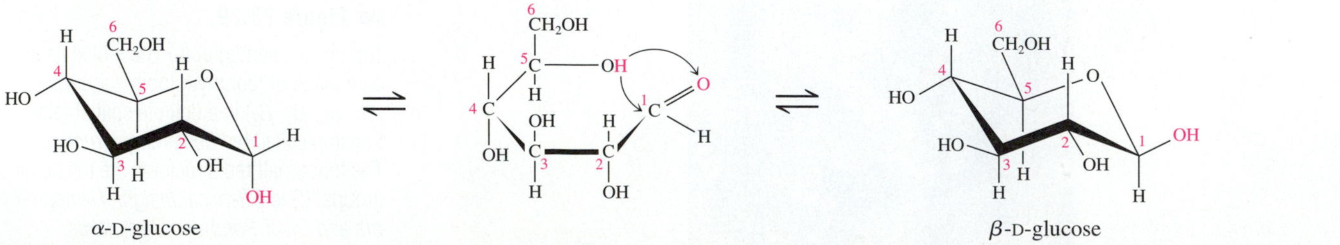

α-D-glucose ⇌ ⇌ β-D-glucose

A **disaccharide** is a condensation product of two cyclic monosaccharides (see Figure 23.32). The oxygen bridge formed by the condensation is called a **glycosidic linkage**; it hydrolyzes in the digestive tract to release the monosaccharides again. *Sucrose*, or table sugar, hydrolyzes to give glucose and fructose. *Lactose* (milk sugar) yields glucose and galactose. *Maltose* (malt sugar) produces two glucose molecules.

The **polysaccharides** *starch, glycogen,* and *cellulose* are large polymers of glucose. Starch, the form in which glucose is stored in plants, is a mixture of the linear polymer *amylose* and the branched polymer *amylopectin* (Figure 23.33). Each of these polymers is a condensation product of α-D-glucose. Their polar OH groups point outward, making the molecules somewhat water soluble. The shape of the glycosidic linkage in starch allows it to interact with the digestive enzymes that hydrolyze it to glucose. Glycogen, the form in which animals store glucose, is similar to amylopectin but has more branching in the polymer chains.

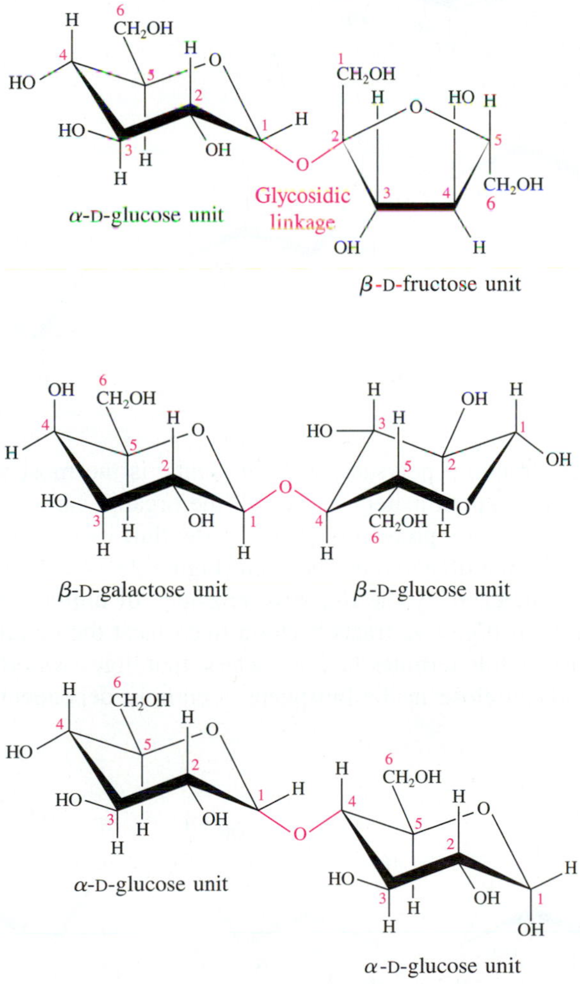

Figure 23.32

Naturally occurring disaccharides. (a) Sucrose (table sugar). (b) Lactose (milk sugar). (c) Maltose (malt sugar).

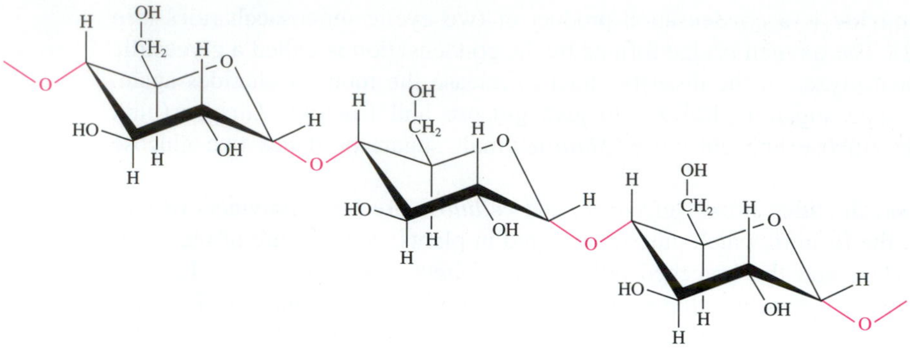

(a)

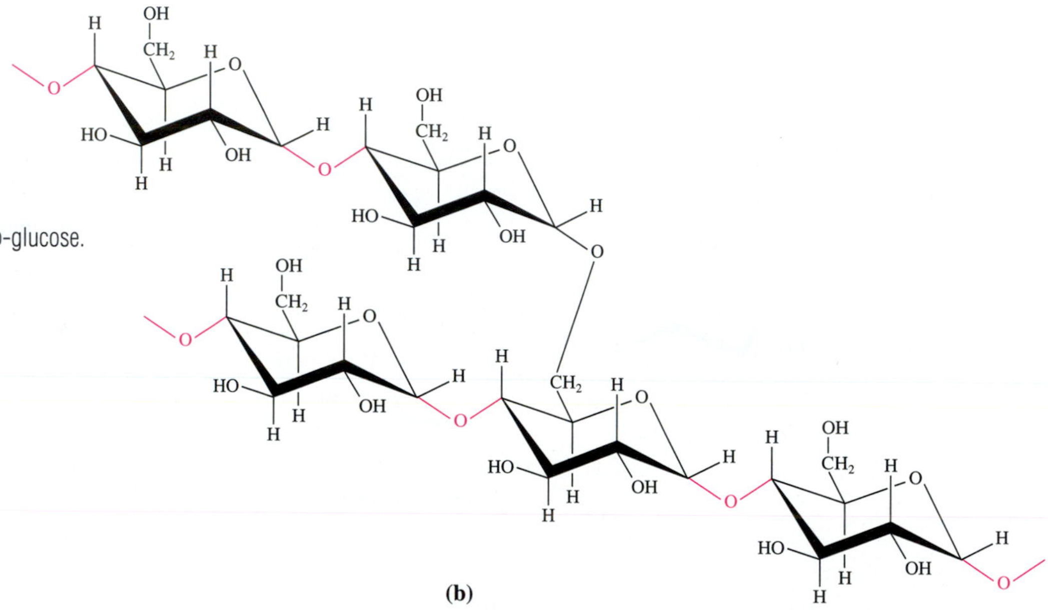

(b)

Figure 23.33
Forms of starch, a polymer of α-D-glucose. (a) Amylose. (b) Amylopectin.

Cellulose, the structural polysaccharide in plants, is the most abundant glucose polymer; it contains approximately half of all the organic carbon in the biosphere. Cellulose is a condensation product of β-D-glucose; thus its glycosidic linkage has a different twist from that of starch or glycogen (Figure 23.34). Cellulose is not water soluble, nor is it attacked by the digestive enzymes of animals. Ruminants (e.g., cows) harbor in their digestive tracts bacteria that digest the cellulose of grass and hay for their hosts, while termites harbor bacteria that digest wood. The continuous biodegradation of cellulose in the biosphere is entirely dependent on microorganisms.

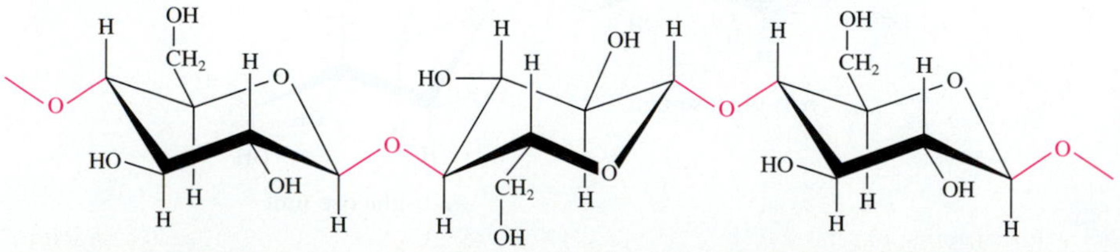

Figure 23.34
Cellulose, a polymer of β-D-glucose.

23.9 NUCLEOTIDES AND NUCLEIC ACIDS

Nucleic acids are the biochemical "ruling class"; they direct and control the life processes of reproduction, growth, and metabolism. The nucleic acid known as deoxyribonucleic acid or DNA, contains the vast amount of inherited information that each cell needs in order to function. Genes, the fundamental units of heredity, are made of DNA. Nucleotides are the monomeric units from which nucleic acids are built. The most important nucleotide is adenosine triphosphate or ATP, the universal currency for energy storage and exchange in living cells.

Nucleotides

A **nucleotide** is a molecule that on hydrolysis yields phosphoric acid (H_3PO_4), a five-carbon sugar that is either ribose or deoxyribose (Figure 23.35), and one of five ***nitrogenous bases***: *adenine, guanine, thymine, cytosine,* or *uracil* (Figure 23.36). These bases are commonly designated by their initial letters: A, G, T, C, and U. The nucleotides involved in energy storage and release are ***adenosine diphosphate*** and

LEARNING HINT

$$\text{nucleotide} \xrightarrow{\text{hydrolysis}} H_3PO_4 + \begin{matrix}\text{ribose}\\ \text{or}\\ \text{deoxyribose}\end{matrix} + \begin{matrix}\text{nitrogeneous}\\ \text{base}\end{matrix}$$

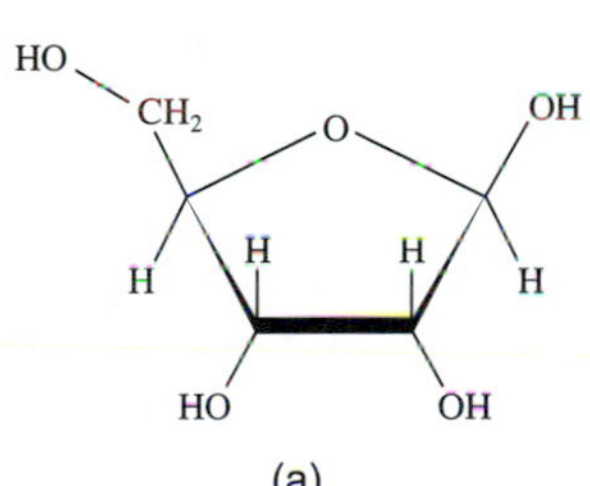

(a)

Figure 23.35

(a) Ribose and (b) deoxyribose.

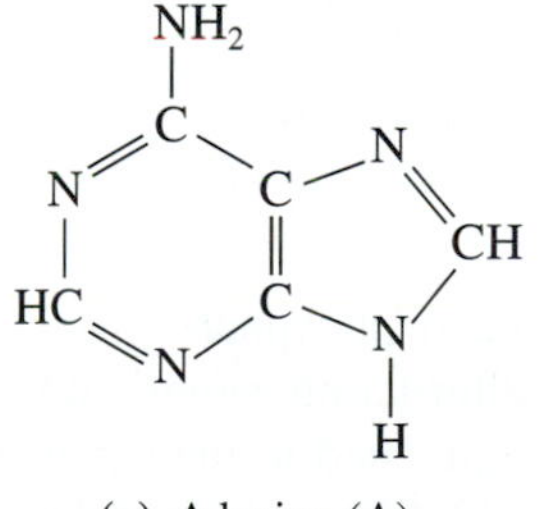

(a) Adenine (A)

(b) Guanine (G)

(c) Thymine (T)

(d) Cytosine (C)

(e) Uracil (U)

Figure 23.36

The nitrogenous bases found in nucleotides and nucleic acids.

Figure 23.37

The structures of ADP and ATP. These nucleotides can be regarded as condensation products of phosphoric acid, the five-carbon sugar *ribose*, and the nitrogeneous base *adenine*. The wavy lines indicate bonds that release energy on hydrolysis.

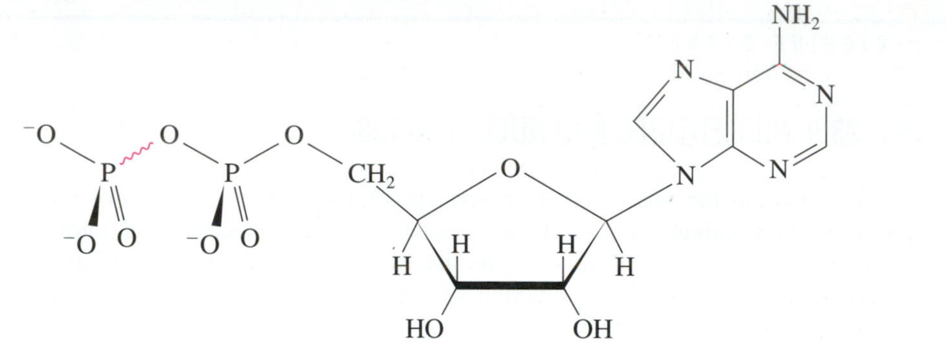

(a) Adenosine diphosphate (ADP)

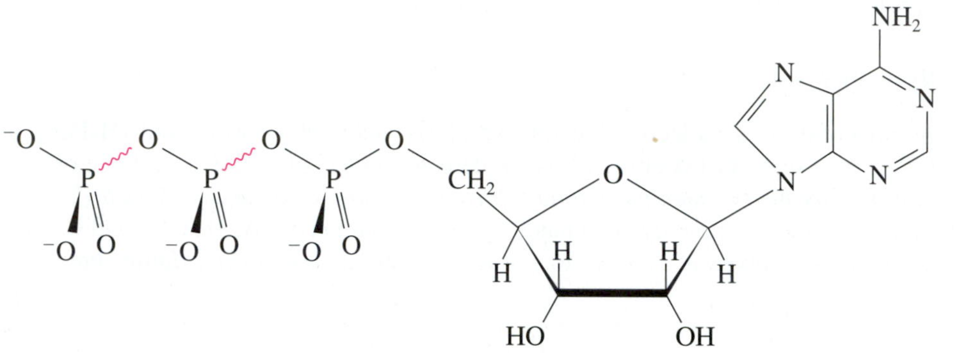

(b) Adenosine triphosphate (ATP)

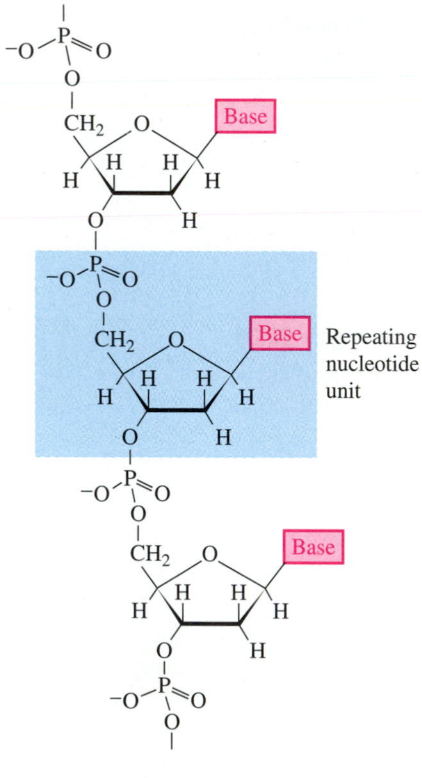

Figure 23.38

Part of the structure of DNA, a nucleic acid. Each base attached to a deoxyribose ring is either adenine (A), guanine (G), thymine (T), or cytosine (C).

Repeating nucleotide unit

adenosine triphosphate, known as **ADP** and **ATP**, respectively (Figure 23.37). All of the body's work, from muscular contraction to transporting molecules through membranes, runs on energy from the hydrolysis of ATP^{4-} to ADP^{3-}:

$$ATP^{4-} + H_2O(l) \rightarrow ADP^{3-} + HPO_4^{2-}(aq) + H^+(aq) \qquad \Delta G° = -31 \text{ kJ}$$

The thermodynamics of this process for storing and releasing energy was discussed in *Chemical Insight: Time's Arrow and Coupled Reactions* (p. 926).

Nucleic Acids

Nucleic acids are nucleotide polymers in which each phosphate group links two ribose units to form a chain or "backbone" of alternating sugar and phosphate groups (Figure 23.38). The nitrogeneous base units attached to the sugar groups are substituents along this chain. **Deoxyribonucleic acid (DNA)** molecules consist of two polymer strands twisted into a **double helix** with the nitrogenous bases pointed inward (Figure 23.39). The two strands are bound to each other by hydrogen bonds between bases that are said to be **complementary**. Adenine is complementary to thymine (A—T), and guanine is complementary to cytosine (G—C) (Figure 23.40). (Adenine is also complementary to uracil (A—U), but DNA does not contain uracil.) Whatever the sequence of bases in one strand, the other strand will have the complementary bases in corresponding order.

The DNA of each individual creature is unique in the sequence of its nucleotide bases. It is this sequence that provides a set of instructions that ultimately directs each cell to synthesize a characteristic set of proteins. When a certain protein is to be synthesized, the DNA molecule sends out strands of **messenger ribonucleic acid (mRNA)** (Figure 23.42a). **RNA** molecules are nucleic acid chains that contain ribose units instead of deoxyribose units, and uracil (U) instead of thymine as a complementary base to adenine. The messenger RNA molecule is assembled on a portion of the DNA molecule that serves as a template; it consists of bases that are complementary to those of the template. This process, which essentially copies the DNA instructions in complementary form (as a photographic print copies a negative), is called **transcription**.

Each of the 20 amino acids used in protein construction is represented by a group of three sequential bases called a **codon**. Put another way, we might say that the "letters" of the **genetic code** are the RNA bases A, G, U, and C, while the "words" are the three-base codons, GCA, AAC, CAU, GUC, and so forth. The four letters or bases can make a total of $4 \times 4 \times 4 = 64$ words or codons (Table 23.9). Since there are only 20 amino acids used in protein construction, many of the codons are "synonyms" for the same amino acid. For example, Table 23.9 shows that the amino acid phenylalanine (Phe) is represented by the codons UUU and UUC. It has been shown that 61 of the 64 codons represent amino acids, and the remaining three are "stop protein synthesis" signals, that is, chain termination signals.

After the newly formed mRNA molecule separates from the DNA, it moves out of the nucleus and becomes attached to several **ribosomes** (Figure 23.42b), which are small cellular particles equipped with the enzymes needed for protein synthesis. Each codon on the messenger RNA now interacts with a molecule of **transfer RNA (tRNA)**. A transfer RNA molecule is a small RNA strand that carries an amino acid and its **anticodon**. If, for example, an amino acid codon on the mRNA is GCC, the code for alanine, it will interact with a tRNA molecule bearing the anticodon CGG and carrying an alanine molecule. The alanine molecule will then be released and incorporated into a polypeptide chain.

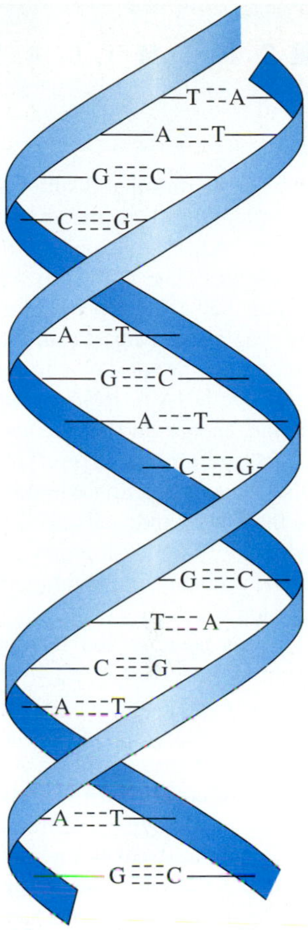

Figure 23.39 (above)

The DNA double helix. The two nucleic acid strands are coiled around a common axis with the bases pointing inward. The strands are joined by hydrogen bonds between pairs of bases.

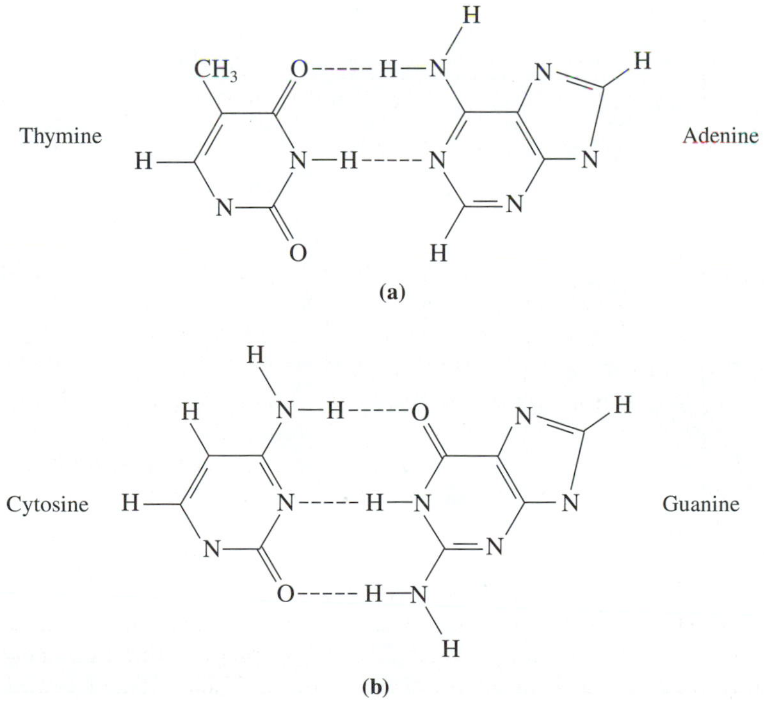

Figure 23.40

The pairing of complementary bases through hydrogen bonding. (a) Thymine pairs with adenine. (b) Cytosine pairs with guanine.

CHEMICAL INSIGHT

THE ORIGIN OF LIFE

The properties that distinguish living from nonliving matter are often listed as metabolism, reproduction, and growth. Nucleic acids have one of these properties: they replicate themselves. The ability to replicate suggests that when the first nucleic acids appeared long ago, by chance or otherwise, life was on the way. Primitive nucleic acids may have been similar to present-day viruses, which essentially consist of extensive DNA or RNA molecules that enter host organisms to find energy and substance; once there, they copy themselves and emerge in greater numbers to infect more cells. DNA cannot function without the materials that cells supply; on the other hand, cells need the DNA. Which came first, the machine or the instruction manual? Proteins and DNA (or RNA) probably evolved together, each supporting the other's development.

It has been argued that nucleic acid molecules are too complex to have formed by mere chance encounter among primitive molecules. However, experiments over the last few decades suggest that nucleic acid molecules have just the sort of structure that might arise from chance alone. The exact composition of earth's early atmosphere is a matter of debate, but all agree that free oxygen was absent, while some combination of gases such as methane, ammonia, hydrogen sulfide, carbon dioxide, and water vapor was present. Energy from the sun's ultraviolet radiation and from lightning continually disrupted these molecules and allowed new ones to form. In 1953 Stanley Miller, then a graduate student at the University of Chicago, was the first to simulate the effect of lightning on a primitive atmosphere. His apparatus is shown in Figure 23.41. Miller and other investigators have since sparked and irradiated a number of different artificial atmospheres composed of the gases listed above, and they found an interesting variety of new molecules in the resulting mixtures. The monomers that make up biomolecules were surprisingly abundant. Aldehydes, acids, and amino acids were found, along with many five- and six-membered rings. Riboses, hexoses, purines, pyrimidines, and even nucleotides have been discovered in such irradiated mixtures. The HCN content of the mixtures was observed to rise at first and then fall as the content of amino acids and nitrogenous bases increased. HCN is an active molecule that will combine with ammonia and aldehydes to give amino acids. It will also polymerize to give cyclic bases such as adenine, which are essential in the formation of DNA and RNA. There is evidence to suggest that certain clays may have provided catalytic surfaces for further polymerization.

Apparently, the development of life was a natural outcome of early conditions rather than the result of some spectacular chain of coincidences that defies probability. Whether the birth of the universe was miraculous, or simply mysterious, life is a natural part of it.

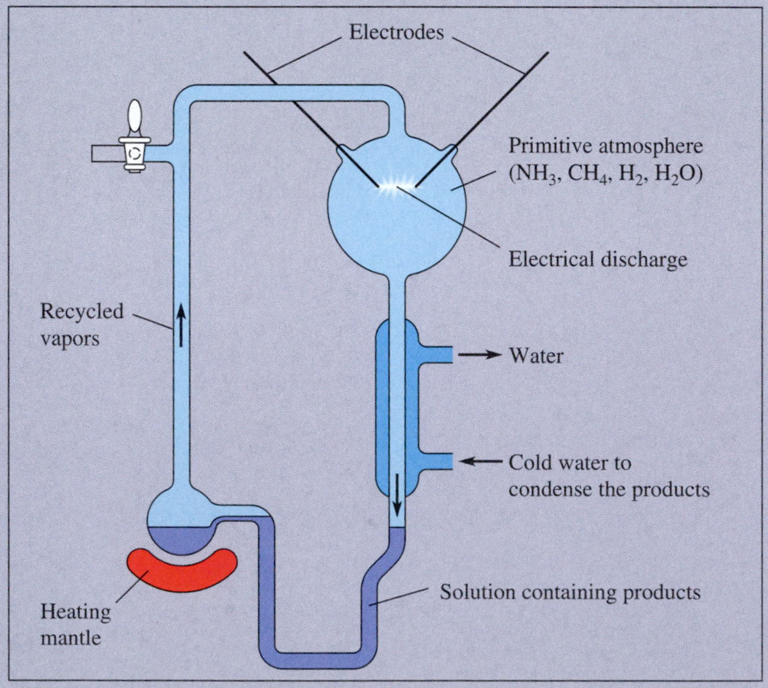

Figure 23.41
The Miller apparatus. Gas mixtures that resemble hypothetical primitive atmospheres are cycled past the electric discharge, which simulates the effect of lightning.

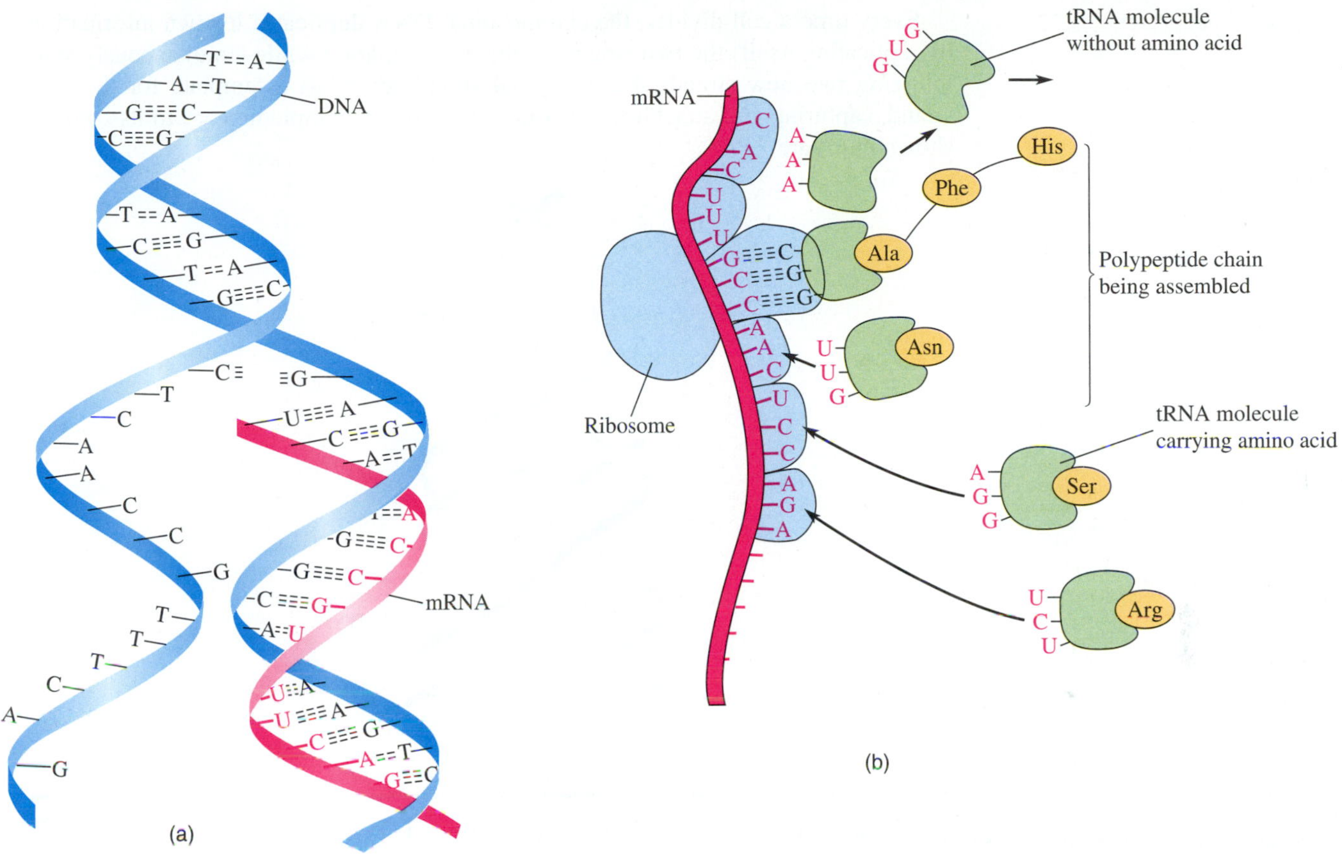

Figure 23.42 (above)

Protein synthesis. (a) mRNA synthesis (transcription). (b) Assembly of polypeptide chain.

TABLE 23.9 The Genetic Code[a]

	U		C		A		G	
U	UUU	Phe	UCU	Ser	UAU	Tyr	UGU	Cys
	UUC	Phe	UCC	Ser	UAC	Tyr	UGC	Cys
	UUA	Leu	UCA	Ser	UAA[b]		UGA[b]	
	UUG	Leu	UCG	Ser	UAG[b]		UGG	Trp
C	CUU	Leu	CCU	Pro	CAU	His	CGU	Arg
	CUC	Leu	CCC	Pro	CAC	His	CGC	Arg
	CUA	Leu	CCA	Pro	CAA	Gln	CGA	Arg
	CUG	Leu	CCG	Pro	CAG	Gln	CGG	Arg
A	AUU	Ile	ACU	Thr	AAU	Asn	AGU	Ser
	AUC	Ile	ACC	Thr	AAC	Asn	AGC	Ser
	AUA	Ile	ACA	Thr	AAA	Lys	AGA	Arg
	AUG	Met	ACG	Thr	AAG	Lys	AGG	Arg
G	GUU	Val	GCU	Ala	GAU	Asp	GGU	Gly
	GUC	Val	GCC	Ala	GAC	Asp	GGC	Gly
	GUA	Val	GCA	Ala	GAA	Glu	GGA	Gly
	GUG	Val	GCG	Ala	GAG	Glu	GGG	Gly

[a]The first base in each codon is given by the row label; the second base is given by the column heading.

[b]This combination serves as a code for the termination of a peptide chain.

Every time a cell divides, the chromosomal DNA duplicates its own information by replicating itself; the two strands of the helix separate while simultaneously synthesizing two new strands. Each original strand serves as a template for the new strand, capturing and attaching nucleotides with bases complementary to its own, as shown in Figure 23.43.

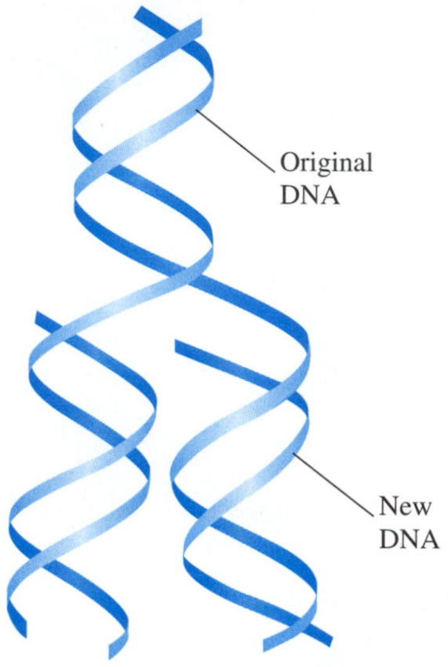

Figure 23.43
DNA replication.

CHAPTER REVIEW

Starred entries are based on the *Digging Deeper* sections.

LEARNING OBJECTIVES BY SECTION

23.1
1. Give the names and structural formulas for straight-chain and branched-chain alkanes with 1–10 carbon atoms.
2. Name alkanes, given their structural formulas; write structural formulas for alkanes, given their names.
3. Give names, structural formulas, and polygon notation for cycloalkanes.
4. Draw the boat and chair configurations of cyclohexane.
5. Draw the basic shape of a steroid molecule.
6. Write equations with structural formulas to illustrate the light-catalyzed reactions of chlorine and bromine with alkanes.

23.2
1. Give names and structural formulas for alkenes and alkynes.
2. Write equations to illustrate the addition of hydrogen, halogens, and hydrogen halides to hydrocarbons with double and triple bonds.
3. Write formulas for polymers formed from ethylene and its derivatives.
4. Given the formula for an addition polymer, identify the monomer.
5. Use equations with structural formulas to show the chain mechanism of addition polymerization.

23.3
1. Give the structural formulas of benzene, naphthalene, and various substituted benzenes.
2. Identify substituents on a benzene ring as *ortho, meta,* or *para.*

23.4
1. Identify the functional groups in a structural formula.
2. Use the structural formula of a compound to identify it as an ether; primary, secondary, or tertiary alcohol; phenol; aldehyde; ketone; or carboxylic acid.
3. Write the structural formulas for all the possible alcohols, aldehydes, ketones, and carboxylic acids with a given carbon skeleton.

4. Give the formulas, properties, and uses of methanol, ethanol, isopropyl alcohol, ethylene glycol, glycerol, and phenol.
5. List some of the sources of methanol and ethanol.
6. Describe the molecular structure of saturated and unsaturated fatty acids.
7. List the oxygen derivatives of a given hydrocarbon in order of the increasing oxidation number of carbon.
8. Explain why glucose is a reducing agent.

23.5 1. Identify an ester from its structural formula.
2. Write the equation for the formation of an ester from a given alcohol and acid.
3. Given the formula of an ester, write the equation for its hydrolysis.
4. Write the formula of the polyester formed from a given dialcohol and diacid.
5. Given the formula of a polyester, write the formulas of its monomers.
6. Write the structural formula of a triglyceride formed from given fatty acids, and predict whether it is a fat or an oil at room temperature.
7. Write an equation to show how soap is made from fat.

23.6 1. Write general formulas for primary, secondary, and tertiary amines.
2. Write equations for the reactions of amines with strong acids.
3. Write the general formula of an α-amino acid.
4. Identify a chiral organic molecule from its formula, and identify the asymmetric carbon atom.

23.7 1. Identify an amide linkage in a structural formula.
2. Write the equation for the formation of an amide from a given acid and amine.
3. Write the formula of the polyamide produced by a given diacid and diamine.
4. Given a polyamide formula, write the formulas of its monomers.
5. Write the structural formula for a polypeptide formed from given amino acids in a specified sequence.
*6. Explain what is meant by primary, secondary, tertiary, and quaternary components of protein structure.
*7. Describe two common types of secondary protein structure.
*8. List and describe the interactions that maintain secondary, tertiary, and quaternary protein structures.

23.8 1. Write the general formula for a monosaccharide.
2. Write molecular formulas, and give examples of a hexose and a pentose.
3. Identify an aldose or ketose from its open-ring structural formula.
4. List four naturally occurring hexoses.
5. Use structural formulas to show the equilibrium between closed- and open-ring forms of glucose.
6. List three natural disaccharides and their monosaccharide components.
7. Describe the glycosidic linkage, and show how two hexoses condense to form a disaccharide.
8. Describe the occurrence and biochemical function of starch, glycogen, and cellulose.
9. Describe the difference between amylose and amylopectin, the components of starch.
10. Explain the differences in the properties of starch and cellulose on the basis of structure.

23.9 *1. List the structural components of a nucleotide molecule.
*2. Discuss the role of ADP and ATP in the storage and release of energy.
*3. Distinguish between nucleotides and nucleic acids.
*4. Describe the structure of DNA and explain what is meant by the genetic code.
*5. Distinguish between mRNA and tRNA, and describe the mechanism of protein synthesis in cells.
*6. Explain how DNA replicates itself.

KEY TERMS BY SECTION

23.1 Alkane
Alkyl group
Boat conformation
Chair conformation
Cycloalkane
Homologous series
Saturated hydrocarbon
Steroid
Substitution reaction

23.2 Addition polymer
Addition reaction
Alkene
Alkyne
Diene
Markovnikov's rule
Monomer
Polymer
Unsaturated hydrocarbon

23.3 Aromatic hydrocarbon
meta Substituents
ortho Substituents
para Substituents
Phenyl group

23.4 Alcohol (primary, secondary, and tertiary)
Aldehyde
Carbonyl group
Carboxyl group
Carboxylic acid
Detergent
Ether
Fatty acid (saturated and unsaturated)
Functional group
Ketone
Phenol
Polyhydroxy alcohol
Soap

23.5 Condensation reaction
Ester
Ester linkage
Fat
Oil
Polyester
Saponification
Triglyceride

23.6 α-Amino acid
Amine (primary, secondary, and tertiary)
Amino acid
Amino group
Asymmetric carbon atom

23.7 *α-Helix
Amide
Amide linkage
Amino acid residue
*β-Pleated sheet
Dipeptide
*Globular protein
Peptide linkage
Polyamide
Polypeptide
*Protein
*Protein structure (primary, secondary, tertiary, and quaternary)

23.8 Aldose
Carbohydrate
Disaccharide
Glycosidic linkage
Hexose
Ketose
Monosaccharide
Pentose
Polysaccharide

23.9 *ADP
*Anticodon
*ATP
*Codon
*Complementary bases
*DNA
*Double helix
*Genetic code
*mRNA
*Nucleic acid
*Nucleotide
*Ribosome
*RNA
*tRNA
*Transcription

FINAL EXERCISES

Answers to exercises with blue numbers are given in Appendix D. Starred exercises are based on the *Digging Deeper* sections.

PART A. QUESTIONS AND PROBLEMS BY SECTION

Alkanes (Section 23.1)

23.1 Define and give examples of each of the following:
(a) general formula
(b) homologous series
(c) saturated hydrocarbon
(d) alkyl group

23.2 List the first 10 straight-chain alkanes by name.

23.3 Sketch the boat and chair conformations of the cyclohexane ring.

23.4 Draw the basic steroid ring structure.

23.5 Is a straight-chain alkane really straight? Explain.

23.6 **(a)** What would be the value of the C—C—C bond angle in a flat ring of six carbon atoms?
(b) Explain why the cyclohexane ring is puckered.
(c) Does puckering make the angles larger or smaller than the value in Part (a)?

23.7 Write a molecular and structural formula for each of the following:
(a) cyclopentane
(b) a six-carbon cycloalkane with a five-membered ring

23.8 Write a molecular and structural formula for each of the following:
(a) a cycloalkane with two six-member rings that share one carbon atom
(b) a cycloalkane with two fused six-member rings (rings that share two carbon atoms)

23.9 Write a structural formula for (a) 5-ethyl-3-methylnonane and (b) 3,3-dimethylpentane.

23.10 Write a structural formula for (a) 2,2,4-trimethylpentane and (b) *n*-heptane. (These compounds are the standards that define the octane rating scale for gasoline).

23.11 Write a structural formula for each of the following compounds. State which of the compounds are structural isomers (see Section 10.3).
(a) 2-methylheptane
(b) 2,4-dimethylhexane
(c) 3-ethylhexane
(d) 2,3,4-trimethylpentane
(e) 3-ethyl-2-methylpentane

23.12 A student named a certain structure 2-propylpropane. Write out the structural formula, and rename it correctly.

23.13 Name each of the following compounds:

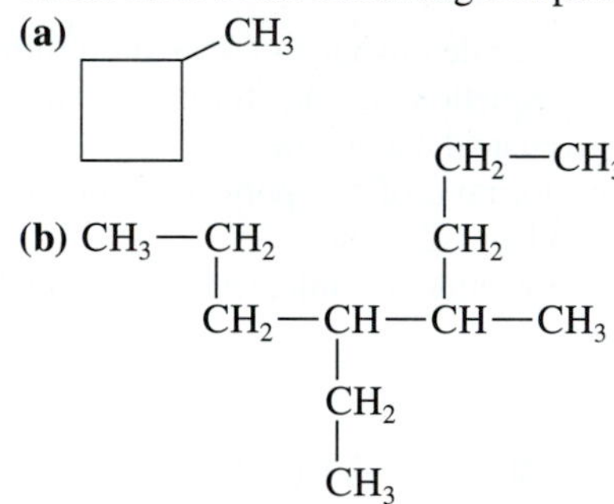

23.14 Name each of the following compounds:

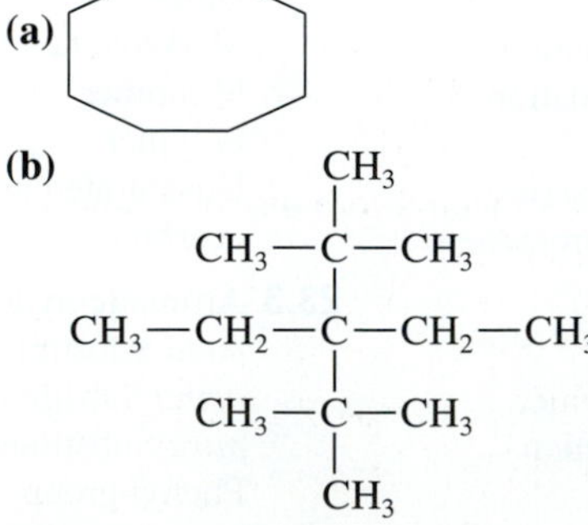

23.15 (a) How many cycloalkane isomers are possible for C_7H_{14}?
(b) Draw each of the isomers in Part (a).

23.16 The compound 1,2-dichlorocyclobutane has *cis* and *trans* isomers. Draw them.

23.17 Write the equation for the complete combustion of butane.

23.18 Write the equation for the complete combustion of cyclopentane.

23.19 Write balanced equations for the successive steps by which a methane molecule is converted to carbon tetrachloride (CCl_4) by the action of light and Cl_2. What is the other product?

23.20 Write formulas for as many as you can of the various halogenated ethanes and propanes produced by the following reactions:

(a) ethane + $Br_2 \xrightarrow{h\nu}$

(b) propane + $Cl_2 \xrightarrow{h\nu}$

Unsaturated Hydrocarbons (Section 23.2)

23.21 Write a general formula for each of the following:
(a) alkene (b) diene (c) triene

23.22 Write a general formula for each of the following:
(a) cycloalkene
(b) cyclodiene
(c) alkyne

23.23 Write a structural formula for each of the following compounds:
(a) 1-butene
(b) 2-butene
(c) acetylene

23.24 Write a structural formula for each of the following compounds:
(a) ethene
(b) cyclobutene
(c) propyne

23.25 Draw a structural formula for each of the following alkenes, and identify those that can have *cis* and *trans* isomers (see Section 10.3).
(a) 1-pentene
(b) 2,3-dimethyl-2-pentene
(c) 2,3-dimethyl-2-butene
(d) 2-pentene
(e) butadiene

23.26 (a) Draw a structural formula for 1,3-heptadiene and 4,4-diethyl-2-hexyne.
(b) Is it possible to have 3,3-diethyl-2-hexyne? Explain why or why not.

23.27 Draw a carbon skeleton for each of the noncyclic structural isomers of C_6H_{12}.

23.28 Draw a structure for (a) a cyclic isomer of *n*-hexene and (b) a noncyclic isomer of decalin:

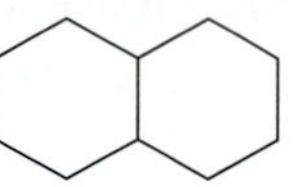

23.29 Name the following alkene:

$$\begin{array}{ccc} CH_3{-}CH_2 & & CH_2{-}CH_2{-}CH_3 \\ & \diagdown\; C{=}C \;\diagup & \\ & \diagup \qquad \diagdown & \\ H & & H \end{array}$$

23.30 Name the following alkene:

$$\begin{array}{ccc} CH_3{-}CH_2 & & H \\ & \diagdown\; C{=}C \;\diagup & \\ & \diagup \qquad \diagdown & \\ H & & CH_2{-}CH_2{-}CH_3 \end{array}$$

23.31 Write equations for the preparation of 2-bromobutane by (a) substitution and (b) addition. Which method is better? Why?

23.32 State Markovnikov's rule. Discuss its application to the preparation of 2-bromobutane in Exercise 23.31.

23.33 Complete the following equations:

(a) $CH_3{-}\underset{}{\overset{\overset{\displaystyle CH_3}{|}}{C}}{=}CH_2 + Br_2 \longrightarrow$

(b) $CH_3{-}CH_2{-}C{\equiv}CH + 2Br_2 \longrightarrow$

(c) $CH_3{-}CH_2{-}C{\equiv}CH + 2HBr \longrightarrow$

23.34 Complete the following equations:

(a) $CH_3{-}\overset{\overset{\displaystyle CH_3}{|}}{C}{=}CH_2 + HBr \longrightarrow$

(b) $CH_3{-}\overset{\overset{\displaystyle CH_3}{|}}{C}{=}CH_2 + H_2 \xrightarrow{Pt}$

(c) $CH_2{=}CH{-}CH{=}CH_2 + 2H_2 \xrightarrow{Pt}$

23.35 Write the structural formula of the principal product in each of the following reactions:

(a) 1-methylcyclohexene (CH_3 on the ring double bond) $+ H_2 \xrightarrow{Pt}$

(b) $CH_3CH{=}CH_2 \longrightarrow$ (polymerize by addition)

23.36 Write the structural formula of the principal product in each of the following reactions:

(a) 1-methylcyclohexene (CH_3 on the ring double bond) $+ HBr \longrightarrow$

(b) benzene ring with $CH{=}CH_2$ $\longrightarrow$ (polymerize by addition)

23.37 Replace each question mark with the structural formula of the reactant. Insert any required catalysts or special conditions over the arrow.

(a) ? + Cl_2 → (1,2-dichlorocyclohexane)

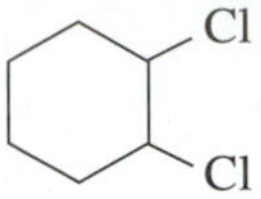

(b) ? + Cl_2 → (chlorocyclohexane) + HCl

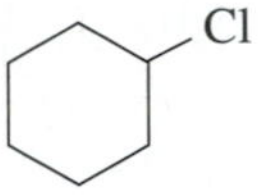

(c) ? + $2H_2$ ⟶ pentane
(d) ? + 2HBr ⟶ 2,2-dibromopentane

23.38 Replace each question mark with the structural formula of the reactant. Insert any required catalysts or special conditions over the arrow.

(a) ? + HCl → (1-chloro-1-methylcyclopentane: CH_3, Cl)

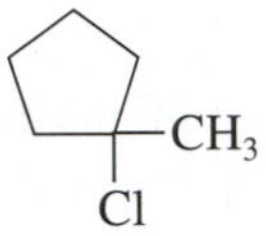

(b) ? + $2H_2$ → (cyclopentane)

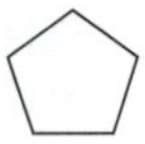

(c) ? + $2Br_2$ ⟶ 1,1,2,2-tetrabromopentane

23.39 Write the formula for the monomer that produces the following addition polymer:

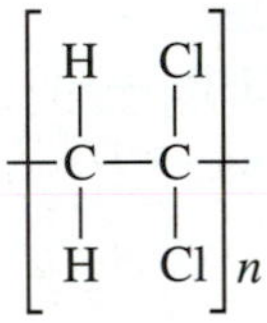

$$\left[\begin{array}{cc} H & Cl \\ | & | \\ -C- & C- \\ | & | \\ H & Cl \end{array}\right]_n$$ polyvinylidene chloride (Saran)

23.40 Write the formula for the monomer that produces the following addition polymer:

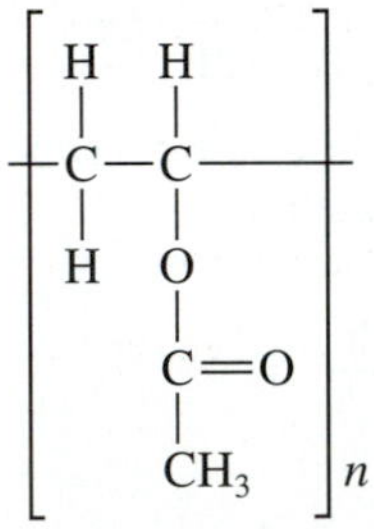

$$\left[\begin{array}{cc} H & H \\ | & | \\ -C- & C- \\ | & | \\ H & O \\ & | \\ & C{=}O \\ & | \\ & CH_3 \end{array}\right]_n$$ polyvinyl acetate (PVA)

Aromatic Hydrocarbons (Section 23.3)

23.41 What bonding feature do aromatic compounds have in common?

23.42 Explain why aromatic compounds, unlike alkenes and alkynes, do not readily participate in addition reactions.

23.43 Give the molecular and structural formulas of each of the following compounds:
(a) ethylbenzene
(b) chlorobenzene
(c) 2-methyl-3-phenylheptane

23.44 Refer to Table 10.1 (p. 428), and give the molecular and structural formulas of each of the following compounds:
(a) anthracene
(b) phenanthrene
(c) 3,4-benzopyrene

23.45 Write structural formulas for the three isomeric xylenes (dimethylbenzenes).

23.46 Write structural formulas for the three isomeric mesitylenes (trimethylbenzenes).

23.47 Name the following cyclic compounds:

(a) 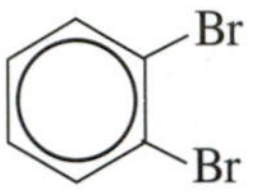 (benzene ring with Br, Br on adjacent positions)

(b) 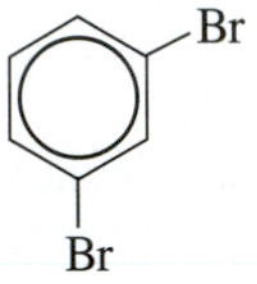 (benzene ring with Br, Br in meta positions)

(c) 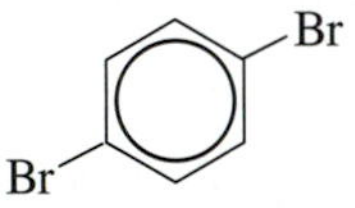 (benzene ring with Br, Br in para positions)

23.48 Name the following cyclic compounds:

(a) 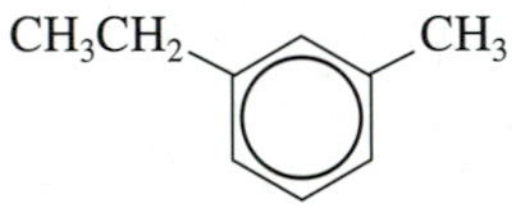 CH_3CH_2–(benzene ring)–CH_3 (meta)

(b) 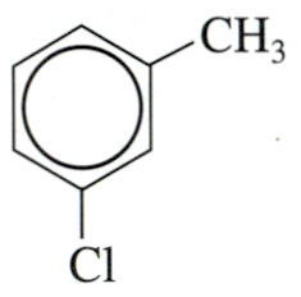 (benzene ring with CH_3 and Cl in meta positions)

(c) 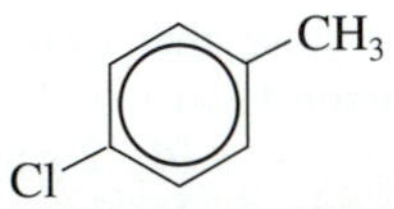 (benzene ring with CH_3 and Cl in para positions)

23.49 Write a balanced equation, including structural formulas, for each of the following reactions:
(a) Ni-catalyzed addition of H_2 to benzene
(b) $FeCl_3$-catalyzed reaction of benzene with Cl_2

23.50 Write a balanced equation, including structural formulas, for each of the following reactions:
(a) formation of TNT from toluene
(b) $AlCl_3$-catalyzed reaction of CH_3Cl with benzene

Functional Groups (Section 23.4)

23.51 What is meant by the term *functional group*? Give some examples of functional groups.

23.52 Explain why a double bond is considered to be a functional group.

23.53 Give the general formula and at least two examples of each of the following compounds. Identify the functional group in each compound.
(a) ether **(c)** primary alcohol
(b) ketone **(d)** carboxylic acid

23.54 Give the general formula and at least two examples of each of the following compounds. Identify the functional group in each compound.
(a) aldehyde **(c)** secondary alcohol
(b) phenol **(d)** alkene

23.55 **(a)** Describe two methods for obtaining ethanol.
(b) Give some practical uses for ethanol.

23.56 **(a)** Describe two methods for obtaining methanol.
(b) Give some practical uses for methanol.

23.57 **(a)** Explain why phenol is more acidic than ethanol.
(b) Write the equation for the reaction of phenol with aqueous sodium hydroxide.

23.58 Explain why the COOH group is generally a better proton donor than the OH group in an alcohol or phenol.

23.59 Define and give specific examples of
(a) a fatty acid
(b) an unsaturated fatty acid
(c) a soap

23.60 Explain the detergent properties of a soluble soap in terms of molecular structure.

23.61 Identify the functional group or groups in the following molecules.

(a) $CH_3{-}\overset{O}{\overset{\|}{C}}{-}CH_3$

(b) $CH_3{-}\overset{OH}{\overset{|}{C}H}{-}CH_3$

(c) $CH_3{-}CH_2{-}\overset{O}{\overset{\|}{C}}{-}OH$

(d) $HO{-}\overset{O}{\overset{\|}{C}}{-}\overset{O}{\overset{\|}{C}}{-}OH$

(e) $CH_3{-}CHBr{-}\overset{O}{\overset{\|}{C}}{-}OH$

(f) $CH_3{-}\overset{O}{\overset{\|}{C}}{-}\overset{O}{\overset{\|}{C}}{-}OH$

(g) $C_6H_4(OH){-}\overset{O}{\overset{\|}{C}}{-}OH$

23.62 The following are three of several known pheromones of the honeybee. Identify the functional groups in each one.

(a) $CH_3\overset{O}{\overset{\|}{C}}CH_2CH_2CH_2CH_2CH_2\overset{H}{\overset{|}{C}}{=}CH{-}\overset{O}{\overset{\|}{C}}{-}OH$

(b) $CH_3\overset{OH}{\overset{|}{C}}HCH_2CH_2CH_2CH_2CH_2\overset{H}{\overset{|}{C}}{=}CH{-}\overset{O}{\overset{\|}{C}}{-}OH$

(c) $H\overset{O}{\overset{\|}{C}}CH{=}\overset{CH_3}{\overset{|}{C}}{-}CH_2CH_2CH{=}\underset{CH_3}{\underset{|}{C}}CH_3$

23.63 Write the structural formulas for all alcohols with the molecular formula $C_4H_{10}O$.

23.64 Write a structural formula and give some uses for (a) formaldehyde and (b) acetone.

23.65 Give the structural formula, uses, and sources for each of the following compounds:
(a) acetic acid **(c)** benzoic acid
(b) oxalic acid **(d)** salicylic acid

23.66 Give the structural formula and sources for each of the following compounds:
(a) linoleic acid
(b) oleic acid
(c) stearic acid

23.67 In which compound of the following pairs would you expect hydrogen bonding be more significant? Explain.
(a) methanol or 1-propanol
(b) butyric acid or formic acid
(c) acetic acid or oxalic acid
(d) 2-propanol or glycerol

23.68 Explain in terms of hydrogen bonding why (a) the solubility of alcohols in water decreases with increasing molar mass, (b) acetone is very volatile and also very soluble in water, and (c) ethylene glycol is added to some ballpoint inks to retard drying.

23.69 Classify each of the following changes as oxidation, reduction, or neither:
(a) methanol $\longrightarrow$ formic acid
(b) $CO_2 \longrightarrow CH_4$
(c) formaldehyde $\longrightarrow$ methanol

23.70 Classify each of the following changes as oxidation, reduction, or neither:
(a) acetone $\longrightarrow$ isopropyl alcohol
(b) isopropyl alcohol $\longrightarrow$ CO_2
(c) acetone $\longrightarrow$ *n*-propane

23.71 Classify each of the following changes as oxidation, reduction, or neither:

(a) $CH_3CHOHCH_3 \longrightarrow CH_3\overset{\overset{\displaystyle O}{\|}}{C}CH_3$

(b) $CH_3CHOHCH_3 \longrightarrow CH_3CH{=}CH_2$

(c) $CH_3\overset{\overset{\displaystyle O}{\|}}{C}-\overset{\overset{\displaystyle O}{\|}}{C}-OH \longrightarrow CH_3\overset{\overset{\displaystyle OH}{|}}{C}H-\overset{\overset{\displaystyle O}{\|}}{C}-OH$

23.72 Classify each of the following changes as oxidation, reduction, or neither:

(a) $H_2\overset{\overset{\displaystyle OH}{|}}{C}-\overset{\overset{\displaystyle O}{\|}}{C}-\overset{\overset{\displaystyle OH}{|}}{C}H_2 \longrightarrow H\overset{\overset{\displaystyle O}{\|}}{C}-\overset{\overset{\displaystyle OH}{|}}{C}H-\overset{\overset{\displaystyle OH}{|}}{C}H_2$

(b) 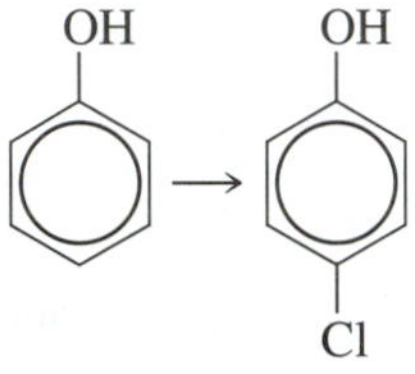 phenol (benzene ring with OH) $\longrightarrow$ benzene ring with OH and para Cl

Esters (Section 23.5)

23.73 Write (a) the formula for the ester linkage and (b) the general formula for an ester.

23.74 Write the general formula for a fat.

23.75 Distinguish between (a) a saturated fat and (b) an unsaturated fat. When is a fat called an oil?

23.76 Explain why fat is more efficient than carbohydrate as a means for storing energy in the body.

23.77 Write a structural formula for each of the following esters:
(a) methyl butyrate
(b) phenyl acetate
(c) glyceryl tripropionate

23.78 Write a balanced equation, including structural formulas, for ester formation from (a) phenol and formic acid, (b) propionic acid and ethanol, and (c) butyric acid and isopropyl alcohol. Name each ester.

23.79 Write formulas for the hydrolysis products of the following compound. Identify each product as an alcohol or an acid.

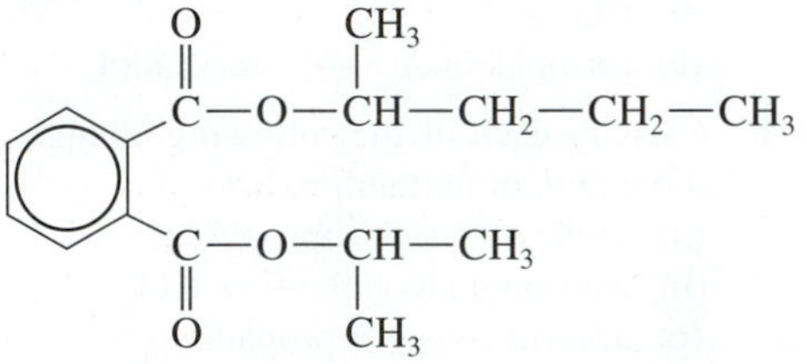

Benzene ring with two adjacent substituents:
$-\overset{\overset{\displaystyle O}{\|}}{C}-O-\overset{\overset{\displaystyle CH_3}{|}}{C}H-CH_2-CH_2-CH_3$
$-\underset{\underset{\displaystyle O}{\|}}{C}-O-\underset{\underset{\displaystyle CH_3}{|}}{C}H-CH_3$

23.80 Write a balanced equation, including structural formulas, for the hydrolysis of aspirin in gastric juice. (*Note*: The pH of gastric juice ranges from 1.0 to 3.0.)

23.81 The simplest possible polyester would be a polymer of ethylene glycol and oxalic acid. Draw a few units of the polymer chain.

23.82 Write the formula of Kevlar, a condensation polymer of terephthalic acid (benzene with *para* carboxyl groups) and phenylenediamine (*p*-diaminobenzene). Kevlar is used to make bulletproof vests.

23.83 **(a)** Write the structural formula for glycerol trioleate.
(b) Write the equation for (1) the hydrolysis of glycerol trioleate in an acid medium and (2) its saponification with KOH to form a soap.

23.84 What is the relationship between the melting point of a fat and its degree of unsaturation? Explain this relationship on the basis of molecular structure. For what purpose are oils sometimes hydrogenated?

Amines, Amino Acids, and Chirality (Section 23.6)

23.85 **(a)** What is the functional group in an amine?
(b) Distinguish between primary, secondary, and tertiary amines, and give an example of each.

23.86 **(a)** Write the general formula for an α-amino acid.
(b) Consult Table 23.8, and give the name and structural formula of an amino acid that (1) has a nonpolar side group, (2) an acidic side group, (3) a basic side group, (4) an aromatic side group, and (5) a sulfur-containing side group.

23.87 What is an asymmetric carbon atom? Give an example of a molecule containing such an atom.

23.88 What type of isomers are associated with the presence of asymmetric carbon atoms? Sketch the isomers for your example in Exercise 23.87.

23.89 Write a structural formula for each of the following compounds:
(a) methylamine
(b) ethylamine
(c) 1,2-diaminoethane (ethylenediamine)

23.90 Write a structural formula for each of the following compounds:
(a) aniline
(b) paraaminobenzoic acid (the sunscreen known as PABA)
(c) 1,6-diaminohexane

23.91 **(a)** Write the formula for the two-carbon amino acid glycine.
(b) Is glycine chiral? If so, sketch its enantiomers.

23.92 Identify the asymmetric carbon in alanine,

$$CH_3\overset{\overset{\displaystyle NH_2}{|}}{C}HCOOH$$

Sketch the enantiomers. Which one is found in nature?

23.93 Which of the following compounds could have enantiomers? Identify each asymmetric carbon atom.
(a) 1-chloropropane
(b) 2-chloropentane
(c) 3-bromopentane
(d) 1,3-dichlorobutane
(e) $C_2H_5CHClCH_3$

23.94 Which of the following properties are the same for both dextrorotatory and levorotatory lactic acid? Which properties differ?
(a) boiling point
(b) specific gravity
(c) solubility in H_2O
(d) rate of reaction with ethyl alcohol
(e) rate of reaction with levorotatory 1-chloro-2-propanol
(f) melting point
(g) effect on polarized light
(h) solubility in dextrorotatory 2-butanol

The Amide Linkage (Section 23.7)

23.95 **(a)** Write the formula for the *amide linkage*.
(b) Write the general formula for an amide.

23.96 **(a)** What is the *peptide linkage*?
(b) Use general formulas to show how amino acids can form a polypeptide chain.

***23.97** Explain what is meant by the primary, secondary, tertiary, and quaternary structure of a protein. Describe the interactions that maintain each type of structure. Which types of structure are found in (a) fibrous proteins, (b) globular proteins, and (c) hemoglobin?

***23.98** **(a)** Describe an α-helix. What keeps a polypeptide α-helix from unwinding?
(b) Describe a β-pleated sheet. What maintains the alignment of the polypeptide chains in such a sheet?

23.99 Draw a structural formula for each of the following compounds:
(a) formamide
(b) acetanilide

23.100 Draw a structural formula for each of the following compounds:
(a) acetamide
(b) acetaminophen (*p*-hydroxyacetanilide)
What is acetaminophen used for?

23.101 Write an equation for the hydrolysis of each compound in
(a) Exercise 23.99
(b) Exercise 23.100

23.102 Using the general formula for an amide, write an equation for its hydrolysis. What type of compound is each product?

23.103 **(a)** Use general formulas to show how a diacid and a diamine can form a condensation polymer (a polyamide).
(b) Write the equation for the formation of Nylon-66.

23.104 Write the equation for the formation of Nylon-46 from adipic acid ($HOOC(CH_2)_4COOH$) and 1,4-diaminobutane. (*Note*: The numbers in a nylon name give the number of carbon atoms in the diamine and diacid, respectively.)

23.105 Write a structural formula for (a) Tyr-Leu and (b) Leu-Tyr.

23.106 Write a structural formula for the pituitary antidiuretic hormone vasopressin, Gly-Arg-Pro-Cys-Asn-Gln-Phe-Tyr-Cys.

***23.107** Refer to Table 23.8, and find amino acid residues that might interact through (a) hydrogen bonding, (b) ionic interactions, and (c) hydrophobic interactions.

***23.108** How does disulfide bridge formation differ from the other interactions that maintain protein structure? Which amino acid residue is involved in disulfide bridge formation.

Carbohydrates (Section 23.8)

23.109 Distinguish between the following monosaccharides:
(a) a hexose and a pentose
(b) an aldose and a ketose
Which of these terms apply to glucose? Which apply to fructose?

23.110 Sketch the open and cyclic forms of glucose.

23.111 Name three disaccharides and identify their hydrolysis products.

23.112 **(a)** Sketch the sucrose molecule.
(b) What is a *glycosidic linkage*? Identify the glycosidic linkage in your sketch of the sucrose molecule.

23.113 What do starch, glycogen, and cellulose have in common? What are their differences? Where is each found in nature?

23.114 Distinguish between amylose and amylopectin, the components of starch.

Nucleotides and Nucleic Acids (Section 23.9)

***23.115** What is the general structure of a nucleotide? What types of molecules do nucleotides produce when hydrolysed?

***23.116** Draw the structures of ADP and ATP. Which ATP bonds release stored energy on hydrolysis?

***23.117** Draw a structural formula for each sugar that may form during the hydrolysis of a nucleotide.

***23.118** Draw a structural formula for each nitrogenous base that may form during the hydrolysis of a nucleotide.

***23.119** What is a nucleic acid? Illustrate your answer by sketching a portion of the DNA molecule.

***23.120** What are the two principal nucleic acids?

***23.121** In what form does DNA hold the instructions for protein synthesis?

***23.122** Describe, with the aid of a diagram, the process of *transcription.*

***23.123** **(a)** What is a codon?
(b) Is each amino acid involved in protein synthesis represented by one codon or by several codons? Explain.
(c) Refer to Table 23.9, and identify the "stop protein synthesis" codons.

***23.124** Describe, with the aid of a diagram, the process by which a polypeptide chain is assembled. What are the functions of mRNA and tRNA in this process?

PART B. MISCELLANEOUS QUESTIONS AND PROBLEMS

23.125 What characteristics of the carbon atom account for the fact that organic compounds are more numerous than other compounds? Explain why the chemistry of silicon, another Group 4A element, is less extensive than the chemistry of carbon.

23.126 The chain mechanism for halogen substitution is initiated by the photodissociation of a halogen molecule. (a) Write initiation, propagation, and termination steps for the reaction of ethane and chlorine. (b) Account for the fact that the final mixture may contain a number of products such as chloroethane, 1,1-dichloroethane, 1,2-dichloroethane, trichloroethane, butane, and chlorobutane.

23.127 Write the steps in the chain mechanism by which polyvinyl chloride, $\text{+}CH_2CHCl\text{+}_n$, forms from vinyl chloride ($H_2C{=}CHCl$). Assume that the chains are initiated by free radicals R·.

.23.128 A saturated hydrocarbon with the molecular formula C_6H_{14} is treated with Br_2, and all the products are identified. Deduce the structure of the original hydrocarbon and give its name if the number of isomeric monobromo derivatives is (a) five, (b) four, (c) three, and (d) two.

23.129 The mass of one mole of a certain hydrocarbon is 72 g, and it forms only one monobromo derivative. Draw the structural formulas of the hydrocarbon and its monobromo derivative.

23.130 A certain compound with the molecular formula C_3H_8O can be converted by mild oxidizing agents to C_3H_6O. Further oxidation can be achieved only with much stronger oxidizing agents. Give the names and structural formulas of the original compound and its oxidation product.

23.131 The *iodine number* of a fat is the number of grams of iodine that will add to the double bonds in 100 g of fat. Calculate the expected iodine number of each of the following fats:
(a) glyceryl trioleate
(b) glyceryl trilinoleate
(c) stearin

23.132 Permanent waving or straightening of hair is achieved by (1) holding the hair in the desired position while treating it with a reducing agent that attacks disulfide bridges, (2) treating the hair with an oxidizing agent, and (3) releasing the hair. Make a sketch showing how this treatment could alter the alignment of the polypeptide chains. What does the oxidizing agent do? Use formulas to show the action of the oxidizing and reducing agents.

23.133 Explain in terms of enzyme active sites why humans cannot digest cellulose.

***23.134** When James Watson and Francis Crick were working out the double helical structure of DNA, one piece of useful evidence was the fact that all DNA samples were found to yield equal molar quantities of adenine and thymine and equal molar quantities of guanine and cytosine. Explain why this observation was significant.

***23.135** Mentally compose a 10-minute talk explaining how protein synthesis is directed and controlled by DNA and RNA. Assume that your audience consists of people who have never heard of DNA or RNA, but who would like to learn about protein synthesis.

APPENDIX A

MATHEMATICAL OPERATIONS

The following sections review scientific (exponential) notation, logarithms, the quadratic formula, and straight-line graphs. If you have never studied these topics, you should ask your instructor to refer you to a more detailed treatment than is given here.

A.1 SCIENTIFIC NOTATION

The numbers encountered in chemistry and other sciences are often very large or very small. For example, there are 50,100,000,000,000,000,000,000 atoms in 1 g of carbon, and the radius of each carbon atom is 0.000,000,000,077 m. Numbers in this form are difficult to read and even more difficult to write. Scientists have found it convenient to express such numbers using powers of 10. You might recall the relations in Table A.1 from your previous mathematics courses. Numbers in the right-hand column of the table are written in *exponential form;* that is, they are written as 10 raised to a whole-number power.

Consider the number 2100. It can also be written as

$$2100 = 2.100 \times 1000 = 2.100 \times 10^3$$

A number written in the exponential form

$$A \times 10^n$$

where A is greater than or equal to 1 but less than 10, and n is a positive or negative integer, is said to be written in *scientific notation*. In the above example, $A = 2.100$ and $n = 3$. Some other examples of scientific notation are

$$365 = 3.65 \times 100 = 3.65 \times 10^2$$

$$0.41 = 4.1 \times 0.1 = 4.1 \times 10^{-1}$$

$$0.00524 = 5.24 \times 0.001 = 5.24 \times 10^{-3}$$

The following rules will help you to write numbers in scientific notation:

1. Find the number A by moving the decimal point from its original location to just after the first nonzero digit. (When the original number is written without a decimal point, for example, 168 or 1750 or 186,000, the decimal point is understood to follow the last digit.) Drop all zeros before the first nonzero digit.
2. Find n, the power of 10, by counting the number of places the decimal point has been moved. If the decimal point is moved to the left, the power of 10 is positive; if it is moved to the right, the power of 10 is negative.

TABLE A.1 Exponential Notation

Number	Exponential Form
1,000,000,000	1×10^{9}
100,000,000	1×10^{8}
10,000,000	1×10^{7}
1,000,000	1×10^{6}
100,000	1×10^{5}
10,000	1×10^{4}
1,000	1×10^{3}
100	1×10^{2}
10	1×10^{1}
1	1×10^{0}
0.1	1×10^{-1}
0.01	1×10^{-2}
0.001	1×10^{-3}
0.0001	1×10^{-4}
0.00001	1×10^{-5}
0.000001	1×10^{-6}
0.0000001	1×10^{-7}
0.00000001	1×10^{-8}
0.000000001	1×10^{-9}

NOTES

1. $10^0 = 1$.
2. When the number is greater than 1, the exponent of 10 is equal to the number of zeros before the decimal point. When the number is less than 1, the exponent of 10 has a negative sign and is one more than the number of zeros after the decimal point.

EXAMPLE A1.1

Write the number 96,500 in scientific notation.

SOLUTION

The decimal point is understood to be after the last zero.

Step 1: Move the decimal point four places to the left so that it is just after the first nonzero digit (between 9 and 6):

$$96\,500. \longrightarrow 9.6\,500$$

There are no zeros to drop, so A is 9.6500.

Step 2: The decimal point was moved four places to the left, so the power of 10 is 4. Hence,

$$96{,}500 = 9.6500 \times 10^{4}$$

EXAMPLE A1.2

Write the radius of a carbon atom in scientific notation.

SOLUTION

The radius of a carbon atom was given above as 0.000,000,000,077 m.

Step 1: Move the decimal point 11 places to the right so that it is just after the first nonzero digit (between 7 and 7):

$$0.000\ 000\ 000\ 077 \longrightarrow 0\ 000\ 000\ 000\ 07.7$$

All zeros before the number 7 are dropped. Hence, A is 7.7.

Step 2: The decimal point was moved 11 places to the right, so the power of 10 is -11. Hence,

$$0.000{,}000{,}000{,}077 \text{ m} = 7.7 \times 10^{-11} \text{ m}$$

EXAMPLE A1.3

Write the number 1225.00 in scientific notation.

SOLUTION

Step 1: Move the decimal point three places to the left so that it is just after the first nonzero digit (between 1 and 2):

$$1225.00 \longrightarrow 1.225\ 00$$

There are no zeros to be dropped, so $A = 1.22500$.

Step 2: The decimal point was moved three places to the left, so the power of 10 is 3. Hence,

$$1225.00 = 1.22500 \times 10^3$$

If you have a scientific calculator, it can be used to express numbers in scientific notation. The buttons to use vary from calculator to calculator but are usually marked EE, EXP, or EEX. For example, to enter the number 6.022×10^{23}, first enter 6.022, push the appropriate button, and then enter 23. The number will be displayed as

$$6.022 \qquad 23$$

The change-sign button, which may be marked CHS or +/−, is used to enter negative exponents. For example, to enter the number 5.4×10^{-8}, first enter 5.4, push the appropriate EE, EXP, or EEX button, push the change-sign button, and then enter 8. The number will be displayed as

$$5.4 \qquad -8$$

Arithmetic Operations

The addition, subtraction, multiplication, and division of numbers written in scientific notation is most easily done with the aid of a scientific calculator. The numbers are entered in the usual way, and the arithmetic operation is carried out according to the instructions that accompany the calculator. Calculators also contain buttons such as X^2 for squaring a number, $1/X$ for finding the reciprocal of a number, $\sqrt{X}$ for finding the square root of a number, and Y^x for finding roots and powers of Y. It is strongly recommended that you familiarize yourself with each of these operations.

If you add, subtract, multiply, or divide without the aid of a calculator, then the following rules should be followed:

Addition and Subtraction. Express each number to the same power of 10, and then add or subtract the decimal numbers. For example, the sum

$$(5.07 \times 10^5) + (4.320 \times 10^6)$$

can be found in two ways:

$$\begin{array}{r} 5.07 \times 10^5 \\ +43.20 \times 10^5 \\ \hline 48.27 \times 10^5 \end{array} \quad \text{or} \quad \begin{array}{r} 0.507 \times 10^6 \\ +4.320 \times 10^6 \\ \hline 4.827 \times 10^6 \end{array}$$

(Remember that if one part of a number is increased by some factor, the other part must be decreased by the same factor. The number 5.07×10^5 was changed to 0.507×10^6 by multiplying 10^5 by 10 and dividing 5.07 by 10. The number 4.320×10^6 was changed to 43.20×10^5 by dividing 10^6 by 10 and multiplying 4.320 by 10.) Note that the first answer to the above calculation (48.27×10^5) is not in scientific notation because 48.27 does not lie between 1 and 10.

Multiplication. When multiplying numbers written in scientific notation, use the relation

$$(A \times 10^m) \times (B \times 10^n) = A \times B \times 10^{m+n}$$

In other words, the coefficients (A and B) are multiplied, but the exponents (m and n) are added. Some examples are:

$$(5.0 \times 10^4) \times (3.2 \times 10^6) = 5.0 \times 3.2 \times 10^{4+6} = 16 \times 10^{10} = 1.6 \times 10^{11}$$

$$(3.1 \times 10^6) \times (2.0 \times 10^{-8}) = 3.1 \times 2.0 \times 10^{6-8} = 6.2 \times 10^{-2}$$

Division. When dividing numbers written in scientific notation, use the relation

$$\frac{A \times 10^m}{B \times 10^n} = \frac{A}{B} \times 10^{m-n}$$

Some examples are

$$\frac{6.4 \times 10^6}{2.0 \times 10^8} = \frac{6.4}{2.0} \times 10^{6-8} = 3.2 \times 10^{-2}$$

$$\frac{8.8 \times 10^2}{4.0 \times 10^{-6}} = \frac{8.8}{4.0} \times 10^{2+6} = 2.2 \times 10^8$$

A.2 LOGARITHMS

A *logarithm* is an exponent. Consider the relation

$$1000 = 10^3$$

The exponent of 10 (3 in this case) is a logarithm. In general, if a number A is written as 10 raised to some power x

$$A = 10^x$$

then x is referred to as the *common logarithm* or the *logarithm to base 10* of A. In symbols, $x = \log_{10} A$ (or simply $\log A$). In other words, the common logarithm of a number is the power to which 10 must be raised in order to produce the number. Using the above example, $\log 1000 = 3$.

The common logarithm of a number can be found from log tables or by using the LOG button on your calculator. Simply enter the number, push the LOG button, and read the resulting display. Some examples are

$$\log (5.0 \times 10^{-6}) = -5.30$$

$$\log (4.55 \times 10^{4}) = 4.658$$

If you are concerned with significant figures (Section 1.4), the following rule will be helpful: *The number of digits to the right of the decimal point in the logarithm of a number is equal to the number of significant figures in the original number.* The number 5.0×10^{-6} has two significant figures; its logarithm should be expressed to two decimal places. The number 4.55×10^4 has three significant figures; its logarithm should be expressed to three decimal places.

The number $e = 2.71828$ is a constant that, like π, occurs in many scientific problems. (A more precise value of e can be obtained by entering 1 into your calculator and pushing the e^x button; try it!) The *natural logarithm* or *logarithm to base e* of a number is the power to which e must be raised to produce the number. If

$$A = e^y$$

then y is the natural logarithm of A. In symbols, $y = \log_e A$ (or $\ln A$). Natural logarithms are found using the LN button on your calculator. The natural logarithms of the numbers given above are

$$\ln (5.0 \times 10^{-6}) = -12.21$$

$$\ln (4.55 \times 10^{4}) = 10.725$$

Common logarithms can be converted to natural logarithms by the relation

$$\ln x = 2.303 \log x$$

The following properties of common and natural logarithms are important and should be memorized:

1. $\log (A \times B) = \log A + \log B$ $\quad\quad$ $\ln (A \times B) = \ln A + \ln B$

2. $\log (A/B) = \log A - \log B$ $\quad\quad$ $\ln (A/B) = \ln A - \ln B$

3. $\log (1/B) = \log 1 - \log B = -\log B$ $\quad\quad$ $\ln (1/B) = \ln 1 - \ln B = -\ln B$

4. $\log A^B = B \times \log A$ $\quad\quad$ $\ln A^B = B \times \ln A$

5. $\log 10^B = B \times \log 10 = B$ $\quad\quad$ $\ln e^B = B \times \ln e = B$

The last relation can be easily demonstrated with a scientific calculator:

$$\log 10^0 = \log 1 = 0$$
$$\log 10^5 = 5$$
$$\log 10^{-4} = -4$$

Antilogarithms

Chemistry students are often expected to solve equations such as

$$\log x = 4.21 \qquad \text{or} \qquad \ln y = -3.821$$

for the unknown quantities x and y. The properties of logarithms (see above) can be used to rearrange these equations into the following forms

$$x = 10^{4.21} \qquad \text{or} \qquad y = e^{-3.821}$$

Because solving for x is the inverse of finding a common logarithm, x is called the *antilog* or *inverse log* of 4.21. Your calculator probably has one of the following buttons for evaluating antilogs: 10^x, y^x, x^y, or INV LOG. Use your calculator to show that

$$x = \text{antilog } 4.21 = 10^{4.21} = 1.6 \times 10^4$$

The value of y is obtained with the e^x button on your calculator. Enter -3.821, push the e^x button, and read the resulting display:

$$y = e^{-3.821} = 2.19 \times 10^{-2}$$

Keep in mind that *the number of significant figures in the antilog of a number is equal to the number of digits to the right of the decimal point in the number.*

A.3 LINEAR EQUATIONS

An equation of the general form

$$y = mx + b$$

is called a *linear equation* because its graph is a straight line (see Figure A.1). In this equation, y and x are variables while m and b are constants characteristic of the line. The constant b, called the *intercept*, is the value of y when $x = 0$. Inspection of Figure A.1 shows that b is the point where the straight line crosses or intercepts the y axis.

The constant m, called the *slope*, is obtained by dividing the vertical distance between any two points on the line, $\Delta y = y_2 - y_1$, by the horizontal distance, $\Delta x = x_2 - x_1$:

$$m = \frac{\Delta y}{\Delta x} = \frac{y_2 - y_1}{x_2 - x_1}$$

The slope may be positive or negative, as shown in Figure A.1.

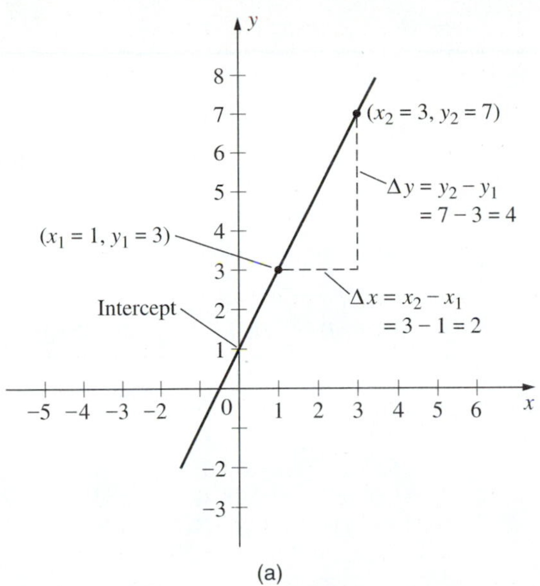

(a)

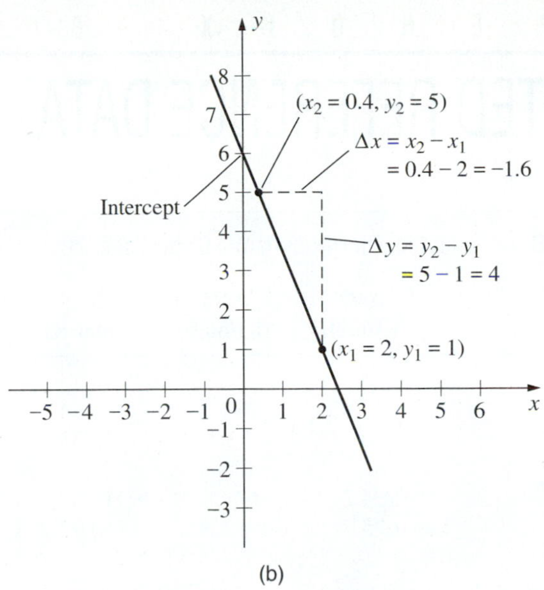

(b)

Figure A1

(a) Graph of the equation $y = 2x + 1$. The slope, $m = \Delta y/\Delta x = 4/2 = 2$, is positive. The intercept $b = 1$.
(b) Graph of the equation $y = -2.5x + 6$. The slope, $m = \Delta y/\Delta x = -4/1.6 = -2.5$, is negative. The intercept $b = 6$.

A.4 QUADRATIC EQUATIONS

A *quadratic equation* has the general form

$$ax^2 + bx + c = 0$$

where x is an unknown quantity, and a, b, and c are known quantities. Some examples are

$$2x^2 + 4x + 1 = 0 \qquad a = 2; b = 4; c = 1$$

$$x^2 - 5x - 3 = 0 \qquad a = 1; b = -5; c = -3$$

Quadratic equations have two roots (two values of x), which are given by the *quadratic formula:*

$$x = \frac{-b + \sqrt{b^2 - 4ac}}{2a} \qquad x = \frac{-b - \sqrt{b^2 - 4ac}}{2a}$$

The values of x for the first example are

$$x = \frac{-4 + \sqrt{16 - 8}}{4} \qquad x = \frac{-4 - \sqrt{16 - 8}}{4}$$

$$= -0.293 \qquad = -1.707$$

It is left as an exercise for you to show that the values of x for the second example are $x = +5.54$ and $x = -0.54$.

A P P E N D I X B

SELECTED REFERENCE DATA

TABLE B.1 Selected Thermodynamic Data[a,b] at 25°C

Substance	ΔH°_f (kJ/mol)	ΔG°_f (kJ/mol)	S° (J/mol·K)
Aluminum			
Al(g)	326.4	285.7	164.54
Al(s)	0	0	28.33
Al^{3+}(aq)	−531	−485	−321.7
$AlCl_3$(s)	−704.2	−628.8	−110.67
Al_2O_3(s)	−1675.7	−1582.3	50.92
Barium			
Ba(g)	180	146	170.243
Ba(s)	0	0	62.8
Ba^{2+}(aq)	−537.64	−560.77	9.6
$BaCl_2$(s)	−858.6	−810.4	123.68
$BaCO_3$(s)	−1216.3	−1137.6	112.1
$BaSO_4$(s)	−1473.2	−1362.2	132.2
Bromine			
Br(g)	111.884	82.396	175.022
Br^-(aq)	−121.55	−103.96	82.4
Br_2(g)	30.907	3.110	245.463
Br_2(l)	0	0	152.231
BrCl(g)	14.64	−0.98	240.1
HBr(g)	−36.40	−53.45	198.695
Calcium			
Ca(g)	178.2	144.3	154.884
Ca(s)	0	0	41.42
Ca^{2+}(aq)	−542.83	−553.58	−53.1
$CaCl_2$(s)	−795.8	−748.1	104.6
$CaCO_3$(s, calcite)	−1206.92	−1128.79	92.9
CaF_2(s)	−1219.6	−1167.3	68.87
CaO(s)	−635.09	−604.03	39.75
$Ca(OH)_2$(s)	−986.09	−898.49	83.39
$CaSO_4$(s)	−1434.11	−1321.79	106.7
Carbon			
C(g)	716.68	671.26	158.10
C(s, diamond)	1.895	2.900	2.377
C(s, graphite)	0	0	5.740
CCl_4(g)	−102.9	−60.59	309.85
CCl_4(l)	−135.44	−65.21	216.40
CH_4(g)	−74.81	−50.72	186.26
C_2H_2(g)	226.73	209.20	200.94
C_2H_4(g)	52.26	68.15	219.56
C_2H_6(g)	−84.68	−32.82	229.60
C_6H_6(l)	49.04	124.42	172.80
CH_3COO^-(aq)	−486.01	−369.31	86.6
CH_3COOH(aq)	−485.76	−396.46	178.7
CH_3COOH(l)	−484.5	−389.9	159.8
CH_3OH(g)	−200.66	−161.96	239.81
CH_3OH(l)	−238.66	−166.27	126.8
C_2H_5OH(g)	−235.10	−168.49	282.70
C_2H_5OH(l)	−277.69	−174.78	160.7
$C_6H_{12}O_6$(s, glucose)	−1274.5	−910.56	212.1
$C_{12}H_{22}O_{11}$(s, sucrose)	−2221.7	−1544.3	360.24
CN^-(aq)	150.6	172.4	94.1
CO(g)	−110.525	−137.168	197.67

[a]Data are based on a standard state pressure of 1 bar, the standard pressure recommended by IUPAC (see Table 19.1). Because 1 atm = 1.01325 bar, calculations based on these data will not significantly differ from those based on a standard pressure of 1 atmosphere, as shown by the following conversion equations:

$$\Delta H_f\,(1\text{ atm}) = \Delta H_f\,(1\text{ bar})$$
$$\Delta G_f\,(1\text{ atm}) = \Delta G_f\,(1\text{ bar}) + \Delta n \times 0.03263\text{ kJ/mol}$$
$$S(1\text{ atm}) = S(1\text{ bar}) \quad \text{(for solids and liquids)}$$
$$S(1\text{ atm}) = S(1\text{ bar}) - 0.1094\text{ J/mol·K} \quad \text{(for gases)}$$

where Δn is the change in the number of moles of gas in the formation equation.

[b]The data for ions in solution is based on the convention that ΔH°_f, ΔG°_f, and S° for H^+(aq) are zero. The basis for this convention is (1) the formation process for an ion in solution is defined in terms of the reaction $M(s) + H^+(aq) \longrightarrow M^+(aq) + \frac{1}{2}H_2(g)$ and (2) E° and ΔG° are defined to be zero for the standard hydrogen electrode. One consequence of this convention is that S° is negative for some aqueous ions.

Substance	ΔH_f° (kJ/mol)	ΔG_f° (kJ/mol)	S° (J/mol·K)
CO_2(aq)	−413.80	−385.98	117.6
CO_2(g)	−393.509	−394.359	213.74
CO_3^{2-}(aq)	−677.14	−527.81	−56.9
$COCl_2$(g)	−218.8	−204.6	283.53
CS_2(l)	89.70	65.27	151.34
HCN(aq)	107.1	119.7	124.7
HCN(g)	135.1	124.7	201.78
$HCOO^-$(aq)	−425.55	−351.0	92
HCO_3^-(aq)	−691.99	−586.77	91.2
HCOOH(aq)	−425.43	−372.3	163
HCOOH(l)	−424.72	−361.35	128.95
Cesium			
Cs(g)	76.065	49.121	175.595
Cs(s)	0	0	85.23
Cs^+(aq)	−258.28	−292.02	133.05
CsBr(s)	−398.978	−391.41	113.05
CsCl(s)	−443.04	−414.53	101.17
CsF(s)	−553.5	−525.5	92.80
CsI(s)	−346.60	−340.58	123.05
Chlorine			
Cl(g)	121.679	105.680	165.198
Cl^-(aq)	−167.159	−131.228	56.5
Cl_2(g)	0	0	223.066
HCl(g)	−92.307	−95.299	186.908
Chromium			
Cr(g)	396.6	351.8	174.50
Cr(s)	0	0	23.77
CrO_4^{2-}(aq)	−881.15	−727.75	50.21
Cr_2O_3(s)	−1139.7	−1058.1	81.2
$Cr_2O_7^{2-}$(aq)	−1490.3	−1301.1	261.9
Cobalt			
Co(g)	424.7	380.3	179.515
Co(s)	0	0	30.04
Co^{2+}(aq)	−58.2	−54.4	−113
Co^{3+}(aq)	92	134	−305
$CoCl_2$(s)	−312.5	−269.8	109.16

Substance	ΔH_f° (kJ/mol)	ΔG_f° (kJ/mol)	S° (J/mol·K)
Copper			
Cu(g)	338.32	298.58	166.38
Cu(s)	0	0	33.150
Cu^+(aq)	71.67	49.98	40.6
Cu^{2+}(aq)	64.77	65.49	−99.6
$CuCl_2$(s)	−220.1	−175.7	108.07
CuO(s)	−157.3	−129.7	42.63
Cu_2O(s)	−168.6	−146.0	93.14
Cu_2S(s)	−79.5	−86.2	120.9
Fluorine			
F(g)	78.99	61.91	158.754
F^-(aq)	−332.63	−278.79	−13.8
F_2(g)	0	0	202.78
HF(g)	−271.1	−273.2	173.779
Hydrogen			
H(g)	217.965	203.247	114.713
H^+(aq)	0	0	0
H_2(g)	0	0	130.684
Iodine			
I(g)	106.838	70.250	180.791
I^-(aq)	−55.19	−51.57	111.3
I_2(g)	62.438	19.327	260.69
I_2(s)	0	0	116.135
I_3^-(aq)	−51.5	−51.4	239.3
HI(g)	26.48	1.70	206.594
Iron			
Fe(g)	416.3	370.7	180.490
Fe(s)	0	0	27.28
Fe^{2+}(aq)	−89.1	−78.90	−137.7
Fe^{3+}(aq)	−48.5	−4.7	−315.9
$FeCl_2$(s)	−341.79	−302.30	117.95
$FeCl_3$(s)	−399.49	−334.00	142.3
Fe_2O_3(s)	−824.2	−742.2	87.40
Fe_3O_4(s)	−1118.4	−1015.4	146.4
FeS_2(s)	−178.2	−166.9	52.93

—Table continues on next page

TABLE B.1 —(Continued)

Substance	ΔH_f° (kJ/mol)	ΔG_f° (kJ/mol)	S° (J/mol·K)
Lead			
Pb(g)	195.0	161.9	175.373
Pb(s)	0	0	64.81
Pb^{2+}(aq)	−1.7	−24.43	10.5
$PbCl_2$(s)	−359.41	−314.10	136.0
$PbCO_3$(s)	−699.1	−625.5	131.0
PbO(s)	−217.32	−187.89	68.70
PbO_2(s)	−277.4	−217.33	68.6
Pb_3O_4(s)	−718.4	−601.2	211.3
PbS(s)	−100.4	−98.7	91.2
$PbSO_4$(s)	−919.94	−813.14	148.57
Lithium			
Li(g)	159.37	126.66	138.77
Li(s)	0	0	29.12
Li^+(aq)	−278.49	−293.31	13.4
LiBr(s)	−351.213	−342.00	74.27
LiCl(s)	−408.61	−384.37	59.33
LiF(s)	−615.97	−587.71	35.65
LiI(s)	−270.41	−270.29	86.78
Magnesium			
Mg(g)	147.70	113.10	148.650
Mg(s)	0	0	32.68
Mg^{2+}(aq)	−466.85	−454.8	−138.1
$MgCl_2$(s)	−641.32	−591.79	89.62
MgO(s)	−601.70	−569.43	26.94
$Mg(OH)_2$(s)	−924.54	−833.51	63.18
$MgSO_4$(s)	−1284.9	−1170.6	91.6
Manganese			
Mn(g)	280.7	238.5	173.70
Mn(s)	0	0	32.01
Mn^{2+}(aq)	−220.75	−228.1	−73.6
MnO_2(s)	−520.3	−465.14	53.05
MnO_4^-(aq)	−541.4	−447.2	191.2
Mercury			
Hg(g)	61.317	31.820	174.96
Hg(l)	0	0	76.02
Hg^{2+}(aq)	171.1	164.40	−32.2
Hg_2^{2+}(aq)	172.4	153.52	84.5
$HgCl_2$(s)	−224.3	−178.6	146.0
Hg_2Cl_2(s)	−265.22	−210.745	192.5
HgO(s, red)	−90.83	−58.539	70.29
HgS(s, red)	−58.2	−50.6	82.4
Nickel			
Ni(g)	429.7	384.5	182.193
Ni(s)	0	0	29.87
Ni^{2+}(aq)	−54.0	−45.	−128.9
NiO(s)	−239.7	−211.7	37.99
NiS(s)	−82.0	−79.5	52.97
Nitrogen			
N(g)	472.704	455.563	153.298
N_2(g)	0	0	191.61
NH_3(aq)	−80.29	−26.50	111.3
NH_3(g)	−46.11	−16.45	192.45
NH_4^+(aq)	−132.51	−79.31	113.4
NH_4Cl(s)	−314.43	−202.87	94.6
NH_4NO_3(s)	−365.56	−183.87	151.08
N_2H_4(g)	95.40	159.35	238.47
NO(g)	90.25	86.55	210.761
NO_2(g)	33.18	51.31	240.06
NO_3^-(aq)	−205.0	−108.74	146.4
N_2O(g)	82.05	104.20	219.85
N_2O_4(g)	9.16	97.89	304.29
NOCl(g)	51.71	66.08	261.69
HNO_3(g)	−135.06	−74.72	266.38
HNO_3(l)	−174.10	−80.71	155.60
Oxygen			
O(g)	249.170	231.731	161.055
O_2(g)	0	0	205.138
O_3(g)	142.7	163.2	238.93
OH^-(aq)	−229.994	−157.244	−10.75
H_2O(g)	−241.818	−228.572	188.825
H_2O(l)	−285.830	−237.129	69.91
H_2O_2(aq)	−191.17	−134.03	143.9
H_2O_2(g)	−136.31	−105.57	232.7
H_2O_2(l)	−187.78	−120.35	109.6

Substance	ΔH_f° (kJ/mol)	ΔG_f° (kJ/mol)	S° (J/mol·K)
Phosphorus			
P(s, red)	−17.6	−12.1	22.80
P_4(g)	58.91	24.44	279.98
P_4(s, white)	0	0	41.09
PCl_3(g)	−287.0	−267.8	311.78
PCl_5(g)	−374.9	−305.0	364.58
PH_3(g)	5.4	13.4	210.23
PO_4^{3-}(aq)	−1277.4	−1018.7	−222
P_4O_{10}(s)	−2984.0	−2697.7	228.86
H_3PO_4(s)	−1260.60	−1119.1	110.50
Potassium			
K(g)	89.24	60.59	160.336
K(s)	0	0	64.18
K^+(aq)	−252.38	−283.27	102.5
KBr(s)	−393.798	−380.66	95.90
KCl(s)	−436.75	−409.14	82.59
$KClO_3$(s)	−397.73	−296.25	143.1
KF(s)	−567.27	−537.75	66.57
KI(s)	−327.900	−324.892	106.32
KNO_3(s)	−494.63	−394.86	133.05
KOH(s)	−424.764	−379.08	78.9
Silicon			
Si(g)	455.6	411.3	167.97
Si(s)	0	0	18.83
SiC(s)	−65.3	−62.8	16.61
$SiCl_4$(l)	−687.0	−619.84	239.7
SiO_2(s, quartz)	−910.94	−856.64	41.84
Silver			
Ag(g)	284.55	245.65	172.997
Ag(s)	0	0	42.55
Ag^+(aq)	105.579	77.107	72.68
AgBr(s)	−100.37	−96.90	107.1
AgCl(s)	−127.07	−109.79	96.2
$AgNO_3$(s)	−124.39	−33.41	140.92
Ag_2O(s)	−31.05	−11.20	121.3
Ag_2S(s)	−32.59	−40.67	144.01
Sodium			
Na(g)	107.32	76.761	153.712
Na(s)	0	0	51.21
Na^+(aq)	−240.12	−261.905	59.0
NaBr(s)	−361.062	−348.983	86.82
NaCl(s)	−411.15	−384.14	72.13
NaF(s)	−573.647	−543.494	51.46
$NaHCO_3$(s)	−950.81	−851.0	101.7
$NaHSO_4$(s)	−1125.5	−992.8	113.0
NaI(s)	−287.78	−286.06	98.53
NaOH(s)	−425.61	−379.49	64.455
Na_2CO_3(s)	−1130.68	−1044.44	134.98
Na_2SO_4(s)	−1387.08	−1270.16	149.58
Sulfur			
S_2(g)	128.37	79.30	228.18
S_8(g)	102.30	49.63	430.98
S_8(s, rhombic)	0	0	31.80
SF_6(g)	−1209	−1105.3	291.82
SO_2(g)	−296.83	−300.19	248.22
SO_3(g)	−395.72	−371.06	256.76
H_2S(g)	−20.63	−33.56	205.79
H_2S(aq)	−39.7	−27.83	121
H_2SO_4(l)	−813.99	−690.00	156.90
Tin			
Sn(g)	302.1	267.3	168.486
Sn(s, white)	0	0	51.55
Sn(s, gray)	−2.09	0.13	44.14
SnO_2(s)	−580.7	−519.6	52.3
Titanium			
Ti(g)	469.9	425.1	180.298
Ti(s)	0	0	30.63
$TiCl_4$(l)	−804.2	−737.2	252.34
TiO_2(s, rutile)	−944.7	−889.5	50.33
Zinc			
Zn(g)	130.73	95.15	160.98
Zn(s)	0	0	41.63
Zn^{2+}(aq)	−153.89	−147.06	−112.1
$ZnCl_2$(s)	−415.05	−369.398	111.46
ZnO(s)	−348.28	−318.30	43.64
ZnS(s, sphalerite)	−205.98	−201.29	57.7

TABLE B.2 Ionization Constants[a] of Weak Acids at 25°C

Acid	Formula	K_a	pK_a
Acetic acid	CH_3COOH	1.76×10^{-5}	4.754
Aluminum ion	$Al(H_2O)_6^{3+}$	1.4×10^{-5}	4.85
Arsenic acid	H_3AsO_4	2.5×10^{-4}	3.60
	$H_2AsO_4^-$	5.6×10^{-8}	7.25
	$HAsO_4^{2-}$	3×10^{-13}	12.5
Ascorbic acid[b]	$H_2C_6H_6O_6$	7.9×10^{-5}	4.10
	$HC_6H_6O_6^-$	1.6×10^{-12}	11.80
Barbituric acid	$HC_4H_3N_2O_3$	9.8×10^{-5}	4.01
Benzoic acid	C_6H_5COOH	6.46×10^{-5}	4.190
Carbonic acid	$CO_2 + H_2O$[c]	4.30×10^{-7}	6.367
	HCO_3^-	5.61×10^{-11}	10.251
Chloroacetic acid	$CH_2ClCOOH$	1.40×10^{-3}	2.85
Chromic acid	H_2CrO_4	1.8×10^{-1}	0.74
	$HCrO_4^-$	3.20×10^{-7}	6.495
Citric acid	$H_3C_6H_5O_7$	7.45×10^{-4}	3.128
	$H_2C_6H_5O_7^-$	1.73×10^{-5}	4.762
	$HC_6H_5O_7^{2-}$	4.02×10^{-7}	6.396
Dichloroacetic acid	$CHCl_2COOH$	3.3×10^{-2}	1.48
Diethylbarbituric acid	$HC_8H_{11}N_2O_3$	3.7×10^{-8}	7.43
Formic acid	$HCOOH$	1.9×10^{-4}	3.72
Hydrazoic acid	HN_3	1.9×10^{-5}	4.72
Hydrocyanic acid	HCN	4.93×10^{-10}	9.307
Hydrofluoric acid	HF	3.53×10^{-4}	3.452
Hydrogen sulfide	H_2S	1.02×10^{-7}	6.991
	HS^-[d]	$\sim 1 \times 10^{-19}$	~19.0
Hypobromous acid	$HOBr$	2.06×10^{-9}	8.686
Hypochlorous acid	$HOCl$	2.8×10^{-8}	7.55
Hypoiodous acid	HOI	2.3×10^{-11}	10.64
Iodic acid	HIO_3	1.69×10^{-1}	0.772
Iron(III) ion	$Fe(H_2O)_6^{3+}$	7.9×10^{-3}	2.10
Lactic acid	$HC_3H_5O_3$	1.37×10^{-4}	3.863
Malonic acid	$H_2C_3H_2O_4$	1.49×10^{-3}	2.827
	$HC_3H_2O_4^-$	2.03×10^{-6}	5.693
Nicotinic acid	$HC_6H_4NO_2$	1.4×10^{-5}	4.85
Nitrous acid	HNO_2	6.0×10^{-4}	3.22
Oxalic acid	$H_2C_2O_4$	5.90×10^{-2}	1.229
	$HC_2O_4^-$	6.40×10^{-5}	4.194
Phenol	C_6H_5OH	1.28×10^{-10}	9.893
Phosphoric acid	H_3PO_4	7.52×10^{-3}	2.124
	$H_2PO_4^-$	6.23×10^{-8}	7.206
	HPO_4^{2-}	4.5×10^{-13}	12.35

—Table continues on next page

TABLE B.2 Ionization Constants[a] of Weak Acids at 25°C—(Continued)

Acid	Formula	K_a	pK_a
o-Phthalic acid	$H_2C_8H_4O_4$	1.3×10^{-3}	2.89
	$HC_8H_4O_4^-$	3.1×10^{-6}	5.51
Propionic acid	$HC_3H_5O_2$	1.34×10^{-5}	4.873
Selenious acid	H_2SeO_3	2.4×10^{-3}	2.62
	$HSeO_3$	5.0×10^{-9}	8.30
Sulfurous acid	H_2SO_3	1.71×10^{-2}	1.767
	HSO_3^-	6.0×10^{-8}	7.22
Sulfuric acid	H_2SO_4	Strong	
	HSO_4^-	1.20×10^{-2}	1.92
Tartaric acid	$H_2C_4H_4O_6$	1.04×10^{-3}	2.983
	$HC_4H_4O_6^-$	4.55×10^{-5}	4.342
Trichloroacetic acid	CCl_3COOH	2×10^{-1}	0.7
Veronal (see Diethylbarbituric acid)			
Zinc ion	$Zn(H_2O)_4^{2+}$	2.5×10^{-10}	9.60

[a]The first ionization constant of a polyprotic acid such as H_3PO_4 is usually symbolized by K_{a_1}, the second by K_{a_2}, and so forth.

[b]Ascorbic acid is the chemical name for vitamin C.

[c]H_2CO_3 is unstable; the first ionization constant is for the reaction $CO_2(aq) + H_2O(l) \rightleftharpoons H^+(aq) + HCO_3^-(aq)$.

[d]The value of K_a for HS^- is controversial. The reported value was taken from R. J. Meyers, *J. of Chem. Ed.*, **63**, 687 (1986).

TABLE B.3 Ionization Constants of Weak Bases at 25°C

Base	Formula	K_b	pK_b
Ammonia	NH_3	1.77×10^{-5}	4.752
Aniline	$C_6H_5NH_2$	4.3×10^{-10}	9.37
Dimethylamine	$(CH_3)_2NH$	5.41×10^{-4}	3.267
Ethylamine	$C_2H_5NH_2$	6.41×10^{-4}	3.193[a]
Hydrazine	H_2NNH_2	8.9×10^{-7}	6.05
Hydroxylamine	NH_2OH	9.1×10^{-9}	8.04
Methylamine	CH_3NH_2	3.70×10^{-4}	3.432
Pyridine	C_5H_5N	1.78×10^{-9}	8.750
Trimethylamine	$(CH_3)_3N$	6.3×10^{-5}	4.20
Urea	H_2NCONH_2	1.2×10^{-14}	13.92[a]

[a]At 20°C.

TABLE B.4 Solubility Product Constants at 25°C

Precipitate	Color	K_{sp}
Acetates		
CH_3COOAg	White	2.5×10^{-3}
Bromides		
AgBr	Cream	5.0×10^{-13}
Hg_2Br_2	Yellow	1.3×10^{-22}
$PbBr_2$	White	2.1×10^{-6}
Carbonates		
Ag_2CO_3	Yellow	8.1×10^{-12}
$BaCO_3$	White	5.0×10^{-9}
$CaCO_3$	White	4.5×10^{-9}
Li_2CO_3	White	1.7×10^{-3}
$MgCO_3$	White	1.6×10^{-6}
$SrCO_3$	White	9.3×10^{-10}
$PbCO_3$	White	7.4×10^{-14}
Chlorides		
AgCl	White	1.8×10^{-10}
CuCl	White	1.7×10^{-7}
Hg_2Cl_2	White	1.3×10^{-18}
$PbCl_2$	White	1.7×10^{-5}
Chromates		
Ag_2CrO_4	Rust red	1.2×10^{-12}
$BaCrO_4$	Yellow	1.2×10^{-10}
$CaCrO_4$	Yellow	7.1×10^{-4}
$PbCrO_4$	Yellow	2.8×10^{-13}
$SrCrO_4$	Yellow	3.6×10^{-5}
Fluorides		
BaF_2	White	1.3×10^{-6}
CaF_2	White	3.9×10^{-11}
MgF_2	White	3.7×10^{-8}
PbF_2	White	3.6×10^{-8}
SrF_2	White	7.9×10^{-10}
Hydroxides		
$Al(OH)_3$	White	3×10^{-34}
$Ba(OH)_2$	White	5×10^{-3}
$Ca(OH)_2$	White	6.5×10^{-6}
$Cr(OH)_2$	Gray-green	6.3×10^{-31}
$Cu(OH)_2$	Pale blue	1.6×10^{-19}
$Fe(OH)_2$	Pale green	7.9×10^{-16}
$Fe(OH)_3$	Rust red	1.6×10^{-39}
$Mg(OH)_2$	White	7.1×10^{-12}
$Mn(OH)_2$	Pink	6×10^{-14}
$Pb(OH)_2$	White	1.2×10^{-15}
$Sr(OH)_2$	White	3.2×10^{-4}
$Zn(OH)_2$	White	4.5×10^{-17}
Iodides		
AgI	Yellow	8.3×10^{-17}
Hg_2I_2	Yellow	4.5×10^{-29}
PbI_2	Yellow	7.9×10^{-9}
Oxalates		
$Ag_2C_2O_4$	White	1.1×10^{-11}
BaC_2O_4	White	1.6×10^{-7}
CaC_2O_4	White	1.3×10^{-9}
MgC_2O_4	White	8.6×10^{-5}
PbC_2O_4	White	8.3×10^{-12}
SrC_2O_4	White	5.6×10^{-8}
Phosphates		
Ag_3PO_4	Yellow	1.8×10^{-18}
$Ba_3(PO_4)_2$	White	6×10^{-39}
$Ca_3(PO_4)_2$	White	1.3×10^{-32}
$Sr_3(PO_4)_2$	White	1×10^{-31}
Sulfates		
Ag_2SO_4	White	1.4×10^{-5}
$BaSO_4$	White	1.1×10^{-10}
$CaSO_4$	White	2.4×10^{-5}
$PbSO_4$	White	6.3×10^{-7}
$SrSO_4$	White	3.2×10^{-7}

TABLE B.5 Formation Constants[a] of Complex Ions at 25°C

Complex Ion	K_f	Complex Ion	K_f
Ammonia complexes		**Halide complexes**	
$Ag(NH_3)_2^+$	1.6×10^7	$AgCl_2^-$	2.5×10^5
$Cd(NH_3)_4^{2+}$	1×10^7	AlF_6^{3-}	6.7×10^{19}
$Cu(NH_3)_4^{2+}$	1.1×10^{13}	$HgCl_4^{2-}$	1.2×10^{15}
$Ni(NH_3)_6^{2+}$	5.5×10^8	HgI_4^{2-}	1.9×10^{30}
$Zn(NH_3)_4^{2+}$	2.9×10^9		
		Hydroxide complexes	
Cyanide complexes		$Al(OH)_4^-$	7.7×10^{33}
$Ag(CN)_2^-$	1×10^{21}	$Zn(OH)_4^{2-}$	2.2×10^{16}
$Au(CN)_2^-$	2×10^{38}		
$Cd(CN)_4^{2-}$	1.2×10^{17}	**Other complexes**	
$Cu(CN)_2^-$	1.0×10^{24}	$Ag(S_2O_3)_2^{3-}$	1.7×10^{13}
$Cu(CN)_3^{2-}$	2×10^{27}	$Fe(C_2O_4)_3^{3-}$	3×10^{20}
$Fe(CN)_6^{4-}$	1.0×10^{24}	$Fe(SCN)^{2+}$	1.2×10^2
$Fe(CN)_6^{3-}$	1.0×10^{31}	$Zn(EDTA)^{2-}$	3.8×10^{16}
$Ni(CN)_4^{2-}$	1×10^{22}		
$Hg(CN)_4^{2-}$	2×10^{41}		
$Zn(CN)_4^{2-}$	6×10^{16}		

[a]The dissociation constant (K_d) is the reciprocal of the formation constant. Formation constants are also called stability constants (K_{stab}); dissociation constants are called **instability constants** (K_{instab}).

TABLE B.6 Standard Reduction Potentials[a] at 25°C

Half-Reaction	$E°$ (V)
$F_2(g) + 2e^- \rightleftharpoons 2F^-(aq)$	2.889
$O_3(g) + 2H^+(aq) + 2e^- \rightleftharpoons O_2(g) + H_2O(l)$	2.075
$Co^{3+}(aq) + e^- \rightleftharpoons Co^{2+}(aq)$	1.95
$H_2O_2(aq) + 2H^+(aq) + 2e^- \rightleftharpoons 2H_2O(l)$	1.763
$Ce^{4+}(aq) + e^- \rightleftharpoons Ce^{3+}(aq)$	1.743
$Au^+(aq) + e^- \rightleftharpoons Au(s)$	1.691
$PbO_2(s) + SO_4{}^{2-}(aq) + 2e^- \rightleftharpoons PbSO_4(s) + 2H_2O(l)$	1.686
$MnO_4{}^-(aq) + 8H^+(aq) + 5e^- \rightleftharpoons Mn^{2+}(aq) + 4H_2O(l)$	1.512
$Au^{3+}(aq) + 3e^- \rightleftharpoons Au(s)$	1.498
$PbO_2(s) + 4H^+(aq) + 2e^- \rightleftharpoons Pb^{2+}(aq) + 2H_2O(l)$	1.458
$Cl_2(g) + 2e^- \rightleftharpoons 2Cl^-(aq)$	1.360
$Cr_2O_7{}^{2-}(aq) + 14H^+(aq) + 6e^- \rightleftharpoons 2Cr^{3+}(aq) + 7H_2O(l)$	1.33
$MnO_2(s) + 4H^+(aq) + 2e^- \rightleftharpoons Mn^{2+}(aq) + 2H_2O(l)$	1.229
$O_2(g) + 4H^+(aq) + 4e^- \rightleftharpoons 2H_2O(l)$	1.229
$2IO_3{}^-(aq) + 12H^+(aq) + 10e^- \rightleftharpoons I_2(s) + 6H_2O(l)$	1.209
$Br_2(l) + 2e^- \rightleftharpoons 2Br^-(aq)$	1.078
$AuCl_4{}^-(aq) + 3e^- \rightleftharpoons Au(s) + 4Cl^-(aq)$	1.00
$NO_3{}^-(aq) + 4H^+(aq) + 3e^- \rightleftharpoons NO(g) + 2H_2O(l)$	0.964
$NO_3{}^-(aq) + 3H^+(aq) + 2e^- \rightleftharpoons HNO_2(aq) + H_2O(l)$	0.928
$2Hg^{2+}(aq) + 2e^- \rightleftharpoons Hg_2{}^{2+}(aq)$	0.908
$ClO^-(aq) + H_2O(l) + 2e^- \rightleftharpoons Cl^-(aq) + 2OH^-(aq)$	0.890
$Hg^{2+}(aq) + 2e^- \rightleftharpoons Hg(l)$	0.852
$Ag^+(aq) + e^- \rightleftharpoons Ag(s)$	0.799
$Hg_2{}^{2+}(aq) + 2e^- \rightleftharpoons 2Hg(l)$	0.796
$Fe^{3+}(aq) + e^- \rightleftharpoons Fe^{2+}(aq)$	0.769
$O_2(g) + 2H^+(aq) + 2e^- \rightleftharpoons H_2O_2(aq)$	0.695
$ClO_2{}^-(aq) + H_2O(l) + 2e^- \rightleftharpoons ClO^-(aq) + 2OH^-(aq)$	0.681
$MnO_4{}^-(aq) + 2H_2O(l) + 3e^- \rightleftharpoons MnO_2(s) + 4OH^-(aq)$	0.597
$I_2(s) + 2e^- \rightleftharpoons 2I^-(aq)$	0.535
$Cu^+(aq) + e^- \rightleftharpoons Cu(s)$	0.518
$O_2(g) + 2H_2O(l) + 4e^- \rightleftharpoons 4OH^-(aq)$	0.401

Stronger Oxidizing Agents ↑ (left side) — Stronger Reducing Agents ↓ (right side)

[a]Each substance is in its standard state; that is, approximately 1 *M* for dissolved species and 1 atm for gases (see Table 19.1).

TABLE B.6 Standard Reduction Potentials[a] at 25°C—(Continued)

Half-Reaction	$E°$ (V)
$ClO_4^-(aq) + H_2O(l) + 2e^- \rightleftharpoons ClO_3^-(aq) + 2OH^-(aq)$	0.398
$Ag(NH_3)_2^+(aq) + e^- \rightleftharpoons Ag(s) + 2NH_3(aq)$	0.372
$Cu^{2+}(aq) + 2e^- \rightleftharpoons Cu(s)$	0.339
$ClO_3^-(aq) + 2H_2O(l) + 2e^- \rightleftharpoons ClO_2^-(aq) + 2OH^-(aq)$	0.271
$Hg_2Cl_2(s) + 2e^- \rightleftharpoons 2Hg(l) + 2Cl^-(aq)$	0.268
$AgCl(s) + e^- \rightleftharpoons Ag(s) + Cl^-(aq)$	0.222
$Cu^{2+}(aq) + e^- \rightleftharpoons Cu^+(aq)$	0.161
$Sn^{4+}(aq) + 2e^- \rightleftharpoons Sn^{2+}(aq)$	0.154
$S(s) + 2H^+(aq) + 2e^- \rightleftharpoons H_2S(aq)$	0.144
$AgBr(s) + e^- \rightleftharpoons Ag(s) + Br^-(aq)$	0.0732
$2H^+(aq) + 2e^- \rightleftharpoons H_2(g)$	0.000
$Pb^{2+}(aq) + 2e^- \rightleftharpoons Pb(s)$	−0.127
$Sn^{2+}(aq) + 2e^- \rightleftharpoons Sn(s)$	−0.14
$AgI(s) + e^- \rightleftharpoons Ag(s) + I^-(aq)$	−0.152
$Ni^{2+}(aq) + 2e^- \rightleftharpoons Ni(s)$	−0.236
$Co^{2+}(aq) + 2e^- \rightleftharpoons Co(s)$	−0.282
$Ag(CN)_2^-(aq) + e^- \rightleftharpoons Ag(s) + 2CN^-(aq)$	−0.31
$PbSO_4(s) + 2e^- \rightleftharpoons Pb(s) + SO_4^{2-}(aq)$	−0.356
$Cd^{2+}(aq) + 2e^- \rightleftharpoons Cd(s)$	−0.402
$Fe^{2+}(aq) + 2e^- \rightleftharpoons Fe(s)$	−0.409
$Cr^{3+}(aq) + 3e^- \rightleftharpoons Cr(s)$	−0.74
$Zn^{2+}(aq) + 2e^- \rightleftharpoons Zn(s)$	−0.762
$2H_2O(l) + 2e^- \rightleftharpoons H_2(g) + 2OH^-(aq)$	−0.828
$SO_4^{2-}(aq) + H_2O(l) + 2e^- \rightleftharpoons SO_3^{2-}(aq) + 2OH^-(aq)$	−0.936
$Zn(NH_3)_4^{2+}(aq) + 2e^- \rightleftharpoons Zn(s) + 4NH_3(aq)$	−1.015
$Mn^{2+}(aq) + 2e^- \rightleftharpoons Mn(s)$	−1.182
$Zn(OH)_4^{2-}(aq) + 2e^- \rightleftharpoons Zn(s) + 4OH^-(aq)$	−1.190
$Al^{3+}(aq) + 3e^- \rightleftharpoons Al(s)$	−1.68
$Mg^{2+}(aq) + 2e^- \rightleftharpoons Mg(s)$	−2.357
$Na^+(aq) + e^- \rightleftharpoons Na(s)$	−2.714
$Ca^{2+}(aq) + 2e^- \rightleftharpoons Ca(s)$	−2.869
$Ba^{2+}(aq) + 2e^- \rightleftharpoons Ba(s)$	−2.906
$K^+(aq) + e^- \rightleftharpoons K(s)$	−2.936
$Li^+(aq) + e^- \rightleftharpoons Li(s)$	−3.040

← Stronger Oxidizing Agents (increasing upward)

Stronger Reducing Agents → (increasing downward)

APPENDIX C

ANSWERS TO PRACTICE EXERCISES

CHAPTER 1

1.3 (a) 2.1 m; (b) 4.10×10^{-7} m; (c) 5.5×10^{-7} m
1.4 (a) 1000 L; (b) 0.018 L; (c) 0.500 L
1.5 (a) 2; (b) 4; (c) 3; (d) 1
1.6 (a) 2; (b) 6; (c) 1; (d) 4
1.7 (a) 78.2 mL; (b) 5.79 cm; (c) 14.25 g
1.8 (a) 0.109; (b) 16; (c) 2.7×10^{-11}
1.9 0.258 kg
1.10 64.3 L
1.11 4.46×10^3 m^2
1.12 44 m/s
1.13 (a) 14.2 g; (b) 1.42×10^4 mg
1.14 (a) 833 g; (b) 333 g
1.15 250 g
1.16 13.58 g/mL
1.17 37.0 cm^3 aluminum; 8.811 cm^3 lead
1.18 4.1×10^2 kJ
1.19 (a) increase; (b) decrease; (c) decrease
1.20 More
1.21 (a) 1.08×10^4 kcal; (b) 4.50×10^4 kJ

CHAPTER 2

2.2 (a) 56; (b) $^{56}_{26}Fe$
2.3 $^{35}_{17}Cl^-$
2.4 7.5% lithium-6, 92.5% lithium-7
2.5 9.6486×10^4 C
2.6 1.009 u
2.7 (a) 24.30 g; (b) 197.0 g
2.8 62.1 g
2.9 1.33×10^{-2} g
2.11 (b) and (d) are pure substances; the others are mixtures.
2.12 (a) 2 C, 4 H, 2 O; (b) 2 C, 7 H, 1 N
2.14 (a) 3.5 mol Ca^{2+}, 7.0 mol Cl^-; (b) 2.1×10^{24} Ca^{2+} ions, 4.2×10^{24} Cl^- ions
2.15 Mg_3N_2
2.16 (a) 58.12 g/mol; (b) 9.651×10^{-23} g/molecule
2.17 (a) 15.61 mol; (b) 9.400×10^{24} molecules
2.18 (a) Na_2CO_3; (b) $MgCl_2$; (c) $FeBr_3$
2.19 (a) potassium ion; (b) calcium ion; (c) gallium ion
2.20 (a) copper(I) ion, copper(II) ion; (b) chromium(II) ion, chromium(III) ion
2.21 (a) bromide ion; (b) nitride ion; (c) oxide ion
2.22 (a) BrO_3^-; (b) BrO^-; (c) BrO_4^-
2.23 (a) barium chloride; (b) calcium sulfide; (c) magnesium nitride; (d) iron(II) sulfate; (e) potassium cyanide; (f) aluminum nitrate
2.24 (a) $FeCl_2$; (b) Na_2S; (c) $K_2Cr_2O_7$; (d) $Mg_3(PO_4)_2$; (e) $Fe_2(SO_4)_3$
2.25 (a) sulfur dioxide; (b) sulfur trioxide; (c) phosphorus trichloride; (d) phosphorus pentachloride; (e) uranium hexafluoride; (f) carbon tetrachloride
2.26 (a) NH_3; (b) CH_4; (c) HBr; (d) BrF_5; (e) CS_2; (f) N_2O_4; (g) IBr
2.27 (a) $CaCO_3$; (b) $NaHCO_3$; (c) CO_2; (d) SiO_2; (e) CaO; (f) N_2O; (g) NaOH

CHAPTER 3

3.1 (a) and (c) chemical; (b) physical
3.2 Methane and oxygen are the reactants; water and carbon dioxide are the products.
3.4 (a) 88.8% O, 11.2% H; (b) 39.3% Na, 60.7% Cl
3.5 24.95% Fe, 46.46% Cr, 28.59% O
3.6 S_8
3.7 H_2S
3.8 (a) CH_2O; (b) $HgNO_3$
3.9 $CHCl_3$
3.10 Al_2S_3
3.11 C_4H_8
3.12 N_4Se_4
3.13 Both formulas are $C_{20}H_{30}O$.

3.14 (a) 88.80% C, 11.18% H; (b) C_4H_6
3.15 $C_2H_4(g) + 3O_2(g) \rightarrow 2CO_2(g) + 2H_2O(g)$
3.16 $2SO_2(g) + O_2(g) \rightarrow 2SO_3(g)$
3.17 $K_2CrO_4(aq) + 2AgNO_3(aq) \rightarrow Ag_2CrO_4(s) + 2KNO_3(aq)$
3.18 (a) $N_2(g) + 3H_2(g) \rightarrow 2NH_3(g)$; (b) $2BaO_2(s) \rightarrow 2BaO(s) + O_2(g)$
3.19 $2F_2(g) + 2H_2O(l) \rightarrow O_2(g) + 4HF(aq)$
3.20 $C_4H_4S(l) + 6O_2(g) \rightarrow 4CO_2(g) + 2H_2O(g) + SO_2(g)$
3.21 Al is oxidized; Fe_2O_3 is reduced.
3.22 8.00 mol
3.23 39.4 kg
3.24 27.3 g
3.25 12.8 g
3.26 (a) O_2; (b) 0.80 mol P_4
3.27 (a) $CaCl_2$ is limiting, Na_3PO_4 is excess; (b) 50.8 g Na_3PO_4
3.28 199.8 g
3.29 40.0%

CHAPTER 4

4.1 (a) $Al_2(SO_4)_3(s) \rightarrow 2Al^{3+}(aq) + 3SO_4^{2-}(aq)$
(b) $HI(aq) \rightarrow H^+(aq) + I^-(aq)$
(c) $HNO_2(aq) \rightleftharpoons H^+(aq) + NO_2^-(aq)$
4.2 $Al_2(SO_4)_3(aq) + 6NaOH(aq) \rightarrow 2Al(OH)_3(s) + 3Na_2SO_4(aq)$
Net ionic: $Al^{3+}(aq) + 3OH^-(aq) \rightarrow Al(OH)_3(s)$
4.3 (a), (c), (d) soluble; (b) sparingly soluble
4.4 (a) $Pb(NO_3)_2(aq) + 2KBr(aq) \rightarrow PbBr_2(s) + 2KNO_3(aq)$
Net ionic: $Pb^{2+}(aq) + 2Br^-(aq) \rightarrow PbBr_2(s)$
(b) $Na_2S(aq) + ZnCl_2(aq) \rightarrow ZnS(s) + 2NaCl(aq)$
Net ionic: $Zn^{2+}(aq) + S^{2-}(aq) \rightarrow ZnS(s)$
4.5 One possibility is
$3AgNO_3(aq) + Na_3PO_4(aq) \rightarrow Ag_3PO_4(s) + 3NaNO_3(aq)$
4.6 30.17% Cl^-
4.7 (a) $Fe(OH)_3(s) + 3HNO_3(aq) \rightarrow Fe(NO_3)_3(aq) + 3H_2O(l)$
(b) $Mg(OH)_2(s) + H_2SO_4(aq) \rightarrow MgSO_4(aq) + 2H_2O(l)$
4.8 (a) $H_3PO_4(aq) + 2NaOH(aq) \rightarrow Na_2HPO_4(aq) + 2H_2O(l)$
(b) $H_2SO_4(aq) + KOH(aq) \rightarrow KHSO_4(aq) + H_2O(l)$
4.9 (a) $NaCN(s) + HCl(aq) \rightarrow HCN(g) + NaCl(aq)$
(b) $NH_4HCO_3(s) + HCl(aq) \rightarrow NH_4Cl(aq) + H_2O(l) + CO_2(g)$
(c) $MgCO_3(s) + 2HCl(aq) \rightarrow MgCl_2(aq) + H_2O(l) + CO_2(g)$
(d) $NaHSO_3(s) + HCl(aq) \rightarrow NaCl(aq) + H_2O(l) + SO_2(g)$
4.10 $Mg(s) + H_2SO_4(aq) \rightarrow MgSO_4(aq) + H_2(g)$
Net ionic: $Mg(s) + 2H^+(aq) \rightarrow Mg^{2+}(aq) + H_2(g)$
4.11 (a) $Mg(s) + 2AgNO_3(aq) \rightarrow Mg(NO_3)_2(aq) + 2Ag(s)$
(b) $Pb(s) + CuSO_4(aq) \rightarrow PbSO_4(s) + Cu(s)$
4.12 232 g
4.13 5.0×10^{-9} g
4.14 1.63 *M*
4.15 0.247 *M*
4.16 16.0 *M*
4.17 167 mL
4.18 30.8 g
4.19 Put 2.65 g Na_2CO_3 into a 100-mL volumetric flask. Add water to dissolve, then dilute to the mark.
4.20 (a) 2.25 mol SO_4^{2-}; (b) 1.50 mol Fe^{3+}
4.21 0.859 *M* Cl^-
4.23 8.3 mL
4.24 1.50 *M*
4.25 49.8 mL
4.26 13.3 mL
4.27 0.4493 *M*
4.28 24.80%

CHAPTER 5

5.1 1.01×10^5 N/m^2
5.2 (a) 790 torr; (b) 0.987 atm
5.3 (a) 120 torr; (b) 0.158 atm; (c) 0.160 bar
5.4 870 torr
5.5 60.0 mL
5.6 217 kPa
5.7 0.628 atm
5.8 (a) increase; (b) by a factor of 780/760
5.9 (a) decrease; (b) by a factor of 4.0/6.0
5.10 (a) 258 K; (b) −100°C
5.11 (a) 245 K; (b) −28°C
5.12 (a) increase; (b) by a factor of 303/190
5.13 34.6 mL
5.14 0.714 g/L
5.15 31.4 L/mol
5.16 266 K = −7°C
5.18 0.764 atm
5.19 88.0 g/mol
5.20 0.537 g/L
5.21 6.01 L
5.22 (a) 22.5 mL; (b) no
5.23 (a) 716 torr; (b) 0.968; (c) 96.8%
5.24 757 torr
5.25 4.42 g
5.26 16.0 g/mol
5.27 1.99×10^4 J
5.29 $H_2 < N_2 < O_2 < CO_2$
5.30 (a) 35.3 atm; (b) 29.6 atm
5.31 (a) $He < H_2 < O_2 < N_2 < CO_2 < HCl$
(b) $He < H_2 < O_2 < N_2 < HCl < CO_2$

CHAPTER 6

6.1 (a) $q = -10$ kJ; (b) $q = +5$ kJ
6.2 (a) $w = -500$ J; (b) $w = +20$ kJ

6.3 (a) $q = +2600$ J; (b) $w = -170$ J; $\Delta E = +2430$ J
6.4 (a) exothermic; (b) endothermic
6.5 (a) negative; (b) positive
6.6 328 kJ evolved
6.7 3.31×10^4 kJ evolved
6.8 (a) $2Fe(s) + \frac{3}{2}O_2(g) \rightarrow Fe_2O_3(s)$ $\Delta H_f^\circ = -824.2$ kJ; (b) 3.69×10^5 kJ
6.9 $\Delta H^\circ = -99$ kJ
6.10 $\Delta H_f^\circ = -484.4$ kJ
6.11 $\Delta H_f^\circ = -1430.4$ kJ
6.12 0.9533 J/g °C
6.13 76.4°C
6.14 65.5 Calories (65.5 kcal)
6.15 (a) no; (b) no; (c) yes
6.16 1.6°C
6.17 $\Delta E = +83.29$ kJ

CHAPTER 7

7.1 192.2 m
7.2 132 nm
7.3 4.38×10^{14} s^{-1}
7.4 The $n = 5$ state
7.5 (a) absorbed; (b) emitted
7.6 -1.36×10^{-19} J; $n = 4$
7.7 656.5 nm
7.8 1.06×10^{-34} m
7.9 Seven orbitals
7.10 (a) no; (b) yes
7.11 Ten states

CHAPTER 8

8.1 (a) the carbon $1s$ electron; (b) the $3s$ electron
8.3 See Table 8.1.
8.4 (a) [Ar]$4s^2 3d^3$; (b) $3d$: [↑ | ↑ | ↑ | |] $4s$: [↑↓]
8.5 (a) 7; (b) 2; (c) 2
8.7 (a) [Xe]; (b) [Xe]$6s^2 4f^{14} 5d^{10}$
8.8 [Xe]
8.9 (a) diamagnetic; (b) and (c) paramagnetic
8.10 (a) I; (b) Si
8.11 (a) Ba; (b) S^{2-}
8.12 (a) Cl; (b) Mg

SURVEY 1

S1.1 $1s^2$
S1.2 (a) $2K(s) + 2H_2O(l) \rightarrow 2KOH(aq) + H_2(g)$
(b) $2Na(s) + Br_2(l) \rightarrow 2NaBr(s)$
(c) $2Li(s) + H_2(g) \rightarrow 2LiH(s)$
S1.3 (a) $Ca(s) + 2H_2O(l) \rightarrow Ca(OH)_2(aq) + H_2(g)$
(b) $Ba(s) + Cl_2(g) \rightarrow BaCl_2(s)$
(c) $Sr(s) + H_2(g) \rightarrow SrH_2(s)$
(d) $3Ca(s) + N_2(g) \rightarrow Ca_3N_2(s)$
S1.4 (a) −268.9°C;
(b) He, Ne, N_2, Ar, O_2, CH_4, Kr, Xe, CO_2, Rn

CHAPTER 9

9.1 (a) :Ṗ· ; (b) :B̤̈r· ; (c) ·Ṡi·
9.2 (a) Mg^{2+}; (b) I^-; (c) S^{2-}
9.3 (a) Both ions obey. (b) Cl^- obeys. (c) Br^- obeys.
9.4 (a) MgF_2; (b) MgO; (c) AlF_3
9.5 $\Delta H_{subl} = +147$ kJ/mol
9.6 (a) HCl; (b) HBr
9.7 $\Delta H^\circ = +83$ kJ/mol

9.8
```
    H
    |
H—C—H
    |
    H
```

9.9 :Ö=S̈—Ö:
9.10 H—N≡C:

9.11 (a)
```
          :O:
           ‖
H—Ö—C—Ö—H
```
(b)
```
[         :O:      ]⁻
[          ‖       ]
[ H—Ö—C—Ö:  ]
```

9.12 (a) Ö=C=Ö no formal charges
(b)
```
          :Ö:(−)
           |
H—Ö—P(+)—Ö—H
           |
          :O:
           |
           H
```
(c)
```
[      H        ]⁺
[      |        ]
[ H—N(+)—H ]
[      |        ]
[      H        ]
```
(d) [:Ö(−)—H]⁻

9.13
```
[   :O:    ]²⁻      [   :Ö:    ]²⁻      [   :Ö:    ]²⁻
[    ‖     ]    ↔   [    |     ]    ↔   [    |     ]
[    C     ]        [    C     ]        [    C     ]
[ :Ö:  :Ö: ]        [ Ö=  :Ö:  ]        [ :Ö:  =Ö  ]
```
Bond lengths are identical, intermediate between C=O and C—O.

9.14 (a)
```
(−1) (+1)        (+1) (−1)       (−2) (+1) (+1)
 N̈=N=Ö         :N≡N—Ö:        :N̈—N≡O:
    I               II              III
```
(b) III

9.15 (a) 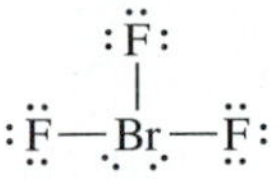 :F̤—Br(:F̤:)—F̤: (b) 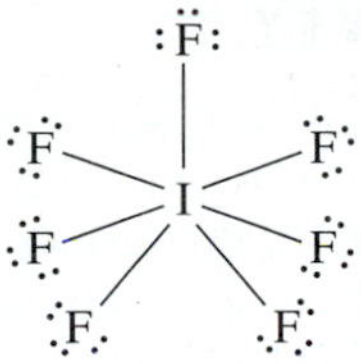 IF_7: I bonded to seven :F̤: atoms

9.16 (a) 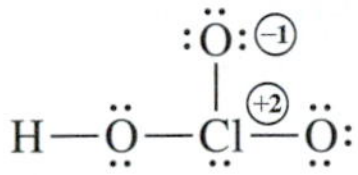 H—Ö—Cl(+2)(—:Ö:(−1))—Ö:(−1)

(b) 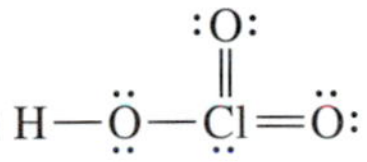 H—Ö—Cl(=O:)=Ö: This contributes more

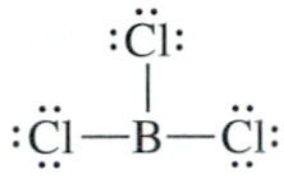

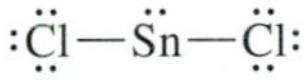

9.17 :F̤—Xe—F̤:

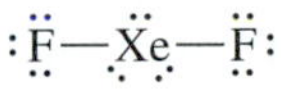

9.18 (a) 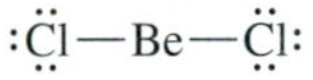 :C̤l—Be—C̤l: (b) :C̤l—B(—:C̤l:)—C̤l:

(c) :C̤l—S̈n—C̤l:

9.19 :Ö=C̈l=Ö: ⟷ :Ö=C̈l—Ö· ⟷ ·Ö—C̈l=Ö:

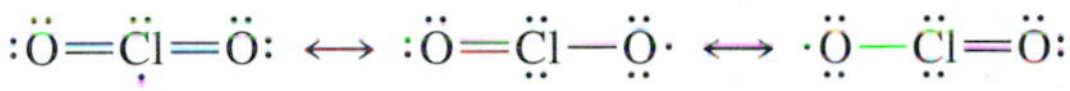

9.20 (a) polar covalent; (b) ionic

CHAPTER 10

10.1 F—O—F (bent)
10.2 (a) 2; (b) 4

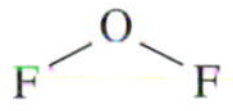

10.3 (a) 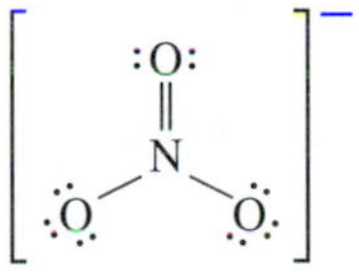 $[NO_3]^-$ (three resonance structures) (b) 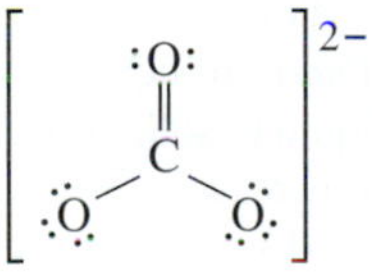 $[CO_3]^{2-}$ (three resonance structures)

10.4 :Ö=Ö—Ö: (2 resonance structures)

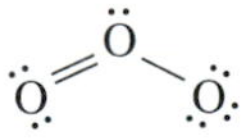

10.5 (a) 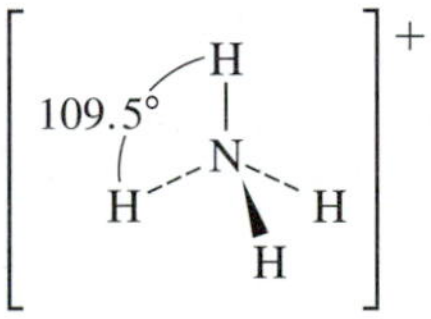 $[NH_4]^+$, 109.5° (b) 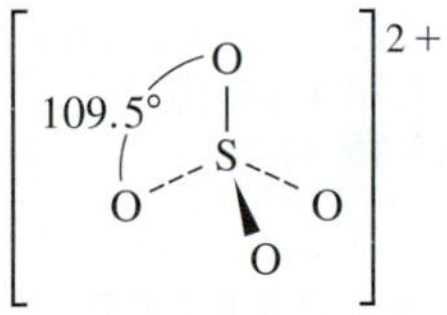 $[SO_4]^{2+}$, 109.5°

(c) 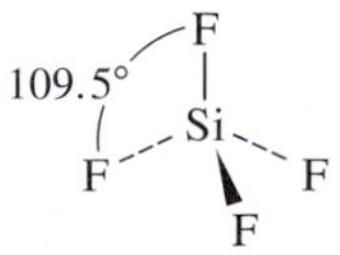 SiF_4, 109.5°

10.6 (a) 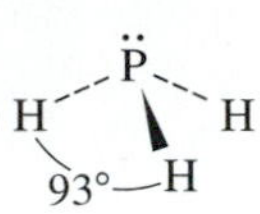 PH_3, 93° (pyramidal) (b) 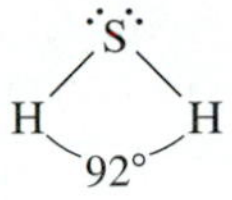 H_2S, 92°

10.7 (a) 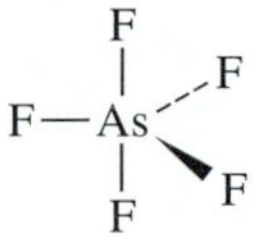 AsF_5 (trigonal bipyramidal) (b) 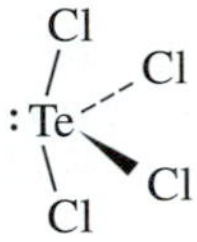 $:TeCl_4$ (bent see-saw)

(c) 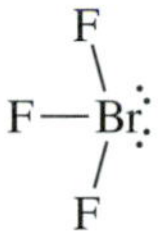 $F_3Br:$ (bent T-shape)

10.8 (a) 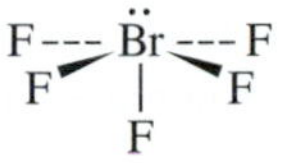 BrF_5 (slightly bent square pyramid) (b) 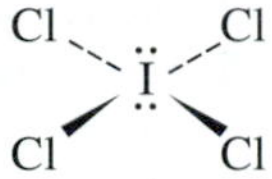 ICl_4 (square planar)

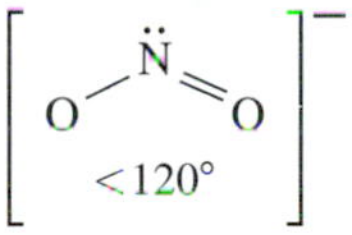

10.9 $[O—N̈=O]^-$ <120°

10.10 NF_3: bond dipole moments; 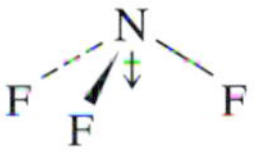 molecular dipole moment

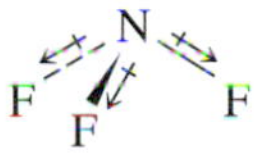

10.11 (a) polar; (b) nonpolar; (c) nonpolar; (d) polar; (e) nonpolar
10.12 (a) and (b) are the same molecule; (c) is a different isomer.
10.13 No

10.14 (a)

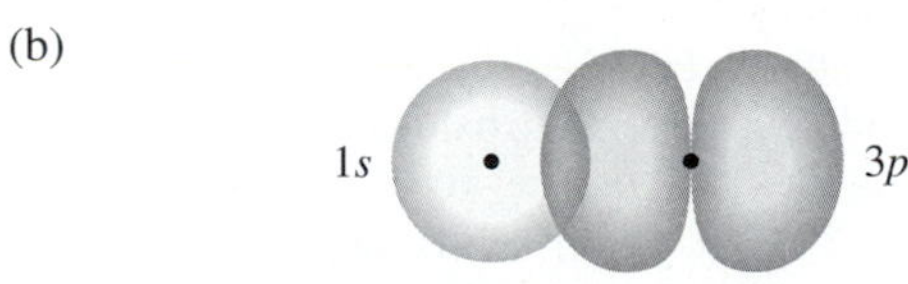

(b)

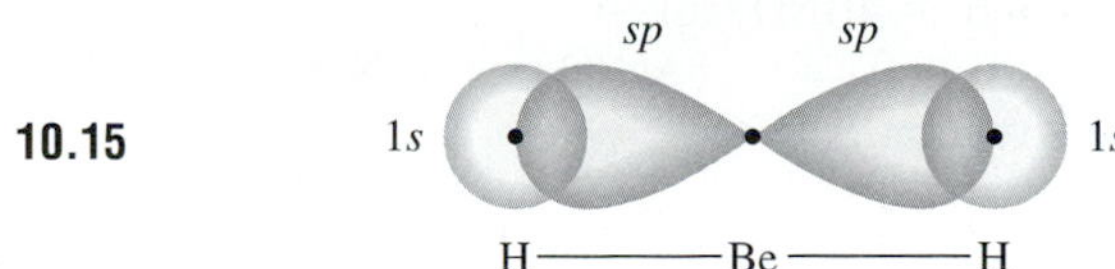

10.15

10.16

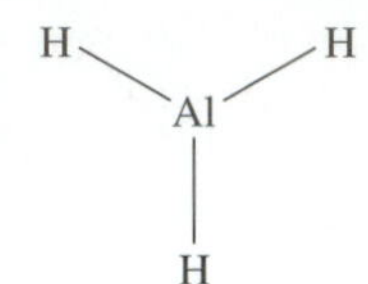

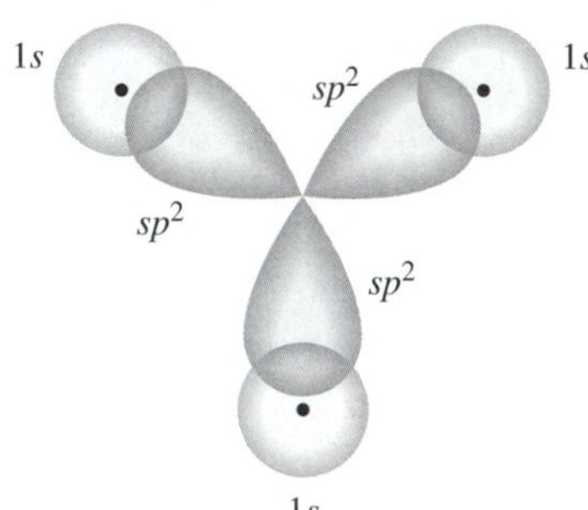

10.17 The 3*s*, $3p_x$, $3p_y$, and $3p_z$ of Si.
10.18 All are about 109°; H—N—H and H—N—C are slightly compressed by lone pair repulsion.
10.19 (a) $\frac{1}{2}$; (b) $\frac{1}{2}$; (c) 0
10.20 (a) $(\sigma_{1s})^2(\sigma^*_{1s})^2(\sigma_{2s})^2(\sigma^*_{2s})^2$; (b) not stable
10.21 (a) $(\sigma_{1s})^2(\sigma^*_{1s})^2(\sigma_{2s})^2(\sigma^*_{2s})^2(\pi_{2p})^4(\sigma_{2p})^2$ (b) no
10.22 (a) O_2^- is paramagnetic; O_2^{2-} is not; (b) $O_2^- > O_2^{2-}$; (c) $O_2^{2-} > O_2^-$
10.23 (a) $(\sigma_{1s})^2$ (b) yes

CHAPTER 11

11.1 (a) $Zn(s) \rightarrow Zn^{2+}(aq) + 2e^-$
$2H^+(aq) + 2e^- \rightarrow H_2(g)$
(b) H^+ is the oxidizing agent; Zn is the reducing agent.
11.2 (a) reduced; (b) becomes oxidized
11.3 Oxidation numbers are Cl, −1; O, −2; C, +4.
11.4 (a) O, −2; N, +5
(b) O, −2; N, +3
(c) N, −3; C, +2
11.5 2.5
11.6 HBr, $HBrO_2$, $HBrO_3$, $HBrO_4$
11.7 Oxidized
11.8 H_2S is oxidized; $FeCl_3$ is reduced.
11.9 SO_3, oxidizing agent; SO_2, oxidizing or reducing agent
11.10 $2IO_3^-(aq) + 12H^+(aq) + 10e^- \rightarrow I_2(s) + 6H_2O(l)$
11.11 (a) $3Ag(s) + NO_3^-(aq) + 4H^+(aq) \rightarrow 3Ag^+(aq) + NO(g) + 2H_2O(l)$
(b) $3Ag(s) + 4HNO_3(aq) \rightarrow 3AgNO_3(aq) + NO(g) + 2H_2O(l)$
11.12 $S_2O_4^{2-}(aq) + 4OH^-(aq) \rightarrow 2SO_3^{2-}(aq) + 2H_2O(l) + 2e^-$
11.13 $2CrO_4^{2-}(aq) + I^-(aq) + 5H_2O(l) \rightarrow 2Cr^{3+}(aq) + IO_3^-(aq) + 10OH^-(aq)$
11.14 $3MnO_4^{2-}(aq) + 4H^+(aq) \rightarrow 2MnO_4^-(aq) + MnO_2(s) + 2H_2O(l)$
11.16 (a) $1MnO_4^-/5Fe^{2+}$ (b) 0.01059 mol $FeSO_4$

SURVEY 2

S2.1 $2I^-(aq) \rightarrow I_2(aq) + 2e^-$
$O_3(g) + H_2O(l) + 2e^- \rightarrow O_2(g) + 2OH^-(aq)$
S2.2 (a) H_2SO_4 (b) $NaHSO_4$ and/or $Na_2SO_4 + H_2O$
S2.3 (a) $H_2O_2(aq) + 2H^+(aq) + 2e^- \rightarrow 2H_2O(l)$
(b) $H_2O_2(aq) + 2e^- \rightarrow 2OH^-(aq)$
(c) $H_2O_2(aq) \rightarrow 2H^+(aq) + O_2(g) + 2e^-$
S2.4 Linear
S2.5 Bent
S2.6 (a) $HNO_2(aq) + H^+(aq) + e^- \rightarrow NO(g) + H_2O(l)$
(b) $NO_2^-(aq) + 2OH^-(aq) \rightarrow NO_3^-(aq) + H_2O(l) + 2e^-$
S2.7 (a) linear; (b) trigonal planar
S2.8 $4HNO_3(aq) \rightarrow 4NO_2(g) + O_2(g) + 2H_2O(l)$
S2.9 (a) $C(s) + 2H_2O(l) \rightarrow CO_2(g) + 4H^+(aq) + 4e^-$
$NO_3^-(aq) + 2H^+(aq) + e^- \rightarrow NO_2(g) + H_2O(l)$
(b) $Cu(s) \rightarrow Cu^{2+}(aq) + 2e^-$
$NO_3^-(aq) + 4H^+(aq) + 3e^- \rightarrow NO(g) + 2H_2O(l)$
S2.10 $2Br^-(aq) \rightarrow Br_2(l) + 2e^-$
$Cr_2O_7^{2-}(aq) + 14H^+(aq) + 6e^- \rightarrow 2Cr^{3+}(aq) + 7H_2O(l)$
S2.11 (a) bent T; (b) bent square pyramid
S2.12 (a) −1; (b) +1; (c) +4; (d) +7
S2.13 (a) bromic acid; (b) periodic acid; (c) sodium hypobromite; (d) sodium iodate
S2.14 (a) $KClO_2$; (b) HOBr; (c) $HBrO_4$; (d) $NaIO_4$
S2.15 (a) bent; (b) trigonal pyramid; (c) tetrahedral

CHAPTER 12

12.1 $q = 22.6$ kJ
12.2 0.091 atm
12.3 (a) liquid; (b) liquid
12.4 (a) liquid to solid to vapor; (b) liquid to vapor
12.5 Both increase.
12.6 (a) no; (b) yes
12.7 (a) H_2S; (b) H_2S
12.8 (a) no; (b) no; (c) yes
12.9 (a) I_2; (b) SiH_4; (c) CH_3CH_3
12.10 Two atoms
12.11 144 pm
12.12 26.8 g/mol; aluminum
12.13 (a) 68.02%; (b) 74.05%
12.14 (a) 4; (b) 4
12.15 12.0°

CHAPTER 13

13.1 1.06×10^{-5} mol/L
13.2 $P_{methanol} = 29.6$ torr; $P_{ethanol} = 29.7$ torr; $P_{total} = 59.3$ torr
13.3 Mole fraction = 0.076
13.4 1.73 mol/kg
13.5 $K_b = 2.54°C{\cdot}kg/mol$

13.6 0.43 mol/kg
13.7 24.5 atm
13.8 385 g/mol
13.9 4.46×10^4 g/mol
13.12 101.11°C

CHAPTER 14

14.1 Eight times faster at 35°C
14.2 0.10 mol/L·s
14.4 1×10^{-4} mol/L·s
14.5 Rate $= k[I^-][S_2O_8^{2-}]$
14.6 Rate $= k[N_2O_5]$
14.7 Rate $= k[S_2O_3^{2-}][H^+]$
14.8 $k = 0.020\ s^{-1}$; 2.0×10^{-4} mol/L
14.9 $k = 0.039\ min^{-1}$
14.10 $k = -m = 0.0373\ min^{-1}$
14.11 Fraction remaining $= 0.19$
14.12 $2.92 \times 10^{-3}\ min^{-1}$
14.13 $1.23 \times 10^2\ min^{-1}$
14.14 91.4 s
14.15 2.98×10^{-3} mol/L
14.16 3.30 s; 6.60 s
14.18 (a) rate $= k[N_2O][O]$; (b) rate $= k[Br_2]$
14.19 $NO_2(g) + F_2(g) \rightarrow NO_2F(g) + F(g)$ (slow)
$NO_2(g) + F(g) \rightarrow NO_2F(g)$ (fast)
14.21 +10 kJ, endothermic
14.22 $0.080\ s^{-1}$
14.23 1.5
14.24 1.14

CHAPTER 15

15.1 (a) $Q_c = \dfrac{[Cl]^2}{[Cl_2]}$
15.3 $K_c = 1.9$
15.4 Forward
15.5 $[Br] = 0.0162$ mol/L
15.6 $[O_2] = 3.04 \times 10^{-9}$ mol/L
15.7 $x = 3.52 \times 10^{-3}$
15.8 (a) $K_c = \dfrac{[HCN][OH^-]}{[CN^-]}$ (b) $K_c = \dfrac{[Be(OH)^+][H^+]}{[Be^{2+}]}$
15.9 0.0013 mol/L
15.10 (a) $P_{PCl_5} = 0.730$ atm; $P_{NO} = 0.270$ atm; $P_{Cl_2} = 0.135$ atm; (b) $K_p = 1.85 \times 10^{-2}$
15.11 1.55×10^{-2} atm
15.12 (a) $\Delta n = +1$; (b) $K_c = 3.50 \times 10^{-7}$
15.13 (a) $K_p = 2.8 \times 10^{-8}$; (b) $K_p = 3.6 \times 10^7$
15.14 $K_p = 1.62 \times 10^{12}$
15.15 (a) $K_c = \dfrac{[NOBr]}{[NO]}$; (b) $K_c = \dfrac{[H^+]^2}{[Sb^{3+}][Cl^-]}$
15.16 $P_{S_2} = 0.152$ atm; $P_{CS_2} = 0.85$ atm
15.17 (a) and (b) more intense.
15.18 The fraction of NO_2 will increase.
15.19 (a) H_2 and F_2 increase, HF decreases; (b) K_p decreases.

CHAPTER 16

16.1 (a) $CH_3COOH(aq) + H_2O(l) \rightleftharpoons H_3O^+(aq) + CH_3COO^-(aq)$
(b) $H_3O^+(aq) + OH^-(aq) \rightleftharpoons 2H_2O(l)$
16.2 $2CH_3NH_2(l) \rightleftharpoons CH_3NH_3^+ + CH_3NH^-$
16.3 (a) $H_2PO_4^-$; (b) O^{2-}
16.4 $HC_2O_4^-$(acid) and $C_2O_4^{2-}$(base); HS^-(base) and H_2S(acid)
16.5 (a) $HClO_4 + H_2O(l) \rightarrow H_3O^+(aq) + ClO_4^-(aq)$
(b) $NH_2^- + H_2O(l) \rightarrow NH_3(g) + OH^-(aq)$
16.6 (a) HS^-; (b) $CH_3NH_3^+$
16.7 (a) $HF(aq) + OH^-(aq) \rightarrow F^-(aq) + H_2O(l)$
(b) $CN^-(aq) + H_3O^+(aq) \rightarrow HCN(aq) + H_2O(l)$
16.8 (a) $HF(aq) + CN^-(aq) \rightleftharpoons F^-(aq) + HCN(aq)$
(b) $NH_3(aq) + CH_3COOH(aq) \rightleftharpoons NH_4^+(aq) + CH_3COO^-(aq)$
16.9 $Cu(H_2O)_4^{2+}(aq) + H_2O(l) \rightleftharpoons Cu(H_2O)_3(OH)^+(aq) + H_3O^+(aq)$
16.10 $H_3PO_4(aq) + H_2O(l) \rightleftharpoons H_3O^+(aq) + H_2PO_4^-(aq)$
$H_2PO_4^-(aq) + H_2O(l) \rightleftharpoons H_3O^+(aq) + HPO_4^{2-}(aq)$
$HPO_4^{2-}(aq) + H_2O(l) \rightleftharpoons H_3O^+(aq) + PO_4^{3-}(aq)$
16.11 $Zn(H_2O)_4^{2+}(aq) + 2OH^-(aq) \rightarrow Zn(H_2O)_2(OH)_2(s) + 2H_2O(l)$
16.12 $[OH^-] = 2.0 \times 10^{-3}\ M$; $[H^+] = 5.0 \times 10^{-12}\ M$
16.13 pH $= 1.82$
16.14 Basic
16.15 pH $= 13.70$; pOH $= 0.30$
16.16 pH $= 7.00$
16.17 $[H^+] = 5 \times 10^{-5}\ M$
16.19 (a) H_2S; (b) O^{2-}
16.20 (a) $CHCl_2COOH$; (b) OI^-
16.21 (a) HBr; (b) OH^-
16.22 $Cr(H_2O)_6^{3+}$
16.23 The acid is $AlCl_3$; the base is Cl^-.

CHAPTER 17

17.1 Hydrofluoric acid
17.2 1.8×10^{-4}
17.3 $[H^+] = [CN^-] = 5.0 \times 10^{-6}\ M$
17.4 (a) 0.100 M: $1.33 \times 10^{-3}\ M$;
$1.00 \times 10^{-4}\ M$: $4.20 \times 10^{-5}\ M$
(b) 0.100 M: $1.32 \times 10^{-3}\ M$;
$1.00 \times 10^{-4}\ M$: $3.41 \times 10^{-5}\ M$
17.5 Trimethylamine
17.6 pH $= 8.90$
17.7 5.65×10^{-10}
17.8 0.940%
17.9 (a) decrease; (b) no change; (c) and (d) increase
17.10 $H_2AsO_4^-$

17.11 (a) pH = 1.46; (b) 2.0×10^{-6} *M*

17.12 $S^{2-}(aq) + H_2O(l) \rightleftharpoons HS^-(aq) + OH^-(aq)$

$$K_b = \frac{[HS^-][OH^-]}{[S^{2-}]}$$

17.13 $CH_3NH_3^+(aq) \rightleftharpoons CH_3NH_2(aq) + H^+(aq)$

$$K_a = \frac{[CH_3NH_2][H^+]}{[CH_3NH_3^+]}$$

17.14 (a) basic; (b) acidic; (c) neutral; (d) slightly acidic

17.15 pH = 9.18

17.16 Basic

17.17 (a) Sodium benzoate (C_6H_5COONa) raises pH.
(b) Aniline hydrochloride ($C_6H_5NH_3Cl$) lowers pH.

17.18 $[OH^-] = 3.70 \times 10^{-4}$ *M*

17.19 pH = 5.69

17.20 pH = 1.20

17.21 pH = 12.60

17.22 $\frac{[HCOO^-]}{[HCOOH]} = 1.7$

17.23 pH = 4.28

17.24 One choice would be formic acid plus sodium formate.

17.25 0.032 mol

17.26 pH = 3.12

17.27 pH = 2.69

17.28 pH = 5.356

17.30 One choice is congo red.

CHAPTER 18

18.1 (a) $K_{sp} = [Ca^{2+}][C_2O_4^{2-}]$; (b) $K_{sp} = [Bi^{3+}]^2[S^{2-}]^3$

18.2 $PbCl_2$

18.3 3.8×10^{-11}

18.4 0.6 g

18.5 (a) less soluble; (b) 0.025 *M*

18.6 No precipitate

18.7 $[Cl^-] = 0.13$ *M*

18.8 $[OH^-] = 8.1 \times 10^{-3}$ *M*

18.9 $[Mg^{2+}] = 1.1 \times 10^{-7}$ *M*, yes

18.10 pH = 3.0 or less

18.11 $[NH_4^+] = 0.15$ *M* or greater

18.12 Only (b)

18.13 $Al_2O_3(s) + 6H^+(aq) \rightarrow 2Al^{3+}(aq) + 3H_2O(l)$
$Al_2O_3(s) + 2OH^-(aq) + 3H_2O(l) \rightarrow 2Al(OH)_4^-(aq)$

18.14 (a) $PbCO_3(s) + 2HNO_3(aq) \rightarrow Pb(NO_3)_2(aq) + H_2O(l) + CO_2(g)$
(b) and (c) do not dissolve.
(d) $PbF_2(s) + 2HNO_3(aq) \rightarrow Pb(NO_3)_2(aq) + 2HF(aq)$

18.15 (a) pH 1.4 or lower; (b) increase the pH

18.16 No

18.17 HgI_4^{2-} is the most stable; $Fe(SCN)^{2+}$ the least stable.

18.18 $[Zn^{2+}] = 6.9 \times 10^{-19}$ *M*

18.19 AgBr will precipitate.

18.20 9.7×10^{-2} mol/L

SURVEY 3

S3.1 Approximately 95°C

S3.2 $Fe_2O_3(s) + 3H_2SO_4(aq) \rightarrow Fe_2(SO_4)_3(aq) + 3H_2O(l)$

S3.3 (a) $SO_4^{2-}(aq) + 10H^+(aq) + 8e^- \rightarrow H_2S(g) + 4H_2O(l)$
(b) $SO_4^{2-}(aq) + 8I^-(aq) + 10H^+(aq) \rightarrow H_2S(g) + 4I_2(s) + 4H_2O(l)$

S3.4 (a) and (b) VSEPR predicts < 109°; the observed angles are larger.

S3.5 $H_3PO_4(aq) + 3NaOH(aq) \rightarrow Na_3PO_4(aq) + 3H_2O(l)$

S3.6 $H_2SO_4(aq) + NaHCO_3(aq) \rightarrow NaHSO_4(aq) + CO_2(g) + H_2O(l)$

S3.7 (a) $[:C{\equiv}C:]^{2-}$; (b) $\ddot{S}{=}C{=}\ddot{S}$; (c) $H{-}C{\equiv}N:$

S3.8 −1

CHAPTER 19

19.1 (a) $\Delta G = 0$; (b) no

19.2 (a) 1636 kJ; (b) 141 kJ evolved

19.3 $\Delta G° = -23.8$ kJ/mol

19.4 (a) $\Delta G° = 8.6$ kJ; (b) water vapor condenses

19.5 $\Delta G° = +40.94$ kJ; $K_p = 7 \times 10^{-8}$

19.6 (a) $\Delta G = +2.15$ kJ/mol; (b) condensation is favored

19.7 (a) increase; (b) decrease; (c) increase

19.8 $\Delta S° = +109$ J/K

19.9 $\Delta S_f° = -99.5$ J/K

19.10 $\Delta S° = +3.363$ J/K; diamond is more ordered

19.11 $\Delta G° = -2872$ kJ/mol

19.12 992 K

CHAPTER 20

20.1

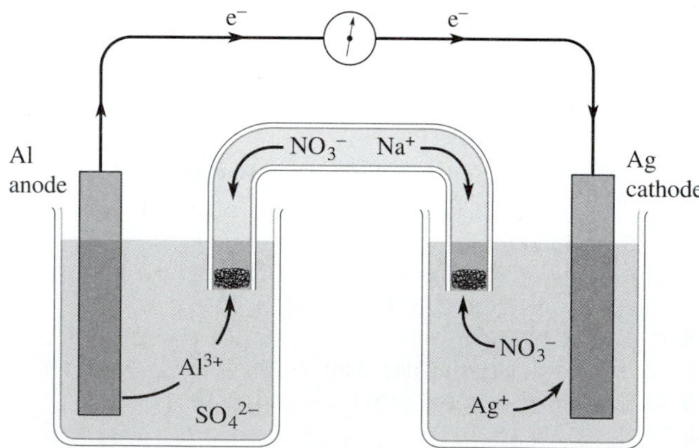

Anode: $Al(s) \rightarrow Al^{3+}(aq) + 3e^-$
Cathode: $Ag^+(aq) + e^- \rightarrow Ag(s)$
Net: $Al(s) + 3Ag^+(aq) \rightarrow Al^{3+}(aq) + 3Ag(s)$

20.2 $Zn(s)|\ Zn^{2+}(aq)\|\ Br^-(aq)|\ Br_2(l)|\ C(gr)$
20.3 $Zn(s)|\ Zn^{2+}(1\ M)\|\ Br^-(1\ M)|\ Br_2(l)|\ C(gr)$
20.4 8.10×10^{-5} J
20.5 $E^\circ = 2.01$ V
20.6 $E^\circ = +2.869$ V
20.7 (a) I^- (b) Co^{2+}
20.8 (a) no; (b) yes
20.9 (a) $MnO_4^-(aq) + 8H^+(aq) + 5e^- \rightarrow Mn^{2+}(aq) + 4H_2O(l)$

$2I^-(aq) \rightarrow I_2(s) + 2e^-$

(b) $E^\circ = 0.977$ V
20.10 $\Delta G^\circ = -88.8$ kJ
20.11 0.39 V
20.12 −0.414 V
20.13 (a) $E^\circ = 0.358$ V; (b) $\Delta G^\circ = -34.5$ kJ
20.14 $K = 3.5 \times 10^8$
20.15 Anode: $2Cl^- \rightarrow Cl_2(g) + 2e^-$

Cathode: $Mg^{2+} + 2e^- \rightarrow Mg(l)$
20.16 3.40 g
20.17 1.03×10^3 C
20.18 834 C
20.19 0.104 L/min

CHAPTER 21

21.1 VCl_3
21.2 (a) In_2O_3; (b) $Fe(OH)_2$
21.3 Zero
21.4 $[Co(NH_3)_4Cl_2]Cl$
21.5 (a) $Ni(CO)_4$: Ni bonded to four CO (tetrahedral) (b) $Ni(CN)_4$: Ni bonded to four CN (square planar)

(tetrahedral) (square planar)

21.6 Three
21.7 (a) Sodium hexanitrocobaltate(III); (b) $Na_2[Pb(EDTA)]$
21.8 (a) and (c)
21.9 (a) has enantiomers; (b) does not
21.10 $Ni(en)_3^{2+}$
21.11 +3

SURVEY 4

S4.1 $Ni_2S_3(s) + 4O_2(g) \rightarrow 2NiO(s) + 3SO_2(g)$
S4.2 Oxygen is the oxidizing agent; gold is the reducing agent.
S4.3 $Al(OH)_3(s)$ and $H_2S(g)$
S4.4 (a) yes; (b) no
S4.5 767°C; cadmium
S4.6 6.9×10^{-7} mol/L
S4.7 $\Delta G^\circ = -34.4$ kJ
S4.8 $Cu(s) + 2NO_3^-(aq) + 4H^+(aq) \rightarrow Cu^{2+}(aq) + 2NO_2(g) + 2H_2O(l)$
S4.9 (a) 7; (b) 6; (c) 8

CHAPTER 22

22.1 (a) mass number = 183; atomic number = 78; ${}^{183}_{78}Pt$

(b) ${}^{28}_{13}Al \rightarrow {}^{28}_{14}Si + {}^{0}_{-1}e$
22.2 ${}^{96}_{42}Mo + {}^{2}_{1}H \rightarrow {}^{97}_{43}Tc + {}^{1}_{0}n$

${}^{209}_{83}Bi + {}^{4}_{2}He \rightarrow {}^{211}_{85}At + 2{}^{1}_{0}n$
22.3 345 h
22.4 17,200 y
22.5 3.93×10^9 y
22.6 Beta emission
22.7 1.04×10^{-4} u
22.8 (a) 0.54365 u; (b) 506.39 MeV; (c) 8.731 MeV/nucleon

CHAPTER 23

23.1 4-Ethyl-2-methylheptane

23.2
```
   H  Cl  Cl  H  H
   |  |   |   |  |
H—C—C—C—C—C—H      2,3-dichloropentane
   |  |   |   |  |
   H  H   H   H  H
```

23.3
```
   H  Cl  Cl  H
   |  |   |   |
H—C—C—C—C—H      2,3-dichlorobutane
   |  |   |   |
   H  H   H   H
```

23.4 (a) Ketone; (b) ether; (c) aldehyde; (d) phenol
23.5 (a) 1:2; (b) 1:2; (c) 3:4; glyceraldehyde to pyruvic acid, and pyruvic acid to carbon dioxide are oxidations.

23.6 $CH_3CH_2CH_2CH_2OH + HO{-}C({=}O){-}CH_2CH_3 \rightarrow CH_3CH_2CH_2CH_2{-}O{-}C({=}O){-}CH_2CH_3$

butyl propionate

23.7 (b) the middle carbon atom

23.8
```
      OH                                            CH3
      |                                             |
   (benzene ring)                 (benzene ring)    S
      |                                 |           |
      |                                 |           CH2
      |                                 |           |
     CH2 O      H  O     H  O          CH2 O       CH2 O
      |  ||     |  ||    |  ||          |  ||       |  ||
H2N-C—C-N-C-C-N-C-C-N-C—C-N-C—C-OH
      |       |  |     |  |       |  |       |  |
      H       H  H     H  H       H  H       H  H
```

A P P E N D I X D

ANSWERS TO SELECTED FINAL EXERCISES

CHAPTER 1

1.7 (a) m/s; (b) m^2; (c) m^3; (d) g/m^3; (e) $kg \cdot m/s^2$; (f) $kg \cdot m^2/s^2$
1.9 (a) m, m^3; (b) 1 m^3 = 1000 L
1.11 (a) 2.2×10^2 cm; (b) 2.2×10^{-3} km; (c) 2.2×10^9 nm
1.13 (a) 1 cm = 10 mm; (b) 1 cm^2 = 100 mm^2
(c) 1 cm^3 = 1000 mm^3
1.15 2.6×10^2 m^2
1.17 (a) 134 cm^3; (b) 0.134 dm^3; (c) 1.34×10^{-4} m^3
1.19 6.88×10^6 pm^3
1.23 (a) and (b) random; (c) systematic
1.25 (b), (c), and (f)
1.27 (a) 5.27×10^6; (b) 4.97×10^6; (c) 7.00×10^6
1.29 (a) 5; (b) 2; (c) 2; (d) 5; (e) 1; (f) 3
1.31 (a) 60; (b) 3.7; (c) 8.6×10^6; (d) 2026
1.33 (a) 3.88×10^5; (b) 5.486×10^{-4}; (c) 1.0078;
(d) 3.4×10^{47}
1.37 (a) 88 km/h; (b) 25 m/s
1.39 14 km/L
1.41 0.18 L/h
1.45 Double- and single-pan balances measure mass. A spring scale measures weight.
1.47 (a) 1.7 m; (b) 65.8 kg; (c) 645 N
1.49 3145 g
1.51 42 g
1.53 39 g
1.55 0.260 g
1.57 8.47 g/mL
1.59 (a) 1.00×10^3 g; (b) 917 g; (c) 0.596 g
1.61 (a) 13.6 kg/dm^3; (b) 849 lb/ft^3
1.63 0.70 L or larger
1.65 (a) 2.06×10^3 g; (b) 2.06 L
1.67 0.7907
1.73 (a) and (c) increase; (b) decrease
1.75 (a) 75.0 J; (b) 17.9 cal
1.77 (a) 2.2×10^2 J; (b) 6.64×10^{-21} J
1.79 4.9×10^5 m/s
1.81 5.40×10^2 cal
1.83 3222°F
1.89 Atoms
1.91 (a) 5.83×10^{-2} mm; (b) 58.3 μm
1.93 (a) 24 mg phenobarbital, 390 mg theophylline, 72 mg ephedrine hydrochloride; (b) 0.37 grains; (c) 1.5 days
1.95 8.0×10^{-2} g/cm^3
1.97 216 g
1.99 28.3%

CHAPTER 2

2.5 (a) increases; (b) remains same
2.9 6.6448×10^{-16} g
2.15 18 groups, 7 periods
2.17 (a) Fe; Hg; Na; N; P; Ca; S; Cu; (b) lead, silver, potassium, gold, neon, zinc, nickel
2.19 (a) hydrogen, helium, nitrogen, oxygen, fluorine, neon, chlorine, argon, krypton, xenon, radon; (b) bromine, mercury
2.21 Zn: A = 64, 30 p, 30 e
I: Z = 53, A = 127, 53 e
Eu: Z = 63, 90 n, 63 e
2.23 (a) 82 p, 126 n, 80 e; (b) 7 p, 7 n, 10 e
2.25 $^{27}_{13}Al^{3+}$
2.27 Chlorine-37: 17 p, 20 n, 17 e; argon-37: 18 p, 19 n, 18 e; sulfur-37: 16 p, 21 n, 16 e
2.29 (a) 12 u, exact; (b) 1 u; (c) 1 u; (d) 4 u
2.35 6 neutrons, 5 neutrons
2.37 2.79 g; both are weighted.
2.39 24.31 u
2.41 75.772% chlorine-35, 24.228% chlorine-37
2.43 64.91 u
2.47 (a) 107.87 u/atom, 107.87 g/mol; (b) 200.59 u/atom, 200.59 g/mol (c) 226.0 u/atom, 226.0 g/mol
2.49 (a) 68.8 g; (b) 4.480×10^{-23} g

2.51 (a) 6.02×10^{22}; (b) 3.06×10^{24}; (c) 1.02×10^{25}
2.53 (a), (b), and (c) 0.0141 moles
2.61 (a) diamond and graphite; (b) oxygen and ozone; (c) white phosphorus and red phosphorus
2.65 (a) 1 Ca, 2 Cl; (b) 1 Mg, 2 O, 2 H; (c) 1 Al, 1 N, 4 H, 2 S, 8 O; (d) 4 C, 10 H
2.67 (a) H_2; (b) O_2; (c) P_4; (d) Ar; (e) Cu; (f) Br_2 (a), (b), (c), and (f) are molecular.
2.69 H_2, He, N_2, O_2, F_2, Ne, Cl_2, Ar, Kr, Xe, Rn
2.71 (a) C_2H_5OH; (b) CH_3COOH; (c) H_2O; (d) H_2O_2; (e) CO; (f) NH_3
2.73 (a) true; (b), (c), and (d) false
2.75 (a) 2.41×10^{24} atoms; (b) 4.00 mol atoms
2.77 (a) 7.50 mol of H atoms; (b) 4.52×10^{24} atoms
2.79 Fe_3O_4
2.81 (a) 180.16 g/mol; (b) 80.07 g/mol
2.83 (a) 40.10 g/mol; (b) 101.96 g/mol; (c) 101.96 g/mol; (d) 60.09 g/mol
2.85 (a) 60.2 g; (b) 658 g
2.87 (a) 1.11 mol; (b) 6.68×10^{23} molecules
2.89 (a) 2.50 mol of each ion; (b) 1.51×10^{24} ions of each
2.91 80.0 g
2.95 (a) CsBr; (b) $SrBr_2$
2.97 (a) $KMnO_4$; (b) $(NH_4)_2CO_3$; (c) Hg_2Cl_2; (d) $Hg(NO_3)_2$; (e) NaSCN; (f) $Mg(ClO_4)_2$
2.99 (a) lead ion; (b) tin(II) ion; (c) iron(III) ion; (d) ammonium ion; (e) copper(II) ion; (f) mercury(I) ion
2.101 (a) potassium dichromate; (b) iron(II) sulfate; (c) potassium permanganate; (d) mercury(I) chloride; (e) potassium oxalate; (f) sodium chlorate
2.103 (a) hydrogen chloride; (b) hydrogen sulfide; (c) water; (d) dinitrogen pentoxide; (e) iodine monobromide; (f) disulfur difluoride
2.109 They all have the same number of protons and the same number of electrons. They may have different numbers of neutrons.
2.113 $10^{-13}\%$
2.115 67.91 u; zinc-68
2.117 (a) 9.65×10^4 C; (b) 1.93×10^5 C
2.121 0.306 mol
2.123 88.9 mL

CHAPTER 3

3.1 (a) and (d) physical; (b) and (c) chemical
3.3 (a) and (d) physical; (b) and (c) chemical
3.7 (a) 0.52 g; (b) 92.6% Hg, 7.4% O
3.9 70.0% Fe, 30.0% O
3.11 CuS contains exactly twice the mass of copper per gram of sulfur as CuS does.
3.13 (a) 26.68% C, 71.08% O, 2.239% H; (b) 75.75% Sn, 24.25% F; (c) 62.04% C, 10.41% H, 27.55% O; (d) 12.14% C, 16.18% O, 71.68% Cl
3.17 (a) NH_2; (b) CH; (c) S
3.19 B_{12}
3.21 (a) As_2S_3; (b) no
3.23 NH_3
3.25 (a) KClO; (b) $KClO_3$; (c) $KClO_4$
3.27 C_4H_8
3.29 $C_6H_{10}O_7$
3.31 (a) Pb_3O_4; (b) no
3.33 (a) 92.24% C, 7.74% H; (b) CH
3.35 $C_2H_8N_2$
3.39 (a) sugar → carbon + water
(b) hydrochloric acid + sodium hydroxide → sodium chloride + water
(c) carbon dioxide + water → glucose + oxygen
3.41 (a) $2C_6H_6(l) + 15O_2(g) \rightarrow 12CO_2(g) + 6H_2O(g)$
(b) $2ZnS(s) + 3O_2(g) \rightarrow 2ZnO(s) + 2SO_2(g)$
(c) $4FeO(s) + O_2(g) \rightarrow 2Fe_2O_3(s)$
(d) $CS_2(l) + 3O_2(g) \rightarrow CO_2(g) + 2SO_2(g)$
3.43 (a) $P_4O_{10}(s) + 6H_2O(l) \rightarrow 4H_3PO_4(aq)$
(b) $XeF_6(s) + 3H_2O(l) \rightarrow XeO_3(s) + 6HF(g)$
(c) $2K(s) + 2H_2O(l) \rightarrow 2KOH(aq) + H_2(g)$
(d) $PCl_5(s) + 4H_2O(l) \rightarrow H_3PO_4(aq) + 5HCl(aq)$
(e) $PBr_3(l) + 3H_2O(l) \rightarrow H_3PO_3(aq) + 3HBr(aq)$
3.45 (a) $C_6H_{12}O_6(aq) \rightarrow 2C_2H_5OH(aq) + 2CO_2(g)$
(b) $6CO_2(g) + 6H_2O(l) \rightarrow C_6H_{12}O_6(aq) + 6O_2(g)$
3.49 CO
3.51 (a) and (c) displacement, (b) decomposition, (d) combination
3.53 (a) $2C_8H_{18}(l) + 25O_2(g) \rightarrow 16CO_2(g) + 18H_2O(g)$
(b) $C_3H_8O(l) + 5O_2(g) \rightarrow 3CO_2(g) + 4H_2O(g)$
(c) $C_6H_{12}O_6(s) + 6O_2(g) \rightarrow 6CO_2(g) + 6H_2O(g)$
(d) $C_3H_6O(l) + 4O_2(g) \rightarrow 3CO_2(g) + 3H_2O(g)$
3.55 (a) $C(s) + O_2(g) \rightarrow CO_2(g)$
(b) $2C(s) + O_2(g) \rightarrow 2CO(g)$
3.57 (a) $GeO_2(s) + 2C(s) \rightarrow Ge(s) + 2CO(g)$
(b) Carbon is oxidized; GeO_2 is reduced.
3.59 (a) $2C(s) + O_2(g) \rightarrow 2CO(g)$
$Fe_2O_3(s) + 3CO(g) \rightarrow 2Fe(l) + 3CO_2(g)$
(b) $SnO_2(s) + 2C(s) \rightarrow Sn(l) + 2CO(g)$
(c) $2C(s) + O_2(g) \rightarrow 2CO(g)$
$Fe_3O_4(s) + 4CO(g) \rightarrow 3Fe(l) + 4CO_2(g)$
(d) $2ZnS(s) + 3O_2(g) \rightarrow 2ZnO(s) + 2SO_2(g)$
$ZnO(s) + C(s) \rightarrow Zn(g) + CO(g)$
3.61 (a) $2KClO_3(s) \rightarrow 2KCl(s) + 3O_2(g)$
(b) 0.50 mol KCl, 0.75 mol O_2
(c) 37 g KCl, 24 g O_2
3.63 176 g O_2
3.65 (a) 0.499 mol CO_2; (b) 9.00 g H_2O
3.67 125 g
3.69 214 g
3.71 89.97%
3.73 (a) Candle is limiting; oxygen is excess. (b) Oxygen is limiting; candle is excess.

3.75 (a) KOH; (b) 1.33 mol; (c) 2.00 mol Cl_2
3.77 41.4 g
3.79 0.499 mol SO_2, 0.499 mol H_2O, 5.50 mol O_2, no H_2S
3.81 (a) 85.6 kg; (b) 64.7%
3.83 34.2 g
3.85 (a) 161 g; (b) 142 g
3.93 (a) HgO; (b) H_2O
3.95 4
3.97 CH_3
3.99 Same
3.101 (a) 2.66 kg; (b) 2.40 kg
3.103 22.2 g
3.105 9.5×10^2 kg
3.107 Hematite
3.109 17.6 g

CHAPTER 4

4.1 (a), (b), (d), (f), homogeneous; (c), (e) heterogeneous
4.5 (a) The crystal will dissolve. (b) No change; the crystal will not dissolve. (c) Excess solute will crystallize.
4.9 (a) $K_2CO_3(s) \rightarrow 2K^+(aq) + CO_3^{2-}(aq)$
(b) $NaHCO_3(s) \rightarrow Na^+(aq) + HCO_3^-(aq)$
(c) $FeSO_4(s) \rightarrow Fe^{2+}(aq) + SO_4^{2-}(aq)$
(d) $Ba(OH)_2(s) \rightarrow Ba^{2+}(aq) + 2OH^-(aq)$
(e) $CaCl_2(s) \rightarrow Ca^{2+}(aq) + 2Cl^-(aq)$
(f) $Na_3PO_4(s) \rightarrow 3Na^+(aq) + PO_4^{3-}(aq)$
4.13 (a), (b), (e) soluble; (c), (d), (f) sparingly soluble
4.15 Yes
4.17 3.578%
4.19 13.14%
4.21 Acids produce $H^+(aq)$; bases produce $OH^-(aq)$; neutralization produces H_2O.
4.23 (a) $H_2SO_3(aq) \rightleftharpoons H^+(aq) + HSO_3^-(aq)$
$HSO_3^-(aq) \rightleftharpoons H^+(aq) + SO_3^{2-}(aq)$
(b) $H_2CO_3(aq) \rightleftharpoons H^+(aq) + HCO_3^-(aq)$
$HCO_3^-(aq) \rightleftharpoons H^+(aq) + CO_3^{2-}(aq)$
(c) $HCN(aq) \rightleftharpoons H^+(aq) + CN^-(aq)$
4.25 (a) $Ba(OH)_2(aq) + 2HNO_3(aq) \rightarrow Ba(NO_3)_2(aq) + 2H_2O(l)$
(b) $3LiOH(aq) + H_3PO_4(aq) \rightarrow Li_3PO_4(aq) + 3H_2O(l)$
(c) $Cd(OH)_2(s) + H_2S(aq) \rightarrow CdS(s) + 2H_2O(l)$
(d) $2Fe(OH)_3(s) + 3H_2SO_4(aq) \rightarrow Fe_2(SO_4)_3(aq) + 6H_2O(l)$
4.27 One reaction each:
(a) $CaCO_3(s) + 2HCl(aq) \rightarrow CaCl_2(aq) + H_2O(l) + CO_2(g)$
$CaCO_3(s) + 2H^+(aq) \rightarrow Ca^{2+}(aq) + H_2O(l) + CO_2(g)$
(b) $Na_2SO_3(aq) + 2HCl(aq) \rightarrow 2NaCl(aq) + H_2O(l) + SO_2(g)$
$SO_3^{2-}(aq) + 2H^+(aq) \rightarrow H_2O(l) + SO_2(g)$
(c) $ZnS(s) + 2HCl(aq) \rightarrow ZnCl_2(aq) + H_2S(g)$
$ZnS(s) + 2H^+(aq) \rightarrow Zn^{2+}(aq) + H_2S(g)$
(d) $NaCN(aq) + HCl(aq) \rightarrow NaCl(aq) + HCN(g)$
$CN^-(aq) + H^+(aq) \rightarrow HCN(g)$
4.29 $NH_4^-(aq) + OH^-(aq) \rightarrow NH_3(g) + H_2O(l)$
4.31 (a) Iron is more active than hydrogen. (b) Aluminum is more active than copper.
4.33 (a) $2Cr(s) + 6HCl(aq) \rightarrow 3H_2(g) + 2CrCl_3(aq)$
(b) $2K(s) + 2H_2O(l) \rightarrow H_2(g) + 2KOH(aq)$
(c) $Zn(s) + CuSO_4(aq) \rightarrow ZnSO_4(aq) + Cu(s)$
(d) $Mg(s) + 2H_2O(g) \rightarrow H_2(g) + Mg(OH)_2(s)$
(e) $2Al(s) + 3H_2SO_4(aq) \rightarrow 3H_2(g) + Al_2(SO_4)_3(aq)$
(f) $Fe(s) + 2AgNO_3(aq) \rightarrow 2Ag(s) + Fe(NO_3)_2(aq)$
4.35 (a) $2Al(s) + 3Pb(NO_3)_2(aq) \rightarrow 2Al(NO_3)_3(aq) + 3Pb(s)$
$2Al(s) + 3Pb^{2+}(aq) \rightarrow 2Al^{3+}(aq) + 3Pb(s)$
(b) No
4.37 (a) 5.0×10^{-4} parts per thousand; (b) 0.50 ppm
4.39 3.0×10^2 mg
4.41 291 g
4.43 (a) 833 g; (b) 733 g; (c) 793 mL
4.45 Moles of solute
4.47 (a) 3.08 *M*; (b) 0.204 *M*; (c) 4.49×10^{-2} *M*; (d) 9.42×10^{-2} *M*
4.49 17.4 *M*
4.51 (a) 4.88 mol; (b) 2.70×10^{-4} mol; (c) 0.162 mol
4.53 (a) 3.50×10^{-2} mol; (b) 3.50×10^{-5} mol; (c) 0.0195 mol
4.57 (a) 2.0×10^3 mL; (b) 55 mL
4.59 0.21 mg
4.61 (a) 5.00 *M;* (b) 2.5×10^2 mmol
4.63 (a) 3.0 *M* K^+, 3.0 *M* NO_3^-; (b) 0.55 *M* Ba^{2+}, 1.1 *M* Cl^-; (c) 0.75 *M* Na^+, 0.75 *M* HCO_3^-; (d) 1.5 *M* Al^{3+}, 2.2 *M* SO_4^{2-} (rounded from 2.25 *M*); (e) 0.75 *M* Na^+, 0.25 *M* PO_4^{3-}; (f) 1.0 *M* Hg_2^{2+}, 2.0 *M* NO_3^-
4.65 (a) 45.0 g; (b) 1.50 *M*
4.67 (a) 0.137 *M*; (b) 0.0857 *M*; (c) 0.249 *M*
4.71 (a) 30 mL; (b) 50 mL; (c) 2.3×10^2 mL
4.73 (a) 12.1 *M*; (b) 10.3 mL
4.79 (a) 5.000×10^{-2} *M*; (b) 0.1000 *M*; (c) 2.500×10^{-2} *M*; (d) 0.1000 *M*
4.81 8.778×10^{-2} *M*
4.83 9.9%
4.85 0.6257 *M*
4.87 (a) 5.9 g; (b) no
4.89 59.9%
4.91 The components of a solution are dispersed as individual molecules or ions. A solution may be colored, but not cloudy.
4.93 460 g solute will crystallize.
4.95 Yes, if sparingly soluble
4.97 Zinc
4.99 True
4.101 (a) $NaHCO_3(s) + HCl(aq) \rightarrow NaCl(aq) + H_2O(l) + CO_2(g)$
$NaHCO_3(s) + H^+(aq) \rightarrow Na^+(aq) + H_2O(l) + CO_2(g)$

(b) $CaCO_3(s) + 2HCl(aq) \rightarrow CaCl_2(aq) + H_2O(l) + CO_2(g)$
$CaCO_3(s) + 2H^+(aq) \rightarrow Ca^{2+}(aq) + H_2O(l) + CO_2(g)$
(c) $Mg(OH)_2(s) + 2HCl(aq) \rightarrow MgCl_2(aq) + 2H_2O(l)$
$Mg(OH)_2(s) + 2H^+(aq) \rightarrow Mg^{2+}(aq) + 2H_2O(l)$

4.103 (a) 5.30 mmol/L; (b) 5.30×10^3 μmol/L
4.105 109 mg/100 mL
4.107 (a) 5×10^{12} g; (b) at $400 per troy ounce, 6×10^{13} dollars
4.109 28.51%
4.113 (a) 25 proof; (b) about 5
4.115 1.73 g KNO_3, 1.45 g $NaNO_3$, 1.27 g KCl
4.117 4.20 g $BaSO_4$; 8.5×10^{-3} M Ba^{2+}; 0.197 M Cl^-; 0.180 M H^+

CHAPTER 5

5.7 (a) greater than; (b) equal to
5.11 (a) 0.980 atm; (b) 9.93×10^4 Pa; (c) 0.993 bar
5.13 (a) 1.97 atm; (b) 1.50×10^3 torr
5.15 4.6 mm
5.17 (a) 591 torr; (b) open side 16.5 cm; closed side 12.3 cm
5.19 P, V, T, and number of moles
5.21 (a) and (b) decreases; (c) increases
5.23 (a) 37.5 mL; (b) 71.2 mL; (c) 50.4 mL; (d) 38.0 mL
5.25 29 mL
5.27 (a) 310 K; (b) 273 K; (c) 373 K; (d) 298 K
5.29 (a) 0.657 L; (b) 0.439 L; (c) 0.352 L; (d) 0.660 L
5.31 (a) 2.7 L; (b) 2.3 L
5.33 (a) 2.5 L; (b) 0.036 mol
5.35 (a) 1.00 g/L; (b) 1.56 g/L
5.43 (a) 990 torr; (b) 258 torr; (c) 623 torr; (d) 1.01×10^3 torr
5.45 962 kPa
5.47 (a) 7.00 L; (b) 6.41 L
5.49 (a) 0.14 mol; (b) 0.12 mol
5.51 (a) 131 g/mol; (b) xenon
5.53 4.00 g/mol
5.55 (a) 6.59 g/L; (b) 1.90 g/L
5.57 (a) 119 g/mol; (b) $CHCl_3$
5.59 5.24×10^3 mL
5.61 8.77 g
5.63 (a) 4.28×10^3 mL; (b) 6.42×10^3 mL
5.65 30 L N_2, 90 L H_2
5.67 Yes
5.73 (a) $X_{He} = 0.316$, $X_{CO_2} = 0.684$; (b) 31.6 mol% He, 68.4 mol% CO_2; (c) 228 torr He, 492 torr CO_2
5.75 (a) 1.135 atm; (b) $X_{NOCl} = 0.643$, $X_{NO} = 0.238$, $X_{Cl_2} = 0.119$
5.77 (a) 0.26 atm H_2, 0.26 atm I_2, 1.8 atm HI; (b) 2.4 atm
5.79 (a) 716 torr; (b) 96.8 mol %
5.81 0.827 g
5.83 O_2, 1.17 times more rapidly
5.85 42.0 g/mol
5.87 (a), (b) greater than
5.91 (a) 425 m/s, 1.82×10^5 m²/s²; 427 m/s
5.95 Increase, by a factor of 1.414
5.103 (a) 7.77 atm; (b) 7.59 atm
5.107 29.0 g/mol
5.109 0.99 torr
5.111 22.4 L
5.113 8 cm
5.115 C_2H_4
5.117 5.22 kg
5.119 0.575 L
5.123 0.40 atm cyclopropane, 0.20 atm O_2, 0.40 atm He; (b) 25 g/mol
5.125 29 g/mol; C_2H_6
5.127 (a) 1.01×10^2 J; (b) 8.29 J/mol·K
5.131 (a) 14.0 atm; (b) 13.8 atm
5.133 2

CHAPTER 6

6.5 q positive, w negative
6.9 (a) $q = +10.6$ J, $w = -5.3$ J, $\Delta E = +5.3$ J;
(b) $q = -25.3$ J, $w = -12.0$ J, $\Delta E = -37.3$ J;
(c) $q = +52.9$ J, $w = +387.0$ J, $\Delta E = +439.9$ J;
(d) $q = -187.5$ J, $w = +10.0$ J, $\Delta E = -177.5$ J
6.11 (a) $q = -2$ J, $w = -15$ J, $\Delta E = -17$ J;
(b) $q = +2$ J, $w = +15$ J, $\Delta E = +17$ J
6.19 Yes
6.21 (a) and (b) positive, endothermic; (c) negative, exothermic
6.23 (a) $C_2H_2(g) + \frac{5}{2}O_2(g) \rightarrow 2CO_2(g) + H_2O(l)$ $\Delta H° = -1299.6$ kJ
(b) $CH_3OH(l) + \frac{3}{2}O_2(g) \rightarrow CO_2(g) + 2H_2O(l)$ $\Delta H° = -726.5$ kJ
6.25 (a) $2Hg(l) + Cl_2(g) \rightarrow Hg_2Cl_2(s)$ $\Delta H_f° = -265.2$ kJ
(b) $Ca(s) + O_2(g) + H_2(g) \rightarrow Ca(OH)_2(s)$ $\Delta H_f° = -986.1$ kJ
(c) $Hg(l) \rightarrow Hg(g)$ $\Delta H_f° = +61.3$ kJ
(d) $S(s) + \frac{3}{2}O_2(g) \rightarrow SO_3(g)$ $\Delta H_f° = -395.72$ kJ
6.27 (b)
6.29 (a) −566.0 J; (b) +283.0 kJ
6.31 1.34×10^3 kJ
6.33 (a)

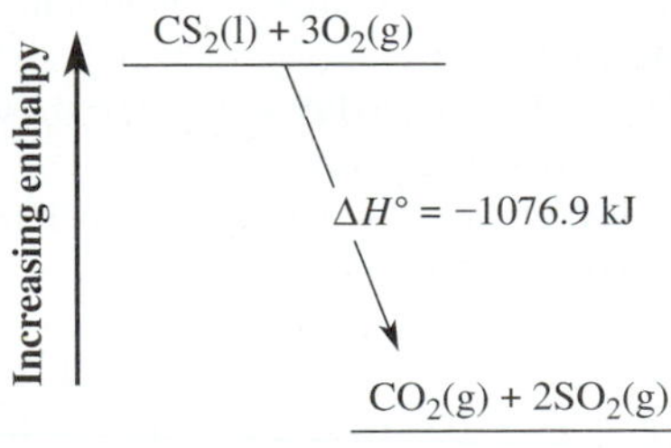

(b) 354 kJ

6.35 47.5 kJ
6.37 1.16×10^3 kJ
6.41 +158 kJ
6.43 −312 kJ
6.45 (a) −283.0 kJ; (b) −92.2 kJ; (c) −285.4 kJ; (d) −78.0 kJ; (e) −128.2 kJ
All are exothermic.
6.47 +90.25 kJ
6.49 (a) −890.3 kJ, 890.4 in Table 6.5; (b) −2218.8 kJ, same in Table 6.5; (c) −5450.4 kJ, −5450.5 kJ in Table 6.5
6.53 (a) 111.4 J/mol·°C; (b) 28.23 J/mol·°C
6.55 (a) 135 J; (b) 367 J
6.57 (a) 8.40 J/g·°C; (b) 8.40×10^2 J/°C
6.59 Gold
6.61 16.4°C
6.63 28.7°C
6.69 −633 kJ/mol
6.71 -5.99×10^3 kJ/mol
6.73 −1790 kJ/mol
6.75 (a) 7.50 kJ/°C; (b) 5.89×10^3 kJ/mol
6.77 (c)
6.79 -5.90×10^{-3} kJ
6.81 +175.8 kJ
6.83 (a) +4.96 kJ; (b) −171.0 kJ
6.89 15.26 kg
6.91 2.1×10^2 g
6.93 9.1 kcal
6.95 (a) 194 kcal; (b) 7.8×10^2 kcal
6.97 1725 kcal/day
6.101 (a) 16°C; (b) 2.3×10^3 J lost; (c) 2.3×10^3 J gained
6.103 (a) $\Delta E = -530$ kJ/mol, $\Delta H = -528$ kJ/mol; (b) +307 kJ
6.107 Constant pressure
6.109 Constant pressure
6.111 (a) −5.94 L·atm; (b) −602 J
6.113 0.16 J

CHAPTER 7

7.1 Radio, microwave, infrared, visible, ultraviolet, x-ray, gamma ray
7.3 (a) gamma ray; (b) radio; (c) gamma ray
7.5 Infrared, less energetic
7.9 2.52×10^3 h
7.11 (a) 4.62×10^{14} s^{-1}, visible; (b) 5.00 nm, where ultraviolet and x-ray meet
7.13 (a) 5.090×10^{14} Hz, 5.085×10^{14} Hz; (b) visible
7.17 (a) decreases; (b) increases
7.19 *Ultraviolet*: 30 nm, 7×10^{-18} J; *radio*: 300 m, 7×10^{-28} J; *green*: 6×10^{14} Hz, 4×10^{-19} J
7.21 (a) 5.085×10^{14} Hz, 3.369×10^{-19} J
7.23 460 nm: 6.52×10^{14} Hz, 4.32×10^{-19} J, 260 kJ
610 nm: 4.92×10^{14} Hz, 3.26×10^{-19} J, 196 kJ
7.25 (a) 1.44×10^{18}; (b) 2.52×10^{18}; (c) 2.3×10^{20}
7.29 No
7.31 1.28×10^{-19} J
7.33 (a) ultraviolet; (b) visible
7.35 (b) -2.179×10^{-18} J; (c) 1
7.37 The lowest energy state, $n = 1$
7.39 One
7.45 (a) -1.362×10^{-19} J; (b) -2.421×10^{-19} J
7.47 (a) absorbed; (b) 3.026×10^{-19} J, 4.567×10^{14} Hz, 656.5 nm; (c) visible
7.49 1.634×10^{-18} J. The other transitions to $n = 1$ are even more energetic.
7.51 (a) -3.405×10^{-20} J, $n = 8$; (b) Pfund
7.53 (b) decreases, decreases
7.57 (a) 1.5×10^{-5} m; (b) 7.9×10^{-9} m; (c) 3.3×10^{-35} m
7.67 3; see Table 7.2
7.69 (a) n; (b) n and l
7.71 3d: $n = 3, l = 2$; 4f: $n = 4, l = 3$
7.73 (a) 3; (b) 5; (c) 1; (d) 49
7.75 (a) and (c) are allowed.
7.77 (a) and (b)
7.79 (a) 9; (b) 5; (c) 5; (d) 9
7.81 Its magnetic field; m_s, $+\frac{1}{2}$ and $-\frac{1}{2}$
7.83

n	l	m_l	m_s
1	0	0	$+\frac{1}{2}$
1	0	0	$-\frac{1}{2}$

7.87 See Figure 7.32.
7.89 See Figure 7.32.
7.91 None
7.95 p_x, in yz plane; d_{xy}, in xz and yz planes
7.103 81
7.105 4.3×10^{28}
7.107 (a) 3.220×10^{-19} J; (b) 616.9 nm, visible
7.111 For transition from n_H (higher) to n_L (lower)

$$\frac{1}{\lambda} = 1.097 \times 10^7 \text{ m}^{-1}\left(\frac{1}{n_L^2} - \frac{1}{n_H^2}\right)$$

7.113 (a) from $n = 2$: 5.448×10^{-19} J; from $n = 3$: 2.421×10^{-19} J; (b) the $n = 2$ electrons; (c) the $n = 3$ electrons
7.115 1.237×10^{-5} m
7.117 No
7.119 4.85 m/s
7.123 (a) $5p \rightarrow 3s$; (b) $5p \rightarrow 2s$; (c) $5p \rightarrow 1s$

CHAPTER 8

8.3 3s, 3p, 3d
8.5 Barium, barium
8.9 3s
8.13 (a) 8, (b) 32, (c) $2n^2$
8.15

(a) 2	0	0	$+\frac{1}{2}$,	2	0	0	$-\frac{1}{2}$
(b) 2	1	−1	$+\frac{1}{2}$,	2	1	−1	$-\frac{1}{2}$
2	1	0	$+\frac{1}{2}$,	2	1	0	$-\frac{1}{2}$
2	1	+1	$+\frac{1}{2}$,	2	1	+1	$-\frac{1}{2}$

8.17 Oxygen: [He]$2s^2 2p^4$; hydrogen: $1s^1$

8.19 $1s\ 2s\ 2p\ 3s\ 3p\ 4s\ 3d\ 4p$

8.23 All are in agreement with Table 8.1.

8.25 (a) Be: [↑↓] [↑↓]; none
$1s$ $2s$

(b) P:

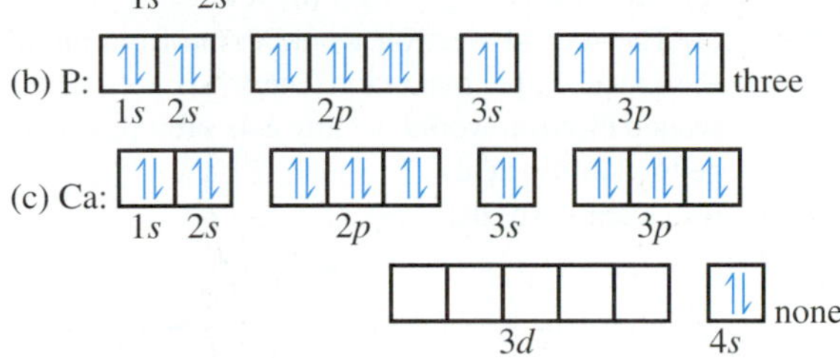

(d) V: [Ar]

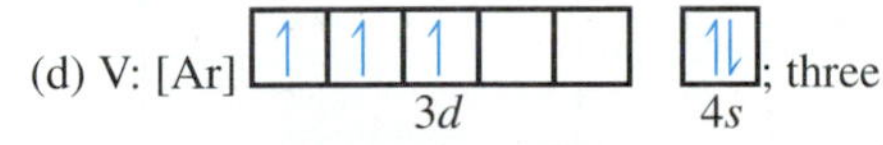

(e) Se: [Ar]

(f) F: [↑↓] [↑↓] [↑↓|↑↓|↑]; one
$1s$ $2s$ $2p$

8.27 Cr: [Ar]$3d^4 4s^2$; Cu: [Ar]$3d^9 4s^2$
These do not agree with Table 8.1.

8.29 (a) and (d)

8.33 (a) $2s$ and $2p$; (b) $3s$ and $3p$; (c) $4s$, $3d$, and $4p$

8.35 (a) s and p; (b) d and f; (c) f; (d) s, p, d, and f; (e) p and s; (f) f

8.37 (a) [He]; (b) [Ar]; (c) [Kr]; (d) [Ne]; (e) [Rn]; (f) [Rn]

8.39 No irregularities

8.41 (a), (b), (c) show irregularities.
(a) [Ar]$3d^4 4s^2$; (b) [Xe]$4f^5 5d^1 6s^2$; (c) [Kr]$4d^4 5s^2$.
(d), (e), (f) agree with Table 8.1.

8.43 (b) [↑↓|↑|↑|↑|↑]
$3d$

(c)

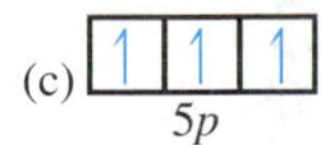

(d)

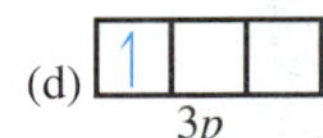

(e)

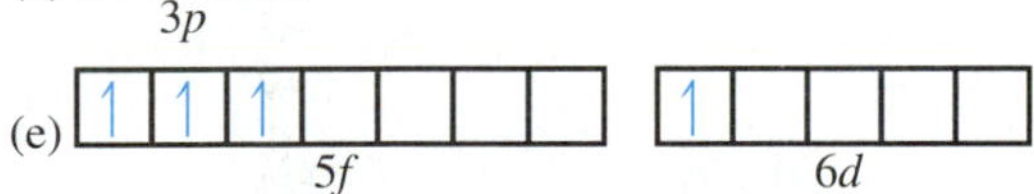

8.45 Cr and Cu show irregularities.

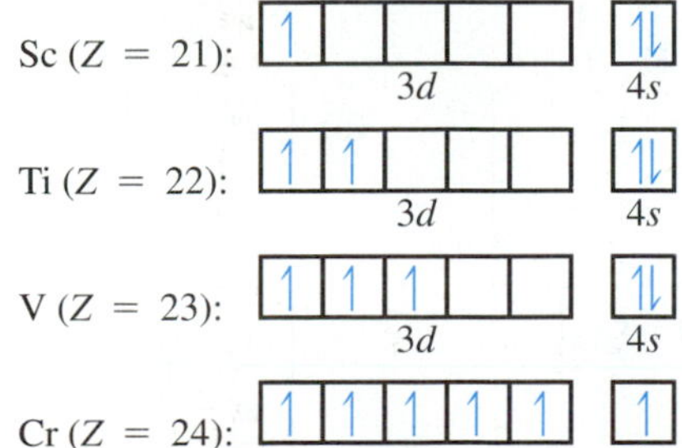

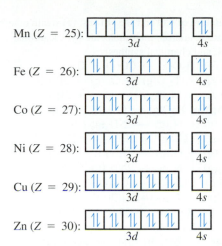

Mn (Z = 25): [↑|↑|↑|↑|↑] $3d$ [↑↓] $4s$

Fe (Z = 26): [↑↓|↑|↑|↑|↑] $3d$ [↑↓] $4s$

Co (Z = 27): [↑↓|↑↓|↑|↑|↑] $3d$ [↑↓] $4s$

Ni (Z = 28): [↑↓|↑↓|↑↓|↑|↑] $3d$ [↑↓] $4s$

Cu (Z = 29): [↑↓|↑↓|↑↓|↑↓|↑↓] $3d$ [↑] $4s$

Zn (Z = 30): [↑↓|↑↓|↑↓|↑↓|↑↓] $3d$ [↑↓] $4s$

8.47 (a) 2; (b) 3; (c) 1; (d) 4

8.49 (a) $3s^1$; (b) $4s^2$; (c) $5s^2 5p^4$; (d) $4s^2 4p^5$

8.51 (a) ns^2np^5; (b) ns^1; (c) ns^2; (d) ns^2; (e) ns^2np^2; (f) $ns^2(n-1)d^{10}$

8.53 (a) [Ar]$3d^{10}$; (b) [Ar]$3d^6$; (c) [Ne]; (d) [Ar]$3d^8$; (e) [Ar]$3d^3$; (f) [Ne]; (g) [Ar]; (h) [Kr]$4d^{10}$

8.59 Fr, H

8.61 He, Cs (of those given in Figure 8.10)

8.71 H, Li, B, C, N, O, and F

8.73 (a) 3; (b) 0; (c) 4; (d) 1

8.75 Co^{3+}, Ni^{2+}, V^{2+}

8.77 Cu: [Ar]$3d^{10}4s^1$; Cu^+: [Ar]$3d^{10}$; Cu^{2+}: [Ar]$3d^9$
Cu and Cu^{2+} each has one unpaired electron.

8.79 (a) S; (b) Ba; (c) Si; (d) Na; (e) C; (f) I

8.81 (a) $Cl < Br < I$; (b) $F < C < Si$; (c) $Mg < Ca < K$

8.83 (a) K; (b) Cl^-; (c) Fe^{2+}; (d) P^{3-}; (e) S^{2-}; (f) I^-

8.85 (a) Ca; (b) F; (c) S; (d) Mg; (e) P; (f) F

8.87 (a) K^+; (b) Cl; (c) Fe^{3+}; (d) P; (e) O^{2-}; (f) Cl^-

8.89 (a) $I < Br < Cl$; (b) $Si < C < F$; (c) $K < Ca < Mg$

8.91 22 each

8.93 87 metals, 22 nonmetals

8.95 Electrical and thermal conductivity, luster, malleability ductility

8.97 6

8.101 (a) 241.3 nm; (b) ultraviolet

8.109 K has a larger atomic radius, a smaller first ionization energy, a smaller electron affinity, more metallic character, and a greater tendency to give up electrons than Br.

8.111 In atomic units (proton charge = 1 unit):
$2s$: $Z^* = 1.26$, $\sigma = 1.74$; $1s$: $Z^* = 2.358$ and $\sigma = 0.642$

SURVEY 1

S1.1 (a) H; (b) H_2; (c) H^+; (d) H^-

S1.3 (a) H^+, H^-; (b) H^-; (c) H^+

S1.5 H and H_2 escape earth's gravity more easily than heavier particles do.

S1.7 (a) a mixture of CO and H_2; (b) yes; (c) steam reacting with carbon and/or small-molecule hydrocarbons; (d) fuel, and chemical synthesis

S1.11 (a) $C(s) + H_2O(g) \rightarrow CO(g) + H_2(g)$
(b) $CH_4(g) + H_2O(g) \rightarrow CO(g) + 3H_2(g)$
(c) $CH_4(g) + H_2O(g) \rightarrow CO(g) + 3H_2(g)$
$CO(g) + H_2O(g) \rightarrow CO_2(g) + H_2(g)$

S1.13 (a) $3H_2(g) + N_2(g) \rightarrow 2NH_3(g)$
(b) $H_2(g) + Cl_2(g) \rightarrow 2HCl(g)$
(c) $2H_2(g) + O_2(g) \rightarrow 2H_2O(l)$

S1.15 (a) $3H_2(g) + N_2(g) \rightarrow 2NH_3(g)$
(b) $CO(g) + 2H_2(g) \rightarrow CH_3OH(g)$
(c) $H_2(g) + Cl_2(g) \rightarrow 2HCl(g)$

S1.17 (a) Na: [Ne]$3s^1$; Mg: [Ne]$3s^2$; (b), (d) Na greater; (c) Mg greater; (e) Na^+, Mg^{2+}

S1.19 (a) K: $4s^1$; Cs: $6s^1$; (b), (d) Cs greater; (c) K greater

S1.21 (a) and (d), the alkali metal; (b) and (c), the alkaline earth metal

S1.23 The crust is mainly Na_2CO_3. Cut a fresh surface to observe the metallic properties.

S1.25 (a) NaCl; (b) $MgCO_3$, $MgCl_2$, $MgSO_4$, dolomite; (c) beryl; (d) limestone, gypsum, fluorite, dolomite

S1.29 (a) Li, Na, K, Rb, Cs, Ca, Sr, Ba; (c) (1) yellow, (2) scarlet, (3) brick red, (4) violet, (5) pale green

S1.33 (a) NaOH; (b), (d) Na_2CO_3; (c), (e) $NaHCO_3$; (f) NaOH and KOH

S1.35 (a) $2KCl(l) \rightarrow 2K(l) + Cl_2(g)$
(b) $CaCl_2(l) \rightarrow Ca(s) + Cl_2(g)$

S1.37 See Table S1.4.

S1.39 (a) $2K(s) + 2H_2O(l) \rightarrow 2KOH(aq) + H_2(g)$
(b) $CaO(s) + 2H_2O(l) \rightarrow Ca(OH)_2(aq)$
(c) $NaH(s) + H_2O(l) \rightarrow NaOH(aq) + H_2(g)$
(d) $Li_3N(s) + 3H_2O(l) \rightarrow NH_3(g) + 3LiOH(aq)$

S1.41 See Tables S1.4 and S1.6.

S1.43 (a) $2KCl(aq) + 2H_2O(l) \rightarrow 2KOH(aq) + H_2(g) + Cl_2(g)$
(b) $2NaHCO_3(s) \rightarrow Na_2CO_3(s) + H_2O(g) + CO_2(g)$
(c) $Na_2CO_3(s) + 2HCl(aq) \rightarrow 2NaCl(aq) + H_2O(l) + CO_2(g)$
(d) $2NaHCO_3(s) + H_2SO_4(aq) \rightarrow Na_2SO_4(aq) + 2H_2O(l) + 2CO_2(g)$

S1.47 Helium, argon, radon

S1.53 (a) See Figure 8.4. (b) He has a complete outer shell.

S1.57 (a) 285.8 kJ/mol; (b) 890.3 kJ/mol; (c) 284.4 kJ/mol

S1.59 (a) 1.46×10^4 kJ; (b) same for both; (c) gas can be piped

S1.61 The electrostatic forces binding the small highly charged Group 2A cation to adjacent anions are stronger than those binding the Group 1A cations.

S1.67 (a) H_2, 0.0824 g/L; He: 0.164 g/L; Ar: 1.63 g/L; Kr: 3.43 g/L; (b) no, no; (c) both decrease

S1.69 4.3 L

S1.71 0.723

CHAPTER 9

9.3 The N—H and N—O bonds are covalent. The bonding between the oppositely charged ions is ionic.

9.5 (a) $\cdot\dot{\underset{\cdot}{C}}\cdot$; (b) $\cdot\dot{\underset{\cdot}{As}}\cdot$; (c) $\cdot\ddot{\underset{\cdot}{N}}\cdot$; (d) K·; (e) $\cdot\dot{Al}\cdot$; (f) $\cdot\ddot{\underset{\cdot\cdot}{O}}\cdot$; (g) ·Mg·; (h) $:\ddot{\underset{\cdot\cdot}{Cl}}\cdot$

9.9 (a) The second electron would come from an underlying shell where the ionization energy is too high. (b) The second electron would occupy a 4*s* orbital, where it would be too loosely held.

9.13 (a) increase; (b) decrease

9.15

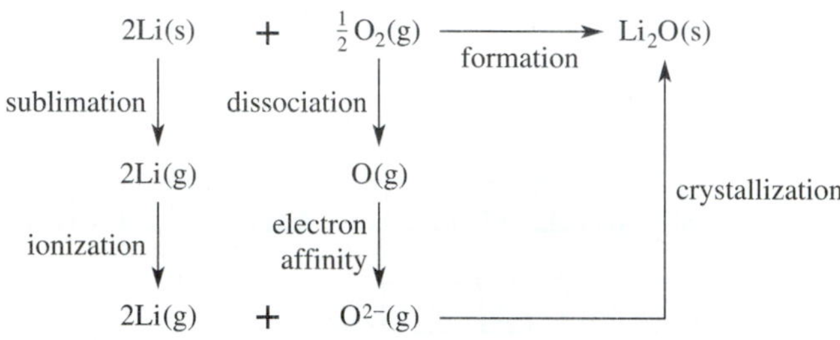

9.17 (a) Se^{2-}; (b) Rb^+; (c) F^-; (d) O^{2-}; (e) P^{3-}; (f) Ba^{2+}

9.19 RbBr

9.21 (a) Sn^{2+}; (b) Fe^{2+}, Fe^{3+}

9.23 (a) $Be(NO_3)_2$; (b) $Ca(OH)_2$

9.25 2255 kJ/mol

9.31 (a) and (b) bond energy increases, bond length decreases.

9.35 The bond length is shorter in Cl_2.

9.37 (a) $H—\ddot{\underset{\cdot\cdot}{O}}—H$ (b) $H—C\equiv C—H$ (↑ triple bond) (c) $H—\ddot{N}—H$ with H bonded below N

9.39 NH_3, NH_3

9.41 (a) H—S; (b) H—Se

9.43 −357 kJ

9.45 163 kJ

9.47 −106 kJ/mol

9.49 H can form only one bond.

9.51 The charge on the molecule or ion

9.53 (a) $H—\ddot{\underset{\cdot\cdot}{S}}—H$ (b) $:\ddot{\underset{\cdot\cdot}{I}}—\ddot{N}—\ddot{\underset{\cdot\cdot}{I}}:$ with $:\ddot{I}:$ bonded above N (c) $H—Si—H$ with H above and H below Si
(d) $H—C—C—H$ with H above and below each C (e) $H—C=\ddot{O}:$ with H above C

9.55 (a) $[H—\ddot{O}—H]^+$ with H above O (b) $[:\ddot{\underset{\cdot\cdot}{O}}—H]^-$
(c) $[:\ddot{\underset{\cdot\cdot}{O}}—\ddot{\underset{\cdot\cdot}{Cl}}:]^-$ (d) $[H—N—H]^+$ with H above and H below N

9.57 No formal charges

9.63, 9.65

(a) 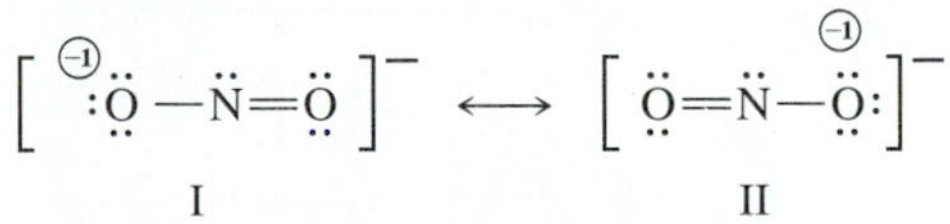 [:Ö(−1)—N̈=Ö]⁻ ⟷ [Ö=N̈—Ö:(−1)]⁻

I II

Both contribute equally.

(b) 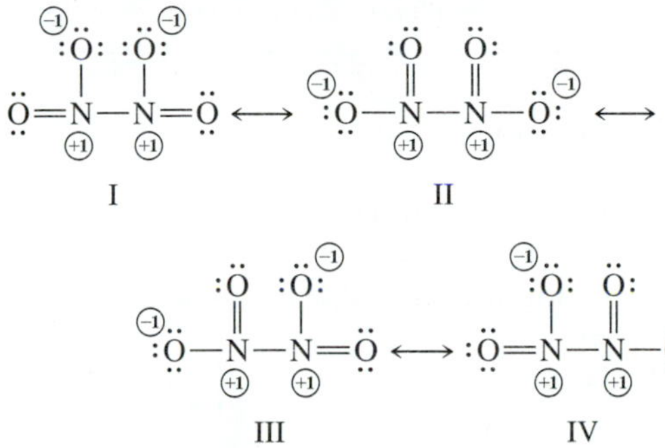 I: Ö=N(+1)(—:Ö:(−1))—N(+1)(—:Ö:(−1))=Ö ⟷ II: :Ö(−1)—N(+1)(=Ö:)—N(+1)(=Ö:)—Ö:(−1) ⟷

III: :Ö(−1)—N(+1)(=Ö)—N(+1)(—:Ö:(−1))=Ö ⟷ IV: :Ö=N(+1)(—:Ö:(−1))—N(+1)(=Ö:)—Ö(−1)

All contribute equally.

(c) 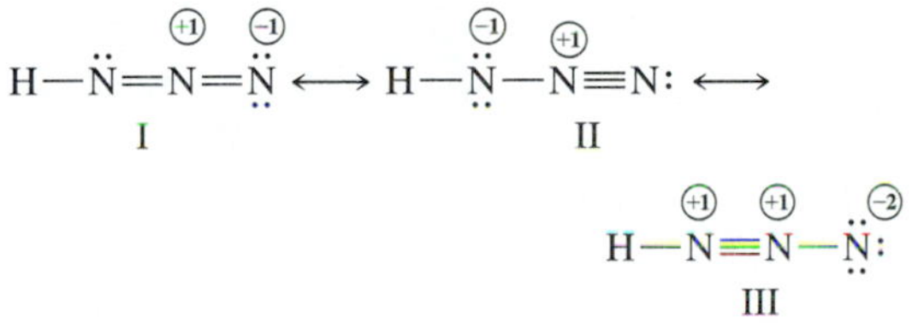 I: H—N̈=N(+1)=N̈(−1) ⟷ II: H—N̈(−1)—N(+1)≡N: ⟷ III: H—N(+1)≡N(+1)—N̈:(−2)

Structures I and II make greater contributions to the hybrid.

(d) 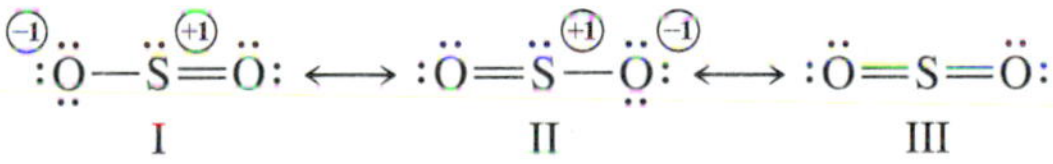 I: :Ö(−1)—S̈(+1)=Ö: ⟷ II: :Ö=S̈(+1)—Ö:(−1) ⟷ III: :Ö=S=Ö:

9.67 Two benzene Kekulé structures (C_6H_6, each C bonded to one H, alternating single and double C—C bonds) ⟷ (double bonds shifted)

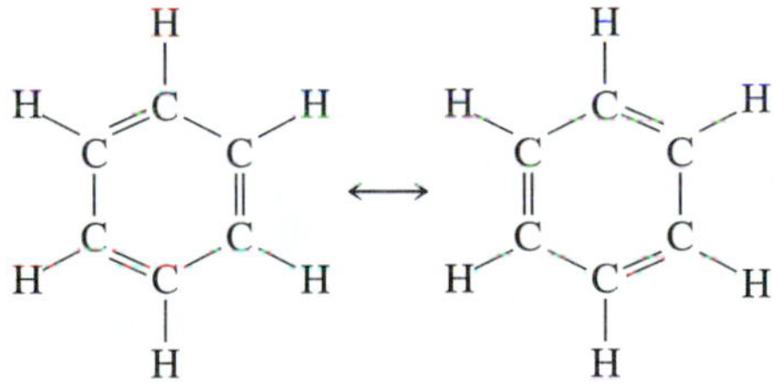

9.73 (a) 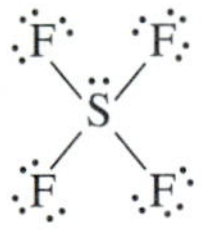 SF_4: S̈ bonded to four :F̈: atoms (b) 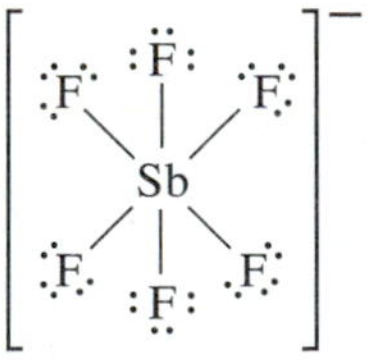 $[SbF_6]^-$: Sb bonded to six :F̈: atoms, in brackets with − charge

9.75 (a) 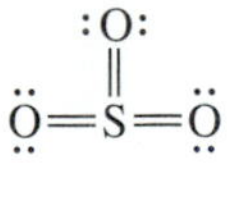 Ö=S(=O:)=Ö (b) 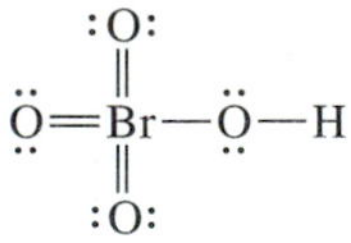 Ö=Br(=:O:)(=:O:)—Ö—H

(c) 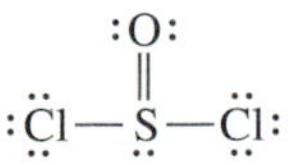 :C̈l—S̈(=:O:)—C̈l: (d) 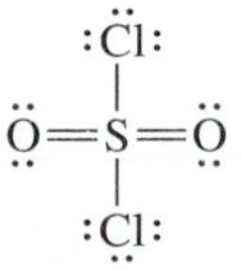 Ö=S(—:C̈l:)(—:C̈l:)=Ö

9.77 (a) 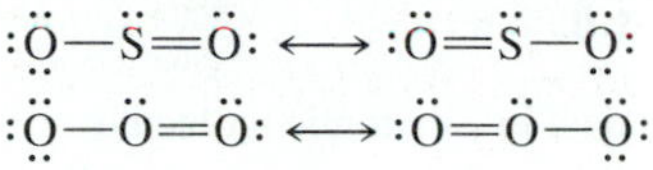 :Ö—S̈=Ö: ⟷ :Ö=S̈—Ö:

:Ö—Ö=Ö: ⟷ :Ö=Ö—Ö:

(b) 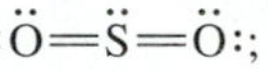 :Ö=S̈=Ö:; no.

9.79 (a) 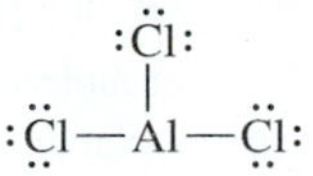 :C̈l—Al(—:C̈l:)—C̈l: (b) Li—Li

(c) 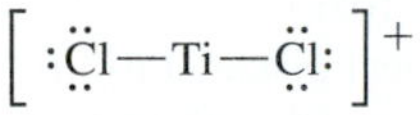 [:C̈l—Ti—C̈l:]⁺

9.81 (a) 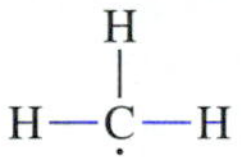 H—C(—H)—H (b) 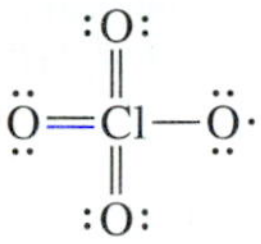 Ö=Cl(=:O:)(=:O:)—Ö· (4 resonance structures)

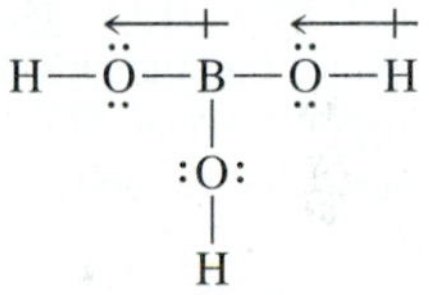

9.91 (a) F; (b) O; (c) Fr

9.93 (a) Cl; (b) O; (c) O; (d) S

9.95 (a) $\overset{\delta+}{N}—\overset{\delta-}{O}$; (b) $\overset{\delta+}{Br}—\overset{\delta-}{Cl}$; (c) $\overset{\delta-}{S}—\overset{\delta+}{H}$

9.97 (a) covalent; (b) and (c) ionic

9.99 (a) 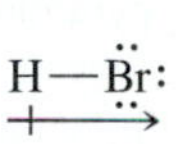 H—B̈r: (dipole arrow ⟶ toward Br) (b) H—Ö—B(—:Ö:—H)—Ö—H (dipole arrows toward O) (c) 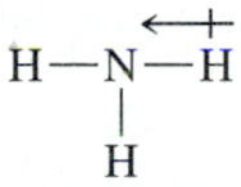 H—N(—H)—H (dipole arrow toward N)

(Arrows are shown for one bond of each type.)

9.101 MgO, greater lattice energy

9.103 −36.4 kJ/mol

9.105 415.9 kJ/mol

9.107 Two Kekulé structures of chlorobenzene (C_6H_5Cl, Cl on one ring carbon, H on the other five, alternating single and double C—C bonds) ⟷ (double bonds shifted)

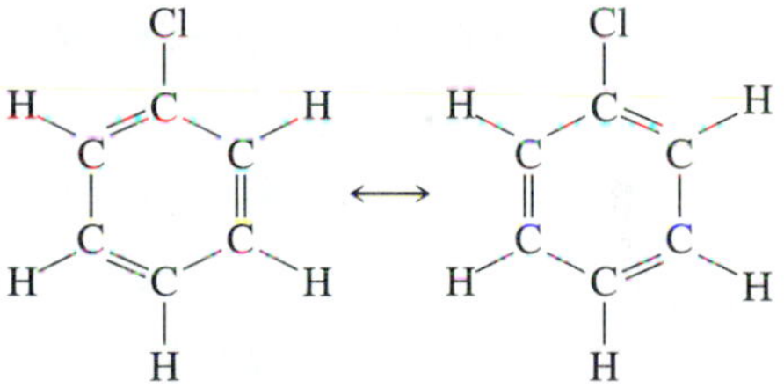

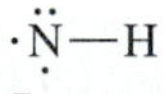

9.109 (a) 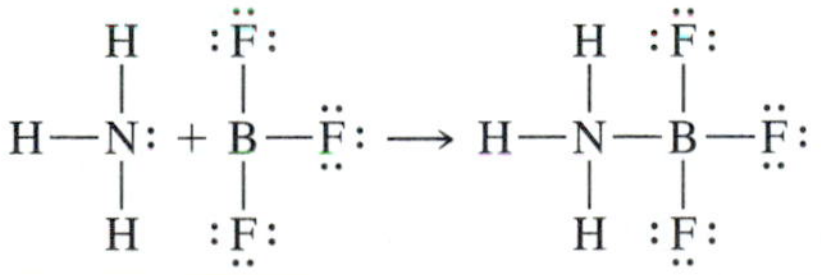 H—N̈(—H)(—H) + B(—:F̈:)(—:F̈:)—F̈: ⟶ H—N(—H)(—H)—B(—:F̈:)(—:F̈:)—F̈:

(b) yes

9.111 (a) 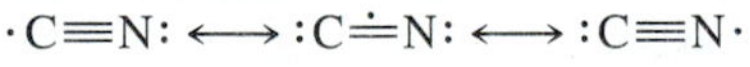 ·C≡N: ⟷ :C≐N: ⟷ :C≡N·

(b) ·N̈—H

(c) [:N≡N·]⁺ ⟷ [:N≐N:]⁺

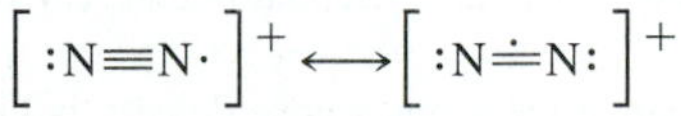

(d) 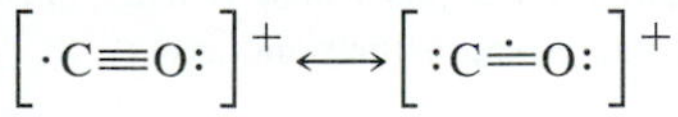 [·C≡O:]⁺ ⟷ [:C≐O:]⁺

(e) 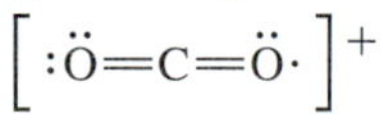 [:Ö=C=Ö·]⁺

9.113 The resonance structures suggest that the center atom tends to be positive relative to the end atoms. The molecule must be bent, because the two bond moments would cancel if it were linear.

CHAPTER 10

10.3 They are identical in a diatomic molecule.

10.5 (a) bent; (b) S, H H, individual (c)

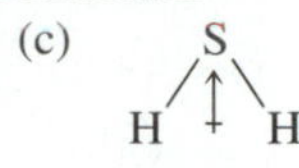

net molecular

10.7 Molecule is symmetrical, i.e., trigonal planar.

10.13 See Figures 10.4 through 10.6.

10.15 (a) linear, 180°; (b) trigonal planar, 120°; (c) tetrahedral, 109.5°; (d) trigonal bipyramidal, 120° and 90°; (e) trigonal pyramidal, $<109.5°$; (f) bent seesaw, $<90°$, $<120°$, $<180°$

10.17 (e) and (f)

10.19 (a) bent seesaw, $<90°$, $<120°$, $<180°$; (b) tetrahedral, 109.5°; (c) and (e) octahedral, 90°; (d) linear, 180°; (f) trigonal bipyramidal, 90°, 120°

10.21 (a) and (b) tetrahedral, 109.5°; (c) bent seesaw; $<90°$, $<120°$, $<180°$; (d) trigonal planar, 120°

10.23 (a) H_2O is bent. H_3O^+ is trigonal pyramidal and has slightly wider bond angles. (b) NO_2^+ linear, NO_2^- bent; (c) SO_3 trigonal planar, SO_3^{2-} trigonal pyramidal

10.29 Conformers can be converted into each other without breaking bonds.

10.31 Tetrahedral, 109.5°

10.33 The double bond causes part of the molecule to be planar.

10.35 See Figure 10.8.

10.37 The carbon skeletons are

```
                                       C
                                       |
1. C—C—C—C—C—C                 2. C—C—C—C—C

     C   C                             C
     |   |                             |
3. C—C—C—C                     4. C—C—C—C
                                       |
                                       C
          C
          |
5. C—C—C—C—C
```

10.39 (a)

```
H3C         CH2—CH3     H3C         H
    C=C                     C=C
H         H             H         CH2—CH3
     cis                     trans
```

(b) no isomers

10.41 See Table 10.1.

10.43 A shared pair of electrons whose major density lies along the bond axis.

10.47 See Figure 10.19.

10.49 (a) 180°, linear; (b) 120°, trigonal planar; (c) 109.5°, tetrahedral

10.51 A shared pair of electrons whose density is bisected by a nodal plane that contains the bond axis.

10.53

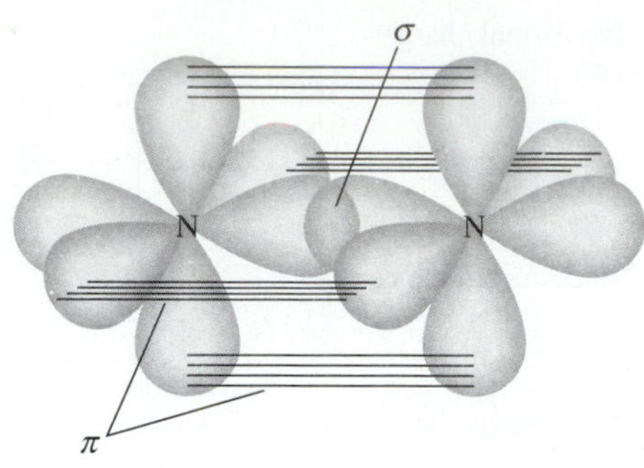

10.55 For Exercise 10.15: (a) sp; (b) sp^2; (c), (e) sp^3; (d), (f) sp^3d

10.57 (a) trigonal pyramidal, sp^3; (b) tetrahedral, sp^3; (c) linear, sp

10.59 (a)

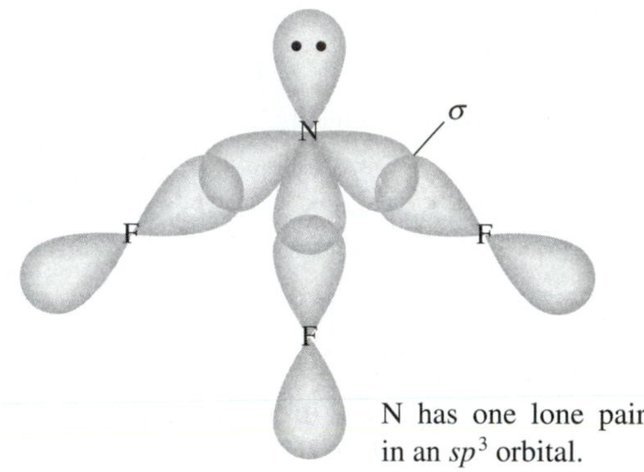

N has one lone pair in an sp^3 orbital.

(b)

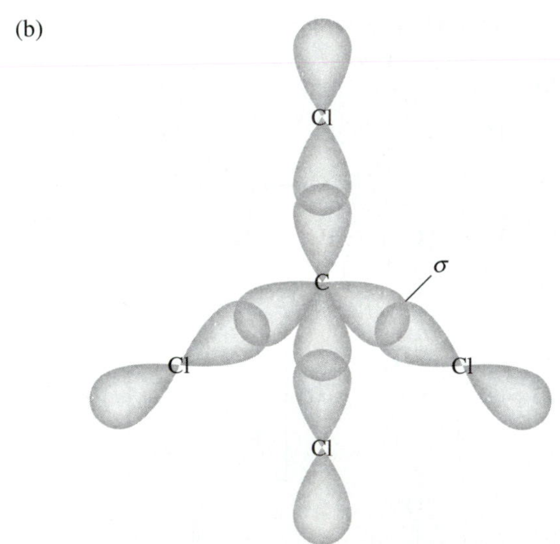

(c)

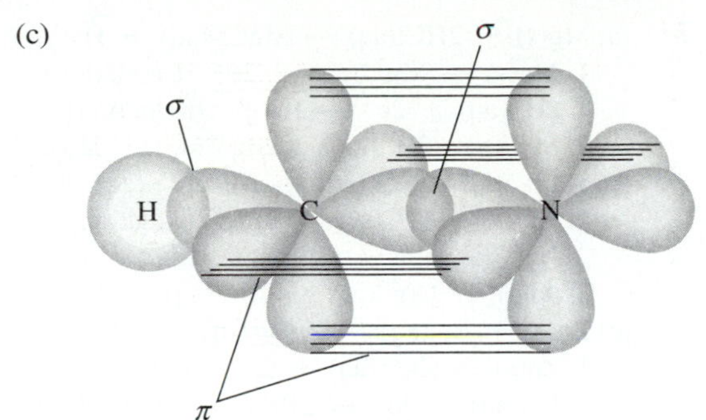

10.61 Oxygen has only four orbitals in its valence shell and thus cannot form more than four covalent bonds.

10.63 (a) See Figures 10.29, 10.30, 10.31; (b) acetylene

10.65 Except where indicated, the hybridization is sp^3 and the angles are approximately 109.5°.

(a) H—Ö—Ö—H (each O with two lone pairs)

(b) H—C(H)(H)—C(H)(H)—Ö—H

(c) H—C(H)(H)—N̈(H)—C(H)(H)—H

(d) H_2C^1=C^2H—CH_3

The angles around C^1 and C^2 are approximately 120°.

10.69 (a) The energy is lower in the bonding orbital. (b) The bonding orbital has more density between the nuclei.

10.71 A sigma bond is a shared pair of electrons. A sigma orbital is a shared orbital that may or may not be occupied by electrons.

10.73 Electrons in bonding orbitals have lower energy than in the separated atoms; electrons in antibonding orbitals have higher energy.

10.75 See Figure 10.38.

10.77 (a) See Table 10.3. (b) Ne_2 does not exist. (c) B_2 is paramagnetic.

10.79 (a) See Table S2.3.
(b) Most stable, O_2^+; least stable O_2^{2-}
(c) O_2, O_2^-, and O_2^+

10.81 (a) $(\sigma_{1s})^2(\sigma^*_{1s})^2(\sigma_{2s})^2(\sigma^*_{2s})^2(\pi_{2p_x})^2(\pi_{2p_y})^2(\sigma_{2p_x})^1$
(b) yes, B_2

10.83 (a) HeH could exist, bond order $\frac{1}{2}$.

10.85 (a) bond order 3:
$(\sigma_{1s})^2(\sigma^*_{1s})^2(\sigma_{2s})^2(\sigma^*_{2s})^2(\pi_{2p_x})^2(\pi_{2p_y})^2(\sigma_{2p_z})^2$
(b) bond order 2: $(\sigma_{1s})^2(\sigma^*_{1s})^2(\sigma_{2s})^2(\sigma^*_{2s})^2(\pi_{2p_x})^2(\pi_{2p_y})^2$

10.89 (a)

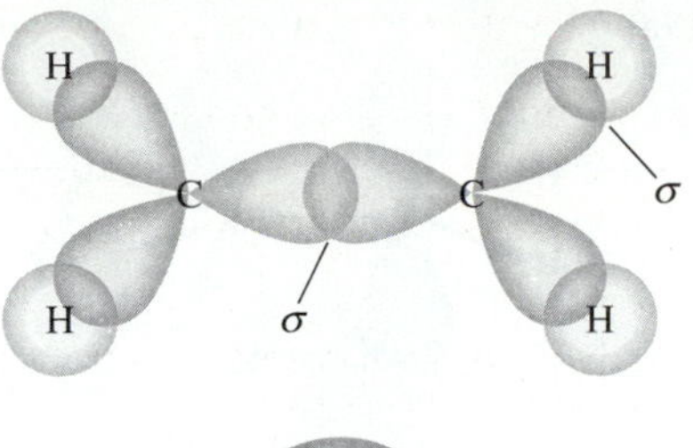

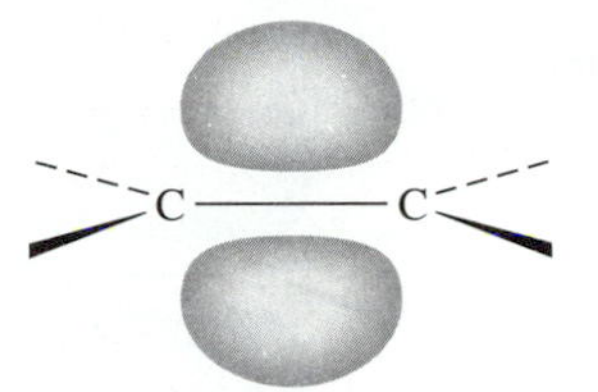

π molecular orbital

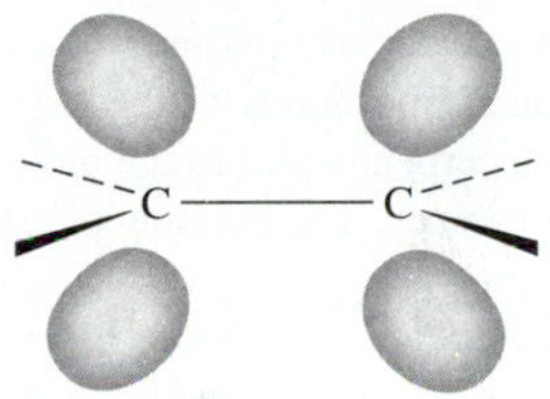

π* molecular orbital

(b)

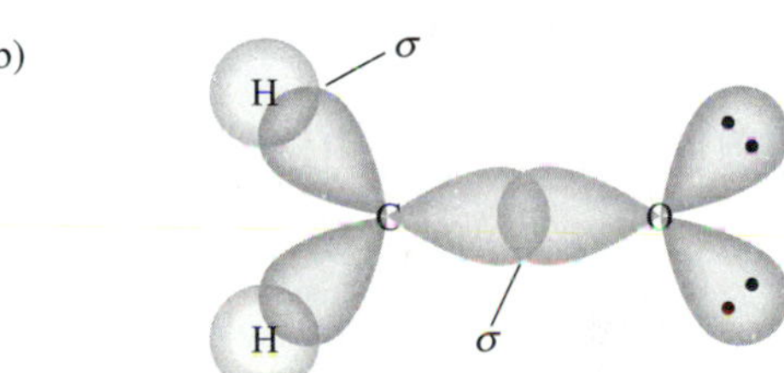

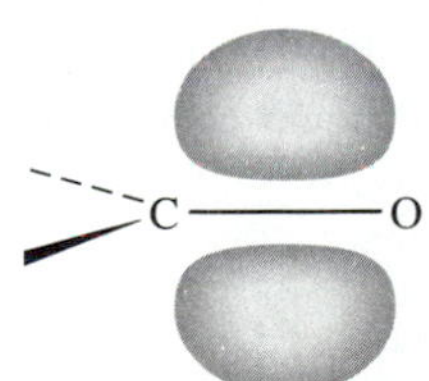

π molecular orbital

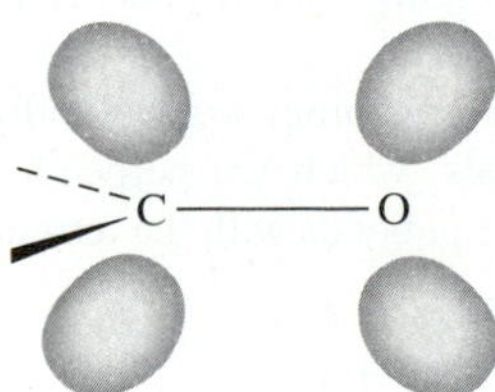

π* molecular orbital

10.91 (a)

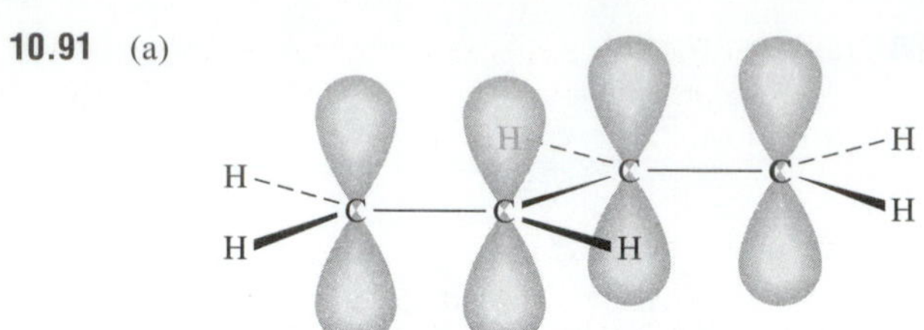

(b) four

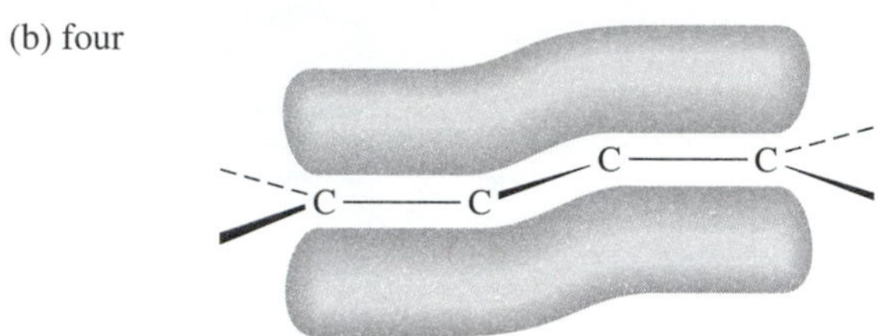

(c) 2, 2, 0, 0

10.93 (a) The center oxygen uses two of its sp^2 hybrid orbitals to form sigma bonds to the two end atoms. The three p orbitals perpendicular to the molecular plane form pi molecular orbitals. The bonding pi orbital contains one pair of electrons.

10.95 Only orthodichlorobenzene has a dipole moment.

10.97 Lone-pair repulsion, multiple bonds, nonidentical substituent atoms

10.99 (b) 72°

10.101 (a)

```
      H   H H   H
       \ /   \ /
        C-----C
   H   /       \   H
    \ /         \ /
     C           C
    / \         / \
   H   \       /   H
        C-----C
       / \   / \
      H   H H   H
```

(b) puckered

10.105 (a) Adding the right number of electrons and bonds to the sigma skeleton gives two resonance structures

```
   \       ..  C            \         C
    C—N                      C=N
   //     \                 /   \
  :O:      H              :Ö:    H
```

The delocalized O—C—N pi bond prevents free rotation about C—N.

(b) Approximately 120°.

10.107 (a) see Figure S3.4; (b) puckered

10.109 (a) H_2; (b) O

10.111 The center C atom is sp hybridized. Its two unhybridized p orbitals, which are perpendicular to each other, form separate pi bonds with the terminal carbon atoms.

CHAPTER 11

11.3 (a) If one species acquires electrons, some other species must lose them. (b) oxygen

11.5 The equation for either the oxidation or the reduction part of a reaction. Electrons appear.

11.7 (a) $Mg(s) + 2HCl(aq) \rightarrow MgCl_2(aq) + H_2(g)$
1. $Mg(s) \rightarrow Mg^{2+}(aq) + 2e^-$ (oxidation);
$2H^+(aq) + 2e^- \rightarrow H_2(g)$ (reduction)
2. $Mg(s) + 2H^+(aq) \rightarrow Mg^{2+}(aq) + H_2(g)$

(b) $Al(s) + 3AgNO_3(aq) \rightarrow Al(NO_3)_3(aq) + 3Ag(s)$
1. $Al(s) \rightarrow Al^{3+}(aq) + 3e^-$ (oxidation);
$Ag^+(aq) + e^- \rightarrow Ag(s)$ (reduction)
2. $Al(s) + 3Ag^+(aq) \rightarrow Al^{3+}(aq) + 3Ag(s)$

(c) $Zn(s) + Br_2(aq) \rightarrow ZnBr_2(aq)$
1. $Zn(s) \rightarrow Zn^{2+}(aq) + 2e^-$ (oxidation);
$Br_2(aq) + 2e^- \rightarrow 2Br^-(aq)$ (reduction)
2. $Zn(s) + Br_2(aq) \rightarrow Zn^{2+}(aq) + 2Br^-(aq)$

11.9 (a) 1. $H^+(aq)$ is the oxidizing agent; Mg(s) is the reducing agent.
2. Mg(s) is oxidized and $H^+(aq)$ is reduced.

(b) 1. $Ag^+(aq)$ is the oxidizing agent; Al(s) is the reducing agent.
2. Al(s) is oxidized and $Ag^+(aq)$ is reduced.

(c) 1. $Br_2(aq)$ is the oxidizing agent; Zn(s) is the reducing agent.
2. Zn(s) is oxidized and $Br_2(aq)$ is reduced.

11.11 $N_2(g) + 3H_2(g) \rightarrow 2NH_3(g)$
H_2 is oxidized and N_2 is reduced.

11.13 (a) A main-group element can lose or share only the electrons in its incomplete outer shell. (b) O, F, Po, At

11.15 An oxidizing agent contains some element that could also be stable in a lower oxidation state; a reducing agent contains some element that could also be stable in a higher oxidation state.

11.17 The O in H_2O_2 and the S in SO_2 could also be stable in both higher and lower oxidation states.

11.19 $N_2O < NO < N_2O_3 < NO_2 = N_2O_4 < N_2O_5$

11.21 In (a) through (i): H is +1, O is −2, K is +1, and F is −1.
(a) Xe, +4; (b) B, +3; (c) I, +3; (d) Br, +3;
(e) Sb, +5; (f) I, +7; (g) Sn, +4; (h) Cl, −1; Cr, +6;
(i) N, −3; Cr, +6

11.23 (a), (b) −1; (c) $-\frac{1}{2}$; (d) +2; (e) +1

11.25 (a) +4; (b) −4; (c) −2; (d) +4; (e) +2; (f) 0

11.27 The oxidizing and reducing agents, respectively, are:
(b) NO_2, NO_2; (c) O_2, SO_2; (d) $K_2Cr_2O_7$, K_2S

11.29 (a) +5; (b) −3

11.31 H_2Se, MnO_2, HI, H_3PO_3, Co^{2+}

11.33 (a) HNO_3, CrO_4^{2-}, $HClO_4$; (b) H_2Se, HI;
(c) MnO_2, H_3PO_3, Co^{2+}

11.37 (a) $2MnO_4^-(aq) + 6H^+(aq) + 5HSO_3^-(aq) \rightarrow 2Mn^{2+}(aq) + 5HSO_4^-(aq) + 3H_2O(l)$

(b) $3CuS(s) + 2NO_3^-(aq) + 8H^+(aq) \rightarrow 3Cu^{2+}(aq) + 3S(s) + 2NO(g) + 4H_2O(l)$

(c) $2Fe^{3+}(aq) + 3H_2S(aq) \rightarrow 2FeS(s) + S(s) + 6H^+(aq)$

(d) $2H^+(aq) + 2NO_3^-(aq) + 3H_2S(aq) \rightarrow 3S(s) + 2NO(g) + 4H_2O(l)$

(e) $Pb(s) + PbO_2(s) + 4H^+(aq) + 2SO_4^{2-}(aq) \rightarrow 2PbSO_4(s) + 2H_2O(l)$

(f) $2MnO_2(s) + 3PbO_2(s) + 4H^+(aq) \rightarrow 2MnO_4^-(aq) + 3Pb^{2+}(aq) + 2H_2O(l)$
(g) $5ClO^-(aq) + I_2(s) + H_2O(l) \rightarrow 5Cl^-(aq) + 2IO_3^-(aq) + 2H^+(aq)$
(h) $Br_2(aq) + SO_2(g) + 2H_2O(l) \rightarrow 2Br^-(aq) + SO_4^{2-}(aq) + 4H^+(aq)$

11.39 (a) $2Al(s) + 2OH^-(aq) + 6H_2O(l) \rightarrow 2Al(OH)_4^-(aq) + 3H_2(g)$
(b) $2MnO_4^-(aq) + 3S^{2-}(aq) + 4H_2O(l) \rightarrow 2MnO_2(s) + 3S(s) + 8OH^-(aq)$
(c) $4Zn(s) + 7OH^-(aq) + 6H_2O(l) + NO_3^-(aq) \rightarrow 4Zn(OH)_4^{2-}(aq) + NH_3(aq)$
(d) $Pb(OH)_3^-(aq) + OCl^-(aq) \rightarrow PbO_2(s) + Cl^-(aq) + OH^-(aq) + H_2O(l)$

11.41 (a) $H_2O_2(aq) + 2H^+(aq) + 2e^- \rightarrow 2H_2O(l)$
(b) $H_2O_2(aq) \rightarrow O_2(g) + 2H^+(aq) + 2e^-$

11.43 (a) $Cr_2O_7^{2-}(aq) + 3H_2O_2(aq) + 8H^+(aq) \rightarrow 2Cr^{3+}(aq) + 3O_2(g) + 7H_2O(l)$
(b) $H_2S(aq) + H_2O_2(aq) \rightarrow S(s) + 2H_2O(l)$
(c) $2MnO_4^-(aq) + 5H_2O_2(aq) + 6H^+(aq) \rightarrow 2Mn^{2+}(aq) + 5O_2(g) + 8H_2O(l)$
(d) $Sn^{2+}(aq) + H_2O_2(aq) + 2H^+(aq) + 6Cl^-(aq) \rightarrow SnCl_6^{2-}(aq) + 2H_2O(l)$
(e) $HAsO_2(aq) + H_2O_2(aq) \rightarrow H_3AsO_4(aq)$

11.45 (a) $3NO_2(g) + H_2O(l) \rightarrow 2NO_3^-(aq) + NO(g) + 2H^+(aq)$
(b) $2ClO_2(g) + 2OH^-(aq) \rightarrow ClO_2^-(aq) + ClO_3^-(aq) + H_2O(l)$

11.47 (a) $3HAsO_3^{2-}(aq) + BrO_3^-(aq) + 6H^+(aq) \rightarrow 3H_3AsO_4(aq) + Br^-(aq)$
(b) $3Na_2HAsO_3(aq) + KBrO_3(aq) + 6HCl(aq) \rightarrow 3H_3AsO_4(aq) + KBr(aq) + 6NaCl(aq)$

11.49 (a) $4Mn(OH)_2(s) + O_2(g) + 2H_2O(l) \rightarrow 4Mn(OH)_3(s)$
(b) $3Ti^{3+}(aq) + RuCl_5^{2-}(aq) + 6OH^-(aq) \rightarrow Ru(s) + 3TiO^{2+}(aq) + 5Cl^- + 3H_2O(l)$
(c) $2H_2S(g) + SO_2(g) \rightarrow 3S(s) + 2H_2O(g)$
(d) $6FeSO_4(aq) + K_2Cr_2O_7(aq) + 8H_2SO_4(aq) \rightarrow 3Fe_2(SO_4)_3(aq) + Cr_2(SO_4)_3(aq) + 2KHSO_4(aq) + 7H_2O(l)$

11.51 1.98 g
11.53 1.06 g
11.55 591 mL
11.57 11.4 g
11.59 0.660 g
11.61 Yes, for monatomic ions
11.65 One
11.67 $2Cu(NH_3)_4^{2+}(aq) + 3CN^-(aq) + 2OH^-(aq) \rightarrow 2CuCN(s) + CNO^-(aq) + 8NH_3(aq) + H_2O(l)$
11.69 $K_4Fe(CN)_6(s) + K_2CO_3(s) \rightarrow 5KCN(s) + KOCN(s) + Fe(s) + CO_2(g)$
11.71 (a) $I_2(s) + 2S_2O_3^{2-}(aq) \rightarrow 2I^-(aq) + S_4O_6^{2-}(aq)$
(b) 2.30 g
11.73 (a) 48.1 g; (b) 0.50 mol SO_2, 1.00 mol H_2O
11.75 0.22 L
11.77 (a) $5Co(NO_2)_6^{3-}(aq) + 11MnO_4^-(aq) + 28H^+(aq) \rightarrow 5Co^{2+}(aq) + 11Mn^{2+}(aq) + 30NO_3^-(aq) + 14H_2O(l)$
(b) 6.31×10^{-4} mol; 49.3 mg

SURVEY 2

S2.5 Pass the gas into starch–iodide solution, and look for blue color.
$2I^-(aq) + O_3(g) + H_2O(l) \rightarrow 2OH^-(aq) + I_2(aq) + O_2(g)$

S2.7 (a) Metals have basic oxides. (b) Nonmetals have acidic oxides. (c) Test with acids and bases to see which it will neutralize.

S2.11 (a) See Figure S2.3. (b) Two lone pairs on each oxygen compress the bond angles to less than the tetrahedral angle.

S2.13 (a) See Table S2.3. (b) O_2 molecules have strong bonds (bond order 2); unpaired electrons cause the molecules to be paramagnetic and chemically reactive.

S2.15 (a) $2H_2O(l) \xrightarrow{\text{electrolysis}} 2H_2(g) + O_2(g)$
(b) $2KClO_3(s) \xrightarrow[\text{heat}]{MnO_2(s)} 2KCl(s) + 3O_2(g)$
(c) $2H_2O_2(l) \longrightarrow 2H_2O(l) + O_2(g)$
(d) $6CO_2(g) + 6H_2O(l) \xrightarrow{h\nu} C_6H_{12}O_6(aq) + 6O_2(g)$

S2.17 (a) $3Fe(s) + 2O_2(g) \rightarrow Fe_3O_4(s)$ or $4Fe(s) + 3O_2(g) \rightarrow 2Fe_2O_3(s)$
(b) $2Zn(s) + O_2(g) \rightarrow 2ZnO(s)$
(c) $S(s) + O_2(g) \rightarrow SO_2(g)$
(d) $P_4(s) + 5O_2(g) \rightarrow P_4O_{10}(s)$
(e) $2SO_2(g) + O_2(g) \rightarrow 2SO_3(g)$
(f) $2C_2H_2(g) + 5O_2(g) \rightarrow 4CO_2(g) + 2H_2O(g)$

S2.19 (a) $Mg(OH)_2(s) \xrightarrow{\text{heat}} MgO(s) + H_2O(g)$
(b) $2LiOH(s) \xrightarrow{\text{heat}} Li_2O(s) + H_2O(g)$
(c) $Fe(OH)_2(s) \xrightarrow{\text{heat}} FeO(s) + H_2O(g)$
(d) $2Fe(OH)_3(s) \xrightarrow{\text{heat}} Fe_2O_3(s) + 3H_2O(g)$

S2.21 (a) $BaO(s) + H_2O(l) \rightarrow Ba(OH)_2(aq)$
(b) $Na_2O(s) + H_2O(l) \rightarrow 2NaOH(aq)$

S2.23 (a) $SO_2(g) + 2NaOH(aq) \rightarrow Na_2SO_3(aq) + H_2O(l)$
$SO_2(g) + Ca(OH)_2(aq) \rightarrow CaSO_3(s) + H_2O(l)$
(b) $SO_3(g) + 2NaOH(aq) \rightarrow Na_2SO_4(aq) + H_2O(l)$
$SO_3(g) + Ca(OH)_2(aq) \rightarrow CaSO_4(s) + H_2O(l)$
(c) $CO_2(g) + 2NaOH(aq) \rightarrow Na_2CO_3(aq) + H_2O(l)$
$CO_2(g) + Ca(OH)_2(aq) \rightarrow CaCO_3(s) + H_2O(l)$

S2.25 $BaO_2(s) + H_2SO_4(aq) \xrightarrow{0^\circ C} BaSO_4(s) + H_2O_2(aq)$

S2.27 See Table S2.4.

S2.29 (a) $PbS(s) + 4H_2O_2(aq) \rightarrow PbSO_4(s) + 4H_2O(l)$
(b) $2Na_2O_2(s) + 2H_2O(l) \rightarrow 4NaOH(aq) + O_2(g)$
(c) $2Na_2O_2(s) + 2CO_2(g) \rightarrow 2Na_2CO_3(s) + O_2(g)$
(d) $4KO_2(s) + 2CO_2(g) \rightarrow 2K_2CO_3(s) + 3O_2(g)$

S2.35 The reactive species are free nitrogen atoms liberated by the discharge.

S2.37 (a) $N_2(g) + 3H_2(g) \xrightarrow[\text{catalyst}]{400°\text{C, 250 atm}} 2NH_3(g)$

(b) $6Li(s) + N_2(g) \xrightarrow{\text{heat}} 2Li_3N(s)$

(c) $3Mg(s) + N_2(g) \xrightarrow{\text{heat}} Mg_3N_2(s)$

S2.39 See answer S2.37(a) above.

S2.41 See Table S2.9.

S2.43 Some of the possible reactions are:

(a) $NH_4NO_3(s) \xrightarrow{\text{heat}} N_2O(g) + 2H_2O(g)$

(b) $N_2(g) + O_2(g) \xrightarrow[\text{electric discharge}]{\text{heat or}} 2NO(g)$

(c) $2NO(g) + O_2(g) \longrightarrow 2NO_2(g)$

(d) $2NO_2(g) \rightleftharpoons N_2O_4(g)$

S2.45 (a) $2HCl(aq) + Ca(NO_2)_2(aq) \xrightarrow{\text{cold}} 2HNO_2(aq) + CaCl_2(aq)$

(b) $3HNO_2(aq) \rightarrow HNO_3(aq) + 2NO(g) + H_2O(l)$

S2.47 (a) $4NH_3(g) + 5O_2(g) \xrightarrow[900°\text{C}]{\text{Pt}} 4NO(g) + 6H_2O(g)$

$2NO(g) + O_2(g) \rightarrow 2NO_2(g)$

$3NO_2(g) + H_2O(l) \rightarrow 2HNO_3(aq) + \underset{\text{recycled}}{NO(g)}$

S2.49 (a) $NO_3^-(aq) + 2H^+(aq) + e^- \rightarrow NO_2(g) + H_2O(l)$

(b) $NO_3^-(aq) + 4H^+(aq) + 3e^- \rightarrow NO(g) + 2H_2O(l)$

(c) $2NO_3^-(aq) + 10H^+(aq) + 8e^- \rightarrow N_2O(g) + 5H_2O(l)$

(d) $2NO_3^-(aq) + 12H^+(aq) + 10e^- \rightarrow N_2(g) + 6H_2O(l)$

(e) $NO_3^-(aq) + 10H^+(aq) + 8e^- \rightarrow NH_4^+(aq) + 3H_2O(l)$

S2.51 (a) $P(s) + 5HNO_3(aq) \rightarrow H_3PO_4(aq) + 5NO_2(g) + H_2O(l)$

(b) $S(s) + 6HNO_3(aq) \rightarrow H_2SO_4(aq) + 6NO_2(g) + 2H_2O(l)$

(c) $I_2(s) + 10HNO_3(aq) \rightarrow 2HIO_3(aq) + 10NO_2(g) + 4H_2O(l)$

S2.53 (a) $Na_2CO_3(s) + 2HNO_3(aq) \rightarrow 2NaNO_3(aq) + CO_2(g) + H_2O(l)$

(b) $Mg(OH)_2(s) + 2HNO_3(aq) \rightarrow Mg(NO_3)_2(aq) + 2H_2O(l)$

(c) $N_2O_5(s) + H_2O(l) \rightarrow 2HNO_3(aq)$

(d) $Li_3N(s) + 3H_2O(l) \rightarrow NH_3(g) + 3LiOH(aq)$

(e) $NaCN(s) + HCl(aq) \rightarrow NaCl(aq) + HCN(g)$

(f) $2NH_3(l) + OCl^-(aq) \rightarrow N_2H_4(aq) + Cl^-(aq) + H_2O(l)$

S2.55 (a) :N̈=Ö: ⟷ :N̈=Ȯ:

(b) :N̈=N=Ö: ⟷ :N≡N—Ö:

(c) O—N=O ⟷ O=N—O ⟷ O=N—O ⟷ O—N=O

(d) O=N(—O)—N=O ⟷ O—N(=O)—N=O

(e) Ö=N—N=Ö (with :Ö: :Ö: above each N) ⟷ :Ö—O—N—Ö: ⟷ :Ö—N—N=Ö ⟷ Ö=N—N—Ö

(f) :Ö—N—Ö—N—Ö: (with :O: above each N) ⟷ Ö=N—Ö—N=Ö (with :Ö: above each N)

S2.57 From S2.55: (b) linear; (c) bent; (d) left side trigonal planar, right side bent, free rotation about center bond; (e) trigonal planar around each nitrogen, free rotation about center bond; (f) trigonal planar around each nitrogen, bent around the central oxygen, free rotation around the single bonds

S2.59 (a) largely as halide ions, X^-; (c) NaCl

S2.61 See Table S2.11.

S2.65 (a) hypoiodous acid; (b) potassium chlorate; (c) perchloric acid; (d) iodic acid; (e) sodium perbromate; (f) magnesium hypochlorite

S2.67 (a) $CaF_2(s) + H_2SO_4(18\ M) \xrightarrow{\text{heat}} 2HF(g) + CaSO_4(s)$

$2HF \text{ (in KF)} \xrightarrow{\text{electrolysis}} H_2(g) + F_2(g)$

(b) $2NaCl(aq) + 2H_2O(l) \xrightarrow{\text{electrolysis}} 2NaOH(aq) + Cl_2(g) + H_2(g)$

(c) $\underset{\text{in seawater}}{2Br^-(aq)} + Cl_2(g) \rightarrow Br_2(l) + 2Cl^-(aq)$

(d) $2I^-(aq) + Cl_2(g) \rightarrow I_2(s) + 2Cl^-(aq)$

S2.69 Selected equations are

(a) $2Na(s) + I_2(g) \rightarrow 2NaI(s)$

(b) $Zn(s) + F_2(g) \rightarrow ZnF_2$

(c) $H_2(g) + Br_2(l) \rightarrow 2HBr(g)$

(d) $Cl_2(g) + H_2O(l) \rightleftharpoons HCl(aq) + HClO(aq)$

S2.71 (a) yes; $Cl_2(g) + 2NaI(aq) \rightarrow 2NaCl(aq) + I_2(s)$

(b) no

S2.73 (a) $CaF_2(s) + H_2SO_4(18\ M) \rightarrow CaSO_4(s) + 2HF(g)$

(b) $NaCl(s) + H_2SO_4(18\ M) \rightarrow NaHSO_4(s) + HCl(g)$

(c) $NaBr(s) + H_3PO_4(15\ M) \rightarrow HBr(g) + NaH_2PO_4(aq)$

(d) $NaI(s) + H_3PO_4(15\ M) \rightarrow HI(g) + NaH_2PO_4(aq)$

S2.75 (a) $Mg(s) + 2HF(g) \rightarrow MgF_2(s) + H_2(g)$
(b) $HF(aq) \rightleftharpoons H^+(aq) + F^-(aq)$
(c) $SiO_2(s) + 4HF(\text{g or aq}) \rightarrow SiF_4(g) + 2H_2O(g)$

S2.77 (a) $HOBr(+1)$; $HBrO_3(+5)$; $HBrO_4(+7)$
(b) $HOBr$; $HBrO_3$; $HBrO_4$
(c) $HBrO_4$; $HBrO_3$; $HOBr$

S2.79 (a) $I_2(s) + 2KOH(aq) \xrightarrow{\text{cool}} KI(aq) + KOI(aq) + H_2O(l)$
(b) $2HOCl(aq) \rightarrow 2HCl(aq) + O_2(g)$
(c) $2KClO_3(s) \xrightarrow[\text{heat}]{MnO_2(s)} 2KCl(s) + 3O_2(g)$

S2.81 (a) H—Ö—Cl—Ö: with :O: below Cl — trigonal pyramid   (b) [:Ö—I—Ö:]⁻ bent
(c) H—Ö—I: bent   (d) [:I—I—I:]⁻ linear
(e) :Ö—Cl—Ö: bent   (f) [ClO₄]⁻ (:O: above and below, :Ö—Cl—Ö: across) tetrahedral

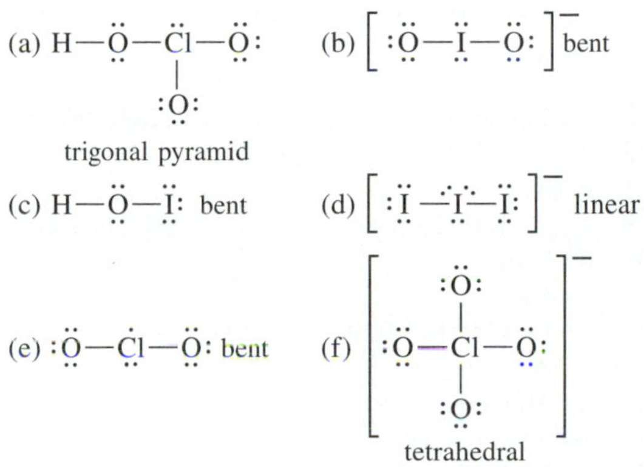

S2.83 (a) dark colored crystalline solid; (b) with metals to form astatides containing the At^- ion; very slightly with water to form a strong acid HAt and a weak acid HAtO; (c) no

S2.85 (a) Ca^{2+} cannot be oxidized further. (b) F_2 reacts with the steel to form a protective metal fluoride coating.

S2.87 (a) dinitrogen oxide; (b) iodine in alcohol solution; (c) concentrated nitric acid and hydrochloric acid in a 1:3 ratio; (d) potassium nitrate; (e) sodium nitrate; (f) calcium fluoride; (g) polytetrafluoroethylene (a polymer); (h) dichlorodifluoromethane; (i) calcium hypochlorite

S2.89 (a) 2.3×10^2 L; (b) 140 L

S2.91 1.3×10^3 kg

S2.93 (a) 14 M; (b) 0.88 M; (c) 0.15 M

S2.95 (a) 22 L

S2.97 642.3 kJ/mol

S2.99 1.1×10^9 kJ

CHAPTER 12

12.1 (a) gas, liquid, solid; (b) solid, liquid, gas

12.7 Vapor pressure increases

12.9 Lower

12.11 In addition to being hot, the steam releases heat as it condenses.

12.13 During any phase change, the particles lose or gain potential energy, but not kinetic energy.

12.15

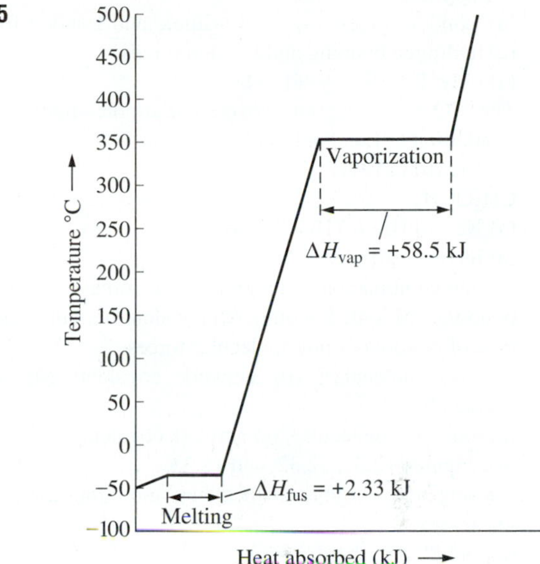

Heating curve for 1 mole of mercury, not drawn to scale

12.17 (a) 334 kJ; (b) 2.26×10^3 kJ

12.19 (a) 54.2 kJ; (b) 13.0 kcal

12.21 68°C

12.23 (a) gas; (b) supercritical fluid

12.25 Increase

12.27 Ice will sublime.

12.29 Yes; yes

12.31

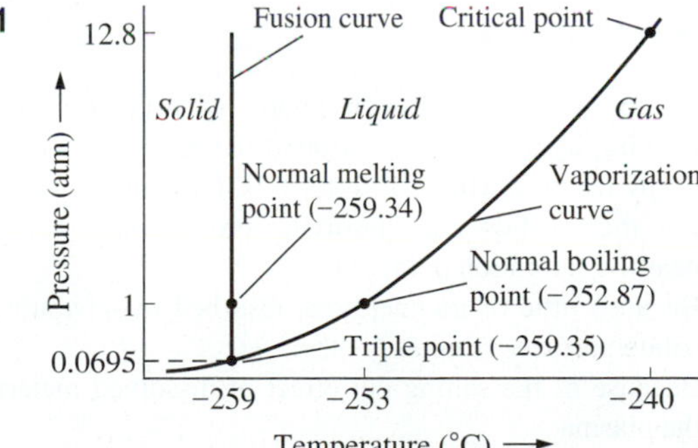

12.33 Greater
12.35 (a) three; (b) no; (c) increases
12.37 Liquid helium I
12.39 (a) graphite, diamond; (b) 4000°C
12.45 Usually linear.
12.49 (a) London forces; (b) dipole–dipole and London forces; (c) hydrogen bonding and London forces
12.51 (a) only; H—F----H—F
12.53 The NO molecules must overcome dipole–dipole forces in addition to London forces.
12.55 (a) and (b) CH_3OH
12.57 CH_3OCH_3
12.59 (a) Xe; (b) HF; (c) $HOCH_2CH_2OH$
12.63 (a) increase; (b) decrease
12.65 (a) any combination of intermolecular forces; (b) metallic bonding; (c) ionic bonding; (d) covalent bonding; (e) any type of bonding or intermolecular forces
12.69 (a), (c) molecular; (b) network covalent; (d) ionic; (e) metallic
12.71 (a) ionic; (b) molecular; (c) network covalent
12.75 See Figures 12.31, 12.32, and 12.33.
12.81 Missing atoms or ions, dislocations, and impurities
12.83 Decreases
12.85 (a) 144.20 pm; (b) 19.29 g/cm^3
12.87 7.93 g/cm^3
12.89 (a) 4, face-centered cubic; (b) 175.03 pm
12.91 (a) see Figure 12.37; (b) 2; (c) 12
12.93 See Figure 12.39. (a) 6; (b) 6
12.95 303 pm
12.101 Butane
12.107 49°C
12.109 (a) 8; (c) no
12.111 2.16 g/cm^3
12.113 A NaCl type of lattice (see Figure 12.39)

CHAPTER 13

13.5 (a) by forming hydrogen bonds with ethanol; (b) by forming hydration layers around the K^+ and NO_3^- ions; (c) by reacting with HBr to form $H^+(aq)$ and $Br^-(aq)$
13.7 For most solids, the dissolving process is endothermic near the saturation point.
13.9 Because little or no energy is absorbed in separating the solute particles from each other.
13.13 Because of the salting-out effect of dissolved material in the plasma
13.15 0.12 mol/L
13.19 (a) Negative deviations from Raoult's law occur when the attractive forces between the component molecules A and B of the solution are greater than the A—A forces or the B—B forces. (b) Positive deviations from Raoult's law occur when the attractive forces between the component molecules A and B of the solution are less than the A—A forces or the B—B forces.
13.21

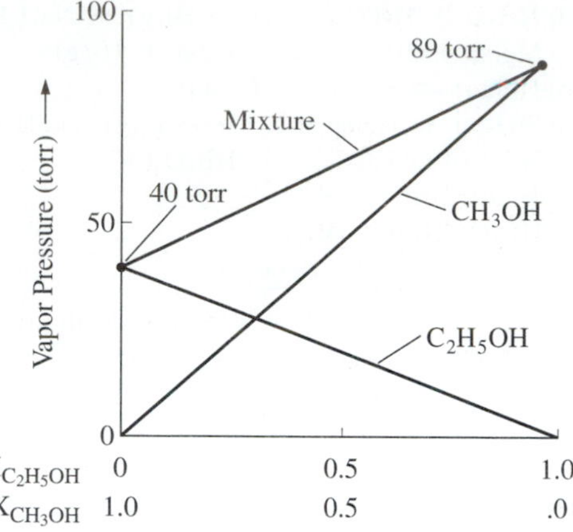

13.23 (a) 209 torr CS_2, 46.3 torr $CH_3COOC_2H_5$; (b) 255 torr
13.25 $X_{water} = 0.9470$; $X_{urea} = 0.0530$
13.27 Positive
13.29 $X_{benzene}$ is 0.528 in the vapor and 0.25 in the solution.
13.33 0.192 *m*
13.35 1166 g
13.37 (a) 1.09 *M*; (b) 0.0200; (c) 1.13 *m*
13.39 2.7 *m*
13.43 (a) All increase. (b) All decrease.
13.45 (a) 0.10 *m* glucose; (b) 0.010 *m* urea
13.47 (a) 5.0°C; (b) −2.81°C
13.49 (a) 55.1 torr; (b) 100.08°C; (c) −0.31°C
13.51 0.56°C
13.53 Chloroform
13.55 (a) 20°C·kg/mol; (b) 1.3×10^2 g/mol
13.57 610 g/mol
13.59 84.2 g/mol
13.61 (a), (d) higher for seawater; (b), (c) lower for seawater
13.63 (a) because of the association of oppositely charged ions; (b) because the degree of association decreases with dilution
13.65 1.0 *m* $CaCl_2$
13.67 (a) 100.256°C; (b) freezing point higher, boiling point lower
13.69 (a) 1.08 mol/kg; (b) 100.553°C
13.71 2.17
13.73 (a) −0.348°C; (b) 100.096°C; (c) 4.53 atm
13.79 See Table 13.5.
13.83 The colloidal particles form hydrogen bonds with the water.
13.85 Adsorbed ions provide all the colloidal particles with charges of the same sign, so they repel each other.
13.87 The heat of solution is endothermic at the saturation point.
13.93 (a) 20.2 mL; (b) 47.3 g
13.95 (a) $X_{benzene} = 0.658$, $X_{toluene} = 0.342$; (b) 891 torr
13.97 (a) 1.1×10^3 g; (b) 108.7°C; (c) no
13.99 2.4 *m*
13.101 (a) 2.9×10^2 g/mol; (b) $C_{19}H_{28}O_2$, 288.43 g/mol

CHAPTER 14

14.9 4

14.15 $\frac{\Delta[H_2O]}{\Delta t} = \frac{6}{4}\frac{\Delta[NO]}{\Delta t}$; $\frac{-\Delta[NH_3]}{\Delta t} = \frac{\Delta[NO]}{\Delta t}$;

and $\frac{-\Delta[O_2]}{\Delta t} = \frac{5}{4}\frac{\Delta[NO]}{\Delta t}$

14.17 (a) 4.06×10^{-4} mol/s; (b) 3.25×10^{-2} mol/L·s
14.19 (a) 2.1×10^{-3} mol/L·min; (b) 4.5×10^{-4} mol/L·min
14.21 1.57×10^{-3} mol/L·min
14.25 The sum of the exponents in the rate law.
14.27 In the rate law, the constant k, determined by experiment.
14.29 (a) s^{-1}; (b) L/mol·s
14.31 Rate $= k[H_2]^2[N_2]$
14.33 (a) rate $= k[A]^2[B]$, third order; (b) 2.0 L^2/mol^2·s
(c) 2.0×10^{-6} mol/L·s
14.35 (a) rate $= k[C_2F_4]^2$, second order; (b) 6.21×10^{-14} L/mmol·s; (c) 2.48×10^{-15} mmol/L·s
14.37 (a) rate $= k[N_2O_4][N_2]$; (b) 1.71×10^6 L/mol·s;
(c) first order, first order
14.43 (b) 3.19×10^{-3} min^{-1}; (c) 70 min
14.45 (a) 0.17 mol; (b) 1.1 atm
14.47 (a) 1.39×10^3 s; (b) 5.43 min
14.49 17 min
14.51 128 min
14.53 (a) 480 min
14.55 (b) 0.0529 L/mol·min; (c) 62.9 min
14.63 (a) rate $= k[Cl_2]$; (b) rate $= k[Cl]^2$
14.65 Rate $= k[ClO_2][F_2]$;
Step 1: $ClO_2(g) + F_2(g) \rightarrow FClO_2(g) + F(g)$ (slow)
Step 2: $F(g) + ClO_2(g) \rightarrow FClO_2(g)$ (fast)
14.67 Rate $= k[H_2][I_2]$

14.69 (a) $2NO(g) + O_2(g) \rightarrow 2NO_2(g)$
N_2O_2 is an intermediate; (b) rate $= k[NO]^2[O_2]$

14.73 (a) Chain initiation step:

$Cl_2(g) \xrightarrow{h\nu} Cl\cdot(g) + Cl\cdot(g)$

Chain propagation steps:

$Cl\cdot(g) + CH_4(g) \rightarrow CH_3\cdot(g) + HCl(g)$
$CH_3\cdot(g) + Cl_2(g) \rightarrow CH_3Cl(g) + Cl\cdot(g)$

Chain termination steps:

$Cl\cdot(g) + Cl\cdot(g) \rightarrow Cl_2(g)$
$CH_3\cdot(g) + CH_3\cdot(g) \rightarrow C_2H_6(g)$
$CH_3\cdot(g) + Cl\cdot(g) \rightarrow CH_3Cl(g)$

(b) The chain propagation steps
14.75 The larger the activation energy, the slower the rate.

14.77

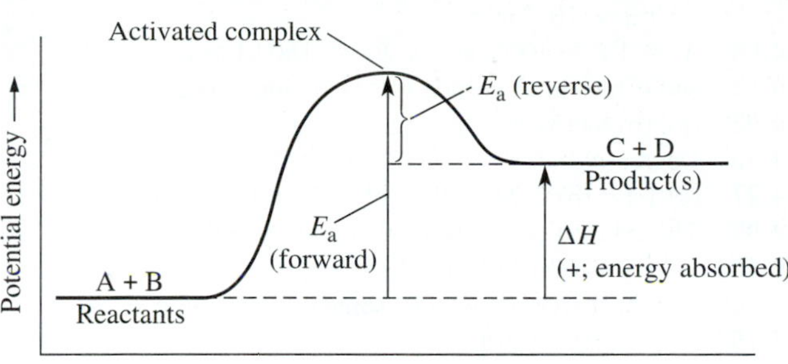

14.79

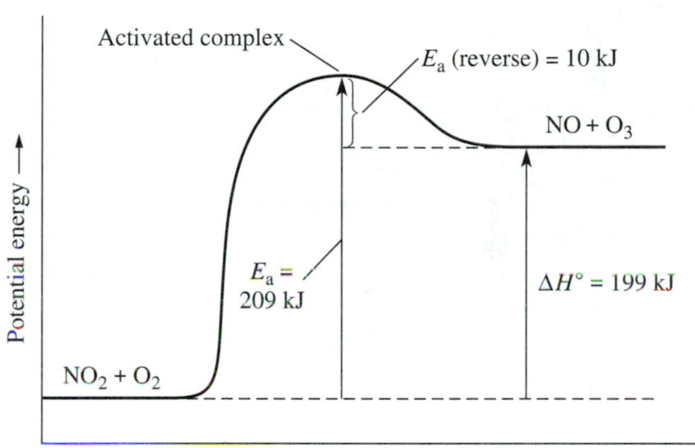

14.81 (a)

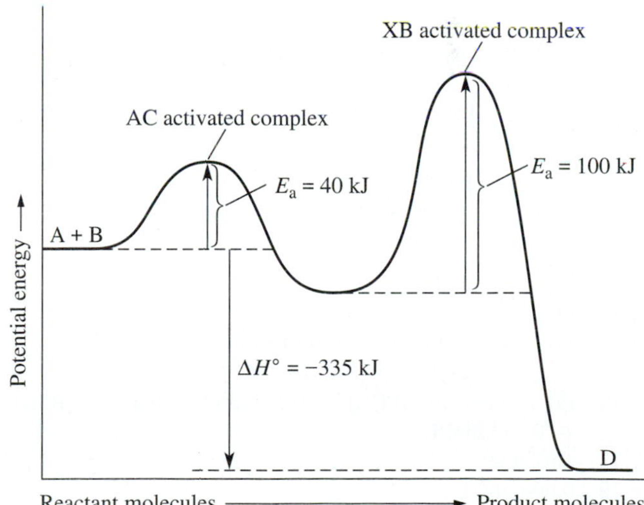

(b) step 2; (c) −335 kJ
14.83 (a) decreases; (b) increases
14.85 53 kJ/mol

14.87 (a) 183 kJ/mol; (b) 1.71×10^{16} min^{-1}; (c) 1.3×10^{-4} min^{-1}
14.89 $A = 1.8 \times 10^8$ L/mol·s, $E_a = 160$ kJ/mol
14.91 Molecules must collide for a reaction to occur.
14.93 (a), (b) increases
14.95 Decreases
14.97 (a) 1.17; (b) 2.29×10^4; (c) 2.68×10^4
14.99 The catalyst provides a lower-energy pathway.
14.103 (a) 119 kJ; (b) 289 kJ
14.105 The discharge dissociates some of the N_2 into atoms.
14.107 2.52×10^{-5} mol/L·s
14.109 15 s
14.111 (a) $[A] = -kt + [A]_0$; (c) $t_{1/2} = [A]_0/2k$
14.113 $\left(\frac{1}{[A]_0 - [B]_0}\right) \ln \frac{[A][B]_0}{[A]_0[B]} = kt$
14.115 Yes
14.117 (a) $CHCl_3(g) + Cl_2(g) \rightarrow CCl_4(g) + HCl(g)$
Cl and CCl_3 are intermediates.
(b) rate $= k[CHCl_3][Cl_2]^{1/2}$
14.119 (a) 0.0099 h^{-1} at 65.0°C, 0.179 h^{-1} at 70.2°C; (b) 540 kJ/mol
14.121 $E_a = 96$ kJ/mol, $A = 2.3 \times 10^{15}$ s^{-1}
14.123 2.4×10^{-2}

CHAPTER 15

15.3 The rates approach equality.
15.5 (a) $Q_c = \frac{[CH_3OH]}{[CO][H_2]^2}$ $K_c = \frac{[CH_3OH]_{eq}}{[CO]_{eq}[H_2]^2_{eq}}$
(b) Q_c is a variable that describes the system at any time. K_c is a constant that applies only to equilibrium concentrations.
15.7 (a) reverse; (b) forward
15.9 (a) $Q_c = \frac{[SO_2]^2[H_2O]^2}{[H_2S]^2[O_2]^3}$
(b) $Q_c = \frac{[Ag^+][CN^-]^2}{[Ag(CN)_2^-]}$
(c) $Q_c = \frac{[I_3^-]}{[I_2][I^-]}$
15.11 O_2
15.13 (a) $[Cl_2] = [PCl_3] = 0.0470$ mol/L, $[PCl_5] = 0.053$ mol/L; (b) 0.042; (c) 0.470
15.15 (a) reverse; (b) $[PCl_5] = 0.127$ mol/L; $[Cl_2] = [PCl_3] = 0.073$ mol/L
15.17 1.87 mol
15.19 0.096 mol/L
15.21 0.090 mol/L, 6.7%
15.23 0.0605 mol/L N_2, 0.182 mol/L H_2, 0.026 mol/L NH_3
15.25 (a) $K_c = \frac{[CrO_4^{2-}]^2}{[Cr_2O_7^{2-}][OH^-]^2}$
(b) $K_c = \frac{[HCO_3^-][OH^-]}{[CO_3^{2-}]}$
(c) $K_c = [H^+][OH^-]$
15.27 9.62×10^{-3} mol/L
15.29 Gas-phase reactions; atm
15.33 (a) $Q_p = P^3_{O_2}/P^2_{O_3}$; (b) $Q_p = P_{C_2H_4}/P_{C_2H_2}P_{H_2}$
15.35 4.14×10^{-2} mol/L
15.37 Increase
15.39 (a) 2.3 atm PCl_5, 2.02 atm PCl_3, 2.02 atm Cl_2; (b) 1.8
15.41 0.212 atm NO_2, 0.394 atm N_2O_4
15.43 (a) 1.67 atm Br_2, 0.654 atm Br; (b) 0.164
15.45 1.49×10^{-2} atm
15.47 (a) $K_c = K_p/RT$; (b) $K_c = K_pRT$
15.49 3.1×10^{14}
15.51 (a) 2.7×10^{-2}; (b) 3.6×10^{-3}
15.53 (a) 1.9×10^6; (b) 1.4×10^3
15.55 3.2×10^{-8}
15.59 (a) $K_p = P_{H_2}P_{CO}/P_{H_2O}$; (b) $K_p = P^4_{H_2O}/P^4_{H_2}$; (c) $K_p = 1/P_{Cl_2}$
15.61 (a) $K_p = P_{O_2}$; (b) $K_p = P_{H_2O}$; (c) $K_p = 1/P_{SO_2}$
15.63 (a) $K_c = [Bi^{3+}]^2[S^{2-}]^3$; (b) $K_c = [H^+][Br^-][HOBr]$
15.65 124.8 atm
15.67 0.11
15.69 1.22×10^{90}
15.73 (a) becomes more yellow; (b) becomes more orange
15.75 (a), (b), (c) color deepens; (d) no change
15.77 endothermic
15.79 (a) increases; (b) decreases
15.81 (a) decreases; (b) increases; (c), (d) no change
15.83 58 kJ/mol
15.85 3.9×10^{-8}
15.87 (a) $Ag(S_2O_3)_2^{3-}$; (b) 1.0 *M* $Ag(NH_3)_2^+$
15.89 (a) forward; (b) $[CH_3COOH] = [CH_3COO^-] = 0.050$ *M*, $[H^+] = 1.76 \times 10^{-5}$ *M*
15.91 16.1 mol %
15.93 0.450 mol/L
15.95 0.0230
15.97 (a) increase; (b), (c) decrease
15.99 0.390 mol/L

CHAPTER 16

16.3 The molecules of Brønsted acids all have protons that can be donated.
16.7 Having the ability to both accept and donate protons. Amphiprotic molecules have protons and lone electron pairs.
16.9 Both are basic with respect to water.
16.11 (a) $HCO_3^-(aq) + HCl(aq) \rightarrow H_2O(l) + CO_2(g) + Cl^-(aq)$
$HCO_3^-(aq) + OH^-(aq) \rightarrow CO_3^{2-}(aq) + H_2O(l)$
(b) $H_2PO_4^- + HCl(aq) \rightarrow H_3PO_4(aq) + Cl^-(aq)$
$H_2PO_4^- + OH^-(aq) \rightarrow HPO_4^{2-}(aq) + H_2O(l)$

16.13 (a) $2CH_3OH(l) \rightleftharpoons CH_3OH_2^+ + CH_3O^-$
(b) $2HCl(l) \rightleftharpoons H_2Cl^+ + Cl^-$

16.19 NH_4^+ is stronger than $C_2H_5NH_3^+$.

16.21 H^+ is the strongest acid; OH^- the strongest base. Stronger acids and bases are completely converted to their conjugate forms.

16.23 Water differentiates between acids weaker than H_3O^+ and bases weaker than OH^-.

16.25 Compare their degree of ionization in aqueous solution.

16.27 Compare their degree of ionization in a very acidic solvent like pure acetic acid.

16.29 (a) HSO_4^-; (b) SO_4^{2-}; (c) PH_3; (d) H_2O; (e) OH^-; (f) $Cr(H_2O)_5(OH)^{2+}$; (g) $C_2H_5O^-$; (h) HS^-

16.31 (a) NH_3 and NH_4^+; HCO_3^- and CO_3^{2-}; (b) CH_3NH_2 and $CH_3NH_3^+$; HCl and Cl^-; (c) H_3O^+ and H_2O; OH^- and H_2O

16.33 OH^- and H_2O; H_3O^+ and H_2O

16.35 (a) $HI(aq) + H_2O(l) \rightarrow H_3O^+(aq) + I^-(aq)$
(b) $NaNH_2(s) + H_2O(l) \rightarrow NaOH(aq) + NH_3$

16.37 (a) $HNO_2(aq) + H_2O(l) \rightleftharpoons H_3O^+(aq) + NO_2^-(aq)$
(b) $HCN(aq) + H_2O(l) \rightleftharpoons H_3O^+(aq) + CN^-(aq)$
(c) $HF(aq) + H_2O(l) \rightleftharpoons H_3O^+(aq) + F^-(aq)$
(d) $CH_3COOH(aq) + H_2O(l) \rightleftharpoons H_3O^+(aq) + CH_3COO^-(aq)$

16.39 (a) $H_2SeO_4(aq) \rightarrow H^+(aq) + HSeO_4^{2-}(aq)$
$HSeO_4^-(aq) \rightleftharpoons H^+(aq) + SeO_4^-(aq)$
(b) $H_2SeO_3(aq) \rightleftharpoons H^+(aq) + HSeO_3^-(aq)$
$HSeO_3^-(aq) \rightleftharpoons H^+(aq) + SeO_3^{2-}(aq)$

16.41 (a) $HCOOH(aq) + H_2O(l) \rightleftharpoons H_3O^+(aq) + HCOO^-(aq)$
(b) $H_2C_2O_4(aq) + H_2O(l) \rightleftharpoons H_3O^+(aq) + HC_2O_4^-(aq)$
$HC_2O_4^-(aq) + H_2O(l) \rightleftharpoons H_3O^+(aq) + C_2O_4^{2-}(aq)$

16.43 (a) $NH_4^+(aq) + H_2O(l) \rightleftharpoons H_3O^+(aq) + NH_3(aq)$
(b) $O^{2-} + H_2O(l) \rightarrow 2OH^-(aq)$
(c) $Cd(H_2O)_4^{2+}(aq) + H_2O(l) \rightleftharpoons H_3O^+(aq) + Cd(H_2O)_3(OH)^+(aq)$
(d) $H_2PO_4^-(aq) + H_2O(l) \rightleftharpoons H_3O^+(aq) + HPO_4^{2-}(aq)$
and $H_2PO_4^-(aq) + H_2O(l) \rightleftharpoons H_3PO_4(aq) + OH^-(aq)$

16.45 (a) $Zn(s) + 2HCl(aq) \rightarrow H_2(g) + Zn^{2+}(aq) + 2Cl^-(aq)$
(b) $Zn(H_2O)_2(OH)_2(s) + 2HNO_3(aq) \rightarrow Zn(H_2O)_4^{2+}(aq) + 2NO_3^-(aq)$
or $Zn(OH)_2(s) + 2HNO_3(aq) \rightarrow Zn(NO_3)_2(aq) + 2H_2O(l)$
(c) $2HClO_4(aq) + Ba(OH)_2(aq) \rightarrow Ba(ClO_4)_2(aq) + 2H_2O(l)$
(d) $NaOH(aq) + H_2C_2O_4(aq) \rightarrow NaHC_2O_4(aq) + H_2O(l)$

16.47 (a) $C_6H_5COOH(aq) + OH^-(aq) \rightarrow C_6H_5COO^-(aq) + H_2O(l)$
(b) $NH_3(aq) + H_3O^+(aq) \rightarrow NH_4^+(aq) + H_2O(l)$
(c) $HCOOH(aq) + F^-(aq) \rightleftharpoons HCOO^-(aq) + HF(aq)$

16.49 +146 kJ

16.51 No

16.55 (a) 8.00×10^{-15} *M;* (b) 2.9×10^{-15} *M;* (c) 2.00×10^{-13} *M*

16.57 (a) 1.82; (b) 0.824; (c) 0.82391; (d) −0.176

16.59 5×10^{-3} *M,* acidic

16.61 1×10^{-1} *M,* basic

16.63 (a) pH −0.097, pOH 14.097; (b) pH −0.53, pOH 14.53; (c) pH 1.301, pOH 12.699

16.65 (a) 6.7×10^{-13} *M,* pOH 12.17; (b), (c) 6.67×10^{-14} *M,* pOH 13.176; (d) 6.67×10^{-15} *M,* pOH 14.176

16.67 Trichloracetic acid; it donates more protons to the solvent.

16.69 0.10 *M* HF: 5.9×10^{-3} *M* H^+, 1.7×10^{-12} *M* OH^-
0.10 *M* trichloracetic acid: 7.4×10^{-2} *M* H^+, 1.4×10^{-13} *M* OH^-

16.71 Acidic, 3×10^{-6} *M*

16.77 (a) PH_4^+ (b) H_2S (c) $H_2PO_4^-$

16.79 (a) The charge on Br^- is more delocalized than on Cl^-, so Br^- is a weaker base. (b) Cl is more electronegative than Br. (c) HNO_3 has an additional electronegative oxygen atom. (d) Al^{3+} has a greater positive charge.

16.81 (a) HI; (b) H_3PO_4; (c) H_2S; (d) HCl; (e) $Cr(H_2O)_6^{3+}$; (f) $HClO_4$; (g) $CH_3CH_3ClCOOH$

16.85 "Electrophile" means "electron-loving." Lewis bases are called "nucleophiles."

16.87 A positive region that will attract an electron pair. BF_3, CO_2, and Al^{3+} are examples of Lewis acids.

16.89

(a) :F̈—B(—:F̈:)(—:F̈:) + :F̈:⁻ ⟶ [:F̈—B(—:F̈:)(—:F̈:)—F̈:]⁻

Lewis acid Lewis base New covalent bond

This bond dissociates

(b) CH_3—N̈(H)(H): + H—C̈l: ⟶ [CH_3—N(H)(H)—H]⁺ + :C̈l:⁻

Lewis base Lewis acid New covalent bond

16.93 No

16.95 (a) $3CaO(s) + 2H_3PO_4(aq) \rightarrow Ca_3(PO_4)_2(s) + 3H_2O(l)$
(b) $12NaOH(aq) + P_4O_{10}(s) \rightarrow 4Na_3PO_4(aq) + 6H_2O(l)$
(c) $CaO(s) + SO_2(g) \rightarrow CaSO_3(s)$
(d) $MgO(s) + 2HNO_3(aq) \rightarrow Mg(NO_3)_2(aq) + H_2O(l)$
(e) $Ba(OH)_2(aq) + SO_2(aq) \rightarrow BaSO_3(aq) + H_2O(l)$
(f) $Cr_2O_3(s) + 3H_2SO_4(aq) \rightarrow Cr_2(SO_4)_3(aq) + 3H_2O(l)$
(g) $2KOH(aq) + CO_2(aq) \rightarrow K_2CO_3(aq) + H_2O(l)$

16.97 13.876

16.99 Moles H^+ per tablet: A, 0.0184; B, 0.0231; C, 0.00999
16.101 343 mL
16.103 (a) 0.390 g; (b) 48.8%

CHAPTER 17

17.1 (b) CCl_3COOH
17.3 Stronger
17.5 (a) 1×10^{-4} M H^+, 1×10^{-10} M OH^-
17.9 1.2×10^{-5}
17.11 1.56×10^{-11}
17.13 (a) 1.1×10^{-3} M; (b) 1.2×10^{-5}
17.15 (a) 2.304, yes; (b) 2.824, no
17.17 (a), (b) decreases; (c), (d), (f) increases; (e) remains same
17.19 (a) 3.7%; (b) 3.8%
17.21 (a) 3.47; (b) 2×10^{-4}
17.23 (a) 4.955; (b) 4.44×10^{-5}
17.25 The negative ionic charge makes proton removal more difficult.
17.27 $[H^+] = [HSO_3^-] = 0.0428$ M; $[SO_3^{2-}] = 6.0 \times 10^{-8}$ M
17.29 (a) 3.92; (b) 1.2×10^{-4} M HCO_3^-, 5.6×10^{-11} CO_3^{2-}
17.35 (a) $Al(NO_3)_3$, acidic; (b) KOBr, basic; (c) $Ca(HCO_3)_2$, basic; (d) NH_4ClO_4, acidic
17.37 KCN is more basic.
17.39 $AlCl_3$ is less acidic.
17.41 (a) 5.65×10^{-10}; (b) 5.04
17.43 11.08
17.45 11.19
17.47 12.97
17.49 (a), (d) decreases; (b), (c), (e) increases; (f) unchanged
17.51 4.00
17.53 10.25
17.55 (a) 2.8×10^{-3} M H^+, 1.6×10^{-12} M $C_6H_6O_6^{2-}$
(b) 0.010 M H^+, 1.3×10^{-13} M $C_6H_6O_6^{2-}$
17.61 No changes
17.63 (a) through (e); the two should be approximately equal.
17.65 One good choice would be acetic acid and sodium acetate.
17.67 (a) 4.01; (b) 3.71
17.69 58.8 g
17.71 (a) 4.3 g; (b) 3.52
17.73 (a) from 4.754 to 1.004; (b) from 4.754 to 4.67
17.75 (a) 8.85; (b) 8.72
17.79 The indicator itself is a weak acid or base.
17.81 11.04
17.83 The pH values are (a) 13.30; (b) 7; (c) 12.70; (d) 10.8; (e) 3.2.
17.85 For Exercise 17.83, any indicator with a transition interval between pH 4 and 10
17.87 (a) Any indicator in Table 17.3 from methyl red through phenolphthalein; (b) phenolphthalein; (c) methyl red
17.89 Points on the curve would be: 1.1% at 0.50 M, 2.5% at 0.10 M, 3.6% at 0.050 M, 7.7% at 0.010 M.
17.93 (a), (b) increases; (c) decreases; (d) unchanged
17.95 0.89
17.97 8.42
17.99 0.6 g
17.101 (a) 0.14 M; (b) 2.99
17.103 $[H_2PO_4^-] = 0.0452$ M, $[HPO_4^{2-}] = 0.0014$ M
17.105 0.79 mol
17.107 0.034 mol
17.109 When the titration involves: (a) weak acid, strong base; (b) strong acid, strong base; (c) weak base, strong acid
17.111 6.56×10^{-2} M
17.113 (a) 9.9×10^{-6} M; (b) 1.9×10^{-6} M HCO_3^-, 5.6×10^{-11} M CO_3^{2-}

CHAPTER 18

18.5 (a) $[Pb^{2+}][SO_4^{2-}]^2$; (b) $[Sb^{3+}]^2[SO_4^{2-}]^3$; (c) $[Fe^{3+}][OH^-]^3$
18.7 (a) 5.3×10^{-6}; (b) 6×10^{-5}; (c) 2.0×10^{-12}
18.9 (a) 3.0×10^{-5} M, 4.4×10^{-4} g/L; (b) 1.3×10^{-3} M, 6.0×10^{-2} g/L
18.11 Ag_2CO_3
18.13 (a) 1.5×10^{-2} mol/L; (b) 4.2×10^{-3} mol/L; (c) 1.4×10^{-3} mol/L
18.15 (a) 6.7×10^{-2} g; (b) 9.0×10^{-6} g
18.17 (a) 3.0×10^{-13}; (b) 8.7×10^{-7} mol/L
18.19 12.38
18.21 9.0×10^{-4}
18.23 (a) equal to; (b) less than; (c) greater than
18.25 (a), (b) no; (c) yes
18.27 (a) $[Ca^{2+}] = [CrO_4^{2-}] = 0.010$ M
$[Cl^-] = [K^+] = 0.020$ M;
(b) $[Ba^{2+}] = 5.0 \times 10^{-3}$ M, $[Cl^-] = 1.0 \times 10^{-2}$ M
$[Na^+] = [F^-] = 3.0 \times 10^{-3}$ M
18.29 4.4×10^{-6}%
18.31 (a) 2.4×10^{-3} M; (b) $[Ba^{2+}] = 0.225$ M, $[Ca^{2+}] = 6.8 \times 10^{-6}$ M
18.33 (a) 0.014 M; (b) $[Ca^{2+}] = 0.050$ M, $[Ba^{2+}] = 8.6 \times 10^{-9}$ M
18.37 (a) $[Fe^{3+}] = 1.6 \times 10^{-18}$ M, $[Pb^{2+}] = 0.12$ M; (b) 3.0
18.39 14.0
18.41 $Mn(OH)_2$ and $Fe(OH)_2$ will precipitate.
18.43 $[NH_4^+]/[NH_3] = 16$ or greater
18.47 (a) $Sn(OH)_2(s) + 2HCl(aq) \rightarrow SnCl_2(aq) + 2H_2O(l)$
(b) $Cr(OH)_3(s) + NaOH(aq) \rightarrow NaCr(OH)_4(aq)$
18.49 Strong base will dissolve $Al(OH)_3$ but not $Fe(OH)_3$.
18.53 CuCN and Cu_2S

18.55 Add HCl(aq).

18.57 (a) $CaC_2O_4(s) + 2HCl(aq) \rightarrow CaCl_2(aq) + H_2C_2O_4(aq)$
(b) $CaCO_3(s) + 2HCl(aq) \rightarrow CaCl_2(aq) + H_2O(l) + CO_2(g)$
(c) $MnS(s) + 2HCl(aq) \rightarrow MnCl_2(aq) + H_2S(g)$

18.59 $2.9 \times 10^{-4}\ M$

18.61 Less soluble

18.63 A K_{spa} is the equilibrium constant for the dissolving of a sparingly soluble salt of a weak acid in an acidic solution.

18.65 $4 \times 10^{-6}\ M$

18.67 CuS will precipitate.

18.69 (a) $3\ M\ H^+$; (b) $[Sn^{2+}] = 0.0010\ M$, $[Pb^{2+}] = 3 \times 10^{-5}\ M$

18.71 (a) Less CdS will precipitate. (b) Some ZnS will precipitate with the CdS.

18.75 $Ag(CN)_2^-$

18.77 1.5×10^{-20}

18.79 (a) $Sn(OH)_2(s) + 2OH^-(aq) \rightarrow Sn(OH)_4^{2-}(aq)$
(b) $AgC_2H_3O_2(s) + 2NH_3(aq) \rightarrow Ag(NH_3)_2^+ + C_2H_3O_2^-(aq)$
(c) $Cu_2O(s) + 6CN^-(aq) + H_2O(l) \rightarrow 2Cu(CN)_3^{2-}(aq) + 2OH^-(aq)$

18.81 $1.0 \times 10^{-3}\ M$

18.83 $4.1 \times 10^{-16}\ M$

18.85 $[Hg^{2+}] = 1.3 \times 10^{-14}\ M$, $[Cl^-] = 0.16\ M$

18.87 $2 \times 10^{-4}\ M$

18.89 $[Zn(NH_3)_4^{2+}] = 0.015\ M$, $[Zn^{2+}] = 1.4 \times 10^{-10}\ M$, $[NH_3] = 0.44\ M$

18.91 $4.5 \times 10^{-5}\ M$

18.93 An equilibrium concentration of $3.8 \times 10^{-2}\ M\ S_2O_3^{2-}$ is needed (plus enough $S_2O_3^{2-}$ to complex all the Ag^+).

18.95 $K_{sp} = 6.25 \times 10^{-5}$; decreases

18.97 $8.8 \times 10^{-11}\ M$

18.99 Na_2CO_3, NaOH, KI, and $Na_2C_2O_4$, et al.

18.101 2.3×10^{-3}

18.103 $4.4 \times 10^{-7}\ M$

18.105 (a) 0.50 mmol; (b) 1.5×10^{-4} mmol AgCl, 0.37 mmol $PbCl_2$

18.107 (a) High pH favors. (b) Precipitation would be initiated somewhere between 1.3×10^{-4} and 2.4×10^{-3} mmol oxalate per day.

18.109 3.4×10^{-2} mol/L

18.111 $1.0 \times 10^{-5}\ M$

18.113 0.0375 mol

SURVEY 3

S3.1 (b) tellurium (c) selenium and tellurium

S3.11 See Figure S3.4. The ring is puckered so that the bond angles can be more nearly tetrahedral.

S3.13 Some of the resonance structures are

(a) O=S–O ↔ O–S=O ↔ O=S=O

(b) O=S(–O)–O (with S–O single bond to top O) ↔ O–S(=O)–O ↔ O–S(–O)=O ↔ O=S(=O)=O

(c) H–O–S(–O)–O–H ↔ H–O–S(=O)–O–H

(d) $[O–S(–O)(–O)–O]^{2-}$ ↔ $[O–S(=O)(=O)–O]^{2-}$

S3.15

(a) SF_6: S bonded to six F atoms

(b) Cl–S–S–Cl

(c) Cl–S–Cl

S3.17 Sulfur can exhibit octet expansion; oxygen cannot.

S3.19 See Table S3.3.

S3.21 (a) +2; (b) +5; (c) +2.5

S3.23 (a) $Na_2SO_3(aq) + 2HCl(aq) \rightarrow 2NaCl(aq) + SO_2(g) + H_2O(l)$
(b) $2S_2O_3^{2-}(aq) + I_2(s) \rightarrow S_4O_6^{2-}(aq) + 2I^-(aq)$
(c) $3S^{2-}(aq) + 2CrO_4^{2-}(aq) + 8H_2O(l) \rightarrow 3S(s) + 2Cr(OH)_4^-(aq) + 8OH^-(aq)$
(d) $Na_2SO_3(aq) + S(s) \xrightarrow{heat} Na_2S_2O_3(aq)$

S3.25 (a) $SO_2(g) + H_2O(l) \rightleftharpoons H_2SO_3(aq)$
(b) $SO_2(g) + OH^-(aq) \rightarrow HSO_3^-(aq)$ (limited base; hydrogen sulfite ion)
$SO_2(g) + 2OH^-(aq) \rightarrow SO_3^{2-}(aq) + H_2O(l)$ (excess base; sulfite ion)
(c) $CaCO_3(s) + SO_2(g) \rightarrow CaSO_3(s) + CO_2(g)$

S3.27 The effects on wood and on $CuSO_4 \cdot 5H_2O$ are due to the dehydrating action of concentrated H_2SO_4. H_2SO_4 oxidizes the Br^- in NaBr to brown Br_2, and oxidizes metallic copper to Cu^{2+}.

S3.29 (a) S has the same number of valence electrons as O. (b) The acid strength is diminished.

S3.31 (b) arsenic and antimony

S3.33 See Figure S3.7.

$$2Ca_3(PO_4)_2(l) + 6SiO_2(l) \xrightarrow{heat} P_4O_{10}(g) + 6CaSiO_3(l)$$

$$P_4O_{10}(g) + 10C(s) \xrightarrow{heat} P_4(g) + 10CO(g)$$

S3.35 See Figure S3.11.

S3.37 See Figures S3.8 and S3.13.

S3.39 Refer to Figure S3.14. In H_3PO_3, the H that is bonded to P is not acidic because the P—H bond has very little polarity.

S3.41 (a) $P_4(s) + 3O_2(g) \xrightarrow{\text{limited oxygen}} P_4O_6(s)$

(b) $P_4(s) + 5O_2(g) \xrightarrow{\text{excess oxygen}} P_4O_{10}(s)$

(c) $P_4(l) + 3S(l) \xrightarrow{heat} P_4S_3(s)$ ($+P_4S_5(s)$, $P_4S_7(s)$, etc.)

(d) $P_4(s) + 12Na(s) \xrightarrow{heat} 4Na_3P(s)$

S3.43 (a) $P_4O_6(s) + 6H_2O(l) \rightarrow 4H_3PO_3(aq)$

(b) $P_4O_{10}(s) + 6H_2O(l) \rightarrow 4H_3PO_4(aq)$

S3.45 (a) 1. $P_4O_{10}(s) + 6H_2O(l \text{ or } g) \rightarrow 4H_3PO_4(l)$

2. $Ca_3(PO_4)_2(s) + 3H_2SO_4(aq) \rightarrow 2H_3PO_4(85\%) + 3CaSO_4(s)$
phosphate rock

(b) Method 1

S3.47 (a) $PCl_5(s) + 4H_2O(l) \rightarrow H_3PO_4(aq) + 5HCl(aq)$

(b) $AsCl_3(s \text{ or } l) + 3H_2O(l) \rightarrow As(OH)_3(aq) + 3HCl(aq)$

S3.49 (a) $Ca_3(PO_4)_2(s) + H_2SO_4(aq) \rightarrow 2CaHPO_4(aq) + CaSO_4(aq)$

(b) $Ca_3(PO_4)_2(s) + 2H_2SO_4(aq) \rightarrow Ca(H_2PO_4)_2(s) + 2CaSO_4(s)$

These products make better fertilizer because they are more soluble than the original phosphate rock.

S3.51 Lower oxidation states are more favored by the heavier members of a main group.

S3.53 (b) and (c) silicon and germanium

S3.55 See Figure 2.18b and Section S3.3.

S3.59 $CaCO_3(s) \xrightarrow{heat} CaO(s) + CO_2(g)$
crushed limestone

$CO_2(g) + NaCl(aq) + NH_3(aq) + H_2O(l) \rightarrow NaHCO_3(s) + NH_4Cl(aq)$
the product

$CaO(s) + 2NH_4Cl(aq) \rightarrow CaCl_2(aq) + H_2O(l) + 2NH_3(g)$
a coproduct; recycled

S3.61 See Figure S3.21.

S3.65 (a) $\ddot{O}{=}C{=}\ddot{O}$ (c) H—C≡N:

(b) H—Ö—C(=:O:)—Ö—H (d) $\ddot{S}{=}C{=}\ddot{S}$

S3.67 (a) $2C(s) + O_2(g) \rightarrow 2CO(g)$

(b) $C(s) + O_2(g) \rightarrow CO_2(g)$

Limited oxygen supply and/or temperatures above 500°C favor the formation of CO.

S3.69 $KCN(aq) + H^+(aq) \rightarrow K^+(aq) + HCN(g)$

HCN(g) is lethal.

S3.71 An acetylide is a carbide containing the C_2^{2-} ion.

(a) $CaO(s) + 3C(s) \xrightarrow{heat} CaC_2(l) + CO(g)$

(b) $CaC_2(s) + 2H_2O(l) \rightarrow Ca(OH)_2(s) + C_2H_2(g)$

S3.75 (a) The bonds are longer and weaker in the Si and Ge lattices (b) Si and Ge atoms are too large to form strong pi bonds.

S3.79 See Figure S3.29. Water glass is an aqueous solution of sodium silicate, mainly the orthosilicate.

S3.81 See Figure S3.29a. Silicones owe their durability to the thermal stability of the Si—O bonds and to the resilience of the long Si—O chains.

S3.85 See Figures S3.31, S3.34, and S3.35.

S3.87 See Figure S3.37.

S3.89

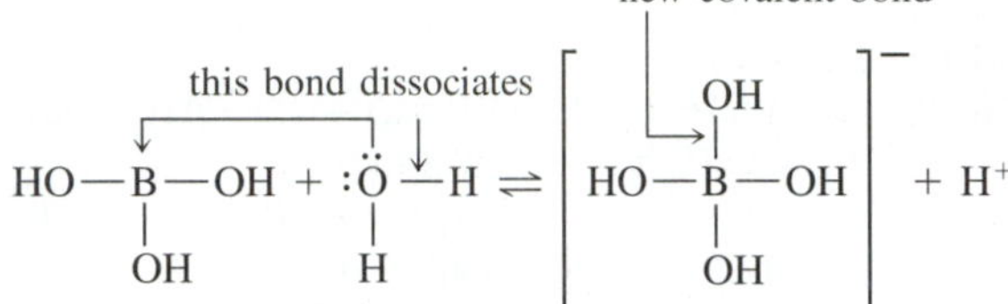

Lewis acid Lewis base

Boric acid is not a Brönsted acid.

S3.91 (a) $Na_2B_4O_7 \cdot 10H_2O(aq) + 2H^+(aq) \rightleftharpoons 4H_3BO_3(aq) + 2Na^+(aq) + 10H_2O(l)$

(b) $B(OH)_3(aq) + H_2O(l) \rightleftharpoons B(OH)_4^-(aq) + H^+(aq)$

(c) $2H_3BO_3(s) \xrightarrow{heat} B_2O_3(s) + 3H_2O(g)$

S3.93 (a) S, sulfur

(b) $MgSO_4 \cdot 7H_2O$, magnesium sulfate heptahydrate

(c) $CaSO_4 \cdot 2H_2O$, calcium sulfate dihydrate

(d) H_2SO_4, sulfuric acid

(e) $H_2S_2O_7$, pyrosulfuric acid

(f) P_4O_{10}, phosphorus pentoxide

(g) SiO_2, silicon dioxide (with some hydration)

(h) $Mg_2(Si_2O_5)_2 \cdot Mg(OH)_2$, a sheet silicate

(i) $Na_2B_4O_7 \cdot 10H_2O$, hydrated sodium tetraborate.

S3.97 (b) 0.082 *M*

S3.99 (a) $Ca(OH)_2(s \text{ or } aq) + SO_2(g) \rightarrow CaSO_3(s) + H_2O(l)$

(b) $MgO(s) + SO_2(g) \rightarrow MgSO_3(s)$

Both sulfites soon oxidize to sulfates in the air.

S3.101 −3640 kJ; propane has a greater enthalpy of combustion and is also more stable than silane.

S3.103 (a) The S_2 configuration involves the third shell of each atom; otherwise it resembles that of O_2.

(b) The S atoms are too large to form strong pi bonds.

S3.105 (a) 53.3 g/mol; (b) B_4H_{10}

S3.107 0.02400 *M*

CHAPTER 19

19.7 (a) 474 kJ; (b) 98 kJ

19.9 $\Delta G = \Delta G^\circ + RT \ln Q$

19.11 (a) decreases; (b) increases

19.13 Only (a) and (b). Most stable: HNO_3. Least stable: N_2O.
19.15 (a) forward; (b) reverse
19.17 0.120 kJ
19.19 Forward
19.21 Same for both
19.25 (a) 23.2 kJ/mol; (b) 18.7 kJ/mol; (c) reverse
19.27 (a) 4×10^{-31}; (b) 1×10^{-58}; (c) 5×10^{-17}
19.29 (a) $K_p = 2.30 \times 10^{-2}$, $\Delta G = +919$ kJ/mol; (b) $K_p = 1$, $\Delta G^\circ = 0$
19.31 (a) +27.1 kJ/mol; (b) −119 kJ/mol; (c) +27.1 kJ/mol
19.35 Decreases
19.39 Negative
19.41 ΔS for system: (a), (e) positive; (b), (c), (d) negative. ΔS for universe: positive (always).
19.43 (a) increases; (b) decreases
19.49 $q_{surr} = -q_{syst}$
19.51 35.3 J/K
19.53 107 J/K
19.55 (a) 24.8 J/K; (b) −138 J/K; (c) −121 J/K
19.57 (a) +2.85 J/K; (b) +99.0 J/K
19.59 (a) −159.74 J/K; (b) −198.76 J/K
19.63 (a), (b) yes; (c) no
19.65 (a) +166.3 kJ; (b) +364.2 kJ; (c) −67.8 kJ
19.67 52°C
19.69 Stable at both temperatures
19.73 (a) Work done reduces the amount of heat evolved. (b) none
19.75 (a) 2; (b) 32; (c) 36 (c) is most likely.
19.79 (a) 1×10^5; (b) −92 kJ/mol; forward
19.83 (a) −75.9 kJ, −27.6 kJ; (b) 2×10^{13}, 140

CHAPTER 20

20.1 (a) negative; (b) positive; (c) from anode to cathode
20.3 The salt bridge allows ions to flow, thus maintaining an unbroken electrical circuit.
20.5 (b)

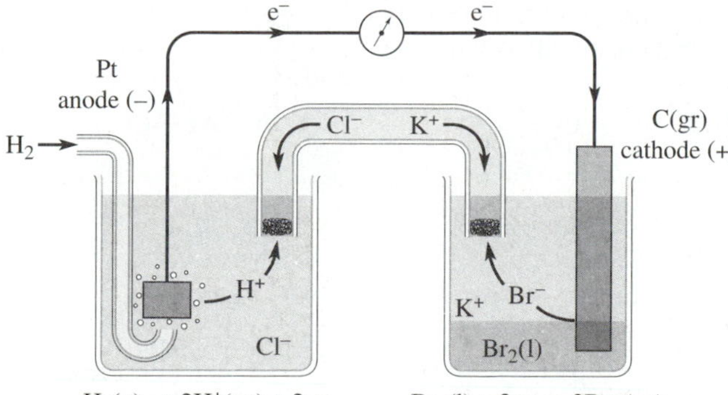

(c)

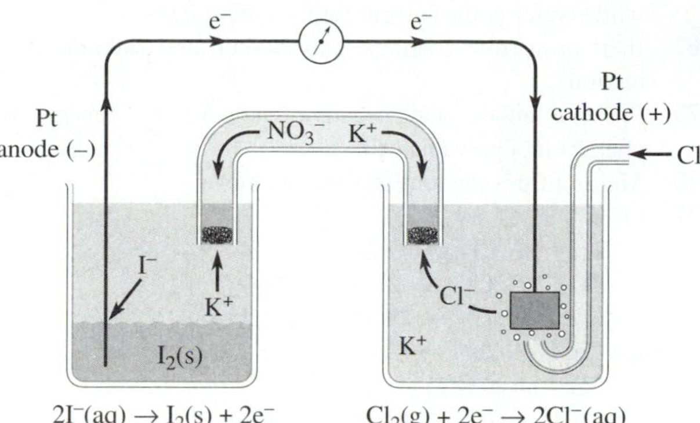

20.7 (a) $Fe(s) \mid Fe^{2+}(aq) \parallel Sn^{2+}(aq) \mid Sn(s)$
(b) $Pt \mid H_2(g) \mid H^+(aq) \parallel Br^-(aq) \mid Br_2(l) \mid C(gr)$
(c) $Pt \mid I_2(s) \mid I^-(aq) \parallel Cl^-(aq) \mid Cl_2(g) \mid Pt$
20.9 (a) $3Ag(s) + NO_3^-(aq) + 4H^+(aq) \rightarrow 3Ag^+(aq) + NO(g) + 2H_2O(l)$
$Ag(s) \mid Ag^+ (1\ M) \parallel H^+ (1\ M), NO_3^- (1\ M) \mid NO (1\ atm) \mid Pt$
(b) The half-reactions are
$Zn(s) + 4NH_3(aq) \rightarrow Zn(NH_3)_4^{2+}(aq) + 2e^-$ anode, oxidation
$Ag(NH_3)_2^+(aq) + e^- \rightarrow Ag(s) + 2NH_3(aq)$ cathode, reduction
$Zn(s) \mid Zn(NH_3)_4^{2+} (1\ M), NH_3 (1\ M) \parallel NH_3 (1\ M), Ag(NH_3)_2^+ (1\ M) \mid Ag(s)$
20.11 (a) coulomb; (b) volts
20.15 −68 kJ
20.17 −52.8 kJ
20.23 (a) Ni^{2+}; (b) Cd
20.25 The acidified permanganate
20.27 $KMnO_4$
20.29 (a) $Zn(s) + 2H^+(aq) \rightarrow Zn^{2+}(aq) + H_2(g)$;
(b) $3Ag(s) + NO_3^-(aq) + 4H^+(aq) \rightarrow 3Ag^+(aq) + NO(g) + 2H_2O(l)$;
(c) no reaction
20.31 (a) 0.27 V, −52 kJ; (b) 1.078 V, −208 kJ; (c) 0.825 V, −159 kJ
20.33 (a) 0.762 V, −147 kJ; (b) 0.165 V, −47.8 kJ; (c) no reaction
20.35 The concentrations are approaching their values at equilibrium, at which point the cell voltage will be zero.
20.39 The 0.10 *M* cell is the anode.
20.41 (a) 0.45 V; (b) 0.077 V
20.43 (a) 1.33 V; (b) 0.815 V
20.45 0.189 V
20.47 0.015 *M*
20.49 (a) 8×10^{46}; (b) 2.8×10^{-9}
20.51 2×10^{-18}
20.53 0.005 *M* Cu^{2+}, 1×10^{-9} Ag^+
20.59 (a) The electrolyte concentrations are lower in the discharged battery. (b) Solid $PbSO_4$ may be dislodged from the electrodes and therefore not be available for reconversion to Pb and PbO during recharging.

20.65 In this galvanic process, oxygen is the oxidizing agent while water is the solvent for the electrolytes.

20.67 Inert protective coatings, galvanizing, and cathodic protection

20.73 Positive anode and negative cathode, the reverse of charges in a galvanic cell

20.75 Melting frees the ions so they can move.

20.77 (a) $K^+ + e^- \rightarrow K(l)$

$2Cl^- \rightarrow Cl_2(g) + 2e^-$

$2K^+ + 2Cl^- \rightarrow 2K(l) + Cl_2(g)$

or $2KCl(l) \rightarrow 2K(l) + Cl_2(g)$

(b) $Ca^{2+} + 2e^- \rightarrow Ca(l)$

$2Cl^- \rightarrow Cl_2(g) + 2e^-$

$Ca^{2+} + 2Cl^- \rightarrow Ca(l) + Cl_2(g)$

or $CaCl_2(l) \rightarrow Ca(l) + Cl_2(g)$

20.79 Water rather than the ion of the active metal would be reduced.

20.83 (a) $2Cl^-(aq) \rightarrow Cl_2(g) + 2e^-$

$2H_2O(l) + 2e^- \rightarrow H_2(g) + 2OH^-(aq)$

$2H_2O(l) + 2Cl^-(aq) \rightarrow Cl_2(g) + H_2(g) + 2OH^-(aq)$

(b) $2Cl^-(aq) \rightarrow Cl_2(g) + 2e^-$

$2H_2O(l) + 2e^- \rightarrow H_2(g) + 2OH^-(aq)$

$2H_2O(l) + 2Cl^-(aq) \rightarrow Cl_2(g) + H_2(g) + 2OH^-(aq)$

20.85 See Figure 20.22. The Na^+ and SO_4^{2-} ions are needed to conduct current.

20.87 (a) 2.80 g; (b) 15.3 L

20.89 0.58 g

20.91 (a) 604 g; (b) 391 L

20.93 Since 2 faradays are used per mole of copper ion, $n = 2$.

20.95 (a) 3.00 *F*; (b) 1.00 *F*

20.97 (a) $+4.59 \times 10^{-2}$ g; (b) $+4.24 \times 10^{-2}$ g

20.99 (a) −0.34 V; (b) All values would be 0.34 V lower; (c) No negative values would become positive. All positive values above 0.34 V would become negative.

20.103 0.090 J

20.105 (a) 0.127 V; (b) −24.5 kJ

20.107 (a) $SO_3^{2-}(aq) + 2OH^-(aq) \rightarrow SO_4^{2-}(aq) + H_2O(l) + 2e^-$

$MnO_4^-(aq) + 2H_2O(l) + 3e^- \rightarrow MnO_2(s) + 4OH^-(aq)$

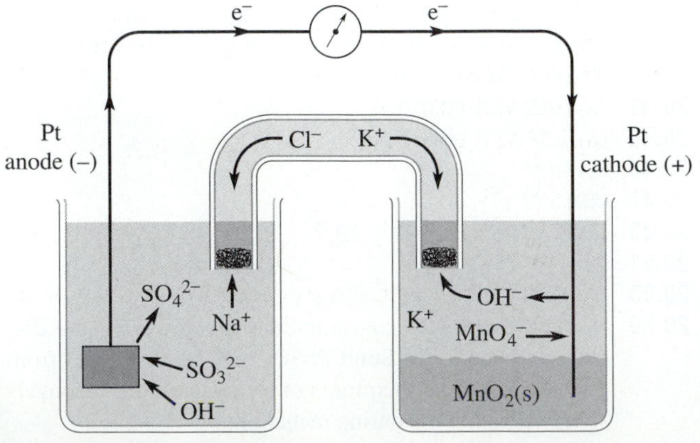

(b) −887 kJ, $K = 10^{155}$

20.109 (a) 2×10^{-10}; (b) 2×10^{-10}

20.111 (a) 3.6×10^{-4} C; (b) 0.37 μg Hg; (c) 4.9×10^{-4} J

20.113 (a) 0.051 g; (b) 1.6×10^{-3} mol; (c) 0.75 watts

CHAPTER 21

21.7 The melting point is expected to decrease because the atomic radius decreases.

21.9 The almost complete lack of negative oxidation states. Most metals form positive ions, ionic bonds, and basic oxides.

21.11 (1) Conversion to separate atoms (endothermic); (2) Ionization (endothermic); (3) Hydration of the ions (exothermic)

21.13 In the *d*, *f*, and *p* blocks. Group 1A metals have only one valence electron.

21.17 TlOH, $Tl(OH)_3$. TlOH is more ionic and thus has a higher melting point.

21.19 (a) above; (b) some above and some below

21.23 −454.8 J, −41 J/K·mol

21.27 (a)

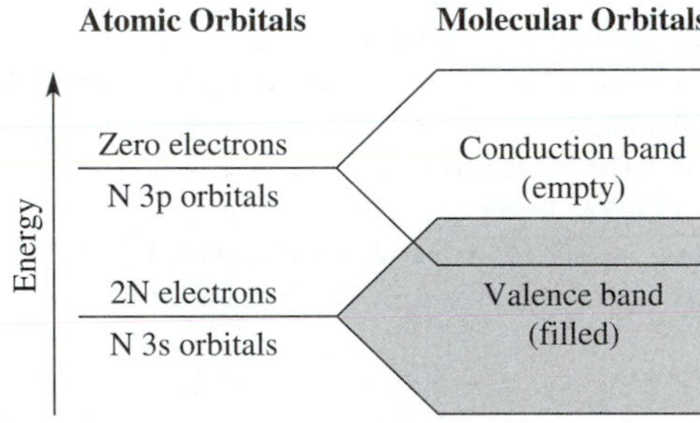

(b) Free migration of valence electrons is possible only because the empty 3*p* band overlaps the filled 3*s* band.

21.29 $[Cr(NH_3)_4Cl_2]Cl(s) \rightarrow Cr(NH_3)_4Cl_2^+(aq) + Cl^-(aq)$

21.31 (a) formula A $[Pt(NH_3)_3(NO_3)]Br$,
formula B $[Pt(NH_3)_3Br]NO_3$;
(b) $[Pt(NH_3)_3(NO_3)]Br(aq) + AgNO_3(aq) \rightarrow AgBr(s) + Pt(NH_3)_3(NO_3)^+(aq) + NO_3^-(aq)$

21.33 (a) $FeCl_6^{3-}$; (b) $Fe(C_2O_4)_3^{3-}$

21.35 (a), (b), (d) +3; (c) +2

21.37 (a) Cl, Cl, Pd, Cl, Cl (square planar) (c) Cl—Ag—Cl

(b) OH, HO, Be, OH, OH (tetrahedral) (d) OCO_2, H_3N, NH_3, Co, H_3N, NH_3, NH_3 (octahedral)

21.39 (a)

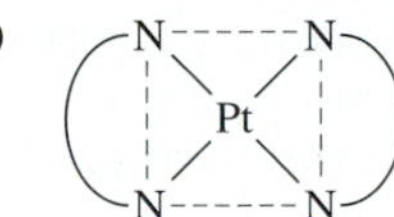

(b)

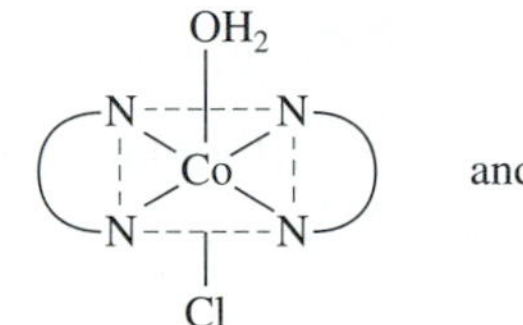

and

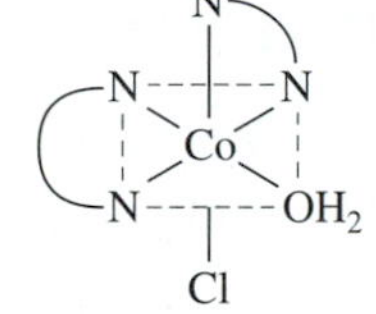

(c)

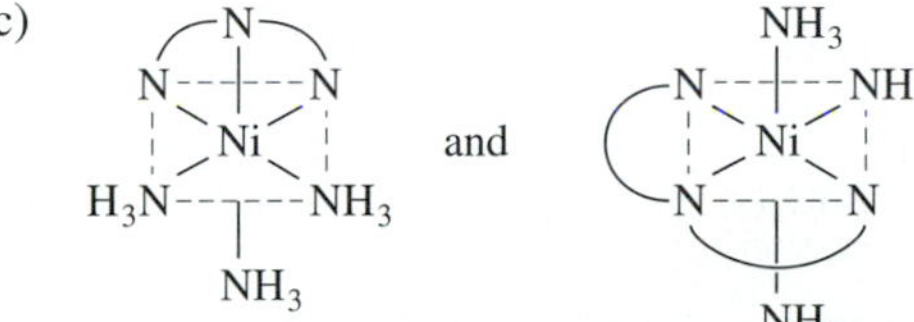

21.41 (a) $Ni(H_2O)_6^{2+}$; (b) $Cu(NH_3)_4^{2+}$; (c) $[Ir(NH_3)_3Cl_3]$; (d) $[Co(NH_3)_4Cl_2]$; (e) $Fe(H_2O)_6^{3+}$

21.43 (a) Exercise 21.35: (a) hexaamminecobalt(III) chloride; (b) potassium hexafluoroferrate(III); (c) potassium hexacyanoferrate(II); (d) potassium hexacyanoferrate(III).
(b) Exercise 21.37: (a) tetrachloropalladate(II) ion; (b) tetrahydroxoberyllate(II) ion; (c) dichloroargentate ion; (d) pentaamminecarbonatocobalt(III) ion.
(c) Exercise 21.39: (a) bis(ethylenediamine)platinum(II) ion; (b) aquachlorobis(ethylenediamine)cobalt(II) ion; (c) triamminediethylenetriaminenickel(II) ion.

21.51 (a)

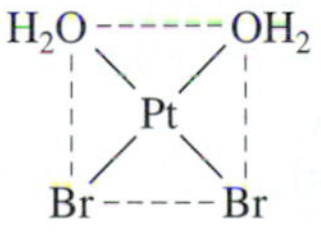

cis 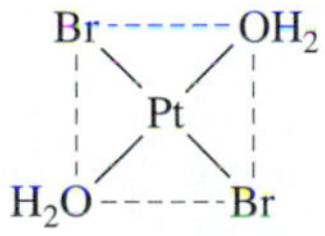 *trans*

(b)

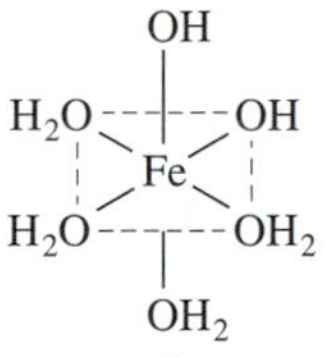

cis 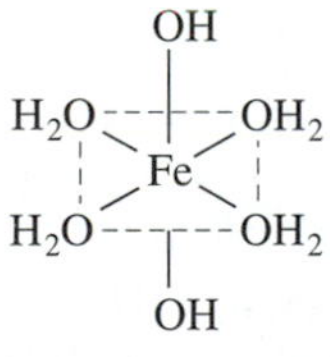 *trans*

21.53 (a)

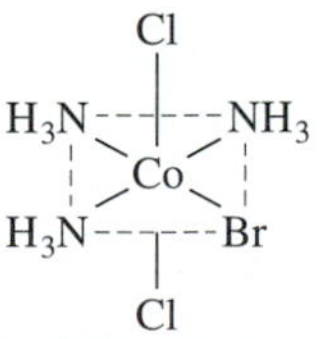

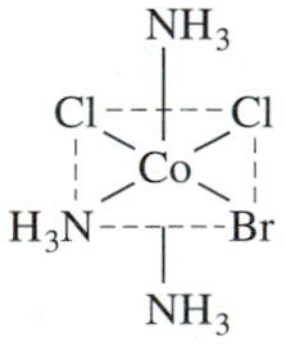

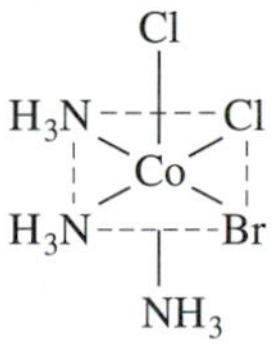

(b)

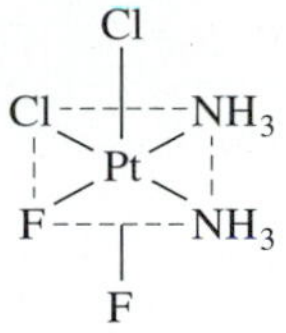

cis 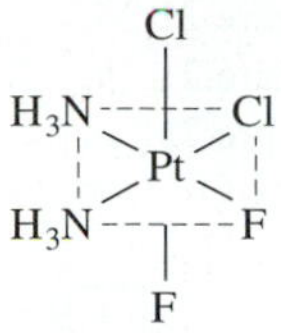 *cis*

Optical isomers

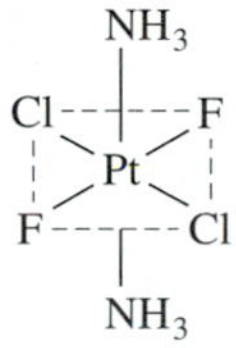

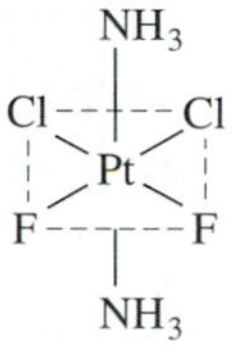

trans

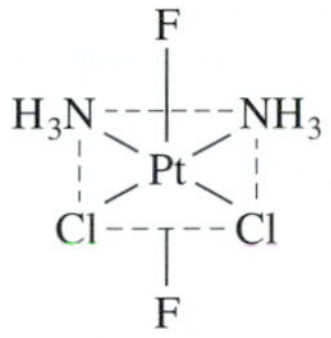

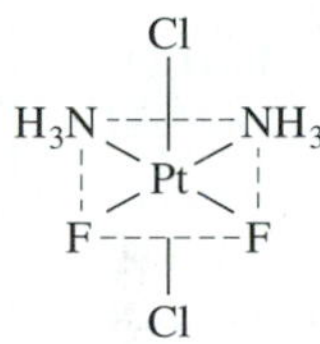

(c)

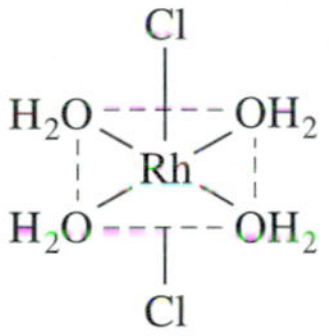

trans 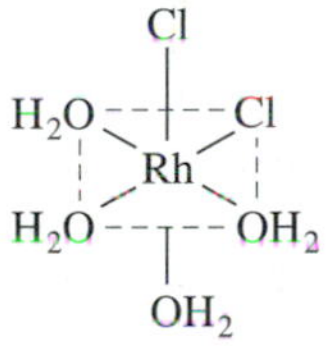 *cis*

(d)

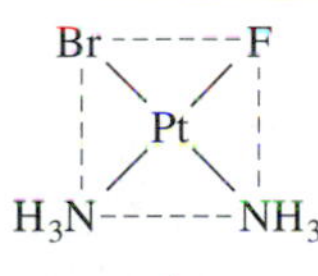

cis 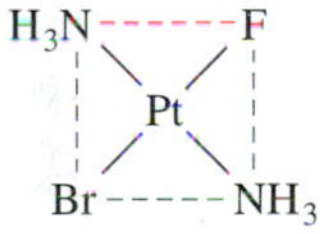 *trans*

21.55 (a)

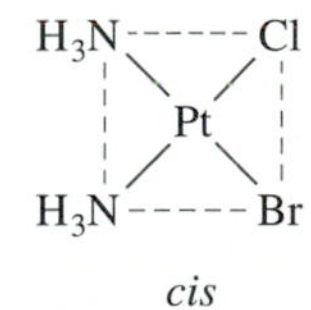

cis 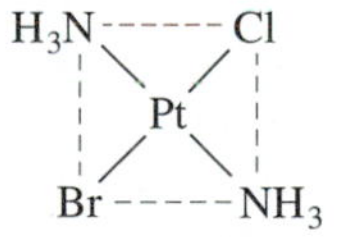 *trans*

Any other sketches can be superimposed on one of these by just rotating through 90° or 180° or by flipping the sketch.

(b)

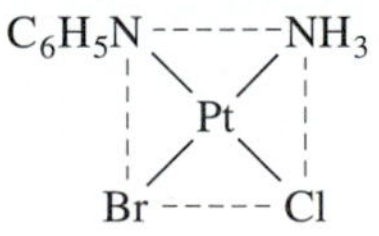

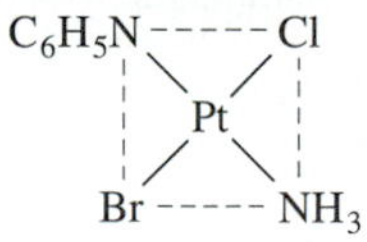

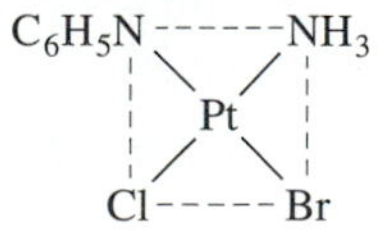

Any other sketches can be superimposed on one of these.

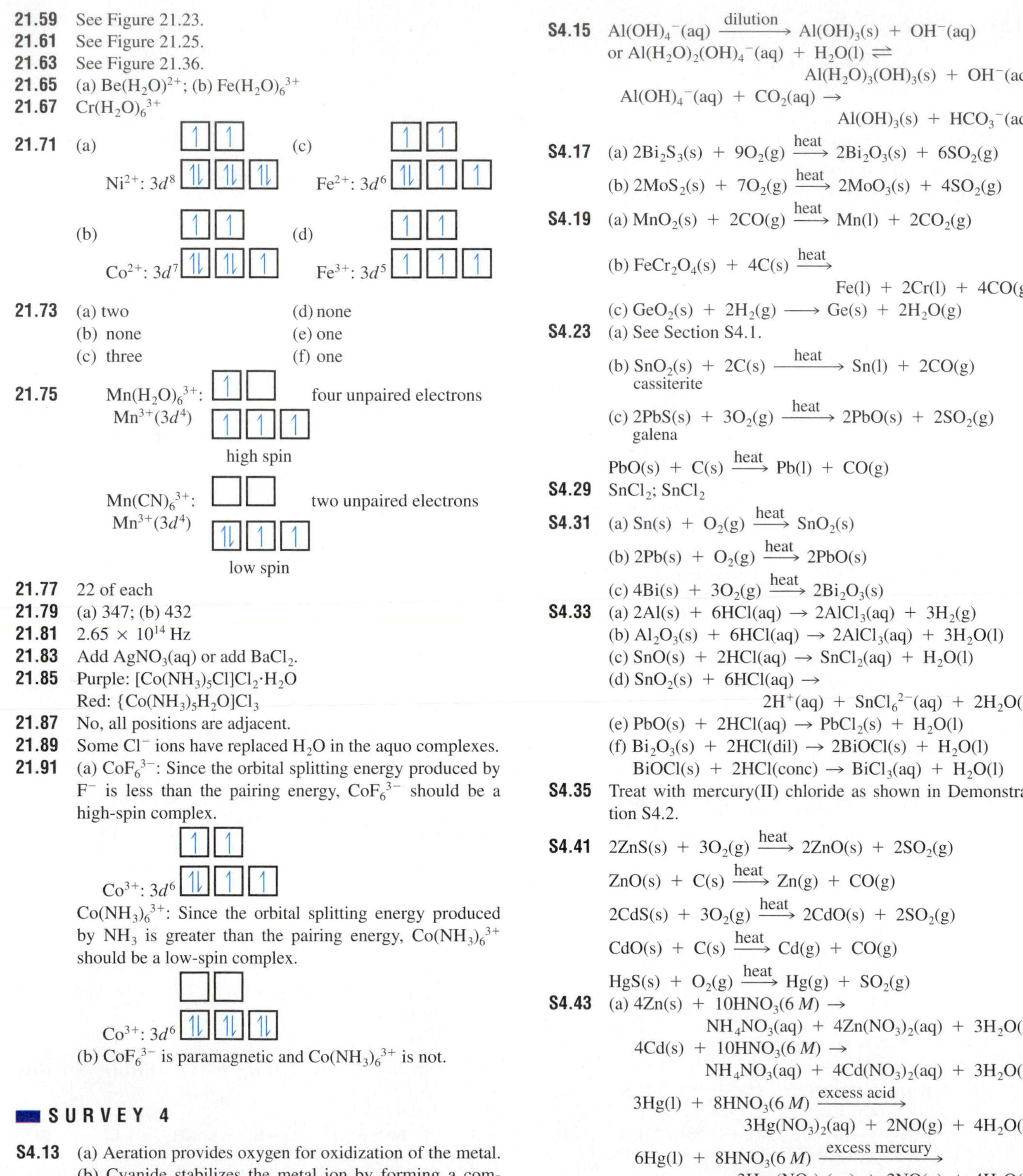

21.59 See Figure 21.23.
21.61 See Figure 21.25.
21.63 See Figure 21.36.
21.65 (a) $Be(H_2O)^{2+}$; (b) $Fe(H_2O)_6^{3+}$
21.67 $Cr(H_2O)_6^{3+}$
21.71 (a) Ni^{2+}: $3d^8$ (b) Co^{2+}: $3d^7$ (c) Fe^{2+}: $3d^6$ (d) Fe^{3+}: $3d^5$
21.73 (a) two (b) none (c) three (d) none (e) one (f) one
21.75 $Mn(H_2O)_6^{3+}$: $Mn^{3+}(3d^4)$ high spin — four unpaired electrons
$Mn(CN)_6^{3+}$: $Mn^{3+}(3d^4)$ low spin — two unpaired electrons
21.77 22 of each
21.79 (a) 347; (b) 432
21.81 2.65×10^{14} Hz
21.83 Add $AgNO_3(aq)$ or add $BaCl_2$.
21.85 Purple: $[Co(NH_3)_5Cl]Cl_2{\cdot}H_2O$
Red: $\{Co(NH_3)_5H_2O]Cl_3$
21.87 No, all positions are adjacent.
21.89 Some Cl^- ions have replaced H_2O in the aquo complexes.
21.91 (a) CoF_6^{3-}: Since the orbital splitting energy produced by F^- is less than the pairing energy, CoF_6^{3-} should be a high-spin complex.
Co^{3+}: $3d^6$
$Co(NH_3)_6^{3+}$: Since the orbital splitting energy produced by NH_3 is greater than the pairing energy, $Co(NH_3)_6^{3+}$ should be a low-spin complex.
Co^{3+}: $3d^6$
(b) CoF_6^{3-} is paramagnetic and $Co(NH_3)_6^{3+}$ is not.

SURVEY 4

S4.13 (a) Aeration provides oxygen for oxidization of the metal. (b) Cyanide stabilizes the metal ion by forming a complex.
S4.15 $Al(OH)_4^-(aq) \xrightarrow{\text{dilution}} Al(OH)_3(s) + OH^-(aq)$
or $Al(H_2O)_2(OH)_4^-(aq) + H_2O(l) \rightleftharpoons Al(H_2O)_3(OH)_3(s) + OH^-(aq)$
$Al(OH)_4^-(aq) + CO_2(aq) \rightarrow Al(OH)_3(s) + HCO_3^-(aq)$
S4.17 (a) $2Bi_2S_3(s) + 9O_2(g) \xrightarrow{\text{heat}} 2Bi_2O_3(s) + 6SO_2(g)$
(b) $2MoS_2(s) + 7O_2(g) \xrightarrow{\text{heat}} 2MoO_3(s) + 4SO_2(g)$
S4.19 (a) $MnO_2(s) + 2CO(g) \xrightarrow{\text{heat}} Mn(l) + 2CO_2(g)$
(b) $FeCr_2O_4(s) + 4C(s) \xrightarrow{\text{heat}} Fe(l) + 2Cr(l) + 4CO(g)$
(c) $GeO_2(s) + 2H_2(g) \longrightarrow Ge(s) + 2H_2O(g)$
S4.23 (a) See Section S4.1.
(b) $\underset{\text{cassiterite}}{SnO_2(s)} + 2C(s) \xrightarrow{\text{heat}} Sn(l) + 2CO(g)$
(c) $\underset{\text{galena}}{2PbS(s)} + 3O_2(g) \xrightarrow{\text{heat}} 2PbO(s) + 2SO_2(g)$
$PbO(s) + C(s) \xrightarrow{\text{heat}} Pb(l) + CO(g)$
S4.29 $SnCl_2$; $SnCl_2$
S4.31 (a) $Sn(s) + O_2(g) \xrightarrow{\text{heat}} SnO_2(s)$
(b) $2Pb(s) + O_2(g) \xrightarrow{\text{heat}} 2PbO(s)$
(c) $4Bi(s) + 3O_2(g) \xrightarrow{\text{heat}} 2Bi_2O_3(s)$
S4.33 (a) $2Al(s) + 6HCl(aq) \rightarrow 2AlCl_3(aq) + 3H_2(g)$
(b) $Al_2O_3(s) + 6HCl(aq) \rightarrow 2AlCl_3(aq) + 3H_2O(l)$
(c) $SnO(s) + 2HCl(aq) \rightarrow SnCl_2(aq) + H_2O(l)$
(d) $SnO_2(s) + 6HCl(aq) \rightarrow 2H^+(aq) + SnCl_6^{2-}(aq) + 2H_2O(l)$
(e) $PbO(s) + 2HCl(aq) \rightarrow PbCl_2(s) + H_2O(l)$
(f) $Bi_2O_3(s) + 2HCl(dil) \rightarrow 2BiOCl(s) + H_2O(l)$
$BiOCl(s) + 2HCl(conc) \rightarrow BiCl_3(aq) + H_2O(l)$
S4.35 Treat with mercury(II) chloride as shown in Demonstration S4.2.
S4.41 $2ZnS(s) + 3O_2(g) \xrightarrow{\text{heat}} 2ZnO(s) + 2SO_2(g)$
$ZnO(s) + C(s) \xrightarrow{\text{heat}} Zn(g) + CO(g)$
$2CdS(s) + 3O_2(g) \xrightarrow{\text{heat}} 2CdO(s) + 2SO_2(g)$
$CdO(s) + C(s) \xrightarrow{\text{heat}} Cd(g) + CO(g)$
$HgS(s) + O_2(g) \xrightarrow{\text{heat}} Hg(g) + SO_2(g)$
S4.43 (a) $4Zn(s) + 10HNO_3(6\,M) \rightarrow NH_4NO_3(aq) + 4Zn(NO_3)_2(aq) + 3H_2O(l)$
$4Cd(s) + 10HNO_3(6\,M) \rightarrow NH_4NO_3(aq) + 4Cd(NO_3)_2(aq) + 3H_2O(l)$
$3Hg(l) + 8HNO_3(6\,M) \xrightarrow{\text{excess acid}} 3Hg(NO_3)_2(aq) + 2NO(g) + 4H_2O(l)$
$6Hg(l) + 8HNO_3(6\,M) \xrightarrow{\text{excess mercury}} 3Hg_2(NO_3)_2(aq) + 2NO(g) + 4H_2O(l)$
(b) $Zn(s) + 4HNO_3(12\,M) \rightarrow Zn(NO_3)_2(aq) + 2NO_2(g) + 2H_2O(l)$

$Cd(s) + 4HNO_3(12\ M) \rightarrow Cd(NO_3)_2(aq) + 2NO_2(g) + 2H_2O(l)$

$3Hg(l) + 8HNO_3(12\ M) \xrightarrow{\text{excess acid}} 3Hg(NO_3)_2(aq) + 2NO(g) + 4H_2O(l)$

$6Hg(l) + 8HNO_3\ (12\ M) \xrightarrow{\text{excess mercury}} 3Hg_2(NO_3)_2(aq) + 2NO(g) + 4H_2O(l)$

S4.45 (a) Add $HNO_3(aq)$.
(b) Treat with concentrated KOH(aq).
(c) Treat with $Cl_2(g)$.
(d) Treat excess mercury with concentrated HNO_3 to get $Hg_2(NO_3)_2(aq)$. Add HCl or NaCl.

S4.49 1. Concentrate the ore.
2. Impurities are converted to slag by roasting and smelting with SiC. A sample equation is:
$2FeS(s) + 3O_2(g) \rightarrow 2FeO(s) + 2SO_2$
3. Air is blown through the molten CuS.
$Cu_2S(l) + O_2(g) \rightarrow 2Cu(l) + SO_2(g)$

S4.53 1. Some of the silver AgBr is sensitized by light.
2. The developer reduces the sensitized AgBr to Ag.
$AgBr(s) + e^-$ (from developer) $\rightarrow Ag(s) + Br^-(aq)$
3. The fixer ($Na_2S_2O_3$) dissolves the unsensitized AgBr.
$AgBr(s) + 2S_2O_3(aq) \rightarrow Ag(S_2O_3)_2{}^{3-}(aq) + Br^-(aq)$

S4.55 (a) $Cu(s) + Cl_2(g) \rightarrow CuCl_2(s)$
$2Ag(s) + Cl_2(g) \rightarrow 2AgCl(s)$
$2Au(s) + 3Cl_2(g) \rightarrow 2AuCl_3(s)$
(b) $2Cu(s) + 2H_2S(g) + O_2(g) \rightarrow 2CuS(s) + 2H_2O(g)$
$4Ag(s) + 2H_2S(g) + O_2(g) \rightarrow 2Ag_2S(s) + 2H_2O(g)$
(c) $Cu(s) + 4H^+(aq) + 2NO_3^-(aq) \rightarrow Cu^{2+}(aq) + 2NO_2(g) + 2H_2O(l)$
$Ag(s) + 2H^+(aq) + NO_3^-(aq) \rightarrow Ag^+(aq) + NO_2(g) + H_2O(l)$

S4.57 (a) $Cu^{2+}(aq) + 2OH^-(aq) \rightarrow Cu(OH)_2(s)$
(b) $2Ag^+(aq) + 2OH^-(aq) \rightarrow Ag_2O(s) + H_2O(l)$

S4.63 $Fe^{3+}(aq) + SCN^-(aq) \rightarrow Fe(SCN)^{2+}(aq)$
deep red
$Co^{2+}(aq) + 7KNO_2(aq) + 2H^+(aq) \rightarrow K_3[Co(NO_2)_6](s) + 4K^+(aq) + NO(g) + H_2O(l)$
(c) Add dimethylglyoxime to the solution. If Ni^{2+} is present, an insoluble bright red complex will form.

S4.65 (a) $3Fe(s) + 2O_2(g) \rightarrow Fe_3O_4(s)$
and/or $4Fe(s) + 3O_2(g) \rightarrow 2Fe_2O_3(s)$
$3Co(s) + 2O_2(g) \rightarrow Co_3O_4(s)$
$2Ni(s) + O_2(g) \rightarrow 2NiO(s)$
(b) $Fe(s) + 2HCl(aq) \rightarrow FeCl_2(aq) + H_2(g)$
$Co(s) + 2HCl(aq) \rightarrow CoCl_2(aq) + H_2(g)$
$Ni(s) + 2HCl(aq) \rightarrow NiCl_2(aq) + H_2(g)$

S4.67 (a) The O—H bonds in the hydrated ion are more polar, and thus stronger proton donors, than those in the solvent H_2O. (b) Fe^{3+}

S4.69 Fe

S4.71 The reaction in acid medium is
$Fe^{2+}(aq) \rightarrow Fe^{3+}(aq) + e^-$ oxidation
$O_2(g) + 4H^+(aq) + 4e^- \rightarrow H_2O(l)$ reduction
$4Fe^{2+}(aq) + O_2(g) + 4H^+(aq) \rightarrow 4Fe^{3+}(aq) + 2H_2O(l)$

S4.75 (a) Both metals have protective oxide films. Both form amphoteric oxides. The $Cr^{3+}(aq)$ and $Al^{3+}(aq)$ ions have similar reactions in solution.
(b) Unlike Al, Cr has several oxidation states, forms many complex ions, and many colored compounds. Cr is denser than Al and has a higher melting point.

S4.77 (a) $FeCr_2O_4(s) + 4C(s) \rightarrow Fe(s) + 2Cr(s) + 4CO(g)$
(b) $MnO_2(s) + 2C(s) \rightarrow Mn(s) + 2CO(g)$
(c) $Cr_2O_3(s) + 2Al(s) \rightarrow 2Cr(l) + Al_2O_3(l)$

S4.79 $Cr^{3+}(aq) + 3OH^-(aq) \rightarrow Cr(OH)_3(s)$
$Cr(OH)_3(s) + OH^-(aq) \rightarrow Cr(OH)_4{}^-(aq)$
$Mn^{2+}(aq) + 2OH^-(aq) \rightarrow Mn(OH)_2(s)$

S4.81 (a) Values for ΔG of formation calculated using Equation 19.7 show that HgO is unstable and that ZnO is stable with respect to the elements at 500°C. (b) Ag

S4.83 (a) −852 kJ (b) 3580 K

S4.85 (a) [1.] $4.1 \times 10^{-4}\ M$; [2.] $6.3 \times 10^{-9}\ M$

S4.87 (a) 7.31×10^8 g; (b) 390 m^3

S4.89 $[Hg—Hg]^{2+}$

S4.91 (a) −29.4 kJ; (b) −44.4 kJ

S4.93 4×10^{39}

CHAPTER 22

22.1 Alpha emission decreases Z by 2, and A by 4. Beta emission increases by Z by 1, does not change A. Gamma emission does not change A or Z.

22.3 Beta particles are less massive than alpha particles.

22.5 See Table 22.2.

22.7 See Figure 22.4.

22.9 (a) ${}^{228}_{90}Th \rightarrow {}^{4}_{2}He + {}^{224}_{88}Ra$ (b) ${}^{28}_{13}Al \rightarrow {}^{0}_{-1}e + {}^{28}_{14}Si$
(c) ${}^{82}_{37}Rb + {}^{0}_{-1}e \rightarrow {}^{82}_{36}Kr$ + x-ray photon
(d) ${}^{11}_{6}C \rightarrow {}^{0}_{1}e + {}^{11}_{5}B$

22.11 There are 14 steps. Equations for the first four are
1. ${}^{238}_{92}U \rightarrow {}^{4}_{2}He + {}^{234}_{90}Th$ 2. ${}^{234}_{90}Th \rightarrow {}^{0}_{-1}e + {}^{234}_{91}Pa$
3. ${}^{234}_{91}Pa \rightarrow {}^{0}_{-1}e + {}^{234}_{92}U$ 4. ${}^{234}_{92}U \rightarrow {}^{4}_{2}He + {}^{230}_{90}Th$

22.13 See Figure 22.7.

22.15 (a) ${}^{10}_{5}B + {}^{4}_{2}He \rightarrow {}^{13}_{7}N + {}^{1}_{0}n$ (b) ${}^{8}_{3}Li + {}^{1}_{0}n \rightarrow {}^{9}_{3}Li$
(c) ${}^{238}_{92}U + {}^{1}_{0}n \rightarrow {}^{239}_{93}Np + {}^{0}_{-1}e$

22.17 (a) ${}^{10}B(\alpha,n){}^{13}N$; (b) ${}^{8}Li(n,0){}^{9}Li$; (c) ${}^{238}U(n,\beta){}^{239}Np$

22.19 ${}^{238}_{92}U + {}^{2}_{1}H \rightarrow {}^{238}_{93}Np + 2{}^{1}_{0}n$
${}^{238}_{93}Np \rightarrow {}^{238}_{94}Pu + {}^{0}_{-1}e$

22.21 ${}^{209}_{83}Bi + {}^{4}_{2}He \rightarrow {}^{211}_{85}At + 2{}^{1}_{0}n$

22.25 No

22.27 45.5 min

22.29 (a) 0.976, 0.781; (b) 0.26

22.31 4.3×10^9 y

22.33 17.4 y

22.35 The rem; about 0.15 rem/year

22.37 110 per min

22.39 (a) 9 days; (b) 2.7 mCi

22.45 Alpha emission, positron emission, electron capture

22.47 Beta emission

22.49 (a), (b), (c) positron emission; (d), (e) beta emission

22.51 (a) ${}^{45}_{22}\text{Ti} \rightarrow {}^{45}_{21}\text{Sc} + {}^{0}_{1}\text{e}$

${}^{45}_{22}\text{Ti} + {}^{0}_{-1}e \rightarrow {}^{45}_{21}\text{Sc}$

(b) 1.05, 1.14; (c) below

22.53 The energy needed to decompose a nucleus into protons and neutrons; iron-56.

22.55 (a) 9.6×10^{10} kJ/mol; (b) 9.6×10^{13} kJ/mol

22.57 (a) 3.958×10^{12} J/atom; (b) 5.955×10^{11} J/g

22.59 ${}^{238}_{92}\text{U} \rightarrow {}^{234}_{90}\text{Th} + {}^{4}_{2}\text{He}$

(a) 931.5 MeV; (b) 1.73×10^{9} J/g

22.61 (a) 6.461 MeV/nucleon; (b) 6.475 MeV/nucleon; (c) 8.46 MeV/nucleon

22.71 (a) ${}^{238}_{92}\text{U} + {}^{1}_{0}\text{n} \rightarrow {}^{239}_{92}\text{U}$

${}^{239}_{92}\text{U} \rightarrow {}^{239}_{93}\text{Np} + {}^{0}_{-1}e$

${}^{239}_{93}\text{Np} \rightarrow {}^{239}_{94}\text{Pu} + {}^{0}_{-1}e$

22.73 (a) ${}^{235}_{92}\text{U} + {}^{1}_{0}\text{n} \rightarrow {}^{90}_{37}\text{Rb} + {}^{144}_{55}\text{Cs} + 2{}^{1}_{0}\text{n}$

(b) ${}^{235}_{92}\text{U} + {}^{1}_{0}\text{n} \rightarrow {}^{90}_{38}\text{Sr} + {}^{143}_{54}\text{Xe} + 3{}^{1}_{0}\text{n}$

(c) ${}^{235}_{92}\text{U} + {}^{1}_{0}\text{n} \rightarrow {}^{90}_{38}\text{Sr} + {}^{144}_{54}\text{Xe} + 2{}^{1}_{0}\text{n}$

(d) ${}^{235}_{92}\text{U} + {}^{1}_{0}\text{n} \rightarrow {}^{97}_{39}\text{Y} + {}^{137}_{53}\text{I} + 2{}^{1}_{0}\text{n}$

(e) ${}^{235}_{92}\text{U} + {}^{1}_{0}\text{n} \rightarrow {}^{137}_{52}\text{Te} + {}^{97}_{40}\text{Zr} + 2{}^{1}_{0}\text{n}$

22.75 32 tons

22.77 (a) 17.60 MeV/atom; (b) 1.698×10^{9} kJ/mol

22.79 At that time, natural radioactivity was the only projectile source.

22.83 7 alpha, 4 beta

22.85

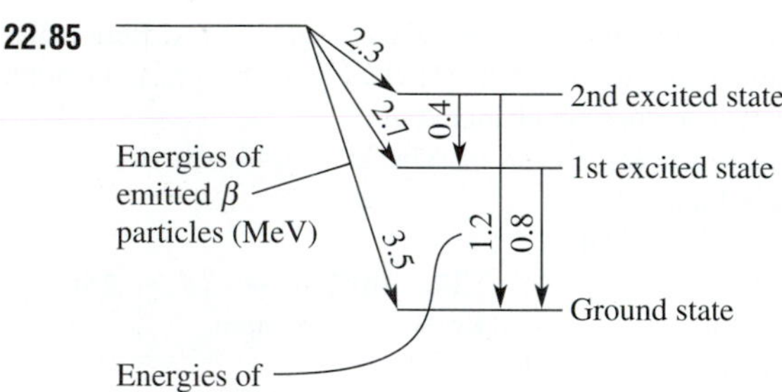

(b) 0.8, 1.2, and 0.4 MeV

22.87 (a) 4.380×10^{-10} g

22.89 22.38 MeV/atom

22.93 6.9×10^{-6} μCi/g

22.95 1.0×10^{9} y

CHAPTER 23

23.3 Refer to Figure 23.2.

23.5 No

23.7 (a) C_5H_{10} (b) C_6H_{12}

—CH_3

23.9 (a)

```
                  H
                  |
          H     H-C-H
          |       |
        H-C-H   H-C-H
  H   H   |   H   |   H   H   H   H
  |   |   |   |   |   |   |   |   |
H-C---C---C---C---C---C---C---C---C-H
  |   |   |   |   |   |   |   |   |
  H   H   H   H   H   H   H   H   H
```

(b)

```
          H
          |
        H-C-H
          |
  H   H   |   H   H
  |   |   |   |   |
H-C---C---C---C---C-H
  |   |   |   |   |
  H   H   |   H   H
          |
        H-C-H
          |
          H
```

23.11 (a)

```
      H
      |
    H-C-H
      |
  H   |   H   H   H   H   H
  |   |   |   |   |   |   |
H-C---C---C---C---C---C---C-H
  |   |   |   |   |   |   |
  H   H   H   H   H   H   H
```

(b)

```
      H       H
      |       |
    H-C-H   H-C-H
      |       |
  H   |   H   |   H   H
  |   |   |   |   |   |
H-C---C---C---C---C---C-H
  |   |   |   |   |   |
  H   H   H   H   H   H
```

(c)

```
          H
          |
        H-C-H
          |
        H-C-H
          |
  H   H   |   H   H   H
  |   |   |   |   |   |
H-C---C---C---C---C---C-H
  |   |   |   |   |   |
  H   H   H   H   H   H
```

(d)
```
            H               H
            |               |
         H—C—H  H   H—C—H
            |    |          |
   H        | H—C—H        |   H
   |        |    |          |   |
H—C———C———C———C—C—H
   |        |    |          |   |
   H        H    H          H   H
```

(e)
```
                     H
                     |
          H      H—C—H
          |          |
       H—C—H  H—C—H
          |          |
   H      |          |     H   H
   |      |          |     |   |
H—C————C—————C———C—C—H
   |      |          |     |   |
   H      H          H     H   H
```

All are isomers of C_8H_{18}.

23.13 (a) methylcyclobutane; (b) 4-ethyl-5-methyloctane

23.15 (a) 17

23.17 $2C_4H_{10}(g) + 13O_2(g) \rightarrow 8CO_2(g) + 10H_2O(g)$

23.19 (a) Step 1. $CH_4 + Cl_2 \xrightarrow{h\nu} CH_3Cl + HCl$

Step 2. $CH_3Cl + Cl_2 \xrightarrow{h\nu} CH_2Cl_2 + HCl$

Step 3. $CH_2Cl_2 + Cl_2 \xrightarrow{h\nu} CHCl_3 + HCl$

Step 4. $CHCl_3 + Cl_2 \xrightarrow{h\nu} CCl_4 + HCl$

(b) The other product is HCl.

23.21 (a) C_nH_{2n}; (b) C_nH_{2n-2}; (c) C_nH_{2n-4}

23.23 (a)
```
H       H   H
|       |   |
C═C—C—C—H
|   |   |   |
H   H   H   H
```
(b)
```
    H   H   H   H
    |   |   |   |
H—C—C═C—C—H
    |           |
    H           H
```
(c) H—C≡C—H

23.25 (a) 1.
```
     H   H   H   H   H
     |   |   |   |   |
H—C═C—C—C—C—H
             |   |   |
             H   H   H
```
2.
```
          H             H
          |             |
       H—C—H     H—C—H
          |             |
   H      |             |   H   H
   |      |             |   |   |
H—C—C═══════C—C—C—H
   |                        |   |
   H                        H   H
```
3.
```
          H             H
          |             |
       H—C—H     H—C—H
          |             |
   H      |             |   H
   |      |             |   |
H—C—C═══════C—C—H
   |                        |
   H                        H
```
4.
```
    H   H   H   H   H
    |   |   |   |   |
H—C—C═C—C—C—H
    |           |   |
    H           H   H
```
5.
```
       H   H   H   H
       |   |   |   |
H—C═C—C═C—H
```
(b) 2-Pentene has *cis–trans* isomers.

23.27
```
C—C—C—C—C═C

C—C—C—C═C—C

C—C—C═C—C—C

          C
          |
C—C—C—C═C

      C
      |
C—C—C═C—C

          C
          |
C—C—C═C—C

      C
      |
C—C═C—C—C

      C
      |
C—C—C═C
      |
      C

              C
              |
C—C—C—C═C

      C
      |
C—C—C—C═C

      C   C
      |   |
C—C—C═C

C—C═C—C
      |   |
      C   C

          C
          |
          C
          |
C—C—C═C
```

23.29 *cis*-3-Heptene

23.31 (a) $CH_3CH_2CH_2CH_3 + Br_2 \xrightarrow{h\nu}$ $CH_3CHBrCH_2CH_3 + HBr$
butane → 2-bromobutane

(b)
```
              H   H
              |   |
CH3CH2C═C—H + HBr ⟶ CH3CH2CHBrCH3
```
butene → 2-bromobutane

(c) the addition method

23.33 (a)
```
              CH3 H
               |   |
⟶ CH3—C—C—H
               |   |
              Br  Br
```

(b) 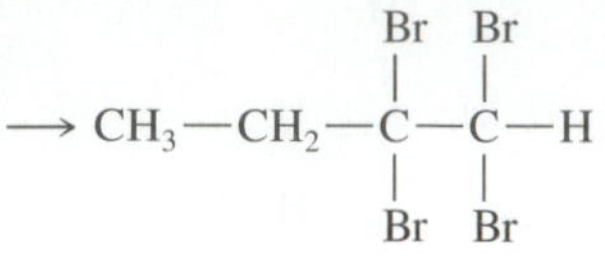 $\longrightarrow CH_3-CH_2-CBr_2-CBr_2-H$

(c) 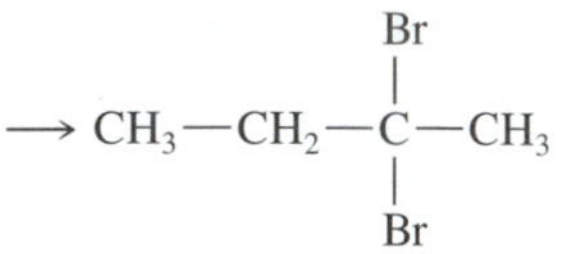 $\longrightarrow CH_3-CH_2-CBr_2-CH_3$

23.35 (a) 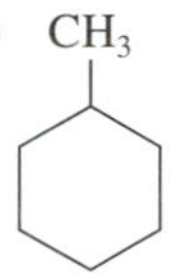 CH_3 (methylcyclohexane)

(b) 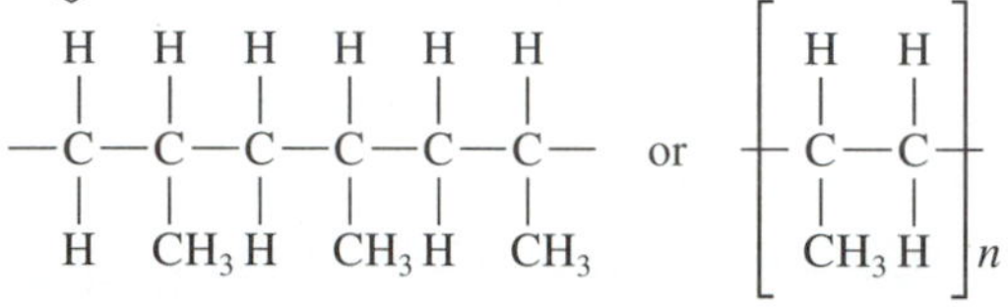 $-CH(CH_3)-CH_2-CH(CH_3)-CH_2-CH(CH_3)-CH_2-$ or $\left[-CH_2-CH(CH_3)-\right]_n$

23.37 (a) 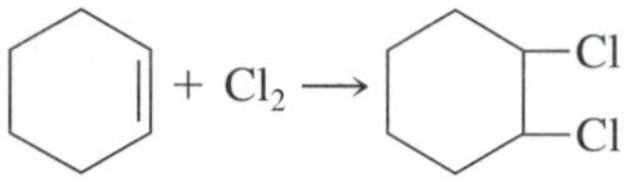 cyclohexene $+ Cl_2 \longrightarrow$ 1,2-dichlorocyclohexane

(b) 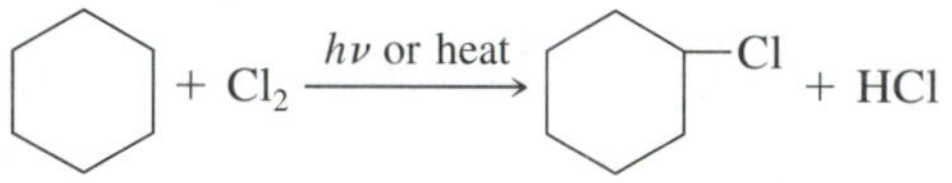 cyclohexane $+ Cl_2 \xrightarrow{h\nu \text{ or heat}}$ chlorocyclohexane $+ HCl$

(c) $CH_3CH_2CH_2-C{\equiv}C-H + 2H_2 \xrightarrow{Pt}$ pentane
or
$H_2C{=}CHCH_2CH{=}CH_2 + 2H_2 \xrightarrow{Pt}$ pentane

(d) $H-C{\equiv}C-CH_2CH_2CH_3 + 2HBr \rightarrow$ 2,2-dibromopentane

23.39 $H_2C{=}CCl_2$

23.43 (a) 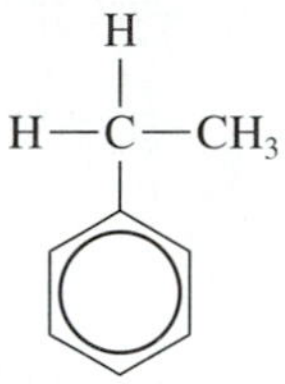 $H-CH(CH_3)-C_6H_5$ (ethylbenzene), C_8H_{10} (b)  C_6H_5Cl (chlorobenzene), C_6H_5Cl

(c) $CH_3-CH_2-C(CH_3)(C_6H_5)-CH_2-CH_2-CH_2-CH_2-CH_3$, $C_{14}H_{22}$

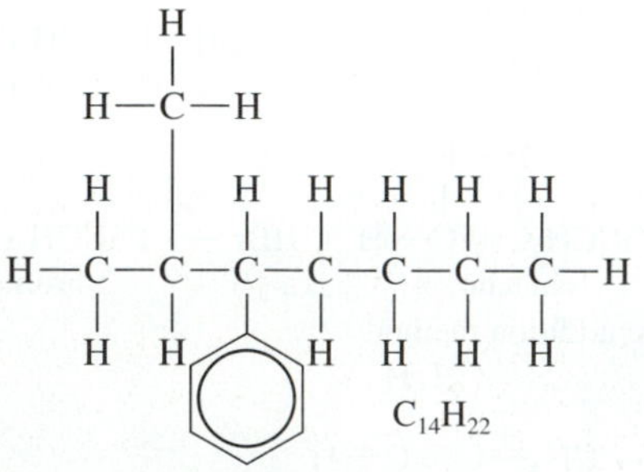

23.45 See Section 23.3.

23.47 (a) 1,2-dibromobenzene or *o*-dibromobenzene;
(b) 1,3-dibromobenzene or *m*-dibromobenzene;
(c) 1,4-dibromobenzene or *p*-dibromobenzene

23.49 (a) $C_6H_6 + 3H_2 \xrightarrow[200°C;\ 25\ atm]{Ni} C_6H_{12}$

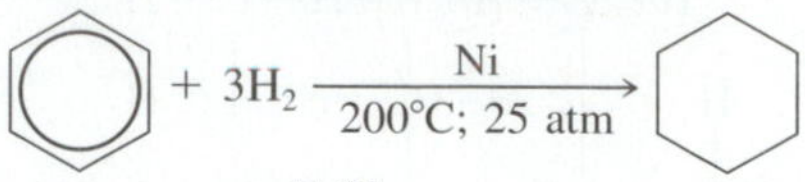

benzene $+ 3H_2 \xrightarrow[200°C;\ 25\ atm]{Ni}$ cyclohexane

(b) $C_6H_6 + Cl_2 \xrightarrow{FeCl_3} C_6H_5Cl + HCl$

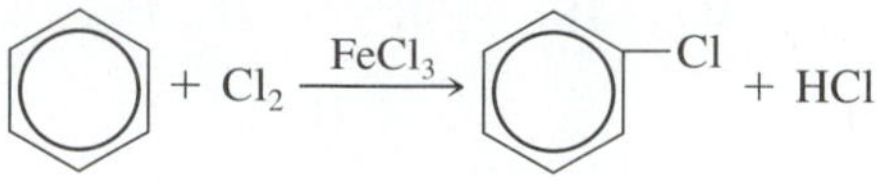

benzene $+ Cl_2 \xrightarrow{FeCl_3}$ chlorobenzene $+ HCl$

23.53 (a) $R-O-R$; (b) $R-C(=O)-R'$; (c) $R-OH$; (d) $R-C(=O)-OH$

23.57 (a) The aromatic ring withdraws electron density from the O—H bond, thus making it more polar.
(b) $C_6H_5OH(l \text{ or } aq) + NaOH(aq) \rightarrow C_6H_5ONa(aq) + H_2O(l)$

23.61 (a) carbonyl group; (b) hydroxyl group; (c) carboxyl group; (d) two carboxyl groups; (e) halide and carboxyl groups; (f) carbonyl and carboxyl groups; (g) hydroxyl and carboxyl groups

23.63 $CH_3-CH_2-CH_2-CH_2-OH$ $CH_3-CH_2-CH(OH)-CH_3$

$(CH_3)_2CH-CH_2-OH$ $(CH_3)_3C-OH$

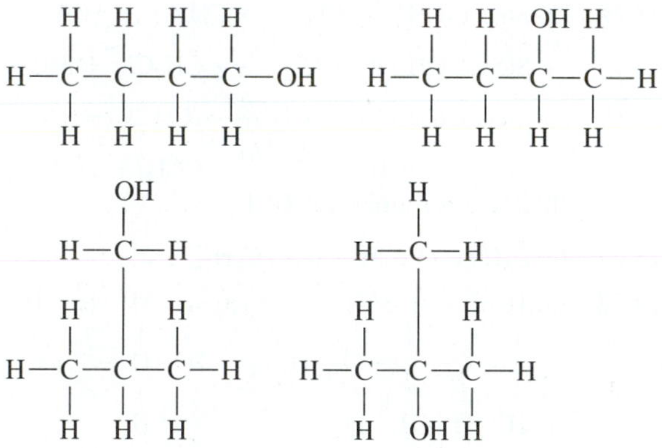

23.65 See Table 23.5.

23.67 (a) methanol; (b) formic acid; (c) oxalic acid; (d) glycerol

23.69 (a) oxidation; (b), (c) reduction

23.71 (a) oxidation; (b) neither; (c) reduction

23.73 (a) $-C(=O)-O$; (b) $R-C(=O)-O-R$

23.77

(a) $CH_3-O-C(=O)-CH_2CH_2CH_3$

(b) 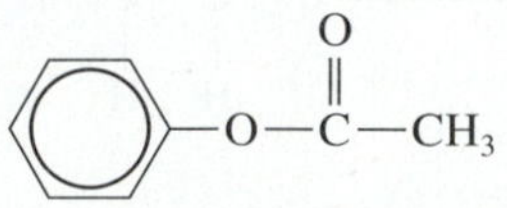 $C_6H_5-O-C(=O)-CH_3$

(c)
```
        O
        ‖
H2C—O—C—CH2CHg
 |
 |      O
 |      ‖
HC—O—C—CH2CH3
 |
H2C—O—C—CH2CH3
        ‖
        O
```

23.79 Acid: (benzene ring with two —C(=O)—OH groups on adjacent carbons; phthalic acid)
```
        O
        ‖
   ⌬ —C—OH
     —C—OH
        ‖
        O
```

Alcohols: $HO—CHCH_2CH_2CH_3$ (with CH_3 on the CH) and $HO—CHCH_3$ (with CH_3 on the CH)

23.81
```
    O  O      H  H      O  O      H  H
    ‖  ‖      |  |      ‖  ‖      |  |
HO—C—C—O—C—C—O—C—C—O—C—C—O—
              |  |                |  |
              H  H                H  H
```

23.83 (a)
```
        O
        ‖
H2C—O—C—(CH2)7CH=CH(CH2)7CH3
 |
 |      O
 |      ‖
HC—O—C—(CH2)7CH=CH(CH2)7CH3
 |
H2C—O—C—(CH2)7CH=CH(CH2)7CH3
        ‖
        O
```

(b) 1.
```
          O
          ‖
  H2C—O—C—C17H33
   |
   |      O
   |      ‖                  acid
  HC—O—C—C17H33 + 3H2O   ⇌
   |                         acid
  H2C—O—C—C17H33
          ‖
          O

H2C—OH
 |
HC—OH + 3C17H33COOH
 |        oleic acid
H2C—OH
glycerol
```

2.
```
        O
        ‖
H2C—O—C—C17H33
 |
 |      O
 |      ‖
HC—O—C—C17H33 + 3KOH ⟶
 |
H2C—O—C—C17H33
        ‖
        O

H2C—OH
 |
HC—OH + 3C17H33COOK
 |        potassium oleate, a soap
H2C—OH
glycerol
```

23.85 (a)
```
    H              H             |
    |              |             |
 —N—H    or    —N—     or    —N—
```

23.89 (a)
```
        H
        |
CH3—N—H
```
(b)
```
             H
             |
CH3CH2—N—H
```
(c) $H_2N—CH_2—CH_2—NH_2$

23.91 (a)
```
   H
   |
H—C—COOH
   |
   NH2
```
(b) no

23.93 (b) $CH_3^*CHClCH_2CH_2CH_3$
(d) $CH_2ClCH_2•CHClCH_3$

23.95 (a)
```
  O  H
  ‖  |
—C—N—
```
(b)
```
   O  H
   ‖  |
R—C—N—R′
```

23.99 (a)
```
   O  H
   ‖  |
H—C—N—H
```
(b)
```
   H  O  H
   |  ‖  |
H—C—C—N—⌬
   |
   H
```

23.127

1.
```
        H  H            H  H
        |  |            |  |
R· + C=C ⟶ R—C—C·        chain initiation
        |  |            |  |
        H  Cl           H  Cl
```

2.
```
   H  H    H  H         H  H  H  H
   |  |    |  |         |  |  |  |
R—C—C· + C=C ⟶ R—C—C—C—C·
   |  |    |  |         |  |  |  |
   H  Cl   H  Cl        H  Cl H  Cl
```
chain propagation

3. Step 2 is repeated over and over again until the chain growth is terminated by combination with another free radical; either R· or another free-radical chain.

23.129
```
     CH3              CH3
     |                |
H3C—C—CH3      H3C—C—CH2Br
     |                |
     CH3              CH3
```

23.131 (a) 86; (b) 173; (c) 0

GLOSSARY OF IMPORTANT TERMS

Note: Parenthetical references at the end of each entry are sections in the text where the term is introduced.

A

Absolute entropy—The entropy of a substance calculated on the assumption that its entropy is zero at zero Kelvin. *(19.4)*

Absolute temperature scale—Synonym for the Kelvin temperature scale. *(5.3)*

Absolute zero—The lower limit of the temperature scale; $-273.15°C = 0$ K. *(5.3)*

Absorption spectrum—The spectrum obtained when light with a continuous spectrum is passed through a substance that absorbs some of the wavelengths of the light. *(7.1)*

Acceptor impurity—An impurity with fewer valence electrons per atom than the semiconductor to which it is added. *(21.3)*

Accuracy—The extent to which a measured value coincides with the true value of the quantity measured. *(1.4)*

Acetylide—A carbide containing the C_2^{2-} ion. *(S3.3)*

Achiral—Interchangeable with its mirror image. *(21.5)*

Acid—A substance that ionizes in water to form H^+ (aq) ions; also see *Arrhenius acid, Brønsted–Lowry acid,* and *Lewis acid. (4.2)*

Acid ionization constant (K_a)—The equilibrium constant for the ionization of a weak acid in water. *(17.1)*

Acid strength—The ability of an acid to give up protons to a base. *(16.2)*

Acidic oxide—An oxide that reacts with water to form an acid or with aqueous bases to form water. Removing the elements of water from an oxo acid produces an acidic oxide. Most nonmetal oxides are acidic. *(S2.1)*

Acidic hydrogen—A hydrogen atom capable of being transferred as a proton from an acid to a base. *(4.2)*

Acidic solution—An aqueous solution in which the hydrogen ion concentration is greater than the hydroxide ion concentration; a solution with a pH below 7.0 at 25°C. *(16.4)*

Actinides—The elements 90 through 103 in the periodic table, immediately following actinium, in which the 5*f* orbitals are being filled. *(8.4)*

Activated complex—The configuration of atoms having the highest energy in the potential energy profile of an elementary step; the configuration of atoms in the transition state. *(14.7)*

Activation energy (E_a)—The minimum energy required for a reaction to occur. Usually only a small fraction of molecules will have this energy at any given time. *(14.7)*

Activity—The ability of a metal to be oxidized; oxidation potential is a measure of metal activity in an aqueous environment. Also the number of atoms disintegrating per second in a radioactive sample. *(4.3, 22.4)*

Activity series—A list of metals and hydrogen in order of their activity. A more active element (higher in the series) will displace a less active element from an ionic solution. *(4.3)*

Actual yield—The amount of product collected at the end of a reaction. *(3.7)*

Addition polymer—A polymer formed from the addition of unsaturated monomers to each other. *(23.2)*

Addition reaction—In organic chemistry, addition of a reactant to a double or triple bond. *(23.2)*

ADP—Adenosine diphosphate. See also *ATP. (23.9)*

Aerosol—A colloidal dispersion of a liquid or solid in a gas. Fog is an example of the former, smoke an example of the latter. *(13.6)*

Alcohol—A compound with the general formula R—OH, where R is a hydrocarbon group; see *primary, secondary,* and *tertiary alcohols. (23.4)*

Aldehyde—A carbonyl compound with the general formula R—CHO. *(23.4)*

Aldose—A *monosaccharide* that contains an aldehyde group, for example, glucose. *(23.8)*

Alkali—Synonym for base, most often used for a strong, water-soluble base. *(4.2)*

Alkali metal—A Group 1A element other than hydrogen. *(S1.2)*

Alkaline earth metal—A Group 2A element. *(S1.3)*

Alkane—A hydrocarbon that contains only single bonds. Straight- or branched-chain alkanes have the general formula C_nH_{2n+2}. *(10.3)*

Alkene—A hydrocarbon with one or more double bonds. Straight- or branched-chain alkenes with one double bond have the general formula C_nH_{2n}. *(10.3)*

Alkyl group—A group formed by removing one hydrogen atom from an alkane. *(23.1)*

Alkyne—A hydrocarbon with one or more triple bonds. Straight- or branched-chain alkynes with one triple bond have the general formula C_nH_{2n-2}. *(10.3)*

Allotropes—Two or more forms of the same element that can exist in the same physical state—for example, O_2 and O_3. *(2.6)*

Alloy—A solid solution composed of metals. *(4.1)*

Alloy steel—Steel that contains elements in addition to iron and carbon. *(S4.1)*

α-Amino acid—See *amino acid.*

Alpha carbon—The carbon atom adjacent to a carboxyl group. *(23.6)*

α-Helix—A spiral that turns in the direction of a right-handed screw. The polypeptide chains of certain proteins have this conformation. *(23.9)*

Alpha particle (α or ^4_2He)—The nucleus of a helium-4 atom, composed of two protons and two neutrons. Alpha particles are emitted by some radionuclides. *(2.2, 22.1)*

Aluminosilicate—A mineral in which aluminum atoms have replaced some of the silicon atoms in a silicate structure. *(S3.4)*

Aluminothermy—A metallurgical process in which aluminum is used to reduce a metal oxide. *(S4.1)*

Amalgam—A solution of a metal in mercury. *(S4.3)*

Amide—A compound that contains the *amide linkage.* The condensation product of a carboxylic acid with ammonia or an amine. *(23.7)*

Amide linkage—The $-\overset{\overset{\displaystyle O}{\|}}{C}-\underset{\underset{\displaystyle H}{|}}{N}-$ group; the functional group of amides. *(23.7)*

Amine—A hydrocarbon derivative of ammonia. Primary, secondary, and tertiary amines have, respectively, one, two, and three of the NH_3 hydrogen atoms replaced by hydrocarbon groups. *(23.6)*

Amino acid—A compound with a carboxyl group and an amino group. In an alpha amino acid, the amino group is on the carbon atom adjacent to the carboxyl group. *(23.6)*

Amino acid residue—In a *polypeptide* chain, a segment that derives from one amino acid. *(23.7)*

Amino group—The NH_2 group; the functional group in primary amines. *(23.6)*

Amontons' law—At constant volume the pressure of a fixed number of moles of gas is directly proportional to the Kelvin temperature; $P = k \times T$. *(5.4)*

Amorphous carbon—Very finely divided carbon such as soot, ashes, and carbon black. The lattice structure in amorphous carbon resembles that of graphite. *(S3.3)*

Amorphous solid—A solid in which the particles do not have a regular three-dimensional arrangement; there is no well-defined crystal lattice. Amorphous solids are also called glassy solids and vitreous solids. *(12.5)*

Ampere (A)—The unit of electric current; 1 A = 1 C/s. *(20.8)*

Amphiprotic—Able to both donate and accept protons. *(16.1)*

Amphoteric—Able to neutralize both acids and bases. Certain oxides and hydroxides are amphoteric and dissolve in both acidic and strongly basic solutions. *(18.3)*

Amphoteric hydroxide—See *amphoteric.*

Anion—A negatively charged ion. *(2.8)*

Anode—The electrode at which oxidation takes place. *(20.1)*

Anode sludge—A sludge that forms around the anode during electrorefining. The sludge contains anode impurities that are not oxidized during the refining process. *(S4.1)*

Anodizing—An electrolytic process by which the film of Al_2O_3 on an aluminum surface is thickened. *(S4.2)*

Antibonding orbital—A molecular orbital that has a higher energy than the atomic orbitals that formed it. *(10.5)*

Anticodon—A sequence of nitrogenous bases complementary to that in a given *codon.* *(23.9)*

Aqua regia—A concentrated mixture of nitric and hydrochloric acids. Aqua regia is a strong oxidizing agent that will dissolve metals such as gold and platinum. *(S2.2)*

Aqueous solution—A solution in which the solvent is water. *(4.1)*

Aquo-complex—A complex ion that contains water molecules as ligands. *(18.5)*

Aromatic hydrocarbon—A hydrocarbon that contains a ring with delocalized pi bonds, usually a benzene ring. *(10.3)*

Arrhenius acid—A substance that produces $H^+(aq)$ in aqueous solution. *(16.1)*

Arrhenius base—A substance that produces $OH^-(aq)$ in aqueous solution. *(16.1)*

Arrhenius equation—$k = Ae^{-E_a/RT}$; this equation describes the observed dependence of the rate constant k on Kelvin temperature for many reactions. *(14.7)*

Asymmetric carbon atom—A carbon atom with four different substituents. Such a carbon is *chiral.* *(23.6)*

Atmosphere (atm)—A unit of pressure; 1 atm $= 1.01325 \times 10^5$ N/m^2 = 760 torr = 1.013 bar. *(5.2)*

Atom—A neutral system of negatively charged electrons moving around a dense, positively charged nucleus. *(2.1)*

Atomic mass unit (u)—A mass equal to one-twelfth the mass of one carbon-12 atom; 1 g $= 6.022 \times 10^{23}$ u. The atomic mass unit is often abbreviated as amu. *(2.3)*

Atomic number (*Z*)—The number of protons in the nucleus of an atom. The atomic number is equal to the number of electrons in a neutral atom. *(2.2)*

Atomic radius—One-half the average distance between centers of two identical, neighboring atoms. *(8.5)*

Atomic weight—See *chemical atomic weight.*

ATP—Adenosine triphosphate. The conversion of ATP to ADP releases energy. *(19.5, 23.9)*

Aufbau procedure—A procedure for building up the ground-state electron configuration of an atom by adding one electron at a time into available orbitals according to an experimentally established filling order. The resulting configuration must satisfy the *Pauli exclusion principle* and *Hund's rule. (8.3)*

Autoionization—The transfer of protons from one molecule to another molecule of the same substance; self-ionization. Amphiprotic solvents undergo autoionization to a limited extent. *(16.1)*

Average reaction rate—The change in concentration (or amount) of a reactant or product divided by the time it takes for the change to occur. *(14.2)*

Avogadro's hypothesis—Equal volumes of different gases at the same temperature and pressure contain the same number of molecules. *(5.3)*

Avogadro's law—At constant temperature and pressure, the volume of a gas is directly proportional to the number of moles; $V = k_a \times n$. *(5.3)*

Avogadro's number (N_A)—6.0221367×10^{23}; the number of atoms in exactly 12 g of carbon-12; the number of units in a mole. *(2.4)*

Axial bond—One of two bonds perpendicular to the center plane of a bipyramidal molecule. *(10.1)*

Azimuthal quantum number (l)—A quantum number determining the shape of a hydrogen-like orbital. For a given value of n, l has integral values ranging from 0 to $n - 1$. *(7.6)*

B

Balance—A device for obtaining the mass of an object by balancing its gravitational attraction against that of known masses. *(1.6)*

Balanced equation—See *chemical equation.*

Balloon picture—An orbital diagram depicting a surface within which there is a large probability of finding the electron (see Figure 7.32). The electron density is the same at all points on the surface depicted. *(7.7)*

Balmer series—The visible series in the emission spectrum of hydrogen, produced by electron transitions from higher levels down to $n = 2$. *(7.3)*

Band—A collection of closely spaced molecular orbitals; see *band theory. (21.3)*

Band theory—A theory that explains the bonding and properties of solids in terms of very closely spaced molecular orbitals, called bands, that extend over the entire crystal. *(21.3)*

Bar—A unit of pressure; 1 bar $= 10^5$ N/m^2. *(5.2)*

Barometer—An instrument for measuring atmospheric pressure. *(5.2)*

Base—A substance that produces OH^-(aq) ions in aqueous solution; also see *Arrhenius base, Brønsted–Lowry base,* and *Lewis base. (4.2)*

Base ionization constant (K_b)—The equilibrium constant for the ionization of a weak base in water. *(17.2)*

Base strength—The ability of a base to capture protons from an acid. *(16.2)*

Basic oxide—An oxide that reacts with water to form an ionic hydroxide or with aqueous acids to form water. Removing the elements of water from a metal hydroxide produces a basic oxide. Most metal oxides are basic oxides. *(S2.1)*

Basic oxygen process—The process by which most steel is made in the United States. *(S4.5)*

Basic solution—An aqueous solution in which the hydroxide ion concentration is greater than the hydrogen ion concentration; a solution with a pH above 7.0 at 25°C. *(16.4)*

Battery—A package of one or more galvanic cells used for the production and storage of electrical energy. *(20.5)*

Bayer process—A process for obtaining pure aluminum oxide from bauxite. *(S4.1)*

Becquerel (Bq)—The SI unit of activity; 1 Bq = 1 disintegration per second. *(22.4)*

Beta particle (β or ${}_{-1}^{0}e$)—A rapidly moving electron emitted by a radionuclide. *(22.1)*

Beta pleated sheet—A side-by-side array of extended polypeptide chains found in proteins. The chains are held in place by hydrogen bonds. *(23.7)*

Bidentate ligand—A ligand with two coordinating atoms. *(21.4)*

Bimolecular step—An elementary step involving the collision of two particles. *(14.5)*

Binary acid—An acid containing only two elements, one of which is hydrogen. HCl is an example of a binary acid. *(16.5)*

Binary compound—A compound containing only two elements; water (H_2O) and ammonia (NH_3) are binary compounds. *(2.9)*

Binding energy—The energy required to decompose a nuclide into protons and neutrons. The energy equivalent of the *mass defect. (22.6)*

Blast furnace—A furnace in which the high-temperature chemical reduction of an ore is carried out. *(S4.1)*

Boat conformation—A boat-shaped cyclohexane conformation. The boat and chair conformations provide tetrahedral angles in a six-membered ring. *(23.1)*

Body-centered cubic unit cell—The unit cell for a lattice built of cubes that have a particle at each corner and a particle in the center. *(12.6)*

Bohr model (of the hydrogen atom)—A planetary model for hydrogen, in which the electron moves in certain definite allowed orbits around the nucleus. *(7.3)*

Boiling point—The temperature at which the vapor pressure of a liquid is equal to the external pressure. *(12.1)*

Boiling point elevation (BPE)—The boiling point of a solution minus the boiling point of the pure solvent; a colligative property proportional to the molality of the solute. *(13.3)*

Boiling point elevation constant (K_b)—The proportionality constant relating boiling point elevation (BPE) to molality (m); $K_b = \text{BPE}/m$. *(13.3)*

Bomb calorimeter—An instrument for measuring constant volume heats of reaction; see Figure 6.13. *(6.6)*

Bond angle—The equilibrium angle between the lines that join two atoms to a third atom in a molecule. *(10.2)*

Bond energy (D)—The average energy required to break a chemical bond in a gaseous molecule, usually expressed in kilojoules per mole of bonds. *(9.3)*

Bond order—$\frac{1}{2}$ × (number of bonding electrons − number of antibonding electrons). *(10.5)*

Bond length—The equilibrium distance between the centers of two bonded atoms. The bond length is the distance at which the attractive and repulsive forces are in balance and the net force on the atoms is zero. *(9.3)*

Bonding orbital—A molecular orbital that has a lower energy than the atomic orbitals that formed it. *(10.5)*

Borane—One of the numerous boron hydrides. Boranes contain multicentered bonds. *(S3.5)*

Born–Haber cycle—A diagram that relates the enthalpy of formation of an ionic solid to the sum of five energy quantities (see Figure 9.5). It is usually used in conjunction with Hess's law to calculate the *lattice energy* or any one of the related energies. *(9.2)*

Bottleneck step—Synonym for *rate-determining step. (14.5)*

Boyle's law—At constant temperature, the volume of a fixed number of moles of gas is inversely proportional to the pressure; $P \times V = k_b$. *(5.3)*

Bragg equation—The equation, $n\lambda = 2d \sin \theta$, relating the wavelength λ, the distance d between atomic layers, and the diffraction angle θ. The Bragg equation is the basis for crystal structure analysis by x-ray diffraction. *(12.6)*

Breeder reactor—A nuclear reactor that produces fissionable nuclear fuel in addition to energy. *(22.7)*

Brønsted–Lowry acid—A substance that can donate protons to a base. *(16.1)*

Brønsted–Lowry base—A substance that can accept protons from an acid. A Brønsted–Lowry base has a pair of electrons that is capable of binding the proton. *(16.1)*

Brownian motion—The spontaneous random motion of finely divided particles suspended in a liquid. *(5.1)*

Buffer—Synonym for *buffer solution.*

Buffer capacity—The number of moles of strong acid or strong base needed to change the pH of 1 L of buffer by one unit. *(17.7)*

Buffer ratio—The ratio of base concentration to acid concentration ([base]/[acid]) in a buffer solution. *(17.7)*

Buffer solution—A solution that resists pH changes. Buffer solutions often consist of a weak acid and its conjugate base. *(17.7)*

Buret—A long narrow glass tube on which volume levels can be read with a high degree of precision. Burets are used in titrations; see Figure 4.14. *(4.6)*

C

Calomel electrode—A reference electrode based on the half-reaction

$$Hg_2Cl_2(s) + 2e^- \rightleftharpoons 2Hg(l) + 2Cl^-(\text{KCl, saturated})$$

Calomel electrodes are used in pH meters. *(20.4)*

Calorie (cal)—A quantity of energy exactly equal to 4.184 J; 1 cal will raise the temperature of 1 g of water by approximately 1°C. *(1.7)*

Calorimeter—An instrument for measuring heats of reaction. Bomb calorimeters are used for measuring constant volume heats of reaction. "Coffee cup" calorimeters, Dewar flasks, and flame calorimeters are used for measuring enthalpy changes. *(6.6)*

Capillary action—The spontaneous rise or fall of liquids in narrow tubes. *(12.4)*

Carbide—A binary compound of carbon and a less electronegative element. Carbides are classified as ionic, interstitial, and covalent. *(S3.3)*

Carbohydrate—A *monosaccharide* or a monosaccharide polymer. *(23.8)*

Carbonyl group—The $>C{=}O$ group; the functional group of aldehydes and ketones. *(23.4)*

Carboxyl group—The COOH group; the functional group of carboxylic acids. *(23.4)*

Carboxylic acid—An organic acid that contains the *carboxyl group. (23.4)*

Cast iron—A fairly brittle iron containing about 4% carbon. *(S4.5)*

Catalyst—A substance that increases the rate of a reaction without being consumed in the reaction. *(14.8)*

Cathode—The electrode at which reduction takes place. *(20.1)*

Cathode ray—An electron beam that travels from cathode to anode in a discharge tube. *(2.1)*

Cathodic protection—A method for preventing corrosion in which a more active metal is put in electrical contact with the metal to be protected. The more active metal will be anodic and corrode; the less active metal will provide the protected cathodic surface on which reduction occurs. *(20.6)*

Cation—A positively charged ion. *(2.8)*

Cell notation—A shorthand notation for describing a galvanic cell. *(20.1)*

Celsius scale—The temperature scale in which the freezing point of water is 0° and the boiling point is 100°. *(1.7)*

Central atom—An atom to which two or more other atoms are bonded. *(10.2)*

Chain initiation step—The first step in a chain mechanism; the initiation step is typically an endothermic dissociation that produces free radicals. *(14.6)*

Chain propagation step—One of the successive steps in a chain mechanism in which one free radical produces another free radical to continue the chain. *(14.6)*

Chain reaction—A reaction in which an initial step leads to a succession of repeating steps that continues indefinitely; chain reactions usually involve free radicals. *(14.6)*

Chain termination step—A step in a chain mechanism in which two free radicals combine, thus removing them from circulation. *(14.6)*

Chair conformation—A chair-shaped cyclohexane conformation; see *boat conformation. (23.1)*

Charge delocalization—The spreading of ionic charge over a large volume. A given charge is more delocalized on a large ion than on a small ion. *(16.5)*

Charles's law—At constant pressure, the volume of a fixed number of moles of gas is directly proportional to the Kelvin temperature; $V = k_c \times T$. *(5.3)*

Chelating agent—A polydentate ligand that coordinates two or more positions of the central metal atom or ion in a complex. *(21.4)*

Chemical atomic weight—The weighted average mass of the isotopes making up an element, obtained by multiplying each isotopic mass by its fractional abundance and adding the products. Chemical atomic weights are expressed in atomic mass units. *(2.3)*

Chemical bond—The strong bond between atoms in a molecule or a crystal. Chemical bonds can be distinguished from intermolecular attractions by the greater energies required to break them, usually between 100 and 1000 kJ/mol. *(9.1)*

Chemical change—Synonym for *chemical reaction.*

Chemical equation—A symbolic form of a chemical reaction in which formulas are used to represent the reactants and products. Chemical equations are balanced in the sense that every atom on the reactant side also appears on the product side. *(3.5)*

Chemical kinetics—The study of chemical reaction rates and mechanisms. *(14.1)*

Chemical property—Any chemical reaction that an element or compound can undergo. *(3.1)*

Chemical reaction—A change in which one or more new elements or compounds are formed. *(3.1)*

Chemisorption—Adsorption that involves some degree of chemical bonding between the adsorbing surface and the adsorbed molecules. *(14.8)*

Chemistry—The study of matter and its transformations. *(1.1)*

Chiral—Having a nonsuperimposable mirror image. A hand is chiral. *(21.5)*

***cis–trans* Isomers**—A form of stereoisomerism in which the isomers differ in the orientation of certain bonds with respect to a molecular axis or a molecular plane. The *cis* isomer has two similar atoms on the same side of the axis or plane; the *trans* isomer has them on opposite sides. *cis–trans* Isomers are also called geometric isomers. *(10.3, 21.5)*

Clausius–Clapeyron equation—An equation relating the temperature and vapor pressure of a liquid: $\ln P = -\Delta H^\circ_{vap}/RT + B$, where ΔH°_{vap} is the molar heat of vaporization and B is a constant for the given liquid. *(12.1)*

Close-packing—An atomic packing arrangement utilizing either hexagonal close-packed or cubic close-packed (face-centered cubic) lattices. Close packing allows the maximum number of identical atoms to be packed into a given volume. *(12.6)*

Codon—A sequence of three nitrogenous bases in a nucleic acid. In the *genetic code,* each amino acid is represented by one or more codons. *(23.9)*

Coefficient (in an equation)—Numbers, usually integers, placed before the formulas in a chemical equation. The coefficients indicate the relative molar quantities involved in the reaction. *(3.5)*

Coinage metal—A Group 1B element; copper, silver, or gold. *(S4.4)*

Coke—Carbon obtained by heating coal to drive off volatile impurities. Coke is often used to chemically reduce oxide ores. *(3.6)*

Colligative effect—See *colligative property.*

Colligative property—Solution properties that depend on the concentration of solute particles rather than on their nature; these properties include vapor pressure lowering, boiling point elevation, freezing point depression, and osmotic pressure. *(13.3)*

Collision frequency (*Z*)—The number of collisions that occur each second in a given sample of matter; see *collision theory of reaction rates. (14.7)*

Collision theory of reaction rates—A theory in which the rate of a bimolecular reaction is assumed to be equal to the collision frequency (Z) multiplied by the fraction of collisions having the activation energy (f_a) and by the fraction having a suitable orientation for reaction (p). *(14.7)*

Colloid—A heterogeneous mixture containing particles with dimensions greater than 1000 pm, but at least one dimension less than 100,000 pm. Colloidal particles are intermediate in size between those of true solutions and suspensions. *(13.6)*

Colloidal dispersion—Synonym for *colloid.*

Combination reaction—A reaction in which a compound forms from simpler substances. *(3.6)*

Combustion analysis—An analytical procedure that involves burning a small sample of the compound in oxygen and weighing the products. *(3.4)*

Combustion reaction—A vigorous self-sustaining reaction in which substances combine and give off heat and light; a reaction in which one substance burns in another. The most familiar combustions are the burning of fuels in oxygen. *(3.6)*

Common ion—An ion that is common to two solutes in a given solution. *(17.6)*

Common ion effect—A shift in equilibrium caused by the addition of an ion common to the equilibrium mixture. *(17.6)*

Complementary bases—A pair of *nitrogenous bases* that hydrogen bond to each other. The pairs in DNA and RNA are adenine–thymine, guanine–cytosine, and adenine–uracil. *(23.9)*

Complex—See *metal complex.*

Complex ion—See *metal complex.*

Compound—A substance made up of two or more elements combined in a fixed ratio. The properties of a compound are different from those of its constituent elements.

Concentration—The amount of solute in a given quantity of solvent or solution. *(4.4)*

Condensation—A phase change from gas to liquid. *(12.1)*

Condensation reaction—A reaction that joins two molecules and eliminates some small molecule such as H_2O. *(23.5)*

Condensed state—The solid or liquid state. *(12.1)*

Conduction band—An empty or partially filled band of molecular orbitals through which electrons are free to move; see *band theory.* *(21.3)*

Conformation—Any one of the various shapes a molecule can get into by twisting and bending. *(10.3)*

Conjugate acid–base pair—A Brønsted acid and base whose formulas differ by a single proton. *(16.2)*

Constant pressure reaction—A reaction occurring under constant external pressure; one in which the pressure of the products is equal to the pressure of the reactants. *(6.2)*

Constant volume reaction—A reaction in which the volume does not change; one in which there is no expansion or contraction. *(6.2)*

Contact catalyst—A catalyst that provides a surface on which the reaction takes place. *(14.1)*

Contact process—An industrial process for preparing sulfuric acid from sulfur. *(S3.1)*

Continuous spectrum—A spectrum containing all the wave lengths for the spectral region of interest. *(7.1)*

Contributing structure—Synonym for *resonance structure.*

Control rods—Neutron-absorbing rods that control the fission rate in a reactor core. Control rods often contain cadmium or boron atoms. *(22.7)*

Conversion factor—A fraction whose numerator and denominator contain the same quantity expressed in different units; used as a multiplying factor to change units. *(1.5)*

Cooling curve—A plot of the temperature of a substance versus heat withdrawn from the substance. *(12.1)*

Coordinate covalent bond—An electron-pair bond in which both electrons come from the same atom. *(21.4)*

Coordinating atom—A ligand atom with a lone electron pair oriented toward the metal atom in a complex. *(21.4)*

Coordination compound—A compound containing a complex. The compound itself may be a complex, or it may contain a complex ion. *(21.4)*

Coordination isomerism—A form of structural isomerism in which the isomers contain the same central metal atoms (two or more) and the same ligands, but differ in how the ligands are distributed between the metal atoms. *(21.5)*

Coordination number—The number of neighboring atoms or ions in contact with a given atom or ion, or the number of ligand atoms in contact with the central metal atom of a complex. *(12.6, 21.4)*

Corrosion—A natural galvanic process during which metals become pitted and eaten away by slow oxidation. The rusting of iron is an example. *(20.6)*

Corrosion inhibitor—A substance that retards the corrosion process by being adsorbed onto the metal surface. *(20.6)*

Cosmic rays—High-energy particles that originate in the stars and travel through space. Cosmic rays consist mostly of electrons, protons, and light nuclides. *(22.4)*

Coulomb (C)—The SI unit of electric charge. *(2.1)*

Coulometer—A device that measures the total charge passed through a circuit. In a silver coulometer, the charge is calculated from the mass of electrolytically deposited silver. *(20.8)*

Covalent bond—One, two, or three pairs of electrons shared by two atoms and attracted to both nuclei. *(9.1)*

Covalent compound—A compound that does not contain ions. *(9.1)*

Critical mass—The minimum mass of fissionable material required to support a self-sustaining nuclear reaction. *(22.7)*

Critical point—The upper endpoint of the vaporization curve in a phase diagram. *(12.2)*

Critical pressure—The pressure at the critical point. *(12.2)*

Critical temperature—The temperature at the critical point; the temperature above which the liquid state cannot exist. *(12.2)*

Crystal lattice—The regular repeating pattern of atoms, molecules, or ions in a crystal. *(12.5)*

Crystalline solid—A solid whose particles are arranged in a crystal lattice. *(12.5)*

Cubic close packing—One of two ways of packing identical spheres into the smallest possible volume; its unit cell is the face-centered cube. *(12.6)*

Cubic decimeter (dm^3)—A unit of volume equal to exactly 1 L. *(1.3)*

Curie (Ci)—A measure of activity; 1 Ci = 3.7×10^{10} disintegrations per second. *(22.4)*

Current (*I*)—The time rate of flow of charge. The unit of current is the *ampere.* *(20.8)*

Cyanide leaching process—A process in which gold or silver is dissolved in an aerated cyanide solution. *(S4.1)*

Cycle—See *wave.*

Cycloalkane—A saturated hydrocarbon in which some or all of the carbon atoms form a ring. *(23.1)*

Cyclotron—An instrument that uses an alternating magnetic field to accelerate charged particles in an ever-widening circular path. *(22.3)*

D

***d* Block elements**—Elements in which *d* orbitals are being filled. *(8.4)*

***d* Orbital**—An atomic orbital with azimuthal quantum number $l = 2$. *(7.6)*

Dalton's law of partial pressures—The total pressure of a gas mixture is equal to the sum of the partial pressures of the individual gases. *(5.6)*

Data—Factual information; obtained by experiment and observation. *(1.2)*

De Broglie wavelength—The wavelength of a matter wave; $\lambda = h/mv$, where *m* and v are the particle's mass and speed, respectively, and *h* is Planck's constant. *(7.4)*

Debye (D)—A unit for measuring molecular dipole moments; 1 D $= 3.33 \times 10^{-30}$ coulomb·meter. *(9.7)*

Decay series—The sequence of steps by which a radioisotope is transformed into a stable isotope. *(22.1)*

Decomposition reaction—A reaction in which a compound breaks down into elements or into simpler compounds. *(3.6)*

Degree Celsius (°C)—One-hundredth of the temperature difference between the freezing and boiling points of water on the Celsius scale. *(1.7)*

Delocalized bond—A chemical bond in which bonding electrons spread out over several atoms, thus helping to bond all of them. The second and third pairs of electrons in multiple bonds are often delocalized. Delocalized bonds can be represented by two or more *resonance structures.* *(9.5)*

Delocalized orbital—A molecular orbital with electron density spread out over three or more atoms. *(10.6)*

Density—The mass per unit volume of a substance; $d = m/V$. *(1.6)*

Deposition—A phase change from gas to solid without passing through the liquid state. *(12.1)*

Derived unit—A unit made by combining two or more other units; the joule (1 J = 1 N·m) is an example of a derived unit. *(1.2)*

Detergent—A cleansing agent with wetting and emulsifying properties. A typical detergent consists of long molecules with one hydrophobic and one hydrophilic end. *(23.4)*

Dextrorotatory—Able to rotate the plane of polarized light to the right; symbolized by (+). *(21.5)*

Diamagnetism—The absence of paramagnetism. *(8.5)*

Diaphragm cell—An industrial cell used for the electrolysis of brine. Aqueous NaOH produced in this cell is contaminated with unreacted sodium chloride. *(20.7)*

Diene—A hydrocarbon with two double bonds. *(23.2)*

Differentiating solvent—A solvent that shows up differences in acid and base strength. Water is a differentiating solvent for acids weaker than $H_3O^+(aq)$ and bases weaker than $OH^-(aq)$. *(16.2)*

Diffraction pattern—The alternating bright and dark regions that result when radiation passes through adjacent slits. A diffraction pattern is obtained when x-rays are passed through a crystal. *(12.6)*

Diffusion—The spontaneous spreading of one substance through another. *(5.1)*

Dilution formula—A formula that relates the volume and concentration of a concentrated solution to the volume and concentration of a solution made from it by dilution; $M_c \times V_c = M_d \times V_d$. *(4.5)*

Dimensional analysis—A problem-solving technique in which the units of each quantity are included in each step of the calculation and are handled (multiplied, canceled, and raised to powers) just like numbers. *(1.5)*

Dipeptide—The condensation product of two amino acids. *(23.7)*

Dipole—Any object whose ends carry equal but opposite charges; HCl is an example of a molecular dipole. *(9.7)*

Dipole–dipole force—The net electrostatic attraction in a collection of polar molecules; often called a dipole–dipole interaction. *(12.3)*

Dipole moment (μ)—The magnitude of the separated charges in a *dipole* multiplied by the distance between their centers. The dipole moment of a molecule is the resultant of the bond dipole moments. *(9.7)*

Diprotic acid—An acid with two acidic hydrogen atoms per molecule, for example, H_2SO_4. *(16.3)*

Direction-focusing mass spectrometer—See *mass spectrometer.*

Directly proportional—A relation between two quantities such that if one changes by a certain factor, the other will change by the same factor. *(5.3)*

Disaccharide—The condensation product of two *monosaccharides.* *(23.8)*

Discharge tube—A sealed glass tube containing two metal electrodes. Discharge tubes were originally used for studying the conduction of electricity in gases. *(2.1)*

Displacement reaction—A reaction in which one element displaces another element from a compound. *(3.6)*

Disproportionation reaction—A redox reaction in which one substance is both oxidized and reduced. *(11.3)*

Dissociation—The process by which ions move apart and spread throughout a solution. *(4.1)*

Dissociation constant (K_d)—The equilibrium constant for the dissociation of a complex ion into a metal ion and its ligands; the reciprocal of the *formation constant. (18.5)*

DNA—Deoxyribonucleic acid. The genetic material in the cell nucleus consists of DNA. *(23.9)*

Donor impurity—An impurity with more valence electrons per atom than the semiconductor to which it is added. *(21.3)*

Doping (of a semiconductor)—Adding carefully selected impurities to enhance the conductivity of a semiconductor. *(21.3)*

Double bond—A covalent bond consisting of two shared pairs of electrons. *(9.3)*

Double displacement reaction—synonym for *metathesis reaction.*

Double helix—The DNA conformation, which consists of two nucleic acid strands twisted together into an α-helix. *(23.9)*

Double replacement reaction—synonym for *metathesis reaction.*

Dow process—A process for obtaining magnesium metal from seawater. *(S4.1)*

Downs cell—An industrial cell used for the electrolysis of molten sodium chloride. *(20.7)*

Ductility—The ability to be drawn out into wires; a property generally associated with metals. *(12.5)*

Dynamic equilibrium—A state in which there is no net change because two opposing processes occur at the same rate. *(4.1)*

E

Effective nuclear charge—The nuclear charge corrected for the shielding effect of intervening electrons. *(8.1)*

Effusion—The passage of gas molecules through small openings. *(5.6)*

Electrical work—Useful work obtained from a galvanic cell; it is the product of the cell voltage E and the number of coulombs Q transferred between the cell terminals. *(20.2)*

Electrochemistry—A branch of chemistry dealing with processes that convert chemical energy into electrical energy, and vice versa. *(20.1)*

Electrode—A conducting substance, usually a metal, through which electrons enter or leave a conducting medium. *(4.1, 20.1)*

Electrode reaction—A chemical change occurring at an electrode in an electrochemical cell. Each electrode reaction is one of the half-reactions that make up the total cell reaction. *(20.1)*

Electrolysis—The use of an electric current to bring about an oxidation–reduction reaction. *(20.7)*

Electrolyte—A solute that provides ions in aqueous solution. *(4.1)*

Electrolytic cell—A cell in which an electrolysis reaction is carried out. *(20.7)*

Electromagnetic radiation—Forms of energy that travel through space setting up oscillating electric and magnetic fields along the way; commonly referred to as "light." The product of the wavelength and frequency of electromagnetic radiation is equal to the speed of light, $c = 2.99792458 \times 10^8$ m/s. *(7.1)*

Electromagnetic spectrum—Electromagnetic waves sorted out and arranged in order of wavelength or frequency. *(7.1)*

Electromotive force (EMF)—The driving force that pushes electrons around a circuit or ions through a solution. *(20.2)*

Electron—A negatively charged fundamental particle found outside the atomic nucleus. See Table 2.1 for its charge and mass. *(2.1)*

Electron affinity (EA)—The enthalpy change accompanying the addition of an electron to a gaseous ground-state atom or ion. *(8.5)*

Electron capture—The capture of an extranuclear electron from the innermost shell by a radionuclide. Electron capture converts a nuclear proton into a neutron. *(22.5)*

Electron cloud—A dot diagram showing how electron density varies with position relative to the nucleus. *(7.7)*

Electron configuration—The distribution of electrons in the orbitals of a many-electron atom or molecule; understood to be the ground-state configuration unless otherwise specified. *(8.3)*

Electron density (ψ^2)—The square of the wave function. The probability of finding an electron in a small volume of space centered about some point is proportional to the electron density at that point. *(7.7)*

Electron pair bond—Synonym for *covalent bond. (9.1)*

Electron spin—The hypothetical spinning motion of an electron that produces its magnetic field. *(7.6)*

Electron volt (eV)—An energy unit; 1 eV $= 1.602 \times 10^{-19}$ J. *(22.6)*

Electronegativity—The tendency of an atom to attract shared electrons. *(9.7)*

Electrophile—An "electron-seeking" species; a *Lewis acid. (16.6)*

Electroplating—The electrolytic deposition of a metal on the surface of another metal. *(20.7)*

Electrorefining—An electrolytic process for refining a crude metal. Less active impurities form an *anode sludge;* more active impurities remain in the electrolyte. *(S4.1)*

Electrostatic model—A theory of metal–ligand bonding in which it is assumed that the ligands of a complex ion are bonded to a central metal ion by simple electrostatic attraction. *(21.6)*

Element—A substance composed of atoms all having the same atomic number. *(2.3)*

Elementary step—A step in a reaction mechanism. The equation for an elementary step describes an event in which one, two, or three particles participate. *(14.5)*

Emission spectrum—The spectrum of any object emitting electromagnetic radiation. *(7.1)*

Empirical formula—A formula whose subscripts represent the simplest whole-number ratio of the atoms in a compound; for example, the empirical formula of benzene (C_6H_6) is CH. *(3.4)*

Emulsifying agent—A hydrophilic substance that is adsorbed by the surface of a hydrophobic colloid particle. Soaps, detergents, and various food additives are examples. Emulsifying agents help stabilize the existence of hydrophobic colloids in water. *(13.6)*

Emulsion—A colloid that consists of one liquid dispersed in another. *(13.6)*

Enantiomers—Isomers that are nonsuperimposable mirror images. *(21.5)*

Endothermic reaction—A reaction that absorbs heat from its surroundings. *(6.2)*

Endpoint—The end of a titration as shown by the color change of an indicator or by some other signal. *(4.6)*

Energy—The ability to do work. *(1.1)*

Energy band—Synonym for *band. (21.3)*

Energy gap—The energy difference between the valence and conduction bands of an atomic solid. *(21.3)*

Energy level diagram—A diagram showing the relative energies of an atom or a molecule. The energy scale is on the vertical axis and a horizontal line is drawn for each allowable energy. *(7.3)*

Enthalpy (*H*)—The sum of the internal energy and the pressure–volume product of a system; $H = E + PV$. *(6.3)*

Enthalpy change (ΔH)—The enthalpy of the products minus the enthalpy of the reactants; equal to the heat of reaction for a constant pressure reaction. *(6.3)*

Enthalpy of combustion—See *standard enthalpy of combustion.*

Enthalpy of formation—See *standard enthalpy of formation.*

Entropy (*S*)—A thermodynamic property that measures the degree of randomness in a system. Entropy is a state function. *(19.3)*

Enzyme—A biological catalyst. *(14.1)*

Equation—See *chemical equation.*

Equation of state—An equation that relates a substance's pressure, volume, temperature, and number of moles. *(5.8)*

Equatorial bond—A bond in the center plane of a bipyramidal molecule. *(10.1)*

Equilibrium constant (*K*)—The value of the reaction quotient at equilibrium. Equilibrium constants expressed in terms of molar concentrations are symbolized by K_c. Constants expressed in terms of partial pressures are symbolized by K_p. *(15.2)*

Equivalence point—The point during a titration at which the reactants have combined in the ratio of their equation coefficients. Neither reactant is in excess. *(17.8)*

Ester—A compound whose molecules contain the *ester linkage.* The condensation product of an alcohol and a carboxylic acid. *(23.5)*

Ester linkage—The $-\overset{\overset{\displaystyle O}{\|}}{C}-O-$ group; the functional group in esters. *(23.5)*

Ether—A compound with the formula $R_1{-}O{-}R_2$, where R_1 and R_2 are hydrocarbon groups. *(23.4)*

Excess reactant—A reactant that is not used up at the end of a reaction. *(3.8)*

Excited state—Any energy state other than the ground state. *(7.3)*

Exothermic reaction—A reaction that gives off heat to its surroundings. *(6.2)*

Expanded octet—More than eight electrons around a central atom in a Lewis structure. Octet expansion is exhibited by elements, especially nonmetals, from the third and following periods; it is never exhibited by first- and second-period elements. *(9.6)*

Experiment—A procedure used to discover facts or test ideas. *(1.1)*

F

***f* Block elements**—Elements in which *f* orbitals are being filled; the *inner transition elements. (8.4)*

Face-centered cubic unit cell—The unit cell for a lattice built of cubes that have a particle at each corner and a particle in the center of each face. *(12.6)*

Fahrenheit scale—The temperature scale in which the freezing point of water is 32° and the boiling point is 212°. *(1.7)*

Faraday (*F*)—The quantity of charge on 1 mol of electrons; 96,485 C. *(20.2)*

Fat—An ester of glycerol with three fatty acid molecules. At room temperature, fats formed from saturated fatty acids tend to be solid while fats formed from unsaturated fatty acids tend to be liquid. *(23.5)*

Fatty acid—A long-chain *carboxylic acid* obtained by hydrolyzing fats or oils. Saturated fatty acids contain only C—C single bonds; unsaturated fatty acids contain one or more double bonds. *(23.4)*

Ferromagnetism—The strong magnetism exhibited by iron and a few other substances; attributable to a cooperative effect between paramagnetic atoms. *(8.5)*

Ferrous alloy—An alloy containing iron. *(S4.1)*

First Bohr radius (a_0)—The radius of the first Bohr orbit for hydrogen; 52.9 pm $= 0.529 \times 10^{-10}$ m. *(7.3)*

First law of thermodynamics—Energy can be converted from one form into another, but it cannot be created or destroyed. *(6.1)*

First-order reaction—A reaction in which the rate is proportional to the first power of a reactant concentration. *(14.3)*

Fission—See *nuclear fission.*

Fission reactor—See *nuclear reactor.*

Flotation—A method for concentrating sulfide ores with the use of oil and a foaming detergent. *(S4.1)*

Fluorocarbons—Compounds containing only carbon and fluorine. *(S2.1)*

Foam—A colloidal dispersion of gas in a liquid or solid. Shaving cream is an example of the former, styrofoam the latter. *(13.6)*

Force—Any push or pull. *(1.6)*

Formal charge—The charge obtained by subtracting the number of electrons assigned to an atom in a Lewis structure from the number of valence electrons contributed by the atom; formal charge = number of valence electrons − number of unshared electrons − $\frac{1}{2}$ × number of bonded electrons. *(9.4)*

Formation constant (K_f)—The equilibrium constant for the formation of a complex ion from a metal ion and its ligands. *(18.5)*

Formula weight—A synonym for the molar mass of a substance. *(2.7)*

Fractional abundance—The percentage abundance divided by 100%. *(2.3)*

Fractional distillation—A process of separating the components of a liquid mixture by gradually raising its temperature and allowing the components to boil off one by one. *(S1.1)*

Fractional precipitation—The use of a precipitating agent to precipitate one ion while leaving another in solution. *(18.2)*

Frasch process—The industrial process by which underground sulfur is forced to the surface with the aid of hot water and air. *(S3.1)*

Free energy—See *Gibbs free energy.*

Free radical—Any atom, molecule, or ion containing at least one unpaired electron. O_2, NO, and NO_2 are among the few free radicals that are stable under ordinary conditions. *(9.2)*

Freezing—A phase change from liquid to solid. *(12.1)*

Freezing point—Synonym for melting point. *(12.1)*

Freezing point depression (FPD)—The freezing point of a pure solvent minus the freezing point of a solution; a colligative property proportional to the molality of the solute. *(13.3)*

Freezing point depression constant (K_f)—The proportionality constant relating freezing point depression (FPD) to molality (m); K_f = FPD/m. *(13.3)*

Fuel cell—A galvanic cell in which reactants are continuously fed in and products are continuously removed. *(20.5)*

Fuel rods—The rods in a nuclear reactor containing enriched uranium-235 fuel in the form of uranium oxide pellets. *(22.7)*

Fuel value—The heat evolved per gram of fuel consumed. *(6.3)*

Functional group—An atom or group of atoms that provides a molecule with a reactive site. *(23.4)*

Fusion—Synonym for melting; also see *nuclear fusion.* *(12.1)*

Fusion curve—On a phase diagram, the curve showing the temperatures and pressures at which the liquid and solid states are in equilibrium. *(12.2)*

G

Galvanic cell—A device that utilizes an oxidation–reduction reaction to produce current and do electrical work. *(20.17)*

Galvanized iron—See *galvanizing.*

Galvanizing—Coating with zinc; galvanized iron is iron coated with zinc. *(20.6)*

Gamma rays (γ)—High-energy photons emitted by radionuclides. *(22.1)*

Gangue—Rocky material, often silicates, found along with the mineral in an ore. *(S4.1)*

Gas constant (R)—The constant R in the ideal gas law; R = 0.0821 L·atm/mol·K = 8.314 J/mol·K. *(5.4)*

Geiger–Müller counter—One type of *ionization counter.* *(22.1)*

Gel—A semirigid colloidal dispersion of a solid in a liquid. Jello and cell protoplasm are examples. *(13.6)*

General formula—A formula that fits all members of a particular group by expressing the subscript in terms of an integer n. For example, C_nH_{2n+2} is the general formula for straight- or branched-chain *alkanes.* *(23.1)*

Genes—The fundamental units of heredity. Genes are made from DNA. *(23.9)*

Genetic code—The code or set of instructions that directs each cell to synthesize a characteristic set of proteins; see *codon.* *(23.9)*

Geometric isomers—See *cis–trans isomers.*

Gibbs free energy (G)—A thermodynamic property analogous to potential energy; it decreases during the course of a spontaneous physical or chemical change. The free energy is defined so that its change, ΔG, is the negative of the maximum useful work that can be obtained from a reaction at constant temperature and pressure. Free energy is a state function. *(19.1)*

Glass—Synonym for *amorphous solid.* *(12.5)*

Glass electrode—An ion-sensitive electrode based on the half-reaction $AgCl(s) + e^- \rightleftharpoons Ag(s) + Cl^-(1\ M)$. Glass electrodes are used in pH meters. *(20.4)*

Globular protein—A protein in which the polypeptide conformation does not follow a repeating pattern. Such protein molecules often have ball-like shapes. *(23.7)*

Glycosidic linkage—The oxygen bridge formed by the condensation of two cyclic *monosaccharides.* *(23.8)*

Gouy balance—A balance that measures the apparent change in weight of a paramagnetic substance in a magnetic field; see Figure 8.5. *(8.5)*

Graham's law of effusion—Under identical conditions, the rates of effusion of different gases are inversely proportional to the square roots of their densities (or molar masses). *(5.6)*

Gram (g)—One-thousandth of a kilogram. *(1.6)*

Gravimetric analysis—An analytical procedure in which the amount of some substance is determined by weighing a reaction product, usually a precipitate. *(4.2)*

Gravitational acceleration (g)—The acceleration imparted to an object by the force of gravity, approximately equal to 9.81 m/s^2 on the earth's surface. *(1.6)*

Gray (Gy)—The SI unit of radiation dose; 1 Gy is the amount of radiation that results in the absorption of 1 J of energy per kilogram of irradiated material. *(22.4)*

Ground state—The lowest energy state for an atom or a molecule. *(7.3)*

Ground-state configuration—The electron configuration of lowest energy. *(8.3)*

Group—A vertical column of elements in the periodic table; sometimes called a family of elements. *(2.3, 8.4)*

H

Haber process—The industrial process in which nitrogen and hydrogen are catalytically combined under pressure to form ammonia. *(S2.2)*

Half-cell—The oxidation or reduction half of a galvanic cell; the compartment and associated electrode of a galvanic cell in which one of the half-reactions occurs. A standard half-cell is half of a standard galvanic cell. *(20.1)*

Half-cell potential—An oxidation or a reduction potential; the contribution of the half-cell to the total cell voltage. *(20.3)*

Half-life ($t_{1/2}$)—The time required for the concentration (or amount) of a reactant to be reduced by half. *(14.4)*

Half-neutralization point—The point in an acid–base titration at which half of the acid (or base) has been neutralized. *(17.8)*

Half-reaction—The oxidation half or the reduction half of an oxidation–reduction reaction. *(11.1)*

Half-reaction method—A method for balancing oxidation–reduction equations; balanced equations are written for the half-reactions and added in such a way that electron gain and loss are equalized. *(11.3)*

Hall–Héroult process—An industrial process for the preparation of aluminum based on the electrolysis of aluminum oxide dissolved in molten cryolite. *(20.7)*

Halogen—A Group 7A element. *(S2.3)*

Heat—A form of energy that flows spontaneously from one object to another because of a temperature difference between the objects. *(1.7)*

Heat capacity—The amount of heat required to raise the temperature of an object by 1°C. *(6.5)*

Heat of combustion—Synonym for enthalpy of combustion; see *standard enthalpy of combustion.*

Heat of formation—Synonym for enthalpy of formation; see *standard enthalpy of formation.*

Heat of fusion—See *molar heat of fusion.*

Heat of reaction—The heat released or absorbed during a chemical reaction. *(6.3)*

Heat of solution ($\Delta H^\circ_{solution}$)—The enthalpy change that occurs when 1 mol of solute dissolves in a given solvent. Most heats of solution are measured at "infinite dilution," that is, when the relative amount of solvent is very large. *(13.1)*

Heat of vaporization—See *molar heat of vaporization.*

Heating curve—A plot of the temperature of a substance versus heat absorbed by the substance. *(12.1)*

Heavy-water reactor—A nuclear reactor using heavy water (D_2O) as the moderator. *(22.7)*

Heisenberg uncertainty principle—It is impossible to make simultaneous and exact measurements of both the position and momentum of a particle. *(7.5)*

Henderson–Hasselbalch equation—An equation that expresses the pH as the sum of the pK_a and the logarithm of the buffer ratio; $pH = pK_a + \log([base]/[acid])$. *(17.7)*

Henry's law—The solubility of a sparingly soluble gas is directly proportional to the pressure of the gas above the solution. *(13.1)*

Hertz (Hz)—Unit of frequency; $1\ Hz = 1\ s^{-1}$. *(7.1)*

Hess's law—The total enthalpy change for a reaction is the same whether the reaction occurs in one or several steps. *(6.4)*

Heterogeneous catalysis—Catalysis in which the catalyst provides a surface on which the reaction takes place. The catalyst, called a *contact catalyst,* is usually a solid. *(14.1)*

Heterogeneous equilibrium—An equilibrium in which the opposing changes occur on the surface of a solid or liquid. *(15.7)*

Heterogeneous mixture—A mixture in which the components are not uniformly mixed. The presence of individual grains or droplets can be detected visually. *(4.1)*

Heterogeneous reaction—A reaction in which the reactants are in two different states; the reaction occurs at the surface where the two states meet. *(14.1)*

Heteronuclear—Composed of different kinds of atoms. *(10.3)*

Hexagonal close-packing—One of two ways of packing identical spheres into the smallest possible volume; three of its unit cells form a hexagonal prism. *(12.6)*

Hexose—A six-carbon *monosaccharide. (23.8)*

High-spin complex—A metal complex in which electrons occupy high-energy *d* orbitals rather than pairing up in low-energy *d* orbitals. *(21.6)*

Homogeneous catalysis—Catalysis in which the catalyst is in solution with the reactants. *(14.1)*

Homogeneous equilibrium—An equilibrium in which all the reacting substances are in the same solution. *(15.7)*

Homogeneous mixture—Synonym for *solution.*

Homogeneous reaction—A reaction in which all substances are dissolved in the same solution. *(14.1)*

Homologous series—Compounds that have the same general formula and are related by regular increments. *(23.1)*

Homonuclear—Composed of identical atoms. *(10.3)*

Hund's rule—Electrons occupy the orbitals of a subshell singly and with parallel spins until each orbital has one electron in it. *(8.3)*

Hybrid orbital—An atomic orbital that is formed by mixing two or more atomic orbitals on the same atom. *(10.4)*

Hybridization—An imaginary mixing process in which the ground-state orbitals of an atom rearrange themselves to form new atomic orbitals called hybrid orbitals. Hybridization provides the bond angles needed for valence bond theory. *(10.4)*

Hydration—Synonym for *solvation* when water is the solvent. *(13.1)*

Hydration energy—Synonym for *solvation energy* when water is the solvent. *(13.1)*

Hydride—A binary hydrogen compound. Ionic hydrides contain the hydride ion, H^-. *(S1.1)*

Hydride ion—H^-. *(S1.1)*

Hydrocarbon—A compound containing only carbon and hydrogen. Methane (CH_4) is the simplest hydrocarbon. *(10.3, 23.1)*

Hydrogen bond—A strong dipole–dipole interaction in which a hydrogen atom comes between two small highly electronegative atoms such as N, O, and F. Hydrogen bonds are the strongest type of intermolecular attraction. *(12.3)*

Hydrogen ion (H^+)—A proton; a hydrogen atom that has lost its one electron. *(2.2)*

Hydrolysis (of a salt)—The reaction of a salt with water to produce an acidic or basic solution. *(17.5)*

Hydrolysis constant—Synonym for the K_a or K_b of an ion. *(17.5)*

Hydronium ion—H_3O^+; a proton in aqueous solution is a hydrated hydronium ion. *(16.1)*

Hydrophilic colloid—A colloid made up of polar molecules arranged so that the polar groups can form hydrogen bonds with water molecules. Starch dispersed in water is an example. *(13.6)*

Hydrophobic colloid—A colloid consisting of particles with little affinity for water. Sulfur dispersed in water is an example. Hydrophobic colloids are stabilized either by the adsorption of charged ions or by the adsorption of an *emulsifying agent.* *(13.6)*

I

Ideal gas—A gas that obeys the ideal gas law; also, a gas whose molecules have no volume of their own and exert no intermolecular forces. *(5.4)*

Ideal gas law—$PV = nRT$; the equation of state for an ideal gas; obeyed approximately by real gases. *(5.4)*

Ideal solution—An imaginary solution in which the component molecules experience forces identical to those in the pure state. The volume of such a solution is the sum of its component volumes, and there is no heat of solution. The vapor pressure of each component in the solution satisfies *Raoult's law,* and the total pressure above the solution is the sum of the partial pressures. *(13.2)*

Indicator—A dye that changes color at the end of a reaction; used for signaling the endpoint of a titration. *(4.6)*

Induced dipole—The dipole that results from the distortion of the electron cloud of an atom or molecule by a neighboring ion or dipole. *(12.3)*

Inert gas—Synonym for *noble gas.*

Infrared radiation—The spectral region immediately adjacent to the visible region on the longer wavelength side. Infrared radiation is usually detected by its heating effects. *(7.1)*

Inner transition elements—Synonym for the *f* block elements; the *lanthanides* and *actinides.* *(8.4)*

Instability constant—A synonym for the *dissociation constant* of a metal complex.

Instantaneous reaction rate—The reaction rate at a given instant of time. *(14.2)*

Insulator—A nonconductor of electricity. *(21.3)*

Integrated rate law—An equation relating reactant concentration to time; obtained by integrating the differential form of the rate law. *(14.4)*

Interhalogen compound—A compound composed of two different halogens. *(S2.3)*

Intermediate (in a reaction)—A species, often unstable, that is produced in one step of a reaction mechanism and used up in a subsequent step. *(14.5)*

Intermolecular attractive forces—Forces other than chemical bonds that cause otherwise independent atoms and molecules to gather together into clusters. Intermolecular forces include dipole–dipole forces, London dispersion forces, and hydrogen bonds; they are usually weaker than chemical bonding forces. *(12.3)*

Internal energy (*E*)—The total energy of a system. *(6.1)*

Interstitial carbide—A nonionic carbide in which the carbon atoms occupy vacant spaces in a metallic lattice. Interstitial carbides are usually formed from carbon and less active metals such as iron. *(S3.3)*

Inversely proportional—A relation between two quantities such that if one increases by a certain factor, the other will decrease by the same factor. *(5.3)*

Ion—An atom or group of atoms bearing an electrical charge. *(2.2)*

Ion–dipole attraction—The attraction between an ion and a polar molecule. *(13.1)*

Ion product—The reaction quotient Q for the dissolving of an ionic solid. In a saturated solution the ion product equals the solubility product constant. *(18.2)*

Ion-sensitive electrode—An electrode sensitive to the concentration changes of some specific ion. *(20.4)*

Ionic bond—The electrostatic attraction between oppositely charged ions. *(9.1)*

Ionic compound—A compound containing positive and negative ions. *(2.8)*

Ionic lattice—The regular three-dimensional pattern in an ionic solid, in which each ion is surrounded by and attracted to neighbors of opposite charge. *(7.1, 12.6)*

Ionic solid—A crystalline solid composed of positive and negative ions. *(12.6)*

Ionic solution—A solution containing dissolved ions. Ionic solutions conduct an electric current. *(4.1)*

Ionization—The formation of ions from neutral atoms or molecules. *(4.1)*

Ionization counter—An instrument that detects and counts ionizing particles and photons by means of the current they produce in a gas. A Geiger counter is an ionization counter. *(22.1)*

Ionization energy (IE)—The energy required to remove an electron from a gaseous ground-state atom or ion; also called the ionization potential. *(8.5)*

Ionization isomerism—A form of structural isomerism exhibited by ionic coordination compounds in which each isomer has a different ion outside the complex. *(21.5)*

Ionization potential—Synonym for *ionization energy.*

Ionizing radiation—Particles and photons with sufficient energy to ionize and fragment molecules and atoms in their path. *(22.1)*

Iron triad—Iron, cobalt, and nickel. *(S4.5)*

Isoelectronic—Having the same number of electrons. Atoms and ions with the same electron configurations are isoelectronic. *(9.2)*

Isomers—Compounds with the same composition but a different arrangement of atoms. *(10.3, 21.3)*

Isothermal—Synonym for constant temperature.

Isotonic solutions—Solutions having the same osmotic pressure. *(13.3)*

Isotopes—Atoms with the same atomic number but different mass numbers. *(2.3)*

J

Joule (J)—The SI unit of energy; 1 J = 1 N·m. *(1.7)*

K

K_a—See *acid ionization constant.*

K_b—See *base ionization constant.*

K_w—See *water constant.*

Kelvin (K)—The temperature unit on the Kelvin scale; the kelvin is equal in magnitude to a Celsius degree. *(5.3)*

Kelvin temperature scale—A temperature scale defined such that each temperature is 273.15 units higher than the corresponding Celsius temperature; all readings on the Kelvin scale are positive. *(5.3)*

Ketone—A compound containing the *carbonyl group* attached to two carbon atoms. *(23.4)*

Ketose—A *monosaccharide* that contains a ketone group—for example, fructose. *(23.8)*

Kilogram (kg)—The SI unit of mass. *(1.6)*

Kinetic energy—The energy an object possesses by virtue of its motion; equal to $mv^2/2$, where m is the mass of the object and v is its speed. *(1.7)*

Kinetic theory—A theory of matter based on the assumption that the particles making up matter are in continuous, rapid, and chaotic motion. *(5.1)*

Kinetics—The study of motion. See *Chemical Kinetics.*

L

Lanthanide contraction—The gradual but marked decrease in size of the lanthanides with increasing atomic number. *(8.5)*

Lanthanides—The elements 58 through 71 in the periodic table, immediately following lanthanum, in which the 4*f* orbitals are being filled. *(8.4)*

Lattice defect—An imperfection in an otherwise regular crystal lattice. Lattice defects include missing or misplaced atoms, impurities, and fragments of extra layers; they often weaken a crystal. *(12.6)*

Lattice energy—The energy required to break up 1 mol of an ionic crystal into independent gaseous ions. The lattice energy is a measure of the strength of the bond in an ionic crystal. *(9.2)*

Law (scientific)—A generalization that summarizes observed facts. *(1.2)*

Law of chemical equilibrium—The equilibrium constant is independent of the initial concentrations and varies only with temperature. *(15.2)*

Law of conservation of energy—Another name for the *first law of thermodynamics. (6.1)*

Law of conservation of mass—There is no measurable change in total mass during a chemical reaction. *(3.2)*

Law of constant composition—The elemental composition by mass of a given compound is the same for all samples of the compound. *(3.2)*

Law of definite proportions—When elements combine to form a compound, they do so in a definite proportion by mass. *(3.2)*

Law of multiple proportions—In different compounds containing the same elements, the masses of one element combined with a fixed mass of the other element are in the ratio of small whole numbers. *(3.2)*

Le Châtelier's principle—When a system in equilibrium is disturbed, the system adjusts to a new equilibrium in a way that partially counteracts the disturbance. Le Châtelier's principle is used to predict the effects of changing conditions and changing concentrations on an equilibrium system. *(15.8)*

Leveling effect—The ability of a solvent to protonate completely all bases stronger than its own conjugate base and deprotonate all acids stronger than its own conjugate acid. As a consequence, all such acids and bases appear equally strong when dissolved in the solvent. *(16.2)*

Levorotatory—Able to rotate the plane of polarized light to the left; symbolized by (−). *(21.5)*

Lewis acid—A substance that accepts electron pairs; an electrophile. *(16.6)*

Lewis base—A substance that donates electron pairs; a nucleophile. *(16.6)*

Lewis structure—A representation of a molecule that shows all the valence electrons. Shared electron pairs are shown as dashes between the bonded atoms and unshared electrons are shown as dots. *(9.3)*

Lewis symbol—A representation of an atom showing valence electrons as dots around the atomic symbol. *(9.1)*

Ligand—One of the molecules or ions that surround the central metal ion or atom of a complex. *(18.5, 21.4)*

Light-water reactor—A nuclear reactor using ordinary water (H_2O) as the moderator. *(22.7)*

Limiting reactant—The reactant that is used up in a reaction, hence the reactant that determines the quantity of product formed. *(3.8)*

Line spectrum—A discontinuous spectrum consisting of distinct spectral lines. *(7.1)*

Linear accelerator—An instrument that uses an alternating electrical field to accelerate charged particles in a linear path. *(22.3)*

Linkage isomerism—A form of structural isomerism in which the isomers differ in the coordinating atom used by one of the ligands. *(21.5)*

Liter (L)—A volume unit equal to 1000 cm^3; the volume of a cube 10 cm on edge. *(1.3)*

Localized bond—A bond that holds two nuclei together. *(9.4)*

London dispersion forces—The net attractive force that exists in a collection of mutually induced dipoles. London forces exist between all atoms and molecules, including those with no permanent polarity. *(12.3)*

Lone pair—In a Lewis structure, an unshared pair of valence electrons; also called a nonbonding pair. *(9.3)*

Lone-pair repulsion—The repulsion exerted by a lone pair of electrons on other pairs of electrons around the atom. The repulsion exerted by a lone pair on the vertex atom of an angle acts to compress the angle. *(10.2)*

Low-spin complex—A metal complex in which electrons pair up in the lower energy *d* orbitals rather than occupying the higher energy *d* orbitals. *(21.6)*

Lyman series—The ultraviolet series in the emission spectrum of hydrogen, produced by electron transitions from higher levels down to $n = 1$. *(7.3)*

M

Magic numbers—Certain numbers of either protons or neutrons that are characteristic of very stable nuclei; the magic numbers are 2, 8, 20, 28, 50, 82, and 126. *(22.5)*

Magnetic quantum number (m_l)—A quantum number determining the orientation of a hydrogenlike orbital in space. For a given value of l, m_l may take on integral values ranging from $-l$ to $+l$, including 0. *(7.6)*

Main-group elements—Synonym for *representative elements.*

Malleability—The ability to be shaped by hammering; a property generally associated with metals. *(12.5)*

Manometer—An instrument consisting of a U-tube partly filled with mercury and used for measuring the pressure inside a closed vessel. *(5.2)*

Many-electron atom—An atom with two or more electrons. *(8.1)*

Markovnikov's rule—When an unsymmetrical reactant adds to a double bond, the more positive part will add to the carbon with more hydrogen atoms. *(23.2)*

Mass—A measure of the quantity of matter in an object; the SI unit of mass is the kilogram. *(1.6)*

Mass defect—The difference between the mass of a nucleus and the sum of the masses of its independent protons and neutrons. *(22.6)*

Mass-energy equation—An equation relating the mass of an object to its energy; $E = mc^2$. *(22.6)*

Mass number (*A*)—The total number of protons and neutrons in the nucleus of an atom. *(2.3)*

Mass spectrometer—An instrument used for measuring precise masses and abundances of atomic and molecular ions. The operating principles of the direction focusing mass spectrometer are described in *Digging Deeper: The Mass Spectrometer. (2.3)*

Mass spectrum—A plot of the intensity of a mass spectrometer signal versus the mass number of the ion giving the signal. *(2.3)*

Matter—Anything that has mass and occupies space. *(1.1)*

Matter wave—A wave associated with a moving particle, see *De Broglie wavelength. (7.4)*

Maxwell–Boltzmann distribution curve—A plot of molecular speed or energy versus the fraction of molecules having that speed or energy. *(5.7)*

Mean free path—The average distance traveled between intermolecular collisions. *(5.8)*

Mean square speed—The average of the squares of the molecular speeds; the sum of the squared speeds divided by the total number of molecules. *(5.7)*

Measurement—A process by which the magnitude of some quantity is found in terms of a previously defined unit that serves as a standard. *(1.2)*

Mechanism (of a reaction)—The series of consecutive steps by which reactants are transformed into products. *(14.5)*

Megaelectron volt (MeV)—An energy unit; 1 MeV $= 1.602 \times 10^{-13}$ J. *(22.1)*

Megaton—A unit of measurement for explosives, the equivalent of 1 million tons of TNT. *(22.7)*

Melting—A phase change from solid to liquid. *(12.1)*

Melting point—The temperature at which the solid and liquid states of a substance are in equilibrium. *(12.1)*

Membrane cell—An industrial cell used for the electrolysis of brine. Aqueous NaOH produced in this cell is less contaminated with NaCl than aqueous NaOH produced in the *diaphragm cell. (20.7)*

Meniscus—The curved boundary between a liquid and a gas. *(12.2)*

Messenger RNA—See RNA.

Meta- (*m*-) substituents—In a benzene ring, substituents that are on carbon atoms separated by one carbon atom. *(23.3)*

Metal—A substance that is usually characterized by a distinctive luster, relatively high electrical and thermal conductivity, malleability, and ductility. Metallic elements are to the left of the stepwise line in the periodic table. *(8.6)*

Metal complex—A molecule or ion consisting of a central metal ion or atom bonded to one or more groups called *ligands;* a charged complex is called a complex ion. *(21.4)*

Metallic bond—A type of chemical bond in which the valence electrons are not confined to individual atoms or bonds, but flow freely throughout the crystal; characteristic of metals. *(9.1, 21.3)*

Metallic conduction—The transport of electric charge by electrons in the conduction band of a metal. Metallic conduction decreases with increasing temperature. *(21.3)*

Metallic luster—The characteristic sheen of a metal surface caused by the absorption and re-emission of visible light by moving electrons. *(21.3)*

Metallic solid—A crystalline solid in which atoms are held together by a metallic bond. *(12.5)*

Metalloid—Synonym for *semimetal.*

Metallurgy—The science of extracting metals from their ores, refining them, and making alloys. *(S4.1)*

Metastable state—An unstable state that is able to persist for an appreciable length of time. In radiochemistry, an excited nucleus that does not immediately decay. Technetium-99m is a metastable form of technetium-99. *(4.1, 22.5)*

Metathesis—A reaction in which atoms or ions exchange partners. *(4.2)*

Meter (m)—The SI unit of length; 1 m is the distance light travels in a vacuum during 1/299,792,458 of a second. *(1.3)*

Method of successive approximations—a method for finding the roots of algebraic equations of order higher than 1. *(15.3)*

Metric system—A decimal system of weights and measures originally based on the meter as the unit of length and the kilogram as the unit of mass. *(1.2)*

Millikan oil drop experiment—An experiment for determining the charge on an electron; see *Digging Deeper: How the Electron Charge and Mass Were Determined. (2.1)*

Mineral—An inorganic substance with a definite crystal structure and a definite range of composition. *(S4.1)*

Miscible—A term used for liquids that mix with each other. *(4.1)*

Mixture—A form of matter containing two or more pure substances (elements or compounds) in which the components retain their identity. Unlike compounds, mixtures may vary in composition, and the components can be separated from each other by physical means. *(2.5)*

Moderator—A substance that controls the neutron speed in a reactor core. Graphite, H_2O, and D_2O are used as moderators. *(22.7)*

Molality (*m*)—A concentration measure equal to the number of moles of solute divided by the number of kilograms of solvent. *(13.3)*

Molar entropy—See *standard molar entropy.*

Molar heat capacity—The heat capacity per mole of substance. *(6.5)*

Molar heat of fusion (ΔH°_{fus})—The enthalpy change associated with melting 1 mol of solid. *(12.1)*

Molar heat of vaporization (ΔH°_{vap})—The enthalpy change associated with vaporizing 1 mol of liquid. *(12.1)*

Molarity (*M*)—The number of moles of solute per liter of solution. *(4.4)*

Molar mass ($\mathcal{M}$)—The mass of 1 mol of any given species. *(2.4)*

Molar volume—The volume of 1 mol of a substance. *(5.3)*

Mole—A counting unit containing the same number of items as there are atoms in exactly 12 g of carbon-12. One mole of any substance contains Avogadro's number of particles. The number of moles is symbolized by *n*. *(2.4)*

Mole of reaction—An Avogadro number of reactions as represented by the chemical equation. *(19.2)*

Mole fraction (*X*)—The number of moles of a component in a mixture divided by the total number of moles; the mole fraction is equal to the molecule fraction. *(5.6)*

Mole percent—100% × mole fraction. *(5.6)*

Molecular equation—An equation that uses complete neutral formulas. The substances involved may or may not be molecular. *(4.2)*

Molecular formula—A formula showing the number of each kind of atom in a molecule. *(2.6)*

Molecular orbital—One of the allowed states for an electron moving in the field of two or more nuclei. *(10.5)*

Molecular orbital theory—A theory of chemical bonding that assumes that atomic orbitals overlap and form new orbitals, called molecular orbitals, whose shapes, energies, and electron density distributions are different from those of the parent orbitals. *(10.5)*

Molecular solid—A crystalline solid consisting of molecules held in place by intermolecular forces. *(12.5)*

Molecular weight—The sum of the atomic weights of all the atoms in a molecule. The terms molecular weight and *molar mass* are often used interchangeably even though they refer to different quantities. *(2.7)*

Molecularity—The number of particles participating in an elementary step. *(14.5)*

Molecule—A distinct, electrically neutral group consisting of a well-defined number of atoms bonded together. *(2.6)*

Momentum—The product of a particle's mass and speed; momentum $= m \times v$. *(7.5)*

Monatomic ion—An atom that bears a charge, formed when an atom loses or gains electrons. *(2.8)*

Monodentate ligand—A ligand with only one coordinating atom. *(21.4)*

Monomer—One of the smaller molecules that combine to form a *polymer. (23.2)*

Monosaccharide—A polyhydroxy aldehyde or ketone with the general formula $(CH_2O)_n$ where n ranges from 3 to 7—for example, glucose or fructose. Monosaccharides are also called simple sugars. *(23.8)*

Most probable distance—The distance from the nucleus corresponding to the maximum radial electron density; the distance at which the electron spends the greatest fraction of its time. *(7.7)*

Most probable speed—The speed possessed by the greatest number of molecules in a collection. *(5.7)*

mRNA—See *RNA.*

N

Native ore—An ore in which an element, usually a metal, is not chemically combined. *(S4.1)*

Negative deviation (from Raoult's law)—Real solutions exhibit negative deviations when their actual vapor pressures are less than those calculated from Raoult's law. Negative deviations occur when the attractive forces in the solution are greater than in the pure components. *(13.2)*

Nernst equation—An equation for determining the cell voltage at concentrations other than standard state concentrations; $E = E° - (RT/nF) \ln Q$. *(20.4)*

Net ionic equation—An equation that does not include spectator ions. *(4.2)*

Network covalent solid—A crystalline solid in which each atom is covalently bonded to its neighbors and in which there are no separate molecules. *(12.5)*

Neutral solution—An aqueous solution in which the hydrogen ion and hydroxide ion concentrations are equal; a solution whose pH is 7.0 at 25°C. *(16.4)*

Neutralization—The reaction between an acid and a base in which their acidic and basic properties are lost, or neutralized. In aqueous solution, a reaction in which an acid and a base are converted to water and a salt. The Arrhenius theory interprets neutralization as the combination of hydrogen and hydroxide ions to form water, the Brønsted–Lowry theory interprets it as proton transfer, and the Lewis theory interprets it as the sharing of an electron pair donated by the base. *(4.2, 16.1, 16.6)*

Neutron—A neutral fundamental particle found inside the atomic nucleus. Its mass (see Table 2.1) is slightly greater than that of the proton. *(2.2)*

Newton (N)—The SI unit of force; $1\ N = 1\ kg{\cdot}m/s^2$. *(1.6)*

Nitride—A compound containing the nitride ion, N^{3-}. *(S2.2)*

Nitrogen fixation—The formation of nitrogen compounds from atmospheric nitrogen. *(S2.2)*

Nitrogenous base—A molecule with a ring of nitrogen and carbon atoms. Adenine, thymine, cytosine, guanine, and uracil are nitrogenous bases. *(23.9)*

***n* + *l* Rule**—Orbitals fill in order of increasing $(n + l)$ value. When two orbitals have the same $(n + l)$ value, the one with the lower n value fills first. *(8.3)*

Noble gas—A Group 8A element; sometimes called a rare gas or inert gas. *(S1.4)*

Noble gas core—The electron configuration of the noble gas preceding an atom in the periodic table. *(8.3)*

Nodal plane—See *node. (7.7)*

Node—A point or surface at which the electron density is zero. For example, the 2*s* and 3*s* orbitals have nodal spherical surfaces (see Figure 7.30) and the *p* and *d* orbitals have nodal planes (see Figures 7.31 and 7.32). *(7.7)*

Nonbonding pair—See *lone pair.*

Nonelectrolyte—A solute that does not form ions in aqueous solution. *(4.1)*

Nonferrous alloy—An alloy that does not contain iron. *(S4.1)*

Nonmetal—A substance that is not lustrous, is a poor conductor of electricity and heat, and whose solid form is brittle; nonmetallic elements are to the right of the stepwise line in the periodic table. *(8.6)*

Nonoxidizing acid—An acid such as HCl in which H^+ is the only oxidizing agent. *(S4.3)*

Nonpolar bond—A covalent bond in which the electron density is equally shared by the bonded atoms. *(9.7)*

Normal boiling point—The boiling point of a liquid under 1 atm pressure. *(12.1)*

Normal melting point—The melting point of a solid under 1 atm pressure. *(12.1)*

***n*-Type semiconductor**—A semiconductor doped with a donor impurity. *(21.3)*

Nuclear atom—A model of the atom in which negative electrons move around a very small, but massive, positive nucleus containing protons and neutrons. *(2.2)*

Nuclear bombardment reaction—A reaction initiated by the collision of a moving particle with a nucleus. *(22.3)*

Nuclear chain reaction—A self-sustaining sequence of fission reactions. *(22.7)*

Nuclear equation—An equation representing a nuclear reaction. The sum of the mass numbers is the same on each side of the equation, and so is the sum of the atomic numbers. *(22.2)*

Nuclear fission—The splitting of a heavy nucleus into two or more lighter nuclei. Energy is released during fission reactions. *(22.6)*

Nuclear force—A very strong attractive force that operates only over very short distances comparable to the dimensions of the atomic nucleus. *(22.5)*

Nuclear fusion—The combination of lighter nuclei to make a heavier nucleus. Energy is released during fusion reactions. *(22.6)*

Nuclear reaction—Any reaction in which a nucleus changes. *(22.2)*

Nuclear reactor—A reactor in which controlled nuclear fission is used to generate power. *(22.7)*

Nucleic acid—A nucleotide polymer. DNA and RNA are nucleic acids. *(23.9)*

Nucleophile—A "nucleus-seeking" species; a *Lewis base. (16.6)*

Nucleotide—A molecule that is a condensation product of phosphoric acid, a five-carbon sugar such as ribose or deoxyribose, and a *nitrogenous base.* ATP and ADP are nucleotides. *(23.9)*

Nucleus—The positively charged core of an atom. The nucleus contains protons and neutrons and accounts for most of the mass of the atom. *(2.2)*

O

Octahedron—A solid geometrical figure with eight triangular faces and six corners. Six VSEPR pairs on an atom will point toward the corners of an octahedron. *(10.2)*

Octet rule—Bonded atoms have noble gas configurations. The octet rule is frequently but not invariably followed in ionic and covalent bonding. *(9.2)*

Oil—A liquid *fat. (23.5)*

Optical activity—The ability of a substance to rotate the plane of polarized light; *enantiomers* are optically active. *(21.5)*

Optical isomers—Synonym for *enantiomers.*

Orbital—A mathematical function, also called a *wave function,* that is unique for each allowable state of an atom or molecule and that can be used to calculate the energy and various properties of an electron in the state; symbolized by ψ. *(7.6)*

Orbital diagram—A version of the electron configuration showing the occupancy of individual orbitals and paired and unpaired electrons. *(8.3)*

Orbital filling order—An experimentally established order in which atomic orbitals fill from one atom to the next in the periodic table; used for writing ground-state electron configurations by the *aufbau procedure. (8.3)*

Orbital splitting energy (Δ)—The energy difference between the high- and low-energy sets of *d* orbitals occupied by the metal *d* electrons in a metal complex. *(21.6)*

Order (of a reaction)—The sum of the concentration exponents in the rate law. *(14.3)*

Ore—A natural mineral deposit from which an element can be profitably extracted. *(3.6, S4.1)*

Organic chemistry—The chemistry of the compounds of carbon. *(23.1)*

Organic compound—A compound that contains carbon and usually hydrogen. *(23.1)*

Orientation factor (*p*)—The fraction of collisions that have the geometry (orientation) necessary for a reaction to take place. *(14.7)*

Ortho- (*o*-) substituents—In a benzene ring, substituents that are on adjacent carbon atoms. *(23.3)*

Osmosis—The net movement of solvent molecules through a semipermeable membrane from a dilute solution to a concentrated solution. *(13.3)*

Osmotic pressure (π)—The pressure required to stop the osmotic flow of solvent into a solution; a colligative effect proportional to the molarity of the solute. *((13.3)*

Ostwald process—The industrial process in which ammonia is converted to nitric acid. *(S2.2)*

Overlap—To share a common region of space. *(10.4)*

Overvoltage—The difference between the actual voltage and the theoretical minimum voltage required to recharge a battery or reverse a spontaneous redox reaction. *(20.5)*

Oxidation—The loss of electrons by an element or compound. For a reaction involving oxygen, the combination of a substance with oxygen. *(3.6, 11.1)*

Oxidation number—The charge calculated for a bonded atom when the electrons in each of its bonds are assigned to the more electronegative atom. *(11.2)*

Oxidation number method—A method for balancing oxidation–reduction equations that is based on identifying and equalizing oxidation number changes. *(11.3)*

Oxidation potential—The contribution of the oxidation half-reaction to the total cell voltage; a high oxidation potential implies a strong tendency to lose electrons. The standard oxidation potential (E°_{oxid}) is the oxidation potential of a standard half-cell. *(20.3)*

Oxidation–reduction reaction—A reaction in which electrons are transferred from one substance to another. *(11.1)*

Oxidation state—Synonym for *oxidation number.*

Oxide—A binary oxygen compound. Ionic oxides contain the oxide ion O^{2-}, the peroxide ion O_2^{2-}, or the superoxide ion O_2^-. *(S2.1)*

Oxidizing acid—An acid such as HNO_3 or H_2SO_4 that has a reducible anion. *(S4.3)*

Oxidizing agent—A substance that accepts electrons in an oxidation–reduction reaction. The oxidizing agent is reduced. *(11.1)*

Oxo acid—An acid containing polar O—H bonds attached to some central atom—for example, H_2SO_4 and HNO_3. *(16.5)*

Oxo anion—The anion formed when an oxo acid loses one or more protons—for example, SO_4^{2-} and NO_3^-. *(16.5)*

P

p Block elements—Elements in which *p* orbitals are being filled; Groups 3A through 8A (excluding helium). *(8.4)*

p Orbital—An atomic orbital with azimuthal quantum number l = 1. *(7.6)*

Paired electrons—Two electrons in the same orbital; paired electrons have opposite (antiparallel) spin. *(8.2)*

Para- (p-) substituents—In a benzene ring, substituents that are on opposite carbon atoms. *(23.3)*

Paramagnetism—The magnetism exhibited by individual atoms, ions, or molecules with unpaired electrons. *(8.5)*

Partial pressure—The pressure that one component of a gas mixture would exert if it were alone in the container and at the same temperature. *(5.6)*

Parts per billion (ppb)—Parts per billion (10^9) parts. *(4.4)*

Parts per million (ppm)—Parts per million (10^6) parts. *(4.4)*

Parts per thousand—Parts per thousand parts. *(4.4)*

Parts per trillion (ppt)—Parts per trillion (10^{12}) parts. *(4.4)*

Pascal (Pa)—The SI unit of pressure; 1 N/m^2. *(5.2)*

Patina—The greenish coating of hydroxide, carbonate, and sulfate that develops on an exposed copper surface. *(S4.4)*

Pauli exclusion principle—Only two electrons can be in one orbital, and these electrons must have opposite spins. An alternative statement is: No two electrons in an atom can have the same four quantum numbers. *(8.2)*

Pauling electronegativity scale—A listing of relative electronegativities for the elements. Pauling electronegativities range from 0 to 4. *(9.7)*

Penetrating ability—The average distance that radiation will travel in a given medium. *(22.1)*

Pentose—A five-carbon *monosaccharide*. *(23.8)*

Peptide linkage—The *amide linkage* when it joins two amino acids. *(23.7)*

Percent—Parts per hundred; 100% × the part divided by the whole.

Percent by mass—100% × (mass of component/total mass of mixture). *(4.4)*

Percent ionization—100% times the fraction of molecules ionized. *(17.3)*

Percent yield—100% × (actual yield/theoretical yield). *(3.9)*

Percentage composition—The elemental composition of a compound expressed as percent of each element by mass. *(3.2)*

Period—A horizontal row of elements in the periodic table. *(2.6)*

Periodic table—A table of the elements arranged in order of increasing atomic number. Each horizontal row or *period* begins with the addition of an *s* electron to an unoccupied shell and ends with the formation of a noble gas atom. Each vertical column or *group* consists of elements with the same outer configuration and similar chemical properties. *(8.4)*

Peroxide ion—O_2^{2-}. *(S1.2)*

pH—A measure of acidity; pH = $-\log [H^+]$. The lower the pH, the more acidic the solution. *(16.4)*

pH meter—An electrochemical cell used for measuring the pH of a solution. It contains an *ion sensitive electrode* whose voltage varies with the hydrogen ion concentration. *(20.4)*

Phase change—A change from one state to another; phase changes include melting, freezing, vaporization, sublimation, condensation, and deposition. *(12.1)*

Phase diagram—A graph that shows the most stable state of a substance at any given temperature and pressure. *(12.2)*

Phenol—A compound containing an OH group attached to an aromatic ring. An aromatic alcohol. *(23.4)*

Phenyl group—The group formed by removing one hydrogen atom from benzene. *(23.2)*

Phosphide—A binary compound of phosphorus and a less electronegative element; ionic phosphides contain the phosphide ion, P^{3-}. *(S3.2)*

Phosphor—A substance that emits a flash of light for each particle or photon of ionizing radiation that strikes it. *(22.1)*

Photoelectric effect—The light-induced emission of electrons from a metal surface. *(7.2)*

Photon—A quantum of electromagnetic radiation; see *Planck's law*. *(7.2)*

Physical change—A change in which each substance retains its identity and no new elements or compounds are formed. *(3.1)*

Physical property—Any property of a substance other than a chemical property. Physical properties include color, odor, density, and melting point. *(3.1)*

Pi bond—A shared pair of electrons whose principal density lies above and below the plane or axis of the molecule. There is no electron density along the line of atomic centers. Pi bonds are formed by the overlap of parallel *p* orbitals. In valence bond theory, a multiple bond consists of a sigma bond and one or two pi bonds. *(10.4)*

Pi orbital—A molecular orbital with regions of electron density above and below the plane or axis of the molecule, and no electron density along the line of atomic centers. Pi orbitals can be bonding or antibonding. *(10.5)*

Pig iron—Iron as it comes from the blast furnace, with about 4% carbon. *(S4.5)*

pK_a—$-\log K_a$ *(17.1)*

pK_b—$-\log K_b$ *(17.2)*

Planck's constant (*h*)—The proportionality constant in Planck's law; $h = 6.6262 \times 10^{-34}$ J·s. *(7.2)*

Planck's law—The energy of a photon is proportional to the radiation frequency; $E = h\nu$. *(7.2)*

Plane of symmetry—An imaginary surface that divides a molecule into two halves, one of which is the mirror image of the other; sometimes called a mirror plane. *(21.5)*

Plane-polarized light—Light whose electric field oscillations are confined to a single plane. The magnetic field oscillations are confined to a plane perpendicular to the plane of the electric field oscillations. *(21.5)*

Plasma—An extremely hot gas consisting of electrons and positive nuclei. *(22.8)*

pOH—A measure of the OH^- concentration; $pOH = -\log [OH^-]$. *(16.4)*

Poisoning (of a catalyst)—The inactivation of a contact catalyst because of accumulated impurities adsorbed on the surface of the catalyst. *(14.8)*

Polar covalent bond—A bond in which the electron density is not equally shared. The bonding electrons spend more time on one of the bonded atoms than on the other. *(9.7)*

Polarimeter—A device that measures *optical activity. (21.5)*

Polyamide—A polymer formed when a *carboxylic acid* and an amine undergo a *condensation reaction.* The units in a polyamide are joined by *amide linkages. (23.7)*

Polyatomic ion—A group of atoms that bears a charge. *(2.8)*

Polydentate ligand—A ligand with two or more coordinating atoms. *(21.4)*

Polyester—A polymer in which the units are joined by *ester linkages. (23.5)*

Polyhydroxy alcohol—A compound whose molecules contain two or more OH groups. *(23.4)*

Polymer—A large molecule formed by combining many similar small molecules; see *monomer. (23.2)*

Polypeptide—A polymer formed when amino acids combine with each other through their amino and carboxyl groups. The units in a polypeptide are joined by *amide linkages. (23.7)*

Polyprotic acid—An acid with two or more acidic hydrogen atoms per molecule; thus, an acid that can donate two or more protons. The ionization of a polyprotic acid occurs in distinct steps. *(16.3)*

Polysaccharide—The condensation product of two or more *monosaccharides. (23.8)*

Positive deviation (from Raoult's law)—Real solutions exhibit positive deviations when their actual vapor pressures are greater than those calculated from Raoult's law. Positive deviations occur when the attractive forces in the solution are less than in the pure components. *(13.2)*

Positive ray—A beam of positive ions that moves toward the cathode in a discharge tube. *(2.2)*

Positron ($^{0}_{+1}e$)—A particle identical to an electron but with a positive charge. *(22.2)*

Potential energy—Stored energy possessed by an object either because of its position or the positions of its parts relative to each other. *(1.7)*

Potential energy profile—A graph of the potential energies of successive configurations formed by reactants in the process of becoming products. *(14.7)*

Potentiometer—A device for obtaining the maximum cell voltage. *(20.2)*

Precipitate—A solid that settles out of solution. *(4.2)*

Precision—The reproducibility of a measurement; also its "fineness" in terms of significant figures. *(1.4)*

Pressure—Force per unit area. *(5.2)*

Pressure–volume work—The work associated with the volume change of a system; the pressure–volume work done on a system during a constant pressure reaction is equal to $P\Delta V$, where P is the pressure and ΔV is the change in volume of the system. *(6.2)*

Primary alcohol—An alcohol in which the OH group is attached to an end carbon atom. *(23.4)*

Primary amine—A derivative of ammonia in which one hydrogen atom has been replaced by a hydrocarbon group. *(23.6)*

Primary protein structure—The sequence of amino acid residues in a protein. *(23.7)*

Principal quantum number (*n*)—The quantum number that determines the energy and the average distance from the nucleus of an electron in a hydrogenlike orbital. *(7.6)*

Probe—A compact galvanic sensor containing a concentration-sensitive electrode and a reference electrode. *(20.4)*

Product (in a reaction)—A substance that forms as a result of a chemical reaction. *(3.1)*

Promotion (of electrons)—The shift of an electron from its original orbital to an orbital that is slightly higher in energy. Electron promotion and rearrangement provide the half-filled orbitals needed for valence bond theory. *(10.4)*

Property—Any characteristic trait possessed by a substance; color, melting point, and density are examples. *(2.6)*

Protein—A polypeptide chain with at least 50 amino acid residues in a definite sequence and with a definite conformation. See *primary, secondary, tertiary,* and *quaternary protein structure. (23.7)*

Proton—A positively charged fundamental particle found inside the atomic nucleus; the nucleus of a hydrogen atom. See Table 2.1 for its charge and mass. *(2.2)*

Proton–proton chain—One of the mechanisms by which helium is produced in stars. *(22.8)*

Proton transfer—Exchange of a proton between two molecules; a Brønsted–Lowry neutralization. *(16.1)*

***p*-Type semiconductor**—A semiconductor doped with an acceptor impurity. *(21.3)*

Pure substance—An element or compound. *(2.5)*

PV work—Abbreviation for *pressure–volume work.*

Q

Quantization—The restriction of an observable quantity such as energy to a discrete set of values. This restriction has led to the belief that energy is composed of particle-like units called *quanta.* *(7.2)*

Quantum—A small particle-like bundle of energy. The quantum of electromagnetic radiation is called a *photon.* *(7.2)*

Quantum mechanics—The science that applies the concept of quantization to atoms and molecules. *(7.6)*

Quantum number—A number, usually an integer, used to describe the allowable states of an atom or a molecule. *(7.3)*

Quaternary protein structure—The spatial relationship between two or more polypeptide chains in a protein. *(23.7)*

Quenching—In the treatment of steel, sudden cooling from a high temperature. *(S4.5)*

R

Racemic mixture—A mixture containing equal amounts of both enantiomers of a chiral substance; a racemic mixture is optically inactive. *(21.5)*

RAD—Abbreviation for "radiation absorbed dose"; one rad is the amount of radiation that results in the absorption of 10^{-2} J of energy per kilogram of irradiated material. *(22.4)*

Radial electron density—The probability of finding an electron at a given distance from the nucleus regardless of direction. *(7.7)*

Radiant energy—Synonym for *electromagnetic radiation.* *(7.1)*

Radioactive decay—The emission of energetic particles and photons from unstable atomic nuclei. Radioactivity can be from natural sources or from artificial radionuclides. *(22.1)*

Radioactivity—Synonym for *radioactive decay.*

Radiocarbon dating—A method that uses the concentration and decay rate of carbon-14 to establish the age of an object. *(22.4)*

Radioisotope—Synonym for *radionuclide.*

Radioisotope dating—Any method that uses the concentration and decay rate of a radionuclide to establish the age of an object. *(22.4)*

Radionuclide—A radioactive nucleus. *(22.1)*

Radiopharmaceutical—A radioactive preparation that can be ingested or injected, usually for therapy or diagnosis. *(22.4)*

Radiotracer—A reagent containing radioactive atoms that can be used to follow the course of a reaction. *(22.4)*

Random error—A small variation in measurement that appears to be caused by chance alone. *(1.4)*

Raoult's law—The vapor pressure P of a component above an ideal solution is proportional to the mole fraction X of the component in the solution; that is, $P = XP°$, where $P°$ is the vapor pressure of the pure substance. *(13.2)*

Rare earth elements—Synonym for the *lanthanides.*

Rare gas—Synonym for *noble gas.*

Rate—Change per unit time. *(14.1)*

Rate constant—The proportionality constant k in the rate law. *(14.3)*

Rate-determining step—The slowest step in a mechanism, often called the bottleneck step. A reaction can go no faster than its slowest step. *(14.5)*

Rate equation—Synonym for *rate law.* *(14.3)*

Rate law—An equation that describes the dependence of rate on concentration for a given reaction. *(14.3)*

Rate of effusion—The amount of gas effusing per second. *(5.6)*

RBE—Abbreviation for "relative biological effectiveness." A factor assigned to each type of radiation to indicate its biological effectiveness. *(22.4)*

Reactant—A substance that undergoes chemical change; one of the original substances in a chemical reaction. *(3.1)*

Reaction mechanism—See *mechanism.*

Reaction quotient (*Q*)—A function of the form product concentrations divided by reactant concentrations. The product concentrations in the numerator are multiplied, the reactant concentrations in the denominator are multiplied, and each concentration is raised to a power equal to its coefficient in the balanced equation. Reaction quotients expressed in terms of molar concentrations are symbolized by Q_c. Quotients expressed in terms of partial pressures are symbolized by Q_p. *(15.2)*

Reaction rate—The time rate of change in the concentration or amount of a reactant or product. *(14.1)*

Reactive collision—A collision that results in product formation, as opposed to a collision that leaves reactant molecules unchanged. *(14.7)*

Reactor core—The part of a nuclear reactor containing the fuel rods, the control rods, and the moderator. *(22.7)*

Reagent—Synonym for *reactant.* Solutions of reactants are often referred to as reagents. *(3.1)*

Real solution—Any actual solution. Real solutions, while never ideal, may approximate ideal behavior. *(13.2)*

Redox reaction—Synonym for *oxidation–reduction reaction.*

Redox titration—A titration involving an *oxidation–reduction reaction.*

Reducing agent—A substance that gives up electrons in an *oxidation–reduction reaction.* The reducing agent is oxidized. *(11.1)*

Reduction—The gain of electrons by an element or compound. For a reaction involving oxygen, the loss of oxygen by a compound. *(3.6, 11.1)*

Reduction potential—The contribution of the reduction half-reaction to the total cell voltage; a high reduction potential implies a strong tendency to gain electrons. The standard reduction potential ($E°_{red}$) is the reduction potential of a standard half-cell. *(20.3)*

Reference electrode—A half-cell, such as the standard hydrogen electrode or the calomel electrode, used as a standard for comparing other half-cells. *(20.4)*

REM—Abbreviation for "radiation equivalent for man." A measure of radiation dosage in terms of biological effectiveness; dose (in rems) = dose (in rads) × RBE. *(22.4)*

Representative elements—Elements in which *s* or *p* orbitals are being filled; elements from Groups 1A through 8A. *(8.4)*

Resonance hybrid—The actual molecule represented by a set of *resonance structures. (9.5)*

Resonance structures—Two or more Lewis structures with the same arrangement of atoms and the same number of electron pairs, but different positions for the electrons. The actual structure of the molecule is intermediate between those depicted by the resonance structures. *(9.5)*

Reverse osmosis—The forced movement of solvent molecules across a semipermeable membrane in a direction opposite to that of osmotic flow. Reverse osmosis, which occurs when the concentrated solution side of a semipermeable membrane is subjected to a pressure greater than its osmotic pressure, has been used to prepare desalinated water from seawater. *(13.4)*

Reversible conditions—Conditions such that the driving force of the forward reaction differs only infinitesimally from the driving force of the reverse reaction. *(19.4)*

Reversible reaction—A reaction that can proceed in the forward and backward directions. All chemical reactions are believed to be reversible as long as the reactants and products remain in contact. *(15.1)*

Ribosome—In a living cell, a small particle equipped with the enzymes needed for protein synthesis. *(23.9)*

RNA—Ribonucleic acid. Messenger RNA (mRNA) molecules bring the directions for protein synthesis to the cell *ribosomes.* Transfer RNA (tRNA) molecules bring the individual amino acids and their anticodons. *(23.9)*

Roasting—Heating strongly in air, often done to convert sulfide and carbonate minerals to oxides. *(S4.1)*

Rock salt structure—An ionic lattice similar to the lattice of sodium chloride. *(12.6)*

Root mean square speed—The square root of the mean square speed; the speed of a molecule possessing the average kinetic energy. *(5.7)*

Rounding off—Finding the number that is closest to a given number but with fewer digits. *(1.4)*

Rydberg constant (R_H)—The constant in the Bohr energy formula for the hydrogen atom; $R_H = 2.179 \times 10^{-18}$ J. *(7.3)*

S

***s* Block elements**—Elements in which *s* orbitals are being filled; Groups 1A and 2A and helium. *(8.4)*

***s* Orbital**—An atomic orbital with azimuthal quantum number $l = 0$. *(7.6)*

Salt—An ionic compound whose ions are neither H^+ nor OH^-. *(4.2)*

Salt bridge—An inverted U-tube filled with an electrolyte solution. Salt bridges are often used to connect the half-cell electrolytes in galvanic cells. *(20.1)*

Salting-in effect—The increase in solubility of a sparingly soluble ionic compound caused by the presence of salts not having an ion in common with the compound. *(13.1)*

Salting-out effect—The decrease in solubility of a nonpolar or slightly polar solute caused by the presence of dissolved salts. *(13.1)*

Saponification—The hydrolysis of an ester in basic solution to produce an alcohol and a salt of the acid. Saponification is used to make soap from fats. *(23.5)*

Saturated fatty acid—See *fatty acid.*

Saturated hydrocarbon—Synonym for *alkane.*

Saturated solution—A solution in which the dissolved solute is in dynamic equilibrium with undissolved solute, or would be in equilibrium if undissolved solute were present. *(4.1)*

Schrödinger wave equation—An equation developed by Erwin Schrödinger for obtaining atomic and molecular orbitals. *(7.1)*

Scientific method—The process of collecting data, formulating scientific laws from the data, and constructing theories to explain the laws. *(1.2)*

Scintillation counter—A device that counts the flashes of light produced by the impact of ionizing radiation on a phosphor such as sodium iodide. Most scintillation counters also measure the intensity of each flash. *(22.1)*

Second law of thermodynamics—All spontaneous changes are accompanied by an increase in the entropy of the universe. *(19.3)*

Secondary alcohol—An alcohol in which the OH group is attached to a carbon atom that is bonded to two hydrocarbon groups. *(23.4)*

Secondary amine—A derivative of ammonia in which two hydrogen atoms have been replaced by hydrocarbon groups. *(23.6)*

Secondary protein structure—A regular repeating protein structure such as an *α-helix* or *beta pleated sheet. (23.7)*

Self-ionization—Synonym for *autoionization.*

Semiconductor—A substance characterized by a relatively low electrical conductivity that increases with increasing temperature. Unlike metals, semiconductors have very few electrons in the conduction band. Unlike insulators, their electrons can be excited into the conduction band. *(21.1)*

Semimetal—An element whose properties are intermediate to those of metals and nonmetals. The semimetals include boron, silicon, germanium, arsenic, antimony, and tellurium, and lie along the stepwise line separating metals from nonmetals in the periodic table. *(S3.1)*

Semipermeable membrane—A membrane that allows only selected materials to pass from one side to the other. *(13.3)*

Shell—A set of atomic orbitals with a given *n* value; each shell contains n^2 orbitals. *(7.6)*

Shielding effect—The decrease in nuclear attraction for an electron due to the presence of other electrons in underlying orbitals, sometimes called the screening effect. *(8.1)*

Short-range order—Orderly arrangements among small groups of molecules, especially in the liquid state. These orderly domains are continuously forming and breaking apart, so that the long-range effect is one of disorder. *(12.4)*

SI units—A set of seven metric units (including the meter, kilogram, and second) that has been recommended for use by the scientific community and that is adequate for all measurements in physical science. *(1.2)*

Sigma bond—A shared pair of electrons whose principal density lies between the nuclei and along the line of atomic centers. In valence bond theory, all single bonds are sigma bonds. *(10.4)*

Sigma orbital—A molecular orbital with regions of electron density along the internuclear axis. Sigma orbitals can be bonding or antibonding. *(10.5)*

Significant figures—The number of digits in a measured value; the number of significant figures is an indication of the precision to which the measurement was made. *(1.4)*

Silane—A silicon–hydrogen compound analogous to a hydrocarbon, for example, SiH_4. *(S3.4)*

Silicone—A compound containing silicon–oxygen chains with hydrocarbon side groups. Silicones combine the flexibility and stability of the Si—O chains with the water-repelling properties of the hydrocarbon groups. *(S3.4)*

Silver coulometer—See *coulometer.*

Simple cubic unit cell—The unit cell for a lattice built of cubes that have a particle at each corner and no other particles. *(12.6)*

Single bond—A covalent bond in which one pair of electrons is shared between two atoms. *(9.3)*

Slag—The mixture, mostly calcium silicate, that floats on top of the molten metal in a blast furnace. *(S4.1)*

Smelting—The high-temperature chemical reduction of a mineral to produce a molten metal. *(S4.1)*

Soap—A salt of a *fatty acid.* (23.4)

Sol—A colloidal dispersion of a solid in a liquid. Paint and milk of magnesia are examples. *(13.6)*

Solid sol—A colloidal dispersion of a solid in a solid. Colored gems are examples. *(13.6)*

Solubility—The amount of solute that dissolves in a given quantity of solvent to form a saturated solution. *(4.1)*

Solubility product constant (K_{sp})—The equilibrium constant for a solid in equilibrium with its ions. *(18.1)*

Solute—Any solution component other than the solvent. *(4.1)*

Solution—A homogeneous mixture in which one or more substances are uniformly dispersed as separate atoms, molecules, or ions throughout another substance. *(4.1)*

Solvation—The step in the solution process in which solvent molecules surround and are attracted by intermolecular forces to solute particles. Solvation helps to dissolve the solute and stabilizes the existence of ions in solution. *(13.1)*

Solvation energy—The energy released during solvation. *(13.1)*

Solvay process—An industrial process for preparing sodium hydrogen carbonate from saltwater and limestone. *(S3.3)*

Solvent—The solution component that determines whether the solution is solid, liquid, or gaseous. *(4.1)*

***sp* Hybrid orbital**—One of the two hybrid atomic orbitals formed from an *s* and a *p* orbital. The two hybrid orbitals make a 180° angle with each other. *(10.4)*

sp^2 Hybrid orbital—One of the three hybrid atomic orbitals formed from one *s* and two *p* orbitals. The three hybrid orbitals point toward the corners of an equilateral triangle. *(10.4)*

sp^3 Hybrid orbital—One of the four hybrid orbitals formed from one *s* and three *p* orbitals. The four hybrid orbitals point toward the corners of a regular tetrahedron. *(10.4)*

sp^3d Hybrid orbital—One of the five hybrid orbitals formed from one *s* orbital, three *p* orbitals, and one *d* orbital. The five hybrid orbitals point toward the corners of a trigonal bipyramid. *(10.4)*

sp^3d^2 Hybrid orbital—One of the six hybrid orbitals formed from one *s* orbital, three *p* orbitals, and two *d* orbitals. The six hybrid orbitals point toward the corners of an octahedron. *(10.4)*

Sparingly soluble—Having very limited solubility, almost insoluble. *(4.2)*

Specific gravity (sp gr)—The ratio of the density of a substance to the density of water. *(1.6)*

Specific heat—The heat capacity per gram of substance. *(6.5)*

Spectator ion—An ion that is present but does not participate in a reaction. *(4.2)*

Spectrochemical series—A list of ligands in order of increasing field strength. *(21.6)*

Spectroscope—An instrument for observing and recording spectra. Entering light is separated into its component wavelengths by passing it through a prism or diffraction grating. *(7.1)*

Spectroscopy—The study of the absorption and emission of radiation by matter. *(7.1)*

Spectrum—A mixture of waves spread out in order of wavelength or frequency. *(7.1)*

Speed of light—See *electromagnetic radiation. (7.1)*

Spherically symmetric orbital—An orbital in which the electron density depends only on the distance from the nucleus; the *s* orbitals are spherically symmetrical. *(7.7)*

Spin quantum number (m_s)—A half-integral quantum number describing the spin state of an electron; $m_s = +\frac{1}{2}$ or $-\frac{1}{2}$. *(7.6)*

Spin state—One of two electron states characterized by different orientations of the electron's magnetic field in an external magnetic field; see *spin quantum number. (7.6)*

Stability band—The narrow band containing stable nuclei in a plot of neutrons versus atomic number. *(22.5)*

Stability constant—Synonym for *formation constant. (18.5)*

Standard cell—A cell in which each reactant and product is in its standard state. *(20.1)*

Standard conditions—The reference conditions for gas volumes and densities; 0°C and 1 atm pressure; see *STP. (5.3)*

Standard electrode—Synonym for *standard half-cell.*

Standard electrode potential—A standard half-cell potential. *(20.3)*

Standard EMF ($E°$)—The maximum voltage produced by a standard cell; $E°$ is equal to the sum of the standard oxidation and reduction potentials for the cell. *(20.3)*

Standard enthalpy change ($\Delta H°$)—The enthalpy change for a reaction in which each reactant and each product is in its standard state. *(6.3)*

Standard enthalpy of combustion—The standard enthalpy change accompanying the combustion of 1 mol of a substance in oxygen. *(6.3)*

Standard enthalpy of formation ($\Delta H_f°$)—The standard enthalpy change accompanying the formation of 1 mol of a substance from its elements, each in its most stable form. *(6.3)*

Standard entropy change ($\Delta S°$)—The entropy change when reactants in their standard states are converted to products in their standard states. *(19.3)*

Standard free-energy change ($\Delta G°$)—The free-energy change when reactants in their standard states are converted to products in their standard states. *(19.2)*

Standard free energy of formation ($\Delta G_f°$)—The standard free-energy change for the formation of 1 mol of a substance from the most stable form of each of its elements. *(19.2)*

Standard half-cell—See *half-cell.*

Standard hydrogen electrode—A reference platinum electrode dipped into 1 *M* HCl. Hydrogen gas at 1 atm partial pressure is bubbled through the HCl and around the platinum. The half-cell reaction is $2H^+(1\ M) + 2e^- \rightleftharpoons H_2(g, 1\ atm)$. *(20.3)*

Standard molar entropy ($S°$)—The absolute entropy of 1 mol of a substance at 1 atm pressure and some specified temperature. *(19.4)*

Standard oxidation potential—See *oxidation potential.*

Standard reduction potential—See *reduction potential.*

Standard solution—A solution whose concentration has been precisely determined. *(4.6)*

Standard state—The state of a solid, liquid, or gas at 1 atm pressure and the temperature of interest, usually 25°C. For dissolved solutes, the standard state is a solution of approximately one molar concentration. *(6.3)*

Standardization—The process of determining the concentration of a solution. *(4.6)*

State function—A property whose change depends only on the initial and final states of the system, for example, internal energy, temperature, pressure, and volume. *(6.1)*

States of matter—Forms in which substances can exist, such as solid, liquid, or gas. *(5.1)*

Steady state—An intermediate is said to be in a steady state when it is used up as fast as it is produced. *(14.5)*

Steady-state approximation—The assumption that the concentration of an intermediate does not change, that its rate of formation is equal to its rate of consumption. The steady-state approximation is useful in deducing reaction mechanisms. *(14.5)*

Steel—A ferrous alloy containing controlled amounts of carbon (0.1–1.5%). Steels usually contain other metals chosen to impart some desired mix of properties. *(S4.5)*

Stereoisomers—Isomers with the same bonds but different bond orientations. Stereoisomers include *geometric isomers* and *enantiomers. (10.3, 21.5)*

Stern–Gerlach experiment—An experiment verifying the existence of two and only two electron spin orientations. *(7.6)*

Steroid—A biologically important group of substances, including cholesterol and various sex hormones, that have the basic steroid structure of three fused six-membered rings and one five-membered ring. *(23.1)*

Stoichiometry—The study of quantitative relationships governing the composition of substances and their reactions. *(3.1)*

STP—Abbreviation for standard temperature and pressure; see *standard conditions.*

Strong acid—Traditionally, an acid that ionizes completely in water. In the Brønsted–Lowry system, a substance with a strong tendency to give up protons; in aqueous solution, an acid that is stronger than hydronium ion. *(4.2, 16.1)*

Strong base—Traditionally, an ionic hydroxide. In the Brønsted–Lowry system, a substance with a strong tendency to accept protons; in aqueous solution, a base that is stronger than hydroxide ion. *(4.2, 16.1)*

Strong electrolyte—An electrolyte that is completely or almost completely ionized in aqueous solution. *(4.1)*

Strong-field ligand—A ligand that produces a large *orbital splitting energy. (21.6)*

Structural isomers—Isomers in which the sequence of atoms or bonds is different. *(10.3)*

Sublimation—A phase change from solid to vapor without passing through the liquid state. *(12.1)*

Sublimation curve—On a phase diagram, the curve showing the temperatures and pressures under which the solid and vapor states are in equilibrium. *(12.2)*

Sublimation temperature—The temperature at which the solid and gaseous states of a substance are in equilibrium. *(12.2)*

Subshell—Within a shell, all atomic orbitals having the same azimuthal quantum number. The number of orbitals in a subshell is $2l + 1$. *(7.6)*

Substance—See *pure substance.*

Substitution reaction—In organic chemistry, a reaction in which one substituent on a carbon atom is replaced by another. *(23.1)*

Supercooled liquid—A metastable liquid existing at a temperature below its freezing point. *(12.1)*

Supercritical fluid—A substance existing at a temperature above its critical temperature. *(12.2)*

Superoxide ion—O_2^-. *(S2.1)*

Supersaturated solution—A solution that contains a greater concentration of solute than would be in equilibrium with undissolved solute. The excess solute precipitates in the presence of "seed" crystals. *(4.1)*

Surface tension—The energy required to stretch the surface of a liquid by some unit amount; a measure of the tendency of a liquid surface to behave like an elastic skin. *(12.4)*

Surfactant—A surface-active agent that lowers the surface tension of a liquid—for example, a detergent. *(12.4)*

Surroundings—Everything not included in a system. *(6.1)*

Suspension—A heterogeneous mixture containing particles with dimensions exceeding 100,000 pm. *(13.6)*

System—A portion of the universe arbitrarily chosen for consideration. *(6.1)*

Systematic error—A consistent error from one measurement to the next caused by an imperfection in the instrument or in the experimental method. *(1.4)*

T

Temperature—The property of a substance that determines the direction of heat flow; hot objects always have higher temperatures than cold ones. *(1.7)*

Temperature scale—The markings on a thermometer that relate temperature to a property that varies with temperature. *(1.7)*

Tempering—In the treatment of steel, moderate heating followed by slow cooling. *(S4.5)*

Termolecular step—An elementary step involving the simultaneous collision of three particles. *(14.5)*

Tertiary alcohol—An alcohol in which the OH group is attached to a carbon atom that is bonded to three hydrocarbon groups. *(23.4)*

Tertiary amine—A derivative of ammonia in which all three hydrogen atoms have been replaced by hydrocarbon groups. *(23.6)*

Tertiary protein structure—Folds and bends that are not part of a repeating *secondary protein structure. (23.7)*

Tetrahedral angle—109.5°; the angle made by lines drawn from the corners of a tetrahedron to its center. *(10.2)*

Tetrahedron—A solid geometrical figure whose faces are four equilateral triangles. Four VSEPR pairs on an atom point toward the corners of a tetrahedron. *(10.2)*

Theoretical yield—The amount of product calculated from the chemical equation. *(3.9)*

Theory—A model based on conjecture that accounts for the laws of nature. *(1.2)*

Thermal equilibrium—A state in which there is no net flow of heat in any direction; objects in thermal equilibrium have the same temperature. *(1.7)*

Thermal motion—The temperature-dependent random motion of atoms and molecules. *(5.1)*

Thermite—A mixture of iron oxides and aluminum powder, sometimes used in welding. *(3.6)*

Thermochemical equation—A chemical equation that includes the enthalpy change. *(6.2)*

Thermochemistry—The study of the heat released or absorbed during chemical reactions. *(6.1)*

Thermodynamic equilibrium constant—An equilibrium constant calculated from thermodynamic data. *(19.2)*

Thermodynamic property—Synonym for *state function.*

Thermodynamics—The branch of science dealing with all forms of energy and their interconversion. *(6.1)*

Thermometer—An instrument for measuring temperature. *(1.7)*

Thermonuclear reaction—A nuclear fusion reaction. *(22.8)*

Third-law entropy—Synonym for *absolute entropy.*

Third law of thermodynamics—The entropy of a pure, perfectly crystalline element or compound may be taken as zero at zero Kelvin. *(19.4)*

Thomson atom—A model of the atom in which negative electrons are embedded in a jellylike sphere of positive electricity; of historical interest only. *(2.2)*

Threshold frequency (ν_0)—The light frequency below which photoelectrons will not be ejected from a metal surface. *(7.2)*

Titration—A procedure for obtaining quantitative information about a reactant by the controlled addition of one solution to another until the reaction is complete. *(4.6)*

Titration curve—A plot of pH versus volume of acid or base added during the course of a titration. *(17.8)*

Torr—A unit of pressure equal to 1/760 atm; 1 torr is equal to the pressure that supports 1 mm of mercury at 0°C. *(5.2)*

Transcription—In biochemistry, the assembling of an RNA molecule with a base sequence complementary to that in a given portion of DNA. The information in the DNA is thus transcribed, or rewritten, in the RNA. *(23.9)*

Transfer RNA—See *RNA.*

Transition elements—The elements between Groups 2A and 3A, in which the *d* and *f* orbitals are being filled. *(8.4, 21.4, 5, 6)*

Transition interval—The pH range over which an indicator changes color. *(17.8)*

Transition series—A horizontal row of transition elements. The third and fourth transition series include the lanthanides and actinides. *(8.4)*

Transition state—The highest point on the potential energy profile for an elementary step. The terms transition state and *activated complex* are often used interchangeably. *(14.7)*

Translational motion—Motion from one point in space to another. *(5.7)*

Transmutation—The changing of one element into another. *(22.3)*

Transuranium element—An element beyond uranium in the periodic table; an element with an atomic number greater than 92. *(22.3)*

Triglyceride—An ester of glycerol with three molecules of fatty acid. Fats and oils are triglycerides. *(23.5)*

Trigonal bipyramid—A five-cornered figure resembling two pyramids sharing a triangular face. Five VSEPR pairs on an atom point toward the corners of a trigonal bipyramid. *(10.2)*

Trigonal planar—An arrangement in which three VSEPR pairs on an atom point toward the corners of an equilateral triangle. *(10.2)*

Triple bond—A covalent bond in which three pairs of electrons are shared between two atoms. *(9.3)*

Triple point—A point on a phase diagram showing the temperature and pressure at which three states can coexist in equilibrium. *(12.3)*

Triprotic acid—A polyprotic acid with three acidic hydrogen atoms per molecule—for example, H_3PO_4. *(16.3)*

tRNA—See *RNA*.

Tyndall effect—The scattering of light by dispersed particles. Colloids exhibit a Tyndall effect, true solutions do not. *(13.6)*

U

Ultraviolet radiation—The shorter wavelength spectral region adjacent to the visible region; ultraviolet radiation produces suntans and sunburn. *(7.1)*

Uncertainty principle—See *Heisenberg uncertainty principle.*

Unimolecular step—An elementary step in which a single atom or molecule breaks down or rearranges itself into other species. *(14.5)*

Unit cell—A three-dimensional arrangement of atoms that repeats itself throughout a crystal lattice. *(12.6)*

Unpaired electron—A single electron in an orbital. *(8.2)*

Unsaturated fatty acid—See *fatty acid.*

Unsaturated hydrocarbon—A hydrocarbon with one or more double or triple bonds. *(23.1)*

Unsaturated solution—A solution that contains less than the equilibrium amount of solute. *(4.1)*

Useful work—Work other than PV work. Useful work includes mechanical work, electrical work, magnetic work, and so forth. *(19.1)*

V

Valence band—A band of molecular orbitals that is completely filled with valence electrons; see *band theory. (21.3)*

Valence bond theory—A theory of chemical bonding that assumes that atomic orbitals overlap when they approach each other. According to this theory, a covalent bond forms because two electrons from the parent orbitals spend most of their time in the overlap region between the nuclei, where the electron density is the highest. Valence bond theory includes the concepts of orbital hybridization and electron promotion. *(10.4)*

Valence electrons—The electrons of an atom that are the principal participants in bond formation. *(8.4)*

Valence shell—An electron shell that contains *valence electrons. (8.4)*

van der Waals constants—The constants a and b in the van der Waals equation; a is a measure of attractive forces and b is a measure of molecular size. *(5.8)*

van der Waals equation—$(P + an^2/V^2)(V - nb) = nRT$; an equation of state for gases that includes a correction for attractive forces and molecular size. *(5.8)*

van der Waals forces—Synonym for *intermolecular attractive forces.*

van't Hoff equation—An equation relating the enthalpy of reaction and the equilibrium constants at two different temperatures (see Equations 15.3 and 15.4). Also an equation relating the osmotic pressure π of a solution to its molarity M; $\pi = MRT$. *(13.3, 15.8)*

van't Hoff factor (*i*)—The effective number of moles of ions produced by 1 mol of solute; i is equal to the observed colligative effect of an ionic solute divided by the calculated effect assuming no dissociation. *(13.5)*

Vapor pressure—The partial pressure of a vapor in equilibrium with its liquid or solid; vapor pressures of water at various temperatures are listed in Table 5.3. *(5.5, 12.1)*

Vapor pressure lowering (VPL)—The vapor pressure of a pure solvent minus the vapor pressure of a solution; a colligative property proportional to the mole fraction of solute. *(13.3)*

Vaporization—A phase change from liquid to gas. *(12.1)*

Vaporization curve—On a phase diagram, the curve showing the temperatures and pressures under which the liquid and vapor states are in equilibrium. *(12.2)*

Volumetric analysis—An analytical procedure in which quantitative information is obtained by measuring the volume of a reacting solution. *(4.6)*

VSEPR—Valence Shell Electron Pair Repulsion Theory. A theory based on the assumption that valence shell electron pairs tend to stay as far apart as possible. VSEPR provides a method for using the number of bonds and lone pairs around a central atom to explain and predict molecular geometry. *(10.2)*

VSEPR pairs—The electron pairs that play a primary role in determining bond angles. The number of VSEPR pairs around a central atom is obtained from the Lewis structure by counting the number of lone pairs on the atom and adding one pair for each single or multiple bond. *(10.2)*

Viscosity—A measure of a liquid's resistance to flow. *(12.4)*

Vitreous solid—Synonym for *amorphous solid.*

Volt (V)—A unit used as a measure of electromotive force; 1 V equals 1 J/C. *(20.2)*

Voltage (*E*)—The potential energy lost by 1 C of charge as it moves between the terminals of a circuit. *(20.2)*

Voltaic cell—Synonym for *galvanic cell.*

W

Water constant (K_w)—The product of the hydrogen and hydroxide concentrations in an aqueous solution; $K_w = 1.00 \times 10^{-14}$ at 25°C. *(16.4)*

Watt—The rate at which energy is produced or consumed; 1 W is equal to 1 J/s. *(6.6)*

Wave—A periodic disturbance or oscillation passing through space or some medium such as water or air. A wave consists of repeating units or *cycles.* The length of each cycle is the *wavelength* (λ), and the number of cycles that pass a given point each second is the *wave frequency* (ν). The product of the wavelength and the wave frequency is the *wave speed. (7.1)*

Wave frequency (ν)—See *wave.*

Wave function—Synonym for *orbital.*

Wave mechanics—A version of quantum mechanics developed by Erwin Schrödinger. *(7.6)*

Wave speed—See *wave.*

Wavelength (λ)—See *wave.*

Weak acid—Traditionally, an acid that is only partially ionized in aqueous solution. In the Brønsted–Lowry system, an acid that does not readily give up protons; in aqueous solution, an acid that is weaker than hydronium ion. *(4.2, 16.1)*

Weak base—Traditionally, a base that is only partially ionized in aqueous solution. In the Brønsted–Lowry system, a base that does not readily accept protons; in aqueous solution, a base that is weaker than hydroxide ion. *(4.2, 16.1)*

Weak electrolyte—A molecular substance that ionizes only to a limited extent in water. Most of a weak electrolyte remains in molecular form. *(4.2)*

Weak-field ligand—A ligand that produces a small *orbital splitting energy. (21.6)*

Weighing—The process of ascertaining the mass of an object. *(1.6)*

Weight—The gravitational force acting on an object, equal to the product of the object's mass m and the gravitational acceleration g. *(1.6)*

White light—Light containing all the visible wavelengths. *(7.1)*

Work—The product of a force f and the distance d through which the force operates; $w = f \times d$. Work is a form of energy. *(1.7)*

X

X-ray diffraction—The passage of x-rays through crystals to obtain diffraction patterns. The analysis of diffraction patterns gives information about crystal structures. *(12.6)*

Z

Zeeman effect—The splitting of spectral lines into sets of three or more lines by a strong magnetic field. *(7.6)*

Zero-order reaction—A reaction in which the rate is not affected by concentration changes. *(14.3)*

Zone refining—A method for preparing ultrapure elements such as silicon and germanium by slowly passing a bar of the solid through a heating ring that melts successive zones. The impurities tend to accumulate in the melted region and are swept toward the end of the bar. *(S3.4)*

INDEX

Page numbers followed by a *t* designate topic coverage in a table. Page numbers followed by an *f* designate topic coverage in a figure.

A

B

C

D

E

F

G

H

I

J

K

L

M

N

O

P

Q

R

S

T

U

V

W

X

Y

Z

Photo Credits

Chapter 1

Chapter Opener, p. xxxiv © Dan McCoy, Rainbow. **1.1a, p. 2** courtesy Oakland University. **1.1b, p. 2** © Joel Gordon 1988. **1.2a, b p. 3** © Anglo-Australian Telescope Board 1987. **1.4a, p. 5** Courtesy of the National Institute of Standards and Technology. **1.4b, p. 5** © Royal Greenwich Observatory/Science Source. **1.6, p. 9** © Joel Gordon 1993. **p. 18** Courtesy of the National Institute of Standards and Technology. **1.9, p. 21** © Joel Gordon 1986. **1.10c, p. 22** © James Prince 1990/Photo Researchers.

Chapter 2

Chapter Opener, p. 34 © Dr. E. R. Degginger. **p. 36** The Granger Collection, New York. **2.7, p. 41** © Joel Gordon 1993. **p. 43** Burndy Library, **p. 50** Courtesy of Finnigan MAT, San Jose, CA. **D2.1, p. 57** © Joel Gordon 1988. **2.14a,b, p. 58** Stock Montage, Chicago Illinois. **2.16b, p. 60** © Joel Gordon 1988. **2.17b, p. 61** © Joel Gordon 1988. **2.18, p. 62** © Joel Gordon 1988. **2.19b, p. 63** A. Singer/© American Museum of Natural History. **2.20a, p. 63** © Joel Gordon 1989. **2.20b, p. 63** © Bruce Coleman. **2.21, p. 65** © Joel Gordon 1993.

Chapter 3

Chapter Opener, p. 82 © Joel Gordon 1993. **3.1, p. 84** © Joel Gordon 1988. **D3.1, p. 86** © Joel Gordon 1986. **p. 87** The Granger Collection, New York. **p. 94** © Dr. E. R. Degginger. **p. 98** Bob Firth/Firth Photobank. **D3.2, p. 100** © Joel Gordon 1986. **D3.3, p. 101** © Joel Gordon 1986. **D3.4, p. 103** © Joel Gordon 1993. **3.6a,b, p. 105** © Joel Gordon 1987.

Chapter 4

Chapter Opener, p. 124 Andrew McClenaghan/Science Photo Library. **4.2, p. 127** S. K. Mittwede/Visuals Unlimited. **4.3, p. 127** © Joel Gordon 1993. **4.4, p. 128** © Joel Gordon 1993. **D4.1, p. 129** © Joel Gordon 1988. **4.7, p. 131** © Joel Gordon 1993. **4.8, p. 134** © Joel Gordon 1987. **4.11, p. 142** © Joel Gordon 1986. **4.12, p. 143** © Joel Gordon 1986. **4.13, p. 149** © Joel Gordon 1986. **D4.2, p. 151** © Joel Gordon 1986. **D4.3, p. 155** © Joel Gordon 1986.

Chapter 5

Chapter Opener, p. 168 © Ray Nelson/Phototake NYC. **5.2, p. 171** © Joel Gordon 1986. **D5.1, p. 174** © Joel Gordon 1986. **p. 176** Stock Montage, Chicago, Illinois. **D5.2, p. 182** © Joel Gordon 1986. **p. 187** Burndy Library. **5.16, p. 201** © Joel Gordon 1993. **5.17, p. 202** Department of Energy. **5.18, p. 203** © Joel Gordon 1986.

Chapter 6

Chapter Opener, p. 222 © Joel Gordon 1988. **D6.1, p. 230** © Joel Gordon 1986. **D6.2, p. 231** © Joel Gordon 1988.

Chapter 7

Chapter Opener, p. 262 © Holt Confer/Phototake NYC. **7.4a, p. 266** United States Naval Observatory. **7.4b, p. 266** © Anglo-Australian Telescope Board, imaging by David Allen. **7.5, p. 267** Runk/Schoenberger from Grant Heilman. **7.6, p. 268** Courtesy of Sargent-Welch Scientific Co., a VWR Company. **7.8a, p. 269** © Dohrn/Science Photo Library. **7.8b, p. 269** © Yoav Levy/Phototake NYC. **7.9, p. 270** Palomar Observatories/Visuals Unlimited. **p. 274** American Institute of Physics. **7.17, p. 280** Education Development Center. **7.18, p. 282** Dr. Mitsuo Ohtsuki/Science Photo Library. **7.19, p. 282** Science VU-IBMRL/Visuals Unlimited. **7.20, p. 283** IBM Corporation, Research Division, Thomas J. Watson Research Center. **7.21, p. 283** IBM Corporation, Research Division, Almaden Research Center. **7.23, p. 284** American Institute of Physics.

Chapter 8

Chapter Opener, p. 302 © Tony Freemen/Photo Edit. **8.3a,b, p. 310** Chemical Heritage Foundation. **8.7, p. 319** © 1973 Fundamental Photographs.

Survey of the Elements 1

Chapter Opener, p. 336 © Cary Wolinski/Stock Boston. **S1.2a, p. 341** © Liane Enkelis/Stock Boston. **S1.2b, p. 341** © Charles M. Falco/Photo Researchers. **S1.3** © Joel Gordon 1986. **DS1.1, p. 343** © Joel Gordon 1986. **S1.4a, b, p. 345** © Joel Gordon 1986. **S1.4c, p. 345** © Joel Gordon 1993. **S1.5, p. 348** John S. White/Smithsonian. **S1.6a, p. 348** John S. White/Smithsonian. **S1.6b, p. 348** © George Gerster/Photo Researchers. **DS1.2, p. 349** © Joel Gordon 1986. **DS1.3, p. 351** © Joel Gordon 1988. **S1.7, p. 352** Betz Laboratories, Inc. **S1.8, p. 353** National Institutes of Health. **S1.9a, b, p. 353** © Joel Gordon 1987. **S1.9c, p. 353** © Joel Gordon 1993. **S1.10a, p. 354** © John D. Cunningham/Visuals Unlimited. **S1.10b, p. 354** © Roy King 1985. **p. 357** E. R. Degginger. **S1.12a,b, p. 358** © Joel Gordon 1993. **S1.12c, p. 358** © Richard Megna, 1988/Fundamental Photographs. **S1.13, p. 359** Argonne National Laboratory.

Chapter 9

Chapter Opener, p. 364 © Thomas Hollyman 1972/Photo Researchers. **9.3a,b, p. 367** Reproduced by permission of the Bancroft Library. **9.8, p. 395** NASA. **9.9, p. 396** Donald Clegg. **D9.1, p. 397** © Joel Gordon 1986. **9.13, p. 400** Bettman Newsphotos.

Chapter 10

Chapter Opener, p. 410 Dr. A. Lesk, Laboratory of Molecular Biology/Science Photo Library. **10.7, p. 422** © Joel Gordon 1988. **10.8, p. 423** © Joel Gordon 1988. **10.9, p. 424** © Joel Gordon 1988. **10.10, p. 425** © Joel Gordon 1988. **10.11, p. 425** © Joel Gordon. **10.13, p. 426** © Joel Gordon 1988. **10.14, p. 427** © Joel Gordon 1988. **10.15, p. 428** © Joel Gordon 1988. **10.16, p. 429** © Joel Gordon 1988.

Chapter 11

Chapter Opener, p. 464 © Richard Megna, 1988/Fundamental Photographs. **D11.1, p. 466** © Joel Gordon 1986. **D11.2, p. 482** © Joel Gordon 1988.

Survey of the Elements 2

Chapter Opener, p. 490 © Tony Stone Worldwide. **p. 493** SIU, Peter Arnold, Inc. **DS2.1, p. 494** © Joel Gordon 1987. **p. 494 (bottom)** E. R. Degginger. **DS2.2, p. 497** © Joel Gordon 1987. **DS2.3, p. 500** © Joel Gordon 1988. **S2.5, p. 505** © John D. Cunningham/Visuals Unlimited. **S2.7, p. 506** © Grant Heilman Photography. **S2.8, p. 509** © A. J. Copley/Visuals Unlimited. **DS2.4, p. 510** © Joel Gordon 1987. **S2.9, p. 511** © Joel Gordon 1988. **S2.10, p. 512** © Joel Gordon 1988. **S2.12, p. 515** © Joel Gordon 1987. **p. 516** © Joel Gordon 1988. **DS2.5, p. 518** © Joel Gordon 1986. **S2.14, p. 523** Courtesy of Johnson Space Center/NASA.

Chapter 12

Chapter Opener, p. 532 © John Koivula/Science Source. **D12.1, p. 539** © Joel Gordon 1988. **12.21, p. 558** © William J. Weber/Visuals Unlimited. **12.27d, p. 563** © Hughes/Visuals Unlimited. **12.28a, p. 564** © Joel Gordon 1988. **12.28b, p. 564** © John D. Cunningham/Visuals Unlimited. **p. 573** Wendy Metzen/Bruce Coleman, Inc. **12.41b, p. 576** Courtesy of Dr. Keith Moffatt, Cornell University, Section of Biochemistry Cell and Molecular Biology.

Chapter 13

Chapter Opener, p. 586 © Peticolas/Megna, 1989/Fundamental Photographs. **D13.1, p. 590** © Joel Gordon 1988. **13.3, p. 591** Yoav Levy/Phototake. **p. 595** Science VU/API/Visuals Unlimited. **D13.2, p. 602** © Joel Gordon 1987. **13.15, p. 608** © David M. Phillips/Visuals Unlimited. **D13.3, p. 609** © Joel Gordon 1988. **13.19, p. 617** © Joel Gordon 1987. **13.22, p. 620** © Joel Gordon 1988. **13.23, p. 620** Visuals Unlimited. **13.27, p. 622** © Omikron/Photo Researchers.

Chapter 14

Chapter Opener, p. 630 © National Geographic Society. **p. 632** Visuals Unlimited. **D14.1, p. 633** © Joel Gordon 1993. **14.1, p. 633** © Joel Gordon 1987. **D14.2, p. 658** © Joel Gordon 1988. **14.19b, p. 675** © National Geographic Society. **14.22, p. 678** A. C. Rochester Division GMC.

Chapter 15

Chapter Opener, p. 690 © Joel Gordon 1993. **15.4, p. 719** © Joel Gordon 1988. **D15.1, p. 720** © Joel Gordon 1988. **15.7, p. 724** © Joel Gordon 1993.

Chapter 16

Chapter Opener, p. 738 © Eric L. Wheater/The Image Bank. **16.1, p. 740** © Joel Gordon 1987. **16.4, p. 750** © Joel Gordon 1988. **16.5, p. 755** Fisher Scientific Company. **16.6a, p. 760** NEW-PCC/Visuals Unlimited. **16.6b, p. 760** © John D. Cunningham/Visuals Unlimited.

Chapter 17

Chapter Opener, p. 774 Cyril Isy-Schwart/The Image Bank. **17.2, p. 788** © Joel Gordon 1988. **D17.1, p. 793** © Joel Gordon 1987. **D17.2, p. 800** © Joel Gordon 1988. **17.6, p. 811** © Joel Gordon 1988. **17.7, p. 812** © Joel Gordon, 1987.

Chapter 18

Chapter Opener, p. 820 Randall Hyman/Stock, Boston. **18.1, p. 826** © Joel Gordon 1988. **18.2, p. 832** © Joel Gordon 1988. **18.4, p. 836** © Joel Gordon 1993. **D18.1, p. 837** © Joel Gordon 1988. **18.5, p. 839** Virginia Runk/Schoenberger/Grant Heilman. **18.6, p. 841** © Joel Gordon 1988.

Survey of the Elements 3

Chapter Opener, p. 854 © Soames Summerhays/Photo Researchers. **S3.1, p. 857** © Vulcain/Explorer/Photo Researchers. **S3.3, p. 859** © Joel Gordon 1988. **DS3.1, p. 861** © Joel Gordon 1988. **DS3.2, p. 864** © Joel Gordon 1988. **DS3.3, p. 865** © Joel Gordon 1988. **DS3.4, p. 866** © Joel Gordon 1988. **DS3.5, p. 867** © Joel Gordon 1988. **S3.9, p. 870** Derby Art Gallery. **S3.10, p. 870** © Joel Gordon 1988. **S3.12a, b, p. 872** © Joel Gordon 1988. **S3.12c, p. 872** © John D. Cunningham/Visuals Unlimited. **S3.16, p. 878** © Joel Gordon 1993. **S3.17a, p. 878** E. R. Degginger. **S3.17b, p. 878** © Joel Gordon 1988. **S3.18, p. 879** © Paul Silverman, 1993/Fundamental Photographs. **S3.20, p. 881** Geoff Tompkinson/Science Photo Library. **DS3.6, p. 882** © Joel Gordon 1993. **S3.22, p. 884** NASA Goddard Institute for Space Studies. **S3.23, p. 886** © Joel Gordon 1982. **S3.25a, p. 887** Runk/Schoenberger/Grant Heilman. **S3.25b, p. 887** © Arthur R. Hill/Visuals Unlimited. **S3.25c,d, p. 887** Runk/Schoenberger/Grant Heilman. **DS3.7, p. 888** © Joel Gordon 1988. **S3.27, p. 890** © A. J. Copley/Visuals Unlimited. **S3.28, p. 890** Chemical Designs, Ltd. **S3.29, p. 891** E. R. Degginger. **S3.33a, p. 894** © Rich Treptow/Visuals Unlimited. **S3.33b, p. 894** © Albert Copley/Visuals Unlimited. **S3.33c, p. 894** John Cunningham/Visuals Unlimited.

Chapter 19

Chapter Opener, p. 904 © Richard Megna/Fundamental Photographs. **19.2, p. 915** © Joel Gordon 1986. **D19.1, p. 916** © Joel Gordon 1988.

Chapter 20

Chapter Opener, p. 936 © Diane Schiumo 1990. **D20.1, p. 938** © Joel Gordon 1988. **20.2, p. 939** © Joel Gordon 1988. **20.3, p. 940** © Joel Gordon 1988. **D20.2, p. 961** © Joel Gordon 1993. **20.14, p. 965** © John D. Cunningham/Visuals Unlimited. **20.16, p. 966** © Joel Gordon 1988. **D20.3, p. 967** © Joel Gordon 1988. **20.18, p. 967** Science VU/Visuals Unlimited. **D20.4, p. 973** © Joel Gordon 1988. **D20.5, p. 975** © Joel Gordon 1988.

Chapter 21

Chapter Opener, p. 986 Mike McNamee/Science Photo Library. **21.27b, p. 1019** © Joel Gordon 1988. **21.29a, b, p. 1020** © Paul Silverman/Fundamental Photographs. **21.30, p. 1021** © Joel Gordon 1988. **21.34, p. 1024** © Joel Gordon 1988. **21.37, p. 1024** © Joel Gordon 1988.

Survey of the Elements 4

Chapter Opener, p. 1032 © Murray Alcosser The Image Bank. **S4.1, p. 1034** Runk/Schoenberger/Grant Heilman. **p. 1034 (bottom)** © Joel Gordon 1988. **DS4.1, p. 1036** © Joel Gordon 1988. **S4.2, p. 1038** © 1988 Dawson Jones/Stock, Boston. **S4.3b, p. 1040** © Don Green/Kennecott. **DS4.2, p. 1041** © Joel Gordon 1988. **S4.4a, p. 1041** © Joel Gordon 1988. **S4.4b, p. 1041** © Dona Lambrecht/Visuals Unlimited. **S4.5, p. 1042** © Joel Gordon 1988. **S4.7, p. 1044** © Joel Gordon 1988. **S4.8, p. 1045** © Joel Gordon 1988. **S4.9, p. 1046** © Joel Gordon 1988. **S4.10, p. 1047** © Joel Gordon 1988. **S4.11a, p. 1048** © John D. Cunningham/Visuals Unlimited. **S4.11b,c, p. 1048** © Joel Gordon 1988. **S4.13, p. 1050** © Jerome Wyckoff/Visuals Unlimited. **S4.14, p. 1051** © AMEX/Visuals Unlimited. **DS4.3, p. 1053** © Joel Gordon 1988. **S4.15, p. 1054** © A.J. Copley/Visuals Unlimited. **S4.16, p. 1055** © Tom McHugh/Photo Researchers. **S4.17, p. 1055** © 1986 Albert Copley/Visuals Unlimited. **S4.18, p. 1056** © Joel Gordon 1988. **S4.19, p. 1056** Science VU/Visuals Unlimited. **S4.20, p. 1056** © Joel Gordon 1988. **DS4.4, p. 1059** © Joel Gordon 1988. **S4.21, p.1059** © Joel Gordon 1988. **S4.22, p. 1060** © Joel Gordon 1988. **S4.23, p. 1061** © Joel Gordon 1988. **S4.24, p. 1062** © Joel Gordon 1988. **DS4.5, p. 1063** © Joel Gordon 1988. **S4.25, p. 1063** © Joel Gordon 1988. **S4.26, p. 1064** © Joel Gordon 1988. **S4.27, p. 1064** Science VU/Visuals Unlimited.

Chapter 22

Chapter Opener, p. 1070 Science VU/Fermilab/Visuals Unlimited. **22.1, p. 1072** © Joel Gordon 1988. **p. 1074** © Joel Gordon. **22.5, p. 1075** © Joel Gordon 1988. **22.8a, p. 1080** Science VU/LBL/Visuals Unlimited. **22.9, p. 1082** The Granger Collection. **22.10, p. 1083** © Porterfield/Chickering/Photo Researchers. **22.11, p. 1086** Science VU/BNL/Visuals Unlimited. **22.12, p. 1087** Science VU/Visuals Unlimited. **22.16, p. 1095** American Institute of Physics. **22.20, p. 1100** Visuals Unlimited. **22.21b, p. 1101** © Dan McCoy/Rainbow. **22.24, p. 1104** Science VU/LNL/Visuals Unlimited. **22.25, p. 1104** Royal Observatory, Edinburg/AATB/Science Photo Library.

Chapter 23

Chapter Opener, p. 1112 Birney Lettick/The Image Bank. **23.1, p. 1117** © Joel Gordon 1988. **D23.1, p. 1119** © Joel Gordon 1988. **D23.2, p. 1121** © Joel Gordon 1988. **23.5, p. 1125** Science VU/API/Visuals Unlimited. **23.6, p. 1129** © Joel Gordon 1988. **23.7, p. 1131** © Joel Gordon 1988. **23.8, p. 1132** © Joel Gordon 1988. **23.9, p. 1132** © Joel Gordon 1988. **23.10, p. 1133** © Joel Gordon 1988. **23.11, p. 1133** © Joel Gordon 1989. **23.13, p. 1134** © Joel Gordon 1988. **D23.3, p. 1137** © Joel Gordon 1988. **23.15, p. 1138** © Joel Gordon 1988. **23.16, p. 1138** © Joel Gordon 1989. **23.17, p. 1140** Science VU/NASA/Visuals Unlimited. **23.19, p. 1142** © Joel Gordon 1988. **23.20, p. 1142** © Joel Gordon 1988. **23.21, p. 1142** © Joel Gordon 1988. **23.22, p. 1144** © Joel Gordon 1988. **23.23, p. 1146** © Joel Gordon 1988. **D23.4, p. 1147** © Joel Gordon 1988. **23.24, p. 1147** © Joel Gordon 1988. **23.29, p. 1152** © Irving Geis.

SOME USEFUL CONVERSION FACTORS

LENGTH

SI unit = meter (m)

1 m = 39.37 in

1 in = 2.54 cm (exactly)

1 mile = 5280 ft = 1.609 km

1 angstrom (Å) = 10^{-10} m

MASS

SI unit = kilogram (kg)

1 kg = 2.205 lb

1 lb = 16 oz = 453.6 g

1 ton (short) = 2000 lb

1 metric ton = 1000 kg = 1.102 tons

1 g = 6.0221×10^{23} atomic mass units (u)

ENERGY

SI unit = joule (J)

1 J = 1 N·m

1 cal = 4.184 J (exactly)

1 L·atm = 101.33 J

1 megaelectron volt (MeV) = 1.602×10^{-13} J

1 u = 931.5 MeV

PRESSURE

SI unit = pascal (Pa)

1 Pa = 1 N/m^2

1 bar = 10^5 Pa

1 atm = 1.01325×10^5 Pa (exactly)

= 1.01325 bar = 760 torr (exactly)

= 760 mm Hg

VOLUME

SI unit = cubic meter (m^3)

1 L = 1000 cm^3 = 1.057 qt (U.S.)

1 gal (U.S.) = 4 qt = 8 pt = 128 fluid ounces

= 3.785 L

MISCELLANEOUS

Charge: 1 F = 96,485 C

Current: 1 ampere (A) = 1 C/s

Force: 1 N = 1 kg·m/s^2

Power: 1 watt = 1 J/s

Time: 1 h = 60 min = 3600 s

Voltage: 1 volt = 1 J/C

PHYSICAL CONSTANTS

Avogadro's number: $N_A = 6.0221 \times 10^{23}$

Elementary charge: $e = 1.6022 \times 10^{-19}$ C

Faraday: $F = 96{,}485$ C

Gas constant: $R = 0.082058$ L·atm/mol·K

= 8.3145 J/mol·K = 1.9872 cal/mol·K

Gravitational acceleration: $g = 9.8066$ m/s^2

Planck's constant: $h = 6.6261 \times 10^{-34}$ J·s

Speed of light in vacuum: $c = 2.9979 \times 10^8$ m/s